MECHANICAL DESIGN AND SYSTEMS HANDBOOK

OTHER McGRAW-HILL HANDBOOKS OF INTEREST

MAGILL, HOLDEN, AND ACKLEY · Air Pollution Handbook
MANAS · National Plumbing Code Handbook
MANTELL · Engineering Materials Handbook
MAYNARD · Industrial Engineering Handbook
MERRITT · Building Construction Handbook
MORROW · Maintenance Engineering Handbook
PERRY · Chemical Engineers' Handbook
PERRY · Engineering Manual
ROSSNAGEL · Handbook of Rigging
SHAND · Glass Engineering Handbook
SOCIETY OF MANUFACTURING ENGINEERS:
Die Design Handbook
Handbook of Fixture Design
Manufacturing planning and Estimating Handbook
Tool Engineers Handbook
STANIAR · Plant Engineering Handbook
STREETER · Handbook of Fluid Dynamics
TOULOUKIAN · Retrieval Guide to Thermophysical Properties Research Literature
TRUXAL · Control Engineers' Handbook

MECHANICAL DESIGN AND SYSTEMS HANDBOOK

HAROLD A. ROTHBART, Editor-in-Chief

Dean, College of Science and Engineering
Fairleigh Dickinson University
Teaneck, New Jersey

McGRAW-HILL BOOK COMPANY

New York St. Louis San Francisco Düsseldorf Johannesburg
Kuala Lumpur London Mexico Montreal New Delhi
Panama Rio de Janeiro Singapore Sydney Toronto

MECHANICAL DESIGN AND SYSTEMS HANDBOOK

 Library of Congress Catalog Card Number: 62-21118

07-054019-5

8 9 – MAMM – 7 5

CONTRIBUTORS

CLAYTON H. ALLEN, Ph.D., *Senior Engineering Scientist, Bolt Beranek and Newman, Inc., Cambridge, Mass.* (Section 7, *Noise Control*)

JOHN W. AXELSON, Ph.D., *Section Chief, Packings Development, Johns Manville Corp., Research Center, Manville, N.J.* (Section 24, *Pressure Components (Seals)*)

W. H. BAIER, Ph.D., *Senior Research Engineer, Armour Research Foundation, Chicago, Ill.* (Section 29, *Belts*)

ANTONIO F. BALDO, Ph.D., *Associate Professor of Mechanical Engineering, The City University of New York, New York, N.Y.* (Section 34, *Dampers*)

PHILIP BARKAN, Ph.D., *Senior Research Engineer, General Electric Company, Switchgear and Control Division, Philadelphia, Pa.* (Section 16, *Impact*)

STANLEY J. BECKER, M.S., *Engineer, Advanced Techniques Re-entry Systems Department, General Electric Corp., Philadelphia, Pa.* (Section 23, *Shrink- and Press-fitted Assemblies*)

KARL BRUNELL, B.M.E., *Research Associate, Princeton Laboratory, American Can Co., Princeton, N.J.* (Section 26, *Power Screws*)

JOHN A. CARLSON, Ph.D., *Engineering Specialist, MRD Division, General American Transportation Corp., Niles, Ill.* (Section 2, *Computers*)

SAUL FENSTER, Ph.D., *Chairman, Department of Mechanical Engineering, Fairleigh Dickinson University, Teaneck, N.J.* (Section 1, *Mathematics*, and Section 18, *Mechanical Design*)

E. T. FORTINI, M.S., *Staff Engineer, Compugraphic Corp., Reading, Mass., and Lecturer in Mechanical Engineering, Lincoln College, Northeastern University, Boston, Mass.* (Section 19, *Fabrication Principles*)

FERDINAND FREUDENSTEIN, Ph.D., *Professor of Mechanical Engineering, Columbia University, New York, N.Y.* (Section 4, *Kinematics of Mechanisms*)

THEODORE GELA, D.Sc., *Professor of Metallurgy, Stevens Institute of Technology, Hoboken, N.J.* (Section 17, *Properties of Engineering Materials*)

HERBERT H. GOULD, M.M.E., *Engineer, Sperry Gyroscope Co., Great Neck, N.Y.* (Section 1, *Mathematics* and Section 18, *Mechanical Design*)

MARSHALL HOLT, Ph.D., *Chief, Mechanical Testing Division, Aluminum Company of America, Alcoa Research Laboratories, New Kensington, Pa.* (Section 22, *Rivets and Riveted Joints*)

Z. J. JANIA, M.S., *Senior Research Engineer, Scientific Laboratory, Ford Motor Company, Dearborn, Mich.* (Section 28, *Friction Clutches and Brakes*)

A. B. JONES, *Consulting Engineer, Newington, Conn.; formerly, Chief Research Engineer, The Fafnir Bearing Company, New Britain, Conn.* (Section 13, *The Mathematical Theory of Rolling-element Bearings*)

WILLIAM R. JOHNSON, B.S., *Assistant Director of Research and Development, Associated Spring Corp., Bristol, Conn.* (Section 33, *Springs*)

S. KAMINSKY, M.M.E., M.S.E.E., *Consulting Engineer, Brooklyn, N.Y.* (Section 8, *System Dynamics*)

GEORGE F. KAPPELT, *Director, Engineering Laboratories, Bell Aerosystems Company, Buffalo, N.Y.* (Section 25, *Welding, Brazing, Soldering, and Adhesive Bonding*)

A. I. KATZ, M.S., M.E., *Director of Training and Education, Electronic Associates Inc., Princeton, N.J.* (Section 2, *Computers*)

K. W. KOHLMEYER, B.M.E., *Senior Engineer, Surface Armament Division, Sperry Gyroscope Company, Great Neck, N.Y.* (Section 9, *Control-system Applications*)

R. P. KOLB, M.S., *Engineering Staff, Technik, Inc., Jericho, Long Island, N.Y.* (Section 15, *Mechanics of Materials*)

JEROME LEIGHT, M.E.E., *Chief Systems Engineering Section, Systems Management Division, Kollsman Instrument Corp., Elmhurst, N.Y.* (Section 35, *Electromechanical Components*)

CARL H. LEVINSON, B.S.M.E., *Reliability Engineer, Sperry Gyroscope Co., Air Armament Division, Central Reliability Engineering Dept., Great Neck, N.Y.* (Section 18, *Mechanical Design*)

V. J. McDONOUGH, M.S., *Engineering Section Head, Surface Armament Division, Sperry Gyroscope Company, Great Neck, N.Y.* (Section 9, *Control-system Applications*)

GEORGE W. MICHALEC, M.S., *Head, Special Products Dept., General Precision, Inc., Pleasantville, N.Y.* (Section 32, *Gearing*)

THOMAS P. MITCHELL, Ph.D., *Associate Professor of Engineering Mechanics, Department of Engineering Mechanics and Materials, Cornell University, Ithaca, N.Y.* (Section 3, *Classical Mechanics*)

F. B. PINTARD, M.S., *Research Engineer, Johns Manville Corp., Research Center, Manville, N.J.* (Section 24, *Pressure Components (Seals)*)

EUGENE I. RADZIMOVSKY, Ph.D., *Professor of Mechanical Engineering, University of Illinois, Urbana, Ill.* (Section 21, *Bolted Joints*)

G. REETHOF, Ph.D., *Manager, Large Jet Engine Department, General Electric Co., Cincinnati, Ohio.* (Section 37, *Hydraulic Components*)

CARL H. RINGE, Dipl. Ing., *Research Engineer, Manager, Patent Dept., R. Hoe and Co., New York, N.Y.* (Section 19, *Fabrication Principles*)

FRIEDRICH O. RINGLEB, Ph.D., *Staff Physicist, Naval Air Engineering Laboratory, Philadelphia, Pa. and Professor of Mechanics, University of Delaware, Newark, Del.* (Section 31, *Cable Dynamics*)

HAROLD A. ROTHBART, D. Eng., *Dean, College of Science and Engineering, Fairleigh Dickinson University, Teaneck, N.J.* (Section 5, *Cam Mechanisms*, and Section 10, *Machine Systems*)

M. SADOWY, Ph.D., *Professor of Mechanical Engineering, Marquette University, Milwaukee, Wis.* (Section 27, *Shafts, Couplings, Keys, Etc.*)

GEORGE N. SANDOR, Eng.Sc.D., *Associate Professor of Mechanical Engineering, Yale University, New Haven, Conn.* (Section 4, *Kinematics of Mechanisms*)

CHARLES B. SCHUDER, M.S.M.E., *Director Systems Analysis Group, Research Dept., Fisher Governor Company, Marshalltown, Iowa.* (Section 36, *Pneumatic Components*)

DAVID SINCLAIR, Ph.D., *Senior Physicist, Johns Manville Corporation, Research Center, Manville, N.Y.* (Section 28, *Friction Clutches and Brakes*)

STANLEY SPORN, M.E.E., *Supervisor, Accelerometer Section, American Bosch Arma Corporation, Arma Division, Garden City, N.Y.* (Section 35, *Electromechanical Components*)

B. STERNLICHT, Ph.D., *Technical Director, Mechanical Technology, Inc.. Schenectady, N.Y.* (Section 12, *Hydrodynamic Lubrication*)

DAVID TABOR, Sc.D., *Laboratory for the Physics and Chemistry of Solids, Department of Physics, Cambridge University, Cambridge, England.* (Section 11, *Friction, Lubrication, and Wear*)

G. V. TORDION, Ph.D., *Director, Mechanical Engineering Department, Université Laval, Faculté des Sciences, Quebec, Canada.* (Section 30, *Power-transmission Chains*)

W. A. TUPLIN, D.Sc., M.I., Mech.E., *Professor of Applied Mechanics, University of Sheffield, Sheffield, England.* (Section 32, *Gearing*)

ERIC E. UNGAR, Eng.Sc.D., *Senior Engineering Scientist, Bolt Beranek ana Newman Inc., Cambridge, Mass.* (Section 6, *Mechanical Vibrations*)

JOHN J. VIEGAS, M.M.E., *Assistant to the Engineering Vice President, R. Hoe and Company, New York, N.Y.* (Section 20, *Standards for Mechanical Elements*)

WALTER WERNITZ, Dr.-Ing., *Institut für Maschinenelemente und Fordertechnik, Technische Hochschule Braunschweig, Braunschweig, West Germany.* (Section 14, *Friction Drives*)

M. ZAID, Sc.D., *President, Technik, Inc., Jericho, Long Island, N.Y.* (Section 15, *Mechanics of Materials*)

PREFACE

This handbook attempts for the first time to provide a systematic and scientific basis for machine design and the dynamic analysis of mechanical systems. It is intended for practising engineers, scientists, and students who have found the older, empirical methods of putting a machine together inadequate to today's needs. The methods of empiricism just cannot cope with demands for faster-moving, more sensitive, and more complex mechanisms.

To meet this need, newer concepts have been formulated of a machine as an assembly of interacting mechanical systems and components, subject to definite methods of mathematical analysis. Thus the approach has been to apply the latest mathematical methods to the analysis and design of mechanical elements and their relationship to one another in mechanical systems. "Practical" data that are still useful for design have not been sacrificed, however.

The system and subsystem concepts that have proved so fruitful in advancing our technology apply not only to equipment but also to corporate structure, to research program planning, and to the organization of thought itself. Actual classification into systems and subsystems is always subject to dispute, so we must start off with an essentially arbitrary definition: *A mechanical system is an entity that includes all interacting material bodies, from the first input to the last output of mechanical energy.*

The smallest element capable of analysis in a given system is termed a *component.* This handbook treats as components the ordinary and necessary hardware of machinery: bolts, keys, washers, screws, and the like. For the purposes of mechanical analysis, the basic types of electromechanical, hydraulic, and pneumatic equipment are also presented on a component basis: we will be concerned only with their performance characteristics and input-output relations.

Assemblies of closely interrelated components are treated as *subsystems;* as discussed here they include such elements as brakes, bearings, gear trains, and linkages. Finally, the *system* definition given above can be limited for practical purposes to consist of the entity under study—a particular machine or a given collection of subsystems.

For example, consider an electric motor driving a gear reducer, belt, pulley, cam, and follower. All are supported on a frame, and with the frame constitute a system. The cam alone is a component; with the follower and the frame it comprises a subsystem. But this mechanical system with its subsystems and components is directly connected to an electrical system, from which the motor obtains its energy input. In turn, the electrical system is powered by a thermal system, which converts chemical energy into useful electricity at the power station.

Sometimes the nature of the situation permits the isolation of a single system from a chain of interconnected systems. Such isolation is an aid in formulating and solving complex problems. Often, as greater understanding is required, it is necessary to include the adjoining systems. Thus the load on a cam caused by the follower may be reflected back into the motor as a fluctuating torque at the motor shaft, and the electrical system will respond with voltage fluctuations.

Other factors—stemming from the effects of friction, heat, or electromagnetic disturbances—often introduce complexities and nonlinearities into the basic differential equations, and therefore have been given emphasis. Consider the example of a spring-loaded friction clutch. The heat produced at the surfaces of contact raises the temperature of the friction materials, shafting, and springs. Two immediate effects of the elevated temperature are that the coefficient of friction of the clutch facing changes, and the modulus of elasticity of the spring material is reduced, changing the spring force. There are other effects as well.

It is important that the broad engineer-scientist use the knowledge at his disposal to evaluate such interrelationships among systems, but it is also important—and highly useful—that he know the approximate magnitude of the various influences so that, when they are very minor, he may safely neglect them. The engineer must always consider the economic relation between means and ends; sometimes overdesign is less costly than extended calculation.

Ultimately, knowledge is most useful when it can be expressed parametrically and is fully substantiated by experiment and experience. Nothing can replace experience, whether in the field or in the laboratory, coupled with the unifying disciplines of mathematics and the physical sciences. "The supreme misfortune," as Leonardo da Vinci observed, "is when theory outstrips performance."

I wish to thank the many in industry who were helpful in providing valuable data, and the various industrial organizations through whose courtesy the data are published. Grateful acknowledgment is due the section contributors, whose efforts and patience have indeed been heartwarming. Finally, special appreciation goes to my colleagues, Dr. S. Fenster and Messrs. S. Kaminsky and C. Ringe, who made many valuable suggestions.

H. A. Rothbart

CONTENTS

PJ MOSCHNER

Section 1

MATHEMATICS

By

SAUL FENSTER, Ph.D., *Chairman, Department of Mechanical Engineering, Fairleigh Dickinson University, Teaneck, N.J.*

HERBERT H. GOULD, M.M.E., *Engineer, Sperry Gyroscope Co., Great Neck, N.Y.*

CONTENTS

1.1. MATHEMATICAL NOTATION

$+$	= plus
$-$	= minus
$\times, \cdot$	= multiplied by
$\div$	= divided by
$=$	= equals
$\neq$	= is not equal to
$\cong, \approx$	= approximately equals
$\equiv$	= is identical to
$>$	= is greater than
$<$	= is less than
$\geqq$	= equals or is greater than
$\leqq$	= equals or is less than
$\propto$	= varies directly as
$:$	= is to (proportion), as in $x:y = u:v$ or $x/y = u/v$
x^n	= x raised to the nth power
$x^{\frac{1}{n}}$	= the nth root of x
$\sqrt[n]{x}$	= the nth root of x
$\sqrt{x}$	= the square root of $x (n = 2)$
$\sqrt[3]{x}$	= the cube root of $x (n = 3)$
$\sqrt{-1}, i, j$	= imaginary unit
$\log, \log_{10}$	= logarithm to the base 10, common logarithm
$\log_e, \ln$	= logarithm to the base e, natural logarithm, Napierian logarithm, hyperbolic logarithm
$\log^{-1} x$	= number whose logarithm is x
e	= base of natural logarithms $e = 2.7182818285$
π	= pi, $\pi = 3.1415926536$
$n!, \lfloor n$	= n factorial, $1 \cdot 2 \cdot 3 \cdot 4 \cdot \cdot \cdot \cdot \cdot (n - 1) \cdot n$
$P(n,r), nPr, P^n r$	= permutations of n different things taken r at a time
$C(n,r), nCr, \binom{n}{r}$	= combinations of n different things taken r at a time
D_r^n	= combinations with repetition of n things taken r at a time
$\angle$	= angle
$\llcorner$	= right angle
$\triangle$	= triangle
$\odot$	= circle
$\square$	= parallelogram
$\perp$	= perpendicular to
$\parallel$	= parallel to
rad	= radian, 57.29578 degrees, 2π rad = 360 degrees
°,deg	= degree
′,min,ft	= minute, feet
″,sec,in.	= second, inches
$u = f(x,y,z)$	= the variable u is a function of x, y, and z
dy	= differential of y
$dy/dx, y'$	= derivative of y taken with respect to x; $y = f(x)$
$d^2y/dx^2, y''$	= second derivative of y taken with respect to x
$d^ny/dx^n, y^{(n)}$	= nth derivative of y taken with respect to x
$\partial y/\partial x, f_x$	= partial derivative of y taken with respect to x; $y = f(x,z, \ldots)$
Δ	= delta, increment of
δ	= delta, variation of
D	= D operator, $Dy = dy/dx$
$\rightarrow$	= approaches as a limit

lim = limit
Σ = summation
$\sum_{i=a}^{b}$ = summation over i from a to b
$\prod_{b}^{a} A_i$ = product taken in order, $A_b \cdot A_{b-1} \cdot A_{b-2} \cdots A_a$
$\int$ = integral of
$\int_a^b$ = definite integral between limits a and b
$\iint$ = double integral
$\mathbf{a}$ = indicates vector quantity
$\mathbf{a} \cdot \mathbf{b}$ = scalar product
$\mathbf{a} \times \mathbf{b}$ = vector product
( ) = parentheses
[ ] = brackets
{ } = braces
∞ = infinity
$|a|$ = absolute value of a
$f(s), \mathcal{L}\{F(t)\}$ = Laplace transformation of $F(t)$
$fs(n), S\{F(x)\}$ = finite sine transform of $F(x)$
$fc(n), C\{F(x)\}$ = finite cosine transform of $F(x)$

1.2. MATHEMATICAL TABLES

Table 1.1. Binomial Coefficients

n	$\binom{n}{0}$	$\binom{n}{1}$	$\binom{n}{2}$	$\binom{n}{3}$	$\binom{n}{4}$	$\binom{n}{5}$	$\binom{n}{6}$	$\binom{n}{7}$	$\binom{n}{8}$	$\binom{n}{9}$	$\binom{n}{10}$
0	1										
1	1	1									
2	1	2	1								
3	1	3	3	1							
4	1	4	6	4	1						
5	1	5	10	10	5	1					
6	1	6	15	20	15	6	1				
7	1	7	21	35	35	21	7	1			
8	1	8	28	56	70	56	28	8	1		
9	1	9	36	84	126	126	84	36	9	1	
10	1	10	45	120	210	252	210	120	45	10	1
11	1	11	55	165	330	462	462	330	165	55	11
12	1	12	66	220	495	792	924	792	495	220	66
13	1	13	78	286	715	1287	1716	1716	1287	715	286
14	1	14	91	364	1001	2002	3003	3432	3003	2002	1001
15	1	15	105	455	1365	3003	5005	6435	6435	5005	3003
16	1	16	120	560	1820	4368	8008	11440	12870	11440	8008
17	1	17	136	680	2380	6188	12376	19448	24310	24310	19448
18	1	18	153	816	3060	8568	18564	31824	43758	48620	43758
19	1	19	171	969	3876	11628	27132	50388	75582	92378	92378
20	1	20	190	1140	4845	15504	38760	77520	125970	167960	184756

Table 1.2. Natural Logarithms*

These two pages give the natural or Napierian logarithms (ln) of numbers between 1 and 10, correct to four places. Moving the decimal point n places to the right [or left] in the number is equivalent to adding n times 2.3026 [or n times $\bar{3}.6974$] to the logarithm. Base e = 2.71828+

n	n (2.3026)	n (0.6974–3)
1	2.3026	0.6974–3
2	4.6052	0.3948–5
3	6.9078	0.0922–7
4	9.2103	0.7897–10
5	11.5129	0.4871–12
6	13.8155	0.1845–14
7	16.1181	0.8819–17
8	18.4207	0.5793–19
9	20.7233	0.2767–21

Number	0	1	2	3	4	5	6	7	8	9	Avg diff
1.0	0.0000	0100	0198	0296	0392	0488	0583	0677	0770	0862	95
1.1	0953	1044	1133	1222	1310	1398	1484	1570	1655	1740	87
1.2	1823	1906	1989	2070	2151	2231	2311	2390	2469	2546	80
1.3	2624	2700	2776	2852	2927	3001	3075	3148	3221	3293	74
1.4	3365	3436	3507	3577	3646	3716	3784	3853	3920	3988	69
1.5	0.4055	4121	4187	4253	4318	4383	4447	4511	4574	4637	65
1.6	4700	4762	4824	4886	4947	5008	5068	5128	5188	5247	61
1.7	5306	5365	5423	5481	5539	5596	5653	5710	5766	5822	57
1.8	5878	5933	5988	6043	6098	6152	6206	6259	6313	6366	54
1.9	6419	6471	6523	6575	6627	6678	6729	6780	6831	6881	51
2.0	0.6931	6981	7031	7080	7129	7178	7227	7275	7324	7372	49
2.1	7419	7467	7514	7561	7608	7655	7701	7747	7793	7839	47
2.2	7885	7930	7975	8020	8065	8109	8154	8198	8242	8286	44
2.3	8329	8372	8416	8459	8502	8544	8587	8629	8671	8713	43
2.4	8755	8796	8838	8879	8920	8961	9002	9042	9083	9123	41
2.5	0.9163	9203	9243	9282	9322	9361	9400	9439	9478	9517	39
2.6	9555	9594	9632	9670	9708	9746	9783	9821	9858	9895	38
2.7	0.9933	9969	*0006	*0043	*0080	*0116	*0152	*0188	*0225	*0260	36
2.8	1.0296	0332	0367	0403	0438	0473	0508	0543	0578	0613	35
2.9	0647	0682	0716	0750	0784	0818	0852	0886	0919	0953	34
3.0	1.0986	1019	1053	1086	1119	1151	1184	1217	1249	1282	33
3.1	1314	1346	1378	1410	1442	1474	1506	1537	1569	1600	32
3.2	1632	1663	1694	1725	1756	1787	1817	1848	1878	1909	31
3.3	1939	1969	2000	2030	2060	2090	2119	2149	2179	2208	30
3.4	2238	2267	2296	2326	2355	2384	2413	2442	2470	2499	29
3.5	1.2528	2556	2585	2613	2641	2669	2698	2726	2754	2782	28
3.6	2809	2837	2865	2892	2920	2947	2975	3002	3029	3056	27
3.7	3083	3110	3137	3164	3191	3218	3244	3271	3297	3324	27
3.8	3350	3376	3403	3429	3455	3481	3507	3533	3558	3584	26
3.9	3610	3635	3661	3686	3712	3737	3762	3788	3813	3838	25
4.0	1.3863	3888	3913	3938	3962	3987	4012	4036	4061	4085	25
4.1	4110	4134	4159	4183	4207	4231	4255	4279	4303	4327	24
4.2	4351	4375	4398	4422	4446	4469	4493	4516	4540	4563	23
4.3	4586	4609	4633	4656	4679	4702	4725	4748	4770	4793	23
4.4	4816	4839	4861	4884	4907	4929	4951	4974	4996	5019	22
4.5	1.5041	5063	5085	5107	5129	5151	5173	5195	5217	5239	22
4.6	5261	5282	5304	5326	5347	5369	5390	5412	5433	5454	21
4.7	5476	5497	5518	5539	5560	5581	5602	5623	5644	5665	21
4.8	5686	5707	5728	5748	5769	5790	5810	5831	5851	5872	20
4.9	5892	5913	5933	5953	5974	5994	6014	6034	6054	6074	20

$\ln x = (2.3026) \log_{10} x \qquad \log_{10} x = (0.4343) \ln x$

where 2.3026 = ln 10 and 0.4343 = $\log_{10} e$

* From Marks' "Mechanical Engineers' Handbook," Theodore Baumeister (ed.), 6th ed., McGraw-Hill Book Company, Inc., 1958.

Table 1.2. Natural Logarithms (Continued)

Number	0	1	2	3	4	5	6	7	8	9	Avg diff
5.0	1.6094	6114	6134	6154	6174	6194	6214	6233	6253	6273	20
5.1	6292	6312	6332	6351	6371	6390	6409	6429	6448	6467	19
5.2	6487	6506	6525	6544	6563	6582	6601	6620	6639	6658	19
5.3	6677	6696	6715	6734	6752	6771	6790	6808	6827	6845	18
5.4	6864	6882	6901	6919	6938	6956	6974	6993	7011	7029	18
5.5	1.7047	7066	7084	7102	7120	7138	7156	7174	7192	7210	18
5.6	7228	7246	7263	7281	7299	7317	7334	7352	7370	7387	18
5.7	7405	7422	7440	7457	7475	7492	7509	7527	7544	7561	17
5.8	7579	7596	7613	7630	7647	7664	7681	7699	7716	7733	17
5.9	7750	7766	7783	7800	7817	7834	7851	7867	7884	7901	17
6.0	1.7918	7934	7951	7967	7984	8001	8017	8034	8050	8066	16
6.1	8083	8099	8116	8132	8148	8165	8181	8197	8213	8229	16
6.2	8245	8262	8278	8294	8310	8326	8342	8358	8374	8390	16
6.3	8405	8421	8437	8453	8469	8485	8500	8516	8532	8547	16
6.4	8563	8579	8594	8610	8625	8641	8656	8672	8687	8703	15
6.5	1.8718	8733	8749	8764	8779	8795	8810	8825	8840	8856	15
6.6	8871	8886	8901	8916	8931	8946	8961	8976	8991	9006	15
6.7	9021	9036	9051	9066	9081	9095	9110	9125	9140	9155	15
6.8	9169	9184	9199	9213	9228	9242	9257	9272	9286	9301	15
6.9	9315	9330	9344	9359	9373	9387	9402	9416	9430	9445	14
7.0	1.9459	9473	9488	9502	9516	9530	9544	9559	9573	9587	14
7.1	9601	9615	9629	9643	9657	9671	9685	9699	9713	9727	14
7.2	9741	9755	9769	9782	9796	9810	9824	9838	9851	9865	14
7.3	1.9879	9892	9906	9920	9933	9947	9961	9974	9988	*0001	13
7.4	2.0015	0028	0042	0055	0069	0082	0096	0109	0122	0136	13
7.5	2.0149	0162	0176	0189	0202	0215	0229	0242	0255	0268	13
7.6	0281	0295	0308	0321	0334	0347	0360	0373	0386	0399	13
7.7	0412	0425	0438	0451	0464	0477	0490	0503	0516	0528	13
7.8	0541	0554	0567	0580	0592	0605	0618	0631	0643	0656	13
7.9	0669	0681	0694	0707	0719	0732	0744	0757	0769	0782	12
8.0	2.0794	0807	0819	0832	0844	0857	0869	0882	0894	0906	12
8.1	0919	0931	0943	0956	0968	0980	0992	1005	1017	1029	12
8.2	1041	1054	1066	1078	1090	1102	1114	1126	1138	1150	12
8.3	1163	1175	1187	1199	1211	1223	1235	1247	1258	1270	12
8.4	1282	1294	1306	1318	1330	1342	1353	1365	1377	1389	12
8.5	2.1401	1412	1424	1436	1448	1459	1471	1483	1494	1506	12
8.6	1518	1529	1541	1552	1564	1576	1587	1599	1610	1622	12
8.7	1633	1645	1656	1668	1679	1691	1702	1713	1725	1736	11
8.8	1748	1759	1770	1782	1793	1804	1815	1827	1838	1849	11
8.9	1861	1872	1883	1894	1905	1917	1928	1939	1950	1961	11
9.0	2.1972	1983	1994	2006	2017	2028	2039	2050	2061	2072	11
9.1	2083	2094	2105	2116	2127	2138	2148	2159	2170	2181	11
9.2	2192	2203	2214	2225	2235	2246	2257	2268	2279	2289	11
9.3	2300	2311	2322	2332	2343	2354	2364	2375	2386	2396	11
9.4	2407	2418	2428	2439	2450	2460	2471	2481	2492	2502	11
9.5	2.2513	2523	2534	2544	2555	2565	2576	2586	2597	2607	10
9.6	2618	2628	2638	2649	2659	2670	2680	2690	2701	2711	10
9.7	2721	2732	2742	2752	2762	2773	2783	2793	2803	2814	10
9.8	2824	2834	2844	2854	2865	2875	2885	2895	2905	2915	10
9.9	2925	2935	2946	2956	2966	2976	2986	2996	3006	3016	10
10.0	2.3026										

Moving the decimal point n places to the right [or left] in the number requires adding n times 2.3026 [or n times (0.6974–3)] in the body of the table. See auxiliary table of multiples on top of the preceding page.

Table 1.3. Trigonometric Functions (at intervals of 10′)†

Annex—10 in columns marked*.

Degrees	Radians	Sines Nat.	Sines Log *	Cosines Nat.	Cosines Log *	Tangents Nat.	Tangents Log *	Cotangents Nat.	Cotangents Log		
0° 00′	0.0000	.0000	∞	1.0000	0.0000	.0000	∞	∞	∞	1.5708	90° 00′
10	0.0029	.0029	7.4637	1.0000	.0000	.0029	7.4637	343.77	2.5363	1.5679	50
20	0.0058	.0058	.7648	1.0000	.0000	.0058	.7648	171.89	.2352	1.5650	40
30	0.0087	.0087	.9408	1.0000	.0000	.0087	.9409	114.59	.0591	1.5621	30
40	0.0116	.0116	8.0658	.9999	.0000	.0116	8.0658	85.940	1.9342	1.5592	20
50	0.0145	.0145	.1627	.9999	.0000	.0145	.1627	68.750	.8373	1.5563	10
1° 00′	0.0175	.0175	8.2419	.9998	9.9999	.0175	8.2419	57.290	1.7581	1.5533	89° 00′
10	0.0204	.0204	.3088	.9998	.9999	.0204	.3089	49.104	.6911	1.5504	50
20	0.0233	.0233	.3668	.9997	.9999	.0233	.3669	42.964	.6331	1.5475	40
30	0.0262	.0262	.4179	.9997	.9999	.0262	.4181	38.188	.5819	1.5446	30
40	0.0291	.0291	.4637	.9996	.9998	.0291	.4638	34.368	.5362	1.5417	20
50	0.0320	.0320	.5050	.9995	.9998	.0320	.5053	31.242	.4947	1.5388	10
2° 00′	0.0349	.0349	8.5428	.9994	9.9997	.0349	8.5431	28.636	1.4569	1.5359	88° 00′
10	0.0378	.0378	.5776	.9993	.9997	.0378	.5779	26.432	.4221	1.5330	50
20	0.0407	.0407	.6097	.9992	.9996	.0407	.6101	24.542	.3899	1.5301	40
30	0.0436	.0436	.6397	.9990	.9996	.0437	.6401	22.904	.3599	1.5272	30
40	0.0465	.0465	.6677	.9989	.9995	.0466	.6682	21.470	.3318	1.5243	20
50	0.0495	.0494	.6940	.9988	.9995	.0495	.6945	20.206	.3055	1.5213	10
3° 00′	0.0524	.0523	8.7188	.9986	9.9994	.0524	8.7194	19.081	1.2806	1.5184	87° 00′
10	0.0553	.0552	.7423	.9985	.9993	.0553	.7429	18.075	.2571	1.5155	50
20	0.0582	.0581	.7645	.9983	.9993	.0582	.7652	17.169	.2348	1.5126	40
30	0.0611	.0610	.7857	.9981	.9992	.0612	.7865	16.350	.2135	1.5097	30
40	0.0640	.0640	.8059	.9980	.9991	.0641	.8067	15.605	.1933	1.5068	20
50	0.0669	.0669	.8251	.9978	.9990	.0670	.8261	14.924	.1739	1.5039	10
4° 00′	0.0698	.0698	8.8436	.9976	9.9989	.0699	8.8446	14.301	1.1554	1.5010	86° 00′
10	0.0727	.0727	.8613	.9974	.9989	.0729	.8624	13.727	.1376	1.4981	50
20	0.0756	.0756	.8783	.9971	.9988	.0758	.8795	13.197	.1205	1.4952	40
30	0.0785	.0785	.8946	.9969	.9987	.0787	.8960	12.706	.1040	1.4923	30
40	0.0814	.0814	.9104	.9967	.9986	.0816	.9118	12.251	.0882	1.4893	20
50	0.0844	.0843	.9256	.9964	.9985	.0846	.9272	11.826	.0728	1.4864	10
5° 00′	0.0873	.0872	8.9403	.9962	9.9983	.0875	8.9420	11.430	1.0580	1.4835	85° 00′
10	0.0902	.0901	.9545	.9959	.9982	.0904	.9563	11.059	.0437	1.4806	50
20	0.0931	.0929	.9682	.9957	.9981	.0934	.9701	10.712	.0299	1.4777	40
30	0.0960	.0958	.9816	.9954	.9980	.0963	.9836	10.385	.0164	1.4748	30
40	0.0989	.0987	.9945	.9951	.9979	.0992	.9966	10.078	.0034	1.4719	20
50	0.1018	.1016	9.0070	.9948	.9977	.1022	9.0093	9.7882	0.9907	1.4690	10
6° 00′	0.1047	.1045	9.0192	.9945	9.9976	.1051	9.0216	9.5144	0.9784	1.4661	84° 00′
10	0.1076	.1074	.0311	.9942	.9975	.1080	.0336	9.2553	.9664	1.4632	50
20	0.1105	.1103	.0426	.9939	.9973	.1110	.0453	9.0098	.9547	1.4603	40
30	0.1134	.1132	.0539	.9936	.9972	.1139	.0567	8.7769	.9433	1.4574	30
40	0.1164	.1161	.0648	.9932	.9971	.1169	.0678	8.5555	.9322	1.4544	20
50	0.1193	.1190	.0755	.9929	.9969	.1198	.0786	8.3450	.9214	1.4515	10
7° 00′	0.1222	.1219	9.0859	.9925	9.9968	.1228	9.0891	8.1443	0.9109	1.4486	83° 00′
10	0.1251	.1248	.0961	.9922	.9966	.1257	.0995	7.9530	.9005	1.4457	50
20	0.1280	.1276	.1060	.9918	.9964	.1287	.1096	7.7704	.8904	1.4428	40
30	0.1309	.1305	.1157	.9914	.9963	.1317	.1194	7.5958	.8806	1.4399	30
40	0.1338	.1334	.1252	.9911	.9961	.1346	.1291	7.4287	.8709	1.4370	20
50	0.1367	.1363	.1345	.9907	.9959	.1376	.1385	7.2687	.8615	1.4341	10
8° 00′	0.1396	.1392	9.1436	.9903	9.9958	.1405	9.1478	7.1154	0.8522	1.4312	82° 00′
10	0.1425	.1421	.1525	.9899	.9956	.1435	.1569	6.9682	.8431	1.4283	50
20	0.1454	.1449	.1612	.9894	.9954	.1465	.1658	6.8269	.8342	1.4254	40
30	0.1484	.1478	.1697	.9890	.9952	.1495	.1745	6.6912	.8255	1.4224	30
40	0.1513	.1507	.1781	.9886	.9950	.1524	.1831	6.5606	.8169	1.4195	20
50	0.1542	.1536	.1863	.9881	.9948	.1554	.1915	6.4348	.8085	1.4166	10
9° 00′	0.1571	.1564	9.1943	.9877	9.9946	.1584	9.1997	6.3138	0.8003	1.4137	81° 00′
		Nat.	Log *	Nat.	Log *	Nat.	Log *	Nat.	Log		
		Cosines		Sines		Cotangents		Tangents		Radians	Degrees

† From Marks' "Mechanical Engineers' Handbook," Theodore Baumeister (ed.), 6th ed., McGraw-Hill Book Company, Inc., 1958.

Table 1.3. Trigonometric Functions (at intervals of 10′) (Continued)

Annex—10 in columns marked*.

Degrees	Radians	Sines Nat.	Sines Log *	Cosines Nat.	Cosines Log *	Tangents Nat.	Tangents Log *	Cotangents Nat.	Cotangents Log		
9° 00′	0.1571	.1564	9.1943	.9877	9.9946	.1584	9.1997	6.3138	0.8003	1.4137	81° 00′
10	0.1600	.1593	.2022	.9872	.9944	.1614	.2078	6.1970	.7922	1.4108	50
20	0.1629	.1622	.2100	.9868	.9942	.1644	.2158	6.0844	.7842	1.4079	40
30	0.1658	.1650	.2176	.9863	.9940	.1673	.2236	5.9758	.7764	1.4050	30
40	0.1687	.1679	.2251	.9858	.9938	.1703	.2313	5.8708	.7687	1.4021	20
50	0.1716	.1708	.2324	.9853	.9936	.1733	.2389	5.7694	.7611	1.3992	10
10° 00′	0.1745	.1736	9.2397	.9848	9.9934	.1763	9.2463	5.6713	0.7537	1.3963	80° 00′
10	0.1774	.1765	.2468	.9843	.9931	.1793	.2536	5.5764	.7464	1.3934	50
20	0.1804	.1794	.2538	.9838	.9929	.1823	.2609	5.4845	.7391	1.3904	40
30	0.1833	.1822	.2606	.9833	.9927	.1853	.2680	5.3955	.7320	1.3875	30
40	0.1862	.1851	.2674	.9827	.9924	.1883	.2750	5.3093	.7250	1.3846	20
50	0.1891	.1880	.2740	.9822	.9922	.1914	.2819	5.2257	.7181	1.3817	10
11° 00′	0.1920	.1908	9.2806	.9816	9.9919	.1944	9.2887	5.1446	0.7113	1.3788	79° 00′
10	0.1949	.1937	.2870	.9811	.9917	.1974	.2953	5.0658	.7047	1.3759	50
20	0.1978	.1965	.2934	.9805	.9914	.2004	.3020	4.9894	.6980	1.3730	40
30	0.2007	.1994	.2997	.9799	.9912	.2035	.3085	4.9152	.6915	1.3701	30
40	0.2036	.2022	.3058	.9793	.9909	.2065	.3149	4.8430	.6851	1.3672	20
50	0.2065	.2051	.3119	.9787	.9907	.2095	.3212	4.7729	.6788	1.3643	10
12° 00′	0.2094	.2079	9.3179	.9781	9.9904	.2126	9.3275	4.7046	0.6725	1.3614	78° 00′
10	0.2123	.2108	.3238	.9775	.9901	.2156	.3336	4.6382	.6664	1.3584	50
20	0.2153	.2136	.3296	.9769	.9899	.2186	.3397	4.5736	.6603	1.3555	40
30	0.2182	.2164	.3353	.9763	.9896	.2217	.3458	4.5107	.6542	1.3526	30
40	0.2211	.2193	.3410	.9757	.9893	.2247	.3517	4.4494	.6483	1.3497	20
50	0.2240	.2221	.3466	.9750	.9890	.2278	.3576	4.3897	.6424	1.3468	10
13° 00′	0.2269	.2250	9.3521	.9744	9.9887	.2309	9.3634	4.3315	0.6366	1.3439	77° 00′
10	0.2298	.2278	.3575	.9737	.9884	.2339	.3691	4.2747	.6309	1.3410	50
20	0.2327	.2306	.3629	.9730	.9881	.2370	.3748	4.2193	.6252	1.3381	40
30	0.2356	.2334	.3682	.9724	.9878	.2401	.3804	4.1653	.6196	1.3352	30
40	0.2385	.2363	.3734	.9717	.9875	.2432	.3859	4.1126	.6141	1.3323	20
50	0.2414	.2391	.3786	.9710	.9872	.2462	.3914	4.0611	.6086	1.3294	10
14° 00′	0.2443	.2419	9.3837	.9703	9.9869	.2493	9.3968	4.0108	0.6032	1.3265	76° 00′
10	0.2473	.2447	.3887	.9696	.9866	.2524	.4021	3.9617	.5979	1.3235	50
20	0.2502	.2476	.3937	.9689	.9863	.2555	.4074	3.9136	.5926	1.3206	40
30	0.2531	.2504	.3986	.9681	.9859	.2586	.4127	3.8667	.5873	1.3177	30
40	0.2560	.2532	.4035	.9674	.9856	.2617	.4178	3.8208	.5822	1.3148	20
50	0.2589	.2560	.4083	.9667	.9853	.2648	.4230	3.7760	.5770	1.3119	10
15° 00′	0.2618	.2588	9.4130	.9659	9.9849	.2679	9.4281	3.7321	0.5719	1.3090	75° 00′
10	0.2647	.2616	.4177	.9652	.9846	.2711	.4331	3.6891	.5669	1.3061	50
20	0.2676	.2644	.4223	.9644	.9843	.2742	.4381	3.6470	.5619	1.3032	40
30	0.2705	.2672	.4269	.9636	.9839	.2773	.4430	3.6059	.5570	1.3003	30
40	0.2734	.2700	.4314	.9628	.9836	.2805	.4479	3.5656	.5521	1.2974	20
50	0.2763	.2728	.4359	.9621	.9832	.2836	.4527	3.5261	.5473	1.2945	10
16° 00′	0.2793	.2756	9.4403	.9613	9.9828	.2867	9.4575	3.4874	0.5425	1.2915	74° 00′
10	0.2822	.2784	.4447	.9605	.9825	.2899	.4622	3.4495	.5378	1.2886	50
20	0.2851	.2812	.4491	.9596	.9821	.2931	.4669	3.4124	.5331	1.2857	40
30	0.2880	.2840	.4533	.9588	.9817	.2962	.4716	3.3759	.5284	1.2828	30
40	0.2909	.2868	.4576	.9580	.9814	.2994	.4762	3.3402	.5238	1.2799	20
50	0.2938	.2896	.4618	.9572	.9810	.3026	.4808	3.3052	.5192	1.2770	10
17° 00′	0.2967	.2924	9.4659	.9563	9.9806	.3057	9.4853	3.2709	0.5147	1.2741	73° 00′
10	0.2996	.2952	.4700	.9555	.9802	.3089	.4898	3.2371	.5102	1.2712	50
20	0.3025	.2979	.4741	.9546	.9798	.3121	.4943	3.2041	.5057	1.2683	40
30	0.3054	.3007	.4781	.9537	.9794	.3153	.4987	3.1716	.5013	1.2654	30
40	0.3083	.3035	.4821	.9528	.9790	.3185	.5031	3.1397	.4969	1.2625	20
50	0.3113	.3062	.4861	.9520	.9786	.3217	.5075	3.1084	.4925	1.2595	10
18° 00′	0.3142	.3090	9.4900	.9511	9.9782	.3249	9.5118	3.0777	0.4882	1.2566	72° 00′
		Nat.	Log *	Nat.	Log *	Nat.	Log *	Nat.	Log		
		Cosines		Sines		Cotangents		Tangents		Radians	Degrees

Table 1.3. Trigonometric Functions (at intervals of 10′) (Continued)

Annex—10 in columns marked*.

Degrees	Radians	Sines		Cosines		Tangents		Cotangents			
		Nat.	Log *	Nat.	Log *	Nat.	Log *	Nat.	Log		
18° 00′	0.3142	.3090	9.4900	.9511	9.9782	.3249	9.5118	3.0777	0.4882	1.2566	72° 00′
10	0.3171	.3118	.4939	.9502	.9778	.3281	.5161	3.0475	.4839	1.2537	50
20	0.3200	.3145	.4977	.9492	.9774	.3314	.5203	3.0178	.4797	1.2508	40
30	0.3229	.3173	.5015	.9483	.9770	.3346	.5245	2.9887	.4755	1.2479	30
40	0.3258	.3201	.5052	.9474	.9765	.3378	.5287	2.9600	.4713	1.2450	20
50	0.3287	.3228	.5090	.9465	.9761	.3411	.5329	2.9319	.4671	1.2421	10
19° 00′	0.3316	.3256	9.5126	.9455	9.9757	.3443	9.5370	2.9042	0.4630	1.2392	71° 00′
10	0.3345	.3283	.5163	.9446	.9752	.3476	.5411	2.8770	.4589	1.2363	50
20	0.3374	.3311	.5199	.9436	.9748	.3508	.5451	2.8502	.4549	1.2334	40
30	0.3403	.3338	.5235	.9426	.9743	.3541	.5491	2.8239	.4509	1.2305	30
40	0.3432	.3365	.5270	.9417	.9739	.3574	.5531	2.7980	.4469	1.2275	20
50	0.3462	.3393	.5306	.9407	.9734	.3607	.5571	2.7725	.4429	1.2246	10
20° 00′	0.3491	.3420	9.5341	.9397	9.9730	.3640	9.5611	2.7475	0.4389	1.2217	70° 00′
10	0.3520	.3448	.5375	.9387	.9725	.3673	.5650	2.7228	.4350	1.2188	50
20	0.3549	.3475	.5409	.9377	.9721	.3706	.5689	2.6985	.4311	1.2159	40
30	0.3578	.3502	.5443	.9367	.9716	.3739	.5727	2.6746	.4273	1.2130	30
40	0.3607	.3529	.5477	.9356	.9711	.3772	.5766	2.6511	.4234	1.2101	20
50	0.3636	.3557	.5510	.9346	.9706	.3805	.5804	2.6279	.4196	1.2072	10
21° 00′	0.3665	.3584	9.5543	.9336	9.9702	.3839	9.5842	2.6051	0.4158	1.2043	69° 00′
10	0.3694	.3611	.5576	.9325	.9697	.3872	.5879	2.5826	.4121	1.2014	50
20	0.3723	.3638	.5609	.9315	.9692	.3906	.5917	2.5605	.4083	1.1985	40
30	0.3752	.3665	.5641	.9304	.9687	.3939	.5954	2.5386	.4046	1.1956	30
40	0.3782	.3692	.5673	.9293	.9682	.3973	.5991	2.5172	.4009	1.1926	20
50	0.3811	.3719	.5704	.9283	.9677	.4006	.6028	2.4960	.3972	1.1897	10
22° 00′	0.3840	.3746	9.5736	.9272	9.9672	.4040	9.6064	2.4751	0.3936	1.1868	68° 00′
10	0.3869	.3773	.5767	.9261	.9667	.4074	.6100	2.4545	.3900	1.1839	50
20	0.3898	.3800	.5798	.9250	.9661	.4108	.6136	2.4342	.3864	1.1810	40
30	0.3927	.3827	.5828	.9239	.9656	.4142	.6172	2.4142	.3828	1.1781	30
40	0.3956	.3854	.5859	.9228	.9651	.4176	.6208	2.3945	.3792	1.1752	20
50	0.3985	.3881	.5889	.9216	.9646	.4210	.6243	2.3750	.3757	1.1723	10
23° 00′	0.4014	.3907	9.5919	.9205	9.9640	.4245	9.6279	2.3559	0.3721	1.1694	67° 00′
10	0.4043	.3934	.5948	.9194	.9635	.4279	.6314	2.3369	.3686	1.1665	50
20	0.4072	.3961	.5978	.9182	.9629	.4314	.6348	2.3183	.3652	1.1636	40
30	0.4102	.3987	.6007	.9171	.9624	.4348	.6383	2.2998	.3617	1.1606	30
40	0.4131	.4014	.6036	.9159	.9618	.4383	.6417	2.2817	.3583	1.1577	20
50	0.4160	.4041	.6065	.9147	.9613	.4417	.6452	2.2637	.3548	1.1548	10
24° 00′	0.4189	.4067	9.6093	.9135	9.9607	.4452	9.6486	2.2460	0.3514	1.1519	66° 00′
10	0.4218	.4094	.6121	.9124	.9602	.4487	.6520	2.2286	.3480	1.1490	50
20	0.4247	.4120	.6149	.9112	.9596	.4522	.6553	2.2113	.3447	1.1461	40
30	0.4276	.4147	.6177	.9100	.9590	.4557	.6587	2.1943	.3413	1.1432	30
40	0.4305	.4173	.6205	.9088	.9584	.4592	.6620	2.1775	.3380	1.1403	20
50	0.4334	.4200	.6232	.9075	.9579	.4628	.6654	2.1609	.3346	1.1374	10
25° 00′	0.4363	.4226	9.6259	.9063	9.9573	.4663	9.6687	2.1445	0.3313	1.1345	65° 00′
10	0.4392	.4253	.6286	.9051	.9567	.4699	.6720	2.1283	.3280	1.1316	50
20	0.4422	.4279	.6313	.9038	.9561	.4734	.6752	2.1123	.3248	1.1286	40
30	0.4451	.4305	.6340	.9026	.9555	.4770	.6785	2.0965	.3215	1.1257	30
40	0.4480	.4331	.6366	.9013	.9549	.4806	.6817	2.0809	.3183	1.1228	20
50	0.4509	.4358	.6392	.9001	.9543	.4841	.6850	2.0655	.3150	1.1199	10
26° 00′	0.4538	.4384	9.6418	.8988	9.9537	.4877	9.6882	2.0503	0.3118	1.1170	64° 00′
10	0.4567	.4410	.6444	.8975	.9530	.4913	.6914	2.0353	.3086	1.1141	50
20	0.4596	.4436	.6470	.8962	.9524	.4950	.6946	2.0204	.3054	1.1112	40
30	0.4625	.4462	.6495	.8949	.9518	.4986	.6977	2.0057	.3023	1.1083	30
40	0.4654	.4488	.6521	.8936	.9512	.5022	.7009	1.9912	.2991	1.1054	20
50	0.4683	.4514	.6546	.8923	.9505	.5059	.7040	1.9768	.2960	1.1025	10
27° 00′	0.4712	.4540	9.6570	.8910	9.9499	.5095	9.7072	1.9626	0.2928	1.0996	63° 00′
		Nat.	Log *	Nat.	Log *	Nat.	Log *	Nat.	Log		
		Cosines		Sines		Cotangents		Tangents		Radians	Degrees

Table 1.3. Trigonometric Functions (at intervals of 10′) (Continued)

Annex—10 in columns marked*.

Degrees	Radians	Sines Nat.	Sines Log *	Cosines Nat.	Cosines Log *	Tangents Nat.	Tangents Log *	Cotangents Nat.	Cotangents Log		
27° 00′	0.4712	.4540	9.6570	.8910	9.9499	.5095	9.7072	1.9626	0.2928	1.0996	63° 00′
10	0.4741	.4566	.6595	.8897	.9492	.5132	.7103	1.9486	.2897	1.0966	50
20	0.4771	.4592	.6620	.8884	.9486	.5169	.7134	1.9347	.2866	1.0937	40
30	0.4800	.4617	.6644	.8870	.9479	.5206	.7165	1.9210	.2835	1.0908	30
40	0.4829	.4643	.6668	.8857	.9473	.5243	.7196	1.9074	.2804	1.0879	20
50	0.4858	.4669	.6692	.8843	.9466	.5280	.7226	1.8940	.2774	1.0850	10
28° 00′	0.4887	.4695	9.6716	.8829	9.9459	.5317	9.7257	1.8807	0.2743	1.0821	62° 00′
10	0.4916	.4720	.6740	.8816	.9453	.5354	.7287	1.8676	.2713	1.0792	50
20	0.4945	.4746	.6763	.8802	.9446	.5392	.7317	1.8546	.2683	1.0763	40
30	0.4974	.4772	.6787	.8788	.9439	.5430	.7348	1.8418	.2652	1.0734	30
40	0.5003	.4797	.6810	.8774	.9432	.5467	.7378	1.8291	.2622	1.0705	20
50	0.5032	.4823	.6833	.8760	.9425	.5505	.7408	1.8165	.2592	1.0676	10
29° 00′	0.5061	.4848	9.6856	.8746	9.9418	.5543	9.7438	1.8040	0.2562	1.0647	61° 00′
10	0.5091	.4874	.6878	.8732	.9411	.5581	.7467	1.7917	.2533	1.0617	50
20	0.5120	.4899	.6901	.8718	.9404	.5619	.7497	1.7796	.2503	1.0588	40
30	0.5149	.4924	.6923	.8704	.9397	.5658	.7526	1.7675	.2474	1.0559	30
40	0.5178	.4950	.6946	.8689	.9390	.5696	.7556	1.7556	.2444	1.0530	20
50	0.5207	.4975	.6968	.8675	.9383	.5735	.7585	1.7437	.2415	1.0501	10
30° 00′	0.5236	.5000	9.6990	.8660	9.9375	.5774	9.7614	1.7321	0.2386	1.0472	60° 00′
10	0.5265	.5025	.7012	.8646	.9368	.5812	.7644	1.7205	.2356	1.0443	50
20	0.5294	.5050	.7033	.8631	.9361	.5851	.7673	1.7090	.2327	1.0414	40
30	0.5323	.5075	.7055	.8616	.9353	.5890	.7701	1.6977	.2299	1.0385	30
40	0.5352	.5100	.7076	.8601	.9346	.5930	.7730	1.6864	.2270	1.0356	20
50	0.5381	.5125	.7097	.8587	.9338	.5969	.7759	1.6753	.2241	1.0327	10
31° 00′	0.5411	.5150	9.7118	.8572	9.9331	.6009	9.7788	1.6643	0.2212	1.0297	59° 00′
10	0.5440	.5175	.7139	.8557	.9323	.6048	.7816	1.6534	.2184	1.0268	50
20	0.5469	.5200	.7160	.8542	.9315	.6088	.7845	1.6426	.2155	1.0239	40
30	0.5498	.5225	.7181	.8526	.9308	.6128	.7873	1.6319	.2127	1.0210	30
40	0.5527	.5250	.7201	.8511	.9300	.6168	.7902	1.6212	.2098	1.0181	20
50	0.5556	.5275	.7222	.8496	.9292	.6208	.7930	1.6107	.2070	1.0152	10
32° 00′	0.5585	.5299	9.7242	.8480	9.9284	.6249	9.7958	1.6003	0.2042	1.0123	58° 00′
10	0.5614	.5324	.7262	.8465	.9276	.6289	.7986	1.5900	.2014	1.0094	50
20	0.5643	.5348	.7282	.8450	.9268	.6330	.8014	1.5798	.1986	1.0065	40
30	0.5672	.5373	.7302	.8434	.9260	.6371	.8042	1.5697	.1958	1.0036	30
40	0.5701	.5398	.7322	.8418	.9252	.6412	.8070	1.5597	.1930	1.0007	20
50	0.5730	.5422	.7342	.8403	.9244	.6453	.8097	1.5497	.1903	0.9977	10
33° 00′	0.5760	.5446	9.7361	.8387	9.9236	.6494	9.8125	1.5399	0.1875	0.9948	57° 00′
10	0.5789	.5471	.7380	.8371	.9228	.6536	.8153	1.5301	.1847	0.9919	50
20	0.5818	.5495	.7400	.8355	.9219	.6577	.8180	1.5204	.1820	0.9890	40
30	0.5847	.5519	.7419	.8339	.9211	.6619	.8208	1.5108	.1792	0.9861	30
40	0.5876	.5544	.7438	.8323	.9203	.6661	.8235	1.5013	.1765	0.9832	20
50	0.5905	.5568	.7457	.8307	.9194	.6703	.8263	1.4919	.1737	0.9803	10
34° 00′	0.5934	.5592	9.7476	.8290	9.9186	.6745	9.8290	1.4826	0.1710	0.9774	56° 00′
10	0.5963	.5616	.7494	.8274	.9177	.6787	.8317	1.4733	.1683	0.9745	50
20	0.5992	.5640	.7513	.8258	.9169	.6830	.8344	1.4641	.1656	0.9716	40
30	0.6021	.5664	.7531	.8241	.9160	.6873	.8371	1.4550	.1629	0.9687	30
40	0.6050	.5688	.7550	.8225	.9151	.6916	.8398	1.4460	.1602	0.9657	20
50	0.6080	.5712	.7568	.8208	.9142	.6959	.8425	1.4370	.1575	0.9628	10
35° 00′	0.6109	.5736	9.7586	.8192	9.9134	.7002	9.8452	1.4281	0.1548	0.9599	55° 00′
10	0.6138	.5760	.7604	.8175	.9125	.7046	.8479	1.4193	.1521	0.9570	50
20	0.6167	.5783	.7622	.8158	.9116	.7089	.8506	1.4106	.1494	0.9541	40
30	0.6196	.5807	.7640	.8141	.9107	.7133	.8533	1.4019	.1467	0.9512	30
40	0.6225	.5831	.7657	.8124	.9098	.7177	.8559	1.3934	.1441	0.9483	20
50	0.6254	.5854	.7675	.8107	.9089	.7221	.8586	1.3848	.1414	0.9454	10
36° 00′	0.6283	.5878	9.7692	.8090	9.9080	.7265	9.8613	1.3764	0.1387	0.9425	54° 00′
		Nat.	Log *	Nat.	Log *	Nat.	Log *	Nat.	Log		
		Cosines		Sines		Cotangents		Tangents		Radians	Degrees

Table 1.3. Trigonometric Functions (at intervals of 10′) (Continued)

Annex—10 in columns marked*.

Degrees	Radians	Sines Nat.	Sines Log *	Cosines Nat.	Cosines Log *	Tangents Nat.	Tangents Log *	Cotangents Nat.	Cotangents Log		
36° 00′	0.6283	.5878	9.7692	.8090	9.9080	.7265	9.8613	1.3764	0.1387	0.9425	54° 00′
10	0.6312	.5901	.7710	.8073	.9070	.7310	.8639	1.3680	.1361	0.9396	50
20	0.6341	.5925	.7727	.8056	.9061	.7355	.8666	1.3597	.1334	0.9367	40
30	0.6370	.5948	.7744	.8039	.9052	.7400	.8692	1.3514	.1308	0.9338	30
40	0.6400	.5972	.7761	.8021	.9042	.7445	.8718	1.3432	.1282	0.9308	20
50	0.6429	.5995	.7778	.8004	.9035	.7490	.8745	1.3351	.1255	0.9279	10
37° 00′	0.6458	.6018	9.7795	.7986	9.9023	.7536	9.8771	1.3270	0.1229	0.9250	53° 00′
10	0.6487	.6041	.7811	.7969	.9014	.7581	.8797	1.3190	.1203	0.9221	50
20	0.6516	.6065	.7828	.7951	.9004	.7627	.8824	1.3111	.1176	0.9192	40
30	0.6545	.6088	.7844	.7934	.8995	.7673	.8850	1.3032	.1150	0.9163	30
40	0.6574	.6111	.7861	.7916	.8985	.7720	.8876	1.2954	.1124	0.9134	20
50	0.6603	.6134	.7877	.7898	.8975	.7766	.8902	1.2876	.1098	0.9105	10
38° 00′	0.6632	.6157	9.7893	.7880	9.8965	.7813	9.8928	1.2799	0.1072	0.9076	52° 00′
10	0.6661	.6180	.7910	.7862	.8955	.7860	.8954	1.2723	.1046	0.9047	50
20	0.6690	.6202	.7926	.7844	.8945	.7907	.8980	1.2647	.1020	0.9018	40
30	0.6720	.6225	.7941	.7826	.8935	.7954	.9006	1.2572	.0994	0.8988	30
40	0.6749	.6248	.7957	.7808	.8925	.8002	.9032	1.2497	.0968	0.8959	20
50	0.6778	.6271	.7973	.7790	.8915	.8050	.9058	1.2423	.0942	0.8930	10
39° 00′	0.6807	.6293	9.7989	.7771	9.8905	.8098	9.9084	1.2349	0.0916	0.8901	51° 00′
10	0.6836	.6316	.8004	.7753	.8895	.8146	.9110	1.2276	.0890	0.8872	50
20	0.6865	.6338	.8020	.7735	.8884	.8195	.9135	1.2203	.0865	0.8843	40
30	0.6894	.6361	.8035	.7716	.8874	.8243	.9161	1.2131	.0839	0.8814	30
40	0.6923	.6383	.8050	.7698	.8864	.8292	.9187	1.2059	.0813	0.8785	20
50	0.6952	.6406	.8066	.7679	.8853	.8342	.9212	1.1988	.0788	0.8756	10
40° 00′	0.6981	.6428	9.8081	.7660	9.8843	.8391	9.9238	1.1918	0.0762	0.8727	50° 00′
10	0.7010	.6450	.8096	.7642	.8832	.8441	.9264	1.1847	.0736	0.8698	50
20	0.7039	.6472	.8111	.7623	.8821	.8491	.9289	1.1778	.0711	0.8668	40
30	0.7069	.6494	.8125	.7604	.8810	.8541	.9315	1.1708	.0685	0.8639	30
40	0.7098	.6517	.8140	.7585	.8800	.8591	.9341	1.1640	.0659	0.8610	20
50	0.7127	.6539	.8155	.7566	.8789	.8642	.9366	1.1571	.0634	0.8581	10
41° 00′	0.7156	.6561	9.8169	.7547	9.8778	.8693	9.9392	1.1504	0.0608	0.8552	49° 00′
10	0.7185	.6583	.8184	.7528	.8767	.8744	.9417	1.1436	.0583	0.8523	50
20	0.7214	.6604	.8198	.7509	.8756	.8796	.9443	1.1369	.0557	0.8494	40
30	0.7243	.6626	.8213	.7490	.8745	.8847	.9468	1.1303	.0532	0.8465	30
40	0.7272	.6648	.8227	.7470	.8733	.8899	.9494	1.1237	.0506	0.8436	20
50	0.7301	.6670	.8241	.7451	.8722	.8952	.9519	1.1171	.0481	0.8407	10
42° 00′	0.7330	.6691	9.8255	.7431	9.8711	.9004	9.9544	1.1106	0.0456	0.8378	48° 00′
10	0.7359	.6713	.8269	.7412	.8699	.9057	.9570	1.1041	.0430	0.8348	50
20	0.7389	.6734	.8283	.7392	.8688	.9110	.9595	1.0977	.0405	0.8319	40
30	0.7418	.6756	.8297	.7373	.8676	.9163	.9621	1.0913	.0379	0.8290	30
40	0.7447	.6777	.8311	.7353	.8665	.9217	.9646	1.0850	.0354	0.8261	20
50	0.7476	.6799	.8324	.7333	.8653	.9271	.9671	1.0786	.0329	0.8232	10
43° 00′	0.7505	.6820	9.8338	.7314	9.8641	.9325	9.9697	1.0724	0.0303	0.8203	47° 00′
10	0.7534	.6841	.8351	.7294	.8629	.9380	.9722	1.0661	.0278	0.8174	50
20	0.7563	.6862	.8365	.7274	.8618	.9435	.9747	1.0599	.0253	0.8145	40
30	0.7592	.6884	.8378	.7254	.8606	.9490	.9772	1.0538	.0228	0.8116	30
40	0.7621	.6905	.8391	.7234	.8594	.9545	.9798	1.0477	.0202	0.8087	20
50	0.7650	.6926	.8405	.7214	.8582	.9601	.9823	1.0416	.0177	0.8058	10
44° 00′	0.7679	.6947	9.8418	.7193	9.8569	.9657	9.9848	1.0355	0.0152	0.8029	46° 00′
10	0.7709	.6967	.8431	.7173	.8557	.9713	.9874	1.0295	.0126	0.7999	50
20	0.7738	.6988	.8444	.7153	.8545	.9770	.9899	1.0235	.0101	0.7970	40
30	0.7767	.7009	.8457	.7133	.8532	.9827	.9924	1.0176	.0076	0.7941	30
40	0.7796	.7030	.8469	.7112	.8520	.9884	.9949	1.0117	.0051	0.7912	20
50	0.7825	.7050	.8482	.7092	.8507	.9942	.9975	1.0058	.0025	0.7883	10
45° 00′	0.7854	.7071	9.8495	.7071	9.8495	1.0000	0.0000	1.0000	0.0000	0.7854	45° 00′
		Cosines Nat.	Cosines Log *	Sines Nat.	Sines Log *	Cotangents Nat.	Cotangents Log *	Tangents Nat.	Tangents Log	Radians	Degrees

Table 1.4. Exponential and Hyperbolic Functions*

x	e^x	e^{-x}	$\sinh x$	$\cosh x$	$\tanh x$	x	e^x	e^{-x}	$\sinh x$	$\cosh x$	$\tanh x$
0.00	**1.0000**	**1.0000**	**0.0000**	**1.0000**	**.00000**	**0.60**	**1.8221**	**.54881**	**0.6367**	**1.1855**	**.53705**
0.01	1.0101	.99005	0.0100	1.0001	.01000	0.61	1.8404	.54335	0.6485	1.1919	.54413
0.02	1.0202	.98020	0.0200	1.0002	.02000	0.62	1.8589	.53794	0.6605	1.1984	.55113
0.03	1.0305	.97045	0.0300	1.0005	.02999	0.63	1.8776	.53259	0.6725	1.2051	.55805
0.04	1.0408	.96079	0.0400	1.0008	.03998	0.64	1.8965	.52729	0.6846	1.2119	.56496
0.05	1.0513	.95123	0.0500	1.0013	.04996	0.65	1.9155	.52205	0.6967	1.2188	.57167
0.06	1.0618	.94176	0.0600	1.0018	.05993	0.66	1.9348	.51685	0.7090	1.2258	.57836
0.07	1.0725	.93239	0.0701	1.0025	.06989	0.67	1.9542	.51171	0.7213	1.2330	.58498
0.08	1.0833	.92312	0.0801	1.0032	.07983	0.68	1.9739	.50662	0.7336	1.2402	.59152
0.09	1.0942	.91393	0.0901	1.0041	.08976	0.69	1.9937	.50158	0.7461	1.2476	.59798
0.10	**1.1052**	**.90484**	**0.1002**	**1.0050**	**.09967**	**0.70**	**2.0138**	**.49659**	**0.7586**	**1.2552**	**.60437**
0.11	1.1163	.89583	0.1102	1.0061	.10956	0.71	2.0340	.49164	0.7712	1.2628	.61068
0.12	1.1275	.88692	0.1203	1.0072	.11943	0.72	2.0544	.48675	0.7838	1.2706	.61691
0.13	1.1388	.87810	0.1304	1.0085	.12927	0.73	2.0751	.48191	0.7966	1.2785	.62307
0.14	1.1503	.86936	0.1405	1.0098	.13909	0.74	2.0959	.47711	0.8094	1.2865	.62915
0.15	1.1618	.86071	0.1506	1.0113	.14889	0.75	2.1170	.47237	0.8223	1.2947	.63515
0.16	1.1735	.85214	0.1607	1.0128	.15865	0.76	2.1383	.46767	0.8353	1.3030	.64108
0.17	1.1853	.84366	0.1708	1.0145	.16838	0.77	2.1598	.46301	0.8484	1.3114	.64693
0.18	1.1972	.83527	0.1810	1.0162	.17808	0.78	2.1815	.45841	0.8615	1.3199	.65271
0.19	1.2092	.82696	0.1911	1.0181	.18775	0.79	2.2034	.45384	0.8748	1.3286	.65841
0.20	**1.2214**	**.81873**	**0.2013**	**1.0201**	**.19738**	**0.80**	**2.2255**	**.44933**	**0.8881**	**1.3374**	**.66404**
0.21	1.2337	.81058	0.2115	1.0221	.20697	0.81	2.2479	.44486	0.9015	1.3464	.66959
0.22	1.2461	.80252	0.2218	1.0243	.21652	0.82	2.2705	.44043	0.9150	1.3555	.67507
0.23	1.2586	.79453	0.2320	1.0266	.22603	0.83	2.2933	.43605	0.9286	1.3647	.68048
0.24	1.2712	.78663	0.2423	1.0289	.23550	0.84	2.3164	.43171	0.9423	1.3740	.68581
0.25	1.2840	.77880	0.2526	1.0314	.24492	0.85	2.3396	.42741	0.9561	1.3835	.69107
0.26	1.2969	.77105	0.2629	1.0340	.25430	0.86	2.3632	.42316	0.9700	1.3932	.69626
0.27	1.3100	.76338	0.2733	1.0367	.26362	0.87	2.3869	.41895	0.9840	1.4029	.70137
0.28	1.3231	.75578	0.2837	1.0395	.27291	0.88	2.4109	.41478	0.9981	1.4128	.70642
0.29	1.3364	.74826	0.2941	1.0423	.28213	0.89	2.4351	.41066	1.0122	1.4229	.71139
0.30	**1.3499**	**.74082**	**0.3045**	**1.0453**	**.29131**	**0.90**	**2.4596**	**.40657**	**1.0265**	**1.4331**	**.71630**
0.31	1.3634	.73345	0.3150	1.0484	.30044	0.91	2.4843	.40252	1.0409	1.4434	.72113
0.32	1.3771	.72615	0.3255	1.0516	.30951	0.92	2.5093	.39852	1.0554	1.4539	.72590
0.33	1.3910	.71892	0.3360	1.0549	.31852	0.93	2.5345	.39455	1.0700	1.4645	.73059
0.34	1.4049	.71177	0.3466	1.0584	.32748	0.94	2.5600	.39063	1.0847	1.4753	.73522
0.35	1.4191	.70469	0.3572	1.0619	.33638	0.95	2.5857	.38674	1.0995	1.4862	.73978
0.36	1.4333	.69768	0.3678	1.0655	.34521	0.96	2.6117	.38289	1.1144	1.4973	.74428
0.37	1.4477	.69073	0.3785	1.0692	.35399	0.97	2.6379	.37908	1.1294	1.5085	.74870
0.38	1.4623	.68386	0.3892	1.0731	.36271	0.98	2.6645	.37531	1.1446	1.5199	.75307
0.39	1.4770	.67706	0.4000	1.0770	.37136	0.99	2.6912	.37158	1.1598	1.5314	.75736
0.40	**1.4918**	**.67032**	**0.1408**	**1.0811**	**.37995**	**1.00**	**2.7183**	**.36788**	**1.1752**	**1.5431**	**.76159**
0.41	1.5068	.66365	0.4216	1.0852	.38847	1.01	2.7456	.36422	1.1907	1.5549	.76576
0.42	1.5220	.65705	0.4325	1.0895	.39693	1.02	2.7732	.36059	1.2063	1.5669	.76987
0.43	1.5373	.65051	0.4434	1.0939	.40532	1.03	2.8011	.35701	1.2220	1.5790	.77391
0.44	1.5527	.64404	0.4543	1.0984	.41364	1.04	2.8292	.35345	1.2379	1.5913	.77789
0.45	1.5683	.63763	0.4653	1.1030	.42190	1.05	2.8577	.34994	1.2539	1.6038	.78181
0.46	1.5841	.63128	0.4764	1.1077	.43008	1.06	2.8864	.34646	1.2700	1.6164	.78566
0.47	1.6000	.62500	0.4875	1.1125	.43820	1.07	2.9154	.34301	1.2862	1.6292	.78946
0.48	1.6161	.61878	0.4986	1.1174	.44624	1.08	2.9447	.33960	1.3025	1.6421	.79320
0.49	1.6323	.61263	0.5098	1.1225	.45422	1.09	2.9743	.33622	1.3190	1.6552	.79688
0.50	**1.6487**	**.60653**	**0.5211**	**1.1276**	**.46212**	**1.10**	**3.0042**	**.33287**	**1.3356**	**1.6685**	**.80050**
0.51	1.6653	.60050	0.5324	1.1329	.46995	1.11	3.0344	.32956	1.3524	1.6820	.80406
0.52	1.6820	.59452	0.5438	1.1383	.47770	1.12	3.0649	.32628	1.3693	1.6956	80757
0.53	1.6989	.58860	0.5552	1.1438	.48538	1.13	3.0957	.32303	1.3863	1.7093	.81102
0.54	1.7160	.58275	0.5666	1.1494	.49299	1.14	3.1268	.31982	1.4035	1.7233	.81441
0.55	1.7333	.57695	0.5782	1.1551	.50052	1.15	3.1582	.31664	1.4208	1.7374	.81775
0.56	1.7507	.57121	0.5897	1.1609	.50798	1.16	3.1899	.31349	1.4382	1.7517	.82104
0.57	1.7683	.56553	0.6014	1.1669	.51536	1.17	3.2220	.31037	1.4558	1.7662	.82427
0.58	1.7860	.55990	0.6131	1.1730	.52267	1.18	3.2544	.30728	1.4735	1.7808	.82745
0.59	1.8040	.55433	0.6248	1.1792	.52990	1.19	3.2871	.30422	1.4914	1.7957	.83058
0.60	**1.8221**	**.54881**	**0.6367**	**1.1855**	**.53705**	**1.20**	**3.3201**	**.30119**	**1.5095**	**1.8107**	**.83365**

* From R. H. Perry, "Engineering Manual," McGraw-Hill Book Company, Inc., New York, 1959.

Table 1.4. Exponential and Hyperbolic Functions (Continued)

x	e^x	e^{-x}	sinh x	cosh x	tanh x	x	e^x	e^{-x}	sinh x	cosh x	tanh x
1.20	**3.3201**	**.30119**	**1.5095**	**1.8107**	**.83365**	**1.80**	**6.0496**	**.16530**	**2.9422**	**3.1075**	**.94681**
1.21	3.3535	.29820	1.5276	1.8258	.83668	1.81	6.1104	.16365	2.9734	3.1371	.94783
1.22	3.3872	.29523	1.5460	1.8412	.83965	1.82	6.1719	.16203	3.0049	3.1669	.94884
1.23	3.4212	.29229	1.5645	1.8568	.84258	1.83	6.2339	.16041	3.0367	3.1972	.94983
1.24	3.4556	.28938	1.5831	1.8725	.84546	1.84	6.2965	.15882	3.0689	3.2277	.95080
1.25	3.4903	.28650	1.6019	1.8884	.84828	1.85	6.3598	.15724	3.1013	3.2585	.95175
1.26	3.5254	.28365	1.6209	1.9045	.85106	1.86	6.4237	.15567	3.1340	3.2897	.95268
1.27	3.5609	.28083	1.6400	1.9208	.85380	1.87	6.4883	.15412	3.1671	3.3212	.95359
1.28	3.5966	.27804	1.6593	1.9373	.85648	1.88	6.5535	.15259	3.2005	3.3530	.95449
1.29	3.6328	.27527	1.6788	1.9540	.85913	1.89	6.6194	.15107	3.2341	3.3852	.95537
1.30	**3.6693**	**.27253**	**1.6984**	**1.9709**	**.86172**	**1.90**	**6.6859**	**.14957**	**3.2682**	**3.4177**	**.95624**
1.31	3.7062	.26982	1.7182	1.9880	.86428	1.91	6.7531	.14808	3.3025	3.4506	.95709
1.32	3.7434	.26714	1.7381	2.0053	.86678	1.92	6.8210	.14661	3.3372	3.4838	.95792
1.33	3.7810	.26448	1.7583	2.0228	.86925	1.93	6.8895	.14515	3.3722	3.5173	.95873
1.34	3.8190	.26185	1.7786	2.0404	.87167	1.94	6.9588	.14370	3.4075	3.5512	.95953
1.35	3.8574	.25924	1.7991	2.0583	.87405	1.95	7.0287	.14227	3.4432	3.5855	.96032
1.36	3.8962	.25666	1.8198	2.0764	.87639	1.96	7.0993	.14086	3.4792	3.6201	.96109
1.37	3.9354	.25411	1.8406	2.0947	.87869	1.97	7.1707	.13946	3.5156	3.6551	.96185
1.38	3.9749	.25158	1.8617	2.1132	.88095	1.98	7.2427	.13807	3.5523	3.6904	.96259
1.39	4.0149	.24908	1.8829	2.1320	.88317	1.99	7.3155	.13670	3.5894	3.7261	.96331
1.40	**4.0552**	**.24660**	**1.9043**	**2.1509**	**.88535**	**2.00**	**7.3891**	**.13534**	**3.6269**	**3.7622**	**.96403**
1.41	4.0960	.24414	1.9259	2.1700	.88749	2.01	7.4633	.13399	3.6647	3.7987	.96473
1.42	4.1371	.24171	1.9477	2.1894	.88960	2.02	7.5383	.13266	3.7028	3.8355	.96541
1.43	4.1787	.23931	1.9697	2.2090	.89167	2.03	7.6141	.13134	3.7414	3.8727	.96609
1.44	4.2207	.23693	1.9919	2.2288	.89370	2.04	7.6906	.13003	3.7803	3.9103	.96675
1.45	4.2631	.23457	2.0143	2.2488	.89569	2.05	7.7679	.12873	3.8196	3.9483	.96740
1.46	4.3060	.23224	2.0369	2.2691	.89765	2.06	7.8460	.12745	3.8593	3.9867	.96803
1.47	4.3492	.22993	2.0597	2.2896	.89958	2.07	7.9248	.12619	3.8993	4.0255	.96865
1.48	4.3929	.22764	2.0827	2.3103	.90147	2.08	8.0045	.12493	3.9398	4.0647	.96926
1.49	4.4371	.22537	2.1059	2.3312	.90332	2.09	8.0849	.12369	3.9806	4.1043	.96986
1.50	**4.4817**	**.22313**	**2.1293**	**2.3524**	**.90515**	**2.10**	**8.1662**	**.12246**	**4.0219**	**4.1443**	**.97045**
1.51	4.5267	.22091	2.1529	2.3738	.90694	2.11	8.2482	.12124	4.0635	4.1847	.97103
1.52	4.5722	.21871	2.1768	2.3955	.90870	2.12	8.3311	.12003	4.1056	4.2256	.97159
1.53	4.6182	.21654	2.2008	2.4174	.91042	2.13	8.4149	.11884	4.1480	4.2669	.97215
1.54	4.6646	.21438	2.2251	2.4395	.91212	2.14	8.4994	.11765	4.1909	4.3085	.97269
1.55	4.7115	.21225	2.2496	2.4619	.91379	2.15	8.5849	.11648	4.2342	4.3507	.97323
1.56	4.7588	.21014	2.2743	2.4845	.91542	2.16	8.6711	.11533	4.2779	4.3932	.97375
1.57	4.8066	.20805	2.2993	2.5073	.91703	2.17	8.7583	.11418	4.3221	4.4362	.97426
1.58	4.8550	.20598	2.3245	2.5305	.91860	2.18	8.8463	.11304	4.3666	4.4797	.97477
1.59	4.9037	.20393	2.3499	2.5538	.92015	2.19	8.9352	.11192	4.4116	4.5236	.97526
1.60	**4.9530**	**.20190**	**2.3756**	**2.5775**	**.92167**	**2.20**	**9.0250**	**.11080**	**4.4571**	**4.5679**	**.97574**
1.61	5.0028	.19989	2.4015	2.6013	.92316	2.21	9.1157	.10970	4.5030	4.6127	.97622
1.62	5.0531	.19790	2.4276	2.6255	.92462	2.22	9.2073	.10861	4.5494	4.6580	.97668
1.63	5.1039	.19593	2.4540	2.6499	.92606	2.23	9.2999	.10753	4.5962	4.7037	.97714
1.64	5.1552	.19398	2.4806	2.6746	.92747	2.24	9.3933	.10646	4.6434	4.7499	.97759
1.65	5.2070	.19205	2.5075	2.6995	.92886	2.25	9.4877	.10540	4.6912	4.7966	.97803
1.66	5.2593	.19014	2.5346	2.7247	.93022	2.26	9.5831	.10435	4.7394	4.8437	.97846
1.67	5.3122	.18825	2.5620	2.7502	.93155	2.27	9.6794	.10331	4.7880	4.8914	.97888
1.68	5.3656	.18637	2.5896	2.7760	.93286	2.28	9.7767	.10228	4.8372	4.9395	.97929
1.69	5.4195	.18452	2.6175	2.8020	.93415	2.29	9.8749	.10127	4.8868	4.9881	.97970
1.70	**5.4739**	**.18268**	**2.6456**	**2.8283**	**.93541**	**2.30**	**9.9742**	**.10026**	**4.9370**	**5.0372**	**.98010**
1.71	5.5290	.18087	2.6740	2.8549	.93665	2.31	10.074	.09926	4.9876	5.0868	.98049
1.72	5.5845	.17907	2.7027	2.8818	.93786	2.32	10.176	.09827	5.0387	5.1370	.98087
1.73	5.6407	.17728	2.7317	2.9090	.93906	2.33	10.278	.09730	5.0903	5.1876	.98124
1.74	5.6973	.17552	2.7609	2.9364	.94023	2.34	10.381	.09633	5.1425	5.2388	.98161
1.75	5.7546	.17377	2.7904	2.9642	.94138	2.35	10.486	.09537	5.1951	5.2905	.98197
1.76	5.8124	.17204	2.8202	2.9922	.94250	2.36	10.591	.09442	5.2483	5.3427	.98233
1.77	5.8709	.17033	2.8503	3.0206	.94361	2.37	10.697	.09348	5.3020	5.3954	.98267
1.78	5.9299	.16864	2.8806	3.0492	.94470	2.38	10.805	.09255	5.3562	5.4487	.98301
1.79	5.9895	.16696	2.9112	3.0782	.94576	2.39	10.913	.09163	5.4109	5.5026	.98335
1.80	**6.0496**	**.16530**	**2.9422**	**3.1075**	**.94681**	**2.40**	**11.023**	**.09072**	**5.4662**	**5.5569**	**.98367**

Table 1.4. Exponential and Hyperbolic Functions (Continued)

x	e^x	e^{-x}	sinh x	cosh x	tanh x
2.40	**11.023**	**.09072**	**5.4662**	**5.5569**	**.98367**
2.41	11.134	.08982	5.5221	5.6119	.98400
2.42	11.246	.08892	5.5785	5.6674	.98431
2.43	11.359	.08804	5.6354	5.7235	.98462
2.44	11.473	.08716	5.6929	5.7801	.98492
2.45	11.588	.08629	5.7510	5.8373	.98522
2.46	11.705	.08543	5.8097	5.8951	.98551
2.47	11.822	.08458	5.8689	5.9535	.98579
2.48	11.941	.08374	5.9288	6.0125	.98607
2.49	12.061	.08291	5.9892	6.0721	.98635
2.50	**12.182**	**.08208**	**6.0502**	**6.1323**	**.98661**
2.51	12.305	.08127	6.1118	6.1931	.98688
2.52	12.429	.08046	6.1741	6.2545	.98714
2.53	12.554	.07966	6.2369	6.3166	.98739
2.54	12.680	.07887	6.3004	6.3793	.98764
2.55	12.807	.07808	6.3645	6.4426	.98788
2.56	12.936	.07730	6.4293	6.5066	.98812
2.57	13.066	.07654	6.4946	6.5712	.98835
2.58	13.197	.07577	6.5607	6.6365	.98858
2.59	13.330	.07502	6.6274	6.7024	.98881
2.60	**13.464**	**.07427**	**6.6947**	**6.7690**	**.98903**
2.61	13.599	.07353	6.7628	6.8363	.98924
2.62	13.736	.07280	6.8315	6.9043	.98946
2.63	13.874	.07208	6.9008	6.9729	.98966
2.64	14.013	.07136	6.9709	7.0423	.98987
2.65	14.154	.07065	7.0417	7.1123	.99007
2.66	14.296	.06995	7.1132	7.1831	.99026
2.67	14.440	.06925	7.1854	7.2546	.99045
2.68	14.585	.06856	7.2583	7.3268	.99064
2.69	14.732	.06788	7.3319	7.3998	.99083
2.70	**14.880**	**.06721**	**7.4063**	**7.4735**	**.99101**
2.71	15.029	.06654	7.4814	7.5479	.99118
2.72	15.180	.06587	7.5572	7.6231	.99136
2.73	15.333	.06522	7.6338	7.6991	.99153
2.74	15.487	.06457	7.7112	7.7758	.99170
2.75	15.643	.06393	7.7894	7.8533	.99186
2.76	15.800	.06329	7.8683	7.9316	.99202
2.77	15.959	.06266	7.9480	8.0106	.99218
2.78	16.119	.06204	8.0285	8.0905	.99233
2.79	16.281	.06142	8.1098	8.1712	.99248
2.80	**16.445**	**.06081**	**8.1919**	**8.2527**	**.99263**
2.81	16.610	.06020	8.2749	8.3351	.99278
2.82	16.777	.05961	8.3586	8.4182	.99292
2.83	16.945	.05901	8.4432	8.5022	.99306
2.84	17.116	.05843	8.5287	8.5871	.99320
2.85	17.288	.05784	8.6150	8.6728	.99333
2.86	17.462	.05727	8.7021	8.7594	.99346
2.87	17.637	.05670	8.7902	8.8469	.99359
2.88	17.814	.05613	8.8791	8.9352	.99372
2.89	17.993	.05558	8.9689	9.0244	.99384
2.90	**18.174**	**.05502**	**9.0596**	**9.1146**	**.99396**
2.91	18.357	.05448	9.1512	9.2056	.99408
2.92	18.541	.05393	9.2437	9.2976	.99420
2.93	18.728	.05340	9.3371	9.3905	.99431
2.94	18.916	.05287	9.4315	9.4844	.99443
2.95	19.106	.05234	9.5268	9.5791	.99454
2.96	19.298	.05182	9.6231	9.6749	.99464
2.97	19.492	.05130	9.7203	9.7716	.99475
2.98	19.688	.05079	9.8185	9.8693	.99485
2.99	19.886	.05029	9.9177	9.9680	.99496
3.00	**20.086**	**.04979**	**10.018**	**10.068**	**.99505**

Table 1.5. Complete Elliptic Integrals, K and E, for Different Values of the Modulus k*

$$K=\int_0^{\frac{\pi}{2}} \frac{dx}{\sqrt{1-k^2\sin^2 x}}; \quad E=\int_0^{\frac{\pi}{2}} \sqrt{1-k^2\sin^2 x}\, dx$$

$\sin^{-1}k$	K	E	$\sin^{-1}k$	K	E	$\sin^{-1}k$	K	E
0°	1.5708	1.5708	50°	1.9356	1.3055	81°.0	3.2553	1.0338
1	1.5709	1.5707	51	1.9539	1.2963	81.2	3.2771	1.0326
2	1.5713	1.5703	52	1.9729	1.2870	81.4	3.2995	1.0314
3	1.5719	1.5697	53	1.9927	1.2776	81.6	3.3223	1.0302
4	1.5727	1.5689	54	2.0133	1.2681	81.8	3.3458	1.0290
5	1.5738	1.5678	55	2.0347	1.2587	82.0	3.3699	1.0278
6	1.5751	1.5665	56	2.0571	1.2492	82.2	3.3946	1.0267
7	1.5767	1.5649	57	2.0804	1.2397	82.4	3.4199	1.0256
8	1.5785	1.5632	58	2.1047	1.2301	82.6	3.4460	1.0245
9	1.5805	1.5611	59	2.1300	1.2206	82.8	3.4728	1.0234
10	1.5828	1.5589	60	2.1565	1.2111	83.0	3.5004	1.0223
11	1.5854	1.5564	61	2.1842	1.2015	83.2	3.5288	1.0213
12	1.5882	1.5537	62	2.2132	1.1920	83.4	3.5581	1.0202
13	1.5913	1.5507	63	2.2435	1.1826	83.6	3.5884	1.0192
14	1.5946	1.5476	64	2.2754	1.1732	83.8	3.6196	1.0182
15	1.5981	1.5442	65	2.3088	1.1638	84.0	3.6519	1.0172
16	1.6020	1.5405	65.5	2.3261	1.1592	84.2	3.6852	1.0163
17	1.6061	1.5367	66.0	2.3439	1.1545	84.4	3.7198	1.0153
18	1.6105	1.5326	66.5	2.3622	1.1499	84.6	3.7557	1.0144
19	1.6151	1.5283	67.0	2.3809	1.1453	84.8	3.7930	1.0135
20	1.6200	1.5238	67.5	2.4001	1.1408	85.0	3.8317	1.0127
21	1.6252	1.5191	68.0	2.4198	1.1362	85.2	3.8721	1.0118
22	1.6307	1.5141	68.5	2.4401	1.1317	85.4	3.9142	1.0110
23	1.6365	1.5090	69.0	2.4610	1.1272	85.6	3.9583	1.0102
24	1.6426	1.5037	69.5	2.4825	1.1228	85.8	4.0044	1.0094
25	1.6490	1.4981	70.0	2.5046	1.1184	86.0	4.0528	1.0086
26	1.6557	1.4924	70.5	2.5273	1.1140	86.2	4.1037	1.0079
27	1.6627	1.4864	71.0	2.5507	1.1096	86.4	4.1574	1.0072
28	1.6701	1.4803	71.5	2.5749	1.1053	86.6	4.2142	1.0065
29	1.6777	1.4740	72.0	2.5998	1.1011	86.8	4.2744	1.0059
30	1.6858	1.4675	72.5	2.6256	1.0968	87.0	4.3387	1.0053
31	1.6941	1.4608	73.0	2.6521	1.0927	87.2	4.4073	1.0047
32	1.7028	1.4539	73.5	2.6796	1.0885	87.4	4.4811	1.0041
33	1.7119	1.4469	74.0	2.7081	1.0844	87.6	4.5609	1.0036
34	1.7214	1.4397	74.5	2.7375	1.0804	87.8	4.6477	1.0031
35	1.7312	1.4323	75.0	2.7681	1.0764	88.0	4.7427	1.0026
36	1.7415	1.4248	75.5	2.7998	1.0725	88.2	4.8478	1.0021
37	1.7522	1.4171	76.0	2.8327	1.0686	88.4	4.9654	1.0017
38	1.7633	1.4092	76.5	2.8669	1.0648	88.6	5.0988	1.0014
39	1.7748	1.4013	77.0	2.9026	1.0611	88.8	5.2527	1.0010
40	1.7868	1.3931	77.5	2.9397	1.0574	89.0	5.4349	1.0008
41	1.7992	1.3849	78.0	2.9786	1.0538	89.1	5.5402	1.0006
42	1.8122	1.3765	78.5	3.0192	1.0502	89.2	5.6579	1.0005
43	1.8256	1.3680	79.0	3.0617	1.0468	89.3	5.7914	1.0004
44	1.8396	1.3594	79.5	3.1064	1.0434	89.4	5.9455	1.0003
45	1.8541	1.3506	80.0	3.1534	1.0401	89.5	6.1278	1.0002
46	1.8691	1.3418	80.2	3.1729	1.0388	89.6	6.3509	1.0001
47	1.8848	1.3329	80.4	3.1928	1.0375	89.7	6.6385	1.0001
48	1.9011	1.3238	80.6	3.2132	1.0363	89.8	7.0440	1.0000
49	1.9180	1.3147	80.8	3.2340	1.0350	89.9	7.7371	1.0000

* From R. S. Burington, "Handbook of Mathematical Tables and Formulas," 3d ed., McGraw-Hill Book Company, Inc., New York, 1948.

Table 1.6. Error Function or Probability Integral,* $\operatorname{erf}(x) = \dfrac{2}{\sqrt{\pi}} \int_0^x e^{-n^2}\, dn$

x	0	1	2	3	4	5	6	7	8	9
0.0		**.01128**	**.02256**	**.03384**	**.04511**	**.05637**	**.06762**	**.07886**	**.09008**	**.10128**
0.1	.11246	.12362	.13476	.14587	.15695	.16800	.17901	.18999	.20094	.21184
0.2	.22270	.23352	.24430	.25502	.26570	.27633	.28690	.29742	.30788	.31828
0.3	.32863	.33891	.34913	.35928	.36936	.37938	.38933	.39921	.40901	.41874
0.4	.42839	.43797	.44747	.45689	.46623	.47548	.48466	.49375	.50275	.51167
0.5	.52050	.52924	.53790	.54646	.55494	.56332	.57162	.57982	.58792	.59594
0.6	.60386	.61168	.61941	.62705	.63459	.64203	.64938	.65663	.66378	.67084
0.7	.67780	.68467	.69143	.69810	.70468	.71116	.71754	.72382	.73001	.73610
0.8	.74210	.74800	.75381	.75952	.76514	.77067	.77610	.78144	.78669	.79184
0.9	.79691	.80188	.80677	.81156	.81627	.82089	.82542	.82987	.83423	.83851
1.0	**.84270**	**.84681**	**.85084**	**.85478**	**.85865**	**.86244**	**.86614**	**.86977**	**.87333**	**.87680**
1.1	.88021	.88353	.88679	.88997	.89308	.89612	.89910	.90200	.90484	.90761
1.2	.91031	.91296	.91553	.91805	.92051	.92290	.92524	.92751	.92973	.93190
1.3	.93401	.93606	.93807	.94002	.94191	.94376	.94556	.94731	.94902	.95067
1.4	.95229	.95385	.95538	.95686	.95830	.95970	.96105	.96237	.96365	.96490
1.5	.96611	.96728	.96841	.96952	.97059	.97162	.97263	.97360	.97455	.97546
1.6	.97635	.97721	.97804	.97884	.97962	.98038	.98110	.98181	.98249	.98315
1.7	.98379	.98441	.98500	.98558	.98613	.98667	.98719	.98769	.98817	.98864
1.8	.98909	.98952	.98994	.99035	.99074	.99111	.99147	.99182	.99216	.99248
1.9	.99279	.99309	.99338	.99366	.99392	.99418	.99443	.99466	.99489	.99511
2.0	**.99532**	**.99552**	**.99572**	**.99591**	**.99609**	**.99626**	**.99642**	**.99658**	**.99673**	**.99688**
2.1	.99702	.99715	.99728	.99741	.99753	.99764	.99775	.99785	.99795	.99805
2.2	.99814	.99822	.99831	.99839	.99846	.99854	.99861	.99867	.99874	.99880
2.3	.99886	.99891	.99897	.99902	.99906	.99911	.99915	.99920	.99924	.99928
2.4	.99931	.99935	.99938	.99941	.99944	.99947	.99950	.99952	.99955	.99957
2.5	.99959	.99961	.99963	.99965	.99967	.99969	.99971	.99972	.99974	.99975
2.6	.99976	.99978	.99979	.99980	.99981	.99982	.99983	.99984	.99985	.99986
2.7	.99987	.99987	.99988	.99989	.99989	.99990	.99991	.99991	.99992	.99992
2.8	.99992	.99993	.99993	.99994	.99994	.99994	.99995	.99995	.99995	.99996
2.9	.99996	.99996	.99996	.99997	.99997	.99997	.99997	.99997	.99997	.99998
3.0	**.99998**									

* From Herbert B. Dwight, "Mathematical Tables of Elementary and Some Higher Mathematical Functions," 3d ed., Dover Publications, Inc., New York, 1961.

1.3. PHYSICAL TABLES

(a) Systems of Units

Force, mass, and acceleration are related by Newton's law which states that the force acting on an object is proportional to the product of its mass and its acceleration, $F \propto ma$. This relation can be made an equality by the introduction of $1/g_c$, a constant of proportionality, i.e., $F = ma/g_c$. By substituting definitions of certain standard masses and forces into the above equation the value of g_c can be determined.

For example, the standard pound mass (lb_m), a quantity of matter defined in terms of the standard kilogram ($1\ lb_m = 0.4536$ kg), in a standard gravitational field of 32.1740 ft/sec² (980.665 cm/sec²) is acted upon by a force of 1 pound (lb_f). The value of g_c is determined as follows:

$$F = ma/g_c \qquad 1\ lb_f = 1\ lb_m \times \frac{32.1740\ ft/sec^2}{g_c}$$

Therefore,

$$g_c = 32.1740\ lb_m\ ft/lb_f\ sec^2$$

When a pound force acts upon a mass of one slug, the resulting acceleration will be 1 ft/sec². Thus

$$1\ lb_f = \frac{1\ slug \times 1\ ft/sec^2}{g_c}$$

Therefore,

$$g_c = 1\ slug\ ft/lb_f\ sec^2$$

When a mass of one pound is accelerated at the rate of 1 ft/sec², the force acting is one poundal. Thus

$$1\ poundal = \frac{1\ lb_m \times 1\ ft/sec^2}{g_c}$$

Therefore,

$$g_c = 1\ lb_m\ ft/poundal\ sec^2$$

In the metric system, a dyne is defined as the force required to accelerate a mass of 1 gram at the rate of 1 cm/sec². Thus 1 dyne $= 1\ \frac{gram\ cm/sec^2}{g_c}$ and, therefore, $g_c = \frac{1\ gram\ cm/sec^2}{dyne}$.

It is important to note that g_c is a constant which depends upon the specific units employed and not upon the magnitude of the local gravitational acceleration g. Table 1.7 summarizes various systems of units.

Table 1.7

Unit	Systems of Units			
	Engineering English	Technical English	Absolute English	Absolute Metric
Mass	Pound mass	Slug	Pound mass	Gram
Force	Pound force	Pound force	Poundal	Dyne
Length	Foot	Foot	Foot	Centimeter
Time	Second	Second	Second	Second
Energy	Foot-lb_f	Foot-lb_f	Foot-poundal	Erg
g_c	$32.1740\ \frac{lb_m\ ft}{lb_f\ sec^2}$	$1\ \frac{slug\ ft}{lb_f\ sec^2}$	$1\ \frac{lb_m\ ft}{poundal\ sec^2}$	$1\ \frac{gm\ cm}{dyne\ sec^2}$

Table 1.8. Conversion Factors

A To convert A into B	B	Multiply A by	C For the converse operation, multiply B by C
Acres	Square feet	4.356×10^4	2.296×10^{-5}
Atmospheres	Dynes per cm^2	1.0132×10^6	0.98697×10^{-6}
Atmospheres	Feet of water **at** 4°C	33.90	2.950×10^{-2}
Atmospheres	Inches of mercury at 0°C	29.92	3.342×10^{-2}
Atmospheres	Kilograms per square meter	1.0332×10^4	9.678×10^{-5}
Atmospheres	Millimeters of mercury at 0°C	760	1.316×10^{-3}
Atmospheres	Newtons per square meter	1.0133×10^5	9.869×10^{-6}
Atmospheres	Pounds per in.^{2}	14.696	6.804×10^{-2}
British thermal units (Btu)	Foot-pounds	778.26	1.2849×10^{-3}
British thermal units (Btu)	Horsepower-hours	3.929×10^{-4}	2,545
British thermal units (Btu)	Joules	1,054.8	9.480×10^{-4}
British thermal units (Btu)	Kilogram-calories	0.2520	3.969
British thermal units (Btu)	Kilowatthours	2.930×10^{-4}	3,413
Btu per minute	Horsepower	0.02358	42.40
Btu per minute	Watts	17.58	0.05688
Btu per square foot per minute	Watts per in.^{2}	0.1221	8.190
Bushels	Cubic feet	1.2445	0.8036
Centimeters	Feet	3.281×10^{-2}	30.48
Centimeters	Inches	0.3937	2.540
Centimeters	Meters	0.010	100.00
Centimeters	Millimeters	10.00	0.10
Centimeters	Mils (10^{-3} in.)	393.7	2.540×10^{-3}
Centimeters per second	Feet per minute	1.969	0.5080
Centimeters per second per second	Feet per second per second	3.281×10^{-2}	30.48
Circular mils	Square centimeters	5.067×10^{-6}	1.973×10^5
Circular mils	Square inches	7.854×10^{-7}	1.273×10^6
Circular mils	Square mils	0.7854	1.273
Cubic inches	Cubic centimeters	16.39	6.102×10^{-2}
Cubic inches	Cubic feet	5.787×10^{-4}	1,728
Cubic inches	Cubic meters	1.639×10^{-5}	6.1023×10^4
Cubic inches	Cubic yards	2.143×10^{-5}	4.6656×10^4
Cubic inches	Gallons	4.329×10^{-3}	231
Cubic inches	Liters	1.639×10^{-2}	61.02
Cubic inches	Quarts (liquid)	0.01732	57.75
Degrees (angular)	Minutes	60	1.667×10^{-2}
Degrees (angular)	Radians	1.745×10^{-2}	57.30
Degrees (angular)	Seconds	3600	2.778×10^{-4}
Degrees Fahrenheit	Degrees centigrade	°C = $\frac{5}{9}$(°F − 32)	°F = $\frac{9}{5}$°C + 32
Dynes	Gram cm per sec^2	1	1
Dynes	Grams at standard gravity	1.020×10^{-3}	980.7
Dynes	Pounds	2.248×10^{-6}	4.448×10^5

Table 1.8. Conversion Factors (Continued)

A To convert A into B	B	Multiply A by	C For the converse operation, multiply B by C
Dynes	Poundals	7.233×10^{-5}	1.3826×10^{4}
Dyne-centimeters	Pound-feet	7.376×10^{-8}	1.356×10^{7}
Ergs	Btu	9.480×10^{-11}	1.054×10^{10}
Ergs	Dyne-centimeters	1	1
Ergs	Foot-pounds	7.376×10^{-8}	1.356×10^{7}
Ergs	Joules	10^{-7}	10^{7}
Feet	Inches	12	8.333×10^{-2}
Feet	Miles	1.894×10^{-4}	5,280
Feet	Yards	0.333	3
Foot-pounds	Horsepower-hours	5.050×10^{-7}	1.98×10^{6}
Foot-pounds	Kilowatthours	3.766×10^{-7}	2.655×10^{6}
Grams	Kilograms	10^{-3}	10^{3}
Grams	Milligrams	10^{3}	10^{-3}
Grams	Ounces (avoir.)	3.527×10^{-2}	28.35
Grams	Ounces (Troy)	3.215×10^{-2}	31.10
Grams per cm^3	Pounds per ft^3	62.43	1.602×10^{-2}
Grams per cm^3	Pounds per $in.^3$	3.613×10^{-2}	27.68
Gram-cm^2	Pound-ft^2	2.37285×10^{-6}	4.21434×10^{5}
Gram-cm^2	Pound-$in.^2$	3.4169×10^{-4}	2.9266×10^{3}
Gram-cm^2	Slug-ft^2	7.37507×10^{-8}	1.3559×10^{7}
Horsepower	Foot-pounds per minute	33,000	3.030×10^{-5}
Horsepower	Foot-pounds per second	550	1.818×10^{-3}
Horsepower	Horsepower (metric)	1.014	0.9863
Horsepower	Kilowatts	0.7457	1.341
Hours	Day	4.167×10^{-2}	24
Hours	Minutes	60	1.667×10^{-2}
Hours	Seconds	3,600	2.778×10^{-4}
Hours	Weeks	5.952×10^{-3}	168
Inches	Centimeters	2.540	0.3937
Inches	Mils	10^{3}	10^{-3}
Joules (int.)	Foot-pounds	0.7376	1.356
Joules (int.)	Watthours	2.778×10^{-4}	3,600
Kilograms	Pounds	2.205	0.4536
Kilograms	Tons (short, 2,000 lb avoir.)	1.102×10^{-3}	907.2
Kilograms	Tons (long, 2,240 lb avoir.)	9.482×10^{-4}	1,016
Kilogram-calories	Foot-pounds	3,088	3.238×10^{-4}
Kilogram-calories	Horsepower-hours	1.560×10^{-3}	641.1
Kilogram-calories	Joules	4,186	2.389×10^{-4}
Kilogram-calories	Kilowatthours	860	1.163×10^{-3}
Kilometers	Feet	3,281	3.048×10^{-4}
Kilometers	Meters	10^{3}	10^{-3}
Kilowatts	Watts	10^{3}	10^{-3}
Knots (speed)	Feet per second	1.688	0.5925
Knots (speed)	Kilometers per hour	1.853	0.5396
Knots (speed)	Miles per hour	1.1508	0.869
Liters	Cubic centimeters	10^{3}	10^{-3}
Liters	Cubic inches	61.02	1.639×10^{-2}
Liters	Gallons (U.S. liquid)	0.2642	3.785
Meters	Yards	1.094	0.9144

Table 1.8. Conversion Factors (Continued)

A To convert A into B	B	Multiply A by	C For the converse operation, multiply B by C
Meters per second	Feet per minute	196.8	5.080×10^{-3}
Miles (nautical)	Feet	6,076.1	1.646×10^{-4}
Miles (statute)	Feet	5,280	1.894×10^{-4}
Miles per hour	Feet per minute	88	1.136×10^{-2}
Millimeters	Meters	10^{-3}	10^{3}
Minutes (angular)	Radians	2.909×10^{-4}	3,438
Minutes (angular)	Seconds	60	1.667×10^{-2}
Newtons	Dynes	10^{5}	10^{-5}
Newton meters	Dyne-centimeters	10^{7}	10^{-7}
Newton meters	Pound-feet	0.7376	1.356
Ounces (avoir.)	Grains	437.5	2.285×10^{-3}
Ounces (avoir.)	Grams	28.35	0.03527
Ounces (avoir.)	Pounds	0.0625	16
Ounces (fluid)	Cubic inches	1.805	0.5540
Ounces (fluid)	Liters	0.02957	33.81
Ounces (fluid)	Quarts	3.125×10^{-2}	32
Ounces (Troy)	Grams	31.10	0.03215
Ounces (Troy)	Pounds (Troy)	0.08333	12
Pints (liquid)	Gallons	0.125	8
Pints (liquid)	Liters	0.4732	2.113
Pints (liquid)	Quarts	0.5000	2
Poundals	Dynes	1.3826×10^{4}	7.233×10^{-5}
Poundals	Pounds	3.108×10^{-2}	32.17
Pounds (avoir.)	Grams	453.6	2.205×10^{-3}
Pounds (avoir.)	Ounces (avoir.)	16	0.0625
Pounds per ft^2	Feet of water at 4°C	1.602×10^{-2}	62.43
Pounds per ft^2	Inches of mercury	1.414×10^{-2}	70.73
Pounds per ft^2	Pounds per in.^{2}	6.944×10^{-3}	144
Radians per second	Revolutions per minute	9.549	0.1047
Radians per second	Revolutions per second	0.1592	6.283
Revolutions per minute per minute	Radians per second per second	1.745×10^{-3}	573.0
Slugs	Pounds (avoir.)	32.174	3.108×10^{-2}
Square centimeters	Square feet	1.076×10^{-3}	929.0
Square centimeters	Square inches	0.1550	6.452
Square centimeters	Square meters	10^{-4}	10^{4}
Square centimeters	Square millimeters	10^{2}	10^{-2}
Square feet	Square inches	144	6.944×10^{-3}
Square feet	Square meters	0.09290	10.76
Square inches	Square mils	10^{6}	10^{-6}
Square millimeters	Circular mils	1.973×10^{3}	5.067×10^{-4}
Watts	Ergs per second	10^{7}	10^{-7}
Watts	Foot-pounds per minute	44.26	2.260×10^{-2}
Watts	Horsepower	1.341×10^{-3}	745.7

1.4. MATHEMATICS

(a) Algebra

Basic. *Fundamental Laws*

Commutative law: $a + b = b + a \qquad ab = ba$

Associative law: $a + (b + c) = (a + b) + c \qquad a(bc) = (ab)c$

Distributive law: $a(b + c) = ab + ac$

Sums of Numbers

The sum of the first n numbers:

$$\sum_{1}^{n} (n) = \frac{n(n+1)}{2}$$

The sum of the squares of the first n numbers:

$$\sum_{1}^{n} (n^2) = \frac{n(n+1)(2n+1)}{6}$$

The sum of the cubes of the first n numbers:

$$\sum_{1}^{n} (n^3) = \frac{n^2(n+1)^2}{4}$$

Progressions. *Arithmetic Progression*

$$a, a + d, a + 2d, a + 3d, \ldots$$

a = first term
d = common difference
n = number of terms
S = sum of n terms
l = last term
$l = a + (n - 1)d$
$S = (n/2)(a + l)$

Arithmetic mean of a and $b = \dfrac{a + b}{2}$

Geometric Progression

$$a, ar, ar^2, ar^3, \ldots$$

a = first term
r = common ratio
n = number of terms
S = sum of n terms
l = last term
$l = ar^{n-1}$

$$S = a\frac{r^n - 1}{r - 1} = \frac{rl - a}{r - 1}; \text{ for } r^2 < 1 \text{ and } n = \infty,\ S = \frac{a}{1 - r}$$

Geometric mean of a and $b = \sqrt{ab}$

Powers and Roots

$$
\begin{aligned}
a^x a^y &= a^{x+y} \\
a^x/a^y &= a^{x-y} \\
(ab)^x &= a^x b^x \\
(a^x)^y &= a^{xy} \\
a^0 &= 1 \text{ if } a \neq 0 \\
a^{-x} &= 1/a^x \\
a^{x/y} &= \sqrt[y]{a^x} \\
a^{1/y} &= \sqrt[y]{a} \\
\sqrt[x]{ab} &= \sqrt[x]{a}\,\sqrt[x]{b} \\
\sqrt[x]{a/b} &= \sqrt[x]{a}/\sqrt[x]{b}
\end{aligned}
$$

Binomial Theorem

$$
(a \pm b)^n = a^n \pm na^{n-1}b + \frac{n(n-1)}{2!}a^{n-2}b^2 \pm \frac{n(n-1)(n-2)}{3!}a^{n-3}b^3 + \cdots + (\pm 1)^m \frac{n(n-1)\cdots(n-m+1)}{m!}a^{n-m}b^m + \cdots
$$

where $m! = \lfloor m = 1 \cdot 2 \cdot 3 \cdots (m-1)m$

The series is finite if n is a positive integer. If n is negative or fractional, the series is infinite and will converge for $|b| < |a|$ only.

Absolute Values. The numerical or absolute value of a number n is denoted by $|n|$ and represents the magnitude of the number without regard to algebraic sign. For example, $|-3| = |+3| = 3$

Logarithms. Definition of a logarithm: If $N = b^x$, the exponent x is the logarithm of N to the base b and is written $x = \log_b N$. The number b must be positive, finite, and different from unity. The base of common, or Briggsian, logarithms is 10. The base of natural, Napierian, or hyperbolic logarithms is $2.7182818 \cdots$ denoted by e.

Laws of Logarithms

$$
\begin{aligned}
\log_b MN &= \log_b M + \log_b N & \log_b 1 &= 0 \\
\log_b M/N &= \log_b M - \log_b N & \log_b b &= 1 \\
\log_b N^m &= m \log_b N & \log_b 0 &= +\infty,\ (0 < b < 1) \\
\log_b \sqrt[r]{N^m} &= m/r \log_b N & \log_b 0 &= -\infty,\ (1 < b < \infty) \\
\log_b N &= \log_a N/\log_a b
\end{aligned}
$$

Important Constants

$$
\begin{aligned}
\log_{10} e &= 0.4342944819 \\
\log_{10} x &= 0.4343 \log_e x = 0.4343 \ln x \\
\ln 10 &= \log_e 10 = 2.3025850930 \\
\ln x &= \log_e x = 2.3026 \log_{10} x
\end{aligned}
$$

Synthetic Division. *Synthetic division* is a process by which a polynomial $f(x)$ is divided by a binomial $(x - a)$. (It is a simplified form of long division.)

The following procedure can be followed:

1. Arrange the coefficients $c_0, c_1, c_2, \ldots, c_n$, including the zeros, of the polynomial $f(x) = c_0x^n + c_x^{n-1} + \cdots + c_n$ on the first line.
2. Write a at the right and c_0 in the first place on the third line.
3. Multiply c_0 by a and place the product (c_0a) on the second line under c_1.
4. Write the sum of c_1 and (c_0a) in the second place on the third line.
5. Multiply this sum by a and add to c_2.
6. Write this sum in the third place on the third line, and so forth.
7. The *remainder* is the last number in the third line. The other numbers on the third line are the coefficients of the powers of x, in descending order, in the quotient.

8. The binomial which is the divisor must be of the form $(x - a)$. If the binomial has the form $(kx - b) = k(x - b/k)$, then divide synthetically by $(x - b/k)$ and then divide the resulting quotient by k.

Example: Divide $(3x^4 - 4x^2 + x)$ by $(x + 2)$

$$\begin{array}{rrrrr|l} 3 & +0 & -4 & +1 & +0 & -2 \\ & -6 & +12 & -16 & +30 & \\ \hline 3 & -6 & +8 & -15 & 30 & \end{array}$$

$$\text{Quotient} = 3x^3 - 6x^2 + 8x - 15$$
$$\text{Remainder} = 30$$

Partial Fractions. A *rational fraction* is the ratio of two polynomials. If the numerator of the algebraic fraction is of lower degree than the denominator, it is termed a *proper fraction* and can be resolved into *partial fractions*. If the numerator is of higher degree than the denominator, division of numerator by denominator will result in the sum of a proper fraction and a polynomial, termed a *mixed fraction*.

A number of commonly found proper fractions are given below together with the form of the partial fractions into which they can be resolved.

1. If the denominator is factorable into real linear factors F_1, F_2, F_3, . . . , $(F = ax + b)$, each different, then corresponding to each factor there exists a partial fraction $A/(ax + b)$. Thus

$$\frac{\text{Numerator}}{F_1 \cdot F_2 \cdot F_3 \cdots} = \frac{\text{numerator}}{(a_1x + b_1)(a_2x + b_2)(a_3x + b_3) \cdots}$$
$$= \frac{A}{a_1x + b_1} + \frac{B}{a_2x + b_2} + \frac{C}{a_3x + b_3} + \cdots$$

where A, B, C, . . . are constants which can be obtained by clearing fractions and equating coefficients of like powers of x.

2. If the denominator is factorable into real linear factors, one or more of which is repeated, then corresponding to each repeated factor there exist the partial fractions

$$\frac{A}{a_1x + b_1} + \frac{B}{(a_1x + b_1)^2} + \cdots + \frac{C}{(a_1x + b_1)^n}$$

where n equals the number of times the factor appears in the denominator.

3. If the denominator is factorable into quadratic factors $ax^2 + bx + c$, none of which is repeated or factorable into real linear factors, then corresponding to each quadratic factor there exists a partial fraction

$$\frac{Ax + B}{ax^2 + bx + c}$$

4. If the denominator is factorable into quadratic factors, one or more of which is repeated and none of which is factorable into real linear factors, then corresponding to each repeated factor there exist the partial fractions

$$\frac{Ax + B}{a_1x^2 + b_1x + c} + \frac{Cx + D}{(a_1x^2 + b_1x + c)^2} + \cdots + \frac{Ex + F}{(a_1x^2 + b_1x + c_1)^n}$$

where n equals the number of times the factor appears in the denominator.

Quadratic Equations. The general form of a quadratic equation is

$$f(x) = ax^2 + bx + c = 0.$$

This equation has two roots, x_1 and x_2:

$$x_1,\ x_2 = \frac{-b \pm \sqrt{b^2 - 4ac}}{2a}$$

where $b^2 - 4ac$ is called the discriminant.

If $b^2 - 4ac > 0$, the roots are real and unequal.
If $b^2 - 4ac = 0$, the roots are real and equal.
If $b^2 - 4ac < 0$, the roots are imaginary.

Cubic Equations. The general form of a cubic equation is

$$f(x) = a_0x^3 + a_1x^2 + a_2x + a_3 = 0$$

It may also be written in the form $f(x) = ax^3 + 3bx^2 + 3cx + d = 0$. Let

$$\alpha = ac - b^2$$
$$\beta = \tfrac{1}{2}(3abc - a^2d) - b^3$$
$$\gamma_1 = (\beta + \sqrt{\alpha^3 + \beta^2})^{1/3}$$
$$\gamma_2 = (\beta - \sqrt{\alpha^3 + \beta^2})^{1/3}$$

The roots are

$$x_1 = \frac{\gamma_1 + \gamma_2 - b}{a}$$

$$x_2 = \frac{1}{a}\left[-\tfrac{1}{2}(\gamma_1 + \gamma_2) + \frac{\sqrt{-3}}{2}(\gamma_1 - \gamma_2) - b\right]$$

$$x_3 = \frac{1}{a}\left[-\tfrac{1}{2}(\gamma_1 + \gamma_2) - \frac{\sqrt{-3}}{2}(\gamma_1 - \gamma_2) - b\right]$$

The above solution for the roots x_1, x_2, x_3 is termed the *algebraic solution.* If, however, $\alpha^3 + \beta^2 < 0$, the above formulas require the finding of the cube roots of complex numbers. In this case, and if desired, for any case, the *trigonometric solution* given below may be employed.

Write the roots as

$$x_1 = \frac{u_1 - b}{a} \qquad x_2 = \frac{u_2 - b}{a} \qquad x_3 = \frac{u_3 - b}{a}$$

and substitute the appropriate value for u_1, u_2, u_3 from Table 1.9. If $\alpha^3 + \beta^2 = 0$, the three roots will be real and at least two will be equal. If $\alpha^3 + \beta^2 > 0$, one root is real and two are complex. If $\alpha^3 + \beta^2 < 0$, the three roots are real.

Combinations and Permutations. *Permutation.* A permutation is an arrangement with regard to order, of all or a part of a set of things. The number of permutations of n different things taken r at a time is

$$P(n,r) = {}_nP_r = P^n{}_r = \frac{n!}{(n - r)!}$$

Combination. A combination is an arrangement without regard to order of all or a part of a set of things. The number of combinations of n different things taken r at a time is

$$C(n,r) = {}_nC_r = \binom{n}{r} = \frac{P(n,r)}{r!} = \frac{n!}{r!(n - r)!}$$

The number of combinations, with repetition of n things taken r at a time, is

$$D^n{}_r = \frac{(n + r - 1)!}{r!(n - 1)!}$$

Table 1.9. For Trigonometric Solution of Cubic Equations

	For negative α		For positive α
	$\alpha^3 + \beta^2 \le 0$	$\alpha^3 + \beta^2 \ge 0$	
u_1	$\pm 2\sqrt{-\alpha}\cos\left(\frac{1}{3}\cos^{-1}\frac{\pm\beta}{\sqrt{-\alpha^3}}\right)$	$\pm 2\sqrt{-\alpha}\cosh\left(\frac{1}{3}\cosh^{-1}\frac{\pm\beta}{\sqrt{-\alpha^3}}\right)$	$\pm 2\sqrt{\alpha}\sinh\left(\frac{1}{3}\sinh^{-1}\frac{\pm\beta}{\sqrt{\alpha^3}}\right)$
u_2	$\pm 2\sqrt{-\alpha}\cos\left(\frac{1}{3}\cos^{-1}\frac{\pm\beta}{\sqrt{-\alpha^3}} + \frac{2\pi}{3}\right)$	$\mp \sqrt{-\alpha}\cosh\left(\frac{1}{3}\cosh^{-1}\frac{\pm\beta}{\sqrt{-\alpha^3}}\right)$ $+ i\sqrt{-3\alpha}\sinh\left(\frac{1}{3}\cosh^{-1}\frac{\pm\beta}{\sqrt{-\alpha^3}}\right)$	$\mp \sqrt{\alpha}\sinh\left(\frac{1}{3}\sinh^{-1}\frac{\pm\beta}{\sqrt{\alpha^3}}\right)$ $+ i\sqrt{3\alpha}\cosh\left(\frac{1}{3}\sinh^{-1}\frac{\pm\beta}{\sqrt{\alpha^3}}\right)$
u_3	$\pm 2\sqrt{-\alpha}\cos\left(\frac{1}{3}\cos^{-1}\frac{\pm\beta}{\sqrt{-\alpha^3}} + \frac{4\pi}{3}\right)$	$\mp \sqrt{-\alpha}\cosh\left(\frac{1}{3}\cosh^{-1}\frac{\pm\beta}{\sqrt{-\alpha^3}}\right)$ $- i\sqrt{-3\alpha}\sinh\left(\frac{1}{3}\cosh^{-1}\frac{\pm\beta}{\sqrt{-\alpha^3}}\right)$	$\mp \sqrt{\alpha}\sinh\left(\frac{1}{3}\sinh^{-1}\frac{\pm\beta}{\sqrt{\alpha^3}}\right)$ $- i\sqrt{3\alpha}\cosh\left(\frac{1}{3}\sinh^{-1}\frac{\pm\beta}{\sqrt{\alpha^3}}\right)$

NOTE: The upper of the alternative algebraic signs above is used when β is positive and the lower when β is negative.

Matrices and Determinants. *Definitions.* A rectangular array of mn elements arranged in m rows and n columns is termed a *matrix* of order $m \times n$. A matrix may be represented in any of the following ways:

$$A \equiv [a_{ij}] \equiv \begin{bmatrix} a_{11} & a_{12} & \cdots & a_{1n} \\ a_{21} & a_{22} & \cdots & a_{2n} \\ \cdots & \cdots & \cdots & \cdots \\ a_{m1} & a_{m2} & \cdots & a_{mn} \end{bmatrix} \equiv \begin{pmatrix} a_{11} & a_{12} & \cdots & a_{1n} \\ a_{21} & a_{22} & \cdots & a_{2n} \\ \cdots & \cdots & \cdots & \cdots \\ a_{m1} & a_{m2} & \cdots & a_{mn} \end{pmatrix}$$

$$\equiv \left\| \begin{matrix} a_{11} & a_{12} & \cdots & a_{1n} \\ a_{21} & a_{22} & \cdots & a_{2n} \\ \cdots & \cdots & \cdots & \cdots \\ a_{m1} & a_{m2} & \cdots & a_{mn} \end{matrix} \right\|$$

where the first subscript of each element designates the row and the second subscript the column.

A *column matrix* consists of elements arranged in a single column, as

$$\begin{bmatrix} a_{11} \\ a_{21} \\ \cdot \\ \cdot \\ \cdot \\ a_{m1} \end{bmatrix}$$

A row matrix consists of elements arranged in a single row, as

$$[a_{11} \quad a_{12} \quad \cdots \quad a_{1n}]$$

If $m = n$, the array is a *square matrix* of order m. A diagonal matrix is a square matrix in which the elements not on the principal diagonal are zero; i.e.,

$$\begin{bmatrix} a_{11} & 0 & \cdots & 0 \\ 0 & a_{22} & \cdots & 0 \\ \cdots & \cdots & \cdots & \cdots \\ 0 & 0 & \cdots & a_{nn} \end{bmatrix}$$

A *scalar matrix* is a diagonal matrix whose elements are equal. A *unit matrix*, denoted by I, is a diagonal matrix whose nonzero elements are unity. A *symmetric matrix* is a square matrix in which $a_{ij} = a_{ji}$. A *skew-symmetric* or *antisymmetric matrix* is one in which $a_{ij} = -a_{ji}$ and $a_{ii} = 0$. If $a_{ii} \neq 0$, the array is a *skew matrix*. A *zero* or *null matrix* is a matrix in which all the elements are zero.

Two matrices are equal only if they have the same number of rows and columns, and if corresponding elements are equal. If a matrix is obtained by interchanging the rows and columns of another matrix, the two matrices are *transposes* or *conjugates* of one another and are represented by the symbols A^T, A', or $\tilde{A}$.

The *determinant* of a square matrix of order n, denoted by the symbols $|A|$, $|a_{ij}|$, or

$$\begin{vmatrix} a_{11} & a_{12} & \cdots & a_{1n} \\ a_{21} & a_{22} & \cdots & a_{2n} \\ \cdots & \cdots & \cdots & \cdots \\ a_{n1} & a_{n2} & \cdots & a_{nn} \end{vmatrix}$$

is the sum of all possible $(n!)$ products $(-1)^k a_{i_1 1} a_{i_2 2} \cdots a_{i_n n}$ obtained by selecting one and only one element from each row and from each column. The algebraic sign of each product is determined by the integer k, which is the number of times the elements a_{ij} in $(-1)^k a_{i_1 1} a_{i_2 2} \cdots a_{i_n n}$ are interchanged in order to arrange the first subscripts $(i_1, i_2, \ldots, i_n)$ in the normal order $1, 2, 3, \ldots, n$. (The interchanges need not be of adjacent elements.) If the determinant $|a_{ij}|$ of a matrix $[a_{ij}]$ is zero, the matrix is *singular*. A matrix is of *rank r* if it contains at least one r-rowed determinant that is not zero, while all its determinants of order higher than r are zero. The *minor* D_{ij} of the element a_{ij} in the square matrix $[a_{ij}]$ is the determinant which remains when the row and the column which contain the element a_{ij} are omitted from $[a_{ij}]$. A minor obtained by omitting the same rows as columns is termed a *principal minor*. The *cofactor* A_{ij} of the element a_{ij} is the minor of a_{ij} multiplied by $(-1)^{i+j}$.

$$A_{ij} = (-1)^{i+j} D_{ij}$$

The *adjoint* of the matrix $[a_{ij}]$ is

$$adj[a_{ij}] = \begin{bmatrix} A_{11} & \cdots & A_{n1} \\ A_{12} & \cdots & A_{n2} \\ \cdots & \cdots & \cdots \\ A_{1n} & \cdots & A_{nn} \end{bmatrix}$$

The *inverse* or *reciprocal* $[a_{ij}]^{-1}$ of the nonsingular square matrix $[a_{ij}]$ is

$$[a_{ij}]^{-1} = \frac{adj[a_{ij}]}{|a_{ij}|} = \begin{bmatrix} \frac{A_{11}}{|a_{ij}|} & \cdots & \frac{A_{n1}}{|a_{ij}|} \\ \cdot & \cdots & \cdot \\ \cdot & \cdots & \cdot \\ \cdot & \cdots & \cdot \\ \frac{A_{1n}}{|a_{ij}|} & \cdots & \frac{A_{nn}}{|a_{ij}|} \end{bmatrix}$$

If the elements of a determinant M, of order k, are common to any k columns and any k rows in a determinant $|a_{ij}|$ of order n, then M is a *kth minor* of $|a_{ij}|$. The elements of $|a_{ij}|$ which are not included in M form a determinant of order $(n - k)$ which is termed the *complementary minor* of M. The complementary minor of M multiplied by $(-1)^{i_1+i_2\cdots+i_k+j_1+j_2+\cdots+j_k}$ where the $i_1 \cdots i_k$ and $j_1 \cdots j_k$ denote the rows and columns of $|a_{ij}|$ which contain M, is termed the *algebraic complement* of the minor M.

Operations with Determinants. The expansion of a second-order determinant is obtained by taking diagonal products with algebraic signs as indicated below:

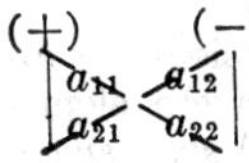

$$= a_{11}a_{22} - a_{12}a_{21}$$

The expansion of a third-order determinant is obtained by taking diagonal products, with the algebraic signs as indicated below (note the use of the repeated columns):

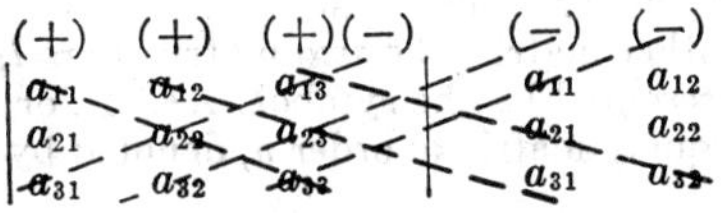

Thus

$$\begin{vmatrix} a_{11} & a_{12} & a_{13} \\ a_{21} & a_{22} & a_{23} \\ a_{31} & a_{32} & a_{33} \end{vmatrix} = a_{11}a_{22}a_{33} + a_{12}a_{23}a_{31} + a_{13}a_{21}a_{32} - a_{13}a_{22}a_{31} - a_{11}a_{23}a_{32} - a_{12}a_{21}a_{33}$$

The method of taking diagonal products as shown above in the expansion of second- and third-order determinants does not apply to the expansion of higher-order determinants. The following rules and methods can be employed in the expansion of higher-order determinants:

1. A determinant is identically zero if all the elements in any row or any column are zero. For example,

$$\begin{vmatrix} 6 & 5 \\ 0 & 0 \end{vmatrix} = 0$$

2. A determinant is identically zero if corresponding elements of two rows or two columns are proportional or equal. For example,

$$\begin{vmatrix} 4 & 6 \\ 2 & 3 \end{vmatrix} = 0 \qquad \text{(row 1 is twice row 2)}$$

3. The value of a determinant remains unchanged if the rows and columns are interchanged in the same order.

4. The value of a determinant remains unchanged if, to each element of a row or column, a constant multiple of the corresponding elements of another row or column is added. For example,

$$\begin{vmatrix} 3 & 1 \\ 4 & 2 \end{vmatrix} = \begin{vmatrix} 3 + (2)(1) & 1 \\ 4 + (2)(2) & 2 \end{vmatrix} = \begin{vmatrix} 5 & 1 \\ 8 & 2 \end{vmatrix} = 2$$

5. The sign of a determinant is changed if any two rows or any two columns are interchanged.

6. The determinant is multiplied by k if all the elements in one row or one column are multiplied by k.

7. A determinant containing a row or column of binomials can be written as the sum of two determinants as follows:

$$\begin{vmatrix} a_{11} & \cdots & (a_{1i} + b_{1i}) & \cdots & a_{1n} \\ a_{21} & \cdots & (a_{2i} + b_{2i}) & \cdots & a_{2n} \\ a_{31} & \cdots & (a_{3i} + b_{3i}) & \cdots & a_{3n} \\ \cdots & \cdots & \cdots & \cdots & \cdots \\ a_{11} & \cdots & (a_{ni} + b_{ni}) & \cdots & a_{nn} \end{vmatrix}$$

$$= \begin{vmatrix} a_{11} & \cdots & a_{1i} & \cdots & a_{1n} \\ a_{21} & \cdots & a_{2i} & \cdots & a_{2n} \\ a_{31} & \cdots & a_{3i} & \cdots & a_{3n} \\ \cdots & \cdots & \cdots & \cdots & \cdots \\ a_{n1} & \cdots & a_{ni} & \cdots & a_{nn} \end{vmatrix} + \begin{vmatrix} a_{11} & \cdots & b_{1i} & \cdots & a_{1n} \\ a_{21} & \cdots & b_{2i} & \cdots & a_{2n} \\ a_{31} & \cdots & b_{3i} & \cdots & a_{3n} \\ \cdots & \cdots & \cdots & \cdots & \cdots \\ a_{n1} & \cdots & b_{ni} & \cdots & a_{nn} \end{vmatrix}$$

8. A determinant equals the sum of the products of the elements of any row or column multiplied by the corresponding cofactors. For example,

$$\begin{vmatrix} 1 & 2 & 3 & 4 \\ 3 & 1 & 1 & 1 \\ 2 & 4 & 5 & 7 \\ 0 & 0 & 2 & 5 \end{vmatrix} = 0 + 0 + 2(-1)^{4+3} \begin{vmatrix} 1 & 2 & 4 \\ 3 & 1 & 1 \\ 2 & 4 & 7 \end{vmatrix} + 5(-1)^{4+4} \begin{vmatrix} 1 & 2 & 3 \\ 3 & 1 & 1 \\ 2 & 4 & 5 \end{vmatrix}$$

by expanding the last row.

9. A determinant can be expanded by choosing any k rows or columns and by summing the products of all the kth minors of $|a_{ij}|$ contained in the k rows or columns multiplied by their algebraic complements. For example, expanding by minors of the first two rows:

$$\begin{vmatrix} 1 & 6 & 1 & 1 \\ 5 & 7 & 1 & 2 \\ 4 & 5 & 0 & 2 \\ 3 & 2 & 5 & 1 \end{vmatrix} = (-1)^{1+2+1+2} \begin{vmatrix} 1 & 6 \\ 5 & 7 \end{vmatrix} \begin{vmatrix} 0 & 2 \\ 5 & 1 \end{vmatrix} + (-1)^{1+2+1+3} \begin{vmatrix} 1 & 1 \\ 5 & 1 \end{vmatrix} \begin{vmatrix} 5 & 2 \\ 2 & 1 \end{vmatrix}$$

$$+ (-1)^{1+2+1+4} \begin{vmatrix} 1 & 1 \\ 5 & 2 \end{vmatrix} \begin{vmatrix} 5 & 0 \\ 2 & 5 \end{vmatrix} + (-1)^{1+2+2+3} \begin{vmatrix} 6 & 1 \\ 7 & 1 \end{vmatrix} \begin{vmatrix} 4 & 2 \\ 3 & 1 \end{vmatrix}$$

$$+ (-1)^{1+2+2+4} \begin{vmatrix} 6 & 1 \\ 7 & 2 \end{vmatrix} \begin{vmatrix} 4 & 0 \\ 3 & 5 \end{vmatrix} + (-1)^{1+2+3+4} \begin{vmatrix} 1 & 1 \\ 1 & 2 \end{vmatrix} \begin{vmatrix} 4 & 5 \\ 3 & 2 \end{vmatrix}$$

$$= \begin{vmatrix} 1 & 6 \\ 5 & 7 \end{vmatrix} \begin{vmatrix} 0 & 2 \\ 5 & 1 \end{vmatrix} - \begin{vmatrix} 1 & 1 \\ 5 & 1 \end{vmatrix} \begin{vmatrix} 5 & 2 \\ 2 & 1 \end{vmatrix} + \begin{vmatrix} 1 & 1 \\ 5 & 2 \end{vmatrix} \begin{vmatrix} 5 & 0 \\ 2 & 5 \end{vmatrix}$$

$$+ \begin{vmatrix} 6 & 1 \\ 7 & 1 \end{vmatrix} \begin{vmatrix} 4 & 2 \\ 3 & 1 \end{vmatrix} - \begin{vmatrix} 6 & 1 \\ 7 & 2 \end{vmatrix} \begin{vmatrix} 4 & 0 \\ 3 & 5 \end{vmatrix} + \begin{vmatrix} 1 & 1 \\ 1 & 2 \end{vmatrix} \begin{vmatrix} 4 & 5 \\ 3 & 2 \end{vmatrix}$$

(The above method is called Laplace's development.)

10. A determinant $|a_{ij}|$ can also be expanded by the following numerical pivotal-element method:

a. Reduce some element a_{ij}, if necessary, to unity by dividing either the ith row or jth column by a_{ij} and then multiplying the determinant by the same factor. This element is the *pivotal element.*

b. Strike out the row and column containing the pivotal element.

c. Subtract from each remaining element in the determinant the product of the two crossed-out elements which are in line, horizontally and vertically, with this remaining element.

d. Multiply the new determinant, which is of order one less than the original determinant by $(-1)^{i+j}$ where i and j are the row and column numbers of the pivotal element. For example,

$$\begin{vmatrix} 4 & 4 & 3 & 8 \\ 3 & 3 & 2 & 9 \\ 2 & 2 & 6 & 2 \\ 0 & 4 & 5 & 2 \end{vmatrix} = 2\begin{vmatrix} 4 & 4 & 3 & 8 \\ 3 & 3 & 2 & 9 \\ 1 & 1 & 3 & 1 \\ 0 & 4 & 5 & 2 \end{vmatrix} = 2(-1)^{3+2}\begin{vmatrix} 4-(4)(1) & 3-(4)(3) & 8-(4)(1) \\ 3-(3)(1) & 2-(3)(3) & 9-(3)(1) \\ 0-(4)(1) & 5-(4)(3) & 2-(4)(1) \end{vmatrix}$$

$$= -2\begin{vmatrix} 0 & -9 & 2 \\ 0 & -4 & 6 \\ -4 & -7 & -2 \end{vmatrix} = (-2)(-4)(-1)^{3+1}\begin{vmatrix} -9 & 2 \\ -4 & 6 \end{vmatrix} = 8(-54+8) = -368$$

Matrix Operations

1. The *sum* (or *difference*) of two $m \times n$ matrices is a matrix, the elements of which are the sums (or differences) of the corresponding elements of the two matrices. Thus,

$$[a_{ij}] \pm [b_{ij}] = [a_{ij} \pm b_{ij}]$$

For example,

$$\begin{bmatrix} 6 & 3 \\ 1 & 0 \end{bmatrix} + \begin{bmatrix} 1 & 5 \\ 2 & 3 \end{bmatrix} = \begin{bmatrix} 7 & 8 \\ 3 & 3 \end{bmatrix}$$

2. If the number of columns of matrix A equals the number of rows of matrix B, the two matrices are *conformable* and can be multiplied in the order $A \times B$. The *product* $A \times B$ is a matrix P in which the ijth element is obtained by selecting the ith row of A and the jth column of B and then summing the products of their corresponding elements:

$$P_{ik} = \sum_{j=1}^{n} a_{ij}b_{jk}$$

For example,

$$\begin{bmatrix} 1 & 2 & 3 \\ 4 & 5 & 6 \end{bmatrix}\begin{bmatrix} 1 & 2 & 4 & 6 \\ 8 & 1 & 1 & 0 \\ 1 & 2 & 1 & 4 \end{bmatrix} = \begin{bmatrix} (1)(1)+(2)(8)+(3)(1) & (1)(2)+(2)(1)+(3)(2)\cdots \\ (4)(1)+(5)(8)+(6)(1) & \cdots\cdots\cdots\cdots \end{bmatrix}$$

Multiplication of matrices is not in general commutative, even when the two matrices are conformable in either order, that is, $AB \neq BA$. Multiplication of matrices is associative. That is, $(AB)C = A(BC)$ if the matrices are conformable. Multiplication of matrices is distributive as long as the matrices being added have the same number of rows and columns, and the matrices being multiplied are conformable. That is, $AB + AC = A(B + C)$.

3. The product of a matrix $[a_{ij}]$ and a scalar S is the matrix $[Sa_{ij}]$. For example

$$5\begin{bmatrix} 4 & 3 \\ 6 & 3 \end{bmatrix} = \begin{bmatrix} (5)(4) & (5)(3) \\ (5)(6) & (5)(3) \end{bmatrix}$$

4. If the product of two or more matrices vanishes, it does not imply that one of the matrices is zero. The factors are termed *divisors of zero.* The product of a col-

umn and a row results in a matrix with proportional rows and proportional columns. The product of a matrix and a column results in a column. The product of a matrix and a row is a row. The product of a row and a column is a scalar. The product of a matrix and its inverse is a unit matrix:

$$AA^{-1} = A^{-1}A = I$$

Solution of n Simultaneous Linear Equations in n Unknowns. A system of linear simultaneous equations is said to be *consistent* if at least one common solution exists. The solution of n unknowns requires n independent equations which are derived from n independent conditions.

In the case of systems of two equations in two unknowns, a graphical solution is possible. The two equations are plotted on a common coordinate system and the coordinates of any point of intersection of the graphs are solutions of the system. (If the graphs intersect at one point, the equations of the system are *independent* and *consistent.* If the graphs coincide, the equations are *dependent* and *consistent.* If the graphs are parallel, the equations are *inconsistent.*)

Given the following system of homogeneous equations:

$$\begin{aligned} a_{11}x_1 + a_{12}x_2 + \cdots\ a_{1n}x_n &= 0 \\ a_{21}x_1 + \cdots\cdots\cdots &= 0 \\ \cdots\cdots\cdots\cdots\cdots\cdots\cdots \\ a_{n1}x_1 + a_{n2}x_2 + \cdots\ a_{nn}x_n &= 0 \end{aligned}$$

The system will have a solution other than the trivial solution $x_1 = x_2 = x_n = 0$ if the determinant of the coefficient matrix $[a_{ij}]$ is zero.

The following system of nonhomogeneous equations, where at least one b is nonzero, has a single solution provided that the determinant of the coefficient matrix $[a_{ij}]$ is not zero.

$$\begin{aligned} a_{11}x_1 + \cdots\ a_{1n}x_n &= b_1 \\ a_{21}x_1 + \cdots\ a_{2n}x_n &= b_2 \\ \cdots\cdots\cdots\cdots\cdots \\ a_{n1}x_1 + \cdots\ a_{nn}x_n &= b_n \end{aligned}$$

Methods of Solution. 1. Cramer's rule. Given the system of equations

$$[a_{ij}]\{x\} = \{b\},\ |a_{ij}| \neq 0$$

then

$$x_i = \frac{|B|}{|a_{ij}|}$$

where $|B|$ is the same determinant as $|a_{ij}|$ except that the column of the coefficients of x_i in $|a_{ij}|$ is replaced by the constants of column $\{b\}$. For example,

$$\begin{bmatrix} 3 & 1 \\ 2 & 1 \end{bmatrix} \begin{Bmatrix} x_1 \\ x_2 \end{Bmatrix} = \begin{Bmatrix} 3 \\ 1 \end{Bmatrix}$$

$$x_1 = \frac{\begin{vmatrix} 3 & 1 \\ 1 & 1 \end{vmatrix}}{\begin{vmatrix} 3 & 1 \\ 2 & 1 \end{vmatrix}} = \frac{3 - 1}{3 - 2} = 2$$

2. Coefficient matrix inversion method. Given

$$[a_{ij}]\{x\} = \{b\}$$

then

$$\{x\} = [a_{ij}]^{-1}\{b\}$$

where

$[a_{ij}]^{-1}$ is the inverse of $[a_{ij}]$

3. Elimination method. Given, for example,

$$\begin{aligned} 2x_1 + 4x_2 + 8x_3 &= 6 \\ x_1 + 3x_2 + 7x_3 &= 10 \\ 3x_1 + 12x_2 + 6x_3 &= 12 \end{aligned}$$

The following solution illustrates the method. Rewrite the array of coefficients and constants as follows:

$$\begin{array}{rrr|r} 2 & 4 & 8 & 6 \\ 1 & 3 & 7 & 10 \\ 3 & 12 & 6 & 12 \end{array}$$

Divide each equation by its leading coefficient. The array thus becomes

$$\begin{array}{rrr|r} 1 & 2 & 4 & 3 \\ 1 & 3 & 7 & 10 \\ 1 & 4 & 2 & 4 \end{array}$$

Subtract the first equation from each of the remaining equations

$$\begin{array}{rrr|r} 1 & 2 & 4 & 3 \\ 0 & 1 & 3 & 7 \\ 0 & 2 & -2 & 1 \end{array}$$

The last two rows constitute a system of equations of one order lower than the original system. Apply the same technique to these (and so on in higher-order systems) to obtain the value of x_3 (or x_n). Thus,

$$\begin{array}{rrr|r} 1 & 2 & 4 & 3 \\ 0 & 1 & 3 & 7 \\ 0 & 1 & -1 & 0.5 \\ \hline 1 & 2 & 4 & 3 \\ 0 & 1 & 3 & 7 \\ 0 & 0 & -4 & -6.5 \\ \hline 1 & 2 & 4 & 3 \\ 0 & 1 & 3 & 7 \\ 0 & 0 & 1 & -6.5/-4 \end{array}$$

Substituting the value of x_3 in the second equation and solving for x_2, and then substituting the values of x_3 and x_2 in the first equation, the solution is obtained.

(b) Mensuration

In the following formulas P = perimeter and A = area.

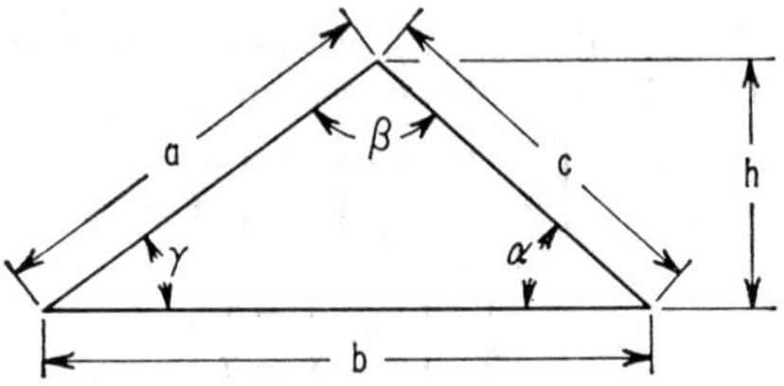

FIG. 1.1

General Triangle (Fig. 1.1)

$$\text{Length of line to side } c \text{ bisecting angle } \gamma = \frac{\sqrt{ab[(a+b)^2 - c^2]}}{a+b}$$

$$\text{Length of median line to side } c = \tfrac{1}{2}\sqrt{2(a^2+b^2) - c^2}$$

$$A = \tfrac{1}{2}bh = \tfrac{1}{2}ba \sin\gamma = \tfrac{1}{2}\frac{a^2 \sin\gamma \sin\beta}{\sin\alpha} = \sqrt{\frac{P}{2}\left(\frac{P}{2} - a\right)\left(\frac{P}{2} - b\right)\left(\frac{P}{2} - c\right)}$$

where $P = a + b + c$

For an *isosceles triangle,* $a = c \quad \gamma = \alpha$
For an *equilateral triangle,* $a = b = c \quad \alpha = \beta = \gamma$

General Quadrilateral or Trapezium (Fig. 1.2)

$$P = a + b + c + d$$
$$A = \tfrac{1}{2} d_1 d_2 \sin \alpha$$

(or the total area may be divided into several simple geometric figures such as two triangles).

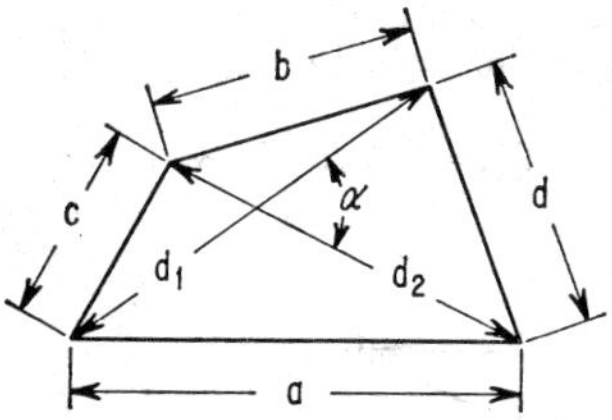

Fig. 1.2

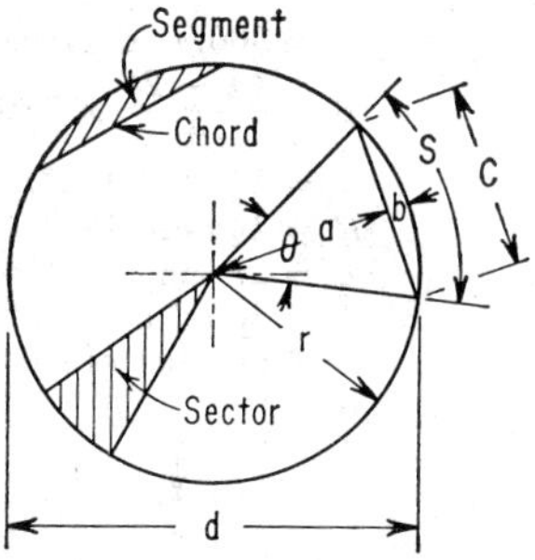

Fig. 1.3

Circle (Fig. 1.3)

$C = 2r \sin (\theta/2) \qquad \theta =$ central angle, radians
$S = r\theta$
$P = 2\pi r = \pi d$
$b = r \mp \sqrt{r^2 - C^2/4} \qquad -$for $\theta \leqq 180°$ (π radians), $+$for $\theta \geqq 180°$ (π radians)
$b = r[1 - \cos (\theta/2)] = r$ versin $(\theta/2)$

$$\text{Area of sector} = \tfrac{1}{2} rS = \tfrac{1}{2} r^2 \theta$$
$$\text{Area of segment} = \tfrac{1}{2} r^2 (\theta - \sin \theta)$$
$$\text{Area of circle} = \tfrac{1}{4} \pi d^2 = \pi r^2$$

Ellipse (Fig. 1.4)

$$P = (4a)E$$

where $E = \int_0^{2\pi} \sqrt{1 - k^2 \sin^2 x}\, dx$, complete elliptic integral (see Table 1.8), and $k = 1 - b^2/a^2$.

$$P = \pi(a + b)(1 + \gamma^2/4 + \gamma^4/64 + \cdots)$$

where $\gamma = (a - b)/(a + b)$.

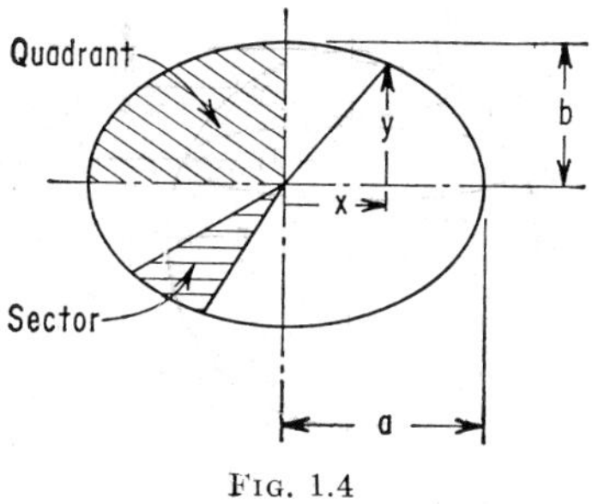

Fig. 1.4

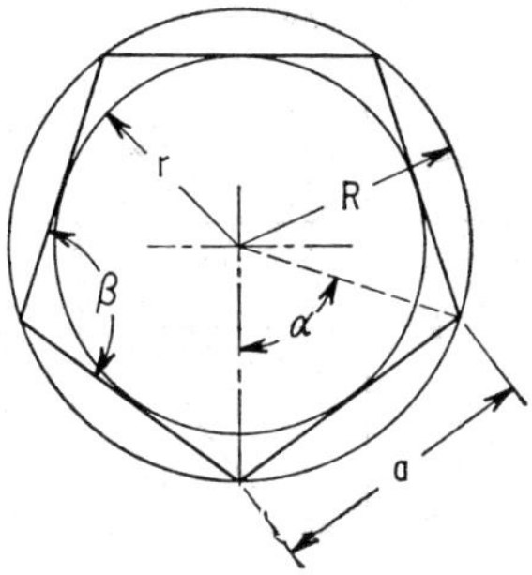

Fig. 1.5

$$P = \pi \frac{a + b}{4} \left[3(1 + \lambda) + \frac{1}{1 - \lambda} \right]$$

where $\lambda = \left[\dfrac{a - b}{2(a - b)} \right]^2$.

$$\text{Area of ellipse} = \pi ab$$

Regular Polygon—Equal Sides; Equal Angles (Fig. 1.5)

$$\alpha = 360/n, \text{ degrees} = 2\pi/n, \text{ radians}$$
$$\beta = \frac{(n - 2)}{n} \times 180, \text{ degrees}$$
$$r = \tfrac{1}{2}a \cot 180°/n = \text{radius of inscribed circle}$$
$$R = \tfrac{1}{2}a \csc 180°/n = \text{radius of circumscribed circle}$$
$$a = 2r \tan(\alpha/2) = 2R \sin(\alpha/2)$$
$$A = \tfrac{1}{4}na^2 \cot 180°/n = nr^2 \tan(\alpha/2) = \tfrac{1}{2}nR^2 \sin \alpha$$
$$n = \text{number of sides}$$

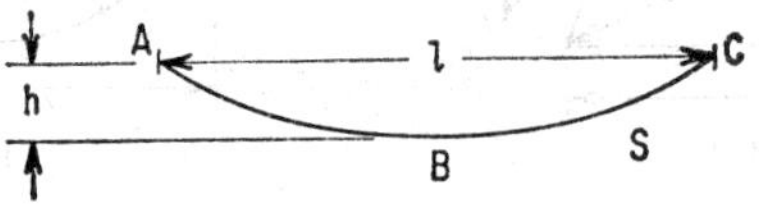

FIG. 1.6

Catenary (Fig. 1.6)

$$S = \text{arc } ABC = l\left[1 + \frac{2}{3}\left(\frac{2h}{l}\right)^2\right] \text{ approximately} \qquad \text{if } h \ll l$$

Parabola (Fig. 1.7)

$$S = \tfrac{1}{2}\sqrt{16H^2 + L^2} + \frac{L^2}{8H} \ln\left(\frac{4H + \sqrt{16H^2 + L^2}}{L}\right)$$
$$= L\left[1 + \frac{2}{3}\left(\frac{2H}{L}\right)^2 - \frac{2}{5}\left(\frac{2H}{L}\right)^4 + \cdots\right]$$
$$h = \frac{H}{L^2}(L^2 - l^2)$$
$$l = L\sqrt{\frac{H - h}{H}}$$
$$A = \tfrac{2}{3}HL$$

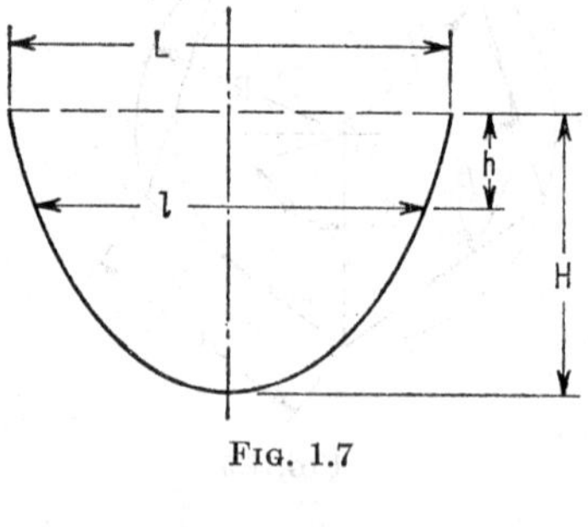

FIG. 1.7

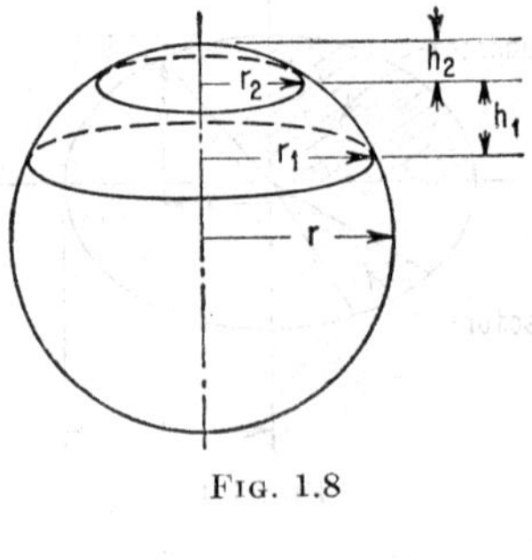

FIG. 1.8

Sphere (Fig. 1.8)

$$\text{Area of sphere} = 4\pi r^2 = \pi d^2$$

$$\text{Volume of spherical segment (one base)} = \tfrac{1}{6}\pi h_2(3r_2^2 + h_2^2) = \tfrac{1}{3}\pi h_2^2(3r - h_2)$$

$$\text{Volume of spherical segment (two bases)} = \tfrac{1}{6}\pi h_1(3r_2^2 + 3r_1^2 + h_1^2)$$

$$\text{Volume of sphere} = \tfrac{4}{3}\pi r^3 = \tfrac{1}{6}\pi d^3$$

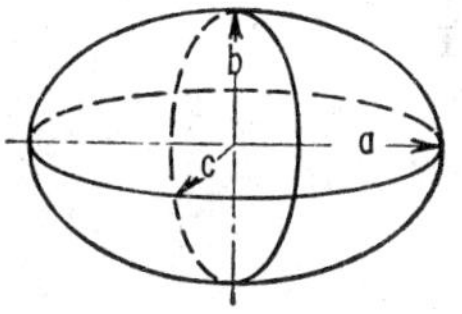

FIG. 1.9

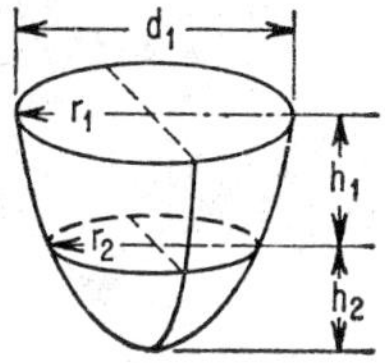

FIG. 1.10

Ellipsoid (Fig. 1.9)

$$V = \tfrac{4}{3}\pi abc$$

Paraboloid of Revolution (Fig. 1.10)

$$A = \frac{2\pi d_1}{3(h_1 + h_2)^2}\left[\left(\frac{d_1^2}{16} + (h_1 + h_2)^2\right)^{3/2} - \left(\frac{d_1}{4}\right)^3\right]$$

$$\text{Volume of paraboloid of revolution (segment of one base)} = \tfrac{1}{2}\pi r_2^2 h_2$$

$$\text{Volume of paraboloid of revolution (segment of two bases)} = \tfrac{1}{2}\pi h_1(r_1^2 + r_2^2)$$

Cone and Pyramid (Figs. 1.11 and 1.12)

$$\text{Lateral area of regular frustum of cone or pyramid} = \tfrac{1}{2}(p_1 + p_2)s_2$$

$$\text{Lateral area of regular figure} = \tfrac{1}{2}\text{ slant height} \times \text{perimeter of base} = \tfrac{1}{2}s_1 p_1$$

$$\text{Volume of frustum of cone or pyramid} = \tfrac{1}{3}(A_1 + A_2 + \sqrt{A_1 A_2})h_2$$

$$\text{Volume of cone or pyramid} = \tfrac{1}{3}\text{ altitude} \times \text{area of base} = \tfrac{1}{3}h_1 A_1$$

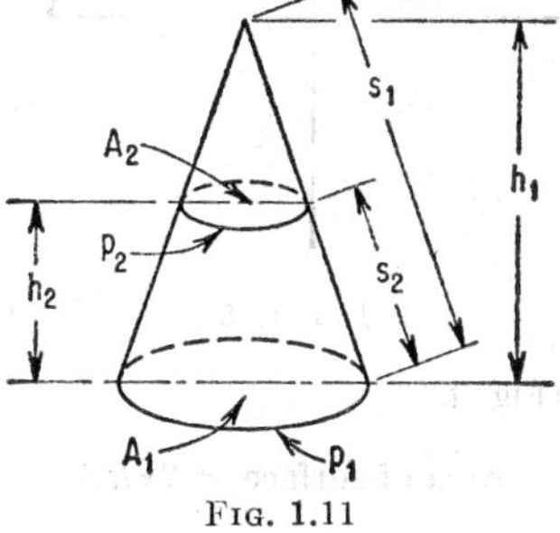

FIG. 1.11

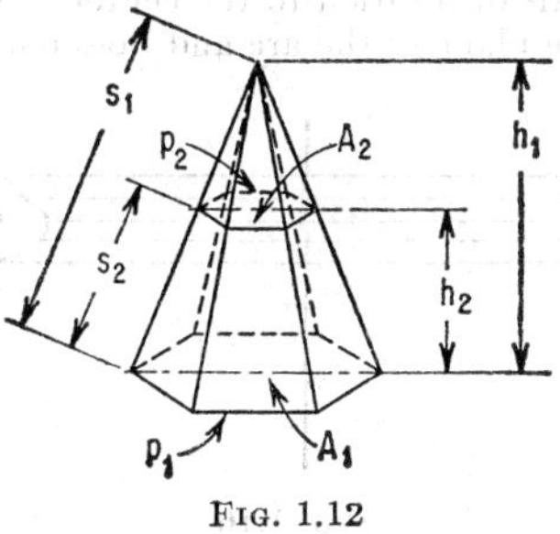

FIG. 1.12

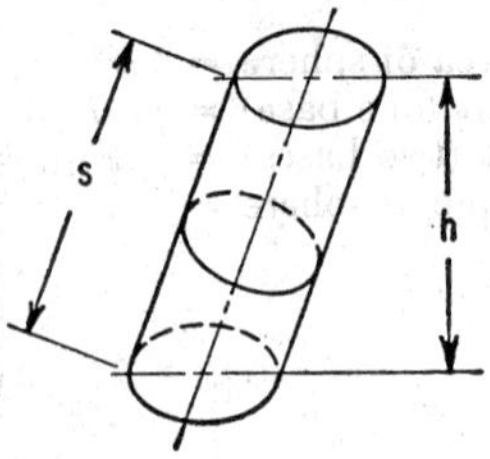

Fig. 1.13

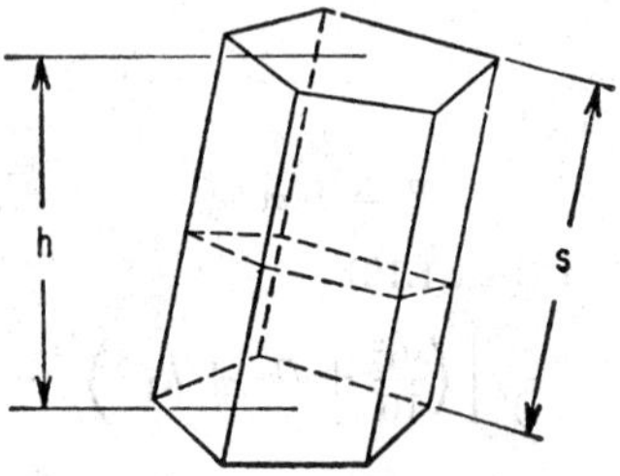

Fig. 1.14

Cylinder and Prism (Figs. 1.13 and 1.14)

$$\text{Lateral area} = \text{lateral edge} \times \text{perimeter of right section}$$
$$V = \text{altitude} \times \text{area of base}$$

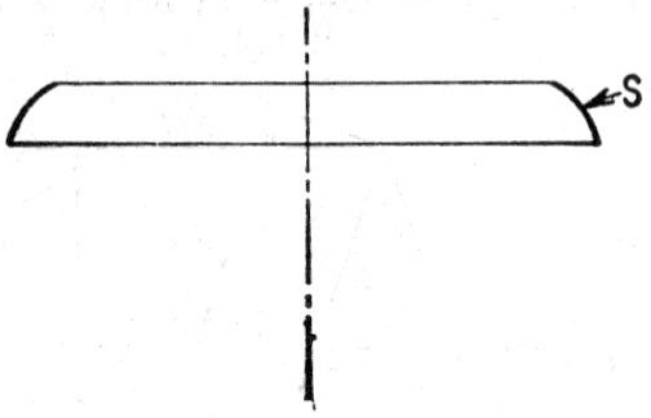

Fig. 1.15

Surface of Revolution (Fig. 1.15)

$$\text{Area of surface} = 2\pi RS$$

where R = distance from axis of rotation to the center of gravity of arc of length S. The axis of rotation is in the plane of the arc and does not cross it.

Fig. 1.16

Volume of Revolution (Fig. 1.16)

$$V = 2\pi RA$$

where R = distance from the axis of rotation to the center of gravity of the area A. The axis of rotation is in the plane of area A and does not cross it.

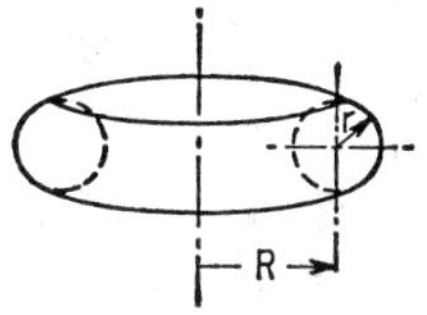

FIG. 1.17

Torus (Fig. 1.17)

$$\text{Area of surface} = 4\pi^2 Rr$$
$$V = 2\pi^2 Rr^2$$

(c) Properties of Plane Sections (see Table 1.10)

The following symbols are used:

A = area
x_c, y_c = coordinates of centroid of section in xy coordinate system
I_{xc}, I_{yc} = moment of inertia about an axis through the centroid parallel to the xy axes
r_{xc}, r_{yc} = radius of gyration of the section with respect to the centroidal axes parallel to the xy axes
$I_{x_c y_c}$ = product of inertia with respect to the centroidal axes parallel to the xy axes
I_x, I_y = moment of inertia with respect to the xy axes shown
r_x, r_y = radius of gyration of the section with respect to the xy axes shown
I_{xy} = product of inertia with respect to the xy axes shown
I_P = polar moment of inertia about an axis passing through the centroid
r_P = radius of gyration of the section about the polar axis passing through the centroid

G marks the centroid.

(d) Properties of Homogeneous Bodies (see Table 1.11)

The following symbols are used:

ρ = mass density
M = mass
x_c, y_c, z_c = coordinates of centroid in xyz coordinate system
I_{xc}, I_{yc}, I_{zc} = moment of inertia about an axis through the centroid parallel to the xyz axes shown
r_{xc}, r_{yc}, r_{zc} = radius of gyration of the body with respect to the centroidal axes parallel to the xyz axes shown
$I_{x_c y_c}, I_{x_c z_c}$, etc. = product of inertia with respect to the centroidal axes parallel to the xyz axes shown
I_x, I_y, I_z = moment of inertia with respect to the xyz axes shown
r_x, r_y, r_z = radius of gyration of the body with respect to the xyz axes shown
I_{xy}, I_{xz}, etc. = product of inertia with respect to the xyz axes shown
I_{AA}, r_{AA} = moments of inertia and radii of gyration with respect to special axes shown

G marks the centroid.

Table 1.10. Properties of Plane Sections*

Figure	Area and centroid	Moment of inertia	r^2	Product of inertia
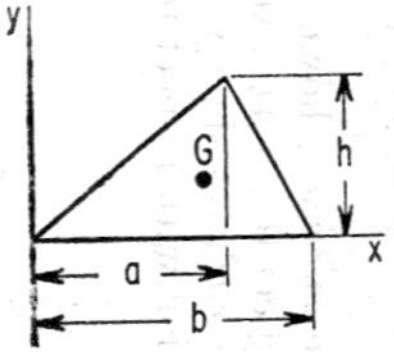 Triangle	$A = \frac{1}{2}bh$ $x_c = \frac{1}{3}(a+b)$ $y_c = \frac{1}{3}h$	$I_{xc} = \frac{bh^3}{36}$ $I_{yc} = \frac{bh}{36}(b^2 - ab + a^2)$ $I_x = \frac{bh^3}{12}$ $I_y = \frac{bh}{12}(b^2 + ab + a^2)$	$r_{xc}^2 = \frac{1}{18}h^2$ $r_{yc}^2 = \frac{1}{18}(b^2 - ab + a^2)$ $r_x^2 = \frac{1}{6}h^2$ $r_y^2 = \frac{1}{6}(b^2 + ab + a^2)$	$I_{xcyc} = \frac{Ah}{36}(2a - b) = \frac{bh^2}{72}(2a - b)$ $I_{xy} = \frac{Ah}{12}(2a + b) = \frac{bh^2}{24}(2a + b)$
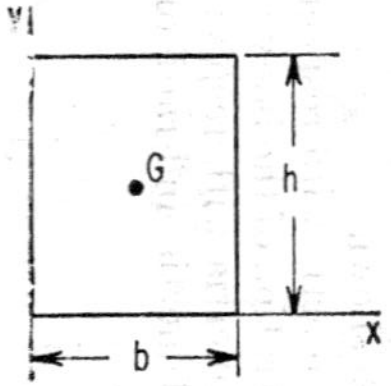 Rectangle	$A = bh$ $x_c = \frac{1}{2}b$ $y_c = \frac{1}{2}h$	$I_{xc} = \frac{bh^3}{12}$ $I_{yc} = \frac{b^3h}{12}$ $I_x = \frac{bh^3}{3}$ $I_y = \frac{b^3h}{3}$ $I_P = \frac{bh}{12}(b^2 + h^2)$	$r_{xc}^2 = \frac{1}{12}h^2$ $r_{yc}^2 = \frac{1}{12}b^2$ $r_x^2 = \frac{1}{3}h^2$ $r_y^2 = \frac{1}{3}b^2$ $r_P^2 = \frac{1}{12}(b^2 + h^2)$	$I_{xcyc} = 0$ $I_{xy} = \frac{A}{4}bh = \frac{b^2h^2}{4}$
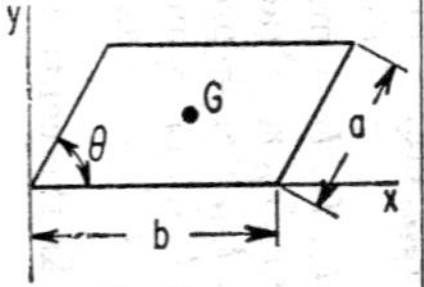 Parallelogram	$A = ab \sin\theta$ $x_c = \frac{1}{2}(b + a\cos\theta)$ $y_c = \frac{1}{2}(a\sin\theta)$	$I_{xc} = \frac{a^3b}{12}\sin^3\theta$ $I_{yc} = \frac{ab}{12}\sin\theta(b^2 + a^2\cos^2\theta)$ $I_x = \frac{a^3b}{3}\sin^3\theta$ $I_y = \frac{ab}{3}\sin\theta\,(b + a\cos\theta)^2 - \frac{a^2b^2}{6}\sin\theta\cos\theta$	$r_{xc}^2 = \frac{1}{12}(a\sin\theta)^2$ $r_{yc}^2 = \frac{1}{12}(b^2 + a^2\cos^2\theta)$ $r_x^2 = \frac{1}{3}(a\sin\theta)^2$ $r_y^2 = \frac{1}{3}(b + a\cos\theta)^2 - \frac{1}{6}(ab\cos\theta)$	$I_{xcyc} = \frac{a^3b}{12}\sin^2\theta\cos\theta$

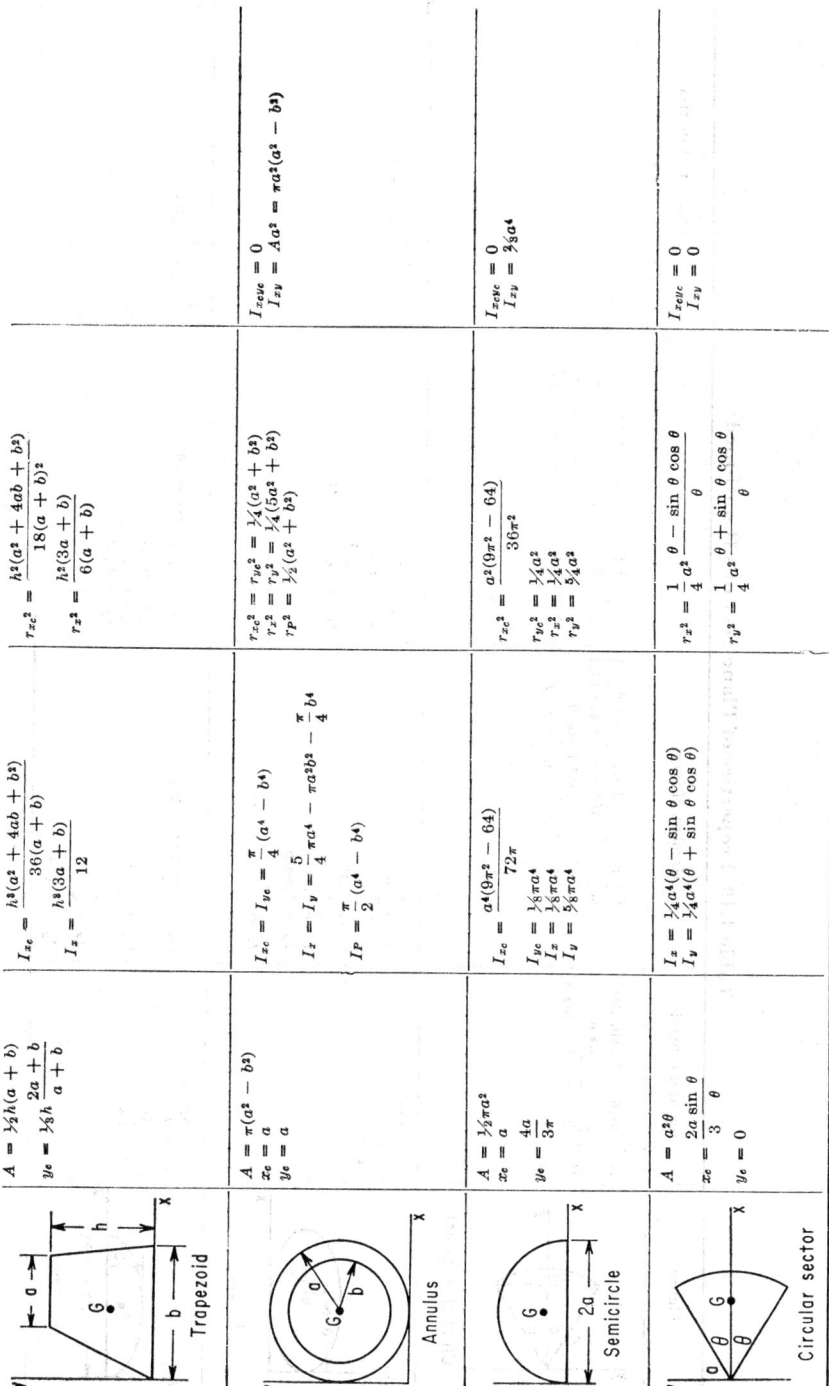

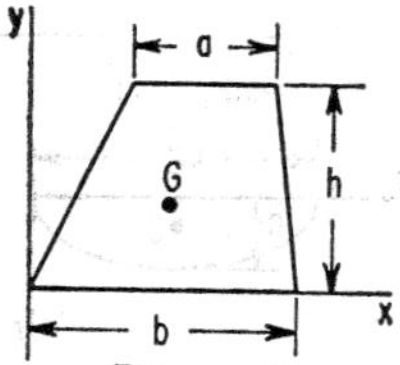 Trapezoid	$A = \frac{1}{2}h(a + b)$ $y_c = \frac{1}{3}h\frac{2a + b}{a + b}$	$I_{x_c} = \frac{h^3(a^2 + 4ab + b^2)}{36(a + b)}$ $I_x = \frac{h^3(3a + b)}{12}$	$r_{x_c}^2 = \frac{h^2(a^2 + 4ab + b^2)}{18(a + b)^2}$ $r_x^2 = \frac{h^2(3a + b)}{6(a + b)}$	
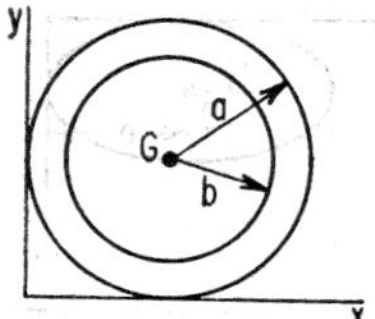 Annulus	$A = \pi(a^2 - b^2)$ $x_c = a$ $y_c = a$	$I_{x_c} = I_{y_c} = \frac{\pi}{4}(a^4 - b^4)$ $I_x = I_y = \frac{5}{4}\pi a^4 - \pi a^2 b^2 - \frac{\pi}{4}b^4$ $I_P = \frac{\pi}{2}(a^4 - b^4)$	$r_{x_c}^2 = r_{y_c}^2 = \frac{1}{4}(a^2 + b^2)$ $r_x^2 = r_y^2 = \frac{1}{4}(5a^2 + b^2)$ $r_P^2 = \frac{1}{2}(a^2 + b^2)$	$I_{x_c y_c} = 0$ $I_{xy} = Aa^2 = \pi a^2(a^2 - b^2)$
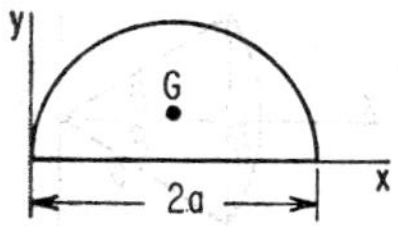 Semicircle	$A = \frac{1}{2}\pi a^2$ $x_c = a$ $y_c = \frac{4a}{3\pi}$	$I_{x_c} = \frac{a^4(9\pi^2 - 64)}{72\pi}$ $I_{y_c} = \frac{1}{8}\pi a^4$ $I_x = \frac{1}{8}\pi a^4$ $I_y = \frac{5}{8}\pi a^4$	$r_{x_c}^2 = \frac{a^2(9\pi^2 - 64)}{36\pi^2}$ $r_{y_c}^2 = \frac{1}{4}a^2$ $r_x^2 = \frac{1}{4}a^2$ $r_y^2 = \frac{5}{4}a^2$	$I_{x_c y_c} = 0$ $I_{xy} = \frac{2}{3}a^4$
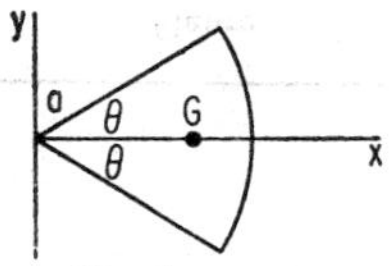 Circular sector	$A = a^2\theta$ $x_c = \frac{2a}{3}\frac{\sin\theta}{\theta}$ $y_c = 0$	$I_x = \frac{1}{4}a^4(\theta - \sin\theta\cos\theta)$ $I_y = \frac{1}{4}a^4(\theta + \sin\theta\cos\theta)$	$r_x^2 = \frac{1}{4}a^2\frac{\theta - \sin\theta\cos\theta}{\theta}$ $r_y^2 = \frac{1}{4}a^2\frac{\theta + \sin\theta\cos\theta}{\theta}$	$I_{x_c y_c} = 0$ $I_{xy} = 0$

Table 1.10. Properties of Plane Sections (Continued)

Figure	Area and centroid	Moment of inertia	r^2	Product of inertia
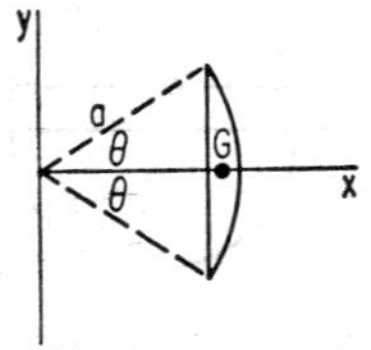 Circular segment	$A = a^2(\theta - \frac{1}{2}\sin 2\theta)$ $x_c = \frac{2a}{3}\,\frac{\sin^3\theta}{\theta - \sin\theta\cos\theta}$ $y_c = 0$	$I_x = \frac{Aa^2}{4}\left[1 - \frac{2\sin^3\theta\cos\theta}{3(\theta - \sin\theta\cos\theta)}\right]$ $I_y = \frac{Aa^2}{4}\left(1 + \frac{2\sin^3\theta\cos\theta}{\theta - \sin\theta\cos\theta}\right)$	$r_x^2 = \frac{a^2}{4}\left[1 - \frac{2\sin^3\theta\cos\theta}{3(\theta - \sin\theta\cos\theta)}\right]$ $r_y^2 = \frac{a^2}{4}\left(1 + \frac{2\sin^3\theta\cos\theta}{\theta - \sin\theta\cos\theta}\right)$	$I_{x_cy_c} = 0$ $I_{xy} = 0$
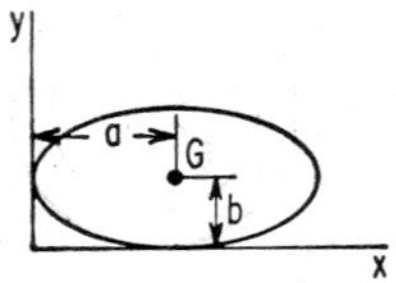 Ellipse	$A = \pi ab$ $x_c = a$ $y_c = b$	$I_{xc} = \frac{\pi}{4}ab^3$ $I_{yc} = \frac{\pi}{4}a^3b$ $I_x = \frac{5}{4}\pi ab^3$ $I_y = \frac{5}{4}\pi a^3b$ $I_P = \frac{\pi ab}{4}(a^2 + b^2)$	$r_{xc}^2 = \frac{1}{4}b^2$ $r_{yc}^2 = \frac{1}{4}a^2$ $r_x^2 = \frac{5}{4}b^2$ $r_y^2 = \frac{5}{4}a^2$ $r_P^2 = \frac{1}{4}(a^2 + b^2)$	$I_{x_cy_c} = 0$ $I_{xy} = Aab = \pi a^2b^2$
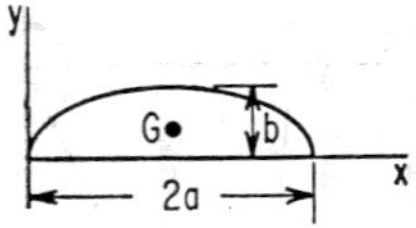 Semiellipse	$A = \frac{1}{2}\pi ab$ $x_c = a$ $y_c = \frac{4b}{3\pi}$	$I_{xc} = \frac{ab^3}{72\pi}(9\pi^2 - 64)$ $I_{yc} = \frac{\pi}{8}a^3b$ $I_x = \frac{\pi}{8}ab^3$ $I_y = \frac{5}{8}\pi a^3b$	$r_{xc}^2 = \frac{b^2}{36\pi^2}(9\pi^2 - 64)$ $r_{yc}^2 = \frac{1}{4}a^2$ $r_x^2 = \frac{1}{4}b^2$ $r_y^2 = \frac{5}{4}a^2$	$I_{x_cy_c} = 0$ $I_{xy} = \frac{2}{3}a^2b^2$

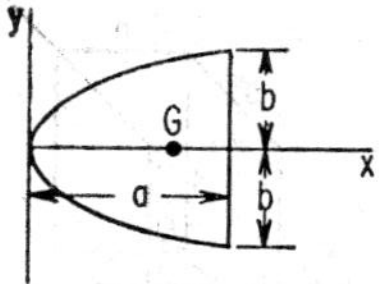 Parabola	$A = \frac{4}{3}ab$ $x_c = \frac{3}{5}a$ $y_c = 0$	$I_{xc} = I_x = \frac{4}{15}ab^3$ $I_{yc} = \frac{16}{175}a^3b$ $I_y = \frac{4}{7}a^3b$	$r_{xc}^2 = r_x^2 = \frac{1}{5}b^2$ $r_{yc}^2 = \frac{12}{175}a^2$ $r_y^2 = \frac{3}{7}a^2$	$I_{xcyc} = 0$ $I_{xy} = 0$
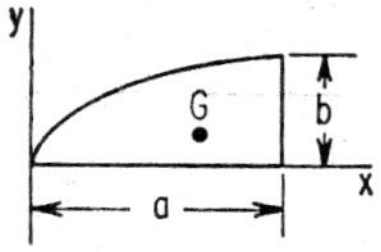 Semiparabola	$A = \frac{2}{3}ab$ $x_c = \frac{3}{5}a$ $y_c = \frac{3}{8}b$	$I_x = \frac{2}{15}ab^3$ $I_y = \frac{2}{7}a^3b$	$r_x^2 = \frac{1}{5}b^2$ $r_y^2 = \frac{3}{7}a^2$	$I_{xy} = \frac{A}{4}ab = \frac{1}{6}a^2b^2$

* From Housner and Hudson, "Applied Mechanics," Vol. II, "Dynamics," copyright 1959, D. Van Nostrand Co., Inc., Princeton, N.J.

Table 1.11. Properties of Homogeneous Bodies*

Body	Mass and centroid	Moment of inertia	r^2	Product of inertia
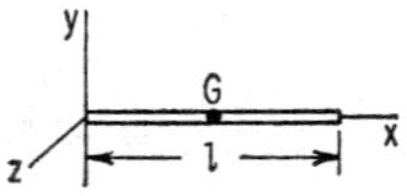 Thin rod	$M = \rho l$ $x_c = \frac{1}{2} l$ $y_c = 0$ $z_c = 0$	$I_x = I_{x_c} = 0$ $I_{y_c} = I_{z_c} = \frac{M}{12} l^2$ $I_y = I_z = \frac{M}{3} l^2$	$r_x^2 = r_{x_c}^2 = 0$ $r_{y_c}^2 = r_{z_c}^2 = \frac{1}{12} l^2$ $r_y^2 = r_z^2 = \frac{1}{3} l^2$	$I_{x_c y_c}$, etc. $= 0$ I_{xy}, etc. $= 0$
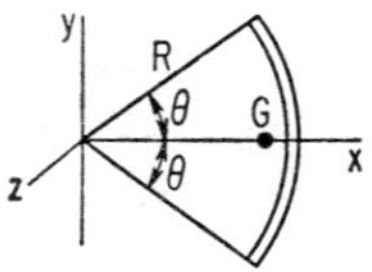 Thin circular rod	$M = 2\rho R\theta$ $x_c = \frac{R \sin \theta}{\theta}$ $y_c = 0$ $z_c = 0$	$I_x = I_{x_c}$ $= \frac{MR^2(\theta - \sin\theta\cos\theta)}{2\theta}$ $I_y = \frac{MR^2(\theta + \sin\theta\cos\theta)}{2\theta}$ $I_z = MR^2$	$r_x^2 = r_{x_c}^2 = \frac{R^2(\theta - \sin\theta\cos\theta)}{2\theta}$ $r_y^2 = \frac{R^2(\theta + \sin\theta\cos\theta)}{2\theta}$ $r_z^2 = R^2$	$I_{x_c y_c}$, etc. $= 0$ I_{xy}, etc. $= 0$
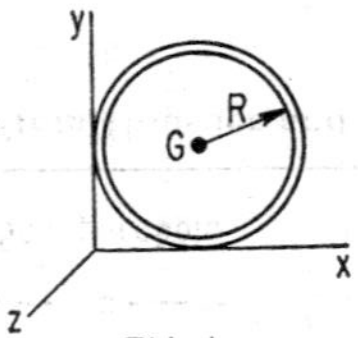 Thin hoop	$M = 2\pi\rho R$ $x_c = R$ $y_c = R$ $z_c = 0$	$I_{x_c} = I_{y_c} = \frac{M}{2} R^2$ $I_{z_c} = MR^2$ $I_x = I_y = \frac{3}{2} MR^2$ $I_z = 3MR^2$	$r_{x_c}^2 = r_{y_c}^2 = \frac{1}{2} R^2$ $r_{z_c}^2 = R^2$ $r_x^2 = r_y^2 = \frac{3}{2} R^2$ $r_z = 3R^2$	$I_{x_c y_c}$, etc. $= 0$ $I_{xy} = MR^2$ $I_{xz} = I_{yz} = 0$
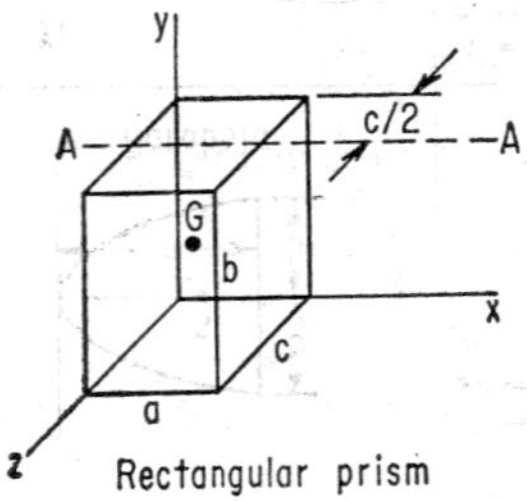 Rectangular prism	$M = \rho abc$ $x_c = \frac{1}{2} a$ $y_c = \frac{1}{2} b$ $z_c = \frac{1}{2} c$	$I_{x_c} = \frac{1}{12} M(b^2 + c_2)$ $I_x = \frac{1}{3} M(b^2 + c^2)$ $I_{AA} = \frac{1}{12} M(4b^2 + c^2)$	$r_{x_c}^2 = \frac{1}{12}(b^2 + c^2)$ $r_x^2 = \frac{1}{3}(b^2 + c^2)$ $r_{AA}^2 = \frac{1}{12}(4b^2 + c^2)$	$I_{x_c y_c}$, etc. $= 0$ $I_{xy} = \frac{1}{4} Mab$ $I_{xz} = \frac{1}{4} Mac$ $I_{yz} = \frac{1}{4} Mbc$

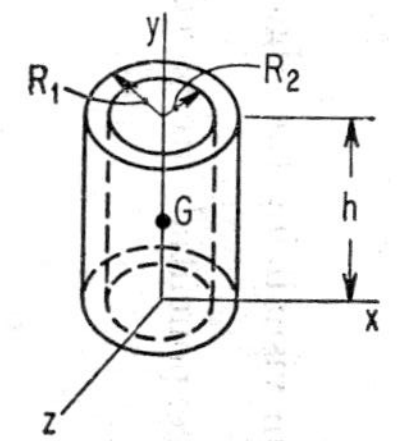 Hollow right circular cylinder	$M = \pi\rho h(R_1^2 - R_2^2)$ $x_c = 0$ $y_c = \frac{1}{2}h$ $z_c = 0$	$I_{x_c} = I_{z_c}$ $= \frac{1}{12}M(3R_1^2 + 3R_2^2 + h^2)$ $I_{y_c} = I_y = \frac{1}{2}M(R_1^2 + R_2^2)$ $I_x = I_z$ $= \frac{1}{12}M(3R_1^2 + 3R_2^2 + 4h^2)$	$r_{x_c}^2 = r_{z_c}^2 = \frac{1}{12}(3R_1^2 + 3R_2^2 + h_2)$ $r_{y_c}^2 = r_y^2 = \frac{1}{2}(R_1^2 + R_2^2)$ $r_x^2 = r_z^2 = \frac{1}{12}(3R_1^2 + 3R_2^2 + 4h^2)$	$I_{x_cy_c}$, etc. $= 0$ I_{xy}, etc. $= 0$
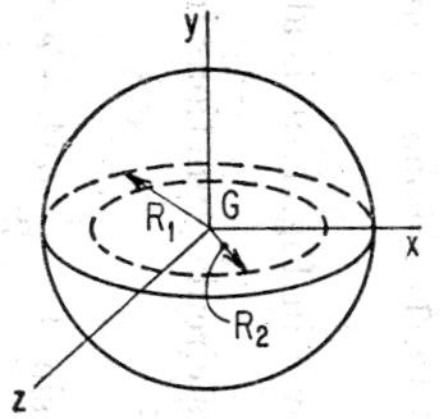 Hollow sphere	$M = \frac{4}{3}\pi\rho(R_1^3 - R_2^3)$ $x_c = 0$ $y_c = 0$ $z_c = 0$	$I_x = I_y = I_z = \frac{2}{5}M\frac{R_1^5 - R_2^5}{R_1^3 - R_2^3}$	$r_x^2 = r_y^2 = r_z^2 = \frac{2}{5}\frac{R_1^5 - R_2^5}{R_1^3 - R_2^3}$	I_{xy}, etc. $= 0$
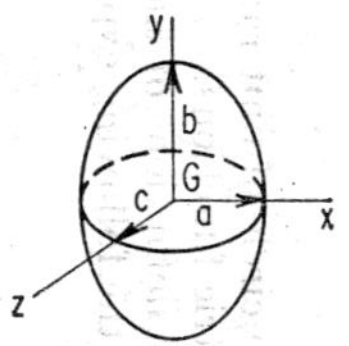 Ellipsoid	$M = \frac{4}{3}\pi\rho abc$ $x_c = 0$ $y_c = 0$ $z_c = 0$	$I_x = \frac{1}{5}M(b^2 + c^2)$ $I_y = \frac{1}{5}M(a^2 + c^2)$ $I_z = \frac{1}{5}M(a^2 + b^2)$	$r_x^2 = \frac{1}{5}(b^2 + c^2)$ $r_y^2 = \frac{1}{5}(a^2 + c^2)$ $r_z^2 = \frac{1}{5}(a^2 + b^2)$	I_{xy}, etc. $= 0$

* From Housner and Hudson, "Applied Mechanics," Vol. II, "Dynamics," copyright 1959, D. Van Nostrand Co., Inc., Princeton, N.J.

(e) Trigonometry

Plane Trigonometry. The following relationships apply to the right triangle shown in Fig. 1.18.

$$\text{sine } A = \sin A = \frac{a}{c}$$

$$\text{cosine } A = \cos A = \frac{b}{c}$$

$$\text{tangent } A = \tan A = \frac{a}{b} = \frac{\sin A}{\cos A} = \frac{1}{\cot A}$$

$$\text{cosecant } A = \csc A = \frac{c}{a} = \frac{1}{\sin A}$$

$$\text{secant } A = \sec A = \frac{c}{b} = \frac{1}{\cos A}$$

$$\text{cotangent } A = \cot A = \frac{b}{a} = \frac{\cos A}{\sin A} = \frac{1}{\tan A}$$

$$\text{exsecant } A = \text{exsec } A = \sec A - 1$$

$$\text{versine } A = \text{vers } A = 1 - \cos A$$

$$\text{coversine } A = \text{covers } A = 1 - \sin A$$

$$\text{haversine } A = \text{hav } A = \tfrac{1}{2} \text{ vers } A$$

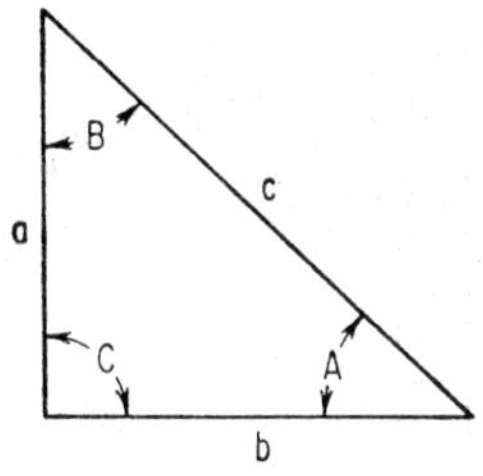

FIG. 1.18. Right triangle.

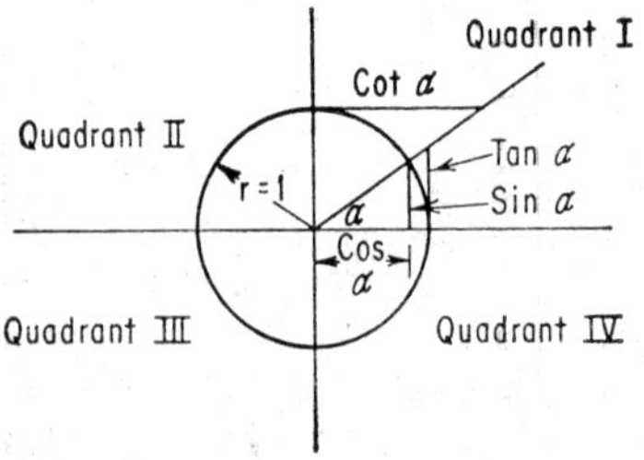

FIG. 1.19. Trigonometric functions.

The circle of unit radius, shown in Fig. 1.19, is also useful in visualizing the trigonometric functions.

The definitions of the various circular functions in terms of exponentials are based upon the following expressions:

$$\sin A = \frac{1}{2i}(e^{iA} - e^{-iA})$$

$$\cos A = \tfrac{1}{2}(e^{iA} + e^{-iA})$$

The sine, cosine, and tangent are graphically represented in Fig. 1.20.

$$\sin^2 x + \cos^2 x = 1$$
$$1 + \tan^2 x = \sec^2 x$$
$$1 + \cot^2 x = \csc^2 x$$

$$\sin x = \cos (90° - x) = \sin (180° - x)$$
$$\cos x = \sin (90° - x) = -\cos (180° - x)$$
$$\tan x = \cot (90° - x) = -\tan (180° - x)$$
$$\cot x = \tan (90° - x) = -\cot (180° - x)$$
$$\csc x = \cot \frac{x}{2} - \cot x$$

$$\sin (x \pm y) = \sin x \cos y \pm \cos x \sin y$$
$$\cos (x \pm y) = \cos x \cos y \mp \sin x \sin y$$
$$\tan (x \pm y) = \frac{\tan x \pm \tan y}{1 \mp \tan x \tan y}$$

$$\sin x + \sin y = 2 \sin \tfrac{1}{2}(x + y) \cos \tfrac{1}{2}(x - y)$$
$$\sin x - \sin y = 2 \cos \tfrac{1}{2}(x + y) \sin \tfrac{1}{2}(x - y)$$
$$\cos x + \cos y = 2 \cos \tfrac{1}{2}(x + y) \cos \tfrac{1}{2}(x - y)$$
$$\cos x - \cos y = -2 \sin \tfrac{1}{2}(x + y) \sin \tfrac{1}{2}(x - y)$$
$$\tan x \pm \tan y = \frac{\sin (x \pm y)}{\cos x \cos y}$$
$$\cot x \pm \cot y = \frac{\pm \sin (x \pm y)}{\sin x \sin y}$$

$$\sin 2x = 2 \sin x \cos x$$
$$\cos 2x = \cos^2 x - \sin^2 x$$
$$= 2 \cos^2 x - 1$$
$$= 1 - 2 \sin^2 x$$
$$\sin 3x = 3 \sin x - 4 \sin^3 x$$
$$\cos 3x = 4 \cos^3 x - 3 \cos x$$
$$\sin 4x = 8 \cos^3 x \sin x - 4 \cos x \sin x$$
$$\cos 4x = 8 \cos^4 x - 8 \cos^2 x + 1$$
$$\sin 5x = 5 \sin x - 20 \sin^3 x + 16 \sin^5 x$$
$$\cos 5x = 16 \cos^5 x - 20 \cos^3 x + 5 \cos x$$
$$\sin 6x = 32 \cos^5 x \sin x - 32 \cos^3 x \sin x + 6 \cos x \sin x$$
$$\cos 6x = 32 \cos^6 x - 48 \cos^4 x + 18 \cos^2 x - 1$$
$$\sin nx = 2 \sin (n - 1)x \cos x - \sin (n - 2)x$$
$$\cos nx = 2 \cos (n - 1)x \cos x - \cos (n - 2)x$$

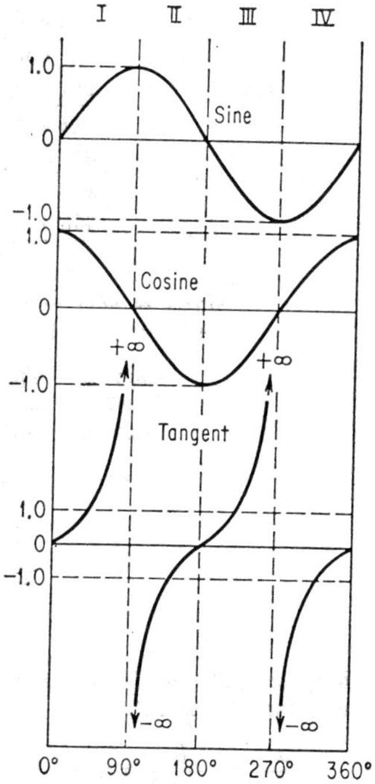

FIG. 1.20. Graphical representation of trigonometric functions.

$$\sin \frac{x}{2} = \pm \sqrt{\frac{1 - \cos x}{2}}$$ positive if $\frac{x}{2}$ in quadrant I or II, negative otherwise

$$\cos \frac{x}{2} = \pm \sqrt{\frac{1 + \cos x}{2}}$$ positive if $\frac{x}{2}$ in quadrant I or IV, negative otherwise

$$\tan \frac{x}{2} = \frac{1 - \cos x}{\sin x} = \frac{\sin x}{1 + \cos x} = \pm \sqrt{\frac{1 - \cos x}{1 + \cos x}}$$

positive if $\frac{x}{2}$ in quadrant I or III, negative otherwise

$$\sin x \sin y = \tfrac{1}{2} \cos (x - y) - \tfrac{1}{2} \cos (x + y)$$
$$\cos x \cos y = \tfrac{1}{2} \cos (x - y) + \tfrac{1}{2} \cos (x + y)$$
$$\sin x \cos y = \tfrac{1}{2} \sin (x + y) + \tfrac{1}{2} \sin (x - y)$$
$$\cos x \sin y = \tfrac{1}{2} \sin (x + y) - \tfrac{1}{2} \sin (x - y)$$

$$\cos nx \cos mx = \tfrac{1}{2} \cos (n - m)x + \tfrac{1}{2} \cos (n + m)x$$
$$\sin nx \sin mx = \tfrac{1}{2} \cos (n - m)x - \tfrac{1}{2} \cos (n + m)x$$
$$\cos nx \sin mx = \tfrac{1}{2} \sin (n + m)x - \tfrac{1}{2} \sin (n - m)x$$

$$\frac{1 + \tan x}{1 - \tan x} = \tan (45° + x)$$

$$\frac{\cot x + 1}{\cot x - 1} = \cot (45° - x)$$

$$\frac{\sin x \pm \sin y}{\cos x + \cos y} = \tan \tfrac{1}{2}(x \pm y)$$

$$\frac{\sin x \pm \sin y}{\cos x - \cos y} = - \cot \tfrac{1}{2}(x \mp y)$$

$$\frac{\sin x + \sin y}{\sin x - \sin y} = \frac{\tan \frac{1}{2}(x + y)}{\tan \frac{1}{2}(x - y)}$$

$$\sin^2 x = \tfrac{1}{2}(1 - \cos 2x) \qquad \sin^3 x = \tfrac{1}{4}(3 \sin x - \sin 3x)$$
$$\cos^2 x = \tfrac{1}{2}(1 + \cos 2x) \qquad \cos^3 x = \tfrac{1}{4}(\cos 3x + 3 \cos x)$$

$$\sin^2 x - \sin^2 y = \sin (x + y) \sin (x - y)$$
$$\cos^2 x - \cos^2 y = - \sin (x + y) \sin (x - y)$$
$$\cos^2 x - \sin^2 y = \cos (x + y) \cos (x - y)$$

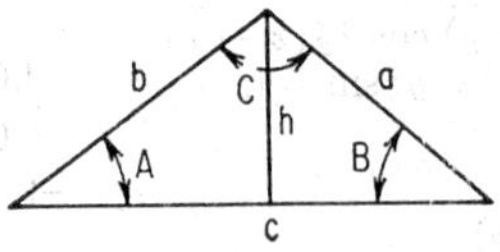

FIG. 1.21. Triangle.

Relations between Sides and Angles of Any Plane Triangle. The following expressions refer to Fig. 1.21.

$$\frac{a}{\sin A} = \frac{b}{\sin B} = \frac{c}{\sin C} = \text{diameter of the circumscribed circle} \qquad \text{law of sines}$$

$$a^2 = b^2 + c^2 - 2bc \cos A \qquad \text{law of cosines}$$

$$\frac{a + b}{a - b} = \frac{\tan \frac{1}{2}(A + B)}{\tan \frac{1}{2}(A - B)} \qquad \text{law of tangents}$$

$$\frac{a + b}{a - b} = \frac{\sin A + \sin B}{\sin A - \sin B} = \frac{\cot \frac{1}{2}C}{\tan \frac{1}{2}(A - B)}$$

$$\tan \frac{A - B}{2} = \frac{a - b}{a + b} \cot \frac{C}{2}$$

$$\sin A = \frac{2}{bc} \sqrt{s(s - a)(s - b)(s - c)} \qquad \text{where } 2s = a + b + c$$

$$\sin \frac{A}{2} = \sqrt{\frac{(s - b)(s - c)}{bc}}$$

$$\cos \frac{A}{2} = \sqrt{\frac{s(s - a)}{bc}}$$

$$\tan \frac{A}{2} = \sqrt{\frac{(s - b)(s - c)}{s(s - a)}}$$

$$h = c \frac{\sin A \sin B}{\sin (A + B)} = \frac{c}{\cot A + \cot B}$$

$$h = c \frac{\sin A \sin B'}{\sin (B' - A)} = \frac{c}{\cot A - \cot B'} \qquad \text{(refer to Fig. 1.22)}$$

Inverse Trigonometric Functions. The solution of $x = \sin y$ is (in radians)

$$y = (-1)^n \sin^{-1} x + n\pi \qquad (-\pi/2 \leqq \sin^{-1} x \leqq \pi/2)$$

where $\sin^{-1} x$ is the principal value of the angle whose sine is x and n is an integer. Likewise for $x = \cos y$,

$$y = \pm \cos^{-1} x + n(2\pi) \qquad (0 \leqq \cos^{-1} x \leqq \pi)$$

and

$$x = \tan y$$
$$y = \tan^{-1} x + n\pi \qquad (-\pi/2 < \tan^{-1} x < \pi/2)$$

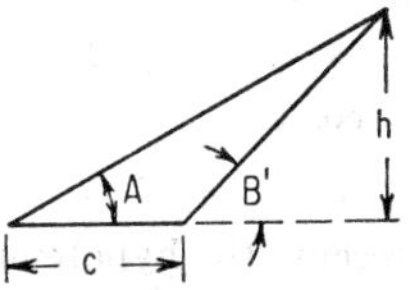

FIG. 1.22. Triangle.

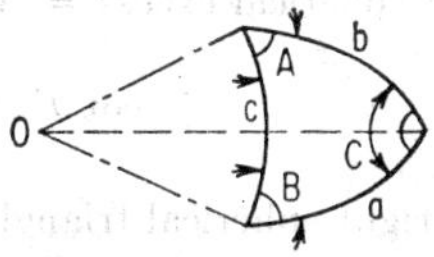

FIG. 1.23. Spherical triangle.

Spherical Trigonometry. The letters a, b, c represent the sides (as measured by the angles subtended at the center of a sphere) of a triangle on the surface of a sphere with center at 0, as shown in Fig. 1.23. A, B, C represent the angles opposite a, b, c, respectively.

$$0° < a + b + c < 360°$$
$$180° < A + B + C < 540°$$

The sides and angles of the polar triangle corresponding to each spherical triangle are $180° - a$, $180° - b$, $180° - c$, and $180° - A$, $180° - B$, $180° - C$, respectively.

General Relationships

$$\frac{\sin A}{\sin a} = \frac{\sin B}{\sin b} = \frac{\sin C}{\sin c} \qquad \text{law of sines}$$

$$\cos a = \cos b \cos c + \sin b \sin c \cos A \qquad \text{law of cosines}$$

$$\cos A = -\cos B \cos C + \sin B \sin C \cos a \qquad \text{law of cosines}$$

$$\tan \frac{A}{2} = \sqrt{\frac{\sin (s-b) \sin (s-c)}{\sin s \sin (s-a)}} \qquad \text{where } s = \tfrac{1}{2}(a+b+c)$$

$$\tan \frac{a}{2} = \sqrt{\frac{-\cos \sigma \cos (\sigma - A)}{\cos (\sigma - B) \cos (\sigma - C)}} \qquad \text{where } \sigma = \tfrac{1}{2}(A+B+C)$$

$$\sin \frac{A}{2} = \sqrt{\frac{\sin (s-b) \sin (s-c)}{\sin b \sin c}}$$

$$\sin \frac{a}{2} = \sqrt{\frac{-\cos \sigma \cos (\sigma - A)}{\sin B \sin C}}$$

$$\cos \frac{A}{2} = \sqrt{\frac{\sin s \sin (s-a)}{\sin b \sin c}}$$

$$\cos \frac{a}{2} = \sqrt{\frac{\cos (\sigma - B) \cos (\sigma - C)}{\sin B \sin C}}$$

$$\sin \tfrac{1}{2}(A+B) \cos \frac{c}{2} = \cos \tfrac{1}{2}(a-b) \cos \tfrac{1}{2}C$$

$$\cos \tfrac{1}{2}(A+B) \cos \frac{c}{2} = \cos \tfrac{1}{2}(a+b) \sin \tfrac{1}{2}C$$

$$\sin \tfrac{1}{2}(A-B) \sin \frac{c}{2} = \sin \tfrac{1}{2}(a-b) \cos \tfrac{1}{2}C$$

$$\cos \tfrac{1}{2}(A-B) \sin \frac{c}{2} = \sin \tfrac{1}{2}(a+b) \sin \tfrac{1}{2}C$$

$$\frac{\sin \frac{1}{2}(A+B)}{\sin \frac{1}{2}(A-B)} = \frac{\tan \frac{1}{2}c}{\tan \frac{1}{2}(a-b)} \qquad \frac{\sin \frac{1}{2}(a+b)}{\sin \frac{1}{2}(a-b)} = \frac{\cot \frac{1}{2}C}{\tan \frac{1}{2}(A-B)}$$

$$\frac{\cos \frac{1}{2}(A+B)}{\cos \frac{1}{2}(A-B)} = \frac{\tan \frac{1}{2}c}{\tan \frac{1}{2}(a+b)} \qquad \frac{\cos \frac{1}{2}(a+b)}{\cos \frac{1}{2}(a-b)} = \frac{\cot \frac{1}{2}C}{\tan \frac{1}{2}(A+B)}$$

$$\tan E/4 = \sqrt{\tan \tfrac{1}{2}s \tan \tfrac{1}{2}(s-a) \tan \tfrac{1}{2}(s-b) \tan \tfrac{1}{2}(s-c)}$$

where E = spherical excess $= A + B + C - 180°$

$$\cot E/2 = \frac{\cot a/2 \cot b/2 + \cos c}{\sin C}$$

For the right spherical triangle, $C = 90°$ and c represents the hypotenuse.

$$\sin a = \sin A \sin c \qquad \sin C = \cot A \cot B$$
$$\sin a = \tan b \cot B \qquad \cos A = \cos a \sin B$$
$$\sin b = \sin B \sin c \qquad \cos A = \tan b \cot C$$
$$\sin b = \tan a \cot A \qquad \cos B = \cos b \sin A$$
$$\sin c = \cos a \cos b \qquad \cos B = \tan a \cot c$$

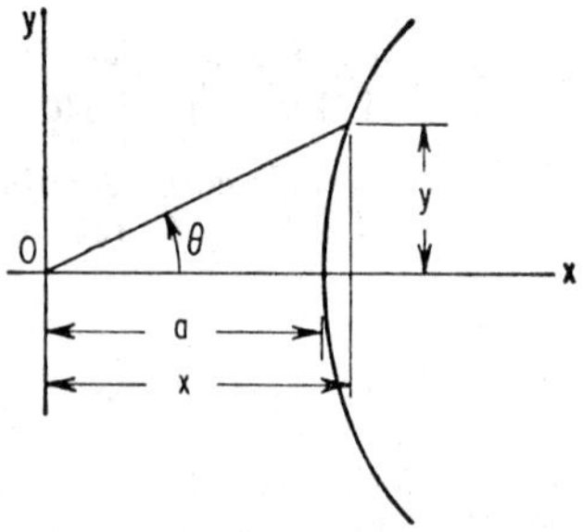

FIG. 1.24. Equilateral hyperbola.

Hyperbolic Trigonometry. The relationships which define the hyperbolic functions refer to the equilateral hyperbola shown in Fig. 1.24.

$$\text{Hyperbolic sine } \theta = \sinh \theta = \frac{y}{a}$$

$$\text{Hyperbolic cosine } \theta = \cosh \theta = \frac{x}{a}$$

$$\text{Hyperbolic tangent } \theta = \tanh \theta = \frac{y}{x} = \frac{\sinh \theta}{\cosh \theta}$$

$$\text{Hyperbolic cotangent } \theta = \coth \theta = \frac{x}{y} = \frac{1}{\tanh \theta}$$

$$\text{Hyperbolic secant } \theta = \operatorname{sech} \theta = \frac{a}{x} = \frac{1}{\cosh \theta}$$

$$\text{Hyperbolic cosecant } \theta = \operatorname{csch} \theta = \frac{a}{y} = \frac{1}{\sinh \theta}$$

As in the case of the circular functions, the hyperbolic functions may be expressed in terms of exponentials.

$$\sinh \theta = \frac{e^{\theta} - e^{-\theta}}{2}$$

$$\cosh \theta = \frac{e^{\theta} + e^{-\theta}}{2}$$

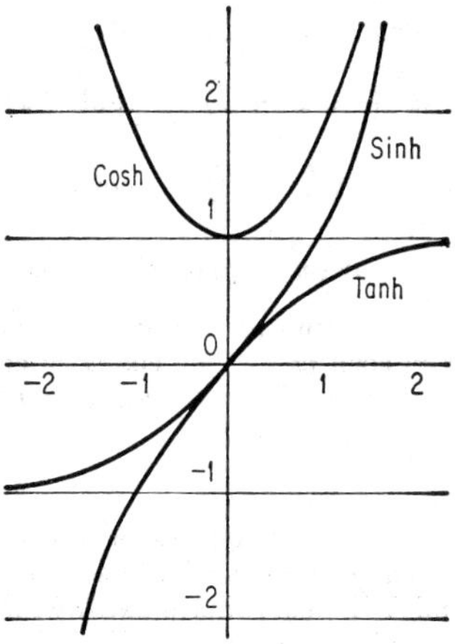

FIG. 1.25. Graphical representation of hyperbolic functions.

Figure 1.25 is a graphical representation of the sinh, cosh, and tanh.

Trigonometric Relationships (Hyperbolic)

$$\cosh^2 x - \sinh^2 x = 1 \qquad \operatorname{csch}^2 x - \coth^2 x = -1$$
$$\operatorname{sech}^2 x + \tanh^2 x = 1 \qquad \cosh^2 x + \sinh^2 x = \cosh 2x$$

$$\sinh (x \pm y) = \sinh x \cosh y \pm \cosh x \sinh y$$
$$\cosh (x \pm y) = \cosh x \cosh y \pm \sinh x \sinh y$$
$$\tanh (x \pm y) = \frac{\tanh x \pm \tanh y}{1 \pm \tanh x \tanh y}$$
$$\coth (x \pm y) = \frac{1 \pm \coth x \coth y}{\coth x \pm \coth y}$$

$$\sinh 2x = 2 \sinh x \cosh x \qquad \sinh^2 x = \tfrac{1}{2} \cosh 2x - \tfrac{1}{2}$$
$$\cosh 2x = \cosh^2 x + \sinh^2 x \qquad \cosh^2 x = \tfrac{1}{2} \cosh 2x + \tfrac{1}{2}$$

$$\sinh x + \sinh y = 2 \sinh \frac{x+y}{2} \cosh \frac{x-y}{2}$$
$$\sinh x - \sinh y = 2 \cosh \frac{x+y}{2} \sinh \frac{x-y}{2}$$
$$\cosh x + \cosh y = 2 \cosh \frac{x+y}{2} \cosh \frac{x-y}{2}$$
$$\cosh x - \cosh y = 2 \sinh \frac{x+y}{2} \sinh \frac{x-y}{2}$$

$$\tanh x \pm \tanh y = \frac{\sinh (x \pm y)}{\cosh x \cosh y}$$
$$(\cosh x \pm \sinh x)^n = \cosh nx \pm \sinh nx$$
$$\sinh \frac{x}{2} = \sqrt{\frac{(\cosh x - 1)}{2}}$$
$$\cosh \frac{x}{2} = \sqrt{\frac{(\cosh x + 1)}{2}}$$

Inverse Hyperbolic Functions

$$\sinh^{-1} x = \ln (x + \sqrt{x^2 + 1}) \qquad \tanh^{-1} x = \tfrac{1}{2} \ln \frac{1 + x}{1 - x}$$

$$\cosh^{-1} x = \ln (x + \sqrt{x^2 - 1}) \qquad \coth^{-1} x = \tfrac{1}{2} \ln \frac{x + 1}{x - 1}$$

Relationships Involving Circular and Hyperbolic Functions

$$\sin x = -i \sinh ix \qquad \sin ix = i \sinh x$$
$$\cos x = \cosh ix \qquad \cos ix = \cosh x$$
$$\tan x = -i \tanh ix \qquad \tan ix = i \tanh x$$

(f) Analytic Geometry

Plane Analytic Geometry. *Rectangular- or Cartesian-coordinate System* (Fig. 1.26). The rectangular- or cartesian-coordinate system is defined by two mutually perpendicular lines $X'X$ and $Y'Y$ (coordinate axes) which intersect at O, the origin The distances along the X axis (abscissa) and Y axis (ordinate) to a point $P(x,y)$ are the coordinates of the point. The positive X direction is to the right of the $Y'Y$ line, and the positive Y direction is above $X'X$.

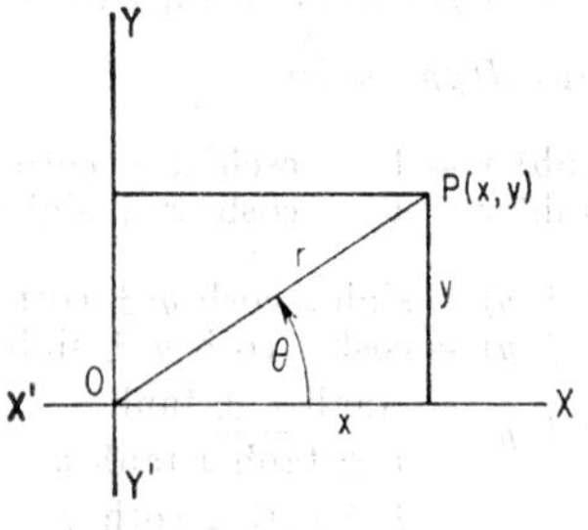

FIG. 1.26. Coordinate systems.

Polar-coordinate System (Fig. 1.26). The position of any point $P(r,\theta)$ may be defined by the coordinates r and θ in a polar-coordinate system. The angle θ is measured positive counterclockwise from the line OX (initial line) and the radius vector r is measured from O, the origin or pole.

Relations between Rectangular- and Polar-coordinate Systems (Fig. 1.26)

$$x = r \cos \theta \qquad r^2 = x^2 + y^2$$
$$y = r \sin \theta \qquad \tan \theta = y/x$$

Transformation of Coordinate Axes. If a rectangular-coordinate system (x,y) is translated (old and new coordinate axes parallel) by amounts h and k (the coordinates of the new origin relative to the old coordinate system are h and k), the new system (x',y') is related to the old by the following:

$$x = h + x' \qquad y = k + y'$$

If the origin of a coordinate system is unchanged, but the coordinate axes rotated through an angle ϕ counterclockwise, the new coordinate system (x',y') is related to the old by

$$x = x' \cos \phi - y' \sin \phi \qquad y = x' \sin \phi + y' \cos \phi$$

Distances and Slopes. The distance d between two points $P_1(x_1y_1)$ and $P_2(x_2y_2)$ is

$$d = \overline{P_1P_2} = \sqrt{(x_2 - x_1)^2 + (y_2 - y_1)^2}$$

The point which divides the line segment $\overline{P_1P_2}$ in the ratio $r_1:r_2$ is

$$\left(\frac{r_1x_2 + r_2x_1}{r_1 + r_2}, \frac{r_1y_2 + r_2y_1}{r_1 + r_2}\right)$$

Thus the mid-point of $\overline{P_1P_2}(1:1)$ is given by

$$\left(\frac{x_1 + x_2}{2}, \frac{y_1 + y_2}{2}\right)$$

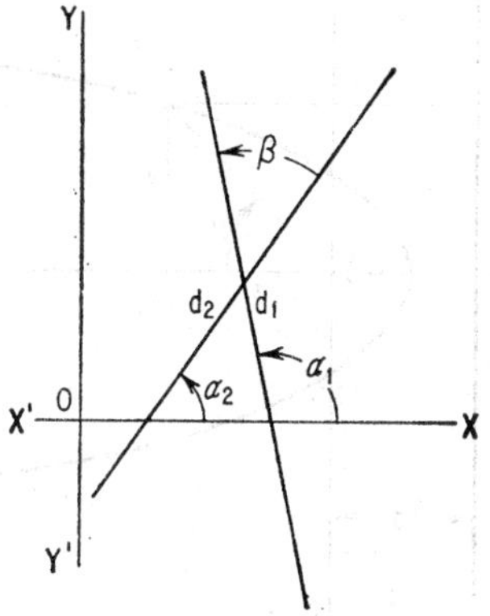

FIG. 1.27. Intersecting lines.

The slope m of a line segment $\overline{P_1P_2}$ (Fig. 1.27)

$$m = \tan \alpha = \frac{y_2 - y_1}{x_2 - x_1}$$

The angle between two lines of slopes m_1 and m_2 (Fig. 1.27)

$$\beta = \tan^{-1} \frac{m_1 - m_2}{1 + m_1m_2}$$

where m_1 denotes the line of greater inclination as shown.

If the line segments are parallel, $m_1 = m_2$ and $\beta = 0$. If the lines are perpendicular $m_1m_2 = -1$ and $\beta = 90°$.

The perpendicular distance from a point $P(x_1y_1)$ to a line $Ax + By + C = 0$ is

$$\left|\frac{Ax_1 + By_1 + C}{\sqrt{A^2 + B^2}}\right|$$

Equations of the Straight Line (Fig. 1.27)

$Ax + By + C = 0$ general form

$y = -(A/B)x - C/B$ slope-intercept form;
slope $= -A/B = m$ $\quad$ y intercept $= b$

or

$y = mx + b$

$y - y_1 = m(x - x_1)$ line through the point x_1y_1 of slope m

$\dfrac{y - y_1}{x - x_1} = \dfrac{y_2 - y_1}{x_2 - x_1}$ line through points x_1y_1 and x_2y_2

$x \cos \alpha + y \sin \alpha = p$ normal form; p is the distance from the origin to the line, and α is the angle which the normal to the line makes with the x axis

Conics (Fig. 1.28)

A conic is the locus of a point P which moves so that its distance from a fixed point F (focus) is related to its distance from a fixed straight line (directrix) by a constant ratio e (eccentricity). See Fig. 1.28.

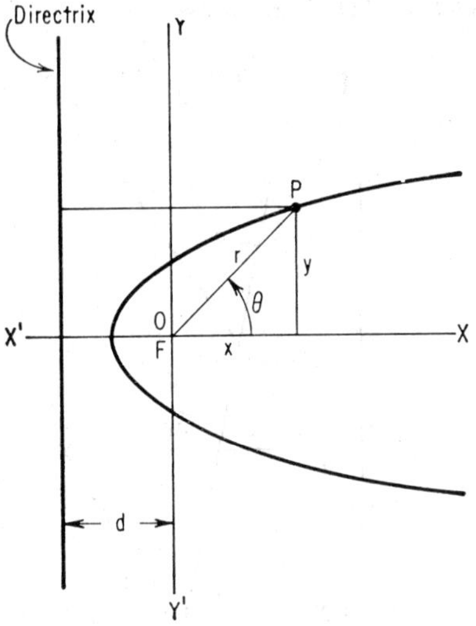

FIG. 1.28. Conic.

$x^2 + y^2 = e^2(d + x)^2$ (d is the distance from focus to directrix and F is at the origin)

$$r = \frac{de}{1 - e \cos \theta}$$

If $e = 1$, the conic is a parabola.
If $e > 1$, the conic is a hyperbola.
If $e < 1$, the conic is an ellipse (of which the circle is a special case).

Equations of the Circle

$x^2 + y^2 + Dx + Ey + F = 0$

general form, center at $-D/2, -E/2$; radius $= \sqrt{(D/2)^2 + (E/2)^2 - F}$

$(x - h)^2 + (y - k)^2 = r^2$ equation of circle of radius r and center at h, k

Equations of the Parabola. The locus of a point such that the distances from a fixed point to a fixed line are equal is called a parabola.

$(y - k)^2 = a(x - h)$ vertex at h, k; axis parallel to x axis (Fig. 1.29)
$(x - h)^2 = a(y - k)$ vertex at h, k; axis parallel to y axis (Fig. 1.30)

where $a = 4$ (distance from vertex to focus) = latus rectum (L,L)

Distance from vertex to focus = distance from vertex to directrix

Equations of the Ellipse. The locus of a point such that the sum of its distances from two fixed points is constant is an ellipse.

$$\frac{(x - h)^2}{a^2} + \frac{(y - k)^2}{b^2} = 1$$ center at h, k; major axis parallel to x axis (Fig. 1.31)

$$\frac{(y - k)^2}{a^2} + \frac{(x - h)^2}{b^2} = 1$$ center at h, k; major axis parallel to y axis (Fig. 1.32)

Major axis $= 2a$, minor axis $= 2b$, eccentricity $= e = \sqrt{a^2 - b^2}/a$.
Distance from center to either focus $= \sqrt{a^2 - b^2}$.
Latus rectum $= 2b^2/a$.
Distance from center to either directrix $= a/e$.
Sum of distances from any point P on the ellipse to foci $= PF_1 + PF_2 = 2a$.

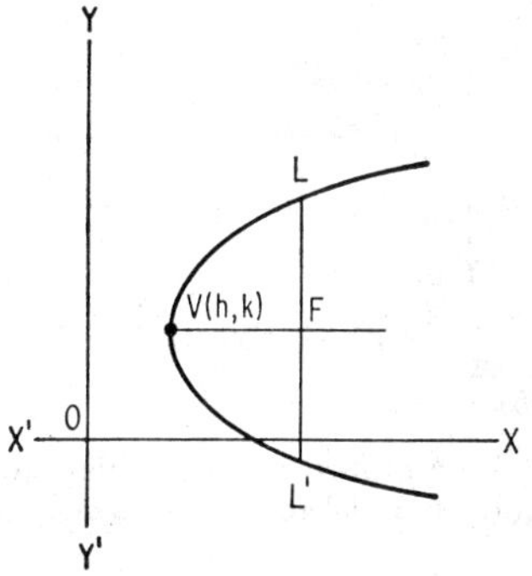

FIG. 1.29. Parabola.

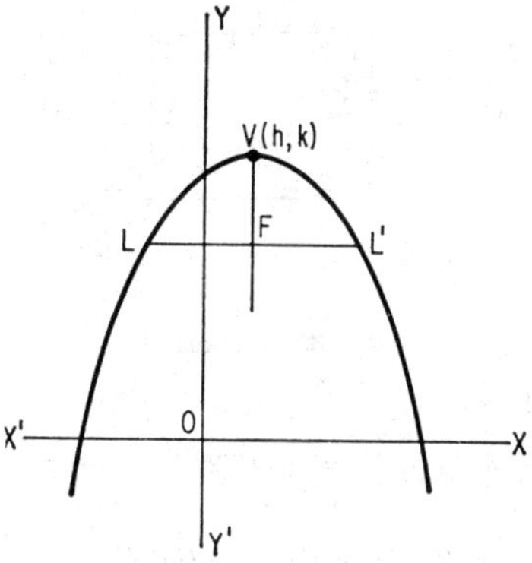

FIG. 1.30. Parabola.

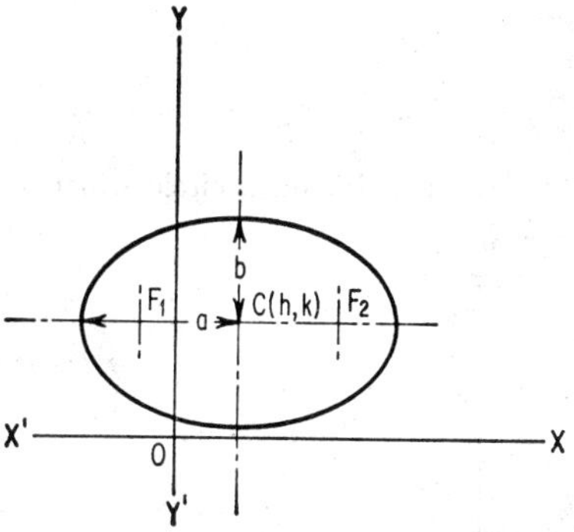

FIG. 1.31. Ellipse.

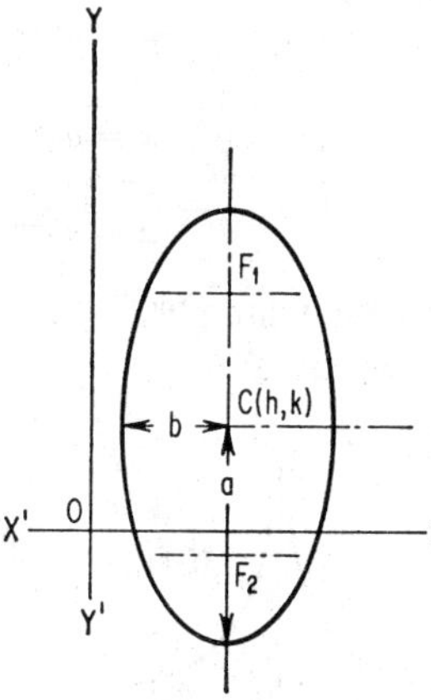

FIG. 1.32. Ellipse.

Equations of the Hyperbola. The locus of a point such that the difference of its distances from two fixed points is constant is a hyperbola.

$$\frac{(x-h)^2}{a^2} - \frac{(y-k)^2}{b^2} = 1$$ center at h, k. Transverse axis parallel to x axis. Slopes of asymptotes $= \pm b/a$ (Fig. 1.33)

$$\frac{(y-k)^2}{a^2} - \frac{(x-h)^2}{b^2} = 1$$ center at h, k. Transverse axis parallel to y axis. Slopes of asymptotes $= \pm a/b$ (Fig. 1.34)

Transverse axis $= 2a$, conjugate axis $= 2b$, eccentricity $= e = \sqrt{a^2 + b^2}/a$.
Latus rectum $= 2b^2/a^2$, distance from center to either focus $= \sqrt{a^2 + b^2}$.
Distance from center to either directrix $= a/e$.

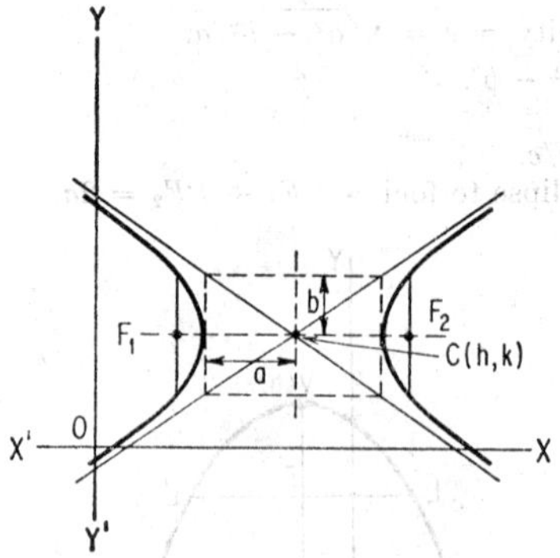

FIG. 1.33. Hyperbola.

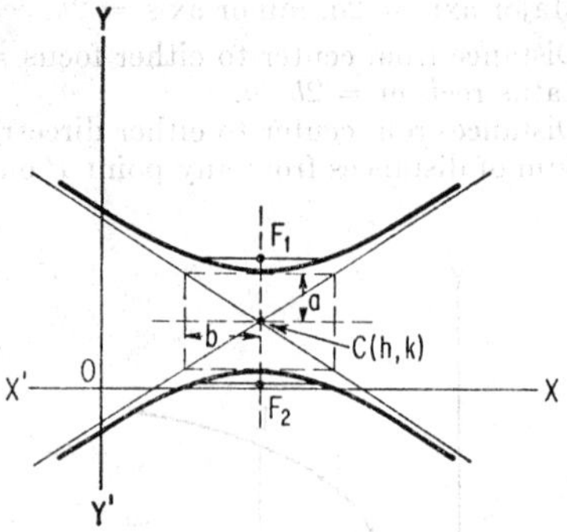

FIG. 1.34. Hyperbola.

Rectangular or Equilateral Hyperbola (Fig. 1.35)

$(x - h)(y - k) = \pm a^2/2$ center at h, k. Asymptotes perpendicular to one another. Asymptotes parallel to x and y axes.

Logarithmic or Exponential Curves

$$y = ab^x \quad \text{or} \quad x = \log_b \frac{y}{a}$$

$$y = ab^{-x} \quad \text{or} \quad x = -\log_b \frac{y}{a}$$

$$x = ab^y \quad \text{or} \quad y = \log_b \frac{x}{a}$$

$$x = ab^{-y} \quad \text{or} \quad y = -\log_b \frac{x}{a}$$

Cycloid (Fig. 1.36). The cycloid is generated by a point on a circle which rolls

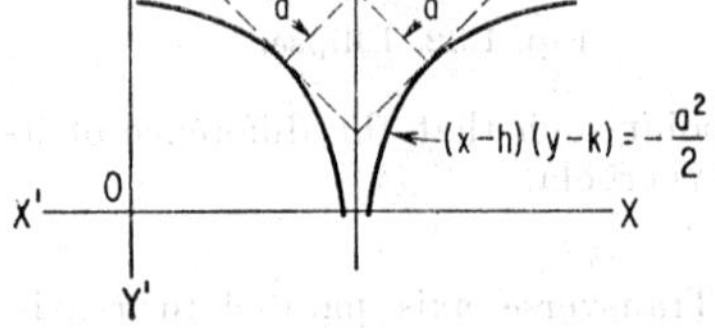

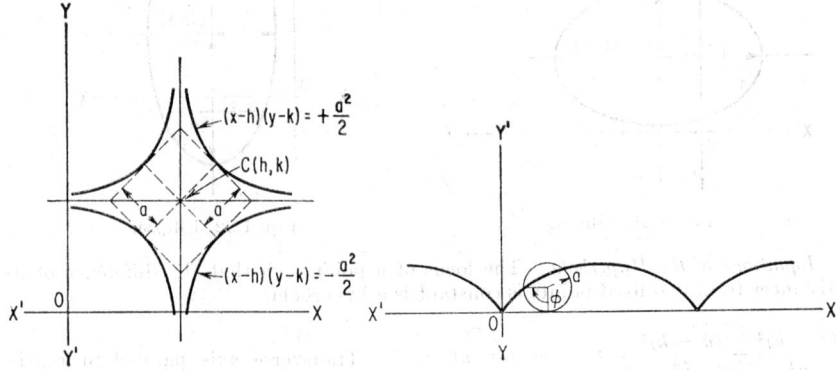

FIG. 1.35. Rectangular or equilateral hyperbola.

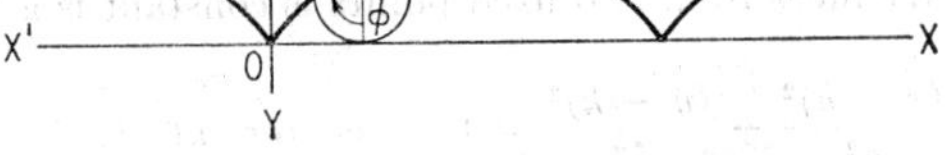

FIG. 1.36. Cycloid.

along a fixed straight line. The equations for the cycloid are $x = a(\phi - \sin \phi)$, $y = a(1 - \cos \phi)$.

Prolate and Curtate Cycloid (Figs. 1.37 and 1.38). These curves are generated by a point on a circle at a distance b from the circle whose radius is a, as the circle rolls along a fixed straight line. The equations for the prolate and curtate cycloid are

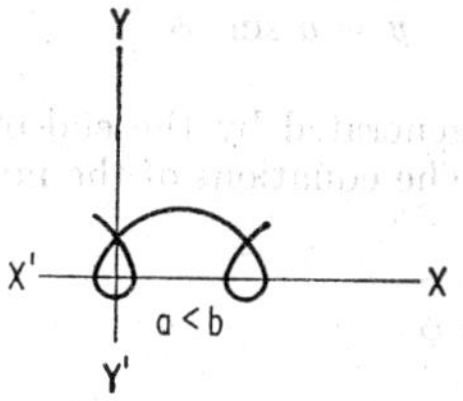

FIG. 1.37. Prolate cycloid.

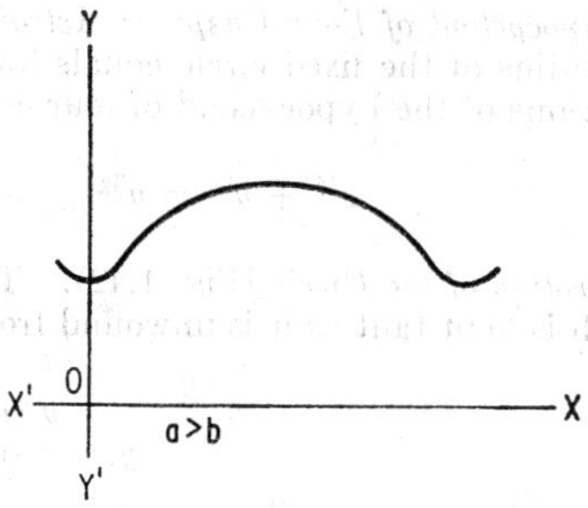

FIG. 1.38. Curtate cycloid.

$x = a\phi - b \sin \phi$, $y = a - b \cos \phi$. For the prolate cycloid $a < b$; for the curtate cycloid $a > b$.

Epicycloid (Fig. 1.39). This curve is generated by a point on a circle which rolls

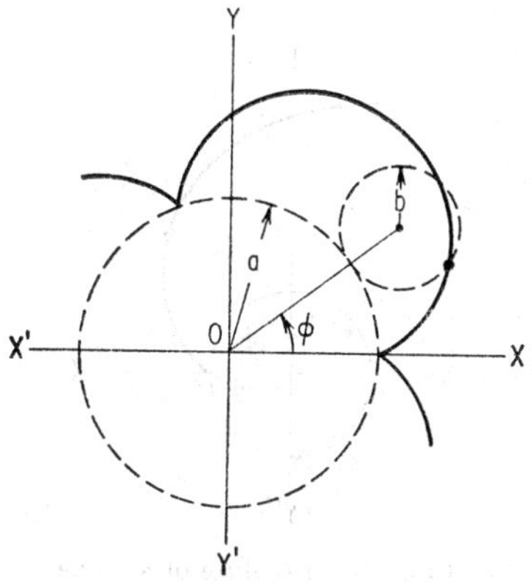

FIG. 1.39. Epicycloid.

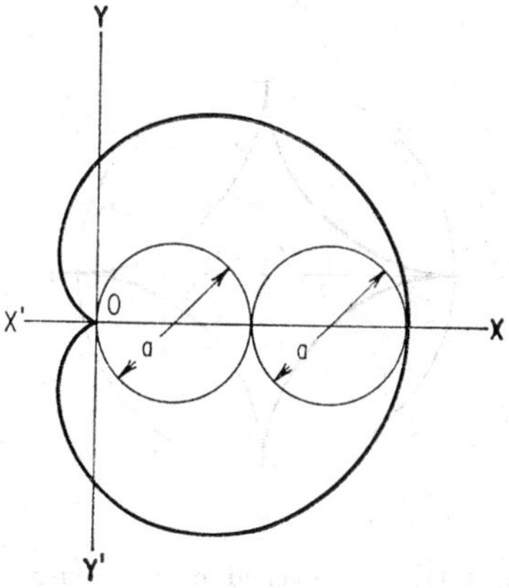

FIG. 1.40. Cardioid.

along the outside of a fixed circle. The equations of the epicycloid are

$$x = (a + b) \cos \phi - b \cos \left(\frac{a + b}{b} \phi \right)$$

$$y = (a + b) \sin \phi - b \sin \left(\frac{a + b}{b} \phi \right)$$

Cardioid. The cardioid is an epicycloid with radii of fixed and rolling circles equal. The equations of the cardioid are

$r = a(1 + \cos \theta)$	Fig. 1.40
$r = a(1 + \sin \theta)$	Fig. 1.40 rotated through +90°
$r = a(1 - \sin \theta)$	Fig. 1.40 rotated through −90°
$r = a(1 - \cos \theta)$	Fig. 1.40 rotated through +180°

Hypocycloid. This curve is generated by a point on a circle which rolls along the inside of a fixed circle. The equations of the hypocycloid are

$$x = (a - b) \cos \phi + b \cos \left(\frac{a - b}{b} \phi \right)$$

$$y = (a - b) \sin \phi - b \sin \left(\frac{a - b}{b} \phi \right)$$

Hypocycloid of Four Cusps or Astroid (Fig. 1.41). This curve is generated when the radius of the fixed circle equals four times the radius of the rolling circle. The equations of the hypocycloid of four cusps are

$$x^{2/3} + y^{2/3} = a^{2/3} \qquad x = a \cos^3 \phi \qquad y = a \sin^3 \phi$$

Involute of the Circle (Fig. 1.42). This curve is generated by the end of a string which is kept taut as it is unwound from a circle. The equations of the involute are

$$x = a \cos \phi + a\phi \sin \phi$$
$$y = a \sin \phi - a\phi \cos \phi$$

Archimedean Spiral. $r = a\theta$ (Fig. 1.43).
Hyperbolic or Reciprocal Spiral. $\rho\theta = a$ (Fig. 1.44).

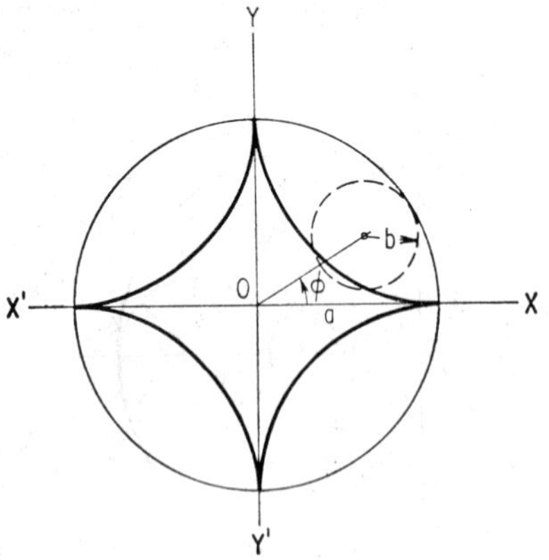

FIG. 1.41. Hypocycloid of four cusps or astroid.

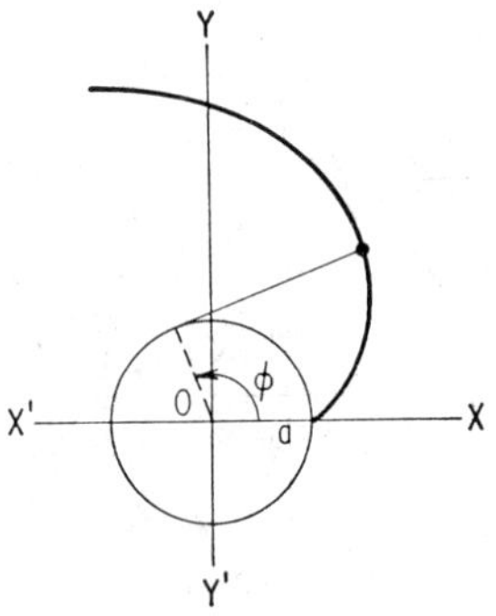

FIG. 1.42. Involute of a circle.

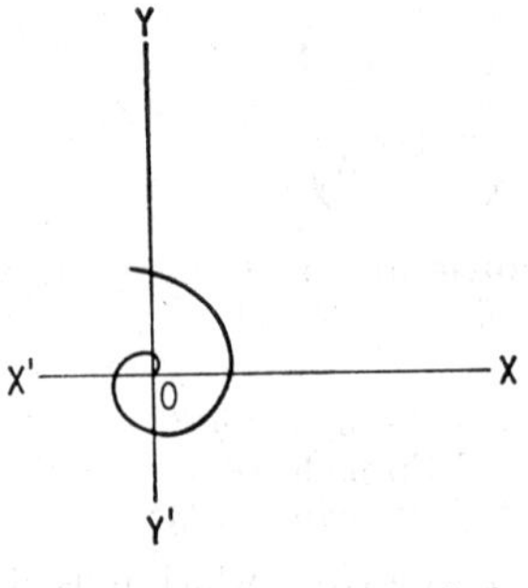

FIG. 1.43. Archimedean spiral.

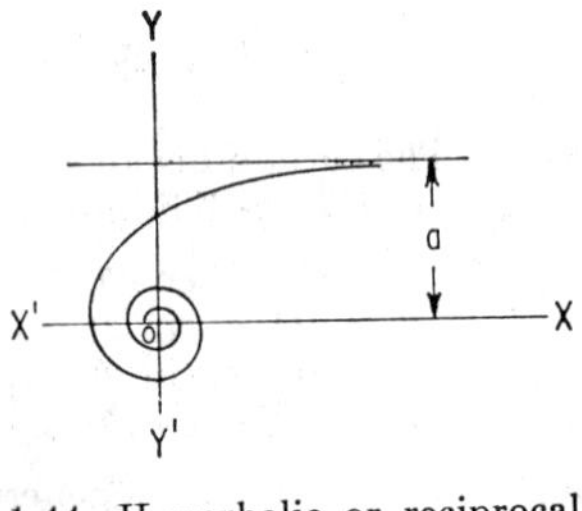

FIG. 1.44. Hyperbolic or reciprocal spiral.

Logarithmic or Equiangular Spiral. $\ln \rho = a\theta$ or $\rho = e^{a\theta}$ (Fig. 1.45).
Parabolic Spiral. $(r - a)^2 = 4aC\theta$ (Fig. 1.46).
Probability Curve. $y = e^{-x^2}$ (Fig. 1.47).
Catenary. $y = a \cosh x/a$ (Fig. 1.48).

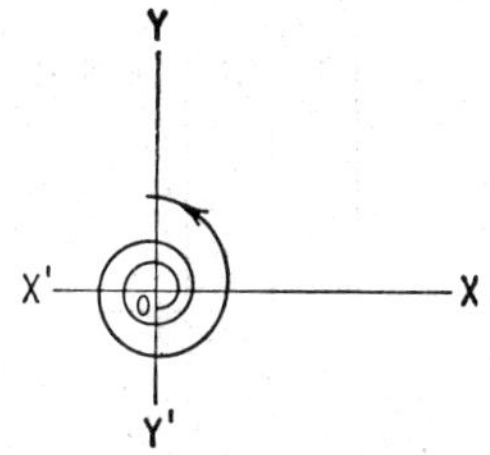

FIG. 1.45. Logarithmic or equiangular spiral.

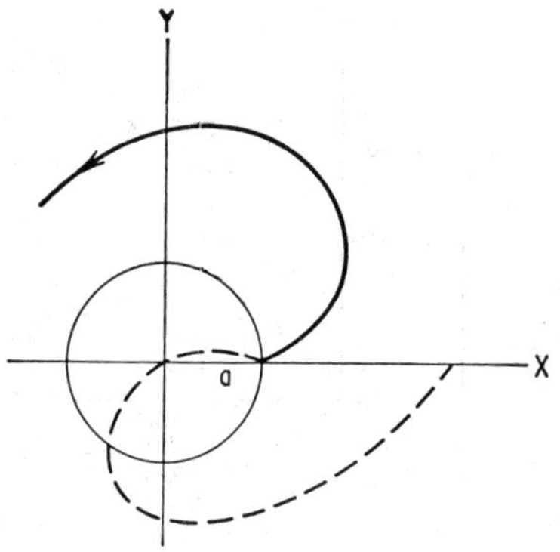

FIG. 1.46. Parabolic spiral.

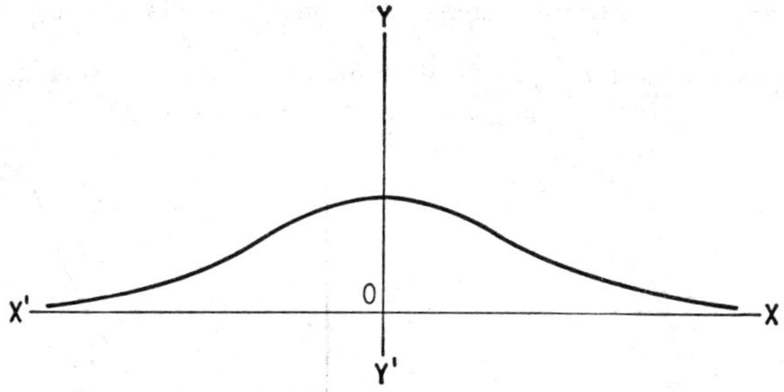

FIG. 1.47. Probability curve.

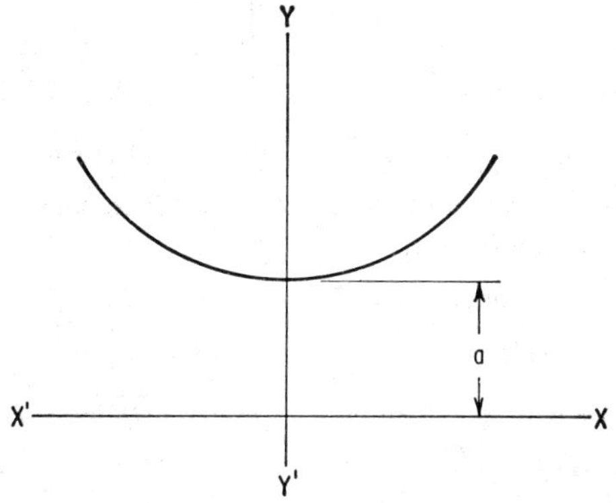

FIG. 1.48. Catenary.

Solid Analytic Geometry. *Rectangular-coordinate System.* The rectangular-coordinate system in space is defined by three mutually perpendicular coordinate axes which intersect at the origin O as shown in Fig. 1.49. The position of a point $P(x,y,z)$ is given by the distances x, y, z from the coordinate planes ZOY, XOZ, and XOY, respectively.

Cylindrical-coordinate System. The position of any point $P(r,\theta,z)$ is given by the polar coordinates r and θ, the projection of P on the XY plane, and by z, the distance from the XY plane to the point (Fig. 1.50).

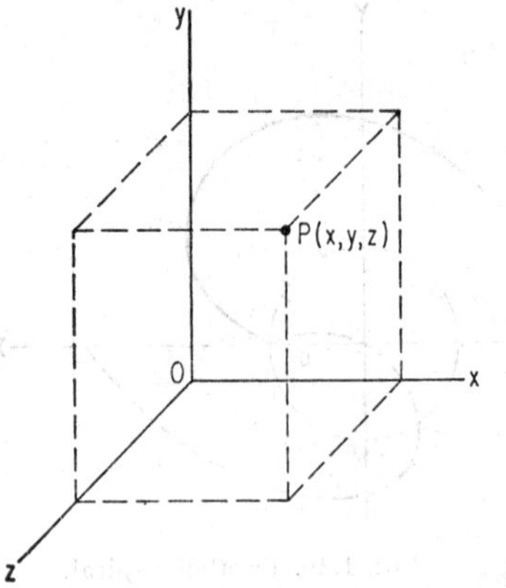

FIG. 1.49. Rectangular-coordinate system.

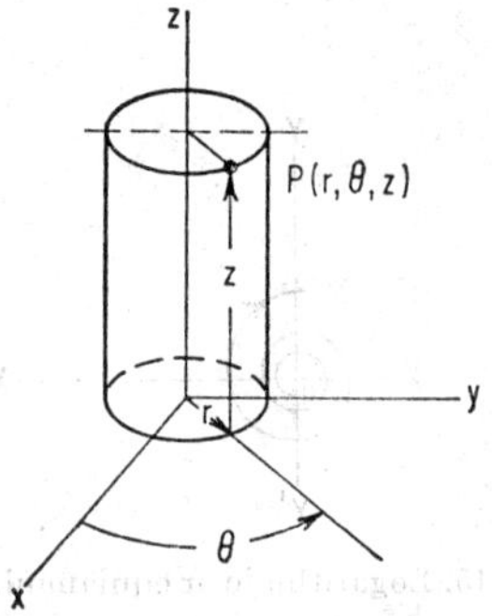

FIG. 1.50. Cylindrical-coordinate system.

Spherical-coordinate System. The position of any point $P(r,\phi,\theta)$ (Fig. 1.51) is given by the distance r $(= \overline{OP})$, the angle ϕ which is formed by the intersection of the

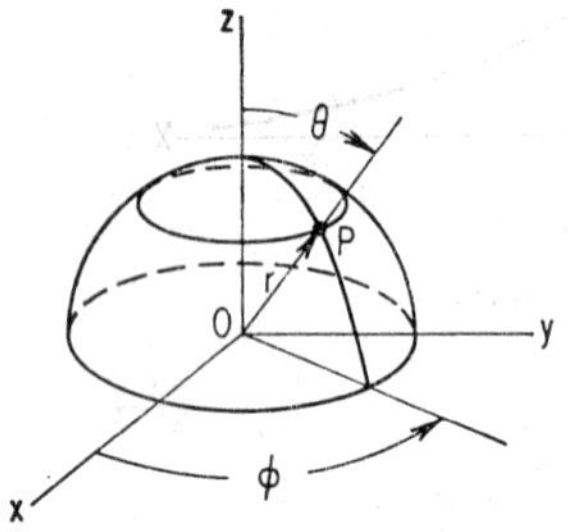

FIG. 1.51. Spherical-coordinate system.

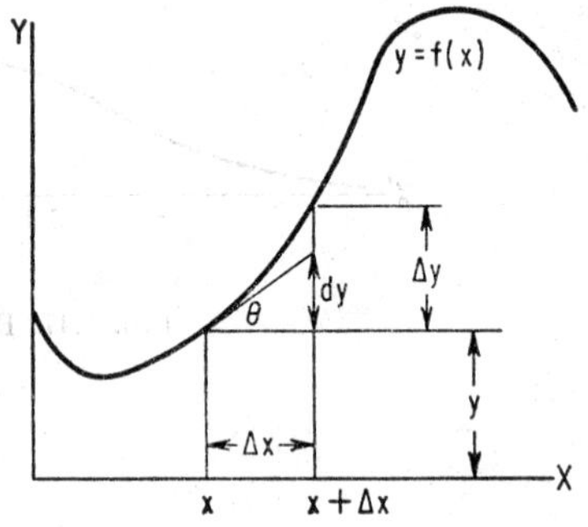

FIG. 1.52. Derivative representation.

X coordinate and the projection of OP on the XY plane, and the angle θ which is formed by $\overline{OP}$ and the coordinate z.

Relations between Coordinate Systems

Rectangular and cylindrical:

$$x = r \cos \theta \qquad y = r \sin \theta \qquad z = z \qquad \overline{OP} = \sqrt{r^2 + z^2}$$

Rectangular and spherical:

$$x = r \sin \theta \cos \phi \qquad y = r \sin \theta \sin \phi \qquad z = r \cos \theta$$

(g) Differential Calculus

Definitions. The derivative of a function $y = f(x)$ of a single variable x is

$$\frac{dy}{dx} = \lim_{\Delta x \to 0} \frac{\Delta y}{\Delta x} = \lim_{\Delta x \to 0} \frac{f(x + \Delta x) - f(x)}{\Delta x} = f'(x)$$

i.e., the derivative of the function y is the limit of the ratio of the increment of the function Δy to the increment of the independent variable Δx as the increment of x

varies and approaches zero as a limit. The derivative at a point can also be shown to equal the slope of the tangent line to the curve at the same point; i.e., $dy/dx = \tan\theta$, as shown in Fig. 1.52. The derivative of $f(x)$ is a function of x and may also be differentiated with respect to x. The first differentiation of the first derivative yields the second derivative of the function d^2y/dx^2 or $f''(x)$. Similarly, the third derivative d^3y/dx^3 or $f'''(x)$ of the function is the first derivative of d^2y/dx^2, and so on.

From Fig. 1.52 it is seen that the function $f(x)$ possesses a maximum value where the derivative is zero and the concavity is downward, and the function possesses a minimum value where the slope is zero and the curve has an upward concavity. If the function $f(x)$ is concave upward, the second derivative will have a positive value; if negative, the curve will be concave downward. If the second derivative equals zero at a point, that point is a point of inflection. More particularly, where the nature of the curve is not well known, Table 1.12 may be used to adjudge the significance of the derivatives.

Table 1.12. Significance of the Derivative

$f'(x)\Big]_{x_0}$	$f''(x)\Big]_{x_0}$	$f'''(x)\Big]_{x_0}$	$f''''(x)\Big]_{x_0}$	Comment
0	<0			x_0 a maximum point
0	>0			x_0 a minimum point
0	0	≠0		x_0 a point of inflection
0	0	0	<0	x_0 a maximum point
0	0	0	>0	x_0 a minimum point

The *partial derivative* of a function $u = u(x,y)$ of two variables, taken with respect to the variable x, is defined by

$$\frac{\partial u}{\partial x} = \lim_{\Delta x \to 0} \frac{u(x + \Delta x, y) - u(x,y)}{\Delta x} = u_x = D_x u$$

The partial derivatives are taken by differentiating with respect to one of the variables only, regarding the remaining variables as momentarily constant. Thus the partial derivative of u with respect to x of the function $u = 2xy^2$ is equal to $2y^2$; similarly $\partial u/\partial y = 4xy$. Higher derivatives are similarly formed. Thus

$$\partial^2 u/\partial y^2 = u_{yy} = 4x \qquad \partial^2 u/\partial x^2 = u_{xx} = 0$$
$$\partial^2 u/\partial x\, \partial y = u_{xy} = 4y \quad \text{and} \quad \partial^2 u/\partial y\, \partial x = 4y$$

The order of differentiation in obtaining the mixed derivatives is immaterial if the derivatives are continuous.

Table 1.13 gives the conditions required to determine maxima, minima, and saddle points for $u = u(x,y)$.

Table 1.13. Conditions for Maxima, Minima, and Saddle Point

$\frac{\partial u}{\partial x}\Big]_{x_0,y_0}$	$\frac{\partial u}{\partial y}\Big]_{x_0,y_0}$	$\frac{\partial u^2}{\partial x^2}\Big]_{x_0,y_0}$	$\frac{\partial^2 u}{\partial y^2}\Big]_{x_0,y_0}$	Comments
0	0	<0	<0	$u(x,y)$ maximum at x_0, y_0 if $(u_{xx})(u_{yy}) - (u_{xy})^2 > 0$
0	0	>0	>0	$u(x,y)$ minimum at x_0, y_0 if $(u_{xx})(u_{yy}) - (u_{xy})^2 > 0$
0	0			Saddle point if u_{xx} and u_{yy} are of different sign

Implicit Functions. If y is an implicit function of x, as, for example,

$$xy - 5x^3y^2 = 3$$

and if it is difficult to solve the equation for y (or x), differentiate the terms as given,

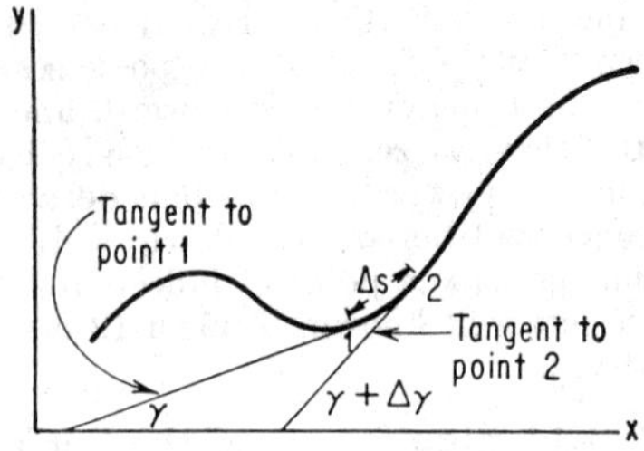

FIG. 1.53. Curvature.

treating y as a function of x and solving for dy/dx. Thus, taking the derivatives of the above expression,

$$(d/dx)(xy) - (d/dx)(5x^3y^2) = (d/dx)(3)$$
$$x(dy/dx) + y - 5x^3 2y(dy/dx) - y^2 15x^2 = 0$$
$$dy/dx = (15x^2y^2 - y)/(x - 10x^3y)$$

Another approach utilizes the relationship $dy/dx = -f_x/f_y$. Thus, in the foregoing example,

$$f_x = y - y^2 15x^2 \qquad f_y = x - 10x^3y$$

Curvature. The curvature K at a point P of a curve $y = f(x)$ (Fig. 1.53)

$$K \equiv \left| \lim_{\Delta s \to 0} \frac{\Delta\gamma}{\Delta s} \right| = \frac{d\gamma}{ds}$$

A working expression for the curvature is

$$K = \frac{y''}{[1 + (y')^2]^{3/2}}$$

where the derivatives are evaluated at the point P. For a curve described in polar coordinates, the corresponding expression for the curvature is

$$K = \frac{\rho^2 + 2\rho'^2 - \rho\rho''}{(\rho^2 + \rho'^2)^{3/2}}$$

where ρ' and ρ'' represent the first and second derivatives of ρ with respect to θ.

The circle of curvature is tangent, on its concave side, to the curve $y = f(x)$ at P. The circle of curvature and the given curve have equal curvature at P. The circle of curvature is described by a radius of curvature R located at the center of curvature.

$$R = \frac{1}{K} = \frac{(1 + y'^2)^{3/2}}{y''}$$

The center of curvature is located at (α,β) given by the following expressions:

$$\alpha = x - \frac{y'(1 + y'^2)}{y''} \qquad \beta = y + \frac{1 + y'^2}{y''}$$

where the expressions are evaluated at the point P.

Table 1.14. Derivatives

u, v, and w are functions of x

$$\frac{d}{dx}(a) = 0 \qquad a = \text{constant}$$

$$\frac{d}{dx}(x) = 1$$

$$\frac{dy}{dx} = \frac{dy}{dv}\frac{dv}{dx} \qquad y = y(v)$$

$$\frac{d}{dx}(au) = a\frac{du}{dx}$$

$$\frac{dy}{dx} = \frac{1}{dx/dy} \qquad \text{if } \frac{dx}{dy} \neq 0$$

$$\frac{d}{dx}(\pm u \pm v \pm \cdots) = \pm\frac{du}{dx} \pm \frac{dv}{dx} \pm \cdots$$

$$\frac{d}{dx}(u^n) = nu^{n-1}\frac{du}{dx}$$

$$\frac{d}{dx}(uv) = u\frac{dv}{dx} + v\frac{du}{dx}$$

$$\frac{d}{dx}\frac{u}{v} = \frac{v\,du/dx - u\,dv/dx}{v^2}$$

$$\frac{d}{dx}(u^v) = vu^{v-1}\frac{du}{dx} + u^v \ln u \frac{dv}{dx}$$

$$\frac{d}{dx}(a^u) = a^u \ln a \frac{du}{dx}$$

$$\frac{d}{dx}(e^u) = e^u\frac{du}{dx}$$

$$\frac{d}{dx}(\ln u) = \frac{1}{u}\frac{du}{dx}$$

$$\frac{d}{dx}(\log_a u) = \frac{\log_a e}{u}\frac{du}{dx}$$

$$\frac{d}{dx}(\sin u) = \cos u \frac{du}{dx}$$

$$\frac{d}{dx}(\cos u) = -\sin u \frac{du}{dx}$$

$$\frac{d}{dx}(\tan u) = \sec^2 u \frac{du}{dx}$$

$$\frac{d}{dx}(\csc u) = -\csc u \cot u \frac{du}{dx}$$

$$\frac{d}{dx}(\sec u) = \sec u \tan u \frac{du}{dx}$$

$$\frac{d}{dx}(\cot u) = -\csc^2 u \frac{du}{dx}$$

$$\frac{d}{dx}(\text{vers } u) = \sin u \frac{du}{dx}$$

$$\frac{d}{dx}\sin^{-1} u = \frac{1}{\sqrt{1-u^2}}\frac{du}{dx} \qquad \frac{-\pi}{2} \leqq \sin^{-1} u \leqq \frac{\pi}{2}$$

$$\frac{d}{dx}\cos^{-1} u = -\frac{1}{\sqrt{1-u^2}}\frac{du}{dx} \qquad 0 \leqq \cos^{-1} u \leqq \pi$$

$$\frac{d}{dx}\tan^{-1} u = \frac{1}{1+u^2}\frac{du}{dx}$$

Table 1.14. Derivatives (Continued)

u, v, and w are functions of x

$$\frac{d}{dx}\sinh^{-1}u = \frac{1}{\sqrt{u^2+1}}\frac{du}{dx}$$

$$\frac{d}{dx}\cosh^{-1}u = \frac{1}{\sqrt{u^2-1}}\frac{du}{dx} \qquad u > 1$$

$$\frac{d}{dx}\tanh^{-1}u = \frac{1}{1-u^2}\frac{du}{dx}$$

$$\frac{d}{dx}\operatorname{csch}^{-1}u = -\frac{1}{u\sqrt{u^2+1}}\frac{du}{dx}$$

$$\frac{d}{dx}\operatorname{sech}^{-1}u = -\frac{1}{u\sqrt{1-u^2}}\frac{du}{dx} \qquad u > 0$$

$$\frac{d}{dx}\coth^{-1}u = \frac{1}{1-u^2}\frac{du}{dx}$$

$$\frac{d}{dx}\csc^{-1}u = -\frac{1}{u\sqrt{u^2-1}}\frac{du}{dx} \qquad -\pi < \csc^{-1}u \leqq -\frac{\pi}{2},\ 0 < \csc^{-1}u \leqq \frac{\pi}{2}$$

$$\frac{d}{dx}\sec^{-1}u = \frac{1}{u\sqrt{u^2-1}}\frac{du}{dx} \qquad -\pi \leqq \sec^{-1}u < -\frac{\pi}{2},\ 0 \leqq \sec^{-1}u < \frac{\pi}{2}$$

$$\frac{d}{dx}\cot^{-1}u = \frac{-1}{1+u^2}\frac{du}{dx}$$

$$\frac{d}{dx}\operatorname{vers}^{-1}u = \frac{1}{\sqrt{2u-u^2}}\frac{du}{dx} \qquad 0 \leqq \operatorname{vers}^{-1}u \leqq \pi$$

$$\frac{d}{dx}\sinh u = \cosh u\frac{du}{dx}$$

$$\frac{d}{dx}\cosh u = \sinh u\frac{du}{dx}$$

$$\frac{d}{dx}\tanh u = \operatorname{sech}^2 u\frac{du}{dx}$$

$$\frac{d}{dx}\operatorname{csch} u = -\operatorname{csch} u\coth u\frac{du}{dx}$$

$$\frac{d}{dx}\operatorname{sech} u = -\operatorname{sech} u\tanh u\frac{du}{dx}$$

$$\frac{d}{dx}\coth u = -\operatorname{csch}^2 u\frac{du}{dx}$$

Differentials. The differential of a function is equal to the derivative of the function multiplied by the differential of the independent variable. Thus

$$dy = dy/dx\,dx = f'(x)\,dx$$

The total differential dz of a function of two variables $z = z(x,y)$ is

$$dz = \partial z/\partial x\,dx + \partial z/\partial y\,dy$$
$$dz/dt = \partial z/\partial x\,dx/dt + \partial z/\partial y\,dy/dt$$

For a function of three variables $u = u(x,y,z)$,

$$du = \partial u/\partial x\,dx + \partial u/\partial y\,dy + \partial u/\partial z\,dz$$
$$du/dt = \partial u/\partial x\,dx/dt + \partial u/\partial y\,dy/dt + \partial u/\partial z\,dz/dt$$

x, y, z being functions of the independent variable t.

In the following relationships for differential of arc in rectangular coordinates ds represents the differential of arc and τ is the angle of the tangent drawn at the point in question; i.e., $\tan \tau$ = slope

$$ds^2 = dx^2 + dy^2$$
$$ds = [1 + (dy/dx)^2]^{1/2}\, dx = [1 + (dx/dy)^2]^{1/2}\, dy$$
$$dx/ds = \cos \tau = \frac{1}{(1 + y'^2)^{1/2}} \qquad dy/ds = \sin \tau = \frac{y'}{(1 + y'^2)^{1/2}}$$

In the following relationships for differential of arc in polar coordinates for the function $\rho = \rho(\theta)$ ds represents the differential of arc:

$$ds = \sqrt{d\rho^2 + \rho^2\, d\theta^2} \qquad ds = [\rho^2 + (d\rho/d\theta)^2]^{1/2}\, d\theta$$

Indeterminate Forms. The function $f(x) = u(x)/v(x)$ has an indeterminate form 0/0 at $x = a$ if $u(x)$ and $v(x)$ each approach zero as x approaches a through values greater than a $(x \to a+)$. The function $f(x)$ is not defined at $x = a$, and therefore, it is often useful to assign a value to $f(a)$. L'Hôpital's rule is readily applied to indeterminacies of the form 0/0:

$$\lim_{x \to a+} u(x)/v(x) = \lim_{x \to a+} u'(x)/v'(x)$$

L'Hôpital's rule may be reapplied as often as necessary, but it is important to remember to differentiate numerator and denominator separately. The above discussion is equally valid if $x \to a-$.

Other indeterminate forms such as ∞/∞, $0 \cdot \infty$, $\infty - \infty$, 0^0, ∞^0, and 1^∞ may also be evaluated by L'Hôpital's rule by changing their forms. For example, in order to evaluate the indeterminate form $0 \cdot \infty$, the function $u(x)v(x)$ may be written $\dfrac{u(x)}{1/v(x)}$ and the same technique employed as before.

(h) Integral Calculus

Indefinite Integrals. The inverse operation of differentiation is integration; i.e., given a differential of a function, the process of integration yields the original function. This process is denoted by the integral sign $\int$ in front of the differential of the function. Thus

$$\int f(x)\, dx = F(x) + C$$

if

$$f(x)\, dx = dF(x)$$

or

$$\frac{dF(x)}{dx} = f(x)$$

C is an arbitrary constant of integration, and the expression $F(x) + C$ is the indefinite integral of $f(x)$ [or $F'(x)$]. The constant of integration is required because any constant term in the function $F(x)$ does not appear in $f(x)$.

Definite Integrals. The difference between the indefinite integral evaluated at b, i.e., the variable equated to b, and the same integral evaluated at a is a definite integral with upper limit b and lower limit a. This is termed *integration between limits.* Since the constant of integration cancels in subtraction, it need not be introduced. Thus

$$\int_a^b f(x)\, dx = F(x)\Big|_a^b = F(b) - F(a)$$

which is always a definite number, can be interpreted geometrically as the area bounded by the x axis, the lines $x = a$ and $x = b$, and the curve $y = f(x)$ provided a and b are finite, $f(x)$ does not cross the x axis, and $f(x)$ is not infinite between a and b. The following fundamental theorems apply to definite integrals:

$$\int_a^b f(x)\,dx = -\int_b^a f(x)\,dx$$

$$\int_a^b f(x)\,dx = \int_a^\phi f(x)\,dx + \int_\phi^b f(x)\,dx$$

$$\int_a^b Kf(x)\,dx = K\int_a^b f(x)\,dx \qquad \text{where } K = \text{a constant}$$

$$\int_a^b [f_1(x) + f_2(x) + \cdots + f_n(x)]\,dx = \int_a^b f_1(x)\,dx + \int_a^b f_2(x)\,dx + \cdots + \int_a^b f_n(x)\,dx$$

$$d/dx \int_a^x f(\xi)\,d\xi = f(x)$$

Improper Integrals

$$\int_a^{+\infty} f(x)\,dx = \lim_{b\to+\infty} \int_a^b f(x)\,dx \qquad \text{if the limit exists}$$

$$\int_{-\infty}^b f(x)\,dx = \lim_{a\to-\infty} \int_a^b f(x)\,dx \qquad \text{if the limit exists}$$

$$\int_a^b f(x)\,dx = \lim_{\epsilon\to 0} \int_{a+\epsilon}^b f(x)\,dx$$

$f(x)$ discontinuous at $x = a$, $a < b$, ϵ positive, and limit exists

$$\int_a^b f(x)\,dx = \lim_{\epsilon\to 0} \int_a^{b-\epsilon} f(x)\,dx$$

$f(x)$ discontinuous at $x = b$, $a < b$, ϵ positive, and limit exists

Elliptic Integrals (See Table 1.5). *Elliptic Integral of the First Kind*

$$F(\phi,k) = \int_0^\phi \frac{d\theta}{\sqrt{1 - k^2 \sin^2\theta}} = \int_0^x \frac{dt}{\sqrt{(1-t^2)(1-k^2t^2)}} \qquad k^2 < 1,\ x = \sin\phi$$

Elliptic Integral of the Second Kind

$$E(\phi,k) = \int_0^\phi \sqrt{1 - k^2\sin^2\theta}\,d\theta = \int_0^x \frac{\sqrt{1-k^2t^2}}{\sqrt{1-t^2}}\,dt \qquad k^2 < 1,\ x = \sin\phi$$

Elliptic Integral of the Third Kind

$$\Pi(\phi,n,k) = \int_0^\phi \frac{d\theta}{(1 + n\sin^2\theta)\sqrt{1 - k^2\sin^2\theta}} = \int_0^x \frac{dt}{(1+nt^2)\sqrt{(1-t^2)(1-k^2t^2)}} \qquad k^2 < 1,\ x = \sin\phi$$

The complete elliptic integrals are

$$K = F\left(\frac{\pi}{2}, k\right) = \frac{\pi}{2}\left[1 + \left(\frac{1}{2}\right)^2 k^2 + \left(\frac{1\times 3}{2\times 4}\right)^2 k^4 + \left(\frac{1\times 3\times 5}{2\times 4\times 6}\right)^2 k^6 + \cdots\right],\ k^2 < 1$$

$$E = E\left(\frac{\pi}{2}, k\right) = \frac{\pi}{2}\left[1 - \left(\frac{1}{2}\right)^2 k^2 - \left(\frac{1\times 3}{2\times 4}\right)^2 \frac{k^4}{3} - \left(\frac{1\times 3\times 5}{2\times 4\times 6}\right)^2 \frac{k^6}{5} - \cdots\right],\ k^2 < 1$$

Approximate Numerical Value of an Elliptic Integral of the First and Second Kind (Landen's Method). Elliptic Integral of the First Kind

$$F(\phi,k) = \int_0^\phi \frac{d\theta}{\sqrt{1 - k^2 \sin^2 \theta}} \qquad (k^2 < 1)$$

$$= (1 + k_1)(1 + k_2)(1 + k_3) \cdots (1 + k_n) \frac{F(k_{n_1}\phi_n)}{2^n}$$

where

$$k_1 = \frac{1 - \sqrt{1 - k^2}}{1 + \sqrt{1 - k^2}}$$

and in general

$$k_m = \frac{1 - \sqrt{1 - k_{m-1}}}{1 + \sqrt{1 - k_{m-1}}}$$

$$\tan (\phi_1 - \phi) = \sqrt{1 - k^2} \tan \phi$$

and in general

$$\tan (\phi_m - \phi_{m-1}) = \sqrt{1 - k^2{}_{m-1}} \tan \phi_{m-1}$$

Also, since $k_n \to 0$ and $F(k_n,\phi_n) \to \phi_n$ as $n \to \infty$, the integral can be approximated by

$$F(\phi,k) \cong (1 + k_1)(1 + k_2)(1 + k_3) \cdots (1 + k_n)\phi_n/2^n$$

Elliptic Integral of the Second Kind

$$E(\phi,k) = \int_0^\phi \sqrt{1 - k^2 \sin^2 \theta}\, d\theta$$

$$= F(\phi,k)[1 - (k^2/2)(1 + k_1/2 + k_1k_2/2^2 + k_1k_2k_3/2^3 + \cdots)]$$

$$+ k\left[\left(\frac{\sqrt{k_1}}{2}\right) \sin \phi_1 + \left(\frac{\sqrt{k_1k_2}}{2^2}\right) \sin \phi_2 + \left(\frac{\sqrt{k_1k_2k_3}}{2^3}\right) \sin \phi_3 + \cdots\right]$$

where the recursion formulas from above apply.

The Gamma Function. The *gamma* or *generalized factorial* function is defined by

$$\Gamma(n) = \int_0^\infty x^{n-1}e^{-x}\, dx \qquad \text{for } n > 0$$

and by means of the recursion formula

$$\Gamma(n) = \Gamma(n + 1)/n$$

for negative values of n.

The gamma function is infinite when n assumes the value of a negative integer or zero (Fig. 1.54). When n is a positive integer, the relation

$$\Gamma(n + 1) = n! \qquad n = 1, 2, 3, \ldots$$

holds. It also follows that

$$\begin{aligned} 0! = \Gamma(1) &= 1 \\ \Gamma(2) &= 1 \\ \Gamma(3) &= 2 \\ \Gamma(4) &= 6, \text{ etc.} \end{aligned}$$

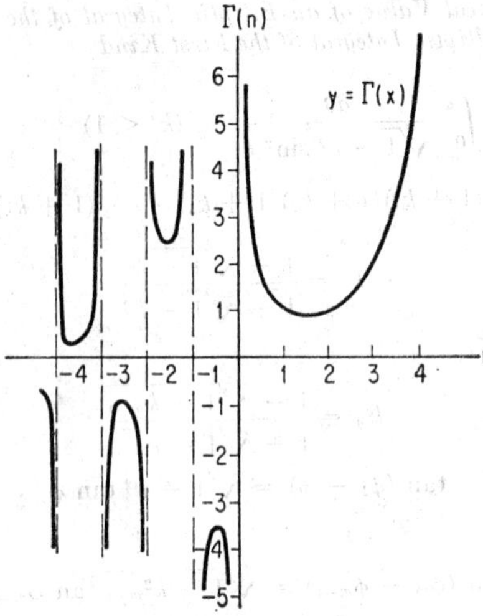

FIG. 1.54. The gamma function.

The Beta Function. The beta function is defined by

$$\beta(m,n) = \int_0^1 x^{m-1}(1 - x)^{n-1}\,dx \qquad m, n > 0$$

or by letting

$$x = \sin^2 \phi$$

$$\beta(m,n) = 2 \int_0^{\frac{\pi}{2}} (\sin \phi)^{2m-1}(\cos \phi)^{2n-1}\,d\phi$$

The beta function and gamma function are connected by the expression

$$\beta(m,n) = \frac{\Gamma(m)\Gamma(n)}{\Gamma(m + n)}$$

The Error Function. The error function is defined by

$$\text{erf}\,(x) = 2/\sqrt{\pi} \int_0^x e^{-n^2}\,dn \qquad \text{(see Table 1.6)}$$

It follows from the definition that

$$\text{erf}\,(0) = 0$$
$$\text{erf}\,(-x) = -\,\text{erf}\,(x)$$
$$\text{erf}\,(\infty) = 1$$
$$\text{erf}\,(iz) = 2i/\sqrt{\pi} \int_0^z e^{n^2}\,dn \qquad \text{where } i = \sqrt{-1}$$

Integration by Parts

$$\int u\,dv = uv - \int v\,du \qquad \text{where } u = u(x) \text{ and } v = v(x)$$

For example, to evaluate $\int x \cos x\,dx$, let $u = x$ and $dv = \cos x\,dx$. Then

$$du = dx \qquad \text{and} \qquad v = \int \cos x\,dx = \sin x$$

Thus $\int x \cos x\,dx = x \sin x - \int \sin x\,dx = x \sin x + \cos x + C$

Also $$\int_a^b u\,dv = uv\Big|_a^b - \int_a^b v\,du$$

Table 1.15. Table of Integrals

Fundamental Indefinite Integrals

1. $\int a\,dx = ax$
2. $\int (u + v + w + \cdots)\,dx = \int u\,dx + \int v\,dx + \int w\,dx + \cdots$
3. $\displaystyle\int x^n\,dx = \frac{x^{n+1}}{n+1} \qquad (n \neq -1)$
4. $\displaystyle\int \frac{dx}{x} = \log_e x + c = \log_e c_1 x \qquad [\log_e x = \log_e\ -x) + (2k+1)\pi i]$
5. $\displaystyle\int e^{ax}\,dx = \frac{1}{a} e^{ax}$
6. $\displaystyle\int a^x\,dx = \frac{a^x}{\log_e a}$
7. $\int a^x \log_e a\,dx = a^x$
8. $\displaystyle\int \sin ax\,dx = -\frac{1}{a}\cos ax$
9. $\displaystyle\int \sin^2 ax\,dx = \frac{1}{2}x - \frac{1}{2a}\sin ax \cos ax$
10. $\displaystyle\int \sin^{-1} ax\,dx = x \sin^{-1} ax + \frac{1}{a}\sqrt{1 - a^2x^2}$
11. $\displaystyle\int \cos ax\,dx = \frac{1}{a}\sin ax$
12. $\displaystyle\int \cos^2 ax\,dx = \frac{1}{2}x + \frac{1}{2a}\sin ax \cos ax$
13. $\int \cos^{-1} ax\,dx = x \cos^{-1} ax - 1/a \sqrt{1 - a^2x^2}$
14. $\int \tan ax\,dx = -1/a \log_e \cos ax$
15. $\int \tan^2 ax\,dx = 1/a \tan ax - x$
16. $\int \tan^{-1} ax\,dx = x \tan^{-1} ax - 1/2a \log_e (1 + a^2x^2)$
17. $\int \cot ax\,dx = 1/a \log_e \sin ax$
18. $\int \cot^2 ax\,dx = -1/a \cot ax - x$
19. $\int \cot^{-1} ax\,dx = x \cot^{-1} ax + 1/2a \log_e (1 + a^2x^2)$
20. $\int \sec ax\,dx = 1/a \log_e (\sec ax + \tan ax)$
21. $\int \sec^2 ax\,dx = 1/a \tan ax$
22. $\int \sec^{-1} ax\,dx = x \sec^{-1} ax - 1/a \log_e (ax + \sqrt{a^2x^2 - 1})$
23. $\displaystyle\int \csc ax\,dx = \frac{1}{a}\log_e (\csc ax - \cot ax)$
24. $\int \csc^2 ax\,dx = -1/a \cot ax$
25. $\int \csc^{-1} ax\,dx = x \csc^{-1} ax + 1/a \log_e (ax + \sqrt{a^2x^2 - 1})$
26. $\displaystyle\int \sinh ax\,dx = \frac{1}{a}\cosh ax$
27. $\displaystyle\int \cosh ax\,dx = \frac{1}{a}\sinh ax$
28. $\displaystyle\int \tanh ax\,dx = \frac{1}{a}\log_e (\cosh ax)$
29. $\displaystyle\int \coth ax\,dx = \frac{1}{a}\log_e (\sinh ax)$
30. $\displaystyle\int \operatorname{sech} ax\,dx = \frac{1}{a}\sin^{-1} (\tanh ax)$
31. $\displaystyle\int \operatorname{csch} ax\,dx = \frac{1}{a}\log_e \left(\tanh \frac{ax}{2}\right)$

Table 1.15. Table of Integrals (Continued)

Fundamental Indefinite Integrals (Continued)

32. $\int \frac{dx}{a^2 + x^2} = \frac{1}{a} \tan^{-1} \frac{x}{a}$

33. $\int \frac{dx}{\sqrt{a^2 - x^2}} = \sin^{-1} \frac{x}{a}$

34. $\int \frac{dx}{\sqrt{2x - x^2}} = \text{versin}^{-1} x$, or $-\text{coversin}^{-1} x$

35. $\int \frac{dx}{x\sqrt{x^2 - 1}} = \sec^{-1} x$, or $-\csc^{-1} x$

36. $\int u \frac{dv}{dx} dx = uv - \int v \frac{du}{dx} dx$

37. $\int u\, dv = uv - \int v\, du$

Integrals Involving $\sin^n ax$ *and* $\sin^{-1} x$

38. $\int \sin ax \sin bx\, dx = \frac{\sin (a - b)x}{2(a - b)} - \frac{\sin (a + b)x}{2(a + b)} \quad (a^2 \neq b^2)$

39. $\int \sin nx \sin^m x\, dx = \frac{1}{m + n}\left[-\cos nx \sin^m x + m \int \cos (n - 1)x \sin^{m-1} x\, dx\right]$

40. $\int \sin ax \sin bx \sin cx\, dx = -\frac{1}{4}\left[\frac{\cos (a - b + c)x}{a - b + c} + \frac{\cos (b + c - a)x}{b + c - a} + \frac{\cos (a + b - c)x}{a + b - c} - \frac{\cos (a + b + c)x}{a + b + c}\right]$

41. $\int (\sin^{-1} x)^2\, dx = x(\sin^{-1} x)^2 - 2x + 2\sqrt{1 - x^2} \sin^{-1} x$

42. $\int x \sin^{-1} x\, dx = \frac{1}{4}[(2x^2 - 1) \sin^{-1} x + x\sqrt{1 - x^2}]$

43. $\int x^n \sin^{-1} x\, dx = \frac{1}{n + 1}\left(x^{n+1} \sin^{-1} x - \int \frac{x^{n+1}\, dx}{\sqrt{1 - x^2}}\right)$

44. $\int \sin^3 ax\, dx = -\frac{1}{a} \cos ax + \frac{1}{3a} \cos^3 ax$

45. $\int \sin^4 ax\, dx = \frac{3}{8} x - \frac{1}{4a} \sin 2ax + \frac{1}{32a} \sin 4ax$

46. $\int \sin^n ax\, dx = -\frac{\sin^{n-1} ax \cos ax}{na} + \frac{n - 1}{n} \int \sin^{n-2} ax\, dx \quad (n = \text{positive integer})$

47. $\int x \sin ax\, dx = \frac{\sin ax}{a^2} - \frac{x \cos ax}{a}$

48. $\int x^2 \sin ax\, dx = \frac{2x}{a^2} \sin ax - \left(\frac{x^2}{a} - \frac{2}{a^3}\right) \cos ax$

49. $\int x^3 \sin ax\, dx = \left(\frac{3x^2}{a^2} - \frac{6}{a^4}\right) \sin ax - \left(\frac{x^3}{a} - \frac{6x}{a^3}\right) \cos ax$

50. $\int x^n \sin ax\, dx = -\frac{x^n}{a} \cos ax + \frac{n}{a} \int x^{n-1} \cos ax\, dx \quad (n > 0)$

51. $\int e^{ax} \sin bx\, dx = \frac{e^{ax}}{a^2 + b^2} (a \sin bx - b \cos bx)$

52. $\int xe^{ax} \sin bx\, dx = \frac{xe^{ax}}{a^2 + b^2} (a \sin bx - b \cos bx) - \frac{e^{ax}}{(a^2 + b^2)^2} [(a^2 - b^2) \sin bx - 2ab \cos bx]$

53. $\int \frac{dx}{\sin^n ax} = -\frac{1}{a(n - 1)} \frac{\cos ax}{\sin^{n-1} ax} + \frac{n - 2}{n - 1} \int \frac{dx}{\sin^{n-2} ax} \quad (n \text{ integer} > 1)$

Table 1.15. Table of Integrals (Continued)

Integrals Involving $\sin^n ax$ *and* $\sin^{-1} x$ *(Continued)*

54. $\int \frac{x\,dx}{\sin^2 ax} = -\frac{x}{a}\cot ax + \frac{1}{a^2}\log_e \sin ax$

55. $\int \frac{dx}{1 + \sin ax} = -\frac{1}{a}\tan\left(\frac{\pi}{4} - \frac{ax}{2}\right)$

56. $\int \frac{dx}{1 - \sin ax} = \frac{1}{a}\cot\left(\frac{\pi}{4} - \frac{ax}{2}\right)$

57. $\int \frac{x\,dx}{1 + \sin ax} = -\frac{x}{a}\tan\left(\frac{\pi}{4} - \frac{ax}{2}\right) + \frac{2}{a^2}\log_e \cos\left(\frac{\pi}{4} - \frac{ax}{2}\right)$

58. $\int \frac{x\,dx}{1 - \sin ax} = \frac{x}{a}\cot\left(\frac{\pi}{4} - \frac{ax}{2}\right) + \frac{2}{a^2}\log_e \sin\left(\frac{\pi}{4} - \frac{ax}{2}\right)$

59. $\int \frac{dx}{b + d\sin ax} = \frac{-2}{a\sqrt{b^2 - d^2}}\tan^{-1}\left[\sqrt{\frac{b - d}{b + d}}\tan\left(\frac{\pi}{4} - \frac{ax}{2}\right)\right] \quad (b^2 > d^2)$

60. $\int \frac{dx}{b + d\sin ax} = \frac{-1}{a\sqrt{d^2 - b^2}}\log_e \frac{d + b\sin ax + \sqrt{d^2 - b^2}\cos ax}{b + d\sin ax} \quad (d^2 > b^2)$

61. $\int \frac{\sin ax}{x^n}\,dx = -\frac{1}{n - 1}\frac{\sin ax}{x^{n-1}} + \frac{a}{n - 1}\int \frac{\cos ax}{x^{n-1}}\,dx$

62. $\int \frac{\sin^n x\,dx}{x^m} = \frac{1}{(m - 1)(m - 2)}\left[-\frac{\sin^{n-1} x((m - 2)\sin x + nx\cos x)}{x^{m-1}}\right.$

$\left. - n^2\int \frac{\sin^n x\,dx}{x^{m-2}} + n(n - 1)\int \frac{\sin^{n-2} x\,dx}{x^{m-2}}\right]$

63. $\int \frac{\sin nx\,dx}{\sin^m x} = 2\int \frac{\cos(n - 1)x\,dx}{\sin^{m-1} x} + \int \frac{\sin(n - 2)x\,dx}{\sin^m x}$

64. $\int \frac{\sin x}{x}\,dx = x - \frac{x^3}{3 \times 3!} + \frac{x^5}{5 \times 5!} - \frac{x^7}{7 \times 7!} + \frac{x^9}{9 \times 9!} \cdots$

65. $\int \frac{\sin x}{x^m}\,dx = -\frac{1}{m - 1}\frac{\sin x}{x^{m-1}} + \frac{1}{m - 1}\int \frac{\cos x}{x^{m-1}}\,dx$

66. $\int \frac{x\,dx}{\sin x} = x + \frac{x^3}{3 \times 3!} + \frac{7x^5}{3 \times 5 \times 5!} + \frac{31x^7}{3 \times 7 \times 7!} + \frac{127x^9}{3 \times 5 \times 9!} + \cdots$

Integrals Involving $\cos^n ax$ *and* $\cos^{-1} x$

67. $\int \cos ax \cos bx\,dx = \frac{\sin(a - b)x}{2(a - b)} + \frac{\sin(a + b)x}{2(a + b)} \quad (a^2 \neq b^2)$

68. $\int \cos nx \cos^m x\,dx = \frac{1}{m + n}\left[\sin nx \cos^m x + m\int \cos(n - 1)x \cos^{m-1} x\,dx\right]$

69. $\int \cos ax \cos bx \cos cx\,dx = \frac{1}{4}\left[\frac{\sin(a + b + c)x}{a + b + c} + \frac{\sin(b + c - a)x}{b + c - a}\right.$

$\left. + \frac{\sin(a - b + c)x}{a - b + c} + \frac{\sin(a + b - c)x}{a + b - c}\right]$

70. $\int (\cos^{-1} x)^2\,dx = x(\cos^{-1} x)^2 - 2x - 2\sqrt{1 - x^2}\cos^{-1} x$

71. $\int x\cos^{-1} x\,dx = \frac{1}{4}[(2x^2 - 1)\cos^{-1} x - x\sqrt{1 - x^2}]$

72. $\int x^n \cos^{-1} x\,dx = \frac{1}{n + 1}\left(x^{n+1}\cos^{-1} x + \int \frac{x^{n+1}\,dx}{\sqrt{1 - x^2}}\right)$

73. $\int \cos^3 ax\,dx = \frac{1}{a}\sin ax - \frac{1}{3a}\sin^3 ax$

74. $\int \cos^4 ax\,dx = \frac{3}{8}x + \frac{1}{4a}\sin 2ax + \frac{1}{32a}\sin 4ax$

Table 1.15. Table of Integrals (Continued)

Integrals Involving $\cos^n ax$ and $\cos^{-1} x$ (Continued)

75. $\int \cos^n ax\,dx = \frac{\cos^{n-1} ax \sin ax}{na} + \frac{n-1}{n}\int \cos^{n-2} ax\,dx$ (n = positive integer)

76. $\int x \cos ax\,dx = \frac{\cos ax}{a^2} + \frac{x \sin ax}{a}$

77. $\int x^2 \cos ax\,dx = \frac{2x}{a^2}\cos ax + \left(\frac{x^2}{a} - \frac{2}{a^3}\right)\sin ax$

78. $\int x^3 \cos ax\,dx = \left(\frac{3x^2}{a^2} - \frac{6}{a^4}\right)\cos ax + \left(\frac{x^3}{a} - \frac{6x}{a^3}\right)\sin ax$

79. $\int x^n \cos ax\,dx = \frac{x^n \sin ax}{a} - \frac{n}{a}\int x^{n-1} \sin ax\,dx \quad (n > 0)$

80. $\int e^{ax} \cos bx\,dx = \frac{e^{ax}}{a^2 + b^2}(a \cos bx + b \sin bx)$

81. $\int xe^{ax} \cos bx\,dx = \frac{xe^{ax}}{a^2 + b^2}(a \cos bx + b \sin bx) - \frac{e^{ax}}{(a^2 + b^2)^2}[(a^2 - b^2) \cos bx + 2ab \sin bx]$

82. $\int \frac{\cos ax}{x^n}\,dx = -\frac{1}{n-1}\frac{\cos ax}{x^{n-1}} - \frac{a}{n-1}\int \frac{\sin ax}{x^{n-1}}\,dx$

83. $\int \frac{dx}{\cos^n ax} = \frac{1}{a(n-1)}\frac{\sin ax}{\cos^{n-1} ax} + \frac{n-2}{n-1}\int \frac{dx}{\cos^{n-2} ax}$ (n integer > 1)

84. $\int \frac{x\,dx}{\cos^2 ax} = \frac{x}{a}\tan ax + \frac{1}{a^2}\log_e \cos ax$

85. $\int \frac{dx}{1 + \cos ax} = \frac{1}{a}\tan\frac{ax}{2}$

86. $\int \frac{dx}{1 - \cos ax} = -\frac{1}{a}\cot\frac{ax}{2}$

87. $\int \frac{x\,dx}{1 + \cos ax} = \frac{x}{a}\tan\frac{ax}{2} + \frac{2}{a^2}\log_e \cos\frac{ax}{2}$

88. $\int \frac{x\,dx}{1 - \cos ax} = -\frac{x}{a}\cot\frac{ax}{2} + \frac{2}{a^2}\log_e \sin\frac{ax}{2}$

89. $\int \frac{dx}{b + d \cos ax} = \frac{2}{a\sqrt{b^2 - d^2}}\tan^{-1}\left(\sqrt{\frac{b-d}{b+d}}\tan\frac{ax}{2}\right) \quad (b^2 > d^2)$

90. $\int \frac{dx}{b + d \cos ax} = \frac{1}{a\sqrt{d^2 - b^2}}\log_e \frac{d + b \cos ax + \sqrt{d^2 - b^2}\sin ax}{b + d \cos ax} \quad (d^2 > b^2)$

91. $\int \frac{\cos^n x\,dx}{x^m} = \frac{1}{(m-1)(m-2)}\left[\frac{\cos^{n-1} x(nx \sin x - (m-2)\cos x)}{x^{m-1}} - n^2\int \frac{\cos^n x\,dx}{x^{m-2}} + n(n-1)\int \frac{\cos^{n-2} x\,dx}{x^{m-2}}\right]$

92. $\int \frac{\sin nx\,dx}{\cos^m x} = 2\int \frac{\sin (n-1)x\,dx}{\cos^{m-1} x} - \int \frac{\sin (n-2)x\,dx}{\cos^m x}$

93. $\int \frac{\cos x}{x}\,dx = \log x - \frac{x^2}{2 \times 2!} + \frac{x^4}{4 \times 4!} - \frac{x^6}{6 \times 6!} + \frac{x^8}{8 \times 8!} \cdots$

94. $\int \frac{\cos x}{x^m}\,dx = -\frac{1}{m-1}\frac{\cos x}{x^{m-1}} - \frac{1}{m-1}\int \frac{\sin x}{x^{m-1}}\,dx$

95. $\int \frac{x\,dx}{\cos x} = \frac{x^2}{2} + \frac{x^4}{4 \times 2!} + \frac{5x^6}{6 \times 4!} + \frac{61x^8}{8 \times 6!} + \frac{1{,}385\,x^{10}}{10 \times 8!} + \cdots$

Table 1.15. Table of Integrals (Continued)

Integrals Involving $\sin^n ax$ *and* $\cos^n ax$

96. $$\int \sin ax \cos bx\, dx = -\frac{1}{2}\left[\frac{\cos (a-b)x}{a-b} + \frac{\cos (a+b)x}{a+b}\right] \quad (a^2 \neq b^2)$$

97. $$\int \sin^n ax \cos ax\, dx = \frac{1}{a(n+1)} \sin^{n+1} ax \quad (n \neq -1)$$

98. $$\int \sin ax \cos^n ax\, dx = -\frac{1}{a(n+1)} \cos^{n+1} ax \quad (n \neq -1)$$

99. $$\int \sin^2 ax \cos^2 ax\, dx = \frac{x}{8} - \frac{\sin 4ax}{32a}$$

100. $$\int \sin^n ax \cos^m ax\, dx = -\frac{\sin^{n-1} ax \cos^{m+1} ax}{a(n+m)} + \frac{n-1}{n+m}\int \sin^{n-2} ax \cos^m ax\, dx$$

$(m, n \text{ positive})$

101. $$\int \frac{dx}{\sin ax \cos ax} = \frac{1}{a} \log_e \tan ax$$

102. $$\int \frac{dx}{b \sin ax + d \cos ax} = \frac{1}{a\sqrt{b^2 + d^2}} \log_e \tan \tfrac{1}{2}\left(ax + \tan^{-1}\frac{d}{b}\right)$$

103. $$\int \frac{\sin ax}{b + d \cos ax}\, dx = -\frac{1}{ad} \log_e (b + d \cos ax)$$

104. $$\int \frac{\cos ax}{b + d \sin ax}\, dx = \frac{1}{ad} \log_e (b + d \sin ax)$$

105. $$\int \frac{\sin^n ax}{\cos^m ax}\, dx = \frac{\sin^{n+1} ax}{a(m-1) \cos^{m-1} ax} - \frac{n-m+2}{m-1}\int \frac{\sin^n ax}{\cos^{m-2} ax}\, dx$$

$(m, n \text{ positive}, m \neq 1)$

106. $$\int \frac{\cos^m ax}{\sin^n ax}\, dx = \frac{-\cos^{m+1} ax}{a(n-1) \sin^{n-1} ax} + \frac{n-m-2}{(n-1)}\int \frac{\cos^m ax}{\sin^{n-2} ax}\, dx$$

$(m, n \text{ positive}, n \neq 1)$

107. $$\int \sin x \cos x\, dx = \tfrac{1}{2} \sin^2 x$$

108. $$\int \frac{dx}{\sin^m x \cos^n x} = \frac{1}{n-1}\,\frac{1}{\sin^{m-1} x \cos^{n-1} x} + \frac{m+n-2}{n-1}\int \frac{dx}{\sin^m x \cos^{n-2} x}$$

$$= -\frac{1}{m-1}\,\frac{1}{\sin^{m-1} x \cos^{n-1} x} + \frac{m+n-2}{m-1}\int \frac{dx}{\sin^{m-2} x \cos^n x}$$

109. $$\int \sin ax \cos bx \cos cx\, dx = -\frac{1}{4}\left[\frac{\cos (a+b+c)x}{a+b+c} - \frac{\cos (b+c-a)x}{b+c-a} + \frac{\cos (a+b-c)x}{a+b-c} + \frac{\cos (a+c-b)x}{a+c-b}\right]$$

110. $$\int \cos ax \sin bx \sin cx\, dx = \frac{1}{4}\left[\frac{\sin (a+b-c)x}{a+b-c} + \frac{\sin (a-b+c)x}{a-b+c} - \frac{\sin (a+b+c)x}{a+b+c} - \frac{\sin (b+c-a)x}{b+c-a}\right]$$

111. $$\int e^{ax} \sin^m x \cos^n x\, dx = \frac{1}{(m+n)^2 + a^2}$$
$$\left[e^{ax} \sin^m x \cos^{n-1} x[a \cos x + (m+n) \sin x] - ma\int e^{ax} \sin^{m-1} x \cos^{n-1} x\, dx + (n-1)(m+n)\int e^{ax} \sin^m x \cos^{n-2} x\, dx\right]$$

Integrals Involving $\tan^n ax$, $\sec^n ax$, $\cot^n ax$, *and* $\csc^n ax$

112. $$\int \tan^n ax\, dx = \frac{1}{a(n-1)} \tan^{n-1} ax - \int \tan^{n-2} ax\, dx \quad (n \text{ integer} > 1)$$

Table 1.15. Table of Integrals (Continued)

Integrals Involving $\tan^n ax$, $\sec^n ax$, $\cot^n ax$, *and* $\csc^n ax$ *(Continued)*

113. $\int \tan ax \sec ax \, dx = \frac{1}{a} \sec ax$

114. $\int \tan^n ax \sec^2 ax \, dx = \frac{1}{a(n+1)} \tan^{n+1} ax \quad (n \neq -1)$

115. $\int \frac{\sec^2 ax \, dx}{\tan ax} = \frac{1}{a} \log_e \tan ax$

116. $\int \cot^n ax \, dx = -\frac{1}{a(n-1)} \cot^{n-1} ax - \int \cot^{n-2} ax \, dx \quad (n \text{ integer} > 1)$

117. $\int \cot ax \csc ax \, dx = -\frac{1}{a} \csc ax$

118. $\int \cot^n ax \csc^2 ax \, dx = -\frac{1}{a(n+1)} \cot^{n+1} ax \quad (n \neq -1)$

119. $\int \frac{\csc^2 ax}{\cot ax} dx = -\frac{1}{a} \log_e \cot ax$

120. $\int \sec^n ax \, dx = \frac{1}{a(n-1)} \frac{\sin ax}{\cos^{n-1} ax} + \frac{n-2}{n-1} \int \sec^{n-2} ax \, dx \quad (n \text{ integer} > 1)$

121. $\int \csc^n ax \, dx = -\frac{1}{a(n-1)} \frac{\cos ax}{\sin^{n-1} ax} + \frac{n-2}{n-1} \int \csc^{n-2} ax \, dx \quad (n \text{ integer} > 1)$

122. $\int \frac{dx}{b + d \tan ax} = \frac{1}{b^2 + d^2} \left[bx + \frac{d}{a} \log_e (b \cos ax + d \sin ax) \right]$

123. $\int \frac{dx}{\sqrt{b + d \tan^2 ax}} = \frac{1}{a\sqrt{b-d}} \sin^{-1} \left(\sqrt{\frac{b-d}{b}} \sin ax \right) \quad (b \text{ positive}, b^2 > d^2)$

124. $\int x \tan^{-1} x \, dx = \frac{1}{2}[(x^2+1) \tan^{-1} x - x]$

125. $\int x \cot^{-1} x \, dx = \frac{1}{2}[(x^2+1) \cot^{-1} x + x]$

126. $\int x \sec^{-1} x \, dx = \frac{1}{2}(x^2 \sec^{-1} x - \sqrt{x^2 - 1})$

127. $\int x \csc^{-1} x \, dx = \frac{1}{2}(x^2 \csc^{-1} x + \sqrt{x^2 - 1})$

Integrals Involving $\sinh^n x$, $\cosh^n x$, *and* $\tanh^n x$

128. $\int \sinh^n x \, dx = \frac{1}{n} \sinh^{n-1} x \cosh x - \frac{n-1}{n} \int \sinh^{n-2} x \, dx$

$$= \frac{1}{n+1} \sinh^{n+1} x \cosh x - \frac{n+2}{n+1} \int \sinh^{n+2} x \, dx$$

129. $\int x \sinh x \, dx = x \cosh x - \sinh x$

130. $\int x^2 \sinh x \, dx = (x^2 + 2) \cosh x - 2x \sinh x$

131. $\int x^n \sinh x \, dx = x^n \cosh x - nx^{n-1} \sinh x + n(n-1) \int x^{n-2} \sinh x \, dx$

132. $\int \cosh^n x \, dx = \frac{1}{n} \sinh x \cosh^{n-1} x + \frac{n-1}{n} \int \cosh^{n-2} x \, dx$

$$= -\frac{1}{n+1} \sinh x \cosh^{n+1} x + \frac{n+2}{n+1} \int \cosh^{n+2} x \, dx$$

133. $\int x \cosh x \, dx = x \sinh x - \cosh x$

134. $\int \sinh x \cosh x \, dx = \frac{1}{4} \cosh (2x)$

135. $\int \sinh (mx) \sinh (nx) \, dx = \frac{1}{m^2 - n^2} [m \sinh (nx) \cosh (mx) - n \cosh (nx) \sinh (mx)]$

136. $\int \cosh (mx) \cosh (nx) \, dx = \frac{1}{m^2 - n^2} [m \sinh (mx) \cosh (nx) - n \sinh (nx) \cosh (mx)]$

137. $\int \cosh (mx) \sinh (nx) \, dx = \frac{1}{m^2 - n^2} [m \sinh (nx) \sinh (mx) - n \cosh (nx) \cosh (mx)]$

Table 1.15. Table of Integrals (Continued)

Integrals Involving $\sinh^n x$, $\cosh^n x$, *and* $\tanh^n x$ *(Continued)*

138. $\int \sinh x \cos x \, dx = \frac{1}{2}(\cosh x \cos x + \sinh x \sin x)$

139. $\int \sinh x \sin x \, dx = \frac{1}{2}(\cosh x \sin x - \sinh x \cos x)$

140. $\int \cosh x \cos x \, dx = \frac{1}{2}(\sinh x \cos x + \cosh x \sin x)$

141. $\int \cosh x \sin x \, dx = \frac{1}{2}(\sinh x \sin x - \cosh x \cos x)$

142. $\int \sinh^{-1} x \, dx = x \sinh^{-1} x - \sqrt{1 + x^2}$

143. $\int \cosh^{-1} x \, dx = x \cosh^{-1} x - \sqrt{x^2 - 1}$

144. $\int \tanh^2 x \, dx = x - \tanh x$

Integrals Involving Algebraic Functions

145. $\int (ax + b)^n \, dx = \frac{1}{a(n + 1)} (ax + b)^{n+1} \quad (n \neq -1)$

146. $\int x(ax + b)^n \, dx = \frac{1}{a^2(n + 2)} (ax + b)^{n+2} - \frac{b}{a^2(n + 1)} (ax + b)^{n+1} \quad (n \neq -1, -2)$

147. $\int \frac{dx}{ax + b} = \frac{1}{a} \log_e (ax + b)$

148. $\int \frac{x \, dx}{ax + b} = \frac{x}{a} - \frac{b}{a^2} \log_e (ax + b)$

149. $\int \frac{x \, dx}{(ax + b)^2} = \frac{b}{a^2(ax + b)} + \frac{1}{a^2} \log_e (ax + b)$

150. $\int \frac{x^2 \, dx}{(ax + b)^2} = \frac{1}{a^3} \left[(ax + b) - 2b \log_e (ax + b) - \frac{b^2}{ax + b} \right]$

151. $\int \frac{x^2 \, dx}{ax + b} = \frac{1}{a^3} \left[\frac{1}{2} (ax + b)^2 - 2b(ax + b) + b^2 \log_e (ax + b) \right]$

152. $\int \frac{dx}{x(ax + b)} = \frac{1}{b} \log_e \frac{x}{ax + b}$

153. $\int \frac{dx}{x(ax + b)^2} = \frac{1}{b(ax + b)} - \frac{1}{b^2} \log_e \frac{ax + b}{x}$

154. $\int (ax^2 + b)^n x \, dx = \frac{1}{2a} \frac{(ax^2 + b)^{n+1}}{n + 1} \quad (n \neq -1)$

155. $\int \frac{dx}{ax^2 + b} = \frac{1}{\sqrt{ab}} \tan^{-1} \left(x \sqrt{\frac{a}{b}} \right) \quad (a \text{ and } b \text{ positive})$

156. $\int \frac{dx}{ax^2 + b} = \frac{1}{2 \sqrt{-ab}} \log_e \frac{x \sqrt{a} - \sqrt{-b}}{x \sqrt{a} + \sqrt{-b}} \quad (a \text{ positive, } b \text{ negative})$

$\quad = \frac{1}{2 \sqrt{-ab}} \log_e \frac{\sqrt{b} + x \sqrt{-a}}{\sqrt{b} - x \sqrt{-a}} \quad (a \text{ negative, } b \text{ positive})$

157. $\int \frac{dx}{(ax^2 + b)^n} = \frac{1}{2(n - 1)b} \frac{x}{(ax^2 + b)^{n-1}} + \frac{2n - 3}{2(n - 1)b} \int \frac{dx}{(ax^2 + b)^{n-1}} \quad (n \text{ integer} > 1)$

158. $\int \frac{x^2 \, dx}{ax^2 + b} = \frac{x}{a} - \frac{b}{a} \int \frac{dx}{ax^2 + b}$

159. $\int \frac{x^2 \, dx}{(ax^2 + b)^n} = - \frac{1}{2(n - 1)a} \frac{x}{(ax^2 + b)^{n-1}} + \frac{1}{2(n - 1)a} \int \frac{dx}{(ax^2 + b)^{n-1}},$
$(n \text{ integer} > 1)$

160. $\int \frac{dx}{ax^2 + bx + d} = \frac{1}{\sqrt{b^2 - 4ad}} \log_e \frac{2ax + b - \sqrt{b^2 - 4ad}}{2ax + b + \sqrt{b^2 - 4ad}} \quad (b^2 > 4ad)$

161. $\int \frac{dx}{ax^2 + bx + d} = \frac{2}{\sqrt{4ad - b^2}} \tan^{-1} \frac{2ax + b}{\sqrt{4ad - b^2}} \quad (b^2 < 4ad)$

Table 1.15. Table of Integrals (Continued)

Integrals Involving Algebraic Functions (Continued)

162. $\int \frac{dx}{ax^2 + bx + d} = -\frac{2}{2ax + b} \quad (b^2 = 4ad)$

Integrals Involving Irrational Algebraic Functions

163. $\int \sqrt{ax + b}\, dx = \frac{2}{3a} \sqrt{(ax + b)^3}$

164. $\int x \sqrt{ax + b}\, dx = -\frac{2(2b - 3ax)\sqrt{(ax + b)^3}}{15a^2}$

165. $\int x^2 \sqrt{ax + b}\, dx = \frac{2(8b^2 - 12abx + 15a^2x^2)\sqrt{(ax + b)^3}}{105a^3}$

166. $\int \frac{\sqrt{ax + b}}{x} dx = 2\sqrt{ax + b} + \sqrt{b} \log_e \frac{\sqrt{ax + b} - \sqrt{b}}{\sqrt{ax + b} + \sqrt{b}}$ (b positive)

167. $\int \frac{\sqrt{ax + b}}{x} dx = 2\sqrt{ax + b} - 2\sqrt{-b} \tan^{-1} \sqrt{\frac{ax + b}{-b}}$ (b negative)

168. $\int \frac{dx}{x\sqrt{ax + b}} = \frac{1}{\sqrt{b}} \log_e \frac{\sqrt{ax + b} - \sqrt{b}}{\sqrt{ax + b} + \sqrt{b}}$ (b positive)

169. $\int \frac{dx}{x\sqrt{ax + b}} = \frac{2}{\sqrt{-b}} \tan^{-1} \sqrt{\frac{ax + b}{-b}}$ (b negative)

170. $\int \frac{dx}{x^2\sqrt{ax + b}} = -\frac{\sqrt{ax + b}}{bx} - \frac{a}{2b\sqrt{b}} \log_e \frac{\sqrt{ax + b} - \sqrt{b}}{\sqrt{ax + b} + \sqrt{b}}$ (b positive)

171. $\int \frac{dx}{x^2\sqrt{ax + b}} = -\frac{\sqrt{ax + b}}{bx} - \frac{a}{b\sqrt{-b}} \tan^{-1} \sqrt{\frac{ax + b}{-b}}$ (b negative)

172. $\int \sqrt{ax^2 + b}\, dx = \frac{x}{2}\sqrt{ax^2 + b} + \frac{b}{2\sqrt{a}} \log_e \frac{x\sqrt{a} + \sqrt{ax^2 + b}}{\sqrt{b}}$ (a positive)

173. $\int \sqrt{ax^2 + b}\, dx = \frac{x}{2}\sqrt{ax^2 + b} + \frac{b}{2\sqrt{-a}} \sin^{-1}\left(x\sqrt{-\frac{a}{b}}\right)$ (a negative)

174. $\int \frac{dx}{\sqrt{ax^2 + b}} = \frac{1}{\sqrt{a}} \log_e (x\sqrt{a} + \sqrt{ax^2 + b})$ (a positive)

175. $\int \frac{dx}{\sqrt{ax^2 + b}} = \frac{1}{\sqrt{-a}} \sin^{-1}\left(x\sqrt{-\frac{a}{b}}\right)$ (a negative)

176. $\int \frac{\sqrt{ax^2 + b}}{x} dx = \sqrt{ax^2 + b} + \sqrt{b} \log_e \frac{\sqrt{ax^2 + b} - \sqrt{b}}{x}$ (b positive)

177. $\int \frac{\sqrt{ax^2 + b}}{x} dx = \sqrt{ax^2 + b} - \sqrt{-b} \tan^{-1} \frac{\sqrt{ax^2 + b}}{\sqrt{-b}}$ (b negative)

178. $\int \frac{x\, dx}{\sqrt{ax^2 + b}} = \frac{1}{a}\sqrt{ax^2 + b}$

179. $\int x\sqrt{ax^2 + b}\, dx = \frac{1}{3a}(ax^2 + b)^{3/2}$

180. $\int \frac{dx}{x\sqrt{ax^n + b}} = \frac{1}{n\sqrt{b}} \log_e \frac{\sqrt{ax^n + b} - \sqrt{b}}{\sqrt{ax^n + b} + \sqrt{b}}$ (b positive)

181. $\int \frac{dx}{x\sqrt{ax^n + b}} = \frac{2}{n\sqrt{-b}} \sec^{-1} \sqrt{-\frac{ax^n}{b}}$ (b negative)

Table 1.15. Table of Integrals (Continued)

Integrals Involving Irrational Algebraic Functions (Continued)

182. $\int \sqrt{ax^2 + bx + d}\, dx = \frac{2ax + b}{4a} \sqrt{ax^2 + bx + d} + \frac{4ad - b^2}{8a} \int \frac{dx}{\sqrt{ax^2 + bx + d}}$

183. $\int x \sqrt{ax^2 + bx + d}\, dx = \frac{(ax^2 + bx + d)^{3/2}}{3a} - \frac{b}{2a} \int \sqrt{ax^2 + bx + d}\, dx$

184. $\int \frac{dx}{\sqrt{ax^2 + bx + d}} = \frac{1}{\sqrt{a}} \log_e (2ax + b + 2\sqrt{a(ax^2 + bx + d)})$ (a positive)

185. $\int \frac{dx}{\sqrt{ax^2 + bx + d}} = \frac{1}{\sqrt{-a}} \sin^{-1} \frac{-2ax - b}{\sqrt{b^2 - 4ad}}$ (a negative)

186. $\int \frac{dx}{x\sqrt{ax^2 + bx + d}} = -\frac{1}{\sqrt{d}} \log_e \left(\frac{\sqrt{ax^2 + bx + d} + \sqrt{d}}{x} + \frac{b}{2\sqrt{d}} \right)$ (d positive)

187. $\int \frac{dx}{x\sqrt{ax^2 + bx + d}} = \frac{1}{\sqrt{-d}} \sin^{-1} \frac{bx + 2d}{x\sqrt{b^2 - 4ad}}$ (d negative)

188. $\int \frac{x\, dx}{ax^2 + bx + d} = \frac{1}{2a} \log_e (ax^2 + bx + d) - \frac{b}{2a} \int \frac{dx}{ax^2 + bx + d}$

189. $\int \frac{x\, dx}{\sqrt{ax^2 + bx + d}} = \frac{\sqrt{ax^2 + bx + d}}{a} - \frac{b}{2a} \int \frac{dx}{\sqrt{ax^2 + bx + d}}$

Integrals Involving e, x^n, a^x, $\ln x$

190. $\int xe^{ax}\, dx = \frac{e^{ax}}{a^2} (ax - 1)$

191. $\int x^m e^{ax}\, dx = \frac{x^m e^{ax}}{a} - \frac{m}{a} \int x^{m-1} e^{ax}\, dx$

192. $\int \frac{e^{ax}}{x^m}\, dx = \frac{1}{m - 1} \left(-\frac{e^{ax}}{x^{m-1}} + a \int \frac{e^{ax}\, dx}{x^{m-1}} \right)$

193. $\int x^n a^x\, dx = \frac{a^x x^n}{\log a} - \frac{na^x x^{n-1}}{(\log a)^2} + \frac{n(n - 1)a^x x^{n-2}}{(\log a)^3} \cdots \pm \frac{n(n - 1)(n - 2) \cdots 2 \cdot 1 a^x}{(\log a)^{n+1}}$

194. $\int \frac{a^x\, dx}{x} = \log x + x \log a + \frac{(x \log a)^2}{2 \times 2!} + \frac{(x \log a)^3}{3 \times 3!} + \cdots$

195. $\int \frac{dx}{1 + e^x} = \log \frac{e^x}{1 + e^x}$

196. $\int \frac{dx}{a + be^{mx}} = \frac{1}{am} [mx - \log (a + be^{mx})]$

197. $\int x^m \log x\, dx = x^{m+1} \left[\frac{\log x}{m + 1} - \frac{1}{(m + 1)^2} \right]$

198. $\int (\log x)^n\, dx = x(\log x)^n - n \int (\log x)^{n-1}\, dx$

199. $\int x^m (\log x)^n\, dx = \frac{x^{m+1} (\log x)^n}{m + 1} - \frac{n}{m + 1} \int x^m (\log x)^{n-1}\, dx$

200. $\int \frac{(\log x)^n\, dx}{x} = \frac{(\log x)^{n+1}}{n + 1}$

201. $\int x^m \log (a + bx)\, dx = \frac{1}{m + 1} \left[x^{m+1} \log (a + bx) - b \int \frac{x^{m+1}\, dx}{a + bx} \right]$

Table 1.15. Table of Integrals (Continued)

Definite Integrals

202. $\int_0^{\frac{\pi}{2}} \sin^n x\,dx = \int_0^{\frac{\pi}{2}} \cos^n x\,dx = \frac{1 \times 3 \times 5 \cdots (n-1)}{2 \times 4 \times 6 \cdots (n)} \frac{\pi}{2}$ if n is an even integer

$= \frac{2 \times 4 \times 6 \cdots (n-1)}{1 \times 3 \times 5 \times 7 \cdots n}$ if n is an odd integer

$= \frac{1}{2}\sqrt{\pi}\, \frac{\Gamma\left(\frac{n+1}{2}\right)}{\Gamma\left(\frac{n}{2}+1\right)}$ for any value of n greater than -1

203. $\int_0^{\infty} \frac{\sin mx\,dx}{x} = \frac{\pi}{2}$, if $m > 0$; 0, if $m = 0$; $-\frac{\pi}{2}$, if $m < 0$

204. $\int_0^{\infty} \frac{\sin x \cos mx\,dx}{x} = 0$, if $m < -1$ or $m > 1$; $\frac{\pi}{4}$, if $m = -1$ or $m = 1$; $\frac{\pi}{2}$, if $-1 < m < 1$

205. $\int_0^{\infty} \frac{\sin^2 x\,dx}{x^2} = \frac{\pi}{2}$

206. $\int_0^{\pi} \sin^2 mx\,dx = \int_0^{\pi} \cos^2 mx\,dx = \frac{\pi}{2}$

207. $\int_0^{\infty} \frac{\cos x\,dx}{\sqrt{x}} = \int_0^{\infty} \frac{\sin x\,dx}{\sqrt{x}} = \sqrt{\frac{\pi}{2}}$

208. $\int_0^{\infty} e^{-ax} \sin mx\,dx = \frac{m}{a^2 + m^2}$, if $a > 0$

209. $\int_0^{\infty} e^{-ax} \cos mx\,dx = \frac{a}{a^2 + m^2}$, if $a > 0$

210. $\int_0^{\frac{\pi}{2}} \log \sin x\,dx = \int_0^{\frac{\pi}{2}} \log \cos x\,dx = -\frac{\pi}{2} \log 2$

211. $\int_0^{\pi} x \log \sin x\,dx = -\frac{\pi^2}{2} \log 2$

212. $\int_0^{\pi} \sin kx \sin mx\,dx = \int_0^{\pi} \cos kx \cos mx\,dx = 0$, if k is different from m

213. $\int_0^{\infty} x^n e^{-ax}\,dx = \frac{\Gamma(n+1)}{a^{n+1}} = \frac{n!}{a^{n+1}}$ $\quad n > -1,\ a > 0$

214. $\int_0^{\infty} x^{2n} e^{-ax^2}\,dx = \frac{1 \times 3 \times 5 \cdots (2n-1)}{2^{n+1} a^n} \sqrt{\frac{\pi}{a}}$

215. $\int_0^{\infty} e^{-a^2x^2}\,dx = \frac{1}{2a}\sqrt{\pi} = \frac{1}{2a}\Gamma(\frac{1}{2})$ $\quad a > 0$

216. $\int_0^{\infty} \frac{e^{-nx}}{\sqrt{x}}\,dx = \sqrt{\frac{\pi}{n}}$

217. $\int_0^{1} \frac{\log x}{1+x}\,dx = -\frac{\pi^2}{12}$

218. $\int_0^{1} \frac{\log x}{1-x}\,dx = -\frac{\pi^2}{6}$

219. $\int_0^{1} \frac{\log x}{1-x^2}\,dx = -\frac{\pi^2}{8}$

Multiple Integrals. Multiple integrals are of the form

$$\int\int f(x,y) \qquad \int\int\int f(x,y,z) \qquad \text{etc.}$$

Two successive integrations, for example, an integration with respect to y holding x constant, and an integration with respect to x between constant limits, will yield the value for the double integral

$$\int_a^b \int_{y_1(x)}^{y_2(x)} f(x,y)\, dy\, dx$$

Similarly a triple integral is evaluated by three successive single integrations. The order of integration can be reversed if the function $f(x,y, \ldots)$ is continuous.

(i) Differential Equations

Introduction. A differential equation describes some fundamental relationship between independent variables $x, y, z, \ldots$ and dependent variables $u, v, w, \ldots$ and some of the derivatives of $u, v, w, \ldots$ with respect to $x, y, z, \ldots$. An ordinary differential equation is one which contains an independent variable, a dependent variable, and various orders of derivatives of the dependent variable. The order of a differential equation is the order of the highest derivative which it contains. A general solution of an ordinary differential equation of nth order, $F(x,y,y', \ldots, y^{(n)}) = 0$, is a family of curves $G(x,y,C_1, \ldots, C_n) = 0$, each curve of which is a solution of the differential equation.

A partial differential equation contains a function of more than one independent variable and its partial derivatives of various order.

The degree of a differential equation is the power to which the derivative of highest order is raised.

Methods of Solution of Ordinary Differential Equations of First Order. *Exact Equations.* An exact differential equation is of the form

$$dy/dx\, N(x,y) + M(x,y) = 0 \qquad \text{or} \qquad M(x,y)\, dx + N(x,y)\, dy = 0$$

for which

$$\partial M/\partial y(x,y) = \partial N/\partial x(x,y)$$

The solution of the exact equation is

$$\int M(x,y)\, dx + \int [N(x,y) - \partial/\partial y \int M(x,y)\, dx]\, dy = C$$

A differential equation, though not initially exact, may be rendered exact by multiplication through by a suitable function $u(x,y)$ called an integrating factor. For example, a differential equation containing the form $(x\, dx + y\, dy)$ may often be made exact by using the integrating factor $(x^2 + y^2)$ or a function of $(x^2 + y^2)$. Likewise, corresponding to the form $(x\, dy + y\, dx)$, try xy or a function of xy, and for $(x\, dy - y\, dx)$, the integrating factors $1/x^2$, $1/y^2$, $1/xy$, $1/(x^2 + y^2)$, or $1/x^2 f(y/x)$ may be tried.

Differential Equations with Variables Separable. If the differential equation is of the form $M(x)\, dx + N(y)\, dy = 0$, it is not only exact but may be integrated directly because the variables are separable. The solution is

$$\int M(x)\, dx + \int N(y)\, dy = 0 \qquad \text{or} \qquad F(x) + G(y) = C$$

Linear Equations. The differential equation of the form $dy/dx + P(x)y = Q(x)$ is a linear equation of first order. If $Q(x) = 0$, the equation is linear homogeneous.

The solution of the general case is

$$y = e^{-\int P dx}\left(\int Q e^{\int P dx}\, dx + c\right)$$

Example: Solve the equation

$$dy/dx + (1/x)y = 2x$$

$$y = e^{-\int dx/x}\left(\int 2x e^{\int dx/x}\, dx + C\right)$$

$$y = \tfrac{2}{3}x^2 + C/x \qquad \text{since} \qquad e^{\ln x} = x$$

Bernoulli's equation is of the form

$$dy/dx + P(x)y = Q(x)y^n$$

By substituting $v = y^{1-n}$, the equation is reducible to the following linear differential equation in v and x:

$$dv/dx + (1 - n)P(x)v = (1 - n)Q(x)$$

Clairaut's equation is of the form $y = xp + f(p)$ where $p = dy/dx$. The solution is $y = Cx + f(C)$, which is the equation of a family of straight lines obtained when an arbitrary constant C is substituted for P, and the curve obtained by eliminating p from the two simultaneous equations $x + df/dx(p) = 0$ and $y = xp + f(p)$.

Ordinary Differential Equations of Higher Than First Order. The equation

$$a_n(x)d^ny/dx^n + a_{n-1}(x)d^{n-1}y/dx^{n-1} + \cdots + a_1(x)dy/dx + a_0(x)y = f(x)$$

is the general representation of a linear differential equation of order n. When $f(x) = 0$, the equation is homogeneous; otherwise it is termed a nonhomogeneous equation. The complete solution of the general differential equation may be written

$$y = y_c + y_p$$

where y_c is the solution of the homogeneous equation corresponding to the general equation, and y_p is a particular solution of the nonhomogeneous equation.

The D-operator Method. This method is frequently employed to solve linear homogeneous differential equations with constant coefficients,

$$a_n\, d^ny/dx^n + a_{n-1}\, d^{n-1}y/dx^{n-1} + \cdots + a_1\, dy/dx + a_0y = 0$$

The operator D is defined by the equation $Dy \equiv dy/dx$ or $D = d/dx$. Rewriting the homogeneous equation in operator form we obtain

$$(a_nD^n + a_{n-1}D^{n-1} + \cdots + a_1D + a_0)y = 0$$

Let $\alpha_n, \alpha_{n-1}, \alpha_{n-2}, \ldots, \alpha_1$ be the roots of the auxiliary algebraic equation

$$a_nD^n + a_{n-1}D^{n-1} + \cdots + a_1D + a_0 = 0$$

The solution of the differential equation if all the roots are real and distinct (none repeated) is

$$y = C_1e^{\alpha_1x} + C_2e^{\alpha_2x} + \cdots C_ne^{\alpha_nx}$$

If q roots are equal, the remaining being real and distinct, the solution is

$$y = e^{\alpha_1x}(C_1 + C_2x + C_3x^2 + \cdots + C_qx^{q-1}) + \cdots + C_{n-1}e^{\alpha_{n-1}x} + C_ne^{\alpha_nx}$$

If two roots are conjugate imaginary,

$$\alpha_1 = \beta + \gamma i \qquad \alpha_2 = \beta - \gamma i$$

the solution is $y = e^{\beta x}(C_1 \cos \gamma x + C_2 \sin \gamma x) + C_3e^{\alpha_3x} + \cdots + C_ne^{\alpha_nx}$. If there are two conjugate imaginary double roots,

$$\alpha_1 = \alpha_2 = \beta + \gamma i \qquad \alpha_3 = \alpha_4 = \beta - \gamma i$$

the solution is $y = e^{\beta x}[(C_1 + C_2x) \cos \gamma x + (C_3 + C_4x) \sin \gamma x]$.

Method of Undetermined Coefficients. This approach to the solution of linear nonhomogeneous differential equations with constant coefficients may be employed where it is possible to guess at the form of the particular solution. The following example will serve to illustrate the method of solution.

Solve the equation

$$d^2y/dx^2 - y = \sin x$$

The solution to the homogeneous equation $y'' - y = 0$ is obtained by the D-operator method; the auxiliary equation being $D^2 - 1 = 0$, $D = \pm 1$; therefore,

$$y_c = C_1 e^x + C_2 e^{-x}$$

To obtain the particular solution, assume the form of solution to be

$$y_p = A \sin x + B \cos x$$

Differentiating y_p and substituting into the original equation, we obtain

$$-2A \sin x - 2B \cos x = \sin x$$

from which

$$A = -\tfrac{1}{2} \qquad \text{and} \qquad B = 0 \qquad y_p = -\tfrac{1}{2} \sin x$$

and the complete solution is

$$y = C_1 e^x + C_2 e^{-x} - \tfrac{1}{2} \sin x$$

In the event the trial particular solution has a component which solves the homogeneous equation, the particular solution should be multiplied by x. If this function also has a component which solves the homogeneous equation, the original trial y_p should be multiplied by x^2, and so on.

If the right-hand side of the differential equation contains terms of more than one type, the trial particular integral will be the sum of trial terms for each type. The trial terms in Table 1.16 are suggested for the type listed.

Table 1.16. Trial Terms

Type of term in $f(x)$	*Form to be included in y_p*
c (constant)	C (constant)
$c \sin x$ or $c \cos x$	$C_1 \sin x + C_2 \cos x$
cx^n	$C_0 x^n + C_1 x^{n-1} + C_2 x^{n-2} + \cdots + C_{n-1} x + C_n$
ce^{kx}	Ce^{kx}
$cx^n e^{kx}$	$e^{kx}(C_0 x^n + C_1 x^{n-1} + \cdots + C_n)$
$e^{kx} \sin \alpha x$ or $e^{kx} \cos \alpha x$	$e^{kx}(C_1 \sin \alpha x + C_2 \cos \alpha x)$
$cx^n e^{kx} \sin \alpha x$ or $cx^n e^{kx} \cos \alpha x$	$(C_0 x^n + \cdots + C_{n-1} x + C_n) e^{kx} \sin \alpha x + (C'_0 x^n + \cdots$ $C'_{n-1} x + C'_n) e^{kx} \cos \alpha x$

Taylor-series Solution. If the solution $y(x)$ of a linear differential equation can be expressed as a Taylor-series expansion valid in some neighborhood of the initial point $x = x_0$, then the first n coefficients $y(x_0)$, $y'(x_0)/1!$, . . . , $y^{n-1}(x_0)/(n-1)!$ of the series $y(x) = y(x_0) + [y'(x_0)/1!](x - x_0) + [y''(x_0)/2!](x - x_0)^2 + \cdots$ are obtained from the n initial conditions and additional coefficients are obtained by differentiating the given differential equation $(n - 1)$ times and solving these simultaneously with the original differential equation. For example, given the equation

$$x\, d^2y/dx^2 + x^2\, dy/dx - y = 0$$

and the initial conditions

$$y(1) = 0 \qquad y'(1) = 1$$

the first four terms of the expansion are required.

The expansion of $y(x)$ is

$$y(x) = y(1) + [y'(1)/1!](x - 1) + [y''(1)/2!](x - 1)^2 + [y'''(1)/3!](x - 1)^3 + \cdots$$

The coefficients $y(1)$ and $y'(1)$ are given by the initial conditions as zero and one, respectively. The remaining two coefficients are obtained from differentiating the differential equation. Thus the equations

$$xy'' + x^2 y' - y = 0$$
$$xy''' + (x^2 + 1)y'' + (2x - 1)y' = 0$$

yield after substituting $x = 1$, $y = 0$, and $y' = 1$

$$y'' = -1 \qquad y''' = 1$$

The first four terms of the expansion are therefore

$$y = 0 + (x - 1) - \tfrac{1}{2}(x - 1)^2 + \tfrac{1}{6}(x - 1)^3 + \cdots$$

Variations of Parameters. This method of solution yields the particular integral of a differential equation of which the coefficients need not be constant. It is required that the complementary solution be known before this approach may be used.

Example: Given the differential equation

$$x^2y'' + xy' - y = 1 + x$$

which has a complementary solution $y = C_1x + C_2/x$ of the corresponding homogeneous equation $x^2y'' + xy' - y = 0$, find the particular integral.

Solution: The arbitrary constants C_1 and C_2 are replaced by functions of x, $A(x)$, and $B(x)$ such that $y = A(x)x + B(x)/x$ is a solution of the given nonhomogeneous equation. Differentiate to obtain

$$y' = A'x + A + B'/x - B/x^2$$

Since there are two functions $A(x)$ and $B(x)$, whose identity must be established, two sets of conditions related to these functions must be written. The first condition is the nonhomogeneous equation itself. The remaining condition (there will be $n - 1$ conditions remaining where there are n functions, in general) may be assigned arbitrarily. Thus we select

$$A'x + B'/x = 0$$

Now differentiate $y = A(x)x + B(x)/x$ to obtain

$$y' = A'x + A + B'/x - B/x^2$$

which together with the above yields

$$y' = A - B/x^2$$

Differentiating, we obtain

$$y'' = A' - B'/x^2 + 2B/x^3$$

which is substituted in the original differential equation

$$x^2(A' - B'/x^2 + 2B/x^3) + x(A - B/x^2) - Ax - B/x = 1 + x$$

or

$$A'x^2 - B' = 1 + x$$

which together with $A'x + B'/x = 0$, yields

$$A' = \tfrac{1}{2}(1/x^2 + 1/x) \qquad A = \tfrac{1}{2}(-1/x + \ln x) + C_3$$
$$B' = -\tfrac{1}{2}(1 + x) \qquad B = -x/2(1 + x/2) + C_4$$

so that

$$y(x) = C_1x + C_2/x + Ax + B/x$$

Systems of Linear Differential Equations with Constant Coefficients. One method of solution of a system of n linear differential equations with constant coefficients is the *symbolic algebraic method.* The system has the form

$$P_1(D)x + Q_1(D)y + \cdots = F_1(t)$$
$$P_2(D)x + Q_2(D)y + \cdots = F_2(t)$$
$$\cdots\cdots\cdots\cdots\cdots\cdots\cdots$$

where $D = d/dt$.

The solution may be found by treating these equations as algebraic equations. If the elimination of $x(t)$ from the first two equations is desired, one may operate on the first equation by $P_2(D)$ and on the second equation by $P_1(D)$. Then by subtraction $x(t)$ can be eliminated. For example, given the two simultaneous equations

$$(D + 1)x + Dy = 0$$
$$(D - 1)x + (D + 1)y = 0$$

We obtain, after operating on the first equation with $(D - 1)$ and on the second equation with $(D + 1)$,

$$(D - 1)(D + 1)x + (D - 1)(D)y = 0$$
$$(D + 1)(D - 1)x + (D + 1)^2y = 0$$

By subtracting the second equation from the first we obtain

$$(3D + 1)y = 0$$

which has only one dependent variable. Solving for $y(t)$ in this equation and substituting back into one of the original equations, $x(t)$ can be found.

Solutions to Important Differential Equations in Engineering. *Legendre's Equation*

$$(1 - x^2)\, d^2y/dx^2 - 2x\, dy/dx + n(n + 1)y = 0$$

where n is a positive integer. The solution is $y(x) = C_1P_n(x) + C_2Q_n(x)$. $P_n(x)$ is a Legendre polynomial of degree n (Legendre function of the first kind). $Q_n(x)$ is the Legendre function of the second kind.

Bessel Differential Equation

$$x^2\, d^2y/dx^2 + x\, dy/dx + (x^2 - n^2)y = 0$$

n may take on fractional, integral, positive, negative, or complex values. The general solution of the Bessel equation is

$$y = C_1y_{\mathrm{I}} + C_2y_{\mathrm{II}}$$

where y_{I} and y_{II} are the Bessel functions of the first and second kind, respectively, and the order of the Bessel functions will depend upon the value of n.

Consider the zero-order Bessel equation obtained by setting $n = 0$,

$$d^2y/dx^2 + (1/x)(dy/dx) + y = 0$$

The solution is

$$y = C_1J_0(x) \text{ and } C_2Y_0(x)$$

where $J_0(x)$ and $Y_0(x)$ are the zero-order Bessel functions of the first and second kinds, respectively.

$$J_0(x) = 1 - \frac{x^2}{2^2} + \frac{x^4}{2^2 \times 4^2} - \frac{x^6}{2^2 \times 4^2 \times 6^2} + \cdots$$

$$Y_0(x) = J_0(x) \ln x + \frac{x^2}{2^2} - \frac{x^4}{2^2 \times 4^2}(1 + \tfrac{1}{2}) + \frac{x^6}{2^2 \times 4^2 \times 6^2}(1 + \tfrac{1}{2} + \tfrac{1}{3}) - \cdots$$

The general solution of an nth-order Bessel equation is

$$y = C_1J_n(x) + C_2Y_n(x) \qquad n = 0, 1, 2, 3, \ldots$$

The general equation for the Bessel functions of the first kind for n equal to zero or a positive integer is

$$J_n(x) = \frac{(x/2)^n}{n!}\left[1 - \frac{x^2}{2(2n + 2)} + \frac{x^4}{2 \times 4(2n + 2)(2n + 4)} + \cdots\right] \qquad n = 0, 1, 2, 3, \ldots$$

For the Bessel function of the second kind we obtain

$$Y_n(x) = J_n(x) \ln x - \frac{1}{2}\sum_{p=0}^{n-1} \frac{(n-p-1)!}{p!}\left(\frac{x}{2}\right)^{2p-n}$$

$$- \frac{1}{2}\sum_{p=0}^{\infty} \frac{(-1)^n}{p!(n+p)!}\left(\frac{x}{2}\right)^{2p+n} [f(P) + f(n+p)]$$

where

$$[f(P) + f(n+p)] = 1/1 + 1/2 + \cdots + 1/p + 1/1 + \cdots + 1/(n+p)$$
$$n = 0, 1, 2, 3, \ldots$$

Recursion Formulas

$$d/dx\, J_0(x) = -J_1(x) \qquad J_1(-x) = -J_1(x)$$
$$d/dx\, J_n(x) = n/x\, J_n(x) - J_{n+1}(x) \qquad J_{-n}(x) = (-1)^n J_n(x) \qquad n = 0, 1, 2, \ldots$$
$$d/dx\, Y_0(x) = -Y_1(x) \qquad Y_{-n}(x) = (-1)^n Y_n(x) \qquad n = 0, 1, 2, \ldots$$
$$J_0(-x) = J_0(x)$$

The Modified Bessel Equation

$$x^2\, d^2y/dx^2 + x\, dy/dx + m^2x^2y = 0 \qquad m = \text{constant}$$

Making the substitution $\xi = mx$ we obtain

$$\xi^2\, d^2y/d\xi^2 + \xi\, dy/d\xi + \xi^2 y = 0$$

whose solution, already discussed, is

$$y = C_1J_0(\xi) + C_2Y_0(\xi) = C_1J_0(mx) + C_2Y_0(mx)$$

The equation

$$x^2\, d^2y/dx^2 + x\, dy/dx - (x^2 + n^2)y = 0$$

can be transformed to the form

$$x^2\, d^2y/dx^2 + x\, dy/dx + (x^2 - n^2)y = 0$$

by making the substitution $x = i\xi$:

$$\xi^2\, d^2y/d\xi^2 + \xi\, dy/d\xi + (\xi^2 - n^2)y = 0$$

For integral n,

$$I_n(x) = i^{-n}J_n(ix)$$
$$K_n(x) = i^{n+1}\,\pi/2\,[J_n(ix) + iY_n(ix)]$$

The solution for the modified Bessel equation is therefore

$$y = C_1I_n(x) + C_2K_n(x) \qquad n = 0, 1, 2, \ldots$$

Two useful relationships are

$$I_{-n}(x) = I_n(x) \qquad n = 0, 1, 2, \ldots$$
$$K_{-n}(x) = K_n(x) \qquad n = 0, 1, 2, \ldots$$

Solution of Partial Differential Equations. The only means of solution which will be discussed is the separation-of-variables method. For the following equation,

$$f_1(x)\, \partial^2z/\partial x^2 + f_2(x)\, \partial z/\partial x + f_3(x)z + g_1(y)\, \partial^2z/\partial y^2 + g_2(y)\, \partial z/\partial y + g_3(y)z = 0$$

the solution of $z = z(x,y)$ will be assumed to have the form of a product of two functions $X(x)$ and $Y(y)$, which are functions of x and y only, respectively,

$$z = z(x,y) = X(x)Y(y)$$

Substituting $z = XY$ into the original differential equation we obtain

$$-1/X[f_1(x)\ \partial^2X/\partial x^2 + f_2(x)\ \partial X/\partial x + f_3(x)X] = 1/Y[g_1(y)\ \partial^2Y/\partial y^2 + g_2(y)\ \partial Y/\partial y + g_3(y)Y]$$

Note that the left-hand side contains the function of x only and that the right-hand side contains functions of y only. Since the right- and left-hand sides are independent of x and y, respectively, they must be equal to a common constant, called a separation constant α; thus

$$f_1(x)\ d^2X/dx^2 + f_2(x)\ dX/dx + [f_3(x) + \alpha]X = 0$$

and

$$g_1(y)\ d^2Y/dy^2 + g_2(y)\ dY/dy + [g_3(y) - \alpha]Y = 0$$

Once the solutions to the above have been obtained, the product solution XY is obtained. The method outlined may be extended to additional variables. A differential equation may be separable, but where the product solution will not satisfy initial and boundary conditions, the foregoing method will not yield a solution.

(j) Numerical Methods

Numerical Differentiation. Numerical differentiation or numerical integration provides means for evaluating the derivative or integral of a function $y = f(x)$ when

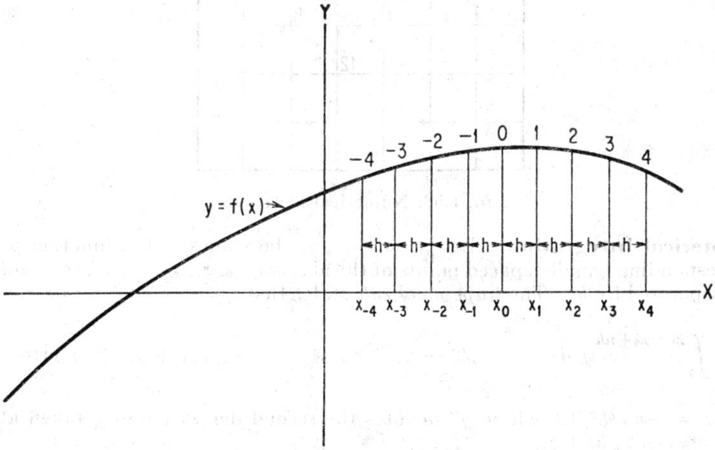

FIG. 1.55. Numerical methods.

the function is described by a series of pairs of values (x_i,y_i) and not by an explicit relationship.

The following expressions refer to Fig. 1.55, which defines the various symbols used

Derivative	*Expression*
$2h(dy/dx)_0$	$y_1 - y_{-1}$
$h^2(d^2y/dx^2)_0$	$y_1 - 2y_0 + y_{-1}$
$2h^3(d^3y/dx^3)_0$	$y_2 - 2y_1 + 2y_{-1} - y_{-2}$
$h^4(d^4y/dx^4)_0$	$y_2 - 4y_1 + 6y_0 - 4y_{-1} + y_{-2}$

The following expressions for the various derivatives of a function of two variables refer to Fig. 1.56, which locates the various points with respect to the point at which the derivative is being taken.

Derivative	*Expression*
$2h(\partial u/\partial x)_0$	$u_1 - u_3$
$2h(\partial u/\partial y)_0$	$u_2 - u_4$
$h^2(\partial^2 u/\partial x^2)_0$	$u_1 - 2u_0 + u_3$
$h^2(\partial^2 u/\partial y^2)_0$	$u_2 - 2u_0 + u_4$
$4h^2(\partial^2 u/\partial x\, \partial y)_0$	$u_5 - u_6 + u_7 - u_8$
$2h^3(\partial^3 u/\partial x^3)_0$	$u_9 - 2u_1 + 2u_3 - u_{11}$
$2h^3(\partial^3 u/\partial y^3)_0$	$u_{10} - 2u_2 + 2u_4 - u_{12}$
$2h^3(\partial^3 u/\partial x^2\, \partial y)_0$	$u_5 - 2u_2 + u_6 - u_8 + 2u_4 - u_7$
$2h^3(\partial^3 u/\partial x\, \partial y^2)_0$	$u_5 - 2u_1 + u_8 - u_6 + 2u_3 - u_7$
$h^4(\partial^4 u/\partial x^4)_0$	$u_9 - 4u_1 + 6u_0 - 4u_3 + u_{11}$
$h^4(\partial^4 u/\partial y^4)_0$	$u_{10} - 4u_2 + 6u_0 - 4u_4 + u_{12}$
$h^4(\partial^4 u/\partial x^2\, \partial y^2)_0$	$u_5 - 2u_2 + u_6 - 2u_1 + 4u_0 - 2u_3 + u_8 - 2u_4 + u_7$

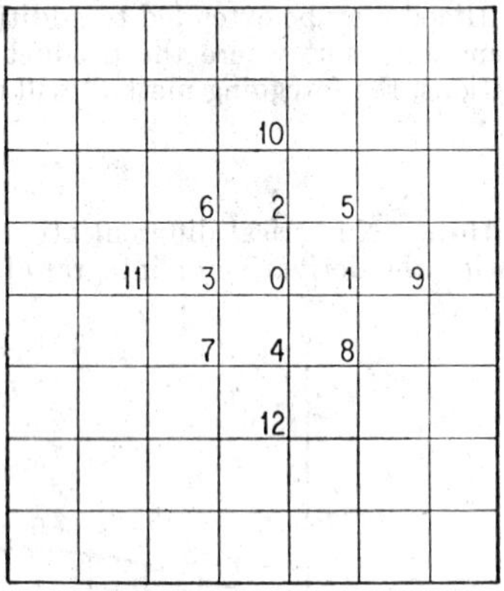

FIG. 1.56. Numerical methods.

Numerical Integration. Let $y_0, y_1, y_2, \ldots$ be values of the function $y = f(x)$ at corresponding equally spaced points of the abscissa $x_0, x_1, x_2, \ldots$, the values of x being separated by h. The *trapezoidal rule* states that

$$\int_{x_0}^{x_n = x_0 + nh} y\, dx = h(y_0/2 + y_1 + y_2 + \cdots + y_{n-1} + y_n/2) + \text{error}$$

Error $= -ny''h^3/12$, where y'' denotes the second derivative of y taken at some point between x_0 and x_n.

Simpson's rule states that

$$\int_{x_0}^{x_{2n} = x_0 + 2nh} y\, dx = h/3(y_0 + 4y_1 + 2y_2 + 4y_3 + 2y_4 + 4y_5 + \cdots + 4y_{2n-1} + y_{2n}) + \text{error}$$

Error $= -ny^{(4)}h^5/90$

Graphical Methods for Determining the Approximate Value of a Real Root. Given the function $y = f(x)$ it is required to find the values of the independent variable x for which the function equals zero. By graphing the function, the approximate values of the real roots can be obtained by noting the values of x corresponding to $y = 0 = f(x)$, i.e., where the graph crosses the x axis.

If the function $y = f(x)$ is written $y = f_1(x) + f_2(x) = y_1(x) + y_2(x)$, the real roots of the function are those which satisfy the equation $f_1(x) = -f_2(x)$. By plotting $f_1(x) = 0$ and $-f_2(x) = 0$, the real roots may be obtained by noting the points of intersection of the two functions.

Numerical Methods for Determining the Value of a Real Root. The graphical methods outlined above provide only approximate values of the real roots. The two methods discussed below permit the solution to any required degree of accuracy.

The Method of False Position (Regula Falsi). Given a real function $y = f(x)$, it is required to obtain the value of the real roots. If the function is continuous between $x = a$ and $x = b$ (Fig. 1.57) and if $y_a = f(a)$ and $y_b = f(b)$ are of opposite sign, then

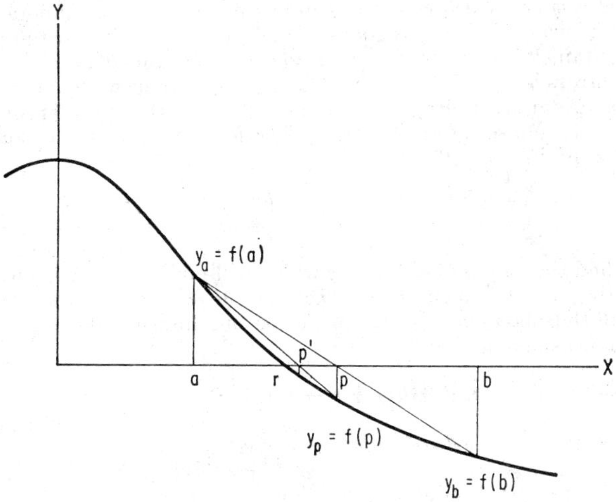

FIG. 1.57. Method of Regula Falsi.

there is at least one real root between $x = a$ and $x = b$. From Fig. 1.57 it can be seen that the function between a and b can be approximated by a straight line provided that the points are close together. From triangles $\overline{ay_ap}$ and $\overline{by_bp}$ it can be shown that

$$p = a + \frac{y_a}{y_a - y_b}(b - a)$$

$f(p)$ must be of sign opposite to $f(a)$ or $f(b)$ since $f(a)$ and $f(b)$ are themselves of opposite sign. Therefore, the method can be reapplied as often as desired. The next approximation to the root is

$$p' = a + \frac{y_a}{y_a - y_p}(p - a)$$

Thus the root is bracketed between p^i and p^j such that $f(p^i)$ and $f(p^j)$ are of opposite sign.

The Newton-Raphson Method. This method may be easily implemented to determine the real roots of many polynomial and transcendental equations provided:

1. The function is smooth.
2. The first derivative of the function is readily obtainable.
3. The first derivative of the function in the neighborhood of the root is not very small (in which case the computation will be slow).
4. The first derivative of the function near the root is not zero (in which case the method fails).

Let a = an approximation of the desired root of the equation $y = f(x) = 0$, and h = a correction which will make the approximate root exact. Thus $x = a + h$ and $f(x) = f(a + h) = 0$. Expanding $f(a + h)$ in a Taylor series and dropping terms in h of degree two and higher, we obtain $f(a) + hf'(a) = 0$. A first calculation of h yields $h_1 = -f(a)/f'(a)$, which leads to a closer approximation of the root: $a_1 = a + h_1$. A new value for h may be computed from a_1: $h_2 = -f(a_1)/f'(a_1)$. This procedure may be repeated as often as necessary to achieve the desired accuracy. The nth calculated approximation is therefore

$$a_n = a_{n-1} - f(a_{n-1})/f'(a_{n-1})$$

The Method of Least Squares. This method is usually employed to determine the equation of a curve which comes as close as possible to a set of points x_i, y_i. It is based upon the principle of least squares, which states that the most probable value which can be obtained from a set of observations is the value for which the sum of the squares of the errors is a minimum, provided all observations were of equal precision.

Assume, for example, that it is required to find the best fit of the curve $y = ax^2 + bx + c$, to a set of data x_i, y_i. Therefore, the sum of the squares of the errors is to be a minimum. Thus

$$\Sigma(ax^2 + bx + c - y)^2 = \text{minimum} = (ax_1^2 + bx_1 + c - y_1)^2 + (ax_2^2 + bx_2 + c - y_2)^2 + \cdots + (ax_n^n + bx_n + c - y_n)^2 = F(a,b,c)$$

In order to find the values of a, b, and c which will minimize $F(a,b,c)$ we obtain the partial derivatives $\partial F/\partial a(a,b,c)$, $\partial F/\partial b(a,b,c)$, and $\partial F/\partial c(a,b,c)$ and set each equal to zero. We shall thus have, in this example, three simultaneous algebraic equations in the three unknowns a, b, and c:

$$(ax_1^2 + bx_1 + c - y_1) + (ax_2^2 + bx_2 + c - y_2) + \cdots + (ax_n^2 + bx_n + c - y_n) = 0 \quad (a)$$

$$x_1(ax_1^2 + bx_1 + c - y_1) + x_2(ax_2^2 + bx_2 + c - y_2) + \cdots + x_n(ax_n^2 + bx_n + c - y_n) = 0 \quad (b)$$

$$x_1^2(ax_1^2 + bx_1 + c - y_1) + x_2^2(ax_2^2 + bx_2 + c - y_2) + \cdots + x_n^2(ax_n^2 + bx_n + c - y_n) = 0 \quad (c)$$

(k) Operational Mathematics

The Laplace Transform. The Laplace transform $f(s)$ of a function $F(t)$ defined for all positive t is

$$\mathfrak{L}\{F(t)\} = f(s) = \int_0^\infty e^{-st}F(t)\,dt$$

In order for a Laplace transform of the function $F(t)$ to exist, the function must be sectionally continuous in every finite interval in $t \geqq 0$ and must be of exponential order. A function $F(t)$ is of exponential order if there exists a constant α such that $e^{-\alpha t}|F(t)|$ is bounded for all t greater than some finite number T.

Given the function $f(s)$, the original function $F(t)$ is obtained from the inversion integral

$$F(t) = \mathfrak{L}^{-1}\{f(s)\} = 1/2\pi i \int_{\gamma-i\infty}^{\gamma+i\infty} e^{st}f(s)\,ds$$

An extremely useful theorem for the solution of differential equations is

$$\mathfrak{L}\{F^{(n)}(t)\} = s^nf(s) - s^{n-1}F(+0) - s^{n-2}F'(+0) - s^{n-3}F''(+0) - \cdots - F^{(n-1)}(+0)$$

In the foregoing equation, the function $F(t)$ has a continuous derivative of order $n - 1$, $F^{(n-1)}(t)$, and a sectionally continuous derivative $F^{(n)}(t)$ in every finite interval $0 \leqq t \leqq T$. In addition $F(t)$, $F'(t) \cdots F^{(n-1)}(t)$ are of exponential order (of order $e^{\alpha t}$) and $s > \alpha$.

Table 1.17. Operations for the Laplace Transformation*

	$F(t)$	$f(s)$
1	$F(t)$	$\int_0^\infty e^{-st}F(t)\,dt$
2	$AF(t) + BG(t)$	$Af(s) + Bg(s)$
3	$F'(t)$	$sf(s) - F(+0)$
4	$F^{(n)}(t)$	$s^n f(s) - s^{n-1}F(+0) - s^{n-2}F'(+0) - \cdots - F^{(n-1)}(+0)$
5	$\int_0^t F(\tau)\,d\tau$	$\frac{1}{s}f(s)$
6	$\int_0^t \int_0^\tau F(\lambda)\,d\lambda\,d\tau$	$\frac{1}{s^2}f(s)$
7	$\int_0^t F_1(t-\tau)F_2(\tau)\,d\tau = F_1 * F_2$	$f_1(s)f_2(s)$
8	$tF(t)$	$-f'(s)$
9	$t^nF(t)$	$(-1)^n f^{(n)}(s)$
10	$\frac{1}{t}F(t)$	$\int_s^\infty f(x)\,dx$
11	$e^{at}F(t)$	$f(s-a)$
12	$F(t-b)$, where $F(t) = 0$ when $t < 0$	$e^{-bs}f(s)$
13	$\frac{1}{c}F\left(\frac{t}{c}\right)$ $(c > 0)$	$f(cs)$
14	$\frac{1}{c}e^{\frac{bt}{c}}F\left(\frac{t}{c}\right)$ $(c > 0)$	$f(cs - b)$
15	$F(t)$, when $F(t+a) = F(t)$	$\dfrac{\int_0^a e^{-st}F(t)\,dt}{1 - e^{-as}}$
16	$F(t)$, when $F(t+a) = -F(t)$	$\dfrac{\int_0^a e^{-st}F(t)\,dt}{1 + e^{-as}}$
17	$F_1(t)$, the half-wave rectification of $F(t)$ in No. 16	$\dfrac{f(s)}{1 - e^{-as}}$
18	$F_2(t)$, the full-wave rectification of $F(t)$ in No. 16	$f(s)\coth\frac{as}{2}$
19	$\sum_1^m \frac{p(a_n)}{q'(a_n)}e^{a_nt}$	$\frac{p(s)}{q(s)}$, $q(s) = (s-a_1)(s-a_2)\cdots(s-a_m)$

* From R. V. Churchill, "Operational Mathematics," 2d ed., McGraw-Hill Book Company, Inc., New York, 1958.

The foregoing theorem in conjunction with Table 1.18 can be used to solve a variety of differential equations encountered in engineering problems. For example, solve the differential equation

$$d^2X/dt^2 + K/mX = 0$$

given the boundary conditions $X(0) = C_1$ and $X'(0) = C_2$. The differential equation may also be written

$$X''(t) + k^2X(t) = 0$$

where $k^2 = K/m$. Applying the foregoing theorem we obtain

$$\begin{aligned} \mathfrak{L}\{X''(t) + k^2X(t)\} &= \mathfrak{L}\{X''(t)\} + k^2\mathfrak{L}\{X(t)\} = 0 \\ \mathfrak{L}\{X''(t)\} &= s^2x(s) - sX(+0) - X'(+0) \\ k^2\mathfrak{L}\{X(t)\} &= k^2x(s) \end{aligned}$$

Substituting, we obtain

$$\begin{aligned} s^2x(s) - sX(+0) - X'(+0) + k^2x(s) &= 0 \\ s^2x(s) - sC_1 - C_2 + k^2x(s) &= 0 \\ x(s) &= \frac{sC_1}{s^2 + k^2} + \frac{C_2}{s^2 + k^2} \end{aligned}$$

Table 1.18. Laplace Transforms†

	$f(s)$	$F(t)$
1	$\frac{1}{s}$	1
2	$\frac{1}{s^2}$	t
3	$\frac{1}{s^n}$ $(n = 1, 2, \ldots)$	$\frac{t^{n-1}}{(n-1)!}$
4	$\frac{1}{\sqrt{s}}$	$\frac{1}{\sqrt{\pi t}}$
5	$s^{-\frac{3}{2}}$	$2\sqrt{\frac{t}{\pi}}$
6	$s^{-(n+\frac{1}{2})}$ $(n = 1, 2, \ldots)$	$\frac{2^n t^{n-\frac{1}{2}}}{1 \times 3 \times 5 \cdots (2n-1)\sqrt{\pi}}$
7	$\frac{\Gamma(k)}{s^k}$ $(k > 0)$	t^{k-1}
8	$\frac{1}{s-a}$	e^{at}
9	$\frac{1}{(s-a)^2}$	te^{at}
10	$\frac{1}{(s-a)^n}$ $(n = 1, 2, \ldots)$	$\frac{1}{(n-1)!} t^{n-1}e^{at}$
11	$\frac{\Gamma(k)}{(s-a)^k}$ $(k > 0)$	$t^{k-1}e^{at}$
12*	$\frac{1}{(s-a)(s-b)}$	$\frac{1}{a-b}(e^{at} - e^{bt})$
13*	$\frac{s}{(s-a)(s-b)}$	$\frac{1}{a-b}(ae^{at} - be^{bt})$
14*	$\frac{1}{(s-a)(s-b)(s-c)}$	$-\frac{(b-c)e^{at} + (c-a)e^{bt} + (a-b)e^{ct}}{(a-b)(b-c)(c-a)}$
15	$\frac{1}{s^2 + a^2}$	$\frac{1}{a}\sin at$
16	$\frac{s}{s^2 + a^2}$	$\cos at$

* Here a, b, and (in 14) c represent distinct constants.

† From R. V. Churchill, "Operational Mathematics," 2d ed., McGraw-Hill Book Company, Inc., New York, 1958.

Table 1.18. Laplace Transforms (Continued)

	$f(s)$	$F(t)$
17	$\dfrac{1}{s^2 - a^2}$	$\dfrac{1}{a} \sinh at$
18	$\dfrac{s}{s^2 - a^2}$	$\cosh at$
19	$\dfrac{1}{s(s^2 + a^2)}$	$\dfrac{1}{a^2} (1 - \cos at)$
20	$\dfrac{1}{s^2(s^2 + a^2)}$	$\dfrac{1}{a^3} (at - \sin at)$
21	$\dfrac{1}{(s^2 + a^2)^2}$	$\dfrac{1}{2a^3} (\sin at - at \cos at)$
22	$\dfrac{s}{(s^2 + a^2)^2}$	$\dfrac{t}{2a} \sin at$
23	$\dfrac{s^2}{(s^2 + a^2)^2}$	$\dfrac{1}{2a} (\sin at + at \cos at)$
24	$\dfrac{s^2 - a^2}{(s^2 + a^2)^2}$	$t \cos at$
25	$\dfrac{s}{(s^2 + a^2)(s^2 + b^2)} \quad (a^2 \neq b^2)$	$\dfrac{\cos at - \cos bt}{b^2 - a^2}$
26	$\dfrac{1}{(s - a)^2 + b^2}$	$\dfrac{1}{b} e^{at} \sin bt$
27	$\dfrac{s - a}{(s - a)^2 + b^2}$	$e^{at} \cos bt$
28	$\dfrac{3a^2}{s^3 + a^3}$	$e^{-at} - e^{at/2} \left(\cos \dfrac{at\sqrt{3}}{2} - \sqrt{3} \sin \dfrac{at\sqrt{3}}{2} \right)$
29	$\dfrac{4a^3}{s^4 + 4a^4}$	$\sin at \cosh at - \cos at \sinh at$
30	$\dfrac{s}{s^4 + 4a^4}$	$\dfrac{1}{2a^2} \sin at \sinh at$
31	$\dfrac{1}{s^4 - a^4}$	$\dfrac{1}{2a^3} (\sinh at - \sin at)$
32	$\dfrac{s}{s^4 - a^4}$	$\dfrac{1}{2a^2} (\cosh at - \cos at)$
33	$\dfrac{8a^3 s^2}{(s^2 + a^2)^3}$	$(1 + a^2 t^2) \sin at - at \cos at$
34*	$\dfrac{1}{s} \left(\dfrac{s - 1}{s} \right)^n$	$L_n(t) = \dfrac{e^t}{n!} \dfrac{d^n}{dt^n} (t^n e^{-t})$
35	$\dfrac{s}{(s - a)^{\frac{3}{2}}}$	$\dfrac{1}{\sqrt{\pi t}} e^{at}(1 + 2at)$
36	$\sqrt{s - a} - \sqrt{s - b}$	$\dfrac{1}{2\sqrt{\pi t^3}} (e^{bt} - e^{at})$

* $L_n(t)$ is the Laguerre polynomial of degree n.

Table 1.18. Laplace Transforms (Continued)

	$f(s)$	$F(t)$
37	$\dfrac{1}{\sqrt{s}+a}$	$\dfrac{1}{\sqrt{\pi t}} - ae^{a^2t}\operatorname{erfc}(a\sqrt{t})$
38	$\dfrac{\sqrt{s}}{s-a^2}$	$\dfrac{1}{\sqrt{\pi t}} + ae^{a^2t}\operatorname{erf}(a\sqrt{t})$
39	$\dfrac{\sqrt{s}}{s+a^2}$	$\dfrac{1}{\sqrt{\pi t}} - \dfrac{2a}{\sqrt{\pi}}e^{-a^2t}\int_0^{a\sqrt{t}} e^{\lambda^2}\,d\lambda$
40	$\dfrac{1}{\sqrt{s}\,(s-a^2)}$	$\dfrac{1}{a}e^{a^2t}\operatorname{erf}(a\sqrt{t})$
41	$\dfrac{1}{\sqrt{s}\,(s+a^2)}$	$\dfrac{2}{a\sqrt{\pi}}e^{-a^2t}\int_0^{a\sqrt{t}} e^{\lambda^2}\,d\lambda$
42	$\dfrac{b^2-a^2}{(s-a^2)(b+\sqrt{s})}$	$e^{a^2t}[b - a\operatorname{erf}(a\sqrt{t})] - be^{b^2t}\operatorname{erfc}(b\sqrt{t})$
43	$\dfrac{1}{\sqrt{s}\,(\sqrt{s}+a)}$	$e^{a^2t}\operatorname{erfc}(a\sqrt{t})$
44	$\dfrac{1}{(s+a)\sqrt{s+b}}$	$\dfrac{1}{\sqrt{b-a}}e^{-at}\operatorname{erf}(\sqrt{b-a}\,\sqrt{t})$
45	$\dfrac{b^2-a^2}{\sqrt{s}\,(s-a^2)(\sqrt{s}+b)}$	$e^{a^2t}\left[\dfrac{b}{a}\operatorname{erf}(a\sqrt{t}) - 1\right] + e^{b^2t}\operatorname{erfc}(b\sqrt{t})$
46*	$\dfrac{(1-s)^n}{s^{n+\frac{1}{2}}}$	$\dfrac{n!}{(2n)!\sqrt{\pi t}}H_{2n}(\sqrt{t})$
47	$\dfrac{(1-s)^n}{s^{n+\frac{3}{2}}}$	$-\dfrac{n!}{\sqrt{\pi}\,(2n+1)!}H_{2n+1}(\sqrt{t})$
48†	$\dfrac{\sqrt{s+2a}}{\sqrt{s}} - 1$	$ae^{-at}[I_1(at) + I_0(at)]$
49	$\dfrac{1}{\sqrt{s+a}\,\sqrt{s+b}}$	$e^{-\frac{1}{2}(a+b)t}I_0\left(\dfrac{a-b}{2}t\right)$
50	$\dfrac{\Gamma(k)}{(s+a)^k(s+b)^k}\ (k>0)$	$\sqrt{\pi}\left(\dfrac{t}{a-b}\right)^{k-\frac{1}{2}}e^{-\frac{1}{2}(a+b)t} \times I_{k-\frac{1}{2}}\left(\dfrac{a-b}{2}t\right)$
51	$\dfrac{1}{(s+a)^{\frac{1}{2}}(s+b)^{\frac{3}{2}}}$	$te^{-\frac{1}{2}(a+b)t}\left[I_0\left(\dfrac{a-b}{2}t\right) + I_1\left(\dfrac{a-b}{2}t\right)\right]$
52	$\dfrac{\sqrt{s+2a}-\sqrt{s}}{\sqrt{s+2a}+\sqrt{s}}$	$\dfrac{1}{t}e^{-at}I_1(at)$

* $H_n(x)$ is the Hermite polynomial, $H_n(x) = e^{x^2}\dfrac{d^n}{dx^n}(e^{-x^2})$.

† $I_n(x) = i^{-n}J_n(ix)$, where J_n is Bessel's function of the first kind.

Table 1.18. Laplace Transforms (Continued)

	$f(s)$	$F(t)$
53	$\dfrac{(a-b)^k}{(\sqrt{s+a}+\sqrt{s+b})^{2k}}\ (k>0)$	$\dfrac{k}{t}e^{-\frac{1}{2}(a+b)t}I_k\left(\dfrac{a-b}{2}t\right)$
54	$\dfrac{(\sqrt{s+a}+\sqrt{s})^{-2\nu}}{\sqrt{s}\sqrt{s+a}}\ (\nu>-1)$	$\dfrac{1}{a^\nu}e^{-\frac{1}{2}at}I_\nu\left(\dfrac{1}{2}at\right)$
55	$\dfrac{1}{\sqrt{s^2+a^2}}$	$J_0(at)$
56	$\dfrac{(\sqrt{s^2+a^2}-s)^\nu}{\sqrt{s^2+a^2}}\ (\nu>-1)$	$a^\nu J_\nu(at)$
57	$\dfrac{1}{(s^2+a^2)^k}\ (k>0)$	$\dfrac{\sqrt{\pi}}{\Gamma(k)}\left(\dfrac{t}{2a}\right)^{k-\frac{1}{2}}J_{k-\frac{1}{2}}(at)$
58	$(\sqrt{s^2+a^2}-s)^k\ (k>0)$	$\dfrac{ka^k}{t}J_k(at)$
59	$\dfrac{(s-\sqrt{s^2-a^2})^\nu}{\sqrt{s^2-a^2}}\ (\nu>-1)$	$a^\nu I_\nu(at)$
60	$\dfrac{1}{(s^2-a^2)^k}\ (k>0)$	$\dfrac{\sqrt{\pi}}{\Gamma(k)}\left(\dfrac{t}{2a}\right)^{k-\frac{1}{2}}I_{k-\frac{1}{2}}(at)$
61	$\dfrac{e^{-ks}}{s}$	$S_k(t)=\begin{cases}0 \text{ when } 0<t<k\\ 1 \text{ when } t>k\end{cases}$
62	$\dfrac{e^{-ks}}{s^2}$	$\begin{cases}0 & \text{when } 0<t<k\\ t-k & \text{when } t>k\end{cases}$
63	$\dfrac{e^{-ks}}{s^\mu}\ (\mu>0)$	$\begin{cases}0 & \text{when } 0<t<k\\ \dfrac{(t-k)^{\mu-1}}{\Gamma(\mu)} & \text{when } t>k\end{cases}$
64	$\dfrac{1-e^{-ks}}{s}$	$\begin{cases}1 \text{ when } 0<t<k\\ 0 \text{ when } t>k\end{cases}$
65	$\dfrac{1}{s(1-e^{-ks})}=\dfrac{1+\coth\frac{1}{2}ks}{2s}$	$1+[t/k]=n$ when $(n-1)k<t<nk$ $(n=1,2,\ldots)$ (Fig. 5)
66	$\dfrac{1}{s(e^{ks}-a)}$	$\begin{cases}0 \text{ when } 0<t<k\\ 1+a+a^2+\cdots+a^{n-1}\\ \quad\text{when } nk<t<(n+1)k\\ \quad(n=1,2,\ldots)\end{cases}$
67	$\dfrac{1}{s}\tanh ks$	$M(2k,t)=(-1)^{n-1}$ when $2k(n-1)<t<2kn$ $(n=1,2,\ldots)$ (Fig. 9)
68	$\dfrac{1}{s(1+e^{-ks})}$	$\dfrac{1}{2}M(k,t)+\dfrac{1}{2}=\dfrac{1-(-1)^n}{2}$ when $(n-1)k<t<nk$
69	$\dfrac{1}{s^2}\tanh ks$	$H(2k,t)$ (Fig. 10)

Table 1.18. Laplace Transforms (Continued)

	$f(s)$	$F(t)$
70	$\dfrac{1}{s \sinh ks}$	$F(t) = 2(n-1)$ when $(2n-3)k < t < (2n-1)k$ $(t > 0)$
71	$\dfrac{1}{s \cosh ks}$	$M(2k, t + 3k) + 1 = 1 + (-1)^n$ when $(2n-3)k < t < (2n-1)k$ $(t > 0)$
72	$\dfrac{1}{s} \coth ks$	$F(t) = 2n - 1$ when $2k(n-1) < t < 2kn$
73	$\dfrac{k}{s^2 + k^2} \coth \dfrac{\pi s}{2k}$	$\|\sin kt\|$
74	$\dfrac{1}{(s^2+1)(1 - e^{-\pi s})}$	$\begin{cases} \sin t \text{ when} & (2n-2)\pi < t < (2n-1)\pi \\ 0 \quad \text{when} & (2n-1)\pi < t < 2n\pi \end{cases}$
75	$\dfrac{1}{s} e^{-(k/s)}$	$J_0(2\sqrt{kt})$
76	$\dfrac{1}{\sqrt{s}} e^{-(k/s)}$	$\dfrac{1}{\sqrt{\pi t}} \cos 2\sqrt{kt}$
77	$\dfrac{1}{\sqrt{s}} e^{k/s}$	$\dfrac{1}{\sqrt{\pi t}} \cosh 2\sqrt{kt}$
78	$\dfrac{1}{s^{\frac{3}{2}}} e^{-(k/s)}$	$\dfrac{1}{\sqrt{\pi k}} \sin 2\sqrt{kt}$
79	$\dfrac{1}{s^{\frac{3}{2}}} e^{k/s}$	$\dfrac{1}{\sqrt{\pi k}} \sinh 2\sqrt{kt}$
80	$\dfrac{1}{s^\mu} e^{-(k/s)} \; (\mu > 0)$	$\left(\dfrac{t}{k}\right)^{(\mu-1)/2} J_{\mu-1}(2\sqrt{kt})$
81	$\dfrac{1}{s^\mu} e^{k/s} \; (\mu > 0)$	$\left(\dfrac{t}{k}\right)^{(\mu-1)/2} I_{\mu-1}(2\sqrt{kt})$
82	$e^{-k\sqrt{s}} \; (k > 0)$	$\dfrac{k}{2\sqrt{\pi t^3}} \exp\left(-\dfrac{k^2}{4t}\right)$
83	$\dfrac{1}{s} e^{-k\sqrt{s}} \; (k \geqq 0)$	$\operatorname{erfc}\left(\dfrac{k}{2\sqrt{t}}\right)$
84	$\dfrac{1}{\sqrt{s}} e^{-k\sqrt{s}} \; (k \geqq 0)$	$\dfrac{1}{\sqrt{\pi t}} \exp\left(-\dfrac{k^2}{4t}\right)$
85	$s^{-\frac{3}{2}} e^{-k\sqrt{s}} \; (k \geqq 0)$	$2\sqrt{\dfrac{t}{\pi}} \exp\left(-\dfrac{k^2}{4t}\right) - k \operatorname{erfc}\left(\dfrac{k}{2\sqrt{t}}\right)$
86	$\dfrac{a e^{-k\sqrt{s}}}{s(a + \sqrt{s})} \; (k \geqq 0)$	$-e^{ak} e^{a^2 t} \operatorname{erfc}\left(a\sqrt{t} + \dfrac{k}{2\sqrt{t}}\right) + \operatorname{erfc}\left(\dfrac{k}{2\sqrt{t}}\right)$

Table 1.18. Laplace Transforms (Continued)

	$f(s)$	$F(t)$
87	$\dfrac{e^{-k\sqrt{s}}}{\sqrt{s}\,(a+\sqrt{s})}\quad (k \geqq 0)$	$e^{ak}e^{a^2t}\operatorname{erfc}\left(a\sqrt{t}+\dfrac{k}{2\sqrt{t}}\right)$
88	$\dfrac{e^{-k\sqrt{s(s+a)}}}{\sqrt{s(s+a)}}$	$\begin{cases} 0 & \text{when } 0 < t < k \\ e^{-\frac{1}{2}at}I_0(\frac{1}{2}a\sqrt{t^2-k^2}) & \text{when } t > k \end{cases}$
89	$\dfrac{e^{-k\sqrt{s^2+a^2}}}{\sqrt{s^2+a^2}}$	$\begin{cases} 0 & \text{when } 0 < t < k \\ J_0(a\sqrt{t^2-k^2}) & \text{when } t > k \end{cases}$
90	$\dfrac{e^{-k\sqrt{s^2-a^2}}}{\sqrt{s^2-a^2}}$	$\begin{cases} 0 & \text{when } 0 < t < k \\ I_0(a\sqrt{t^2-k^2}) & \text{when } t > k \end{cases}$
91	$\dfrac{e^{-k(\sqrt{s^2+a^2}-s)}}{\sqrt{s^2+a^2}}\quad (k \geqq 0)$	$J_0(a\sqrt{t^2+2kt})$
92	$e^{-ks} - e^{-k\sqrt{s^2+a^2}}$	$\begin{cases} 0 & \text{when } 0 < t < k \\ \dfrac{ak}{\sqrt{t^2-k^2}}J_1(a\sqrt{t^2-k^2}) & \text{when } t > k \end{cases}$
93	$e^{-k\sqrt{s^2-a^2}} - e^{-ks}$	$\begin{cases} 0 & \text{when } 0 < t < k \\ \dfrac{ak}{\sqrt{t^2-k^2}}I_1(a\sqrt{t^2-k^2}) & \text{when } t > k \end{cases}$
94	$\dfrac{a^\nu e^{-k\sqrt{s^2+a^2}}}{\sqrt{s^2+a^2}\,(\sqrt{s^2+a^2}+s)^\nu}\quad (\nu > -1)$	$\begin{cases} 0 & \text{when } 0 < t < k \\ \left(\dfrac{t-k}{t+k}\right)^{\frac{1}{2}\nu} J_\nu(a\sqrt{t^2-k^2}) & \text{when } t > k \end{cases}$
95	$\dfrac{1}{s}\log s$	$\Gamma'(1) - \log t \quad [\Gamma'(1) = -0.5772]$
96	$\dfrac{1}{s^k}\log s\ (k > 0)$	$t^{k-1}\left\{\dfrac{\Gamma'(k)}{[\Gamma(k)]^2} - \dfrac{\log t}{\Gamma(k)}\right\}$
97*	$\dfrac{\log s}{s-a}\quad (a > 0)$	$e^{at}[\log a - \operatorname{Ei}(-at)]$
98†	$\dfrac{\log s}{s^2+1}$	$\cos t \operatorname{Si} t - \sin t \operatorname{Ci} t$
99	$\dfrac{s\log s}{s^2+1}$	$-\sin t \operatorname{Si} t - \cos t \operatorname{Ci} t$
100	$\dfrac{1}{s}\log(1+ks)\ (k > 0)$	$-\operatorname{Ei}\left(-\dfrac{t}{k}\right)$

*The exponential-integral function $\operatorname{Ei}(-t) = -\int_t^\infty \frac{e^{-x}}{x}\,dx\ (t > 0)$ is a tabulated function.

†The cosine-integral function is defined as $Ci(t) = -\int_t^\infty (\cos x/x)\,dx$ $Si(t) = \int_0^t (\sin x/x)\,dx$.

Table 1.18. Laplace Transforms (Continued)

	$f(s)$	$F(t)$
101	$\log \dfrac{s-a}{s-b}$	$\dfrac{1}{t}(e^{bt} - e^{at})$
102	$\dfrac{1}{s}\log(1 + k^2s^2)$	$-2\,\mathrm{Ci}\left(\dfrac{t}{k}\right)$
103	$\dfrac{1}{s}\log(s^2 + a^2)\ (a > 0)$	$2\log a - 2\,\mathrm{Ci}(at)$
104	$\dfrac{1}{s^2}\log(s^2 + a^2)\ (a > 0)$	$\dfrac{2}{a}[at\log a + \sin at - at\,\mathrm{Ci}(at)]$
105	$\log \dfrac{s^2 + a^2}{s^2}$	$\dfrac{2}{t}(1 - \cos at)$
106	$\log \dfrac{s^2 - a^2}{s^2}$	$\dfrac{2}{t}(1 - \cosh at)$
107	$\arctan \dfrac{k}{s}$	$\dfrac{1}{t}\sin kt$
108	$\dfrac{1}{s}\arctan \dfrac{k}{s}$	$\mathrm{Si}(kt)$
109	$e^{k^2s^2}\,\mathrm{erfc}(ks)\ (k > 0)$	$\dfrac{1}{k\sqrt{\pi}}\exp\left(-\dfrac{t^2}{4k^2}\right)$
110	$\dfrac{1}{s}e^{k^2s^2}\,\mathrm{erfc}(ks)\ (k > 0)$	$\mathrm{erf}\left(\dfrac{t}{2k}\right)$
111	$e^{ks}\,\mathrm{erfc}\sqrt{ks}\ \ (k > 0)$	$\dfrac{\sqrt{k}}{\pi\sqrt{t}\,(t + k)}$
112	$\dfrac{1}{\sqrt{s}}\,\mathrm{erfc}(\sqrt{ks})$	$\begin{cases} 0 & \text{when } 0 < t < k \\ (\pi t)^{-\frac{1}{2}} & \text{when } t > k \end{cases}$
113	$\dfrac{1}{\sqrt{s}}e^{ks}\,\mathrm{erfc}(\sqrt{ks})\ (k > 0)$	$\dfrac{1}{\sqrt{\pi(t + k)}}$
114	$\mathrm{erf}\left(\dfrac{k}{\sqrt{s}}\right)$	$\dfrac{1}{\pi t}\sin(2k\sqrt{t})$
115	$\dfrac{1}{\sqrt{s}}e^{k^2/s}\,\mathrm{erfc}\left(\dfrac{k}{\sqrt{s}}\right)$	$\dfrac{1}{\sqrt{\pi t}}e^{-2k\sqrt{t}}$
116*	$K_0(ks)$	$\begin{cases} 0 & \text{when } 0 < t < k \\ (t^2 - k^2)^{-\frac{1}{2}} & \text{when } t > k \end{cases}$
117	$K_0(k\sqrt{s})$	$\dfrac{1}{2t}\exp\left(-\dfrac{k^2}{4t}\right)$
118	$\dfrac{1}{s}e^{ks}K_1(ks)$	$\dfrac{1}{k}\sqrt{t(t + 2k)}$
119	$\dfrac{1}{\sqrt{s}}K_1(k\sqrt{s})$	$\dfrac{1}{k}\exp\left(-\dfrac{k^2}{4t}\right)$
120	$\dfrac{1}{\sqrt{s}}e^{k/s}K_0\left(\dfrac{k}{s}\right)$	$\dfrac{2}{\sqrt{\pi t}}K_0(2\sqrt{2kt})$

* $K_n(x)$ is Bessel's function of the second kind for the imaginary argument.

Table 1.18. Laplace Transforms (Continued)

	$f(s)$	$F(t)$
121	$\pi e^{-ks} I_0(ks)$	$\begin{cases} [t(2k-t)]^{-\frac{1}{2}} & \text{when } 0 < t < 2k \\ 0 & \text{when } t > 2k \end{cases}$
122	$e^{-ks} I_1(ks)$	$\begin{cases} \dfrac{k-t}{\pi k \sqrt{t(2k-t)}} & \text{when } 0 < t < 2k \\ 0 & \text{when } t > 2k \end{cases}$
123	$-e^{as} \operatorname{Ei}(-as)$	$\dfrac{1}{t+a} \quad (a > 0)$
124	$\dfrac{1}{a} + se^{as} \operatorname{Ei}(-as)$	$\dfrac{1}{(t+a)^2} \quad (a > 0)$
125	$\left(\dfrac{\pi}{2} - \operatorname{Si} s\right) \cos s + \operatorname{Ci} s \sin s$	$\dfrac{1}{t^2+1}$

From Table 1.18 we find that

$$\mathcal{L}^{-1}\left\{\frac{s}{s^2+k^2}\right\} = \cos kt \qquad \text{and} \qquad \mathcal{L}^{-1}\left\{\frac{1}{s^2+k^2}\right\} = \frac{1}{k} \sin kt$$

The solution is therefore

$$X(t) = C_1 \cos kt + (C_2/k) \sin kt$$

In the case of a nonhomogeneous equation, the Laplace transform of both sides of the equation is obtained, and the procedure is identical with that indicated above.

Often the function $f(s)$ is such that it is adaptable to the inversions of Table 1.18 by use of partial fractions. Thus

$$f(s) = \frac{s+4}{s^2+2s} = \frac{s+4}{s(s+2)} = \frac{A}{s} + \frac{B}{s+2}$$

Multiplying through by $(s^2 + 2s)$ we obtain $s + 4 = (A + B)s + 2A$. Equating coefficients of like powers of s, the constants are evaluated: $A = 2$, $B = -1$. Therefore, $f(s) = 2/s - 1/(s + 2)$. From Table 1.18 we find that $F(t) = 2 - e^{-2t}$. For a more complete discussion of partial fractions, see Art. 1.4(a).

The *convolution* of two functions $F_1(t)$ and $F_2(t)$ is denoted $F_1 * F_2$ and is defined as follows:

$$F_1 * F_2 = \int_0^t F_1(t-\tau) F_2(\tau)\, d\tau$$

The correspondence between the transforms of two functions $f_1(s) \cdot f_2(s)$ and the convolution of the functions $F_1(t)$ and $F_2(t)$ is given by

$$f_1(s) \cdot f_2(s) = \mathcal{L}\{F_1(t) * F_2(t)\}$$

Thus $$\mathcal{L}^{-1}\{f_1(s) \cdot f_2(s)\} = F_1(t) * F_2(t) = \int_0^t F_1(t-\tau) F_2(\tau)\, d\tau$$

Finite Fourier Sine Transforms (See Table 1.19). The finite Fourier sine transform of a sectionally continuous function $F(x)$ is defined by the following relationship:

$$f_s(n) = S\{F(x)\} = \int_0^\pi F(x) \sin nx\, dx \qquad (n = 1, 2, \ldots) \qquad (0 < x < \pi)$$

Table 1.19. Finite Sine Fourier Transforms*

	$f_s(n)$	$F(x)$
1	$f_s(n) = \int_0^\pi F(x) \sin nx\, dx \; (n = 1, 2, \ldots)$	$F(x) \quad (0 < x < \pi)$
2	$(-1)^{n+1} f_s(n)$	$F(\pi - x)$
3	$\dfrac{1}{n}$	$\dfrac{\pi - x}{\pi}$
4	$\dfrac{(-1)^{n+1}}{n}$	$\dfrac{x}{\pi}$
5	$\dfrac{1 - (-1)^n}{n}$	1
6	$\dfrac{\pi}{n^2} \sin nc \quad (0 < c < \pi)$	$\begin{cases} (\pi - c)x & (x \leqq c) \\ c(\pi - x) & (x \geqq c) \end{cases}$
7	$\dfrac{\pi}{n} \cos nc \quad (0 \leqq c \leqq \pi)$	$\begin{cases} -x & (x < c) \\ \pi - x & (x > c) \end{cases}$
8	$\dfrac{(-1)^{n+1}}{n^3}$	$\dfrac{x(\pi^2 - x^2)}{6\pi}$
9	$\dfrac{1 - (-1)^n}{n^3}$	$\dfrac{x(\pi - x)}{2}$
10	$\dfrac{\pi^2(-1)^{n-1}}{n} - \dfrac{2[1 - (-1)^n]}{n^3}$	x^2
11	$\pi(-1)^n \left(\dfrac{6}{n^3} - \dfrac{\pi^2}{n}\right)$	x^3
12	$\dfrac{n}{n^2 + c^2} [1 - (-1)^n e^{c\pi}]$	e^{cx}
13	$\dfrac{n}{n^2 + c^2}$	$\dfrac{\sinh c(\pi - x)}{\sinh c\pi}$
14	$\dfrac{n}{n^2 - k^2} \quad (\lvert k \rvert \neq 0, 1, 2, \ldots)$	$\dfrac{\sin k(\pi - x)}{\sin k\pi}$
15	$0 (n \neq m); f_s(m) = \dfrac{\pi}{2}$	$\sin mx \quad (m = 1, 2, \ldots)$
16	$\dfrac{n}{n^2 - k^2} [1 - (-1)^n \cos k\pi]$	$\cos kx \quad (\lvert k \rvert \neq 1, 2, \ldots)$
17	$\dfrac{n}{n^2 - m^2} [1 - (-1)^{n+m}], \; (n \neq m);$ $f_s(m) = 0$	$\cos mx \quad (m = 1, 2, \ldots)$
18	$\dfrac{n}{(n^2 - k^2)^2} \quad (\lvert k \rvert \neq 0, 1, 2 \ldots)$	$\dfrac{\pi \sin kx}{2k \sin^2 k\pi} - \dfrac{x \cos k(\pi - x)}{2k \sin k\pi}$
19	$\dfrac{b^n}{n} \quad (\lvert b \rvert \leqq 1)$	$\dfrac{2}{\pi} \arctan \dfrac{b \sin x}{1 - b \cos x}$
20	$\dfrac{1 - (-1)^n}{n} b^n \quad (\lvert b \rvert \leqq 1)$	$\dfrac{2}{\pi} \arctan \dfrac{2b \sin x}{1 - b^2}$

* From R. V. Churchill, "Operational Mathematics," 2d ed:, McGraw-Hill Book Company, Inc., New York, 1958.

If $F'(x)$ is sectionally continuous and the function $F(x)$ is defined at each point of discontinuity by

$$F(x_0) = \tfrac{1}{2}[F(x_0 + 0) + F(x_0 - 0)] \qquad (0 < x_0 < \pi)$$

then the inversion relationship for the finite sine transform is

$$S^{-1}\{f_s(n)\} = F(x) = \frac{2}{\pi}\sum_{1}^{\infty} f_s(n) \sin nx \qquad (0 < x < \pi)$$

If $F(x)$ for $0 < x < \pi$ has a sectionally continuous derivative of order $2\nu(\nu = 1, 2, \ldots)$ and a continuous derivative of order $2\nu - 1$, then $S\{F^{(2\nu)}(x)\}$ is given by

$$S\{F^{(2\nu)}(x)\} = (-n^2)^\nu f_s(n) - (-1)^\nu n^{2\nu-1}[F(0) - (-1)^n F(\pi)] - (-1)^{\nu-1} n^{2\nu-3}[F''(0) - (-1)^n F''(\pi)] - \cdots + n[F^{(2\nu-2)}(0) - (-1)^n F^{(2\nu-2)}(\pi)]$$

Finite Fourier Cosine Transforms (See Table 1.20). The finite Fourier cosine transform of a sectionally continuous function $F(x)$ is defined by the following relationship:

$$f_c(n) = C\{F(x)\} = \int_0^\pi F(x) \cos nx \, dx \qquad (n = 1, 2, \ldots) \qquad (0 < x < \pi)$$

If $F(x)$ and $F'(x)$ are sectionally continuous, then the inversion relationship for the finite cosine transform is

$$C^{-1}\{f_c(n)\} = F(x) = (1/\pi)f_c(0) + 2/\pi \sum_{1}^{\infty} f_c(n) \cos nx$$

If $F(x)$ for $0 < x < \pi$ has a sectionally continuous derivative of order $2\nu(\nu = 1, 2, \ldots)$ and a continuous derivative of order $2\nu - 1$, then $C\{F^{(2\nu)}(x)\}$ is given by

$$C\{F^{(2\nu)}(x)\} = (-n^2)^\nu f_c(n) - (-1)^{\nu-1} n^{2\nu-2}[F'(0) - (-1)^n F'(\pi)] - (-1)^{\nu-2} n^{\nu-4}[F''(0) - (-1)^n F'''(\pi)] - \cdots - [F^{(2\nu-1)}(0) - (-1)^n F^{(2\nu-1)}(\pi)]$$

(l) Complex Variables

Complex Numbers. A complex number z consists of a real part x and an imaginary part y and is represented as

$$z = x + iy \qquad \text{where } i = \sqrt{-1} \; (i^2 = -1)$$

The conjugate $\bar{z}$ of a complex number is defined as

$$\bar{z} = x - iy$$

Two complex numbers are equal only if their real parts are equal and their imaginary parts are equal; i.e.,

$$x_1 + iy_1 = x_2 + iy_2$$

only if $x_1 = x_2$ and $y_1 = y_2$

Also $x + iy = 0$

only if $x = 0$ and $y = 0$

Complex numbers satisfy the distributive, associative, and commutative laws of algebra.

Complex numbers may be graphically represented on the $z(x - y)$ plane or in polar (r,θ) coordinates. The polar coordinates of a complex number are

$$r = |z| = \sqrt{x^2 + y^2} = \text{modulus or absolute value of } z = \text{mod } z \qquad (r \geqq 0)$$

and θ = argument or amplitude of z = arg z = amp z = $\tan^{-1} y/x$

arg z is multiple-valued, but for an angular interval of range 2π there is only one value of θ for a given z.

Table 1.20. Finite Cosine Fourier Transforms*

	$f_c(n)$	$F(x)$
1	$f_c(n) = \int_0^\pi F(x) \cos nx\, dx \quad (n = 0, 1, 2, \ldots)$	$F(x) \quad (0 < x < \pi)$
2	$(-1)^n f_c(n)$	$F(\pi - x)$
3	0 when $n = 1, 2, \ldots ; f_c(0) = \pi$	1
4	$\frac{2}{n} \sin nc;\ f_c(0) = 2c - \pi$	$\begin{cases} 1 & (0 < x < c) \\ -1 & (c < x < \pi) \end{cases}$
5	$-\frac{1 - (-1)^n}{n^2};\ f_c(0) = \frac{\pi^2}{2}$	x
6	$\frac{(-1)^n}{n^2};\ f_c(0) = \frac{\pi^2}{6}$	$\frac{x^2}{2\pi}$
7	$\frac{1}{n^2};\ f_c(0) = 0$	$\frac{(\pi - x)^2}{2\pi} - \frac{\pi}{6}$
8	$3\pi^2 \frac{(-1)^n}{n^2} - 6 \frac{1 - (-1)^n}{n^4};\ f_c(0) = \frac{\pi^4}{4}$	x^3
9	$\frac{(-1)^n e^{c\pi} - 1}{n^2 + c^2}$	$\frac{1}{c} e^{cx}$
10	$\frac{1}{n^2 + c^2}$	$\frac{\cosh c(\pi - x)}{c \sinh c\pi}$
11	$\frac{(-1)^n \cos k\pi - 1}{n^2 - k^2} \quad (\lvert k \rvert \neq 0, 1, 2, \ldots)$	$\frac{1}{k} \sin kx$
12	$\frac{(-1)^{n+m} - 1}{n^2 - m^2};\ f_c(m) = 0 \quad (m = 1, 2, \ldots)$	$\frac{1}{m} \sin mx$
13	$\frac{1}{n^2 - k^2} \quad (\lvert k \rvert \neq 0, 1, 2, \ldots)$	$-\frac{\cos k(\pi - x)}{k \sin k\pi}$
14	$0 (n \neq m);\ f_c(m) = \frac{\pi}{2} \quad (m = 1, 2, \ldots)$	$\cos mx$
15	$b^n (n \neq 0),\ f_c(0) = 0,\ (\lvert b \rvert < 1)$	$\frac{2b}{\pi} \frac{\cos x - b}{1 - 2b \cos x + b^2}$

* From R. V. Churchill, "Operational Mathematics," 2d ed., McGraw-Hill Book Company, Inc., New York, 1958.

The complex number $z = x + iy$ is written in polar coordinates as

$$z = r(\cos \theta + i \sin \theta) \qquad \text{or} \qquad z = re^{i\theta}$$

Useful Formulas

$$z_1 \pm z_2 = (x_1 \pm x_2) + i(y_1 \pm y_2)$$
$$z_1 z_2 = r_1 r_2[\cos (\theta_1 + \theta_2) + i \sin (\theta_1 + \theta_2)] = r_1 r_2 e^{i(\theta_1+\theta_2)}$$
$$z_1/z_2 = r_1/r_2[\cos (\theta_1 - \theta_2) + i \sin (\theta_1 - \theta_2)] = r_1/r_2 e^{i(\theta_1-\theta_2)}$$
$$iz = r[\cos (\theta + \pi/2) + i \sin (\theta + \pi/2)]$$
$$(\cos \theta + i \sin \theta)^n = \cos n\theta + i \sin n\theta \qquad n = \pm 1, \pm 2, \ldots$$
$$z^n = r^n(\cos n\theta + i \sin n\theta) \qquad n = \pm 1, \pm 2, \ldots$$
$$e^z = e^{x+iy} = e^x(\cos y + i \sin y)$$
$$e^{(z+2\pi i)} = e^z$$
$$\overline{z_1 + z_2} = \bar{z}_1 + \bar{z}_2$$
$$\overline{z_1 z_2} = \bar{z}_1 \bar{z}_2$$
$$\overline{(z_1/z_2)} = \bar{z}_1/\bar{z}_2$$

$$z + \bar{z} = 2x = 2R(z) = (2)\times \text{(real part of } z)$$
$$z - \bar{z} = 2iy = 2iI(z) = (2i)\times \text{(imaginary part of } z)$$
$$z\bar{z} = x^2 + y^2 = |z|^2$$
$$|\bar{z}| = |z|$$
$$|z_1 z_2| = |z_1| \cdot |z_2|$$
$$|z_1 z_2|^2 = |z_1|^2 \cdot |z_2|^2$$
$$|z_1/z_2| = |z_1|/|z_2|$$
$$|z^n| = |z|^n$$
$$|z_1 + z_2| \leqq |z_1| + |z_2|$$
$$|z_2 - z_1| \geqq ||z_1| - |z_2||$$
$$\cos z = \frac{e^{iz} + e^{-iz}}{2} = \cos x \cosh y - i \sin x \sinh y$$
$$\sin z = \frac{e^{iz} - e^{-iz}}{2i} = \sin x \cosh y + i \cos x \sinh y$$
$$\cos (iz) = \cosh z$$
$$\sin (iz) = i \sinh z$$
$$\cosh (iz) = \cos z$$
$$\sinh (iz) = i \sin z$$
$$\sinh (x + iy) = \sinh x \cos y + i \cosh x \sin y$$
$$\cosh (x + iy) = \cosh x \cos y + i \sinh x \sin y$$

Complex Variables. *Continuity.* The single-valued function $f(z)$ is continuous at a point z_0 if

$$f(z_0) \text{ and } \lim_{z\to z_0} f(z) \text{ exist and } \lim_{z\to z_0} f(z) = f(z_0)$$

Derivative. The derivative of the function $f(z)$ at z_0 is

$$f'(z_0) = \lim_{\Delta z \to 0} \frac{f(z_0 + \Delta z) - f(z_0)}{\Delta z}$$

and the differentiation formulas of functions of real variables apply to the suitably defined functions of complex variables.

The Cauchy-Riemann Conditions. The function

$$f(z) = u(x,y) + iv(x,y)$$

is said to satisfy the Cauchy-Riemann conditions if

$$\partial u/\partial x = \partial v/\partial y \qquad \text{and} \qquad \partial u/\partial y = -\partial v/\partial x$$

or in the polar form

$$\frac{\partial u}{\partial r} = \frac{1}{r}\frac{\partial v}{\partial \theta} \qquad \text{and} \qquad \frac{\partial v}{\partial r} = -\frac{1}{r}\frac{\partial u}{\partial \theta}$$

Analytic Function. Given $f(z) = u(x,y) + iv(x,y)$, where $u(x,y)$, $v(x,y)$ and their first-order partial derivatives are single-valued, continuous, and satisfy the Cauchy-Riemann conditions in a given region. Then $f(z)$ is analytic at all points in the region.

Singularity of the Function. If the function $f(z)$ is not analytic at a point z_0 but is analytic at some point in every neighborhood of z_0, then z_0 is a *singular point.*

Harmonic Functions. If $f(z) = u(x,y) + iv(x,y)$ is analytic in a region, then u and v are harmonic; i.e.,

$$\partial u^2/\partial x^2 + \partial u^2/\partial y^2 = 0 \qquad \text{and} \qquad \partial^2 v/\partial x^2 + \partial^2 v/\partial y^2 = 0$$

and u and v are termed *conjugate harmonic functions.*

Simply Connected Region. A region is *simply* connected if every closed curve within the region will enclose points of the given region only.

Integrals of Analytic Functions. If in a simply connected region

$$f(z) = dF(z)/dz$$

Then

$$F(z) = \int f(z)\,dz$$

and $\int f(z)\,dz$ is analytic in the region. If the path of integration is within the simply connected region and the functions under consideration are analytic in the region then

$$\int_{z_1}^{z_0} f(z)\,dz = -\int_{z_0}^{z_1} f(z)\,dz$$

$$\int_{z_1}^{z_2} f(z)\,dz = \int_{z_0}^{z_1} f(z)\,dz + \int_{z_1}^{z_2} f(z)\,dz$$

$$\int_{z_0}^{z_1} [a\,f_1(z) + b\,f_2(z)]\,dz = a\int_{z_0}^{z_1} f_1(z)\,dz + b\int_{z_0}^{z_1} f_2(z)\,dz$$

The Cauchy-Goursat Theorem. If a function $f(z)$ is single-valued and analytic within and on a simple closed contour C, then

$$\int_C f(z)\,dz = 0$$

The Cauchy Integral Formula. If $f(z)$ is single-valued and analytic within and on a closed curve C and if z_0 is any point within C, then

$$f(z_0) = \frac{1}{2\pi i}\int_C \frac{f(z)}{z - z_0}\,dz$$

and

$$f^{(n)}(z_0) = \frac{n!}{2\pi i}\int_C \frac{f(z)}{(z - z_0)^{n+1}}\,dz \qquad n = 1, 2, \ldots$$

Laurent's Series. If the function $f(z)$ is analytic on two concentric circles C_1 and C_2, of radii R_1 and R_2, respectively, and within the ring between C_1 and C_2, then at every point z between C_1 and C_2, $f(z)$ can be represented by the convergent series

$$f(z) = \sum_{n=-\infty}^{\infty} C_n(z - z_0)^n$$

where z_0 is the center of the two concentric circles and $R_2 < |z - z_0| < R_1$ and

$$C_n = \frac{1}{2\pi i}\int_C \frac{f(\xi)\,d\xi}{(\xi - z_0)^{n+1}} \qquad n = 0, \pm 1, \pm 2, \pm 3, \ldots$$

where C is any closed curve within the annulus encircling C_2, and each integral is taken in the counterclockwise direction.

Residues. If $f(z)$, which is single-valued and analytic everywhere in a region except at the *isolated singular point* z_0, is expanded in a Laurent series in the neighborhood of z_0, then the *residue* of $f(z)$ at z_0 is

$$C_{-1} = \frac{1}{2\pi i}\int_C f(z)\,dz$$

where the integral is taken counterclockwise enclosing z_0 and no other singular point. Thus the coefficient of $(z - z_0)^{-1}$ in the Laurent expansion is the residue of $f(z)$ at z_0.

Residue Theorem. If $f(z)$ is analytic within and on the closed curve C except for a finite number of singular points $z_1, z_2, z_3, \ldots, z_n$ within C_1 then

$$\int_C f(z)\,dz = 2\pi i(m_1 + m_2 + m_3 + \cdots + m_n)$$

when $m_1, m_2, m_3, \ldots, m_n$ are the residues of $f(z)$ at $z_1, z_2, z_3, \ldots, z_n$ and the integral is taken counterclockwise.

(m) Vector Analysis

Introduction. A quantity (such as displacement, velocity, acceleration, and force) completely specified by both magnitude and direction is termed a *vector*. The vector **V** (denoted by boldface type as shown or by $\overrightarrow{V}$) may be graphically represented by a directed line segment $\overrightarrow{OA}$ (Fig. 1.58*a*), where the length of the segment signifies the magnitude of the vector, and the arrow indicates its direction. Two vectors are equal if they have the same magnitude and direction regardless of the position of their starting points. Line segment $\overrightarrow{AO}$ represents the vector $-\mathbf{V}$.

A *scalar* quantity is one which is completely described by its magnitude only (such as mass, time, and any real number). Scalars are represented by ordinary type. Scalar operations are the same as those in elementary algebra.

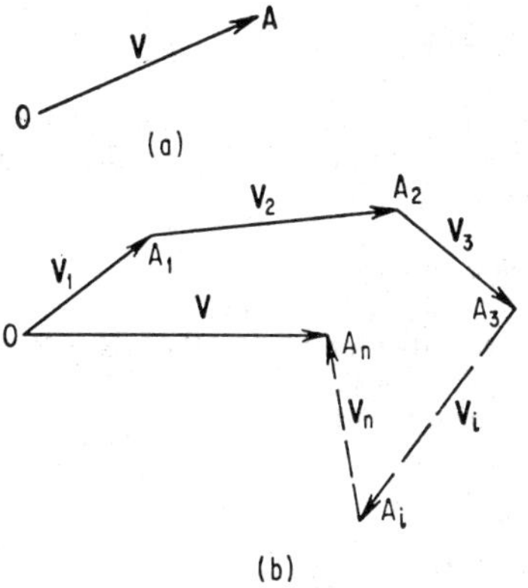

FIG. 1.58. Vectors. (*a*) Representation. (*b*) Addition.

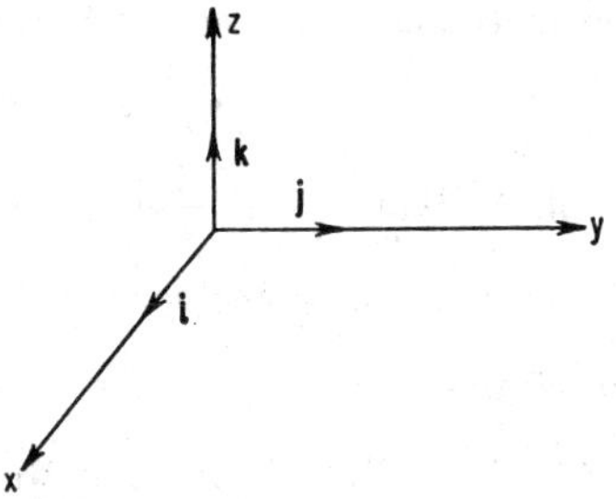

FIG. 1.59. Unit vectors in rectangular-coordinate system.

The graphical sum or resultant or n vectors (Fig. 1.58*b*) is formed by drawing the first vector $\mathbf{V}_1 = \overrightarrow{OA_1}$ from any point O, then drawing $\mathbf{V}_2$ at the end of $\mathbf{V}_1$, and continuing in this manner making the initial point of each vector the end point of the preceding one until the last vector $\mathbf{V}_n$ is added. The resultant

$$\mathbf{V} = \mathbf{V}_1 + \mathbf{V}_2 + \mathbf{V}_3 + \cdots + \mathbf{V}_n$$

is the vector joining the starting point of $\mathbf{V}_1$ with the end point of $\mathbf{V}_n$.

Unit Vectors. Vectors of unit length, generally of the same sense as the coordinate axes defining a coordinate system, are termed unit vectors. Thus the unit vectors **i**, **j**, **k** of a rectangular-coordinate system have unit lengths and are positive in the positive x, y, z directions, respectively. Figure 1.59 shows the unit vectors of a right-handed rectangular-coordinate system.

A right-handed coordinate system is one in which a right-hand screw rotated through 90° from Ox to Oy will advance in the direction of positive z.

Vector Components. The projection of a vector on a coordinate axis is called the component of the vector along that axis. Thus the vector **R**, having components along the x, y, and z axes, may be represented as the vector sum of these components

$$\mathbf{R} = R_x\mathbf{i} + R_y\mathbf{j} + R_z\mathbf{k}$$

where R_x, R_y, and R_z are the magnitudes of the respective components. The magnitude of the vector **R** is $\sqrt{R_x^2 + R_y^2 + R_z^2}$.

Vector Algebra. Whereas, by employing proper definitions, the scalar operations of addition, subtraction, and multiplication may be extended to vectors, division by a vector is not defined.

The product of a scalar m and a vector **R** is a vector $m\mathbf{R}$ whose direction is the same as or opposite to that of **R** (depending upon the sign of m) and whose magnitude is m times the magnitude of the vector **R**.

In the following laws of vector algebra, **R**, $\mathbf{R}_1$, $\mathbf{R}_2$, and $\mathbf{R}_3$ represent vectors, and m, m_1, m_2 represent scalar quantities.

Commutative law	$m\mathbf{R} = \mathbf{R}m$
Associative law	$m_1(m_2\mathbf{R}) = m_1m_2\mathbf{R}$
Distributive law	$m(\mathbf{R}_1 + \mathbf{R}_2) = m\mathbf{R}_1 + m\mathbf{R}_2$
Distributive law	$(m_1 + m_2)\mathbf{R} = m_1\mathbf{R} + m_2\mathbf{R}$
Commutative law	$\mathbf{R}_1 + \mathbf{R}_2 = \mathbf{R}_2 + \mathbf{R}_1$
Associative law	$\mathbf{R}_1 + (\mathbf{R}_2 + \mathbf{R}_3) = (\mathbf{R}_1 + \mathbf{R}_2) + \mathbf{R}_3$

The *scalar* or *dot product* of two vectors $\mathbf{R}_1$ and $\mathbf{R}_2$, a scalar quantity, a real number, is equal to the product of the magnitudes of the vectors multiplied by the cosine of the included angle between the vectors, thus:

$$\mathbf{R}_1 \cdot \mathbf{R}_2 = |\mathbf{R}_1|\,|\mathbf{R}_2| \cos\theta = \mathbf{R}_2 \cdot \mathbf{R}_1$$

It follows that $\mathbf{R}_1 \cdot \mathbf{R}_1 = |\mathbf{R}_1|^2$. If $\mathbf{R}_1$ is perpendicular to $\mathbf{R}_2$, then $\mathbf{R}_1 \cdot \mathbf{R}_2 = 0$. Also, $\mathbf{i}\cdot\mathbf{i} = \mathbf{j}\cdot\mathbf{j} = \mathbf{k}\cdot\mathbf{k} = 1$ and $\mathbf{i}\cdot\mathbf{j} = \mathbf{j}\cdot\mathbf{k} = \mathbf{k}\cdot\mathbf{i} = 0$.

The following additional laws apply to scalar or dot products:

Distributive law

$$\mathbf{R}_1 \cdot (\mathbf{R}_2 + \mathbf{R}_3) = \mathbf{R}_1 \cdot \mathbf{R}_2 + \mathbf{R}_1 \cdot \mathbf{R}_3$$

$$m(\mathbf{R}_1 \cdot \mathbf{R}_2) = (m\mathbf{R}_1) \cdot \mathbf{R}_2 = (\mathbf{R}_1 \cdot \mathbf{R}_2)m = \mathbf{R}_1 \cdot (m\mathbf{R}_2)$$

If $\mathbf{R}_1 = R_{1x}\mathbf{i} + R_{1y}\mathbf{j} + R_{1z}\mathbf{k}$ and $\mathbf{R}_2 = R_{2x}\mathbf{i} + R_{2y}\mathbf{j} + R_{2z}\mathbf{k}$

it follows that

$$\mathbf{R}_1 \cdot \mathbf{R}_1 = |\mathbf{R}_1|^2 = R_{1x}^2 + R_{1y}^2 + R_{1z}^2 \qquad \mathbf{R}_2 \cdot \mathbf{R}_2 = |\mathbf{R}_2|^2 = R_{2x}^2 + R_{2y}^2 + R_{2z}^2$$

and

$$\mathbf{R}_1 \cdot \mathbf{R}_2 = R_{1x}R_{2x} + R_{1y}R_{2y} + R_{1z}R_{2z}$$

The *vector* or *cross product* of two vectors $\mathbf{R}_1$ and $\mathbf{R}_2$ is a vector ($\mathbf{R} = \mathbf{R}_1 \times \mathbf{R}_2$) normal to the plane containing $\mathbf{R}_1$ and $\mathbf{R}_2$. The direction of **R** is such that a right-handed rotation about **R** through an angle equal to or less than π radians will carry $\mathbf{R}_1$ into $\mathbf{R}_2$. The magnitude of the vector product is $|\mathbf{R}| = |\mathbf{R}_1|\,|\mathbf{R}_2| \sin\theta$ where θ is the included angle between $\mathbf{R}_1$ and $\mathbf{R}_2$. The vector $\mathbf{R} = |\mathbf{R}_1|\,|\mathbf{R}_2| \sin\theta\mathbf{u}$ where **u** is a unit vector normal to the plane containing $\mathbf{R}_1$ and $\mathbf{R}_2$.

The following laws apply to vector products:

$$\mathbf{R}_1 \times \mathbf{R}_2 = -\mathbf{R}_2 \times \mathbf{R}_1$$

$$\mathbf{R}_1 \times (\mathbf{R}_2 + \mathbf{R}_3) = \mathbf{R}_1 \times \mathbf{R}_2 + \mathbf{R}_1 \times \mathbf{R}_3$$

$$m(\mathbf{R}_1 \times \mathbf{R}_2) = (\mathbf{R}_1 \times \mathbf{R}_2)m = \mathbf{R}_1 \times (m\mathbf{R}_2) = (m\mathbf{R}_1) \times \mathbf{R}_2$$

$$\mathbf{R}_1 \times \mathbf{R}_2 = \begin{vmatrix} \mathbf{i} & \mathbf{j} & \mathbf{k} \\ R_{1x} & R_{1y} & R_{1z} \\ R_{2x} & R_{2y} & R_{2z} \end{vmatrix}$$

$$\mathbf{i} \times \mathbf{i} = \mathbf{j} \times \mathbf{j} = \mathbf{k} \times \mathbf{k} = 0$$

$$\mathbf{i} \times \mathbf{j} = \mathbf{k} = -\mathbf{j} \times \mathbf{i}$$

$$\mathbf{j} \times \mathbf{k} = \mathbf{i} = -\mathbf{k} \times \mathbf{j}$$

$$\mathbf{k} \times \mathbf{i} = \mathbf{j} = -\mathbf{i} \times \mathbf{k}$$

Triple Products

$$\mathbf{V}_1 \cdot (\mathbf{V}_2 \times \mathbf{V}_3) = (\mathbf{V}_1 \times \mathbf{V}_2) \cdot \mathbf{V}_3 = \mathbf{V}_3 \cdot (\mathbf{V}_1 \times \mathbf{V}_2) = \mathbf{V}_2 \cdot (\mathbf{V}_3 \times \mathbf{V}_1)$$
$$\mathbf{V}_1 \times (\mathbf{V}_2 \times \mathbf{V}_3) = (\mathbf{V}_1 \cdot \mathbf{V}_3)\mathbf{V}_2 - (\mathbf{V}_1 \cdot \mathbf{V}_2)\mathbf{V}_3$$
$$(\mathbf{V}_1 \times \mathbf{V}_2) \cdot (\mathbf{V}_3 \times \mathbf{V}_4) = (\mathbf{V}_1 \cdot \mathbf{V}_3)(\mathbf{V}_2 \cdot \mathbf{V}_4) - (\mathbf{V}_1 \cdot \mathbf{V}_4)(\mathbf{V}_2 \cdot \mathbf{V}_3)$$
$$(\mathbf{V}_1 \times \mathbf{V}_2) \times (\mathbf{V}_3 \times \mathbf{V}_4) = (\mathbf{V}_1 \times \mathbf{V}_2 \cdot \mathbf{V}_4)\mathbf{V}_3 - (\mathbf{V}_1 \times \mathbf{V}_2 \cdot \mathbf{V}_3)\mathbf{V}_4$$
$$\mathbf{V}_1 \cdot (\mathbf{V}_2 \times \mathbf{V}_3) = \begin{vmatrix} V_{1x} & V_{1y} & V_{1z} \\ V_{2x} & V_{2y} & V_{2z} \\ V_{3x} & V_{3y} & V_{3z} \end{vmatrix}$$

Scalar Field. A scalar field ϕ is defined in a region if corresponding to each point (x,y,z) in the region, a scalar $\phi(x,y,z)$ exists. ϕ is termed a *scalar point function* or *scalar function of position.*

Vector Field. A vector field $\mathbf{V}$ is defined in a region if, corresponding to each point (x,y,z) in the region, a vector $\mathbf{V}(x,y,z)$ exists. $\mathbf{V}$ is termed a *vector point function* or *vector function of position.*

Vector Differentiation. When $\mathbf{V}$, $\mathbf{V}_1$, $\mathbf{V}_2$, and $\mathbf{V}_3$ are vector functions of the scalar x and ϕ is a scalar function of x, the following differentiation formulas apply:

$$\frac{d}{dx}(\mathbf{V}_1 + \mathbf{V}_2) = \frac{d\mathbf{V}_1}{dx} + \frac{d\mathbf{V}_2}{dx}$$
$$\frac{d}{dx}(\phi\mathbf{V}) = \phi\frac{d\mathbf{V}}{dx} + \frac{d\phi}{dx}\mathbf{V}$$
$$\frac{d}{dx}(\mathbf{V}_1 \cdot \mathbf{V}_2) = \mathbf{V}_1 \cdot \frac{d\mathbf{V}_2}{dx} + \frac{d\mathbf{V}_1}{dx} \cdot \mathbf{V}_2$$
$$\frac{d}{dx}(\mathbf{V}_1 \times \mathbf{V}_2) = \mathbf{V}_1 \times \frac{d\mathbf{V}_2}{dx} + \frac{d\mathbf{V}_1}{dx} \times \mathbf{V}_2 = \mathbf{V}_1 \times \frac{d\mathbf{V}_2}{dx} - \mathbf{V}_2 \times \frac{d\mathbf{V}_1}{dx}$$
$$\frac{d}{dx}(\mathbf{V}_1 \cdot \mathbf{V}_2 \times \mathbf{V}_3) = \mathbf{V}_1 \cdot \mathbf{V}_2 \times \frac{d\mathbf{V}_3}{dx} + \mathbf{V}_1 \cdot \frac{d\mathbf{V}_2}{dx} \times \mathbf{V}_3 + \frac{d\mathbf{V}_1}{dx} \cdot \mathbf{V}_2 \times \mathbf{V}_3$$
$$\frac{d}{dx}[\mathbf{V}_1 \times (\mathbf{V}_2 \times \mathbf{V}_3)] = \mathbf{V}_1 \times \left(\mathbf{V}_2 \times \frac{d\mathbf{V}_3}{dx}\right) + \mathbf{V}_1 \times \left(\frac{d\mathbf{V}_2}{dx} \times \mathbf{V}_3\right) + \frac{d\mathbf{V}_1}{dx} \times (\mathbf{V}_2 \times \mathbf{V}_3)$$

Differential Operators. *Cartesian Coordinates.* The vector differential operator *del:*

$$\nabla = \mathbf{i}\frac{\partial}{\partial x} + \mathbf{j}\frac{\partial}{\partial y} + \mathbf{k}\frac{\partial}{\partial z}$$

The *gradient* of a *scalar field:*

$$\nabla\phi = \text{grad } \phi = \left(\mathbf{i}\frac{\partial}{\partial x} + \mathbf{j}\frac{\partial}{\partial y} + \mathbf{k}\frac{\partial}{\partial z}\right)\phi = \frac{\partial\phi}{\partial x}\mathbf{i} + \frac{\partial\phi}{\partial y}\mathbf{j} + \frac{\partial\phi}{\partial z}\mathbf{k}$$

The *divergence* of a *vector field:*

$$\nabla \cdot \mathbf{V} = \text{div } \mathbf{V} = \left(\mathbf{i}\frac{\partial}{\partial x} + \mathbf{j}\frac{\partial}{\partial y} + \mathbf{k}\frac{\partial}{\partial z}\right) \cdot (V_x\mathbf{i} + V_y\mathbf{j} + V_z\mathbf{k})$$
$$= \frac{\partial V_x}{\partial x} + \frac{\partial V_y}{\partial y} + \frac{\partial V_z}{\partial z}$$

The *curl* of a *vector field:*

$$\nabla \times \mathbf{V} = \text{curl } \mathbf{V} = \left(\mathbf{i}\frac{\partial}{\partial x} + \mathbf{j}\frac{\partial}{\partial y} + \mathbf{k}\frac{\partial}{\partial z}\right) \times (V_x\mathbf{i} + V_y\mathbf{j} + V_z\mathbf{k})$$
$$= \begin{vmatrix} \mathbf{i} & \mathbf{j} & \mathbf{k} \\ \frac{\partial}{\partial x} & \frac{\partial}{\partial y} & \frac{\partial}{\partial z} \\ V_x & V_y & V_z \end{vmatrix} = \left(\frac{\partial V_z}{\partial y} - \frac{\partial V_y}{\partial z}\right)\mathbf{i} + \left(\frac{\partial V_x}{\partial z} - \frac{\partial V_z}{\partial x}\right)\mathbf{j} + \left(\frac{\partial V_y}{\partial x} - \frac{\partial V_x}{\partial y}\right)\mathbf{k}$$

Important Relationships Involving the Differential Operators. In the following relations $\mathbf{V}_1$ and $\mathbf{V}_2$, ϕ and ψ represent vector and scalar point functions.

$$\nabla(\phi + \psi) = \nabla\phi + \nabla\psi$$
$$\nabla(\phi\psi) = \phi\nabla\psi + \psi\nabla\phi$$
$$\nabla \cdot (\phi\mathbf{V}_1) = (\nabla\phi) \cdot \mathbf{V}_1 + \phi(\nabla \cdot \mathbf{V}_1)$$
$$\nabla \times (\phi\mathbf{V}_1) = (\nabla\phi) \times \mathbf{V}_1 + \phi(\nabla \times \mathbf{V}_1)$$
$$\nabla \cdot (\nabla\phi) = \nabla^2\phi = \frac{\partial^2\phi}{\partial x^2} + \frac{\partial^2\phi}{\partial y^2} + \frac{\partial^2\phi}{\partial z^2}$$

$$\nabla^2 = \frac{\partial^2}{\partial x^2} + \frac{\partial^2}{\partial y^2} + \frac{\partial^2}{\partial z^2} = \text{Laplacian operator (in cartesian coordinates)}$$

$$\nabla \times (\nabla\phi) = 0$$
$$\nabla \cdot (\mathbf{V}_1 + \mathbf{V}_2) = \nabla \cdot \mathbf{V}_1 + \nabla \cdot \mathbf{V}_2$$
$$\nabla \times (\mathbf{V}_1 + \mathbf{V}_2) = \nabla \times \mathbf{V}_1 + \nabla \times \mathbf{V}_2$$
$$\nabla \cdot (\mathbf{V}_1 \times \mathbf{V}_2) = \mathbf{V}_2 \cdot (\nabla \times \mathbf{V}_1) - \mathbf{V}_1 \cdot (\nabla \times \mathbf{V}_2)$$
$$\nabla \times (\mathbf{V}_1 \times \mathbf{V}_2) = (\mathbf{V}_2 \cdot \nabla)\mathbf{V}_1 - \mathbf{V}_2(\nabla \cdot \mathbf{V}_1) - (\mathbf{V}_1 \cdot \nabla)\mathbf{V}_2 + \mathbf{V}_1(\nabla \cdot \mathbf{V}_2)$$
$$\nabla (\mathbf{V}_1 \cdot \mathbf{V}_2) = (\mathbf{V}_2 \cdot \nabla)\mathbf{V}_1 + (\mathbf{V}_1 \cdot \nabla)\mathbf{V}_2 + \mathbf{V}_2 \times (\nabla \times \mathbf{V}_1) + \mathbf{V}_1 \times (\nabla \times \mathbf{V}_2)$$
$$\nabla \cdot (\nabla \times \mathbf{V}_1) = 0$$
$$\nabla \times (\nabla \times \mathbf{V}_1) = \nabla(\nabla \cdot \mathbf{V}_1) - \nabla^2\mathbf{V}_1$$

Cylindrical Coordinates (ρ, θ, z with Unit Vectors $\mathbf{k}_\rho$, $\mathbf{k}_\theta$, and $\mathbf{k}_z$, Respectively)

$$\nabla\psi = \text{grad }\psi = \frac{\partial\psi}{\partial\rho}\mathbf{k}_\rho + \frac{1}{\rho}\frac{\partial\psi}{\partial\theta}\mathbf{k}_\theta + \frac{\partial\psi}{\partial z}\mathbf{k}_z$$

$$\text{div }\mathbf{V} = \nabla \cdot \mathbf{V} = \frac{1}{\rho}\frac{\partial}{\partial\rho}(\rho V_\rho) + \frac{1}{\rho}\left(\frac{\partial V_\theta}{\partial\theta}\right) + \frac{\partial V_z}{\partial z}$$

$$\text{curl }\mathbf{V} = \nabla \times \mathbf{V} = \left(\frac{1}{\rho}\frac{\partial V_z}{\partial\theta} - \frac{\partial V_\theta}{\partial z}\right)\mathbf{k}_\rho + \left(\frac{\partial V_\rho}{\partial z} - \frac{\partial V_z}{\partial\rho}\right)\mathbf{k}_\theta + \left[\frac{1}{\rho}\frac{\partial}{\partial\rho}(\rho V_\theta) - \frac{1}{\rho}\frac{\partial V_\rho}{\partial\theta}\right]\mathbf{k}_z$$

$$\nabla^2\psi = \frac{1}{\rho}\frac{\partial}{\partial\rho}\left(\rho\frac{\partial\psi}{\partial\rho}\right) + \frac{1}{\rho^2}\frac{\partial^2\psi}{\partial\theta^2} + \frac{\partial^2\psi}{\partial z^2}$$

Spherical Coordinates (r, θ, ϕ with Unit Vectors $\mathbf{k}_r$, $\mathbf{k}_\theta$, and $\mathbf{k}_\phi$, Respectively)

$$\text{grad }\psi = \nabla\psi = \frac{\partial\psi}{\partial r}\mathbf{k}_r + \frac{1}{r}\frac{\partial\psi}{\partial\theta}\mathbf{k}_\theta + \frac{1}{r\sin\theta}\frac{\partial\psi}{\partial\theta}\mathbf{k}_\phi$$

$$\text{div }\mathbf{V} = \nabla \cdot \mathbf{V} = \frac{1}{r^2}\frac{\partial}{\partial r}(r^2V_r) + \frac{1}{r\sin\theta}\frac{\partial}{\partial\theta}(V_\theta\sin\theta) + \frac{1}{r\sin\theta}\frac{\partial V_\phi}{\partial\phi}$$

$$\text{curl }\mathbf{V} = \nabla \times \mathbf{V} = \frac{1}{r\sin\theta}\left[\frac{\partial}{\partial\theta}(V_\phi\sin\theta) - \frac{\partial V_\theta}{\partial\phi}\right]\mathbf{k}_r + \frac{1}{r}\left[\frac{1}{\sin\theta}\frac{\partial Vr}{\partial\phi} - \frac{\partial}{\partial r}(rV_\phi)\right]\mathbf{k}_\theta + \frac{1}{r}\left[\frac{\partial}{\partial r}(rV_\theta) - \frac{\partial V_r}{\partial\theta}\right]\mathbf{k}_\phi$$

$$\nabla^2\psi = \frac{1}{r^2}\frac{\partial}{\partial r}\left(r^2\frac{\partial\psi}{\partial r}\right) + \frac{1}{r^2\sin\theta}\frac{\partial}{\partial\theta}\left(\sin\theta\frac{\partial\psi}{\partial\theta}\right) + \frac{1}{r^2\sin^2\theta}\frac{\partial^2\psi}{\partial\phi^2}$$

Line Integrals. The line integral $\int_a^b \mathbf{V} \cdot d\mathbf{r}$ denotes the scalar integral from a to b of the tangential component of the vector function $\mathbf{V}(x,y,z)$ defined at each point P along the path C (Fig. 1.60). $d\mathbf{r}$ represents a differential increment of the position vector $\mathbf{r}$. If s is a distance along C to point P, and $\mathbf{t}$ is a unit vector tangent to C at P then

$$\int_a^b \mathbf{V} \cdot d\mathbf{r} = \int_a^b \mathbf{V} \cdot \mathbf{t}\, ds = \int_a^b V_t\, ds$$

In general the value of the line integral depends upon the path. If, however, $\mathbf{V} = \nabla\phi$, where $\phi(x,y,z)$ has continuous derivatives and is single-valued in the region, then the value of the integral depends only upon the end points. Thus

$$\int_a^b \mathbf{V} \cdot d\mathbf{r} = \phi(b) - \phi(a)$$

and the integral around a closed curve vanishes. It can also be stated that

$$\oint \mathbf{V} \cdot d\mathbf{r} = 0$$

if

$$\nabla \times \mathbf{V} = 0$$

Surface Integrals. The surface integral $\int_s \mathbf{V} \cdot \mathbf{n}\, da$ denotes the scalar integral of the normal components of the vector function $\mathbf{V}(x,y,z)$ over the surface s. $\mathbf{n}$ represents a unit vector normal to the positive side of the surface (positive being arbitrarily defined).

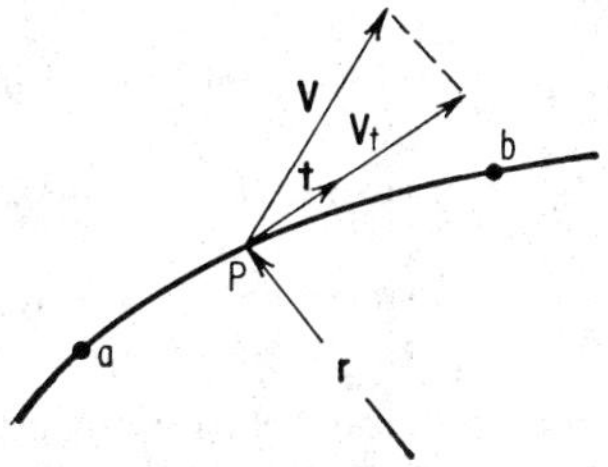

FIG. 1.60. Line integral.

Volume Integrals. The volume integral $\int_v \mathbf{V}\, dv$ denotes the vector integral of the vector function $\mathbf{V}(x,y,z)$ over the volume v.

Integration Theorems and Formulas

$$\int_c \mathbf{V} \cdot d\mathbf{r} = \int_s \mathbf{n} \cdot \nabla \times \mathbf{V}\, da \qquad \text{(Stokes's theorem)}$$

$$\int_s \mathbf{V} \cdot \mathbf{n}\, da = \int_v \nabla \cdot \mathbf{V}\, dv \qquad \text{(divergence theorem)}$$

$$\int_s \mathbf{n} \times \mathbf{V}\, da = \int_v \nabla \times \mathbf{V}\, dv$$

$$\int_c d\mathbf{r} \times \mathbf{V} = \int_s (\mathbf{n} \times \nabla) \times \mathbf{V}\, da$$

$$\int_s f\mathbf{n}\, da = \int_v \nabla f\, dv$$

$$\int_c f\, d\mathbf{r} = \int_s \mathbf{n} \times \nabla f\, da$$

(n) Fourier Series and Harmonic Analysis

Fourier Series. A function $f(x)$ can be represented by the series

$$f(x) = \tfrac{1}{2}a_0 + \sum_{n=1}^{\infty} (a_n \cos nx + b_n \sin nx) \qquad (a)$$

where

$$a_n = 1/\pi \int_{-\pi}^{\pi} f(x) \cos nx\, dx \qquad n = 0, 1, 2, \ldots \qquad (b)$$

and

$$b_n = 1/\pi \int_{-\pi}^{\pi} f(x) \sin nx\, dx \qquad n = 1, 2, 3, \ldots \qquad (c)$$

when

1. $f(x)$ is a single-valued bounded function defined in the interval $-\pi \leq x \leq \pi$.
2. $f(x)$ is periodic; i.e., $f(x + 2\pi) = f(x)$ outside the interval $-\pi \leq x \leq \pi$.
3. $f(x)$ has a finite number of points of finite discontinuity in the interval $-\pi \leq x \leq \pi$.
4. $f(x)$ has a finite number of maxima and minima in the interval $-\pi \leq x \leq \pi$.

It follows from the above conditions that the Fourier series $f(x)$ converges to $f(x)$ at all points where the function is continuous, and converges to the average of the right- and left-hand limits of $f(x)$ at each point where the function is discontinuous.

If the interval $\alpha \leq x \leq \alpha + 2\pi$, where α is any real number, is more convenient than the interval $-\pi \leq x \leq \pi$ then

$$a_n = 1/\pi \int_\alpha^{\alpha+2\pi} f(x) \cos nx \, dx \qquad n = 0, 1, 2, \ldots$$

and

$$b_n = 1/\pi \int_\alpha^{\alpha+2\pi} f(x) \sin nx \, dx \qquad n = 1, 2, 3, \ldots$$

When $\alpha = 0$, which corresponds to the frequently used interval $0 \leq x \leq 2\pi$,

$$a_n = 1/\pi \int_0^{2\pi} f(x) \cos nx \, dx \qquad n = 0, 1, 2, \ldots$$

$$b_n = 1/\pi \int_0^{2\pi} f(x) \sin nx \, dx \qquad n = 1, 2, 3, \ldots$$

When the period under consideration is of arbitrary length $2l$ rather than 2π, then for the interval $\alpha \leq x \leq \alpha + 2l$, the series assumes the general form

$$f(x) = \tfrac{1}{2}a_0 + \sum_{n=1}^{\infty} [a_n \cos (n\pi/l)x + b_n \sin (n\pi/l)x]$$

where

$$a_n = 1/l \int_\alpha^{\alpha+2l} f(x) \cos (n\pi/l)x \, dx \qquad n = 0, 1, 2, \ldots$$

and

$$b_n = 1/l \int_\alpha^{\alpha+2l} f(x) \sin (n\pi/l)x \, dx \qquad n = 1, 2, 3, \ldots$$

If $f(x)$ is an even function, i.e., $f(x) = f(-x)$, of period $2l$ in the interval $-l \leq x \leq l$ then

$$f(x) = \tfrac{1}{2}a_0 + \sum_{n=1}^{\infty} a_n \cos (n\pi/l)x$$

where

$$a_n = 2/l \int_0^l f(x) \cos (n\pi/l)x \, dx \qquad n = 0, 1, 2, \ldots$$

If $f(x)$ is an odd function, i.e., $f(-x) = -f(x)$, of period $2l$ in the interval $-l \leq x \leq l$ then

$$f(x) = \sum_{n=1}^{\infty} b_n \sin (n\pi/l)x$$

where

$$b_n = 2/l \int_0^l f(x) \sin (n\pi/l)x \, dx \qquad n = 1, 2, 3, \ldots$$

If $f(x)$ is defined in the half interval $0 \leq x \leq l$, its expansion is obtained by prolonging $f(x)$ arbitrarily into the interval $-l \leq x \leq 0$. It is generally desirable to prolong the function as either an even or an odd function. However, no matter how the function is extended into the interval $-l \leq x \leq 0$, the expansion will represent the original function in the intervals $0 \leq x \leq l$, $2l \leq x \leq 3l$, etc.

The Fourier series can also be expressed in exponential form as

$$f(x) = \sum_{n=-\infty}^{\infty} c_n e^{in\pi x/l}$$

where

$$c_n = 1/2l \int_{\alpha}^{\alpha+2l} f(x)e^{-in\pi x/l}\,dx \qquad n = \ldots, -3, -2, -1, 0, 1, 2, 3, \ldots$$

Harmonic Analysis. If the function $f(x)$ to be analyzed is given in tabulated form or by a graph rather than analytically, or if the integrals in the expressions for a_n and b_n are too difficult for exact evaluation, then methods of harmonic analysis

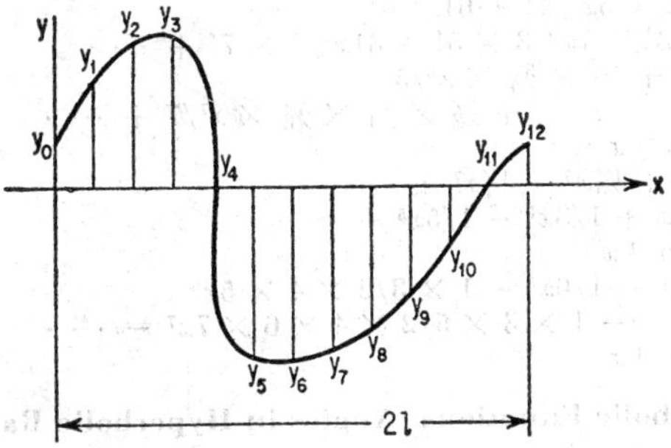

FIG. 1.61. Harmonic analysis.

can be employed to approximate a_n and b_n. The 12-ordinate scheme carried to obtain the first 6 harmonics is given below.

Let the values of the function $y = f(x)$ be given at intervals of one-twelfth of a period. (Note that $y_0 = y_{12}$.)

The y's are recorded in the following fashion, and additions and subtractions are performed (Fig. 1.61):

	y_0	y_1	y_2	y_3	y_4	y_5	y_6
		y_{11}	y_{10}	y_9	y_8	y_7	
Sums:	s_0	s_1	s_2	s_3	s_4	s_5	s_6
Differences:		d_1	d_2	d_3	d_4	d_5	

Similarly, the s_i and d_i are operated on:

	s_0	s_1	s_2	s_3
	s_6	s_5	s_4	
Sums:	S_0	S_1	S_2	S_3
Differences:	D_0	D_1	D_2	

	d_1	d_2	d_3
	d_5	d_4	
Sums:	$\bar{S}_1$	$\bar{S}_2$	$\bar{S}_3$
Differences:	$\bar{D}_1$	$\bar{D}_2$	

The coefficients of the first six harmonics are given below:

$$\begin{aligned}
\tfrac{1}{2}a_0 &= \tfrac{1}{12}(S_0 + S_1 + S_2 + S_3)\\
a_1 &= \tfrac{1}{6}(D_0 + \sqrt{3}/2D_1 + \tfrac{1}{2}D_2)\\
a_2 &= \tfrac{1}{6}[(S_0 - S_3) + \tfrac{1}{2}(S_1 - S_2)]\\
a_3 &= \tfrac{1}{6}(D_0 - D_2)\\
a_4 &= \tfrac{1}{6}[(S_0 + S_3) - \tfrac{1}{2}(S_1 + S_2)]\\
a_5 &= \tfrac{1}{6}(D_0 - \sqrt{3}/2D_1 + \tfrac{1}{2}D_2)\\
a_6 &= \tfrac{1}{6}(S_0 - S_1 + S_2 - S_3)\\
b_1 &= \tfrac{1}{6}(\tfrac{1}{2}\bar{S}_1 + \sqrt{3}/2\bar{S}_2 + \bar{S}_3)\\
b_2 &= \sqrt{3}/12(\bar{D}_1 + \bar{D}_2)\\
b_3 &= \tfrac{1}{6}(\bar{S}_1 - \bar{S}_3)\\
b_4 &= \sqrt{3}/12(\bar{D}_1 - \bar{D}_2)\\
b_5 &= \tfrac{1}{6}(\tfrac{1}{2}\bar{S}_1 - \sqrt{3}/2\bar{S}_2 + \bar{S}_3)\\
b_6 &= 0
\end{aligned}$$

and the truncated series is

$$f(x) = \tfrac{1}{2}a_0 + \sum_{n=1}^{6} [a_n \cos (n\pi/l)x + b_n \sin (n\pi/l)x]$$

(o) Series

Series for Trigonometric Functions (Angles in Radians)

$\sin x = x - x^3/3! + x^5/5! - x^7/7! + \cdots \qquad -\infty < x < \infty$

$\cos x = 1 - x^2/2! + x^4/4! - x^6/6! + \cdots \qquad -\infty < x < \infty$

$\tan x = x + x^3/3 + 2x^5/15 + 17x^7/315 + 62x^9/2{,}835 + \cdots \qquad -\pi/2 < x < \pi/2$

$\cot x = 1/x - x/3 - x^3/45 - 2x^5/945 - x^7/4{,}725 - \cdots \qquad -\pi < x < \pi$

$\sec x = 1 + x^2/2! + 5x^4/4! + 61x^6/6! + \cdots \qquad -\pi/2 < x < \pi/2$

$\csc x = 1/x + x/3! + 7x^3/3 \times 5! + 31x^5/3 \times 7! + \cdots \qquad -\pi < x < \pi$

$\sin^{-1} x = x + x^3/6 + \tfrac{1}{2} \times \tfrac{3}{4} \times x^5/5 + \tfrac{1}{2} \times \tfrac{3}{4} \times \tfrac{5}{6} \times x^7/7 + \cdots \qquad -1 \leqq x \leqq 1$

$\cos^{-1} x = \tfrac{1}{2}\pi - \sin^{-1} x$

$\tan^{-1} x = x - \tfrac{1}{3}x^3 + \tfrac{1}{5}x^5 - \tfrac{1}{7}x^7 + \cdots \qquad -1 \leqq x \leqq 1$

$\tan^{-1} x = \pi/2 - 1/x + 1/3x^3 - 1/5x^5 + \cdots \qquad x^2 > 1$

$\cot^{-1} x = \tfrac{1}{2}\pi - \tan^{-1} x$

$\sec^{-1} x = \pi/2 - 1/x - 1/6x^3 - 1 \times 3/2 \times 4 \times 5x^5 - 1 \times 3 \times 5/2 \times 4 \times 6 \times 7x^7 - \cdots \qquad x^2 > 1$

$\csc^{-1} x = \pi/2 - \sec^{-1} x$

Series for Hyperbolic Functions (Angles in Hyperbolic Radians)

$\sinh x = x + x^3/3! + x^5/5! + x^7/7! + \cdots \qquad -\infty < x < \infty$

$\cosh x = 1 + x^2/2! + x^4/4! + x^6/6! + \cdots \qquad -\infty < x < \infty$

$\tanh x = x - x^3/3 + 2x^5/15 - 17x^7/315 + \cdots \qquad \pi/2 < x < \pi/2$

$\coth x = 1/x + x/3 - x^3/45 + 2x^5/945 - x^7/4{,}725 + \cdots \qquad -\pi < x < \pi$

$\operatorname{sech} x = 1 - x^2/2! + 5x^4/4! - 61x^6/6! + 1{,}385x^8/8! - \cdots \qquad -\pi/2 < x < \pi/2$

$\operatorname{csch} x = 1/x - x/6 + 7x^3/360 - 31x^5/15{,}120 + \cdots \qquad -\pi < x < \pi$

$\sinh^{-1} x = x - \tfrac{1}{2} \times x^3/3 + \tfrac{1}{2} \times \tfrac{3}{4} \times x^5/5 - \tfrac{1}{2} \times \tfrac{3}{4} \times \tfrac{5}{6} \times x^7/7 + \cdots \qquad x^2 < 1$

$\sinh^{-1} x = \ln 2x + 1/2 \times 2x^2 - 1 \times 3/2 \times 4 \times 4x^4 + 1 \times 3 \times 5/2 \times 4 \times 6 \times 6x^6 - \cdots \qquad x^2 > 1$

$\cosh^{-1} x = \pm(\ln 2x - 1/2 \times 2x^2 - 1 \times 3/2 \times 4 \times 4x^4 - 1 \times 3 \times 5/2 \times 4 \times 6 \times 6x^6 - \cdots) \qquad x > 1$

$\tanh^{-1} x = x + x^3/3 + x^5/5 + x^7/7 + \cdots \qquad -1 < x < 1$

$\coth^{-1} x = 1/x + 1/3x^3 + 1/5x^5 + 1/7x^7 + \cdots \qquad x^2 > 1$

$\operatorname{sech}^{-1} x = \pm[\log (2/x) - (1/2 \times 2)x^2 - (1 \times 3/2 \times 4 \times 4)x^4 - (1 \times 3 \times 5/2 \times 4 \times 6 \times 6)x^6 - \cdots] \qquad 0 < x < 1$

$\operatorname{csch}^{-1} x = 1/x - 1/2 \times 3x^3 + 1 \times 3/2 \times 4 \times 5x^5 - 1 \times 3 \times 5/2 \times 4 \times 6 \times 7x^7 + \cdots \qquad x^2 > 1$

Series for Exponential and Logarithmic Functions

$$e = 1 + 1/1! + 1/2! + 1/3! + 1/4! + \cdots$$

$$e^x = 1 + x + x^2/2! + x^3/3! + x^4/4! + \cdots \qquad -\infty < x < \infty$$

$$a^x = e^{x \ln a} = 1 + x \ln a + (x \ln a)^2/2! + (x \ln a)^3/3! + \cdots \qquad a > 0,\ -\infty < x < \infty$$

$$e^{-x^2} = 1 - x^2 + x^4/2! - x^6/3! + x^8/4! - \cdots \qquad -\infty < x < \infty$$

$$\int_0^x e^{-x^2}\, dx = x - \tfrac{1}{3}x^3 + x^5/5 \times 2! - x^7/7 \times 3! + \cdots \qquad -\infty < x < \infty$$

$$\ln x = (x - 1) - \tfrac{1}{2}(x - 1)^2 + \tfrac{1}{3}(x - 1)^3 - \cdots \qquad 0 < x \leqq 2$$

$$\ln x = \frac{x-1}{x} + \frac{1}{2}\left(\frac{x-1}{x}\right)^2 + \frac{1}{3}\left(\frac{x-1}{x}\right)^3 + \cdots \qquad x > 1/2$$

$$\ln x = 2\left[\frac{x-1}{x+1} + \frac{1}{3}\left(\frac{x-1}{x+1}\right)^3 + \frac{1}{5}\left(\frac{x-1}{x+1}\right)^5 + \cdots\right] \qquad x > 0$$

$$\ln (1+x) = x - x^2/2 + x^3/3 - x^4/4 + x^5/5 - \cdots \qquad -1 < x \le 1$$

$$\ln (a+x) = \ln a + 2\left[\frac{x}{2a+x} + \frac{1}{3}\left(\frac{x}{2a+x}\right)^3 + \frac{1}{5}\left(\frac{x}{2a+x}\right)^5 + \cdots\right] \qquad a > 0,\ -a < x < \infty$$

$$\ln (1-x) = -x - x^2/2 - x^3/3 - x^4/4 - x^5/5 - \cdots \qquad -1 < x < +1$$

$$\ln\left(\frac{1+x}{1-x}\right) = 2\left(x + \frac{x^3}{3} + \frac{x^5}{5} + \frac{x^7}{7} + \cdots\right) \qquad -1 < x < +1$$

$$\ln\left(\frac{x+1}{x-1}\right) = 2\left(\frac{1}{x} + \frac{1}{3x^3} + \frac{1}{5x^5} + \frac{1}{7x^7} + \cdots\right) \qquad x < -1 \text{ or } +1 < x$$

$$\ln\left(\frac{x+1}{x}\right) = 2\left[\frac{1}{2x+1} + \frac{1}{3(2x+1)^3} + \frac{1}{5(2x+1)^5} + \cdots\right] \qquad x > 0$$

$$\ln (x + \sqrt{1+x^2}) = x - \tfrac{1}{2}x^3/3 + \tfrac{1}{2} \times \tfrac{3}{4} \times x^5/5 - \tfrac{1}{2} \times \tfrac{3}{4} \times \tfrac{5}{6} \times x^7/7 + \cdots \qquad x^2 < 1$$

$$\ln \sin x = \ln x - x^2/6 - x^4/180 - x^6/2{,}835 - \cdots \qquad x^2 < \pi^2$$

$$\ln \cos x = -x^2/2 - x^4/12 - x^6/45 - 17x^8/2{,}520 - \cdots \qquad x^2 < \pi^2/4$$

$$\ln \tan x = \ln x + x^2/3 + 7x^4/90 + 62x^6/2{,}835 + \cdots \qquad x^2 < \pi^2/4$$

$$e^{\sin x} = 1 + x + x^2/2! - 3x^4/4! - 8x^5/5! - 3x^6/6! + 56x^7/7! + \cdots \qquad \infty < x < \infty$$

$$e^{\cos x} = e(1 - x^2/2! + 4x^4/4! - 31x^6/6! + \cdots) \qquad \infty < x < \infty$$

$$e^{\tan x} = 1 + x + x^2/2! + 3x^3/3! + 9x^4/4! + 37x^5/5! + \cdots \qquad -\pi/2 < x < \pi/2$$

Maclaurin's Series.

$$f(x) = f(0) + \frac{x}{1!}f'(0) + \frac{x^2}{2!}f''(0) + \cdots + \frac{x^{n-1}}{(n-1)!}f^{(n-1)}(0) + \cdots$$

where the remainder R_n after n terms $= \dfrac{f^{(n)}(\epsilon x)}{n!}x^n$, $0 < \epsilon < 1$, and $R_n \to 0$ as $n \to \infty$.

Taylor's Series.

$$f(x) = f(x_0) + \frac{f'(x_0)}{1!}(x - x_0) + \frac{f''(x_0)}{2!}(x - x_0)^2 + \cdots + \frac{f^{(n-1)}(x_0)}{(n-1)!}(x - x_0)^{n-1} + \cdots$$

where the remainder R_n after n terms $= \dfrac{f^{(n)}(\zeta)}{n!}(x - x_0)^n$, $\zeta = x_0 + \epsilon(x - x_0), 0 < \epsilon < 1$.

Also $f(x + h) = f(x) + \dfrac{f'(x)}{1!}h + \dfrac{f''(x)}{2!}h^2 + \cdots + \dfrac{f^{(n-1)}(x)}{(n-1)!}h^{n-1} + \cdots$ where the remainder R_n after n terms $= \dfrac{f^{(n)}(x + \epsilon h)}{n!}h^n$, $0 < \epsilon < 1$, and $R_n \to 0$ as $n \to \infty$.

Taylor's Series for Two Variables.

$$f(x + h_1, y + h_2) = f(x,y) + \frac{1}{1!}\left(h_1\frac{\partial}{\partial x} + h_2\frac{\partial}{\partial y}\right)f(x,y) + \frac{1}{2!}\left(h_1\frac{\partial}{\partial x} + h_2\frac{\partial}{\partial y}\right)^2 f(x,y) + \cdots + \frac{1}{(n-1)!}\left(h_1\frac{\partial}{\partial x} + h_2\frac{\partial}{\partial y}\right)^{n-1} f(x,y) + \cdots$$

where the remainder R_n after n terms $= \dfrac{1}{n!}\left(h_1\dfrac{\partial}{\partial x} + h_2\dfrac{\partial}{\partial y}\right)^n f(x + \epsilon h_1, y + \epsilon h_2)$, $0 < \epsilon < 1$, and $R_n \to 0$ as $n \to \infty$.

Useful Approximate Relationships

$$\frac{1}{1+x} = 1 - x + \qquad \text{(for } 0 < x < 1\text{, error } < x^2\text{)}$$

$$\frac{1}{1+x} = 1 - x + x^2 - \qquad \text{(for } 0 < x < 1\text{, error } < x^3\text{)}$$

$$\frac{1}{1-x} = 1 + x + \qquad \text{(for } 0 < x < \tfrac{1}{2}\text{, error } < x^2 + 2x^3\text{)}$$

$$\frac{1}{1-x} = 1 + x + x^3 + \qquad \text{(for } 0 < x < \tfrac{1}{2}\text{, error } < x^3 + 2x^4\text{)}$$

(p) Probability and Statistics

Definitions. If an event may succeed in s ways and fail in f ways, each having the same likelihood of occurrence, the probability of success in a single trial is

$$p = \frac{s}{s + f}$$

and the probability of failure in a single trial is

$$q = \frac{f}{s + f}$$

so that $p + q = 1$.

If two independent events E_1 and E_2 have probabilities of occurrence p_1 and p_2, respectively, then the probability that both events will occur simultaneously or in order is p_1p_2.

If two independent events E_1 and E_2 have probabilities of occurrence p_1 and p_2, respectively, then the probability that either event will occur is $p_1 + p_2$.

The probability that an event will occur exactly r times in n trials is

$$\binom{n}{r} p^r q^{n-r} \qquad \text{(binomial formula, Bernoulli distribution)}$$

The probability that an event will occur at least r times in n trials is

$$\sum_{i=r}^{i=n} \binom{n}{i} p^i q^{n-i}$$

The probability that an event will occur at most r times in n trials is

$$\sum_{i=0}^{i=r} \binom{n}{i} p^i q^{n-i}$$

The average or arithmetic mean of a set of observations $r_1, r_2, \ldots, r_n$ is

$$\bar{r} = \mu_1(r) = \frac{r_1 + r_2 + \cdots + r_n}{n}$$

where n is finite.

The kth moment is defined as the average of the kth powers of the observations. Thus

$$\mu_k(r) = \overline{r^k} = \frac{r_1^k + r_2^k + \cdots + r_n^k}{n}$$

Corresponding to a set of observations there is a set of deviations $d_1, d_2, \ldots, d_n$.

$$\begin{aligned} d_1 &= r_1 - \bar{r} = r_1 - \mu_1(x) \\ d_2 &= r_2 - \bar{r} \\ d_n &= r_n - \bar{r} \end{aligned}$$

The first moment of the deviations, $\mu_1(d) = 0 = \bar{d}$. The second moment of the deviations is the mean-square deviation

$$\begin{aligned} \mu_2(d) &= \mu_2(r) - [\mu_1(r)]^2 = \overline{d^2} = \text{variance} \\ &= \overline{r^2} - (\bar{r})^2 \end{aligned}$$

$\sigma = \sqrt{\mu_2(d)}$ = root-mean-square value of the deviation or the standard deviation

Normal Distribution. The distribution of the deviations may often be described by the normal-probability curve (Fig. 1.62)

$$y = \frac{1}{\sqrt{2\pi}\,\sigma} e^{-x^2/2\sigma^2}$$

where x is the magnitude of a deviation and y is a function such that the area under

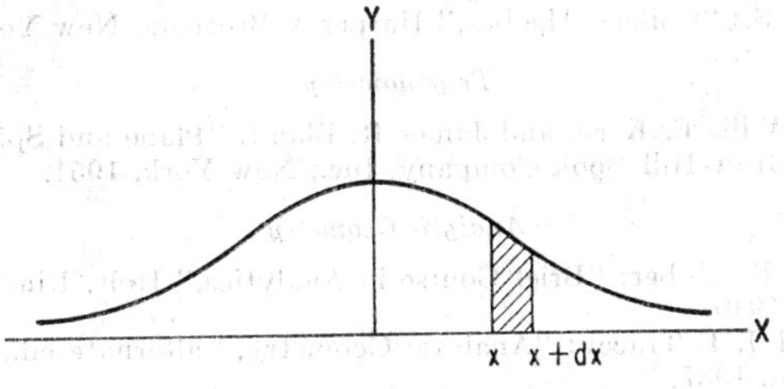

FIG. 1.62. Normal-probability curve.

the curve between x and $x + dx$ is equal to the probability that a deviation will lie in that range. Thus the probability P_{ab} that a deviation lies between a and b is

$$P_{ab} = \frac{1}{\sqrt{2\pi}\,\sigma} \int_a^b e^{-x^2/2\sigma^2}\, dx$$

The probability P_{ab} may also be expressed in terms of the error function (erf) (see Table 1.6)

$$P_{ab} = \tfrac{1}{2}(\text{erf } hb - \text{erf } ha)$$

where h is the measure of precision.

The second moment $\mu_2(x)$ of a distribution described by the normal curve is $\mu_2(x) = 1/2h^2$, but $\sigma = \sqrt{\mu_2(x)}$, so that $\mu_2(x) = \sigma^2 = 1/2h^2$ or $\sigma = 1/\sqrt{2}\,h$.

The percentage of the observations related to the spread in σ units is given below.

	Per cent
$\pm 0.674\sigma$	50
$\pm 1\sigma$	68.27
$\pm 2\sigma$	95.45
$\pm 3\sigma$	99.73

The mean absolute error for the normal-distribution curve is

$$\mu_1(|x|) = \frac{1}{h\sqrt{\pi}} = \mu$$

The probable error e is the deviation for which the probability is ½. Thus

$$\tfrac{1}{2} = P_{-e,e} = \operatorname{erf} he \qquad e = 0.4769/h = 0.6745\sigma = 0.8453\mu$$

Poisson's Law. Poisson's law, which applies to many physical situations, represents a simple approximation to the binomial formula and may be written as follows:

$$\frac{(np)^r}{r!}e^{-np}$$

The Poisson approximation to the binomial formula improves as n becomes larger and p diminishes.

Bibliography

Mathematical Tables

Jahnke, Eugene, and Fritz Emde: "Tables of Functions with Formulae and Curves," 4th ed., Dover Publications, Inc., New York, 1945.

Algebra

Frazer, R. A., W. J. Duncan, and A. R. Collar: "Elementary Matrices," Cambridge University Press, New York, 1952.

Peterson, Thurman S.: "College Algebra," Harper & Brothers, New York, 1947.

Trigonometry

Kells, Lyman M., Willis F. Kern, and James R. Bland: "Plane and Spherical Trigonometry," 3d ed., McGraw-Hill Book Company, Inc., New York, 1951.

Analytic Geometry

Hill, M. A., and J. B. Linber: "Brief Course in Analytics," Holt, Rinehart and Winston, Inc., New York, 1940.

Wilson, W. A., and J. I. Tracey: "Analytic Geometry," alternate ed., D. C. Heath and Company, Boston, 1937.

Differential and Integral Calculus

Byerly, W. E.: "Elements of the Integral Calculus," 2d ed. (1888), reprinted by G. E. Stechert & Company, New York, 1941.

Franklin, P.: "Methods of Advanced Calculus," McGraw-Hill Book Company, Inc., New York, 1944.

Granville, W. A., P. F. Smith, and W. R. Longley: "Elements of the Differential and Integral Calculus," rev. ed., Ginn and Company, Boston, 1941.

Peirce, B. O.: "A Short Table of Integrals," 3d rev. ed., Ginn and Company, Boston, 1929.

Numerical Methods

Allen, D. N. de G.: "Relaxation Methods," McGraw-Hill Book Company, Inc., New York, 1954.

Booth, Andrew D.: "Numerical Methods," 2d ed., Academic Press Inc., Butterworth Scientific Publications, New York, 1957.
Milne, W. E.: "Numerical Calculus," Princeton University Press, Princeton, N.J., 1949.
Salvadori, Mario G., and Melvin L. Baron: "Numerical Methods in Engineering," Prentice-Hall, Inc., Englewood Cliffs, N.J., 1952.
Scarborough, J. D.: "Numerical Mathematical Analysis," The Johns Hopkins Press, Baltimore, 1930.
Shaw, F. S.: "An Introduction to Relaxation Methods," Dover Publications, Inc., New York, 1953.

Differential Equations

Agnew, Ralph Palmer: "Differential Equations," 2d ed., McGraw-Hill Book Company, Inc., New York, 1960.
Bowman, Frank: "Introduction to Bessel Functions," Dover Publications, Inc., New York, 1958.
Forsyth, Andrew Russell: "Theory of Differential Equations," vols. 1–6, Dover Publications, Inc., New York, 1959.
Golomb, Michael, and Merrill Shanks: "Elements of Ordinary Differential Equations," McGraw-Hill Book Company, Inc., New York, 1950.
Miller, K. S.: "Partial Differential Equations in Engineering Problems," Prentice-Hall, Inc., Englewood Cliffs, N.J., 1953.
Phillips, H. B.: "Differential Equations," 3d ed., John Wiley & Sons, Inc., New York, and Chapman & Hall, Ltd., London, 1946.
Salvadori, M. G., and R. J. Schwarz: "Differential Equations in Engineering Problems," Prentice-Hall, Inc., Englewood Cliffs, N.J., 1954.
Sommerfeld, Arnold: "Partial Differential Equations in Physics," Academic Press Inc., New York, 1949.
Webster, A. G.: "Partial Differential Equations of Mathematical Physics," 2d ed., Dover Publications, Inc., New York, 1955.

Operational Mathematics

Churchill, R. V.: "Operational Mathematics," 2d ed., McGraw-Hill Book Company, New York, 1958.

Complex Variables

Churchill, R. V.: "Complex Variables and Applications," 2d ed., McGraw-Hill Book Company, Inc., New York, 1960.

Vector Analysis

Phillips, H. B.: "Vector Analysis," John Wiley & Sons, Inc., New York, 1957.

Fourier Analysis and Harmonic Analysis

Wylie, C. R., Jr.: "Advanced Engineering Mathematics," 2d ed., McGraw-Hill Book Company, Inc., New York, 1960.

Probability and Statistics

Anderson, R. L., and T. A. Bancroft: "Statistical Theory in Research," McGraw-Hill Book Company, Inc., New York, 1952.
Cramér, H.: "Mathematical Methods of Statistics," Princeton University Press, Princeton, N.J., 1946.
Parzen, E.: "Modern Probability Theory and Its Applications," John Wiley & Sons, Inc., New York, 1960.

Section 2

COMPUTERS

By

A. I. KATZ, M.S., M.E., *Director of Training and Education, Electronic Associates Inc., Princeton, N.J.* (*Analog Computers*)

JOHN A. CARLSON, Ph.D., *Engineering Specialist, MRD Division, General American Transportation Corp., Niles, Ill.* (*Digital Computers*)

CONTENTS

ANALOG COMPUTERS

DIGITAL COMPUTERS

ANALOG COMPUTERS*

2.1. INTRODUCTION

The analog computer can be an effective tool in either of two categories: model building—an inductive process, and model analysis—a deductive process. In model building, an analytical relation between variables is hypothesized to describe the physical system of interest. Forcing functions identical to those in the physical system can then be applied to the hypothetical model. Because of the ease with which parameter variation and model changes can be accomplished, the analog computer is useful in conducting many "trial-and-error" experiments on the model to obtain the best fit to the physical system. It is also possible to mechanize the computer so that it will automatically seek the best fit once a desired optimum is quantitatively defined.

On the other hand, in model analysis, it is necessary to have a mathematical statement available which describes the physical system to be studied. These mathematical statements (equations) are often supplemented by graphical information as well as logic statements. A definite range of values for the parameters is assigned for the study. Experiments are then performed, varying the inputs and the parameters which describe the system, to obtain finally an optimum response for the system or to develop a better understanding of the intrinsic nature of the system by studying input-output relations.

Only the principles of operation and application from the point of view of model analysis will be discussed here because the use of the computer in analysis is straightforward and more readily defined than in model building. References 1 through 4 yield more information on analog computers.

2.2. CONCEPT OF THE ANALOG COMPUTER

An analog computer is a collection of operational devices which are capable of performing basic mathematical operations. By interconnecting components which can integrate, add, multiply by -1, multiply by a constant, and multiply variables, systems described by ordinary differential equations can be analyzed.

The dynamic equilibrium for the simple spring-mass-dashpot combination is described by (Sec. 6)

$$m\, d^2x/dt^2 + c\, dx/dt + kx = P_0 \sin \omega t \tag{2.1}$$

Here we are interested in investigating the motion of the mass m subjected to an external exciting force $P_0 \sin \omega t$. x represents the motion of the mass. The dependent variable x as a function of the independent variable time t [i.e., $x = f(t)$] must be found. Rewriting Eq. (2.1),

$$m\, d^2x/dt^2 = -c\, dx/dt - kx + P_0 \sin t \tag{2.2}$$

If $m\, d^2x/dt^2$ is known, double integration and multiplication by $1/m$ will yield the dependent variable x. With dx/dt available as the first integral of d^2x/dt^2, x avail-

* By A. I. Katz.

able from the integral of dx/dt, and $P_0 \sin \omega t$ a known function of time, we can readily produce the three terms on the right-hand side of Eq. (2.2) and then sum them as shown in the mathematical block diagram of Fig. 2.1. To force equality of $m\, d^2x/dt^2$ with the right-hand side of Eq. (2.2), we connect point a to point b, as shown by the dotted line in Fig. 2.1.

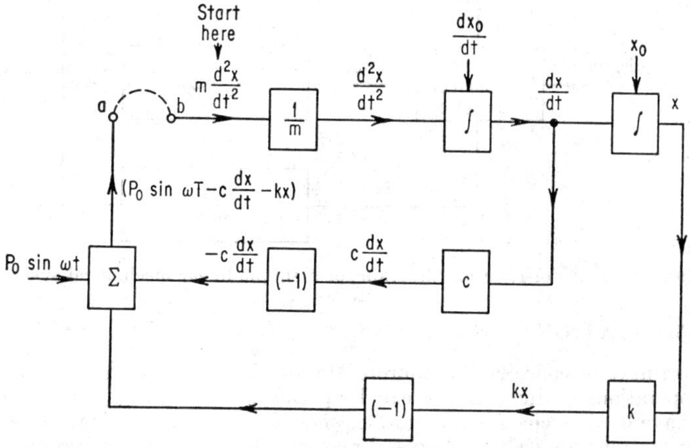

FIG. 2.1. Mathematical block diagram for the solution of Eq. (2.2).

In the physical system, the mass could have an initial velocity dx_0/dt and an initial deflection x_0. This is introduced mathematically by allowing for an initial condition input to the integrating devices. With initial conditions applied, the output of the integrator is now the definite integral of the input between the limits $t = 0$ and $t = t_1$.

2.3. DYNAMIC MODEL OF PHYSICAL SYSTEM

The mechanical elements of a simple linear mechanical system have transfer functions (input-output relationships) which can be used to describe the system in a manner analogous to the mathematical-block-diagram approach of Fig. 2.1. The mass *motion* (velocity) is *due to* an applied *force,* and *spring* or *dashpot force* is the *result of* a relative motion across the element. Then we can picture a mass as an integrator (of forces) and multiplier (by the constant $1/m$), and a spring as an integrator (of motion) and multiplier (by k). The dashpot simply multiplies (velocity) by a constant c.

From Newton's second law the forces $F_k = kx$, $F_c = c\, dx/dt$, and $P_0 \sin \omega t$ acting on mass m produce the inertial force $m\, d^2x/dt^2$ (acting in the plus direction—positive upward). From Eq. (2.2), the resulting motion of m is

$$dx_m/dt = 1/m \int F_m\, dt \tag{2.3}$$

where $F_m = -F_k - F_c + P_0 \sin \omega t$

Since the spring and dashpot each have one end attached to the mass, while the other end is fixed to the earth (see Sec. 6), the motion across these elements is also dx_m/dt. For positive motion of the mass, there is a negative force exerted by the spring and dashpot on the mass. For the dashpot, the force is $-c\, dx_m/dt$, and for the spring it is $-kx_m$ where x_m is the instantaneous deflection of the spring. For dynamic equilibrium we must connect a to b.

Initial kinetic energy of the system is defined by the initial velocity of the mass dx_0/dt. Initial potential energy is introduced by an initial deflection x_0 of the spring.

Comparison of Figs. 2.1 and 2.2 reveals that the interconnection of mathematical computing components produces a diagram similar to the one obtained by considering the dynamic equilibrium for the physical system.

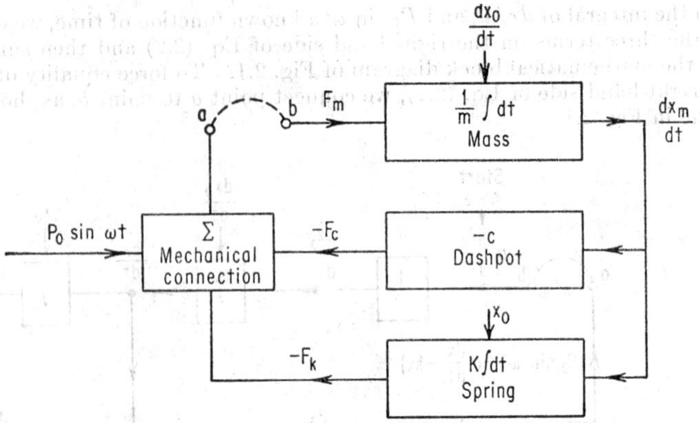

FIG. 2.2. Transfer-function diagram for a simple linear mechanical system.

2.4. COMPUTATIONAL ELEMENTS

We have next to consider the appropriate physical medium in which the mathematical operations outlined above can be performed. Although analog computers can and have used mechanical, hydraulic, pneumatic, and other elements, an electrical medium is most desirable for several reasons: (1) wide dynamic range, (2) ease of interconnection, (3) availability of recording devices activated by electrical signals, (4) availability of precision components, and (5) highly developed analytical techniques. Therefore, discussion here will be limited to the electronic general-purpose analog computer or electronic differential analyzer in which the problem variables and their derivatives appear as voltages.

(a) Attenuator

The attenuator is the simplest computing element. Its output is a voltage which is the product of the input voltage and a constant, say ρ $(0 \leq \rho \leq 1.0)$. The symbols and input-output relations for an attenuator are shown in Fig. 2.3. Attenuators are usually precision potentiometers with calibrated dials.

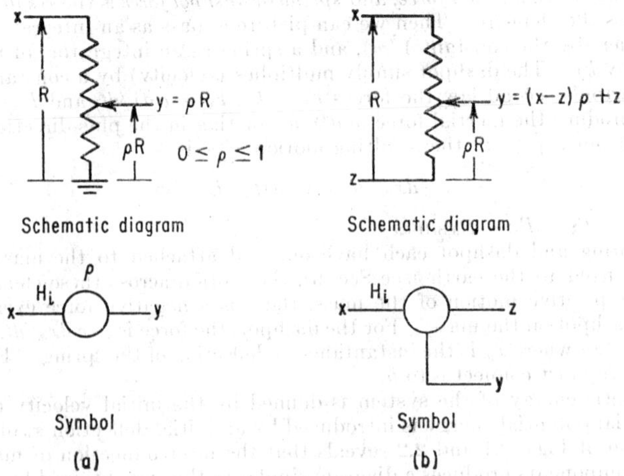

FIG. 2.3. Attenuator. (a) Grounded. (b) Ungrounded.

(b) Operational Amplifier

The operational amplifier is the basic component of the modern electronic analog computer (Fig. 2.4*a*). The triangular symbol represents a high-gain direct-coupled amplifier having a wide frequency range extending from zero to beyond 25 kc, and effectively zero grid current. Z_f and Z_i are passive electrical components with impedance (voltage to current ratio) matched to better than 0.01 per cent, an accuracy

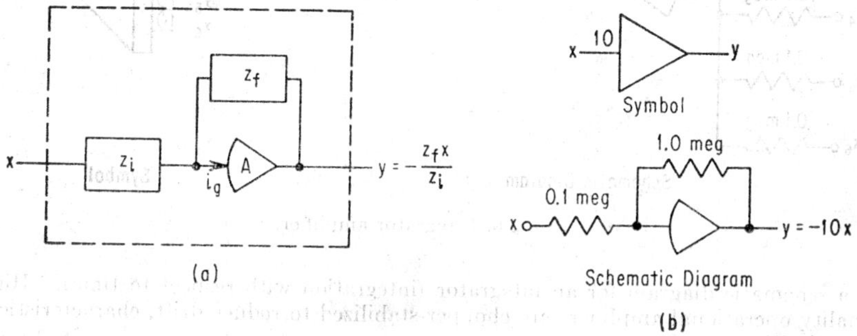

FIG. 2.4. Operational amplifier. (*a*) Block diagram for general operational amplifier. (*b*) Multiplication by −10 with an operational amplifier.

more than adequate for the majority of engineering calculations. The general relationship between the output y and the input x for the operational amplifier is

$$y = -(Z_f/Z_i)x \tag{2.4}$$

The amplifier inherently performs the operation of mathematical inversion (multiplication by −1) while Z_f and Z_i establish the specific mathematical operation of the device. For example, if Z_f is 1 megohm (10^6 ohms) and Z_i 0.1 megohm (10^5 ohms) (Fig. 2.4*b*) the input-output relation is

$$y = -(10^6/10^5)x = -10x$$

By introducing additional input paths to the d-c amplifier (Fig. 2.5), summation can be performed with the same device.

In addition to inversion, multiplication by a constant, and summation, integration can also be performed using a feedback capacitor with the d-c amplifier. Figure 2.6

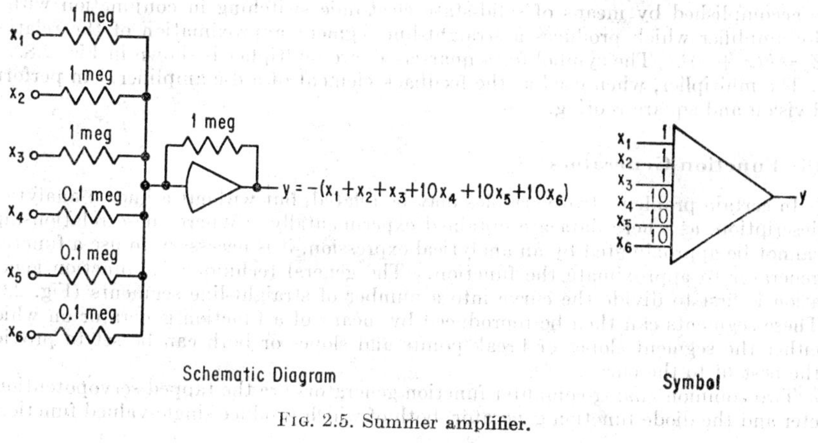

FIG. 2.5. Summer amplifier.

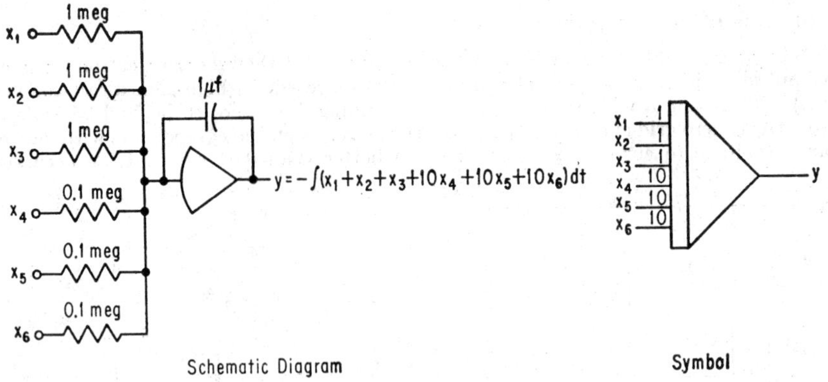

FIG. 2.6. Integrator amplifier.

is a schematic diagram for an integrator (integration with respect to time). High-quality operational amplifiers are chopper-stabilized to reduce drift, characteristic of d-c amplifiers.

(c) Multiplier

In linear problems with time-varying coefficients, as well as in nonlinear problems, it is also necessary to produce the product of two variables. Although there are several methods for multiplying two variables, the most common techniques are the servomultiplier and the quarter-square multiplier.

In the servomechanical multiplier (Fig. 2.7), a potentiometer wiper is positioned as a function of one variable, say x, by means of a position-feedback control loop. The voltage applied to the potentiometer is a function of a second variable, say y, and thus the output of the wiper will be proportional to the product of the two variables. By ganging several potentiometers, it is possible to obtain simultaneously products of x and the variables applied to the additional potentiometers.

The quarter-square multiplier is based upon the relation

$$xy = \tfrac{1}{4}[(x + y)^2 - (x - y)^2] \tag{2.5}$$

which reduces multiplication to the operations of summation and squaring. Squaring is accomplished by means of solid-state electronic switching in conjunction with a d-c amplifier which produces a straight-line-segment approximation of the relation $Z = k(x + y)^2$. The symbol for a quarter-square multiplier is shown in Fig. 2.8.

The multiplier, when used as the feedback element of a d-c amplifier, can perform division and square rooting.

(d) Function Generators

In certain problems two variables may be related, but without a known analytical description, as where data are obtained experimentally. Where these relationships cannot be approximated by an analytical expression, it is necessary to use a function generator to approximate the function. The general technique for function generation is first to divide the curve into a number of straight-line segments (Fig. 2.9). These segments can then be reproduced by means of a function generator on which either the segment slopes or break points and slopes or both can be set to provide the best fit to the curve.

Two common analog-computer function generators are the tapped servopotentiometer and the diode function generator, both of which produce single-valued functions.

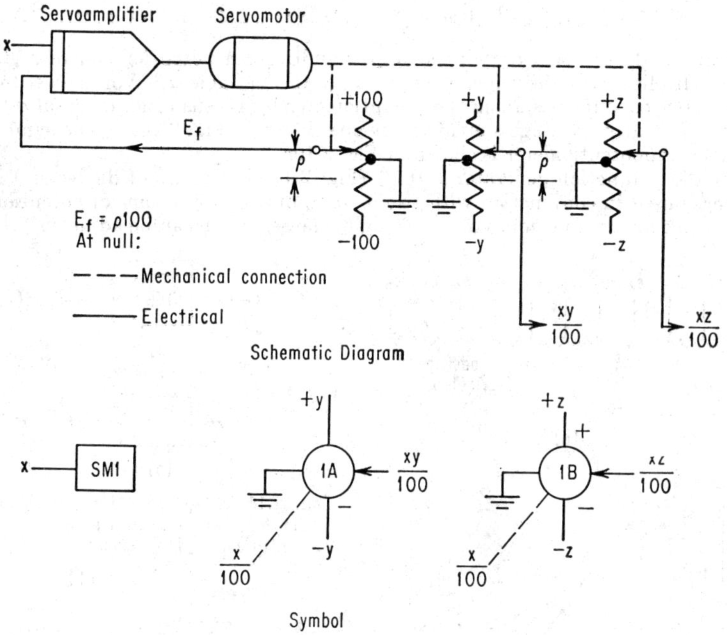

FIG. 2.7. Servomultiplier.

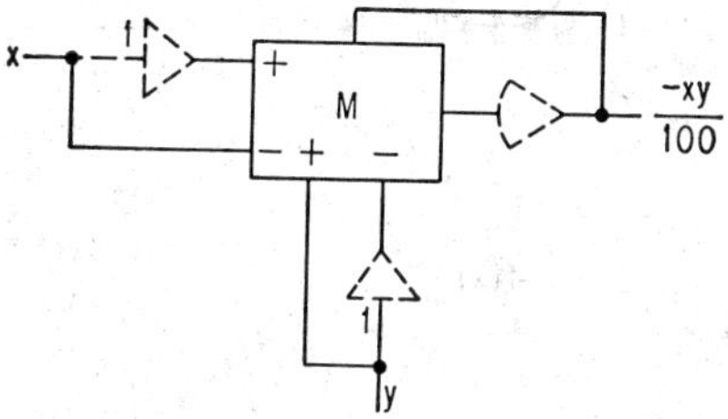

FIG. 2.8. Quarter-square multiplier. Amplifiers are shown dotted since they are not normally fixed to a specific multiplier.

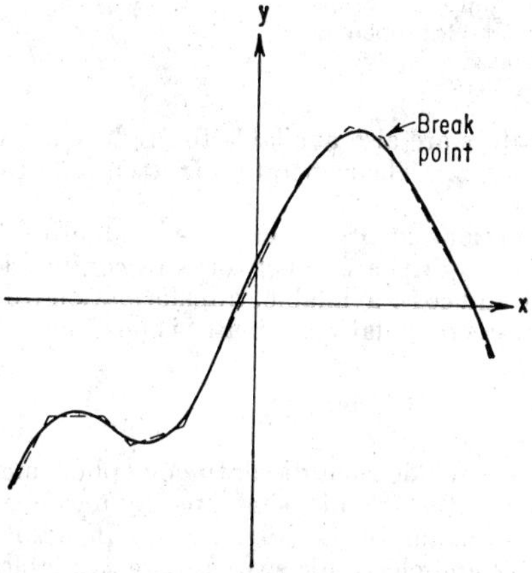

FIG. 2.9. Straight-line approximation of a nonanalytic function.

Figure 2.10*a* is a schematic of a tapped multiplying potentiometer of a standard servomultiplier with the wiper positioned by the variable x. Voltages e_1 to e_n are forced at fixed intervals along the potentiometer by means of an external source, to produce an output voltage similar to the one shown in Fig. 2.10*b*. The symbol for a tapped servopotentiometer is shown in Fig. 2.10*c*.

The diode function generator (DFG) (Fig. 2.11) is a series of diodes and resistor networks used in conjunction with two d-c amplifiers. By means of potentiometers, both break points and slopes of the straight-line segments can be adjusted.

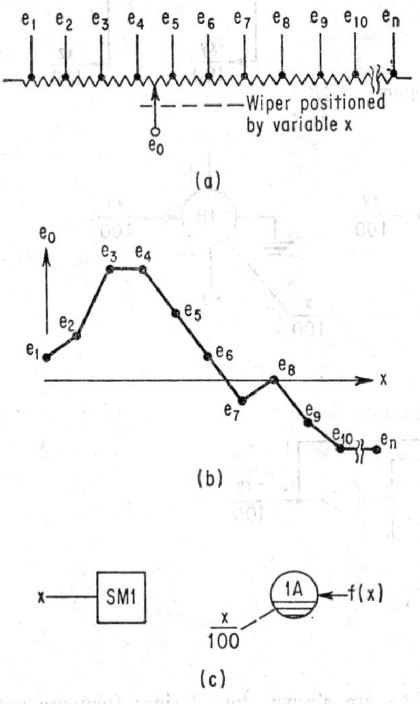

FIG. 2.10. Tapped servopotentiometer function generator. (*a*) Schematic of tapped potentiometer. (*b*) Output of tapped potentiometer as a function of X. (*c*) Symbol.

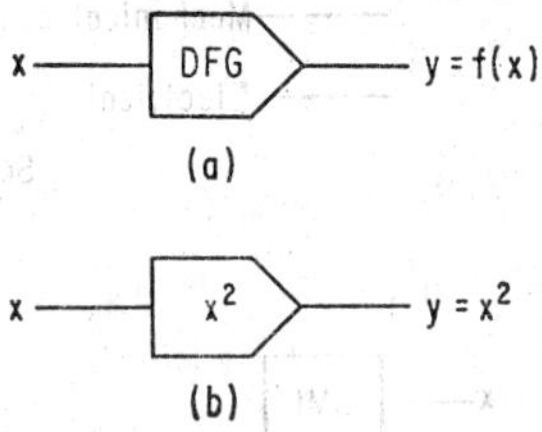

FIG. 2.11. Diode function generators. (*a*) Variable diode function generator. (*b*) X^2 diode function generator.

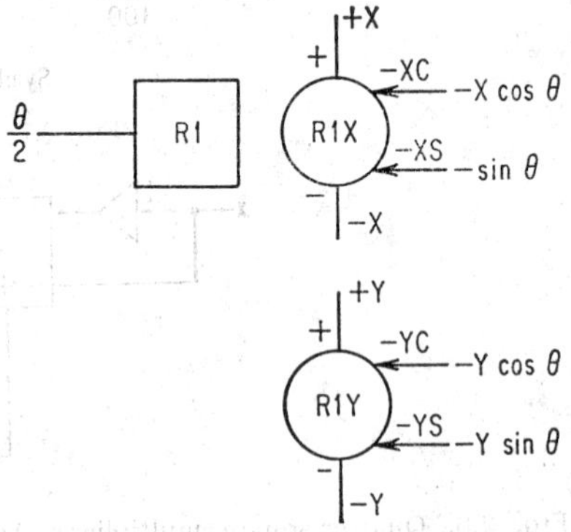

FIG. 2.12. Servoresolver.

Fixed function generators are also available to produce a given function of a variable such as x^2, x^4, and log x. Their operation is essentially the same as the variable DFG.

The trigonometric functions sin x and cos x are produced by using special potentiometers on servomultipliers or shaping networks in conjunction with d-c amplifiers (Fig. 2.12). With sin θ and cos θ available, transformation from rectangular-to-polar coordinates and polar-to-rectangular coordinates is possible.

(e) Special Devices

It is often necessary to include such discontinuous phenomena as limits, backlash, and dead zone. These functions can be simulated by means of biased diodes used in conjunction with high-gain amplifiers (Fig. 2.13). In addition, high-speed relays (comparators) (Fig. 2.14) and electronic switches are available to perform additional logic operations.

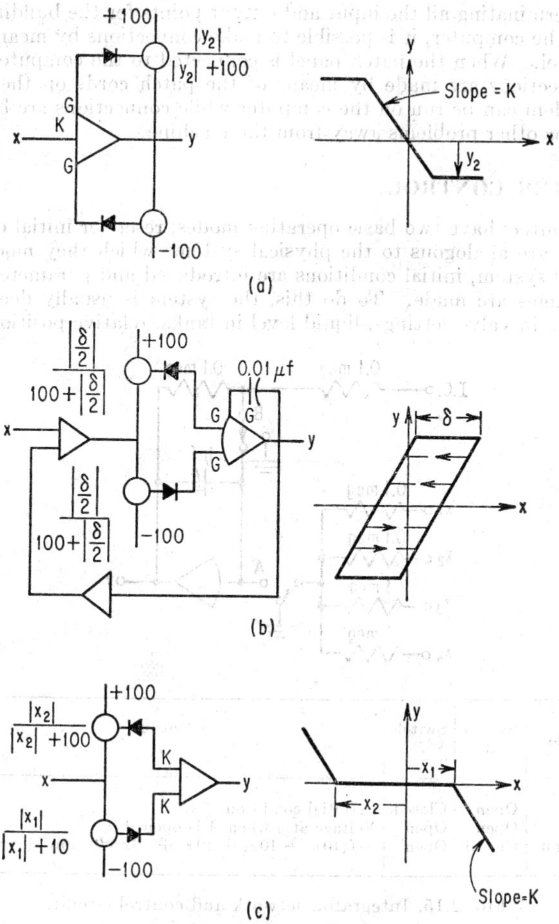

FIG. 2.13. Simple diode circuits to represent discontinuous functions. *G* indicates connection to grid of amplifier. (*a*) Limiter. (*b*) Backlash. (*c*) Dead zone.

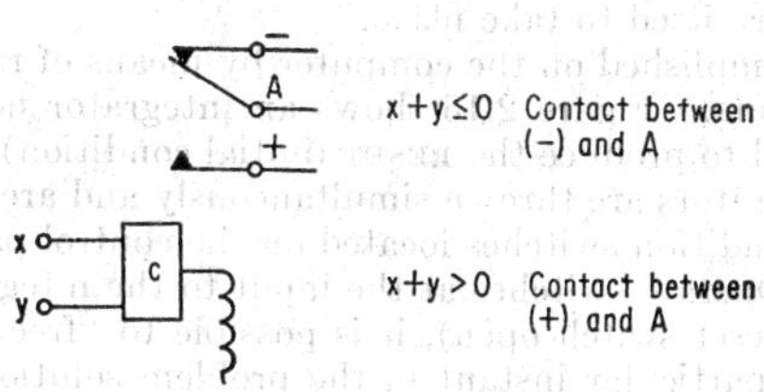

FIG. 2.14. Comparator.

2.5. CONNECTING COMPONENTS—PATCHING

Once an assemblage of computer building blocks is available, it is then necessary to provide means to interconnect the output of one unit to the input of one or more other units. For small computers this is usually accomplished by connecting directly from the output of one device into the input of another by means of a cord or wire.

However, by terminating all the input and output points for the building blocks at a single area on the computer, it is possible to make connections by means of a removable patch panel. When the patch panel is connected to the computer, the component interconnections are made by means of the patch cords on the patch panel. Thus one problem can be run on the computer while connections are being made on patch panels for other problems away from the machine.

2.6. COMPUTER CONTROL

Analog computers have two basic operating modes, reset (or initial condition) and operate, which are analogous to the physical systems which they model. In reset, on the physical system, initial conditions are introduced and parameter adjustments or system changes are made. To do this, the system is usually deenergized, permitting changes in valve settings, liquid level in tanks, relative position of elements,

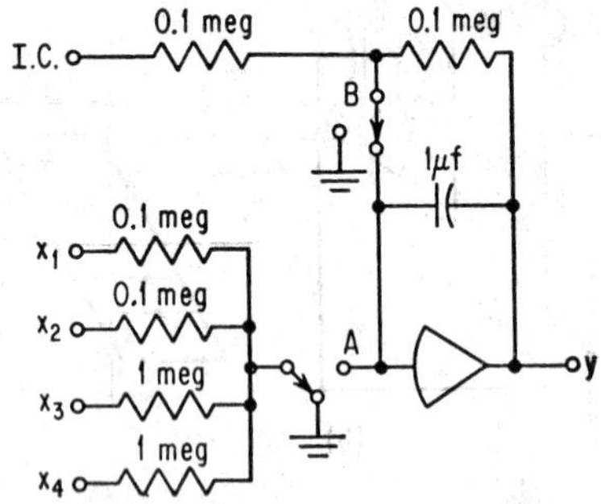

Mode	Switch *A*	Switch *B*	Output *y*
RESET	Open	Closed	−Initial condition
HOLD	Open	Open	Voltage at *y* when *A* is opened
OPERATE	Closed	Open	$-\int(10x_1 + 10x_2 + 10x_3 + 10x_4)dt$ −initial condition

FIG. 2.15. Integrator network and control circuit.

deflection of springs, and replacement of weights. In operate, power is applied to the system and the dynamic response (or operation), under imposed initial conditions or forcing functions, is permitted to take place.

Mode control is accomplished on the computer by means of relay switching associated with the integrators. Figure 2.15 shows an integrator network and amplifier and the two relays used to produce the RESET (initial condition) and OPERATE modes. The relays for all integrators are thrown simultaneously and are controlled by master Operate and Initial Condition switches located on the control panel of the computer.

By opening all the Operate switches at the input to the integrators during a problem run (leaving the reset switch open), it is possible to "freeze" the variables and their derivatives at a particular instant in the problem solution. This is the HOLD mode of the computer. When the computer is returned to the OPERATE mode it will have, in effect, a new set of initial conditions: the conditions existing when the computer was placed in the HOLD mode (or modified during HOLD by the programmer).

2.7. COMPUTER READOUT

Several methods are available for displaying the results obtained in a computer solution. A digital voltmeter–printer combination allows monitoring and subsequent print-out of all voltages in a problem at a particular instant—or when the voltage reaches a fixed value. Readout is available to 0.01 per cent.

In addition to the digital-voltmeter readout, a continuous record can be made of problem variables as functions of time. Strip-chart recorders, either single or multi-channel, are available for this task. If it is desirable to plot one variable vs. another as in the case of a phase-plane plot, a rectangular XY plotter can be used. By means of these devices, dynamic-response data for the problem can be recorded.

2.8. PROGRAMMING

Computer programming will be demonstrated by preparing a computer diagram to solve Eq. (2.1). The following routine will be used as a programming procedure:

1. Separate the highest-order derivative(s) (nth order) to the left-hand side of the equation.
2. Assuming that the highest derivative(s) is available, integrate n times to produce the dependent variable.
3. Using the variable(s) and the $(n - 1)$th-order derivatives produced by integration, generate the terms on the right-hand side of the equation(s) (Fig. 2.16).

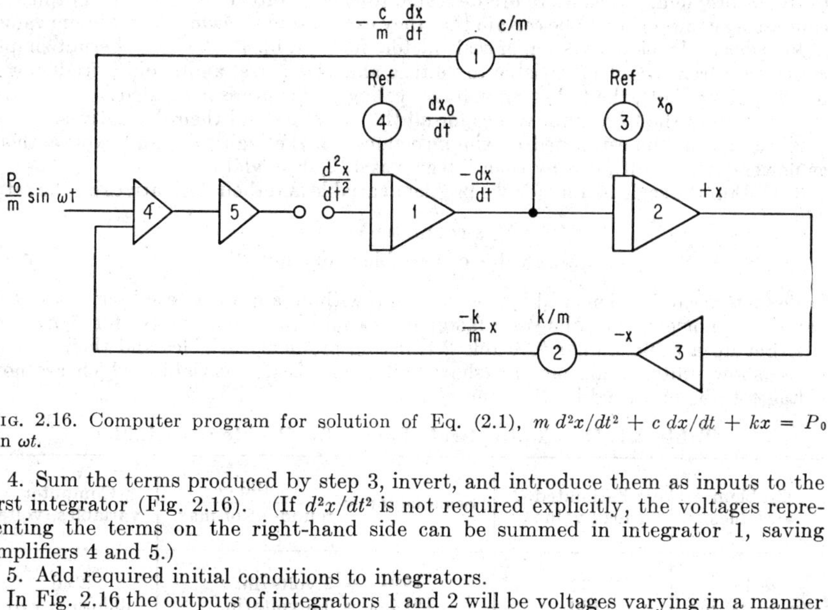

FIG. 2.16. Computer program for solution of Eq. (2.1), $m\, d^2x/dt^2 + c\, dx/dt + kx = P_0 \sin \omega t$.

4. Sum the terms produced by step 3, invert, and introduce them as inputs to the first integrator (Fig. 2.16). (If d^2x/dt^2 is not required explicitly, the voltages representing the terms on the right-hand side can be summed in integrator 1, saving amplifiers 4 and 5.)
5. Add required initial conditions to integrators.

In Fig. 2.16 the outputs of integrators 1 and 2 will be voltages varying in a manner analogous to the problem variables dx/dt and x, respectively. Problem parameters are adjusted by potentiometers 1 and 2. By changing the initial-condition potentiometers (3 and 4) on the integrators we can change initial conditions in the problem.

2.9. SCALING

On the analog computer, voltage is the dependent variable; the independent variable is time. To fit a problem to the computer it is necessary to provide definite relations between the problem variables and the computer voltage, and also between the problem-independent variable and time on the computer.

(a) Voltage Scaling

On the basis of amplifier design there is a voltage range in which the computer can operate to produce the mathematical functions described previously. On modern general-purpose analog computers, this voltage is usually ±10 volts or ±100 volts. The latter will be used as the basis for further discussions of scaling.

In a physical system the maximum value of the dependent variable can be quite small ($\ll 1.0$) (e.g., motions of intricate mechanisms). On the other hand, acceleration forces caused by shock blast result in a very large variable ($\gg 10^3$ lb). In order to transform the problem variable(s) into a computer voltage and assure that the maximum allowable voltage on the computer will not be exceeded when the maximum value of the variable in the problem is reached, a set of scaled relations between problem variables and computer voltages must be developed.

In relating problem variables to voltage, scale factors are required of the form volts per inch, volts per degree, volts per foot per second, etc. To obtain these scale factors, the maximum values of the variables which will occur in the problem must first be estimated. Errors in estimating maximum values are not serious because these errors are quickly indicated by overloads of computing elements after the problem is put on the computer. (Most analog computers have both an audible and a visual alarm to indicate overload of the computing elements.) In addition to not exceeding the voltage limitation of the computing elements, proper scaling also requires that the voltages be greater than some minimum voltage level at some time during a problem. It is therefore desirable to check voltage levels at the outputs of computing components to be certain that they exceed some reasonable minimum value.

In order to develop a system of scale factors for a problem which will permit simple conversion from problem variable to voltage and vice versa, and avoid introducing a new set of symbols, the following voltage-scaling procedure is suggested:

1. Estimate the maximum values for all the variables and their derivatives.
2. Round off the estimated maximum values (on the high side) and express them as powers of 10 multiplied by such integers as 1, 2, 4, 5, and 8.
3. Obtain the scale factors for the problem variable and their derivatives by the ratio

$$\frac{\text{Max computer voltage}}{\text{Max value of variable [from step 2]}}$$

Scale factors can be identified by the symbol S with an appropriate subscript for each variable. For example, the scale factor for x could be designated S_x, for $\dot{x}$, $S_{\dot{x}}$, etc.

4. Set up a table similar to Table 2.1, listing problem variables and their derivatives (show units), scale factors (show units), and scaled variables which are now voltages to be produced by the computer.

Table 2.1. Developing Scale Factor for Problem Variables

Problem variables	Estimated max value	Scale factor $= \dfrac{100}{\text{round-off of max}}$	Computer variables, volts
x_1 in.	32	$100/50 = 2$ volts/in.	$[2x_1]$
$(x_1 - x_2)$ in.	190	$100/200 = 0.5$ volts/in.	$[0.5(x_1 - x_2)]$
e^{-kx} lb	10^6	$\dfrac{001}{10^6} = 10^{-4}$ volts/lb	$[10^2 e^{-kx}]$
sin θ	1	$100/1 = 100$	$[100 \sin \theta]$

To prepare the computer patching diagram use scaled voltage equations which can be obtained from the original problem equations and scaled voltage variables by the following procedure:

1. Use the problem equation(s) in the form where the highest derivative(s) of the dependent variable(s) is alone on the left-hand side of the equation.
2. Multiply the equation through by the scale factor of the highest derivative (or the scale factor of the $(n-1)$th derivative if the highest derivative is not required explicitly as an output voltage of a computing element). This converts each of the terms in the equation from units of the problem variable to a voltage.

3. On the right-hand side of the equation, replace the problem variables by the scaled variables.

4. In order to maintain equality in the equation, it is necessary to multiply each term by the reciprocal of the scale factor. Each term now consists of two parts: a voltage and a coefficient.

5. Factor the coefficient into a potentiometer setting (a number less than 1 for all values of the parameters) and an amplifier gain. The magnitudes of the amplifier gain developed in this step are an indication of the possible need for time-scaling the problem.

(b) Time Scaling

Whether or not time scaling is required in a problem is determined by several factors:

1. Limitation of dynamic response of computing elements
2. Limitation of dynamic response of recording elements
3. Length of time required for problem solution

Time scaling can be reduced to a straightforward procedure which will not affect the voltage scale factors previously established for the dependent variables. To increase or decrease the time required for a phenomenon to occur, it is necessary to increase or decrease the rate at which the phenomenon occurs. Since the inputs to all integrators are rates, to increase or decrease the rate of problem solution on the computer, increase or decrease the gain of the integrators.

The time τ required for physical problem solution can be related to time t by $\tau = \beta t$. To increase the time required for solution, β will be a number greater than 1. For example, if the physical system takes 10 sec to reach steady state, on the computer the same solution would require 100 sec for $\beta = 10$. The relation between τ and t can be mechanized by having the integrator inputs modified by the factor $1/\beta$. Thus for β equal to 10, $1/\beta$ will be a number less than 1, the rate to the integrators is decreased, and the solution time on the computer increased. For β less than 1, $1/\beta$ will be a number greater than 1 and the solution time on the computer will be reduced because of the increase in integrator gain (rate).

In the preliminary analysis of a problem, it is often possible to estimate frequencies (or time constants) that will occur in the system. If these frequencies are either less than 0.01 cps or greater than 5 cps, time scaling may be necessary. During programming and voltage scaling, the resulting amplifier gains are an excellent indication as to whether or not time scaling is required. If large integrator gains are required, it is desirable to slow the problem solution down by introducing large values for β. On the other hand, if small potentiometer settings, say less than 0.01, are required throughout a problem, time scaling is appropriate to increase problem-solution speed and obtain suitable settings.

If time scaling is accomplished by changing the gain of the integrators, it is not necessary to rescale the voltage scale factors. One simply has to note on the computer diagram that β is other than 1 and take this into consideration when evaluating response data as a function of the independent variable.

2.10. PROBLEM CHECKING

Once the computer program has been completed it is of utmost importance to ascertain that the final computer program is actually the model of the original problem equations. This can be accomplished as follows:

1. Select an arbitrary set of initial conditions for all the derivatives and variables on the right-hand side of the equation. (Initial conditions selected should not result in zero terms.)

2. With the initial conditions chosen in step 1 and appropriate values selected for the parameters, calculate the voltages which should exist at each point in the computer diagram as well as the inputs to all integrators.

3. Substitute the same initial conditions and parameters into the *original* problem equations and solve for the highest derivatives of the unknowns. This should be related to the calculated voltage from step 2 representing the highest derivative by the following expression:

$$d^nx/dt^n \text{ (from equation)} = \left(\frac{\beta}{\text{scale factor for } dx^n/dt^n}\right) \times \text{(voltage calculated from computer diagram for } d^nx/dt^n)$$

This verifies that the computer diagram is the correct mechanization of the original equation.

When the problem is finally set up on the computer, the problem-check initial condition and parameters are programmed as a static check run. The voltage existing at the input to the integrator (whose input is analogous to d^nx/dt^n) should equal the voltage calculation in step 2 above. This final check verifies computer patching, pot settings, amplifier gains, and functioning of the nonlinear computing elements.

2.11. PARTIAL DIFFERENTIAL EQUATIONS

In general, the analog computer can handle differential equations with only one independent variable. However, partial differential equations can be solved if they can be replaced by a set of simultaneous ordinary differential equations. Consider Eq. (2.6) which describes the transverse vibration of a beam:

$$\rho\, \partial^2 y/\partial t^2 = -\partial^2/\partial x^2(EI\, \partial^2 y/\partial x^2) \tag{2.6}$$

The right-hand side of the equation can be written as four first-order derivatives using the following relations:

$$\begin{aligned} S \text{ (shear force)} &= \partial/\partial x(-EI\, \partial^2 y/\partial x^2) = \partial M/\partial x \\ M \text{ (bending moment)} &= -EI\, \partial^2 y/\partial x^2 = -EI\, \partial\theta/\partial y \\ \theta \text{ (slope)} &= \partial y/\partial x \end{aligned}$$

to yield

$$\begin{aligned} \rho\, \partial^2 y/\partial t^2 &= \partial S/\partial x \\ S &= \partial M/\partial x \\ M &= -EI\, \partial\theta/\partial y \\ \theta &= \partial y/\partial x \end{aligned}$$

Replacing the first-order derivatives by a finite-difference approximation, a set of N simultaneous ordinary differential equations is obtained of the form

$$\rho \frac{d^2 y_i}{dt^2} = \frac{S_{i+\frac{1}{2}} - S_{i-\frac{1}{2}}}{\Delta x}$$

where

$$\begin{aligned} S_{i+\frac{1}{2}} &= \frac{M_{i+1} - M_i}{\Delta x} \\ M_i &= -E_i I_i \left(\frac{\theta_{i+\frac{1}{2}} - \theta_{i-\frac{1}{2}}}{\Delta x}\right) \\ \theta_{i+\frac{1}{2}} &= \frac{y_{i+1} - y_i}{\Delta x} \end{aligned} \tag{2.7}$$

These equations can now be handled in the same manner as ordinary differential equations.

DIGITAL COMPUTERS*

2.12. WHAT A DIGITAL COMPUTER IS

The digital computer should be considered as just another tool to aid in solving problems of design and analysis more efficiently. The digital computer permits the utilization of many of the mathematical techniques of applied mechanics in machine and mechanical-systems design. Time, cost, and errors are kept to a minimum.

A computer system consists of three basic components: input devices, computer, and output devices. Input is usually in the form of punch cards, paper tape, and/or magnetic tape, together with manual operation of keyboards and switches. In addition to cards and tapes, output may be printed page and visual displays of the computer contents.

The computer itself may be broken down into three functional parts. A *control unit* interprets the instructions presented to it and initiates their execution. Actual numerical manipulation and arithmetic operations are accomplished in the *arithmetic unit,* which consists of an accumulator and an operand (i.e., addition, subtrahend, multiplicand, divisor) register. Probably the most distinctive part of the computer is the *storage,* in which the machine's instructions and the data needed to carry out instructions are placed. Several forms of storage media are in use, such as magnetic cores, magnetic drums and disks, and magnetic tape.

The storage or memory is divided into groups of bits, each group being designated as a word and identified in some way by an address. Each word or location may contain a piece of data, a number, or an instruction. A location may contain more than one instruction in some computers. Among the systems used in computers to represent numbers are binary, biquinary, and decimal. However, in most modern computers input and output are decimal so that the engineer is usually not concerned with the internal system used. Hence, in the following discussion all numbers will be considered as decimal.

2.13. WHAT A DIGITAL COMPUTER DOES

A digital computer might be called an automated high-speed desk calculator. Its arithmetic unit contains an accumulator with a capacity of $2N$ decimal digits, just as does the desk calculator, an operand of N decimal digits, corresponding to the calculator keyboard, and a multiplier register of N decimal digits (a portion of the accumulator may be used for this purpose) supplied as a separate keyboard on the desk calculator. In the computer all the above numbers are also supplied with algebraic sign, plus or minus. Once the data have been entered into the arithmetic unit the operation specified by the instruction may be carried out. Data as required by certain operations are entered before execution, automatically.

A list of operations is furnished with each computer. The basic types of computer operations are arithmetic and logical.

The most common arithmetic operations are addition, subtraction, multiplication, and division. In addition to these, most digital computers provide several operations for manipulating the contents of the arithmetic unit. These may include shifting the contents of the accumulator to the left or right, rounding off the number in the accumulator, and storing the contents of the arithmetic unit in the memory unit. In addition, certain store operations may be provided to facilitate modifying instructions. A certain value of a function stored in the memory as a table may be searched for by a table-look-up operation.

Logical operations are those in which the computer must make a decision based on the results of previous arithmetic operations. The decision is usually *yes* or *no,* with the computer subsequently following one of two sequences of operations offered. Common questions asked are: "Is the accumulator positive or negative?" "Is the accumulator (or part of the accumulator, such as the remainder from a divide oper-

* By John A. Carlson.

ation) not zero?" "Has the accumulator overflowed (from trying to add in too many numbers, such that the number of digits exceeds $2N$)?" The combination of logical and arithmetic operations gives practically unlimited range to the calculations which digital computers may undertake.

Other operations which may not be classed as either arithmetic or logical are *control input* and *output*, or *stop the computer*.

2.14. OUTPUT FORMS AND THEIR USES

The different forms of output were mentioned previously. Thus far, an on-line page printer for a computer has either been very expensive or very slow. In addition, the printed page has quite limited usefulness since it is not easily converted to other forms. For these reasons, direct printed-page output from computers is practical only if the output volume is small compared with the amount of computations or if large volumes of printed-page output are consistently desired. Sometimes it may be desired to print the results in an order different from that in which it is computed. In this case cards are the forms which are most suitable for sorting outside the computer. Cards may be used in conjunction with a tabulating machine to yield great flexibility of printed form. Some curve-plotting output devices are available which may work either directly from the computer or from other output media.

Punched cards or tapes from the computer may be used directly in numerically controlled machines. These can also be used for telegraph transmission of data and results to distant points.

While magnetic tape offers a high-speed input and output to the computer it is usually not a convenient medium outside the computer. It always requires a reader for interpretation, and magnetic-tape readers are costly. Small amounts of output are not conveniently handled by magnetic tape since it must always be handled on reels. Usually magnetic-tape output is converted to other forms outside the computer and may offer an advantage where computer time is at a premium.

2.15. HOW TO PUT A DIGITAL COMPUTER TO WORK

A digital-computer program is an assembly of arithmetic and logical operations set in a sequence so as to obtain a desired result from input data. Computer programming demands exacting care.

(a) Step 1. Formulation of the Problem

The general digital-computer problem is the solution of n equations in n unknowns. These equations may be algebraic, differential or partial differential, linear or nonlinear. It is usually necessary to solve these equations for different values of variable parameters, such as coefficients and/or initial and boundary conditions. Reduction of the problem to such a form is the first step in programming. The resulting equations may be the same as those which would be necessary to analyze the problem without the computer. However, numerical techniques, such as the Monte Carlo, possible only through the use of a computer, may be employed which require different equations. After deriving the problem formulas it may be asked whether it is more economical to proceed to program for the computer or to carry out the computations by hand. Only the simplest, least repetitive calculations of nonrecurring problems are usually excluded by this test. Many times a particular problem program or parts of it may be used for other problems and programs, even if the same problem, with different data, does not recur.

(b) Step 2. Logic Block Diagram

Formulation, step 1, results in an array of algebraic equations which can be solved to yield the answer to the problem. If these are to be solved numerically, one must decide where to start the computations and in which order the various formulas should be used. Provision must be made for the computer to yield the results in some form of output and for the computer to stop when all the desired results have been obtained. Complex formulas must be broken down into relatively simple groups of arithmetic operations.

Intermediate results may be obtained which may require different treatment, depending on their magnitude and/or sign, to yield the final results. For these, alternative groups of operations must be provided together with criteria for deciding which group should be used. The point at which such a decision must be made is called a *branch point.* An alternative sequence of operations may be obtained by modifying certain instructions of a given group. After instructions are modified and used, it may be necessary to return them to their original form.

A block diagram is more easily understood if the different types of operations, arithmetic, input, output, branching, program modification, etc., are indicated by blocks of different shapes. Each block may require many computer operations for its execution; e.g., one or more complicated formulas may be involved in calculating certain results which will be needed for further calculations or which may be the output. Input or output operations involve the transfer of information from one storage medium to another, one of them usually being the main computer memory or storage unit.

Branching operations include the formulation of the question as well as the actual machine branch operation. Most questions are answered yes or no; however, some computers provide branch operations in which more than two alternatives are possible. Program modification usually follows a branch operation or may be present in a closed loop of a repetitive program, as when an independent variable assumes a series of values.

A working program should be free of manual operations except to start perhaps. Generally, starting may be made automatic by providing for it in the computer-loading operation. It may be desirable to provide temporary manual operations for the purpose of checking out the program the first time it is run. Using these operations at certain points will allow the checking of intermediate results obtained in calculating. However, most computers have facilities provided on the consoles for stopping the program at any point, looking into the memory and arithmetic units, and restarting at the same point.

The different blocks are connected by flow lines with arrows indicating the sequence of operation. If it is inconvenient or confusing to draw a flow line from one block to another, a connection may be indicated by a matching pair of circled numbers, one circle adjacent to the one block and the other circle adjacent to the following block. One number may be used as an exit from any number of blocks but may be used to enter only one block. This leads to a basic rule of computer logic: No ambiguity is permitted!

(c) Step 3. Programming

Programming is the process of converting the logical calculations and procedures indicated by the block diagram into a series of machine instructions which the computer can understand and execute. The computer operating language is numerical, and operations and storage locations are coded numerically. Knowing the numerical language of the machine, it is possible to write a series of instructions and enter them into the computer directly through one of its inputs. The collection of numbers which represent the machine operations is called the *machine code* and is different for each make and model of computer. Memorization and retention of these codes by a programmer are possible if the same computer is used constantly; however, for a person using a computer occasionally this is not practical.

Thus alphabetic codes have been developed for several computers in which two or three letters are used to represent the operation, chosen so that the code has a high mnemonic value. Such a code is much easier to memorize and retain. Computer programs are used to translate the alphabetic language into numerical machine language. In SOAP II, a system developed for the IBM 650 computer, storage locations as well as operations may be alphabetically coded and be referred to by names. In the systems described thus far each machine step must be programmed.

One disadvantage of computer programming is the multiplicity of operation codes and instruction forms. Considerable effort has been made to develop a more or less universal computer language. The common language of computing is mathematics, more specifically, algebra. Several manufacturers have devised programs for their machines to convert symbolically coded algebraic statements into step-by-step computer operations. In the statements, variables may be represented by words or letters and operations indicated by symbols, such as, "+" for add, "−" for subtract, "/" for divide, "*" for multiply ("×" not used because of possible confusion with letter X), etc. From here, the systems differ somewhat in form. The logic is, of course, similar. In addition to mathematical statements, control statements may be made for input and output, looping, logical decisions, and setting up tables. The algebraic programming systems are faster to program than the step-by-step and permit the use of the computer for shorter runs of routine calculations. Processing of the algebraic program results in a step-by-step program in machine language which in turn is loaded into the computer.

At this point, step-by-step programming has one advantage in that, if difficulty arises in running, the programmer can more easily trace the trouble and correct it directly if he has a knowledge of the computer steps.

(d) Use of Subroutines—Library

In most engineering programs, operations and functions arise which are outside the range of simple operations built into the computer. These include square roots, cube roots, trigonometric functions, exponential and logarithmic functions, and Bessel functions. Computer subroutines utilizing numerical techniques have been developed for each computer for the more commonly used operations and functions and are part of libraries available to the computer user. These subroutines are written so that they may be used in the object program at any number of points. Only one region in the memory must be reserved for each of the subroutines required, regardless of the number of times it is used. In the program a subroutine usually requires that the argument of the desired function be placed in the arithmetic unit (say, in the accumulator) and that the locations of the start of the subroutine and the next program step following the subroutine be specified.

Many problems which arise frequently in engineering have been programmed and entered in the computer library as routines. The solution of simultaneous algebraic and differential equations, Laplace's equation, roots of polynomials, and statistical analysis fall into this category. Much time may be saved by using available programs. A private library may be built up by the individual computer installation, and programs should be written with an eye toward their inclusion in a library, together with instructions for their use, called a *write-up*.

(e) Decimal Point—Fixed, Floating

In most digital computers every number is treated as a whole number just as on a desk calculator or slide rule. Fractional numbers are obtained by scaling, that is, by considering every number to be of the form $N(10)^n$, where N and n are integers and may be independently positive or negative. For example, $\pi = 3141592653 \times 10^{-9}$. The computer deals only with the number N. The scale factor 10^n for each input and output variable must be fixed and applied by the programmer when preparing the input data or interpreting the output data. Since n is fixed for each variable, the arithmetic system is called *fixed-point* and a computer operating on numbers in the above form is said to be operating in the fixed-point mode. Inside the computer, N may be in decimal or binary form.

Scaling must be considered as the problem is programmed, and certain rules must be observed for arithmetic operations:

1. The scale factors of two numbers being added or subtracted must be the same. In $A \pm B = C$, where $A = N_A(10)^{n_A}$, $B = N_B(10)^{n_B}$, $C = N_C(10)^{n_C}$; $n_A = n_B = n_C$.

2. When two numbers are multiplied together, their scale factors must also be multiplied together. In $AB = C$, $n_C = n_A + n_B$.

3. When one number is divided by another, their scale factors must also be divided. In $A/B = C$, $n_C = n_A - n_B$.

4. In division, the scale factors must be such that the quotient $N_C(= N_A/N_B)$ neither exceeds the capacity of the computer nor possesses insufficient digits for the desired accuracy.

In some problems, it is difficult or impossible to apply all the above rules of scaling, such as when variables vary greatly in magnitude. To handle these, floating-point subroutines or interpretive systems have been developed in which the exponent of the scale factor n is operated upon as well as the characteristic N. A floating-point number is in the form Na or aN in the computer, where $N = X.XXXXX \cdots$ (or $.XXXXXX \cdots$), and $a = n + b$. b is chosen so that a is positive for $10^{-b} \leq N(10)^n < 10^b$ [or $10^{-b-1} \leq N(10)^n < 10^{b+1}$]. As an example,

$$-0.000123456 = -1.23456(10)^{-4}$$

appears in the computer as $-1.2345600\underline{46}$ or $-\underline{46}1.2345600$, where $n = -4$ and $b = 50$. Thus the signs of the number and of the exponent can be considered independently while assigning only one sign, that of the number, to the memory location containing the floating-point number. The first digit of N is always nonzero unless the number is identically zero. A few computers have floating-point arithmetic units built in or available as accessories, in which cases the memory is not used up by a subroutine.

Floating-point operation is usually slower than fixed-point and should for this reason not be used indiscriminately.

2.16. NUMERICAL-ANALYSIS TECHNIQUES APPLICABLE TO DIGITAL COMPUTERS

Mathematical operations such as interpolation, differentiation, integration, and transformations must be reduced to basic arithmetic operations in the formulation of a problem for a digital computer. The following collection of formulas is used extensively in computer programming.

(a) Interpolation

Differences. Given a table of values of a variable, differences of any order n may be calculated. These differences may be horizontal or backward differences, diagonal or forward differences, or central differences, depending upon the use which is to be made of them. Tables 2.2 to 2.4 illustrate differences of the various types with distinguishing notation.

Table 2.2. Horizontal or Backward Differences

i	y_i	∇y_i	$\nabla^2 y_i$	$\nabla^3 y_i \cdots$
0	y_0			
1	y_1	∇y_1		
2	y_2	∇y_2	$\nabla^2 y_2$	
3	y_3	∇y_3	$\nabla^2 y_3$	$\nabla^3 y_3$
$\cdots$	$\cdots$	$\cdots$	$\cdots$	$\cdots$

Table 2.3. Diagonal or Forward Differences

i	y_i	Δy_i	$\Delta^2 y_i$	$\Delta^3 y_i$ ···
0	y_0	Δy_0	$\Delta^2 y_0$	$\Delta^3 y_0$ ···
1	y_1	Δy_1	$\Delta^2 y_1$	···
2	y_2	Δy_2	···	
3	y_3	···		
···	···			

Table 2.4. Central Differences

i	y_i	δy_i	$\delta^2 y_i$	$\delta^3 y_i$ ···
···	···			
−3	y_{-3}	···		
		$\delta y_{-2\frac{1}{2}}$	···	
−2	y_{-2}		$\delta^2 y_{-2}$	···
		$\delta y_{-1\frac{1}{2}}$		$\delta^3 y_{-1\frac{1}{2}}$
−1	y_{-1}		$\delta^2 y_{-1}$	
		$\delta y_{-\frac{1}{2}}$		$\delta^3 y_{-\frac{1}{2}}$
0	y_0		$\delta^2 y_0$	
		$\delta y_{\frac{1}{2}}$		$\delta^3 y_{\frac{1}{2}}$
1	y_1		$\delta^2 y_1$	
		$\delta y_{1\frac{1}{2}}$		$\delta^3 y_{1\frac{1}{2}}$
2	y_2		$\delta^2 y_2$	···
		$\delta y_{2\frac{1}{2}}$	···	
3	y_3	···		
···	···			

The terms in the tables are defined by the following formulas:

$$\nabla^n y_i = \nabla^{n-1} y_i - \nabla^{n-1} y_{i-1} \tag{2.8}$$

$$\Delta^n y_i = \Delta^{n-1} y_{i+1} - \Delta^{n-1} y_i \tag{2.9}$$

$$\delta^n y_i = \delta^{n-1} y_{i+\frac{1}{2}} - \delta^{n-1} y_{i-\frac{1}{2}} \tag{2.10}$$

$$\nabla^0 y_i = \Delta^0 y_i = \delta^0 y_i = y_i \tag{2.11}$$

Gregory-Newton Formulas. Given

$$y_i = f(x_i) \qquad \text{at } x_i = x_0 + ih,\ i = 0, 1, 2, \cdots, n$$

Forward Interpolation

$$(x - x_0)/h = u \qquad \text{or} \qquad x = x_0 + hu$$

$$y(x) = y_0 + u\,\Delta y_0 + \frac{u(u-1)}{2!}\Delta^2 y_0 + \frac{u(u-1)(u-2)}{3!}\Delta^3 y_0 + \cdots + \frac{u(u-1)(u-2)\cdots(u-n+1)}{n!}\Delta^n y_0 \tag{2.12}$$

Backward Interpolation

$$(x - x_n)/h = u \qquad \text{or} \qquad x = x_n + hu$$

$$y(x) = y_n + u\,\nabla y_n + \frac{u(u+1)}{2!}\nabla^2 y_n + \frac{u(u+1)(u+2)}{3!}\nabla^3 y_n + \cdots + \frac{u(u+1)(u+2)\cdots(u+n-1)}{n!}\nabla^n y_n \tag{2.13}$$

Equation (2.12) is used to interpolate near the beginning of a table or to extrapolate before a table. Equation (2.13) is used near the end of or beyond a table.

Central-difference Formulas. *Stirling's Formula*

$$(x - x_0)/h = u$$

$$y(x) = y_0 + u\frac{\delta y_{-1/2} + \delta y_{1/2}}{2} + \frac{u^2}{2!}\delta^2 y_0 + \frac{u(u^2 - 1^2)}{3!}\frac{\delta^3 y_{-1/2} + \delta^3 y_{1/2}}{2}$$
$$+ \frac{u^2(u^2 - 1^2)}{4!}\delta^4 y_0 + \frac{u(u^2 - 1^2)(u^2 - 2^2)}{5!}\frac{\delta^5 y_{-1/2} + \delta^5 y_{1/2}}{2} + \frac{u^2(u^2 - 1^2)(u^2 - 2^2)}{6!}\delta^6 y_0$$
$$+ \cdots + \frac{u(u^2 - 1^2)(u^2 - 2^2) \cdots [u^2 - (n-1)^2]}{(2n-1)!}\frac{\delta^{2n-1} y_{-1/2} + \delta^{2n-1} y_{1/2}}{2}$$
$$+ \frac{u^2(u^2 - 1^2)(u^2 - 2^2) \cdots [u^2 - (n-1)^2]}{(2n)!}\delta^{2n} y_0 \qquad (2.14)$$

Bessel's Formula

$$\frac{x - x_0}{h} = u \qquad x \text{ between } x_0 \text{ and } x_1$$

$$y(x) = y_0 + u\,\delta y_{1/2} + \frac{u(u-1)}{2}\frac{\delta^2 y_0 + \delta^2 y_1}{2} + \frac{(u - \frac{1}{2})u(u-1)}{3!}\delta^3 y_{1/2}$$
$$+ \frac{u(u-1)(u+1)(u-2)}{4!}\frac{\delta^4 y_0 + \delta^4 y_1}{2} + \frac{(u - \frac{1}{2})u(u-1)(u+1)(u-2)}{5!}\delta^5 y_{1/2}$$
$$+ \cdots + \frac{u(u-1)(u+1)(u-2)(u+2) \cdots (u-n)(u+n-1)}{(2n)!}\frac{\delta^{2n} y_0 + \delta^{2n} y_1}{2}$$
$$+ \frac{(u - \frac{1}{2})u(u-1)(u+1) \cdots (u-n)(u+n-1)}{(2n+1)!}\delta^{2n+1} y_{1/2} \qquad (2.15)$$

Use Stirling's formula (2.14) for $-0.25 < u < 0.25$ and Bessel's formula (2.15) for $0.25 < |u| < 0.75$ for greatest accuracy in most cases.

Lagrange's Formula. Given values of $y_i = f(x_i)$ for $i = 0, 1, 2, \ldots, n$, intermediate values of $y(x)$ may be found by approximating $f(x)$ by a polynomial.

$$y(x) \approx \sum_{j=0}^{n} \frac{\prod_{i=0}^{n} (x - x_i)}{(x - x_j)\, d/dx \prod_{i=0}^{n} (x - x_i)\Big|_{x=x_j}} y_j \qquad (2.16)$$

Lagrange's formula may be used equally well to obtain $x(y)$ by interchanging x and y in (2.16). The intervals of x or y need not be equidistant. Equation (2.16) should not be used where (2.12), (2.13), (2.14), or (2.15) are applicable.

Inverse Interpolation. Lagrange's formula given above may be used.

Reversion of Series. If the Gregory-Newton, Stirling, and Bessel formulas are written in the form of a power series

$$y(x) = \sum_{i=0}^{\infty} a_i x^i \qquad (2.17)$$

this series may be reverted to obtain

$$x = \sum_{i=0}^{\infty} c_i (y - a_0/a_1)^{i+1} \qquad (2.18)$$

where $c_0 = 1$, $c_1 = -a_2/a_1$, $c_2 = -(a_3/a_1) + 2(a_2/a_1)^2$
$c_3 = -a_4/a_1 + 5(a_2a_3/a_1^2) - 5(a_2/a_1)^3$
$c_4 = -a_5/a_1 + 6(a_2a_4/a_1^2) + 3(a_3/a_1)^2 - 21(a_2^2a_3/a_1^3) + 14(a_2/a_1)^4, \ldots$

(b) Roots of Equations

Newton-Raphson Method. See Sec. 1. The Newton-Raphson method may be used for simultaneous equations

$$f_j(x_1,x_2, \ldots ,x_N) = 0 \qquad j = 1, 2, \ldots , N \tag{2.19}$$

The nth calculated approximation

$$a_i^n = a_i^{n-1} + h_i^n \qquad i = 1, 2, \ldots , N \tag{2.20}$$

In this case the corrections h_i^n are found as the solution of the set of linear simultaneous algebraic equations

$$\sum_{i=1}^{N} h_i^n(\partial f_j/\partial x_i)_{x_i=a_i^{n-1}} = -f_j(a_1^{n-1},a_2^{n-1}, \ldots ,a_N^{n-1}) \qquad j = 1, 2, \ldots , N \tag{2.21}$$

Method of Iteration. When $f(x) = 0$ can be expressed in the form $x = \phi(x)$, the real roots can be found by application of the following iteration formula: $a^n = \phi(a^{n-1})$, provided $d\phi/dx < 1$ in the vicinity of the root.

This method can also be applied to simultaneous equations which can be written in the form

$$x_j = \phi_j(x_1,x_2, \ldots ,x_N) \qquad j = 1, 2, \ldots , N$$

The iteration formulas become

$$a_j^n = \phi_j(a_1^n,a_2^n, \ldots ,a_{j-1}^n,a_j^{n-1}, \ldots ,a_N^{n-1}) \qquad j = 1, 2, \ldots , N \tag{2.22}$$

For convergence it is necessary that the following conditions be satisfied in the neighborhood of the root:

$$\sum_{j=1}^{N} \left| \frac{\partial \phi^j}{\partial x_i} \right| < 1 \qquad i = 1, 2, \ldots , N \tag{2.23}$$

Complex Roots of Algebraic Equations. Numerical techniques for finding both real and complex roots of algebraic equations may be found in refs. 8, 9, 10, 11, and 12. Library programs are available for several computers for solving algebraic equations.

Linear Simultaneous Algebraic Equations. Programs are available in the libraries for most computers for solving linear simultaneous equations. These programs may also perform matrix inversion. The methods used in the programs are the same as those used for manual computation (see Matrices and Determinants, Sec. 1).

(c) Numerical Differentiation

Derivatives of tabulated functions may be found by differentiating the appropriate interpolation formula, Eqs. (2.12), (2.13), (2.14), (2.15). Use may be made of the relation

$$\frac{dy}{dx} = \frac{1}{h}\frac{dy}{du}$$

(d) Numerical Integration—Simpson's Rule

See Sec. 1

(e) First-order Differential Equations

Gregory-Newton Backward-interpolation Formula

$$\frac{dy}{dx} = f(x,y) \tag{2.24}$$

Integrating,

$$y = \int_{x_0}^{x} f(x,y)\,dx + y_0 \tag{2.25}$$

or

$$y_n = \int_{x_0}^{x_n} f(x,y)\,dx + y_0 = I_0^n + y_0 \tag{2.26}$$

where $x_n = x_0 + nh$.

$$y_{n+1} = I_0^n + I_n^{n+1} + y_0 \tag{2.27}$$

I_n^{n+1} may be found by substituting the Gregory-Newton backward-interpolation formula [Eq. (2.13)] into

$$I_n^{n+1} = \int_{x_n}^{x_n+h} f(x,y)\,dx = h\int_0^1 f(x_n,y_n)\,du \tag{2.28}$$

yielding

$$I_n^{n+1} = h(f_n + \tfrac{1}{2}\nabla f_n + \tfrac{5}{12}\nabla^2 f_n + \tfrac{3}{8}\nabla^3 f_n + \tfrac{251}{720}\nabla^4 f_n) \tag{2.29}$$

The equation

$$y_n^{i+1} = I_0^{n-1} + {}^iI_{n-1}^n + y_0 \qquad i = 1, 2, \ldots \tag{2.30}$$

may be used to apply a correction to the value of y_{n+1} obtained from (2.27) and (2.29), if y_{n+1} is taken as y_n^i in

$$\begin{aligned} {}^iI_{n-1}^n &= \int_{x_n-h}^{x_n} f(x,y)\,dx = h\int_{n-1}^{0} f(x_n,y_n^i)\,du \\ &= h(f_n^i - \tfrac{1}{2}\nabla f_n^i - \tfrac{1}{12}\nabla^2 f_n^i - \tfrac{1}{24}\nabla^3 f_n^i - \tfrac{19}{720}\nabla^4 f_n^i) \end{aligned} \tag{2.31}$$

The corrected value y_n^{i+1} is obtained from (2.30) by using the value of ${}^iI_{n-1}^n$ from (2.31). This procedure may be repeated as often as necessary to minimize the error; i.e., $y_n^{i+1} - y_n^i$ may be made as small as necessary.

Euler's Formula, Modified. Euler's formula for finding successive values of the dependent variable y in Eq. (2.24) is

$$y_{i+1} = y_i + (dy/dx)_i h \qquad i = 1, 2, \ldots \tag{2.32}$$

where $(dy/dx)_i = f(x_i,y_i)$ and $h = x_{i+1} - x_i$. This formula is modified as follows:

$$y^{k+1}_{i+1} = y_i + \frac{(dy/dx)_i + (dy/dx)^k_{i+1}}{2}h \tag{2.33}$$

where $(dy/dx)^k_{i+1} = f(x_{i+1},y^k_{i+1})$, y^1_{i+1} being the value of y_{i+1} obtained from Eq. (2.32). Equation (2.33) may be repeatedly applied by increasing k to obtain a stable value of y_{i+1}.

Milne's Method. Taking $y'_n = (dy/dx)_n = f(x_n,y_n)$, the solution to Eq. (2.24) may be found for $n > 3$ from

$$y^{(1)}_{n+1} = y_{n-3} + (4h/3)(2y'_{n-2} - y'_{n-1} + 2y'_n) \tag{2.34}$$

and

$$y^{(2)}_{n+1} = y_{n-1} + (h/3)(y'_{n-1} + 4y'_n + y'_{n+1}) \tag{2.35}$$

Equation (2.34) is used to find a first value of y_{n+1}, $y^{(1)}_{n+1}$. This value is used in Eq. (2.35) to obtain a corrected value of y_{n+1}, $y^{(2)}_{n+1}$. If the error

$$E_2 = \tfrac{1}{29}(y^{(2)}_{n+1} - y^{(1)}_{n+1}) \tag{2.36}$$

is larger than the desired accuracy in y, smaller intervals of x must be chosen; i.e., h must be reduced.[5]

Runge-Kutta Method. Successive values of the dependent variable in Eq. (2.24) are given by

$$y_{n+1} = y_n + \Delta y \qquad (2.37)$$

where $\Delta y = \frac{1}{6}(k_1 + 2k_2 + 2k_3 + k_4)$ and the values of k_1, k_2, k_3, k_4 are given by[8] $k_1 = f(x_n,y_n)h$, $k_2 = f(x_n + h/2, \quad y_n + k_1/2)h$, $k_3 = f(x_n + h/2, \quad y_n + k_2/2)h$, $k_4 = f(x_n + h, y_n + k_3)h$. $x_n = x_0 + nh$.

Runge-Fox Method for Linear Equations. Given a linear first-order differential equation

$$dy/dx = f(x)y + g(x) \qquad (2.38)$$

Taking $x_n = x_0 + nh$, $f(x_n) = f_n$, $g(x_n) = g_n$, the Runge-Fox recurrence equation is

$$y_{n+1}^{(i)} = \frac{1}{1 - (h/2)f_{n+1}} \left[\left(1 + \frac{h}{2} f_n\right) y_n^{(i)} + \frac{h}{2}(g_n + g_{n+1}) + \epsilon_{n+1}^{(i-1)} \right] \qquad (2.39)$$

where

$$\epsilon_{n+1}^{(i)} = -(\tfrac{1}{12}\delta^3 - \tfrac{1}{120}\delta^5 + \tfrac{1}{840}\delta^7 - \cdots)y_{n+\frac{1}{2}}^{(i)} \qquad (2.40)$$

$\epsilon_{n+1}^{(0)}$ is taken as zero, and the values of $y_n^{(i)}$ are found using Eq. (2.39). This solution is equivalent to the stable values of y_n found with the modified Euler's formula. The correction $\epsilon_{n+1}^{(i-1)}$ is found using Eq. (2.40) where the central differences are calculated using the values $y_n^{(i-1)}$ and then substituted in Eq. (2.39) to obtain the corrected values $y_n^{(i)}$. This process is done for $i = 2, 3, \ldots$ until $\epsilon_{n+1}^{(i-1)}$ is found to be the same as $\epsilon_{n+1}^{(i-2)}$. This method yields an error approximately an order of magnitude smaller than the modified Euler formula.[9]

(f) Second-order Differential Equations

The general second-order differential equation is written

$$d^2y/dx^2 = f(x,y, dy/dx) \qquad (2.41)$$

In the *Runge-Kutta method* Eq. (2.41) may be integrated step by step by applying the following formulas:

$$\begin{aligned} k_1 &= hf(x_n,y_n,y_n') \\ k_2 &= hf[x_n + h/2,\ y_n + (h/2)y_n' + (h/8)k_1,\ y_n' + k_1/2] \\ k_3 &= hf[x_n + h/2,\ y_n + (h/2)y_n' + (h/8)k_1,\ y_n' + k_2/2] \\ k_4 &= hf[x_n + h,\ y_n + hy_n' + (h/2)k_3,\ y_n' + k_3] \\ \Delta y &= h[y_n' + \tfrac{1}{6}(k_1 + k_2 + k_3)] \\ \Delta y' &= \tfrac{1}{6}(k_1 + 2k_2 + 2k_3 + k_4) \end{aligned}$$

where

$$\begin{aligned} x_n &= x_0 + nh \\ y_{n+1} &= y_n + \Delta y \\ y'_{n+1} &= y'_n + \Delta y' \end{aligned}$$

For the special second-order equation,

$$y'' = f(x,y) \qquad (2.42)$$

note that $k_3 = k_2$, and k_3 may be eliminated.[8]

The *Adams-Störmer* recurrence formula for solving Eq. (2.41) is

$$y_{n+1} = -y_{n-1} + 2y_n + h^2(1 + \tfrac{1}{12}\nabla^2 + \tfrac{1}{12}\nabla^3 + \tfrac{19}{240}\nabla^4 + \tfrac{3}{40}\nabla^5 + \cdots)f_n \qquad (2.43)$$

which is used together with

$$y'_{n+1} = y'_n + h(1 + \tfrac{1}{2}\nabla + \tfrac{5}{12}\nabla^2 + \tfrac{3}{8}\nabla^3 + \tfrac{251}{720}\nabla^4 + \tfrac{95}{288}\nabla^5 + \cdots)f_n \qquad (2.44)$$

For the solution of Eq. (2.42) using this method, only Eq. (2.43) is needed.

The linear second-order differential equation is written

$$y'' + f(x)y' + g(x)y = F(x) \qquad (2.45)$$

Taking $x_n = x_0 + nh$, Fox's formula for this equation is

$$y_{n+1}^{(i)} = \frac{1}{1 + (h/2)f_n} - \left[\left(1 - \frac{h}{2} f_n\right) y_{n-1}^{(i)} + (2 - h^2 g_n) y_n^{(i)} + h^2 F_n + \epsilon_{n+1}^{(i-1)}\right] \tag{2.46}$$

where the correction is given by

$$\epsilon_{n+1}^{(i-1)} = (\delta^4/12 - \delta^6/90 + \cdots) y_n^{(i-1)} + hf_n\mu(\delta^3/6 - \delta^5/30 + \cdots) y_n^{(i-1)} \tag{2.47}$$

The operator μ indicates that the odd differences are averages of the differences at $n - \frac{1}{2}$ and $n + \frac{1}{2}$; thus $\mu\delta^3 y_n = \frac{1}{2}(\delta^3 y_{n-\frac{1}{2}} + \delta^3 y_{n+\frac{1}{2}})$.

Taking $\epsilon_{n+1}^{(0)} = 0$, $y_{n+1}^{(1)}$ is computed using Eq. (2.46) for $n = 1, 2, \ldots$. The value of y_1 must be determined by some other method such as a Taylor-series expansion about x_0,

$$y_1 = y_0 + hy_0' + (h^2/2)y_0'' + (h^3/6)y_0''' + (h^4/24)y_0^{\mathrm{IV}} + (h^5/120)y_0^{\mathrm{V}} + \cdots \tag{2.48}$$

Corrections are calculated using Eq. (2.47) and substituted into Eq. (2.46) for $i = 2, 3, \ldots$, stopping when $\epsilon_{n+1}^{(i-1)} = \epsilon_{n+1}^{(i-2)}$.[9]

(g) Partial Differential Equations

Finite-difference Quotients. Partial differential equations are solved numerically by expressing them as finite-difference equations. The field covered by the partial differential equation is divided into a lattice of finite points, and the function is evaluated at these points. The lattice is arranged in a coordinate system, usually rectangular, corresponding to that of the differential equation

$$\begin{aligned} x &= x_0 + lh & l &= 0, 1, 2, \ldots \\ y &= y_0 + mh & m &= 0, 1, 2, \ldots \\ z &= z_0 + nh & n &= 0, 1, 2, \ldots \end{aligned}$$

The differentials of a function $F(x,y,z)$ may be written about the point (x,y,z) in terms of values of F at surrounding points as difference quotients. Thus the forward first difference with respect to x may be written

$$F_x\left(x + \frac{h}{2}, y, z\right) = \frac{F(x + h, y, z) - F(x,y,z)}{h} \tag{2.49}$$

and the backward first difference is

$$F_{\bar{x}}\left(x - \frac{h}{2}, y, z\right) = \frac{F(x,y,z) - F(x - h, y, z)}{h} \tag{2.50}$$

The second-difference quotient with respect to x

$$F_{\bar{x}x}(x,y,z) = \frac{F_x - F_{\bar{x}}}{h} = \frac{F(x + h, y, x) - 2F(x,y,z) + F(x - h, y, z)}{h^2} \tag{2.51}$$

Similar expressions may be written for difference quotients with respect to the y and z coordinates. Higher-order difference quotients may be obtained using additional points.

The difference quotient with respect to time

$$F_t = \frac{F(x, y, z, t + \Delta t) - F(x,y,z,t)}{\Delta t} \tag{2.52}$$

Method of Iteration. Boundary-value problems such as those arising from Poisson's or Laplace's equations are solved by expressing the partial differential equation as a difference equation and assuming an approximate solution which satisfies the boundary conditions but not the differential equation. New values of the dependent function are found by solving the difference equation for the value of the function at the central point $F(x,y,z)$ in terms of the values at the surrounding points $F(x + h, y, z)$,

$F(x, y + h, z)$, etc. This is done for all points in the lattice, thus replacing the approximate solution by a new solution. By repeating this process a solution may be found which does not differ significantly from the previous solution.

The inherent error in the final solution obtained above is of the order h^2 for second-order equations. An approximate value of the inherent error is given by

$$E_2 = \tfrac{1}{3}(F_2 - F_1)$$

where F_1 is the solution found when $h = h_1$ and F_2 is the solution found when $h = h_2 = \tfrac{1}{2}h_1$ (see ref. 8, p. 338).

For Laplace's equation in two dimensions

$$\partial^2F/\partial x^2 + \partial^2F/\partial y^2 = 0 \tag{2.53}$$

the difference equation is

$$\frac{F(x + h, y) - 2F(x,y) + F(x - h, y)}{h^2} + \frac{F(x, y + h) - 2F(x,y) + F(x, y - h)}{h^2} = 0 \tag{2.54}$$

from which the iteration formula is

$$F(x,y) = \tfrac{1}{4}[F(x + h, y) + F(x - h, y) + F(x, y + h) + F(x, y - h)] \tag{2.55}$$

or

$$F_{lm} = F(x_l,y_m) = \tfrac{1}{4}(F_{l+1,m} + F_{l-1,m} + F_{l,m+1} + F_{l,m-1})$$

Programs are available in computer libraries for certain partial differential equations which arise frequently.

(h) Curve Fitting

Given a set of data x_i, y_i, $i = 1, 2, \ldots, m$, programs are available for fitting a polynomial curve to the data using the method of least squares (see Sec. 1). These programs involve a subprogram which solves the set of simultaneous algebraic equations. The order of the polynomial and the number of points which may be handled are limited by the capacity of the computer used.

References

1. Fifer, Stanley: "Analogue Computation," vols. I–IV, McGraw-Hill Book Company, Inc., New York, 1960.
2. Jackson, Albert S.: "Analog Computation," McGraw-Hill Book Company, Inc., New York, 1960.
3. Johnson, C. L.: "Analog Computer Techniques," McGraw-Hill Book Company, Inc., New York, 1956.
4. Rogers, A. E. and T. W. Connolly: "Analog Computation in Engineering Design," McGraw-Hill Book Company, Inc., New York, 1960.
5. Livesley, R. K.: "Automatic Digital Computers," Cambridge University Press, New York, 1957.
6. McCormick, E. M.: "Digital Computer Primer," McGraw-Hill Book Company, Inc., New York, 1959.
7. Ivall, T. E.: "Electronic Computers, Principles and Applications," Philosophical Library, Inc., New York, 1960.
8. Scarborough, J. B.: "Numerical Mathematical Analysis," The Johns Hopkins Press, Baltimore, 1958.
9. Salvadori, M. G., and M. L. Baron: "Numerical Methods in Engineering," Prentice-Hall, Inc., Englewood Cliffs, N.J., 1952.
10. Stibitz, G. R., and J. A. Larrivee: "Mathematics and Computers," McGraw-Hill Book Company, Inc., New York, 1956.
11. Lin, S. N.: A Method of Successive Approximations of Evaluating the Real and Complex Roots of Cubic and Higher Order Equations, *J. Math. and Phys.*, vol. 20, p. 153, 1941.
12. Friedman, B.: Note on Approximating Complex Zeros of a Polynomial, *Communs. Pure Appl. Math.*, vol. II, June–September, 1949.
13 Todd, John: "Survey of Numerical Analysis," McGraw-Hill Book Company, Inc., New York, 1962.

Section 3

CLASSICAL MECHANICS

By

THOMAS P. MITCHELL, Ph.D., *Associate Professor of Engineering Mechanics, Department of Engineering Mechanics and Materials, Cornell University, Ithaca, N.Y.*

CONTENTS

The aim of this section is to present the concepts and results of Newtonian dynamics which are required in a discussion of rigid-body motion. The detailed analysis of particular rigid-body motions is not included. The chapter contains a few topics which while not directly needed in the discussion either serve to round out the presentation or are required elsewhere in this handbook.

INTRODUCTION

The study of classical dynamics is founded on Newton's three laws of motion and on the accompanying assumptions of the existence of absolute space and absolute time. In addition, in problems in which gravitational effects are of importance, Newton's law of gravitation is adopted. The object of the study is to enable one to predict, being given the initial conditions and the forces which act, the evolution in time of a mechanical system or, being given the motion, to determine the forces which produce it.

The mathematical formulation and development of the subject can be approached in two ways. The vectorial method, that used by Newton, emphasizes the vector quantities force and acceleration. The analytical method, which is largely due to Lagrange, utilizes the scalar quantities work and energy. The former method is the more physical and generally possesses the advantage in situations in which dissipative forces are present. The latter is more mathematical and accordingly is very useful in developing powerful general results.

3.1. THE BASIC LAWS OF DYNAMICS

The *first law of motion* states that a body which is under the action of no force remains at rest or continues in uniform motion in a straight line. This statement is also known as the law of inertia, inertia being that property of a body which demands that a force is necessary to change its motion. Inertial mass is the numerical measure of inertia. The conditions under which an experimental proof of this law could be carried out are clearly not attainable.

In order to investigate the motion of a system it is necessary to choose a frame of reference, which is assumed to be rigid, relative to which the displacement, velocity, etc., of the system are to be measured. The law of inertia immediately classifies the possible frames of reference into two types. For, suppose that in a certain frame S the law is found to be true; then it must also be true in any frame which has a constant velocity vector relative to S. However, the law is found not to be true in any frame which is in accelerated motion relative to S. A frame of reference in which the law of inertia is valid is called an inertial frame, and any frame in accelerated motion relative to it is said to be noninertial. Any one of the infinity of inertial frames can claim to be at rest while all others are in motion relative to it. Hence it is not possible to distinguish, by observation, between a state of rest and one of uniform motion in a straight line. The transformation rules by which the observations relative to two inertial frames are correlated can be deduced from the second law of motion.

Newton's *second law of motion* states that in an inertial frame the force acting on a mass is equal to the time rate of change of its linear momentum. Linear momentum, a vector, is defined to be the product of the inertial mass and the velocity. The law can be expressed in the form

$$d/dt(m\mathbf{v}) = \mathbf{F} \qquad (3.1)$$

which, in the many cases in which the mass m is constant, reduces to

$$m\mathbf{a} = \mathbf{F} \qquad (3.2)$$

where $\mathbf{a}$ is the acceleration of the mass.

The *third law, the law of action and reaction,* states that the force with which a mass m_i acts on a mass m_j is equal in magnitude and opposite in direction to the force which m_j exerts on m_i. The additional assumption that these forces are collinear is needed in some applications, e.g., in the development of the equations governing the motion of a rigid body.

The law of gravitation asserts that the force of attraction between two point masses is proportional to the product of the masses and inversely proportional to the square of the distance between them. The masses involved in this formula are the gravitational masses. The fact that falling bodies possess identical accelerations leads, in conjunction with Eq. (3.2), to the proportionality of the inertial mass of a body to its gravitational mass. The results of very precise experiments by Eotvös show that inertial mass is, in fact, equal to gravitational mass. In the future the word mass will be used without either qualifying adjective.

(a) The Rules of Transformation

If a mass in motion possesses the position vectors $\mathbf{r}_1$ and $\mathbf{r}_2$ relative to the origins of two inertial frames S_1 and S_2, respectively, and if further S_1 and S_2 have a relative velocity $\mathbf{v}$, then it follows from Eq. (3.2) that

$$\begin{aligned} \mathbf{r}_1 &= \mathbf{r}_2 + \mathbf{V}t_2 + \text{const} \\ t_1 &= t_2 + \text{const} \end{aligned} \tag{3.3}$$

in which t_1 and t_2 are the times measured in S_1 and S_2. The transformation rules (3.3), in which the constants depend merely upon the choice of origin, are called *Galilean transformations.* It is clear that acceleration is an invariant under such transformations.

The rules of transformation between an inertial frame and a noninertial frame are considerably more complicated than Eq. (3.3). Their derivation is facilitated by the application of the following theorem: A frame S_1 possesses relative to a frame S an angular velocity $\boldsymbol{\omega}$ passing through the common origin of the two frames. The time rate of change of any vector $\mathbf{A}$ as measured in S is related to that measured in S_1 by the formula

$$(d\mathbf{A}/dt)_S = (d\mathbf{A}/dt)_{S_1} + \boldsymbol{\omega} \times \mathbf{A} \tag{3.4}$$

The interpretation of Eq. (3.4) is clear. The first term on the right-hand side accounts for the change in the magnitude of $\mathbf{A}$ while the second corresponds to its change in direction.

If S is an inertial frame and S_1 is a frame rotating relative to it, as explained in the statement of the theorem, S_1 being therefore noninertial, the substitution of the position vector $\mathbf{r}$ for $\mathbf{A}$ in Eq. (3.4) produces the result

$$\mathbf{v}_{\text{abs}} = \mathbf{v}_{\text{rel}} + \boldsymbol{\omega} \times \mathbf{r} \tag{3.5}$$

In Eq. (3.5) $\mathbf{v}_{\text{abs}}$ represents the velocity measured relative to S, $\mathbf{v}_{\text{rel}}$ the velocity relative to S_1, and $\boldsymbol{\omega} \times \mathbf{r}$ is the transport velocity of a point rigidly attached to S_1. The law of transformation of acceleration is found on a second application of Eq. (3.4) in which $\mathbf{A}$ is replaced by $\mathbf{v}_{\text{abs}}$. The result of this substitution leads directly to

$$(d^2\mathbf{r}/dt^2)_S = (d^2\mathbf{r}/dt^2)_{S_1} + \boldsymbol{\omega} \times (\boldsymbol{\omega} \times \mathbf{r}) + \dot{\boldsymbol{\omega}} \times \mathbf{r} + 2\boldsymbol{\omega} \times \mathbf{v}_{\text{rel}} \tag{3.6}$$

in which $\dot{\boldsymbol{\omega}}$ is the time derivative, in either frame, of $\boldsymbol{\omega}$. The physical interpretation of Eq. (3.6) can be shown in the form

$$\mathbf{a}_{\text{abs}} = \mathbf{a}_{\text{rel}} + \mathbf{a}_{\text{trans}} + \mathbf{a}_{\text{cor}} \tag{3.7}$$

where $\mathbf{a}_{\text{cor}}$ represents the Coriolis acceleration $2\boldsymbol{\omega} \times \mathbf{v}_{\text{rel}}$. The results, Eqs. (3.5) and (3.7), constitute the rules of transformation between an inertial and a noninertial frame. Equation (3.7) shows in addition that in a noninertial frame the second law of motion takes the form

$$m\mathbf{a}_{\text{rel}} = \mathbf{F}_{\text{abs}} - m\mathbf{a}_{\text{cor}} - m\mathbf{a}_{\text{trans}} \tag{3.8}$$

The modifications required in the above formulas for the case in which S_1 is translating as well as rotating relative to S are easily made. For, if $\mathbf{D}(t)$ is the position vector of the origin of the S_1 frame relative to that of S, Eq. (3.5) is replaced by

$$\mathbf{V}_{\text{abs}} = (d\mathbf{D}/dt)_S + \mathbf{v}_{\text{rel}} + \boldsymbol{\omega} \times \mathbf{r}$$

and consequently, Eq. (3.7) by

$$\mathbf{a}_{\text{abs}} = (d^2\mathbf{D}/dt^2)_S + \mathbf{a}_{\text{rel}} + \mathbf{a}_{\text{trans}} + \mathbf{a}_{\text{cor}}$$

In practice the decision as to what constitutes an inertial frame of reference depends upon the accuracy sought in the contemplated analysis. In many cases a set of axes rigidly attached to the earth's surface is sufficient even though such a frame is noninertial to the extent of its taking part in the daily rotation of the earth about its axis and also its yearly rotation about the sun. When more precise results are required, a set of axes fixed at the center of the earth may be used. Such a set of axes is subject only to the orbital motion of the earth. In still more demanding circumstances an inertial frame is taken to be one whose orientation relative to the fixed stars is constant.

3.2. THE DYNAMICS OF A SYSTEM OF MASSES

The problem of locating a system in space involves the determination of a certain number of variables as functions of time. This basic number, which cannot be reduced without the imposition of constraints, is characteristic of the system and is known as its number of degrees of freedom. A point mass free to move in space has three degrees of freedom. The system of two point masses free to move in space but subject to the constraint that the distance between them remains constant possesses five degrees of freedom. It is clear that the presence of constraints reduces the number of degrees of freedom of a system.

Three possibilities arise in the analysis of the motion of mass systems. Firstly the system may consist of a small number of masses and hence its number of degrees of freedom is small. Secondly, there may be a very large number of masses in the system but the constraints which are imposed on it reduce the degrees of freedom to a small number; this happens in the case of a rigid body. Finally, it may be that the constraints acting on a system which contains a large number of masses do not provide an appreciable reduction in the number of degrees of freedom. This third case is treated in statistical mechanics, the degrees of freedom being reduced by statistical methods.

In the following paragraphs the fundamental results relating to the dynamics of mass systems are derived. The system is assumed to consist of n constant masses m_i $(i = 1, 2, \ldots, n)$. The position vector of m_i, relative to the origin O of an inertial frame is denoted by $\mathbf{r}_i$. The force acting on m_i is represented in the form

$$\mathbf{F}_i = \mathbf{F}_i^e + \sum_{j=1}^{n} \mathbf{F}_{ij} \tag{3.9}$$

in which $\mathbf{F}_i^e$ is the external force acting on m_i, $\mathbf{F}_{ij}$ is the force exerted on m_i by m_j, and $\mathbf{F}_{ii}$ is zero.

(a) The Motion of the Center of Mass

The motion of m_i relative to the inertial frame is determined from the equation

$$\mathbf{F}_i^e + \sum_{j=1}^{n} \mathbf{F}_{ij} = m_i \, d\mathbf{v}_i/dt \tag{3.10}$$

On summing the n equations of this type one finds

$$\mathbf{F}^e + \sum_{i=1}^{n} \sum_{j=1}^{n} \mathbf{F}_{ij} = \sum_{1}^{n} m_i \, d\mathbf{v}_i/dt \tag{3.11}$$

where $\mathbf{F}^e$ is the resultant of all the external forces which act on the system. But Newton's third law states that

$$\mathbf{F}_{ij} = -\mathbf{F}_{ji}$$

and hence the double sum in Eq. (3.11) vanishes. Further, the position vector $\mathbf{r}_c$ of the center of mass of the system relative to O is defined by the relation

$$\mathfrak{M}\mathbf{r}_c = \sum_{1}^{n} m_i \mathbf{r}_i \tag{3.12}$$

in which $\mathfrak{M}$ denotes the total mass of the system. It follows from Eq. (3.12) that

$$\mathfrak{M}\mathbf{v}_c = \sum_1^n m_i\mathbf{v}_i \tag{3.13}$$

and therefore from Eq. (3.11) that

$$\mathbf{F}^e = \mathfrak{M}\, d^2\mathbf{r}_c/dt^2 \tag{3.14}$$

which proves the theorem:

The center of mass moves as if the entire mass of the system were concentrated there and the resultant of the external forces acted there.

Two first integrals of Eq. (3.14) provide useful results [Eqs. (3.15) and (3.16)]:

$$\int_{t_1}^{t_2} \mathbf{F}^e\, dt = \mathfrak{M}\mathbf{v}_c(t_2) - \mathfrak{M}\mathbf{v}_c(t_1) \tag{3.15}$$

The integral on the left-hand side is called the *impulse of the external force.* Equation (3.15) shows that the change in linear momentum of the center of mass is equal to the impulse of the external force. This leads to the conservation of linear momentum theorem:

The linear momentum of the center of mass is constant if no resultant external force acts on the system or, in view of Eq. (3.13), *the total linear momentum of the system is constant if no resultant external force acts.*

$$\int_1^2 \mathbf{F}^e \cdot d\mathbf{r}_c = \tfrac{1}{2}\mathfrak{M}\mathbf{v}_c^2\Big]_1^2 \tag{3.16}$$

which constitutes the work-energy theorem:

The work done by the resultant external force acting at the center of mass is equal to the change in the kinetic energy of the center of mass.

In certain cases the external force $\mathbf{F}_i^e$ may be the gradient of a scalar quantity V which is a function of position only. Then

$$\mathbf{F}^e = -\partial V/\partial \mathbf{r}_c$$

and Eq. (3.16) takes the form

$$[\tfrac{1}{2}\mathfrak{M}\mathbf{v}_c^2 + V]_1^2 = 0 \tag{3.17}$$

If such a function V exists the force field is said to be conservative and Eq. (3.17) provides the conservation of energy theorem.

(b) The Kinetic Energy of a System

The total kinetic energy of a system is the sum of the kinetic energies of the individual masses. However, it is possible to cast this sum into a form which frequently makes the calculation of the kinetic energy less difficult. The total kinetic energy of the masses in their motion relative to O is

$$T = \tfrac{1}{2}\sum_{i=1}^n m_i\mathbf{v}_i^2$$

but

$$\mathbf{r}_i = \mathbf{r}_c + \boldsymbol{\delta}_i$$

where $\boldsymbol{\sigma}_i$ is the position vector of m_i relative to the system center of mass C (see Fig. 3.1).

Hence

$$T = \tfrac{1}{2}\sum_{i=1}^{n} m_i\dot{\mathbf{r}}_c^2 + \sum_{1}^{n} m_i\dot{\mathbf{r}}_c\cdot\dot{\boldsymbol{\sigma}}_i + \tfrac{1}{2}\sum_{i=1}^{n} m_i\dot{\boldsymbol{\sigma}}_i^2$$

but

$$\sum_{i=1}^{n} m_i\boldsymbol{\sigma}_i = 0$$

by definition, and so

$$T = \tfrac{1}{2}\mathfrak{M}\dot{\mathbf{r}}_c^2 + \tfrac{1}{2}\sum_{i=1}^{n} m_i\dot{\boldsymbol{\sigma}}_i^2 \tag{3.18}$$

which proves the theorem:

The total kinetic energy of a system is equal to the kinetic energy of the center of mass plus the kinetic energy of the motion relative to the center of mass.

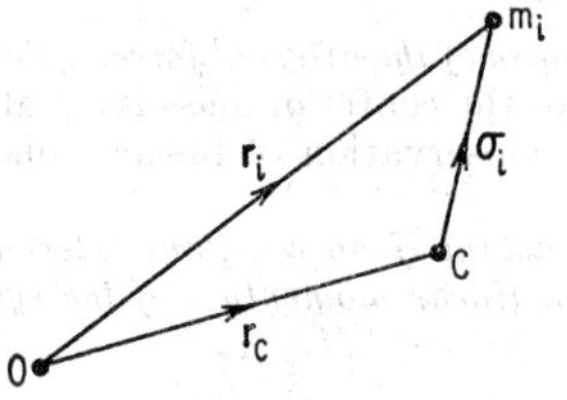

FIG. 3.1

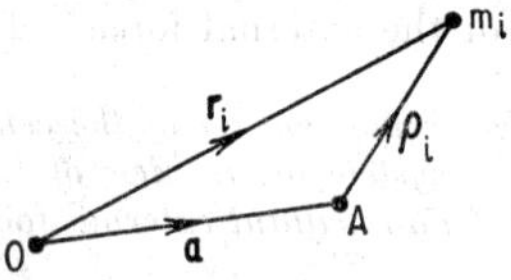

FIG. 3.2

(c) Angular Momentum of a System (Moment of Momentum)

Each mass m_i of the system has associated with it a linear momentum vector $m_i\mathbf{v}_i$. The moment of this momentum about the point O is $\mathbf{r}_i \times m_i\mathbf{v}_i$. The moment of momentum of the motion of the system relative to O, about O, is

$$\mathbf{H}(O) = \sum_{i=1}^{n} \mathbf{r}_i \times m_i\mathbf{v}_i$$

It follows that

$$(d/dt)\mathbf{H}(O) = \sum_{i=i}^{n} \mathbf{r}_i \times m_i(d^2\mathbf{r}_i/dt^2)$$

which, by Eq. (3.10), is equivalent to

$$(d/dt)\mathbf{H}(O) = \sum_{i=i}^{n} \mathbf{r}_i \times \mathbf{F}_i^e + \sum_{i=i}^{n} \mathbf{r}_i \times \sum_{j=1}^{n} \mathbf{F}_{ij} \tag{3.19}$$

It is now assumed that, in addition to the validity of Newton's third law, the force $\mathbf{F}_{ij}$ is collinear with $\mathbf{F}_{ji}$ and acts along the line joining m_i to m_j; i.e., the internal forces are central forces. Consequently, the double sum in Eq. (3.19) vanishes and

$$(d/dt)\mathbf{H}(O) = \sum_{i=i}^{n} \mathbf{r}_i \times \mathbf{F}_i^e = \mathbf{M}(O) \tag{3.20}$$

where $\mathbf{M}(O)$ represents the moment of the external forces about the point O. The following extension of this result to certain noninertial points is useful.

Let A be an arbitrary point with position vector $\mathbf{a}$ relative to the inertial point O (see Fig. 3.2). If $\boldsymbol{\rho}_i$ is the position vector of m_i relative to A, then in the notation already developed

$$\mathbf{H}(A) = \sum_{i=i}^{n} \boldsymbol{\varrho}_i \times m_i(d\mathbf{r}_i/dt)$$

$$= \sum_{i=i}^{n} (\mathbf{r}_i - \mathbf{a}) \times m_i(d\mathbf{r}_i/dt)$$

$$= \mathbf{H}(O) - \mathbf{a} \times \mathfrak{M}\mathbf{v}_c$$

Thus $$(d/dt)\mathbf{H}(A) = (d/dt)\mathbf{H}(O) - \dot{\mathbf{a}} \times \mathfrak{M}\mathbf{v}_c - \mathbf{a} \times \mathfrak{M}(d\mathbf{v}_c/dt)$$

which reduces on application of Eqs. (3.14) and (3.20) to

$$(d/dt)\mathbf{H}(A) = \mathbf{M}(A) - \dot{\mathbf{a}} \times \mathfrak{M}\mathbf{v}_c$$

The validity of the result

$$(d/dt)\mathbf{H}(A) = \mathbf{M}(A) \tag{3.21}$$

is assured if the point A satisfies either of the conditions

1. $\dot{\mathbf{a}} = O$; i.e., the point A is fixed relative to O.
2. $\dot{\mathbf{a}}$ is parallel to $\mathbf{v}_c$; i.e., the point A is moving parallel to the center of mass of the system.

A particular, and very useful, case of condition 2 is that in which the point A is the center of mass. The preceding results [Eqs. (3.20) and (3.21)] are contained in the theorem:

The time rate of change of the moment of momentum about a point is equal to the moment of the external forces about that point if the point is inertial, is moving parallel to the center of mass, or is the center of mass.

As a corollary to the foregoing one can state that the moment of momentum of a system about a point satisfying the conditions of the theorem is conserved if the moment of the external forces about that point is zero.

The moment of momentum about an arbitrary point A of the motion relative to A is

$$\mathbf{H}_{\text{rel}}(A) = \sum_{i=1}^{n} \boldsymbol{\varrho}_i \times m_i(d\boldsymbol{\varrho}_i/dt)$$

$$= \sum_{i=i}^{n} \boldsymbol{\varrho}_i \times m_i(\dot{\mathbf{r}}_i - \dot{\mathbf{a}})$$

$$= \mathbf{H}(A) + \dot{\mathbf{a}} \times \sum_{i=i}^{n} m_i\boldsymbol{\varrho}_i \tag{3.22}$$

If the point A is the center of mass C of the system Eq. (3.22) reduces to

$$\mathbf{H}_{\text{rel}}(C) = \mathbf{H}(C) \tag{3.23}$$

which frequently simplifies the calculation of $\mathbf{H}(C)$.

Further general theorems of the type derived above are available in the literature. The present discussion is limited to the more commonly applicable results.

3.3. THE MOTION OF A RIGID BODY

As was mentioned earlier, a rigid body is a dynamical system which, although it can be considered to consist of a very large number of point masses, possesses a small number of degrees of freedom. The rigidity constraint reduces the degrees of freedom to six in the most general case, which is that in which the body is translating and rotating in space. This can be seen as follows: The position of a rigid body in space is determined once the positions of three noncollinear points in it are known.

These three points have nine coordinates, among which the rigidity constraint prescribes three relationships. Hence only six of the coordinates are independent. The same result can be obtained otherwise.

Rather than view the body as a system of point masses, it is convenient to consider it to have a mass density per unit volume. In this way the formulas developed in the analysis of the motion of mass systems continue to be applicable if the sums are replaced by integrals.

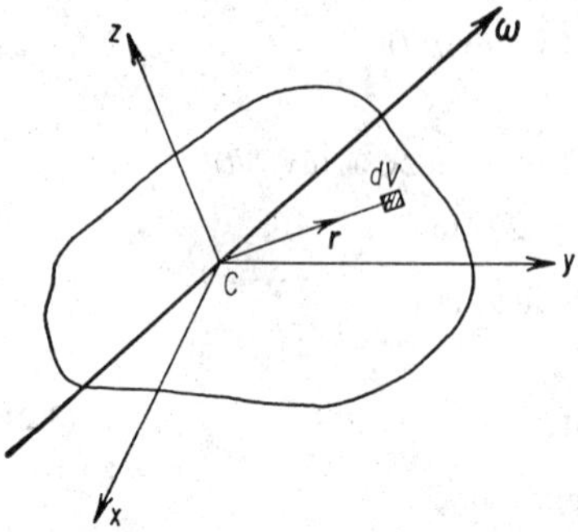

FIG. 3.3

The six degrees of freedom demand six equations of motion for the determination of six variables. Three of these equations are provided by Eq. (3.14), which describes the motion of the center of mass, and the remaining three are found from moment-of-momentum considerations, e.g., Eq. (3.21). It is assumed, therefore, in what follows that the motion of the center of mass is known, and the discussion is limited to the rotational motion of the rigid body about its center of mass C.*

Let $\boldsymbol{\omega}$ be the angular velocity of the body. Then the moment of momentum about C is, by Eq. (3.23),

$$\mathbf{H}(C) = \int_V \mathbf{r} \times (\boldsymbol{\omega} \times \mathbf{r})\rho \, dV \tag{3.24}$$

where $\mathbf{r}$ is now the position vector of the element of volume dV relative to C (see Fig. 3.3), ρ is the density of the body, and the integral is taken over the volume of the body. By a direct expansion one finds

$$\begin{aligned} \mathbf{r} \times (\boldsymbol{\omega} \times \mathbf{r}) &= \mathbf{r}^2\boldsymbol{\omega} - \mathbf{r}(\mathbf{r} \cdot \boldsymbol{\omega}) \\ &= \mathbf{r}^2\boldsymbol{\omega} - \mathbf{rr} \cdot \boldsymbol{\omega} \\ &= \mathbf{r}^2\mathbf{I} \cdot \boldsymbol{\omega} - \mathbf{rr} \cdot \boldsymbol{\omega} \\ &= (\mathbf{r}^2\mathbf{I} - \mathbf{rr}) \cdot \boldsymbol{\omega} \end{aligned}$$

and hence

$$\mathbf{H}(C) = \mathbf{I}(C) \cdot \boldsymbol{\omega} \tag{3.25}$$

where

$$\mathbf{I}(C) = \int_V \rho(\mathbf{r}^2\mathbf{I} - \mathbf{rr}) \, dV \tag{3.26}$$

is the inertia tensor of the body about C.

In Eq. (3.26), $\mathbf{I}$ denotes the identity tensor. The inertia tensor can be evaluated once the value of ρ and the shape of the body are prescribed. A short digression is made here to discuss the structure and properties of $\mathbf{I}(C)$.

For definiteness let xyz be an orthogonal set of cartesian axes with origin at C (see Fig. 3.3). Then in matrix notation

$$\mathbf{I}(C) = \begin{pmatrix} I_{xx} & -I_{xy} & -I_{xz} \\ -I_{yx} & I_{yy} & -I_{yz} \\ -I_{zx} & -I_{zy} & I_{zz} \end{pmatrix}$$

where

$$I_{xx} = \int_v \rho(y^2 + z^2) \, dV$$

$$I_{xy} = \int_v \rho xy \, dV \qquad \text{etc.}$$

It is clear that

1. The tensor is second-order symmetric with real elements.
2. The elements are the usual moments and products of inertia.
3. The moment of inertia about a line through C defined by a unit vector $\mathbf{e}$ is

$$\mathbf{e} \cdot \mathbf{I}(C) \cdot \mathbf{e}$$

* Rotational motion about any *fixed* point of the body is treated in a similar way.

4. Because of the property expressed in condition 1 it is always possible to determine at C a set of mutually perpendicular axes relative to which $\mathbf{I}(C)$ is diagonalized.

Returning to the analysis of the rotational motion one sees that the inertia tensor $\mathbf{I}(C)$ is time-dependent unless it is referred to a set of axes which rotate with the body. For simplicity the set of axes S_1 which rotates with the body is chosen to be the orthogonal set in which $\mathbf{I}(C)$ is diagonalized. A space-fixed frame of reference with origin at C is represented by S. Accordingly, from Eqs. (3.4) and (3.21),

$$[(d/dt)\mathbf{H}(C)]_S = [(d/dt)\mathbf{H}(C)]_{S_1} + \boldsymbol{\omega} \times \mathbf{H}(C) = \mathbf{M}(C) \tag{3.27}$$

which, by Eq. (3.25), reduces to

$$\mathbf{I}(C)(d\boldsymbol{\omega}/dt) + \boldsymbol{\omega} \times \mathbf{I}(C) \cdot \boldsymbol{\omega} = \mathbf{M}(C) \tag{3.28}$$

where

$$\mathbf{H}(C) = \mathbf{i}I_{xx}\omega_x + \mathbf{j}I_{yy}\omega_y + \mathbf{k}I_{zz}\omega_z \tag{3.29}$$

In Eq. (3.29) the xyz axes are those for which

$$\mathbf{I}(C) = \begin{pmatrix} I_{xx} & 0 & 0 \\ 0 & I_{yy} & 0 \\ 0 & 0 & I_{zz} \end{pmatrix}$$

and $\mathbf{i}$, $\mathbf{j}$, $\mathbf{k}$ are the conventional unit vectors. Equation (3.28) in scalar form supplies the three equations needed to determine the rotational motion of the body. These equations—the Euler equations—are

$$\begin{aligned} I_{xx}(d\omega_x/dt) + \omega_y\omega_z(I_{zz} - I_{yy}) &= M_x \\ I_{yy}(d\omega_y/dt) + \omega_z\omega_x(I_{xx} - I_{zz}) &= M_y \\ I_{zz}(d\omega_z/dt) + \omega_x\omega_y(I_{yy} - I_{xx}) &= M_z \end{aligned} \tag{3.30}$$

The analytical integration of the Euler equations in the general case defines a problem of classical difficulty. However, in special cases solutions can be found. The sources of the simplifications in these cases are the symmetry of the body and the absence of some components of the external moment. Since discussion of the various possibilities lies outside the scope of this chapter reference is made to refs. 1 and 2 and for a survey of the recent work to ref. 3. Of course, in situations in which energy or moment of momentum, or perhaps both, are conserved first integrals of the motion can be written down without employing the Euler equations. To do so it is convenient to have an expression for the kinetic energy T of the rotating body. This expression is readily found in the following manner.

The kinetic energy is

$$\begin{aligned} T &= \tfrac{1}{2} \int_V \rho(\boldsymbol{\omega} \times \mathbf{r})^2 \, dV \\ &= \tfrac{1}{2} \int_V \rho\boldsymbol{\omega} \cdot [\mathbf{r} \times (\boldsymbol{\omega} \times \mathbf{r})] \, dV \end{aligned}$$

which by Eqs. (3.24), (3.25), and (3.26) is

$$T = \tfrac{1}{2}\boldsymbol{\omega} \cdot \mathbf{I}(C) \cdot \boldsymbol{\omega} \tag{3.31}$$

or, in matrix notation,

$$2T = (\omega_x\omega_y\omega_z) \begin{pmatrix} I_{xx} & 0 & 0 \\ 0 & I_{yy} & 0 \\ 0 & 0 & I_{zz} \end{pmatrix} \begin{pmatrix} \omega_x \\ \omega_y \\ \omega_z \end{pmatrix}$$

Equation (3.31) can be put in a simpler form by writing

$$T = \tfrac{1}{2}\omega^2(\boldsymbol{\omega}/\omega) \cdot \mathbf{I}(C) \cdot (\boldsymbol{\omega}/\omega)$$

and hence

$$T = \tfrac{1}{2}I_{\omega\omega}\omega^2 \tag{3.32}$$

In Eq. (3.32) $I_{\omega\omega}$ is the moment of inertia of the body about the axis of the angular velocity vector $\boldsymbol{\omega}$.

3.4. ANALYTICAL DYNAMICS

The knowledge of the time dependence of the position vectors $\mathbf{r}_i(t)$ which locate an n-mass system relative to a frame of reference can be attained indirectly by determining the dependence upon time of some parameters q_j $(j = 1 \cdots m)$ if the functional relationships

$$\mathbf{r}_i = \mathbf{r}_i(q_j,t) \qquad i = 1 \cdots n;\ j = 1 \cdots m \tag{3.33}$$

are known. The parameters q_j which completely determine the position of the system in space are called *generalized coordinates*. Any m quantities can be used as generalized coordinates on condition that they uniquely specify the positions of the masses. Frequently the q_j are the coordinates of an appropriate curvilinear system.

It is convenient to define two types of mechanical systems:

1. A holonomic system is one for which the generalized coordinates and the time may be arbitrarily and independently varied without violating the constraints.
2. A nonholonomic system is such that the generalized coordinates and the time may not be arbitrarily and independently varied because of some (say s) nonintegrable constraints of the form

$$\sum_{i=1}^{m} A_{ji}\,dq_i + A_j\,dt = 0 \qquad j = 1, 2, \ldots s \tag{3.34}$$

In the constraint equations (3.37) the A_{ji} and A_j represent functions of the q_k and t. Holonomic and nonholonomic systems are further classified as rheonomic or scleronomic depending upon whether the time t is explicitly present or absent, respectively, in the constraint equations.

(a) Generalized Forces and d'Alembert's Principle

A virtual displacement of the system is denoted by the set of vectors $\delta\mathbf{r}_i$. The work done by the forces in this displacement is

$$\delta W = \sum_{i=1}^{n} \mathbf{F}_i \cdot \delta\mathbf{r}_i \tag{3.35}$$

If the force $\mathbf{F}_i$, acting on the mass m_i, is separable in the sense that

$$\mathbf{F}_i = \mathbf{F}_i{}^a + \mathbf{F}_i{}^c \tag{3.36}$$

in which the first term is the applied force and the second the force of constraint, then

$$\delta W = \sum_{i=1}^{n} (\mathbf{F}_i{}^a + \mathbf{F}_i{}^c) \cdot \left[\sum_{j=1}^{m} (\partial\mathbf{r}_i/\partial q_j)\,\delta q_j + (\partial\mathbf{r}_i/\partial t)\,\delta t \right] \tag{3.37}$$

The generalized applied forces and the generalized forces of constraint are defined by

$$Q_j{}^a = \sum_{i=1}^{n} \mathbf{F}_i{}^a \cdot \partial\mathbf{r}_i/\partial q_j \tag{3.38}$$

and

$$Q_j{}^c = \sum_{i=1}^{n} \mathbf{F}_i{}^c \cdot \partial\mathbf{r}_i/\partial q_j \tag{3.39}$$

respectively. Hence, Eq. (3.37) assumes the form

$$\delta W = \sum_{j=1}^{m} Q_j{}^a\,\delta q_j + \sum_{j=1}^{m} Q_j{}^c\,\delta q_j + \sum_{i=1}^{n} (\mathbf{F}_i{}^a + \mathbf{F}_i{}^c) \cdot (\partial\mathbf{r}_i/\partial t)\,\delta t \tag{3.40}$$

If the virtual displacement is compatible with the instantaneous constraints $\delta t = 0$, and if in such a displacement the forces of constraint do no work, e.g., sliding friction is absent, then

$$\delta W = \sum_{j=1}^{m} Q_j{}^a \, \delta q_j \tag{3.41}$$

The assumption that a function $V(q_j,t)$ exists such that

$$Q_j{}^a = -\partial V/\partial q_j$$

leads to the result

$$\delta W = -\delta V \tag{3.42}$$

In Eq. (3.42) $V(q_j,t)$ is called the potential or work function.

The first step in the introduction of the kinetic energy of the system is taken by using d'Alembert's principle. The equations of motion (3.10) can be written

$$\mathbf{F}_i - m_i\ddot{\mathbf{r}}_i = 0$$

and consequently

$$\sum_{i=1}^{n} (\mathbf{F}_i - m_i\ddot{\mathbf{r}}_i) \cdot \delta\mathbf{r}_i = 0 \tag{3.43}$$

The principle embodied in Eq. (3.43) constitutes the extension of the principle of virtual work to dynamical systems and is named after d'Alembert. When attention is confined to $\delta\mathbf{r}_i$ which represent virtual displacements compatible with the instantaneous constraints and to forces $\mathbf{F}_i$ which satisfy Eqs. (3.36) and (3.41), the principle states that

$$\sum_{j=1}^{m} Q_j{}^a \, \delta q_j = \sum_{i=1}^{n} m_i\ddot{\mathbf{r}}_i \cdot \delta\mathbf{r}_i \tag{3.44}$$

(b) The Lagrange Equations

The central equations of analytical mechanics can now be derived. These equations, which were developed by Lagrange, are presented here for the general case of a rheonomic nonholonomic system consisting of n masses m_i, m generalized coordinates q_j, and s constraint equations

$$\sum_{j=1}^{m} A_{kj} \, dq_j + A_k \, dt = 0 \qquad k = 1, 2, \ldots s \tag{3.45}$$

The equations are found by writing the acceleration terms in d'Alembert's principle (3.43) in terms of the kinetic energy T and the generalized coordinates. By definition

$$T = \tfrac{1}{2} \sum_{1}^{n} m_i\dot{\mathbf{r}}_i{}^2$$

where
$$\dot{\mathbf{r}}_i = \sum_{j=1}^{m} (\partial\mathbf{r}_i/\partial q_j)(dq_j/dt) + \frac{\partial\mathbf{r}_i}{dt} \qquad i = 1, 2, \ldots n$$

Thus
$$\partial\dot{\mathbf{r}}_i/\partial\dot{q}_j = \partial\mathbf{r}_i/\partial q_j$$
$$\partial\dot{\mathbf{r}}_i/\partial q_j = (d/dt)(\partial\mathbf{r}_i/\partial q_j)$$
$$\partial T/\partial q_j = \sum_{i=1}^{n} m_i\dot{\mathbf{r}}_i \cdot (d/dt)(\partial\mathbf{r}_i/\partial q_j)$$

and
$$\partial T/\partial\dot{q}_j = \sum_{i=1}^{n} m_i\dot{\mathbf{r}}_i \cdot (\partial\mathbf{r}_i/\partial q_j)$$

Accordingly,

$$(d/dt)(\partial T/\partial \dot{q}_j) - \partial T/\partial q_j = \sum_{i=1}^{n} m_i \ddot{\mathbf{r}}_i \cdot (\partial \mathbf{r}_i/\partial q_j) \qquad j = 1, 2, \ldots m \tag{3.46}$$

and by summing over all values of j one finds

$$\sum_{j=1}^{m} [(d/dt)(\partial T/\partial \dot{q}_j) - \partial T/\partial q_j]\,\delta q_j = \sum_{i=1}^{n} m_i \ddot{\mathbf{r}}_i \cdot \delta \mathbf{r}_i \tag{3.47}$$

because

$$\delta \mathbf{r}_i = \sum_{j=1}^{m} (\partial \mathbf{r}_i/\partial q_j)\,\delta q_j$$

for instantaneous displacements. From Eqs. (3.44) and (3.47) it follows that

$$\sum_{j=1}^{m} [(d/dt)(\partial T/\partial \dot{q}_j) - \partial T/\partial q_j - Q_j^a]\,\delta q_j = 0 \tag{3.48}$$

The δq_j which appear in (3.48) are not independent but must satisfy the instantaneous constraint equations

$$\sum_{j=1}^{m} A_{kj}\,\delta q_j = 0 \qquad k = 1, 2, \ldots s \tag{3.49}$$

The "elimination" of s of the δq_j between (3.48) and (3.49) is effected, in the usual way, by the introduction of s Lagrange multipliers λ_k ($k = 1, 2, \ldots, s$). This step leads directly to the equations

$$(d/dt)(\partial T/\partial \dot{q}_j) - \partial T/\partial q_j = Q_j^a - \sum_{k=1}^{s} \lambda_k A_{kj} \qquad j = 1, 2, \ldots m \tag{3.50}$$

These m second-order ordinary differential equations are the Lagrange equations of the system. The general solution of the equations is not available.* For a holonomic system with n degrees of freedom Eq. (3.50) reduces to

$$(d/dt)(\partial T/\partial \dot{q}_j) - \partial T/\partial q_j = Q_j^a \qquad j = 1, \cdots n \tag{3.51}$$

In the presence of a function V such that

$$Q_j^a = -\partial V/\partial q_j$$

and

$$\partial V/\partial \dot{q}_j = 0$$

Eqs. (3.51) can be written in the form

$$(d/dt)(\partial \mathcal{L}/\partial \dot{q}_j) - \partial \mathcal{L}/\partial q_j = 0 \qquad j = 1, 2, \cdots n \tag{3.52}$$

in which

$$\mathcal{L} = T - V$$

The scalar function $\mathcal{L}$—the Lagrangian—which is the difference between the kinetic and potential energies is all that need be known to write the Lagrange equations in this case.

The major factor which contributes to the solving of Eqs. (3.52) is the presence of ignorable coordinates. In fact, in dynamics problems generally the possibility of finding analytical representations of the motion depends crucially on there being ignorable coordinates. A coordinate, say q_l, is said to be ignorable if it does not appear explicitly in the Lagrangian, i.e., if

$$\partial \mathcal{L}/\partial q_l = 0 \tag{3.53}$$

* Nonholonomic problems are frequently more tractable by vectorial than by Lagrangian methods.[6]

If Eq. (3.53) is valid then Eq. (3.52) leads to

$$\partial \mathfrak{L}/\partial \dot{q}_l = \text{const} = c_l$$

and hence a first integral of the motion is available. Clearly the more ignorable coordinates that exist in the Lagrangian, the better. This being so, considerable effort has been directed toward developing systematic means of generating ignorable coordinates by transforming from one set of generalized coordinates to another more suitable set. This transformation theory of dynamics, while extensively developed, is not generally of practical value in engineering problems.

(c) The Euler Angles

To use Lagrangian methods in analyzing the motion of a rigid body one must choose a set of generalized coordinates which uniquely determines the position of the body relative to a frame of reference fixed in space. It suffices to examine the motion of a body rotating about its center of mass.

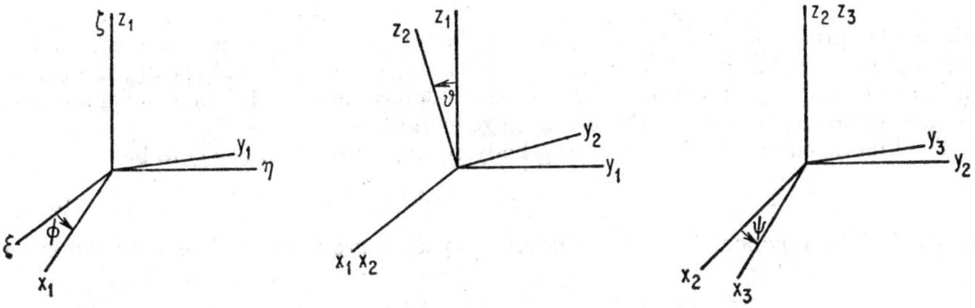

FIG. 3.4

An inertial set of orthogonal axes $\xi\eta\zeta$ with origin at the center of mass and a noninertial set xyz fixed relative to the body with the same origin are adopted. The required generalized coordinates are those which specify the position of the xyz axes relative to the $\xi\eta\zeta$ axes. More than one set of coordinates which achieves this purpose can be found. The most generally useful one, viz., the Euler angles, is used here.

The frame $\xi\eta\zeta$ can be brought into coincidence with the frame xyz by three finite rigid-body rotations through angles ϕ, ϑ, and ψ,* in that order, defined as follows (see Fig. 3.4):

1. A rotation about the ζ axis through an angle ϕ to produce the frame x_1, y_1, z_1
2. A rotation about the x_1 axis through an angle ϑ to produce the frame x_2, y_2, z_2
3. A rotation about the z_2 axis through an angle ψ to produce the frame x_3, y_3, z_3

which coincides with the frame x, y, z.

Each rotation can be represented by an orthogonal matrix operation so that the process of getting from the inertial to the noninertial frame is

$$\begin{pmatrix} x_1 \\ y_1 \\ z_1 \end{pmatrix} = \begin{pmatrix} \cos\phi & \sin\phi & 0 \\ -\sin\phi & \cos\phi & 0 \\ 0 & 0 & 1 \end{pmatrix} \begin{pmatrix} \xi \\ \eta \\ \zeta \end{pmatrix} = A \begin{pmatrix} \xi \\ \eta \\ \zeta \end{pmatrix} \qquad (a)$$

$$\begin{pmatrix} x_2 \\ y_2 \\ z_2 \end{pmatrix} = \begin{pmatrix} 1 & 0 & 0 \\ 0 & \cos\vartheta & \sin\vartheta \\ 0 & -\sin\vartheta & \cos\vartheta \end{pmatrix} \begin{pmatrix} x_1 \\ y_1 \\ z_1 \end{pmatrix} = B \begin{pmatrix} x_1 \\ y_1 \\ z_1 \end{pmatrix} \qquad (b)$$

$$\begin{pmatrix} x_3 \\ y_3 \\ z_3 \end{pmatrix} = \begin{pmatrix} \cos\psi & \sin\psi & 0 \\ -\sin\psi & \cos\psi & 0 \\ 0 & 0 & 1 \end{pmatrix} \begin{pmatrix} x_2 \\ y_2 \\ z_2 \end{pmatrix} = C \begin{pmatrix} x_2 \\ y_2 \\ z_2 \end{pmatrix} \qquad (c)$$

* This notation is not universally adopted. See ref. 5 for discussion.

Consequently

$$\begin{pmatrix} x \\ y \\ z \end{pmatrix} = CBA \begin{pmatrix} \xi \\ \eta \\ \zeta \end{pmatrix} = D \begin{pmatrix} \xi \\ \eta \\ \zeta \end{pmatrix} \tag{3.54}$$

where

$$D = CBA = \begin{pmatrix} \cos\psi\cos\phi - \cos\vartheta\sin\phi\sin\psi & \cos\psi\sin\phi + \cos\vartheta\cos\phi\sin\psi & \sin\psi\sin\vartheta \\ -\sin\psi\cos\phi - \cos\vartheta\sin\phi\cos\psi & -\sin\psi\sin\phi + \cos\vartheta\cos\phi\cos\psi & \cos\psi\sin\vartheta \\ \sin\vartheta\sin\phi & -\sin\vartheta\cos\phi & \cos\vartheta \end{pmatrix} \tag{3.55}$$

Since A, B, and C are orthogonal matrices, it follows from Eq. (3.54) that

$$\begin{pmatrix} \xi \\ \eta \\ \zeta \end{pmatrix} = D^{-1} \begin{pmatrix} x \\ y \\ z \end{pmatrix} = D' \begin{pmatrix} x \\ y \\ z \end{pmatrix} \tag{3.56}$$

where the prime denotes the transpose of the matrix. From Eqs. (3.54) and (3.55) one sees that, if the time dependence of the three angles ϕ, ϑ, ψ is known, the orientation of the xyz axes relative to the $\xi\eta\zeta$ axes is determined. This time dependence is sought by attempting to solve the Lagrange equations.

The kinetic energy T of the rotating body is found from Eq. (3.31) to be

$$2T = I_{xx}\omega_x^2 + I_{yy}\omega_y^2 + I_{zz}\omega_z^2 \tag{3.57}$$

in which the components of the angular velocity ω are provided by the matrix equation

$$\begin{pmatrix} \omega_x \\ \omega_y \\ \omega_z \end{pmatrix} = CB \begin{pmatrix} 0 \\ 0 \\ \dot{\phi} \end{pmatrix} + C \begin{pmatrix} \dot{\vartheta} \\ 0 \\ 0 \end{pmatrix} + \begin{pmatrix} 0 \\ 0 \\ \dot{\psi} \end{pmatrix} \tag{3.58}$$

It is to be noted that if

$$I_{xx} \neq I_{yy} \neq I_{zz} \tag{3.59}$$

none of the angles is ignorable. Hence considerable difficulty is to be expected in attempting to solve the Lagrange equations if this inequality (3.59) holds. A similar inference could be made on examining Eqs. (3.30). The possibility of there being ignorable coordinates in the problem arises if the body has axial, or so-called kinetic, symmetry about (say) the z axis. Then

$$I_{xx} = I_{yy} = I$$

and, from Eq. (3.57),

$$2T = I(\dot{\phi}^2\sin^2\vartheta + \dot{\vartheta}^2) + I_{zz}(\dot{\phi}\cos\vartheta + \dot{\psi})^2 \tag{3.60}$$

The angles ϕ and ψ do not occur in Eq. (3.60). Whether or not they are ignorable depends on the potential energy $V(\phi,\vartheta,\psi)$.

(d) Small Oscillations of a System near Equilibrium

The Lagrange equations are particularly useful in examining the motion of a system near a position of equilibrium. Let the generalized coordinates $q_1, q_2, \ldots, q_n$—the explicit appearance of time being ruled out—represent the configuration of the system. It is not restrictive to assume the equilibrium position at

$$q_1 = q_2 = \cdots = q_n = 0$$

and, since motion near this position is being considered, the q_i and $\dot{q}_i$ may be taken to be small.

The potential energy can be expanded in a Taylor series about the equilibrium point in the form

$$V(q_1 \cdots q_n) = V(0) + \sum_{i=1}^{n} (\partial V/\partial q_i)_0 q_i + \tfrac{1}{2} \sum_i \sum_j (\partial^2 V/\partial q_i\, \partial q_j)_0 q_i q_j + \cdots \quad (3.61)$$

In Eq. (3.61) the first term can be neglected because it merely changes the potential energy by a constant and the second term vanishes because $\partial V/\partial q_i$ is zero at the equilibrium point. Thus, retaining only quadratic terms in q_i, one finds

$$V(q_1 \cdots q_n) = \tfrac{1}{2} \sum_i \sum_j V_{ij} q_i q_j \quad (3.62)$$

in which

$$V_{ij} = (\partial^2 V/\partial q_i\, \partial q_j)_0 = V_{ji} \quad (3.63)$$

are real constants.

The kinetic energy T of the system is representable by an analogous Taylor series

$$T(\dot{q}_i \cdots \dot{q}_n) = \tfrac{1}{2} \sum_i \sum_j T_{ij} \dot{q}_i \dot{q}_j \quad (3.64)$$

where

$$T_{ij} = T_{ji} \quad (3.65)$$

are real constants. The quadratic forms (3.62) and (3.64) in matrix notation, a prime denoting a transpose, are

$$V = \tfrac{1}{2} q' \mathcal{V} q \quad (3.66)$$

and

$$T = \tfrac{1}{2} \dot{q}' \mathcal{T} \dot{q} \quad (3.67)$$

In these expressions $\mathcal{V}$ and $\mathcal{T}$ represent the matrices with elements V_{ij} and T_{ij}, respectively, and q the column vector $(q_1 \cdots q_n)$. The form (3.67) is necessarily positive definite owing to the nature of kinetic energy. Rather than create the Lagrange equations in terms of the coordinates q_i a new set of generalized coordinates ρ_i is introduced in terms of which the energies are simultaneously expressible as quadratic forms without product terms. That the transformation to such coordinates is possible can be seen by considering the equations

$$\mathcal{V} b_j = \lambda_j \mathcal{T} b_j \qquad j = 1, 2, \ldots n \quad (3.68)$$

in which λ_j, the roots of the equation

$$|\mathcal{V} - \lambda \mathcal{T}| = 0$$

are the eigenvalues—assumed distinct—and b_j are the corresponding eigenvectors. The matrix of eigenvectors b_j is symbolized by B and the diagonal matrix of eigenvalues λ_j by Λ. One can write

$$b_k' \mathcal{V} b_j = \lambda_j b_k' \mathcal{T} b_j$$

and

$$b_k' \mathcal{V} b_j = \lambda_k b_k' \mathcal{T} b_j$$

because of the symmetry of $\mathcal{V}$ and $\mathcal{T}$. Thus, if $\lambda_j \neq \lambda_k$, it follows that

$$b_k' \mathcal{T} b_j = 0 \qquad k \neq j$$

and, since the eigenvectors of (3.68) are each undetermined to within an arbitrary multiplying constant, one can always normalize the vectors so that

$$b_i' \mathcal{T} b_i = 1$$

Hence

$$B' \mathcal{T} B = I \quad (3.69)$$

where I is the unit matrix. But

$$\mathcal{V} B = \mathcal{T} B \Lambda \quad (3.70)$$

and so

$$B' \mathcal{V} B = B' \mathcal{T} B \Lambda = \Lambda \quad (3.71)$$

Furthermore, denoting the complex conjugate by a superscript bar, one has

$$\mathcal{V}\bar{b}_j = \lambda_j \mathcal{T}\bar{b}_j$$

and

$$b_j'\mathcal{V}\bar{b}_j = \lambda_j b_j'\mathcal{T}\bar{b}_j \tag{3.72}$$

since $\mathcal{V}$ and $\mathcal{T}$ are real. However,

$$b_j'\mathcal{V}\bar{b}_j = \lambda_j b_j'\mathcal{T}\bar{b}_j \tag{3.73}$$

because $\mathcal{V}$ and $\mathcal{T}$ are symmetric. From (3.72) and (3.73) it follows that

$$(\lambda_j - \lambda_j)b_j'\mathcal{T}\bar{b}_j = 0 \tag{3.74}$$

The symmetry and positive definiteness of $\mathcal{T}$ ensure that the form $b_j'\mathcal{T}\bar{b}_j$ is real and positive definite. Consequently the eigenvalues λ_j, and eigenvectors b_j, are real. Finally one can solve (3.68) for the eigenvalues in the form

$$\lambda_j = b_j'\mathcal{V}b_j/b_j'\mathcal{T}b_j \tag{3.75}$$

The transformation from the q_i to the p_i coordinates can now be made by writing

$$q = Bp$$

from which

$$V = \tfrac{1}{2}q'\mathcal{V}q = \tfrac{1}{2}p'B'\mathcal{V}Bp = \tfrac{1}{2}p'\Lambda p \tag{3.76}$$

and

$$T = \tfrac{1}{2}\dot{q}'\mathcal{T}\dot{q} = \tfrac{1}{2}\dot{p}'B'\mathcal{T}B\dot{p} = \tfrac{1}{2}\dot{p}'I\dot{p} \tag{3.77}$$

It is seen from Eqs. (3.76) and (3.77) that V and T have the desired forms and that the corresponding Lagrange equations (3.52) are

$$d^2p_i/dt^2 + \omega_i^2 p_i = 0 \qquad i = 1 \cdots n \tag{3.78}$$

where $\omega_i^2 = \lambda_i$. If the equilibrium position about which the motion takes place is stable the ω_i^2 are positive. The eigenvalues λ_i must then be positive and Eq. (3.75) shows that $\mathcal{V}$ is positive definite. In other words, the potential energy is a minimum at a position of stable equilibrium. In this case the motion of the system can be analyzed in terms of its normal modes—the n harmonic oscillators (3.78). If the matrix $\mathcal{V}$ is not positive definite Eq. (3.75) indicates that negative eigenvalues may exist and hence Eqs. (3.78) may have hyperbolic solutions. The equilibrium is then unstable. Regardless of the nature of the equilibrium the Lagrange equations (3.78) can always be arrived at because it is possible to diagonalize simultaneously two quadratic forms one of which (the kinetic-energy matrix) is positive definite.

(e) Hamilton's Principle

In conclusion it is remarked that the Lagrange equations of motion can be arrived at by methods other than that presented above. The point of departure adopted here is Hamilton's principle, the statement of which for holonomic systems is:

Provided the initial (t_1) and final (t_2) configurations are prescribed, the motion of the system from time t_1 to time t_2 occurs in such a way that the line integral

$$\int_{t_1}^{t_2} \mathcal{L}\, dt = \text{extremum}$$

where $\mathcal{L} = T - V$. That the Lagrange equations (3.52) can be derived from this principle is shown here for the case of a single-mass one-degree-of-freedom system. The generalization of the proof to include an n-degree-of-freedom system is made without difficulty.

The Lagrangian is

$$\mathcal{L}(q,\dot{q},t) = T - V$$

in which q is the generalized coordinate and $q(t)$ describes the motion which actually occurs. Any other motion can be represented by

$$\bar{q}(t) = q(t) + \epsilon f(t) \tag{3.79}$$

in which $f(t)$ is an arbitrary differentiable function such that $f(t_1) = f(t_2) = 0$ and ϵ is a parameter defining the family of curves $\bar{q}(t)$. The condition

$$\int_{t_1}^{t_2} \mathfrak{L}(q_1,\dot{q}_1,t)\, dt = \text{extremum}$$

is tantamount to

$$\partial/\partial\epsilon \int_{t_1}^{t_2} \mathfrak{L}(\bar{q}_1,\dot{\bar{q}}_1,t)\, dt = 0 \qquad \epsilon = 0 \tag{3.80}$$

for all $f(t)$. But

$$\partial/\partial\epsilon \int_{t_1}^{t_2} \mathfrak{L}(\bar{q}_1,\dot{\bar{q}}_1,t)\, dt = \int_{t_1}^{t_2} [(\partial\mathfrak{L}/\partial\bar{q})(\partial\bar{q}/\partial\epsilon) + (\partial\mathfrak{L}/\partial\dot{\bar{q}})(\partial\dot{\bar{q}}/\partial\epsilon)]\, dt$$

which, by Eq. (3.79), is

$$= \int_{t_1}^{t_2} [f(t)(\partial\mathfrak{L}/\partial\bar{q}) + \dot{f}(t)(\partial\mathfrak{L}/\partial\dot{\bar{q}})]\, dt \tag{3.81}$$

Its second term having been integrated by parts, Eq. (3.81) reduces, because $f(t_1) = f(t_2) = 0$, to

$$\partial/\partial\epsilon \int_{t_1}^{t_2} \mathfrak{L}(\bar{q},\dot{\bar{q}},t)\, dt = \int_{t_1}^{t_2} f(t)[\partial\mathfrak{L}/\partial q - (d/dt)(\partial\mathfrak{L}/\partial\dot{\bar{q}})]\, dt$$

Hence Eq. (3.80) is equivalent to

$$\int_{t_1}^{t_2} f(t)[\partial\mathfrak{L}/\partial q - (d/dt)(\partial\mathfrak{L}/\partial\dot{q})]\, dt = 0 \tag{3.82}$$

for all $f(t)$. Equation (3.82) can hold for all $f(t)$ only if

$$(d/dt)(\partial\mathfrak{L}/\partial\dot{q}) - \partial\mathfrak{L}/\partial q = 0$$

which is the Lagrange equation of the system.

The extension to an n-degree-of-freedom system is made by employing n arbitrary differentiable functions $f_k(t)$, $k = 1 \cdots n$ such that $f_k(t_1) = f_k(t_2) = 0$. For the generalizations of Hamilton's principle which are necessary in treating nonholonomic systems the references should be consulted.

It is noteworthy that the application of Hamilton's principle provides no information regarding a dynamical system which is not obtainable by the direct use of Newton's laws of motion or of Lagrange's equations.

References

1. Routh, E. J.: "Advanced Dynamics of a System of Rigid Bodies," 6th ed., Dover Publications, Inc., New York, 1955.
2. Whittaker, E. T.: "A Treatise on Analytical Dynamics," 4th ed., Dover Publications, Inc., New York, 1944.
3. Leimanis, E., and N. Minorsky: "Dynamics and Nonlinear Mechanics," John Wiley & Sons, Inc., New York, 1958.
4. Corben, H. C., and P. Stehle: "Classical Mechanics," 2d ed., John Wiley & Sons, Inc., New York, 1960.
5. Goldstein, H.: "Classical Mechanics," Addison-Wesley Publishing Company, Inc., Reading, Mass., 1957.
6. Milne, E. A.: "Vectorial Mechanics," Methuen & Co., Ltd., London, 1948.
7. Scarborough, J. B.: "The Gyroscope," Interscience Publishers, Inc., New York, 1958.
8. Synge, J. L., and B. A. Griffith: "Principles of Mechanics," 3d ed., McGraw-Hill Book Company, Inc., New York, 1959.
9. Lanczos, C.: "The Variational Principles of Mechanics," University of Toronto Press, Toronto, Canada, 1949.
10. Synge, J. L.: "Classical Dynamics," Handbuch Der Physik Bd III/1, Springer Verlag, Berlin, 1960.

Section 4

KINEMATICS OF MECHANISMS

By

FERDINAND FREUDENSTEIN, Ph.D., *Professor of Mechanical Engineering, Columbia University, New York, N.Y.*

GEORGE N. SANDOR, Eng.Sc.D., *Associate Professor of Engineering and Applied Science, Yale University, New Haven, Conn.; Consulting Engineer, Huck Design Corp., New York, N. Y.*

CONTENTS

4.1 DESIGN—USE OF THE MECHANISMS SECTION

The design process involves intuition, invention, synthesis, and analysis. Although no arbitrary rules can be given, the following design procedure is suggested:

1. Define the problem in terms of inputs, outputs, their time-displacement curves, sequencing, and interlocks.

2. Select a suitable mechanism, either from experience or with the help of the several available compilations of mechanisms, mechanical movements, and components (Art. 4.8).

3. To aid systematic selection, investigate degrees of freedom, kinematic inversions, equivalent mechanisms, and if necessary, modifications of the initial selection (Arts. 4.2 and 4.6).

4. Develop a first approximation to the mechanism proportions from known design requirements, layouts, geometry, velocity and acceleration analysis, and path-curvature considerations (Arts. 4.3 and 4.4).

5. Obtain a more precise dimensional synthesis, such as outlined in Art. 4.5, possibly with the aid of computer programs, charts, diagrams, tables, and atlases (Arts. 4.5, 4.6, 4.7, and 4.9).

6. Complete the design by the methods outlined in Art. 4.6 and check end results. Note that cams and gears are treated in Secs. 5 and 32, respectively.

4.2. BASIC CONCEPTS

(a) Kinematic Elements

Mechanisms are often studied as though made up of rigid-body members, or *links*, connected to each other by rigid *kinematic elements* or *element pairs*. The nature and arrangement of the kinematic links and elements determine the kinematic properties of the mechanism.

If two mating elements are in surface contact, they are said to form a *lower pair;* element pairs with line or point contact form *higher pairs*. Three types of lower pairs

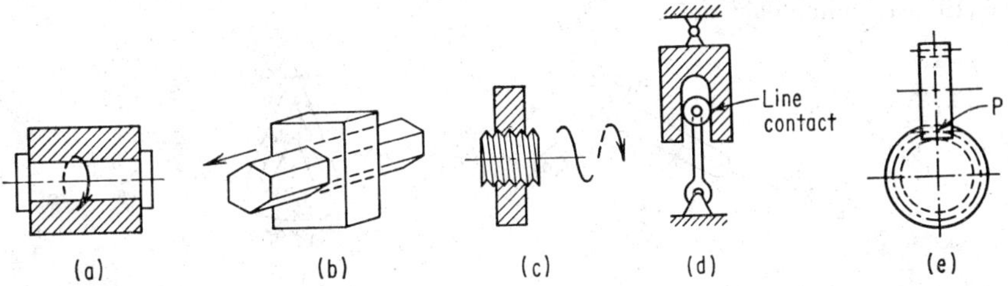

FIG. 4.1. Examples of kinematic-element pairs: lower pairs *a*, *b*, *c*, and higher pairs *d* and *e*. (*a*) Turning pair. (*b*) Sliding pair. (*c*) Screw pair. (*d*) Roller in slot. (*e*) Helical gears at right angles.

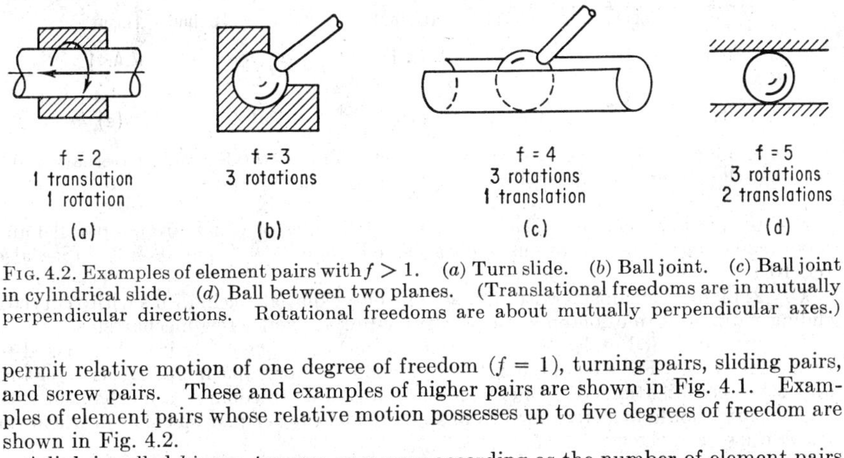

FIG. 4.2. Examples of element pairs with $f > 1$. (*a*) Turn slide. (*b*) Ball joint. (*c*) Ball joint in cylindrical slide. (*d*) Ball between two planes. (Translational freedoms are in mutually perpendicular directions. Rotational freedoms are about mutually perpendicular axes.)

permit relative motion of one degree of freedom ($f = 1$), turning pairs, sliding pairs, and screw pairs. These and examples of higher pairs are shown in Fig. 4.1. Examples of element pairs whose relative motion possesses up to five degrees of freedom are shown in Fig. 4.2.

A link is called *binary*, *ternary*, or *n-nary* according as the number of element pairs connected to it is 2, 3, or n. A ternary link, pivoted as in Fig. 4.3*a* and *b*, is often called a *rocker* or a *bell crank*, according as A is obtuse or acute.

A ternary link having three parallel turning-pair connections with coplanar axes, one of which is fixed, is called a *lever* when used to overcome a weight or resistance (Fig. 4.3*c*, *d*, and *e*). A link without fixed elements is called a *floating link*. Mechanisms consisting of a chain of rigid links (one of which, *the frame*, is considered fixed) are said to be closed by *chain closure* if all element pairs are constrained by material boundaries. All others, such as may involve springs or body forces for chain closure, are said to be closed by means of *force closure*. In the latter nonrigid elements may be included in the chain.

(b) Degrees of Freedom[6,9,10,13,79,86,122,189,301]

Let F = degree of freedom of mechanism
l = total number of links, including fixed link
j = total number of joints
f_i = degree of freedom of relative motion between element pairs of ith joint.

Then, *in general,*

$$F = \lambda(l - j - 1) + \sum_{i=1}^{j} f_i \tag{4.1}$$

where λ is an integer whose value is determined as follows:

$\lambda = 3$: plane mechanisms with turning pairs, or turning and sliding pairs; spatial mechanisms with turning pairs only (motion on sphere); spatial mechanisms with rectilinear sliding pairs only.

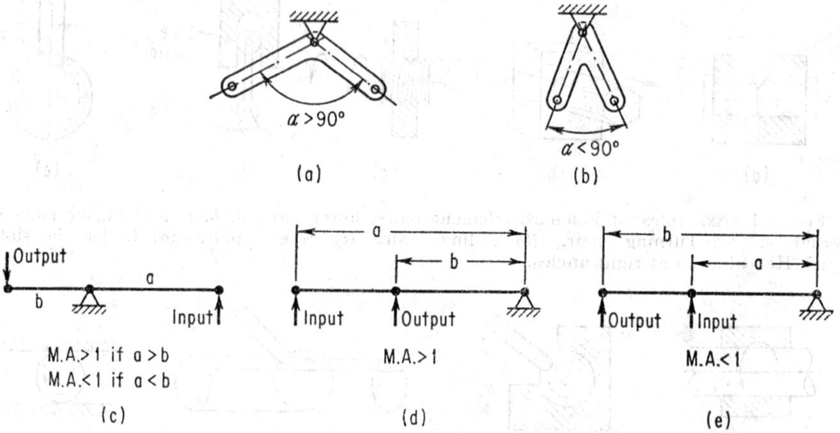

FIG. 4.3. Links and levers. (*a*) Rocker (ternary link). (*b*) Bell-crank (ternary link). (*c*) First-class lever. (*d*) Second-class lever. (*e*) Third-class lever.

$\lambda = 6$: spatial mechanisms with lower pairs, the axes of which are nonparallel and nonintersecting; note exceptions such as listed under $\lambda = 2$ and $\lambda = 3$. (See also ref. 10.)

$\lambda = 2$: plane mechanisms with sliding pairs only; spatial mechanisms with "curved" sliding pairs only (motion on a sphere); three-link coaxial screw mechanisms.

Although included under Eq. (4.1), the motions on a sphere are usually referred to as special cases. For a comprehensive discussion and formulas including screw chains and other combinations of elements, see ref. 13. The freedom of a mechanism with higher pairs should be determined from an equivalent lower-pair mechanism whenever feasible (see Sec. 4.2).

Mechanism Characteristics Depending on Degree of Freedom Only. For plane mechanisms with turning pairs only and one degree of freedom,

$$2j - 3l + 4 = 0, \quad \text{except in special cases.} \tag{4.2}$$

Furthermore, if this equation is valid,

1. The number of links is even;
2. The minimum number of binary links is four;
3. The maximum number of joints in a single link cannot exceed one-half the number of links;
4. If one joint connects m links, the joint is counted as $(m - 1)$-fold.

In addition, for nondegenerate plane mechanisms with turning and sliding pairs and one degree of freedom,

1. If a link has only sliding elements, they cannot all be parallel;
2. Except for the three-link chain, binary links having sliding pairs only cannot, in general, be directly connected;
3. No closed nonrigid loop can contain less than two turning pairs.

For plane mechanisms, having any combination of higher and/or lower pairs, and with one degree of freedom,

1. The number of links may be odd;
2. The maximum number of elements in a link may exceed one-half the number of links, but an upper bound can be determined;[122,301]
3. If a link has only higher-pair connections, it must possess at least three elements.

For constrained spatial mechanisms in which Eq. (4.1) applies with $\lambda = 6$, the sum of the degrees of freedom of all joints must add up to seven whenever the number of links is equal to the number of joints.

Special Cases. F can exceed the value predicted by Eq. (4.1) in certain special cases. These occur, generally, when a sufficient number of links are parallel in plane motion (Fig. 4.4*a*) or, in spatial motions, when the axes of the joints intersect (Fig. 4.4*b*—motion on a sphere, considered special in the sense that $\lambda \neq 6$).

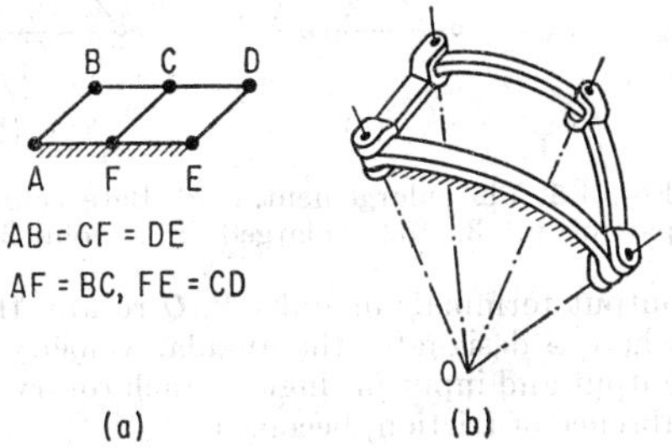

FIG. 4.4. Special cases which are exceptions to Eq. (4.1). (*a*) Parallelogram motion, $F = 1$. (*b*) Spherical four-bar mechanism, $F = 1$; axes of four turning joints intersect at O.

The existence of these special cases or "critical forms" can sometimes also be detected by multigeneration effects involving pantographs, inversors, or mechanisms derived from these (see Art. 4.6 and ref. 122). In the general case, the critical form is associated with the singularity of the functional matrix of the differential displacement equations of the coordinates;[103] this singularity is usually difficult to ascertain, however, especially when higher pairs are involved. Known cases are summarized in ref. 122. For two-degree-of-freedom systems, additional results are listed in refs. 86 and 189.

(c) Kinematic Inversion

Kinematic inversion refers to the process of considering different links as the frame in a given kinematic chain. Thereby different and possibly useful mechanisms can be obtained. The slider crank, the turning-block, and the swinging-block mechanism are mutual inversions, as are also drag-link and "crank-and-rocker" mechanisms.

(d) Pin Enlargement

Another method for developing different mechanisms from a base configuration involves enlarging the joints, illustrated in Fig. 4.5.

(e) Mechanical Advantage

Neglecting friction and dynamic effects, the instantaneous power input and output of a mechanism must be equal and, in the absence of branching (one input, one output, connected by a single "path"), equal to the "power flow" through any other point of the mechanism.

In a single-degree-of-freedom mechanism without branches, the power flow at any point J is the product of the force F_j at J and the velocity V_j at J in the direction of the force. Hence, for any point in such a mechanism,

$$F_iV_i = \text{constant} \qquad \text{(neglecting friction and dynamic effects)} \tag{4.3}$$

For the point of input P and the point of output Q of such a mechanism, the mechanical advantage is defined as

$$\text{MA} = F_Q/F_P \tag{4.4}$$

(f) Velocity Ratio

The *linear velocity ratio* for the motion of two points P, Q representing the input and output members or "terminals" of a mechanism is defined as V_Q/V_P. If input and

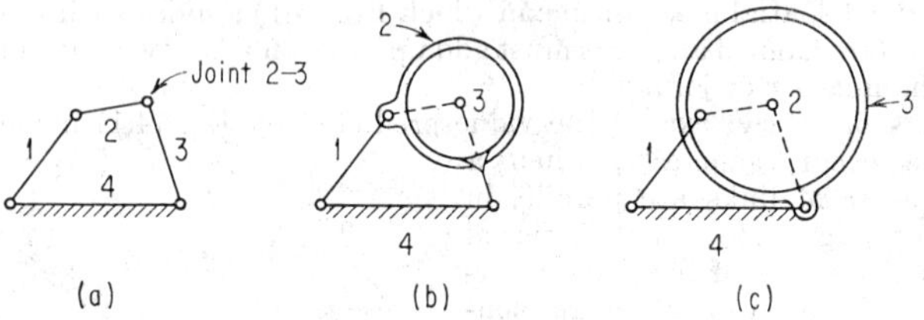

FIG. 4.5. Pin enlargement. (*a*) Base configuration. (*b*) Enlarged pin at joint 2-3; pin part of link 3. (*c*) Enlarged pin at joint 2-3; pin takes place of link 2.

output terminals or links P, Q rotate, the *angular velocity ratio* is defined as ω_Q/ω_P, where ω designates the angular velocity of the link. If T_Q and T_P refer to torque output and input in single-branch rotary mechanisms, the power-flow equation, in the absence of friction, becomes

$$T_P\omega_P = T_Q\omega_Q \tag{4.5}$$

(g) Conservation of Energy

Neglecting friction and dynamic effects, the product of the mechanical advantage and the linear velocity ratio is unity for all points in a single-degree-of-freedom mechanism without branch points, since $F_QV_Q/F_PV_P = 1$.

(h) Toggle

Toggle mechanisms are characterized by sudden snap or overcenter action, such as in Fig. 4.6*a* and *b*, schematics of a crushing mechanism and a light switch. The

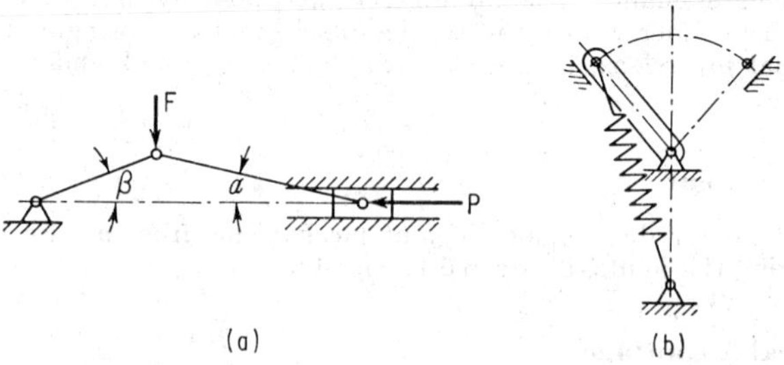

FIG. 4.6. Toggle actions. (*a*) $P/F = (\tan\alpha + \tan\beta)^{-1}$ (neglecting friction). (*b*) Schematic of a light switch.

mechanical advantage, as in Fig. 4.6*a*, can become very high. Hence toggles are often used in operations such as clamping, crushing, and coining.

(i) Transmission Angle[15,125,126,132,133,135,142,165] (see Arts. 4.6 and 4.9)

The transmission angle μ is used as a geometrical indication of the ease of motion of a mechanism under static conditions, excluding friction. It is defined by the ratio

$$\tan \mu = \frac{\text{force component tending to move driven link}}{\text{force component tending to apply pressure on driven-link bearing or guide}} \tag{4.6}$$

where μ is the transmission angle

In four-link mechanisms, μ is the angle between the coupler and the driven link (or the supplement of this angle) (Fig. 4.7) and has been used in optimizing linkage pro-

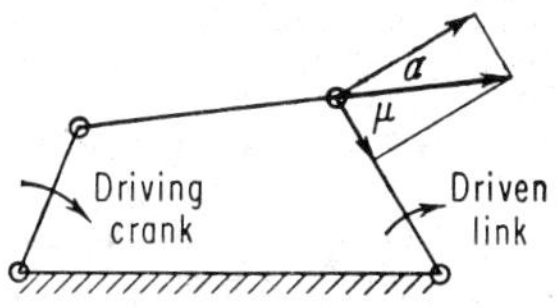

FIG. 4.7. Transmission angle μ and pressure angle α in a four-link mechanism.

portions (Arts. 4.6 and 4.9). Its ideal value is 90°; in practice it may deviate from this value by 30° and possibly more.

(j) Pressure Angle

In cam and gear systems, it is customary to refer to the complement of the transmission angle, called the pressure angle α, defined by the ratio

$$\tan \alpha = \frac{\text{force component tending to put pressure on follower bearing or guide}}{\text{force component tending to move follower}} \tag{4.7}$$

The ideal value of the pressure angle is zero; in practice it is frequently held to within 30° (Fig. 4.8). To ensure movability of the output member the ultimate criterion is to preserve a sufficiently large value of the ratio of driving force (or torque) to friction force (or torque) on the driven link. For a link in pure sliding (Fig. 4.8*c*) the motion will lock if the pressure angle and the friction angle add up to or exceed 90°.

A mechanism, the output link of which is shown in Fig. 4.8*d*, will lock if the ratio of p, the distance of the line of action of the force F from the fixed pivot axis, to the bearing radius r_b is less than or equal to the coefficient of friction, f, i.e., if the line of action of the force F cuts the *friction circle* of radius fr_b, concentric with the bearing.[137]

(k) Kinematic Equivalence[125,148,229,231,281,308] (see Art. 4.6)

This term, when applied to two mechanisms, refers to equivalence in motion, the precise nature of which must be defined in each case.

The motion of joint C in Fig. 4.9*a* and *b* is entirely equivalent if the quadrilaterals $ABCD$ are identical; the motion of C as a function of the rotation of link AB is also equivalent throughout the range allowed by the slot. In Fig. 4.9*c*, B and C are the centers of curvature of the contacting surfaces at N; $ABCD$ is one equivalent four-bar mechanism in the sense that, if AB is integral with body 1, the angular velocity and angular acceleration of link CD and body 2 are the same in the position shown, but not necessarily elsewhere.

Equivalence is used in design to obtain alternate mechanisms, which may be mechanically more desirable than the original. If, as in Fig. 4.9*d*, A_1A_2 and B_1B_2 are conjugate point pairs (see Art. 4.4), with A_1B_1 fixed on roll curve 1, which is in rolling contact with roll curve 2 (A_2B_2 are fixed on roll curve 2), then the path of E

on link A_2B_2 and of the coincident point on the body of roll curve 2 will have the same path tangent and path curvature in the position shown, but not generally elsewhere.

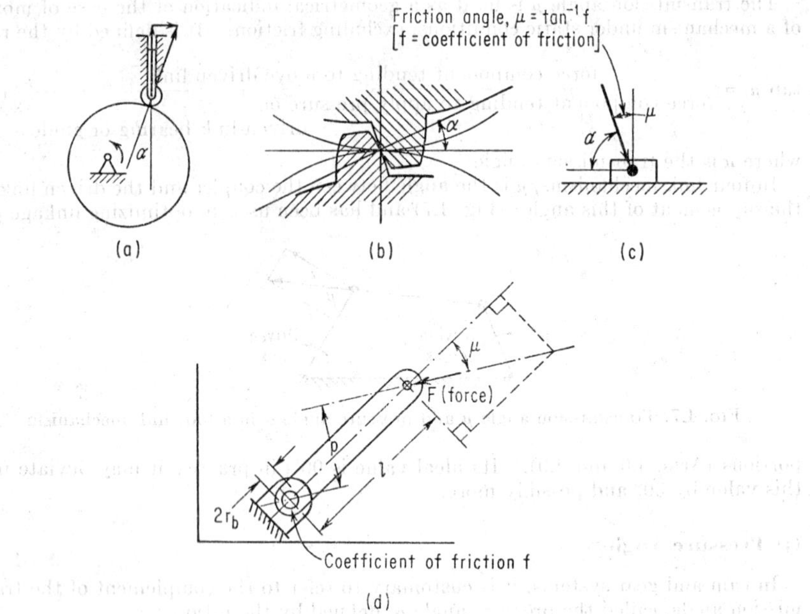

FIG. 4.8. Pressure angle. (*a*) Cam and follower. (*b*) Gear teeth in mesh. (*c*) Link in sliding motion; condition of locking by friction $(\alpha + \mu) \geq 90°$. (*d*) Condition for locking by friction of a rotating link: $\sin \mu \leq fr_b/l$.

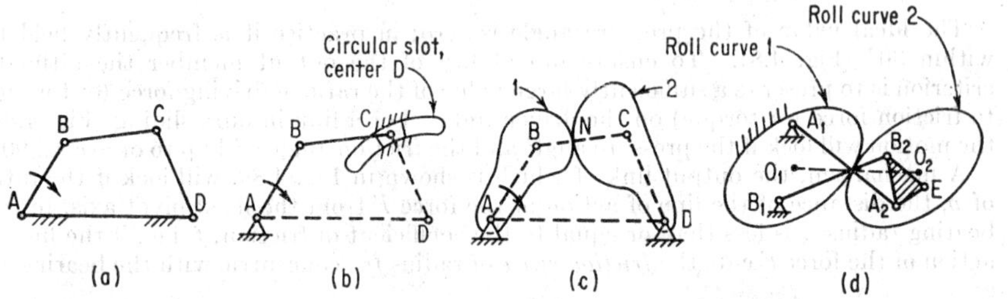

FIG. 4.9. Kinematic equivalence. (*a*), (*b*), (*c*) for four-bar motion. (*d*) illustrates rolling motion and an equivalent mechanism: when O_1, O_2 are fixed, curves are in rolling contact; when roll curve 1 is fixed and rolling contact is maintained, O_2 generates circle with center O_1.

(l) The Instant Center

At any instant in the plane motion of a link, the velocities of all points on the link are proportional to their distance from a particular point P, called the instant center. The velocity of each point is perpendicular to the line joining that point to P (Fig. 4.10). Regarded as a point on the link, P has an instantaneous velocity of zero. In pure translation, P is at infinity.

The instant center is defined in terms of velocities and is not the center of path curvature for the points on the moving link in the instant shown, except in special cases, e.g., points on common tangent between centrodes (see Sec. 4.4).

An extension of this concept to the "instantaneous screw axis" in spatial motions has been described.[32]

(m) Centrodes, Polodes, Pole Curves

Relative plane motion of two links can be obtained from the pure rolling of two curves, the *fixed* and *movable centrodes* (*polodes* or *pole curves*), respectively, which can be constructed as illustrated in the following example.

As shown in Fig. 4.11, the intersections of path normals locate successive instant centers $P, P', P'', \ldots,$ whose locus constitutes the fixed centrode. The movable centrode can be obtained either by inversion (i.e., keeping AB fixed, moving the guide, and constructing the centrode as before) or by *direct construction:* superposing triangles $A'B'P', A''B''P'', \ldots,$ on AB so that A' covers A and B' covers B, etc. The new locations thus found for $P', P'', \ldots,$ marked $\pi', \pi'', \ldots,$ then constitute points on the movable centrode, which rolls without slip on the fixed centrode and carries AB with it, duplicating the original motion. Thus, for the motion of AB, the centrode-rolling motion is kinematically equivalent to the original guided motion

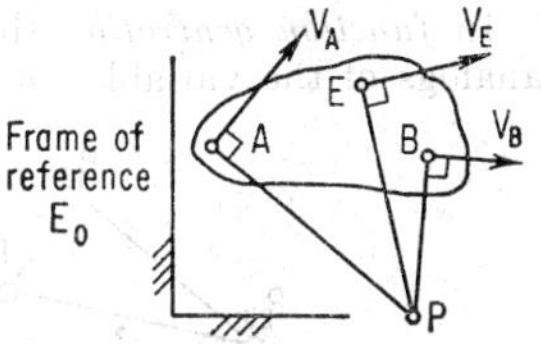

FIG. 4.10. Instant center, P. $V_E/V_B = EP/BP$, $V_E \perp EP$, etc.

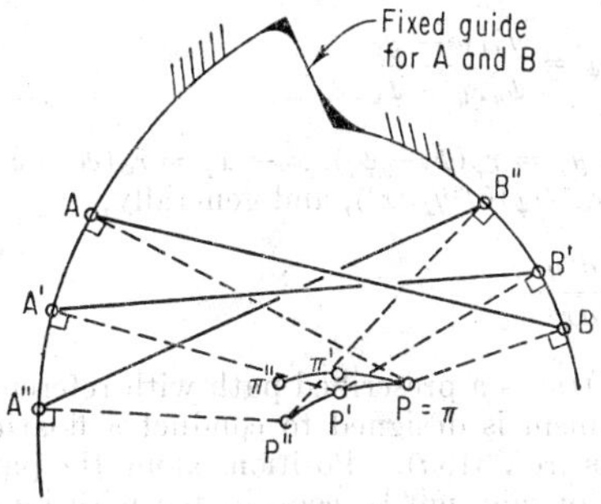

FIG. 4.11. Construction of fixed and movable centrodes. Link AB in plane motion, guided at both ends; $PP' = P\pi'$; $\pi'\pi'' = P'P''$, etc.

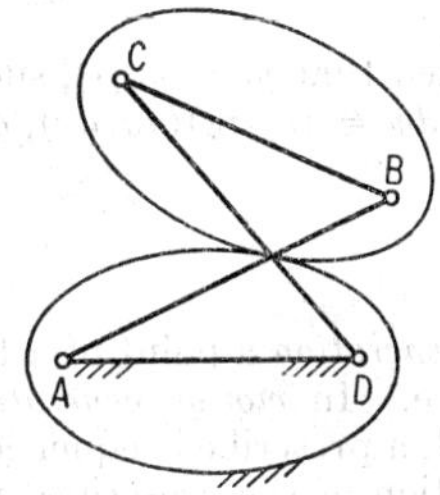

FIG. 4.12. Antiparallel equal-crank linkage; rolling ellipses, foci at A, D, B, C; $AD < AB$.

In the antiparallel equal-crank linkage, with the shortest link fixed, the centrodes for the coupler motion are identical ellipses with foci at the link pivots (Fig. 4.12); if the longer link AB were held fixed, the centrodes for the coupler motion of CD would be identical hyperbolas with foci at A, B and C, D, respectively.

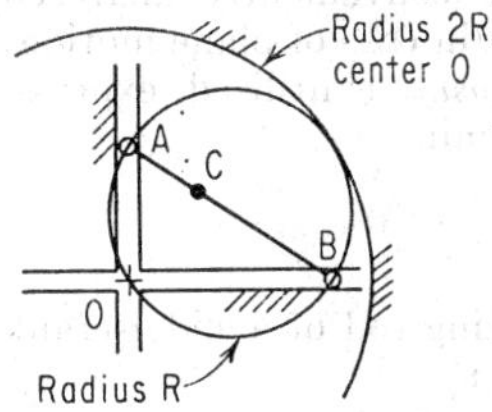

FIG. 4.13. Cardanic motion of the elliptic trammel, so called because any point C of AB describes an ellipse; midpoint of AB describes circle, center O (point C need not be collinear with AB).

In the elliptic trammel motion (Fig. 4.13) the centrodes are two circles, the smaller rolling inside the larger, twice its size. Known as *Cardanic motion,* it is used in press drives, resolvers, and straight-line guidance.

Apart from their use in kinematic analysis, the centrodes are used to obtain alternate, kinematically equivalent mechanisms, and sometimes to guide the original mechanism past the "in-line" or "dead-center" positions.[167]

(n) The Theorem of Three Centers

Also known as Kennedy's or the Aronhold-Kennedy theorem, this theorem states that, for any three bodies i, j, k in plane motion, the relative instant centers P_{ij}, P_{jk}, P_{ki} are collinear; here P_{ij}, for instance, refers to the instant center of the motion of link i relative to link j, or vice versa. Figure 4.14 illustrates the theorem with respect to four-bar motion. It is used in determining the location of instant centers and in curvature investigations.

(o) Function, Path, and Motion Generation

In *function generation* the input and output motions of a mechanism are linear analogs of the variables of a function $F(x,y, \ldots) = 0$. The number of degrees of freedom of the mechanism is equal to the number of independent variables.

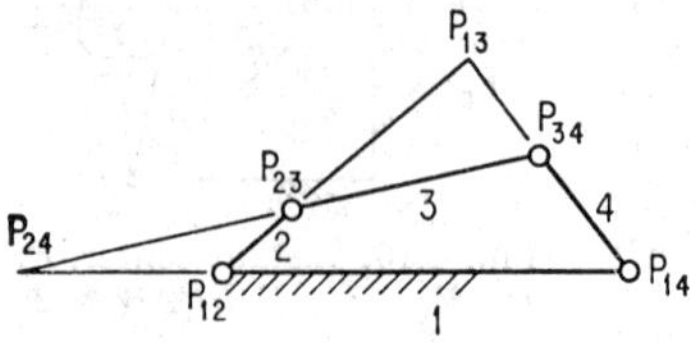

FIG. 4.14. Instant centers in four-bar motion.

For example, let ϕ and ψ, the linear or rotary motions of the input and output links or "terminals," be linear analogs of x and y, where $y = f(x)$ within the range $x_0 \leq x \leq x_{n+1}$, $y_0 \leq y \leq y_{n+1}$. Let the *input* values ϕ_0, ϕ_j, ϕ_{n+1} and the *output* values ψ_0, ψ_j, ψ_{n+1} correspond to the values x_0, x_j, x_{n+1} and y_0, y_j, y_{n+1}, of x and y respectively, where the subscripts 0, j, and $(n + 1)$ designate starting, jth intermediate, and terminal values. Scale factors r_ϕ, r_ψ are defined by

$$r_\phi = \frac{x_{n+1} - x_0}{\phi_{n+1} - \phi_0} \qquad r_\psi = \frac{y_{n+1} - y_0}{\psi_{n+1} - \psi_0}$$

(it is assumed that $y_0 \neq y_{n+1}$), such that $y - y_j = r_\psi(\psi - \psi_j)$, $x - x_j = r_\phi(\phi - \phi_j)$, whence $d\psi/d\phi = (r_\phi/r_\psi)(dy/dx)$, $d^2\psi/d\phi^2 = (r_\phi{}^2/r_\psi)(d^2y/dx^2)$, and generally,

$$\frac{d^n\psi}{d\phi^n} = \frac{r_\phi{}^n}{r_\psi}\frac{d^ny}{dx^n}.$$

In *path generation* a point of a floating link traces a prescribed path with reference to the frame. In *motion generation* a mechanism is designed to conduct a floating link through a prescribed sequence of positions (ref. 313*a*). Positions along the path or specification of the prescribed motion may or may not be coordinated with input displacements.

4.3. PRELIMINARY DESIGN ANALYSIS: DISPLACEMENTS, VELOCITIES, AND ACCELERATIONS[35,50,53,54,81,90,91,102,113,138,147,156,171,204,219,239,243,250,295,315,322,366] (see Art. 4.9)

Displacements in mechanisms are obtained graphically (from scale drawings) or analytically or both. Velocities and accelerations can be conveniently analyzed graphically by the "vector-polygon" method or analytically (in case of plane motion) via complex numbers. In all cases, the *vector equation of closure* is utilized, expressing the fact that the mechanism forms a closed kinematic chain.

(a) Velocity Analysis: Vector-polygon Method

The method is illustrated using a point D on the connecting rod of a slider-crank mechanism (Fig. 4.15). The vector-velocity equation for C is

$$\mathbf{V}_C = \mathbf{V}_B + \mathbf{V}^n{}_{C/B} + \mathbf{V}^t{}_{C/B} = \text{a vector parallel to line } AX$$

where $\mathbf{V}_C$ = velocity of C (Fig. 4.15)
$\mathbf{V}_B$ = velocity of B
$\mathbf{V}^n{}_{C/B}$ = normal component of velocity of C relative to B = component of relative velocity along BC = zero (owing to the rigidity of the connecting rod)
$\mathbf{V}^t{}_{C/B}$ = tangential component of velocity of C relative to B, value $(\omega_{BC})\overline{BC}$, perpendicular to BC

The velocity equation is now "drawn" by means of a vector polygon as follows:

1. Choose an arbitrary origin o (Fig. 4.16).
2. Label terminals of velocity vectors with lowercase letters, such that absolute velocities start at o and terminate with the letter corresponding to the point whose velocity is designated. Thus $\mathbf{V}_B = \mathbf{ob}$, $\mathbf{V}_C = \mathbf{oc}$, to a certain scale.
3. Draw $\mathbf{ob} = (\omega_{AB})\overline{AB} \times k_v$, where k_v is the velocity scale factor, say, in. per in./sec
4. Draw $\mathbf{bc} \perp BC$ and $\mathbf{oc} \| AX$ to determine intersection c.
5. Then $\mathbf{V}_C = (\mathbf{oc})/k_v$; absolute velocities always start at o.
6. Relative velocities $\mathbf{V}_{C/B}$, etc., connect the terminals of absolute velocities. Thus $\mathbf{V}_{C/B} = (\mathbf{bc})/k_v$. Note the reversal of order in C/B and $\mathbf{bc}$.
7. To determine the velocity of D, one way is to write the appropriate velocity-vector equation and draw it on the polygon: $\mathbf{V}_D = \mathbf{V}_C + \mathbf{V}^n_{D/C} + \mathbf{V}^t_{D/C}$; the second is to utilize the *principle of the velocity image.* This principle states that Δbcd in the velocity polygon is similar to ΔBCD in the mechanism, and the sense $b \to c \to d$ is the same as that of $B \to C \to D$. This *image construction* applies to any three points on a rigid link in plane motion. It has been used in Fig. 4.16 to locate d, whence $\mathbf{V}_D = (\mathbf{od})/k_v$.

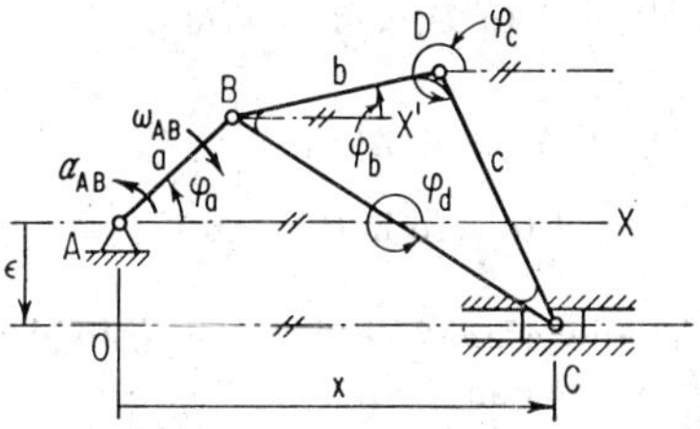

FIG. 4.15. Offset slider-crank mechanism.

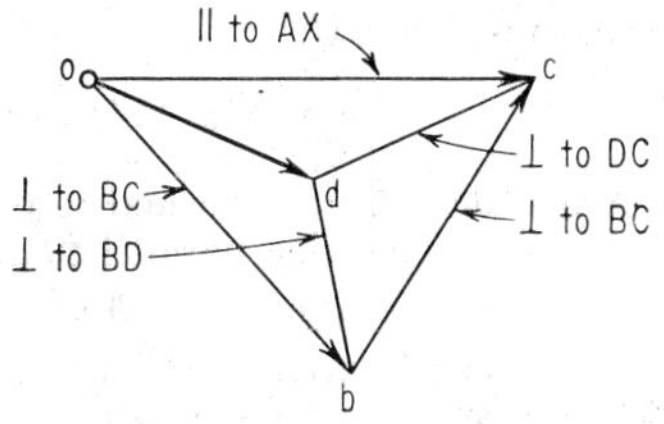

FIG. 4.16. Velocity polygon for slider-crank mechanism of Fig. 4.15.

8. The angular velocity ω_{BC} of the coupler can now be determined from

$$|\omega_{BC}| = \frac{|\mathbf{V}_{B/C}|}{\overline{BC}} = \frac{|(\mathbf{cb})/k_v|}{\overline{BC}}$$

9. Note that to determine the velocity of D it is easier to proceed in steps, to determine the velocity of C first and thereafter to use the image-construction method.

(b) Velocity Analysis: Complex-number Method

Using the slider crank of Fig. 4.15 once more as an illustration with x axis along the center line of the guide, and recalling that $i^2 = -1$, we write the complex-number equations as follows, with the equivalent vector equation below each:

Displacement:
$$ae^{i\varphi_a} + be^{i\varphi_b} + ce^{i\varphi_c} = x + i\epsilon \tag{4.8}$$
$$\mathbf{AB} + \mathbf{BD} + \mathbf{DC} = \mathbf{AC}$$

Velocity:
$$iae^{i\varphi_a}\omega_{AB} + (ibe^{i\varphi_b} + ice^{i\varphi_c})\omega_{BC} = dx/dt \qquad (t = \text{time}) \tag{4.9}$$
$$\mathbf{V}_B + \mathbf{V}_{D/B} + \mathbf{V}_{C/D} = \mathbf{V}_C$$

Note that $\omega_{AB} = d\varphi_a/dt$ is positive when counterclockwise and negative when clockwise; in this problem ω_{AB} is negative.

The complex conjugate of Eq. (4.9)

$$-iae^{-i\varphi_a}\omega_{AB} - (ibe^{-i\varphi_b} + ice^{-i\varphi_c})\omega_{BC} = dx/dt \tag{4.10}$$

From Eqs. (4.9) and (4.10), regarded as simultaneous equations:

$$\frac{\omega_{BC}}{\omega_{AB}} = \frac{ia(e^{i\varphi_a} + e^{-i\varphi_a})}{-ib(e^{i\varphi_b} + e^{-i\varphi_b}) - ic(e^{i\varphi_c} + e^{-i\varphi_c})} = \frac{a \cos \varphi_a}{-(b \cos \varphi_b + c \cos \varphi_c)}$$

$$\mathbf{V}_D = \mathbf{V}_B + \mathbf{V}_{D/B} = iae^{i\varphi_a}\omega_{AB} + ibe^{i\varphi_b}\omega_{BC}$$

The quantities φ_a, φ_b, φ_c are obtained from a scale drawing or by trigonometry.

Both the vector-polygon and the complex-number methods can be readily extended to accelerations, and the latter also to the higher accelerations.

(c) Acceleration Analysis: Vector-polygon Method

We continue with the slider crank of Fig. 4.15. After solving for the velocities via the velocity polygon, write out and "draw" the acceleration equations. Again proceed in order of increasing difficulty: from B to C to D, and determine first the acceleration of point C:

$$\mathbf{A}_C = \mathbf{A}^n{}_C + \mathbf{A}^t{}_C = \mathbf{A}^n{}_B + \mathbf{A}^t{}_B + \mathbf{A}^n{}_{C/B} + \mathbf{A}^t{}_{C/B}$$

where $\mathbf{A}^n{}_C$ = acceleration normal to path of C (equal to zero in this case)
$\mathbf{A}^t{}_C$ = acceleration parallel to path of C
$\mathbf{A}^n{}_B$ = acceleration normal to path of B, value $\omega^2{}_{AB}(\overline{AB})$, direction B to A
$\mathbf{A}^t{}_B$ = acceleration parallel to path of B, value $\alpha_{AB}(\overline{AB})$, $\perp AB$
$\mathbf{A}^n{}_{C/B}$ = acceleration component of C relative to B, in the direction C to B, value $(\overline{BC})\omega^2{}_{BC}$
$\mathbf{A}^t{}_{C/B}$ = acceleration component of C relative to B, $\perp\overline{BC}$, value $\alpha_{BC}(\overline{BC})$. Since α_{BC} is unknown, so is the magnitude of $\mathbf{A}^t{}_{C/B}$

The acceleration polygon is now drawn as follows (Fig. 4.17):

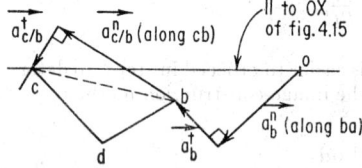

1. Choose an arbitrary origin o, as before.
2. Draw each acceleration to scale k_a [in. per in./sec.^{2}], and label the appropriate vector terminals with the lowercase letter corresponding to the point whose acceleration is designated, e.g., $\mathbf{A}_B = (\mathbf{ob})/k_a$. Draw $\mathbf{A}^n{}_B$, $\mathbf{A}^t{}_B$, and $\mathbf{A}^n{}_{C/B}$.
3. Knowing the direction of $\mathbf{A}^t{}_{C/B}(\perp\overline{BC})$, and also of $\mathbf{A}_C$ (along the slide), locate c at the intersection of a line through o, parallel to AX, and the line representing $\mathbf{A}^t{}_{C/B}$. $\mathbf{A}_C = (\mathbf{oc})/k_a$.

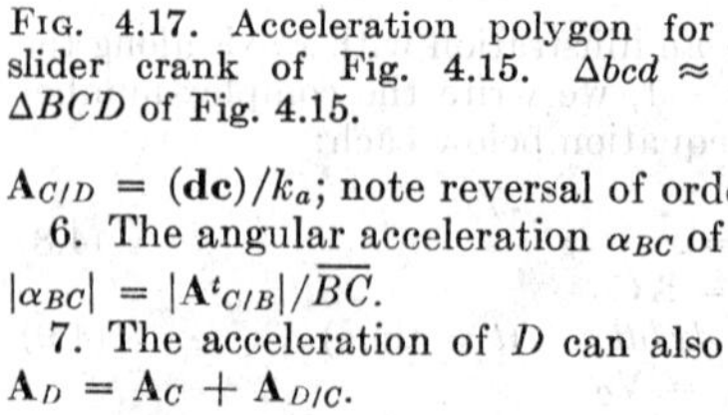

FIG. 4.17. Acceleration polygon for slider crank of Fig. 4.15. $\Delta bcd \approx \Delta BCD$ of Fig. 4.15.

4. The acceleration of D is obtained using the *principle of the acceleration image*, which states that, for any three points on a rigid body, such as link BCD, in plane motion, Δbcd and ΔBCD are similar, and the sense $b \rightarrow c \rightarrow d$ is the same as that of $B \rightarrow C \rightarrow D$. $\mathbf{A}_D = (\mathbf{od})/k_a$.
5. Relative accelerations can also be found from the polygon. For instance, $\mathbf{A}_{C/D} = (\mathbf{dc})/k_a$; note reversal of order of the letters C and D.
6. The angular acceleration α_{BC} of the connecting rod can now be determined from $|\alpha_{BC}| = |\mathbf{A}^t{}_{C/B}|/\overline{BC}$.
7. The acceleration of D can also be obtained by direct drawing of the equation $\mathbf{A}_D = \mathbf{A}_C + \mathbf{A}_{D/C}$.

(d) Acceleration Analysis: Complex-number Method (see Fig. 4.15)

Differentiating Eq. (4.9), obtain the acceleration equation of the slider-crank mechanism:

$$ae^{i\varphi_a}(i\alpha_{AB} - \omega^2{}_{AB}) + (be^{i\varphi_b} + ce^{i\varphi_c})(i\alpha_{BC} - \omega^2{}_{BC}) = d^2x/dt^2 \qquad (4.11)$$

This is equivalent to the vector equation

$$\mathbf{A}^t{}_B + \mathbf{A}^n{}_B + \mathbf{A}^t{}_{D/B} + \mathbf{A}^n{}_{D/B} + \mathbf{A}^t{}_{C/D} + \mathbf{A}^n{}_{C/D} = \mathbf{A}^n{}_C + \mathbf{A}^t{}_C$$

Combining Eq. (4.11) and its complex conjugate, eliminate d^2x/dt^2 and solve for α_{BC}. Substitute the value of α_{BC} in the following equation for $\mathbf{A}_D$:

$$\mathbf{A}_D = \mathbf{A}_B + \mathbf{A}_{D/B} = ae^{i\varphi_a}(i\alpha_{AB} - \omega^2{}_{AB}) + be^{i\varphi_b}(i\alpha_{BC} - \omega^2{}_{BC})$$

The above complex-number approach also lends itself to the analysis of motions involving Coriolis acceleration. The latter is encountered in the determination of the relative acceleration of two instantaneously coincident points on different links (refs. 137, 315). The general complex-number method is discussed more fully in ref. 313. An alternate approach, using the acceleration center, is described in Art. 4.4. The accelerations in certain specific mechanisms are discussed in Art. 4.9.

(e) Higher Accelerations (see also Art. 4.4)

The second acceleration (time derivative of acceleration), also known as *shock, jerk,* or *pulse,* is significant in the design of high-speed mechanisms and has been investigated in several ways.[35,53,54,239,313,315,366] It can be determined by direct differentiation of the complex-number acceleration equation.[313]

Basic equations:

Shock of B relative to A (where A,B represents two points on one link whose angular velocity is ω_p; $\alpha_p = d\omega_p/dt$):[239]

Component along AB:

$$-3\alpha_p\omega_p(AB)$$

Component perpendicular to AB:

$$(AB)[(d\alpha_p/dt) - \omega_p{}^3] \qquad \text{(in direction of } \boldsymbol{\omega}_p \times \mathbf{AB})$$

Absolute shock:[239]

Component along path tangent (in direction of $\boldsymbol{\omega}_p \times \mathbf{AB}$):

$$d^2v/dt^2 - v^3/\rho^2$$

where v = the velocity of B

ρ = the radius of curvature of path of B

Component directed toward the center of curvature:

$$(v/\rho)[3(dv/dt) - (v/\rho)(d\rho/dt)]$$

Absolute shock with reference to rolling centrodes (Fig. 4.18, Sec. 4.4)[239][l,m as in Eq. (4.22)]

Component along AP:

$$-3\omega_p{}^3\delta\left[\delta\sin\psi\left(\frac{1}{m}+\frac{1}{g}\right)+\delta\cos\psi\left(\frac{1}{l}-\frac{1}{\delta}\right)-\frac{r}{g}\right], \qquad g = \frac{\omega_p{}^2\delta}{\alpha_p}$$

Component perpendicular to AP in direction of $\boldsymbol{\omega}_p \times \mathbf{PA}$

$$r\left(\frac{d\alpha_p}{dt} - \omega_p{}^3\right) + 3\delta^2\omega_p{}^3\left[\cos\psi\left(\frac{1}{m}+\frac{1}{g}\right) - \sin\psi\left(\frac{1}{l}-\frac{1}{g}\right)\right]$$

(f) Acceleration in Complex Mechanisms

When the number of real unknowns in the complex-number or vector equations is greater than two, several methods can be used.[113,250] These are applicable to mechanisms with more than four links.

(g) Finite Differences in Velocity and Acceleration Analysis[171,307,322]

When the time-displacement curve of a point in a mechanism is known, the calculus of finite differences can be used for the calculation of velocities and accelerations. The data can be numerical or analytical. The method is useful also in ascertaining the existence of local fluctuations in velocities and accelerations, such as occur in cam-follower systems, for instance:[171]

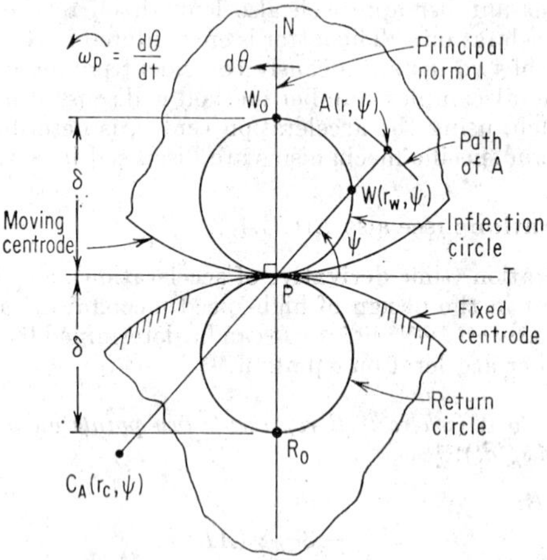

FIG. 4.18. Notation for the Euler-Savary equation.

Let a time-displacement curve be subdivided into equal time intervals Δt and define the *i*th, the general interval, as $t_i \leq t \leq t_{i+1}$, such that $\Delta t = t_{i+1} - t_i$. The *central-difference* formulas then give the following approximate values for velocities dy/dt, accelerations d^2y/dt^2, and shock d^3y/dt^3, where y_i denotes the displacement y at the time $t = t_i$:

Velocity at $t = t_i + \frac{1}{2}\Delta t$:
$$\frac{dy}{dt} = \frac{y_{i+1} - y_i}{\Delta t} \tag{4.12}$$

Acceleration at $t = t_i$:
$$\frac{d^2y}{dt^2} = \frac{y_{i+1} - 2y_i + y_{i-1}}{(\Delta t)^2} \tag{4.13}$$

Shock at $t = t_i + \frac{1}{2}\Delta t$:
$$\frac{d^3y}{dt^3} = \frac{y_{i+2} - 3y_{i+1} + 3y_i - y_{i-1}}{(\Delta t)^3} \tag{4.14}$$

If the values of the displacements y_i are known with absolute precision (no error), the values for velocities, accelerations, and shock in the above equations become increasingly accurate as Δt approaches zero, provided the curve is smooth. If, however, the displacements y_i are known only within a given tolerance, say $\pm\epsilon y$, then the accuracy of the computations will be high only if the interval Δt is sufficiently small and, in addition, if

$$2(\epsilon y/\Delta t) \ll dy/dt \quad \text{for velocities}$$
$$4[\epsilon y/(\Delta t)^2] \ll d^2y/dt^2 \quad \text{for accelerations}$$
$$8[\epsilon y/(\Delta t)^3] \ll d^3y/dt^3 \quad \text{for shock}$$

and provided also that these requirements are mutually compatible.

Further estimates of errors resulting from the use of Eqs. (4.12), (4.13), and (4.14), as well as alternate formulations involving "forward" and "backward" differences, are found in texts on numerical mathematics (e.g., ref. 155, pp. 94–97 and 110–112, with a discussion of truncation and round-off errors).

The above equations are particularly useful when the displacement-time curve is given in the form of a numerical table, as frequently happens in checking an existing design and in redesigning.

Some current computer programs in displacement, velocity, and acceleration analysis are listed in ref. 102; the kinematic properties of specific mechanisms, including spatial mechanisms, are summarized in Art. 4.9.[102]

4.4. PRELIMINARY DESIGN ANALYSIS: PATH CURVATURE

The following principles apply to the analysis of a mechanism in a given position, as well as to synthesis when motion characteristics are prescribed in the vicinity of a particular position. The technique can be used to obtain a quick "first approximation" to mechanism proportions which can be refined at a later stage.

(a) Polar-coordinate Convention

Angles are measured counterclockwise from a directed line segment, the *pole tangent* PT, origin at P (see Fig. 4.18); the polar coordinates (r,ψ) of a point A are either $r = |PA|$, $\psi = \angle TPA$ or $r = -|PA|$, $\psi = \angle TPA \pm 180°$. For example, in Fig. 4.18 r is positive, but r_c is negative.

(b) The Euler-Savary Equation (Fig. 4.18)

PT = common tangent of fixed and moving centrodes at point of contact P (the instant center).

PN = principal normal at P; $\angle TPN = 90°$

PA = line or ray through P.

$C_A(r_c,\psi)$ = center of curvature of path of $A(r,\psi)$ in position shown.

A and C_A are called *conjugate points.*

θ = angle of rotation of moving centrode, positive counterclockwise.

s = arc length along fixed centrode, measured from P, positive toward T.

The Euler-Savary equation is valid under the following assumptions:

1. During an infinitesimal displacement from the position shown, $d\theta/ds$ is finite and different from zero.
2. Point A does not coincide with P.
3. AP is finite.

Under these conditions, the curvature of the path of A in the position shown can be determined from the following "Euler-Savary" equations:

$$[(1/r) - (1/r_c)] \sin \psi = -d\theta/ds = -\omega_p/v_p \qquad (4.15)$$

where ω_p = angular velocity of moving centrode = $d\theta/dt$, t = time

v_p = corresponding velocity of point of contact between centrodes along the fixed centrode = ds/dt

Let r_w = polar coordinate of point W on ray PA, such that radius of curvature of path of W is infinite in the position shown; then W *is called the inflection point on ray* PA, and

$$1/r - 1/r_c = 1/r_w \qquad (4.16)$$

The locus of all inflection points W in the moving centrode is the *inflection circle,* tangent to PT at P, of diameter $PW_0 = \delta = -ds/d\theta$, where W_0, the *inflection pole,* is the inflection point on the principal normal ray. Hence,

$$[(1/r) - (1/r_c)] \sin \psi = 1/\delta \qquad (4.17)$$

The centers of path curvature of all points at infinity in the moving centrode are on the *return circle,* also of diameter δ, and obtained as the reflection of the inflection circle about line PT. The reflection of W_0 is known as the *return pole* R_0. For the pole velocity we have

$$v_p = ds/dt = -\omega_p\delta \qquad (4.18)$$

The curvatures of the paths of all points on a given ray are concave toward the inflection point on that ray.

For the diameter of the inflection and return circles we have

$$\delta = r_p r_\pi/(r_\pi - r_p) \tag{4.19}$$

where r_p and r_π are the polar coordinates of the centers of curvature of the moving and fixed centrodes, respectively, at P. Let $\rho = (r_c - r)$ be the instantaneous value of the radius of curvature of the path of A, and $w = \overline{AW}$, then

$$r^2 = \rho w \tag{4.20}$$

which is known as the *quadratic form* of the Euler-Savary equation.

Conjugate points in the planes of the moving and fixed centrodes are related by a "quadratic transformation."[26] When the above assumptions 1, 2, and 3, establishing the validity of the Euler-Savary equations, are not satisfied, see R. Mehmke;[222] for a further curvature theorem, useful in relative motions, see ref. 23.

Example: Cylinder of radius 2 in., rolling inside a fixed cylinder of radius 3 in.; common tangent horizontal, both cylinders above the tangent; δ = 6 in., $W_0(6{,}90°)$; for point $A_1(\sqrt{2}{,}45°)$: $r_{c1} = 1.5\sqrt{2}$, $C_{A1}(1.5\sqrt{2}{,}45°)$, $\rho_{A1} = 0.5\sqrt{2}$, $r_{w1} = 3\sqrt{2}$, $v_p = -6\omega_p$; for point $A_2(-\sqrt{2}{,}135°)$: $r_{c2} = -0.75\sqrt{2}$, $C_{A2}(-0.75\sqrt{2}{,}135°)$, $\rho_{A2} = 0.25\sqrt{2}$, $r_{w2} = 3\sqrt{2}$.

(c) Generating Curves and Envelopes[301]

Let g-g be a smooth curve attached to the moving centrode and e-e be the curve in the fixed centrode enveloping the successive positions of g-g during the rolling of the centrodes. Then g-g is called a *generating curve* and e-e its *envelope* (Fig. 4.19).

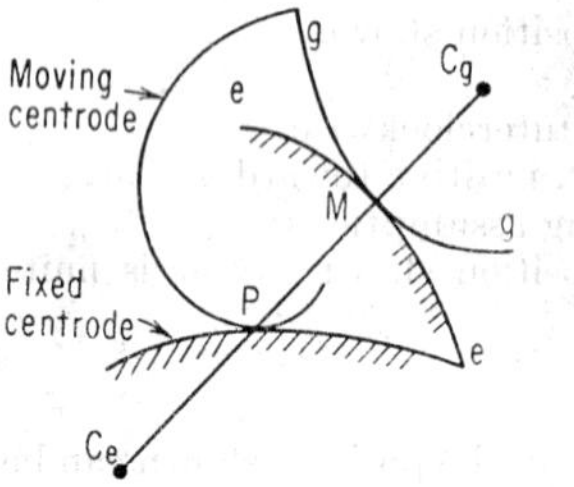

FIG. 4.19. Generating curve and envelope.

If C_g is the center of curvature of g-g and C_e that of e-e (at M):

1. C_e, P, M, and C_g are collinear (M being the point of contact between g-g and e-e).
2. C_e and C_g are conjugate points; i.e., if C_g is considered a point of the moving centrode, the center of curvature of its path lies at C_e; interchanging the fixed and moving centrodes will invert this relationship.
3. Aronhold's first theorem: The return circle is the locus of the centers of curvature of all envelopes whose generating curves are straight lines.
4. If a straight line in the moving plane always passes through a fixed point by sliding through it and rotating about it, that point is on the return circle.
5. Aronhold's second theorem: The inflection circle is the locus of the centers of curvature for all generating curves whose envelopes are straight lines.

Example (utilizing paragraph 4 above): In the swinging-block mechanism of Fig. 4.20, point C is on the return circle, and the center of curvature of the path of C as a point of link BD is therefore at C_c, halfway between C and P. Thus $ABCC_c$ constitutes a four-bar mechanism, with C_c as a fixed pivot, equivalent to the original mechanism in the position shown with reference to path tangents and path curvatures of points in the plane of link BD.

(d) Bobillier's Theorem

Consider two separate rays, 1 and 2 (Fig. 4.21), with a pair of distinct conjugate points on each, A_1, C_1 and A_2, C_2. Let $Q_{A_1A_2}$ be the intersection of A_1A_2 and C_1C_2. Then the line through $PQ_{A_1A_2}$ is called the *collineation axis*, unique for the pair of rays 1 and 2, regardless of the choice of conjugate point pairs on these rays. Bobillier's theorem states that the angle between the common tangent of the centrodes and one ray is equal to the angle between the other ray and the collineation axis, both angles being described in the same sense.[301]

The collineation axis is parallel to the line joining the inflection points on the two rays.

Bobillier's construction for determining the curvature of point-path trajectories is illustrated for two types of mechanisms in Figs. 4.22 and 4.23.

Another method for finding centers of path curvature is Hartmann's construction, described in ref. 68.

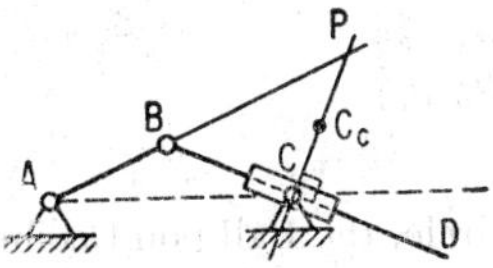

FIG. 4.20. Swinging-block mechanism: $CC_c = C_cP$.

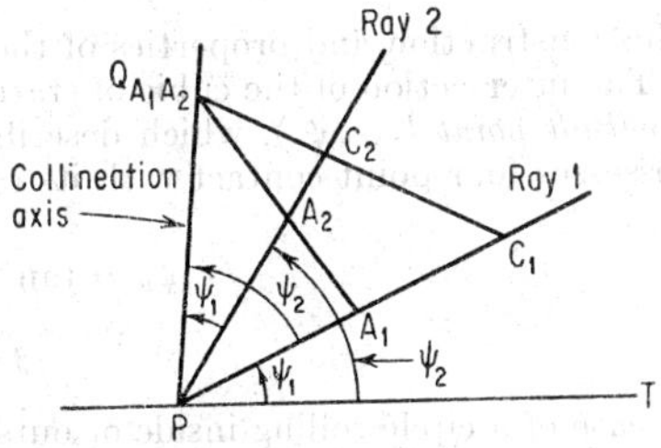

FIG. 4.21. Bobillier's construction.

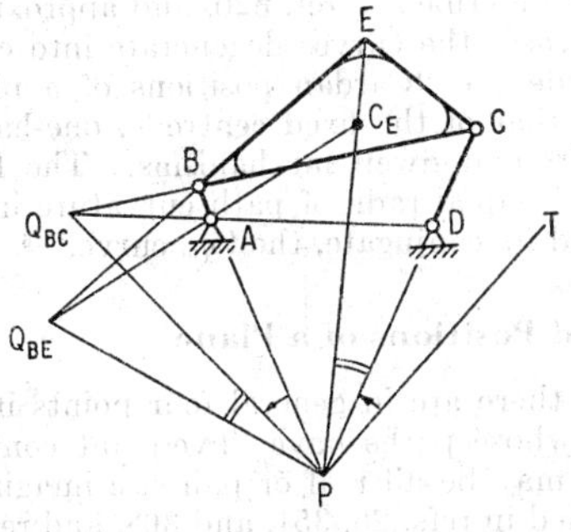

FIG. 4.22. Bobillier's construction for the center of curvature C_E of path of E on coupler of four-bar mechanism in position shown.

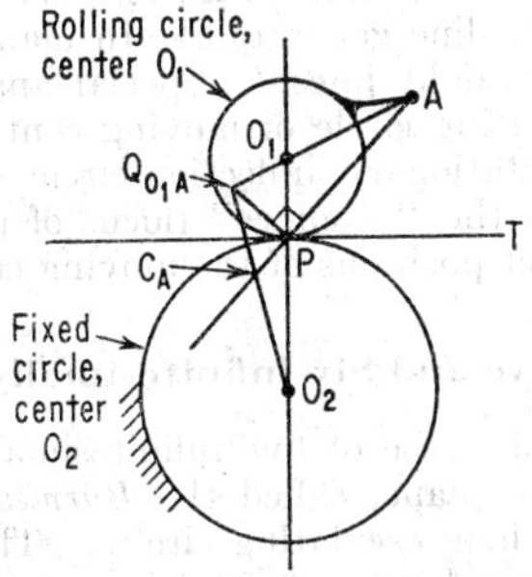

FIG. 4.23. Bobillier's construction for cycloidal motion. Determination of C_A, the center of curvature of the path of A, attached to the rolling circle (in position shown).

Occasionally, especially in the design of linkages with a dwell (temporary rest of output link), one may also use the *sextic of constant curvature*, known also as the ρ curve,[26] the locus of all points in the moving centrode whose paths at a given instant have the same numerical value of the radius of curvature.

(e) The Cubic of Stationary Curvature (the k_u Curve)

The "k_u curve" is defined as the locus of all points in the moving centrode whose rate of change of path curvature in a given position is zero: $d\rho/ds = 0$. Paths of points on this curve possess "four-point contact" with their osculating circles. Under the same assumptions as in Art. 4.4(a), the following is the equation of the k_u curve:

$$(\sin\psi\cos\psi)/r = \sin\psi/m + \cos\psi/l \tag{4.21}$$

where (r,ψ) = polar coordinates of a point on the k_u curve,

$$m = -3\delta/(d\delta/ds)$$
$$l = 3r_pr_\pi/(2r_\pi - r_p) \tag{4.22}$$

In cartesian coordinates (xy axes $PT - PN$),

$$(x^2 + y^2)(mx + ly) - lmxy = 0 \tag{4.23}$$

The locus of the centers of curvature of all points on the k_u curve is known as the *cubic of centers of stationary curvature*, or the "k_a curve." Its equation is

$$(x^2 + y^2)(mx + l^*y) - l^*mxy = 0 \tag{4.24}$$

where

$$(1/l) - (1/l^*) = 1/\delta \tag{4.25}$$

The construction and properties of these curves are discussed in refs. 26 and 256.

The intersection of the cubic of stationary curvature and the inflection circle yields the *Ball point* $U(r_u,\psi_u)$, which describes an approximate straight line; i.e., its path possesses four-point contact with its tangent. The coordinates of the Ball point are

$$\psi_u = \tan^{-1} \frac{2r_p - r_\pi}{(r_\pi - r_p)(d\delta/ds)} \tag{4.26}$$

$$r_u = \delta \sin \psi_u \tag{4.27}$$

In case of a circle rolling inside or outside a fixed circle, the Ball point coincides with the inflection pole.

Technical applications of the cubic of stationary curvature, other than design analysis in general, include the generation of n-sided polygons,[26] the design of intermittent-motion mechanisms such as the type described in ref. 320, and approximate straight-line generation. In many of these cases the curves degenerate into circles and straight lines.[26] Special analyses include the "Cardan positions of a plane" (osculating circle of moving centrode inside that of the fixed centrode, one-half its size; stationary inflection-circle diameter)[43,99] and dwell mechanisms. The latter utilize the "q_1 curve" (locus of points having equal radii of path curvature in two distinct positions of the moving centrode) and its conjugate, the "q_m curve."[26]

(f) Five and Six Infinitesimally Separated Positions of a Plane

In the case of five infinitesimal positions, there are in general four points in the moving plane, called the *Burmester points*, whose paths have "five-point contact" with their osculating circles. These points may be all real or pairwise imaginary. Their application to four-bar motion is outlined in refs. 26, 351, and 368, and related computer programs are listed in ref. 102, the latter also summarizing the applicable results of six-position theory, in so far as they pertain to four-bar motion. Burmester points and points on the cubic of stationary curvature have been used in a variety of six-link dwell mechanisms.[26,125]

(g) Application of Curvature Theory to Accelerations

1. The acceleration $\mathbf{A}_p$ of the instant center (as a point of the moving centrode) is given by $\mathbf{A}_p = \omega_p^2(\mathbf{PW}_0)$; it is the only point of the moving centrode whose acceleration is independent of the angular acceleration α_p.

2. The inflection circle (also called the *de la Hire circle* in this connection), is the locus of points having zero acceleration normal to their paths.

3. The locus of all points on the moving centrode, whose tangential acceleration (i.e., acceleration along path) is zero, is another circle, the *Bresse circle*, tangent to the principal normal at P, with diameter equal to $-\omega_p^2\delta/\alpha_P$ where α_P is the angular acceleration of the moving centrode, the positive sense of which is the same as that of θ.

4. The intersection of these circles, other than P, determines the point Γ, with zero total acceleration, known as the *acceleration center*. It is located at the intersection of the inflection circle and a ray of angle γ, where

$$\gamma = \angle W_0P\Gamma = \tan^{-1}(\alpha_p/\omega_p^2) \qquad 0 \leq |\gamma| \leq 90°$$

measured in the direction of the angular acceleration.

5. The acceleration $\mathbf{A}_B$ of any point B in the moving system is proportional to its distance from the acceleration center:

$$\mathbf{A}_B = (\mathbf{B\Gamma})(e^{-i\gamma})|(\omega_p{}^4 + \alpha_p{}^2)^{\frac{1}{2}}| \tag{4.28}$$

6. The acceleration vector $\mathbf{A}_B$ of any point B makes an angle γ with the line joining it to the acceleration center [see Eq. (4.28)] where γ is measured from $\mathbf{A}_B$ in the direction of angular acceleration.

7. When the acceleration vectors of two points (V,U) on one link, other than the pole, are known, the location of the acceleration center can be determined from item 6 and the equation

$$|\tan\gamma| = \frac{|\mathbf{A}^t{}_{U/V}|}{|A^n{}_{U/V}|}$$

8. The concept of acceleration centers and images can be extended also to the higher accelerations[35] (see also Sec. 4.3).

(h) Examples of Mechanism Design and Analysis Based on Curvature Theory

1. Mechanism used in guiding the grinding tool in large gear generators (Fig. 4.24): The radius of path curvature ρ_m of M at the instant shown: $\rho_m = (W_1W_2)/(2\tan^3\theta)$, at which instant M is on the cubic of stationary curvature belonging to link W_1W_2; ρ_m is arbitrarily large if θ is sufficiently small.

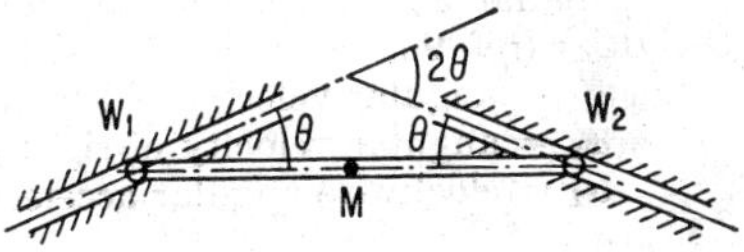

Fig. 4.24. Mechanism used in guiding the grinding tool in large gear generators; due to A. H. Candee, Rochester, N. Y. $MW_1 = MW_2$; link W_1W_2 constrained by straight-line guides for W_1 and W_2.

2. Machining of radii on tensile test specimens[141,367] (Fig. 4.25): C lies on cubic of stationary curvature; AB is the diameter of the inflection circle for the motion of link ABC; radius of curvature of path of C in the position shown:

$$\rho_c = (AC)^2/(BC)$$

3. Pendulum with large period of oscillation, yet limited size[224,326] (Fig. 4.26), as used in recording ship's vibrations: $AB = a$, $AC = b$, $CS = s$, r_t = radius of gyration

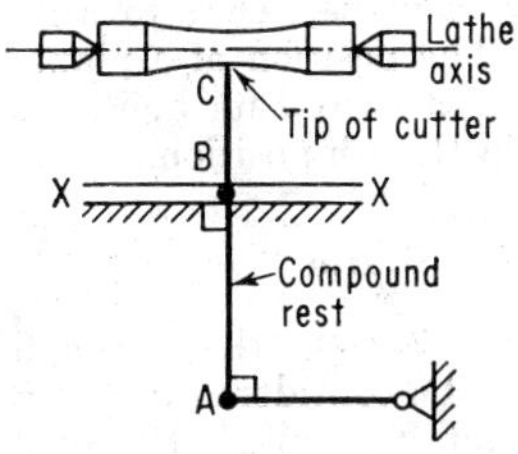

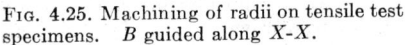

Fig. 4.25. Machining of radii on tensile test specimens. B guided along X-X.

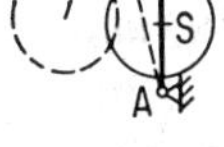

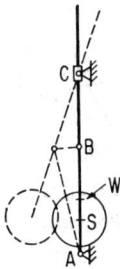

Fig. 4.26. Pendulum with large period of oscillation.

of the heavy mass S about its center of gravity. If the mass other than S and friction are negligible, the length l of the equivalent simple pendulum is given by

$$l = (r_t{}^2 + s^2)/(b/a)(b-a) - s$$

where the distance CW is equal to $(b/a)(b-a)$. The location of S is slightly below the inflection point W, in order for the oscillation to be stable and slow.

4. Modified geneva drive in high-speed bread wrapper[309] (Fig. 4.27): The driving pin of the geneva motion can be located at or near the Ball point of the pinion motion; the path of the Ball point, approximately square, can be used to give better kinematic characteristics to a four-station geneva than the regular crankpin design, by reducing peak velocities and accelerations.

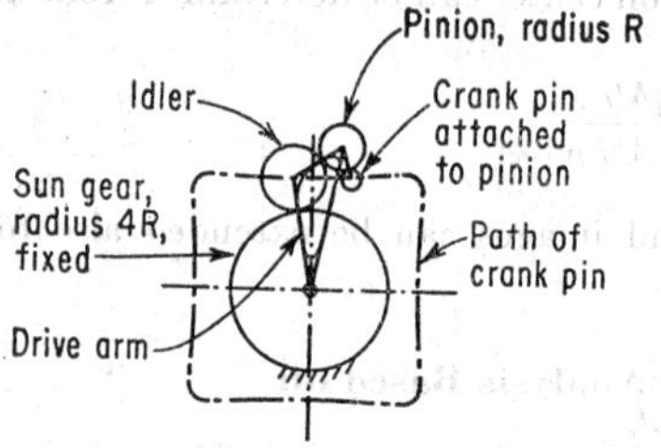

FIG. 4.27. Modified geneva drive in high-speed bread wrapper.

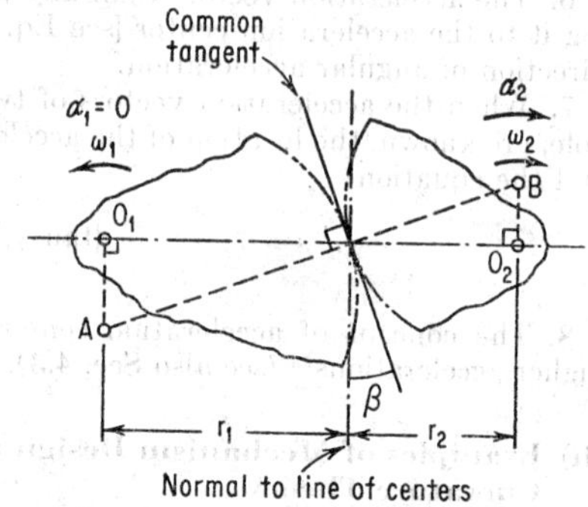

FIG. 4.28. Angular acceleration diagram for noncircular gears.

5. Angular acceleration of noncircular gears (obtainable from equivalent linkage O_1ABO_2) (ref. 90, discussion by A. H. Candee) (Fig. 4.28):

Let ω_1 = angular velocity of left gear, assumed constant, counterclockwise
ω_2 = angular velocity of right gear, clockwise
α_2 = clockwise angular acceleration of right gear

then
$$\alpha_2 = \frac{r_1(r_1 + r_2)}{r_2^2} (\tan \beta)\omega_1^2$$

4.5. DIMENSIONAL SYNTHESIS: PATH, FUNCTION, AND MOTION GENERATION

In the design of automatic machinery, it is often required to guide a part through a sequence of prescribed positions. Such motions can be mechanized by dimensional synthesis based on the kinematic geometry of distinct positions of a plane. In plane motion, a kinematic plane, hereafter called a plane, refers to a rigid body, arbitrary in extent. The position of a plane is determined by the location of two of its points, A and B, designated as A_i, B_i in the ith position.

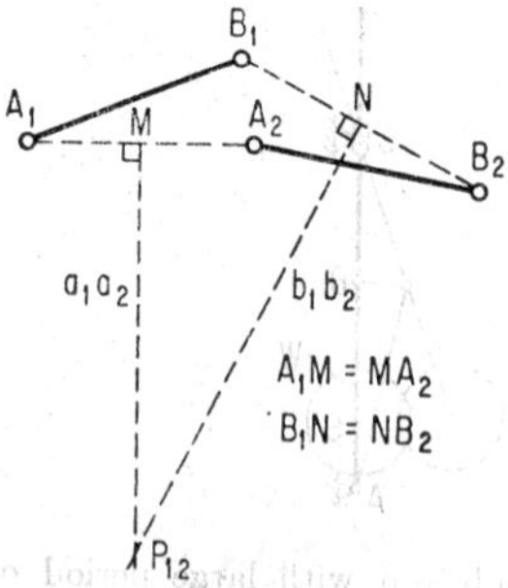

FIG. 4.29. Two positions of a plane. Pole $P_{12} = a_1a_2 \times b_1b_2$.

(a) Two Positions of a Plane

According to *Chasles's theorem*, the motion from A_1B_1 to A_2B_2 (Fig. 4.29) can be considered *as though* it were a rotation about a point P_{12}, called the pole, which is the intersection of the perpendicular bisectors a_1a_2, b_1b_2 of A_1A_2 and B_1B_2, respectively. $A_1, A_2, \ldots$, are called *corresponding positions* of point A; $B_1, B_2, \ldots$, those of point B; $A_1B_1, A_2B_2, \ldots$, those of the plane AB.

A similar construction applies to the *relative motion of two planes* (Fig. 4.30) AB and CD (positions A_iB_i and C_iD_i, $i = 1, 2$). The *relative pole* Q_{12} is constructed by transferring the figure $A_2B_2C_2D_2$ as a rigid body to bring A_2 and B_2 into coincidence with A_1 and B_1, respectively, and denoting the new positions of C_2, D_2, by C_2^1, D_2^1, respectively. Then Q_{12} is obtained from C_1D_1 and $C_2^1D_2^1$ as in Fig. 4.29.

1. The motion of A_1B_1 to A_2B_2 in Fig. 4.29 can be carried out by four-link mechanisms in which A and B are coupler-hinge pivots and the fixed-link pivots A_0, B_0 are located on the perpendicular bisectors a_1a_2, b_1b_2, respectively.

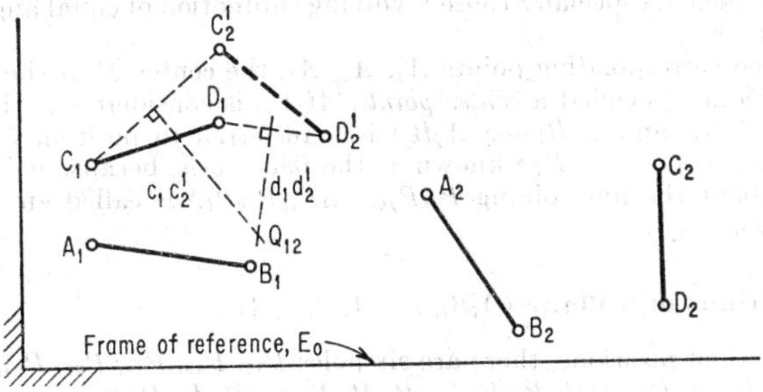

Fig. 4.30. Relative motion of two planes, AB, CD. Relative pole, $Q_{12} = c_1c_2^1 \times d_1d_2^1$.

2. To construct a four-bar mechanism A_0ABB_0 when the corresponding angles of rotation of the two cranks are prescribed (in Fig. 4.31 the construction is illustrated with ϕ_{12} clockwise for A_0A and ψ_{12} clockwise for B_0B):

a. From line A_0B_0X, lay off angles $\frac{1}{2}\phi_{12}$ and $\frac{1}{2}\psi_{12}$ opposite to desired direction of rotation of the cranks, locating Q_{12} as shown.

b. Draw any two straight lines L_1 and L_2 through Q_{12}, such that

$$\angle L_1Q_{12}L_2 = \angle A_0Q_{12}B_0$$

in magnitude and sense.

c. A_1 can be located on L_1, B_1 on L_2, and when A_0A_1 rotates clockwise by ϕ_{12}, B_0B_1 will rotate clockwise by ψ_{12}. Care must be taken, however, to ensure that the mechanism will not lock in an intermediate position.

(b) Three Positions of a Plane (A_iB_i, $i = 1, 2, 3$)

In this case there are three poles P_{12}, P_{23}, P_{31} and three associated rotations ϕ_{12}, ϕ_{23}, ϕ_{31}, where $\phi_{ij} = \angle A_iP_{ij}A_j = \angle B_iB_{ij}B_j$. The three poles form the vertices of the *pole triangle* (Fig. 4.32). Note that $P_{ij} = P_{ji}$, and $\phi_{ij} = -\phi_{ji}$.

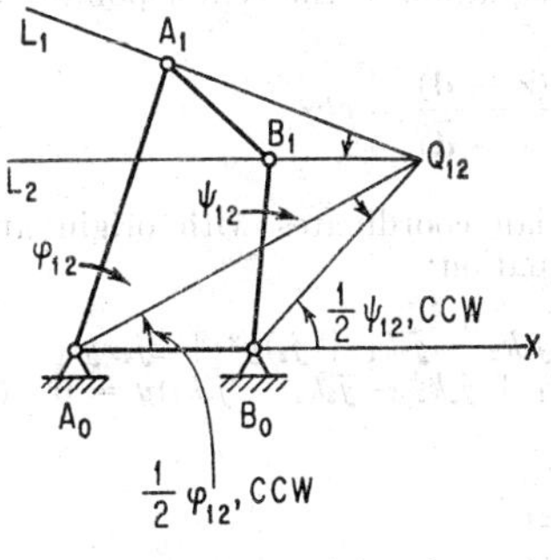

Fig. 4.31. Construction of four-bar mechanism $A_0A_1B_1B_0$ in position 1, for prescribed rotations ϕ_{12} vs. ψ_{12}, both clockwise in this case.

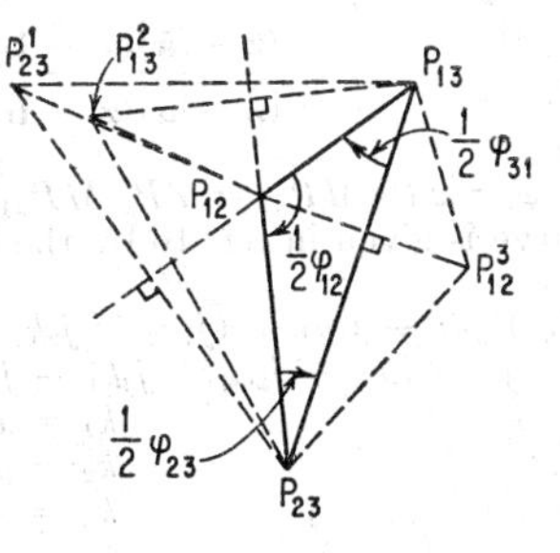

Fig. 4.32. Pole triangle for three positions of a plane. Pole triangle $P_{12}P_{23}P_{13}$ for three positions of a plane; image poles P_{12}^3, P_{23}^1, P_{31}^2; subtended angles $\frac{1}{2}\phi_{12}$, $\frac{1}{2}\phi_{23}$, $\frac{1}{2}\phi_{31}$.

Theorem of the Pole Triangle. The internal angles of the pole triangle, corresponding to three distinct positions of a plane, are equal to the corresponding halves of the associated angles of rotation ϕ_{ij} which are connected by the equation

$$\tfrac{1}{2}\phi_{12} + \tfrac{1}{2}\phi_{23} + \tfrac{1}{2}\phi_{31} = 180^\circ \qquad \angle\tfrac{1}{2}\phi_{ij} = \angle P_{ik}P_{ij}P_{jk}$$

Further developments, especially those involving subtention of equal angles, are found in the literature.[26]

For any three corresponding points A_1, A_2, A_3, the center M of the circle passing through these points is called a *center point*. If P_{ij} is considered as though fixed to link A_iB_i (or A_jB_j) and A_iB_i (or A_jB_j) is transferred to position k (A_kB_k), then P_{ij} moves to a new position $P_{ij}{}^k$ known as the *image pole*, because it is the image of P_{ij} reflected about the line joining $P_{ik}P_{jk}$. $\Delta P_{ik}P_{jk}P_{ij}{}^k$ is called an *image-pole triangle* (Fig. 4.32).

(c) Four Positions of a Plane (A_iB_i, $i = 1, 2, 3, 4$)

With four distinct positions, there are six poles P_{12}, P_{13}, P_{14}, P_{23}, P_{24}, P_{34} and four pole triangles ($P_{12}P_{23}P_{13}$), ($P_{12}P_{24}P_{14}$), ($P_{13}P_{34}P_{14}$), ($P_{23}P_{34}P_{24}$).

Any two poles, whose subscripts are all different, are called *complementary poles*. e.g., $P_{23}P_{14}$, or generally $P_{ij}P_{kl}$, where i, j, k, l represents any permutation of the numbers 1, 2, 3, 4. Two complementary-pole pairs constitute the two diagonals of a *complementary-pole quadrilateral*, of which there are three: ($P_{12}P_{24}P_{34}P_{13}$), ($P_{13}P_{32}P_{24}P_{14}$), and ($P_{14}P_{43}P_{32}P_{12}$).

Also associated with four positions are six further points Π_{ik} found by intersections of opposite sides of complementary-pole quadrilaterals, or their extensions, as follows: $\Pi_{ik} = P_{il}P_{kl} \times P_{ij}P_{kj}$.

(d) The Center-point Curve or Pole Curve[16,26,56,100]

For three positions, a center point corresponds to *any* set of corresponding points; for four corresponding points to have a common center point, point A_1 can no longer be located arbitrarily in plane AB. However, a curve exists in the frame of reference called the *center-point curve* or pole curve, which is the locus of centers of circles, each of which passes through *four* corresponding points of the plane AB. The center-point curve may be obtained from any complementary-pole quadrilateral; if associated with positions i, j, k, l, the center-point curve will be denoted by m_{ijkl}. Using complex numbers, let $\mathbf{OP}_{13} = \mathbf{a}$, $\mathbf{OP}_{23} = \mathbf{b}$, $\mathbf{OP}_{14} = \mathbf{c}$, $\mathbf{OP}_{24} = \mathbf{d}$, and $\mathbf{OM} = \mathbf{z} = x + iy$, where $\mathbf{OM}$ represents the vector from an arbitrary origin O to a point M on the center-point curve. The equation of the center-point[100] curve is given by

$$\frac{(\bar{\mathbf{z}} - \bar{\mathbf{a}})(\mathbf{z} - \mathbf{b})}{(\mathbf{z} - \mathbf{a})(\bar{\mathbf{z}} - \bar{\mathbf{b}})} = \frac{(\bar{\mathbf{z}} - \bar{\mathbf{c}})(\mathbf{z} - \mathbf{d})}{(\mathbf{z} - \mathbf{c})(\bar{\mathbf{z}} - \bar{\mathbf{d}})} = e^{2i\varphi} \tag{4.29}$$

where $\varphi = \angle P_{13}MP_{23} = \angle P_{14}MP_{24}$. In cartesian coordinates with origin at P_{12}, this curve is given in ref. 16 by the following equation:

$$\begin{aligned}(x^2 + y^2)(j_2x - j_1y) + (j_1k_2 - j_2k_1 - j_3)x^2 + (j_1k_2 - j_2k_1 + j_3)y^2 + 2j_4xy \\ + (-j_1k_3 + j_2k_4 + j_3k_1 - j_4k_2)x + (j_1k_4 + j_2k_3 - j_3k_2 - j_4k_1)y = 0\end{aligned} \tag{4.30}$$

where

$$\begin{aligned} k_1 &= x_{13} + x_{24} \\ k_2 &= y_{13} + y_{24} \\ k_3 &= x_{13}y_{24} + y_{13}x_{24} \\ k_4 &= x_{13}x_{24} - y_{13}y_{24} \\ j_1 &= x_{23} + x_{14} - k_1 \\ j_2 &= y_{23} + y_{14} - k_2 \\ j_3 &= x_{23}y_{14} + x_{14}y_{23} - k_3 \\ j_4 &= x_{23}x_{14} - y_{23}y_{14} - k_4 \end{aligned} \tag{4.31}$$

and (x_{ij}, y_{ij}) are the cartesian coordinates of pole P_{ij}. Equation (4.30) represents a third-degree algebraic curve, passing through the six poles P_{ij} and the six points

Π_{ij}.[26] Furthermore, any point M on the center-point curve subtends equal angles, or angles differing by two right angles, at opposite sides $(P_{ij}P_{jl})$ and $(P_{ik}P_{kl})$ of a complementary-pole quadrilateral, provided the sense of rotation of subtended angles is preserved:

$$\angle P_{ij}MP_{jl} = \angle P_{ik}MP_{kl} \cdots \qquad (4.32)$$

Construction of the Center-point Curve m_{ijkl}.[26] When the four positions of a plane are known (A_iB_i, $i = 1, 2, 3, 4$), the poles P_{ij} are constructed first; thereafter, the center-point curve is found as follows:

A chord $P_{ij}P_{jk}$ of a circle, center O, radius

$$R = \overline{P_{ij}P_{jk}}/2 \sin \theta$$

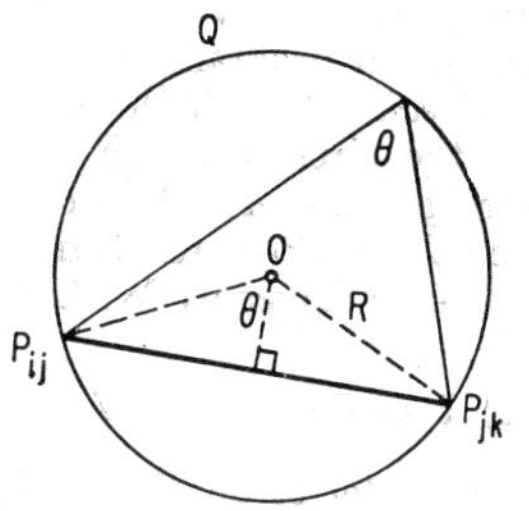

FIG. 4.33. Subtention of equal angles.

(Fig. 4.33) subtends the angle θ (mod π) at any point on its circumference. For any value of θ, $-180° \leq \theta \leq 180°$, two corresponding circles can be drawn following Fig. 4.33, using as chords the opposite sides $P_{ij}P_{jk}$ and $P_{il}P_{kl}$ of a complementary-pole quadrilateral; intersections of such corresponding circles are points (M) on the center-point curve, provided Eq. (4.32) is satisfied.

As a check, it is useful to keep in mind the following angular equalities:

$$\tfrac{1}{2}\angle A_iMA_l = \angle P_{ij}MP_{jl} = \angle P_{ik}MP_{kl}$$

Use of the Center-point Curve. Given four positions of a plane $A_iB_i (i = 1, 2, 3, 4)$ in a coplanar motion-transfer process, we can mechanize the motion by selecting points on the center-point curve as fixed pivots.

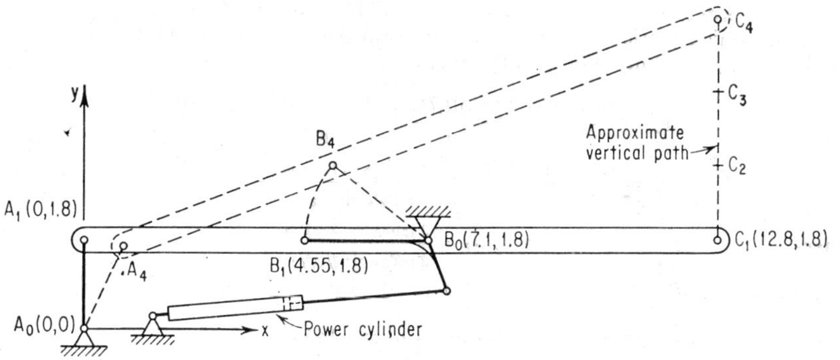

FIG. 4.34. Stacker conveyor drive.

Example[76] (Fig. 4.34): A stacker conveyor for corrugated boxes is based on the design shown schematically in Fig. 4.34. The path of C should be as nearly vertical as possible; if A_0, A_1, AC, $C_1C_2C_3C_4$ are chosen to suit the specifications, B_0 should be chosen on the center-point curve determined from A_iC_i, $i = 1, 2, 3, 4$; B_1 is then readily determined by inversion, i.e., by drawing the motion of B_0 relative to A_1C_1 and locating B_1 at the center of the circle thus described by B_0 (also see next paragraphs).[125]

(e) The Circle-point Curve

The circle-point curve is the kinematical inverse of the center-point curve. It is the locus of all points K in the moving plane whose four corresponding positions lie on one circle. If the circle-point curve is to be determined for position i of the plane AB, Eqs. (4.29), (4.30), and (4.31) would remain unchanged, except that P_{jk}, P_{kl}, and P_{jl} would be replaced by the image poles $P_{jk}{}^i$, $P_{kl}{}^i$, and $P_{jl}{}^i$, respectively.

The center-point curve lies in the frame or reference plane; the circle-point curve lies in the moving plane. In the above example, point B_1 is on the circle-point curve for plane AC in position 1. The example can be solved also by selecting B_1 on the circle-point curve in A_1C_1; B_0 is then the center of the circle through $B_1B_2B_3B_4$. A computer program for the center-point and circle-point curves is outlined in ref. 314.

Special Case: If the corresponding points $A_1A_2A_3$ lie on a straight line, A_1 must lie on the circle through $P_{12}P_{13}P_{23}{}^1$; for four corresponding points $A_1A_2A_3A_4$ on one straight line, A_1 is located at intersection, other than P_{12}, of circles through $P_{12}P_{13}P_{23}{}^1$ and $P_{12}P_{14}P_{24}{}^1$, respectively. Applied to straight-line guidance in slider-crank and four-bar drives by W. Lichtenheldt;[196] see also R. Beyer.[26]

(f) Five Positions of a Plane (A_iB_i, $i = 1, 2, 3, 4, 5$)

In order to obtain accurate motions, it is desirable to specify as many positions as possible; at the same time the design process becomes more involved, and the number of "solutions" becomes more restricted. Frequently four or five positions are the most that can be economically prescribed.

Associated with five positions of a plane are four sets of points $K_u{}^{(i)}$ ($u = 1, 2, 3, 4$ and i is the position index as before) whose corresponding five positions lie on one circle; to each of these circles, moreover, corresponds a center point M_u. These circle points $K_u{}^{(1)}$ and corresponding center points M_u are called *Burmester point pairs.* These four point pairs may be all real or pairwise imaginary (all real, two-point pairs real and two point pairs imaginary, or all point pairs imaginary).[100] Note the difference, due to historical reasons, between the above definition and that given in Art. 4.4(f) for infinitesimal motion. The location of the center points, M_u, can be obtained as the intersections of two center-point curves, such as m_{1234} and m_{1235}.

A complex-number derivation of their location[100] as well as a computer program for simultaneous determination of the coordinates of both M_u and $K_u{}^{(i)}$ is available.[100,312]

An algebraic equation for the coordinates (x_u, y_u) of M_u is given in ref. 16 as follows. Origin at P_{12}, coordinates of P_{ij} are x_{ij}, y_{ij}.

$$x_u = \frac{(u - \tan \tfrac{1}{2}\theta_{12})[l_1(k_2 - k_3u) - l_2(e_2 - e_3u)]}{p_1u^2 + p_2u + p_3} \qquad (4.33)$$

$$y_u = \frac{(u - \tan \tfrac{1}{2}\theta_{12})[l_1(k_4 + k_1u) - l_2(e_4 + e_1u)]}{p_1u^2 + p_2u + p_3}$$

where
$$\tan \tfrac{1}{2}\theta_{12} = \frac{x_{13}y_{23} - x_{23}y_{13}}{x_{13}x_{23} + y_{13}y_{23}} \qquad (4.34)$$

and u is a root of $\quad m_4u^4 + m_3u^3 + m_2u^2 + m_1u + m_0 = 0$

wherein
$$\begin{aligned}
m_0 &= p_3(q_1 + l_3p_3)\\
m_1 &= p_2(q_1 + 2l_3p_3)p_2 + q_2p_3 - q_3 \tan \tfrac{1}{2}\theta_{12}\\
m_2 &= q_0p_3 + q_2p_2 + q_1p_1 + l_3(p_2^2 + 2p_1p_3) - q_5 \tan \tfrac{1}{2}\theta_{12} + q_3\\
m_3 &= q_0p_2 + p_1(q_2 + 2l_3p_2) - q_4 \tan \tfrac{1}{2}\theta_{12} + q_5\\
m_4 &= p_1(q_0 + l_3p_1) + q_4
\end{aligned} \qquad (4.35)$$

$$\begin{aligned}
q_0 &= d_1h_3 - d_3h_1 & h_1 &= k_1l_1 - e_1l_2\\
q_1 &= d_2h_4 - d_4h_2 & h_2 &= k_2l_1 - e_2l_2\\
q_2 &= -d_1h_2 + d_2h_1 - d_3h_4 + d_4h_3 & h_3 &= k_3l_1 - e_3l_2\\
q_3 &= h_2^2 + h_4^2 & h_4 &= k_4l_1 - e_4l_2\\
q_4 &= h_1^2 + h_3^2 & p_1 &= k_3e_1 - k_1e_3\\
q_5 &= 2(h_1h_4 - h_2h_3) & p_2 &= k_3e_4 + k_1e_2 - k_2e_1 - k_4e_3\\
d_1 &= x_{15} + x_{25} & p_3 &= k_4e_2 - k_2e_4\\
d_2 &= x_{15} - x_{25} & e_1 &= d_1 - x_{13} - x_{23}\\
d_3 &= y_{15} + y_{25} & e_2 &= d_2 - x_{13} + x_{23}\\
d_4 &= y_{15} - y_{25} & e_3 &= d_3 - y_{13} - y_{23}\\
k_1 &= d_1 - x_{14} - x_{24} & e_4 &= d_4 - y_{13} + y_{23}\\
k_2 &= d_2 - x_{14} + x_{24} & l_1 &= x_{13}x_{23} + y_{13}y_{23} - l_3\\
k_3 &= d_3 - y_{14} - y_{24} & l_2 &= x_{14}x_{24} + y_{14}y_{24} - l_3\\
k_4 &= d_4 - y_{14} + y_{24} & l_3 &= x_{15}x_{25} + y_{15}y_{25}
\end{aligned} \qquad (4.36)$$

The Burmester point pairs are discussed in refs. 16, 56, and 100 and extensions of the theory in ref. 313*a*. It is suggested that, except in special cases, their determination warrants programmed computation.

Use of the Burmester Point Pairs. As in the example of Art. 4.5(d), the Burmester point pairs frequently serve as convenient pivot points in the design of linked mechanisms. Thus, in the stacker of Art. 4.5(d), five positions of C_i could have been specified in order to obtain a more accurately vertical path for C; the choice of locations of B_0 and B_1 would then have been limited to at most two Burmester point pairs (since A_0A_1 and $C_1C_0^\infty$, prescribed, are also Burmester point pairs).

(g) Point-position Reduction[2,125,156]

"Point-position reduction" refers to a construction for simplifying design procedures involving several positions of a plane. For five positions, graphical methods would involve the construction of two center-point curves or their equivalent. In point-position reduction, a fixed-pivot location, for instance, would be chosen so that one or more poles coincide with it. In the relative motion of the fixed pivot with reference to the moving plane, therefore, one or more of the corresponding positions coincide, thereby reducing the problem to four or fewer positions of the pivot point; the center-point curves, therefore, may not have to be drawn. The reduction in complexity of construction is accompanied, however, by increased restrictions in the choice of mechanism proportions. An exhaustive discussion of this useful tool is found in ref. 125.

(h) Complex-number Methods[96,304,305,312,313,315,327]

Burmester-point theory has been applied to function generation as well as to path generation and combined path and function generation.[100,312] The most general approach to path and function generation in plane motion utilizes complex numbers. The vector closure equations are used for each independent loop of the mechanism for every prescribed position and are differentiated once or several times if velocities, accelerations, and higher rates of change are prescribed. The equations are then solved for the unknown mechanism proportions. This method has been applied to four-bar path and function generators[96,100,304,312,315] (the former with prescribed crank rotations), as well as to a variety of other mechanisms. The so-called *path-increment* and *path-increment-ratio* techniques (see below) simplify the mathematics in so far as this is possible. In addition to path and function specification, these methods can take into account prescribed transmission angles, mechanical advantages, velocity ratios, accelerations, etc., and combinations of these.

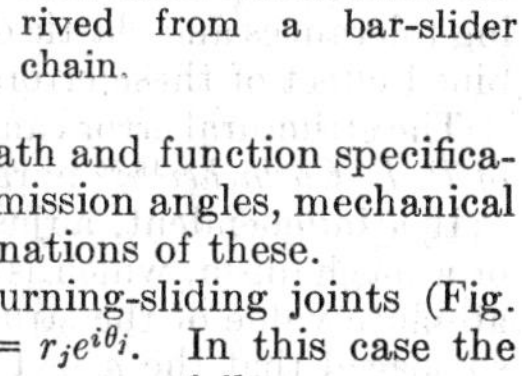

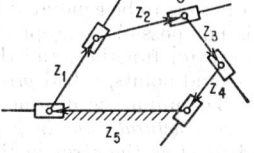

FIG. 4.35. Mechanism derived from a bar-slider chain.

Consider, for instance, a chain of links connected by turning-sliding joints (Fig. 4.35). Each bar slider is represented by the vector $z_j = r_j e^{i\theta_j}$. In this case the closure equations for the position shown, and its derivatives, are as follows:

Closure: $$\sum_{j=1}^{5} z_j = 0$$

Velocity: $$d/dt \sum_{j=1}^{5} z_j = 0 \quad \text{or} \quad \sum_{j=1}^{5} \lambda_j z_j = 0$$

where $$\lambda_j = (1/r_j)(dr_j/dt) + i(d\theta_j/dt) \qquad (t = \text{time})$$

Acceleration: $$d/dt \sum_{j=1}^{5} \lambda_j z_j = 0 \quad \text{or} \quad \sum_{j=1}^{5} \lambda_j \mu_j z_j = 0$$

where $$\mu_j = \lambda_j + (1/\lambda_j)(d\lambda_j/dt)$$

Similar equations hold for other positions. After suitable constraints are applied on the bar-slider chain (i.e., on r_j, θ_j) in accordance with the properties of the particular type of mechanism under consideration, the equations are solved for the z_j vectors, i.e., for the "initial" mechanism configuration.

If the path of a point such as C in Fig. 4.35 (although not necessarily a joint in the actual mechanism represented by the schematic or "general" chain) is specified for a number of positions by means of vectors $\delta_1, \delta_2, \ldots, \delta_k$, the *path increments* measured from the initial position are $(\delta_j - \delta_1)$, $j = 2, 3, \ldots, k$, and the *path increment ratios* are $(\delta_j - \delta_1)/(\delta_2 - \delta_1)$, $j = 3, 4, \ldots, k$. By working with these quantities, only moving links or their ratios are involved in the computations. The solution of these equations of synthesis usually involves the prior solution of nonlinear *compatibility equations,* obtained from matrix considerations. Additional details are covered in the above-mentioned references. A number of related computer programs for the synthesis of linked mechanisms are described in ref. 102. Numerical methods suitable for such syntheses are described in ref. 305.

4.6. DESIGN REFINEMENT

After the mechanism is selected and its approximate dimensions determined, it may be necessary to refine the design by means of relatively small changes in the proportions, based on more precise design considerations. Equivalent mechanisms may also present improvements.

(a) Optimization of Proportions for Generating Prescribed Motions with Minimum Error

Whenever mechanisms possess a limited number of independent dimensions, only a finite number of independent conditions can be imposed on their motion. Thus, if a path is to be generated by a point on a linkage (rather than, say, a cam follower), it is not possible—except in special cases—to generate the curve exactly. A desired path (or function) and the actual, or generated, path (or function) may coïncide at several points, called *precision points*; between these, the curves differ.

The minimum distance from a point on the ideal path to the actual path is called the *structural error in path generation.* The *structural error in function generation* is defined as the error in the ordinate (dependent variable y) for a given value of the abscissa (independent variable x). Structural errors exist independent of manufacturing tolerances and elastic deformations and are thus inherent in the design. The combined effect of these errors should not exceed the maximum tolerable error.

The structural error can be minimized by the application of the fundamental *Theorem of P. L. Chebichev*[16,36] phrased nonrigorously for mechanisms as follows:

If n independent, adjustable proportions (parameters) are involved in the design of a mechanism, which is to generate a prescribed path or function, then the largest absolute value of the structural error is minimized when there are n precision points so spaced that the $n + 1$ maximum values of the structural error between each pair of adjacent precision points—as well as between terminals and the nearest precision points—are numerically equal with successive alterations in sign.

In Fig. 4.36 (applied to function generation) the maximum structural error in each "region," such as 01, 12, 23, and 34, is shown as ϵ_{01}, ϵ_{12}, ϵ_{23}, and ϵ_{34}, respectively, which represent vertical distances between ideal and generated functions having three precision points. In general, the mechanism proportions and the structural error will vary with the choice of precision points. The spacing of precision points which yields least maximum structural error is called *optimal spacing.* Other definitions and concepts, useful in this connection, are the following:

n-point approximation: generated path (or function) has n precision points.

nth-order approximation: limiting case of n-point approximation, as the spacing between precision points approaches zero. In the limit, one precision point is retained, at which point, however, the first $(n - 1)$ derivatives, or rates of change of the generated path (or function), have the same values as those of the ideal path (or function).

The following paragraphs apply both to function generation and to planar path generation, provided (in the latter case) that x is interpreted as the arc length along the ideal curve and the structural error ϵ_{ij} refers to the distance between generated and ideal curves.

Chebichev Spacing.[95] For an n-point approximation to $y = f(x)$, within the range $x_0 \leq x \leq x_{n+1}$, Chebichev spacing of the n precision points x_j is given by

$$x_j = \tfrac{1}{2}(x_0 + x_{n+1}) - \tfrac{1}{2}(x_{n+1} - x_0)\cos\{[(2j-1)\pi]/2n\} \qquad j = 1, 2, \ldots, n$$

Though not generally optimum for finite ranges, Chebichev spacing often represents a good first approximation to optimal spacing.

The process of respacing the precision points, so as to minimize the maximum structural error, is carried out numerically[95] unless an algebraic solution is feasible.[36]

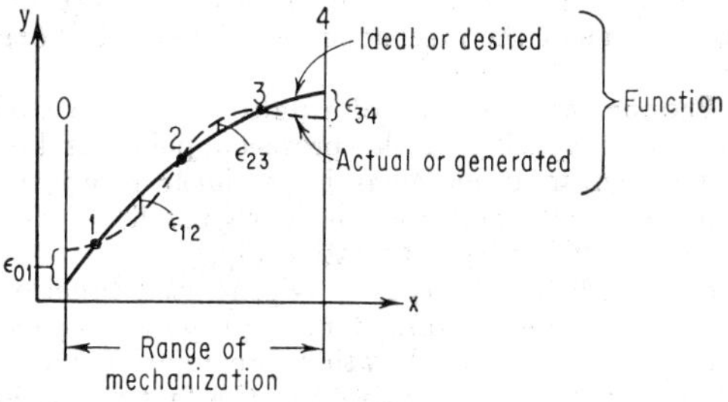

FIG. 4.36. Precision points 1, 2, and 3 and "regions" 01, 12, 23, and 34, in function generation.

Respacing of Precision Points to Reduce Structural Error via Successive Approximations. Let $x_{ij}^{(1)} = x_j^{(1)} - x_i^{(1)}$, where $j = i+1$, and let $x_i^{(1)} (i = 1, 2, \ldots, n)$ represent precision-point locations in a first approximation as indicated by the superscript (1). Let $\epsilon_{ij}^{(1)}$ represent the maximum structural error between points $x_i^{(1)}$, $x_j^{(1)}$, in the first approximation with terminal values x_0, x_{n+1}. Then a second spacing $x_{ij}^{(2)} = x_j^{(2)} - x_i^{(2)}$ is sought for which $\epsilon_{ij}^{(2)}$ values are intended to be closer to optimum (i.e., more nearly equal); it is obtained from

$$x_{ij}^{(2)} = \frac{x_{ij}^{(1)}(x_{n+1} - x_0)}{[\epsilon_{ij}^{(1)}]^m \sum_{i=0}^{n} \{x_{ij}^{(1)}/[\epsilon_{ij}^{(1)}]^m\}} \tag{4.37}$$

The value of the exponent m generally lies between 1 and 3. Errors can be minimized also according to other criteria, for instance, according to least squares.[194]

Estimate of Least Possible Maximum Structural Error ϵ_{opt}. In case of an n-point Chebichev spacing in the range $x_0 \leq x \leq x_{n+1}$ with maximum structural errors, ϵ_{ij}, $j = i + 1$, $i = 0, 1, \ldots, n$,

$$\epsilon^2_{\text{opt (estimate)}} = (1/2n)[\epsilon^2_{01} + \epsilon^2_{n(n+1)}] + (1/n)[\epsilon^2_{12} + \epsilon^2_{23} + \cdots + \epsilon^2_{(n-1)n}] \tag{4.38}$$

In other spacings different estimates should be used; in the absence of more refined evaluations, the root-mean-square value can be used in the general case. These estimates may show whether a refinement of precision-point spacing is worthwhile.

Chebichev Polynomials. Concerning the effects of increasing the number of precision points or changing the range, some degree of information may be gained from an examination of the *Chebichev polynomials.* The Chebichev polynomial $T_n(t)$ is that nth-degree polynomial in t (with leading coefficient unity) which deviates least from zero within the interval $\alpha \le t \le \beta$. It can be obtained from the following differential-equation identity by equating to zero coefficients of like powers of t:

$$2[t^2 - (\alpha + \beta)t + \alpha\beta]T_n''(t) + [2t - (\alpha + \beta)]T_n'(t) - 2n^2T_n(t) \equiv 0$$

where the primes refer to differentiation with respect to t. The maximum deviation from zero, L_n, is given by

$$L_n = (\alpha - \beta)^n/2^{2n-1}$$

For the interval $-1 \le t \le 1$, for instance,

$$T_n(t) = (1/2^{n-1}) \cos (n \cos^{-1} t) \qquad L_n = 1/2^{n-1}$$
$$T_1(t) = t$$
$$T_2(t) = t^2 - \tfrac{1}{2}$$
$$T_3(t) = t^3 - (\tfrac{3}{4})t$$
$$T_4(t) = t^4 - t^2 + \tfrac{1}{8} \qquad \text{etc.}$$

Chebichev polynomials can be used directly in algebraic synthesis, provided the motion and proportions of the mechanism can be suitably expressed in terms of such polynomials.[36]

Adjusting the Proportions of a Mechanism for Given Respacing of Precision Points. Once the respacing of the precision points is known, it is possible to recompute the mechanism proportions by a linear computation[95,140,194,336] provided the changes in the proportions are sufficiently small.

Let $f(x)$ = ideal or desired functional relationship,

$g(x) = g(x, p_0^{(1)}, p_1^{(1)}, \ldots, p_{n-1}^{(1)})$ = generated functional relationship in terms of mechanism parameters or proportions $p_j^{(1)}$, where $p_j^{(k)}$ refers to the jth parameter in the kth approximation,

$\epsilon^{(1)}(x_i^{(2)})$ = value of structural error at $x_i^{(2)}$ in the first approximation, where $x_i^{(2)}$ is a new or respaced location of a precision point, such that ideally $\epsilon^{(2)}(x_i^{(2)}) = 0$ (where $\epsilon = f\text{-}g$).

Then the new values of the parameters $p_j^{(2)}$ can be computed from the equations

$$\epsilon^{(1)}(x_i^{(2)}) = \sum_{j=0}^{n-1} [\partial g(x_i^{(2)})/\partial p_j^{(1)}](p_j^{(2)} - p_j^{(1)}) \qquad i = 1, 2, \ldots, n \tag{4.39}$$

These are n linear equations, one each at the n "precision" points $x_i^{(2)}$ in the n unknowns $p_j^{(2)}$. The convergence of this procedure depends on the appropriateness of neglecting higher-order terms in Eq. (4.39); this in turn depends on the functional relationship and the mechanism and cannot in general be predicted. For related investigations, see refs. 104 and 150; for respacing via automatic computation and for accuracy obtainable in four-bar function generators, see ref. 95.

(b) Tolerances and Precision[17,115,124,140,179,190,363]

After the structural error is minimized, the effects of manufacturing errors still remain.

The accuracy of a motion is frequently expressed as a percentage defined as the maximum output error divided by total output travel (range).

For a general discussion of the various types of errors, see ref. 363.

Machining errors may cause changes in link dimensions, as well as clearances and backlash. Correct tolerancing requires the investigation of both. If the errors in link dimensions are small compared with the link lengths, their effect on displacements, velocities, and accelerations can be determined by a linear computation, using only first-order terms.

The effects of clearances in the joints and of backlash are more complicated and, in addition to kinematic effects, are likely to affect adversely the dynamic behavior of the mechanism.[115] The kinematic effect manifests itself as an uncertainty in displacements, velocities, accelerations, etc., which, in the absence of load reversal, can be computed as though due to a change in link length, equivalent to the clearance or backlash involved.

Since the effect of tolerances will depend on the mechanism and on the "location" of the tolerance in the mechanism, each tolerance should be specified in accordance with the magnitude of its effect on the pertinent kinematic behavior.

(c) Harmonic Analysis (see also Art. 4.9 and bibliography in ref. 371)

It is sometimes desirable to express the motion of a machine part as a Fourier series in terms of driving motion, in order to analyze dynamic characteristics and to ensure satisfactory performance at high speeds. Harmonic analysis, for example, is used in computing the inertia forces in slider-crank mechanisms in internal-combustion engines[19,33,300] and also in other mechanisms.[101,227,230,371]

Generally, two types of investigations arise:

1. Determination of the "harmonics" in the motion of a given mechanism as a check on inertial loads and critical speeds
2. Proportioning to minimize higher harmonics[101]

(d) Transmission Angles [see also Art. 4.2(i)][378]

In a mechanism with varying transmission angle μ the optimum condition is frequently attained by equalizing the extreme deviation of μ from a right angle: $\mu_{\max} = 90° + \nu$, $\mu_{\min} = 90° - \nu$. In crank-and-rocker mechanisms, for instance (lengths: crank $= a$, coupler $= b$, rocker $= c$, fixed link $= d$), this is achieved when $a^2 + b^2 = c^2 + d^2$, in which case $\sin \nu = ab/cd$ (see also Art. 4.9).

(e) Design Charts

To save labor in the design process, charts and atlases are useful when available. Among these are refs. 161 and 169 in four-link motion; the VDI-Richtlinien Duesseldorf (obtainable through Beuth-Vertrieb Gmbh, Berlin), such as 2131, 2132 on the offset turning block and the offset slider crank, and 2125, 2126, 2130, 2136 on the offset slider-crank and crank-and-rocker mechanisms; 2123, 2124 on four-bar mechanisms; 2137 on the in-line swinging block; and data sheets in the technical press.

(f) Equivalent and "Substitute" Mechanisms[87]

Kinematic equivalence is explained in Art. 4.2(k). Ways of obtaining equivalent mechanisms include (1) pin enlargement, (2) kinematic inversion, (3) use of centrodes, (4) use of curvature constructions, (5) use of pantograph devices, (6) use of multigeneration properties, (7) substitution of tapes, racks, and chains for rigid links[69,83,125,152,226] and other ways depending on the inventiveness of the designer.* Of these, 5 and 6 require additional explanation.

The *pantograph* can be used to reproduce a given motion, unchanged, enlarged, reduced, or rotated. It is based on "Sylvester's plagiograph," shown in Fig. 4.37.

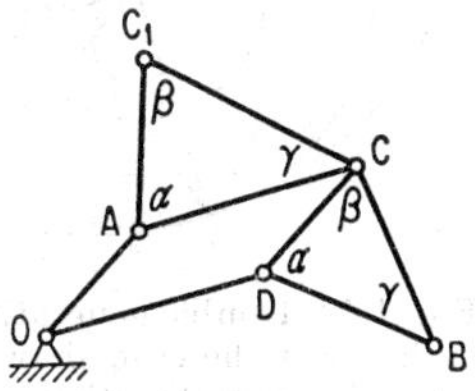

FIG. 4.37. Sylvester's plagiograph or skew pantograph.

AODC is a parallelogram linkage with point O fixed with two similar triangles ACC_1, DBC, attached as shown. Points B and C_1 will trace similar curves, altered in the ratio $OC_1/OB = AC_1/AC$ and rotated relative to each other by an amount equal to the angle α. The ordinary pantograph is the special case obtained when B, D, C, and C, A, C_1 are collinear. It is used in engraving machines and other motion-copying devices.

* Investigation of enumeration of mechanisms based on degree-of-freedom requirements are found in refs. 125 and 128 to 131 with application to clamping devices, tools, jigs, fixtures, and vice jaws.

Roberts's theorem[26,148,229,281] states that there are three different but related four-bar mechanisms generating the same coupler curve (Fig. 4.38): the "original" *ABCDE*, the "right cognate" *LKGDE*, and the "left cognate" *LHFAE*. Similarly, slider-crank mechanisms have one cognate each.[148]

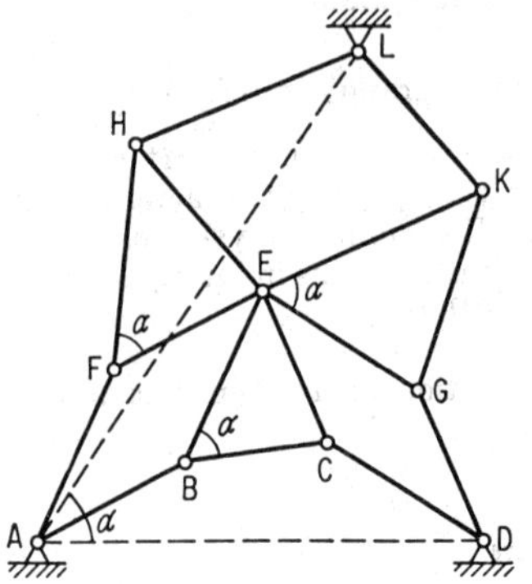

FIG. 4.38. Roberts's theorem. $\Delta BEC \approx \Delta FHE \approx \Delta EKG \approx \Delta ALD \approx \Delta AHC \approx \Delta BKD \approx \Delta FLG$; *AFEB*, *EGDC*, *HLKE* are parallelograms.

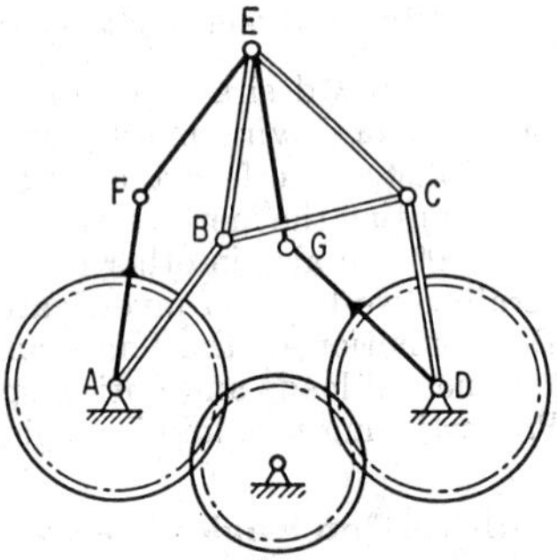

FIG. 4.39. Four-bar linkage *ABCDE* and equivalent 1:1 geared five-bar mechanism *AFEGD*; *AFEB* and *DGEC* are parallelograms.

If the "original" linkage has poor proportions, a cognate may be preferable. When Grashof's inequality is obeyed (Art. 4.9) and the original is a double rocker, the cognates are crank-and-rocker mechanisms; if the original is a drag link, so are the cognates; if the original does not obey Grashof's inequality, neither do the cognates, and all three are either double-rockers or folding linkages. Several well-known straight-line guidance devices (Watt and Evans mechanisms) are cognates.

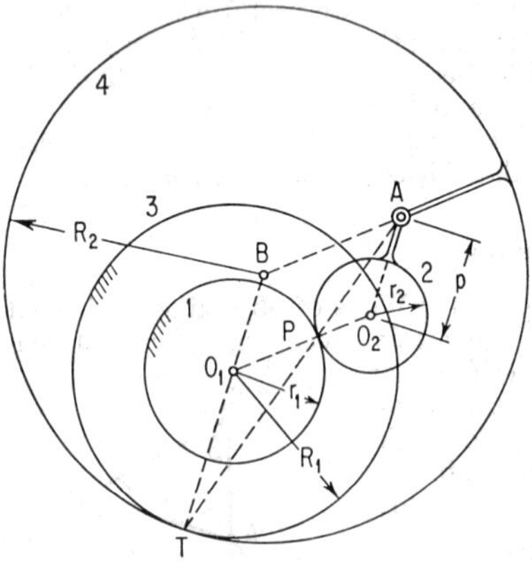

FIG. 4.40. Double generation of a cycloidal path. For the case shown O_1O_2 and AO_2 rotate in same direction.

$$R_2/R_1 = r_2/r_1 + 1; \qquad R_1 = p(r_1/r_2);$$
$$R_2 = p\,[1 + (r_1/r_2)]$$

Radius ratios are considered positive or negative depending on whether gearing is internal or external.

Geared five-bar mechanisms[80,93,93a,281,305] may also be used to generate the coupler curve of a four-bar mechanism, possibly with better transmission angles and proportions, as, for instance, in the drive of a deep-draw press. The gear ratio in this case is 1:1 (Fig. 4.39), where *ABCDE* is the four-bar linkage and *AFEGD* is the five-bar mechanism with links *AF* and *GD* geared to each other by 1:1 gearing. The path of *E* is identical in both mechanisms.

In Fig. 4.38 each cognate has one such derived geared five-bar mechanism (as in Fig. 4.39), thus giving a choice of six different mechanisms for the generation of any one coupler curve.

Double Generation of Cycloidal Curves.[256] A given cycloidal motion can be obtained by two different pairs of rolling circles (Fig. 4.40). Circle 2 rolls on fixed circle 1 and point *A*, attached to circle 2, describes a cycloidal curve. If O_1, O_2 are centers of circles 1 and 2, *P* their point of contact, and O_1O_2AB a parallelogram, circle 3, which is also fixed, has center O_1 and radius O_1T, where *T* is the intersection of extensions of O_1B and *AP*; circle 4 has center B, radius BT, and rolls on circle 3. If point A is now rigidly attached to circle 4, its path will be the same as before. Dimensional relationships are given in the caption of Fig. 4.40. For analysis of cycloidal motions see ref. 377.

Equivalent mechanisms obtained by multigeneration theory may yield patentable devices by producing "unexpected" results, which constitutes one criterion of patentability. In one application, cycloidal path generation has been used in a speed reducer.[42,320,373] Another form of "cycloidal equivalence" involves adding an idler gear to convert from, say, internal to external gearing; applied to resolver mechanism in ref. 291.

4.7. THREE-DIMENSIONAL MECHANISMS[5,21,29,161a] (Art. 4.9)

Also called *spatial mechanisms*, points on these move on three-dimensional curves. The basic three-dimensional mechanisms are the *spherical four-bar mechanism* (Fig. 4.41) and the *offset* or *spatial four-bar mechanism* (Fig. 4.42).

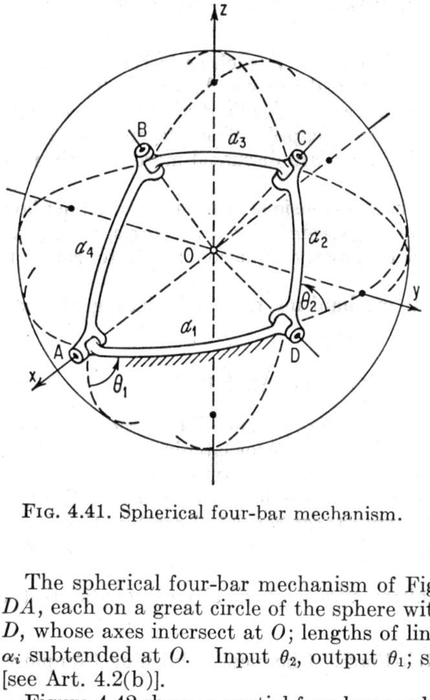

FIG. 4.41. Spherical four-bar mechanism.

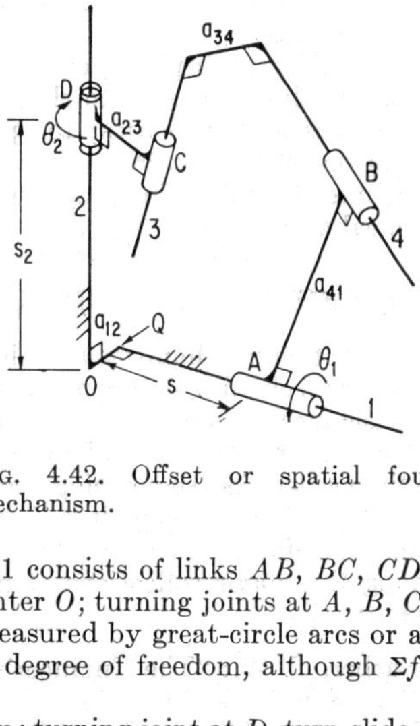

FIG. 4.42. Offset or spatial four-bar mechanism.

The spherical four-bar mechanism of Fig. 4.41 consists of links AB, BC, CD, and DA, each on a great circle of the sphere with center O; turning joints at A, B, C, and D, whose axes intersect at O; lengths of links measured by great-circle arcs or angles α_i subtended at O. Input θ_2, output θ_1; single degree of freedom, although $\Sigma f_i = 4$ [see Art. 4.2(b)].

Figure 4.42 shows a spatial four-bar mechanism; turning joint at D, turn-slide joints at B, C, and D; a_{ij} denote minimum distances between axes of joints; input θ_2 at D; output at A consists of translation s and rotation θ_1; $\Sigma f_i = 7$; freedom, $F = 1$.

Three-dimensional mechanisms used in practice are usually special cases of the above two mechanisms. Among these are Hooke's joint (a spherical four-bar, with $\alpha_2 = \alpha_3 = \alpha_4 = 90°$, $90° < \alpha_1 \leq 180°$), the wobble plates ($\alpha_4 < 90°$, $\alpha_2 = \alpha_3 = \alpha_1 = 90°$), the space crank,[268] the spherical slider crank,[245] and other mechanisms, whose analysis is outlined in Art. 4.9.

The analysis and synthesis of spatial mechanisms require special tools to reduce their complexity. The analysis of displacements, velocities, and accelerations of the general spatial chain (Fig. 4.42) is conveniently accomplished with the aid of dual numbers, dual vectors, matrices, quaternions, and Cayley-Klein parameters.[70,72,372] The spherical four-bar (Fig. 4.41) can be analyzed the same way, or by spherical trigonometry.[28] A computer program by J. Denavit and R. S. Hartenberg[102] is available for the analysis and synthesis of a spatial four-link mechanism whose terminal axes are nonparallel and nonintersecting, and whose two moving pivots are ball joints. See also ref. 372 for additional spatial computer programs. For the simpler problems, for verification of computations and for visualization, graphical layouts are useful.[21,27,30,32,346]

Applications of three-dimensional mechanisms involve these motions:

1. Combined translation and rotation (e.g., door openers to lift and slide simultaneously[3])
2. Compound motions, such as in paint shakers, mixers, dough-kneading machines and filing[8,29,30,245]
3. Motions in shaft couplings, such as universal and constant-velocity joints[4,6,21,29,203] (see Art. 4.9)
4. Motions around corners and in limited space, such as in aircraft, certain wobble-plate engines, and lawn mowers.[52,251,268]
5. Complex motions, such as in aircraft landing gear and remote-control handling devices[60,211]

When the motion is constrained ($F = 1$), but $\Sigma f_i < 7$ (such as in the mechanism shown in Fig. 4.41), any elastic deformation will tend to cause binding. This is not the case when $\Sigma f_i = 7$, as in Fig. 4.42, for instance. Under light-load low-speed conditions, however, the former may represent no handicap.[10] The "degenerate" cases, usually associated with parallel or intersecting axes, are discussed more fully in refs. 3, 10, 111, and 369.

In the analysis of displacements and velocities, extensions of the ideas used in plane kinematic analysis have led to the notions of the *instantaneous screw axis*,[32] valid for displacements and velocities; to spatial Euler-Savary equations; and to concepts involving line geometry.[175]

Care must be taken in designing spatial mechanisms to avoid binding and low mechanical advantages.

4.8. CLASSIFICATION AND SELECTION OF MECHANISMS

In this section, mechanisms and their components are grouped into three categories:

A. Basic mechanism components, such as those adapted for latching, fastening, etc.

B. Basic mechanisms: the building blocks in most mechanism complexes.

C. Groups or assemblies of mechanisms, characterized by one or more displacement-time schedules, sequencing, interlocks, etc.; these consist of combinations from categories A and B and constitute important mechanism units or independent portions of entire machines.

Among the major collections of mechanisms and mechanical movements are the following:

1M: Barber, T. W.: "The Engineer's Sketch-Book," Chemical Publishing Company, Inc., New York, 1940.

2M: Beggs, J. S.: "Mechanism," McGraw-Hill Book Company, Inc., New York, 1955.

3M: Hain, K.: "Die Feinwerktechnik," Fachbuch-Verlag, Dr. Pfanneberg & Co., Giessen, Germany, 1953.

4M: "Ingenious Mechanisms for Designers and Inventors," vols. 1–2, edited by F. D. Jones; vol. 3, edited by H. L. Horton, The Industrial Press, New York, 1930–1951.

5M: Rauh, K.: "Praktische Getriebelehre," Springer-Verlag OHG, Berlin, vol. I, 1951; vol. II, 1954.

There are, in addition, numerous others, as well as more special compilations, the vast amount of information in the technical press, the AWF publications,[350] and (as a useful reference in depth), the Engineering Index. For some mechanisms, especially the more elementary types involving fewer than six links, a systematic enumeration of kinematic chains based on degrees of freedom may be worthwhile,[128–131] particularly if questions of patentability are involved. Mechanisms are derived from the kinematic chains by holding one link fixed and possibly by using equivalent and substitute mechanisms (Sec. 4.6f). The present state of the art is summarized in ref. 125.

In the following list of mechanisms and components, each item is classified according to category (A, B, or C) and is accompanied by references, denoting one or more of the above five sources, or those at the end of this section. In using this listing, it is to be remembered that a mechanism used in one application may frequently be employed in a completely different one, and sometimes combinations of several mechanisms may be useful.

The categories A, B, C, or their combinations are approximate in some cases, since it is often difficult to determine a precise classification.

Adjustments, fine (A,1M)(A,2M)(A,3M)
Adjustments, to a moving mechanism (A,2M)(AB,1M); see also Transfer, power
Airplane instruments and linkages (C,3M)[276]
Analog computing mechanisms;[247] see also Computing mechanisms
Anchoring devices (A,1M)
Automatic machinery, special-purpose;[127] automatic handling[297]
Ball bearings, guides and slides (A,3M)
Ball-and-socket joints (A,1M); see also Joints
Band drives (B,3M); see also Tapes
Bearings (A,3M)(A,1M); jewel;[192] for oscillating motion[48]
Belt gearing (B,1M)
Bolts (A,1M)
Brakes (B,1M)
Business machines, bookkeeping and records (C,3M)
Calculating devices (C,3M);[218] see also Mathematical instruments
Cameras (C,3M); see also Photographic devices
Cam-link mechanisms (BC,1M)(BC,5M)
Cams and cam drives (BC,4M)(BC,5M)(BC,1M)[307]
Carriages and cars (BC,1M)
Centrifugal devices (BC,1M)
Chain drives (B,1M)(B,5M)[206]
Chucks, clamps, grips, holders (A,1M)
Circular-motion devices (B,1M)
Clock mechanisms;[18] see also Escapements
Clutches, overrunning (BC,1M)(C,4M); see also Couplings and clutches
Computing mechanisms (BC,5M)[66,212,272,336]
Couplings and clutches (B,1M)(B,3M)(B,5M);[107,108,176,202,270,332,343] see also Joints
Covers and doors (A,1M)
Cranes (AC,1M)[65]
Crank and eccentric gear devices (BC,1M)
Crushing and grinding devices (BC,1M)
Curve-drawing devices (BC,1M); see also Writing instruments and Mathematical instruments
Cushioning devices (AC,1M)
Cutting devices (A,1M)[265,288]
Detents (A,3M)
Differential motions (C,1M)(C,4M)[151]
Differentials (B,5M)[4]
Dovetail slides (A,3M)
Drilling and boring devices (AC,1M)
Driving mechanisms for reciprocating parts (C,3M)
Duplicating and copying devices (C,3M)
Dwell linkages (C,4M)
Ejecting mechanisms for power presses (C,4M)

Spanners (A,1M)
Speed-changing mechanisms (C,4M); also see Transmissions
Spindles and centers (A,1M)
Springs (A,1M)(A,3M); and devices;[67,326] fastening of[277]
Springs and mechanisms (BC,5M)
Steering mechanisms (BC,5M)[199,337,373]
Stop mechanisms (C,4M)
Stops (A,2M)[277]
Straight-line motions, guides, parallel motions and devices (B,1M)(BC, 4M);[61,118,123,357] Art. 4.9
Struts and ties (A,1M)
Substitute mechanisms[87]
Swivels (A,1M)
Tape drives and devices (B,3M)(B,5M)[21,152]
Threads[277]
Three-dimensional drives;[8,245] Art. 4.9
Time-measuring devices (C,3M); timers[215]
Toggles[106,112,216,321,375]
Torsion devices[109]
Toys, mechanisms used in[253]
Tracks and rails (A,1M)
Transducers (AB,2M)(C,3M)[1,46,59,85,358]
Transfer, of parts, or station advance (B,2M)
Transfer, power to moving mechanisms (AB,2M)
Transmissions and speed changers (BC,5M);[4,7,329] see also Variable mechanical advantage and Speed-changing mechanisms
Transmissions, special (C,4M)[145]
Tripping mechanisms (C,4M)
Typewriting devices (C,3M)[154,207]
Universal joints[21,203,205,311,335]
Valve gear (BC,1M)
Valves (A,1M); design of nonlinear[269,306]
Variable mechanical advantage and power devices (A,1M);[209,333,376] see also Transmissions and speed changers
Washing devices (A,1M)
Wedge devices (A,3M)(B,5M)
Weighing devices (AB,1M)
Weights, for compensation and balancing (A,1M)
Wheels (A,1M)
Wheels, elastic (A,1M)
Windmill and feathering devices (A,1M)
Window-regulating mechanisms[110]
Woodworking machines[214]
Writing instruments (C,3M); see also Curve-drawing devices and Mathematical instruments

4.9. KINEMATIC PROPERTIES OF MECHANISMS

For more complete literature survey see refs. 1 to 6, cited in ref. 97, and the Engineering Index; currently available computer programs are listed in ref. 102.

(a) The General Slider-crank Chain (Fig. 4.43)

Nomenclature

A = crankshaft axis	t = time
B = crankpin axis	Block at C = slider
C = wrist-pin axis	s = stroke
FD = guide	θ = crank angle

$AF = e =$ offset, $\perp FD$
$AB = r =$ crank
$BC = l =$ connecting rod
$x =$ displacement of C in direction of guide, measured from F
$\phi =$ angle between connecting rod and slide, pressure angle
$\tau = \angle ABC$
$\eta = \angle BGP =$ auxiliary angle
$PG =$ collineation axis; $CP \perp FD$

The following mechanisms are derivable from the general slider-crank chain:

1. The *slider-crank mechanism;* guide fixed; if $e \neq 0$, called "offset," if $e = 0$, called "in-line"; $\lambda = r/l$; in case of the in-line slider crank, if $\lambda < 1$, AB rotates; if $\lambda > 1$, AB oscillates.
2. *Swinging-block mechanism;* connecting rod fixed; "offset" or "in-line" as in 1.
3. *Turning-block mechanism;* crank fixed; exact kinematic equivalent of 2; see Fig. 4.44.
4. The *standard geneva mechanism* is derivable from the special case, $e = 0$ (Fig. 4.45*b*).
5. Several *variations* of the *geneva mechanism* and other pin-and-slot or block-and-slot drives.

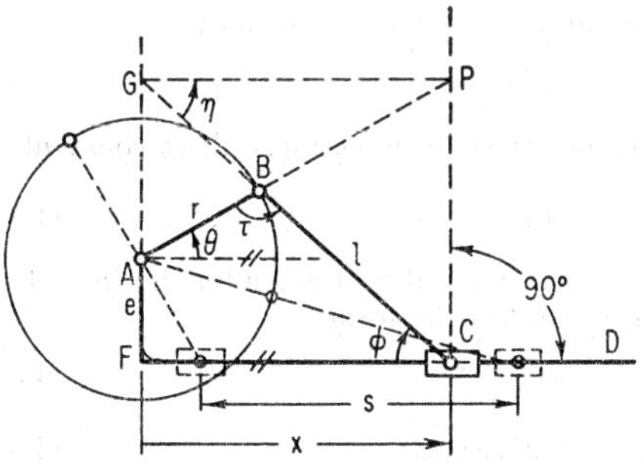

FIG. 4.43. General slider-crank chain.

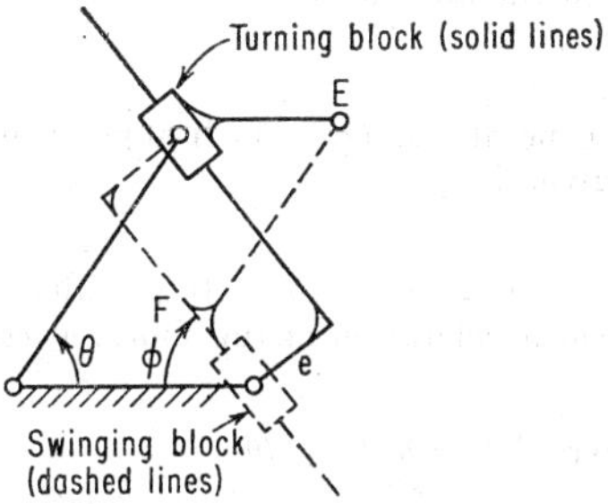

FIG. 4.44. Kinematic equivalence of the swinging-block and turning-block mechanisms, shown by redundant connection *EF*.

(b) The Offset Slider-crank Mechanism (see Fig. 4.43 with *AFD* stationary[33,36,84,148,237,324,378])

Let

$$\lambda = r/l \qquad \epsilon = e/l \tag{4.40}$$

where l is the length of the connecting rod, then

$$s = l[(1 + \lambda)^2 - \epsilon^2]^{1/2} - l[(1 - \lambda)^2 - \epsilon^2]^{1/2} \tag{4.41}$$

$$\sin \phi = \epsilon + \lambda \sin \theta \tag{4.42}$$

$$\tau = \pi - \phi - \theta \tag{4.43}$$

and

$$x = r \cos \theta + l \cos \phi \tag{4.44}$$

Let the angular velocity of the crank be $d\theta/dt = \omega$; then the slider velocity is given by

$$\frac{dx}{dt} = r\omega \frac{-\sin(\theta + \phi)}{\cos \phi} \tag{4.45}$$

Extreme value of dx/dt occurs when the auxiliary angle $\eta = 90°$.[90]
Slider acceleration ($\omega =$ constant):

$$\frac{d^2x}{dt^2} = r\omega^2 \left[\frac{-\cos(\theta + \phi)}{\cos \phi} - \lambda \frac{\cos^2 \theta}{\cos^3 \phi} \right] \tag{4.46}$$

Slider shock (ω = constant):

$$\frac{d^3x}{dt^3} = r\omega^3 \left[\frac{\sin(\theta + \phi)}{\cos\phi} + \frac{3\lambda\cos\theta}{\cos^5\phi}(\sin\theta\cos^2\phi - \lambda\sin\phi\cos^2\theta) \right] \tag{4.47}$$

For the angular motion of the connecting rod, let the angular velocity ratio,

$$m_1 = d\phi/d\theta = \lambda(\cos\theta/\cos\phi) \tag{4.48}$$

Then the angular velocity of the connecting rod

$$d\phi/dt = m_1\omega \tag{4.49}$$

Let

$$m_2 = d^2\phi/d\theta^2 = m_1(m_1\tan\phi - \tan\theta) \tag{4.50}$$

Then the angular acceleration of the connecting rod, at constant ω, is given by

$$d^2\phi/dt^2 = m_2\omega^2 \tag{4.51}$$

In addition, let

$$m_3 = d^3\phi/d\theta^3 = 2m_1m_2\tan\phi - m_2\tan\theta + m_1^3\sec^2\phi - m_1\sec^2\theta \tag{4.52}$$

Then the angular shock of the connecting rod, at constant ω, becomes

$$d^3\phi/dt^3 = m_3\omega^3$$

In general, the $(n - 1)$th angular acceleration of the connecting rod, at constant ω, is given by

$$d^n\phi/dt^n = m_n\omega^n \tag{4.53}$$

where $m_n = dm_{n-1}/d\theta$. In a similar manner, the general expression for the $(n - 1)$th linear acceleration of the slider, at constant ω, takes the form

$$d^nx/dt^n = r\omega^n M_n \tag{4.54}$$

where $M_n = dM_{n-1}/d\theta$

$$M_1 = -\sin(\theta + \phi)/\cos\phi \tag{4.55}$$

and M_2 and M_3 are the bracketed expressions in Eqs. (4.46) and (4.47), respectively.

Kinematic characteristics governed by Eqs. (4.40) to (4.55). Examples for path and function generation, ref. 324. Harmonic analyses, refs. 33 and 237. Coupler curves, ref. 84. Cognates, ref. 148. Offset slider-crank mechanism can be used to reduce the friction of the slider in the guide during the "working" stroke; transmission-angle charts, ref. 378.

Amplitudes of the harmonics are slightly higher than for the in-line slider-crank with same λ value.

For a nearly constant slider velocity $(1/\omega)(dx/dt) = k$ over a portion of the motion cycle, the proportions[36]

$$12k = 3e \pm \sqrt{9e^2 - 8(l^2 - 9r^2)}$$

may be useful.

(c) The In-line Slider-crank Mechanism ($e = 0$)[21,26,33,62,84,90,102,148,237,353,378]

If $\omega \neq$ constant, see ref. 21. In general, see Eqs. (4.44) to (4.55).

Equations (4.56) and (4.57) give approximate values when $\lambda < 1$, and with ω = constant. (For nomenclature refer to Fig. 4.43, with $e = 0$, and guide fixed.)

Slider velocity: $$dx/dt = r\omega(-\sin\theta - \tfrac{1}{2}\lambda\sin 2\theta) \tag{4.56}$$

Slider acceleration: $$d^2x/dt^2 = r\omega^2(-\cos\theta - \lambda\cos 2\theta) \tag{4.57}$$

Extreme Values. $(dx/dt)_{max}$ occurs when the auxiliary angle $\eta = 90°$. For a prescribed extreme value, $(1/r\omega)(dx/dt)_{max}$, λ is obtainable from Eq. (22) of ref. 90.

At extended dead center: $$d^2x/dt^2 = -r\omega^2(1 + \lambda) \tag{4.58}$$

At folded dead center: $$d^2x/dt^2 = r\omega^2(1 - \lambda) \tag{4.59}$$

Equations (4.58) and (4.59) yield exact extreme values whenever $0.264 < \lambda < 0.88$.[353]

Computations. See computer programs in ref. 102, and also Kent's "Mechanical Engineers Handbook," 1956 ed., Sec. II, Power, Sec. 14, pp. 14-61 to 14-63, for displacements, velocities, and accelerations vs. λ and θ; similar tables including also kinematics of connecting rod in ref. 62 for $0.2 \leq \lambda \leq 0.7$ in increments of 0.1.

Harmonic Analysis[33]

$$x/r = A_0 + \cos\theta + \tfrac{1}{4}A_2\cos 2\theta - \tfrac{1}{16}A_4\cos 4\theta + \tfrac{1}{36}A_6\cos 6\theta - \cdots \tag{4.60}$$

If ω = constant,

$$-(1/r\omega^2)(d^2x/dt^2) = \cos\theta + A_2\cos 2\theta - A_4\cos 4\theta + A_6\cos 6\theta - \cdots \tag{4.61}$$

where A_j are given in Table 4.1.[33]

Table 4.1. Values of A_j*

l/r	A_2	A_4	A_6
2.5	0.4173	0.0182	0.0009
3.0	0.3431	0.0101	0.0003
3.5	0.2918	0.0062	0.0001
4.0	0.2540	0.0041	0.0001
4.5	0.2250	0.0028	0.0000
5.0	0.2020	0.0021	
5.5	0.1833	0.0015	
6.0	0.1678	0.0012	

* Biezeno and Grammel, "Engineering Dynamics," vol. 4, Blackie and Son, Ltd., Glasgow, 1954.

For harmonic analysis of $\phi(\theta)$, and for inclusion of terms for $\omega \neq$ constant, see ref. 33; coupler curves (described by a point in the plane of the connecting rod) in refs. 26 and 84; "cognate" slider-crank mechanism (i.e., one, a point of which describes the same coupler curve as the original slider-crank mechanism), ref. 148; straight-line coupler-curve guidance, see VDI.—Richtlinien No. 2136.

(d) Miscellaneous Mechanisms Based on the Slider-crank Chain[19,33,36,84,230,232,233,238,240,242,283,300,324,328]

1. In-line swinging-block mechanism
2. In-line turning-block mechanism
3. External geneva motion
4. Shaper drive
5. Offset swinging-block mechanism
6. Offset turning-block mechanism
7. Elliptic slider-crank drive

In-line swinging-block and *in-line turning-block* mechanisms (Fig. 4.45*a* and *b*). The following applies to both mechanisms. $\lambda = r/a$; θ is considered as input, with ω_{AB} = constant.

Displacement:
$$\phi = \tan^{-1}\frac{\lambda\sin\theta}{1 - \lambda\cos\theta} \tag{4.62}$$

Angular velocities (positive clockwise):
$$d\phi/dt = \omega_{AB}\frac{\lambda\cos\theta - \lambda^2}{1 + \lambda^2 - 2\lambda\cos\theta} \tag{4.63}$$

$$(1/\omega_{AB})|d\phi/dt|_{\min,(\theta=180°)} = -\lambda/(\lambda + 1) \tag{4.64}$$

$$(1/\omega_{AB})(d\phi/dt)_{\max,(\theta=0°)} = \lambda/(1 - \lambda) \tag{4.65}$$

Angular acceleration:
$$\alpha_{BD} = \frac{d^2\phi}{dt^2} = \frac{(\lambda^3 - \lambda)\sin\theta}{(1 + \lambda^2 - 2\lambda\cos\theta)^2}\omega^2_{AB} \tag{4.66}$$

Extreme value of α_{BD} occurs when $\theta = \theta_{max}$, where

$$\cos\theta_{max} = -G + (G^2 + 2)^{1/2} \qquad (4.67)$$

and

$$G = \tfrac{1}{4}(\lambda + 1/\lambda) \qquad (4.68)$$

Angular velocity ratio ω_{BD}/ω_{AB} and the ratio $\alpha_{BD}/\omega^2_{AB}$ are found from Eqs. (4.63) and (4.66), respectively, where $\omega_{BD} = d\phi/dt$. See also section 4.9g.

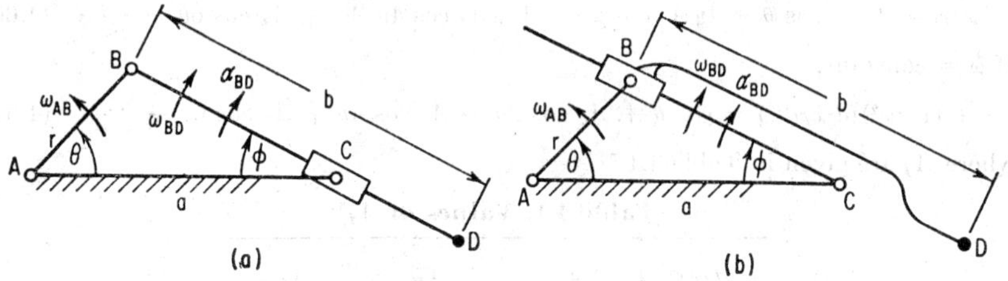

FIG. 4.45. (*a*) In-line swinging-block mechanism. (*b*) In-line turning-block mechanism.

Straight-line Guidance.[36,378] Point *D* (see Fig. 4.45*a* and *b*) will generate a close point-approximation to a straight line for a portion of its (bread-shaped) path, when

$$b = 3a - r + \sqrt{8a(a - r)} \qquad (4.69)$$

Approximate Circular Arc (for a portion of motion cycle).[36] Point *D* (Fig. 4.45*a* and *b*) will generate an approximately circular arc whose center is at a distance *c* to the right of *A* (along *AC*) when

$$[b(a + c) - c(a - r)]^2 = 4bc(c - a)(a + r)$$

with $b > 0$ and $|c| > a > r$.

Proportions can be used in intermittent drive by attachment of two additional links (*VDI-Berichte*, vol. 29, p. 28, 1958).[36,354]

Harmonic Analysis[33,227,230,242] (see Fig. 4.45*a* and *b*).

Case 1, $\lambda < 1$

$$\phi = \sum_{n=1}^{\infty} (\lambda^n \sin n\theta)/n \qquad (4.70)$$

$$\omega_{BD}/\omega_{AB} = d\phi/d\theta = \sum_{n=1}^{\infty} \lambda^n \cos n\theta \qquad (4.71)$$

Case 2, $\lambda > 1$

$$\phi = \pi - \theta - \sum_{n=1}^{\infty} (\sin n\theta)/n\lambda^n \qquad (4.72)$$

$$d\phi/d\theta = -1 - \sum_{n=1}^{\infty} (\cos n\theta)/\lambda^n \qquad (4.73)$$

Note that in case 1 *AB* rotates and *BC* oscillates, while in case 2 both links perform full rotations.

External geneva motion. Equations (4.62) to (4.68) apply. For more extensive data, including tables of third derivatives and various numerical values, see Intermittent-motion Mechanisms in Art. 4.9 and refs. 197, 198, and 286.

An analysis of the *shaper drive* involving the turning-block mechanism is described in ref. 230, part 2; see also ref. 36.

Offset swinging-block and *offset turning-block* mechanisms.[242] See Fig. 4.44. Synthesis of offset turning-block mechanisms for path and function generation described in ref. 324; see also ref. 328 for velocities and accelerations; extreme values of angular velocity ratio $d\phi/d\theta = q$ are related by the equation $q_{max}^{-1} + q_{min}^{-1} = -2$; these occur for the same position of the driving link or for those whose crank angles add up to two right angles, depending on whether the driving link swings (oscillates) or rotates, respectively[328]; for graphical analysis of accelerations involving relative motion between two (instantaneously) coincident points on two moving links, use Coriolis's acceleration, or complex numbers in analytical approach.

For the "*elliptic slider-crank drive*" see refs. 234, 236, and 238.

(e) Four-bar Linkages (Plane)[2,12,16,26,36,50,53,54,84,89,92,95,96,98,100,101,102,125,126,132,136,139,142,146,156,161,162,165,167,169,174,194,220,225,239,241,244,248,256,312,324,327,338,342,352,354,357,364,368,370,378]

See Fig. 4.46. Four-bar mechanism, $ABCD$, E on coupler; AB = crank b; BC = coupler c; CD = crank or link d; AD = fixed link a; AB is assumed to be the driving link.

Grashof's Inequality. Length of longest link + length of shortest link < sum of lengths of two intermediate links.

Types of Mechanisms

1. If Grashof's inequality is satisfied and b or d is the shortest link, the linkage is a *crank and rocker;* the shortest link is the "crank," and the opposite link is the "rocker."
2. If Grashof's inequality is satisfied and the fixed link is the shortest link, the linkage is a *drag linkage;* both cranks can make complete rotations.
3. All other cases except 4: the linkage is a *double-rocker* mechanism (cranks can only oscillate); this will be the case, for instance, whenever the coupler is the smallest link.
4. Special cases: where the equal sign applies in Grashof's inequality. These involve "folding" linkages and "branch positions," at which the motion is not positive. Examples: parallelogram linkage; antiparallel equal-crank linkage ($AB = CD$, $BC = AD$, but AB is not parallel to CD).[167]

Angular Displacement. In Fig. 4.46, $\psi = \psi_1 + \psi_2$ (a minus sign would occur in front of ψ_2 when a mechanism lies entirely on one side of diagonal BD).

$$\psi = \cos^{-1}\frac{h^2 + a^2 - b^2}{2ah} + \cos^{-1}\frac{h^2 + d^2 - c^2}{2hd} \tag{4.74}$$

$$h^2 = a^2 + b^2 + 2ab\cos\phi \tag{4.75}$$

For alternative equation between tan $\tfrac{1}{2}\psi$ and tan $\tfrac{1}{2}\varphi$ (useful for automatic computation) see F. M. Dimentberg.[72]

The general closure equation:[89]

$$R_1\cos\phi - R_2\cos\psi + R_3 = \cos(\phi - \psi) \tag{4.76}$$

where $\quad R_1 = a/d \qquad R_2 = a/b \qquad R_3 = (a^2 + b^2 - c^2 + d^2)/2bd \quad$ (4.77)

The ϕ, τ equation:

$$p_1\cos\phi - p_2\cos\tau + p_3 = \cos(\phi - \tau) \tag{4.78}$$

where $\quad p_1 = b/c \qquad p_2 = b/a \qquad p_3 = (a^2 + b^2 + c^2 - d^2)/2ac \quad$ (4.79)

$$\theta = \tau - \phi = \angle AQB \tag{4.80}$$

Extreme rocker-angle values in a crank and rocker:

$$\psi_{max} = \cos^{-1}\frac{a^2 + d^2 - (b + c)^2}{2ad} \tag{4.81}$$

$$\psi_{min} = \cos^{-1}\frac{a^2 + d^2 - (c - b)^2}{2ad} \tag{4.82}$$

Total range of rocker: $\qquad \Delta\psi = \psi_{max} - \psi_{min}$

To determine inclination of the coupler $\angle AQB = \theta$, determine length of AC:

$$\overline{AC}^2 = a^2 + d^2 - 2ad\cos\psi = k^2 \text{ (say)} \tag{4.83}$$

then compute $\angle ABC = \tau$ from

$$\cos \tau = (b^2 + c^2 - k^2)/2bc \tag{4.84}$$

and use Eq. (4.80). See also refs. 248 and 338 for angular displacements; for extreme positions see ref. 225; for geometrical construction of proportions for given ranges and extreme positions, see ref. 139.

Velocities

Angular velocity ratio: $\omega_{CD}/\omega_{AB} = QA/QD$ (4.85)

Velocity ratio:

$$V_C/V_B = PC/PB \qquad P = AB \times CD \qquad V_p \text{ (} P \text{ on coupler)} = 0 \tag{4.86}$$

Velocity ratio of tracer point E: $V_E/V_B = PE/PB$ (4.87)

Angular velocity ratio of coupler to input link:

$$\omega_{BC}/\omega_{AB} = BA/BP \tag{4.88}$$

When cranks are parallel, B and C have the same linear velocity, and $\omega_{BC} = 0$.

When coupler and fixed link are parallel, $\omega_{CD}/\omega_{AB} = 1$.

At an extreme value of angular velocity ratio, $\lambda = 90°$.[90,325] When ω_{BC}/ω_{AB} is at a maximum or minimum, $QP \perp CD$.

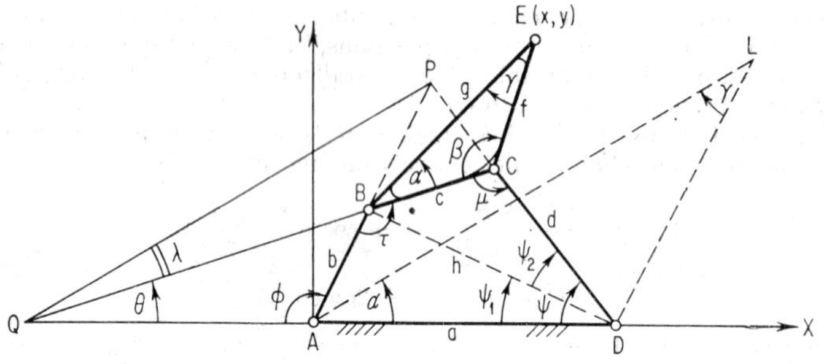

FIG. 4.46. Four-bar mechanism.

Angular velocity ratio of output to input link is also obtainable by differentiation of Eq. (4.76):

$$m_1 = \frac{\omega_{CD}}{\omega_{AB}} = \frac{d\psi}{d\phi} = \frac{\sin(\phi - \psi) - R_1 \sin \phi}{\sin(\phi - \psi) - R_2 \sin \psi} \tag{4.89}$$

Accelerations (ω_{AB} = constant, t = time)

$$m_2 = \frac{d^2\psi}{d\phi^2} = \frac{1}{\omega_{AB}^2}\frac{d^2\psi}{dt^2} = \frac{(1 - m_1)^2 \cos(\phi - \psi) - R_1 \cos \phi + m_1^2 R_2 \cos \psi}{\sin(\phi - \psi) - R_2 \sin \psi} \tag{4.90}$$

Alternate formulation

$$\frac{1}{\omega_{AB}^2}\frac{d^2\psi}{dt^2} = m_1(1 - m_1) \cot \lambda \tag{4.91}$$

(useful when $m_1 \neq 1, 0$, and $\lambda \neq 0°, 180°$).

On extreme values, see ref. 90; velocities, accelerations, and point-path curvature discussed via complex numbers in ref. 50; computer programs in ref. 102.

Second acceleration or shock (ω_{AB} = constant)

$$\frac{1}{\omega_{AB}{}^3}\frac{d^3\psi}{dt^3} = \frac{d^3\psi}{d\phi^3} = m_3$$

$$= \frac{R_1 \sin\phi - (m_1{}^3 \sin\psi - 3m_1m_2\cos\psi)R_2 - 3m_2(1-m_1)\cos(\phi-\psi) - (1-m_1)^3\sin(\phi-\psi)}{\sin(\phi-\psi) - R_2\sin\psi} \tag{4.92}$$

Coupler Motion. Angular velocity of coupler:

$$d\theta/dt = (n_1 - 1)\omega_{AB}$$

where

$$n_1 = 1 + \frac{d\theta}{d\phi} = \frac{\sin(\phi-\tau) - p_1\sin\phi}{\sin(\phi-\tau) - p_2\sin\tau} \qquad (p_1 \text{ and } p_2 \text{ as before [Eq. (4.79)]}) \tag{4.93}$$

Let

$$n_2 = \frac{d^2\theta}{d\phi^2} = \frac{(1-n_1)^2\cos(\phi-\tau) - p_1\cos\phi + n_1{}^2p_2\cos\tau}{\sin(\phi-\tau) - p_2\sin\tau} \tag{4.94}$$

Then the angular acceleration of the coupler:

$$d^2\theta/dt^2 = n_2\omega_{AB}{}^2 \qquad (\omega_{AB} = \text{const}) \tag{4.95}$$

If

$$n_3 = \frac{d^3\theta}{d\phi^3} = \frac{p_1\sin\phi - (n_1{}^3\sin\tau - 3n_1n_2\cos\tau)p_2 - 3n_2(1-n_1)\cos(\phi-\tau) - (1-n_1)^3\sin(\phi-\tau)}{\sin(\phi-\tau) - p_2\sin\tau} \tag{4.96}$$

the angular shock of the coupler $d^3\theta/dt^3$ at ω_{AB} = constant is given by

$$d^3\theta/dt^3 = n_3\omega_{AB}{}^3 \tag{4.97}$$

See also refs 53 and 54 for angular acceleration and shock of coupler; for shock of points on the coupler see ref. 239.

Harmonic Analysis (ψ vs. ϕ). Literature survey in ref. 371. General equations for crank and rocker in ref. 98. Formulas for special crank-and-rocker mechanisms designed to minimize higher harmonics:[101] Choose ν: $0 \ll \nu \ll 90°$, and let

$$AB = \tan\tfrac{1}{2}\nu, \qquad BC = (1/\sqrt{2})\sec\tfrac{1}{2}\nu = CD, \qquad AD = 1$$
$$\mu_{\max} = 90° + \nu, \qquad \mu_{\min} = 90° - \nu$$

$$\psi = \text{const.} + \sum_{m=1}^{\infty}\frac{(-\tan\tfrac{1}{2}\nu)^m}{-m}\sin m\phi - \frac{C_0}{4}\sin\nu\cos\phi - \sum_{m=1}^{\infty}\frac{\sin\nu}{4m}(C_{m-1} - C_{m+1})\cos m\phi$$

where

$$\sin\nu = p, \qquad a_2 = \tfrac{1}{4}p^2, \qquad a_4 = \tfrac{3}{64}p^4, \qquad a_6 = \tfrac{5}{512}p^6, \qquad a_8 = \frac{35}{128^2}p^8$$

$$C_m\ (m \text{ odd}) = 0, \qquad C_8 = a_8 + \cdots, \qquad C_6 = a_6 + 8C_8 + \cdots$$
$$C_4 = a_4 + 6C_6 - 20C_8 + \cdots, \qquad C_2 = a_2 + 4C_4 - 9C_6 + 16C_8 + \cdots$$
$$C_0 = 1 + C_2 - C_4 + C_6 - C_8 + \cdots$$

For numerical tables see ref. 101. For four-bar linkages with adjacent equal links (driven crank = coupler), as in ref. 101, see also refs. 39, 241, and 244.

Three-point Function Synthesis. To find mechanism proportions when (ϕ_i,ψ_i) are prescribed for $i = 1, 2, 3$ (see Fig. 4.46).

$$a = 1 \qquad b = \frac{w_2w_3 - w_1w_4}{w_1w_6 - w_2w_5} \qquad d = \frac{w_2w_3 - w_1w_4}{w_3w_6 - w_4w_5}$$

$$c^2 = 1 + b^2 + d^2 - 2bd \cos(\phi_i - \psi_i) - 2d \cos \psi_i + 2b \cos \phi_i \qquad i = 1, 2, 3$$

where $w_1 = \cos \phi_1 - \cos \phi_2 \qquad w_2 = \cos \phi_1 - \cos \phi_3$

$$w_3 = \cos \psi_1 - \cos \psi_2 \qquad w_4 = \cos \psi_1 - \cos \psi_3$$

$$w_5 = \cos(\phi_1 - \psi_1) - \cos(\phi_2 - \psi_2) \qquad w_6 = \cos(\phi_1 - \psi_1) - \cos(\phi_3 - \psi_3)$$

Four- and Five-point Synthesis. For maximum accuracy, use five points; for greater flexibility in choice of proportions and transmission-angle control, choose four points.

Four-point Path and Function Generation. Path generation together with prescribed crank rotations in refs. 96 and 304. Function generation in ref. 313.

Five-point Path and Function Generation. See refs. 96 and 312; the latter reference usable for five-point path vs. prescribed crank rotations, for Burmester point-pair determinations pertaining to five distinct positions of a plane, and for function generation with the aid of ref. 100; additional references include 313, 327, and others at beginning of section; minimization of structural error in refs. 16, 95, and 194, the latter with least squares; see refs. 92, 95, and 156 for minimum-error function generators such as log x, sin x, tan x, e^x, x^n, tanh x; infinitesimal motions, Burmester points in refs. 351 and 368.

General. Atlases for path generation (ref. 161) and for function generation via "trace deviation" (refs. 169 and 352); point-position-reduction discussed in refs. 2, 125, 156, and Art. 4.5(g); nine-point path generation in ref. 305.

Coupler Curve.[26,84,256] Traced by point E, in cartesian system with origin at A, and xy axes as in Fig. 4.46:

$$U = f[(x - a) \cos \gamma + y \sin \gamma](x^2 + y^2 + g^2 - b^2) - gx[(x - a)^2 + y^2 + f^2 - d^2]$$

$$V = f[(x - a) \sin \gamma - y \cos \gamma](x^2 + y^2 + g^2 - b^2) + gy[(x - a)^2 + y^2 + f^2 - d^2]$$

$$W = 2gf \sin \gamma[x(x - a) + y^2 - ay \cot \gamma]$$

With these
$$U^2 + V^2 = W^2 \tag{4.98}$$

Equation (4.98) is a tricircular, trinodal, sextic, algebraic curve. Any intersection of this curve with circle through ADL (Fig. 4.46) is a double point, in special cases a cusp; coupler curves may possess up to three real double points or cusps (excluding curves traced by points on folding linkages); construction of coupler curves with cusps and application to instrument design (dwells, noiseless motion reversal, etc.) described in refs. 26, 125, and 220, theory in ref. 55; detailed discussion of curves, including Watt straight-line motion and equality of two adjacent links, in ref. 84; instant center (at intersection of cranks, produced if necessary) describes a cusp.

Radius of Path Curvature R for Point E (Fig. 4.46). In this case, as in other linkages, analytical determination of R is readily performed *parametrically*. Parametric equations of the coupler curve:

$$x = x(\phi) = -b \cos \phi + g \cos(\theta + \alpha) \tag{4.99}$$

$$y = y(\phi) = b \sin \phi + g \sin(\theta + \alpha) \tag{4.100}$$

where $\theta = \tau - \phi$, $\tau = \tau(\phi)$ is obtainable from Eqs. (4.74), (4.75), (4.83), and (4.84) and $\cos \alpha = (g^2 - f^2 + c^2)/2gc$.

$$x' = dx/d\phi = b \sin \phi - g(n_1 - 1) \sin(\theta + \alpha) \tag{4.101}$$

$$y' = dy/d\phi = b \cos \phi + g(n_1 - 1) \cos(\theta + \alpha) \tag{4.102}$$

$$x'' = d^2x/d\phi^2 = b \cos \phi - gn_2 \sin(\theta + \alpha) - g(n_1 - 1)^2 \cos(\theta + \alpha) \tag{4.103}$$

$$y'' = d^2y/d\phi^2 = -b \sin \phi + gn_2 \cos(\theta + \alpha) - g(n_1 - 1)^2 \sin(\theta + \alpha) \tag{4.104}$$

where n_1 and n_2 are given in Eqs. (4.93) and (4.94).

$$R = \frac{(x'^2 + y'^2)^{3/2}}{x'y'' - y'x''} \tag{4.105}$$

Equivalent or "Cognate" Four-bar Linkages. For Robert's theorem, see Art. 4.6, Fig. 4.38. Proportions of the cognates are as follows (Figs. 4.38 and 4.46):

Left Cognate

$$\mathbf{AF} = \mathbf{BC}z \qquad \mathbf{HF} = \mathbf{AB}z \qquad \mathbf{HL} = \mathbf{CD}z \qquad \mathbf{AL} = \mathbf{AD}z$$

where $z = (g/c)e^{i\alpha}$ $\quad \alpha = \angle CBE$

and where **AF,** etc., represent the complex-number form of the vector $\overrightarrow{AF}$, etc.

Right Cognate

$$\mathbf{GD} = \mathbf{BC}u \qquad \mathbf{GK} = \mathbf{CD}u \qquad \mathbf{LK} = \mathbf{AB}u \qquad \mathbf{LD} = \mathbf{AD}u$$

where $u = (f/c)e^{-i\beta}$ $\quad \beta = \angle ECB$

Same construction can be studied systematically with the "Cayley diagram."[26]

Symmetrical Coupler Curves. Coupler curves with an axis of symmetry are obtained when $BC = CD = EC$ (Fig. 4.46); also by cognates of such linkages; used by K. Hunt for path of driving pin in geneva motions;[162] also for dwells and straight-line guidance (see Art. 4.8, 5M). Symmetrical coupler-curve equation[36] for equal-crank linkage, traced by mid-point E of coupler in Fig. 4.47.

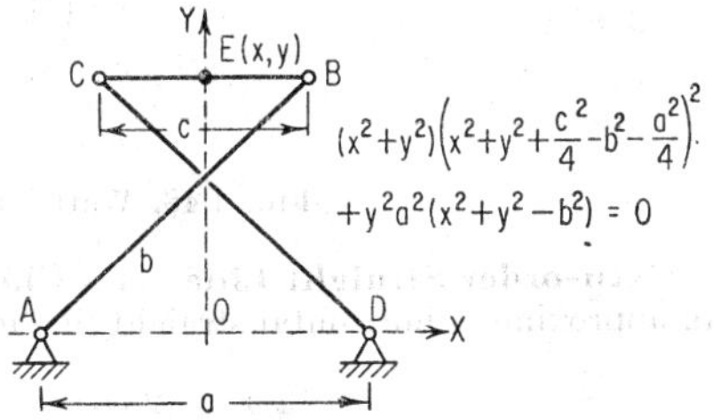

FIG. 4.47. Equal-crank linkage showing equation of symmetric coupler curve generated by point E, midway between C and B.

Transmission Angles (μ, Fig. 4.46). μ should be as close to 90° as possible; nontrivial extreme values occur when AB and AD are parallel or antiparallel ($\phi = 0°, 180°$). Generally

$$\cos \mu = [(c^2 + d^2 - a^2 - b^2)/2cd] - (2ab/2cd) \cos \phi \tag{4.106}$$

$\cos \mu_{\max}$ occurs when $\phi = 180°$, $\cos \phi = -1$; $\cos \mu_{\min}$ when $\phi = 0°$, $\cos \phi = 1$. Good crank-and-rocker proportions given in Art. 4.6:

$$b^2 + a^2 = c^2 + d^2 \qquad |\mu_{\min} - 90°| = |\mu_{\max} - 90°| = \nu \qquad \sin \nu = ab/cd$$

A computer program for path generation with optimum transmission angles and proportions is described in ref. 304. Charts for optimum transmission-angle designs as follows: drag links in ref. 126; double rockers in ref. 132; general, in refs. 136 and 165. See also VDI charts in ref. 378.

Approximately Constant Angular-velocity Ratio of Cranks (m_1) [Eq. (4.89)] **Over a Portion of Crank Rotation** (see also ref. 36). In Fig. 4.46, if $d = 1$, a three-point approximation is obtained when

$$a^2 = c^2 \frac{(1 - 2m_1)(m_1 - 2)}{9m_1} + b^2 \frac{(1 - m_1)(2 - m_1)}{m_1(m_1 + 1)} + \frac{(1 - m_1)(1 - 2m_1)}{(m_1 + 1)}$$

Useful only for limited crank rotations, possibly involving connection of distant shafts, high loads.

Straight-line Mechanisms. Survey in refs. 61, 168; modern and special applications in refs. 174, 354, 357, theory and classical straight-line mechanisms in ref. 36; see also below; order-approximation theory in ref. 342.

Fifth-order Approximate Straight Line via a Watt Mechanism.[36] "Straight" line of length $2l$, generated by M on coupler, such that $y \cong kx$ (Fig. 4.48).

Choose k, l, r; let $\beta = l(1 + k^2)^{-1/2}$; then maximum error from straight-line path $\cong 0.038(1 + k^2)^3\beta^6$. To compute d and c:

$$\begin{aligned}(d^2 - c^2) &= [r^4 + 6(7 - 4\sqrt{3})l^4 + 3(3 - 2\sqrt{3})l^2r^2]^{1/2} \\ p_2 &= 3(3 - 2\sqrt{3})\beta \\ 4k^2d^2 &= 2(1 + k^2)(d^2 - c^2 + r^2) + p_2(1 + k^2)^2\end{aligned}$$

For less than fifth-order approximation, proportions can be simpler: $AB = CD$, $BM = MC$.

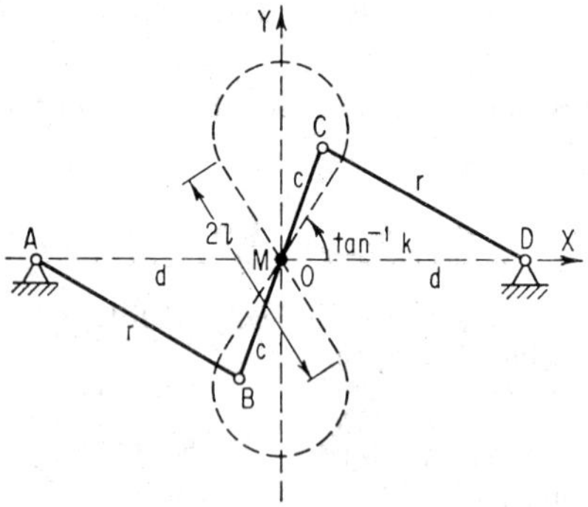

FIG. 4.48. Watt straight-line mechanism.

Sixth-order Straight Line via a Chebichev Mechanism.[36,370] M will describe an approximate horizontal straight line in the position shown in Fig. 4.49, when

$$\begin{aligned}a/r &= (2\cos^2\phi\cos 2\phi)/\cos 3\phi \\ b/r &= -\sin^2 2\phi/\cos 3\phi \\ c/r &= (\cos^2\phi\cos 2\phi\tan 3\phi)/\cos 3\phi\end{aligned}$$

when $\phi = 60°$, $NM = 0$.

Lambda Mechanism[12,36] **and a Related Motion.** The four-bar lambda mechanism of Fig. 4.50 consists of crank $AC' = r$, fixed link $CC' = d$, coupler AB, driven

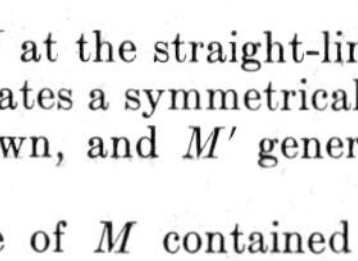

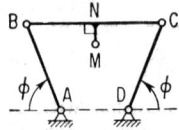

FIG. 4.49. Chebichev straight-line mechanism. $BN = NC = \frac{1}{2}a$ $AD = b$, $AB = CD = r$ $NM = c$ (+ downward). In general $\phi = 90° - \theta$, where θ is the angle between axis of symmetry and crank in symmetry position.

link BC, with generating point M at the straight-line extension of the coupler, where $BC = MB = BA = 1$. M generates a symmetrical curve. In a related mechanism, $M'B = BA$, $\omega = \angle M'BA$ as shown, and M' generates another symmetrical coupler curve.

Case 1. Entire coupler curve of M contained between two concentric circles, center O_1, O_1M_0C collinear. $M = M_0$ when $AC'C$ are collinear as shown.

Let ψ'' be a parameter, $0 \leq \psi'' \leq 45°$. Then a six-point approximate circle is generated by M with least maximum structural error when

$$\begin{aligned}r &= \frac{2\sin\psi''\sin 2\psi''\sqrt{2\cos 2\psi''}}{\sin 3\psi''} \\ d &= \sin 2\psi''/\sin 3\psi'' \\ O_1C &= 2\cos^2\psi''/\sin 3\psi''\end{aligned}$$

Radius, R, of generated circle (at precision points):

$$R = r \cot \psi''$$

Maximum radial (structural) error:

$$2 \cos 2\psi''/\sin 3\psi''$$

For table of numerical values see ref. 12.

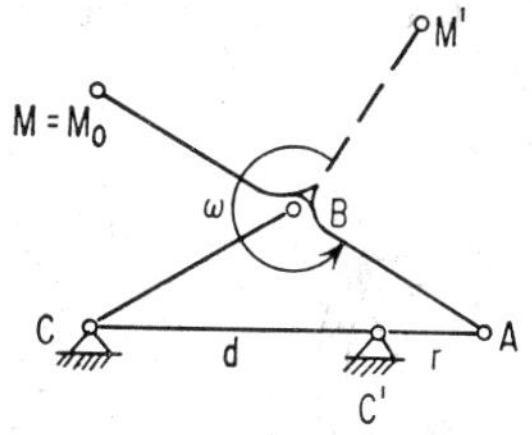

FIG. 4.50. Lambda mechanism.

Case 2. Entire coupler curve contained between two straight lines (six-point approximation of straight line with least maximum structural error). In the equations above, M' generates this curve when $\angle M'BA = \omega = \pi + 2\psi''$. Maximum deviation from straight line:[12]

$$\frac{2 \sin 2\psi'' \sqrt{2 \cos^3 2\psi''}}{\sin 3\psi''}$$

Case 3. Six-point straight line for a portion of the coupler curve of M':

$$r = \tfrac{1}{4} \qquad d = \tfrac{3}{4}$$

Case 4. Approximate circle for a *portion* of the coupler curve of M. Any proportions for r and d give reasonably good approximation to *some* circle because of symmetry. Exact proportions are shown in refs. 12 and 36.

Balancing of Four-bar Linkages for High-speed Operation.[338] Make links as light as possible; if necessary, counterbalance cranks, including appropriate fraction of coupler on each.

Multilink Planar Mechanisms. Geared five-bar mechanisms;[80,93,93a] six-link mechanisms.[282]

Design Charts.[378] See also section on transmission angles.

(f) Three-dimensional Mechanisms[3,5,6,8,9,10,21,27,28,29,30,31,52,63,70,71,164,175,177,195,203,235,236,243,245,246,251,263,268,295,346,347,348,369,371,373]

Spherical Four-bar Mechanism[70] (θ_1, output vs. θ_2, input) (Fig. 4.51)

$$A \sin \theta_1 + B \cos \theta_1 = C \qquad \text{where} \qquad A = \sin \alpha_2 \sin \alpha_4 \sin \theta_2 \tag{4.107}$$

$$B = -\sin \alpha_4(\sin \alpha_1 \cos \alpha_2 + \cos \alpha_1 \sin \alpha_2 \cos \theta_2) \tag{4.108}$$

$$C = \cos \alpha_3 - \cos \alpha_4(\cos \alpha_1 \cos \alpha_2 - \sin \alpha_1 \sin \alpha_2 \cos \theta_2) \tag{4.109}$$

Other relations given in ref. 70. Convenient equations between $\tan \tfrac{1}{2}\theta_1$ and $\tan \tfrac{1}{2}\theta_2$ given in ref. 72.

Maximum angular velocity ratio $d\theta_1/d\theta_2$ occurs when $\lambda = 90°$.[243]

Types of mechanisms: Assume $\alpha_i (i = 1, 2, 3, 4) < 180°$ and apply Grashof's rule (p. 4-41) to equivalent mechanism with identical axes of turning joints, such that all links except possibly the coupler, $<90°$.

Harmonic analysis: see ref. 371.

Special Cases of the Spherical Four-bar Mechanism

1. *Hooke's joint:* $\alpha_2 = \alpha_3 = \alpha_4 = 90°$, $90° < \alpha_1 \leq 180°$; in practice, if $\alpha_1 = 180° - \beta$, then $0 \leq \beta \leq 37\tfrac{1}{2}°$. If angles $\theta, \varphi(\theta = \theta_1 - 90°, \varphi = 180° - \theta_2)$ are measured from

a starting position (shown in Fig. 4.52) in which the planes ABO and OCD are perpendicular, then

$$\tan \varphi / \tan \theta = \cos \beta \qquad (\beta = 180° - \alpha_1)$$

Angular velocity ratio: $$\frac{\omega_2}{\omega_1} = \frac{\cos \beta}{1 - \sin^2 \theta \sin^2 \beta}$$

Maximum value, $(\cos \beta)^{-1}$, occurs at $\theta = 90°, 270°$; minimum value, $\cos \beta$, occurs at $\theta = 0°, 180°$.

A graph of θ vs. φ will show two "waves" per revolution.

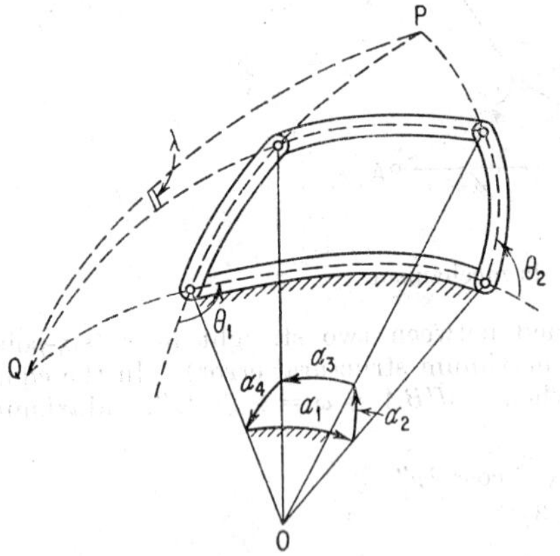

FIG. 4.51. Spherical four-bar mechanism.

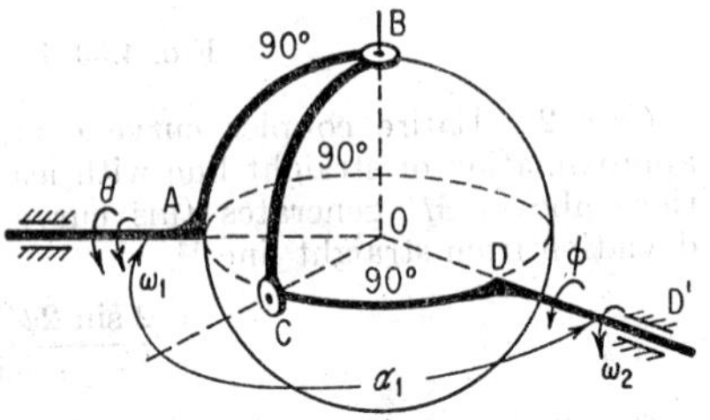

FIG. 4.52. Hooke's joint.

If ω_1 = constant, the angular-acceleration ratio of shaft DD' is given by

$$\frac{1}{\omega_1^2}\frac{d^2\varphi}{dt^2} = \frac{\cos \beta \sin^2 \beta \sin 2\theta}{(1 - \sin^2 \beta \sin^2 \theta)^2}; \qquad \text{maximum at } \theta = \theta_{max}$$

where $$\cos 2\theta_{max} = G - (G^2 + 2)^{1/2} \tag{4.110}$$

and $$G = \frac{2 - \sin^2 \beta}{2 \sin^2 \beta} \tag{4.111}$$

For $d^2\varphi/dt^2$ as a function of β, see ref 205.

Harmonic Analysis of Hooke's Joint.[205,371,373] If $\alpha_1 = 180° - \beta$, the amplitude ϵ_m of the mth harmonic, in the expression $\varphi(\theta)$, is given by $\epsilon_m = 0$, m odd; and $\epsilon_m = (2/m)(\tan \frac{1}{2}\beta)^m$, m even. Two Hooke's joints in series[81,263] can be used to transmit constant (1:1) angular velocity ratio between two intersecting or nonintersecting shafts 1 and 3, provided that the angles between shafts 1 and 3 and the intermediate shaft are the same (Fig. 4.53) and that when fork 1 lies in the plane of shafts 1 and 2, fork 2 lies in the plane of shafts 2 and 3; thus in case shafts 1 and 3 intersect, forks 1 and 2 are coplanar; see also refs. 70 and 263 when $\alpha \neq \alpha'$, which may arise due to misalignment or the effect of manufacturing tolerances, or may be intentional for use as a vibration-excitation drive.

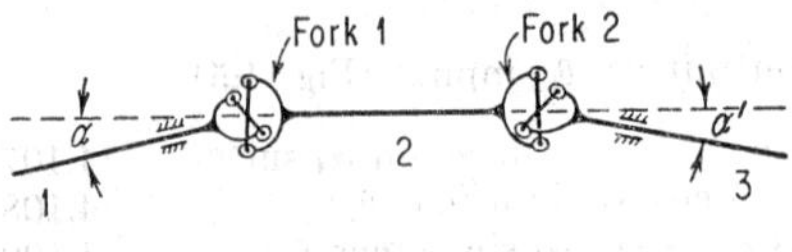

FIG. 4.53. Two Hooke's joints in series.

2. *Other Special Cases of the Spherical Four-bar*

Wobble-plate Mechanism. $\alpha_4 < 90°$, $\alpha_2 = \alpha_3 = \alpha_1 = 90°$; ref. 29 gives displacements, velocities, and equation of "coupler curve." See also refs. 30, 235, and 371, the latter giving harmonic analysis.

Two Angles 90°, Two Angles Arbitrary[30,236,246]

Fixed Link = 90°. Reference 268 includes applications to universal joints, gives velocities and accelerations ("space crank").

3. *Spherical Four-bar Mechanisms with Dwell*: see ref. 29.

4. *Spherical Slider-crank Mechanism.* Three turning pairs, one moving joint a turn slide, input pair and adjacent pair at right angles: see ref. 245 for displacements; if input and output axes intersect at right angles, obtain "skewed Hooke's joint."[235,245]

Spatial Four-bar Mechanisms (see Fig. 4.42). To any spatial four-bar mechanism a corresponding spherical four-bar can be assigned as follows: Through O (Fig. 4.41) draw four radii, parallel to the axes of joints A, B, C, and D in Fig. 4.42 to intersect the surface of a sphere in four points corresponding to the joints of the spherical four-bar. The rotations of the spatial four-bar (Fig. 4.42) are the same as those of the corresponding spherical four-bar and are independent of the offsets a_{ij}, the minimum distances between the (noninteresting) axes i and j, (ij = 12, 23, 34, and 41) of the spatial four-bar (Fig. 4.42). The input and output angles θ_2, θ_1, of the spatial four-bar, can be measured as in Fig. 4.41 for the corresponding spherical four-bar; for s, the sliding at the output joint, measured from A to Q in Fig. 4.42, the general displacement equations[70,71] are (for rotations see Fig. 4.51 and accompanying equations)

$$s = \frac{A_1 \sin\theta_1 + B_1 \cos\theta_1 - C_1}{-A\cos\theta_1 + B\sin\theta_1} \tag{4.112}$$

where A and B are given in Eqs. (4.107) and (4.108) and

$$A_1 = (a_2 \cos\alpha_2 \sin\alpha_4 + a_4 \sin\alpha_2 \cos\alpha_4)\sin\theta_2 + s_2 \sin\alpha_2 \sin\alpha_4 \cos\theta_2 \tag{4.113}$$

$$\begin{aligned} B_1 = {} & -a_4 \cos\alpha_4(\sin\alpha_1 \cos\alpha_2 + \cos\alpha_1 \sin\alpha_2 \cos\theta_2) - a_1 \sin\alpha_4(\cos\alpha_1 \cos\alpha_2 \\ & - \sin\alpha_1 \sin\alpha_2 \cos\theta_2) + a_2 \sin\alpha_4(\sin\alpha_1 \sin\alpha_2 - \cos\alpha_1 \cos\alpha_2 \cos\theta_2) \\ & + s_2 \sin\alpha_4 \cos\alpha_1 \sin\alpha_2 \sin\theta_2 \end{aligned} \tag{4.114}$$

$$\begin{aligned} C_1 = {} & -a_3 \sin\alpha_3 + a_4 \sin\alpha_4(\cos\alpha_1 \cos\alpha_2 - \sin\alpha_1 \sin\alpha_2 \cos\theta_2) \\ & + a_1 \cos\alpha_4(\sin\alpha_1 \cos\alpha_2 + \cos\alpha_1 \sin\alpha_2 \cos\theta_2) \\ & + a_2 \cos\alpha_4(\cos\alpha_1 \sin\alpha_2 + \sin\alpha_1 \cos\alpha_2 \cos\theta_2) \\ & - s_2 \cos\alpha_4 \sin\alpha_1 \sin\alpha_2 \sin\theta_2 \end{aligned} \tag{4.115}$$

where $a_1 = a_{12}$ of Fig. 4.42, and similarly $a_2 = a_{23}$, $a_3 = a_{34}$, $a_4 = a_{41}$, $\alpha_1 = \angle$ between axes 1 and 2, $\alpha_2 = \angle 2$ and 3, $\alpha_3 = \angle 3$ and 4, and $\alpha_4 = \angle 4$ and 1 (Figs. 4.41 and 4.42). For complete nomenclature and sign convention, see ref. 372. These equations are used principally in special cases, in which they simplify.

Special Cases of the Spatial Four-bar and Related Three-dimensional Four-bar Mechanisms.[8]

1. *Spatial Four-bar with Two Ball Joints on Coupler and Two Turning Joints.* Displacements and velocities,[52,162,251,295] synthesis for function generation,[71,195,295] computer programs for displacements and synthesis according to ref. 71 are listed in ref. 102; forces and torques.[52]
2. Spatial four-bar: three angles 90°.[28,29]
3. Spatial four-bar with one ball joint and two turn slides (three links).[29]
4. Spatial four-bar with one ball joint, one turn slide, two turning joints.[346]
5. The "3-D crank slide," one ball joint, one turn slide, intersecting axes; used for agitators.[245]
6. "Degenerate" mechanisms, wherein $F = 1$, $\Sigma f_i < 7$; conditions for,[369] practical constructions in;[3] see also refs. 111, 156, and 348.
7. Spherical geneva.[31,286]
8. Spatial five-link mechanisms.[6,78]
9. Spatial six-link mechanisms.[9,78]

Harmonic Analysis. Rotations θ_1 vs. θ_2 in ref. 371, which also includes special cases, such as Hooke's joint, wobble plates, and spherical-crank drive.

8. Special applications.[3,29,30,235,245,268]

9. Miscellaneous special shaft couplings;[29,235] Cayley-Klein parameters and dual vectors;[70] screw axes and graphical methods.[30,119,158,346]

(g) Intermittent-motion Mechanisms[11,16,21,22,38,116,117,157,162,164,172,173,178,197,198,248,285,286,289,309,318,319,320,334,339,340,341,349,350]

1. The external geneva.[197,198,286] In Fig. 4.54,

a = center distance
α = angle of driver, radians; in Fig. 4.54, $|\alpha| = \alpha_0$
β = angle of driven or geneva wheel, radians (α and β measured from center-line in the direction of motion); in Fig. 4.54, $|\beta| = \beta_0$

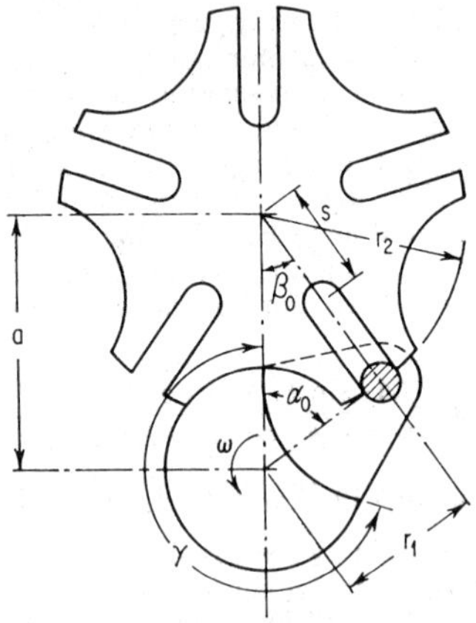

FIG. 4.54. External geneva mechanism in starting position.

r_1 = radius to center of driving pin
γ = locking angle of driver, radians
n = number of equally spaced slots in geneva (≥ 3) at start: $|\alpha| = \alpha_0$,

$$|\beta| = \beta_0 = \frac{\pi}{2} - \alpha_0$$

so that driving pin can enter slot tangentially to reduce shock
r_2 = radius of geneva $= a \cos \beta_0$
r_2' = outside radius of geneva wheel, with correction for finite pin diameter[285] $= r_2 \sqrt{1 + r_p^2/r_2^2}$; r_p = pin radius

$$\epsilon = \text{gear ratio} = \frac{\text{angle moved by driver during motion}}{\text{angle moved by geneva during motion}} = \frac{n-2}{2}$$

μ = radius ratio $= r_2/r_1 = \cot \beta_0$
s = distance of center of semicircular end of slot from center of geneva $\leq a(1 - \sin \beta_0)$
γ = angle of locking action $= (\pi/n)(n + 2)$ radians. Note that classical locking action shown is subject to play in practice. Better constructions in 4M, Art. 4.8
ν = ratio, time of motion of geneva wheel to time for one revolution of driver $= [(n - 2)/2n]$ $(< \frac{1}{2})$
$r_1 = a \sin \beta_0$, $\beta_0 = \pi/n$

Let ω = angular velocity of driving wheel, assumed constant, t = time.

Displacement (β vs. α). Let $r_1/a = \lambda$; then

$$\beta = \tan^{-1} \frac{\lambda \sin \alpha}{1 - \lambda \cos \alpha} \tag{4.116}$$

Velocities

$$\frac{1}{\omega}\frac{d\beta}{dt} = \frac{\lambda \cos \alpha - \lambda^2}{1 - 2\lambda \cos \alpha + \lambda^2} \tag{4.117}$$

$$\frac{1}{\omega}\left(\frac{d\beta}{dt}\right)_{\max} (\text{at } \alpha = 0) = \frac{\lambda}{1 - \lambda} \tag{4.118}$$

Accelerations

$$\frac{1}{\omega^2}\frac{d^2\beta}{dt^2} = \frac{(\lambda^3 - \lambda)\sin\alpha}{(1 + \lambda^2 - 2\lambda\cos\alpha)^2} \tag{4.119}$$

$$(1/\omega^2)(d^2\beta/dt^2)_{\text{initial}\,\alpha=-\alpha_0} = \tan\beta_0 = r_1/r_2 \tag{4.120}$$

Maximum acceleration occurs at $\alpha = \alpha_{\max}$, where

$$\cos\alpha_{\max} = -G + (G^2 + 2)^{1/2} \qquad G = \tfrac{1}{4}(\lambda + 1/\lambda) \tag{4.121}$$

Second Acceleration or Shock

$$\frac{1}{\omega^3}\frac{d^3\beta}{dt^3} = \frac{\lambda(\lambda^2 - 1)[2\lambda\cos^2\alpha + (1 + \lambda^2)\cos\alpha - 4\lambda]}{(1 + \lambda^2 - 2\lambda\cos\alpha)^3} \tag{4.122}$$

$$\frac{1}{\omega^3}\frac{d^3\beta}{dt^3}_{(\alpha=0)} = \frac{\lambda(\lambda + 1)}{(\lambda - 1)^3} \tag{4.123}$$

Starting Shock ($|\beta| = \beta_0$)

$$\frac{1}{\omega^3}\frac{d^3\beta}{dt^3} = \frac{3\lambda^2}{1 - \lambda^2}$$

Table 4.2. External Geneva Characteristics[198]*

n	β_0	α_0	ϵ	(r_1/a)	(r_2/a)	μ	$s_{\min}/a$	γ	ν	$(d\beta/d\alpha)_{\max}$
3	60°	30°	0.5	0.8660	0.5000	0.5774	0.13397	300°	0.1667	6.46
4	45°	45°	1	0.7071	0.7071	1.0000	0.2929	270°	0.2500	2.41
5	36°	54°	1.5	0.5878	0.8090	1.3764	0.4122	252°	0.3000	1.43
6	30°	60°	2	0.5000	0.8660	1.7320	0.5000	240°	0.3333	1.00
7	25°43′	64°17′	2.5	0.4339	0.9009	2.0765	0.5661	231°26′	0.3571	0.766
8	22°30′	67°30′	3	0.3827	0.9239	2.4142	0.6173	225°	0.3750	0.620
9	20°	70°	3.5	0.3420	0.9397	2.7475	0.6580	220°	0.3889	0.520
10	18°	72°	4	0.3090	0.9511	3.0777	0.6910	216°	0.4000	0.447
∞	0°	90°	∞	0	1	∞	1	180°	0.5000	0

* O. Lichtwitz, Getriebe fuer Aussetzende Bewegung, Springer-Verlag OHG, Berlin, 1953.

Table 4.3. External Geneva Characteristics[198]

n	$(d^2\beta/d\alpha^2)_{\text{initial}}$ $\alpha=-\alpha_0$	$\alpha_{\max}$	$(d^2\beta/d\alpha^2)_{\max}$	$(d^3\beta/d\alpha^3)_{\alpha=0}$
3	1.732	4°46′	31.44	−672
4	1.000	11°24′	5.409	− 48.04
5	0.7265	17°34′	2.299	− 13.32
6	0.5774	22°54′	1.350	− 6.000
7	0.4816	27°33′	0.9284	− 3.429
8	0.4142	31°38′	0.6998	− 2.249
9	0.3640	35°16′	0.5591	− 1.611
10	0.3249	38°30′	0.4648	− 1.236
∞	0	90°	0	0

* O. Lichtwitz, Getriebe fuer Aussetzende Bewegung, Springer-Verlag OHG, Berlin, 1953.

Design Procedure (ω = const)

1. Select number of stations ($n \geq 3$).
2. Select center distance a.
3. Compute: $r_1 = a\sin\beta_0$; $r_2'^2 = (a\cos\beta_0)^2 + r_p^2$; $s \leq a(1 - \sin\beta_0)$;

$$|\alpha_0| = \left(\frac{\pi}{2}\right)\left[\frac{(n-2)}{n}\right] \qquad |\beta_0| = \frac{\pi}{2} - \alpha_0 \qquad \text{radians}$$

$$\gamma = (\pi/n)(n + 2) \qquad \text{radians}$$

4. Determine kinematic characteristics from tables, including maximum velocity, acceleration, and shock: $d^n\beta/dt^n = (\omega^n)(d^n\beta/d\alpha^n)$, taking the last fraction from the tables.

Check for resulting forces, stresses, and vibrations.

Modifications of Standard External Geneva. More than one driving pin[197,198]; pins not equally spaced[16,197,198]; designs for small indexing mechanism[289]; for high-speed indexing[172]; mounting driving pin on a planet pinion to reduce peak loading[11,198,309]; pin guided on four-bar coupler[162] (see also Art. 4.8, ref. 5M); double rollers and different entrance and exit slots, especially for starwheels[178,349]; eccentric gear drive for pin[164].

Internal Genevas.[197,198,286] Used when $\nu > \frac{1}{2}$; better kinematic characteristics, but more expensive.

Star Wheels.[178,197,198,349] Both internal and external are used; permits considerable freedom in choice of ν, which can equal unity, in contrast to genevas. Kinematic properties of external star wheels are better or worse than of external geneva with same n, according as the number of stations (or shoes), n, is less than six or greater than five, respectively.

Special Intermittent and/or Dwell Linkages. The three-gear drive[21,88a,157,173,318,334] cardioid drive (slotted link driven by pin on planetary pinion)[286,319,320]; link-gear (and/or) -cam mechanisms to produce dwell, reversal, or intermittent motions[22,339,340,341] include link-dwell mechanisms[116]; eccentric-gear mechanisms[117]. These special motions may be required when control of rest, reversal, and kinematic characteristics exceeds that possible with the standard genevas.

(h) Noncircular Cylindrical Gearing and Rolling-contact Mechanisms[16,37,64,143,200,201,233,255,258,259,262,275,284,290,323,330,364]

Most of the data for this article are based on refs. 37 and 258. Noncircular gears can be used for producing positive unidirectional motion; if the pitch curves are closed curves, unlimited rotations may be possible; only externally meshed, plane spur-type gearing will be considered; point of contact between pitch surfaces must lie on the line of centers.

A pair of roll curves may serve as pitch curves for noncircular gears (see Table 4.4) C = center distance; β = angle between the common normal to roll-curves at contact and the line of centers. Angular velocities ω_1 and ω_2 measured in opposite directions; polar-coordinate equations of curves; $R_1 = R_1(\theta_1)$, $R_2 = R_2(\theta_2)$, such that the points $R_1(\theta_1 = 0)$, $R_2(\theta_2 = 0)$ are in mutual contact, where θ_1 and θ_2 are respective oppositely directed rotations from a starting position. Centers O_1, O_2 and contact point Q are collinear.

Twin rolling curves. Mating or pure rolling of two identical curves, e.g., two ellipses when pinned at foci.

Mirror Rolling Curves. Curve mates with mirror image.

Theorems

1. Every mating curve to a mirror (twin) rolling curve is itself a mirror (twin) rolling curve.
2. All mating curves to a given mirror (twin) rolling curve will mate with each other.
3. A closed roll curve can generally mate with an entire set of different closed roll curves at varying center distances, depending upon the value of the "average gear ratio."

Average Gear Ratio. For mating closed roll curves: ratio of the total number of teeth on each gear.

Rolling Ellipses and Derived Forms. If an ellipse, pivoted at the focus, mates with a roll curve so that the average gear ratio n (ratio of number of teeth on mating curve to number of teeth on ellipse) is integral, the mating curve is called an nth-*order ellipse*. The case $n = 1$ represents an identical (twin) ellipse; second-order ellipses are oval-shaped and appear similar to ordinary ellipses; third-order ellipses appear pear-shaped with three lobes; fourth-order ellipses appear nearly square; nth-order ellipses appear approximately like n-sided polygons. Equations for several of these are found in ref. 37. Characteristics for five noncircular gear systems are given in Table 4.4.[37]

Design Data[37]. Data usually given in one of three ways:
1. Given $R_1 = R_1(\theta_1)$, C. Find R_2 in parametric form: $R_2 = R_2(\theta_1)$; $\theta_2 = \theta_2(\theta_1)$.

$$\theta_2 = -\theta_1 + C \int_0^{\theta_1} \frac{d\theta_1}{C - R_1(\theta_1)} \qquad R_2 = C - R_1(\theta_1)$$

2. Given $\theta_2 = f(\theta_1)$, C. Find $R_1 = R_1(\theta_1)$, $R_2 = R_2(\theta_2)$.

$$R_1 = \frac{(df/d\theta_1)C}{1 + df/d\theta_1} \qquad R_2 = C - R_1$$

3. Given $\omega_2/\omega_1 = g(\theta_1)$, C. Find $R_1 = R_1(\theta_1)$, $R_2 = R_2(\theta_2)$.

$$R_1 = \frac{Cg(\theta_1)}{1 + g(\theta_1)} \qquad R_2 = C - R_1 \qquad \theta_2 = \int_0^{\theta_1} g(\theta_1)\, d\theta_1$$

4. *Checking for closed curves.* Let $R_1 = R_1(\theta_1)$ be a single-turn closed curve; then $R_2 = R_2(\theta_2)$ will be a single-turn closed curve also, if and only if C is determined from

$$4\pi = C \int_0^{2\pi} \frac{d\theta_1}{C - R_1(\theta_1)}$$

When the average gear ratio is not unity, see ref. 37.

5. *Checking angle* β, also called the angle of obliquity.

$$\beta = \tan^{-1} [(1/R_i)(dR_i/d\theta_i)] \qquad i = 1 \text{ or } 2.$$

Values of β between 0 and 45° are generally considered reasonable.

6. *Check for tooth undercut.* Let ρ = radius of curvature of pitch curve (roll curve)

$$\rho = \frac{[R_i^2 + (dR_i/d\theta_i)^2]^{3/2}}{R_i^2 - R_i(d^2R_i/d\theta_i^2) + 2(dR_i/d\theta_i)^2} \qquad i = 1 \text{ or } 2$$

Number of teeth: $T_{\min}$ = 32 for 14½° pressure angle cutting tool
= 18 for 20° pressure angle cutting tool

Condition to avoid undercut in noncircular gears:

$$\rho > \frac{T_{\min}}{2 \times \text{diametral pitch}}$$

7. *Determining length S of roll curves.* $S = \int_0^{2\pi} [R^2 + (dR/d\theta)^2]^{1/2}\, d\theta$; best computed automatically by numerical integration or determined graphically by large-scale layout.

8. *Check on number of teeth.* For closed single-turn curves,

$$\text{Number of teeth} = \frac{S(\text{diametral pitch})}{\pi}$$

Diametral pitch should be integral but may vary by a few percentage points. For symmetrical twin curves, use odd number of teeth for proper meshing following identical machining.

Manufacturing information in ref. 37.

Special Topics in Noncircular Gearing. Survey[275]; elliptic gears[233,284,290]; noncircular cams and rolling-contact mechanisms such as in shears and recording instruments[143,201,255]; noncircular bevel gears[259]; algebraic properties of roll curves[364]; miscellaneous[200,262,330].

Table 4.4. Characteristics of Five Noncircular Gear Systems[30]

Type	Comments	Basic equations	Velocity equation
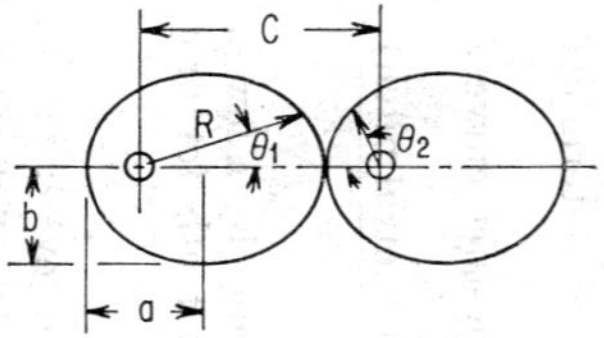 Two ellipses pivoted at foci	Gears are identical, easy to manufacture. Used for quick-return mechanisms, printing presses, automatic machinery.	$R = \dfrac{b^2}{a(1 + \epsilon \cos \theta)}$ assuming $\theta = 0$ at $R = R_{\min}$ ϵ = eccentricity $= \sqrt{1 - (b/a)^2}$ $a = \frac{1}{2}$ major axis $b = \frac{1}{2}$ minor axis	$\omega_2 = \omega_1 \left[\dfrac{r^2 + 1 + (r^2 - 1) \cos \theta_2}{2r} \right]$ where $r = \dfrac{R_{max}}{R_{min}}$ 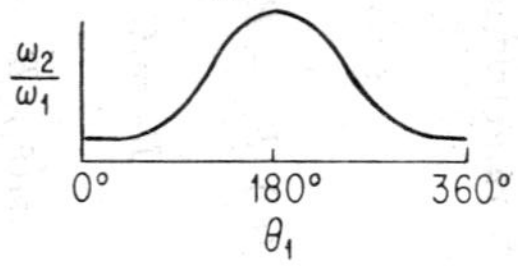
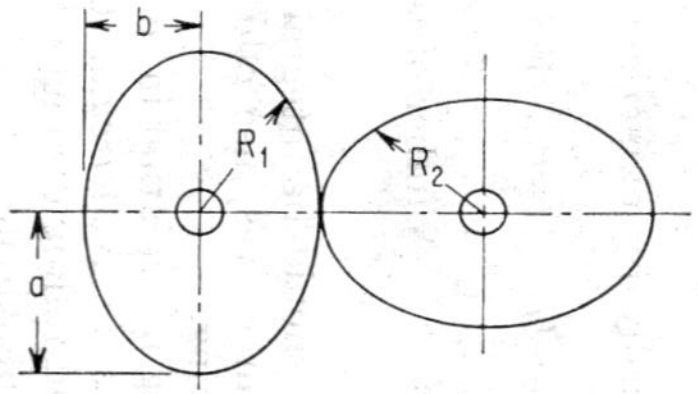 Second-order ellipses rotated about their geometric centers	Gears are identical. Well-known geometry. Better balance than true elliptical. Two complete speed cycles in one revolution.	$R = \dfrac{2ab}{(a + b) - (a - b) \cos 2\theta}$ assuming $\theta = 0$ at $R = R_{\min}$ $C = a + b$ a = max radius b = min radius	$\omega_2 = \omega_1 \left[\dfrac{r + 1 + (r^2 - 1) \cos 2\theta_2}{2r} \right]$ where $r = \dfrac{a}{b}$ 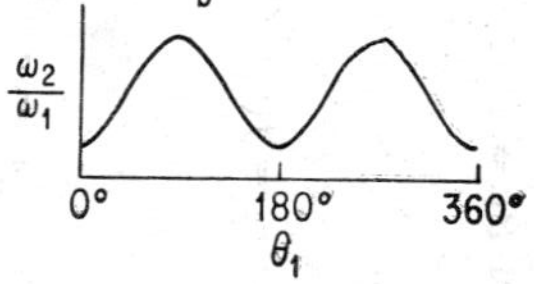

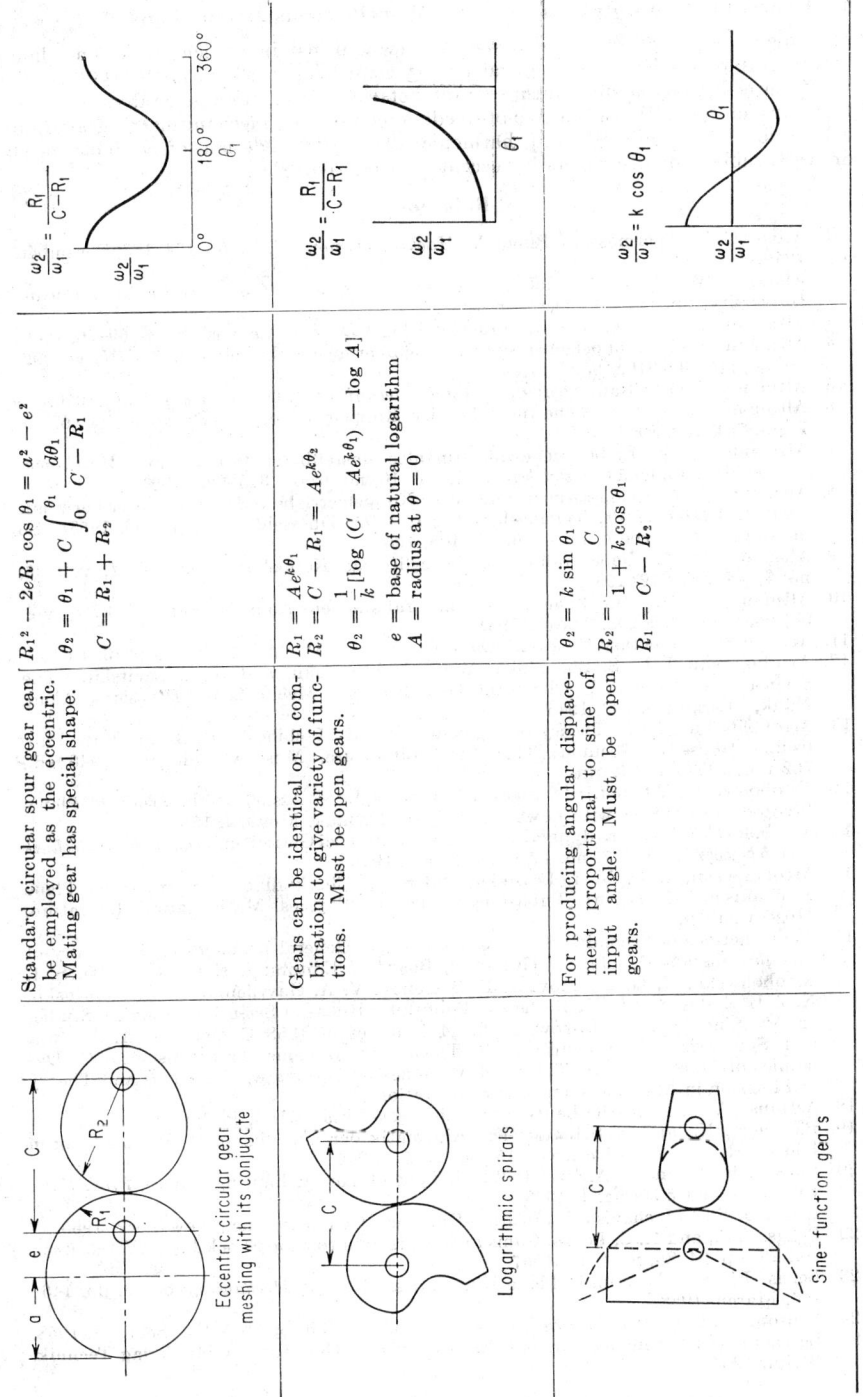

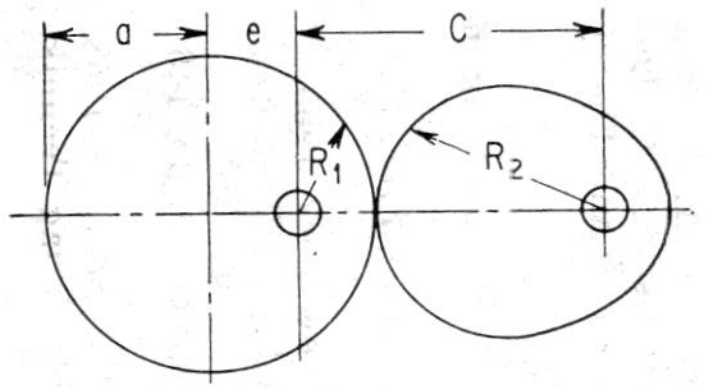 Eccentric circular gear meshing with its conjugcte	Standard circular spur gear can be employed as the eccentric. Mating gear has special shape.	$R_1^2 - 2eR_1 \cos\theta_1 = a^2 - e^2$ $\theta_2 = \theta_1 + C\int_0^{\theta_1} \frac{d\theta_1}{C - R_1}$ $C = R_1 + R_2$	$\frac{\omega_2}{\omega_1} = \frac{R_1}{C - R_1}$ 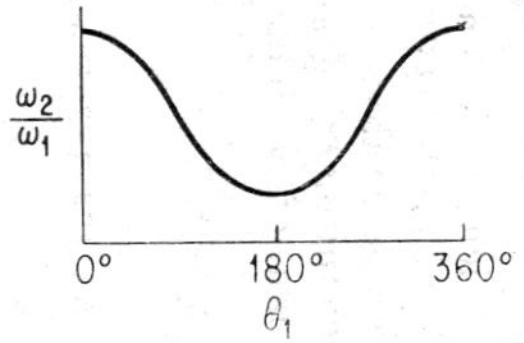
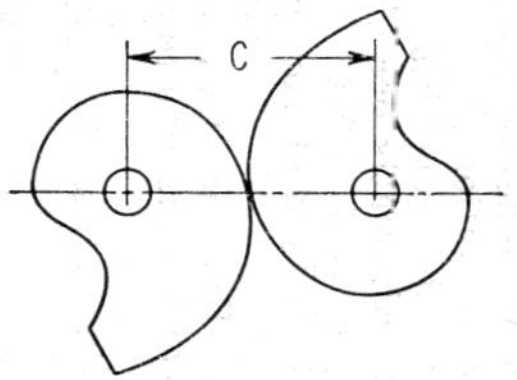 Logarithmic spirals	Gears can be identical or in combinations to give variety of functions. Must be open gears.	$R_1 = Ae^{k\theta_1}$ $R_2 = C - R_1 = Ae^{k\theta_2}$ $\theta_2 = \frac{1}{k}[\log (C - Ae^{k\theta_1}) - \log A]$ e = base of natural logarithm A = radius at $\theta = 0$	$\frac{\omega_2}{\omega_1} = \frac{R_1}{C - R_1}$ 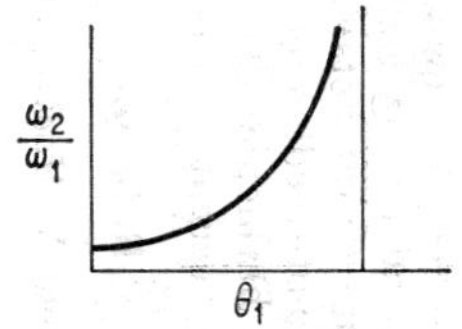
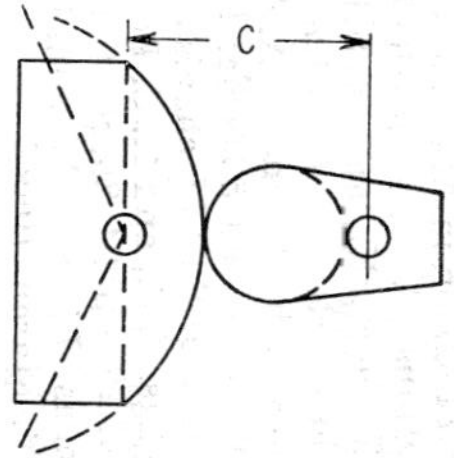 Sine-function gears	For producing angular displacement proportional to sine of input angle. Must be open gears.	$\theta_2 = k \sin\theta_1$ $R_2 = \frac{C}{1 + k\cos\theta_1}$ $R_1 = C - R_2$	$\frac{\omega_2}{\omega_1} = k \cos\theta_1$ 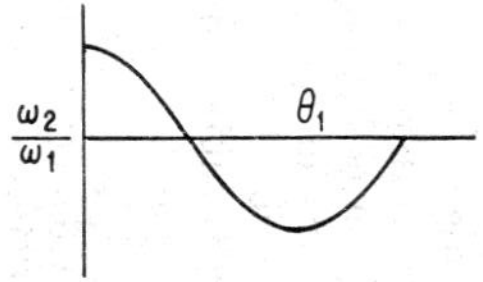

(i) Gear-link-cam Combinations and Miscellaneous Mechanisms

Two-gear drives[228,355,356]; *slider-crank* mechanisms in which rack on slide drives output gear[226,266]; mechanical analog computing mechanisms[21,34,105,247,264,272,331,336]; three-link screw mechanisms[293]; ratchets[18,296,303]; function generators with two four-bars in series[193]; two-degree-of-freedom computing mechanisms[264]; gear-train calculations[21,24,144,221,223,294]; the harmonic drive[57,99,257]; design of variable-speed drives[25]; rubber-covered rollers[313a]; eccentric-gear drives[120].

References

1. Alban, C. F.: Thermostatic Bimetals, *Machine Design*, vol. 18, pp. 124–128, December, 1946.
2. Allen, C. W.: Point Position Reduction, *Trans. Fifth Conf. Mechanisms*, Purdue University, pp. 181–193, October, 1958.
3. Altmann, F. G.: Ausgewaehlte Raumgetriebe, *VDI-Berichte*, vol. 12, pp. 69–76, 1956.
4. Altmann, F. G.: Koppelgetriebe fuer gleichfoermige Uebersetzung, *Z.VDI*, vol. 92, no. 33, pp. 909–916, 1950.
5. Altmann, F. G.: "Raumgetriebe," *Feinwerktechnik*, vol. 60, no. 3, pp. 1–10, 1956.
6. Altmann, F. G.: Raümliche fuenfgliedrige Koppelgetriebe, *Konstruktion*, vol. 6, no. 7, pp. 254–259, 1954.
7. Altmann, F. G.: Reibgetriebe mit stufenlos einstellbarer Übersetzung, *Maschinenbautechnik*, supplement *Getriebetechnik*, vol. 10, no. 6, pp. 313–322, 1961.
8. Altmann, F. G.: Sonderformen raümlicher Koppelgetriebe und Grenzen ihrer Verwendbarkeit, *VDI-Berichte*, Tagungsheft no. 1, *VDI*, Duesseldorf, pp. 51–68, 1953; also in *Konstruktion*, vol. 4, no. 4, pp. 97–106, 1952.
9. Altmann, F. G.: Ueber raümliche sechsgliedrige Koppelgetriebe, *Z.VDI*, vol. 96, no. 8, pp. 245–249, 1954.
10. Altmann, F. G.: Zur Zahlsynthese der raümlichen Koppelgetriebe, *Z.VDI*, vol. 93, no. 9, pp. 205–208, March, 1951.
11. Arnesen, L.: Planetary Gears, *Machine Design*, vol. 31, pp. 135–139, Sept. 3, 1959.
12. Artobolevskii, I. I., S. Sh. Blokh, V. V. Dobrovolskii, and N. I. Levitskii: "The Scientific Works of P. L. Chebichev II—Theory of Mechanisms" (Russian), Akad. Nauk, Moscow, p. 192, 1945.
13. Artobolevskii, I. I.: "Theory of Mechanisms and Machines" (Russian), State Publishing House of Technical-Theoretical Literature, Moscow-Leningrad, 1940 ed., 762 pp.; 1953 ed., 712 pp.
14. Artobolevskii, I. I. (ed.): "Collected Works (of L. V. Assur) on the Study of Plane, Pivoted Mechanisms with Lower Pairs," Akad. Nauk, Moscow, 1952.
15. Artobolevskii, I. I.: Das dynamische Laufkriterium bei Getrieben, *Maschinenbautechnik (Getriebetechnik)*, vol. 7, no. 12, pp. 663–667, 1958.
16. Artobolevskii, I. I., N. I. Levitskii, and S. A. Cherkudinov: "Synthesis of Plane Mechanisms" (Russian), Publishing House of Physical-Mathematical Literature, Moscow, 1959.
17. "Transactions of Second U.S.S.R. Conference on General Problems in the Theory of Mechanisms and Machines" (Russian), Board of Editors: I. I. Artobolevskii, S. I. Artobolevskii, G. G. Baranov, A. P. Bessonov, V. A. Gavrilenko, A. E. Kobrinskii, N. I. Levitskii, and L. N. Reshetov, Publishing House of Scientific Technical Studies on Machine Design, Moscow, 1960. 4 Volumes of 1958 Conference: A. Analysis and Synthesis of Mechanisms. B. Theory of Machine Transmissions. C. Dynamics of Machines. D. Theory of Machines in Automatic Operations and Theory of Precision in Machinery and Instrumentation.
18. Assmus, F.: "Technische Laufwerke," Springer-Verlag OHG, Berlin, 1958.
19. Bammert, K., and A. Schmidt: Die Kinematik der mittelbaren Pleuelanlenkung in Fourier-Reihen, *Ing.-Arch.*, vol. 15, pp. 27–51, 1944.
20. Bareiss, R. A., and P. A. Brand: Which Type of Torque-limiting Device, *Prod. Eng.*, vol. 29, pp. 50–53, Aug. 4, 1958.
21. Beggs, J. S.: "Mechanism," McGraw-Hill Book Company, Inc., New York, 1955.
22. Beggs, J. S., and R. S. Beggs: Cams and Gears Join to Stop Shock Loads, *Prod. Eng.*, vol. 28, pp. 84–85, Sept. 16, 1957.
23. Beggs, J. S.: A Theorem in Plane Kinematics, *J. Appl. Mechanics*, vol. 25, pp. 145–146, March, 1958.
24. Benson, A.: Gear-train Ratios, *Machine Design*, vol. 30, pp. 167–172, Sept. 18, 1958.
25. Berthold, H.: "Stufenlos verstellbare mechanische Getriebe," VEB Verlag Technik, Berlin, 1954.

26. Beyer, R.: "Kinematische Getriebesynthese," Springer-Verlag OHG, Berlin, 1953.
27. Beyer, R., and E. Schoerner: "Raumkinematische Grundlagen," Johann Ambrosius Barth, Munich, 1953.
28. Beyer, R.: Zur Geometrie und Synthese eigentlicher Raumkurbelgetriebe, *VDI-Berichte*, vol. 5, pp. 5–10, 1955.
29. Beyer, R.: Zur Synthese und Analyse von Raumkurbelgetrieben, *VDI-Berichte*, vol. 12, pp. 5–20, 1956.
30. Beyer, R.: Wissenschaftliche Hilfsmittel und Verfahren zur Untersuchung räumlicher Gelenkgetriebe, *VDI Zeitschrift*, vol. 99, pt. I, no. 6, pp. 224–231, February, 1957; pt. 2, no. 7, pp. 285–290, March, 1957.
31. Beyer, R.: Räumliche Malteserkreutzgetriebe, *VDI-Forschungsheft*, no. 461, pp. 32–36, 1957 (see ref. 120).
32. Beyer, R.: Space Mechanisms, *Trans. Fifth Conf. Mechanisms*, Purdue University, Oct. 13–14, 1958, pp. 141–163.
33. Biezeno, C. B., and R. Grammel: "Engineering Dynamics" (translated by M. P. White), vol. 4, Blackie & Son, Ltd., London, 1954.
34. Billings, J. H.: Review of Fundamental Computer Mechanisms, *Machine Design*, vol. 27, pp. 213–216, March, 1955.
35. Blaschke, W., and H. R. Mueller: "Ebene Kinematik," R. Oldenbourg KG, Munich, 1956.
36. Blokh, S. Sh.: "Angenaeherte Synthese von Mechanismen," VEB Verlag Technik, Berlin, 1951.
37. Bloomfield, B.: Non-circular Gears, *Prod. Eng.*, vol. 31, pp. 59–66, Mar. 14, 1960.
38. Bogardus, F. J.: A Survey of Intermittent-motion Mechanisms, *Machine Design*, vol. 28, pp. 124–131, Sept. 20, 1956.
39. Bogdan, R. C., and C. Pelecudi: Contributions to the Kinematics of Chebichev's Dyad, *Rev. Mecan. Appliquee*, vol. V, no. 2, pp. 229–240, 1960.
40. Bolz, W.: Gaskets in Design, *Machine Design*, vol. 17, pp. 151–156, March, 1945.
41. Bolz, R. W., J. W. Greve, K. R. Harnar, and R. E. Denega: High Speeds in Design, *Machine Design*, vol. 22, pp. 147–194, April, 1950.
42. Botstiber, D. W., and L. Kingston: Cycloid Speed Reducer, *Machine Design*, vol. 28, pp. 65–69, June 28, 1956.
43. Bottema, O.: On Cardan Positions for the Plane Motion of a Rigid Body, *Koninkl. Ned. Akad. Wetenschap. Proc.*, pp. 643–651, June, 1949.
44. Bottema, O.: On Gruebler's Formulae for Mechanisms, *Appl. Sci. Research*, vol. A2, pp. 162–164, 1950.
45. Brandenberger, H.: "Kinematische Getriebemodelle," Schweizer, Druck & Verlagshaus, AG, Zurich, 1955.
46. Breunich, T. R.: Sensing Small Motions, *Prod. Eng.*, vol. 21, pp. 113–116, July, 1950.
47. Buffenmyer, W. L.: Pressure Gages, *Machine Design*, vol. 31, p. 119, July 23, 1959.
48. Bush, R. R.: Bearings for Intermittent Oscillatory Motions, *Machine Design*, vol. 22, pp. 117–119, January, 1950.
49. Campbell, J. A.: Pneumatic Power—3, *Machine Design*, vol. 23, pp. 149–154, July, 1951.
50. Capitaine, D.: "Zur Analyse und Synthese ebener Vier- und Siebengelenkgetriebe in komplexer Behandlungsweise," doctoral dissertation, Technische Hochschule, Munich, Germany (1955–1960).
51. Carnetti, B., and A. J. Mei: Hermetic Motor Pumps for Sealed Systems, *Mech. Eng.*, vol. 77, pp. 488–494, June, 1955.
52. Carter, B. G.: Analytical Treatment of Linked Levers and Allied Mechanisms, *J. Roy. Aeron. Soc.*, vol. 54, pp. 247–252, October, 1950.
53. Carter, W. J.: Kinematic Analysis and Synthesis Using Collineation-axis Equations, *Trans. ASME*, vol. 79, pp. 1305–1312, 1957.
54. Carter, W. J.: Second Acceleration in Four-bar Mechanisms as Related to Rotopole Motion, *J. Appl. Mechanics*, vol. 25, pp. 293–294, June, 1958.
55. Cayley, A.: On Three-bar Motion, *Proc. London Math. Soc.*, vol. 7, pp. 135–166, 1875–1876.
56. Cherkudinov, S. A.: "Synthesis of Plane, Hinged, Link Mechanisms" (Russian), Akad. Nauk, Moscow, 1959.
57. Chironis, N.: Harmonic Drive, *Prod. Eng.*, vol. 31, pp. 47–51, Feb. 8, 1960.
58. Chow, H.: Linkage Joints, *Prod. Eng.*, vol. 29, pp. 87–91, Dec. 8, 1958.
59. Conklin, R. M., and H. M. Morgan: Transducers, *Prod. Eng.*, vol. 25, pp. 158–162, December, 1954.
60. Conway, H. G.: "Landing-gear Design," Chapman & Hall, Ltd., London, 1958.
61. Conway, H. G.: Straight-line Linkages, *Machine Design*, vol. 22, p. 90, 1950.

62. Creech, M. D.: Dynamic Analysis of Slider-crank Mechanisms, *Prod. Eng.*, vol. 33, pp. 58–65, Oct. 29, 1962.
63. Crossley, F. R. E., and F. W. Keator: Three Dimensional Mechanisms, *Machine Design*, vol. 27, no. 8, pp. 175–179; no. 9, pp. 204–209, 1955.
64. Cunningham, F. W.: Non-circular Gears, *Machine Design*, vol. 31, pp. 161–164, Feb. 19, 1959.
65. Skoda Harbor Cranes for the Port of Madras, India, *Czechoslovak Heavy Industry*, February, 1961, pp. 21–28.
66. Deimel, R. F., and W. A. Black: Ball Integrator Averages Sextant Readings, *Machine Design*, vol. 18, pp. 116–120, March, 1946.
67. Deist, D. H.: Air Springs, *Machine Design*, vol. 30, pp. 141–146, Jan. 6, 1958.
68. de Jonge, A. E. R.: A Brief Account of Modern Kinematics, *Trans. ASME*, vol. 65, pp. 663–683, 1943.
69. de Jonge, A. E. R.: Kinematic Synthesis of Mechanisms, *Mech. Eng.*, vol. 62, pp. 537–542, 1940. See also refs. 167, 168.
70. Denavit, J.: Displacement Analysis of Mechanisms Based on 2 × 2 Matrices of Dual Numbers (in English), *VDI-Berichte*, vol. 29, pp. 81–88, 1958.
71. Denavit, J., and R. S. Hartenberg: Approximate Synthesis of Spatial Linkages, *J. Appl. Mechanics*, vol. 27, pp. 201–206, March, 1960.
72. Dimentberg, F. M.: Determination of the Motions of Spatial Mechanisms, Akad. Nauk, Moscow, 142 p. 1950. See also ref. 78.
73. Double-throw Crankshaft Imparts Vertical Motion to Worktable, *Design News*, Apr. 14, 1958, p. 26.
74. "Sky-lift"—Tractor and Loader with Vertical Lift Attachment, *Design News*, Aug. 3, 1959, pp. 24–25.
75. Hydraulic Link Senses Changes in Force to Control Movement in Lower Limb, *Design News*, Oct. 26, 1959, pp. 26–27.
76. Four-bar Linkage Moves Stacker Tip in Vertical Straight Line, *Design News*, Oct. 26, 1959, pp. 32–33.
77. Linkage Balances Forces in Thrust Reverser on Jet Star, *Design News*, Feb. 15, 1960, pp. 30–31.
78. Dimentberg, F. M.: A General Method for the Investigation of Finite Displacements of Spatial Mechanisms and Certain Cases of Passive Constraints (Russian), Akad. Nauk, Moscow, *Trudi Sem. Teor. Mash. Mekh.*, vol. 5, no. 17, pp. 5–39, 1948. See also ref. 72.
79. Dobrovolskii, V. V.: "Theory of Mechanisms" (Russian), State Publishing House of Technical Scientific Literature, Machine Construction Division, Moscow, p. 464, 1951.
80. Dobrovolskii, V. V.: "Trajectories of Five-Link Mechanisms," *Trans. Moscow Machine-tool Construc. Inst.*, vol. 1, 1937.
81. Doughtie, V. L., and W. H. James: "Elements of Mechanism," John Wiley & Sons, Inc., New York, 1954.
82. Dudley, D. W.: "Gear Handbook," McGraw-Hill Book Company, Inc., New York, 1962.
83. Eastman, F. S.: Flexure Pivots to Replace Knife-edges and Ball Bearings, *Univ. Wash. Eng. Expt. Sta. Bull.* 86, November, 1935.
84. Ebner, F.: "Leitfaden der Technischen Wichtigen Kurven," B. G. Teubner Verlagsgesellschaft, GmbH, Leipzig, 1906.
85. Ellis, A. H., and J. H. Howard: What to Consider When Selecting a Metallic Bellows, *Prod. Eng.*, vol. 21, pp. 86–89, July, 1950.
86. Federhofer, K.: "Graphische Kinematik und Kinetostatik," Springer-Verlag OHG, Berlin, 1932.
87. Franke, R.: "Vom Aufbau der Getriebe," vol. I, Beuth-Vertrieb, GmbH, Berlin, 1948; vol. II, VDI Verlag, Duesseldorf, Germany, 1951.
88. Nonmetallic Gaskets, based on studies by E. C. Frazier, *Machine Design*, vol. 26, pp. 157–188, November, 1954.
88*a*. Freudenstein, F.: Design of Four-link Mechanisms, doctoral dissertation, Columbia University, University microfilms, Ann Arbor, Mich., 1954.
89. Freudenstein, F.: Approximate Synthesis of Four-bar Linkages, *Trans. ASME*, vol. 77, pp. 853–861, August, 1955.
90. Freudenstein, F.: On the Maximum and Minimum Velocities and the Accelerations in Four-link Mechanisms, *Trans. ASME*, vol. 78, pp. 779–787, 1956.
91. Freudenstein, F.: Ungleichfoermigkeitsanalyse der Grundtypen ebener Getriebe, *VDI-Forschungsheft*, no. 461, series B, vol. 23, pp. 6–10, 1957.
92. Freudenstein, F.: Four-bar Function Generators *Machine Design*, vol 30, pp 119–123, Nov. 27, 1958.

93. Freudenstein, F., and E. J. F. Primrose: Geared Five-bar Motion I—Gear Ratio Minus One, *J. Appl. Mech.*, vol. 30; *Trans. ASME, ser. E*, vol. 85, pp. 161–169, June, 1963.

93*a*. Freudenstein, F. (E. J. F. Primrose and): Geared Five-bar Motion II—Arbitrary Commensurate Gear Ratio, *ibid.*, pp. 170–175. See also ref. 268*a*.

94. Freudenstein, F., and B. Roth: Numerical Solution of Systems of Nonlinear Equations, *J. Assoc. Computing Machinery*, October, 1963.

95. Freudenstein, F.: Structural Error Analysis in Plane Kinematic Synthesis, *J. Eng. Ind., Trans. ASME*, vol. 81B, pp. 15–22, February, 1959.

96. Freudenstein, F., and G. N. Sandor: Synthesis of Path-generating Mechanisms by Means of a Programmed Digital Computer, *J. Eng. Ind., Trans. ASME*, vol. 81B, pp. 159–168, 1959.

97. Freudenstein, F.: Trends in the Kinematics of Mechanisms, *Appl. Mechanics Revs.*, vol. 12, no. 9, September, 1959, survey article.

98. Freudenstein, F.: Harmonic Analysis of Crank-and-rocker Mechanisms with Application, *J. Appl. Mechanics*, vol. 26, pp. 673–675, December, 1959.

99. Freudenstein, F.: The Cardan Positions of a Plane, *Trans. Sixth Conf. Mechanisms* Purdue University, October, 1960, pp. 129–133.

100. Freudenstein, F., and G. N. Sandor: On the Burmester Points of a Plane, *J. Appl. Mechanics*, vol. 28, pp. 41–49, March, 1961; discussion, September, 1961, pp. 473–475.

101. Freudenstein, F., and K. Mohan: Harmonic Analysis, *Prod. Eng.*, vol. 32, pp. 47–50, Mar. 6, 1961.

102. Freudenstein, F.: "Automatic Computation in Mechanisms and Mechanical Networks and a Note on Curvature Theory," presented at the International Conference on Mechanisms at Yale University, March, 1961—Shoestring Press, Inc., New Haven, Conn., 1961, pp. 43–62.

103. Freudenstein, F.: "On the Variety of Motions Generated by Mechanisms," *Trans. ASME*, vol. 81B, *J. Eng. Ind.*, February, 1962, pp. 156–160.

104. Freudenstein, F.: Bi-variate, Rectangular, Optimum-interval Interpolation, *Mathematics of Computation*, vol. 15, no. 75, pp. 288–291, July, 1961.

105. Fry, M.: Designing Computing Mechanisms, *Machine Design*, Vol. 17–18, 1945–1946; I, August, pp. 103–108; II, September, pp. 113–120; III, October, pp. 123–128; IV, November, pp. 141–145; V, December, pp. 123–126; VI, January, pp. 115–118; VII, February, pp. 137–140.

106. Fry, M.: When Will a Toggle Snap Open, *Machine Design*, vol. 21, p. 126, August, 1949.

107. Gagne, A. F., Jr.: One-way Clutches, *Machine Design*, vol. 22, pp. 120–128, April, 1950.

108. Gagne, A. F., Jr.: Clutches, *Machine Design*, vol. 24, pp. 123–158, August, 1952.

109. Geary, P. J.: Torsion Devices, Part 3, Survey of Instrument Parts, British Scientific Instrument Research Association, *Research Rept.* R. 249, 1960, South Hill, Chislehurst, Kent, England; see also Rept. 1, on Flexure Devices; 2 on Knife-edge Bearings, all by the same author.

110. Auto Window Regulator Mechanism, *Gen. Motors Eng. J.*, January, February, March, 1961, pp. 40–42.

111. Goldberg, M.: New Five-bar and Six-bar Linkages in Three Dimensions, *Trans. ASME*, vol. 65, pp. 649–661, 1943.

112. Goodman, T. P.: Toggle Linkage Applications in Different Mechanisms, *Prod. Eng.*, vol. 22, no. 11, pp. 172–173, 1951.

113. Goodman, T. P.: An Indirect Method for Determining Accelerations in Complex Mechanisms, *Trans. ASME*, vol. 80, pp. 1676–1682, November, 1958.

114. Goodman, T. P.: Four Cornerstones of Kinematic Design, *Trans. Sixth Conf. Mechanisms*, Purdue University, Oct. 10–11, 1961, pp. 4–30; also published in *Machine Design*, 1960–1961.

115. Goodman, T. P.: Dynamic Effects of Backlash, *Trans. Seventh Conf. Mechanisms*, Purdue University, pp. 128–138, 1962.

116. Grodzenskaya, L. S.: Computational Methods of Designing Linked Mechanisms with Dwell (Russian), *Trudi Inst. Mashinoved.*, Akad. Nauk, Moscow, vol. 19, no. 76, pp. 34–45, 1959; see also vol. 71, pp. 69–90, 1958.

117. Grodzinski, P.: Eccentric Gear Mechanisms, *Machine Design*, vol. 25, pp. 141–150, May, 1953.

118. Grodzinski, P.: Straight-line Motion, *Machine Design*, vol. 23, pp. 125–127, June, 1951.

119. Grodzinski, P., and E. M'Ewen: Link Mechanisms in Modern Kinematics, *Proc. Inst. Mech. Engrs. (London)*, vol. 168, no. 37, pp. 877–896, 1954.

120. Grodzinski, P.: Applying Eccentric Gearing, *Machine Design*, vol. 26, pp. 147-151, July, 1954.
121. Grodzinski, P.: "A Practical Theory of Mechanisms," Emmett & Co., Ltd., Manchester, England, 1947.
122. Gruebler, M.: "Getriebelehre," Springer-Verlag OHG, Berlin, 1917/21.
123. Hagedorn, L.: "Getriebetechnik und Ihre praktische Anwendung," *Konstruktion*, vol. 10, no. 1, pp. 1-10, 1958.
124. Hain, K.: Der Einfluss der Toleranzen bei Gelenkrechengetrieben, *Die Messtechnik*, vol. 20, pp. 1-6, 1944.
125. Hain, K.: "Angewandte Getriebelehre," 2d ed., VDI Verlag, Duesseldorf, 1961, 587 pp., English translation, McGraw-Hill Book Company, Inc., New York, in press.
126. Hain, K.: Uebertragungsguenstige unsymmetrische Doppelkurbelgetriebe, *VDI-Forschungsheft*, no. 461, supplement to *Forsch. Gebiete Ingenieurw.*, series B, vol. 23, pp. 23-25, 1957.
127. Hain, K.: Selbsttaetige Getriebegruppen zur Automatisierung von Arbeitsvorgaengen, *Feinwerktechnik*, vol. 61, no. 9, pp. 327-329, September, 1957.
128. Hain, K.: Achtgliedrige kinematische Ketten mit dem Freiheitsgrad F = −1 fuer gegebene Kraefteverhaeltnisse, *Das Industrieblatt*, vol. 62, no. 6, pp. 331-337, June, 1962.
129. Hain, K.: Beispiele zur Systematik von Spannvorrichtungen aus sechsgliedrigen kinematischen Ketten mit dem Freiheitsgrad F = −1, *Das Industrieblatt*, vol. 61, no. 12, pp. 779-784, December, 1961.
130. Hain, K.: Die Entwicklung von Spannvorrichtungen mit mehreren Spannstellen aus kinematischen Ketten, *Das Industrieblatt*, vol. 59, no. 11, pp. 559-564, November, 1959.
131. Hain, K.: Entwurf viergliedriger kraftverstaerkender Zangen fuer gegebene Kraefteverhaeltnisse, *Das Industrieblatt*, pp. 70-73, February, 1962.
132. Hain, K.: Der Entwurf Uebertragungsguenstiger Kurbelgetriebe mit Hilfe von Kurventafeln, *VDI-Berichte*, Getriebetechnik und Ihre praktische Anwendung, vol. 29, pp. 121-128, 1958.
133. Hain, K.: Drag Link Mechanisms, *Machine Design*, vol. 30, pp. 104-113, June 26, 1958.
134. Hain, K.: Hydraulische Schubkolbenantriebe fuer schwierige Bewegungen, *Oelhydraulik & Pneumatik*, vol. 2, no. 6, pp. 193-199, September, 1958.
135. Hain, K., and G. Marx: How to Replace Gears by Mechanisms, *Trans. ASME*, vol. 81, pp. 126-130, May, 1959.
136. Hain, K.: Mechanisms, a 9-step Refresher Course (translated by F. R. E. Crossley), *Prod. Eng.*, vol. 32, 1961 (Jan. 2, 1961-Feb. 27, 1961, in 9 parts).
137. Ham, C. W., E. J. Crane, and W. L. Rogers: "Mechanics of Machinery," 4th ed., McGraw-Hill Book Company, Inc., New York, 1958.
138. Hall, A. S., and E. S. Ault: How Acceleration Analysis Can Be Improved, *Machine Design*, vol. 15, part I, pp. 100-102, February, 1943; part II, pp. 90-92, March, 1943.
139. Hall, A. S.: Mechanism Properties, *Machine Design*, vol. 20, pp. 111-115, February, 1948.
140. Hall, A. S., and D. C. Tao: Linkage Design—a Note on One Method, *Trans. ASME*, vol. 76, no. 4, pp. 633-637, 1954.
141. Hall, A. S.: A Novel Linkage Design Technique, *Machine Design*, vol. 31, pp. 144-151, July 9, 1959.
142. Hall, A. S.: "Kinematics and Linkage Design," Prentice-Hall, Inc., Englewood Cliffs, N.J., 1961.
143. Hannula, F. W.: Designing Non-circular Surfaces, *Machine Design*, vol. 23, pp. 111-114, 190, 192, July, 1951.
144. Handy, H. W.: "Compound Change Gear and Indexing Problems," The Machinery Publishing Co., Ltd., London.
145. Harnar, R. R.: Automatic Drives, *Machine Design*, vol. 22, pp. 136-141, April, 1950.
146. Harrisberger, L.: "Mechanization of Motion," John Wiley & Sons, Inc., New York, 1961.
147. Hartenberg, R. S.: Complex Numbers and Four-bar Linkages, *Machine Design*, vol. 30, pp. 156-163, Mar. 20, 1958.
148. Hartenberg, R. S., and J. Denavit: Cognate Linkages, *Machine Design*, vol. 31, pp. 149-152, Apr. 16, 1959.
149. Hartenberg, R. S.: Die Modellsprache in der Getriebetechnik, *VDI-Berichte*, vol. 29, pp. 109-113, 1958.
150. Hastings, C., Jr., J. T. Hayward, and J. P. Wong, Jr.: "Approximations for Digital Computers," Princeton University Press, Princeton, N.J., 1955.

151. Heidler, G. R.: Spring-loaded Differential Drive (for tensioning), *Machine Design*, vol. 30, p. 140, Apr. 3, 1958.
152. Hekeler, C. B.: Flexible Metal Tapes, *Prod. Eng.*, vol. 32, pp. 65–69, Feb. 20, 1961.
153. Herst, R.: Servomechanisms (types of), *Elec. Mfg.*, May, 1950, pp. 90–95.
154. Hildebrand, S.: Moderne Schreibmaschinenantriebe und Ihre Bewegungsvorgänge, *Getriebetechnik, VDI-Berichte*, vol. 5, pp. 21–29, 1955.
155. Hildebrand, F. B.: "Introduction to Numerical Analysis," McGraw-Hill Book Company, Inc., New York, 1956.
156. Hinkle, R. T.: "Kinematics of Machines," 2d ed., Prentice-Hall, Inc., Englewood Cliffs, N.J., 1960.
157. Hirschhorn, J.: New Equations Locate Dwell Position of Three-gear Drive, *Prod. Eng.*, vol. 30, pp. 80–81, June 8, 1959.
158. Hohenberg, F.: "Konstruktive Geometrie in der Technik," 2d ed., Springer-Verlag OHG, Vienna, 1961.
159. Hornsteiner, M.: Getriebetechnische Fragen bei der Faltschachtel-Fertigung, *Getriebetechnik, VDI-Berichte*, vol. 5, pp. 69–74, 1955.
160. Hotchkiss, C., Jr.: Flexible Shafts, *Prod. Eng.*, vol. 26, pp. 168–177, February, 1955.
161. Hrones, J. A., and G. L. Nelson: "Analysis of the Four-bar Linkage," The Technology Press of the Massachusetts Institute of Technology, Cambridge, Mass., and John Wiley & Sons, Inc., New York, 1951.
162. Hunt, K. H.: "Mechanisms and Motion," John Wiley & Sons, Inc., New York, N.Y., 1959, p. 108; see also *Proc. Inst. Mech. Eng. (London)*, vol. 174, no. 21, pp. 643–668, 1960.
163. Compensation Practice, *Instruments and Control Systems*, August, 1959, p. 1185.
164. Jahr, W., and P. Knechtel: "Grundzuege der Getriebelehre," Fachbuch Verlag, Leipzig, vol. 1, 1955; vol. II, 1956.
165. Jensen, P. W.: Four-bar Mechanisms, *Machine Design*, vol. 33, pp. 173–176, June 22, 1961.
166. Jones, H. B., Jr.: Recording Systems, *Prod. Eng.*, vol. 26, pp. 180–185, March, 1955.
167. de Jonge, A. E. R.: Analytical Determination of Poles in the Coincidence Position of Links in Four-bar Mechanisms Required for Valves Correctly Apportioning Three Fluids in a Chemical Apparatus, *Trans. ASME*, 84B; *J. Eng. Ind.*, pp. 359–372, August, 1962.
168. de Jonge, A. E. R.: The Correlation of Hinged Four-bar Straight-line Motion Devices by Means of the Roberts Theorem and a New Proof of the Latter, *Ann. N.Y. Acad. Sci.*, vol. 84, pp. 75–145, 1960. See also refs. 68, 69.
168*a*. Johnson, C.: Dynamic Sealing with O-rings, *Machine Design*, vol. 27, pp. 183–188, August, 1955.
169. Johnson, H. L.: "Synthesis of the Four-bar Linkage," M.S. dissertation, Georgia Institute of Technology, Atlanta, Ga., June, 1958.
170. Johnson, R. C.: Geneva Mechanisms, *Machine Design*, vol. 28, pp. 107–111, Mar. 22, 1956.
171. Johnson, R. C.: Method of Finite Differences in Cam Design—Accuracy—Applications, *Machine Design*, vol. 29, pp. 159–161, Nov. 14, 1957.
172. Johnson, R. C.: Development of a High-speed Indexing Mechanism, *Machine Design*, vol. 30, pp. 134–138, Sept. 4, 1958.
173. Kaplan, J., and H. North: Cyclic Three-Gear Drives, *Machine Design*, vol. 31, pp. 185–188, Mar. 19, 1959.
174. Kearny, W. R., and M. G. Wright: Straight-line Mechanisms, *Trans. Second Conf. Mechanisms*, Purdue University, October, 1954, pp. 209–216.
175. Keler, M. K.: Analyse und Synthese der Raumkurbelgetriebe mittels Raumliniengeometrie und dualer Groessen, *Forsch. Gebiete Ingenieurw.*, vol. 75, pp. 26–63, 1959.
176. Kinsman, F. W.: "Controlled-acceleration Single-revolution Drives," *Trans. Seventh Conf. Mechanisms*, Purdue University, pp. 229–233, 1962.
177. Kislitsin, S. G.: General Tensor Methods in the Theory of Space Mechanisms (Russian), Akad. Nauk, Moscow, *Trudi Sem. Teor. Mash. Mekh.*, vol. 14, no. 54, 1954.
178. Kist, K. E.: Modified Starwheels, *Trans. Third Conf. Mechanisms*, Purdue University, May, 1956, pp. 16–20.
179. Kobrinskii, A. E.: On the Kinetostatic Calculation of Mechanisms with Passive Constraints and with Play, *Trudi Sem. Teor. Mash. Mekh.*, vol. V, no. 20, pp. 5–53, Akad. Nauk, Moscow, 1948.
180. Kobrinskii, A. E., M. G. Breido, V. S. Gurfinkel, E. P. Polyan, Y. L. Slavitskii, A. Y. Sysin, M. L. Zetlin, and Y. S. Yacobson: On the Investigation of Creating a Bioelectric Control System, *Trudi Inst. Mashinoved.* Akad. Nauk, Moscow, vol. 20, no. 77, pp. 39–50, 1959.

181. Kovacs, J. P., and R. Wolk: Filters, *Machine Design*, vol. 27, pp. 167–178, January, 1955.
182. Kraus, C. E.: Chuting and Orientation in Automatic Handling, *Machine Design*, vol. 21, pp. 95–98, September, 1949.
183. Kraus, C. E.: Automatic Positioning and Inserting Devices, *Machine Design*, vol. 21, pp. 125–128, October, 1949.
184. Kraus, C. E.: Elements of Automatic Handling, *Machine Design*, vol. 23, pp. 142–145, November, 1951.
185. Kraus, R.: "Getriebelehre," vol. I, VEB Verlag Technik, Berlin, 1954.
186. Kuhlenkamp, A.: Linkage Layouts, *Prod. Eng.*, vol. 26, no. 8, pp. 165–170, August, 1955.
187. Kuhn, H. S.: Rotating Joints, *Prod. Eng.*, vol. 27, pp. 200–204, August, 1956.
188. Kupfrian, W. J.: Flexible Shafts, *Machine Design*, vol. 26, pp. 164–173, October, 1954.
189. Kutzbach, K.: Mechanische Leitungsverzweigung, ihre Gesetze und Anwendungen, Maschinenbau, vol. 8, 1929, pp. 710–716; see also *Z.VDI*, vol. 77, p. 1168, 1933.
190. Langosch, O.: Der Einfluss der Toleranzen auf die Genauigkeit von periodischen Getrieben, *Konstruktion*, vol. 12, no. 1, p. 35, 1960.
191. Laughner, V. H., and A. D. Hargan: "Handbook of Fastening and Joining Metal Parts," McGraw-Hill Book Company, Inc., New York, 1956.
192. Lawson, A. C.: Jewel Bearing Systems, *Machine Design*, vol. 26, pp. 132–137, April, 1954.
193. Levitskii, N. I.: On the Synthesis of Plane, Hinged, Six-link Mechanisms, pp. 98–104. See item *A*, ref. 17, in this section of book.
194. Levitskii, N. I.: "Design of Plane Mechanisms with Lower Pairs" (Russian), Akad. Nauk, Moscow-Leningrad, 1950.
195. Levitskii, N. I., and Sh. Shakvasian: Synthesis of Spatial Four-link Mechanisms with Lower Pairs, Akad. Nauk, Moscow, *Trudi Sem. Teor. Mash. Mekh.*, vol. 14, no. 54, pp. 5–24, 1954.
196. Lichtenheldt, W.: "Konstruktionslehre der Getriebe," Akademie Verlag, Berlin, 1961.
197. Lichtwitz, O.: Mechanisms for Intermittent Motion, *Machine Design*, vol. 23–24, 1951–1952; I, December, pp. 134–148; II, January pp. 127–141; III, February, pp. 146–155; IV, March, pp. 147–155.
198. Lichtwitz, O.: "Getriebe fuer Aussetzende Bewegung," Springer-Verlag OHG, Berlin, 1953.
199. Lincoln, C. W.: A Summary of Major Developments in the Steering Mechanisms of American Automobiles, *Gen. Motors Eng. J.*, March–April, 1955, pp. 2–7.
200. Litvin, F. L.: Design of Non-circular Gears and Their Application to Machine Design, Akad. Nauk, Moscow, *Trudi Inst. Machinoved.*, vol. 14, no. 55, pp. 20–48, 1954.
201. Lockenvitz, A. E., J. B. Oliphint, W. C. Wilde, and J. M. Young: Non-circular Cams and Gears, *Machine Design*, vol. 24, pp. 141–145, May, 1952.
202. Lundquist, I.: Miniature Mechanical Clutches, *Machine Design*, vol. 28, pp. 124–133, Oct. 18, 1956.
203. Mabie, H. H.: Constant Velocity Universal Joints, *Machine Design*, vol. 20, pp. 101–105, May, 1948.
204. Mabie, H. H., and F. W. Ocvirk: "Mechanisms and Dynamics of Machinery," John Wiley & Sons, Inc., New York, 1957.
205. Velocity and Acceleration Analysis of Universal Joints, *Machine Design*, vol. 14 (data sheet), November, 1942, pp. 93–94.
206. Chain Drive, *Machine Design*, vol. 23, pp. 137–138, June, 1951.
207. Constant Force Action (manual typewriters), *Machine Design*, vol. 28, p. 98, Sept. 20, 1956.
208. *Machine Design*, vol. 30, p. 134, Feb. 20, 1958.
209. Constant-force Output, *Machine Design*, vol. 30, p. 115, Apr. 17, 1958.
210. Taut-band Suspension, *Machine Design*, vol. 30, p. 152, July 24, 1958. Intermittent-motion Gear Drive, *Machine Design*, vol. 30, p. 151, July 24, 1958.
211. Almost Human Engineering, *Machine Design*, vol. 31, pp. 22–26, Apr. 30, 1959.
212. Aircraft All-mechanical Computer, *Machine Design*, vol. 31, pp. 186–187, May 14, 1959.
213. Virtual Hinge Pivot, *Machine Design*, vol. 33, p. 119, July 6, 1961.
214. Mansfield, J. H.: Woodworking Machinery, *Mech. Eng.*, vol. 74, pp. 983–995, December, 1952.
215. Marich, F.: Mechanical Timers, *Prod. Eng.*, vol. 32, pp. 54–56, July 17, 1961.
216. Marker, R. C.· Determining Toggle-jaw Force, *Machine Design*, vol. 22, pp. 104–107, March, 1950

217. Martin, G. H., and M. F. Spotts: An Application of Complex Geometry to Relative Velocities and Accelerations in Mechanisms, *Trans. ASME*, vol. 79, pp. 687–693, April, 1957.
218. Mathi, W. E., and C. K. Studley, Jr.: Developing a Counting Mechanism, *Machine Design*, vol. 22, pp. 117–122, June, 1950.
219. Maxwell, R. L.: "Kinematics and Dynamics of Machinery," Prentice-Hall, Inc., Englewood Cliffs, N.J., 1960.
220. Mayer, A. E.: Koppelkurven mit drei Spitzen und Spezielle Koppelkurvenbueschel, *Z. Math. Phys.*, vol. 43, p. 389, 1937.
221. McComb, G. T., and W. N. Matson: Four Ways to Select Change Gears—and the Faster Fifth Way, *Prod. Eng.*, vol. 31, pp. 64–67, Feb. 15, 1960.
222. Mehmke, R.: Ueber die Bewegung eines Starren ebenen Systems in seiner Ebene, *Z. Math. Phys.*, vol. 35, pp. 1–23, 65–81, 1890.
223. Merritt, H. E.: "Gear Trains," Sir Isaac Pitman & Sons, Ltd., London, 1947.
224. Meyer zur Capellen, W.: Getriebependel, *Z. Instrumentenk.*, vol. 55, 1935; pt. I, October, pp. 393–408; pt. II, November, pp. 437–447.
225. Meyer zur Capellen, W.: Die Totlagen des ebenen Gelenkvierecks in analytischer Darstellung, *Forsch. Ing. Wes.*, vol. 22, no. 2, pp. 42–50, 1956.
226. Meyer zur Capellen, W.: Der einfache Zahnstangen-Kurbeltrieb und das entsprechende Bandgetriebe, *Werkstatt u. Betrieb*, vol. 89, no. 2, pp. 67–74, 1956.
227. Meyer zur Capellen, W.: Harmonische Analyse bei der Kurbelschleife, *Z. angew. Math. u. Mechanik*, vol. 36, pp. 151–152, March–April, 1956.
228. Meyer zur Capellen, W.: Kinematik des Einfachen Koppelraedertriebes, *Werkstatt u. Betrieb*, vol. 89, no. 5, pp. 263–266, 1956.
229. Meyer zur Capellen, W.: Bemerkung zum Satz von Roberts über die Dreifache Erzeugung der Koppelkurve, *Konstruktion*, vol. 8, no. 7, pp. 268–270, 1956.
230. Meyer zur Capellen, W.: Kinematik und Dynamik der Kurbelschleife, *Werkstatt u. Betrieb*, pt. I, vol. 89, no. 10, pp. 581–584, 1956; pt. II, vol. 89, no. 12, pp. 677–683, 1956.
231. Meyer zur Capellen, W.: Ueber gleichwertige periodische Getriebe, *Fette, Seifen, Anstrichmittel*, vol. 59, no. 4, pp. 257–266, 1957.
232. Meyer zur Capellen, W.: Die Kurbelschleife Zweiter Art, *Werkstatt u. Betrieb*, vol. 90, no. 5, pp. 306–308, 1957.
233. Meyer zur Capellen, W.: Die elliptische Zahnraeder und die Kurbelschleife, *Werkstatt u. Betrieb*, vol. 91, no. 1, pp. 41–45, 1958.
234. Meyer zur Capellen, W.: Die harmonische Analyse bei elliptischen Kurbelschleifen, *Z. angew. Math. Mechanik*, vol. 38, no. 1/2, pp. 43–55, 1958.
235. Meyer zur Capellen, W.: Das Kreuzgelenk als periodisches Getriebe, *Werkstatt u. Betrieb*, vol. 91, no. 7, pp. 435–444, 1958.
236. Meyer zur Capellen, W.: Ueber elliptische Kurbelschleifen, *Werkstatt u. Betrieb*, vol. 91, no. 12, pp. 723–729, 1958.
237. Meyer zur Capellen, W.: Bewegungsverhaeltnisse an der geschraenkten Schubkurbel, *Forchungsberichte des Landes Nordrhein-Westfalen, West-Deutscher Verlag*, Cologne, no. 449, 1958.
238. Meyer zur Capellen, W.: Eine Getriebegruppe mit stationaerem Geschwindigkeitsverlauf (Elliptic Slider-crank Drive) (in German), *Forschungsberichte des Landes Nordrhein-Westfalen, West-Deutscher Verlag, Cologne*, no. 606, 1958.
239. Meyer zur Capellen, W.: Die Beschleunigungsaenderung, *Ing.-Arch.*, vol. 27, pt. I, no. 1, pp. 53–65; pt. II, no. 2, pp. 73–87, 1959.
240. Meyer zur Capellen, W.: Die geschraenkte Kurbelschleife zweiter Art, *Werkstatt u. Betrieb*, vol. 92, no. 10, pp. 773–777, 1959.
241. Meyer zur Capellen, W.: Harmonische Analyse bei Kurbeltrieben, *Forschungsberichte des Landes Nordrhein-Westfalen, West-Deutscher Verlag*, Cologne, pt. I, no. 676, 1959; pt. II, no. 803, 1960.
242. Meyer zur Capellen, W.: Die geschraenkte Kurbelschleife, *Forschungsberichte des Landes Nordrhein-Westfalen, West-Deutscher Verlag*, Cologne, pt. I, no. 718, 1959; pt. II, no. 804, 1960.
243. Meyer zur Capellen, W.: Die Extrema der Uebersetzungen in ebenen und sphaerischen Kurbeltrieben, *Ing. Arch.*, vol. 27, no. 5, pp. 352–364, 1960.
244. Meyer zur Capellen, W.: Die gleichschenklige zentrische Kurbelschwinge, *Z. prakt. Metallbearbeitung*, vol. 54, no. 7, pp. 305–310, 1960.
245 Meyer zur Capellen, W.: Three-dimensional Drives, *Prod. Eng.*, vol. 31, pp. 76–80, June 30, 1960.
246. Meyer zur Capellen, W.: Kinematik der sphaerischen Schubkurbel, *Forschungsberichte des Landes Nordrhein-Westfalen, West-Deutscher Verlag*, Cologne, no. 873, 1960.

247. Michalec, G. W.: Analog Computing Mechanisms, *Machine Design*, vol. 31, pp. 157–179, Mar. 19, 1959.
248. Miller, H.: Analysis of Quadric-chain Mechanisms, *Prod. Eng.*, vol. 22, pp. 109–113, February, 1951.
249. Miller, W. S.: Packaged Speed Reducers and Gearmotors, *Machine Design*, vol. 29, pp. 121–149, Mar. 21, 1957.
250. Modrey, J.: Analysis of Complex Kinematic Chains with Influence Coefficients, *J. Appl. Mechanics*, vol. 26; *Trans. ASME*, vol. 81, E., pp. 184–188, June, 1959; discussion, *J. Appl. Mechanics*, vol. 27, pp. 215–216, March, 1960.
251. Moore, J. W., and M. V. Braunagel: Space Linkages, *Trans. Seventh Conf. Mech.*, Purdue University, pp. 114–122, 1962.
252. Moreinis, I. Sh.: Biomechanical Studies of Some Aspects of Walking on a Prosthetic Device (Russian), *Trudi Inst. Machinoved.*, Akad. Nauk, Moscow, vol. 21, no. 81–82, pp. 119–131, 1960.
253. Morgan, P.: Mechanisms for Moppets, *Machine Design*, vol. 34, pp. 105–109, Dec. 20, 1962.
254. Moroshkin, Y. F.: General Analytical Theory of Mechanisms (Russian), Akad. Nauk, Moscow, *Trudi Sem. Teor. Mash. Mekh.*, vol. 14, no. 54, pp. 25–50, 1954.
255. Morrison, R. A.: Rolling-surface Mechanisms, *Machine Design*, vol. 30, pp. 119–123, Dec. 11, 1958.
256. Mueller R.: "Einfuehrung in die theoretische Kinematik," Springer-Verlag. OHG, Berlin, 1932.
257. Musser, C. W.: Mechanics Is Not a Closed Book, *Trans. Sixth Conf. Mechanisms*, Purdue University, October, 1960, pp. 31–43.
258. Olsson, V.: Non-circular Cylindrical Gears, *Acta Polytech., Mech. Eng. Ser.*, vol. 2, no. 10, Stockholm, 1953.
259. Olsson, V.: Non-circular Bevel Gears, *Acta Polytech., Mech. Eng., Ser.* 5, Stockholm, 1959.
260. Paul, B.: A Unified Criterion for the Degree of Constraint of Plane Kinematic Chains, *J. Appl. Mechanics*, vol. 27; *Trans. ASME*, ser. E., vol. 82, pp. 196–200, March, 1960.
261. Peek, H. L.: Trip-free Mechanisms, *Mech. Eng.*, vol. 81, pp. 193–199, March, 1959.
262. Peyrebrune, H. E.: Application and Design of Non-circular Gears, *Trans. First Conf. Mechanisms*, Purdue University, 1953, pp. 13–21.
263. Philipp, R. E.: "Kinematics of a General Arrangement of Two Hooke's Joints," ASME paper 60-WA-37, 1960.
264. Pike, E. W., and T. R. Silverberg: Designing Mechanical Computers, *Machine Design*, vol. 24, pt. I, pp. 131–137, July, 1952; pt. II, pp. 159–163, August, 1952.
265. Pollitt, E. P.: High-speed Web-cutting, *Machine Design*, vol. 27, pp. 155–160, December, 1955.
266. Pollitt, E. P.: Motion Characteristics of Slider-crank Linkages, *Machine Design*, vol. 30, pp. 136–142, May 15, 1958.
267. Pollitt, E. P.: Some Applications of the Cycloid in Machine Design, *Trans. ASME, ser. B, J. Eng. Ind.*, vol. 82, no. 4, pp. 407–414, November, 1960.
268. Predale, J. O., and A. B. Hulse, Jr.: The Space Crank, *Prod. Eng.*, vol. 30, pp. 50–53, Mar. 2, 1959.
268*a*. Primrose, E. J. F., and F. Freudenstein: Geared Five-bar Motion II—Arbitrary Commensurate Gear Ratio, *J. Appl. Mech.*, vol. 30; *Trans. ASME, ser. E*, vol. 85, June, 1963, pp. 170–175.
269. Procopi, J.: Control Valves, *Machine Design*, vol. 22, pp. 153–155, September, 1950.
270. Proctor, J.: Selecting Clutches for Mechanical Drives, *Prod. Eng.*, vol. 32, pp. 43–58, June 19, 1961.
271. Mechanisms Actuated by Air or Hydraulic Cylinders, *Prod. Eng.*, vol. 20, pp. 128–129, December, 1949.
272. Computing Mechanisms, *Prod. Eng.*, vol. 27, I, p. 200, March; II, pp. 180–181, April, 1956.
273. High-speed Electrostatic Clutch, *Prod. Eng.*, vol. 28, pp. 189–191, February, 1957.
274. Linkage Keeps Table Flat, *Prod. Eng.*, vol. 29, p. 63, Feb. 3, 1958.
275. *Prod. Eng.*, vol. 30, pp. 64–65, Mar. 30, 1959.
276. Down to Earth with a Four-bar Linkage, *Prod. Eng.*, vol. 31, p. 71, June 22, 1959.
277. Design Work Sheets, no. 14, *Prod. Eng.*
278. Radcliffe, C. W.: Prosthetic Mechanisms for Leg Amputees, *Trans. Sixth Conf. Mechanisms*, Purdue University, October, 1960, pp. 143–151.
279. Radcliffe, C. W.: "Biomechanical Design of a Lower-extremity Prosthesis," ASME paper 60-WA-305, 1960.
280. Rainey, R. S.: Which Shaft Seal, *Prod. Eng.*, vol. 21, pp. 142–147, May, 1950.

281. Rankers, H.: Vier genau gleichwertige Gelenkgetriebe für die gleiche Koppelkurve, *Das Industrieblatt*, January, 1959, pp. 17–21.
282. Rankers, H.: Anwendungen von sechsgliedrigen Kurbelgetrieben mit Antrieb an einem Koppelpunkt, *Das Industrieblatt*, pp. 78–83, February, 1962.
283. Rankers, H.: Bewegungsverhaeltnisse an der Schubkurbel mit angeschlossenem Kreuzschieber, *Das Industrieblatt*, pp. 790–796, December, 1961.
284. Rantsch, E. J.: Elliptic Gears Depend on Accurate Layout, *Machine Design*, vol. 9, pp. 43–44, March, 1937.
285. Rappaport, S.: A Neglected Design Detail, *Machine Design*, vol. 20, p. 140, September, 1948.
286. Rappaport, S.: "Kinematics of Intermittent Mechanisms," *Prod. Eng.*, vols. 20–21, 1949–1950, I: The External Geneva Wheel, July, pp. 110–112; II, The Internal Geneva Wheel, August, pp. 109–112; III, The Spherical Geneva Wheel, October, pp. 137–139; IV, The Three-gear Drive, January, 1950, pp. 120–123; V, The Cardioid Drive, 1950, pp. 133–134.
287. Rappaport, S.: Crank-and-Slot Drive, *Prod. Eng.*, vol. 21, pp. 136–138, July, 1950.
288. Rappaport, S.: Shearing Moving Webs, *Machine Design*, vol. 28, pp. 101–104, May 3, 1956.
289. Rappaport, S.: Small Indexing Mechanisms, *Machine Design*, vol 29, pp. 161–163, Apr. 18, 1957.
290. Rappaport, S.: Elliptical Gears for Cyclic Speed Variation, *Prod. Eng.*, vol. 31, pp. 68–70, Mar. 28, 1960.
291. Rappaport, S.: Intermittent Motions and Special Mechanisms, *Trans. Conf. Mechanisms*, Yale University, Shoestring Press, New Haven, Conn., pp. 91–122, 1961.
292. Rappaport, S.: Review of Mechanical Integrators, *Trans. Seventh Conf. Mechanics*, Purdue University, pp. 234–240, 1962.
293. Rasche, W. H.: Design Formulas for Three-link Screw Mechanisms, *Machine Design*, vol. 17, pp. 147–149, August, 1945.
294. Rasche, W. H.: Gear Train Design, *Virginia Polytechnic Inst. Eng. Expt. Sta. Bull.* 14, 1933.
295. Raven, F. H.: Velocity and Acceleration Analysis of Plane and Space Mechanisms by Means of Independent-position Equations, *J. Appl. Mechanics*, vol. 25, March, 1958; *Trans. ASME*, vol. 80, pp. 1–6.
296. Reuleaux, F.: "The Constructor" (translated by H. H. Suplee), D. Van Nostrand Company, Inc., Princeton, N.J., 1893.
297. Richardson, I. H.: Trend toward Automation in Automatic Weighing and Bulk Materials Handling, *Mech. Eng.*, vol. 75, pp. 865–870, November, 1953.
298. Ring, F.: Remote Control Handling Devices, *Mech. Eng.*, vol. 78, pp. 828–831, September, 1956.
299. Roemer, R. L.: Flight-control Linkages, *Mech. Eng.*, vol. 80, pp. 56–60, June, 1958.
300. Root, R. E., Jr.: "Dynamics of Engine and Shaft," John Wiley & Sons, Inc., New York, 1932.
301. Rosenauer, N., and A. H. Willis: "Kinematics of Mechanisms," Associated General Publications, Pty. Ltd., Sydney, Australia, 1953.
302. Rosenauer, N.: Some Fundamentals of Space Mechanisms, *Mathematical Gazette*, vol. 40, no. 334, pp. 256–259, December, 1956.
303. Rossner, E. E.: Ratchet Layout, *Prod. Eng.*, vol. 29, pp. 89–91, Jan. 20, 1958.
304. Roth, B., F. Freudenstein, and G. N. Sandor: Synthesis of Four-link Path Generating Mechanisms with Optimum Transmission Characteristics, *Trans. Seventh Conf. Mech.*, Purdue University, pp. 44–48, October, 1962 (available from *Machine Design*, Penton Bldg., Cleveland).
305. Roth, B., and F. Freudenstein: Synthesis of Path generating Mechanisms by Numerical Methods, *Trans. ASME*, 85B, *J. Eng. Ind.*, pp. 298–306, August, 1963.
306. Roth, G. L.: Modifying Valve Characteristics, *Prod. Eng.*, Annual Handbook, pp. 16–18, 1958.
307. Rothbart, H. A.: "Cams," John Wiley & Sons, Inc., New York. 1956.
308. Rothbart, H. A.: Equivalent Mechanisms for Cams, *Machine Design*, vol. 30, pp. 175–180, Mar. 20, 1958.
309. Rumsey, R. D.: Redesigned for Higher Speed, *Machine Design*, vol. 23, pp. 123–129, April, 1951.
310. Rzeppa, A. H.: Universal Joint Drives, *Machine Design*, vol. 25, pp. 162–170, April, 1953.
311. Saari, O.: Universal Joints, *Machine Design*, vol. 26, pp. 175–178, October, 1954.
312. Sandor, G. N., and F. Freudenstein: "Kinematic Synthesis of Path-generating Mechanisms by Means of the IBM 650 Computer," Program 9.5.003, IBM Library, Applied

Programming Publications, IBM, 590 Madison Avenue, New York 22, N.Y., 1958.

313. Sandor, G. N.: "A General Complex-number Method for Plane Kinematic Synthesis with Applications," doctoral dissertation, Columbia University, University Microfilms, Ann Arbor, Mich., 1959.

313*a*. Sandor, G. N.: On the Kinematics of Rubber-Covered Cylinders Rolling on a Hard Surface, ASME paper 61-SA-67, abstr. *Mech. Engrg.*, Vol. 83, No. 10, October, 1961, p. 84.

314. Sandor, G. N.: On Computer-aided Graphical Kinematics Synthesis, Technical Seminar Series, *Rept.* 4, Princeton University, Dept. of Graphics and Engineering Drawing, 1962.

315. Sandor, G. N.: On the Loop Equations in Kinematics, *Trans. Seventh Conf. Mechanisms*, Purdue University, pp. 49–56, 1962.

315*a*. Sandor, G. N.: On the Existence of a Cycloidal Burmester Theory in Planar Kinematics, *J. Appl. Mechanics*, vol. 31, *Trans. ASME*, vol. 86E, 1964.

316. Saxon, A. F.: Multistage Sealing, *Machine Design*, vol. 25, pp. 170–172, March, 1953.

317. Soled, J.: Industrial Fasteners, *Machine Design*, vol. 28, pp. 105–136, Aug. 23, 1956.

318. Schashkin, A. S.: Study of an Epicyclic Mechanism with Dwell (ref. 17-B), pp. 117–132.

319. Schmidt, E. H.: Cyclic Variations in Speed, *Machine Design*, vol. 19, pp. 108–111, March, 1947.

320. Schmidt, E. H.: Cycloidal-crank Mechanisms, *Machine Design*, vol. 31, pp. 111–114, Apr. 2, 1959.

321. Schulze, E. F. C.: Designing Snap-action Toggles, *Prod. Eng.*, vol. 26, pp 168–170, November, 1955.

322. Shaffer, B. W., and I. Krause: Refinement of Finite Difference Calculations in Kinematic Analysis, *Trans. ASME*, vol. 82B, no. 4, pp. 377–381, November, 1960.

323. Sheppard, W. H.: Rolling Curves and Non-circular Gears, *Mech. World*, pp. 5–11, January, 1960.

324. Sieker, K. H.: Kurbelgetriebe-Rechnerische Verfahren, *VDI-Bildungswerk*, no. 077 (probably 1960–1961).

325. Sieker, K. H.: Extremwerte der Winkelgeschwindigkeiten in Symmetrischen Doppelkurbeln, *Konstruktion*, vol. 13, no. 9, pp. 351–353, 1961.

326. Sieker, K. H.: "Getriebe mit Energiespeichern," C. F. Winter'sche Verlagshandlung, Fussen, 1954.

327. Sieker, K. H.: Zur algebraischen Mass-Synthese ebener Kurbelgetriebe, *Ing. Arch.*, vol. 24, pt. I, no. 3, pp. 188–215; pt. II, no. 4, pp. 233–257, 1956.

328. Sieker, K. H.: Winkelgeschwindigkeiten und Winkelbeschleunigungen in Kurbelschleifen, *Feinwerktechnik*, vol. 64, no. 6, pp. 1–9, 1960.

329. Simonis, F. W.: "Stufenlos verstellbare Getriebe," Werkstattbuecher no. 96, Springer-Verlag OHG, Berlin, 1949.

330. Sloane, W. W.: Utilizing Irregular Gears for Inertia Control, *Trans. First Conf. Mechanisms*, Purdue University, 1953, pp. 21–24.

331. Karplus, W. J., and W. J. Soroka: "Analog Methods in Computation and Simulation," 2d ed., McGraw-Hill Book Company, Inc., New York, 1959.

332. Spector, L. F.: Flexible Couplings, *Machine Design*, vol. 30, pp. 101–128, Oct. 30, 1958.

333. Spector, L. F.: Mechanical Adjustable-speed Drives, *Machine Design*, vol. 27, I, April, pp. 163–196; II, June, pp. 178–189, 1955.

334. Spotts, M. F.: Kinematic Properties of the Three-gear Drive, *J. Franklin Inst.*, vol. 268, no. 6, pp. 464–473, December, 1959.

335. Strasser, F.: Ten Universal Shaft-couplings, *Prod. Eng.*, vol. 29, pp. 80–81, Aug. 18, 1958.

336. Svoboda, A.: "Computing Mechanisms and Linkages," MIT Radiation Laboratory Series, vol. 27, McGraw-Hill Book Company, Inc., New York, 1948.

337. Taborek, J. J.: Mechanics of Vehicles 3, Steering Forces and Stability, *Machine Design*, vol. 29, pp. 92–100, June 27, 1957.

338. Talbourdet, G. J.: Mathematical Solution of Four-bar Linkages, *Machine Design*, vol. 13, I, II, no. 5, pp. 65–68; III, no. 6, pp. 81–82; IV, no. 7, pp. 73–77, 1941.

339. Talbourdet, G. J.: Intermittent Mechanisms (data sheets), *Machine Design*, vol. 20, pt. I, September, pp. 159–162; pt. II, October, pp. 135–138, 1948.

340. Talbourdet, G. J.: Motion Analysis of Several Intermittent Variable-speed Drives, *Trans. ASME*, vol. 71, pp. 83–96, 1949.

341. Talbourdet, G. J.: Intermittent Mechanisms, *Machine Design*, vol. 22, pt. I, September, pp. 141–146; pt. II, October, pp. 121–125, 1950.

342. Tesar, D.: "Translations of Papers (by R. Mueller) on Geometrical Theory of Motion

Applied to Approximate Straight-line-Motion," *Kansas State Univ. Eng. Exp. Sta., Spec. Rept.* 21, 1962.

343. Thearle, E. L.: A Non-reversing Coupling, *Machine Design*, vol. 23, pp. 181–184, April, 1951.

344. Thumin, C.: Designing Quick-acting Latch Releases, *Machine Design*, vol. 19, pp. 110–115, September, 1947.

345. Tolle, M.: Regelung der Kraftmaschinen, 3d ed., Springer-Verlag, OHG, Berlin, 1921.

346. Trinkl, F.: Analytische und zeichnerische Verfahren zur Untersuchung eigentlicher Raumkurbelgetriebe, *Konstruktion*, vol. 11, no. 9, pp. 349–359, 1959.

347. Uhing, J.: Einfache Raumgetriebe für ungleichfoermige Dreh- und Schwingbewegung, *Konstruktion*, vol. 9, no. 1, pp. 18–21, 1957.

348. Uicker, J. J., Jr.: "Displacement Analysis of Spatial Mechanisms by an Iterative Method Based on 4 × 4 Matrices," M.S. dissertation, Northwestern University, Evanston, Ill., 1963.

349. Vandeman, J. E., and J. R. Wood: Modifying Starwheel Mechanisms, *Machine Design*, vol. 25, pp. 255–261, April, 1953.

350. "Sperrgetriebe," AWF-VDMA-VDI Getriebehefte, Ausschuss f. Wirtschaftliche Fertigung, Berlin, no. 6061 pub. 1955, nos. 6062, 6071 pub. 1956, no. 6063 pub. 1957.

351. Veldkamp, G. R.: "Curvature Theory in Plane Kinematics," J. B. Wolters, Groningen, 1963.

352. Vidosic, J. P., and H. L. Johnson: Synthesis of Four-bar Function Generators, *Trans. Sixth Conf. Mechanisms*, Purdue University, October, 1960, pp. 82–86.

353. Vogel, W. F.: Crank Mechanism Motions, *Prod. Eng.*, Vol. 12, pt. I, June, pp. 301–305; II, July, pp. 374–379; III, pp. 423–428, August; IV, September, pp. 490–493, 1941.

354. Volmer, J.: Konstruktion eines Gelenkgetriebes fuer eine Geradfuehrung, *VDI-Berichte*, vol. 12, pp. 175–183, 1956.

355. Volmer, J.: Systematik, Kinematik des Zweiradgetriebes, *Maschinenbautechnik*, vol. 5, no. 11, pp. 583–589, 1956.

356. Volmer, J.: Raederkurbelgetriebe, *VDI-Forschungsheft* no. 461, pp. 52–55, 1957.

357. Volmer, J.: "Gelenkgetriebe zur Geradfuhrung einer Ebene" *Z. prakt. Metallbearbeitung*, vol. 53, no. 5, pp. 169–174, 1959.

358. Wallace, W. B., Jr.: Pressure Switches, *Machine Design*, vol. 29, pp. 106–114, Aug. 22, 1957.

359. Weise, H.: Bewegungsverhaeltnisse an Filmschaltgetriebe, Getriebetechnik, *VDI-Berichte*, vol. 5, 1955, pp. 99–106.

360. Weise, H.: Getriebe in photographischen und kinematographischen Geraeten, *VDI-Berichte*, vol. 12, pp. 131–137, 1956.

361. "Stability of a Lifting Rig," Engineering Problems, vol. II, Westinghouse Electric Corp., Pittsburgh, Pa., approx. 1960.

362. Weyth, N. C., and A. F. Gagne: Mechanical Torque-limiting Devices, *Machine Design*, vol. 18, pp. 127–130, May, 1946.

363. Whitehead, T. N.: "Instruments and Accurate Mechanisms," Dover Publications, Inc., New York, 1954.

364. Wieleitner, H.: "Spezielle ebene Kurven," G. J. Goeschen Verlag, Leipzig, 1908.

365. Wittel, O., and D. C. Haefele: A Non-intermittent Film Projector, *Mech. Eng.*, vol. 79, pp. 345–347, April, 1957.

366. Wolford, J. C., and A. S. Hall: "Second-acceleration Analyses of Plane Mechanisms," ASME paper 57-A-52, 1957.

367. Wolford, J. C., and D. C. Haack: Applying the Inflection Circle Concept, *Trans. Fifth Conf. Mechanisms*, Purdue University, 1958, pp. 232–239.

368. Wolford, J. C.: An Analytical Method for Locating the Burmester Points for Five Infinitesimally Separated Positions of the Coupler Plane of a Four-bar Mechanism, *J. Appl. Mechanics*, vol. 27, *Trans. ASME*, vol. 82E, pp. 182–186, March, 1960.

369. Woerle, H.: Sonderformen zwangläufiger viergelenkiger Raumkurbelgetriebe, *VDI-Berichte*, vol. 12, pp. 21–28, 1956.

370. Wunderlich, W.: Zur angenaeherten Geradfuehrung durch symmetrische Gelenkvierecke, *Z. angew Math. Mechanik*, vol. 36, no. 3/4, pp. 103–110, 1956.

371. Yang, A. T.: Harmonic Analysis of Spherical Four-bar Mechanisms, *J. Appl. Mechanics*, vol. 29, no. 4, *Trans. ASME*, vol. 84E, pp. 683–688, December, 1962.

372. Yang, A. T.: "Application of Quaternion Algebra and Dual Numbers to the Analysis of Spatial Mechanisms," doctoral dissertation, Columbia University, New York, 1963. See also *J. Appl. Mech.*, 1964.

373. Yudin, V. A.: General Theory of Planetary-cycloidal Speed Reducer, Akad. Nauk, Moscow, *Trudi Sem. Teor. Mash. Mekh.*, vol. 4, no. 13, pp. 42–77, 1948.

373*a*. "Computer Explosion in Truck Engineering," a collection of seven papers, *SAE Publ.* SP240, December, 1962.
374. "Mechanical Design of Spacecraft," Jet Propulsion Laboratory, Seminar Proceedings, Pasadena, Calif., August, 1962.
375. Guide to Limit Switches, *Prod. Eng.* vol. 33, pp. 84–101, Nov. 12, 1962.
376. Mechanical Power Amplifier, *Machine Design*, vol. 32, p. 104, Dec. 22, 1960.
377. Wunderlich, W.: Hoehere Radlinien, *Österr. Ing.-Arch.*, vol. 1, pp. 277–296, 1947.
378. VDI Richtlinien, VDI Duesseldorf; for transmission-angle charts, refer to (*a*) four-bars, *VDI* 2123, 2124, August, 1959; (*b*) slider cranks, *VDI* 2125, August, 1959. For straight-line generation, refer to (*a*) in-line swinging-blocks, *VDI* 2137, August, 1959; (*b*) in-line slider-cranks, *VDI* 2136, August, 1959. (*c*) Planar four-bar, *VDI* 2130-2135, August, 1959.

Section 5

CAM MECHANISMS

By

HAROLD A. ROTHBART, D. Eng., *Dean, College of Science and Engineering, Fairleigh Dickinson University, Teaneck, N.J.*

CONTENTS

NOTATIONS AND DEFINITIONS

a, b = distance, in.
A = follower overhang, in.
B = follower bearing length, in.
C = a constant
F = force normal to cam profile, lb
g = gravitational constant
h = maximum displacement of follower, in.
m = equivalent mass of follower, lb-sec²/in.
n = any number
N_1, N_2 = forces normal to translating follower stem, lb
r = radius from cam center to center of curvature, in.
r_c = radius from cam center to center of roller follower, in.
R_a = radius of prime circle (smallest circle to the roller center), in

R_p = radius of pitch circle (circle drawn to roller center having the maximum pressure angle), in.
R_r = radius of follower roller, in.
t = time for cam to rotate angle θ, sec
T = time for cam to rotate angle θ_0, sec
y = displacement of translating follower, in.
$\dot{y}$ = velocity of follower, ips
$\ddot{y}$ = acceleration of follower, in./sec^2
α, β, φ = angles, rad
ψ = pressure angle, rad
θ_0 = ωt-cam-angle rotation for rise h, rad
θ = ωt-cam-angle rotation for displacement y, rad
μ = coefficient of friction
ρ = radius of curvature of pitch curve (path of roller center), in.
τ = angle between radius of curvature and follower motion, rad
ω = cam angular velocity, rad/sec

5.1. INTRODUCTION[1,2,3,16]

A cam is a mechanical member for transmitting a desired motion to a follower by direct contact. The driver is called a *cam* and the driven member is called the *follower*. Either may remain stationary, translate, oscillate, or rotate. The motion is given by $y = f(\theta)$.

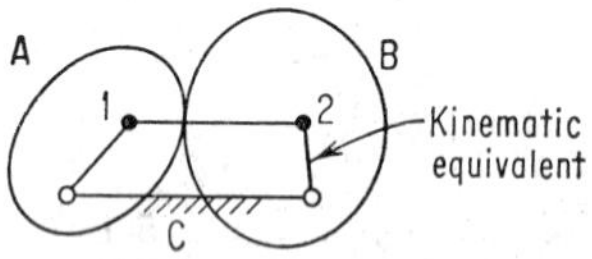

FIG. 5.1. Basic cam mechanism and its kinematic equivalent (points 1 and 2 are centers of curvature) of the contact point. (*Courtesy of John Wiley & Sons, Inc.*[1])

Kinematically speaking, in its general form the plane cam mechanism (Fig. 5.1) consists of two shaped members A and B connected by a fixed third body C. Either body A or body B may be the driver with the other the follower. We may at each instant replace these shaped bodies by an equivalent mechanism. These are pin-jointed at the instantaneous centers of curvature, 1 and 2, of the contacting surfaces. In general, at any other instant the points 1 and 2 are shifted and the links of the equivalent mechanism have different lengths. Figure 5.2 shows the two most popular cams.

Cam synthesis may be accomplished by (1) shaping the cam body to some known curve such as involute, spirals, parabolas, or circular arcs, (2) mathematically controlling and establishing the follower motion and forming the cam by tabulated data of this action, (3) establishing the cam contour in parametric form, and (4) layout of cam profile by eye. This last method is acceptable only for low speeds in which a smooth "bumpless" curve may fulfill the requirements. But as either loads, mass, speed, or elasticity of the members increases, a detailed study must be made of both the dynamic aspects of the cam curve and the accuracy of cam fabrication.

A constant-speed cam composed of an eccentric circle and a flat-faced follower has the following characteristics[1,4,5] (Fig. 5.3*a*):

$$y = r(1 - \cos\theta) \tag{5.1}$$
$$\dot{y} = r\omega \sin\theta \tag{5.2}$$
$$\ddot{y} = r\omega^2 \cos\theta \tag{5.3}$$

and if a roller follower is employed (Fig. 5.3*b*)

$$y = r - \rho - r\cos\theta + \rho\cos\tau \tag{5.4}$$
$$\dot{y} = r\omega(\sin\theta - \cos\theta\tan\tau) \tag{5.5}$$
$$\ddot{y} = r\omega^2[\cos\theta - (r\cos^2\theta/\rho\cos^3\tau) + \sin\theta\tan\tau] \tag{5.6}$$

Cams composed of blended circular arcs have different values and different equivalent mechanisms for every arc. Complex cam curves can be investigated in the same manner by approximating the radius of curvature at each position. Reference 18

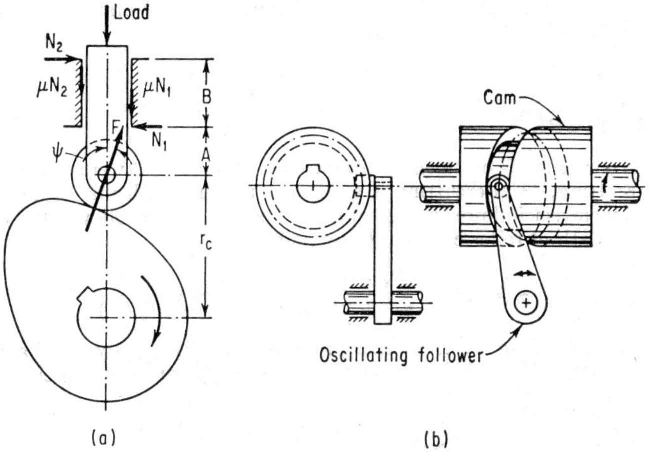

FIG. 5.2. Popular cams. (*a*) Radial cam—translating roller follower (open cam). (*b*) Cylindrical cam—oscillating roller follower (closed cam). (*Courtesy of John Wiley & Sons, Inc.*[1])

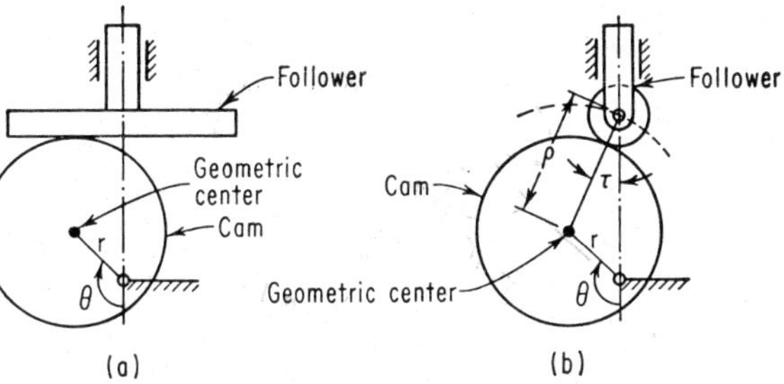

FIG. 5.3. Eccentric-circle cams translating follower. (*a*) Flat-faced follower. (*b*) Roller follower. (*Courtesy of John Wiley & Sons, Inc.*[1])

is the most comprehensive investigation of constant-diameter positive-drive cams. References 1, 20, and 21 cover computing cam mechanisms of all sorts.

5.2. PRESSURE ANGLE[1,6]

The pressure angle is the angle between the normal to the cam profile and the instantaneous direction of the follower. It is related to the follower force distribution. With flat-faced followers, the pressure-angle force distribution rarely is of concern. With oscillating roller followers, the distance from the cam center to follower pivot center and the length of the follower arm are pertinent to this study. A radial translating roller follower having rigid members has a theoretical limiting pressure angle based on forces shown in Fig. 5.2*a*.

$$\psi_m < \tan^{-1} \frac{B}{N(2A + B)} \tag{5.7}$$

To be practical, the pressure angle is generally limited to 30° for the translating roller follower because of the elasticity of the follower stem and the clearance between the stem in its guide bearing and the friction. We can show that the pressure angle

$\tan \psi = \ddot{y}/r_c\omega$. Symmetrical cam curves have the point of maximum pressure angle approximately at the mid-point of the rise

$$\tan \psi_m = \dot{y}_m/R_p\omega \tag{5.8}$$

5.3. CURVATURE[1,7]

The cam curvature is pertinent in establishing the cam follower surface stresses, life, and undercutting. Undercutting is a phenomenon in which inadequate cam curvature yields incorrect follower movement. For a translating follower the curvature of pitch curve (path of roller center) at any point

$$\rho = \frac{[(R_a + y)^2 + (\dot{y}/\omega)^2]^{3/2}}{(R_a + y)^2 + 2(\dot{y}/\omega)^2 - (R_a + y)(\ddot{y}/\omega^2)} \tag{5.9}$$

If ρ is positive we have a convex cam profile and if ρ is negative we have a concave cam profile (Fig. 5.4). For a convex curved cam to prevent undercutting, which

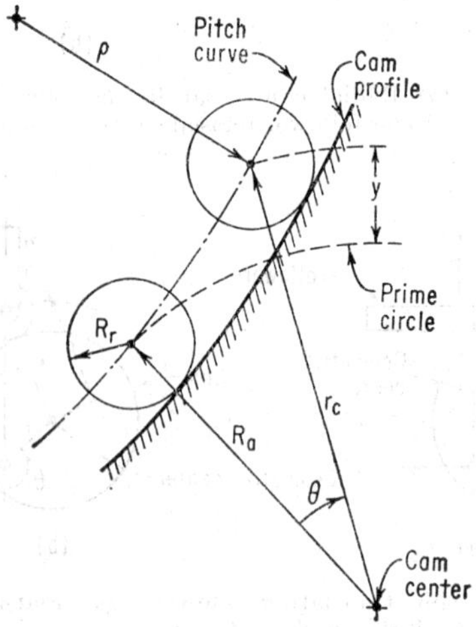

FIG. 5.4. Curvature cam profile terminology. (*Courtesy of John Wiley & Sons, Inc.*[1])

generally occurs in the point of maximum negative acceleration, $\rho > R_r$. In all cams the minimum radius of curvature should be determined to establish that the cam-follower system does not have excessive stresses or undercutting.

5.4. COMPLEX NOTATION

Complex numbers[17,19] can be employed to obtain the curvature, profile, cutter location, etc., for any kind of cam and follower. Use of a high-speed computer will aid in its use. Let us take the disk cam with an oscillating roller follower as an example (Fig. 5.5). The angular displacement of the follower from initial position β_0

$$\beta = \beta_0 + f(\theta)$$

The relations to the roller or cutter center

$$\mathbf{r}_c = r_c e^{i\theta_c} = re^{i\alpha} + \rho e^{i\varphi} = a + ib + le^{i\beta} \tag{5.10}$$

The equation of the cam contour measured from the initial position

$$r_c e^{i(\theta+\theta_c)} = (a + ib + le^{i\beta})e^{i\theta} \tag{5.11}$$

Solving for r_c and θ_c yields the information necessary to construct the cam profile. Separating Eq. (5.10) into real and imaginary parts

$$r \cos \alpha + \rho \cos \varphi = a + l \cos \beta \tag{5.12}$$
$$r \sin \alpha + \rho \sin \varphi = b + l \sin \beta \tag{5.13}$$

differentiating we obtain the radius of curvature

$$\rho = \frac{(E^2 + D^2)^{3/2}}{(E^2 + D^2)[1 + f'(\theta)] - (GE + bD)f'(\theta) + (G \sin \beta - b \cos \beta)lf''(\theta)} \tag{5.14}$$

where

$$E = [1 + f'(\theta)]l \cos \beta + a \tag{5.15}$$
$$D = [1 + f'(\theta)]l \sin \beta + b \tag{5.16}$$

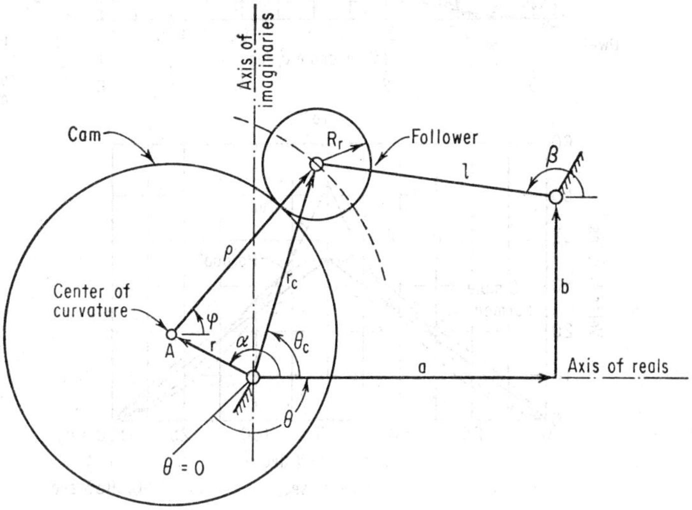

FIG. 5.5. Complex notation for oscillating roller follower.

to avoid undercutting $\rho > R_r$.

5.5. BASIC DWELL-RISE-DWELL CURVES

The most popular method in establishing the cam shape is by first choosing a simple curve which is easy to construct and analyze. Conventionally they are of two families: the simple polynomial, i.e., parabolic and cubic, and the trigonometric, i.e., simple harmonic, cycloidal, etc. The polynomial family

$$y = C_0 + C_1\theta + C_2\theta^2 + C_3\theta^3 + \cdots + C_n\theta^n \tag{5.17}$$

of which the basic curve is generally with a low integer

$$y = C_n\theta^n \tag{5.18}$$

The trigonometric family

$$y = C_0 + C_1 \sin \omega t + C_2 \cos \omega t + C_3 \sin^2 \omega t + C_4 \cos^2 \omega t + \cdots \tag{5.19}$$

are reduced to the basic curves using only the lowest-order terms. The trigonometric are better since they yield smaller cams, lower follower side thrust, cheaper manufacturing costs, and easier layout and duplication. A summary of equations and curves is shown in Fig. 5.6 and Tables 5.1 and 5.2.

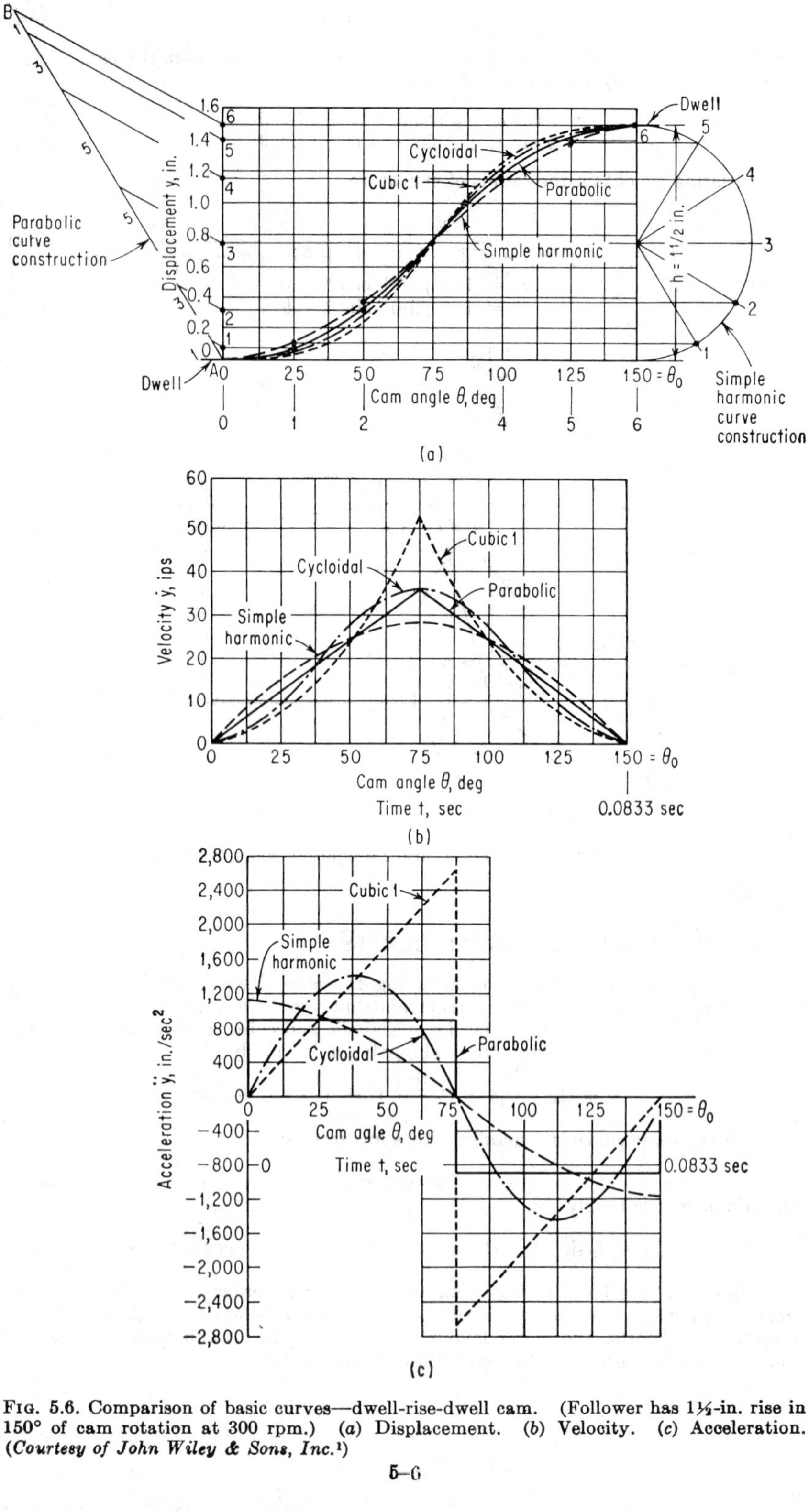

Fig. 5.6. Comparison of basic curves—dwell-rise-dwell cam. (Follower has 1½-in. rise in 150° of cam rotation at 300 rpm.) (*a*) Displacement. (*b*) Velocity. (*c*) Acceleration. (*Courtesy of John Wiley & Sons, Inc.*[1])

Table 5.1. Characteristic Equations of Basic Curves*

Curves		Displacement y, in.	Velocity $\dot{y}$, ips	Acceleration $\ddot{y}$, in./sec²
Straight line		$h\theta/\theta_0$	$\omega h/\theta_0$	0
Simple harmonic motion (SHM)		$h/2[1 - \cos(\pi\theta/\theta_0)]$	$(h\pi\omega/2\theta_0)\sin(\pi\theta/\theta_0)$	$(h/2)(\pi\omega/\theta_0)^2\cos(\pi\theta/\theta_0)$
Cycloidal		$h/\pi[\pi\theta/\theta_0 - \frac{1}{2}\sin(2\pi\theta/\theta_0)]$	$h\omega/\theta_0[1 - \cos(2\pi\theta/\theta_0)]$	$2h\pi\omega^2/\theta_0^2\sin(2\pi\theta/\theta_0)$
Parabolic or constant acceleration	$\theta/\theta_0 \lesssim 0.5$	$2h(\theta/\theta_0)^2$	$4h\omega\theta/\theta_0^2$	$4h(\omega^2/\theta_0^2)$
	$\theta/\theta_0 \gtrsim 0.5$	$h[1 - 2(1 - \theta/\theta_0)^2]$	$(4h\omega/\theta_0)(1 - \theta/\theta_0)$	$-4h(\omega^2/\theta_0^2)$
Cubic 1 or constant pulse 1	$\theta/\theta_0 \lesssim 0.5$	$4h(\theta/\theta_0)^3$	$(12h\omega/\theta_0)(\theta/\theta_0)^2$	$(24h\omega^2/\theta_0^2)(\theta/\theta_0)$
	$\theta/\theta_0 \gtrsim 0.5$	$h[1 - 4(1 - \theta/\theta_0)^3]$	$(12h\omega/\theta_0)(1 - \theta/\theta_0)^2$	$-(24h\omega^2/\theta_0^2)(1 - \theta/\theta_0)$

where h = maximum rise of follower, in.
θ_0 = cam angle of rotation to give rise h, rad
ω = cam angular velocity, rad/sec
θ = cam-angle rotation for follower displacement y, rad

* Courtesy of John Wiley & Sons, Inc.,[1]

The parabolic curve is constructed (Fig. 5.6*a*) by dividing any line *OB* into odd incremental spacings, i.e., 1, 3, 5, 5, 3, 1 the same number as the abscissa spacing. This line is projected to rise *h* and then to spacings. The simple harmonic curve is constructed by dividing the *h* diameter into equal parts and projecting.

5.6. CAM LAYOUT

The fundamental basis for cam layout is one of inversion in which the cam profile is developed by fixing the cam moving the follower to its respective relative positions.

Example: A radial cam rotating at 180 rpm is driving a ¾-in.-diameter translating roller follower. Construct the cam profile with the pressure angle limited to 20° on the rise. The motion is as follows: (1) rise of 1 in. with simple harmonic motion in 150° of cam rotation, (2) dwell for 60°, (3) fall of 1 in. with simple harmonic motion in 120° of cam rotation, (4) dwell for the remaining 30°.

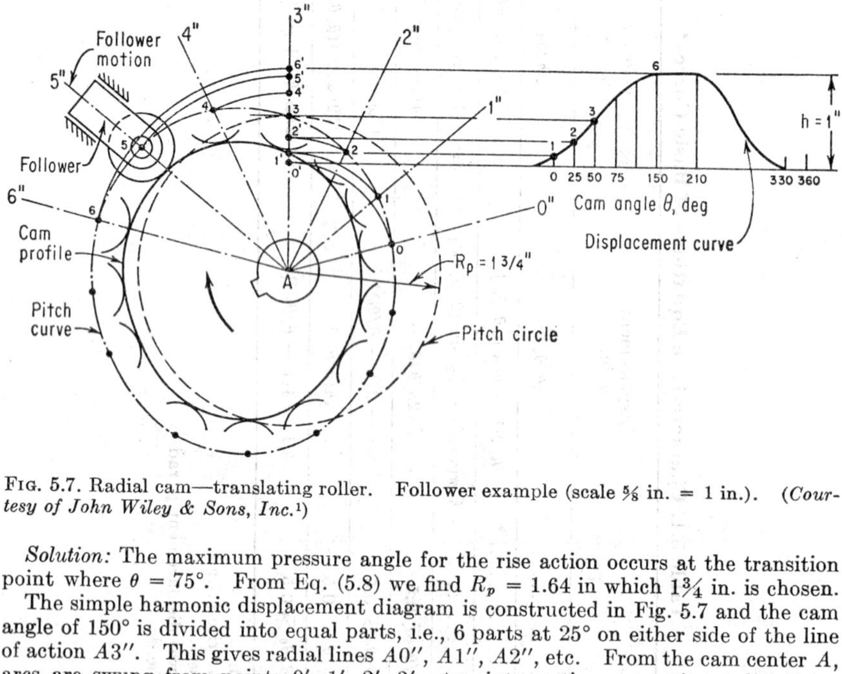

FIG. 5.7. Radial cam—translating roller. Follower example (scale ⅝ in. = 1 in.). (*Courtesy of John Wiley & Sons, Inc.*[1])

Solution: The maximum pressure angle for the rise action occurs at the transition point where $\theta = 75°$. From Eq. (5.8) we find $R_p = 1.64$ in which 1¾ in. is chosen. The simple harmonic displacement diagram is constructed in Fig. 5.7 and the cam angle of 150° is divided into equal parts, i.e., 6 parts at 25° on either side of the line of action *A*3″. This gives radial lines *A*0″, *A*1″, *A*2″, etc. From the cam center *A*, arcs are swung from points 0′, 1′, 2′, 3′, etc., intersecting respective radial lines, locating points 0, 1, 2, 3, etc. Next we draw the smooth pitch curve, roller circles, and cam profile tangent to the rollers. We continue in the same manner to complete the cam.

5.7. ADVANCED CURVES[15]

Special curve shapes are obtained by (1) employing polynomial or trigonometric relations previously mentioned, (2) numerical procedure by the method of finite differences, (3) blending portions of curves, and (4) trial and error employing of a high-speed computer starting with the fourth-derivative (d^4y/dt^4) curve.

Rise or fall portions of any curve are each composed of a positive acceleration period and a negative acceleration period. For any smooth curve, we know that $\int_0^{\theta/\omega} \ddot{y}\,dt = 0$. Therefore, for any rise or fall curve, the area under the positive-acceleration portion equals the area under the negative-acceleration portion for each curve.

(a) Polynomial and Trigonometric

Equations (5.17) and (5.19) show the general expressions. The symmetrical dwell-rise-dwell cam curve can be developed from a Fourier series with only odd multiples of the fundamental frequency[14]

$$\ddot{y} = \sum_{k \text{ odd}}^{n} C_k \sin \omega_k t = C_1 \sin \omega_1 t + C_3 \sin \omega_3 t + C_5 \sin \omega_5 t + \cdots \qquad (5.20)$$

where $\omega_k = 2\pi k/T$ and C_r's are coefficients.

If $k = 1$ it yields the well-known cycloidal curve, if $k = \infty$ we obtain the parabolic curve.

For $n > 1$ we can progressively flatten curves for any shape desired.

(b) Finite Differences[1,8]

In the application of finite differences the displacement y is known for pivotal points equally spaced by Δt. The derivation of formulas by Taylor-series expansion of $y(t + \Delta t)$ about t is

$$y(t + \Delta t) = y(t) + \Delta t\, \dot{y}(t) + [(\Delta t)^2/2!]\ddot{y}(t) + [(\Delta t)^3/3!]\dddot{y}(t) + \cdots \qquad (5.21)$$

For simplicity we take a three-point approximation which consists of passing a parabola through the three pivotal points. At any interval i

$$y_{i+1} \simeq y_i + \Delta t\, \dot{y}_i + (\Delta t)^2/2!\ddot{y}_i$$
$$y_{i-1} \simeq y_i - \Delta t\, \dot{y}_i + (\Delta t)^2/2!\ddot{y}_i$$

Solving for the velocity and acceleration, respectively,

$$\dot{y}_i \simeq (y_{i+1} - y_{i-1})/2\Delta t \qquad (5.22)$$
$$\ddot{y}_i \simeq (y_{i+1} + y_{i-1} - 2y_i)/(\Delta t)^2 \qquad (5.23)$$

Equations (5.22) and (5.23) can be used to determine the numerical relationships between y, $\dot{y}$, $\ddot{y}$ characteristics curves for incremental steps to include the cam accuracy if desired.

(c) Blending Portions of Curves[1]

The primary condition for these combination curves is that (1) there must be no discontinuities in the acceleration curve and (2) the acceleration curve has a smooth shape. Therefore, weaker harmonics result. Table 5.2 gives a comparison of all the basic curves and some combination curves.

As an illustration of an advanced curve choice, automobile and aircraft designers are employing an unsymmetrical cam acceleration curve in which the maximum positive acceleration is much greater than the maximum negative acceleration (Fig. 5.8). This is a high-speed cam giving minimum spring size, maximum cam radius of curvature, and longer surface life. For cam dynamics the reader is referred to Secs. 6 and 11, and refs. 1 and 9 through 14.

Example: Derive the relationship for the excellent dwell-rise-fall-dwell cam curve shown in Fig. 5.9 having equal maximum acceleration values. Portions I and III are harmonic; portions II and IV are horizontal straight lines.

Solution:

Let the θ's and β's be the angles for each portion shown. Note for velocity and acceleration one should multiply values by ω and ω^2, respectively, and the boundary conditions are $y(0) = 0$, $\dot{y}(0) = 0$, and $\dot{y}(\beta_4) = 0$ and $y(\beta_4) =$ total rise h.

Table 5.2. Comparison of Symmetrical Curves*

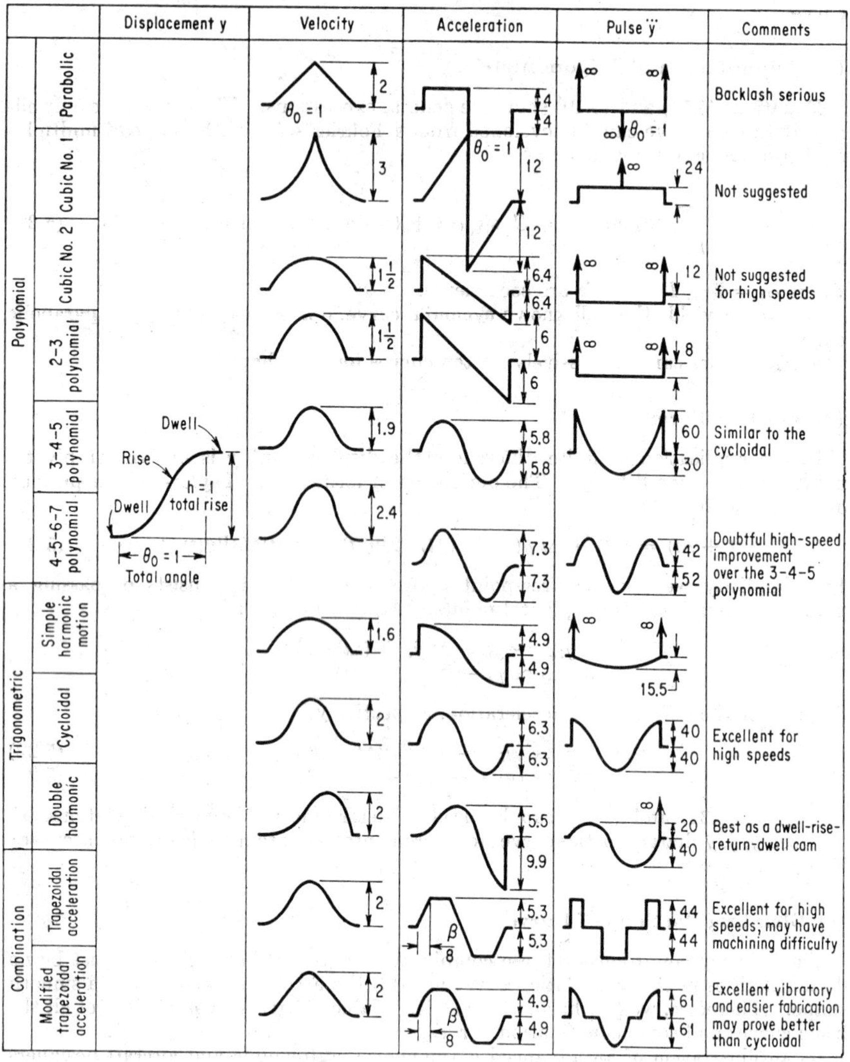
Displacement y
Velocity
Acceleration
Pulse $\dddot{y}$
Comments
Polynomial
Parabolic
Cubic No. 1
Cubic No. 2
2–3 polynomial
3–4–5 polynomial
4–5–6–7 polynomial
Trigonometric
Simple harmonic motion
Cycloidal
Double harmonic
Combination
Trapezoidal acceleration
Modified trapezoidal acceleration
Dwell
Rise
Dwell
h = 1 total rise
θ_0 = 1
Total angle
Backlash serious
Not suggested
Not suggested for high speeds
Similar to the cycloidal
Doubtful high-speed improvement over the 3-4-5 polynomial
Excellent for high speeds
Best as a dwell-rise-return-dwell cam
Excellent for high speeds; may have machining difficulty
Excellent vibratory and easier fabrication may prove better than cycloidal

* Courtesy of John Wiley & Sons, Inc.[1]

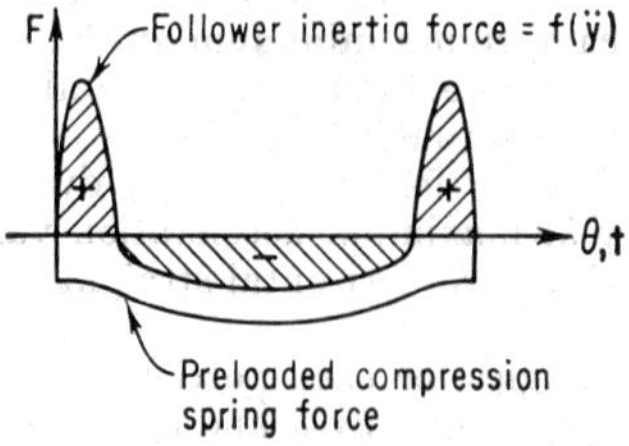

FIG. 5.8. Typical high-speed cam curve used for automotive or aircraft engine valve system.

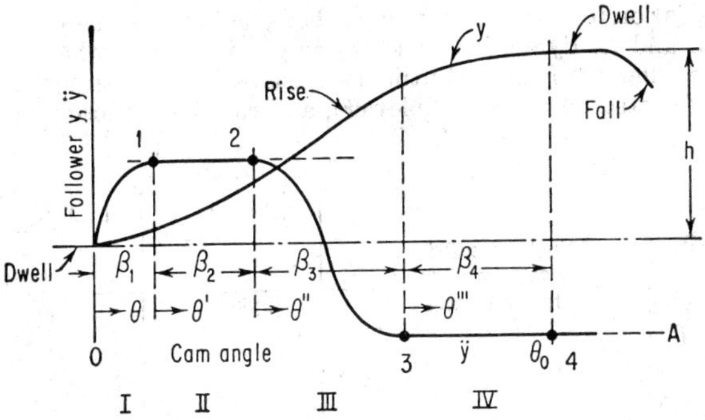

FIG. 5.9. Example of symmetrical dwell-rise-fall-dwell cam curve.

Portion I

$$\begin{aligned}
\ddot{y} &= A \sin (\pi\theta/2\beta_1) \\
\dot{y} &= (2\beta_1 A/\pi)[1 - \cos (\pi\theta/2\beta_1)] \\
y &= -A(2\beta_1/\pi)^2 \sin (\pi\theta/2\beta_1) + (2A\beta_1/\pi)\theta \\
\dot{y}_1 &= 2A\beta_1/\pi \\
y_1 &= (2A\beta_1^2/\pi)(1 - 2/\pi)
\end{aligned}$$

Portion II

$$\begin{aligned}
\ddot{y} &= A \\
\dot{y} &= A\theta' + \dot{y}_1 \\
y &= A(\theta')^2/2 + \dot{y}_1\theta' + y_1 \\
\dot{y}_2 &= A\beta_2 + 2A\beta_1/\pi \\
y_2 &= A\beta_2^2/2 + 2A\beta_1\beta_2/\pi + 2A\beta_1^2/\pi(1 - 2/\pi)
\end{aligned}$$

Portion III

$$\begin{aligned}
\ddot{y} &= A \cos (\pi\theta''/\beta_3) \\
\dot{y} &= A\beta_3/\pi \sin (\pi\theta''/\beta_3) + \dot{y}_2 \\
y &= -A(\beta_3/\pi)^2[1 - \cos (\pi\theta''/\beta_3)] + \dot{y}_2\theta'' + y_2 \\
\dot{y}_3 &= \dot{y}_2 \\
y_3 &= 2A(\beta_3/\pi)^2 + A\beta_2\beta_3 + 2A\beta_1\beta_3/\pi + A\beta_2^2/2 + 2A\beta_1\beta_2/\pi + (2A\beta_1^2/\pi)(1 - 2/\pi)
\end{aligned}$$

Portion IV

$$\begin{aligned}
\ddot{y} &= -A \\
\dot{y} &= -A\theta''' + \dot{y}_3 \\
y &= -[A(\theta''')^2/2] + \dot{y}_3\theta''' + y_3 \\
\dot{y}_4 &= -A\beta_4 + \dot{y}_3 = -A\beta_4 + A\beta_2 + 2A\beta_1/\pi = 0 \\
\beta_4 &= \beta_2 + 2\beta_1/\pi
\end{aligned}$$

Total Rise

$$h = y_4 = -(A\beta_4^2/2) + (A\beta_2 + 2A\beta_1/\pi)\beta_4 + y_3$$

Substituting

$$A = \frac{h}{-\beta_4^2/2 + \beta_2\beta_4 + 2\beta_1\beta_4/\pi + (\beta_3/\pi)^2 + \beta_2\beta_3 + 2\beta_1\beta_3/\pi + \beta_2^2/2 + 2\beta_1\beta_2/\pi + (2\beta_1^2/\pi)(1 - 2/\pi)}$$

For a given total rise h for angle θ_0 and any two of the β angles given one can solve for all angles and all values of the derivative curves.

5.8. ACCURACY[1]

A surface may appear smooth to the eye and yet have poor dynamic properties since the determination of whether a cam is satisfactory is based on the acceleration curve. The acceleration being the second derivative of lift, the large magnification of error will appear on it. There are two methods for establishing the accuracy of the cam fabrication and its dynamic effect: (1) employing the mathematical numerical method of finite differences, and (2) utilizing electronic instrumentation of an accelerometer pickup (located on the cam follower), and cathode-ray oscilloscope.

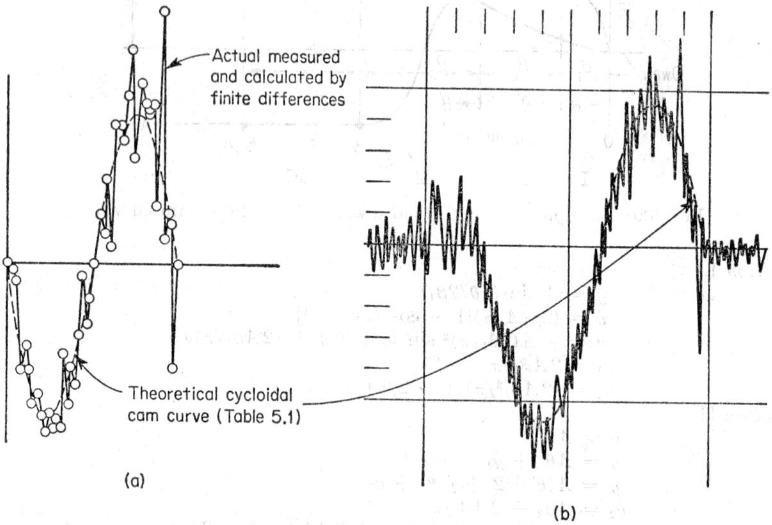

FIG. 5.10. Accuracy investigation of a cam turning 300 rpm, rise = ¾ in. in 45°. (a) Finite difference calculations. (b) Experimental with accelerometer and oscilloscope.

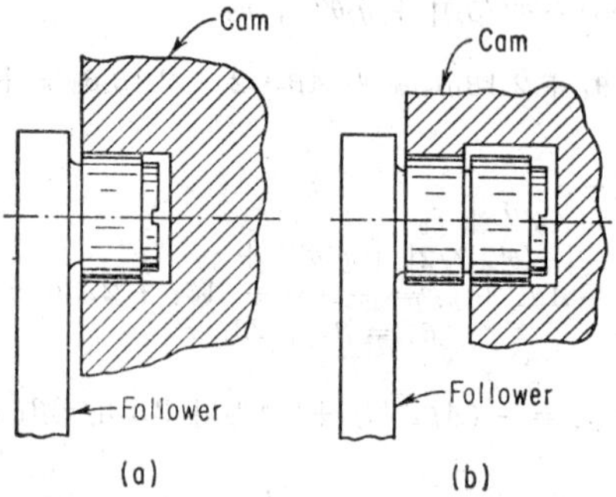

FIG. 5.11. Some roller and groove designs. (a) Single roller. (b) Double roller, undercut groove. (*Courtesy of John Wiley & Sons, Inc.*[1])

Many types of errors were investigated on measured shapes. All cams employed the basic cycloidal curve. In Fig. 5.10 we see a cam investigated at 300 rpm. The greatest error in fabrication exists at about the 40° cam angle. Correlation is observed between the finite difference and the experimental results. The theoretical maximum acceleration was calculated from the cycloidal formula to be 19.5 g's and the maximum acceleration of cam with errors is about 34 g's. On a broad average the maximum acceleration difference for the error only is 12 g's per 0.001-in. error at a speed of 300 rpm.

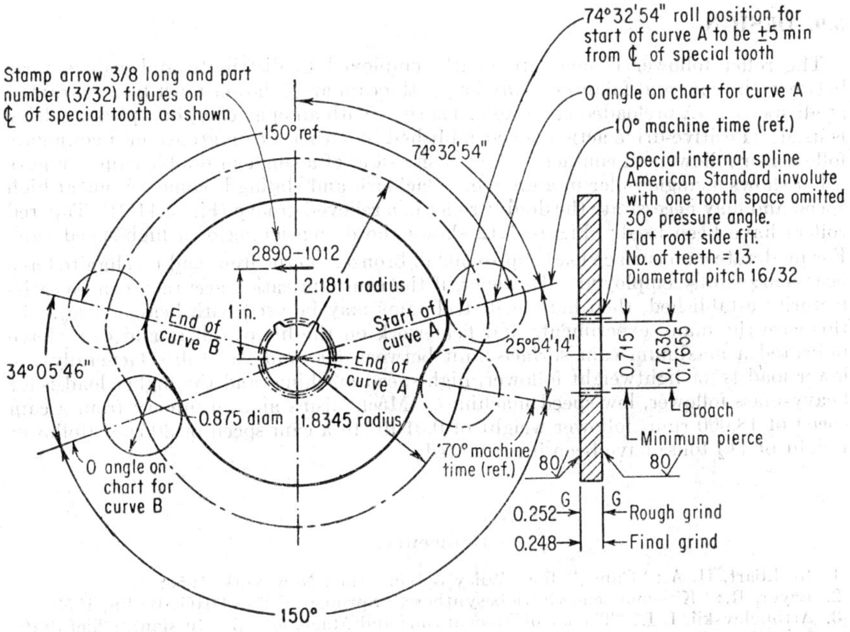

One required, No. A4140 steel 5⁄16 in. thick. Blank and pierce allow 3⁄64 in. on contour for finishing cam to be flat within ±0.010 after blanking. Heat-treat before machining and after blanking to 30 to 34 Rockwell C.

Induction-harden cam periphery 50 to 55 Rockwell C. Tolerance on cam curve must not vary in excess of ±0.001 from true curve, profile surface 32 μin.

Cam profile (roller center)

Points for curve A		Points for curve B	
Angle	Radius	Angle	Radius
0° 0′ 0″	2.1811	0° 0′ 0″	1.8345
0° 30′ 0″	2.1809	0° 22′ 30″	1.8345
1° 0′ 0″	2.1801	0° 45′ 2″	1.8345
1° 30′ 0″	2.1788	1° 07′ 38″	1.8346
2° 0′ 0″	2.1769	1° 30′ 18″	1.8349
2° 30′ 0″	2.1746	1° 53′ 5″	1.8353
3° 0′ 0″	2.1707	2° 16′ 1″	1.8359
Etc.	Etc.	Etc.	Etc.
23° 34′ 14″	1.8388	32° 35′ 28″	2.1807
24° 24′ 14″	1.8369	32° 58′ 8″	2.1810
24° 54′ 14″	1.8355	33° 20′ 44″	2.1810
25° 24′ 14″	1.8347	33° 43′ 15″	2.1811
25° 54′ 14″	1.8345	34° 35′ 46″	2.1811
150° from end of curve A to start of curve B at 1.8345 rad		150° from end of curve B to start of curve A at 2.1811 rad	

FIG. 5.12. Cam for loom. (*Courtesy of Warner & Swasey Co., Cleveland, Ohio.*)

5.9. DESIGN

The roller follower is most frequently employed to distribute and reduce wear between the cam and follower. Obviously the cam and follower must be constrained at all speeds. A preloaded compression spring (with an open cam) or a positive drive is used. Positive-drive action is accomplished by either a cam groove or a conjugate follower or followers in contact with opposite sides of a single or double cam. Figure 5.11*a* shows a single roller in a groove. Backlash and sliding become serious at high speed and may necessitate the double roller in a relieved groove (Fig. 5.11*b*). Tapered rollers have been used. Figure 5.12 shows the dimensioning of a high-speed cam. For moderate cam surface loads up to 500 lb bronze, ductile iron, and meehanite have been successfully employed. However, if the cam-fabrication accuracy can be satisfactorily established, then hardened-steel cams may be used with heavier load. In this case the many experiments and field work on the life of rollers and cams have indicated a maximum cam surface limit between 500 and 2,000 lb. Generally the lower load is for lightweight follower, high-speed machines and the higher load is for heavy-mass follower, low-speed machines. Mechanisms and machinery from a cam speed of 18,000 rpm (follower weight of 0.01 lb) to a cam speed of 30 rpm (follower weight of 1½ tons) have been investigated.

References

1. Rothbart, H. A.: ' Cams," John Wiley & Sons, Inc., New York, 1956.
2. Beyer, R.: "Kinematische Getriebesynthese," Springer-Verlag OHG, Berlin, 1953.
3. Artobolevskii, I. I.: "Theory of Mechanisms and Machines" (in Russian), Chief State Publishing House for Technical Theoretical Literature, Moscow, U.S.S.R., 1953.
4. Spotts, M F.: Straight Line Follower Motion Obtained with Circular Arc Cams, *Prod. Eng.*, vol. 21, pp. 110–113, August, 1950.
5. Thomson, W. R.: "The Simple Cams Having Profiles of Circular Arcs and Straight Lines," Machinery Publishing Co., Ltd., London.
6. Kloomok, M., and R. V. Muffley: Determination of Pressure Angles for Swinging-follower Cam Systems, *Trans. ASME*, vol. 70, p. 473, 1948.
7 Kloomok, M., and R. V. Muffley: "Determination of Radius of Curvature for Radial and Swinging Follower Cam Systems," ASME paper 55-SA-89, 1955.
8. Johnson, R. C.: Method of Finite Differences for Cam Design, *Machine Design*, vol. 27, p. 195, November, 1955.
9. Jehle, F., and W. R. Spiller: Idiosyncrasies of Valve Mechanisms and Their Causes, *Trans. SAE*, vol. 24, p. 197, 1929.
10. Turkish, M. C.: "Valve Gear Design," Eaton Mfg. Co., Detroit, Mich., 1946.
11. Dudley, W. M.: New Methods in Valve Cam Design, *Trans. SAE*, vol. 2, p. 19, January, 1948.
12. Mitchell, D. B.: Tests on Dynamic Response of Cam-follower Systems, *Mech. Eng.*, vol. 72, p. 467, June, 1950.
13. Hrones, J. A.: Analysis of Dynamic Forces in a Cam-driven System, *Trans. ASME*, vol. 70, p. 473, 1948.
14. Freudenstein, F.: On the Dynamics of High-speed Cam Profiles, *Intern. J. Mech. Sci.*, vol. 1, pp. 342–349, 1960.
15. Rothbart, H. A.: Cam Dynamics, *Trans. First Intern. Conf. Mechanisms*, Yale University, March 27, 1961.
16. Beggs, J. S.: "Mechanism," pp. 139–141, McGraw-Hill Book Company, Inc., New York, 1955.
17. Raven, F. H.: Analytical Design of Disk Cams and Three-dimensional Cams by Independent Position Equations, *Trans. ASME*, vol. 26, series E, no. 1, pp. 18–24, March, 1959.
18. Brunell, K.: "Constant Diameter Cams," ASME paper 61-SA-2, June, 1961.
19. Hinkle, R.: "Kinematics of Machines," pp. 148–161, Prentice-Hall, Inc., Englewood Cliffs, N. J., 1960.
20. Rothbart, H.: Mechanical Control Cams, *Control Eng.*, vol. 7, no. 6, pp. 118–122, June, 1960.
21. Rothbart, H.: Cams in Control Systems, *Control Eng.*, vol. 8, no. 11, pp. 97–101, November, 1961.

Section 6

MECHANICAL VIBRATIONS

By

ERIC E. UNGAR, Eng.Sc.D., *Senior Engineering Scientist, Bolt Beranek and Newman Inc., Cambridge, Mass.*

CONTENTS

INTRODUCTION

The field of dynamics deals essentially with the interrelation between the motions of objects and the forces causing them. The words shock and vibration imply particular forces and motions; hence this chapter concerns itself essentially with a subfield of dynamics. However, oscillatory phenomena occur also in nonmechanical systems; e.g., electric circuits, and many of the methods and some of the nomenclature used for mechanical systems are derived from nonmechanical systems.

Mechanical vibrations may be caused by forces whose magnitudes and/or directions and/or points of application vary with time. Typical forces may be due to rotating unbalanced masses, to impacts, to sinusoidal pressures (as in a sound field), or to random pressures (as in a turbulent boundary layer). In some cases the resulting vibrations may be of no consequence; in others they may be disastrous. Vibrations may be undesirable because they can result in deflections of sufficient magnitude to lead to malfunction, in high stresses which may lead to decreased life by increasing material fatigue, in unwanted noise, or in human discomfort.

Article 6.1 serves to delineate the concepts, phenomena, and analytical methods associated with the motions of systems having a single degree of freedom and to introduce the nomenclature and ideas discussed in the subsequent sections. Article 6.2 deals similarly with systems having a finite number of degrees of freedom, and Art. 6.3 with continuous systems (having an infinite number of degrees of freedom). Mechanical shocks are discussed in Art. 6.4, and in Art. 6.5 appears additional information concerning design considerations, vibration-control techniques, and rotating machinery, as well as charts and tables of natural frequencies, spring constants, and material properties. The appended references substantiate and amplify the presented material.

Complete coverage of mechanical vibrations and associated fields is clearly impossible within the allotted space. However, it was attempted to present enough information so that an engineer who is not a specialist in this field can solve the most prevalent problems with a minimum amount of reference to other publications.

6.1. SYSTEMS WITH A SINGLE DEGREE OF FREEDOM

A system with a single degree of freedom is one whose configuration at any instant can be described by a single number. A mass constrained to move without rotation along a given path is an example of such a system; its position is completely specified when one specifies its distance from a reference point, as measured along the path. Single-degree-of-freedom systems can be analyzed more readily than more complicated ones; therefore, actual systems are often approximated by systems with a single degree of freedom, and many concepts are derived from such simple systems and then enlarged to apply also to systems with many degrees of freedom.

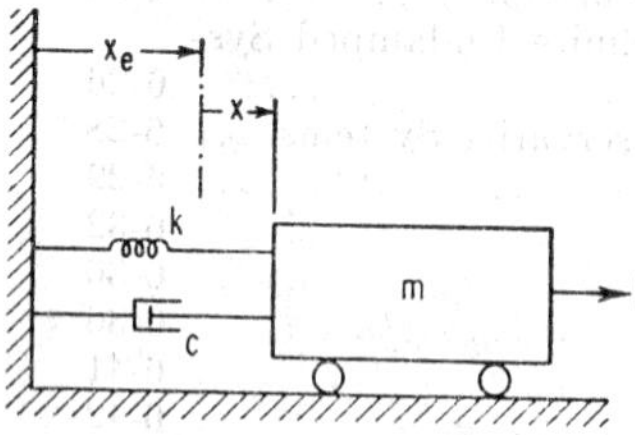

FIG. 6.1. A system with a single degree of freedom.

Figure 6.1 may serve as a model for all single-degree-of-freedom systems. This model consists of a pure inertia component (mass m supported on rollers which are devoid of friction and inertia), a pure restoring component (massless spring k), a pure energy-dissipation component (massless dashpot c), and a driving component (external force F). The inertia component limits acceleration. The restoring component opposes system deformation from equilibrium and tends to return the system to its equilibrium configuration in absence of other forces.

(a) Linear Single-degree-of-freedom Systems[6,17,35,42,59,62]

If the spring supplies a restoring force proportional to its elongation and the dashpot provides a force which opposes motion of the mass proportionally to its velocity, then the system response is proportional to the excitation, and the system is said to be linear. If the position x_e indicated in Fig. 6.1 corresponds to the equilibrium position of the mass and if x denotes displacement from equilibrium, then the spring force may be written as $-kx$ and the dashpot force as $-c\,dx/dt$ (where the displacement x and all forces are taken as positive in the same coordinate direction). The equation of motion of the system then is

$$m\, d^2x/dt^2 + c\, dx/dt + kx = F \tag{6.1}$$

Free Vibrations. In absence of a driving force F and of damping c, i.e., with $F = c = 0$, Eq. (6.1) has a general solution which may be expressed in any of the following ways:

$$\begin{aligned} x &= A \cos (\omega_n t - \phi) = (A \cos \phi) \cos \omega_n t + (A \sin \phi) \sin \omega_n t \\ &= A \sin (\omega_n t - \phi + \pi/2) = A \operatorname{Re} \{e^{i(\omega_n t - \phi)}\} \end{aligned} \tag{6.2}$$

A and ϕ are constants which may in general be evaluated from initial conditions. A is the maximum displacement of the mass from its equilibrium position and is called the *displacement amplitude;* ϕ is called the *phase angle.* The quantities ω_n and f_n, given by

$$\omega_n = \sqrt{k/m} \qquad f_n = \omega_n/2\pi$$

are known as the *undamped natural frequencies;* the first is in terms of "circular frequency" and is expressed in radians per unit time, the second is in terms of cyclic frequency and is expressed in cycles per unit time.

If damping is present, $c \neq 0$, one may recognize three separate cases depending on the value of the damping factor $\zeta = c/c_c$, where c_c denotes the critical damping coefficient (the smallest value of c for which the motion of the system will not be oscillatory). The critical damping coefficient and the damping factor are given by

$$c_c = 2\sqrt{km} = 2m\omega_n \qquad \zeta = c/c_c = c/2\sqrt{km} = c/2m\omega_n$$

The following general solutions of Eq. (6.1) apply when $F = 0$:

$$c > c_c,\ (\zeta > 1): \quad x = Be^{(-\zeta+\sqrt{\zeta^2-1})\omega_n t} + Ce^{(-\zeta-\sqrt{\zeta^2-1})\omega_n t} \tag{6.3a}$$
$$c = c_c,\ (\zeta = 1): \quad x = (B + Ct)e^{-\omega_n t} \tag{6.3b}$$
$$c < c_c,\ (\zeta < 1): \quad x = Be^{-\zeta\omega_n t} \cos (\omega_d t + \phi) = B \operatorname{Re} \{e^{i(\omega_N t+\phi)}\} \tag{6.3c}$$

where $\qquad \omega_d \equiv \omega_n \sqrt{1 - \zeta^2} \qquad i\omega_N = -\zeta\omega_n + i\omega_d$

denote, respectively, the "undamped natural frequency" and the "complex natural frequency," and the B, C, ϕ are constants that must be evaluated from initial conditions in each case.

Equation (6.3a) represents an extremely highly damped system; it contains two decaying exponential terms. Equation (6.3c) applies to a lightly damped system and is essentially a sinusoid with exponentially decaying amplitude. Equation (6.3b) pertains to a critically damped system and may be considered as the dividing line between highly and lightly damped systems.

Figure 6.2 compares the motions of systems (initially displaced from equilibrium and released with zero velocity) having several values of the damping factor ζ.

In all cases where $c > 0$ the displacement x approaches zero with increasing time. The damped natural frequency ω_d is generally only slightly lower than the undamped natural frequency ω_n; for $\zeta \leq 0.5$, $\omega_d \geq 0.87\ \omega_n$.

The static deflection x_{st} of the spring k due to the weight mg of the mass m (where g denotes the acceleration of gravity) is related to natural frequency as

$$x_{st} = mg/k = g/\omega_n^2 = g/(2\pi f_n)^2$$

This relation provides a quick means for computing the undamped natural frequency (or for approximating the damped natural frequency) of a system from its static deflection. For x in inches and f_n in cps it becomes

$$f_n^2 \text{ (cps)} = 9.80/x_{st} \text{ (in.)}$$

which is plotted in Fig. 6.3.

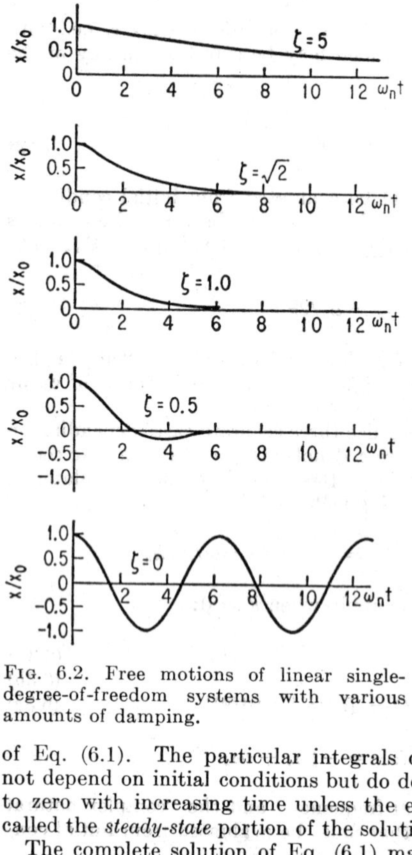

FIG. 6.2. Free motions of linear single-degree-of-freedom systems with various amounts of damping.

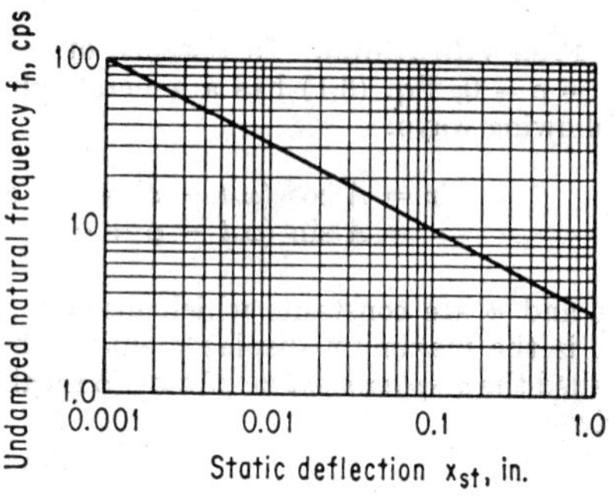

FIG. 6.3. Relation between natural frequency and static deflection of linear undamped single-degree-of-freedom system.

Forced Vibrations. The previous section dealt with cases where the forcing function F of Eq. (6.1) was zero. The solutions obtained were the so-called "general solution of the homogeneous equation" corresponding to Eq. (6.1). Since these solutions vanish with increasing time (for $c > 0$), they are sometimes also called the "transient solutions." For $F \neq 0$ the solutions of Eq. (6.1) are made up of the aforementioned general solution (which incorporates constants of integration that depend on the initial conditions) plus a "particular integral" of Eq. (6.1). The particular integrals contain no constants of integration and do not depend on initial conditions but do depend on the excitation. They do not tend to zero with increasing time unless the excitation tends to zero and hence are often called the *steady-state* portion of the solution.

The complete solution of Eq. (6.1) may be expressed as the sum of the general (transient) solution of the homogeneous equation and a particular (steady-state) solution of the nonhomogeneous equation. The associated general solutions have already been discussed; hence the present discussion will be concerned primarily with the steady-state solutions.

The steady-state solutions corresponding to a given excitation $F(t)$ may be obtained from the differential equation (6.1) by use of various standard mathematical techniques[11,19,28] without a great deal of difficulty. Table 6.1 gives the steady-state responses x_{ss} to some common forcing functions $F(t)$.

Superposition. Since the governing differential equation is linear, the response corresponding to a sum of excitations is equal to the sum of the individual responses; or, if

$$F(t) = A_1F_1(t) + A_2F_2(t) + A_3F_3(t) + \cdots$$

where $A_1, A_2, \ldots$ are constants, and if $x_{ss_1}, x_{ss_2}, \ldots$ are solutions corresponding, respectively, to $F_1(t), F_2(t), \ldots$, then the steady-state response to $F(t)$ is

$$x_{ss} = A_1x_{ss_1} + A_2x_{ss_2} + A_3x_{ss_3} + \cdots$$

Superposition permits one to determine the response of a linear system to any time-dependent force $F(t)$ if one knows the system's impulse response $h(t)$. This impulse response is the response of the system to a Dirac function $\delta(t)$ of force; also $h(t) = \dot{u}(t)$, where $u(t)$ is the system response to a unit step function of force $[F(t) = 0$ for $t < 0$, $F(t) = 1$ for $t > 0]$. In the determination of $h(t)$ and $u(t)$ the system is taken as at rest and at equilibrium at $t = 0$.

The motion of the system may be found from

$$x_{ss}(t) = \int_0^t F(\tau)h(t - \tau)\,d\tau \tag{6.4}$$

in conjunction with the proper "transient" solution expression, the constants in which must be adjusted to agree with specified initial conditions. For single-degree-of-freedom systems,

$$h(t) = \frac{\omega_n}{2k\sqrt{\zeta^2 - 1}}\left[e^{(-\zeta+\sqrt{\zeta^2-1})\omega_n t} + e^{(-\zeta-\sqrt{\zeta^2-1})\omega_n t}\right] \qquad \text{for } \zeta > 1$$

$$h(t) = \frac{\omega_n{}^2 t}{k}\,e^{-\omega_n t} \qquad \text{for } \zeta = 1$$

$$h(t) = \frac{\omega_n}{k\sqrt{1 - \zeta^2}}\,e^{-\zeta\omega_n t}\sin \omega_d t \qquad \text{for } \zeta < 1$$

$$h(t) = \frac{\omega_n}{k}\sin \omega_n t \qquad \text{for } \zeta = 0$$

Table 6.1. Steady-state Responses of Linear Single-degree-of-freedom Systems

$F(t)$	x_{ss}
1	$1/k$
t	$t/k - c/k^2$
t^2	$t^2/k - 2ct/k^2 + 2c^2/k^3 - 2m/k^2$
$e^{\pm\alpha t}$*	$h(\pm\alpha)$
$te^{\pm\alpha t}$*	$h(\pm\alpha)e^{\pm\alpha t} - h^2(\pm\alpha)(c \pm 2m\alpha)te^{\pm\alpha t}$
$e^{i\omega t}$	$(k - m\omega^2 + ic\omega)^{-1}$
$\sin \beta t$†	$g(\beta)[(k - m\beta^2)\sin \beta t - c\beta\cos \beta t]$
$\cos \beta t$†	$g(\beta)[c\beta \sin \beta t + (k - m\beta^2)\cos \beta t]$
$e^{\pm\alpha t}\sin \beta t$‡	$(A\cos \beta t + B\sin \beta t)e^{-\alpha t}/D$
$e^{\pm\alpha t}\cos \beta t$‡	$(A\sin \beta t - B\cos \beta t)e^{-\alpha t}/D$

* $h(\pm\alpha) = (m\alpha^2 \pm c\alpha + k)^{-1}$
† $g(\beta) = [(k - m\beta^2)^2 + (c\beta)^2]^{-1}$
‡ $A = k + m(\alpha^2 - \beta^2) \pm \alpha c$
$B = \beta(c \pm 2\alpha m)$
$D = A^2 + B^2$

Sinusoidal (Harmonic) Excitation. With an excitation

$$F(t) = F_0 \sin \omega t$$

one obtains a response which may be expressed as

$$x_{ss} = X_0 \sin(\omega t - \phi)$$

where
$$X_0 = \frac{F_0}{\sqrt{(k - m\omega^2)^2 + (c\omega)^2}} \qquad \tan\phi = \frac{c\omega}{k - m\omega^2} \tag{6.5}$$

The ratio

$$H_s(\omega) = \frac{X_0}{F_0/k} = \left\{\left[1 - \left(\frac{\omega}{\omega_n}\right)^2\right]^2 + \left(2\zeta\frac{\omega}{\omega_n}\right)^2\right\}^{-1/2} \tag{6.6}$$

is called the frequency response or the magnification factor. As the latter name implies, this ratio compares the displacement amplitude X_0 with the displacement F_0/k that a force F_0 would produce if it were applied statically. $H_s(\omega)$ is plotted in Fig. 6.4.

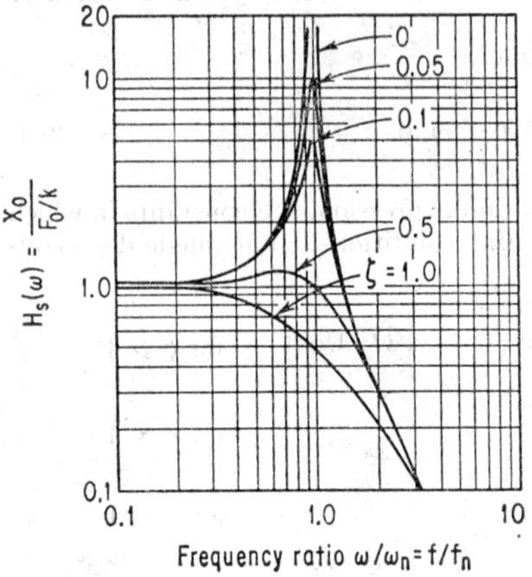

FIG. 6.4. Frequency response (magnification factor) of linear single-degree-of-freedom system.

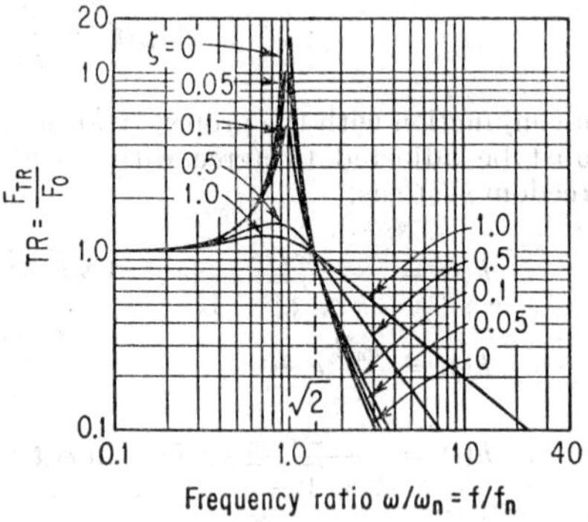

FIG. 6.5. Transmissibility of linear single-degree-of-freedom system.

Complex notation is convenient for representing general sinusoids.* Corresponding to a sinusoidal force

$$F(t) = F_0 e^{i\omega t}$$

one obtains a displacement

$$x_{ss} = X_0 e^{i\omega t}$$

where

$$\frac{X_0}{F_0/k} = H(\omega) = \left[1 - \left(\frac{\omega}{\omega_n}\right)^2 + 2i\zeta\left(\frac{\omega}{\omega_n}\right)\right]^{-1} \tag{6.7}$$

$H(\omega)$ is called the complex frequency response, or the complex magnification factor,† and is related to that of Eq. (6.6) as

$$H_s(\omega) = |H(\omega)|$$

From the model of Fig. 6.1 one may determine that the force F_{TR} exerted on the wall at any instant is given by

$$F_{TR} = kx + c\dot{x}$$

The ratio of the amplitude of this transmitted force to the amplitude of the sinusoidal applied force is called the transmissibility TR_s and obeys

$$TR_s = \frac{F_{TR}}{F_0} = \sqrt{\frac{1 + [2\zeta(\omega/\omega_n)]^2}{[1 - (\omega/\omega_n)^2]^2 + [2\zeta(\omega/\omega_n)]^2}} \tag{6.8}$$

Transmissibility $TR_s(\omega)$ is plotted in Fig. 6.5.

* In complex notation[42] it is usually implied, though it may not be explicitly stated, that only the *real parts* of excitations and responses represent the physical situation. Thus the complex form $Ae^{i\omega t}$ (where the coefficient $A = a + ib$ is also complex in general) implies the oscillation given by

$$\text{Re}\{Ae^{i\omega t}\} = \text{Re}\{(a + ib)(\cos\omega t + i\sin\omega t)\} = a\cos\omega t - b\sin\omega t$$

† An alternate formulation in terms of mechanical impedance is discussed in Art. 6.2(f).

In complex notation

$$TR = \frac{1 + 2i\zeta(\omega/\omega_n)}{1 - (\omega/\omega_n)^2 + 2i\zeta(\omega/\omega_n)} \qquad TR_s(\omega) = |TR(\omega)|$$

As evident from Figs. 6.4 and 6.5,

$$|H(\omega)| \approx |TR| \approx 1 \qquad \text{for } \omega \ll \omega_n$$
$$|H(\omega)| \approx (\omega_n/\omega)^2 \qquad |TR| \approx 2\zeta(\omega_n/\omega) \qquad \text{for } \omega \gg \omega_n$$

Increased damping ζ always reduces the frequency response H. For $\omega/\omega_n < \sqrt{2}$ increased damping also decreases TR, but for $\omega/\omega_n > \sqrt{2}$ increased damping increases TR.

The frequencies at which the maximum transmissibility and amplification factor occur for a given damping ratio are shown in Fig. 6.6; the magnitudes of these maxima are shown in Fig. 6.7. For small damping ($\zeta < 0.3$, which applies to many practical problems), the maximum transmissibility $|TR|_{max}$ and maximum amplification factor $|H|_{max}$ both occur at $\omega_d \approx \omega_n$, and

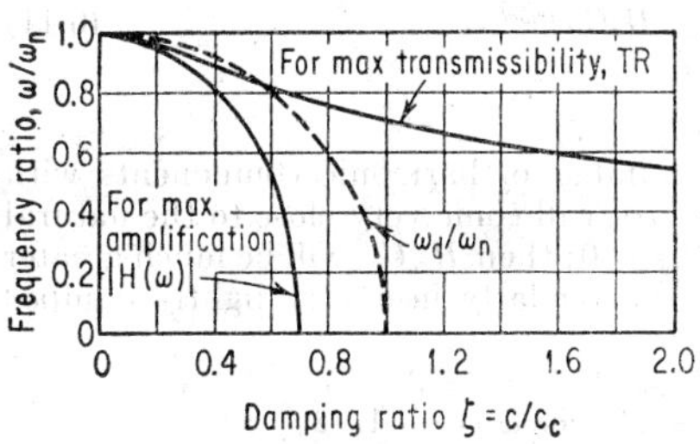

FIG. 6.6. Frequencies at which magnification and transmissibility maxima occur for given damping ratio.

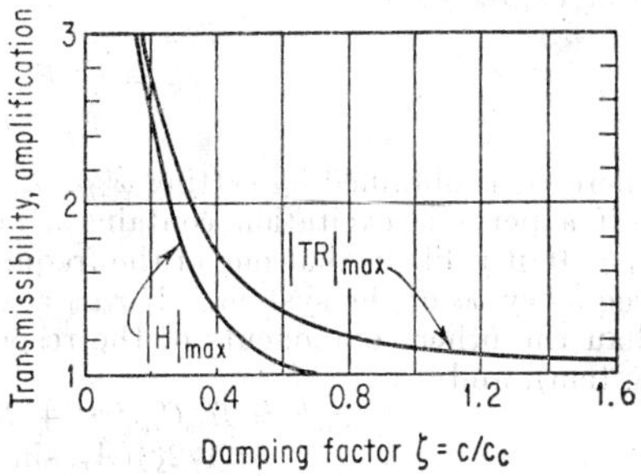

FIG. 6.7. Maximum values of magnification and transmissibility.

$$|TR|_{max} \approx |H|_{max} \approx (2\zeta)^{-1}$$

The quantity $(2\zeta)^{-1}$ is often given the symbol Q, termed the "quality factor" of the system. The frequency at which the greatest amplification occurs is called the resonant frequency; the system is then said to be in resonance. For lightly damped systems the resonant frequency is practically equal to the natural frequency, and often no distinction is made between the two. Thus, for lightly damped systems resonance (i.e., maximum amplification) occurs essentially when the exciting frequency ω is equal to the natural frequency ω_n.

Equation (6.6) shows that

$$X_0/F_0 \approx 1/k \text{ (system is stiffness-controlled) for } \omega \ll \omega_n$$
$$\approx 1/2k\zeta \text{ (system is damping-controlled) for } \omega \approx \omega_n (\zeta \ll 1)$$
$$\approx 1/m\omega^2 \text{ (system is mass-controlled) for } \omega \gg \omega_n$$

General Periodic Excitation. Any periodic excitation may be expressed in terms of a Fourier series (i.e., a series of sinusoids) and any aperiodic excitation may be expressed in terms of a Fourier integral, which is an extension of the Fourier-series concept. In view of the superposition principle applicable to linear systems the response can then be obtained in terms of a corresponding series or integral.

A periodic excitation with period T may be expanded in a Fourier series as

$$F(t) = A_0/2 + \sum_{r=1}^{\infty} (A_r \cos r\omega_0 t + B_r \sin r\omega_0 t) = \sum_{r=-\infty}^{\infty} C_r e^{ir\omega_0 t} \tag{6.9}$$

where the period T and fundamental frequency ω_0 are related by

$$\omega_0 T = 2\pi$$

The Fourier coefficients A_r, B_r, C_r may be computed from

$$A_r = (2/T) \int_t^{t+T} F(t) \cos (r\omega_0 t)\, dt \qquad B_r = (2/T) \int_t^{t+T} F(t) \sin (r\omega_0 t)\, dt$$

$$C_r = \tfrac{1}{2}(A_r - iB_r) = (1/T) \int_t^{t+T} F(t) e^{-ir\omega_0 t}\, dt \tag{6.10}$$

Superposition permits the steady-state response to the excitation given by Eq. (6.9) to be expressed as

$$x_{ss} = (1/k) \sum_{r=-\infty}^{\infty} H_r C_r e^{ir\omega_0 t} \tag{6.11}$$

where H_r is obtained by setting $\omega = r\omega_0$ in Eq. (6.7).

If a periodic excitation contains a large number of harmonic components with $C_r \neq 0$, it is likely that one of the frequencies $r\omega_0$ will come very close to the natural frequency ω_n of the system. If $r_0\omega_0 \approx \omega_n$, $C_{r_0} \neq 0$; then $H_{r_0}C_{r_0}$ will be much greater than the other components of the response (particularly in a very lightly damped system), and

$$x_{ss}k \approx H_{r_0}C_{r_0}e^{i\omega_n t} + H_{-r_0}C_{-r_0}e^{-i\omega_n t} + A_0/2$$
$$\approx (1/2\zeta)(A_{r_0} \sin \omega_n t - B_{r_0} \cos \omega_n t) + A_0/2$$

General Nonperiodic Excitation.[3,12,16,42,48] The response of linear systems to any well-behaved* forcing function may be determined from the impulse response as discussed in conjunction with Eq. (6.4) or by application of Fourier integrals. The latter may be visualized as generalizations of Fourier series applicable for functions with infinite period.

A "well-behaved"* forcing function $F(t)$ may be expressed as†

$$F(t) = (1/2\pi) \int_{-\infty}^{\infty} \Phi(\omega) e^{i\omega t}\, d\omega \tag{6.12}$$

where

$$\Phi(\omega) = \int_{-\infty}^{\infty} F(t) e^{-i\omega t}\, dt \tag{6.13}$$

[These are analogous to Eqs. (6.9) and (6.10).] With the ratio $H(\omega)$ of displacement to force as given by Eq. (6.7), the displacement-response transform then is

$$X(\omega) = (1/k)H(\omega)\Phi(\omega)$$

and, analogously to Eq. (6.11), one finds a displacement given by

$$x_{ss}(t) = (1/2\pi) \int_{-\infty}^{\infty} X(\omega)e^{i\omega t}\, d\omega = (1/2\pi k) \int_{-\infty}^{\infty} H(\omega)\Phi(\omega)e^{i\omega t}\, d\omega$$
$$= (1/2\pi k) \int_{-\infty}^{\infty} H(\omega) \left[\int_{-\infty}^{\infty} F(t)e^{-i\omega t}\, dt \right] e^{i\omega t}\, d\omega$$

* "Well-behaved" means that $|F(t)|$ is integrable and $F(t)$ has bounded variation.

† Other commonly used forms of the integral transforms can be obtained by substituting $j = -i$. Since $j^2 = i^2 = -1$, all the developments still hold. Fourier transforms are also variously defined as regards the coefficients. For example, instead of $1/2\pi$ in Eq. (6.12) there often appears a $1/\sqrt{2\pi}$; then a $1/\sqrt{2\pi}$ factor is added in Eq. (6.13) also. In all cases the product of the coefficients for a complete cycle of transformations is $1/2\pi$.

One may expect the components of the excitation with frequencies nearest the natural frequency of a system to make the most significant contributions to the response. For lightly damped systems one may assume that these most significant components are contained in a small frequency band containing the natural frequency. Usually one uses a "resonance bandwidth" $\Delta\omega = 2\zeta\omega_n$, thus effectively assuming that the most significant components are those with frequencies between $\omega_n(1 - \zeta)$ and $\omega_n(1 + \zeta)$. (At these two limiting frequencies, commonly called the half-power points, the rate of energy dissipation is one-half of that at resonance. The amplitude of the response at these frequencies is $1/\sqrt{2} \approx 0.707$ times the amplitude at resonance.) Noting that the largest values of the complex amplification factor $H(\omega)$ occur for $\omega \approx \pm\omega_d \approx \pm\omega_n$, one may write

$$2\pi x_{ss}k \approx -i\omega_n e^{-\zeta\omega_n t}[\Phi(\omega_n)e^{i\omega_n t} + \Phi(-\omega_n)e^{-i\omega_n t}]$$

Random Vibrations. Mean Values, Spectra, Spectral Densities.[3,12,16,48] In many cases one is interested only in some mean value as a characterization of response. The time average of a variable $y(t)$ may be defined as

$$\bar{y} = \lim_{\tau\to\infty} (1/\tau) \int_0^\tau y(t)\, dt \tag{6.14}$$

where it is assumed that the limit exists. For periodic $y(t)$ one may take τ equal to a period and omit the limiting process.

The mean-square value of $y(t)$ thus is given by

$$\overline{y^2} = \lim_{\tau\to\infty} (1/\tau) \int_0^\tau y^2(t)\, dt$$

and the root-mean-square value by $y_{\text{rms}} = \sqrt{\overline{y^2}}$. For a sinusoid $x = \text{Re}\,\{Ae^{i\omega t}\}$ one finds $\overline{x^2} = \frac{1}{2}|A|^2 = \frac{1}{2}AA^*$ where A^* is the complex conjugate of A.

The mean-square response $\overline{x^2}$ of a single-degree-of-freedom system with frequency response $H(\omega)$ [Eq. (6.7)] to a sinusoidal excitation of the form $F(t) = \text{Re}\,\{F_0e^{i\omega t}\}$ is given by

$$k^2\overline{x^2} = H(\omega)F_0H^*(\omega)F_0{}^*/2 = |H(\omega)|^2\overline{F^2}$$

Similarly, the mean-square value of a general periodic function $F(t)$, expressed in Fourier-series form as

$$F(t) = \sum_{r=-\infty}^{\infty} C_r e^{ir\omega_0 t}$$

is

$$\overline{F^2} = \sum_{r=-\infty}^{\infty} C_rC_r{}^*/2 = \frac{1}{2} \sum_{r=-\infty}^{\infty} |C_r|^2 \tag{6.15}$$

The mean-square displacement of a single-degree-of-freedom system in response to the aforementioned periodic excitation is given by

$$k^2\overline{x^2} = \frac{1}{2} \sum_{r=-\infty}^{\infty} |C_r|^2|H(r\omega_0)|^2$$

where convergence of all the foregoing infinite series is assumed.

If one were to plot the cumulative value of (the sum representing) the mean-square value of a periodic variable as a function of frequency, starting from zero, one would obtain a diagram somewhat like Fig. 6.8. This graph shows how much each frequency (or "spectral component") adds to the total mean-square value. Such a graph* is called the spectrum (or possibly more properly the integrated spectrum) of $F(t)$. It is generally of relatively little interest for periodic functions but is extremely useful for aperiodic (including random) functions.

The derivative of the (integrated) spectrum with respect to ω is called the *mean-square spectral density* (or power spectral density) of F. Thus the power spectral density S_F of F is defined as*

$$S_F(\omega) = 2\pi d(\overline{F^2})/d\omega \qquad (6.16)$$

where $\overline{F^2}$ is interpreted as a function of ω as in Fig. 6.8. The mean-square value of F is related to power spectral density as

$$\overline{F^2} = (1/2\pi) \int_0^\infty S_F(\omega)\, d\omega \qquad (6.17)$$

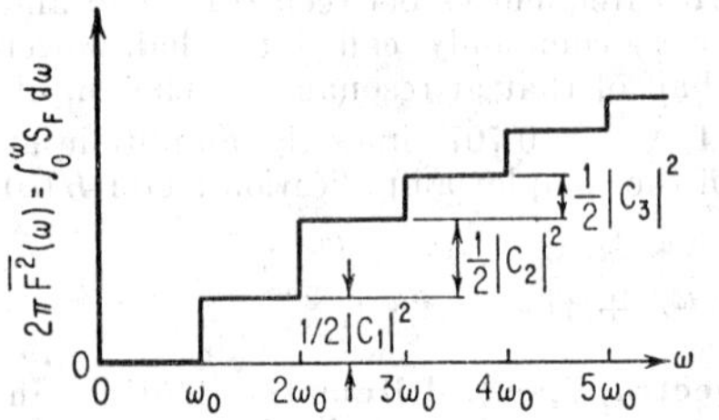

FIG. 6.8. (Integrated) spectrum of periodic function $F(t)$.

This integral over all frequencies is analogous to the infinite sum of Eq. (6.15). From Fig. 6.8 and Eq. (6.17) one may visualize that power spectral density is a convenient means for expressing the contributions to the mean-square value in any frequency range.

For nonperiodic functions one obtains contributions to the mean-square value over a continuum of frequencies instead of at discrete frequencies, as in Fig. 6.8. The (integrated) spectrum and the power spectral density then are continuous curves. The relations governing the mean responses to nonperiodic excitation can be obtained by the same limiting processes which permit one to proceed from the Fourier series to Fourier integrals. However, the results are presented here in a slightly more general form so that they can be applied also to systems with random excitation.†

Response to Random Excitation; Autocorrelation Functions. For stationary ergodic random processes‡ whose sample functions are $F(t)$ or for completely specified functions $F(t)$ one may define an autocorrelation function $R_F(\tau)$ as

$$R_F(\tau) = \lim_{T\to\infty} (1/2T) \int_{-T}^{T} F(t)F(t+\tau)\, dt \qquad (6.18)$$

* The factor 2π appearing in Fig. 6.8 and Eqs. (6.16) and (6.17) is a matter of definition. Different constants are sometimes used in the literature, and one must use care in comparing results from different sources.

† In the previous discussion the excitation was described as some known function of time and the responses were computed as other completely defined time functions; in each case the values at each instant were specified or could be found. Often the stimuli cannot be defined so precisely; only some statistical information about them may be available. Then, of course, one may only obtain some similar statistical information about the responses.

‡ A "random process" is a mathematical model useful for representing randomly varying physical quantities. Such a process is determined not by its values at various instants but by certain average and spectral properties. One sacrifices precision in the description of the variable for the sake of tractability.

One may envision a large number of sample functions (such as force vs. time records obtained on aircraft landing gears, with time datum at the instant of landing). One may compute an average value of these functions at any given time instant; such an average is called a *statistical average* and generally varies with the instant selected. On the other hand, one may also compute the time average of any given sample function over a long interval. The statistical average will be equal to the time average of almost every sample function provided that the sample process is both stationary and ergodic.[12]

A random process is stationary essentially if the statistical average of the sample functions is independent of time, i.e., if the ensemble appears unchanged if the time origin is changed. Ergodicity essentially requires that almost every sample function is "typical" of the entire group. General mathematical results are to a large degree available only for stationary ergodic random processes; hence the following discussion is limited to such processes.

This function has the properties

$$R_F(0) = \overline{F^2} \geq R_F(\tau) \qquad R_F(-\tau) = R_F(\tau)$$

For many physical random processes the values of F observed at widely separated intervals are uncorrelated, that is

$$\lim_{\tau \to \infty} R_F(\tau) = (\bar{F})^2$$

or R_F approaches the square of the mean value (not the mean-square value!) of F for large time separation τ. (Many authors define variables measured from a mean value; if such variables are uncorrelated, their $R_F \to 0$ for large τ.) One may generally find some value of τ beyond which R_F does not differ "significantly" from $(\bar{F})^2$. This value of τ is known as the "scale" of the correlation.

The power spectral density of F is given by*,†

$$S_F(\omega) = 2 \int_{-\infty}^{\infty} R_F(\tau) e^{-i\omega\tau}\, d\tau \tag{6.19}$$

and is equal to twice* the Fourier transform of the autocorrelation function R_F. Inversion of this transform gives*,†

$$R_F(\tau) = (1/4\pi) \int_{-\infty}^{\infty} S_F(\omega) e^{i\omega\tau}\, d\tau$$

whence*,†

$$R_F(0) = \overline{F^2} = (1/4\pi) \int_{-\infty}^{\infty} S_F(\omega)\, d\omega \tag{6.20}$$

The last of these relations agrees with Eq. (6.17).

System Response to Random Excitation. For a system with a complex frequency response $H(\omega)$ as given by Eq. (6.7) one finds that the power spectral density S_x of the response is related to the power spectral density S_F of the exciting force according to

$$k^2 S_x(\omega) = |H(\omega)|^2 S_F(\omega) \tag{6.21}$$

In order to compute the mean-square response of a system to aperiodic (or stationary ergodic random) excitation one may proceed as follows:

1. Calculate $R_F(\tau)$ from Eq. (6.18).
2. Find $S_F(\omega)$ from Eq. (6.19).
3. Determine $S_x(\omega)$ from Eq. (6.21).
4. Find $\overline{x^2}$ from Eq. (6.20) (with F subscripts replaced by x).

"White noise" is a term commonly applied to functions whose power spectral density is constant for all frequencies. Although such functions are not realizable physically, it is possible to obtain power spectra that remain virtually constant over a frequency region of interest in a particular problem (particularly in the neighborhood of the resonance of the system considered, where the response contributes most to the total).

The mean-square displacement of a single-degree-of-freedom system to a (real) white-noise excitation F, having the power spectral density $S_F(\omega) = S_0$, is given by

$$\overline{x^2} = \omega_n S_0 / 8\zeta k^2 = S_0 / 4ck$$

Probability Distributions of Excitation and Response.[3,12,16,48] For most practical purposes it is sufficient to define the probability of an event as the fraction

* Definitions involving different numerical coefficients are also in general use.

† For real $F(t)$ one may multiply the coefficient shown here by 2 and replace the lower limit of integration by zero.

of the number of "trials" in which the event occurs, provided that a large number of trials are made. (In throwing an unbiased die a large number of times one expects to obtain a given number, say 2, one-sixth of the time. The probability of the number 2 here is $1/6$.)

If a given variable can assume a continuum of values (unlike the die for which the variable, i.e., the number of spots, can assume only a finite number of discrete values) it makes generally little sense to speak of the probability of any given value. Instead, one may profitably apply the concepts of probability distribution and probability density functions. Consider a continuous random variable x and a certain value x_0 of that variable. The probability distribution function P_{dis} then is defined as a function expressing the probability P that the variable $x \leq x_0$. Symbolically,

$$P_{\text{dis}}(x_0) = P(x \leq x_0)$$

The probability density function P_{dens} is defined by

$$P_{\text{dens}}(x_0) = \frac{dP_{\text{dis}}(x_0)}{dx_0}$$

so that the probability of x occurring between x_0 and $x_0 + dx_0$ is

$$P(x_0 < x \leq x + dx_0) = P_{\text{dens}}(x_0)\,dx_0$$

Among the most widely studied distributions are the Gaussian (or normal) distri bution, for which

$$P_{\text{dens}}(x) = [1/\sigma \sqrt{2\pi}]e^{-(x-M)^2/2\sigma^2} \tag{6.22}$$

and the Rayleigh distribution, for which

$$P_{\text{dens}}(x) = (x/\sigma^2)e^{-x^2/2\sigma^2} \tag{6.23}$$

In the foregoing, M denotes the statistical mean value, defined by

$$M = \int_{-\infty}^{\infty} xP_{\text{dens}}(x)\,dx$$

and σ^2 denotes the variance of x and is defined by

$$\sigma^2 = \int_{-\infty}^{\infty} (x - M)^2 P_{\text{dens}}(x)\,dx$$

σ is called the *standard deviation* of the distribution. For a stationary ergodic random process

$$\sigma^2 + M^2 = R_x(0) = \overline{x^2}$$

The Gaussian distribution is by far the most important, since it represents many physical conditions relatively well and permits mathematical analysis to be carried out relatively simply. With some qualifications, the "central limit theorem" states that any random process, each of whose sample functions is constructed from the sum of a large number of sample functions selected independently from some other random process, will tend to become Gaussian as the number of sample functions added tends to infinity. A stationary Gaussian random process with zero mean is completely characterized by either its autocorrelation function or its power spectral density. If the excitation of a linear system is a Gaussian random process, then so is the system response.

For a Gaussian random process $x(t)$ with autocorrelation function $R_x(\tau)$ and power spectral density $S_x(\omega)$ one may find the *average* number of times N_0 that $x(t)$ passes through zero in unit time from

$$(\pi N_0)^2 = -(d^2R_x/d\tau^2)\Big|_{\tau=0} [R_x(0)]^{-1} = \left[\int_0^\infty \omega^2 S_x(\omega)\,d\omega\right]\left[\int_0^\infty S_x(\omega)\,d\omega\right]^{-1}$$

The quantity N_0 gives an indication of the "apparent frequency" of $x(t)$. A sinusoid crosses zero twice per cycle and has $f = N_0/2$. This relation may be taken as the definition of apparent frequency for a random process.

The average number of times N_α that the aforementioned Gaussian $x(t)$ crosses the value $x = \alpha$ per unit time is given by

$$N_\alpha = N_0 e^{-\alpha^2/2R_x(0)}$$

The average number of times per unit time that $x(t)$ passes through α with positive slope is half the foregoing value. The average number of peaks* of $x(t)$ occurring per unit time between $x = \alpha$ and $x = \alpha + d\alpha$ is

$$N_{\alpha,\alpha+d\alpha} = \frac{N_0 \alpha\, d\alpha}{2R_x(0)} e^{-\alpha^2/2R_x(0)}$$

For a linear single-degree-of-freedom system with natural circular frequency ω_n subject to white noise of power spectral density $S_F(\omega) = S_0$, one finds

$$N_0 = \omega_n/\pi \qquad f_{\text{apparent}} = \omega_n/2\pi = f_n$$

The displacement vs. time curve representing the response of a lightly damped system to broad-band excitation has the appearance of a sinusoid with the system natural frequency, but with randomly varying amplitude and phase. The average number of peaks per unit time occurring between α and $\alpha + d\alpha$ in such an oscillation is given by

$$(2\pi/\omega_n)N_{\alpha,\alpha+d\alpha} = [\alpha\, d\alpha/R_x(0)]e^{-\alpha^2/2R_x(0)}$$

The term on the right-hand side is, except for the $d\alpha$, the Rayleigh probability density of Eq. (6.23).

(b) Nonlinear Single-degree-of-freedom Systems[15,55,58]

The previous discussion dealt with systems whose equations of motion can be expressed as linear differential equations (with constant coefficients), for which solutions can always be found. The present section deals with systems having equations of motion for which solutions cannot be found so readily. Approximate analytical solutions can occasionally be found, but these generally require insight and/or a considerable amount of algebraic manipulation. Numerical or analog computations or graphical methods appear to be the only ones of general applicability.

Practical Solution of General Equations of Motion. After one sets up the equations of motion of a system one wishes to analyze, one should determine whether solutions of these are available by referring to texts on differential equations and compendia like ref. 27. (The latter reference also describes methods of general utility for obtaining approximate solutions, such as that involving series expansion of the variables.) If these approaches fail one is generally reduced to the use of numerical or graphical methods. In the following pages two generally useful methods of these types are outlined. Methods and results applicable to some special cases are discussed in subsequent sections.

A Numerical Method. The equation of motion of a single-degree-of-freedom system can generally be expressed in the forms

$$m\ddot{x} + G(x,\dot{x},t) = 0 \qquad \ddot{x} + f(x,\dot{x},t) = 0 \tag{6.24}$$

where f includes all nonlinear and nonconstant coefficient effects. (f may occasionally also depend on higher time derivatives. These are not considered here, but the method discussed here may be readily extended to account for them.) It is assumed that f is a known function, given in graphical, tabular, or analytic form.

* Actually average excess of peaks over troughs, but for $\alpha \gg x_{\text{rms}}$ the probability of troughs in the interval becomes very small.

In order to integrate Eq. (6.24) numerically as simply as possible one assumes that f remains virtually constant in a small time interval Δt. Then one may proceed by the following steps:

1*a*. Determine $f_0 = f(x_0,\dot{x}_0,0)$, the initial value of f, from the specified initial displacement x_0 and initial velocity $\dot{x}_0$. Then the initial acceleration is $\ddot{x}_0 = -f_0$.

b. Calculate the velocity $\dot{x}_1$ at the end of a conveniently chosen small time interval $\Delta t_{0\text{-}1}$, and the average velocity $\bar{\dot{x}}_{0\text{-}1}$ during the interval from

$$\dot{x}_1 = \dot{x}_0 + \ddot{x}_0\,\Delta t_{0\text{-}1} \qquad \bar{\dot{x}}_{0\text{-}1} = \tfrac{1}{2}(\dot{x}_0 + \dot{x}_1)$$

c. Calculate the displacement x_1 at the end of the interval $\Delta t_{0\text{-}1}$ and the average displacement $\bar{x}_{0\text{-}1}$ during the interval from

$$x_1 = x_0 + \bar{\dot{x}}_{0\text{-}1}\,\Delta t_{0\text{-}1} \qquad \bar{x}_{0\text{-}1} = \tfrac{1}{2}(x_0 + x_1)$$

d. Compute a better approximation* to the average f and $\ddot{x}$ during the interval $\Delta t_{0\text{-}1}$ by using $x = \bar{x}_{0\text{-}1}$, $\dot{x} = \bar{\dot{x}}_{0\text{-}1}$, $t = \tfrac{1}{2}\Delta t_{0\text{-}1}$ in the determination of f.

2. Repeat steps 1*b* to 1*d*, beginning with the new approximation to f, until no further changes in f occur (to the desired accuracy).

3. Select a second time interval $\Delta t_{1\text{-}2}$ (not necessarily of the same magnitude as $\Delta t_{0\text{-}1}$) and continue to

a. Find $\ddot{x}_1 = -f_1 = -f(x_1,\dot{x}_1,t_1)$.

b. Calculate the velocity $\dot{x}_2$ at the end of the interval $\Delta t_{1\text{-}2}$, and the average velocity $\bar{\dot{x}}_{1\text{-}2}$ during the interval, from

$$\dot{x}_2 = \dot{x}_1 + \ddot{x}_1\,\Delta t_{1\text{-}2} \qquad \bar{\dot{x}}_{1\text{-}2} = \tfrac{1}{2}(\dot{x}_1 + \dot{x}_2)$$

c. Similarly, find the final and average displacements for the $\Delta t_{1\text{-}2}$ interval from

$$x_2 = x_1 + \bar{\dot{x}}_{1\text{-}2}\,\Delta t_{1\text{-}2} \qquad \bar{x}_{1\text{-}2} = \tfrac{1}{2}(x_1 + x_2)$$

d. Compute a better approximation to the average f and $\ddot{x}$ during the interval $\Delta t_{1\text{-}2}$ by using $x = \bar{x}_{1\text{-}2}$, $\dot{x} = \bar{\dot{x}}_{1\text{-}2}$, $t = \Delta t_{0\text{-}1} + \tfrac{1}{2}\Delta t_{1\text{-}2}$ in the determination of f.

4. Repeat steps 3*b* to 3*d*, starting with the better value of f, until no changes in f occur to within the desired accuracy.

5. One may then continue by essentially repeating steps 3 and 4 for additional time intervals until one has determined the motion for the desired total time of interest.

Generally, the smaller the time intervals selected, the greater will be the accuracy of the results (regardless of the f-averaging method used). Use of smaller time intervals naturally leads to a considerable increase in computational labor. If high accuracy is required, one may generally benefit by employing one of the many available more sophisticated numerical-integration schemes.[30,53] In many practical instances the labor of carrying out the required calculations by "hand" becomes prohibitive, however, and use of a digital computer is indicated.

A Semigraphical Method. One may avoid some of the tedium of the foregoing numerical-solution method and gain some insight into a problem by using the "phase-plane delta" method[26,35] discussed here. This method, like the foregoing numerical one, is essentially a stepwise integration for small time increments. It is based on rewriting the equation of motion (6.24) as

$$\ddot{x} + \omega_0^2(x + \delta) = 0 \qquad \delta = -x + f(x,\dot{x},t)/\omega_0^2 \tag{6.25}$$

where ω_0 is any convenient constant circular frequency. (Any value may be chosen for ω_0, but it is usually useful to select one with some physical meaning, e.g., $\omega_0 = \sqrt{k_0/m_0}$ where k_0 and m_0 are values of stiffness and mass for small x, $\dot{x}$, t.) If

* It should be noted that evaluation of f at the average values of the variables involved is only one of many possible ways of obtaining an average f for the interval considered. Other averages, for example, can be obtained from $\sqrt{f_0 f_1}$, $\tfrac{1}{2}(f_0 + f_1)$. One can rarely predict which average will produce the most accurate results in a given case.

one introduces into Eq. (6.25) a reduced velocity v given by

$$v = \dot{x}/\omega_0$$

and assumes that $\delta(x,\dot{x},t)$ remains essentially constant in a short time interval one may integrate the resulting equation to obtain

$$v^2 + (x + \delta)^2 = R^2 = \text{const}$$

Thus for small time increments the solutions of (6.25) are represented in the xv plane (Fig. 6.9) by short arcs of circles whose centers are at $x = -\delta$, $v = 0$.

The angle $\Delta\theta$ subtended by the aforementioned circular arc is related to the time interval Δt according to

$$\Delta t \approx \Delta x / v\omega_0 \approx \Delta\theta/\omega_0 \qquad (6.26)$$

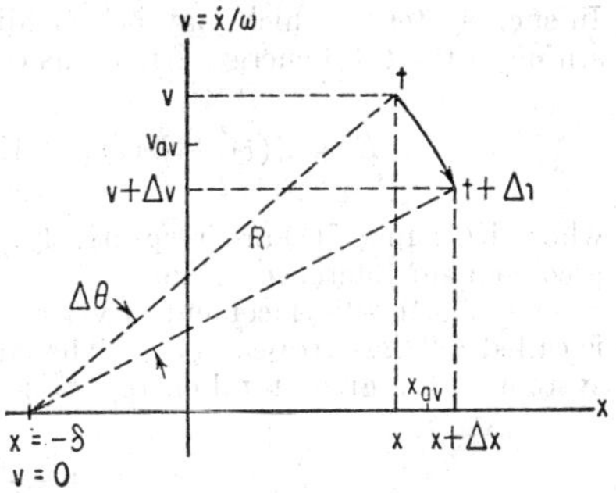

FIG. 6.9. Phase-plane delta method.

On the basis of the foregoing discussion one may thus proceed as follows:

1. Calculate $\delta(x_0,\dot{x}_0,t_0)$ from Eq. (6.25) using the given initial conditions.
2. Locate the circle center $(-\delta,0)$ and the initial point (x_0,v_0) on the xv plane; draw a small clockwise arc.
3. At the end of this arc is the point x_1, v_1 corresponding to the end of the first time interval.
4. Measure or calculate (in radians) the angle $\Delta\theta$ subtended by the arc; calculate the length of the time increment from Eq. (6.26).
5. Calculate $\delta(x_1,\dot{x}_1,t_1)$, and continue as before.
6. Repeat this process until the desired information is obtained.
7. Plots, such as those of x, $\dot{x}$, or $\ddot{x}$, against time, may then be readily obtained from the xv curve and the computed time information.

If increased accuracy is desired, particularly where δ changes rapidly, δ should be evaluated from average conditions $(x_{av},\dot{x}_{av},t_{av})$ during the time increment instead of conditions at the beginning of this increment (see Fig. 6.9). If δ depends on only one variable, a plot of δ against this variable may generally be used to advantage, particularly if it is superposed onto the vx plane.

Mathematical-approximation Methods. An analytical expression is usually preferable to a series of numerical solutions, since it generally permits greater insight into a given problem. If exact analytical solutions cannot be found, approximate ones may be the next best approach.

Series expansion of the dependent in terms of the independent variable is often a useful expedient. Power series and Fourier series are most commonly used, but occasionally series of other functions may be employed. The approach consists essentially of writing the dependent variable in terms of a series with unknown coefficients, substituting this into the differential equation, and then solving for the coefficients. However, in many cases these solutions may be difficult, or the series may converge slowly or not at all.

Other methods attempt to obtain solutions by separating the governing equations into a linear part (for which a simple solution can be found) and a nonlinear part. The solution of the linear part is then applied to the nonlinear part in some way so as to give a first correction to the solution. The correction process is then repeated until a second better approximation is obtained, and the process is continued. Such methods include:

1. Perturbation,[15,42] which is particularly useful where the nonlinearities (deviations from linearity) are small.
2. Reversion,[15] which is a special treatment of the perturbation method.

3. Variation of parameters,[15] useful where nonlinearities do not result in additive terms.
4. Averaging methods, based on error minimization.
   a. Galerkin's method.[15]
   b. Ritz method.[15]

Conservative Systems; the Phase Plane.[15,55,58] A conservative system is one whose equation of motion can be written

$$m\ddot{x} + f(x) = 0 \tag{6.27}$$

In such systems (which may be visualized as masses attached to springs of variable stiffness) the total energy E remains constant; that is

$$E = V(\dot{x}) + U(x) \qquad V(\dot{x}) = \tfrac{1}{2}m\dot{x}^2,\ U(x) = \int_{x_0}^{x} f(x)\,dx$$

where $V(\dot{x})$ and $U(x)$ are, respectively, the kinetic and the potential energies and x_0 is a convenient reference value.

The velocity-displacement ($\dot{x}$ vs. x) plane is called the "phase plane"; a curve in it is called a "phase trajectory." The equation of a phase trajectory of a conservative system with a given total energy E is

$$\dot{x}^2 = \frac{2}{m}[E - U(x)] \tag{6.28}$$

the time interval $(t - t_1)$ in which a change of displacement from x_1 to x occurs, is given by

$$t - t_1 = \int_{x_1}^{x} \frac{dx}{\dot{x}} = \int_{x_1}^{x} \frac{dx}{\sqrt{\dfrac{2}{m}[E - U(x)]}} \tag{6.29}$$

Some understanding of the geometry of phase trajectories may be obtained with the aid of Fig. 6.10, which shows the dependence of phase trajectories on total energy for a hypothetical potential energy function $U(x)$. For $E = E_1$ the motion is periodic; zero velocity and velocity reversal occur where $U(x) = E_1$. With $E = E_2$ periodic oscillations are possible about two points; the initial conditions applicable in a given case dictate which type of oscillation occurs in that case. For $E = E_3$ only a single periodic motion is possible; for $E = E_4$ the motion is aperiodic. For $E = E_u$ there exists an instability at x_u; there the mass may move either in increasing or decreasing x direction. (The arrows on the phase trajectories point in the direction of increasing time.)

For a linear undamped system $f(x) = kx$, $U(x) = \tfrac{1}{2}kx^2$, and phase trajectories are ellipses with semiaxes $(2E/k)^{1/2}$, $(2E/m)^{1/2}$.

The following facts may be summarized for conservative systems:

1. Oscillatory motions occur about minima in $U(x)$.
2. Phase trajectories are symmetric about x axis and cross x axis perpendicularly.
3. If $f(x)$ is single-valued, the phase trajectories for different energies E do not intersect.
4. All finite motions are periodic.

For nonconservative systems the phase trajectories tend to cross the constant-energy trajectories for the corresponding conservative systems. For damped systems the trajectories tend toward lower energy; i.e., they spiral in to a point of stability. For excited systems the trajectories spiral outward, either toward a "limit-cycle" trajectory or indefinitely.[58]

The period T of an oscillation of a conservative system occurring with maximum

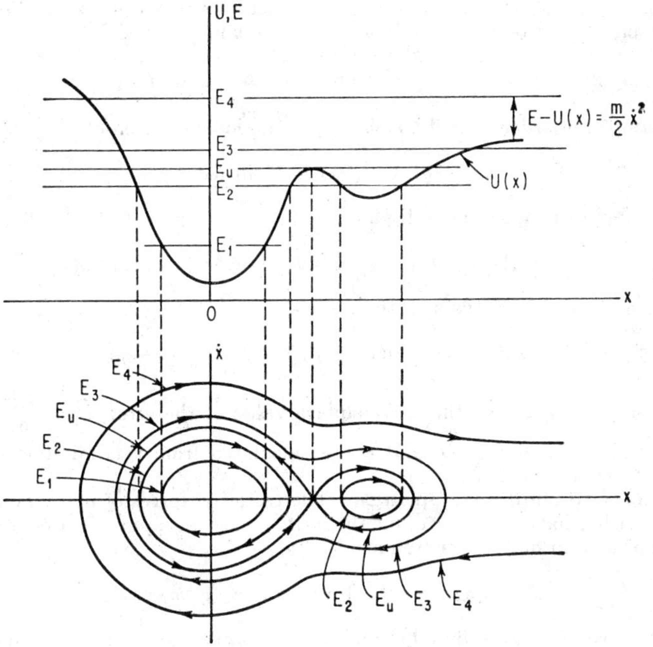

FIG. 6.10. Dependence of phase trajectories on energy.

displacement (amplitude) x_{max} may be computed from

$$T = \sqrt{2m} \int_{x_{min}}^{x_{max}} \left[\int_{x}^{x_{max}} f(x)\, dx \right]^{-1/2} dx \tag{6.30}$$

where x_{max} and x_{min} are the largest and smallest values of x (algebraically) for which zero velocity $\dot{x}$ occurs. The frequency f may then be obtained from $f = 1/T$.

For a linear spring-mass system with clearance, as shown in Fig. 6.11, the frequency is given by[17,35]

$$f = \frac{\sqrt{k/m}}{2\left[\pi + \dfrac{2}{(x_{max}/a) - 1}\right]}$$

where x_{max} is the maximum excursion of the mass from its middle position.

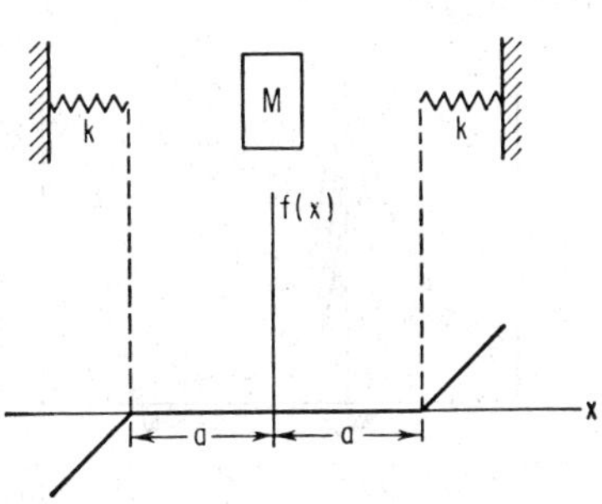

FIG. 6.11. Spring-mass system with clearance.

For a system governed by Eq. (6.27) with $f(x) = kx|x|^{b-1}$ [or $f(x) = kx^b$, if b is odd], the frequency is given by[35]

$$f = \sqrt{\frac{(b+1)k}{8\pi m} x_{max}{}^{b-1}} \; \frac{\Gamma[1/(b+1) + 1/2]}{\Gamma[1/(b+1)]}$$

in terms of the gamma function Γ, values of which are available in many tables.

Steady-state Periodic Responses. In many cases, particularly in steady-state analyses of periodically forced systems, periodic oscillatory solutions are of primary interest. A number of mathematical approaches are available to deal with these problems. Most of these, including the well-known methods of Stoker[58] and Schwesinger,[35,54] are based on the idea of "harmonic balance."[15] They essentially assume a Fourier expansion of the solution and then require the coefficients to be adjusted so that relevant conditions on the lowest few harmonic components are satisfied.

For example, in order to find a steady-state periodic solution of

$$m\ddot{x} + g(\dot{x}) + f(x) = F \sin (\omega t + \phi)$$

one may substitute an assumed displacement

$$x = A_1 \sin \omega t + A_2 \sin 2\omega t + \cdots + A_n \sin n\omega t$$

and impose certain restrictions on the error ϵ,

$$\epsilon(t) = m\ddot{x} + g(\dot{x}) + f(x) - F \sin (\omega t + \phi)$$

In Schwesinger's method[54] the mean-square value of the error, $\overline{\epsilon^2} = \int_0^{2\pi} \epsilon^2(t)d(\omega t)$, is minimized, and values of F and ϕ are calculated from this minimization corresponding to an assumed A_1.

Systems with Nonlinear Springs. The restoring forces of many systems (particularly with small amounts of nonlinearity) may be approximated so that the equation of motion may be written as

$$\ddot{x} + 2\zeta\omega_0\dot{x} + \omega_0^2 x + (a/m)x^3 = (F/m) \cos \omega t \tag{6.31}$$

in the presence of viscous damping and a sinusoidal force. ζ is the damping factor and ω_0 the natural frequency of a corresponding undamped linear system (i.e., for $a = 0$). For $a > 0$ the spring becomes stiffer with increasing deflection and is called "hard;" for $a < 0$ the spring becomes less stiff and is called "soft."

Figure 6.12 compares the responses of linear and nonlinear lightly damped spring systems. The responses are essentially of the form $x = A \cos \omega t$; curves of response amplitude A vs. forcing frequency ω are sketched for several values of forcing amplitude F, for constant damping ζ. For a linear system the frequency of free oscillations ($F = 0$) is independent of amplitude; for a hard system it increases; for a soft system it decreases with increasing amplitude. The response curves of the nonlinear spring systems may be visualized as "bent-over" forms of the corresponding curves for the linear systems.

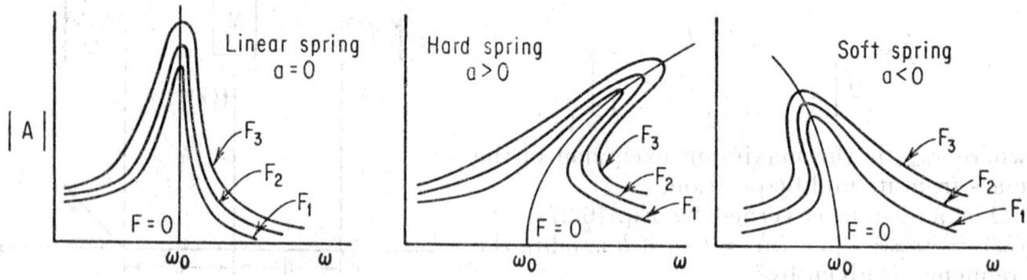

FIG. 6.12. Comparison of frequency responses of linear and nonlinear systems.

As apparent from Fig. 6.12, the response curves of the nonlinear spring systems are triple-valued for some frequencies. This fact leads to "jump" phenomena, as sketched in Figs. 6.13 and 6.14. If a given force amplitude is maintained as forcing frequency is changed slowly, then the response amplitude follows the usual response curve until point 1 of Fig. 6.13 is reached. The hatched regions between points 1 and 3 correspond to unstable conditions; an increase in ω above point 1 causes the amplitude to jump to that corresponding to point 2. A similar condition occurs when frequency is slowly decreased; the jump then occurs between points 3 and 4.

As also evident from Fig. 6.12, a curve of response amplitude vs. force amplitude at constant frequency is also triple-valued in some regions of frequency. Thus amplitude jumps occur also when one changes the forcing amplitude slowly at constant frequency ω_c. This condition is sketched in Fig. 6.14. For a hard spring this can occur only at frequencies above ω_0, for soft springs below ω_0. The equations characterizing a lightly damped nonlinear spring system and its jumps are summarized in Fig. 6.15.

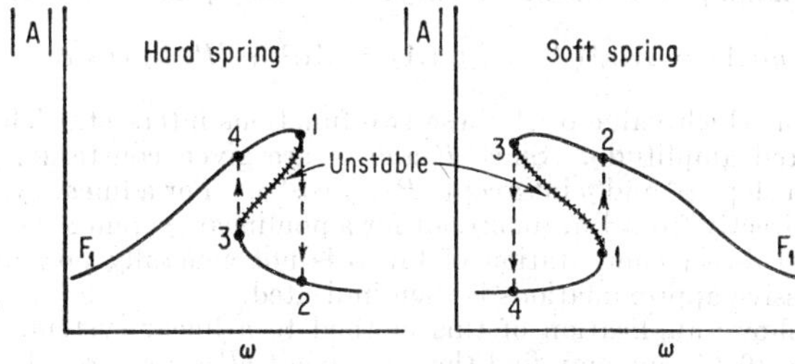

FIG. 6.13. Jump phenomena with variable frequency and constant force.

The previous discussion deals with the system response as if it were a pure sinusoid $x = A \cos \omega t$. However, in nonlinear systems there occur also harmonic components (at frequencies $n\omega$; where n is an integer) and subharmonic components (frequencies ω/n). In addition, components occur at frequencies which are integral multiples of the subharmonic frequencies. The various harmonic and subharmonic components tend to be small for small amounts of nonlinearity, and damping tends to limit the

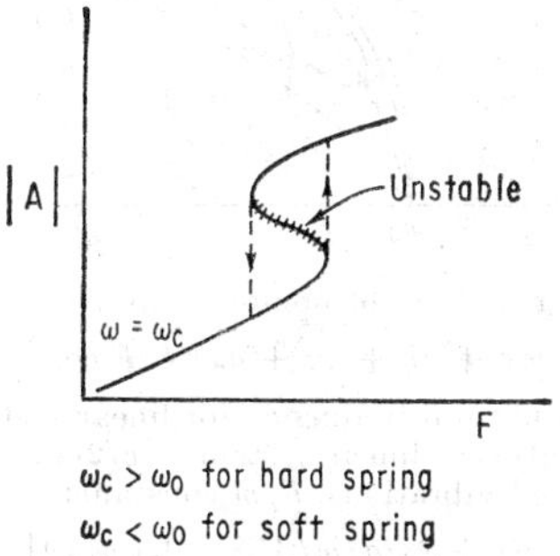

FIG. 6.14. Jump phenomena with variable force and constant frequency.

occurrence of subharmonics. The amplitude of the component with frequency 3ω (the lowest harmonic above the fundamental with finite amplitude for an undamped system) is given by $aA^3/36m\omega^2$, where A is the amplitude of the fundamental response, in view of Eq. (6.31). For more complete discussions see refs. 15 and 18.

Graphical Determination of Response Amplitudes. A relatively easily applied method for approximating response amplitudes was developed by Martienssen[38] and improved by Mahalingam.[37] It is based on the often observed fact that the response to sinusoidal excitation is essentially sinusoidal. The method is here first explained for a linear system, then illustrated for nonlinear ones.

In order to obtain the steady-state response of a linear system one substitutes an assumed trial solution

$$x = A \cos (\omega t - \phi)$$

into the equation of motion

$$m\ddot{x} + c\dot{x} + kx = P \cos \omega t$$

By equating coefficients of corresponding terms on the two sides of the resulting equation one obtains a pair of equations which may be solved to yield

$$\omega_n{}^2 A = \omega^2 A + (P/m) \cos \phi \qquad \tan \phi = \frac{c\omega/m}{(\omega_n/\omega)^2 - 1} \tag{6.32}$$

where ω_n is the undamped natural frequency. One may plot the functions

$$y_1(A) = A\omega_n{}^2(A) \qquad y_2(A) = A\omega^2 + (P/m) \cos \phi$$

and determine for which value of A these two functions intersect. This value of A then is the desired amplitude. Since P and m are given constants, y_2 plots as a straight line with slope ω^2 and y intercept $(P/m) \cos \phi$. For a linear system one may compute $\tan \phi$ directly from Eq. (6.32) but for a nonlinear system $\omega_n{}^2$ depends on the amplitude A and direct computation of $\tan \phi$ is not generally possible. Use of a method of successive approximations is then indicated.

Figure 6.16*a* shows application of this method to a linear system. After calculating ϕ from Eq. (6.32) one may find the y intercept $(P/m) \cos \phi$. For a given fre-

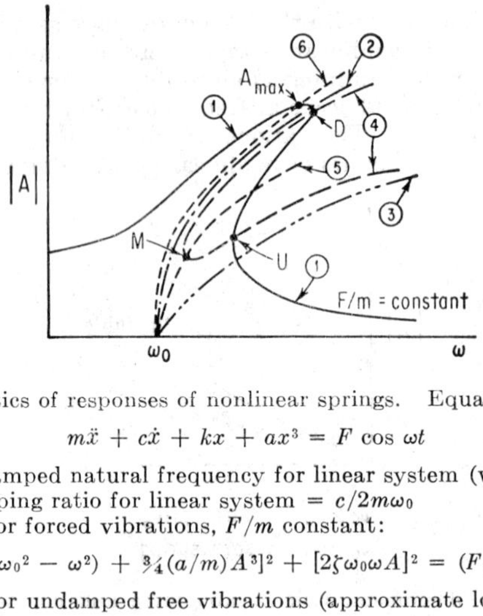

FIG. 6.15. Characteristics of responses of nonlinear springs. Equation of motion:

$$m\ddot{x} + c\dot{x} + kx + ax^3 = F \cos \omega t$$

ω_0 = undamped natural frequency for linear system (with $a = 0$)
ζ = damping ratio for linear system $= c/2m\omega_0$

1. Response curve for forced vibrations, F/m constant:

$$[A(\omega_0{}^2 - \omega^2) + \tfrac{3}{4}(a/m)A^3]^2 + [2\zeta\omega_0\omega A]^2 = (F/m)^2$$

2. Response curve for undamped free vibrations (approximate locus of downward jump points D, and of A_{max}):

$$\omega^2 = \omega_0{}^2 + \tfrac{3}{4}(a/m)A^2$$

3. Locus of upward jump points U with zero damping; approximate locus of same with finite damping:

$$\omega^2 = \omega_0{}^2 + \tfrac{9}{4}(a/m)A^2$$

4. Locus of upward and downward jump points U and D with finite damping:

$$[\omega_0{}^2 - \omega^2 + \tfrac{3}{4}(a/m)A^2][\omega_0{}^2 - \omega^2 + \tfrac{9}{4}(a/m)A^2] + (2\zeta\omega_0\omega)^2 = 0$$

5. Locus of points M below which no jumps occur:

$$\omega^2 = \omega_0{}^2 + \tfrac{9}{8}(a/m)A^2$$

6. Locus of maximum amplitudes A_{max}:

$$\omega^2 - \tfrac{3}{4}(a/m)A^2 = \omega_0{}^2(1 - 2\zeta^2)$$

quency ω one may then draw a line of slope ω^2 through that intercept to represent the function y_2. For a linear system y_1 is a straight line with slope ω_n^2 and passing through the origin. The amplitude of the steady-state oscillation may then be determined as the value A_0 of A where the two lines intersect.

Figure 6.16*b* shows a diagram analogous to Fig. 6.16*a* but for an arbitrary nonlinear system. The function $y_1 = A\omega_n^2$ is not a straight line in general since ω_n generally is a function of A. This function may be determined from the restoring function $f(x)$ by use of Eq. (6.30). The possible amplitudes corresponding to a given driving frequency ω and force amplitude P are determined here, as before, by the intersection of the y_1 and y_2 curves. As shown in the figure, more than one amplitude may correspond to a given frequency—a condition often encountered in nonlinear systems.

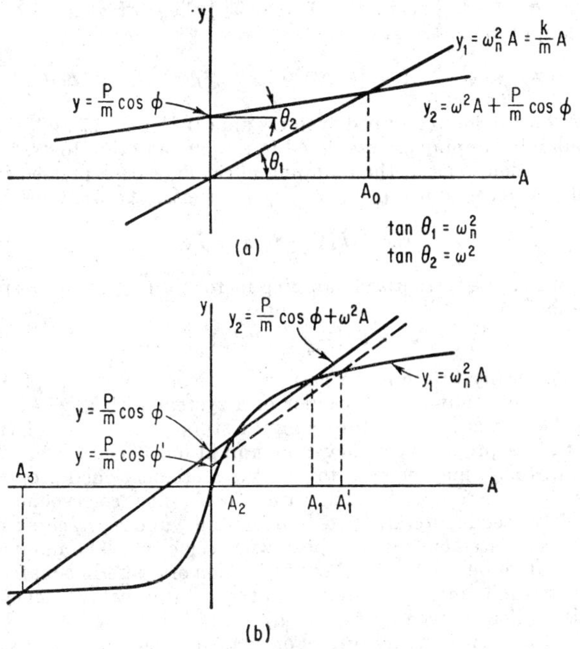

FIG. 6.16. Graphical determination of amplitude. (*a*) Linear system. (*b*) Nonlinear system.

In applying the previously outlined method to nonlinear systems one can generally not find the correct value of ϕ at once from Eq. (6.32), since ω_n depends on the amplitude A, as has been pointed out. Instead one may assume any value of ϕ, such as ϕ', and determine a first approximation A_1' to A_1. Using this approximate amplitude one may then determine better values of $\omega_n(A)$ and ϕ from Eq. (6.32) and then use these better values to obtain a better approximation to A_1. This process may be repeated until ϕ and A_1 have been found to the desired degree of accuracy. A separate iteration process of this sort is generally required for each of the possible amplitudes. (Different values of ϕ correspond to the different amplitudes A_1, A_2, A_3.)

Systems with Nonlinear Damping. The governing equation in this case may be written

$$m\ddot{x} + C(x,\dot{x}) + kx = F \sin \omega t \tag{6.33}$$

where $C(x,\dot{x})$ represents the effect of damping. Exact or reasonably good approximate solutions are available only for relatively few cases. However, for many practical cases where the damping is not too great, the system response is essentially sinusoidal. One may then use an "equivalent" viscous-damping term $c_e\dot{x}$ instead of $C(x,\dot{x})$, so that the equivalent damping does the same work per cycle as the original nonlinear damping. For a response of the form $x = A \sin \omega t$ the equivalent viscous damping may be computed from

$$c_e = (1/\pi\omega A) \int_0^{2\pi} C(A \sin \omega t, A\omega \cos \omega t) d(\omega t) \tag{6.34}$$

In contrast to the usual linear case, c_e is generally a function of frequency and amplitude. Once c_e has been found the response amplitude may then be computed from Eq. (6.6); successive approximations must be used if c_e is amplitude-dependent.

For "dry" or Coulomb friction, where the friction force is $\pm\mu$ (constant in magnitude but always directed opposite to the velocity) the equivalent viscous damping c_e and response amplitude A are given by

$$c_e = 4\mu/\pi A\omega \qquad A = [F/k(1 - \omega^2/\omega_n^2)] \sqrt{1 - [(4/\pi)(\mu/F)]^2}$$

Further details on Coulomb damped systems appear in refs. 17 and 62. Free vibrations with Coulomb damping are discussed in ref. 59 and other texts.

If complex notation is used, the damping effect may be expressed in terms of an imaginary stiffness term, and Eq. (6.33) may alternately be written

$$m\ddot{x} + k(1 + i\gamma)x = Fe^{i\omega t}$$

where γ is known as the "structural damping factor."[12,21,50] For a response given by $x = Ae^{i\omega t}$ one finds

$$|A| = F/|-m\omega^2 + k(1 + i\gamma)| \tag{6.35}$$

This reduces identically to the linear case with viscous damping c, if γ is defined so that $k\gamma = \omega c$. The steady-state behavior of a system with any reasonable type of damping may be represented by this complex stiffness concept, provided that γ is prescribed with the proper frequency and amplitude dependence. The foregoing relation and Eq. (6.34) may be used to find the aforementioned proper dependences for a given resisting force $C(x,\dot{x})$. The case of constant γ corresponds to "structural damping" (widely used in aircraft flutter calculations) and represents a damping force proportional to displacement but in phase with velocity. The damping factor γ is particularly useful for describing the damping action of rubberlike materials, for which the damping is virtually independent of amplitude (but not of frequency),[50] since then the response is explicitly given by Eq. (6.35).

Response to Random Excitation. For quantitative results the reader is referred to refs. 2, 7, 33, and 34.

Qualitatively, the response of a linear system to random excitation is essentially a sinusoid at the system's natural frequency. The amplitudes (i.e., the envelope of this sinusoid) vary slowly and have a Rayleigh distribution. Compared with a linear system a system with a "hard" spring has a higher natural frequency, a lower probability of large excursions, and waves with flattened peaks. ("Soft" spring systems exhibit opposite characteristics.) The effects of the nonlinearities on frequency and on the wave shape are generally very small.

Self-excited Systems. If the damping coefficient c of a linear system is negative, the system tends to oscillate with ever-increasing amplitude. Positive damping extracts energy from the system; negative damping contributes energy to it. A system (such as one with negative damping) for which the energy-contributing forces are controlled by the system motion is called self-excited. A source of energy must be available if a system is to be self-excited. The steady-state amplitude of a self-excited oscillation may generally be determined from energy considerations, i.e., by requiring the total energy dissipated per cycle to equal the total energy supplied per cycle.

The chatter of cutting tools, screeching of hinges or locomotive wheels, and the chatter of clutches are due to self-excited oscillations associated with friction forces which decrease with increasing relative velocity. The larger friction forces at lower relative velocities add energy to the system; smaller friction forces at higher velocities (more slippage) remove energy. While the oscillations build up the energy added is greater than that removed; at steady state in each cycle the added energy is equal to the extracted energy.

More detailed discussions of self-excited systems may be found in refs. 15 and 58.

6.2. SYSTEMS WITH A FINITE NUMBER OF DEGREES OF FREEDOM

The instantaneous configurations of many physical systems can be specified by means of a finite number of coordinates. Continuous systems, which have an infinite number of degrees of freedom, can be approximated for many purposes by systems with only a finite number of degrees of freedom, by "lumping" of stiffnesses, masses, and distributed forces.

(a) Systematic Determination of Equations of Motion

Generalized Coordinates; Constraints.[52,55,63] A set of n quantities $q_i (i = 1, 2, \ldots, n)$ which at any time t completely specify the configuration of a system are called *generalized coordinates* of the system. The quantities may or may not be usual space coordinates.

If one selects more generalized coordinates than the minimum number necessary to describe a given system fully, then one finds some interdependence of the selected coordinates dictated by the geometry of the system. This interdependence may be expressed as

$$G(q_1,q_2, \ldots ,q_n;\dot{q}_1,\dot{q}_2, \ldots ,\dot{q}_n;t) = 0 \qquad (6.36)$$

Relations like those of Eq. (6.36) are known as *equations of constraint;* if no such equations can be formulated for a given set of generalized coordinates, the set is known as *kinematically independent.*

The constraints of a set of generalized coordinates are said to be integrable if all equations like (6.36) either contain no derivatives $\dot{q}_i$ or if such $\dot{q}_i$ that do appear can be eliminated by integration. If to the set of n generalized coordinates there correspond m constraints, all of which are integrable, then one may find a new set of $(n - m)$ generalized coordinates which are subject to no constraints. This new system is called *holonomic,* and $(n - m)$ is the number of degrees of freedom of the system (that is, the smallest number of quantities necessary to describe the system configuration at any time). In practice one may often be able to select a holonomic system of generalized coordinates by inspection. Henceforth the discussion will be limited to holonomic systems.

Lagrangean Equations of Motion. The equations governing the motion of any holonomic system may be obtained by application of Lagrange's equation

$$(d/dt)(\partial T/\partial \dot{q}_i) - \partial T/\partial q_i + \partial U/\partial q_i + \partial F/\partial \dot{q}_i = Q_i \qquad i = 1, 2, \ldots, n \qquad (6.37)$$

where T denotes the kinetic energy, U the potential energy of the entire dynamical system, F is a dissipation function, and Q_i the generalized force associated with the generalized coordinate q_i.

The generalized force Q_i may be obtained from

$$\delta W = Q_i \, \delta q_i \qquad (6.38)$$

where δW is the total work done on the system by all external forces not contributing to U when the single coordinate q_i is changed to $q_i + \delta q_i$. The potential energy U accounts only for forces which are "conservative" (that is, for those forces for which the work done in a displacement of the system is a function of only the initial and final configurations). The dissipation function F represents half the rate at which energy is lost from the system; it accounts for the dissipative forces that appear in the equations of motion.

Lagrange's equations can be applied to nonlinear as well as linear system , but little can be said about solving the resulting equations of motion if they are nonlinear, except that small oscillations of nonlinear systems about equilibrium can always be approximated by linear equations. Methods of solving linear sets of equations of motion are available and are discussed subsequently.

(b) Matrix Methods for Linear Systems—Formalism

The kinetic energy T, potential energy U, and dissipation function F of any *linear* holonomic system with n degrees of freedom may be written as

$$
\begin{aligned}
T &= \tfrac{1}{2} \sum_{i=1}^{n} \sum_{j=1}^{n} a_{ij}\dot{q}_i\dot{q}_j \qquad a_{ij} = a_{ji} \\
U &= \tfrac{1}{2} \sum_{i=1}^{n} \sum_{j=1}^{n} c_{ij}q_iq_j \qquad c_{ij} = c_{ji} \\
F &= \tfrac{1}{2} \sum_{i=1}^{n} \sum_{j=1}^{n} b_{ij}\dot{q}_i\dot{q}_j \qquad b_{ij} = b_{ji}
\end{aligned}
\tag{6.39}
$$

The equations of motion may then readily be determined by use of Eqs. (6.37). They may be expressed in matrix form as

$$A\{\ddot{q}\} + B\{\dot{q}\} + C\{q\} = \{Q(t)\} \tag{6.40}$$

where A, B, C are symmetric square matrices with n rows and n columns whose elements are the coefficients appearing in Eqs. (6.39) and $\{q\}$, $\{\dot{q}\}$, $\{\ddot{q}\}$, $\{Q(t)\}$ are n-dimensional column vectors.* The elements of $\{q\}$ are the coordinates q_i, the elements of $\{\dot{q}\}$ are the first time derivatives of q_i (i.e., the generalized velocities $\dot{q}_i$); those of $\{\ddot{q}_i\}$ are the generalized accelerations $\ddot{q}_i$; those of $\{Q\}$ are the generalized forces Q_i.

Free Vibrations—General System. If all the generalized forces Q_i are zero, then Eq. (6.40) reduces to a set of homogeneous linear differential equations. To solve it one may postulate a time dependence given by

$$\{q\} = \{r\}e^{st} \qquad s = \sigma + i\omega$$

which, when introduced into Eq. (6.40), results in

$$(s^2A + sB + C)\{r\} = 0 \tag{6.41}$$

One may generally find n nontrivial solutions $\{r^{(j)}\}$, $j = 1, 2, \ldots, n$, each corresponding to a specific value $s_{(j)}$ of s. The general solution of the homogeneous equation may then be expressed as

$$\{q\} = \sum_{j=1}^{n} \alpha_j\{r^{(j)}\}e^{s_{(j)}t} = R\{p\} \tag{6.42}$$

* A is called the inertia matrix, B the damping matrix, C the elastic or the stiffness matrix. For example,

$$
A = \begin{bmatrix} a_{11} & a_{12} & \cdots & a_{1n} \\ a_{21} & a_{22} & \cdots & a_{2n} \\ \cdot & \cdot & & \cdot \\ \cdot & \cdot & & \cdot \\ \cdot & \cdot & & \cdot \\ a_{n1} & a_{n2} & & a_{nn} \end{bmatrix}
\qquad
\{q\} = \begin{bmatrix} q_1 \\ q_2 \\ \cdot \\ \cdot \\ \cdot \\ q_n \end{bmatrix}
\qquad
\{\dot{q}\} = \begin{bmatrix} \dot{q}_1 \\ \dot{q}_2 \\ \cdot \\ \cdot \\ \cdot \\ \dot{q}_n \end{bmatrix}
\qquad
\{Q\} = \begin{bmatrix} Q_1 \\ Q_2 \\ \cdot \\ \cdot \\ \cdot \\ Q_n \end{bmatrix}
$$

in terms of complex constants α_j and the newly defined

$$R = \begin{bmatrix} r_1^{(1)} & r_1^{(2)} & \cdots & r_1^{(n)} \\ r_2^{(1)} & r_2^{(2)} & \cdots & r_2^{(n)} \\ \cdot & & & \\ \cdot & & & \\ \cdot & & & \\ r_n^{(1)} & r_n^{(2)} & \cdots & r_n^{(n)} \end{bmatrix} \qquad \{p(t)\} = \begin{bmatrix} \alpha_1 & e^{s(1)t} \\ \alpha_2 & e^{s(2)t} \\ \cdot & \\ \cdot & \\ \cdot & \\ \alpha_n & e^{s(n)t} \end{bmatrix} \tag{6.43}$$

Initial conditions, e.g., $\{q(0)\}$, $\{\dot{q}(0)\}$ may be introduced into Eq. (6.42) to evaluate the constants α_j.

Forced Vibrations—General System. The forced motion may be described in terms of the sum of two motions, one satisfying the homogeneous equation (with all $Q_i = 0$) and including all the constants of integration, the other satisfying the complete Eq. (6.40) and containing no integration constants. (The constants must be evaluated so that the total solution satisfies the prescribed initial conditions.) The latter constant-free solution is often called the *steady-state solution.*

Steady-state Solution for Periodic Generalized Forces. If the generalized forces are harmonic with frequency ω_0 one may set

$$\{Q\} = \{\bar{Q}\}e^{i\omega_0 t} \qquad \{q\} = \{\bar{q}\}e^{i\omega_0 t} \tag{6.44}$$

in Eq. (6.40) and obtain

$$(-\omega_0^2 A + i\omega_0 B + C)\{\bar{q}\} = \{\bar{Q}\}$$

from which $\{\bar{q}\}$ may be determined. Equation (6.44) then gives the steady-state solutions.

If the Q_i are periodic with period T, but not harmonic, one may expand them and the components q_i of the steady-state solution in Fourier series:

$$\{Q\} = \sum_{N=-\infty}^{\infty} \{\bar{Q}^{(N)}\}e^{iN\omega_0 t} \qquad \{q\} = \sum_{N=-\infty}^{\infty} \{\bar{q}^{(N)}\}e^{iN\omega_0 t} \qquad \omega_0 = T/2\pi$$

The Fourier components $\{\bar{q}^{(N)}\}$ may be evaluated from

$$(-N^2\omega_0^2 A + iN\omega_0 B + C)\{\bar{q}^{(N)}\} = \{\bar{Q}^{(N)}\}$$

Steady-state Solution for Aperiodic Generalized Forces. For general $\{Q(t)\}$ one may write the solutions of Eq. (6.40) as

$$q_i(t) = \sum_{j=1}^{n} \int_0^t h_i^{(j)}(t - \tau)Q_j(\tau)\, d\tau \tag{6.45}$$

where $h_i^{(j)}(t)$ is the response of coordinate q_i to a unit impulse acting in place of Q_j. The "impulse response function" $h_i^{(j)}$ may be found from

$$h_i^{(j)}(t) = (d/dt)u_i^{(j)}(t)$$

where $u_i^{(j)}$ is the response of q_i to a unit step function acting in place of Q_j. (A unit step function is zero for $t < 0$, unity for $t > 0$.)

If one defines a square matrix $[H(t)]$ whose elements are $h_i^{(j)}$, one may write Eq. (6.45) alternately as

$$\{q(t)\} = \int_0^t [H(t - \tau)]\{Q(\tau)\}\, d\tau$$

where $\quad [H(t)] = (1/2\pi i) \int_{c+i\infty}^{c-i\infty} [T(s)]e^{st}\, ds \qquad [T(s)] = (s^2A + sB + C)^{-1}$

Undamped Systems. For undamped systems all elements of the B matrix of Eq. (6.40) are zero, and Eq. (6.41) may be rewritten in the classical eigenvalue form

$$E\{r\} = \omega^2\{r\} \qquad E = A^{-1}C \tag{6.46}$$

The eigenvalues $\omega_{(j)}$, i.e., the values for which nonzero solutions $r^{(j)}$ exist, are real. They are the natural frequencies; the corresponding solution vectors $r^{(j)}$ describe the mode shapes.

One may then find a set of principal coordinates ψ_i, in terms of which the equations of motion are uncoupled and may be written in the following forms:

$$\{\ddot{\psi}\} + \Omega\{\psi\} = R^{-1}A^{-1}\{Q\} \qquad \text{or} \qquad M\{\ddot{\psi}\} + K\{\psi\} = \Phi(t)$$

Here $$M = \bar{R}AR \qquad K = \bar{R}CR \qquad \{\Phi(t)\} = \bar{R}\{Q(t)\}$$

Ω is the diagonal matrix of natural frequencies

$$\Omega = \begin{bmatrix} \omega^2{}_{(1)} & 0 & \cdots & 0 \\ 0 & \omega^2{}_{(2)} & \cdots & 0 \\ \vdots & \vdots & & \vdots \\ 0 & 0 & & \omega^2{}_{(n)} \end{bmatrix}$$

R is given by Eq. (6.43), and $\bar{R}$ denotes the transpose of R.

The principal coordinates ψ_i are related to the original coordinates q_i by

$$\{q\} = R\{\psi\} \qquad \{\psi\} = R^{-1}\{q\} \tag{6.47}$$

The response of the system to any forcing function $\{Q(t)\}$ may be determined in terms of the principal coordinates from

$$\psi_j(t) = m_{jj}k_{jj} \int_0^t \Phi_j(\tau) \sin [\omega_{(j)}(t - \tau)]\, d\tau + \psi_j(0) \cos [\omega_{(j)}t] + [1/\omega_{(j)}]\dot{\psi}_j(0) \sin [\omega_{(j)}t] \tag{6.48}$$

The response in terms of the original coordinates q_i may then be obtained by substitution of the results of Eq. (6.48) into Eq. (6.47).

(c) Matrix Iteration Solution of Positive-definite Undamped Systems

Positive-definite Systems; Influence Coefficient and Dynamic Matrixes. A system is *positive-definite* if its potential energy U, as given by Eq. (6.39), is greater than zero for any $\{q\} \neq \{0\}$. Systems connected to a fixed frame are positive-definite; systems capable of motion (changes in the coordinates q_i) without increasing U are called *semidefinite*.[63] The latter motions occur without energy storage in the elastic elements and are called *rigid-body* motions or *zero modes.* (They imply zero natural frequency.)

Rigid-body motions are generally of no interest in vibration study. They may be eliminated by proper choice of the generalized coordinates or by introducing additional relations (constraints) among an arbitrarily chosen system of generalized coordinates by applying conservation-of-momentum concepts. Thus any system of generalized coordinates can be reduced to a positive-definite one.

For positive-definite linear systems C^{-1}, the inverse of the elastic matrix, is known as the influence coefficient matrix D. The elements of D are the influence coefficients; the typical element d_{ij} is the change in coordinate q_i due to a unit generalized force Q_j (applied statically), with all other Q's equal to zero. Since these influence coefficients can be determined from statics, one generally need not find C at all. It should be noted that for systems that are not positive-definite one cannot compute the influence coefficients from statics alone.

For iteration purposes it is useful to rewrite Eq. (6.46), the system equation of free sinusoidal motion, as

$$G\{r\} = (1/\omega^2)\{r\} \tag{6.49}$$

where $$\{q\} = \{r\}e^{i\omega t} \qquad G = C^{-1}A = DA = E^{-1} \tag{6.50}$$

The matrix G is called the "dynamic matrix" and is defined, as above, as the product of the influence coefficient matrix D and the inertia matrix A.

Iteration for Lower Modes. In order to solve Eq. (6.49), which is a standard eigenvalue matrix equation, numerically for the lowest mode one may proceed as follows: Assume any vector $\{r_{(1)}\}$; then compute $G\{r_{(1)}\} = \alpha_{(1)}\{r_{(2)}\}$, where $\alpha_{(1)}$ is a constant chosen so that one element (say, the first) of $\{r_{(2)}\}$ is equal to the corresponding element of $\{r_{(1)}\}$. Then find $G\{r_{(2)}\} = \alpha_{(2)}\{r_{(3)}\}$, with $\alpha_{(2)}$ chosen like $\alpha_{(1)}$ before. Repeat this process until $\{r_{(n+1)}\} \approx \{r_{(n)}\}$ to the desired degree of accuracy. The corresponding constant $\alpha_{(n)}$ which satisfies $G\{r_{(n)}\} = \alpha_{(n)}\{r_{(n+1)}\}$ then yields the lowest natural frequency ω_1 of the system and $\{r_{(n)}\}$ describes the shape of the corresponding (first) mode $\{r^{(1)}\}$. In view of Eq. (6.49)

$$\omega_1^2 = 1/\alpha_{(n)}$$

The second mode $\{r^{(2)}\}$ must satisfy the orthogonality relation

$$\{\overline{r^{(2)}}\}A\{r^{(1)}\} = 0 \qquad \text{or} \qquad \sum_{i=1}^{n}\sum_{j=1}^{n} r_i^{(2)}a_{ij}r_j^{(1)} = 0 \tag{6.51}$$

In order to obtain a vector that satisfies Eq. (6.51) from an arbitrary vector $\{r\}$ one may select $(n - 1)$ components of $\{r^{(2)}\}$ as equal to the corresponding components of $\{r\}$ and then compute the nth from Eq. (6.51). This process may be expressed as

$$\{r^{(2)}\} = S_1\{r\}$$

where S_1 is called the *first sweeping matrix*. S_1 appears like the identity matrix in n dimensions, except for one row which describes the interrelation Eq. (6.51). If A is diagonal one may take, for example,

$$S_1 = \begin{bmatrix} 0 & -\dfrac{a_{22}r_2^{(1)}}{a_{11}r_1^{(1)}} & -\dfrac{a_{33}r_3^{(1)}}{a_{11}r_1^{(1)}} & \cdots & \dfrac{a_{nn}r_n^{(1)}}{a_{11}r_1^{(1)}} \\ 0 & 1 & 0 & \cdots & 0 \\ 0 & 0 & 1 & \cdots & 0 \\ \cdot & \cdot & \cdot & & \\ \cdot & \cdot & \cdot & & \\ \cdot & \cdot & \cdot & & \\ 0 & 0 & 0 & \cdots & 1 \end{bmatrix}$$

To obtain the second lowest mode shape $\{r^{(2)}\}$ and the second lowest natural frequency ω_2 one may form $H_1 = GS_1$, and solve

$$H_1\{r\} = (1/\omega^2)\{r\} \tag{6.52}$$

by iteration. Since Eq. (6.52) is of the same form as Eq. (6.49) one may proceed here as previously discussed, i.e., by assuming a trial vector $\{r_{(1)}\}$, forming $H_1\{r_{(1)}\} = \alpha_{(1)}\{r_{(2)}\}$, so that one element of $\{r_{(2)}\}$ is equal to the corresponding element of $\{r_{(1)}\}$, then forming $H_1\{r_{(2)}\} = \alpha_{(2)}\{r_{(3)}\}$, etc. This process converges to $\{r^{(2)}\}$ and $\alpha = 1/\omega_2^2$.

The third mode $\{r^{(3)}\}$ similarly must satisfy

$$\{\overline{r^{(3)}}\}A\{r^{(1)}\} = 0 \qquad \{\overline{r^{(3)}}\}A\{r^{(2)}\} = 0$$

$$\text{or} \quad \sum_{i=1}^{n}\sum_{j=1}^{n} r_i^{(3)}a_{ij}r_j^{(1)} = 0 \qquad \sum_{i=1}^{n}\sum_{j=1}^{n} r_i^{(3)}a_{ij}r_j^{(2)} = 0 \tag{6.53}$$

One may thus select $n - 2$ components of $\{r^{(3)}\}$ as equal to the corresponding components of an arbitrary vector $\{r\}$ and adjust the remaining two components to satisfy Eq. (6.53). The matrix S_2 expressing this operation, or

$$\{r^{(3)}\} = S_2\{r\}$$

is called the *second sweeping matrix*. For diagonal A one possible form of S_2 is

$$S_2 = \begin{bmatrix} 0 & -\dfrac{a_{22}r_2^{(1)}}{a_{11}r_1^{(1)}} & -\dfrac{a_{33}r_3^{(1)}}{a_{11}r_1^{(1)}} & \cdots & -\dfrac{a_{nn}r_n^{(1)}}{a_{11}r_1^{(1)}} \\ 0 & -\dfrac{a_{22}r_2^{(2)}}{a_{11}r_1^{(2)}} & -\dfrac{a_{33}r_3^{(2)}}{a_{11}r_1^{(2)}} & \cdots & -\dfrac{a_{nn}r_n^{(2)}}{a_{11}r_1^{(2)}} \\ 0 & 0 & 1 & \cdots & 0 \\ \cdot & & & & \cdot \\ \cdot & & & & \cdot \\ \cdot & & & & \cdot \\ 0 & 0 & 0 & \cdots & 1 \end{bmatrix}$$

Then one may form $H_2 = GS_2$ and solve

$$H_2\{r\} = (1/\omega^2)\{r\}$$

by iteration. The process here converges to $\{r^{(3)}\}$ and $\alpha = 1/\omega_3{}^2$.

Higher modes may be treated similarly; each mode must be orthogonal to all the lower ones, so that $p - 1$ relations like Eq. (6.51) must be utilized to find the $(p - 1)$st sweeping matrix. Iteration on $H_{(p-1)} = GS_{(p-1)}$ then converges to the pth mode.

Iteration for the Higher Modes. The previously outlined process begins with the lowest natural frequency and works toward the highest. It is not very useful for the highest few modes because of the tedium and of the accumulation of rounding-off errors. Results for the higher modes can be obtained more simply and accurately by starting with the highest frequency and working toward lower ones.

The highest mode may be obtained by solving Eq. (6.46) directly by iteration. This is accomplished by assuming any trial vector $\{r_{(1)}\}$, forming $E\{r_{(1)}\} = \beta_{(1)}\{r_{(2)}\}$ with $\beta_{(1)}$ chosen so that one element of the result $\{r_{(2)}\}$ is equal to the corresponding element of $\{r_{(1)}\}$. Then one may form $E\{r_{(2)}\} = \beta_{(2)}\{r_{(3)}\}$ similarly, and continue until $\{r_{(p+1)}\} \approx \{r_{(p)}\}$ to within the required accuracy. Then $\{r_{(p)}\} = \{r^{(n)}\}$ and $\beta_{(p)} = \omega_n{}^2$.

The next-to-highest $[(n - 1)\text{st}]$ mode may be found by writing

$$\{r^{(n-1)}\} = T_1\{r\}$$

where T_1 is a sweeping matrix that "sweeps out" the nth mode. T_1 appears like the identity matrix, except for one row, which expresses the orthogonality relation

$$\overline{\{r^{(n-1)}\}}A\{r^{(n)}\} = 0 \qquad \text{or} \qquad \sum_{i=1}^{n}\sum_{j=1}^{n} r_i{}^{(n-1)}a_{ij}r_j{}^{(n)} = 0$$

Iterative solution of

$$J_1\{r\} = \omega^2\{r\} \qquad \text{where} \qquad J_1 = ET_1$$

then converges to $\{r^{(n-1)}\}$ and $\omega^2{}_{(n-1)}$.

The next lower modes may be obtained similarly, using other sweeping matrixes embodying additional orthogonality relations in complete analogy to Art. 6.2(c).

(d) Approximate Natural Frequencies of Conservative Systems

A conservative system is one which executes free oscillations without dissipating energy. The potential energy $\hat{U}$ that the system has at an instant when its velocity (and hence its kinetic energy) is zero must therefore be exactly equal to the kinetic energy $\hat{T}$ of the system when it occupies its equilibrium position (zero potential energy) during its oscillation.

For sinusoidal oscillations the (generalized) coordinates obey $q_j = \bar{q}_j \sin \omega t$ where

$\bar{q}_j$ is a constant (i.e., the amplitude of q_j). For linear holonomic systems, in view of Eq. (6.39),

$$\hat{T} = \tfrac{1}{2}\omega^2 \sum_{i=1}^{n} \sum_{j=1}^{n} a_{ij}\bar{q}_i\bar{q}_j \qquad \hat{U} = \tfrac{1}{2} \sum_{i=1}^{n} \sum_{j=1}^{n} c_{ij}\bar{q}_i\bar{q}_j$$

Rayleigh's Quotient. Rayleigh's quotient RQ, defined as[63]

$$RQ = \frac{\sum_{i=1}^{n} \sum_{j=1}^{n} c_{ij}\bar{q}_i\bar{q}_j}{\sum_{i=1}^{n} \sum_{j=1}^{n} a_{ij}\bar{q}_i\bar{q}_j} \tag{6.54}$$

is a function of $\{\bar{q}_1,\bar{q}_2, \ldots ,\bar{q}_n\}$. However, multiplication of each $\bar{q}_i$ by the same number does not change the value of RQ. Rayleigh's quotient has the following properties:

1. The value of RQ one obtains with any $\{\bar{q}_1,\bar{q}_2, \ldots ,\bar{q}_n\}$ always equals or exceeds the square of the lowest natural frequency of the system; $RQ \geq \omega_1^2$.
2. $RQ = \omega_n^2$ if $\{\bar{q}_i,\bar{q}_2, \ldots ,\bar{q}_n\}$ corresponds to the nth mode shape (eigenvector) of the system, but even fairly rough approximations to the eigenvector generally result in good approximations to ω_n^2.

For systems whose influence coefficients d_{ij} are known, one may substitute an arbitrary vector $\{\bar{q}\} = \{\bar{q}_1,\bar{q}_2, \ldots ,\bar{q}_n\}$ into the right-hand side of

$$\{\bar{q}\}_1 = \alpha G\{\bar{q}\}$$

where $G = DA$ is the dynamic matrix of Eq. (6.50) and α is an arbitrary constant. [This relation follows directly from Eq. (6.49).] The resulting vector $\{\bar{q}\}_1$, when substituted into Eq. (6.54), results in a value for RQ which is nearer to ω_1^2 than the value obtained by direct substitution of the arbitrary vector $\{\bar{q}\}$. Usually one obtains good results rapidly if one assumes $\{\bar{q}\}$ initially so as to correspond to the deflection of the system due to gravity (i.e., the "static" deflection).

Rayleigh-Ritz Procedure. An alternate method useful for obtaining improved approximations to ω_1^2 from Rayleigh's quotient is the so-called "Rayleigh-Ritz" procedure. It consists of computing RQ from Eq. (6.54) for a trial vector $\{\bar{q}\}$ made up of a linear combination of arbitrarily selected vectors

$$\{\bar{q}\} = \alpha_1\{\bar{q}\}_1 + \alpha_2\{\bar{q}\}_2 + \cdots$$

then minimizing RQ with respect to the coefficients α of the selected vectors. The resulting minimum value of RQ is approximately equal to ω_1^2. That is,

$$RQ \text{ (evaluated so that } \partial RQ/\partial\alpha_1 = \partial RQ/\partial\alpha_2 = \cdots = 0) \approx \omega_1^2$$

Dunkerley's Equation.[59] This equation states that

$$1/\omega_1^2 + 1/\omega_2^2 + \cdots + 1/\omega_n^2 = 1/\Omega_1^2 + 1/\Omega_2^2 + \cdots + 1/\Omega_n^2 \tag{6.55}$$

where ω_i denotes the ith natural frequency of an n-degree-of-freedom system, and Ω_i denotes the natural frequency that the ith inertia element would have if all others were removed from the system. Usually $\omega_n^2 \gg \cdots \gg \omega_2^2 \gg \omega_1^2$, so that the left-hand side of Eq. (6.55) is approximately equal to $1/\omega_1^2$ and Eq. (6.55) may be used directly for estimation of the fundamental frequency ω_1. In many cases the Ω_1 are obtainable almost by inspection, or by use of Table 6.8.

(e) Chain Systems

A chain system is one in which the inertia elements are arranged in series, so that each is directly connected only to the one preceding and the one following it. Shafts carrying a number of disks (or other rotational inertia elements) are the most common example and are discussed in more detail subsequently. Translational chain systems, as sketched in Fig. 6.17, may be treated completely analogously and hence will not be discussed separately.

Sinusoidal Steady-state Forced Motion. If the torque acting on the sth disk is $T_s e^{i\omega t}$, where T_s is a known complex number, then the equations of motion of the system may be written as

$$\begin{aligned}
(k_{12} - I_1\omega^2)\theta_1 - k_{12}\theta_2 &= T_1 \\
-k_{12}\theta_1 + (k_{12} + k_{23} - I_2\omega^2)\theta_2 - k_{23}\theta_3 &= T_2 \\
-k_{23}\theta_2 + (k_{23} + k_{34} - I_3\omega^2)\theta_3 - k_{34}\theta_4 &= T_3 \\
&\cdot \\
&\cdot \\
&\cdot \\
-k_{n-1,n}\theta_{n-1} + (k_{n-1,n} - I_n\omega^2)\theta_n &= T_n
\end{aligned} \tag{6.56}$$

where $\theta_s e^{i\omega t}$ describes the angular motion of the sth disk, as measured from equilibrium. [Damping in the system may be taken into account by assigning complex values to the k's, as in the last portion of Art. 6.1(b).]

One may solve this set of equations simply by using each equation in turn to eliminate one of the θ's, so that one may finally solve for the last remaining θ, then obtain the others by substitution of the determined value into the given equations. This procedure becomes prohibitively tedious if more than a few disks are involved.

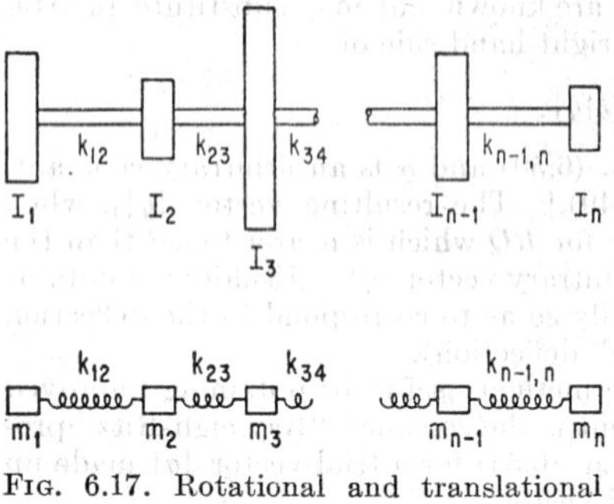

FIG. 6.17. Rotational and translational chain systems.

If all T's and k's are real (i.e., if the driving torques are in phase and if damping is neglected), one may assume a real value for θ_1, then calculate the corresponding value of θ_2 from the first of Eqs. (6.56). Then one may find θ_3 from the second equation, θ_4 from the third, and so on. Finally, one may compute T_n from the last (nth) equation and compare it with the given value. This process may be repeated with different initially assumed θ_1 values until the computed value of T_n comes out sufficiently close to the specified one. After a few computations one may often make good use of a plot of computed T_n vs. assumed θ_1 for determining by interpolation or extrapolation a good approximation to the correct value of θ_1.

If all T's are zero, except T_n, one may proceed as before. But, since θ_1 is proportional to T_n in this case, the correct value of θ_1 may be computed directly after a single complete calculation by use of the proportionality

$$\theta_{1,\text{correct}} = (\theta_{1,\text{assumed}})\left(\frac{T_{n,\text{specified}}}{T_{n,\text{calculated}}}\right)$$

The latter approach may be used also for damped systems (i.e., with complex k's).

Natural Frequencies; Holzer's Method. Free oscillations of the system considered obey Eqs. (6.56), but with all $T_s = 0$. To obtain the natural frequencies one may proceed by assuming a value of ω and setting $\theta_1 = 1$, then calculating θ_2 from the first equation, thereafter θ_3 from the second, etc. Finally one may compute T_n from the last equation. If this T_n comes out zero, as required, the assumed frequency is a natural frequency. By repeating this calculation for a number of assumed

Table 6.2. Mechanical Impedances of Simple Systems

SYSTEM	DIAGRAM	IMPEDANCE Z	FREQUENCY DEPENDENCE
MASS	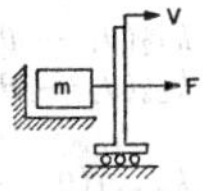	$i\omega m$	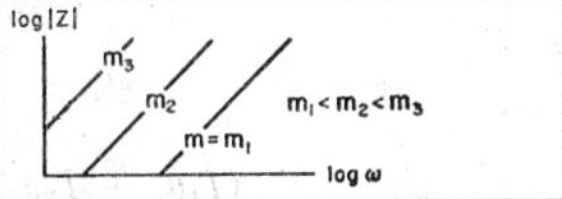
VISCOUS DASHPOT	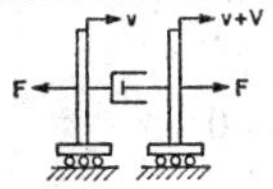	c	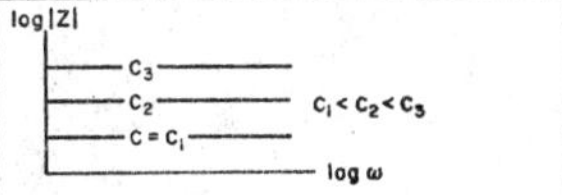
SPRING	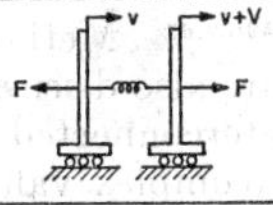	$k/i\omega$	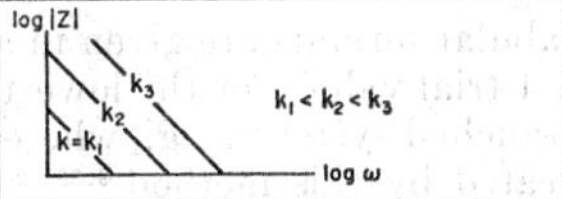
DRIVEN MASS ON SPRING 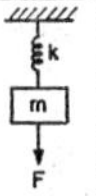	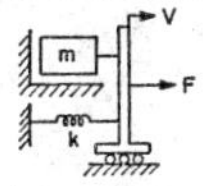	$\frac{k}{i\omega} + i\omega m$	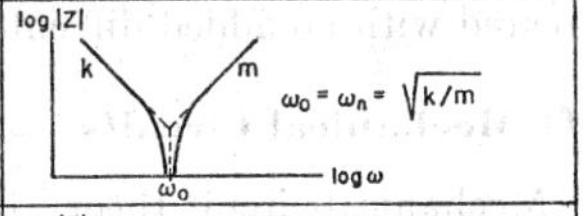
DAMPED MASS 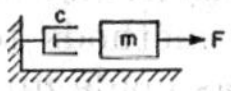	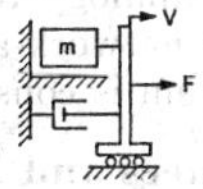	$c + i\omega m$	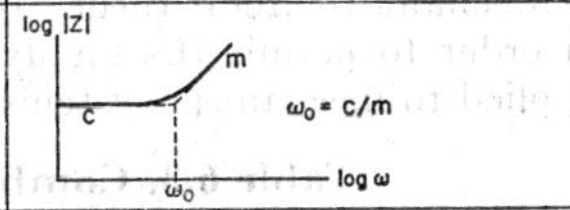
SPRING AND DASHPOT IN PARALLEL	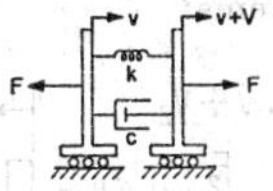	$c + k/i\omega$	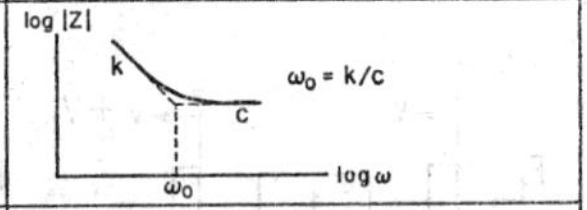
DAMPED SINGLE DEGREE OF FREEDOM SYSTEM	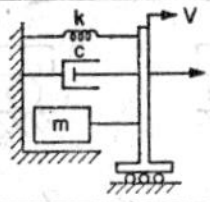	$i\omega m + c + \frac{k}{i\omega}$	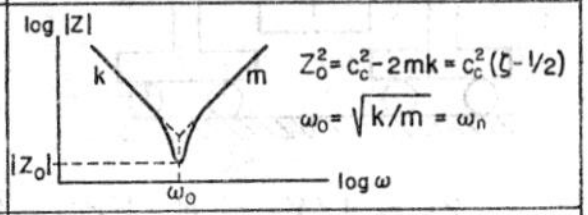
MASS ON DRIVEN SPRING 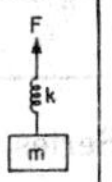	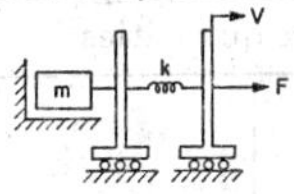	$\frac{1}{\frac{i\omega}{k} + \frac{1}{i\omega m}}$	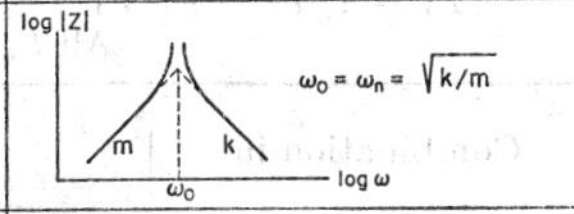
MASS DRIVEN THROUGH DASHPOT 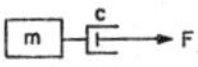	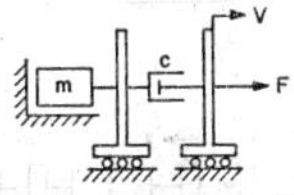	$\frac{1}{\frac{1}{c} + \frac{1}{i\omega m}}$	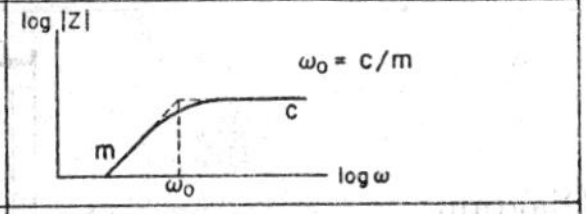
SPRING AND DASHPOT IN SERIES	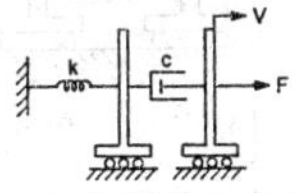	$\frac{1}{\frac{1}{c} + \frac{i\omega}{k}}$	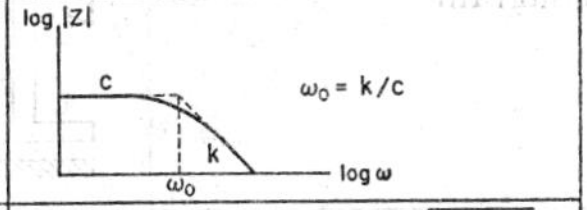
DAMPED MASS DRIVEN THROUGH SPRING 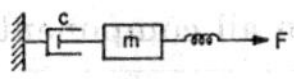	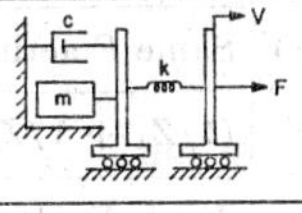	$\frac{1}{\frac{1}{c+i\omega m} + \frac{i\omega}{k}} = \frac{c_c\left[\zeta + \frac{i}{2}\frac{\omega}{\omega_n}\right]}{1 - \left(\frac{\omega}{\omega_n}\right)^2 + 2i\zeta\frac{\omega}{\omega_n}}$	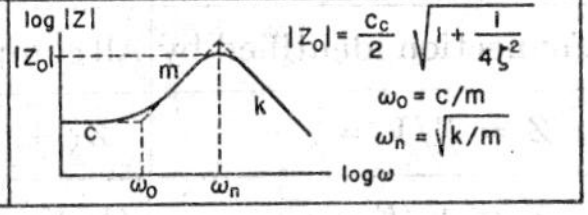

values of ω one may arrive at a plot of T_n vs. ω, which will aid in the estimation of subsequent trial values of ω. (Natural frequencies are obtained where this curve crosses the ω axis.) One should keep in mind that a system composed of n disks has n natural frequencies. The mode shape (i.e., a set of values of θ's that satisfy the equation of motion) is also obtained in the course of the calculations.

Convenient tabular calculation methods (Holzer tables) may be set up on the basis of the equations of motion Eq. (6.56) rewritten in the following form:

$$
\begin{aligned}
I_1\theta_1\omega^2 &= k_{12}(\theta_1 - \theta_2) \\
(I_1\theta_1 + I_2\theta_2)\omega^2 &= k_{23}(\theta_2 - \theta_3) \\
&\cdots \\
(I_1\theta_1 + I_2\theta_2 + \cdots + I_s\theta_s)\omega^2 &= k_{s,s+1}(\theta_s - \theta_{s+1}) \\
&\cdots \\
\sum_{s=1}^{n} (I_s\theta_s)\omega^2 &= 0
\end{aligned}
\tag{6.57}
$$

Tabular formats are given in a number of texts.[17,59,62,63] Methods for obtaining good first trial values for the lowest natural frequency are also discussed in refs. 17 and 59. Branched systems, e.g., where several shafts are interconnected by gears, may also be treated by this method.[9,59,62] Damped systems (complex values of k) may also be treated with no added difficulty in principle.[59]

(f) Mechanical Circuits

Mechanical-circuit theory is developed in direct analogy to electric-circuit theory in order to permit the highly developed electrical-network-analysis methods to be applied to mechanical systems. A mechanical system is considered as made up of a

Table 6.3. Combination of Impedances and Mobilities

Elements or subsystems:

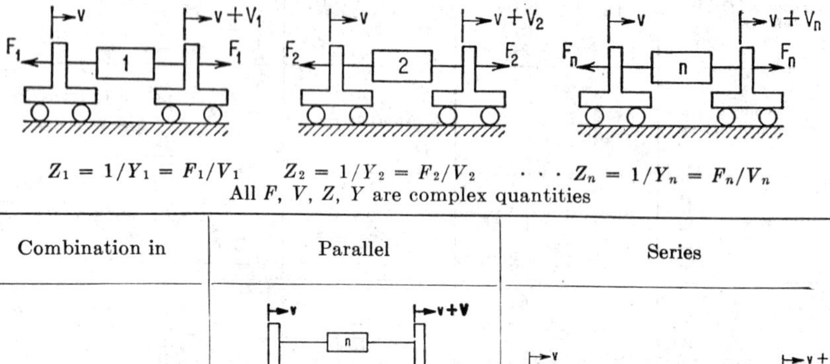

$Z_1 = 1/Y_1 = F_1/V_1 \qquad Z_2 = 1/Y_2 = F_2/V_2 \qquad \cdots \; Z_n = 1/Y_n = F_n/V_n$

All F, V, Z, Y are complex quantities

Combination in	Parallel	Series
Diagram	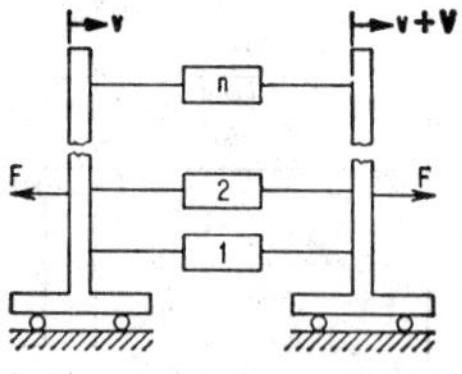	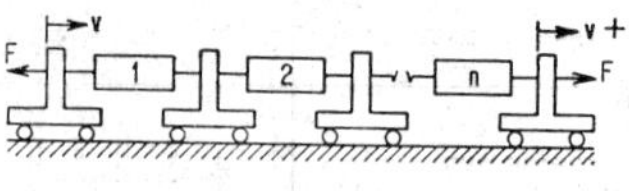
Connection identified by	All components having same V	Same F acting on all components
$Z = F/V =$	$Z_1 + Z_2 + \cdots + Z_n$	$(1/Z_1 + 1/Z_2 + \cdots + 1/Z_n)^{-1}$
$Y = V/F =$	$(1/Y_1 + 1/Y_2 + \cdots + 1/Y_n)^{-1}$	$Y_1 + Y_2 + \cdots + Y_n$
Z's combine like electrical resistances in	Series	Parallel
Y's combine like electrical resistances in	Parallel	Series

Table 6.4. Mechanical-electrical Analogies (Lumped Systems)[1]

Mechanical System		Electrical Systems	
Translational	Rotational	Force-voltage (classical) analog	Force-current (mobility) analog
F Force	T Torque	V Voltage	I Current
u Velocity	Ω Angular velocity	I Current	V Voltage
x Displacement	θ Angular displacement	q Charge	ϕ Magnetic flux
m Mass	J Moment of inertia	L Inductance	C Capacitance
c Damping	c_r Rotational damping	R Resistance	$G = \frac{1}{R}$ Conductance
k Spring rate	k_r Torsional spring rate	$S = \frac{1}{C}$ Elastance	$\Gamma = \frac{1}{L}$ Inverse inductance
$Z = \frac{F}{u}$ Mechanical impedance	$Z_r = \frac{T}{\Omega}$ Rotational impedance	$Z = \frac{V}{I}$ Impedance	$Y = \frac{I}{V}$ Admittance
$Y = \frac{u}{F}$ Mobility	$Y_r = \frac{\Omega}{T}$ Rotational mobility	$Y = \frac{I}{V}$ Admittance	$Z = \frac{V}{I}$ Impedance

number of mechanical-circuit elements (e.g., masses, springs, force generators) connected in series or parallel, in much the same way that an electrical network is considered to be made up of a number of interconnected electrical elements (e.g., resistances, capacitances, voltage sources). From known behavior of the elements one may then, by proper combination according to established rules, determine the system responses to given excitations.

Mechanical-circuit concepts are useful for determination of the equations of motion (which may then be solved by classical or transform techniques,[11,23,29]) for analyzing and visualizing the effects of system interconnections, for dealing with electromechanical systems, and for the construction of electrical analogs by means of which one may evaluate the responses by measurement.

Mechanical Impedance and Mobility. As in electric-circuit theory, the sinusoidal steady state is assumed, and complex notation is used in basic mechanical-impedance analysis. That is, forces F and relative velocities V are expressed as

$$F = F_0 e^{i\omega t} \qquad V = V_0 e^{i\omega t}$$

where F_0 and V_0 are complex in the most general case.

Table 6.4. Mechanical-electrical Analogies (Lumped Systems)[1] (Continued)

Mechanical Systems		Electrical Systems	
Translational	Rotational	Force-voltage (classical) analog	Force-current (mobility) analog
Kinetic energy = $\frac{1}{2}mu^2$	Kinetic energy = $\frac{1}{2}I\Omega^2$	Magnetic energy = $\frac{1}{2}LI^2$	Electrical energy = $\frac{1}{2}CV^2$
Potential energy = $\frac{1}{2}kx^2$	Potential energy = $\frac{1}{2}k_r\theta^2$	Electrical energy = $\frac{1}{2}\frac{1}{C}q^2$	Magnetic energy = $\frac{1}{2}\frac{1}{L}\phi^2$
Power loss = u^2c	Power loss = Ω^2c_r	Power loss = I^2R	Power loss = V^2G
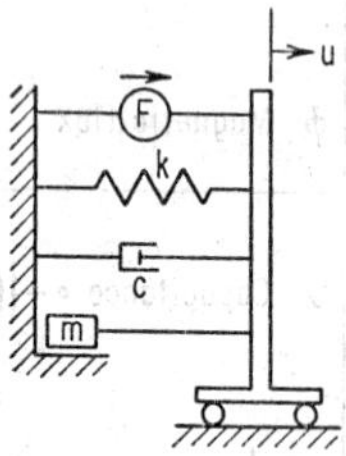	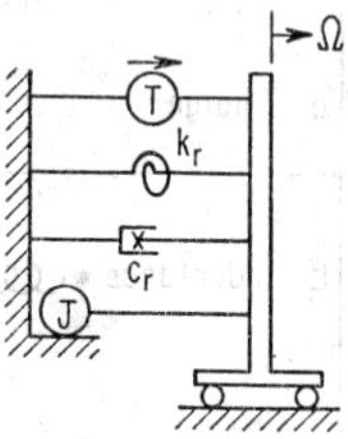	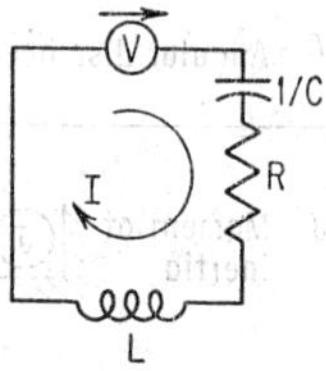	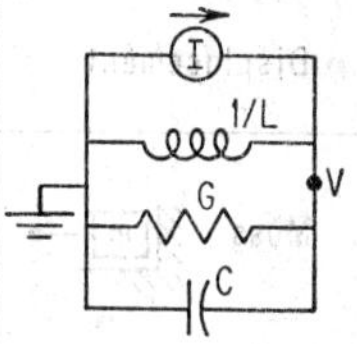
$m\dot{u}+cu+k\int u\,dt = F(t)$	$J\dot{\Omega}+c_r\Omega+k_r\int\Omega\,dt = T(t)$	$L\dot{I}+RI+\frac{1}{C}\int I\,dt = V(t)$	$C\dot{V}+RV+\frac{1}{L}\int V\,dt = I(t)$
Elements connected to same node have same velocity u	Elements connected to same node have same angular velocity Ω	Elements in same loop have same current I	Elements connected to same node have same voltage v
Element connected to ground has one end at zero velocity	Element connected to ground has one end at zero velocity	Element in one loop has only one loop current through it	Element connected to ground has one end at zero (reference) voltage
Elements between two nodes	Elements between two nodes	Elements in two loops	Elements between two nodes
Elements in parallel	Elements in parallel	Elements in series	Elements in parallel

Mechanical impedance Z and mobility Y are complex quantities defined by

$$Z = F_0/V_0 \qquad Y = 1/Z = V_0/F_0 \tag{6.58}$$

If the force and velocity refer to an element or system, the corresponding impedance is called the *impedance of the element* or system; if F and V refer to quantities at the same point of a mechanical network, then Z is called the *driving-point impedance* at that point. If F and V refer to different points, the corresponding Z is called the *transfer impedance* between those points.

Mechanical impedances (and mobilities) of elements can be combined exactly like electrical impedances (and admittances), and the driving-point impedances of composites can easily be obtained. From a knowledge of the elemental impedances and of how they combine one may calculate (and often estimate quickly) the behavior of composite systems.

Basic Impedances, Combination Laws, Analogies. Table 6.2 shows the basic mechanical-circuit elements and their impedances and summarizes the impedances of some simple systems.

In Table 6.3 are indicated the combination laws for mechanical impedances and mobilities. Table 6.4 is a summary of analogies between translational and rotational

Table 6.4. Mechanical-electrical Analogies (Lumped Systems)[1] (Continued)

Mechanical Systems		Electrical Systems	
Translational	Rotational	Force-voltage (classical) analog	Force-current (mobility) analog
Newton's law: sum of forces on node = 0	Newton's law: sum of torques on node = 0	Kirchhoff's voltage law: sum of voltages around loop = 0	Kirchhoff's current law: sum of currents into node = 0
Ideal lever 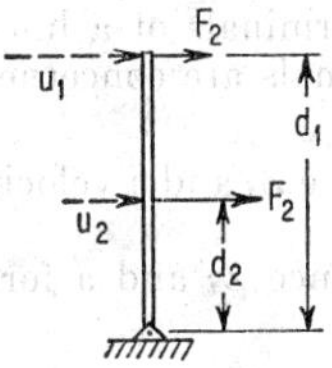	Ideal gear train 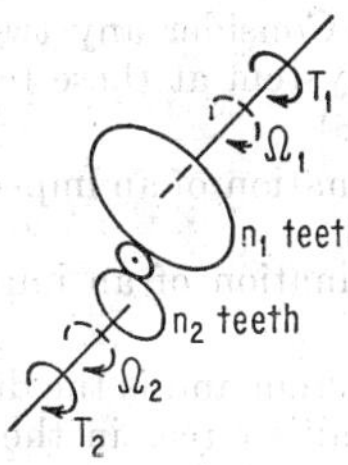	Ideal voltage transformer 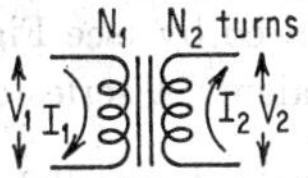	Ideal current transformer 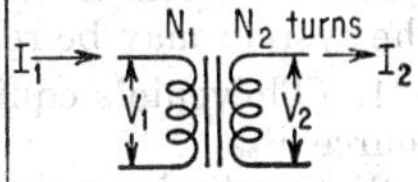
$\frac{u_1}{u_2} = \frac{F_2}{F_1} = \frac{d_1}{d_2}$	$\frac{\Omega_1}{\Omega_2} = \frac{T_2}{T_1} = \frac{n_2}{n_1}$	$\frac{I_1}{I_2} = \frac{V_2}{V_1} = \frac{N_2}{N_1}$	$\frac{V_1}{V_2} = \frac{I_2}{I_1} = \frac{N_1}{N_2}$

mechanical systems and electrical networks. The impedances of some distributed mechanical systems are discussed in Art. 6.3(b) and summarized in Tables 6.6 and 6.7.

Systems which are a combination of rotational and translational elements or electromechanical systems may be treated as either all-mechanical systems of a single type or as all-electrical systems by suitable substitution of analogous elements and variables.[23] In systems where both rotation and translation of a single mass occur, this single mass may have to be represented by two or more mass elements, and the concept of mutual mass (analogous to mutual inductance) may have to be introduced.[23]

Some Results from Electrical-network Theory. Resonances. Resonances occur at those frequencies for which the system impedances are minimum; antiresonances occur when the impedances are maximum.

Force and Velocity Sources. An ideal velocity generator supplies a prescribed relative velocity amplitude regardless of the force amplitude. A force generator supplies a prescribed force amplitude regardless of the velocity. When a velocity source

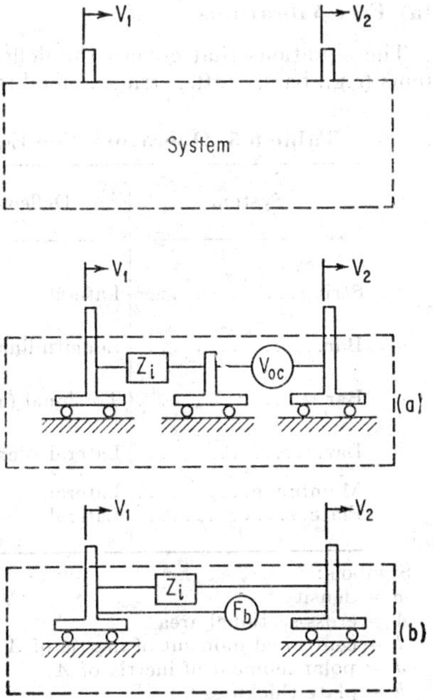

FIG. 6.18. (*a*) Thévenin's and (*b*) Norton's equivalent networks.

is "turned off," $V = 0$, it acts like a rigid connection between its terminals. When a force generator is turned off, $F = 0$, it acts like no connection between the terminals.

Reciprocity. The transfer impedance Z_{ij} (the force in the jth branch divided by the relative velocity of a generator in the ith branch) is equal to the transfer impedance Z_{ji} (force in the ith branch divided by relative velocity of generator in jth branch).

Foster's Reactance Theorem. For a general undamped system the driving-point impedance can be written $Z = if(\omega)$, where $f(\omega)$ is a real function of (real) ω. The function $f(\omega)$ always has $df/d\omega > 0$ and has a pole or zero at $\omega = 0$ and at $\omega = \infty$. All poles and zeros are simple (not repeated), poles and zeros alternate (i.e., there is always a zero between two poles), and $f(\omega)$ is determined within a multiplicative factor by its poles and zeros.

Thévenin's and Norton's Theorems. Consider any two terminals of a linear system. Then, as far as the effects of the system at these terminals are concerned, the system may be replaced by (see Fig. 6.18)

1. (Thévenin's equivalent) A series combination of an impedance Z_i and a velocity source V_{oc}
2. (Norton's equivalent) A parallel combination of an impedance Z_i and a force source F_b

Z_i is called the *internal impedance* of the system and is the driving-point impedance obtained at the terminals considered when all sources in the system are turned off (velocity generators replaced by rigid links, force generators replaced by disconnections). V_{oc} is the "open-circuit" velocity, i.e., the velocity occurring between the terminals considered (with all generators active). F_b is the "blocked force," i.e., the force transmitted through a rigid link inserted between the terminals considered, with all generators active.

6.3. CONTINUOUS LINEAR SYSTEMS

(a) Free Vibrations

The equations that govern the deflection $u(x,y,t)$ of many continuous linear systems (e.g., bars, shafts, strings, membranes, plates) in absence of external forces may

Table 6.5. Operators [See Eq. (6.59)] for Some Elastic Systems

System	Deflection	$\mathfrak{M} = \mu$	$\mathcal{K}$
String	Lateral	$\rho A(x)$	$T \dfrac{\partial^2}{\partial x^2}$
Bar	Longitudinal	$\rho A(x)$	$-\dfrac{\partial}{\partial x}\left(EA\dfrac{\partial}{\partial x}\right)$
Bar	Torsional (angular)	$\rho J(x)$	$-\dfrac{\partial}{\partial x}\left(KG\dfrac{\partial}{\partial x}\right)$
Bar	Lateral (flexure)	$\rho A(x)$	$\dfrac{\partial^2}{\partial x^2}\left(EI\dfrac{\partial^2}{\partial x^2}\right)$
Membrane	Lateral	$\rho h(x,y)$	$-S\nabla^2$
Plate	Lateral	$\rho h(x,y)$	$\nabla^2(D\nabla^2)$

Symbols:

ρ = density
A = cross-sectional area
I = centroidal moment of inertia of A
J = polar moment of inertia of A
h = plate thickness
T = tensile force
S = tension/unit length
E = Young's modulus
G = shear modulus
K = torsional constant of A (= J for circular sections)
$D = Eh^3/12(1 - \nu^2)$
ν = Poisson's ratio

be expressed as

$$\mathfrak{M}\ddot{u} + \mathfrak{K}u = 0 \tag{6.59}$$

where $\mathfrak{M}$ and $\mathfrak{K}$ are linear differential operators involving the coordinate variables only. Table 6.5 lists $\mathfrak{M}$ and $\mathfrak{K}$ for a number of common systems. Solutions of Eq. (6.59) may be expressed in terms of series composed of terms of the form $\phi(x,y)e^{i\omega t}$, where ϕ satisfies

$$(\mathfrak{K} - \mathfrak{M}\omega^2)\phi = 0 \tag{6.60}$$

in addition to the boundary conditions of a given problem. In solving the differential equation (6.60) by use of standard methods and introducing the boundary conditions applicable in a given case one finds that solutions ϕ that are not identically zero exist only for certain frequencies. These frequencies are called the *natural frequencies of the system;* the equation that the natural frequencies must satisfy for a given system is called the *frequency equation of the system;* the functions ϕ that satisfy Eq. (6.60) in conjunction with the natural frequencies are called the *eigenfunctions* or *mode shapes* of the system. One-dimensional systems (strings, bars) have an infinite number of natural frequencies ω_n and eigenfunctions ϕ_n; $n = 1, 2, \ldots$. Two-dimensional systems (membranes, plates) have a doubly infinite set of natural frequencies ω_{mn}, and eigenfunctions ϕ_{mn}; $m, n = 1, 2, \ldots$. Table 6.6 lists eigenfunctions and frequency equations for some common systems.

The eigenfunctions of flexural systems where all edges are either free, built-in, or pinned are "orthogonal," that is,

$$\begin{aligned}\int_0^L \mu(x)\phi_n(x)\phi_{n'}(x)\,dx &= \begin{cases}\bar{\mu}L\Phi_n & \text{for } n = n' \\ 0 & \text{for } n \neq n'\end{cases} \\ \iint_{A_s} \mu(x,y)\phi_{mn}(x,y)\phi_{m'n'}(x,y)\,dA_s &= \begin{cases}\bar{\mu}A_s\Phi_{mn} & \text{for } m = m',\ n = n' \\ 0 & \text{otherwise}\end{cases}\end{aligned} \tag{6.61}$$

Table 6.6a. Modal Properties for Some One-dimensional Systems[5]

System	Wave velocity c	Boundary conditions $x = 0$	Boundary conditions $x = L$	Mode shape† $\phi_n(x)$	Natural frequency $\omega_n = 2\pi f_n$	$\frac{\lambda}{L} = \frac{2\pi}{Lk_n}$
String.	$\sqrt{T/\rho A}$	Fixed	Fixed	$\sin\frac{n\pi x}{L}$	$\frac{n\pi}{L}c$	$\frac{2}{n}$
Uniform shaft* in torsion.	$\sqrt{\frac{GK}{\rho J}}$	Clamped	Free	$\sin\frac{(2n-1)\pi x}{2L}$	$\frac{(2n-1)\pi}{2L}c$	$\frac{4}{2n-1}$
		Free	Free	$\cos\frac{n\pi x}{L}$	$\frac{n\pi}{L}c$	$\frac{2}{n}$
Uniform bar* in longitudinal vibration	$\sqrt{E/\rho}$	Clamped	Clamped	$\sin\frac{n\pi x}{L}$	$\frac{n\pi}{L}c$	$\frac{2}{n}$

Symbols:

A = cross-sectional area
ρ = material density
T = tensile force
E = Young's modulus
G = shear modulus
J = polar moment of inertia of A
K = torsional constant of A (= J for circular sections)
GK = torsional rigidity
λ_n = wavelength
k_n = wave number
n = mode number

* The same modal properties, but with different values of c, apply for uniform shafts vibrating torsionally and uniform bars vibrating longitudinally.

† $\Phi_n = (1/L)\int_0^L \phi_n^2\,dx = \frac{1}{2}$

Table 6.6b. Modal Properties for Flexural Vibrations of Uniform Beams[5]

Natural frequency $\omega_n = 2\pi f_n = k_n^2 c_L r = k_n^2 \sqrt{EI/\rho A}$

Wavelength $\lambda_n = 2\pi/k_n$

Wave velocity $c_n = \omega_n/k_n = k_n c_L r = k_n \sqrt{EI/\rho A} = \sqrt{\omega_n c_L r}$

For mode shapes $\phi_n(x)$ listed below, $\Phi_n = (1/L)\int_0^L \phi_n^2\,dx = 1$

Boundary conditions $x = 0$	Boundary conditions $x = L$	Mode shape $\phi_n(x)$	σ_n	Frequency equation	k_1L	k_2L	k_3L	k_4L	k_5L	$k_nL,\ n > 5$
Pinned	Pinned	$\sqrt{2}\sin(k_nx)$		$\sin(k_nL) = 0$	π	2π	3π	4π	5π	$n\pi$
Clamped	Clamped	$\cosh(k_nx) - \cos(k_nx) - \sigma_n[\sinh(k_nx) - \sin(k_nx)]$	$\dfrac{\cosh(k_nL) - \cos(k_nL)}{\sinh(k_nL) - \sin(k_nL)}$	$\cos(k_nL)\cosh(k_nL) = 1$	4.73	7.85	11.00	14.14	17.29	$\approx \dfrac{2n+1}{2}\pi$
Free	Free	$\cosh(k_nx) + \cos(k_nx) - \sigma_n[\sinh(k_nx) + \sin(k_nx)]$								
Clamped	Free	$\cosh(k_nx) - \cos(k_nx) - \sigma_n[\sinh(k_nx) - \sin(k_nx)]$	$\dfrac{\sinh(k_nL) - \sin(k_nL)}{\cosh(k_nL) + \cos(k_nL)}$	$\cos(k_nL)\cosh(k_nL) = -1$	1.875	4.69	7.85	11.00	14.14	$\approx \dfrac{2n-1}{2}\pi$
Clamped	Pinned	$\cosh(k_nx) - \cos(k_nx) - \sigma_n[\sinh(k_nx) - \sin(k_nx)]$	$\cot(k_nL)$	$\tan(k_nL) = \tanh(k_nL)$	3.93	7.07	10.21	13.35	16.49	$\approx \dfrac{4n+1}{4}\pi$
Free	Pinned	$\cosh(k_nx) + \cos(k_nx) - \sigma_n[\sinh(k_nx) - \sin(k_nx)]$								

Symbols:

k_n = wave number
E = Young's modulus
ρ = material density
$c_L = \sqrt{E/\rho}$ = longitudinal wave velocity
A = cross-section area
I = moment of inertia of A
$r = \sqrt{I/A}$ = radius of gyration
L = beam length

Table 6.6c. Modal Properties for Some Plates[42]

Frequency	$\omega_{mn} = 2\pi f_{mn} = k_{mn}{}^2 \sqrt{D/\rho h} \approx k_{mn}{}^2 c_L r$	$D = Eh^3/12(1-\nu^2)$
Wavelength	$\lambda_{mn} = 2\pi/k_{mn}$	$r = h/\sqrt{12}$
Wave velocity	$c_{mn} = \omega_{mn}/k_{mn} = k_{mn}\sqrt{D/\rho h} \approx k_{mn} c_L r$	$c_L = \sqrt{E/\rho}$

Rectangular, on simple supports; $0 < x < a$, $0 < y < b$:

$$\phi_{mn}(x,y) = \sin\frac{m\pi x}{a}\sin\frac{n\pi y}{b} \qquad \overline{\phi_{mn}} = 1/4$$

$$k_{mn} = \pi\sqrt{m^2/a^2 + n^2/b^2}$$

Circular,* clamped at circumference; $0 < r < a$, $0 < \theta < 2\pi$:

$$\phi_{mn_e}(r,\theta) = \cos(m\theta)F_{mn}(r)$$
$$\phi_{mn_o}(r,\theta) = \sin(m\theta)F_{mn}(r)$$
$$F_{mn}(r) = J_m(k_{mn}r) - \frac{J_m(k_{mn}a)}{I_m(k_{mn}a)} I_m(k_{mn}r)$$

Same orthogonality holds as for membranes (see Table 6.6*d*)

Frequency equation:

$$I_m(k_{mn}a)\left[\frac{d}{dr}J_m(k_{mn}r)\right]_{r=a} = J_m(k_{mn}a)\left[\frac{d}{dr}I_m(k_{mn}r)\right]_{r=a}$$

Table of $k_{mn}a/\pi$ for Clamped Circular Plate

		n			
		1	2	3	>3
m	0	1.015	2.007	3.000	$\approx n + \frac{m}{2}$
	1	1.468	2.483	3.490	
	2	1.879	2.992	4.000	

* For circular plates one finds two sets of ϕ_{mn}, one even (subscript e), one odd (subscript o).

where L is the total length of a (one-dimensional) system, A_s is the total surface area of a (two-dimensional) system, μ is a weighting function, and $\bar{\mu}$ is the mean value of μ for the system. (The weighting functions of the common systems tabulated in Table 6.5 are equal to the mass operator $\mathfrak{M}$.)

$$\bar{\mu}L = \int_0^L \mu(x)\,dx \qquad \text{for one-dimensional systems}$$

$$\bar{\mu}A_s = \iint_{A_s} \mu(x,y)\,dA_s \qquad \text{for two-dimensional systems}$$

For uniform systems $\mu = \bar{\mu}$ is a constant and may be canceled from Eqs. (6.61).*

The free motions (or "transient" responses) of two-dimensional† systems with

* Note that for the systems of Table 6.6 $\mathfrak{M} = \mu$ is a multiplicative factor, not an operator in the more general sense. However, for other systems $\mathfrak{M}$ may be a more general operator (e.g., for lateral vibrations of beams when rotatory inertia is not neglected) and may differ from μ. The distinction between $\mathfrak{M}$ and μ is maintained throughout the subsequent discussion to permit application of the results to systems that are more complicated than those of Table 6.6.

† These equations apply also for one-dimensional systems if all m subscripts and y dependences are deleted, if A_s is replaced by L, and if the double integrations over A_s are replaced by a single integration from 0 to L.

Table 6.6d. Modal Properties for Some Membranes[42]

Wave velocity $c = \sqrt{S/\rho h}$

Rectangular; $0 < x < a, \quad 0 < y < b$:

$$\phi_{mn}(x,y) = \sin \frac{m\pi x}{a} \sin \frac{n\pi y}{b} \qquad \Phi_{mn} = \tfrac{1}{4}$$

$$\omega_{mn} = 2\pi f_{mn} = \pi c \sqrt{(m/a)^2 + (n/b)^2} \qquad k_{mn} = 2\pi/\lambda_{mn} = \omega_{mn}/c$$

Circular,* $0 < r < a, \quad 0 < \theta < 2\pi$:

$$\phi_{mn_e}(r,\theta) = \cos(m\theta) J_n(\omega_{mn} r/c)$$
$$\phi_{mn_o}(r,\theta) = \sin(m\theta) J_n(\omega_{mn} r/c)$$

$$\iint_{A_s} \phi_{mn_j} \phi_{m'n'j'}\, dA_s = \begin{cases} \pi a^2 \Phi_{mn} & \text{for } m' = m,\ n' = n,\ j' = j \\ 0 & \text{otherwise} \end{cases}$$

$$\Phi_{mn} = \begin{cases} [J_1(\omega_{mn} a/c)]^2 & \text{for } m = 0 \\ \tfrac{1}{2}[J_{m-1}(\omega_{mn} a/c)]^2 & \text{for } m > 0 \end{cases}$$

Frequency equation:

$$J_m(\omega a/c) = 0$$

Table of $\omega_{mn}a/\pi c$ for Circular Membrane

		n: 1	2	3	>3
m	0	0.7655	1.7571	2.7546	$\approx n + \frac{m}{2} - \frac{1}{4}$
	1	1.2197	2.2330	3.2383	
	2	1.6347	2.6793	3.6987	

Table of Φ_{mn} for Circular Membrane

		n: 1	2	3	4
m	0	0.2695	0.1241	0.07371	0.05404
	1	0.08112	0.04503	0.02534	0.02385
	2	0.05770	0.03682	0.02701	0.02134

Symbols for Tables 6.6c and *d*:

S = tension force/unit length
μ = mass/unit area
D = flexural rigidity $= Eh^3/12(1 - \nu^2)$
k_{mn} = wave number
E = Young's modulus
ν = Poisson's ratio
c_L = longitudinal wave velocity $= \sqrt{E/\rho}$
r = radius of gyration of section
h = plate thickness
ρ = material density
J_m = mth-order Bessel function
I_m = mth-order hyperbolic Bessel function

* For circular membranes one finds two sets of ϕ_{mn}, one even (subscript e), one odd (subscript o).

initial displacement $u(x,y,0)$ from equilibrium and initial velocity $\dot{u}(x,y,0)$ are given by

$$\begin{aligned} u_t(x,y,t) &= \sum_{m=1}^{\infty} \sum_{n=1}^{\infty} \phi_{mn}(B_{mn} \cos \omega_{mn}t + C_{mn} \sin \omega_{mn}t) \\ B_{mn} &= (1/\bar{\mu}A_s\Phi_{mn}) \iint_{A_s} u(x,y,0)\mu(x,y)\phi_{mn}(x,y)\, dA_s \\ C_{mn} &= (1/\bar{\mu}A_s\Phi_{mn}\omega_{mn}) \iint_{A_s} \dot{u}(x,y,0)\mu(x,y)\phi_{mn}(x,y)\, dA_s \end{aligned} \tag{6.62}$$

(b) Forced Vibrations

Forced vibrations of distributed systems are governed by

$$\mathfrak{M}\ddot{u} + \mathfrak{K}u = F(x,y,t)$$

where the right-hand side represents the applied load distribution in space and time.

General Response; Modal Displacements, Forcing Functions, Masses. The displacement $u(x,y,t)$ of a system* may in general be expressed in a modal series as

$$u(x,y,t) = \sum_{m=1}^{\infty} \sum_{n=1}^{\infty} U_{mn}(t)\phi_{mn}(x,y) \tag{6.63}$$

where the modal displacement $U_{mn}(t)$ is given by

$$U_{mn}(t) = (1/\omega_{mn}M_{mn}) \int_0^t G_{mn}(\tau) \sin \omega_{mn}(t - \tau)\, d\tau \tag{6.64}$$

The modal forcing function $G_{mn}(t)$ and the modal mass M_{mn} are given by

$$G_{mn}(t) = \iint_{A_s} F(x,y,t)\phi_{mn}(x,y)\, dA_s \qquad M_{mn} = \iint_{A_s} \phi_{mn}(x,y)\mathfrak{M}\phi_{mn}(x,y)\, dA_s \tag{6.65}$$

For the special case of a uniform system with $\mathfrak{M} = \mu = \bar{\mu} =$ constant,

$$M_{mn} = \mu A_s \Phi_{mn}$$

where Φ_{mn} is defined as in Eq. (6.61).

The "steady-state" response given by (6.63) and (6.64) must be combined with the "transient" response given by (6.62) if one desires the complete solution.

Sinusoidal Response; Input Impedance. For a sinusoidal forcing function $F(x,y,t) = e^{i\omega t}F_0(x,y)$ one finds a steady-state response given by

$$u_{ss}(x,y,t) = e^{i\omega t} \sum_{m=1}^{\infty} \sum_{n=1}^{\infty} U_{mn}\phi_{mn}(x,y)$$

with

$$U_{mn} = \frac{G_{mn}}{(\omega_{mn}^2 - \omega^2)M_{mn}} \qquad G_{mn} = \iint_{A_s} F_0(x,y)\phi_{mn}(x,y)\, dA_s$$

and M_{mn} given by (6.65).

For a point force $F_1e^{i\omega t}$ applied at $x = x_0$, $y = y_0$, one finds $G_{mn} = F_1\phi_{mn}(x_0,y_0)$. The input impedance $Z(x_0,y_0,\omega)$ of the system at $x = x_0$, $y = y_0$ for frequency ω then

* These equations apply also for one-dimensional systems if all m subscripts and y dependences are deleted, if A_s is replaced by L, and if the double integrations over A_s are replaced by a single integration from 0 to L.

may be found from

$$\frac{1}{i\omega Z(x_0,y_0,\omega)} = \sum_{m=1}^{\infty} \sum_{n=1}^{\infty} \frac{\phi_{mn}^2(x_0,y_0)}{(\omega_{mn}^2 - \omega^2)M_{mn}} \tag{6.66}$$

Point-impulse Response; Alternate Formulation of General Response. The response of a system to a point force applied at $x = x_0$, $y = y_0$ and varying like a Dirac impulse function (of unit magnitude) with time is given by

$$u_\delta(x,y;\, x_0,y_0;\, t) = \sum_{m=1}^{\infty} \sum_{n=1}^{\infty} \frac{\phi_{mn}(x_0,y_0)\phi_{mn}(x,y)}{\omega_{mn}M_{mn}} \sin \omega_{mn}t$$

An alternate expression for the steady-state response of a system to a general distributed force $F(x,y,t)$ may then be written as

$$u_{ss}(x,y,t) = \iint_{A_s} \left[\int_0^t F(x_0,y_0,\tau)u_\delta(x,y;\, x_0,y_0;\, t - \tau)\, d\tau \right] dx_0\, dy_0$$

This expression is entirely analogous to the result one obtains by combining Eqs. (6.63) and (6.64).

(c) Approximation Methods

Finite-difference Equations, Lumped-parameter Approximations. The most widely applicable numerical method, particularly if digital-computation equipment is available, consists of replacing the applicable differential equations by finite-difference equations[13,51,63] which may then be solved numerically.

A second method consists of replacing the continuous (infinite-degrees-of-freedom) system by one made up of a finite number of suitably interconnected masses, then applying methods developed for systems with a finite number of degrees of freedom, as outlined in Art. 6.2. These "lumped-parameter" approximations may be obtained, for example, by dividing a beam or plate to be analyzed into arbitrary segments and assuming the mass of each segment concentrated at its center of gravity or "lumping point." If one then establishes the influence coefficients between the various lumping points, one has enough information to apply directly the methods outlined in Art. 6.2. Alternately, one may often replace the structure between lumping points by equivalent springs. [For example, for beams and plates by springs with stiffness equal to that of a clamped-clamped beam extending from one lumping point to the next, width equal to the beam width (or distance between lumping points, for plates), thickness equal to the thickness of the structure. Suitable boundary conditions for the springs may be taken where the original structure has its boundaries.] Then one obtains a new system analogous to the continuous one; a beam is replaced by a linear array, a plate by a two-dimensional network of masses interconnected by springs. One may then proceed by determining the equations of motion of the new systems and by solving these as discussed in Art. 6.2.

Fundamental Frequencies*—Rayleigh's Quotient. RQ for an arbitrary deflection function $u(x,y)$ of a two-dimensional continuous system is given by[63]

$$RQ = \frac{\iint_{A_s} u\mathcal{K}(u)\, dA_s}{\iint_{A_s} u\mathfrak{M}(u)\, dA_s}$$

*These equations apply also for one-dimensional systems if all m subscripts and y dependences are deleted, if A_s is replaced by L, and if the double integrations over A_s are replaced by a single integration from 0 to L.

For $u = \phi_{mn}$ (the mn mode shape) RQ takes on the value ω_{mn}^2. For any function u that satisfies the boundary conditions of the given system

$$RQ \geq \omega_{11}^2$$

Thus RQ produces an estimate of the fundamental frequency ω_{11} which is always too high.

To obtain a better estimate one may use the Rayleigh-Ritz procedure. In this procedure one forms RQ for a linear combination of any convenient number of functions $u_i(x,y)$ that satisfy the boundary conditions of the problem; that is, one forms RQ from

$$u = \alpha_1 u_1 + \alpha_2 u_2 + \cdots$$

where the α are constants. A good approximation to ω_{11}^2 is then obtained in general by minimizing RQ with respect to the various α's.

$$RQ \text{ (evaluated so that } \partial RQ/\partial\alpha_1 = \partial RQ/\partial\alpha_2 = \cdots = 0) \approx \omega_{11}^2$$

Special Methods for Lateral Vibrations of Nonuniform Beams. In addition to the foregoing methods a number of others are available that have been developed specially for dealing with the vibrations of beams and shafts. In all these methods the beam mass is replaced by a number of masses concentrated at lumping points.

The Stodola method[9,52,59,62] is related to both the Rayleigh and matrix-iteration methods. In it one may proceed as follows:

1. Assume a deflection curve that satisfies the boundary conditions. (Usually the static-deflection curve gives good results.)
2. Determine a first approximation to ω_1 using Rayleigh's quotient, from

$$\omega_1^2 \approx RQ = \frac{g\Sigma m_i u_i}{\Sigma m_i u_i^2}$$

where g denotes the acceleration of gravity and u_i the assumed deflection of the mass m_i.

3. Calculate the deflection of the beam as if inertia forces $(-m_i u_i \omega_1^2)$ were applied statically. (This is usually done best by graphical or numerical means.) Use these new deflections instead of the original u_i in the foregoing equation. Repeat this process until no further changes in RQ result to the degree of accuracy desired. Then $RQ = \omega_1^2$ to within the desired accuracy.

The Myklestad method[43] is essentially the same as Holzer's method but considerably simpler to use for flexural vibrations. Extensions of Myklestad's original method also apply to coupled bending-torsion vibrations and to vibrations in centrifugal fields.[43] Some simplifications of Myklestad's method have been developed by Thomson.[56,60] Because of the details necessary for a sufficient discussion of these procedures the reader is referred to the original sources.

(d) Systems of Infinite Extent

Truly infinite or semi-infinite systems do not occur in reality. However, as far as the local response to a local excitation is concerned, finite systems behave like infinite ones if the ends are far (many wavelengths) removed from the excitation and if there is enough dissipation in the system or at the ends so that little effect of reflected waves is felt near the driving point.

The velocities with which waves travel in infinite systems are listed under the heading of "wave velocity" in Table 6.6, but in infinite systems the frequencies and wave numbers are not restricted, as they are in finite ones. In all cases the wave velocity c is related to wavelength λ, wave number k, frequency f (cycles/time), and circular frequency ω (radians/time) as

$$c = f\lambda = \omega\lambda/2\pi = \omega/k$$

Table 6.7. Driving-point Impedances of Some Infinite Uniform Systems*[,24]

Beams

Loading	Extent		
	Infinite	Semi-infinite	
Axial force..........	$Z = 2A\rho c_L$	$Z = A\rho c_L$	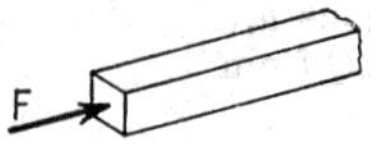
Lateral force.........	$Z = 2mc_B(1 + i)$	$Z = \frac{mc_B}{2}(1 + i)$	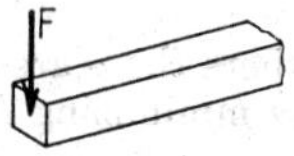
Moment............	$Z_M = \frac{2mc_B}{k^2}(1 + i)$	$Z_M = \frac{mc_B}{2k^2}(1 + i)$	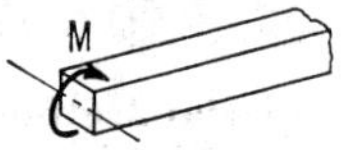
Torsion.............	$Z_M = 2J\rho c_T$	$Z_M = J\rho c_T$	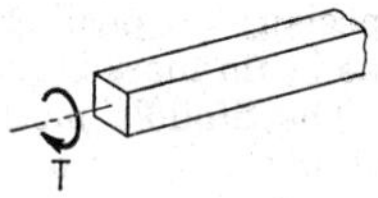

Symbols:

A = cross-section area
ρ = material density
E = Young's modulus
J = polar moment of inertia of A
K = torsional constant
G = shear modulus
$r = \sqrt{I/A}$

$c_L = \sqrt{E/\rho}$ = longitudinal wave velocity
$c_T = \sqrt{GK/\rho J}$ = torsional wave velocity
$c_B = \sqrt{\omega r c_L}$ = flexural wave velocity
$k = \sqrt{\omega / r c_L}$ = flexural wave number
$\frac{c_B}{k^2} = \sqrt{\frac{(r c_L)^3}{\omega}}$

Plates

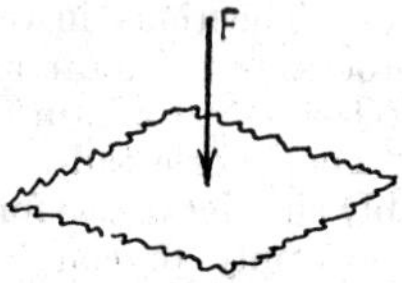

Infinite isotropic: $Z = 8\sqrt{D\rho h} \approx 2.3h^2\sqrt{\frac{E\rho}{1-\nu^2}}$

Infinite orthotropic: $Z \approx 8\sqrt[4]{D_x D_y}\sqrt{\rho h}$

Isotropic, infinite in x direction, on simple supports at $y = 0, L$, forced at $(0, y_0)$:

$$\frac{1}{Z} = \frac{i}{2\sqrt{D\rho h}}\sum_{n=1}^{\infty}[(n^2\pi^2 - K)^{-\frac{1}{2}} - (n^2\pi^2 + K)^{-\frac{1}{2}}]\sin^2\left(\frac{n\pi y_0}{L}\right)$$

$$K = (kL)^2 = L^2\omega\sqrt{\frac{\rho h}{D}}$$

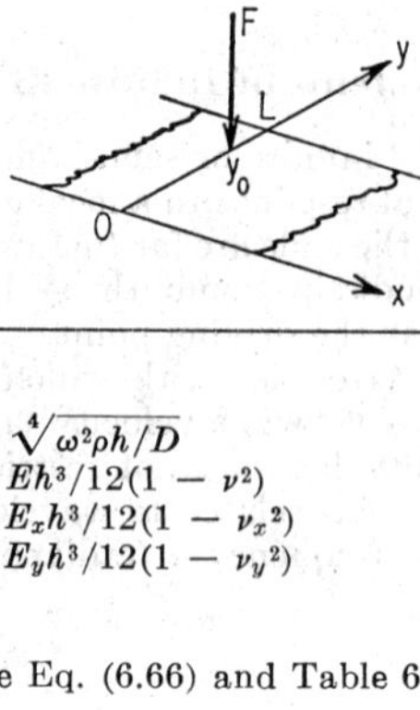

Symbols:

ρ = density of plate material
h = plate thickness
E = Young's modulus of plate material
E_x, E_y = Young's modulus in principal directions
ν = Poisson's ratio of plate material
ν_x, ν_y = Poisson's ratio in principal directions

$k = \sqrt[4]{\omega^2\rho h/D}$
$D = Eh^3/12(1 - \nu^2)$
$D_x = E_x h^3/12(1 - \nu_x{}^2)$
$D_y = E_y h^3/12(1 - \nu_y{}^2)$

* Systems are assumed undamped. For finite systems see Eq. (6.66) and Table 6.6.

Input impedances of infinite structures are useful for estimation of the responses of mechanical systems that are composed of or connected to one or more structures, if the responses of the latter may be approximated by those of corresponding infinite structures in the light of the first paragraph of this article. These impedances may be used precisely like previously discussed impedances of systems with only a few degrees of freedom.

If a force $F_0e^{i\omega t}$ gives rise to a velocity $V_0e^{i\omega t}$ at its point of application (where F_0 and V_0 may be complex in general), then the driving-point impedance is defined as $Z = F_0/V_0$. Similarly, the driving-point moment impedance is defined by $Z_M = M_0/\Omega_0$, where M_0 denotes the amplitude of a driving moment $M_0e^{i\omega t}$ and where $\Omega_0e^{i\omega t}$ is the angular velocity at the driving point. Table 6.7 lists the driving-point impedances of some infinite and semi-infinite systems.

6.4. MECHANICAL SHOCKS

By a mechanical shock one generally means a relatively suddenly applied transient force or support acceleration. The responses of mechanical systems to shocks may be computed by direct application of the previously discussed methods for determination of transient responses, provided that the forcing functions are known. If the forcing functions are not known precisely, one may approximate them by some idealized functions or else describe them in some rough way, for example, in terms of the subsequently discussed shock spectra.

(a) Idealized Forcing Functions

Among the most widely studied idealized shocks are those associated with sudden support displacements or velocity changes, or with suddenly applied forces. The responses of systems with one or two degrees of freedom to idealized shocks have been studied in considerable detail, since for such systems solutions may be obtained relatively simply by analytical or analog means.

Sudden Support Displacement. Sudden (vertical) support displacements occur, for example, when an automobile hits a sudden change in level of the roadbed. The responses of simple systems to such shocks have been studied in considerable detail. Families of curves describing system responses to a certain type of rounded step-function displacement are given in ref. 56. It is found among other things that the responses are generally of an oscillatory nature, except for peaks in acceleration that occur during the "rise" time of the displacement.

Sudden Support Velocity Change. Velocity shocks, i.e., those associated with instantaneous velocity changes, have been studied considerably. They approximate a number of physical situations where impact occurs; for example, when a piece of equipment is dropped the velocity of its outer parts changes from some finite value to zero at the instant these parts hit the ground. Reference 40 discusses in detail how one may compute the velocity shock responses of single-degree-of-freedom systems with linear and various nonlinear springs. It also presents charts for the computation of the important parameters of responses of simple systems attached to the aforementioned velocity-shock-excited systems. That is, if one system is mounted inside another, one may first compute the motion of the outer system in response to a velocity shock, then the response of the inner system to this motion (assuming the motion of the inner system has little effect on the shock response of the outer). This permits one to estimate, for example, what happens to an electronic component (inner system) when a chassis containing it (outer system) is dropped. Damping effects are generally neglected in ref. 40, but detailed curves for lightly damped linear two-degree-of-freedom systems appear in refs. 45 and 69. Some discussion appears also in ref. 14.

Suddenly Applied Forces. If a force is suddenly applied to a linear system with a single degree of freedom, then this force causes at most twice the displacement and twice the stress that this same force would cause if it were applied statically. This rule of thumb can lead to considerable error for systems with more than one degree of freedom.[46] One generally does well to carry out the necessary calculations in detail for such systems.

(b) Shock Spectra

The shock-spectrum concept is useful for describing shocks and the responses of simple systems exposed to them, particularly where the shocks cannot be described precisely and where they cannot be reasonably approximated by one of the idealized shocks of Art. 6.4(a).

Physical Interpretation. Assume that a given (support acceleration) shock is applied to an undamped linear single-degree-of-freedom system whose natural frequency is f_1. The maximum displacement of the system (relative to its supports) in response to this shock then gives one point on a plot of displacement vs. frequency. Repetition with the same shock applied to systems with different natural frequencies gives more points which, when joined in a curve, make up the "displacement shock spectrum" corresponding to the given shock. If instead of the maximum displacement one had noted the maximum velocity or acceleration for each test system, one would have obtained the velocity or acceleration shock spectrum.

Definitions.[1,22] If a force $F(t)$ is applied to an undamped single-degree-of-freedom system* with natural frequency ω and mass m, then the displacement, velocity, and acceleration of the system (assumed to be initially at rest and at equilibrium) are given by

$$x(t,\omega) = (1/m\omega) \int_0^t F(\tau) \sin \omega(t - \tau)\, d\tau$$

$$\dot{x}(t,\omega) = (1/m) \int_0^t F(\tau) \cos \omega(t - \tau)\, d\tau$$

$$\ddot{x}(t,\omega) = F(t)/m - \omega^2 x(t,\omega)$$

The various shock spectra of the force shock $F(t)$ are then defined as follows:

$$\begin{aligned} D_+'(\omega) &= \text{timewise maximum of } x(t,\omega) = \text{positive-displacement spectrum} \\ D_-'(\omega) &= \text{timewise minimum of } x(t,\omega) = \text{negative-displacement spectrum} \\ D'(\omega) &= \text{timewise maximum of } |x(t,\omega)| = \text{displacement spectrum} \end{aligned}$$

Similarly, for example,

$$\begin{aligned} V'(\omega) &= \text{timewise maximum of } |\dot{x}(t,\omega)| = \text{velocity spectrum} \\ A'(\omega) &= \text{timewise maximum of } |\ddot{x}(t,\omega)| = \text{acceleration spectrum} \end{aligned}$$

Similarly, if the previously discussed test system is exposed to a support acceleration $\ddot{s}(t)$, then the displacement, velocity, and acceleration of the system relative to its supports are given by

$$y(t,\omega) = -(1/\omega) \int_0^t \ddot{s}(\tau) \sin \omega(t - \tau)\, d\tau = x - s$$

$$\dot{y}(t,\omega) = -\int_0^t \ddot{s}(\tau) \cos \omega(t - \tau)\, d\tau = \dot{x} - \dot{s}$$

$$\ddot{y}(t,\omega) = -\ddot{s}(t) - \omega^2 y(t,\omega) = \ddot{x} - \ddot{s}$$

where $y = x - s$ denotes the relative displacement, x the absolute displacement of the mass, and s the displacement of the support.

* The shock spectra defined here are essentially descriptions of the shock in the frequency domain. Shock spectra are related to the Fourier transforms, the latter being lower bounds to the former.[64] Inclusion of damping appears unnecessary for purposes of describing the shock, but most authors[1,22] include damping in their definitions since they tend to be more concerned with descriptions of system responses to shocks than with descriptions of shocks. The term "shock spectra of a system" is often applied to descriptions of responses of a specified system to specified shocks.[61]

The various shock spectra of the acceleration shock $\ddot{s}(t)$ are defined as follows:

$$\begin{aligned} D(\omega) &= \text{timewise maximum of } |y(t,\omega)| = \text{relative-displacement spectrum} \\ V(\omega) &= \text{timewise maximum of } |\dot{y}(t,\omega)| = \text{relative-velocity spectrum} \\ A(\omega) &= \text{timewise maximum of } |\ddot{y}(t,\omega)| = \text{relative-acceleration spectrum} \\ D_a(\omega) &= \text{timewise maximum of } |x(t,\omega)| = \text{absolute-displacement spectrum} \\ V_a(\omega) &= \text{timewise maximum of } |\dot{x}(t,\omega)| = \text{absolute-velocity spectrum} \\ A_a(\omega) &= \text{timewise maximum of } |\ddot{x}(t,\omega)| = \text{absolute-acceleration spectrum} \end{aligned}$$

$$\tilde{V}(\omega) = \omega\, D(\omega) = \text{pseudo-velocity spectrum}$$

For any $s(t)$ it is true that

$$A_a(\omega) = \omega^2 D(\omega)$$

but $\tilde{V}(\omega) \neq V(\omega)$, $\tilde{V}(\omega) \neq V_a(\omega)$ in general.

One often distinguishes also between system responses during the action of a shock and responses after the shock action has ceased. The spectra associated with system motion during the shock action are called *primary spectra;* those associated with system motion after the shock are called *residual spectra.*

Shock spectra are generally reduced to dimensionless "amplification spectra," e.g., by dividing $D'(\omega)$ by the static displacement due to the maximum value of $F(t)$, or $A(\omega)$ by the maximum value of $\ddot{s}(t)$. Frequency is also usually reduced to dimensionless form by division by some suitable frequencylike parameter.

Simple Shocks. A simple shock is one like that sketched in Fig. 6.19; it is generally nonoscillatory, has a unique absolute maximum reached within a finite "rise time" t_m, and a finite duration t_0. Spectra of simple shocks are usually presented in terms of ft_m or ωt_m (and/or ft_0 or ωt_0) or simple multiples of these dimensionless quantities.

Shock spectra may, by virtue of their definitions, be used directly to determine the maximum responses of single-degree-of-freedom systems to the shocks to which the spectra pertain.

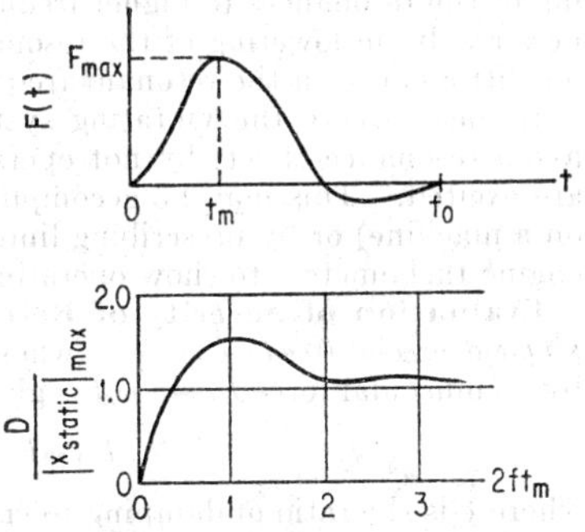

FIG. 6.19. Typical simple shock and dimensionless amplification spectrum.

The main utility of the shock-spectrum concept, however, is due to the fact that the amplification spectra of roughly similar shocks, when presented in terms of the foregoing dimensionless-frequency parameters, tend to coincide very nearly; i.e., amplification spectra are relatively insensitive to details of the pulse form. (Good coincidence of spectra of similar shocks is obtained for small values of f if the spectra are plotted against ft_0, for large values of f if spectra are plotted against ft_m.)

The effect of damping is to reduce system responses in general. The greatest reduction usually occurs near the peaks of the amplification spectra; near relative minima of these spectra the reduction due to damping is generally small.

The amplification spectra of simple shocks have the following properties:

They pass through the origin.

They do not exceed 2.0.

They approach 1.0 for large ft_m.

They generally reach their maxima between $2\ ft_m = 0.7$ and 2.0, but most often near $2\ ft_m = 1$.

Simple shocks occur in drop tests, aircraft landing impact, hammer impact, gun recoil, explosion blasts, and ground shock due to explosive detonations. Details of many spectra appear in refs. 1, 20, 22, 66, and 68. Good discussions appear also in ref. 44.

6.5. DESIGN CONSIDERATIONS

(a) Design Approach

Since vibration is generally considered an undesirable side effect, it seldom controls the primary design of a machine or structure. Items usually are designed first to fulfill their main function, then analyzed from a vibration viewpoint in regard to possible equipment damage or malfunction, structural fatigue failure, noise, or human discomfort or annoyance.

The most severe effects of vibration generally occur at resonance; therefore, one usually is concerned first with determination of the resonant frequencies of the preliminary design. (Damping is usually neglected in the pertinent calculations for all but the simplest systems, unless a prominent damping effect is anticipated.) If resonant frequencies are found to lie within the intended range of driving frequencies, one should attempt a redesign to shift the resonances out of the driving-frequency range.

If resonances cannot be avoided reasonably, the designer must determine the severity of these resonances. Damping must then be considered in the pertinent calculations, since it is primarily damping that limits response at resonance. If resonant responses are too severe, one must reduce the excitation and/or incorporate increased damping in the system or structure.

Shifting of Resonances. If resonances are found to occur within the range of excitation frequencies, one should try to redesign the system or structure to change its resonant frequencies. Added stiffness with little addition of mass results in shifting of the resonances to higher frequencies. Added mass with little addition of stiffness results in lowering of the resonant frequencies. Addition of damping generally has little effect on the resonant frequencies.

In cases where the vibrating system cannot be modified satisfactorily, one may avoid resonance effects by not operating at excitation frequencies where resonances are excited. This may be accomplished by automatic controls (e.g., speed controls on a machine) or by prescribing limitations on use of the system (e.g., "red lines" on engine tachometers to show operating speeds to be avoided).

Evaluation of Severity of Resonances. The displacement amplitude X_0 of a *single-degree-of-freedom system* (whose natural frequency is ω_n) excited at resonance by a sinusoidal force $F_0 \sin \omega_n t$ is given by

$$kX_0/F_0 = 1/2\zeta = Q = 1/\eta$$

where ζ is the ratio of damping to critical damping, Q the "quality factor," and η the loss factor of the system. The velocity amplitude V_0 and acceleration amplitude A_0 are given by

$$A_0 = \omega_n V_0 = \omega_n^2 X_0$$

The maximum force exerted by the system's spring (stiffness k) is

$$F_s = kX_0$$

The maximum spring stress may readily be calculated from the foregoing spring force.

Multiple-degree-of-freedom systems generally require detailed analysis in accordance with methods outlined in Art. 6.2. The previous expressions pertaining to single-degree-of-freedom systems hold also for systems with a number of degrees of freedom if X_0 is taken as the amplitude of a generalized (principal) coordinate that is independent of the other coordinates, and if F_0 is taken as the corresponding generalized force. Then $\eta = 2\zeta = Q^{-1}$ must describe the effective damping for that coordinate.

Stresses in the various members may be determined from the mode shapes (vectors) corresponding to the resonant mode; i.e., from the maximum displacements or forces to which the elements are subjected.

For uniform distributed systems* the modal displacement U_{mn} at resonance of the

* The summary here is presented for two-dimensional systems (plates), and double subscripts are used for the various modal parameters. However, the identical expressions apply also for one-dimensional systems (beams, strings) if the n subscripts are deleted. See also Art. 6.3.

m, n mode due to a modal excitation $G_{mn} \sin \omega_{mn}t$ is given by

$$|U_{mn}| = |G_{mn}|/M_{mn}\omega_{mn}^2\eta(\omega_{mn})$$

and is related to modal velocity and acceleration amplitudes $\dot{U}_{mn}$, $\ddot{U}_{mn}$, according to

$$|\ddot{U}_{mn}| = \omega_{mn}^2|U_{mn}| = \omega_{mn}|\dot{U}_{mn}|$$

$\eta(\omega)$ denotes the system loss factor, which varies with frequency depending on the damping mechanism present. For beams and plates in flexure[65] η is related to the usual viscous-damping coefficient c and flexural rigidity D as*

$$\eta(\omega) = c\omega/D$$

The maximum stress $\sigma_{\max}$ and maximum strain $\epsilon_{\max}$ that occur at resonance in bending of beams and plates may be approximated by

$$\sigma_{\max} \approx \frac{C}{r}|U_{mn}|\omega_{mn}\rho c_L g_{mn} \approx \epsilon_{\max}\rho c_L{}^2$$

where $c_L = \sqrt{E/\rho}$ denotes the velocity of sound in the material (E is Young's modulus, ρ the material density) and C denotes the distance from the neutral to the outermost fiber, r the radius of gyration of the cross section. For beams with rectangular cross sections and for plates, $C/r = \sqrt{3}$. For plates $M_{mn} = \rho h$; for beams $M_m = \rho A$, where h denotes plate thickness, A beam cross-section area.

The factor g_{mn} accounts for boundary conditions and takes on the following approximate values:

Boundary conditions on pair of opposite edges	g_{mn}
Pinned-pinned or clamped-free	1.00
Clamped-clamped or clamped-pinned	1.33
Free-free or pinned-free	0.80

For conservative design if the boundary conditions are not known one should use $g_{mn} = 1.33$. Otherwise, one should use the largest value of g_{mn} pertinent to the existing boundary conditions.

Reduction of Severity of Resonances. If redesign to avoid resonances is not feasible, one can generally do little to modify mass and stiffness and has only two means for reducing resonant amplitudes: (1) reduction of modal excitation and (2) increase of damping.

Excitation reduction may take the form of running a machine at reduced power, isolating the resonating system from the source of excitation, or shielding the system from exciting pressures. Increased damping may be obtained by addition of energy-dissipating devices or structures. For example, one might use metals with high internal damping for the primary structure, or else attach coatings or sandwich media with large energy-dissipation capacities to a primary structure of common materials.[50] Alternately, one might rely on structural joints or shaft bearings to absorb energy by friction, or else attempt to extract energy by means of viscous friction or acoustic radiation in fluids in contact with the resonant system. Simple dashpots, localized friction pads, or magnetically actuated eddy-current-damping devices may also be used, particularly for systems with a few degrees of freedom; however, such localized devices may not be effective for higher modes of distributed systems.

(b) Source-path-receiver Concept

If one is called upon to analyze or modify an existing or projected system from the vibration viewpoint, one may find it useful to examine the system in regard to

* For plates $D = Eh^3/12(1 - \nu^2)$; for beams $D = EI$.

1. Sources of vibration (e.g., reciprocating engines, unbalanced rotating masses, fluctuating air pressures)
2. Paths connecting sources to critical items (e.g., substructures, vibration mounts)
3. Receivers of vibratory energy; i.e., critical items that malfunction if exposed to too much vibration (e.g., electronic components)

The problem of limiting the effects of vibration may then be attacked by any or all of the following means:

1. Vibration elimination at the source (e.g., designing machines in opposed pairs so that inertia forces cancel, balancing all rotating items, smoothing or deflecting oscillating air flows)
2. Modification of paths (e.g., changing substructures so as to transmit fewer vibrations, introducing vibration mounts)
3. Decreasing vibration sensitivity of critical items (e.g., changing orientation of components with respect to excitation, using more rugged components, using more fatigue-resistant materials, designing more damping into components in order to decrease the effects of internal resonances)

Vibration Reduction at the Source; Vibration Absorbers. The strength of vibration sources may often be reduced by proper design. Such design may include balancing, use of the lightest possible reciprocating parts, arranging components so that inertia forces cancel, or attaching vibration absorbers.

A vibration absorber* is essentially a mass m attached by means of a spring k to a primary mass M in order to reduce the response of M to a sinusoidal force acting directly on M (or to a sinusoidal support acceleration when the M is attached to the support by a spring K). If an absorber for which $k/m = \omega_0^2$ is attached to M, then M will experience no excursion if the excitation occurs at a frequency ω_0. (The displacement amplitude of m will be F/k, where F is the force amplitude.)

For driving frequencies very near ω_0 attachment of the vibration absorber results in small amplitudes for M, but for other frequencies it generally results in amplitudes that are greater than those obtained without an absorber. Vibration absorbers are very frequency-sensitive and hence should be used only where there exists essentially a single relatively accurately known driving frequency. The addition of damping to an absorber[17,62] extends to some extent the frequency range over which the absorber results in vibration reduction of the primary mass, but increased damping also results in more motion of M at the optimum frequency ω_0.

Path Modification; Vibration Mounts. The prime means for modifying paths traversed by vibratory energy consists of the addition of vibration mounts. If a mass M is excited by a vibratory force of frequency ω one may reduce the vibratory force that the mass exerts on a rigid support to which it is attached by inserting a spring of stiffness k between mass and support, such that $k/M < \omega^2/2$. Further reduction in k/M reduces the transmitted force further; added damping with k/M held constant increases the transmitted force [see also Art. 7.1(a)].

If a mass M is attached by means of a spring k to a support that is vibrating at frequency ω, then M can be made to vibrate less than the support if k is chosen so that $k/M < \omega^2/2$. Further reduction of k/M further reduces the motion of the mass; addition of damping increases this motion.

Fuller discussions of vibration and shock mounting are to be found in refs. 14 and 35. Data on commercial mounts may be found in manufacturers' literature.

Modification of Critical Items. Critical items may often be made less sensitive to vibration by redesigning them so that all their internal component resonance frequencies fall outside the excitation frequency range. Occasionally this can be done merely by reorienting a critical item with respect to the direction of excitation; more often it requires stiffening the components (or possibly adding mass). If fatigue rather than malfunction is a problem, one may obtain improved parts by careful

* The discussion is presented here only for translational systems. It may be extended by analogy to apply also to rotational systems.

redesign to eliminate stress concentrations and/or by using materials with greater fatigue resistance. If internal resonances cannot be avoided, one may reduce their effect by designing damping into the resonant components.

(c) Rotating Machinery

A considerable amount of information has been amassed in relation to reciprocating and turbine machines. Comprehensive treatments of the associated torsional vibrations may be found in refs. 44 and 68; less detailed discussions of these appear in standard texts.[17,35,59,62] A few of the most important items pertaining to rotating machinery are outlined subsequently.

Vibrations of rotating machines are caused by the following factors; reduction of these generally serves to reduce vibrations:

1. Unbalance of rotating components
2. Reciprocating components
3. Whirling of shafts
4. Gas forces
5. Instabilities, such as those due to slip-stick phenomena

Balancing. A rotor mounted on frictionless bearings so that its axis of rotation is horizontal remains motionless (if subject only to gravity) in any angular position,

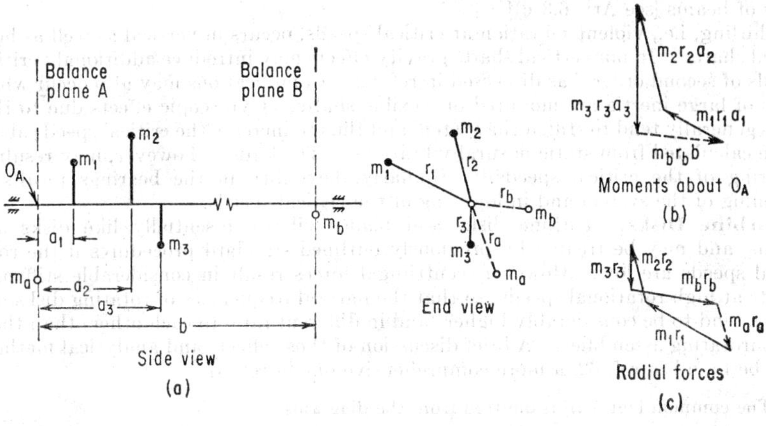

FIG. 6.20. Balancing of rigid rotors.

if the rotor is *statically* balanced. Static balance implies an even mass distribution around the rotational axis. However, even statically balanced rotors may be dynamically unbalanced. Dynamic unbalance occurs if the centrifugal forces set up during rotation result in a nonzero couple. (Static balance assures only that the centrifugal forces result in zero net radial force.)

Rigid rotors can always be balanced statically by addition of a single weight in any arbitrarily chosen plane, or dynamically (and statically) by addition of two weights in two arbitrarily chosen planes. The procedure, illustrated in Fig. 6.20, is as follows[9]:

1. Divide the rotor into a convenient number of sections by passing planes perpendicular to the rotational axis.

2. Determine the mass and center-of-gravity position for each section.

3. Draw a diagram like Fig. 6.20*a* where each section is represented by its mass located at the center-of-gravity position.

4. Select balance planes, i.e., planes in which weights are to be attached for balance.

5. Draw a diagram (Fig. 6.20*b*) of centrifugal force moments* about point O_A (where rotational axis intersects balance plane A). Each moment is represented by a vector of length $m_i r_i a_i$ (to some suitable scale) parallel to r_i in the end view. The vector required to close the diagram then is $m_b r_b b$; its direction gives the direction of r_b in the end view; $m_b r_b$ may be calculated and either m_b or r_b may be selected arbitrarily.

6. Draw a diagram (Fig. 6.20*c*) of centrifugal forces* $m_i r_i$, including $m_b r_b$. Each force vector is drawn parallel to r_i in the end view; the vector required to close the diagram is $m_a r_a$ and defines the mass (and its location) to be added in plane A. Again, either m_a or r_a may be selected arbitrarily; r_a in the end view must be parallel to the $m_a r_a$ vector, however.

Balancing of flexible rotors or of rotors on flexible shafts (particularly when operating above critical speeds) can generally be accomplished for only one particular speed or for none at all.[17,62] This problem is discussed in refs. 10, 17, and 62.

Balancing machines, their principles and use, and field balancing procedures are discussed in refs. 17, 41, 43, 59, and 62, among others.

Balancing of reciprocating engines is treated in ref. 17 and in a number of texts on dynamics of machines, such as ref. 25.

Whirling of Shafts; Critical Speeds. Rotating shafts become unstable at certain speeds, and large vibrations are likely to develop. These speeds are known as *critical speeds*. At a critical speed the number of revolutions per second is generally very nearly equal to a natural frequency (in cycles per second) of the shaft considered as a nonrotating beam vibrating laterally. Thus critical speeds of shafts may be found by any of the means for calculating the natural frequencies of lateral vibrations of beams [see Art. 6.3(c)].

Whirling, i.e., violent vibration at critical speeds, occurs in vertical as well as horizontal shafts. In nonvertical shafts gravity effects may introduce additional "critical speeds of second order," as discussed in ref. 62. Complications may also occur where disks of large inertia are mounted on flexible shafts. Gyroscopic effects due to thin disks generally tend to stiffen the system and thus to increase the critical speeds above those calculated from static flexural vibrations. Thick disks, however, may result in lowering of the critical speeds.[17] Similarly, flexibility in the bearings results in softening of the system and in lowering of the critical speeds.

Turbine Disks. Turbine disks and blades vibrate essentially like disks and beams, and may be treated by previously outlined standard procedures if the rotational speeds are low. However, centrifugal forces result in considerable stiffening effects at high rotational speeds; so that the natural frequencies of rotating disks and blades tend to be considerably higher (and in different ratio to each other) than those of nonrotating assemblies. A brief discussion of those effects and analytical methods may be found in ref. 62, a more comprehensive one in ref. 57.

* The common factor ω^2 is omitted from the diagrams.

(d) Charts and Tables

This article and Art. 6.3 contain tabulated data on natural frequencies, spring constants, and material properties, as follows:

Natural frequencies	*Tables* 6.8 *and* 6.6
Simple translational systems	6.8*a*
Simple torsional systems	6.8*b*
Beams, bars, shafts (uniform section unless noted otherwise):	
Uniform and variable section, in flexure	6.8*c*
Flexure; mode shapes	6.6*b*
Torsion; mode shapes	6.6*a*
Longitudinal vibration; mode shapes	6.6*a*
On multiple evenly spaced supports	6.8*d*
Free-free, on elastic foundation	6.8*i*
Strings, uniform; mode shapes	6.6*a*
Membranes, uniform:	
Circular	6.8*i*
Rectangular and circular; mode shapes	6.6*d*
Plates, uniform:	
Square, various boundary conditions	6.8*e*
Rectangular, various boundary conditions	6.8*g*
Circular, various boundary conditions	6.8*h*
Rectangular, on simple supports; mode shapes	6.6*c*
Circular, clamped; mode shapes	6.6*c*
Cantilever, various shapes	6.6*f*
Rings	6.8*j*
Mass free to rotate and translate in plane	6.8*j*
Mass on spring of finite mass	6.8*j*

Spring constants	*Table* 6.9
Combinations of springs	6.9*a*
Round-wire helical springs	6.9*b*
Beams with force or moment inputs	6.9*c*
Torsion springs	6.9*d*
Shafts in torsion	6.9*d,e*
Plates loaded at centers	6.9*f*

Sound velocity and K_m *(for Table* 6.8) *for engineering materials*	*Table* 6.10

Table 6.8a. Natural Frequencies of Simple Translational Systems[17,39]

System	$\omega^2 =$	Remarks
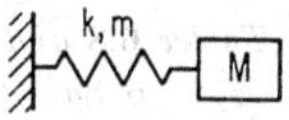	$\dfrac{k}{M+0\cdot33m}$	
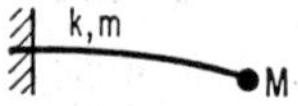	$\dfrac{k}{M+0\cdot23m}$	Point mass on cantilever beam
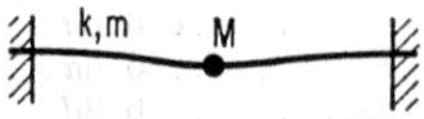	$\dfrac{k}{M+0\cdot375m}$	Point mass at center of clamped beam — See table 6.9c for k
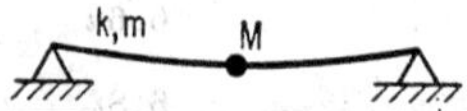	$\dfrac{k}{M+0\cdot50m}$	Point mass at center of simply supported beam
	$k\left[\dfrac{1}{M_1}+\dfrac{1}{M_2}\right]$	
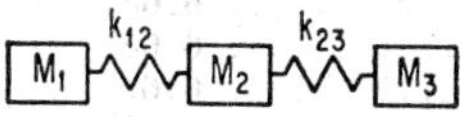		$2B=\dfrac{k_{12}}{M_1}+\dfrac{k_{23}}{M_3}+\dfrac{k_{12}+k_{23}}{M_2}$ $C=\dfrac{k_{12}+k_{23}}{M_1\,M_2\,M_3}\,(M_1+M_2+M_3)$
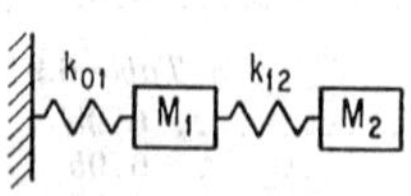	$B\pm\sqrt{B^2-C}$	$2B=\dfrac{k_{01}+k_{12}}{M_1}+\dfrac{k_{12}}{M_2}$ $C=\dfrac{k_{01}\,k_{12}}{M_1\,M_2}$
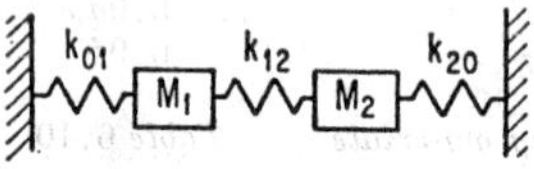		$2B=\dfrac{k_{01}+k_{12}}{M_1}+\dfrac{k_{12}+k_{20}}{M_2}$ $C=\dfrac{1}{M_1M_2}\,(k_{01}k_{12}+k_{12}k_{20}+k_{20}k_{01})$
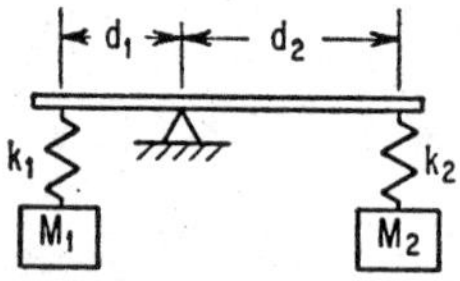	$\dfrac{\dfrac{1}{M_1}+\dfrac{1}{R^2M_2}}{\dfrac{1}{k_1}+\dfrac{1}{R^2k_2}}$	Inertia of lever negligible $R=\dfrac{d_2}{d_1}$

Symbols:

ω = circular natural frequency.
M = mass.
m = total mass of spring element.
k = spring constant.

Table 6.8b. Natural Frequencies of Simple Torsional Systems[17,39]

System	$\omega^2 =$	Remarks
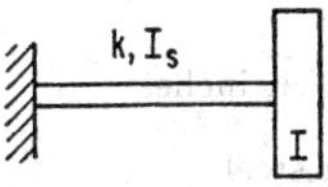	$\dfrac{k}{I + I_s/3}$	
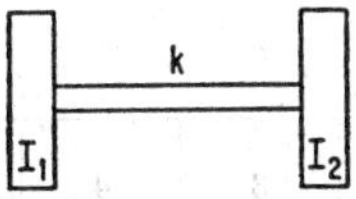	$k\left(\dfrac{1}{I_1} + \dfrac{1}{I_2}\right)$	
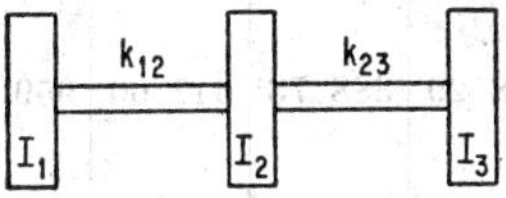	$B \pm \sqrt{B^2 - C}$	$2B = \dfrac{k_{12}}{I_1} + \dfrac{k_{23}}{I_3} + \dfrac{k_{12} + k_{23}}{I_2}$ $C = \dfrac{k_{12}\,k_{23}}{I_1\,I_2\,I_3}(I_1 + I_2 + I_3)$
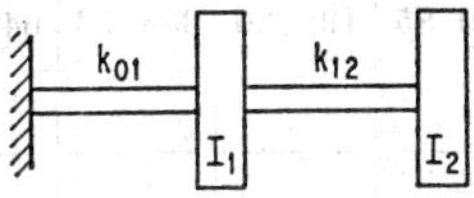	$B \pm \sqrt{B^2 - C}$	$2B = \dfrac{k_{01} + k_{12}}{I_2} + \dfrac{k_{12}}{I_2}$ $C = \dfrac{k_{01}\,k_{12}}{I_1\,I_2}$
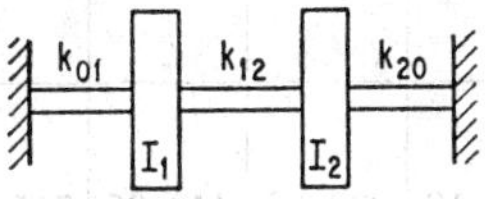	$B \pm \sqrt{B^2 - C}$	$2B = \dfrac{k_{01} + k_{12}}{I_1} + \dfrac{k_{12} + k_{20}}{I_2}$ $C = \dfrac{1}{I_1\,I_2}(k_{01}k_{12} + k_{12}k_{20} + k_{20}k_{01})$
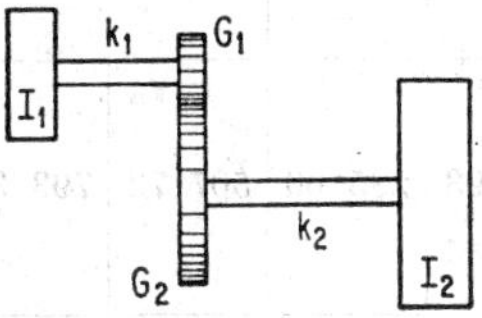	$\dfrac{\dfrac{1}{I_1} + \dfrac{1}{R^2 I_2}}{\dfrac{1}{k_1} + \dfrac{1}{R^2 k_2}}$	Inertia of gears G_1, G_2 assumed negligible $R = \dfrac{\text{number of teeth on gear 1}}{\text{number of teeth on gear 2}}$ $= \dfrac{\text{rpm of shaft 2}}{\text{rpm of shaft 1}}$

Symbols:

ω = circular natural frequency.
k = torsional spring constant; see Table 6.9d.
I = polar mass moment of inertia.
I_s = polar mass moment of inertia of entire shaft.

Table 6.8c. Natural Frequencies of Beams in Flexure[36]

$$f_n = C_n \frac{r}{L^2} \times 10^4 \times K_m$$

f_n = nth natural frequency, cps
C_n = frequency constant listed in these tables
r = radius of gyration of cross section = $\sqrt{I/A}$, inches
L = beam length, inches
K_m = material constant (Table 6.10) = 1.00 for steel

Uniform-section beams		n = 1	2	3	4	5
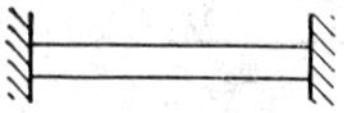 Clamped–Clamped	 Free–Free	71.95	198.29	388.73	642.60	959.94
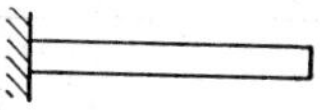 Clamped–Free		11.30	70.85	198.30	388.73	642.60
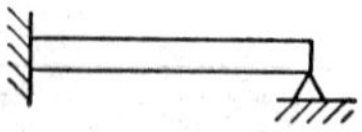 Clamped–Hinged	 Free–Hinged	49.57	160.65	335.17	573.20	874.65
 Clamped–Guided	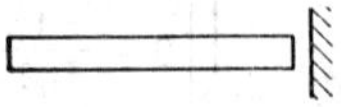 Free–Guided	17.98	97.18	239.98	446.25	715.98
 Hinged–Hinged	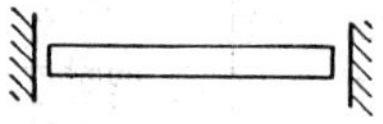 Guided–Guided	31.73	126.93	285.60	507.73	793.33
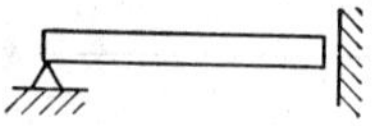 Hinged–Guided		7.93	71.40	198.33	388.73	642.60

Table 6.8c. Natural Frequencies of Beams in Flexure (Continued)

Variable-section beams (Use maximum r to calculate f_n)	Shape		n		
	b/b_0	h/h_0	1	2	3
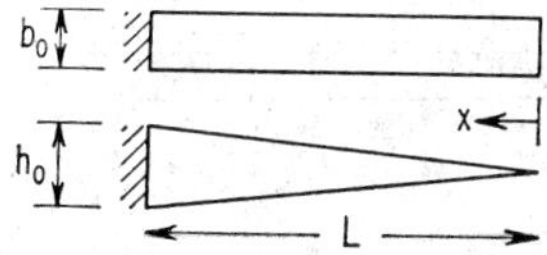	1	x/L	17.09	48.89	96.57
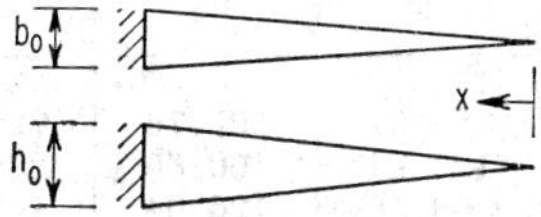	x/L	x/L	26.08	68.08	123.64
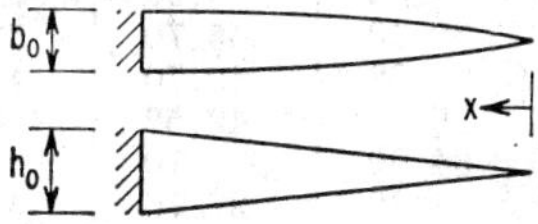	$\sqrt{x/L}$	x/L	22.30	58.18	109.90
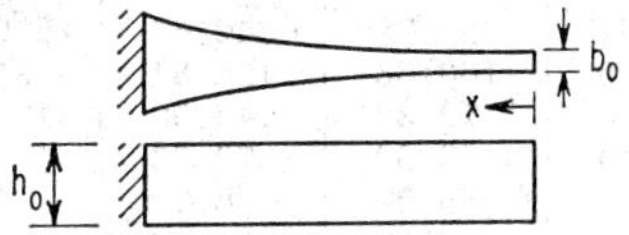	exp x/L	1	15.23	77.78	206.07
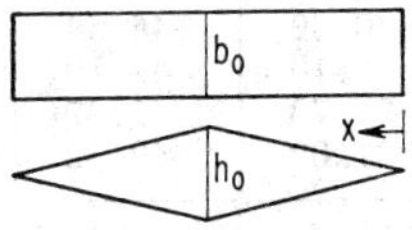	1	x/L	21.21* 35.05†	56.97*	
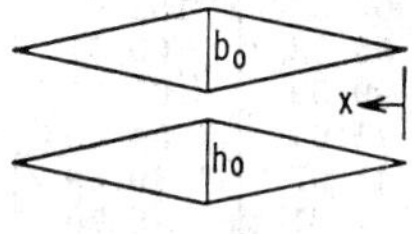	x/L	x/L	32.73* 49.50†	76.57*	
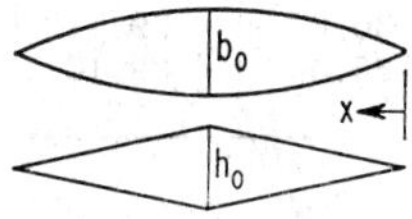	$\sqrt{x/L}$	x/L	25.66* 42.02†	66.06*	

* Symmetric mode.
† Antisymmetric mode.

Table 6.8d. Natural Frequencies of Uniform Beams on Multiple Equally Spaced Supports[36]

$$f_n = C_n \frac{r}{L^2} \times 10^4 \times K_m$$

f_n = nth natural frequency, cps
C_n = frequency constant listed in these tables
r = radius of gyration of cross section = $\sqrt{I/A}$, inches
L = span length, inches
K_m = material constant (Table 6.10) = 1.00 for steel

	Number of spans	n = 1	2	3	4	5
 Ends simply supported	1	31.73	126.94	285.61	507.76	793.37
	2	31.73	49.59	126.94	160.66	285.61
	3	31.73	40.52	59.56	126.94	143.98
	4	31.73	37.02	49.59	63.99	126.94
	5	31.73	34.99	44.19	55.29	66.72
	6	31.73	34.32	40.52	49.59	59.56
	7	31.73	33.67	38.40	45.70	53.63
	8	31.73	33.02	37.02	42.70	49.59
	9	31.73	33.02	35.66	40.52	46.46
	10	31.73	33.02	34.99	39.10	44.19
	11	31.73	32.37	34.32	37.70	41.97
	12	31.73	32.37	34.32	37.02	40.52
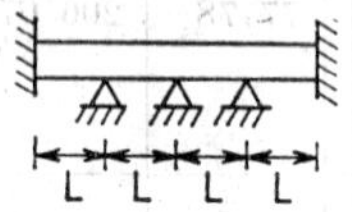 Ends clamped	1	72.36	198.34	388.75	642.63	959.98
	2	49.59	72.36	160.66	198.34	335.20
	3	40.52	59.56	72.36	143.98	178.25
	4	37.02	49.59	63.99	72.36	137.30
	5	34.99	44.19	55.29	66.72	72.36
	6	34.32	40.52	49.59	59.56	67.65
	7	33.67	38.40	45.70	53.63	62.20
	8	33.02	37.02	42.70	49.59	56.98
	9	33.02	35.66	40.52	46.46	52.81
	10	33.02	34.99	39.10	44.19	49.59
	11	32.37	34.32	37.70	41.97	47.23
	12	32.37	34.32	37.02	40.52	44.94
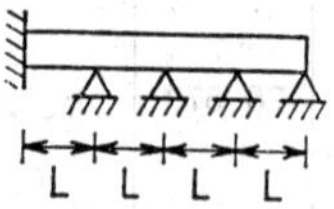 Ends clamped-supported	1	49.59	160.66	335.2	573.21	874.69
	2	37.02	63.99	137.30	185.85	301.05
	3	34.32	49.59	67.65	132.07	160.66
	4	33.02	42.70	56.98	69.51	129 49
	5	33.02	39.10	49.59	61.31	70.45
	6	32.37	37.02	44.94	54.46	63.99
	7	32.37	35.66	41.97	49.59	57.84
	8	32.37	34.99	39.81	45.70	53.63
	9	31.73	34.32	38.40	43.44	49.59
	10	31.73	33.67	37.02	41.24	46.46
	11	31.73	33.67	36.33	39.81	44.19
	12	31.73	33.02	35.66	39.10	42.70

Table 6.8e. Natural Frequencies of Square Plates[36]

$$f_n = C_n \frac{h}{a^2} \times 10^4 \times K_m$$

f_n = nth natural frequency, cps
C_n = frequency constant listed in table
h = plate thickness, inches
a = plate edge length, inches
K_m = material constant (Table 6.10) = 1.00 for steel

Boundary conditions F = free C = clamped S = simply supported	n					
	1	2	3	4	5	6
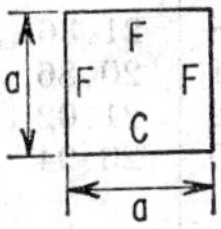	3.40	8.32	20.86	26.71	30.32	
	6.77	23.43	26.07	46.75	61.44	
	13.72	19.99	23.26	34.98	59.93	63.47
	19.20	48.00	76.82	96.01	124.82	163.25
	23.01	50.28	57.06	83.79	97.58	110.13
	28.16	53.26	67.44	92.02	99.43	125.60
	35.01	71.42	105.36	128.03	128.71	160.72

Table 6.8f. Natural Frequencies of Cantilever Plates[36,70]

$$f_n = C_n \frac{h}{a^2} \times 10^4 \times K_m$$

f_n = nth natural frequency, cps
C_n = frequency constant listed in table
h = plate thickness, inches
a = plate dimension, as shown, inches
K_m = material constant (Table 6.10) = 1.00 for steel

Boundary conditions F = free C = clamped S = simply supported		n = 1	2	3	4	5
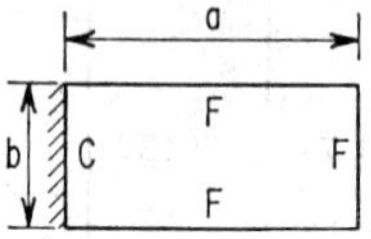	$a/b = \frac{1}{2}$	3.41	5.23	9.98	21.36	24.18
	$a/b = 1$	3.40	8.32	26.71	20.86	30.32
	$a/b = 2$	3.38	14.52	91.92	21.02	47.39
	$a/b = 5$	3.36	33.79	548.60	20.94	103.03
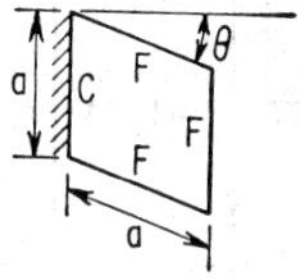	$\theta = 15°$	3.50	8.63			
	$\theta = 30°$	3.85	9.91			
	$\theta = 45°$	4.69	13.38			
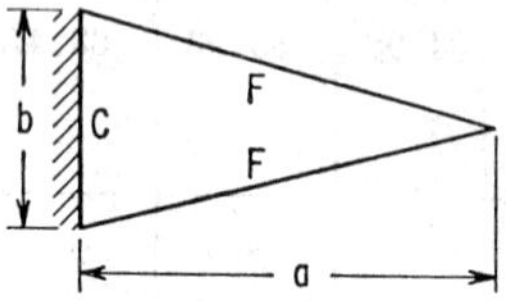	$a/b = 2$	6.7	28.6	56.6	137	
	$a/b = 4$	6.6	28.5	83.5	240	
	$a/b = 8$	6.6	28.4	146.1	457	
	$a/b = 14$	6.6	28.4	246	790	
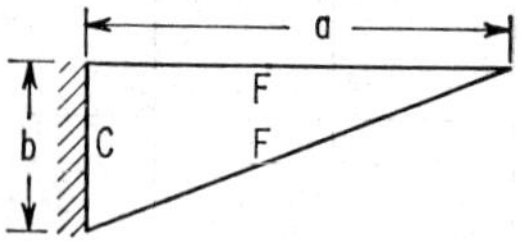	$a/b = 2$	5.5	23.6			
	$a/b = 4$	6.2	26.7			
	$a/b = 7$	6.4	28.1			

Table 6.8g. Natural Frequencies of Rectangular Plates[36,72]

$$f = C\frac{h}{a^2} \times 10^4 \times K_m$$

f = lowest natural frequency, cps
C = frequency constant given in table
a = plate edge length, inches, as shown
h = plate thickness, inches
K_m = material constant (Table 6.10) = 1.00 for steel

Boundary conditions S = simply supported C = clamped	b/a						a/b					
	1.0	1.5	2.0	2.5	3.0	∞	1.0	1.5	2.0	2.5	3.0	∞
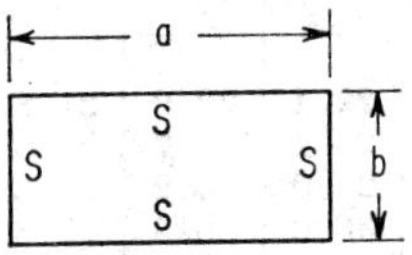	19.20	13.87	12.00	11.14	10.67	9.60	19.20	13.87	12.00	11.14	10.67	9.60
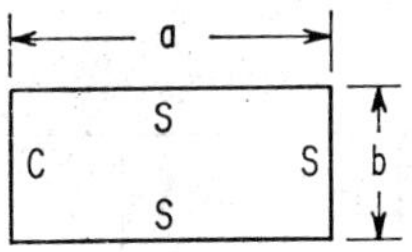	23.01	18.39	16.86	16.18	15.82	15.01	23.01	15.15	12.57	11.43	10.84	9.60
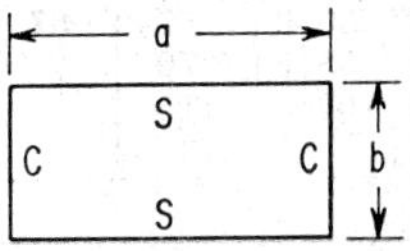	28.16	24.37	23.17	22.64	22.37	21.76	28.16	16.90	13.32	11.80	11.05	9.60

Table 6.8g. Natural Frequencies of Rectangular Plates (Continued)

$$f_n = C_n \frac{h}{a^2} \times 10^4 \times K_m$$

f_n = nth natural frequency, cps
C_n = frequency constant listed in these tables
h = plate thickness, inches
a = plate edge length, inches
K_m = material constant (Table 6.10) = 1.00 for steel

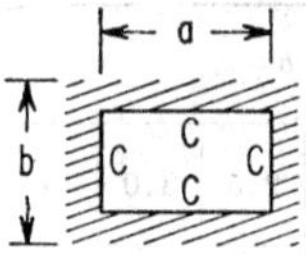

		b/a									
		1.0	0.9	0.8	0.7	0.6	0.5	0.4	0.3	0.2	0.1
n	1	35.0	31.8	29.1	26.9	25.1	23.9	23.0	22.4	22.0	21.8
	2	71.4	60.6	51.1	42.9	36.2	30.9	27.1	24.4	22.7	22.0
	3	105.3	68.8	66.6	64.8	55.4	43.6	33.5	28.0	24.2	22.3
	4	128.0	95.4	86.7	69.8	63.5	61.6	45.4	33.6	26.3	22.8

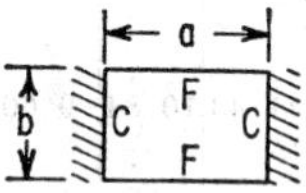

		b/a									
		1.0	0.9	0.8	0.7	0.6	0.5	0.4	0.3	0.2	0.1
n	1	21.5	21.4	21.4	21.4	21.3	21.3	21.2	21.2	21	21
	2	25.6	26.7	27.6	29.2	31.5	34.8	40.4	50.0	57	57
	3	42.3	47.4	54.2	59.2	59.0	59.0	58.6	58.5	70	
	4	59.4	59.5	59.2	64.4	74.2	79.7	89.3	106.5	115	

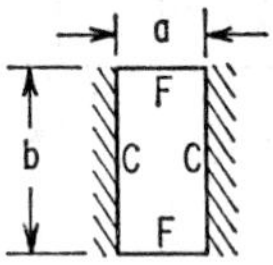

		a/b									
		1.0	0.9	0.8	0.7	0.6	0.5	0.4	0.3	0.2	0.1
n	1	21.5	21.5	21.5	21.5	21.5	21.5	21.6	21.6	21.6	21.6
	2	25.6	24.9	24.2	23.6	23.1	22.7	22.3	22.0	21.8	21.8
	3	42.3	38.3	34.8	31.6	28.9	26.7	24.8	23.4	22.4	22.0
	4	59.4	59.4	56.5	47.9	40.6	34.5	29.7	26.0	23.5	22.4

Table 6.8h. Natural Frequencies of Circular Plates[36]

$$f = C\frac{h}{r^2} \times 10^4 \times K_m$$

f = natural frequency, cps
C = frequency constant listed in table
h = plate thickness, inches
r = plate radius, inches
K_m = material constant (Table 6.10) = 1.00 for steel

Boundary conditions		Number of nodal circles	Number of nodal diameters			
			0	1	2	3
Clamped at circumference		0	9.936	20.651	33.906	
		1	38.713			
		2	86.516			
Free	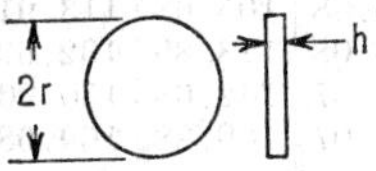	0			5.110	11.902
		1	8.832	19.970	34.295	51.491
		2	37.487	58.255		
Clamped at center	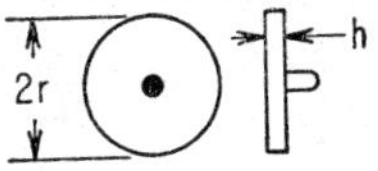	0	3.649			
		1	20.349			
		2	59.053			
		3	116.490			
Simply supported at circumference	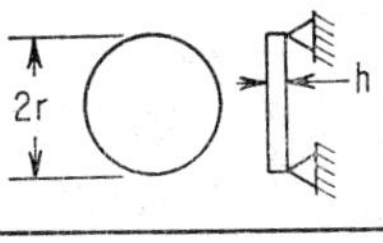	0	5.0			

Table 6.8i. Natural Frequencies of Circular Membranes[36]

$$f = \frac{C}{r}\sqrt{\frac{S}{h}}\,K_m$$

f = natural frequency, cps
C = frequency constant listed in table
r = membrane radius, inches
h = membrane thickness, inches
S = tension at circumference, lb/in.
K_m = material constant (Table 6.10) = 1.00 for steel

Boundary conditions	Number of nodal circles	Number of nodal diameters					
		0	1	2	3	4	5
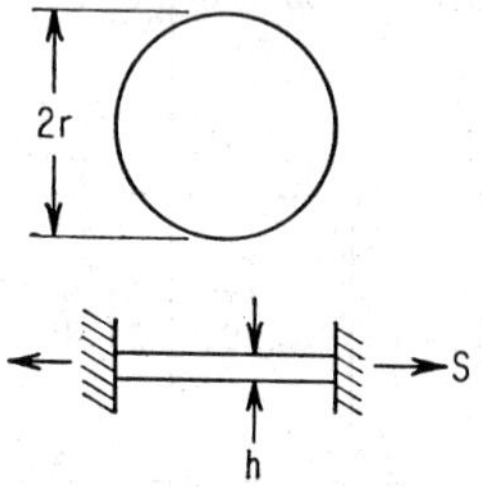	1	14.09	22.49	30.12	37.46	44.56	51.55
	2	32.41	41.22	49.44	57.30	64.94	72.22
	3	50.79	59.71	64.94	76.45	84.55	92.18
	4	69.28	78.09	86.90	95.12	103.34	111.56
	5	87.48	96.88	105.68	113.91	122.13	130.35
	6	106.27	115.08	123.89	132.69	140.91	149.13
	7	124.47	133.87	142.68	150.90	159.70	167.92
	8	143.26	152.07	160.88	169.68	178.49	186.71

Table 6.8j. Natural Frequencies of Miscellaneous Systems

Mass Free to Rotate and Translate in Plane[59]

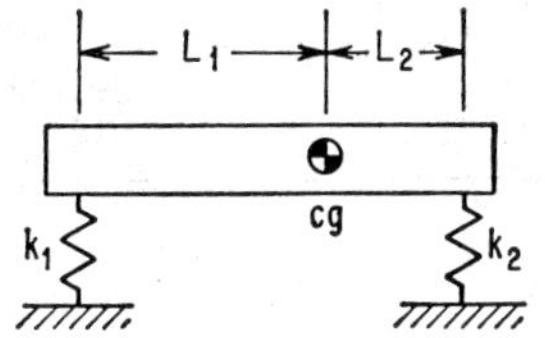

Mass $= M$, moment of inertia about c.g. $= r^2M$

$$2\omega^2 = a + c \pm \sqrt{(a - c)^2 + (2b/r)^2}$$

$$a = \frac{k_1 + k_2}{M} \qquad b = \frac{L_2k_2 - L_1k_1}{M}$$

$$c = \frac{L_1^2k_1 + L_2^2k_2}{r^2M}$$

Thin Rings[17]

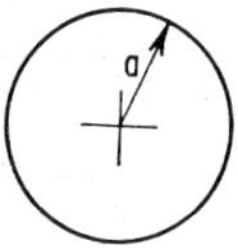

Extension (pure radial mode): $\omega = \dfrac{1}{a} c_L$

Bending in its own plane: $\omega_n = \dfrac{n(n^2 - 1)}{\sqrt{n^2 + 1}} \dfrac{r}{a^2} c_L$

$n = 2, 3, \ldots$

$r = \sqrt{I/A}$

I = moment of inertia of ring section area A for bending in its own plane
For out-of-plane bending and partial rings, see Den Hartog[17]

Free-free Beam on Elastic Foundation[39]

Same mode shapes as without support, but frequencies increased:

$$\omega^2_{n,\text{ on support}} = \omega^2_{n,\text{ unsupported}} + \frac{k'}{\rho A}$$

$$\frac{k'}{\rho A} = \frac{\text{support spring constant/unit length}}{\text{beam mass/unit length}} = \omega^2_{\text{rigid beam on elastic foundation}}$$

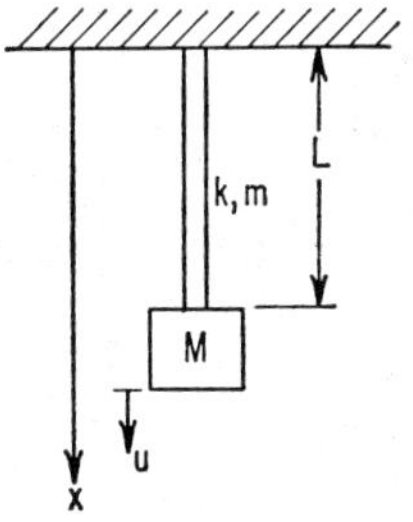

Mass on Rod or Spring of Static Stiffness k, Spring Total Mass m Uniformly Distributed[63]

Frequency equation:

$$(M/m)(\omega \sqrt{m/k}) = \cot (\omega \sqrt{m/k})$$

$$\phi_n(x) = \sin (\omega_n x/L \sqrt{m/k})$$

Weighting function $\mu(x) = m/L + M\delta(x - L)$
[see Art 6.3]

M/m	$\omega_1 \sqrt{m/k}$	$\omega_2 \sqrt{m/k}$	$\omega_3 \sqrt{m/k}$	$\omega_4 \sqrt{m/k}$	$\omega_5 \sqrt{m/k}$
0*	$\pi/2$	$3\pi/2$	$5\pi/2$	$7\pi/2$	$9\pi/2$
½	1.077	3.644	6.579	9.630	12.722
1	0.960	3.435	6.437	9.426	12.645
∞ †	0	π	2π	3π	4π

* Corresponds to spring free at $x = L$.
† Corresponds to spring clamped at $x = L$.

Table 6.9a. Combination of Spring Constants

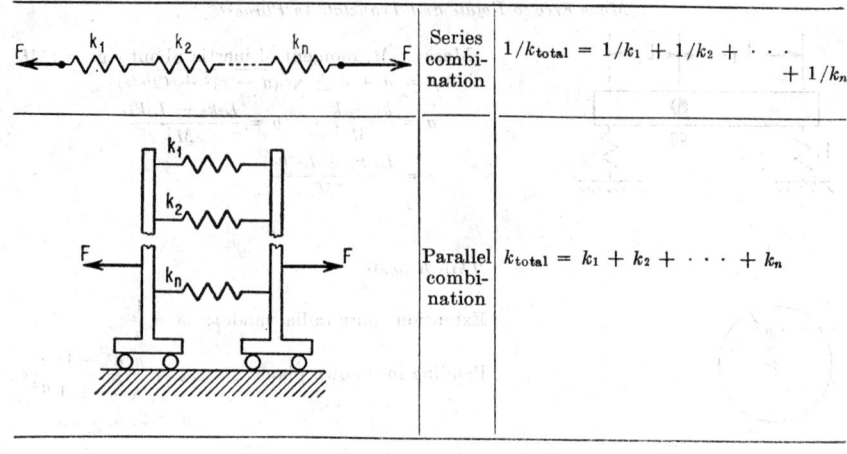

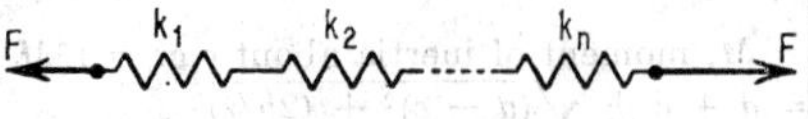	Series combination	$1/k_{\text{total}} = 1/k_1 + 1/k_2 + \cdots + 1/k_n$
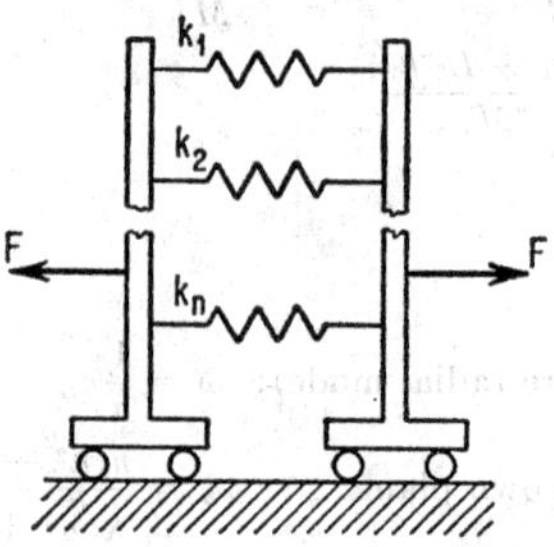	Parallel combination	$k_{\text{total}} = k_1 + k_2 + \cdots + k_n$

Table 6.9b. Spring Constants of Round-wire Helical Springs[17]

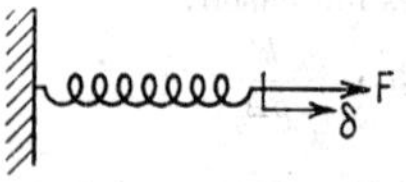	Axial loading	$\frac{F}{\delta} = k = \frac{Gd^4}{8nD^3}$
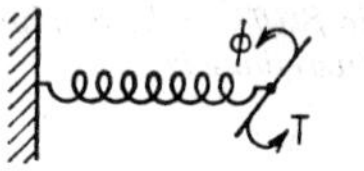	Torsion	$\frac{T}{\phi} = k_r = \frac{Ed^4}{64nD}$
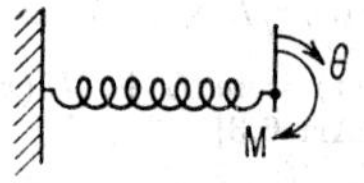	Bending	$\frac{M}{\theta} = k_b = \frac{2}{2 + \nu} k_r$

Symbols:

F = axial force
δ = axial deflection
T = torque
ϕ = torsion angle
M = bending moment
θ = flexure angle
G = shear modulus
E = elastic modulus
ν = Poisson's ratio
d = wire diameter
D = mean coil diameter
n = number of coils

Table 6.9c. Spring Constants of Beams[49]

	Translational (force F)	Rotational (moment M)
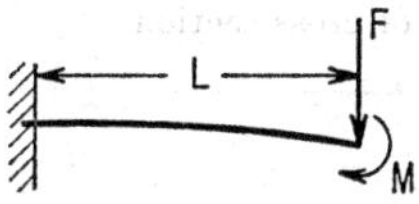	$k = \frac{3EI}{L^3}$	$k_r = \frac{EI}{L}$
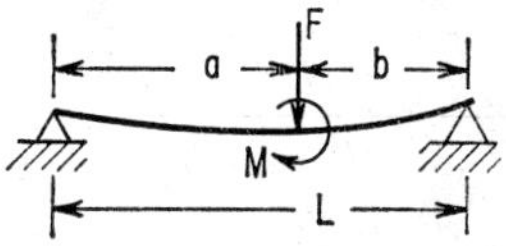	$k = \frac{3EIL}{(ab)^2}$	$k_r = \frac{3EIL}{L^2 - 3ab}$
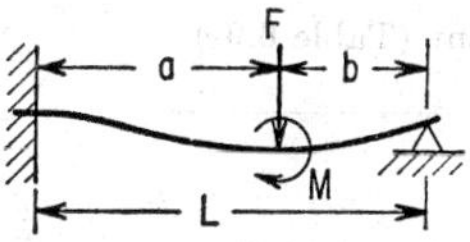	$k = \frac{96EI}{b(5b^2 - 3L^2)}$	$k_r = \frac{4EIL^3}{b[4L^3 - 3b(L + a)(L + b)]}$
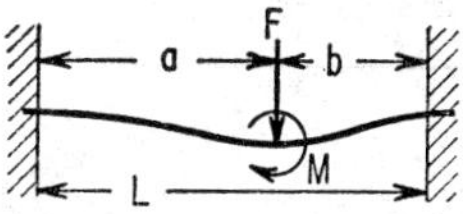	$k = \frac{3EIL^3}{(ab)^3}$	$k_r = \frac{EIL^3}{ab(L^2 - 3ab)}$
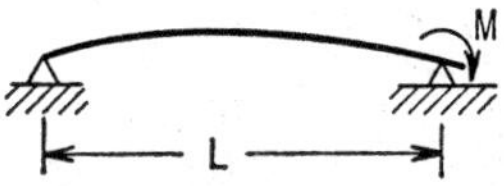		$k_r = \frac{3EI}{L}$
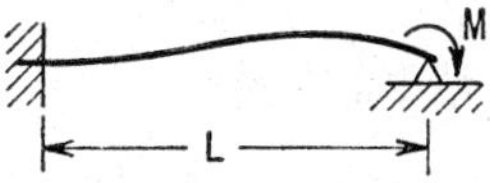		$k_r = \frac{4EI}{L}$

Symbols:

E = modulus of elasticity
I = centroidal moment of inertia of cross section
F = force
M = bending moment

$k = F/\delta$, translational spring constant
$k_r = M/\theta$, rotational spring constant
δ = lateral deflection at F
θ = flexural angle at M

Table 6.9d. Spring Constants of Torsion Springs

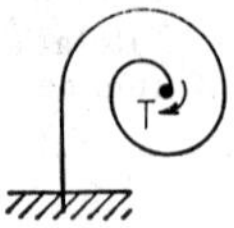	Spiral spring[17] $k_r = T/\phi = EI/L$ E = modulus of elasticity L = total spring length I = moment of inertia of cross section
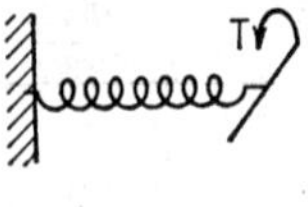	Helical spring See Table 6.9*b*
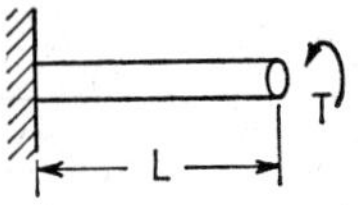	Uniform shaft $k_r = GJ/L$ G = shear modulus L = length J = torsional constant (Table 6.9*e*)
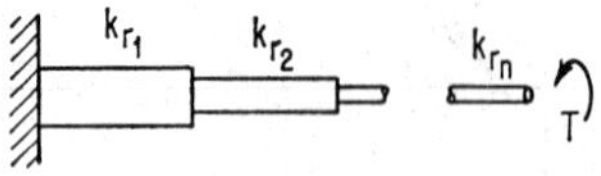	Stepped shaft $1/k_r = 1/k_{r_1} + 1/k_{r_2} + \cdots 1/k_{r_n}$ $k_{r_j} = k_r$ of jth uniform part by itself

Table 6.9e. Torsional Constants J of Common Sections[49]

Circle	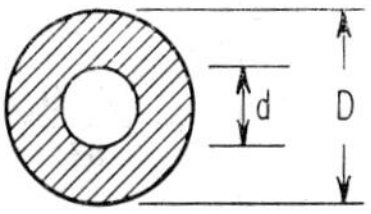	$J = \frac{\pi}{32}(D^4 - d^4)$	
Ellipse	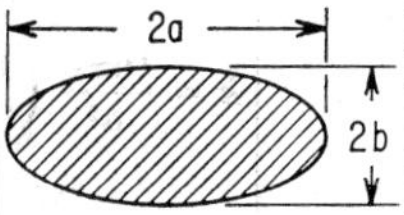	$J = \frac{\pi a^3 b^3}{a^2 + b^2}$	
Square	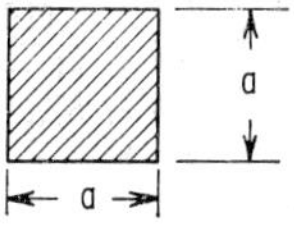	$J = 0.1406\, a^4$	
Rectangle	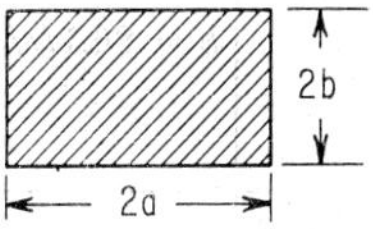	$J = ab^3\left[\frac{16}{3} - 3.36\frac{b}{a}\left(1 - \frac{b^4}{12a^4}\right)\right]$	
Equilateral triangle	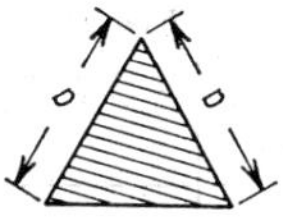	$J = \frac{\sqrt{3}\, a^4}{80}$	
Any solid compact section without reentrant angles	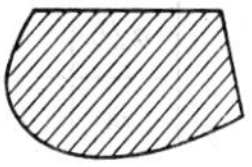	$J \approx \frac{A^4}{40I}$	
Any thin closed tube of uniform thickness t	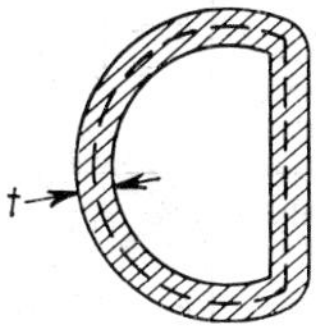	$J \approx \frac{4A^2 t}{U}$	A = cross-section area U = mean circumferential length (length of dotted lines shown) I = polar moment of inertia of section about its centroid
Any thin open tube of uniform thickness t	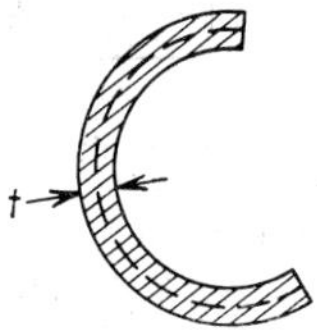	$J \approx \frac{Ut^3}{3}$	

Table 6.9f. Spring Constants of Centrally Loaded Plates[49]

Circular	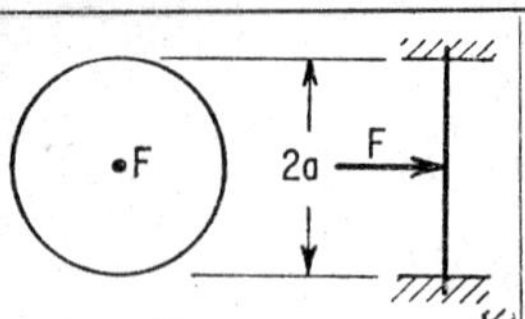	Clamped	$k = \dfrac{16\pi D}{a^2}$
	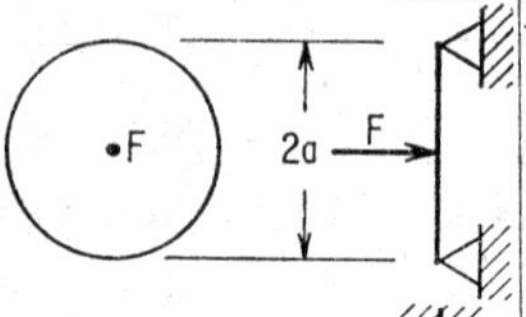	Simply supported	$k = \dfrac{16\pi D}{a^2}\left(\dfrac{1+\nu}{3+\nu}\right)$
	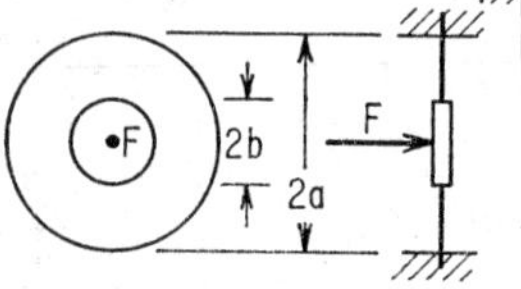	Clamped, with concentric rigid insert	$k = \dfrac{(16\pi D/a^2)(1-c^2)}{(1-c^2)^2 - 4c^2(\ln c)^2}$ $c = b/a$
Square	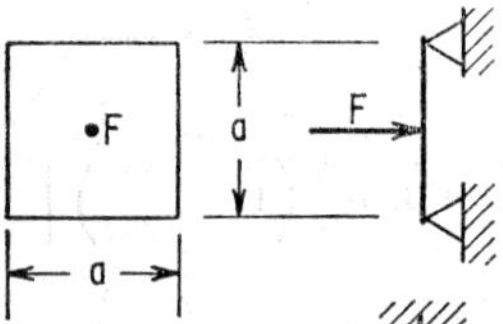	All edges simply supported	$k = 86.1D/a^2$
	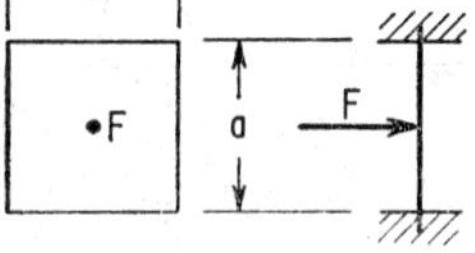	All edges clamped	$k = 192.2D/a^2$
Rectangular	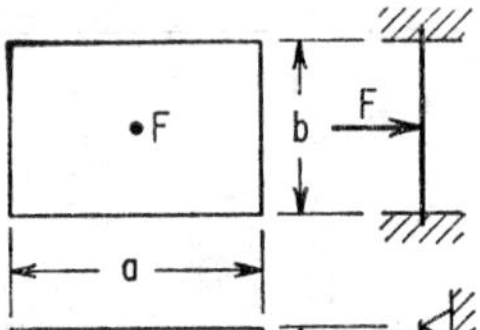	All edges clamped	$k = D\alpha/b^2$ a/b: 4, 2, 1 α: 167, 147, 192
	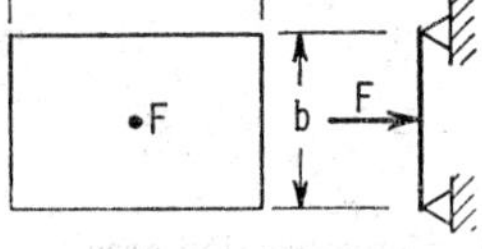	All edges simply supported	$k = 59.2(1 + 0.462c^4)D/b^2$ $c = b/a < 1$
Equilateral triangular	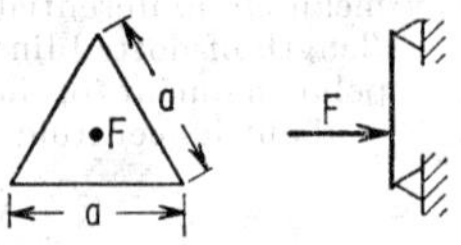	All edges simply supported	$k = 175D/a^2$

Symbols:

k = force/deflection, spring constant

$D = \dfrac{Eh^3}{12(1-\nu^2)}$

E = modulus of elasticity

ν = Poisson's ratio

h = plate thickness

Table 6.10. Sound Velocity and K_m for Table 6.8 for Engineering Materials

Materials	Temp.,* °C	Sound velocity[71] $c_L = \sqrt{E/\rho}$, ft/sec	$K_m = \frac{(c_L)_{material}}{(c_L)_{steel}}$
Aluminum	...	16,740	0.985
Brass, bronze	...	11,480	0.680
Copper	...	11,670	0.682
Copper	100	10,080	0.598
Copper	200	9,690	0.575
Iron, average steel	...	16,820	1.000
Iron	100	17,390	1.032
Iron	200	15,480	0.919
Cast steel	...	16,360	0.972
Cast steel	200	15,710	0.933
Lead	...	4,026	0.239
Magnesium	...	15,100	0.896
Monel metal	...	14,700	0.872
Nickel	...	10,340	0.615
Silver	...	8,550	0.508
Silver	100	8,660	0.515
Tin	...	8,200	0.488
Titanium	...	16,300–16,700	0.97–0.99
Titanium	90	15,800–16,300	0.94–0.97
Titanium	200	15,300–15,700	0.91–0.93
Titanium	325	14,000–14,600	0.83–0.87
Titanium	540	13,100–14,000	0.78–0.83
Zinc	...	12,140	0.722
Concrete	...	12,000–15,000	0.71–0.89
Cork	...	1,640	0.0974
Granite	...	19,680	1.169
Marble	...	12,500	0.743
Glass	...	16,400–19,700	0.97–1.17
Plywood	...	≈7,000	≈0.42
Woods:			
Along fibers	...	11,000–16,000	0.65–0.95
Across fibers	...	4,000– 4,500	0.24–0.27

* Room temperature is implied where no entry appears.

References

1. Barton, M. V. (ed.): "Shock and Structural Response," American Society of Mechanical Engineers, New York, 1960.
2. Beaquiére, A.: "Mecanique non linéaire," Gauthier-Villars, Paris, 1960.
3. Bendat, J. S.: "Principles and Applications of Random Noise Theory," John Wiley & Sons, Inc., New York, 1958.
4. Biezeno, C. B., and R. Grammel: "Engineering Dynamics," vol. III, "Steam Turbines," Blackie & Son, Ltd., London, 1954.
5. Bishop, R. E. D., and D. C. Johnson: "Vibration Analysis Tables," Cambridge University Press, Cambridge, 1956.
6. Burton, R.: "Vibration and Impact," Addison-Wesley Publishing Company, Inc., Reading, Mass., 1958.
7. Caughey, T. K.: Response of Van der Pol's Oscillator to Random Excitation, *J. Appl. Mech.*, vol. 81, no. 3, pp. 345–348, September, 1959.
8. Chenea, P. F.: "On the Application of the Impedance Method to Continuous Systems," paper 52-A-28, presented at ASME annual meeting, 1952.
9. Church, A. H.: "Mechanical Vibrations," John Wiley & Sons, Inc., New York, 1957.
10. Church, A.H ., and R. Plunkett: Balancing Flexible Rotors, *Trans. ASME* (*Ser. B*), vol. 83, pp. 383–389, November, 1961.

11. Churchill, R. V.: "Operational Mathematics," 2d ed., McGraw-Hill Book Company, Inc., New York, 1958.
12. Crandall, S. H.: "Random Vibration," The Technology Press of the Massachusetts Institute of Technology, Cambridge, Mass., 1958.
13. Crandall, S. H.: "Engineering Analysis," McGraw-Hill Book Company, Inc., New York, 1956.
14. Crede, C. E.: "Vibration and Shock Isolation," John Wiley & Sons, Inc., New York, 1951.
15. Cunningham, W. J.: "Introduction to Nonlinear Analysis," McGraw-Hill Book Company, Inc., New York, 1958.
16. Davenport, W. B., and W. L. Root: "Introduction to Random Signals and Noise," McGraw-Hill Book Company, Inc., New York, 1958.
17. Den Hartog, J. P.: "Mechanical Vibrations," 4th ed., McGraw-Hill Book Company, Inc., New York, 1956.
18. Felgar, R. P.: Formulas for Integrals Containing Characteristic Function of a Vibrating Beam, *Univ. Texas, Bur. Eng. Res., Cir.* 14, 1950.
19. Ford, L. R.: "Differential Equations," 2d ed., McGraw-Hill Book Company, Inc., New York, 1955.
20. Frankland, J. M.: Effects of Impact on Simple Elastic Structures, *Proc. Soc. Exptl. Stress Anal.*, vol. 6, pp. 7–27, 1948.
21. Fung, Y. C.: "An Introduction to the Theory of Aeroelasticity," John Wiley & Sons, Inc., New York, 1955.
22. Fung, Y. C., and M. V. Barton: Some Shock Spectra Characteristics and Uses, *J. Appl. Mech.*, September, 1958, pp. 365–372.
23. Gardner, M. F., and J. L. Barnes: "Transients in Linear Systems," vol. I, John Wiley & Sons, Inc., New York, 1942.
24. Heckl, M. A.: Compendium of Impedance Formulas, Bolt Beranek and Newman Inc., *Rept.* 774, submitted to Office of Naval Research, Code 411, May 26, 1961.
25. Holowenko, R.: "Dynamics of Machinery," John Wiley & Sons, Inc., New York, 1955.
26. Jacobsen, L. S., and R. S. Ayre, "Engineering Vibrations," McGraw-Hill Book Company, Inc., New York, 1958.
27. Kamke, E.: "Differentialgleichungen, Lösungsmethoden und Lösungen," 3d ed., Chelsea Publishing Company, New York, 1948.
28. Kaplan, W.: "Ordinary Differential Equations," Addison-Wesley Publishing Company, Inc., Reading, Mass., 1958.
29. LePage, W. R., and S. Seely: "General Network Analysis," McGraw-Hill Book Company, Inc., New York, 1952.
30. Levy, H., and E. A. Baggott: "Numerical Solutions of Differential Equations," Dover Publications, Inc., New York, 1950.
31. Lowe, R., and R. D. Cavanaugh: Correlation of Shock Spectra and Pulse Shape with Shock Environment, *Environ. Eng.*, February, 1959.
32. Lukasik, S., and A. W. Nolle (eds.): "Handbook of Acoustic Noise Control," vol. I, Supplement 1, Wright Air Development Center Report, WADC TR 52-204, April, 1955.
33. Lyon, R. H.: On the Vibration Statistics of a Randomly Excited Hard-spring Oscillator, *J. Acoust. Soc. Am.*, vol. 32, pp. 716–719, June, 1960.
34. Lyon, R. H.: Equivalent Linearization of the Hard Spring Oscillator, *J. Acoust. Soc. Am.*, vol. 32, pp. 1161–1162, September, 1960.
35. Macduff, J. N., and J. R. Curreri: "Vibration Control," McGraw-Hill Book Company, Inc., New York, 1958.
36. Macduff, J. N., and R. P. Felgar: Vibration Design Charts, *Trans. ASME*, vol. 79, pp. 1459–1475, 1957; also Vibration Frequency Charts, *Machine Design*, Feb. 7, 1957.
37. Mahalingam, S.: Forced Vibration of Systems with Non-linear Non-symmetrical Characteristics, *J. Appl. Mech.*, vol. 24, pp. 435–439, September, 1957.
38. Martienssen, O.: Über neue Resonanzerscheinungen in Wechselstromkreisen, *Physik. Z.*, vol. 11, pp. 448–460, 1910.
39. McGoldrick, R. T.: "A Vibration Manual for Engineers," 2d ed., David Taylor Model Basin Report R-189, December, 1957.
40. Mindlin, R. D.: Dynamics of Package Cushioning, *Bell System Tech. J.*, vol. 24, nos. 3, 4, pp. 353–461, July–October, 1945.
41. Morrill, B.: "Mechanical Vibrations," The Ronald Press Company, New York, 1957.
42. Morse, P. M.: "Vibration and Sound," 2d ed., McGraw-Hill Book Company, Inc., New York, 1948.
43. Myklestad, N. O.: "Fundamentals of Vibration Analysis," McGraw-Hill Book Company, Inc., New York, 1956.
44. Nestroides, E. J. (ed.): "Bicera: Handbook on Torsional Vibrations," Cambridge

University Press, New York, 1958 (British Internal Combustion Engine Research Association).
45. Ostergren, S. M.: "Shock Response of a Two-degree-of-freedom System," Rome Air Development Center Report, RADC TN 58-251.
46. Pistiner, J. S., and H. Reisman: Dynamic Amplification Factor of a Two-degree-of-freedom System, *J. Environ. Sci.*, October, 1960, pp. 4–8.
47. Plunkett, R. (ed.): "Mechanical Impedance Methods for Mechanical Vibrations," American Society of Mechanical Engineers, New York, 1958.
48. Rice, S. O.: Mathematical Analysis of Random Noise, *Bell System Tech. J.*, vol. 23, pp. 282–332, 1944; vol. 24, pp. 46–156, 1945. Also reprinted in Wax, N.: "Selected Papers on Noise and Stochastic Processes," Dover Publications, Inc., New York, 1954.
49. Roark, R. J.: "Formulas for Stress and Strain," 3d ed., McGraw-Hill Book Company, Inc., New York, 1954.
50. Ruzicka, J. (ed.): "Structural Damping," American Society of Mechanical Engineers, New York, 1959.
51. Salvadori, M. G., and M. L. Baron: "Numerical Methods in Engineering," Prentice-Hall, Inc., Englewood Cliffs, N.J., 1952.
52. Scanlan, R. H., and R. Rosenbaum: "Introduction to the Study of Aircraft Vibration and Flutter," The Macmillan Company, New York, 1951.
53. Scarborough, F. B.: "Numerical Mathematical Analysis," The Johns Hopkins Press, Baltimore, 1930.
54. Schwesinger, G.: On One-term Approximations of Forced Nonharmonic Vibrations, *J. Appl. Mech.*, vol. 17, no. 2, pp. 202–208, June, 1950.
55. Slater, J. C., and N. H. Frank: "Mechanics," McGraw-Hill Book Company, Inc., New York, 1947.
56. Snowdon, J. C., and G. G. Parfitt: Isolation from Mechanical Shock with One and Two-stage Mounting Systems, *J. Acoust. Soc. Am.* vol. 31, pp. 967–976, July, 1959.
57. Stodola, A. (Translated by L. C. Lowenstein): "Steam and Gas Turbines," McGraw-Hill Book Company, Inc., New York, 1927.
58. Stoker, J. J.: "Nonlinear Vibrations," Interscience Publishers, Inc., New York, 1950.
59. Thomson, W. T.: "Mechanical Vibrations," 2d ed., Prentice-Hall, Inc., Englewood Cliffs, N.J., 1953.
60. Thomson, W. T.: A Note on Tabular Methods for Flexural Vibration, *J. Aeron. Sci.*, January, 1953, pp. 62–63.
61. Thomson, W. T.: Shock Spectra of a Nonlinear System, *J. Appl. Mech.*, vol. 27, pp. 528–534, September, 1960.
62. Timoshenko, S.: "Vibration Problems in Engineering," 2d ed., D. Van Nostrand Company, Inc., Princeton, N.J., 1937.
63. Tong, K. N.: "Theory of Mechanical Vibration," John Wiley & Sons, Inc., New York, 1960.
64. Trent, H. M.: Physical Equivalents of Spectral Notions, *J. Acoust. Soc. Am.*, vol. 32, pp. 348–351, March, 1960.
65. Ungar, E. E.: Maximum Stresses in Beams and Plates Vibrating at Resonance, *Trans. ASME* (*Ser. B*), vol. 84, pp. 149–155, February, 1962.
66. Vigness, I.: Fundamental Nature of Shock and Vibrations, *Elec. Mfg.*, vol. 63, pp. 89–108, June, 1959.
67. Walsh, J. P., and R. E. Blake: The Equivalent Static Accelerations of Shock Motions, *Proc. Soc. Exptl. Stress Anal.*, vol. 6, pp. 7–27, 1948.
68. Wilson, W. K.: "Practical Solution of Torsional Vibration Problems," John Wiley & Sons, Inc., New York, 1956.
69. Zorowski, C. F.: Shock-damping Calculations, *Machine Design*, May, 1961.
70. Anderson, B. W.: Vibration of Triangular Cantilever Plates by the Ritz Method, *Trans. ASME*, vol. 76, pp. 365–370, 1954.
71. Hodgman, C. D. (ed.): "A Handbook of Chemistry and Physics," 31st ed., Chemical Rubber Publishing Company, Cleveland, Ohio, 1949.
72. Claassen, R. W., and C. J. Thorne: "Transverse Vibrations of Thin Rectangular Isotropic Plates," NAVWEPS Report 7016, U.S. Naval Ordnance Test Station, China Lake, Calif., Aug. 18, 1960.

Section 7

NOISE CONTROL

By

CLAYTON H. ALLEN, Ph.D., *Senior Engineering Scientist, Bolt Beranek and Newman Inc., Cambridge, Mass.*

CONTENTS

7.1. INTRODUCTION

Increasing demands for high speed and compactness in new machines and the greater awareness of noise as an annoyance and potential hazard have focused attention on the problem of noise control. Higher speed generally increases the power employed in machine systems and, consequently, the noise. Compactness frequently requires some sacrifice of the space needed for noise and vibration isolation. Thus, noise control has become a major consideration in machine design.

Noise control is the technology of obtaining an acceptable noise environment which is consistent with economic and operational considerations. Noise control differs from noise reduction in that satisfactory noise control may result from a change in the character of the noise even without any reduction in the overall amount of noise. Under some circumstances, satisfactory noise control may be obtained by adding some masking noise to obscure objectionable noise components.

Noise control is of vital economic importance in machine design. This is particularly obvious in consumer products where two of the prime factors governing acceptance are the amount and quality of noise. The consumer often judges the quality of a machine, its power, and even its life expectancy, consciously or unconsciously, by the sound of its operation.

Noise control in machine design is closely related to vibration control, since a large portion of machine noise results from the vibrations set up within the solid structure. However, it is essential to differentiate clearly between vibrations in the solid structure and noise actually generated in air. Noise, as treated in this chapter, refers only to airborne sound, but the reduction of airborne noise may involve some operations on solid objects which vibrate.

Noise is a subtle thing which defies exact definition. Practically, *noise is any unwanted sound.* As arbitrary as this definition may be, it is probably the most exact and useful definition, for it is a fact that any sound may or may not be noise, depending upon the activity and mental attitude of the listener, the motivation for his use of a noisy device, or the relationship between the noise and his general welfare.

It is not practical, economical, or even desirable to eliminate all noise from mechanical systems. In fact, a noiseless machine may suffer by comparison with a noisy one because of its apparent lack of power or quality. To aid in evaluating the need for noise control in any particular application and to enable the designer to direct his noise reduction efforts and limit them at an appropriate level, this chapter first reviews the physics of sound, then presents criteria for acceptable noise levels in various applications, and finally deals with noise-reduction techniques from the standpoint of practical application.

Several modern texts and other references on acoustics and noise control are available (refs. 1 to 11) describing both practical applications and theoretical studies of noise control.

7.2. PHYSICS OF NOISE

(a) Wave Motion

Noise is a *wave motion* in air. A wave motion is generated in an elastic medium such as air when the medium is disturbed in any way by a driving force, a *source.* Since the density and elasticity are distributed throughout the medium, a motion at one point sets up a motion at adjacent points; therefore, a disturbance at one region is coupled to adjacent regions and spreads throughout the medium (see Sec. 8).

The speed of sound in a gas

$$c = \sqrt{\gamma P/\rho} \tag{7.1}$$

where γ = ratio of specific heats
P = ambient or atmospheric pressure
ρ = density of air at pressure P and temperature T

Substitution of the ideal-gas law and appropriate constants for air into Eq. (7.1) yields

$$c = 49.03\sqrt{T} \quad \text{ft/sec} \tag{7.2}$$

where T = absolute temperature, degrees Rankine (degrees Rankine = degrees Fahrenheit + 460)

Most sounds which are of interest in noise control are *periodic;* they are generated by a source which vibrates more or less continuously at some frequency f. Nonperiodic sounds are comprised of a large number of periodic sounds of various frequencies, all having sinusoidal wave shape. Since sound waves do not interact with each other, their physical characteristics are additive, and the properties of noise can be understood most directly by studying the characteristics of simple sinusoidal sound waves and then considering the effects of their addition.

Generally the frequencies of sounds important in noise control lie in the *audible range* (between 20 and 20,000 cps). The performance of many of the devices employed in noise control is related to the *wavelength* λ of sound; it is therefore important to consider the length of the wave at the frequencies of interest in relation to the dimensions of the mechanical device which may generate the noise or be used to reduce the noise. The wavelength

$$\lambda = c/f \tag{7.3}$$

where f = frequency

Sound waves displace air a small distance and return it to its original position when they have passed. The *excess pressure* above ambient which causes the displacement of the particles of air is called the *sound pressure*, the to-and-fro displacement of the air is called the *particle displacement*, and the velocity with which the particles move is called *particle velocity*. In general, there is no net flow of air. However, just as surface waves can travel on a flowing stream, so sound waves pass through a current of air and are substantially unaffected by the air flow.

(b) Sound Sources

There are two important kinds of sources, *simple sources*, or more correctly *monopole sources*, and *dipole sources*, illustrated in Fig. 7.1. A monopole source has the same *phase* over its entire surface, i.e., it expands and contracts in such a way that all its surfaces move in and out together. A dipole consists of a pair of monopoles of equal strength, but opposite in phase, located close to each other. For a dipole source, half the surface moves out while the other half moves in.

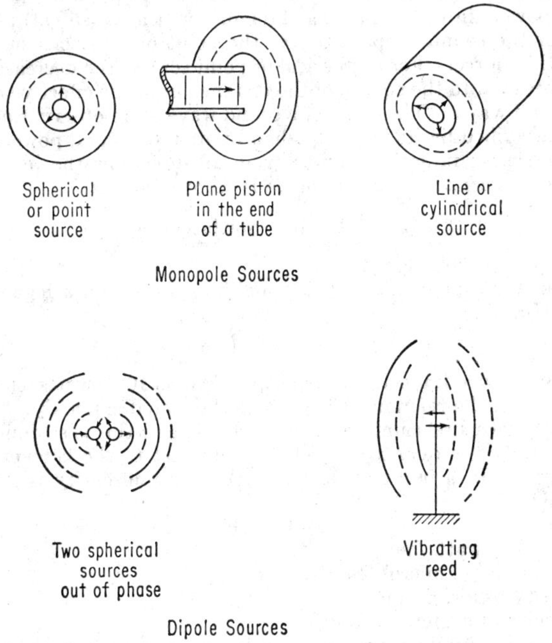

FIG. 7.1. Sources of sound.

A monopole, which is small compared with a wavelength, radiates in all directions equally and is said to be a *nondirectional* source. A dipole radiates best in the directions of its motion and very little at right angles regardless of how small it is. Such a source is said to be *directional.* Real machinery generally has a combination of many monopole and dipole sources and may have a complicated *directivity pattern.*

(c) Sound Energy

A wave carries energy away from the source. The amount of energy carried away per unit time is defined as sound power radiated by the source. Since the amount of power radiated from common sources varies over a wide range, it is convenient to express sound power on a logarithmic scale in terms of *sound-power level* (PWL), decibels (db)

$$\text{PWL} = 10 \log W/W_{\text{ref}} \tag{7.4}$$

where W = radiated sound power, watts

W_{ref} = a reference power, frequently taken as 10^{-13} watt but sometimes taken as 10^{-12} watt (denoted "re 10^{-12} watt")

Since the numerical value of the sound level depends upon the reference power, the reference must be stated explicitly.

The amount of sound power radiated by a source is a characteristic of the source itself and its mode of operation and is affected only slightly by its surroundings. Scales of sound power and sound-power level are compared in Fig. 7.2, and the relative positions of several sources of interest in noise-control problems are indicated on these scales. Generally, a sound which is 6 to 10 db higher than another similar sound is found to be twice as loud on a subjective scale.

As a sound wave propagates away from a source, little sound is absorbed by the air; so that the total sound power in the wave remains almost constant even for large distances, a hundred feet or more. As the wave travels, the sound power generally spreads out over an increasingly large area. Therefore, the sound intensity, which is defined as the power (in watts) passing through a unit area (sq cm) perpendicular to the direction of the sound propagation at the point of observation, decreases with distance from the source. For a spherical or point source, the decrease in intensity is inversely proportional to the square of the distance from the effective center of divergence of the sound wave. The effect of a sound wave at an observation point is more appropriately described by its intensity than by its total sound power.

For a simple source radiating uniformly in all directions the intensity over any spherical surface concentric with the source is uniform and is

$$I = W/S \quad \text{watts/cm}^2 \tag{7.5}$$

where S = area of the sphere, cm^2

Like sound power, sound intensity is usually expressed on a logarithmic scale as *intensity level* (IL), db,

$$\text{IL} = 10 \log I/I_{\text{ref}} \tag{7.6}$$

where I_{ref} = a reference intensity, frequently taken as 10^{-16} watts/cm²

As with power level, the reference value should be stated explicitly.

It is difficult to measure sound intensity directly, and this is seldom done. Sound pressure is generally the quantity which is of interest to the listener and most frequently measured. In a free progressive wave, sound intensity is related to sound pressure by

$$I = p^2/\rho c \tag{7.7}$$

where I = intensity, watts/cm²

p = rms pressure, microbars (dynes/cm²)

ρ = density of air, g/cm³

c = velocity of sound, cm/sec

Sound pressure is usually expressed on a logarithmic scale as *sound-pressure level* (SPL), which is defined as an extension of intensity level. By combining Eqs. (7.6)

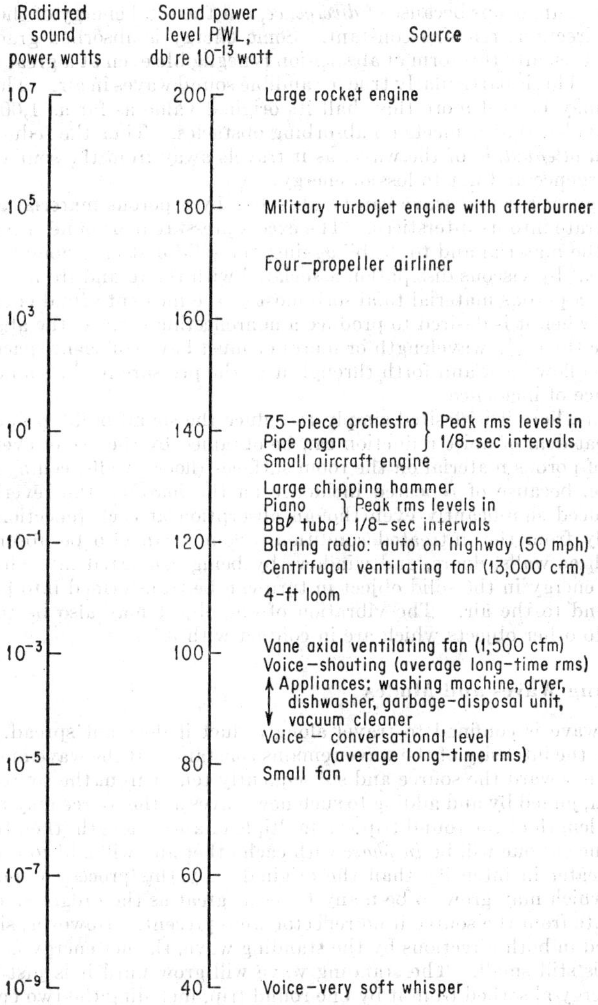

FIG. 7.2. Sound-power levels of some common sound sources.

and (7.7) the intensity level becomes

$$\text{IL} = 10 \log \frac{p^2/\rho c}{(p^2/\rho c)_{\text{ref}}} \quad \text{db} \tag{7.8}$$

which reduces to the following equation defining sound-pressure level when ρc is assumed constant:

$$\text{SPL} = 20 \log p/p_{\text{ref}} \quad \text{db} \tag{7.9}$$

where p_{ref} = reference pressure, usually taken as 0.0002 microbar (this reference pressure makes the intensity level and the sound-pressure level equal in a free progressive sound wave at room temperature and pressure)

The reference pressure should be stated explicitly when sound-pressure levels are reported. As a free progressive sound wave spreads, its intensity and consequently its sound pressure drops because of *divergence,* but the total energy of the wave radiating in all directions remains constant. Some energy is absorbed gradually by the medium, but usually this form of absorption is negligible even at a great distance from the source. This is particularly true for audible sound waves in air. The total energy of a wave may be still more than half its original value as far as 1,000 ft from the source* provided that it meets no absorbing obstacles. Thus the reduction in intensity, termed *attenuation* of the wave, as it travels away from the source, is primarily due to divergence and not to loss of energy.

The energy of a sound wave can be absorbed by a porous material which permits air to penetrate into its interstices. The excess pressure in a sound wave forces air to move into the material and to "rub" against the solid matter. Sound energy is converted to heat by viscous dissipation associated with the to-and-fro motion of the air. In order for a porous material to absorb most of the incident sound energy, as would be required when it is desired to produce a nonreflecting surface, the absorbing material must be thick (¼ wavelength or more) or must have sufficient space behind it to permit air to flow back and forth through it as the pressure in the sound wave varies at the surface of incidence.

In a room where it is desired simply to reduce the sound buildup due to reverberation, a great deal of noise reduction can be obtained by the use of even a relatively thin layer of porous material on the room surfaces (floors, walls, ceilings, furnishings, etc.). Here, because of repeated incidence on the material the reverberant sound will be reduced significantly even though absorption at each reflection is increased only slightly from the untreated condition. Sound can also be absorbed by solid objects such as walls, doors, and windows by being converted into vibration. The vibrational energy in the solid object in turn can be transformed into heat or reradiated as sound to the air. The vibration of one object may also be transmitted as vibrations to other objects which are in contact with it.

(d) Standing Waves and Modes

When a wave is confined to travel along a duct it does not spread. Aside from small losses, the intensity of the wave remains constant. If the wave meets a reflector it will return toward the source and subsequently reflect from the source to begin its course again, joined by and adding to such new waves as the source may be generating. If the total length of the round trip is a multiple of a wavelength, then the new sound wave and the old one will be *in phase* with each other and will add to produce a wave which is greater in intensity than the original. By this process a *standing wave* is generated which may grow to be many times as great as the progressive wave which would radiate from the source if no reflector were present. However, since energy is being carried in both directions by the standing wave, the net energy flow away from the source is still small. The standing wave will grow until it is just large enough that the energy absorbed from it by one round trip, including the two end reflections, equals the amount of energy supplied by the source. This is an important concept in noise control. It is the buildup of standing waves which makes it difficult to quiet a noisy device by confining it in an enclosure.

Sounds which are only partially confined between parallel reflecting surfaces will act in the same way as if they were in a duct when the reflecting surfaces are large compared with a wavelength and the distance between reflectors is a multiple of one-half of the wavelength.

In rooms or other hard-walled enclosures, more complicated standing sound waves called *room modes* or simply *modes* will form. The frequency for the lowest mode in a

* Attenuation of sound in still air depends upon humidity and frequency. Whereas the effect of air attenuation is negligible upon the noise produced by machines in ordinary rooms, it may become significant and should be considered in setting up noise-level criteria for large machines operating outdoors far away from residential communities. For such situations, prevailing winds, ground cover, and general climatic conditions must also be considered (ref. 2, pp. 185*ff.*).

one-dimensional room (the equivalent of a long tube) is

$$f_1 = c/2d \tag{7.10}$$

where f_1 = frequency of the lowest mode, cps
c = speed of sound, fps
d = distance separating two reflecting surfaces, ft

In general, modes will exist at frequencies f_n, where $f_n = nf_1$ with n being an integer 1, 2, 3, ...

When a sound wave strikes an isolated object, the reflected sound radiates very much as if the reflector were the source of the sound. Such a reflector is sometimes referred to as a *secondary source.* An object which is small compared with a wavelength reflects very little sound and acts as a weak point source. A large plane surface acts like a mirror and redirects the wave as if it came from a mirror image of the true source.

Perhaps one of the most important secondary sources is a hole in a wall or in a barrier of any sort. The hole will transmit sound with little, if any, attenuation in sound pressure. Therefore, the intensity of the sound over the area of the hole is substantially undiminished while the intensity over the rest of the area of the wall or barrier may be diminished by several decibels depending upon the construction. If the wall gives a moderate transmission loss, say 40 db, the amount of sound power which passes through a small hole will equal that which passes through an area of the wall which is 10,000 times larger. The effect of a hole is increasingly serious as the quality of the wall, i.e., its transmission loss, is increased by improved construction.

(e) Acoustic Impedances

Air has a characteristic impedance (rayls, dynes-sec/cm³)

$$Z_c \equiv p/u \tag{7.11}$$

where p = *instantaneous excess pressure,* or briefly the *sound pressure* in microbars (10^{-6} atm)
u = instantaneous *particle velocity,* cm/sec

The characteristic impedance of air (40.6 rayls at room temperature and pressure)

$$Z_c = \rho c = (\rho\gamma P)^{1/2} \tag{7.12}$$

where ρ = ambient density of air (0.0012 gm/cm³)
c = velocity of sound in air (34,400 cm/sec)
γ = ratio of specific heats (1.4 for air)
P = ambient pressure (10^6 microbars or dynes/cm² at 1 atm)

A sound source has a radiation impedance also, called the *specific acoustic impedance* (rayls), defined as follows:

$$Z_s = p/u_{\text{source}} \tag{7.13}$$

where u_{source} = instantaneous source velocity, cm/sec

When the source is large compared with a wavelength and has substantially no air leaks around it, the pressure generated by its motion is radiated into the air as sound pressure, and the radiation impedance of the source equals the characteristic impedance of air. When the source is small compared with a wavelength, air slips around it; the radiation impedance decreases, and little sound is radiated. In this case the pressure and particle velocity are no longer directly in phase with the motion of the source, and the impedance is therefore complex, $Z_s = R_s + iX_s$. R_s is the *resistive* or *real* component in which the pressure and velocity are in phase; it is responsible for the generation of sound. X_s is the *reactive* or *imaginary* component in which the velocity and pressure are 90° out of phase; it results only in a circulation of air back and forth around the source.

The resistive and reactive components of the radiation impedance are plotted in Fig. 7.3 for four important sources, a sphere, a circular piston in a wall or baffle, a circular piston in a long unflanged pipe, and a circular disk vibrating perpendicular to its own plane. The impedances have been normalized by dividing them by the characteristic impedance ρc of air. The efficiency of sound radiation is a maximum when the resistive component of the normalized impedance approaches unity and the reactive component is small.

It can be observed that the resistive component of impedance for each of these sources is close to unity when the perimeter of the sources equals or is larger than a wavelength, i.e., when $ka \equiv 2\pi a/\lambda \geq 1$. The radiation resistance drops rapidly if the source is made smaller. For sources which are not round, an approximate equivalent radius may be used to obtain qualitative results. It should be noted that the radiation resistance of the sphere and the piston behave similarly for small radii.

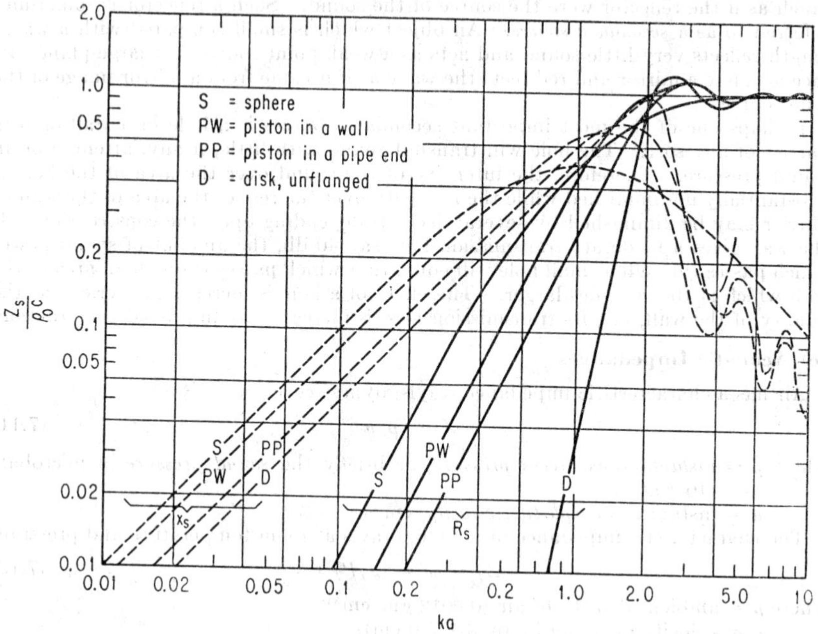

FIG. 7.3. Real and imaginary parts of radiation impedance normalized against the characteristic impedance of air. Note: $Z_s/\rho_0 c = m/\rho_0 c = Z_a S/\rho_0 c$ where S = the radiating area, $ka = 2\pi a/\lambda$, a = radius. S = sphere, PW = piston in a wall, PP = piston in a pipe end, D = disk, unflanged.

Each of these is a simple source since the entire radiating surface has the same phase. In contrast, the radiation resistance of the disk drops twice as rapidly with decreasing size as that of either of the other sources. This is because the disk acts as a dipole since one side is out of phase with the other—air can move around it from front to back (whereas it can only be pushed aside by the other sources). This fact is of significance in noise control since any means for permitting the flow of air from front to back of a small vibrating member will decrease its sound radiation significantly. This control measure will be useful, however, only when the perimeter of the vibrating member is small compared with a wavelength, or when the source can be pierced with holes large enough and close enough together to reduce the effective size of the source between holes to the equivalent of a disk which is small compared with a wavelength.

(f) Mechanical Impedance

Most solid sources have a high internal or *mechanical impedance*, defined as the ratio of the driving force to the velocity amplitude of the driven member. Because solid objects are usually dense compared with air, the forces required to move the object are large compared with those required to move the air with the same acceleration. Thus the additional load caused by the air in contact with the object causes only a negligible decrease in its motion (if it is driven with a constant force amplitude) or a negligible increase in its driving force (if it is constrained to move with a given displacement amplitude). It is this high internal impedance which is characteristic of most mechanical sources of noise.

Not all mechanical sources have a high internal impedance, however. A vibrating panel or diaphragm is an example of a mechanical source having a fairly low impedance. The driving force is usually applied at one or more mounting points and spreads throughout the panel as a vibrational wave. The air loading may add significantly to the driving force required to maintain the panel vibration at a panel resonance frequency. In particular, the air load becomes great when the vibrational bending wave velocity in a panel equals or exceeds the velocity of sound in air. In that case there will be some angle ϕ of radiation for which the bending wavelength in the panel just matches the wave trace in air along the panel; i.e., the bending wavelength λ_B in the panel equals the wavelength in air λ divided by sin ϕ. A large amount of sound will be radiated at this angle from such a panel. Likewise, as shown in Fig. 7.4*a*, such a panel struck by sound waves incident at this same angle will vibrate easily and will transmit sound from one side to the other almost as if the panel were not there. This effect is called the *coincidence effect.*

The angle of coincidence varies with frequency because the velocity of the bending wave in a panel increases as the square root of frequency, while the speed of sound in air is independent of frequency. At some frequency the speed of sound in the panel just equals that in air, and at this frequency coincidence occurs only at grazing incidence. Below this frequency, coincidence cannot occur, and the panel acts like a limp mass, radiating sound very poorly and transmitting incident sound poorly also.

The coincidence frequency of most panels and wall structures occurs in the low audible-frequency range; sound radiation and transmission will therefore occur for such barriers over a wide frequency range. The coincidence frequency can be raised by reducing the panel stiffness; therefore an ideal wall is a limp wall, e.g., a dense, nonporous blanket with zero stiffness. Some specially designed panel structures[15] use a controlled stiffness that raises the coincidence frequency to the high audible-frequency range and greatly increases the sound isolation.

(g) Transmission Loss and Noise Reduction

The effectiveness of a sound barrier is measured by its *transmission loss* or by the *noise reduction* it produces. Transmission loss and noise reduction are related and are both measured in decibels, but they are not the same quantity. Transmission loss is defined as ten times the logarithm of the ratio of the sound intensity incident upon a wall to the intensity transmitted through it. Noise reduction is the difference between the sound-pressure level measured under two different conditions; the two conditions, for example, may be two positions on opposite sides of a wall or may be the same position with and without the wall in place. It is apparent that transmission loss is a quantity characteristic of the barrier. Noise reduction, by contrast, involves the barrier, the acoustic environment, and any other conditions which may be changed between the two measurements being compared.

The transmission loss of a limp wall increases with frequency and increases with the surface weight. This relation, often referred to in acoustics as the *mass law,* is presented graphically in Fig. 7.4*b*.

The relation between noise reduction and transmission loss depends upon so many factors that texts on acoustics should be consulted for details, but some special cases of interest may be noted here:

1. If a wall separates a reverberant noisy area from a quiet, *anechoic* space, i.e., a region without reflecting surfaces, and if the noise reduction is defined as the difference between the noise level in the reverberant space and the noise level just outside the wall, the noise reduction will be 6 db greater than the transmission loss. The

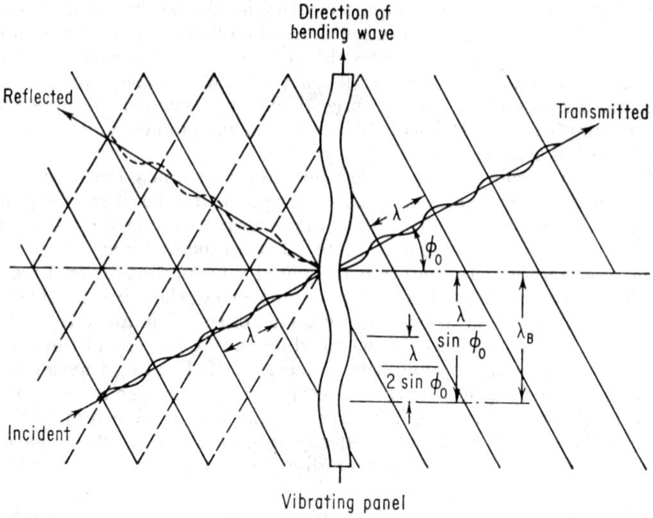

(a)

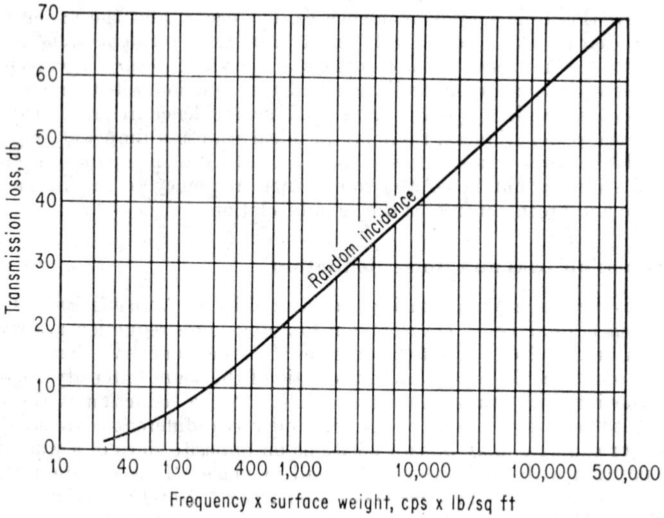

(b)

FIG. 7.4. Sound reflection and transmission. (a) Wave coincidence. The wavelength of the bending wave in the panel is λ_B. A sound wave in air, whose wavelength is λ, impinges on a plate at the angle ϕ_0. When $\lambda/\sin \phi_0$ is equal to λ_B, the intensity of the transmitted wave approaches the intensity of the incident wave. The frequency for which $\lambda = \lambda_B$ is called the critical frequency.

(b) Transmission loss (TL) for solid damped partitions.[2] The average transmission loss may be determined from this graph by assuming a frequency of 500 cps. For undamped walls subtract about 5 db from the average transmission loss. For plane sound waves incident normal to the face of the panel, the transmission loss increases 6 db for a doubling of the abscissa rather than 5 db, as is shown here for randomly incident sound.

reason is that within the reverberant region the average intensity level of the sound striking a wall is 6 db less than the average sound-pressure level measured in the reverberant space.

2. If a noise source is surrounded by an enclosure and if noise reduction is defined as the difference between the noise level at a listening position before and after adding the enclosure, then the noise reduction will generally be less than the transmission loss of the enclosure walls. The reason is that the sound within the enclosure will reflect from the walls many times and will increase in intensity until a balance is reached between the sound generated by the source and the dissipation of sound as heat within the enclosure, as vibration in solid structures, and as reradiated sound outside the enclosure. When the dissipation of sound within the enclosure as vibration and heat is small, the sound level within the enclosure will increase by a number of decibels approaching (or perhaps exceeding) the value of the transmission loss and the observed noise reduction may be nearly zero. This fact points out the necessity of adding as much sound-absorbing material as is practicable within an enclosure. When the walls are perfectly nonreflecting, the noise reduction equals the transmission loss.

(h) Sound Spectrum

A sound spectrum is a measure of the distribution of sound energy throughout the band of frequencies being considered. When a wave is sinusoidal, it contains only the single *fundamental* frequency characteristics of the *period* of the wave. Since all the energy of such a sound is concentrated at a single frequency, it is said to have a *line spectrum* located at the fundamental frequency. The hum of a transformer has a line spectrum. However, even transformer hum generally contains *harmonic frequencies,* and therefore the spectrum consists of a series of lines at frequencies which are multiples of the fundamental. Harmonics are more evident in the whine of gears or the shrill tone of a siren. The fundamental and its harmonics still form a periodic but nonsinusoidal wave. Such spectra are characteristic of rotating or other periodically moving machinery and are generally found to be the most disturbing components of their noise.

In contrast to line spectra, *continuous noise spectra* are generated by *random processes* such as the falling of water, which has no inherent periodicity. Continuous spectra may be made up of sound generated by numerous random pulses. Such a spectrum contains all frequencies, though some portions of the spectrum may carry more energy than others and may therefore have greater intensity. In general, machine noise is made up of a combination of a continuous or *broad-band* spectrum and a line spectrum.

Noise control is generally directed toward reducing broad-band spectra to an acceptable level for speech-communication requirements and reshaping the broad-band spectrum to produce a more natural or agreeable sound, then reducing the line spectrum so that it is inaudible in the presence of the broad-band spectrum.

The techniques of measurement and analysis of noise as required for the purpose of noise control are adequately discussed in refs. 1, 2, 3, and 8. However, a brief description of the basic process is necessary to the understanding of published information on noise criteria and noise-reduction procedures.

Noise is measured by means of a pressure-sensitive microphone that produces an electrical-signal replica of the sound pressure. This electrical signal then can be operated upon to determine all the essential characteristics of the noise. The electrical signal and the noise itself are so intimately identified with each other that the electrical signal is frequently looked upon as if it were the noise. Thus, such an expression as "passing the noise through a filter" really means "passing the electrical signal through an electrical filter."

Viewing the shape of the electrical wave on an oscilloscope gives a quick and qualitative picture of the sound wave, but such a presentation is of limited value for noise-control purposes. In general the electrical signal must be passed through some form of electrical filtering system and presented to a meter which has been calibrated to read directly the equivalent SPL in decibels. The narrower the filter bandwidth, the more detailed the information which can be obtained about the sound.

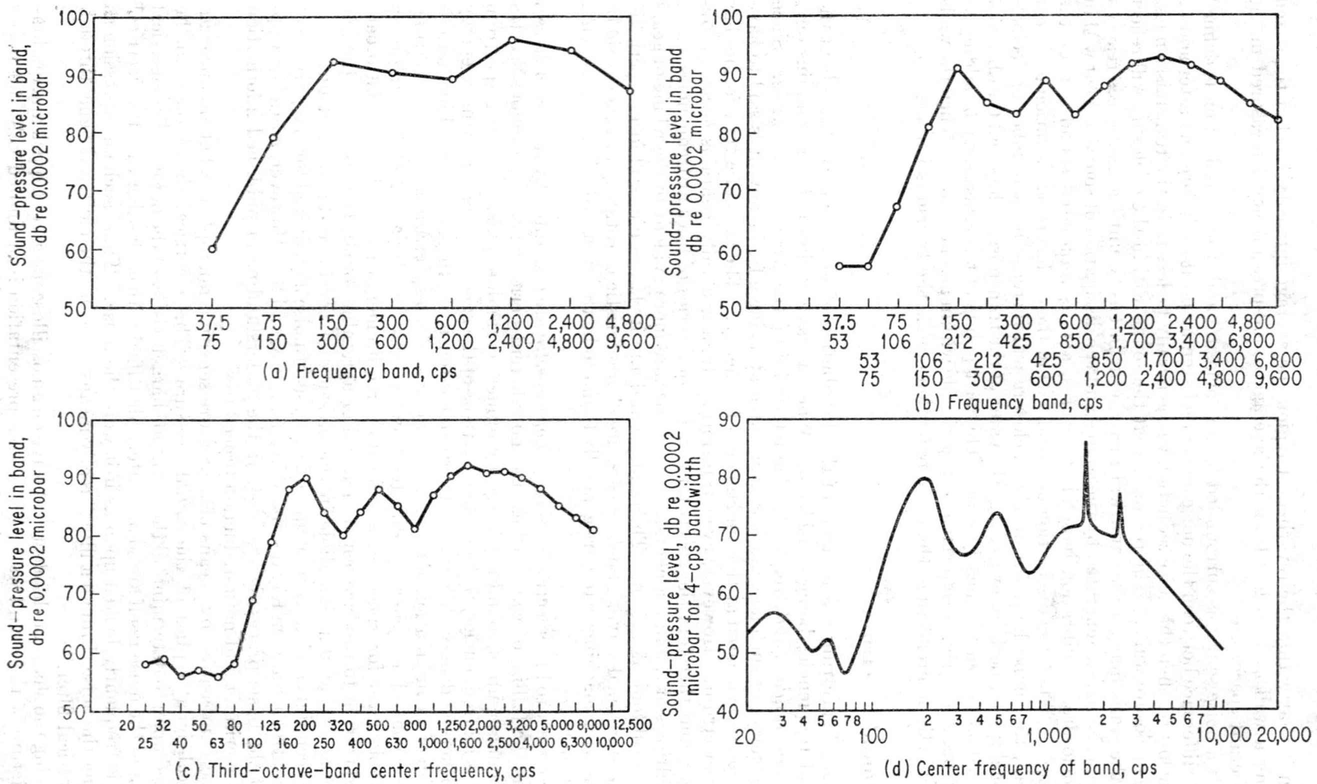

FIG. 7.5. Plots of a noise spectrum with a broad-band background and two pure tones at 15 and 2,500 cps, as measured on four different frequency analyzers. (*a*) Octave-band analyzer. (*b*) One-half-octave-band analyzer. (*c*) One-third-octave-band analyzer. (*d*) Constant-bandwidth (4-cps) analyzer. See ref. 2.

When the unfiltered signal is presented to meter, the reading is the *overall* SPL, which is of little value except where the spectral shape is known by some other means. Commercial sound-level meters usually have three weighting networks referred to as the *A*, *B*, and *C* scales, which permit a single number reading of overall noise levels approximating the overall response of an ear. The *A* scale is used for low sound levels, where the ear sensitivity varies widely with frequency; the *B* scale for the medium range; and the *C* scale for the high range, where the ear sensitivity is nearly flat.

The *A*, *B*, and *C* scale measurements are now generally supplanted by *octave-band* measurements using the eight filters that cover the ranges of 20 to 75, 75 to 150, 150 to 300, 300 to 600, 600 to 1,200, 1,200 to 2,400, 2,400 to 4,800, and 4,800 to 10,000 cps. The sound-pressure levels measured in octave bands give a sufficiently detailed picture of a broad-band spectrum for most noise-control purposes. Where pure tones are present or sharply peaked spectra are involved, closer inspection is necessary and a set of one-half- or one-third-octave filters or a single narrow-band tunable filter may be used.

A noise spectrum containing broad-band noise and two pure tones as measured by four different filtering systems is shown in Fig. 7.5. The pure tones are seen to be completely obscured, and even the irregular nature of the broad-band spectrum is not apparent with octave-band filtering. Half-octave and one-third-octave filters show improved resolution of the spectral shape, but a narrow-band filter is generally required to accurately display any pure tones which may be present.

It is important to note from Fig. 7.5 that the SPL measured with a wide-band filter is higher than that for a narrower-band filter. The difference in decibels will be nearly the value of 10 $\times$ log of bandwidth ratio. When the strength of a single pure tone dominates the signal passed by two filters of different bandwidths, the SPL measured will be nearly the same for both.

7.3. CRITERIA FOR NOISE CONTROL

Generally sound is objectionable and classed as noise requiring control when it is annoying, interferes with speech, causes hearing damage, or results in fatigue or malfunction of equipment and structures.

(a) Annoyance

Little is known in a quantitative way about annoyance caused by noise, because annoyance is so subjective a response. Broad-band noises, e.g., rustle of leaves, are not unpleasant and generally will be tolerated without annoyance even at relatively high levels. These noises are such a common part of normal experience that they are generally not noticed unless attention is drawn to their existence. Sounds of this type are useful in masking more objectionable noises which otherwise would be irritating.

Noises which are intermittent, shrill, excessively loud, or which contain a concentration of energy in pure tones or narrow bands are among the more annoying. Annoyance is generally reduced by reducing loudness, particularly by reducing the intensity of the high-frequency components and by avoiding intermittent changes in level. Noises which startle or those which convey some form of information are inordinately annoying because they induce fright or intrude upon privacy. Noises which have monotonous repetition may cause "psychological amplification," as in the case of a dripping faucet.

Rasping noise or rattle within a machine connotes excessive wear, improper assembly, or maladjustment and thereby generally evokes annoyance far beyond that which might be expected from the overall loudness or spectral character. Such noises have not submitted to the precise evaluation required for the development of numerical criteria; however, they are of vital importance in the consumer acceptance of household and personal products as well as of office and laboratory equipment. As a general rule it is safe to permit such noises only when they are 5 db or more below the level of other more acceptable masking noises which accompany the normal use of the device in question.

Some success has been obtained in predicting community reaction to noise.[12] This information is particularly useful in noise-control efforts required in the design of large manufacturing machinery, aircraft, transportation vehicles, and earth-moving or construction equipment.

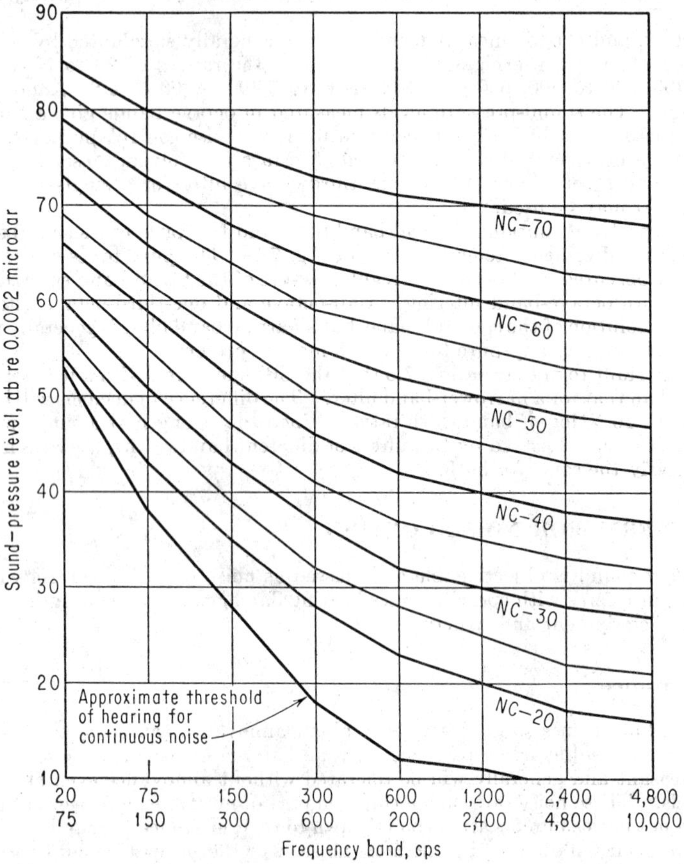

FIG. 7.6. Criterion curves for use with Tables 7.1 and 7.2 in determining the permissible (or desirable) SPL's in eight octave bands. These curves are recommended for specifications wherever a favorable relation between the low-frequency and the high-frequency portion of the spectrum is desired (see text).

Recent work with community response to jet and propeller aircraft[13] has led to the development of a new subjective measure of noise called *perceived-noise level* (units = PNdb) which is found to be particularly well suited to the evaluation of air-traffic noise problems. The perceived-noise level (PNL) of a particular noise in PNdb is defined as the sound-pressure level of a band of noise, from 910 to 1,090 cps, that sounds equally "noisy."

(b) Speech Interference

Human response to noise in buildings is governed by three subjective factors: the ease of speech communication, annoyance, and the desire for privacy. The ease of speech communication generally dominates. It is controlled primarily by the noise in the three octave bands 600 to 1,200 cps, 1,200 to 2,400 cps, and 2,400 to 4,800 cps. It has been found experimentally that the noise levels in these three octave bands are almost equally important and that a speech-interference level (SIL) obtained by averaging the decibel levels in these three bands predicts the ease of speech communication for a wide variety of spectra. Annoyance is controlled not only by the

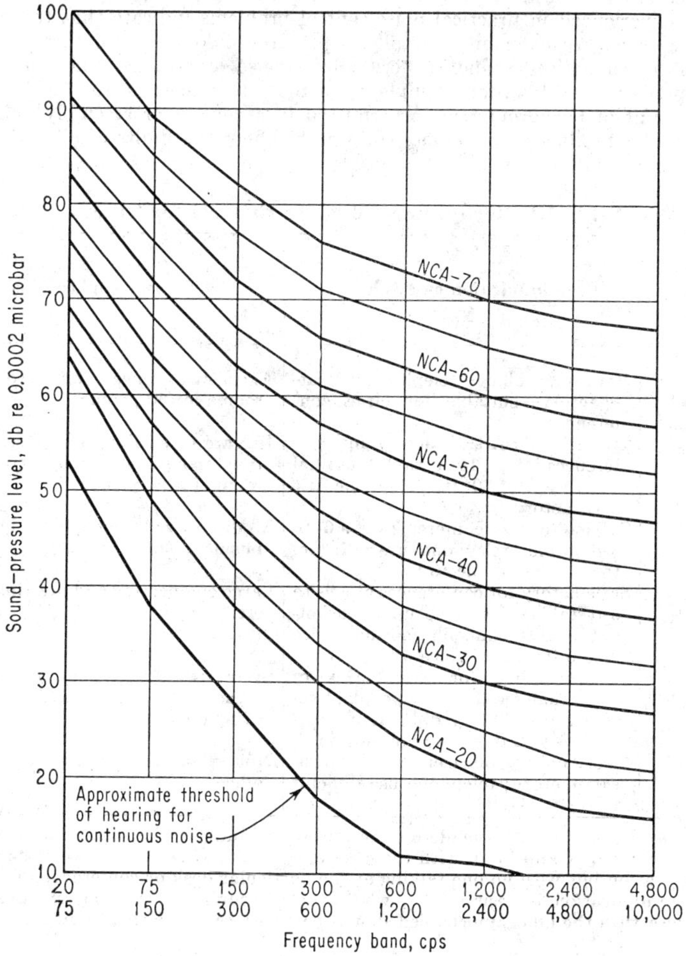

FIG. 7.7. Same as for Fig. 7.6 except that these curves may be used in place of those of Fig. 7.6 wherever economy dictates a maximum compromise and where, in addition, the noise is steady and free from beats between low-frequency pure-tone components.

SIL but by the values of noise outside the 600- to 4,800-cps region. The addition of noise outside the 600- to 4,800-cps region has little effect on speech intelligibility or the ease of speech communication.

Where a minimum of annoyance is desired consistent with a predetermined SIL value, the noise levels in each of the eight octave bands in the audible-frequency range should be maintained below the noise criteria (NC curves) given in Fig. 7.6. When a maximum permissible annoyance consistent with a preestablished SIL value must be permitted because of economic or other considerations, the octave-band noise levels should not be permitted to rise above the noise criteria (NCA curves) given in

Fig. 7.7. The NCA curves may be used as alternates for the NC curves only in instances where extreme economy is needed and where the noise is steady and free from beats between low-frequency pure-tone components.

The NC and NCA criteria have been found to be applicable to offices and other rooms where voice communication and associated listening activities predominate. The NC curves recommended for various office conditions are given in Table 7.1. For spaces other than offices, detailed studies are not available. However, general reactions experienced in practical noise-control problems indicate that the NC and NCA criteria curves have broad application. It should be noted that individual conditions affect attitudes and the critical response of people. The quality of a sound-reproducing system, for example, may allow as much as 5 to 15 db variation in the amount of background noise permitted in a concert or assembly hall. The noise normally produced in a home will greatly affect the limits of allowable noise from consumer products.

Table 7.1. Recommended Noise Criteria for Offices

NC curve of Fig. 7.6, NC units	Communication environment	Typical applications
20–30	Very quiet office—telephone use satisfactory—suitable for large conferences	Executive offices and conference rooms for 50 people
30–35	"Quiet" office: satisfactory for conferences at a 15-ft table; normal voice 10 to 30 ft; telephone use satisfactory	Private or semiprivate offices, reception rooms, and small conference rooms for 20 people
35–40	Satisfactory for conferences at a 6- to 8-ft table: telephone use satisfactory: normal voice 6 to 12 ft	Medium-sized offices and industrial business offices
40–50	Satisfactory for conferences at a 4- to 5-ft table; telephone use occasionally slightly difficult; normal voice 3 to 6 ft; raised voice 6 to 12 ft	Large engineering and drafting rooms, etc.
50–55	Unsatisfactory for conferences of more than two or three people; telephone use slightly difficult; normal voice 1 to 2 ft; raised voice 3 to 6 ft	Secretarial areas (typing) accounting areas (business machines), blueprint rooms, etc.
Above 55	"Very noisy"; office environment unsatisfactory; telephone use difficult	Not recommended for any type of office

NOTE: Noise measurements made for the purpose of judging the satisfactoriness of the noise in an office by comparison with these criteria should be performed with the office in normal operation but with no one talking at the particular desk or conference table where speech communication is desired (i.e., where the measurement is being made). Background noise with the office unoccupied should be lower, say by 5 to 10 units.

Table 7.2 presents values of the NC curves which provide background noise levels recommended for a variety of interior spaces under average conditions.

In the design of mechanical equipment, distinctive noises should be kept 3 to 5 db below the most critical background noise level in which the device is intended to operate. Where pure tones or narrow bands of noise exist, and particularly if these noises fluctuate in level, their peak values should be 10 to 15 db below the octave band levels of the applicable NC or NCA criteria.

(c) Privacy

Privacy criteria have been developed primarily in connection with speech transmission from one room to another. Similar criteria might be found to apply for mechanical devices such as water closets, electric razors, typewriters, air conditioners, and lawn mowers. Such items can be masked by steady broad-band background noise (resembling wind noise) such as diffuser noise from ventilation systems, but a machine designer has little control over these noise sources and should take into account the environment in which the device under design may be expected to operate.

It has been found as a criterion,[14] that speech privacy is obtained when the long time average* sound level of speech is at least 5 to 10 db below the background noise levels when compared in octave bands. With this criterion, approximately 1 per cent of the speech sound-level peaks† will exceed the background levels by 5 db. These peaks convey such a small proportion of the speech information that privacy is main-

Table 7.2. Recommended Noise Criteria for Rooms

Type of space	Recommended NC curve of Fig. 7.6 or 7.7, NC units	Computed equivalent SLM readings,* weighting scale A, dba
Broadcast studios	15–20	25–30
Concert halls	15–20	25–30
Legitimate theaters (500 seats, no amplification)	20–25	30–35
Music rooms	25	35
Schoolrooms (no amplification)	25	35
Television studios	25	35
Apartments and hotels	25–30	35–40
Assembly halls (amplification)	25–35	35–40
Homes (sleeping areas)	25–35†	35–45†
Motion-picture theaters	30	40
Hospitals	30	40
Churches (no amplification)	25	35
Courtrooms (no amplification)	25	30–35
Libraries	30	40–45
Restaurants	45	55
Coliseums for sports only (amplification)	50	60

NOTE: Noise levels are to be measured in unoccupied rooms. Each noise-criterion curve is a code for specifying permissible sound-pressure levels in eight octave bands. It is intended that in no one frequency band should the specified level be exceeded. The computed equivalent dba numbers in the right-hand column are presented for information only and are not recommended for use in specifications. Ventilating systems should be operating, and outside noise sources, traffic conditions, etc., should be normal when measurements are made.

* If there were relatively less noise in the low-frequency bands than indicated by the recommended noise-criterion curve, the dba numbers would be lower by about 5 dba. All numbers in this column should be dropped by 5 dba if they are to be used to estimate the compliance of a normally encountered noise level with an NC criterion; these numbers should not be used for specification purposes. (dba = decibels "a" scale.)

† Room air conditioners manufactured prior to 1957 commonly produce levels of 40 to 55 dba in sleeping areas.

tained. This criterion also appears to ensure that the received speech sounds generally will not be considered annoying. On this basis it can be assumed that the same criterion applied to the recognizable noises from mechanical devices will be conservatively safe.

* The "long time average" level of speech is determined by averaging the measured sound-pressure level over a period of several seconds of speaking time.

† Important speech sounds are generally distributed throughout a range of 30 db in level regardless of whether the conversation takes place at a low voice level in the quiet or at raised voice level in a noisy location. It is found that 1 per cent of the peak sound levels in speech exceed a value 12 db above the long term average. This level represents the upper edge of the 30-db range of speech sound levels.

(d) Hearing Conservation

Exposure to high noise levels will cause permanent loss of hearing acuity. Because the loss is gradual and cumulative, it generally is not recognized until it has reached serious proportions. Although the relevant criteria for determining upper limits for safe values of noise exposure are complicated because they depend upon the type of noise, the length of exposure, and the age and health and personal susceptibility of individuals, some useful criteria have been established.

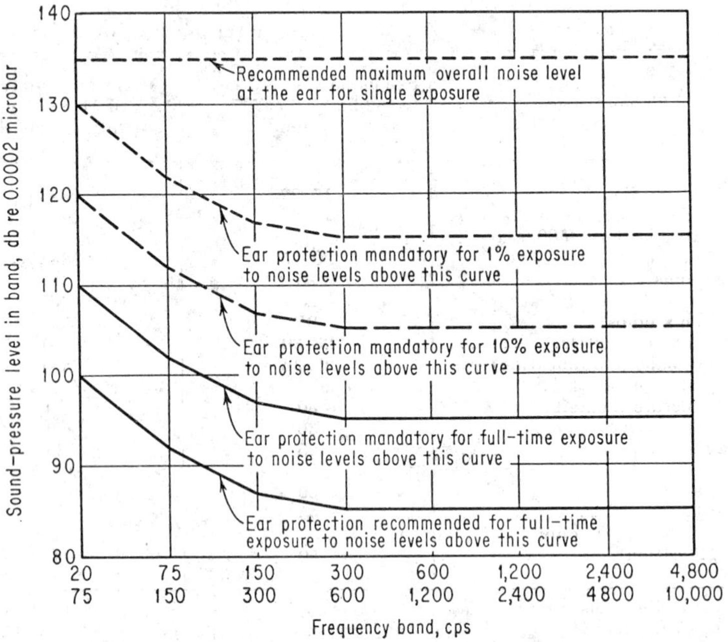

FIG. 7.8. Broad-band noise-level exposures for which conservation-of-hearing measures are recommended or mandatory. The parameter of the upper three curves is percentage of each working day. Exposure over a period of years is assumed. (*After Beranek,*[2] *p.* 572, *Fig.* 22.1.)

Pure tones and narrow bands of noise are more damaging than a broad band of noise of the same overall intensity. For a sufficiently narrow band of noise the damaging effectiveness seems to be indistinguishable from that of a pure tone of the same intensity. The largest band for which this relation holds is called a *critical band* for hearing-damage risk; it is approximately ⅓ octave wide. Bands of noise which are wider than ⅓ octave are potentially less damaging than the pure tones or critical bands when equal in overall noise level.

The maximum safe noise levels to which the average population may be subjected are summarized in Figs. 7.8 and 7.9. For pure-tone or critical bands the noise levels should be decreased 6 to 10 db from those shown. The criteria of Figs. 7.8 and 7.9 are consistent with the current issue of Air Force Regulation 160-3. Legal measures are continually being developed to enforce hearing conservation in noisy industries.

(e) Fatigue and Malfunction

In the intense fields generated by jet and rocket engines or even by large fans, compressors, or stationary engines, it is found that auxiliary mechanical devices such as control mechanisms or communications gear may be damaged or caused to malfunction by the sound received from the primary noise sources.

The noise levels which cause fatigue or failure in mechanical devices are highly dependent upon the mechanical construction of the device, its resonance frequencies, internal damping, and mounting. Therefore, no single criterion for the safe operation of equipment can be presented. It is possible to cause fatigue and malfunction of delicate equipment in noise fields having sound-pressure levels greater than about 120 db re 0.0002 microbar. Generally, equipment used in aircraft and missiles is expected to withstand extended exposures to sound-pressure levels in the range from 140 to 150 db while exterior structures encounter levels 20 db higher. Rocket devices generate levels up to 190 db over large regions near the jet stream, but since the

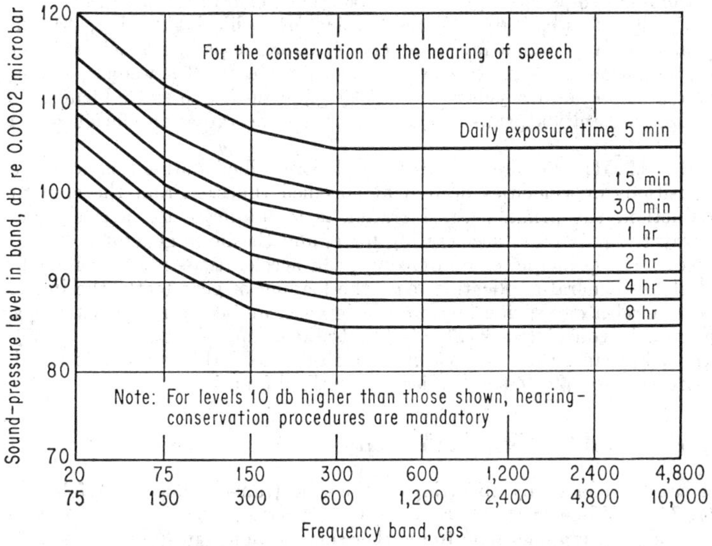

FIG. 7.9. Hearing-conservation criteria for use in noise-control design for broad-band noise levels of various average daily exposure times.[2] These curves are built on the lowest curve of Fig. 7.10, assuming that a halving of the daily exposure permits 3 db higher noise levels.

exhaust is located at the tail and the overall structure is large, the sound intensities (145 to 155 db re 0.0002 microbar) at the surfaces of the missile itself are not so high as those encountered on some aircraft where a jet engine is located directly beneath the wing or close to the side of the fuselage.

7.4. NOISE - CONTROL PRINCIPLES

(a) General Approach

For many applications in machine-generated noise, it is helpful to characterize the noise problem by a *vibration source, vibration path, noise sources, noise path, and a receiver*. The primary source of noise may be a vibration generated somewhere within the mechanism, not necessarily at the point primarily responsible for the radiation of noise. Separating the vibration source and the noise source helps show that noise control in machine design involves not only the control of airborne sound external to the machine but also the control of solid-borne vibration within the machine itself. From this standpoint there must be close correlation between use of the present section and Sec. 6 on Mechanical Vibrations. We shall deal here primarily with airborne sound and its control but shall consider vibration to the extent that it is directly responsible for the production of airborne sound.

(b) Noise Control at the Vibration Source

The most effective means of reducing noise is to alter the source. This includes the substitution of a quiet operation for a noisy one when such a substitution is feasible at the design stage. The prime mover for the source of airborne sound may be and frequently is a vibration generated at some remote location; power from the prime mover is transmitted to the radiating source of sound by means of solid-borne vibrations. Once the prime mover is located, the driving force can be reduced by one or more of the following steps: (1) Improve balancing of rotating parts, counterbalance eccentric loads; (2) substitute uniform rotation for oscillating motion; (3) reduce the moving mass of any oscillating member; (4) avoid impact, replace hammering by pressing; (5) replace dogs and latches by smooth-acting toggles, wedges, clutches, etc.; (6) reduce acceleration in cyclical motions (e.g., attention to curve blending in cams to minimize acceleration peaks); (7) reduce the sound-radiation efficiency of moving parts.

Where a cyclical motion is cam-governed, the prescribed motion is generally detailed as the positions or velocities desired at specified times. From the standpoint of noise reduction, the primary consideration is to minimize the acceleration peaks. In this way the maximum accelerating force, and the noise generated by it, will be minimized. The principle of minimizing peak accelerations is not confined to cam design but has general application, e.g., the loading of gear teeth, the impacting of relays, or the release of spring tension. Tolerances and clearances between parts and linkages must be considered as contributing to the flexing of the system. They should be minimized because they increase the response time in the system and thus increase the peak forces required to accomplish a prescribed function.

(c) Noise Control in the Vibration Path

Noise can be reduced by operating on the solid transmission in the following ways (see Fig. 7.10): When distances between the source and the isolated point are large (as, for example, from one room to another several rooms away in a building), vibrations are generally reduced by spreading through the structure. Because of resonances within the structure, this separation is not always successful, unless the distances involved are very large and a significant amount of vibration damping occurs in the intervening sections.

A vibration break is generally the most effective and least expensive means of vibration isolation. Ideally, a vibration break is a physical break in the solid structure across which vibrational forces cannot be transmitted. Practically, a vibration break must be filled or bridged by some material which will hold or locate the two parts with respect to each other. This bridging material is effective when it is dynamically soft compared with the solid structure. Many considerations enter into the choice of a material used to bridge a vibration break: The bridge must be as soft as possible in order to prevent the transmission of vibrational forces. The bridge must be stiff enough to give the required alignment of parts or to transmit the required steady or low-frequency forces for functional operation. The bridging material generally should be resistive as well as resilient, so that it will not create a system which will resonate and produce violent, and perhaps destructive, vibrations which are greater than the unisolated vibrations at some critical frequencies. Springs may be used if proper attention is given to damping. Sometimes a series combination of a spring for low frequencies and a feltlike material for high frequencies provides a wide range of vibration isolation.

A vibration break is effective only when *all* solid connections between two isolated parts are broken. No solid structure must be permitted to come in contact with both sides of a vibration break. Such a solid contact is as effective in conducting a vibration across a vibration break as a single wire is effective in shorting across an otherwise well insulated electric circuit.

Where a vibrating member cannot be separated physically from the remainder of the structure, a massive structure may be attached between the vibrating member and the remaining structure. This mass must be large and relatively immovable compared with the vibrating member. It serves to reflect vibrations back away from the portion of the structure to be isolated. A vibration block is particularly effective if it can be preceded or followed by a vibration break or by a relatively compliant structural section. In general, a vibration block will be ineffective if the vibrational energy reflected from it returns to the source and adds in phase to the new vibrations being generated. The vibration block relies on the existence of sufficient damping between the source and the block to prevent excessive buildup of vibrational energy.

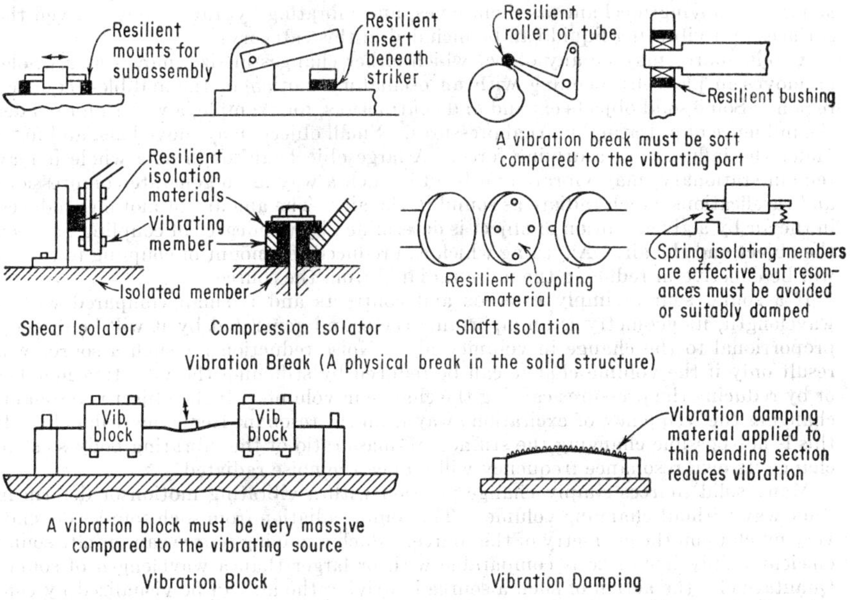

FIG. 7.10. Vibration isolation.

Where the source and observation point are relatively far apart, and where the intervening structure is of light construction, significant reduction of vibration can be obtained by application of vibration-damping materials. Damping materials are particularly useful in reducing vibrations in highly resonant members such as panels, webs, springs, and reeds. Vibration breaks and blocks are made more effective by use of damping materials on the vibrating members. Since the vibration break and block work by reflecting vibrational energy, large vibrations tend to build up in the isolated source member unless means are available to absorb this energy.

The choice of vibration-damping materials depends upon many things, including type of structure, the frequency of vibration, the operating temperature, and the particular function of the vibrating part. Commonly used materials include lead, sand, asphalt, impregnated felt, and mastic. Several new specially designed damping materials and composite damping structures are commercially available (see Sec. 34 and ref. 2).

Damping materials should have a stiffness which is generally comparable with the stiffness of the material being damped so as to provide an effective resistance to bending motion of the vibrating part. Operating temperatures must be considered in the choice of damping materials. Usually the temperature range of optimum performance is limited, but there is a wide choice of materials in most temperature ranges. Damping materials should be applied on the relatively thin sections of a vibrating member where the maximum deformation occurs.

(d) Noise Control at the Acoustic Source

Solid Sources. Having considered the source of vibration and means of treatment of solid-borne vibration paths, consider now the actual sources of airborne sound of importance in machine design: solid sources, aerodynamic sources, and secondary sources.

Solid sources are those most intimately associated with machinery and with the solid-borne vibration just considered. These sources include all vibrating areas such as panels, moving mechanisms themselves, and vibrating hydraulic lines. Even the surfaces of a vibrating liquid can be included in this category.

A solid source may be any object which either changes in size, moves as a whole, or moves in parts by bending with an oscillatory motion in the audible-frequency region. Some solid objects expand and contract (as, for example, a water pipe) under the influence of pulsating internal pressures. Small objects may move back and forth under the influence of a driving force. A large object, although as a whole it may remain stationary, may vibrate and bend in such a way as to generate compressions and rarefications which radiate as sound in the air. The amount of motion produced in the air by a given motion of object is determined by the degree of coupling between the object and the air. Anything which can reduce the amount of coupling to the air will be effective in reducing the noise radiated from the source.

If a sound source simply expands and contracts and is small compared with a wavelength, its geometry is unimportant; the sound generated by it will be directly proportional to the change in volume only. Noise reduction for such a source will result only if the volume change can be reduced by stiffening the vibrating member or by reducing the pressures causing the change in volume. If the object is resonant, changing the frequency of excitation away from the resonant frequency will help. If this is not possible changing the stiffness-to-mass ratio of the vibrating body so as to change its own resonance frequency will reduce the noise radiated.

Many solid sources simply change position with a vibrating motion or deform in some way without changing volume. The sound radiation from such sources depends very much upon the geometry of the source. Such a moving source can radiate sound efficiently only if its size is comparable with or larger than a wavelength of sound. Qualitatively, the action of such a source in driving the air can be visualized by considering the motion of the air produced first by a very small moving object and then by a successively larger object. As a small object such as a bead or a wire moves, the air flows around it easily without causing any appreciable pressure or disturbance. If the moving object is slightly larger, there may be a little turbulence caused by its motion, but still little resistance to its motion is encountered. With a larger object the air encounters difficulty in flowing around and an appreciable pressure is created in front of the moving object. This pressure causes air to flow around the object but also causes some air to compress ahead of the object. The slight compression thus produced is the source of a sound wave which radiates in all directions and travels away at the characteristic speed of sound.

If the solid object approaches a wavelength of sound in its linear dimensions, little air can flow around, and most of the pressure which is created as the object moves back and forth results in the creation of sound.

From this discussion it is evident that, if a vibrating object is to be kept from radiating sound at the frequency of vibration, the linear dimensions of the object must be kept small, of the order of one-tenth of the wavelength of the undesired sound in air. The undesired wavelength may be that of the fundamental frequency of vibration or may be some higher harmonic which may fall in a more important region of the audible-frequency spectrum.

When the overall dimensions of a vibrating object are large because of the necessary mechanical operations being performed, there are several ways in which its effective size as a sound generator can be reduced: large surfaces may be pierced with holes, frames may be made more slender by using stronger or stiffer materials, webbing may be removed or pierced between stiffeners. Also, of course, large radiating surfaces might be disconnected from the source of vibration by use of a vibration break as indicated in an earlier article.

When a panel can be neither isolated by a vibration break nor made smaller or pierced with holes, its radiation efficiency might still be reduced by several means: (1) Its surface density might be increased by adding mass, thus making it more difficult to drive and reducing its amplitude. Simply increasing the thickness may not be successful because the coincidence effect (see below) might in some *instances* increase objectionable noise radiation. (2) Damping material may be added to the surface; this adds mass as well as damping with very little increase in stiffness.[2] (3) A special construction can be used which can result in a panel which acts statically stiff but which is sufficiently limp in vibration to raise the coincidence to a high frequency, out of the normal vibrational range.[15]

Aerodynamic Sources. Sound from aerodynamic sources is caused by a motion of the air itself, e.g., fans, jets, stationary objects in a moving air stream such as grilles or dampers in ducts, or objects toward which high-speed air streams may be directed. In all these sources anything which can reduce turbulence will reduce noise radiation.

The internal impedance of aerodynamic sources is generally low because the forces associated with the turbulence are of the same order of magnitude as the forces required to generate sound waves in air. Confining the aerodynamic sources usually decreases noise radiation, unless the confinement increases the speed of air flow.

The noise generated by aerodynamic sources increases with relative air velocity. The relation between sound pressure and air velocity is not generally sharply defined. It may vary from V^5 to V^8, depending upon the velocity range and the flow geometry (ref. 2, pp. 644–666). The most effective way of reducing aerodynamic noise is to reduce the flow velocity. This is particularly evident in air-moving devices such as fans[16] but also applies for ducts, dampers, ventilating louvers, and terminal devices used in air-conditioning or cooling systems. Where size and flow velocity are fixed by other considerations, careful attention to the streamlining of all solid objects contacting the moving air will yield significant reduction in noise.

Turbulence-induced noise is sometimes augmented by resonant cavities or the presence of parallel surfaces which reflect sound and reinforce certain frequencies. Such sound reinforcement should be eliminated by filling or closing resonant cavities and by avoiding parallel reflecting surfaces where practical.

Other ways in which aerodynamic noise may be reduced include addition of a diffusing section to exhaust openings, removal of obstacles from the air path (especially sharp or angular obstacles), and streamlining all air conduits and all objects which necessarily must remain in the air stream.

Secondary Sources. Secondary sources, such as a hole or a crack in a wall or the opening in the end of a duct, act by radiating sound from a noisy region into a quiet one. Although the opening is not the source of the sound, it has all the characteristics of a source as observed from the quiet region.

Airborne sound from a secondary source consisting of an opening in a wall, duct, or other sound barrier can be reduced by the following changes:

Reduction of the area of opening, preferably by providing an airtight closure.

Directing the opening away from the listener. This helps only for high frequencies, i.e., frequencies for which the perimeter of the opening is larger than a wavelength.

Addition of a sound-attenuating muffler ahead of the opening. This may take the form of an acoustical lining for a duct or conduit, or it may be an acoustically lined labyrinth. For a hole in a wall, even a simple baffle which is large compared with the opening and large compared with a wavelength of the sound considered gives some reduction, although this will vary greatly with the design.

Generally, decreasing the size of an opening is effective in decreasing sound radiated. Cracks between panels or cover plates are particularly severe sources of noise leaks. Wherever noise isolation is required, such cracks should be gasketed and made airtight; even paper-thin cracks around large panels will amount to a large total effective open area. It should be noted, however, that where a crack or opening is the principal exit for the sound from a reverberant enclosure, the reduction in size of the opening may result only in an increase of reverberation within the enclosure. A corresponding increase in the sound level in the enclosure will, in turn, counterbalance the decrease in hole size and result in little decrease in radiated sound. Sound-absorbing material within the enclosure is necessary to prevent an increase in reverberation as the hole is reduced. The addition of sound-absorbing material within the enclosure may in fact be very effective in decreasing the radiated sound even when an opening cannot be reduced in size.

An acoustically lined channel or duct can reduce the noise from an opening in an enclosure. At least one bend (frequently more) should be used to prevent direct line-of-sight transmission of sound. The more bends, the more noise reduction will result; however, a bend will be of little value at frequencies below that for which the duct is less than ¼ wavelength across in the direction of the bend. Walls of the duct or baffle should be massive and nonporous so as to provide transmission loss against sound passing directly through the duct or baffle wall.

Sound in air can drive mechanical parts such as panels and walls to such an extent that they, in turn, will generate serious vibrational forces which might travel through solid structure and reradiate sound at a remote region, which may otherwise be vibrationally isolated from the prime source of vibrational energy.

Vibrations caused by the reaction of sound waves in air on solid parts or walls can be reduced by avoiding the conversion of air pressure into mechanical forces. The means for preventing the sound from causing vibrations in solid members are the same as for preventing vibrations from generating sound. They may be reiterated as follows:

Perforate the solid surface to admit a relatively free flow of air so that the pressure on opposite sides may equalize. For best results, holes in the surface should be as large as practicable and closer than ¼ wavelength (in air) of the sound considered. Perforation of a panel is effective in reducing the conversion of sound energy into solid-borne vibrations, but this technique cannot be used where the panel must provide sound-transmission loss from one region to another.

Make flexible sections more massive. This applies particularly to enclosures where perforations would defeat the purpose of the enclosure.

Add damping materials to the surface to eliminate resonances.

(e) Noise Control in the Acoustic Path

Some operations are intrinsically noisy, and the noise cannot be reduced at the source. Some processes make too much noise even when the best practicable vibration isolation has been employed.

Reduction of noise from such sources must be accomplished after the noise has been generated in the air. This requires some form of isolation or attenuation to be inserted in the air path between the source and the receiver. Isolation techniques for airborne sound differ from those for solid-borne vibrations.

The methods discussed below of isolating a sound source from a receiver are all of practical importance in machine design.

A source and receiver may be separated by increasing the physical distance between them. This is effective if the sound can spread out (so as to reduce the pressure fluctuations) as it travels from the source to the receiver. Sound will spread out if it encounters no obstacles such as walls and other hard surfaces which will reflect it.

Where hard walls confine the sound, as in a room, the sound pressure remains high throughout the enclosure. Separating the source and receiver in such a space is not effective in reducing sound at the receiver, except when the source and receiver are initially very close together. Lining a hard-walled room with acoustical absorbing material reduces sound reflection and thereby lowers the pressure fluctuations throughout the enclosed space. Separation of source and receiver is effective over greater distances in a room with absorbing material. Where source and receiver are contained in a corridor-shaped enclosure, such as a duct or pipe, the effectiveness of a separation between the source and receiver can be greatly increased by use of acoustical absorbing material on the inner surfaces of the corridor. For some types of ducts, some very effective acoustical treatments are available commercially.[2]

When the distance between the source and listener cannot be increased, noise can be reduced by using a barrier. A total airtight enclosure using massive walls is the most effective type of barrier. Figure 7.11 illustrates the performance of typical enclosures.

The noise reduction of an enclosure increases with decrease in area of openings, increase in mass of the wall, increase in frequency of the sound, and decrease in area of the enclosure.

The noise reduction of a good enclosure can be destroyed by even a small air leak (refer to the previous article on Secondary Sources). Therefore, design of an enclosure is limited by the total noise escaping from any openings. Good enclosure design first minimizes the area of all openings and reduces the noise from these by suitable acoustic treatment, then provides a wall structure to equal or surpass (by a small margin) the isolation provided by the openings. The design of wall sections which are much better than the opening isolation would be wasteful and would result in little added noise reduction.

The performance of a good enclosure is generally improved by the addition of acoustical absorptive treatment inside to prevent the buildup of sound due to multiple reflections from the walls of the enclosure. However, the addition of absorptive material to the outside surfaces of an enclosure is almost useless except for the mass it adds to the walls and the possible vibration damping it may achieve. A wall of porous "sound-absorbing" material by itself gives practically no acoustic isolation. Any solid connection between the source and the enclosure will destroy the acoustic isolation and may greatly increase the noise output over that produced with no enclosure. The reason for this is that a direct vibrational path is made to the radiating area of the enclosure, which is generally much greater than that of the mechanical device. The enclosure therefore provides a better impedance match to the air and radiates more sound than the machine itself.

Partial enclosures[8] are useful where only a small noise reduction is required in a high-frequency range. A partial enclosure relies on the directionality of high-frequency sounds. It is effective when its dimensions are large compared with a wavelength. It is particularly effective if it forces the sound to make one or more reflections from an absorbing surface before reaching the receiver. A baffle is a simplified partial enclosure usually consisting of a single wall preventing direct line-of-sight transmission of sound from a source to a receiver. A baffle provides its greatest effectiveness when it is close to the source and/or receiver, so that the path length between the two is greatly increased.

(f) Noise Control at the Receiver

The receiver is the object most directly concerned with the results of any noise-control efforts. For example, the operator of a noisy machine may be relocated by moving the controls with respect to the principal noise sources.

Changing the spectral shape or the addition of a more acceptable noise to mask the objectionable noise are techniques which may be more effective than noise reduction. Sometimes advertising the qualities (power, response, and utility, etc.) of a mechanism which are implied by its sound might gain more consumer acceptance then a detailed and successful noise-control modification. Noise control in machine design is directed toward obtaining a satisfying noise environment, not necessarily toward the reduction of noise level.

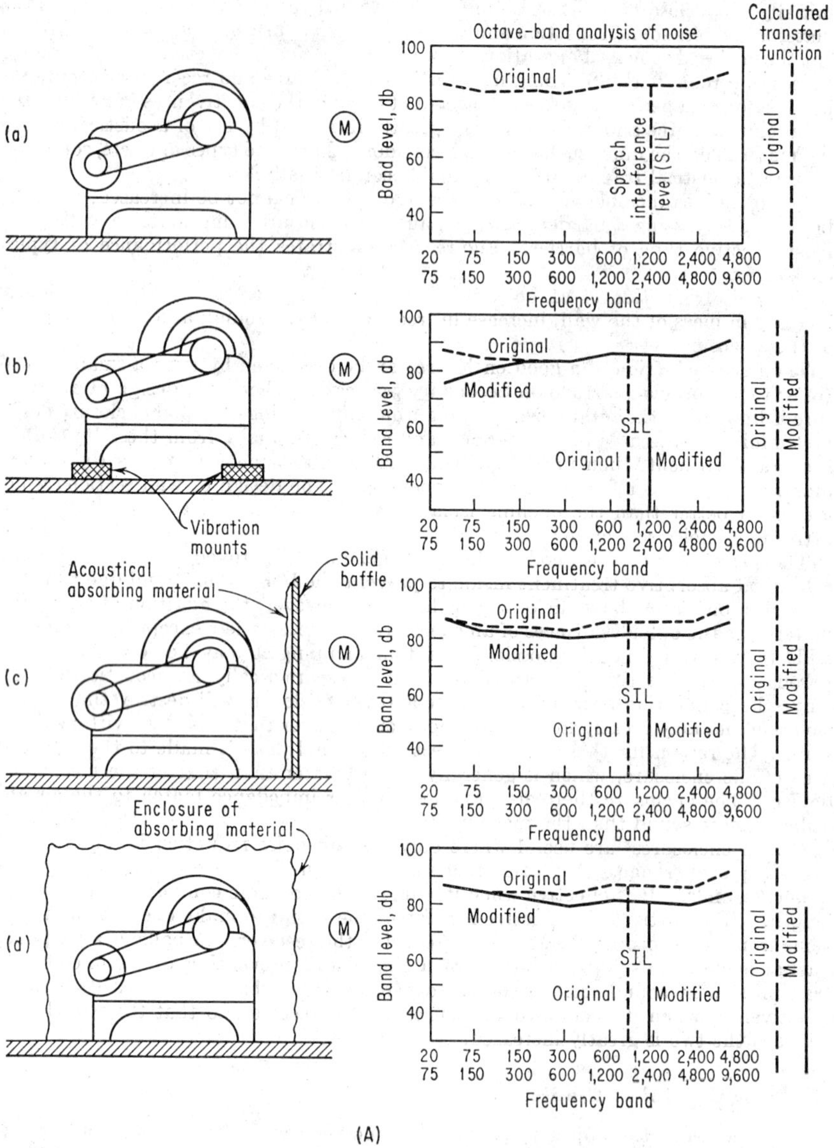

FIG. 7.11. Typical enclosures. Hypothetical examples of noise reduction showing relative values of vibration mounts, barriers, enclosures, and acoustic materials used in various combinations. The speech-interference level is shown by the vertical line marked SIL. The calculated transfer function (loudness) before and after modification is shown by the vertical lines to the right of the graphs. (*After Beranek,*[3] *pp.* 336 *and* 337, *Fig.* 11.1.)

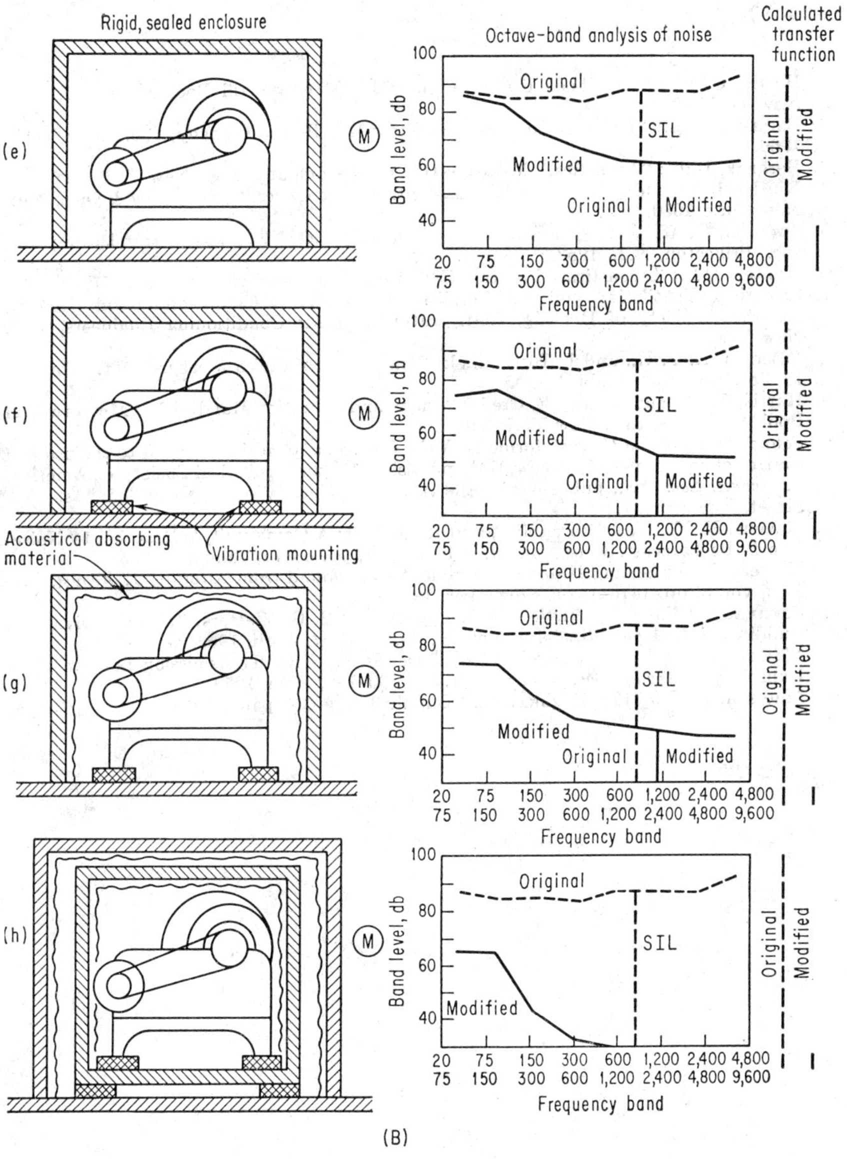

FIG. 7.11. Continued.

References

1. Harris, C. M.: "Handbook of Noise Control," McGraw-Hill Book Company, Inc. New York, 1957.
2. Beranek, L. L.: "Noise Reduction," McGraw-Hill Book Company, Inc., New York, 1960.
3. Beranek, L. L.: "Acoustics," McGraw-Hill Book Company, Inc., New York, 1954.
4. Kinsler, L. E., and A. R. Frey: "Fundamentals of Acoustics," John Wiley & Sons, Inc., New York, 1950.
5. Swenson, G. W., Jr.: "Principles of Modern Acoustics," D. Van Nostrand Company, Inc., Princeton, N.J., 1953.
6. Hunter, J. L.: "Acoustics," Prentice-Hall, Inc., Englewood Cliffs, N.J., 1957.
7. "Heating, Ventilating and Air Conditioning Guide," Chapter on Sound Control, American Society of Heating, Refrigerating and Air Conditioning Engineers, Inc., New York.
8. Peterson, A. P. G., and L. L. Beranek: "Handbook of Noise Measurement," General Radio Company, Cambridge, Mass.
9. Harris, C. M., and Charles E. Crede: "Shock and Vibration Handbook," McGraw-Hill Book Company, Inc., New York, 1961.
10. *Sound, Its Uses and Control*, formerly *Noise Control*, a magazine devoted to the use and control of sound noise and vibration, published by the Acoustical Society of America.
11. *J. Acoust. Soc. Am.*, published by the Acoustical Society of America, 335 East 45th Street, New York 17, N.Y.
12. Stevens, K. N., W. A. Rosenblith, and R. H. Bolt: A Community's Reaction to Noise: Can It Be Forecast, *Noise Control*, vol. 1, no. 1, pp. 63–71, January, 1955.
13. Kryter, K. D.: Scaling Human Reactions to the Sound from Aircraft, *J. Acoust. Soc. Am.*, vol. 3, pp. 1415–1429, November, 1959.
14. Cavanaugh, W., P. W. Hirtle, W. R. Farrell, and B. G. Watters: "Speech Privacy in Buildings," *J. Acoust. Soc. Am.*, vol. 34, pp. 475–492, April, 1962.
15. Watters, B., and G. Kurtze: New Wall Design for High Transmission Loss on High Damping, *J. Acoust. Soc. Am.*, vol. 31, pp. 739–748, June, 1959.
16. Madison, R. D.: "Fan Engineering," Buffalo Forge Company, 1949.

Section 8

SYSTEM DYNAMICS

By

S. KAMINSKY, M.M.E., M.S.E.E., *Consulting Engineer, Brooklyn, N.Y.*

CONTENTS

8.1. INTRODUCTION—PRELIMINARY CONCEPTS

A physical system undergoing a time-varying interchange or dissipation of energy among or within its elementary storage or dissipative devices is said to be in a *dynamic state.* The elements are in general inductive, capacitative, or resistive—the first two being capable of storing energy while the latter is dissipative. All are called *passive;*

i.e., they are incapable of generating net energy. A system composed of a finite number or a denumerable infinity of storage elements is said to be *lumped* or *discrete*, while a system containing elements which are dense in physical space is called *continuous*. The mathematical description of the dynamics for the discrete case is a set of ordinary differential equations, while for the continuous case it is a set of partial differential equations.

The mathematical formulation depends upon the constraints (e.g., kinematic or geometric) and the physical laws governing the behavior of the system. For example, the motion of a single point mass obeys $\mathbf{F} = m(d\mathbf{v}/dt)$ in accordance with Newton's second law of motion. Analogously, the voltage drop across a perfect coil of self-inductance L is $V = L(di/dt)$, a consequence of Faraday's law. In the first case the energy-storage element is the mass, which stores $mv^2/2$ units of kinetic energy while

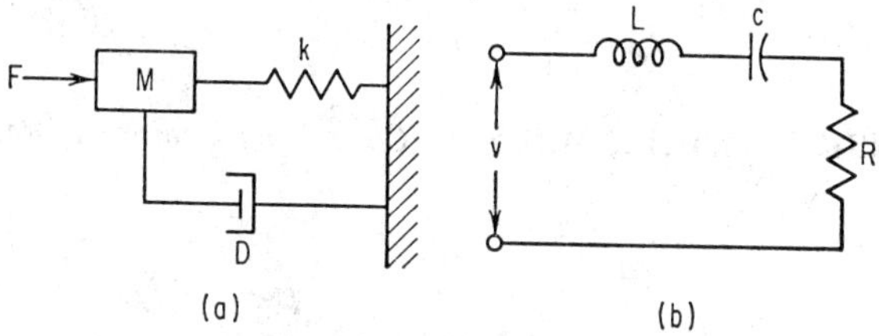

FIG. 8.1. Second-order systems. (*a*) Mechanical system. (*b*) Electrical analog.

the inductance L stores $Li^2/2$ units of energy in the second case. A spring-mass system and its electrical analog, an inductive-capacitive series circuit, represent higher-order discrete systems. The unbalanced force acting on the mass is $F - kx$. Thus

$$F = kx + m\ddot{x} \qquad m, k > 0 \tag{8.1}$$

Analogously for the electrical case,

$$V = L\ddot{q} + q/c \qquad L, c > 0$$

following Kirchhoff's voltage-drop law (i.e., the sum of voltage drop around a closed loop is zero). To show that Eq. (8.1) expresses the dynamic exchange of energy, multiply Eq. (8.1) by $\dot{x}\,dt$ (which is equal to dx) and integrate.

$$\int_0^t F\dot{x}\,dt = \int_{x=x_0}^{x} F\,dx = \int_0^t m\dot{x}\ddot{x}\,dt + \int_0^t k\dot{x}x\,dt$$

$$\text{Work input} = \underbrace{m\dot{x}^2/2\Big]_0^t}_{\Delta\text{ K.E.}} + \underbrace{kx^2/2\Big]_0^t}_{\Delta\text{ P.E.}} = m\dot{x}^2/2 - m\dot{x}_0^2/2 + kx^2/2 - kx_0^2/2$$

which is a statement of the law of conservation of energy. This illustrates that work input is divided into two parts, one part increasing the kinetic energy, the remainder increasing the potential energy. The actual partition between the two energy sources at any instant is time-varying, depending on the solution to Eq. (8.1).

If a viscous damping element is added to the system the force equation becomes (see Fig. 8.1*a*)

$$m\ddot{x} + c\dot{x} + kx = F \qquad c > 0$$

and performing the same operation of multiplying by $\dot{x}\,dt$, (dx) and integrating we obtain

$$\int_0^t m\dot{x}\ddot{x}\,dt + \int_0^t c\dot{x}^2\,dt + \int_0^t kx\dot{x}\,dt = \int_{x_0}^{x} F\,dx \tag{8.2a}$$

$$m\dot{x}^2/2\Big]_0^t + \int_0^t c\dot{x}^2\,dt + kx^2/2\Big]_0^t = \int_{x_1}^{x} F\,dx \tag{8.2b}$$

again expressing the energy-conservation law. Note that the integrand $c\dot{x}^2 \geq 0$ and that the integral in Eq. (8.2b) is thus a monotonically increasing function of time. This condition assures that, for $F = 0$, the free (homogeneous) system must eventually come to rest since under this condition Eq. (8.2b) becomes

$$m\dot{x}^2/2 + kx^2/2 + \int_0^t c\dot{x}^2\,dt = \text{const} = m\dot{x}_0^2/2 + kx_0^2/2 \tag{8.3}$$

which again is an expression of the law of energy conservation. The first two terms are positive since they contain the squared factors $\dot{x}^2$ and x^2, while the third term, as noted above, increases with time. It follows that the sum of the first two must decrease monotonically in order to satisfy Eq. (8.3); moreover, neither term can be greater than the sum. It follows that, as $t \to \infty$, $x \to 0$ and $\dot{x} \to 0$.

Formulation of the foregoing simple problems was based upon fundamental physical laws. The derivation by LaGrange equations, which in this simple case offers little advantage, is (Sec. 3)

$$L = T - V \qquad T = \tfrac{1}{2}m\dot{x}^2 \qquad V = kx^2/2$$

For conservative systems (e.g., spring-mass),

$$(d/dt)(\partial L/\partial\dot{x}) - \partial L/\partial x = F = m\ddot{x} + kx$$

For nonconservative systems with dissipation function $\mathfrak{F}$,

$$\begin{aligned}\mathfrak{F} &= c\dot{x}^2/2\\ (d/dt)(\partial L/\partial\dot{x}) - \partial L/\partial x &= -\partial\mathfrak{F}/\partial\dot{x} + F\\ m\ddot{x} + kx &= -c\dot{x} + F\\ m\ddot{x} + c\dot{x} + kx &= F\end{aligned}$$

Precisely the same form is deducible from a LaGrange statement of the electrical equivalent (Fig. 8.1b).

(a) Degrees of Freedom

Thus far it has been observed that one independent variable x was employed to describe the system dynamics. In general, however, several variables $x_1, x_2, \ldots, x_n$ are necessary to describe the motion of a complex system. The minimum number

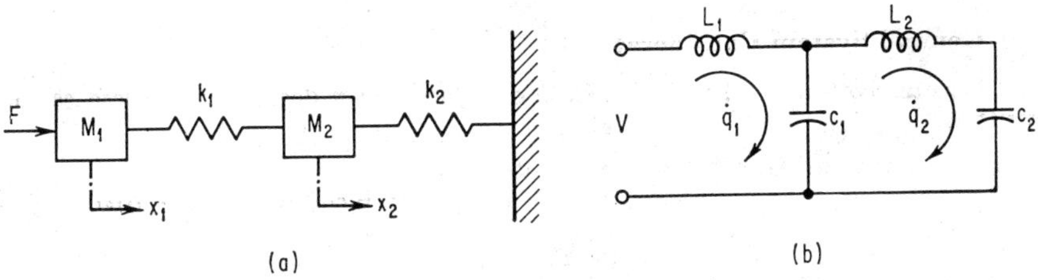

FIG. 8.2. Two-degree-of-freedom systems. (a) Mechanical. (b) Electrical analog.

of coordinates that are so required is defined as the number of degrees of freedom of the system. Simple examples of two-degree-of-freedom systems are shown in Fig. 8.2. The respective equations of motion are

Mechanical:
$$\begin{aligned}m_1\ddot{x}_1 + k_1(x_1 - x_2) &= F\\ m_2\ddot{x}_2 + k_2x_2 + k_1(x_2 - x_1) &= 0\end{aligned} \tag{8.4a}$$

Electrical:
$$\begin{aligned}L_1\ddot{q}_1 + (q_1 - q_2)/c_1 &= V\\ L_2\ddot{q}_2 + q_2/c_2 + (q_2 - q_1)/c_1 &= 0\end{aligned} \tag{8.4b}$$

derivable from force and loop volt-drop considerations.

Another example of a two-degree-of-freedom system is shown in Fig. 8.3, a compound pendulum constrained to move in a plane. While the system may at first appear to have four degrees of freedom with the positions of m_1 and m_2 given by r_1, θ_1, and r_1, θ_1, r_2, θ_2, respectively, two seemingly trivial expressions of constraint r_1 = constant and r_2 = constant show that the motion is describable in terms of θ_1 and θ_2 only. If a spring were interposed between m_1 and the pivot r_1, then r_1 would no longer be a constant and the motion would involve r_1, θ_1, θ_2, or three independent variables, resulting in a three-degree-of-freedom system.

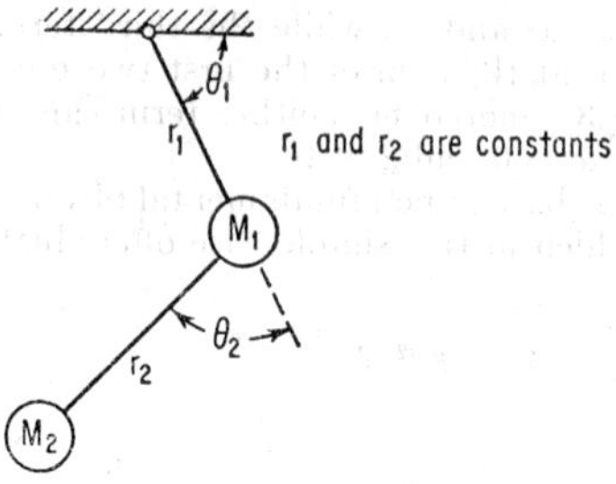

FIG. 8.3. Two-degree-of-freedom system (compound pendulum).

(b) Coupled and Uncoupled Systems

Equations (8.4*a*) or (8.4*b*) also illustrate a coupled system. The term *coupled* is a consequence of having more than one independent variable present in each equation of a set. In Eq. (8.4*a*), x_1 and x_2 and/or their derivatives appear in each of the two dynamic equations, implying that motion of one mass excites motion in the other mass. Only in conservative linear systems is it always possible to uncouple the system by a linear transformation.

An n-degree-of-freedom system requires for description n independent equations, usually of second order or lower. It is sometimes convenient to make changes in variables to facilitate the analysis of complex systems, or indeed to express the motion in terms of parameters, which are more accessible. In any case this amounts to having

$$x_i = x_i(q_1,q_2, \ldots ,q_m) \qquad i = 1, 2, \ldots , n$$

The q's, called generalized coordinates, when judiciously chosen play a useful role in the analysis of complex systems. The q's need not be independent. This implies $m > n$ and the existence of $m - n$ equations that connect the q's, since the motion must involve only n independent equations. The case of Fig. 8.3 is an example in which $m = 4$ and $n = 2$ with $m - n = 2$ or two constraint equations, namely,

$$r_1 = \text{const} \qquad r_2 = \text{const}$$

(c) General System Considerations

Discrete Systems. The equations for a system of n degrees of freedom can be written

$$(b_{11}p^2 + c_{11}p + d_{11})x_1 + (b_{12}p^2 + c_{12}p + d_{12})x_2 + \cdots + (b_{1n}p^2 + c_{1n}p + d_{1n})x_n = f_1(t)$$

$$\vdots$$

$$(b_{n1}p^2 + c_{n1}p + d_{n1})x_1 + \cdots + (b_{nn}p^2 + c_{nn}p + d_{nn})x_n = f_n(t) \qquad (8.5a)$$

where $p = d/dt$, $p^2 = d^2/dt^2$. Or, more concisely,

$$\sum_{j=1}^{n} (b_{ij}p^2 + c_{ij}p + d_{ij})x_j = f_i(t) \qquad i = 1, \ldots , n \qquad (8.5b)$$

where b_{ij}, c_{ij}, d_{ij} are in general functions of x_k, $\dot{x}_k$, $\ddot{x}_k$, $k = 1, \ldots , n$, and time. In terms of generalized coordinates,

$$\sum_{j=1}^{m} (b'_{ij}p^2 + c'_{ij}p + d'_{ij})q_j = Q_i(t) \qquad i = 1, \ldots , m \qquad m \geq n \qquad (8.5c)$$

where the Q_i's are the generalized forces (see Sec. 3). The number of degrees of freedom appears to have increased in Eq. (8.5c) for $m > n$, but this really is not the case, because of the existence of $m - n$ constraint equations connecting the q's.

The general form depicted by Eq. (8.5) is nonlinear in view of the b_{ij}, c_{ij}, and d_{ij} dependence on x_k and its time derivatives. Removal of this dependence yields the linear form of Eq. (8.5). Elimination of the time dependence in these coefficients yields the linear constant-coefficient form, which is of greatest engineering interest because it is the only one yielding completely to analysis and because a large class of systems can be approximated by this form. This is in contradistinction to the nonlinear and linear time-variable cases for which analytic solutions are in general not obtainable and not obtainable in closed form, respectively.

The initial state of each coordinate of Eq. (8.5) must be known [i.e., $x_i(0)$, $\dot{x}_i(0)$, $i = 1, 2, \ldots, n$], before the general solution is possible; hence $2n$ initial conditions are available which coincide with the maximum order of the differential equation obtained by eliminating $n - 1$ variables in Eq. (8.5). If the order is less than $2n$, then some of the initial conditions are not independent.

Continuous Systems. In passing from the description of discrete to that of continuous systems, the ordinary differential equation of n degrees of freedom becomes the set of partial differential equations as $n \to \infty$; i.e., the storage and dissipative elements become densely packed. The initial conditions are similar to those of the ordinary differential equation case which required initial velocity and position coordinates of each elementary mass particle; for now, in the limit of continuous systems, the initial displacement from equilibrium $\mathbf{u}(x,y,z,0)$ and the displacement velocity $(d\mathbf{u}/dt)(x,y,z,0)$ as well as the conditions $\mathbf{u}(x_b,y_b,z_b,t)$ that bound the system (where x_b, y_b, z_b are the continuous coordinates that bound the unperturbed system) are essential. As an example, consider the propagation of a pressure wave moving longitudinally in an infinite elastic dissipationless medium of small cross section. The equation of motion is derived by considering the elemental width dx having a stress $\sigma_{(x)}$ at the position x. Newton's law of motion applied to the element of mass of unit cross section is written

$$\begin{aligned} \{\sigma - [\sigma + (\partial\sigma/\partial x)\,dx]\} &= m\,dx\,(\partial^2 u/\partial t^2) \\ -\,\partial\sigma/\partial x &= m(\partial^2 u/\partial t^2) \end{aligned} \tag{8.6}$$

where m = mass density
σ = compressive stress

The displacement from equilibrium u results in strain (compressive)

$$\epsilon = -\partial u/\partial x$$

and by Hooke's law,

$$\sigma = -Y(\partial u/\partial x) \qquad \text{(positive } \sigma \text{ is compressive)} \tag{8.7}$$

where Y is Young's modulus for solids and a proportionality constant for other elastic media. Substitution of Eq. (8.7) into Eq. (8.6) yields

$$(Y/m)(\partial^2 u/\partial x^2) = \partial^2 u/\partial t^2 \tag{8.8}$$

which is the simple one-dimensional wave equation.

8.2. SYSTEMS OF LINEAR PARTIAL DIFFERENTIAL EQUATIONS[1,2,3,4,5]

(a) Elastic Systems

That class of systems characterized by interchange of kinetic and elastic energy is termed *elastic*. The formulation of the nondissipative (conservative) type leads to the simple wave equation

$$c^2\nabla^2 u = \partial^2 u/\partial t^2 \tag{8.9}$$

where ∇^2 = Laplace operator (three-dimensional in general)

Examples of such systems are as follows:

Hydrodynamics and Acoustics. Applying Newton's second law to an elementary particle yields

$$\rho(dx\,dy\,dz)(d\mathbf{v}/dt) = (dx\,dy\,dz)\mathbf{F} + \Sigma\mathbf{f} \tag{8.10}$$

where **F** represents the external forces (body forces) acting per unit volume on the element (e.g., gravity or inertia, using d'Alembert's principle), $\Sigma\mathbf{f}$ the sum of forces acting on the surfaces, and ρ the mass density. We can write

$$\Sigma\mathbf{f} = -(\partial p/\partial x)\,dx\,dy\,dz\mathbf{i} - (\partial p/\partial y)\,dy\,dx\,dz\mathbf{j} - (\partial p/\partial z)\,dz\,dx\,dy\mathbf{k} \tag{8.11}$$

$$\Sigma\mathbf{f} = -\nabla p\,dx\,dy\,dz \tag{8.12}$$

where ∇ is the "del" or gradient operator and p is the pressure.

Substituting Eq. (8.12) into Eq. (8.10) yields

$$\rho(d\mathbf{v}/dt) = \mathbf{F} - \nabla p \tag{8.13}$$

Expanding the left-hand side of Eq. (8.13), we obtain

$$\rho(\partial\mathbf{v}/\partial t + \underbrace{V_x\,\partial\mathbf{v}/\partial x + V_y\,\partial\mathbf{v}/\partial y + V_z\,\partial\mathbf{v}/\partial z}_{(\mathbf{v}\cdot\nabla)\mathbf{v}}) = \mathbf{F} - \nabla p \tag{8.14}$$

where $\mathbf{v} = V_x\mathbf{i} + V_y\mathbf{j} + V_z\mathbf{k}$.

From continuity (conservation of mass),

$$\partial\rho/\partial t + \text{div}\,(\rho\mathbf{v}) = 0 \tag{8.15}$$

$$\partial\rho/\partial t + \mathbf{v}\cdot\nabla\rho + \rho\,\text{div}\,\mathbf{v} = 0 \tag{8.16}$$

Equation (8.15) states that the rate of mass increase in elementary volume $dx\,dy\,dz$ $(\partial\rho/\partial t)$ equals the rate of flow into the same volume, $-dx\,dy\,dz\,\text{div}\,(\rho\mathbf{v})$. If $|\partial\mathbf{v}/\partial t| \gg |(\mathbf{v}\cdot\nabla)\mathbf{v}|$ then to a good approximation of Eq. (8.14)

$$\rho\,\partial\mathbf{v}/\partial t \approx \mathbf{F} - \nabla p \tag{8.17}$$

Let the density be given by

$$\rho = \rho_0(1 + \epsilon) \qquad \rho_0 = \text{const} \tag{8.18}$$

Taking a first differential of Eq. (8.18) we obtain

$$d\rho/\rho_0 = d\epsilon \tag{8.19}$$

Elimination of ρ in Eq. (8.16) yields

$$\partial\epsilon/\partial t + \text{div}\,\mathbf{v} + \mathbf{v}\cdot\nabla\epsilon \approx 0 \tag{8.20}$$

where it is assumed that the variation of density about ρ_0 is small (i.e., $|\epsilon| \ll 1$). As a further consequence the third term of Eq. (8.20), involving space derivatives of ϵ which are of higher order, is accordingly dropped leaving to a good approximation

$$\partial\epsilon/\partial t + \text{div}\,\mathbf{v} \approx 0 \tag{8.21}$$

which together with Eq. (8.17) in rearranged form

$$\partial\mathbf{v}/\partial t - \mathbf{F}/\rho_0 + \nabla p/\rho_0 \approx 0 \tag{8.22}$$

provide two of the three essential relationships for small perturbation analysis; the remaining expression is the equation of state (e.g., $dp = k\,d\epsilon$).

Substituting $\nabla p = k\nabla\epsilon$ (where k is a constant) into Eq. (8.22), the following is obtained:

$$\partial\mathbf{v}/\partial t - \mathbf{F}/\rho_0 + k(\nabla\epsilon/\rho_0) \approx 0 \tag{8.23}$$

which together with Eq. (8.21) forms a fundamental set.

Calling u the particle displacement, $\mathbf{v} = \partial\mathbf{u}/\partial t$. Substituting for $\mathbf{v}$ in Eq. (8.23) and Eq. (8.21) yields

$$\partial^2\mathbf{u}/\partial t^2 - \mathbf{F}/\rho_0 + k\,\nabla\epsilon/\rho_0 = 0 \tag{8.24}$$

$$\partial\epsilon/\partial t + \partial/\partial t\,(\text{div}\,\mathbf{u}) = 0 \tag{8.25}$$

and

$$\epsilon + \text{div}\,\mathbf{u} = 0 \tag{8.26}$$

Taking the divergence of Eq. (8.24) yields

$$-(\partial^2/\partial t^2)\,\text{div}\,\mathbf{u} - \text{div}\,\mathbf{F}/\rho_0 + k/\rho_0\,\text{div grad}\,\epsilon = 0 \tag{8.27}$$

Next, substituting ϵ for $-$ div $\mathbf{u}$ results in

$$-(\partial^2\epsilon/\partial t^2) - \text{div}\,(\mathbf{F}/\rho_0) + (k/\rho_0)\nabla^2\epsilon = 0 \tag{8.28}$$

For the case div $\mathbf{F}/\rho_0 = 0$, Eq. (8.28) becomes

$$\partial^2\epsilon/\partial t^2 = (k/\rho_0)\nabla^2\epsilon$$

which is the three-dimensional wave equation in ϵ.

If, in addition, the velocity is derivable from a scalar potential ϕ, i.e.,

$$\mathbf{v} = \text{grad}\,\phi = \partial\mathbf{u}/\partial t \tag{8.29}$$

then substitution in Eq. (8.24) gives

$$(\partial/\partial t)\,\text{grad}\,\phi = -(k/\rho_0)\,\text{grad}\,\epsilon + \mathbf{F}/\rho_0 \tag{8.30}$$

Differentiating with respect to time and assuming $\mathbf{F}$ time-independent

$$\text{grad}\,[(\partial^2/\partial t^2)\phi - (k/\rho_0)\nabla^2\phi] = 0$$

From Eqs. (8.26) and (8.29),

$$\partial\epsilon/\partial t = -\,\text{div grad}\,\phi = -\nabla^2\phi$$

whence, by a suitable choice of ϕ,

$$(\partial^2/\partial t^2)\phi - (k/\rho_0)\nabla^2\phi = 0$$

i.e., the velocity-potential function is also of the wave type. For the special case of one-dimensional propagation with $\mathbf{F} = 0$

$$V_x\mathbf{i} = (\partial u_x/\partial t)\mathbf{i} = \text{grad}\,\phi$$

Differentiating with respect to time and substituting from Eqs. (8.30) and (8.26),

$$\partial^2 u_x/\partial t^2 = (k/\rho_0)(\partial^2 u_x/\partial x^2) \tag{8.31}$$

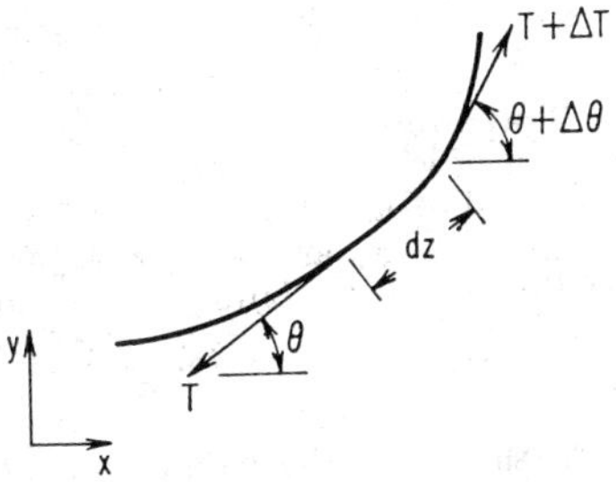

FIG. 8.4. String under tension.

Transverse Motion of an Elastic String Due to a Slight Perturbation. Consider a string under uniform tension T_0 initially stretched in a horizontal (x) direction (Fig. 8.4). Assuming that the weight of the string is negligible compared with inertia forces, and the elongation is negligible, the force balance in the y direction for an elementary section of length dz is

$$[\partial(T\sin\theta)/\partial z]\,dz = m\,dz\,\partial^2 y/\partial t^2 \tag{8.32a}$$

where $\sin\theta = \partial y/\partial z$. Considering only small displacements from the unperturbed position, i.e., $|\partial y/\partial z| \ll 1$ and

$$dz = dx\,[1 + (\partial y/\partial x)^2]^{1/2} \qquad \partial x/\partial z \approx 1$$

$$\frac{\partial[T(\partial y/\partial z)]}{\partial z} = \frac{\partial\{T[(\partial y/\partial x)(\partial x/\partial z)]\}}{\partial x}\frac{\partial x}{\partial z} \approx \frac{\partial[T(\partial y/\partial x)]}{\partial x} \tag{8.32b}$$

Assuming tension T is essentially constant and that additive elongation is negligible,

$$T = T_0(1 + \epsilon) \qquad |\epsilon| \ll 1$$

then to a first approximation,

$$\frac{\partial[T(\partial y/\partial x)]}{\partial x} \approx T_0 \frac{\partial^2 y}{\partial x^2}$$

and Eq. (8.32*a*) becomes, after canceling dz, the one-dimensional wave equation.

$$(T_0/m)(\partial^2 y/\partial x^2) = \partial^2 y/\partial t^2$$

Transverse Vibration of Stretched Membrane. Consider the stretched membrane of circular cross section (see Fig. 8.5). The transverse motion under a pressure p is found by forming the equation of motion on an elementary annulus of width dr and again ignoring the membrane weight,

$$dr\,(\partial/\partial r)(T 2\pi r \sin\theta) = 2\pi r\,dr\,p + 2\pi\rho r\,dr\,(\partial^2 y/\partial t^2)$$

where $\sin\theta = \partial y/\partial r$
T = tension
ρ = density

whence, for T essentially constant

$$[T = T_0(1 + \epsilon),\ |\epsilon| \ll 1]$$

$$(T_0/\rho)(1/r)(\partial/\partial r)[r(\partial y/\partial r)] \approx \partial^2 y/\partial t^2 + p/\rho$$

$$(T_0/\rho)\nabla^2 y = \partial^2 y/\partial t^2 + p/\rho \quad \text{(inhomogeneous wave equation)}$$

where ∇^2 = Laplacian operator.

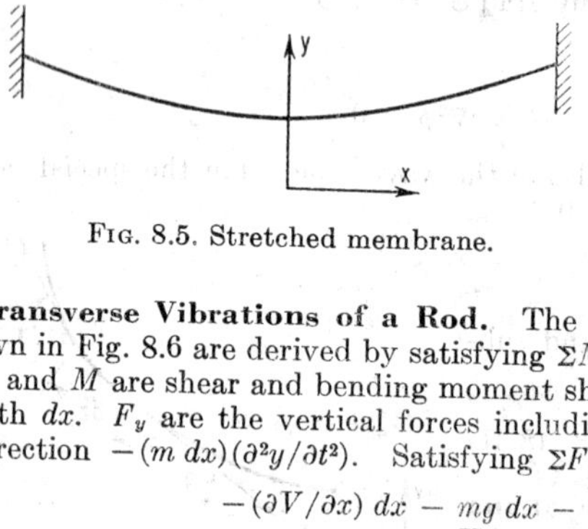

FIG. 8.5. Stretched membrane.

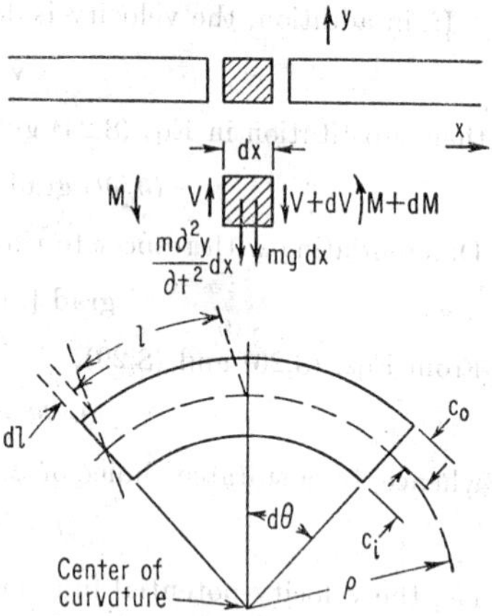

FIG. 8.6. Bending of a bar.

Transverse Vibrations of a Rod. The dynamics of motion of a uniform bar shown in Fig. 8.6 are derived by satisfying $\Sigma M = 0$ and $\Sigma F_y = 0$.

V and M are shear and bending moment shown acting on the elemental section of length dx. F_y are the vertical forces including the d'Alembert inertia force in the y direction $-(m\,dx)(\partial^2 y/\partial t^2)$. Satisfying $\Sigma F_y = 0$ in the positive y direction yields

$$-(\partial V/\partial x)\,dx - mg\,dx - (m\,dx)(\partial^2 y/\partial t^2) = 0$$
$$\partial V/\partial x + mg + m\,\partial^2 y/\partial t^2 = 0 \qquad (8.33a)$$

and satisfying the moment equation about the center of mass of the elementary section results in

$$\partial M/\partial x = V \qquad (8.33b)$$

Now a physical relation exists between M and y which is derivable by considering the bent section which is compressed on the inner fiber and stretched on the upper fiber with a "neutral axis," unstressed at the initial length (Fig. 8.6).

From geometric considerations,

$$dl = c_0\,d\theta \qquad \rho \gg c_0$$
$$d\theta = l/\rho \qquad \rho \gg c_i \qquad (8.34a)$$

where ρ is the radius of curvature, c_0 and c_i distances from the neutral axis to the outer and inner fiber, respectively, l the half width of the elementary section and $d\theta$ the half angle subtended by the section under stressed conditions.

ρ is further expressed by (from elementary calculus)

$$\frac{1}{\rho} = \frac{\partial^2 y/\partial x^2}{[1 + (\partial y/\partial x)^2]^{3/2}} \approx \partial^2 y/\partial x^2 \quad \text{for } \partial y/\partial x \ll 1 \tag{8.34b}$$

The strain at the outer fiber is $\epsilon_0 \approx dl/l$. From the geometry, the strain at any other point is $\epsilon_0(y/c_0)$ where y is the position measured from the neutral axis. From Eq. (8.34a), $dl/l = c_0/\rho$ and therefore the strain at y is

$$\epsilon = \epsilon_0(y/c_0) = (dl/l)(y/c_0) = (c_0/\rho)(y/c_0) = y/\rho$$

The stress, following Hooke's law, is

$$\sigma = E\epsilon = Ey/\rho$$

where E is the modulus of elasticity.

The bending moment about the neutral axis is expressed by

$$M = \int_{-c_i}^{c_0} y\sigma b\, dy$$

where b is the depth and $b\, dy$ is the elementary cross-sectional area. Substituting for σ the above expression becomes

$$\int_{-c_i}^{c_0} y(Ey/\rho)b\, dy = E/\rho \int_{-c_i}^{c_0} y^2 b\, dy$$

The integral on the right is I, the area moment of inertia about the neutral axis, a geometric property.

Thus

$$M = EI/\rho$$

and from Eq. (8.34b)

$$M/EI = \partial^2 y/\partial x^2 \tag{8.35}$$

Taking two derivatives of Eq. (8.35) with respect to x, one derivative of Eq. (8.33b), and substituting for $\partial y/\partial x$ in Eq. (8.33a) yields

$$EI\, \partial^4 y/\partial x^4 + m(g + \partial^2 y/\partial t^2) = 0$$

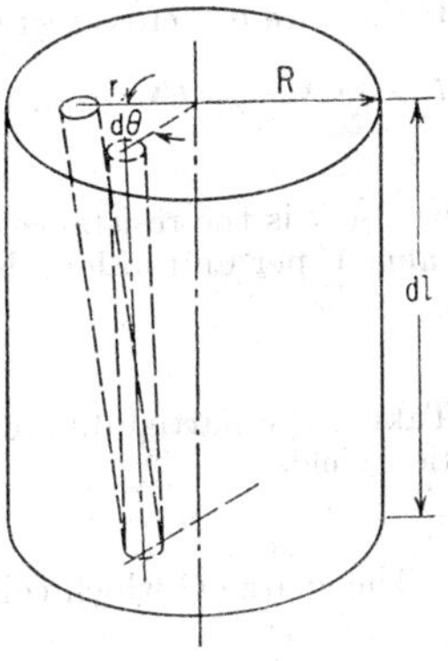

FIG. 8.7. Torsion in a rod.

Torsional Motion of a Rod. Consider an elemental cylindrical section of length dl and a twist angle $d\theta$ (see Fig. 8.7). The strain on an elemental area da is $r\, d\theta/dl$ and the associated stress is

$$\sigma = Gr(d\theta/dl) \tag{8.36}$$

where G is the shear modulus and r is the radius to the point in question. The total torque

$$T = \int_A r\sigma\, da = \int_A Gr(d\theta/dl)\, r\, da = G(d\theta/dl) \int_A r^2\, da$$

$$T = G(d\theta/dl)J$$

where J = polar moment of inertia $= \int_A r^2\, da = \int_0^R r^2 2\pi r\, dr$.

The expression for torsional oscillations is obtained from Newton's second law:

$$dl(\partial T/\partial l) = I(\partial^2\theta/\partial t^2)\, dl \tag{8.37}$$
$$JG(\partial^2\theta/\partial l^2) = I(\partial^2\theta/\partial t^2)$$

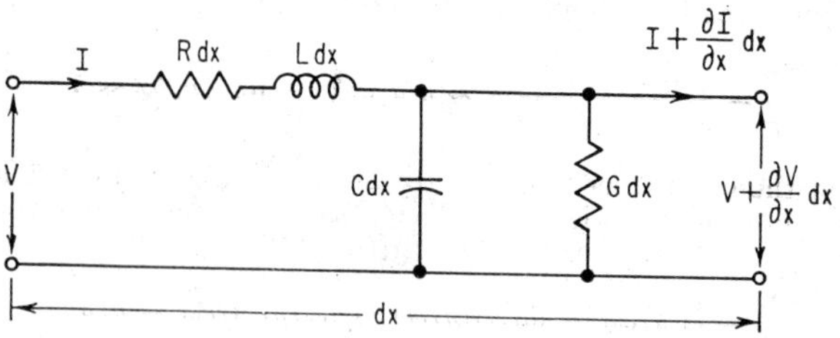

FIG. 8.8. Electric transmission line.

where I = mass moment of inertia per unit length. For homogeneous media $I = \rho J$ and Eq. (8.37) becomes

$$(G/\rho)(\partial^2\theta/\partial l^2) = \partial^2\theta/\partial t^2 \tag{8.38}$$

Electric-transmission-line Equation for Low-frequency Operation. Consider a section of length dx as shown in Fig. 8.8. From Ohm's law the current density is

$$\mathbf{i} = -k \text{ grad } V$$

If the wire has cross section **A**, the total current I is

$$I = \mathbf{i} \cdot \mathbf{A} = -k\mathbf{A} \cdot \text{grad } V = -(1/R) \text{ grad } V$$
$$= -(1/R)(\partial V/\partial x); \quad -\partial V/\partial x = IR \tag{8.39}$$

where R is the resistance per unit length. The wire also acts as a distributed capacitance C per unit of length; following Faraday's law:

$$dQ/Cdx = dV \tag{8.40}$$

Taking the partial differential of Eq. (8.40) with respect to time for the elemental section yields

$$(1/C\,dx)(\partial Q/\partial t) = \partial V/\partial t \tag{8.41}$$

The charge Q which collects within the dx section is

$$Q = \int_0^t I\,dt - \int_0^t [I + (\partial I/\partial x)\,dx]\,dt \tag{8.42}$$

and therefore

$$\partial Q/\partial t = -(\partial I/\partial x)\,dx$$

Substitution in Eq. (8.41) yields

$$(1/C)(\partial I/\partial x) = -\partial V/\partial t \tag{8.43}$$

If in addition some current leaks off and is proportional to V, then Eq. (8.42) should be modified as follows:

$$Q = \int_0^t I\,dt - \int [I + (\partial I/\partial x)\,dx]\,dt - \int (GV\,dx)\,dt \tag{8.44}$$

where G is the leakage conductance per unit length.

Taking the partial derivative of Eq. (8.44) with respect to time

$$\partial Q/\partial t = -(\partial I/\partial x)\,dx - GV\,dx$$

and replacing $\partial Q/\partial t$ in Eq. (8.41) leads to the modified equation

$$-\partial I/\partial x = GV + C(\partial V/\partial t) \tag{8.45}$$

Also, the inductance along the wire owing to Faraday's law introduces an additional voltage drop to modify Eq. (8.39) to read

$$-(\partial V/\partial x)\,dx = (RI)\,dx + L(\partial I/\partial t)\,dx \tag{8.46}$$

where L is the inductance per unit length

$$-\partial V/\partial x = RI + L(\partial I/\partial t)$$

Combining Eqs. (8.45) and (8.46) results in

$$CL(\partial^2 I/\partial t^2) + (RC + GL)(\partial I/\partial t) + RGI = \partial^2 I/\partial x^2 \tag{8.47}$$

and the identical form in V, i.e.,

$$CL(\partial^2 V/\partial t^2) + (RC + GL)(\partial V/\partial t) + RGV = \partial^2 V/\partial x^2 \tag{8.48}$$

Equations (8.47) and (8.48) are the telegrapher's equation which was first reported by Kirchhoff. Note that, if R and G are zero, they reduce to the simple wave equation.

(b) Inelastic Systems

Flow of Heat, Electricity, and Fluid. The flow of heat across a boundary as given by Fourier is

$$-k \text{ grad } T = \mathbf{Q} \tag{8.49}$$

where $\mathbf{Q}$ is the heat flux and T is the temperature.

For electricity, Ohm's law is analogous to Fourier's law; thus

$$-k \text{ grad } V = \mathbf{i} \tag{8.50}$$

where V = voltage and $\mathbf{i}$ = current flow density.

Following Fick's law[6] for flow of incompressible fluid through finely divided porous media,

$$-k \text{ grad } p = \mathbf{v} \tag{8.51}$$

where p = pressure
$\mathbf{v}$ = flow rate per unit area

Conservation laws applied to Eqs. (8.49), (8.50), and (8.51) yield the following expressions:

For Eq. (8.49), conservation of thermal energy implies

$$\rho c(\partial T/\partial t) = -\text{ div } \mathbf{Q} \tag{8.52}$$

where c = specific heat per unit mass
ρ = mass density

Similarly, for Eq. (8.50), conservation of charge, and Eq. (8.51), conversation of mass, are implied respectively by

$$\partial q/\partial t = -\text{ div } \mathbf{i} \tag{8.53}$$

where q = charge density

$$\partial \rho/\partial t = -\text{ div } (\rho \mathbf{v}) \tag{8.54}$$

where ρ = mass density

$\mathbf{Q}$ is eliminated between Eqs. (8.49) and (8.52) by taking div of Eq. (8.49):

$$\rho c(\partial T/\partial t) = -(-\text{div } k \text{ grad } T) = k\,\nabla^2 T \tag{8.55}$$

Similarly, for Eqs. (8.50) and (8.53),

$$\partial q/\partial t = k\,\nabla^2 V \tag{8.56}$$

And, for Eqs. (8.51) and (8.54),

$$\partial \rho/\partial t = k \text{ div } (\rho \text{ grad } p) \tag{8.57}$$

If ρ = constant, Eq. (8.57) reduces to

$$\nabla^2 p = 0 \qquad \text{(Laplace's equation)} \tag{8.58}$$

If Eq. (8.52), (8.53), or (8.54) had volume sources at the points of investigation, for example,

$$\rho c(\partial T/\partial t) = -\operatorname{div} \mathbf{Q} + S$$

Then Eqs. (8.55), (8.56), and (8.58) would read

$$k\,\nabla^2 T = \rho c(\partial T/\partial t) - S \tag{8.59}$$
$$k\,\nabla^2 V = \partial q/\partial t - S \tag{8.60}$$
$$k\rho\,\nabla^2 p = -S \tag{8.61}$$

In the absence of time-varying potentials, Eqs. (8.59) and (8.60) reduce to the Poisson form of Eq. (8.61), and where no source is present all reduce to the form of Laplace's equation (8.58). Electrostatic phenomena are closely related to the above developments. The electrostatic field **E** is given by

$$\mathbf{E} = -\operatorname{grad} V \tag{8.62}$$

and the flux **D** is linearly related to **E** by

$$\mathbf{D} = \epsilon \mathbf{E} \tag{8.63}$$
$$\epsilon = \text{dielectric constant}$$

By Gauss's law, which follows from Coulomb's law of forces,

$$\operatorname{div} \mathbf{D} = \rho \tag{8.64}$$
$$\rho = \text{charge density}$$

Eliminating **D** and **E** among Eqs. (8.62), (8.63), and (8.64) yields

$$\rho = \operatorname{div} \mathbf{D} = -\epsilon \operatorname{div} \operatorname{grad} V$$
$$= -\epsilon\,\nabla^2 V$$
$$\nabla^2 V = -\rho/\epsilon \tag{8.65}$$

which is Poisson's equation, degenerating to Laplace's equation in the absence of sources (i.e., $\rho = 0$).

8.3. SYSTEMS OF ORDINARY DIFFERENTIAL EQUATIONS

(a) Fundamentals

All systems which occur in nature are nonlinear and distributed. To an excellent approximation, many systems can be "lumped," permitting vast simplifications of the mathematical model. For example, the lumped spring-mass-damping system is, strictly speaking, a distributed system with the "mass" composed of an infinity of densely packed elementary springs and masses and damping elements arranged in some uncertain order. Because of the theoretical difficulties encountered in formulating an accurate mathematical model which fits the actual system and the analytical difficulties in attacking the complex problem, the engineer (with experimental justification) makes the "mass" a point mass which cannot be deformed, and the spring a massless spring without damping. If damping is present an element called the *damper* is isolated so that the "lumped" system is composed of discrete elements. Having settled on an equivalent lumped physical model, the equations describing system behavior are next formulated on the basis of known physical laws.

The equations thus derived constitute a set of ordinary differential equations, generally nonlinear, implying the existence of one or more lumped elements which do not behave in a "linear" fashion, e.g., nonlinearity of load vs. deflection of a spring.

In mathematical terms it is easier to define a nonlinear set by first defining what constitutes a linear set and then using the exclusion principle as follows.

A set of ordinary differential equations is linear if terms containing the dependent variable(s) or their time derivatives appear to the first degree only. The physical system it characterizes is termed *linear*. All other systems are nonlinear and the physical systems they define are nonlinear.

An example of a linear system is the set

$$\begin{aligned} t^2(d^2x_1/dt^2) + t(d^2x_2/dt^2) + d^2x_1/dt^2 + (\sin t)^2x_2 + G &= 0 \\ dx_1/dt + d^2x_2/dt^2 &= 0 \end{aligned}$$

where it is noted that the factors containing functions of t, the independent variable, are ignored in determining linearity, and each term containing the dependent variable x_1, x_2 or their derivatives is of the first degree.

Two examples of nonlinear systems are

$$(dx_1/dt)^2 + 2x_1 = 0 \tag{8.66a}$$

$$\left\{\begin{aligned} x_1(d^3x_2/dt^2) + x_1x_2 &= 0 \\ x_2 + d^2x_1/dt^2 + d^3x_2/dt^3 &= 0 \end{aligned}\right\} \tag{8.66b}$$

In the first, the square of the first derivative immediately rules it as nonlinear. The first equation of the second set is nonlinear on two counts: first, by virtue of the product of x_1, and a time derivative d^3x_2/dt^2, and second, because of the term containing the product of two dependent variables x_1, x_2. Despite the linearity of the second equation in the set, the overall set [Eq. (8.66*b*)] is nonlinear.

In general, and with few exceptions, the nonlinear equation does not yield to analysis, so that machine, numerical, or graphical methods must be employed. Wherever possible and under very special circumstances approximations are made to "linearize" a nonlinear system in order to make the problem amenable to analysis.

(b) Introduction to Systems of Nonlinear Differential Equations[7,8,9]

Perhaps the simplest classic example of a nonlinear system is the undamped free pendulum, the equation of motion of which is

$$\ddot{\theta} + g/l \sin \theta = 0 \tag{8.67}$$

This belongs to a class of elastic systems containing nonlinear restoring forces. Here $\sin \theta$ is clearly the nonlinear term. For small displacements of θ, $\sin \theta \approx \theta$ and Eq. (8.67) becomes

$$\ddot{\theta} + g/l\theta = 0 \tag{8.68}$$

The general solution of Eq. (8.68) is

$$\theta = A \sin \sqrt{g/l}\, t + B \cos \sqrt{g/l}\, t \tag{8.69}$$

A and B are constants of integration depending upon initial conditions. As θ gets large Eq. (8.68) no longer holds, and therefore Eq. (8.69) is an invalid approximation to Eq. (8.67). Under this condition Eq. (8.67) cannot be "linearized." Other nonlinear restoring forces are characterized as hard and soft springs whose force (F) vs. deflection (x) characteristics are given by $F = ax + bx^3$, $a > 0$ where $b < 0$ for soft springs, $b > 0$ for hard springs, and $b = 0$ for linear springs (see Sec. 6). The degree of nonlinearity is measured by the relative magnitudes of bx^3 and ax and implies some knowledge of x. Linearization of the spring-mass system given by

$$\ddot{x} + ax + bx^3 = 0 \tag{8.70}$$

is possible if

$$|bx^3| \ll |ax|$$

for all x experienced, yielding the approximation

$$\ddot{x} + ax \approx 0$$

An electric analog of this system of Eq. (8.70) exists for an LC circuit where C depends upon q in accordance with $1/C = \alpha + \beta q^2$.

From $q/C + L(d^2q/dt^2) = 0$ the following is derived after substitution for $1/C$:

$$L\ddot{q} + \alpha q + \beta q^3 = 0$$

Another analog derives from the nonlinear dependence of flux ϕ on current i in an LC circuit given by $i = \alpha\phi + \beta\phi^3$, which when substituted in the first time derivative of the loop-drop equation

$$L d\phi/dt + q/C = 0$$

yields

$$L d^2\phi/dt^2 + i/c = L d^2\phi/dt^2 + (\alpha\phi + \beta\phi^3)/C \tag{8.71}$$

Expressed in generalized form, the foregoing nonlinear spring-mass (capacitance-inductance) systems are given by

$$\ddot{x} + f(x) = 0 \tag{8.72}$$

Multiplying Eq. (8.72) by $\dot{x}\,dt$ and integrating we have

$$\int_0^t \dot{x}\ddot{x}\,dt + \int_0^t f(x)\dot{x}\,dt = \dot{x}^2/2\Big]_0^t + \int_{x(0)}^{x(t)} f(x)\,dx$$

$$\dot{x}^2/2 + \int_{x(0)}^{x(t)} f(x)\,dx = \dot{x}^2_{(0)}/2 \tag{8.73}$$

which is a statement expressing energy conservation. If $V(x)$ is the indefinite integral

$$\int f(x)\,dx = +V(x)$$

Then Eq. (8.73) becomes

$$\dot{x}^2/2 - x^2(0)/2 + V(x) - V[x(0)] = 0$$
$$\dot{x}^2/2 + V(x) = \dot{x}^2(0)/2 + V[x(0)] \triangleq E \tag{8.74}$$

$V(x)$ is the potential-energy function which represents stored energy from some arbitrary reference level, and E, a constant, is defined to be the "total energy" at any time. Solving Eq. (8.74),

$$\dot{x} = \sqrt{2[E - V(x)]} \tag{8.75}$$

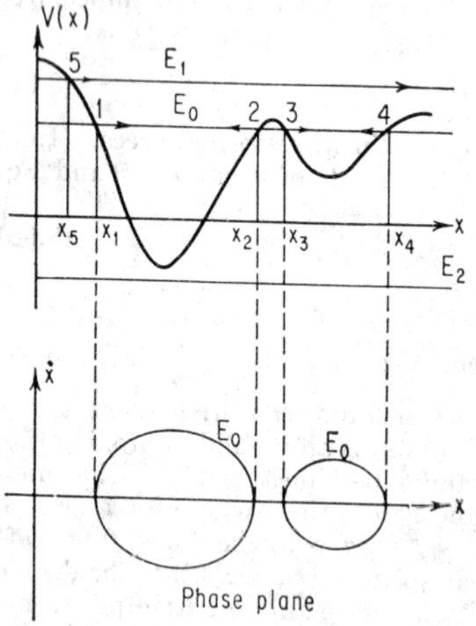

FIG. 8.9. Qualitative behavior of second-order free system.

Qualitative Behavior of the Conservative Free System. From Eq. (8.75) it is evident that physically realizable motion demands that $E \geq V(x)$ for all possible x. Consider a possible graph of $V(x)$ (Fig. 8.9) with E_0 drawn intersecting at points 1, 2, 3, 4, which points correspond to $E_0 = V(x)$, and from Eq. (8.75), $\dot{x} = 0$. Since $f(x) = (dV/dx)$, the slopes of the curve at these points give the spring force $f(x)$. From Eq. (8.72)

$$\ddot{x} = -f(x) = -dV/dx \tag{8.76}$$

Consequently acceleration corresponds to the direction of arrows shown for the two possible states of motion in Fig. 8.9, implying periodic motion between x_1 and x_2 in one case, and between x_3 and x_4 in the other. To find the period for case 1, for example,

$$\tau = \int_{x_1}^{x_2} dx/\dot{x} + \int_{x_2}^{x_1} dx/\dot{x} = \oint dx/\dot{x} \tag{8.77}$$

where integration is around a cycle loop in a phase-plane plot shown in Fig. 8.9, where $\dot{x}$ is plotted as a function of x, and the sign of $\dot{x}$ equals the sign of dx.

For E_1 as an initial-energy level shown in Fig. 8.9, motion is possible when $E_1 \geq V(x)$; it is seen that, for an initial negative velocity, the system will come to rest at point 5 and then, from Eq. (8.76), since the acceleration at that point is positive, motion would start to the right. Since $E_1 > V$ for $x > x_5$, it is impossible for x to reach zero again, and hence motion would continue in the positive direction without bound.

If $E = E_2$ as shown in Fig. 8.9, $E_2 < V(x)$ for all x; this cannot correspond to a physical system, a consequence of Eq. (8.75).

As an example of the above, the energy of the simple undamped pendulum is found from Eqs. (8.67) and (8.74)

$$\dot{\theta}^2/2 - g/l \cos \theta = E$$

where motion is indicated between θ_1 and θ_2 for $E = E_0$ (Fig. 8.10). For $E = E_1$ motion continues in a single direction, which physically amounts to putting in more energy than that required to bring the pendulum into the position where it is vertically above its support.

Graphical Analysis of Second-order Nonlinear Autonomous Differential Equations. Consider the following form of a free second-order equation with time-invariant coefficients

$$\ddot{x} + f(x,\dot{x}) = 0 \tag{8.78}$$

It is possible to analyze this very restrictive equation by a graphical method called *phase-plane analysis.* Equation (8.78) is first rewritten as

$$\begin{aligned} \dot{x}(d\dot{x}/dx) &= -f(x,\dot{x}) \\ d\dot{x}/dx &= -f(x,\dot{x})/\dot{x} \end{aligned} \tag{8.79}$$

A plot is next made of $\dot{x}$ as a function of x (phase-plane plot). At every point Eq. (8.79) states that the slope is $-f(x,\dot{x})/\dot{x}$.

The initial conditions $\dot{x}(0)$, $x(0)$ place the origin of the system in the phase plane. An arc with slope equal to $\dfrac{-f[x(0),\dot{x}(0)]}{\dot{x}(0)}$ is laid off extended over a small length terminating at $x(1)$, $\dot{x}(1)$. The process is continued until either a stable point is reached, or a limit cycle is manifest, or indications show the growth without bound of the system parameters x or $\dot{x}$. To find x as a function of time, $t = \int dx/\dot{x}$.

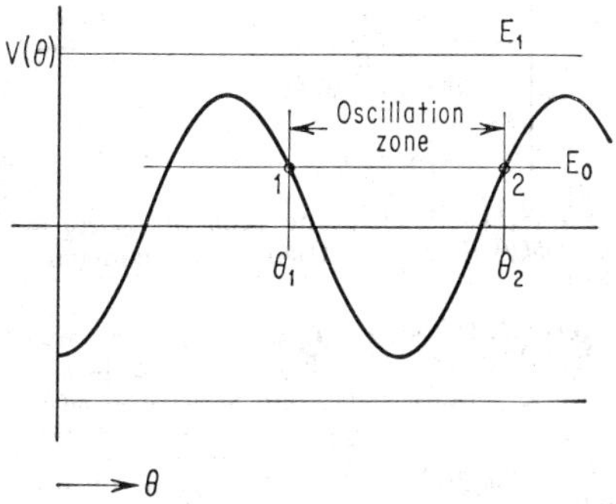

FIG. 8.10. Potential function for undamped pendulum $V(\theta) = \int \sin \theta \, d\theta = -\cos \theta$.

Several convenient techniques are available to facilitate procedures (e.g., the isocline method), the essentials of which were described above. Special cases of phase-plane analyses are given below.

Special Case 1. *Linear Spring-mass System*

$$\ddot{x} + kx = 0$$
$$\dot{x}^2/2 + kx^2/2 = \dot{x}(0)^2/2 + kx^2(0)/2 = E$$

The equation is of an ellipse in the phase plane $\dot{x}$ vs. x, or if we make the following changes of variable

$$y = x/\sqrt{k} \qquad \tau = \sqrt{k}\,t$$

and substitute in the above we obtain

$$\begin{aligned} (dy/d\tau)^2 + y^2 &= 2E/k^2 \\ \dot{y}^2 + y^2 &= 2E/k^2 \end{aligned} \tag{8.80}$$

a circle of radius $= \sqrt{2E/k^2}$ about the origin in the phase plane of $\dot{y}$ vs. y.

Special Case 2. *Spring-mass-damper System*

$$\ddot{x} + c\dot{x} + kx = 0$$
$$\dot{x}^2/2 + \int_0^t c\dot{x}^2 \, dt + kx^2/2 = E \qquad c > 0$$

Writing the energy form, where the damping integral is greater than zero as shown earlier,

$$\dot{x}^2/2 + kx^2/2 = E - \int_0^t c\dot{x}^2\,dt$$

$$k^2\dot{y}^2/2 + k^2y^2/2 = E - k^2/k^{1/2}\int_0^\tau c\dot{y}^2\,d\tau$$

The right side decreases in time, so that in the phase-plane plot the locus must lie on a continuously decreasing radius from the origin as time increases, until the origin is reached. The actual path is a logarithmic spiral. Other systems of the form $\ddot{x} + f(\dot{x}) + kx = 0$ are, by suitable changes of variable, shown equivalent to

$$\ddot{y} + \phi(\dot{y}) + y = 0$$

whence

$$\frac{d\dot{y}}{dy} = \frac{-[\phi(\dot{y}) + y]}{\dot{y}} \tag{8.81}$$

The phase-plane plot of Eq. (8.81) is obtainable by a neat method due to Liénard, described as follows: In Fig. 8.11, first $-\phi(\dot{y})$ is drawn. Then for any point of state, say P_1, the locus has a center of curvature in the phase plane located on the y axis shown by dotted construction. The slope must be that given by Eq. (8.81), tan ω. From geometry,

$$\tan\omega = -\frac{1}{\tan\theta} = \frac{-[\phi(\dot{y}) + y]}{\dot{y}}$$

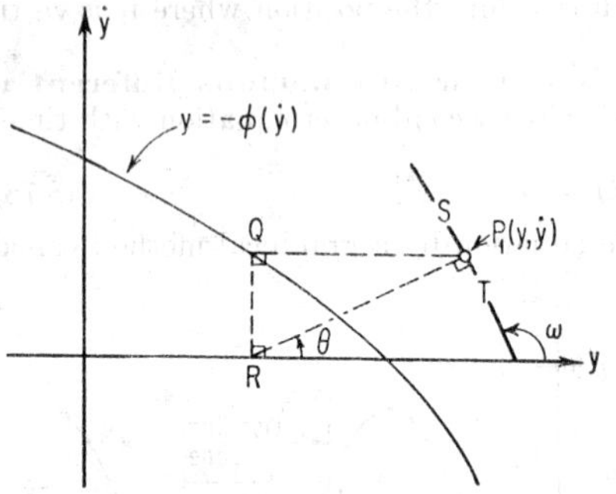

FIG. 8.11. Phase-plane plot of $d\dot{y}/dy = \{-[\phi(\dot{y}) + y]\}/\dot{y}$ (Liénard's construction).

Method: $QR = \dot{y}$
$QP = y - [-\phi(\dot{y})] = y + \phi(\dot{y})$
$\tan\theta = QR/QP = \dot{y}/[y + \phi(\dot{y})]$
$= -1/(d\dot{y}/dy)$

Hence it follows that the slope of line ST, perpendicular to line RP_1 at P_1, is $d\dot{y}/dy$.

Special Case 3. *Coulomb Damping* (*Dry Friction*). *Second-order System*

$$\ddot{x} + c\,\text{sgn}\,\dot{x} + x = 0 \tag{8.82}$$

where sgn = sign of. The phase-plane plot in Fig. 8.12 is accomplished, following Liénard's method, by first plotting $-c$ sgn $\dot{x}$ and then following in accordance with the above description. The plot consists of arcs of two circles centered at 1 for $\dot{x} > 0$ and 2 for $\dot{x} < 0$. This is shown for two different initial conditions corresponding to p_1 and p_1' in Fig. 8.12. Note that motion stops at a position corresponding to 4 since $\dot{x} = 0$ and the spring force is less than the impending damping force, thus preventing motion. This can also be shown analytically by considering two regions $\dot{x} < 0$ and $\dot{x} > 0$. Rewriting Eq. (8.82), we obtain

$$\dot{x}(d\dot{x}/dx) + (x + c\,\text{sgn}\,\dot{x}) = 0$$

Multiplying Eq. (8.82) by dx and integrating for the two regions,

$$\dot{x}^2/2 + x^2/2 + cx = E_1 \rightarrow \dot{x}^2 + (x + c)^2 = 2E_1 + c^2 \qquad \dot{x} > 0$$

Similarly,

$$\dot{x}^2 + (x - c)^2 = 2E_2 + c^2 \qquad \dot{x} < 0$$

Limit Cycles and Sustained Oscillations. Consider the system governed by $\ddot{x} + f(x,\dot{x}) + x = 0$.

If

$$\dot{x}f(x,\dot{x}) < 0 \qquad |x| < \delta \tag{8.83a}$$

$$\dot{x}f(x,\dot{x}) > 0 \qquad |x| > \delta \tag{8.83b}$$

where δ is some positive constant, the system will exhibit a limit cycle which corresponds to a closed curve in the phase plane. When Eq. (8.83a) holds, there is a net increase in the system energy $\mathcal{E}$.

$$\dot{x}^2/2 + x^2/2 + \int_0^t f(x,\dot{x})\dot{x}\,dt = E = \dot{x}^2(0)/2 + x^2(0)/2 = \mathcal{E}(0)$$

$$\mathcal{E} = \dot{x}^2/2 + x^2/2 = E - \int_0^t f(\dot{x},x)\dot{x}\,dt \tag{8.84}$$

given by the initial state E minus the integral. Since $2\mathcal{E}$ is the radius squared from the center to the point of state in the phase plane, there is a time rate of increase of radius every time the motion falls within the shaded zone (Fig. 8.13) and a decrease for motion corresponding to points outside the shaded zone. The type of oscillation is self-sustained and will start of its own accord for any initial condition. The van der Pol equation is an example of this type:

$$\ddot{x} - \epsilon\dot{x} + \beta x^2\dot{x} + x = 0 \qquad \epsilon > 0 \tag{8.85}$$
$$\dot{x}f(x,\dot{x}) = \dot{x}^2(-\epsilon + \beta x^2)$$

The term $f(x,\dot{x})$ changes sign when

$$-\epsilon + \beta x^2 = 0$$
$$x = \pm\sqrt{\epsilon/\beta} = \pm\delta$$

Limit cycles for higher-order systems are conceptually depicted by closed curves in multidimensional space, which is the generalization of single-degree-of-freedom systems.

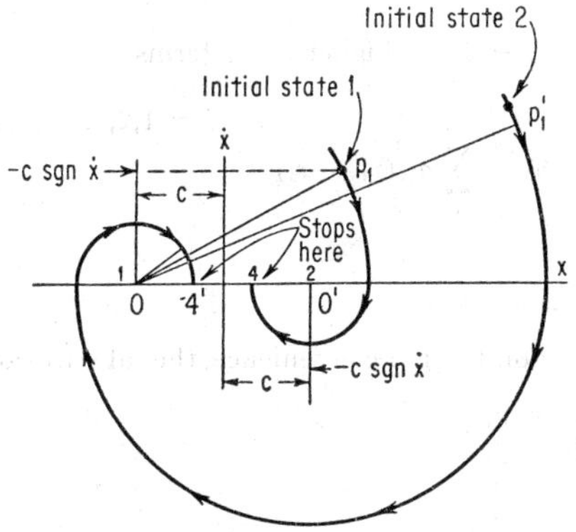

FIG. 8.12. Phase-plane plot for Coulomb damping of spring-mass system.

$$\ddot{x} + c \operatorname{sgn} \dot{x} + x = 0$$

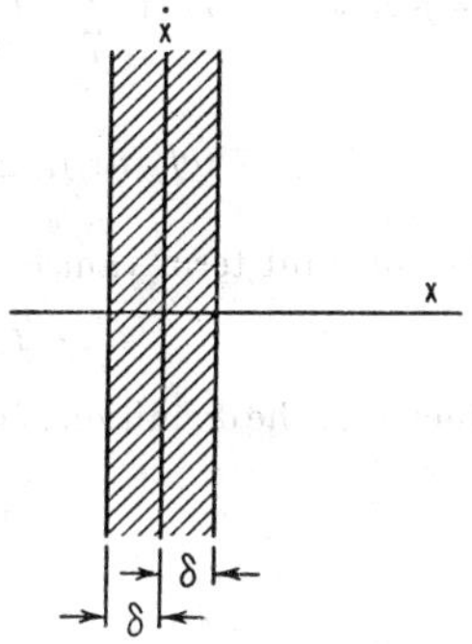

FIG. 8.13. Limit cycles and sustained oscillations.

Singular Points and Stability. It can be shown[7] that any autonomous set of nonlinear differential equations can be represented by

$$\begin{aligned} \dot{x}_1 &= f_1 \ (x_1 \ \cdots \ x_n) \\ \dot{x}_2 &= f_2 \ (x_1 \ \cdots \ x_n) \\ &\cdots \\ \dot{x}_n &= f_n \ (x_1 \ \cdots \ x_n) \end{aligned} \tag{8.86}$$

The equilibrium positions are given by the roots of

$$\begin{aligned} f_1 \ (x_1 \ \cdots \ x_n) &= 0 \\ f_2 \ (x_1 \ \cdots \ x_n) &= 0 \\ &\cdots \\ f_n \ (x_1 \ \cdots \ x_n) &= 0 \end{aligned} \tag{8.87}$$

The roots $x_1, \ldots, x_n$ of this set are called *singular equilibrium points* where all the time derivatives $\dot{x}_1, \dot{x}_2, \ldots, \dot{x}_n$ are equal to zero.

The algebraic solution to Eq. (8.87) gives in general one or more sets of singular points, e.g.,

$$\begin{array}{llll} x_1^{(0)} & x_2^{(0)} & \cdots & x_n^{(0)} \quad \text{one set} \\ x_1^{(1)} & x_2^{(1)} & \cdots & x_n^{(1)} \quad \text{second set, etc.} \end{array}$$

If the motion at any time corresponds to one of these points, the system is at rest. If left undisturbed, from Eq. (8.86) $\dot{x}_1 = \dot{x}_2 = \cdots \dot{x}_n = 0$, the point (in phase space), and therefore the corresponding motion, does not change in time; i.e., the system remains at rest. If, however, the point is disturbed from its equilibrium position (perturbation), it is of interest from the stability point of view as to whether or not it will return to the point. If for a small perturbation from equilibrium the system tends to return to the same equilibrium position as $t \to \infty$, the system is said to be *asymptotically stable*. If, however, the system diverges from the equilibrium point, it is said to be in a state of *unstable equilibrium*, or the point is unstable. Special points in which neither of these events occurs are said to display neutral stability and are exceptional. To test stability, the nonlinear system is "linearized" in the neighborhood of the equilibrium point $\bar{x}_1, \bar{x}_2, \bar{x}_3, \ldots, \bar{x}_n$ by performing a Taylor's-series expansion about the point and ignoring terms higher than the first power of x_i. The typical expansion is

$$f_j = f_j(\bar{x}_1,\bar{x}_2, \ldots ,\bar{x}_n) + \sum_i (\partial f_j/\partial x_i)_{x_k=\bar{x}_k} (x_i - \bar{x}_i) + \text{higher-order terms}$$

$$j = 1,2, \ldots , n$$

Defining $\quad (\partial f_j/\partial x_i)_{x_k=\bar{x}_k} = a_{ji} \qquad f_j \approx \sum_i a_{ji}(x_i - \bar{x}_i)$

where the constant term vanishes, i.e.,

$$f_j(\bar{x}_1,\bar{x}_2, \ldots ,\bar{x}_n) = 0$$

a consequence of the definition of equilibrium point. For convenience, the substitution

$$\begin{aligned} x_1 - \bar{x}_1 &= y_1 \\ x_2 - \bar{x}_2 &= y_2 \\ &\vdots \\ x_n - \bar{x}_n &= y_n \end{aligned}$$

placed in Eq. (8.86) yields

$$\begin{aligned} \dot{y}_1 &= a_{11}y_1 + a_{12}y_2 + \cdots a_{1n}y_n \\ &\vdots \\ \dot{y}_n &= a_{n1}y_1 + \cdots\cdots a_{nn}y_n \end{aligned} \tag{8.88}$$

This is the well-known linear set whose solution is of the exponential type. Assuming a solution,

$$y_j = A_j e^{\lambda t}$$

and making this substitution in Eq. (8.88) yields

$$\begin{aligned} 0 &= (a_{11} - \lambda)A_1 + a_{12}A_2 + \cdots + a_{1n}A_n \\ 0 &= a_{n1}A_1 + \cdots\cdots + (a_{nn} - \lambda)A_n \end{aligned} \tag{8.89}$$

From linear theory, the necessary condition for Eq. (8.89) to have nontrivial solutions is

$$\begin{vmatrix} a_{11} - \lambda & a_{12} & a_{13} & \cdots & a_{1n} \\ a_{21} & a_{22} - \lambda & \cdot & \cdots & \cdot \\ \cdot & \cdot & \cdot & \cdots & \cdot \\ \cdot & \cdot & \cdot & \cdots & \cdot \\ a_{n1} & \cdot & \cdot & \cdots & a_{nn} - \lambda \end{vmatrix} = 0$$

which when expanded leads to an nth-order algebraic equation

$$b_n\lambda^n + b_{n-1}\lambda^{n-1} + \cdots + b_1x + b_0 = 0 \tag{8.90}$$

which has n roots for λ, the characteristic roots of the matrix

$$\begin{bmatrix} a_{11} & a_{12} & \cdots & a_{1n} \\ \cdot & & & \\ \cdot & & & \\ \cdot & & & \\ a_{n1} & \cdots & \cdots & a_{nn} \end{bmatrix}$$

Each root corresponds to a solution $y_j = A_j e^{\lambda_i t}$. If λ_i has a real part greater than zero, y_j will grow without bound. Hence the necessary and sufficient condition for stability at the equilibrium point is that Re $\lambda_i < 0$ for all roots, $i = 1, \ldots, n$. As an example, consider the second-order van der Pol equation (8.85) in the form

$$\dot{x} = v$$
$$\dot{v} = \epsilon v - \beta x^2 v - x$$

The only equilibrium point is $x = v = 0$ obtained after invoking Eq. (8.87).

Expanding about $x = 0$, $v = 0$, carrying linear terms

$$\dot{x} = v$$
$$\dot{v} \approx -x + \epsilon v$$

The characteristic roots are found from

$$\begin{vmatrix} -\lambda & 1 \\ -1 & \epsilon - \lambda \end{vmatrix} = 0$$

$$-\lambda(\epsilon - \lambda) + 1 = 0 \qquad \lambda = \frac{\epsilon \pm \sqrt{\epsilon^2 - 4}}{2}$$

It is evident that one or more roots must satisfy Re $\lambda > 0$; therefore, the system is unstable about the equilibrium point as observed previously. Note that, if ϵ is negative, the system is stable.

8.4. SYSTEMS OF ORDINARY LINEAR DIFFERENTIAL EQUATIONS[11,12,13,14]

Formulation of the linearized form of a lumped dynamic system leads to a set of linear differential equations. The question of validity in assuming linearity is, in general, complicated. One method (though not conclusive) is to assume a linear form, solve the set, cast the solution into the original form to measure deviations from linearity, and finally, on this basis, render a decision on validity. An example is the simple pendulum undergoing a forced vibration where the steady-state solution is of interest:

$$\ddot{\theta} + \omega_0^2 \sin\theta = A \sin\omega t \qquad \omega_0^2 = g/l$$

The linearized form and its characteristic solution are

$$\ddot{\theta} + \omega_0^2\theta = A \sin\omega t$$
$$\theta = \frac{A}{\omega_0^2 - \omega^2} \sin\omega t$$

And therefore θ is bounded by

$$|\theta| \leq \left|\frac{A}{\omega_0^2 - \omega^2}\right|$$

The linear form is valid if $(\sin\theta - \theta)/\theta$ remains small for all motion, i.e.,

$$|(\sin\theta - \theta)/\theta| \ll 1$$

But

$$\sin\theta = \theta - \theta^3/3! + \theta^5/5! - \cdots$$

Therefore

$$\sin\theta/\theta = 1 - \theta^2/3! + \theta^4/5! - \cdots; \qquad |(\sin\theta - \theta)/\theta| \leq \theta^2/3!$$

and the necessary condition for validity becomes

$$|\theta^2/3!| \ll 1$$

Linear systems are classified as time-variant or time-invariant and in the former derive from a system containing time-variable parameters leading to product terms in independent and dependent variables, e.g.,

$$\sin \alpha t(d^3x/dt^3) \qquad t^2x$$

whereas the time-invariant case shows no parametric dependence on time.

An example of a time-variable linear system is that of a rocket propelled in free unidirectional flight by a jet-exhaust thrust $C(t)$. Fuel expenditure results in a rocket mass loss. The equation of motion is

$$m(t)(dv/dt) = C(t) \tag{8.91}$$

with $m(t)$ and $C(t)$, the mass and thrust, connected under some broad assumptions by the linear differential equation

$$C(t) = -k[dm(t)/dt] \tag{8.92}$$

The solution is

$$v - v(0) = k \ln [m(0)/m]$$

This is an exceptional case since a closed solution is obtainable. More generally, however, time-variable systems are practically invulnerable to analytic attack in contrast to their time-invariant counterparts whose solutions are completely known.

Properties of Linear Differential Equations. The general properties of linear differential equations are stated as follows:

1. The general homogeneous linear equation is expressed in operator form

$$D(Y) = f(t)$$

with initial conditions $Y(0) = a_0$.

$$\dot{Y}(0) = a_1 \qquad (d^{n-1}/dt^{n-1})Y(0) = a_{n-1} \tag{8.93}$$

where D is the linear operator of nth order:

$$D = \sum_{m=0}^{n} P_m(t)(d^m/dt^m)$$

where the coefficient (function of time in general) of the nth derivative term is $P_m(t)$

2. The linear operator has the properties

$$D(\alpha y) = \alpha\, Dy \qquad (\alpha = \text{const})$$
$$D(y_1 + y_2) = Dy_1 + Dy_2$$

Therefore,

$$D(\alpha y_1 + \beta y_2) = \alpha\, Dy_1 + \beta\, Dy_2$$

3. The complete solution to the homogeneous equation $D(Z) = 0$ of the nth degree is the sum of n linearly independent solutions, viz.,

$$Z = b_1z_1 + b_2z_2 + \cdots + b_nz_n$$

where $b_1 \cdots b_n$ are constants which can be adjusted to satisfy n initial conditions of the problem.

4. The general solution to Eq. (8.93) is the sum of two solutions $Y = Z + X$ where Z is the total solution to the homogeneous equation

$$D(Z) = 0 \tag{8.94}$$

and X is *any* solution to $D(X) = f(t)$ regardless of initial conditions. It should be emphasized, and is implied, that X is not unique, containing any number of solutions to the homogeneous equation, e.g., $(c_1z_1 + \cdots + c_mz_m)$.

5. If the solution to $D(X) = f(t)$ is confined to be the asymptotic solution (i.e., the solution as $t \to \infty$), then this solution $X \triangleq X_p$ is called the particular solution and the remainder $Z = b_1z_1 + b_2z_2 + \cdots + b_nz_n \triangleq Z_c$ is called the *complementary solution* and is called the transient solution if it vanishes with time. $b_1 \cdots b_n$ are chosen such that all initial conditions of Eq. (8.93) are satisfied. Then since the solution to Eq. (8.93) is unique[11]

$$Y = X_p + Z_c = X + Z$$

6. Another important specialized partitioning of X and Z restricts X to satisfy zero initial conditions, i.e., the solution to $D(x) = f(t)$ for

$$X(0) = \dot{X}(0) = \cdots = d^{n-1}/dt^{n-1}X(0) = 0$$

which is

$$X_{(1)} = \int_0^t W(t,T)f(T)\,dT$$

where $W(t,T)$ is the solution to

$$D(W) = \delta(t - T)$$

where $W(t,T) \equiv 0$ for $t \leq T$ and $\delta(t - T) =$ Dirac delta function with the remaining part of the solution

$$Z_{(1)} = c_1z_1 + c_2z_2 + \cdots + c_nz_n$$

where the c's are chosen to satisfy the initial conditions of the problem which are absorbed in the z's alone.

$$Y(0) = Z_{(1)}(0) = \sum_{i=1}^{n} c_iz_i(0)$$

$$Y'(0) = Z_{(1)}'(0) = \sum_{i=1}^{n} c_iz_i'(0)$$

$$Y''(0) = Z_{(1)}''(0) = \sum_{i=1}^{n} c_iz_i''(0)$$

$$\vdots \qquad \vdots$$

$$(d^{n-1}/dt^{n-1})Y(0) = d^{n-1}Z_{(1)}(0)/dt^{n-1} = \sum_{i=1}^{n} c_i(d^{n-1}/dt^{n-1})z_{(i)}(0)$$

and

$$Y = X_{(1)} + Z_{(1)} = X + Z$$

7. As a direct consequence of property 6, the principle of superposition follows For a system initially inert, i.e., for zero initial conditions,

y_1 is the response to a forcing function $f_1(t)$
y_2 is the response to a forcing function $f_2(t)$

The response y to forcing function $f(t) = \alpha f_1(t) + \beta f_2(t)$ is $y = \alpha y_1 + \beta y_2$.

8. The necessary and sufficient condition that the solution y be bounded in Eq.

(8.93) for any bounded input $f(t)$ is that

$$\int_0^\infty |W(t,T)|\, dT < M < \infty$$

where M is some arbitrarily large positive number. Systems satisfying this criterion are said to be stable.

9. The fact that n solutions to Eq. (8.94) are linearly independent requires that the Wronskian be different from zero at any point in the interval, $t_1 < t < t_2$ where Eq. (8.94) is valid.

$$W_r(z_1, z_2, \cdots, z_n) = \det \begin{bmatrix} z_1 & z_2 & \cdots & z_n \\ dz_1/dt & dz_2/dt & \cdots & dz_n/dt \\ \vdots & \vdots & & \vdots \\ d^{n-1}z_1/dt^{n-1} & d^{n-1}z_2/dt^{n-1} & \cdots & d^{n-1}z_n/dt^{n-1} \end{bmatrix}$$

The Wronskian of the solutions to

$$d^nz/dt^n + Q_{(n-1)}(t)(dz^{n-1}/dt)^{n-1}Q_1(t)(dz/dt) + Q_0(t)z = 0$$

is given by

$$W_r(z_1 \cdots z_n) = W_r[z_1(\tau), z_2(\tau), \cdots, z_n(\tau)] \exp \int_\tau^t Q_{n-1}(x)\, dx \qquad (8.95)$$

As a direct consequence of Eq. (8.95) the Wronskian of a set satisfying Eq. (8.94) either does not vanish at all or vanishes identically since the exponential term cannot vanish.

(a) Introduction to Matrix Analysis of Differential Equations[15]

The nth-order differential equation

$$d^ny_1/dt^n + Q_{n-1}(d^{n-1}y_1/dt^{n-1}) + \cdots + Q_0y_1 = F \qquad (8.96)$$

can be written as n first-order differential equations

$$\begin{aligned} dy_1/dt &= y_2 \\ dy_2/dt &= y_3 \\ &\vdots \\ dy_{n-1}/dt^{n-1} &= y_n \\ dy_n/dt &= -Q_{n-1}y_n - Q_{n-2}y_{n-1} - \cdots - Q_0y_1 + F \end{aligned} \qquad (8.97)$$

which in matrix form is written

$$d\mathbf{y}/dt = A\mathbf{y} + \mathbf{f} \qquad y = \begin{vmatrix} y_1 \\ y_2 \\ y_3 \\ \cdot \\ \cdot \\ \cdot \\ y_n \end{vmatrix} \qquad (8.98)$$

In Eq. (8.97),

$$A = \begin{bmatrix} 0 & 1 & 0 & 0 & 0 & \cdot & \cdot & \cdot & 0 \\ 0 & 0 & 1 & 0 & 0 & \cdot & \cdot & \cdot & 0 \\ 0 & 0 & 0 & 1 & 0 & \cdot & \cdot & \cdot & 0 \\ 0 & 0 & 0 & 0 & 1 & \cdot & \cdot & \cdot & \cdot \\ \cdot & \cdot & \cdot & 0 & 0 & \cdot & \cdot & 0 & 0 \\ \cdot & \cdot & \cdot & \cdot & 0 & \cdot & \cdot & 1 & 0 \\ \cdot & \cdot & \cdot & \cdot & \cdot & \cdot & \cdot & 0 & 1 \\ -Q_0 & -Q_1 & \cdot & \cdot & \cdot & \cdot & \cdot & -Q_{n-2} & -Q_{n-1} \end{bmatrix} \qquad \mathbf{f} = \begin{bmatrix} 0 \\ 0 \\ \cdot \\ \cdot \\ \cdot \\ F \end{bmatrix}$$

Equation (8.98) is linear if matrix $A = A(t)$ and $\mathbf{f} = \mathbf{f}(t)$ and linear time-invariant if A does not depend on t.

Following classical methods, the total solution to

$$d\mathbf{y}/dt = A(t)\mathbf{y} + \mathbf{f}(t)$$

is the sum of two solutions, one to the homogeneous equation

$$d\mathbf{z}/dt = A(t)\mathbf{z} \tag{8.99}$$

plus any solution to

$$d\mathbf{x}/dt = A(t)\mathbf{x} + \mathbf{f}(t) \tag{8.100}$$

where

$$\mathbf{y} = \mathbf{x} + \mathbf{z}$$

The general solution chosen here will let the homogeneous solution satisfy the initial conditions of the problem, i.e.,

$$\mathbf{y}(0) = \mathbf{z}(0) = \mathbf{c}$$

and the remaining solution satisfy the null-vector condition at $t = 0$,

$$\mathbf{x}(0) = \mathbf{0}$$

so that

$$\mathbf{y}(0) = \mathbf{z}(0) + \mathbf{x}(0) = \mathbf{y}(0) + \mathbf{0} = \mathbf{y}(0) = \mathbf{c}$$

as required.

If the solution to Eq. (8.99) is known, the solution to Eq. (8.100) can be obtained by Lagrange's method of variation of parameters as follows: Consider the matrix differential equation

$$dZ(t)/dt = A(t)Z(t) \qquad Z(0) = I \text{ (initial conditions)} \tag{8.100a}$$

Postmultiplication by $\mathbf{c}$ yields

$$d(Z\mathbf{c})/dt = A(Z\mathbf{c})$$

From this and Eq. (8.99) it follows that

$$\mathbf{z} = Z(t)\mathbf{c} = Z(t)\mathbf{z}(\mathbf{0})$$

Now let

$$\mathbf{y} \triangleq Z(t)\mathbf{u} \tag{8.100b}$$

Substituting for $\mathbf{y}$ in Eq. (8.98),

$$d\mathbf{y}/dt = A(t)[Z(t)\mathbf{u}] + \mathbf{f} = Z(t)(d\mathbf{u}/dt) + [dZ(t)/dt]\mathbf{u} = Z(t)(d\mathbf{u}/dt) + A(t)Z(t)\mathbf{u}$$

whence $\mathbf{f} = Z d\mathbf{u}/dt$. Premultiplying by Z^{-1} yields $Z^{-1}\mathbf{f} = d\mathbf{u}/dt$, and integrating after separation of variables gives $\mathbf{u} = \mathbf{u}(0) + \int_0^t Z^{-1}(\tau)\mathbf{f}(\tau)\,d\tau$. Premultiply by Z to give

$$\mathbf{y} = Z(t)\mathbf{u} = Z(t)\mathbf{c} + \int_0^t Z(t)Z^{-1}(\tau)\mathbf{f}(\tau)\,d\tau \tag{8.101}$$

where $\mathbf{u}(0) = \mathbf{y}(0) = \mathbf{c}$ from Eq. (8.100b) and $Z(0) = I$.

For the time-invariant case,

$$Z(t)Z^{-1}(\tau) = Z(t - \tau)$$

and Eq. (8.101) takes the simpler form

$$\mathbf{y} = Z(t)\mathbf{c} + \int_0^t Z(t - \tau)\mathbf{f}(\tau)\,d\tau \tag{8.102}$$

Eqs. (8.101) and (8.102) hinge on the solution to the homogeneous matrix equation

$$dZ/dt = A(t)Z \tag{8.103}$$

However, for $A(t)$, the time-variable case, a solution is rarely possible. Turning to the time-invariant case,

$$d\mathcal{Z}(t)/dt = A\mathcal{Z}(t) \qquad \mathcal{Z}(0) = I \tag{8.104}$$

The formal solution is

$$\mathcal{Z}(t) = e^{At}$$

which must be defined as

$$e^{At} \triangleq \sum_{n=0}^{\infty} \frac{A^n t^n}{n!}$$

and

$$\mathcal{Z}(t + \tau) = e^{A(t+\tau)}$$

Equation (8.102) then becomes

$$\mathbf{y} = e^{At}\mathbf{c} + \int_0^t e^{A(t-\tau)}\mathbf{f}(\tau)\, d\tau \tag{8.105}$$

Equation (8.105) is in a form that is not useful for quantitative analysis.

It is desirable to obtain $\mathcal{Z}$ in closed form. If $\mathcal{Z}$ is represented by column vectors

$$\mathcal{Z} = \begin{bmatrix} z_1^{(1)} & \cdots & z_1^{(n)} \\ z_2^{(1)} & & \\ \cdot & & \\ \cdot & & \\ \cdot & & \\ z_n^{(1)} & \cdots & z_n^{(n)} \end{bmatrix} \triangleq \mathbf{z}^{(1)}\mathbf{z}^{(2)}\mathbf{z}^{(3)} \cdots \mathbf{z}^{(n)}$$

then

$$d\mathcal{Z}/dt = A\mathcal{Z} \qquad \mathcal{Z}(0) = I = \begin{bmatrix} 1 & 0 & \cdot & \cdot & \cdot & \cdot & \cdot & 0 \\ 0 & 1 & \cdot & \cdot & \cdot & \cdot & \cdot & 0 \\ 0 & 0 & 1 & \cdot & \cdot & \cdot & \cdot & 0 \\ 0 & 0 & 0 & 1 & \cdot & \cdot & \cdot & 0 \\ \cdot & \cdot & \cdot & \cdot & \cdot & \cdot & \cdot & \cdot \\ 0 & \cdot & \cdot & \cdot & \cdot & \cdot & 1 & 0 \\ 0 & \cdot & \cdot & \cdot & \cdot & \cdot & \cdot & 1 \end{bmatrix}$$

is equivalent to n equations

$$\begin{aligned} d\mathbf{z}^{(1)}/dt &= A\mathbf{z}^{(1)} & \mathbf{z}^{(1)}(0) &= \begin{bmatrix} 1 \\ 0 \\ 0 \\ \cdot \\ \cdot \\ 0 \end{bmatrix} \\ &\vdots & &\vdots \\ d\mathbf{z}^{(n)}/dt &= A\mathbf{z}^{(n)} & \mathbf{z}^{(n)}(0) &= \begin{bmatrix} 0 \\ 0 \\ \cdot \\ \cdot \\ \cdot \\ 1 \end{bmatrix} \end{aligned} \tag{8.106}$$

The homogeneous matrix equation possesses n possible solution vectors. Assuming one such vector solution

$$\mathbf{z}^{(k)} = \mathbf{c}^{(k)}e^{\lambda_j t} \qquad j = 1, 2, \ldots, n$$

and entering this into Eq. (8.106) yields

$$\begin{aligned} \lambda_j \mathbf{c}^k &= A\mathbf{c}^{(k)} \\ (A - \lambda_j I)\mathbf{c}^k &= 0 \end{aligned} \tag{8.107}$$

which is the eigenvector equation that must satisfy

$$|A - \lambda_j I| = 0$$

an nth-order equation yielding n roots for λ_j. The discussion here will be limited to distinct roots.

For each λ_j so found there exists a column vector formed from the cofactors of any row of $(A - \lambda_j I)$; let the matrix formed by the n column vectors thus formed be called B:

$$B = \mathbf{c}^{k(1)}\mathbf{c}^{k(2)} \cdots \mathbf{c}^{k(n)}$$

Equation (8.107) can be represented as

$$AB = B\Lambda \qquad \Lambda = \begin{bmatrix} \lambda_1 & 0 & \cdots & 0 \\ 0 & \lambda_2 & \cdots & 0 \\ 0 & 0 & \cdots & \lambda_n \end{bmatrix}$$

Premultiplying by B^{-1} yields

$$B^{-1}AB = \Lambda$$

which shows that A is diagonalized by a linear transformation.

From Eq. (8.100a),

$$dZ/dt = AZ \qquad z(0) = I$$

Let W be introduced by defining the transformation

$$Z = BW$$

Substituting in Eq. (8.100a),

$$\begin{aligned} d(BW)/dt &= A(BW) \\ B(dW/dt) &= ABW \end{aligned}$$

Premultiplying by B^{-1},

$$dW/dt = B^{-1}ABW = \Lambda W = \begin{bmatrix} \lambda_1 & 0 & & 0 \\ 0 & \lambda_2 & & 0 \\ & & \ddots & \\ 0 & 0 & & \lambda_n \end{bmatrix} W$$

The solution for W is easily verified to be

$$W = \begin{bmatrix} e^{\lambda_1 t} & 0 & & 0 \\ 0 & e^{\lambda_2 t} & & 0 \\ & & \ddots & \\ 0 & 0 & & e^{\lambda_n t} \end{bmatrix} B^{-1} \qquad W(0) = B^{-1}$$

which satisfies its differential equation and the initial conditions, viz.,

$$dW/dt = \Lambda W$$

$$\begin{bmatrix} \lambda_1 e^{\lambda_1 t} & 0 & \cdots & 0 \\ 0 & \lambda_2 e^{\lambda_2 t} & \cdots & 0 \\ \cdot & \cdot & & \cdot \\ \cdot & \cdot & & \cdot \\ \cdot & \cdot & & \cdot \\ 0 & 0 & \cdots & \lambda_n e^{\lambda_n t} \end{bmatrix} B^{-1} = \begin{bmatrix} \lambda_1 & 0 & \cdots & 0 \\ 0 & \lambda_2 & \cdots & 0 \\ \cdot & \cdot & & \cdot \\ \cdot & \cdot & & \cdot \\ \cdot & \cdot & & \cdot \\ 0 & 0 & \cdots & \lambda_n \end{bmatrix} \begin{bmatrix} e^{\lambda_1 t} & 0 & \cdots & 0 \\ 0 & e^{\lambda_2 t} & \cdots & 0 \\ \cdot & \cdot & & \cdot \\ \cdot & \cdot & & \cdot \\ \cdot & \cdot & & \cdot \\ 0 & 0 & \cdots & e^{\lambda_n t} \end{bmatrix} B^{-1}$$

$$Z(0) = BW(0) = BB^{-1} = I \qquad \text{as required}$$

Linear Time-invariant Systems.[16,17,18,19,20,21,22,23,24] The important property that distinguishes these systems from the time-variable systems is:

1. If the input $f(t)$ yields the response $y(t)$, the input $f(t + T)$ yields the response $y(t + T)$ (where all initial conditions are zero). As a consequence of property 1 and the superposition property we have:

2. The response to the derivative of an arbitrary input is equal to the derivative of the response. If $f(t) \rightarrow y(t)$

then
$$\frac{f(t+\epsilon) - f(t)}{\epsilon} \rightarrow \frac{y(t+\epsilon) - y(t)}{\epsilon}$$
$$\epsilon \rightarrow 0 \qquad f'(t) \rightarrow y'(y)$$

or, vectorially, $\mathbf{f}(t) \rightarrow \phi[\mathbf{y}(t)]$ where ϕ is a linear time-invariant operator. Then

$$\frac{\mathbf{f}(t+\epsilon) - \mathbf{f}(t)}{\epsilon} \rightarrow \frac{\phi[\mathbf{y}(t+\epsilon) - \mathbf{y}(t)]}{\epsilon} = \phi[\mathbf{y}'(t)]$$
$$\mathbf{f}'(t) \rightarrow \phi[\mathbf{y}'(t)] \qquad \epsilon \rightarrow 0$$

3. The solution to a free (homogeneous) time-invariant system of equations is composed of exponential terms, there being as many terms as the highest degree of the differential equation obtained in one dependent variable. These terms must be linearly independent, satisfying $W_r \neq 0$.

The general form of the coupled time-invariant system of n degrees of freedom is

$$\begin{array}{c} a_{11}(p)y_1 + a_{12}(p)y_2 \cdots + a_{1n}(p)y_n = f_1(t) \\ \vdots \\ a_{n1}(p)y_1 + \cdots\cdots + a_{nn}(p)y_n = f_n(t) \end{array} \tag{8.108}$$

In matrix-operator form,

$$A(p)\mathbf{y} = \mathbf{f}(t) \qquad A(p) = \begin{bmatrix} a_{11}(p) & \cdots & a_{1n}(p) \\ \vdots & & \vdots \\ a_{n1}(p) & \cdots & a_{nn}(p) \end{bmatrix} \qquad \mathbf{f} = \begin{bmatrix} f_1 \\ f_2 \\ \vdots \\ f_n \end{bmatrix} \tag{8.108a}$$

where p is the differential-integral operator defined by

$$p \triangleq d(\ )/dt \qquad 1/p \triangleq \int_0^t (\ )\,dt$$

For linear passive systems (i.e., containing inductance/inertia, capacitance/spring, or resistance/damping), each coefficient takes the form

$$a_{ij}(p) = L_{ij}p + 1/C_{ij}p + R_{ij}$$

where L_{ij}, C_{ij}, and R_{ij} are constants.

Recall that the general solution to Eq. (8.108) is composed of two solutions, one to the homogeneous system

$$A(p)\mathbf{z} = 0 \qquad \begin{array}{c} a_{11}(p)z_1 + a_{12}(p)z_2 + \cdots + a_{1n}(p)z_n = 0 \\ \vdots \\ a_{n1}(p)z_1 + \cdots\cdots + a_{nn}(p)z_n = 0 \end{array} \tag{8.109}$$

and one to the inhomogeneous equation.

$$A(p)\mathbf{x} = \mathbf{f}(t)$$

Thus
$$\mathbf{y} = \mathbf{x} + \mathbf{z} \text{ is the total solution} \tag{8.110}$$

To find $\mathbf{z}$ assume, as before, the exponential form [similar to Eq. (8.107)]

$$\mathbf{z} = \begin{vmatrix} z_1 = c_1^{(i)} & e^{\lambda t} \\ z_2 = c_2^{(i)} & e^{\lambda t} \\ \cdot & \cdot \\ \cdot & \cdot \\ \cdot & \cdot \\ z_n = c_n^{(i)} & e^{\lambda t} \end{vmatrix}$$

Substitution in Eq. (8.109) yields

$$\begin{aligned} a_{11}(\lambda)c_1 + a_{12}(\lambda)c_2 + \cdots + a_{1n}(\lambda)c_n &= 0 \\ \cdot \qquad\qquad\qquad & \cdot \\ \cdot \qquad\qquad\qquad & \cdot \\ \cdot \qquad\qquad\qquad & \cdot \\ a_{n1}(\lambda)c_1 + \cdots\cdots + a_{nn}(\lambda)c_n &= 0 \end{aligned} \qquad A(\lambda)\mathbf{c} = 0 \qquad (8.111)$$

From linear theory a solution c_1, c_2, c_n different from zero (the trivial case) can exist if and only if the determinant vanishes, thus

$$\det [A(\lambda)] = 0$$

$$\begin{vmatrix} a_{11}(\lambda) & a_{12}(\lambda) & a_{13}(\lambda) & \cdots & a_{1n}(\lambda) \\ a_{21}(\lambda) & a_{22}(\lambda) & & & \cdot \\ \cdot & \cdot & & & \cdot \\ \cdot & \cdot & & & \cdot \\ a_{n1}(\lambda) & \cdots & & & a_{nn}(\lambda) \end{vmatrix} = 0$$

which leads to an mth-degree algebraic equation in λ called the characteristic equation of the matrix A. In general, $m \neq n$ and $m \leqq 2n$ for passive systems.

The mth-degree equation yields m roots $\lambda_1, \lambda_2, \ldots, \lambda_m$ each of which satisfies Eq. (8.111). If the roots are distinct the corresponding $\mathbf{c}$ column vector

$$\mathbf{c} = \begin{bmatrix} c_1 \\ c_2 \\ \cdot \\ \cdot \\ \cdot \\ c_n \end{bmatrix}$$

can be found for each λ_i, being the cofactors of any row of $A(\lambda_i)$ in order from left to right: $c_1 = \operatorname{cof}(a_{j1})$; $c_2 = \operatorname{cof}(a_{j2})$; $c_n = \operatorname{cof}(a_{jn})$ for any j. The solution to Eq. (8.109) is then

$$\mathbf{z} = \sum_{i=1}^{n} \mathbf{c}^{(i)} e^{\lambda_i t}$$

or

$$\begin{aligned} z_1 &= c_1^{(1)} e^{\lambda_1 t} + c_1^{(2)} e^{\lambda_2 t} + \cdots + c_1^{(m)} e^{\lambda_m t} \\ \cdot \quad & \quad \cdot \\ \cdot \quad & \quad \cdot \\ \cdot \quad & \quad \cdot \\ z_n &= c_n^{(1)} e^{\lambda_1 t} + \cdots\cdots + c_n^{(m)} e^{\lambda_m t} \end{aligned}$$

If some of the roots are repeated, then the above fails and these roots of multiplicity ν_j, $(\nu_j > 1)$ yield solution

$$\Sigma[\mathbf{c}^{(j)} + \mathbf{d}^{(j)} t + \cdots + \mathbf{h}^{(j)} t^{\lambda_j - 1}] e^{\lambda_j t}$$

The total solution is

$$\mathbf{z} = \sum_{i=l+1}^{m-q+l} \mathbf{c}^{(i)} e^{\lambda_i t} + \sum_{j=1}^{l} [\mathbf{c}^{(j)} + \mathbf{d}^{(j)} t + \cdots + \mathbf{h}^{j} t^{\nu_j - 1}] e^{\lambda_j t} \qquad q = \Sigma \nu_j$$

Alternatively and if interest is focused on one of the dependent variables in Eq. (8.109), then all other variables can be eliminated to yield

$$\det \begin{bmatrix} a_{11}(p) & \cdots & a_{1n}(p) \\ \cdot & & \cdot \\ \cdot & & \cdot \\ \cdot & & \cdot \\ a_{n1}(p) & \cdots & a_{nn}(p)z \end{bmatrix} z_i = 0$$

$$|A(p)|z_i = 0 \qquad (i = 1,2, \ldots, n)$$

giving the identical homogeneous equation for each of the dependent variables. If, as before, the exponential form $z_1 = c_1e^{\lambda t}$ is assumed, substitution gives $|A(\lambda)|c_1e^{\lambda t} = 0$ where for $c_1 \neq 0$, $|A(\lambda)| = 0$, giving the same characteristic equation for the exponential constants as before. The inhomogeneous reduced equations from Eq. (8.108) are formally obtained by purely algebraic considerations as

$$|A(p)|y_j = \sum_{l=1}^{n} M_{ij}(p)f_i \qquad j = 1, 2, \ldots, n$$

where M_{ij} is the cofactor of the element in the ith row, jth column of A.

Let

$$|A(p)| \triangleq D(p)$$

$$D(p)y_j = \sum_{i=1}^{n} M_{ij}(p)f_i(t) \qquad j = 1, \ldots, n$$

From the previous considerations,

$$D(p)z_j = 0 \tag{8.112a}$$

$$D(p)x_j = \sum_{i=1}^{n} M_{ij}(p)f_i(t) \qquad j = 1, \ldots, n \tag{8.112b}$$

$$y_j = x_j + z_j$$

To conclude, in general the solution x_j may be determined without regard for the initial conditions. It can be obtained in many ways depending on the character of the f_i's. If the f_i's are known as a finite power series, the method of undetermined coefficients will be expeditious; if it has more general behavior, it may be convenient to use the method of variation of parameters; if f_i's are exponential (including sin, cos, sinh, cosh) then an assumed exponential solution for each exponent will yield the answer; if the f_i's are periodic, by Fourier analysis these can be reconstructed as exponential functions and solved as outlined above; if f_i's are not periodic, having certain restrictive integral-convergence behavior, Fourier integral methods can be utilized. Last and most powerful is the Laplace-transform method, which not only has the widest range of applicability but can be utilized to obtain y (the total solution) directly.

(b) Fourier-series Analysis

If $f(t)$ is real and periodic of period T, with few restrictions, it can be approximately expressed as a linear sum of sine and cosine terms (Fourier series) or exponential terms. That is, if

$$f(t) = f(t + T)$$

then

$$f(t) = a_0/T + \sum_{n=1}^{\infty} a_n \cos \omega_0 nt + b_n \sin \omega_0 nt$$

where

$$\omega_0 = 2\pi/T \qquad \text{and } a_n,\ b_n \text{ are real constants}$$

or alternatively

$$f(t) = \sum_{n=-\infty}^{+\infty} c_n e^{i\omega_0 nt} \tag{8.113}$$

$$\text{where } a_0 = \int_t^{t+T} f(t)\, dt$$

$$a_n = 2/T \int_t^{t+T} f(t) \cos n\omega_0 t\, dt$$

$$b_n = 2/T \int_t^{t+T} f(t) \sin n\omega_0 t\, dt$$

$$c_n = 1/T \int_t^{t+T} f(t) e^{-jn\omega_0 t}\, dt$$

$$2c_n = a_n - jb_n$$
$$2c_{-n} = a_n + jb_n$$

Let $e^{j\omega t}$ be the input f_i in Eq. (8.112). Assuming a response $X_{ij}^{(1)} = H_{ij}(j\omega)e^{j\omega t}$ it is required to find $H_{ij}(j\omega)$. From Eq. (8.112) it is clear that all operations on $e^{j\omega t}$ are equivalent to replacing p by $(j\omega)$ and Eq. (8 112) becomes, for $f_i = e^{j\omega t}$,

$$D(j\omega)H_{ij}(j\omega)e^{j\omega t} = M_{ij}(j\omega)e^{j\omega t}$$

yielding
$$X_{ij}^{(1)} = H_{ij}(j\omega)e^{j\omega t} = \frac{M_{ij}(j\omega)}{D(j\omega)} e^{j\omega t}$$
$$= [P(j\omega) + Q(j\omega)]e^{j\omega t}$$
$$= R(j\omega)e^{j\phi}e^{j\omega t} \qquad \phi = \tan^{-1} \frac{-jQ(j\omega)}{P(j\omega)}$$
$$= R(j\omega)e^{j(\omega t+\phi)} \qquad R(j\omega) = [P^2(j\omega) + Q^2(j\omega)]^{1/2} \tag{8.114}$$

where $P(j\omega)$ and $R(j\omega)$ are real and therefore even functions of $(j\omega)$, and $Q(j\omega)$ is imaginary and an odd function of $(j\omega)$ and the following properties apply:

$$Q(j\omega) = -Q(-j\omega) \qquad \text{odd function}$$
$$\left.\begin{aligned} P(j\omega) &= P(-j\omega) \\ R(j\omega) &= R(-j\omega) \end{aligned}\right\} \qquad \text{even functions}$$

Similarly for an input $e^{-j\omega t}$, the output is

$$X_{ij}^{(2)} = H_{ij}(-j\omega)e^{-j\omega t} = [P(-j\omega) + Q(-j\omega)]e^{-j\omega t}$$
$$= Re^{j\phi'}e^{-j\omega t}$$

Since
$$\phi' = \tan^{-1} \frac{-jQ(-j\omega)}{P(j\omega)} = \tan^{-1} \frac{jQ(j\omega)}{P(j\omega)} = -\phi$$

the response $X_{ij}^{(2)}$ is

$$X_{ij}^{(2)} = H_{ij}(-j\omega)e^{-j\omega t} = Re^{-j(\omega t+\phi)} \tag{8.115}$$

The sum of the responses is

$$X_{ij}^{(1)} + X_{ij}^{(2)} = H_{ij}(-j\omega)e^{-j\omega t} + H_{ij}(j\omega)e^{j\omega t} = Re^{j(\omega t+\phi)} + Re^{-j(\omega t+\phi)}$$
$$X_{ij}^{(1)} + X_{ij}^{(2)} = 2R \cos (\omega t + \phi)$$

which is just twice the real part of either response or

$$X_{ij}^{(1)} + X_{ij}^{(2)} = 2 \operatorname{Re} X_{ij}^{(1)} = 2 \operatorname{Re} X_{ij}^{(2)}$$

showing that the total input, $e^{j\omega t} + e^{-j\omega t} = 2 \cos \omega t$, results in an output of different phase and amplitude.

$$H_{ij}(j\omega) = M_{ij}(j\omega)/D(j\omega) = R(j\omega)e^{j\phi(j\omega)}$$

is called the *transfer function* for sinusoidal inputs (real frequency) containing both amplitude and phase information.

It follows readily from superposition that the periodic responses to forcing functions having Fourier-series representations are available as a sum of responses of the form

$$X = \sum_{n=0}^{\infty} Ra_n \sin (\omega_0 nt + \phi_n) + Rb_n \cos (\omega_0 nt + \phi_n)$$

$$\phi_n = \phi(j\omega_0 n) \qquad \phi_{-n} = \phi(-j\omega_0 n)$$

or

$$X = \sum_{-\infty}^{+\infty} Rc_n e^{j(\omega_0 nt + \phi_n)}$$

(c) Complex Frequency-domain Analysis[16]

It is often convenient to cast the linear system from its time-domain representation into a frequency-domain form in order to simplify analysis or exhibit more clearly certain of its important properties (e.g., spectrum, stability). The Fourier- and Laplace-transform methods are most prominent in this regard.

Fourier-transform Method. If the input forcing function(s) are not periodic functions of time, the Fourier-transform method may be employed to solve Eq. (8.112) for each input f_i whenever

$$\int_{-\infty}^{+\infty} |f_i(t)|\, dt < \infty \tag{8.116}$$

That is, the absolute convergence of the infinite integral is a sufficient condition for Fourier transformability. Examples of functions not satisfying Eq. (8.116) are the step function,* sinusoid, rising exponentials, functions containing t to positive exponents, e.g., ramp function (αt). Examples of functions satisfying Eq. (8.116) are pulses of finite duration. The Fourier-integral theorem asserts

$$f(t) = 1/2\pi \int_{-\infty}^{+\infty} d\omega \int_{-\infty}^{+\infty} f(T) e^{j\omega(t-T)}\, dT \tag{8.117}$$

If $f(t)$ has a finite discontinuity at any point, then this integration will yield the average value of $f(t)$ at the discontinuity. From Eq. (8.117) the Fourier-transform pair is obtained:

$$f(t) = 1/2\pi \int_{-\infty}^{+\infty} F(j\omega) e^{+j\omega t}\, d\omega = \int_{-\infty}^{+\infty} F(j2\pi f) e^{j2\pi ft}\, df \tag{8.118a}$$

$$F(j\omega) = \int_{-\infty}^{+\infty} f(t) e^{-j\omega t}\, dt = \int_{-\infty}^{+\infty} f(t) e^{-j2\pi ft}\, dt \tag{8.118b}$$

where $F(j\omega)$ is in general complex and is denoted as the complex spectrum of $f(t)$. Equation (8.118a) can be imagined to express $f(t)$ as the infinite sum of Fourier components $F(j\omega)e^{j\omega t}\, d\omega/2\pi$. From the superposition principle, the total response is made up of the sum of each of the responses $H_{ij}(j\omega)F(j\omega)e^{j\omega t}\, d\omega/2\pi$ where $H_{ij}(j\omega)$ was defined as the real frequency-transfer function and the sum is expressed (since it is continuous in ω) as

$$x(t) = 1/2\pi \int_{-\infty}^{+\infty} H_{ij}(j\omega) F(j\omega) e^{j\omega t}\, d\omega \tag{8.119}$$

But $x(t)$ has a transform representation from Eq. (8.118)

$$x(t) = 1/2\pi \int_{-\infty}^{+\infty} X(j\omega) e^{j\omega t}\, d\omega \tag{8.120}$$

The integrals Eqs. (8.119) and (8.120) are evidently identical. Hence

$$X(j\omega) = H_{ij}(j\omega) F(j\omega) \tag{8.121}$$

which gives the *important property* that the product of the transfer function (at real frequency) and the Fourier transform of the driving function yields the Fourier transform of the response.

* These have Fourier representations despite violation of Eq. (8.116). Note that Eq. (8.116) is only a sufficient condition.

Consider the linear time-invariant differential equation

$$d^nx/dt^n + Q_{n-1}(d^{n-1}x/dt^{n-1}) + \cdots + Q_0x = f(t) \qquad f(t) = 1/2\pi \int_{-\infty}^{+\infty} F(j\omega)e^{j\omega t}\, d\omega$$

$$1/2\pi \int_{-\infty}^{+\infty} [(j\omega)^n + Q_{n-1}(j\omega)^{n-1} + \cdots + Q_0]X_{(j\omega)}e^{j\omega t}\, dt = 1/2\pi \int_{-\infty}^{+\infty} F(j\omega)e^{j\omega t}\, d\omega$$

$$x(t) = 1/2\pi \int_{-\infty}^{+\infty} X(j\omega)e^{j\omega t}\, d\omega$$

whence
$$X(j\omega) = \frac{F(j\omega)}{(j\omega)^n + Q_{n-1}(j\omega)^{n-1} + \cdots + Q_1(j\omega) + Q_0}$$

The Fourier transform, in terms of its real and imaginary parts, is

$$X(j\omega) = M'(j\omega) + N'(j\omega) = M(\omega) + jN(\omega)$$

where M is an even function of ω and N an odd function of ω.

Then

$$\begin{aligned} x(t) &= 1/2\pi \int_{-\infty}^{+\infty} [M(\omega) + jN(\omega)]e^{j\omega t}\, d\omega \\ &= 1/2\pi \int_{-\infty}^{+\infty} [M(\omega) \cos \omega t - N(\omega) \sin \omega t]\, d\omega \\ &= 1/\pi \int_{-0}^{+\infty} (M \cos \omega t - N \sin \omega t)\, d\omega \end{aligned} \tag{8.122}$$

where only the even parts of the integrands can contribute because integration of the odd terms vanishes over the infinite limits. Now for the system (causal)

$$x(t) = 0 \qquad \text{for } t < 0$$

which is mathematically equivalent to

$$x(-t) = 0 \qquad \text{for } t > 0$$

Substitution in Eq. (8.122) yields

$$x(-t) = 0 = 1/\pi \int_0^{\infty} (M \cos \omega t + N \sin \omega t)\, d\omega \qquad t > 0 \tag{8.123}$$

Adding Eqs. (8.122) and (8.123) we obtain

$$x(t) = 2/\pi \int_0^{\infty} M \cos \omega t\, d\omega = 1/\pi \int_{-\infty}^{+\infty} M \cos \omega t\, d\omega \qquad t > 0 \tag{8.124a}$$

Subtracting Eq. (8.123) from Eq. (8.122),

$$x(t) = -2/\pi \int_0^{\infty} N \sin \omega t\, d\omega = -1/\pi \int_{-\infty}^{+\infty} N \sin \omega t\, d\omega \qquad t > 0 \tag{8.124b}$$

Since $x(t)$ has two integral representations they are equal:

$$2/\pi \int_0^{\infty} M \cos \omega t\, d\omega = -2/\pi \int_0^{\infty} N \sin \omega t\, d\omega$$

$$\int_0^{\infty} M(\omega) \cos \omega t\, d\omega = -\int_0^{\infty} N(\omega) \sin \omega t\, d\omega$$

$$M(\omega) = \int_{-\infty}^{+\infty} x(t) \cos \omega t\, dt \tag{8.125a}$$

$$N(\omega) = -\int_{-\infty}^{\infty} x(t) \sin \omega t\, dt \tag{8.125b}$$

which are the real Fourier-transform coefficients and together with Eq. (8.124) constitute the real Fourier-transform pair. Properties of Fourier transforms are given in Table 8.1.

Table 8.1. Properties of Fourier-transform Pairs

Property	Fourier transform	Time function
Basic pairs......	$F(\omega)$ $F(\omega) = \int_{-\infty}^{+\infty} f(t)e^{-j\omega t}\,dt$	$f(t)$ $f(t) = 1/2\pi \int_{-\infty}^{+\infty} F(\omega)e^{j\omega t}\,d\omega$
Linearity, a_1 and a_2 constants...	$a_1F_1(\omega) + a_2F_2(\omega)$	$a_1f_1(t) + a_2f_2(t)$
Time multiplication, a real constant	$\frac{1}{\lvert a\rvert} F(\omega/a)$	$f(at)$
Time shift.......	$F(\omega)e^{-j\omega t_1}$	$f(t - t_1)$
Frequency multiplication, a real constant	$F(a\omega)$	$\frac{f(t/a)}{\lvert a\rvert}$
Frequency shift..	$F(\omega - \omega_1)$	$f(t)e^{j\omega_1 t}$
Time differentiation.........	$(j\omega)^n F(\omega)$	$(d^n/dt^n)f(t)$
Integration......	$(1/j\omega)F(\omega)$	$\int_{-\infty}^{t} f(x)\,dx$
Frequency differentiation.....	$j^n(d^n/d\omega^n)F(\omega)$	$t^n f(t)$
Convolution time domain	$F_1(\omega)F_2(\omega)$	$\int_{-\infty}^{+\infty} f_1(t - T)f_2(T)\,dT \triangleq f_1(t) * f_2(t)$
Convolution frequency domain	$F_1(\omega) * F_2(\omega) = \frac{1}{2\pi}\int_{-\infty}^{+\infty} F_1(\omega - x)F_2(\omega)\,dx$	$f_1(t)f_2(t)$
Forms for real $f(t)$	$F(\omega) = R(\omega) + jX(\omega)$ $R(\omega) = \int_{-\infty}^{+\infty} f(t)\cos\omega t\,dt$ $X(\omega) = -\int_{-\infty}^{+\infty} f(t)\sin\omega t\,dt$	$f(t) = 1/\pi \int_0^{\infty} [R(\omega)\cos\omega t\,d\omega - X(\omega)\sin\omega t]\,d\omega$ $= 1/\pi \int_0^{\infty} \lvert F(\omega)\rvert \cos(\omega t + \phi)\,d\omega$ $\phi = \tan^{-1} X(\omega)/R(\omega)$
Forms for real even $f(t)$	$X(\omega) = 0$ $R(\omega) = -2\int_0^{\infty} f(t)\sin\omega t\,dt$	$f(t) = -1/\pi \int_0^{\infty} R(\omega)\cos\omega t\,d\omega$
Forms for real odd $f(t)$	$R(\omega) = 0$ $X(\omega) = -2\int_0^{\infty} f(t)\sin\omega t\,dt$	$f(t) = -1/\pi \int_0^{\infty} X(\omega)\sin\omega t\,d\omega$
$f(t)$ causal, i.e., $f(t) = 0$ for $t < 0$	$F(\omega) = \int_0^{\infty} f(t)e^{-j\omega t}\,dt$ $= R(\omega) + jX(\omega)$	$f(t) = 2/\pi \int_0^{\infty} R(\omega)\cos\omega t\,d\omega$ $= -2/\pi \int_0^{\infty} X(\omega)\sin\omega t\,d\omega \quad t > 0$
Forms for periodic function $f(t) = f(t + T)$	$2\pi \sum_{n=-\infty}^{+\infty} a_n\delta(\omega - \omega_n)$ $a_n = 1/T \int_0^{T} f(t)e^{-j\omega_n t}\,dt$	$f(t) = f(t + T)$ $= \sum_{n=-\infty}^{+\infty} a_n e^{j\omega_n t}$ $\omega_n = 2\pi n/T$

Laplace-transform Method. When Eq. (8.116) does not hold for a forcing function $f(t)$, recourse may be taken to the unilateral Laplace-transform method provided that $f(t) \equiv 0$ for $t < 0$ and there exists a positive number c (in most physical problems of interest one exists) such that

$$\int_{-\infty}^{+\infty} |f(t)|e^{-ct}\,dt < \infty \tag{8.126}$$

The minimum value of c for which Eq. (8.126) holds is designated as the abscissa of convergence, equal to c_1. Equation (8.126) ensures that the Laplace transform of $f(t)$, written $\mathfrak{L}\{f(t)\}$ and defined by

$$\mathfrak{L}\{f(t)\} = \int_0^\infty f(t)e^{-st}\,dt = F(s) \tag{8.127}$$

will converge to a function of s. s is a complex variable given by

$$s = c + j\omega \qquad c \geq c_1$$

Examples of c_1 are: for $f(t) = \sin \omega t$, $c_1 > 0$; for unit step, $c_1 > 0$; for $t^n e^{dt}$, $c_1 > d$, n finite, d real. An example of a function where c cannot be found to satisfy Eq. (8.126) is $f(t) = \exp(t^n)$ for $n > 1$, and Laplace methods will accordingly fail. Equation (8.127) looks like the Fourier transform of

$$f(t)e^{-ct} \qquad \text{if } f(t) = 0 \text{ for } t < 0$$
$$\mathfrak{L}\{f(t)\} = \mathfrak{F}[f(t)e^{-ct}] \qquad \mathfrak{F} \triangleq \text{Fourier transform}$$

Invoking the Fourier-transform theorem [Eq. (8.117)] and manipulating as follows

$$f(t)e^{-ct} = 1/2\pi \int_{-\infty}^{+\infty} d\omega \int_{-\infty}^{+\infty} f(T)e^{-cT}e^{j\omega(t-T)}\,dT$$
$$= 1/2\pi \int_{-\infty}^{+\infty} e^{j\omega t}\,d\omega \int_{-\infty}^{+\infty} f(T)e^{-(c+j\omega)T}\,dT$$
$$f(t) = 1/2\pi \int_{-j\infty}^{+j\infty} e^{(c+j\omega)t}[d(j\omega)/j] \int_0^\infty f(T)e^{-(c+j\omega)T}\,dT \qquad f(T) = 0 \qquad T < 0$$

Since $s = c + j\omega$ and c is a constant, $ds = d(j\omega)$ and the above becomes

$$f(t) = 1/2\pi j \int_{c-j\infty}^{c+j\infty} e^{st}\,ds \int_0^\infty f(T)e^{-sT}dT = 1/2\pi j \int_{c-j\infty}^{c+j\infty} [\mathfrak{L}f(t)]e^{st}\,ds$$
$$f(t) = 1/2\pi j \int_{c-j\infty}^{c+j\infty} F(s)e^{st}\,ds \tag{8.128}$$

Equation (8.128) is the inversion form for going from $\mathfrak{L}\{f(t)\}$ to $f(t)$. Equations (8.127) and (8.128) constitute the Laplace-transform pair.

From Eq. (8.114), the real-frequency transfer function

$$H_{ij}(j\omega) = M_{ij}(j\omega)/D(j\omega)$$

was deduced for sinusoidal inputs. Similarly, consider the response [Eq. (8.112)] x_{ij} due to f_i.

$$D(p)x_{ij}(t) = M_{ij}(p)f_i(t)$$

where $D(p)$ and $M_{ij}(p)$ are linear differential operators. For zero initial conditions the Laplace transformation of both sides yields

$$D(s)X_{ij}(s) = M_{ij}(s)F_i(s) \tag{8.129}$$

Transfer Function. From Eq. (8.129), dropping all subscripts for clarity, $X(s)/F(s) = M(s)/D(s) = H(s)$, which is, by definition, the transfer function where s replaces $j\omega$ in the argument $H(j\omega)$, the real-frequency transfer function for sinusoidal input. $H(s)$ in itself has no physical significance; it contains, however, the complete characterization of the system. This is in contrast with $H(j\omega)$ which gives the steady-state response to a sinusoidal input, its amplitude being the gain and its argument the phase difference between output and input.

It follows that, if $H(j\omega)$ is a known analytic function of $j\omega$, then $H(s)$ is immediately available (by analytic continuation) for a complete system description.

The total response $y(t)$ satisfying the equation

$$\begin{aligned} D(p)y(t) &= M(p)f(t) \\ f(t) &= 0 \qquad \text{for } t < 0 \end{aligned} \tag{8.130}$$

can be obtained directly by taking the Laplace transform [including initial conditions which result in the polynomial $L(s)$ of lower order than $D(s)$] as follows:

$$\begin{aligned} D(s)Y(s) - L(s) &= M(s)F(s) \\ Y(s) &= M(s)F(s)/D(s) + L(s)/D(s) \\ Y(s) &= H(s)F(s) + L(s)/D(s) \end{aligned} \tag{8.130a}$$

By inversion:

$$y(t) = \underbrace{\frac{1}{2\pi j}\int_{c-j\infty}^{c+j\infty} H(s)F(s)e^{st}\,ds}_{x(t) = \int_0^t W(t-T)f(T)\,dT} + \underbrace{\frac{1}{2\pi j}\int_{c-j\infty}^{c+j\infty} \frac{L(s)}{D(s)}e^{st}\,ds}_{z(t)} \tag{8.131}$$

The first integral in Eq. (8.131), $x(t)$, is the solution to Eq. (8.112*b*) for zero initial conditions; the second, $z(t)$, is the solution to the homogeneous form Eq. (8.112*a*) which satisfies the initial conditions of Eq. (8.130). $W(t - T)$ is the response $x(t)$ at time t to a unit impulse input $f = \delta(t - T)$, the system being initially at rest.

Inversion. The transformation (inversion) from the complex frequency representation to the time domain is given by

$$f(t) = 1/2\pi j \int_{c-j\infty}^{c+j\infty} F(s)e^{st}\,dt$$

which is a line integral along the line Re $s = c$ in the complex s plane where $c > c_1$, and c_1 is defined as the abscissa of convergence. For the case $|F(s)| \to 0$ as $|s| \to \infty$ the line integral is most readily evaluated by forming a contour inc uding this line and an infinite semicircle connected on the left and considering the contour integral

$$\oint F(s)e^{st}\,ds$$

Since $|F(s)| \to 0$ as $|s| \to \infty$ the line integral around the semicircular portion of this contour vanishes as a consequence of Jordan's lemma.[16] This leads to the equality of the contour integral with the inversion integral, viz.,

$$f(t) = 1/2\pi j \int_{c-j\infty}^{c+j\infty} F(s)e^{st}\,ds = 1/2\pi j \oint F(s)e^{st}\,ds \qquad t > 0$$

From Cauchy's residue theorem, the right-hand side equals the sum of the residues of $F(s)e^{st}$ enclosed. The residues are evaluated at each simple pole s_k by

$$\underset{s \to s_k}{(s - s_k)F(s)e^{st}} = R_{s_k} \qquad \text{(residue at } s_k)$$

If $F(s)$ is a fraction, $F(s) = A(s)/B(s)$ where $A(s)$ and $B(s)$ are analytic functions of s inside the contour (excluding poles at infinity) then the poles of $F(s)$ are clearly the zeros of $B(s)$. A pole of multiplicity m is equal to the excess of zeros of $B(s)$ over $A(s)$ at the pole. For simple poles of $F(s)$, i.e., where $m = 1$, the residue at $s = s_k$ is simply $(s - s_k)F(s)e^{st} = A(s_k)e^{s_kt}/B'(s_k) \quad s \to s_k$. The residue at pole s_j of multiplicity m is

$$\frac{1}{(m-1)!}\frac{d^{m-1}}{ds^{m-1}}\left[(s - s_j)^m F(s)e^{st}\right]_{s=s_j}$$

If there are no zeros of $A(s)$ at the point $s = s_j$, then an alternate form of this is

$$\frac{md^{m-1}}{ds_j^{m-1}}\left[\frac{A(s_j)e^{s_jt}}{d^mB(s_j)/ds_j^m}\right]$$

Table 8.2. Properties of Laplace-transform Pairs for Causal Time Function
$f(t) = 0 \quad t < 0$

Property	Laplace transform	Time function
Basic pairs	$F(s)$ $F(s) = \int_0^{\infty} f(t)e^{-st}\,dt$ $= \int_{-\infty}^{+\infty} f(t)e^{-st}\,dt$	$f(t)$ $f(t) = 1/2\pi j \int_{c-j\infty}^{c+j\infty} F(s)e^{st}\,ds$
Linearity a_1 and a_2 constants	$a_1F_1(s)t + a_2F_2(s)$	$a_1f_1(t) + a_2f_2(t)$
Time scale multiplication a positive real constant	$\frac{F(s/a)}{a}$	$f(at)$
Time shift	$F(s)e^{-st_1}$	$f(t - t_1)$
Complex frequency multiplication a positive real constant	$F(as)$	$\frac{f(t/a)}{a}$
Complex frequency shift	$F(s - s_1)$	$f(t)e^{s_1t}$
Time differentiation	$s^nF(s) - \sum_{k=1}^{n} f^{k-1}(0)s^{n-k}$	$\frac{d^n}{dt^n} f(t)$
Integration	$\frac{F(s)}{s} + \frac{f^{-1}(0)}{s}$	$\int_0^t f(t)\,dt$
Complex frequency differentiation	$(-1)^n \frac{d^n}{ds^n} F(s)$	$t^nf(t)$
Convolution time domain	$F_1(s)F_2(s)$	$f_1(t) * f_2(t) = \int_0^{\infty} f_1(t-T)f_2(T)\,dT$ $= \int_0^t f_1(t - T)f_2(T)\,dT$
Convolution frequency domain	$1/2\pi j \int_{c-j\infty}^{c+j\infty} F_1(s - x)F_2(x)\,dx$ $= (1/2\pi j)F_1(s) * F_2(s)$	$f_1(t)f_2(t)$
Forms for periodic function $f_k(t) = f_k(t + T)$	$\frac{1/2\pi j \int_0^T f_k(t)e^{-st}\,dt}{1 - e^{-sT}}$	$f_k(t) = f_k(t + T)$
Initial-value theorem [$f(t)$ and $f'(t)$ are Laplace transformable]	$\lim_{s\to\infty} sF(s)$	$\lim_{t\to 0} f(t)$
Final-value theorem $f(t)$, $f'(t)$ Laplace transformable and $sF(s)$ analytic $Re\ s \geq 0$	$\lim_{s\to 0} sF(s)$	$\lim_{t\to\infty} f(t)$

Some important properties of $H(s)$ for passive systems of differential equations are.

1. $H(s)$ is a rational function with real coefficients.
2. The degree of $D(s)$ is equal to or greater than $M_{ij}(s)$.
3. As a consequence of property 1 the complex zeros and poles occur in conjugate pairs.
4. All poles of $H(s)$ lie in the closed left half plane, a consequence of passivity.

Table 8.2 has properties of Laplace-transform pairs.

(d) Time-domain Analysis

The general solution to Eq. (8.130) given by Eq. (8.131) is

$$y(t) = \int_0^t W(t - T)f(T)\,dT + z(t) \tag{8.132}$$

where the integral expression alone satisfies the inhomogeneous equation with zero initial conditions, and $z(t)$ contains the linearly independent solutions to the homogeneous form fulfilling the initial conditions on $y(t)$, i.e.,

$$y(0) = z(0)$$

$W(t)$ is the inverse transform of the transfer function $H(s)$ which physically is the response to the delta function $\delta(t)$. The Dirac delta function $\delta(t - T)$ (defined as a pulse of infinite height at $t = T$, with unit area) has the following properties:

$$\int_{-\infty}^{+\infty} \delta(t - T)\,dt = 1$$

and

$$\int_{-\infty}^{+\infty} \delta(t - T)A(t)\,dt = A(T)$$

Examples of the delta function are shown in Fig. 8.14. To find the response $y(t)$ for input $\delta(t - T)$ and zero initial conditions, we first evaluate

$$F(s) = \mathcal{L}[\delta(t - T)] = e^{-sT}$$

and substitute in Eq. (8.130a)

$$Y(s) = H(s)F(s) = H(s)e^{-sT}$$

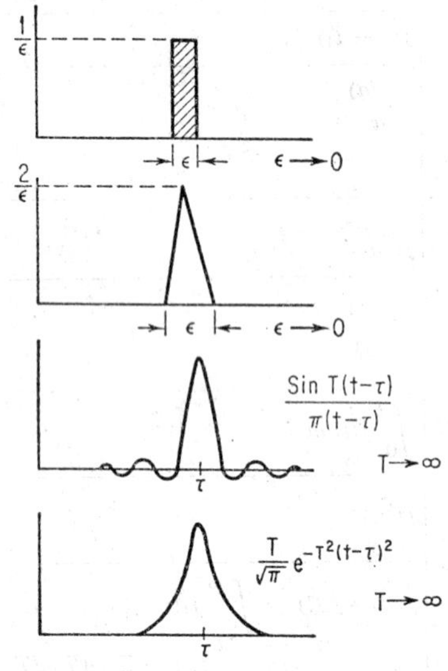

FIG. 8.14. Examples of delta functions, $\delta(t - T)$.

where $L(s) = 0$ is a consequence of zero initial conditions. Transforming to the time domain, we have

$$y(t) = 1/2\pi j \int_{c-j\infty}^{c+j\infty} H(s)e^{-sT}e^{st}\,ds = 1/2\pi j \int_{c-j\infty}^{c+j\infty} H(s)e^{s(t-T)}\,ds$$

$$y(t) = W(t - T) \qquad t < T$$

Any forcing function can be approximated by an infinite number of "delta" functions of strength $f(T)\,\Delta T$ with the responses at time t the sum of the responses $W(t - T)$ per unit impulse for each of these pulses which occurred $(t - T)$ seconds previous to the time of inspection, t. Figure 8.15 shows a graphical construction with equal-duration rectangular pulses, ΔT wide and height $f(T)$. Each pulse has an area $f(T)\,\Delta T$ so that the function can be approximated by

$$\sum_{i=0}^{i=t/\Delta T} f(T)\,\Delta T\,\delta(t - i\,\Delta T)$$

The transient produced at time t due to the pulse $f(T)\,\Delta T\,\delta(t-T)$ is

$$W(t-T)f(T)\,\Delta T$$

The total effect of all pulses is by superposition

$$y(t) = \sum_{T=0}^{t} f(T)\,\Delta T W(t-T) = \sum_{i=0}^{i=t/\Delta T} f(i\,\Delta T)\,\Delta T\,W(t-i\,\Delta t)$$

which in the limit $\Delta T \to 0$ is

$$\int_0^t f(T)W(t-T)\,dT$$

$$W(t-T) = 0 \qquad \begin{cases} t \leqq T \\ t \leq 0 \end{cases} \qquad (8.133)$$

$$f(T) = 0 \qquad T < 0$$

consistent with Eq. (8.132).

In view of the restrictions on $W(t-T)$ and $f(t)$ the limits on Eq. (8.133) can be changed to any of the following:

$$y(t) = \int_0^t = \int_{-\infty}^{+\infty} = \int_{-\infty}^{t} = \int_0^{\infty}$$

By a change of variable, y can be represented as

$$y(t) = \int_0^t f(t-T)W(T)\,dT$$

and also is obtainable directly as an alternate form from the convolution integral in going from complex to the real time domain.

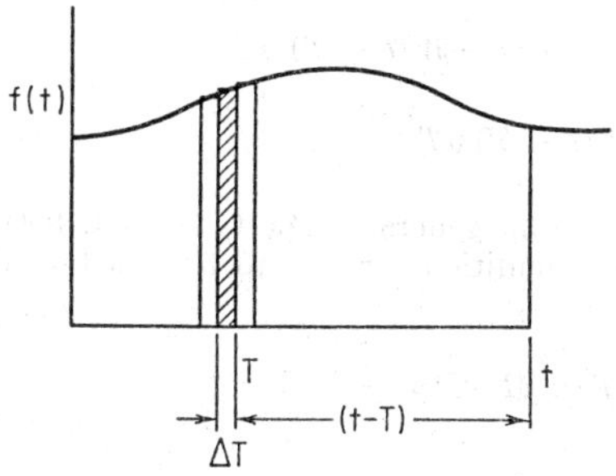

FIG. 8.15. Pulse synthesis for time-convolution theorem.

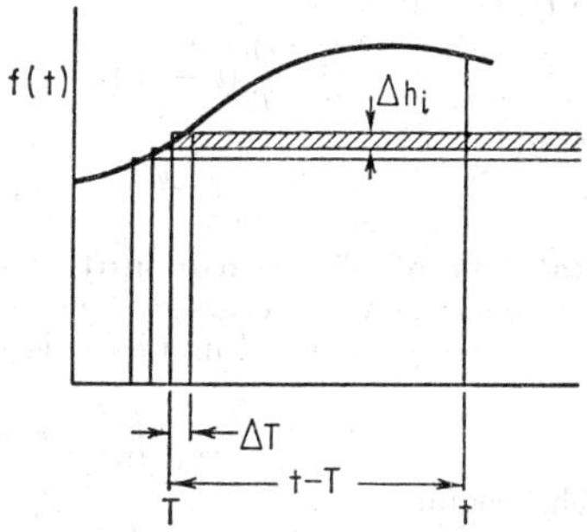

FIG. 8.16. Staircase synthesis.

Staircase Development. Another manner of depicting the forcing function $f(t)$ is shown in Fig. 8.16 as the synthesis of step functions.

Since the unit step function is the integral of the delta function

$$U(t-T) = \int_{-\infty}^{+\infty} \delta(t-T)\,dt \qquad \text{and} \qquad \frac{dU(t-T)}{dt} = \delta(t-T)$$

From property 2 of time-invariant systems previously discussed, if the response to the unit step is

$$\text{Input} = U(t-T) \to Q(t-T) = \text{response}$$

then the response to the derivative $[\partial U(t-T)]/\partial t = \delta(t-T)$ is $[\partial Q(t-T)]/\partial t$. But the response to $\delta(t - T)$ has already been shown to be $W(t - T)$. Therefore,

$$W(t - T) = (\partial/\partial t)Q(t - T)$$

The elementary step functions are of varying amplitudes Δh_i for fixed ΔT. From geometrical considerations

$$\Delta h_i \approx f'(T)\,\Delta T \qquad T = i\,\Delta T$$

and

$$\Delta h_i\, U(t - T) = f'(T)\,\Delta T U(t - T)$$

The response to the elementary step $\Delta h_i U(t - T)$ is

$$f'(T)\,\Delta T\, Q(t - T)$$

The total response is therefore

$$f(0)Q(t) + \sum_{i=0}^{i=T\Delta T} f'(T)\,\Delta T Q(t - T) = \sum_{i=0}^{i=t/\Delta T} f'(i\,\Delta T)\,\Delta T Q(t - i\,\Delta T) + f(0)Q(t)$$

As $\Delta T \to 0$ this becomes in the limit

$$y(t) = f(0)Q(t) + \int_0^t f'(T)Q(t - T)\,dT \tag{8.134}$$

which is identical to Eq. (8.133). This is shown by integrating Eq. (8.134) by parts.

$$\begin{aligned} y(t) &= f(0)Q(t) + f(T)Q(t - T)\Big|_{T=0}^{T=t} - \int_0^t f(T)\frac{\partial Q(t - T)}{\partial T}\,dT \\ &= f(0)Q(t) + f(t)Q(0) - f(0)Q(t) + \int_0^t f(T)\frac{\partial Q}{\partial t}(t - T)\,dT \end{aligned}$$

since $Q(0) = 0$ and

$$\frac{\partial Q}{\partial T}(t - T) = \frac{-\partial Q}{\partial t}(t - T) = -W(t - T)$$

$$y(t) = \int_0^t f(T)W(t - T)\,dT$$

Stability of Time-invariant System. From general property 8 (Art. 8.4) of linear systems the necessary and sufficient condition for $y(t)$ to have a bounded output for any bounded input $f(t)$ is

$$\int_0^\infty |W(t,T)|\,dT < M < \infty$$

which becomes

$$\int_0^\infty W(t - T)\,dT < M < \infty \tag{8.135}$$

for the time-invariant case.

By definition, $W(t - T)$, the weighting function, is the response to the unit delta function $\delta(t - T)$, and it has been shown that $W(t)$ is the inverse of the system transfer function

$$W(t) = H^{-1}(s) = [M(s)/D(s)]^{-1}$$

From the inversion theorem,

$$W(t) = \sum_{k=1}^{m} \sum_{n=1}^{\nu_k} a_{kn} t^{\nu_k - n} e^{s_k t}$$

where ν_k = multiplicity of roots s_k
s_k = zeros of $D(s)$
m = number of different poles of $H(s)$

Invoking Eq. (8.135) for a typical term $at^q e^{s_k t}$, q = integer,

$$\int_0^\infty |W(t - T)|\, dt = \int_0^\infty at^q e^{\mathrm{Re}\, s_k t}\, dt < M < \infty$$

which can hold if and only if Re s_k is negative. Hence the criterion for stability for the time-invariant case is simply that all the zeros of $D(s)$ lie in the left half complex plane excluding the imaginary axis. Translated to the time domain, this is equivalent to stating that the real parts of the exponent of each solution to the homogeneous equation $D(p)z = 0$ must be negative so that as $t \to \infty$ they all tend to vanish. Real systems composed only of passive elements are necessarily stable since it can be shown that the poles of their transfer function are restricted to the left-hand plane. The systematic investigation of locating the zeros of $D(s)$ has been motivated by control theory and is discussed subsequently.

8.5. BLOCK DIAGRAMS AND THE TRANSFER FUNCTION

(a) General

A convenient and descriptive way of viewing a system is by use of a block diagram. While a block diagram has little practical value for a simple system, in a complex array of coupled systems it suggests the flow of signals and facilitates analysis. The basic "block" essentially defines the system by giving a description of the physical processes which occur. Specifically for an input i the block gives information on the output o. An example of a block expressing an algebraic relationship is shown in Fig. 8.17a. A block of a more general operator relationship which includes the nonlinear differential-integral operator Φ is shown in Fig. 8.17b.

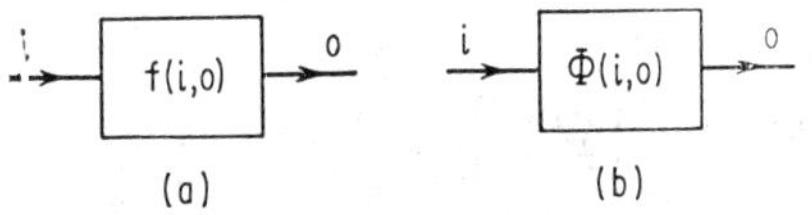

FIG. 8.17. Basic block diagrams. (a) $f(i,O)$ Algebraic relationship. (b) $\phi(i,O)$ General operator relationship.

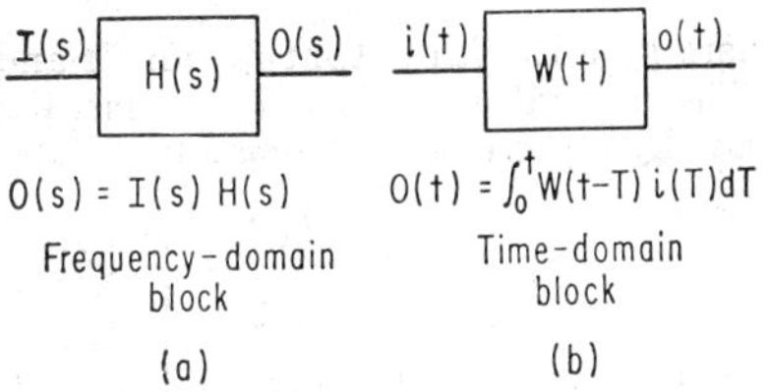

FIG. 8.18. Transfer blocks for linear time-invariant system.

(b) Linear Time-invariant Systems

Single-degree-of-freedom Case. The case of the linear time-invariant system (Fig. 8.18) is of special interest because the block-diagram characterization can be very simply shown in terms of the transfer function $H(s)$ in the frequency domain.

The input and output $I(s)$, $O(s)$ are Laplace transforms of the input and output signals. From Fig. 8.18a we have

$$O(s)/I(s) = H(s) \qquad i(t) = o(t) = 0 \qquad \text{for } t < 0$$

The multiplication property $O(s) = H(s)I(s)$ is of fundamental importance when dealing with linear cascaded systems such as occur in control theory. Moreover, the complete analysis of such systems can be made in this domain without translating to real time. An alternative representation in the time domain is shown in Fig. 8.18b where $W(t)$ is the weighting function implying the convolution relation

$$o(t) = \int_0^t W(t - T)i(T)\, d(T)$$

Returning to the frequency-domain representation, the response to cascaded systems is simply

$$O(s) = H_1(s)H_2(s)H_3(s) \cdots H_n(s)I(s)$$

Some care must be exercised in implementing this formula since the $H_1(s)$ (e.g., network systems) sometimes displays a loading effect, i.e., their individual free transfer functions differ from the transfer functions in cascade.

Consider the second-order system

$$d^2x/dt^2 + c\,dx/dt + kx = f(t) \tag{8.136}$$

$$A(p)x = f(t) \qquad A(p) = (p^2 + cp + k)$$

The system transfer function is obtained by taking the Laplace transform of Eq. (8.136) with zero initial conditions, thus:

$$H(s) = 1/A(s) = 1/(s^2 + cs + k) = X(s)/F(s)$$

Multi-degree-of-freedom Case. For the more general system of n forcing functions f_i, $i = 1, \ldots, n$, e.g., Eq. (8.108) with n outputs $y_1, \ldots, y_n$, and assuming A^{-1} exists, i.e., A is a nonsingular matrix operator, consider

Laplace form

$$A(p)\mathbf{y} = \mathbf{f}(t) \qquad A(s)\mathbf{Y}(s) = \mathbf{F}(s)$$

$$y_j = \sum_{i=1}^{n} \frac{M_{ij}(p)f_i(t)}{D(p)} \qquad Y_j(s) = \sum_{i=1}^{n} \frac{M_{ij}(s)}{D(s)} F_i(s)$$

$$y_{ij} = M_{ij}(p)f_i(t)/D(p) \qquad Y_{ij}(s) = \frac{M_{ij}(s)}{D(s)} F_i(s)$$

where y_j is the jth output for all f's and y_{ij} is a component of y_j produced by f_i.

The natural extension of the one-dimensional case to this n-dimensional case is made by representing the matrix as follows:

$$\mathbf{Y}(s) = [H(s)]\mathbf{F}(s)$$

$$\mathbf{Y}(s) = \begin{bmatrix} Y_1(s) \\ Y_2(s) \\ \cdot \\ \cdot \\ \cdot \\ Y_n(s) \end{bmatrix} \qquad \mathbf{F}(s) = \begin{bmatrix} F_1(s) \\ F_2(s) \\ \cdot \\ \cdot \\ \cdot \\ F_n(s) \end{bmatrix}$$

$$H(s) = A^{-1}(s) = \begin{bmatrix} h_{11}(s) & \cdots & h_{1n}(s) \\ \cdot & & \cdot \\ \cdot & & \cdot \\ \cdot & & \cdot \\ h_{n1}(s) & \cdots & h_{nn}(s) \end{bmatrix}$$

which formally is identical to Fig. 8.18, the one-dimensional case.

(c) Feedback Control-system Dynamics[10,25,26,27,28,29,30]

In the feedback control system one or more dependent variables (output) of a dynamic process is controlled. To this end, the difference(s) (error) between the desired value (input) and output is measured and functionally operated on to obtain a correcting signal (e.g., force) which is imparted to the basic system for the purpose of driving the output to correspondence with the input. The following descriptions imply the basic blocks which distinguish a control system.

1. The plant—the uncontrolled system
2. Sensor—to detect the output

3. Transmittor—a device to transmit the output or input signals to the comparator
4. Comparator—a device to detect differences between output and input
5. Controller—a device which takes some useful function of the input and output to correct the "error"

An overall description of the controlled process will yield forms outlined in the foregoing articles which dealt with single systems, where emphasis was placed upon passive types. Stability of these types of systems was assured without recourse to mathematical analysis. On the other hand, the control-system equations relating input to output (overall transfer) are not in general "passive" because the control system contains energy sources such as power amplifiers. The basic problem of control is the synthesis of an optimum control system which exhibits absolute as well as relative stability.

The general equations for control and the corresponding block-diagram representations are as follows:

1. The plant

$$\omega = \phi(v,d,t) \qquad \begin{matrix} d \rightarrow \\ v \rightarrow \end{matrix} \boxed{\phi(v,d,t)} \rightarrow \omega$$

where ϕ = general operator
v = input
d = disturbance
t = time

2. Sensor

$$x = \Omega(\omega) \qquad \xrightarrow{\omega} \boxed{\Omega(\omega)} \xrightarrow{x}$$

3. Transmitter

$$y = \theta(x) \qquad \xrightarrow{x} \boxed{\theta(x)} \xrightarrow{y}$$

4. Error device (comparator)

$$\epsilon = z - y \qquad z \rightarrow \otimes \rightarrow \epsilon = (z - y), \quad \uparrow y$$

5. Controller

$$v = \psi(\epsilon) \qquad \xrightarrow{\epsilon} \boxed{\psi(\epsilon)} \xrightarrow{v} \tag{8.137}$$

The total feedback control system for the control of output w is shown in Fig. 8.19. Here ϕ, ψ, Ω, θ are in general nonlinear differential operators. Because of the extraordinary complexity of systems involving nonlinear operators, only the linear time-invariant case in which the operators in the set of control equations are linear with constant coefficients will be presented.

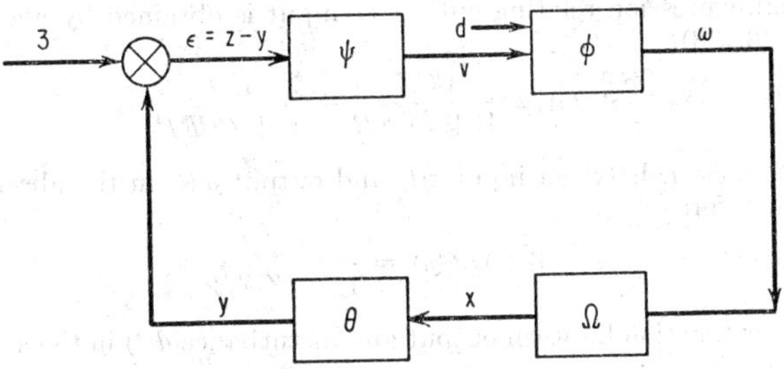

FIG. 8.19. Closed-loop feedback control system.

(d) The Linear Time-invariant Control System

Linearity and time invariance admit to simplifying techniques of frequency-domain methods of analyses. In block-diagram form the system transfer functions and the Laplace transforms of the signals entering and leaving are given for each block. The linear time-invariant control-system representation of Eq. (8.137) is depicted in Fig.

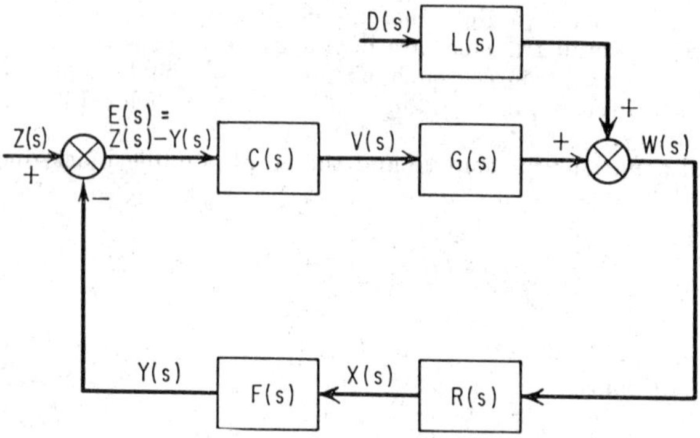

FIG. 8.20. Closed-loop control system for linear time-invariant system.

8.20. The disturbance and prescribed input to the plant are, in the linear case, connected by the differential equation

$$A(p)w(t) = B(p)v(t) + K(p)\,d(t)$$

whence the transfer function derives

$$\begin{aligned} A(s)W(s) &= B(s)V(s) + K(s)D(s) \\ W(s) &= [B(s)/A(s)]V(s) + [K(s)/A(s)]D(s) \\ W(s) &= G(s)V(s) + L(s)D(s) \end{aligned}$$

where $G(s) = B(s)/A(s)$ and $L(s) = K(s)/A(s)$ (8.138)

The equivalent frequency-domain forms of Eq. (8.137) are

$$W(s) = G(s)V(s) + L(s)D(s) \qquad (8.138a)$$
$$E(s) = Z(s) - Y(s) \qquad (8.138b)$$
$$V(s) = C(s)E(s) \qquad (8.138c)$$
$$X(s) = R(s)W(s) \qquad (8.138d)$$
$$Y(s) = F(s)X(s) \qquad (8.138e)$$

The functional expression relating output to input is obtained by algebraic manipulation of Eq. (8.138).

$$W(s) = \frac{CGZ}{1 + CGFR} + \frac{LD}{1 + CGFR} \qquad (8.139)$$

The transfer function between input $z(t)$ and output $w(t)$ in the absence of disturbance $d(t)$ is therefore

$$W(s)/Z(s) = \frac{CG}{1 + CGFR} \qquad (8.140)$$

and the transfer function between output and disturbance $d(t)$ in the absence of signal $z(t)$ is

$$W(s)/D(s) = L/(1 + CGFR) \qquad (8.141)$$

The matrix generalization for the simultaneous control of many variables is obtained by considering the transfer functions to be transfer matrices from which

$$\mathbf{W}(s) = [1 + CGFR]^{-1}CG\mathbf{Z}(s) + [1 + CGFR]^{-1}L\mathbf{D}(s)$$

where the column $\mathbf{W}(s)$ implies the separate controlled variables, and $\mathbf{Z}(s)$ and $\mathbf{D}(s)$, the input and disturbance vectors.[28]

Multiloop Systems. In general, control systems are more complicated than the one shown in Fig. 8.20, being composed of many loops, as shown, for example, in Fig. 8.21.

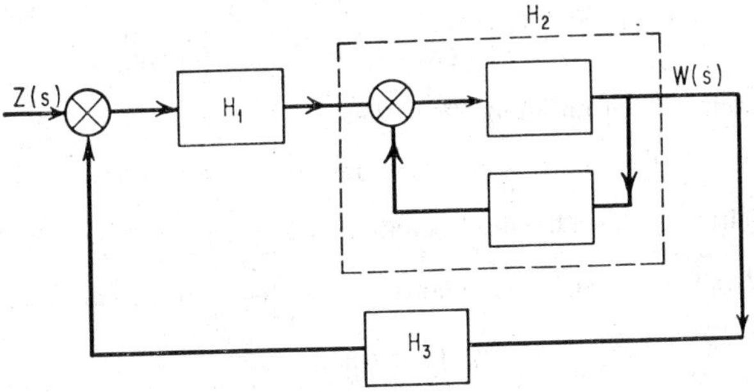

FIG. 8.21. Multiloop control system.

If the inner loop has a transfer function (dotted) H_2 then the overall transfer function is accordingly

$$W(s)/Z(s) = H_1H_2/(1 + H_1H_2H_3) \tag{8.142}$$

(e) Analysis of Control System

The transfer-function representations, e.g., Eqs. (8.140) and (8.142), permit the evaluation of three important properties of the control system without transformation to the time domain.

Equation (8.140) or (8.142) can be represented conveniently as

$$W(s)/Z(s) = Y_1/(1 + Y_1Y_2) \tag{8.143}$$

The error-to-input transfer function is obtained by subtracting both sides of the above from unity to give

$$\frac{E(s)}{Z(s)} = \frac{1 + Y_1Y_2 - Y_1}{1 + Y_1Y_2} \tag{8.144}$$

Properties Deducible from Frequency-domain Representation. *Property* 1. The steady-state error is

$$\epsilon(t)_{t\to\infty} = sE(s)_{s\to 0} = s\frac{1 + Y_1(s)Y_2(s) - Y_1(s)}{1 + Y_1(s)Y_2(s)}Z(s) \tag{8.145}$$

If the input $z(t)$ is the unit step $Z(s) = 1/s$, substitution in Eq. (8.145) yields

$$\epsilon(t)_{t\to\infty} = s\frac{1 + Y_1(0)Y_2(0) - Y_1(0)}{1 + Y_1(0)Y_2(0)}\frac{1}{s}\Bigg|_{s\to 0} \triangleq \frac{1}{1 + K_p} \tag{8.146}$$

If the input $z(t)$ is the ramp t, $Z(s) = 1/s^2$ Eq. (8.145) becomes

$$\epsilon(t)_{t\to\infty} = s\frac{1 + Y_1(s)Y_2(s) - Y_1(s)}{1 + Y_1(s)Y_2(s)}\frac{1}{s^2} = \frac{1 + Y_1(s)Y_2(s) - Y_1(s)}{sY_1(s)Y_2(s)}\Bigg|_{s\to 0} \triangleq \frac{1}{K_v} \tag{8.147}$$

If the input $Z(t)$ is t^2, $Z(s) = 2/s^3$ and Eq. (8.145) becomes

$$\underset{t\to\infty}{\epsilon(t)} = \underset{s\to 0}{2\,\frac{1 + Y_1(s)Y_2(s) - Y_1(s)}{s^2 Y_1(s) Y_2(s)}} \triangleq \frac{1}{K_a} \tag{8.148}$$

where the steady-state errors due to a step, ramp, and acceleration of the input are developed in Eqs. (8.146) through (8.148) assuming that they exist in each case.

If $E(s)/Z(s)$ can be represented by a Maclaurin series and is stable,

$$E(s)/Z(s) = (a_0 + a_1 s + a_2 s^2 + \cdots)$$
$$E(s) = a_0 Z(s) + a_1 s Z(s) + \cdots$$

the steady state $e(t)$ for an input $z(t)$ is then

$$e(t) \to a_0 z(t) + a_1 z'(t) + a_2 z''(t) + \cdots \qquad t \to \infty$$

which implicitly ignores all initial conditions, i.e., assumes the initial disturbances vanish as $t \to \infty$.

Property 2. The rms error is a quantitative measure of the effectiveness of control,

$$\bar{\varepsilon}^2 = 1/T \int_0^T |e(t)|^2\, dt \qquad T \to \infty$$

can be obtained from[10,16,27]

$$\bar{\varepsilon}^2 = \tfrac{1}{2} \int_{-\infty}^{+\infty} G_e(f)\, df = \int_0^{\infty} G_e(f)\, df$$

where $G_e(f)$ is the spectral density of the error $e(t)$. G_z for the input $z(t)$ is defined by

$$G_z(j\omega) = 1/2T \left| \int_{-T}^{+T} z(t) e^{-j\omega t}\, dt \right|^2$$

For the input forms which are bounded, G_z exists and the overall G_e is given by

$$G_e = \left| \frac{E(j\omega)}{Z(j\omega)} \right|^2 G_z(j\omega) \qquad T \to \infty$$

The spectral density G_z for random-type inputs and disturbances is also obtainable by statistical methods; then the property becomes the "expected rms error" owing to the nonspecific character of input.

Property 3. The stability of the system, however complex, is completely ascertained by the locations of the poles of the right-hand member of Eq. (8.143), namely,

$$\frac{Y_1(s)}{1 + Y_1(s) Y_2(s)} \tag{8.149}$$

The system is stable if no poles of this function lie in the right half s plane, the imaginary axis included. Otherwise it is unstable.

There are three prominent methods for determining whether or not poles of the function lie in the right-hand plane: (1) Routh-Horowitz, (2) root locus, and (3) Nyquist criterion.

Routh-Horowitz Method. The method is applicable to rational fractions

$$Y_1(s) = A(s)/B(s) \qquad Y_2(s) = C(s)/D(s)$$

First the fractions are cleared, leaving

$$\frac{A/B}{1 + (A/B)(C/D)} = \frac{AD}{AC + BD}$$

assuming the fraction in the lowest form $AC + BD$ cannot coincide with zeros of its numerator. The poles of Eq. (8.149) correspond to the zeros of the denominator. The denominator polynomial can be written

$$AC + BD = a_n s^n + a_{n-1}s^{n-1} + \cdots + a_1 s + a_0$$

System stability is therefore governed by the location of zeros of this polynomial.

Stability—Routh-Horowitz Criterion. Given the general nth-order equation

$$a_n s^n + a_{n-1}s^{n-1} + \cdots + a_1 s + a_0 = 0 \tag{8.150}$$

with real coefficients.

The following statements apply to the roots:

1. The roots occur in conjugate complex pairs.
2. A necessary condition for the real parts of all roots to be negative is that all coefficients have the same sign, and hence is a necessary condition for stability.
3. A necessary condition for all real parts to be nonpositive is that all coefficients $a_0, a_1, a_2, \ldots, a_n$ be different from zero, i.e.,

$$a_0, a_1, a_2, \ldots, a_n \neq 0$$

and is a necessary condition for stability.

If conditions 2 or 3 fail, the system is unstable. If, on the other hand, Eq. (8.150) meets conditions 2 and 3, further tests (Routh-Horowitz) must be made to determine stability.

The procedure is first to arrange the coefficients in two rows as shown followed by a third row developed from the first two rows, viz.,

$$a_n \quad a_{n-2} a_{n-4} \cdots \tag{a}$$
$$a_{n-1} a_{n-3} a_{n-5} \cdots \tag{b}$$
$$b_{n-1} b_{n-3} b_{n-5} \tag{c}$$

$$b_{n-1} \triangleq \frac{-1}{a_{n-1}} \begin{vmatrix} a_n & a_{n-2} \\ a_{n-1} & a_{n-3} \end{vmatrix}$$

$$b_{n-3} \triangleq \frac{-1}{a_{n-1}} \begin{vmatrix} a_n & a_{n-4} \\ a_{n-1} & a_{n-5} \end{vmatrix} \quad \text{etc.}$$

where the bars indicate the determinant of the enclosed array. In a like fashion form a fourth row $c_{n-1}c_{n-3}c_{n-5}$ developed from rows (b) and (c).

$$c_{n-1} = \frac{-1}{b_{n-1}} \begin{vmatrix} a_{n-1} & a_{n-3} \\ b_{n-1} & b_{n-3} \end{vmatrix}$$

$$c_{n-3} = \frac{-1}{b_{n-1}} \begin{vmatrix} a_{n-1} & a_{n-5} \\ b_{n-1} & b_{n-5} \end{vmatrix}$$

Continue this procedure of forming a new row from the two preceding rows until zeros are obtained; $(n + 1)$ rows will result.

The Routh-Horowitz criterion states that the number of roots with positive real parts equals the number of changes of sign in the first column. Since only one root with positive real part is sufficient to cause instability, the following stability criterion may be stated. A system whose characteristic equation is Eq. (8.150) is stable if and only if the elements formed in the first column $a_n, a_{n-1}, b_{n-1}, \ldots$ are all of the same algebraic sign.

Root-locus Method. This method, due to Evans, takes the denominator of Eq. (8.149) and factors $Y_1 Y_2$ into zeros and poles.

$$1 + Y_1(s)Y_2(s) = 1 + \frac{K\Pi_i(s - z_i)}{\Pi_j(s - p_j)}$$

where Π denotes product.

One first explores the zeros of $1 + Y_1Y_2$—which must satisfy two conditions—the amplitude and the phase.

$$1 + \frac{K\Pi_i(s - z_i)}{\Pi_j(s - p_j)} = 0$$

$$K\left|\frac{\Pi_i(s - z_i)}{\Pi_j(s - p_j)}\right| = 1 \qquad \text{amplitude condition}$$

$$\arg \frac{\Pi_i(s - z_i)}{\Pi_i(s - p_j)} = \pi(2n + 1) \qquad n \text{ an integer} \qquad \text{Phase condition} \qquad (8.151)$$

where arg = argument.

The location of all possible s that satisfy the phase condition (8.151) is drawn in the s plane; each corresponds to a K satisfying the amplitude condition. Figure 8.22 shows a typical example where the locus is drawn in solid lines.

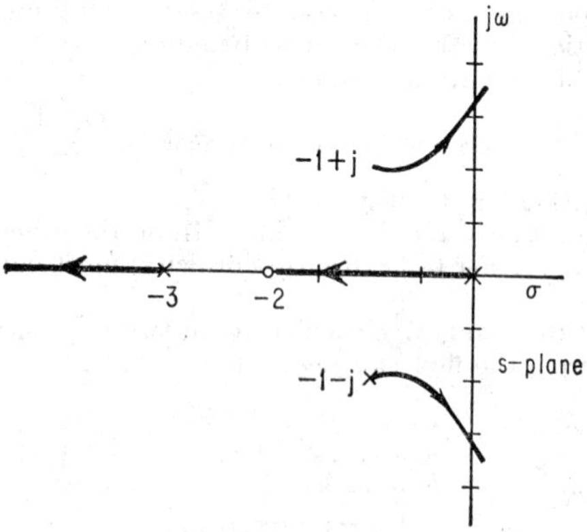

FIG. 8.22. Typical root-locus plot for $H(s)$.

$$H(s) = \frac{(s + 2)K}{s(s + 3)(s^2 + 2s + 2)}$$

Stability for the system is ascertained by the points on the locus (of roots) that apply for the specific K in question. If the locus in question lies in the left half plane the system is stable; otherwise it is unstable. Relative stability is judged by the proximity to the imaginary axis.

Complex-function Theory for Nyquist Criterion. The Nyquist criterion utilizes complex-function theory and in particular the so-called "argument principle." It is stated and proved as follows: Given a function $F(s)$ regular in a closed region R (except for a finite number of poles) bounded by the closed curve C, then the curve C maps into the curve C' in the plane, $w = F(s)$. The theorem asserts that the number of times the curve C' encircles the origin in the w plane is equal to the difference between the number of zeros and poles of $F(s)$ included within the region R, $N = Z - P$, including the multiplicity of the zeros and poles. In proof:

Near a zero inside R, say near $s = z_i$,

$$F(s) = (s - z_i)^{\gamma_i}[\psi(s)]$$

Near a pole inside R at p_i,

$$F(s) = \frac{1}{(s - p_i)^{\beta_i}}[\Omega(s)]$$

In general, therefore,

$$F(s) = \frac{(s - z_1)^{\gamma_1}(s - z_2)^{\gamma_2} \cdots (s - z_n)^{\gamma_n}}{(s - p_1)^{\beta_1}(s - p_2)^{\beta_2} \cdots (s - p_m)^{\beta_m}}[\phi(s)]$$

where $\phi(s)$ has no zeros or poles inside the region R. Taking the logarithmic derivative of $F(s)$,

$$(d/ds) \ln F(s) = F'(s)/F(s) = \gamma_1/(s - z_1) + \gamma_2/(s - z_1) + \cdots$$
$$- \beta_1/(s - p_1) - \beta_2/(s - p_2) \cdots + \phi'(s)/\phi(s)$$

Integrating around the closed contour C in the s plane and employing Cauchy's residue theorem we get

$$\oint_c d[\ln F(s)] = \ln F(s) \Big|_c = 2\pi j \left(\sum_i \gamma_i - \sum_k \beta_k \right) + 0 = \ln w \Big|_{\text{contour } c'} = 2\pi j N$$

$$N = \sum_i \gamma_i - \sum_k \beta_k$$

where it is noted that the evaluation of $\ln w$ over the closed curve c' in the w plane yields the change of argument times j or $2\pi jN$ where N is the number of encirclements of the origin with due regard for sign. Also, since ϕ has no zeros or poles in region R, ϕ'/ϕ is regular inside R and its contour integral vanishes over the closed path, viz.,

$$\oint (\phi'/\phi) \, ds = 0$$

Nyquist Criterion. This principle is now applied to the transfer function

$$\frac{Y_1(s)}{1 + Y_1(s)Y_2(s)}$$

to determine the zeros of $1 + Y_1(s)Y_2(s)$ in the right half s plane. The scanning contour in the s plane is the region bounded by the imaginary axis and the right-hand infinite semicircle. Instead of examining the number of encirclements of $1 + Y_1(s)Y_2(s)$ around the origin of w, by a shift of axis one unit, it is exactly equivalent to examining the contour of $Y_1(s)Y_2(s)$ mapped into w with reference to the -1 point. Now the number of encirclements of $Y_1(s)Y_2(s)$ around -1 is given by

$$N = Z - P \qquad Z = N + P \tag{8.151a}$$

If the number of right half plane poles P of $Y_1(s)Y_2(s)$ are known (none for passive blocks, but for multiloop systems they may be present) then the number of right half plane zeros Z can be determined from Eq. (8.151a) since N, the number of turns, can be counted. If Z has a value different from zero it implies that there are zeros of $1 + Y_1(s)Y_2(s)$ in the right-hand s plane establishing the case of instability. Otherwise the system is stable. The necessary and sufficient condition for stability is $Z \equiv 0$.

The actual plotting of the contour $Y_1(s)Y_2(s)$ (for rational fractions) in the w plane requires only one half imaginary axis, say $s = j\omega$, since rational functions of s display real axis symmetry as follows:

$$Y_1(s)Y_2(s) \to Y_1(j\omega)Y_2(j\omega) = M(j\omega) + jN(j\omega) \qquad s = j\omega$$
$$Y_1(s)Y_2(s) \to Y_1(-j\omega)Y_2(-j\omega) = M(j\omega) - jN(j\omega) \qquad s = -j\omega$$

where M and N are real even and odd functions, respectively, of ω. Hence the contour mapping $Y_1(j\omega)Y_2(j\omega)$ is symmetric with $Y_1(-j\omega)Y_2(-j\omega)$ with respect to the real w axis. In plotting the infinite semicircles

$$s = Re^{j\phi} \qquad R \to \infty$$

the terms in the highest power of the numerator and denominator are retained for evaluation. Consider, for example,

$$Y_1(s)Y_2(s) = \frac{as^3 + bs^2 + cs + d}{es^5 + fs^4 + gs^3 + hs^2 + ns + m} \to \frac{as^3}{es^5} = \frac{a}{es^2} \to \frac{a}{eR^2e^{2\phi j}}$$
$$s = Re^{j\phi} \qquad R \to \infty$$

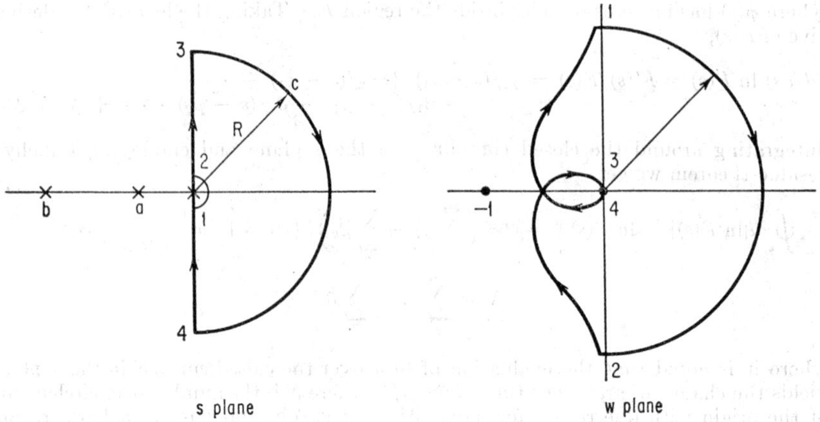

FIG. 8.23. Nyquist plot. $Y_1(s)Y_2(s) = \dfrac{K}{s(s+a)(s+b)}$.

A Nyquist plot is shown for a third-order system with a pole at the origin in Fig. 8.23.

Relative Stability. In addition to providing information on stability a measure of relative stability is provided by noting the proximity of the graphical plot to the -1 point in the w plane.

In quantitative terms, relative stability is determined by the relative gain of the open-loop transfer function $Y_1(j\omega)Y_2(j\omega)$ (Nyquist plot) 180° out of phase with the input, shown with amplitude e. The ratio $1/e$ is called the *gain margin*, implying that an increase of gain by this factor would make the system unstable. If $e > 1$, then the system is unstable. Similarly, the angle ϕ at unit distance is called the *phase margin* and indicates the additional amount of phase lag necessary to destabilize a stable system (see Fig. 8.24).

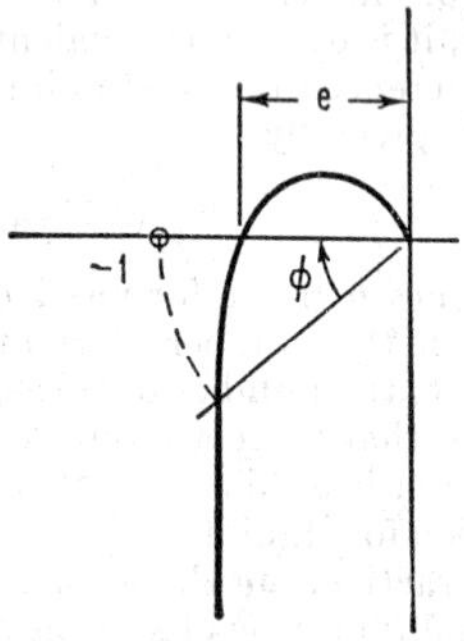

FIG. 8.24. Gain and phase margin. Phase margin $= \phi$. Gain margin $= 1/e$.

Systems with Feedback Time Lag. Interest is often centered on the destabilizing effect of time delays in the feedback path of a system which is otherwise stable. In some applications they are intentionally introduced for such an effect (oscillator).

The basic system is shown in Fig. 8.25 with an overall transfer function $Y_1/(1 + Y_1Y_2)$. Introduction of the time-delay block $\boxed{D}$ which is e^{-sT} changes the transfer function to $Y_1/(1 + Y_1Y_2e^{-sT})$. The system stability is readily evaluated by making the basic system Nyquist plot of Y_1Y_2 which is stable (by hypothesis) and incorporating the e^{-sT} by rotating $e - \omega T$ at each point along the existing $Y_1(j\omega)Y_2(j\omega)$ basic plot. The new plot provides the required stability picture.

A method due to Satch, applied to a first-order differential equation with a time lag T, is shown below for the first-order system.

$$dx/dt + \alpha x + \beta x(t - T) = 0$$

The transfer function is

$$\frac{1}{s + \alpha + \beta e^{-sT}} \tag{8.152}$$

which in order to exhibit stability must be free of right-hand plane poles. The equation for poles is

$$\frac{-(s+\alpha)}{\beta} = e^{-sT} \tag{8.153}$$

Let each side map the right-hand s plane, shown superimposed in Fig. 8.26*b* and *c* for two different cases. The right side maps into the unit circle, the left side the half plane displaced by $-\alpha/\beta$, shown crosshatched. The two closed curves are shown intersected in Fig. 8.26*b* for $\alpha/\beta < 1$, and therefore the included region corresponds to points in the s plane which satisfy Eq. (8.153) and consequently the system may be unstable. Further investigation would be required to ascertain stability or instability.

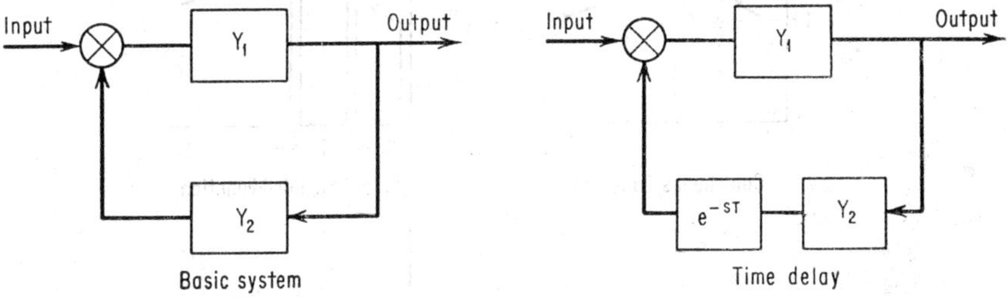

FIG. 8.25. System with feedback time delay.

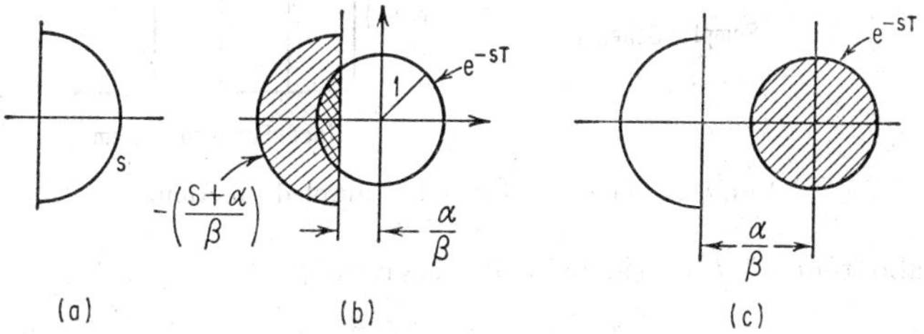

FIG. 8.26. Satch diagram for $dx/dt + \alpha x + \beta x(t - T) = 0$. Intersection $\alpha/\beta < 1$. No intersection (unconditionally stable) $\alpha/\beta > 1$.

If the regions intersected contain in each case one or more coincident points of the right half s plane the system is unstable; otherwise it is stable. The situation depicted in Fig. 8.26*c*, $\alpha/\beta > 1$, reveals no intersection and illustrates the unconditionally stable system.

(f) The Problem of Synthesis[31,32,33]

The problem of synthesis is to realize an overall transfer function within a set of specifications which often requires optimization of several conflicting requirements. Consider Fig. 8.20. The designer usually has little control over any block except $C(s)$, the "controller." Hence synthesis involves realization of transfer functions in cascade with fixed elements to produce the desired overall transfer function.

(g) Linear Discontinuous Control—Sampled Data[34,25]

If at one or more points in a linear control system the signal is interrupted intermittently at a prescribed rate, the resultant system is discontinuous and linear. If the rate is constant, the system is called a linear sampled-data control system. In so far as analysis is concerned, this merely introduces another building block called the "sampler" at each sampling point.

The sampler shown schematically in Fig. 8.27 has the property of taking the input $e(t)$ and periodically sampling it for time durations such that the area under each pulse (strength) is proportional to the instantaneous input. If the proportionality constant is made unity and the sampling pulse duration is small compared with the sampling period T, then to an excellent approximation which offers considerable analytic advantages, the output is assumed to be a train of impulses of strength $e(t)$ at each sampling "instant." The actual and ideal outputs are shown in Fig. 8.27.

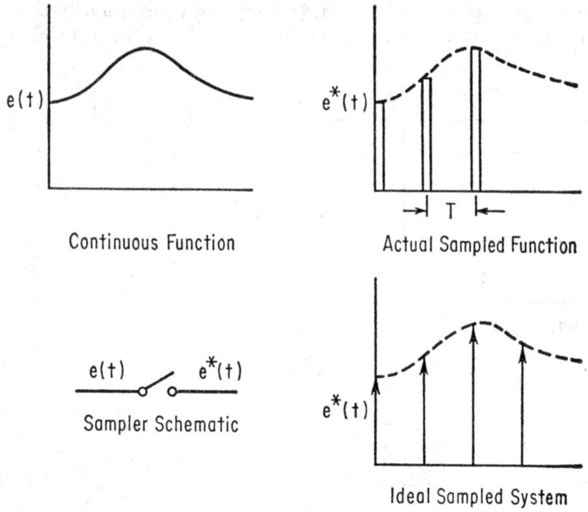

FIG. 8.27. The sampler and a sampled function.

The ideal output $e * (t)$ considered below is given by

$$e * (t) = e(t) \sum_{n=0}^{\infty} \delta(t - nT) = \sum_{n=0}^{\infty} e(nT)\delta(t - nT) \tag{8.154}$$

$$\delta(x) = \text{Dirac delta function}$$

Following continuous theory it is desirable to obtain the frequency-domain behavior of the "sampling" block. To this end consider the Laplace transform of the sampled signal

$$\mathcal{L}\{e * (t)\} \triangleq E * (s) = \mathcal{L}\left[e(t) \sum_{n=0}^{\infty} \delta(t - nT)\right] \tag{8.155}$$

From the convolution theorem,

$$\mathcal{L}[x(t)y(t)] = X(s) * Y(s)$$

Eq. (8.155) becomes

$$E * (s) = \mathcal{L}[e * (t)] = E(s) * \mathcal{L} \sum_{n=0}^{\infty} \delta(t - nT)$$

$$= E(s) * \sum_{n=0}^{\infty} e^{-nTs} = E(s) * I(s) \tag{8.156}$$

$$I(s) = \sum_{n=0}^{\infty} e^{-nTs}$$

where $I(s)$ is defined as the Laplace transform of the impulse train

$$\sum_{n=0}^{\infty} \delta(t - nT)$$

An alternative and less useful form is derived from the right-side representation of Eq. (8.154),

$$E^*(s) = \mathcal{L}e^*(t) = \mathcal{L} \sum_{n=0}^{\infty} e(nT)\delta(t - nT) = \sum_{n=0}^{\infty} e(nT)e^{-nTs}$$

Using the closed-form representation of $I(s)$

$$I(s) = \sum_{n=0}^{\infty} e^{-nTs} = \frac{1}{1 - e^{-sT}}$$

Eq. (8.156) is evaluated by a closed-contour integration as follows:

$$E^*(s) = \frac{1}{2\pi j} \int_{c-j\infty}^{c+j\infty} E(w)I(s - w)\, dw = \frac{1}{2\pi j} \oint \frac{E(w)\, dw}{1 - e^{-(s-w)T}} \tag{8.157}$$

The abscissa c is chosen so that the poles of $E(w)$ have real parts $<c$ and s is defined for Re s (real of s) $> c$ where the contour integral shown in Fig. 8.28 is employed since the integral over the infinite right-hand semicircle vanishes. The infinity of poles of the integrand inside this contour are then the zeros of $1 - e^{-(s-w)T}$, the latter corresponding to $s - w = \pm 2n\pi j/T$.

$$w = s \pm 2n\pi j/T \qquad \text{poles}$$
$$2\pi/T = \omega_s \qquad \text{sampling frequency}$$

which are an infinity of simple poles. Equation (8.157) is evaluated by Cauchy's residue theorem, taking the residues of the infinity of poles yielding

$$E^*(s) = \sum_{n=-\infty}^{+\infty} \frac{E(s + jn\omega_s)}{T} \tag{8.158}$$

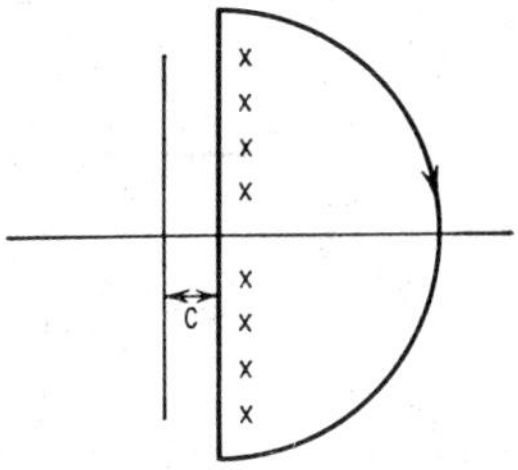

FIG. 8.28. W-plane contour for complex convolution.

If $E(s)$ has no right half plane poles then Eq. (8.158) is defined for the entire right half s plane. Equation (8.158) is clearly a periodic function of s having the complex period $j\omega_s$ as shown

$$E^*(s + j\omega_s) = \sum_{n=-\infty}^{+\infty} \frac{E[s + j(n+1)\omega_s]}{T} = \sum_{-\infty}^{+\infty} \frac{E(s + jn\omega_s)}{T} = E^*(s) \tag{8.159}$$

From the periodic character of $E(s)$ it follows that if $E^*(s)$ is known in any strip in the complex s plane bounded by

$$jx < \text{Im } s < j(x + \omega_s) \tag{8.160}$$

it is known everywhere in the s plane. The transfer function at real frequency $s = j\omega$ is found directly from Eq. (8.158). Its amplitude spectrum is sketched (Fig. 8.29) for $\omega_s/2 > \omega_0$ where ω_0 is the cutoff frequency of $E(j\omega)$, i.e., for the sampling frequency greater than twice the highest frequency component of $e(t)$. Note that $|E^*(j\omega)|$ yields the infinitely repeated spectrum of $|E(j\omega)|$ attenuated by $1/T$. If $\omega_s/2 < \omega_0$, there is overlapping and resultant distortion of the input signal. Returning to the case $\omega_s/2 > \omega_0$, practically all the input $e(t)$ information is stored in $E^*(j\omega)$ over the frequency range $0 < \omega < \omega_0$. By ideal low-pass filtering of the signal $E^*(j\omega)$, spectral components greater than ω_0 can be eliminated, leaving the fundamental signal shape. The resultant system would then be the equivalent of the continuous system with an attenuator $1/T$ placed after the input signal. In practice, however, deviations from ideality of this filter introduce severe stability problems. A smoothing device is utilized as a compromise between high degree of filtering and its concomitant stability problems.

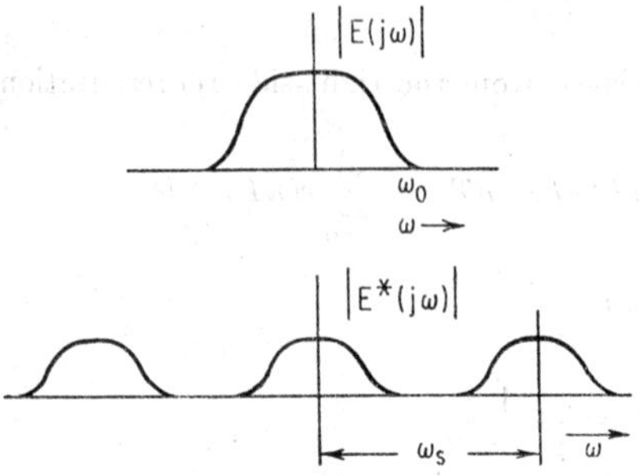

FIG. 8.29. Frequency spectrum for function and for sampled function.

One such smoothing device is the holding circuit whose transfer function is

$$(1 - e^{-Ts})/s \tag{8.161}$$

which for any impulse input $\delta(t - nT)$ yields the output $u(t - nT) - u[t - (n + 1)T]$, a pulse of unit height starting at $t = nT$ and of duration T.

Stability Investigation of Sampled-data Control Systems. Consider the sampled-data control system shown in Fig. 8.30. The overall transfer function is derived from the basic properties of transfer functions.

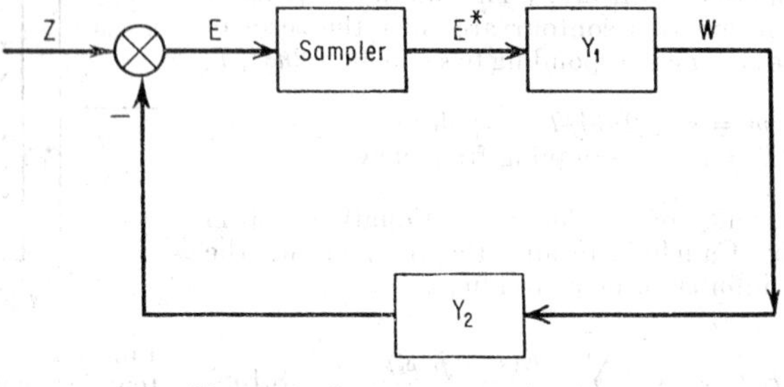

FIG. 8.30. Sampled-data system.

$$E = Z - Y_2W \qquad W = Y_1E^* \tag{8.162}$$

Elimination of W in Eq. (8.162) gives

$$E = Z - Y_1Y_2E^* \tag{8.163}$$

Now a fundamental property of sampling is stated and proved as follows:

Given: $A = BC^*$
Then: $A^* = B^*C^*$

Proof [utilizing Eq. (8.158)]:

$$X^*(s) = 1/T \sum_{n=-\infty}^{+\infty} X(s + jnw_s)$$

$$A^*(s) = 1/T \sum_{n=-\infty}^{+\infty} A(s + jnw_s) = 1/T \sum_{n=-\infty}^{+\infty} B(s + jnw_s)C^*(s + jnw_s)$$

But from Eq. (8.159)

$$C^*(s + jnw_s) = C^*(s)$$

and therefore

$$A^*(s) = C^*(s)1/T \sum_{n=-\infty}^{+\infty} B(s + jnw_s)$$
$$= C^*(s)B^*(s) = B^*(s)C^*(s)$$

Sampling Eq. (8.163) and utilizing the results of this theorem yields

$$E^* = Z^* - (Y_1Y_2)^*E^* \tag{8.163a}$$

and

$$E^* = \frac{Z^*}{1 + (Y_1Y_2)^*} \qquad W = Y_1E^* = \frac{Z^*Y_1}{1 + (Y_1Y_2)^*}$$

Following previous work, the stability of the system rests with the location of the poles of $1/[1 + (Y_1Y_2)^*]$ or more specifically the zeros of $1 + (Y_1Y_2)^*$.

The transcendental form of $(Y_1Y_2)^*$ makes this evaluation using methods cited earlier extremely difficult to apply, per se. The task is simplified, however, owing to the periodic property of sampled functions embodied in Eq. (8.159) which implies that if one investigates the zeros of $(Y_1Y_2)^* + 1$ bounded by a strip given in Eq.

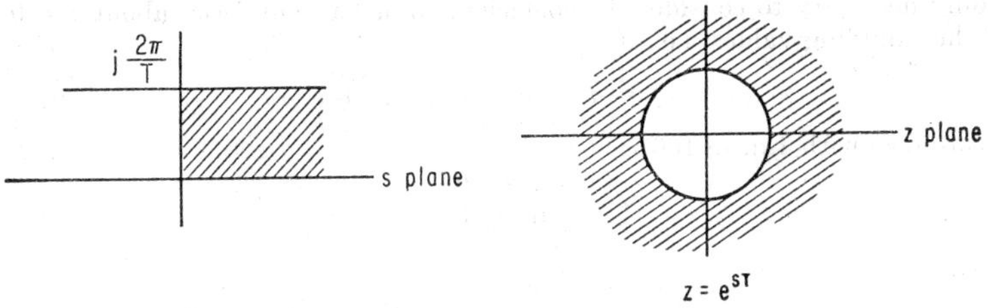

FIG. 8.31. Mapping the strip onto the z plane.

(8.160) in region Re $s > 0$, then this effectively gives the zero configuration in the entire right half plane. In practice the strip chosen is $0 < \text{Im } s < j\omega_s = j2\pi/T$, and Re $s > 0$ shown crosshatched in Fig. 8.31. Now a function z is defined by

$$z = e^{sT} \tag{8.164}$$
$$s = \ln z/T \tag{8.165}$$

with s defined only in the whole strip

$$0 < \text{Im } s < j2\pi/T$$

so that s and z are single-valued analytic functions of each other (except for a branch cut on the real z axis). Equation (8.164) implies a mapping of the whole strip in the s plane onto the entire z plane with the shaded portion mapping outside the unit circle as shown in Fig. 8.31. The location of zeros of $[1 + (Y_1Y_2)^*]$ in the right half strip of s corresponds to the location of zeros outside the unit circle of the z plane, and their presence or absence is translated as unstable or stable conditions, respectively.

***z* Transform.** If the transformation Eq. (8.164) is applied to the Laplace transformation of a sampled function $x^*(t)$ as follows:

$$X(z) \triangleq X^*(s) \qquad z = e^{sT}$$

then $X(z)$ is defined as the z transform of $x(t)$. $X(z)$ is valid only at the sampling instants despite its general continuity properties. Note that the z transform of $x^*(t)$ is the same as that of $x(t)$ since the sampled $x^*(t)$ is indeed $x^*(t)$. The z transformation of any function is obtained by first sampling the function, then taking its Laplace transform and finally making the substitution [Eq. (8.165)] to eliminate s in favor of z. From this definition,

$$E(z) = E^*(s) = \sum_{n=0}^{\infty} e(nT)\, e^{-nTs} = \sum_{n=0}^{\infty} e(nT) z^{-n} \tag{8.166}$$

or from the convolution form [Eq. (8.157)],

$$E^*(s) = \frac{1}{2\pi j} \int_{c-j\infty}^{c+j\infty} E(w) \frac{dw}{1 - e^{-st}e^{wT}} = \frac{1}{2\pi j} \int_{c-j\infty}^{c+j\infty} E(w) \frac{dw}{1 - e^{wT}z^{-1}} = E(z)$$

$$\text{Re } s > c$$

If this integration is performed over the contour enclosing the left-hand infinite semicircle in contradistinction to the contour used previously so that the zeros of $1 - e^{wt}z^{-1}$ are not contained inside the contour, then application of the residue theorem yields

$$E(z) = \sum \text{Residues of} \left[E(w) \frac{1}{1 - e^{wT}z^{-1}} \right]$$

for poles of $E(w)$ only.

Inversion of $E(z)$ to Time Domain. To go from the z domain to the time domain, it is only necessary to consider the coefficients of a Laurent series about $z = 0$ which yield the sampling values, i.e., if

$$F(z) = a_0 + a_1/z + a_2/z^2 + \cdots$$

in accordance with Eq. (8.166)

$$a_0 = e(0)$$
$$a_1 = e(T)$$
$$\cdot$$
$$\cdot$$
$$\cdot$$
$$a_n = e(nT)$$

Table 8.3. Properties of z Transforms of Causal Time Function
$f(t) = 0 \quad t < 0$

Property	z transform	Time function
Basic pairs	$F(z) = \sum_{n=0}^{\infty} f(nT)z^{-n}$	$f(t)$ $f(nT) \quad n = 0,1, \ldots, \infty$ $f(nT) = 1/2\pi j \oint F(z)z^{n-1}\,dz$
Linearity	$a_1F_1(z) + a_2F_2(z)$	$a_1f_1(t) + a_2f_2(t)$
Time shift	$z[F(z) - f(0)]$	$f(t + T)$
Initial-value theorem	$\lim_{z \to \infty} F(z)$	$\lim_{t \to 0} f(t)$
Final-value theorem	$\lim_{z \to 1} (z - 1)F(z)$	$\lim_{t \to \infty} f(t)$

Formally using any closed contour around $z = 0$, this is equivalent to the contour integral

$$e(nT) = 1/2\pi j \oint E(z) z^{n-1}\, dz = \text{Residues of } E(z) z^{n-1}$$

Table 8.3 shows some z transforms and their properties.

Example of Stability Investigation. Consider the sampled-proportional-level control system shown schematically and in block form in Fig. 8.32 with the constant

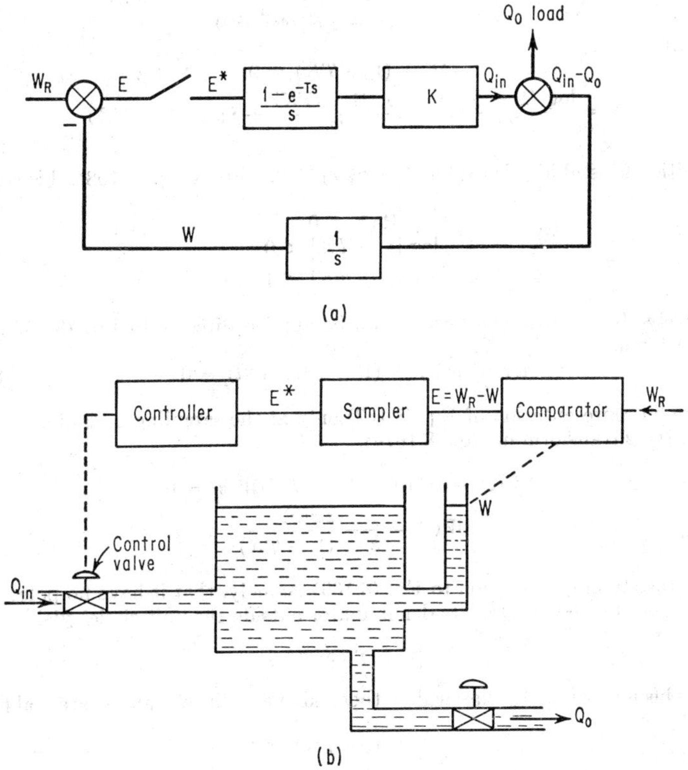

FIG. 8.32. Sampled-level control system. (a) Control-block diagram. (b) System.

input w_R (desired level). The system is described by

1. $q_0(t)$ is an arbitrary-rate flow of effluent.
2. The replenishment rate is proportional to the error existing one sample time prior to

$$q_{in} = -K(w[t/T] - w_R)$$

where $[x]$ is defined as the smallest integral value of x.

Writing the equation conserving mass yields

$$dw/dt = -K(w[t/T] - w_R) - q_0(t)$$

Integrating over sampling times nT and $(n + 1)\,T$ yields the equation

$$w\{(n + 1)T\} - w(nT) = TK\{w_R - w(nT)\} - \int_{nT}^{(n+1)T} q_0\, dt$$

$$w\{(n + 1)T\} - (1 - TK)w(nT) = TKw_R - \int_{nT}^{(n+1)T} q_0\, dt$$

which is the inhomogeneous difference equation whose theory parallels that of ordinary differential equations. The stability is a function of the solutions to the homogeneous equation

$$w_1[(n+1)T] - (1 - TK)w_1(nT) = 0 \tag{8.167}$$

A solution is found by assuming the exponential form

$$w_1(t) = e^{\lambda t}$$

whence

$$e^{\lambda(n+1)T} - (1 - TK)e^{n\lambda T} = 0 \tag{8.168}$$

from which

$$e^{\lambda T} - (1 - TK) = 0$$

$$\lambda = \frac{\log (1 - TK)}{T} = \frac{\log |1 - TK|}{T} + j \arg (1 - TK)$$

The condition for stability is that as $t \to \infty$, $w_1 \to 0$. From Eq. (8.168), this is met by requiring

$$\text{Re}\,\lambda < 0$$
$$\log |1 - TK| < 0$$
$$|1 - TK| < 1$$

Alternatively the z-transform method can be applied directly to Eq. (8.167), which is just written as

$$w_1^*(t + T) - (1 - TK)w_1^*(t) = 0 \tag{8.167a}$$

and is a valid representation of Eq. (8.167) only at the sampling instants.

Taking the z transform of Eq. (8.167a),

$$z[W(z) - w(0)] - (1 - TK)W(z) = 0$$

$$W(z) = \frac{zw(0)}{z - (1 - TK)} \tag{8.169}$$

Applying the stability condition to Eq. (8.169), namely, that it have no poles outside the unit circle (i.e., no zeros of its denominator outside the unit circle) yields

$$|1 - TK| < 1$$

which has been obtained by classical methods above. Since T and K are real the condition is

$$-1 < (1 - TK) < 1$$

Finally, and more generally, direct consideration of the control block (Fig. 8.32) and Eq. (8.163a) yields

$$E^* = \frac{W_R^*}{1 + K[(1 - e^{-Ts})/s^2]^*} + \frac{[Q_0(s)/s]^*}{1 + K[(1 - e^{-Ts})/s^2]^*} \tag{8.170}$$

From tables in ref. 10

$$\left[\frac{1 - e^{-Ts}}{s^2}\right]^* = (1 - z^{-1})\frac{Tz}{(z-1)^2} = \frac{T}{z - 1}$$

The transform of Eq. (8.170) is

$$E(z) = \frac{W_R(z)}{1 + KT/(z - 1)} + \frac{[Q_0(s)/s]^*}{1 + KT/(z - 1)} \qquad s = \ln \frac{z}{T}$$

No poles of the numerator are envisioned for practical systems, so that the zeros of the denominator give all $E(z)$ poles

$$z = 1 - TK$$

which for stability demands $|z| < 1$ or once again

$$-1 < 1 - TK < 1$$

(h) Nonlinear Control Systems[36,37,38,39]

The treatment of control systems containing nonlinearities (as defined under general nonlinear systems) is for the most part so formidable that all known analytic methods fail. The very special case of the second-order autonomous (time-invariant coefficients) can be handled most conveniently by graphical methods in the phase plane. Also under very special conditions it is possible for a higher-degree nonlinear

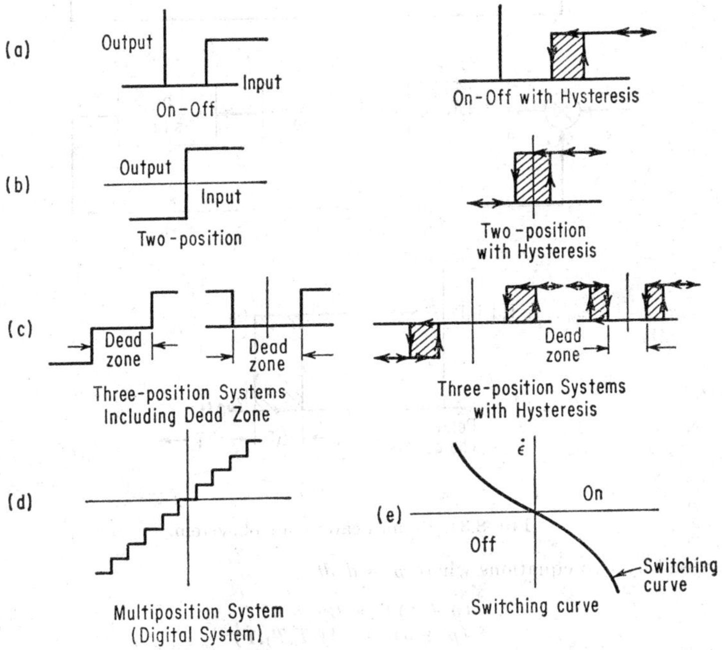

FIG. 8.33. Switching functions.

system to be analyzed by a "describing-function" technique. Systems which contain switching-function nonlinearities which are otherwise linear can be analyzed (with great difficulty) by linear methods over each (linear) regime of operation satisfying boundaries between regimes. However, as the order of the equation goes beyond three, the difficulty in matching boundaries becomes prohibitive. Often the second-order case is handled most conveniently by phase-plane graphical methods.

Linear Systems with Discontinuous Switching. These systems are characterized by switching operations. If switching occurs at a constant rate then the system is the linear sampled-data system described above. If on the other hand the switching operation occurs whenever the signal (e.g., error) reaches a prescribed level of some function of the output, then the system is of the relay type and is in general nonlinear. Examples of this type are shown in Fig. 8.33. More generally, the switching points may be mixed functions of the input variable and its derivative; ϵ and $\dot{\epsilon}$ have a phase-plane representation as shown for the example in Fig. 8.33e. In addition to the behavior as a function of input, hysteresis, and dead zone, there are time lags inherent in operation because of inertia and inductance.

An illustration of a linear system with switching is a room heater whose block diagram is drawn in Fig. 8.34 with relay characteristics shown in Fig. 8.34*c* indicating that the furnace goes on whenever $T < T_1$ and off whenever $T > T_2$. The furnace and room are two first-order systems connected in tandem. Analytically, this result

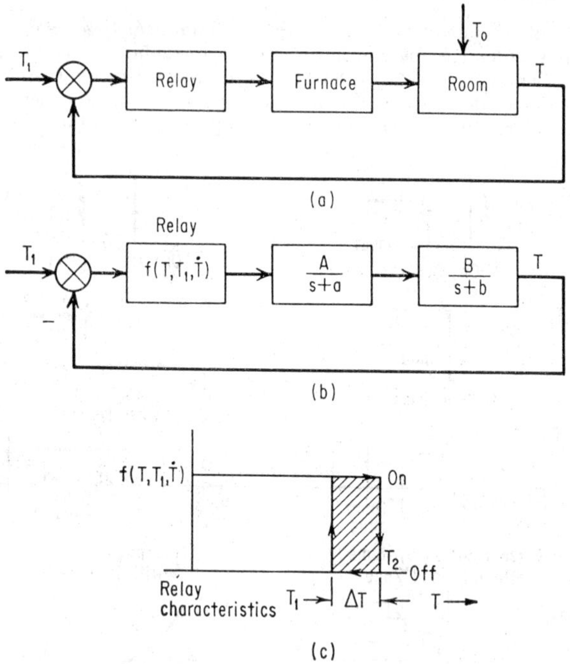

FIG. 8.34. Room-heater control system.

is expressed by two equations where $p = d/dt$

$$\begin{aligned}(p + b)T &= By + bT_o \\ (p + a)y &= Af(T,T_1,\dot{T})\end{aligned} \tag{8.171}$$

where y is the furnace output, T_o the outside temperature, and $f(T,T_1,\dot{T})$ the switching function. Elimination of y in Eq. (8.171) yields

$$(p + a)(p + b)T = ABf(T,T_1,\dot{T}) + abT_o \tag{8.171a}$$

Note that the function f not only depends upon T and T_1 but also on the sign of $\dot{T}$ as follows:

$$\begin{aligned} f(T,T_1,\dot{T}) &= Q \qquad T < T_1 \\ &= Q \qquad T_1 < T < T_2 \qquad \dot{T} > 0 \\ &= 0 \qquad T_1 < T < T_2 \qquad \dot{T} < 0 \\ &= 0 \qquad T > T_2 \end{aligned} \tag{8.172}$$

which constitute four possible regimes of operation. Accordingly Eq. (8.172) must be solved for the heating cycle $f(T,T_1,\dot{T}) = Q$ and the cooling cycle $f(T,T_1,\dot{T}) = 0$. For heating,

$$T \text{ heating} = \alpha e^{-at} + \beta e^{-bt} + T_\infty$$

where T_∞ is the asymptotic temperature for uncontrolled continuous heating. For cooling,

$$T \text{ cooling} = \gamma e^{-at} + \delta e^{-bt} + T_o$$

The four constants of integration, α, β, γ, δ, must be determined by matching conditions at the transitions of any two regimes where continuity of T and its derivative $\dot{T}$ must be preserved. This becomes a most laborious procedure, since these constants change repeatedly, regime after regime, cycle after cycle. Only under constant load T_o will a cycle that is repetitive be eventually reached (limit cycle).

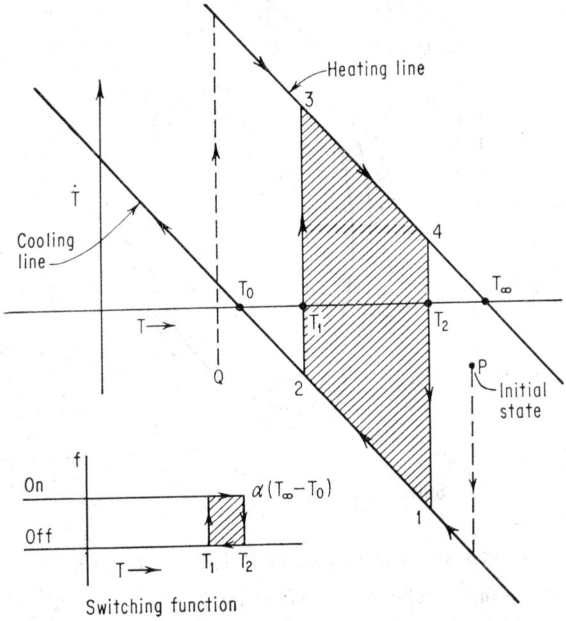

FIG. 8.35. Phase-plane plot for first-order temperature-control system $\dot{T} + \alpha T = f + \alpha T_0$. f = switching function.

The prohibitive analytic method is seldom justified for second-order systems whose complete graphical solution in the phase plane can be easily generated. As an introduction to the method, consider the first-order temperature-control system

$$\dot{T} + \alpha T = f(T,T_1,\dot{T}) + \alpha T_o \tag{8.173}$$

with its characteristic switching function f. The two modes of operation for Eq. (8.173) are

Heating: $\dot{T} + \alpha T = Q + \alpha T_o \triangleq \alpha T_\infty$

Cooling: $\dot{T} + \alpha T = \alpha T_o$

In the phase plane for both cases (taking one time derivative) assuming T_o constant,

$$\dot{T}(d\dot{T}/dT)\ \ddot{T} = \ddot{T} = -\alpha\dot{T}$$
$$d\dot{T}/dT = -\alpha$$

The corresponding locus of phase-plane operation therefore consists of two parallel lines of operation of slope $-\alpha$ as drawn in Fig. 8.35 identified by their T intercepts. The limit cycle 1-2-3-4 is shown crosshatched. Jump action occurs at points 2 and 4, the switching points. For any initial $\dot{T}$, T to the left of $T = T_1$ shown for point Q, the point will jump vertically to the heating line and proceed to 3-4-1-2-3, closing the cycle. Similarly for point P at $T > T_2$, the point inside the zone $T_1 < T < T_2$ goes

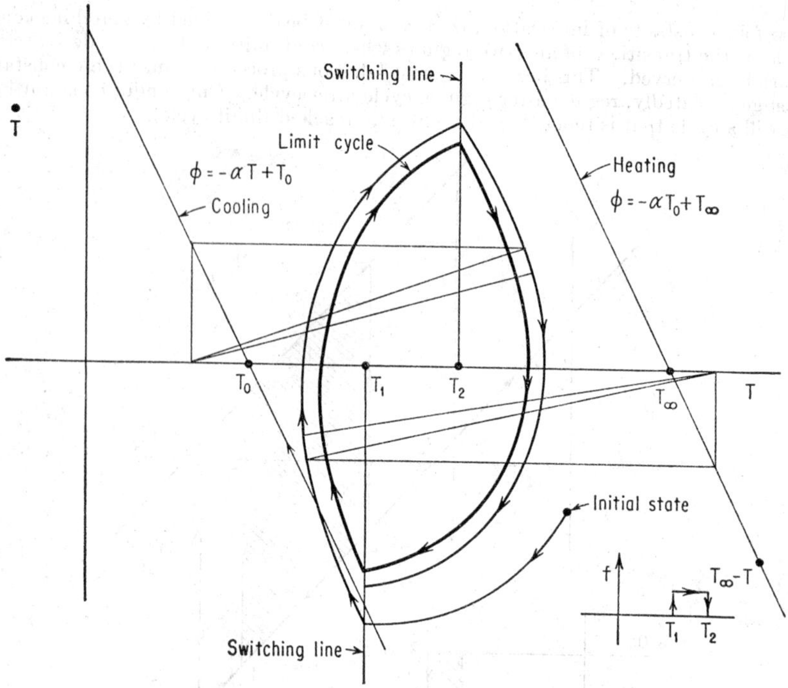

FIG. 8.36. Phase-plane plot for $\ddot{T} + \dot{T}\alpha + T = f + T_0$.

to the cooling curve first. The closed-cycle time is

$$\oint dT/\dot{T} = \int_{T_1}^{T_2} dT/\dot{T} + \int_{T_2}^{T_1} dT/\dot{T} \tag{8.174}$$

Returning to the second-order temperature system defined by Eq. (8.171*a*)

$$\ddot{T} + (a + b)\dot{T} + abT = ABf(T,T_1,\dot{T}) + abT_o$$

By a suitable change of variable, Eq. (8.171) can always be represented as

$$\ddot{T}' + \alpha'\dot{T}' + T' = f'(T',T_1,\dot{T}') + T_o'$$

where primes have been appended to imply the transformation. Dropping these primes for convenience leaves

$$\ddot{T} + \alpha\dot{T} + T = f(T,T_1,\dot{T}) + T_o$$

whose representation for the phase-plane plot has been shown to be

$$\frac{d\dot{T}}{dT} = \frac{-\alpha\dot{T} - T + f + T_o}{\dot{T}}$$

$$= \frac{-\alpha\dot{T} - T + T_\infty}{\dot{T}} \quad \text{heating cycle}$$

$$= \frac{-\alpha\dot{T} - T + T_o}{\dot{T}} \quad \text{cooling cycle}$$

The phase-plane plot of this system is determined by first drawing the switching lines $T = T_1$ and $T = T_2$, and executing the Liénard construction for each of the two operating regimes. A limit cycle is reached as shown in Fig. 8.36.

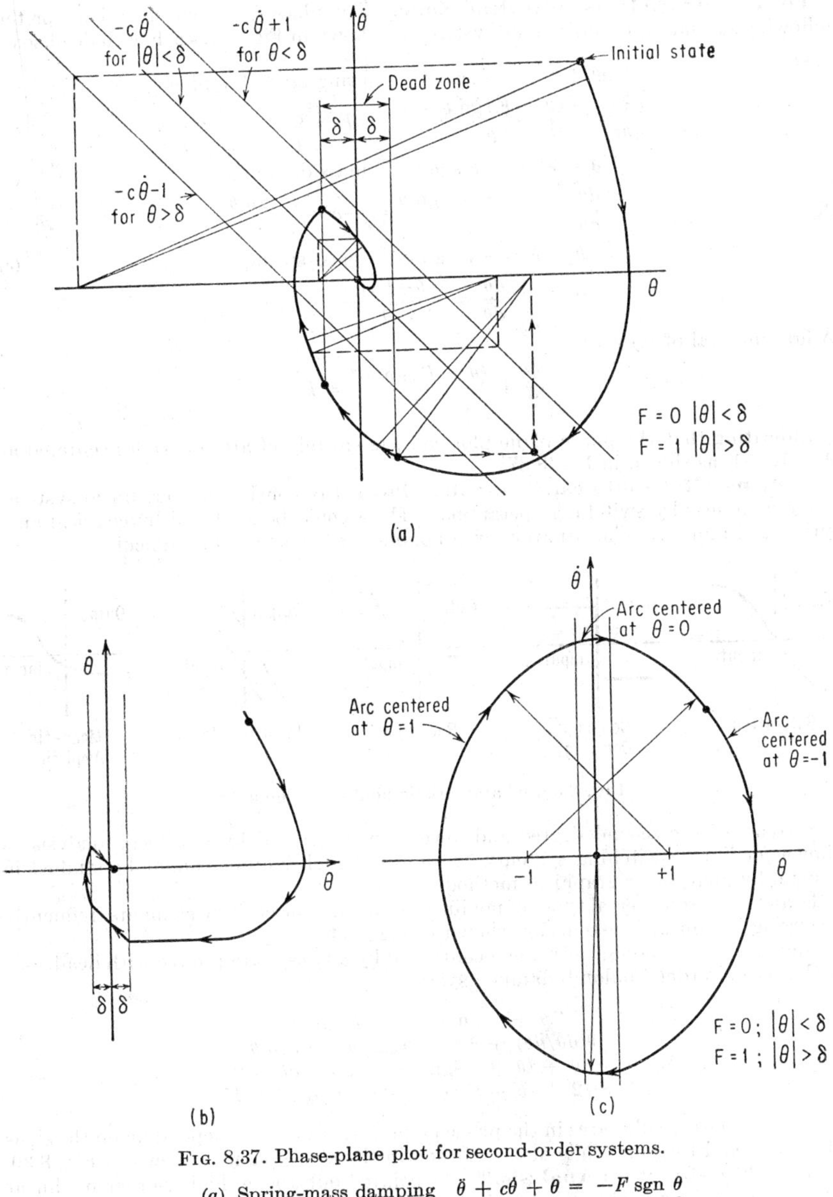

FIG. 8.37. Phase-plane plot for second-order systems.

(a) Spring-mass damping $\ddot{\theta} + c\dot{\theta} + \theta = -F \operatorname{sgn} \theta$
(b) Mass damping $\ddot{\theta} + c\dot{\theta} = -F \operatorname{sgn} \theta$
(c) Mass spring $\ddot{\theta} + \theta = -F \operatorname{sgn} \theta$

Positioning Systems with Dead Zone. The phase-plane constructions for the following second-order positioning systems are shown in Fig. 8.37 each with dead zone.

$$\ddot{\theta} + c\dot{\theta} + \theta = -F \operatorname{sgn} \theta \qquad \text{spring-mass damping} \tag{a}$$

$$\frac{d\dot{\theta}}{d\theta} = \frac{-c\dot{\theta} - F \operatorname{sgn} \theta - \theta}{\dot{\theta}} \qquad \begin{aligned} F &= 0 \quad |\theta| < \delta \\ &= 1 \quad |\theta| > \delta \end{aligned}$$

$$\ddot{\theta} + c\dot{\theta} = -F \operatorname{sgn} \theta \qquad \text{mass damping} \tag{b}$$

$$\frac{d\dot{\theta}}{d\theta} = \frac{-c\dot{\theta} - F \operatorname{sgn} \theta}{\dot{\theta}} = -c - \frac{F \operatorname{sgn} \theta}{\dot{\theta}}$$

$$\ddot{\theta} + \theta = -F \operatorname{sgn} \theta \qquad \text{mass spring} \tag{c}$$

$$\frac{d\dot{\theta}}{d\theta} = \frac{-\theta - F \operatorname{sgn} \theta}{\dot{\theta}}$$

A first integral of (*c*) gives

$$\frac{\dot{\theta}^2}{2} + \frac{(\theta + F \operatorname{sgn} \theta)^2}{2} = k_0^2$$

Indicating that the phase-plane plot consists entirely of arcs of circles centered at 0, −1, +1, as shown in Fig. 8.37.

Systems with Nonlinear Elements. Previously considered were linear systems made nonlinear by switching operations. These could be analyzed by classical analytical techniques. The situation with nonlinear systems is more difficult.

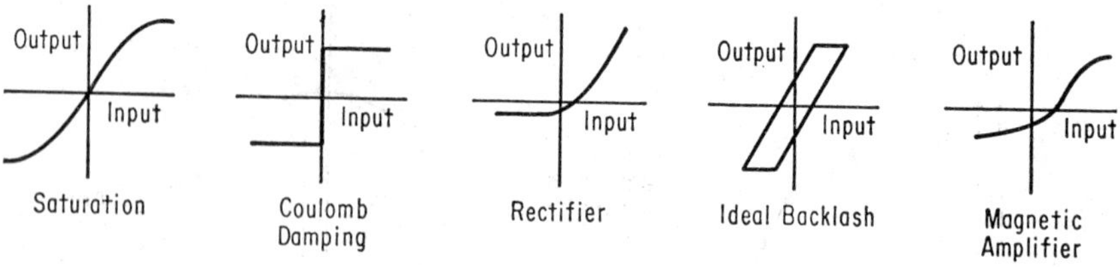

FIG. 8.38. Characteristic nonlinear elements.

Systems of the second degree and lower can be treated by graphical analysis as shown for linear switching systems. Systems of higher order cannot be studied in general by analytic or graphical methods.

Examples of some usual types of nonlinear frequency-insensitive elements frequently occurring in "linear" systems are shown in Fig. 8.38.

An example of a nonlinear system controlled by a two-position force with dead zone is the second-order Coulomb damped system

$$\begin{aligned} \ddot{\theta} + c \operatorname{sgn} \dot{\theta} + \theta &= F \operatorname{sgn} \theta \\ \dot{\theta}(d\dot{\theta}/d\theta) + \theta &= -b \operatorname{sgn} \theta - c \operatorname{sgn} \dot{\theta} \\ \dot{\theta}\, d\dot{\theta} + (\theta - F \operatorname{sgn} \theta + c \operatorname{sgn} \dot{\theta})\, d\theta &= 0 \\ \dot{\theta}^2/2 + (\theta - F \operatorname{sgn} \theta + c \operatorname{sgn} \dot{\theta})^2/2 &= k^2 \end{aligned}$$

which describes circular arcs in the phase plane with six centers depending on the signs of θ and $\dot{\theta}$ and the amplitude of θ. Motion in the phase plane is shown in Fig. 8.39.

Describing-function Analysis.[10] As pointed out above, higher-order nonlinear systems are not amenable to graphical analysis. A method of analysis has evolved which is valid under very restrictive conditions. It employs linear concepts in an attempt to simplify the complex nonlinear problem and bring it within the realm of analysis.

The analysis is limited to systems containing one nonlinearity or where many can be grouped to yield effectively one nonlinear and time-invariant block. For a sinusoidal input, the resultant output will be composed of the fundamental plus higher harmonics. The essence of the analysis is to ignore all harmonics other than the fundamental. This is the most restrictive assumption and can often be justified for slight nonlinearities where the higher harmonics are small to begin with; these are further attenuated since most systems are usually natural low-pass filters.

Implementing the foregoing description, the object is to obtain for a fundamental input of amplitude A and frequency ω a Fourier series whose fundamental amplitude

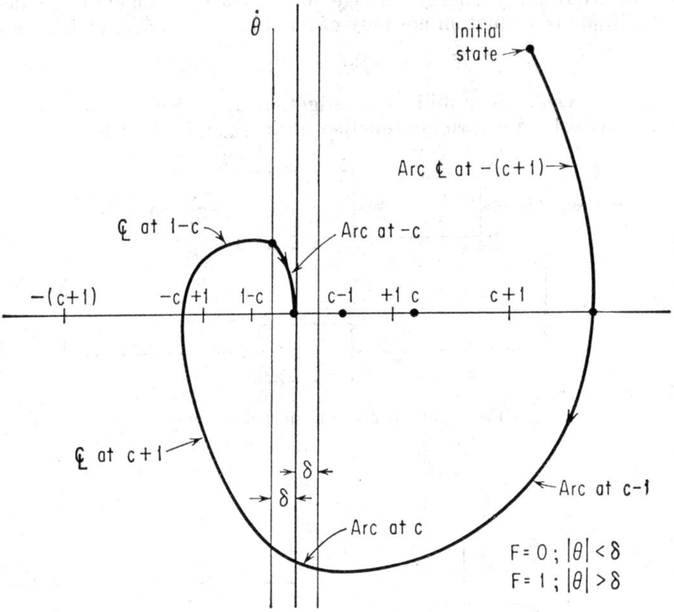

FIG. 8.39. Phase-plane plot for $\ddot{\theta} + c \operatorname{sgn} \dot{\theta} + \theta = -F \operatorname{sgn} \theta$.

is $B(A,\omega)$ and phase $\phi(A,\omega)$. If a functional relation connects the input with the output, say

$$x_2 = f(x_1)$$

then for an input

$$x_1 = A \sin \omega t$$

the output is

$$x_2(A \sin \omega t) = f(A \sin \omega t)$$

which shows that x_2 is a periodic function of time, with period $2\pi/\omega$. It therefore has a Fourier series development ($\phi = \omega t$)

$$\begin{aligned} x_2(A \sin \phi) &= f(A \sin \phi) \\ x_2(\phi) &= a_1 \sin \phi + a_2 \sin 2\phi + \cdots + a_n \sin n\phi \\ &\quad + b_1 \cos \phi + b_2 \cos 2\phi + \cdots \end{aligned}$$

For an input $A \cos \phi$, the output would be obtained by adding $\pi/2$ to ϕ

$$x_2(\cos \phi) = x_2 [\sin (\phi + \pi/2)] = a_1 \cos \phi + \cdots - b_1 \sin \phi + \cdots$$

The amplitude ratio of the fundamental output to input is

$$|N| = \frac{(a_1^2 + b_1^2)^{1/2}}{A} \qquad \begin{aligned} a_1 &= a_1(A) \\ b_1 &= b_1(A) \end{aligned}$$

and the phase

$$\theta = \tan^{-1} b_1/a_1$$

By definition, the describing function

$$N = \frac{(a_1^2 + b_1^2)^{1/2}}{A} < \theta$$

is the complex ratio of the fundamental component of output to input. For nonlinear systems containing energy storage or dissipative elements the describing-function amplitude is a function not only of amplitude but of frequency as well, i.e.,

$$N = N(A,\omega) \qquad <\theta(A,\omega)$$

Stability Analysis. A stability investigation of a system with one nonlinear block characterized by a describing function utilizes graphical techniques.

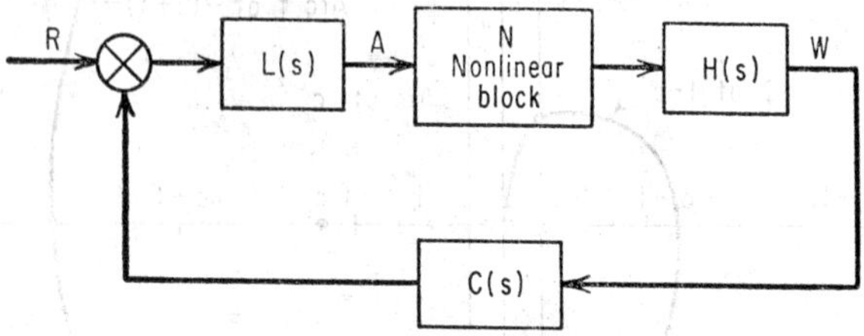

FIG. 8.40. Nonlinear control system.

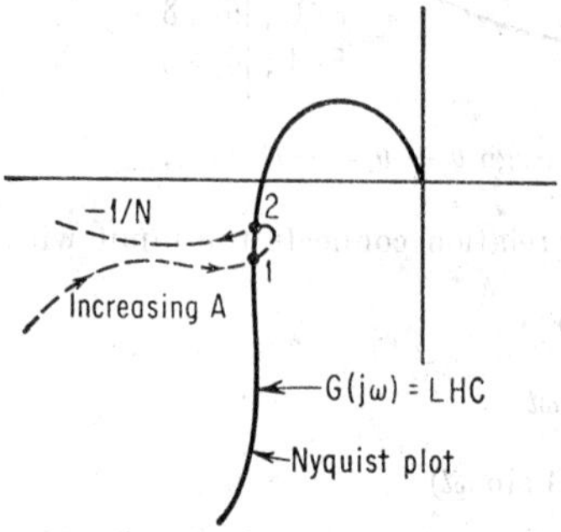

FIG. 8.41. Nyquist plot for nonlinear system.

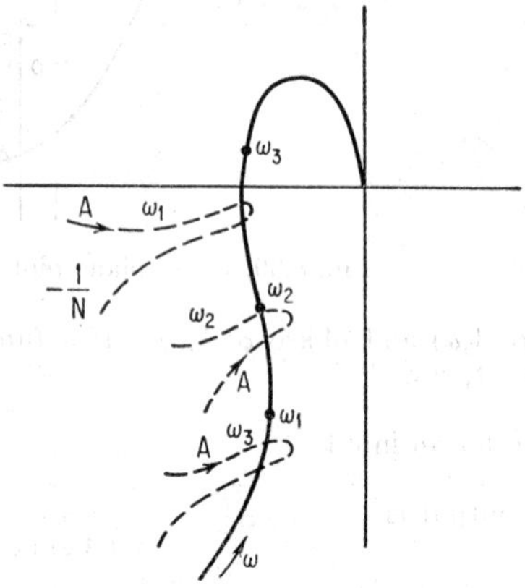

FIG. 8.42. Nyquist plot for nonlinear system where N is a function of frequency and amplitude.

Consider, for example, the control system shown in Fig. 8.40. Assume a signal whose fundamental amplitude A impinges on the input to the nonlinear block. The transfer functions of the linear blocks have their arguments s replaced by $j\omega$ to obtain overall characteristics for real frequency. It should be noted here that right-half plane poles no longer have meaning in the usual sense and interest must be necessarily restricted to sinusoidal signals owing to the describing-function definition. The loop gain at frequency ω is obtained by starting a signal amplitude A at zero phase at the input to the nonlinear device, and going completely around the loop giving

$$LHCN \times A$$

If this signal is 180° out of phase and greater than A, the amplitude grows; if it is less than A in amplitude, the signal decays. Grouping the linear blocks $LHC = G$ the condition for a net increase of signal is:

Amplitude $|G(j\omega)N(A,\omega)| > 1$
Argument $G(j\omega)N(A,\omega) = (2n + 1)\pi \quad n = \text{integer}$

A convenient way of illustrating this is indicated in Fig. 8.41, where first $G(j\omega)$ is plotted in a usual Nyquist plot. On the same set $-1/N$ is plotted (frequency-independent case) with the arrow in the direction of increasing amplitude A. Intersection for this case corresponds to sustained oscillations. At point 1 the frequency corresponding to the plot $G(j\omega)$ is an unstable point, since if perturbed in the direction where $-1/N > G(j\omega)$, A will decay. If perturbed in the opposite direction $-1/N < G(j\omega)$, the condition for increased growth of A, it will proceed to point 2, the stable oscillation point. If the $-1/N$ curve lies entirely within the $G(j\omega)$ plot, the amplitude will grow without bound, since $-1/N < G(j\omega)$ for all A; if outside, the system is unconditionally stable.

If N is a function of ω as well, then in the plot of $G(j\omega)$ and N as shown in Fig. 8.42 N is drawn for constant ω, e.g., ω_1, ω_2, ω_3. A qualitative analysis similar to the foregoing can be inferred. Stable or unstable points of intersection take on meaning only where there is correspondence of ω as well as amplitude as shown at ω_2.

References

1. Miller, F. H.: "Partial Differential Equations," John Wiley & Sons, Inc., New York, 1941.
2. Courant, R., and D. Hilberg: "Methods of Mathematical Physics," vol. I, Interscience Publishers, Inc., New York, 1953.
3. Frank, P. L., and R. von Mises: "Differential und Integral Gleichungen der Mechanik und Physik," Friedr. Vieweg & Sohn, Germany, 1935.
4. Sagan, H.: "Boundary and Eigenvalue Problems in Mathematical Physics," John Wiley & Sons, Inc., New York, 1961.
5. Webster, A. G.: "Partial Differential Equations," S. B. Teubner, Leipzig, 1927.
6. Lamb, H.: "Hydrodynamics," Cambridge University Press, New York, 1932.
7. Struble, R. A: "Nonlinear Differential Equations," McGraw-Hill Book Company, Inc., New York, 1962.
8. Cunningham, W. J.: "Introduction to Nonlinear Analysis," McGraw-Hill Book Company, Inc., New York, 1958.
9. Stoker, J. J.: "Nonlinear Vibrations," Interscience Publishers, Inc., New York, 1950.
10. Truxal, J. G.: "Automatic Feedback Control System Synthesis," McGraw-Hill Book Company, Inc., New York, 1955.
11. Coddington, E. A., and N. Levinson: "Theory of Ordinary Differential Equations," McGraw-Hill Book Company, Inc., New York, 1955.
12. Kumke, E.: "Differential Gleichungen reeler Funktionen," Akademische Verlagsgesellschaft Geest & Portig KG, Leipzig, 1930.
13. Lepschetz, S.: "Lectures on Differential Equations," Princeton University Press, Princeton, N.J., 1946.
14. Kaplan, W.: "Ordinary Differential Equations," Addison-Wesley Publishing Company, Inc., Reading, Mass., 1958.
15. Bellman, R. E.: "Stability Theory of Differential Equations," McGraw-Hill Book Company, Inc., New York, 1953.
16. A Papoulis: "The Fourier Integral and Its Applications to Linear Systems," McGraw-Hill Book Company ,Inc., New York, 1962.
17. Carslaw, H. S.: "Introduction to the Theory of Fourier Series and Integrals," 3d ed., Dover Publications, Inc., New York, 1930.
18. Churchill, R. V.: "Operational Mathematics," 2d ed., McGraw-Hill Book Company, Inc., New York, 1944.
19. Weber, E.: "Linear Transient Analysis," vol. II, John Wiley & Sons, Inc., New York, 1956.
20. Gardner, M. F., and J. S. Barnes: "Transients in Linear Systems," vol. I, John Wiley & Sons, Inc., New York, 1942.
21. Goldman, S.: "Transformation Calculus and Electrical Transients," Prentice-Hall, Inc., Englewood Cliffs, N.J., 1949.

22. Cheng, D. L.: "Analysis of Linear Systems," Addison-Wesley Publishing Company, Inc., Reading, Mass., 1959.
23. Pipes, L. A.: "Applied Mathematics for Engineers and Physicists," 2d ed., McGraw-Hill Book Company, Inc., New York, 1958.
24. Pfeiffer, P. E.: "Linear Systems Analysis," McGraw-Hill Book Company, Inc., New York, 1961.
25. Scott, E. J.: "Transform Calculus with an Introduction to Complex Variables," Harper & Row, Publishers, Incorporated, New York, 1955.
26. Evans, W. R.: "Control-system Dynamics," McGraw-Hill Book Company, Inc., New York, 1954.
27. Solodovnickoff, V. V.: "Introduction to the Statistical Dynamics of Automatic Control Systems," Dover Publications, Inc., New York, 1960.
28. Tsien, H. S.: "Engineering Cybernetics," McGraw-Hill Book Company, Inc., New York, 1954.
29. Seifert, W. W., and C. W. Steeg: "Control Systems Engineering," McGraw-Hill Book Company, Inc., New York, 1960.
30. Smith, O. J. M.: "Feedback Control Systems," McGraw-Hill Book Company, Inc., New York, 1958.
31. Chang, S. S.: "Synthesis of Optimum Control Systems," McGraw-Hill Book Company, Inc., New York, 1961.
32. Kipiniak, W.: "Dynamic Optimization Use Control," John Wiley & Sons, Inc., New York, 1961.
33. Gibson, J. E. (ed.): "Proceedings of Dynamic Programming Workshop," Purdue University, 1961.
34. Ragazzini, J. R., and Gene F. Franklin: "Sampled-data Control Systems," McGraw-Hill Book Company, Inc., New York, 1958.
35. Jury, E. J.: "Sampled Data Control Systems," John Wiley & Sons, Inc., New York, 1958.
36. McRuer, Graham D.: "Analysis of Nonlinear Control Systems," John Wiley & Sons, Inc., New York, 1961.
37. Cosgriff, R. L.: "Nonlinear Control Systems," McGraw-Hill Book Company, Inc., New York, 1958.
38. Ku, Y. H.: "Analysis and Control of Nonlinear Systems," The Ronald Press Company, New York, 1958.
39. Setov, J.: "Stability in Nonlinear Control Systems," Princeton University Press, Princeton, N.J., 1961.

Section 9

CONTROL-SYSTEM APPLICATIONS

By

K. W. KOHLMEYER, B.M.E., *Senior Engineer, Surface Armament Division, Sperry Gyroscope Company, Great Neck, N.Y.*

V. J. McDONOUGH, M.S., *Engineering Section Head, Surface Armament Division, Sperry Gyroscope Company, Great Neck, N.Y.*

CONTENTS

9.1. INTRODUCTION

This section presents various methods of combining components in designing closed-loop control systems.

The initial concern of the controls engineer is the physical characteristics of the load (e.g., inertia); static, Coulomb, and viscous friction; and the required maximum velocity and acceleration. From these considerations, the maximum torque and power requirements can be ascertained. Next, the static and steady-state dynamic-error coefficients or open-loop gain of the servo are determined. It is important at this point to investigate the overall system complex in which the servo is to be employed and to make certain that the static- and dynamic-error requirements are neither too stringent nor too loose. Unnecessary accuracy can be costly when selecting components.

Upon completion of the preliminary studies, consideration is directed toward methods of mechanization, where the factors of prime concern are cost, weight, and volume. Once the method of mechanization has been established, a theoretical analysis of system performance can begin.

9.2. ERROR ANALYSIS

The two errors of primary concern in design analysis are static error and dynamic error. In some servo applications, the error associated with noise is also of great importance.

(a) Static Error

The static error of a servo is the error which exists when the servo is at rest. This error results from error-detector resolution capabilities, torque-to-error gain of the

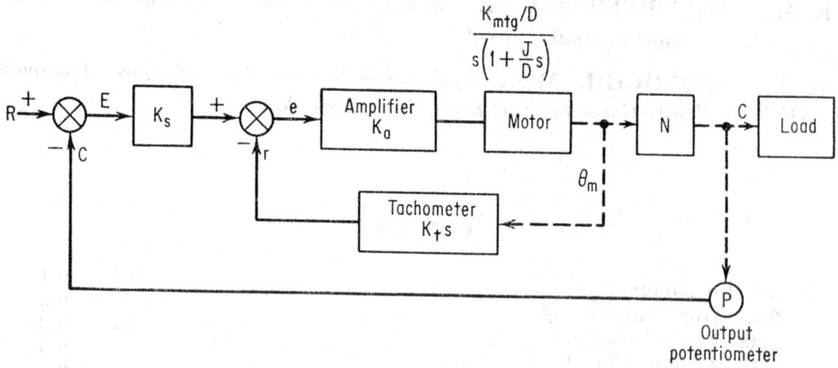

FIG. 9.1. General type 1 servo block diagram.

forward loop, and gearing inaccuracies. For very stringent static-accuracy requirements, error-detector resolution and gearing errors can be minimized by selection of precision components. The degree of precision is dictated by the limitations imposed on the torque-to-error gain of the servo. The maximum available torque at the controlled shaft is limited by the size of the prime mover and the gear ratio. The maximum available torque for a particular error is also limited by the gain of the amplifier driving the controller, because for linear servo operation, the amplifier must operate unsaturated in the range of anticipated variation of the reference or input command signal.

Static accuracy resulting from torque-to-error gain can be determined as follows, referring to Fig. 9.1:

The magnitude of the error E necessary to deliver sufficient torque to cause the load to move is to be determined. The following expression relates static error to the ratio of load-torque to system-torque gain:

$$E = \frac{T_L}{K_s K_a K_{mtg} N} \tag{9.1}$$

where T_L = total load torque (including friction torque and any steady load torque independent of speed)

K_s = loop-error-detector sensitivity, volt/rad of controlled shaft displacement relative to reference input

K_{mtg} = motor-torque gradient, delivered torque per volt of error on control field
K_a = amplifier gain, volts/volt
E = error, rad
N = gear ratio, $N \gg 1$

(b) Dynamic Error

When determining the dynamic accuracy of a servo, the frequency-sensitive elements must be considered in addition to the static-gain constants. Again referring to Fig. 9.1, an expression can be obtained relating the error E to the input or reference signal R and its derivatives. The following procedure may be used to determine the error coefficients for all the derivatives, and thus the total dynamic error of a closed-loop servo system:

$$\text{Error } E = R - C \tag{9.2}$$

where R = reference input signal, rad
C = output position, rad

Dividing Eq. (9.2) by R,

$$E/R = 1 - C/R \tag{9.3}$$

and

$$C/R = G/(1 + G) \tag{9.4}$$

where G = major open-loop transfer function when feedback is unity. Substituting Eq. (9.4) into Eq. (9.3),

$$E = R[1/(1 + G)] \tag{9.5}$$

where

$$G = G'/(1 + G'H)$$

where G' = transfer function of minor loop
H = transfer function of minor-loop feedback

That is,

$$G' = \frac{K_a K_m}{s(1 + \tau_m s)} \tag{9.6}$$

$$H = K_t s \tag{9.7}$$

where K_m = motor gain, volts per rad/sec = K_{mtg}/D
s = Laplace-transform variable
τ_m = motor time constant, sec = J/D
K_t = tachometer sensitivity, volts per rad/sec
D = motor viscous-damping coefficient, torque per rad/sec, obtainable from motor torque-speed curves

Substitution of Eqs. (9.6) and (9.7) into Eq. (9.5) and multiplying by $1/N$ and K_s yields

$$G = \frac{K_s K_a K_m}{N(K_a K_t K_m + 1)} \frac{1}{s[1 + (\tau_m/1 + K_a K_t K_m)s]} \tag{9.8}$$

Setting

$$K = \frac{K_s K_a K_m}{N(K_a K_t K_m + 1)} \quad \text{and} \quad \tau = \frac{\tau_m}{1 + K_a K_t K_m}$$

in Eq. (9.8), we obtain

$$G = \frac{K}{s(1 + \tau s)} \tag{9.9}$$

Substituting Eq. (9.9) for G into Eq. (9.5) yields

$$E = \frac{R}{K} \frac{\tau s^2 + s}{(\tau/K)s^2 + s/K + 1} \tag{9.10}$$

Dividing numerator by denominator results in the following Maclaurin expansion:

$$E = (1/K)Rs + [(K\tau - 1)/K^2]Rs^2 + (1/K^3)Rs^3 \tag{9.11}$$

Returning to the time domain,

$$E = (1/K)(dR/dt) + [(K_T - 1)/K^2](d^2R/dt^2) + (1/K^3)(d^3R/dt^3) \quad (9.12)$$

where K = velocity-error coefficient, rad/sec per rad = K_v
$K^2/(K_T - 1)$ = acceleration-error coefficient, rad/sec² per rad = K_a
K^3 = error coefficient of third derivative

Equation (9.12) is representative of a type 1 or velocity servo.

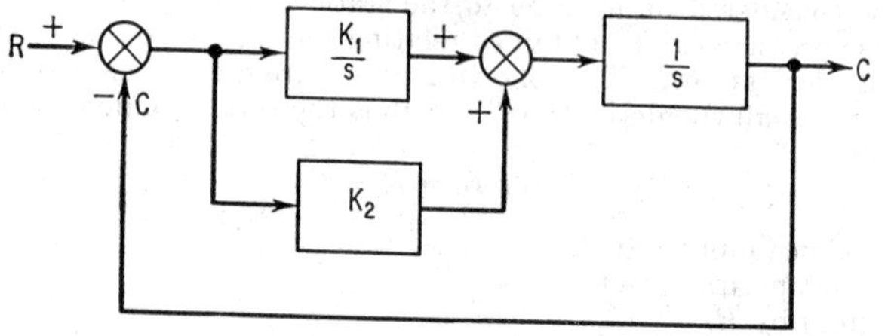

FIG. 9.2. Type 2 servo simplified block diagram.

Figure 9.2 is a simplified block diagram of a type 2 or acceleration servo. The open-loop transfer function

$$G = (K_1/s + K_2)(1/s) = \frac{K_1[1 + (K_2/K_1)s]}{s^2} \quad (9.13)$$

The equations relating error to input are

$$E = \frac{Rs^2}{s^2 + K_2 s + K_1} \quad (9.14)$$

and

$$E = (1/K_1)(d^2R/dt^2) - (K_2/K_1^2)(d^3R/dt^3) \quad (9.15)$$

where K_1 = acceleration-error coefficient = K_a
K_1^2/K_2 = error coefficient of third derivative

Note that position error is eliminated in Eq. (9.12) and the velocity error is eliminated in Eq. (9.15). The advantage, therefore, of a type 2 servo is that the steady-state dynamic error is reduced because of the elimination of the error contribution of the first derivative.

9.3. INSTRUMENT SERVOS

This type of servo is primarily employed in analog-computing devices to drive either mechanical or electrical components. The static and dynamic requirements for this servo are usually very stringent, thus requiring very accurate components and low velocity- and acceleration-error coefficients. In order to achieve these requirements, careful consideration must be given motor selection, reflected inertia of gearing and load to the motor shaft, and friction of the gear train. Some applications of instrument servos are discussed below.

(a) Servo Integrators

The servo integrator is frequently employed in analog computers to perform integration with respect to time. Its effectiveness over the desired frequency range is essentially determined by the steady state and dynamic performance. In the steady state, for a constant input reference voltage R, the output or controlled shaft C should rotate at some constant angular velocity. Under dynamic conditions, the phase shift of the output relative to the input should be 90° throughout the desired band of frequencies.

Tachometer Integrator. Figure 9.3 illustrates a motor-tachometer integrator system. The relationship connecting the frequency-sensitive elements of the loop is

$$E = R - K_t Cs \tag{9.16}$$

$$E = \frac{Cs(1 + \tau_1 s)}{K_a K_m} \tag{9.17}$$

where $\tau_1 = J_T/D_m$
J_T = total inertia at motor shaft
T_m = motor torque
N_m = motor speed
D_m = motor viscous-damping coefficient (e.g., in.-oz per rad/sec)

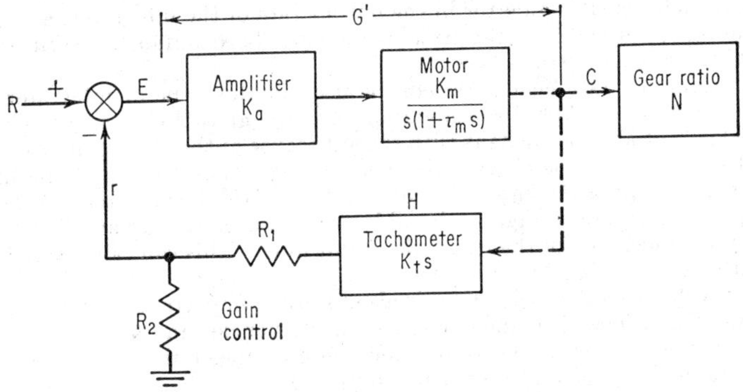

FIG. 9.3. Block diagram of tachometer integrator.

Substituting Eq. (9.17) into Eq. (9.16) yields

$$R = Cs\frac{1 + \tau_1 s}{K_a K_m} + K_t \tag{9.18}$$

and

$$\frac{C}{R} = \frac{K}{s(1 + \tau s)} \tag{9.19}$$

where

$$\tau = \frac{\tau_1}{1 + K_t K_a K_m} = \frac{1}{\omega}$$

$$K = \frac{K_a K_m}{1 + K_t K_a K_m}$$

In practice, it is the objective to make $K_t K_a K_m \gg 1$ and $\tau_1 \ll 1$, so that Eq. (9.19) can be written

$$C/R = (1/s)(1/K_t) \tag{9.20}$$

where $1/K_t = K$ = gain constant, rad/sec of output per volt of input

A more general approach can be illustrated as follows:

Let

$$\frac{C}{E} = G' = \frac{K_a K_m}{s(1 + \tau_m s)} \tag{9.21}$$

and

$$r/C = H = K_t s \tag{9.22}$$

then

$$C/R = G = G'/(1 + G'H) \tag{9.23}$$

Rearranging Eq. (9.23) as follows: $G = (G'H/1 + G'H)(1/H)$, it follows that, when $G'H \gg 1$, the system response follows $1/H$, and when $G'H \ll 1$, the system response follows G'.[1,9]

Inspection of Eq. (9.19) indicates the effect of K, τ, on the bandwidth, when K_a, K_t, and τ_1 are increased or decreased. As mentioned previously, the primary consideration is to design the integrator so that ω contributes negligible phase shift over the band of frequencies to be integrated in order to validate Eq. (9.20). This is the limiting parameter when determining the fidelity or effectiveness of integration. As the feedback voltage K_t is reduced, the gain of C/R increases, thus increasing the bandwidth and the response of the system to changes in R. However, ω decreases, thus contributing additional phase shift. In practice the gain is determined by the nature of the input signal and the scale factor of integration required.

The accuracy of an integrator of the type described is very much dependent upon the accuracy of the tachometer. In systems requiring extreme accuracy, linearity of the tachometer gain must be maintained under all operational environments, and therefore it is often necessary to provide temperature compensation. Nonlinearity of the tachometer creates an error in the output rate of the integrator, and therefore the operating speed range of the tachometer must be maintained within the linear portion of the characteristic curve.

Either a-c or d-c rate generators can be used in this type of integrator. When using a-c tachometers careful consideration must be given to phasing. The output phase angle is not always constant with speed because of the change in leakage resistance and reactance. This characteristic can decrease the fidelity of the integrator because of the additional phase-shift contribution of the tachometer. Another consideration is quadrature voltage produced by the presence of third and fifth harmonics created by nonlinearities of the magnetic-circuit iron. This can cause amplifier saturation without producing useful results.[3]

The null voltage of the tachometer is also of prime importance because these voltages contain both harmonic and fundamental frequencies and are present when the tachometer speed is zero. The harmonic components of the null voltage can be filtered, but the fundamental cannot be filtered.[3]

A d-c rate generator has several advantages and disadvantages when compared with an a-c rate generator. The phasing and nulling problems associated with alternating current are eliminated in the d-c tachometer. One significant disadvantage of direct current is the friction introduced by the brushes. This factor is very important when the integrator is required to rotate at very slow speeds. Brush jump is another disadvantage which becomes evident at high velocities.[3]

Ball-and-disk Integrator. This mechanical integrator is frequently used in analog computers to perform integration and differentiation.[4]

The transfer function of the ball and disk integrator is

$$\phi(s)/C(s) = K_i/s \tag{9.24}$$

where $\phi(s)$ = integrator output
$C(s)$ = input to integrator
K_i = scale factor, rad/sec per in.

Figure 9.4 illustrates a method of mechanizing the ball-and-disk integrator to perform differentiation. While the minor tachometer loop is identical with the one described previously, the tachometer in this case is used for stabilization rather than computation, and therefore need not be of high precision.

Bandwidth is an important consideration when designing a differentiator. The integrator has inherent smoothing whereas the amount of smoothing obtained in a differentiation is purely a function of design. The initial consideration is that the bandwidth must be sufficiently broad to differentiate effectively over the expected frequency range of the input. A perfect differentiator has a 90° phase lead over the frequency spectrum from zero to infinity. Thus input noise at high frequencies is greatly amplified until, at infinite frequency, the amplitude of the output is also infinite. Therefore, the bandwidth must be limited to the band of frequencies over which differentiation is required. This requirement is dictated by the input information rates or dynamics.

Referring to Fig. 9.4, the initial objective is to determine the transfer function of the output of the servo C to the input R. Transfer functions are often determined experimentally. A steady-state measurement is made at the output and compared with a sinusoidal input. The output of the mechanical differential positions the rotor of the rotary transformer as a function of the difference E between the angular motion of the differential inputs R and ϕ. A servo error e is generated as a result of the displacement of the transformer rotor with respect to the stator. The servo positions the input C of the ball-and-disk integrator and the stator of the transformer until the errors E and e are driven to zero.

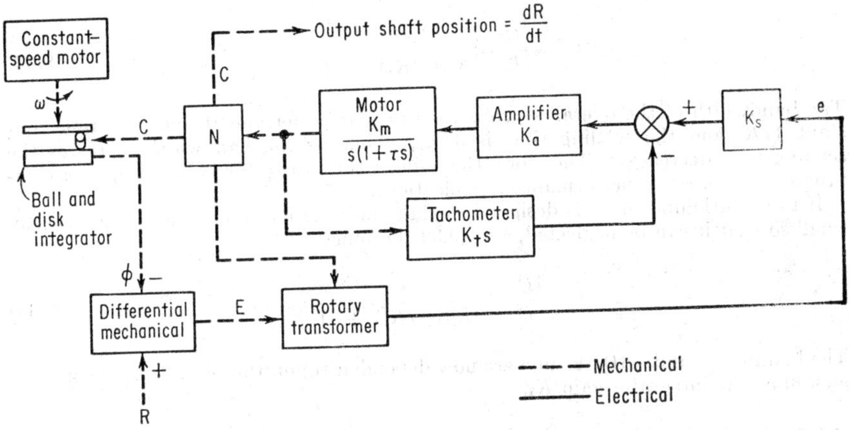

FIG. 9.4. Block diagram of servo differentiator.

The derivation of the transfer function relating C/R or the closed-loop response is as follows:

$$C/e = NKK_s/s(1 + \tau_2 s) = G_1 \tag{9.25}$$

For the closed loop (minor loop),

$$\frac{C}{E} = \frac{G_1}{1 + G_1 H_1} = G_2 \tag{9.26}$$

where $H_1 = 1$. Substituting for G_1 and H_1 in Eq. (9.26) yields

$$G_2 = \frac{1}{(\tau_2/KK_sN)s^2 + s/KK_sN + 1} \tag{9.27}$$

Referring to Eq. (9.19) to obtain the exact expression for τ_2 and K and substituting in Eq. (9.27) results in the following:

$$G_2 = \frac{1}{(\tau_1/K_AK_mK_sN)s^2 + (K_tK_AK_m/K_AK_mK_sN)s + 1} \tag{9.28}$$

Employing the approximations $\tau_1 \ll 1$, $K_AK_mK_sN \gg 1$, Eq. (9.28) becomes

$$G_2 = 1/(\tau_3 s + 1) \tag{9.29}$$

where $\tau_3 = K_t/K_sN$. Finally, the closed loop of the complete system can be expressed as follows:

$$C/R = G_2/(1 + G_2H_2) \tag{9.30}$$

where $H_2 = \phi(s)/C(s) = K_i/s$. Making the necessary substitutions in Eq. (9.30) and simplifying,

$$\frac{C}{R} = \frac{s}{K_i[(K_t/K_sK_iN)s^2 + s/K_i + 1]} \tag{9.31}$$

Thus the output shaft position of the servo is the derivative of the input.

Referring to Eq. (9.29), if the servo parameters are chosen such that the time constant τ_3 is much smaller than $1/K_i$, the positioning servo transfer function becomes essentially equal to unity and Eq. (9.31) reduces to

$$\frac{C}{R} = \frac{s}{K_i[(s/K_i) + 1]} \tag{9.32}$$

The bandwidth of the overall servo is then primarily dependent upon the time constant $1/K_i$ and the settling time is $3/K_i$. Thus as the bandwidth increases the settling time decreases. Therefore, the gain of the integrator is based upon a compromise of these two performance specifications.

If the positioning servo is designed so that the time constant τ_3 is not sufficiently small so that it can be neglected, Eq. (9.32) becomes

$$\frac{C}{R} = \frac{s}{K_i[(\tau_3 s/K_i) + 1]} \tag{9.33}$$

The bandwidth and settling time are now dependent upon the servo time constant τ_3 as well as the integrator gain K_i.

(b) Inertially Damped Resolver Servo

Resolver servos are used extensively in analog computers to perform coordinate transformations and rotation of coordinate axes. The system shown in Fig. 9.5 illustrates a method for converting from rectilinear to polar coordinates. The system utilizes an inertially damped servomotor whose transfer function $\dfrac{K_N(1 + s/\omega_2)}{s(1 + s/\omega_1)(1 + s/\omega_3)}$ is written

$$\frac{\theta_0}{E} = \frac{(K/D_m)[(J_{fw}/D_{fw})s + 1]}{s[(JJ_{fw}/D_mD_{fw})s^2 + (J_{fw}/D_{fw} + J/D_m + J_{fw}/D_m)s + 1]} \tag{9.34}$$

where J = total inertia at the motor shaft (includes motor rotor, reflected gearing, and load)
J_{fw} = flywheel inertia
D_m = motor viscous-damping coefficient
D_{fw} = flywheel damping coefficient
K = motor torque gradient
E = control-signal input

The transfer function given by Eq. (9.34) can be approximated by examining the behavior of the system at low and high input frequencies. At low frequencies the flywheel θ' follows the motor shaft θ_0. The first corner frequency then becomes

$$\omega_1 = D_m/(J + J_{fw}) \tag{9.35}$$

At high frequencies, θ' can no longer follow θ_0 and can therefore be considered at rest.[3] Therefore,

$$\omega_3 = (D_m + D_{fw})/J \tag{9.36}$$

Equation (9.34) can then be expressed

$$\frac{\theta_0}{E} = \frac{K_m(1 + s/\omega_2)}{s(1 + s/\omega_1)(1 + s/\omega_3)} \tag{9.37}$$

where $K_m = K/D_m$

$\omega_2 = D_{fw}/J_{fw}$

Referring to Fig. 9.5, the servo drives one output from the resolver to zero. When this is accomplished, the shaft position of the servo represents θ and the voltage from the other output of the resolver represents the magnitude of R.

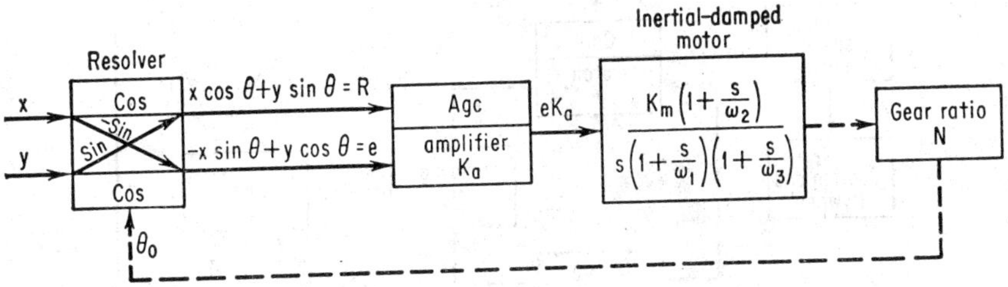

FIG. 9.5. Resolver servo employing inertial-damped motor.

The equations representing the resolver outputs during steady state are[3,5]

$$R = x \cos \theta + y \sin \theta \tag{9.38}$$
$$-x \sin \theta + y \cos \theta = 0 \tag{9.39}$$

Consideration must be given resolver sensitivity. It is desirable to evaluate the angular error for small perturbations of θ about the zero position. If it is assumed that the servo shaft is rotated by $\Delta\theta$, the resolver equations are

$$e = -x \sin (\theta + \Delta\theta) + y \cos (\theta + \Delta\theta) \tag{9.40}$$
$$R = x \cos (\theta + \Delta\theta) + y \sin (\theta + \Delta\theta) \tag{9.41}$$

Expanding Eqs. (9.40) and (9.41) and making the small angle approximation yields

$$e = -x \sin \theta + y \cos \theta - \Delta\theta(x \cos \theta + y \sin \theta) \tag{9.42}$$
$$R = x \cos \theta + y \sin \theta + \Delta\theta(-x \sin \theta + y \cos \theta) \tag{9.43}$$

Substituting in Eqs. (9.38) and (9.39) results in

$$e = -\Delta\theta R \qquad \text{or} \qquad e/\theta = -R \tag{9.44}$$
$$R = x \cos \theta + y \sin \theta \tag{9.45}$$

The condition described by Eq. (9.44) is undesirable. As R increases, loop gain increases and the servo has good response. As R decreases, however, the loop gain decreases and the servo becomes very sluggish. This condition is remedied by incorporating an automatic gain control (AGC) circuit in the servoamplifier. The voltage output representing R from the resolver varies the amplifier gain so that it is inversely proportional to R. Thus the amplifier output has constant sensitivity (volts/rad).

It is important to note that, because a resolver is a transformer device, a-c voltages must be used and careful consideration be given to phase shift of the a-c carrier in order to prevent generation of error and quadrature noise. For example, if two a-c signals are added with one slightly phase-shifted by an amount ϕ,

$$E = E_1 \sin \omega t + E_2 \sin (\omega t + \phi) \tag{9.46}$$

Making small angle approximations yields

$$E = (E_1 + E_2) \sin \omega t + E_2\phi \cos \omega t \tag{9.47}$$

The quadrature component $E_2\phi \cos \omega t$, if large enough, may cause amplifier saturation. In addition, if subsequent phase shifting of this signal should occur, an in-phase signal will be obtained, causing system error.

(c) Clutch Servo[1]

This type of servo employs a clutch as the actuator instead of a conventional servomotor. The system prime mover is an electric motor operating continuously

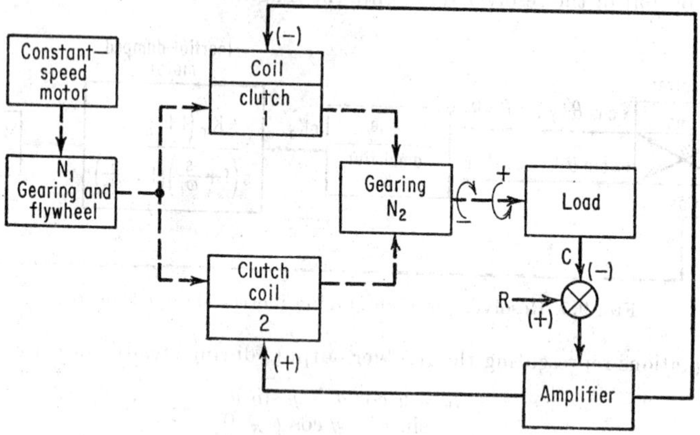

Fig. 9.6. Schematic of clutch servo.

at constant speed. The clutch acts as a torque transmitter and controls the coupling between the constant-speed drive motor and the load. A servoamplifier controls the clutch, and the degree of coupling between the input and output shafts is proportional to the magnitude of the error signal. Figure 9.6 is a general block diagram of a clutch-operated servo. The function of the drive motor is to supply power to the system upon demand, thus making the servo completely independent of the frequency-sensitive characteristics of the motor.

A clutch servo can be either a linear (continuous control) or nonlinear (on-off control) system, depending upon the type of clutch employed and the method of operation. For linear operation the clutch coupling must vary directly with the magnitude of the control signal. Figure 9.7 shows the relationship between control current and torque for a magnetic-particle clutch. A nonlinear clutch (friction disk) can be linearized by intentionally introducing dither. Improved linearity of a clutch (eddy-current, hysteresis, magnetic-fluid, and magnetic-particle type) can be achieved by push-pull operation and biasing. Biasing is accomplished by maintaining a small quiescent current in the clutch coils when there is zero control signal at the input to the control amplifier. Figure 9.8 illustrates the operation of a pair of biased magnetic-particle clutches. The current in the coils produces a bidirectional closed-loop system; the clutch outputs are in opposition. Therefore, the algebraic sum of the torques is zero and the load experiences no motion. However, when an error signal appears at the input to the

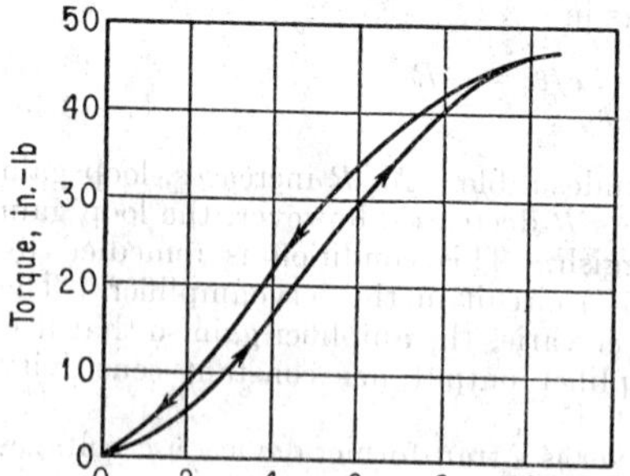

Fig. 9.7. Torque vs. current for single-magnetic-particle clutch showing hysteresis.

amplifier, it causes a current unbalance in the coils, which creates an unbalanced torque. The direction and speed of load rotation depend upon the polarity and magnitude of the error signal. This method of push-pull clutch operation essentially eliminates gearing backlash. A disadvantage of this type of mechanization is that the clutches dissipate energy under no-load conditions.[6]

After the inertia, friction, and maximum desired speed and acceleration characteristics of the load are determined, equations can be formulated to determine the power requirements of the prime mover and clutch, and also the optimum gear ratio between the clutch output member and the load.

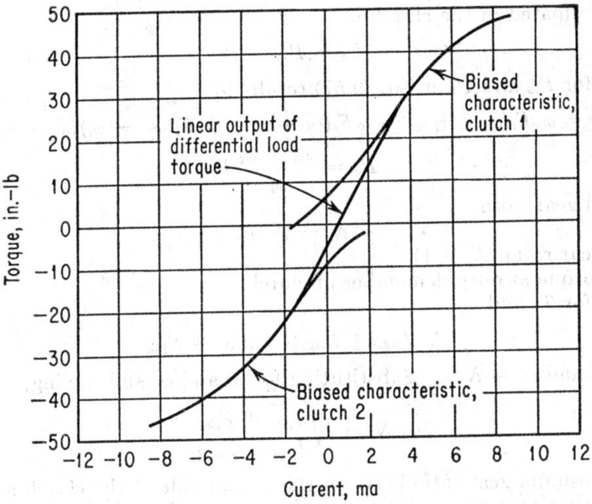

FIG. 9.8. Biased operation of pair of clutches showing net torque is linearly proportional to differential current.

Power Required to Drive Load. The power requirements for overcoming friction may be expressed

$$P_f = T_f\omega_l \tag{9.48}$$

where T_f = effective friction torque at load
ω_l = load velocity

The power required to accelerate the load

$$P_a = T_a\omega_l \tag{9.49}$$

where T_a = torque required at load shaft = $\alpha_l J_l$
α_l = load acceleration
J_l = total inertia at load shaft

The total power required to drive the load is then

$$P_l = P_f + P_a \tag{9.50}$$

Substituting for P_f and P_a,

$$P_l = T_f\omega_l + T_a\omega_l \tag{9.51}$$

where $T_f = F\omega_l$. Note that the friction torque coefficient F must include all viscous and kinetic friction in the system. Substituting for T_f and T_a in Eq. (9.51),

$$P_l = F\omega_l^2 + J_l\alpha_l\omega_l = F_l\omega_l(\omega_l + t_l\alpha_l) \tag{9.52}$$

where $\tau_l = J_l/F_l$ = time constant of the load.

Energy Transfer from Motor to Load. The torque on the input side of the clutch T_i and the total torque required by the system T_0 are

$$T_i = P_m/\omega_m \tag{9.53}$$
$$T_0 = F_l\omega_l + J_l\alpha_l \tag{9.54}$$

where P_m = power delivered by the motor
ω_m = motor speed

Substituting,

$$P_m/\omega_m = F_l\omega_l + J_l\alpha_l \tag{9.55}$$
$$P_m = \omega_m(F_l\omega_l + J_l\alpha_l) = \omega_m F_l(\omega_l + t_l\alpha_l) \tag{9.56}$$

The power dissipated in the clutch is

$$P_c = P_m - P_l \tag{9.57}$$

Substituting for P_m and P_l in Eq. (9.57) results in

$$P_c = \omega_m F_l(\omega_l + \tau_l\alpha_l) - \omega_l F_l(\omega_l + \tau_l\alpha_l) = F_l(\omega_m - \omega_l)(\omega_l + \tau_l\alpha_l) \tag{9.58}$$
$$P_c = \frac{\omega_m - \omega_l}{\omega_l} P_l \tag{9.59}$$

The optimum gear ratio

$$N = T_l/T_c \tag{9.60}$$

where N = gear ratio ($N > 1$)
T_c = torque at output member of clutch

Substituting for T_0 and T_l,

$$N(J_c\alpha_c + F_c\omega_c) = J_l\alpha_l + F_l\omega_l \tag{9.61}$$

Let $\alpha_c = N\alpha_l$ and $\omega_c = N\omega_l$. Substituting for α_c and ω_c and solving,

$$N = \sqrt{\frac{J_l\alpha_l + F_l\omega_l}{J_c\alpha_l + F_c\omega_l}} \tag{9.62}$$

This is the optimum gear ratio between the output side of the clutch and the load.[6]

After the prime mover and clutch have been selected on the basis of power requirements, an analysis of the overall servo loop can be made. For a pure inertia load, referring to Eq. (9.62), the maximum load acceleration is achieved by making the reflected inertia of the load (J_l/N^2) equal to the inertia of the output side of the clutch.

Consideration must also be given the input side of the clutch. The inertia of the drive motor and flywheel added to the reflected inertia of the input clutch member must be large compared with $2J_T$, where J_T is the inertia of the clutch output member plus the reflected inertia of gearing and load. In addition, the motor and flywheel should have sufficient reserve kinetic energy to drive the load during peak-demand transients. The maximum desired speed of the output and the allowable maximum speed of the input clutch member must also be carefully investigated when choosing the gear ratios.

The foregoing equations are general and can be applied to all clutches. Some modifications are necessary when considering pure inertia loads and negligible loads such as those encountered in most instrument servos.

(d) Magnetic-particle Clutch

Clutch Operation. The magnetic-particle clutch has numerous instrument and medium-power servo applications.

The clutch-coupling medium is a magnetic-particle mixture contained within the gap between a continuously rotating clutch body and a low-inertia output member. Energizing the clutch with direct current causes the magnetic particles to span the gap with multiple-particle chains. The shear strength of the coupling is proportional to the excitation current.

Under conditions of slip, slippage occurs only at the center of the gap, an immobile layer of particles protects each clutch surface. Because the clutch is designed for intermittent loads and not for continuous slip, the frictional heat generated during continuous slipping should be carefully considered.

Typical torque-speed characteristics of the magnetic-particle clutch are shown in Fig. 9.9. The inherent damping compared with the conventional servomotor is negligible and exhibited only by a slight downward slope from the initial point of maximum torque. At each value of differential input current ΔI, the applied torque is almost maximum until the limiting velocity is attained; then the accelerating torque disappears quickly. This characteristic makes the clutch servo similar in operation to a hydraulic servo.[7]

The magnetic-particle clutch offers extreme smoothness of operation over the entire range of magnetization.

Below the point of saturation of the magnetic current, torque output is a linear function of excitation except in the vicinity of zero excitation.

Output torque is essentially independent of relative clutch slippage.

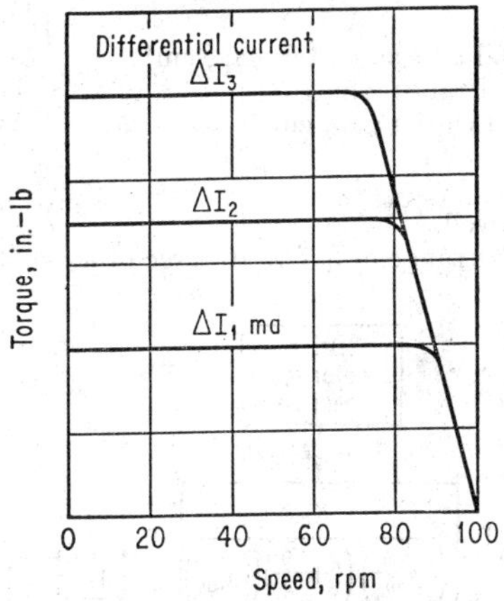

FIG. 9.9. Torque-speed characteristics for various differential currents.

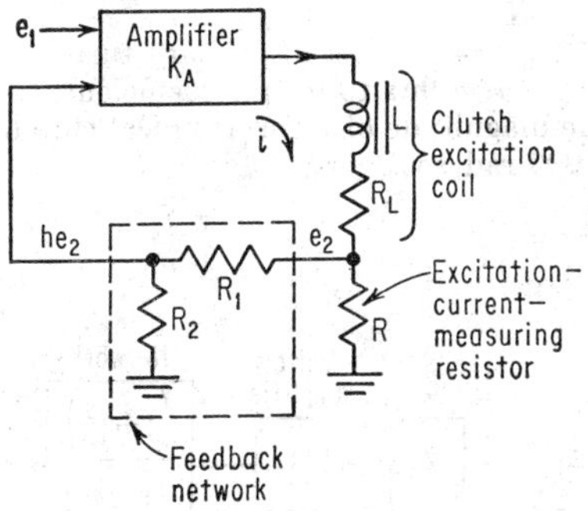

FIG. 9.10. Equivalent circuit for coil time-lag compensation.

Clutch Transfer Function. Figure 9.10 shows the equivalent circuit for the clutch-coil time-lag compensation. Without compensation the time lag is proportional to L/R_L. The function of the compensation circuit is to reduce this time constant sufficiently so that it is far removed from the servo operating frequency and therefore does not degrade the phase margin appreciably. The transfer function of the overall compensation circuit is then[8]

$$i/e = K_1/[R_L + R(1 + K_1h)] \tag{9.63}$$

Simplifying the above expression,

$$i/e = K_A/(1 + \tau_a s) \tag{9.64}$$

where R_L = internal coil resistance
K_1 = gain of voltage amplifier
L = excitation-coil self-inductance
$\tau_a = L/[R_L + R(1 + K_1h)] = 1/\omega_a$
$K_A = K_1/[R_L + R(1 + K_1h)]$
h = gain of current feedback network
R = excitation current measuring resistance

If the feedback term K_1h is large, the time constant τ_a for the coil and amplifier is small. R should be made as small as possible in order to minimize input power requirements. If pentodes with high plate resistance are used to drive the coils, the time constant

$$\tau_a = L/(R_p + R_L)$$

where R_p = plate resistance of pentode. τ_a can be of the order of 1.6×10^{-3} sec ($\omega_a = 628$ rad/sec). If a triode or transistor amplifier is used, the current feedback circuit shown in Fig. 9.10 helps to provide a stiff current source, thus maintaining τ_a small.

The second time lag τ_c results from the induced secondary current in the clutch rotor. The interaction of the driving current and induced secondary current produces the flux time lag

$$T/i = K_c/(1 + \tau_c s) \tag{9.65}$$

where T = shaft output torque = $\nu\phi$

$\tau_c = \dfrac{0.4\pi\mu A_g \times 10^{-8}}{l_1 R_s}$

ϕ = magnetic flux, maxwells

μ = permeability of clutch gap

A_g = magnetic circuit mean cross-sectional area of clutch gap, cm²

R_s = single-turn secondary equivalent resistance, ohms

l_1 = mean flux path length through total clutch gap, cm

$K_c = 0.4\pi\mu N A_g \nu / l_1$

N = number of primary turns

ν = flux to torque coefficient, ft-lb/maxwell

The magnitude of τ_c for a typical clutch is in the range of 3.5×10^{-3} sec or $\omega_c = 333$ to 200 rad/sec.

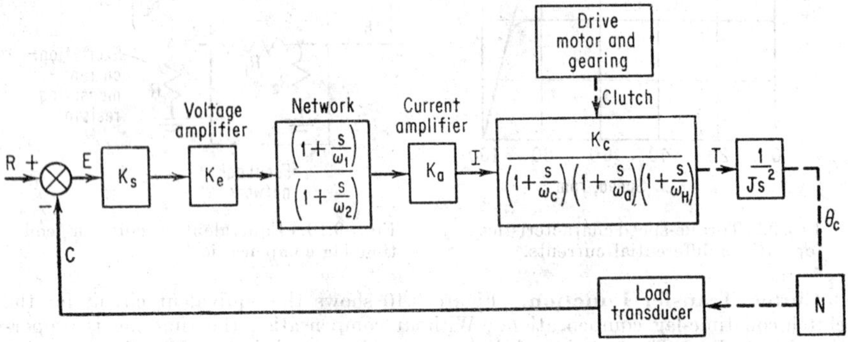

FIG. 9.11. Servo block diagram, magnetic-particle clutch.

The third lag to be considered results from the magnetic hysteresis in the magnetic particles. At low frequencies, hysteresis contributes a small fixed phase lag (10°) without attenuating the torque output. This lag must be added, however, when considering the phase margin. At high frequencies, it approximates a constant time delay. A typical value for the constant time delay τ_H is 1.5×10^{-3} sec. A more descriptive discussion of the effects of hysteresis on magnitude and phase angle can be found in ref. 6.

The overall clutch and amplifier transfer function describing the output torque for an input voltage is, from Fig. 9.11,

$$\frac{T}{e} = \frac{K_a K_c}{(1 + \tau_a s)(1 + \tau_c s)(1 + \tau_H s)} \tag{9.66}$$

Since $T = J_c s^2 \theta_c$, the transfer function relating output motion to voltage input to the amplifier is

$$\frac{\theta_c}{E} = \frac{K_e K_A K_c / J_c}{s^2(1 + \tau_c s)(1 + \tau_a s)(1 + \tau_H s)} \tag{9.67}$$

where J_c = inertia of clutch output member

K_c = clutch gain, e.g., in.-lb/amp

The gain of the clutch (K_c/J_c, rad/sec² per amp) is very high. For instrument-type clutches, torque-inertia ratios of 3.5×10^5 rad/sec² and gains of 4×10^6 rad/sec² per amp are typical. The gain is reduced by any reflected inertia from gearing and load. However, the load on an instrument servo is usually negligible and consists of the error sensor or some data-transmitting component. Because the load is small, small-pitch gears can be used which also reduce the reflected inertia. The major reduction in gain results from the required gear ratio between clutch and error sensor.

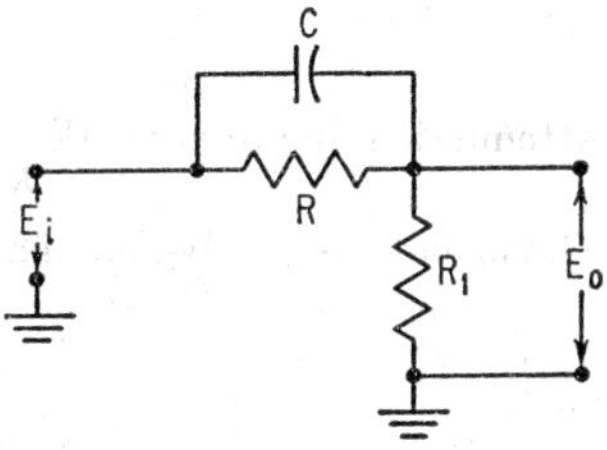

FIG. 9.12. Compensating network.

Servo Application and Stability Analysis. In order to employ the clutch previously described in a closed-loop system, some means of stabilization must be employed. The method used will depend upon the static and dynamic requirements placed on the servo. The clutch can be stabilized with the conventional tachometer feedback or with a series compensation network placed in the error channel. The method of stabilization discussed here will be a compensating network in series with the error signal. Figure 9.12 is a schematic of the network which has the following transfer function:

$$E_i = I\left[\frac{(1/Cs)R}{(1/Cs) + R} + R_1\right] \tag{9.68}$$

Simplifying,

$$\frac{E_0}{E_i}(s) = \frac{R_1}{R + R_1}\,\frac{1 + RCs}{1 + [R_1/(R + R_1)]Cs} \tag{9.69}$$

where $E_0 = IR_1$. In order to obtain maximum effectiveness from the lead term, the lag time constant $[R_1/(R + R_1)]Cs$ must be made as small as possible so that it contributes negligible phase shift at the crossover frequency. This is not achieved without limitations. The limitation is the attenuating characteristic of the network $R_1/(R + R_1)$. As the spread between the lead and lag is increased, attenuation of the loop gain is also increased. In addition, it is also desirable to assign reasonable values to the capacitor and resistors.

Example: Using the following clutch characteristics, a Bode plot can be constructed to determine the necessary time constants of the network: $T_c/J_c = 350{,}000$ rad/sec²; T_c max $= 1.5$ in.-lb; $J_c = 4.25 \times 10^{-6}$ in.-lb-sec²;

$$\omega_c = \text{maximum clutch speed} = 2{,}000 \text{ rpm}$$

i_c = control current for maximum torque = 85 ma; $K_c = \dfrac{T_c/J_c}{i_c} = 4.2 \times 10^6$ rad/sec² per amp; $\omega_a = 628$ rad/sec; $\omega_H = 670$ rad/sec.

Assuming the following characteristics for the servo: maximum operating acceleration of output shaft = 3 rad/sec²; minimum acceleration-error coefficient

$$K_a = 600 \text{ rad/sec per rad}$$

maximum operating velocity of output shaft of gearing = 5 rad/sec; minimum bandwidth = 8 cps. It is desirable to have the lead occur at least an octave before the crossover frequency of 8 cps (50.3 rad/sec). Using this information, in addition to the desired K_a, a Bode plot can be constructed to determine more accurately where the lead must be placed to obtain the desired bandwidth and K_a. One additional approximation can be made: a 50:1 spread can be assumed between the lead and lag break frequencies. Figure 9.13 shows a Bode plot of the servo. The gain margin, i.e., the magnitude by which the open-loop gain can increase before the phase lag is 180° (0° phase margin), is approximately 9 db. From the Bode plot it can be seen that lead time constant $\tau_1 = RC = 0.0715$ sec; lead break frequency $\omega_1 = 1/\tau_1 = 14$ rad/

sec; lag time constant $\tau_2 = [R_1/(R_1 + R)]C = 0.00143$ sec; lag break frequency $\omega_2 = 700$ rad/sec.

Values for R, R_1, and C can now be obtained. Let $C = 0.25\ \mu f$

$$\tau_1 = 0.25R = 0.0715 \text{ sec}$$
$$R = 286{,}000 \text{ ohms}$$
$$R_1 = R/50$$
$$R_1 = 5{,}730 \text{ ohms}$$

Attenuation due to network

$$K_{nw} = R_1/(R + R_1) = 0.0196$$

The open-loop steady-state gain is

$$K_a = \frac{K_e K_a K_s K_c K_{nw}}{N} \tag{9.70}$$

$$\text{Gear ratio } N = \omega_c/\omega_{os}$$

where ω_c = maximum clutch speed = 2,000 rpm = 209 rad/sec
ω_{os} = maximum desired output shaft speed = 5 rad/sec

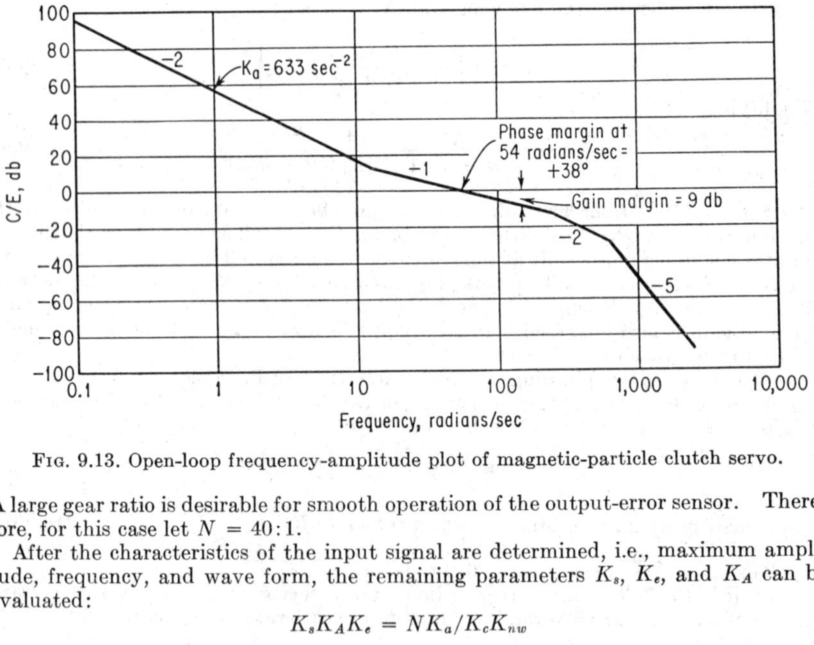

FIG. 9.13. Open-loop frequency-amplitude plot of magnetic-particle clutch servo.

A large gear ratio is desirable for smooth operation of the output-error sensor. Therefore, for this case let $N = 40{:}1$.

After the characteristics of the input signal are determined, i.e., maximum amplitude, frequency, and wave form, the remaining parameters K_s, K_e, and K_A can be evaluated:

$$K_s K_A K_e = N K_a / K_c K_{nw}$$

The amplifier should be designed so that it does not saturate before the maximum anticipated error-signal amplitude is reached. The sensitivity voltage is chosen so that the minimum error-signal amplitude is sufficiently high so as to be out of the noise. The overall open-loop transfer function for this servo is

$$\frac{C}{E} = \frac{633 \text{ sec}^{-2}(1 + s/14)}{s^2(1 + s/250)(1 + s/628)(1 + s/667)(1 + s/700)} \tag{9.71}$$

The closed loop can be approximated by

$$\frac{C}{R} = \frac{1 + s/14}{s^3/250K_a + s^2/K_a + s/14 + 1} \tag{9.72}$$

By substituting j_w for s, the following expression is obtained:

$$\frac{C}{R}(j\omega) = \frac{1 + j\omega/14}{-j\omega^3/250K_a + \omega^2/K_a + j\omega/14 + 1} \tag{9.73}$$

At the lower frequencies the first two terms in the denominator can be neglected and the amplitude ratio is thus unity. The ratio and phase shift can be evaluated at any critical frequency to determine the magnitude of peaking.

9.4. POWER SERVOS

Power servos operate in the medium and high bands of the power spectrum. In contradistinction to instrument servos, the load magnitude is of prime consideration in the design of power servos.

(a) Rotating-amplifier Servo

The rotating-amplifier servo is classified under three different designations: Ward-Leonard system, Rototrol or Regulex generator, and the amplidyne generator. The

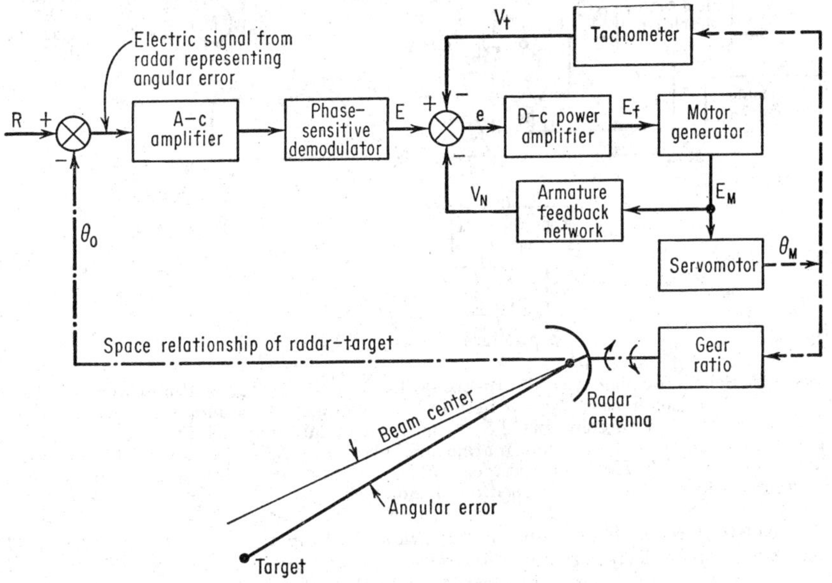

FIG. 9.14. Block diagram of rotating-amplifier antenna angle-drive servo.

basic operation of these systems is very similar. Each type consists of an induction motor-generator set operating at 3,600 rpm (1,800 rpm for the larger size), d-c armature-controlled servomotor with constant current applied to the field, and an electronic or magnetic power amplifier to drive the field winding of the generator. The amplification characteristics of these systems result from the large ratio of generator armature current to generator field control current; a small current flowing in the field winding of the generator field produces a large current in the generator armature. The d-c servomotor armature is connected in series with the generator armature; therefore, a small field current produces a large torque at the motor output shaft. When employing this type of drive to move very large loads, such as exceptionally large radar antennas subjected to large wind loads, it sometimes is necessary to employ more than one drive. Elimination of backlash can also be accomplished by the push-pull configuration which was previously outlined in the discussion on the clutch servo.

Employing the rotating amplifier in a closed-loop system can be accomplished using conventional error-sensing devices. Stabilization of the loop can be achieved with either electrical networks, tachometer feedback, or a combination thereof. Figure 9.14 shows a functional block diagram of a rotating-amplifier closed-loop servosystem. References 1, 3, 4, and 6 provide additional design information for the Amplidyne and Rototrol systems.

Ward-Leonard System Analysis. This system is a motor driven by a separately excited generator. Figure 9.15 is a schematic of the amplifier system. The control field power can vary from 0.5 to 10 per cent of the required load power. Compensating windings are usually used to cancel the internal magnetic fields by the load current, thus reducing the armature inductance. Sufficient reduction in the armature

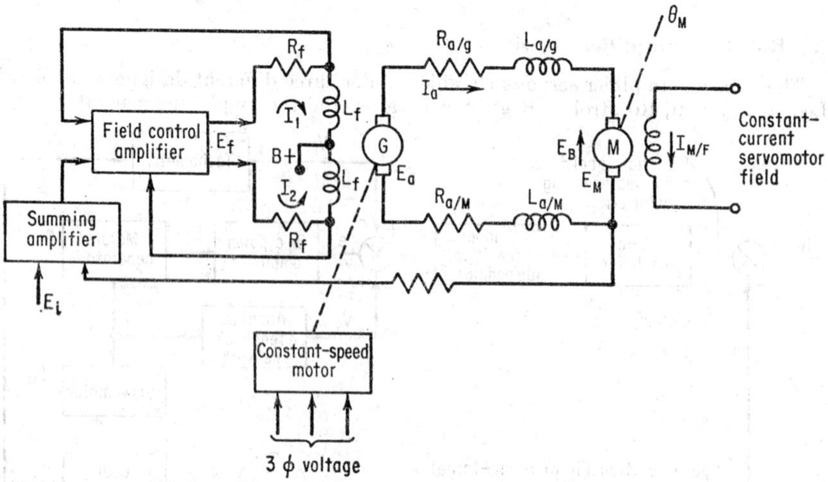

FIG. 9.15. Schematic diagram of Ward-Leonard system. E_a = generator armature voltage; E_B = back emf of motor; E_f = generator field voltage; E_m = motor armature voltage; I_a = generator armature current; $I_{m/f}$ = motor field current (constant); $L_{a/g}$ = generator armature inductance; $L_{a/m}$ = motor armature inductance; L_f = generator field inductance; $R_{a/g}$ = generator armature resistance; $R_{a/m}$ = motor armature resistance; R_f = d-c generator field resistance; θ_0 = position of motor shaft.

time constant L_a/R_a can be obtained so that stable operation can be achieved. The field time constant L_f/R_f can be decreased by adding resistance in series with the field winding in addition to field current feedback. However, this reduction is limited because as the series resistor (sometimes referred to as a forcing resistor) is increased the control power requirements increase; thus a decrease in power gain results. Increasing the feedback also reduces the gain. Lags caused by eddy currents in the iron pole structure are, for all practical purposes, eliminated by a laminated magnetic circuit. This lag can be reduced to from 0.001 to 0.002 sec.[6]

The transfer function for the overall Ward-Leonard system, without feedback, is (referring to Fig. 9.15) derived as follows:

$$E_f = I_f R_f + L_f I_f s \tag{9.74}$$

$$E_a = I_a R_{a/g} + L_{a/g} I_a s + E_m \tag{9.75}$$

where E_m = servomotor load

$$E_a = K_g I_f \tag{9.76}$$

where K_g = gain relationship between field current and armature volts
The equation for the load (servomotor) is

$$E_m = E_B + I_a R_{a/m} + L_{a/m} I_a s \tag{9.77}$$
$$E_B = K_B \theta_m s = \text{motor back emf} \tag{9.78}$$
$$I_a = T/K_B \tag{9.79}$$

where $K_B = K_\phi$ = proportionality constant
ϕ = flux
T = developed torque
Substituting for E_a in Eq. (9.76) and E_B in Eq. (9.77), then solving Eqs. (9.74) and (9.77) simultaneously and simplifying, results in

$$E_f = (R_f + L_f s)[I_a(R_a + L_a s) + K_B \theta_m s] \tag{9.80}$$

where $R_a = R_{a/g} + R_{a/m}$
$L_a = L_{a/g} + L_{a/m}$
The motor equation for a mechanical load on the motor consisting of inertia and viscous friction[6]

$$T = J_m \theta_m s^2 + f\theta_m s \tag{9.81}$$

where θ_m = motor-shaft position, rad
J_m = motor gearing and load inertia at the motor shaft
f = coefficient of viscous friction
Substituting Eq. (9.79) for I_a and Eq. (9.81) for T in Eq. (9.80) and simplifying results in an expression relating motor-output-shaft position to amplifier input voltage:

$$\frac{\theta_m}{E_i} = \frac{K_a K_B K_g}{R_f(fR_a + K_B^2)} \frac{1}{s[1 + (L_f/R_f)s][(J_m L_a/fR_a + K_B^2)s^2 + (fL_a + J_m R_a/fR_a + K_B^2)s + 1]} \tag{9.82}$$

where $K_a = E_f/E_i$ = electronic amplifier gain.

The loops shown in Fig. 9.16*a* can be simplified and reduced to the block diagram of Fig. 9.16*b*. The transfer functions can be expressed as follows:

Field, motor, armature, and overall transfer functions, respectively,

$$G_1 = \frac{K_1 K_2}{(1 + K_2 K_f)[1 + (\tau_f/K_2 K_f)s](1 + \tau_e s)} \tag{9.83}$$

$$G_4 = \frac{G_2}{1 + G_2 G_3 K_B N} \tag{9.84}$$

$$G_5 = \frac{G_1 G_4}{1 + G_1 G_4 k_a N} \tag{9.85}$$

$$\frac{\theta_L}{E} = \frac{G_5 G_3 N}{s} \tag{9.86}$$

(b) Eddy-current Clutch Servo

The eddy-current clutch has numerous applications in both closed- and open-loop-type systems and is frequently employed in angular drives for large high-precision radio-telescope antennas. The eddy-current clutch can also be used in an opposing-drive configuration as discussed previously.

The eddy clutch operates on the same basic principle as the induction motor. The input member consists of an electromagnetic yoke which is driven at a constant speed by the drive motor. A metal disk is mounted on the output member which is connected to the load. When the electromagnetic yoke is energized by the control signal, eddy currents are induced in the metal disk as a result of the relative motion between the yolk and the disk. Coupling between the input and output shafts results from the electromagnetic forces set up between the eddy currents in the disk and the magnetic field of the yolk. The amount of torque transmitted by the eddy current is proportional to the ratio of input to output shaft speeds (slip). Maximum torque transfer occurs when the speed of the output shaft is one-half the speed of the input shaft. The magnitude of slip is proportional to the control current.

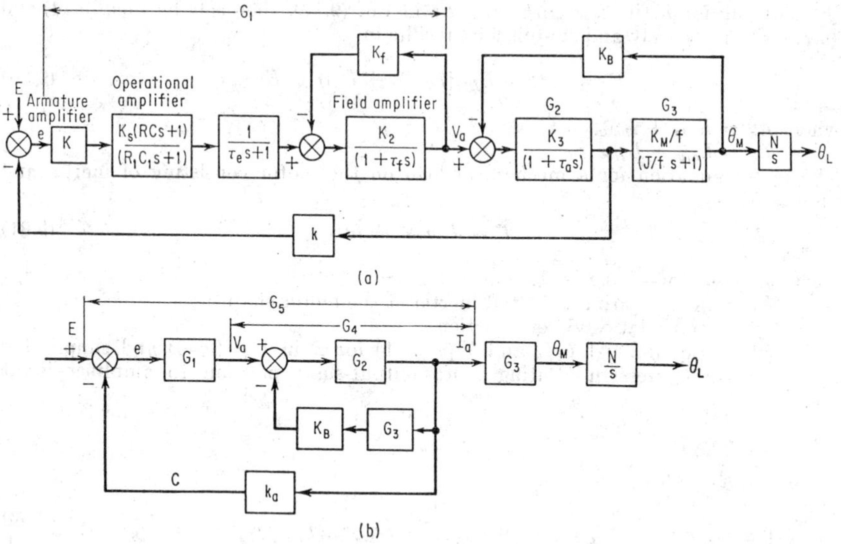

FIG. 9.16. Block diagrams of motor generator. (*a*) General. (*b*) Simplified. f = viscous friction, ft-lb per rad/sec; J = total inertia at motor; K_1 = gain of sum amp, volts/volt; K_2 = gain of field amp, volts/amp; $K_3 = 1/R$, amp/volt; K_a = armature feedback, volts/amp; K_B = motor back emf, volts per rad/sec; K_f = field feedback constant, volts/amp; K_M = motor gain, ft-lb/amp; N = gear ratio; $RC = 0.1$ sec; $R_1C_1 = 0.6$ sec; τ_a = armature lag; τ_e = eddy-current lag; θ_L = load position.

The major disadvantage of this type of clutch is the power loss resulting from slippage. When used in a system requiring a wide range of speed control and high starting torque, the clutch suffers a reduction in efficiency. One definite advantage of an eddy-current clutch servo relative to a rotating amplifier is the lower reliability of the latter because of brush and commutator wear associated with continuous operation.

Although the clutches and motor for a high-powered system are heavy and bulky, the rotating-amplifier combinations are heavier, bulkier, and much more expensive. Additional advantages are that the clutch control generates no r-f noise and can be controlled by an a-c amplifier, thus eliminating any drift problem; provides smooth control at all speeds; and has inherent linear characteristics, making it ideal for continuous control application. Stabilizing the rotating amplifier is much more difficult because of the additional time lags involved in the generator field and armature circuit; therefore, an additional feedback loop is required between the armature and the input of the servoamplifier.[6]

Transfer-function Analysis. Similar to many other electromechanical conversion machines, the eddy-current clutch has two time constants associated with it, an electrical and a mechanical. The electrical time constant results from the field inductance and resistance and can be derived as follows:

$$E(s) = I(s)Z(s) \tag{9.87}$$

where $E(s)$ = voltage drop across impedance
$Z(s)$ = clutch-coil impedance
$I(s)$ = current through impedance
s = Laplace operation

$$Z(s) = R_c + L_c(s) \tag{9.88}$$

where R_c = clutch-winding equivalent d-c resistance
L_c = clutch-winding inductance

Substituting for $Z(s)$ in Eq. (9.87) and solving for current output for a given voltage input, the transfer function of the clutch is found to be equal to the admittance of the clutch winding

$$I(s)/E(s) = K_a/(R_c + L_cs) \tag{9.89}$$

$$I(s)/E(s) = K_a/[R_c(1 + t_cs)] \tag{9.90}$$

where K_a = control-amplifier gain
$t_c = L_c/R_c$ = electrical time constant

The relationship between the clutch input current and the torque output to the load is derived in the manner shown below. The torque at the output side of clutch is directly proportional to current in the clutch:

$$T_c/I_c = K_t \tag{9.91}$$

where K_t = torque constant, ft-lb/amp
J_t = total inertia at clutch output side
F_t = total reflected viscous friction at clutch output

$$T_c = J_t(d^2\theta_c/dt^2) + F_t(d\theta_c/dt) \tag{9.92}$$

Transforming Eq. (9.92) yields the torque equation in the frequency domain:

$$T_c(s) = J_ts^2\theta_c + F_ts\theta_c \tag{9.93}$$

Simplifying,

$$T_c(s) = F_ts\theta_c(1 + \tau_ms) \tag{9.94}$$

where τ_m = mechanical time constant = J_t/F_t.

Closed-loop Servo Analysis. A two-speed fine-coarse synchro feedback loop is used to illustrate the closed-loop operation. Figure 9.17 is a block diagram of the overall system. When the shafts of control-transmitter synchros CX in the remote unit are rotated through an angle, the rotor excitation voltage induces voltages into the three-stator windings $S_1S_2S_3$ the magnitude and polarity of which define the angular position of the shaft with respect to the stator. These stator voltages appear on the stator windings of the control transformer CT in the servo loop. A voltage proportional to the rotation of the CX synchro is induced into the rotor of the CT. This voltage appears as the error signal at the servo input and causes the prime mover to drive until the CT synchros are rotated through the commanded angle. The sensitivity of the synchro error is 1 volt/degree.

The fine and coarse synchros are separated by a 36:1 gear ratio, which is one of the standard separations for this type of operation; some other ratios are 25:1, 16:1, and 20:1. Fine-coarse operation provides for a greater degree of accuracy when operating off the fine or high-speed synchro. The errors are reduced because of the higher sensitivity and also by a factor of 36 due to the gearing.

The fine-coarse switching block shown in Fig. 9.17 is required to switch from the fine synchro output to the coarse synchro output when the error exceeds a set threshold. Likewise, as a large error is diminished it switches from coarse to fine operation. (For more detailed discussion on synchros, see refs. 3 and 6.)

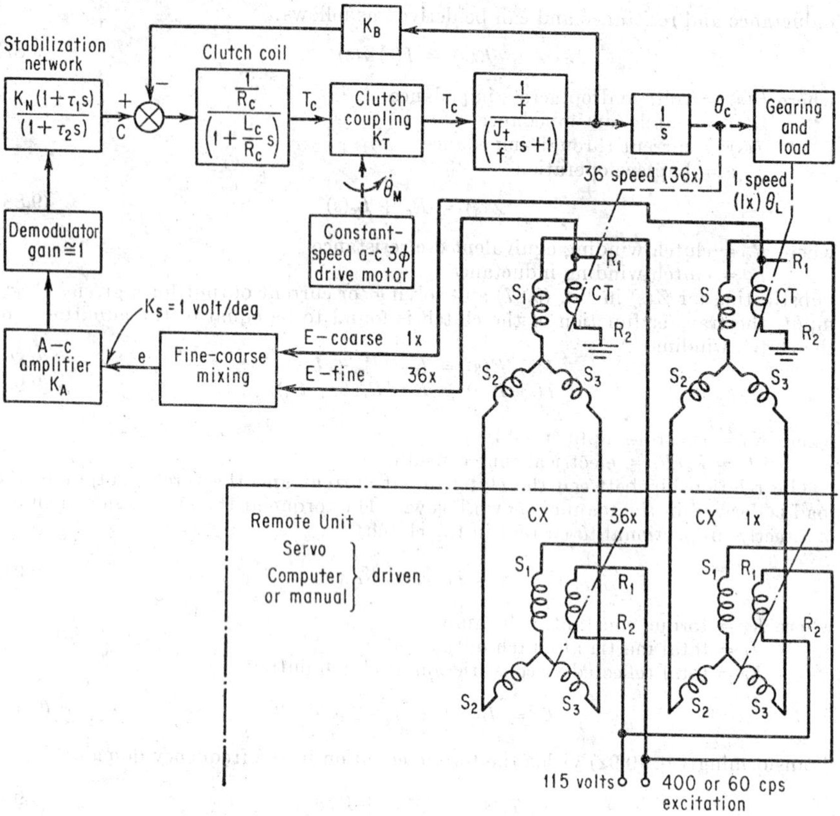

FIG. 9.17. Block diagram eddy-current clutch servo.

The transfer function of shaft position output to clutch input is:

$$\frac{\theta_L}{C} = \left(\frac{1}{\dfrac{R_c f}{K_T} + K_B}\right)\left\{\frac{1}{s\left[\dfrac{L_c J_t s^2}{R_c f + K_T K_B} + \dfrac{f L_c + R_c J_t}{R_c f + K_T K_B} s + 1\right]}\right\} \tag{9.95}$$

where K_T = clutch proportionality constant, ft-lb/amp
K_B = back emf, volts per rad/sec

(c) Hydraulic Servomechanisms

While the hydraulic servo has been here classed as a power servo, it is not limited to large-load applications. They are utilized throughout the power spectrum from small aircraft and missile instrument servos to large gun-turret systems.

The closed-loop hydraulic servo consists of an electronic amplifier, hydraulic amplifier, prime mover, and any one of the conventional error-sensing devices. In addition, auxiliary equipment such as relief and reducing valves, accumulators, filters, and heat exchangers may also be required. These components are generally not considered when investigating loop stability, however.

Electronic amplification is usually required to amplify the small error voltage from the error sensor. The amplified voltage is in turn applied to the controls of the hydraulic amplifier. The prime mover controlled by the hydraulic amplifier can be either a rotary actuator, a fixed-displacement motor, a piston, or a variety of turbines. For hydraulic components see Sec. 37.

Of the many possible systems, two will be here described, one employing pump control and one utilizing valve control.

Hydraulic-valve Control System. Hydraulic-valve control systems can be divided into two categories, flow control and pressure control. Only the pressure control system will be discussed.

This system consists of a pressure-control valve, a pressure-compensated pump, and a fixed-displacement motor. The pressure-control valve is connected directly to a fixed-displacement hydraulic motor. The valve has four hydraulic openings. Two output ports are connected to the motor, a third is connected to a pressure-compensated hydraulic pump, and the fourth is returned to an oil sump. The differential current ΔI across the coils of the valve is proportional to the servo error signal

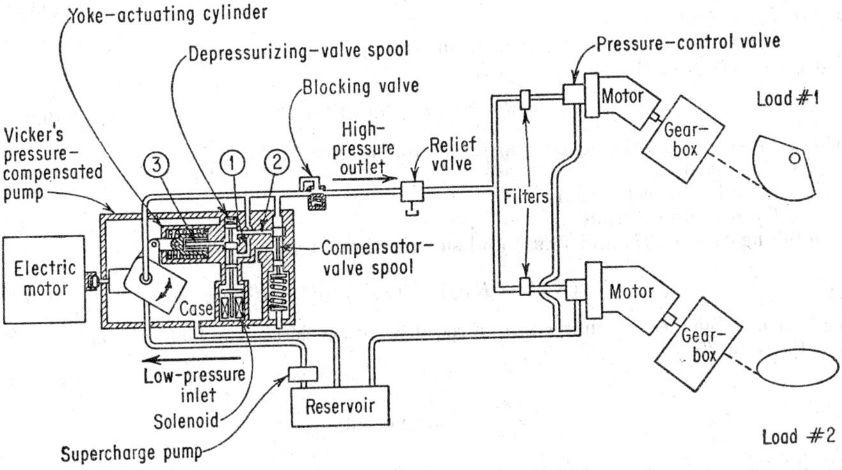

FIG. 9.18. Vickers electrical depressurized pressure-compensated pump.

and controls the differential pressure across the motor. The differential pressure is substantially independent of flow, but a slight loss of pressure with flow does occur.

The pressure-compensated pump shown in Fig. 9.18 is a variable-delivery positive-displacement hydraulic pump. It maintains substantially constant pressure in the system by automatically varying the pump yoke position as a change in flow is required and is thus especially suited for applications characterized by intermittent or changing flow demands. In the pump type shown in Fig. 9.18, system pressure is reduced to a low value (300 to 600 psi) during standby operation, thus greatly reducing losses and increasing pump life. When the solenoid is not energized, the depressurizing-valve spool is held down by its spring, so that the yoke-actuating cylinder line (3) is connected to the compensator line (2) but is closed to the outlet-pressure line (1). In this mode, the pump operates as a basic pressure-compensating unit. When the solenoid is actuated (by an automatic or manual switch) the depressurizing spool is forced upward. This connects the yoke-actuating cylinder line (3) to the outlet-pressure line (1) but blocks the compensator line (2). With outlet pressure acting on its actuating cylinder, the yoke is forced to a position very near center. Actually there is still a small displacement, in order to provide enough outlet pressure (300 to 600 psi) to hold the yoke in its depressurized position and to supply enough leakage for pump lubrication.

Since the driving speed of the pump is continuous and unaffected by depressurization, the outlet pressure builds up very rapidly when the solenoid circuit is opened. The solenoid valve is displaced downward and pressure regulation resumes.

A blocking valve may be used in order to isolate the pump from the rest of the system during depressurized periods. The valve is set to close at a value slightly above the idling pressure, and it opens again only when the solenoid is deenergized.

Motor and Valve Transfer Functions. The relationship between differential pressure and flow is as follows

$$P_d = K_P I - K_f Q \tag{9.96}$$

where P_d = differential pressure, psi
K_P = pressure gain, psi/ma
I = differential current, ma
Q = flow out of servo valve, gpm
K_f = servo-valve pressure drop (e.g., psi/gpm)

The output torque at the motor shaft is

$$T_M = D_M P_d \tag{9.97}$$

where T_M = motor torque
D_M = motor displacement (e.g., in.³/rad)

Torque reflected to the motor shaft is

$$T_r = (J_T/N^2)(d^2\theta_L/dt^2)N + T_F \tag{9.98}$$

where J_T = total inertia load (gearing, motor) at output shaft
N = gear ratio, $N \gg 1$
$d^2\theta_L/dt^2$ = load acceleration
T_F = friction torque

Combining Eqs. (9.97) and (9.98) and substituting for P_d results in

$$D_M(K_P I - K_f Q) = (J_T/N)(d^2\theta_L/dt^2) \tag{9.99}$$

T_F can be neglected assuming torque required to accelerate the load inertia is much greater than T_F

$$Q = D_M N(d\theta_L/dt) \tag{9.100}$$

where $\dfrac{d\theta_L}{dt}$ = load velocity

Combining Eqs. (9.99) and (9.100), and transforming, results in the following transfer function:

$$\frac{\theta_L}{I} = \frac{K_P}{K_f N D_M} \frac{1}{s[(J_T/K_f N^2 D_M^2)s + 1]} \tag{9.101}$$

The velocity coefficient K_V of the system

$$K_V = K_P K_A K_s / K_f N D_M \tag{9.102}$$

where K_A = amplifier gain, ma/volt
K_s = error sensitivity, volts/rad

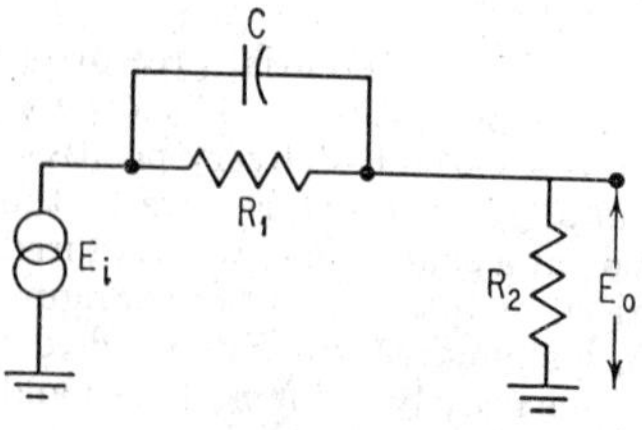

FIG. 9.19. Stabilization circuit.

Valve-coil and Stabilization Network. An additional lag is due to the valve-control-coil time constant (L/R), however, the time constant can be made small. Addition of a lead-lag network is usually required to achieve both the desired open-loop gain and stability. The stabilization circuit is shown in Fig. 9.19, from which

$$\frac{E_o}{E_i(s)} = \frac{R_1}{R_1 + R_2} \frac{\tau_1 s + 1}{\tau_2 s + 1} \tag{9.103}$$

where $\tau_1 = R_1 C$
$\tau_2 = [R_1/(R_1 + R_2)]C$

A system block diagram is shown in Fig. 9.20. The open-loop transfer function is

$$\frac{\theta_L}{E} = \frac{K_V(1 + \tau_1 s)}{s(\tau_M s + 1)(\tau_2 s + 1)(\tau_V s + 1)} \tag{9.104}$$

where $K_V = \dfrac{K_A K_s K_P K_N}{K_f N D_M (K_N R_1 + 1)}$

τ_V = valve time constant
$= L_V/R_V$

$\tau_M = \dfrac{J_T}{K_f N^2 D_M^2}$

$K_N = \dfrac{R_1}{R_1 + R_2}$

Power Requirements. The maximum power requirements for the hydraulic motor can be calculated as follows:

$$\text{hp} = \frac{P_d D_M N \omega_{\max}}{6{,}600} \qquad (N \gg 1) \tag{9.105}$$

where $\omega_{\max}$ = maximum required load angular velocity.

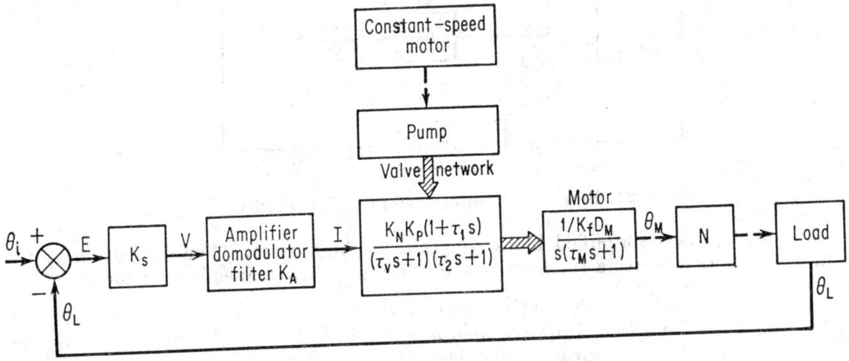

FIG. 9.20. Simplified system block diagram.

Pump Flow Requirement. The maximum flow requirement of the pump can be calculated as follows:

Motor and valve:

$$Q = N D_M V_{\max} \qquad \text{in.}^3/\text{sec} \qquad (N \gg 1) \tag{9.106}$$

Leakage, provided by manufacturer

Total flow required

$$Qt = Q \text{ motor} + Q \text{ valve} + Q \text{ leakage} \tag{9.107}$$

Pump delivery

$$Q_P = (D_P S_P/60) = Q_T \tag{9.108}$$

where Q_P = maximum pump delivery, in.³/sec
D_P = pump displacement, in.³/rev
S_P = pump speed, rpm

Load-isolation Analysis. One of the inherent features of this type of system is the capability of isolating the load from external disturbances. Unlike other hydraulic or electrical servo systems, the pressure-control valve decouples the load from the disturbances on the platform by offering little resistance to the inertia of the load. Consider a broadside-looking antenna mounted on a rolling, moving ship. The disturbance occurs in the elevation axis. As the ship rolls, the inertia of the load tends to maintain the load positioned in space. The pressure-control valve allows the gear train between the deck and the load to rotate as a function of the disturbance input. The magnitude of the ratio of the gearing and motor inertia (reflected to the load shaft) to the total inertia determines the response of the servo to disturbances of a high frequency.

For the derivation of the system transfer function under these conditions, it is assumed that the load is balanced and that the servo is stabilized by a single lead network. Ideally, if the center of gravity of the load is exactly at the center of rotation, the system is balanced. For any other condition a static error is required to supply the torque necessary to eliminate the unbalance. The system is assumed frictionless.

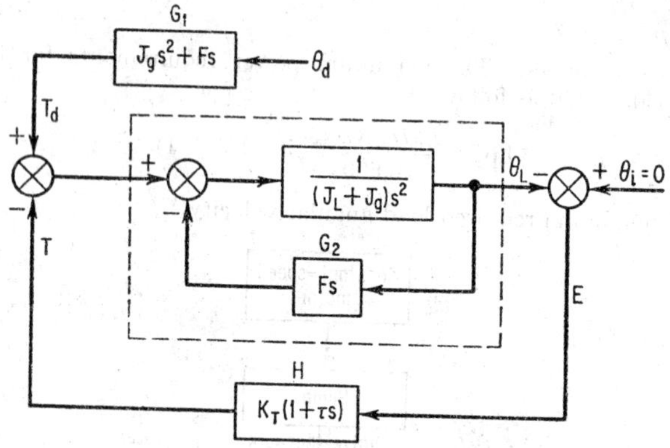

FIG. 9.21. Block diagram of system with disturbance input.

Figure 9.21 is a simplified block diagram of the system. The derivation of the transfer function relating the load position to input disturbance is as follows:

$$\text{Disturbance torque } T_d = J_g s^2\theta_d + Fs\theta_d \tag{9.109}$$

$$\text{Reaction torque of load } T_L = (J_L + J_g)s^2\theta_L + Fs\theta_L \tag{9.110}$$

$$\text{Error torque assisting load } T_E = K_T\theta_L(1 + \tau_1 s) \tag{9.111}$$

where $\theta_L = E$ in Fig. 9.21. When $\theta_i = 0$,

$$T = T_d - T_L - T_E = 0 \tag{9.112}$$

Combining Eqs. (9.109), (9.110), and (9.111),

$$\theta_L[K_T(1 + \tau_1 s) + (J_L + J_g)s^2 + Fs] = \theta_d(J_g s^2 + Fs) \tag{9.113}$$

Rearranging terms and simplifying,

$$\frac{\theta_L}{\theta_d} = \frac{Fs[(J_g/F)s + 1]}{K_T\{[(J_L + J_g)/K_T]s^2 + [(F + K_T\tau_1)/K_T]s + 1\}} = \frac{G_1G_2}{1 + G_2H} \tag{9.114}$$

where F = total viscous damping due to load and valve pressure drop, ft-lb per rad/sec
K_T = system torque gain, ft-lb/rad
τ_1 = lead time constant, sec
J_L = load inertia at load shaft, ft-lb-sec²
J_g = gearing and hydraulic motor inertia at load shaft, ft-lb-sec²

The value of F can be calculated as follows:

$$F = \frac{K_f T_{M\max} N D_M}{P_{d\max}} + F_L \tag{9.115}$$

Figure 9.22 is a general system open-loop Bode plot and Fig. 9.23 is a general plot of the response of the load to a disturbance θ_L/θ_d. It shows the effects of the various system parameters in Eq. (9.114) on the isolation capability of the system.

The advantages of a pressure-controlled system are as follows: System gain is not limited by length of lines because valve is integrated with motor; thus extremely low dynamic errors can be achieved. Pressure control provides isolation of load from external disturbances. More than one hydraulic motor can be driven from only one pump as shown in Fig. 9.18.

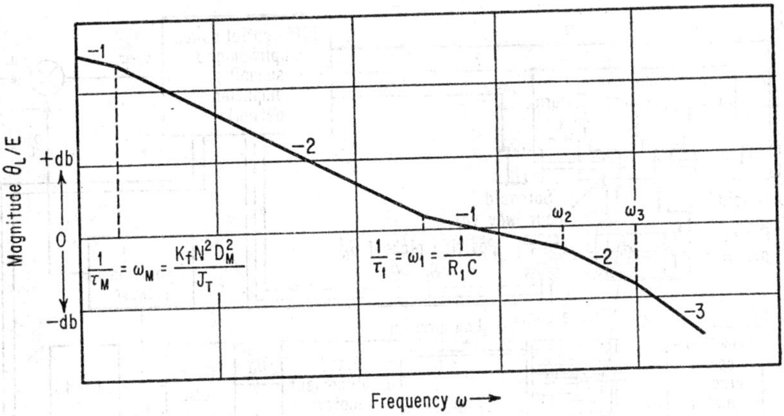

FIG. 9.22. Open-loop frequency response.

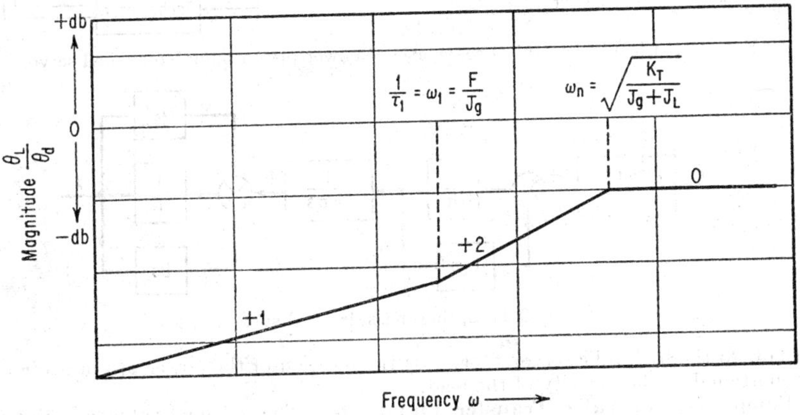

FIG. 9.23. System response to input disturbance.

Disadvantages of this system include the following: Maximum pressure is required all the time from zero to maximum load speed, thus reducing efficiency. Continuous operation creates fluid-heating problem.

Variable-displacement Pump and Power Piston. This type of hydraulic system employs a variable-displacement pump which is somewhat similar to the previously described pressure-compensated pump. The primary difference is that the angle of the yoke is continuously varied about the center position depending on the demands of the load. The yoke is displaced by a power piston driven by a hydraulic control valve connected to a constant pressure source. The valve piston is driven by a solenoid which is driven by an electronic amplifier. A fixed-displacement hydraulic motor identical to the one previously described is connected to the output of the pump.

Referring to the overall functional diagram (Fig. 9.24), the error E represents the difference in position of the load θ_L and commanded input θ_i. The polarity and magnitude of the error determine the direction and distance the stroking piston moves. The piston is moved until the position feedback θ_P from the piston pickoff nulls the error E. The pump yoke is rotated through an angle proportional to the linear displacement of the power piston. This angle determines the magnitude and direction

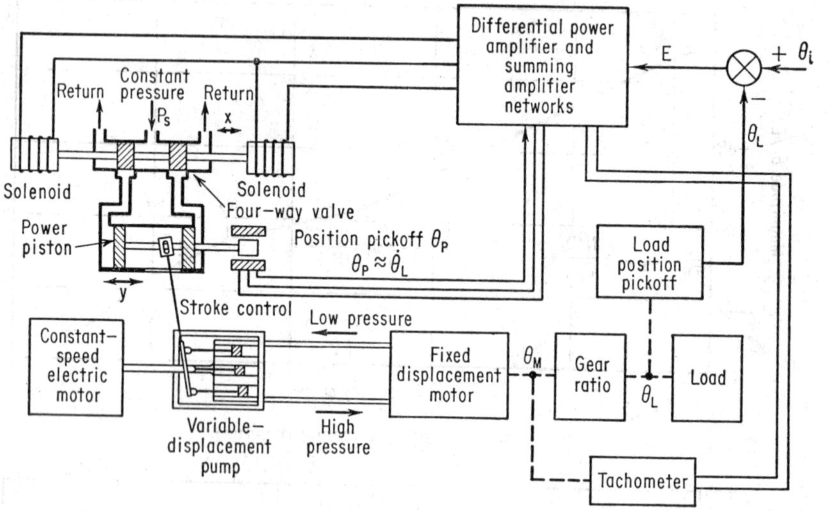

FIG. 9.24. Electromechanical schematic of power-piston pump-controlled servo.

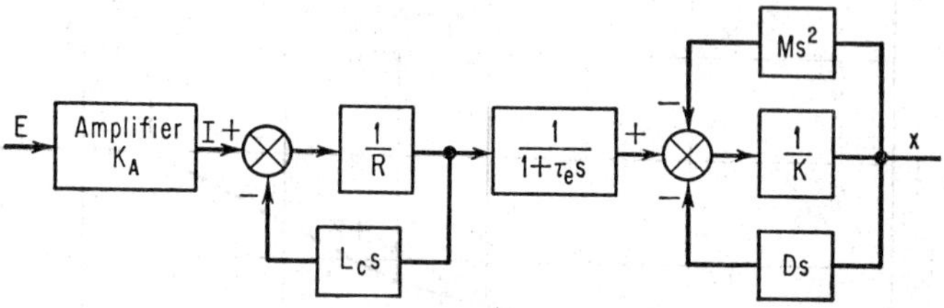

FIG. 9.25. Spring-mass damped system.

of flow to the hydraulic servomotor; thus the position of the power piston is directly proportional to the velocity of the load.

Solenoids and Valve Transfer Functions. The solenoid-valve combination has both electrical and mechanical lags associated with it. The electrical time constants result from the inductance and resistance of the winding and also eddy currents. The mechanical lags result because the solenoid armature and control valve are a spring-mass damped system. This system can be represented by the simplified block diagram shown in Fig. 9.25:

$$\frac{x}{E} = \left(\frac{K_A}{RK}\right) \frac{1}{(1 + \tau_e s)[1 + (L_c/R)s][Ms^2/K + (D/K)s + 1]} \qquad (9.116)$$

where x = valve travel, in.
E = error voltage
K_A = amplifier gain, ma/volt

R = solenoid and amplifier resistance, ohms
L_c = coil inductance, henrys
τ_e = time constant due to eddy currents, sec
M = solenoid armature control valve and effective spring mass, lb/in./sec
K = solenoid spring gradient, lb/in.
D = viscous damping of solenoid armature and control valve, lb/in./sec

It is very difficult to obtain the transfer function of the solenoid without actually performing a frequency response because the value τ_e varies as a function of frequency. In addition, the solenoid has inherent nonlinearities. The time constant L_c/R quite often can be neglected unless the solenoid is driven from a low-impedance source such as a magnetic amplifier. One additional characteristic of the solenoid-valve combination is the relatively high static friction, which results in a dead band. This condition can be minimized by introducing a mechanical or electrical dither into the system.[6]

Valve-power Stroke Piston. The transfer function relating valve displacement to power-piston (yoke) motion is derived as follows:[6] The relationship between valve displacement and load (power-piston and yoke) velocity is

$$X_1 = Ys/C_1 \tag{9.117}$$

where X_1 = valve displacement producing load velocity, in.
Y = load and piston travel, in.
C_1 = velocity gradient, in./sec/in., obtained from valve-flow gradient at zero differential pressure Q_0; for an inertia load maximum velocity occurs at zero differential pressure

$$C_1 = Q_0/A \tag{9.118}$$

The valve displacement including viscous friction of piston and load is

$$X_1 = (1/C_1 + \alpha/C_2)Ys \tag{9.119}$$

where C_2 = force gradient, lb/in.
α = viscous friction of load and piston, lb/in./sec

$$C_2 = PA \tag{9.120}$$

P = differential pressure gradient, e.g., psi/in.
A = piston area

Force required to accelerate an inertia load is

$$F = Ms^2Y \tag{9.121}$$

and the relationship between valve displacement and valve-piston force gradient is

$$X_2 = (M/C_2)s^2Y \tag{9.122}$$

A rate of change of control pressure produces a rate of change of load acceleration. The fluid is compressible; thus the valve is required to supply the compressibility flow in order to produce this change of pressure. The amount of flow depends on the entrained-fluid volume and its bulk modulus. For one side of valve and piston,

$$Q_0X_3 = (V_1/B)sP_1 \tag{9.123}$$

where V_1 = entrained fluid volume on one side of circuit
B = bulk modulus of fluid, psi
P_1 = change of control pressure on one side of piston

If the system is symmetrical, there is an equal flow from the other side of the circuit:

$$Q_0X_3 = (V_2/B)sP_2 \tag{9.124}$$

The change in differential pressure is the sum of the change in each control pressure:

$$Q_0X_3 = (V/2B)sH \tag{9.125}$$

where H = increment of differential pressure, psi

Dividing Eq. (9.125) by A,

$$(Q_0/A)X_3 = (V/2BA)sH \tag{9.126}$$

The valve compressibility flow occurs at minimum differential pressure and thus is the same as Q_0. Equation (9.126) can be simplified by substituting C_1 from Eq. (9.118)

$$C_1X_3 = (V/2BA)sH \tag{9.127}$$

The differential pressure H can be expressed in terms of load motion y by considering force exerted by the piston to provide load acceleration and overcome viscous friction. The resulting relationship is

$$H = (M/A)s^2y + (\alpha/A)sy \tag{9.128}$$

Combining Eqs. (9.127) and (9.128),

$$X_3 = (VM/2A^2C_1B)s^3y + (V\alpha/2A^2C_1B)s^2y \tag{9.129}$$

Thus the overall valve displacement is

$$X = X_1 + X_2 + X_3 \tag{9.130}$$

Substituting for X_1, X_2, and X_3 in Eq. (9.130) and solving for the relationship of valve displacement X to piston stroke y results in the following transfer function:

$$\frac{y}{X} = \frac{C_1/[1 + \alpha(C_1/C_2)]}{s\left\{\dfrac{VM}{2BA^2[1 + \alpha(C_1/C_2)]}s^2 + \dfrac{C_1M/C_2 + V\alpha/2BA^2}{1 + \alpha(C_1/C_2)}s + 1\right\}} \tag{9.131}$$

$$\omega_n = \sqrt{(2BA^2/VM)[1 + \alpha(C_1/C_2)]} = \text{natural frequency} \tag{9.132}$$

and

$$\delta = \frac{C_1M\omega_n}{2C_2}\,\frac{1 + \alpha(VC_2/\alpha C_1BA^2M)}{1 + \alpha(C_1/C_2)} = \text{damping coefficient} \tag{9.133}$$

or

$$\frac{y}{X} = \frac{C_1/[1 + \alpha(C_1/C_2)]}{s[s^2/\omega_n^2 + (2\delta/\omega_n)s + 1]} \tag{9.134}$$

Variable-displacement Pump and Motor. The transfer function relating piston (yoke) motion y to motor angle θm is derived as follows:[6]

$$Q_P = Q_M + Q_L + Q_C \tag{9.135}$$
$$Q_M = Q_P - Q_L - Q_C \tag{9.136}$$

where Q_P = volume rate of oil from pump, in.³/sec
Q_M = flow of oil through motor, in.³/sec
Q_L = leakage flow around motor, in.³/sec, due to pressure drop across pump pistons
Q_C = compressibility flow, in.³/sec

$$Q_P = [nA_P(\dot\theta_P/2\pi)]y = K_Py \tag{9.137}$$

where n = number of pistons in pump
A_P = area of piston, in.²
$\dot\theta_P/2\pi$ = pump speed, rps
K_P = flow from pump per unit displacement of y, in.³/sec/in.

$$Q_M = d_m\dot\theta_M \tag{9.138}$$

where $\dot\theta_M$ = motor speed, rad/sec
d_m = motor volumetric displacement, in.³/rad

Equating mechanical power developed by the motor $T_M\dot\theta_M$ to the hydraulic power PQ_M,

$$PQ_M = T_M\dot\theta_M \tag{9.139}$$

Combining Eqs. (9.138) and (9.139) results in a relation between pump discharge pressure and motor torque:

$$P = T_M/d_m \tag{9.140}$$

$$\text{Leakage flow } Q_L = PK_L \tag{9.141}$$

where K_L = leakage coefficient, in.³/sec per psi
Combining Eqs. (9.140) and (9.141),

$$Q_L = (K_L/d_m)T_M \tag{9.142}$$

The volume rate of flow delivered to the motor is also decreased because of the compressibility of the oil. The effect of compressibility is a function of the quantity of fluid in the system and pressure. If the system is operating at a high pressure and the hydraulic lines between the pump and motor are long, compressibility has a great bearing on the gain and stability of the system. The effect of compressibility on the overall transfer function is as follows:

$$\Delta V = (V/B)\,\Delta P \tag{9.143}$$

where ΔV = change in volume
The compressibility flow

$$Q_C = \lim_{\Delta t \to 0} \Delta V/\Delta t = (V/B)(dP/dt) \tag{9.144}$$

Combining Eqs. (9.140) and (9.144),

$$Q_C = (V/Bd_m)(dT_M/dt) = (K_C/d_m)(dT_M/dt) \tag{9.145}$$

where K_C = compressibility coefficient
Substituting of Eqs. (9.137), (9.138), (9.141), and (9.145) into Eq. (9.136) yields

$$d_m\dot{\theta}_M = K_P y - (K_L/d_m)T_M - (K_C/d_m)(dT_M/dt) \tag{9.146}$$

If the load at the motor shaft consists of inertia and viscous damping, then

$$T_M = J\ddot{\theta}_M + F\dot{\theta}_M \tag{9.147}$$

where J = total inertia at motor shaft $(J_M + J_L/N^2)$
F = viscous damping coefficient
Substituting for T_M in Eq. (9.146) and transforming results in the transfer function relating yoke displacement y and motor shaft angle θ_M as follows:

$$\frac{\theta_M}{y} = \frac{K_P[1/(K_LF + d_m^2)]}{s\left(\dfrac{K_CJ}{K_LF + d_m^2}s^2 + \dfrac{K_LJ + K_CF}{K_LF + d_m^2}s + 1\right)} \tag{9.148}$$

$$\omega_n = \sqrt{(K_LF + d_m^2)/K_CJ} = \text{resonant frequency} \tag{9.149}$$

$$\zeta = \frac{1}{2}\frac{K_LJ + K_CF}{K_LF + d_m^2}\,\omega_n = \text{damping ratio} \tag{9.150}$$

System Application and Analysis. Figure 9.26 is a block diagram of a pump-controlled servosystem employing tachometer feedback through a stabilizing lead-lag network. Stroke-position feedback is used around the solenoid, valve, and power piston; thus the position of the piston represents the velocity of the load.

Solenoid-valve Loop. The transfer function of this minor loop

$$\frac{y}{C} = \frac{K_2K_aG_1G_2}{1 + K_2K_aG_1G_2} \tag{9.151}$$

where $G_1 = \dfrac{K_3}{(1 + \tau_e s)[1 + (L_C/R)s](Ms^2/K + Ds/K + 1)}$

$G_2 = \dfrac{C_1/[1 + \alpha(C_1/C_2)]}{s[s^2/\omega_n^2 + (2\zeta/\omega_n)s + 1]}$

K_2 = error sensitivity

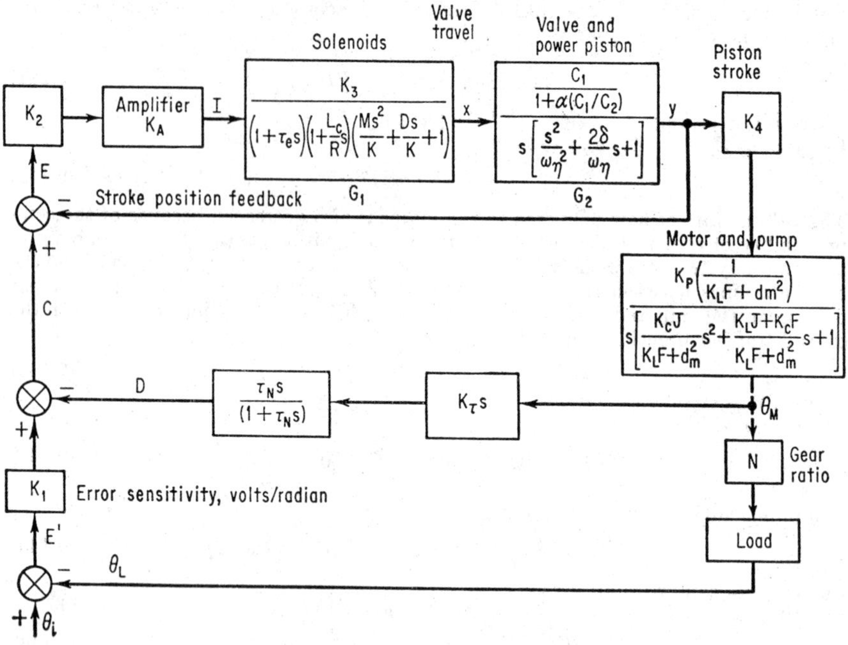

FIG. 9.26. Pump-controlled servosystem block diagram.

Values of ω_n and ζ are given in Eqs. (9.145) and (9.146), respectively. Substituting for G_1 and G_2 in Eq. (9.151) and letting $C_1/[1 + 2(C_1/C_2)] = K_{PV}$ results in

$$\frac{y}{C} = \frac{K_2K_3K_{PV}K_a}{s\left(\frac{s^2}{\omega_n{}^2} + \frac{2\zeta}{\omega_n}s + 1\right)(1 + \tau_e s)\left(1 + \frac{L_C}{R}s\right)\left(\frac{Ms^2}{K} + \frac{Ds}{K} + 1\right) + K_2K_3K_{PV}K_a} \tag{9.152}$$

The gain of this loop can usually be made sufficiently large and the time constants sufficiently small so that, when considered as part of the tachometer loop, $y/C \approx 1$.

Motor-tachometer Loop. The open-loop (D/E) transfer function of this loop must be investigated for stability. Thus

$$\frac{D}{E} = GH = \underbrace{\frac{K_4K_P\left(\frac{1}{K_LF + dm^2}\right)}{s\left[\frac{K_cJ}{K_LF + dm^2}s^2 + \frac{K_LJ + K_cF}{K_LF + dm^2}s + 1\right]}}_{G} \times \underbrace{\frac{\tau_N K_t s^2}{1 + \tau_N s}}_{H} \tag{9.153}$$

where K_4 = sensitivity of stroke mechanism.

$$GH = \frac{Ks}{[s^2/\omega_n{}^2 + (2\zeta/\omega_n)s + 1](1 + s/\omega_n)} \tag{9.154}$$

where $K = \tau_N K_4 K_P[1/(K_LF + dm^2)]$

$\omega_n = \sqrt{(K_LF + dm^2)/K_CJ}$

$\zeta = \frac{1}{2}\left(\frac{K_LJ + K_CF}{K_LF + dm^2}\right)\omega_n$

If a high-performance system is required, i.e., high acceleration- and velocity-error coefficients, the tachometer open response GH requires the addition of network in order to achieve stabilization (refer to the discussion devoted to the Ward-Leonard system). Figure 9.27 is a plot showing the configuration of the open-loop response of the motor-tachometer loop.

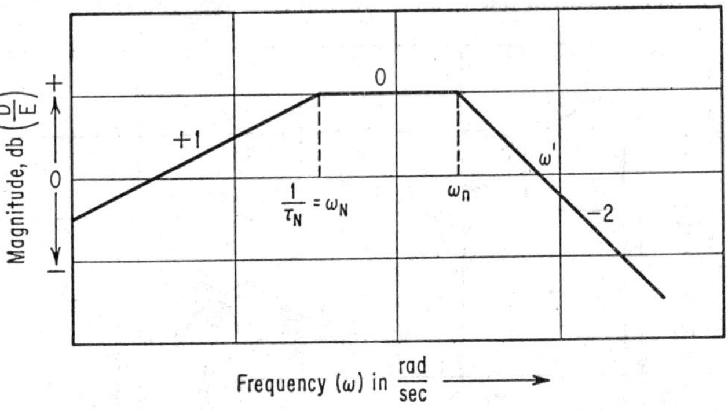

FIG. 9.27. Open-loop response of tachometer-motor loop.

The closed-loop response of the tachometer loop is

$$\theta_L/E' = G'/(1 + G'H) \tag{9.155}$$

where $G' = \dfrac{K_1K_4K_P[1/(K_LF + dm^2)]/N}{s[s^2/\omega_n{}^2 + (2\zeta/\omega_n)s + 1]}$

$H = \dfrac{NK\tau_N s^2}{K_1(1 + \tau_N)}$

$$\text{Velocity coefficient} = K_V = \frac{K_1K_4K_P[1/(K_LF + dm^2)]}{N} \tag{9.156}$$

where K_1 = error sensitivity.

$$\frac{\theta_L}{E'} \cong \frac{K_V(1 + s/\omega_n)}{s(1 + s/\omega_1)(1 + s/\omega_2)^2} \tag{9.157}$$

Figure 9.28 is a plot showing the configuration of the tachometer-motor closed-loop response. The magnitude of the peaking at ω_n depends upon the values of ω_1 and ω_2 in the quadratic in the denominator of the pump-motor transfer function [Eq. (9.148)].[1] This plot was obtained by first plotting G and then $1/H$ where the dotted line between ω_1 and ω_2 represents $1/H$.

If the response curves of the pump-controlled system and valve-controlled system are compared, the effect of the quadratic, appearing in the pump-controlled system, on the loop gain can be evaluated. The absence of this term in the valve-controlled system provides for better dynamic performance.

(d) Pneumatic Servosystem

The primary source of power for operating a pneumatic system is any gaseous fluid often readily available in aircraft and missile applications as a result of combustion processes or from storage devices. When operating in severe environmental conditions, a hot-gas type of pneumatic servo should be considered. Quite often it is not feasible to build an electromechanical system which can withstand a severe environment. Hydraulic systems may be affected by variation in fluid viscosity or may, under certain conditions, present a fire hazard. When comparing hydraulic and pneumatic systems, the hydraulic system has the advantage of operating with an incompressible fluid; thus a more positive action results in smaller time lags. However, this disadvantage in the pneumatic system can be minimized by higher operating pressures.[4] For gas components see Sec. 36.

Figure 9.29 is a schematic of a single-stage flapper-valve pneumatic system. The flapper-valve configuration has definite advantages over the spool valve in that it is

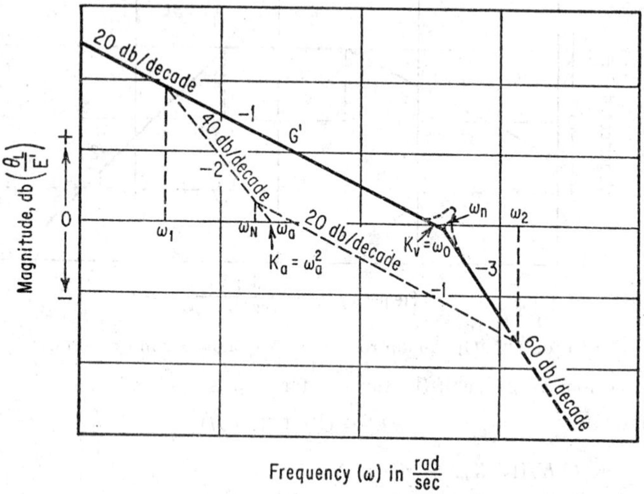

FIG. 9.28. Closed-loop response of tachometer-motor loop.

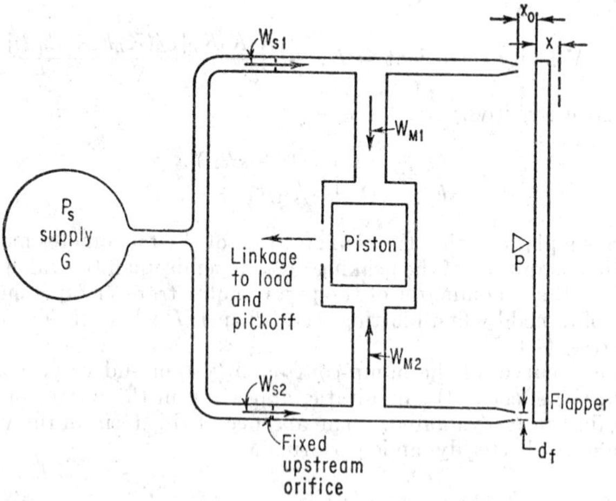

FIG. 9.29. Flapper valve and piston.

less costly because of less stringent tolerances. In addition they are less sensitive to particles of dirt, which in the case of a spool valve may cause it to become inoperative.

The system shown in Fig. 9.29 is operated by moving the flapper about the pivot point P. This is accomplished by means of coils which are energized by an error signal coming from an electronic amplifier. When no error signal is present, the gas passes from the constant-pressure-source gas generator G through fixed upstream orifices and escapes through the variable orifices. The flow through the two paths is equal; there is therefore no differential pressure across the piston. When an error signal exists, an imbalanced current flows in the coils; thus the flow through one path is increased and the flow through the other path is decreased. Because of the change in pressure drop across the orifices, a differential pressure exists at the piston to cause the load to move. A pickoff at the load provides an electrical signal proportional to load motion, which is fed back into the amplifier to close the loop.

Transfer-function Derivation. *Valve.* The describing function for the valve can be derived from the general weight-flow equation as follows:

$$W = C_1 A (P/\sqrt{T})(A'/A) \tag{9.158}$$

where T = temperature
A = area
A' = critical area

$$C_1 = g\sqrt{\frac{k}{R[(k+1)/2](k+1)/(k-1)}}$$

where g = gravitational constant
R = specific gas constant
k = ratio of specific heat

The weight-flow equation relating flow through upstream nozzle to sum of motor flow and flow through downstream nozzle is as follows:

$$W_{s_1} = C_1 A_\mu (P_s/\sqrt{T_s})(A'/A) \tag{9.159}$$

where P_s = supply pressure
W_{s_1} = weight of flow through upstream orifice off side one
T_s = absolute temperature of gas supplied

The downstream nozzle is always choked; thus A'/A has a constant value of 1 and

$$W_{f_1} = C_1 A_D (P_1/\sqrt{T_1}) \tag{9.160}$$

where W_{f_1} = weight of flow through downstream orifice
A_D = downstream nozzle area
P_1 = pressure on one side

and

$$A_D = \pi d_f x_0 (1 - x/x_0) \tag{9.161}$$

where $A_D = A_{D_0}(1 - x/x_0)$
d_f = diameter of downstream nozzle
x_0 = quiescent nozzle gap
x = armature displacement
A_{D_0} = quiescent downstream nozzle area

therefore

$$W_{f_1} = C_1 A_{D_0} (P_1/\sqrt{T_1})(1 - x/x_0) \tag{9.162}$$

The weight flow can be converted to volumetric flow using the following relationship:

$$Q_0 = W_{f_0}(RT_1/gP_1) \tag{9.163}$$

where Q_0 = quiescent volumetric flow.

At quiescent conditions $A_D = A_{D_0}$ and $x/x_0 = 0$. The motor weight flow can be obtained by subtracting Eq. (9.162) from Eq. (9.159) and multiplying the result by the specific volume.

$$Q_{M_1} = W_{M_1}(RT_1/gP_1) = (RT/gP_1)(W_{s_1} - W_{f_1}) \tag{9.164}$$

$$Q_{M_1} = Q_0[(C/a)(P_s/P_1)(A'/A)_1 - (1 - x/x_0)] \tag{9.165}$$

where $C = \sqrt{T_1/T_s} \approx \sqrt{T_2/T_s}$

a = ratio of downstream nozzle area to upstream orifice area at quiescence

and
$$Q_{M_2} = Q_0[(C/a)(P_s/P_2)(A'/A)_2 - (1 + x/x_0)] \tag{9.166}$$

where $Q_M = Q_{M_1} = -Q_{M_2}$.

If Eqs. (9.166) and (9.167) are plotted, they can be added graphically to produce a curve relating motor flow Q_m and ratio of load to supply pressure P_1/P_s. The choice of a and the choice of gas constant are the two parameters which must be defined before curves can be produced. The static servo performance can be obtained from information on these curves.

The load rates can be obtained from the following relationship:

$$\dot{\theta}_L = Q_M r/A_R \tag{9.167}$$

where $\dot{\theta}_L$ = load rate, rad/sec

A_R = piston area

r = rocker-arm radius driving load

The maximum servo torque capability is

$$T = P_L A_R r \tag{9.168}$$

where P_L = load pressure.

The foregoing equations provide a means for determining motor flow to the cylinders. Under dynamic conditions this flow is divided into compressible flow and ram flow, i.e., the flow necessary to fill swept-out volume created by piston motion, thus

$$Q_{M_1} = Q_{C_1} + A_R\dot{Y} \tag{9.169}$$

where $\dot{Y}$ = piston rate

Q_{C_1} = compressible flow

The expression for compressible flow Q_{C_1} is obtained from the general expression for isentropic flow; thus

$$P_1V_1^K = \text{const}$$

Differentiating Eq. (9.169) results in

$$KP_1V_1^{K-1}(dV_1/dP_1) + V_1^K = 0 \tag{9.170}$$
$$Q_{C_1} = -dV_1/dt = -(V_1/KP_1)(dP_1/dt) \tag{9.171}$$

If the operation is considered about the quiescent point, then

$$Q_{C_1} = (V_0/KP_0)(dP_1/dt) \tag{9.172}$$

Combining Eqs. (9.165), (9.169), and (9.171) results in

$$(V_0/KP_0)(dP_1/dt) + A_R\dot{Y}_1 = Q_0[(C/a)(P_s/P_1)(A'/A)_1 - (1 - x/x_0)] \tag{9.173}$$

and for side two:

$$(V_0/KP_0)(dP_2/dt) + A_R\dot{Y}_2 = Q_0[(C/a)(P_s/P_2)(A'/A)_2 - (1 + x/x_0)] \tag{9.174}$$

When operating about the quiescent point the following linearizations can be made:

$$(A'/A)_1 \approx (A'/A)_2 \approx 1$$
$$P_1 = P_0 + \Delta P_1$$
$$P_2 = P_0 + \Delta P_2$$
$$Y_1 = -Y_2 = Y_0$$

Equations (9.168) and (9.169) can now be combined to yield

$$\frac{V_0}{KP_0}\frac{dP_L}{dt} + 2A_R\dot{Y} = \frac{Q_0CP_s}{2}\frac{-P_2}{P_0^2 + \Delta P_1P_0 + \Delta P_2P_0} + 2Q_0\frac{x}{x_0} \tag{9.175}$$

From Eq. (9.165) or (9.166)

$$P_0 = CP_s/a \tag{9.176}$$

For small perturbations about the quiescent point, $\Delta P_1 P_0 + \Delta P_2 P_0$ can be neglected. The describing function for the valve is

$$Q_0(x/x_0) = (V_0 a/2KCP_s)(dP_2/dt) + A_R Y + (Q_0 a/2CP_s)P_L \qquad (9.177)$$

Consider the load for the dynamic analysis to be a simple spring-mass system, thus

$$P_L A_R = m\ddot{Y} + f\dot{Y} + KY \qquad (9.178)$$

where m = load mass
f = viscous-damping coefficient
K = spring constant

Torque Motor. The torque motor has a single lag which is sufficiently high in frequency so that it can usually be neglected. However, the flapper spring system cannot be neglected and is expressed as follows:

$$K_T \Delta i = Kx + P_L A_N \qquad (9.179)$$

where Δi = differential coil current
K_T = torque motor constant
K = spring rate of torque motor
A_N = effective force area

Amplifier and Feedback Transducer

$$\Delta i = K(e - K_f Y) \qquad (9.180)$$

where e = error signal
K_f = feedback sensitivity

Transforming Eqs. (9.175), (9.178), (9.179), and (9.180) results in the block diagram of Fig. 9.30. The lead-lag network is for compensation stabilization. Additional information can be found in refs. 2, 4, and 6.

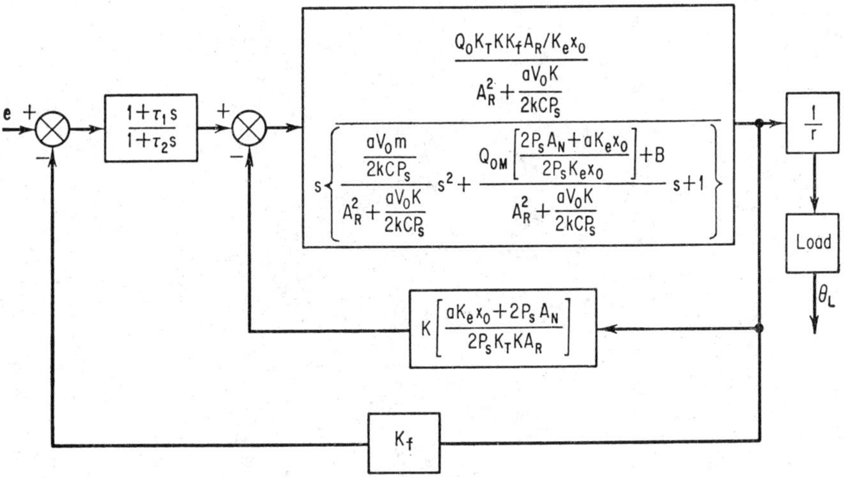

FIG. 9.30. Block diagram of pneumatic servo.

References

1. Chestnut, H., and R. W. Mayer: "Servomechanisms and Regulating System Design," vol. 1, John Wiley & Sons, Inc., New York, 1951.
2. Ordinance Corps: "Ordinance Engineering Design Handbook—Servomechanisms," Sec. 1, Theory, Office of the Chief of Ordinance, Washington, D.C., August, 1959.
3. Ahrendt, W. R., and C. J. Savant, Jr.: "Servomechanism Practice," 2d ed., McGraw-Hill Book Company, Inc., New York, 1960.

4. Raven, F. R.: "Automatic Control Engineering," McGraw-Hill Book Company, Inc., New York, 1961.
5. Korn, G. A., and T. M. Korn: "Electronic Analog Computers," 2d ed., McGraw-Hill Book Company, Inc., New York, 1956.
6. Truxal, J. C.: "Control Engineers' Handbook," Sec. 14, McGraw-Hill Book Company, Inc., New York, 1958.
7. Graham, Dunstan: Magnetic Clutches Add Muscle to Electronic Servos, *Space/Aeronautics*, vol. 31, no. 4, pp. 139, 141, April, 1959.
8. Hruby, R. J.: Optimum Compensation of a Position Servo with a Magnetic Clutch Actuator, *IRE Trans. Auto. Control*, vol. A.C.-5, no. 3, p. 220, August, 1960.
9. Del Toro, V., and S. Parker: "Principles of Control System Engineering," McGraw-Hill Book Company, Inc., New York, 1960.

Section 10

MACHINE SYSTEMS

By

HAROLD A. ROTHBART, D. Eng., *Dean, College of Science and Engineering Fairleigh Dickinson University, Teaneck, N.J.*

CONTENTS

10.1. SYSTEMS IN GENERAL

(a) Introduction

A *machine* shall be defined herein as a system of mechanical elements whose purpose it is to translate, transmit, or transfer a force, motion, or energy input at one or more elements (input) into a prescribed force, motion, or energy at one or more other elements (output). Its geometry is called the *mechanism*.

The primary purpose of dynamic analysis is to predict the nature and magnitude of forces and motion existing at foundations and machine elements so that stress or motion analysis can be evaluated and function can be sustained. The reader is referred to ref. 19 for more information on foundations of machines.

Degrees of Freedom. The minimum number of coordinates which must be specified to define the spatial configuration of a system is termed the *number of degrees of freedom of the system*. Examples of a single-degree-of-freedom system are the ideal slider crank or four-bar planer mechanisms, the angular position of any linkage uniquely determining the space configuration of the entire mechanism. Owing to flexibility (elastic behavior) either by design (by deliberate use of springs) or by virtue of deflection of members under loading, all real systems are of the multi-degree-of-freedom type. Backlash (i.e., free motion) also increases the number of degrees of freedom. Consequently the single-degree-of-freedom system is an idealization of a system connected by rigid elements (links) in the absence of backlash. All other systems are of a higher degree of freedom. Before performing analysis, interest is centered first on the number of degrees of freedom, since this provides some measure of the complexity.

(b) General Approach to the Machine System

Any system, however complex, can be analyzed as follows: Frame the physical model of the system to render it for analysis, choose the method of dynamic analysis, formulate the problem by writing the equations of dynamics and constraints, and then solve the problem.

Framing the Physical Model. This is the most important part of the overall analysis. It involves a comprehensive study of the system under static as well as dynamic conditions. Since all components have continuously distributed mass, springs, and damping, formulating an exact physical model would involve a prohibitive amount of labor in addition to the theoretical problems associated with, in particular, the elemental damping behavior, under dynamic conditions. Moreover, even if such a model were available, the mathematical difficulties could be insurmountable. The analyst must usually make a decision, therefore, to simplify the problem in order to render it mathematically manageable but at the same time of sufficient accuracy to yield reliable results. Herein lies the heart of analysis. The model can often be simplified by lumping physical properties at discrete stations, as, for example, in torsional systems and in translational vibrating systems where judicious choices of lumping stations are made to retain sufficient accuracy. In mechanisms where deflections are small, linkages can often be considered rigid, leading to considerable simplification in analysis.

Successive simplifications of the model often involve reduction in the number of degrees of freedom. Indeed the simplest, but not necessarily reliable, system is the single-degree-of-freedom system, whose omnipresence especially in mechanism analysis is all too well known.

Method of Analysis. Having established the physical model the method of analysis based upon laws derivable from Newtonian mechanics must be selected. The optimum method from the standpoint of minimizing labor depends upon the physical model. The most useful methods are given below together with their corresponding areas of application. A description and some of the foundations of their development are given later in this section.

Method	*Area of application*
1. Lagrange equations of motion	Most powerful method. Can be used for all systems. Recommended for all systems with multiconstraints where friction is either absent or of the viscous type
2. Direct application of Newton's laws with d'Alembert's principle	Can be used for all systems. Recommended for all one-degree-of-freedom systems with motion prescribed in time. Multi-degree-of-freedom systems with no constraints or linear constraints; e.g., linear vibrating systems

Method	*Area of application*
3. Euler's equations of rigid-body dynamics	Systems of one degree of freedom containing rigid bodies—motion prescribed in time
4. Virtual work	One-degree-of-freedom conservative or nonconservative systems with viscous friction. Used to determine end-point (reaction) forces

Formulation of the Problem. The formulation involves writing the equations of motion and of constraints imposed by the geometry. Note that the equations of Lagrange may contain n generalized coordinates, in order to describe the dynamics. If the number of degrees of freedom of the system is m, the system has only m *independent* generalized coordinates. Therefore, $n - m$ equations of constraint must be found to complete the formulation.

Method of Solution. The complexity of the system (which depends upon the number of degrees of freedom, linearity or nonlinearity), and the constraints, usually dictate the method. In all but a few cases, numerical or digital-computer methods must be employed owing primarily to the general nonlinear nature of most systems. Such methods are discussed in Secs. 2, 6, and 8. The exceptional case of linear multi-degree-of-freedom systems with no constraints or linear constraints can be handled by noncomputer techniques. This is the case of linear vibrating systems treated in Sec. 6. Note that linear constraints means that the generalized variables are connected by algebraic equations of the first degree only.

The special case of one degree of freedom with motion given in time is amenable to graphical as well as analytical analysis (see Sec. 4).

The general case of multidegree of freedom with nonlinear friction, constraints, and forces will always yield to analysis. The method employing Lagrange's equations stands out as the most powerful method of dealing with this type of situation. If the friction is linear (viscous), depending upon the velocity difference of two generalized coordinates, the equation of Lagrange is straightforward. Under the more general conditions of nonlinear friction, the friction terms can be treated as external forces of unknown magnitude. By starting with an assumed frictional force (usually zero) a first solution to the equations of motion will yield another frictional force. This force(s) is now introduced into the equation of motion and the system solved once again, yielding another refinement. This iterative process, which lends itself to computer methods, is carried out until there is an insignificant difference in two successive results.

The infinity of possible machine systems and types cannot of course be covered in a single section or indeed an entire book; accordingly this section will treat some of the more important systems as regards practicability and exemplification of principles.

(c) Review of Classical Dynamics

Newtonian Dynamics. The following summarizes the more important concepts drawn from Sec. 3 for application to the greatest variety of practical systems of machinery. In exceptional cases resort can be taken to the fundamental developments of that section.

In general the center of mass of a system moves as if the entire mass were located at the center of mass (CM) and all the external forces were acting at that point, thus,

$$M d^2\mathbf{R}_{cm}/dt^2 = \Sigma \mathbf{F}_i^e = \mathbf{F}^e \tag{10.1}$$

where $\mathbf{F}^e$ represents the sum of external forces $\mathbf{F}_i^e$.

The time rate of change of angular momentum is equal to the external torque when referred to any inertial frame of reference, to the center of mass (CM), or to an axis moving parallel to the center of mass.

$$(d/dt)\mathbf{H}(C) = \Sigma \mathbf{r}_i \times \mathbf{F}_i^e = \mathbf{N}(C) \tag{10.2}$$

where $\mathbf{H}(C)$ and $\mathbf{N}(C)$ are the angular momentum and torque, respectively, referred to axis C, or in most practical cases,

$$(d\mathbf{H}/dt)(\mathrm{CM}) = \Sigma \mathbf{r}_j \times \mathbf{F}_j^e = \mathbf{N}(\mathrm{CM}) \tag{10.3}$$

when referred to CM axes.

Rigid-body Dynamics—Euler. The angular momentum of a rigid body is expressed in matrix form by

$$\mathbf{H}(C) = \begin{bmatrix} I_{xx} & -I_{xy} & -I_{xz} \\ -I_{yx} & I_{yy} & -I_{yz} \\ -I_{zx} & -I_{zy} & I_{zz} \end{bmatrix} \begin{bmatrix} \omega_x \\ \omega_y \\ \omega_z \end{bmatrix} = [I]\boldsymbol{\omega} \tag{10.4}$$

when referred to a set of mutually perpendicular axes at a point C where it is understood that the elements of the column $\mathbf{H}(C)$ are the three components of $\mathbf{H}(C)$ in the x, y, z directions in order from top to bottom. It is always possible to find another set of axes oriented at the same origin to reduce the matrix to diagonal form if the origin is at the center of mass.

$$\mathbf{H}'(C^*) = \begin{bmatrix} I_{xx}' & 0 & 0 \\ 0 & I_{yy}' & 0 \\ 0 & 0 & I_{zz}' \end{bmatrix} \begin{bmatrix} \omega_x' \\ \omega_y' \\ \omega_z' \end{bmatrix} = \begin{bmatrix} I_{xx}' & \omega_x' \\ I_{yy}' & \omega_y' \\ I_{zz}' & \omega_z' \end{bmatrix} \tag{10.5}$$

where the primes denote a change of coordinate system, the new axes x', y', z' are the *principal axes* of motion, and the asterisk applies to the principal axis.

Equation (10.5) can be expressed in more compact form as

$$\mathbf{H}(C) = [I]\boldsymbol{\omega} \tag{10.5a}$$

The time rate of change of any vector $\mathbf{A}$ as measured in a system S is related to that measured in another frame S_1 by

$$(d\mathbf{A}/dt)_S = (d\mathbf{A}/dt)_{S_1} + \boldsymbol{\omega} \times \mathbf{A} \tag{10.6}$$

where $\boldsymbol{\omega}$ is the angular velocity of S_1 relative to S. Two cases of widest interest are where S_1 is fixed to a rigid body with the axis at the CM:

Case 1. S is fixed in space.

Case 2. S is a reference frame whose origin is fixed to the CM with direction of axes remaining fixed.

Substituting the vector $\mathbf{H}(C)$ for $\mathbf{A}$ in Eq. (10.6) yields

$$(d\mathbf{H}(C)/dt)_S = \begin{bmatrix} I_{xx} & -I_{xy} & -I_{xz} \\ -I_{yx} & I_{yy} & -I_{yz} \\ -I_{zx} & -I_{zy} & I_{zz} \end{bmatrix} \begin{bmatrix} \dot\omega_x \\ \dot\omega_y \\ \dot\omega_z \end{bmatrix} + \begin{bmatrix} 0 & -\omega_z & \omega_y \\ \omega_z & 0 & -\omega_x \\ -\omega_y & \omega_x & 0 \end{bmatrix} [I] \begin{bmatrix} \omega_x \\ \omega_y \\ \omega_z \end{bmatrix} = (\mathbf{N}(C))_S \tag{10.7}$$

For the two cases above,

Case 1 $\qquad d\mathbf{H}(C)/dt = [I]\dot{\boldsymbol{\omega}} + \boldsymbol{\omega} \times \mathbf{H}(C) = \mathbf{N}(C) \qquad (10.8)$

Case 2 $\qquad d\mathbf{H}(\mathrm{CM})/dt = [I]\dot{\boldsymbol{\omega}} + \boldsymbol{\omega} \times \mathbf{H}(\mathrm{CM}) = \mathbf{N}(\mathrm{CM}) \qquad (10.9)$

Virtual Work. The concept of virtual work is essential in the classical development of Lagrange and is useful in limited engineering applications where end-point forces are desired. It assumes an instantaneous infinitesimal displacement of the system in zero time which is necessary for systems with moving constraints which are prescribed in time. In the absence of these constraints the zero-time condition can be abandoned and these so-called "virtual" displacements become real. The literature does not, however, distinguish, using "virtual" for both types of system. The fundamental statement of the theorem is: The virtual work accompanying a virtual displacement is zero:

$$0 = \Sigma \mathbf{F}_i^a \cdot \delta \mathbf{r}_i + \Sigma \mathbf{F}_i^c \cdot \delta \mathbf{r}_i = \Sigma \mathbf{F}_i \cdot \delta \mathbf{r}_i \tag{10.10}$$

where the first term represents the work of the external + inertia forces $\mathbf{F}_i^a$ and the second term the work of the constraint forces $\mathbf{F}_i^c$. If generalized coordinates are used,

$$0 = \Sigma Q_j^a \, \delta q_j + \Sigma Q_j^c \, \delta q_j \tag{10.11}$$

where

$$\mathbf{r}_i = \mathbf{r}_i(q_1, q_2, \ldots, q_n)$$

$$Q_j^a = \sum_i (\partial \mathbf{r}_i/\partial q_j) \cdot \mathbf{F}_i^a \qquad Q_j^c = \sum_i (\partial \mathbf{r}_i/\partial q_j) \cdot \mathbf{F}_i^c$$

where the vectorial representation is eliminated. If the system is *conservative*, the second term on the right-hand side of Eq. (10.11) vanishes since motion is then transverse to the surface of constraint. This then affords great simplification and is really the great contribution of Lagrange's equations of motion, which are remarkably independent of the explicit knowledge of these forces. On the other hand, to determine the dissipative system, both terms on the right-hand side of Eq. (10.11) are required, which in general adds complication to Lagrange formulations.

10.2. SINGLE-DEGREE-OF-FREEDOM SYSTEMS (RIGID-BODY ELEMENTS)

(a) Introduction

This is the simplest model possible. It is a valid model for machines which are free of springs, containing members which are connected with negligible backlash and which experience insignificant deflection and store negligible energy under dynamic conditions. Ideally the members are termed *rigid* to imply constancy of geometry or total absence of strain. The single-degree-of-freedom system under general loading conditions, despite its relative simplicity compared with the actual multi-degree-of-freedom system, presents a formidable problem of analysis.

(b) System Motion Prescribed in Time

If motion of one of the members is given as a function of time (and therefore all motion can be obtained) vast simplifications in analysis are possible. In this case the usual problems associated with interdependence of forces and motion are eliminated; i.e., in general, forces exist as nonlinear differential functions of the motion. Since in this case the motion is known, these differential functions become explicit functions of time. For motion prescribed in time the superposition property holds; i.e., at any instant of time the constraint forces can be obtained by summing the effects of individual applied forces or d'Alembert forces acting alone.

This special case is a fair approximation to the "steady-state" operation of machines containing heavy inertial elements (e.g., flywheels) since in the "steady state," the flywheel maintains essentially uniform rotational velocity. It is important to note that "steady state" is a state of approximate uniform motion since energy is constantly absorbed by and delivered to the flywheel at unequal rates; only the average rates over a cycle are equal. Furthermore, the transient problem of starting the system from zero to "steady state" clearly does not fall under the above classification but must be treated with more generality.

(c) Three-dimensional System—Motion Prescribed in Time

It is possible to subdivide the general three-dimensional system into "free bodies" and apply the rigid-body equations of Euler to each free body. It is common practice to choose whole elements as "free bodies," each member treated as a rigid body. As shown in Fig. 10.1, the forces acting on any member are the externally applied force(s) and the forces of constraint imposed by adjoining members.

Without loss of generality the applied force can be resolved into a force passing through the center of mass of the member, and a couple. The unqualified term *force* shall refer to the force vector made up of three components of force and three components of moment given by either the column vector or usual vector notation as follows:

$$\mathbf{F} = \begin{bmatrix} f_x \\ f_y \\ f_z \\ N_x \\ N_y \\ N_z \end{bmatrix} = \begin{bmatrix} \mathbf{f} \\ \mathbf{N} \end{bmatrix} \qquad \mathbf{f} = \begin{bmatrix} f_x \\ f_y \\ f_z \end{bmatrix} \quad \mathbf{N} = \begin{bmatrix} N_x \\ N_y \\ N_z \end{bmatrix} \tag{10.12}$$

where, for example, f_x is a force directed along the positive x axis and N_y is a moment about the y axis using the usual right-hand rule. The origin of the body-fixed axes will, unless otherwise specified, be fixed at the center of mass of the linkage with the directions always fixed.

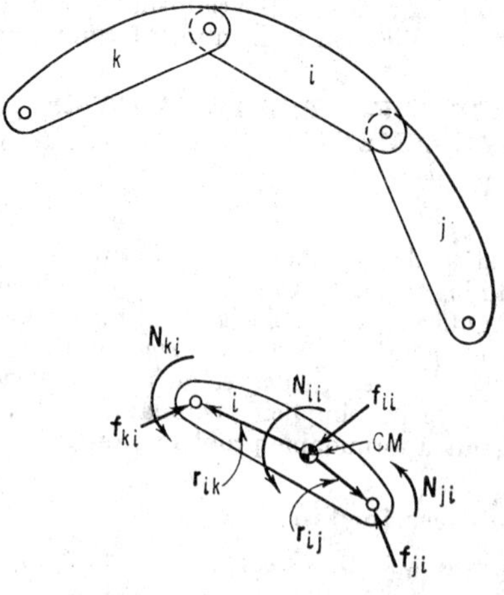

FIG. 10.1. General mechanism.

The unknown forces are the internal $\mathbf{f}_{lm}$ as well as the torque on the input (output) element. Their evaluation is the essence of the dynamics problem. If there are N links, there are N junctions and hence N internal force vectors to be evaluated in addition to the input force. If $\mathbf{f}_{lm}$ is the internal force of the lth member on the mth member then the reciprocity relation expressing equality of action and reaction at a point is

$$\mathbf{f}_{lm} = -\mathbf{f}_{ml}$$

A free-body analysis for each link i yields the unbalanced force $\mathbf{F}_i$.

$$\mathbf{F}_i = \begin{bmatrix} \mathbf{f}_i \\ \mathbf{N}_i \end{bmatrix} \begin{aligned} &= \mathbf{f}_{ji} + \mathbf{f}_{ki} + \mathbf{f}_{ii}^{\text{ext}} \\ &= \mathbf{r}_{ij} \times \mathbf{f}_{ji} + \mathbf{r}_{ik} \times \mathbf{f}_{ki} + \mathbf{N}_{ji} + \mathbf{N}_{ki} + \mathbf{N}_{ii}^{\text{ext}} \end{aligned} \tag{10.13}$$

The right-hand side of Eq. (10.3) can be equated to $\mathbf{N}_i$ while the left-hand side of Eq. (10.1) is equated to $\mathbf{f}_i$.

$$(Md^2/dt^2)\mathbf{R}_{cm(i)} = \mathbf{f}_i$$

$$d\mathbf{H}(C)_i/dt = [I]\begin{bmatrix} \dot{\omega}_x \\ \dot{\omega}_y \\ \dot{\omega}_z \end{bmatrix} + \begin{bmatrix} 0 & -\omega_z & \omega_y \\ \omega_z & 0 & -\omega_x \\ -\omega_y & \omega_x & 0 \end{bmatrix} [I] \begin{bmatrix} \omega_x \\ \omega_y \\ \omega_z \end{bmatrix} = \mathbf{N}_i \tag{10.14}$$

Since the motion is known, the left-hand side of Eq. (10.14) is known as well as the external force $\mathbf{f}_{ii}$; $\mathbf{N}_{ii} = \mathbf{N}_{ii}^{\text{ext}}$ + input + torque. Since there are N links, there are N force vectors and an input force which are unknown, and if a slider is present, the position of the reaction force is also unknown. Further, if one of the links is grounded to earth, its equation of motion is trivially useless, and $N - 1$ nontrivial relations of the type Eq. (10.14) remain. The interacting moments are often linear functions of the corresponding $\mathbf{f}_{mn}$ and opposite to the motion (Coulomb law), in which case

$$\mathbf{N}_{nm} \propto |\mathbf{f}_{mn} \times \boldsymbol{\omega}_{nm}|[\boldsymbol{\omega}_{nm}/|\boldsymbol{\omega}_{nm}|^2] \tag{10.15}$$

where $\boldsymbol{\omega}_{nm} = \boldsymbol{\omega}_n - \boldsymbol{\omega}_m$; or for viscous friction takes the form

$$\mathbf{N}_{nm} \propto \boldsymbol{\omega}_{nm} \tag{10.16}$$

Under these general conditions, there are $3N$ unknown reaction forces plus the input force vector containing six unknown components with $6(N - 1)$ equations. Equating number of unknowns to equations,

$$\begin{aligned} 6(N - 1) &= 3N + 6 \\ N &= 4 \end{aligned} \tag{10.17}$$

which shows that four members yield a compatible system.

Considerable care must usually be exercised in the formulation since there is often an excess of equations over unknowns. This is because the several equations are not always independent.

(d) Motion with Unidirectional Angular Velocity

Under these conditions the set Eq. (10.14) reduces to

$$\frac{M\, d^2\mathbf{R}_{cm}}{dt^2} = \begin{bmatrix} m\ddot{x}_c \\ m\ddot{y}_c \\ m\ddot{z}_c \end{bmatrix} \begin{matrix} = \\ = \\ = \end{matrix} \begin{bmatrix} f_x \\ f_y \\ f_z \end{bmatrix} \tag{10.18a}$$

$$\frac{d\mathbf{H}(C)}{dt} = \begin{bmatrix} -I_{xz}\dot{\omega} & +I_{yz}\omega^2 \\ -I_{yz}\dot{\omega} & -I_{xz}\omega^2 \\ +I_{zz}\dot{\omega} & +I_{zz}\dot{\omega} \end{bmatrix} \begin{matrix} = \\ = \\ = \end{matrix} \begin{bmatrix} N_x \\ N_y \\ N_z \end{bmatrix} \tag{10.18b}$$

$$\boldsymbol{\omega} = \mathbf{k}\omega_z \tag{10.18c}$$

where $\boldsymbol{\omega}$ is the angular velocity having only a component normal to the planes of motion.

(e) Principles of Statics Applied to Dynamic Systems with Motion Prescribed in Time

A system in equilibrium or static equilibrium implies

$$\frac{d\mathbf{H}(C)}{dt} = \mathbf{N} = \begin{bmatrix} N_x \\ N_y \\ N_z \end{bmatrix} = 0 = \begin{bmatrix} 0 \\ 0 \\ 0 \end{bmatrix} \tag{10.19}$$

$$\frac{d}{dt}\frac{M\, d\mathbf{R}_{cm}}{dt} = \frac{M\, d^2\mathbf{R}_{cm}}{dt^2} = M \begin{bmatrix} \ddot{x} \\ \ddot{y} \\ \ddot{z} \end{bmatrix} = \begin{bmatrix} f_x \\ f_y \\ f_z \end{bmatrix} = \begin{bmatrix} 0 \\ 0 \\ 0 \end{bmatrix} \tag{10.20}$$

i.e., if the angular momentum and linear momentum are constants of the motion, the system is in a state of equilibrium. For convenience $H(C)$ is referred to CM axes, as will be the usual case in dealing with machine systems.

For a dynamic system, Newton's laws take the form

$$d\mathbf{H}(C)/dt = \mathbf{N} \qquad (M\, d^2\mathbf{R}_{cm}/dt^2) = \mathbf{f} \tag{10.21}$$

By simple rearrangement of these equations,

$$0 = \mathbf{N} - d\mathbf{H}(C)/dt \tag{10.22}$$

$$0 = \mathbf{f} - M\, d^2\mathbf{R}_{cm}/dt^2 \tag{10.23}$$

If $-d\mathbf{H}(C)/dt$ and $-M\, d^2\mathbf{R}_{cm}/dt^2$ are considered forces, the simple substitution

$$\mathbf{N}' = \mathbf{N} - d\mathbf{H}(C)/dt \tag{10.24}$$
$$\mathbf{f}' = \mathbf{f} - M\, d^2\mathbf{R}_{cm}/dt^2 \tag{10.25}$$

yields the seemingly trivial results

$$\mathbf{N}' = 0 \qquad \mathbf{f}' = 0 \tag{10.26}$$

which is formally the same as Eqs. (10.19) and (10.20). The forces $-d\mathbf{H}(C)/dt$ and $-d^2\mathbf{R}_{cm}/dt^2$ are known as d'Alembert or inertia forces. Hence the dynamic system can be treated as a static one by imposing a linear force $-M\, d^2\mathbf{R}_{cm}/dt^2$ at the CM and a moment $-d\mathbf{H}(C)/dt$ about the CM axes, to be considered in addition to the actual forces present and then invoking Eq. (10.26), the equation of statics.

(f) Reaction Forces—Motion Prescribed in Time

It is not unusual to be interested in reaction or end-point forces to the exclusion of all other internal forces. In this case the "free body" can be taken as the ensemble

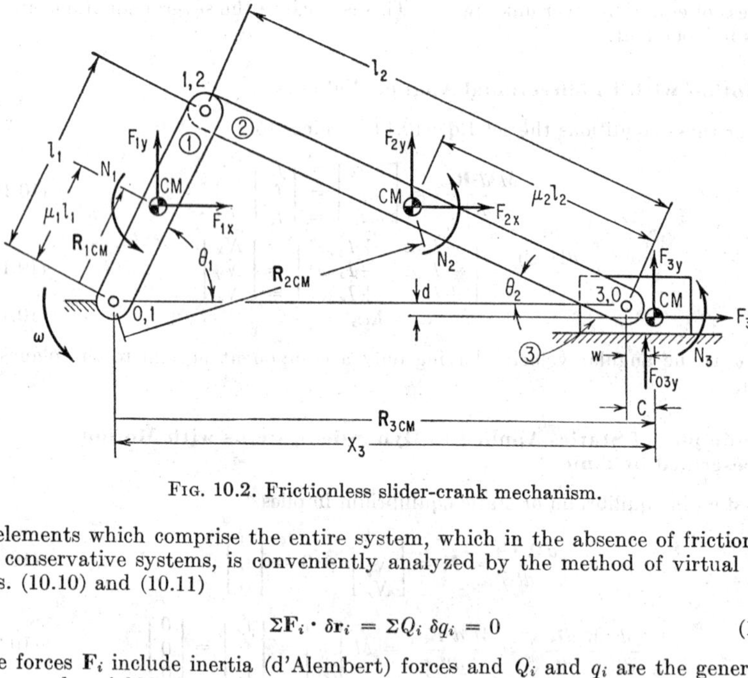

FIG. 10.2. Frictionless slider-crank mechanism.

of elements which comprise the entire system, which in the absence of friction, i.e., for conservative systems, is conveniently analyzed by the method of virtual work: Eqs. (10.10) and (10.11)

$$\Sigma\mathbf{F}_i \cdot \delta\mathbf{r}_i = \Sigma Q_i\, \delta q_i = 0 \tag{10.27}$$

The forces $\mathbf{F}_i$ include inertia (d'Alembert) forces and Q_i and q_i are the generalized forces and variables, respectively.

As an example consider the frictionless slider-crank mechanism of Fig. 10.2. Generalized coordinates and forces will be used. From the known motion of one link, say the crank, kinematic analysis yields the entire motion for any mass. The motion of the CM and the angular momentum about the CM are

$$\mathbf{R}_{i(cm)} = \mathbf{R}_{i(cm)}(\theta) \qquad \mathbf{H}_i = \mathbf{H}_{iz} = \mathbf{H}_i(\theta,\dot\theta) = \dot\theta f(\theta) \tag{10.28}$$

whence the d'Alembert forces can be determined by taking derivatives.

$$\begin{aligned} -M_i\ddot{\mathbf{R}}_{i(cm)} &= -M_i\{\ddot\theta[d\mathbf{R}_i(\theta)/d\theta] + [d^2\mathbf{R}_i(\theta)/d\theta^2]\dot\theta^2\} \\ -d\mathbf{H}_{iz}/dt &= -\{\ddot\theta f(\theta) + \dot\theta^2[df(\theta)/d\theta]\} \end{aligned} \tag{10.29}$$

In terms of generalized coordinates, the d'Alembert forces

$$f_{jx} = -M_j\ddot{x}_{j(cm)} = -M_j\{\ddot{\theta}[dx_j(\theta)/d\theta] + \dot{\theta}^2[d^2x_j(\theta)/d\theta^2]\} \tag{10.30}$$
$$f_{jy} = -M_j\ddot{y}_{j(cm)} = -M_j\{\ddot{\theta}[dy_j(\theta)/d\theta] + \dot{\theta}^2[d^2y_j(\theta)/d\theta^2]\} \tag{10.31}$$
$$f_{jz} = -dH_j/dt = -\ddot{\theta}f_j(\theta) + \dot{\theta}^2(df_j\theta/d\theta) \tag{10.32}$$

The external forces and inertia forces are now all known with the exception of N_1 or F_{3x}. As stated earlier, these forces can always be replaced by a linear force passing through the CM of the member plus a moment about the CM; so that without loss of generality, the force structure is that shown in Eq. (10.2).

$$\Sigma\mathbf{Q}_i \cdot \delta\mathbf{q}_i = N_1\,\delta\theta_1 + F_{1x}\,\delta X_1 + F_{1y}\,\delta Y_1 + N_2\,\delta\theta_2 + F_{2x}\,\delta X_2 + F_{2y}\,\delta Y_2 + F_{3x}\,\delta X_3 + F_{3y}(0) = 0 \tag{10.33}$$

where the subscripts 1, 2, 3 refer to the crank, connecting rod, and piston, respectively; the displacements x, y, z the motion of the CM; and θ_i the angular motion of the members. Included among the forces are those of inertia obtained in the foregoing.

The generalized coordinates θ_1, θ_2, X_1, Y_1, X_2, Y_2, X_3 are connected by six equations of constraint:

$$l_1 \cos\theta_1 + l_2 \cos\theta_2 = X_3 - C \tag{10.34}$$
$$l_1 \sin\theta_1 = l_2 \sin\theta_2 \tag{10.35}$$
$$X_1 = \mu_1 l_1 \cos\theta_1 \tag{10.36}$$
$$Y_1 = \mu_1 l_1 \sin\theta_1 \tag{10.37}$$
$$X_2 = l_1\cos\theta_1 + (1 - \mu_2)l_2\cos\theta_2 \tag{10.38}$$
$$Y_2 = \mu_2 l_2 \sin\theta_2 \tag{10.39}$$

Assuming that all forces except N_1 are given, Eqs. (10.33) and (10.34) to (10.39) yield N_1, the total input moment. The force structure has now been evaluated except for the reaction forces. These are readily obtainable by the usual methods, i.e., static analysis about convenient points, in order to satisfy

$$\mathbf{N}' = 0 \qquad \Sigma\mathbf{f}' = 0 \tag{10.40}$$

which could be obtained by virtual-work considerations.

Consider the simple well-known frictionless case where gravity and inertia forces are negligible and the only external force F_{3x} is given. Then Eq. (10.33) becomes

$$N_1\,\delta\theta_1 + F_{3x}\,\delta X_3 = 0 \tag{10.41}$$

whence $N_1 = -F_{3x}(dX_3/d\theta_1)$. It remains only to find $dX_3/d\theta_1$. From the geometry,

$$(X_3 - l_1\cos\theta_1)^2 + (l_1\sin\theta_1)^2 = l_2^2 \tag{10.42}$$

Differentiating and simplifying yields

$$\frac{dX_3}{d\theta_1} = \frac{l_1X_3\sin\theta_1}{l_1\cos\theta_1 - X_3} \tag{10.43}$$

which permits an explicit evaluation of N_1. The subscript 0 represents the ground. F_{03Y} is obtained by taking moments about the junction 01 after establishing $\omega = 0$ from a free-body analysis of the slider.

$$F_{03Y}(X_3 - C + w) + N_1 + F_{3X}d = 0 \tag{10.44}$$
$$F_{03Y}(X_3 - C) + N_1 + F_{3X}d = 0$$

From
$$F_{01Y} = -F_{03Y}$$
$$F_{01X} = -F_{3X}$$

If friction of the viscous type is present, the equation of virtual work can still be used. The transverse force of friction f_{nm} is a function of velocity difference of two generalized coordinates q_n and q_m only, thus

$$f_{nm} \propto (\dot{q}_n - \dot{q}_m) \tag{10.45}$$

Eq. (10.33) is then modified to read

$$N_1\,\delta\theta_1 + \underset{\sim\sim\sim}{N_{01}\,\delta\theta_1} + \underset{\sim\sim\sim}{N_{21}\,\delta\theta_1} + F_{1X}\,\delta X_1 + F_{1Y}\,\delta Y_1 + N_2\,\delta\theta_2 + \underset{\sim\sim\sim}{N_{12}\,\delta\theta_2} + \underset{\sim\sim\sim}{N_{32}\,\delta\theta_2} + F_{2X}\,\delta X_2 + F_{2Y}\,\delta Y_2 + \underset{\sim\sim\sim}{F_{03X}\,\delta X_3} + F_{3X}\,\delta X_3 = 0 \quad (10.46)$$

where the wavy line indicates the additional frictional terms.

(g) Examples—System Motion Prescribed in Time

Example 1: Consider the simple machine system in Fig. 10.3 consisting of a link rotating about the z axis with angular velocity $\boldsymbol{\omega} = \mathbf{k}\omega$ and accelerating with $\dot{\boldsymbol{\omega}} = \mathbf{k}\dot{\omega}$. It is required to find the reaction force $\mathbf{f}_{01}$ and the input torque $\mathbf{N}_{ext}$.

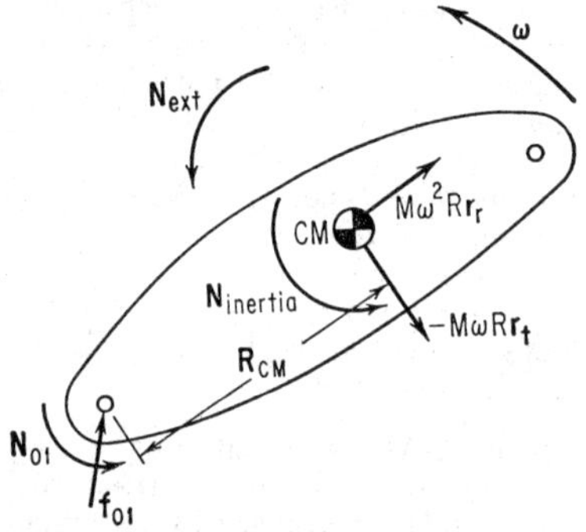

FIG. 10.3. Simple machine system.

Friction Absent. The inertia force is found using center-of-mass axes. The CM moves in a circle of radius R with acceleration

$$\ddot{\mathbf{R}}_{cm} = -\omega^2 R\mathbf{r}_r + \dot{\omega} R\mathbf{r}_t \quad (10.47)$$

where the vectors $\mathbf{r}_r$ and $\mathbf{r}_t$ are the unit radial and tangent vectors. The time rate of change of angular momentum is found using the form of Eq. (10.18) for motion in parallel planes:

$$\frac{d\mathbf{H}(\mathrm{CM})}{dt} = \begin{bmatrix} -I_{xz}\dot{\omega} & +I_{yz}\omega^2 \\ -I_{yz}\dot{\omega} & -I_{xz}\omega^2 \\ +I_{zz}\dot{\omega} & \end{bmatrix} = \begin{bmatrix} N_x \\ N_y \\ N_z \end{bmatrix} \quad (10.48)$$

The inertia force and moment are then

$$\mathbf{f}_{\text{inertia}} = -M\ddot{\mathbf{R}}_{cm} = M\omega^2 R\mathbf{r}_r - M\dot{\omega} R\mathbf{r}_t \quad (10.49)$$

$$\mathbf{N}_{\text{inertia}} = -\frac{d\mathbf{H}(\mathrm{CM})}{dt} = \begin{bmatrix} I_{xz}\dot{\omega} & -I_{yz}\omega^2 \\ I_{yz}\dot{\omega} & +I_{xz}\omega^2 \\ -I_{zz}\dot{\omega} & \end{bmatrix} \quad (10.50)$$

Invoking Eq. (10.26) yields

$$\mathbf{f}_{01} + \mathbf{f}_{\text{inertia}} = 0 \quad (10.51)$$

$$\mathbf{N}_{\text{ext}} + \mathbf{N}_{\text{inertia}} - \mathbf{R}_{cm} \times \mathbf{f}_{01} + \mathbf{N}_{01} = 0 \quad (10.52)$$

$$\mathbf{f}_{01} = -\mathbf{f}_{\text{inertia}} = -M\omega^2 R\mathbf{r}_r + M\dot{\omega} R\mathbf{r}_t \quad (10.53)$$

$$\mathbf{N}_{\text{ext}} = -\mathbf{N}_{\text{inertia}} - \mathbf{R}_{cm} \times \mathbf{f}_{\text{inertia}} = \begin{bmatrix} -I_{xz}\dot{\omega} & +I_{yz}\omega^2 \\ -I_{yz}\dot{\omega} & -I_{xz}\omega^2 \\ +I_{zz}\dot{\omega} & \end{bmatrix} + \begin{bmatrix} 0 \\ 0 \\ +MR^2\dot{\omega} \end{bmatrix} \quad (10.54)$$

$$\mathbf{N}_{\text{ext}} = \begin{bmatrix} -I_{xz}\dot{\omega} & +I_{yz}\omega^2 \\ -I_{yz}\dot{\omega} & -I_{xz}\omega^2 \\ +I_{zz}\dot{\omega} & +MR^2\dot{\omega} \end{bmatrix} \qquad \mathbf{N}_{01} = 0 \quad (10.55)$$

Friction Present. If friction is present in the viscous form

$$\mathbf{N}_{\text{friction}} = -\mu\omega \tag{10.56}$$

where μ is a proportionality constant, then

$$\mathbf{N}_{01} = \begin{bmatrix} 0 \\ 0 \\ -\mu\omega \end{bmatrix}$$

If the friction is of the Coulomb type then

$$\mathbf{N}_{01} = -\mu|\mathbf{f}_{01}| \frac{\omega}{|\omega|} = \begin{bmatrix} 0 \\ 0 \\ -\mu|\mathbf{f}_{01}| \end{bmatrix} \tag{10.57}$$

and Eq. (10.55) would be modified to read

$$\mathbf{N}_{\text{ext}} = \begin{bmatrix} -I_{xz}\dot{\omega} & +I_{yz}\omega^2 \\ -I_{yz}\dot{\omega} & -I_{xz}\omega^2 \\ I_{zz}\dot{\omega} & +MR^2\dot{\omega} \end{bmatrix} - \mathbf{N}_{01} \tag{10.55a}$$

Example 2. The Slider-crank Chain—Solution by "Direct Method" without Friction: First, the inertia forces are found which together with the externally applied forces yield the general force structure (in accordance with the foregoing) shown in Fig. 10.4, i.e., forces through the CM and moments about CM. Invoking laws of static equilibrium,

$$\mathbf{N}' = 0$$
$$\mathbf{f} = 0$$

where moments are taken about respective CM axis and the number of equations are implied,

$$\mathbf{r}_{10} \times \mathbf{f}_{01} + \mathbf{r}_{12} \times \mathbf{f}_{21} + \mathbf{N}_1 = 0 \qquad +1 \tag{10.58}$$
$$\mathbf{f}_{01} + \mathbf{f}_{11} + \mathbf{f}_{21} = 0 \qquad +2 \tag{10.59}$$
$$\mathbf{r}_{21} \times \mathbf{f}_{12} + \mathbf{r}_{23} \times \mathbf{f}_{32} + \mathbf{N}_2 = 0 \qquad +1 \tag{10.60}$$
$$\mathbf{f}_{12} + \mathbf{f}_{22} + \mathbf{f}_{32} = 0 \qquad +2 \tag{10.61}$$
$$\mathbf{r}_{30} \times \mathbf{f}_{03} + \mathbf{r}_{32} \times \mathbf{f}_{23} + \mathbf{N}_3 = 0 \qquad +1 \tag{10.62}$$
$$\mathbf{f}_{23} + \mathbf{f}_{33} + \mathbf{f}_{03} = 0 \qquad +2 \tag{10.63}$$

Note that the $\mathbf{r}_{ij}$ in this case are measured from the center of gravity. In the previous example (Fig. 10.3) the vectors were measured from the fixed point. Also the reciprocity relation

$$\mathbf{f}_{nm} = -\mathbf{f}_{mn} \tag{10.64}$$

must hold at every junction. The above is equivalent to nine equations in the nine unknown quantities f_{x01}, f_{y01}, f_{x12}, f_{y12}, f_{x23}, f_{y23}, f_{x03}, f_{y03}, N.

Coulomb Friction in Bearings and Slider. Solution by "Direct Method": To account for Coulomb friction at the link connections, these forms must be changed slightly to include terms of the type N_{nm} interacting moments. These are obtainable from elementary analysis of bearing friction.[12]

$$\hat{\mathbf{f}}_{nm} = \mathbf{f}_{nm} + \mu_{nm}\mathbf{f}_{nm}^* \tag{10.65}$$

N_{nm} is found from

$$|\mathbf{N}_{nm}| = |\mathbf{r}_0 \times \mu_{nm}\mathbf{f}_{nm}| = \frac{r_0\mu_{nm}}{\sqrt{1 + \mu_{nm}^2}} |\mathbf{f}_{nm}| \tag{10.66}$$

where the asterisk indicates a vector rotated 90° from the original direction and r_0 is the radius of the bearing ($r_0\mu_{nm}$ is approximately equal to the radius of the friction

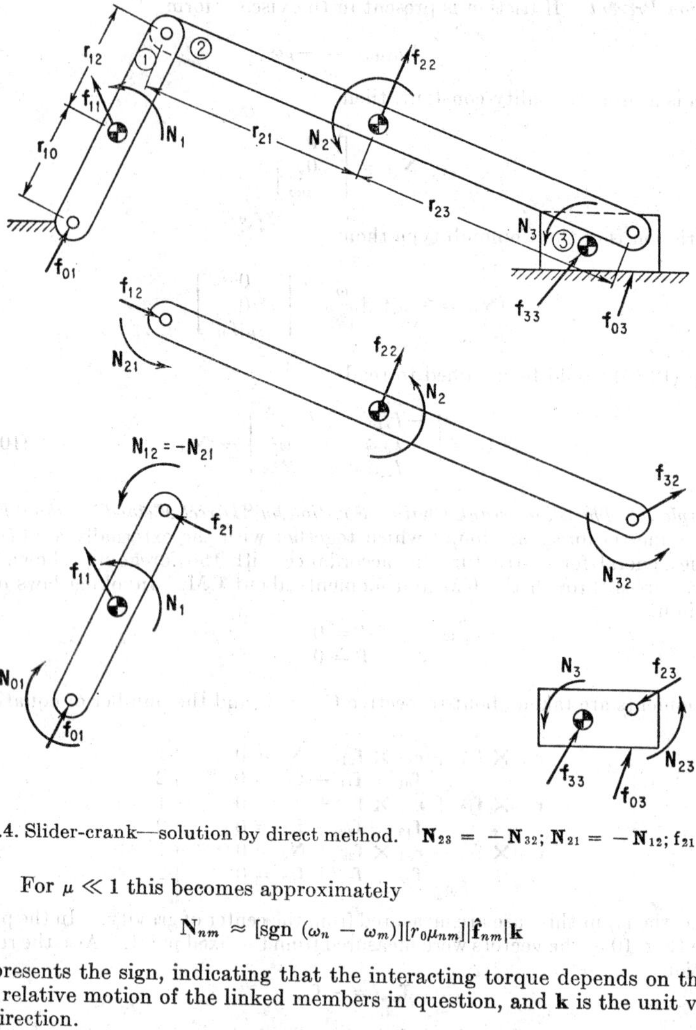

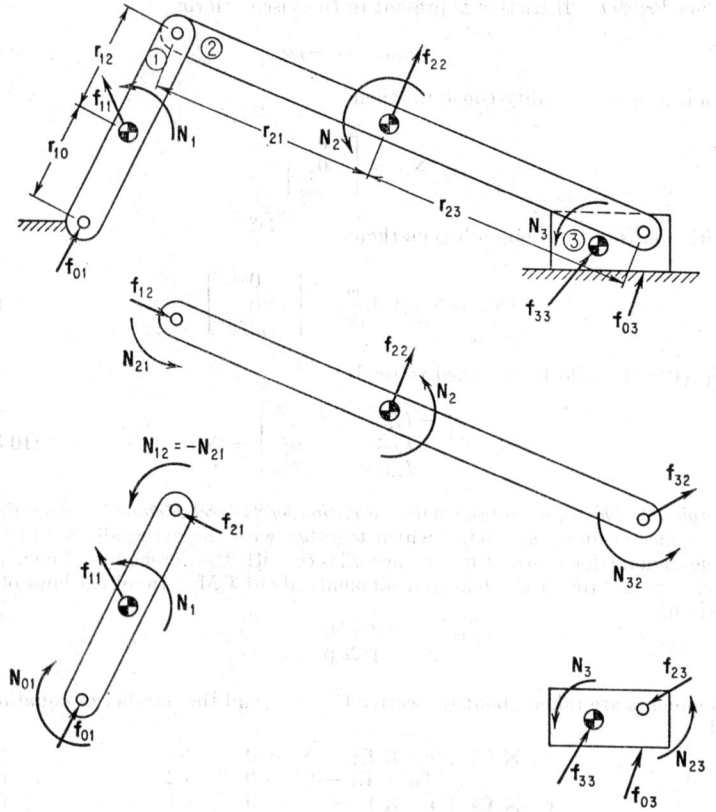

FIG. 10.4. Slider-crank—solution by direct method. $\mathbf{N}_{23} = -\mathbf{N}_{32}$; $\mathbf{N}_{21} = -\mathbf{N}_{12}$; $\mathbf{f}_{21} = -\mathbf{f}_{12}$.

circle). For $\mu \ll 1$ this becomes approximately

$$\mathbf{N}_{nm} \approx [\text{sgn}\ (\omega_n - \omega_m)][r_0\mu_{nm}]|\hat{\mathbf{f}}_{nm}|\mathbf{k} \tag{10.67}$$

sgn represents the sign, indicating that the interacting torque depends on the direction of relative motion of the linked members in question, and $\mathbf{k}$ is the unit vector in the z direction.

The equations can now be written to include Coulomb friction

$$\mathbf{r}_{10} \times \hat{\mathbf{f}}_{01} + \mathbf{r}_{12} \times \hat{\mathbf{f}}_{21} + \mathbf{N}_1 + \mathbf{N}_{01} + \mathbf{N}_{21} = 0 \tag{10.68}$$

$$\hat{\mathbf{f}}_{01} + \hat{\mathbf{f}}_{11} + \hat{\mathbf{f}}_{21} = 0 \tag{10.69}$$

$$\mathbf{r}_{21} \times \hat{\mathbf{f}}_{12} + \mathbf{r}_{23} \times \hat{\mathbf{f}}_{32} + \mathbf{N}_2 + \mathbf{N}_{12} + \mathbf{N}_{32} = 0 \tag{10.70}$$

$$\hat{\mathbf{f}}_{12} + \hat{\mathbf{f}}_{22} + \hat{\mathbf{f}}_{32} = 0 \tag{10.71}$$

$$\mathbf{r}_{30} \times \hat{\mathbf{f}}_{03} + \mathbf{r}_{32} \times \hat{\mathbf{f}}_{23} + \mathbf{N}_3 + \mathbf{N}_{23} + \mathbf{N}_{03} = 0 \tag{10.72}$$

$$\hat{\mathbf{f}}_{23} + \hat{\mathbf{f}}_{33} + \hat{\mathbf{f}}_{03} = 0 \qquad f_{nm} = \hat{f}_{nm} \tag{10.73}$$

$$\mathbf{N}_{nm} = \text{sgn}\ (\omega_n - \omega_m) r_0 \mu_{nm} |\mathbf{f}_{nm}|\mathbf{k} \tag{10.74}$$

Coulomb friction in the slider is given by

$$\hat{\mathbf{f}}_{x03} = -\mu_{03}\hat{\mathbf{f}}_{y03}\ (\text{sgn}\ \dot{x}_3) \tag{10.75}$$

Note that there are now 10 unknown quantities with 10 equations.

Solution by Lagrange's equations. Case without Friction

$$T = \sum_{i=1}^{2} (M_i/2)(\dot{x}_i^2 + \dot{y}_i^2) + \tfrac{1}{2} I_i \dot{\theta}_i^2 + (M_3/2)\dot{x}_3^2 \tag{10.76}$$

$$V = M_1 g y_1 + M_2 g y_2 \tag{10.77}$$

Lagrange's equations of motion are

$$\Sigma[(d/dt)(\partial L/\partial \dot{q}_i) - \partial L/\partial q_i - Q_i]\,\delta q_i = 0 \tag{10.78}$$

which lead to

$$\begin{aligned}(M_1\ddot{X}_1 - f_{1x})\,\delta X_1 + (M_1\ddot{Y}_1 - f_{1y} + M_1 g)\,\delta Y_1 + (I_1\ddot{\theta}_1 - N_1')\,\delta\theta_1 \\ + (I_2\ddot{\theta}_2 - N_2')\,\delta\theta_2 + (M_2\ddot{X}_2 - f_{2x})\,\delta X_2 + (M_2\ddot{Y}_2 - f_{2y} + M_2 g)\,\delta Y_2 \\ + (M_3\ddot{X}_3 - F_{3x})\,\delta X_3 = 0\end{aligned} \tag{10.79}$$

where the f_{nm} and N_m' are actual external forces excluding inertia forces. Note that a simple rearrangement of this form is identically the equation of virtual work [Eq. (10.33)] where terms like $-M_2\ddot{X}_2$ were considered inertia forces.

Six constraint equations connect the generalized coordinates as in Eqs. (10.34) to (10.39). It is possible to eliminate all but one unknown at a time between Eqs. (10.34) to (10.39) and (10.79) leading to solutions. Alternatively the method of Lagrange's multipliers proves more expeditious in this regard, as described below.

A first variation followed by multiplication of an arbitrary factor in each equation of (10.34) to (10.39) gives

$$\lambda_1 l_1 \sin\theta_1\,\delta\theta_1 + \lambda_1 l_2 \sin\theta_2\,\delta\theta_2 + \lambda_1\,\delta X_3 = 0 \tag{10.80}$$
$$\lambda_2 l_1 \cos\theta_1\,\delta\theta_1 - \lambda_2 l_2 \cos\theta_2\,\delta\theta_2 = 0 \tag{10.81}$$
$$\lambda_3 \mu_1 l_1 \sin\theta_1\,\delta\theta_1 + \lambda_3\,\delta X_1 = 0 \tag{10.82}$$
$$-\lambda_4 \mu_1 l_1 \cos\theta_1\,\delta\theta_1 + \lambda_4\,\delta Y_1 = 0 \tag{10.83}$$
$$\lambda_5 l_1 \sin\theta_1\,\delta\theta_1 + l_2\lambda_5(l - \mu_2) l_2 \sin\theta_2\,\delta\theta_2 + \lambda_5\,\delta X_2 = 0 \tag{10.84}$$
$$-\lambda_6 \mu_2 l_2 \cos\theta_2\,\delta\theta_2 + \lambda_6\,\delta Y_2 = 0 \tag{10.85}$$

Subtracting the sum of this set from Eq. (10.79) and setting each factor of δq_i to zero in accordance with principles of the Lagrange-multiplier method yields

$$M_1\ddot{X}_1 - f_{1x} - \lambda_3 = 0 \tag{10.86}$$
$$M_1\ddot{Y}_1 + M_1 g - f_{1y} - \lambda_4 = 0 \tag{10.87}$$
$$M_2\ddot{X}_2 - f_{2x} - \lambda_5 = 0 \tag{10.88}$$
$$M_2\ddot{Y}_2 + M_2 g - f_{2y} - \lambda_6 = 0 \tag{10.89}$$
$$M_3\ddot{X}_3 - f_{3x} - \lambda_1 = 0 \tag{10.90}$$
$$\begin{aligned}I_1\ddot{\theta}_1 - N_1 - \lambda_1 l_1 \sin\theta_1 - \lambda_2 l_2 \cos\theta_1 - \lambda_3\mu_1 l_1 \sin\theta_1 \\ + \lambda_4\mu_1 l_1 \cos\theta_1 - \lambda_5 l_1 \sin\theta_1 = 0\end{aligned} \tag{10.91}$$
$$I_2\ddot{\theta}_2 - N_2 - \lambda_1 l_2 \sin\theta_2 + \lambda_2 l_2 \cos\theta_2 - \lambda_5(1 - N_2) l_2 \sin\theta_2 = 0 \tag{10.92}$$

Equations (10.86) to (10.90) are recognized as equivalent to one-dimensional cases. Since motion is prescribed, these values of λ_1, λ_3, λ_4, λ_5, λ_6 are evaluated directly.

The forces of "constraint" can be due only to internal forces at the joints. Hence,

$$\lambda_3 = f_{x01} + f_{x21} = F_{x1} \tag{10.93}$$
$$\lambda_4 = f_{y01} + f_{y21} = F_{y1} \tag{10.94}$$
$$\lambda_5 = f_{x12} + f_{x32} = F_{x2} \tag{10.95}$$
$$\lambda_6 = f_{y12} + f_{y32} = F_{y2} \tag{10.96}$$
$$\lambda_1 = f_{x03} + f_{x23} = F_{x3} \tag{10.97}$$
$$\mathbf{f}_{nm} = \mathbf{i} f_{xnm} + \mathbf{j} f_{ynm}$$

From Eqs. (10.91) and (10.92) λ_2 and the only unknown N_1 is calculated. The other unknowns are obtained most readily using the forms of the "direct method" after making the identifications of the λ's with total forces (including inertia) at the center of mass of each member as shown in Fig. 10.5. For the cases of Coulomb friction, iterative techniques are required when using the Lagrange method.

For the case of motion prescribed in time, the direct method of analysis is far superior to Lagrange's equation of motion. However, when the force structure is given, the Lagrange method is the better choice in so far as determining the motion. The problem is: Given all the external forces. To find the internal forces and/or motion. The two methods will be discussed for the two cases of interest.

Direct Method. Force Structure Given. Frictionless and Coulomb Friction Cases: The direct method is embodied in the set of equations (10.58) to (10.64). Now the

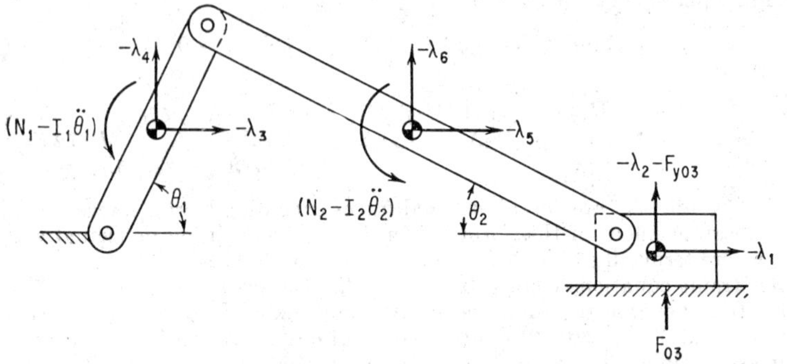

FIG. 10.5. Slider-crank chain—Lagrange multiplier.

unknowns are not only the internal forces $\mathbf{f}_{nm}$ but the forces $\mathbf{f}_{mm}$ at the CM since they include the inertia forces whose values depend on the motion $\theta_1 = \theta_1(t)$ which is not known.

In contrast with the case of known motion the input torque and $\theta_1(t)$ have exchanged roles as unknowns, but the complexity has increased by an order of magnitude and now becomes essentially a set of differential equations in θ, of the form $i = 1, 2, 3$, etc.

$$\mathbf{r}_{10}(\theta_1) \times \mathbf{f}_{01} + \mathbf{r}_{12}(\theta_1) \times \mathbf{f}_{21} + N_1(\dot{\theta}_1,\ddot{\theta}_1,\theta_1) = 0 \tag{10.98}$$

$$\mathbf{f}_{01} + \mathbf{f}_1(\ddot{\theta}_1,\dot{\theta}_1,\theta_1) + \mathbf{f}_{21} = 0 \qquad \text{etc.} \tag{10.99}$$

The solution to this set presents formidable difficulties notwithstanding the use of the digital computer. The case of Coulomb friction is of a somewhat higher order of difficulty owing to the nonlinear dependence of frictional force and motion.

Lagrange's Equation of Motion. Force Structure Given. Frictionless. The basic formulation has been given in Eq. (10.79), which, together with Eqs. (10.34) to (10.39), defines the system. Now six unknowns can be eliminated in favor of the seventh θ_1 which after substitution in Eq. (10.79) yields

$$[F(\ddot{\theta}_1,\dot{\theta}_1,\theta_1)]\ \delta\theta_1 = 0 \tag{10.100}$$

Since $\delta\theta_1$ is a free virtual displacement in order for Eq. (10.100) to hold,

$$F(\ddot{\theta}_1,\dot{\theta}_1,\theta_1) = 0$$

This ordinary differential equation is nonlinear but can always be solved directly by machine methods. A solution is tantamount to finding the motion.

To obtain the internal forces the "direct methods" of Lagrange's equations of motion given in Sec. 10.9(c) have proved effective.

Force Structure Given. Coulomb Friction Present. The presence of Coulomb friction can be accounted for by iterative techniques outlined as follows: First (to first-order approximation) the system forces are determined as for the frictionless case which leads to the motion whence the internal forces are next obtained. The first-order approximation of internal motion yields friction forces which are used as part of the force structure to be entered into the second approximation, etc. The process is repeated until sufficient accuracy is obtained. The case of viscous friction is readily and directly admitted by the Lagrange method, with the damping term being expressed as the gradient (in velocity coordinates) of a Raleigh dissipation function F where

$$\mathcal{F} = \tfrac{1}{2} \sum_{i,j} e_{ij}\dot{q}_i\dot{q}_j \qquad (10.101)$$

And the Lagrange's equations are modified to become

$$\sum_{qi} [(d/dt)(\partial L/\partial \dot{q}_i) - \partial L/\partial q_i + \partial \mathcal{F}/\partial \dot{q}_i - Q_i]\,\delta q_i = 0 \qquad (10.102)$$

(h) Summary of Direct vs. Lagrange Method of Solution

For single-degree-of-freedom systems: When motion is given, and constraints are complex or multiconstraints, use the direct method; when motion is given or unknown and constraints are none or simple, method depends on the system although the direct method is usually suitable; when motion is unknown, and constraints are complex or multiconstraints, use Lagrange's method to find motion, and then the direct method to evaluate forces.

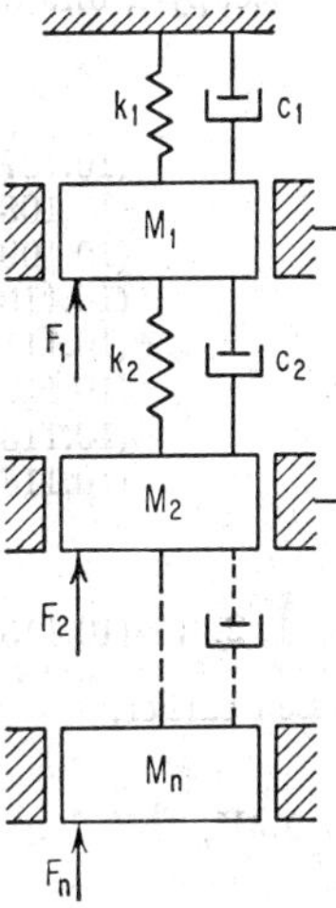

FIG. 10.6. Multi-degree-of-freedom system.

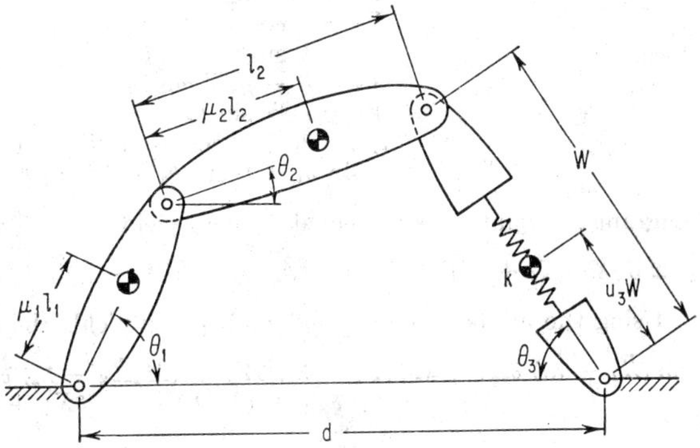

FIG. 10.7. Quadric chain with one flexible link. (l_3 is the unextended length of link 3.)

10.3. MULTI-DEGREE-OF-FREEDOM SYSTEMS (ELASTIC BODY)

(a) Introduction

The simplest case of a multi-degree-of-freedom system is a vibrating system with the translational mode of vibration of the type depicted in Fig. 10.6. Direct application of Newton's laws or Lagrange's equations yields the set of n equations

$$\begin{aligned} M_m\ddot{X}_m + c_m(\dot{X}_m - \dot{X}_{m-1}) + k_m(X_m - X_{m-1}) - c_{m+1}(\dot{X}_{n+1} - \dot{X}_n) \\ - k_{m+1}(X_{m+1} - X_m) = F_m \qquad m = 1, 2, \ldots, n \end{aligned} \qquad (10.103)$$

The solution to this set is fairly straightforward for two reasons, namely:

1. The assumed linearity of the friction and spring-force terms.
2. The absence of constraints; i.e., there is no algebraic interdependence of coordinates X_m.

(b) Examples

Example 1: In order to visualize a more complex situation (number of degrees of freedom only two) consider the quadric chain with two rigid links and one elastic link (Fig. 10.7). It is clear that for this system input motion θ_1 does not determine the motion of the entire system. Hence in no practical case can motion be given in time (a most desirable situation for analysis). Under conditions of this system, Lagrange's equations appear to be the most useful tool of analysis and will therefore be used to formulate the equations of motion. Reactions and internal forces can then be found using a formulation similar to Eq. (10.103).

The Lagrange formulation is, in the absence of gravity,

$$L = T - V \qquad T = \Sigma M_i/2(\dot{X}_i^2 + \dot{Y}_i^2) + \Sigma \tfrac{1}{2} I_{ii}\dot{\theta}_i^2 \tag{10.104}$$

$$V = (k/2)(W - l_3)^2 \tag{10.105}$$

$$\sum_i [(d/dt)(\partial L/\partial \dot{q}_i) - \partial L/\partial q_i - Q_i]\,\delta q_1 = 0 \tag{10.106}$$

where W is the instantaneous length of the elastic member and the 10 generalized coordinates are X_1, Y_1, X_2, Y_2, X_3, Y_3, θ_1, θ_2, θ_3, and W. Eight equations of constraint must now be found since the system has two degrees of freedom; this follows from the fact that, if W and θ_1 were specified, the system would occupy a unique position in space.

The eight constraint equations are obtained from the geometry:

$$X_1 = \mu_1 l_1 \cos\theta_1 \tag{10.107}$$

$$Y_1 = \mu_1 l_1 \sin\theta_1 \tag{10.108}$$

$$X_2 = l_1 \cos\theta_1 + \mu_2 l_2 \cos\theta_2 \tag{10.109}$$

$$Y_2 = l_1 \sin\theta_1 + \mu_2 l_2 \sin\theta_2 \tag{10.110}$$

$$X_3 = d - \mu_3 W \cos\theta_3 \tag{10.111}$$

$$Y_3 = \mu_3 W \sin\theta_3 \tag{10.112}$$

$$l_1 \cos\theta_1 + l_2 \cos\theta_2 + W\cos\theta_3 = d \tag{10.113}$$

$$l_1 \sin\theta_1 + l_2 \sin\theta_2 - W\sin\theta_3 = 0 \tag{10.114}$$

Lagrange's equation is next evaluated to yield

$$\Sigma M_i\ddot{X}_i\,\delta X_i + M_i\ddot{Y}_i\,\delta Y_i + I_{ii}\ddot{\theta}_i\,\delta\theta_i + k(W - l_3)\,\delta W = 0 \qquad i = 1, 2, 3 \tag{10.115}$$

Using the method of Lagrange multipliers yields the form discussed earlier,

$$\begin{aligned}(M_1\ddot{X}_1 - \lambda_1)\,\delta X_1 &+ (M_1\ddot{Y}_1 - \lambda_2)\,\delta Y_1 + (M_2\ddot{X}_2 - \lambda_3)\,\delta X_2 \\ &+ (M_2\ddot{Y}_2 - \lambda_4)\,\delta Y_2 + (M_3\ddot{X}_3 - \lambda_5)\,\delta X_3 + (M_3\ddot{Y}_3 - \lambda_6)\,\delta Y_3 \\ &+ \left[(I_{11}\ddot{\theta}_1) - \sum_{i=1}^{8} \lambda_i f_i(\theta_1,\theta_2,\theta_3,W)\right]\delta\theta_1 \\ &+ [I_{22}\ddot{\theta}_2 - \Sigma\lambda_i\Omega_i(\phi_i\dot{\theta}_i,\theta_2,\dot{\theta}_3,\ldots,W)]\,\delta\theta_2 \\ &+ [k(\omega - l_3) - \Sigma\lambda_i\psi_i(\theta_1{}^n,\theta_2{}^n,\theta_3{}^n,W)] \\ &\qquad + [I_{33}\ddot{\theta}_3 - \Sigma\lambda_L\phi_i(\dot{\theta}_1,\theta_2,\theta_3,\ldots,W)]\,\delta\theta_3 = 0\end{aligned} \tag{10.116}$$

where f_1, Ω_1, ψ_1, and ϕ_i are in general transcendental functions of the coordinates. Each coefficient of q_i is separately equated to zero as follows:

$$\begin{aligned} M_1\ddot{X}_1 - \lambda_1 &= 0 \\ M_1\ddot{Y}_1 - \lambda_2 &= 0 \qquad \text{etc.}\end{aligned} \tag{10.117}$$

λ_1 through λ_6 can now be found in terms of θ_1, θ_2, θ_3, and W. For example, λ_2 is obtained by taking two time deviations of Eq. (10.110) and substituting for $\ddot{Y}$ in the above, leaving

$$\lambda_2 = M_1\ddot{Y}_1 = M_1\mu_1 l_1(\ddot{\theta}_1 \cos\theta_1 - \dot{\theta}_1^2 \sin\theta_1) \tag{10.118}$$

By further elimination of W in Eqs. (10.113) and (10.114), etc., one can get the forms

$$\begin{aligned} \phi(\theta_1,\theta_2,\theta_3) &= 0 \\ F_1(\ddot{\theta}_1,\ddot{\theta}_2,\ddot{\theta}_3,\dot{\theta}_1,\dot{\theta}_2,\dot{\theta}_3,\theta_1,\theta_2,\theta_3) &= 0 \\ F_2(\ddot{\theta}_1,\ddot{\theta}_2,\ddot{\theta}_3,\dot{\theta}_1,\dot{\theta}_2,\dot{\theta}_3,\theta_1,\theta_2,\theta_3) &= 0 \\ F_3(\ddot{\theta}_1,\ddot{\theta}_2,\ddot{\theta}_3,\dot{\theta}_1,\dot{\theta}_2,\dot{\theta}_3,\theta_1,\theta_2,\theta_3) &= 0 \end{aligned} \tag{10.119}$$

or any other two differential equations in two other generalized coordinates. Computer methods are available for solution to this set, which yields the motion. From the motion [and Eq. (10.119)] all the reaction and internal forces can be found.

A similar analysis can be performed if θ_1 is specified and the input torque N_1 is unknown.

Example 2: Consider next the system shown in Fig. 10.8 consisting of a heavy inertial element driven with a known input torque $T_1(t)$; the output angular displacement is θ_1. A cam is connected to the flywheel through a shaft of torsional stiffness k_{12}. The cam follower is connected to the output by an effective spring k_{34}

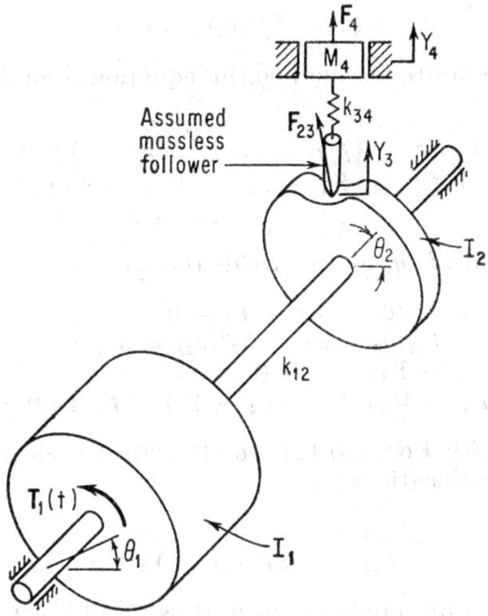

Fig. 10.8. Cam system—three degrees of freedom.

and the output force is, in general, a function of the displacement y_4 and time t; i.e., $F = F(y_4,t)$. The problem is to find the motion and the internal forces in the connecting shaft and the cam surface whose contour is given by $y_3 = f(\theta_2)$.

The system contains three degrees of freedom, with the convenient generalized coordinates θ_1, θ_2, Y_3, Y_4. The one equation of constraint is the cam contour, namely,

$$Y_3 = f(\theta_2) \tag{10.120}$$

Formulation of the Problem

By classical method:

$T_1 - k_{12}(\theta_1 - \theta_2) = I_1\ddot{\theta}_1$	moment balance on I_1	(10.121)
$F_{y23}\,\delta Y_3 + T_{32}\,\delta\theta_2 = 0$	virtual work	(10.122)
$k_{12}(\theta_1 - \theta_2) + T_{32} = I_2\ddot{\theta}_2$	moment balance on I_2	(10.123)
$k_{34}(Y_3 - Y_4) + F_4(Y_4,t) - M_4g = M_4\ddot{Y}_4$	force balance on M_4	(10.124)
$Y_3 = f(\theta_2)$	cam contour	(10.125)
$\delta y_3 = f'(\theta_2)\,\delta\theta_2$	differential	(10.126)
$F_{y23}f'(\theta_2)\,\delta\theta_2 + T_{32}\,\delta\theta_2 = 0$	$F_{y23}f'(\theta_2) + T_{32} = 0$	(10.127)

$\mathbf{F}_{32} = -\mathbf{F}_{23}$ is the vertical force component of the follower on the surface and T_{32} is its turning moment about the axis.

$$T_{32} = Y_3F_{x32} \tag{10.128}$$

By Lagrange's equation:

$$T = \tfrac{1}{2}(I_1\dot{\theta}_1^2 + I_2\dot{\theta}_2^2 + M_4\dot{Y}_4^2) \tag{10.129}$$
$$V = M_4 g Y_4 + (k_{34}/2)(Y_3 - Y_4)^2 + (k_{12}/2)(\theta_2 - \theta_1)^2 \tag{10.130}$$
$$L = T - V$$
$$\Sigma[(d/dt)(\partial L/\partial \dot{q}_i) - \partial L/\partial q_i - Q_i]q_i = 0 \tag{10.131}$$
$$[I_1\ddot{\theta}_1 + k_{12}(\theta_1 - \theta_2) - T_1]\,\delta\theta_1 + [I_2\ddot{\theta}_2 + k_{12}(\theta_2 - \theta_1)]\,\delta\theta_2 + [M_4\ddot{Y}_4 + k_{34}(Y_4 - Y_3) + M_4 g - F_4(Y_4,t)]\,\delta Y_4 + k_{34}(Y_3 - Y_4)\,\delta Y_3 = 0 \tag{10.132}$$

From the constraint $Y_3 = f(\theta_2)$, the Lagrange multiplier method yields

$$\lambda\,\delta y_3 - \lambda f'(\theta_2)\,\delta\theta_2 = 0 \tag{10.133}$$

Following the Lagrange-multiplier method this equation is combined with Eq. (10.131) to give

$$[I_1\ddot{\theta}_1 + k_{12}(\theta_1 - \theta_2) - T_1]\,\delta\theta_1 + [I_2\ddot{\theta}_2 - k_{12}(\theta_1 - \theta_2) + \lambda f'(\theta_2)]\,\delta\theta_2 + [M_4\ddot{Y}_4 + k_{34}(Y_4 - Y_3) + M_4 g - F_4(Y_4,t)]\,\delta Y_4 + [k_{34}(Y_3 - Y_4) - \lambda]\,\delta Y_3 = 0 \tag{10.134}$$

Equating each coefficient of δq_i to zero yields the set

$$I_1\ddot{\theta}_1 + k_{12}(\theta_1 - \theta_2) - T_1 = 0 \tag{10.135}$$
$$I_2\ddot{\theta}_2 + k_{12}(\theta_2 - \theta_1) + \lambda f'(\theta_2) = 0 \tag{10.136}$$
$$k_{34}(Y_3 - Y_4) - \lambda = 0 \tag{10.137}$$
$$M_4\ddot{Y}_4 + M_4 g + k_{34}(Y_4 - Y_3) - F_4(Y_4,t) = 0 \tag{10.138}$$

Comparing this set with Eqs. (10.121) to (10.126) it is clear that they are exactly equivalent with the identifications

$$F_{y23} = \lambda \tag{10.139}$$
$$T_{32} = -\lambda f'(\theta_2) = Y_3 F_{x32} \tag{10.140}$$

The solution to the generally nonlinear form [Eqs. (10.133) through (10.137)] can be obtained by machine methods to yield the motion and forces required.

If a linkage system (gearbox) of single degree is interposed between the flywheel and the cam, the problem can be treated in essentially the same way with the Lagrange-equation-of-motion method. The kinetic-energy term would be modified by the addition of energy stored in the linkage which depends upon the input position and velocity. The modified kinetic energy is then

$$T' = T + \phi(\dot{\theta}_1,\theta_1) \tag{10.141}$$

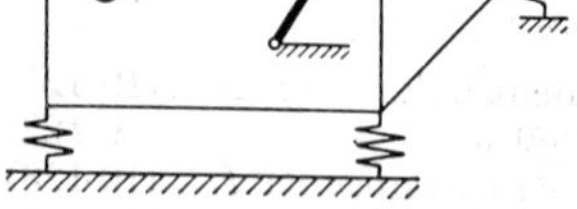

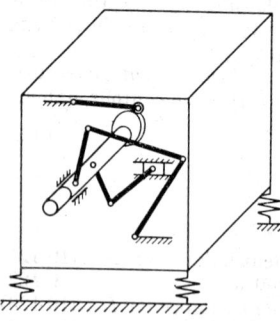

FIG. 10.9. Multi-degree-of-freedom system—operating in parallel planes.

Example 3: Consider the multi-degree-of-freedom problem presented by a parallel series of linkages resting on springs as shown in Fig. 10.9.

As a first approximation assume that the supports are rigid, leading to the single-degree-of-freedom problem similar to that discussed in Example 2. If the input motion is known, all motion can be determined. First the inertia forces are found and added to the external forces in the manner shown before, i.e., a linear force at the CM and a rotational force (moment) about each member. More specifically, for each member the inertia forces are, from Eqs. (10.28) and (10.29),

$$-dH/dt = -\ddot{\theta}(\partial H/\partial\dot{\theta})(\dot{\theta},\theta) - \dot{\theta}[\partial H(\dot{\theta},\theta)/\partial\theta] \quad \text{(inertia moment about CM)} \tag{10.142}$$

$$-M\,d^2R/dt^2 = M\{-\ddot{\theta}[dR(\theta)/d\theta] - \dot{\theta}^2[d^2R(\theta)/d\theta^2]\} \quad \text{(inertia force through CM)} \tag{10.143}$$

where the angular momentum $\mathbf{H}(\theta,\dot\theta)$ and position at CM $\mathbf{R}_{cm}(\theta)$ can be obtained easily. All forces of reaction and internal force can be found by formulation similar to Eq. (10.13).

If motion is not prescribed, the method of Lagrange's equations can be employed as shown previously to determine the motion and the internal forces.

Consideration of *elastic supports* presents the case of the multi-degree-of-freedom problem first posed. It has seven degrees of freedom. The unconstrained generalized coordinates are therefore seven in number: θ, ψ_x, ψ_y, ψ_z, $\bar{X}$, $\bar{Y}$, $\bar{Z}$, which are, respectively, the relative angular position of the input rotation, the rotations of the structure about the three coordinates, and the position of the center of mass of the structure excluding the linkages.

The problem is formulated using Lagrange's equation of motion. First, the kinetic energy is

$$T = \sum_i \tfrac{1}{2}M_i[(\dot{X}_i + \dot{\bar{X}})^2 + (\dot{Y}_i + \dot{\bar{Y}})^2 + (\dot{Z}_i + \dot{\bar{Z}})^2] + \tfrac{1}{2}\sum_i \boldsymbol{\omega}_i I_i \boldsymbol{\omega}_i + (M_i/2)(\dot{\bar{X}}^2 + \dot{\bar{Y}}^2 + \dot{\bar{Z}}^2) \qquad (10.144)$$

where the subscript i refers to the ith member, $\boldsymbol{\omega}_i$ the three vectors using fixed coordinates, and I_i the inertia tensor which is a function of θ, ψ_x, ψ_y, ψ_z. For the usually assumed small rotational motion of the structure, the sinusoidal terms can be approximated by

$$\sin\psi \approx \psi \qquad (10.145)$$
$$\cos\psi \approx 1 \qquad (10.146)$$

from which it follows that the orientation of the body is independent of the order of rotation of ψ_x, ψ_y, ψ_z to terms of the second order in these variables. The inertia tensor is referred to fixed axes about CM of each element.

$$I_i = U^{-1}[I(\theta_i)]U \qquad (10.147)$$

where

$$U \approx \begin{bmatrix} 1 & \psi_z & \psi_y \\ -\psi_z & 1 & -\psi_x \\ -\psi_y & \psi_x & 1 \end{bmatrix} \quad \text{to first order } \psi\text{'s} \qquad (10.148)$$

$$U^{-1} \approx \mathbf{U} \qquad (10.149)$$

$$I(\theta)_i = \begin{bmatrix} \cos\theta_i & \sin\theta_i & 0 \\ -\sin\theta_i & \cos\theta_i & 0 \\ 0 & 0 & 1 \end{bmatrix} [I(0)] \begin{bmatrix} \cos\theta_i & -\sin\theta_i & 0 \\ \sin\theta_i & \cos\theta_i & 0 \\ 0 & 0 & 1 \end{bmatrix} \qquad \theta_i = F(\theta) \qquad (10.150)$$

The potential energy is assumed to be of the form

$$V = \tfrac{1}{2}K_x\bar{X}^2 + \tfrac{1}{2}K_y\bar{Y}^2 + \tfrac{1}{2}K_z\bar{Z}^2 + \tfrac{1}{2}K_{\psi_x}\psi_x^2 + \tfrac{1}{2}K_{\psi_y}\psi_y^2 + \tfrac{1}{2}K_{\psi_z}\psi_z^2 \qquad (10.151)$$

where interacting terms usually present are assumed negligible. Writing the equations of constraints we have

$$X_i = X_i(\theta) \qquad \dot{X}_i = \dot\theta[dX_i(\theta)/d\theta] \qquad (10.152)$$
$$Y_i = Y_i(\theta) \qquad \dot{Y}_i = \dot\theta[dY_i(\theta)/d\theta] \qquad (10.153)$$

The Lagrangian is formulated as follows with the substitutions above borne in mind:

$$M\ddot{\bar{X}} + \sum_i M_i(\ddot{X}_i + \ddot{\bar{X}}) + K_xX = 0 \qquad (10.154)$$

$$M\ddot{\bar{Y}} + \sum_i M_i(\ddot{Y}_i + \ddot{\bar{Y}}) + K_yY = 0 \qquad (10.155)$$

$$M\ddot{\bar{Z}} + \sum_i M_i(\ddot{Z}_i + \ddot{\bar{Z}}) + K_zZ = 0 \qquad (10.156)$$

$$(d/dt)\left((\partial/\partial\dot{\theta})\left\{\sum_i \tfrac{1}{2}\omega_i[I_i](\theta,\psi_x,\psi_y,\psi_z)\omega_i + (\tfrac{1}{2}m_i\dot{\theta}X_i'\theta + \dot{\mathrm{X}})^2\right\}\right)$$

$$- (\partial/\partial\theta)\left\{\sum_i \tfrac{1}{2}\omega_i[I_i](\theta,\psi_x,\psi_y,\psi_z)\omega_i + \sum \tfrac{1}{2}m_i[\dot{\theta}X_i'(\theta) + \dot{X}^2]\right\} - N_0 = 0 \quad (10.157)$$

$$\begin{aligned}
(d/dt)[(\partial/\partial\dot{\psi}_x)\Sigma(\tfrac{1}{2}\omega_i[I_i]\omega_i)] - (\partial/\partial\psi_x)\Sigma\tfrac{1}{2}\omega_i[I_i]\omega_i &= 0\\
(d/dt)[(\partial/\partial\dot{\psi}_y)\Sigma(\tfrac{1}{2}\omega_i[I_i]\omega_i)] - (\partial/\partial\psi_y)\Sigma\tfrac{1}{2}\omega_i[I_i]\omega_i &= 0\\
(d/dt)[(\partial/\partial\dot{\psi}_z)\Sigma(\tfrac{1}{2}\omega_i[I_i]\omega_i)] - (\partial/\partial\psi_z)\Sigma\tfrac{1}{2}\omega_i[I_i]\omega_i &= 0
\end{aligned}$$

where N_0 is the input torque. Equations (10.154) to (10.157) can be solved to yield the motion. From the motion all inertia forces can be evaluated, from which all forces can be obtained. References 15 and 16 show some interesting applications of the principles discussed.

(c) Synthesis of Mechanical Filters

The basic concepts of filter analysis and design can be applied to mechanical systems on the basis of the electric-circuit configurations for which filter theory finds its widest applications. While every possible combination of the three linear electrical elements—resistance, inductance, and capacitance—may not be transformable into a physically realizable mechanical network, it is possible to construct direct mechanical analogs to a wide variety of electric circuits.

Design in the electrical domain permits the direct application of the systematic methods and techniques of circuit analysis. While these techniques are similarly applicable to mechanical systems under certain simplifying assumptions, most of the available literature is written in electrical terms.

Filter theory is generally applied to linear systems. These techniques have recently been applied to nonlinear mechanical systems with success. The theoretical aspects of filter design and synthesis are shown in refs. 7 to 11. Table 6.4 presents analog symbols.

A filter is a two-terminal-pair network which is placed in a system for the purpose of preventing one or more of the harmonic components of the driving function from delivering power to the driven load. This is accomplished by designing the filter such that the filter-load combination presents an imaginary impedance to the driving source at the frequencies which are to be attenuated while presenting a real impedance to the source at the frequencies which are to be transmitted.

The terms "pass band" and "stop band" are used to designate the bands of frequencies over which transmission and attenuation respectively occur. The "cutoff frequencies" are located at the points of transition from pass-band operation to stop-band characteristics.

Although it is possible to construct electrical filters of the low-pass, high-pass, and band-elimination variety, the practical limitations such as stress levels on mechanical networks will probably limit the use of heavy-duty mechanical filters primarily to the low-pass type. In fact, mechanical-system requirements rarely dictate the need for other than low-pass or band-elimination filtering.

Figure 10.10 shows the lumped equivalent electrical-mechanical networks to guide the designer. Note that the flywheel on an elastic shaft is a low-pass filter and is composed of a "split-T" network of inductance and capacitance.

10.4. SIMPLIFIED APPROACHES TO MACHINE ANALYSIS (LUMPED SYSTEMS)

(a) Introduction

To simplify analysis it is necessary to establish simpler systems which are approximately dynamically equivalent. In machine analysis, the determination of *equivalent masses* is fundamental. In elastic systems simplification also demands the establishment of *equivalent springs*.

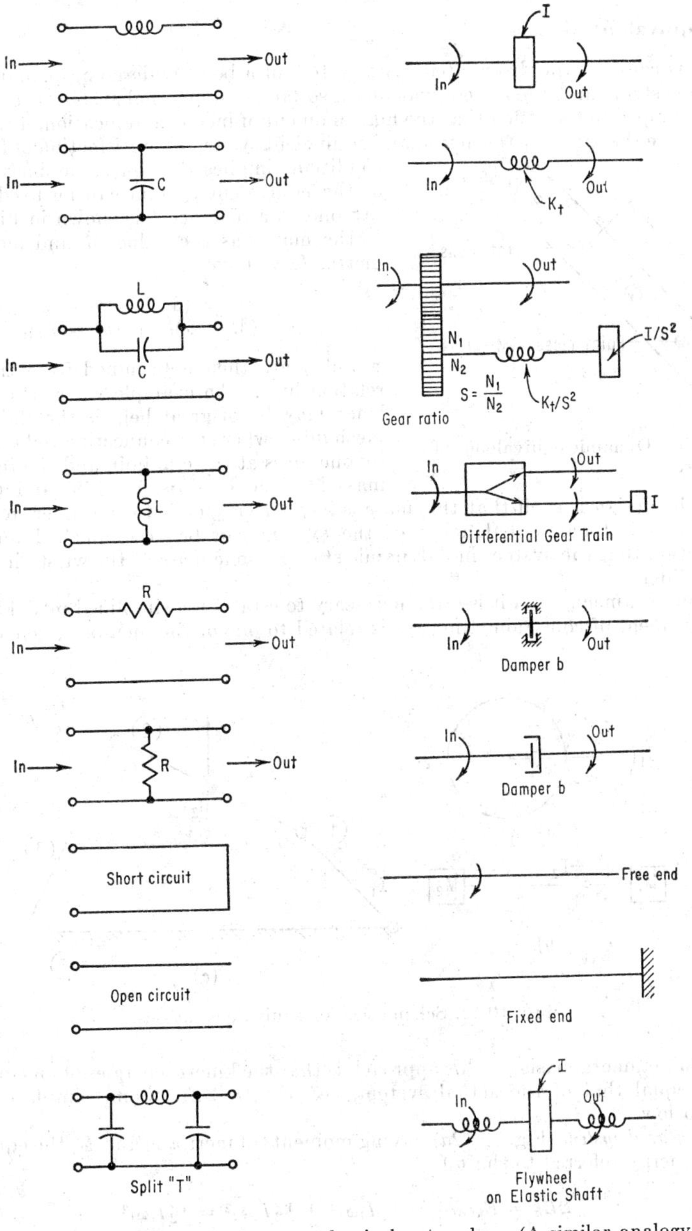

FIG. 10.10. Lumped equivalent electromechanical networks. (A similar analogy may be made for translational motion.) I = moment of inertia; b = damper; k_t = torsional spring constant; L = inductance; C = capacitance; R = resistance.

(b) Equivalent Masses

Let us consider the distributed mass system of a body undergoing planar motion. Such a system can be made equivalent (in so far as an external system is concerned) to two lumped masses such that the mass moment of inertia and location of the center of mass are the same for the actual and equivalent systems. Satisfaction of these two conditions implies that only one mass position of the equivalent system can be fixed at will, say one at a distance l as shown in Fig. 10.11. If the mass has the value M and moment of inertia I, we have

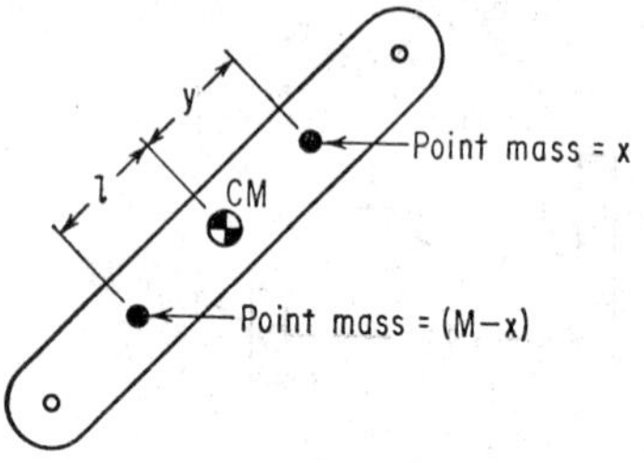

FIG. 10.11. Dynamic equivalence of a member.

$$(M - x)y = lx \tag{10.158}$$

$$(M - x)l^2 + xy^2 = I \tag{10.159}$$

x and y are then determined from these two relationships. An example where this equivalence may be of great help is the slider-crank mechanism where the connecting rod is replaced by one mass at the crankpin and the remaining mass is given by Eqs. (10.158) and (10.159). Often the design is such that this mass is located very close to the slider so that the analysis and dynamic balancing of the system can be accomplished under the assumption that the system in a dynamic state has one mass at the wrist pin and one at the slider.

In complex machinery, it is often necessary to establish a simplified model in which the movement of constrained masses is related to one of the members, generally the input.

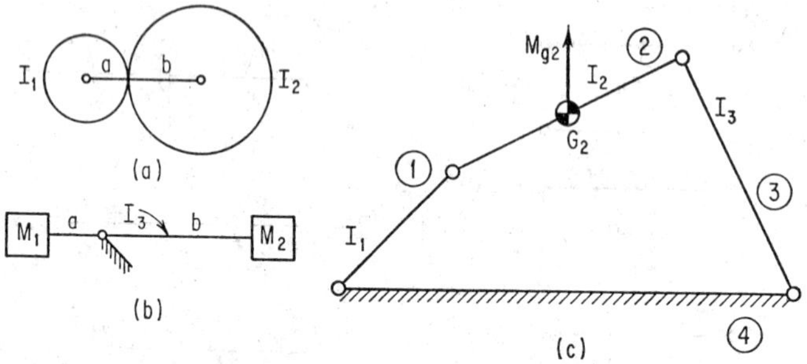

FIG. 10.12. Schematics for equivalent masses.

The fundamental basis for this approach is that the kinetic energies of an equivalent system equal that of the actual system, i.e., $E_{k(e)} = \Sigma E_k$. Some simple cases are shown below.

For a *geared system* (Fig. 10.12*a*) having moments of inertia I_1 and I_2, the equivalent kinetic energy referred to shaft 1

$$\Sigma E_k = E_{k(e)(1)} = \tfrac{1}{2} I_1\omega_1^2 + \tfrac{1}{2} I_2\omega_2^2 = \tfrac{1}{2} I_e\omega_1^2 \tag{10.160}$$

and its equivalent moment of inertia

$$I_{e(1)} = I_1 + (a/b)^2 I_2 = \sum_{i=1}^{n} R_i^2 I_i \tag{10.161}$$

where $R = a/b$, the gear ratio.

For a *lever system* having masses at 1 and 2 (Fig. 10.12*b*) and a moment of inertia I_3 about the pivot point, the equivalent kinetic energy referred to mass 1 is

$$E_{ke(1)} = \tfrac{1}{2}M_1V_1^2 + \tfrac{1}{2}I_3\omega_3^2 + \tfrac{1}{2}M_2V_2^2 = \tfrac{1}{2}M_{e(1)}V_1^2 \tag{10.162}$$

and the equivalent mass

$$M_{e(1)} = M_1 + (b/a)^2M_2 + I_3/a^2 \tag{10.163}$$

For *mechanisms* one member generally has both rotation and translation. The equivalent kinetic energy referred to crank 1 for the quadric mechanism (Fig. 10.12*c*)

$$E_{ke(1)} = \tfrac{1}{2}(I_1\omega_1^2 + I_g{}^2\omega_2^2 + M_2V_{g2}^2 + I_3\omega_3^2) = \tfrac{1}{2}I_{e(1)}\omega_1^2 \tag{10.164}$$

and its equivalent moment of inertia

$$I_{e(1)} = I_1 + I_2(\omega_2/\omega_1)^2 + M_2(V_g{}^2/\omega_1)^2 + I_3(\omega_3/\omega_1)^2 \tag{10.165}$$

(c) Equivalent Springs

All springs are nonlinear. In most instances the degree of nonlinearity is small for the range of deflection under consideration, and the spring may be assumed approximately linear. Refer to Sec. 6 for load-deflection curves for "hard" springs such as belts and chains, "soft" springs, and linear springs. Belt and chain spring nonlinearity is quite significant. Reference 14 presents the solution of an undamped single-degree-of-freedom system containing a cubic spring.

Equivalent springs are calculated on the basis of the equivalence of potential energy (elastic energy in deflected members). The potential energy of springs ($i = 1 \cdots n$) loaded by direct force and moment is, respectively,

$$\text{PE} = \int F\,dx = \sum_{i=1}^{n} F_i^2/2k_i \tag{10.166}$$

$$\text{PE} = \int T\,d\theta = \sum_{i=1}^{n} T_i^2/2k_{ti} \tag{10.167}$$

where F = load (force), lb
$k = \Delta R/\Delta x$ = spring constant
x = linear deflection, in.
T = load (torque), lb-in.
θ = angular deflection, rad
$k_t = \Delta T/\Delta\theta$ = torsional spring constant, lb-in./rad

Table 6.9 presents commonly employed equivalent spring constants, while refs. 17 and 18 have more information.

For calculating the equivalent spring constant for belts or chains one should consider the portion wrapped around the pulley and its reduced loading. In general, installed belts and chains may be equated to nonlinear springs plus backlash elements; the latter effect is due to slack side-belt curvature.

Elasticity considerations in torsional cam-driven systems lead to a nonlinear equation (Hill's).[14] Elastic mechanisms reduced to equivalent systems contain both equivalent masses and springs which are not constant during a cycle of operation, which also lead to nonlinear equations.[1]

10.5. DYNAMIC BALANCING

(a) Introduction

Dynamic balance is the elimination or counteraction of free inertia forces and torques in steady-state operation.

Consider machines operating in parallel xy planes with a connection parallel to the z axis. Steady-state dynamic-balance investigations which may be made include:

1. Inertia balance—balancing against inertia forces through the x and y axes (shaking forces) and their moments about the x and y axes.
2. Torque balance—balancing against torque variation in rotating shafts parallel to the z axis. External loads should also be included where present.
3. Torque reaction—the effect of torques about the z axis acting on stationary members (i.e., machine frame). This phenomenon can rarely be improved by means of balancing. Improvement can be realized, however, by shifting the center of mass of the system and isolating the driving couple by employing resilient supports (vibration isolators).

Overriding all methods of balancing is the general design rule that members be as light in weight as possible. Rigidity, however, should not be sacrificed for weight.

(b) Inertia Balancing of Rotating Systems

Consider a rigid system rotating about some fixed axis AB of Fig. 10.13. The rotating shaft is supported in bearings at A and B. The system is said to be dynamically balanced if for all possible values and time derivatives of angular rotation, where

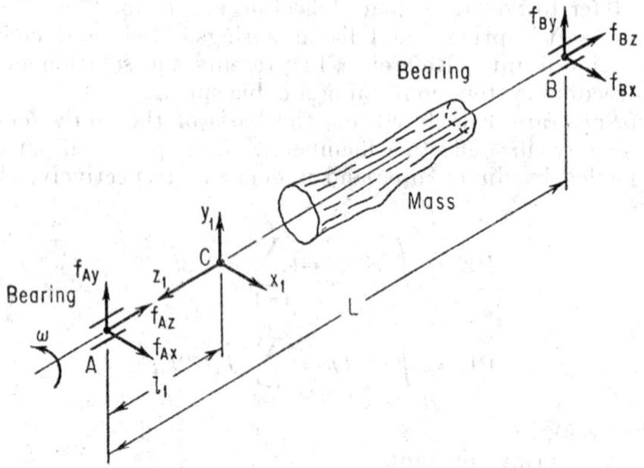

FIG. 10.13. Dynamic balancing.

rotation is confined to the fixed AB axis, the reaction forces at the bearings are zero. The implications of this statement follow from consideration of Euler's equation (10.18) for the rigid body with unidirectional angular velocity. For axes with origin at point A, the Euler equation becomes

$$\begin{matrix} -I_{xz}\dot{\omega} + I_{yz}\omega^2 = \\ -I_{yz}\dot{\omega} - I_{xz}\omega^2 = \end{matrix} \begin{bmatrix} N_x \\ N_y \end{bmatrix} = \begin{bmatrix} f_{By}L \\ f_{Bx}L \end{bmatrix} = \mathbf{r} \times \mathbf{f}_B \qquad (10.168)$$

Invoking one condition for dynamic balance,

$$\mathbf{f}_B = 0 \qquad (10.169)$$

which implies from Eq. (10.168) that

$$I_{xz} = I_{yz} = 0 \qquad (10.170)$$

must hold since no restrictions are placed on ω and $\dot{\omega}$.

The remaining condition for balance, $\mathbf{F}_A = 0$, must be explored to find another compatible condition. To this end the coordinate axis can be shifted to another point C on the shaft axis located arbitrarily a distance l_1 from A. Now the system dynamics referred to this axis become

$$\begin{matrix} -I_{xz}'\dot{\omega} + I_{yz}'\omega^2 \\ -I_{yz}'\dot{\omega} - I_{xz}\omega^2 \end{matrix} = \begin{bmatrix} N_x' \\ N_y' \end{bmatrix} = \mathbf{r}_a \times \mathbf{f}_a + \mathbf{r}_b \times \mathbf{f}_B = \begin{bmatrix} f_{By}(L - l_1) - f_{Ay}l_1 \\ f_{Bx}(L - l_1) - f_{Ax}l_1 \end{bmatrix} \tag{10.171}$$

Now invoking the balancing condition

$$\mathbf{f}_A = \mathbf{f}_B = 0 \tag{10.172}$$

leads to

$$I_{xy}' = I_{yz}' = 0 \tag{10.173}$$

and the general condition for dynamic balance is that the products of inertia I_{xy} and I_{yz} vanish when referred to two (and therefore for all) different coordinate systems separated in the z direction. Thus, from the definition of products of inertia (for continuous systems the summation sign is replaced by the integral),

$$I_{xz} = \sum_i M_i X_i Z_i = \sum M_i X_i' Z_i' = I_{xz}' = 0 \tag{10.174}$$

$$I_{yz} = \sum M_i Y_i Z_i = \sum M_i Y_i' Z_i' = I_{yz}' = 0 \tag{10.175}$$

$$Z_i' = l_1 + Z_i \tag{10.176}$$

$$X_i = X_i' \tag{10.177}$$

$$Y_i = Y_i' \tag{10.178}$$

where the primed and unprimed coordinates are referred to the axes at A and at C, respectively.

Rewriting Eqs. (10.174) to (10.178) more compactly in vector form,

$$\Sigma M_i Z_i (X_i \mathbf{i} + Y_i \mathbf{j}) = 0 \tag{10.179}$$

$$\Sigma M_i Z_i (X_i' \mathbf{i} + Y_i' \mathbf{j}) = 0 \tag{10.180}$$

After substitution of unprimed and primed coordinates,

$$\Sigma M_i Z_i (X_i \mathbf{i} + Y_i \mathbf{j}) = 0 \tag{10.181}$$

$$\Sigma M_i (Z_i + l_i)(X_i \mathbf{i} + Y_i \mathbf{j}) = 0 \tag{10.182}$$

Subtracting Eq. (10.181) from (10.182) yields

$$\Sigma M_i l_1 (X_i \mathbf{i} + Y_i \mathbf{j}) = 0 = l_1 \Sigma M_i (X_i \mathbf{i} + Y_i \mathbf{j}) \tag{10.183}$$

whence

$$\Sigma M_i X_i = M_i Y_i = 0 \tag{10.184}$$

Since l_1 is arbitrary this states that the center of mass lies on the axis of rotation. This is the condition of so-called "static balance."

It is now easily shown that this condition (10.181) and (10.184) are both necessary and sufficient to establish dynamic balance. To show this, rewrite Eqs. (10.181) and (10.184)

$$\Sigma M_i Z_i (X_i \mathbf{i} + Y_i \mathbf{j}) = 0$$

$$\Sigma M_i (X_i \mathbf{i} + Y_i \mathbf{j}) = 0$$

Multiply Eq. (10.184) by b and add Eq. (10.181) to obtain

$$\Sigma M_i (Z_i + b)(X_i \mathbf{i} + Y_i \mathbf{j}) = 0 \tag{10.185}$$

But this is the statement that the products of inertia equal zero when referred to the axis at b removed from the original axis, and since b is arbitrary it holds for two (and for all) axes which have been shown to establish the condition for dynamic balance.

A system of the foregoing type in a state of dynamic imbalance can always be balanced in general by at most two "point" masses ΔM_1 and ΔM_2 located at radial positions $\mathbf{r}_1$ and $\mathbf{r}_2$. As a simple consequence of the above, Eqs. (10.181) and (10.184) become

$$\Sigma M_i(X_i\mathbf{i} + Y_i\mathbf{j}) + \Delta M_1\mathbf{r}_1 + \Delta M_2\mathbf{r}_2 = 0 \tag{10.186}$$
$$\Sigma M_i Z_i(X_i\mathbf{i} + Y_i\mathbf{j}) + \Delta M_1 Z_1\mathbf{r}_1 + \Delta M_2 Z_2\mathbf{r}_2 = 0 \tag{10.187}$$
$$\mathbf{r}_1 = X\mathbf{i} + Y_1\mathbf{j} + Z_1\mathbf{k} \qquad \mathbf{r}_2 = X_2\mathbf{i} + Y_2\mathbf{j} + Z_2\mathbf{k} \tag{10.188}$$

Note that this pair of vector equations [(10.186) and (10.187)] is equivalent to four independent equations. Since the number of parameters ΔM_1, ΔM_2, X_1, Y_1, X_2, Y_2, Z_1, and Z_2 that could be chosen to satisfy this set is eight, four are free to be chosen at will. Usually some are constrained by geometry and other design considerations. Often the radial distance $|r_1| = |r_2|$ and the balancing planes z_1 and z_2 are specified, leaving ΔM_1 and ΔM_2 and the angular positions θ_1 and θ_2 to be uniquely determined by Eqs. (10.186) and (10.187). Reference 6 provides criteria for allowable unbalance.

(c) Inertia Balancing of Slider-crank Chain

In most mechanisms, motion, forces, and moments cannot be represented by explicit mathematical expressions and are usually developed in power or Fourier series with

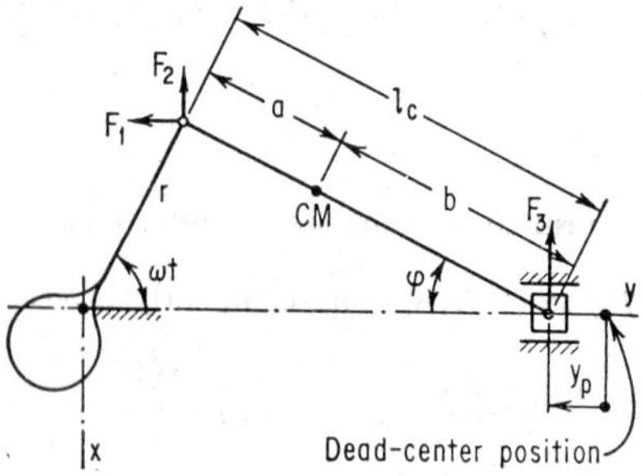

FIG. 10.14. Slider-crank mechanism—inertia balancing.

the higher-order terms neglected. In this section we shall discuss the slider-crank chain. For the cam and quadric mechanisms, see refs. 2 to 4. The displacement, velocity, and acceleration[1] of the slider-crank piston shown in Fig. 10.14 are

$$\begin{aligned} y_p &= r\cos\omega t + l_c\cos\varphi \\ &= r + r^2/4l_c - r[\cos\omega t + (r/4l_c)\cos 2\omega t] \end{aligned} \tag{10.189}$$
$$\dot{y}_p = r\omega[\sin\omega t + (r/2l_c)\sin 2\omega t] \tag{10.190}$$
$$\ddot{y}_p = r\omega^2[\cos\omega t + (r/l_c)\cos 2\omega t] \tag{10.191}$$

where r = crank radius, l_c = connecting-rod length, and ω = crank speed.

The crank components are

$$y_c = r(1 - \cos\omega t) \qquad X_c = -r\sin\omega t \tag{10.192}$$
$$\dot{y}_c = r\omega\sin\omega t \qquad \dot{X}_c = -r\omega\cos\omega t \tag{10.193}$$
$$\ddot{y}_c = r\omega^2\cos\omega t \qquad \ddot{X}_c = r\omega^2\sin\omega t \tag{10.194}$$

It is possible to divide the connecting-rod mass into two masses, one reciprocating on the piston M_{re} and one rotating on the crankpin M_{ro}, and obtain total forces.

$$\begin{aligned} F_y &= -M_{re}\ddot{Y}_p + M_{ro}\ddot{Y}_c \\ &= (M_{re} + M_{ro})r\omega^2\cos\omega t + M_{re}(r^2/l^2)(\omega^2\cos\omega t) \end{aligned} \tag{10.195}$$
$$F_x = M_{ro}\ddot{X}_c = M_{ro}r\omega^2\sin\omega t \tag{10.196}$$

The force alternates periodically and can be divided into a first harmonic or inertia force at the same frequency as the speed of rotation and a second harmonic at twice the frequency.

Because of connecting-rod inertia only, the *slider-crank mechanism* torque acting on the frame about the shaft center O:

$$T_f = F_3(l_c \cos \omega t + \varphi r \cos \omega t) + \tfrac{1}{2}M_{re}\omega^2 r^2 \left\{ \left[\frac{(r^2 + 8l_c^2)(k^2 - ab)}{4ral_c^2} + \frac{r}{2l_c} \right] \sin \omega t - \left(\frac{ab - k^2}{al_c} + 1 \right) \sin 2\omega t - \left[\frac{3r(k^2 - ab)}{4al_c^2} + \frac{3r}{2l} \right] \sin 3\omega t \right\} \quad (10.197)$$

where k is the radius of gyration of the rod $mk^2 = I_G$, and F_3 is force on the piston shown.

Only because of inertia forces, the slider-crank mechanism has a torque in the shaft.

$$T_s = -F_1 r \sin \omega t - F_2 r \cos \omega t \quad (10.198)$$

$$T_s = \tfrac{1}{2}M_{re}\omega^2 r^2 \left[\frac{r}{2l_c} \sin \omega t - \left(1 + \frac{ab - k^2}{al_c} \right) \sin 2\omega t - \frac{3r}{2l_c} \sin 3\omega t \right] \quad (10.199)$$

where F_2 and F_1 are forces shown. Note that the inertia torques on the shaft and frame are not equal. They differ by the moment of the inertia force of the connecting rod.

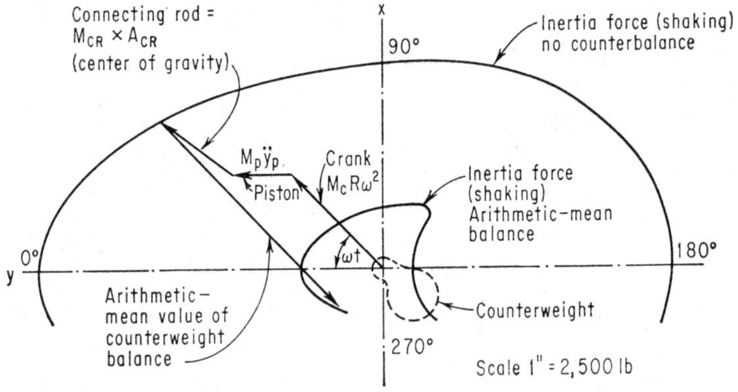

FIG. 10.15. Slider-crank mechanism—polar plot. 3,200 rpm piston = 2.14 lb; connecting rod = 1.80 lb; i = 1 in.; center of gravity of connecting rod to wrist-pin axis = 6 in.; radius of gyration of connecting rod about center of gravity = 2.90 in.; crank unbalance at crankpin center = 3.13 lb.

M_{CR} = mass of connecting rod; A_{CR} = acceleration of center of gravity of connecting rod; M_p = mass of piston $\ddot{y}$; M_C = mass of crank R; $\ddot{y}_p$ = acceleration of piston; $\ddot{y}$ = acceleration of crank.

Balancing of secondary forces is complicated because a counterweight rotating at 2ω is required. Balancing generally consists of adding a counterweight to the crank. The rotating masses can be counterbalanced but the reciprocating masses (which are appreciable) in the single cylinder engine cannot. A polar plot, called a *shaking-force diagram*, is shown in Fig. 10.15. If the center of gravity of the connecting rod and piston were at the crankpin, the system could be perfectly balanced. Other balancing can be accomplished by employing dummy mechanisms (180° mirror images), two mechanisms in line, or using rotating masses at primary and secondary frequencies. Figure 10.16 shows a mechanism completely balanced for forces and moments. The torque, however, is doubled. References 1 and 2 have more on slider-crank chain balancing.

(d) Torque Balancing of Shafts (Inertia and External Loads)

The first step in torque balancing is the determination of the total torque curve (composed of inertia and working loads) for the complete cycle. This curve is made up of a number of harmonic components any of which can be dangerous from the point of view of vibration excitation. The object of torque balancing is to reduce torque fluctuation at its source and at the same time reduce the cyclic peak torque. The machine will usually run at a more uniform speed.

Reversal of torque is undesirable in most dynamic systems. It is particularly detrimental in geared, belt-driven, and chain-driven systems where inherent backlash

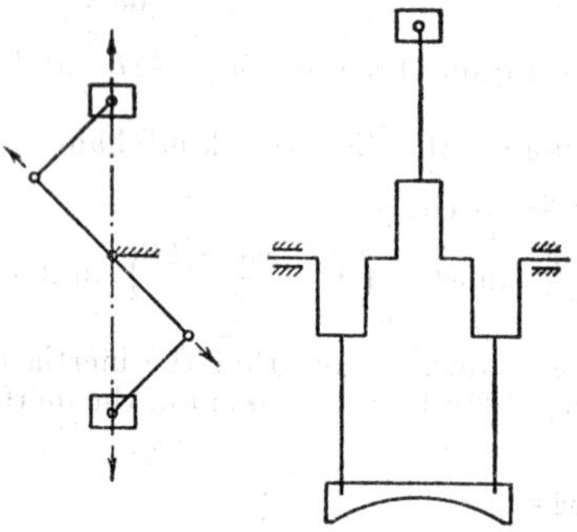

FIG. 10.16. Slider-crank mechanism and dummy mechanism. Completely balanced for forces and moments.

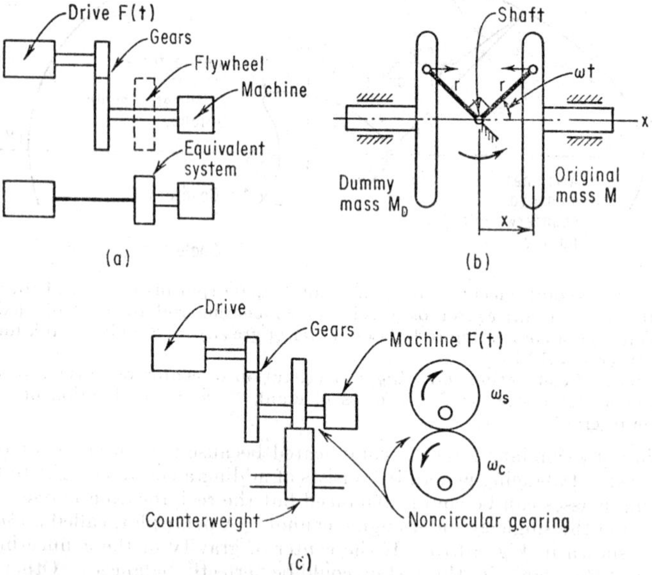

FIG. 10.17. Some torque-balancing methods. (a) Flywheel. (b) Dummy mechanism (Scotch-yoke mechanism shown). (c) Noncircular-gearing counterweight.

exists; shocks and surges would occur. Thus an additional objective of torque balancing is the elimination of torque reversal. Torque balancing can be accomplished by (1) adding a flywheel, (2) employing an out-of-phase dummy mechanism, (3) having rotating masses driven by noncircular gears, (4) using a feedback-controlled torque system. Multiple units out of phase can also be employed to reduce the torque.

Flywheel. A *flywheel* is a high-moment-of-inertia element incorporated between the driving and driven portions of a machine (Fig. 10.17*a*). The discussion which follows considers rigid systems only. Figure 10.17*a* shows the general case in which both input and output torques vary.

The net change in kinetic energy during a complete cycle is $E_k = \int_0^{2\pi} T\,d\theta = 0$.

The speed will fluctuate, however, during the cycle; a positive acceleration will occur in areas of positive torque where the input torque exceeds the output torque. The largest area will cause the largest speed fluctuation.

The change of kinetic energy during the time interval *ab*:

$$\Delta E_k = \int_a^b T\,d\theta = (I_0/2)(\omega_b^2 - \omega_a^2) \approx \Delta(I_0/2)\omega^2 \approx I_0\omega\,\Delta\omega \tag{10.200}$$

This equals the energy change in the rotating mass.

$$\Delta E = I_0 C\omega^2 \tag{10.201}$$

where $I_0 = I_f + I_p$, total moment of inertia
I_f = moment of inertia of flywheel
I_p = moment of inertia of other moving parts subject to energy change
$\omega = \dfrac{\omega_b + \omega_a}{2}$ = mean speed
$C = \dfrac{\omega_b - \omega_a}{\omega}$ = coefficient of regulation

The reciprocal $1/C$ is often called the *coefficient of steadiness* whose values depend upon the nature of service, i.e., permissible speed variation. For electrical machinery, $C = 0.003$; for machine tools, pumps, and textile machinery, $C = 0.03$; and for hammering machinery, $C = 0.2$.

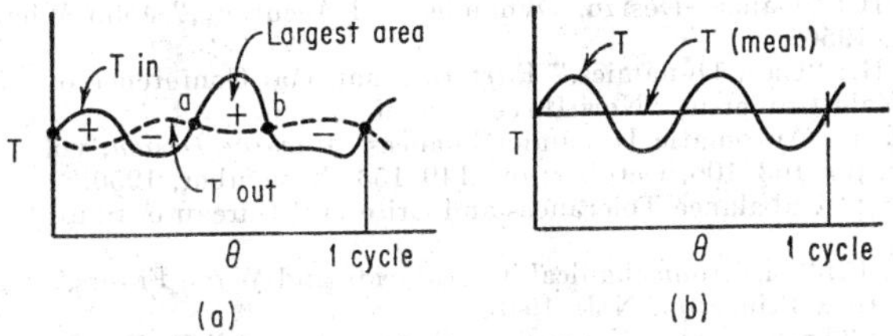

FIG. 10.18. Torque curves. (*a*) General case. (*b*) Either input or output torque fluctuating and *T* (mean) = constant.

Generally, either the input or output is a constant, at a value the mean of the output or input respectively (Fig. 10.18). Note that the flywheel experiences torsional vibration with angular acceleration $\ddot{\theta} = T/I_0$.

Dummy Mechanisms. The use of a *dummy mechanism* is generally impractical because of the larger machine which results. For example, the Scotch-yoke mechanism is torsionally balanced if the dummy mass is made equal to the original mass, with crank 90° offset (Fig. 10.17*b*).

The inertia force of the mass *M*

$$F = M\ddot{x} = +Mr\omega^2 \cos \omega t \tag{10.202}$$

and the torque on the shaft

$$T = Fr \sin \omega t = (Mr^2\omega^2/2) \sin 2\omega t \tag{10.203}$$

In a similar manner M_D, the dummy-mass torque

$$T = -\frac{M_D r^2\omega^2}{2} \sin 2\omega t \tag{10.204}$$

Furthermore, under this condition the inertia force in the shaft is eliminated. Note in the mechanism of Fig. 10.16 the torque is doubled by use of a dummy mechanism to eliminate the inertia forces.

Noncircular-gearing Counterweight. Let us consider masses rotating on *noncircular gearing* at some frequency (Fig. 10.17*c*). Both gears are radially balanced. It can be shown that the ratio ω_s/ω_c is continuously changing. This provides a fluctuating torque due to the counterweight: $T = I(d\omega/dt)$. The shape of the gears and the mass of the counterweight are chosen so that, when they are added to the system torque diagram, the total curve is smoother. The inertia forces in the shafts resulting from addition of the counterweight may be eliminated by using a duplicate pair of noncircular-gear counterweights.

Feedback-torque-controlled System. Such a system is applied to an indexing mechanism, for example, a compressed-air servosystem which provides energy which is released during the negative-acceleration period of the action.

(e) Automatic Balancing

Reference 5 offers an excellent survey of automatic dynamic balances, such as those used in centrifugal extractions of domestic laundry machines. In it are described the LeBlanc-type (liquid) mechanical-type self-balancer and the autobalancer. The latter is effective over the entire speed range.

References

1. Den Hartog, J. P.: "Mechanical Vibrations," 4th ed., McGraw-Hill Book Company, Inc., New York, 1956.
2. Talbourdet, G. J.: Mathematical Solution of 4-bar Linkage, *Machine Design*, May, pp. 65–68; June, pp. 81–82; July, pp. 73–77, 1941.
3. Rothbart, H.: "Cams—Design, Dynamics and Accuracy," John Wiley & Sons, Inc., New York, 1956.
4. Rothbart, H.: "Cam Dynamics," First International Conference on Mechanics, Mar. 27, 1961, Yale University, New Haven, Conn.
5. Thearle, E. L.: Automatic Dynamic Balances, *Machine Design*, vol. 27, pp. 119–124, September; pp. 103–106, October; pp. 149–153, November, 1950.
6. Feldman, S.: "Unbalance Tolerances and Criteria," Bureau of Ships Code 7, Washington, D.C.
7. Mason, W. P.: "Electromechanical Transducers and Wave Filters," D. Van Nostrand Company, Inc., Princeton, N.J., 1948.
8. Shea, T. E.: "Transmission Networks and Wave Filters," D. Van Nostrand Company, Inc., Princeton, N.J., 1948.
9. Van Volkenburg, M. E.: "Network Analysis," Prentice-Hall, Inc., Englewood Cliffs, N.J., 1955.
10. Plunkett, R.: "Mechanical Impedance Methods," ASME Annual Meeting, Dec. 2, 1958.
11. Guillemin, E. A.: "Introductory Circuit Theory," John Wiley & Sons, Inc., New York, 1953.
12. Holowenko, A. R.: "Dynamics of Machinery," John Wiley & Sons, Inc., New York, 1955.
13. Shigley, J. E.: "Dynamic Analysis of Machines," McGraw-Hill Book Company, Inc., New York, 1961.
14. Cunningham, W. J.: "Introduction to Nonlinear Analysis," McGraw-Hill Book Company, Inc., New York, 1958.
15. Barkan, P.: Dynamics of High Capacity Outdoor Oil Breakers, *AIEE Trans.*, vol. 74, part III, pp. 671–676, 1955.
16. Barkan, P., and E. J. Tuohy: "The Effect of Linkage Flexibility on Dynamics of High Capacity Outdoor Circuit Breakers," AIEE paper CP 8–943, summer meeting, Buffalo, N.Y., June 22–25, 1958.
17. Tuplin, W. A.: "Torsional Vibration," John Wiley & Sons, Inc., New York, 1934.
18. Wilson, W.: "Practical Solution of Torsional Vibration Problems," vols. 1 and 2, John Wiley & Sons, Inc., New York, 1956.
19. Kozesnik, J.: "Dynamics of Machines" (in English), E. P Noordhoff Ltd., Groningen, Czechoslovakia, 1962.

Section 11

FRICTION, LUBRICATION, AND WEAR

By

DAVID TABOR, Sc.D., *Laboratory for the Physics and Chemistry of Solids, Department of Physics, Cambridge University, Cambridge, England*

CONTENTS

11.1. INTRODUCTION

Sliding friction is primarily a surface phenomenon. Consequently it depends very markedly on surface conditions such as roughness, degree of work hardening, type of oxide film, and surface cleanliness. In general the roughness has only a secondary effect but surface contamination can have a profound influence on the friction (and wear), particularly with surfaces that are nominally clean. Because of this the account given here concentrates mainly on the mechanisms involved in friction so that the main factors may be assessed in any particular situation. Tables of friction values are given but they must be used with caution. Very wide differences in friction may be obtained under *apparently* similar conditions, especially with unlubricated surfaces.

11.2. DEFINITIONS AND LAWS OF FRICTION

(a) Definition

The friction between two bodies is generally defined as the force which acts between them at their surface of contact so as to resist their sliding on one another. The frictional force F is the force required to initiate or maintain motion. If W is the normal reaction of one body on the other, the coefficient of friction

$$\mu = F/W \tag{11.1}$$

(b) Static and Kinetic Friction

If the force to initiate motion of one of the bodies is F_s and the force to maintain its motion at a given speed is F_k, there is a corresponding coefficient of static friction $\mu_s = F_s/W$ and a coefficient of kinetic friction $\mu_k = F_k/W$. In some cases these coefficients are approximately equal; in most cases $\mu_s > \mu_k$.

(c) Basic Laws of Friction

The two basic laws of friction, which are valid over a wide range of experimental conditions, state that: (1) The frictional force F between solid bodies is proportional to the normal force between the surfaces; i.e., μ is independent of W. (2) The frictional force F is independent of the apparent area of contact.

These two laws of friction are reasonably well obeyed for sliding metals whether clean or lubricated. With polymeric solids (plastics) the laws are not so well obeyed; in particular the coefficient of friction usually decreases with increasing load.

11.3. FRICTION OF CLEAN METALS

(a) Theory of Metallic Friction

When clean metal surfaces are placed in contact they do not usually touch over the whole of their apparent area of contact. In general they are supported by the surface irregularities which are present even on the most carefully prepared surfaces. The smallest loads produce plastic flow at these regions and the asperities crush down until they are large enough to support the load. Under these conditions the area of real contact is approximately proportional to the load and independent of the size and geometry of the surfaces. Metallic junctions are formed at the regions of real contact by a process of cold welding, and these junctions must be sheared if sliding is to take place. Thus the friction may be written

$$F = As + P \tag{11.2}$$

where A is the real area of contact, s the specific shear strength of the interface, and P a deformation or plowing term which arises if a harder surface slides over a softer one. In general the adhesion term As is very much larger than the deformation term P, so that $F \approx As$. Thus the friction is directly proportional to the load and independent of the size of the bodies. This mechanism thus explains the two laws of friction and the type of damage which occurs during sliding.[13]

If metal surfaces are thoroughly cleaned in a vacuum it is almost impossible to slide them on one another.[8,9,11] Any attempt to do so causes further deformation at the regions of contact. The surfaces, being clean, adhere wherever they touch; the resistance to motion increases the harder the surfaces are pulled, and complete seizure occurs. The coefficient of friction is of the order $\mu = 5$ to 100. The smallest trace of gas or vapor produces a profound reduction in the friction to about $\mu = 1$ by inhibiting the growth of strong metallic junctions. With most metals prepared in the atmosphere, the surface oxide films serve a similar role and the friction is of the order $\mu_s = 0.5$ to 1.3 (Table 11.1).

Table 11.1. Coefficient of Static Friction μ_s of Unlubricated Metals Prepared Grease-free*

Pure Metals on Themselves

Metal	Ag	Al	Cd	Cu	Cr	Fe	In	Mg	Mo	Ni	Pb	Pt
μ_s	1.4	1.3	0.5	1.3	0.4	1.0	2	0.5	0.9	0.7	1.5	1.3

Metals on Steel†

Metal	Ag	Al	Cu	Cr	In	Mo	Ni	Pb	Sn
μ_s	0.5	0.5	0.8	0.5	2	0.5	0.5	1.2	0.9

Alloys on Steel†

Alloy	μ_s	Alloy	μ_s
Copper-lead (dendritic: Pb 20)	0.22	Phosphor-bronze	0.35
Copper-lead (nondendritic: Pb 27)	0.22	Aluminum-bronze	0.45
White metal (tin base: Sb 6.4, Cu 4.2, Ni 0.1, Sn 89.2)	0.8	Brass (Cu 70, Zn 30)	0.5
White metal (lead base: Sb 15, Cu 0.5, Sn 6, Pb 78.5)	0.55	Constantan	0.4
Wood's alloy	0.7	Steel (0.13 C, 3.42 Ni)	0.8
		Cast iron	0.4

* The actual values depend crucially on the state of surface cleanliness and on the degree to which the surface oxide is ruptured. The values should therefore be considered as representative rather than absolute. The results are for spectroscopically pure metals. Small amounts of impurity do not have a marked effect on the friction if they do not produce a second phase and if they do not appreciably modify the nature of the oxide film normally present.

† (0.13 per cent C, 3.42 per cent Ni, normalized.) Mild steel gives essentially the same results.

With repeated traversals of the same surface, pickup of the softer metal occurs and in time the sliding becomes characteristic of the softer metal sliding on itself.

(b) Microdisplacements before Sliding Occurs

When surfaces are placed in contact under a normal load and a tangential force F is applied the combined stresses produce further flow in the junctions long before gross sliding occurs. On a microscopic scale the surfaces sink together, increasing the area of contact, and at the same time a minute tangential displacement occurs. As the tangential force is increased this process continues until a stage is reached at which the applied shear stress is greater than the strength of the interface. Junction growth comes to an end and gross sliding occurs.[29] The tangential force has its critical value F_s. The tangential displacements before sliding are always very small. The values given below correspond to the stage where F has reached 90 per cent of the value necessary to produce gross sliding ($F = 0.9F_s$). The results are for a hemispherically tipped conical slider on a flat finely abraded surface of the same metal.[14] As a crude approximation these tangential displacements are roughly proportional to the square root of the normal load (Table 11.2).

Table 11.2. Microdisplacements before Gross Sliding

Surfaces	Vickers hardness, kg/mm²	Tangential displacements at $F = 0.9F_s$, units in 10^{-4} cm		
		1-g load	100-g load	10,000-g load
Indium	1	30	100	
Tin	7	1	20	200
Gold	19	0.5	5	
Platinum	117	...	≈1	60
Mild steel	280	...	1	8

(c) Breakdown of Oxide Films

During the sliding process the breakdown and penetration of surface oxide films may have a very great effect on the frictional behavior. If the surface deformation is sufficiently small the surface oxide may not be ruptured so that all the sliding may occur within the oxide film itself. Because the junctions formed in the oxide film are often weaker than purely metallic junctions the friction may be appreciably less than when the oxide is ruptured. Since the shearing process occurs within the oxide film

Table 11.3. Breakdown of Oxide Film Produced during Sliding as Shown by Electrical-conductance Measurements*

Metal	Vickers hardness, kg/mm²		Load, g, at which appreciable metallic contact occurs
	Metal	Oxide	
Gold	20	.....	0
Silver	26	.....	0.003
Tin	5	1,650	0.02
Aluminum	15	1,800	0.2
Zinc	35	200	0.5
Copper	40	130	1
Iron	120	150	10
Chromium plate	800	.....	500

* The results are for a spherical slider on a flat electrolytically polished surface. The actual breakdown loads will depend on the geometry of the surfaces and on the thickness of the oxide film. The values given in this table provide a relative measure of the protective properties of the oxide film normally present on metals prepared by electrolytic polishing.[13]

the surface damage and wear are always considerably reduced. The criterion for "survival" of the oxide film is that it should be sufficiently soft or ductile compared with the substrate metal itself so that it deforms with it and is not easily ruptured or fractured. Thus the oxide normally present on copper is not easily penetrated, whereas aluminum oxide, being a hard oxide on a soft substrate, is readily shattered during sliding, and even at the smallest loads there is some metallic interaction (Table 11.3). Thicker oxide films often provide more effective protection to the surfaces. Thus, with anodically oxidized surfaces of aluminum or aluminum alloys, the sliding may be entirely restricted to the oxide layer. Similarly, with very hard metal substrates such as chromium, the surface deformation may be so small that the oxide is never ruptured.

(d) Effect of Hardness and Structure on Static Friction

The friction of metals depends little on hardness (Tables 11.4 and 11.5). This is well explained by the adhesion theory of friction. With a softer metal, for example, the area of contact A for a given load is relatively large; but the interface is weaker so that s is small. Consequently the product $F = As$ is scarcely affected by the hardness. However, the friction of hard metals is, on the whole, somewhat less than with softer metals. This is partly because of the reduced ductility of the metal junctions, which restricts junction growth, but mainly because the harder substrate provides greater support to the surface oxide film (see results for copper-beryllium alloy[23] in Table 11.4). The friction also clearly depends on the nature and strength of the oxide film itself. With ferrous alloys the homogeneity of the alloy is at least as important a factor, since heterogeneous materials will give weakened junctions.

Table 11.4. Effect of Hardness and Structure on Static-friction Coefficient μ_s

Material	Composition or designation	Hardness, Vickers, kg/mm²	μ_s
Hard-steel slider (Vickers 464) on copper-beryllium alloy	Beryllium alloy (Be 2, Co 0.25)		
	As quenched	120	1
	Fully aged condition	410	0.4
	Overaged condition	200	0.9–1
Ferrous materials sliding on themselves:			
Pure iron	Cold rolled	≈ 150	1–1.2
Austenitic steel	Cr 18, Ni 8	200	1
Normalized steel	C 0.13, Ni 3.42	170	0.7–0.8
Ball-race steel	....	900	0.6–0.7
Chromium plate	Hard bright	≈1,000	0.6–0.7
Cast iron (gray)	Piston ring, pearlitic with fine dispersed graphite flakes	200	0.3–0.4
Tool steel	Heterogeneous, containing carbides	900	0.3–0.4

(e) Friction of Thin Metallic Films

If soft metal films of suitable thickness are plated on to hard metal substrates the hard substrate supports the load, giving a small area of contact A. If the film is not penetrated shearing occurs in the soft film itself, giving a low value of s. Conse-

Table 11.5. Static-friction Coefficient μ_s of Very Hard Solids

Bonded Tungsten Carbide (Cobalt Binder)*

Metal	μ_s
Tungsten carbide	≈0.2
Copper	≈0.4
Cadmium	0.8–1
Iron	0.4–0.8
Cobalt	0.3

Unbonded Carbides Sliding on Themselves

Carbide	Coefficient of friction μ_s			
	In air	Outgassed and measured in vacuo		
	20°C	20°C	1000°C	Comments
Tungsten carbide	0.15	0.4	0.3	Rises above 1000°C
Titanium carbide	0.15	0.5	0.3	Rises above 1200°C
Boron carbide	0.1	0.6	0.5	Rises above 1800°C

Diamond and Sapphire[11]

Surfaces and condition		μ_s
Diamond on self	In air at 20°C, clean	≈0.05–0.1
	Outgassed at 1000°C, measured in vacuo at 20°C	0.5
Sapphire on self	In air at 20°C, clean	0.2
	Outgassed at 1000°C, measured in vacuo at 20°C	0.9
Diamond on steel	In air at 20°C, clean and lubricated	0.1–0.15
Sapphire on steel	In air at 20°C, clean and lubricated	0.15–0.2

* Friction of curved slider of carbide on flat specimen of other metal. Unlubricated.[26]

Table 11.6. Coefficient of Static Friction of Thin Metallic Films

Load, g	Coefficient of static friction, μ_s			
	Indium film on steel	Indium film on silver	Lead film on copper	Copper film on steel
4,000	0.08	0.1	0.18	0.3
8,000	0.04	0.07	0.12	0.2

quently the resulting frictional force $F = As$ may be small and the low friction generally persists up to the melting point of the surface film.[12] Copper lead-bearing alloys function in this way. Table 11.6 gives typical friction values for thin films of indium, lead, and copper (10^{-3} to 10^{-4} cm thick) deposited on various metal substrates. The other sliding member is a steel sphere ¼ in. in diameter.

Another very useful combination consists of a thin flash of gold on plated rhodium. This appears not only to have favorable friction and wear properties: it also gives very satisfactory electrical contact.

(f) Kinetic Friction of Metals

The main effect of speed of sliding is the generation of high local temperatures; these are due to frictional heating at the points of real contact and may profoundly modify the frictional behavior. For example, local hot spots may produce phase changes or alloy formation at and near the sliding interface, they may produce local melting, and they may greatly change the rate of surface oxidation. Two main groups of results are given in Table 11.7 for ordinary "engineering" speeds of a few fps and at very high sliding speeds up to 1,500 fps where surface melting is the predominant factor.

Table 11.7. Coefficient of Kinetic Friction μ_k for Sliders Rubbing on a Mild Steel Disk[6] at Speeds of a Few fps

Slider	μ_k
Nickel, mild steel	0.55–0.6
Aluminum, brass (70:30), cadmium, chromium, magnesium, hard steel, tin-base alloy, brake-band material	.04–0.5
Copper, lead-base alloy, phosphor bronze, garnet	0.3–0.4
Carbon, constantan, copper-lead alloy, bakelite	0.2

(g) Friction at Very High Sliding Speeds

The results given here are for a rapidly rotating sphere of ball-bearing steel sliding against another surface in a moderate vacuum ($\approx 10^{-4}$ mm Hg). In general the friction falls off at very high sliding speeds because of the formation of a very thin molten surface layer which acts as a lubricant film. With diamond the friction first diminishes and then increases at the highest speed; this is due to transfer of steel from the ball to the diamond so that the sliding becomes representative of metal-metal friction. In some cases the metal may fragment at these very high speeds. The results are roughly independent of load over the load range of 10 to 500 g used in this work.[7] The friction depends critically on the duration of sliding time since the frictional

Table 11.8. Coefficient of Friction of Hard-steel Ball on Various Solid Surfaces at Various Speeds

Surface	Coefficient of friction μ_k			
	30 fps	150 fps	750 fps	1,500 fps
Bismuth	0.25	0.1	0.05	
Lead	0.8	0.6	0.2	0.12
Cadmium	0.3	0.25	0.15	0.1
Copper	1	1	0.8	0.25
Molybdenum	1	0.8	0.3	0.2
Tungsten	0.5	0.4	0.2	0.2
Diamond	≈0.05	≈0.05	0.01	≈0.1

heating may profoundly affect the surface conditions. The results given in Table 11.8 are for sliding times of the order of 1 to 10 sec.

11.4. FRICTION OF POLYMERS

(a) Mechanism of Polymer Friction

The friction of thermoplastic polymers is fairly adequately explained in terms of the adhesion theory of friction.[20] There are, however, two main differences from the performance of metals. First the deformation at the regions of real contact depends on the geometry and time of loading as well as on the load itself. In general the true area of contact A increases less rapidly than the load W. The shear strength s of the interface may increase slightly with pressure but this is not a large effect. Thus the resultant friction $F = As$ increases less rapidly than the load so that μ_s decreases with W according to a relation $\mu_s = cW^{-\beta}$. For curved surfaces, approximating to a single region of contact, $\beta = 0.2$ to 0.3; for flat surfaces giving multiple-point contact, $\beta \approx 0.1$. The friction also tends to be larger if the geometric area is increased. Further, if the surfaces are left in contact the area of true contact increases with time because of creep and the starting friction may be correspondingly larger. The second difference is that the plowing or deformation term P is larger than for metals and may constitute a relatively large part of the observed friction. In many cases the deformation may be essentially elastic; in that case the deformation term corresponds to internal friction or hysteresis losses within the polymer itself (Table 11.9). Thus the plowing term becomes important when the adhesion component is small as with lubricated polymers[17] or with Teflon (Fluon) or when the material has large hysteresis losses as in wood[4] (Table 11.10).

Table 11.9. Static Coefficient of Friction of Unlubricated Polymers on Themselves[28]

Surfaces	Coefficient of friction μ_s	
	Bulk specimens, $\approx$100-g load	Crossed fibers, $\approx 10^{-2}$-g load
Nylon	0.3	0.8
Perspex (Plexiglass)	0.4–0.5	
Polyvinyl chloride	0.4–0.5	
Polystyrene	0.4–0.5	
Polythene	0.6	1.5
P.T.F.E. (Fluon, Teflon)	0.1	1.2

Table 11.10. Static Coefficient of Friction of Woods on Themselves*

Wood	Coefficient of friction μ_s	
	Degreased	Natural state
Soft wood	0.5	0.2–0.5
Lignum vitae	0.4–0.5	0.1

* This depends on moisture content and particularly on the natural waxes present.

(b) Effect of Speed on Friction of Polymers

The effect of speed is again mainly due to frictional heating. In some cases the friction is diminished. In other cases surface softening may cause increased interfacial adhesion and seizure. With P.T.F.E. (Teflon, Fluon) there is a marked increase in friction, possibly as a result of thermal degradation of the polymer.

More recent work suggests that the friction of polymers, of rubber, and of other viscoelastic materials varies with speed and temperature in a manner which reflects the viscoelastic properties of the solid itself [McLaren, K. G., and D. Tabor: Friction of Polymers: Influence of Speed and Temperature, *Nature (London)*, vol. 197, pp. 856–8, 1963, and Grosch, K. A.: Relation between Friction and Viscoelastic Properties of Rubber, *Nature (London)*, vol. 197, pp. 858–9, 1963.]

11.5. ROLLING FRICTION

(a) Mechanism of Rolling Friction

In the past it has been considered that rolling friction is primarily due to minute slip at the interface. Recent work, however, shows that in free rolling, although microslip may occur, the main source of friction is deformation losses as the sphere or cylinder traverses the surface.[27] In rolling over soft metals the deformation is plastic. For a sphere of diameter D (centimeters) rolling under a load W (grams) on a metal of yield pressure or hardness p (g/cm^2) the rolling friction F (grams) for the initial traversal is approximately equal to

$$F = 0.7(W^{3/2}/D)(1/p^{1/2}) \qquad (11.3)$$

Since this relation is balanced dimensionally the relation holds if F is in pounds, W in pounds, D in inches, and p in psi.

In rolling on elastic solids the rolling resistance arises primarily from internal friction or hysteresis losses in the solid. For a hard sphere on a flat surface the relation is

$$F = k(W^{4/3}/D^{2/3})(1/E^{1/3}) \qquad (11.4)$$

where E is Young's modulus (in g/cm^2 for metric units, or in psi if D is in inches and W and F are in pounds), and k is a numerical factor depending on the hysteresis-loss factor of the solid. Typical values for k are 0.005 to 0.01 for natural rubber, 0.05 to 0.1 for cross-linked rubber, 0.2 for a typical black-tread rubber, 0.2 to 0.3 for synthetic butyl rubber, and 0.05 to 0.2 for most grades of plastics. With these materials k will depend on speed of rolling and on temperature. For metals typical values of k are 0.005 for ball-race steel and 0.01 to 0.02 for mild steel, duralumin, and phosphor bronze. In these cases there is far less dependence on speed and temperature, but if the deformation exceeds the elastic limit a large increase in F will occur.

11.6. BOUNDARY LUBRICATION

(a) Definition

When moving surfaces are separated by a relatively thick film of lubricant the resistance to motion is due entirely to the viscosity of the interposed layer. The friction is extremely low ($\mu = 0.001$ to 0.0001) and there is no wear of the solid surfaces. These are the conditions of hydrodynamic lubrication under which bearings operate in the ideal case. If the pressures are too high or the sliding speeds too low the hydrodynamic film becomes so thin that it may be less than the height of the surface irregularities. The asperities then rub on one another and are separated by films only one or two molecular layers thick. The friction under these conditions ($\mu = 0.05$ to 0.15) is much higher than for ideal hydrodynamic lubrication, and some wear of the surfaces occurs. This type of lubricated sliding is called *boundary* lubrication. The friction depends not on the viscosity of the lubricant but on a more elusive property sometimes called *oiliness.* Under boundary conditions as for unlubricated surfaces the frictional resistance is proportional to the load and independent of the size of the surfaces.

(b) Theory of Boundary Lubrication

Boundary lubricants function by interposing between the sliding surfaces a thin film which can reduce metallic interaction and which is, in itself, easily sheared. The latter criterion restricts boundary lubricants almost exclusively to long-chain organic compounds, e.g., paraffins, alcohols, esters, fatty acids, and waxes. Radioactive-tracer experiments show that while a good boundary lubricant may reduce the friction by a factor of about 20 (from $\mu_s \approx 1$ to 0.05) it may reduce the metallic transfer

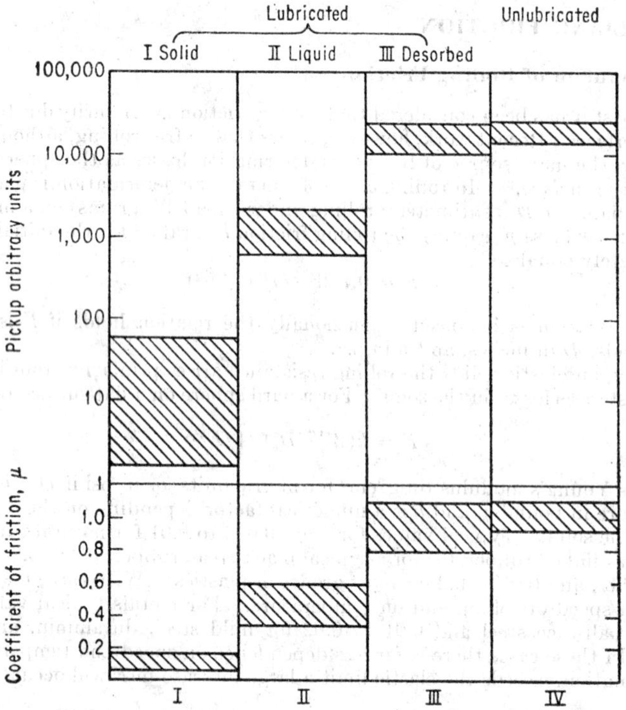

FIG. 11.1. Diagram in schematic form showing friction and metallic transfer for a typical reactive metal when the lubricant film is (I) solid, (II) liquid, (III) desorbed or mobile. The behavior in the desorbed or mobile state is essentially the same as for unlubricated surfaces (IV).

by a factor of 20,000 or more. Under these conditions the metallic junctions contribute very little to the frictional resistance; the friction is due almost entirely to the force required to shear the lubricant film itself. For this reason two good boundary lubricants may give indistinguishable coefficients of friction, but one may easily give twenty times as much metallic transfer (i.e., wear) as the other. Thus with good boundary lubricants the friction may be an inadequate indication of the effectiveness of the lubricant.

Boundary films are most effective when they are solid (Fig. 11.1, column I). If the temperature is raised there is a marked deterioration in the lubrication at the melting point of the film. The friction increases by a factor of 5 to 10 while the metallic transfer increases by a factor of 100 to 1,000 (Fig. 11.1, column II). At this stage the lubricant molecules have lost the strong orientation to the surface which exists in the solid state, but they remain attached to the surface. At a somewhat higher temperature a further deterioration occurs. The friction increases by a small factor but the transfer increases by another factor of 20 to 100. The friction and surface damage are now the same as those observed in the complete absence of lubricant (Fig. 11.1, columns III and IV). The reason is that the lubricant molecules are now mobile or desorbed; so that wherever the surfaces come together the lubricant is pushed away and there is no reduction in metallic interaction. These changes are due to changes in the physical state of the lubricant film and are reversible on cooling. These effects explain (at least in part) the superiority of fatty acids over paraffin and alcohols. The fatty acid usually reacts with the metal to form a metallic soap which has a higher melting point than the parent acid. Consequently the first breakdown temperature is raised and there is also an increase in the temperature of desorption. The absolute value of the breakdown temperature depends to some extent on the operating conditions, especially the speed. At higher sliding speeds the lubricant can act in a quasi-hydrodynamic manner so that effective lubrication may survive to a higher temperature.

With more protracted heating, oxidation of the lubricant occurs and the behavior is now determined by the properties of the oxidation products themselves. In the early stages these may be beneficial but later they lead to polymerization, gumming, and the formation of other deleterious products.

(c) Lubricating Properties of Long-chain Hydrocarbons

Single monolayers of long-chain paraffins, alcohols, or fatty acids can produce a marked reduction in the friction of metals. The coefficient of friction decreases with increasing chain length but there is a limit to this situation.[19] For paraffins this is reached at a chain length of about 20 carbon atoms in the molecule, for paraffins about 15, for fatty acids on reactive metal at about 12. The lower value of the static friction is about $\mu_s = 0.1$. The kinetic friction is generally somewhat smaller. Thus for steel surfaces lubricated with a long-chain fatty acid $\mu_k \approx 0.07$.

(d) Lubrication of Steel Surfaces by Various Lubricants

We give here Tables 11.11 and 11.12. The first describes the lubrication of steel on steel by various lubricants, the second the behavior of various metals on steel in the presence of four different lubricants.

Table 11.11. Lubrication of Steel Surfaces. Static Coefficient of Friction μ_s[15]

Lubricant	Static friction μ_s		Lubricant	Static friction μ_s	
	20°C	100°C		20°C	100°C
None	0.58		Mineral oils:		
Vegetable oils:			Light machine	0.16	0.19
Castor	0.095	0.105	Thick gear	0.125	0.15
Rape	0.105	0.105	Solvent refined	0.15	0.2
Olive	0.105	0.105	Heavy motor	0.195	0.205
Coconut	0.08	0.08	B.P. paraffin	0.18	0.22
			Extreme pressure	0.09.0.1	0.09–0.1
Animal oils:			Graphited oil	0.13	0.15
Sperm	0.10	0.10	Oleic acid	0.08	0.08
Pale whale	0.095	0.095	Trichloroethylene	0.33	
Neat's-foot	0.095	0.095	Alcohol	0.43	
Lard	0.085	0.085	Benzene	0.48	
			Glycerin	0.2	0.25

Table 11.12. Lubrication of Metals on Steel. Static Coefficient of Friction μ_s

Bearing surface	Rape oil, μ_s	Castor oil, μ_s	Mineral oil, μ_s	Long-chain fatty acids, μ_s
Hard steel (axle steel)	0.14	0.12	0.16	0.09
Cast iron	0.10	0.13	0.21	
Gun metal	0.15	0.16	0.21	
Bronze	0.12	0.12	0.16	
Pure lead	....	....	0.05	0.22
Lead-base white metal (Sb 15, Cu 0.5, Sn 6, Pb 78.5)	....	....	0.1	0.08
Pure tin	....	....	0.6	0.21
Tin-base white metal (Sb 6.5, Cu 4.2, Ni 0.1, Sn 89.2)	....	....	0.11	0.07
Sintered bronze	....	....	0.13	
Brass (Cu 70, Zn 30)	....	0.11	0.19	0.13

11.7. EXTREME-PRESSURE LUBRICATION

(a) General Information

Even the best boundary lubricants (e.g., long-chain acids or soaps) cease to provide any lubrication above about 200°C. Since localized hot spots of very much higher temperature are often reached in running mechanisms it is necessary to use surface films that have a high melting point and which, as far as possible, possess a low shear strength. Such materials are called *extreme-pressure* (EP) agents, though it would be more correct to refer to them as extreme-temperature agents. One obvious way of obtaining such lubrication is to coat the metal with a thin film of a softer metal. These films are effective up to their melting point but are gradually worn away with repeated sliding. Materials which are very effective are listed in Table 11.13.

Table 11.13. Coefficient of Static Friction μ_s of Mild Steel Lubricated with Certain Protective Films

Protective film	Static friction μ_s	Temperature up to which lubrication is effective
Teflon, Fluon (P.T.F.E.)	0.05	≈320°C
Graphite	0.07–0.13	≈600°C
Molybdenum disulfide	0.07–0.1	≈800°C

(b) Sulfide and Chloride Films

Another approach is to form a protective film *in situ* by chemical attack, a small quantity of a suitable reactive compound being added to the lubricating oil. The most common materials are additives containing sulfur or chlorine or both; phosphates are also used. The additive must not be too reactive; otherwise excessive corrosion will occur. Only when there is danger of incipient seizure should chemical reaction take place, the metal sulfide or metal chloride that is formed acting as a protective film. The detailed behavior is specific to the metal concerned, but in general, sulfides

Table 11.14. Effect of Sulfide and Chloride Films on Coefficient of Static Friction μ_s of Metals[16,18,21]

Metal	Static friction μ_s				
	Clean	Sulfide films		Chloride films	
		Dry	Covered with lubricating oil	Dry	Covered with lubricating oil
Cadmium on cadmium	0.5	...	....	0.3	0.15
Copper on copper	1.4	0.3	0.2	0.3	0.25
Silver on silver	1.4	0.4	0.2		
Steel on steel (0.13 C, 3.42 Ni)	0.8	0.2	0.05	0.15	0.05

give a higher friction than chlorides. However, sulfides do not decompose in the presence of water and are usually stable up to higher temperatures (Table 11.14).

11.8. WEAR

(a) Laws of Wear

Although the laws of friction are fairly well substantiated there are no satisfactory laws of wear. In general it is safe to say that wear increases with time of running and that with hard surfaces the wear is less than with softer surfaces. But there are many exceptions and the dependence of wear on load, nominal area of contact, speed, etc., is even less generally agreed upon. This is because there are many factors involved in wear and a slight change in conditions may completely alter the importance of individual factors or change their mode of interaction.

(b) Mild and Severe Wear

One of the most general characteristics of wear, both for clean and for lubricated surfaces, is that below a certain load the wear is small (mild wear); above this load it rises catastrophically to values that may be 1,000 or 10,000 times greater (severe wear). In severe wear the wear is mainly due to the shearing of intermetallic junctions. If the interface is weaker than the mating metals the junctions shear near the interface and the wear may be relatively low; this often occurs in the sliding of dissimilar metals. If on the other hand the interface is strong the shearing takes place a short distance from the interface and the wear is extremely heavy. This occurs in the sliding of identical metals since the junctions at the interface are highly work hardened. Thus even in the range of severe wear the wear may vary by a very large factor, say a factor of 1:100, whereas the friction may be substantially the same in both cases. This is shown in Table 11.15 where the wear per centimeter of sliding was determined using radioactive-tracer techniques.[24]

If the wear process, in mild or severe wear, is primarily due to the shearing of interfacial junctions the wear will be proportional to the load W and independent of the apparent area of the solids exactly as in the laws of friction. In addition it is not greatly dependent on the speed of sliding if this does not produce excessive frictional heating. However, although all the junctions contribute to the friction not all of them contribute to the wear. If K represents the fraction of the friction junctions which produce wear it may be shown on a very simple model[2] that the wear volume Z (cm^2/cm) per unit distance of travel is given by

$$Z = K(W/3p) \tag{11.5}$$

Table 11.15. "Severe" Wear of Unlubricated Surfaces (Sphere Sliding on a Flat, load 2,000 g., Sliding Speed 0.01 cm/sec)*

Surfaces	Vickers hardness, kg/mm²	Coefficient of static friction μ_s	Transfer from slider, 10^{-6} g/cm
Similar metals:			
Cadmium on cadmium	20	0.8	50
Zinc on zinc	38	0.5	200
Silver on silver	43	0.8	20
Copper on copper	95	1.3	20
Platinum on platinum	138	0.6	40
Mild steel (1 Cr)	158	0.8	15
Stainless steel (18 Cr)	217	0.8	5
Dissimilar metals:			
Cadmium on mild steel	...	0.5	0.3
Copper on mild steel	...	0.4	1
Platinum on mild steel	...	0.4	1.5
Mild steel on copper	...	0.6	0.3

* Note that repeated transfer from a soft metal to a hard may change the wear behavior to that of similar metals.

Table 11.16. Wear Rates of Various Combinations of Materials

Surfaces	Hardness, 10^6 g/cm²	K (calc.)
Similar metals, 0.01 cm/sec:		
Cadmium	2	10^{-2}
Zinc	3.8	10^{-1}
Silver	4.3	10^{-2}
Mild steel	15.8	10^{-2}
Dissimilar metals, 0.01 cm/sec:		
Cadmium on mild steel	.....	10^{-4}
Copper on mild steel	.....	10^{-3}
Platinum on mild steel	.....	10^{-3}
Mild steel on copper	.....	4×10^{-4}
Similar metals, 180 cm/sec:		
Mild steel	18.6	7×10^{-3}
Hardened tool steel	85	10^{-4}
Sintered tungsten carbide	130	10^{-6}
On hard tool steel, 180 cm/sec:		
Brass 60:40	9.5	6×10^{-4}
Silver steel	32	6×10^{-5}
Beryllium copper	21	4×10^{-5}
Stellite I	69	5×10^{-5}
Nonmetals on hard steel, 180 cm/sec:		
P.T.F.E. (Teflon, Fluon)	0.5	2×10^{-5}
Perspex (Plexiglass)	2	7×10^{-6}
Bakelite (poor)	2.5	7×10^{-6}
Bakelite (good)	3	7×10^{-7}
Polythene	0.2	1×10^{-7}

where W (grams) is the load and p(g/cm²) the hardness or yield pressure of the softer of the two sliding metals.

Typical results are given for unlubricated sliding in Table 11.16. The first set of data are from Table 11.15, for a sliding speed of 0.01 cm/sec. The second portion are for a sliding speed of 180 cm/sec at a load of 400 g.

(c) Effect of Environment

Wear is essentially a surface phenomenon and is greatly affected by surface films. In general lubricants reduce the wear, but if they contain very reactive EP additives corrosive wear may become relatively heavy. Surface oxide films generally reduce wear.

The surrounding atmosphere can have a marked effect, and in many cases air or oxygen or water vapor reduce the wear rate. However, this is not always the case. If, for example, the metal oxide is hard and the conditions favor abrasive wear the continuous formation of oxidized wear fragments may lead to a large increase in wear rate. With ferrous materials in air the atmospheric nitrogen may play an important part. Frictional heating and rapid cooling can produce martensite but with low-carbon steels surface hardening can still occur by reaction with nitrogen. When these hard surface films are formed the wear generally decreases.[30]

(d) Effect of Speed

The main effect of speed arises from increased surface temperatures. Four of the most important consequences are: (1) High hot-spot temperatures increase reactivity of the surfaces and the wear fragments with the environment. (2) Rapid heating and cooling of asperity contacts can lead to metallurgical changes which can change the wear process. (3) High temperatures may greatly increase the ease of interdiffusion and alloy formation. It has been suggested that low wear will occur only if no alloy formation occurs or if the alloy is relatively brittle.[25] (4) Surface melting may occur. In some cases, if melting is restricted to theo utermost surface layers, the friction and wear may become very low.

(e) Wear by Abrasives

Wear by hard abrasive particles is very common in running machinery. Measurements of the wear rate Z on abrasive papers show that the abrasion resistance $(1/Z)$

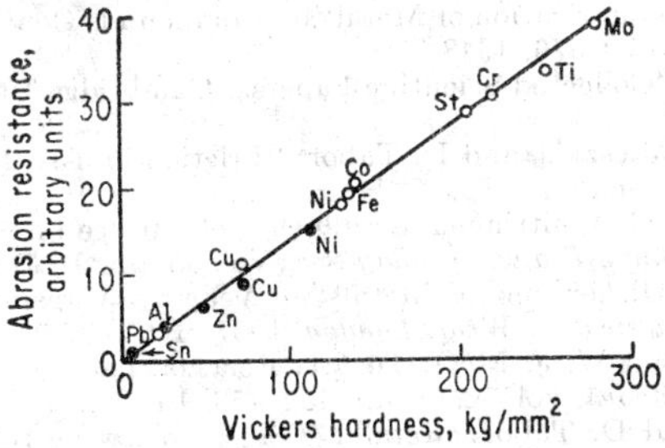

FIG. 11.2. Variation of abrasion resistance (reciprocal of wear rate) as a function of hardness for metals rubbed under standard conditions on dry adhesive paper. (St = 1.2% carbon steel.)

increases almost proportionally with the hardness of the metal.[22] This is shown in Fig. 11.2. The surfaces are grooved by the abrasive particles, but the wear is mainly in the form of fine shavings. It has been estimated that the amount of wear corresponds to about 10 per cent of the volume of the material displaced in the grooves.[5]

References

1. Part of the material given here is based on an article by the author in "Metals Reference Book" edited by Dr. C. J. Smithells, Butterworth & Co. (Publishers), Ltd., London, 1961.

2. Archard, J. F.: Contact and Rubbing of Flat Surfaces, *J. Appl. Phys.*, vol. 24, pp. 981–988, 1953.
3. Archard, J. F., and W. Hirst: The Wear of Metals under Unlubricated Conditions, *Proc. Roy. Soc.* (*London*), vol. A236, pp. 397–410, 1956.
4. Atack, D., and D. Tabor: The Friction of Wood, *Proc. Roy. Soc.* (*London*), vol. A246, pp. 539–555, 1958.
5. Avient, B. W. E., J. Goddard, and H. Wilman: An Experimental Study of Friction and Wear during Abrasion of Metals, *Proc. Roy. Soc.* (*London*), vol. A258, pp. 159–180, 1960.
6. Beare, W. G., and F. P. Bowden: Physical Properties of Surfaces: I, Kinetic Friction, *Phil. Trans. Roy. Soc. London*, vol. A234, pp. 329–354, 1935.
7. Bowden, F. P., and E. H. Freitag: The Friction of Solids at Very High Speeds: I, Metal on Metal; II, Metal on Diamond, *Proc. Roy. Soc.* (*London*), vol. A248, pp. 350–367, 1958.
8. Bowden, F. P., and T. P. Hughes: The Friction of Clean Metals and the Influence of Absorbed Gases: The Temperature Coefficient of Friction, *Proc. Roy. Soc.* (*London*), vol. A172, pp. 263–279, 1939.
9. Bowden, F. P., and G. W. Rowe: The Adhesion of Clean Metals, *Proc. Roy. Soc.* (*London*), vol. A233, pp. 429–442, 1956.
10. Bowden, F. P., and J. E. Young: Friction of Clean Metals and the Influence of Adsorbed Films, *Proc. Roy. Soc.* (*London*), vol. A208, pp. 311–325, 1951.
11. Bowden, F. P., J. E. Young, and G .W. Rowe: Friction of Diamond, Graphite and Carbon: The Influence of Adsorbed Films, *Proc. Roy. Soc.* (*London*), vol. A212, pp. 485–488, 1952.
12. Bowden, F. P., and D. Tabor: Lubrication by Thin Films and the Action of Beaming Metals, *J. Appl. Phys.*, vol. 14, pp. 141–151, 1943.
13. Bowden, F. P., and D. Tabor: "Friction and Lubrication of Solids," Oxford University Press, London, part I, 1954; part II, 1963.
14. Courtney-Pratt, J. S., and E. Eisner: The Effect of a Tangential Force on the Contact of Metallic Bodies, *Proc. Roy. Soc.* (*London*), vol. A238, pp. 529–550, 1957.
15. Fogg, A., and S. A. Hunswick: The Static Friction of Lubricated Surfaces, *J. Inst. Petrol.*, vol. 26, pp. 1–18, 1940.
16. Greenhill, E. B.: The Lubrication of Metals by Compound Containing Sulphur, *J. Inst. Petrol.*, vol. 34, pp. 659–669, 1948.
17. Greenwood, J. A., and D. Tabor: The Friction of Hard Sliders on Lubricated Rubber: The Importance of Deformation Losses, *Proc. Phys. Soc. London*, vol. 71, pp. 989–1001, 1957.
18. Gregory, J. N.: The Lubrication of Metals by Compounds Containing Chlorine, *J. Inst. Petrol.*, vol. 34, pp. 670–676, 1948.
19. Hardy, Sir W. B.: "Collected Scientific Papers," Cambridge University Press, London, 1936.
20. Howell, H. G., K. Mieszkis, and D. Tabor: "Friction in Textiles," Butterworth & Co. (Publishers), Ltd., London, 1959.
21. Hughes, T. P., and G. Whittingham: Influence of Surface Films on the Dry and Lubricated Sliding of Metals, *Trans. Faraday Soc.*, vol. 38, pp. 9–27, 1942.
22. Kruschov, M. M.: Resistance of Metals to Wear by Abrasion, as Related to Hardness, *Proc. Conf. Lubrication Wear, London*, 1957, p. 655.
23. Moore, A. J. W., and W. J. McG. Tegart: Relation between Friction and Hardness, *Proc. Roy. Soc.* (*London*), vol. A212, pp. 452–458, 1952.
24. Rabinowicz, E., and D. Tabor: Metallic Transfer between Sliding Metals: an Autoradiographic Study, *Proc. Roy. Soc.* (*London*), vol. A208, pp. 455–475, 1950.
25. Roach, A. E., C. L. Goodzeit, and R. P. Hunnicutt: Scoring Characteristics of Thirty-eight Different Elemental Metals in High Speed Sliding Contact with Steel, *Trans. ASME*, vol. 78, p. 1659, 1956.
26. Shooter, K. V.: Frictional Properties of Tungsten Carbide and of Bonded Carbides, *Research* (*London*), vol. 4, pp. 136–139, 1951.
27. Tabor, D.: The Mechanism of Rolling Friction: II, The Elastic Range, *Proc. Roy. Soc.* (*London*), vol. A229, pp. 198–220, 1955.
28. Tabor, D.: Friction, Lubrication and Wear of Synthetic Fibres, *Wear*, vol. 1, pp. 5–24, 1957.
29. Tabor, D.: Junction Growth in Metallic Friction: The Role of Combined Stresses and Surface Contamination, *Proc. Roy. Soc.* (*London*), vol. A251, pp. 378–393, 1959.
30. Welsh, N. C.: Frictional Heating and Its Influence on the Wear of Steel, *J. Appl. Phys.*, vol. 28, 960–968, 1957.

Section 12

HYDRODYNAMIC LUBRICATION

By

B. STERNLICHT, Ph.D., *Technical Director, Mechanical Technology, Inc., Schenectady, N.Y.*

CONTENTS

List of Symbols

B = breadth, width (parallel to direction of motion)
C = radial clearance
D = diameter
E = energy, Young's modulus
F = frictional force, force
G = weight flow rate
H = power, work rate
I = moment of inertia
J = mechanical equivalent of heat
K = spring constant
L = bearing length (normal to direction of motion)
M = mass, moment
N = revolutions per unit time
P = unit loading = W/LD
Q = volume flow rate
R = bearing radius
$\mathcal{R}$ = perfect-gas constant
Re = Reynolds number
S = Sommerfeld number = $(\mu N/P)(R/C)^2$
T = temperature
U = linear velocity
V = velocity, volume
W = load
Y = dimensionless parameter
a = ratio of inlet to outlet film thickness
b = damping coefficient
c = specific heat
e = eccentricity
f = coefficient of friction, dimensionless force = $1/S$
g = constant of gravitational acceleration
h = film thickness
k = ratio of specific heats, thermal conductivity, spring constant
m = mass, ellipticity
n = polytropic constant
p = pressure
q = heat flow rate, volume flow rate per unit length
$\bar{q}$ = dimensionless flow coefficient = $Q/\pi RCNL$
t = time, thickness
u, v, w = linear-velocity components
$\bar{\bar{x}}$ = center of pressure
x, y, z = rectangular coordinates

α = dimensionless taper, load angle
β = angular span of bearing arc or sector
δ = amount of taper = $(h_1 - h_2)$, ellipticity ratio, recess depth
ϵ = eccentricity ratio = e/C
λ = feeding parameter
Λ = bearing number = $6\mu UB/p_a h_2^2$ or $(6\mu\omega/p_a)(R/C)^2$
μ = absolute viscosity
ν = kinematic viscosity
ρ = density
τ = shear stress
ϕ = attitude angle
ω = angular velocity
η, ξ = coordinates

Subscripts

D = discharge coefficient
H = horizontal
L = laminar incompressible, load
P = load
R = radial
T = tangential, turbulent incompressible
V = vertical, volume
a = ambient
b = bearing
c = common, critical
j = journal
l = laminar compressible
r = radial
s = slider, supply
t = tangential, turbulent compressible
GR = groove
avg = average
max = maximum
min = minimum
opt = optimum
red = reduced
0 = point of maximum pressure, equilibrium
1 = beginning, inlet, inner
2 = end, outlet, outer
(1) = first-order perturbation

Superscripts

- = dimensionless ($\bar{x} = x/B$)
. = first derivative with respect to time
.. = second derivative with respect to time
... = third derivative with respect to time
* = dynamic bearing number $\Lambda^* = (1 - 2\alpha')\Lambda$
' = first derivative with respect to $1/\omega\, dt$

12.1. INTRODUCTION TO BEARINGS

Mechanical requirements, environmental conditions, and relative cost are primary factors in determining the type of bearing to be used. The advantages and disadvantages of slider-type bearings as compared with rolling-contact bearings are summarized in Table 12.1.

Slider bearings, in which the principal motion is the relative sliding of two parts, include journal bearings and thrust bearings which, respectively, prevent movement

Table 12.1. Selection of Bearing Type[50]

Service factors	Characteristic	Sliding	Rolling
Mechanical requirements	Load: Unidirectional Cyclic Starting Unbalance Shock Emergency	Excellent Excellent Poor Excellent Excellent Fair	Excellent Excellent Excellent Excellent Good Fair
	Speed limited by	Turbulence Temperature rise	Centrifugal loading Dynamic effects
	Misalignment tolerance	Fair	Poor in ball bearings except where designed for at sacrifice of load capacity. Good in spherical roller bearings. Poor in cylindrical roller bearings
	Starting friction	Poor	Good
	Space requirements (radial bearing) Radial dimension Axial dimension	 Small $\frac{1}{4}$ to $L/D = 2$ for gas bearings times the shaft diam	 Large $\frac{1}{5}$ to $\frac{1}{2}$ the shaft diam
	Type of failure	Often permits limited emergency operation after failure	Limited operation may continue after fatigue failure but not after lubricant failure
	Damping	Excellent	Poor
	Type of lubricant	Oil or other liquids, grease, dry lubricants, or gas	Oil or grease
	Lubrication, quantity required	Large, except in low-speed boundary-lubrication types	Very small, except where large amounts of heat must be removed
	Noise	Quiet	May be noisy, depending upon quality of bearing and resonance of mounting
	Power consumption	Varies as $\frac{N^2D^3L}{C}$	Varies widely depending upon type of lubrication. Varies directly as speed. Usually lower than slider bearing
Environmental conditions	Low-temp starting	Poor	Good
	High-temp operation	Limited by lubricant	Limited by lubricant
Economics	Life	Unlimited, except for cyclic loading	Limited by fatigue properties of bearing metal
	Maintenance	Clean lubricant required	Clean lubricant required. Only occasional attention with grease
	Cost	Very small in mass-production quantities, or simple types	Intermediate, but standardized, varying little with quantity
	Ease of replacement	Function of design and installation	Function of type of installation. Usually shaft need not be replaced

N = rpm D = diam L = length C = diametrical clearance in the radial and axial directions of shafts or other moving machine parts. In addition, guide bearings which guide machine elements in lengthwise motion are another type of slider bearing.

Many types of slider bearing are in use. These bearings are primarily classified by method of lubrication and design. Two major slider-bearing types are: (1) the hydrodynamic or self-acting bearing in which the load is supported by fluid pressures

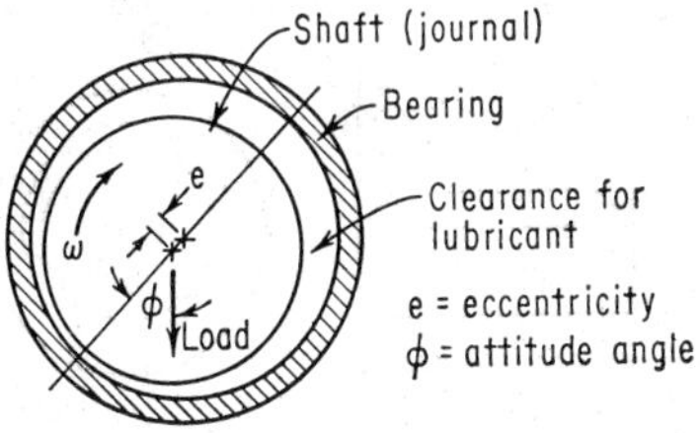

Fig. 12.1. Journal bearing.

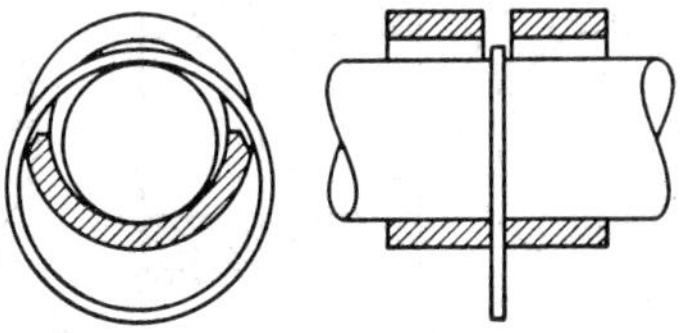

Fig. 12.2. Sketch showing construction principle of ring-oiled bearing.[50]

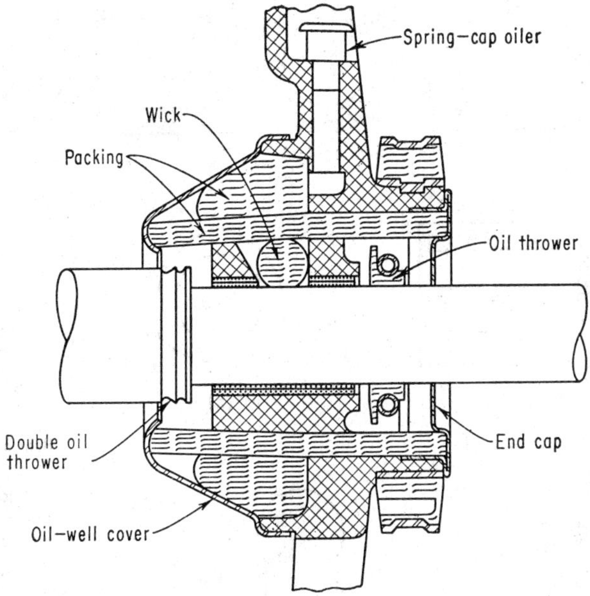

Fig. 12.3. A typical wick-oiled bearing installation for fractional-horsepower electric motors.[50]

generated in the lubricant film and (2) the hybrid in which the load is supported by the combination of hydrodynamic and hydrostatic effects.

(a) Journal Bearings

In its simplest form a journal bearing may be a reamed hole in a machine or a bushing, supporting a rotating shaft as shown in Fig. 12.1.

The hydrodynamic journal bearing includes bushings, oil-ring (Fig. 12.2), and wick-fed (Fig. 12.3) bearings, all of which are of the nonpressurized lubricant-feed type and are used primarily at low speeds and moderate loads. Bushings may be lubricated with grease or oil, and in some cases these may run dry. Sintered materials impregnated with oil ("Oilite bronze") are often used for bushings. As a rule, bushings operate in the boundary lubrication regime rather than in the hydrodynamic regime.

For horizontal shafts an oil ring which is rotated by contact with the shaft transfers oil from an oil reservoir beneath the shaft to the top of the shaft. Grooves are used to spread the oil along the bearing.

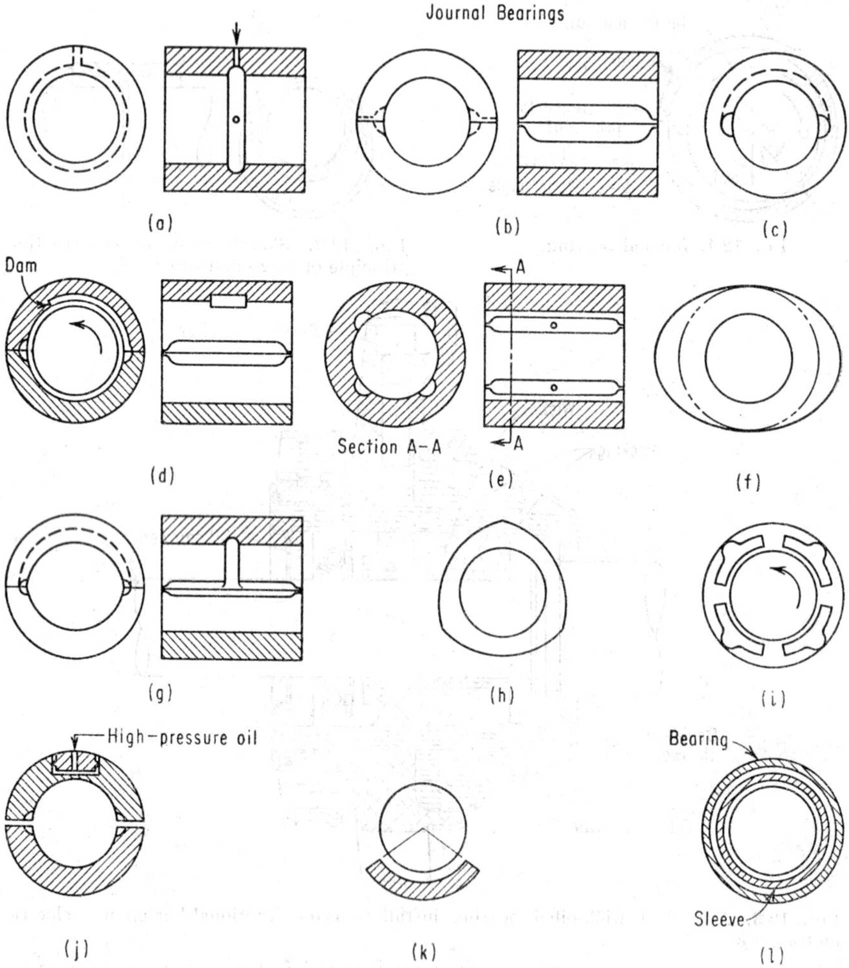

FIG. 12.4. Sliding bearings. Sketches showing shapes of several types of pressure-fed bearings. (*a*) Circumferential-groove. (*b*) Cylindrical. (*c*) Cylindrical overshot groove. (*d*) Pressure pad. (*e*) Multiple-groove. (*f*) Elliptical. (*g*) Elliptical overshot groove. (*h*) Three-hole. (*i*) Pivoted-shoe (tilting pad). (*j*) Nutcracker. (*k*) Partial arc. (*l*) Floating sleeve.[50]

In wick-oiled bearings an oil-saturated wick which rubs against the shaft supplies oil to the bearing. Seals retain the oil within the bearing and the oil is returned to the wick by gravity.

Hydrodynamic journal bearings in which the lubricant is pressure-fed are shown in Fig. 12.4.

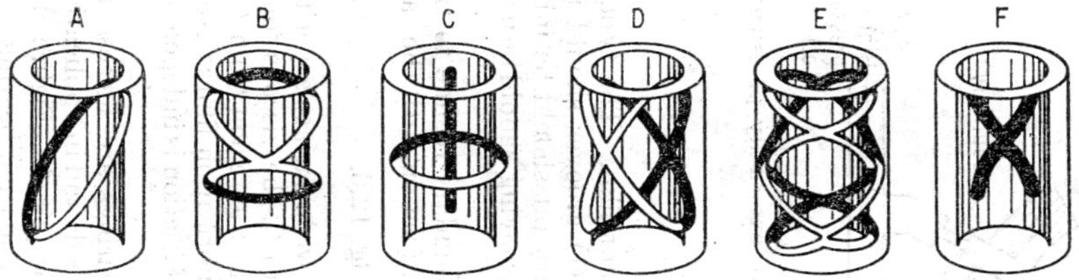

Application	Type of bearing	Bearing support	Direction of		Method of grooving
			Load	Rotation	
General-purpose	Solid or two-part	Fixed in stationary housing	Unidirectional	Unidirectional	Straight groove, stopping short of each end
General-purpose	Solid or two-part. Lubrication provided through shaft	Fixed in stationary housing	Unidirectional	Unidirectional or changing	Circular groove to coincide with oil exit in shaft and straight groove stopping short of each end (*C*)
General-purpose	Solid	Fixed in moving housing (pulleys, gears, etc.)	Unidirectional or changing	Unidirectional or changing	Oval-type groove (*A*)
General-purpose	Solid or two-part	Fixed in stationary housing	Unidirectional	Changing	Double spiral groove short of bearing ends (*F*)
General-purpose	Solid	Fixed in stationary housing	Unidirectional	Changing	Complete figure-eight type of groove (*B*)
General-purpose	Two-part	Fixed in stationary or rotating housing	Unidirectional or changing	Unidirectional or changing	Longitudinal or circular grooves and/or chamfers on each side at parting line
Slow-speed, grease-lubricated	...	...	...	...	Grooved as shown in (*D*) and (*E*). Recommended only for grease or graphite lubrication

FIG. 12.5. Fundamental types of oil grooving.[51]

A pressure-fed system may include supply and return lines for the oil, a storage tank, filter, cooler, and temperature and pressure regulators.

Where the bearing load always acts in one direction, the bearing need not extend for 360°, and partial bearings may be used (Fig. 12.4*k*).

In addition, lubrication may be achieved by (1) hand oiling from an oil can, (2) cup oiling, (3) bath lubrication, and (4) splash lubrication.

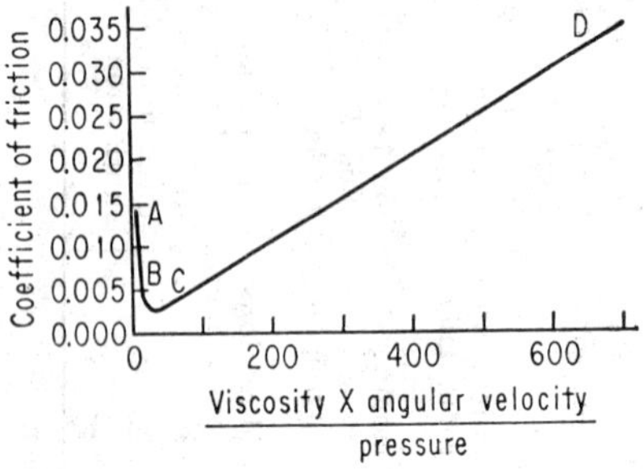

FIG. 12.6. Friction coefficient.[52] $\frac{\text{Bearing length}}{\text{Diameter}} = 1. \quad \frac{\text{Clearance}}{\text{Diameter}} = 0.001.$

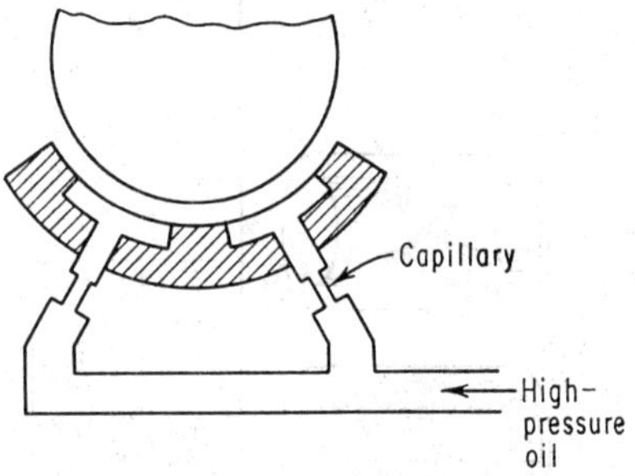

FIG. 12.7. Sketch of externally pressurized bearing.[50]

Grooves are used primarily to maintain an efficient lubricant film. The type of groove to be used depends on the application. Generally, (1) grooves should not connect high-pressure areas to low-pressure areas, (2) the lubricant should be introduced in a low-pressure region, and (3) a minimum amount of grooving should be used. Fundamental types of oil grooving are shown in Fig. 12.5.

The bearings described above operate from the thin-film boundary regime to true hydrodynamic lubrication. The variation in the coefficient of friction with the ratio $\frac{\text{viscosity} \times \text{angular velocity}}{\text{pressure}}$ for a given bearing is shown in Fig. 12.6.

In the thin-film boundary region from *A* to *B* the rubbing between surface asperities causes most of the friction. In the transition region from *B* to *C*, the viscous forces begin to prevail and rubbing between asperities diminishes. In the region from *C* to *D* a true hydrodynamic lubrication film is established and the coefficient of friction is independent of surface conditions.

Hydrostatic journal bearings are successfully used at very low or zero speeds and in applications where low torques are required. Figure 12.7 shows a partial hydrostatic journal bearing.

This type of bearing requires a high-pressure lubricant supply.

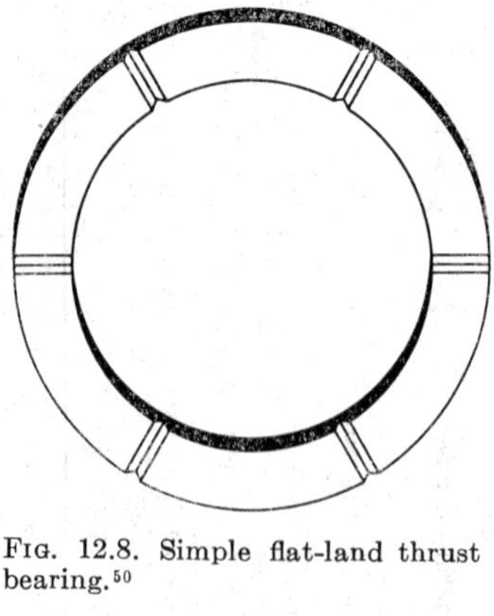

FIG. 12.8. Simple flat-land thrust bearing.[50]

(b) Thrust Bearings

Thrust bearings include types which operate in the boundary-film lubrication regime and in the hydrodynamic regime and types which are externally pressurized.

The flat-land thrust bearing (Fig. 12.8) used for low loads consists of a collar on the shaft which bears against a flat stationary surface. Oil is admitted to the center of the bearing and flows out over the bearing surface in oil grooves.

The tapered-land thrust bearing (Fig. 12.9*a*) is a unidirectional bearing, while Fig. 12.9*b* is a double-acting thrust bearing. With adequate oil-film thickness, this type of bearing can sustain high loads.

Pivoted-shoe thrust bearings are shown in Figs. 12.10*a* and *b*.

In the case of Kingsbury-type bearings each of the shoes is centrally pivoted and the load is equalized through a linkage mechanism. The middle is pivoted on a central line. In the case of a spring-supported bearing each shoe is elastically supported. The shoe profile adjusts itself for each condition of speed and load. These bearings have high load capacities in both directions and will tolerate a high degree of misalignment.

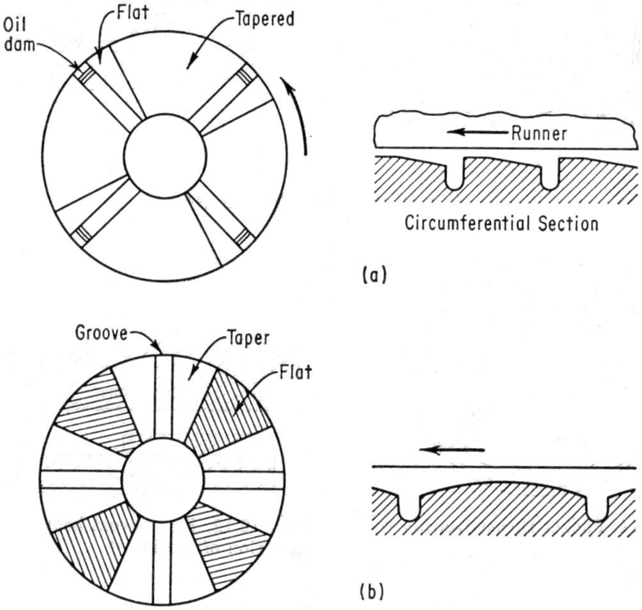

FIG. 12.9. (*a*) Sketch of tapered-land thrust bearing. (*b*) Double taper.[50]

Other types of thrust bearings are the pocket thrust bearing, the step thrust bearing, and the externally pressurized bearing, which are shown in Figs. 12.11, 12.12, and 12.13, respectively.

The groove-design considerations for journal bearings apply for thrust bearings.

(c) Lubricants (also see Sec. 37)

Oils, greases, water, liquid metals, and gases may be employed as lubricants in slider bearings. The available means of lubricant application, environmental conditions, and the bearing design will govern the selection of the lubricant.[49] Gas as a lubricant has the following limitations as compared with a liquid: (1) load capacity is low, (2) closer machining tolerances on bearing parts are required, and (3) there is a greater tendency to instability. Nevertheless, gas-lubricated bearings have found many new applications in the following areas:

1. Extreme-temperature devices. Gas bearings can satisfactorily operate at temperatures even higher than 3000°F or as low as −454°F. The upper limit for liquid lubricants (other than liquid glasses or metals) is approximately 600°F, lower limit approximately −120°F. Gases have the added property of increasing viscosity with increases in temperature. Liquids have the reverse property.

2. High-speed devices. Relatively low viscosity of gases means that the power consumed is much lower than for liquid-lubricated bearings.

3. Clean operation. Absence of possible oil drips and leakage.

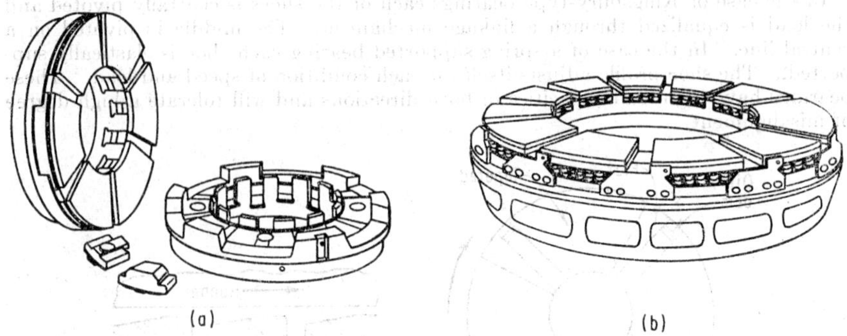

FIG. 12.10. Thrust bearings.[50]

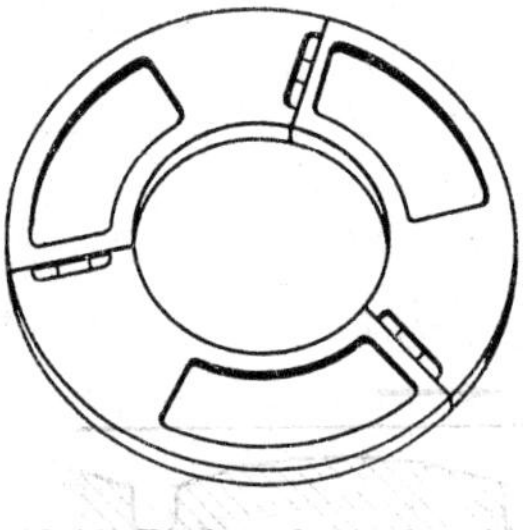

FIG. 12.11. Pocket thrust bearing.[50]

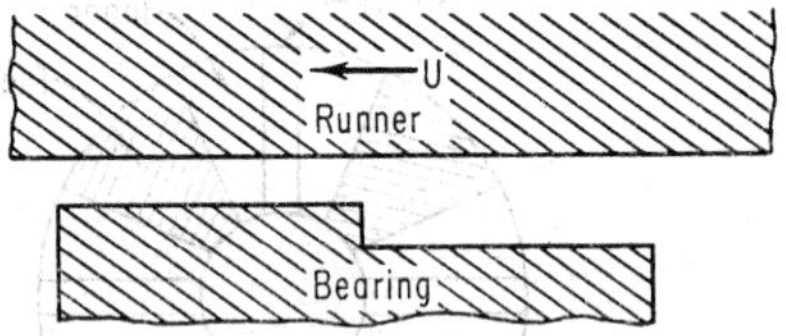

FIG. 12.12. Step thrust bearing—section in direction of motion.[50]

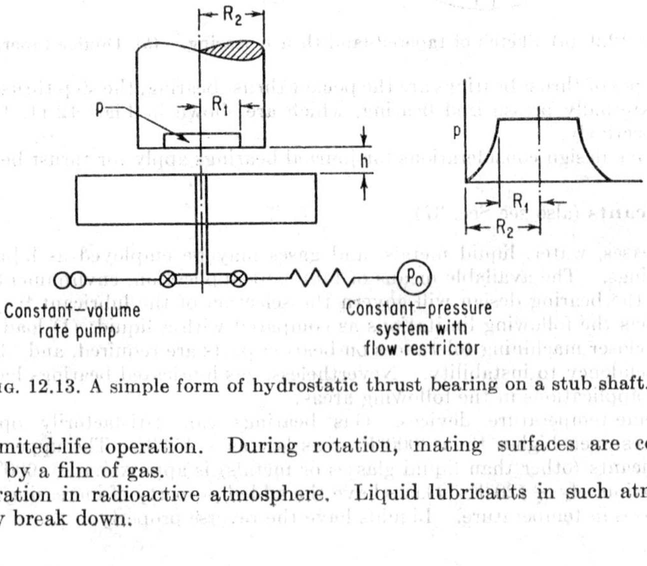

FIG. 12.13. A simple form of hydrostatic thrust bearing on a stub shaft.[50]

4. Unlimited-life operation. During rotation, mating surfaces are completely separated by a film of gas.

5. Operation in radioactive atmosphere. Liquid lubricants in such atmospheres frequently break down.

Typical properties of commercial petroleum oils are given in Table 12.2.

(d) Bearing Materials

Tables 12.3, 12.4, and 12.5 give some data on common bearing materials.

12.2. INTRODUCTION TO HYDRODYNAMIC LUBRICATION

The study of hydrodynamic lubrication is, from a mathematical standpoint, the study of a particular form of the Navier-Stokes equations. The Reynolds equation can be deduced either from the Navier-Stokes equations or from equations of continuity. The Reynolds equation contains viscosity, density, and film thickness and time as parameters. These parameters both determine and depend on the temperature and pressure fields and on the elastic behavior of the bearing surfaces. Thus, to get a complete and accurate representation of the hydrodynamics of the lubricating film, it is oftentimes necessary to consider simultaneously the Reynolds equation, the energy equation, the elasticity equation, and the equation of state.

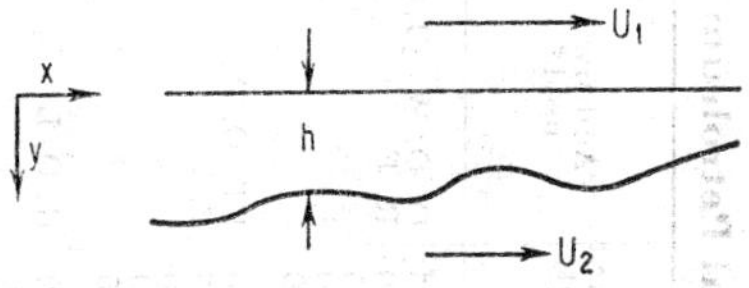

Fig. 12.14. The fluid film.

(a) The Generalized Reynolds Equation

The assumptions involved in reducing Navier-Stokes equations to the Reynolds equation are, referring to the fluid film in Fig. 12.14, as follows:

1. The height of the fluid film y is very small compared with the span and length x, z. This permits us to ignore the curvature of the fluid film, such as in the case of journal bearings, and to replace rotational by translational velocities.
2. No variation of pressure occurs across the fluid film. Thus

$$\partial p/\partial y = 0$$

3. The flow is laminar; no vortex flow and no turbulence occur anywhere in the film.
4. No external forces act on the film. Thus

$$X = Y = Z = 0$$

5. Fluid inertia is small compared with the viscous shear. These inertia forces consist of acceleration of the fluid, centrifugal forces acting in curved films, and fluid gravity. Thus

$$Du/dt = Dv/dt = Dw/dt = 0$$

6. No slip occurs at the bearing surfaces.
7. Compared with the two velocity gradients $\partial u/\partial y$ and $\partial w/\partial y$, all other velocity gradients are considered negligible. Since u and, to a lesser degree, w are the predominant velocities and y is a dimension much smaller than either x or z, the above assumption is valid. The two velocity gradients $\partial u/\partial y$ and $\partial w/\partial y$ can be considered shears, while all others are acceleration terms, and the simplification is also in line with assumption 5. Thus any derivatives of terms other than $\partial u/\partial y$ and $\partial w/\partial y$ will be of a much higher order and negligible. All derivatives with the exception of $\partial^2 u/\partial y^2$ and $\partial^2 w/\partial y^2$ can thus be omitted.

Employing the above assumptions, the Reynolds equation can be derived.[1]

$$\frac{\partial}{\partial x}\left(\frac{\rho h^3}{\mu}\frac{\partial p}{\partial x}\right) + \frac{\partial}{\partial z}\left(\frac{\rho h^3}{\mu}\frac{\partial p}{\partial z}\right) = 6(U_1 - U_2)\frac{\partial(\rho h)}{\partial x} + 6\rho h\frac{\partial}{\partial x}(U_1 + U_2) + 12\rho V + 12h\frac{\partial \rho}{\partial t} \quad (12.1)$$

Table 12.2. Typical Properties of Commercial Petroleum Oils[50]

Oil type	Viscosity				Viscosity index	Density, g/cm^2 at 60°F	Flash point, °F	Pour point, °F	Additives used*	Uses
	Centistokes		Saybolt sec							
	100°F	210°F	100°F	210°F						
Automotive:										
SAE 10W	41	6.0	190	46	102	0.870	410	−15	O, D, W, VI, P, R, F	Automobile, truck, and marine reciprocating engines. Very-heavy-duty types used in diesel engines
20W	71	8.5	330	54	96	0.885	440	−10		
30	114	11.3	530	64	92	0.891	460	−5		
40	173	14.8	800	77	90	0.890	475	0		
50	270	19.7	1,250	97	90	0.902	490	10		
Industrial gear:										
SAE 75	47	7.3	220	50	121	0.900	380	−10	EP, O, R, P, F	Enclosed reduction gearsets, pinion stands, and combination pinion-reduction gear units
80	69	7.9	320	52	78	0.934	365	−25		
90	287	20.4	1,330	100	91	0.930	450	−10		
140	725	34	3,350	160	82	0.937	500	0		
250	1,220	47	5,660	220	83	.....	490	+5		
Turbine:										
Light	32	5.4	150	44	109	0.872	410	0	R, O, L	Direct-connected turbine, electric motors.
Medium	65	8.2	300	53	105	0.877	455	10		Land-geared turbines, electric motors.
Heavy	99	10.8	460	62	100	0.885	470	10		Marine-propulsion geared turbines
Hydraulic:										
Light	32	4.8	150	42	64	0.887	370	−45	R, O	Hydraulic fluids for most indoor industrial hydraulic equipment
Medium	67	7.3	310	50	66	0.895	405	−15	R, O	
Heavy	196	14.0	910	74	70	0.901	495	10	R, O	Heavier loads, higher temps
Extra low temp	14	5.2	74	43	226	0.844	230	−80	R, O, VI, W	Aircraft hydraulic systems
Wide temp	56	10.5	260	61	149	0.871	310	−45		Misc. wide temperature and outdoor use
General-purpose industrial	22	3.9	105	39	80	0.881	350	25	None	All general-purpose lubrication. Machine tools
	44	6.0	205	46	80	0.898	390	25		
	66	7.0	305	49	83	0.915	365	10		
	110	9.9	510	59	25	0.915	390	5		
	200	15.5	930	80	80	0.890	455	15		
Aviation	5	1.6	43	33	87	0.858	232	−85	None	Turbojet engines
	10	2.5	59	35	79	0.864	295	−80	None	Turbojet engines
	76	9.3	350	57	106	0.876	420	0	F	Various reciprocating aircraft engines
	111	11.3	514	64	96	0.884	435	0	F	Various reciprocating aircraft engines
	179	15.5	829	80	95	0.887	450	0	F	Various reciprocating aircraft engines
	268	20.1	1,240	99	95	0.892	480	0	F	Various reciprocating aircraft engines
	369	25.0	1,711	120	96	0.892	505	20	F	Various reciprocating aircraft engines
Steam cylinder	390	27	1,800	130	101	0.895	500	55	L	Cylinder applications for railroad and stationary steam engines, some enclosed gears
	810	45	3,750	210	107	0.910	600	35		
	1,400	64	6,470	300	103	0.904	650	60		
Refrigerator	14	2.9	72	36	53	0.895	295	−50	P	Ammonia compressors
	42	5.1	195	43	22	0.898	330	−35		
	51	5.7	235	45	34	0.909	360	−20		
	72	7.0	335	49	35	0.902	375	−10		

* Commercially available oils frequently include the indicated additives where O = oxidation inhibitor, D = detergent, VI = viscosity-index improver, R = rust inhibitor, P = pour depressant, F = antifoam, EP = extreme-pressure agent, W = wear preventive, and L = load-carrying "oiliness" agent.

Table 12.3. Properties of Bearing Alloys[50]

Bearing material	Hardness, room temp, Brinell	Hardness 300°F, Brinell	Min shaft hardness, Brinell	Load-carrying capacity, psi	Max operating temp, °F	Compatibility*	Conformability and embeddability*	Corrosion resistance*	Fatigue strength*
Tin-base babbitt	20–30	6–12	150 or less	800–1,500	300	1	1	1	5
Lead-base babbitt	15–20	6–12	150 or less	800–1,200	300	1	1	3	5
Alkali-hardened lead	22–26	11–17	200–250	1,200–1,500	500	2	1	5	5
Cadmium base	30–40	15	200–250	1,500–2,000	500	1	2	5	4
Copper-lead	20–30	20–23	300	1,500–2,500	350	2	2	5	3
Tin bronze	60–80	60–70	300–400	4,000+	500+	3	5	2	1
Lead bronze	40–80	40–60	300	3,000–4,500	450–500	3	4	4	2
Aluminum alloy	45–50	40–45	300	4,000+	225–300	5	3	1	2
Silver (overplated)	25	25	300	4,000+	500	2	3	1	1
Three-component bearings babbitt surfaced	.....	.....	230 or less	2,000–4,000+	225–300	1	2	2	3

* Arbitrary scale with 1 being the best material and 5 the worst.

Table 12.4. Typical Journal-bearing Materials[50]

Bearing material	Typical composition, %	Compatibility-test data,* moderate load, boundary lubrication: Shaft material†	Shaft damage: Wear	Shaft damage: Pickup	Shaft damage: Scoring	Dry coef of friction	Physical properties‡: Hardness: Brinell	Hardness: Rockwell	Ultimate tensile strength, psi	Modulus of elasticity, psi × 10^6	Specific gravity	Uses and remarks
Aluminum alloys	Cu 1.0, Sn 6.5, Ni 1.5, Al 91.0	4140	3	3	2	0.33	45	......	22,000	10.3	2.86	Used in diesels and other high-temperature high-load service. Requires good lubrication and hardened shaft
Babbitts:												
Tin base	Cu 8.3, Sb 8.3, Sn 83.4	4140	2	3	2	0.28	27	......	11,200	7.6	7.46	Used in automotive and diesel engines as well as in large steam turbines and motors
Lead base	Sn 10.0, Sb 15.0, Pb 75.0	4140	2	1	1	0.34	23	......	10,000	4.2	9.69	Used for automobiles and electrical equipment
Cadmium base	Ni 1.4, Cd 98.6	4140	2	2	3	0.34	.....	......	....	....	8.6	Good for applications where lubrication is intermittent
Copper alloys:												
Clock brass	Pb 3.0, Zn 35.5, Cu 61.5	416	3	1	2	....	54–142	B49–75	54,000–64,000	15	8.4	Suited to light loading applications
Bronze, high-lead	Sn 4.0, Pb 14.0, Zn 1.5 Ni 1.0 max, Cu 79.5	4140	3	2	1	0.15	45	......	20,000	....	......	Used in poorly lubricated applications with moderately heavy loads
Bronze, high-lead	Sn 16.0, Pb 14.0, Cu 70.0	4140	2	1	2	0.37	.....	......	....	....	......	Same as above. Can withstand higher loads
Bronze, leaded tin	Sn 8.0, Pb 3.5, Zn 3.5, Cu 85.0	4140	4	2	1	0.26	53	......	30,000 min	....	8.4	Moderately heavy duty
Bronze, 80–10–10	Sn 10.0, Pb 10.0, Cu 80.0	4140	3	2	1	0.15	65	......	25,000 min	11	8.86	General-duty bearing bronze
Bronze, nickel tin	Sn 10.0, Ni 3.5, Pb 2.5, Cu 84.0	4140	3	2	1	0.37	90	......	45,000	....	......	Used in medium-heavy-duty application requiring good strength

Bronze, aluminum	Al 10.5, Fe 3.5, Cu 86.0	4140	2	1	1	0.52	202	....	100,000	16	7.60	Applicable in heavy-duty bearings requiring high strength and good impact resistance
Bronze, zinc	Al 1.0, Si 0.8, Mn 2.5, Zn 37.5, Cu 58.2	C1020 carburized 55 Rockwell C	2	1	2	0.39	......	B80–92	70,000 min	15	8.09	Heavy-duty impact loadings on hardened shafts
Iron-base:												
Gray cast iron	C 3.5, Si 2.5, Fe 94.0	4140	3	1	1	0.37	180	....	30,000 min	27	7.2	High-load low-speed applications with good lubrication. Used in refrigerator compressors
Sintered iron	Cu 7.5, Fe 92.5	4140	2	1	3	0.30	......	....	....	....	....	Impregnated with oil will give good results in low-speed applications
Graphite:												
Carbon graphite	C + binder	4140	1	0	0	0.15	......	Shore scleroscope 75	750/2,500	....	1.63–1.86	Particularly suited to high-temperature applications where lubrication is difficult
Carbon graphite and metal	C + Cu + binder	4140	2	0	1	0.17	......	Shore scleroscope 80	3,000/6,000	....	2.9/3.8	Same as above—higher strength
Cemented carbide	Tungsten carbide 97.0, Co 3.0	Carboloy 20% Co	0	0	0	0.20	......	C80	815,000 (compressive)	97.5	15.1	Can withstand extreme loading and high speeds. Must have perfect alignment and good lubrication. Used in high-speed precision grinders
Plastics:												
Nylon	Polyamide	4140	2	2	2	0.86	......	M90	10,000	0.325	1.44	Makes exceptional use of a little lubricant. Used in many household appliances and other lightly loaded applications
Teflon	Polytetrafluoroethylene	4140	2	0	0	0.17	......	....	3,000	0.060	2.2	Useful in corrosive conditions
Textolite 2001	Phenolic, graphite, and cotton cloth	4140	0	0	0	0.18	......	M100	10,000	0.63–0.92	1.36	Exceptional low wear and good compatibility characteristics

* 250 psi load, 700 fpm surface speed. Shaft damage rated on arbitrary scale with 0 indicating the best results and 3 the poorest.

† Condition of shaft material—AISI 4140, 25 Rockwell C hardness, F3 finish; AISI 416, 98 to 100 Rockwell B hardness, F1 finish; AISI C1020, carburized and surface hardened to 55 Rockwell C F1 finish.

‡ Average properties shown except where noted.

Table 12.5. Application Limits for Semilubricated Sintered-metal and Nonmetallic Bearings

Material	Load capacity, psi	Max temp, °F	Max speed, fpm	*PV* limit P = psi load, V = surface fpm
Porous metals	4,000	150	1,500	20,000
Rubber	50	150	1,000	15,000
Graphite materials	600	700	2,500	15,000 dry 150,000 lubricated
Phenolics	5,000	200	2,500	15,000
Nylon	1,000	200	500	2,500
Wood (maple and lignum vitae)	2,000	150	2,000	15,000
Teflon	500	500	250	1,000

Neglecting higher-order terms the generalized Reynolds equation becomes

$$\frac{\partial}{\partial x}\left(\frac{\rho h^3}{\mu}\frac{\partial p}{\partial x}\right) + \frac{\partial}{\partial z}\left(\frac{\rho h^3}{\mu}\frac{\partial p}{\partial z}\right) = 6(U_1 - U_2)\frac{\partial(\rho h)}{\partial x} + 12\rho V + 12h\frac{\partial p}{\partial t} \tag{12.2}$$

The first right-hand term, $6(U_1 - U_2)\partial(\rho h)/\partial x$, is the contribution of the bearing velocities along the oil film, while the term $12\rho V$ is due to the relative velocity of the bearing surfaces in a direction normal to the fluid film. It is of interest to note that the effect of the term $6(U_1 - U_2)\partial(\rho h)/\partial x$ depends on whether the bearing surfaces have translational or angular velocities. For a thrust slider if $U_1 = U_2$, the first right-hand term of Eq. (12.2) disappears and—since in the absence of any normal movement of bearing surfaces, $V = 0$—such a bearing has zero load capacity; conversely, if $U_1 = -U_2$, the load capacity is doubled. However, in a journal bearing if $U_1 = U_2$, the first right-hand term disappears, but the angular motion of the two surfaces introduces both tangential and normal components of velocity. These velocities are

$$\text{Tangential velocity} = U \cos\alpha \approx U$$
$$\text{Normal velocity} = U \sin\alpha \approx U \tan\alpha = U(\partial h/\partial x) = V$$

Thus the right-hand side of Eq. (12.2) becomes $2U\partial h/\partial x$ and the load capacity is doubled; conversely, if the bearing surface rotates in the opposite direction, i.e., if $U_1 = -U_2$, the first right-hand term becomes $2U\,\partial h/\partial x$, but then the sign of V is reversed, giving $-2U\,\partial h/\partial x$, and the net result is zero load capacity.

In journal bearings, therefore, when both bearing and journal rotate in the same direction, the velocities are additive; in thrust bearings they are subtractive. However, if the center of curvature of the journal bearing surface does not coincide with the center of rotation, counterrotation at identical speeds will produce hydrodynamic forces. Thus bearings with a noncircular cross section will yield a load capacity even under the above conditions.

Equation (12.2) holds for both compressible and incompressible lubricants. By setting ρ = constant, the Reynolds equation for incompressible fluids is obtained:

$$\frac{\partial}{\partial x}\left(\frac{h^3}{\mu}\frac{\partial p}{\partial x}\right) + \frac{\partial}{\partial z}\left(\frac{h^3}{\mu}\frac{\partial p}{\partial z}\right) = 6(U_1 - U_2)\frac{\partial h}{\partial x} + 12V \tag{12.3}$$

In Eq. (12.2) the viscosity μ is still treated as a variable, being a function of both the x and z coordinates. The film thickness h, too, is general enough and can be a function of both coordinates.

In most practical cases, the bearing is stationary and only the runner in thrust bearings and the shaft in journal bearings are moving. In that case, Eq. (12.2) reduces to

$$\frac{\partial}{\partial x}\left(\frac{\rho h^3}{\mu}\frac{\partial p}{\partial x}\right) + \frac{\partial}{\partial z}\left(\frac{\rho h^3}{\mu}\frac{\partial p}{\partial z}\right) = 6U\frac{\partial(\rho h)}{\partial x} + 12\rho V_0 + 12h\frac{\partial p}{\partial t} \tag{12.4}$$

with U the sliding velocity of either runner or journal and V_0 representing the squeeze-film velocity. For steady loading ($V_0 = 0$) and incompressible lubricants

$$(\rho = \text{constant})$$

Eq. (12.4) becomes

$$\frac{\partial}{\partial x}\left(\frac{h^3}{\mu}\frac{\partial p}{\partial x}\right) + \frac{\partial}{\partial z}\left(\frac{h^3}{\mu}\frac{\partial p}{\partial z}\right) = 6U\frac{\partial h}{\partial x} \tag{12.5}$$

which is the most commonly encountered form of the Reynolds equation.
In cylindrical coordinates the above equation becomes

$$\frac{\partial}{\partial r}\left(\frac{rh^3}{\mu}\frac{\partial p}{\partial r}\right) + \frac{1}{r}\frac{\partial}{\partial \theta}\left(\frac{h^3}{\mu}\frac{\partial p}{\partial \theta}\right) = 6U\frac{\partial h}{\partial \theta} \tag{12.6}$$

(b) Flow and Shear Equations

Several important expressions have been formulated in the process of deriving the Reynolds equation. These are the flow and shear equations of lubrication. The velocity equations when integrated between the two bearing surfaces provide the lubricant flow at any given section:

$$q_x = -(h^3/12\mu)(\partial p/\partial x) + (h/2)(U_1 + U_2) \tag{12.7}$$
$$q_z = -(h^3/12\mu)(\partial p/\partial z) \tag{12.8}$$

In polar coordinates Eq. (12.6) is

$$q_\theta = -(h^3/12r\mu)(\partial p/\partial \theta) + r(\omega_1 + \omega_2)h/2 \tag{12.9}$$
$$q_r = -(h^3/12\mu)(\partial p/\partial r) \tag{12.10}$$

The shear stress from the definition of a Newtonian fluid is given by

$$\tau_x = \mu(\partial u/\partial y) \qquad \tau_z = \mu(\partial w/\partial y)$$

Since the total frictional force is given by integrating τ over the bearing surface, we have

$$F = \iint \tau\, dA \qquad F = \int_0^z \int_0^x \{[\pm\tfrac{1}{2}(\partial p/\partial x)h + (\mu/h)(U_2 - U_1)]\, dx\}\, dz \tag{12.11}$$

In polar coordinates the above equation is, by expressing Eq. (12.11) in terms of torque rather than force,

$$M = \int_0^r \int_0^\theta [\pm(h/2r)(\partial p/\partial \theta) + \mu r(\omega_2 - \omega_1)/h]r^2\, dr\, d\theta \tag{12.12}$$

(c) Derivation of Energy Equation

In rigorous bearing analysis the variation of viscosity with temperature must be considered. As the fluid is sheared, work is being done on it and there is a temperature rise which in turn reduces the viscosity of incompressible fluids and raises the viscosity of compressible fluids. This variation of viscosity must be included in the solution of the Reynolds equation. Likewise, from the standpoint of heat transfer and thermal distortion, it is desirable to determine the temperature gradients that exist in the bearing. From the first law of thermodynamics we have

$$E_i + H_{do} = E_s + E_o + H_{db}$$

where E_i = energy transported into the control volume
E_o = energy transported out of the control volume
E_s = energy stored transiently in the control volume
H_{do} = work done on the fluid volume by the surroundings
H_{db} = work done by the fluid volume on the surroundings

When steady-state conditions are assumed, the above equation becomes

$$E_o - E_i = H_{do} - H_{db} \tag{12.13}$$

There are two modes in which energy may be transported into and out of control volumes: by conduction according to Fourier's law and by convection of intrinsic energy, i.e., transport of fluid possessing kinetic energy and internal energy. A possible third mode, radiation, is neglected.

The other energies involved in the energy balance are the mechanical works done by the surface stresses and body forces through an incremental distance in an increment of time. For the lubrication problem at hand, body forces, such as gravity, are neglected.

By equating the expressions for E and H according to Eq. (12.13),

$$\left[\frac{\partial(\rho ue)}{\partial x} + \frac{\partial(\rho ve)}{\partial y} + \frac{\partial(\rho we)}{\partial z}\right] - J\left[\frac{\partial}{\partial x}\left(k\frac{\partial T}{\partial x}\right) + \frac{\partial}{\partial y}\left(k\frac{\partial T}{\partial y}\right) + \frac{\partial}{\partial z}\left(k\frac{\partial T}{\partial z}\right)\right]$$
$$= \frac{\partial}{\partial x}(u\sigma_x + v\tau_{yx} + w\tau_{zx}) + \frac{\partial}{\partial y}(u\tau_{xy} + v\sigma_y + w\tau_{zy}) + \frac{\partial}{\partial z}(u\tau_{zx} + v\tau_{yz} + w\sigma_z) \tag{12.14}$$

The above equation may be reduced to the energy equation[2]

$$J\left\{\left[\left(\frac{\rho Uh}{2} - \frac{h^3}{12\nu}\frac{\partial p}{\partial x}\right)\frac{\partial(c_vT)}{\partial x} - \frac{h^3}{12\nu}\frac{\partial p}{\partial z}\frac{\partial(c_vT)}{\partial z}\right] - \left[\frac{\partial}{\partial x}\left(hk\frac{\partial T}{\partial x}\right) + \frac{\partial}{\partial z}\left(hk\frac{\partial T}{\partial z}\right)\right] + K_T(T - T_w)\right\} = \rho\frac{\nu U^2}{h} + \frac{h^3}{12\rho\nu}\left[\left(\frac{\partial p}{\partial x}\right)^2 + \left(\frac{\partial p}{\partial z}\right)^2\right] \tag{12.15}$$

For ordinary lubricants, the characteristic values of the parameters are such that conduction is a minor mode of heat transfer. Further, the heat-transfer coefficient at the fluid boundaries as reported in ref. 2 is very small. Since it is expected that the fluid-temperature rise through a bearing will not be large, the specific heat of the fluid will essentially be constant. With the above-mentioned fluid-parameter characteristics, Eq. (12.15) becomes

$$6JU\rho c_v h\left[\left(1 - \frac{h^2}{6\mu U}\frac{\partial p}{\partial x}\right)\frac{\partial T}{\partial x} - \frac{h^2}{6\mu U}\frac{\partial p}{\partial z}\frac{\partial T}{\partial z}\right] = \frac{12\mu U^2}{h}\left\{1 + \frac{h^4}{12\mu^2U^2}\left[\left(\frac{\partial p}{\partial x}\right)^2 + \left(\frac{\partial p}{\partial z}\right)^2\right]\right\} \tag{12.16}$$

The above equation neglects conduction.

Equation (12.16) can be made more convenient for numerical work by reducing it to nondimensional form. By setting $\bar{x} = x/B$, $\bar{z} = z/B$, $\bar{h} = h/B$, $\bar{\mu} = \mu/\mu_1$, $\bar{\rho} = \rho/\rho_1$, $\bar{p} = pB/6\mu_1U$, $\bar{T} = T\rho_1Jc_vB/\mu_1U$, where B is a representative length in the x direction, Eq. (12.16) becomes

$$\bar{\rho}\bar{h}\left(1 - \frac{\bar{h}^2}{\bar{\mu}}\frac{\partial\bar{p}}{\partial\bar{x}}\right)\frac{\partial\bar{T}}{\partial\bar{x}} - \frac{\bar{\rho}\bar{h}^3}{\bar{\mu}}\frac{\partial\bar{p}}{\partial\bar{z}}\frac{\partial\bar{T}}{\partial\bar{z}} = 2\frac{\bar{\mu}}{\bar{h}}\left\{1 + \frac{3\bar{h}^4}{\bar{\mu}^2}\left[\left(\frac{\partial\bar{p}}{\partial\bar{x}}\right)^2 + \left(\frac{\partial\bar{p}}{\partial\bar{z}}\right)^2\right]\right\} \tag{12.17}$$

Likewise, the Reynolds equation becomes

$$\frac{\partial}{\partial\bar{x}}\left[\bar{\rho}\bar{h}\left(1 - \frac{\bar{h}^2}{\bar{\mu}}\frac{\partial\bar{p}}{\partial\bar{x}}\right)\right] - \frac{\partial}{\partial\bar{z}}\left(\frac{\bar{\rho}\bar{h}^3}{\bar{\mu}}\frac{\partial\bar{p}}{\partial\bar{z}}\right) = 0 \tag{12.18}$$

(d) Equation of State

In Eq. (12.16), μ is a known function of p and T and h a known function of x and z. One more equation is needed because there are three unknowns (p, ρ, and T), and it is provided by the equation of state given by

$$pv = \mathcal{R}T \tag{12.19}$$

For lubrication with a gas, the assumption that the gas obeys the perfect-gas law will be adequate. Since $\bar{p}$ and $\bar{T}$ are quantities of practical importance, the procedure might be to eliminate $\bar{\rho}$ from Eqs. (12.17) and (12.18) by the substitution

$$\bar{p} = 1/6(n - 1)\bar{\rho}\bar{T}$$

For lubrication with a liquid film the choice of an equation of state is more difficult. Even for simple liquids the equations proposed are a modified van der Waals type of considerable algebraic complexity; so that their introduction into Eqs. (12.17) and (12.18) would increase the complications to such an extent that the solution would probably be a major computing operation.

If it is sufficiently accurate to ignore the variation of $\bar{\rho}$, $\bar{\mu}$, etc., with temperature, or if the variation with temperature can be replaced by a variation, known a priori, with $\bar{x}$ and $\bar{z}$, then Eq. (12.18) becomes an equation for $\bar{p}$ only. The solution thus obtained can be inserted into Eq. (12.17), which then becomes an equation in $\bar{T}$ only. This situation, which also arises in other branches of applied mechanics, enables the equations to be solved successively instead of simultaneously, but only in the order (12.18) to (12.17). However, as soon as it becomes necessary to take variation with temperature into account, the equations become interlocked and must be solved simultaneously.

12.3. INCOMPRESSIBLE LUBRICATION

(a) One-dimensional Journal Bearings

Infinitely Long Bearing. The earliest solution of the infinitely long full journal bearing is due to Sommerfeld.[3] Equation (12.5) becomes, for μ constant and $x = R\theta$:

$$(d/d\theta)[h^3(dp/d\theta)] = 6\mu UR(dh/d\theta) \tag{12.20}$$

Integrating Eq. (12.20) and using the proper boundary conditions,

$$p = p_a + \frac{6\mu UR\epsilon}{C^2}\frac{(2 + \epsilon\cos\theta)\sin\theta}{(2 + \epsilon^2)(1 + \epsilon\cos\theta)^2} \tag{12.21}$$

where p_a is the pressure at $\theta = 0$.

The pressure distribution resulting from Eq. (12.21) is always antisymmetrical about $\theta = \pi$ and $p = p_a$.

The vertical load component is derived by integrating the pressure over the bearing surface:

$$W\sin\phi = \frac{12\pi\mu UL(R/C)^2\epsilon}{(2 + \epsilon^2)(1 - \epsilon^2)^{1/2}} \tag{12.22}$$

The attitude angle is

$$\phi = \pi/2 \tag{12.23}$$

and the displacement of the shaft is always at right angles to $W\sin\phi$. Moreover, since there is no load component at right angles to $W\sin\phi$, $W\sin\phi = W$ is the total resultant load. This unrealistic result is a consequence of including the negative pressures in the integration for load capacity. Thus we can rewrite Eq. (12.22) as

$$S = \frac{\mu N}{P}\left(\frac{R}{C}\right)^2 = \frac{(2 + \epsilon^2)(1 - \epsilon^2)^{1/2}}{12\pi^2\epsilon} \tag{12.24}$$

and the Sommerfeld number is seen to be a function of ϵ only.

The frictional force on the journal is

$$F_j = \mu UL \frac{R}{C} \frac{4\pi(1 + 2\epsilon^2)}{(2 + \epsilon^2)(1 - \epsilon^2)^{1/2}} \tag{12.25}$$

The friction factor defined as $f = F/W$ is then

$$f = \frac{F_j}{W} = \frac{C}{R} \frac{1 + 2\epsilon^2}{3} \tag{12.26}$$

At the bearing surface, the friction force is

$$F_b = \frac{\mu ULR}{C} \frac{4\pi(1 - \epsilon^2)^{1/2}}{2 + \epsilon^2}$$

The difference in the journal and bearing torques is balanced by the external load W, which exerts a moment through its eccentricity e:

$$RF_j = RF_b + We \tag{12.27}$$

The friction in a concentric journal bearing when $\epsilon = dp/d\theta = 0$ is often referred to as Petroff's equation and is given simply from Eq. (12.25) by $2\pi\mu ULR/C$.

The major shortcomings of the foregoing analysis can be eliminated by imposing a more realistic boundary condition at the trailing end of the pressure wave as given below:

$$\begin{aligned} p &= 0 \qquad \text{at } \theta = 0 \\ dp/d\theta &= 0 \qquad \text{at } \theta = \theta_2 \\ p &= 0 \qquad \text{at } \theta = \theta_2 \end{aligned} \tag{12.28}$$

Employing Sommerfeld's substitutions and using the first two boundary conditions of Eq. (12.28),

$$p = \frac{6\mu UR}{C^2(1 - \epsilon^2)^{3/2}} \left\{ \psi - \epsilon \sin \psi - \frac{(2 + \epsilon^2)\psi - 4\epsilon \sin \psi + \epsilon^2 \sin \psi \cos \psi}{2[1 + \epsilon \cos (\psi_2 - \pi)]} \right\} \tag{12.29}$$

where $\cos \psi = (\epsilon + \cos \theta)/(1 + \epsilon \cos \theta)$ and ψ_2 corresponds to θ_2. By using the last condition, namely, $p = 0$ at $\psi = \psi_2$, we obtain from Eq. (12.29)

$$\epsilon[\sin (\psi_2 - \pi) \cos (\psi_2 - \pi) - \psi_2] + 2[\psi_2 \cos (\psi_2 - \pi) - \sin (\psi_2 - \pi)] = 0$$

which determines ψ_2 and thus θ_2. Equation (12.29) with ψ_2 determined from the above equation gives a pressure profile satisfying all the conditions of Eq. (12.28).

For the two load components, by writing $\psi_2' = \psi_2 - \pi$,

$$W \cos \phi = -\frac{3\mu UL\epsilon(R/C)^2(1 + \cos \psi_2')^2}{(1 - \epsilon^2)(1 + \epsilon \cos \psi_2')}$$

$$W \sin \phi = \frac{6\mu UL(R/C)^2(\psi_2 \cos \psi_2' - \sin \psi_2')}{(1 - \epsilon^2)^{1/2}(1 + \epsilon \cos \psi_2')}$$

$$W = \frac{3\mu UL(R/C)^2}{(1 - \epsilon^2)^{1/2}(1 + \epsilon \cos \psi_2')} \left[\frac{\epsilon^2(1 + \cos \psi_2')^4}{1 - \epsilon^2} + 4(\psi_2 \cos \psi_2' - \sin \psi_2')^2 \right]^{1/2} \tag{12.30}$$

$$\tan \phi = -\frac{2(1 - \epsilon^2)^{1/2}(\sin \psi_2' - \psi_2 \cos \psi_2')}{\epsilon(1 + \cos \psi_2')^2} \tag{12.31}$$

$$\frac{R}{C} f = \frac{\epsilon \sin \phi}{2} + \frac{2\pi^2 S}{(1 - \epsilon^2)^{1/2}} \tag{12.32}$$

The use of conditions (12.28) results in the elimination of the region of negative pressures and the derivation of a journal locus [Eq. (12.31)], which conforms with experimental evidence. Numerical results of Eqs. (12.30) to (12.32) are given in Table 12.6, where they are tabulated together with the solutions of finite bearings.

Infinitely Short Bearings.[4] For aligned journals $h = f(x)$ and $p = 0$ at $+L/2$, we can integrate the Reynolds equation for infinitely narrow journal bearings to get

$$p(\theta,z) = \frac{3\mu U}{RC^2}\left(\frac{L^2}{4} - z^2\right)\frac{\epsilon \sin\theta}{(1+\epsilon\cos\theta)^3} \tag{12.33}$$

Here the problem of negative pressures is dealt with by deleting the region $\pi \leq \theta \leq 2\pi$ where the negative pressures occur. By summing forces only over the interval $0 \leq \theta \leq \pi$,

$$W_x = \frac{\mu UL^3}{C^2}\frac{\epsilon^2}{(1-\epsilon^2)^2}$$

$$W_y = \frac{\mu UL^3}{4C^2}\frac{\pi\epsilon}{(1-\epsilon^2)^{3/2}}$$

The total load capacity is then given by

$$W = \frac{\mu UL^3}{4C^2}\frac{\epsilon}{(1-\epsilon^2)}[\pi^2(1-\epsilon^2)+16\epsilon^2]^{1/2} \tag{12.34}$$

or

$$\frac{\mu N}{P}\left(\frac{L}{C}\right)^2 = S\left(\frac{L}{C}\right)^2 = \frac{(1-\epsilon^2)^2}{\pi\epsilon[\pi^2(1-\epsilon^2)+16\epsilon^2]^{1/2}} \tag{12.35}$$

The effect of diameter in Eq. (12.35) enters through C, which is usually a function of bearing diameter. The attitude angle is given by

$$\tan\phi = \frac{\pi}{4}\frac{(1-\epsilon^2)^{1/2}}{\epsilon} \tag{12.36}$$

Since there is no pressure-induced shear,

$$\tau = \frac{\mu U}{h}$$

and

$$F = \int_0^{2\pi} \mu\frac{U}{h} LR\,d\theta = \frac{\mu ULR}{C}\frac{2\pi}{(1-\epsilon^2)^{1/2}} \tag{12.37}$$

with the friction coefficient given by

$$\frac{R}{C}f = \frac{2\pi^2 S}{(1-\epsilon^2)^{1/2}} \tag{12.38}$$

The lubricant flow out the sides of the bearing is

$$Q_z = 2\int_0^{\pi} \frac{Rh^3}{12\mu}\frac{dp}{dz}\bigg|_{\pm L/2} d\theta = \epsilon ULC \tag{12.39}$$

The parameters at the point of maximum pressure are

$$\cos\theta_0 = \frac{1-(1+24\epsilon^2)^{1/2}}{4\epsilon}$$

$$h_0 = \frac{C}{4}[5-(1+24\epsilon^2)^{1/2}]$$

$$p_0 = \frac{3\mu UL^2}{4RC^2}\frac{\epsilon\sin\theta_0}{(1+\epsilon\cos\theta_0)^3}$$

This treatment yields a fair approximation to the performance of narrow bearings at low eccentricities.

(b) Finite Journal Bearings

Numerical Methods. Employing the substitutions

$$\bar{x} = x/D \qquad \bar{z} = z/L \qquad \bar{h} = h/2C \qquad \bar{p} = (p/\mu N)(C/R)^2$$

the dimensionless Reynolds equation may be written

$$(\partial/\partial\bar{x})[\bar{h}^3(\partial\bar{p}/\partial\bar{x})] + (D/L)^2(\partial/\partial\bar{z})[\bar{h}^3(\partial\bar{p}/\partial\bar{z})] = 6\pi(\partial\bar{h}/\partial\bar{x}) \tag{12.40}$$

This equation can be written in finite-difference form and solved for pressure ($p_{i,j}$) at every point in the grid.

$$6\pi\frac{h_{i,j-\frac{1}{2}} - h_{i,j+\frac{1}{2}}}{\Delta x} + \left(\frac{D}{L}\right)^2\left[h^3_{i+\frac{1}{2},j}\frac{P_{i+1,j}}{\Delta z^2} + h^3_{i-\frac{1}{2},j}\frac{P_{i-1,j}}{\Delta z^2}\right]$$
$$+\,h^3_{i,j+\frac{1}{2}}\frac{P_{i,j+1}}{\Delta x^2} + h^3_{i,j-\frac{1}{2}}\frac{P_{i,j-1}}{\Delta x^2}$$
$$+\,p_{i,j} = \overline{\left(\frac{D}{L}\right)^2\frac{h^3_{i+\frac{1}{2},j} + h^3_{i-\frac{1}{2},j}}{\Delta z^2} + \frac{h^3_{i,j+\frac{1}{2}} + h^3_{i,i-\frac{1}{2}}}{\Delta x^2}} \tag{12.41}$$

Equation (12.41) is of the form

$$p_{i,j} = a_0 + a_1p_{i+1,j} + a_2p_{i-1,j} + a_3p_{i,j+1} + a_4p_{i,j-1}$$

with a_0, a_1, a_2, a_3, and a_4 given constants for each point (i,j) of the mesh and the pressure $p_{i,j}$ a function of these constants and the four surrounding pressures. For $n \times m$ points in the mesh there will be $n \times m$ simultaneous equations which can be solved either in matrix form or by an iteration process.

If the latter is used, the iterative process is repeated until an error smaller than the prescribed value A is reached. An error may be defined by

$$\frac{\sum_{j=1}^{m}\sum_{i=1}^{n}|(p_{i,j})^k - (p_{i,j})^{k-1}|}{\sum_{j=1}^{m}\sum_{i=1}^{n}(p_{i,j})^k} < A$$

where k is the number of iterations performed.

Integration of Eq. (12.41) gives load-carrying capacity. Other quantities such as attitude angle, flow, and horsepower loss can also be computed quite readily.[5]

Eccentrically Loaded Partial Bearings. The results given in Table 12.6 are the basic data for any bearing analysis. They give the resulting hydrodynamic forces as a function of any arbitrary combination of parameters β, L/D, ϵ, and ϕ. The solutions presented in the following sections were prepared from this basic information by a proper summation of the force vectors resulting for the various bearing elements. As they stand, the data constitute solutions for partial bearings in which the load is at any arbitrary position with respect to the bearing arc, or with α/β as a variable.

Table 12.6. Eccentrically Loaded Partial Journal Bearings

$\beta = 150°$

L/D	ϵ	α/β	ϕ	S	$\bar{q}_{in}$	$\bar{q}_z$
1½	0.2	0.200	20	4.16	......	0.020
		0.266	35	1.80	......	0.038
		0.380	48	0.782	......	0.105
		0.493	61	0.477	......	0.186
		0.546	83	0.510	1.09	0.206
		0.612	103	0.738	1.14	0.169
		0.859	156	32.7	1.18	0.0098
	0.3	0.273	34	0.91	......	0.061
	0.4	0.286	32	0.520	0.735	0.085
		0.453	47	0.244	0.882	0.234
		0.600	65	0.231	1.14	0.407
		0.646	98	0.389	1.26	0.347
		0.866	155	24.6		
	0.5	0.100	10	6.58	......	0.0049
		0.213	23	0.614	......	0.046
		0.293	31	0.308	......	0.103
		0.353	37	0.224	......	0.105
		0.420	42	0.164	......	0.248
		0.567	50	0.133	......	0.417
	0.6	0.300	30	0.176	0.553	0.118
		0.443	38	0.114	0.708	0.266
		0.660	66	0.125	1.17	0.607
		0.683	93	0.259	1.35	0.534
		0.860	156	23.5	1.56	0.030
	0.7	0.100	10	1.22	......	0.006
		0.220	22	0.160	......	0.059
		0.320	28	0.101	......	0.131
		0.397	30	0.0793	......	0.212
		0.467	35	0.0690	......	0.308
		0.520	37	0.0630	......	0.386
	0.8	0.213	23	0.0695	0.287	0.061
		0.334	25	0.0497	......	0.142
		0.725	86	0.183	1.40	0.722
		0.800	156	24.1	1.76	0.040
	0.9	0.253	17	0.0225	......	0.0628
		0.368	20	0.0187	......	0.146
		0.453	22	0.0172	......	0.236
		0.540	24	0.0161	......	0.361
	0.95	0.396	15	0.0077	......	0.147
		0.493	16	0.0071	......	0.247
1	0.2	0.100	10	71.2	......	0.002
		0.200	25	4.92	......	0.031
		0.273	34	2.15	......	0.069
		0.380	48	1.05	......	0.148
		0.487	62	0.756	......	0.241
		0.552	72	0.780	0.408	0.274
		0.620	102	1.08	0.367	0.230
		0.860	156	36.8	0.0659	0.0145

Table 12.6. Eccentrically Loaded Partial Journal Bearings (Continued)

$\beta = 150°$

L/D	ϵ	α/β	ϕ	S	$\bar{q}_{in}$	$\bar{q}_z$
1	0.4	0.207	24	1.32	......	0.0565
		0.360	41	0.435	0.671	0.215
		0.500	52	0.286	1.00	0.385
		0.610	74	0.327	0.810	0.545
		0.655	97	0.568	0.732	0.467
		0.866	155	27.2	0.115	0.029
	0.5	0.100	10	7.02	......	0.007
		0.207	24	0.732	......	0.081
		0.293	31	0.368	......	0.155
		0.360	36	0.277	......	0.231
		0.427	41	0.228	......	0.342
		0.500	46	0.209	0.94	0.425
	0.6	0.214	23	0.383	0.497	0.079
		0.394	36	0.163	0.733	0.303
		0.500	41	0.137	0.865	0.450
		0.672	64	0.176	1.19	0.810
		0.694	91	0.371	1.10	0.720
		0.807	155	26.3	0.130	0.044
	0.7	0.100	10	1.31	......	0.009
		0.220	22	0.181	......	0.088
		0.317	27	0.116	......	0.197
		0.340	29	0.111	......	0.232
		0.390	31	0.0972	......	0.307
		0.407	35	0.0840	......	0.445
		0.500	36	0.835	0.776	0.461
		0.550	38	0.0788	......	0.600
	0.8	0.233	20	0.0797	0.316	0.090
		0.333	25	0.0581	0.428	0.202
		0.523	32	0.0468	1.04	0.500
		0.754	52	0.0785	1.55	1.07
		0.740	84	0.255	1.17	0.97
		0.866	155	27.3	0.285	0.059
	0.9	0.360	21	0.0213	......	0.218
		0.423	21½	0.01905	......	0.308
		0.486	22	0.01875	......	0.410
		0.814	43	0.0414	1.63	1.16
	0.95	0.394	16	0.00825	......	0.222
		0.420	17	0.00910	......	0.264
		0.474	19	0.00980	......	0.364
		0.534	20	0.0105	......	0.478
½	0.2	0.180	28	9.80	......	0.049
		0.240	39	4.77	......	0.103
		0.346	53	2.85	......	0.202
		0.574	59	2.05	......	0.339
		0.567	80	2.17	1.19	0.359
		0.643	99	2.89	1.20	0.307
		0.827	161	54.0	1.19	0.0252

Table 12.6. Eccentrically Loaded Partial Journal Bearings (Continued)

$\beta = 150°$

L/D	ϵ	α/β	ϕ	S	$\bar{q}_{in}$	$\bar{q}_z$
½	0.3	0.246	38	2.63	......	0.120
	0.4	0.160	21	2.68	0.67	0.060
		0.266	35	1.34	0.812	0.202
		0.480	53	0.807	......	0.55
		0.526	56	0.794	......	0.60
		0.632	70	0.905	1.365	0.707
		0.686	72	1.46	1.42	0.614
		0.866	155	41.0	1.38	0.054
	0.5	0.0932	11	8.77	......	0.016
		0.196	25	1.26	......	0.120
		0.266	35	0.798	......	0.241
		0.333	40	0.642	......	0.365
		0.406	44	0.559	......	0.505
		0.560	51	0.521	......	0.752
	0.6	0.500	41	0.303	1.03	0.622
		0.706	59	0.420	1.21	1.05
		0.734	85	0.897	1.09	0.422
		0.876	154	39.7	1.18	0.081
	0.7	0.0932	11	1.685	......	0.021
		0.213	23	0.319	......	0.158
		0.300	30	0.232	......	0.376
		0.376	33	0.210	......	0.485
		0.460	36	0.189	......	0.693
		0.580	38	0.1815	......	0.940
	0.8	0.180	18	0.159	0.331	0.105
		0.327	26	0.105	......	0.360
		0.500	30	0.0912	0.976	0.752
		0.786	47	0.156	1.56	1.37
		0.776	79	0.567	1.75	1.25
		0.866	155	41.0	0.256	0.109
	0.9	0.253	17	0.0226	......	0.063
		0.306	20	0.0321	......	0.386
		0.460	21	0.0307	......	0.588
		0.540	24	0.0161	......	0.362
		0.814	43	0.0417	......	1.15
	0.95	0.334	15	0.01135	......	0.402
		0.500	15	0.01065	......	0.618
¼	0.2	0.156	31	25.1	......	0.068
		0.220	42	15.3	......	0.125
		0.326	56	9.62	......	0.243
		0.453	67	8.00	......	0.335
		0.500	70	7.75	......	0.360
		0.580	78	7.70	1.21	0.398
		0.670	95	9.80	1.22	0.346
		0.870	155	115	1.19	0.042
	0.4	0.173	29	6.40	......	0.136
		0.240	39	4.18	......	0.256
		0.370	49	3.09	......	0.482

Table 12.6. Eccentrically Loaded Partial Journal Bearings (Continued)

$\beta = 150°$

L/D	ϵ	α/β	ϕ	S	$\bar{q}_{in}$	$\bar{q}_z$
¼	0.4	0.480	63	2.84	......	0.680
		0.653	67	3.03	1.42	0.795
		0.720	87	4.07	1.44	0.687
		0.873	154	87.0	1.40	0.086
	0.6	0.187	27	1.74	......	0.201
		0.273	34	1.265	......	0.388
		0.426	41	1.075	......	0.730
		0.487	47	1.065	......	0.882
		0.735	55	1.27	1.61	1.16
		0.767	80	2.70	1.67	1.03
	0.8	0.0866	12	0.954	......	0.04
		0.220	22	0.319	......	0.257
		0.326	26	0.277	......	0.496
		0.506	29	0.257	......	0.947
		0.830	42	0.398	1.66	1.40
		0.820	73	1.51	1.85	1.34
		0.875	154	86.0	1.80	0.173
	0.9	0.100	10	0.154	......	0.042
		0.246	18	0.0836	......	0.271
		0.370	19.5	0.0788	......	0.515
		0.526	21	0.0720	......	0.983
		0.870	35	0.159	1.90	1.74

$\beta = 120°$

L/D	ϵ	α/β	ϕ	S	$\bar{q}_{in}$	$\bar{q}_z$	$(R/C)f$
1	0.1	0	0	∞	0.90	0	∞
		0.190	17	36.4	0.910	0.00514	268
		0.360	37	4.70	0.945	0.0350	33.5
		0.438	48	2.85	0.97	0.0595	20.1
		0.482	72	2.25	1.00	0.0809	15.5
		0.510	79	2.12	1.03	0.0930	14.2
		0.532	96	2.27	1.06	0.0940	14.8
		0.554	114	2.74	1.07	0.0830	17.4
		0.568	122	3.18	1.08	0.0735	19.9
		0.586	130	3.85	1.08	0.0618	23.9
		0.615	136	4.98	1.08	0.0487	30.6
		0.651	142	6.88	1.07	0.0368	42.1
	0.20	0	0	∞	0.800	0	∞
		0.190	17	13.8	0.818	0.0100	108
		0.367	36	1.89	0.888	0.0685	14.8
		0.451	47	1.20	0.936	0.116	9.19
		0.502	60	0.998	1.00	0.160	7.32
		0.532	76	1.00	1.05	0.184	6.98
		0.552	94	1.15	1.11	0.189	7.54
		0.570	111	1.48	1.15	0.170	9.18
		0.581	120	1.77	1.16	0.150	10.7
		0.597	130	2.20	1.17	0.126	12.9
		0.622	135	2.90	1.17	0.100	16.8
		0.656	142	4.09	1.15	0.0750	23.4

Table 12.6. Eccentrically Loaded Partial Journal Bearings (Continued)

$\beta = 120°$

L/D	ϵ	α/β	ϕ	S	$\bar{q}_{in}$	$\bar{q}_z$	$(R/C)f$
1	0.4	0	0	∞	0.600	0	∞
		0.192	18	3.33	0.635	0.0192	32.3
		0.383	34	0.574	0.762	0.129	5.68
		0.484	41	0.401	0.855	0.219	3.79
		0.547	54	0.369	0.971	0.310	3.32
		0.579	70	0.426	1.09	0.368	3.33
		0.594	88	0.569	1.20	0.382	3.90
		0.602	108	0.844	1.29	0.345	5.10
		0.607	117	1.07	1.32	0.310	6.09
		0.616	127	1.39	1.34	0.263	7.58
		0.634	135	1.91	1.34	0.208	10.0
		0.664	140	2.80	1.30	0.154	14.2
	0.6	0	0	∞	0.402	0	∞
		0.196	17	0.837	0.45	0.0282	11.0
		0.408	31	0.200	0.615	0.178	2.77
		0.526	37	0.155	0.747	0.306	2.03
		0.605	47	0.160	0.911	0.445	1.84
		0.634	64	0.219	1.10	0.547	2.08
		0.637	84	0.355	1.28	0.580	2.68
		0.631	105	0.620	1.42	0.530	3.76
		0.630	115	0.834	1.47	0.478	4.62
		0.633	124	1.14	1.50	0.407	5.86
		0.646	132	1.63	1.51	0.322	7.86
		0.670	139	2.46	1.46	0.238	11.3
	0.7	0	0	∞	0.300	0	∞
		0.199	14	0.376	0.352	0.0311	6.10
		0.425	29	0.112	0.531	0.178	1.96
		0.553	34	0.0925	0.682	0.342	1.52
		0.643	47	0.0998	0.870	0.505	1.41
		0.667	79	0.155	1.08	0.630	1.69
		0.660	81	0.287	1.30	0.678	2.32
		0.646	103	0.550	1.48	0.627	3.39
		0.641	113	0.766	1.54	0.564	4.21
		0.641	123	1.08	1.57	0.477	5.40
		0.651	132	1.57	1.59	0.379	7.28
		0.673	139	2.40	1.54	0.294	10.6
	0.8	0	0	∞	0.200	0	∞
		0.204	16	0.142	0.255	0.0333	3.11
		0.450	26	0.0563	0.443	0.212	1.37
		0.587	29	0.0494	0.610	0.376	1.12
		0.691	37	0.0559	0.820	0.668	1.05
		0.706	56	0.104	1.17	0.790	1.38
		0.685	78	0.230	1.33	0.785	2.04
		0.660	101	0.495	1.54	0.725	3.11
		0.651	112	0.714	1.61	0.652	3.91
		0.648	123	1.03	1.67	0.560	5.06
		0.655	132	1.53	1.68	0.442	6.88
		0.676	140	2.38	1.62	0.326	10.0
	0.9	0	0	∞	0.100	0	∞
		0.216	15	0.0360	0.151	0.0336	0.130
		0.490	21	0.0210	0.346	0.224	0.861

Table 12.6. Eccentrically Loaded Partial Journal Bearings (Continued)

$\beta = 120°$

L/D	ϵ	α/β	ϕ	S	$\bar{q}_{in}$	$\bar{q}_z$	$(R/C)f$
1	0.9	0.640	24	0.0197	0.529	0.404	0.756
		0.762	39	0.0232	0.766	0.630	0.697
		0.755	50	0.0626	1.05	0.815	1.09
		0.712	75	0.182	1.35	0.892	1.81
		0.675	99	0.449	1.60	0.800	2.89
		0.662	110	0.673	1.69	0.890	3.69
		0.656	122	1.00	1.73	0.945	4.81
		0.660	131	1.51	1.76	0.975	6.58
		0.679	138	2.38	1.70	0.905	9.67
	0.97	0	0	∞	0.0300	0	∞
		0.243	11	0.00594	0.0706	0.0311	0.500
		0.552	14	0.00493	0.270	0.227	0.455
		0.712	15	0.00487	0.462	0.417	0.428
		0.854	19	0.00588	0.715	0.665	0.374
		0.803	44	0.0372	1.02	0.872	0.878
		0.733	73	0.152	1.36	0.965	1.66
		0.685	98	0.420	1.62	0.890	2.77
		0.669	110	0.649	1.73	0.807	3.56
		0.660	121	0.982	1.79	0.687	4.68
		0.663	130	1.50	1.82	0.547	6.42
		0.681	138	2.39	1.75	0.403	9.47
½ *	0.10	0	0	∞	0.900	0	∞
		0.177	19	60.0	0.915	0.0098	427
		0.324	41	10.7	0.96	0.0534	76.7
		0.400	52	7.22	1.00	0.0864	50.7
		0.467	63	5.78	1.03	0.116	39.5
		0.510	79	5.46	1.06	0.131	36.5
		0.544	94	5.86	1.08	0.131	38.1
		0.578	111	7.10	1.16	0.122	44.9
		0.629	125	9.74	1.10	0.0877	60.3
		0.691	137	15.9	1.08	0.0557	97.3
	0.20	0	0	∞	0.802	0	∞
		0.178	19	22.0	0.825	0.0195	171
		0.333	40	4.19	0.916	0.107	33.5
		0.416	51	3.02	0.98	0.169	22.9
		0.492	61	2.52	1.05	0.231	18.4
		0.537	76	2.56	1.11	0.264	17.6
		0.569	92	2.95	1.15	0.262	19.2
		0.599	108	3.82	1.18	0.233	23.4
		0.641	123	5.55	1.19	0.176	32.5
		0.698	136	9.48	1.15	0.111	54.2
	0.4	0	0	∞	0.600	0	∞
		0.180	18	5.22	0.650	0.0380	50.5
		0.355	38	1.26	0.825	0.209	12.3
		0.456	46	0.977	0.95	0.342	9.05
		0.548	54	0.902	1.08	0.450	7.63
		0.595	69	1.04	1.21	0.522	7.89

* The data for L/D = ½ and L/D = ¼ of $\beta = 120°$ are taken from as yet unpublished results by J. Boyd and A. A. Raimondi, Westinghouse Research Laboratories.

Table 12.6. Eccentrically Loaded Partial Journal Bearings (Continued)

$\beta = 120°$

L/D	ϵ	α/β	ϕ	S	$\bar{q}_{in}$	$\bar{q}_z$	$(R/C)f$
½	0.4	0.619	86	1.41	1.31	0.532	9.38
		0.636	104	2.11	1.36	0.470	12.5
		0.664	120	3.47	1.38	0.358	18.6
		0.709	135	6.50	1.31	0.228	32.9
	0.6	0	0	∞	0.402	0	∞
		0.185	18	1.28	0.468	0.0544	16.8
		0.389	31	0.414	0.723	0.300	5.56
		0.509	39	0.354	0.909	0.776	4.38
		0.618	46	0.359	1.11	0.663	3.85
		0.661	63	0.499	1.30	0.775	4.38
		0.671	80	0.834	1.46	0.800	5.87
		0.672	100	1.50	1.55	0.716	8.71
		0.684	118	2.79	1.57	0.550	14.0
		0.718	134	5.73	1.46	0.348	26.1
	0.70	0	0	∞	0.300	0	∞
		0.190	17	0.562	0.375	0.061	9.07
		0.412	31	0.2207	0.344	0.177	3.69
		0.544	36	0.197	0.88	0.555	3.02
		0.660	41	0.210	1.12	0.757	2.68
		0.697	56	0.332	1.35	0.922	3.27
		0.697	77	0.646	1.53	1.04	4.78
		0.689	97	1.31	1.63	0.838	7.57
		0.693	116	2.61	1.66	0.646	12.6
		0.721	134	5.58	1.53	0.406	24.3
	0.8	0	0	∞	0.199	0	∞
		0.197	17	0.202	0.281	0.062	4.42
		0.444	27	0.101	0.602	0.280	2.32
		0.587	29	0.0942	0.85	0.628	1.96
		0.712	35	0.105	1.13	0.658	1.75
		0.739	52	0.205	1.36	1.02	2.38
		0.724	74	0.494	1.60	1.07	3.91
		0.705	95	1.14	1.72	0.961	6.68
		0.701	116	2.47	1.75	0.742	11.7
		0.725	134	5.50	1.61	0.467	23.0
	0.9	0	0	∞	0.100	0	∞
		0.212	15	0.0482	0.18	0.0675	1.72
		0.493	21	0.0323	0.53	0.41	1.24
		0.648	22	0.0315	0.81	0.688	1.09
		0.784	25	0.0371	1.13	0.973	0.965
		0.789	45	0.11	1.43	1.17	1.63
		0.754	70	0.369	1.66	1.20	3.20
		0.722	94	1.00	1.81	1.085	5.97
		0.709	115	2.38	1.84	0.837	10.9
		0.729	133	5.53	1.69	0.528	22.1
¼	0.1	0	0	∞	0.90	0	∞
		0.157	22	139	0.92	0.015	994
		0.291	45	33.2	0.97	0.0675	237
		0.368	53	23.9	1.01	0.100	169
		0.451	66	19.5	1.03	0.133	134
		0.511	79	18.5	1.07	0.152	124

Table 12.6. Eccentrically Loaded Partial Journal Bearings (Continued)

$\beta = 120°$

L/D	ϵ	α/β	ϕ	S	$\bar{q}_{in}$	$\bar{q}_z$	$(R/C)f$
¼	0.1	0.556	94	19.8	1.09	0.152	128
		0.603	108	23.9	1.10	0.132	150
		0.665	120	32.0	1.10	0.102	197
		0.727	133	48.8	1.08	0.0682	298
	0.2	0	0	∞	0.805	0	∞
		0.158	21	50.6	0.835	0.029	395
		0.301	44	13.3	0.94	0.135	104
		0.388	53	10.1	1.01	0.201	76.6
		0.480	62	8.61	1.18	0.290	62.3
		0.543	74	8.61	1.14	0.305	59.2
		0.567	92	9.85	1.18	0.303	63.7
		0.627	105	12.6	1.20	0.268	77.5
		0.680	118	18.0	1.19	0.203	105
		0.734	133	29.0	1.12	0.132	165
	0.40	0	0	∞	0.600	0	∞
		0.162	20	11.9	0.665	0.0595	115
		0.329	41	3.9	0.885	0.271	37.9
		0.434	47	3.24	1.01	0.402	29.6
		0.548	58	3.00	1.15	0.530	25.0
		0.612	67	3.42	1.27	0.605	25.3
		0.646	82	4.55	1.35	0.605	29.7
		0.670	100	6.83	1.39	0.532	39.8
		0.706	115	10.9	1.37	0.412	58.3
		0.747	130	19.6	1.31	0.277	99.0
	0.60	0	0	∞	0.40	0	∞
		0.169	20	2.87	0.494	0.087	37.5
		0.370	36	1.24	0.81	0.398	16.4
		0.498	40	1.14	1.02	0.602	13.7
		0.681	45	1.12	1.21	0.782	11.6
		0.689	57	1.51	1.40	0.900	12.7
		0.707	75	2.53	1.52	0.905	17.2
		0.713	95	4.64	1.59	0.800	26.3
		0.729	113	8.60	1.56	0.622	42.5
		0.757	129	17.0	1.46	0.418	77.5
	0.7	0	0	∞	0.300	0	∞
		0.175	19	1.28	0.408	0.100	19.9
		0.399	32	0.638	0.775	0.462	10.4
		0.539	35	0.611	1.01	0.688	8.92
		0.678	38	0.618	1.26	0.922	7.44
		0.729	53	0.950	1.46	1.05	8.78
		0.735	73	1.87	1.61	1.05	13.2
		0.731	93	3.94	1.69	0.935	22.1
		0.739	113	8.0	1.65	0.719	38.1
		0.762	129	16.6	1.54	0.486	71.6
	0.8	0	0	∞	0.200	0	∞
		0.185	18	0.433	0.322	0.112	9.25
		0.438	28	0.277	0.74	0.522	6.02
		0.588	29	0.266	1.01	0.800	5.15

Table 12.6. Eccentrically Loaded Partial Journal Bearings (Continued)

$\beta = 120°$

L/D	ϵ	α/β	ϕ	S	$\bar{q}_{in}$	$\bar{q}_z$	$(R/C)f$
¼	0.8	0.734	33	0.282	1.29	1.04	4.29
		0.773	48	0.54	1.51	1.19	5.72
		0.764	69	1.36	1.70	1.20	10.1
		0.748	90	3.36	1.79	1.06	18.8
		0.748	110	7.40	1.76	0.835	34.2
	0.9	0	0	∞	0.10	0	∞
		0.205	16	0.0935	0.228	0.12	3.23
		0.496	20	0.0750	0.70	0.578	2.64
		0.657	22	0.0745	1.01	0.880	2.34
		0.806	24	0.0825	1.32	1.17	1.88
		0.823	42	0.253	1.59	1.35	3.33
		0.794	65	0.955	1.79	1.37	7.57
		0.766	88	2.87	1.88	1.21	16.0
		0.756	110	6.98	1.85	0.95	31.3
		0.770	127	16.1	1.69	0.63	63.9

$\beta = 100°$

L/D	ϵ	α/β	ϕ	S	$\bar{q}_{in}$	$\bar{q}_z$
1	0.2	0.11	9	142	....	0.0016
		0.28	22	6.85	0.803	0.0222
		0.43	23	1.95	0.889	0.0752
		0.51	64	1.32	0.997	0.142
		0.52	68	1.32	0.218	0.148
		0.54	86	1.42	1.08	0.159
		0.56	114	2.05	1.16	0.141
		0.65	135	5.00	1.18	0.0716
		0.70	156	37.6	1.18	0.0151
		0.79	171	42.5	1.19	0.0107
		0.92	168	83.5	1.20	0.0012
	0.3	0.11	9	69.0	0.709	0.00235
		0.28	22	3.29	0.743	0.0328
		0.44	36	1.02	0.841	0.111
		0.53	57	0.745	0.966	0.201
		0.56	84	0.925	1.09	0.238
		0.57	113	1.49	1.23	0.213
		0.65	135	3.94	1.27	0.111
		0.95	165	645	1.29	0.0019
	0.4	0.53	47	0.476	....	0.278
		0.55	55	0.485	....	0.263
	0.6	0.12	8	7.1	0.424	0.00475
		0.29	21	0.472	....	0.351
		0.52	33	0.191	0.657	0.234
		0.54	36	0.184	0.695	0.265
		0.56	39	0.180	0.738	0.295

Table 12.6. Eccentrically Loaded Partial Journal Bearings (Continued)

$\beta = 100°$

L/D	ϵ	α/β	ϕ	S	$\bar{q}_{in}$	$\bar{q}_z$
1	0.6	0.62	78	0.40	1.40	0.487
		0.92	113	3.79	1.38	0.193
		0.93	107	614	1.59	0.0038
	0.8	0.12	8	0.827	0.231	0.0059
		0.31	19	0.097	0.300	0.0727
		0.42	23	0.0688	0.368	0.143
		0.48	37	0.0595	0.430	0.205
		0.54	26	0.0572	0.470	0.245
		0.67	63	0.245	1.22	0.652
		0.63	107	0.782	1.61	0.605
		0.64	122	1.40	1.71	0.48
		0.66	129	2.20	1.72	0.363
		0.95	165	710	1.77	0.0051
	0.9	0.13	7	0.133	0.136	0.00604
		0.36	16	0.0279	0.203	0.0734
		0.50	20	0.0219	0.304	0.179
		0.55	20	0.0211	0.342	0.212
		0.70	70	0.187	1.22	0.736
		0.64	106	0.725	1.66	0.685
		0.65	125	1.72	1.80	0.475
		0.66	139	2.19	1.81	0.412
		0.95	165	710	1.88	0.0057
	0.95	0.36	14	0.0101	0.150	0.0685
		0.45	15	0.00945	0.198	0.116
		0.50	15	0.00898	0.224	0.142
		0.52	15	0.00895	0.235	0.153
		0.65	105	0.704	1.71	0.728
		0.64	121	1.33	1.72	0.470
		0.65	122	1.47	1.79	0.542
½	0.2	0.13	7	183	0.808	0.0033
		0.16	24	12.1	0.809	0.00395
		0.40	40	4.31	0.926	0.121
		0.52	68	2.98	1.07	0.224
		0.53	72	3.02	1.08	0.232
		0.55	85	3.27	1.14	0.246
		0.58	112	4.76	1.20	0.214
		0.66	134	10.00	1.20	0.119
		0.79	141	56.5	1.20	0.0225
		0.85	130	127	1.19	0.0137
		0.91	169	945	1.08	0.0027
	0.3	0.11	9	87.7	0.710	0.0048
		0.27	23	5.75	0.769	0.059
		0.41	39	2.22	0.89	0.18
		0.48	62	1.70	1.02	0.312
		0.57	83	2.10	1.20	0.368
		0.59	111	3.40	1.30	0.324
		0.67	133	7.76	1.18	0.181
		0.78	152	32.3	1.30	0.057
		0.92	192	835	1.29	0.0041

Table 12.6. Eccentrically Loaded Partial Journal Bearings (Continued)

$\beta = 100°$

L/D	ϵ	α/β	ϕ	S	$\bar{q}_{in}$	$\bar{q}_z$
½	0.4	0.11	9	40.5	0.617	0.0064
		0.26	24	2.92	0.692	0.0775
		0.34	31	1.75	0.701	0.146
		0.43	37	1.25	0.851	0.236
		0.52	48	1.04	0.981	0.356
		0.54	51	1.04	....	0.352
		0.55	55	1.06	1.06	0.413
		0.59	81	1.50	1.27	0.492
		0.61	109	2.73	1.60	0.435
		0.63	122	4.04	1.42	0.350
		0.67	133	6.85	....	0.273
		0.74	146	16.4	1.40	0.120
		0.77	148	20.9	....	0.096
		0.78	152	29.2	1.40	0.075
		0.98	198	740	1.40	0.0054
	0.5	0.27	23	1.57	0.59	0.095
		0.58	52	0.677	1.00	0.503
		0.62	108	2.34	1.42	0.535
		0.77	153	28.7	1.47	0.090
	0.6	0.11	9	8.62	0.429	0.0094
		0.20	15	1.75	0.44	0.045
		0.28	22	0.795	0.53	0.111
		0.50	35	0.392	0.805	0.383
		0.53	37	0.382	0.843	0.428
		0.56	39	0.376	0.906	0.484
		0.61	49	0.414	1.00	0.596
		0.64	76	0.855	1.38	0.725
		0.65	106	2.05	1.59	0.656
		0.71	139	9.17	1.605	0.265
		0.73	142	11.75	1.60	0.221
		0.78	152	27.7	1.56	0.113
	0.7	0.29	21	0.37	0.42	0.125
		0.665	114	1.70	1.68	0.87
	0.8	0.22	8	1.0	0.250	0.0121
		0.305	20	0.146	0.364	0.136
		0.41	24	0.113	0.489	0.261
		0.49	26	0.103	0.529	0.301
		0.53	27	0.100	0.656	0.428
		0.70	40	0.14	0.98	0.78
		0.70	71	0.488	1.51	0.99
		0.66	104	1.69	1.785	0.87
		0.67	118	3.021	1.845	0.715
		0.68	127	4.593	1.76	0.555
		0.69	131	5.724	....	0.496
		0.79	151	29.6	1.75	0.157
	0.9	0.12	8	0.159	0.132	0.0126
		0.22	13	0.0561	0.163	0.055
		0.33	17	0.0394	0.271	0.14
		0.50	15	0.0330	....	0.317
		0.55	20	0.0327	0.505	0.375
		0.76	54	0.0575	0.985	0.844

Table 12.6. Eccentrically Loaded Partial Journal Bearings (Continued)

$\beta = 100°$

L/D	ϵ	α/β	ϕ	S	$\bar{q}_{in}$	$\bar{q}_z$
½	0.9	0.70	70	0.392	1.56	1.12
		0.70	85	0.785	1.61	1.08
		0.68	102	1.55	1.90	1.00
		0.78	152	27.7	1.84	0.171
		0.92	192	745	1.89	0.0123
	0.95	0.36	14	0.0131	0.214	0.131
		0.45	15	0.0125	0.287	0.204
		0.50	15	0.0122	0.331	0.248
		0.52	15	0.0121	0.349	0.266
		0.71	84	0.721	1.80	1.145
		0.70	100	2.14	1.915	1.03
		0.67	113	2.32	1.99	0.926
		0.68	117	2.91	1.985	0.835
		0.68	119	3.17	1.98	0.80
		0.79	151	29.8	1.89	0.18
¼	0.2	0.05	105	1180	1.19	0.005
		0.10	10	294	0.811	0.006
		0.23	27	30.1	0.511	0.055
		0.36	44	12.9	0.953	0.148
		0.51	71	9.54	1.10	0.271
		0.52	73	9.60	1.11	0.278
		0.56	84	10.1	1.16	0.286
		0.61	109	14.4	1.21	0.250
		0.69	131	25.2	1.31	0.147
		0.83	157	161	1.21	0.032
	0.3	0.09	11	138	0.719	0.009
		0.23	27	14.4	0.715	0.052
		0.38	42	6.7	0.732	0.222
		0.59	81	6.31	0.853	0.428
		0.63	107	10.3	0.977	0.375
		0.81	119	14.5	1.25	0.322
		0.91	169	944	1.30	0.0078
	0.4	0.20	10	65.8	0.622	0.0118
		0.24	26	7.50	0.723	0.108
		0.31	34	4.85	0.809	0.194
		0.395	40.5	3.78	0.905	0.290
		0.53	52	3.28	1.085	0.47
		0.55	55	3.25	1.115	0.49
		0.63	77	4.47	1.31	0.57
		0.65	105	8.27	1.42	0.50
		0.67	118	12.1	1.425	0.405
		0.71	129	19.1	1.42	0.296
		0.78	147	52.0	1.36	0.137
		0.79	151	65.4	1.41	0.110
		0.91	169	942	1.40	0.0105
	0.6	0.20	10	13.9	0.438	0.0177
		0.35	25	1.75	0.58	0.16
		0.52	38	1.17	0.955	0.595
		0.56	39	1.12	0.995	0.59
		0.68	72	2.42	1.46	0.86
		0.68	102	6.00	1.025	0.75
		0.74	136	25.1	1.66	0.38

Table 12.6. Eccentrically Loaded Partial Journal Bearings (Continued)

$\beta = 100°$

L/D	ϵ	α/β	ϕ	S	$\bar{q}_{in}$	$\bar{q}_z$
¼	0.6	0.75	140	30.1	....	0.29
		0.91	169	945	1.60	0.016
	0.8	0.11	21	1.57	....	0.0255
		0.29	21	0.349	0.931	0.205
		0.40	25	0.285	0.596	0.37
		0.48	27	0.276	0.726	0.50
		0.52	28	0.270	0.690	0.464
		0.74	66	1.27	1.50	1.13
		0.71	99	4.74	1.525	1.00
		0.72	114	8.61	1.856	0.816
		0.73	123	13.2	1.857	0.677
		0.91	191	945	1.80	0.021
	0.9	0.11	9	0.24	0.155	0.0238
		0.33	17	0.081	0.352	0.221
		0.50	20	0.0933	0.589	0.458
		0.55	30	0.0754	0.680	0.550
		0.78	62	0.862	1.66	1.27
		0.73	97	0.428	1.94	1.13
		0.95	195	1,010	1.89	0.023
	0.95	0.37	13	0.0234	0.285	0.203
		0.46	14	0.0224	0.407	0.326
		0.49	16	0.0229	0.452	0.371
		0.73	99	4.08	1.99	1.19
		0.73	107	6.50	2.015	1.05
		0.72	113	8.32	2.035	0.985

$\beta = 75°$

L/D	ϵ	α/β	ϕ	S	$\bar{q}_{in}$	$\bar{q}_z$
1½	0.2	0.167	10	91.0	0.00155	0.00155
		0.207	12	41.2	0.0290	0.0290
		0.514	59	1.87	0.0933	0.0548
		0.54	73	1.87	0.147	0.0655
		0.56	78	1.95	0.110	0.0665
		0.687	151	20.6	0.0628	0.01485
		0.715	154	27.1	0.0907	0.01175
	0.3	0.136	8	101	0.0111	0.0111
		0.179	10	41.0	0.023	0.023
		0.515	49	15.8	0.0853	0.0755
		0.535	67.5	1.18	0.203	0.0915
		0.540	72	1.23	0.205	0.0945
		0.660	148	11.6	0.111	0.0294
		0.686	151	15.8	0.096	0.0230
	0.5	0.143	7.5	23.1	0.00182	0.00182
		0.515	34	6.67	0.6365	0.0885
		0.527	38	0.312	0.174	0.103
		0.553	46	0.407	0.219	0.120
		0.567	65	0.481	0.320	0.101

Table 12.6. Eccentrically Loaded Partial Journal Bearings (Continued)

$\beta = 75°$

L/D	ϵ	α/β	ϕ	S	$\bar{q}_{in}$	$\bar{q}_z$
1½	0.5	0.580	125	2.15	0.342	0.124
		0.590	135	2.81	0.326	0.118
		0.615	152	6.67	0.215	0.0012
	0.7	0.540	27	0.125	0.129	0.099
		0.580	26	0.126	0.202	0.133
		0.587	114	1.35	0.414	0.219
		0.590	124	1.35	0.495	0.200
	0.9	0.467	12	0.0254		
		0.480	17	0.0266	0.0842	0.0842
		0.527	18	0.0245	0.1065	0.1065
		0.620	101	1.04	0.757	0.430
1	0.2	0.167	10	62.5	0.00244	0.00244
		0.207	12	46.5	0.00425	0.00425
		0.512	59	2.29	0.1055	0.055
		0.520	74	2.32	0.1625	0.094
		0.527	78	2.40	0.163	0.0956
		0.687	151	21.20	0.0685	0.0230
		0.713	154	32.30	0.0557	0.0167
	0.3	0.127	8	108	0.00175	0.00175
		0.107	10	44.2	0.0036	0.0036
		0.512	49	1.31	0.875	0.111
		0.534	68	1.36	0.227	0.136
		0.540	72	1.44	0.237	0.140
		0.660	148	12.71	0.121	0.044
		0.686	151	17.4	0.103	0.0346
	0.5	0.512	34	0.487	0.670	0.129
		0.560	56	0.550	0.310	0.201
		0.567	60	0.578	0.337	0.215
		0.621	141	6.03	0.270	0.1145
		0.654	145	8.07	0.107	0.0927
	0.7	0.528	28	0.158	0.160	0.145
		0.554	34	0.147	0.305	0.194
		0.567	113	1.14	0.562	0.308
		0.570	123	2.40	0.529	0.275
	0.9	0.474	17	0.0265	0.0842	0.0842
		0.527	18	0.0245	0.1065	0.1065
		0.620	101	1.03	0.720	0.430
		0.635	155	0.84	0.719	0.442
½	0.2	0.200	12.5	60.8	0.0083	0.0083
		0.495	54	4.28	0.162	0.1335
		0.505	56	4.44	0.168	0.136
		0.514	64	4.20	0.185	0.1495
		0.526	78	4.48	0.207	0.161
		0.586	139	12.30	0.145	0.0965
		0.640	145	15.80	0.109	0.0666
		0.727	153	44.20	0.064	0.0324
	0.3	0.126	8	131	0.0034	0.0034
		0.140	12	62.1	0.0067	0.0066

Table 12.6. Eccentrically Loaded Partial Journal Bearings (Continued)

$\beta = 75°$

L/D	ϵ	α/β	ϕ	S	$\bar{q}_{in}$	$\bar{q}_z$
½	0.3	0.515	49	2.47	0.204	0.184
		0.541	67	2.64	0.279	0.223
		0.548	72	2.90	0.281	0.222
		0.70	90	28.2	....	0.061
		0.676	147	21.7	0.132	0.076
	0.5	0.515	34	0.890	0.242	0.221
		0.535	43	0.895	0.295	0.269
		0.560	51	0.935	0.324	0.320
		0.575	60	1.065	0.398	0.359
		0.606	137	8.25	0.354	0.239
		0.654	145	12.2	0.284	0.178
	0.7	0.527	28	0.273	0.253	0.245
		0.560	30	0.260	0.314	0.292
		0.580	36	0.272	0.360	0.328
		0.620	111	3.26	0.507	0.500
		0.615	117	3.90	0.647	0.476
		0.615	122	4.85	0.600	0.430
	0.9	0.475	17	0.0365	0.161	0.161
		0.581	19	0.0342	0.439	0.248
		0.648	94	1.50	0.942	0.736
		0.654	104	2.46	....	0.662
		0.915	174	3,090	0.00414	0.00414
¼	0.2	0.154	11	211	....	0.0082
		0.180	14	123	....	0.0135
		0.500	60	12.4	....	0.184
		0.527	73	12.3	....	0.206
		0.530	78	12.6	....	0.209
		0.714	149	77.0	....	0.061
		0.741	152	99.0	....	0.0487
	0.3	0.126	8	202	....	0.0062
		0.154	11	95	....	0.0123
		0.500	50	6.95	....	0.243
		0.540	67	7.15	....	0.298
		0.547	72	7.50	....	0.307
		0.694	146	50.2	....	0.113
		0.714	149	63.4	....	0.092
	0.5	0.515	34	2.46	....	0.300
		0.573	55	2.69	....	0.455
		0.580	59	2.90	....	0.203
		0.660	138	2.82	....	0.225
	0.7	0.515	34	0.680		
		0.573	36	0.683		
		0.646	110	0.885		
		0.650	120	12.5		
	0.9	0.460	18	0.785	....	0.251
		0.520	19	0.762	....	0.305
		0.672	92	4.06	....	0.93
		0.666	98	5.10	....	0.905
		0.880	171	1,085	....	0.0158

(c) Finite Thrust Bearings

The analysis of thrust bearings is made easier by the simplicity of the expressions for film shape and by the simple boundary conditions.

$$p(0) = p(B) = p(L/2) = p(-L/2) = 0$$

By writing $h = \alpha x$ for the film shape, we have, from Eq. (12.7),

$$\partial^2 p/\partial x^2 + (3/x)(\partial p/\partial x) + \partial^2 p/\partial z^2 - 6\mu U/\alpha^2 x^3 = 0 \tag{12.42}$$

The solution of Eq. (12.42) can be written as follows:[6]
For small nx,

$$p(x,z) = \sum_{1,3,5}^{\infty} (\sin nz/nx)\{A_n I_1(nx) + B_n K_1(nx) - (24\mu U/\pi\alpha^2)[1 + (nx)^2/3 + (nx)^4/5 \times 3^2 + (nx)^6/7 \times 5^2 3^2 + \cdots]\} \tag{12.43a}$$

For large nx,

$$p(x,z) = \sum_{1,3,5}^{\infty} \sin nz/nx \{A_n' I_1(nx) + B_n' K_1(nx) - 24\mu U/\pi\alpha^2(nx)^2[1 + 3/(nx)^2 + 5 \times 3^2/(nx)^4 + 7 \times 5^2 3^2/(nx)^6 + \cdots]\} \tag{12.43b}$$

where I_1 and K_1 are the Bessel functions of the first and second kind.

The coefficients A_n and B_n must be determined from $p(0) = p(\pi) = 0$ for all values of z. This is done by making the terms in the bracket, which is independent of z, zero. However, their evaluation by analytical means is not easily obtained. The solution of these equations by numerical means is given later in the text.

The Step Bearing. A complete analytical solution of the step bearing can be obtained by essentially solving Laplace's equation with one nonzero boundary condition.[7] For the plane slider, the film thickness is given by h = constant, and Eq. (12.7) transforms into

$$\partial^2 p/\partial x^2 + \partial^2 p/\partial z^2 = 0 \tag{12.44}$$

Employing boundary conditions,

$$p(0,z) = p(x,0) = p(x,L) = 0 \quad \text{and} \quad p = \sum_{1,3,5}^{\infty} p_n \sin (n\pi z/L) \quad \text{at } x = B_2$$

The solution is

$$p(x,z) = \sum_{1,3,5}^{\infty} \frac{p_n}{\sinh (n\pi B_2/L)} \sin \frac{n\pi z}{L} \sinh \frac{n\pi x}{L} \tag{12.45a}$$

For region I,

$$p(x,z) = -\sum_{1,3,5}^{\infty} \frac{p_n}{\sinh (n\pi B_1/L)} \sin \frac{n\pi z}{L} \sinh \frac{n\pi x}{L} \tag{12.45b}$$

where

$$p_n = \frac{24\mu UL(h_1 - h_2)}{n^2\pi^2[h_1^3 \coth (n\pi B_1/L) + h_2^3 \coth (n\pi B_2/L)]}$$

Thus Eq. (12.45) with the value of p_n as given above provides the pressure distribution. The load capacity is

$$W = \frac{48\mu UL^3(h_1 - h_2)}{\pi^4} \sum_{1,3,5}^{\infty} \frac{1}{n^4} \frac{\tanh (n\pi B_1/2L) + \tanh (n\pi B_2/2L)}{h_1^3 \coth (n\pi B_1/L) + h_2^3 \coth (n\pi B_2/L)} \tag{12.46}$$

The frictional force is

$$F = \mu UL\left(\frac{B_1}{h_1} + \frac{B_2}{h_2}\right) + \frac{24\mu UL^2(h_1 - h_2)^2}{\pi^3} \sum_{1,3,5}^{\infty} \frac{1}{n^3[h_1^3 \coth(n\pi B_1/L) + h_2^3 \coth(n\pi B_2/L)]} \tag{12.47}$$

Flow in at the leading edge is

$$Q_1 = \frac{Uh_1L}{2} - \frac{4ULh_1^3(h_1 - h_2)}{\pi^2} \sum_{1,3,5}^{\infty} \frac{1}{n^2 \sinh(n\pi B_1/L)} \frac{1}{h_1^3 \coth(n\pi B_1/L) + h_2^3 \coth(n\pi B_2/L_1)} \tag{12.48a}$$

Flow out at the trailing edge is

$$Q_2 = \frac{Uh_2L}{2} + \frac{4ULh_2^3(h_1 - h_2)}{\pi^2} \sum_{1,3,5}^{\infty} \frac{1}{n^2 \sinh(n\pi B_2/L)} \frac{1}{h_1^3 \coth(n\pi B_1/L) + h_2^3 \coth(n\pi B_1/L)} \tag{12.48b}$$

The side leakage is then

$$Q_z = Q_1 - Q_2$$

For a sectorial step bearing with θ_1 and θ_2 replacing B_1 and B_2, Laplace's equation is

$$\partial^2 p/\partial r^2 + (1/r^2)(\partial^2 p/\partial \theta^2) + (1/r)(\partial p/\partial r) = 0$$

By using

$$r = R_1 e^{\rho}$$
$$r = R_1 \qquad \text{for } \rho = 0$$
$$r = R_2 \qquad \text{for } \rho = \rho_2 = \ln R_2/R_1$$

The equation above becomes

$$\partial^2 p/\partial \rho^2 + \partial^2 p/\partial \theta^2 = 0$$

The solutions to this equation are thus identical with those for a rectangular shape.[4]

$$p(r,\theta) = \sum_{1}^{\infty} \frac{p_n}{\sinh[n\pi\theta_2/\ln(R_2/R_1)]} \sin\frac{n\pi \ln(r/R_1)}{\ln(R_2/R_1)} \sinh\frac{n\pi\theta}{\ln(R_2/R_1)} \tag{12.49a}$$

and for region I

$$p(r,\theta) = -\sum_{1}^{\infty} \frac{p_n}{\sinh[n\pi\theta_1/\ln(R_2/R_1)]} \sin\frac{n\pi \ln(r/R_1)}{\ln(R_2/R_1)} \sinh\frac{n\pi\theta}{\ln(R_2/R_1)} \tag{12.49b}$$

where

$$p_n = \frac{12\mu R_1^2\omega \ln(R_2/R_1)[(-1)^{n+1}(R_2/R_1)^2 + 1](h_1 - h_2)}{\{(n\pi)^2 + [2\ln(R_2/R_1)]^2\}\{h_1^3 \coth[n\pi\theta_1/\ln(R_2/R_1)] + h_2^3 \coth[n\pi\theta_2/\ln(R_2/R_1)]\}}$$

The total load capacity is

$$W = 12\mu\omega \left(\ln \frac{R_2}{R_1}\right)^3 (h_1 - h_2)$$

$$\sum_{n=1}^{\infty} \left[\frac{(-1)^{n+1}R_2^2 + R_1^2}{(n\pi)^2 + [2 \ln (R_2/R_1)]^2}\right]^2 \frac{\tanh [n\pi\theta_1/\ln (R_2/R_1)^2] + \tanh [n\pi\theta_2/\ln (R_2/R_1)^2]}{h_1^3 \coth [n\pi\theta_1/\ln (R_2/R_1)] + h_2^3 \coth [n\pi\theta_2/\ln (R_2/R_1)]} \tag{12.50}$$

The torque required is

$$M = \frac{\mu\omega(R_2^4 - R_1^4)}{4}\left(\frac{\theta_1}{h_1} + \frac{\theta_2}{h_2}\right) + 6\pi(h_1 - h_2)^2\mu\omega\left(\ln \frac{R_2}{R_1}\right)^2$$

$$\sum_{1,3,5}^{\infty} \left[\frac{(-1)^{n+1}R_2^2 + R_1^2}{(n\pi)^2 + [2 \ln (R_2/R_1)]^2}\right]^2 \frac{1}{h_1^3 \coth [n\pi\theta_1/\ln (R_2/R_1)] + h_2^3 \coth [n\pi\theta_2/\ln (R_2/R_1)]} \tag{12.51}$$

Flow becomes

$$Q_1 = \frac{\omega h_1(R_2^2 - R_1^2)}{4} - 2\omega\left(\ln \frac{R_2}{R_1}\right) h_1^3(h_1 - h_2)(R_2^2 - R_1^2)$$

$$\sum_{n=1,3,5}^{\infty} \frac{1}{\{(n\pi)^2 + [2 \ln (R_2/R_1)]^2\} \sin h[n\pi\theta_1/\ln (R_2/R_1)]}$$

$$\left[h_1^3 \coth \frac{n\pi\theta_1}{\ln (R_2/R_1)} + h_2^3 \coth \frac{n\pi\theta_2}{\ln (R_2/R_1)}\right] \tag{12.52a}$$

$$Q_2 = \frac{\omega h_2(R_2^2 - R_1^2)}{4} + 2\omega \ln \left(\frac{R_2}{R_1}\right)^2 h_2^3(h_1 - h_2)(R_2^2 - R_1^2)$$

$$\sum_{n=1,3,5}^{\infty} \frac{1}{\{(n\pi)^2 + [2 \ln (R_2/R_1)]^2\} \sin h\, [n\pi\theta_2/\ln (R_2/R_1)]}$$

$$\left[h_1^3 \coth \frac{n\pi\theta_1}{\ln (R_2/R_1)} + h_2^3 \coth \frac{n\pi\theta_2}{\ln (R_2/R_1)}\right] \tag{12.52b}$$

Sector Pad; Computer Solutions. The solution of a sectorial thrust bearing can be obtained by methods similar to those used on journal bearings.[8] However, since the underlying differential equation and the geometry of the pad are different, new dimensionless ratios must be used.

$$\bar{r} = r/R_2 \qquad U = 2\pi r N = 2\pi R_2 \bar{r} U$$
$$\bar{h} = h/\delta \qquad \bar{p} = (p/\mu N)(\delta/L)^2$$

where $\delta = h_1 - h_2$.

The dimensionless load factor $\bar{p} = (p/\mu N)(\delta/L)^2$ for thrust bearings and can be considered the equivalent of the Sommerfeld number for journal bearings. By writing these substitutions in the Reynolds equation, we have

$$(\partial/\partial\bar{r})[\bar{r}\bar{h}(\partial\bar{p}/\partial\bar{r})] + (1/r)(\partial/\partial\theta)[h^3(\partial\bar{p}/\partial\theta)] = 12\pi\bar{r}(R_2/L)^2(\partial h/\partial\theta) \tag{12.53}$$

By expressing the derivatives in finite-difference form and using them in Eq. (12.53) we have, for any point in the grid,

$$\bar{p}_{i,j} = \frac{12\pi\bar{r}_{i,j}\left(\frac{R}{L}\right)^2 \frac{\bar{h}_{i,j-\frac{1}{2}} - \bar{h}_{i,j+\frac{1}{2}}}{\Delta\theta} + \bar{h}^3_{i+\frac{1}{2},j}\frac{(\bar{r}\bar{p})_{i+1,j}}{(\Delta\bar{r})^2} + \bar{h}^3_{i-\frac{1}{2},j}\frac{(\bar{r}\bar{p})_{i-1,j}}{(\Delta\bar{r})^2} + \bar{h}^3_{i,j+\frac{1}{2}}\frac{\bar{p}_{i,j+1}}{\bar{r}_{i,j}\,\Delta\theta} + \bar{h}^3_{i,j-\frac{1}{2}}\frac{\bar{p}_{i,j-1}}{\bar{r}_{i,j}\,\Delta\theta}}{\frac{\bar{h}^3_{i+\frac{1}{2},j}\bar{r}_{i+1,j}}{(\Delta\bar{r})^2} + \frac{\bar{h}^3_{i-\frac{1}{2},j}\bar{r}_{i-1,j}}{(\Delta\bar{r})^2} + \frac{1}{\bar{r}_{i,j}\,\Delta\theta}(\bar{h}^3_{i,j-\frac{1}{2}} + \bar{h}^3_{i,j-\frac{1}{2}})} \tag{12.54}$$

The expression for the load capacity is

$$W = \int_{R_1}^{R_2}\int_0^\beta pr\,d\theta\,dr = [\mu N/(L/\delta)^2]R_2{}^2\,\Delta\bar{r}\,\Delta\theta \sum_{j=1}^{m}\sum_{i=1}^{n} \bar{p}_{i,j}\bar{r}_{i,j} \tag{12.55}$$

The dissipated power is

$$\begin{aligned} H &= \int_0^\beta \int_{R_1}^{R_2} \mu r\,[2\pi r N/h + (h/2\mu)(\partial p/r\partial\theta)]\,r\,dr\,d\theta \\ &= (\pi\mu N^2 R_2{}^4/\delta)\left\{\pi[1 - (1 - L/R_2)^4]\beta \ln(1 + \delta/h_2) \right. \\ &\quad \left. + (L/R_2)^2\,\Delta\bar{r}\,\Delta\theta \sum_{j=1}^{m}\sum_{i=1}^{n} \bar{h}_{i,j}\bar{r}_{i,j}(\partial\bar{p}_{i,j}/\partial\theta)\right\} = j(\pi\mu N^2 R_2{}^4/\delta) \end{aligned} \tag{12.56}$$

where j is the term in the brackets.

The flow of lubricant is

$$\begin{aligned} Q_z &= \int_0^\beta (h^3/12\mu)(\partial p/\partial z)\,dx = \bar{q}_r \pi R_2 N L\delta \\ Q_x &= \int_{R_1}^{R_2} [Uh/2 - (h^3/12)(\partial p/\partial x)]\,dz = (\pi N L h_2/2)(R_1 + R_2) + \bar{q}_1 \pi R_2 N L\delta \end{aligned} \tag{12.57}$$

where

$$\bar{q}_r = (1/12\pi)(L/R_2)\sum_j^n \bar{h}^3\bar{r}(\partial\bar{p}/\partial\bar{r})\Big|_{R_1,R_2}\Delta\theta$$

$$\bar{q}_1 = (1/12\pi)(L/R_2)\sum_i^m (\bar{h}^3/\bar{r})(\partial\bar{p}/\partial\theta)\Big|_{\theta=0}\Delta\bar{r}$$

For a uniform taper in the circumferential direction the film shape can be expressed by

$$h = h_2 + \delta(1 - \theta/\beta) \tag{12.58}$$

or in dimensionless form

$$\bar{h} = h_2/\delta + (1 - \theta/\beta)$$

The results for the sector thrust bearing are given in Table 12.7. When the quantity $(\mu N/P)(L/\delta)^2$ is plotted vs. the angular span β, there is a minimum in the curve indicating the optimum number of pads for a given set of conditions, and these are tabulated in Table 12.8. Table 12.9 gives a comparison of results for an inclined slider, a slider with exponential film, and for a sector with a circumferential taper only.

Table 12.7. Performance of Sector Thrust Bearings

L/R_2	h_2/δ	β, deg	$\frac{\mu N}{P}\left(\frac{L}{\delta}\right)^2$	$\bar{q}_z$ At R_1	$\bar{q}_z$ At R_2	$\bar{q}_{in}$	Center of pressure* $\bar{\bar{\theta}}$	Center of pressure* $\bar{\bar{r}}$	j
⅓	1	80	1.423	0.34	0.46	0.87	0.64	0.37	2.44
		55	1.108	0.32	0.44	0.84	0.625	0.45	1.685
		40	0.947	0.28	0.395	0.81	0.61	0.49	1.26
		30	0.870	0.235	0.35	0.75	0.605	0.51	0.95
	½	80	0.321	0.35	0.47	0.87	0.71	0.37	3.94
		55	0.257	0.32	0.44	0.84	0.69	0.47	2.70
		40	0.225	0.28	0.40	0.79	0.67	0.50	2.00
		30	0.211	0.24	0.36	0.74	0.66	0.51	1.57
	¼	80	0.0855	0.35	0.47	0.87	0.78	0.41	5.96
		55	0.0714	0.32	0.44	0.83	0.76	0.45	4.25
		40	0.0652	0.29	0.41	0.78	0.74	0.505	3.23
		30	0.0635	0.245	0.36	0.70	0.73	0.52	2.54
	⅛	80	0.0278	0.36	0.48	0.85	0.83	0.465	8.51
		55	0.0247	0.33	0.45	0.81	0.815	0.50	6.23
		40	0.0238	0.29	0.41	0.75	0.795	0.51	4.86
		30	0.0242	0.25	0.37	0.67	0.78	0.565	3.91
½	1	80	1.72	0.23	0.405	0.75	0.62	0.48	2.90
		55	1.494	0.19	0.36	0.69	0.61	0.51	1.96
		40	1.435	0.145	0.31	0.61	0.60	0.53	1.47
		30	1.489	0.11	0.26	0.57	0.59	0.55	1.13
	½	80	0.402	0.23	0.41	0.74	0.685	0.46	4.72
		55	0.3585	0.19	0.33	0.61	0.67	0.52	3.33
		40	0.352	0.15	0.31	0.60	0.655	0.53	2.49
		30	0.370	0.11	0.26	0.53	0.65	0.55	1.92
	¼	80	0.1138	0.24	0.42	0.72	0.755	0.48	7.32
		55	0.1062	0.20	0.27	0.65	0.735	0.52	5.29
		40	0.1080	0.15	0.32	0.56	0.72	0.54	4.065
		30	0.1103	0.11	0.27	0.49	0.71	0.56	3.18
	⅛	80	0.0402	0.25	0.42	0.70	0.81	0.50	10.81
		55	0.0399	0.20	0.28	0.62	0.78	0.53	8.06
		40	0.0423	0.16	0.32	0.53	0.77	0.55	6.30
		30	0.0470	0.11	0.27	0.44	0.765	0.57	5.01
⅔	1	80	2.240	0.12	0.35	0.60	0.61	0.50	3.06
		55	2.185	0.082	0.295	0.53	0.60	0.55	2.12
		40	2.320	0.052	0.245	0.48	0.59	0.58	1.57
		30	2.590	0.033	0.200	0.44	0.59	0.61	1.20
	½	80	0.538	0.13	0.35	0.58	0.67	0.51	5.07
		55	0.537	0.084	0.30	0.51	0.66	0.56	3.59
		40	0.578	0.0535	0.25	0.45	0.65	0.59	2.70
		30	0.653	0.034	0.20	0.40	0.645	0.61	2.07
	¼	80	0.1598	0.13	0.36	0.56	0.735	0.53	8.00
		55	0.1655	0.087	0.30	0.46	0.72	0.57	5.79
		40	0.1820	0.055	0.25	0.40	0.71	0.60	4.43
		30	0.2085	0.035	0.21	0.36	0.705	0.62	3.46
	⅛	80	0.0599	0.14	0.365	0.53	0.79	0.55	12.07
		55	0.0649	0.09	0.31	0.44	0.78	0.58	8.98
		40	0.0737	0.056	0.25	0.35	0.765	0.61	6.94
		30	0.0861	0.036	0.21	0.29	0.76	0.63	5.47

* $\bar{\bar{\theta}} = \theta/\beta \qquad \bar{\bar{r}} = (r - R_1)/L$.

Table 12.8. Optimum Geometry in Tapered-land Thrust Bearing

L/R_2	h_2/δ	β, deg	Number of pads
⅓	1	<30	>10
	½	<30	>10
	¼	35	9
	⅛	40	8
½	1	40	8
	½	45	7
	¼	50	6
	⅛	60	5
⅔	1	50	6
	½	60	5
	¼	80	4
	⅛	>80	4

Table 12.9. Performance of Thrust Bearings with Various Film Configurations

a	Plane slider*	Exponential slider†	Sector pad‡
	$\bar{P} = PL^2h_2^2/\mu\omega R_2^4$		
2.00	0.0810	0.0819	0.0826
2.50	0.113	0.1137	0.106
2.85	0.135	0.135	0.125
	$\bar{F} = Fh_2/\mu\omega R_2^4$		
2.00	0.66	0.81	0.78
2.50	0.74	0.875	0.825
3.04	0.84	0.95	0.88

* $h = \alpha x$.
† $h = k_1 e^{k_2 x}$.
‡ $h = h_2 + \delta(1 - \theta/\beta)$.

12.4. HYDRODYNAMIC GAS BEARINGS

Infinitely Long Slider. Under isothermal conditions, viscosity may be considered constant and the normalized equation describing pressure distribution may then be written as

$$\bar{p}\bar{h}^3(\partial\bar{p}/\partial\bar{x}) = \Lambda(\bar{p}\bar{h} - C_1) \tag{12.59}$$

where $\Lambda = 6\mu UB/h_2^2 p_a$, $\bar{p} = p/p_a$, $\bar{h} = h/h_2$, $\bar{x} = x/B$, and C_1 is a new integration constant and represents the product of pressure and the film thickness at the point where the maximum pressure occurs.[9]

The integration of Eq. (12.59) is simplified by expressing $\bar{h}$ as the independent variable. Since $(a - 1)(\partial/\partial\bar{h}) = -(\partial/\partial\bar{x})$, the above equation may be written as

$$-\bar{p}\bar{h}^3(\partial\bar{p}/\partial\bar{h}) = [\Lambda/(a - 1)](\bar{p}\bar{h} - C_1)$$

As before we choose a new dependent variable $\xi = ph$. For the boundary conditions we set $\bar{p}(0) = 1$ and $\bar{p}(1) = \bar{p}_2$. Therefore, at the leading edge $\bar{x} = 0$, $\xi = a$, while at the trailing edge $\bar{x} = 1$, $\xi = \bar{p}_2$. For the plane slider bearing, $\bar{p}_2 = 1$. The same methods apply for $\bar{p}_2 \neq 1$. Setting $\Lambda/a - 1 = \bar{\Lambda}$:

$$\xi(\xi^2 - \bar{\Lambda}\xi + \bar{\Lambda}C_1)^{-1}\, d\xi = d\bar{h}/\bar{h}$$

Integration yields

$$\xi^2 - \bar{\Lambda}\xi + \bar{\Lambda}C_1 = C_2\bar{h}^2\psi(\xi) \tag{12.60}$$

in which C_2 is a constant of integration and $\psi(\xi)$ assumes values as follows:

$$\psi_1(\xi) = \exp^- \left\{\frac{2\bar{\Lambda}}{[\bar{\Lambda}(4C_1 - \bar{\Lambda})]^{1/2}} \tan^{-1} \frac{2\xi - \bar{\Lambda}}{[\bar{\Lambda}(4C_1 - \bar{\Lambda})]^{1/2}}\right\} \qquad \text{for } \bar{\Lambda} < 4C_1$$

$$\psi_c(\xi) = \exp \frac{8C_1}{2\xi - \bar{\Lambda}} \qquad \text{for } \bar{\Lambda}_c = 4C_1$$

$$\psi_2(\xi) = \left\{\frac{2\xi - \bar{\Lambda} + [\bar{\Lambda}(4C_1 - \bar{\Lambda})]^{1/2}}{2\xi - \bar{\Lambda} - [\bar{\Lambda}(4C_1 - \bar{\Lambda})]^{1/2}}\right\}^{\bar{\Lambda}/[\bar{\Lambda}(4C_1 - \bar{\Lambda})]^{1/2}} \qquad \text{for } \bar{\Lambda} > 4C_1$$

The load-carrying capacity may be obtained by integration of Eq. (12.60):

$$\bar{W} = (a - 1)^{-1}\{\bar{\Lambda} \ln a + (\Lambda - C_1) \ln [\psi(a)/\psi(\bar{p}_2)]\} \qquad \text{for } \bar{\Lambda} < 4C_1$$

$$\bar{W} = (a - 1)^{-1}\left\{\bar{\Lambda}_c \ln \frac{2a - \bar{\Lambda}_c}{2\bar{p}_2 - \Lambda_c} - \frac{\bar{\Lambda}_c}{2}[(2a - \bar{\Lambda}_c)^{-1} - (2\bar{p}_2 - \Lambda_c)^{-1}]\right\} \qquad \text{for } \bar{\Lambda}_c = 4C_1$$

$$\bar{W} = (a - 1)^{-1}\{\bar{\Lambda} \ln a + C_1 \ln [\psi(a)/\psi(\bar{p}_2)]\} \qquad \text{for } \Lambda > 4C_1$$

Table 12.10 gives the isothermal load capacity and center of pressure for infinitely long slider bearings. For bearing numbers approaching zero the load-carrying capacity may be expressed as

$$\bar{W}_{\Lambda\to 0} = \Lambda(a^2 - 1)^{-1}[\ln a - 2(a - 1)(a + 1)^{-1}] - 1$$

For large bearing numbers with adiabatic conditions the bearing load is given by

$$\bar{W}_{\Lambda\to\infty} = \frac{a(a^{n-1} - 1)}{(n - 1)(a - 1)} - 1$$

For isothermal conditions

$$\bar{W}_{\Lambda\to\infty} = \frac{a}{a - 1} - 1$$

The first moment may be obtained by evaluation of Eq. (12.60). For limiting conditions, the center of pressure may be expressed as

$$\bar{x}_{\Lambda\to 0} = (a - 1)^{-5} \frac{a(a + 2) \ln a - (5a - 1)(a - 1)}{2[(a^2 - 1) \ln a - 4(a - 1)^2]}$$

When the bearing number approaches infinity and the gas path is polytropic, the center of pressure may be expressed by

$$\bar{x}_{\Lambda\to 0} = (a - 1)^{-1}\left[a - \frac{(1 - a^{2-n})(n - 2)^{-1}a^n + \frac{1}{2}(1 - a^2)}{(1 - a^{1-n})(n - 1)^{-1}a^n + 1 - a}\right]$$

Table 12.10. Performance of Infinitely Long Sliders under Isothermal Conditions[9]

a	Λ	W/LBp_a	$\bar{\bar{x}}$
1.5	0.5	0.01091	0.5456
	1.0	0.02172	0.552
	5.0	0.0957	0.5861
	10.0	0.1486	0.6168
	25.0	0.1942	0.6585
2.0	0.5	0.01323	0.5724
	1.0	0.02640	0.5761
	5.0	0.1234	0.6008
	10.0	0.2124	0.6235
	30.0	0.3367	0.6705
	50.0	0.3639	0.6870
3.0	0.5	0.01232	0.6095
	1.0	0.02063	0.6116
	5.0	0.1201	0.6264
	10.0	0.2252	0.6402
	50.0	0.5618	0.6950
4.0	0.5	0.01034	0.6349
	1.0	0.02068	0.6364
	5.0	0.1023	0.6464
	10.0	0.1980	0.6560
	25.0	0.4264	0.6756
	50.0	0.6424	0.6976
6.0	0.5	0.007241	0.6688
	1.0	0.01448	0.6696
	5.0	0.07243	0.6753
	10.0	0.1435	0.6809
	50.0	0.5943	0.7057
	100.0	0.9021	0.7255

Finite Slider Bearings. Tables 12.11 and 12.12 give the dimensionless coefficients of load, center of pressure, and friction force for various ratios of inlet to outlet film thickness a, length-to-breadth ratio, and bearing numbers Λ for the case of plain inclined sliders under isothermal conditions. These results were obtained by solving the compressible Reynolds equation numerically.[10] Using these coefficients it is possible to calculate such bearing characteristics as

$$\text{Load} = W = (p_a BL\Lambda/6)\bar{W}$$

where $\bar{W} = (6/\Lambda)(P/p_a)$.

$$\text{Friction force} = F = BL\iint_0^1 \tau_x \,\partial\bar{x}\,\partial\bar{z} = (h_2 p_a L\Lambda/6)\bar{F}$$

$$\text{Center of pressure } \bar{\bar{x}} = \frac{\iint_1^0 (\bar{p}-1)\bar{x}\,\partial\bar{x}\,\partial\bar{z}}{\iint_0^1 (\bar{p}-1)\,\partial\bar{x}\,\partial\bar{z}}$$

Table 12.11. $\bar{W}$, $\bar{\bar{x}}$, $\bar{F}$ Coefficients for Various Values of a and L/B

a	$\Lambda = 10$ $L/B = 0.25$			$\Lambda = 10$ $L/B = 0.40$		
	$\bar{\bar{W}}$	$\bar{\bar{x}}$	$\bar{F}$	$\bar{\bar{W}}$	$\bar{\bar{x}}$	$\bar{F}$
1.0	0	0.50000	1.00000	0	0.50000	1.00000
1.5	0.00705	0.58987	0.81269	0.01532	0.59445	0.81476
1.8	0.00859	0.61923	0.73817	0.01873	0.61688	0.74222
2.0	0.00917	0.63585	0.69773	0.01996	0.62977	0.70313
2.2	0.00945	0.64749	0.66272	0.02071	0.64125	0.66947
2.5	0.00980	0.66634	0.61812	0.02126	0.65644	0.62680
2.8	0.00994	0.68309	0.58096	0.02137	0.66988	0.59124
3.0	0.00996	0.69311	0.55926	0.02128	0.67816	0.57058
4.0	0.00968	0.72726	0.47662	0.02004	0.70989	0.49216
5.0	0.00921	0.75199	0.42078	0.01827	0.73362	0.43890
6.0	0.00855	0.77289	0.37972	0.01655	0.75163	0.39974
7.0	0.00792	0.78869	0.34808	0.01502	0.76585	0.36938

a	$\Lambda = 10$ $L/B = 0.50$			$\Lambda = 10$ $L/B = 0.75$		
1.0	0	0.50000	1.00000	0	0.50000	1.00000
1.5	0.02150	0.59864	0.81630	0.03564	0.60591	0.81984
1.8	0.02623	0.61762	0.74523	0.04382	0.61928	0.75226
2.0	0.02792	0.62870	0.70711	0.04672	0.62742	0.71651
2.2	0.02891	0.63872	0.67440	0.04836	0.63489	0.68606
2.5	0.02958	0.65194	0.63305	0.04940	0.64496	0.64791
2.8	0.02965	0.66384	0.59869	0.04933	0.65444	0.61641
3.0	0.02945	0.67121	0.57875	0.04880	0.66042	0.59810
4.0	0.02722	0.70107	0.50293	0.04405	0.68539	0.52817
5.0	0.02447	0.72302	0.45130	0.03885	0.70432	0.48005
6.0	0.02195	0.73983	0.41323	0.03422	0.71944	0.44390
7.0	0.01975	0.75325	0.38357	0.03032	0.73174	0.41528

a	$\Lambda = 10$ $L/B = 1.00$			$\Lambda = 10$ $L/B = 1.50$		
1.0	0	0.50000	1.00000	0	0.50000	1.00000
1.5	0.04630	0.60964	0.82250	0.05938	0.61272	0.82577
1.8	0.05768	0.62006	0.75780	0.07531	0.62044	0.76486
2.0	0.06172	0.62662	0.72401	0.08115	0.62551	0.73372
2.2	0.06401	0.63278	0.69546	0.08448	0.63045	0.70774
2.5	0.06545	0.64118	0.65995	0.08654	0.63734	0.67576
2.8	0.06532	0.64923	0.63080	0.08646	0.64399	0.64982
3.0	0.06452	0.65438	0.61382	0.08533	0.64840	0.63463
4.0	0.05768	0.67642	0.54862	0.07572	0.66775	0.57567
5.0	0.05034	0.69372	0.50305	0.06542	0.68353	0.53320
6.0	0.04393	0.70777	0.46817	0.05651	0.69663	0.49964
7.0	0.03856	0.71945	0.44001	0.04918	0.70764	0.47185

Table 12.12. $\bar{W}$, $\bar{\bar{x}}$, $\bar{F}$ Coefficients for Various Values of a and L/B

a	$\Lambda = 0.01$ $L/B = 1.00$			$\Lambda = 0.01$ $L/B = \infty$		
	$\bar{W}$	$\bar{\bar{x}}$	$\bar{F}$	$\bar{W}$	$\bar{\bar{x}}$	$\bar{F}$
1.0	0	0.50000	1.00000	0	0.50000	1.00000
1.5	0.04651	0.55401	0.82256	0.10912	0.54560	0.83821
1.8	0.06116	0.57199	0.75920	0.13881	0.56068	0.79026
2.0	0.06597	0.58310	0.72613	0.14821	0.56984	0.76725
2.2	0.06838	0.59336	0.69808	0.15222	0.57835	0.74838
2.5	0.06925	0.60731	0.66280	0.15206	0.58995	0.72490
2.8	0.06823	0.61966	0.63341	0.14790	0.60036	0.70512
3.0	0.06699	0.62727	0.61629	0.14400	0.60669	0.69330
4.0	0.06031	0.65446	0.55257	0.14868	0.61839	0.68511
5.0	0.05344	0.67453	0.50924	0.11263	0.64362	0.62762
6.0	0.04620	0.69213	0.47386	0.09134	0.66156	0.58670
7.0	0.03992	0.70737	0.44409	0.07630	0.67566	0.55322
	$\Lambda = 1.00$ $L/B = 1.00$			$\Lambda = 1.00$ $L/B = \infty$		
1.0	0	0.50000	1.00000	0	0.50000	1.00000
1.5	0.05167	0.55815	0.82385	0.12007	0.55260	0.84095
1.8	0.06301	0.57675	0.75994	0.14245	0.56790	0.79171
2.0	0.06733	0.58758	0.72681	0.14976	0.57663	0.67803
2.2	0.06960	0.59733	0.69881	0.15326	0.58450	0.74900
2.5	0.07042	0.61045	0.66368	0.15304	0.59516	0.72564
2.8	0.06943	0.62212	0.63449	0.14892	0.60475	0.70604
3.0	0.06820	0.62920	0.61751	0.14505	0.61064	0.69436
4.0	0.06159	0.65470	0.55448	0.12531	0.63346	0.65006
5.0	0.05471	0.67369	0.51178	0.10517	0.65145	0.61269
6.0	0.04746	0.69021	0.47701	0.08905	0.66557	0.58098
7.0	0.04118	0.70434	0.44787	0.07668	0.67667	0.55434
	$\Lambda = 40$ $L/B = 1.00$			$\Lambda = 40$ $L/B = \infty$		
1.0	0	0.50000	1.00000	0	0.50000	1.00000
1.5	0.02253	0.66171	0.81656	0.02938	0.67002	0.81828
1.8	0.03119	0.66922	0.74721	0.04286	0.67580	0.75188
2.0	0.03550	0.67358	0.71090	0.05051	0.67865	0.71840
2.2	0.03886	0.67757	0.68037	0.05722	0.68086	0.69138
2.5	0.04246	0.68299	0.64270	0.06562	0.68324	0.66008
2.8	0.04469	0.68787	0.61223	0.07212	0.68486	0.63692
3.0	0.04566	0.69091	0.59497	0.07541	0.68572	0.62472
4.0	0.04630	0.70416	0.53155	0.08162	0.68999	0.58452
5.0	0.04319	0.71585	0.48875	0.07861	0.69502	0.55958
6.0	0.03931	0.72591	0.45662	0.07153	0.70146	0.53718
7.0	0.03549	0.73475	0.43078	0.06402	0.70800	0.51638

(a) Infinitely Long Journal Bearings

Journal Bearing with Inertia Considered. The differential equation which governs steady-state flow without side leakage and with inertia effects considered is

$$\mu(d^2u/dy^2) = \partial p/\partial x + \rho(Du/dt) = \partial p/\partial x + \rho[u(\partial u/\partial x) + v(\partial u/\partial y)] \quad (12.61)$$

The continuity equation, for isothermal compressible fluids with p varying in x direction only, becomes

$$\partial u/\partial x + \partial v/\partial y + (u/p)(\partial p/\partial x) = 0 \quad (12.62)$$

By averaging the inertia of the fluid across the film height, integrating Eq. (12.61) twice, and substituting the proper boundary conditions we get

$$\frac{d\bar{p}}{d\theta} - \frac{6}{(1+\epsilon\cos\theta)^2} + \frac{12A}{\bar{p}(1+\epsilon\cos\theta)^3} = \frac{2B^2}{15}\left\{\frac{\bar{p}\epsilon\sin\theta}{1+\epsilon\cos\theta} - \frac{9A^2\epsilon\sin\theta}{\bar{p}(1+\epsilon\cos\theta)^3} + \frac{d\bar{p}}{d\theta}\left[8 + \frac{9A^2}{\bar{p}^2(1+\epsilon\cos\theta)^2} - \frac{18}{1+\epsilon\cos\theta}\right]\right\} \quad (12.63)$$

where $\bar{p} = (p/\mu\omega)(C/R)^2$, $A = G'C\mathcal{R}T/\mu R^3\omega^2$, $B = R\omega/\sqrt{g\mathcal{R}T}$, $h = C(1 + \epsilon\cos\theta)$, and G' is the flow per unit length.

$$G' \equiv \rho g \int_0^h u\,dy = \rho g[Uh/2 - f(x)h^3/12]$$

The quantity A in this equation, and the constant which will arise when the equation is integrated for pressure, can be evaluated by the pressures specified at the beginning and the end of the bearing arc. Equation (12.63) is difficult to integrate exactly, but it may be solved numerically by using the Runge-Kutta method of integration.[11]

Table 12.13. Effect of Lubricant Inertia on Performance of Journal Bearings

ϵ	Speed, rpm	W/L, lb/in.		ϕ	
		Inertia considered	Inertia neglected	Inertia considered	Inertia neglected
0.2	25,000	6.65	6.65	32.3	32.0
	50,000	8.33	8.65	27.2	26.5
	100,000	8.66	9.15	25.7	22.0
	150,000	8.66	9.58	26.3	20.8
	200,000	8.66	9.60	28.0	20.3
0.4	25,000	16.6	16.7	28.0	27.1
	50,000	20.0	20.8	24.0	21.8
	100,000	21.6	22.9	23.1	18.4
	150,000	21.6	23.3	24.5	18.0
	200,000	21.2	23.5	26.8	17.5
0.6	25,000	33.3	33.4	23.2	22.0
	50,000	40.0	41.7	20.7	17.5
	100,000	44.1	46.2	21.2	15.3
	150,000	44.3	47.9	23.5	14.5
	200,000	43.3	48.8	26.0	14.0

$D = 2$ in., $C/R = 0.001$. At $\theta_1 = 45°$ and $\theta_2 = 225°$, $p = p_a$. θ_1 = angle between line of centers and inlet edge.

Journal Bearing with Inertia Neglected. The right-hand side of Eq. (12.63) is a measure of the lubricant inertia. By neglecting these terms and replacing $\bar{p}$ and A by their dimensional parameters, we obtain

$$\frac{dp}{d\theta} = \frac{6\mu UR}{C^2(1 + \epsilon \cos \theta)^2}\left[1 - \frac{C_1}{C(1 + \epsilon \cos \theta)p}\right] \tag{12.64}$$

At high speeds the inertia effect can be significant for gas-lubricated journal bearings. Table 12.13 compares some results obtained from both Eqs. (12.63) and (12.64) and shows the effects of inertia on the operation of a partial 180° gas-lubricated journal bearing in laminar flow.

(b) Finite Journal Bearings

Linearized *ph* Solution. An improved analytical solution which largely eliminates the defects of first-order perturbation is accomplished by linearizing the differential equation by setting the product ph of pressure and film thickness as the dependent variable. The resulting solution is called the "linearized ph" solution.[12] This analysis gives for the pressure

$$p = \frac{p_a}{1 + \epsilon \cos \theta}\left[1 + \epsilon \frac{\Lambda}{1 + \Lambda^2}(g_1\zeta \sin \theta + g_2\zeta \cos \theta)\right] \tag{12.65a}$$

where
$$g_1\zeta = 1 - A \sinh \alpha\zeta \sin \beta\zeta + B \cosh \alpha\zeta \cos \beta\zeta$$
$$g_2\zeta = 1/\Lambda + A \cosh \alpha\zeta \cos \beta\zeta + B \sinh \alpha\zeta \sin \beta\zeta$$

$$A = \frac{\Lambda \cosh \alpha(L/D) \cos \beta(L/D) + \sinh \alpha(L/D) \sin \beta(L/D)}{\sinh^2 \alpha(L/D) + \cos^2 \beta(L/D)}$$

$$B = \frac{\Lambda \cosh \alpha(L/D) \sin \beta(L/D) - \cosh \alpha(L/D) \cos \beta(L/D)}{\sinh^2 \alpha(L/D) + \cos^2 \beta(L/D)}$$

$$\alpha^2 = (\sqrt{1 + \Lambda^2} + 1)/2$$
$$\beta^2 = (\sqrt{1 + \Lambda^2} - 1)/2$$

or

$$p = \frac{p^{(1)} + p_a\epsilon \cos \theta}{1 + \epsilon \cos \theta} \tag{12.65b}$$

where $p^{(1)}$ is the first-order perturbation pressure solution.

The load components W_R and W_T are

$$W_R = W_R^{(1)} \frac{2}{\epsilon^2}\left(\frac{1 - \sqrt{1 - \epsilon^2}}{\sqrt{1 - \epsilon^2}}\right)$$

$$W_T = W_T^{(1)} \frac{2}{\epsilon^2}(1 - \sqrt{1 - \epsilon^2})$$

where $W_R^{(1)}$ and $W_T^{(1)}$ are the first-order perturbation results. The total load and the attitude angle are

$$W = W^{(1)} \frac{2}{\epsilon^2}\frac{1 - \sqrt{1 - \epsilon^2}}{\sqrt{1 - \epsilon^2}}\sqrt{1 - \epsilon^2 \sin^2 \phi^{(1)}} \tag{12.66}$$

$$\tan \phi = 1 - \epsilon^2 \tan \phi^{(1)} \tag{12.67}$$

where $W^{(1)}$ and $\phi^{(1)}$ are the first-order perturbation results. Figure 12.15 shows the $W^{(1)}$ and $\phi^{(1)}$ as functions of bearing parameter Λ for various values of L/D.

It can be shown that the quasi-steady-state Reynolds equation, with squeeze-film velocity terms present, can be solved by perturbation techniques.[13] When such an analysis is performed, the results show that correspondence relations exist between

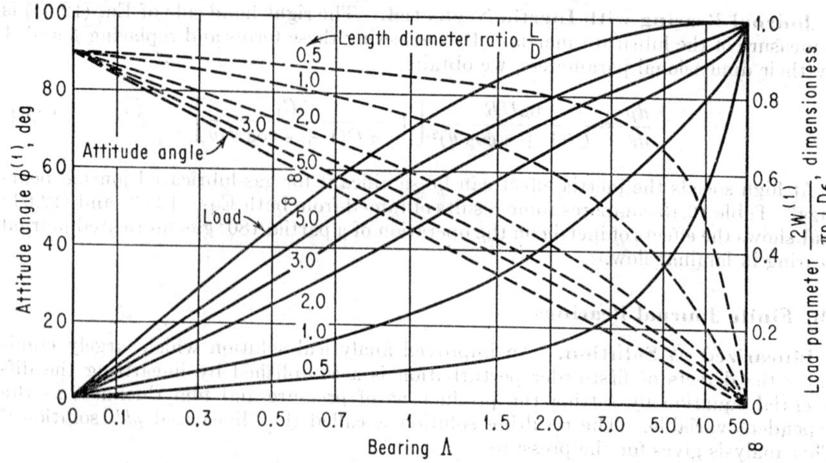

FIG. 12.15. Isothermal bearing load and attitude angle vs. bearing number for first-order perturbation solution.

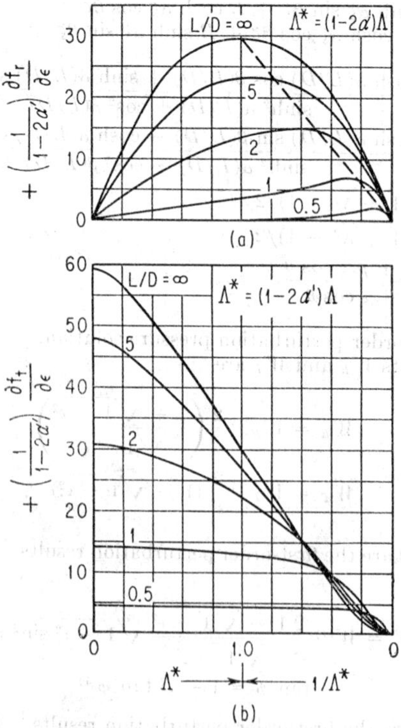

FIG. 12.16*a*. Dimensionless radial stiffness vs. dynamic compressibility number (from perturbation solution).

FIG. 12.16*b*. Dimensionless tangential stiffness vs. dynamic compressibility number (from perturbation solution).

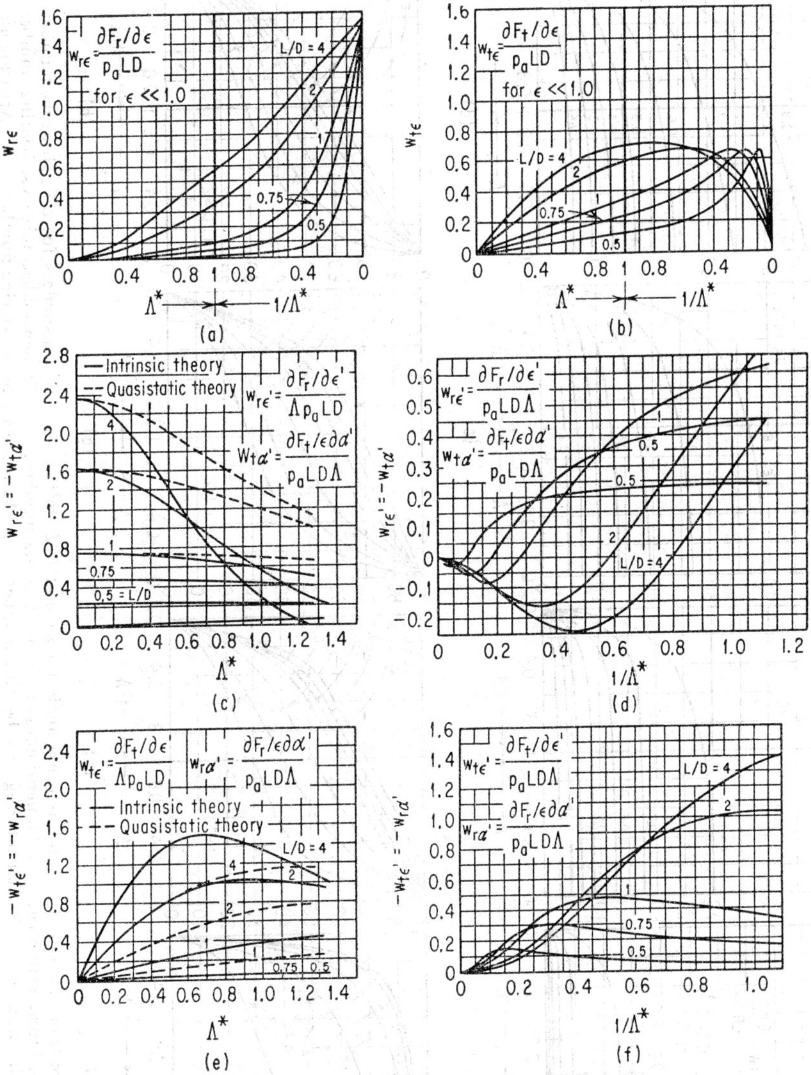

FIG. 12.17. Dimensionless radial-tangential stiffness and damping vs. dynamic compressibility number.

the derivatives of force with respect to displacement and velocity which are

$$\partial W_R/\partial \epsilon' = [2/(1 - 2\alpha')](\partial W_T/\partial \epsilon)_D = 2(\partial W_T/\partial \epsilon)_S \qquad (12.68a)$$
$$\partial W_T/\partial \epsilon' = -[2/(1 - 2\alpha')](\partial W_R/\partial \epsilon)_D = -2(\partial W_R/\partial \epsilon)_S \qquad (12.68b)$$

where $\epsilon' = d\epsilon/\omega\, dt$
$\alpha' = d\alpha/\omega\, dt$

Subscripts D and S correspond to dynamic and static cases, respectively.

These derivatives are necessary in dynamic analysis. Figure 12.16 shows the dimensionless radial and tangential stiffness vs. dynamic bearing number $\Lambda^* = (1 - 2\alpha')\Lambda$

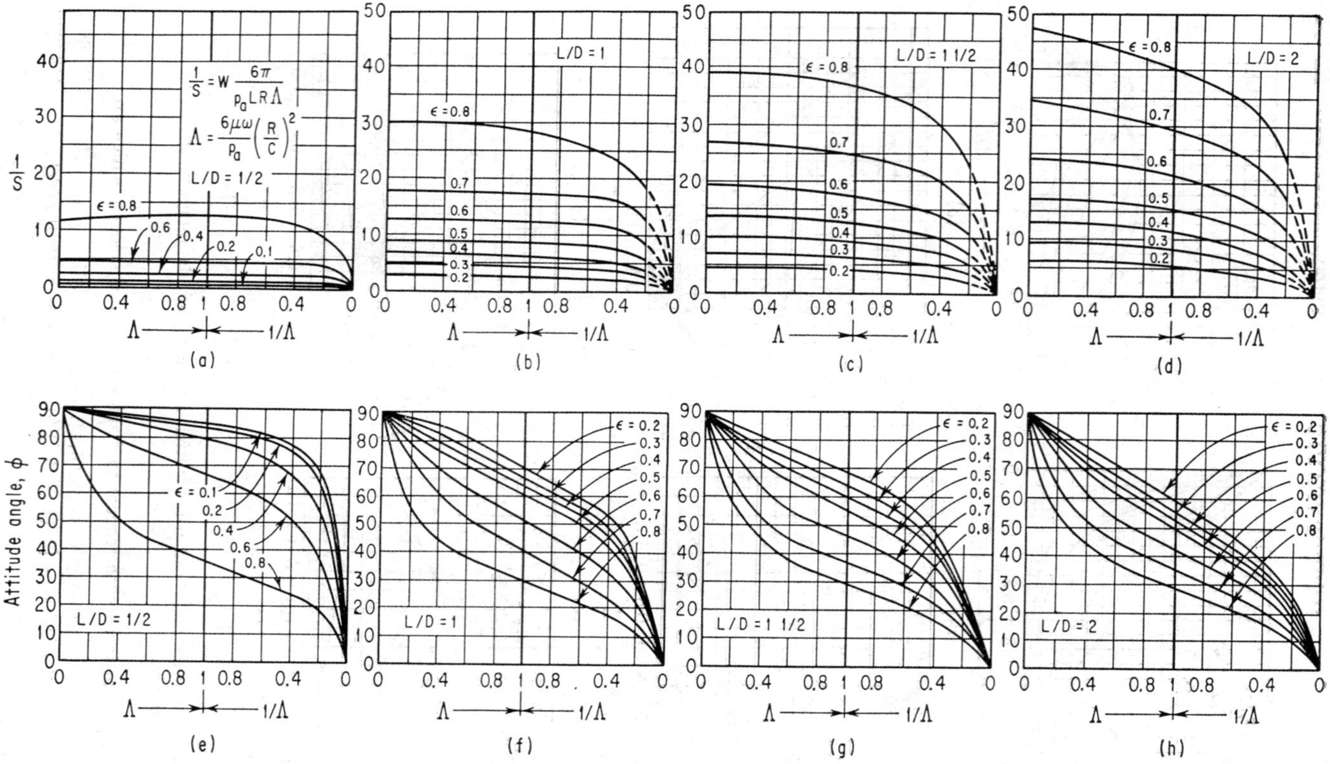

FIG. 12.18. Solutions of dimensionless load and attitude angle for several L/D ratios. (*a*) Dimensionless force vs. bearing number. (*b*) Dimensionless force vs. bearing number. (*c*) Dimensionless force vs. bearing number. (*d*) Dimensionless force vs. bearing number. (*e*) Attitude angle vs. bearing number. (*f*) Attitude angle vs. bearing number. (*g*) Attitude angle vs. bearing number. (*h*) Attitude angle vs. bearing number.

obtained from the first-order perturbation of ϵ and ϵ'. In the absence of whirling, $\Lambda^* = \Lambda$ and the same figures are still applicable. By use of Eqs. (12.68*a*) and (12.68*b*) the other two derivatives with respect to velocity can be obtained.

Using intrinsic time dependence and linearized ph the dynamic Reynolds equation can be solved and the coefficients for stiffness and looping can be obtained.[14] These results are given in Fig. 12.17 and the intrinsic and quasi-static result are compared. These coefficients are required for dynamic analysis of critical speed, synchronous whirl, half-frequency whirl, etc.

Numerical Solution. The Reynolds equation for steady load and isothermal compressible fluid can be solved by the use of the difference equation and iterative procedure[15] in a similar manner to that of Art. 12.3(*b*). Figure 12.18*a* to *h* represents graphically solutions of dimensionless load and attitude angle for several L/D ratios, eccentricity ratios, and bearing number Λ, while Fig. 12.19 gives friction factor vs. Λ for several L/D ratios and eccentricity ratios. Figure 12.20*a* to *d* gives results of radial and tangential stiffness and damping vs. eccentricity ratio for several L/D ratios and bearing numbers. Also see the discussion of stability for stiffness and damping functions.

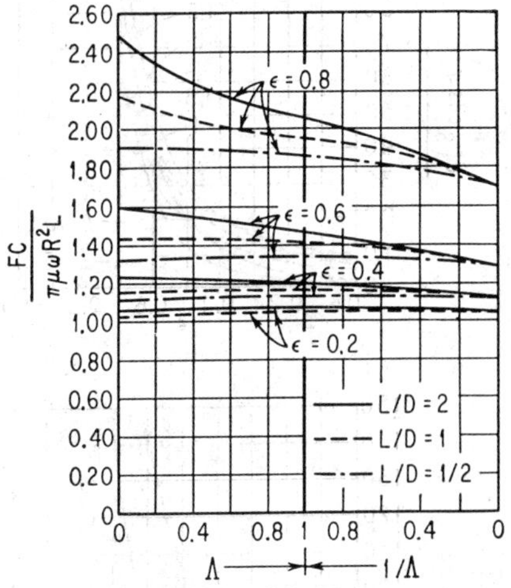

FIG. 12.19. Friction force vs. bearing number.

12.5. HYDROSTATIC BEARINGS

The externally pressurized or hydrostatic bearing, in contrast to the hydrodynamic bearing, demands external fluid pressure to support a load. This system of lubrication offers some distinct advantages not found in hydrodynamic bearings. Among the most important of these characteristics are extreme rigidity, a load capacity independent of velocity, and very small frictional drag.

(a) Plain Journal Bearings[16–18]

Incompressible Lubrication. Laminar Feeding. For the nonrotating bearing the pressure distribution may be expressed by setting $U = 0$ in Eq. (12.7) as

$$(\partial/\partial x)[h^3(\partial p/\partial x)] + (\partial/\partial z)[h^3(\partial p/\partial z)] = 0 \tag{12.69a}$$

or

$$\partial^2 p/\partial z^2 + \partial^2 p/\partial x^2 + (3/h)(\partial h/\partial x)(\partial p/\partial x) = 0 \tag{12.69b}$$

The boundary conditions for this problem are

$$p = p_a \qquad \text{when } z = \pm L/2$$
$$p = p_i(x) \qquad \text{when } z = 0$$

where subscript i represents inlet to the bearing clearance and

$$q_{\text{in}} = 2[-(h^3/12\mu)(\partial p/\partial z)_i] = (\pi a^4 n/8\mu l)(p_s - p_i)$$

where q is the flow per unit length, l is the length and a the radius of the feeding tube, and n the number of feeding tubes per unit length in circumferential direction.

Since $\partial p/\partial x \to 0$ as $\epsilon \to 0$, for small ϵ, $\partial p/\partial x$ is of the same order of magnitude as ϵ. This means that $(3/h)(\partial h/\partial x)(\partial p/\partial x)$ is of the order ϵ^2 for $\epsilon \ll 1.0$, which allows Eq.

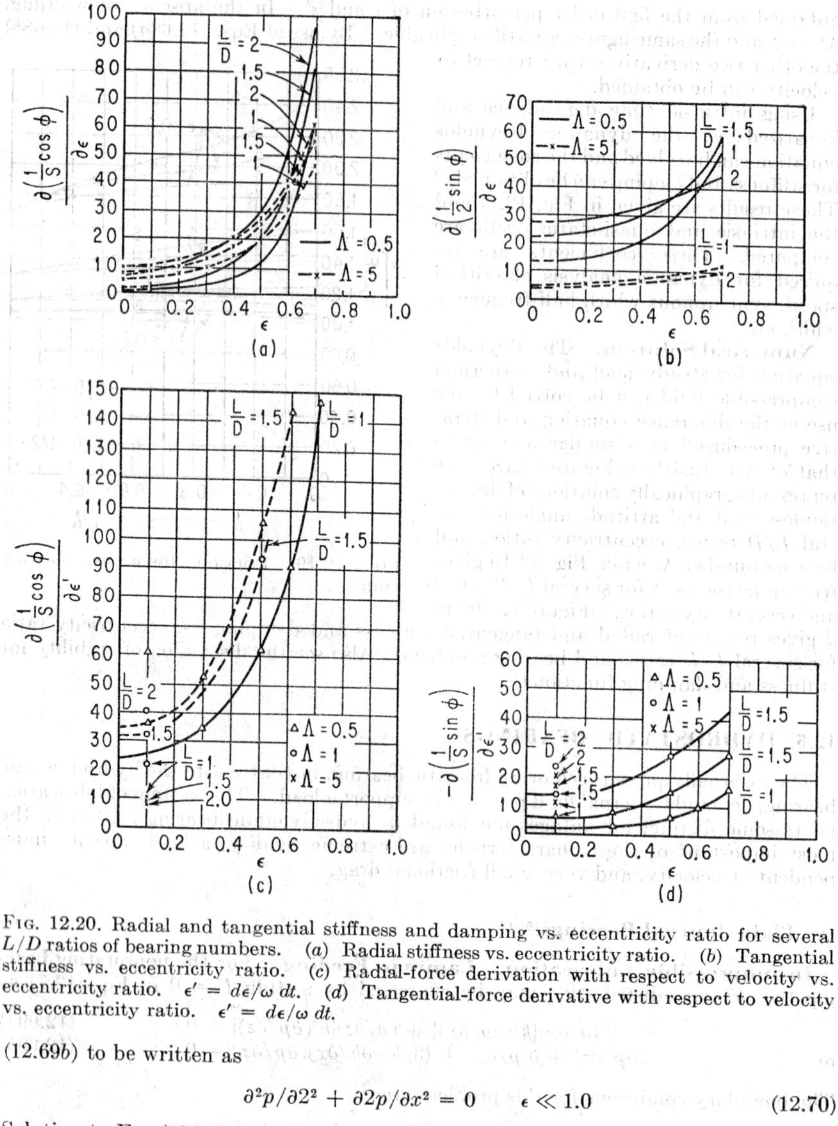

FIG. 12.20. Radial and tangential stiffness and damping vs. eccentricity ratio for several L/D ratios of bearing numbers. (*a*) Radial stiffness vs. eccentricity ratio. (*b*) Tangential stiffness vs. eccentricity ratio. (*c*) Radial-force derivation with respect to velocity vs. eccentricity ratio. $\epsilon' = d\epsilon/\omega\, dt$. (*d*) Tangential-force derivative with respect to velocity vs. eccentricity ratio. $\epsilon' = d\epsilon/\omega\, dt$.

(12.69*b*) to be written as

$$\partial^2 p/\partial 2^2 + \partial 2p/\partial x^2 = 0 \qquad \epsilon \ll 1.0 \tag{12.70}$$

Solution to Eq. (12.70) gives for the average pressure

$$P = \frac{W}{LD} = \frac{3\pi\Lambda_L R^2(p_s - p_a)[\cosh\,(L/2R) - 1]\epsilon}{L(1 + \Lambda_L L/2)[\cosh\,(L/2R) + \Lambda_L R\,\sinh\,(L/2R)]} \tag{12.71}$$

The power required to pump the fluid through the feeding line into the bearing is

$$H = (p_s - p_a)\int_0^{2\pi} q_{\text{in}} R\, d\theta = -(p_s - p_a)\int_0^{2\pi} (h^3/\partial u)(\partial p/\partial 2)_i R\, d\theta$$

$$H_L = \frac{4C^3P^2}{27\pi\mu\epsilon^2}\left(\frac{L}{2R}\right)^2 \frac{(1 + \Lambda_L L/2)\,[\cosh\,(L/2R) + \Lambda_L R\,\sinh\,(L/2R)]^2}{\Lambda_L R[\cosh\,(L/2R) - 1]^2} \tag{12.72}$$

Minimizing H with respect to Λ_L reveals

$$\Lambda_{L\,\text{opt}} = \frac{1}{2L}\left\{\left[1 + \frac{4L/R}{\tanh (L/2R)}\right]^{1/2} - 1\right\}$$

The above equation can be rewritten as

$$\left(\frac{Na^4}{lC^3}\right)_{\text{opt}} = \frac{\left[1 + \dfrac{8L/D}{\tanh (L/D)}\right]^{1/2} - 1}{1.5L/D}$$

By letting the minimum pumping power for an arbitrary L/D ratio be defined as the reduced power, we have

$$H_{\text{red}} = \frac{H_L}{C^3P^2/108\pi\mu\epsilon^2} = \frac{L}{D}\left[3 + \left(1 + \frac{8L/D}{\tanh L/D}\right)^{1/2}\right]$$
$$\frac{\left\{4(L/D)\cosh (L/D) + \left[\left(1 + \dfrac{8L/D}{\tanh L/D}\right)^{1/2} - 1\right]\sinh (L/D)\right\}^2}{\left[\left(1 + \dfrac{8L/D}{\tanh L/D}\right)^{1/2} - 1\right][\cosh (L/D) - 1]^2} \tag{12.73}$$

By taking the derivative of H_{red} with respect to the L/D ratio, we obtain a minimum value H_{red} of 721 at $L/D = 1.1$. At this point Eq. (12.73) yields

$$H_{L\min} = 2.15(C^3P^2/\mu\epsilon^2)$$

The optimum supply pressure for $L/D = 1.1$ is

$$(p_s - p_a)_{\text{opt}} = 2.43(P/\epsilon)$$

while the frictional force is

$$F = -2\pi\mu RLU/C \tag{12.74}$$

Turbulent feeding for turbulent flow in a pipe

$$\Lambda_T = (12\pi a^2\mu n/C^3)\sqrt{a/fl\rho}$$

and in an orifice

$$\Lambda_T = (12\pi a^2 n\mu C_d/C^3)\sqrt{1/2m\rho}$$

Let

$$\Lambda_T\frac{L}{2} = \lambda_T\sqrt{\frac{p_s - p_a}{1 + \lambda_T}}$$

The pressure distribution becomes

$$p = p_a + \frac{\lambda_T(p_s - p_a)}{1 + \lambda_T}\left(1 - \frac{2z}{L}\right) - \frac{3\epsilon\lambda_T(e^{z/R} - e^{(L-z)/R})(p_s - p_a)\cos (x/R)}{2e^{L/D}(1 + \lambda_T)[(L/D)\cosh (L/D) + (\lambda_T/2)\sinh (L/D)]} \tag{12.75}$$

By comparing this result with the result previously obtained for laminar feeding where $\lambda_L = \Lambda_L L/2$, we have

$$p = p_a + \frac{\lambda_L(p_s - p_a)}{1 + \lambda_L}\left(1 - \frac{2z}{L}\right) - \frac{3\epsilon\lambda_L(e^{z/R} - e^{(L-z)/R})(p_s - p_a)\cos (x/R)}{2e^{L/D}(1 + \lambda_L)[(L/D)\cosh (L/D) + \lambda_L\sinh (L/D)]} \tag{12.76}$$

The last expression differs only in form by the coefficient of sinh (L/D), which is $0.5\lambda_T$ for turbulent flow and λ_L for laminar flow. The dimensionless quantities λ_T and λ_L have, however, different values. For laminar flow,

$$\lambda_L = 3\pi a^4 L_n / 8C^3 l$$

For turbulent capillary flow,

$$\lambda_T = \frac{18\pi^2 a^5 L^2 \mu^2 n^2}{flC^6 \rho (p_s - p_a)} \left\{1 + \left[1 + \frac{flC^6 \rho (p_s - p_a)}{9\pi^2 a^5 L^2 \mu^2 n^2}\right]^{1/2}\right\}$$

and for series orifice flow,

$$\lambda_T = \frac{9\pi^2 C_d^2 a^4 L^2 \mu^2 n^2}{mC^6 \rho (p_s - p_a)} \left\{1 + \left[1 + \frac{mC^6 \rho (p_s - p_a)}{4.5\pi^2 a^4 L^2 \mu^2 n^2 C_d^2}\right]^{1/2}\right\}$$

The load-carrying capacity is obtained by integrating Eq. (12.75),

$$P = \frac{3\pi\lambda_T[\cosh (L/D) - 1](p_s - p_a)\epsilon}{2L/D(1 + \lambda_T)[(L/D) \cosh (L/D) + \frac{1}{2}\lambda_T \sinh (L/D)]} \tag{12.77}$$

By defining

$$\bar{P} = P/p_a\epsilon \qquad \bar{p}_s = p_s/p_a$$

we get for turbulent feeding

$$\bar{P} = \frac{3\pi\lambda_T[\cosh (L/D) - 1](\bar{p}_s - 1)}{(2L/D)(1 + \lambda_T)[(L/D) \cosh (L/D) + \frac{1}{2}\lambda_T \sinh (L/D)]} \tag{12.78}$$

and for laminar feeding

$$\bar{P} = \frac{3\pi\lambda_L[\cosh (L/D) - 1](\bar{p}_s - 1)}{(2L/D)(1 + \lambda_L)[(L/D) \cosh (L/D) + \lambda_L \sinh (L/D)]} \tag{12.79}$$

The power necessary to pump the fluid through the bearing is

$$H_T = \frac{\pi C^3 D}{3\mu L}\left(\frac{\lambda_T}{1 + \lambda_T}\right)(p_s - p_a)^2$$

Let the dimensionless power be defined by

$$\bar{H}_T = \frac{H_T \mu}{C^3 p_a^2} = \frac{\pi\lambda_T D(\bar{p}_s - 1)^2}{3L(1 + \lambda_T)}$$

then elimination of $\bar{p}_s - 1$ between Eq. (12.79) and the last expression yields

$$\bar{H}_T = \frac{4}{27\pi} \frac{\bar{P}^2 L(\lambda_T + 1)[(L/D) \cosh (L/D) + \frac{1}{2}\lambda_T \sinh (L/D)]^2}{D\lambda_T[\cosh (L/D) - 1]^2} \tag{12.80}$$

As a comparison the dimensionless power for laminar feeding is

$$\bar{H}_L = \frac{4}{27\pi} \frac{\bar{P}^2 L(\lambda_L + 1)[(L/D) \cosh (L/D) + \lambda_L \sinh (L/D)]^2}{D\lambda_L[\cosh (L/D) - 1]^2} \tag{12.81a}$$

Minimizing $\bar{H}$ with respect to λ, that is, setting $\partial\bar{H}/\partial\lambda = 0$, gives the optimum values for λ:

$$\lambda_{T\,\text{opt}} = \frac{1}{4}\left\{\left[1 + \frac{16L}{D \tanh (L/D)}\right]^{1/2} - 1\right\}$$

$$\lambda_{L\,\text{opt}} = \frac{1}{4}\left\{\left[1 + \frac{8L}{D \tanh (L/D)}\right]^{1/2} - 1\right\}$$

Insertion of these values in Eqs. (12.78) and (12.79) yields

$$\bar{H}_T = \frac{4}{27\pi}\bar{P}^2 \frac{\frac{L}{D}\left\{\left[1 + \frac{16L}{D\tanh(L/D)}\right]^{1/2} + 3\right\}\left(\frac{L}{D}\cosh\frac{L}{D} + \frac{1}{8}\left\{\left[1 + \frac{16L}{D\tanh(L/D)}\right]^{1/2} - 1\right\}\sinh\frac{L}{D}\right)^2}{\left\{\left[1 + \frac{16L}{D\tanh(L/D)}\right]^{1/2} - 1\right\}\left(\cosh\frac{L}{D} - 1\right)^2} \tag{12.81b}$$

$$\bar{H}_L = \frac{4}{27\pi}\bar{P}^2 \frac{\frac{L}{D}\left\{\left[1 + \frac{8L}{D\tanh(L/D)}\right]^{1/2} + 3\right\}\left(\frac{L}{D}\cosh\frac{L}{D} + \frac{1}{4}\left\{\left[1 + \frac{8L}{D\tanh(L/D)}\right]^{1/2} - 1\right\}\sinh\frac{L}{D}\right)^2}{\left\{\left[1 + \frac{8L}{D\tanh(L/D)}\right]^{1/2} - 1\right\}\left(\cosh\frac{L}{D} - 1\right)^2} \tag{12.81c}$$

Evaluation of above equations gives the following optimum parameters:

$$\begin{aligned} (L/D)_{T\,\text{opt}} &= 1.04 & \bar{H}_{T\,\text{min}} &= 1.45\bar{P}^2 \\ (L/D)_{L\,\text{opt}} &= 1.10 & H_{L\,\text{min}} &= 2.15\bar{P}^2 \\ (\lambda_T)_{\text{opt}} &= 0.934 & (\bar{p}_s)_{T\,\text{opt}} &= 1 + 1.725\bar{P} \\ (\lambda_L)_{\text{opt}} &= 0.615 & (\bar{p}_s)_{L\,\text{opt}} &= 1 + 2.43\bar{P} \end{aligned}$$

From this information it can be shown that for turbulent capillary flow

$$(a^5N^2\mu^2/flC^6\rho p_a)_{\text{opt}} = 0.02\bar{P}$$

and for turbulent orifice flow

$$(a^4N^2\mu^2C_d{}^2/mC^6\rho p_a)_{\text{opt}} = 0.04\bar{P}$$

while for laminar flow

$$(a^4N/lC^3)_{\text{opt}} = 1.495$$

Compressible Lubrication. Laminar Feeding. By using the perfect-gas relation for isothermal flow, $p/\rho g$ = constant, and assuming that $V = U_1 = U_2 = 0$ the following may be written:

$$h^3(\partial/\partial z)[p(\partial p/\partial z)] + (\partial/\partial x)[ph^3(\partial p/\partial x)] = 0 \tag{12.82}$$

After expansion of Eq. (12.82), setting $\epsilon(\partial p/\partial x) \approx 0$ and defining a new dependent variable p such that

$$p = \sqrt{\hat{p}}$$

Equation (12.82) becomes

$$\partial^2\hat{p}/\partial x^2 + \partial^2\hat{p}/\partial z^2 = 0 \tag{12.83}$$

which is the same differential equation as for the incompressible case. If we put $\hat{p}_a = p_a{}^2$, the first boundary condition of our problem reads: For $z = L/2$ (from capillary exit to gap) $\hat{p} = \hat{p}_a$. This is again the same condition as for incompressible flow, with $\hat{p}_a$ replaced by $\hat{p}$. Therefore, Eq. (12.76) is a solution to Eq. (12.83) except that p is replaced by $\hat{p}$. Thus

$$\hat{p} = \hat{p}_a + \frac{\lambda_l(\hat{p}_s - \hat{p}_a)}{1 + \lambda_l}\left(1 - \frac{2z}{L}\right) - \frac{3\epsilon\lambda_l(e^{z/R} - e^{(L-z)/R})(\hat{p}_s - \hat{p}_a)\cos(x/R)}{2e^{L/D}(1 + \lambda_l)[(L/D)\cosh(L/D) + \lambda_l \sinh(L/D)]} \tag{12.84}$$

By substitution of the relation between p and $\hat{p}$, Eq. (12.84) may be written (neglecting higher-order terms in ϵ) as

$$p = p_a\left[\frac{1 + \lambda_l\bar{p}_s^2 - \lambda_l(\bar{p}_s^2 - 1)2z/L}{1 + \lambda_l}\right]^{1/2} \left\{1 - \frac{3\lambda_l\epsilon(\bar{p}_s^2 - 1)(e^{z/R} - e^{(L-z)/R})\cos\theta}{4e^{L/D}[1 + \bar{p}_s^2\lambda_l - (\bar{p}_s^2 - 1)\lambda_l 2z/L][(L/D)\cosh(L/D) + \lambda_l\sinh(L/D)]}\right\} \tag{12.85}$$

The load capacity is

$$P = \frac{3\pi\lambda_l\epsilon p_a(\bar{p}_s^2 - 1)}{4L\sqrt{1 + \lambda_l}\, e^{L/D}[(L/D)\cosh(L/D) + \lambda_l\sinh(L/D)]} \int_0^{L/2} \frac{(e^{(L-z)/R} - e^{z/R})\,dz}{[1 + \lambda_l\bar{p}_s^2 - \lambda_l(\bar{p}_s^2 - 1)2z/L]^{1/2}} \tag{12.86}$$

This equation may be integrated by using the transform

$$t = \left\{\frac{(L/D)[1 + \lambda_l\bar{p}_s^2 - \lambda_l(\bar{p}_s^2 - 1)2z/L]}{\lambda_l(\bar{p}_s^2 - 1)}\right\}^{1/2}$$

This integral can be resolved into the Gaussian error integrals

$$\Phi(z) = (2/\sqrt{\pi})\int_0^z e^{-t^2}\,dt \qquad \text{and} \qquad \psi(z) = \int_0^z e^{+t^2}\,dt$$

both of which are tabulated. Thus from Eq. (12.86) we get

$$\bar{P} = \frac{3\pi}{4Y}\left\{\frac{e^{-Y^2}[\psi(\sqrt{Y^2 + L/D}) - \psi(Y)] - (\sqrt{\pi}/2)e^{Y^2}[\Phi(\sqrt{Y^2 + L/D}) - \Phi(Y)]}{L/D\cosh(L/D) + \lambda_l\sinh(L/D)}\right\} \tag{12.87}$$

where

$$Y = \left[\frac{(L/D)(1 + \lambda_l)}{\lambda_l(\bar{p}_s^2 - 1)}\right]^{1/2}$$

Neglecting higher-order terms in ϵ, the laminar pumping power is given by

$$H_l = \frac{\pi\lambda_l p_a^2 C^3(\bar{p}_s^2 - 1)\ln\bar{p}_s}{L/D\; 6\mu(1 + \lambda_l)}$$

or in dimensionless form

$$\bar{H}_l = \frac{\pi}{6}\,\frac{\lambda_l D(\bar{p}_s^2 - 1)\ln\bar{p}_s}{L(1 + \lambda_l)}$$

In terms of Y the above equation reads

$$\bar{H}_l = \frac{\pi}{12Y^2}\ln\left[1 + \frac{(L/D)(1 + \lambda_l)}{\lambda_l Y^2}\right] \tag{12.88}$$

If for brevity we put Eq. (12.87) into the form

$$F(Y,L/D) = 3\pi D/4YL\{e^{-Y^2}[\psi(\sqrt{Y^2+L/D}) - \psi(Y)] - \sqrt{\pi}/2\,e^{Y^2}[\Phi(\sqrt{Y^2+L/D}) - \Phi(Y)]\}$$

then Eq. (12.87) can be written as

$$\bar{P} = \frac{F(Y,L/D)}{\cosh(L/D) + \lambda_l D/L \sinh(L/D)} \tag{12.89}$$

By eliminating λ_l between Eq. (12.88) and the last expression, we obtain

$$\bar{H}_l = \frac{\pi}{12Y^2}\ln\left[\frac{1}{Y^2}\frac{\bar{P}\sinh(L/D)}{F(Y,L/D) - \bar{P}\cosh(L/D)} + \frac{L}{DY^2} + 1\right] \tag{12.90}$$

Figure 12.21*a* shows a plot of $F(Y,L/D)$ against Y for various values of L/D. $(\lambda_l)_{opt}$ can be obtained also from Eq. (12.89), and it takes the following form:

$$(\lambda_l)_{opt} = \left(\frac{L}{D}\right)_{opt}\frac{F[Y_{opt},(L/D)_{opt}] - \bar{P}\cosh(L/D)_{opt}}{\bar{P}\sinh(L/D)_{opt}} \tag{12.91}$$

Finally, $(\bar{p}_s)_{opt}$ can also be obtained from Eq. (12.86), and it yields

$$(\bar{p}_s)_{opt} = \left[1 + \left(\frac{L}{D}\right)_{opt}\frac{1+(\lambda_l)_{opt}}{Y^2_{opt}(\lambda_l)_{opt}}\right]^{1/2}$$

By using the above equation, Fig. 12.21*b* was plotted; it compares the performance characteristics of laminar-fed bearings using compressible and incompressible fluids.

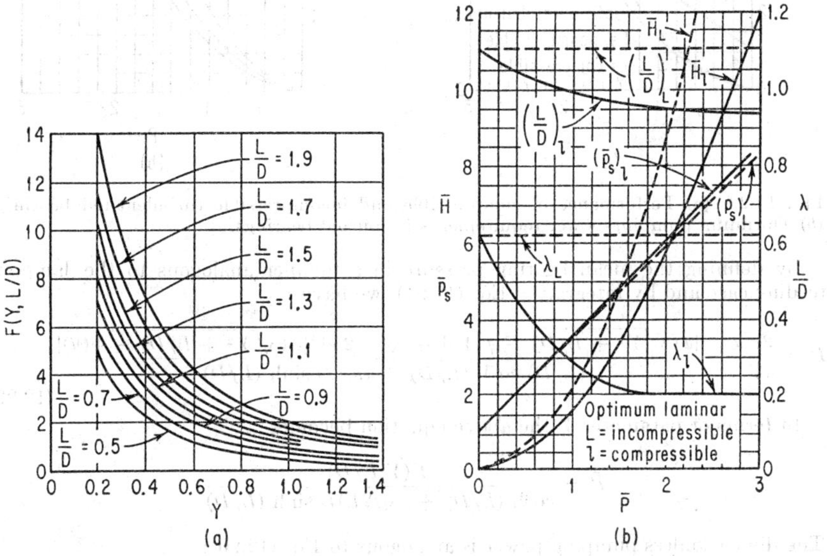

FIG. 12.21. Performance of laminar-fed journal bearings. (*a*) Performance of laminar-fed compressible journal bearings. (*b*) Performance of compressible and incompressible laminar-fed journal bearings.

Turbulent Feeding. The pressure distribution has the same form as in the laminar-feeding case, namely, Eq. (12.85),

$$p = p_a \left[\frac{1 + \lambda_t \bar{p}_s^2 - \lambda_t(\bar{p}_s^2 - 1)2z/L}{1 + \lambda_t}\right]^{1/2}$$

$$\left\{1 - \frac{3\lambda_t \epsilon(\bar{p}_s^2 - 1)(e^{z/R} - e^{(L-z)/R})\cos\theta}{4e^{L/D}[1 + \bar{p}_s^2\lambda_t - (\bar{p}_s^2 - 1)\lambda_t 2z/L][(L/D)\cosh(L/D) + (\lambda_t/2)\sinh(L/D)]}\right\}$$

where

$$\Lambda_t = \frac{12\mu\pi C_d a^2 n}{C^3}\left(\frac{Kg}{m}\right)^{1/2} = \frac{2\lambda_t}{L}\sqrt{\frac{\hat{p}_s - \hat{p}_a}{1 + \lambda.}}$$

Subscript t stands for turbulence in compressible fluids and $K = \mathcal{R}T$.

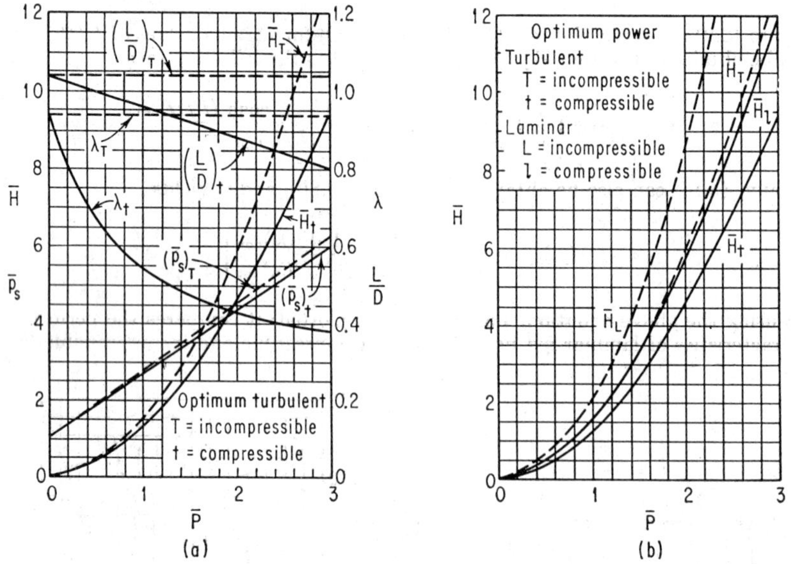

FIG. 12.22. (*a*) Performance of compressible and incompressible turbulent-fed bearings. (*b*) Optimum pumping-power requirements in journal bearings.

By defining the mean bearing pressure in a manner analogous to the laminar-feeding case and by integrating Eq. (12.91), we have

$$\bar{P} = \frac{3\pi}{4Y}\,\frac{e^{-Y^2}[\psi(\sqrt{Y^2 + L/D}) - \psi(Y)] - \sqrt{\pi/2}\,e^{Y^2}[\phi(\sqrt{Y^2 + L/D}) - \Phi(y)]}{(L/D)\cosh(L/D) + (\lambda_t/2)\sinh(L/D)} \tag{12.92}$$

In terms of parameter Y the above equation becomes

$$\bar{P} = \frac{F(Y,L/D)}{\cosh(L/D) + (\lambda_t/2L)D\sinh(L/D)}$$

The dimensionless pumping power is analogous to Eq. (12.90):

$$\bar{H}_t = \frac{\pi}{12Y^2}\ln\left[\frac{1}{2Y^2}\,\frac{\bar{P}\sinh(L/D)}{F(Y,L/D) - \bar{P}\cosh(L/D)} = \frac{L}{DY^2} + 1\right] \tag{12.93}$$

For a given value of the quantity $\bar{P}$ we can find a minimum for $\bar{H}_t$ with respect to the variables Y and L/D. In this way the values of $(L/D)_{opt}$ and Y_{opt} are obtained as functions of $\bar{P}$; the related values of $\bar{H}_{min}$ are also obtained. $(\lambda_t)_{opt}$ then follows from Eq. (12.92):

$$(\lambda_t)_{opt} = 2\left(\frac{L}{D}\right)_{opt} \frac{F(Y_{opt},(L/D)_{opt}) - \bar{P}\cosh (L/D)_{opt}}{\bar{P}\sinh (L/D)_{opt}}$$

$$(\bar{p}_s)_{opt} = \left[1 + \left(\frac{L}{D}\right)_{opt} \frac{1 + (\lambda_t)_{opt}}{Y^2{}_{opt}(\lambda_t)_{opt}}\right]^{1/2}$$

Figure 12.22*a* compares the bearing performance characteristics of the compressible and incompressible turbulent-fed bearing; Fig. 12.22*b* compares dimensionless pump-

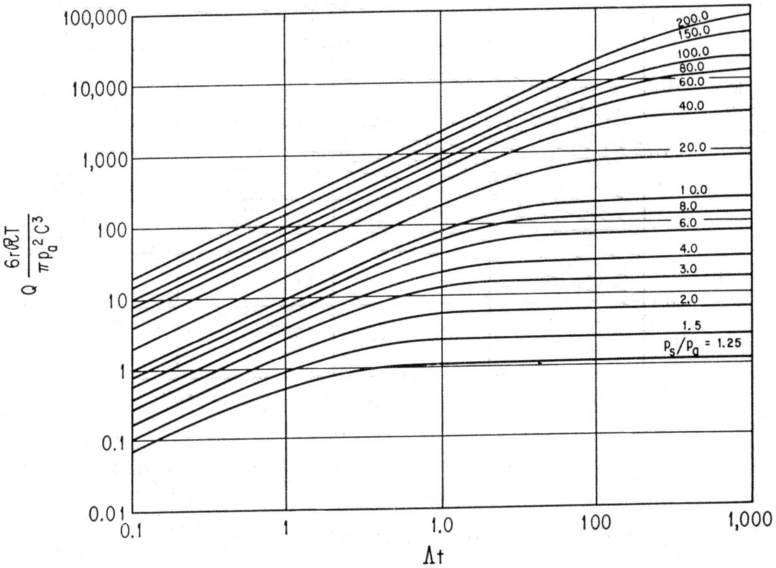

FIG. 12.23. Dimensionless flow vs. restrictor characteristics, $L/D = \frac{1}{2}$.

$$\Lambda_t = \frac{6\mu n C_d A^2}{p_a C^3}\sqrt{\frac{R_g T}{1+\delta}}$$

μ = viscosity, lb-sec/in.^{2}
n = number of feeding holes around circumference
C_d = orifice coefficient $\begin{pmatrix}\sim 0.6 \text{ orifice compensation} \\ \sim 0.7 \text{ inherent compensation}\end{pmatrix}$
a = orifice radius, in.
p_a = ambient pressure, psia
C = radial clearance
R = gas content, in./°R
g = acceleration, in./sec^2
T = total temperature, °R
δ = a^2/dC
d = diameter of feeding hole
Q = flow, lb/sec

ing power for the various cases studied. The friction force is similar for all four cases and is expressed by Eq. (12.74).

Design curves for mass flow and load-carrying capacity for hydrostatic plain cylindrical journal bearings with one row of feeders are given in Figs. 12.23 to 12.30. The flow through the feeders is assumed to be turbulent in these design charts.

(b) Step Thrust Bearing—Isothermal Operation

Compressible Lubrication. In a simple hydrostatic bearing, the gas is introduced at the center of the pad and is distributed out at a radius R_0 by means of grooves

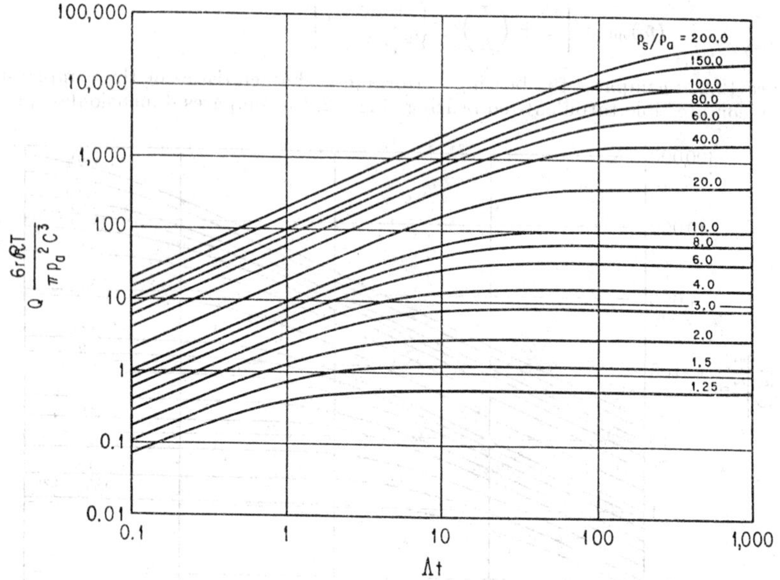

FIG. 12.24. Dimensionless flow vs. restrictor characteristics, $L/D = 1$.

or reliefs. The pressure distribution is given by

$$p = p_0 \left[1 - \frac{1 - (p_a/p_0)^2}{\ln (R/R_0)} \left(\ln \frac{r}{R_0} \right) \right]^{1/2} \tag{12.94}$$

and integration by parts gives for the load capacity in terms of error function

$$W = \pi p_0 R_0^2 \left[1 - \frac{p_a}{p_0} \frac{R^2}{R_0^2} + \frac{e^{2/C_1} \sqrt{2C_1}}{2} \left(xe^{-x^2} - \int e^{-x^2}\, dx \right)_\alpha^\beta \right] \tag{12.95}$$

Evaluation of Eq. (12.94) from R_0 to R indicates that the pressure is nearly linear. Thus it is possible to simplify Eq. (12.95) by assuming linearity and integrating the pressure over the area over which it acts.[20] We then have

$$p = p_0 - p_a \qquad 0 \le r \le R_0$$

$$p = (p_0 - p_a) \frac{R - r}{R - R_0} \qquad R_0 \le r \le R$$

and

$$W = \frac{\pi}{3} (p_0 - p_a) \frac{R^3 - R_0^3}{R - R_0} \tag{12.96}$$

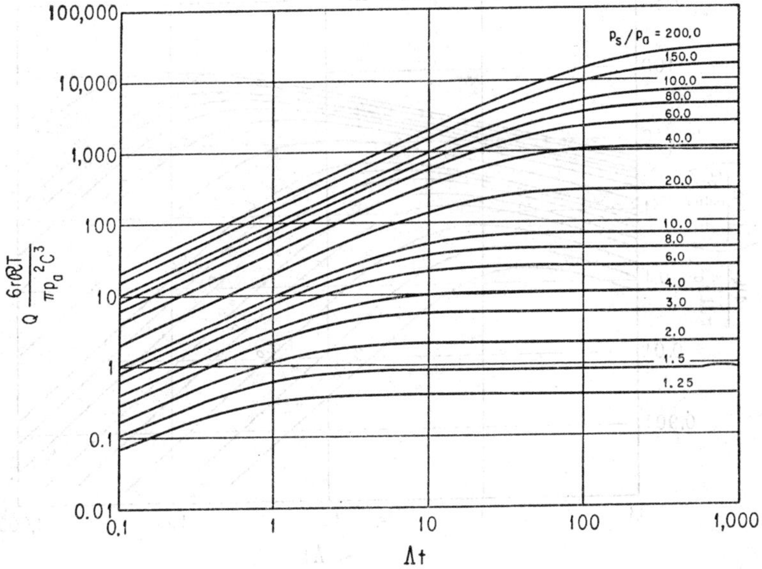

FIG. 12.25. Dimensionless flow vs. restrictor characteristics, $L/D = 1\frac{1}{2}$.

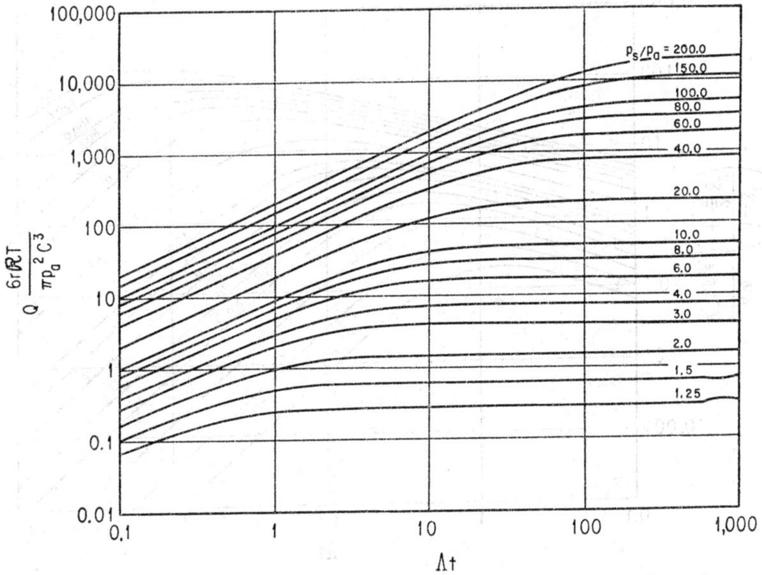

FIG. 12.26. Dimensionless flow vs. restrictor characteristics, $L/D = 2$.

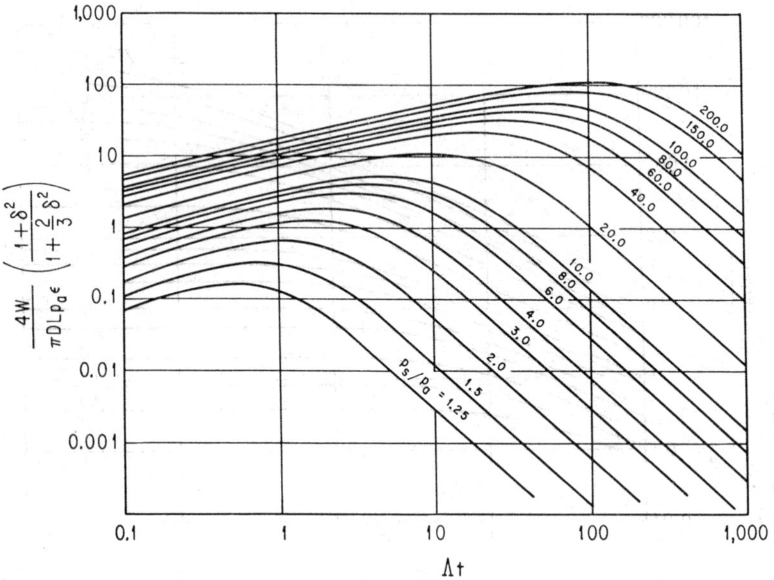

FIG. 12.27. Dimensionless flow vs. restrictor characteristics, $L/D = \frac{1}{2}$.

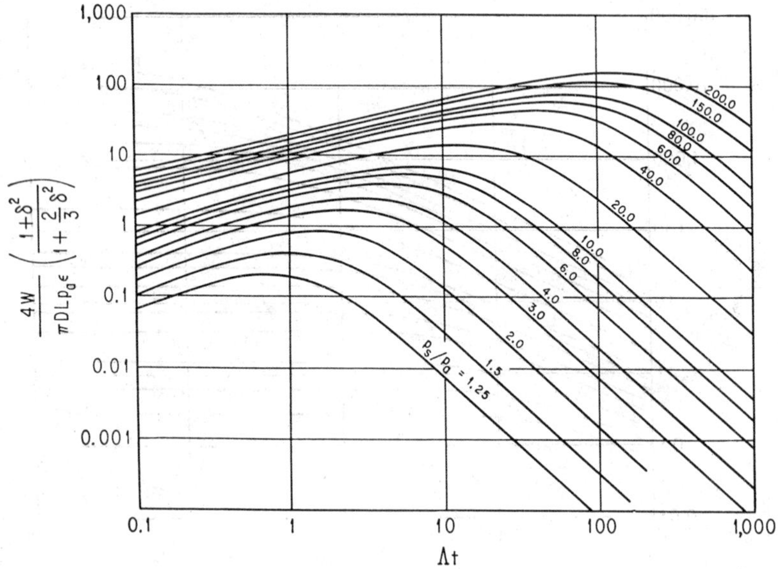

FIG. 12.28. Dimensionless flow vs. restrictor characteristics, $L/D = 1$.

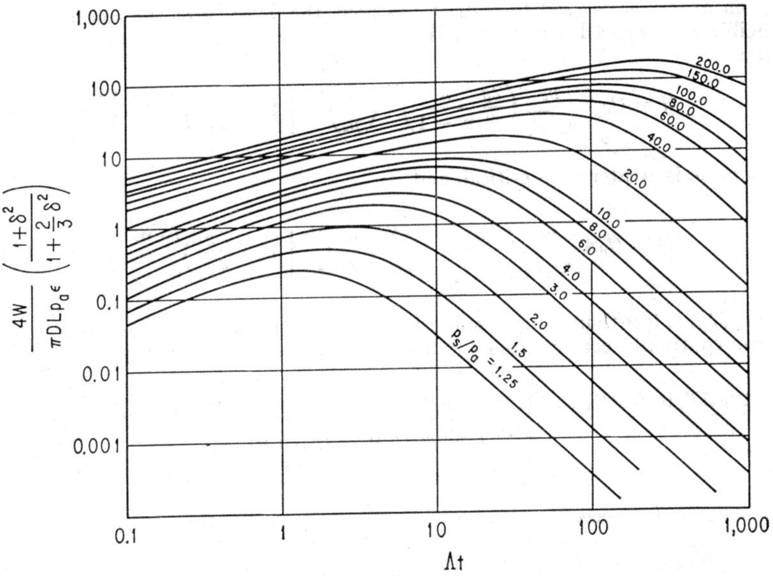

FIG. 12.29. Dimensionless flow vs. restrictor characteristics, $L/D = 1\frac{1}{2}$.

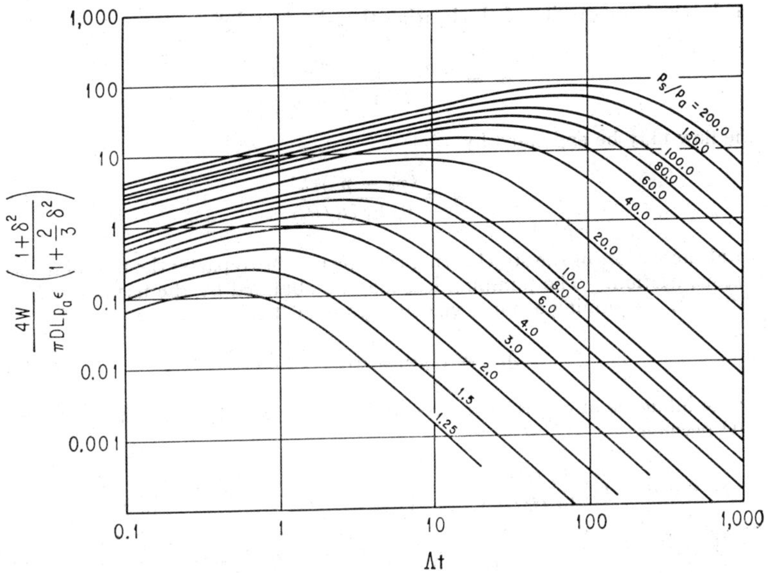

FIG. 12.30. Dimensionless flow vs. restrictor characteristics, $L/D = 2$.

When an orifice is used to restrict the flow, the velocity of the gas upstream is negligible as compared with that at the throat of the orifice. The following relation is true:

$$G = C_d A \frac{p_s}{\sqrt{T_s}} \left(\frac{2g}{\mathcal{R}} \frac{n}{n-1}\right)^{1/2} \left(\frac{p_0}{p_s}\right)^{1/n} \left[1 - \left(\frac{p_0}{p_s}\right)^{(n-1)/n}\right]^{1/2} \tag{12.97a}$$

where p_0/p_s is greater than critical, and

$$G = C_d A \frac{p_s}{\sqrt{T_s}} \left(\frac{2g}{\mathcal{R}} \frac{n}{n-1}\right)^{1/2} \left(\frac{p_0}{p_s}\right)_c^{1/n} \left[1 - \left(\frac{p_0}{p_s}\right)_c^{(n-1)/n}\right]^{1/2} \tag{12.97b}$$

where p_0/p_s is less than critical; subscript c refers to critical pressure ratio.

From continuity considerations we get

$$h^3 = \frac{12\mu \mathcal{R} T_s \ln (R/R_0) C_d A \{(2g/\mathcal{R})[n/(n-1)]\}^{1/2} (p_0/p_s)^{1/n} [1 - (p_0/p_s)^{(n-1)/n}]^{1/2}}{\pi \sqrt{T_s} p_s [(p_0/p_s)^2 - (p_a/p_s)^2]} \tag{12.98}$$

Design curves for mass flow, load, and bearing stiffness for the case of strictly radial flow (Fig. 12.31) are given in Figs. 12.32 to 12.40.[19] Note that these curves are very similar to those given for the longitudinal flow with parameters somewhat modified.

Incompressible Lubrication. For viscous fluid flow the pressure distribution may be expressed by

$$p - p_a = (p_0 - p_a) \frac{\ln (R/r)}{\ln (R/R_0)} \tag{12.99}$$

The load-carrying capacity of the bearing then is given by

$$W = \frac{\pi (p_0 - p_a)(R^2 - R_0^2)}{2 \ln (R/R_0)} \tag{12.100}$$

and the flow may be expressed by

$$Q = \frac{(p_0 - p_a)\pi h^3}{6\mu \ln (R/R_0)} \tag{12.101}$$

The total energy loss of the step bearing is made up of two parts: the pumping loss and the viscous friction. The pumping loss may be obtained by the use of Eqs. (12.99) and (12.101). Assuming a linear velocity gradient, the viscous torque M^* is

$$M^* = (\pi \mu \omega / 2h)(R^4 - R_0^4) \tag{12.102}$$

Under the influence of centrifugal force and squeeze film, Eqs. (12.99), (12.100), and (12.101) become

$$p - p_a = (p_0 - p_a) \frac{\ln (R/\nu)}{\ln (R/R_0)} + C_1 \left[(\nu^2 - R^2) + (R^2 - R_0^2) \frac{\ln R/\nu}{\ln R/R_0} \right] \tag{12.103}$$

$$W = \frac{\pi}{2} (R^2 - R_0^2) \left\{ \frac{p_0 - p_a}{\ln (R/R_0)} - C_1 \left[(R^2 + R_0^2) - \frac{R^2 - R_0^2}{\ln R/R_0} \right] \right\} \tag{12.104}$$

$$Q = \frac{\pi h^3}{6\mu \ln (R/R_0)} \left[(p_0 - p_a) + C_1 (R^2 - R_0^2) - \frac{6\mu \nu^2 h \ln (R/R_0)}{h^3} \right] \tag{12.105}$$

where

$$C_1 = \tfrac{3}{2}(\rho \omega^2 / g + 2\mu \dot{h} / h^3)$$

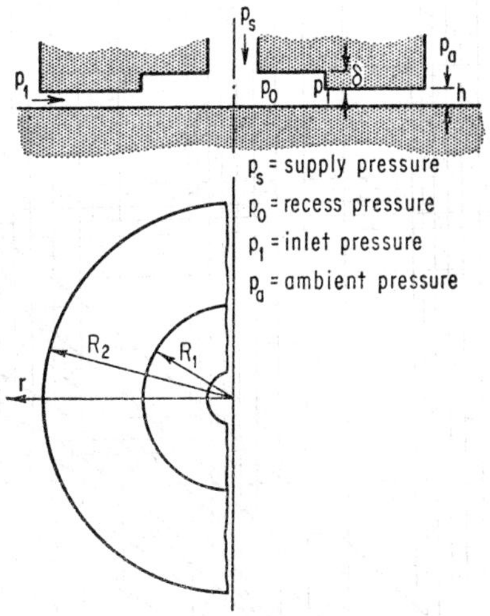

FIG. 12.31. Thrust bearing with a recess.

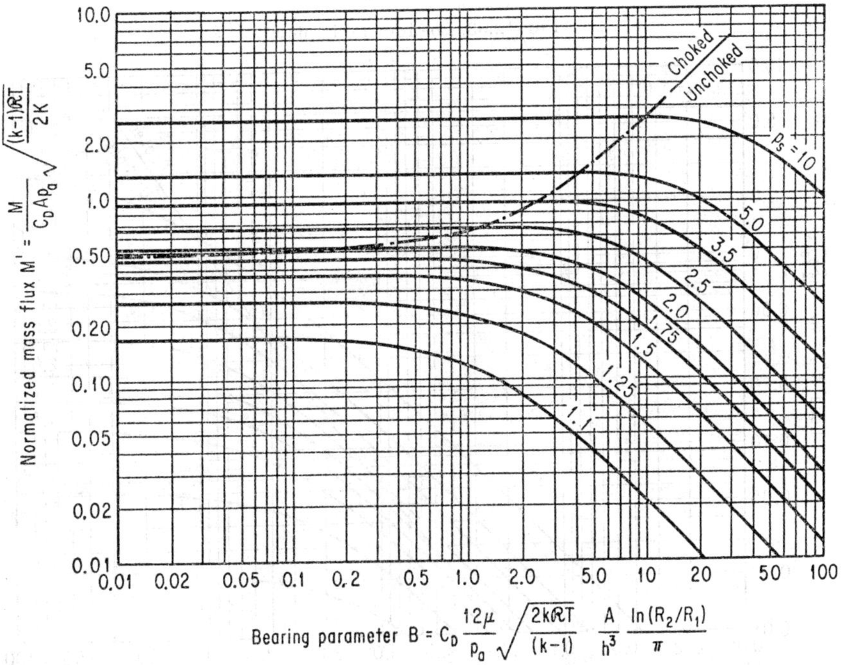

FIG. 12.32. Flow rate.

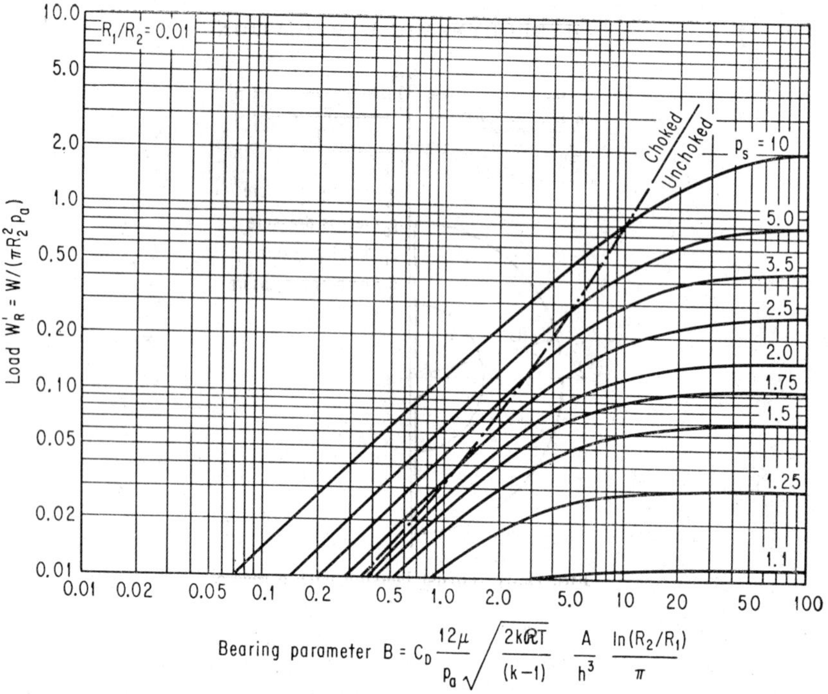

FIG. 12.33. Load for radial flow.

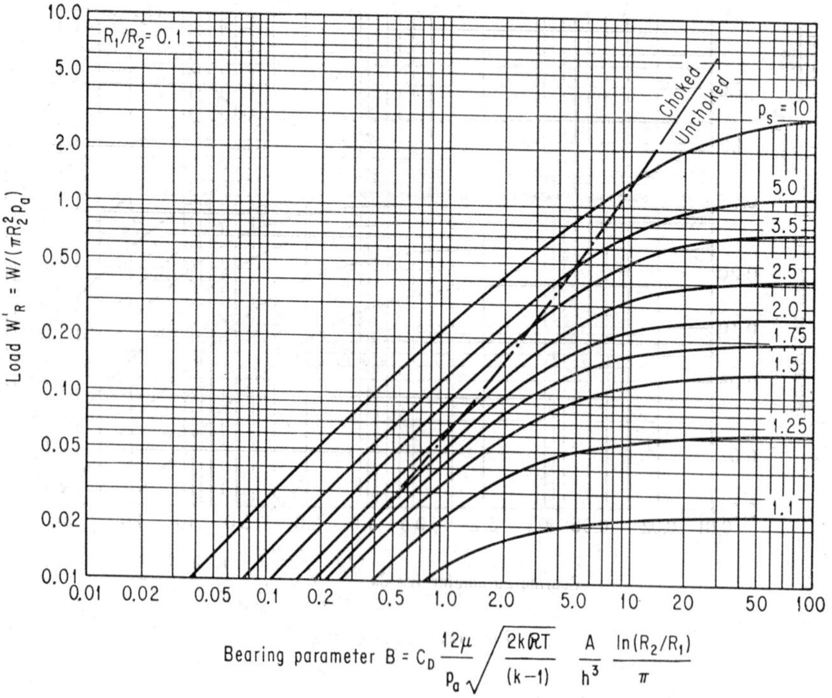

FIG. 12.34. Load for radial flow.

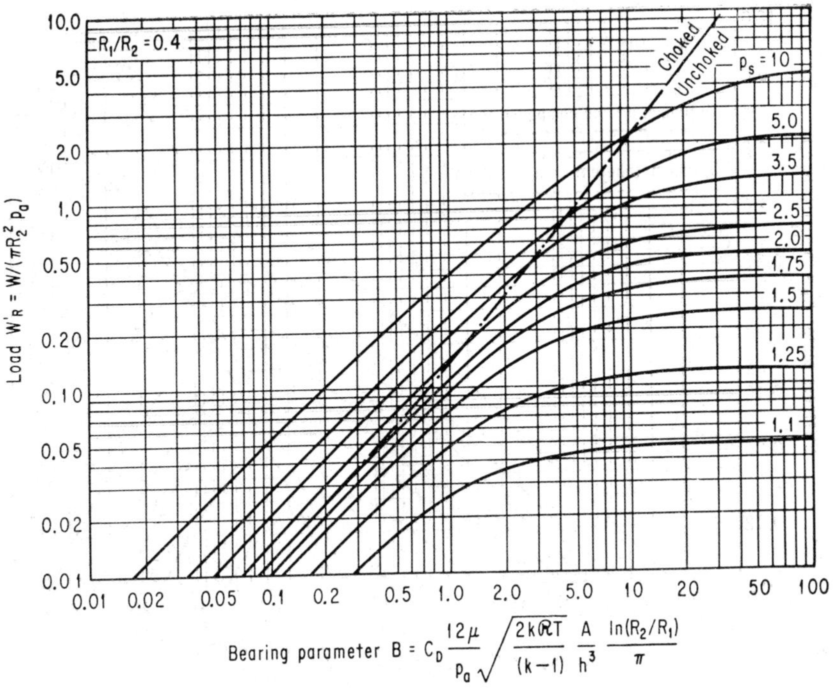

FIG. 12.35. Load for radial flow.

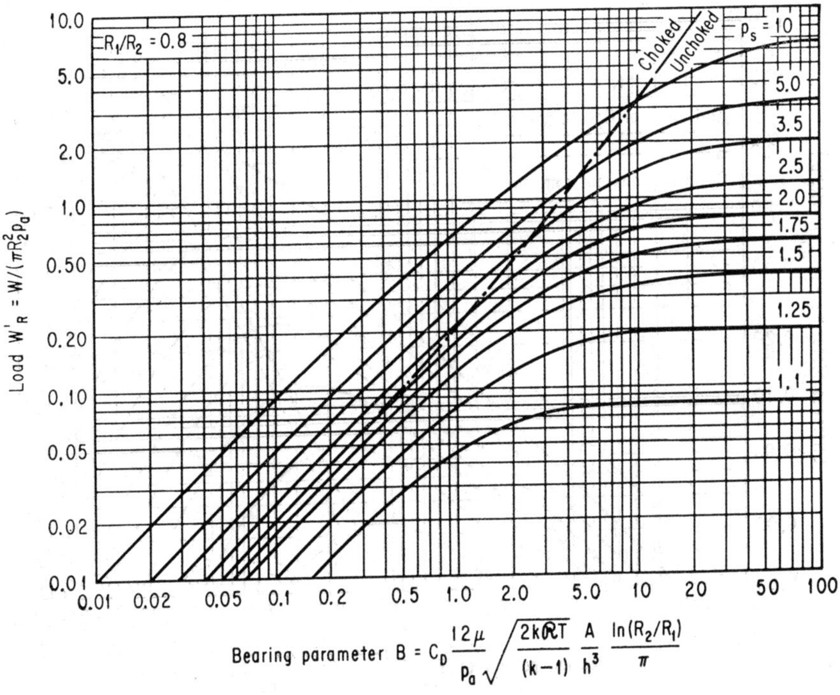

FIG. 12.36. Load for radial flow.

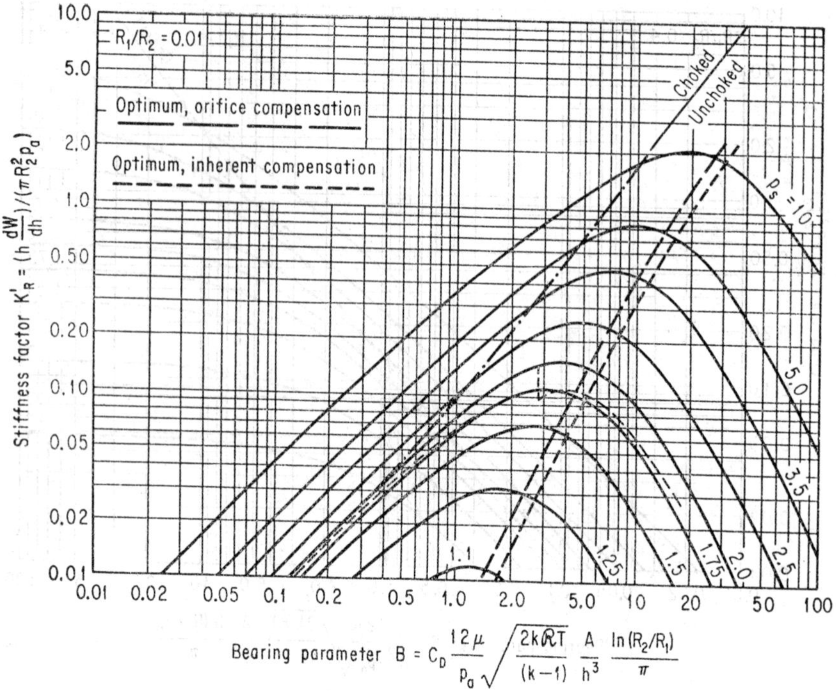

FIG. 12.37. Stiffness factor for radial flow.

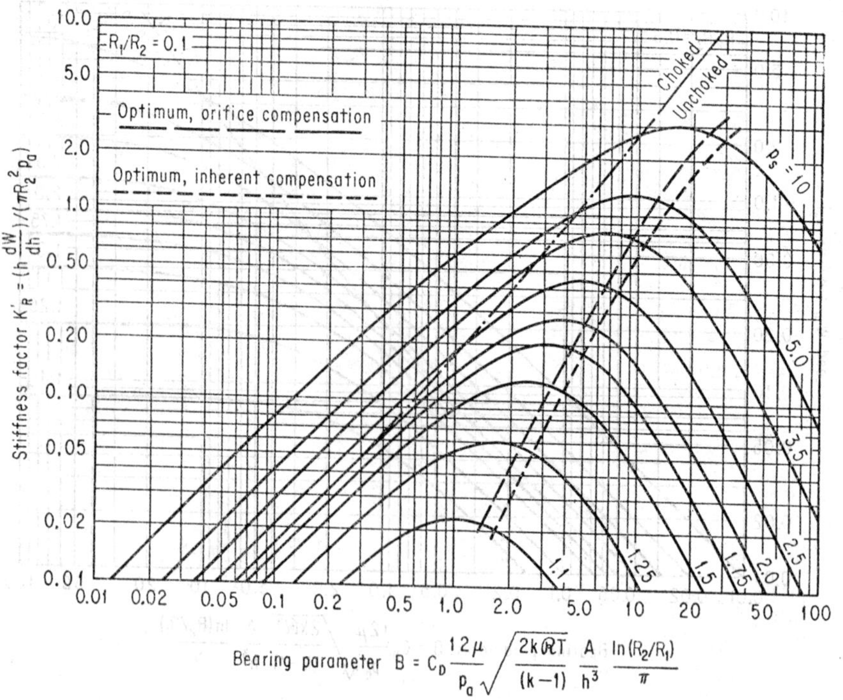

FIG. 12.38. Stiffness factor for radial flow.

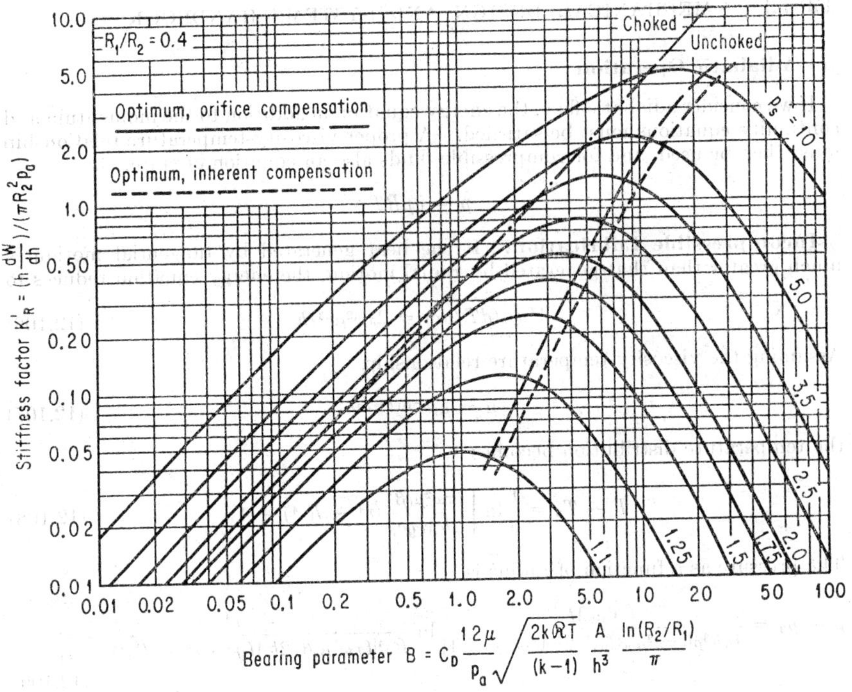

FIG. 12.39. Stiffness factor for radial flow.

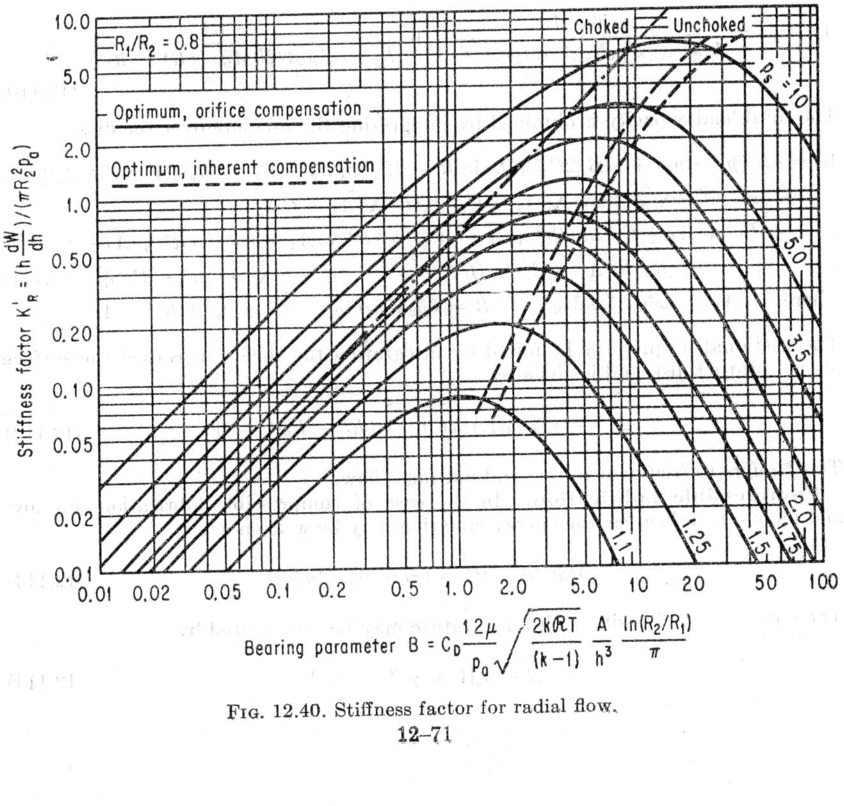

FIG. 12.40. Stiffness factor for radial flow.

(c) Adiabatic Operation

If we consider adiabatic flow, the energy equation in addition to the momentum and continuity equations must be satisfied. A proper viscosity-temperature relationship must then be used, and for compressible fluids also an equation of state.

$$p = g\rho RT$$

Incompressible Lubrication. If the heat generated by tangential motion is much greater than that generated by radial motion, the energy equation reduces to

$$Mgc_v(dT/dr) = 2\pi\omega^2\mu r^3/h \tag{12.106}$$

Assuming the viscosity-temperature relationship

$$\mu = \mu_0 e^{-\beta(T-T_0)} \tag{12.107}$$

the temperature distribution becomes

$$T - T_0 = \frac{1}{\beta}\ln\left[\frac{\pi\omega^2\mu_0\beta}{2hMgc_v}(r^4 - R_0^4) + 1\right] \tag{12.108}$$

The pressure as a function of radius is

$$p - p_0 = \frac{6\mu_0 M}{4\pi h^3\rho[(\pi\omega^2\mu_0\beta R_0^4/2hMgc_v) - 1]}\ln\frac{r^4}{R_0^4[(\pi\omega^2\mu_0\beta/2hMgc_v)(r^4 - R_0^4) + 1]} \tag{12.109}$$

By using the boundary conditions $p = p_a$ at $r = R$, M can be evaluated

$$p_a - p_0 = \frac{6\mu_0 M}{4\pi h^3\rho[(\pi\omega^2\mu_0 R_0^4/2hMgc_v) - 1]}\ln\frac{R^4}{R_0^4[(\pi\omega^2\mu_0\beta/2hMgc_v)(R^4 - R_0^4) + 1]} \tag{12.110}$$

The total load capacity is obtained by integrating the pressure over the disk:

$$\begin{aligned} W = {} & \pi R^2(p_0 - p_a) + (\pi B/2C_1^2)[2R^2\ln R - 2R_0^2\ln R_0 - (R^2 - R_0^2)(1 + 2\ln R_0)] \\ & - (\pi B/4C_1^2\sqrt{A_1})[(R^2\sqrt{A_1} + C_1)\ln(R^2\sqrt{A_1} + C_1) \\ & - (R_0^2\sqrt{A_1} + C_1)\ln(R_0^2\sqrt{A_1} + C_1) + (R^2\sqrt{A_1} - C_1)\ln(R^2\sqrt{A_1} - C_1) \\ & - (R_0^2\sqrt{A_1} - C_1)\ln(R_0^2\sqrt{A_1} - C_1) - 2\sqrt{A_1}(R^2 - R_0^2)] \end{aligned} \tag{12.111}$$

where $A_1 = \pi\omega^2\mu_0\beta/2hMgc_v$ $\quad B = 6\mu_0 M/\pi h^3\rho$ $\quad C_1 = \sqrt{A_1R_0^4 - 1}$

The frictional torque may be found by integrating the shear stress over the surface of the rotating disk and is given by

$$M^* = (2\pi\omega\mu_0/A_1h)\ln[A_1(R^4 - R_0^4) + 1] \tag{12.112}$$

The * refers to frictional torque and not mass flow.

Compressible Lubrication. In the case of compressible lubrication, for low and moderate velocities, the energy equation may be written as

$$Mgc_v(dT/dr) = (\pi h^3 r/6\mu)(dp/dr)^2 \tag{12.113}$$

The variation of viscosity with temperature may be represented by

$$\mu = \mu_0[1 + \gamma(T - T_0)] \tag{12.114}$$

By assuming that the pressure distribution under adiabatic conditions is similar to the isothermal case and varies nearly linearly with radius, we obtain

$$\frac{dp}{dr} = \frac{p_0 - p_a}{R - R_0} = K$$

By combining Eqs. (12.113) with the above equation and integrating we obtain

$$T - T_0 = (1/\gamma)\{[1 + (\pi h^3 K^2/6\mu_0 M g c_v)\gamma(r^2 - R_0^2)]^{1/2} - 1\} \tag{12.115}$$

By rewriting Eq. (12.114) and substituting the new equation for μ and integrating, we obtain a more accurate expression for pressure distribution than by assuming that it is isothermal.

If $\gamma D_1 R_0^2 < 1$ the pressure distribution is given by

$$\begin{aligned}\frac{p_0^2 - p^2}{2E} &= (1 - \gamma D_1 R_0^2) \ln \frac{r}{R_0} + \frac{\gamma D_1}{2}(r^2 - R_0^2) \\ &\quad - (\gamma T_0 - 1)\left\{1 - \sqrt{1 + \gamma D_1(r^2 - R_0^2)}\right. \\ &\quad \left. + \sqrt{1 - \gamma D_1 R_0^2} \ln \frac{R_0[\sqrt{1 - \gamma D_1 R_0^2} + \sqrt{1 + \gamma D_1(r^2 + R_0^2)}]}{r(\sqrt{1 - \gamma D_1 R_0^2} + 1)}\right\}\end{aligned} \tag{12.116a}$$

If $\gamma D_1 R_0^2 > 1$,

$$\begin{aligned}\frac{p_0^2 - p^2}{2E} &= (1 - \gamma D_1 R_0^2) \ln \frac{r}{R_0} + \frac{\gamma D_1}{2}(r^2 + R_0^2) \\ &\quad - (\gamma T_0 - 1)\left[1 - \sqrt{1 + \gamma D_1(r^2 - R_0^2)} + \sqrt{\gamma D_1 R_0^2 - 1}\right. \\ &\quad \left. \sec^{-1} r\sqrt{\frac{\gamma D_1}{\gamma D_1 R_0^2 - 1}} - \sqrt{\gamma D_1 R_0^2 - 1}\,\sec^{-1} R_0\sqrt{\frac{\gamma D_1}{\gamma D_1 R_0^2 - 1}}\right]\end{aligned} \tag{12.116b}$$

where

$$E = 6\mathcal{R} M g \mu_0/\pi h^3 \gamma$$
$$D_1 = \pi h^3 K^2/6\mu_0 M g c_v$$

The frictional torque is given by

$$\begin{aligned}M^* = [2\pi\mu_0\omega/15h(\gamma D_1)^2]\{[\gamma D_1(3R^2 + 2R_0^2) - 2][1 + \gamma D_1(R^2 - R_0^2)]^{3/2} \\ - (5D_1\gamma R_0^2 - 2)\}\end{aligned} \tag{12.117}$$

12.6. SQUEEZE FILM AND DYNAMIC LOADING

(a) Self-excited Vibrations in Gas-lubricated Step Thrust Bearings

This discussion deals with the problem of self-excited vibration in step thrust bearings using compressible fluids.[22,23] The stability analysis is based on a number of simplifying assumptions:

1. At equilibrium the recess area is subjected to a uniform pressure p_0 and the pressure drop from the edge of the recess to the bearing periphery is linear.
2. For small deviations from the equilibrium point, this type of pressure distribution is preserved.
3. Changes in gas density are due primarily to pressure variations; therefore, the relationship $p/\rho = g\mathcal{R}T_0$ is considered valid.
4. External damping may be neglected.
5. The motion is purely vertical.

Based on the assumptions made, the equation of motion using linear pressure gradients is

$$m_1\ddot{h} = p\pi\left[R^2 - \frac{2R^3 - R_0^3 - 3R^2R_0}{3(R - R_0)}\right] = p\pi R_e^2 = pA_e \tag{12.118}$$

where

$$R_e^2 = R^2 - \frac{2R^3 + R_0^3 - 3R^2R_0}{3(R - R_0)}$$

$$A_e = \pi R_e^2$$

m_1 = mass of the upper plate

For small deviations from the equilibrium point (p and H) there are corresponding variations in inflow and outflow which, to a first degree of approximation, can be written respectively as

$$\begin{aligned} M_{11} &= (dM_1/dp)_0 p = -\alpha p \\ M_{22} &= (\partial M_2/\partial p)_0 p + (\partial M_2/\partial h)_0 H = \beta p + \theta H \end{aligned} \tag{12.119}$$

where M is rate of mass and $H = h - h_0$ variation in film thickness. The time rate of change of the bearing gas mass content then becomes

$$M = M_{11} - M_{22} = -(\alpha + \beta)p - \theta H \tag{12.120}$$

where α, β, and θ are all positive.

The gas mass m_2 contained between the bearing surfaces is

$$m_2 = (1/gRT_0)[hp_rA_e + \Delta p_r\pi R_0^2 + hp_a(\pi R^2 - A_e)] \tag{12.121}$$

where Δ is the recess depth. The time rate of change of the bearing gas content $\dot{M}_2$ is evidently equal to the difference between inflow and outflow M and corresponds to the time rates of small deviations from the equilibrium point ($\dot{p}$ and $\dot{H}$).

$$\dot{m}_2 = (\partial m_2/\partial p)_0\dot{p} + (\partial m_2/\partial h)_0\dot{H} = q\dot{p} + s\dot{H} \tag{12.122}$$

where, by differentiation of Eq. (12.121),

$$\begin{aligned} q &= \left(\frac{\partial m_2}{\partial p}\right)_0 = \frac{A_eH_0 + \Delta\pi R_0^2}{g\mathcal{R}T_0} \\ s &= \left(\frac{\partial m_2}{\partial h}\right)_0 = \frac{A_e(p_0 - p_a) + \pi R^2 p_a}{g\mathcal{R}T_0} \\ \frac{q}{s} &= \frac{A_eh_0 + \Delta\pi R_0^2}{A_e(p_0 - p_a) + \pi R^2p_a} \end{aligned} \tag{12.123}$$

From Eqs. (12.120) and (12.122) we have

$$q\dot{p} + s\dot{H} + (\alpha + \beta)p + \theta H = 0 \tag{12.124}$$

and from Eq. (12.118),

$$\begin{aligned} p &= (m_1/A_e)\ddot{H} \\ \dot{p} &= (m_1/A_e)\dddot{H} \end{aligned} \tag{12.125}$$

By elimination of p and $\dot{p}$ between Eqs. (12.124) and (12.125) the following differential equation is obtained:

$$\dddot{H} + \frac{\alpha + \beta}{q}\ddot{H} + \frac{SA_e}{m_1q}\dot{H} + \frac{\theta A_e}{m_1q}H = 0 \tag{12.126}$$

Equation (12.126) is of the form

$$\dddot{H} + C_2\ddot{H} + C_1\dot{H} + C_0H = 0 \tag{12.127}$$

where all coefficients C are positive. Applying Routh's stability criteria to Eq. (12.127), the following inequality must be satisfied in order to achieve stability:

$$C_1C_2 > C_0 \tag{12.128}$$

Therefore, a stability criterion is given by the inequality

$$\frac{\alpha + \beta}{\theta} > \frac{q}{s} \tag{12.129}$$

A study of Fig. 12.41 indicates that a high ratio of $\alpha + \beta/\theta$ corresponds to large values of the recess pressure p_0 and small values of the annulus height h_0 (where the subscript 0 pertains to equilibrium conditions). For a given supply pressure p_s, a favorable condition results if the maximum possible load is being supported within the safety limits of a minimum annular height h_0. Under those conditions the ratio of α/θ has a large value, though β is small.

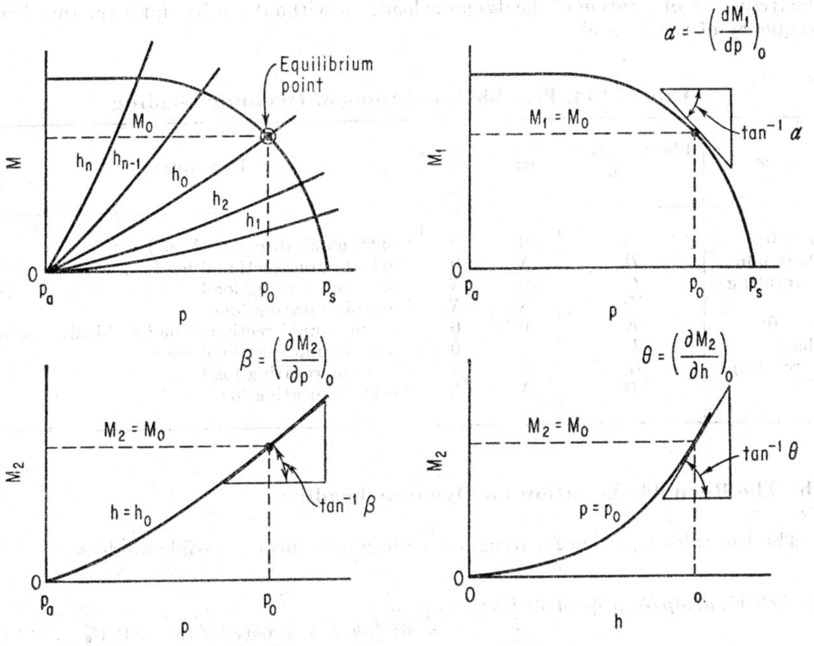

FIG. 12.41. Rates of change of mass flow about equilibrium point.

Equation (12.123) shows that the value of q/s is proportional to the recess depth Δ and the annulus height h_0 and inversely proportional to the recess pressure p_0. It can thus be noted that the values of P_0 and h_0 have an opposite effect on the magnitude of the ratios forming the two sides of the equality (12.129).

For purposes of evaluating the recess depth the following two equations may be written:

$$\left(\frac{R_e}{R_0}\right)^2 \left[\frac{a_n K(r_1^{2/k} - r_1^{(k+1)/k})^{1/2}}{p_0 r_1(1 - r_2^2)}\right]^{1/3} \left(\left\{1 + r_2\left[\left(\frac{R}{R_e}\right)^2 - 1\right]\right\} \left[\frac{2}{3(1 - r_2^2)} + \frac{(k+1)r_1^{(k+1)/k} - 2r_1^{2/k}}{6k(r_1^{2/k} - r_1)^{(k+1)/k}}\right] - 1\right) > \Delta \quad \text{for } r_1 > 0.528 \tag{12.130a}$$

$$\left(\frac{R_e}{R_0}\right)^2 \left[\frac{a_n K_1}{p_0 r_1(1 - r_2^2)}\right]^{1/3} \left(\frac{2}{3(1 - r_2^2)} \left\{1 + r_2\left[\left(\frac{R}{R_e}\right)^2 - 1\right]\right\} - 1\right) > \Delta \quad \text{for } r_1 < 0.528 \tag{12.130b}$$

where a_n = cross-sectional area of nozzle

$$K = \left(\frac{2gk\mathfrak{R}T}{k-1}\right)^{1/2} \frac{12\mu \ln (R/R_0)}{\pi}$$

$$K_1 = 0.532\mathfrak{R}T^{1/2} \frac{12\mu \ln (R/R_0)}{\pi}$$

$$r_1 = p_0/p_s$$
$$r_2 = p_a/p_0$$

Under dynamic loading the shaft may or may not rotate, and the load may vary in both magnitude and direction; thus there are eight different conditions of dynamic loading, and these are listed in Table 12.14. In this table, as throughout the discussion, ω_P denotes the frequency of oscillation of a unidirectional load, while ω_L is the frequency of rotation of the dynamic load. ω without a subscript is the rotational frequency of the journal.

Table 12.14. Possible Conditions of Dynamic Loading

ω	Identification letter	ω_P	ω_L	Description
$\omega = 0$; shaft non-rotating	A	0	0	Constant unidirectional load
	B	X	0	Variable unidirectional load
	C	0	X	Constant rotating load
	D	X	X	Variable rotating load
$\omega \neq 0$; shaft rotating	E	0	0	Constant unidirectional load suddenly applied
	F	X	0	Variable unidirectional load
	G	0	X	Constant rotating load
	H	X	X	Variable rotating load

(b) The Reynolds Equation for Dynamic Loading

The Reynolds equation for dynamic loading with incompressible fluids is

$$(\partial/\partial x)[(h^3/\mu)(\partial p/\partial x)] + (\partial/\partial z)[(h^3/\mu)(\partial p/\partial z)] = 6U(\partial h/\partial x) + 6h(\partial U/\partial x) + 12V_0 \quad (12.131)$$

The right-hand side of Eq. (12.131) contains three terms; each of them contributes to the hydrodynamic forces in the bearing. The first term, $6U\, \partial h/\partial x$, represents the action of the journal rotating with a velocity U over a wedge-shaped fluid film given by $h(x)$. In order for this term to generate positive pressures, it must be negative, since a wedge-shaped film implies that $\partial h/\partial x < 0$. The second term, $6h\, \partial U/\partial x$, implies a variation of tangential velocity along the bearing surface, and in order that this term may contribute to the positive pressures, $\partial U/\partial x$ must be negative, i.e., the velocity must decrease along the fluid film. The last term is the expression for the velocity of shaft center and is responsible for the squeeze-film action. Since $V_0 = dh/dt$, it can be seen that, when V_0 acts in the same direction as the applied load, the film will decrease ($dh/dt < 0$) and the velocity will contribute to the load capacity.

By integrating the Reynolds equation for an infinitely long bearing between 0 and 2π with the conditions

$$p(0) = p(2\pi) = p_a$$

we obtain

$$p - p_a = 6\mu\left(\frac{R}{C}\right)^2\left[\frac{2 + \epsilon\cos\theta}{(1 + \epsilon\cos\theta)^2}\left(\omega - 2\omega_L - 2\frac{d\phi}{dt}\right)\frac{\epsilon}{2 + \epsilon^2}\sin\theta + \frac{1}{\epsilon}\frac{1}{(1 + \epsilon\cos\theta)^2} - \frac{1}{(1 + \epsilon)^2}\frac{d\epsilon}{dt}\right] \tag{12.132}$$

If $\omega_L = d\phi/dt = d\epsilon/dt = 0$, Eq. (12.132) reduces to the Sommerfeld solution.

Equation (12.132) integrated along and at right angles to the line of centers yields

$$\frac{\sin\phi}{12\pi^2 S} = \frac{\epsilon}{(2 + \epsilon^2)(1 - \epsilon^2)^{1/2}}\frac{1}{\omega}\left(\omega - 2\omega_L - 2\frac{d\phi}{dt}\right) \tag{12.133a}$$

$$\frac{\cos\phi}{12\pi^2 S} = \frac{\epsilon}{(1 - \epsilon^2)^{3/2}}\frac{1}{\omega}\frac{d\epsilon}{dt} \tag{12.133b}$$

where Eq. (12.133*a*) is the load capacity due to the wedge action and Eq. (12.133*b*) is the load capacity due to the squeeze film. This set of equations provides a relation between ϵ and ϕ, the phase angle between load and line of centers. For the infinitely short bearing by again neglecting terms of order higher than C/R we have

$$p - p_a = 6\mu\left(\frac{R}{C}\right)^2\left[\left(\frac{L}{D}\right)^2 - \left(\frac{z}{R}\right)^2\right]\frac{1}{(1 + \epsilon\cos\theta)^3}\left[\frac{\epsilon}{2}\left(\omega - 2\omega_L - 2\frac{d\phi}{dt}\right)\sin\theta - \frac{d\epsilon}{dt}\cos\theta\right] \tag{12.134}$$

Integration over 2π using the conditions $p(0) = p(2\pi) = p(\pm L/2) = p_a$ yields

$$\frac{\sin\phi}{4\pi^2 S} = \left(\frac{L}{D}\right)^2\frac{\epsilon}{2(1 - \epsilon^2)^{3/2}}\frac{1}{\omega}\left(\omega - 2\omega_L - 2\frac{d\phi}{dt}\right) \tag{12.135a}$$

$$\frac{\cos\phi}{4\pi^2 S} = \left(\frac{L}{D}\right)^2\frac{1 + 2\epsilon^2}{(1 + \epsilon^2)^{5/2}}\frac{1}{\omega}\frac{d\epsilon}{dt} \tag{12.135b}$$

Constant Loads (Type A of Table 12.14). If P = constant, Eq. (12.133) yields

$$\frac{\epsilon}{(1 - \epsilon^2)^{1/2}} - \frac{\epsilon_1}{(1 - \epsilon_1^2)^{1/2}} = \frac{P}{6\pi(R/C)^2\mu}(t - t_1) \tag{12.136}$$

Equation (12.136) determines the time required for a journal to move a distance $\epsilon - \epsilon_1$ when subjected to a constant load P.

Alternating Loads (Type B of Table 12.14). If the load is given by

$$P = P_0 \sin\omega_P t$$

using $\epsilon = \epsilon_1$ at $\omega_P t = \pi/2$, we have, from Eq. (12.133),

$$\frac{\epsilon}{(1 - \epsilon^2)^{1/2}} - \frac{\epsilon_1}{(1 - \epsilon_1)^{1/2}} = -\frac{\cos\omega_P t}{12\pi^2 S'} \tag{12.137}$$

where

$$S' = (R/C)^2(\mu\omega_P/2\pi P_0)$$

Equation (12.137) describes a periodic motion which is not necessarily symmetrical about ϵ_1. If $\epsilon_1 = 0$, the oscillation is symmetrical about the bearing center and the maximum eccentricity is given by setting $\cos\omega_P t = 1$ or

$$\epsilon_{max}/(1 - \epsilon^2_{max})^{1/2} = 1/12\pi^2 S' \tag{12.138}$$

The displacement as seen from Eq. (12.137) always lags the load vector by $\pi/2$.

For square-wave loading,

$$P = +P_0 \text{ for the first half cycle } \pi/\omega_P$$
$$P = -P_0 \text{ for the second half cycle}$$

Since the displacement lags the load vector by $\pi/2$, the maximum eccentricity will occur one-fourth cycle later, i.e., at $\pi/2\omega_P$. By use of $\epsilon_1 = 0$ at $t = 0$, Eq. (12.137) yields at $t = \pi/2\omega_P$

$$\epsilon_{max}/(1 - \epsilon^2_{max})^{1/2} = 1/24\pi S' \qquad (12.139)$$

Rotating Loads (Type C of Table 12.14). In a symmetrical bearing with the load constant and rotating at a uniform frequency, it is reasonable to expect a fixed relation between load vector and displacement as well as a constant eccentricity. For this type of motion

$$\omega = d\phi/dt = d\epsilon/dt = 0$$

and thus by Eq. (12.133)

$$\frac{P \sin \phi}{6\pi(R/C)^2\mu} = \frac{\epsilon}{(2 + \epsilon^2)(1 - \epsilon^2)^{1/2}}(-2\omega_L)$$
$$\frac{P \cos \phi}{12\pi^2(R/C)^2\mu} = 0$$

These two equations yield

$$\phi = \pi/2 \qquad P = -12\pi\mu\left(\frac{R}{C}\right)^2 \omega_L \frac{\epsilon}{(2 + \epsilon^2)(1 - \epsilon^2)^{1/2}} \qquad (12.140)$$

Equation (12.140) gives the load carried by a nonrotating journal subject to a constant load rotating at an angular velocity ω_L. This expression is similar to the expression for a steady load carried by a journal rotating at $\omega = \omega_L$ except for the sign, which indicates that the two resultants are always 180° out of phase.

When the equations for the infinitely short bearing are used, the load capacity is

$$P = -2\pi(R/C)^2(L/D)^2\mu\epsilon\omega_L/(1 - \epsilon^2)^{3/2} \qquad (12.141)$$

and $\phi = \pi/2$. The comments made about the infinitely long solution apply also to this last expression.

(c) Noncyclic Squeeze Films

This discussion is concerned with the time required for two bodies, separated by a thin fluid film, to approach each other over a certain distance when one of them is subjected to a velocity V. The velocity V is usually imparted by a load W which will be considered throughout this treatment as constant in magnitude and direction and symmetrically placed with respect to the boundaries of the system. One such expression is given by Eq. (12.136), which determines the time it takes a journal in a full sleeve bearing to travel the distance $\epsilon - \epsilon_1$.

Journal Bearings. For $U = 0$ we have

$$Q = -(Lh^3/12\mu R)(dp/d\theta)$$

The flow at any θ due to the velocity V is

$$Q = VLR \sin \theta$$

The two equations yield

$$p = \frac{12\mu V}{(C/R)^3R}\left[\frac{1}{2\epsilon(1 - \epsilon \cos \theta)^2} + C_1\right] \qquad (12.142)$$

which, when integrated for the case of a full bearing with V replaced by $C\,d\epsilon/dt$, yields

$$W = \frac{12\pi\mu L}{(C/R)^3(1-\epsilon^2)^{3/2}}\,C\,\frac{d\epsilon}{dt} \tag{12.143}$$

Integrating again, we obtain

$$\Delta t = \frac{12\pi\mu LR}{(C/R)^2W}\left[\frac{\epsilon_2}{(1-\epsilon_2{}^2)^{1/2}} - \frac{\epsilon_1}{(1-\epsilon_1{}^2)^{1/2}}\right] \tag{12.144}$$

the same expression as Eq. (12.136).

If Eq. (12.142) is applied to a 180° partial bearing, the boundary conditions are $p(\pm\pi/2) = 0$:

$$p = \frac{6\mu V}{(C/R)^3R\epsilon}\left[\frac{1}{(1-\epsilon\cos\theta)^2} - 1\right] \tag{12.145}$$

This expression integrated between 0 and π yields

$$W = 2LR\int_0^\pi p\cos\theta\,d\theta$$

$$= \frac{12\mu VL}{(C/R)^3}\left[\frac{\epsilon}{(1-\epsilon^2)} + \frac{2}{(1-\epsilon^2)^{3/2}}\tan^{-1}\left(\frac{1+\epsilon}{1-\epsilon}\right)^{1/2}\right] \tag{12.146}$$

By replacing V by $C\,d\epsilon/dt$ and integrating, we obtain

$$\Delta t = \frac{24\mu LR}{(C/R)^2W}\left[\frac{\epsilon_2}{(1-\epsilon_2{}^2)^{1/2}}\tan^{-1}\left(\frac{1+\epsilon_2}{1-\epsilon_2{}^2}\right)^{1/2} - \frac{\epsilon_1}{(1-\epsilon_1{}^2)^{1/2}}\tan^{-1}\left(\frac{1+\epsilon_1}{1+\epsilon_1}\right)^{1/2}\right] \tag{12.147}$$

(d) Dynamic Loading of Journal Bearings

Constant Unidirectional Loads. In steady-state hydrodynamics a constant nonrotating load yields journal attitudes which are fixed with respect to the bearing. However, if the more general approach of an arbitrary initial shaft position is taken, then the resultant solution is one of the shaft center moving in a cyclic orbit. The locus will depend on the applied load and initial shaft position and will have as its pole the steady-state eccentricity corresponding to the applied load.

This case, therefore, corresponds to type E in Table 12.14 and from Eq. (12.133) by setting $\omega_L = 0$, we have

$$\frac{\epsilon}{(2+\epsilon^2)(1-\epsilon^2)^{1/2}}\left(1 - \frac{2}{\omega}\frac{d\phi}{dt}\right) = \frac{\sin\phi}{12\pi^2S}$$

$$\frac{1}{(1-\epsilon^2)^{3/2}}\,\frac{d\epsilon/dt}{\omega} = \frac{\cos\phi}{12\pi^2S}$$

With $d\phi/dt = d\epsilon/dt = 0$, these two equations yield the steady Sommerfeld solutions

$$\frac{\epsilon_0}{(2+\epsilon_0{}^2)(1-\epsilon_0{}^2)} = \frac{1}{12\pi^2S} \qquad \phi = \frac{\pi}{2} \tag{12.148}$$

where ϵ_0 is now the pole of the general orbit traveled by the shaft center. This orbit is obtained by eliminating dt between the two equations above and integrating the resulting expression in ϵ and ϕ. The result of the integration is

$$\sin\phi = \frac{12\pi^2S}{5\epsilon(1-\epsilon^2)^{1/2}} + K\,\frac{(1-\epsilon^2)^{3/4}}{\epsilon} \tag{12.149}$$

which gives the cyclic locus of shaft center. This locus depends upon the value of the constant K, which is determined by the initial position of shaft center. Equation (12.149) represents a family of curves with one extreme being a point locus as given by Eq. (12.148) and the other a circle of radius C corresponding to an initial shaft position of $\epsilon = 1$.[25]

The periodicity of this motion is given from the starting equations by $1 - d\phi/dt/\omega$. The ratio of vibration frequency to the rotational speed is

$$\frac{d\phi/dt}{\omega} = \frac{1}{2}\left[1 - \frac{\sin\phi/12\pi^2 S}{\epsilon/(2+\epsilon^2)(1-\epsilon^2)^{1/2}}\right] \tag{12.150}$$

which is seen to depend on the instantaneous position of the shaft center as well as on the pole of the orbit. When this expression is examined for the particular case of $\epsilon = 1$, the average value of $(d\phi/dt)/\omega$ is always one-half. For all orbits that have $\epsilon < 1$ but enclose the bearing center and for all very small orbits, this ratio is less than one-half.

For an infinitely short bearing, the same conditions yield, from Eqs. (12.135*a*) and (12.135*b*),

$$\frac{\epsilon}{2(1-\epsilon^2)^{3/2}}\left(1 - 2\frac{d\phi/dt}{\omega}\right) = \frac{\sin\phi}{4\pi^2(L/D)^2 S}$$

$$\frac{1+2\epsilon^2}{(1+\epsilon^2)^{5/2}}\,\frac{d\epsilon/dt}{\omega} = \frac{\cos\phi}{4\pi^2(L/D)^2 S}$$

By elimination of dt between the two equations and integration,

$$\sin\phi = \frac{\pi^2(L/D)^2 S\epsilon}{(1-\epsilon^2)^{3/2}} - K^2\frac{(1-\epsilon^2)^{3/2}S}{\epsilon} \tag{12.151}$$

This yields a family of curves similar to that obtained from the infinitely long case. By using the expression of Eq. (12.151), we can obtain from the preceding equation the motion of the shaft center in terms of ϵ and K as a parameter. This yields

$$2\pi^4\left(\frac{L}{D}\right)^4 S^2\frac{\epsilon^2}{(1-\epsilon^2)^3} = \left[1 + 4\pi^2K^2S^2\left(\frac{L}{D}\right)^2\right]^{1/2}\sin\frac{\omega(t-t_0)}{2} + \left[1 + 2\pi^2K^2S^2\left(\frac{L}{D}\right)^2\right] \tag{12.152}$$

The frequency of shaft motion is from the term $\sin 1/2\omega\,\Delta t$ seen to equal $1/2\omega$, or exactly one-half of journal rotation. The pole for each orbit is given by setting $d\phi/dt = d\epsilon/dt = 0$ or

$$\frac{1}{S} = 2\pi^2\left(\frac{L}{D}\right)^2\frac{\epsilon_0}{(1-\epsilon_0^2)^{3/2}} \tag{12.153}$$

Variable Unidirectional Loads. For the case in which the frequency of both the applied load and journal rotation may vary let us define by

α = angular displacement of journal
ψ = angular displacement of load
$P = P_0\bar{P}(\omega_P t) = P_0\bar{P}(\tau)$, where P_0 is the maximum load amplitude and ω_P its frequency
$\zeta = \dfrac{\epsilon}{(1-\epsilon^2)^{1/2}} = \tan(\sin^{-1}\epsilon)$
$S_0 = (R/C)^2(\mu N/P_0)$

From the above, $\omega = d\alpha/dt$ and $\omega_P = d\psi/dt$. By assuming that ω is in most cases related to ω_P, by a multiple of its frequency we have, from Eq. (12.133),

$$\frac{1+\zeta^2}{2+3\zeta^2}\zeta\frac{\omega_P}{\omega}\frac{d}{d\tau}[\alpha(\tau) - 2\psi(\tau) - 2\phi] = \frac{\bar{P}(\tau)\sin\phi}{12\pi^2 S_0} \tag{12.154a}$$

$$\frac{\omega_P}{\omega}\frac{d\zeta}{d\tau} = \frac{\bar{P}(\tau)\cos\phi}{12\pi^2 S_0} \tag{12.154b}$$

A plot $[(1+\zeta^2)/(2+3\zeta^2)]\zeta$ vs. ζ will show that, beginning with $\zeta > 0.7$ or $\epsilon > 0.58$, the expression $[(1+\zeta^2)/(2+3\zeta^2)]\zeta$ can be approximated by a straight line, the deviation there being no more than 1½ per cent. Thus, except for lightly loaded bearings, $[(1+\zeta^2)/(2+3\zeta^2)]\zeta$ can be replaced by $a\zeta + b$, and a and b are constants. Writing now

$$x = 12\pi^2 S_0(\omega_P/\omega)(a\zeta + b)$$

we have, from Eq. (12.154),

$$x(d/d\tau)(\alpha - 2\psi - 2\phi) = \bar{P}(\tau)\sin\phi \qquad (1/a)(dx/d\tau) = \bar{P}(\tau)\cos\phi$$

These last equations give x (that is, ϵ) and ϕ as functions of $\tau = \omega_P t$. For unidirectional load, $d\psi/dt = 0$, and the equations become

$$x(d\alpha/d\tau - 2d\phi/d\tau) = \bar{P}(\tau)\sin\phi \tag{12.155a}$$
$$(1/a)(dx/d\tau) = \bar{P}(\tau)\cos\phi \tag{12.155b}$$

The solutions of these equations will, of course, depend on the form of loading $\bar{P}(\tau)$. In the equation above the value of ϵ_{max} which should be used as a basis of comparison of dynamic and steady-state loading is given by

$$12\pi^2 S_0\left[a\frac{\epsilon_{max}}{(1-\epsilon^2_{max})^{1/2}} + b\right] = \frac{x_{max}}{\omega_P/\omega} \tag{12.156}$$

A similar approach to the infinitely short bearing yields equations corresponding to those of Eqs. (12.154a) and (12.154b) in the form

$$\frac{\xi}{2}\frac{\omega_P}{\omega}\frac{d}{d\tau}[\alpha(\tau) - 2\beta(\tau) - 2\phi] = \frac{\bar{P}(\tau)\sin\phi}{4\pi^2(L/D)^2 S_0} \tag{12.157a}$$

$$\frac{\omega_P}{\omega}\frac{d\xi}{d\tau} = \frac{\bar{P}(\tau)\cos\phi}{4\pi^2(L/D)^2 S_0} \tag{12.157b}$$

where $\xi = \epsilon/(1-\epsilon^2)^{3/2}$.

$$x' = 2\pi^2 S_0(L/D)^2(\omega_P/\omega)\xi$$

the two equations become

$$x'(d/d\tau)(\alpha - 2\psi - 2\phi) = \bar{P}(\tau)\sin\phi \qquad 2(dx'/d\tau) = \bar{P}(\tau)\cos\phi$$

and with $d\psi/dt = 0$

$$x'[d\alpha/d\tau - 2(d\phi/dt)] = \bar{P}(\tau)\sin\phi \tag{12.158a}$$
$$2(dx'/dt) = \bar{P}(\tau)\cos\phi \tag{12.158b}$$

Equations (12.158a) and (12.158b) have the advantage of not containing the approximation of a linearized ξ which restricts the infinitely long solution to moderate and heavy loads ($\epsilon > 0.55$).

Sinusoidal Loading

$$\bar{P}(\tau) = \sin\omega_P t$$

If the journal speed is constant, we have, from Eqs. (12.55a) and (12.55b),

$$d\alpha/d\tau = \omega/\omega_P$$

and thus

$$x[\omega/\omega_P - 2(d\phi/d\tau)] = \sin\tau\sin\phi \tag{12.159a}$$
$$(1/a)(dx/d\tau) = \sin\tau\cos\phi \tag{12.159b}$$

The solutions may be obtained by numerical methods involving a trial-and-error procedure. The calculations are summarized in Fig. 12.42, where the ordinate gives the ratio of x_{max} and thus ϵ_{max} of a sinusoidally loaded bearing to that of a steady-state bearing in terms of frequency ω_P/ω. At $\omega_P/\omega = 0$ the ratio is simply the steady-state solution. At $\omega_P/\omega = \frac{1}{2}$ the load capacity is zero. As the load frequency ω_P rises, the load capacity also rises, and as $\omega_P/\omega \to \infty$ the ratio $(\omega_P/\omega/x_{max})$ approaches the line $3\omega_P/\omega$.

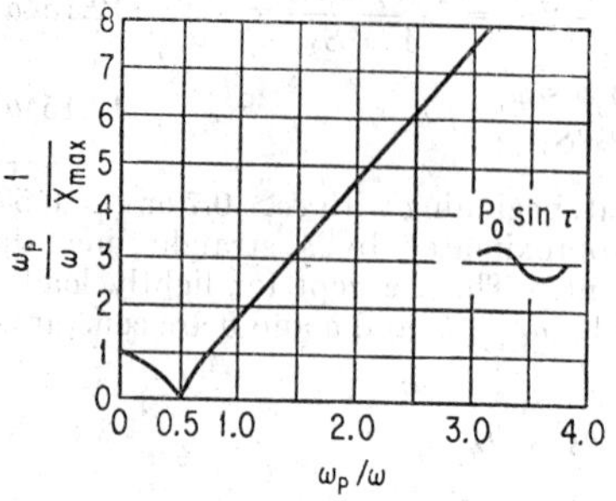

FIG. 12.42. Load capacity for sinusoidal loads relative to constant loading.

The locus of shaft center changes its form depending upon the frequency ratio ω_P/ω.[26]

Square-wave Loading. In this type of loading

$$\bar{P}(\tau) = +1 \qquad \text{for } 0 < \tau < \pi$$
$$\bar{P}(\tau) = -1 \qquad \text{for } \pi < \tau < 2\pi$$

The solutions here are the sum of two solutions for a fixed load acting in opposite directions over the two respective half circles.

By denoting by ϵ_m the eccentricity corresponding to either $\phi = 0$ or $\phi = \pi$, we have, from Eq. (12.151),

$$\pi(L/D)[\epsilon_m/(1 - \epsilon_m)^{3/2}] = K \tag{12.160}$$

This K substituted in Eq. (12.152) gives the time at which the shaft center is at these points, namely, the two values of t resulting from

$$t - t_0 = \frac{2}{\omega} \sin^{-1} \frac{-1}{1 + 4\pi^2 S_0^2 (L/D)^4 [\epsilon_m^2/(1 - \epsilon_m^2)^{3/2}]}$$

The difference between the two roots is the time required for the shaft center to complete one-half cycle which in turn must equal one-half the load cycle. This, therefore, gives a relation between ϵ_m and ω_P:

$$\tan(\pi/4)(\omega/\omega_P) = 2\pi^2 S_0 (L/D)^2 [\epsilon_m/(1 - \epsilon_m^2)^{3/2}] \tag{12.161}$$

It is seen that, when $\omega > 1/2\omega_P$, this ϵ_m is also the maximum eccentricity; however, when $\omega < 1/2\omega_P$, the maximum eccentricity occurs at right angles to ϵ_m. This can be found, then, by setting $\sin\phi = \pm 1$ in Eq. (12.151).

$$\frac{\epsilon^2_m}{(1 - \epsilon^2_m)^3} = \frac{\epsilon^2_{max}}{(1 - \epsilon^2_{max})^3} - \frac{1}{\pi^2 S_0 (L/D)^2} \frac{\epsilon_{max}}{(1 - \epsilon^2_{max})^{3/2}} \tag{12.162}$$

Thus, by combining the relation between ω_P and ϵ_m of Eq. (12.161) with the relations between ϵ_m and ϵ_{max}, we obtain

$$2\pi^2 (L/D)^2 S_0 [\epsilon_{max}/(1 - \epsilon^2_{max})^{3/2}] = 1 - \sec(\pi/4)(\omega/\omega_P) \qquad \tfrac{1}{4} < \omega_P/\omega < \tfrac{1}{2} \tag{12.163a}$$
$$2\pi^2 (L/D)^2 S_0 [\epsilon_{max}/(1 - \epsilon^2_{max})^{3/2}] = \tan(\pi/4)(\omega/\omega_P) \qquad \tfrac{1}{2} < \omega_P/\omega \tag{12.163b}$$

By comparing with Eq. (12.153), it is seen that for a constant load the quantity

$$\frac{W_{\omega_P} \neq 0}{W_{\omega_P} = 0} = \frac{1}{2\pi^2 S_0 (L/D)^2 [\epsilon_{max}/(1 - \epsilon^2_{max})^{3/2}]} \tag{12.164}$$

equals unity. This ratio thus represents the relative load capacity referred to the h_{min} of dynamic and steady loading. This ratio is plotted in Fig. 12.43. The plot is not valid below $\omega/\omega_P < 0.2$; at high values of ω_P/ω, it approaches the value of $(4/\pi)(\omega_P/\omega)$.

A comparison of the relative load capacity of bearings subjected to various forms of an alternating load is given in Table 12.15.

Table 12.15. ω_P/ω

Form of load	0	0.5	1	2	3
$P = P_0$	1	0	1	3	5
$P = P_0 \sin \omega_P t$	1	0	1.74	4.72	7.90
$P = \pm P_0$	1	0	1.25	3.27	5.24

Constant Rotating Loads. For a constant load rotating with a frequency ω_L and by assuming the phase and amplitude of the orbit of shaft center to be constant, we have, from Eq. (12.133),

$$\frac{\epsilon}{(2+\epsilon^2)(1-\epsilon^2)^{1/2}}\left(1-2\frac{\omega_L}{\omega}\right) = \pm\frac{1}{2\pi^2 S} \qquad (12.165)$$

$$\phi = \pm\pi/2$$

The resultant of Eq. (12.165) is seen to depend on the value of $(1 - 2\omega_L/\omega)$, being 1 at $\omega_L/\omega = 0$ (steady-state solution), becomes zero at $\omega_L = 1/2\omega$, and then rising

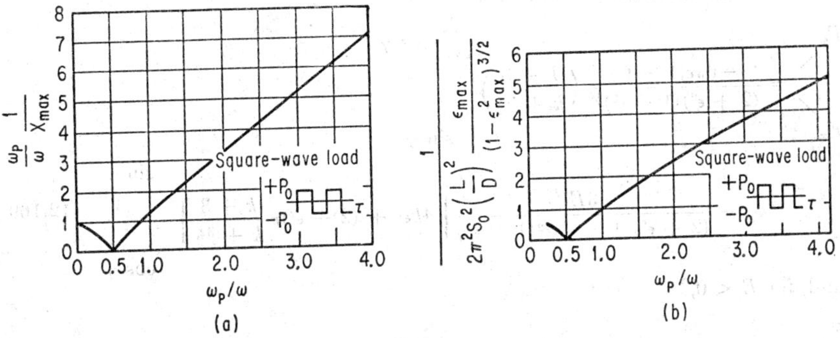

FIG. 12.43. Load capacity for square-wave loading relative to constant loading. (*a*) Infinitely long bearing. (*b*) Infinitely short bearing.

continuously with a further rise in ω_L. To determine the nature of the orbit of vibration, we can for a moment discard the assumption of $d\epsilon/dt = d\phi/dt = 0$. Then from Eq. (12.33) we have

$$\frac{\epsilon}{(2+\epsilon^2)(1-\epsilon^2)^{1/2}}\left(1-2\frac{\omega_L}{\omega}-\frac{2}{\omega}\frac{d\phi}{dt}\right) = \frac{\sin\phi}{12\pi^2 S}$$

$$\frac{1}{(1-\epsilon^2)^{3/2}}\frac{1}{\omega}\frac{d\epsilon}{dt} = \frac{\cos\phi}{12\pi^2 S}$$

and, by eliminating dt between the two equations and integrating, we obtain

$$\sin\phi = \frac{12\pi^2 S(1-2\omega_L/\omega)}{5\epsilon(1-\epsilon^2)^{1/2}} + K\frac{(1-\epsilon^2)^{3/4}}{\epsilon} \qquad (12.166)$$

This last equation is similar in form to Eq. (12.149) except for the term $1 - 2\omega_L/\omega$.
For the infinitely short bearing a similar treatment yields

$$\sin\phi = \frac{\pi^2(L/D)S(1-2\omega_L/\omega)}{(1-\epsilon^2)^{3/2}} - K^2\frac{(1-\epsilon^2)^{3/2}}{\epsilon} \qquad (12.167)$$

Both the infinitely short and infinitely long solutions show that the line of centers is always at right angles to the load and that the eccentricity of the shaft is the same as if the load were fixed provided its magnitude is modified by the factor $(1 - 2\omega_L/\omega)$.

(e) Dynamic Loading of Journal Bearings with No Negative Pressures

Equation (12.132) was integrated over the entire region 2π and, as in the case of the full Sommerfeld solution, it includes negative pressures of a magnitude that cannot possibly occur. The chief reason for presenting these results is that they permit easy comparison between the various operations of dynamically loaded journal bearings. Here we deal with the more realistic fluid forces which are obtained by integration of the dynamic Reynolds equation over the positive pressure regions and by setting the pressure in the remaining portions equal to zero. This involves finding the zero points, i.e., the angles at which the pressure wave begins and ends.[26]

$$P_y = -\frac{1}{2}\frac{\mu\omega}{(C/R)^2}\int \Pi(\theta)\sin(\theta+\gamma)\,d\theta = \frac{1}{2}\frac{\mu\omega}{(C/R)^2}F_y \tag{12.168a}$$

$$P_x = -\frac{1}{2}\frac{\mu\omega}{(C/R)^2}\int \Pi(\theta)\cos(\theta+\gamma)\,d\theta = \frac{1}{2}\frac{\mu\omega}{(C/R)^2}F_x \tag{12.168b}$$

Evaluation of these integrals gives, for $E > 0$,

$$\begin{Bmatrix} F_y \\ F_x \end{Bmatrix} = \frac{-6\pi\epsilon(1-G)}{(2+\epsilon^2)(1-\epsilon^2)^{1/2}}\left(\frac{k+3}{k+3/2}\right)\begin{Bmatrix} -\cos\gamma \\ \sin\gamma \end{Bmatrix} - \frac{3E}{(2+\epsilon^2)(1-\epsilon^2)^{1/2}(1-\epsilon^2)}\left[4k\epsilon^2 + (2+\epsilon^2)\pi\frac{k+3}{k+3/2}\right]\begin{Bmatrix} \sin\gamma \\ \cos\gamma \end{Bmatrix} \tag{12.169}$$

and, for $E < 0$,

$$\begin{Bmatrix} F_y \\ F_x \end{Bmatrix} = -\frac{6\pi\epsilon(1-G)}{(2+\epsilon^2)(1-\epsilon^2)^{1/2}}\left(\frac{k}{k+3/2}\right)\begin{Bmatrix} -\cos\gamma \\ \sin\gamma \end{Bmatrix} + \frac{3E}{(2+\epsilon^2)(1-\epsilon^2)^{1/2}(1-\epsilon^2)}\left[4k\epsilon^2 - (2+\epsilon^2)\pi\frac{k}{k+3/2}\right]\begin{Bmatrix} \sin\gamma \\ \cos\gamma \end{Bmatrix} \tag{12.170}$$

where $k = (1 - \epsilon^2)^{1/2}\left[\left(\frac{1-G}{E}\right)^2 + \frac{1}{\epsilon^2}\right]^{1/2} \qquad E = \frac{2}{\omega}\frac{d\epsilon}{dt} \qquad G = \frac{2}{\omega}\frac{d\gamma}{dt}$

By setting $E = G = 0$, the above equations can be reduced to the standard Sommerfeld solution for a $0 - \pi$ positive-pressure wave, or

$$F_y = \frac{6\pi\epsilon}{(2+\epsilon^2)(1-\epsilon^2)^{1/2}} \qquad F_x = \frac{12\epsilon^2}{(2+\epsilon^2)(1-\epsilon^2)} \tag{12.171}$$

and the attitude angle is

$$\phi = \tan^{-1}\frac{2\epsilon}{\pi(1-\epsilon^2)^{1/2}} \tag{12.172}$$

Solutions for Prescribed Loci. **Circular Locus of Uniform Velocity.** If the locus is circular, ϵ = constant, and if the velocity of travel of shaft center is uniform, $\gamma = \omega t$ and $d\gamma/dt$ = constant = ω. We thus have

$$E = 0 \qquad G = 2(\omega_L/\omega)$$

$$\begin{Bmatrix}F_y\\ F_x\end{Bmatrix} = -\left(1 - 2\frac{\omega_L}{\omega}\right)\frac{6\pi\epsilon}{(2+\epsilon^2)(1-\epsilon^2)^{1/2}}\begin{Bmatrix}-\cos\omega t\\ \sin\omega t\end{Bmatrix} - \left|\left(1 - 2\frac{\omega_L}{\omega}\right)\right|\frac{12\epsilon^2}{(2+\epsilon^2)(1-\epsilon^2)}\begin{Bmatrix}\sin\omega t\\ \cos\omega t\end{Bmatrix} \qquad (12.173)$$

or, for the resultant load capacity,

$$P = \mu\left(\frac{R}{C}\right)^2 |\omega - 2\omega_L| \frac{6\epsilon[\pi^2 - \epsilon^2(\pi^2 - 4)]^{1/2}}{(2+\epsilon^2)(1-\epsilon^2)} \qquad (12.174)$$

Unidirectional Sinusoidal Locus. Let the equation of the locus be given by

$$\epsilon = \epsilon_0 + a \sin \omega_P t$$

where the motion takes place along a straight line with γ equal to either 0 or π. From these conditions we have

$$E = (2/\omega)a\omega_P \cos \omega_P t \quad G = 0$$

With γ equal to either zero or π, $\sin \gamma = 0$ and Eqs. (12.169) and (12.170) become, for $E > 0$,

$$\begin{aligned} F_y &= \frac{6\pi\epsilon}{(2+\epsilon^2)(1-\epsilon^2)^{1/2}} \frac{k+3}{k+3/2}\cos\gamma \\ F_x &= -\frac{3E}{(2+\epsilon^2)(1-\epsilon^2)^{3/2}}\left[4k\epsilon^2 + (2+\epsilon^2)\pi\frac{k+3}{k+3/2}\right]\cos\gamma \end{aligned} \qquad (12.175)$$

and, for $E < 0$,

$$\begin{aligned} F_y &= \frac{6\pi\epsilon}{(2+\epsilon^2)(1-\epsilon^2)^{1/2}} \frac{k}{k+3/2}\cos\gamma \\ F_x &= \frac{3E}{(2+\epsilon^2)(1-\epsilon^2)^{3/2}}\left[4k\epsilon^2 - (2+\epsilon^2)\pi\frac{k}{k+3/2}\right]\cos\gamma \end{aligned} \qquad (12.176)$$

$$k = \left[\left(\frac{1}{E^2} + \frac{1}{\epsilon^2}\right)(1-\epsilon^2)\right]^{1/2}$$

Elliptical Orbits. The parametric equations of an ellipse are

$$\epsilon \sin \gamma = \sigma b \sin \omega t \qquad \epsilon \cos \gamma = a \cos \omega t$$

where a and b are, respectively, the major and minor semiaxes and σ is ± 1.

Here too the load frequency is assumed to be that of shaft rotation. The ellipse, of course, is nothing more than a simultaneous sinusoidal motion in two directions, and we can distinguish two cases: those in which the orbit progresses either in the same direction as the load or in the opposite direction. By assigning to σ the value of either $+1$ or -1, this can be taken into account in a single expression. From the expressions for ϵ and γ given above by noting that

$$\tan \gamma = \sigma(b/a) \tan \omega t \qquad (1/\cos^2\gamma)(d\gamma/dt) = \sigma(b/a)(\omega/\cos^2\omega t)$$

we have

$$E = (b^2 - a^2)\frac{\sin 2\omega t}{\epsilon} \qquad G = \sigma\frac{2ab}{\epsilon^2} \qquad q = \frac{\epsilon^2 + 2\sigma ab}{(b^2 - a^2)\epsilon \sin 2\omega t}$$

These can now be used in Eqs. (12.169) and (12.170) to obtain the vertical and horizontal load components for the respective positions of the shaft center as given by the various values of ωt. The loads for the apexes A and B can be easily obtained from Eq. (12.173) by setting

For A: $\gamma = 0$, $\epsilon = a$

$$F_y = \frac{6\pi(a - 2\sigma b)}{(2 + a^2)(1 - a^2)^{1/2}} \qquad F_x = \frac{12a|(a - 2\sigma b)|}{(2 + a^2)(1 - a^2)}$$

For B: $\gamma = \pi/2$, $\epsilon = b$

$$F_y = -\frac{12b|b - 2\sigma a|}{(2 + b^2)(1 - b^2)} \qquad F_x = -\frac{6\pi(b - 2\sigma a)}{(2 + b^2)(1 - b^2)^{1/2}}$$

For the points opposite to A and B the same expressions apply with their signs reversed.

(f) Solutions for Finite Journal Bearings

Rotor-bearing Dynamics.[27] For an incompressible fluid the dynamic Reynolds equation may be written

$$(\partial/\partial x)[(h^3/\mu)(\partial P/\partial x)] + (\partial/\partial z)[(h^3/\mu)(\partial P/\partial z)] = 6R\omega[1 - 2(\dot\alpha/\omega)](\partial h/\partial x) + 12\,\dot e \cos\theta \qquad (12.177a)$$

Introducing the following dimensionless parameter

$$x = Dx' \qquad z = Lz' \qquad h = 2Ch' \qquad e = C\epsilon$$
$$P = (\mu\omega/2\pi)(R/C)^3P'$$

and assuming constant viscosity, the Reynolds equation in dimensionless form becomes

$$(\partial/\partial x')[h'^3(\partial P'/\partial x')] + (D/L)^2(\partial/\partial z')[h'^3(\partial P'/\partial z')] = 6\pi(\partial h'/\partial x') + 12\pi\frac{\dot\epsilon/\omega}{1 - 2(\dot\alpha/\omega)}\cos\theta \qquad (12.177b)$$

where $h' = \frac{1}{2}(1 + \epsilon\cos\theta)$. The resulting fluid-film forces in radial and tangential directions are

$$F_r = -\lambda\omega\left(1 - 2\frac{\dot\alpha}{\omega}\right)\iint P'\cos\theta\,dx'\,dz' = \lambda\omega\left(1 - 2\frac{\dot\alpha}{\omega}\right)f_r\left(\epsilon, \frac{\dot\epsilon/\omega}{1 - 2(\dot\alpha/\omega)}, \frac{L}{D}\right)$$
$$F_t = \lambda\omega\left(1 - 2\frac{\dot\alpha}{\omega}\right)\iint P'\sin\theta\,dx'\,dz' = \lambda\omega\left(1 - 2\frac{\dot\alpha}{\omega}\right)f_t\left(\epsilon, \frac{\dot\epsilon/\omega}{1 - 2(\dot\alpha/\omega)}, \frac{L}{D}\right)$$

where $\lambda = (\mu RL/\pi)(R/C)^2$. For the rotor-bearing dynamic analysis these forces are linearized with respect to displacement and velocity, i.e.,

$$dF = \lambda\omega\left(1 - 2\frac{\dot\alpha}{\omega}\right)\left[\frac{\partial f}{\partial\epsilon}d\epsilon + \frac{\partial f}{\partial(\dot\epsilon/\omega)}d\left(\frac{\dot\epsilon}{\omega}\right) + \frac{\partial f}{\partial(\dot\alpha/\omega)}d\left(\frac{\dot\alpha}{\omega}\right) - \frac{2f}{1 - 2(\dot\alpha/\omega)}d\left(\frac{\dot\alpha}{\omega}\right)\right]$$

Taking as reference for this Taylor expansion the steady-state equilibrium position it follows that $\dot\epsilon = \dot\alpha = 0$ and thus $[\partial f/\partial(\dot\alpha/\omega)] = 0$, i.e.,

$$dF = \lambda\omega\left[\frac{\partial f}{\partial\epsilon}d\epsilon + \frac{\partial f}{\partial(\dot\epsilon/\omega)}d\left(\frac{\dot\epsilon}{\omega}\right) - \frac{2f}{\epsilon}\frac{1}{\omega}\epsilon\,d\dot\alpha\right]$$

Introducing rectangular coordinates with the x axis vertical downward and the y axis horizontal,

$$x = C\epsilon\cos\alpha \qquad y = C\epsilon\sin\alpha$$

it is found that

$$dF_x = \frac{1}{C}\lambda\omega\left\{-\left[\frac{\partial f_r}{\partial\epsilon}\cos^2\alpha + \frac{f_r}{\epsilon}\sin^2\alpha + \left(-\frac{f_t}{\epsilon}+\frac{\partial f_t}{\partial\epsilon}\right)\cos\alpha\sin\alpha\right]dx\right.$$
$$-\left[\frac{\partial f_r}{\partial(\dot{\epsilon}/\omega)}\cos^2\alpha + \frac{2f_t}{\epsilon}\sin^2\alpha + \left(\frac{2f_r}{\epsilon}+\frac{\partial f_t}{\partial(\dot{\epsilon}/\omega)}\right)\cos\alpha\sin\alpha\right]\frac{1}{\omega}(\dot{dx})$$
$$+\left[-\frac{f_t}{\epsilon}\cos^2\alpha - \frac{\partial f_t}{\partial\epsilon}\sin^2\alpha + \left(\frac{f_r}{\epsilon}-\frac{\partial f_r}{\partial\epsilon}\right)\cos\alpha\sin\alpha\right]dy$$
$$\left.+\left[\frac{2f_r}{\epsilon}\cos^2\alpha - \frac{\partial f_t}{\partial(\dot{\epsilon}/\omega)}\sin^2\alpha + \left(\frac{2f_t}{\epsilon}-\frac{\partial f_r}{\partial(\dot{\epsilon}/\omega)}\right)\cos\alpha\sin\alpha\right]\frac{1}{\omega}(\dot{dy})\right\} \quad (12.178)$$

and similarly for dF_y.

For use in the rotor analysis this is written

$$\begin{aligned} dF_x &= -K_{xx}\,dx - C_{xx}(\dot{dx}) + K_{xy}\,dy + C_{xy}(\dot{dy}) \\ dF_y &= K_{yx}\,dx + C_{yx}(\dot{dx}) - K_{yy}\,dy - C_{yy}(\dot{dy}) \end{aligned} \quad (12.179)$$

The above forces and derivatives are calculated numerically and summarized in Tables 12.16 to 12.19 for cylindrical, four-axial-groove, and elliptical bearings,

Table 12.16. Plain Cylindrical Bearing Computer Results

L/D	ϵ	$\dot{\epsilon}/\omega$	f_v	α	f_r	f_t
	0.2	0	0.4653	75.772	0.11436	0.45102
	0.2	0.03	....	......	0.28010	0.48125
	0.2	−0.03	....	......	−0.02311	0.41915
	0.22	0	....	......	0.14035	0.50170
	0.18	0	....	......	0.09149	0.40187
	0.5	0	1.862	54.763	1.0744	1.5210
	0.5	0.03	....	......	1.4528	1.6485
½	0.5	−0.03	....	......	0.7538	1.3971
	0.52	0	....	......	1.2172	1.6367
	0.48	0	....	......	0.9508	1.4192
	0.7	0	5.160	40.432	3.9274	3.3462
	0.7	0.03	....	......	5.1132	3.6797
	0.7	−0.03	....	......	3.0057	3.0218
	0.715	0	....	......	4.3861	3.5934
	0.685	0	....	......	3.5368	3.1349
	0.2	0	1.504	76.836	0.34257	1.4647
	0.2	0.03	....	......	0.86908	1.5517
	0.2	−0.03	....	......	−0.10182	1.3768
	0.22	0	....	......	0.41810	1.6240
	0.18	0	....	......	0.27556	1.3089
	0.5	0	5.299	57.502	2.8470	4.4693
	0.5	0.03	....	......	3.9082	4.7825
1	0.5	−0.03	....	......	1.9853	4.1527
	0.52	0	....	......	3.1858	4.7621
	0.48	0	....	......	2.5455	4.2090
	0.7	0	12.34	43.280	8.9832	8.4594
	0.7	0.03	....	......	11.455	9.1654
	0.7	−0.03	....	......	6.805	7.7610
	0.715	0	....	......	9.8687	8.9177
	0.685	0	....	......	8.1932	8.0365

Table 12.17. Four-axial-groove Bearing Computer Results

L/D	ϵ	α	$\dot{\epsilon}/\omega$	f_v	f_h
	0.2	80.000	0	0.2711	0.007144
	0.2	81.647	0	0.26946	0.000218
	0.2	81.696	0	0.26941	0
	0.2	81.696	0.03	0.29197	−0.042655
	0.2	81.696	−0.03	0.25277	0.022276
	0.22	81.696	0	0.29980	−0.004059
	0.18	81.696	0	0.23986	0.003245
	0.5	61.000	0	1.1700	−0.004287
	0.5	60.413	0	1.1789	−0.000642
	0.5	60.315	0	1.1805	0
	0.5	60.315	0.03	1.2164	−0.010655
	0.5	60.315	−0.03	1.1482	0.005458
	0.52	60.315	0	1.2847	−0.019631
	0.48	60.315	0	1.0848	0.016799
½	0.7	36.000	0	4.3096	−0.041049
	0.7	35.181	0	4.3135	−0.002017
	0.7	35.139	0	4.3137	0
	0.7	35.139	0.03	4.4712	0.010973
	0.7	35.139	−0.03	4.1559	−0.010862
	0.715	35.139	0	4.7899	−0.073303
	0.685	35.139	0	3.8962	0.059837
	0.2	80.000	0	0.52918	0.023913
	0.2	83.881	0	0.51908	−0.002508
	0.2	83.531	0	0.52004	0
	0.2	83.531	0.03	0.55702	−0.074742
	0.2	83.531	−0.03	0.49585	0.038249
	0.22	83.531	0	0.57700	−0.005625
	0.18	83.531	0	0.46426	0.004520
	0.5	61.000	0	2.2295	0.039473
	0.5	64.123	0	2.1370	0.003321
	0.5	64.431	0	2.1279	0
	0.5	64.431	0.03	2.1846	−0.022831
	0.5	64.431	−0.03	2.0730	0.018699
1	0.52	64.431	0	2.2948	−0.030393
	0.48	64.431	0	1.9714	0.026434
	0.7	36.000	0	7.7047	−0.006564
	0.7	35.918	0	7.7060	−0.000638
	0.7	35.909	0	7.7061	0
	0.7	35.909	0.03	7.9922	0.001840
	0.7	35.909	−0.03	7.4203	−0.001750
	0.715	35.909	0	8.4906	−0.12758
	0.685	35.909	0	7.0118	0.10588

respectively.[28] The resulting spring and damping coefficients as calculated from Eq. (12.178) are shown in Table 12.20. These results can be used directly when calculating the vibrations of the rotor. However, the usual calculation procedure allows for only four coefficients, one spring and damping coefficient in the vertical and in the horizontal direction. No provision is made for taking into account the four cross-coupling terms K_{xy}, C_{xy}, K_{yx}, and C_{yx}. Therefore, it becomes important to eliminate them and to reduce the original eight coefficients to four equivalent coefficients. Because of the nonsymmetry of the cross-coupling terms they do not disappear by the introduction of the principal axis. Instead the eight coefficients may be combined to four by coupling the bearing with the rotor in such a way that the resulting motion remains the same. Consider the dynamic response of a symmetrical rotor with one degree of freedom.

Let 0 be the steady-state position of the journal center (i.e., at zero unbalance), A is the actual journal center, B is the shaft center at mid-span, and G is the center of gravity of the rotor. A force balance gives

$$\begin{aligned} M\ddot{x}_b + k(x_b - x_a) &= M\,\delta\dot{\omega}^2 \cos \omega t \\ k(x_b - x_a) &= 2K_{xx}x_a + 2C_{xx}\dot{x}_a - 2K_{xy}y_a - 2C_{xy}\dot{y}_a \\ M\ddot{y}_b + k(y_b - y_a) &= M\,\delta\omega^2 \sin \omega t \\ k(y_b - y_a) &= -2K_{yx}x_a - 2C_{yx}\dot{x}_a + 2K_{yy}y_c + 2C_{yy}\dot{y}_a \end{aligned} \tag{12.180}$$

The following parameters are introduced:

$$\omega_c^2 = k/M \tag{12.181a}$$

$$\varkappa = \tfrac{1}{2}k \frac{\omega^2}{\omega_c^2 - \omega^2} \tag{12.181b}$$

In dimensionless form Eq. (12.181b) becomes

$$\bar{\varkappa} = \frac{\tfrac{1}{2}k}{(1/C)\lambda\omega_c} \frac{\omega/\omega_c}{1 - (\omega/\omega_c)^2}$$

when $\dfrac{\tfrac{1}{2}k}{(1/C)\lambda\omega_c} = 5$ the above equation is identified by $\bar{\varkappa} = S \dfrac{(\omega/\omega_c)^1}{1 - (\omega/\omega_c)^{12}}$. Furthermore the solution is taken in the form

$$\begin{aligned} x_a &= A \cos \omega t + B \sin \omega t \\ x_b &= \frac{A\omega_c^2 + \delta\omega^2}{\omega_c^2 - \omega^2} \cos \omega t + \frac{B\omega_c^2}{\omega_c^2 - \omega^2} \sin \omega t \\ y_a &= E \cos \omega t + F \sin \omega t \\ y_b &= \frac{E\omega_c^2}{\omega_c^2 - \omega^2} \cos \omega t + \frac{F\omega_c^2 + \delta\omega^2}{\omega_c^2 - \omega^2} \sin \omega t \end{aligned} \tag{12.182}$$

Substituting Eqs. (12.181a), (12.181b), and (12.182) into Eq. (12.180) yields a matrix in terms of the eight coefficients which can be reduced to a matrix of the same form as the one for a rotor without cross-coupling terms. Such a rotor has only four spring and damping coefficients, which are denoted K_x, B_x, K_y, and B_y. The reduced matrix is

$A/\varkappa\delta$	$B/\varkappa\delta$	$E/\varkappa\delta$	$F/\varkappa\delta$	
$(K_x - \varkappa)$	ωB_x	0	0	1
$-\omega B_x$	$(K_x - \varkappa)$	0	0	0
0	0	$(K_y - \varkappa)$	ωB_y	0
0	0	$-\omega B_y$	$(K_y - \varkappa)$	1

(12.183)

Table 12.18. Elliptical Bearing Computer Results

	ϵ	α	f_v	f_h	ϵ_1	α_1	$\dot{\epsilon}_1/\omega$	f_{v1}	f_{h1}	ϵ_2	α_2	$\dot{\epsilon}_2/\omega$	f_{v2}	f_{h2}
	0.2	103.797	0.51882	0.00002	0.28045	43.833	0	0.61757	0.12873	0.35546	33.122	0	0.09874	0.12871
	0.2	100	0.56567	0.03009	0.29178	42.457	0	0.64848	0.14102	0.34622	34.674	0	0.08281	0.11093
	0.2	105	0.50359	−0.00944	0.27680	44.261	0	0.60762	0.12512	0.35830	32.627	0	0.10403	0.13456
	0.2	103.797			0.28045	43.833	0.03	0.79123	0.06618	0.35546	32.122	0.03	0.19696	0.22105
	0.2	103.797			0.28045	43.833	−0.03	0.47744	0.16161	0.35546	33.122	−0.03	0.04035	0.06220
	0.22	103.797	0.57692	−0.01623	0.29098	47.245	0	0.67017	0.10981	0.37031	35.236	0	0.09326	0.12604
	0.18	103.797	0.46239	0.01538	0.27099	40.170	0	0.56563	0.14603	0.34112	30.827	0	0.10323	0.13065
	0.5	55.861	4.3933	−0.00220	0.67291	37.953	0	4.3933	−0.00220	0.41497	85.771	0	0	0
	0.5	54	4.5213	0.05936	0.67783	36.640	0	4.5213	0.05936	0.40688	83.811	0	0	0
	0.5	57	4.3114	−0.03537	0.66982	38.759	0	4.3114	−0.03537	0.41993	86.958	0	0	0
$L/D = \frac{1}{2}$, $m = 0.25$	0.5	55.861			0.67291	37.953	0.03	5.3265	−0.35063	0.41497	85.771	0.03	0	0
	0.5	55.861			0.67291	37.953	−0.03	3.5767	0.24659	0.41497	85.771	−0.03	0	0
	0.52	55.861	4.9324	−0.12343	0.69196	38.461	0	4.9324	−0.12343	0.43242	84.453	0	0	0
	0.48	55.861	3.9271	0.09314	0.65390	37.414	0	3.9271	0.09314	0.39776	87.216	0	0	0
	0.7	24.931	55.096	−0.24058	0.93268	18.443	0	55.262	−0.00608	0.48489	37.484	0	0.16607	0.23450
	0.7	23	56.591	1.3452	0.93524	17.005	0	56.895	1.6153	0.47992	34.744	0	0.20402	0.27002
	0.7	29	49.981	−2.4458	0.92662	21.484	0	50.093	−2.2712	0.49637	43.133	0	0.11239	0.17457
	0.7	24.931			0.93268	18.443	0.03	78.739	−4.3878	0.48489	37.484	0.03	0.30661	0.38396
	0.7	24.931			0.93268	18.443	−0.03	37.607	2.5599	0.48489	37.484	−0.03	0.07941	0.12701
	0.71	24.931	68.015	−1.5541	0.94261	18.512	0	68.197	−1.2984	0.49465	37.231	0	0.18257	0.25573
	0.69	24.931	45.601	0.61758	0.92274	18.373	0	45.752	−0.8332	0.47513	37.745	0	0.15167	0.21566

	0.15	89.728	0.90262	0.00002	0.52270	16.676	0	1.6100	0.66137	0.52133	16.722	0	0.70742	0.66135
	0.15	88	0.96191	0.03285	0.52701	16.526	0	1.6459	0.67508	0.51698	16.856	0	0.68394	0.64223
	0.15	91	0.85855	−0.02429	0.51950	16.780	0	1.5839	0.65151	0.52452	16.615	0	0.72537	0.67580
	0.15	89.728			0.52270	16.676	0.03	2.0538	0.68903	0.52133	16.722	0.03	1.0166	0.85694
	0.15	89.728			0.52270	16.676	−0.03	1.2394	0.61299	0.52133	16.722	−0.03	0.46709	0.48771
	0.17	89.728	1.0342	−0.00490	0.52887	18.750	0	1.7035	0.64976	0.52735	18.806	0	0.66925	0.65466
	0.13	89.728	0.77582	0.00535	0.51722	14.557	0	1.5187	0.67130	0.51603	14.591	0	0.74285	0.66595
	0.3	84.491	2.4814	0.00067	0.60729	29.454	0	2.8966	0.52047	0.55785	32.364	0	0.41515	0.51979
	0.3	80	2.8740	0.15687	0.62618	28.153	0	3.2077	0.58385	0.53657	33.410	0	0.33363	0.42698
	0.3	85	2.4382	−0.01778	0.60510	29.597	0	2.8636	0.51359	0.56023	32.240	0	0.42546	0.53136
$L/D = \frac{1}{2}$ $m = 0.50$	0.3	84.491			0.60729	29.454	0.03	3.5861	0.39014	0.55785	32.364	0.03	0.66512	0.75307
	0.3	84.491			0.60729	29.454	−0.03	2.3463	0.57677	0.55785	32.364	−0.03	0.22776	0.32351
	0.32	84.491	2.7378	−0.04439	0.61897	30.971	0	3.1320	0.46528	0.56717	34.167	0	0.39426	0.50968
	0.28	84.491	2.2696	0.03571	0.59605	27.878	0	2.7007	0.56154	0.54911	30.501	0	0.43117	0.52583
	0.5	20.678	298.93	−0.15595	0.98376	10.339	0	298.93	−0.15595	0.17947	79.667	0	0	0
	0.5	20	290.39	5.5707	0.98481	10.000	0	290.39	5.5707	0.17365	80.000	0	0	0
	0.5	25	239.49	−8.4445	0.97630	12.500	0	239.49	−8.4445	0.21644	77.500	0	0	0
	0.5	20.678			0.98376	10.339	0.03	700.79	−56.134	0.17947	79.667	0.03	0.00056	0.00137
	0.5	20.678			0.98376	10.339	−0.03	145.05	14.503	0.17947	79.667	−0.03	0	0
	0.51	20.678	775.40	−59.809	0.99360	10.442	0	775.40	−59.809	0.18153	82.776	0	0	0
	0.49	20.678	182.28	6.9082	0.97393	10.233	0	182.28	6.9082	0.17795	76.496	0	0	0

Table 12.18. Elliptical Bearing Computer Results (Continued)

	ϵ	α	f_v	f_h	ϵ_1	α_1	$\dot{\epsilon}/\omega$	f_{v1}	f_{h1}	ϵ_2	α_2	$\dot{\epsilon}_2/\omega$	f_{v2}	f_{h2}
	0.2	105.652	1.4664	−0.00048	0.27481	44.491	0	1.6448	0.21708	0.35983	32.358	0	0.17845	0.21755
	0.2	105	1.4862	0.00993	0.27680	44.261	0	1.6581	0.22110	0.35830	32.627	0	0.17197	0.21117
	0.2	110	1.3309	−0.06822	0.26134	45.983	0	1.5541	0.19212	0.36973	30.551	0	0.22319	0.26034
	0.2	105.652			0.27481	44.491	0.03	2.1236	0.05678	0.35983	32.358	0.03	0.39306	0.40089
	0.2	105.652			0.27481	44.491	−0.03	1.2333	0.31543	0.35983	32.358	−0.03	0.06671	0.09808
	0.22	105.652	1.6282	−0.04982	0.28500	48.015	0	1.7953	0.16166	0.37494	34.403	0	0.16709	0.21147
	0.18	105.652	1.3012	0.03769	0.26574	40.710	0	1.4977	0.26527	0.34523	30.137	0	0.19654	0.22758
	0.5	54.288	10.010	−0.00007	0.67707	36.842	0	10.010	−0.00007	0.40813	84.113	0	0	0
	0.5	51	10.301	0.29239	0.68544	34.534	0	10.301	0.29239	0.39392	80.553	0	0	0
	0.5	57	9.7394	−0.23160	0.66982	38.759	0	9.7394	−0.23160	0.41993	86.958	0	0	0
$L/D = 1$	0.5	54.288			0.67707	36.842	0.03	12.159	−0.73831	0.40813	84.113	0.03	0	0
$m = 0.25$	0.5	54.288			0.67707	36.842	−0.03	8.1820	0.50890	0.40813	84.113	−0.03	0	0
	0.52	54.288	11.101	−0.23287	0.69618	37.336	0	11.101	−0.23287	0.42560	82.773	0	0	0
	0.48	54.288	9.0474	0.18671	0.65802	36.320	0	9.0474	0.18671	0.39091	85.571	0	0	0
	0.7	25.628	80.156	0.00028	0.93170	18.964	0	80.401	0.33508	0.48676	38.464	0	0.24576	0.33480
	0.7	25	81.039	0.58077	0.93258	18.495	0	81.305	0.93516	0.48507	37.581	0	0.26580	0.35439
	0.7	27.5	76.911	−1.4131	0.92895	20.362	0	77.103	−1.1332	0.49198	41.070	0	0.19230	0.27996
	0.7	25.628			0.93170	18.964	0.03	109.66	−4.7174	0.48676	38.464	0.03	0.49286	0.58856
	0.7	25.628			0.93170	18.964	−0.03	53.209	4.7403	0.48676	38.464	−0.03	0.10437	0.16437
	0.71	25.628	96.227	−1.8917	0.94164	19.034	0	96.497	−1.5264	0.49651	38.207	0	0.27005	0.36528
	0.69	25.628	68.075	1.2639	0.92177	18.891	0	68.299	1.5704	0.47701	38.731	0	0.22330	0.30652

	0.15	94.263	2.0616	0.00001	0.51122	17.014	0	3.5147	1.2483	0.53259	16.312	0	1.4531	1.2483
	0.15	93	2.1490	0.04401	0.51444	16.930	0	3.5673	1.2661	0.52948	16.434	0	1.4183	1.2221
	0.15	95	2.0101	−0.02575	0.50934	17.060	0	3.4840	1.2381	0.53439	16.238	0	1.4739	1.2638
	0.15	94.263			0.51122	17.014	0.03	4.5497	1.2912	0.53259	16.312	0.03	2.1909	1.6634
	0.15	94.263			0.51122	17.014	−0.03	2.6536	1.1529	0.53259	16.312	−0.03	0.91724	0.89050
	0.17	94.263	2.3271	−0.2857	0.51601	19.180	0	3.7276	1.2161	0.53994	18.299	0	1.4004	1.2447
	0.13	94.263	1.7678	0.01590	0.50719	14.810	0	3.3031	1.2773	0.52589	14.271	0	1.5353	1.2614
	0.3	87.235	5.5986	−0.00009	0.59538	30.218	0	6.3631	0.90295	0.57055	31.682	0	0.76451	0.90304
	0.3	84	6.1060	0.20133	0.60939	29.314	0	6.7522	0.98053	0.55556	32.483	0	0.64621	0.77920
	0.3	90	5.1412	−0.17954	0.58310	30.964	0	6.0241	0.84342	0.58310	30.964	0	0.88294	1.02296
$L/D = 1$ $m = 0.50$	0.3	87.235			0.59538	30.218	0.03	7.8472	0.60341	0.57055	31.682	0.03	1.2947	1.3640
	0.3	87.235			0.59538	30.218	−0.03	5.0667	1.0708	0.57055	31.682	−0.03	0.42279	0.55739
	0.32	87.235	6.0708	−0.10044	0.60650	31.803	0	6.8003	0.78855	0.58049	33.409	0	0.72948	0.88899
	0.28	87.235	5.1252	0.07084	0.58473	28.574	0	5.9506	1.0032	0.56115	29.893	0	0.82538	0.93231
	0.5	21.256	361.66	−0.61222	0.98285	10.628	0	361.66	−0.61222	0.18443	79.367	0	0	0
	0.5	20	349.96	9.8135	0.98481	10.000	0	349.96	9.8135	0.17365	80.000	0	0	0
	0.5	25	294.90	−6.5638	0.97630	12.500	0	294.90	−6.5638	0.21644	77.500	0	0	0
	0.5	21.256			0.98285	10.628	0.03	759.62	−54.350	0.18443	79.367	0.03	0.00067	0.00161
	0.5	21.256			0.98285	10.628	−0.03	165.36	19.725	0.18443	79.367	−0.03	0	0
	0.51	21.256	904.50	−73.916	0.99268	10.735	0	904.50	−73.916	0.18654	82.388	0	0	0
	0.49	21.256	227.18	9.3925	0.97302	10.519	0	227.18	9.3925	0.18285	76.287	0	0	0

Table 12.19. Computed on Basis of Tables 12.16, 12.17, and 12.18

Bearing type	$\frac{L}{D}$	m	ϵ	α	$\frac{\partial f_v}{\partial \epsilon}$	$\frac{\partial f_v}{\epsilon\, \partial \alpha}$	$\frac{\partial f_v}{\partial(\dot{\epsilon}/\omega)}$	$\frac{2f_v}{\epsilon}$	$\frac{\partial f_h}{\partial \epsilon}$	$\frac{\partial f_h}{\epsilon\, \partial \alpha}$	$\frac{\partial f_h}{\partial(\dot{\epsilon}/\omega)}$	$\frac{2f_h}{\epsilon}$
Plain cylindrical bearing	½		0.2	75.772	2.719	0	2.245	4.791	−0.571	−2.396	−4.644	0
			0.5	54.763	8.285	0	10.14	7.449	−2.304	−3.725	−7.098	0
			0.7	40.432	31.46	0	33.85	14.74	−6.726	−7.371	−14.43	0
	1		0.2	76.836	8.483	0	6.524	15.04	−1.676	−7.521	−15.09	0
			0.5	57.502	20.26	0	26.07	21.20	−6.073	−10.60	−21.39	0
			0.7	43.280	60.79	0	72.47	32.26	−16.91	−17.63	−36.09	0
Four-axial-groove bearing	½		0.2	81.696	1.499	−0.270	0.653	2.694	−0.183	−1.286	−0.108	0
			0.5	60.315	4.998	−1.753	1.137	4.722	−0.911	−0.723	−0.269	0
			0.7	35.139	29.79	−0.390	5.255	12.33	−4.438	−3.901	0.364	0
	1		0.2	83.531	2.819	−0.781	1.020	5.200	−0.254	−2.044	−1.883	0
			0.5	64.431	8.085	−3.382	1.860	8.512	−1.421	−1.267	−0.692	0
			0.7	35.909	49.29	−1.240	9.532	22.02	−7.782	−5.916	0.060	0
Elliptical bearing	½	0.25	0.2	103.799	2.863	−3.606	−2.722	8.921	−0.7902	−2.256	−3.030	1.419
			0.5	55.789	25.13	−8.085	23.73	21.39	−5.414	−3.531	−9.471	−3.061
			0.7	24.609	1,121	−75.17	670.1	198.2	−108.6	−62.32	−112.2	−9.234
		0.50	0.15	89.729	6.460	−13.18	−1.928	24.33	−0.2563	−7.281	−1.361	8.570
			0.3	84.510	11.70	−16.28	2.056	29.57	−2.003	−6.896	−4.758	5.662
			0.5	20.656	29,660	794.2	9,057	1,961	−3,336	−818.5	−1,158	−211.2
	1	0.25	0.2	105.622	8.175	−8.709	−6.706	23.15	−2.188	−4.571	−6.406	1.261
			0.5	54.287	51.35	−11.41	54.37	48.08	−10.49	−10.04	−19.83	−6.233
			0.7	25.628	1,408	−121.8	918.2	286.0	−157.8	−72.17	−152.3	−11.82
		0.50	0.15	94.263	13.98	−26.61	−8.788	55.65	−1.112	−13.33	−4.391	17.03
			0.3	87.234	23.64	−31.49	2.613	64.52	−4.282	−12.26	−10.51	8.412
			0.5	21.155	33,870	52.03	9,598	2,550	−4,165	−717.9	−1,214	−227.7

Bearing type	$\frac{L}{D}$	ϵ	$\frac{\partial f_r}{\partial \epsilon}$	$\frac{\partial f_r}{\partial(\dot{\epsilon}/\omega)}$	$\frac{2f_r}{\epsilon}$	$\frac{\partial f_t}{\partial \epsilon}$	$\frac{\partial f_t}{\partial(\dot{\epsilon}/\omega)}$	$\frac{2f_t}{\epsilon}$
Plain cylindrical bearing	½	0.2	1.222	5.054	1.144	2.496	1.035	4.653
		0.5	6.662	11.65	4.298	5.438	4.190	6.084
		0.7	28.31	35.13	11.22	15.28	10.96	9.561
	1	0.2	3.564	16.18	3.426	7.878	2.916	14.65
		0.5	16.01	32.05	11.39	13.83	10.50	17.88
		0.7	55.85	77.50	25.67	29.37	23.41	24.17

Bearing type	$\frac{L}{D}$	m	ϵ	$\frac{\partial f_{v1}}{\partial(\dot{\epsilon}_1/\omega)}$	$\frac{\partial f_{v2}}{\partial(\dot{\epsilon}_2/\omega)}$	$\frac{2f_{v1}}{\epsilon_1}$	$\frac{2f_{v2}}{\epsilon_2}$	$\frac{\partial f_{h1}}{\partial(\dot{\epsilon}_1/\omega)}$	$\frac{\partial f_{h2}}{\partial(\dot{\epsilon}_2/\omega)}$	$\frac{2f_{h1}}{\epsilon_1}$	$\frac{2f_{h2}}{\epsilon_2}$
Elliptical bearing	½	0.25	0.2	5.230	2.610	4.404	0.5556	−1.591	2.647	0.9180	0.7242
			0.5	29.16	0	13.06	0	−9.954	0	0	0
			0.7	685.5	3.787	118.5	0.6850	−115.8	4.283	0.5412	1.043
		0.50	0.15	13.57	9.158	6.161	2.714	1.267	6.154	2.531	2.537
			0.3	20.66	7.289	9.539	1.488	−3.111	7.159	1.714	1.864
			0.5	9,262	0.0094	303.9	0	−1,177	0.0228	0	0
	1	0.25	0.2	14.84	5.439	11.97	0.9918	−4.311	5.047	1.580	1.209
			0.5	66.28	0	29.57	0	−20.79	0	0	0
			0.7	940.9	6.475	172.6	1.010	−157.6	7.070	0.7193	1.376
		0.50	0.15	31.60	21.23	13.75	5.457	2.305	12.88	4.883	4.688
			0.3	46.34	14.53	21.38	2.680	−7.789	13.44	3.033	3.165
			0.5	9,904	0.01117	735.9	0	−1,235	0.0269	0	0

NOTE: For the elliptical bearing $\frac{\partial f_v}{\partial(\dot{\epsilon}/\omega)}$, $\frac{2f_v}{\epsilon}$, $\frac{\partial f_h}{\partial(\dot{\epsilon}/\omega)}$, and $\frac{2f_h}{\epsilon}$ are not computed directly but calculated from Eq. (12.16).

Table 12.20. Computed from Table 12.19

Bearing type	$\frac{L}{D}$	m	ϵ	$\frac{K_{xx}}{(1/C)\lambda\omega}$	$\frac{\omega C_{xx}}{(1/C)\lambda\omega}$	$\frac{K_{xy}}{(1/C)\lambda\omega}$	$\frac{\omega C_{xy}}{(1/C)\lambda\omega}$	$\frac{K_{yx}}{(1/C)\lambda\omega}$	$\frac{\omega C_{yx}}{(1/C)\lambda\omega}$	$\frac{K_{yy}}{(1/C)\lambda\omega}$	$\frac{\omega C_{yy}}{(1/C)\lambda\omega}$
Plain cylindrical bearing	½		0.2	0.651	5.20	−2.64	−0.999	2.18	−1.11	1.14	4.51
			0.5	4.78	11.94	−6.77	−3.99	1.71	−4.10	4.03	5.80
			.7	23.95	35.32	−20.40	−10.73	−0.339	−10.99	9.97	9.36
	1		0.2	1.93	16.13	−8.26	−2.93	6.94	−3.44	3.34	14.70
			0.5	10.89	31.88	−17.09	−10.60	5.68	−11.49	10.82	18.04
			0.7	44.26	76.93	−41.68	−24.02	−0.233	−26.28	24.42	24.74
Four-axial-groove bearing	½		0.2	0.484	2.76	−1.44	−0.257	1.25	−0.156	0.366	1.07
			0.5	4.00	4.67	−3.47	1.35	0.177	−0.133	1.15	0.233
			0.7	24.59	11.39	−16.83	7.05	−1.38	0.298	5.74	−0.209
	1		0.2	1.09	5.28	−2.71	−0.428	2.00	−0.212	0.482	1.87
			0.5	6.54	8.48	−5.83	2.00	0.530	−0.299	1.83	0.624
			0.7	40.65	20.63	−27.90	12.24	−2.83	0.048	9.36	−0.035
Elliptical bearing	½	0.25	0.2	2.819	9.313	−3.640	0.5157	2.379	2.101	0.2293	2.604
			0.5	20.82	31.03	−16.24	−7.598	−0.1239	−7.856	6.462	6.111
			0.7	1,050	691.8	−398.5	−98.85	−72.78	−105.9	101.9	38.33
		0.50	0.15	13.21	24.32	−6.398	2.043	7.280	8.563	0.2907	1.402
			0.3	17.32	29.63	−10.09	0.7824	6.673	5.181	2.654	5.278
			0.5	27,470	9,166	−11,210	−1,360	−2,833	−1,158	1,943	210.9
	1	0.25	0.2	6.186	24.10	−10.22	0.2242	4.991	2.939	0.8763	5.830
			0.5	39.24	70.78	−35.03	−16.08	2.029	−16.64	14.38	12.46
			0.7	1,322	951.6	−499.2	−139.3	−111.1	−142.4	133.3	55.22
		0.50	0.15	25.50	56.15	−15.92	4.627	13.38	17.31	0.1181	3.113
			0.3	32.59	64.57	−22.09	0.5038	12.04	7.895	4.869	10.90
			0.5	31,570	9,871	−12,270	−1,086	−3,625	−1,214	2,173	225.8

The equivalent four coefficients can be expressed in terms of the eight coefficients as

$$\begin{aligned} K_x &= K_{xx} - (1/\psi_x)[\zeta K_{xy} + \eta(\omega C_{xy})] \\ \omega B_x &= \omega C_{xx} + (1/\psi_x)[\eta K_{xy} - \zeta(\omega C_{xy})] \\ K_y &= K_{yy} - (1/\psi_y)[\zeta K_{yx} - \eta(\omega C_{yx})] \\ \omega B_y &= \omega C_{yy} - (1/\psi_y)[\eta K_{yx} + \zeta(\omega C_{yx})] \end{aligned} \tag{12.184}$$

where

$$\begin{aligned} \psi_x &= (K_{yy} - \varkappa + \omega C_{xy})^2 + (K_{xy} - \omega C_{yy})^2 \\ \psi_y &= (K_{xx} - \varkappa - \omega C_{yx})^2 + (K_{yx} + \omega C_{xx})^2 \\ \zeta &= (K_{xx} - \varkappa - \omega C_{yx})(K_{xy} - \omega C_{yy}) + (K_{yy} - \varkappa + \omega C_{xy})(K_{yx} + \omega C_{xx}) \\ \eta &= (K_{xx} - \varkappa - \omega C_{yx})(K_{yy} - \varkappa + \omega C_{xy}) - (K_{xy} - \omega C_{yy})(K_{yx} + \omega C_{xx}) \end{aligned} \tag{12.185}$$

Since Eqs. (12.183) are linear they may also be solved for the amplitudes:

$$\begin{aligned} \frac{A}{\delta} &= \frac{\varkappa(K_x - \varkappa)}{(K_x - \varkappa)^2 + (\omega B_x)^2} \\ \frac{B}{\delta} &= \frac{\varkappa(\omega B_x)}{(K_x - \varkappa)^2 + (\omega B_x)^2} \\ \frac{E}{\delta} &= \frac{-\varkappa(\omega B_y)}{(K_y - \varkappa)^2 + (\omega B_y)^2} \\ \frac{F}{\delta} &= \frac{\varkappa(K_y - \varkappa)}{(K_y - \varkappa)^2 + (\omega B_y)^2} \end{aligned} \tag{12.186}$$

The force transmitted to the bearing pedestal is given by

$$\begin{aligned} P_x &= K_x x_a + B_x \dot{x}_a \\ P_y &= K_y y_a + B_y \dot{y}_a \end{aligned} \tag{12.187}$$

Combining Eqs. (12.186) and (12.187) yields

$$\begin{aligned} \frac{P_x}{\delta} &= \varkappa \sqrt{\frac{K_x^2 + (\omega B_x)^2}{(K_x - \varkappa)^2 + (\omega B_x)^2}} \cos(\omega t - \varphi_x + \gamma_x) \\ \tan\varphi_x &= \frac{\omega B_x}{K_x - \varkappa} \qquad \tan\gamma_x = \frac{\omega B_x}{K_x} \\ \frac{P_y}{\delta} &= \varkappa \sqrt{\frac{K_y^2 + (\omega B_y)^2}{(K_y - \varkappa)^2 + (\omega B_y)^2}} \sin(\omega t - \varphi_y + \gamma_y) \\ \tan\varphi_y &= \frac{\omega B_y}{K_y - \varkappa} \qquad \tan\gamma_y = \frac{\omega B_y}{K_y} \end{aligned} \tag{12.188}$$

Instead of expressing the rotor amplitude in x and y coordinates a better physical picture is obtained by finding the corresponding elliptical path of the journal center. Combining the first and the third of Eq. (12.182), we get

$$\begin{aligned} a/\delta &= \sqrt{\tfrac{1}{2}[(A/\delta)^2 + (B/\delta)^2 + (E/\delta)^2 + (F/\delta)^2] + \tfrac{1}{2}\sqrt{[(A/\delta)^2 + (B/\delta)^2 + (E/\delta)^2 + (F/\delta)^2]^2 - [(A/\delta)(F/\delta) - (B/\delta)(E/\delta)]^2}} \\ b/\delta &= \sqrt{\tfrac{1}{2}[(A/\delta)^2 + (B/\delta)^2 + (E/\delta)^2 + (F/\delta)^2] - \tfrac{1}{2}\sqrt{[(A/\delta)^2 + (B/\delta)^2 + (E/\delta)^2 + (F/\delta)^2]^2 - [(A/\delta)(F/\delta) - (B/\delta)(E/\delta)]^2}} \\ \tan 2\beta &= \frac{2[(A/\delta)(E/\delta) + (B/\delta)(F/\delta)]}{[(A/\delta)^2 + (B/\delta)^2 - (E/\delta)^2 - (F/\delta)^2]} \end{aligned} \tag{12.189}$$

where a is the major axis of the ellipse, b is the minor axis, and β is the angle between the x axis and the major axis. Rotor resonance may be defined as the speed where the

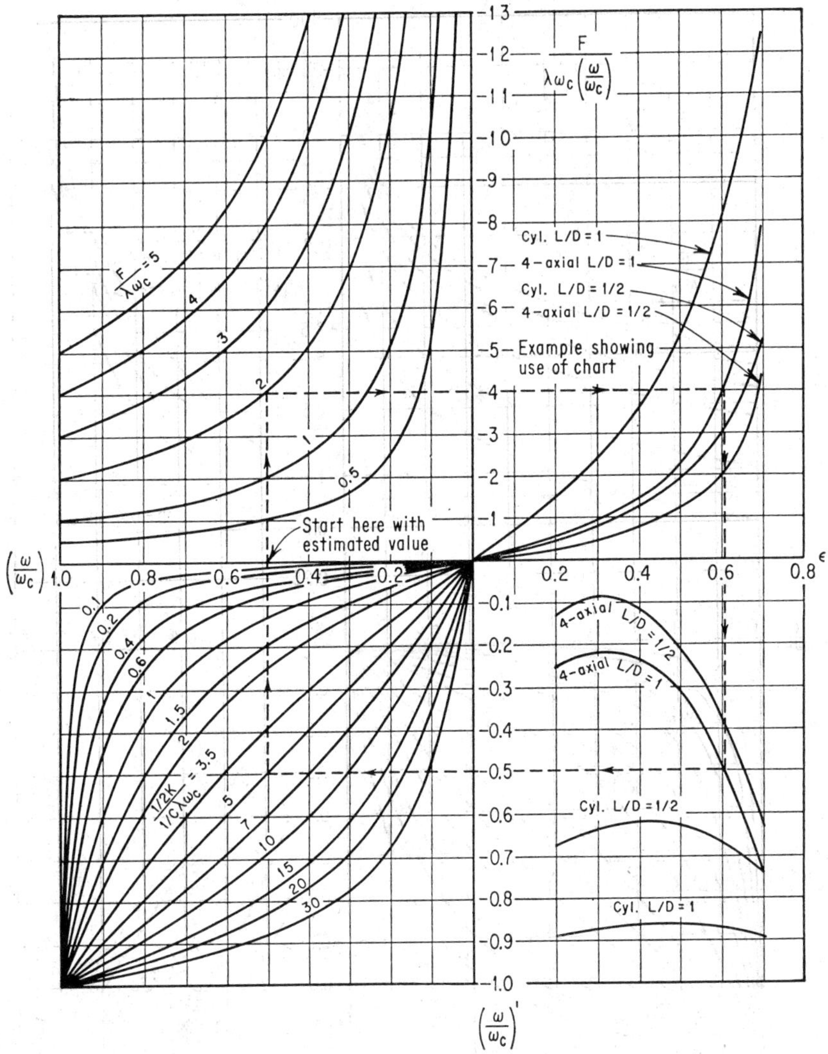

FIG. 12.44. Design chart for critical-speed calculation. Cylindrical and four-axial-groove bearing.

major axis is a maximum. This maximum is found by plotting the major axis as a function of $\varkappa$. The results are shown in Figs. 12.44 and 12.45. From this graph the rotor critical speed can be found directly for a given rotor by a trial-and-error process.

12.7. HYDRODYNAMIC INSTABILITY

Threshold of Half-frequency Whirl. This is a vibration of a journal or a bearing which may occur at any rotational speed and at a frequency of one-half or nearly one-half of the speed of the rotating member. The motion of the journal or bearing center is in the same direction as the direction of the rotating member. In general a

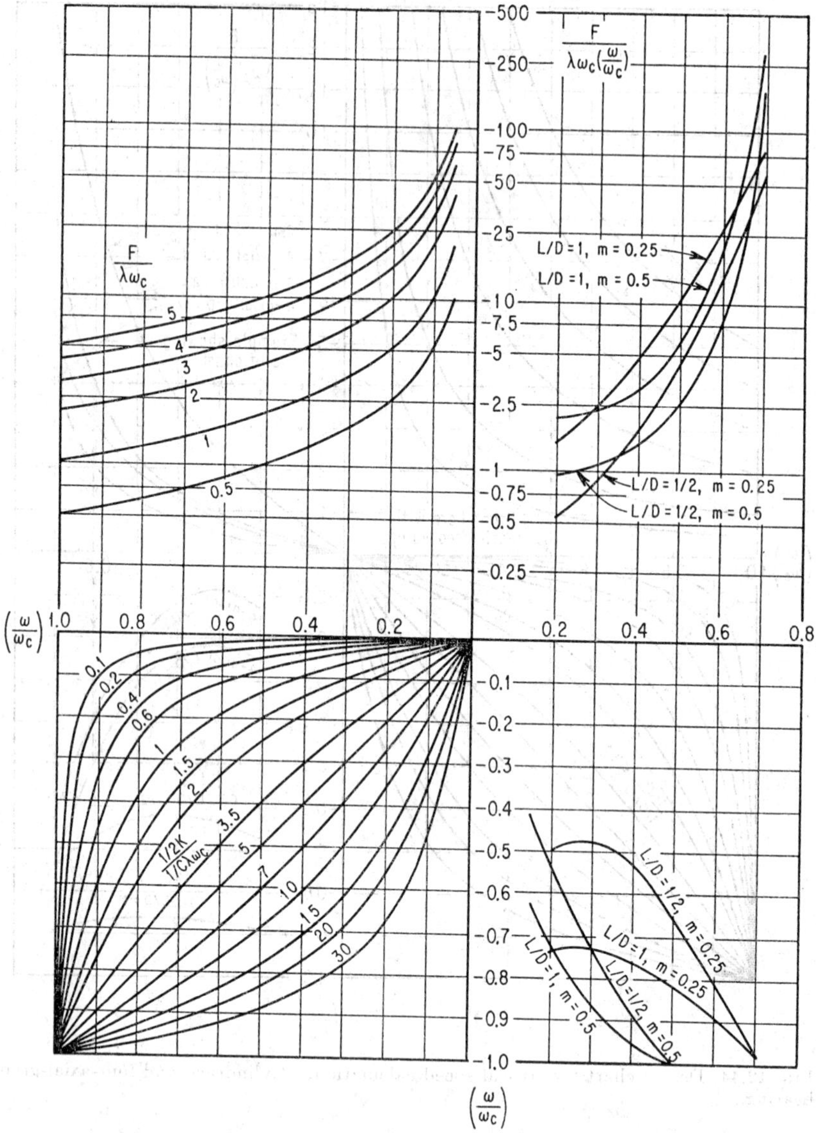

FIG. 12.45. Design chart for critical-speed calculation. Elliptical bearing.

rotor-bearing system has two modes of half-frequency whirl which depend on the rotor mass and the rotor inertia:

1. Translatory (cylindrical)—where the axis of the whirling member remains parallel to the axis of the other member.
2. Conical—where the center of gravity remains fixed in space and an axis through the center of gravity whirls in a conical mode.

By employing the hydrodynamic forces generated within the fluid film it can be shown that the threshold of translatory (cylindrical) instability of a symmetrical system[29] may be represented by

$$M\omega^2(1/k + 1/K_1) < 4 \tag{12.190a}$$

where $1/K_1$ is the dynamic radial fluid-film resilience and $1/k$ is the rotor resilience. This relation applies to rotors supported on either incompressible or compressible fluid bearings. However, in the case of compressible fluid bearings the rotor resilience in general can be neglected, being considerably lower than the fluid-film resilience; thus Eq. (12.190a) becomes

$$\omega^2 < 4K_1/M \tag{12.190b}$$

the threshold for the conical whirl becomes

$$\omega^2 < L^2K_2/I \tag{12.190c}$$

where K_2 is dynamic conical fluid-film stiffness, L is the span between center line of bearings, and I is the difference between transverse and polar inertia of the rotor.

Stability and Variational Equation of Motion. The variational equations are[30]

$$\begin{aligned}(b/g)\delta\ddot{u} + (f_{r\epsilon'}/C\omega f_0)\delta\dot{u} + [f_{r\epsilon}/Cf_0 - (b/g)(\omega_0\alpha_0')^2]\delta u \\ + [f_{r\alpha'}/\omega f_0 - (2b/g)\epsilon_0\omega_0\alpha_0']\delta\phi + (f_t/f_0 + \cos\phi_0)\delta 0 = -(f_r/f_t)_\omega\delta\omega\end{aligned} \tag{12.191a}$$

$$\begin{aligned}[(2b/g)\omega_0\alpha_0' - f_{t\epsilon'}/Cf_0\omega]\delta\dot{u} + [(b/g)(\omega\alpha')_0 - f_{t\epsilon}/Cf_0]\delta u + (\epsilon_0 b/g)\delta\ddot{\phi} \\ + [(2b/g)C\epsilon_0'\omega_0 - f_{t\alpha'}/\omega f_0]\delta\dot{\phi} + (f_r/f_0 - \sin\phi_0)\delta\phi = (f_t/f_0)_\omega\delta\omega\end{aligned} \tag{12.191b}$$

where α_0', ϵ_0' correspond to the solution (ξ_0, ϕ_0).

The solutions of the above equations depend on the coefficients of the dependent variables. The case under consideration is the "equilibrium" (quasi-static) case, where the shaft center is fixed in space but the journal is rotating about its own center; this case makes the coefficient of the δu's and $\delta\phi$'s constant and Eqs. (12.191) are linear differential equations with constant coefficients. The homogeneous part of the above equations admits solutions of the form

$$\delta u = Ae^{\nu t} \qquad C\delta\phi = Be^{\nu t} \tag{12.192}$$

The characteristic equation of the homogeneous system is

$$a_0\nu^4 + a_1\nu^3 + a_2\nu^2 + a_3\nu + a_4 = 0 \tag{12.193}$$

where

$$\begin{aligned}a_0 &= b^2\epsilon_0/g^2 \\ a_1 &= (b/C\omega f_0 g)(\epsilon_0 f_{r\epsilon'} - f_{t\alpha'}) \\ a_2 &= (b/Cf_0 g)(f_r + \epsilon_0 f_{r\epsilon} - f_0 \sin\phi_0) + (1/C^2\omega^2 f_0^2)(f_{t\epsilon'}f_{r\alpha'} - f_{r\epsilon'}f_{t\alpha'}) \\ &\quad + (2\omega b/C\omega f_0 g)[\epsilon_0' f_{r\epsilon} - (\epsilon_0\alpha_0'/C)f_{t\epsilon} - \alpha_0' f_{r\alpha'}] + (\epsilon_0 b^2\alpha_0'^2\omega^2/Cg^2)(4 - C) \\ a_3 &= (1/C^2\omega f_0)[(1/f_0)(f_t f_{r\epsilon'} + f_t f_{t\epsilon'} + f_{t\epsilon}f_{r\alpha'} - f_{r\epsilon}f_{r\alpha'}) + f_{t\epsilon}\cos\phi_0 \\ &\quad - f_{r\epsilon'}\sin\phi_0] + (b/C\omega f_0 g)\left[(\alpha_0'\omega_0)^2 f_{t\alpha'} - \frac{0}{(\alpha'\omega)_0} f_{r\alpha'}\right] + (2\omega b/Cf_0 g)[\epsilon_0' f_{r\epsilon} - \alpha_0' f_t \\ &\quad - (\epsilon_0\alpha_0'/C)f_{t\epsilon} - (\alpha_0'/C)\cos\phi_0 + (\epsilon_0 b/g)\alpha_0' \frac{0}{(\omega\alpha')_0} - C(\omega_0\alpha_0')^2] \\ a_4 &= (1/C^2 f_0)[(1/f_0)(f_r f_{r\epsilon} + f_t f_{t\epsilon}) + (f_{t\epsilon}\cos\phi_0) - (f_{r\epsilon}\sin\phi_0)] \\ &\quad - (\omega\alpha_0' b/Cg)[(\omega\alpha_0')(f_r/f_0 - \sin\phi_0) + (f_t/f_0 + \cos\phi_0)]\end{aligned} \tag{12.194}$$

The roots of Eq. (12.193) determine the nature of solutions (12.192). These roots are functions of the coefficients of Eq. (12.193) and consequently of the parameters of the bearing.

In order to assure stability, it is necessary that all the roots of Eq. (12.193) have negative real parts or, at the worst, be pure imaginary numbers. These requirements are fulfilled if the following conditions are satisfied:

If $\quad a_4 > 0$

then $\quad a_3 > 0 \qquad a_2 a_3 - a_1 a_4 > 0$

$$a_1 a_2 a_3 - a_1^2 a_4 - a_0 a_3^2 \geq 0 \tag{12.195}$$

The a_j's are defined by Eqs. (12.194). Thus, for any motion that satisfies the conditions of this analysis, we seek the maximum angular velocity ω for which all the conditions for stability are satisfied.

$$\epsilon_0' = \alpha_0' = 0 \qquad \omega \neq 0 \tag{12.196}$$

Under these conditions Eqs. (12.194) reduce to

$$\begin{aligned} a_0 &= b^2\epsilon_0/g^2 \\ a_1 &= (b/C\omega f_0 g)a_1{}^* \\ a_2 &= (1/Cf)[(1/C\omega^2 f_0)a_{22}{}^* + (b/g)a_{21}{}^*] \\ a_3 &= (1/C^2\omega f_0)a_3{}^* \\ a_4 &= (1/C^2 f_0)a_4{}^* \end{aligned} \tag{12.197}$$

where the $a_j{}^*$'s are defined as follows:

$$\begin{aligned} a_1{}^* &= \epsilon_0 f_{r\epsilon'} - f_{t\alpha'} \\ a_{21}{}^* &= f_r + \epsilon_0 f_{r\epsilon} - f_0 \sin\phi_0 \\ a_{22}{}^* &= f_{t\epsilon'}f_{r\alpha'} - f_{r\epsilon'}f_{t\alpha'} \\ a_3{}^* &= (1/f_0)(f_r f_{r\epsilon'} + f_t f_{t\epsilon'} + f_{t\epsilon}f_{r\alpha'} - f_{r\epsilon}f_{t\alpha'}) + (f_{t\epsilon'}\cos\phi_0 - f_{r\epsilon'}\sin\phi_0) \\ a_4{}^* &= (1/f_0)(f_r f_{r\epsilon} + f_t f_{t\epsilon}) + (f_{t\epsilon}\cos\phi_0 f_{r\epsilon}\sin\phi_0) \end{aligned}$$

The stability conditions [Eq. (12.195)] become

If
$$a_4{}^* > 0 \tag{12.198a}$$
then
$$a_3{}^* > 0 \tag{12.198b}$$
$$C\omega^2 \lessgtr q_{01}/b = q_1 \qquad \text{for } a_{22}{}^* \gtrless 0 \tag{12.198c}$$
$$C\omega^2 \lessgtr q_{02}/b = q_2 \qquad \text{for } a_{22}{}^* \gtrless 0 \tag{12.198d}$$

where

$$q_{01} = \frac{ga_3{}^*a_{22}{}^*}{f_0(a_1{}^*a_4{}^* - a_3{}^*a_{21}{}^*)}$$

$$q_{02} = \frac{ga_1{}^*a_3{}^*a_{22}{}^*}{f_0{}^2\epsilon_0 a_3{}^{*2} + f_0 a_1{}^*(a_1{}^*a_4{}^* - a_3{}^*a_{21}{}^*)}$$

and, as previously defined, $b = mg/W$.

It should be noted here, again, that the a_j's and $a_j{}^*$'s depend on f_r, f_t, and their derivatives which in turn depend on ϵ_0 and Λ (for the case of compressible fluid). The f_r, f_t, and derivatives are given in Figs. 12.46 through 12.53 as functions of Λ and ϵ_0.

Intersection of the family of curves

$$C^*\omega^{*2} = q_j(\Lambda,\epsilon_0) \tag{12.199a}$$
$$\omega^*/C^{*2} = \Lambda \tag{12.199b}$$

where

$$\begin{aligned} \omega^* &= (p_a/6\mu R^2)^{-1/5}\omega \\ C^* &= (p_a/6\mu R^2)^{2/5}C \end{aligned}$$

gives the upper bound of the stability region in C^*, ω^* plane. Figure 12.54*a* gives stability curves for $L/D = 2$.

Perturbed fluid-film forces can be combined with the perturbed dynamical equation to yield characteristic equations for ν in terms of the parameters $(\epsilon_0,\Lambda,L/D,mC\omega^2/p_aLD)$. The system is unstable if any root of the characteristic equation has a positive real part. In the case that the characteristic equation is a polynomial in ν, stability may be verified conveniently by Hurwitz-Routh criteria. Using the perturbed quasistatic linearized PH results given in Figs. 12-17(*a-f*) it is also possible to perform such a stability analysis. Even though this uncoupled method lacks rigorous justification it is extremely simple and agrees well with practice. Using the quasistatic linearized PH forces stability analysis was performed and the results are given for a range of L/D ratios and one commonly used value of $P/p_a = 0.2$. (Refer to Fig. 12-54*b*.) The effect of mass to weight ratio is shown in Fig. 12-54*c*.

Synchronous Whirl. This is a whirling motion of the journal or bearing at a frequency equal to the rotational frequency. The motion of the journal or bearing center is in the same direction as the direction of the rotating member, and the force leads the point of h_{min} by an angle $\leq 90°$.

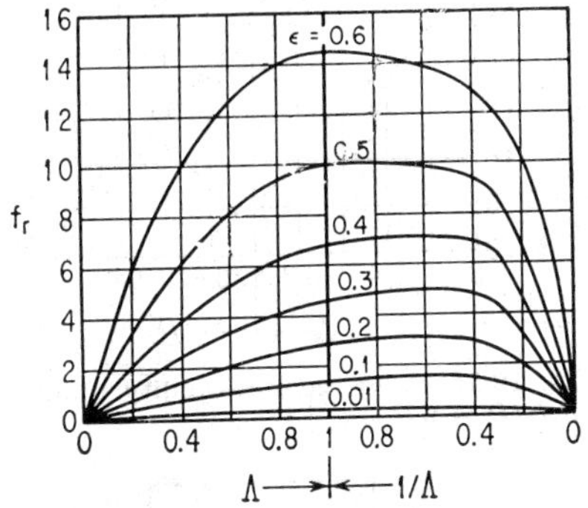

FIG. 12.46. Dimensionless radial force vs. compressibility number for $L/D = 2$.

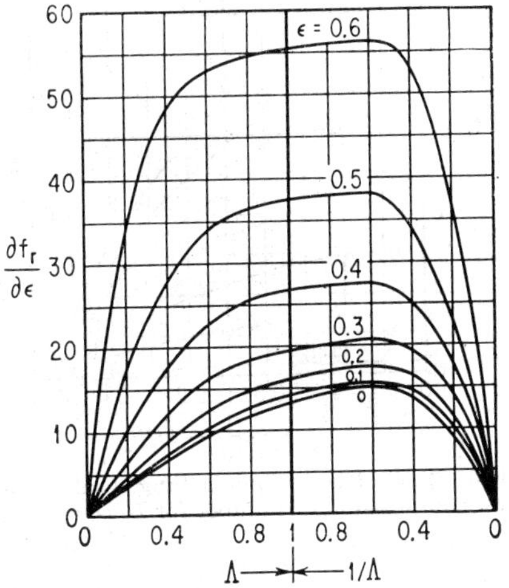

FIG. 12.47. $f_{r\epsilon}$ vs. compressibility number for $L/D = 2$.

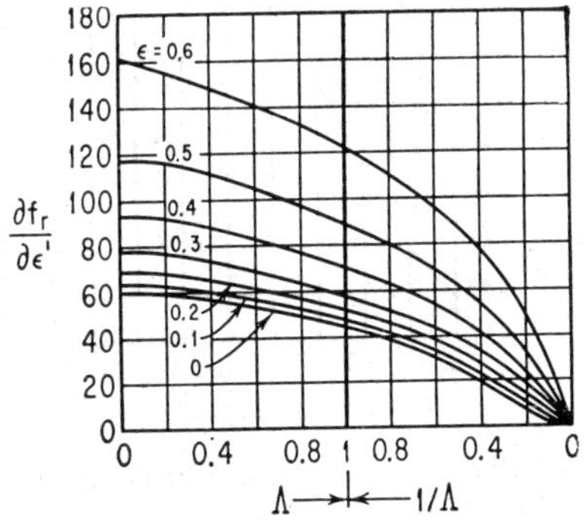

FIG. 12.48. $f_{r\epsilon}$ vs. compressibility number for $L/D = 2$.

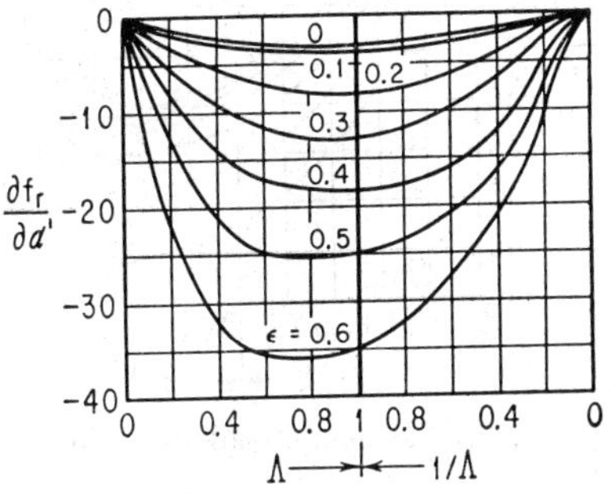

FIG. 12.49. $f_{r\alpha}'$ vs. compressibility number for $L/D = 2$.

An example of the synchronous whirl is the case of a rotor mass M which has a slight unbalance so that the center of gravity is separated from the geometric shaft center by a small distance δ. The shaft has an instantaneous eccentricity e from the bearing center and the planes of maximum film thickness and the mass unbalance form, respectively, the instantaneous angles α and β from a stationary reference plane. The rotor has an angular speed ω. Choosing the bearing center as the polar axis, then the motion of the shaft center is described by the radial velocity $\dot{e}$ and angular velocity $\dot{\alpha}$.

For the vertical rotor in plain cylindrical journal bearings, since axisymmetry prevails, the steady-state condition is described by a circular orbit of the shaft center with a whirl angular speed $\dot{\alpha} = \omega =$ constant.

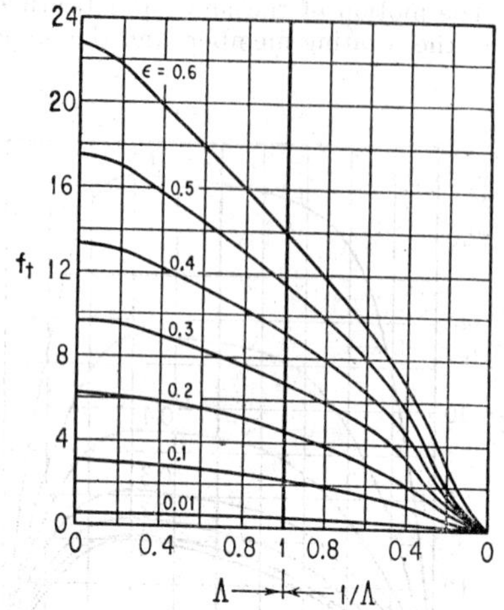

Fig. 12.50. Dimensionless tangential force vs. compressibility number for $L/D = 2$.

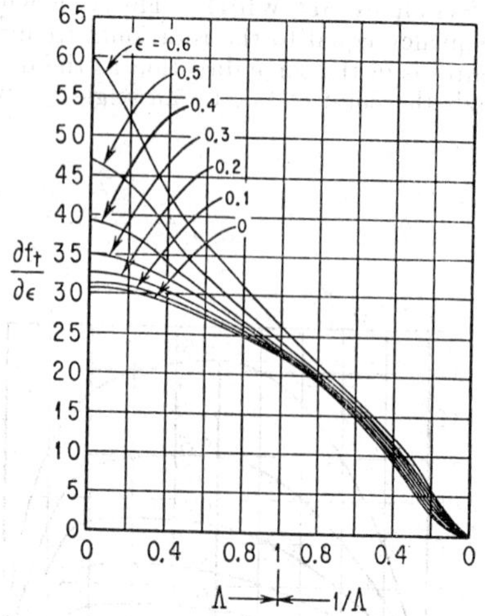

Fig. 12.51. $f_{t\epsilon}$ vs. compressibility number for $L/D = 2$.

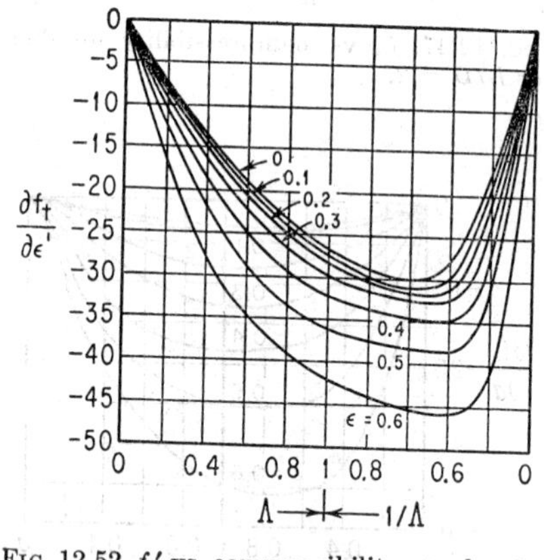

Fig. 12.52. f_t' vs. compressibility number for $L/D = 2$.

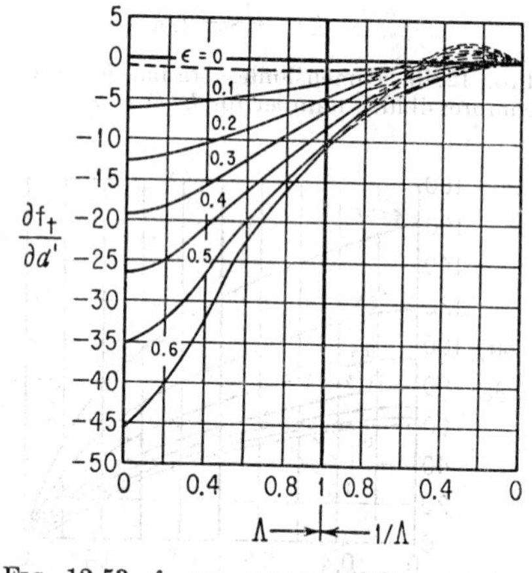

Fig. 12.53. $f_{t\alpha}$ vs. compressibility number for $L/D = 2$.

Under dynamical conditions, either e or $\dot{\alpha}$ or both may not be constant. The condition of dynamical equilibrium gives rise to the following equations:

$$F_r = -M\{\delta[\omega^2 \cos(\beta - \alpha) + \dot{\omega} \sin(\beta - \alpha)] + \ddot{e} - e(\dot{\alpha})^2\} \qquad (12.200a)$$

$$F_t = M\{\delta[\omega^2 \sin(\beta - \alpha) - \dot{\omega} \cos(\beta - \alpha)] + e\ddot{\alpha} + 2\dot{e}\dot{\alpha}\} \qquad (12.200b)$$

If the shaft undergoes a steady whirl motion both e and $\dot{\alpha}$ would be constant. The results of the steady-whirl analysis are applicable when the above situation does not hold precisely, provided the following conditions are satisfied:

$$\dot{e}/e\dot{\alpha} \ll 1 \qquad \ddot{\alpha}/(\dot{\alpha})^2 \ll 1$$

Under these conditions, it is advantageous to employ rotating coordinates which are stationary with respect to the maximum film thickness. Thus the geometrical boundary conditions are now independent of time, a condition especially important for the compressible Reynolds equation.[31] Under the steady-whirl condition the

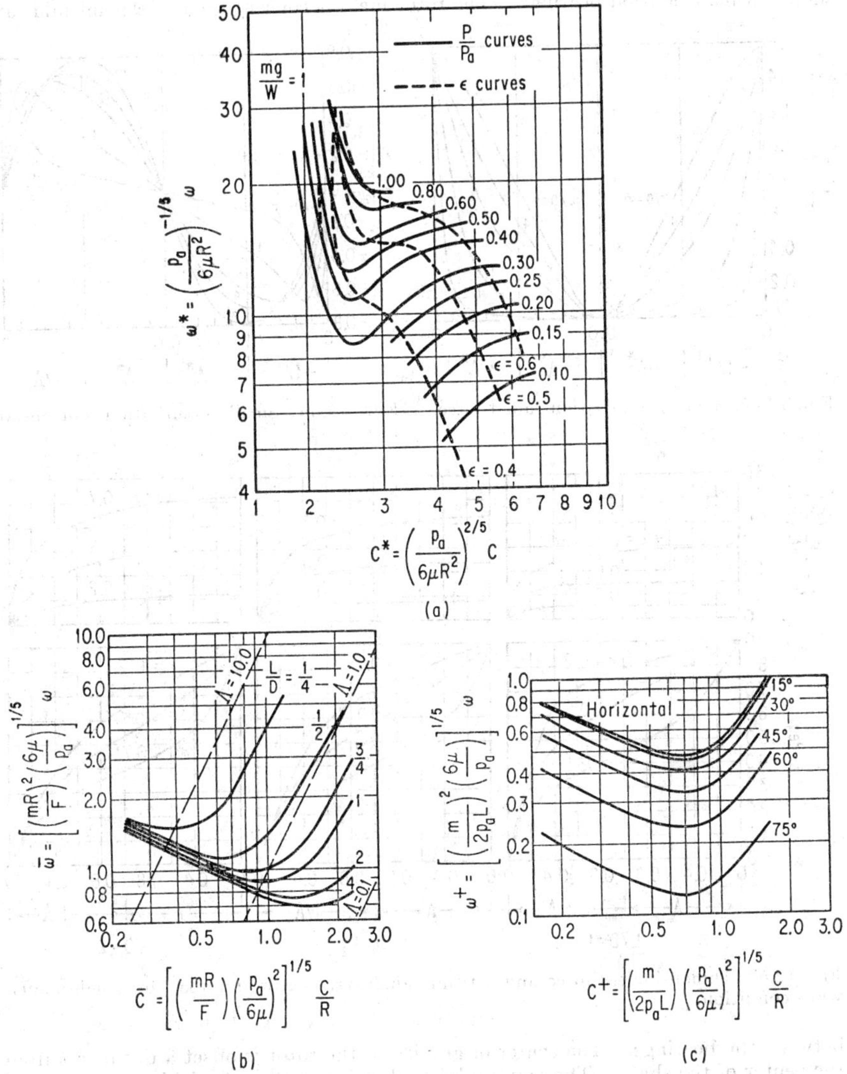

FIG. 12.54. Stability curves, C^* vs. ω^* for $b = 1$.

forces are dependent on only three parameters ϵ, $(1 - 2\dot{\alpha}/\omega)\Lambda$, and L/D. Figures 12.55 and 12.56 give the radial and tangential forces as a function of

$$\Lambda^* = (1 - 2\dot{\alpha}/\omega)\Lambda$$

and L/D using small perturbation for the solution of the dynamic Reynolds equation. At synchronous whirl $\dot{\alpha}/\omega = 1$; thus $\Lambda^* = -\Lambda$. Figure 12.57 gives design curves for the dimensionless force and phase angle for three L/D ratios and eccentricity ratios obtained by iterative solution.[32] Note that these figures are quite similar to Fig. 12.18a to h for steady-load conditions except that load now leads the point of h_{min}.

Forced Vibration of Vertical Rotor. Consider a vertical rotor supported by two identical plain journal bearings. The rotor mass is concentrated at a point midway

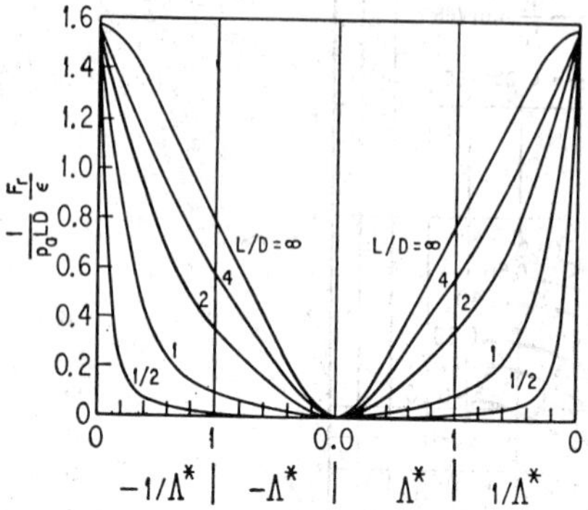

FIG. 12.55. Small-perturbation radial force.

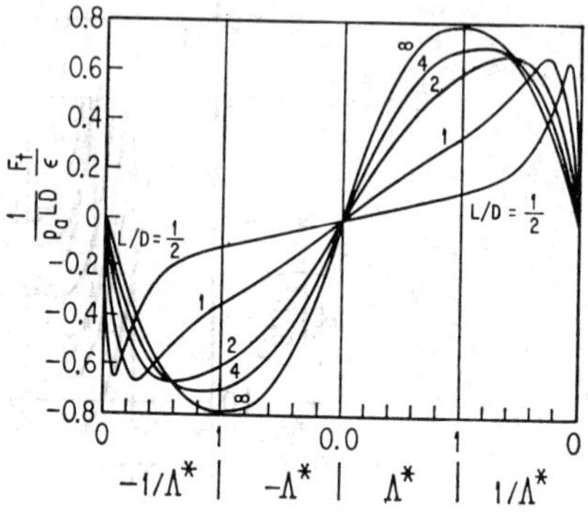

FIG. 12.56. Small-perturbation tangential force.

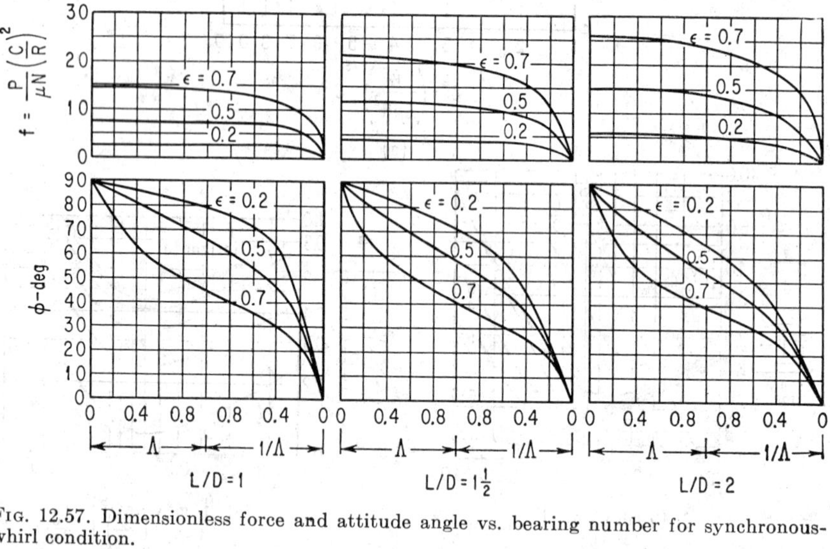

FIG. 12.57. Dimensionless force and attitude angle vs. bearing number for synchronous-whirl condition.

between the bearings. The center of gravity of the rotor is offset a distance δ from the center of the shaft. The amplitude of vibration of the rotor is the vector sum of e_2 and δ.

The centrifugal force acting on the rotor will be given by the following expression:

$$CF = M\omega^2(e_2 + \delta) \qquad (12.201)$$

If we consider the disturbing force as the unbalance force, then it rotates with the same angular velocity as the journal, that is, $\dot{\alpha} = \omega$. If use is made of the incompressible fluid-film force f (Table 12.16) or the compressible-fluid force given in Fig. 12.57 and the rotation is taken into account, the force CF will be given by the following expression:

$$CF/2 = (1/2\pi)\mu\omega LD(R/C)^2 fe^{i(\omega t+\phi)} \tag{12.202}$$

For convenience, the attitude angle ϕ has been chosen to be positive when the force on the fluid film leads the displacement. Here, f and ϕ are a function of the eccentricity ratio ϵ as given in Table 12.16 or in Fig. 12.57.

Let us introduce the symbol β for the phase-angle difference between vectors δ and $e_2 + \delta$. By resolving the displacement vectors in the x and y direction, we arrive at the following two equations:

$$e \sin \phi = \delta \sin \beta \tag{12.203}$$

$$\delta + e_2 = e \cos \phi + \delta \cos \beta + y \tag{12.204}$$

The deflection of the shaft is obtained from simple-beam theory and is given by

$$y = CFl^3/48EI \tag{12.205}$$

If use is made of Eqs. (12.201), (12.202), and (12.205), Eq. (12.204) can now be written as

$$(1/\pi)\mu LD(R/C)^2 f(1/M\omega - l^3\omega/48EI) = e \cos \phi + \delta \cos \beta \tag{12.206}$$

The unknown β can be eliminated between Eqs. (12.203) and (12.206), and as a result we obtain the following relation between the frequency and the amplitude of whirl:

$$[(1/\pi)\mu LD(R/C)^2 f(1/M\omega - l^3\omega/48EI)]^2 \\ - (2/\pi)\mu LD(R/C)^2 f(1/M\omega - l^3\omega/48EI)e \cos \phi + e^2 = \delta^2 \tag{12.207}$$

It will now be convenient to introduce into the above equation a new variable defined by

$$\gamma = 1/M\omega - l^3\omega/48EI \tag{12.208}$$

This leads to a simple quadratic equation, the solution of which is

$$\gamma = (\pi/f\mu LD)(C/R)^2\{e \cos \phi \pm [e^2 \cos^2 \phi + (\delta^2 - e^2)]^{1/2}\} \tag{12.209}$$

By making use of Table 12.16 or Fig. 12.57 and the above equation, values can be found for γ as a function of any eccentricity e.

The frequency ω is defined in terms of γ by Eq. (12.208). By rearranging and solving for ω, we get the following formula:

$$\omega = -24EI\gamma/l^3 \pm [(24EI\gamma/l^3)^2 + 48EI/l^3M]^{1/2} \tag{12.210}$$

Fluid-film and Rotor Resonance. Fluid-film resonance is a consequence of the quasi-elastic properties of the fluid film. The phenomenon may therefore be studied by assuming that the rotor is rigid. The conditions of a rigid rotor may be simulated by setting $l^3/EI = 0$ in Eq. (12.208). Thus we obtain

$$\gamma = 1/M\omega$$

By substituting the expression for γ in Eq. (12.209) we obtain the following equation for ω:

$$\omega = \frac{\mu LD(R/C)^2 f}{\pi M\{e \cos \phi \pm [e^2 \cos^2 \phi + (\delta^2 - e^2)]^{1/2}\}} \tag{12.211}$$

By considering the forces acting on the journal, it is possible to derive the frequency-response characteristics for the system and predict the variation in amplitude of "fluid resonance" with speed. Denoting the disturbing force by vector P, the instantaneous position of the journal by the vector e, and the force which the fluid film exerts on the journal by F, the inertia force acting on the journal will be given by $M\ddot{e}$. If the journal is vertical and there are no other external forces acting, we can express the conditions of force equilibrium for the journal by the following vector equation:

$$M\ddot{e} + F = F_2 \tag{12.212}$$

The disturbing force acting on the rotor is

$$CF = F_2 = m\,\delta\omega^2 e^i(\beta + \omega t) \tag{12.213}$$

If we assume that the journal center moves in a circular path around the bearing center, we can write

$$e = e_0 e^{i\omega t}$$

The inertia force acting on the journal can be expressed by

$$M\ddot{e} = -M\omega^2 e_0 e^{i\omega t} \tag{12.214}$$

If we assume that a single bearing is taking the total load, Eq. (12.211) becomes

$$F = (1/2\pi)\mu\omega LD(R/C)^2 f e^{i(\omega t+\phi)} \tag{12.215}$$

By making use of Eqs. (12.213) to (12.215) in Eq. (12.212), we obtain for ω

$$\omega^2 - \frac{Me_0\mu LD(R/C)^2 f\cos\phi}{\pi[(Me_0)^2 - (m\delta)^2]}\,\omega + \frac{[\mu LD(R/C)^2 f]^2}{4\pi^2[(Me_0)^2 - (m\delta)^2]} = 0 \tag{12.216}$$

The condition for resonance is that the two roots of the above equation coincide. This will lead to the following relation:

$$\cos^2\phi = 1 - (m\delta/Me_0)^2 \tag{12.217}$$

and the resonance frequency will be given by

$$\omega_{\text{res}} = \frac{Me_0\mu LD(R/C)^2 f\cos\phi}{2\pi[(Me_0)^2 - (m\delta)^2]} \tag{12.218}$$

Resonant Whip. This is a vibration of a flexible rotor which is excited by the system first critical frequency and occurs at rotational speeds about twice the actual system first critical. The frequency of vibration is approximately equal to the system first critical regardless of running speed.

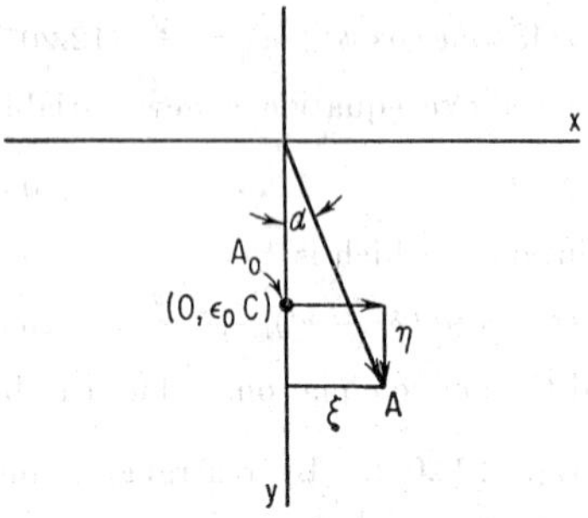

FIG. 12.58. Coordinates for small displacement from equilibrium.

Equations of Small Oscillations.[33] Let the components of the displacement along the x, y direction be ξ, η and let α be measured from the y axis. Then for small ξ, η there results from Fig. 12.58

$$d\epsilon = \eta/C \qquad d\alpha = \xi/\epsilon_0 C \tag{12.219}$$
$$d\epsilon' = d\dot{\epsilon}/\omega = \dot{\eta}/C\omega \qquad d\dot{\alpha} = \dot{\xi}/\epsilon_0 C$$

$$\begin{aligned} X &= \lambda[-\omega f_r\xi/\epsilon_0 C - 2f_t\dot{\xi}/\epsilon_0 C + (\omega\,\partial f_t/\partial\epsilon)(\eta/C) + (\partial f_t/\partial\epsilon')(\dot{\eta}/C)] \\ &= (M/2)(\ddot{x} + \ddot{\xi}) = -kx/2 \\ Y &= \lambda[-\omega f_t\xi/\epsilon_0 C + 2f_r\dot{\xi}/\epsilon_0 C - (\omega\,\partial f_r/\partial\epsilon)(\eta/C) - (\partial f_r/\partial\epsilon')(\dot{\eta}/C)] \\ &= (M/2)(\ddot{y} + \ddot{\eta}) = -ky/2 \end{aligned} \tag{12.220}$$

Here the functions f_r, f_t, and their derivatives with respect to ϵ, ϵ' are all evaluated at the equilibrium eccentricity ratio, and for $\epsilon' = 0$. They are given in Fig. 12.59*a* to *d* for the incompressible cases.

The differential equations (12.220) are linear in the variables ξ, η; x, y, and their solutions contain time as an exponential. These may be expressed in $e^{\nu\tau}$

where $$\tau = \omega_0 t \qquad \omega_0 = \sqrt{k/M} \tag{12.221}$$

Here ω_0 is the critical speed of the simply supported shaft-rotor system whose mass is M and whose stiffness is k.

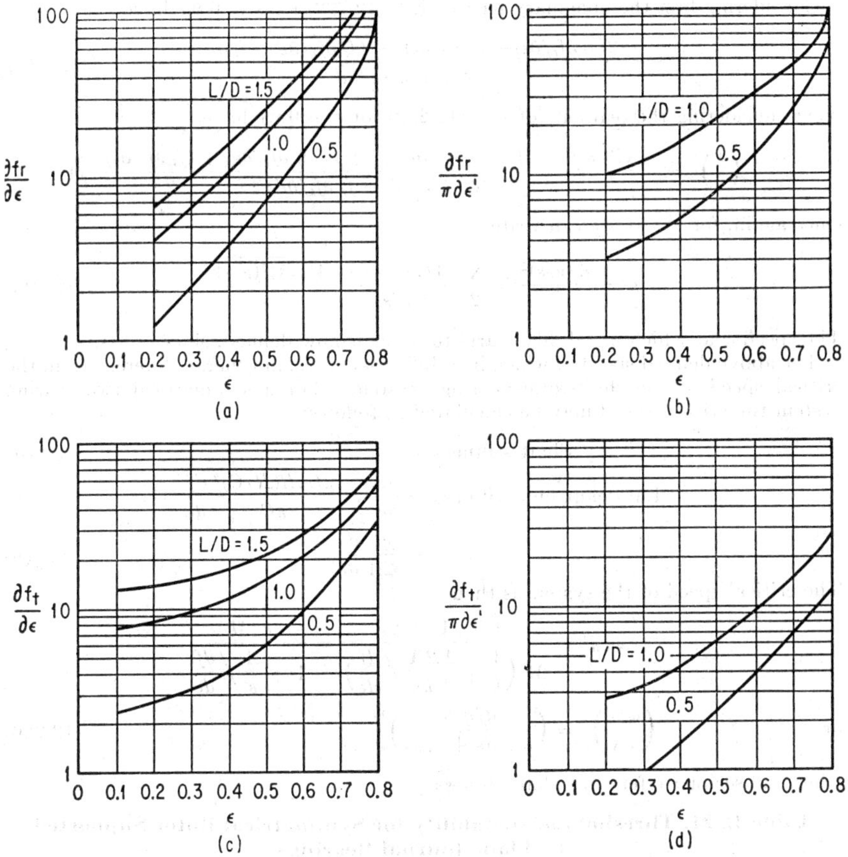

FIG. 12.59. Derivatives of f_r and f_t with respect to ϵ and ϵ'. (*a*) Radial stiffness for incompressible plain bearings. (*b*) Derivative of radial force with respect to velocity for incompressible plain bearings. (*c*) Tangential stiffness for incompressible plain bearings. (*d*) Derivative of tangential force with respect to velocity for incompressible plain bearings.

If we introduce the dimensionless ratio

$$s = \omega/\omega_0$$

where ω is the angular speed at the threshold of instability and

$$\zeta = A\nu^2/(1 + \nu^2)$$

where

$$A = kC^3\pi/2\mu LR^3\omega_0$$

Eq. (12.220) leads to the detrimental equation

$$\omega_0^2 \begin{vmatrix} sf_r + 2\nu f_t + \epsilon_0\zeta & s(\partial f_t/\partial\epsilon) + \nu(\partial f_t/\partial\epsilon') \\ -sf_t + 2\nu f_r & s(\partial f_r/\partial\epsilon) + \nu(\partial f_r/\partial\epsilon') + \zeta \end{vmatrix} = 0 \qquad (12.222)$$

If the system is dynamically stable, the real part of the complex number ν is negative. Conversely, if the system is dynamically unstable, the real part of ν is positive. Thus, at the threshold of instability, ν will be a pure imaginary number.

We now solve Eq. (12.222) for the condition where ν is wholly imaginary in order to obtain the value of ω at the onset of instability.

Considering first the imaginary part of Eq. (12.222) for $s \neq 0$ we have

$$\frac{\zeta}{s} = \frac{-2(f_t\,\partial f_r/\partial\epsilon - f_r\,\partial f_t/\partial\epsilon) - (f_r\,\partial f_r/\partial\epsilon' + f_t\,\partial f_t/\partial\epsilon')}{2f_t + \epsilon\,\partial f_r/\partial\epsilon'} \tag{12.223}$$

Next considering the real part of Eq. (12.222) for $s \neq 0$ we have

$$\left(\frac{\nu}{s}\right)^2 = \frac{-\epsilon(\zeta/s)^2 - (f_r + \epsilon\,\partial f_r/\partial\epsilon)(\zeta/s) - (f_r\,\partial f_r/\partial\epsilon) + (f_t\,\partial f_t/\partial\epsilon)}{2(f_t\,\partial f_r/\partial\epsilon' - f_r\,\partial f_t/\partial\epsilon')} \tag{12.224}$$

Once again, for $s \neq 0$, we can write

$$s = \frac{A(\nu/s)^2 \pm \sqrt{[A(\nu/s)^2]^2 - 4(\zeta/s)^2(\nu/s)^2}}{2(\zeta/s)(\nu/s)^2} \tag{12.225}$$

The speed ω at which instability starts to occur is now defined, since $\omega = s\omega_0$.

The above defined speed at which instability sets in is, in general, different from the critical speed of the shaft-rotor-bearings system. For a symmetrical two-bearing system the critical speed may be calculated as follows:

$$\text{Shaft stiffness} = k \tag{a}$$

$$\text{Lubricant-film stiffness} = \frac{dF}{de} = \frac{\mu L\omega R(R/C)^2}{\pi C}\frac{df}{d\epsilon} = \frac{sk}{2A}\frac{df}{d\epsilon} \tag{b}$$

The critical speed of the system is then

$$\omega_{CR}{}^2 = \frac{1}{M\left(\dfrac{1}{k} + \dfrac{1}{2}\dfrac{2A}{ks}\Big/\dfrac{df}{d\epsilon}\right)} = \frac{k/M}{1 + \dfrac{A}{s}\Big/\dfrac{df}{d\epsilon}}$$

or

$$\left(\frac{\omega_{CR}}{\omega_0}\right)_r = \left(\frac{df_r/d\epsilon}{df_r/d\epsilon + A/s}\right)^{1/2} \tag{12.226}$$

where subscript r refers to radial stiffness.

Table 12.21. Threshold of Instability for Symmetrical Rotor Supported by Plain Journal Bearings

L/D	ϵ	A	ζ/s	$(\nu/s)^2$	$\frac{1}{i}\frac{\nu}{s}$	s	$\frac{(\omega_{CR})_r}{\omega_0}$	$\frac{\omega}{(\omega_{CR})_r}$
½	0.2	0.1	− 0.9429	−0.1411	0.3756	2.6096	0.9852	2.6488
	0.5	0.1	− 3.1968	−0.1252	0.3538	2.8104	0.9974	2.8177
	0.8	0.1	−13.4452	−0.05543	0.2354	4.2438	0.9998	4.2446
½	0.2	100.0	− 0.9429	−0.1411	0.3756	0.06678	0.02910	2.2948
	0.5	100.0	− 3.1968	−0.1252	0.3538	0.2533	0.1315	1.9262
	0.8	100.0	−13.4452	−0.05543	0.2354	1.9267	0.7751	2.4857
1	0.2	0.1	− 2.6761	−0.1377	0.37	2.6763	0.9948	2.6903
	0.3	0.1	− 4.324	−0.1338	0.37	2.7224	0.9971	2.7303
	0.7	0.1	−18.484	−0.0846	0.29	3.4179	0.9998	3.4186
	0.8	0.1	−29.2005	−0.04898	0.22	4.5169	0.9999	4.5173
1	0.2	100.0	− 2.6761	−0.1377	0.37	0.1951	0.0834	2.3420
	0.3	100.0	− 4.324	−0.1338	0.37	0.3198	0.1390	2.300
	0.7	100.0	−18.483	−0.0846	0.29	1.6553	0.7000	2.3647
	0.8	100.0	−29.2005	−0.04898	0.22	3.1199	0.8998	3.4675

The dimensionless number A is a function of bearing geometry, shaft stiffness, and fluid viscosity. Calculations for the threshold of instability in which A was varied from 0.1 to 100 for $0.1 \leq \epsilon \leq 0.8$ and $L/D = 0.5$ and 1 are given in Table 12.21.

12.8. ADIABATIC SOLUTIONS

(a) One-dimensional Solutions

The Geometric and Thermal Wedge. By employing the dimensionless parameters

$$\bar{x} = x/B \qquad \bar{h} = h/B \qquad \bar{\mu} = \mu/\mu_1 \qquad \bar{\rho} = \rho/\rho_1 \qquad \bar{p} = pB/6\mu_1 U \qquad \bar{T} = T\rho gcB/\mu_1 U$$

Table 12.22. Hydrodynamic Equations of the Geometric and Thermal Wedges[2]

	Geometric wedge	Thermal wedge
Equations	$\dfrac{d\bar{p}}{d\bar{x}} = \dfrac{1}{\bar{h}^2}\left(1 - \dfrac{\bar{h}_0}{\bar{h}}\right)$ $\dfrac{d\bar{T}}{d\bar{x}} = \dfrac{8}{\bar{h}_0\bar{h}}\left[1 - \dfrac{3}{2}\dfrac{\bar{h}_0}{\bar{h}} + \dfrac{3}{4}\left(\dfrac{\bar{h}_0}{\bar{h}}\right)^2\right]$	$\dfrac{d\bar{p}}{d\bar{x}} = \dfrac{1}{\bar{h}_2^2}\left(1 - \dfrac{\bar{\rho}_0}{\bar{\rho}}\right)$ $\dfrac{d\bar{T}}{d\bar{x}} = \dfrac{8}{\bar{h}_2^2}\left[1 - \dfrac{3}{2}\dfrac{\bar{\rho}_0}{\bar{\rho}} + \dfrac{3}{4}\left(\dfrac{\bar{\rho}_0}{\bar{\rho}}\right)^2\right]$
Boundary conditions	$\bar{x} = 0,\ \bar{h} = \bar{h}_1,\ \bar{p} = 0,\ \bar{T} = \bar{T}_1$ $\bar{x} = 1,\ \bar{h} = \bar{h}_2,\ \bar{p} = 0,\ \bar{T} = \bar{T}_2$	$\bar{x} = 0,\ \bar{\rho} = 1,\ \bar{p} = 0,\ \bar{T} = \bar{T}_1$ $\bar{x} = 1,\ \bar{\rho} = \bar{\rho}_2,\ \bar{p} = 0,\ \bar{T} = \bar{T}_2$
Definition of a	$\bar{h}_1/\bar{h}_2$	$1/\bar{\rho}_2$
Law of variation	$\bar{h} = a\bar{h}_2\left(1 - \dfrac{a-1}{a}\bar{x}\right)$	$\rho = 1 - \dfrac{(a-1)\bar{x}}{a}$
$\bar{h}_0$ or $\bar{\rho}_0$	$2a\bar{h}_2/(a+1)$	$(a-1)/a \ln a$
$\bar{p}$	$\dfrac{(\bar{h}_1 - \bar{h})(\bar{h} - \bar{h}_2)}{(a^2-1)(\bar{h}_2{}^2\bar{h}^2)}$	$\left[\dfrac{a}{a-1}(1-\bar{\rho}) - \dfrac{\ln(1/\bar{\rho})}{\ln a}\right] \Big/ \bar{h}_2{}^2$
$\bar{P}$	$\dfrac{6}{(a-1)^2}\left[\ln a - \dfrac{2(a-1)}{a+1}\right]$	$3\left(\dfrac{a+1}{a-1} - \dfrac{2}{\ln a}\right)$
$\bar{F}$	$\dfrac{2}{a-1}\left[2\ln a - \dfrac{3(a-1)}{a+1}\right]$	1
$\Delta\bar{T}$	$\dfrac{2}{a}\left[2\dfrac{(a+1)}{a-1}\ln a - 3\right]$	$6\dfrac{(a-1)^2}{a(\ln a)^2} - 4$

Load criterion $\qquad \bar{P} = \dfrac{\text{load per unit area}}{\mu_1 U/B}\left(\dfrac{h_2}{B}\right)^2 = 6\bar{h}_2{}^2 \displaystyle\int_0^1 \bar{p}\, d\bar{x}$

Drag criterion $\qquad \bar{F} = \dfrac{\text{frictional drag per unit area}}{\mu_1 U/B}\left(\dfrac{h_2}{B}\right)$

$$= 4\bar{h}_2 \int_0^1 \frac{1}{\bar{h}}\left(1 - \frac{3\bar{\rho}_0\bar{h}_0}{4\bar{\rho}\bar{h}}\right) d\bar{x}$$

Temperature criterion $\Delta\bar{T} = \dfrac{\text{temperature rise through bearing}}{\mu_1 U/g\rho_1 cB}\, h_2{}^2 = (\bar{T}_2 - \bar{T}_1)\bar{h}_2{}^2$

$$a = \frac{\text{inlet film thickness}}{\text{"effective film thickness" at point of closest approach}}$$

When both geometrical and thermal wedges are present,

Table 12.23. Effect of the Geometric and Thermal Wedges on the Performance of Plane Sliders

a	h_2/B	α	β	$\bar{P}$	F	$F/\bar{P}$	$\Delta\bar{T}$
$^{10}/_9$	10^{-3}	0	0	47×10^{-3}	0.95	20	1.81
$^{10}/_9$	10^{-3}	0	−1.5	32×10^{-3}	0.63	20	1.18
$^{10}/_9$	10^{-3}	0	−3	24×10^{-3}	0.48	20	0.91
1.25	10^{-3}	0	0	88×10^{-3}	0.90	10	1.63
1.27	10^{-3}	10^{-3}	0	95×10^{-3}	0.90	9.5	1.65
1.25	10^{-3}	0	−1.5	61×10^{-3}	0.61	10	1.09
1.26	10^{-3}	10^{-3}	−1.5	64×10^{-3}	0.61	9.6	1.10
1.25	10^{-3}	0	−3	48×10^{-3}	0.48	10	0.84
1.26	10^{-3}	10^{-3}	−3	49×10^{-3}	0.48	9.6	0.85
1.5	10^{-3}	0	0	131×10^{-3}	0.84	6.4	1.41
1.5	10^{-3}	0	−1.5	95×10^{-3}	0.60	6.3	0.98
1.5	10^{-3}	0	−3	75×10^{-3}	0.47	6.3	0.77
1.75	10^{-3}	0	0	151×10^{-3}	0.80	5.3	1.26
1.76	10^{-3}	0	0	154×10^{-3}	0.80	5.2	1.27
1.75	10^{-3}	0	−1.5	112×10^{-3}	0.58	5.2	0.89
1.76	10^{-3}	10^{-3}	−1.5	115×10^{-3}	0.58	5.1	0.90
1.75	10^{-3}	0	−3	92×10^{-3}	0.47	5.1	0.71
2	10^{-3}	0	0	159×10^{-3}	0.77	4.9	1.16
2	10^{-3}	0	−1.5	122×10^{-3}	0.57	4.7	0.84
2.01	10^{-3}	10^{-3}	−1.5	123×10^{-3}	0.57	4.7	0.84
2	10^{-3}	0	−3	101×10^{-3}	0.47	4.6	0.67
1.02	10^{-3}	10^{-3}	0	11×10^{-3}	1.00	92	2.02
1.01	10^{-3}	10^{-3}	−1.5	4×10^{-3}	0.64	147	1.29
1.01	10^{-3}	10^{-3}	−3	3×10^{-3}	0.48	196	0.97
1.06	6.32×10^{-4}	10^{-3}	0	28×10^{-3}	1.00	36	2.06
1.02	$\sqrt{40} \times 10^{-4}$	10^{-3}	−1.5	6×10^{-3}	0.47	79	0.95
1.02	$\sqrt{40} \times 10^{-4}$	10^{-3}	−3	3×10^{-3}	0.32	120	0.64
1.13	4.47×10^{-4}	10^{-3}	0	60×10^{-3}	1.00	16.7	2.13
1.04	$\sqrt{20} \times 10^{-4}$	10^{-3}	−1.5	7×10^{-3}	0.35	52	0.72
1.02	$\sqrt{20} \times 10^{-4}$	10^{-3}	−3	3×10^{-3}	0.22	88	0.44
1.33	3.16×10^{-4}	10^{-3}	0	142×10^{-3}	1.00	7.0	2.35
1.06	$\sqrt{10} \times 10^{-4}$	10^{-3}	−1.5	7×10^{-3}	0.26	36	0.53
1.03	$\sqrt{10} \times 10^{-4}$	10^{-3}	−3	2×10^{-3}	0.15	67	0.30

α is the coefficient of cubical expansion of the oil with temperature:

$$\rho = \rho_1[1 + \alpha(T - T_1)]$$

β is the index of T in a power-law variation of viscosity with temperature:

$$\mu = \mu_1(T/T_1)^\beta$$

a is the ratio of inlet to minimum wedge thickness

$$a = \frac{h_1}{h_2} + \frac{K\,\Delta\bar{T}}{1 - K\,\Delta\bar{T}}$$

where K is written for $\alpha\mu_1 UB/g\rho_1 c h_2{}^2$

The integrated energy and Reynolds equations for infinitely long bearings may be written as

$$d\bar{T}/d\bar{x} = (8/\bar{p}_0\bar{h}_0)(\bar{\mu}/\bar{h})[1 - \tfrac{3}{2}\bar{p}_0\bar{h}_0/\bar{p}\bar{h} + \tfrac{3}{4}(\bar{p}_0\bar{h}_0/\bar{p}\bar{h})^2] \qquad (12.227a)$$

$$d\bar{p}/d\bar{x} = (\bar{\mu}/\bar{h}^2)(1 - \bar{p}_0\bar{h}_0/\bar{p}\bar{h}) \qquad (12.227b)$$

where $\bar{p}_0\bar{h}_0$ is the constant of integration.

Solutions of Eq. (12.227b) for the two simple cases, namely, $\bar{h}$ a linear function of $\bar{x}$, the pure geometric wedges, and $\bar{p}$ a linear function of $\bar{T}$, the pure thermal wedge, are given in Table 12.22.

$$a = \bar{h}_1/\bar{h}_2 + 1/\bar{p}_2 - 1$$
$$= \frac{\bar{h}_1}{\bar{h}_2} + \frac{A(\bar{T}_2 - \bar{T}_1)}{1 - A(\bar{T}_2 - \bar{T}_1)}$$
$$= \frac{\bar{h}_1}{\bar{h}_2} + \frac{(A/\bar{h}_2^2)\,\Delta\bar{T}}{1 - (A/\bar{h}_2^2)\,\Delta\bar{T}}$$

From this definition the value of a has been calculated. The solutions of Eqs. (12.227) under the conditions stated above are given in Table 12.23.

Parallel Slider with $\rho = f(T)$ and $\mu = f(p,T)$.[34] The Reynolds and the energy equation with the density, viscosity, and internal energy represent a system of simultaneous equations in pressure and temperature as the two unknowns.

For the case of infinitely long slider if we represent the density relation by

$$\rho_0/\rho = e^{\lambda(T-T_0)} = 1 + \lambda(T - T_0) + \lambda^2/2(T - T_0)^2 + \cdots$$

and $e^{\alpha p} = k_1 =$ constant the energy equation can be integrated, yielding

$$T = (1/\beta)\ln(1 + k_1kx) \qquad (12.228)$$

where $k = 2U\mu_1\beta/h^2c\rho_0g$ and β is the temperature coefficient of viscosity.

Note that $T = 0$ when $x = 0$. Substituting Eq. (12.228) into the Reynolds equation and integrating yields for pressure

$$p = p_0 - \frac{1}{\alpha}\ln\left[1 + \frac{3\alpha\lambda c\rho_0 g}{2\beta^2k_1}e^{\alpha p_0}\left(\ln\frac{1 + k_1kx}{1 + k_1kx_0}\right)^2\right] \qquad (12.229)$$

where $p = p_0$ when $x = x_0$.

$$x_0 = (h^2c\rho_1g/2U\mu_1\beta k_1)e^{-\lambda T_0}(e^{\beta T_0} - 1)$$
$$p_0 = -(1/\alpha)\ln[1 - (3\alpha\lambda c\rho_1g/2k_1)e^{-\lambda T_0}T_0^2]$$
$$T_0 = (1/2\beta)\ln(1 + k_1k\beta)$$

12.9. ELASTICITY CONSIDERATIONS

(a) One-dimensional Solutions

The Perfectly Elastic Journal Bearing. We shall first consider a limiting case in which the bearing is entirely devoid of rigidity, by which is meant that the bearing consists of an extremely flexible foil stretched around the circumference of the journal such as shown in Fig. 12.60.[35] The film separating the foil from the journal will consist of the following parts:

1. The parallel part, extending over half the circumference of the journal
2. The leading part, at the intake to the parallel part
3. The trailing part, at the outflow of the parallel part

The Reynolds equation

$$dp/dx = 6\mu U[(h - h_0)/h^3] \qquad (12.230)$$

is valid for all three parts of the film. For the parallel film, $h = h_0$ and consequently $dp/dx = 0$; hence the pressure is constant. For sufficiently low values of h_0, say, for $h_0/R < 10^{-2}$, the pressures are generated mainly in the region immediately before the beginning of the parallel film, and the wedge formed between a plane surface and a cylinder may be replaced by a parabolic wedge. Thus

$$h = h_0 + x^2/2R \qquad (12.231)$$

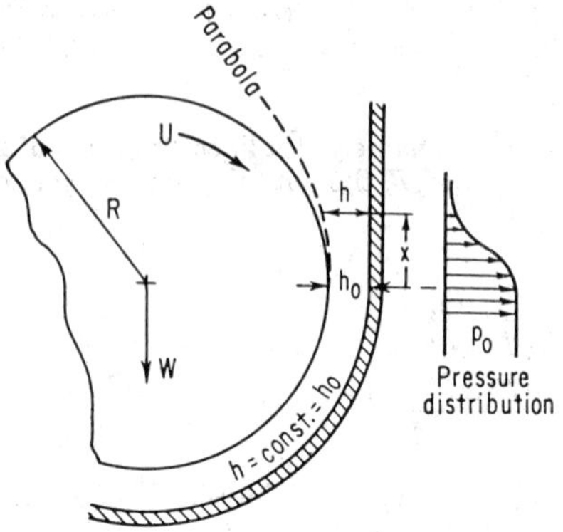

FIG. 12.60. Pressure distribution of foil bearing.

By using the boundary conditions

$$p = 0 \qquad \text{for } x = \infty$$

and

$$p = p_0 \qquad \text{for } x = 0$$

Eq. (12.230) integrated gives

$$h_0/R = 4.78(\mu N/p_0)^{2/3} \qquad (12.232)$$

where p_0 represents the specific pressure, which may be written $p_0 \cong s/R$ where s is the tension in the foil.

For the purpose of comparing the operational reliability of the foil bearing with that of the conventional bearing, the value of h_{min} for a rigid bearing with film thickness as given by Eq. (12.231) is

$$h_{min}/r = 2.45 L\mu U/W \qquad (12.233)$$

where $1/r = 1/R - 1/(R + C) \approx C/R^2$ (if C/R, the clearance ratio, is sufficiently small). Thus Eq. (12.233) becomes

$$h_{min}/R = 7.70(R/C)(\mu N/p_0) \qquad (12.234)$$

Eqs. (12.232) and (12.234) enable us to compare the minimum film thickness of the two bearings at high loads. To this end the ratio h_0/h_{min} is determined, and it is found that

$$h_0/h_{min} > 1 \qquad \text{for } \mu N/p_0 < 0.24(C/R)^3$$

It follows that for a clearance ratio C/R of, say, 10^{-3} the foil bearing gives greater film thickness than the classical bearing when $10^8 \mu N/p_0 < 0.024$, which is below the acceptable limits of safe bearing operation. From this point of view, then, the foil bearing has no advantage over the conventional bearing.

Throughout the parallel part of the film the pressure is constant and, consequently, the velocity distribution is linear across the film. For the parallel part the friction exerted on the foil may be written

$$F_1 = 2\pi^2 L\mu N R(R/h_0) \qquad (12.235a)$$

The friction force in the leading and trailing part is

$$F_2 = 2\pi^2 L \sqrt{2}\, \mu R(R/h_0)^{1/2} \qquad (12.235b)$$

The coefficient of friction may be written

$$f = 2.05(\mu N/p_0)^{1/3} + 6.30(\mu N/p_0)^{2/3} \qquad (12.236)$$

Spring-supported Thrust Bearing. A solution is obtained by starting with the solution for a rigid bearing and then computing a "correction."[36]

The Rigid-bearing Case. For a slider bearing with the bearing rigid the film-thickness variation is given by

$$h_0 = h_2 \left[1 + (a - 1)\frac{B - x}{B}\right] \qquad (12\,237)$$

The zero subscript will be used to denote the fact that the bearing is being considered rigid. The pressure is

$$p_0 = \frac{6\mu UB}{h_2^2} \frac{a-1}{a+1} \frac{(x/B)(1-x/B)}{[1+(a-1)(1-x/B)]^2} \tag{12.238a}$$

From Eq. (12.238*a*) the dimensionless pressure distribution for a bearing is given by

$$\bar{p}_0 = 6\left(\frac{a-1}{a+1}\right) \frac{\bar{x}(1-\bar{x})}{[1+(a-1)\bar{x}]^2} \tag{12.238b}$$

where $\bar{p}_0 = \dfrac{p_0 h_2^2}{\mu UB}$ and $\bar{x} = \dfrac{B-x}{B}$

and the load capacity is

$$W_0 = \frac{6\mu ULB^2}{h_2^2} \frac{1}{(a-1)^2} \left(\ln a - 2\frac{a-1}{a+1}\right) \tag{12.239a}$$

The dimensionless load capacity from Eq. (12.239*a*) reduces to

$$\bar{W}_0 = \frac{6}{(a-1)^2} \left(\ln a - 2\frac{a-1}{a+1}\right) \tag{12.239b}$$

where $\bar{W}_0 = W_0 h_2^2 / L\mu UB^2$

The Flexible-bearing Case. For the slider bearing with a flexible bearing plate, $h = h_0 + h_c$, $p = p_0 + p_c$, and $W = W_0 + W_c$, where h_c, p_c, and W_c are small quantities which account for the flexibility of the bearing.

$$h_c = \frac{3\beta^3 \mu U}{kBh_2^2} \left(\frac{a-1}{a+1}\right) [A_1(B-x) + A_2(B-x)^2] \tag{12.240}$$

where $A_1 = 2(BI_{22} - I_{32})$ and $A_2 = (I_{22} - BI_{12})$.

$$I_{mn} = \int_0^B \frac{(B-s)^m}{[1+(a-1)(B-s)/B]^n}\, ds$$

$\beta^4 = k/4D$, where D is the flexural rigidity of the plate per unit width given by $D = Et^3/12(1-\nu_p^2)$, E is the modulus of elasticity, ν_p is Poisson's ratio, and t is the plate thickness.

The dimensionless film-thickness correction is

$$\bar{h}_c = h_c E h_2^2 / \mu UB^2$$

The pressure correction is given by

$$p_c = \frac{36\mu^2\beta^3 U^2}{kB^2 h_2^5} \frac{a^2(a-1)}{(a+1)^2} \left[2A_1(I_{03}J_{13} - I_{13}J_{03}) - 6A_1 \frac{a}{a+1} (I_{03}J_{14} - I_{14}J_{03}) \right.$$
$$\left. + 2A_2(I_{03}J_{23} - I_{23}J_{03}) - 6A_2 \frac{a}{a+1} (I_{03}J_{24} - I_{24}J_{03}) \right] \tag{12.241}$$

where the J's are indefinite integrals defined by

$$J_{mn} = \int_0^x \frac{(B-x)^m}{[1+(a-1)(B-x)/B]^n}\, dx$$

The dimensionless pressure and load corrections are, respectively,

$$\bar{p}_c = p_c E h_2^5 / \mu^2 U^2 B^3$$
$$\bar{W}_c = W_c E h_2^5 / \mu^2 U^2 L B^4$$

Note that in this dimensionless form the load capacity and load-capacity corrections depend only on a. The above equations are evaluated and plotted in Figs. 12.61 to 12.64.

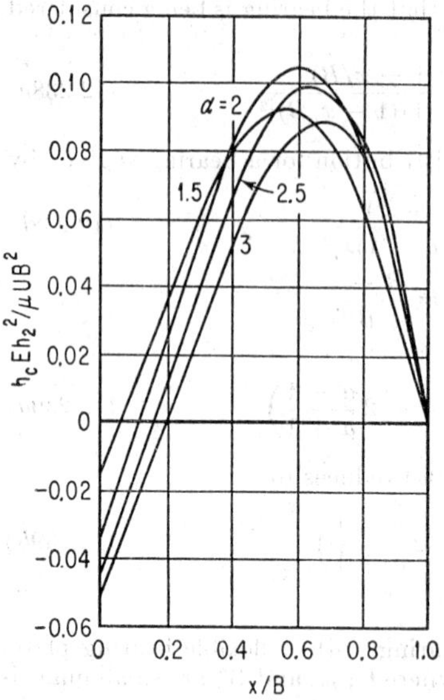

FIG. 12.61. Film-thickness correction along the slider.

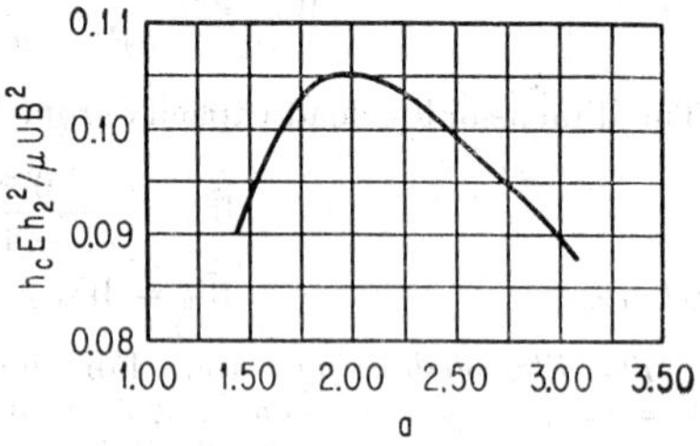

FIG. 12.62. Maximum film-thickness correction.

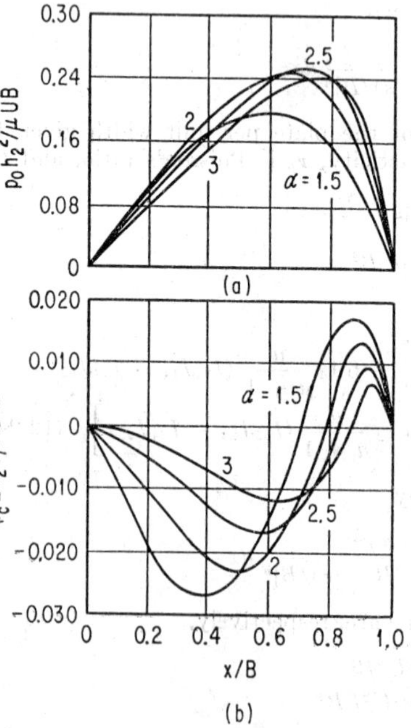

FIG. 12.63. Pressure distribution in plane sliders. (*a*) Rigid surface. (*b*) elastic-surface correction.

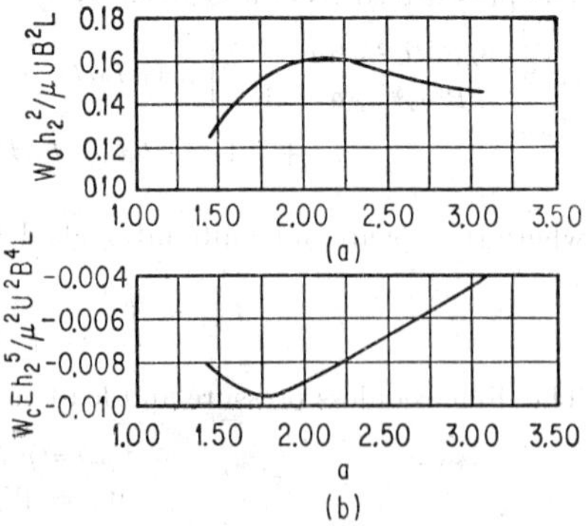

FIG. 12.64. Load capacity of elastic slider bearings. (*a*) Rigid surface, (*b*) Elastic-surface correction.

(b) Two-dimensional Solutions of Centrally Pivoted Sectors

A centrally pivoted-sector thrust bearing can carry load and maintain equilibrium of moments by virtue of either of two processes: a thermal wedge or elastic deformation. Hydrodynamic pressures can also be generated and equilibrium of moments satisfied in a pivoted-sector thrust bearing by offsetting the pivot. This design yields optimum characteristics; however, it can be used only in applications where one direction of rotation is present. The analysis presented here employs numerical methods for the simultaneous solution of the Reynolds, energy, and elasticity equations.[35] In Art. 12.8 the simultaneous solution of the first two equations was discussed; thus here only the added factors will be considered.

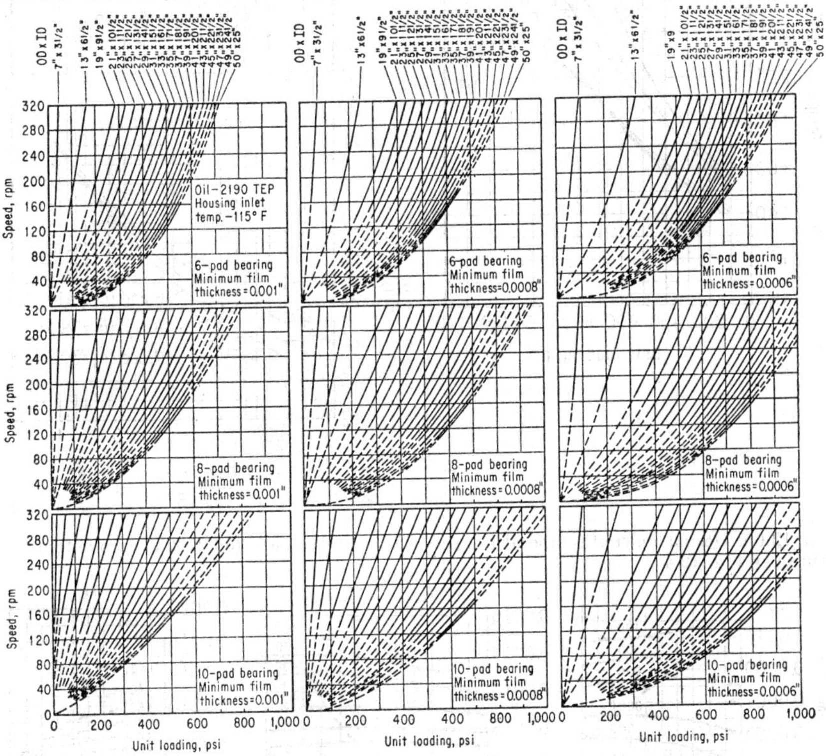

Fig. 12.65. Design chart of minimum film thickness.

The elastic equation may be represented by

$$\nabla^2(\nabla^2\delta) = \frac{12(1-\nu^2)p}{Et^3} + \frac{1+\nu}{t}\nabla^2(\alpha\,\Delta T) \tag{12.242}$$

where $\nabla^2 = [\partial^2/\partial r^2 + \tfrac{1}{2}\partial/\partial r + (1/r^2)(\partial/\partial\theta^2)]$

(For analysis including thermal gradients refer to ref. 38.)

In the analysis that follows the distortion due to thermal gradients is neglected and it is further assumed that, under load, the initially flat surfaces of the pads become slightly convex, assuming a slope that can be represented by part of a sphere with large radius of curvature. It can then be shown that the curvature of the pads under load is approximated by the following relation:

$$1/2R_c = 0.225(W/t^3_{av}E) \tag{12.243}$$

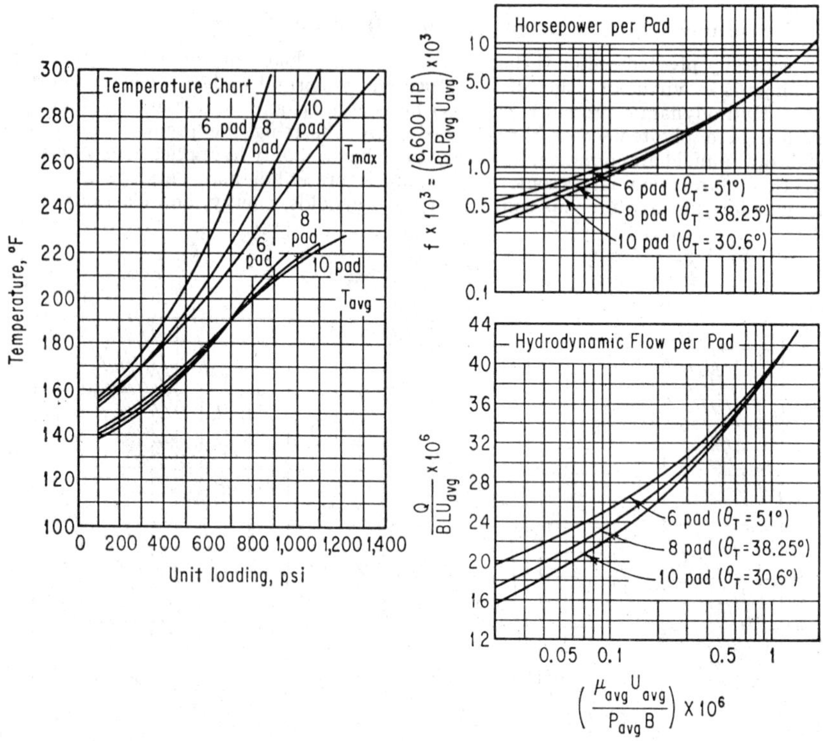

FIG. 12.66. Design chart of temperature, power loss, and oil flow (oil 2190T, groove temperature per Fig. 12.4).

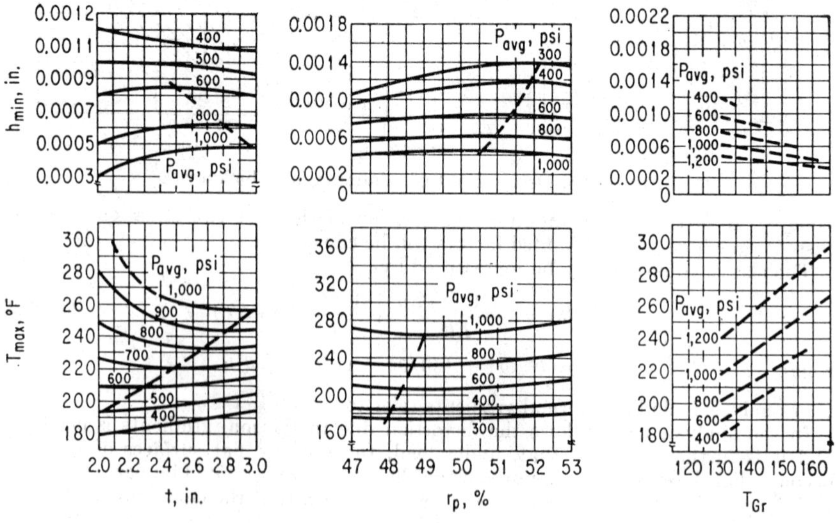

FIG. 12.67. Effects of pad thickness, pivot location, and groove temperature. 31 in. outside diameter by 15½ in. inside diameter, 1 pad, 320 rpm.

The general equation for the film shape in polar coordinates corresponding to this curvature and to inclinations m_r and m_θ is

$$h = h_a + m_\theta[r_a \sin(\theta_a - \beta/2) - r\sin(\theta - \beta/2)] - m_r[r_a \cos(\theta_a - \beta/2) - r\cos(\theta - \beta/2)] + (1/2R_c)[r^2 - r_a^2 - 2r\bar{\bar{r}}\cos(\theta - \bar{\bar{\theta}}) + 2r_a\bar{\bar{r}}\cos(\theta_a - \bar{\bar{\theta}})] \tag{12.244}$$

Eq. (12.244) can also be used to describe the film shape for flat pads. In such cases, R_c is infinite, thus eliminating the fourth term on the right-hand side of the equation.

At each operating point, the pad deformation has to be related to the pad load in accordance with Eq. (12.243). The film shape which depends on this deformation and on the inclinations of the pad has to be such that the resulting center of pressure passes through the pivot. Figures 12.65 to 12.67 give design charts obtained numerically for bearing sizes ranging from 7 to 50 in. in diameter. These bearings are all geometrically similar, having the following properties:

$$L/R_2 = 0.5 \qquad t_{av}/R_2 = 0.154 \qquad \bar{\bar{r}}\% = \bar{\bar{\theta}}\% = 50$$

with 15 per cent of the bearing surface taken up by oil-inlet grooves.

12.10. HYDRODYNAMICS OF ROLLING ELEMENTS

(a) Fluid Film with Rigid Surfaces

Solutions with Constant Viscosity.[39] For conditions of pure rolling found at the gear pitch line, the theory is the same as for a roller bearing without slip. The film of a roller above a flat plate may be expressed by

$$(2R - h + h_{min})(h - h_{min}) = x^2$$

The pressure is given by

$$p/12\mu = (U\sqrt{2Rh_{min}}/h^2_{min})[\bar{x}/2 + \sin 2\bar{x}/4 - \pi/4 - C_1(3\bar{x}/8 - 3\pi/16 + \sin 2\bar{x}/4 + \sin 4\bar{x}/32)] \tag{12.245}$$

where C_1 is an integration constant $U_1 = U_2 = U$ and $\tan\bar{x} = x/\sqrt{2Rh_{min}}$.

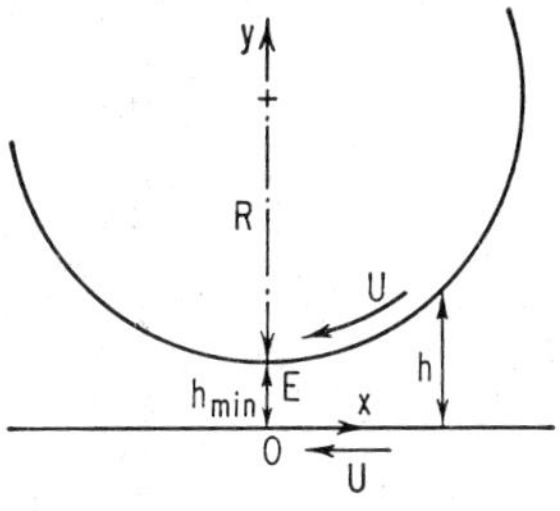

FIG. 12.68. Hydrodynamics of cylinder and plane surface.

To determine C_1, we note that, when the pinion tooth begins to engage the rack tooth, the pressure is zero when x is zero. From the pressure-generation standpoint this is an unfavorable condition, as compared with the case represented by Fig. 12.68, where the region of pressure extends also to the left of OE. Taking it as one possible case, however, we note that, when $x = 0$, $\bar{x}$ is also zero, so that putting $p = 0$ in Eq. (12.69) with this value for x gives $C_1 = 4/3$. Thus

$$p/12\mu = (U/h^2_{min})[(2Rh_{min})^{1/2}/12](\sin 2\bar{x} + \tfrac{1}{2}\sin 4\bar{x}) = -U(2Rh_{min})^{1/2}\phi/h^2_{min}$$

The above function is evaluated in Table 12.24.

$$W = 4.896(\mu URL/h_{min}) \tag{12.246}$$

The maximum pressure is attained when

$$x_0 = 0.47517\sqrt{2Rh_{min}}$$

and its value is

$$p_0 = 1.521(\mu U/h^2_{min})\sqrt{2Rh_{min}}$$

Table 12.24. Pressure-distribution Functions

$\frac{x}{\sqrt{2Rh_{\min}}} = \tan \bar{x}$	$\phi = \frac{1}{12}\left(\sin 2\bar{x} + \frac{\sin 4\bar{x}}{2}\right)$	$\theta = \frac{1}{8}\left(\frac{\pi}{2} - \bar{x} + \frac{\sin 4\bar{x}}{4}\right)$	$\psi = 0.67725\phi + 0.32275\theta$; $\psi = 0$ when $\frac{x}{\sqrt{2Rh_{\min}}} = -0.47517$
−0.5	−0.10667	0.22431	0.000155
−0.4	−0.09908	0.21270	0.00154
−0.3	−0.08417	0.20406	0.00885
−0.2	−0.06163	0.19883	0.02244
−0.1	−0.03266	0.19668	0.04135
0.0	0.00000	0.19635	0.06337
0.1	0.03266	0.19602	0.08538
0.2	0.06163	0.19387	0.10430
0.3	0.08417	0.18864	0.11788
0.4	0.09908	0.18000	0.12519
0.5	0.10667	0.16839	0.12658
0.6	0.10813	0.15474	0.12317
0.7	0.10510	0.14011	0.11639
0.8	0.09915	0.12540	0.10762
0.9	0.09158	0.11129	0.09794
1.0	0.08333	0.09818	0.08812
1.5	0.04734	0.05132	0.04862
2.0	0.02667	0.02796	0.02708
2.5	0.01585	0.01635	0.01601
3	0.01000	0.01022	0.01077
4	0.00589	0.00467	0.00549
5	0.00246	0.00223	0.00239
6	0.00146	0.00147	0.00146
7	0.00093	0.00094	0.00936
8	0.00063	0.00063	0.00632
9	0.00045	0.00045	0.00445
10	0.00028	0.00033	0.00029

These expressions apply to a pinion gearing with a rack. If, however, the pinion meshes with a gear, we must replace R by ρ, where

$$1/\rho = 1/R_1 + 1/R_2$$

In this expression R_1 denotes the radius of curvature of the pinion tooth and R_2 denotes that of the gear tooth. The work done on both acting surfaces is

$$F = \int_{\bar{x}}^{\infty} dF = 2U\mu L \left.\frac{du}{dy}\right|_0 dx = \frac{12U^2\mu L\sqrt{2Rh_{\min}}}{h_{\min}}\left[\frac{\pi}{2} - \bar{x} + C_1\left(\frac{\bar{x}}{2} + \frac{\sin 2\bar{x}}{4} - \frac{\pi}{4}\right)\right] \tag{12.247}$$

Viscosity as a Function of Pressure.[40] If the variation of viscosity with pressure is represented by the form $\mu = \mu_0 e^{\alpha p}$ then

$$dp/dx = \mu_0 e^{\alpha p}(\partial^2 u/\partial y^2) \tag{12.248}$$

The right-hand side of Eq. (12.245) will be the same; the left-hand side will be $1 - e^{-\alpha p}/\alpha$ instead of p

$$\frac{1 - e^{-\alpha p}}{\alpha} = 3(U_1 + U_2)\mu_0 \frac{\sqrt{2Rh_{\min}}}{h^2_{\min}} [(1 + \tfrac{3}{4} \sec^2 \bar{x}_0)(\pi/2 + \bar{x} + \sin \bar{x} \cos \bar{x}) - \tfrac{1}{2} \sec^2 \bar{x}_0 \sin \bar{x} \cos^3 \bar{x}]$$

and
$$1 - e^{-\alpha p_0} = 0.76\mu_0\alpha(U_1 + U_2)(2R/h^3_{\min})^{1/2}$$

We see that p_0 will be infinite when the right-hand side goes to unity.

The obvious criticism of this analysis is that an infinite pressure will produce a number of other phenomena neglected in this simple treatment: an infinite pressure will cause infinite viscosity, and at that point infinite heat generation, which will reduce μ_0 and α; and the disks will deform and alter the shape of the oil film, which in turn will change the pressure curve.

(b) Fluid Film with Elastic Deformation

Combine the hydrodynamic theory with a theory of elasticity for two conditions, constant viscosity μ_0 and a viscosity given by $\mu = \mu_0 e^{\alpha p}$. The displacements s of the boundary points for a given normal load W are determined by the equation

$$\frac{ds}{dx} = -\frac{2(1 - \nu^2)}{E} \frac{1}{\pi} \int_{-\infty}^{x} [P(\xi) - p(2x - \xi)] \frac{1}{x - \xi} d\xi$$

If we select a pressure distribution p as a function of x, we can then determine ds/dx and, from it, by further integration, s. The change of the gap becomes then $s_s = s_1 + s_2 = 2s$.

The film thickness with deformation then becomes

$$h = h_e + s_s + \frac{x^2 - x_e^2}{2\rho}$$

Constant Viscosity. For constant viscosity ref. 41 gives the minimum film thickness which was obtained by using series expansion of specially defined elementary pressure and deformation functions

$$h_{\min} = \frac{RL\mu(U_1 + U_2)}{W} \frac{6.5}{2.7 - s} \tag{12.249}$$

where the "deformation parameter"

$$s = \frac{2}{1 + \dfrac{5L^2R\mu E(U_1 + U_2)}{(1 - \nu^2)W^2}}$$

This is applicable to $s < 1.7$.

Viscosity a Function of Pressure. Another analysis[42] which contains many approximations but which includes effects of pressure on viscosity and deformation gives for the minimum film thickness the following relation:

$$h_{\min} = \frac{1.13[\ln \alpha(U_1 + U_2)\mu_0]^{0.727}\rho^{0.364}}{(W\theta/L)^{0.091}} \tag{12.250}$$

where
$$\theta = \frac{1}{\pi}\left(\frac{1 - \nu_1^2}{E_1} + \frac{1 - \nu_2^2}{E_2}\right)$$

$$\mu = \mu_0 e^{\alpha p}$$

For a good approximation, a simplified relation that may be used to determine the coefficient of friction is given by

$$f = \frac{\mu|p_0(U_1 - U_2)}{h_{\min}p_0} \tag{12.251}$$

The minimum pressure p_0 in Eq. (12.251) may be estimated from the Hertzian theory.

12.11. INERTIA AND TURBULENCE EFFECTS

(a) Introduction

Two of the assumptions underlying the derivation of the Reynolds equation are that the inertia forces of the lubricant are negligible and that the flow is laminar. The familiar Reynolds number of fluid dynamics, given by $\rho UD/\mu$, expresses the ratio of the inertia forces to the viscous forces; in bearings a similar Reynolds number can be formulated by writing $\text{Re} = \rho Uh/\mu$, where h is some representative film thickness. When this Reynolds number becomes sufficiently high, the two assumptions referred to above no longer hold. The inertia forces become of the same order of magnitude as the shearing forces, and laminar flow may give way to a turbulent or semiturbulent state. Under these conditions, the Reynolds equation as formulated in Art. 12.1 no longer represents the true state of the lubricant. Inertia effects can be accounted for by including those terms of the Navier-Stokes equations which were originally dropped because of our assumption of negligible inertia.

(b) Effects of Fluid Inertia

The inertia terms will be of the same order of magnitude as the viscous forces when[43]

$$\mu/\rho\omega h^2 = 1 \tag{12.252}$$

This can be rewritten as

$$\rho R\omega h/\mu \equiv \rho Uh/\mu \equiv \text{Re} = R/h$$

which gives the magnitudes of speed, bearing size, and viscosity necessary to produce noticeable inertia effects in hydrodynamic bearings. This, however, does not mean that the contribution of inertia to the load capacity, or to any other bearing characteristic, is of the same order as the shearing forces. In fact, it will be seen later that, even when bearings operate in the range given by Eq. (12.252), the contribution of the inertia forces to the dynamics of bearings is still very small.

For one-dimensional bearings the equation governing the flow of lubricant with inertia effects included is given by

$$\rho[\partial u/\partial t + u(\partial u/\partial x) + v(\partial u/\partial y)] = -\partial p/\partial x + \mu(\partial^2 u/\partial y^2) \tag{12.253}$$

If only steady-state conditions are considered, $\partial u/\partial t = 0$.

Using the iteration method the above equation neglecting $\partial u/\partial t$ is solved by first calculating with the inertia forces neglected. Thus the left-hand side of Eq. (12.253) becomes a known function of x and y and the remaining unknowns are p and u on the right-hand side. Denoting by the subscript v the inertialess solutions, we have

$$\rho[u_v(\partial u_v/\partial x) + v_v(\partial u_v/\partial y)] = -\partial p/\partial x + \mu(\partial^2 u/\partial y^2) \tag{12.254}$$

where

$$u = u_v + u_c \qquad p = p_v + p_c \tag{12.255}$$

u_c and p_c above are the correction terms due to inertia, and u and p without subscripts are the solutions for a bearing in which both viscous and inertia forces are present. If a further refinement is desired, the values of u and v can later be substituted into the left-hand side of Eq. (12.253) and a new set of u and p values obtained. By putting Eq. (12.255) into Eq. (12.254) we have, since $-dp_v/dx + \mu\, \partial^2 u_v/\partial y^2 = 0$, the equation

$$\rho[u_v(\partial u_v/\partial x) + v_v(\partial u_v/\partial y)] = -dp_c/dx + \mu(\partial^2 u_c/\partial y^2) \tag{12.256}$$

Slider with Inertia Considered.[44] For a plane slider with the coordinate axis at the leading edge, we have

$$h = h_1 - ax$$

The pressure correction is

$$p_c(x) = \rho U^2 \left\{ \frac{1}{7} \frac{(k-1)^2}{2k-1} \left[\frac{3}{5} \frac{1}{2k-1} + \ln \frac{k}{k-1} \right] \left[1 - \left(\frac{1}{1-\bar{x}/k} \right)^2 \right] - \frac{3}{35} \frac{k-1}{2k-1} \left(1 - \frac{1}{1-\bar{x}/k} \right) + \frac{1}{7} \ln \frac{1}{1-\bar{x}/k} \right\} \tag{12.257}$$

where $\bar{x} = x/B \qquad k = h_1/\delta$

It is seen that $p_c \propto \rho U^2$ and that it is independent of the inclination of the slider or film height but depends only on the ratio h_1/δ. The value of this p_c is always positive, thus adding to the total load capacity; its shape is similar to p_v. By writing for a modified Reynolds number

$$\text{Re}^* = (UB/\nu)(h_1/B)^2$$

we have for the total pressure

$$p(x) = p_v + p_c = (\mu UB/h_1^2)(\bar{p}_v + \text{Re}^* \bar{p}_c) \tag{12.258}$$

where $\bar{p}_v = \dfrac{6k^2}{2k-1} \dfrac{\bar{x}(1-\bar{x})}{(k-\bar{x})^2}$

is the solution for a slider without inertia and $\bar{p}_c$ is the expression given in the brackets of Eq. (12.257). By integrating for load capacity, we have

$$W_c = LB \int_0^1 p_c(\bar{x})\, d\bar{x} = (\mu UB^2L/h_1^2)\,\text{Re}^*\, \bar{W}_c = \rho U^2 BL\bar{W}_c$$

where $$\bar{W}_c = \frac{1}{7} - \frac{6}{35} \frac{k^2-k}{(2k-1)^2} - \frac{1}{5} \frac{k^2-k}{2k-1} \ln \frac{k}{k-1} \tag{12.259}$$

The total load capacity is

$$W = W_v + W_c = (\mu UB^2L/h_1^2)(\bar{W}_v + \text{Re}^* \bar{W}_c) \tag{12.260}$$

where $\bar{W}_v = 6k^2 \left(\ln \dfrac{k}{k-1} - \dfrac{2}{2k-1} \right)$

Values for $\bar{W}_v$ and $\bar{W}_c$ are given in Table 12.25. For friction,

$$\tau_c = \mu(\partial u_c/\partial y)\Big|_{y=0}$$

and upon integration

$$F_c = -BL \int_0^1 \tau_c(\bar{x})\, d\bar{x} = (\mu ULB/h_1)\,\text{Re}^*\, \bar{F}_c = \rho U^2 h_1 L\bar{F}_c$$

or $$F = (\mu ULB/h_1)(\bar{F}_v + \text{Re}^* \bar{F}_c) \tag{12.261}$$

$$\bar{F}_c = \frac{18}{35} \frac{k-1}{(2k-1)^2} - \frac{1}{35} \frac{1}{k} \tag{12.262}$$

where $\bar{F}_v = 2k \left(2 \ln \dfrac{k}{k-1} - \dfrac{3}{2k-1} \right)$

and the coefficient of friction is

$$f = F/W = f_v(1 + \text{Re}^* f_c) \tag{12.263}$$

where $f_c = \dfrac{\bar{F}_c}{\bar{F}_v} - \dfrac{\bar{W}_c}{\bar{W}_v}$ and $f_v = \dfrac{h_1}{B} \dfrac{1}{3k} \dfrac{2 \ln [k/(k-1)] - 3/(2k-1)}{\ln [k/(k-1)] - 2/(2k-1)}$

Values of $\bar{F}_v$, $\bar{F}_c$, $\bar{f}_v$, and $\bar{f}_c$ are given in Table 12.25. In general the friction is increased except at very low values of k.

Journal Bearing with Inertia Considered. For the inertialess component of the final solutions, we shall here employ the so-called "half-Sommerfeld expressions," i.e., bearing characteristics based on integrations over the range 0 to π, with the negative pressure of the region π to 2π set equal to zero. These expressions are from previous considerations

$$p_v = \frac{\mu UR}{C^2}\bar{p}_v \qquad \text{where} \qquad \bar{p}_v = \frac{6\epsilon}{2+\epsilon^2}\frac{(2+\epsilon\cos\theta)\sin\theta}{(1+\epsilon\cos\theta)^2}$$

$$\frac{1}{S_v}\cos\phi_v = \frac{12\pi\epsilon^2}{(1-\epsilon^2)(2+\epsilon^2)} \qquad \frac{1}{S_v}\sin\phi_v = \frac{6\pi^2\epsilon}{(1-\epsilon^2)^{1/2}(2+\epsilon^2)}$$

$$\tan\phi_v = \frac{\pi}{2}\left(\frac{1-\epsilon^2}{\epsilon^2}\right)^{1/2}$$

$$F_v = \frac{\mu URL}{C}\bar{F}_v \qquad \text{where} \qquad \bar{F}_v = \frac{2\pi}{(1-\epsilon^2)^{1/2}}\frac{1+2\epsilon^2}{2+\epsilon^2}$$

$$f_v = \frac{C}{R}\frac{\pi}{3}\bar{f}_v \qquad \text{where} \qquad \bar{f}_v = \frac{(1+2\epsilon^2)(1-\epsilon^2)^{1/2}}{\epsilon[\pi^2(1-\epsilon^2)+4\epsilon^2]^{1/2}}$$

Table 12.25. Inertia Effects in Thrust Bearings

$k = \frac{h_1}{h_1 - h_2}$	$\bar{W}_v$	$\bar{W}_c$	$\bar{F}_v$	$\bar{F}_c$	$\bar{f}_v$	$\bar{f}_c$
1.0	∞	0.1429	∞	−0.0286	0.820	
1.1	5.330	0.0857	5.050	0.0097	0.948	−0.0148
1.2	3.130	0.0588	3.460	0.0287	1.105	−0.0105
1.3	2.190	0.0449	2.760	0.0384	1.260	−0.0066
1.4	1.680	0.0349	2.348	0.0431	1.400	−0.0029
1.5	1.330	0.0283	2.095	0.0452	1.572	0.0003
1.6	1.110	0.0237	1.920	0.0462	1.730	0.0026
1.7	0.940	0.0198	1.789	0.0458	1.901	0.0045
1.8	0.818	0.0167	1.685	0.0451	2.080	0.0063
1.9	0.719	0.0142	1.604	0.0441	2.260	0.0074
2.0	0.634	0.0123	1.543	0.0428	2.440	0.0082
2.5	0.405	0.0069	1.360	0.0368	3.360	0.0099
3.0	0.302	0.0044	1.272	0.0316	4.290	0.0095
∞	0	0	1	0	∞	

and h, of course, is given by $h = C(1 + \epsilon\cos\theta)$. The pressure correction after integration is

$$p_c = \left(\frac{\mu UR}{C^2}\right)\text{Re}^{**}\,\bar{p}_c = \rho U^2\bar{p}_c \tag{12.264}$$

$$\text{where} \quad \bar{p}_c = \frac{216}{35}\frac{\epsilon}{(2+\epsilon^2)^2}\left(\frac{I_1}{\pi}-\frac{I_2}{2}\right) - \frac{12}{35}\frac{\epsilon}{2+\epsilon^2}\left(\frac{I_1}{\pi}-\frac{I_3}{2}\right) - \frac{2}{7}\ln\frac{1}{\lambda}\frac{I_1}{\pi} + \frac{1}{7}\ln\frac{1+\epsilon}{1+\epsilon\cos\theta}$$

$$I_1 = \tan^{-1}\left(\sqrt{\lambda}\tan\frac{\theta}{2}\right) - \frac{3}{2}\epsilon\frac{(1-\epsilon^2)^{1/2}}{2+\epsilon^2}\frac{\sin\theta}{1+\epsilon\cos\theta} - \frac{\epsilon}{2}\frac{(1-\epsilon^2)^{3/2}}{2+\epsilon^2}\frac{\sin\theta}{(1+\epsilon\cos\theta)^2}$$

$$I_2 = \frac{\epsilon}{[1+\lambda\tan^2(\theta/2)]^2} - \frac{1+\epsilon}{1+\lambda\tan^2(\theta/2)} + 1$$

$$I_3 = 1 - \frac{1}{1+\lambda\tan^2(\theta/2)} \qquad \lambda = \frac{1-\epsilon}{1+\epsilon}$$

and Re** has been redefined as

$$\mathrm{Re}^{**} = UR/\nu(C/R)^2 \tag{12.265}$$

Thus the total pressure in a journal bearing with both viscous and inertia terms considered is

$$p = (\mu UR/C^2)(\bar{p}_v + \mathrm{Re}^{**}\,\bar{p}_c) \tag{12.266}$$

In general, the correction $\bar{p}_c$ is positive except at low eccentricities and high values of θ, when it becomes negative. For the two load components, we have

$$(1/S)\cos\phi = (1/S_v)\cos\phi_v + \mathrm{Re}^{**}\,(1/\pi S_{c1}) \tag{12.267a}$$

where, after integration of Eq. (12.266),

$$\frac{1}{\pi S_{c1}} = \frac{432}{35\pi}\frac{(1+\epsilon^2)^{1/2}}{(2+\epsilon^2)^3} - \left(27\pi + \frac{24}{\pi}\right)\frac{1}{35}\frac{\epsilon(1-\epsilon^2)^{1/2}}{(2+\epsilon^2)^2} + \frac{3\pi}{35}\frac{(1-\epsilon^2)^{1/2} - (1-\epsilon^2)}{\epsilon(2+\epsilon^2)} + \frac{\pi}{7}\frac{1-(1-\epsilon^2)^{1/2}}{\epsilon} - \frac{4}{7\pi}\frac{(1-\epsilon^2)^{1/2}}{(2+\epsilon^2)}\ln\frac{1+\epsilon}{1-\epsilon}$$

and

$$(1/S)\sin\phi = (1/S_v)\sin\phi_v + \mathrm{Re}^{**}(1/\pi S_c2) \tag{12.267b}$$

where

$$\frac{1}{\pi S_{c2}} = -\frac{324}{35}\frac{\epsilon^2}{(2+\epsilon^2)^3} + \frac{18}{5}\frac{\epsilon^2}{(2+\epsilon^2)^2} - \frac{6}{35}\frac{1}{2+\epsilon^2} + \frac{2}{7} - \frac{1}{5}\frac{1-\epsilon^2}{\epsilon(2+\epsilon^2)}\ln\frac{1+\epsilon}{1-\epsilon}$$

The total load capacity by neglecting terms of higher order is

$$1/S = 1/S_v + \mathrm{Re}^{**}(1/\pi S_c) \tag{12.268}$$

with

$$\frac{1}{\pi S_c} = \frac{(1/S_v)\sin\phi_v\,(1/\pi S_{c2}) + (1/S_v)\cos\phi_v\,(1/S_{c1})}{1/S_v}$$

The new attitude angle is

$$\phi = \phi_v + \mathrm{Re}^{**}\phi_c \tag{12.269a}$$

where

$$\phi_c = \frac{(1/S_v)\cos\phi_v\,(1/\pi S_{c2}) - (1/S_v)\sin\phi_v\,(1/\pi S_{c1})}{(1/S_v)^2} \tag{12.269b}$$

Values for all the dimensionless constants are given in Table 12.26. It can be seen that the effect of the inertia forces on the total load capacity and attitude angle of a journal bearing is quite small.

The correction factor for friction is

$$F_c = (\mu URL/C)\,\mathrm{Re}^{**}\bar{F}_c = \rho U^2 LC\bar{F}_c \tag{12.270}$$

where

$$\bar{F}_c = \epsilon\left[\frac{108}{35}\frac{1-\epsilon^2}{(2+\epsilon^2)^3} - \frac{18}{35}\frac{1-\epsilon^2}{2+\epsilon^2} - \frac{2}{35}\right]$$

and the friction coefficient

$$f = f_v(1 + \mathrm{Re}^{**}\bar{f}_c) \tag{12.271}$$

with

$$\bar{f}_c = \frac{\bar{F}_c}{\bar{F}_v} - \frac{1/S_c}{1/S_v}$$

Method of Averaged Inertia. Since the inertia forces are small and the fluid film is very thin, a reasonable approach is to average out the inertia effects across the film. We can thus rewrite Eq. (12.253) in the following manner:

$$\rho\left[\frac{1}{h}\int_0^h\left(u\frac{\partial u}{\partial x} + v\frac{\partial u}{\partial y}\right)dy\right] = -\frac{\partial p}{\partial x} + \mu\frac{\partial^2 u}{\partial y^2} \tag{12.272}$$

Table 12.26. Inertia Effects in Journal Bearings

ϵ	$\frac{1}{\pi S_v}\cos\phi_v$	$1/\pi S_{c1}$	$\frac{1}{\pi S_v}\sin\phi_v$	$1/\pi S_{c2}$	$1/\pi S_v$	$1/\pi S_c$	ϕ_v, deg	ϕ_c, deg
0	0	0	0	0	0	0	90	0
0.1	0.0603	−0.0075	0.9425	0.0003	0.945	−0.0002	86.34	0.46
0.2	0.2451	−0.0107	1.8861	0.0019	1.902	0.0005	82.67	0.33
0.3	0.5679	−0.0138	2.8361	0.0069	2.893	0.0041	78.68	0.26
0.4	1.0582	−0.0125	3.8087	0.0183	3.954	0.0169	74.47	0.22
0.5	1.7778	−0.0055	4.8369	0.0377	5.154	0.0334	69.82	0.19
0.6	2.8602	0.0104	5.9903	0.0669	6.638	0.0648	64.48	0.17
0.7	4.6303	0.0387	7.4206	0.1061	8.748	0.1105	58.04	0.16
0.8	8.0808	0.0847	9.5200	0.1545	12.490	0.1726	49.67	0.16
0.85	11.4760	0.1216	11.2741	0.1820	16.090	0.2142	44.49	0.15
0.9	18.2056	0.1626	13.8500	0.2118	22.880	0.2576	37.26	0.15
0.95	38.2694	0.2288	19.7583	0.2452	43.070	0.3158	27.31	0.15
1.0	∞	0.4488	∞	0.2857	∞	0.3700	0	0.15

The left-hand side of Eq. (12.272) after integration is a function of x alone, and the expression can thus be integrated for y in the manner of previous solutions. By writing

$$\frac{\rho}{\mu h}\int_0^h\left(u\frac{\partial u}{\partial x}+v\frac{\partial u}{\partial y}\right)dy+\frac{1}{\mu}\frac{dp}{dx}=f(x)$$

we have, from Eq. (12.272),

$$\partial^2u/\partial y^2 = f(x) \tag{12.273}$$

$$dp/dx = \mu f(x) + \rho[-\tfrac{1}{6}UU' - \tfrac{1}{5}hf(x)V + \tfrac{1}{6}U^2(h'/h) - \tfrac{1}{60}f(x)U'h^2 + \tfrac{1}{10}f(x)Uhh' + \tfrac{1}{120}f^2(x)h^3h'] \tag{12.274}$$

Equation (12.274) can be evaluated when U, V, and the function h are given. This expression will now be applied to thrust and journal bearings starting with the simple case of squeeze-film action between two plates.

Squeeze Films. Considering two infinitely long parallel plates of span B approaching each other with a relative velocity V, we have

$$U=0 \qquad h'=0 \qquad V=\text{const}$$

and so, from Eq. (12.274),

$$dp/dx = (\mu + h\rho V/5)f(x)$$

$$p(x) = (12Vx/h^3)(\mu + \rho hV/5)(B-x) \tag{12.275}$$

Journal Bearing. For ordinary bearing operation, expression (12.275) can be somewhat simplified. Under conditions of steady loading and constant linear velocity, we have $V = 0$ and $U' = 0$, and Eq. (12.275) becomes

$$dp/dx = \mu f(x) + \rho[\tfrac{1}{6}U^2(h'/h) + \tfrac{1}{10}f(x)Uhh' + \tfrac{1}{120}f^2(x)h^3h'] \tag{12.276}$$

$$\bar{p} = \bar{p}_v + \bar{p}_c \qquad h_x = h_0 + h_c$$

The values of $\bar{p}_v$ and h_0 are the known Sommerfeld solutions:

$$\bar{p}_v = \frac{\epsilon\sin\theta(2+\epsilon\cos\theta)}{(2+\epsilon^2)(1+\epsilon\cos\theta)^2} \qquad h_0 = 2C\frac{1-\epsilon^2}{2+\epsilon^2} \tag{12.277}$$

By using the same Sommerfeld substitutions and the proper boundary conditions on p_c, we obtain

$$\bar{p}_c = \frac{h_c}{C(1-\epsilon^2)^{5/2}}\left[\frac{2+\epsilon^2}{2}\cos^{-1}\frac{\epsilon+\cos\theta}{1+\cos\theta} + \frac{\epsilon(1-\epsilon^2)^{1/2}(\epsilon^2-3\epsilon\cos\theta-4)}{2(1+\epsilon\cos\theta)^2}\sin\theta\right]$$
$$+\frac{3}{5}\left(\frac{1-\epsilon^2}{2+\epsilon^2}\right)^2\left[\frac{1}{(1+\epsilon)^2}-\frac{1}{(1+\epsilon\cos\theta)^2}\right]+\frac{2}{15}\ln\frac{1+\epsilon}{1+\epsilon\cos\theta} \tag{12.278}$$

where $$h_c = \frac{4C(1-\epsilon^2)^{5/2}}{5(2+\epsilon^2)}\left[\frac{6\epsilon}{(2+\epsilon^2)^2}-\frac{1}{3}\ln\frac{1+\epsilon}{1-\epsilon}\right]$$

The ratio of the pressures obtained from the purely viscous and the viscous-inertia solutions is given by

$$p/p_v = 1 + (\mathrm{Re}/6)(C/R)(p_c/p_v) \tag{12.279}$$

From the criterion for turbulence, the maximum value of Re at which laminar flow still prevails is $\mathrm{Re} = 41.1\,(R/C)^{1/2}$. Thus Eq. (12.279) can be written

$$(p/p_v)_{\max} = 1 + (p_c/6p_v)41.1(C/R)^{1/2}$$

By using a representative value of 0.001 for C/R we get for a maximum possible correction

$$(p_c/p_v)41.1(C/R)^{1/2} = 0.01 \times 41.1/\sqrt{1{,}000} = 1.3 \text{ per cent}$$

which, as previously mentioned, is relatively small.

Slider Bearing. By using for the slider-film thickness the exponential function

$$h = h_{\min} r^{-x/B}$$

where $r = h_{\min}/h_1$, we get for pressure

$$\bar{p}_v = \frac{r^2-1}{2\ln r}\left(\frac{1-r^{3\bar{x}}}{1-r^3}-\frac{1-r^{2\bar{x}}}{1-r^2}\right)$$
$$\bar{p}_c = \frac{27\ln r}{40}\left(\frac{1-r^2}{1-r^3}\right)^2\bar{p}_v + \frac{2\ln r}{15}\left(\bar{x}-\frac{1-r^{3\bar{x}}}{1-r^3}\right) \tag{12.280}$$

and for the load capacity

$$\bar{P} = Ph^2_{\min}/6\mu UB = \int_0^B p_v\,dx + \int_0^B p_c\,dx = \bar{P}_v + \bar{P}_c$$

we have, after integrating p_v and p_c,

$$\bar{P}_v = \frac{r^2-1}{2\ln r}\left\{\frac{1-[(r^3-1)/3\ln r]}{1-r^3}-\frac{1-[(r^2-1)/2\ln r]}{1-r^2}\right\}$$
$$\bar{P}_c = \frac{27\ln r}{40}\left(\frac{1-r^2}{1-r^3}\right)^2\bar{P}_v + \frac{\ln r}{15}\left[1-2\frac{1-(r^3-1)/(3\ln r)}{1-r^3}\right] \tag{12.281}$$

and the ratio of the load capacities is

$$P/P_v = 1 + K(\rho U h^2_{\min}/6\mu B) = 1 + K(h_{\min}/6B)\,\mathrm{Re} \tag{12.282}$$

where K is the ratio of $\bar{P}_c/\bar{P}_v$ and is given by

$$K = \frac{27\ln r}{40}\left(\frac{1-r^2}{1-r^3}\right)^2 + \frac{2\ln^2 r}{15(r^2-1)}\,\frac{1-2\dfrac{1-(r^3-1)/(3\ln r)}{1-r^3}}{\dfrac{1-(r^3-1)/(3\ln r)}{1-r^3}-\dfrac{1-(r^2-1)/(2\ln r)}{1-r^2}}$$

which is seen to be a function of r only. Here again by numerical examples it can be shown that even at high linear speeds the inertia effects on load capacity are of the order of 1 per cent or less.

(c) Acceleration Effects in Bearings

Tangential acceleration of journal or thrust runner always occurs in starting and stopping of engines, and some machine components such as piston rings or spur gears experience a cyclic sequence of acceleration and deceleration. The following is a simplified examination of the effects such accelerations have on the performance of bearings. We shall in Eq. (12.253) ignore the inertia of the lubricant and consider only the inertia term due to the unsteady linear velocity of the bearing surface.[46]

When integrated over 2π with $p = p_a$ at $\theta = 0$, we get

$$p - p_a = \frac{6\mu UR}{C^2}\,\frac{\epsilon \sin\theta(2 + \epsilon\cos\theta)}{(1 + \epsilon\cos\theta)^2(2 + \epsilon^2)} + \frac{\epsilon\sin\theta(\epsilon^2 - 4 - 3\epsilon\cos\theta)(1 - \epsilon^2)^{1/2}}{2(1 + \epsilon\cos\theta)^2(2 + \epsilon^2)}\,\rho R\,\frac{dU}{dt} + \cos^{-1}\frac{\cos\theta + \epsilon}{1 + \epsilon\cos\theta}\,\frac{\rho R}{2}\,\frac{dU}{dt} - \frac{\rho R}{2}\,\frac{dU}{dt}\,\theta \tag{12.283}$$

$$W = \frac{2\pi^2\epsilon\mu R^3 LN}{5C^2(2 + \epsilon^2)(1 - \epsilon^2)^{1/2}} - \frac{\pi^2\epsilon R^3 L\rho}{10(2 + \epsilon^2)}\,\frac{dN}{dt} \tag{12.284}$$

The percentage decrease in load capacity due to an acceleration dN/dt is, from Eq. (12.284),

$$\Delta W = -\frac{1}{\mu/\rho}\,\frac{dN/dt}{N}\,\frac{C^2(1 - \epsilon^2)^{1/2}}{4}$$

which shows that the reduction is proportional to the relative acceleration $dN/dt/N$.

For slider bearings with a film given by the equation[47]

$$h = h_2(a - a\bar{x} + \bar{x})$$

By use of the boundary conditions $p(0) = p(1) = 0$, the integrated pressure is

$$p(x) = \frac{6\mu UB}{h_2^2}\,\frac{\bar{x}(a - 1)(1 - \bar{x})}{(a + 1)(a - a\bar{x} + \bar{x})^2} + \frac{\rho B}{2(a^2 - 1)}\,\frac{dU}{dt}\left[\frac{a^2}{(a - a\bar{x} + \bar{x})^2} + (\bar{x} - a^2\bar{x} + 1)\right] \tag{12.285}$$

The second right-hand term is always negative; thus a positive acceleration will yield lower pressure than for steady-state conditions.

The load capacity by integrating Eq. (12.285) is

$$W = \frac{6\mu UB^2L}{h_2^2(a - 1)^2}\left[\ln a - \frac{2(a - 1)}{a + 1}\right] - \frac{LB^2\rho(a - 1)}{4(a + 1)}\,\frac{dU}{dt} \tag{12.286}$$

and the relative reduction in W due to the acceleration is

$$\frac{\Delta W}{W} = -\frac{1}{\mu/\rho}\,\frac{dU/dt}{U}\,\frac{h_2^2(a - 1)^3}{24(a + 1)\ln a - 48(a - 1)}$$

which here too is seen to depend on the relative acceleration $dU/dt/U$.

(d) Effects of Turbulence

When the bearings are operated at sufficiently high speeds or high clearances or when the viscosity of the lubricant is sufficiently low, the flow will change from laminar to turbulent. One can visualize the inception of turbulence as that instance in which the centrifugal forces become high enough to overcome viscous resistance and thus, instead of the laminar streamlets, circulation and vortexes are set up. A turbulent fluid film affects the performance of bearings in a number of ways: it raises the power losses, lowers the flow, and alters the locus of the shaft center. Before a study of turbulence in bearings can be undertaken, it is necessary first to understand the conditions that promote turbulence and to determine the range in which either turbulent, intermediate, or laminar flow prevails.

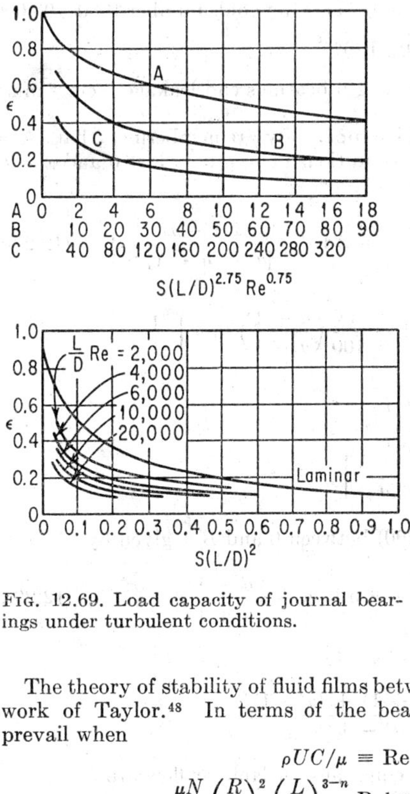

FIG. 12.69. Load capacity of journal bearings under turbulent conditions.

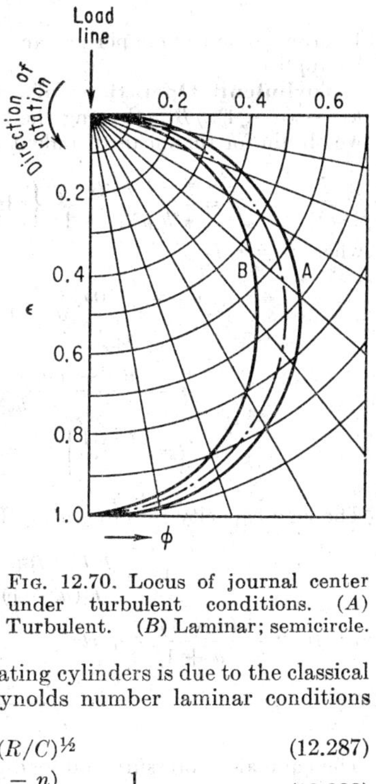

FIG. 12.70. Locus of journal center under turbulent conditions. (A) Turbulent. (B) Laminar; semicircle.

The theory of stability of fluid films between rotating cylinders is due to the classical work of Taylor.[48] In terms of the bearing Reynolds number laminar conditions prevail when

$$\rho UC/\mu \equiv \text{Re} < 41.1(R/C)^{1/2} \qquad (12.287)$$

$$\frac{\mu N}{P}\left(\frac{R}{C}\right)^2\left(\frac{L}{D}\right)^{3-n}\text{Re}^{1-n} = \frac{16(4-n)}{f_1\pi\epsilon^{2-n}}\,\frac{1}{(A^2+B^2)^{1/2}} \qquad (12.288)$$

The attitude angle is of course given by

$$\phi = \arctan B/A \qquad (12.289)$$

It will be seen that here, in addition to the Sommerfeld number and the L/D ratio, the Reynolds number is also a parameter; i.e., the attitude of the shaft is determined by the dimensionless grouping

$$S(L/D)^{3-n}\,\text{Re}^{1-n} = f(\epsilon)$$

For laminar flow $n = 1$, and thus this grouping will be reduced to the standard form obtained previously. For laminar flow $K = 96$ and then $A = 2/(1-\epsilon^2)^2$, $B = (\pi/2)(1-\epsilon^2)^{-3/2}$, and Eqs. (12.288) to (12.289) will reduce to those of laminar flow. Figures 12.69 and 12.70 show the characteristics of turbulent flow as a function of grouping $S(L/D)^{2.75}\,\text{Re}^{0.75}$. It is seen that turbulence in the fluid film causes an increase in load capacity. The position of shaft center, however, as seen from Fig. 12.70, is to the right of the laminar locus, yielding higher values of ϕ for a given eccentricity.

Two simple relations for load-carrying capacity and friction have been established for the infinitely long journal bearing of 360° and 180°.[49]

$$\left[\frac{W}{\mu NLD}\left(\frac{R}{C}\right)^2\right]_T = 0.0415\left[\frac{W}{\mu NLD}\left(\frac{R}{C}\right)^2\right]_L (\mathrm{Re})^{0.6}$$

valid for eccentricity ratios in the range 0.1 to 0.6 and Reynolds number 1000–100,000

$$f_T = 0.039 f_L (\mathrm{Re})^{0.57}$$

In order to determine performance of finite length bearings end leakage factors must be applied.

Turbulent Operation of Slider Bearings. By remembering that $h = h_2a - (a-1)x/B$ and using the boundary conditions of $p = 0$ at $x = 0$ and $x = b$, we obtain for the mean turbulent pressure

$$\bar{p} = \frac{k^2U^2B}{\pi^2h_2a^2(a-1)}\left\{r \ln\left(1 - \frac{a-1}{a}\bar{x}\right) + s\frac{\bar{x}(1-t\bar{x})}{[a-(a-1)\bar{x}]^2}\right\} \tag{12.290}$$

where $\bar{x} = x/B$

$$C_3 = -10h_2\frac{1}{a+1}\left\{1 - \left[1 - \frac{48}{100}\left(\frac{a+1}{a-1}\right)\ln a\right]^{1/2}\right\}$$

$$r = 24a^2$$

$$s = \frac{a(a-1)C_3(C_3 + 10ah_2)}{h_2^2}$$

$$t = \frac{a-1}{a}\left[1 - \frac{C_3}{2(C_3 + 10ah_2)}\right]$$

The load capacity by integrating Eq. (12.290) between 0 and B is given by

$$W = \frac{k^2U^2LB^2(a+1)}{h_2(a-1)^3\pi^2a^2}\left(m + n \ln a + s\frac{a-1}{a+1}\right) \tag{12.291}$$

where $m = \dfrac{(a-1)^2}{a+1}r + st$

$$n = -\frac{s(2at - a + 1) + (a-1)^2r}{a^2 - 1}$$

The equivalent pressure and load-capacity equations for laminar flow are

$$p = \frac{6\mu UB(a-1)}{h_2^2(a+1)^2}\frac{\bar{x}(1-\bar{x})}{[a-(a-1)\bar{x}]^2}$$

$$W = \frac{6\mu ULB^2}{h_2^2(a-1)^2}\left(\ln a - 2\frac{a-1}{a+1}\right)$$

The frictional force in the slider is, by integrating the shear stress in the viscous sublayer,

$$F = \frac{k^2U^2BL}{\pi^2h^2(a-1)}\left[3C_3 \ln a + \frac{36h_2^2a + C_3^2}{4h_2a}(a-1)\right] \tag{12.292}$$

with the corresponding drag in the laminar flow given by

$$F = \frac{\mu UBL}{h_2(a-1)}\left[4 \ln a - \frac{6(a-1)}{a+1}\right]$$

Because of the original assumption of $\mu \partial^2 \bar{u}/\partial y^2 = 0$, the turbulent equations for pressure and load are independent of viscosity. Instead, they depend on k^2, which has the dimensions of density and is an experimental quantity. Even for low values of k^2 the load capacity of turbulent flow is higher than that of laminar flow. A numerical example for $B = 2$, $U = 2{,}500$, $h = 0.002$, $a = 2$, $\mu = 0.145$, and Re = 2,410 would yield the following values:

k	W/L	p_0	H/L, hp
0	2.30×10^2	1.81×10^2	0.212
0.01	3.30×10^4	3.08×10^4	3.93×10
0.1	3.30×10^6	3.08×10^6	3.93×10^3
1.0	3.30×10^8	3.08×10^8	3.93×10^5

References

1. Reynolds, O.: On the Theory of Lubrication and Its Application to Mr. Beauchamp Tower's Experiments, *Phil. Trans. Roy. Soc. London*, vol. 177, part 1, 1886.
2. Cope, W. F.: The Hydrodynamical Theory of Film Lubrication, *Proc. Roy. Soc. (London)*, vol. A197, p. 201, 1949.
3. Sommerfeld, A.: Zur hydrodynamischen Theorie der Schmiermittelreibung, *Z. Math. Physik*, vol. 50, p. 97, 1904.
4. DuBois, G. B., and F. W. Ocvirk: Analytical Derivation and Experimental Evaluation of Short Bearing Approximation for Full Journal Bearings, *NACA Rept.* 1157, 1953.
5. Pinkus, O., and B. Sternlicht: "Theory of Hydrodynamic Lubrication," McGraw-Hill Book Company, Inc., New York, 1961.
6. Michell, A. G. M.: The Lubrication of Plane Surfaces, *Z. Math. Physik*, vol. 132, p. 123, 1905.
7. Archibald, F. R.: A Simple Hydrodynamic Thrust Bearing, *Trans. ASME*, vol. 72, May, 1950.
8. Pinkus, O.: Solution of the Tapered-land Sector Thrust Bearing, *Trans. ASME*, vol. 80, October, 1958.
9. Gross, W. A.: Compressible Lubrication of Infinitely Long Slider and Journal Bearings, *IBM Notes*, June, 1958.
10. Brody, S.: "Solution of Reynolds Equation for a Plain Slider Bearing of Finite Width with an Isothermal Gas Flow," ASLE paper 60AM5A-4.
11. Osterle, J. F., and W. J. Hughes: High Speed Effects in Pneumodynamic Journal Bearing Lubrication, *Appl. Sci. Res., Sect. A*, vol. 7, 1958.
12. Ausman, J. S.: "An Improved Analytical Solution for Self-acting Gas-lubricated Journal Bearings of Finite Length," ASME paper 60-LUB-9.
13. Sternlicht, B.: Gas Lubricated Cylindrical Journal Bearings of Finite Length, Part II, Dynamic Loading, *ONR Rept*, Contract NONR 2844(00), Sept. 9, 1960.
14. Pan, C. H. T.: On the Time Dependent Effects of Self-acting Gas Journal Bearings, February, 1962, Contract NONR 3730(00), *ONR Rept.*
15. Sternlicht, B., and R. C. Elwell: Theoretical and Experimental Analysis of Hydrodynamic Gas Lubricated Journal Bearings, *Trans. ASME*, vol. 80, June, 1958.
16. Heinrich, G.: Über Strömungslager, und Das zylindrische Strömungslager, *Maschinenbau u. Warmewirtsch., Wien*, vol. 4, no. 11, pp. 176–179, 1949; vol. 5, no. 8, pp. 136–143, 1950.
17. Heinrich, G.: Das Strömungs-Spurlager, *Maschinenbau u. Warmewirtsch., Wien*, vol. 6, pp. 57–60, 78–87, 1951.
18. Heinrich, G.: The Aerodynamic Bearing, part I, Laminar Flow, *Maschinenbau u. Warmewirtsch., Wien*, vol. 7, nos. 7, 8, pp. 117–120, 129–135, 1952.
19. Gross, W. A., and I. C. Tang: Analysis and Design of Externally Pressurized Gas Bearings, *IBM Res. Rept.*, RJ-191, Apr. 17, 1961.
20. Weber, R. R.: The Analysis and Design of Hydrodynamic Gas Bearings, *North Am. Rept.*, AL 699, 1949.
21. Hughes, W. F., and J. F. Osterle: "Temperature Effects in Hydrodynamic Thrust Bearing Lubrication," ASME paper 55-LUB-11.

22. Licht, L., D. D. Fuller, and B. Sternlicht: Self-excited Vibrations of an Air-lubricated Thrust Bearing, *Trans. ASME*, vol. 80, 1958.
23. Licht, L., and H. G. Elrod, Jr.: An Analytical and Experimental Study of the Stability of Externally-pressurized, Gas-lubricated Thrust Bearings, *Franklin Inst. Interim Rept.*, I-A2049-12, February, 1961.
24. Archibald, F.: Load Capacity and Time Relations for Squeeze Films, *Trans. ASME*, vol. 78, pp. 24–35, January, 1956.
25. Burwell, J. T.: The Calculated Performance of Dynamically Loaded Sleeve Bearings, *J. Appl. Mech.*, vol. 69, pp. A231–A245, 1947; vol. 71, pp. 358–360, 1949.
26. Ott, H. H.: "Zylindrische Gleitlager bei instationärer Belastung," Verlag A. G. Leemann, Zurich, 1948.
27. Lund, J. W., and B. Sternlicht: "Rotor-bearing Dynamics with Emphasis on Attenuation," ASME paper, 1961 annual meeting.
28. Lund, J. W., and B. Sternlicht: Bearing Attenuation, *Bureau of Ships Rept.*, Contract NObs 78930, Apr. 28, 1961.
29. Boeker, G. F., and B. Sternlicht: Investigation of Translatory Fluid Whirl in Vertical Machines, *Trans. ASME*, vol. 78, pp. 13–20, 1956.
30. Rentzepis, G. M., and B. Sternlicht: On the Stability of Rotors in Cylindrical Journal Bearings, *ONR Rept.*, Contract NONR 2844(00), May 15, 1961.
31. Pan, C. H. T., and B. Sternlicht: "On the Translatory Whirl Motion of a Vertical Rotor in Plain Cylindrical Gas Dynamic Journal Bearings," ASME paper, October, 1961.
32. Sternlicht, B., and R. C. Elwell: Synchronous Whirl in Plain Journal Bearings, *ONR Rept.*, Contract NONR 2844(00), Jan. 30, 1961.
33. Sternlicht, B., H. Poritsky, and E. Arwas: Dynamic Stability Aspects of Compressible and Incompressible Cylindrical Journal Bearings, First International Gas Bearing Symposium, 1959 (sponsored by ONR).
34. Cameron, A., and W. L. Wood: Parallel Surface Thrust Bearing, *Proc. Sixth Intern. Congr. Appl. Mech.*, 1946.
35. Blok, H., and J. J. van Rosaus: The "Foil Bearing": A New Departure in Hydrodynamic Lubrication, *Delft Publ.* 140, Dec. 18, 1952.
36. Osterle, F., and E. Saibel: The Spring Supported Thrust Bearing, *Trans. ASME*, vol. 79, February, 1957.
37. Sternlicht, B., and E. Arwas: "Propeller Shaft Thrust Bearing Analysis," Phase I, Bureau of Ships, U.S. Navy, 1959.
38. Sternlicht, B., G. K. Carter, and E. B. Arwas: "Adiabatic Analysis of Elastic Centrally Pivoted Sector, Thrust Bearing Pads," ASME paper 60-WA-104.
39. Martin, H. M.: The Lubrication of Gear Teeth, *Engineering*, vol. 102, p. 119, 1916.
40. Cameron, A.: Hydrodynamic Lubrication of Rotating Disks in Pure Sliding: A New Type of Oil Film Formation, *J. Inst. Petrol*, vol. 37, p. 471, 1951.
41. Weber, C., and K. Saalfeld: *Z. Angew. Math. Mechanik*, vol. 34, no. 1/2, pp. 54–64, 1954.
42. Grubin, A. N.: Investigation of Contact of Machine Components, *Tsentral Nauchn., Issled. Inst. Tekhnol. i. Mashinostr., Kniga*, vol. 30, 1949.
43. Brand, R. S.: Inertia Forces in Lubricating Films, *Trans. ASME*, vol. 77, p. 363, 1955.
44. Osterle, F., and E. A. Saibel: On the Effect of Lubricant Inertia in Hydrodynamic Lubrication, *Z. Angew. Math. Phys.*, vol. 6, p. 334, 1955.
45. Osterle, F., Y. T. Chou, and E. A. Saibel: "The Effect of Lubricant Inertia in Journal Bearing Lubrication," ASME paper 57APM-37.
46. Ladanyi, D. J.: Effects of Temporal Tangential Bearing Acceleration on Performance Characteristics of Slider and Journal Bearings, *NACA Tech. Note* 1730, 1948.
47. Lyman, F. A., and E. A. Saibel: Transient Lubrication of an Accelerated Infinite Slider Bearing, ASME-ASLE Joint Lubrication Conference, Boston, 1960.
48. Taylor, G. I.: Stability of a Viscous Liquid Contained between Two Rotating Cylinders, *Phil. Trans. Roy. Soc. London*, vol. A223, pp. 289–343, 1923.
49. Sternlicht, B.: *Prod. Eng.*, Aug. 21, 1961, p. 49.
50. Wilcock, D. F., and E. R. Booser: "Bearing Design and Application," McGraw-Hill Book Company, Inc., New York, 1957.
51. Vallance, A., and V. L. Doughtie: "Design of Machine Members," 3d ed., McGraw-Hill Book Company, Inc., New York, 1951.
52. Maleev, V. L., and J. B. Hartman: *Machine Design*, International Textbook Company, Scranton, Pa., 1954.

Section 13

THE MATHEMATICAL THEORY OF ROLLING-ELEMENT BEARINGS

By

A. B. JONES, *Consulting Engineer, Newington, Conn., Formerly Chief Research Engineer, Fafnir Bearing Company.*

CONTENTS

13.1. INTRODUCTION

Ball and roller bearings are produced in many types and sizes which are well described in manufacturers' catalogues and literature. Their application, installation, and lubrication are thoroughly discussed in the manufacturers' publications and in a number of excellent texts and articles.

This section is concerned solely with the mathematical analysis of rolling-element bearings as elastic systems and with the prediction of their performance and life.

Ball bearings are generally the first choice among rolling-element bearings. A ball-bearing assembly usually consists of four parts: an inner race, an outer race, the balls, and a cage or separator. The separator maintains even ball spacing. Figure 13.1 shows typical ball-bearing types.

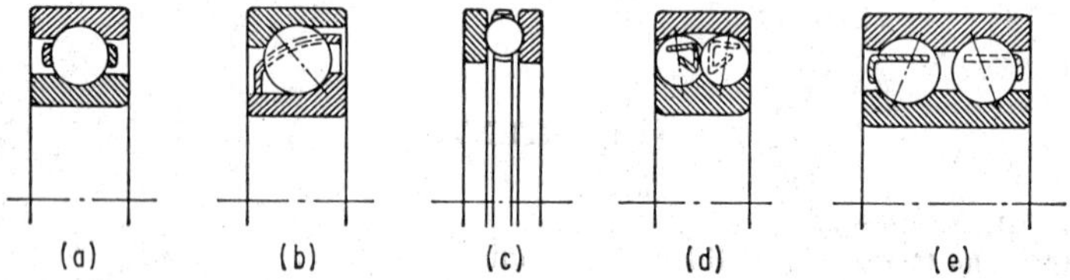

FIG. 13.1. Ball bearings. (*a*) Single-row radial—for radial loads. (*b*) Single-row angular contact—for radial and axial loads. (*c*) Axial thrust. (*d*) Double row—for heavier radial loads. (*e*) Self-aligning—for radial and axial loads and large amounts of angular misalignment.

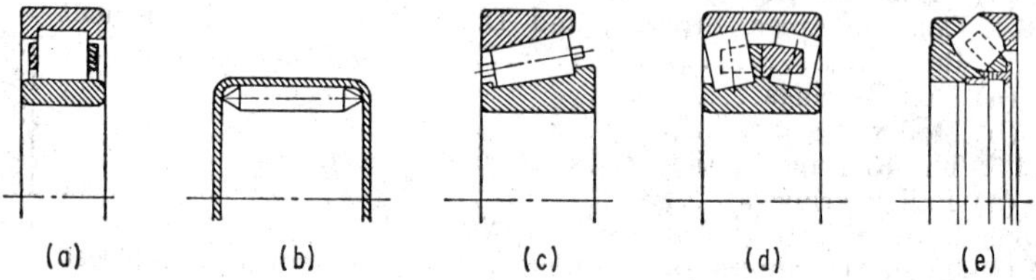

FIG. 13.2. Roller bearings. (*a*) Cylindrical roller bearings. (*b*) Needle bearing—for low speeds, intermittent loads. (*c*) Tapered roller bearing—for heavy axial loads. (*d*) Spherical roller bearing—for thrust loads and large amounts of angular misalignment. (*e*) Spherical thrust roller bearing.

Where shock or heavy loading exists, *roller bearings* are used (Fig. 13.2). Roller bearings can be applied where angular misalignment caused by shaft deflection is slight (except for type Fig. 13.2*d*).

For a rotating shaft, relative rotation between shaft and bearing is usually prevented by mounting the inner race with a press fit and securing it with a nut threaded on the shaft with a washer. The outer race is mounted less tightly but is not loose in the housing. When two ball bearings are mounted on the same shaft, the outer race on one of them should be permitted to shift axially in order to correct for relative motion of shaft and housing (see Fig. 13.3). In bearing design, the shaft should be designed to be as rigid as possible.

The balls and races are subjected to compressive stresses of 200,000 to 300,000 psi. They are made of high-carbon chrome steel SAE 52100, 60 to 65 Rockwell C. The coefficient of friction of ball bearings generally ranges from 0.001 to 0.002. For roller bearings the range is 0.002 to 0.005.

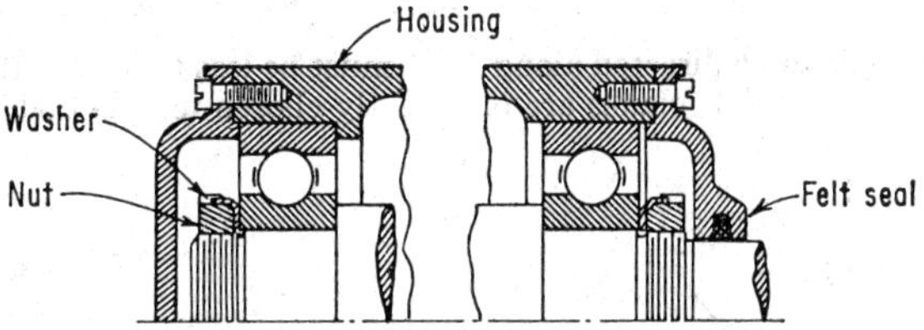

FIG. 13.3. Ball-bearing mounting.

A lubricant must be used for rolling-element bearings. The functions of the lubricant are to dissipate heat, reduce friction, protect the surface from corrosion, and exclude dust. Mineral oil or grease should be used in minimum controlled quantities. Grease is used for slow speeds only. References 11 through 15 give design information.

13.2. THE CONTACT OF SOLID ELASTIC BODIES

Rolling-element bearings derive their load-carrying ability and high efficiency from the use of hardened bodies of revolution in rolling contact.

In all rolling-element bearings, the initial contact between the load-carrying bodies is a mathematical point or line depending on the design geometry. Some types of bearings employ point contact at one raceway and line contact at the other.

When the bodies are pressed together with a force, a flattening occurs in the region of contact. A small area is formed as the result of elastic deformation and the force is distributed over this area, but not uniformly.

The size and shape of the pressure area and the distribution of the normal stress over it were first found by Hertz (see Sec. 15). Later investigators have determined the nature of stress fields surrounding the pressure area. The treatment used here is due to N. M. Belyayev.[1]

Hertz assumed that the pressure area is flat, that its dimensions are small compared with the radii of curvature of the contacting bodies, that the materials are isotropic, that the elastic limit is not exceeded, and that only normal stress acts at the interface.

While none of these assumptions is completely valid in practice, theoretical calculations are quite accurate, even with the close conformities of the contacting bodies found in ball bearings.

(a) Elliptical Pressure Area

Figure 13.4 shows two solid elastic bodies, 1 and 2, which are to be brought into contact. Each body is referred to a separate, orthogonal, XYZ coordinate system such that planes XZ and YZ contain the principal radii of curvature of the bodies.

Let the two bodies be brought into contact so that the Z axes of both coordinate systems coincide. At this time the XZ planes of the two coordinate systems do not necessarily coincide and there exists the angle ω between X_1 and X_2 as shown in Fig. 13.5.

If the two bodies are now pressed together by forces acting along Z, an elliptical pressure area will be produced whose major axis is directed along $\bar{X}$ in Fig. 13.5. Figure 13.6 shows the dimensions of the pressure area.

The curvatures of the bodies are

$$\begin{aligned} \rho_{1_x} &= 1/R_{1_x} \qquad & \rho_{1_y} &= 1/R_{1_y} \\ \rho_{2_x} &= 1/R_{2_x} \qquad & \rho_{2_y} &= 1/R_{2_y} \end{aligned} \tag{13.1}$$

ρ is positive if the body is convex in that plane. The orientation of the pressure area is found from

$$\tan 2\omega_1 = \frac{-(\rho_{2_x} - \rho_{2_y}) \sin 2\omega}{(\rho_{1_x} - \rho_{1_y}) + (\rho_{2_x} - \rho_{2_y}) \cos 2\omega} \tag{13.2}$$

Let

$$\omega_2 = \omega_1 + \omega \tag{13.3}$$

$$A = \tfrac{1}{2}(\rho_{1_x} \cos^2 \omega_1 + \rho_{1_y} \sin^2 \omega_1 + \rho_{2_x} \cos^2 \omega_2 + \rho_{2_y} \sin^2 \omega_2) \tag{13.4}$$

$$B = \tfrac{1}{2}(\rho_{1_x} \sin^2 \omega_1 + \rho_{1_y} \cos^2 \omega_1 + \rho_{2_x} \sin^2 \omega_2 + \rho_{2_y} \cos^2 \omega_2) \tag{13.5}$$

For the major axis to be directed along $\bar{X}$, A must be less than B. If this condition is not satisfied the X and Y planes are incorrectly chosen for the bodies.

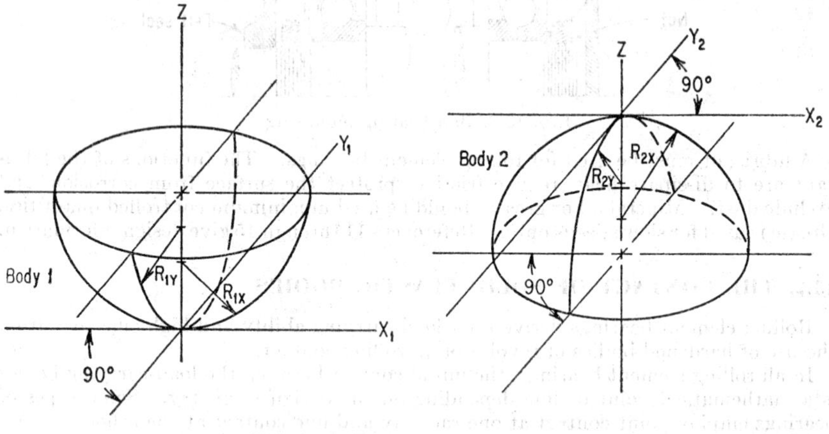

FIG. 13.4. Two solid elastic bodies.

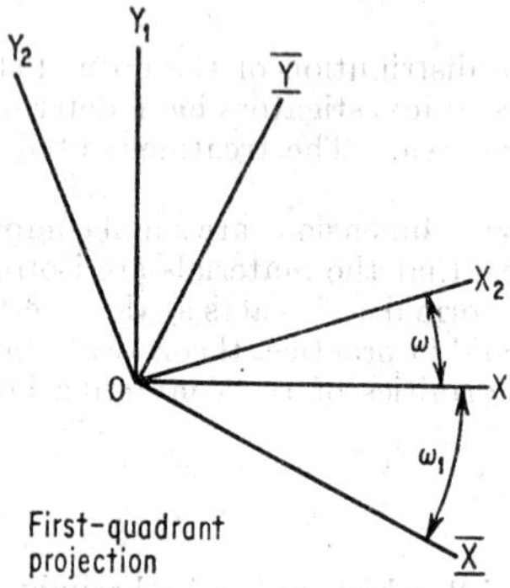

FIG. 13.5. First-quadrant XY projection.

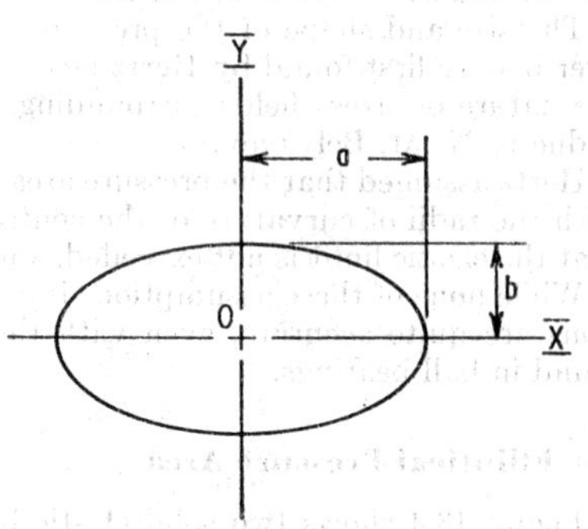

FIG. 13.6. Dimension of the pressure area.

If the X and Y axes of the two coordinate system are parallel, ω is zero and the major axis is directed along X. This is the case with all rolling-element bearings.

When ω is zero

$$A = \tfrac{1}{2}(\rho_{1_x} + \rho_{2_x}) \tag{13.6}$$

$$B = \tfrac{1}{2}(\rho_{1_y} + \rho_{2_y}) \tag{13.7}$$

As before, $A < B$.

The shape of the pressure area depends only on A and B and is independent of the compressive load.

The eccentricity of the pressure ellipse, denoted by e, is found from

$$\frac{A}{B} = \frac{(1 - e^2)(\bar{K} - \bar{E})}{\bar{E} - (1 - e^2)\bar{K}} \tag{13.8}$$

$\bar{K}$ and $\bar{E}$ are the complete elliptic integrals formed with the modulus e.

$$\frac{\tau_{xy}}{\delta_0} = \frac{\mu}{\lambda + \mu}\frac{\beta^2\bar{x}\bar{y}}{e^3}\left(\frac{1}{\bar{x}}\tanh^{-1}\frac{\bar{x}\beta e}{1+\beta\gamma} - \frac{1}{\bar{y}}\tan^{-1}\frac{\bar{y}e}{\gamma+\beta}\right) \tag{13.23}$$

$$\bar{x} = x/b \tag{13.24}$$

$$\bar{y} = y/b \tag{13.25}$$

$$\bar{z} = z/b \tag{13.26}$$

$$\beta = b/a \tag{13.27}$$

$$\gamma = \sqrt{1 - (x^2/a^2) - (y^2/b^2)} \tag{13.28}$$

$$e = \sqrt{1 - \beta^2} \tag{13.29}$$

$$\lambda = \frac{Em}{(1+m)(1-2m)} \tag{13.30}$$

$$\mu = \frac{E}{2(1+m)} \tag{13.31}$$

The subsurface stresses are

$$\frac{\delta_x}{\delta_0} = Q\left[\frac{t\bar{x}}{(1/\beta^2) + t^2}\right]^2 + (1-2m)N_x - 2(1-m)\frac{\bar{z}M_x}{t} + \frac{2m\bar{z}M_z}{t} \tag{13.32}$$

$$\frac{\delta_y}{\delta_0} = Q\left(\frac{t\bar{y}}{1+t^2}\right)^2 + (1-2m)N_y - 2(1-m)\frac{\bar{z}M_y}{t} + \frac{2m\bar{z}M_z}{t} \tag{13.33}$$

$$\delta_z/\delta_0 = Q(\bar{z}/t)^2 \tag{13.34}$$

$$\frac{\tau_{xy}}{\delta_0} = Q\frac{t\bar{x}}{(1/\beta^2) + t^2}\frac{t\bar{y}}{1+t^2} - (1-2m)N \tag{13.35}$$

$$\tau_{yz}/\delta_0 = Q\bar{z}\bar{y}/(1+t^2) \tag{13.36}$$

$$\frac{\tau_{zx}}{\delta_0} = \frac{Q\bar{z}\bar{x}}{(1/\beta^2) + t^2} \tag{13.37}$$

where t is the largest nonnegative root of

$$\frac{\bar{x}^2}{(1/\beta^2) + t^2} + \frac{\bar{y}^2}{1+t^2} + \frac{\bar{z}^2}{t^2} - 1 = 0 \tag{13.38}$$

t^2 can be found by iteration of Eq. (13.38) using a starting value $t^2 > 1$.

$$t^2_{i+1} = t_i^2 - \frac{(t_i^2)^3 + [1 + (1/\beta^2) - \bar{x}^2 - \bar{y}^2 - \bar{z}^2](t_i^2)^2 - [\bar{x}^2 + \bar{z}^2 + (1/\beta^2)(\bar{y}^2 + \bar{z}^2 - 1)]t_i^2 - \bar{z}^2/\beta^2}{3(t_i^2)^2 + 2[1 + (1/\beta^2) - \bar{x}^2 - \bar{y}^2 - \bar{z}^2]t_i^2 - [\bar{x}^2 + \bar{z}^2 + (1/\beta^2)(\bar{y}^2 + \bar{z}^2 - 1)]} \tag{13.39}$$

or by solving the cubic in t^2.

Also

$$Q = \frac{\bar{z}}{t\beta\{[(1/\beta^2) + t^2](1+t^2)\}^{1/2}\left\{\left[\frac{t\bar{x}}{(1/\beta^2) + t^2}\right]^2 + \left(\frac{t\bar{y}}{1+t^2}\right)^2 + \left(\frac{\bar{z}}{t}\right)^2\right\}} \tag{13.40}$$

$$M_x = \frac{t[F(\varphi,k) - E(\varphi,k)]}{(1/\beta^2) - 1} \tag{13.41}$$

$$M_y = \frac{t[E(\varphi,k)/\beta^2 - F(\varphi,k)]}{(1/\beta^2) - 1} - \frac{t^2}{\beta\{[(1/\beta^2) + t^2](1+t^2)\}^{1/2}} \tag{13.42}$$

$$M_z = \frac{1}{\beta}\frac{1+t^2}{(1/\beta^2) + t^2} - tE(\varphi,k) \tag{13.43}$$

$F(\varphi,k)$ and $E(\varphi,k)$ are elliptic integrals of the first and second class formed with the modulus $k^2 = 1 - \beta^2$ and the amplitude $\varphi = \tan^{-1} 1/\beta t$.

$$N_x = \frac{\beta}{1-\beta^2}\left\{1 - \frac{\bar{z}}{t}\left[\frac{1+t^2}{(1/\beta^2) + t^2}\right]^{1/2} - \bar{x}\Omega_x - \bar{y}\Omega_y\right\} \tag{13.44}$$

$$N_y = \frac{\beta}{1-\beta^2}\left\{\frac{\bar{z}}{t}\left[\frac{(1/\beta^2)+t^2}{1+t^2}\right]^{1/2} + \bar{x}\Omega_x + \bar{y}\Omega_y - 1\right\} \tag{13.45}$$

$$N = \frac{\beta}{1-\beta^2}(\bar{x}\Omega_y - \bar{y}\Omega_x) \tag{13.46}$$

$$\Omega_x = \frac{\beta}{(1-\beta^2)^{1/2}}\tan^{-1}\frac{\bar{x}(1/\beta^2-1)^{1/2}}{(1/\beta^2)+t^2+(\bar{z}/l)\{[(1/\beta^2)+t^2](1+t^2)\}^{1/2}}$$

$$\Omega_y = \frac{\beta}{(1-\beta^2)^{1/2}}\tan^{-1}\frac{\bar{y}[(1/\beta^2)-1]^{1/2}}{1+t^2+(\bar{z}/t)\{[(1/\beta^2)+t^2](1+t^2)\}^{1/2}} \tag{13.47}$$

The stress conditions along the Z axis are of particular interest for it is here that the maximum shear stresses occur. The maximum shear stress T in the plane of two normal stresses depends upon the difference in the normal stresses and acts at 45° to each.

$$T_{yx} = (\delta_y - \delta_x)/2 \tag{13.48}$$
$$T_{zy} = (\delta_z - \delta_y)/2 \tag{13.49}$$
$$T_{zx} = (\delta_z - \delta_x)/2 \tag{13.50}$$

$$T_{yx} = \frac{1}{2\beta\{[(1/\beta^2)+\bar{z}^2](1+\bar{z}^2)\}^{1/2}} \left\{\frac{\beta^2(1-2m)[\sqrt{(1/\beta^2)+\bar{z}^2}-\sqrt{1+\bar{z}^2}]^2}{1-\beta^2} + 2(1-m)\bar{z}^2\right\} - \frac{(1-m)\bar{z}\{[(1/\beta^2)+1]E(\varphi,k)-2F(\varphi,k)\}}{(1/\beta^2)-1} \tag{13.51}$$

$$T_{zy} = \frac{1-2(1-m)\bar{z}^2}{2\beta\{[(1/\beta^2)+\bar{z}^2](1+\bar{z}^2)\}^{1/2}} + \frac{(1-m)\bar{z}}{(1/\beta^2)-1}\left[\frac{E(\varphi,k)}{\beta^2}-F(\varphi,k)\right] - \frac{m}{\beta}\left[\frac{1+\bar{z}^2}{(1/\beta^2)+\bar{z}^2}\right]^{1/2} - \frac{(1-2m)\beta}{2(1-\beta^2)}\left\{\left[\frac{(1/\beta^2)+\bar{z}^2}{1+\bar{z}^2}\right]^{1/2}-1\right\} + m\bar{z}E(\varphi,k) \tag{13.52}$$

$$T_{zx} = \frac{1}{2\beta\{[(1/\beta^2)+\bar{z}^2](1+\bar{z}^2)\}^{1/2}} + \frac{(1-m)\bar{z}[F(\varphi,k)-E(\varphi,k)]}{(1/\beta^2)-1} - \frac{m}{\beta}\left[\frac{1+\bar{z}^2}{(1/\beta^2)+\bar{z}^2}\right]^{1/2} - \frac{(1-2m)\beta}{2(1-\beta^2)}\left\{1-\left[\frac{1+\bar{z}^2}{(1/\beta^2)+\bar{z}^2}\right]^{1/2}\right\} + m\bar{z}E(\varphi,k) \tag{13.53}$$

The greatest value of maximum shear stress occurs in the ZY plane some distance below the surface of the pressure area. Figure 13.10 shows the variation in $T_{zy\,\max}$

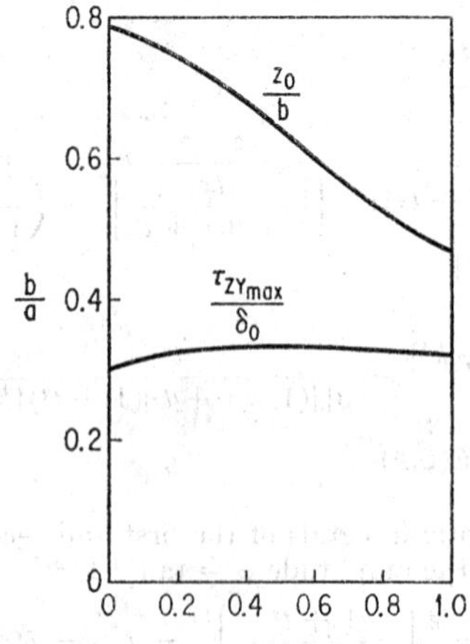

FIG. 13.10. Variation in $T_{zy\,\max}$ and Z_0.

and Z_0, the depth to the point of maximum shear stress, with the shape of the pressure ellipse.

(b) Line Contact

Roller bearings employing cylindrical, convex, or concave rollers having initial contact with a raceway along a line extending the whole length of the roller provide examples of line-contact conditions. In some instances, one or both of the bodies in contact are crowned to provide lower stress at the roller ends than occurs when the rolling elements contact the raceway along the full length of the roller under no load. The degree of crowning or relief is specified by manufacturers to account for the range of loads normally encountered in service.

The stress state in the region of the contact of two cylinders of infinite length and parallel axes is given for the case of uniform loading per unit length. The initial contact is a line of zero width. After loading is applied the line is transformed into a narrow band bounded by parallel lines.

The stress state for such contact regions was obtained by Belyayev by considering the limiting value of the stress state in an elliptical region of contact for that particular instance for which eccentricity of the contact ellipse $e \to 1$. This is equivalent to the statement that the major axis $2a$ of the contact ellipse increases without bound and that the elliptic integral of the first kind becomes ln ∞, while the elliptic integral of the second kind becomes unity.

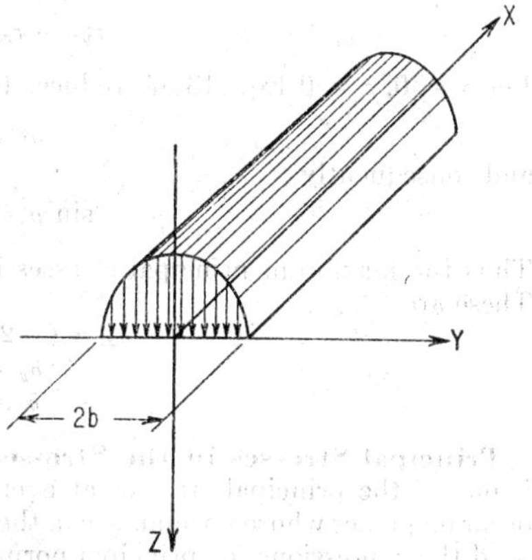

FIG. 13.11. Elliptical coordinate system

Belyayev used the coordinate system shown in Fig. 13.11. The X axis extends along the length of the contact band; Y originates at the center of the contact band and extends to the right in the original unstrained surface normal to X. The axis Z is directed into the body, normal to X and Y. Inasmuch as infinite length of contact has been assumed the origin of X can be set arbitrarily, and stress states show no variation with the X coordinate.

The stresses at points on the surfaces in contact and in the material in the vicinity of the contact region are expressed in terms of the transformations

$$Y = b \cosh \alpha \cos \rho \tag{13.54}$$

$$Z = b \sinh \alpha \sin \rho \tag{13.55}$$

which are used to define each point X, Y determined by intersection of ellipses

$$Y^2/(b^2 \cosh^2 \alpha) + Z^2/(b^2 \sinh^2 \alpha) = 1 \tag{13.56}$$

and the hyperbolas

$$Y^2/(b^2 \cos^2 \rho) - Z^2/(b^2 \sin^2 \rho) = 1 \tag{13.57}$$

The points at the bounding lines of the contact-region width correspond to $\alpha = 0$ and $\rho = 0$; those on the Z axis to $\rho = \pi/2$. The contact stress formulas given below are in terms of p, the load per unit length, and b, the half width of the contact band. The expression for b is given in a later section of this paragraph.

Stresses on *XYZ* Planes. The stresses on planes perpendicular to the coordinate axes are

$$\delta_x = \frac{-2p}{\pi b} \frac{\lambda e^{-\alpha}}{\lambda + \mu} \sin \rho \tag{13.58}$$

$$\delta_y = \frac{-2p}{\pi b} e^{-\alpha} \sin \rho + \frac{2p}{\pi b} \sin \rho \sinh \alpha \left(1 - \frac{\sinh 2\alpha}{\cosh 2\alpha - \cos 2\rho}\right) \tag{13.59}$$

$$\delta_z = \frac{-2p}{\pi b} e^{-\alpha} \sin \rho - \frac{2p}{\pi b} \sin \rho \sinh \alpha \left(1 - \frac{\sinh 2\alpha}{\cosh 2\alpha - \cos 2\rho}\right) \tag{13.60}$$

$$\tau_{yz} = \frac{-2p}{\pi b} \sinh \alpha \sin \rho \frac{\sin 2\rho}{\cosh 2\alpha - \cos 2\rho} \tag{13.61}$$

$$\tau_{xy} = \tau_{zx} = 0 \tag{13.62}$$

Stresses at Contact Surface. For points in the contact region

$$\delta_x = \frac{-2p}{\pi b} \frac{\lambda}{\lambda + \mu} \sin \rho \tag{13.63}$$

$$\delta_y = \frac{-2p}{\pi b} \sin \rho \tag{13.64}$$

$$\delta_z = \frac{-2p}{\pi b} \sin \rho \tag{13.65}$$

$$\tau_{yz} = \tau_{zx} = 0 \tag{13.66}$$

For $\alpha = 0$, $z = 0$ Eq. (13.57) reduces to

$$y^2 = b^2 \cos^2 \rho \tag{13.67}$$

and consequently

$$\sin \rho = \sqrt{1 - y^2/b^2} \tag{13.68}$$

Thus the maximum principal stresses in the contact surface occur at $y = 0$, $z = 0$. These are

$$\delta_x = (-2p/\pi b)[\lambda/(\lambda + \mu)] \tag{13.69}$$

$$\delta_y = -2p/\pi b \tag{13.70}$$

$$\delta_z = -2p/\pi b \tag{13.71}$$

Principal Stresses in the Stressed Volume. Noting from Eq. (13.62) that δ is one of the principal stresses at every point, the remaining two principal stresses occur on planes whose normals are in the XZ plane. For their determination, Belyayev used the expressions for principal normal stresses

$$\delta_N = \tfrac{1}{2}(\delta_y + \delta_z) \pm \sqrt{(\delta_y - \delta_z)^2 + 4\tau_{yz}^2} \tag{13.72}$$

The maximum shear stress corresponding to the principal stresses given by Eq. (13.72) is

$$\tau_{\max} = \tfrac{1}{2} \sqrt{(\delta_y - \delta_z)^2 + 4\tau_{yz}^2} \tag{13.73}$$

These maximum shear stresses occur on planes whose normals are ±45° from the normals of the planes of principal normal stresses.

Substituting Eqs. (13.59), (13.60), and (13.61) into Eq. (13.72) there results

$$\delta_N = \frac{-2p}{\pi b} e^{-\alpha} \sin \rho \left(1 \mp \frac{\sinh \alpha}{\sin^2 \rho + \sinh^2 \alpha}\right) \tag{13.74}$$

and into Eq. (13.73)

$$\tau_{\max} = \frac{2pe^{-\alpha} \sin \rho \sinh \alpha}{\pi b \sqrt{\sin^2 \rho + \sinh^2 \alpha}} \tag{13.75}$$

From Eq. (13.75) it is apparent that the maximum shear stress at a point in the contact region corresponds to α = constant and $\sin \rho = 1$. Hence $\tau_{\max}$ is encountered along the Z axis for which $\rho = \pi/2$. The condition for maximum or minimum of $\tau_{\max}$ is

$$(\partial/\partial\alpha)\tau_{\max} = 0 \tag{13.76}$$

Substituting Eq. (13.75) into Eq. (13.76),

$$\frac{\partial}{\partial\alpha} \frac{2pe^{-\alpha} \sin \rho \sinh \alpha}{\pi b \sqrt{\sin^2 \rho + \sinh^2 \alpha}} = 0 \tag{13.77}$$

which yields the following expression in α:

$$(e^{-\alpha}/\cosh \alpha)(-\sinh \alpha + 1/\cosh \alpha) = 0 \tag{13.78}$$

having solutions

$$e^{-\alpha} = 0 \qquad \text{therefore } \alpha_1 = \infty \tag{13.79}$$
$$1/\cosh \alpha = 0 \qquad \text{therefore } \alpha_2 = \infty \tag{13.80}$$
$$\sinh 2\alpha = 2 \qquad \text{therefore } \alpha_3 = 0.722 \tag{13.81}$$

Substituting Eq. (13.81) into Eq. (13.75) gives

$$(\tau_{\max})_{\max} = (2p/\pi b)e^{-0.722} \tanh 0.722 = 0.304(2p/\pi b) \tag{13.82}$$

This shear-stress maximum occurs at the point $Y = 0$ along the Z axis at a depth given by Eq. (13.55), keeping in mind that $\rho = \pi/2$ as discussed above. This yields

$$Z/b = \sinh \alpha_3 = 0.78 \tag{13.83}$$

Thus the maximum shear stress which occurs at a depth $Z/b = 0.78$ is 0.304 times the maximum normal stress given by Eq. (13.71) occurring at the center of the contact width. It is this maximum shear stress which is critical in static-load applications. The maximum shear stress is attained monotonically as the rolling element approaches and recedes and similarly as the roller recedes. However, as will be shown below, there is an alternating shear stress which occurs with passage of the roller which is critical in endurance conditions.

Maximum Alternating Shear Stress. As it is commonly accepted that fatigue damage is a function of maximum range of shear or normal stress encountered during a loading cycle, it is of interest to investigate the range of alternating shear stress τ_{YE} experienced during passage of a rolling body. τ_{YE} reverses sign during passage of the rolling body; hence the shear-stress variation of interest will be $2|\tau_{YE}|_{\max}$. Thus, recalling Eq. (13.61),

$$\tau_{yz} = \frac{-2p}{\pi b} \sinh \alpha \sin \rho \left(\frac{\sin 2p}{\cosh 2\alpha - \cos 2p}\right)$$

Requirements for τ_{yz} to be a maximum or a minimum are

$$\partial\tau_{yz}/\partial\alpha = 0 \tag{13.84}$$
$$\partial\tau_{yz}/\partial\rho = 0 \tag{13.85}$$

From Eq. (13.85) there results

$$2 \sinh^2 \alpha = 1 - \cos 2\rho \tag{13.86}$$

and from Eq. (13.85)

$$5 \sinh^2 \alpha = 2 \tag{13.87}$$

which have as solutions

$$\alpha = 34°22' \tag{13.88}$$
$$\beta = 39°32' \tag{13.89}$$

Thus

$$\tau_{yz}\Big|_{\max} = -0.242(2p/\pi b) \tag{13.90}$$

occurring at

$$Y = 0.915b \tag{13.91}$$
$$Z = 0.4b \tag{13.92}$$

The maximum of τ_{yz} given above corresponds to a range of variation of shear stress which is

$$2|\tau_{yz}|_{\max} = 0.484(2p/\pi b) \tag{13.93}$$

which exceeds the maximum value of shear stress along the Z axis given by Eq. (13.82) whose range is $0.304(2p/\pi b)$.

Contact Width and Deflections. The half width of the contact band for two cylindrical bodies having contact initially along a line subjected to uniform load p

per unit length can be expressed in terms of previously designated quantities as

$$b = \sqrt{\frac{4pR_1R_2(\vartheta_1 + \vartheta_2)}{R_1 + R_2}} \tag{13.94}$$

where p = force per unit length
$\vartheta = (1 - m^2)/E$
m = Poisson's ratio
E = modulus of elasticity
R_1, R_2 = arc radii of curvature, positive for convex surfaces, negative for concave

For contact of cylinder and plane, $R_2 = \infty$ and the half width of the contact area is

$$b = \sqrt{4pR_1(\vartheta_1 + \vartheta_2)} \tag{13.95}$$

Using Eq. (13.94) for the half width of the contact area between bodies having initial contact along a line, it is possible to obtain the maximum compressive stresses on the contact surface from Eqs. (13.94) and (13.71).

$$\delta_0 = -\frac{2p}{\pi b} = -\sqrt{\frac{p(1/R_1 + 1/R_2)}{\pi^2(\vartheta_1 + \vartheta_2)}} \tag{13.96}$$

The deformation at the contacts, normally taken as the reduction in distance between distant points in the two bodies in contact, cannot be obtained from the theory for line contact discussed above. An approximation of reduction of distance between distant points is generally obtained by treating the raceway as an elastic half space, assuming a pressure distribution over a contact zone which approximates that due to a finite-length roller, calculating the displacement of the surface of the half space relative to its unstrained location, and then adding the radial displacement of the contact surface of the cylindrical roller with respect to its own axis.

Lundberg, in 1939, calculated the displacement of the surface of an elastic half space under the action of a pressure distributed ellipsoidally in the narrow dimension and constant in the finite-length dimension. For this circumstance, he found that

$$\Delta_1 = [2P(1 - m^2)/El\pi][1.1932 + \ln(b/l)] \tag{13.97}$$

in which Δ_1 = depression of center-of-pressure zone
E = modulus of elasticity
l = length of contact
m = Poisson's ratio
b = half width of contact area
P = total load distributed over the area

Kovalsky, in 1940, considered the deformation of a circular cylinder of finite length, loaded from two sides by pressure distributed across the width of contact according to an elliptical distribution. He considered both the local-contact deformation and the over-all deformation of the cylinder. The change in diameter parallel to the direction of the applied force was found to be

$$2\Delta_2 = [4P(1 - m^2)/El\pi][\ln(2R/b) + 0.407] \tag{13.98}$$

where Δ_2 = reduction in distance to contact zone from roller axis
R = roller radius

and all other quantities are as defined above.

Dinnik, in an earlier paper, had calculated the reduction in diameter 2Δ, assuming a parabolic law for distribution of pressure over the width of contact·

$$2\Delta = [4P(1 - m^2)/El\pi][\ln(2R/b) + \tfrac{1}{3}] \tag{13.99}$$

This result differs only slightly from that of Kovalsky.

In the following, we shall accept Kovalsky's and Lundberg's results and write the approach of the distant point in the half plane to the axis of the cylinder as the sum of Δ_1 and Δ_2 per Eqs. (13.97) and (13.98). For the roller and raceway of the same material,

$$\Delta = \Delta_1 + \Delta_2 = [2P(1 - m^2)El\pi][\ln (2Rl/b^2) + 1.6002] \qquad (13.100)$$

Noting that b can be conveniently written as

$$b = \sqrt{8PR(1 - m^2)/El\pi} \qquad (13.101)$$

the approach of the distant points in the elastic half space to the axis of the cylinder can be written as

$$\Delta = [2P(1 - m^2)/El\pi]\{\ln [El^2\pi/P(1 - m^2)] + 0.2139\} \qquad (13.102)$$

13.3. BEARING GEOMETRY

(a) Ball Bearings

Figure 13.12*a* is a cross section of a typical ball bearing and defines some of the terminology to be used. The transverse radius of curvature of the toric race is

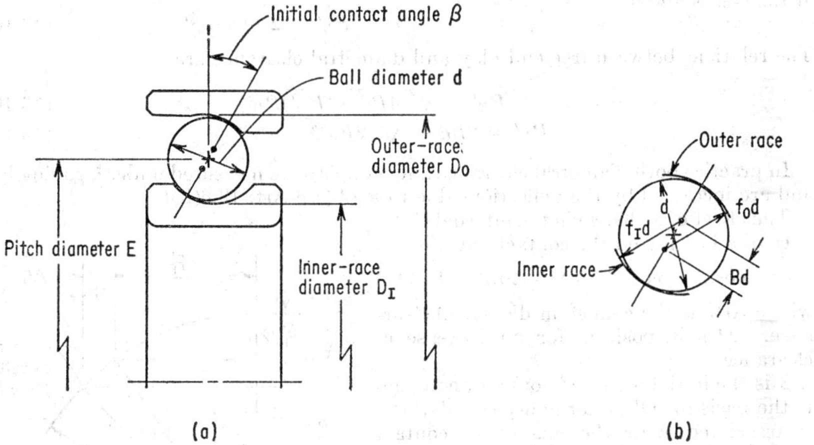

FIG. 13.12. Bearing geometry. (*a*) Terminology. (*b*) Initial contact.

always somewhat greater than the ball radius. The degree of osculation in the transverse plane is called the curvature of the race. Although mathematically a misnomer, curvature is defined as

$$f = r/d \qquad (13.103)$$

When a ball is in initial contact with two races, as in Fig. 13.12*b*, the distance Bd is defined by the ball diameter and the race curvatures. B is known as the total curvature

$$B = f_O + f_I - 1 \qquad (13.104)$$

If the two races are moved in any manner in the plane of the paper so that the ball is compressed between them, the distance Bd is increased by the amount of the deflections at the race contacts as in Fig. 13.13. Δ_O and Δ_I are the elastic deflections at outer- and inner-race contacts, respectively. The total approach Δ of outer and inner races is the sum of the contact deformations.

$$\Delta = Bd(D - 1) \qquad (13.105)$$

For a ball to be under compression D must be greater than unity.

Ball bearings are generally manufactured with internal looseness, though for special purposes the internal looseness may be completely removed. The degree of internal looseness determines the bearing's contact angle and hence its ability to sustain thrust load.

The internal fitup of the bearing may be greatly changed by ring expansions due to interference-mounting fits and temperature effects. Thus it is necessary to distinguish between unmounted and mounted conditions. In the following, the primed items refer to unmounted conditions and the unprimed items to mounted conditions.

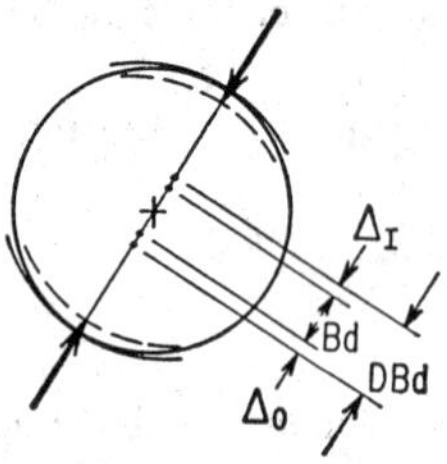

FIG. 13.13. Deflections at the race contacts.

The initial contact angle β' is dependent on the diametral clearance P_D' initially built into the bearing at manufacture.

$$\cos \beta' = (2Bd - P_D')/2Bd \tag{13.106}$$

P_D' is positive for loose bearings and is defined by

$$P_D' = D_O - D_I - 2d \tag{13.107}$$

P_D' is negative for internally preloaded or tight bearings, and the contact angle as defined by Eq. (13.106) does not exist.

The free end play P_E' in a loose bearing is

$$P_E' = 2Bd \sin \beta' \tag{13.108}$$

The relations between free end play and diametral clearance are

$$P_E' = \sqrt{(4Bd - P_D')P_D'} \tag{13.109}$$

$$P_D' = 2Bd - \sqrt{(2Bd)^2 - P_E'^2} \tag{13.110}$$

In practice both diametral clearance and end play are measured under a gaging load and are increased by the deflections due to load [see Art. 13.6(h)].

The effect of a change in the internal clearance of a bearing on the contact angle is

$$\cos \beta = \cos \beta' - \Delta P_D/2Bd \tag{13.111}$$

where ΔP_D is the change in diametral clearance. ΔP_D is positive for an increase in clearance.

β is the initial mounted contact angle and is the basis for all performance calculations.

Associated with the change in contact angle resulting from change in internal clearance is a relative axial displacement ΔS of the inner ring with respect to the outer, which must be considered where axial preloading is involved.

$$\Delta S = Bd(\sin \beta - \sin \beta') \tag{13.112}$$

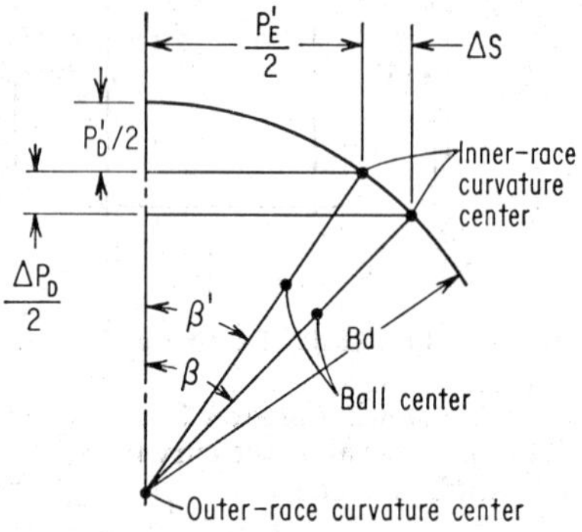

FIG. 13.14. Geometric interpretation.

A positive value of ΔS denotes a reduction in the overall width of the bearing from thrust face of inner to thrust face of outer and is associated with a positive value of ΔP_D. The geometric interpretation of the foregoing is shown in Fig. 13.14.

(b) Roller Bearings

The most common type of roller bearing is the cylindrical radial roller bearing. However, for purposes of developing a general theory a cylindrical angular-contact roller bearing (Fig. 13.15) is used.

The contact angle in a roller bearing is unaffected by clearance variations. Nevertheless changes in ring diameters which provide the equivalent of ΔP_D in a ball bearing produce an axial shift of the roller-bearing inner ring with respect to the outer as in a ball bearing. For a roller bearing

$$\Delta S = (\Delta P_D/2) \cot \beta \qquad (13.113)$$

A tapered roller bearing can be treated as a cylindrical angular-contact roller bearing whose contact angle is equal to the outer-race or cup contact angle of the tapered roller bearing and whose roll diameter is equal to the mean diameter of the tapered roller as in Fig. 13.16.

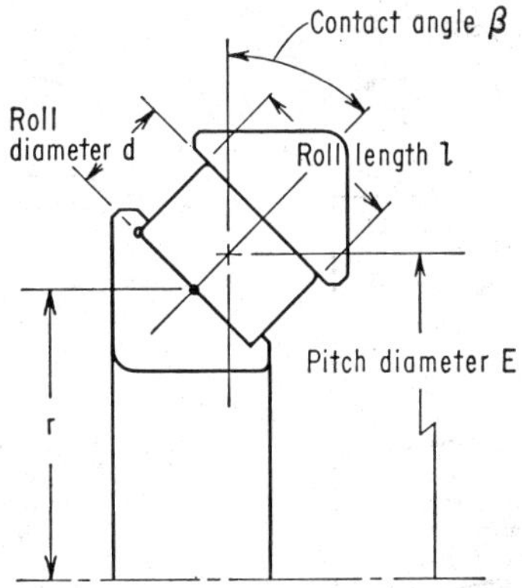

FIG. 13.15. Cylindrical angular-contact roller bearing.

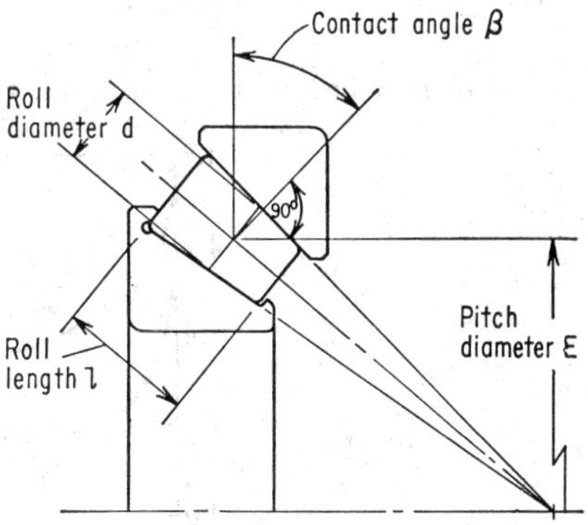

FIG. 13.16. Tapered roller bearing.

The spherical roller bearing, which has substantially line contact at the inner race and point contact at the outer, is treated similarly to the tapered roller bearing in that the contact angle is determined by the outer-race contact.

13.4. CONTACT AREAS, STRESSES, AND DEFLECTIONS IN ROLLING-ELEMENT BEARINGS

(a) Ball Bearings

Figure 13.17 shows a ball compressed between outer and inner races by the force P. Small elliptical pressure areas with semiaxes a and b are formed at the contact points because of elastic deformations under the load P. With toric races of conventional curvature, the semiminor axes lie in the rolling direction.

For purposes of calculating the pressure-area dimensions, let the plane of the paper be the XZ plane. The YZ plane contains the line of action of the load P and the rolling direction. The XY plane is the tangent plane at a contact. The Z axis coincides with the line of action of P.

In accordance with Art. 13.2(b) let the ball be body 1 and a race body 2.

$$\rho_{1_x} = 2/d \qquad \rho_{2_x} = -1/fd \qquad (13.114)$$

$$\rho_{1_y} = 2/d \qquad \rho_{2_y} = -2\gamma/(1+\gamma)d \qquad (13.115)$$

where

$$\gamma = \pm d \cos \beta / E \qquad (13.116)$$

The upper sign in Eq. (13.116) applies to an outer-race contact and the lower to an inner-race contact.

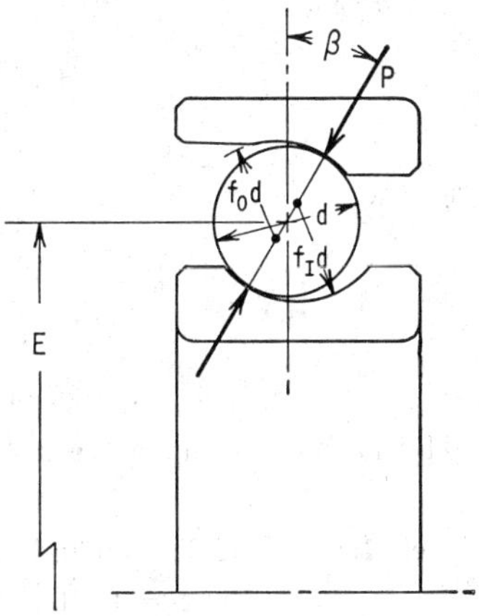

FIG. 13.17. Ball compressed between outer and inner races.

Then

$$A = (1/d)(1 - 1/2f) \quad (13.117)$$
$$B = 1/d(1 + \gamma) \quad (13.118)$$
$$A/B = (1 - 1/2f)(1 + \gamma) \quad (13.119)$$

For ball bearings of conventional materials with $E = 29 \times 10^6$ psi and $m = 0.25$ the dimensions of the pressure area are

$$a = 0.003137K_a[2fdP/(2f - 1)]^{1/3} \quad (13.120)$$
$$b = 0.003137K_b[2fdP/(2f - 1)]^{1/3} \quad (13.121)$$

where

$$K_a = [(\bar{K} - \bar{E})/e^2]^{1/3} \quad (13.122)$$
$$K_b = K_a(1 - e^2)^{1/2} \quad (13.123)$$

K_a and K_b may be obtained from Figs. 13.18 and 13.19.

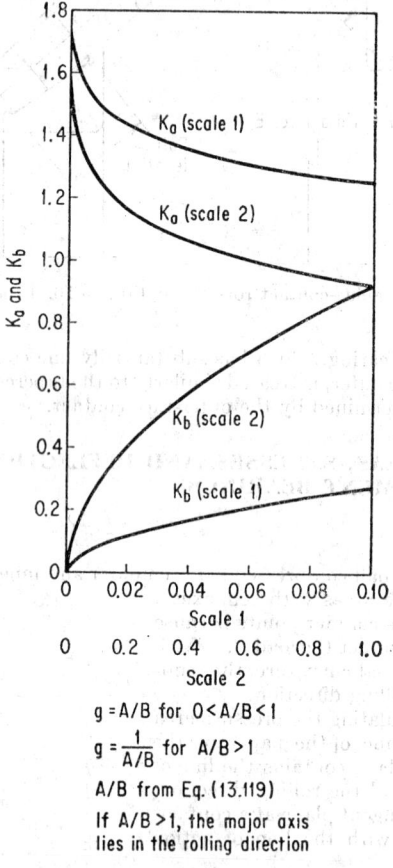

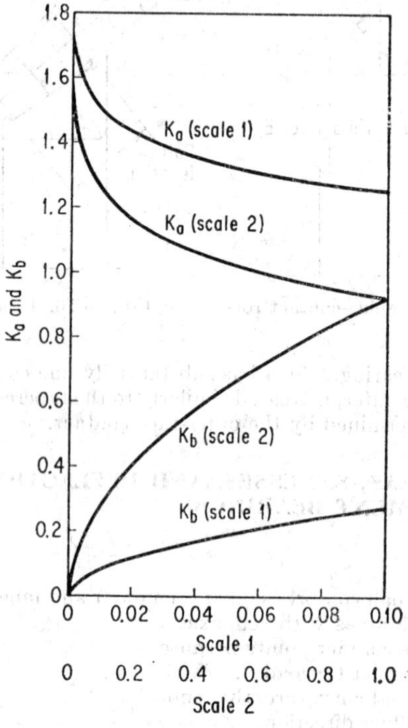

FIG. 13.18. Values of K_a and K_b.

The maximum compressive stress at the pressure-area center is

$$\delta_0 = -3P/2\pi ab \quad (13.124)$$

The elastic approach of ball and race is

$$\Delta = 1.3104(\vartheta_1 + \vartheta_2)^{2/3}C_\Delta[(2f - 1)P^2/2fd]^{1/3} \quad (13.125)$$

where

$$C_\Delta = \bar{K}e^{2/3}/(\bar{K} - \bar{E})^{1/3} \quad (13.126)$$

Values of C_Δ are given in Fig. 13.19

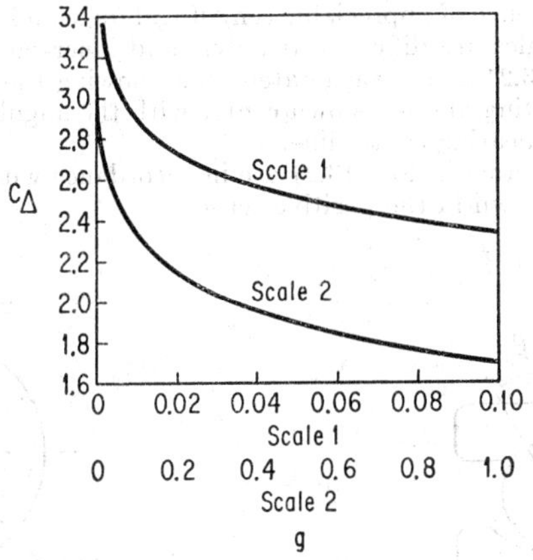

FIG. 13.19. Values of C_a.

For balls and rings with $E = 29 \times 10^6$ psi and $m = 0.25$

$$\Delta = 9.8408 \times 10^{-6} C_\Delta[(2f - 1)P^2/2fd]^{1/3} \tag{13.127}$$

(b) Roller Bearings

The principal radii of curvature in a cylindrical angular-contact roller bearing are

$$R_{1y} = d/2 \tag{13.128}$$
$$R_{2y} = -(1 + \gamma)d/2\gamma \tag{13.129}$$

where

$$\gamma = \pm d \cos \beta / E \tag{13.130}$$

The upper sign in Eq. (13.130) applies to an outer-race contact and the lower to an inner-race contact.

In accordance with Art. 13.2(b) the semiwidth b of the pressure area is

$$b = \left[\frac{2Pd(1 + \gamma)(\vartheta_1 + \vartheta_2)}{l_e}\right]^{1/2} \tag{13.131}$$

For rollers and races with $E = 29 \times 10^6$ psi and $m = 0.25$

$$b = 0.0002029[Pd(1 + \gamma)/l_e]^{1/2} \tag{13.132}$$

The elastic approach of outer and inner race along the line defined by the contact angle is

$$\Delta = 0.90094 \times 10^{-6}(P^{0.9}/l_e^{0.8}) \tag{13.133}$$

l_e is the effective length of the roller and depends on the type of crowning and also, to some extent, on the loading. For high-duty precision roller bearings such as are used in aircraft turbines l_e can be taken as approximately 90 per cent of the actual roller length.

13.5. THE RELATIVE MOTIONS OF THE ROLLING ELEMENTS

(a) Ball Bearings

The most general case of ball motion is that occurring in a high-speed angular-contact bearing.

Because of the existence of appreciable centrifugal force and gyroscopic moment the operating contact angles are different at outer- and inner-race contacts as shown in Fig. 13.20. Figure 13.21 is an exaggerated view showing the ball fixed in the plane of the paper and rotating about its own center with the angular velocity ω_B directed at the angle α to the bearing center line.

Angular-velocity vectors in Fig. 13.21 are in accordance with the right-hand screw convention and are shown in the positive sense.

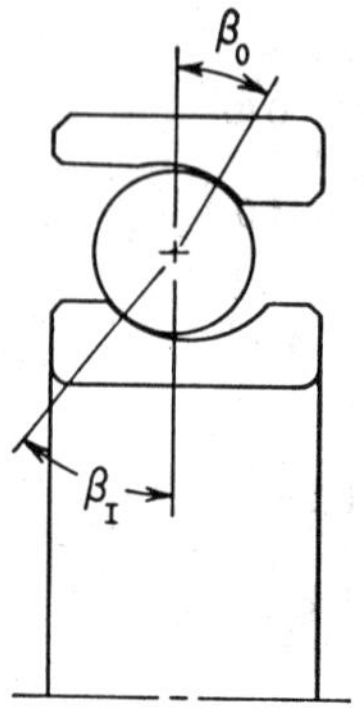

FIG. 13.20. Operating contact angle of ball bearing.

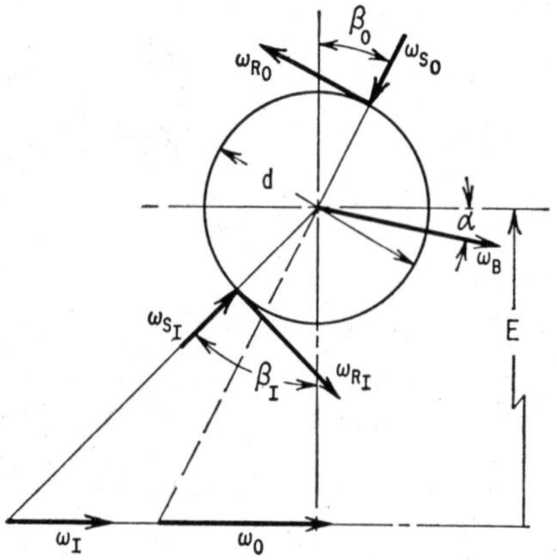

FIG. 13.21. Angular velocities of ball.

If the effects of contact-area deformation on the rolling radii are neglected,

$$\omega_B = (E/d)[(1 + \gamma_O)/\cos(\beta_O - \alpha)]\omega_O \tag{13.134}$$
$$\omega_B = (-E/d)[(1 - \gamma_I)/\cos(\beta_I - \alpha)]\omega_I \tag{13.135}$$

For the outer race to be stationary the ball must orbit with the angular velocity Ω_E.

$$\Omega_E = -\omega_O \tag{13.136}$$

Then the absolute angular velocity of the inner race is Ω_I.

$$\Omega_I = \omega_I - \omega_O \tag{13.137}$$

For stationary outer race and rotating inner race there is found

$$\omega_B = \frac{-E\Omega_I}{d}\frac{(1+\gamma_O)(1-\gamma_I)}{(1+\gamma_O)\cos(\beta_I-\alpha)+(1-\gamma_I)\cos(\beta_O-\alpha)} \tag{13.138}$$

$$\Omega_E = \Omega_I\frac{(1-\gamma_I)\cos(\beta_O-\alpha)}{(1+\gamma_O)\cos(\beta_I-\alpha)+(1-\gamma_I)\cos(\beta_O-\alpha)} \tag{13.139}$$

where

$$\gamma_O = d\cos\beta_O/E \tag{13.140}$$
$$\gamma_I = d\cos\beta_I/E \tag{13.141}$$

Similarly for stationary inner race and rotating outer

$$\omega_B = \frac{E\Omega_O}{d}\frac{(1+\gamma_O)(1-\gamma_I)}{(1+\gamma_O)\cos(\beta_I-\alpha)+(1-\gamma_I)\cos(\beta_O-\alpha)} \tag{13.142}$$

$$\Omega_E = \Omega_O\frac{(1+\gamma_O)\cos(\beta_I-\alpha)}{(1+\gamma_O)\cos(\beta_I-\alpha)+(1-\gamma_I)\cos(\beta_O-\alpha)} \tag{13.143}$$

For simultaneous rotation of outer and inner race

$$\omega_B = \frac{E(\Omega_O - \Omega_I)}{d} \frac{(1+\gamma_O)(1-\gamma_I)}{(1+\gamma_O)\cos(\beta_I - \alpha) + (1-\gamma_I)\cos(\beta_O - \alpha)} \qquad (13.144)$$

$$\Omega_E = \frac{\Omega_O(1+\gamma_O)\cos(\beta_I - \alpha) + \Omega_I(1-\gamma_I)\cos(\beta_O - \alpha)}{(1+\gamma_O)\cos(\beta_I - \alpha) + (1-\gamma_I)\cos(\beta_O - \alpha)} \qquad (13.145)$$

If α is arbitrarily chosen there will, in general, be a spin at both race contacts. This is equivalent to a twisting of the race body with respect to the ball about the normal to the center of the pressure area and which is superimposed on the rolling motion. Spinning torque is the major source of contact-area friction.

From Fig. 13.21 the angular velocities of spin are

$$\omega_{S_O} = -\omega_O \sin\beta_O + \omega_B \sin(\beta_O - \alpha) \qquad (13.146)$$

$$\omega_{S_I} = \omega_I \sin\beta_I - \omega_B \sin(\beta_I - \alpha) \qquad (13.147)$$

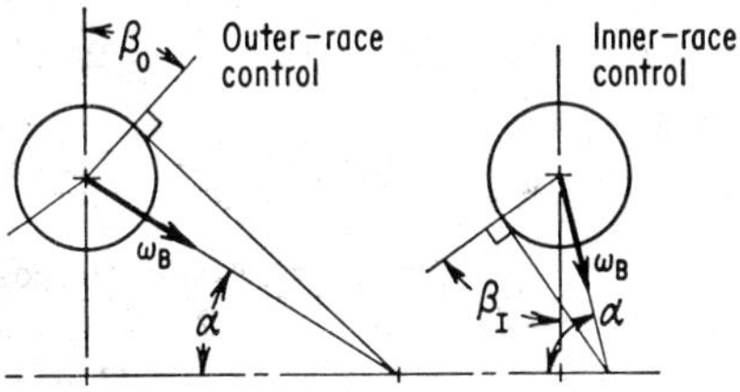

FIG. 13.22. Geometric configuration of ball.

The attitude α of the ball's angular-velocity vector is not arbitrary but is well defined in a properly operating bearing.

High-speed photography of ball motion in gas-turbine thrust bearings confirms the fact that the ball rolls on one race without spin while all spin occurs at the other race contact.

If no spin occurs at the outer-race contact, "outer-race control" exists. Similarly "inner-race control" can occur. The type of control determines the value of α and greatly influences the ball's motion. If Q_{S_O} and Q_{S_I} are the spinning moments at outer- and inner-race contacts, respectively, outer-race control is definite if

$$Q_O \cos(\beta_I - \beta_O) > Q_I \qquad (13.148)$$

From Eq. (13.146) with ω_{S_O} zero the value of α corresponding to outer-race control is

$$\tan\alpha = \sin\beta_O \cos\beta_O/(\cos^2\beta_O + \gamma_O) \qquad (13.149)$$

and from Eq. (13.147) for inner-race control

$$\tan\alpha = \sin\beta_I \cos\beta_I/(\cos^2\beta_I - \gamma_I) \qquad (13.150)$$

The geometric significance of outer- and inner-race control is shown in Fig. 13.22. The angular velocities with which the races roll relative to the ball are

$$\omega_{R_O} = -\omega_O \cos\beta_O + \omega_B \cos(\beta_O - \alpha) \qquad (13.151)$$

$$\omega_{R_I} = \omega_I \cos\beta_I - \omega_B \cos(\beta_I - \alpha) \qquad (13.152)$$

In terms of the absolute angular velocities of outer and inner races

$$\omega_{S_O} = \frac{(1-\gamma_I)(\Omega_O - \Omega_I)\cos(\beta_O - \alpha)\cos\beta_O}{\gamma_O} \frac{-\gamma_O \tan\beta_O + (1+\gamma_O)\tan(\beta_O + \alpha)}{(1+\gamma_O)\cos(\beta_I - \alpha) + (1-\gamma_I)\cos(\beta_O - \alpha)} \qquad (13.153)$$

$$\omega_{S_I} = \frac{-(1+\gamma_O)(\Omega_O - \Omega_I)\cos(\beta_I - \alpha)\cos\beta_I}{\gamma_I} \frac{\gamma_I \tan\beta_I + (1-\gamma_I)\tan(\beta_I - \alpha)}{(1+\gamma_O)\cos(\beta_I - \alpha) + (1-\gamma_I)\cos(\beta_O - \alpha)} \qquad (13.154)$$

$$\omega_{R_O} = \frac{(1-\gamma_I)(\Omega_O - \Omega_I)\cos(\beta_O - \alpha)\cos\beta_O}{\gamma_O[(1+\gamma_O)\cos(\beta_I - \alpha) + (1-\gamma_I)\cos(\beta_O - \alpha)]} \qquad (13.155)$$

$$\omega_{R_I} = \frac{-(1+\gamma_O)(\Omega_O - \Omega_I)\cos(\beta_I - \alpha)\cos\beta_I}{\gamma_I[(1+\gamma_O)\cos(\beta_I - \alpha) + (1-\gamma_I)\cos(\beta_O - \alpha)]} \qquad (13.156)$$

The spin/roll ratio is required in fatigue-life calculations:

$$|\omega_{SO}/\omega_{RO}| = |(1 + \gamma_O) \tan (\beta_O - \alpha) - \gamma_O \tan \beta_O| \qquad (13.157)$$
$$|\omega_{SI}/\omega_{RI}| = |(1 - \gamma_I) \tan (\beta_I - \alpha) + \gamma_I \tan \beta_I| \qquad (13.158)$$

If centrifugal force and gyroscopic moment are negligible, β_O and β_I are identical, as are γ_O and γ_I. Equations (13.159) through (13.176) apply to this situation

$$\gamma = \gamma_O = \gamma_I = d \cos \beta / E \qquad (13.159)$$
$$\beta = \beta_O = \beta_I \qquad (13.160)$$

and

$$\omega_B = \frac{(\Omega_O - \Omega_I)(1 - \gamma^2) \cos \beta}{2\gamma \cos (\beta - \gamma)} \qquad (13.161)$$
$$\Omega_E = \frac{\Omega_O(1 + \gamma) + \Omega_I(1 - \gamma)}{2} \qquad (13.162)$$
$$\omega_{SO} = \frac{(1 - \gamma)(\Omega_O - \Omega_I) \cos \beta}{2\gamma} [-\gamma \tan \beta + (1 + \gamma) \tan (\beta - \alpha)] \qquad (13.163)$$
$$\omega_{SI} = \frac{-(1 + \gamma)(\Omega_O - \Omega_I) \cos \beta}{2\gamma} [\gamma \tan \beta + (1 - \gamma) \tan (\beta - \alpha)] \qquad (13.164)$$
$$\omega_{RO} = \frac{(1 - \gamma)(\Omega_O - \Omega_I) \cos \beta}{2\gamma} \qquad (13.165)$$
$$\omega_{RI} = \frac{-(1 + \gamma)(\Omega_O - \Omega_I) \cos \beta}{2\gamma} \qquad (13.166)$$
$$|\omega_{SO}/\omega_{RO}| = |(1 + \gamma) \tan (\beta - \alpha) - \gamma \tan \beta| \qquad (13.167)$$
$$|\omega_{SI}/\omega_{RI}| = |(1 - \gamma) \tan (\beta - \alpha) + \gamma \tan \beta| \qquad (13.168)$$

For outer-race control

$$\omega_{SO} = 0 \qquad (13.169)$$
$$\omega_{SI} = -(\Omega_O - \Omega_I) \sin \beta \qquad (13.170)$$
$$|\omega_{SO}/\omega_{RO}| = 0 \qquad (13.171)$$
$$|\omega_{SI}/\omega_{RI}| = |2\gamma \tan \beta/(1 + \gamma)| \qquad (13.172)$$

For inner-race control

$$\omega_{SO} = -(\Omega_O - \Omega_I) \sin \beta \qquad (13.173)$$
$$\omega_{SI} = 0 \qquad (13.174)$$
$$|\omega_{SO}/\omega_{RO}| = |2\gamma \tan \beta/(1 - \gamma)| \qquad (13.175)$$
$$|\omega_{SI}/\omega_{RI}| = 0 \qquad (13.176)$$

13.6. INTERNAL ROLLING-ELEMENT LOAD DISTRIBUTION

Until very recently the internal rolling-element load distribution could be calculated from the applied load for only a few simple cases of loading. Such solutions as existed generally contained simplifying assumptions which were introduced to facilitate hand computation.

A completely general solution has now been obtained which enables the proper assessment of all initial and operating conditions. The solution is prohibitive for hand-computation means but is easily accomplished with a high-speed digital computer. Although not discussed in detail here, solutions have also been found which consider structural elastic effects other than those occurring at the rolling elements' contacts.

In Art. 13.6(a) the basic theory of the high-speed ball and radial roller bearing is developed. Structural deflections are not considered but initial conditions such as preloads or misalignments are.

In this treatment the entire grouping of bearings is considered as a single nonlinear elastic system in five degrees of freedom.

In Art. 13.6(b) a similar concept is employed except that dynamical effects are neglected.

In Art. 13.6(c) certain simplifying assumptions are introduced, the number of degrees of freedom is reduced to three and a somewhat tedious hand solution is attained.

In Art. 13.6(d) the special case of two angular-contact bearings is solved in five degrees of freedom using some of the techniques of Art. 13.6(c).

Finally, in Art. 13.6(e) the previously known solutions for a single bearing with not more than two degrees of freedom are derived directly from Art. 13.6(c).

(a) Load Distribution and Deflections in a System of High-speed Ball and Roller Bearings

The High-speed Ball Bearing. Figure 13.23 shows an angular-contact ball bearing referred to an orthogonal XYZ coordinate system. The outer ring is fixed but the inner ring may move with respect to the coordinate system. Both rings are free to rotate about their axes.

Three linear displacements δ', δ_2', and δ_3' and two angular displacements δ_4' and δ_5' are required to define the spatial position and attitude of the inner ring when it is displaced from its initial position. For purposes of derivation the initial situation is that existing when the bearing's end play is just taken up in the thrust direction. Figure 13.24a shows these displacements in the positive sense. Figure 13.24b shows the forces and moments which result from the displacements and the dynamical effects of the bearing's rotation at high speed. These are shown in the positive sense and act on the shaft so as to oppose the displacements.

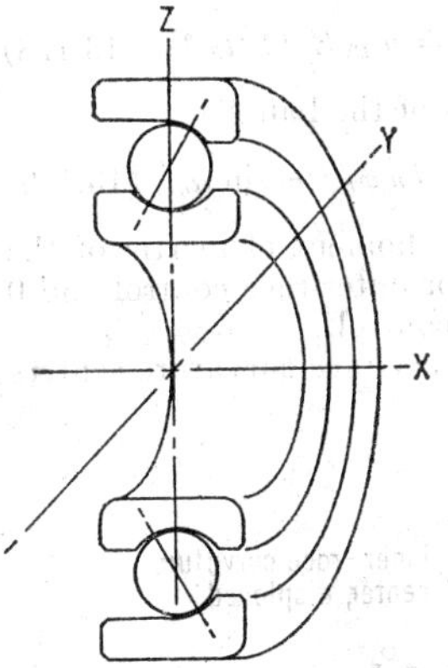

FIG. 13.23. Angular contact of ball bearing.

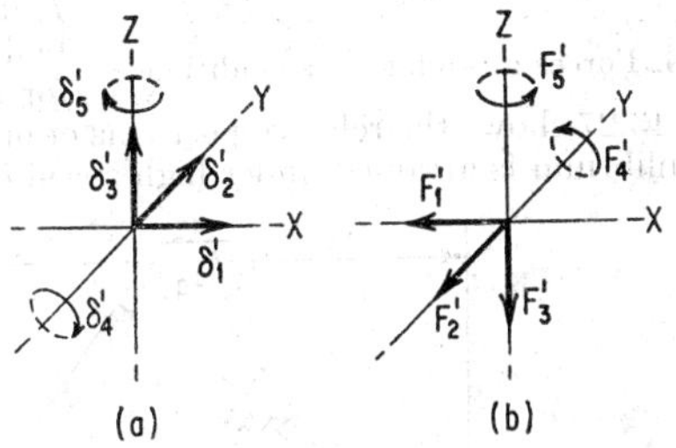

FIG. 13.24. High-speed ball bearing. (a) Displacements. (b) Forces and moments.

Figure 13.25 shows some important dimensions and establishes the convention of the ball-position index q. The contact angle β is the initial mounted contact angle

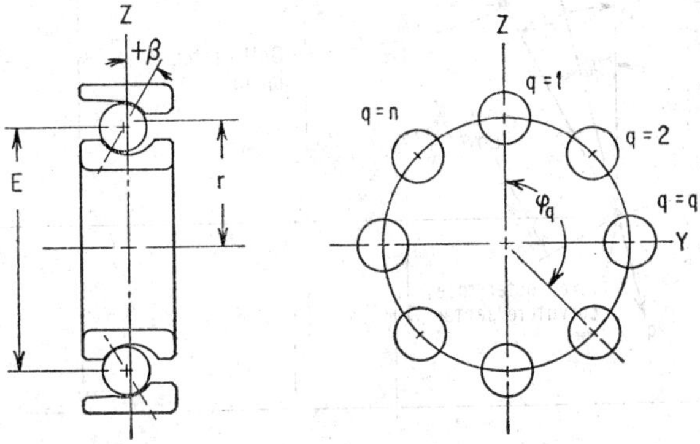

FIG. 13.25. Dimensions and convention of ball-position index q.

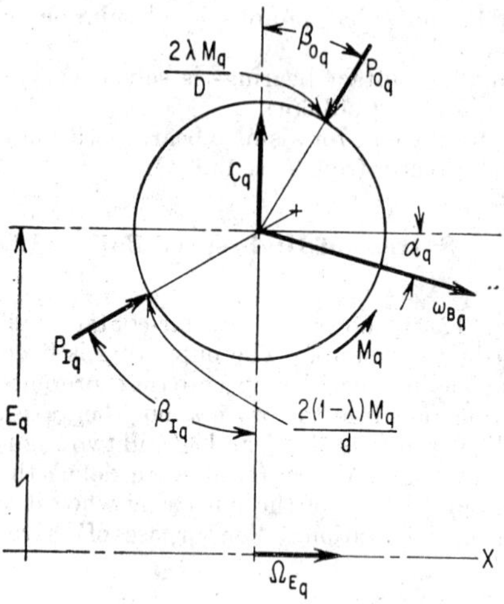

FIG. 13.26. Forces at high-speed conditions.

and is shown in the positive sense. Note that the YZ plane does not contain the ball centers but does contain the locus of the inner-race curvature centers.

The distance r is

$$r = E/2 + (f_I - 0.5)d \cos \beta \quad (13.177)$$

Under high-speed conditions a ball is acted upon by forces at the contacts and by body forces resulting from the ball's motion as in Fig. 13.26. C_q and M_q are, respectively, the centrifugal force and the gyroscopic moment

$$C_q = m_B(E_q/2)\Omega_{E_q}^{\ 2} \quad (13.178)$$

m_B is the mass of the ball.

$$M_q = I_{B_q}\omega_{B_q}\Omega_{E_q} \sin \alpha_q \quad (13.179)$$

I_B is the mass moment of inertia of the ball. λ is 1 for outer-race control and 0 for inner-race control.

Figure 13.27 shows the relative positions of outer- and inner-race curvature centers when equilibrium is attained under high-speed conditions.

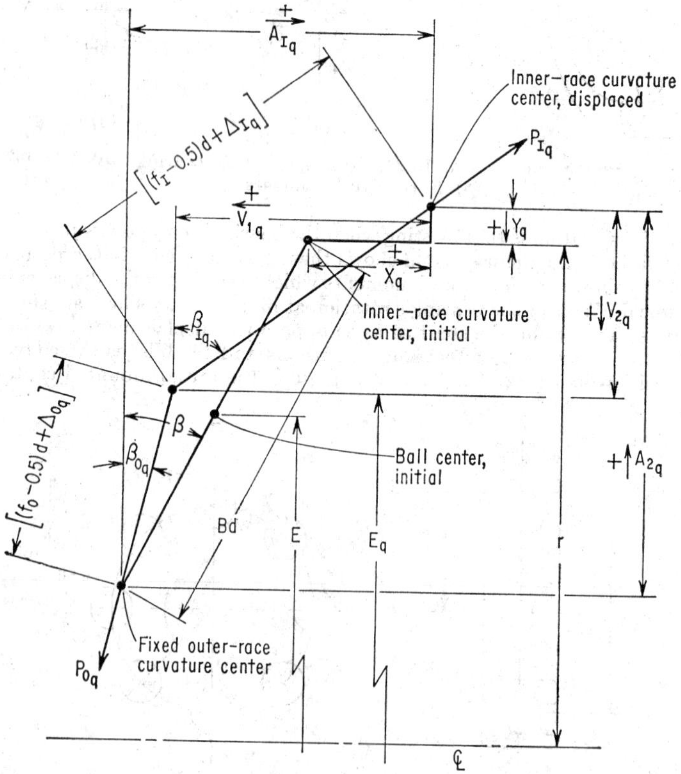

FIG. 13.27. Relative positions of outer- and inner-race curvature centers at high-speed equilibrium.

x_q and y_q are related to the displacements δ_j' at the bearing center

$$x_q = \delta_1' + r(\delta_4' \cos \varphi_q + \delta_5' \sin \varphi_q) \quad (13.180)$$
$$y_q = \delta_2' \sin \varphi_q + \delta_3' \cos \varphi_q - P_D/2 \quad (13.181)$$

P_D is any change in diametral clearance and is positive for an increase in clearance.

$$A_{1_q} = Bd \sin \beta + \delta_1' + r(\delta_4' \cos \varphi_q + \delta_5' \sin \varphi_q) \quad (13.182)$$
$$A_{2_q} = Bd \cos \beta + \delta_2' \sin \varphi_q + \delta_3' \cos \varphi_q - P_D/2 \quad (13.183)$$

and

$$\varphi_q = 2\pi(q-1)/n \quad (13.184)$$

The elastic deflections at the race contacts are

$$\Delta_{O_q} = \sqrt{(A_{1_q} - V_{1_q})^2 + (A_{2_q} - V_{2_q})^2} - (f_O - 0.5)d \geq 0 \quad (13.185)$$
$$\Delta_{I_q} = \sqrt{V_{1_q}{}^2 + V_{2_q}{}^2} - (f_I - 0.5)d \quad (13.186)$$

The normal loads at the race contacts are

$$P_{O_q} = K_{O_q}\Delta_{O_q}{}^{3/2} \quad (13.187)$$
$$P_{I_q} = K_{I_q}\Delta_{I_q}{}^{3/2} \quad (13.188)$$

K_{O_q} and K_{I_q} are elastic coefficients derived from Eqs. (13.187) and (13.188)

$$K_{O_q} = \frac{1}{(\vartheta_1 + \vartheta_2)(1.3104 C_{\Delta_{O_q}})^{3/2}} \left(\frac{2f_O d}{2f_O - 1}\right)^{1/2} \quad (13.189)$$

$$K_{I_q} = \frac{1}{(\vartheta_1 + \vartheta_2)(1.3104 C_{\Delta_{I_q}})^{3/2}} \left(\frac{2f_I d}{2f_I - 1}\right)^{1/2} \quad (13.190)$$

The operating contact angles are found from

$$\tan \beta_{O_q} = (A_{1_q} - V_{1_q})/(A_{2_q} - V_{2_q}) \quad (13.191)$$
$$\tan \beta_{I_q} = V_{1_q}/V_{2_q} \quad (13.192)$$

The instantaneous pitch diameter is

$$E_q = E + 2\{[(f_O - 0.5)d + \Delta_{O_q}] \cos \beta_{O_q} - (f_O - 0.5)d \cos \beta\} \quad (13.193)$$

Under certain conditions a ball may be completely out of contact with the inner race. Loss of inner-race contact occurs when

$$A_{1_q}{}^2 + [A_{2_q} - (f_O - 0.5)d - (C_q/K_{O_q})^{2/3}]^2 \leq [(f_I - 0.5)d]^2 \quad (13.194)$$

At such times the outer-race contact angle β_{O_q}, the inner-race contact force, and the gyroscopic moment are zero. The force at the outer-race contact is C_q and can be taken approximately as the average of the centrifugal forces of all balls which have maintained inner-race contact.

Equilibrium of the ball (Fig. 13.26) requires that

$$P_{O_q} \sin \beta_{O_q} - P_{I_q} \sin \beta_{I_q} - (2M_q/d)[\lambda \cos \beta_{O_q} - (1-\lambda) \cos \beta_{I_q}] = 0 = \psi_{1_q} \quad (13.195)$$
$$P_{O_q} \cos \beta_{O_q} - P_{I_q} \cos \beta_{I_q} + (2M_q/d)[\lambda \sin \beta_{O_q} - (1-\lambda) \sin \beta_{I_q}] - C_q = 0 = \psi_2 \quad (13.196)$$

These equations are nonlinear in the variables V_{1_q} and V_{2_q} and can be solved by iteration of

$${}_2V_{jq} = {}_1V_{jq} - (\Delta V_{jq}) \quad (13.197)$$

${}_1V_{jq}$ are current estimates and ${}_2V_{jq}$ are improved values. The ΔV_{jq} are calculated at current estimates and are found from

$$\Delta V_{1_q} = \frac{\begin{vmatrix} \psi_{1_q} & \partial\psi_{1_q}/\partial V_{2_q} \\ \psi_{2_q} & \partial\psi_{2_q}/\partial V_{2_q} \end{vmatrix}}{\begin{vmatrix} \partial\psi_{1_q}/\partial V_{1_q} & \partial\psi_{1_q}/\partial V_{2_q} \\ \partial\psi_{2_q}/\partial V_{1_q} & \partial\psi_{2_q}/\partial V_{2_q} \end{vmatrix}} \tag{13.198}$$

$$\Delta V_{2_q} = \frac{\begin{vmatrix} \partial\psi_{1_q}/\partial V_{1_q} & \psi_{1_q} \\ \partial\psi_{2_q}/\partial V_{1_q} & \psi_{2_q} \end{vmatrix}}{\begin{vmatrix} \partial\psi_{1_q}/\partial V_{1_q} & \partial\psi_{1_q}/\partial V_{2_q} \\ \partial\psi_{2_q}/\partial V_{1_q} & \partial\psi_{1_q}/\partial V_{2_q} \end{vmatrix}} \tag{13.199}$$

ψ_{1q} and ψ_{2q} are the residuals in Eqs. (13.195) and (13.196) calculated at current estimates. The partial derivatives are

$$\begin{aligned} \frac{\partial\psi_{1q}}{\partial V_{jq}} = {} & [P_{O_q}\cos\beta_{O_q} + (2\lambda M_q/d)\sin\beta_{O_q}](\partial\beta_{O_q}/\partial V_{jq}) - \{P_{I_q}\cos\beta_{I_q} \\ & + [2(1-\lambda)M_q/d]\sin\beta_{I_q}\}(\partial\beta_{I_q}/\partial V_{jq}) + (\partial P_{O_q}/\partial V_{jq})\sin\beta_{O_q} \\ & - (\partial P_{I_q}/\partial V_{jq})\sin\beta_{I_q} - (2/d)[\lambda\cos\beta_{O_q} - (1-\lambda)\cos\beta_{I_q}](\partial M_q/\partial V_{jq}) \end{aligned} \tag{13.200}$$

$$\begin{aligned} \frac{\partial\psi_{2q}}{\partial V_{jq}} = {} & -[P_{O_q}\sin\beta_{O_q} - (2\lambda M_q/d)\cos\beta_{O_q}](\partial\beta_{O_q}/\partial V_{jq}) + \{P_{I_q}\sin\beta_{I_q} \\ & - [2(1-\lambda)M_q/d]\cos\beta_{I_q}\}(\partial\beta_{I_q}/\partial V_{jq}) + (\partial P_{O_q}/\partial V_{jq})\cos\beta_{O_q} \\ & - (\partial P_{I_q}/\partial V_{jq})\cos\beta_{I_q} + (2/d)[\lambda\sin\beta_{O_q} - (1-\lambda)\sin\beta_{I_q}](\partial M_q/\partial V_{jq}) - (\partial C_q/\partial V_{jq}) \end{aligned} \tag{13.201}$$

The partial derivatives of β_{O_q}, β_{I_q}, P_{O_q}, and P_{I_q} required in the foregoing are found in Table 13.1. In addition, if z is any variable such as V_{1q}, V_{2q}, A_{1q}, or A_{2q}

$$\partial C_q/\partial z = m_B[E_q\Omega_{E_q}(\partial\Omega_{E_q}/\partial z) + (\Omega_{E_q}{}^2/2)(\partial E_q/\partial z)] \tag{13.202}$$

$$\partial M_q/\partial z = I_B\{[\omega_{B_q}(\partial\Omega_{E_q}/\partial z) + \Omega_{E_q}(\partial\omega_{B_q}/\partial z)]\sin\alpha_q + \omega_{B_q}\Omega_{E_q}\cos\alpha_q(\partial\alpha_q/\partial z)\} \tag{13.203}$$

$$\frac{\partial\alpha_q}{\partial z} = \frac{[\cos^2\beta_{O_q} \mp \gamma_{O_q}(\cos^2\beta_{O_q} - \sin^2\beta_{O_q})](\partial\beta_{O_q}/\partial z) \mp \sin\beta_{O_q}\cos\beta_{O_q}(\partial\gamma_{O_q}/\partial z)}{(\cos^2\beta_{O_q} \pm \gamma_{O_q})^2}\cos^2\alpha_q \tag{13.204}$$

In Eq. (13.204) the upper sign applies with outer-race control and the lower sign with inner-race control.

$$\begin{aligned} \frac{\partial\omega_{B_q}}{\partial z} = {} & \frac{E_q(\Omega_O - \Omega_I)}{d} \\ & \left[\frac{\begin{gathered}[(1+\gamma_{O_q})\cos(\beta_{I_q} - \alpha_q) + (1-\gamma_{I_q})\cos(\beta_{O_q} - \alpha_q)][-(1+\gamma_{O_q})(\partial\gamma_{I_q}/\partial z) \\ + (1-\gamma_{I_q})(\partial\gamma_{O_q}/\partial z)] - (1+\gamma_{O_q})(1-\gamma_{I_q})[-(1+\gamma_{O_q}) \\ \times \sin(\beta_{I_q} - \alpha_q)(\partial\beta_{I_q}/\partial z - \partial\alpha_q/\partial z) + \cos(\beta_{I_q} - \alpha_q)(\partial\gamma_{O_q}/\partial z) - (1-\gamma_{I_q}) \\ \times \sin(\beta_{O_q} - \alpha_q)(\partial\beta_{O_q}/\partial z - \partial\alpha_q/\partial z) - \cos(\beta_{O_q} - \alpha_q)(\partial\gamma_{I_q}/\partial z)]\end{gathered}}{[(1+\gamma_{O_q})\cos(\beta_{I_q} - \alpha_q) + (1-\gamma_{I_q})\cos(\beta_{O_q} - \alpha_q)]^2} \right] \\ & + \frac{\omega_{B_q}}{E_q}\frac{\partial E_q}{\partial z} \end{aligned} \tag{13.205}$$

$$\begin{aligned} \frac{\partial\Omega_{E_q}}{\partial z} = {} & \\ & \left[\frac{\begin{gathered}[(1+\gamma_{O_q})\cos(\beta_{I_q} - \alpha_q) + (1-\gamma_{I_q})\cos(\beta_{O_q} - \alpha_q)]\{\Omega_O[-(1+\gamma_{O_q}) \\ \times \sin(\beta_{I_q} - \alpha_q)(\partial\beta_{I_q}/\partial z - \partial\alpha_q/\partial z) + \cos(\beta_{I_q} - \alpha_q)(\partial\gamma_{O_q}/\partial z)] \\ - \Omega_I[(1-\gamma_{I_q})\sin(\beta_{O_q} - \alpha_q)(\partial\beta_{O_q}/\partial z - \partial\alpha_q/\partial z) + \cos(\beta_{O_q} - \alpha_q)(\partial\gamma_{I_q}/\partial z)]\} \\ - [\Omega_O(1+\gamma_{O_q})\cos(\beta_{I_q} - \alpha_q) + \Omega_I(1-\gamma_{I_q})\cos(\beta_{O_q} - \alpha_q)] \\ [-(1+\gamma_{O_q})\sin(\beta_{I_q} - \alpha_q)(\partial\beta_{I_q}/\partial z - \partial\alpha_q/\partial z) + \cos(\beta_{I_q} - \partial_q)(\partial\gamma_{O_q}/\partial z) \\ - (1-\gamma_{I_q})\sin(\beta_{O_q} - \alpha_q)(\partial\beta_{O_q}/\partial z - \partial\alpha_q/\partial z) - \cos(\beta_{O_q} - \alpha_q)(\partial\gamma_{I_q}/\partial z)]\end{gathered}}{[(1+\gamma_{O_q})\cos(\beta_{I_q} - \alpha_q) + (1-\gamma_{I_q})\cos(\beta_{O_q} - \alpha_q)]^2} \right] \end{aligned} \tag{13.206}$$

Table 13.1

Γ	$\dfrac{\partial\Gamma}{\partial V_{1q}}$	$\dfrac{\partial\Gamma}{\partial V_{2q}}$	$\dfrac{\partial\Gamma}{\partial A_{1q}}$	$\dfrac{\partial\Gamma}{\partial A_{2q}}$
β_{Oq}	$-\dfrac{\cos^2\beta_{Oq}}{A_{2q}-V_{2q}}$	$\dfrac{\sin^2\beta_{Oq}}{A_{1q}-V_{1q}}$	$\dfrac{\cos^2\beta_{Oq}}{A_{2q}-V_{2q}}$	$-\dfrac{\sin^2\beta_{Oq}}{A_{1q}-V_{1q}}$
β_{Iq}	$\dfrac{\cos^2\beta_{Iq}}{V_{2q}}$	$-\dfrac{\sin^2\beta_{Iq}}{V_{1q}}$	0	0
P_{Oq}	$-\frac{3}{2}K_{Oq}\Delta_{Oq}{}^{1/2}\sin\beta_{Oq}$	$-\frac{3}{2}K_{Oq}\Delta_{Oq}{}^{1/2}\cos\beta_{Oq}$	$\frac{3}{2}K_{Oq}\Delta_{Oq}{}^{1/2}\sin\beta_{Oq}$	$\frac{3}{2}K_{Oq}\Delta_{Oq}{}^{1/2}\cos\beta_{Oq}$
P_{Iq}	$\frac{3}{2}K_{Iq}\Delta_{Iq}{}^{1/2}\sin\beta_{Iq}$	$\frac{3}{2}K_{Iq}\Delta_{Iq}{}^{1/2}\cos\beta_{Iq}$	0	0
E_q	0	-2	0	2
γ_{Oq}	$\dfrac{\gamma_{Oq}\sin\beta_{Oq}\cos\beta_{Oq}}{A_{2q}-V_{2q}}$	$\gamma_{Oq}\left[\dfrac{-\sin^3\beta_{Oq}}{(A_{1q}-V_{1q})\cos\beta_{Oq}}+\dfrac{2}{E_q}\right]$	$\dfrac{-\gamma_{Oq}\sin\beta_{Oq}\cos\beta_{Oq}}{A_{2q}-V_{2q}}$	$\gamma_{Oq}\left[\dfrac{\sin^3\beta_{Oq}}{(A_{1q}-V_{1q})\cos\beta_{Oq}}-\dfrac{2}{E_q}\right]$
γ_{Iq}	$\dfrac{-\gamma_{Iq}\sin\beta_{Iq}\cos\beta_{Iq}}{V_{2q}}$	$\gamma_{Iq}\left(\dfrac{\sin^3\beta_{Iq}}{V_{Iq}\cos\beta_{Iq}}+\dfrac{2}{E_q}\right)$	0	$\dfrac{-2\gamma_I}{E_q}$

The forces and moments with which the inner ring acts on the shaft at its center are

$$F_1' = \sum_{q=1}^{n} \{P_{I_q} \sin \beta_{I_q} - [2(1 - \lambda)M_q]/d \cos \beta_{I_q}\} \tag{13.207}$$

$$F_2' = \sum_{q=1}^{n} \{P_{I_q} \cos \beta_{I_q} + [2(1 - \lambda)M_q/d] \sin \beta_{I_q}\} \sin \varphi_q \tag{13.208}$$

$$F_3' = \sum_{q=1}^{n} \{P_{I_q} \cos \beta_{I_q} + [2(1 - \lambda)M_q/d] \sin \beta_{I_q}\} \cos \varphi_q \tag{13.209}$$

$$F_4' = \sum_{q=1}^{n} \{rP_{I_q} \sin \beta_{I_q} - [2(1 - \lambda)M_q/d](r \cos \beta_{I_q} - f_I d)\} \cos \varphi_q \tag{13.210}$$

$$F_5' = \sum_{q=1}^{n} \{rP_{I_q} \sin \beta_{I_q} - [2(1 - \lambda)M_q/d](r \cos \beta_{I_q} - f_I d)\} \sin \varphi_q \tag{13.211}$$

(b) The High-speed Roller Bearing

This discussion is limited to the radial-type cylindrical roller bearing. All other types of roller bearings are limited to lower-speed operation.

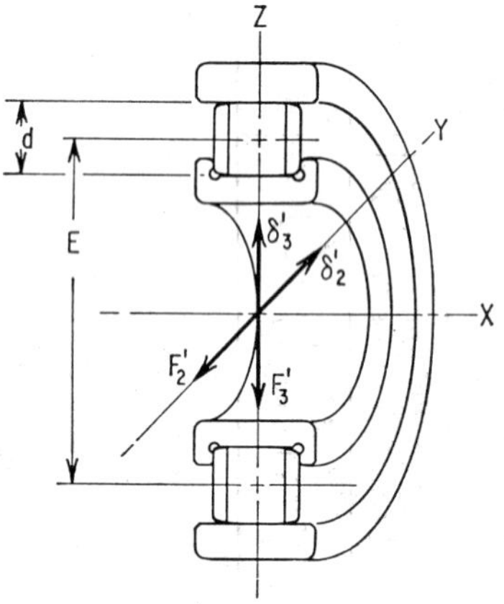

Fig. 13.28. Cylindrical roller bearing.

Figure 13.28 shows a cylindrical roller bearing referred to the XYZ coordinate system. q is the roll position index and has the same significance as with a ball bearing.

If misalignments are small, a radial roller bearing can be loaded only through the displacements δ_2' and δ_3'. The elastic approach of the roll body to a raceway is related to the contact load through

$$P_{O_q} = K_O'\Delta_{O_q}^{10/9} \tag{13.212}$$

$$P_{I_q} = K_I'\,\Delta_{I_q}^{10/9} \tag{13.213}$$

In certain instances a roller may be in elastic contact with both outer and inner races. At other times the roller may be forced against the outer race by centrifugal force alone and be out of contact with the inner race. The centrifugal force on a roller is

$$C_q = m_R(E/2)\Omega_E^2 \tag{13.214}$$

The orbital angular velocity is obtained from Eq. (13.145) with $\beta_O = \beta_I = 0$ and $\gamma_O = \gamma_I = \gamma$ where

$$\gamma = d/E \tag{13.215}$$

$$\Omega_E = \frac{\Omega_O(1 + \gamma) + \Omega_I(1 - \gamma)}{2} \tag{13.216}$$

Loss of inner-race contact will occur if

$$\delta_2' \sin \varphi_q + \delta_3' \cos \varphi_q - (C_q/K_O)^{9/10} - P_D/2 < 0 \tag{13.217}$$

where P_D is the mounted diametral clearance of the roller bearing. C_q is constant for all rolling elements in the bearing.

If the roll is in contact with both races the total of the elastic deformations at outer- and inner-race contacts is

$$\Delta_q = \delta_2' \sin \varphi_q + \delta_3' \cos \varphi_q - P_D/2 \geq 0 \tag{13.218}$$

and is related to the deflections at the individual race contacts through

$$\Delta_q = \Delta_{O_q} + \Delta_{I_q} \tag{13.219}$$

From the equilibrium of forces on a roll

$$K_O'(\Delta_q - \Delta_{I_q})^{10/9} - K_I'\,\Delta_{I_q}{}^{10/9} - C_q = 0 = \xi \tag{13.220}$$

Equation (13.220) is solved for Δ_{I_q} by iteration of

$$_2\Delta_{I_q} = {}_1\Delta_{I_q} + \frac{9\xi}{10[K_o'(\Delta_q - {}_1\Delta_{I_q})^{1/9} + K_I'\,{}_1\Delta_{I_q}{}^{1/9}]} \tag{13.221}$$

${}_1\Delta_{I_q}$ is a current value and ${}_2\Delta_{I_q}$ an improved value of Δ_{I_q}. ξ is also calculated at current estimates. For all practical purposes $K_O' = K_I' = K'$. K' is calculated in accordance with Eq. (13.222) for bearings of conventional steel.

$$K' = 11.254 \times 10^6 l_e{}^{8/9} \tag{13.222}$$

where l_e is the effective length of the roll.

The forces and moments with which the bearing acts on the shaft at its center are

$$F_1' = 0 \tag{13.223}$$

$$F_2' = K' \sum_{q=1}^{n} \Delta_{I_q}{}^{10/9} \sin \varphi_q \tag{13.224}$$

$$F_3' = K' \sum_{q=1}^{n} \Delta_{I_q}{}^{10/9} \cos \varphi_q \tag{13.225}$$

$$F_4' = 0 \tag{13.226}$$

$$F_5' = 0 \tag{13.227}$$

(c) The Solution of a System of High-speed Ball and Roller Bearings under Static Loading

Figure 13.29 shows a system of p ball bearings and t radial roller bearings on a common shaft. The number of bearings is unrestricted.

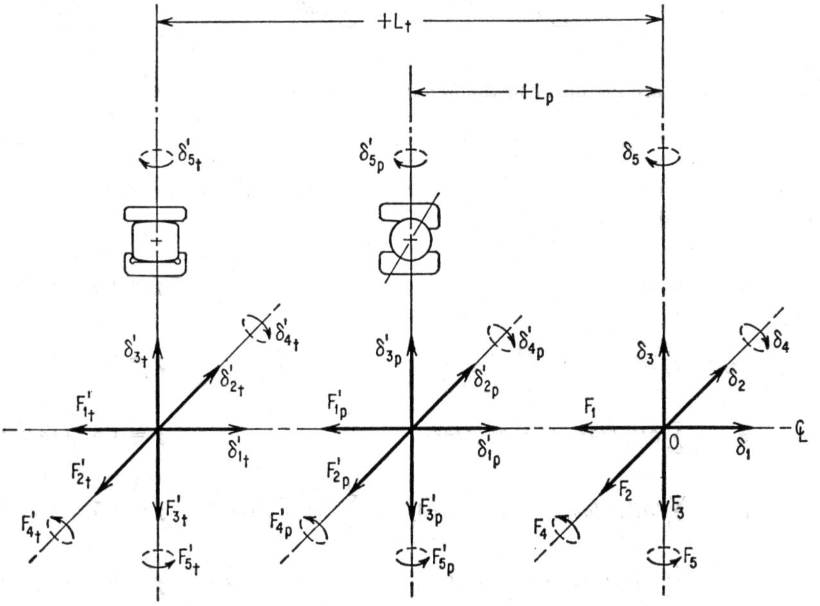

FIG. 13.29. System of p ball bearings.

All external loads acting anywhere along the shaft can be reduced to a set of three forces and two moments acting at point o, which can be arbitrarily chosen. These forces and moments are the F_i.

As the result of the F_i there will occur the five displacements δ_j at point o.

At each bearing position there will be the δ_j' which are linearly dependent on the δ_j and the F_i' which are nonlinearly related to the δ_j at point o.

The displacements δ_j' at a bearing center are related to the δ_j at point o through the following:

$$\delta_1' = \delta_1 + \delta_1'' \tag{13.228}$$

$$\delta_2' = \delta_2 + L\delta_5 + \delta_2'' \tag{13.229}$$

$$\delta_3' = \delta_3 + L\delta_4 + \delta_3'' \tag{13.230}$$

$$\delta_4' = \delta_4 + \delta_4'' \tag{13.231}$$

$$\delta_5' = \delta_5 + \delta_5'' \tag{13.232}$$

where the L is chosen as appropriate.

The δ_j'' are initial displacements in any degree of freedom at a particular bearing location as, for example, axial preload deflection, misalignment, eccentricity, etc. The equilibrium of the system is defined at point o by

$$F_1 + \sum_1^p F_{1_p}' = 0 = \Lambda_1 \tag{13.233}$$

$$F_2 + \sum_1^p F_{2_p}' + \sum_1^t F_{2_t}' = 0 = \Lambda_2 \tag{13.234}$$

$$F_3 + \sum_1^p F_{3_p}' + \sum_1^t F_{3_t}' = 0 = \Lambda_3 \tag{13.235}$$

$$F_4 + \sum_1^p (F_{4_p}' + L_pF_{3_p}') + \sum_1^t L_tF_{3_t}' = 0 = \Lambda_4 \tag{13.236}$$

$$F_5 + \sum_1^p (F_{5_p}' + L_pF_{2_p}') + \sum_1^t L_tF_{2_t}' = 0 = \Lambda_5 \tag{13.237}$$

These equations are nonlinear in the δ_j and are solved by iteration of

$$_2\delta_i = {}_1\delta_i - [a_{ij}]^{-1}\{\Lambda_j\} \qquad i = 1,5 \qquad j = 1,5 \tag{13.238}$$

The elements of the coefficient matrix are the $\partial\Lambda_i/\partial\delta_j$ evaluated at current estimates as is the constant vector.

$$a_{1j} = \sum_1^p \partial F_{1_p}'/\partial\delta_j \qquad j = 1,5 \tag{13.239}$$

$$a_{2j} = \sum_1^p \partial F_{2_p}'/\partial\delta_j + \sum_1^t \partial F_{2_t}'/\partial\delta_j \qquad j = 1,5 \tag{13.240}$$

$$a_{3j} = \sum_1^p \partial F_{3_p}'/\partial\delta_j + \sum_1^t \partial F_{3_t}'/\partial\delta_j \qquad j = 1,5 \tag{13.241}$$

$$a_{4j} = \sum_1^p [\partial F_{4_p}'/\partial\delta_j + L_p(\partial F_{3_p}'/\partial\delta_j)] + \sum_1^t L_t(\partial F_{3_t}'/\partial\delta_j) \qquad j = 1,5 \tag{13.242}$$

$$a_{5j} = \sum_1^p [\partial F_{5_p}'/\partial\delta_j + L_p(\partial F_{2_p}'/\partial\delta_j)] + \sum_1^t L_t(\partial F_{2_t}'/\partial\delta_j) \qquad j = 1,5 \tag{13.243}$$

The Calculation of $\partial F_{i_p}'/\partial\delta_j$ for Ball Bearings. Differentiation of Eqs. (13.207) through (13.211) with respect to the δ_j yields, for a ball bearing denoted by p,

$$\partial F_{1p}'/\partial\delta_j = \sum_{q=1}^{n} (\{P_{I_q}\cos\beta_{I_q} + [2(1-\lambda)M_q/d]\sin\beta_{I_q}\}(\partial\beta_{I_q}/\partial\delta_j) + (\partial P_{I_q}/\partial\delta_j)\sin\beta_{I_q} - [2(1-\lambda)/d](\partial M_q/\partial\delta_j)\cos\beta_{I_q}) \quad (13.244)$$

$$\partial F_{2p}'/\partial\delta_j = \sum_{q=1}^{n} (-\{P_{I_q}\sin\beta_{I_q} - [2(1-\lambda)M_q/d]\cos\beta_{I_q}\}(\partial\beta_{I_q}/\partial\delta_j) + (\partial P_{I_q}/\partial\delta_j)\cos\beta_{I_q} + [2(1-\lambda)/d](\partial M_q/\partial\delta_j)\sin\beta_{I_q})\sin\varphi_q \quad (13.245)$$

$$\partial F_{3p}'/\partial\delta_j = \sum_{q=1}^{n} (-\{P_{I_q}\sin\beta_{I_q} - [2(1-\lambda)M_q/d]\cos\beta_{I_q}\}(\partial\beta_{I_q}/\partial\delta_j) + (\partial P_{I_q}/\partial\delta_j)\cos\beta_{I_q} + [2(1-\lambda)/d](\partial M_q/\partial\delta_j)\cos\beta_{I_q})\cos\varphi_q \quad (13.246)$$

$$\partial F_{4p}'/\partial\delta_j = \sum_{q=1}^{n} (r\{P_{I_q}\cos\beta_{I_q} + [2(1-\lambda)M_q/d]\sin\beta_{I_q}\}(\partial\beta_{I_q}/\partial\delta_j) + r(\partial P_{I_q}/\partial\delta_j)\sin\beta_{I_q} - [2(1-\lambda)/d](r\cos\beta_{I_q} - f_I d)(\partial M_q/\partial\delta_j))\cos\varphi_q \quad (13.247)$$

$$\partial F_{5p}'/\partial\delta_j = \sum_{q=1}^{n} (r\{P_{I_q}\cos\beta_{I_q} + [2(1-\lambda)M_q/d]\sin\beta_{I_q}\}(\partial\beta_{I_q}/\partial\delta_j) + r(\partial P_{I_q}/\partial\delta_j)\sin\beta_{I_q} - [2(1-\lambda)/d](r\cos\beta_{I_q} - f_I d)(\partial M_q/\partial\delta_j))\sin\varphi_q \quad (13.248)$$

The partial derivatives required above are given in Eqs. (13.249) and (13.250) as are other derivatives that are needed. The values of $\partial A_{1q}/\partial\delta_j$ and $\partial A_{2q}/\partial\delta_j$ are given in Table 13.2, subscript p.

$$(\partial/\partial\delta_j)(\beta_{I_q},P_{I_q}) = (\partial/\partial V_{1q})(\beta_{I_q},P_{I_q})(\partial V_{1q}/\partial\delta_j) + (\partial/\partial V_{2q})(\beta_{I_q},P_{I_q})(\partial V_{2q}/\partial\delta_j) \quad (13.249)$$

$$(\partial/\partial\delta_j)(\beta_{O_q},P_{O_q},M_q,C_q) = (\partial/\partial V_{1q})(\beta_{O_q},P_{O_q},M_q,C_q)(\partial V_{1q}/\partial\delta_j) + (\partial/\partial V_{2q})(\beta_{O_q},P_{O_q},M_q,C_q)(\partial V_{2q}/\partial\delta_j) + (\partial/\partial A_{1q})(\beta_{O_q},P_{O_q},M_q,C_q)(\partial A_{1q}/\partial\delta_j) + (\partial/\partial A_{2q})(\beta_{O_q},P_{O_q},M_q,C_q)(\partial A_{2q}/\partial\delta_j) \quad (13.250)$$

If Eqs. (13.195) and (13.196) are set equal to zero and are differentiated with respect to the δ_j there result two simultaneous equations which are linear in $\partial V_{1q}/\partial\delta_j$ and $\partial V_{2q}/\partial\delta_j$ and from which these can be found. The solution is

$$\frac{\partial V_{1q}}{\partial\delta_j} = \frac{\begin{vmatrix} b_{1q} & \partial\psi_{1q}/\partial V_{2q} \\ b_{2q} & \partial\psi_{2q}/\partial V_{2q}\end{vmatrix}}{\begin{vmatrix} \partial\psi_{1q}/\partial V_{1q} & \partial\psi_{1q}/\partial V_{2q} \\ \partial\psi_{2q}/\partial V_{1q} & \partial\psi_{2q}/\partial V_{2q}\end{vmatrix}} \quad (13.251)$$

$$\frac{\partial V_{2q}}{\partial\delta_j} = \frac{\begin{vmatrix} \partial\psi_{1q}/\partial V_{1q} & b_{1q} \\ \partial\psi_{2q}/\partial V_{1q} & b_{2q}\end{vmatrix}}{\begin{vmatrix} \partial\psi_{1q}/\partial V_{1q} & \partial\psi_{1q}/\partial V_{2q} \\ \partial\psi_{2q}/\partial V_{1q} & \partial\psi_{2q}/\partial V_{2q}\end{vmatrix}} \quad (13.252)$$

The partials of ψ_1 and ψ_2 with respect to V_{1q} and V_{2q} are given by Eqs. (13.200) to (13.202) with the aid of Table 13.1.

The constant vector is obtained from

$$b_{1q} = -[(\partial P_{O_q}/\partial A_{1q})(\partial A_{1q}/\partial\delta_j) + (\partial P_{O_q}/\partial A_{2q})(\partial A_{2q}/\partial\delta_j)]\sin\beta_{O_q} - [(\partial\beta_{O_q}/\partial A_{1q})(\partial A_{1q}/\partial\delta_j) + (\partial\beta_{O_q}/\partial A_{2q})(\partial A_{2q}/\partial\delta_j)][P_{O_q}\cos\beta_{O_q} + (2\lambda M_q/d)\sin\beta_{O_q}] + (2/d)[\lambda\cos\beta_{O_q} - (1-\lambda)\cos\beta_{I_q}][(\partial M_q/\partial A_{1q})(\partial A_{1q}/\partial\delta_j) + (\partial M_q/\partial A_{2q})(\partial A_{2q}/\partial\delta_j)] \quad (13.253)$$

$$b_{2q} = -[(\partial P_{O_q}/\partial A_{1q})(\partial A_{1q}/\partial\delta_j) + (\partial P_{O_q}/\partial A_{2q})(\partial A_{2q}/\partial\delta_j)]\cos\beta_{O_q} + [(\partial\beta_{O_q}/\partial A_{1q})(\partial A_{1q}/\partial\delta_j) + (\partial\beta_{O_q}/\partial A_{2q})(\partial A_{2q}/\partial\delta_j)][P_{O_q}\sin\beta_{O_q} - (2\lambda M_q/d)\cos\beta_{O_q}] - (2/d)[\lambda\sin\beta_{O_q} - (1-\lambda)\sin\beta_{I_q}][(\partial M_q/\partial A_{1q})(\partial A_{1q}/\partial\delta_j) + (\partial M_q/\partial A_{2q})(\partial A_{2q}/\partial\delta_j)] + (\partial C_q/\partial A_{1q})(\partial A_{1q}/\partial\delta_j) + (\partial C_q/\partial A_{2q})(\partial A_{2q}/\partial\delta_j) \quad (13.254)$$

The Calculation of $\partial F_{it}'/\partial\delta_j$ for Roller Bearings. Differentiation of Eqs. (13.223) through (13.227) gives, for roller bearings,

$$\partial(F_1',F_4',F_5')/\partial\delta_j = 0 \tag{13.255}$$

$$\partial(F_2',F_3')/\partial\delta_1 = 0 \tag{13.256}$$

$$\frac{\partial F_2'}{\partial(\delta_2,\delta_3,\delta_4,\delta_5)} = \frac{10}{9} K_I' \sum_{q=1}^{n} \Delta_{I_q}^{1/9} \sin\varphi_q \frac{\partial\Delta_{I_q}}{\partial(\delta_2,\delta_3,\delta_4,\delta_5)} \tag{13.257}$$

$$\frac{\partial F_3'}{\partial(\delta_2,\delta_3,\delta_4,\delta_5)} = \frac{10}{9} K_I' \sum_{q=1}^{n} \Delta_{I_q}^{1/9} \cos\varphi_q \frac{\partial\Delta_{I_q}}{\partial(\delta_2,\delta_3,\delta_4,\delta_5)} \tag{13.258}$$

The derivatives of Δ_{I_q} required in the right member of Eq. (13.258) are found as follows:

Let

$$f(\Delta) = K_O'(\Delta_q - \Delta_{I_q})^{10/9} - K_I'\,\Delta_{I_q}^{10/9} - C_q \tag{13.259}$$

$$\partial f(\Delta)/\partial\,\Delta_{I_q} = -{}^{10}\!/_9 K_O'(\Delta_q - \Delta_{I_q})^{1/9} - {}^{10}\!/_9 K_I'\,\Delta_{I_q}^{1/9} \tag{13.260}$$

$$\partial f(\Delta)/\partial\,\Delta_q = {}^{10}\!/_9 K_O'(\Delta_q - \Delta_{I_q})^{1/9} \tag{13.261}$$

$$\frac{\partial\Delta_{I_q}}{\partial\Delta_q} = -\frac{\partial f(\Delta)/\partial\Delta_q}{\partial f\Delta/\partial\Delta_{I_q}} = \left[1 + \frac{K_I'}{K_O'}\left(\frac{\Delta_{I_q}}{\Delta_q - \Delta_{I_q}}\right)^{1/9}\right]^{-1} \tag{13.262}$$

$$\partial\Delta_{I_q}/\partial\delta_j = (\partial\Delta_{I_q}/\partial\Delta_q)[(\partial\delta_2'/\partial\delta_j)\sin\varphi_q + (\partial\delta_3'/\partial\delta_j)\cos\varphi_q] \tag{13.263}$$

Values of $\partial\delta_2'/\partial\delta_j$ and $\partial\delta_3'/\partial\delta_j$ are found in Table 13.2, subscript t. Then

$$\partial\Delta_{I_q}/\partial\delta_1 = 0 \tag{13.264}$$

Table 13.2

Γ	$\frac{\partial\Gamma}{\partial\delta_1}$	$\frac{\partial\Gamma}{\partial\delta_2}$	$\frac{\partial\Gamma}{\partial\delta_3}$	$\frac{\partial\Gamma}{\partial\delta_4}$	$\frac{\partial\Gamma}{\partial\delta_5}$
A_{1qp}	1	0	0	$r_p \cos\varphi_q$	$r_p \sin\varphi_q$
A_{2qp}	0	$\sin\varphi_q$	$\cos\varphi_q$	$L_p \cos\varphi_q$	$L_p \sin\varphi_q$
δ_{2t}'	0	1	0	0	L_t
δ_{3t}'	0	0	1	L_t	0

$$\frac{\partial\Delta_{I_q}}{\partial\delta_2} = \left[1 + \frac{K_I'}{K_O'}\left(\frac{\Delta_{I_q}}{\Delta_q - \Delta_{I_q}}\right)^{1/9}\right]^{-1} \sin\varphi_q \tag{13.265}$$

$$\frac{\partial\Delta_{I_q}}{\partial\delta_3} = \left[1 + \frac{K_I'}{K_O'}\left(\frac{\Delta_{I_q}}{\Delta_q - \Delta_{I_q}}\right)^{1/9}\right]^{-1} \cos\varphi_q \tag{13.266}$$

$$\frac{\partial\Delta_{I_q}}{\partial\delta_4} = \left[1 + \frac{K_I'}{K_O'}\left(\frac{\Delta_{I_q}}{\Delta_q - \Delta_{I_q}}\right)^{1/9}\right]^{-1} L_t \cos\varphi_q \tag{13.267}$$

$$\frac{\partial\Delta_{I_q}}{\partial\delta_5} = \left[1 + \frac{K_I'}{K_O'}\left(\frac{\Delta_{I_q}}{\Delta_q - \Delta_{I_q}}\right)^{1/9}\right]^{-1} L_t \sin\varphi_q \tag{13.268}$$

The coefficient matrix required in Eq. (13.238) can now be completed and the δ_j found.

(d) Load Distribution in a System of Ball and Roller Bearings under Static Loading

The Ball Bearing under Static Loading. If the body forces acting on a ball can be neglected considerable simplification in the solution can be made. The ball center then lies on the line joining the outer- and inner-race curvature centers, and the contact loads and operating contact angles are the same at both contacts.

The sum Δ_q of the elastic displacement at the outer- and inner-race contacts is simply

$$\Delta_q = Bd(D_q - 1) \geq 0 \tag{13.269}$$

and D_q is related to the displacements δ_j at a bearing center by

$$D_q = \{[\sin\beta + \delta_1'/Bd + (r/Bd)(\delta_4' \cos\varphi_q + \delta_5' \sin\varphi_q)]^2 + [\cos\beta + (\delta_2'/Bd)\sin\varphi_q + (\delta_3'/Bd)\cos\varphi_q - \Delta P_D/2Bd]^2\}^{1/2} \geq 1 \quad (13.270)$$

If $D_q < 1$ the ball is completely unloaded. The operating contact angle is

$$\tan\beta_q = \frac{\sin\beta + \delta_1'/Bd + (r/Bd)(\delta_4' \cos\varphi_q + \delta_5' \sin\varphi_q)}{\cos\beta + (\delta_2'/Bd)\sin\varphi + (\delta_3'/Bd)\cos\varphi_q - \Delta P_D/2Bd} \quad (13.271)$$

$+\Delta P_D$ represents any increase in the bearing's diametral clearance. The compressive load on a ball is

$$P_q = K\Delta_q^{3/2} \geq 0 \quad (13.272)$$

where

$$K = (1/K_O^{2/3} + 1/K_I^{2/3})^{-3/2}$$

K_O and K_I are found from Eqs. (13.189) and (13.190). K varies insignificantly with change of contact angle.

The forces with which the bearing acts on the shaft at its center are

$$F_1' = \sum_{q=1}^{n} P_q \sin\beta_q \quad (13.273)$$

$$F_2' = \sum_{q=1}^{n} P_q \cos\beta_q \sin\varphi_q \quad (13.274)$$

$$F_3' = \sum_{q=1}^{n} P_q \cos\beta_q \cos\varphi_q \quad (13.275)$$

$$F_4' = r \sum_{q=1}^{n} P_q \sin\beta_q \cos\varphi_q \quad (13.276)$$

$$F_5' = r \sum_{q=1}^{n} P_q \sin\beta_q \sin\varphi_q \quad (13.277)$$

The Roller Bearing under Static Load. While the discussion of high-speed roller bearings was limited to the radial-type cylindrical roller bearing, a more general type can be considered where operating speeds are moderate. For purposes of developing a theory applicable to several types of bearings the cylindrical angular-contact roller bearing is chosen (see Fig. 13.15).

The sum of the elastic deflections at outer- and inner-race contacts in an angular-contact roller bearing under static load is

$$\Delta_q = [\delta_1' + r(\delta_4' \cos\varphi_q + \delta_5' \sin\varphi_q)]\sin\beta + (\delta_2' \sin\varphi_q + \delta_3' \cos\varphi_q - \Delta P_D/2)\cos\beta \geq 0 \quad (13.278)$$

where ΔP_D is any change in a ring diameter after the bearing is mounted. It is positive if it tends to loosen the fitup of the bearing.

The compressive loading of the roller is

$$P = K'\Delta_q^{10/9}$$

where $K' = 5.21 \times 10^6 l_e^{8/9}$ for bearings of conventional steel and l_e is the effective length of the roller.

The contact angle of a roller bearing does not change with load and is β at all times.

The forces with which the roller bearing acts on the shaft at its center are

$$F_1' = \sum_{q=1}^{n} P_q \sin\beta \quad (13.279)$$

$$F_2' = \sum_{q=1}^{n} P_q \cos\beta \sin\varphi_q \quad (13.280)$$

$$F_3' = \sum_{q=1}^{n} P_q \cos\beta \cos\varphi_q \tag{13.281}$$

$$F_4' = r \sum_{q=1}^{n} P_q \sin\beta \cos\varphi_q \tag{13.282}$$

$$F_5' = r \sum_{q=1}^{n} P_q \sin\beta \sin\varphi_q \tag{13.283}$$

The Solution of a System of Ball and Roller Bearings under Static Loading. Consider a system of p ball bearings and t roller bearings as in Fig. 13.29 except that the radial roller bearing is now replaced by an angular-contact roller bearing of the type shown in Fig. 13.15. By setting the contact angle of the angular-contact roller bearing equal to zero it becomes a radial-type roller bearing.

The equilibrium of forces and moments at point o on the shaft requires that

$$F_1 + \sum_1^p F_{1p}' + \sum_1^t F_{1t}' = \Lambda_1 = 0 \tag{13.284}$$

$$F_2 + \sum_1^p F_{2p}' + \sum_1^t F_{2t}' = \Lambda_2 = 0 \tag{13.285}$$

$$F_3 + \sum_1^p F_{3p}' + \sum_1^t F_{3t}' = \Lambda_3 = 0 \tag{13.286}$$

$$F_4 + \sum_1^p (F_{4p}' + L_pF_{3p}') + \sum_1^t (F_{4t}' + L_tF_{3t}') = \Lambda_4 = 0 \tag{13.287}$$

$$F_5 + \sum_1^p (F_{5p}' + L_pF_{2p}') + \sum_1^t (F_{5t}' + L_tF_{2t}') = \Lambda_5 = 0 \tag{13.288}$$

The solution of the displacements δ_j at point o is accomplished by iteration of

$$_2\delta_i = {_1\delta_i} - [a_{ij}]^{-1}\{\Lambda_j\} \tag{13.289}$$

The elements of the coefficient matrix are easily found to be

$$a_{11} = \sum_1^{p,t} \sum_{q=1}^{n} {_1G_{p,t}} \tag{13.290}$$

$$a_{12} = a_{21} = \sum_1^{p,t} \sum_{q=1}^{n} {_2G_{p,t}} \sin\varphi_q \tag{13.291}$$

$$a_{13} = a_{31} = \sum_1^{p,t} \sum_{q=1}^{n} {_2G_{p,t}} \cos\varphi_q \tag{13.292}$$

$$a_{14} = a_{41} = \sum_1^{p,t} \sum_{q=1}^{n} {_4G_{p,t}} \cos\varphi_q \tag{13.293}$$

$$a_{15} = a_{51} = \sum_1^{p,t} \sum_{q=1}^{n} {_4G_{p,t}} \sin\varphi_q \tag{13.294}$$

$$a_{22} = \sum_1^{p,t} \sum_{q=1}^{n} {_3G_{p,t}} \sin^2\varphi_q \tag{13.295}$$

$$a_{23} = a_{32} = \sum_1^{p,t} \sum_{q=1}^{n} {_3G_{p,t}} \sin\varphi_q \cos\varphi_q \tag{13.296}$$

$$a_{24} = a_{42} = \sum_{1}^{p,t} \sum_{q=1}^{n} {}_5G_{p,t} \sin \varphi_q \cos \varphi_q \tag{13.297}$$

$$a_{25} = a_{52} = \sum_{1}^{p,t} \sum_{q=1}^{n} {}_5G_{p,t} \sin^2 \varphi_q \tag{13.298}$$

$$a_{33} = \sum_{1}^{p,t} \sum_{q=1}^{n} {}_3G_{p,t} \cos^2 \varphi_q \tag{13.299}$$

$$a_{34} = a_{43} = \sum_{1}^{p,t} \sum_{q=1}^{n} {}_5G_{p,t} \cos^2 \varphi_q \tag{13.300}$$

$$a_{35} = a_{53} = \sum_{1}^{p,t} \sum_{q=1}^{n} {}_5G_{p,t} \sin \varphi_q \cos \varphi_q \tag{13.301}$$

$$a_{44} = \sum_{1}^{p,t} \sum_{q=1}^{n} {}_6G_{p,t} \cos^2 \varphi_q \tag{13.302}$$

$$a_{45} = a_{54} = \sum_{1}^{p,t} \sum_{q=1}^{n} {}_6G_{p,t} \sin \varphi_q \cos \varphi_q \tag{13.303}$$

$$a_{55} = \sum_{1}^{p,t} \sum_{q=1}^{n} {}_6G_{p,t} \sin^2 \varphi_q \tag{13.304}$$

where, for ball bearings only

$${}_1G_p = K\Delta_q^{1/2}(3 \sin^2 \beta_q/2 + \Delta_q \cos^2 \beta_q/D_qBd) \tag{13.305}$$
$${}_2G_p = K\Delta_q^{1/2}(3/2 - \Delta_q/D_qBd) \sin \beta_q \cos \beta_q \tag{13.306}$$
$${}_3G_p = K\Delta_q^{1/2}(3 \cos^2 \beta_q/2 + \Delta_q \sin^2 \beta_q/D_qBd) \tag{13.307}$$

For roller bearings only

$${}_1G_t = {}^{10}\!/_9 K'\Delta_q^{1/9} \sin^2 \beta \tag{13.308}$$
$${}_2G_t = {}^{10}\!/_9 K'\Delta_q^{1/9} \sin \beta \cos \beta \tag{13.309}$$
$${}_3G_t = {}^{10}\!/_9 K'\Delta_q^{1/9} \cos^2 \beta \tag{13.310}$$

For ball and roller bearings

$${}_4G_{p,t} = r_{p,t}\, {}_1G_{p,t} + L_{p,t}\, {}_2G_{p,t} \tag{13.311}$$
$${}_5G_{p,t} = r_{p,t}\, {}_2G_{p,t} + L_{p,t}\, {}_3G_{p,t} \tag{13.312}$$
$${}_6G_{p,t} = r_{p,t}\, {}_4G_{p,t} + L_{p,t}\, {}_5G_{p,t} \tag{13.313}$$

(e) An Approximate Solution of the Load Distribution in a System of Ball and Roller Bearings under Static Loading

The foregoing solutions of the load-distribution problem are exact but impractical for hand solution and require programming for a digital computer.

Except for the case of pure axial displacement an exact solution suitable for hand-computation means cannot be attained for ball bearings.

If the bearing system is lightly loaded the displacements are small and the change of contact angle occurring in a ball bearing can be neglected.

In order to simplify the computations further the number of degrees of freedom is reduced to three. That is, in Fig. 13.29, only the displacements δ_1, δ_3, and δ_4 and the forces F_1, F_3, and F_4 are considered.

Load Distribution in a Ball or Roller Bearing under Static Load in Three Degrees of Freedom—Approximate Solution. If the change in contact angle occurring in a ball bearing is neglected the sum of the elastic deflections at a rolling element's contacts can be stated by a single expression for both ball and roller bearings. In terms of the displacements δ_1', δ_3', and δ_4' at the bearing center

$$\Delta_q = \delta_1' \sin \beta + (\delta_3' \cos \beta + r\delta_4' \sin \beta) \cos \varphi_q - (\Delta P_D/2) \cos \beta \tag{13.314}$$

ΔP_D represents any increase in mounted diametral clearance.

The compressive loading of a rolling element is

$$P_q = K\Delta_q^{\alpha} \tag{13.315}$$

where K and α are chosen as appropriate for the type of bearing concerned.

The forces with which a bearing acts on the shaft at its center are

$$F_1' = K \sin \beta \sum_{q=1}^{n} (A + B \cos \varphi_q)^{\alpha} \tag{13.316}$$

$$F_2' = K \cos \beta \sum_{q=1}^{n} (A + B \cos \varphi_q)^{\alpha} \cos \varphi_q \tag{13.317}$$

$$F_3' = Kr \sin \beta \sum_{q=1}^{n} (A + B \cos \varphi_q)^{\alpha} \cos \varphi_q \tag{13.318}$$

The most heavily loaded rolling element may lie in either the upper or lower half of the bearing. Its location is determined by the sign of B.

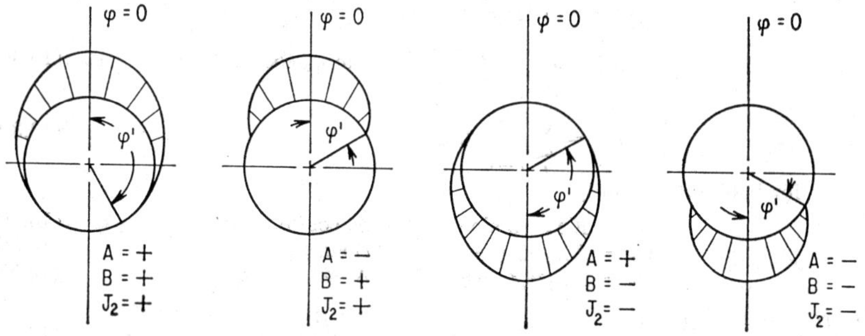

FIG. 13.30. Orientation of the loaded zone and significance of φ'.

Figure 13.30 shows the orientation of the loaded zone and the significance of φ', where

$$A = \delta_1' \sin \beta - (\Delta P_D/2) \cos \beta \tag{13.319}$$

$$B = \delta_3' \cos \beta + \delta_4' r \sin \beta \tag{13.320}$$

Letting

$$J_1 = 1/n \sum_{q=1}^{n} [A/|B| + (B/|B|) \cos \varphi_q]^{\alpha} \approx 1/\pi \int_0^{\varphi'} [A/|B| + (B/|B|) \cos \varphi]^{\alpha}\, d\varphi \tag{13.321}$$

$$J_2 = 1/n \sum_{q=1}^{n} [A/|B| + (B/|B|) \cos \varphi_q]^{\alpha} \cos \varphi_q$$

$$\approx 1/\pi \int_0^{\varphi'} [A/|B| + (B/|B|) \cos \varphi]^{\alpha} \cos \varphi\, d\varphi \tag{13.322}$$

then

$$F_1' = nK|B|^{\alpha} J_1 \sin \beta \tag{13.323}$$

$$F_3' = nK|B|^{\alpha} J_2 \cos \beta \tag{13.324}$$

$$F_4' = rnK|B|^{\alpha} J_2 \sin \beta \tag{13.325}$$

φ' is the extent of the loaded zone and is related to A and B by

$$\cos \varphi' = -A/B \tag{13.326}$$

Values of J_1 and J_2 have been calculated for point and line contact, $\alpha = 3/2$ and $\alpha = 10/9$, respectively, and are given in Fig. 13.31.

In reading values of J_2 from the tables it should be noted that J_2 always takes the sign of B.

Consider a system of u ball and roller bearings as in Fig. 13.29 with the radial roller bearing replaced by the angular-contact type.

For external forces F_1, F_3, and F_4 applied at point o the conditions of equilibrium are

$$F_1 + \sum_1^u n_u K_u |B_u|^{\alpha_u} J_{1_u} \sin \beta_u = \Gamma_1 \doteq 0 \tag{13.327}$$

$$F_3 + \sum_1^u n_u K_u |B_u|^{\alpha_u} J_{2_u} \cos \beta_u = \Gamma_3 \doteq 0 \tag{13.328}$$

$$F_4 + \sum_1^u n_u K_u |B_u|^{\alpha_u} J_{2_u} (r_u \sin \beta_u + L_u \cos \beta_u) = \Gamma_4 \doteq 0 \tag{13.329}$$

Since

$$\delta_{1u}' = \delta_1 + \delta_{1u}'' \tag{13.330}$$

$$\delta_{3u}' = \delta_3 + L_u \delta_4 \tag{13.331}$$

$$\delta_4' = \delta_4 \tag{13.332}$$

A_u and B_u for use with Eqs. (13.327) through (13.329) should be written

$$A_u = (\delta_1 + \delta_{1u}'') \sin \beta_u - (\Delta P_{Du}/2) \cos \beta_u \tag{13.333}$$

$$B_u = \delta_3 \cos \beta_u + \delta_4 (r_u \sin \beta_u + L_u \cos \beta_u) \tag{13.334}$$

Equations (13.327) through (13.329) are nonlinear in the variables δ_1, δ_3, and δ_4. Iteration of Eq. (13.335) leads to a solution

$$_2\delta_i = {}_1\delta_i - \Delta\delta_i \qquad i = 1, 3, 4 \tag{13.335}$$

The corrections are

$$\begin{Bmatrix} \Delta\delta_1 \\ \Delta\delta_3 \\ \Delta\delta_4 \end{Bmatrix} = \frac{1}{D} \begin{bmatrix} (a_{22}a_{33} - a_{23}^2) & -(a_{12}a_{33} - a_{13}a_{23}) & (a_{12}a_{23} - a_{13}a_{22}) \\ -(a_{12}a_{33} - a_{13}a_{23}) & (a_{11}a_{33} - a_{13}^2) & -(a_{11}a_{23} - a_{12}a_{13}) \\ (a_{12}a_{23} - a_{13}a_{22}) & -(a_{11}a_{23} - a_{12}a_{13}) & (a_{11}a_{22} - a_{12}^2) \end{bmatrix} \begin{Bmatrix} \Gamma_1 \\ \Gamma_3 \\ \Gamma_4 \end{Bmatrix} \tag{13.336}$$

where $D = a_{11}(a_{22}a_{33} - a_{23}^2) - a_{12}(a_{12}a_{33} - a_{13}a_{23}) + a_{13}(a_{12}a_{23} - a_{13}a_{22})$ (13.337)

Note that the matrix is symmetric and

$$a_{11} = \sum_1^u n_u K_u |B_u|^{\alpha-1} J_{3u} \sin^2 \beta_u \tag{13.338}$$

$$a_{12} = \sum_1^u n_u K_u |B_u|^{\alpha-1} J_{4u} \sin \beta_u \cos \beta_u \tag{13.339}$$

$$a_{13} = \sum_1^u n_u K_u |B_u|^{\alpha-1} J_{4u} (r_u \sin \beta_u + L_u \cos \beta_u) \sin \beta_u \tag{13.340}$$

$$a_{22} = \sum_1^u n_u K_u |B_u|^{\alpha-1} J_{5u} \cos^2 \beta_u \tag{13.341}$$

$$a_{23} = \sum_1^u n_u K_u |B_u|^{\alpha-1} J_{5u} (r_u \sin \beta_u + L_u \cos \beta_u) \cos \beta_u \tag{13.342}$$

$$a_{33} = \sum_1^u n_u K_u |B_u|^{\alpha-1} J_{5u} (r_u \sin \beta_u + L_u \cos \beta_u)^2 \tag{13.343}$$

J_3, J_4, and J_5 are obtained by differentiation of J_1 and J_2 with respect to A and B and are

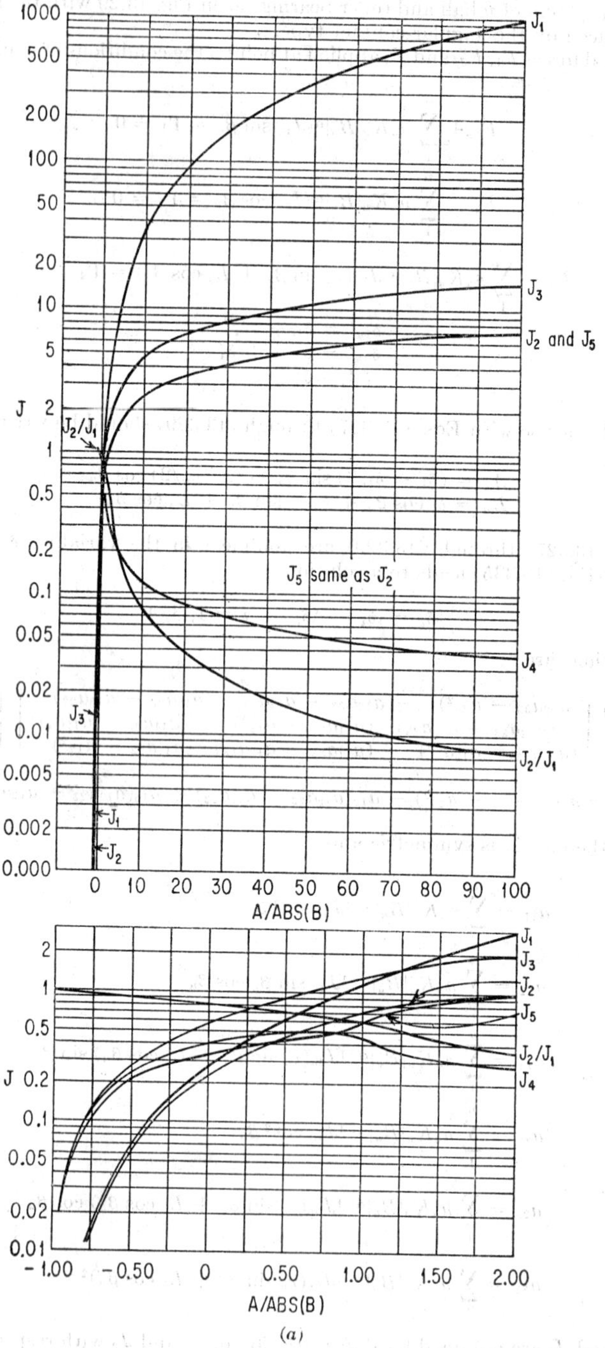

(a)

FIG. 13.31. Integrals for load

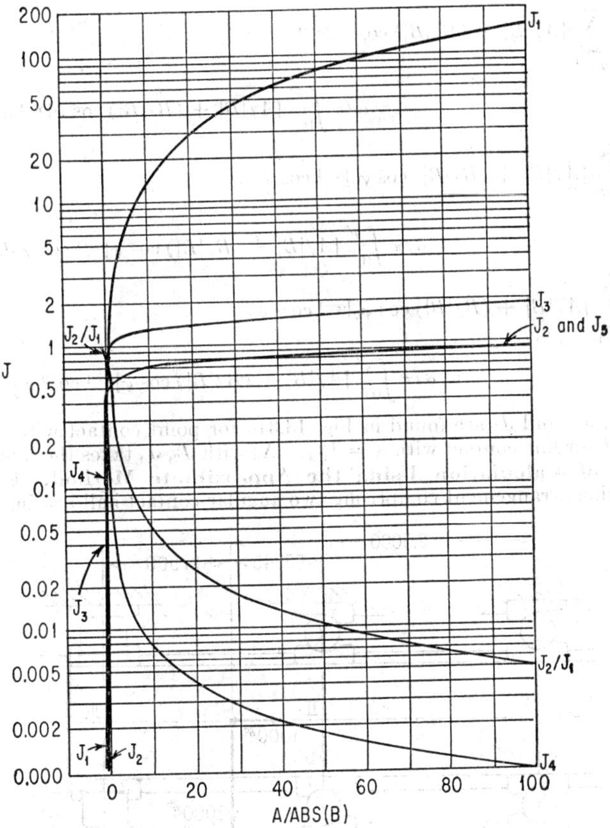

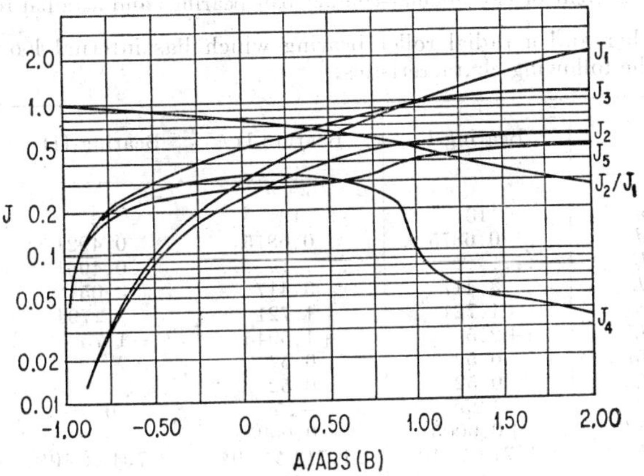

(*b*)

distribution. (*a*) Point contact. (*b*) Line contact.

$$J_3 = (\alpha/n) \sum_{q=1}^{n} [A/|B| + (B/|B|) \cos \varphi_q]^{\alpha-1}$$

$$\approx \alpha/\pi \int_0^{\varphi'} [A/|B| + (B/|B|) \cos \varphi]^{\alpha-1}\, d\varphi \quad (13.344)$$

$$J_4 = \alpha/n \sum_{q=1}^{n} [A/|B| + (B/|B|) \cos \varphi_q]^{\alpha-1} \cos \varphi_q$$

$$\approx \alpha/\pi \int_0^{\varphi'} [A/|B| + (B/|B|) \cos \varphi]^{\alpha-1} \cos \varphi\, d\varphi \quad (13.345)$$

$$J_5 = \alpha/n \sum_{q=1}^{n} [A/|B| + (B/|B|) \cos \varphi_q]^{\alpha-1} \cos^2 \varphi_q$$

$$\approx \alpha/\pi \int_0^{\varphi'} [A/|B| + (B/|B|) \cos \varphi]^{\alpha-1} \cos^2 \varphi\, d\varphi \quad (13.346)$$

Values of J_3, J_4, and J_5 are found in Fig. 13.31*a* for point contact with $\alpha = 3/2$ and in Fig. 13.31*b* for line contact with $\alpha = 10/9$. As with J_2, J_4 takes the same sign as B.

Example of Calculation Using the Approximate Method. Figure 13.32 shows a bearing arrangement comprising two angular-contact ball bearings which are

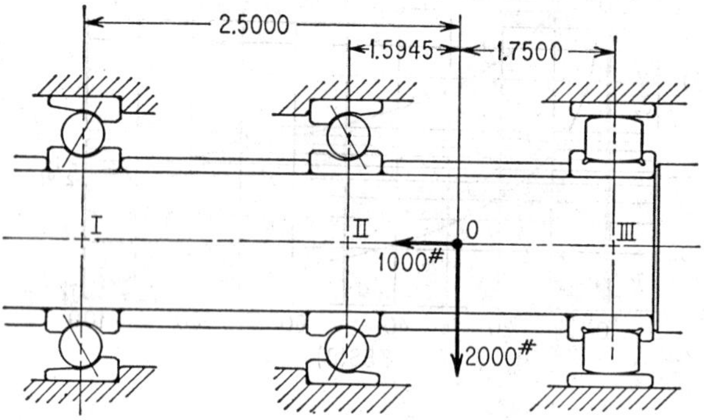

FIG. 13.32. Arrangement of two angular-contact ball bearings and a radial roller bearing.

preloaded together and a radial roller bearing which has internal looseness. The bearings have the following characteristics:

	Bearing I	Bearing II	Bearing III
η	13	13	16
d	0.6875	0.6875	0.4921
l_e	. . . .	. . . .	0.4961
E	3.417	3.417	3.051
r	1.721	1.721	1.2794
L	+2.5	+1.5945	−1.75
f_O	0.52	0.52	
f_I	0.52	0.52	
β	+25°	−25°	0°
δ_1''	+0.0005	−0.0005	
K	11.712×10^6	11.712×10^6	2.794×10^6
α	3/2	3/2	10/9
ΔP_D	. . . .	. . . .	+0.001

δ_1'' indicates that the inner ring of each bearing is axially displaced with respect to its outer ring so that each ball bearing has an initial preload deflection of 0.0005. ΔP_D indicates that the roller bearing has a mounted diametral clearance of 0.001. Initial estimates of the deflections at point o must be made. These are chosen as $\delta_1 = -0.003$, $\delta_3 = -0.002$, $\delta_4 = 0$.

The first iteration proceeds as follows:

	Bearing I	Bearing II	Bearing III
A [Eq. (13.333)]	−0.001057	0.001479	−0.0005
B [Eq. (13.334)]	−0.001813	−0.001813	−0.002
$nK\|B\|^{\alpha}$	11,753.6	11,753.6	44,822.0
$nK\|B\|^{\alpha-1}$	6.4830×10^6	6.4830×10^6	22.4110×10^6
$A/\|B\|$	−0.583012	0.815774	−0.2500
J_1	0.0470	0.9708	0.1907
J_2	−0.0436	−0.6301	−0.1619
J_3	0.2273	1.1578	0.4203
J_4	−0.2030	−0.5117	−0.3169
J_5	0.1838	0.5277	0.2599

The external forces are $F_1 = +1{,}000$ lb, $F_3 = +2{,}000$ lb, $F_4 = 0$ lb-in.

Equations (13.327) through (13.329) yield $\Gamma_1 = -3{,}588.78$, $\Gamma_3 = -12{,}433.19$, $\Gamma_4 = +5{,}850.01$.

The a_{ij} [Eqs. (13.338) through (13.343)] are

$$
\begin{aligned}
a_{11} &= 1.6038 \times 10^6 & a_{22} &= 9.6134 \times 10^6\\
a_{12} &= 0.7665 \times 10^6 & a_{23} &= -4.7352 \times 10^6\\
a_{13} &= -0.6584 \times 10^6 & a_{33} &= 30.2755 \times 10^6
\end{aligned}
$$

The matrix solution for the corrections $\Delta\delta_j$ is

$$
\begin{Bmatrix}\Delta\delta_1\\ \Delta\delta_3\\ \Delta\delta_4\end{Bmatrix} = \frac{10^{-6}}{413.65}\begin{bmatrix}268.63 & -20.09 & 2.70\\ -20.09 & 48.12 & 7.09\\ 2.70 & 7.09 & 14.83\end{bmatrix}\begin{Bmatrix}-3{,}588.78\\ -12{,}433.19\\ 5{,}850.01\end{Bmatrix} = \begin{Bmatrix}-0.001689\\ -0.001172\\ -0.000027\end{Bmatrix}
$$

Improved values of the displacements are, from Eq. (13.335),

$$
\begin{aligned}
\delta_1 &= -0.003 + 0.001689 = -0.001311\\
\delta_3 &= -0.002 + 0.001172 = -0.000828\\
\delta_4 &= 0 + 0.000027 = +0.000027
\end{aligned}
$$

Subsequent iterations give

Iteration	δ_1	δ_3	δ_4	Γ_1	Γ_3	Γ_5
2	−0.001311	−0.000828	+0.000027	−536.43	−1,406.00	112.07
3	−0.000901	−0.000636	+0.000066	−32.93	−84.51	−26.70
4	−0.000871	−0.000622	+0.000071	+0.24	−3.71	+8.39
5	−0.000871	−0.000622	+0.000071			

The internal load distribution may be calculated using Eqs. (13.330) through (13.332) and Eqs. (13.314) and (13.315). Table 13.3 shows the results of these calculations. Also shown in Table 13.3 are results from an exact calculation of the same problem using an IBM-7090 computer.

Table 13.3

	Bearing I		Bearing II		Bearing III	
	Approx	Exact	Approx	Exact	Approx	Exact
q	P_q	......	P_q			
1	....	......	6.28	3.05		
2	....	......	16.3	11.6		
3	....	......	57.1	49.6		
4	....	......	137.7	127.4		
5	....	1.148	245.8	233.1		
6	12.7	24.368	350.0	335.6		
7	29.2	44.424	414.0	398.7	23.9	26.9
8	29.2	44.424	414.0	398.7	203.9	208.7
9	12.7	24.368	350.0	335.6	273.0	278.3
10	....	1.148	245.8	233.1	203.9	208.7
11	....	......	137.7	127.4	23.9	26.9
12	....	......	57.1	49.6		
13	....	......	16.3	11.6		
14						
15						
16						

(f) An Approximate Solution for the Reactions of Two Opposed Angular-contact Bearings with Loadings in Five Degrees of Freedom

If the change of contact angle occurring with ball bearings is neglected and the system comprises only two opposed angular-contact ball and/or roller bearings, considerable simplification in the determination of the bearing's reactions can be made.

Consideration of Eqs. (13.324) and (13.325) shows that a constant ratio exists between a bearing's radial and moment reactions at its center.

$$F_4'/F_3' = r \tan \beta \tag{13.347}$$

This implies that, if the bearings are spaced with their centers at A and B as in Fig. 13.33, radial reactions calculated as though the bearings were located at A' and B' automatically establish the equilibrium of the bearings' moment reactions as well.

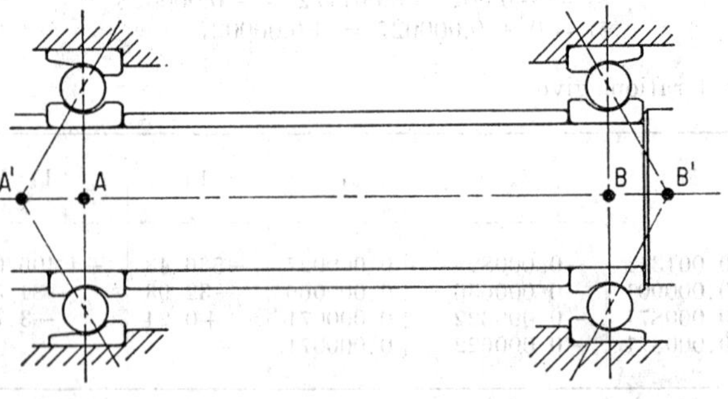

FIG. 13.33. Illustration of contact angles meeting at a point.

Figure 13.33 illustrates a "back-to-back" arrangement in which the contact angles converge to points on the shaft outside the bearings. A "face-to-face" arrangement, in which the contact angles are reversed, is treated in exactly the same manner.

Figure 13.34 shows a system consisting of two angular-contact ball bearings in "back-to-back" arrangement and loaded by thrust, radial, and moment loads at point o.

In accordance with the convention established in Art. 13.1 and illustrated in Fig. 13.29, L is positive when a bearing lies to the left of the loading point and negative when it lies to the right.

In Fig. 13.34 as in any two-bearing system, it is easy to determine by inspection which bearing provides the greater thrust reaction. Call this bearing bearing I and the other bearing II.

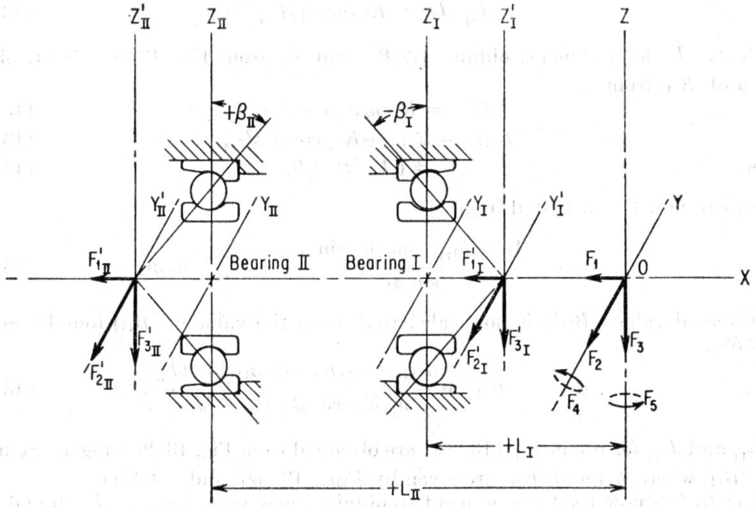

FIG. 13.34. Angular-contact ball bearings loaded by thrust, radial, and moment loads.

The radial reactions at the apices of the contact angles are

$$F_{2_{\mathrm{I}}}' = \frac{-F_2(L_{\mathrm{II}} + r_{\mathrm{II}} \tan \beta_{\mathrm{II}}) + F_5}{(L_{\mathrm{II}} + r_{\mathrm{II}} \tan \beta_{\mathrm{II}}) - (L_{\mathrm{I}} + r_{\mathrm{I}} \tan \beta_{\mathrm{I}})} \tag{13.348}$$

$$F_{2_{\mathrm{II}}}' = \frac{F_2(L_{\mathrm{I}} + r_{\mathrm{I}} \tan \beta_{\mathrm{I}}) - F_5}{(L_{\mathrm{II}} + r_{\mathrm{II}} \tan \beta_{\mathrm{II}}) - (L_{\mathrm{I}} + r_{\mathrm{I}} \tan \beta_{\mathrm{I}})} \tag{13.349}$$

$$F_{3_{\mathrm{I}}}' = \frac{-F_3(L_{\mathrm{II}} + r_{\mathrm{II}} \tan \beta_{\mathrm{II}}) + F_4}{(L_{\mathrm{II}} + r_{\mathrm{II}} \tan \beta_{\mathrm{II}}) - (L_{\mathrm{I}} + r_{\mathrm{I}} \tan \beta_{\mathrm{I}})} \tag{13.350}$$

$$F_{3_{\mathrm{II}}}' = \frac{F_3(L_{\mathrm{I}} + r_{\mathrm{I}} \tan \beta_{\mathrm{I}}) - F_4}{(L_{\mathrm{II}} + r_{\mathrm{II}} \tan \beta_{\mathrm{II}}) - (L_{\mathrm{I}} + r_{\mathrm{I}} \tan \beta_{\mathrm{I}})} \tag{13.351}$$

The signs of the contact angles and their trigonometric functions must be preserved.

The radial reactions at bearings I and II are R_{I} and R_{II}. These will not necessarily lie in the same plane, but this is not important.

$$R_{\mathrm{I}} = \sqrt{(F_{2\mathrm{I}}')^2 + (F_{3_{\mathrm{I}}}')^2} \tag{13.352}$$

$$R_{\mathrm{II}} = \sqrt{(F_{2_{\mathrm{II}}}')^2 + (F_{3_{\mathrm{II}}}')^2} \tag{13.353}$$

An assumption is now made of the extent of the loaded zone in bearing II, which has the lesser thrust reaction, and an appropriate choice of $A_{\mathrm{II}}/|B_{\mathrm{II}}|$ is made. $J_{2\mathrm{I}}/J_{1\mathrm{II}}$ and $J_{2\mathrm{II}}$ are now found from Fig. 13.31 for point or line contact, respectively.

The thrust reaction of bearing II is now calculated from Eq. (13.354) and takes the sign of β_{II}.

$$F_{1_{II}}' = \frac{R_{II} \tan \beta_{II}}{J_{2_{II}}/J_{1_{II}}} \tag{13.354}$$

The thrust reaction of bearing I is

$$F_{1_I}' = -F_1 - F_{1_{II}}' \tag{13.355}$$

Next, the value of J_{2_I}/J_{1_I} is found from

$$J_{2_I}/J_{1_I} = R_I \tan \beta_I / F_{1_I}' \tag{13.356}$$

With J_{2_I}/J_{1_I} as argument, obtain $A_I/|B_I|$ and J_{2_I} from Fig. 13.31. Next, obtain $|B_I|$ and $|B_{II}|$ from

$$|B_I|^\alpha = R_I/n_I K_I \cos \beta_I J_{2_I} \tag{13.357}$$

$$|B_{II}|^\alpha = R_{II}/n_{II} K_{II} \cos \beta_{II} J_{2_{II}} \tag{13.358}$$

Then

$$A_I = (A_I/|B_I|)|B_I| \tag{13.359}$$

A_{II} can now be calculated from

$$A_{II} = \frac{(A_I - \delta_{1_{II}}'' \sin \beta_I) \sin \beta_{II}}{\sin \beta_I} + \delta_{1_{II}}'' \sin \beta_{II} \tag{13.360}$$

A corrected value, $|B_{II}'|$, is now calculated using the value of $|B_{II}|$ found from Eq. (13.358).

$$|B_{II}'| = |B_{II}| + \frac{R_{II} - n_{II} K_{II} \cos \beta_{II} |B_{II}|^\alpha J_{2_{II}}}{n_{II} K_{II} \cos \beta_{II} |B_{II}|^{\alpha-1} J_{5_{II}}} \tag{13.361}$$

$J_{2_{II}}$ and $J_{5_{II}}$ for use in Eq. (13.361) are obtained from Fig. 13.28 using as argument $A_{II}/|B_{II}|$ where A_{II} and $|B_{II}|$ are given by Eqs. (13.360) and (13.361).

$A_{II}/|B_{II}'|$ is now used as argument to obtain a new value for $J_{2_{II}}/J_{1_{II}}$ found from the tables. The process is repeated, entering the solution at Eq. (13.354), until converged.

The internal load distributions within the bearings are now in equilibrium with the external force system and life can be calculated by the methods of Art. 13.7. The following example illustrates the procedure. Figure 13.35 shows, schematically, a system of two angular-contact ball bearings with loads applied to the shaft at point o such as might result from a spiral bevel pinion.

The bearings have the following characteristics:

	Bearing I	Bearing II		Bearing I	Bearing II
n	13	13	r	1.721	1.721
d	0.6875	0.6875	K	11.712×10^6	11.712×10^6
β	$-25°$	$+25°$	L	$+1.000$	$+3.062$
δ_1''	-0.0002	$+0.0002$	$nK \cos \beta$	137.9908×10^6	137.9908×10^6
α	$3/2$	$3/2$			

The values of δ_1'' indicate that each bearing is preloaded.

The external load system is

$$F_1 = +600$$
$$F_2 = +800$$
$$F_3 = +200$$
$$F_4 = +900$$
$$F_5 = 0$$

The radial reactions are

$$F_{2_I}' = \frac{800(3.062 + 1.721 \times 0.466307) - 0}{(1 - 1.721 \times 0.466307) - (3.062 + 1.721 \times 0.466307)} = -843.084 \text{ lb}$$

$$F_{2_{II}}' = \frac{-800(1 - 1.721 \times 0.466307) + 0}{(1 - 1.721 \times 0.466307) - (3.062 + 1.721 \times 0.466307)} = 43.084 \text{ lb}$$

$$F_{3_I}' = \frac{200(3.062 + 1.721 \times 0.466307) - 900}{(1 - 1.721 \times 0.466307) - (3.062 + 1.721 \times 0.466307)} = 34.659 \text{ lb}$$

$$F_{3_{II}}' = \frac{-200(1 - 1.721 \times 0.466307) + 900}{(1 - 1.721 \times 0.466307) - (3.062 + 1.721 \times 0.466307)} = -234.659 \text{ lb}$$

$$R_I = \sqrt{(-843.084)^2 + (34.659)^2} = 843.796 \text{ lb}$$

$$R_{II} = \sqrt{(43.084)^2 + (-234.659)^2} = 238.581 \text{ lb}$$

Assume $A_{II}/|B_{II}| = 0$. From Fig. 13.31a for point contact

$$J_{2_{II}} = 0.228828 \qquad J_{5_{II}} = 0.343242$$

$$J_{2_{II}}/J_{1_{II}} = 0.822503$$

$$F_{1_{II}}' = \frac{238.581 \times 0.466307}{0.822503} = +135.260 \leftarrow$$

$$F_{1_I}' = -600 - 135.260 = -735.260 \rightarrow$$

$$\frac{J_{2_I}}{J_{1_I}} = \frac{843.796 \times (-0.466307)}{-735.260} = 0.535141$$

From Fig. 13.31a

$$A_I/|B_I| = 1.205290$$

$$J_{2_I} = 0.803100$$

$$|B_I|^{3/2} = \frac{843.796}{137.9908 \times 10^6 \times 0.803100} = 7.614085 \times 10^{-6}$$

$$|B_I| = 0.00038703$$

$$|B_{II}|^{3/2} = \frac{238.581}{137.9908 \times 10^6 \times 0.228828} = 7.555732 \times 10^{-6}$$

$$|B_{II}|^{1/2} = 0.019623$$

$$|B_{II}| = 0.00038505$$

$$A_I = 1.205290 \times 0.00038703 = 0.00046648$$

$$A_{II} = \frac{[0.00046648 - (-0.0002)(-0.422618)] \times 0.422618}{-0.422618} + 0.0002 \times 0.422618$$

$$A_{II} = -0.00029743$$

$$\frac{A_{II}}{|B_{II}|} = \frac{-0.00029743}{0.00038505} = -0.772445$$

$$J_{2_{II}} = 0.013356$$

$$J_{5_{II}} = 0.109111$$

$$|B_{II}| = 0.00038505 + \frac{238.581 - 137.9908 \times 10^6 \times 7.555732 \times 10^{-6} \times 0.013356}{137.9908 \times 10^6 \times 0.019623 \times 0.109111}$$

$$|B_{II}| = 0.00114543$$

$$|B_{II}|^{1/2} = 0.033844$$

$$|B_{II}|^{3/2} = 38.766164 \times 10^{-6}$$

Subsequent iterations yield the following:

Iteration	$F_{1_{II}}'$	F_{1_I}'	A_{II}	$\lvert B_{II}\rvert$	A_I	$\lvert B_I\rvert$
2	127.721	−727.721	−0.00029310	0.00079180	0.00046215	0.00038921
3	124.847	−724.847	−0.00029143	0.00073088	0.00046048	0.00039006
4	124.132	−724.132	−0.00029102	0.00072807	0.00046007	0.00039028
5	124.107	−724.107	−0.00029100	0.00072806	0.00046005	0.00039028

The internal load distribution can now be obtained from the converged values of A and $|B|$.

(g) Particular Solutions for the Internal Rolling-element Load Distribution and Displacements in Single Bearings

There are very few cases of loading of single ball or roller bearings that can be precisely evaluated by simple hand computation. However, methods can be prepared which enable the solution of single-bearing load distribution and deflection problems for a number of types of loading. The following shows examples of such methods for selected loadings.

Axial Deflection and Change of Contact Angle in Ball Bearings under Thrust Load. If axial displacement δ_1' only is considered, the total deflection of inner- and outer-race contacts is, in accordance with Eq. (13.268),

$$\Delta = Bd(D - 1) \geq 0 \tag{13.362}$$

and D, which is equal at all ball positions, is found from Eq. (13.269) as

$$D = [(\sin \beta + \delta_1'/Bd)^2 + \cos^2 \beta]^{1/2} \geq 1 \tag{13.363}$$

The convention of force and contact-angle signs is to be disregarded in this discussion. All angles are to be considered as first-quadrant angles.

From Eq. (13.271) the operating contact angle is

$$\tan \beta_1 = \frac{\sin \beta + \delta_1'/Bd}{\cos \beta} \tag{13.364}$$

The compressive load on a ball is

$$P = K\Delta^{3/2} \tag{13.365}$$

where K is found from Eq. (13.273).

The thrust reaction of the bearing on the shaft is, from Eq. (13.273),

$$F_1' = nP \sin \beta_1 \tag{13.366}$$
$$F_1' = nK(Bd)^{3/2} \sin \beta_1[(\cos \beta/\cos \beta_1) - 1]^{3/2} \tag{13.367}$$

The axial deflection of the bearing under thrust load is found from Eq. (13.364).

$$\delta_1' = Bd(\tan \beta_1 \cos \beta - \sin \beta) \tag{13.368}$$

The operating contact angle is easily found by iteration of

$$\cos \beta_1' = \frac{\cos \beta}{1 + (1/Bd)(F_1'/nK \sin \beta_1)^{2/3}} \tag{13.369}$$

where β_1' is an improved estimate of β_1 and $\sin \beta_1$ is in accordance with the current estimate.

$$P = F_1'/n \sin \beta_1 \tag{13.370}$$

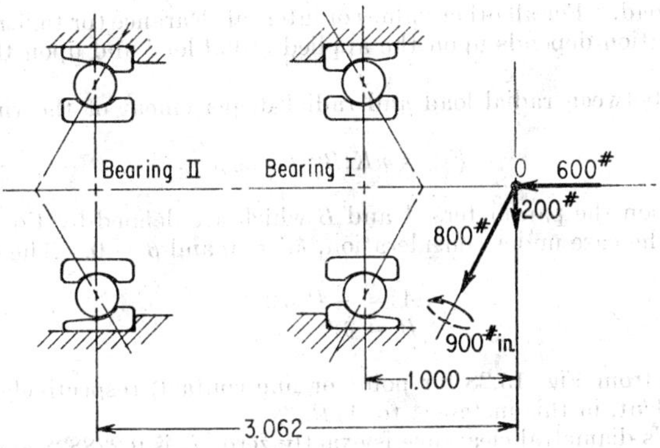

FIG. 13.35. Two angular-contact ball bearings with loads applied to shaft.

As an example the operating contact angle and axial deflection of bearing I, Fig. 13.35, are calculated for 1,600 lb pure thrust load. For this bearing

$$
\begin{aligned}
F_1' &= 1{,}600 \text{ lb} \\
\beta &= 25^\circ \\
nK &= 13 \times 11.712 \times 10^6 = 152.256 \times 10^6 \\
\cos\beta &= 0.906308 \\
Bd &= 0.0275
\end{aligned}
$$

To start the solution assume β_1 somewhat greater than β, say 30° in this case. Then,

$$\cos\beta_1' = \frac{0.906308}{1 + (1/0.0275)(1{,}600/152.256 \times 10^6 \times 0.5)^{2/3}}$$

First iteration $\cos\beta_1' = 0.881885$
Second iteration $\cos\beta_1' = 0.880937$
Third iteration $\cos\beta_1' = 0.880998$
Fourth iteration $\cos\beta_1' = 0.880994$
$\beta_1 = 28.2375^\circ$

The axial displacement is

$$\delta_1' = 0.0275(0.537039 \times 0.906308 - 0.422618) = 0.001763 \text{ in.}$$

and the rolling-element load is

$$P = \frac{1{,}600}{13 \times 0.473128} = 260 \text{ lb}$$

Axial Deflection in Angular-contact Roller Bearings. Since the contact angle of a roller bearing does not change with thrust load, the axial deflection can be found directly.

$$\delta_1 = (F_1'/nK)^{9/10} \sin^{-19/10}\beta \tag{13.371}$$

where K is chosen for line contact. $K = 5.21 \times 10^{-6} l_e^{8/9}$ for roller bearings of conventional materials.

The load on a rolling element is

$$P = T/n \sin\beta \tag{13.372}$$

Radial Deflection in Radial Ball and Roller Bearings under Radial Load. A radial bearing is here defined as one in which the loci of outer- and inner-race curvature centers are constrained to motion in the same plane.

Only in the case of a bearing with no internal clearance, i.e., $P_D = 0$, does the compressive load on the most heavily loaded rolling element bear a fixed ratio to the external radial load. For all other values of internal clearance (or tightness) the internal load distribution depends upon the applied radial load and upon the fitup of the bearing.

The relation between radial load and radial displacement of the rings is implied in Eq. (13.327).

$$F_3' = nK|B|^\alpha J_2 \cos\beta \tag{13.373}$$

J_2 depends upon the parameters A and B which are defined by Eqs. (13.319) and (13.320). For the case under consideration, $\delta_1' = 0$ and $\beta = 0$. Then

$$A = -P_D/2 \tag{13.374}$$
$$B = \delta_3' \tag{13.375}$$

and J_2 is found from Fig. 13.28 for point or line contact, respectively, as functions of $A/|B|$ equivalent, in this instance, to A/B.

If the bearing's diametral clearance is exactly zero, J_2 is 0.228828 and 0.244799 for point and line contact, respectively, and the radial displacement is directly, for point contact,

$$\delta_{1_3}' = 0.3741(F_3'/nK)^{2/3} \tag{13.376}$$

and, for line contact,

$$\delta_{1_3}' = 0.2818(F_3'/nK)^{9/10} \tag{13.377}$$

Also, for the case of zero diametral clearance, the load on the most heavily loaded rolling element is, for point contact,

$$P = 4.3701F_3'/n \tag{13.378}$$

and, for line contact,

$$P = 4.0850F_3'/n \tag{13.379}$$

With internal clearance ($+P_D$) or preload ($-P_D$) the radial displacement cannot be found directly but can readily be calculated as follows:

An estimate is made of the radial displacement δ_3' and $A/|B|$ is calculated from

$$A/|B| = -P_D/2\delta_3' \tag{13.380}$$

and J_2 and J_5 are obtained from Fig. 13.31 for line or point contact.

If δ_3' is an estimate an improved value is ${}_1\delta_3'$.

$${}_1\delta_3' = \delta_3' + \frac{R - nK\delta_3'^{\alpha} J_2}{nK\delta_3'^{\alpha-1} J_5} \tag{13.381}$$

Iteration of Eq. (13.381) yields δ_3' to any desired accuracy.

The load on the most heavily loaded rolling element is related to the external radial load through

$$P = \frac{(1 - P_D/2\delta_3')^\alpha}{nJ_2} F_3' \tag{13.382}$$

δ_3' is always the total relative radial displacement of outer- and inner-ring axes. With bearings which are initially loose the elastic displacement due to load alone is $\delta_3' - P_D/2$. With initially tight bearings δ_3' is entirely due to load. When solving for δ_3' by the iterative process [Eq. (13.381)] the initial guess of δ_3' must always be in excess of $P_D/2$ when diametral clearance is present. An overestimate of δ_3' is desirable in any case. The following example illustrates the process.

An aircraft radial roller bearing is operated under a radial load of 1,800 lb and has a mounted diametral clearance of 0.0010. Determine the radial deflection and the load on the most heavily loaded roller.

The bearing has a complement of 15 rollers, 11 mm diameter × 13 mm overall length. The contact angle is zero.

The effective length may be taken as 90 per cent of the overall length.

$$l_e = 0.9 \times 13 \times 0.03937 = 0.4606 \text{ in.}$$
$$K = 5.21 \times 10^6(0.4606)^{8/9} = 2.6156 \times 10^6$$
$$nK = 15 \times 2.615 \times 10^6 = 39.234 \times 10^6$$

Assume

$$\delta_3' = 0.002$$
$$P_D = +0.0010$$
$$\frac{A}{|B|} = -\frac{0.0010}{2 \times 0.002} = -0.25$$

From Fig. 13.31*b*

$$J_2 = 0.161940 \qquad J_5 = 0.259172$$
$${}_1\delta_3' = 0.002 + \frac{1{,}800 - 39.234 \times 10^6(0.002)^{10/9} \times 0.161940}{39.234 \times 10^6 \times (0.002)^{1/9} \times 0.259172}$$
$${}_1\delta_3' = 0.002 - 0.000897 = 0.001103$$

Subsequent iterations yield

Iteration	$A/\|B\|$	J_2	J_5	${}_1\delta_3'$
2	−0.453309	0.100967	0.2389	0.001046
3	−0.478011	0.094116	0.235560	0.001046

The last two values of δ_3' are in agreement to 10^{-6} in. The load on the most heavily loaded roller is

$$P = \frac{(1 - 0.0010/2 \times 0.001046)^{10/9} \times 1{,}800}{15 \times 0.094116} = 619.2 \text{ lb}$$

The extent of the loaded zone, in accordance with Eq. (13.326), is

$$\varphi' = \cos^{-1} 0.478011$$
$$\varphi' = \pm 61.44°$$

13.7. FRICTIONAL EFFECTS IN ROLLING-ELEMENT BEARINGS

One of the most difficult items to assess with rolling-contact bearings is the energy dissipated as heat as the result of the bearing's operation under service conditions.

In the majority of applications, and in particular in those which operate at high speed and at relatively low load, the major sources of friction are due to causes not directly associated with the rolling process. More important under such conditions are the resistances arising from the shearing and churning of the lubricant and especially those occurring as a result of the rolling element's confinement in the pocket of the cage or retainer and the friction associated with the guidance of the cage by a piloting surface, if any.

Because of the endless variety of rolling-element configurations, the many types of solid, semifluid, and fluid lubricants, and the various ways in which they can be applied, comprehensive rules or formulas for rolling-element power loss cannot be established.

Energy losses resulting from the substantially elastic contact of a rolling element and a raceway are of two types. First, there is the energy which is transformed into heat as the result of material hysteresis occurring from repeated distor-

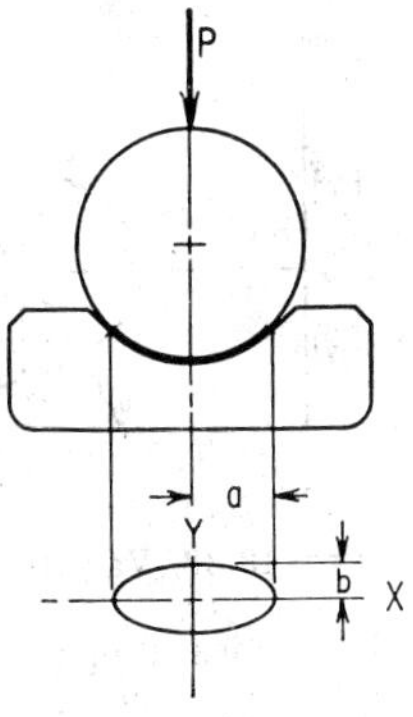

FIG. 13.36. Grooved raceway.

tions of the stressed volume of material at the rolling element's passages. In properly designed roller bearings and in certain types of ball bearings in which the pressure areas are substantially flat hysteresis is probably the main source of contact-area friction. Its effect has not yet been completely evaluated.

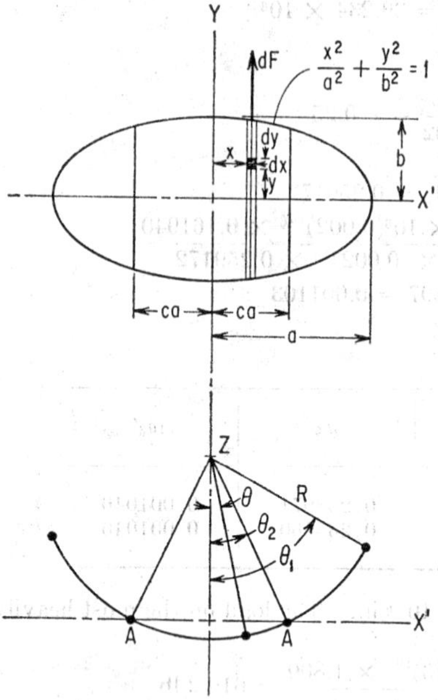

FIG. 13.37. Plan and elevation of pressure area.

Second, in conventional ball-bearing types operating under substantial loads and in which rather elongated pressure ellipses occur with their minor axes in the rolling direction, sliding friction due to relative motions of the contacting bodies within the bounds of the pressure areas account for the major energy loss due to the rolling action. The solution for the latter case is given in Art. 13.7(a) and the results are in good agreement with experimental data.

(a) Sliding Friction in Doubly Curved, Elliptical Pressure Areas

Rolling without Spin. Figure 13.36 illustrates a ball in a grooved raceway such as the inner race of a radial ball bearing under radial load. The ball is assumed to be rolling without spin.

Under the load P a doubly curved pressure ellipse of dimensions a and b is formed which may subtend an appreciable portion of the race-contour arc.

Considering only the curvature of the pressure ellipse in the plane of the paper it is clear that true rolling motion cannot exist at all points within the pressure area and some sliding must exist. Figure 13.37 shows an enlarged plan and elevation of the pressure area.

There can be, at most, two lines of true rolling along which the linear velocities of ball and raceway are the same. Let these lines be as at A in Fig. 13.37. Between

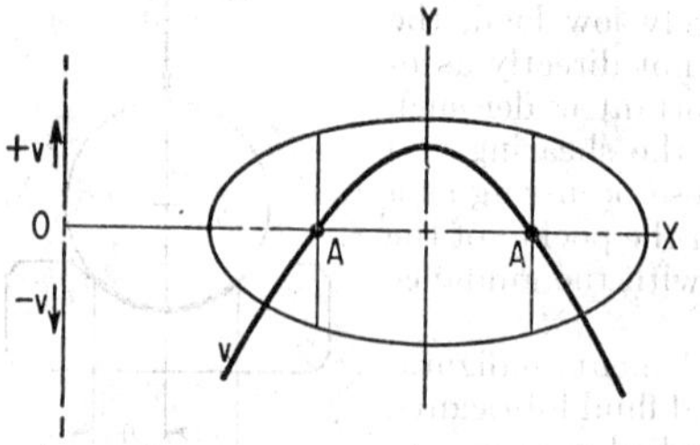

FIG. 13.38. Variation of sliding velocity.

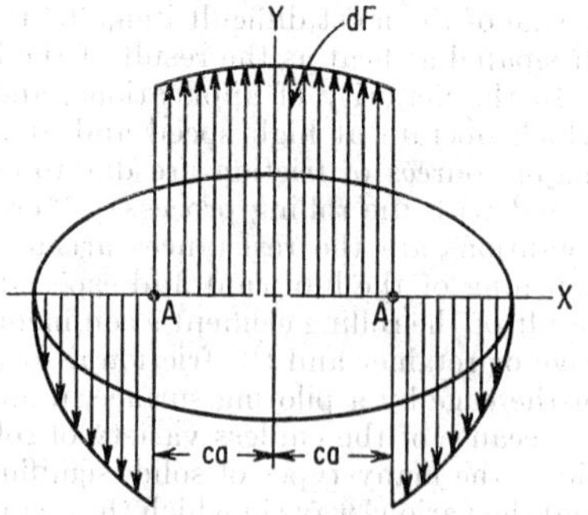

FIG. 13.39. Frictional-force lines.

these lines slippage occurs in one direction, and without the lines the direction of sliding is reversed. Figure 13.38 illustrates the variation in sliding velocity.

Figure 13.39 illustrates the orientations of the frictional forces resulting from interfacial slip.

The lines of frictional force shown in Fig. 13.39 correspond to the friction force dF on an elementary strip in Fig. 13.37.

Considering the ellipse as a whole, the effect of the friction forces is to produce a moment about the X axis which can be imagined as piercing the doubly curved pressure area at A.

For an arbitrary location of the lines of true rolling there will be a net force in or opposed to the rolling direction depending upon the value of c.

The compressive loading is distributed over the pressure surface in accordance with

$$|\delta_z| = (3P/2\pi ab)\sqrt{1 - x^2/a^2 - y^2/b^2} \qquad (13.383)$$

If μ is the coefficient of sliding friction, taken constant over the ellipse, the total friction force in the rolling direction is

$$F = (6P\mu/\pi ab)\left(\int_0^{ca}\int_0^{b\sqrt{1-x^2/a^2}} \sqrt{1 - x^2/a^2 - y^2/b^2}\,dy\,dx - \int_{ca}^{a}\int_0^{b\sqrt{1-x^2/a^2}} \sqrt{1 - x^2/a^2 - y^2/b^2}\,dy\,dx\right) \qquad (13.384)$$

$$F = \mu P(3c - c^3 - 1) \qquad (13.385)$$

This force is analogous to a tractive or braking effort, depending on its direction relative to the ball's motion, and is directed along the Y axis in Fig. 13.37.

In a properly operating ball bearing the net tractive or braking force on a ball is negligible and $F \doteq 0$, corresponding to $c = 0.3473+$. The lines of true rolling then lie at $x = \pm 0.3473a$ in agreement with observed locations of the "nonslip bands" sometimes found in radial-bearing raceways in which well-defined load paths have developed after long-time operation under heavy and constant load. Similar "nonslip bands" are frequently found on one race only of angular-contact bearings run under constant thrust load. The latter observation further substantiates the hypothesis that a ball in an angular-contact bearing normally rolls without spin on one race while all spin occurs at the other race contact.

The rigid-body type of interfacial slip discussed above is sometimes referred to as "Heathcoate slip" and neglects the effects of tangential elastic compliance at the pressure-surface interface.

Consider a point on the ball and a point on the raceway which move in the same plane. Let these points come into engagement but at slightly different velocities. If these points become locked at the instant of engagement, tangential forces immediately begin to build up at the interface. These forces result from tangential distortions due to the different translational velocities of the two points viewed as points on rigid bodies. At some instant during the engagement the tangential force may become equal to the friction force and at that time slip occurs. Thus there cannot be slip over the entire area. However, in conventional bearings with commonly used race contours the region of elastic compliance is confined to a relatively small portion of the pressure ellipse and in a region where the surface compression is low.

Returning to Fig. 13.37, it is clear that the moment of the elementary force dF about the X' axis is

$$dm = R\cos\theta\,dF \qquad (13.386)$$

R is the radius of curvature of the deformed pressure ellipse in the plane containing its major axis.

$$R = 2fd/(2f + 1) \qquad (13.387)$$

Integration over the ellipse yields

$$M = 2R\left(\int_0^{\theta_2}\cos\theta\,dF - \int_{\theta_2}^{\theta_1}\cos\theta\,dF\right) \qquad (13.388)$$

from which

$$M = \frac{3\mu PR}{4 \sin \theta_1}\left[2 \sin \theta_2 \cos \theta_2 - \sin \theta_1 \cos \theta_1 - \frac{\sin 4\theta_1 - 2 \sin 4\theta_2}{16 \sin^2 \theta_1} + (\theta_1 - 2\theta_2)\left(\frac{1}{4 \sin^2 \theta_1} - 1\right)\right] \quad (13.389)$$

From Fig. 13.37

$$\theta_1 = \sin^{-1}(a/R) \quad (13.390)$$
$$\theta_2 = \sin^{-1}(ca/R) \quad (13.391)$$

For the tractive or braking force to vanish, $c = 0.3473+$, and for this condition an approximate expression is obtained which is valid for a wide range of (a/R):

$$M = \frac{0.0796\mu a^{1.986}P}{R^{0.986}} \quad (13.392)$$

For bearings of conventional materials

$$M = \frac{6.784 \times 10^{-7}\mu(2f+1)^{0.986}K_R P^{1.662}}{(fd)^{0.324}(2f-1)^{0.662}} \quad (13.393)$$

Values of K_R are given in Fig. 13.40 for various race curvatures.

It is clear from the above that the assumption of a constant coefficient of *rolling* friction, such that $M \sim P$, is untenable.

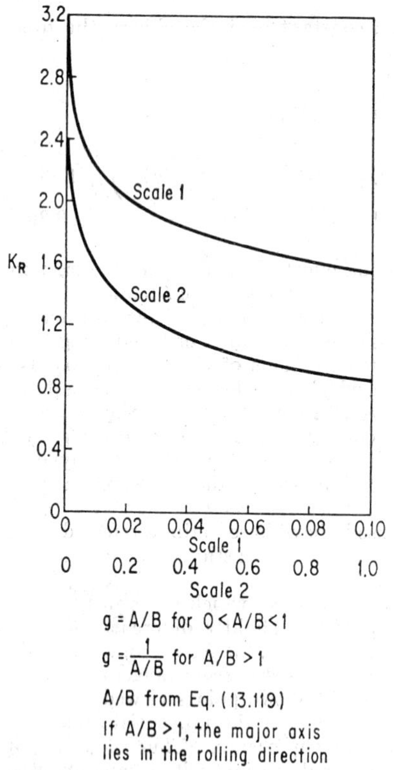

FIG. 13.40. Values of K_R for race curvature.

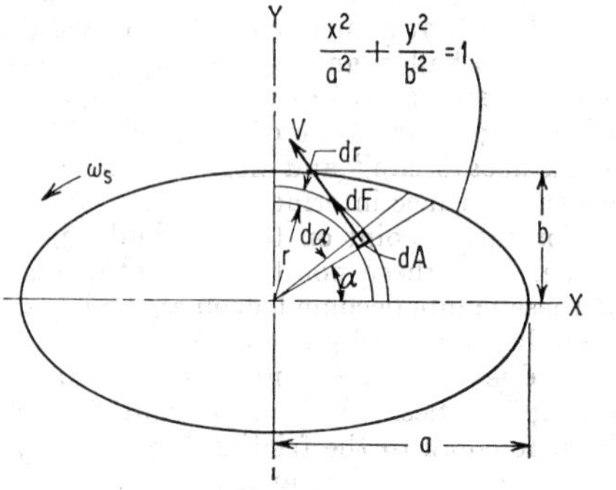

FIG. 13.41. Elliptical pressure areas between two solid elastic bodies.

Good agreement with experimental results is obtained by Eq. (13.392) with $\mu = 0.075$ for running friction and with $\mu = 0.100$ for starting friction.

Spinning Torque. Figure 13.41 illustrates an elliptical pressure area between two solid elastic bodies which are spinning with respect to each other about an axis normal to the pressure area at its center with the angular velocity ω_S. The linear slip velocity of an element of area dA is

$$V = r\omega_S \quad (13.394)$$

The force of friction on the elementary area is dF and is directed along V. In polar coordinates,

$$dF = 3\mu P/2\pi ab \sqrt{1 - (r^2/b^2)(1 - e^2 \cos^2 \alpha)}\; r\,d\alpha\, dr \quad (13.395)$$

where e, eccentricity of the pressure ellipse, is written

$$e = \sqrt{1 - b^2/a^2} \tag{13.396}$$

The moment of dF about the origin is

$$dQ_S = r\,dF \tag{13.397}$$

Integration over the ellipse yields the spinning torque Q_S.

$$Q_S = (3\mu aP/8)E(e) \tag{13.398}$$

where $E(e)$ is the complete elliptic integral of the second kind formed with the modulus e.

For bearings of conventional materials,

$$Q_S = 1.482 \times 10^{-3}\mu K_S[f\,d/(2f-1)]^{1/3}P^{4/3} \tag{13.399}$$

Values of K_S are given in Fig. 13.42 for various race curvatures.

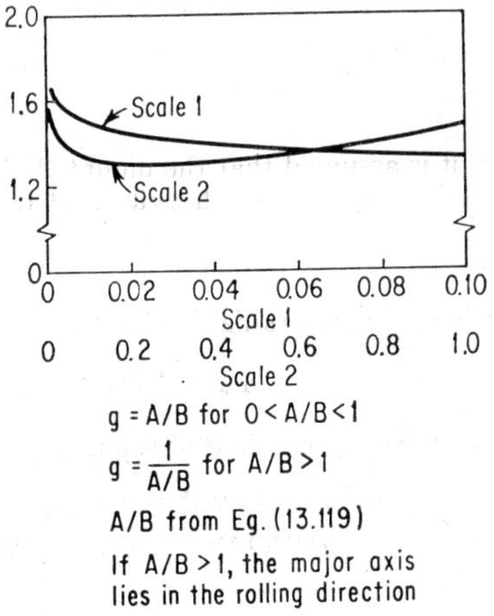

FIG. 13.42. Values of K_S for race curvature.

(b) Friction Torque in Ball Bearings

General Case of Dynamic Loading. For a ball bearing operating under complex loading at high speed, the frictional power loss due to interfacial slip at all contacts is

$$H = 1/6{,}600 \sum_{q=1}^{n} \left(Q_{SO_q}|\omega_{SO_q}| + Q_{SI_q}|\omega_{SI}{}^q| + M_{O_q}|\omega_{RO_q}| + M_{I_q}|\omega_{RI_q}|\right) \tag{13.400}$$

where H is the power loss in horsepower and the ω's are the angular velocities of spin and roll in radians per second. ω_{SO_q} or ω_{SI_q} will be zero depending on whether outer- or inner-race control exists.

The torque required to rotate the bearing under these conditions is $\bar{Q}$.

$$\bar{Q} = \frac{63{,}025H}{|N_O - N_I|} \tag{13.401}$$

where $\bar{Q}$ is the torque in lb-in. and N_O and N_I the angular velocities of outer and inner races in rpm.

General Case of Static Loading. When body forces arising from the balls' motions are neglected the power loss due to interfacial slip is

$$H = \frac{|N_O - N_I|}{63{,}025} \sum_{q=1}^{n} \left\{ |(Q_{SO_q} + Q_{SI_q}) \sin \beta_q| + \frac{E}{2d}[M_{O_q}(1 - \gamma_q) + M_{I_q}(1 + \gamma_q)] \right\} \tag{13.402}$$

and the torque $\bar{Q}$ in lb-in. is

$$\bar{Q} = \sum_{q=1}^{n} \{|(Q_{SO_q} + Q_{SI_q}) \sin \beta_q| + (E/2d)[M_{O_q}(1 - \gamma_q) + M_{I_q}(1 + \gamma_q)]\} \tag{13.403}$$

where

$$\gamma_q = d \cos \beta_q / E \tag{13.404}$$

Q_{SO_q} or Q_{SI_q} is to be set equal to zero depending on whether outer- or inner-race control exists.

For a radial ball bearing under radial load the spinning velocity is zero at each race contact. Then Eq. (13.403) can be written

$$\bar{Q} = (E/2d) \sum_{q=1}^{n} [M_{O_q}(1 - \gamma) + M_{I_q}(1 + \gamma)] \tag{13.405}$$

With Eq. (13.393)

$$\bar{Q} = \frac{3.392 \times 10^{-7}\mu E}{d^{1.324}} \left[\frac{(2f_O + 1)^{0.986} K_{RO}(1 - \gamma)}{f_O^{0.324}(2f_O - 1)^{0.662}} + \frac{(2f_I + 1)^{0.986} K_{RI}(1 + \gamma)}{f_I^{0.324}(2f_I - 1)^{0.662}} \right] \sum_{q=1}^{n} P_q^{1.662} \tag{13.406}$$

If it is assumed that the mounted diametral clearance of the bearing is just zero, the load P_0 on the most heavily loaded ball is

$$P_0 = 4.37\bar{R}/n \tag{13.407}$$

where $\bar{R}$ is the radial load on the bearing. Also, the load P_q on any other ball is

$$P_q = P_0 \cos^{3/2}(2\pi q/n) \qquad -\pi/2 \leq 2\pi q/n \leq \pi/2 \tag{13.408}$$

since exactly one-half of the bearing is loaded. Then

$$\bar{Q} = \frac{1.253 \times 10^{-6}\mu E(\bar{R})^{1.662}}{n^{0.662}d^{1.324}} \left[\frac{(2f_O + 1)^{0.986} K_{RO}(1 - \gamma)}{f_O^{0.324}(2f_O - 1)^{0.662}} + \frac{(2f_I + 1)^{0.986} K_{RI}(1 + \gamma)}{f_I^{0.324}(2f_I - 1)^{0.662}} \right] \int_0^{\pi/2} \cos^{2.493}\varphi \, d\varphi \tag{13.409}$$

$$\bar{Q} = \frac{9.015 \times 10^{-7}\mu E(\bar{R})^{1.662}}{n^{0.662}d^{1.324}} \left[\frac{(2f_O + 1)^{0.986} K_{RO}(1 - \gamma)}{f_O^{0.324}(2f_O - 1)^{0.662}} + \frac{(2f_I + 1)^{0.986} K_{RI}(1 + \gamma)}{f_I^{0.324}(2f_I - 1)^{0.662}} \right] \tag{13.410}$$

Under pure thrust load, all balls in an angular-contact ball bearing are equally loaded and the ball load P may be written

$$P = T/n \sin \beta_1 \tag{13.411}$$

The friction torque of the bearing is then

$$\bar{Q} = \frac{3.392 \times 10^{-7}\mu E T^{1.662}}{n^{0.662}d^{1.324}\sin^{1.662}\beta_1} \left[\frac{(2f_O + 1)^{0.986} K_{RO}(1 - \gamma)}{f_O^{0.324}(2f_O - 1)^{0.662}} + \frac{(2f_I + 1)^{0.986} K_{RI}(1 + \gamma)}{f_I^{0.324}(2f_I - 1)^{0.662}} \right] + \frac{1.482 \times 10^{-3}\mu\, d^{1/3}T^{4/3}}{n^{1/3}\sin^{1/3}\beta} \left[\left(\frac{f_O}{2f_O - 1}\right)^{1/3} K_{SO} + \left(\frac{f_I}{2f_I - 1}\right)^{1/3} K_{SI} \right] \tag{13.412}$$

K_{SO} or K_{SI} is set equal to zero depending on whether the bearing has outer- or inner-race control.

13.8. THE FATIGUE LIFE OF BALL AND ROLLER BEARINGS

Ball and roller bearings, properly mounted and lubricated, in surroundings free from the harmful effects of contaminants and extremes of temperature and not overloaded, show little or no measurable wear even after years of continuous operation.

Rolling-contact bearings do, however, have a definite and statistically predictable life whose end is signaled by the occurrence of a small pit or spall on a rolling surface.

If the operation of the bearing is continued after the appearance of the small initial spall, further disintegration of the load-carrying surfaces rapidly occurs and the bearing is rendered useless.

The life of an individual bearing cannot be predicted. If a substantial number of seemingly identical bearings are run under carefully controlled test conditions a wide variation in individual lives will be found.

If the individual failures are ranked in the order of increasing life and plotted, a definite pattern is found. Figure 13.43 shows such a curve obtained from extensive laboratory tests of deep-groove radial ball bearings under radial load. Approximately the same curve is obtained for all types of rolling-contact bearings under normal test conditions.

Because of the wide dispersion in individual lives, it is necessary to treat the problem of rolling-contact fatigue life statistically. By this means one can establish the fatigue life of a bearing in a given application corresponding to an assigned probability of survival to that life. In practice, fatigue-life calculations presume a 90 per cent probability of survival.

Actually, at this time, very little is known about the precise mechanism of rolling-contact fatigue and the reason for such wide variations in bearing lives. It is known, however, that the basic problem is one of material rather than design or extreme dimensional precision. Intensive basic research is already initiated to enable a better understanding of the rolling-contact fatigue process.

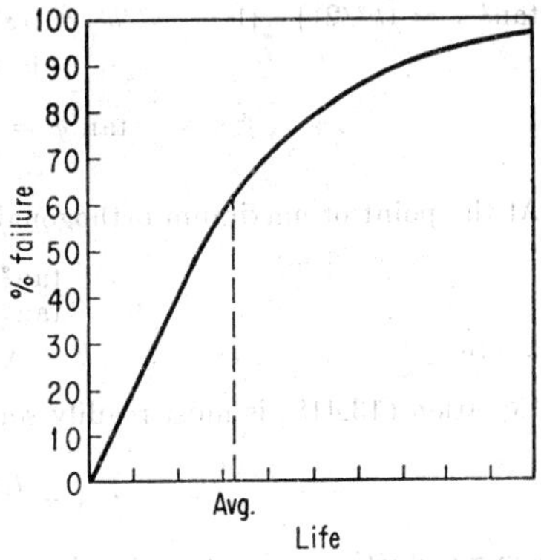

FIG. 13.43. Failure vs. life for deep-groove radial ball bearings under radial load.

In spite of the lack of knowledge concerning the rolling-contact fatigue mechanism it has been possible to relate the material stressings in conventional rolling-element bearings to fatigue life and to obtain surprisingly good agreement with laboratory tests. We are indebted to Dr. Arvid Palmgren and Prof. Gustav Lundberg of SKF (Sweden) who in 1947 applied Weibull's statistical theory of material strength to the rolling-contact problem. A brief discussion of their work follows.

(a) The Decisive Stress and Its Magnitude

Observations of failed ball and roller bearings have shown that the initial crack, which eventually leads to a spall, occurs below the rolling surface in well-lubricated bearings operating under normal conditions. Poor lubrication, excessive load, or surface defects can cause the initial crack to occur close to, or at, the surface. This discussion is restricted to the idealized, subsurface failure.

Fatigue cracks initiate at weak points in the material such as inclusions or in regions of high stress concentration resulting from defects in the crystal structure.

Under repeated stressings of high magnitude, local plastic distortions and slippages occur in the highly stressed region. If a weak point of the material occurs in a region of severe plastic deformation, the stress concentration at the defect is severe and a crack is eventually initiated. Under continued running, the crack propagates to the surface and a spall appears.

In the Lundberg-Palmgren treatment of the rolling-contact fatigue problem the orthogonal shear-stress amplitude $2|\tau_{ZY}|$ is taken as the stress mechanism causing the cumulative damage to the material. The orthogonal shear stress occurs on planes parallel to the rolling surface and attains a maximum some distance below the surface and at a depth where the greatest material damage has been observed. We are concerned only with the orthogonal shear stress lying in the central plane of the rolling path because it is here that the maximum values are found.

The stress system is referred to an orthogonal coordinate system XYZ, as in Fig. 13.16. The Z axis is normal to the pressure surface at its center. The Y axis is directed in the rolling direction and, in conventional rolling-element bearings, contains the semiminor axis of the pressure area. The X axis coincides with the longer dimension of the pressure area.

For an arbitrary point $(0,y,z)$ in the central rolling plane

$$\tau_{zy} = k\delta_0 \frac{\cos^2 \varphi \sin \varphi \sin \gamma}{\tan^2 \gamma + k^2 \cos^2 \varphi} \tag{13.413}$$

where k is the ratio b/a and φ and γ are found from

$$\tan^2 \gamma = (k^2/2)\{-[1 - (y^2/b^2) - (z^2/b^2)] + \sqrt{[1 - (y^2/b^2) - (z^2/b^2)]^2 + 4(z^2/b^2)}\} \tag{13.414}$$

$$\tan \varphi = \frac{(y/b) \tan \gamma}{(z/b)(k^2 + \tan^2 \gamma)^{1/2}} \tag{13.415}$$

At the point of maximum orthogonal shear-stress set

$$\tan^2 \varphi = t \tag{13.416}$$

$$\tan^2 \gamma = t - 1 \tag{13.417}$$

where

$$k = \sqrt{(t^2 - 1)(2t - 1)} \tag{13.418}$$

Equation (13.418) is most readily solved for t by iteration of

$$t_{i+1} = \frac{t_i^2(4t_i - 1) - 1 + k^2}{2t_i(3t_i - 1) - 2} \tag{13.419}$$

using a starting value $t_1 = 1$.

The magnitude of the maximum orthogonal shear stress is

$$\frac{\tau_0}{\delta_0} = \frac{\pm \sqrt{2t - 1}}{2t(t + 1)} \tag{13.420}$$

The locations of the points of maximum $|\tau_{zy}|$ are $(0, \pm y_0, z_0)$ where

$$\frac{z_0}{b} = \frac{1}{(t + 1)\sqrt{2t - 1}} \tag{13.421}$$

$$\frac{y_0}{b} = \frac{t}{t + 1}\left(\frac{2t + 1}{2t - 1}\right)^{1/2} \tag{13.422}$$

At $k = 2.2$ the orthogonal shear-stress amplitude $\pm|\tau_{zy}|$ is equal to the maximum shear stress in the plane at right angles to the rolling direction. For $k > 2.2$ it is less and the orthogonal shear-stress amplitude can no longer be considered decisive in producing cumulative damage of the material.

Since all conventional rolling-contact bearings are such that $0 \leq k < 2.2$ the orthogonal shear-stress amplitude in the rolling direction can be considered decisive in fatigue-life calculations.

(b) The Statistical Nature of the Fatigue Problem

Since fatigue cracks first appear in the neighborhood of weak points in the material the probability of a fatigue failure occurring anywhere in a stressed member must be related to the dispersion of weak points throughout the material volume.

If a uniform dispersion of weak points is assumed to exist throughout the material, the magnitude of the stressed volume of material can be taken as a measure of the number of weak points in the neighborhood of which microslips are apt to occur and at which cracks can initiate.

For treatment of rolling-contact fatigue problems, the magnitude of the stressed volume is taken as

$$V \sim az_0l \tag{13.423}$$

where l is the length of the rolling path.

A consideration in rolling-contact fatigue is that the stress-field gradient is such that, at points away from the greatest stress amplitude, the likelihood of failure is greatly reduced. Thus only the maximum stress amplitude need be considered, that is, the orthogonal shear-stress amplitude at the depth z_0.

In the Lundberg-Palmgren theory, the probability of survival of a volume of material stressed in rolling contact is shown to be dependent upon the maximum orthogonal shear-stress amplitude, the depth from the surface at which it occurs, and the magnitude of the stressed volume.

In this manner a general formula is found which is applicable to rolling-contact fatigue phenomena:

$$\log 1/S \sim \frac{\tau_0{}^c N^e al}{z_0{}^{h-1}} \tag{13.424}$$

where S is the probability of survival of the stressed volume of material to N millions of repetitions of the orthogonal shear-stress amplitude. These relations are continuous for $0 \le k \le \infty$. For $k > 1$ the major and minor axes of the pressure ellipse are inverted and the orthogonal shear stresses correspond to those in the plane of the semimajor axis.

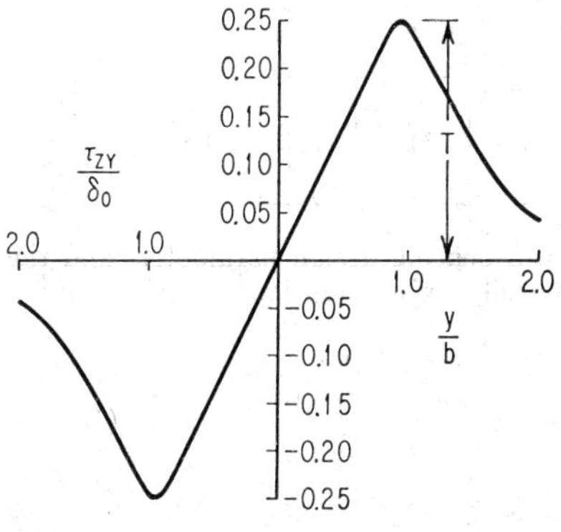

FIG. 13.44. Variation of τ_{zy} for a rectangular pressure area.

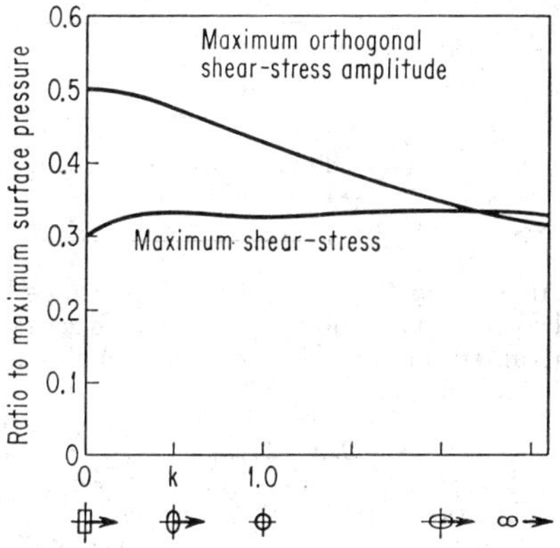

FIG. 13.45. Ratio of maximum surface pressure vs. K.

Figure 13.44 shows the variation in τ_{zy} at the depth where it is a maximum as a rolling element approaches and recedes from the point $(0,0,z_0)$ for a rectangular pressure area $k = 0$.

The maximum shear stress T_{zy}, which is one-half the difference in the principal stresses in the plane of the minor axis and which occurs along the Z axis, attains a maximum value which is considerably greater than $|\tau_0|$ and at a depth somewhat greater than z_0.

Figure 13.45 compares the maximum orthogonal shear-stress amplitude in the rolling direction with the maximum shear stress for various pressure-area shapes as defined by k.

For $0 \le k \le 1$ both shear stresses lie in the rolling direction.

For $k \ge 1$ the orthogonal shear-stress amplitude τ_{zy} is still that in the rolling direction. However, the maximum shear stress τ_{zy} is now a maximum in the plane at right angles to the rolling direction. The orthogonal shear stress in the latter plane does not reverse direction with the passage of a rolling element but oscillates between 0 and $|\tau_{zy}|$. Hence, in the latter plane, the maximum shear stress is considerably greater than the orthogonal shear-stress amplitude.

Equation (13.424) is applicable to point contact. It can be made to apply to line contact as well if the latter is considered the limiting case of point contact.

Line contact is normally associated with the contact of two cylinders along their generatrices. If a roller of finite length is assumed to have a large, but finite, radius of curvature in the plane of its axis and is pressed against, say, a flat surface which is parallel to the roller axis, point contact will initially exist. At some higher load the major axis of the pressure ellipse becomes equal to the roller length. If the load is further increased the extremities of the ellipse are nonexistent and the unit pressure increases more rapidly toward the ends of the roller than at the middle. When the theoretical length of a fully contained pressure ellipse is 1.5 times the roller length ($2a = 1.5l_e$) the pressure distribution along the roller length is substantially uniform. At this time the initial point contact has transformed into line contact.

For line contact the probability of survival can be written

$$\log 1/S \sim \tau_0{}^c N^e l_e l / z_0{}^{h-1} \tag{13.425}$$

Exhaustive endurance tests of ball and roller bearings have enabled the determination of the exponents c, e, and h for conventional bearing materials as follows:

	c	e	h
Point contact..............	$31/3$	$10/9$	$7/3$
Line contact...............	$31/3$	$9/8$	$7/3$

Introducing the functional relationships between bearing design parameters and the decisive stress conditions and with certain simplifying assumptions, expressions are obtained for the capacity of a race contact Q_c.

For point contact

$$Q_{c_p} = A_p \lambda_p \left(\frac{2f}{2f-1}\right)^{0.41} \frac{(1 \mp d \cos\beta/E)^{1.39}}{n^{1/3}} \left(\frac{d}{E}\right)^{0.3} d^{1.8} \tag{13.426}$$

For line contact

$$Q_{cl} = A_l \lambda_R \frac{(1 \mp d \cos\beta/E)^{29/27}}{n^{1/4}} \left(\frac{d}{E}\right)^{2/9} l_e{}^{7/9} d^{29/27} \tag{13.427}$$

The upper signs refer to an inner-ring contact and the lower to an outer.

For conventional materials $A_p = 5{,}667$ and $A_l = 41{,}624$.

λ_p is a factor to account for the reduction in fatigue life due to contact-area spin. At the present time this effect has not been completely evaluated but λ_p appears to have the following form:

$$\lambda_p = 1 - C_1(\omega_S/\omega_R)^{c_2} \tag{13.428}$$

where ω_S/ω_R is the spin/roll ratio of the race contact in question in accordance with Art. 13.5.

Until the effect of the spin/roll ratio can be evaluated take

$$\lambda_p = 1 - \tfrac{1}{3} \sin\beta \tag{13.429}$$

In the case of roller bearings λ_l is a factor to account for stress concentration at the roller's ends and in exactly centered loads on the rollers.

Acceptable values for λ_l are

$$\lambda_l = 0.45 \text{ for uncrowned cylindrical rollers}$$
$$\lambda_l = 0.61 \text{ for crowned rollers}$$

Q_c is the rolling-element load for the basic dynamic capacity of the raceway. It is that load which, if applied equally at each rolling-element passage, would yield one million revolutions of the ring considered, with 90 per cent probability of survival.

The fatigue life of a rolling-element bearing is a nonlinear function of the load. For point contact, life varies inversely as the cube of the load. For line contact, life varies inversely as the fourth power of the load.

Only with centric thrust loading is the rolling-element load constant. With all other types of bearing load, the rolling-element load varies with the rolling element's position. It is therefore necessary to find an equivalent rolling-element load which, if applied equally at each rolling-element passage, would result in the same raceway life as the actual varying loading.

For a raceway which rotates with respect to an externally applied load the stressing of a particular point varies cyclically with time and the equivalent rolling-element load is, for point contact,

$$P_E = \left[(1/2\pi) \int_0^{2\pi} P_q^3 \, d\varphi \right]^{1/3} \approx \left[(1/n) \sum_{q=1}^{n} P_q^3 \right]^{1/3} \qquad (13.430)$$

and, for line contact,

$$P_E = \left[(1/2\pi) \int_0^{2\pi} P_q^4 \, d\varphi \right]^{1/4} \approx \left[(1/n) \sum_{q=1}^{n} P_q^4 \right]^{1/4} \qquad (13.431)$$

A point on a raceway which is stationary with respect to an externally applied load is subjected to the same stressing at every rolling-element passage. The equivalent rolling-element load is derived from the fact that the probability of survival of the raceway as a whole is the product of the probabilities of survival of its parts.

For point contact

$$P_E = \left[(1/2\pi) \int_0^{2\pi} P_q^{10/3} \, d\varphi \right]^{3/10} \approx \left[(1/n) \sum_{q=1}^{n} P_q^{10/3} \right]^{3/10} \qquad (13.432)$$

and for line contact

$$P_E = \left[(1/n) \int_0^{2\pi} P_q^{9/2} \, d\varphi \right]^{2/9} \approx \left[(1/n) \sum_{q=1}^{n} P_q^{9/2} \right]^{2/9} \qquad (13.433)$$

The finite sums in Eqs. (13.430) and (13.433) may be used to evaluate raceway life when the rolling-element load distribution has been determined by the methods outlined in Art. 13.6.

In the particular case of the three-degree-of-freedom solution (Art. 13.3) the rolling-element load distribution is symmetric in the XZ plane and one obtains the following relations.

For a raceway which rotates with respect to the externally applied load and with point contact

$$P_E = K|B|^{3/2} \left\{ (1/\pi) \int_0^{\varphi'} [A/|B| + (B/|B|) \cos \varphi]^{9/2} \, d\varphi \right\}^{1/3} \qquad (13.434)$$

or

$$P_E = K|B|^{3/2} J_6 \qquad (13.435)$$

With line contact

$$P_E = K|B|^{10/9} \left\{ (1/\pi) \int_0^{\varphi'} [A/|B| + (B/|B|) \cos \varphi]^{40/9} \, d\varphi \right\}^{1/4} \qquad (13.436)$$

or

$$P_E = K|B|^{10/9} J_6$$

For a raceway which is stationary with respect to the externally applied load and with point contact

$$P_E = K|B|^{3/2} \left\{ (1/\pi) \int_0^{\varphi'} [A/|B| + (B/|B|) \cos \varphi]^5 \, d\varphi \right\}^{3/10} \qquad (13.437)$$

or

$$P_E = K|B|^{3/2} J_7 \qquad (13.438)$$

With line contact

$$P_E = K|B|^{10/9} \left\{(1/\pi) \int_0^{\varphi'} [A/|B| + (B/|B|) \cos \varphi]^5 \, d\varphi\right\}^{2/9} \qquad (13.439)$$

or

$$P_E = K|B|^{10/9} J_7 \qquad (13.440)$$

values of J_6 and J_7 are given in Fig. 13.46 for point and line contact.

The life of a raceway in hours for a 90 per cent probability of survival is

$$L = (Q_c/P_E)^p (10^6/60|\Omega - \Omega_E|) \qquad (13.441)$$

where p is 3 for point contact and 4 for line contact. Q_c, P_E, and Ω are chosen as appropriate for the race concerned.

Having determined the life of the individual raceways, the life of the complete bearing is found from

$$L_B = [(1/L_O)^e + (1/L_I)^e]^{-1/e} \qquad (13.442)$$

where e is $10/9$ for point contact and $9/8$ for line contact. For bearings which have point contact at one race and line contact at the other take e as 1.12. If two or more bearings which operate simultaneously are to be considered as a unit, as, for example, double-row bearings, the life of the unit is

$$L = \left[\sum_1^i (1/L_{B_i})^e\right]^{-1/e} \qquad (13.443)$$

(c) The AFBMA Method

Prior to 1950, each bearing manufacturer had his own method for establishing load ratings for his product. In many instances such ratings were inadequately supported by technical or experimental data and there was always the commercially expedient temptation to show higher ratings than a competitor. Load-rating formulas were closely guarded trade secrets and there was little interchange of technical information. The resulting confusion was certainly undesirable from the bearing users' standpoint.

In the mid-1940s the Anti-Friction Bearing Manufacturers' Association (AFBMA) undertook the study of the load-rating problem in an attempt to find a method of rating rolling-element bearings that would be both accurate and equitable. A number of theories were examined and discarded.

In 1947, Lundberg and Palmgren published the results of their studies of the statistical nature of the rolling-element-bearing fatigue problem, and this work came under the scrutiny of the AFBMA technical committees.

After 3 years of intensive study and minor modifications to meet the needs of American industry the AFBMA Standard Method of Evaluating the Load Ratings of Annular Ball Bearings was established substantially in agreement with the original theory of Lundberg and Palmgren. Minor revisions have been made from time to time to keep abreast of the advancing technology.

The validity of the constants and exponents used to relate bearing design parameters to service life was the subject of an exhaustive study by the Bureau of Standards of all the available test data regarding single-row radial ball bearings run under radial load in controlled laboratory conditions. This was a cooperative effort on the part of the American ball-bearing industry which made all pertinent test data available. The Bureau of Standards' study confirmed the validity of the rating method.

The standardization of ball-bearing ratings was followed at a later date by a similar standard applicable to roller bearings.

Both the ball-bearing and the roller-bearing rating methods established by AFBMA are today incorporated into the standards of the American Standards Association.

Both standards discuss the static as well as the dynamic carrying capacities of the respective bearing types.

In Art. 13.6 the complexity of the internal rolling-element load-distribution problem has been clearly outlined To arrive at the true load distribution with consideration of ball-bearing contact-angle change is a task that requires the services of a high-speed computer. Even the simplified solution is time-consuming for hand computation.

In order for a standard to be widely applicable it must be fairly simple and concise and not prohibitive for hand-computation means. Therefore, certain simplifying assumptions and approximations, mainly concerning the relationship between the external bearing load and the internal load distribution, had to be made to render the method practical for hand calculation.

The simplifying assumptions introduced into the more complete theory are

1. The inner and outer raceways of a bearing are constrained to parallel motion, that is, to thrust and radial displacements only.
2. The contact angle of a ball bearing changes with thrust load but is unaffected by radial load.
3. A combined thrust and radial load can be replaced, for purposes of fatigue-life calculations, by an equivalent radial load which is a linear combination of the thrust and radial-load components.
4. Body forces due to the rolling elements' motions are neglected.
5. Initial conditions, such as preload displacements, are not considered.

In spite of these necessary assumptions, the AFBMA methods provide satisfactory solutions to the majority of bearing-application problems. The more complex problems require the analytical treatments described previously.

Since most bearing manufacturers now calculate their catalogue ratings in accordance with the AFBMA standards, the confusion previously resulting from the use of different, and sometimes questionable, rating methods is largely eliminated.

Both AFBMA ball-bearing and roller-bearing rating methods are reproduced here verbatim, except that the nomenclature has been changed to agree with that used here and certain descriptive material has been eliminated.

Each method assumes that the thrust and radial-load components acting on a bearing have been established.

Method of Evaluating Dynamic Load Ratings for Radial Ball Bearings. The "life" of an individual ball bearing is defined as the number of revolutions (or hours at some given constant speed) which the bearing runs before the first evidence of fatigue develops in the material of either ring or of any of the rolling elements.

The "rating life" of a group of apparently identical ball bearings is defined as the number of revolutions (or hours at some given constant speed) that 90 per cent of a group of bearings will complete or exceed before the first evidence of fatigue develops. As presently determined, the life which 50 per cent of the group of ball bearings will complete or exceed is approximately five times this rating life.

Calculation of Basic Load Rating, Rating Life, and Equivalent Load. The "basic load rating" is that constant stationary radial load which a group of apparently identical ball bearings with stationary outer ring can endure for a rating life of one million revolutions of the inner ring. In single-row angular-contact ball bearings, the basic load rating relates to the radial component of the load, which results in a purely radial displacement of the bearing rings in relation to each other.

Load ratings, if given for specific speeds, are to be based on a rating life of 500 hr.

The "equivalent load" is defined as that constant stationary radial load which, if applied to a bearing with rotating inner ring and stationary outer ring, would give the same life as that which the bearing will attain under the actual conditions of load and rotation.

The magnitude of the basic load rating C, for radial and angular-contact ball bearings, except filling-slot bearings, with balls not larger than 25.4 mm or 1 in. in diameter, is

$$C = f_c(i \cos \beta)^{0.7} n^{2/3} d^{1.8}$$

With balls larger than 25.4 mm in diameter when kg and mm units are used,

$$C = 3.647 f_c(i \cos \beta)^{0.7} n^{2/3} d^{1.4}$$

With balls larger than 1 in. in diameter when pound and inch units are used,

$$C = f_c(i \cos \beta)^{0.7} n^{2/3} d^{1.4}$$

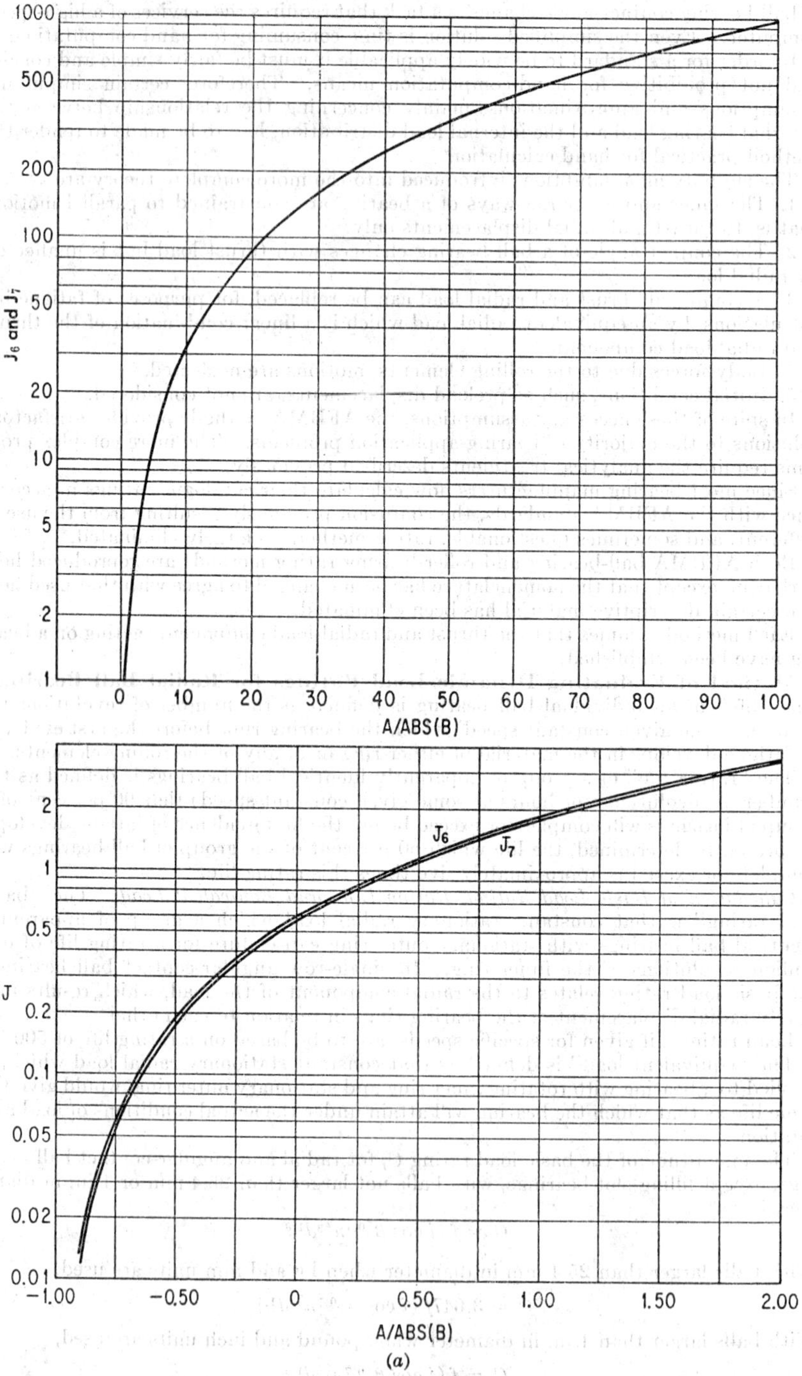

FIG. 13.46. Integrals for life

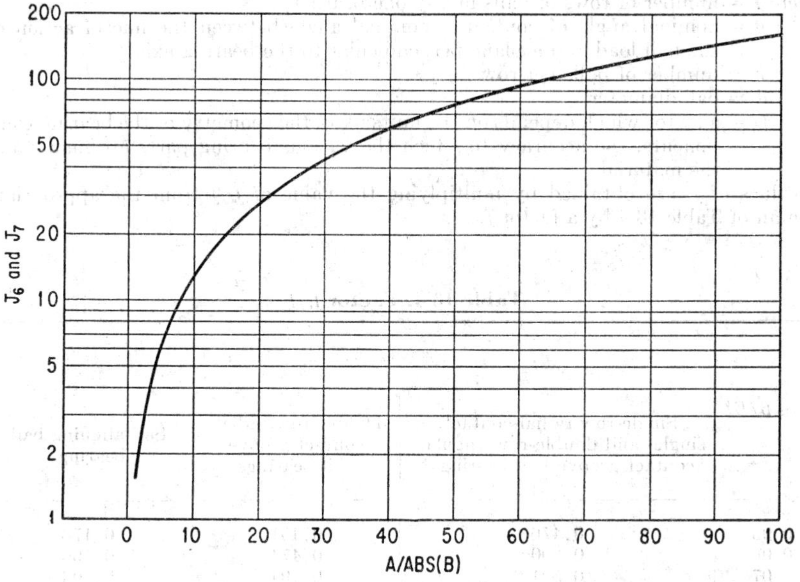

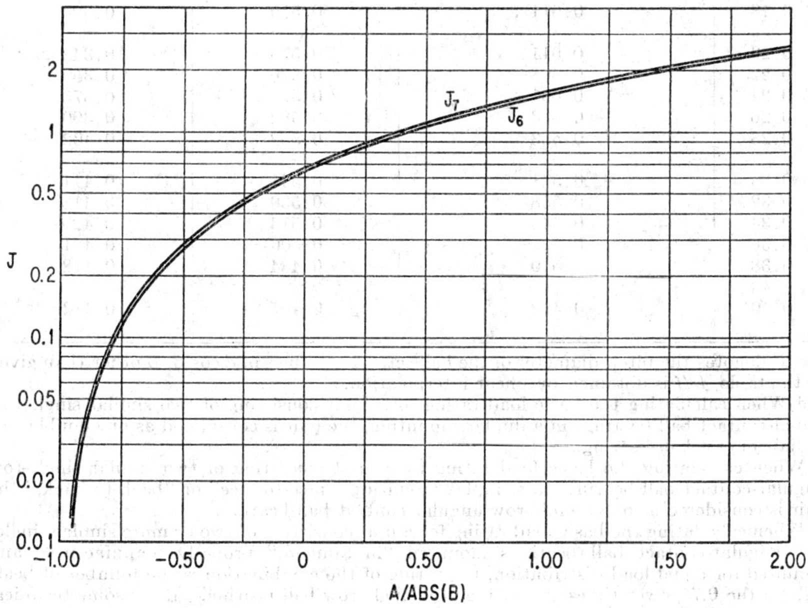

(b)

calculation. (a) Point contact. (b) Line contact.

where i = number of rows of balls in any one bearing
β = nominal angle of contact = nominal angle between the line of action of the ball load and a plane perpendicular to the bearing axis
n = number of balls per row
d = ball diameter
f_c = a factor which depends on the units used, the geometry of the bearing components, the accuracy to which the various bearing parts are made, and the material

Values of f_c are obtained by multiplying the value of f_c/f from the appropriate column of Table 13.4 by a factor f.

Table 13.4. Factor f_c/f

$d \cos \beta/E$*	f_c/f		
	Single-row radial-contact, single- and double-row angular-contact groove ball bearings†	Double-row radial-contact groove ball bearings	Self-aligning ball bearings
0.05	0.476	0.451	0.176
0.06	0.500	0.474	0.190
0.07	0.521	0.494	0.203
0.08	0.539	0.511	0.215
0.09	0.554	0.524	0.227
0.10	0.566	0.537	0.238
0.12	0.586	0.555	0.261
0.14	0.600	0.568	0.282
0.16	0.608	0.576	0.303
0.18	0.611	0.579	0.323
0.20	0.611	0.579	0.342
0.22	0.608	0.576	0.359
0.24	0.601	0.570	0.375
0.26	0.593	0.562	0.390
0.28	0.583	0.552	0.402
0.30	0.571	0.541	0.411
0.32	0.558	0.530	0.418
0.34	0.543	0.515	0.420
0.36	0.527	0.500	0.421
0.38	0.510	0.484	0.418
0.40	0.492	0.467	0.412

* E denotes the pitch diameter of the ball set. For values of $d \cos \beta/E$ other than given in the table, f_c/f is obtained by linear interpolation.

† When calculating the basic load rating for a unit consisting of two similar single-row radial-contact ball bearings in a duplex mounting, the pair is considered as one double-row radial-contact ball bearing.

When calculating the basic load rating for a unit consisting of two similar single-row angular-contact ball bearings in a duplex mounting, "face-to-face" or "back-to-back," the pair is considered as one double-row angular-contact ball bearing.

When calculating the basic load rating for a unit consisting of two or more similar single-row angular-contact ball bearings mounted "in tandem," properly manufactured and mounted for equal load distribution, the rating of the combination is the number of bearings to the 0.7 power times the rating of a single-row ball bearing. If for some technical reason the unit may be treated as a number of individually interchangeable single-row bearings, this footnote does not apply.

A recommended value of the factor f based on current tests of ball bearings of good-quality hardened ball-bearing steel is

$$f = 10 \text{ when kg and mm units are used}$$
$$f = 7{,}450 \text{ when pound and inch units are used}$$

The approximate magnitude of the rating life L, for ball bearings, except filling-slot bearings, is

$$L = (C/R_E)^3 \qquad \text{million revolutions}$$

where R_E = equivalent load

The magnitude of the equivalent load R_E, for radial and angular-contact ball bearings of conventional types, except filling-slot bearings, under combined constant radial and constant thrust loads, is

$$R_E = XVF_r + YF_a$$

where X = a radial factor
V = a rotation factor
Y = a thrust factor
F_r = radial load
F_a = thrust load

Values of X, V, and Y are given in Table 13.5. The factor V, because of lack of sufficient experimental evidence, is used as a matter of precaution.

This standard is limited to bearings whose ring raceways have a cross-sectional radius not larger than:

In deep-groove and angular-contact ball-bearing inner rings, 52 per cent of the ball diameter

In deep-groove and angular-contact ball-bearing outer rings, 53 per cent of the ball diameter

In self-aligning ball-bearing inner rings, 53 per cent of the ball diameter

The basic load rating is not increased by the use of smaller groove radii but is reduced by the use of larger radii than those given above.

Method of Evaluating Dynamic Load Ratings for Thrust Ball Bearings. The "life" of an individual thrust ball bearing is defined as the number of revolutions (or hours at some given constant speed) which the bearing runs before the first evidence of fatigue develops in the material of either washer or of any of the rolling elements.

The "rating life" of a group of apparently identical thrust ball bearings is defined as the number of revolutions (or hours at some given constant speed) that 90 per cent of a group of bearings will complete or exceed before the first evidence of fatigue develops.

The "basic load rating" is that constant, central thrust load which a group of apparently identical thrust ball bearings can endure for a rating life of one million revolutions of one of the bearing washers.

Load ratings, if given for specific speeds, are to be based on a rating life of 500 hr.

The "equivalent thrust load" is defined as that constant central, purely axial load which, if applied to a bearing with rotating shaft washer and stationary housing washer, would give the same life as that which the bearing will attain under the actual conditions of load and rotation.

Calculation of Basic Load Rating, Rating Life, and Equivalent Thrust Load. The magnitude of the basic load rating C_a for single-row single- and double-direction thrust ball bearings with balls not larger than 25.4 mm or 1 in. in diameter, is

For $\beta = 90°$ $\qquad C_a = f_c n^{2/3} d^{1.8}$

For $\beta \neq 90°$ $\qquad C_a = f_c \cos^{0.7} \beta \tan \beta n^{2/3} d^{1.8}$

with balls larger than 25.4 mm in diameter when kg and mm units are used:

For $\beta = 90°$ $\qquad C_a = 3.647 f_c n^{2/3} d^{1.4}$

For $\beta \neq 90°$ $\qquad C_a = 3.647 f_c \cos^{0.7} \beta \tan \beta n^{2/3} d^{1.4}$

Table 13.5. Factors X, V, and Y*

Bearing type		In relation to the load the inner ring is: Rotating	Stationary	Single-row bearings† $F_a/VF_r > e$		Double-row bearings‡ $F_a/VF_r \leq e$		Double-row bearings‡ $F_a/VF_r > e$		e
		V	V	X	Y	X	Y	X	Y	
Radial-contact groove ball bearings	$\frac{F_a}{ind^2}$ Units lb, in.									
	25				2.30				2.30	0.19
	50				1.99				1.99	0.22
	100				1.71				1.71	0.26
	150	1	1.2	0.56	1.55	1	0	0.56	1.55	0.28
	200				1.45				1.45	0.30
	300				1.31				1.31	0.34
	500				1.15				1.15	0.38
	750				1.04				1.04	0.42
	1,000				1.00				1.00	0.44
Angular-contact groove ball bearings with contact angle: 5°	$\frac{F_a}{nd^3}$ Units lb, in.			For this type use the X, Y, and e values applicable to single-row radial-contact bearings						
	25						2.78		3.74	0.23
	50						2.40		3.23	0.26
	100						2.07		2.78	0.30
	150	1	1.2			1	1.87	0.78	2.52	0.34
	200						1.75		2.36	0.36
	300						1.58		2.13	0.40
	500						1.39		1.87	0.45
	750						1.26		1.69	0.50
	1,000						1.21		1.63	0.52
	25				1.88		2.18		3.06	0.29
	50				1.71		1.98		2.78	0.32
	100				1.52		1.76		2.47	0.36
	150				1.41		1.63		2.29	0.38
10°	200	1	1.2	0.46	1.34	1	1.55	0.75	2.18	0.40
	300				1.23		1.42		2.00	0.44
	500				1.10		1.27		1.79	0.49
	750				1.01		1.17		1.64	0.54
	1,000				1.00		1.16		1.63	0.54
	25				1.47		1.65		2.39	0.38
	50				1.40		1.57		2.28	0.40
	100				1.30		1.46		2.11	0.43
15°	150				1.23		1.38		2.00	0.46
	200	1	1.2	0.44	1.19	1	1.34	0.72	1.93	0.47
	300				1.12		1.26		1.82	0.50
	500				1.02		1.14		1.66	0.55
	750				1.00		1.12		1.63	0.56
	1,000				1.00		1.12		1.63	0.56
20°		1	1.2	0.43	1.00	1	1.09	0.70	1.63	0.57
25°		1	1.2	0.41	0.87	1	0.92	0.67	1.41	0.68
30°		1	1.2	0.39	0.76	1	0.78	0.63	1.24	0.80
35°		1	1.2	0.37	0.66	1	0.66	0.60	1.07	0.95
40°		1	1.2	0.35	0.57	1	0.55	0.57	0.93	1.14
Self-aligning ball bearings		1	1	0.40	0.4 cot β	1	0.42 cot β	0.65	0.65 cot β	1.5 tan β

For footnotes to table see top of next page.

Table 13.5. Factors X, V, and Y (Continued)

* Values of X, Y, and e for a load or contact angle other than shown in Table 13.5 are obtained by linear interpolation.

† For single-row bearings, when $F_a/VF_r \leq e$, use $X = 1$ and $Y = 0$. Two similar single-row angular-contact ball bearings mounted "face-to-face" or "back-to-back" are considered as one double-row angular-contact bearing.

For two or more similar single-row ball bearings mounted "in tandem," use the values of X, Y, and e which apply to one single-row ball bearing. When β is smaller than 20°, F_r and F_a are not the total loads but the loads per single-row bearing. i refers to one single-row bearing.

‡ Double-row bearings are presumed to be symmetrical.

---

with balls larger than 1 in. in diameter when pound and inch units are used:

For $\beta = 90°$ $\qquad C_a = f_c n^{2/3} d^{1.4}$

For $\beta \neq 90°$ $\qquad C_a = f_c \cos^{0.7}\beta \tan\beta n^{2/3} d^{1.4}$

where β = nominal angle of contact = nominal angle between the line of action of the ball load and a plane perpendicular to the bearing axis

n = number of balls in a single-row single-direction bearing

d = ball diameter

f_c = a factor which depends on the units used, the geometry of the bearing components, the accuracy to which the various bearing parts are made, and the material

Values of f_c are obtained by multiplying the value of f_c/f from Table 13.6 by a factor f.

Table 13.6. Factor f_c/f

d/E*	f_c/f, $\beta = 90°$	$\frac{d \cos \beta}{E}$*	f_c/f, $\beta = 45°$	f_c/f, $\beta = 60°$	f_c/f, $\beta = 75°$
0.01	0.374	0.01	0.429	0.399	0.381
0.02	0.461	0.02	0.527	0.490	0.468
0.03	0.521	0.03	0.594	0.553	0.527
0.04	0.568	0.04	0.645	0.600	0.572
0.05	0.607	0.05	0.686	0.639	0.609
0.06	0.641	0.06	0.720	0.670	0.639
0.07	0.671	0.07	0.749	0.697	0.665
0.08	0.699	0.08	0.774	0.720	0.687
0.09	0.724	0.09	0.795	0.740	0.705
0.10	0.747	0.10	0.812	0.756	0.721
0.12	0.789	0.12	0.840	0.782	
0.14	0.827	0.14	0.858	0.798	
0.16	0.860	0.16	0.868	0.808	
0.18	0.891	0.18	0.872	0.812	
0.20	0.920	0.20	0.871	0.811	
0.22	0.947	0.22	0.866		
0.24	0.972	0.24	0.856		
0.26	0.995	0.26	0.844		
0.28	1.02	0.28	0.829		
0.30	1.04	0.30	0.811		
0.32	1.06				
0.34	1.08				

* E = pitch diameter of the ball set.

A recommended value of the factor f based on current tests of ball bearings of good-quality hardened ball-bearing steel is

f = 10 when kg and mm units are used
f = 7,450 when pound and inch units are used

The magnitude of the basic load rating C_a for thrust ball bearings with two or more rows of similar balls carrying load in the same direction, is

$$C_a = (n_1 + n_2 + \cdots + n_i)[(n_1/C_{a_1})^{10/3} + (n_2/C_{a_2})^{10/3} + \cdots + (n_i/C_{a_i})^{10/3}]^{-3/10}$$

where $n_1, n_2, \ldots, n_i$ = number of balls in respective rows of a one-direction multirow bearing
$C_{a_1}, C_{a_2}, \ldots, C_{a_i}$ = basic load rating per row of a one-direction multirow thrust ball bearing, each calculated as a single-row bearing with $n_1, n_2, \ldots, n_i$ balls, respectively

The approximate magnitude of the rating life L is

$$L = (C_a/T_e)^3 \qquad \text{million revolutions}$$

where T_e = equivalent thrust load

The magnitude of the equivalent thrust load T_e for thrust ball bearings with $\beta \neq 90°$ under combined constant thrust and constant radial loads is

$$T_e = XF_r + YF_a$$

X = a radial factor
Y = a thrust factor
F_r = radial load
F_a = thrust load

Values of X and Y are given in Table 13.7.

Table 13.7. Factors X and Y

Bearing type	Single-direction bearings		Double-direction bearings*				e
	$F_a/F_r > e$		$F_a/F_r \leq e$		$F_a/F_r > e$		
	X	Y	X	Y	X	Y	
Thrust ball bearings† with contact angle:							
$\beta = 45°$	0.66	1	1.18	0.59	0.66	1	1.25
$\beta = 60°$	0.92	1	1.90	0.54	0.92	1	2.17
$\beta = 75°$	1.66	1	3.89	0.52	1.66	1	4.67

* Double-direction bearings are presumed to be symmetrical.
† For $\beta = 90°$, $F_r = 0$ and $Y = 1$.

This standard is limited to bearings whose washer raceways have a cross-sectional radius not larger than 54 per cent of the ball diameter.

The basic load rating is not increased by the use of smaller groove radii but is reduced by the use of larger radii than that given above.

Method of Evaluating Static-load Ratings for Radial Ball Bearings. The "static load" is defined as a load acting on a nonrotating bearing.

Permanent deformations appear in balls and raceways under static load of moderate magnitude and increase gradually with increasing load. The permissible static load is therefore dependent upon the permissible magnitude of permanent deformation.

Experience shows that a total permanent deformation of 0.0001 of the ball diameter, occurring at the most heavily loaded ball and race contact, can be tolerated in most bearing applications without impairment of bearing operation.

In certain applications where subsequent rotation of the bearing is slow and where smoothness and friction requirements are not too exacting, a much greater total permanent deformation can be permitted. Likewise, where extreme smoothness is required or friction requirements are critical, less total permanent deformation may be tolerated.

For purposes of establishing comparative ratings, the "basic static-load rating" is therefore defined as that static radial load which corresponds to a total permanent deformation of ball and race at the most heavily stressed contact of 0.0001 of the ball diameter.

In single-row angular-contact ball bearings the basic static-load rating relates to the radial component of that load which causes a purely radial displacement of the bearing rings in relation to each other.

The "static equivalent load" is defined as that static radial load which, if applied, would cause the same total permanent deformation at the most heavily stressed ball and race contact as that which occurs under the actual condition of loading.

Calculation of Basic Static Load Rating and Static Equivalent Load. The magnitude of the basic static-load rating C_0 is

$$C_0 = f_0 i n d^2 \cos \beta$$

where i = number of rows of balls in any one bearing
β = nominal angle of contact = nominal angle between the line of action of the ball load and a plane perpendicular to the bearing axis
n = number of balls per row
d = ball diameter

Values of the factor f_0 for different kinds of bearings as commonly designed and manufactured and made of hardened steel are given in Table 13.8.

Table 13.8. Factor f_0

Bearing type	f_0	
	Units kg, mm	Units lb, in.
Self-aligning ball bearings	0.34	484
Radial- and angular-contact groove ball bearings	1.25	1,780

The magnitude of the static equivalent load R_0 for radial bearings under combined radial and thrust loads is the greater of

$$R_0 = X_0 F_r + Y_0 F_a$$
$$R_0 = F_r$$

where X_0 = a radial factor
Y_0 = a thrust factor
F_r = radial load
F_a = thrust load

Values of X_0 and Y_0 are given in Table 13.9.

This standard is limited to bearings whose ring raceways have a cross-sectional radius not larger than:

In deep-groove and angular-contact ball-bearing inner rings, 52 per cent of ball diameter

In deep-groove and angular-contact ball-bearing outer rings, 53 per cent of ball diameter

In self-aligning ball-bearing inner rings, 53 per cent of ball diameter

Table 13.9. Factors X_0 and Y_0

Bearing type	Single-row bearings†		Double-row bearings‡	
	X_0	Y_0	X_0	Y_0
Radial-contact groove ball bearings*,†	0.6	0.5	0.6	0.5
Angular-contact groove ball bearings:§				
$\beta = 20°$	0.5	0.42	1	0.84
$\beta = 25°$	0.5	0.38	1	0.76
$\beta = 30°$	0.5	0.33	1	0.66
$\beta = 35°$	0.5	0.29	1	0.58
$\beta = 40°$	0.5	0.26	1	0.52
Self-aligning ball bearings	0.5	0.22 cot β	1	0.44 cot β

* Permissible maximum value of F_a/C_0 depends on the bearing design (groove depth and internal clearance).

† R_0 is always $\geq F_r$.

‡ Double-row bearings are presumed to be symmetrical.

§ For two similar single-row angular-contact ball bearings mounted "face-to-face" or "back-to-back," use values of X_0 and Y_0 which apply to a double-row angular-contact ball bearing. For two or more similar single-row angular-contact ball bearings mounted "in tandem," use the values of X_0 and Y_0 which apply to a single-row angular-contact ball bearing.

The basic static-load rating is reduced by the use of larger radii than those given above.

Method of Evaluating Static Load Ratings for Thrust Ball Bearings. The "static load" is defined as a load acting on a nonrotating bearing.

Permanent deformations appear in balls and raceways under static load of moderate magnitude and increase gradually with increasing load. The permissible static load is therefore dependent upon the permissible magnitude of permanent deformation.

Experience shows that a total permanent deformation of 0.0001 of the ball diameter, occurring at the most heavily loaded ball and race contact, can be tolerated in most bearing applications without impairment of bearing operation.

In certain applications where subsequent rotation of the bearing is slow and where smoothness and friction requirements are not too exacting, a much greater total permanent deformation can be permitted. Likewise, where extreme smoothness is required or friction requirements are critical, less total permanent deformation may be tolerated.

For purposes of establishing comparative ratings, the "basic static-load rating" is therefore defined as that static central thrust load which corresponds to a total permanent deformation of ball and race at the most heavily stressed contact of 0.001 of the ball diameter.

The "static equivalent load" is defined as that static, central, purely axial load which, if applied, would cause the same total permanent deformation at the most heavily stressed ball and race contact as that which occurs under the actual condition of loading.

Calculation of Basic Static Load Rating and Static Equivalent Load. The magnitude of the basic static-load rating C_{0_a} is:

$$C_{0_a} = f_{0_a} n d^2 \sin \beta$$

where β = nominal angle of contact = nominal angle between the line of action of the ball load and a plane perpendicular to the bearing axis

n = number of balls carrying thrust in one direction

d = ball diameter

f_{0_a} = 5 when kg and mm units are used

f_{0_a} = 7,100 when pound and inch units are used.

These values of f_{0a} are valid for bearings as commonly designed and manufactured and made of hardened steel.

The magnitude of the static equivalent load T_0, for thrust bearings with contact angle $\beta \neq 90°$ under the combined radial and thrust loads, is

$$T_0 = F_a + 2.3F_r \tan \beta$$

where F_a = thrust load
F_r = radial load

The formula is valid for $F_r \leq 0.44F_a \cot \beta$.

This standard is limited to bearings whose washer raceways have a cross-sectional radius not larger than 54 per cent of the ball diameter.

The basic static-load rating is reduced by the use of larger radii than given above.

Method of Evaluating Dynamic Load Ratings of Radial Roller Bearings. The "life" of an individual roller bearing is defined as the number of revolutions (or hours at some given constant speed) which the bearing runs before the first evidence of fatigue develops in the material of either ring or of any of the rolling elements.

The "rating life" of a group of apparently identical roller bearings is defined as the number of revolutions (or hours at some given constant speed) that 90 per cent of a group of bearings will complete or exceed before the first evidence of fatigue develops.

The "basic load rating" is that constant stationary radial load which a group of apparently identical roller bearings with stationary outer ring can endure for a rating life of one million revolutions of the inner ring. In single-row angular-contact roller bearings the basic load rating relates to the radial component of the load which causes a purely radial displacement of the bearing rings in relation to each other.

Load ratings, if given for specific speeds, are to be based on a rating life of 500 hr.

The "equivalent load" is defined as that constant stationary radial load which, if applied to a bearing with rotating inner ring and stationary outer ring, would give the same life as that which the bearing will attain under the actual conditions of load and rotation.

Calculation of Basic Load Rating, Rating Life, and Equivalent Load. The magnitude of the basic load rating C, for radial roller bearings, is

$$C = f_c n^{3/4} d^{29/27} (il_e \cos \beta)^{7/9}$$

where i = number of rows of rollers in any one bearing
β = angle of contact = angle between the line of action of the roller resultant load and a plane perpendicular to the bearing axis
n = number of rollers per row
d = roller diameter (mean diameter of tapered rollers)
l_e = effective length of contact between one roller and that ring where the contact is the shortest (overall roller length minus roller chamfers, or minus grinding undercuts)
f_c = a factor which depends on the units used, the exact geometrical shape of the load-carrying surfaces of the rollers and rings, the accuracy to which the various bearing parts are made, and the material

Values of f_c are obtained by multiplying f_c/f by a factor f.

A recommended value of the factor f based on current tests of roller bearings of good quality hardened roller-bearing steel is

$$f = 56.2 \text{ when kg and mm units are used}$$
$$f = 49{,}500 \text{ when pound and inch units are used}$$

Roller bearings vary considerably in design and execution. Small differences in relative shape of contacting surfaces may account for distinct differences in load-carrying capacity. It is therefore not possible to cover all design variations adequately.

Generally a bearing of good quality and made by a reputable bearing manufacturer must be expected to have a capacity lower than that obtained by using a value of f_c/f taken from column a in Table 13.10, if, under load, a local stress concentration is present in some part of the roller contact. This may be the case if the rollers are not accurately guided. Stress concentration is also present at the end of a line contact or in the center of a point contact even though the rollers are well guided. On the other hand, the capacity may be expected to be higher than that obtained by using a value of f_c/f taken from column a in Table 13.10 if even stress distribution over the whole roller length is automatically assured. For no bearing type or execution will the factors f_c/f be higher than those taken from column k in Table 13.10.

This general evaluation of f_c/f is applied to some specific bearing designs listed in Table 13.11. The last column in this table indicates from which column in Table 13.10 the value of f_c/f should be selected for the respective bearing execution.

Table 13.10. Values of f_c/f

$d \cos \beta/E$*	f_c/f					
	Column					
	j	h	b	a	g	k
0.01	0.061	0.069	0.077	0.083	0.089	0.095
0.02	0.072	0.081	0.090	0.097	0.104	0.111
0.03	0.078	0.088	0.099	0.106	0.113	0.120
0.04	0.083	0.094	0.105	0.113	0.120	0.128
0.05	0.087	0.099	0.110	0.118	0.126	0.134
0.06	0.091	0.102	0.114	0.123	0.131	0.139
0.07	0.093	0.105	0.118	0.126	0.135	0.144
0.08	0.096	0.108	0.121	0.130	0.138	0.147
0.09	0.098	0.110	0.123	0.132	0.141	0.150
0.10	0.099	0.112	0.125	0.134	0.143	0.152
0.12	0.102	0.115	0.128	0.138	0.147	0.156
0.14	0.103	0.117	0.130	0.140	0.149	0.159
0.16	0.104	0.118	0.131	0.141	0.151	0.161
0.18	0.105	0.118	0.132	0.142	0.151	0.161
0.20	0.105	0.118	0.132	0.142	0.151	0.161
0.22	0.104	0.117	0.131	0.141	0.150	0.160
0.24	0.103	0.116	0.130	0.140	0.149	0.159
0.26	0.102	0.115	0.128	0.138	0.147	0.156
0.28	0.100	0.113	0.126	0.136	0.145	0.154
0.30	0.099	0.112	0.124	0.134	0.143	0.152

* E = mean pitch diameter of the roller set.

The approximate magnitude of the rating life L is found from the formula

$$L = (C/R_E)^{10/3} \qquad \text{million revolutions}$$

where R_E = equivalent load

When the values for f_c/f listed in Table 13.10 are applied to non-self-aligning bearings with line contact or modified line contact at both rings, the C values obtained refer to mountings so designed and executed that uniform load distribution over the roller length is assured. If misalignment is present a reduction in the C value should be made before the rating life is estimated.

The magnitude of the equivalent load R_E is found for tapered and self-aligning radial roller bearings of conventional types under combined constant radial and constant thrust loads, from the formula

$$R_E = XVF_r + YF_a$$

Table 13.11. Selection of f_c/f for Various Bearing Designs

Item	Type of bearing	f_c/f selected from the following column in Table 13.10
1	Cylindrical roller bearings with line contact* between rollers and both rings, the rollers not accurately guided	*j*
2	Cylindrical roller bearings with line contact* between rollers and both rings, the rollers accurately guided by a machined cage or by uninterrupted guide surfaces at each end of the rollers, both of these surfaces associated with one of the bearing rings	*h*
3	Cylindrical roller bearings with line contact* between rollers and one ring and point contact† with the other, the rollers accurately guided by a machined cage or by uninterrupted guide surfaces at each end of the rollers, both of these surfaces associated with one of the bearing rings	*b*
4	Cylindrical roller bearings with modified‡ line contact between rollers and both rings, the rollers guided by a cage or by two snap rings, or one snap ring and one integral rib	*b*
5	Cylindrical roller bearings with modified line contact‡ between rollers and both rings, the rollers accurately guided by uninterrupted surfaces at each end of the rollers, both of these surfaces associated with one of the bearing rings (when the rollers are longer than 2.3 times their diameter, use column *b* in Table 13.10)	*k*
6	Tapered roller bearings with line contact* between rollers and both rings, the rollers accurately guided by one integral rib	*h*
7	Tapered roller bearings with line contact* between rollers and one ring with point contact† with the other, the rollers accurately guided by one integral rib	*b*
8	Tapered roller bearings with modified line contact‡ between rollers and both rings, the rollers accurately guided by one integral rib	*g*

* The term "line contact" refers to rollers and raceways so formed that under no load and when in good alignment they contact along the full length of their basic form.

† The term "point contact" refers to rollers and raceways so formed that under no load and when in good alignment they contact at a point located approximately at the middle of the rollers, and that under a bearing load of about one-quarter of the basic load rating C a contact ellipse at the most heavily loaded roller is formed whose length is equal to the length of their basic form.

‡ The term "modified line contact" refers to such departure from the basic form of the rollers and/or ring raceways toward their ends that, under a bearing load of about one-half of the basic load rating C, the material stress at the ends of the contact of the most heavily loaded roller with the ring is approximately the same as in the rest of this contact.

where X = a radial factor
Y = a thrust factor
F_r = radial load
F_a = thrust load
V = a rotation factor = 1.0 for inner ring rotating in relation to the load and 1.2 for inner ring stationary in relation to the load

Values of X and Y are given in Table 13.12.

The factor V, because of lack of sufficient experimental evidence, is used as a matter of precaution.

Method of Evaluating Dynamic Load Ratings of Thrust Roller Bearings. The "life" of an individual thrust roller bearing is defined as the number of revolutions (or hours at some given constant speed) which the bearing runs before the first evidence of fatigue develops in the material of either washer or of any of the rolling elements.

The "rating life" of a group of apparently identical thrust roller bearings is defined as the number of revolutions (or hours at some given constant speed) that 90 per cent of a group of bearings will complete or exceed before the first evidence of fatigue develops.

The "basic load rating" is that constant, central thrust load, which a group of apparently identical thrust roller bearings can endure for a rating life of one million revolutions of one of the bearing washers.

Load ratings, if given for specific speeds, are to be based on a rating life of 500 hr.

The "equivalent thrust load" is defined as that constant, central, purely axial load which, if applied to a bearing with rotating shaft washer and stationary housing washer, would give the same life as that which the bearing will attain under the actual conditions of load and rotation.

Table 13.12. Factors X and Y

Bearing type	$F_a/VF_r \leq e$		$F_a/VF_r < e$		e
	X	Y	X	Y	
Self-aligning and tapered roller bearings, $\beta \neq 0°$:*					
Single-row bearing	1	0	0.4	0.4 cot β	1.5 tan β
Double-row bearings†	1	0.45 to β	0.67	0.65 cot β	1.5 tan β

* For $\beta = 0°$, $F_a = 0$ and $X = 1$.
† Double-row bearings are presumed to be symmetrical.

Calculation of Basic Load Rating, Rating Life, and Equivalent Thrust Load. The magnitude of the basic load rating C_a for single-row single- and double-direction thrust roller bearings is

For $\beta = 90°$ $$C_a = f_c n^{3/4} l_e^{7/9} d^{29/27}$$
For $\beta \neq 90°$ $$C_a = f_c n^{3/4} (l_e \cos \beta)^{7/9} d^{29/27} \tan \beta$$

where β = angle of contact = angle between the line of action of the roller resultant load and a plane perpendicular to the bearing axis
n = number of rollers* in a single-row† single-direction bearing
d = roller diameter (mean diameter of tapered rollers)
l_e = effective length of contact between one roller and that washer where the contact is the shortest (overall roller length minus roller chamfers, or minus grinding undercuts)
f_c = a factor which depends on the units used, the exact geometrical shape of the load-carrying surfaces of the rollers and washers, the accuracy to which the various bearing parts are made, and the material

Values of f_c are obtained by multiplying f_c/f by a factor f.

A recommended value of the factor f based on current tests of thrust roller bearings of good-quality hardened roller-bearing steel is

$$f = 56.2 \text{ when kg and mm units are used}$$
$$f = 49{,}500 \text{ when pound and inch units are used}$$

* In case the bearing is so designed that several rollers are located on one common roller axis, these rollers are considered as one roller of a length equal to the total effective contact length of the several rollers.

† "Rollers" as defined or portions thereof which contact the same washer raceway area belong to one row.

Roller bearings vary considerably in design and execution. Small differences in the relative shape of contacting surfaces may account for distinct differences in load-carrying capacity. It is therefore not possible to give detailed information about the exact basic load rating of a certain general type of thrust roller bearing.

The approximate basic load rating of a thrust roller bearing of good design in all details and made by a reputable roller-bearing manufacturer may be obtained by the use of f_c/f values given in Table 13.13.

Actual basic load ratings of different bearing types and executions may differ from the values indicated. A bearing must be expected to have a lower carrying capacity when, under load, more or less accentuated stress concentrations in the roller contacts are present (point contact) or are not effectively prevented (edge loading at the end of a line contact). The sliding which occurs in the roller contacts in cylindrical thrust roller bearings may also account for a reduction in capacity. For no bearing type or execution will the factor f_c/f exceed the value indicated by more than 14 per cent.

The magnitude of the basic load rating C_a, for thrust roller bearings with two or more rows of rollers carrying load in the same direction, is

$$C_a = \left(\sum_{i=1}^{I} n_i l_{e_i} \right) \left[\sum_{i=1}^{I} \left(\frac{n_i l_{e_i}}{C_{a_i}} \right)^{9/2} \right]^{-2/9}$$

where $n_1, n_2, \ldots, n_I$ = number of rollers in respective rows of a one-direction multirow bearing

$C_{a_1}, C_{a_2}, \ldots, C_{a_I}$ = basic load rating per row of a one-direction multirow thrust roller bearing, each calculated as a single-row bearing with $n_1, n_2, \ldots, n_I$ rollers, respectively

Table 13.13

$d \cos \beta / E$	f_c/f, $45° < \beta \leq 62°$	f_c/f, $62° < \beta \leq 85°$	d/E	f_c/f, $\beta = 90°$
0.01	0.185	0.170	0.01	0.179
0.02	0.215	0.208	0.02	0.208
0.03	0.235	0.227	0.03	0.228
0.04	0.250	0.241	0.04	0.243
0.05	0.262	0.253	0.05	0.255
0.06	0.271	0.262	0.06	0.266
0.07	0.279	0.270	0.07	0.275
0.08	0.286	0.276	0.08	0.284
0.09	0.291	0.281	0.09	0.291
0.10	0.296	0.286	0.10	0.298
0.12	0.303	0.293	0.12	0.310
0.14	0.307	0.297	0.14	0.321
0.16	0.310	0.299	0.16	0.331
0.18	0.310	0.300	0.18	0.340
0.20	0.310	0.299	0.20	0.348
0.22	0.308	0.297	0.22	0.355
0.24	0.305	0.295	0.24	0.362
0.26	0.301	0.291	0.26	0.369
0.28	0.297	0.287	0.28	0.375
0.30	0.292	0.282	0.30	0.380

When the values for f_c/f listed in Table 13.13 are applied to non-self-aligning bearings, the C_a values obtained refer to mountings so designed and executed that uniform load distribution among the rollers is assured and maintained. If misalignment may be present a reduction in the C_a value should be made before the rating life is estimated.

E = mean pitch diameter of the roller set.

The approximate magnitude of the rating life L is

$$L = (C_a/T_E)^{10/3} \quad \text{million revolutions}$$

where T_E = equivalent thrust load

The magnitude of the equivalent thrust load T_E, for thrust roller bearings with $\beta = 90°$ under combined constant thrust and constant radial loads, is

$$T_E = XF_r + YF_a$$

where X = a radial factor
Y = a thrust factor
F_r = radial load
F_a = thrust load

Values of X and Y are given in Table 13.14.

Table 13.14. Factors *X* and *Y*

Bearing type	Single-direction bearings		Double-direction bearings*				e
	$F_a/F_r > e$		$F_a/F_r \leq e$		$F_a/F_r > e$		
	X	Y	X	Y	X	Y	
Self-aligning and tapered thrust roller bearings, $\beta \neq 90°$†	$\tan\beta$	1	$1.5\tan\beta$	0.67	$\tan\beta$	1	$1.5\tan\beta$

* Double-direction bearings are presumed to be symmetrical.
† For $\beta = 90°$, $F_r = 0$ and $Y = 1$.

Method of Evaluating Static-load Ratings of Radial Roller Bearings. The "static load" is defined as a load, acting on a nonrotating bearing.

Permanent deformations appear in rollers and raceways under static load of moderate magnitude and increase gradually with increasing load. Their permissible static load is therefore dependent upon the permissible magnitude of permanent deformation.

Experience shows that a total permanent deformation of 0.0001 of the roller diameter, occurring at the most heavily loaded roller and race contact, can be tolerated in most bearing applications without impairment of bearing operation.

In certain applications where subsequent rotation of the bearing is slow and where smoothness and friction requirements are not too exacting, a much greater total permanent deformation can be permitted. Likewise, where extreme smoothness is required or friction requirements are critical, less total permanent deformation may be tolerated.

For purposes of establishing comparative ratings the "basic static-load rating" is therefore defined as that static radial load which corresponds to a total permanent deformation of roller and race at the most heavily stressed contact of 0.0001 of the roller diameter.

In single-row angular-contact roller bearings the basic static-load rating relates to the radial component of that load which causes a purely radial displacement of the bearing rings in relation to each other.

The "static equivalent load" is defined as that static radial load which, if applied, would cause the same total permanent deformation at the most heavily stressed roller and race contact as that which occurs under the actual condition of loading.

Calculation of Basic Static Load Rating and Static Equivalent Load. The magnitude of the basic static-load rating C_s is

$$C_s = f_s inl_e d \cos \beta$$

where i = number of rows of rollers in any one bearing
β = angle of contact = angle between the line of action of the roller resultant load and a plane perpendicular to the bearing axis
n = number of rollers per row
d = roller diameter (mean diameter of tapered rollers)
l_e = effective length of contact between one roller and that ring where the contact is the shortest (overall roller length minus roller chamfers or minus grinding undercuts)
f_s = 2.2 when kg and mm units are used
f_s = 3,130 when pound and inch units are used

These values of f_s are valid for bearings as commonly designed and manufactured and made of hardened steel.

The magnitude of the static equivalent load R_s, for radial bearings under combined radial and thrust loads, is the greater of

$$R_s = X_s F_r + Y_s F_a$$
$$R_s = F_r$$

where X_s = a radial factor
Y_s = a thrust factor
F_r = radial load
F_a = thrust load

Values of X_s and Y_s are given in Table 13.15.

Table 13.15

Bearing type	Single-row bearings*		Double-row bearings†	
	X_s	Y_s	X_s	Y_s
Self-aligning and tapered roller bearings, $\beta \neq 0°$	0.5	0.22 cot β	1	0.44 cot β

* R_s is always $\geq F_r$.
† Double-row bearings are presumed to be symmetrical.

Method of Evaluating Static-load Ratings of Thrust Roller Bearings. The "static load" is defined as a load acting on a nonrotating bearing.

Permanent deformations appear in rollers and raceways under static load of moderate magnitude and increase gradually with increasing load.

Experience shows that a total permanent deformation of 0.0001 of the roller diameter, occurring at the most heavily loaded roller and race contact, can be tolerated in most bearing applications without impairment of bearing operation.

In certain applications where subsequent rotation of the bearing is slow and where smoothness and friction requirements are not too exacting, a much greater total permanent deformation may be tolerated.

For purposes of establishing comparative ratings the "basic static-load rating" is therefore defined as that static central thrust load which corresponds to a total permanent deformation of roller and race at the most heavily stressed contact of 0.0001 of the roller diameter.

The "static equivalent load" is defined as that static, central, purely axial load which, if applied, would cause the same total permanent deformation at the most heavily stressed roller and race contact as that which occurs under the actual condition of loading. The magnitude of the basic static-load rating C_{sa} is

$$C_{sa} = f_{sa} n l_e d \sin \beta$$

where β = angle of contact = angle between the line of action of the roller resultant load and a plane perpendicular to the bearing axis
n = number of rollers carrying thrust in one direction*
d = roller diameter (mean diameter of tapered rollers)
l_e = effective length of contact between one roller and that washer where the contact is the shortest (overall roller length minus roller chamfers or minus grinding undercuts)*
f_{sa} = 10 when kg and mm units are used
f_{sa} = 14.220 when pound and inch units are used

These values of f_{sa} are valid for bearings as commonly designed and manufactured and made of hardened steel.

The magnitude of the static equivalent load T_s, for thrust bearings with contact angle $\beta \neq 90°$, under combined radial and thrust loads, is

$$T_s = F_a + 2.3F_r \tan \beta$$

where F_a = thrust load
F_r = radial load

The accuracy of the formula decreases in the case of single-direction bearings when $F_r > 0.44F_a \cot \beta$.

References

1. Belyayev, N. M.: "Work in the Theory of Elasticity and Plasticity," State Press for Technical Theoretical Literature, Moscow, 1957.
2. Jones, A. B.: "Analysis of Stresses & Deflections," vols. I and II, New Departure Division, G.M.C., Bristol, Conn., 1946.
3. Lundberg, G., and A. Palmgren: "Dynamic Capacity of Roller Bearings," Acta Polytechnica, Mechanical Engineering Series, vol. I, no. 3, Stockholm, 1947.
4. Lundberg, G., and A. Palmgren: "Dynamic Capacity of Roller Bearings," Acta Polytechnica, Mechanical Engineering Series, vol. II, no. 4, Stockholm, 1952.
5. Jones, A. B.: A General Theory for Elastically Constrained Ball and Radial Roller Bearings under Arbitrary Load and Speed Conditions, *Trans. ASME, J. Bas. Eng.*, June, 1960.
6. Shevchenko, R. P., and P. Bolan: "A Visual Study of Ball Motion—A High-speed Thrust Bearing," SAE Annual Meeting January 14th through 18th, 1957.
7. Jones, A. B.: "Ball Motion and Sliding Friction in Ballbearing." *Trans. ASME J. Bas. Eng.*, March, 1959.
8. Palmgren, A.: "Ball and Roller Bearing Engineering," SKF Industries, Philadelphia, 1959.
9. "Method of Evaluating Load Ratings for Ballbearings," AFBMA Standard, Section No. 9, Revision No. 4, October, 1960 (The Anti-Friction Bearing Manufacturers Association, Inc., N.Y.).
10. "Method of Evaluating Load Ratings for Roller Bearings," AFBMA Standard, Section No. 11, July, 1960 (The Anti-Friction Bearing Manufacturers Association, Inc., N.Y.).
11. Lawson, A. C.: Design Factors for Jewel Bearing Systems, *Machine Design*, pp. 132–137, April, 1954.
12. Wilcock, D. F., and Booser, E. R.: Why Bearings Fail, *Prod. Eng.*, pp. 167–182, October, 1956.
13. Williams, D. L.: Eleven Ways to Oil Lubricate Ball Bearings, pp. F18–F19, in "Annual Handbook of Product Design," Product Engineering, New York, 1953.
14. Rounds, T. E.: Lubrication Systems for High-performance Ball Bearings, *Machine Design*, pp. 114–120, Sept. 20, 1956.
15. Timmerman, W. L.: Miniature Ball Bearings, *Elec. Mfg.*, pp. 106–113, February, 1954.

* In case rollers of different lengths are used to carry load in one direction nl_e is taken as the total effective length of all the rollers over which the load is distributed.

Section 14

FRICTION DRIVES

By

WALTER WERNITZ, Dr.-Ing., *Institut für Maschinenelemente und Fordertechnik, Technische Hochschule Braunschweig, Braunschweig, West Germany*

CONTENTS

14.1. INTRODUCTION

This chapter deals with drives, called *friction drives,* which transmit circumferential forces and velocities primarily by friction. The simplest case of such a drive is two cylinders held together by normal force and transmitting a torque. The principal effect is friction, but with soft materials, a part of the torque may be attributable to an asymmetrical force distribution.[35]

A distinction is made between continuously operating friction drives (constant or infinitely variable speed ratios) and those of intermittent operation such as ratchets and free-wheeling mechanisms. The designs of the latter mechanisms vary greatly and will not be treated here.[1,2]

14.2. FRICTION WHEELS FOR ALMOST CONSTANT SPEED RATIOS

Friction-wheel drives with quasi-constant speed ratios are feasible in the cylindrical, plane, or conical forms and provide a simple means for transmitting motion and power. In these forms, a quasi-rectangular Hertzian area exists (line contact). For such applications, metals or soft materials are often used. Among the metals, hardened steel and cast iron find widespread application. However, their low coefficients of friction, particularly under conditions of lubrication, result in high bearing forces.

This generally limits the success of hardened steel for constant speed ratios, even though the product of coefficient of friction and normal force is highest because of the high allowable Hertzian pressure. On the other hand, the use of hardened steel for infinitely variable drives has been successful in many devices.

Friction wheels, particularly cylindrical, fabricated of soft materials (rubber and rubberlike substances and special reinforced plastics) are widely used. Cylindrical wheels experience tangential rolling and therefore pure tangential creep. The development of special rubber with high allowable coefficients of friction (0.7 to 0.8, when dry) and good durability, together with modern design techniques with automatically controlled contact forces, have considerably furthered the application of friction drives.

An increase in the coefficient of friction by a factor of 2 to 4 over previously used friction wheels made of nonferrous hard materials has been achieved by the pairing of rubber and iron. The development of rubberlike materials with good thermal, wear, and kneading characteristics has contributed largely to the development of friction drives and offers the following advantages: (1) considerable weight and volume reduction, (2) flexible power transmission, (3) smooth, quiet operation with little or no

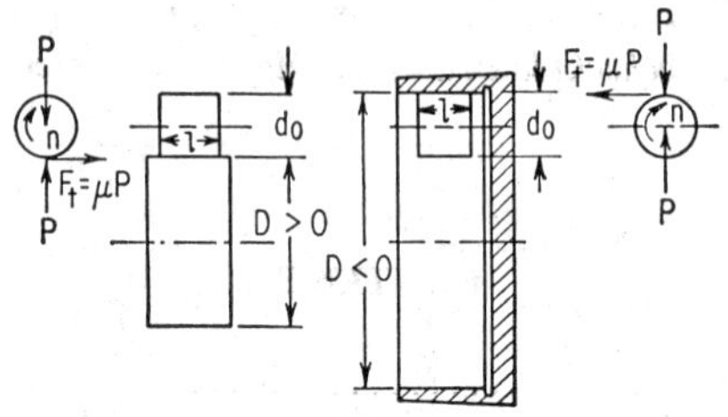

FIG. 14.1. Friction wheels. (*a*) External drive. (*b*) Internal drive.

vibration, (4) little or no maintenance required if carefully assembled, and (5) ability to serve as an overload clutch if suitably designed. A maximum of approximately 100 hp may be transmitted by this type of drive. Apart from limited power transmission and peripheral velocity ($v < 1{,}200$ in./sec), certain other disadvantages exist when compared with gear drives. These are associated with the more rugged bearings required, and with the differences between force-locked and form-locked (positive-drive) drives which are related to slip (a function of peripheral load to a certain degree).

Referring to Fig. 14.1, the torque transmitted is

$$T = F_t d_0/2 = \mu P d_0/2 \tag{14.1}$$

where coefficient of friction $\mu \leq \mu_{allow} = \mu_{max}/S_r$

S_r = a factor to ensure against slip
P = load
F_t = tangential force

A useful empirical stress parameter K is defined as follows:

$$K = P/ld_m \tag{14.2}$$

where $1/d_m = 1/d_0 \pm 1/D$, the negative sign applying to concave curvature (Fig. 14.2).

For laminated wood, reinforced fabrics, and plastics running dry against cast iron at approximately 150 in./sec, $\mu_{allow} = 0.3$ to 0.4 ($S_r = 1.5$) and $K_{allow} = 70$ to 140 psi.[4]

Allowable values for continuous-duty oil-lubricated metal pairs with no tangential loading are given in Table 14.1.[5]

Table 14.2[5] presents allowable values for unlubricated metal pairs, with $\mu = 0.15$ and tangential loading present.

Allowable values for oil-lubricated metal pairs with $\mu = 0.05$ and tangential loading present are given in Table 14.3.[6] Also see ref. 12.

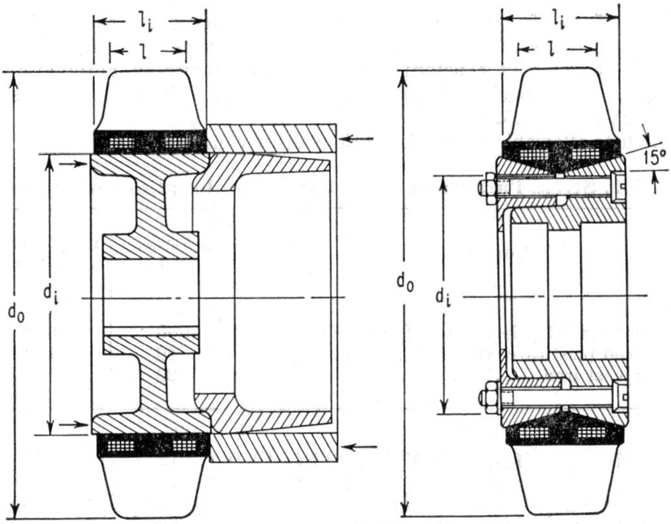

FIG. 14.2. Rubber friction rings (with steel-wire stiffeners).[9]

Table 14.1. Stress Parameter K for Various Metals with Lubrication

Combination	Stress parameter K_{allow}, psi	Brinell hardness number
Cast iron/steel	570	230
Cast iron/steel	425	150
High-carbon steel/steel	850	230
Hardened steel/hardened steel	7,800	650
Aluminum/steel	780	200

Table 14.2. Stress Parameter K for Unlubricated Metal Pairs $\mu = 0.15$

Combination	K_{allow}, psi
Cast iron/steel	300–600
AISI 1045/AISI 1065	500–1000
AISI 1075/AISI 1065	350

Table 14.3. Stress Parameter K for Lubricated Metal Pairs $\mu = 0.05$

Combination	Brinell hardness number	Modulus of elasticity E, psi	K_{allow}, psi
Hardened steel/hardened steel	650	30,000,000	$\dfrac{9{,}250}{1 + v/200}$

where v = peripheral velocity, in./sec.

Because the effects of peripheral velocity and generated heat have not been taken into account, the (μK_{allow}) product should be used only as a general guide for soft materials. In the case of constant contact pressure (without automatic pressure compensation), only 75 per cent of the above values should be used, and S_r should be somewhat greater than 1.5.

Table 14.4 lists allowable values for various other combinations.[7]

Table 14.4. Stress Parameter K for Other Combinations of Materials

Combination	μK_{allow}, psi	max K_{allow} ($\mu \to 0$), psi
Hardened steel/hardened steel (oil-lubricated)....	200	6,000
Reinforced plastics/cast iron....................	40	150
Rubber/cast iron..........................	20	50

14.3. RUBBER WHEELS AGAINST STEEL AND CAST IRON

The proportions given below are for special types of rubber (80 to 90 Shore) particularly resistant to abrasion, kneading, heat, and aging, vulcanized on cast-iron hubs with $d_0 = 1.5$ to 6.25 in. Referring to Fig. 14.2,

$$d_i = 0.625 d_0$$

The width of the diameter of vulcanization

$$l_i = d_0/4$$

which for a taper angle of 12° results in a friction face width

$$l = 0.7 l_i = d_0/5.7$$

Allowable Rolling Pressure. An experimentally determined expression for the allowable rolling pressure follows (for $40 < v < 1{,}200$ in./sec):

$$K_{0,\text{allow}} = \frac{P_{\text{allow}}}{(l)(d_0)} = \frac{K_0^*}{\sqrt[3]{1 + d_0/D}} \left(\frac{40}{v}\right)^{3/4} \quad \text{psi} \tag{14.3}$$

$$K^* = 60 \text{ to } 80 \text{ psi}$$

where the correction for diameter D, (d_0/D), applies only to convex curvature, and where for concave curvature (internal drive), the term $\sqrt[3]{1 + d_0/D}$ is set equal to unity.

The above expression for $K_{0,\text{allow}}$ takes into account the effect of generated heat and also the increased kneading work attributable to indentation. For $v \geq 40$ in./sec,

$$K_{0,\text{allow}} = \frac{K^*}{\sqrt[3]{1 + d_0/D}} \tag{14.4}$$

Coefficient of Friction. For dry continuous operation, $\mu_{allow} = 0.7$; intermittent operation, $\mu_{allow} = 0.5$; and for operation under humid conditions, $\mu_{allow} = 0.1$ to 0.3.

Slip. Slip should generally not exceed 2 to 3 per cent under conditions of continuous operation. Short-time slip of 4 per cent or higher using higher friction coefficients has been experienced in smooth intermittent service without detrimental effects.

The phenomenon of slip has not been finally formulated. Some tests have shown that slip remains almost constant for a constant product of friction coefficient raised to a power and the contact force. The writer's research has shown that, for constant velocity, the following expression is valid:

$$\mu^k K_0 = \text{constant for constant slip} \qquad (14.5)$$

where $k = 1.5$ for a specific case of dry friction for which

$K_0 = K_{0,\text{allow}}$, $\mu = 0.2$, and slip $= 3$ per cent

$K_0 = K_{0,\text{allow}}$, $\mu = 0.3$, and slip $= 5$ per cent

Other research has shown that the value of k can exceed 3 as rolling velocity is increased. It is therefore recommended that, in order to reduce slip and improve the transfer of power with good durability, moderate rolling velocities be considered $v = 100$ to 400 in./sec. In Fig. 14.3 it is shown that the product of contact force and rolling velocity remains constant from $v = 120$ to 600 in./sec, dropping off at higher velocities. The maximum power which can be transmitted by a friction wheel under conditions of dry continuous operation, can be assumed to be:

$$\text{for } d_0 = 1.5 \text{ to } 6.25 \text{ in., maximum power} = 0.15 \text{ to } 2.5 \text{ hp}$$

Several friction wheels may be arranged side by side to transmit proportionately greater power.

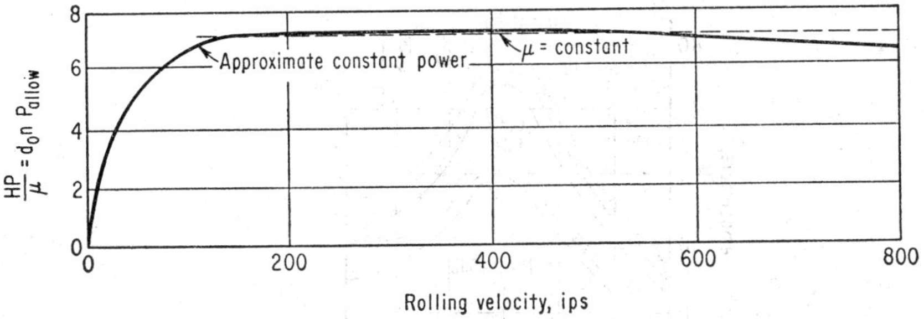

FIG. 14.3. Horsepower vs. rolling velocity for friction ring $d_0 = 10$ in., $L_i = 2$ in., $V = (\pi/60)d_0 n$, in./sec.

Heat Generation.[8,9] The temperature rise at the friction surface of wheels made of soft materials should not exceed 150°F. The heat generated is attributable primarily to kneading work. In the absence of tangential loading, the maximum temperature occurs at the center of rubber shown in Fig. 14.4*c*. The kneading work is primarily a function of velocity, indentation depth, and material properties. The temperature is further influenced by frictional work which manifests itself primarily as a surface temperature rise.

14.4. RUBBER FRICTION RINGS AGAINST STEEL AND CAST IRON[9]

Friction rings made of special rubber operating against steel or cast iron are similar to those previously described except that they are pressed onto a cylindrical metal hub or screw fastened onto a tapered section having an angle of 15° (see Fig. 14.2). The important dimensions of such rings are

$$d_0 = 7 \text{ to } 30 \text{ in.}$$
$$d_i = 0.5d_0 \text{ to } 0.85d_0$$
$$l_i = 0.10d_0 \text{ to } 0.33d_0$$
$$2l_i/(d_0 - d_i) = 1.3 \text{ to } 1.875$$
$$l = l_i - (d_0 - d_i) \sin 15°$$

The allowable rolling pressure probably cannot be precisely described because of the variable ratios of friction surface width to rubber thickness. There is a somewhat greater velocity dependence than in the case of vulcanized rubber. A characteristic

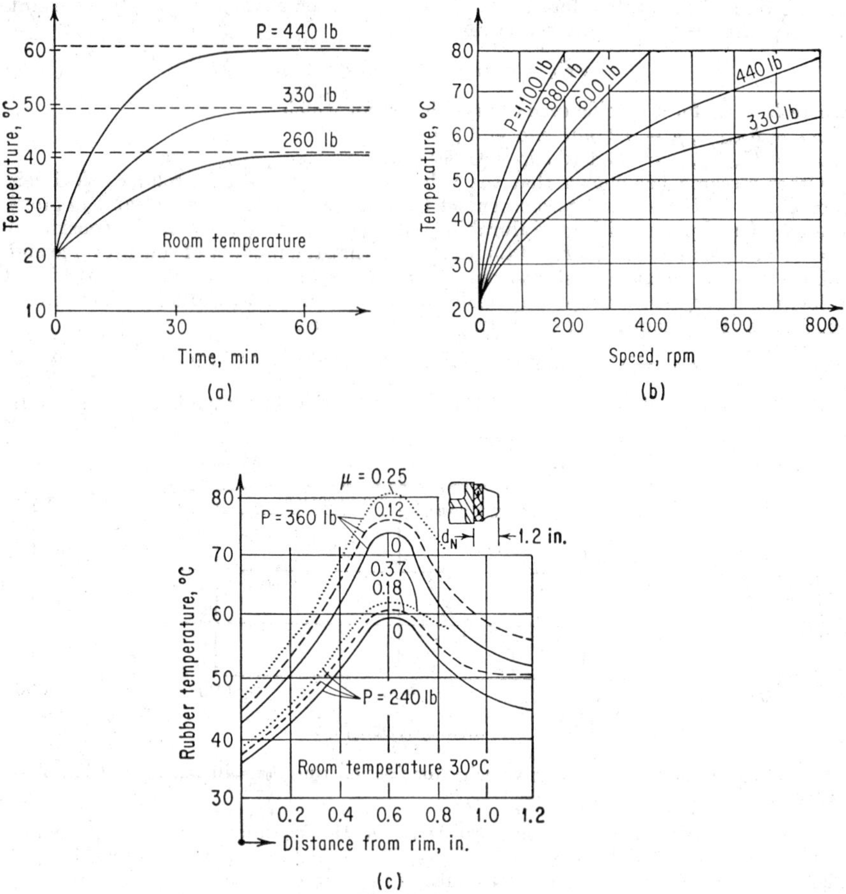

Fig. 14.4. Friction-ring heat.[9] (*a*) Heating of friction surface. (*b*) Steady-state temperature vs. speed. (*c*) Radial temperature distribution.

relationship is presented in Fig. 14.3, wherein the product of contact force and velocity is plotted as a function of velocity.

In the case of low speed, $v \leq 8$ in./sec, the following relationship may be used for constant contact pressures:

$$K_0 = \frac{P}{l_i d_0} = \frac{1}{\sqrt[3]{1 + d_0/D}} (K - md_0) \tag{14.6}$$

where $D > 0$ for convex contact; for concave curvature $D < 0$, set $\sqrt[3]{1 + d_0/D}$ equal to unity.

While the limits of K and m are, respectively, 60 to 80 psi and 0.12 to 0.17 lb/in., representative values are

$$K_0 = 70 \text{ psi} \qquad m = 0.17 \text{ lb/in.}$$

Maximum horsepower transmission is obtained at higher rolling velocities. Note that Fig. 14.3 shows a flat maximum. A constant power transmission can therefore be expected between 120 and 600 in./sec, approximately. Because of low slip, moderate peripheral velocities (120 to 400 in./sec) are recommended.

The maximum allowable peripheral velocity may be assumed to be 600 in./sec for smaller friction rings, and 1,200 in./sec for large rings ($d_0 \geq 14$ in.).

For this type of friction ring, the maximum transmitted horsepower can be estimated from the following relationship, which applies for $120 < v < 600$ in./sec:

$$\text{hp} = \mu l_i \lambda \log \frac{d_0}{d_c} \tag{14.7}$$

where representative values of λ and d_c are 10.45 hp/in. and 4.6 in., respectively, and 7 in. $< d_0 <$ 30 in. The same values of maximum allowable coefficients of friction, slip, and temperature rise are valid as for vulcanized friction wheels discussed previously. Figure 14.4 presents temperature-rise data for friction rings.[9]

14.5. FRICTION-DRIVE DESIGN

In machines operating continuously at constant load, the rolling drive may be safely pressed against the driven wheel with a calculated contact force. During a rest period, an indentation occurs which causes rough running at startup. This usually disappears after several revolutions. Most machines operate, however, under conditions of variable and shock loading. It is desirable under such operating conditions to control contact pressure in order to increase life and efficiency and to prevent detrimental overloads.

Figure 14.5 shows several designs which maintain a well-controlled contact pressure by means of spring-loaded pivoted drives. A control angle α of 35 to 38° is used in external drives; for internal drives, α = 38 to 40°.

For the arrangement shown in Fig. 14.5*a*,

$$P_0 = \frac{Sb - Gw}{e} \tag{14.8}$$

where P_0 = preload.

$P_0 \cong 0.1 P_{\text{allow}}$ for dry running ($\mu_{\text{allow}} = 0.7$)

$P_0 \cong 0.33 P_{\text{allow}}$ for frequent starting and shock loading ($\mu_{\text{allow}} = 0.5$)

$P_0 \cong 0.6 P_{\text{allow}}$ for humid operating conditions ($\mu_{\text{allow}} = 0.3$)

When the tangential force $F_t = \mu P$ is applied, the coefficient of friction may be expressed

$$\mu = (1 - P_0/P) \tan \alpha \qquad \text{where } \mu < \tan \alpha \tag{14.9}$$

In order to avoid excessive contact forces caused by temporary line-of-contact overloads, values of $\mu = 0.7$ may generally be used and values of α selected accordingly.

The spring load for no rotation:

$$S_0 = \frac{Gw + P_0 e}{b} \tag{14.10}$$

The recommended spring rate[9]

$$K = S/\delta = 25F_{t\max}/d_0 = 25\mu P_{\text{allow}}/d_0 \qquad \delta = \text{deflection} \tag{14.11}$$

from which the spring deflection

$$S_0 = \frac{Gw + P_0 e}{P_{\text{allow}} b} \frac{d_0}{25\mu} \tag{14.12}$$

Basically, either the small or larger wheel may be rubber-clad. Experience has shown, however, that it is best to use the small wheel as the rubber wheel for ratios up to 1:8, and the large wheel for ratios up to approximately 1:18. The mating wheel should be constructed of steel or cast iron. Figure 14*b* is an internal drive arrangement similar in concept to Fig. 14*a*.

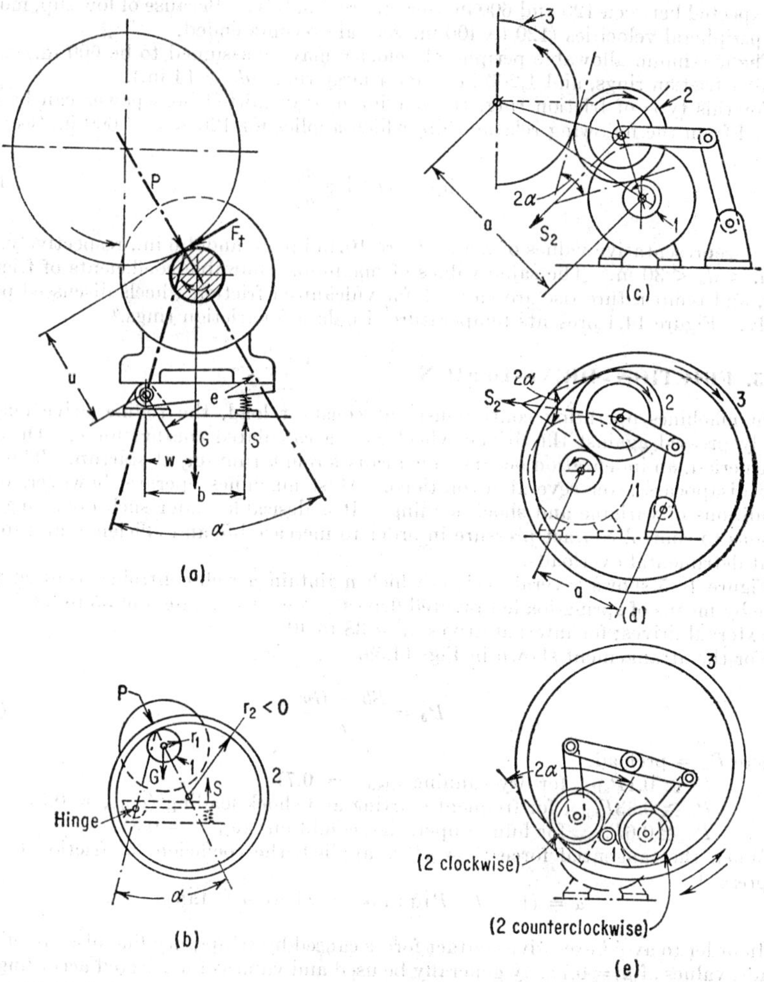

FIG. 14.5. Friction drives. (*a*) Device for automatically controlling contact force (external drive). (*b*) Device for automatically controlling contact force (internal drive). (*c*) External drive. (*d*) Internal drive. (*e*) Reversible internal drive.

14.6. FRICTION WHEELS AS IDLERS

As shown in Fig. 14.5*c*, *d*, *e*, frictions wheels may be used as idlers for large center distances. If larger transmission ratios are required (e.g., where there is direct use of motor shafts), or if the drive shaft or the driven shaft cannot be pivotally mounted, the control angle α must be maintained. This is possible only if the center distance and the diameters of the three wheels bear a particular relationship with one another, and if the angle formed at the center of the idler by lines at the contact points drawn through the nonidler wheel center is $(180 - 2\alpha)$ degrees.

The radius of rubber idler r_2 is given by

$$r_2 = -\left(\frac{r_1 + r_3}{2}\right) + \sqrt{\frac{2a^2 - (r_1 - r_3)^2(1 - \cos 2\alpha)}{4(1 + \cos 2\alpha)}} \tag{14.13}$$

Assuming $\alpha = 35°$,

$$r_2 = -\left(\frac{r_1 + r_3}{2}\right) + \sqrt{0.372a^2 - 0.123(r_1 - r_3)^2} \tag{14.14}$$

There is also experimental evidence[10] to indicate that a smaller value of α may be used:

$$\tan \alpha \approx 0.7\mu_{max} \approx 0.5 \qquad \alpha \approx 27°$$

The value of α is governed by the following relationship:

$$\cos 2\alpha = \frac{(r_2 + r_1)^2 + (r_2 + r_3)^2 - a^2}{2(r_2 + r_1)(r_2 + r_3)} \tag{14.15}$$

The intermediate idler preload P_0 can be applied by a spring. The resulting force should act in the direction of the lines bisecting the angle 2α between the two tangential forces.

The characteristics of the friction-wheel drive are such as to make it a force-limiting device suitable for use as an elastic coupling or even an overload clutch. However, the contact force and the bearing load cannot be arbitrarily increased because of the limited wear resistance of the friction materials. An arrangement with provision for shutoff may therefore be preferred for rough working conditions as in stone-crushing equipment.

14.7. METAL-PAIR FRICTION WHEELS

Table 14.4 indicates that the peripheral force can be substantially increased by matching hardened steel with hardened steel. This combination can be applied only

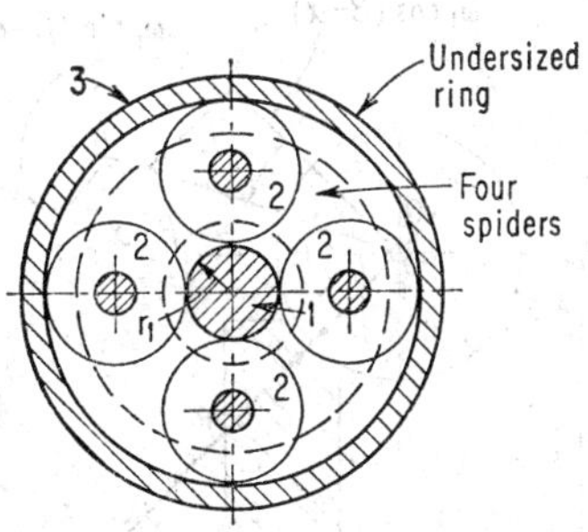

FIG. 14.6. Planetary friction drive.

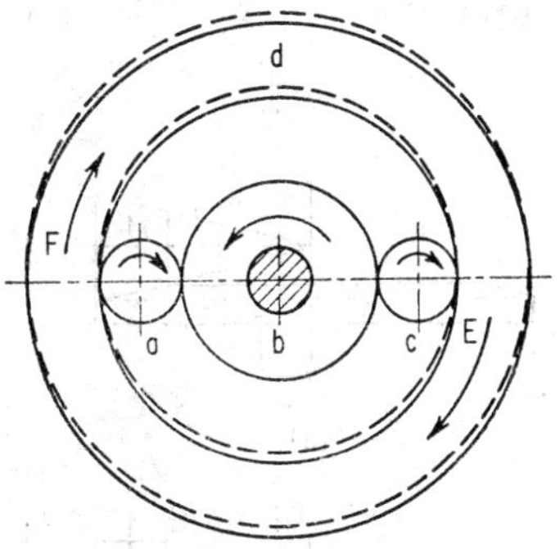

FIG. 14.7. Krupp-Garrard drive.

where the high contact force necessitated by the low coefficient of friction does not result in excessive bearing loads.

Talbourdet[12] investigated, with respect to endurance, rolling pairs having line contact. It was found that combinations which are most suitable in this regard are those involving the hardest materials operating against hardened tool steel. These include a carbon steel hardened to 58 Rockwell C and an induction-hardened SAE 4340 steel hardened to 50 to 55 Rockwell C. Both withstood a Hertzian pressure of 230,000 psi over a period of 10^8 cycles under oil lubrication and no slip ($v = 130$ in./sec). The allowable pressure decreases to approximately 190,000 psi with 9 per cent slip. A somewhat improved performance may be expected by increasing the hardness to 64 Rockwell C.

The methods of wind-up and self-winding have been frequently applied in metallic-friction pairs in order to decrease bearing load.

Figure 14.6 shows a simple elastic wind-up of cylindrical friction wheels (2) in a planetary drive by means of an undersized ring (3).

The Krupp-Garrard drive (Fig. 14.7), which uses an external elastic ring, manifests a self-winding effect which is a function of torque. The free-floating ring d, which is slightly self-winding, is shifted upward when a or c is driving at points E and F below the center line. A self-adjusting pressure which is torque-dependent results from the shifting of the elastic ring. The transmissible torque decreases with increasing numbers of revolutions.[13,14]

14.8. FRICTION DRIVES WITH BORING FRICTION

In order to diminish the bearing load or attain infinite speed variation an additional boring action at the contacting surfaces may be acceptable.

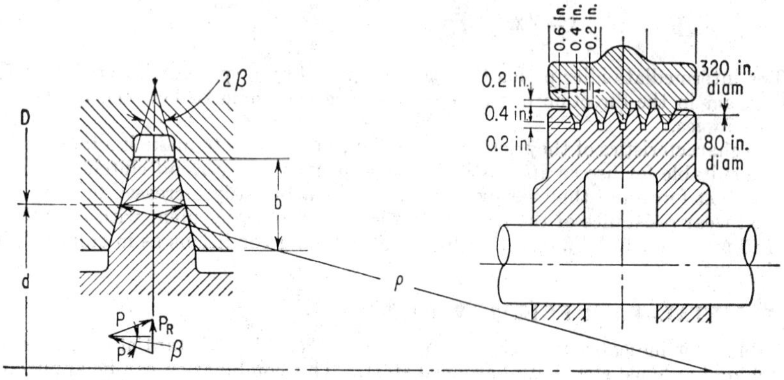

FIG. 14.8. Cast-iron V-groove friction wheels. Peripheral force = 10 to 40 lb per groove.

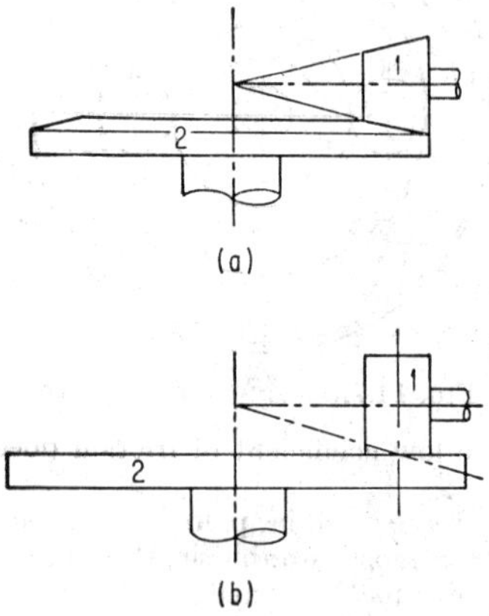

FIG. 14.9. Friction drives. (a) Without spin friction. (b) With spin friction (used for variable-speed drive).

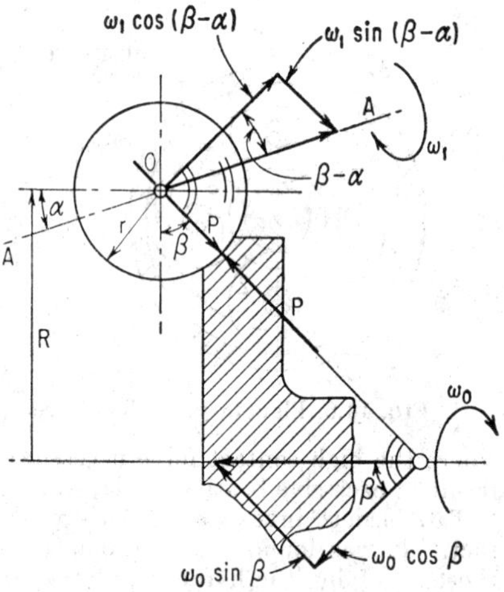

FIG. 14.10. Spin and roll vectors at elements of Cleveland speed variator.[31]

V-groove Friction Wheels.[13] Referring to Fig. 14.8, a radial force P_R results in a tangential force $(\mu/\sin\beta)P_R$. The most frequently used value of β is 15°. Because of the higher values of the effective coefficient of friction $(\mu/\sin\beta)$, the use of hardened steel against cast iron is practical.

The depth b of the V-grooves is made as small as possible in order to reduce the sliding-boring action over the contact area, which results in wear.

Design Characteristics of Infinitely Variable Drives. The difference between the motion which causes boring friction and that which does not may be seen in Fig.

14.9. The circumferential velocities of the bevel-wheel pair are equal at every contact point. Compare this with the infinitely variable disk drive in which a relative boring velocity exists at the contact area.

Figure 14.10 indicates the boring motion, using as an example the steel elements of the Cleveland[31] speed variator:

$$\omega_b = \omega_{b1} - \omega_{b2} = \omega_0 \sin \beta + \omega_1 \sin (\beta - \alpha) \tag{14.16}$$

where ω_b is the boring angular velocity and where it is assumed that the ball can rotate about adjustment shaft A-A (adjustment angle α). Similar boring action occurs in all such drives, primarily variable, in which the mantle lines of the contact areas do not intersect at the common point of intersection of the axes of rotation.

The speed ratio of the Graham[30] drive, shown in Fig. 14.11, depends upon the ratio of diameters of a fixed-diameter wheel to the diameter at the point of contact of a conical roller.

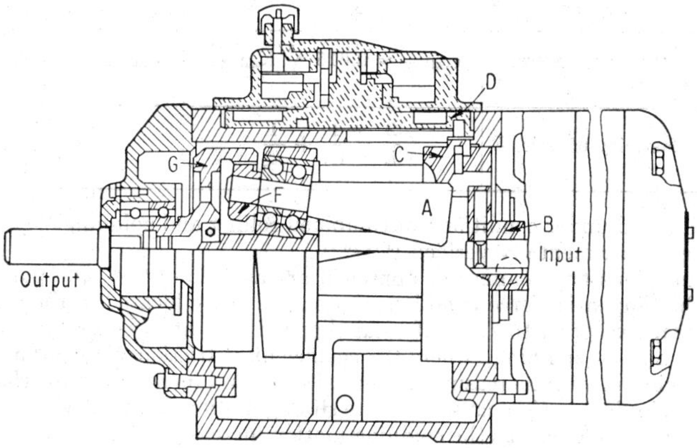

FIG. 14.11. The Graham drive.[30] (*A*) Conical rollers. (*C*) Fixed but axially adjustable ring. (*D*) Adjustment device. (*F*) Planetary bevel gear. (*G*) Driven drum.

The double Hayes drive is a good example of a drive with counterbalanced axial loads which serves to limit bearing loads.[14,15]

The Graham and Hayes drives transmit power by steel on steel contact. Other drives utilize materials having a higher coefficient of friction but poorer wear characteristics. In these cases materials such as laminated plastic and special types of rubber are used in combination with cast iron. For other drives, see refs. 16 and 17.

Approximate Calculation of Transmitted Power. Torque and tangential force can be calculated making the following assumptions, and neglecting the boring action:

1. Force is applied at the center of the Hertzian area.
2. The coefficient of friction is divided by a safety factor S_r (1.3 to 2) in order to provide a margin of safety against slippage.

The circumferential force

$$F_t = \mu P / S_r \tag{14.17}$$

Table 14.5 presents formulas for allowable values of P and p_0 as defined by Stribeck and Hertz.[32,33]

The allowable pressures are lower for line contact than for point contact. However, the length of the line of contact can be easily made longer, thus increasing the contact area. It is probably more difficult to increase the contact area in point-contact designs because of other design considerations. Increasing the contact area introduces larger power losses and increases the amount of heat generated in the case of boring friction, however. These consequences will now be taken into consideration.

Friction with Combined Rolling and Spinning Motion. A solution of the problem of slip, which presents many problems in the case of rolling without spinning,

Table 14.5. Contact Pressure Formulas

	Point contact (elliptic Hertzian area)	Line contact (quasi-rectangular Hertzian area)
Relationship for K_{allow}	$K_{allow} = \dfrac{P_{allow}}{d_0^2}$	$K_{allow} = \dfrac{P_{allow}}{2ad_0}$ $2a$ = length of mantle line
Definition of d_0 (d_0 negative for concave curvature)	$\dfrac{2}{d_0} = \dfrac{1}{d_1} + \dfrac{1}{d_2} + \dfrac{1}{d_3} + \dfrac{1}{d_4}$	$\dfrac{1}{d_0} = \dfrac{1}{d_1} + \dfrac{1}{d_2}$
Maximum circumferential force, F_c	$F_c = \dfrac{\mu}{S_r} K_{allow}\, d_0^2$	$F_c = \dfrac{\mu}{S_r} K_{allow}\, d_0\, 2a$
Maximum Hertzian pressure, $p_{0,allow}$	$p_{0,allow} = 0.615 \sqrt[3]{K_{allow} E^2}$	$p_{0,allow} = 0.592 \sqrt[2]{K_{allow} E}$
Modulus of elasticity of pair, E	$\dfrac{2}{E} = \dfrac{1}{E_1} + \dfrac{1}{E_2}$	$\dfrac{2}{E} = \dfrac{1}{E_1} + \dfrac{1}{E_2}$

has been obtained under conditions of combined rolling and spinning.[14,18] This solution has been experimentally verified.[14,19,20,21]

The assumptions and manner of approach are as follows:[18,19] The point of application of the tangential force is no longer assumed to be located in the center of the Hertzian area, except in the case of total slippage. Contrary to rolling with a tangential force and no spinning, where the coefficient of friction varies from essentially zero to a maximum with increasing tangential force and increasing slip, the combined rolling and spinning motion causes definite sliding motion in the Hertzian area which becomes more pronounced as the spinning motion increases. Thus the effective coefficient of friction may be defined by the velocity of sliding in the Hertzian area. It has been demonstrated theoretically and experimentally that the average coefficient of friction may be assumed constant within the entire Hertzian area.

Figure 14.12 shows for a long narrow Hertzian area (characteristic, for example, of a Prym-type drive) the location (force pole) and magnitude of the tangential force expressed as a function of the center of the twisting or spinning motion (spin pole). For this case the reduction ratio is

$$i = \frac{l_2 - l}{l_1 - l} > i_0 = \frac{l_2}{l_1}$$

and the efficiency at friction contact is

$$\eta_a = \frac{M_2}{M_1} \cdot \frac{\omega_2}{\omega_1} = \frac{l_2 + l_N - l}{l_1 + l_N - l} \cdot \frac{l_1 - l}{l_2 - l}$$

where l_N = distance from the force pole to the spin pole (offset of the poles)
l = distance from the twist pole to the center of pressure of the Hertzian area (offset of spin pole)

Figure 14.13 presents additional examples of combined spinning and rolling friction. Here l_2 is positive if opposite of l_1, and l_2 is negative if in the same side as l_1, and $|l_2| > |l_1|$. The slip is

$$S = 1 - \frac{i_0}{i} = \frac{l/l_1 + l/l_2}{1 + l/l_2}$$

The torque loss is

$$\frac{\Delta M}{\Delta M_1} = 1 - \frac{M_2}{M_1} = \frac{(l_N - l)/l_1 + (l_N - l)/l_2}{1 + (l_N - l)/l_1}$$

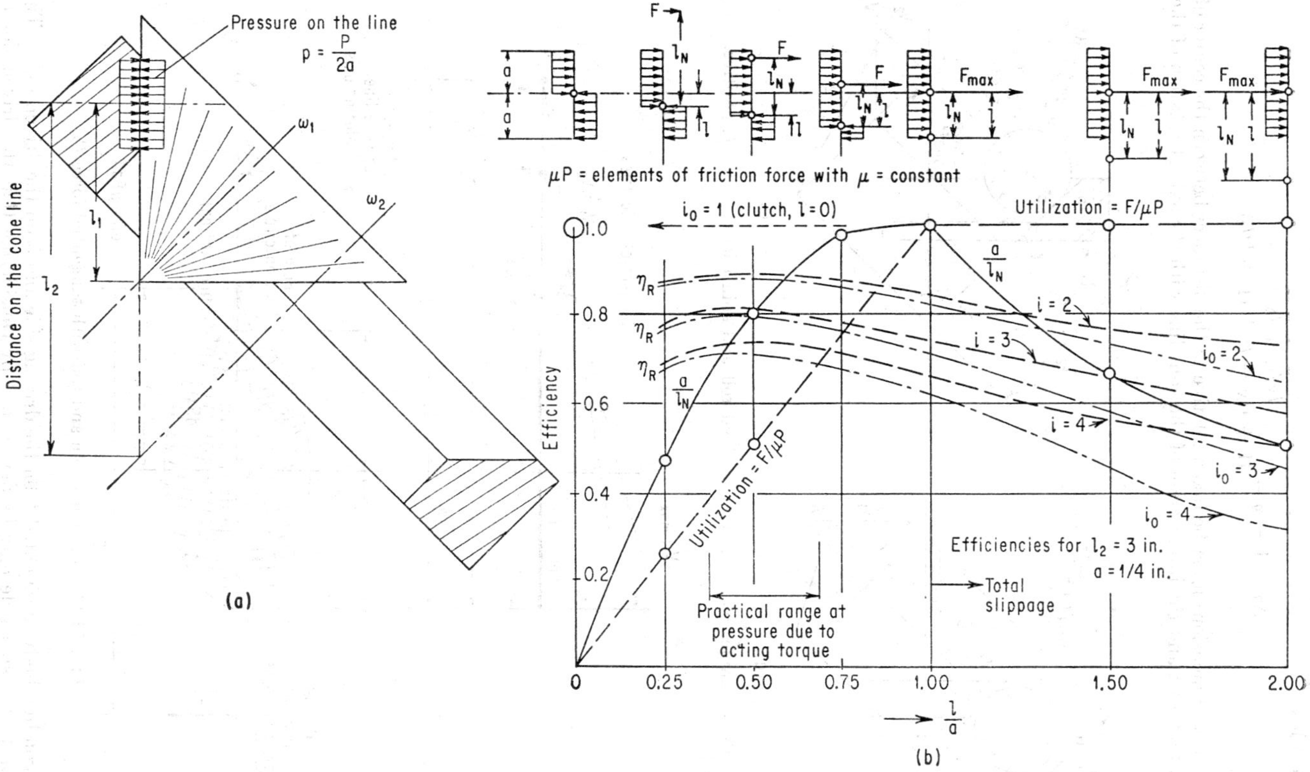

FIG. 14.12. Combined rolling and spinning at very long and narrow contact area. (*a*) Structure. (*b*) Efficiency curves.

efficiency loss

$$\Delta\eta = 1 - \eta = \frac{l_N/l_1 + l_N/l_2}{[1 + (l_N - l)/l_1](1 + l/l_2)}$$

Indicated dimensions must be taken relative to the distances between the center of the contact area and the intersection of the cone line with the axis of rotation of the

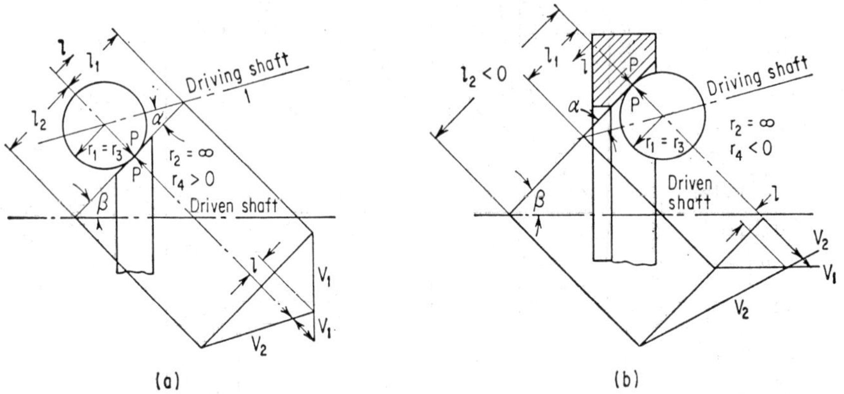

FIG. 14.13. Examples of losses for combined roll and spin friction. (*a*) l_2 positive. (*b*) l_2 negative.

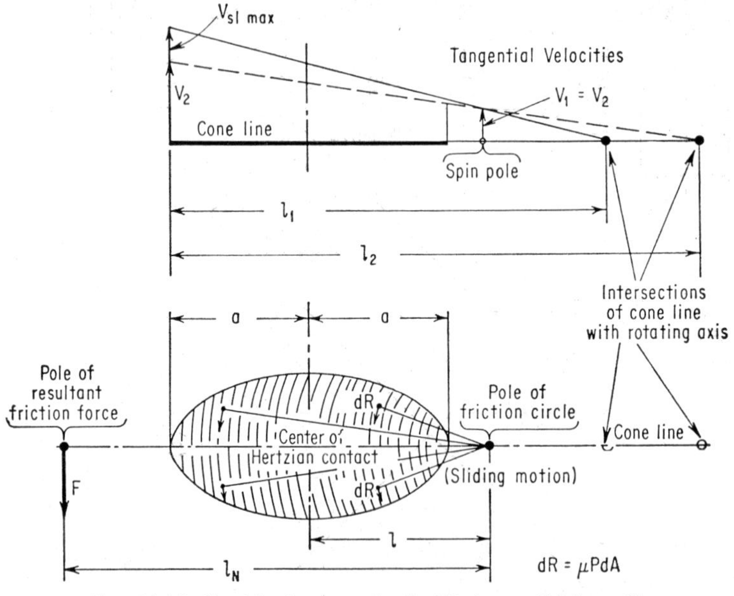

FIG. 14.14. Combined spin and roll with tangential force *F*.

corresponding body, designated l_1 on the driving side and l_2 on the driven side. The location of the spin pole may be determined by considering that the driven side at the center of the contact area has a lower speed than the driving side.

Employing these assumptions (see Fig. 14.14), mathematical analyses for rectangular[20,21] and elliptical[19] Hertzian areas have been performed. The results are presented

Table 14.6. Elliptical Hertzian Areas. Offsets of Force Pole and Spin Pole for Calculation of Friction Losses

cos θ		+0.843		+0.600		+0.539		+0.481		+0.424		+0.368		+0.264		+0.140		±0.0	
a/b		0.2		0.4		0.45		0.50		0.55		0.60		0.70		0.827		1.0	
$\frac{\mu P}{T} = S_R$	$\frac{T}{\mu P}$	$l_N/\sqrt{ab}$	$l/\sqrt{ab}$	$l_N/\sqrt{ab}$	$l/\sqrt{ab}$	$l_N/\sqrt{ab}$	$l/\sqrt{ab}$	$l_N/\sqrt{ab}$	$l/\sqrt{ab}$	$l_N/\sqrt{ab}$	$l/\sqrt{ab}$	$l_N/\sqrt{ab}$	$l/\sqrt{ab}$	$l_N/\sqrt{ab}$	$l/\sqrt{ab}$	$l_N/\sqrt{ab}$	$l/\sqrt{ab}$	$l_N/\sqrt{ab}$	$l/\sqrt{ab}$
1.25	0.8	1.880	1.130	1.395	0.840	1.325	0.825	1.290	0.800	1.246	0.780	1.222	0.773	1.188	0.760	1.173	0.772	1.186	0.800
1.4	0.715	1.720	0.825	1.373	0.648	1.303	0.633	1.278	0.629	1.250	0.630	1.225	0.634	1.202	0.638	1.193	0.655	1.197	0.680
1.6	0.625	1.754	0.617	1.424	0.517	1.368	0.514	1.330	0.515	1.303	0.522	1.285	0.535	1.260	0.542	1.257	0.550	1.250	0.575
1.8	0.556	1.867	0.490	1.507	0.435	1.455	0.442	1.404	0.444	1.383	0.450	1.365	0.460	1.333	0.472	1.327	0.477	1.316	0.500
2.0	0.5	2.000	0.405	1.600	0.378	1.555	0.387	1.488	0.392	1.467	0.396	1.450	0.402	1.422	0.415	1.412	0.424	1.409	0.437
2.2	0.455	2.170	0.343	1.720	0.332	1.650	0.345	1.585	0.352	1.560	0.355	1.540	0.356	1.520	0.370	1.480	0.380	1.504	0.390
2.4	0.417	2.377	0.295	1.854	0.294	1.765	0.310	1.703	0.317	1.676	0.320	1.657	0.325	1.635	0.330	1.627	0.340	1.593	0.353
2.6	0.385	2.532	0.262	1.968	0.268	1.880	0.280	1.818	0.285	1.788	0.290	1.769	0.295	1.732	0.302	1.725	0.312	1.710	0.320
3.33	0.3	3.33	0.300	2.49	0.195	2.40	0.205	2.34	0.213	2.28	0.215	2.24	0.218	2.16	0.225	2.13	0.236	2.095	0.242
4	0.25	4.08	0.245	3.04	0.157	2.97	0.164	2.82	0.170	2.73	0.175	2.67	0.179	2.56	0.186	2.51	0.193	2.49	0.198

cos θ		−0.133		−0.25		−0.340		−0.415		−0.480		−0.532		−0.5785		−0.617	
a/b		1.2		1.4		1.6		1.8		2.0		2.2		2.4		2.6	
$\frac{\mu P}{T} = S_R$	$\frac{T}{\mu P}$	$l_N/\sqrt{ab}$	$l/\sqrt{ab}$	$l_N/\sqrt{ab}$	$l/\sqrt{ab}$	$l_N/\sqrt{ab}$	$l/\sqrt{ab}$	$l_N/\sqrt{ab}$	$l/\sqrt{ab}$	$l_N/\sqrt{ab}$	$l/\sqrt{ab}$	$l_N/\sqrt{ab}$	$l/\sqrt{ab}$	$l_N/\sqrt{ab}$	$l/\sqrt{ab}$	$l_N/\sqrt{ab}$	$l/\sqrt{ab}$
1.25	0.8	1.202	0.824	1.227	0.865	1.263	0.897	1.280	0.936	1.325	0.977	1.360	1.013	1.388	1.047	1.433	1.079
1.4	0.715	1.220	0.711	1.245	0.743	1.275	0.772	1.292	0.802	1.325	0.835	1.370	0.862	1.399	0.889	1.437	0.923
1.6	0.625	1.270	0.603	1.298	0.630	1.333	0.652	1.366	0.673	1.389	0.700	1.424	0.720	1.460	0.743	1.488	0.773
1.8	0.556	1.346	0.527	1.385	0.547	1.420	0.565	1.447	0.581	1.486	0.600	1.500	0.620	1.545	0.640	1.570	0.669
2.0	0.5	1.435	0.465	1.482	0.483	1.524	0.495	1.550	0.511	1.585	0.525	1.613	0.543	1.640	0.568	1.675	0.586
2.2	0.455	1.527	0.418	1.580	0.434	1.620	0.447	1.653	0.458	1.687	0.472	1.715	0.487	1.751	0.505	1.785	0.527
2.4	0.417	1.643	0.375	1.680	0.392	1.730	0.402	1.760	0.415	1.795	0.427	1.831	0.440	1.870	0.457	1.906	0.477
2.6	0.385	1.748	0.343	1.787	0.357	1.835	0.367	1.865	0.379	1.905	0.390	1.940	0.405	1.980	0.420	2.039	0.437
3.33	0.3	2.15	0.259	2.18	0.269	2.24	0.276	2.32	0.282	2.35	0.296	2.39	0.307	2.46	0.318	2.50	0.334
4.0	0.25	2.50	0.212	2.56	0.218	2.62	0.225	2.65	0.236	2.76	0.244	2.84	0.252	2.92	0.260	2.94	0.277

Table 14.7. Rectangular Hertzian Areas. Offsets of Force Pole and Spin Pole for Calculation of Friction Losses

a/b		0.2		0.3		0.4		0.5		0.6		0.7		0.8		1.0	
$\frac{\mu P}{T} = S_R$	$\frac{T}{\mu P}$	$l_N/\sqrt{ab}$	$l/\sqrt{ab}$	$l_N/\sqrt{ab}$	$l/\sqrt{ab}$	$l_N/\sqrt{ab}$	$l/\sqrt{ab}$	$l_N/\sqrt{ab}$	$l/\sqrt{ab}$	$l_N/\sqrt{ab}$	$l/\sqrt{ab}$	$l_N/\sqrt{ab}$	$l/\sqrt{ab}$	$l_N/\sqrt{ab}$	$l/\sqrt{ab}$	$l_N/\sqrt{ab}$	$l/\sqrt{ab}$
1.25	0.8	2.105	1.270	1.780	1.080	1.600	0.985	1.478	0.940	1.417	0.920	1.388	0.915	1.387	0.920	1.398	0.955
1.4	0.715	1.980	0.985	1.705	0.835	1.555	0.780	1.460	0.765	1.403	0.760	1.392	0.770	1.398	0.785	1.421	0.830
1.6	0.625	1.980	0.730	1.725	0.650	1.577	0.625	1.500	0.625	1.478	0.630	1.473	0.650	1.480	0.665	1.516	0.710
1.8	0.556	2.105	0.585	1.797	0.540	1.651	0.535	1.597	0.535	1.595	0.545	1.590	0.565	1.596	0.580	1.613	0.625
2.0	0.5	2.14	0.505	1.871	0.465	1.770	0.465	1.745	0.470	1.735	0.480	1.728	0.500	1.732	0.515	1.744	0.550
2.2	0.455	2.17	0.440	1.945	0.420	1.905	0.420	1.875	0.430	1.865	0.440	1.858	0.455	1.851	0.465	1.868	0.500
2.4	0.417	2.22	0.390	2.12	0.375	2.07	0.380	2.04	0.395	2.02	0.400	2.00	0.415	2.00	0.430	2.01	0.455
2.6	0.385	2.29	0.360	2.27	0.335	2.25	0.350	2.22	0.365	2.20	0.370	2.18	0.385	2.17	0.390	2.14	0.420
3.33	0.3	2.85	0.275	2.79	0.265	2.75	0.275	2.72	0.280	2.70	0.285	2.66	0.295	2.65	0.300	2.67	0.325
4.0	0.25	3.23	0.225	3.19	0.220	3.17	0.230	3.15	0.235	3.12	0.240	3.08	0.250	3.05	0.255	3.05	0.275

a/b		1.2		1.4		1.6		1.8		2.0		2.2		2.4		2.6	
$\frac{\mu P}{T} = S_R$	$\frac{T}{\mu P}$	$l_N/\sqrt{ab}$	$l/\sqrt{ab}$	$l_N/\sqrt{ab}$	$l/\sqrt{ab}$	$l_N/\sqrt{ab}$	$l/\sqrt{ab}$	$l_N/\sqrt{ab}$	$l/\sqrt{ab}$	$l_N/\sqrt{ab}$	$l/\sqrt{ab}$	$l_N/\sqrt{ab}$	$l/\sqrt{ab}$	$l_N/\sqrt{ab}$	$l/\sqrt{ab}$	$l_N/\sqrt{ab}$	$l/\sqrt{ab}$
1.25	0.8	1.430	0.995	1.472	1.040	1.505	1.090	1.563	1.135	1.618	1.185	1.672	1.235	1.728	1.275	1.774	1.320
1.4	0.715	1.486	0.870	1.511	0.920	1.559	0.965	1.613	1.010	1.666	1.055	1.719	1.100	1.765	1.150	1.804	1.195
1.6	0.625	1.562	0.755	1.610	0.790	1.660	0.830	1.711	0.875	1.762	0.910	1.813	0.960	1.863	1.005	1.905	1.055
1.8	0.556	1.648	0.665	1.699	0.700	1.751	0.730	1.804	0.775	1.858	0.810	1.918	0.855	1.978	0.895	2.035	0.935
2.0	0.5	1.770	0.590	1.807	0.625	1.861	0.655	1.918	0.695	1.980	0.730	2.04	0.765	2.105	0.800	2.155	0.830
2.2	0.455	1.899	0.330	1.943	0.565	1.993	0.595	2.048	0.630	2.10	0.660	2.17	0.695	2.24	0.730	2.31	0.760
2.4	0.417	2.03	0.485	2.07	0.510	2.12	0.545	2.18	0.575	2.25	0.605	2.32	0.635	2.39	0.670	2.46	0.695
2.6	0.385	2.17	0.445	2.22	0.475	2.28	0.495	2.34	0.530	2.41	0.555	2.48	0.585	1.56	0.615	2.64	0.645
3.33	0.3	2.69	0.345	2.75	0.365	2.85	0.385	2.89	0.410	2.98	0.430	3.04	0.455	3.12	0.480	3.15	0.510
4.0	0.25	3.12	0.290	3.23	0.310	3.31	0.325	3.40	0.340	3.51	0.360	3.58	0.380	3.71	0.400	3.77	0.420

in Figs. 14.15 and 14.16 as well as Tables 14.6 and 14.7. The Hertzian relationships are given in Table 14.8, where a = one-half the length of the contact area in the direction of the mantle line of the cone, b = one-half the length of the contact area in the direction of rolling, r_1 and r_2 = radii of curvature in the direction of rolling perpendicular to the contact area, r_3 and r_4 = radii of curvature in the direction of the mantle line of the cone perpendicular to the contact area.

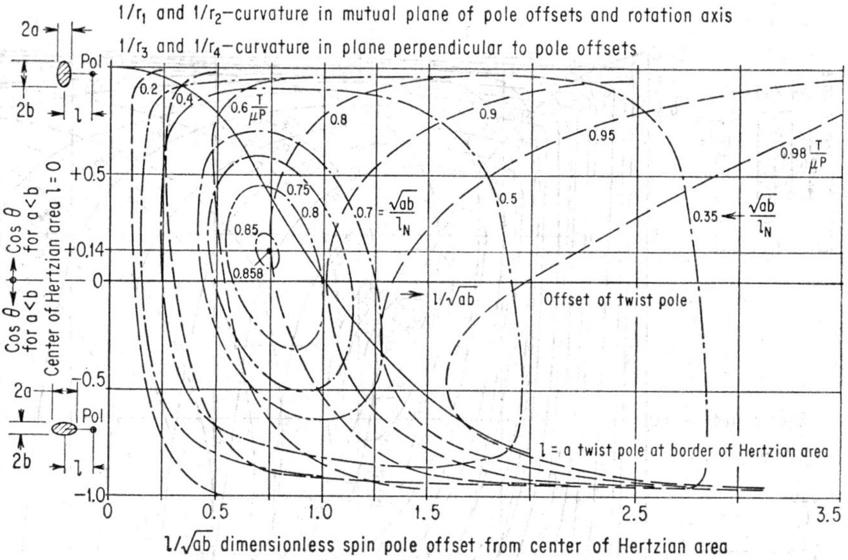

FIG. 14.15. Elliptical Hertzian areas, friction situation for roll and spin.

In the graphs and tables, the offset of the spin or twist pole $l/\sqrt{ab}$ is given as a function of a/b and $\cos\theta$, respectively. The coefficient of friction μ used in defining the utilization $F_t/\mu P$ is the mean coefficient resulting from sliding velocity in the Hertzian area. Even if the coefficient of friction decreases with the sliding velocity (see Figs. 14.18 and 14.19), the maximum sliding velocity in the contact area can be safely assumed to be

$$v_{\text{sliding}} = (l + a)\omega_b \qquad (14.18)$$

for estimating a coefficient of friction. With these values, $l_N/\sqrt{ab}$ can be determined.

Because the quantity $l_N/\sqrt{ab}$ is directly proportional to the total loss in efficiency at the contact area, the maximum reciprocal value $\sqrt{ab}/l_N$ indicates the condition of the lengths l_1 and l_2, and as seen in Fig. 14.13, the losses decrease with increasing l_1

Table 14.8. Hertzian Relationships

Elliptical Hertzian area	*Rectangular Hertzian area*
$ab = 3P/2\pi p_0$	$ab = P/\pi P_0$
$\dfrac{a}{b}$ = function of $\cos\theta$ or curvatures	$b^2 = \dfrac{4Pv_0(1 - v^2)}{\pi Ea}$
$\cos\theta = \dfrac{(1/r_1 + 1/r_2) - (1/r_3 + 1/r_4)}{1/r_1 + 1/r_2 + 1/r_3 + 1/r_4}$	$1 = \dfrac{1}{r_1} + \dfrac{1}{r_2}$
$\cos\theta = 1$, circular area	$a/b = 1$, square area
	$r < 0$, concave curvature

and l_2. Because it is difficult to use Fig. 14.16 for very low values of b/a, another presentation of the offsets l and $(l_N - l)$ is given in Fig. 14.17. The case shown in Fig. 14.12 is included in Fig. 14.17 as $a/b = \infty$.

As explained in ref. 21 in greater detail, elliptical Hertzian areas result in lower losses than corresponding rectangular areas. These conditions are difficult to produce with soft materials; however, in practice one member of the pair is always rubbing on the same ring path, and therefore the rectangular contact area wears to an

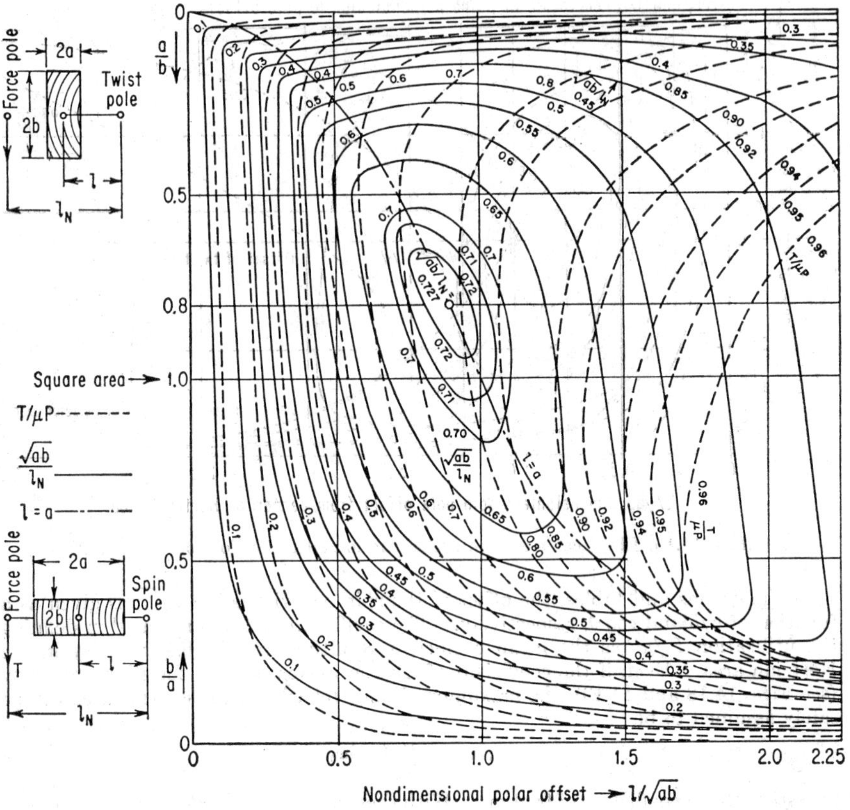

FIG. 14.16. Rectangular Hertzian areas. Friction for roll and spin.

$$ab = P/\pi p_0 \qquad b^2 = 4Pr_0(l - v^2)/\pi Ea$$

l due to slip, P = normal force to Hertzian area, $F/\mu P$ = uilization, l_N due to efficiency, P_0 = maximum of Hertzian pressure, μ = coefficient of friction.

elliptical area. Hard materials such as hardened steel are generally fabricated with curvatures so as to produce elliptical Hertzian areas.

Although soft materials running dry with high coefficients of friction vary widely, it is sufficient to use the values given previously for the various drives of constant transmission ratios. For lubricated pairs made of hardened steel, the friction depends upon several factors but does not vary so much as it does for relatively soft materials. The lubricant serves as a coolant and provides a thin layer between contacting bodies which results in an advantageous distribution of Hertzian pressure. Usual surfaces are not honed but ground, and hydrodynamic support is low but not insignificant at high relative velocities. For high Hertzian pressures of 70,000 to 280,000 psi, the lubricant viscosity should be high in order to transmit the necessary high tangential

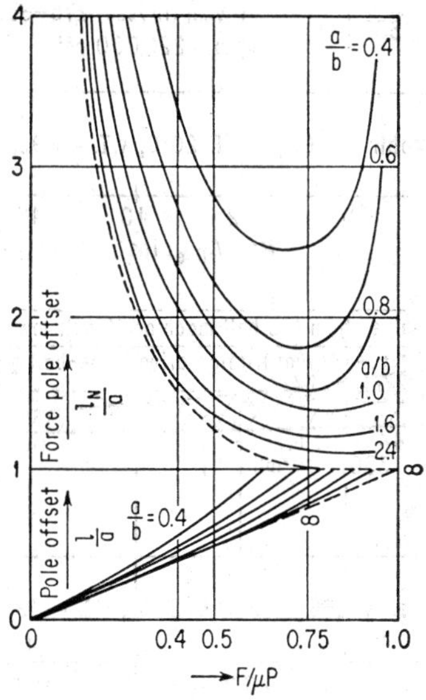

FIG. 14.17. Roll and spin for rectangular Hertzian areas.

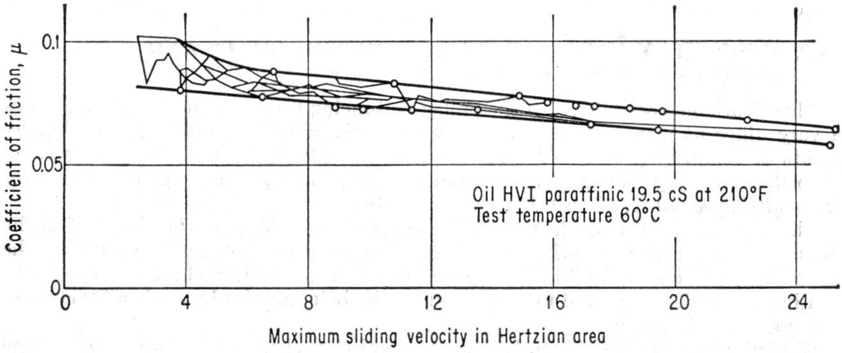

FIG. 14.18. Sliding friction from two ball tests.[19,22] P_0 454,000 psi (low rolling speed).

forces with relatively high friction. Research and practice[14,19,21,22] with many infinitely variable drives indicate that satisfactory lubricants are presently available (see Figs. 14.18 and 14.19).

The allowable Hertzian pressure with respect to endurance and tractive capacity is not yet precisely defined. For surface hardnesses of 62 to 64 Rockwell C and elliptic Hertzian areas, the approximate limit is 425,000 psi; for rectangular areas the limit is 280,000 psi. Two-thirds of these limiting values are realized in presently existing drives.

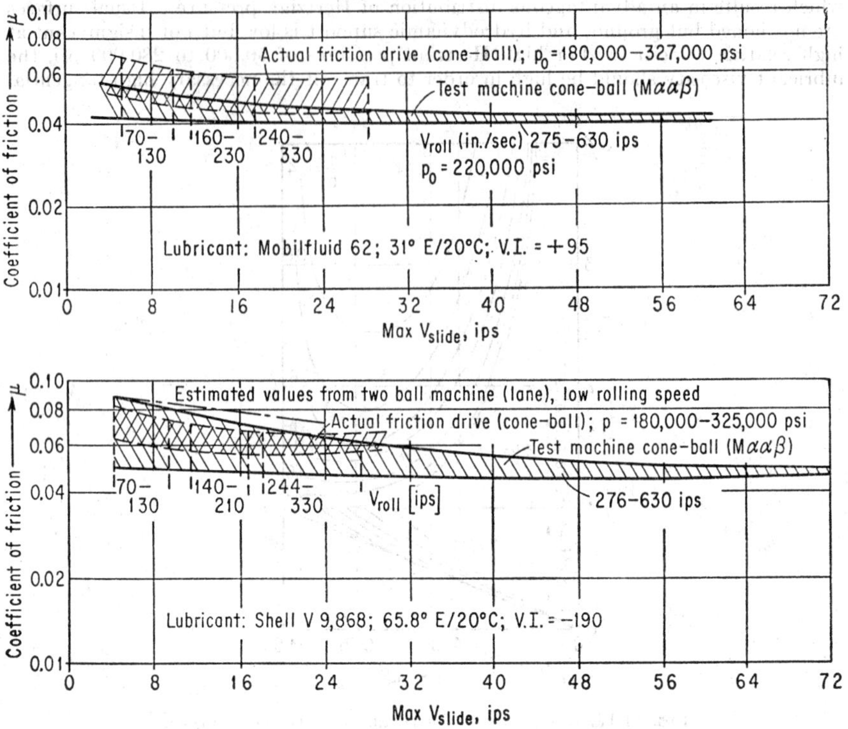

FIG. 14.19. Combined roll and twist comparison of measured friction in different test stands.

14.9. INFINITELY VARIABLE DRIVES OF THE CHAIN AND BELT TYPES[23,24,25,27,28]

These variable-speed drives of the pulley-embracing type operate by friction, but in a very different manner from those previously described. In these drives, the elements of the driving and driven members are all conical. The pulleys are wedge-shaped, formed by two axially displaceable pulley halves forced against the power transmitting member. For V-belt drives, see Sec. 29.

The adjustment of a single V-belt can be accomplished by moving one half of the pulley axially toward the other. Most drive slides maintain proper spacing by the action of spring force (Fig. 14.20). In this way a range of adjustability of about 10:1 can be achieved. The axial force exerted on the belt can also be made a function of the transmitted torque.

Chain-type Drives. If steel power-transmission elements are used, the wedge angle can be substantially reduced (approximately 12° total angle as compared with 24 to 34° with V-belt drives). The forerunner of these steel-on-steel wedge drives is the PIV drive system with laminated links. This drive experiences very little slip because of a form-locked connection between the toothed pulley and a link which is carried by the axially slidable laminations, as shown in Fig. 14.21*a*.

A later development is the PIV double roller chain shown in Fig. 14.21*b*, which is force-locked by Hertzian pressure on smooth-sided pulleys. This drive is suitable for lower powers but higher velocities. The newest development for larger power requirements (up to 100 hp) is the PIV ring rolling chain shown in Fig. 14.21*c*, which because of its larger ring diameter permits the transmission of larger forces. More power can be transmitted by a two-chain arrangement shown in Fig. 14.21*d*.

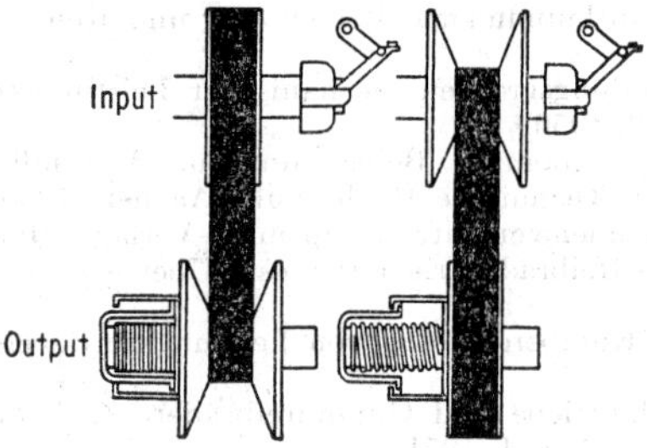

FIG. 14.20. Wide V-belt positions at high speed (left) and low speed (right).

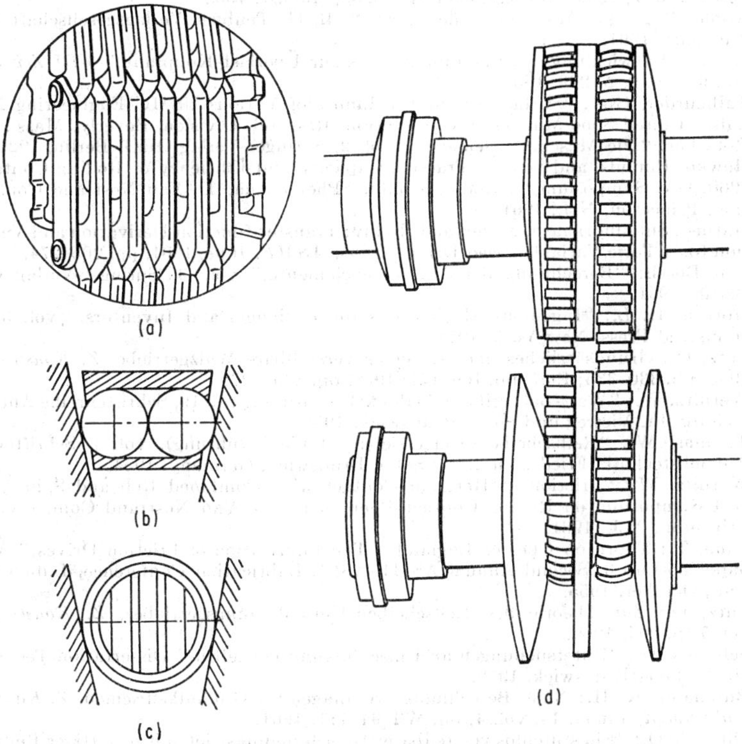

FIG. 14.21. PIV chain drives. (*a*) Laminated chain. (*b*) Double roller chain. (*c*) Ring rolling chain. (*d*) Double chain.

In ring rolling chain drives very often an arrangement is used by which the pulleys are forced against the load-transmitting element with a force which is dependent upon the magnitude of the transmitted torque. In this way the contact force is automatically reduced as the transmitted torque is reduced, and therefore the efficiencies of operation are more favorable during part-load operation.

For further information on this section the reader is referred to refs. 3, 4, 10, 11, 22, 26, 31, and 34.

References

1. v. Thüngen: "Stufenlose Getriebe, insbes. Schaltwerksgetriebe" in Bussien, "Automobiltechnisches Handbuch," Bd. I, Verlag Herbert Cram, Berlin, 1953.

2. Stolzle, Hart: "Freilaufkupplungen, Berechnung und Konstruktion," Springer-Verlag OHG, Berlin, 1961.
3. Thomas, W.: Anwendungsgrenzen mechanischer Leistungsgetriebe, *VDI Zeitschrift*, vol. 92, no. 33, p. 902, 1950.
4. Vieregge, "Energieübertragung, Berechnung u. Anwendbarkeit van Reibradgetrieben," Dissertation Technische Hochschule, Aachen, 1950.
5. Niemann, G.: "Maschinenelemente I," Springer-Verlag OHG, Berlin, 1950.
6. Peppler: Zweiachsige Reibradantriebe für feste Übersetzungen, *Z. Konstruktion*, 1949, pp. 289, 336.
7. Richter, Ohlendorf: Kurzberechnung von Leistungsgetrieben, *Z. Konstruktion*, 1959, p. 424.
8. Feighofen: "Reibradantriebe und Gummireibräder, *Z. Industriekurier, Technik und Forschung*, no. 50 (12), Apr. 4, 1951.
9. Continental Gummi Werke AG, Hannover: "Informationen über Antriebstechnik," Series 2, 6, 7, 1959, and pamphlet WT 5235, August, 1958.
10. Kohler-Rognitz: "Maschinenteile," Part 2, B. G. Teubner Verlagsgesellschaft mbH, Stuttgart, 1961.
11. Kroner: Entwicklung des Reibradantriebes zur Überlast-Kupplung," *VDI Zeitschrift*, vol. 93, no. 9, p. 229, 1951.
12. Talbourdet, Guy: "Surface Endurance Limits of Various USMC Engineering Materials," United Shoe Machinery Corporation, Research Division, Beverly, Mass., 1957.
13. Rotscher: "Die Maschinenelemente," vol. 2, Springer-Verlag OHG, Berlin, 1929.
14. Hewko, Rounds, and Scott: "Tractive Capacity and Efficiency of Rolling Contacts," 1960, GM Symposium on Rolling Contact Phenomena, D. Van Nostrand Company, Inc., Princeton, N.J., 1961.
15. Harned, Subhindranath, Miller, and Roddy: Transfer-function Derivation and Verification for a Toric Variable-speed Drive, *Trans. ASME*, June, 1961, pp. 265–274.
16. Ten Bosch: "Berechnung der Maschinenelemente," vol. 3, Springer-Verlag OHG, Berlin, 1951.
17. Horton, H. L.: "Ingenious Mechanisms for Designers and Inventors," vol. 3, The Industrial Press, New York, 1951.
18. Lutz, O.: Grundsätzliches über stufenlos verstellbare Wälzgetriebe, *Z. Konstruktion*, 1955, pp. 330–335; 1957, pp. 169–171; 1958, pp. 425–427.
19. Wernitz, W.: "Wälz-Bohrreibung" (Punktberührung), vol. 19, Schriftenreihe Antriebstechnik, Fr. Vieweg und Sohn, Brunswick, 1958.
20. Thomas, W.: "Reibscheibenregelgetriebe" (Linienberührung), vol. 4, Schriftenreihe Antriebstechnik, Friedr. Vieweg & Sohn, Brunswick, Germany, 1954.
21. Wernitz, W.: "Friction at Hertzian Contact with Combined Roll and Spin," 1960, GM Symposium on Rolling Contact Phenomena, D. Van Nostrand Company, Inc., Princeton, N.J., 1961.
22. Lane, T.: Thornton Chester England: "The Lubrication of Friction Drives," ASME paper 55-Lub 3, Second Annual ASME-ASLE Lubrication Conference, Indianapolis, Ind., October, 1955.
23. Lutz, O.: Zur Theorie des Keilscheiben-Umschlingungsgetriebes, *Z. Konstruktion*, vol. 7, p. 265, 1960.
24. Schlums, D.: "Untersuchungen an Umschlingungsgetrieben," Dissertation Technische Hochschule, Brunswick, 1959.
25. Bubmann, K. H.: Neue Berechnungsgrundlagen für Gummikeilriemen, *Z. Kautschuk und Gummi*, annual 14, vol. 4, pp. WT 94–111, 1961.
26. Dittrich, O.: "Ein stufenlos verstellbarer Umschlingungstrieb mit neuartiger Reibungskette" in "Getriebe-Kupplungen-Antriebeselemente," vol. 18, p. 75, Schriftenreihe Antriebstechnik, Friedr. Vieweg & Sohn, Brunswick, Germany, 1957.
27. Worley, W. Sp.: "Designing Adjustable-speed V-belt Drives for Farm Implements," SAE National Tractor Meeting, Milwaukee, Wis., Sept. 16, 1954.
28. Marco, S. M., W. L. Starkey, and K. G. Hornung: "A Quantitative Investigation of the Factors Which Influence the Fatigue Life of a V-belt," ASME paper 59-SA-18.
29. Despins, M. P.: "The Mechanical Variable Speed Drive as a Final Control Element in Process Control Systems," Instrument Society of America paper 4-59, 14th Annual Instrument Automation Conference, Chicago, Ill., Sept. 21–25, 1959.
30. Graham Transmissions, Inc., Menomonie Falls, Wis.
31. Cleveland Worm and Gear Co., Cleveland, Ohio.
32. Hertz, H.: "Gesammelte Werke," Springer, Berlin, 1895.
33. Stribeck: Kugellager für beliebige Belostungen, VDI, 1901, vol. 45, p. 473 (*Trans. ASME*, vol. 29, pp. 420–463).
34. Anon.: "Friction and Wear in Machinery," ASME, vol. 14, 1960, New York, pp. 183–201, ***Friction and Wear in Friction Transmission of the Toroidal-Spherical Type.***
35. **Dvorak, D. Z.: Mechanical Drive, *Prod. Eng.*, Dec. 23, 1963, p. 63; Feb. 3, 1964 p. 59**

Section 15

MECHANICS OF MATERIALS

By

M. ZAID, Sc.D., *President, Technik, Inc., Jericho, Long Island, N.Y.*

R. P. KOLB, M.S., *Engineering Staff, Technik, Inc., Jericho, Long Island, N Y.*

CONTENTS

15.1. INTRODUCTION

Mechanics of materials deals with the internal reaction of a body to the application of external loads. This reaction, called *stress* (force per unit area), tends to produce *strain* (deformation per unit length) at the same point where the stress is measured. The relationship between stress and strain, at a point, is called the *stress-strain* law. The stress-strain law is linear in small-deformation theory of elasticity, nonlinear but reversible in large-deformation theory of elasticity, and nonlinear and nonreversible in the theory of plasticity.

In general it can be shown from *d'Alembert's principle* that any dynamic problem can be reduced to a static-problem formulation by the inclusion of inertia forces as part of the loading system. Accordingly this section will be restricted to statics.

ELASTIC THEORY

15.2. STRESS

(a) Definition[2]

Stress is defined as the force per unit area acting on an "elemental" plane in the body. Engineering units of stress are generally pounds per square inch. If the force is normal to the plane the stress is termed *tensile* or *compressive* depending upon whether the force tends to extend or shorten the element. If the force acts parallel to the elemental plane the stress is termed *shear*. Shear tends to deform by causing neighboring elements to slide relative to one another.

(b) Components of Stress[2]

A complete description of the internal forces (stress distributions) requires that stress be defined on three perpendicular faces of an interior element of a structure. In Fig. 15.1 a small element is shown, and, omitting higher-order effects, the stress resultant on any face can be considered as acting at the center of the area.

The direction and type of stress at a point are described by subscripts to the stress symbol σ or τ. The first subscript defines the plane on which the stress acts and the second indicates the direction in which it acts. The plane on which the stress acts is indicated by the normal axis to that plane; e.g., the x plane is normal to the x axis. Conventional notation omits the second subscript for the normal stress and replaces the σ by a τ for the shear stresses. The *stress components* can thus be represented:

Normal stress:

$$\sigma_{xx} \equiv \sigma_x \qquad \sigma_{yy} \equiv \sigma_y \qquad \sigma_{zz} \equiv \sigma_z \tag{15.1}$$

Shear stress:

$$\begin{aligned} \sigma_{xy} &\equiv \tau_{xy} & \sigma_{yz} &\equiv \tau_{yz} \\ \sigma_{xz} &\equiv \tau_{xz} & \sigma_{zx} &\equiv \tau_{zx} \\ \sigma_{yx} &\equiv \tau_{yx} & \sigma_{zy} &\equiv \tau_{zy} \end{aligned} \tag{15.2}$$

In tensor notation, the stress components are

$$\sigma_{ij} = \begin{pmatrix} \sigma_x & \tau_{xy} & \tau_{xz} \\ \tau_{yx} & \sigma_y & \tau_{yz} \\ \tau_{zx} & \tau_{zy} & \sigma_z \end{pmatrix} \tag{15.3}$$

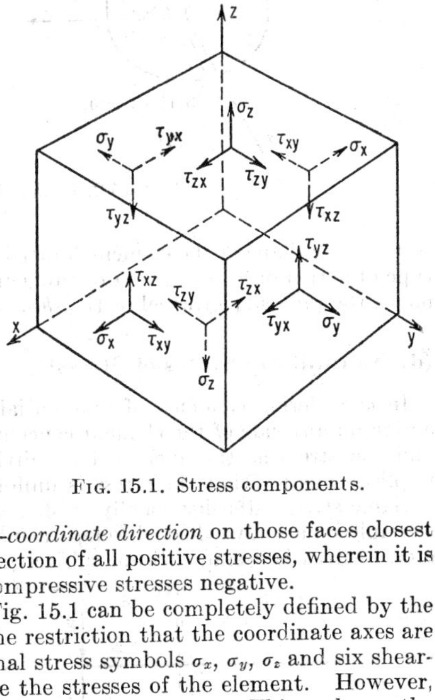

FIG. 15.1. Stress components.

Stress is *positive* if it acts in the *positive-coordinate direction* on those element faces farthest from the origin, and in the *negative-coordinate direction* on those faces closest to the origin. Figure 15.1 indicates the direction of all positive stresses, wherein it is seen that tensile stresses are positive and compressive stresses negative.

The total load acting on the element of Fig. 15.1 can be completely defined by the stress components shown, subject only to the restriction that the coordinate axes are mutually orthogonal. Thus the three normal stress symbols σ_x, σ_y, σ_z and six shear-stress symbols τ_{xy}, τ_{xz}, τ_{yx}, τ_{yz}, τ_{zx}, τ_{zy} define the stresses of the element. However, from equilibrium considerations, $\tau_{xy} = \tau_{yx}$, $\tau_{yz} = \tau_{zy}$, $\tau_{xz} = \tau_{zx}$. This reduces the necessary number of symbols required to define the stress state to σ_x, σ_y, σ_z, τ_{xy}, τ_{xz}, τ_{yz}

(c) Simple Uniaxial States of Stress[1]

Consider a simple bar subjected to axial loads only. The forces acting at a transverse section are all directed normal to the section. The uniaxial normal stress at the section is obtained from

$$\sigma = P/A \tag{15.4}$$

where P = total force, A = cross-sectional area.

Uniaxial shear occurs in a circular cylinder, loaded as in Fig. 15.2*a*, with a radius which is large compared to the wall thickness. This member is subjected to a torque distributed about the upper edge:

$$T = \Sigma P r \tag{15.5}$$

Now consider a surface element (assumed plane) and examine the stresses acting. The stresses τ which act on surfaces *a-a* and *b-b* in Fig. 15.2*b* tend to distort the original

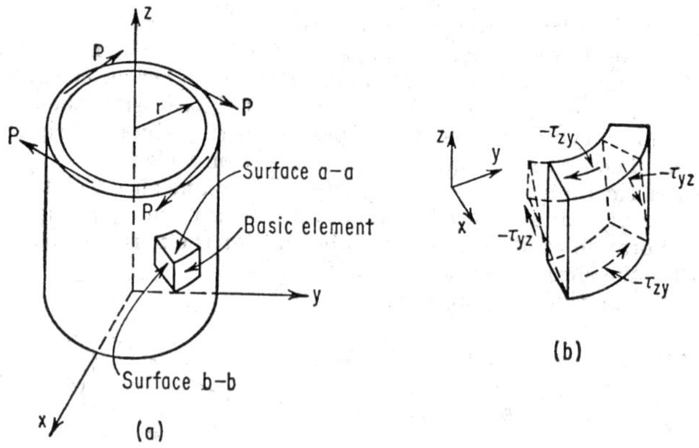

FIG. 15.2. Uniaxial shear basic element.

rectangular shape of the element (dotted shape) into the parallelogram shown. This type of action of a force along or tangent to a surface produces shear within the element, the intensity of which is the *shear stress.*

(d) Nonuniform States of Stress[1]

In considering elements of differential size, it is permissible to assume the force acting on any side of the element concentrated at the center of the area of that side, and the stress as the average force divided by the side area. Hence it has been implied thus far that the stress is uniform. In members of finite size, however, a variable stress intensity usually exists across any given surface of the member. An example of a body which develops a distributed stress pattern across a transverse cross section is a simple beam subjected to a bending load as shown in Fig. 15.3*a.*

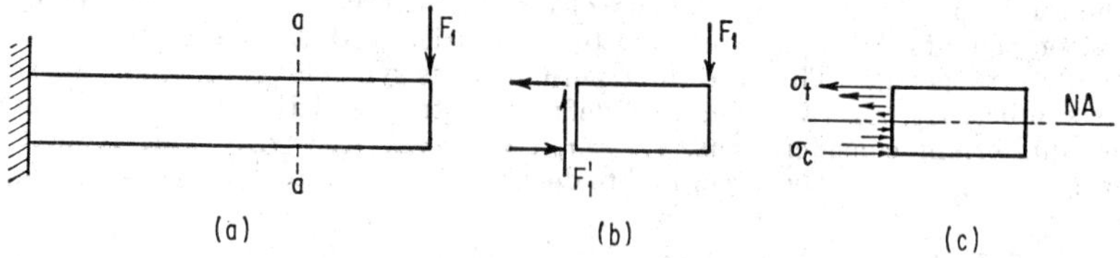

FIG. 15.3. Distributed stress on a simple beam subjected to a bending load.

If a section is then taken at *a-a*, F_1' must be the internal force acting along *a-a* to maintain equilibrium. Forces F_1 and F_1' constitute a couple which tends to rotate the element in a clockwise direction, and therefore a resisting couple must be developed at *a-a* (see Fig. 15.3*b*). The internal effect at *a-a* is a stress distribution with the upper portion of the beam in tension and the lower portion in compression, as in Fig. 15.3*c.* The line of zero stress on the transverse cross section is the *neutral axis* and passes through the centroid of the area.

(e) Combined States of Stress

Tension-Torsion. A body loaded simultaneously in direct tension and torsion, such as a rotating vertical shaft, is subject to a combined state of stress. Figure 15.4*a* depicts such a shaft with end load W, and constant torque T applied to maintain uniform rotational velocity. With reference to *a-a*, considering each load separately, a force system as shown in Fig. 15.4*b* and *c* is developed at the internal surface

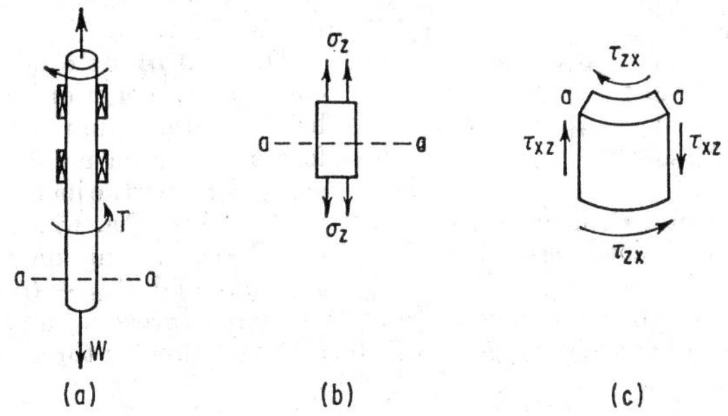

FIG. 15.4. Body loaded in direct tension and torsion.

a-a for the weight load and torque, respectively. These two stress patterns may be superposed to determine the *combined* stress situation for a shaft element.

Flexure-Torsion. If in the above case the load W were horizontal instead of vertical, the combined stress picture would be altered. From previous considerations of a simple beam, the stress distribution varies linearly across section *a-a* of the shaft of Fig. 15.5*a*. The stress pattern due to flexure then depends upon the location of the

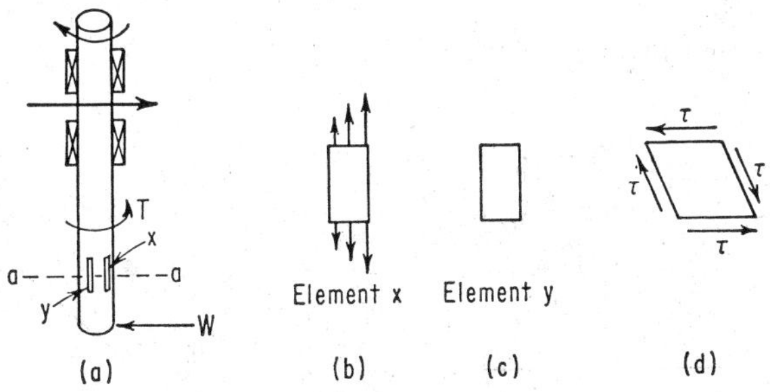

FIG. 15.5. Body loaded in flexure and torsion.

element in question; e.g., if the element is at the outside (element x) then it is undergoing maximum tensile stress (Fig. 15.5*b*), and the tensile stress is zero if the element is located on the horizontal center line (element y) (Fig. 15.5*c*). The shearing stress is still constant at a given element, as before (Fig. 15.5*d*). Thus the *combined* or *superposed* stress state for this condition of loading varies across the entire transverse cross section.

(f) Stress Equilibrium

Equilibrium relations must be satisfied by each element in a structure. These are satisfied if the resultant of all forces acting on each element equals zero in each of three mutually orthogonal directions on that element. The above applies to all situations of *static equilibrium*. In the event that some elements are in motion an inertia term must be added to the equilibrium equation. The inertia term is the elemental mass multiplied by the absolute acceleration taken along each of the mutually perpendicular axes. The equations which specify this latter case are called *dynamic-equilibrium equations* (see Sec. 3).

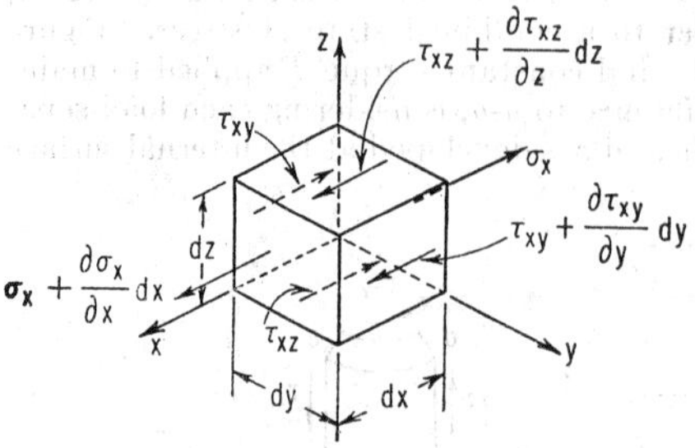

FIG. 15.6. Incremental element (*dx, dy, dz*) with incremental variation of stress.

Three-dimensional Case.[5] The equilibrium equations can be derived by separately summing all x, y, and z forces acting on a differential element accounting for the incremental variation of stress (see Fig. 15.6). Thus the normal forces acting on areas $dz\,dy$ are $\sigma_x\,dz\,dy$ and $[\sigma_x + (\partial\sigma_x/\partial x)\,dx]\,dz\,dy$. Writing x force-equilibrium equations, and by a similar process y and z force-equilibrium equations, and canceling *higher-order* terms, the following three *cartesian equilibrium equations* result:

$$\partial\sigma_x/\partial x + \partial\tau_{xy}/\partial y + \partial\tau_{xz}/\partial z = 0 \qquad (15.6)$$
$$\partial\sigma_y/\partial y + \partial\tau_{yz}/\partial z + \partial\tau_{yx}/\partial x = 0 \qquad (15.7)$$
$$\partial\sigma_z/\partial z + \partial\tau_{zx}/\partial x + \partial\tau_{zy}/\partial y = 0 \qquad (15.8)$$

or, in cartesian stress-tensor notation,

$$\sigma_{ij,j} = 0 \qquad i, j = x, y, z \qquad (15.9)$$

and, in general tensor form,

$$g^{jk}\sigma_{ij,k} = 0 \qquad (15.10)$$

where g^{jk} is the contravariant metric tensor.

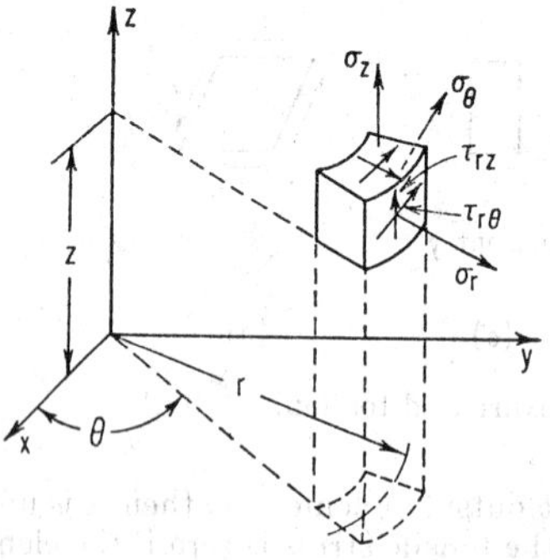

FIG. 15.7. Stresses on a cylindrical element.

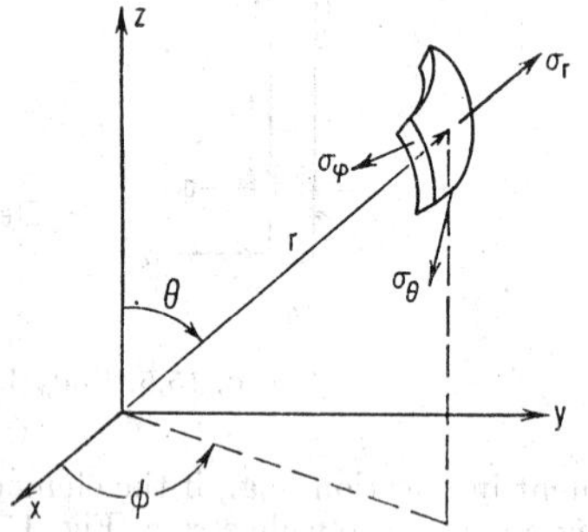

FIG. 15.8. Stresses on a spherical element.

Cylindrical-coordinate equilibrium considerations lead to the following set of equations (see Fig. 15.7):

$$\partial\sigma_r/\partial r + (1/r)(\partial\tau_{r\theta}/\partial\theta) + \partial\tau_{rz}/\partial z + (\sigma_r - \sigma_\theta)/r = 0 \qquad (15.11)$$
$$\partial\tau_{r\theta}/\partial r + (1/r)(\partial\sigma_\theta/\partial\theta) + \partial\tau_{\theta z}/\partial z + 2\tau_{r\theta}/r = 0 \qquad (15.12)$$
$$\partial\tau_{rz}/\partial r + (1/r)(\partial\tau_{\theta z}/\partial\theta) + \partial\sigma_z/\partial z + \tau_{rz}/r = 0 \qquad (15.13)$$

The corresponding *spherical polar-coordinate* equilibrium equations are (see Fig. 15.8)

$$\partial\sigma_r/\partial r + (1/r)(\partial\tau_{r\theta}/\partial\theta) + (1/r\sin\theta)(\partial\tau_{r\phi}/\partial\phi) + (1/r)(2\sigma_r - \sigma_\theta - \sigma_\phi + \tau_{r\theta}\cot\theta) = 0 \quad (15.14)$$

$$\partial\tau_{r\theta}/\partial r + (1/r)(\partial\sigma_\theta/\partial\theta) + (1/r\sin\theta)(\partial\tau_{\theta\phi}/\partial\phi) + (1/r)[(\sigma_\theta - \sigma_\phi)\cot\theta + 3\tau_{r\theta}] = 0 \quad (15.15)$$

$$\partial\tau_{r\phi}/\partial r + (1/r)(\partial\tau_{\theta\phi}/\partial\theta) + (1/r\sin\theta)(\partial\sigma_\phi/\partial\phi) + (1/r)(3\tau_{r\phi} + 2\tau_{\theta\phi}\cot\theta) = 0 \quad (15.16)$$

The general orthogonal curvilinear-coordinate equilibrium equations are

$$h_1h_2h_3[(\partial/\partial\alpha)(\sigma_\alpha/h_2h_3) + (\partial/\partial\beta)(\tau_{\alpha\beta}/h_3h_1) + (\partial/\partial\gamma)(\tau_{\gamma\alpha}/h_1h_2)] + \tau_{\alpha\beta}h_1h_2(\partial/\partial\beta)(1/h_1) + \tau_{\gamma\alpha}h_1h_3(\partial/\partial\gamma)(1/h_1) - \sigma_\beta h_1h_2(\partial/\partial\alpha)(1/h_2) - \sigma_\gamma h_1h_3(\partial/\partial\alpha)(1/h_3) = 0 \quad (15.17)$$

$$h_1h_2h_3[(\partial/\partial\beta)(\sigma_\beta/h_3h_1) + (\partial/\partial\gamma)(\tau_{\beta\gamma}/h_1h_2) + (\partial/\partial\alpha)(\tau_{\alpha\beta}/h_2h_3)] + \tau_{\beta\gamma}h_2h_3(\partial/\partial\gamma)(1/h_2) + \tau_{\alpha\beta}h_2h_1(\partial/\partial\alpha)(1/h_2) - \sigma_\gamma h_2h_3(\partial/\partial\beta)(1/h_3) - \sigma_\alpha h_2h_1(\partial/\partial\beta)(1/h_1) = 0 \quad (15.18)$$

$$h_1h_2h_3[(\partial/\partial\gamma)(\sigma_\gamma/h_1h_2) + (\partial/\partial\alpha)(\tau_{\gamma\alpha}/h_2h_3) + (\partial/\partial\beta)(\tau_{\beta\gamma}/h_3h_1)] + \tau_{\gamma\alpha}h_3h_1(\partial/\partial\alpha)(1/h_3) + \tau_{\beta\gamma}h_3h_2(\partial/\partial\beta)(1/h_3) - \sigma_\alpha h_3h_1(\partial/\partial\gamma)(1/h_1) - \sigma_\beta h_3h_2(\partial/\partial\gamma)(1/h_2) = 0 \quad (15.19)$$

where the α, β, γ specify the coordinates of a point and the distance between two coordinate points ds is specified by

$$(ds)^2 = (d\alpha/h_1)^2 + (d\beta/h_2)^2 + (d\gamma/h_3)^2 \quad (15.20)$$

which allows the determination of h_1, h_2, and h_3 in any specific case.

Thus, in cylindrical coordinates,

$$(ds)^2 = (dr)^2 + (r\,d\theta)^2 + (dz)^2 \quad (15.21)$$

so that

$$\alpha = r \qquad h_1 = 1$$
$$\beta = \theta \qquad h_2 = 1/r$$
$$\gamma = z \qquad h_3 = 1$$

In spherical polar coordinates

$$(ds)^2 = (dr)^2 + (r\,d\theta)^2 + (r\sin\theta\,d\phi)^2 \quad (15.22)$$

so that

$$\alpha = r \qquad h_1 = 1$$
$$\beta = \theta \qquad h_2 = 1/r$$
$$\gamma = \phi \qquad h_3 = 1/(r\sin\theta)$$

All the above equilibrium equations define the conditions which must be satisfied by each interior element of a body. In addition, these stresses must satisfy all surface-stress-boundary conditions. In addition to the cartesian-, cylindrical-, and spherical-coordinate systems, others may be found in the current literature or obtained by reduction from the general curvilinear-coordinate equations given above.

In many applications it is useful to integrate the stresses over a finite thickness and express the resultant in terms of zero or nonzero force or moment resultants as in the beam, plate, or shell theories.

Two-dimensional Case—Plane Stress.[2] In the special but useful case where the stresses in one of the coordinate directions are negligibly small ($\sigma_z = \tau_{xz} = \tau_{yz} = 0$) the general cartesian-coordinate equilibrium equations reduce to

$$\partial\sigma_x/\partial x + \partial\tau_{xy}/\partial y = 0 \quad (15.23)$$
$$\partial\sigma_y/\partial y + \partial\tau_{yx}/\partial x = 0 \quad (15.24)$$

The corresponding cylindrical-coordinate equilibrium equations become

$$\partial\sigma_r/\partial r + (1/r)(\partial\tau_{r\theta}/\partial\theta) + (\sigma_r - \sigma_\theta)/r = 0 \quad (15.25)$$
$$\partial\tau_{r\theta}/\partial r + (1/r)(\partial\sigma_\theta/\partial\theta) + 2(\tau_{r\theta}/r) = 0 \quad (15.26)$$

This situation arises in *thin slabs*, as indicated in Fig. 15.9, which are essentially two-dimensional problems. Because these equations are used in formulations which allow only stresses in the *plane* of the slab, they are classified as *plane-stress* equations.

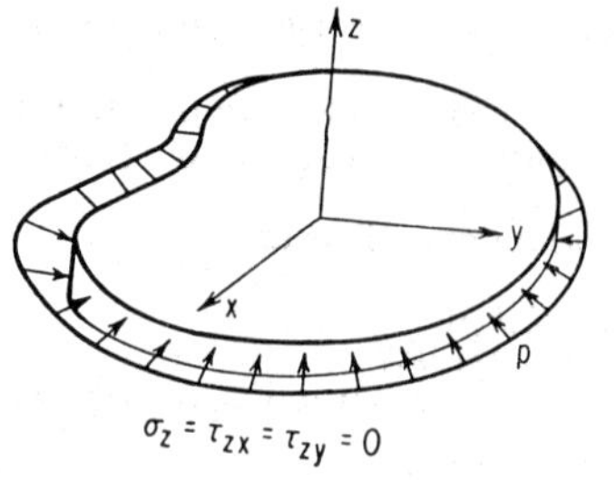

FIG. 15.9. Plane stress on a thin slab.

(g) Stress Transformation—Three-dimensional Case[4,5]

It is frequently necessary to determine the stresses at a point in an element which is rotated with respect to the x, y, z coordinate system; i.e., in an orthogonal x', y', z' system. Using equilibrium concepts and measuring the angle between any specific original and rotated coordinate by the direction cosines (cosine of the angle between the two axes) the following transformation equations result:

$$\begin{aligned}\sigma_{x'} = {} & [\sigma_x \cos(x'x) + \tau_{xy}\cos(x'y) + \tau_{zx}\cos(x'z)]\cos(x'x) \\ & + [\tau_{xy}\cos(x'x) + \sigma_y\cos(x'y) + \tau_{yz}\cos(x'z)]\cos(x'y) \\ & + [\tau_{zx}\cos(x'x) + \tau_{yz}\cos(x'y) + \sigma_z\cos(x'z)]\cos(x'z) \qquad (15.27)\end{aligned}$$

$$\begin{aligned}\sigma_{y'} = {} & [\sigma_x \cos(y'x) + \tau_{xy}\cos(y'y) + \tau_{zx}\cos(y'z)]\cos(y'x) \\ & + [\tau_{xy}\cos(y'x) + \sigma_y\cos(y'y) + \tau_{yz}\cos(y'z)]\cos(y'y) \\ & + [\tau_{zx}\cos(y'x) + \tau_{yz}\cos(y'y) + \sigma_z\cos(y'z)]\cos(y'z) \qquad (15.28)\end{aligned}$$

$$\begin{aligned}\sigma_{z'} = {} & [\sigma_x \cos(z'x) + \tau_{xy}\cos(z'y) + \tau_{zx}\cos(z'z)]\cos(z'x) \\ & + [\tau_{xy}\cos(z'x) + \sigma_y\cos(z'y) + \tau_{yz}\cos(z'z)]\cos(z'y) \\ & + [\tau_{zx}\cos(z'x) + \tau_{yz}\cos(z'y) + \sigma_z\cos(z'z)]\cos(z'z) \qquad (15.29)\end{aligned}$$

$$\begin{aligned}\tau_{x'y'} = {} & [\sigma_x \cos(y'x) + \tau_{xy}\cos(y'y) + \tau_{zx}\cos(y'z)]\cos(x'x) \\ & + [\tau_{xy}\cos(y'x) + \sigma_y\cos(y'y) + \tau_{yz}\cos(y'z)]\cos(x'y) \\ & + [\tau_{zx}\cos(y'x) + \tau_{yz}\cos(y'y) + \sigma_z\cos(y'z)]\cos(x'z) \qquad (15.30)\end{aligned}$$

$$\begin{aligned}\tau_{y'z'} = {} & [\sigma_x \cos(z'x) + \tau_{xy}\cos(z'y) + \tau_{zx}\cos(z'z)]\cos(y'x) \\ & + [\tau_{xy}\cos(z'x) + \sigma_y\cos(z'y) + \tau_{yz}\cos(z'z)]\cos(y'y) \\ & + [\tau_{zx}\cos(z'x) + \tau_{yz}\cos(z'y) + \sigma_z\cos(z'z)]\cos(y'z) \qquad (15.31)\end{aligned}$$

$$\begin{aligned}\tau_{z'x'} = {} & [\sigma_x \cos(x'x) + \tau_{xy}\cos(x'y) + \tau_{zx}\cos(x'z)]\cos(z'x) \\ & + [\tau_{xy}\cos(x'x) + \sigma_y\cos(x'y) + \tau_{yz}\cos(x'z)]\cos(z'y) \\ & + [\tau_{zx}\cos(x'x) + \tau_{yz}\cos(x'y) + \sigma_z\cos(x'z)]\cos(z'z) \qquad (15.32)\end{aligned}$$

In tensor notation these can be abbreviated:

$$\tau_{k'l'} = A_{l'n}A_{k'm}\tau_{mn} \qquad (15.33)$$

where $A_{ij} = \cos(ij) \qquad m, n \rightarrow x, y, z \qquad k', l' \rightarrow x', y', z'$

A special but very useful coordinate rotation occurs when the direction cosines are so selected that all the shear stresses vanish. The remaining mutually perpendicular *normal stresses* are called *principal stresses*.

The magnitudes of the principal stresses σ_x, σ_y, σ_z are the three roots of the cubic equations associated with the determinant:

$$\begin{vmatrix} (\sigma_x - \sigma) & \tau_{xy} & \tau_{zx} \\ \tau_{xy} & (\sigma_y - \sigma) & \tau_{yz} \\ \tau_{zx} & \tau_{yz} & (\sigma_z - \sigma) \end{vmatrix} = 0 \qquad (15.34)$$

where $\sigma_x \cdots$, $\tau_{xy} \cdots$ are the general nonprincipal stresses which exist on an element.

The direction cosines of the principal axes x', y', z' with respect to the x, y, z axes are obtained from the simultaneous solution of the following three equations considering separately the cases where $n = x', y', z'$:

$$\tau_{xy} \cos(xn) + (\sigma_y - \sigma_n) \cos(yn) + \tau_{yz} \cos(zn) = 0 \quad (15.35)$$
$$\tau_{zx} \cos(xn) + \tau_{yz} \cos(yn) + (\sigma_z - \sigma_n) \cos(zn) = 0 \quad (15.36)$$
$$\cos^2(xn) + \cos^2(yn) + \cos^2(zn) = 1 \quad (15.37)$$

(h) Stress Transformation—Two-dimensional Case[2,4]

Selecting an arbitrary coordinate direction in which the stress components vanish, it can be shown, either by equilibrium considerations or by general transformation formulas, that the two-dimensional stress-transformation equations become

$$\sigma_n = \frac{\sigma_x + \sigma_y}{2} + \frac{\sigma_x - \sigma_y}{2} \cos 2\alpha + \tau_{xy} \sin 2\alpha \quad (15.38)$$

$$\tau_{nt} = \frac{\sigma_x - \sigma_y}{2} \sin 2\alpha - \tau_{xy} \cos 2\alpha \quad (15.39)$$

where the directions are defined in Figs. 15.10 and 15.12 ($\tau_{xy} = -\tau_{nt}$, $\alpha = 0$).

The principal directions are obtained from the condition that

$$\tau_{nt} = 0 \qquad \text{or} \qquad \tan 2\alpha = \frac{2\tau_{xy}}{\sigma_x - \sigma_y} \quad (15.40)$$

where the two lowest roots of (first and second quadrants) are taken. It can be easily seen that the first and second principal directions differ by 90°. It can be shown that the *principal* stresses are also the *maximum or minimum normal stresses.*

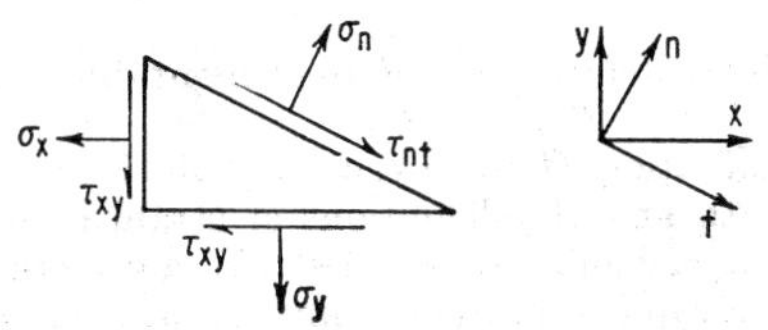

FIG. 15.10. Two-dimensional plane stress.

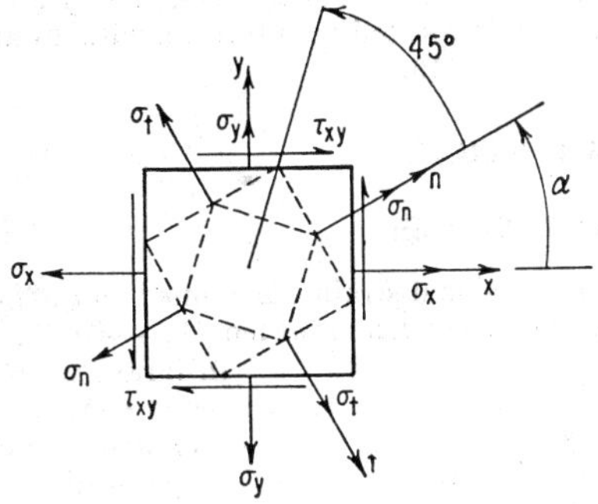

FIG. 15.11. Plane of maximum shear.

The *plane of maximum shear* is defined by

$$\tan 2\alpha = -\frac{\sigma_x - \sigma_y}{2\tau_{xy}} \quad (15.41)$$

These are also represented by planes which are 90° apart and are displaced from the principal stress planes by 45° (Fig. 15.11).

(i) Mohr's Circle

Mohr's circle is a convenient representation of the previously indicated transformation equations. Considering the x, y directions as positive in Fig. 15.11, the stress condition on any elemental plane can be represented as a point in the *Mohr diagram* (clockwise shear taken positive). The Mohr's circle is constructed by connecting the two stress points and drawing a circle through them with center on the σ axis. The stress state of any basic element can be represented by the stress coordinates at the intersection of the circle with an arbitrarily directed line through the circle center. Note that point x for positive τ_{xy} is below the σ axis and vice versa. The element is

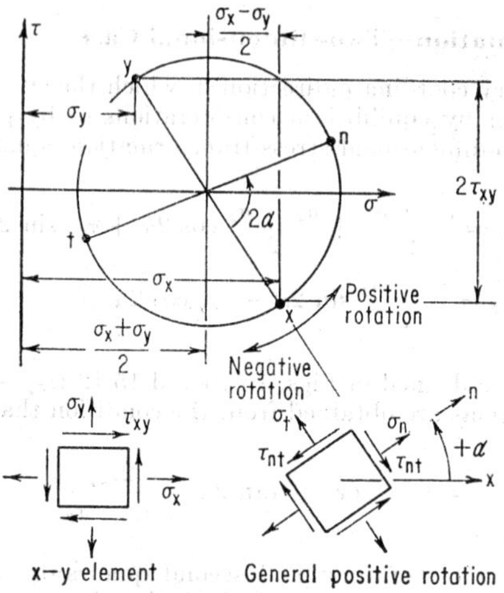

FIG. 15.12. Stress state of basic element.

taken as rotated counterclockwise by an angle α with respect to the x-y element when the line is rotated counterclockwise an angle 2α with respect to the x-y line, and vice versa (Fig. 15.12).

15.3. STRAIN

(a) Definition[2]

Extensional strain ϵ is defined as the extensional deformation of an element divided by the basic elemental length, $\epsilon = u/l_0$.

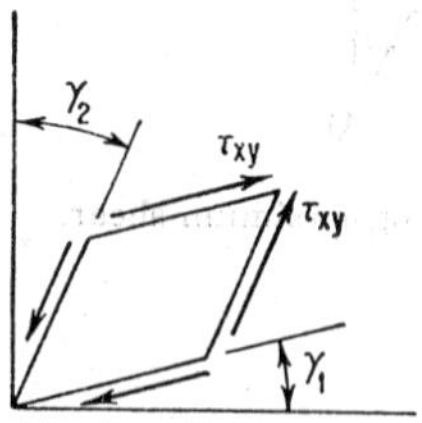

FIG. 15.13. Shear strain deformed element.

In large-strain considerations, l_0 must represent the instantaneous elemental length and the definitions of strain must be given in incremental fashion. In small strain considerations, to which the following discussion is limited, it is only necessary to consider the original elemental length l_0 and its change of length u. Extensional strain is taken positive or negative depending on whether the element increases or decreases in extent. The units of strain are dimensionless (inches/inch).

Shear strain γ is defined as the angular distortion of an original right-angle element. The direction of positive shear strain is taken to correspond to that produced by a positive shear stress (and vice versa) (see Fig. 15.13). Shear strain γ is equal to $\gamma_1 + \gamma_2$. The *units* of shear strain are dimensionless (radians).

(b) Components of Strain[2]

A complete description of strain requires the establishment of three orthogonal extensional and shear strains. Following the cartesian stress nomenclature, the strain components are

Extensional strain:

$$\begin{aligned} \epsilon_{xx} &\equiv \epsilon_x \\ \epsilon_{yy} &\equiv \epsilon_y \\ \epsilon_{zz} &\equiv \epsilon_z \end{aligned} \tag{15.42}$$

Shear strain:

$$\begin{aligned} \epsilon_{xy} &= \epsilon_{yx} \equiv \tfrac{1}{2}\gamma_{xy} \\ \epsilon_{yz} &= \epsilon_{zy} \equiv \tfrac{1}{2}\gamma_{yz} \\ \epsilon_{zx} &= \epsilon_{xz} \equiv \tfrac{1}{2}\gamma_{zx} \end{aligned} \tag{15.43}$$

where positive ϵ_x, ϵ_y, or ϵ_z corresponds to a positive stretching in the x, y, z directions and positive γ_{xy}, γ_{yz}, γ_{zx} refers to positive shearing displacements in the xy, yz, and zx planes. In tensor notation the strain components are

$$\epsilon_{ij} = \begin{pmatrix} \epsilon_x & \tfrac{1}{2}\gamma_{xy} & \tfrac{1}{2}\gamma_{zx} \\ \tfrac{1}{2}\gamma_{xy} & \epsilon_y & \tfrac{1}{2}\gamma_{yz} \\ \tfrac{1}{2}\gamma_{zx} & \tfrac{1}{2}\gamma_{yz} & \epsilon_z \end{pmatrix} \tag{15.44}$$

(c) Simple and Nonuniform States of Strain[2]

Corresponding to each of the stress states previously illustrated there exists either a simple or nonuniform strain state.

In addition to these, a state of *uniform dilatation* exists when the shear strain vanishes and all the extensional strains are equal in sign and magnitude. Dilatation is defined as

$$\Delta = \epsilon_x + \epsilon_y + \epsilon_z \tag{15.45}$$

and represents the change of volume per increment volume.

In uniform dilatation,

$$\Delta = 3\epsilon_x = 3\epsilon_y = 3\epsilon_z \tag{15.46}$$

(d) Strain-Displacement Relationships[4,5]

Considering only small strain, and the previous definitions, it is possible to express the strain components at a point in terms of the associated displacements and their derivatives in the coordinate directions (e.g., u, v, w are displacements in the x, y, z coordinate system).

Thus in a *cartesian system* (x,y,z),

$$\begin{aligned} \epsilon_x &= \partial u/\partial x & \gamma_{xy} &= \partial v/\partial x + \partial u/\partial y \\ \epsilon_y &= \partial v/\partial y & \gamma_{yz} &= \partial w/\partial y + \partial v/\partial z \\ \epsilon_z &= \partial w/\partial z & \gamma_{zx} &= \partial u/\partial z + \partial w/\partial x \end{aligned} \tag{15.47}$$

or, in stress-tensor notation,

$$2\epsilon_{ij} = u_{i,j} + u_{j,i} \qquad i, j \rightarrow x, y, z \tag{15.48}$$

In addition the dilatation

$$\Delta = \partial u/\partial x + \partial v/\partial y + \partial w/\partial z \tag{15.49}$$

or in tensor form

$$\Delta = u_{i,i} \qquad i \rightarrow x, y, z \tag{15.50}$$

Finally, all incremental displacements can be composed of a *pure strain* involving all the above components, plus "rigid-body" rotational components. That is, in general

$$U = \epsilon_x X + \tfrac{1}{2}\gamma_{xy} Y + \tfrac{1}{2}\gamma_{zx} Z - \bar{\omega}_z Y + \bar{\omega}_y Z \tag{15.51}$$

$$V = \tfrac{1}{2}\gamma_{xy} X + \epsilon_y Y + \tfrac{1}{2}\gamma_{yz} Z - \bar{\omega}_x Z + \bar{\omega}_z X \tag{15.52}$$

$$W = \tfrac{1}{2}\gamma_{zx} X + \tfrac{1}{2}\gamma_{yz} Y + \epsilon_z Z - \bar{\omega}_y X + \bar{\omega}_x Y \tag{15.53}$$

where U, V, W represent the incremental displacement of the point $x + X$, $y + Y$, $z + Z$ in excess of that of the point x, y, z where XYZ are taken as the sides of the incremental element. The rotational components are given by

$$\begin{aligned} 2\bar{\omega}_x &= \partial w/\partial y - \partial v/\partial z \\ 2\bar{\omega}_y &= \partial u/\partial z - \partial w/\partial x \\ 2\bar{\omega}_z &= \partial v/\partial x - \partial u/\partial y \end{aligned} \tag{15.54}$$

or, in tensor notation,

$$2\bar{\omega}_{ij} = u_{i,j} - u_{j,i} \qquad i, j = x, y, z \tag{15.55}$$
$$\bar{\omega}_{zy} \equiv \bar{\omega}_x, \ \bar{\omega}_{xz} \equiv \bar{\omega}_y, \ \bar{\omega}_{yx} \equiv \bar{\omega}_z$$

In *cylindrical coordinates,*

$$\begin{aligned} \epsilon_r &= \partial u_r/\partial r & \gamma_{\theta z} &= (1/r)(\partial u_z/\partial \theta) + \partial u_\theta/\partial z \\ \epsilon_\theta &= (1/r)(\partial u_\theta/\partial \theta) + u_r/r & \gamma_{zr} &= \partial u_r/\partial z + \partial u_z/\partial r \\ \epsilon_z &= \partial u_z/\partial z & \gamma_{r\theta} &= \partial u_\theta/\partial r - u_\theta/r + (1/r)(\partial u_r/\partial \theta) \end{aligned} \tag{15.56}$$

The dilatation is

$$\Delta = (1/r)(\partial/\partial r)(ru_r) + (1/r)(\partial u_\theta/\partial \theta) + \partial u_z/\partial z \tag{15.57}$$

and the rotation components are

$$\begin{aligned} 2\bar{\omega}_r &= (1/r)(\partial u_z/\partial \theta) - \partial u_\theta/\partial z \\ 2\bar{\omega}_\theta &= \partial u_r/\partial z - \partial u_z/\partial r \\ 2\bar{\omega}_z &= (1/r)(\partial/\partial r)(ru_\theta) - (1/r)(\partial u_r/\partial \theta) \end{aligned} \tag{15.58}$$

In *spherical polar coordinates,*

$$\begin{aligned} \epsilon_r &= \partial u_r/\partial r & \gamma_{\theta\phi} &= (1/r)(\partial u_\phi/\partial \theta - u_\phi \cot \theta) + (1/r \sin \theta)(\partial u_\theta/\partial \phi) \\ \epsilon_\theta &= (1/r)(\partial u_\theta/\partial \theta) + u_r/r & \gamma_{\phi r} &= (1/r \sin \theta)(\partial u_r/\partial \phi) + \partial u_\phi/\partial r - u_\phi/r \\ \epsilon_\phi &= (1/r \sin \theta)(\partial u_\phi/\partial \phi) + (u_\theta/r)\cot \theta + u_r/r & \gamma_{r\theta} &= \partial u_\theta/\partial r - u_\theta/r + (1/r)(\partial u_r/\partial \theta) \end{aligned} \tag{15.59}$$

The dilatation is

$$\Delta = (1/r^2 \sin \theta)[(\partial/\partial r)(r^2 u_r \sin \theta) + (\partial/\partial \theta)(ru_\theta \sin \theta) + (\partial/\partial \phi)(ru_\phi)] \tag{15.60}$$

The rotation components are

$$\begin{aligned} 2\bar{\omega}_r &= (1/r^2 \sin \theta)[(\partial/\partial \theta)(ru_\phi \sin \theta) - (\partial/\partial \phi)(ru_\theta)] \\ 2\bar{\omega}_\theta &= (1/r \sin \theta)[\partial u_r/\partial \phi - (\partial/\partial r)(ru_\phi \sin \theta)] \\ 2\bar{\omega}_\phi &= (1/r)[(\partial/\partial r)(ru_\theta) - \partial u_r/\partial \theta] \end{aligned} \tag{15.61}$$

In general *orthogonal curvilinear coordinates,*

$$\begin{aligned} \epsilon_\alpha &= h_1(\partial u_\alpha/\partial \alpha) + h_1 h_2 u_\beta(\partial/\partial \beta)(1/h_1) + h_3 h_1 u_\gamma(\partial/\partial \gamma)(1/h_1) \\ \epsilon_\beta &= h_2(\partial u_\beta/\partial \beta) + h_2 h_3 u_\gamma(\partial/\partial \gamma)(1/h_2) + h_1 h_2 u_\alpha(\partial/\partial \alpha)(1/h_2) \\ \epsilon_\gamma &= h_3(\partial u_\gamma/\partial \gamma) + h_3 h_1 u_\alpha(\partial/\partial \alpha)(1/h_3) + h_2 h_3 u_\beta(\partial/\partial \beta)(1/h_3) \\ \gamma_{\beta\gamma} &= (h_2/h_3)(\partial/\partial \beta)(h_3 u_\gamma) + (h_3/h_2)(\partial/\partial \gamma)(h_2 u_\beta) \\ \gamma_{\gamma\alpha} &= (h_3/h_1)(\partial/\partial \gamma)(h_1 u_\alpha) + (h_1/h_3)(\partial/\partial \alpha)(h_3 u_\gamma) \\ \gamma_{\alpha\beta} &= (h_1/h_2)(\partial/\partial \alpha)(h_2 u_\beta) + (h_2/h_1)(\partial/\partial \beta)(h_1 u_\alpha) \end{aligned} \tag{15.62}$$

$$\Delta = h_1 h_2 h_3[(\partial/\partial \alpha)(u_\alpha/h_2 h_3) + (\partial/\partial \beta)(u_\beta/h_3 h_1) + (\partial/\partial \gamma)(u_\gamma/h_1 h_2)] \tag{15.63}$$

$$\begin{aligned} 2\bar{\omega}_\alpha &= h_2 h_3[(\partial/\partial \beta)(u_\gamma/h_3) - (\partial/\partial \gamma)(u_\beta/h_2)] \\ 2\bar{\omega}_\beta &= h_3 h_1[(\partial/\partial \gamma)(u_\alpha/h_1) - (\partial/\partial \alpha)(u_\gamma/h_3)] \\ 2\bar{\omega}_\gamma &= h_1 h_2[(\partial/\partial \alpha)(u_\beta/h_2) - (\partial/\partial \beta)(u_\alpha/h_1)] \end{aligned} \tag{15.64}$$

where the quantities h_1, h_2, h_3 have been discussed with reference to the equilibrium equations.

In the event that one deflection (i.e., w) is constant or zero and the displacements are a function of x, y only, a special and useful class of problems arises termed *plane strain,* which are analogous to the *plane-stress* problems. A typical case of plane strain occurs in slabs rigidly clamped on their faces so as to restrict all axial deformation. Although all the stresses may be nonzero, and the general equilibrium equations apply, it can be shown that, after combining all the necessary stress and strain relationships, both classes of *plane* problems yield the same form of equations. From this, one solution suffices for both the related *plane-stress* and *plane-strain* problems, provided that the elasticity constants are suitably modified. In particular the applicable strain-displacement relationships reduce in *cartesian coordinates* to

$$\begin{aligned} \epsilon_x &= \partial u/\partial x \\ \epsilon_y &= \partial v/\partial y \\ \gamma_{xy} &= \partial v/\partial x + \partial u/\partial y \end{aligned} \tag{15.65}$$

and in *cylindrical coordinates* to

$$\begin{aligned} \epsilon_r &= \partial u_r/\partial r \\ \epsilon_\theta &= (1/r)(\partial u_\theta/\partial \theta) + u_r/r \\ \gamma_{r\theta} &= \partial u_\theta/\partial r - u_\theta/r + (1/r)(\partial u_r/\partial \theta) \end{aligned} \tag{15.66}$$

(e) Compatibility Relationships[2,4,5]

In the event that a single-valued continuous-displacement field (u,v,w) is not explicitly specified, it becomes necessary to ensure its existence in solution of the stress, strain, and stress-strain relationships. By writing the strain-displacement relationships and manipulating them to eliminate displacements it can be shown that the following six equations are both necessary and sufficient to ensure compatibility:

$$\begin{aligned} \partial^2\epsilon_y/\partial z^2 + \partial^2\epsilon_z/\partial y^2 &= \partial^2\gamma_{yz}/\partial y\,\partial z \\ 2(\partial^2\epsilon_x/\partial y\,\partial z) &= (\partial/\partial x)(-\partial\gamma_{yz}/\partial x + \partial\gamma_{zx}/\partial y + \partial\gamma_{xy}/\partial z) \end{aligned} \tag{15.67}$$

$$\begin{aligned} \partial^2\epsilon_z/\partial x^2 + \partial^2\epsilon_x/\partial z^2 &= \partial^2\gamma_{zx}/\partial x\,\partial z \\ 2(\partial^2\epsilon_y/\partial z\,\partial x) &= (\partial/\partial y)(+\partial\gamma_{yz}/\partial x - \partial\gamma_{zx}/\partial y + \partial\gamma_{xy}/\partial z) \end{aligned} \tag{15.68}$$

$$\begin{aligned} \partial^2\epsilon_x/\partial y^2 + \partial^2\epsilon_y/\partial x^2 &= \partial^2\gamma_{xy}/\partial x\,\partial y \\ 2(\partial^2\epsilon_z/\partial x\,\partial y) &= (\partial/\partial z)(+\partial\gamma_{yz}/\partial x + \partial\gamma_{zx}/\partial y - \partial\gamma_{xy}/\partial z) \end{aligned} \tag{15.69}$$

In tensor notation the most general compatibility equations are

$$\epsilon_{ij,kl} + \epsilon_{kl,ij} - \epsilon_{ik,jl} - \epsilon_{jl,ik} = 0 \qquad i,\ j,\ k,\ l = x,\ y,\ z \tag{15.70}$$

which represents 81 equations. Only the above six equations are essential.

In addition to satisfying these conditions everywhere in the body under consideration, it is also necessary that all *surface strain* or *displacement boundary conditions* be satisfied.

(f) Strain Transformation[4,5]

As with stress, it is frequently necessary to refer strains to a rotated orthogonal coordinate system (x',y',z'). In this event it can be shown that the stress and strain tensors transform in an identical manner.

$$\begin{array}{llll} \sigma_{x'} \to \epsilon_{x'} & \sigma_x \to \epsilon_x & \tau_{x'y'} \to \tfrac{1}{2}\gamma_{x'y'} & \tau_{xy} \to \tfrac{1}{2}\gamma_{xy} \\ \sigma_{y'} \to \epsilon_{y'} & \sigma_y \to \epsilon_y & \tau_{y'z'} \to \tfrac{1}{2}\gamma_{y'z'} & \tau_{yz} \to \tfrac{1}{2}\gamma_{yz} \\ \sigma_{z'} \to \epsilon_{z'} & \sigma_z \to \epsilon_z & \tau_{z'x'} \to \tfrac{1}{2}\gamma_{z'x'} & \tau_{zx} \to \tfrac{1}{2}\gamma_{zx} \end{array}$$

In tensor notation the strain transformation can be written

$$\epsilon_{k'l'} = A_{l'n}A_{k'm}\epsilon_{mn} \qquad \begin{array}{l} m, n \to x,\ y,\ z \\ l'k' \to x',\ y',\ z' \end{array} \tag{15.71}$$

As a result the stress and strain principal directions are coincident; so that all remarks made for the principal stress and maximum shear components and their directions apply equally well to strain tensor components. Note that in the use of Mohr's circle in the two-dimensional case one must be careful to substitute $\frac{1}{2}\gamma$ for τ in the *ordinate* and ϵ for σ in the abscissa (Fig. 15.14).

15.4. STRESS - STRAIN RELATIONSHIPS

(a) Introduction[2]

It can be experimentally demonstrated that a one-to-one relationship exists between uniaxial stress and strain during a single loading. Further, if the material is always loaded within its *elastic* or *reversible* range, a one-to-one relationship exists for all loading and unloading cycles.

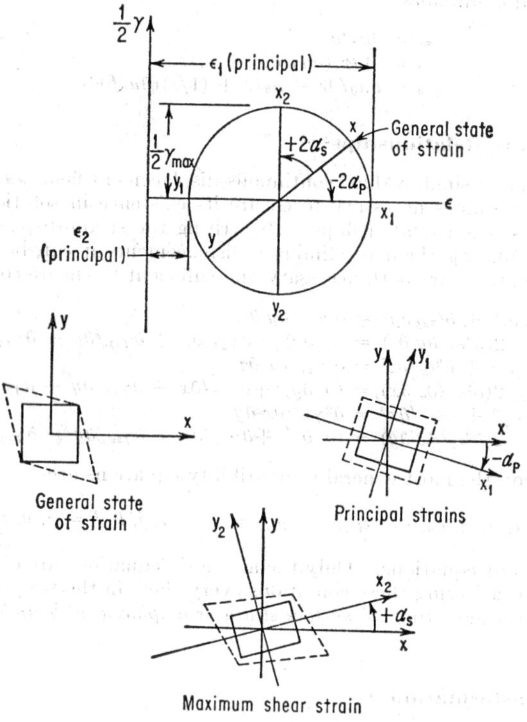

FIG. 15.14. Strain transformation.

For stresses below a certain characteristic value termed the *proportional limit,* the stress-strain relationship is very nearly linear. The stress beyond which the stress-strain relationship is no longer reversible is called the *elastic limit.* In most materials the proportional and elastic limits are identical. Because the departure from linearity is very gradual it is often necessary to prescribe arbitrarily an *apparent or offset elastic limit.* This is obtained as the intersection of the stress-strain curve with a line parallel to the linear stress-strain curve, but offset by a prescribed amount, e.g., 0.02 per cent (see Fig. 15.15*a*). The *yield point* is the value of stress at which continued deformation of the bar takes place with little or no further increase in load, and the *ultimate limit* is the maximum stress that the specimen can withstand.

Note that some materials may show no clear difference between the apparent elastic, inelastic, and proportional limits or may not show clearly defined yield points (Fig. 15.15*b*).

The concept that a useful linear range exists for most materials and that a simple mathematical law can be formulated to describe the relationship between stress and strain in this range is termed *Hooke's law.* It is an essential starting point in the *small-strain theory of elasticity* and the associated mechanics of materials. In the above-described tensile specimen, the law is expressed as

$$\sigma = E\epsilon \tag{15.72}$$

as in the analogous torsional specimen

$$\tau = G\gamma \tag{15.73}$$

where E and G are the slope of the appropriate stress-strain diagrams and are called the *Young's modulus* and the *shear modulus* of elasticity, respectively.

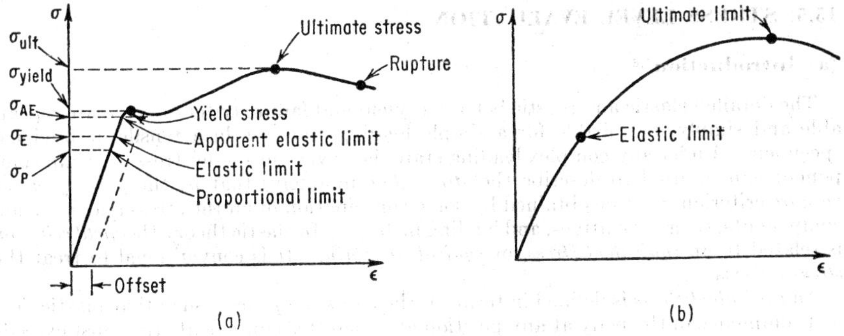

FIG. 15.15. Stress-strain relationship.

(b) General Stress-Strain Relationship[2,4,5]

The one-dimensional concepts discussed above can be generalized for both small and large strain and elastic and nonelastic materials. The following discussion will be limited to small-strain elastic materials consistent with much engineering design. Based upon the above, Hooke's law is expressed as

$$\begin{aligned} \epsilon_x &= (1/E)[\sigma_x - \nu(\sigma_y + \sigma_z)] \qquad & \gamma_{xy} &= \tau_{xy}/G \\ \epsilon_y &= (1/E)[\sigma_y - \nu(\sigma_z + \sigma_x)] \qquad & \gamma_{yz} &= \tau_{yz}/G \\ \epsilon_z &= (1/E)[\sigma_z - \nu(\sigma_x + \sigma_y)] \qquad & \gamma_{zx} &= \tau_{zx}/G \end{aligned} \tag{15.74}$$

where ν is *Poisson's ratio,* the ratio between longitudinal strain and lateral contraction in a simple tensile test.

In cartesian *tensor form* Eq. (15.74) is expressed

$$\epsilon_{ij} = [(1 + \nu)/E]\sigma_{ij} - (\nu/E)\delta_{ij}\sigma_{kk} \qquad (i,j,k = x,y,z) \tag{15.75}$$

where
$$\delta_{ij} = \begin{cases} 0, & (i \neq j) \\ 1, & (i = j) \end{cases}$$

The stress-strain laws appear in inverted form as

$$\begin{aligned} \sigma_x &= 2G\epsilon_x + \lambda\Delta \\ \sigma_y &= 2G\epsilon_y + \lambda\Delta \\ \sigma_z &= 2G\epsilon_z + \lambda\Delta \\ \tau_{xy} &= G\gamma_{xy} \\ \tau_{yz} &= G\gamma_{yz} \\ \tau_{zx} &= G\gamma_{zx} \end{aligned} \tag{15.76}$$

where

$$\lambda = \frac{\nu E}{(1+\nu)(1-2\nu)} \qquad \Delta = \epsilon_x + \epsilon_y + \epsilon_z$$

$$G = \frac{E}{2(1+\nu)}$$

In *cartesian tensor form* Eq. (15.76) is written

$$\sigma_{ij} = 2G\epsilon_{ij} + \lambda\,\Delta\delta_{ij} \qquad (i,j = x,y,z) \tag{15.77}$$

and general tensor form

$$\sigma_{ij} = 2G\epsilon_{ij} + \lambda\,\Delta g_{ij} \tag{15.78}$$

where g_{ij} is the *covariant metric tensor* and these coefficients (stress modulus) are often referred to as *Lamé's constants,* and $\Delta = g^{mn}\epsilon_{mn}$.

15.5. STRESS-LEVEL EVALUATION

(a) Introduction[1,6]

The detailed elastic and plastic behavior, yield and failure criterion, etc., are repeatable and simply describable for a simple loading state, as in a tensile or torsional specimen. Under any complex loading state, however, no single stress or strain component can be used to describe the *stress state* uniquely; that is, the *yield, flow,* or *rupture* criterion must be obtained by some combination of all the stress and/or strain components, their derivatives, and loading history. In elastic theory the *yield criterion* is related to an *equivalent stress,* or *equivalent strain.* It is conventional to treat the *stress* criteria.

An *equivalent stress* is defined in terms of the *stress components* such that plastic flow will commence in the body at any position at which this equivalent stress just exceeds the one-dimensional yield-stress value, for the material under consideration. That is, yielding commences when

$$\sigma_{\text{equivalent}} \geq \sigma_E$$

The *elastic safety factor* at a point is defined as the ratio of the one-dimensional yield stress to the equivalent stress at that position, i.e.,

$$n_i = \frac{\sigma_E}{\sigma_{\text{equivalent}}} \tag{15.79}$$

and the elastic safety factor for the entire structure under any specific loading state is taken as the lowest safety factor of consequence that exists anywhere in the structure.

The *margin of safety,* defined as $n - 1$, is another measure of the proximity of any structure to yielding. When $n > 1$ the structure has a *positive margin of safety* and will not yield. When $n = 1$, the margin of safety is zero and the structure just yields. When $n < 1$, the margin of safety is negative and the structure is considered unsafe. Note that highly localized yielding is often permitted in ductile materials if it is of such nature as to redistribute stresses without failure, building up a *residual stress state* which allows all subsequent loadings to be accomplished with elastic-stress states. This is the basic concept used in the *autofrettage process* in the strengthening of gun tubes, and it also explains why ductile materials often have low *notch sensitivity* (see Secs. 15 and 18).

(b) Effective Stress[1,6]

The concept of effective stress is very closely connected with yield criteria. Geometrically it can be shown that a unique surface can be constructed in stress space in terms of principal stresses ($\sigma_1,\sigma_2,\sigma_3$) such that all nonyielding states of stress lie within that surface, and yielding states lie on or outside the surface. For *ductile materials* the yield surface may be taken as an infinitely long right cylinder having a center axis defined by $\sigma_1 = \sigma_2 = \sigma_3$. Because the yield criterion is represented by a right cylinder it is adequate to define the yield curve, which is the intersection with a plane normal to the axis of the cylinder. Many yield criteria exist; among these, the *Mises criterion* takes this curve as a circle, and the *Tresca criterion* as a regular hexagon (see Fig. 15.16). The former is often referred to by names such as *Hencky, Hencky-Mises,* or *distortion-energy criterion* and the latter by *shear criterion.*

Considering any one of the principal stresses as constant, the *yield locus* can be represented by the intersection of the plane σ_i = constant with the cylindrical yield

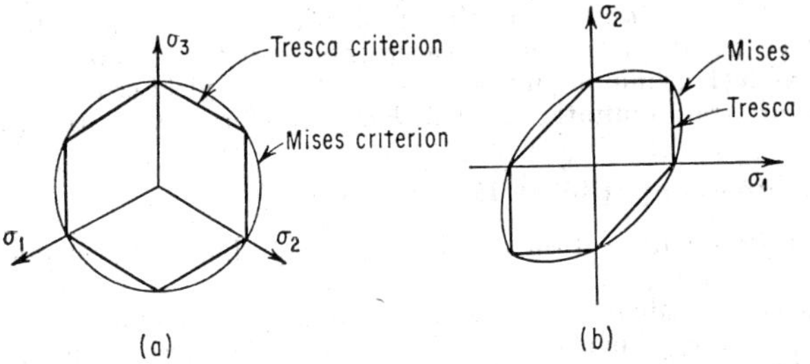

FIG. 15.16. Tresca and Mises criteria. (*a*) General yield criteria. (*b*) One principal stress constant.

surface, which is represented as an ellipse for the Mises criterion and an elongated hexagon for the Tresca criterion (Fig. 15.16*b*).

In general the yield criteria indicate, upon experimental evidence for a ductile material, that yielding is essentially independent of *hydrostatic compression* or tension for the loadings usually considered in engineering problems. That is, yielding depends only on the *deviatoric* stress component. In general this can be represented as

$$\sigma_k' = \sigma_k - \tfrac{1}{3}(\sigma_1 + \sigma_2 + \sigma_3) \tag{15.80}$$

where σ' is the principal deviatoric stress component and σ the actual principal stress.

In tensor form

$$\sigma_{ij}' = \sigma_{ij} - \tfrac{1}{3}\sigma_{kk}\delta_{ij} \tag{15.81}$$

The analytical representation of the yield criteria can be shown to be a function of the *deviatoric stress-tensor invariants.* In component form these can be expressed as follows, where σ_0 is the yield stress in simple tension,

Mises:

$$\sigma_0 = \frac{1}{\sqrt{2}} \sqrt{(\sigma_1 - \sigma_2)^2 + (\sigma_2 - \sigma_3)^2 + (\sigma_3 - \sigma_1)^2} \tag{15.82}$$

where σ_1, σ_2, σ_3 are principal stresses.

$$\sigma_0 = 1/\sqrt{2}\,\sqrt{(\sigma_x - \sigma_y)^2 + (\sigma_y - \sigma_z)^2 + (\sigma_z - \sigma_x)^2 + 6(\tau_{xy}^2 + \tau_{yz}^2 + \tau_{zx}^2)} \tag{15.83}$$

or, in tensor form,

$$\tfrac{2}{3}\sigma_0^2 = \sigma_{ij}'\sigma_{ij}' \tag{15.84}$$

Tresca:

$$\sigma_0 = \sigma_1 - \sigma_3 \tag{15.85}$$

where $\sigma_1 > \sigma_2 > \sigma_3$, or, in general symmetric terms,

$$[(\sigma_1 - \sigma_3)^2 - \sigma_0^2][(\sigma_2 - \sigma_1)^2 - \sigma_0^2][(\sigma_3 - \sigma_2)^2 - \sigma_0^2] = 0 \tag{15.86}$$

15.6. FORMULATION OF GENERAL MECHANICS-OF-MATERIAL PROBLEM

(a) Introduction[2,4,5]

Generally the mechanics-of-material problem is stated as follows: Given a prescribed structural configuration, and surface tractions and/or displacements, find the stresses and/or displacements at any, or all, positions in the body. Additionally it is often desired to use the derived stress information to determine the *maximum load-carrying capacity* of the structure, prior to yielding. This is usually referred to as the problem of *analysis*. Alternatively the problem may be inverted and stated: Given a set of surface tractions and/or displacements, find the geometrical configuration for a constraint such as minimum weight, subject to the yield criterion (or some other general stress or strain limitation). This latter is referred to as the *design* problem.

(b) Equilibrium-Compatibility Formulations[2,4,5]

The classical formulation of the equation for the problem of mechanics of materials is as follows:

It is necessary to evaluate the six stress components σ_{ij}, six strain components ϵ_{ij}, and three displacement quantities u_i which satisfy the three equilibrium equations, six compatibility equations and six stress-strain relationships, all subject to the appropriate stress and/or displacement boundary conditions. It is apparent that these supply an adequate number of equations to evaluate all the unknowns, and as such furnish a complete formulation of the general problem. Certain assumptions generally reduce the number of variables and associated equations.

Based upon the above discussion and the previous derivations, the most general three-dimensional formulation in cartesian coordinates is

$$\begin{aligned}\partial\sigma_x/\partial x + \partial\tau_{xy}/\partial y + \partial\tau_{xz}/\partial z &= 0\\ \partial\sigma_y/\partial y + \partial\tau_{yz}/\partial z + \partial\tau_{yx}/\partial x &= 0\\ \partial\sigma_z/\partial z + \partial\tau_{zx}/\partial x + \partial\tau_{zy}/\partial y &= 0\end{aligned} \qquad \text{equilibrium} \qquad (15.87)$$

$$\begin{aligned}\partial^2\epsilon_x/\partial y^2 + \partial^2\epsilon_y/\partial x^2 &= \partial^2\gamma_{xy}/\partial x\,\partial y\\ \partial^2\epsilon_y/\partial z^2 + \partial^2\epsilon_z/\partial y^2 &= \partial^2\gamma_{yz}/\partial y\,\partial z\\ \partial^2\epsilon_z/\partial x^2 + \partial^2\epsilon_x/\partial z^2 &= \partial^2\gamma_{zx}/\partial z\,\partial x\\ 2(\partial^2\epsilon_x/\partial y\,\partial z) &= (\partial/\partial x)(-\partial\gamma_{yz}/\partial x + \partial\gamma_{zx}/\partial y + \partial\gamma_{xy}/\partial z)\\ 2(\partial^2\epsilon_y/\partial z\,\partial x) &= (\partial/\partial y)(\partial\gamma_{yz}/\partial x - \partial\gamma_{zx}/\partial y + \partial\gamma_{xy}/\partial z)\\ 2(\partial^2\epsilon_z/\partial x\,\partial y) &= (\partial/\partial z)(\partial\gamma_{yz}/\partial x + \partial\gamma_{zx}/\partial y - \partial\gamma_{xy}/\partial z)\end{aligned} \qquad \text{compatibility} \qquad (15.88)$$

$$\begin{aligned}\epsilon_x &= (1/E)[\sigma_x - \nu(\sigma_y + \sigma_z)]\\ \epsilon_y &= (1/E)[\sigma_y - \nu(\sigma_z + \sigma_x)]\\ \epsilon_z &= (1/E)[\sigma_z - \nu(\sigma_x + \sigma_y)]\\ \gamma_{xy} &= (1/G)\tau_{xy}\\ \gamma_{yz} &= (1/G)\tau_{yz}\\ \gamma_{zx} &= (1/G)\tau_{zx}\end{aligned} \qquad \text{stress-strain relationships} \qquad (15.89)$$

In cartesian tensor form these appear as

$$\sigma_{ij,j} = 0 \qquad \text{equilibrium} \qquad (15.90)$$

$$\epsilon_{ij,kl} + \epsilon_{kl,ij} - \epsilon_{ik,jl} - \epsilon_{jl,ik} = 0 \qquad \text{compatibility} \qquad (15.91)$$

$$\epsilon_{ij} = \frac{1+\nu}{E}\sigma_{ij} - \frac{\nu}{E}\delta_{ij}\sigma_{kk} \qquad \text{stress-strain} \qquad (15.92)$$

All are subject to appropriate boundary conditions.

It is possible to simplify the above set of 15 equations considerably by combining and eliminating many of the unknowns. One such reduction is obtained by eliminating stress and strain:

$$\begin{aligned} \nabla^2 u + \frac{1}{1-2\nu}\frac{\partial \Delta}{\partial x} &= 0 \\ \nabla^2 v + \frac{1}{1-2\nu}\frac{\partial \Delta}{\partial y} &= 0 \\ \nabla^2 w + \frac{1}{1-2\nu}\frac{\partial \Delta}{\partial z} &= 0 \end{aligned} \qquad (15.93)$$

where ∇^2 is the Laplacian operator which in cartesian coordinates is $\partial^2/\partial x^2 + \partial^2/\partial y^2 + \partial^2/\partial z^2$; and Δ is the dilatation, which in cartesian coordinates is $\partial u/\partial x + \partial v/\partial y + \partial w/\partial z$.

Using the above general principles it is possible to formulate completely many of the technical problems of mechanics of materials which appear under special classifications such as "beam theory" and "shell theory." These formulations and their solutions will be treated under Special Applications.

(c) Energy Formulations[2,4,5]

Alternative useful approaches exist for the problem of mechanics of materials. These are referred to as *energy, extremum,* or *variational* formulations. From a strictly formalistic point of view these could be obtained by establishing the analogous integral equations, subject to various restrictions, such that they reduce to a minimum. This is not the usual approach; instead energy functions U, W are established so that the stress-strain laws are replaced by

$$\begin{aligned} \sigma_x &= \partial U/\partial \epsilon_x & \tau_{xy} &= \partial U/\partial \gamma_{xy} \\ \sigma_y &= \partial U/\partial \epsilon_y & \tau_{yz} &= \partial U/\partial \gamma_{yz} \\ \sigma_z &= \partial U/\partial \epsilon_z & \tau_{zx} &= \partial U/\partial \gamma_{zx} \end{aligned}$$

or

$$\sigma_{ij} = \partial U/\partial \epsilon_{ij}$$

and

$$\begin{aligned} \epsilon_x &= \partial W/\partial \sigma_x & \gamma_{xy} &= \partial W/\partial \tau_{xy} \\ \epsilon_y &= \partial W/\partial \sigma_y & \gamma_{yz} &= \partial W/\partial \tau_{yz} \\ \epsilon_z &= \partial W/\partial \sigma_z & \gamma_{zx} &= \partial W/\partial \tau_{zx} \end{aligned}$$

or

$$\epsilon_{ij} = \partial W/\partial \sigma_{ij}$$

The energy functions are given by

$$U = \tfrac{1}{2}[2G(\epsilon_x^2 + \epsilon_y^2 + \epsilon_z^2) + \lambda(\epsilon_x + \epsilon_y + \epsilon_z)^2 + G(\gamma_{xy}^2 + \gamma_{yz}^2 + \gamma_{zx}^2)] \qquad (15.94)$$

$$W = \tfrac{1}{2}[(1/E)(\sigma_x^2 + \sigma_y^2 + \sigma_z^2) - (2\nu/E)(\sigma_x\sigma_y + \sigma_y\sigma_z + \sigma_z\sigma_x) + (1/G)(\tau_{xy}^2 + \tau_{yz}^2 + \tau_{zx}^2)] \qquad (15.95)$$

The variational principle for strains, or theorem of minimum potential energy, is stated as follows: Among all states of strain which satisfy the strain-displacement relationships and displacement boundary conditions the associated stress state, derivable through the stress-strain relationships, which also satisfies the equilibrium equations, is determined by the minimization of Π where

$$\Pi = \int_{\text{volume}} U\, dV - \int_{\text{surface}} (\bar{p}_x u + \bar{p}_y v + \bar{p}_z w)\, dS \qquad (15.96)$$

where $\bar{p}_x$, $\bar{p}_y$, $\bar{p}_z$ are the x, y, z components of any prescribed surface stresses.

The analogous variational principle for stresses, or principle of least work, is: Among all the states of stress which satisfy the equilibrium equations and stress boundary conditions, the associated strain state, derivable through the stress-strain relationships, which also satisfies the compatibility equations is determined by the minimization of I, where

$$I = \int_{\text{volume}} W\, dV - \int_{\text{surface}} (p_x\bar{u} + p_y\bar{v} + p_z\bar{w})\, dS \qquad (15.97)$$

where $\bar{u}$, $\bar{v}$, $\bar{w}$ are the x, y, z components of any prescribed surface displacements and p_x, p_y, p_z are the surface stresses.

In the above theorems $\Pi_{\min}$ and $I_{\min}$ replace the equilibrium and compatibility relationships, respectively. Their most powerful advantage arises in obtaining approximate solutions to problems which are generally intractable by exact techniques. In this, one usually introduces a limited class of assumed stress or displacement functions for minimization, which in themselves satisfy all other requirements imposed in the statement of the respective theorems. Then with the use of these theorems it is possible to find the best solution in that limited class which provides the best minimum to the associated Π or I function. This in reality does not satisfy the missing equilibrium or compatibility equation, but it does it as well as possible for the class of function assumed to describe the stress or strain in the body, within the framework of the principle established above. It has been shown that most reasonable assumptions, regardless of their simplicity, provide useful solutions to most problems of mechanics of materials.

(d) Example. Energy Techniques[2,4,5]

It can be shown that for beams the variational principle for strains reduces to

$$\Pi_{\min} = \left\{ \int_0^L [\tfrac{1}{2}EI(y'')^2 - qy]\,dx - \sum P_i y_i \right\}_{\min} \tag{15.98}$$

where EI is the flexural rigidity of the beam at any position x, I, the moment of inertia of the beam, y is the deflection of the beam, the y' refers to x derivative of y, q is the distributed loading, the P_i's represent concentrated loads, and L is the span length.

If the minimization is carried out, subject to the restrictions of the variational principle for strains, the beam equation results. However, it is both useful and instructive to utilize the above principle to obtain two approximate solutions to a specific problem and then compare these with the exact solutions obtained by other means.

First a centrally loaded, simple-support beam problem will be examined. The function of minimization becomes

$$\Pi = \int_0^{L/2} EI(y'')^2\,dx - Py_{L/2} \tag{15.99}$$

Select the class of displacement functions described by

$$y = Ax(\tfrac{3}{4}L^2 - x^2) \qquad 0 \le x \le L/2 \tag{15.100}$$

This satisfies the boundary conditions

$$y(0) = y''(0) = y'(L/2) = 0$$

In this A is an arbitrary parameter to be determined from the minimization of Π.

Properly introducing the value of y, y'' into the expression for Π and integrating, then minimizing Π with respect to the open parameter by setting

$$\partial\Pi/\partial A = 0$$

yields

$$y = \frac{Px}{12EI}(\tfrac{3}{4}L^2 - x^2) \qquad 0 \le x \le L/2 \tag{15.101}$$

It is coincidental that this is the exact solution to the above problem.

A second class of deflection function is now selected

$$y = A \sin(\pi x/L) \qquad 0 \le x \le L \tag{15.102}$$

which satisfies the boundary conditions

$$y(0) = y''(0) = y(L) = y''(L) = 0$$

which is intuitively the expected deflection shape. Additionally, $y(L/2) = A$. Introducing the above information into the expression for Π and minimizing as before yields

$$y = (PL^3/EI)[(2/\pi^4) \sin (\pi x/L)] \tag{15.103}$$

The ratio of the approximate to the exact central deflection is 0.9855, which indicates that the approximation is of sufficient accuracy for most applications.

15.7. FORMULATION OF GENERAL THERMOELASTIC PROBLEM[2,9]

A nonuniform temperature distribution or a nonuniform material distribution with uniform temperature change introduces additional stresses and/or strains, even in the absence of external tractions.

Within the confines of the linear theory of elasticity and neglecting small coupling effects between the *temperature-distribution problem* and the *thermoelastic problem* it is possible to solve the general mechanics-of-material problem as the superposition of the previously defined mechanics-of-materials problem and an initially traction-free thermoelastic problem.

Taking the same consistent definition of stress and strain as previously presented it can be shown that the strain-displacement, stress-equilibrium, and compatibility relationships remain unchanged in the thermoelastic problem. However, because a structural material can change its size even in the absence of stress, it is necessary to modify the stress-strain laws to account for the additional strain due to temperature (αT). Thus Hooke's law is modified as follows:

$$\begin{aligned} \epsilon_x &= (1/E)[\sigma_x - \nu(\sigma_y + \sigma_z)] + \alpha T \\ \epsilon_y &= (1/E)[\sigma_y - \nu(\sigma_z + \sigma_x)] + \alpha T \\ \epsilon_z &= (1/E)[\sigma_z - \nu(\sigma_x + \sigma_y)] + \alpha T \end{aligned} \tag{15.104}$$

The shear strain-stress relationships remain unchanged. α is the coefficient of thermal expansion and T the temperature rise above the ambient stress-free state. In uniform, nonconstrained structures this ambient base temperature is arbitrary, but in problems associated with nonuniform material or constraint this base temperature is quite important.

Expressed in cartesian tensor form the stress-strain relationships become

$$\epsilon_{ij} = [(1 + \nu)/E]\sigma_{ij} - (\nu/E)\delta_{ij}\sigma_{kk} + \alpha T\delta_{ij} \tag{15.105}$$

In inverted form the modified stress-strain relationships are

$$\begin{aligned} \sigma_x &= 2G\epsilon_x + \lambda\Delta - (3\lambda + 2G)\alpha T \\ \sigma_y &= 2G\epsilon_y + \lambda\Delta - (3\lambda + 2G)\alpha T \\ \sigma_z &= 2G\epsilon_z + \lambda\Delta - (3\lambda + 2G)\alpha T \end{aligned} \tag{15.106}$$

or in cartesian tensor form

$$\sigma_{ij} = 2G\epsilon_{ij} + \lambda\Delta\delta_{ij} - (3\lambda + 2G)\alpha T\delta_{ij} \tag{15.107}$$

Considering the equilibrium compatibility formulations it can be shown that the analogous thermoelastic displacement formulations result in

$$\begin{aligned} (\lambda + G)(\partial\Delta/\partial x) + G\nabla^2 u - (3\lambda + 2G)\alpha(\partial T/\partial x) &= 0 \\ (\lambda + G)(\partial\Delta/\partial y) + G\nabla^2 v - (3\lambda + 2G)\alpha(\partial T/\partial y) &= 0 \\ (\lambda + G)(\partial\Delta/\partial z) + G\nabla^2 w - (3\lambda + 2G)\alpha(\partial T/\partial z) &= 0 \end{aligned} \tag{15.108}$$

A useful alternate stress formulation is

$$
\begin{aligned}
&(1+\nu)\nabla^2\sigma_x + \frac{\partial^2\Theta}{\partial x^2} + \alpha E\left(\frac{1+\nu}{1-\nu}\nabla^2 T + \frac{\partial^2 T}{\partial x^2}\right) = 0\\
&(1+\nu)\nabla^2\sigma_y + \frac{\partial^2\Theta}{\partial y^2} + \alpha E\left(\frac{1+\nu}{1-\nu}\nabla^2 T + \frac{\partial^2 T}{\partial y^2}\right) = 0\\
&(1+\nu)\nabla^2\sigma_z + \frac{\partial^2\Theta}{\partial z^2} + \alpha E\left(\frac{1+\nu}{1-\nu}\nabla^2 T + \frac{\partial^2 T}{\partial z^2}\right) = 0\\
&(1+\nu)\nabla^2\tau_{xy} + \frac{\partial^2\Theta}{\partial x\,\partial y} + \alpha E\frac{\partial^2 T}{\partial x\,\partial y} = 0\\
&(1+\nu)\nabla^2\tau_{yz} + \frac{\partial^2\Theta}{\partial y\,\partial z} + \alpha E\frac{\partial^2 T}{\partial y\,\partial z} = 0\\
&(1+\nu)\nabla^2\tau_{zx} + \frac{\partial^2\Theta}{\partial z\,\partial x} + \alpha E\frac{\partial^2 T}{\partial z\,\partial x} = 0
\end{aligned}
\qquad (15.109)
$$

where $\Theta = \sigma_x + \sigma_y + \sigma_z$

15.8. CLASSIFICATION OF PROBLEM TYPES

In mechanics of materials it is frequently desirable to classify problems in terms of their geometric configurations and/or assumptions that will permit their codification and ease of solution. As a result there exist problems in plane stress or strain, beam theory, curved-beam theory, plates, shells, etc. Although the defining equations can be obtained directly from the general theory together with the associated assumptions, it is often instructive and convenient to obtain them directly from physical considerations. The difference between these two approaches marks one of the principal distinguishing differences between the *theory of elasticity* and *mechanics of materials.*

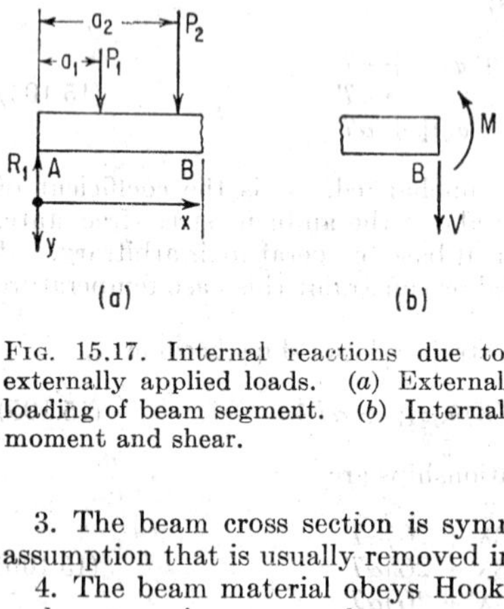

FIG. 15.17. Internal reactions due to externally applied loads. (*a*) External loading of beam segment. (*b*) Internal moment and shear.

15.9. BEAM THEORY

(a) Mechanics of Materials Approach[1]

The following assumptions are basic in the development of elementary beam theory:

1. Beam sections, originally plane, remain plane and normal to the "neutral axis."
2. The beam is originally straight and all bending displacements are small.
3. The beam cross section is symmetrical with respect to the loading plane; an assumption that is usually removed in the general theory.
4. The beam material obeys Hooke's law and the moduli of elasticity in tension and compression are equal.

Consider the beam portion loaded as shown in Fig. 15.17*a*.

For static equilibrium, the internal actions required at section B which are supplied by the immediately adjacent section to the right must consist of a vertical shearing force V and an internal moment M, as shown in Fig. 15.17*b*.

The evaluation of the shear V is accomplished by noting, from equilibrium $\Sigma F_y = 0$,

$$V = R - P_1 - P_2 \qquad \text{(for this example)} \qquad (15.110)$$

The algebraic sum of all the shearing forces at one side of the section is called the shearing force at that section. The moment M is obtained from $\Sigma M = 0$:

$$M = R_1 x - P_1(x - a_1) - P_2(x - a_2) \qquad (15.111)$$

The algebraic sum of the moments of all external loads to one side of the section is called the bending moment at the section.

Note the sign conventions employed thus far:

1. Shearing force is positive if the right portion of the beam tends to shear downward with respect to the left.
2. Bending moment is positive if it produces bending of the beam concave upward.
3. Loading w is positive if it acts in the positive direction of the y axis.

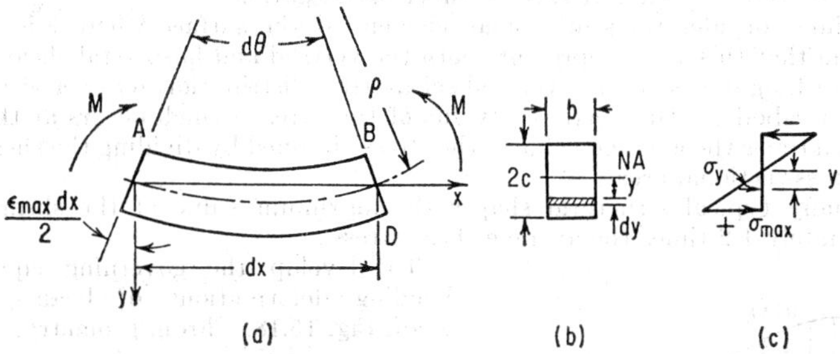

FIG. 15.18. Beam bending with externally applied load. (*a*) Beam element. (*b*) Cross section. (*c*) Bending-stress pattern at section *B-D*.

In Fig. 15.18*a* a portion of one of the beams previously discussed is shown with the bending moment M applied to the element.

Equilibrium conditions require that the sum of the normal stresses σ on a cross section must equal zero, a condition satisfied only if the "neutral axis," defined as the plane or axis of zero normal stress, *is also the centroidal axis* of the cross section.

$$\int_{-c}^{+c} \sigma b\,dy = (\sigma/y)\int_{-c}^{+c} by\,dy = 0 \qquad (15.112)$$

where $\sigma/y = (\epsilon/y)E = (\epsilon_{max}/y_{max})E = \text{const}$

Further, if the moments of the stresses acting on the element dy of the figure are summed over the height of the beam,

$$M = \int_{-c}^{+c} \sigma b y\,dy = (\sigma/y)\int_{-c}^{+c} by^2\,dy = (\sigma/y)I \qquad (15.113)$$

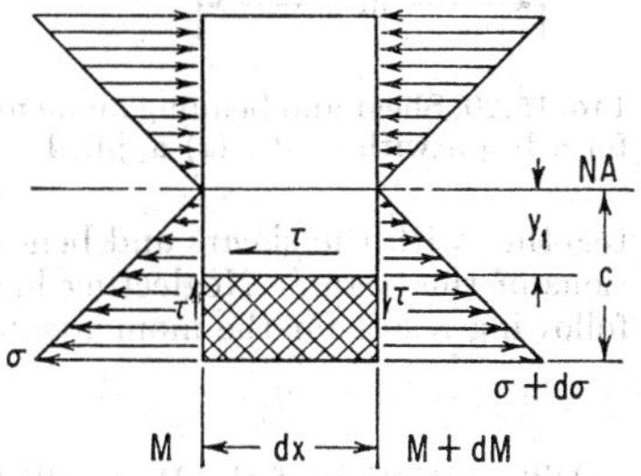

FIG. 15.19. Shear-stress diagram for beam subjected to varying bending moment.

where y = distance from neutral axis to point on cross section being investigated,

and $I = \int_{-c}^{+c} by^2\,dy$ = area moment of inertia about the centroidal axis of the cross section. Equation (15.113) defines the flexural stress in a beam subject to moment M:

$$\sigma = My/I \qquad (15.114)$$

Thus

$$\sigma_{max} = Mc/I \qquad (15.115)$$

To develop the equations for shear stress τ, the general case of the element of the beam subjected to a varying bending moment is taken as in Fig. 15.19.

Applying axial-equilibrium conditions to the shaded area of Fig. 15.19 yields the following general expression for the horizontal shear stress at the lower surface of the shaded area:

$$\tau = (dM/dx)(1/Ib)\int_{y_1}^{c} y\,dA \qquad (15.116)$$

or, in familiar terms,

$$\tau = (V/Ib) \int_{y_1}^{c} y\,dA = (V/Ib)Q \qquad (15.117)$$

where Q = moment of area of cross section about neutral axis for the shaded area above the surface under investigation
V = net vertical shearing force
b = width of beam at surface under investigation

Equilibrium considerations of a small element at the surface where τ is computed will reveal that this value represents both the vertical and horizontal shear.

For a rectangular beam, the vertical shear-stress distribution across a section of the beam is parabolic. The maximum value of this stress (which occurs at the neutral axis) is 1.5 times the average value of the stress obtained by dividing the shear force V by the cross-sectional area.

For many typical structural shapes the maximum value of the shear stress is approximately 1.2 times the average shear stress.

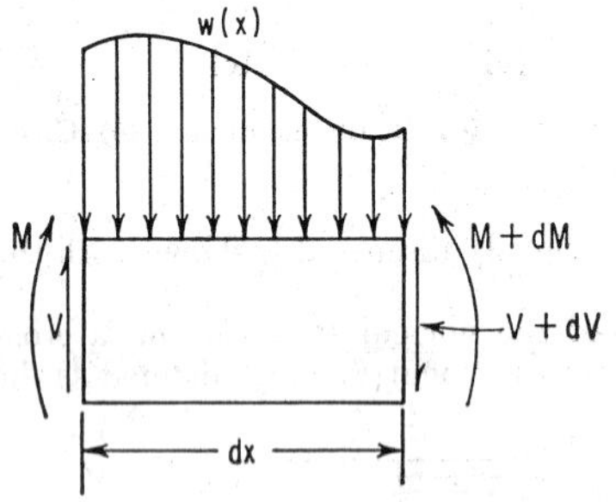

FIG. 15.20. Shear and bending moments for a beam with load $w(x)$ applied.

To develop the governing equation for bending deformations of beams, consider again Fig. 15.18. From geometry,

$$\frac{(\epsilon/2)\,dx}{y} = \frac{dx/2}{\rho} \qquad (15.118)$$

Combining Eqs. (15.118), (15.114), and (15.72) yields

$$1/\rho = M/EI \qquad (15.119)$$

Since $1/\rho \approx -d^2y/dx^2 = -y''$ (15.120)

Therefore $y'' = -M/EI$

(Bernoulli-Euler equation) (15.121)

In Fig. 15.20 the element of the beam subjected to an arbitrary load $w(x)$ is shown together with the shears and bending moments as applied by the adjacent cross sections of the beam. Neglecting higher-order terms, moment summation leads to the following result for the moments acting on the element:

$$dM/dx = V \qquad (15.122)$$

Differentiation of the Bernoulli-Euler equation yields

$$y''' = -V/EI \qquad (15.123)$$

In similar manner, the summation of transverse forces in equilibrium yields

$$dV/dx = -w(x) \qquad (15.124)$$

or

$$y^{\mathrm{IV}} = \frac{w(x)}{EI} \qquad (15.125)$$

where due attention has been given to the proper sign convention.

Table 15.1 presents typical shear, moment, and deflection formulas for beams.

(b) Energy Considerations

The total strain energy of bending is

$$U_b = \int_0^L (M^2/2EI)\,dx \qquad (15.126)$$

The strain energy due to shear is

$$U_s = \int_0^L (V^2/2GA)\,dx \tag{15.127}$$

In calculating the deflections by the energy techniques, shear-strain contributions need not be included unless the beam is short and deep.

The deflections can then be obtained by the application of Castigliano's theorem, of which a general statement is: The partial derivative of the total strain energy of any structure with respect to any one generalized load is equal to the generalized deflection at the point of application of the load, and is in the direction of the load. The generalized loads can be forces or moments and the associated generalized deflections are displacements or rotations:

$$Y_a = \partial U/\partial P_a \tag{15.128}$$

$$\theta_a = \partial U/\partial M_a \tag{15.129}$$

where U = total strain energy of bending of the beam
P_a = load at point a
M_a = moment at point a
Y_a = deflection of beam at point a
θ_a = rotation of beam at point a

Thus

$$Y_a = \partial U/\partial P_a = (\partial/\partial P_a)\int_0^L M^2\,dx/2EI = \int_0^L (M/EI)(\partial M/\partial P_a)\,dx \tag{15.130}$$

$$\theta_a = \partial U/\partial M_a = (\partial/\partial M_a)\int_0^L M^2\,dx/2EI = \int_0^L (M/EI)(\partial M/\partial M_a)\,dx \tag{15.131}$$

An important restriction on the use of this theorem is that the deflection of the beam or structure must be a linear function of the load; i.e., geometrical changes and other nonlinear effects must be neglected.

A second theorem of Castigliano states that

$$P_a = \partial U/\partial Y_a \tag{15.132}$$

$$M_a = \partial U/\partial \theta_a \tag{15.133}$$

and is just the inverse of the first theorem. Because it does not have a "linearity" requirement it is quite useful in special problems.

To illustrate, the deflection y at the center of wire of length $2L$ due to a central load P will be found.

From geometry, the extension δ of each half of the wire is, for small deflections,

$$\delta \approx y^2/2L \tag{15.134}$$

The strain energy absorbed in the system is

$$U = 2\tfrac{1}{2}(AE/L)\,\delta^2 = (AE/4L^3)y^4 \tag{15.135}$$

Then, by the second theorem,

$$P = \partial U/\partial y = (AE/L^3)y^3 \tag{15.136}$$

or the deflection is

$$y = L\sqrt[3]{P/AE} \tag{15.137}$$

Among the other useful energy theorems are:

Theorem of Virtual Work. If a beam which is in equilibrium under a system of external loads is given a small deformation (*virtual deformation*), the work done by the load system during this deformation is equal to the increase in internal strain energy.

Principle of Least Work. For beams with statically indeterminate reactions, the partial derivative of the total strain energy with respect to the unknown reactions must be zero.

$$\partial U/\partial P_i = 0 \qquad \partial U/\partial M_i = 0 \tag{15.138}$$

depending on the type of support. (This follows directly from Castigliano's theorems.) The magnitudes of the reactions thus determined are such as to minimize the strain energy of the system.

(c) Elasticity Approach[2]

In developing the conventional equations for beam theory from the basic equations of elastic theory (i.e., stress equilibrium, strain compatibility, and stress-strain relations) the beam problem is considered a plane-stress problem. The equilibrium equations for plane stress are

$$\partial\sigma_x/\partial x + \partial\tau_{xy}/\partial y = 0 \tag{15.139}$$
$$\partial\sigma_y/\partial y + \partial\tau_{xy}/\partial x = 0 \tag{15.140}$$

By using an *Airy stress function* ψ defined as follows:

$$\sigma_x = \partial^2\psi/\partial y^2 \qquad \sigma_y = \partial^2\psi/\partial x^2 \qquad \tau_{xy} = -\partial^2\psi/\partial x\,\partial y \tag{15.141}$$

and the compatibility equation for strain, as set forth previously, the governing equations for beams can be developed.

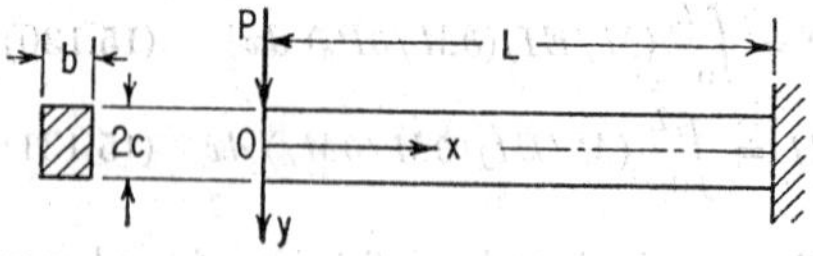

FIG. 15.21. Cantilever beam with end load P.

The only compatibility equation not identically satisfied in this case is

$$\partial^2\epsilon_x/\partial y^2 + \partial^2\epsilon_y/\partial x^2 = \partial^2\gamma_{xy}/\partial x\,\partial y \tag{15.142}$$

Substituting the stress-strain relationships into the compatibility equations and introducing the Airy stress function yields

$$\partial^4\psi/\partial x^4 + 2\partial^4\psi/\partial x^2\,\partial y^2 + \partial^4\psi/\partial y^4 = \nabla^4\psi = 0 \tag{15.143}$$

which is the *biharmonic* equation where ∇^2 is the Laplace operator.

To illustrate the utility of this equation consider a uniform-thickness cantilever beam (Fig. 15.21) with end load P. The boundary conditions are $\sigma_y = \tau_{xy} = 0$ on the surfaces $y = \pm c$, and the summation of shearing forces must be equal to the external load P at the loaded end, $\int_{-c}^{+c} \tau_{xy} b\,dy = P$. The solution for σ_x is

$$\sigma_x = \partial^2\psi/\partial y^2 = cxy \tag{15.144}$$

Introducing $b(2c)^3/12 = I$, the final expressions for the stress components are

$$\sigma_x = -Pxy/I = -\frac{My}{I}$$
$$\sigma_y = 0 \tag{15.145}$$
$$\tau_{xy} = -P(c^2 - y^2)/2I$$

To extend the theory further to determine the displacements of the beam, the definitions of the strain components are

$$\epsilon_x = \partial u/\partial x = \sigma_x/E = -Pxy/EI$$
$$\epsilon_y = \partial v/\partial y = -\nu\sigma_x/E = \nu Pxy/EI \tag{15.146}$$
$$\gamma_{xy} = \partial u/\partial y + \partial v/\partial x = [2(1+\nu)/E]\tau_{xy} = [(1+\nu)P/EI](c^2 - y^2)$$

Solving explicitly for the u and v subject to the boundary conditions

$$u = v = \partial u/\partial x = 0 \qquad \text{at } x = L \text{ and } y = 0$$

there results

$$v = \nu Pxy^2/2EI + Px^3/6EI - PL^2x/2EI + PL^3/3EI \tag{15.147}$$

The equation of the deflection curve at $y = 0$ is

$$(v)_{y=0} = (P/6EI)(x^3 - 3L^2x + 2L^3) \tag{15.148}$$

The curvature of the deflection curve is therefore the Bernoulli-Euler equation

$$1/\rho \approx -(\partial^2 v/\partial^2 x)_{y=0} = -Px/EI = M/EI = -y'' \tag{15.149}$$

Example 1: The moment at any point x along a simply supported uniformly loaded beam (w lb/ft) of span L is

$$M = wLx/2 - wx^2/2 \tag{15.150}$$

Integrating Eq. (15.121) and employing the boundary conditions $y(0) = y(L) = 0$, the solution for the elastic or deflection curve becomes

$$y = (wL^4/24EI)(x/L)[1 - 2(x/L)^2 + (x/L)^3] \tag{15.151}$$

Example 2: In order to obtain the general deflection curve a fictitious load P_a is placed at a distance a from the left support of the previously described uniformly loaded beam.

$$\begin{aligned} M &= -wx^2/2 + wLx/2 + [P_a x(L-a)]/L \qquad 0 < x < a \\ M &= -wx^2/2 + wLx/2 + [P_a a(L-x)]/L \qquad a < x < L \end{aligned} \tag{15.152}$$

From Castigliano's theorem,

$$y_a = \partial U/\partial P_a = \int_0^L (M/EI)(\partial M/\partial P_a)\,dx \tag{15.153}$$

and therefore

$$y_a = (1/EI)(\tfrac{1}{24}wa^4 - \tfrac{1}{12}wLa^3 + \tfrac{1}{24}waL^3) \tag{15.154}$$

The elastic curve of the beam is obtained by substituting x for a in Eq. (15.154), resulting in the same expression as obtained by the double-integration technique.

Example 3: It can be shown, considering stresses away from the beam ends, and essentially considering temperature variations only in the direction perpendicular to the beam axis (or in two or more directions by superposition) that the traction-free thermoelastic stress distribution is given by

$$\sigma_x = -\alpha ET + (1/2c)\int_{-c}^{+c} \alpha ET\,dy + (3y/2c^3)\int_{-c}^{+c} \alpha ETy\,dy \tag{15.155}$$

Since this stress distribution results in a net axial force-free, and moment-free, distribution but nonzero bending displacements and slopes, it would be necessary to superpose any additional stresses associated with actual boundary constraints.

Because the stress distribution is independent of any linear temperature gradient, it is always possible to add an arbitrary linear distribution T', such that $T_{\text{tot}} = T + T'$ and $\int T_{\text{tot}}\,dy = \int T_{\text{tot}}y\,dy = 0$.

Thus

$$\sigma_x = -\alpha ET_{\text{tot}} \tag{15.156}$$

In general T' can be chosen with sufficient accuracy by *visual examination* of the temperature distribution to make the *total-temperature integral* and its *first moment* equal zero. Thus the simplified formula together with its interpretation presents a useful graphical thermoelastic solution for beam and slab problems.

For the beam with temperature distribution $T = a(c^2 - y^2)$ and a is an arbitrary constant, superimpose a temperature distribution T' such that $A_T = -A_{T'}$, where A_T refers to the area under the temperature distribution curve. Evidently

$$T' = -\tfrac{2}{3}ac^2$$

and

$$T_{\text{tot}} = T + T' = a(c^2 - y^2) - \tfrac{2}{3}ac^2$$

and $\int T_{\text{tot}}\,dy = \int T_{\text{tot}}y\,dy = 0$, evident from the selection of T' and symmetry.

The stress is therefore

$$\sigma = -\alpha E T_{\text{tot}} = -\alpha E[a(c^2 - y^2) - \tfrac{2}{3}ac^2] \tag{15.157}$$

15.10. CURVED-BEAM THEORY[1]

A curved beam (see Fig. 15.22) is defined as a beam in which the line joining the centroid of the cross sections (hereafter referred to as the *center line*) is a curve. In the standard developments of the equations for the stresses and deflections for curved beams the following *assumptions* are usually made and represent the *restrictions* on the applicability of curved-beam theory:

The sections of the beam originally plane and normal to the center line of the beam remain so after bending.

All cross sections have an axis of symmetry in the plane of the center line.

The beam is subjected to forces and moments acting in the plane of symmetry.

(a) Equilibrium Approach

In Fig. 15.22 a *positive bending* moment is taken as one which tends to *decrease* the *curvature* of the beam. If R denotes the curvature of the beam at the centroid of a section, then it can be shown that the neutral axis is displaced from R a distance

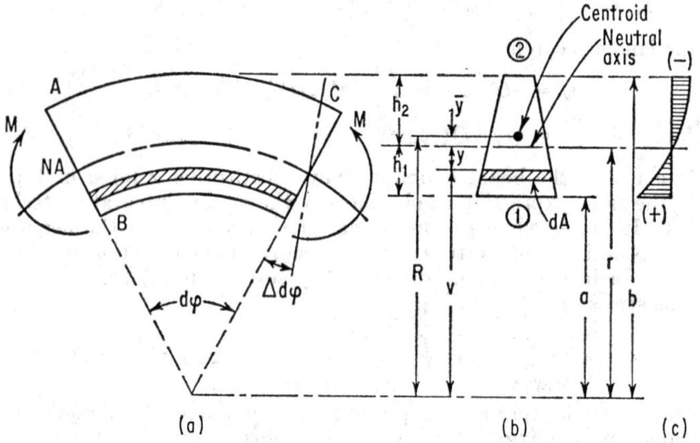

FIG. 15.22. Bending of curved beam element. (*a*) Beam element. (*b*) Cross section. (*c*) Bending-stress pattern at section *C-D*.

toward the center of curvature. As with straight beams, the *neutral axis* is defined as that axis about which the integrated tangential force is zero, when the external traction is restricted to a bending moment. This distance $\bar{y}$ may be computed from

$$\bar{y} = R - \frac{A}{\int dA/v} \tag{15.158}$$

where A = cross-section area of beam, v the distance from the center of curvature to the incremental area; and the integration $\int dA/v$ is carried out over the entire section of the beam.

The flexural stress at any point a distance y from the neutral axis is

$$\sigma = E\epsilon = \frac{Ey(\Delta\, d\phi)}{(r - y)\, d\phi} \tag{15.159}$$

which shows that the normal stress distribution over a cross section is not linear, as would be the case in simple beam theory, but hyperbolic in shape. Equating the sum of the moments of each of these segments of normal stress to the applied bending moment on the element,

$$\int \sigma y \, dy = \frac{E(\Delta \, d\phi)}{d\phi} \int \frac{y^2 \, dA}{r - y} = M \tag{15.160}$$

yields the resulting expression for the stress

$$\sigma = \frac{My}{A\bar{y}(r - y)} \tag{15.161}$$

with the stresses at the extreme fibers, points 1 and 2, expressed by

$$\sigma_1 = \frac{Mh_1}{A\bar{y}a} \quad \text{(tensile)} \qquad \sigma_2 = \frac{Mh_2}{A\bar{y}b} \quad \text{(compressive)} \tag{15.162}$$

The above-described stresses result from pure bending only. If a more general loading condition is given, it must be reduced to the statically equivalent couple and a normal force through the centroid of the section in question.* Then the extensional stresses resulting from the normal force are superposed on the flexural stresses due to the couple.

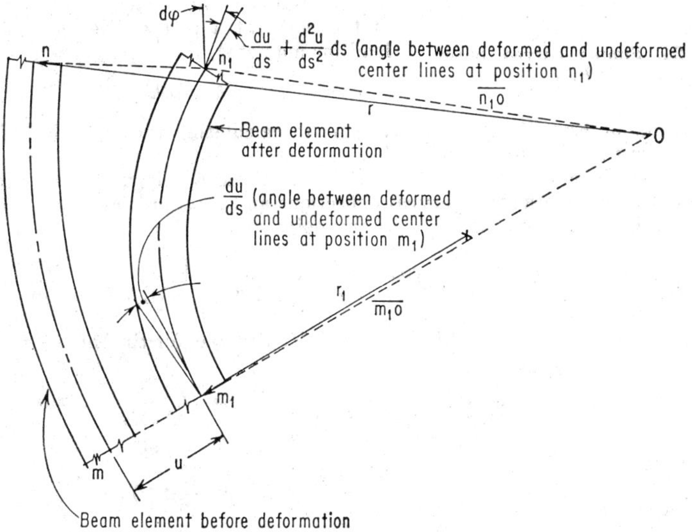

FIG. 15.23. Curved beam with circular center line.

The development of the governing equation for the displacement of a curved beam, for small deflections with a circular center line, depends on the differential quantities shown in Fig. 15.23, where $1/r$ and $1/r_1$ are the respective curvatures of the undeflected and the deflected beam

Note that the radial displacement u is taken positive inward in these relations. A comparison of the changes in length $\Delta \, ds$ and central angle $\Delta \, d\phi$ due to deformation

* This also applies in the general case when a shearing force also may be acting on this section.

leads to

$$1/r_1 - 1/r = u/r^2 + d^2u/ds^2 \tag{15.163}$$

If the thickness of the beam is small compared with the curvature; then

$$1/r_1 - 1/r = -M/EI \tag{15.164}$$

and the differential equation for the deflection curve of a curved beam, which is entirely analogous to that of simple-beam theory, becomes

$$d^2u/ds^2 + u/r^2 = -M/EI \tag{15.165}$$

For an infinitely large r this reduces to the Bernoulli-Euler equation for simple beams.

(b) Energy Approach

A second and more powerful approach, which does not require that the center line of the beam be circular, is essentially the application of Castigliano's theorem. The expression for the strain energy in bending of a curved beam is similar to that of simple-beam theory,

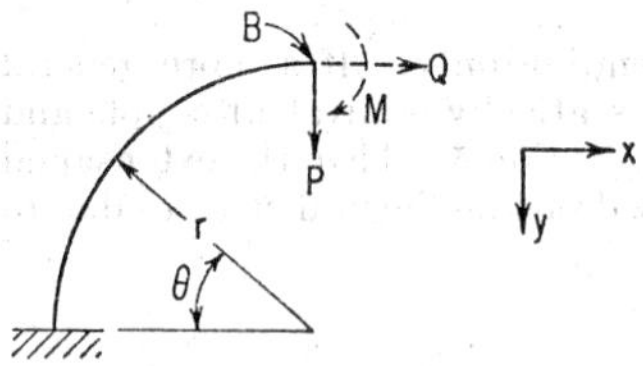

FIG. 15.24. Simple curved beam with vertical load P.

$$U = \int_0^s \frac{M^2\,ds}{2EI} \tag{15.166}$$

where the integration is over the entire length of the beam. The deflection (or rotation) of the beam at a point under a concentrated load (or moment), is

$$\delta_i = \partial U/\partial P_i \tag{15.167}$$

The utility of Eqs. (15.166) and (15.167) in calculating the deflection of curved beams can best be shown by example.

Example 1: The curved beam shown in Fig. 15.24 has a uniform cross section. To determine the horizontal and vertical deflections and rotations, at point B fictitious load Q and M are assumed to act. At any point s on the beam,

$$M_s = -[Pr\cos\theta + Qr(1 - \sin\theta) + M] \tag{15.168}$$

$$\partial M_s/\partial P = -r\cos\theta \qquad \partial M_s/\partial Q = -r(1 - \sin\theta) \qquad \partial M_s/\partial M = -1 \tag{15.169}$$

Substituting $Q = 0$, $M = 0$ in the expression M_s we have, for the three deformations with $ds = r\,d\theta$,

$$\delta_y = \left.\frac{\partial U}{\partial P}\right|_{\substack{Q\to 0\\ M\to 0}} = \int_0^s \frac{M_s}{EI}\frac{\partial M_s}{\partial P}\,ds = \int_0^{\pi/2} \frac{Pr^3}{EI}\cos^2\theta\,d\theta = \frac{\pi}{4}\frac{Pr^3}{EI} \tag{15.170}$$

$$\delta_x = \left.\frac{\partial U}{\partial Q}\right|_{\substack{Q\to 0\\ M\to 0}} = \int_0^s \frac{M_s}{EI}\frac{\partial M_s}{\partial Q}\,ds = \int_0^{\pi/2} \frac{Pr^3}{EI}\cos\theta\,(1 - \sin\theta)\,d\theta = \frac{Pr^3}{2EI} \tag{15.171}$$

$$\theta = \left.\frac{\partial U}{\partial M}\right|_{\substack{Q\to 0\\ M\to 0}} = \int_0^s \frac{M_s}{EI}\frac{\partial M_s}{\partial M}\,ds = \int_0^{\pi/2} \frac{Pr^2}{EI}\cos\theta\,d\theta = \frac{Pr^2}{EI} \tag{15.172}$$

Example 2. *Circular Ring–Energy Approach:* The circular ring is subjected to equal and opposite forces P as shown in Fig. 15.25. From symmetry considerations an equivalent model may be constructed where the load on the horizontal section is denoted by moment M_A and force $P/2$. From symmetry, there is no rotation of the horizontal section at point A. Therefore, by Castigliano's theorem,

$$\theta_A = \partial U/\partial M_A = 0 \tag{15.173}$$

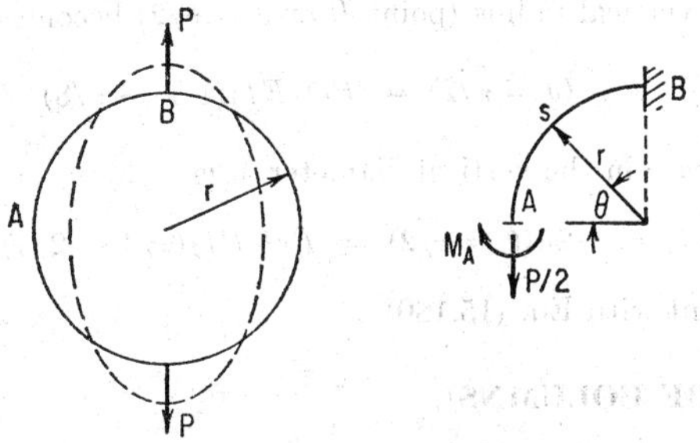

FIG. 15.25. Circular ring.

The moment at any point s is given by

$$M_s = M_A - (P/2)r(1 - \cos\theta) \qquad \text{and} \qquad \partial M_s/\partial M_A = 1 \tag{15.174}$$

Substituting in Eq. (15.173) and imposing the condition of zero rotation,

$$\theta_A = 0 = \partial U/\partial M_A = \int_0^s (M_s/EI)(\partial M_s/\partial M_A)\, ds$$

$$= \int_0^{\pi/2} (1/EI)[M_A - (Pr/2)(1 - \cos\theta)]r\, d\theta \tag{15.175}$$

which leads to

$$M_A = (Pr/2)(1 - 2/\pi) \tag{15.176}$$

which yields the moment expression

$$M_s = (Pr/2)(\cos\theta - 2/\pi) \tag{15.177}$$

The total strain energy for the entire ring is four times that of the quadrant considered. To obtain the total increase in the vertical diameter, the following steps are taken:

$$U = 4\int_0^{\pi/2} \frac{M_s^2}{2EI} r\, d\theta \tag{15.178}$$

$$\delta_y = \partial U/\partial P = 4/EI \int_0^{\pi/2} M_s(\partial M_s/\partial P) r\, d\theta$$

$$= Pr^3/EI \int_0^{\pi/2} (\cos\theta - 2/\pi)^2\, d\theta \tag{15.179}$$

$$\delta_y = (Pr^3/EI)(\pi/4 - 2/\pi) \tag{15.180}$$

Equilibrium Approach: The basic equation for this problem is

$$d^2u/ds^2 + u/r^2 = -M_s/EI \tag{15.181}$$

Substituting Eq. (15.177) yields

$$r^2 d^2u/ds^2 + u = (Pr^3/2EI)(2/\pi - \cos\theta) \tag{15.182}$$

The boundary conditions, derived from the symmetry of the ring, are

$$u'(\theta = 0) = u'(\theta = \pi/2) = 0$$

The general solution of Eq. (15.182) is

$$u = A\cos\theta + B\sin\theta + Pr^3/EI\pi - (Pr^3/4EI)\,\theta\sin\theta \tag{15.183}$$

$$A = -Pr^3/4EI \qquad \text{and} \qquad B = 0 \tag{15.184}$$

Then $\quad u = Pr^3/EI\pi - (Pr^3/4EI)\cos\theta - (Pr^3/4EI)\theta\sin\theta$

The increase in the vertical radius (point B at $\theta = \pi/2$) becomes

$$u = (\theta = \pi/2) = (Pr^3/EI)(1/\pi - \pi/8) \tag{15.185}$$

Thus the total increase in the vertical diameter δ_v is

$$\delta_v = -2u(\theta = \pi/2) = (Pr^3/EI)(\pi/4 - 2/\pi) \tag{15.186}$$

which is in agreement with Eq. (15.180).

15.11. THEORY OF COLUMNS[1]

The equilibrium approach for slender symmetrical columns is based on two sets of assumptions:

1. All the basic assumptions inherent in the derivation of the Bernoulli-Euler equation apply to columns also.

2. The transverse deflection of the column at the point of load application is *not* small when compared with the eccentricity of the applied load.

This theory is best described with the aid of a typical column, taken with a built-in support and subjected to an eccentric load, as illustrated in Fig. 15.26.

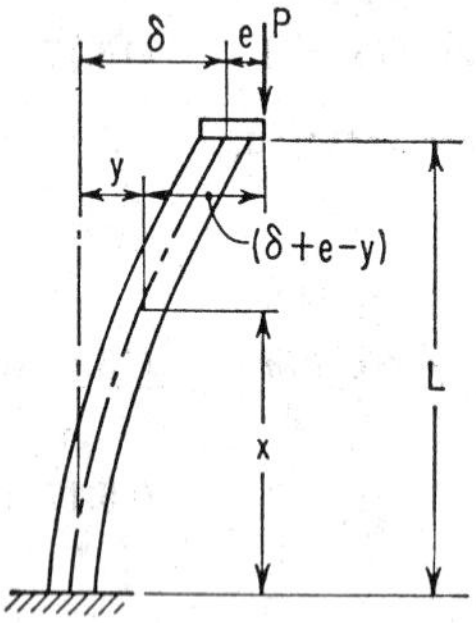

FIG. 15.26. Column subjected to an eccentric load.

The moment at any section a distance x from the base is

$$M = -P(\delta + e - y) \tag{15.187}$$

where the negative sign is in accordance with the sign convention for simple beams. Writing the Bernoulli-Euler equation for the bending deflection of this member with the aid of the substitution $p^2 = P/EI$ gives

$$y'' + p^2 y = p^2(\delta + e) \tag{15.188}$$

The general solution of Eq. (15.188) is

$$y = A \sin px + B \cos px + \delta + e \tag{15.189}$$

From the boundary conditions for a built-in end

$$y(0) = y'(0) = 0$$

the equation for the deflection curve is

$$y = (\delta + e)(1 - \cos px) \tag{15.190}$$

The deflection at the end of the column at $x = L$ is seen to be

$$\delta = \frac{e(1 - \cos pL)}{\cos pL} \tag{15.191}$$

The complete description of the deflection curve for any point in the column thus becomes

$$y = \frac{e(1 - \cos px)}{\cos pL} \tag{15.192}$$

These results can easily be extended to cover a column hinged at both ends by redefining terms as indicated in Fig. 15.27. Thus, from symmetry, the relation of the deflection at the mid-span can be written directly from the previous results:

$$\delta = \frac{e[1 - \cos (pL/2)]}{\cos p(L/2)} \tag{15.193}$$

In applying these equations note that the deflection is not proportional to the compressive load P. Hence the method of superposing a compressive deflection due to P and a bending deflection due to the couple Pe cannot be used for column action.

In considering column action the concept of *critical load* is of fundamental importance. As the argument of the cosine in the equations for the maximum deflection approaches a value of $\pi/2$ the deflection δ increases without bound and in actual practice the column will fail in a buckling *regardless of the eccentricity e.* Substituting the value $pL = \pi/2$ and $pL/2 = \pi/2$ back into $p^2 = P/EI$ will yield

$P_{CR} = \pi^2EI/4L^2$ for the single built-in support (15.194)
$P_{CR} = \pi^2EI/L^2$ for the hinged ends (15.195)
$P_{CR} = 4\pi^2EI/L^2$ for both ends built-in (15.196)

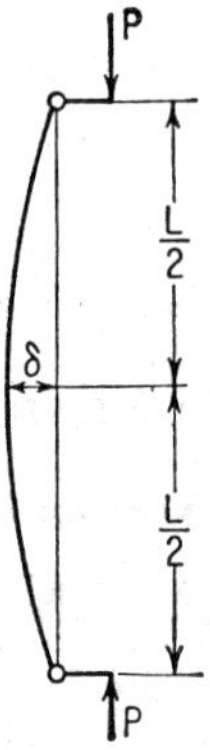

FIG. 15.27. Column hinged at both ends.

The above equations for the critical load (Euler's loads) depend only on the dimensions of the column (I/L^2) and the modulus of the material E. If the moment of inertia is written $I = Ak^2$ where k is the radius of gyration, the critical-load expressions take the form

$$P_{CR} = \frac{C_1\pi^2AE}{(L/k)^2} \tag{15.197}$$

where the dependence is now on the material and a slenderness ratio L/k.

The *critical stress* for a column with hinged ends is given by

$$\sigma_{CR} = \frac{P_{CR}}{A} = \frac{\pi^2E}{(L/k)^2} \tag{15.198}$$

A plot of (σ_{CR} vs. L/k) is hyperbolic in form, as shown in Fig. 15.28.

The horizontal line in Fig. 15.28 indicates the compressive yield stress of a typical structural steel. For analysis purposes, one uses the compressive yield stress as the design criterion for *small* slenderness ratios and the Euler curve for higher ratios.

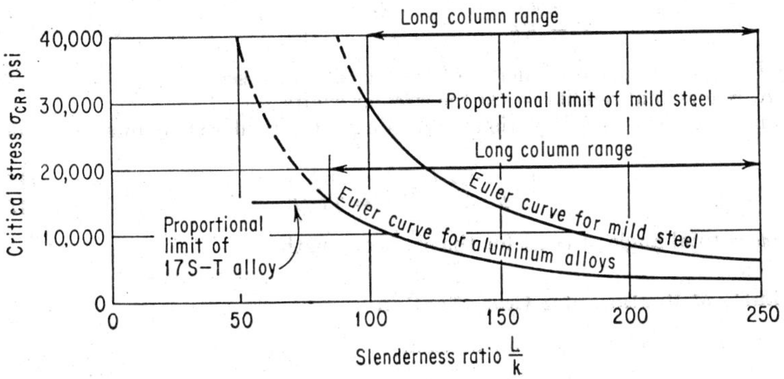

FIG. 15.28. Curve of critical stress vs. slenderness ratio for a column with hinged ends.

Much column design, especially in heavy structural engineering, is accomplished by means of the application of empirical formulas developed as a result of experimental work and practical experience. Several of these formulas are presented below, for hinged bars. Straight-line formulas for structural-steel bars:

$$\sigma_{CR} = P_{CR}/A = 48{,}000 - 210(L/k) \tag{15.199}$$

Parabolic formula for structural steels:

$$\sigma_{CR} = 40{,}000 - 1.35(L/k)^2 \tag{15.200}$$

Gordon-Rankine formula for main members with $120 < L/k < 200$:

$$\sigma_w = \frac{18{,}000}{1 + L^2/18{,}000k^2}\left(1.6 - \frac{L}{200k}\right) \tag{15.201}$$

where σ_w is a working stress.

15.12. SHAFTS, TORSION, AND COMBINED STRESS[1]

(a) Torsion of Solid Circular Shafts

When a solid circular shaft is subjected to a pure torsional load the reaction of the shaft for small angles of twist is assumed to be subject to the following restraints:

1. Circular cross sections remain circular and their diameters remain unchanged.
2. The axial distances between adjacent cross sections do not change.
3. A lateral surface element of the shaft is in a state of pure shear.

The shearing stresses acting on an element a distance r from the axis are

$$\tau = Gr\theta \tag{15.202}$$

where θ = angle of twist per unit length of shaft

G = modulus of rigidity $= \dfrac{E}{2(1+\nu)}$

The shear stress, maximum at the surface is

$$\tau_{\max} = \tfrac{1}{2}G\theta D \tag{15.203}$$

where D = diameter of the shaft.

The total torque acting on the shaft can be expressed as

$$T = \int_A r(\tau\, dA) = \int_A G\theta r^2\, dA = G\theta J \tag{15.204}$$

$$T = K\theta \tag{15.205}$$

where J = polar moment of inertia of the circular cross section

K = torsional rigidity $= GJ$ for circular shafts

Therefore, the most useful relations in dealing with circular-shaft torsional problems are

$$\tau_{\max} = TD/2J = 16T/\pi D^3 \tag{15.206}$$

$$\phi = TL/GJ = TL/K \tag{15.207}$$

where ϕ = total angle of twist for the shaft of length L.

(b) Shafts of Rectangular Cross Section

For a shaft of rectangular cross section,

$$\tau_{\max} = (T/ht^2)[3 + 1.8(t/h)] \qquad (h > t) \tag{15.208}$$

$$\varphi = TL/K \tag{15.209}$$

where $K = \beta h t^3 G$ and the constant β is 0.141, 0.249, and 0.333 for various ratios of h/t of 1.0, 2.5, and ∞, respectively.

(c) Single-cell Tubular-section Shaft

In Fig. 15.29 a general tubular shape is subjected to a pure torque load; the resultant angular rotations will be about point 0. Considering a slice of the tube of length L, the *shear flow* q can be shown to be constant around the tube. In thin-walled-tube problems the shear is expressed in terms of force per unit length and is thought of as "flowing" from *source* to *sink* in much the same manner as in hydrodynamics problems. Thus it can be shown that

$$q = T/2A \tag{15.210}$$

where A = area enclosed by the median line of the section

$A = \pi r^2$ for a circular tubular section of mean radius r

The total strain energy and total angle of twist (with no warping) for the tube is

$$U = \oint (q^2L/2tG)\, ds \tag{15.211}$$

$$\phi = \partial U/\partial T = (TL/4A^2G) \oint ds/t = (L/2AG) \oint (q/t)\, ds \tag{15.212}$$

The unit angle of twist is

$$\theta = \phi/L = (1/2AG) \oint (q/t)\, ds \tag{15.213}$$

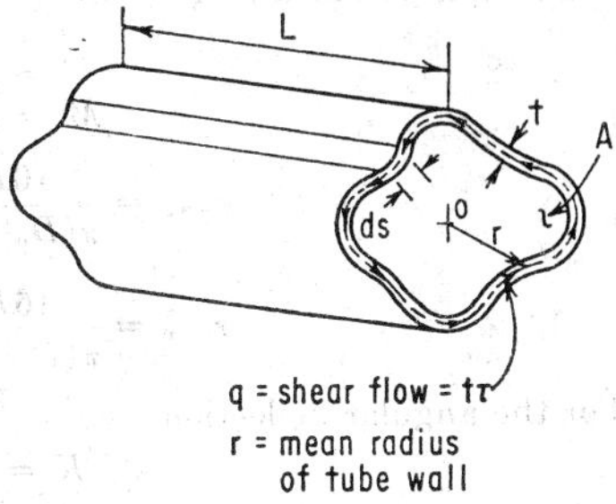

FIG. 15.29. Shaft with general tubular shape.

The torsional rigidity of the tubular section may then be defined as

$$K = \frac{4A^2}{\oint (ds/t)} G \tag{15.214}$$

For a circular pipe of mean diameter D_m the value of the rigidity becomes

$$K = (\pi D_m{}^3 t/4)G \tag{15.215}$$

(d) Combined Stresses

Frequently problems in shafts involve combined torsion and bending. If the weight of the shaft is neglected relative to load P, there are two major components contributing to the maximum stress:

1. Torsional stress τ, which is a maximum at any point on the surface of the shaft
2. Flexural stress σ, which is a maximum at the built-in end on an element most remote from the shaft axis

The direct stresses due to shear are not significant, since they are a maximum at the axis of the shaft where the flexural stress is zero. The loads on the element at the built-in end are $T = Pr$ and $M = -PL$, and the stresses may be evaluated as follows:

$$\tau = TD/2J = PrD/2J \tag{15.216}$$

$$\sigma = Mc/I = \pm PLD/2I \tag{15.217}$$

where r = load offset distance

D = shaft diameter

L = shaft length

The maximum principle stress on the element in tension is

$$\sigma_{max} = \sigma/2 + \tfrac{1}{2}\sqrt{\sigma^2 + 4\tau^2} \tag{15.218}$$

Noting that for a circular shaft the polar moment is twice the area moment about a diameter ($J = 2I$):

$$\sigma_{max} = D/4I(M + \sqrt{M^2 + T^2}) = (PLD/4I)[1 + \sqrt{1 + (r/L)^2}] \quad (15.219)$$

where

$$I = \pi D^4/64$$

The maximum shear stress on the element is

$$\tau_{max} = D/4I \sqrt{M^2 + T^2} = (PLD/4I) \sqrt{1 + (r/L)^2} \quad (15.220)$$

Example 1: For a hollow circular shaft the loads and stresses are

$$T = Pr \qquad \tau = PrD/2J$$
$$M = PL \qquad \sigma = PLD/2I$$

$$\sigma_{max} = \frac{16D_0PL}{\pi(D_0^4 - D_i^4)}\left[1 + \sqrt{1 + \left(\frac{r}{L}\right)^2}\right] \quad (15.221)$$

$$\tau_{max} = \frac{16D_0PL}{\pi(D_0^4 - D_i^4)}\left[\sqrt{1 + \left(\frac{r}{L}\right)^2}\right]$$

For the angular deflection,

$$K = \pi D^3tG/4$$
$$\theta = T/K = 4Pr/\pi D^3tG \quad (15.222)$$
$$\phi = \theta L = 4PrL/\pi D^3tG$$

where $D = \dfrac{D_0 + D_i}{2}$

Example 2: The shear stresses and angular deflection of a box as shown in Fig. 15.30 will now be investigated.

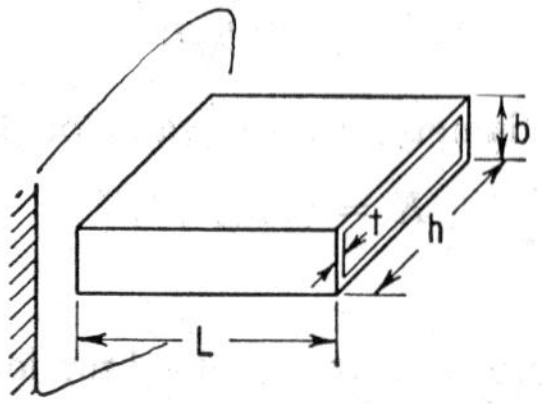

FIG. 15.30. Rectangular thin-walled section.

The shear flow is given by

$$q = T/2A \quad (15.223)$$

where A = plane area $(b - t)(h - t)$. The shear stress

$$\tau = \frac{q}{t} = \frac{T}{2t(b - t)(h - t)} \quad (15.224)$$

The steps in obtaining the torsional rigidity are

$$\oint ds/t \approx 1/t[2(b - t) + 2(h - t)] \quad (15.225)$$

$$K = \frac{4A^2G}{\oint ds/t} = \frac{2Gt[(b - t)(h - t)]^2}{(b - t) + (h - t)} \quad (15.226)$$

$$\phi = \theta L = \frac{TL}{K} = \frac{TL[(b - t) + (h - t)]}{2Gt[(b - t)(h - t)]^2} \quad (15.227)$$

Table 15.1. Shear, Moment, and Deflection Formulas for Beams[1,12]

Notation: W = load (lb); w = unit load (lb/linear in.). M is positive when clockwise; V is positive when upward; y is positive when upward. Constraining moments, applied couples, loads, and reactions are positive when acting as shown. All forces are in pounds, all moments in inch-pounds; all deflections and dimensions in inches. θ is in radians and $\tan\theta = \theta$

Loading, support, and reference number	Reactions R_1 and R_2, vertical shear V	Deflection y, maximum deflection, and end slope θ
End supports Uniform load 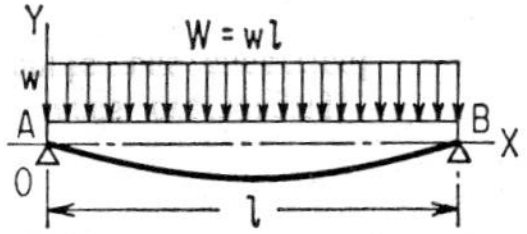 (1)	$R_1 = +\frac{1}{2}W \quad R_2 = +\frac{1}{2}W$ $V = \frac{1}{2}W\left(1 - \frac{2x}{l}\right)$	$y = -\frac{1}{24}\frac{Wx}{EIl}(l^3 - 2lx^2 + x^3)$ Max $y = -\frac{5}{384}\frac{Wl^3}{EI}$ at $x = \frac{1}{2}l$ $\theta = -\frac{1}{24}\frac{Wl^2}{EI}$ at A $\quad \theta = +\frac{1}{24}\frac{Wl^2}{EI}$ at B
End supports Intermediate load 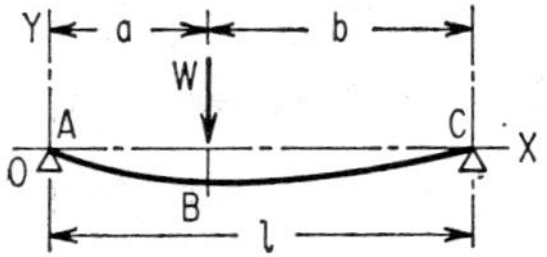 (2)	$R_1 = +W\frac{b}{l} \quad R_2 = +W\frac{a}{l}$ (A to B) $V = +W\frac{b}{l}$ (B to C) $V = -W\frac{a}{l}$	(A to B) $y = -\frac{Wbx}{6EIl}[2l(l-x) - b^2 - (l-x)^2]$ (B to C) $y = -\frac{Wa(l-x)}{6EIl}[2lb - b^2 - (l-x)^2]$ Max $y = -\frac{Wab}{27EIl}(a+2b)\sqrt{3a(a+2b)}$ at $x = \sqrt{\frac{1}{3}a(a+2b)}$ when $a > b$ $\theta = -\frac{1}{6}\frac{W}{EI}\left(bl - \frac{b^3}{l}\right)$ at A; $\quad \theta = +\frac{1}{6}\frac{W}{EI}\left(2bl + \frac{b^3}{l} - 3b^2\right)$ at C

Table 15.1. Shear, Moment, and Deflection Formulas for Beams (Continued)

Loading, support, and reference number	Reactions R_1 and R_2, vertical shear V	Deflection y, maximum deflection, and end slope θ
Supported at both ends Two symmetrical loads 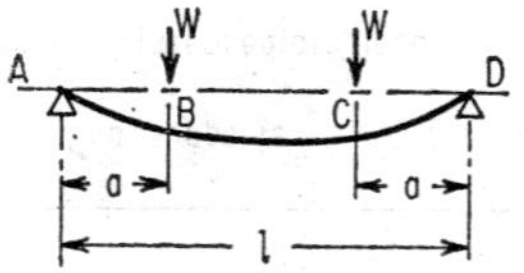 (3)	$R_1 = R_2 = W$ $(A$ to $B)$ and $(C$ to $D)$ $V = W$ $(B$ to $C)$ $V = 0$	$(A$ to $B)$ and $(C$ to $D)$ $y = -\frac{Wx}{6EI}(3\,la - 3a^2 - x^2)$ $(B$ to $C)$ $y = -\frac{Wa}{6EI}(3lx - 3x^2 - a^2)$ Max $y = -\frac{Wa}{24EI}(3l^2 - 4a^2)$ $\theta_A = -\frac{Pa}{2EI}(l - a)$
Both ends overhanging, supports unsymmetrical Uniform load 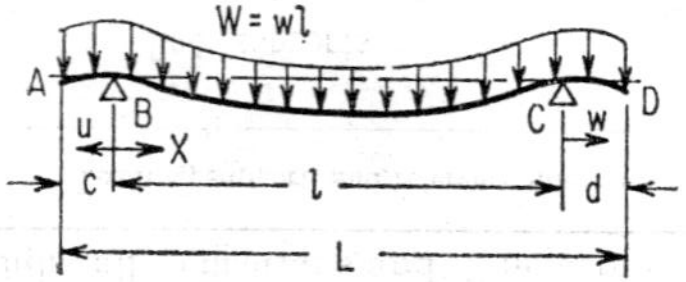 (4)	$R_1 = \frac{W}{2l}(l^2 - d^2 + c^2)$ $R_2 = \frac{W}{2l}(l^2 + d^2 - c^2)$	$(A$ to $B)$ $y = -\frac{Wu}{24EIl}[2l(d^2 + 2c^2) + 6c^2u - u^2(4c - u) - l^3]$ $(B$ to $C)$ $y = -\frac{Wx(l - x)}{24EIl}\left\{x(l - x) + l^2 - 2(d^2 + c^2) - \frac{2}{l}[d^2x + c^2(l - x)]\right\}$ Deflection at end: $y = -\frac{Wc}{24EIl}[2l(d^2 + 2c^2) + 3c^3 - l^3]$

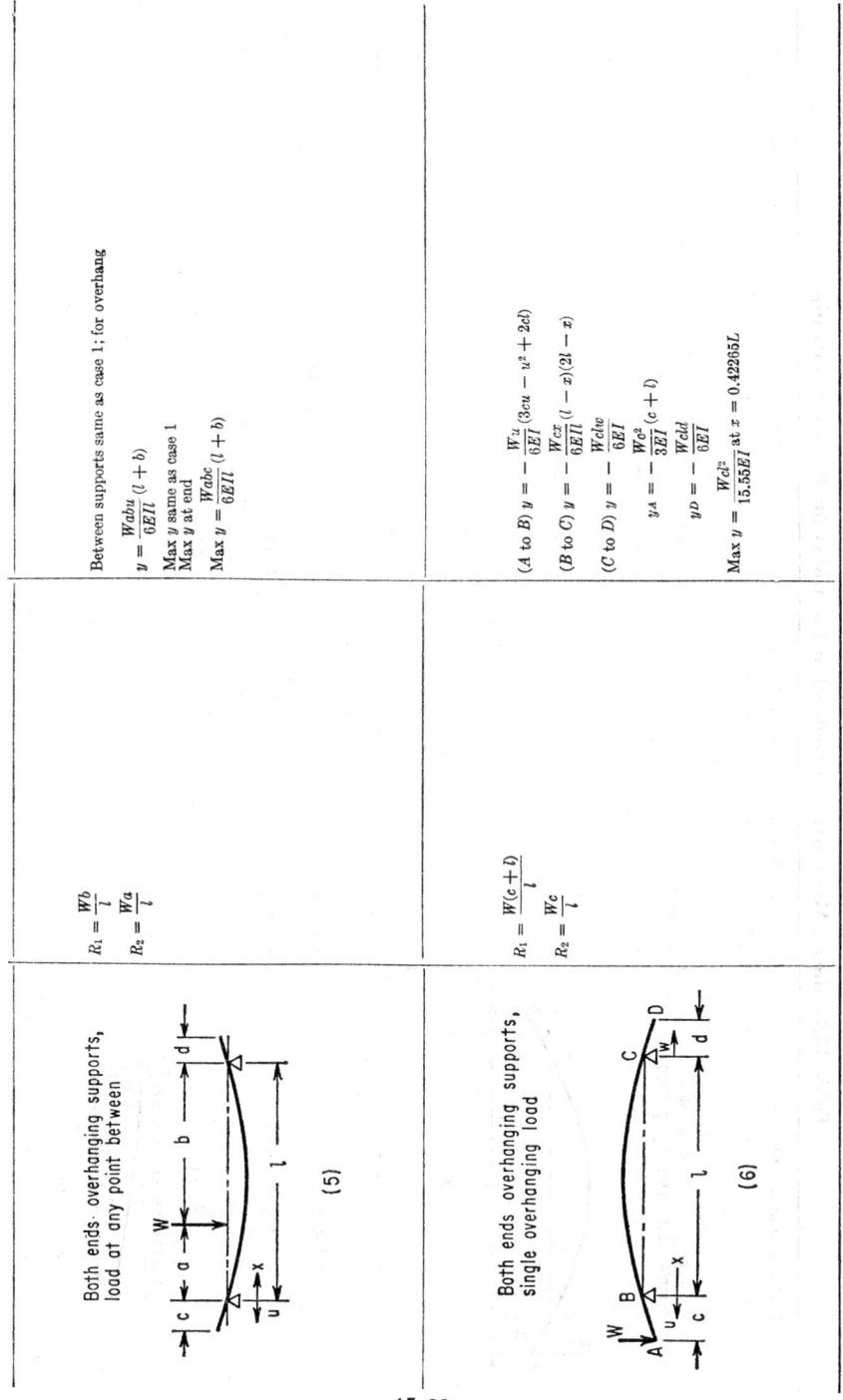

Both ends overhanging supports, load at any point between 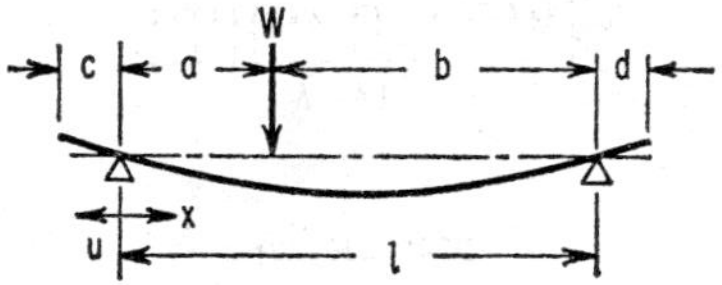 (5)	$R_1 = \frac{Wb}{l}$ $R_2 = \frac{Wa}{l}$	Between supports same as case 1; for overhang $y = \frac{Wabu}{6EIl}(l + b)$ Max y same as case 1 Max y at end Max $y = \frac{Wabc}{6EIl}(l + b)$
Both ends overhanging supports, single overhanging load 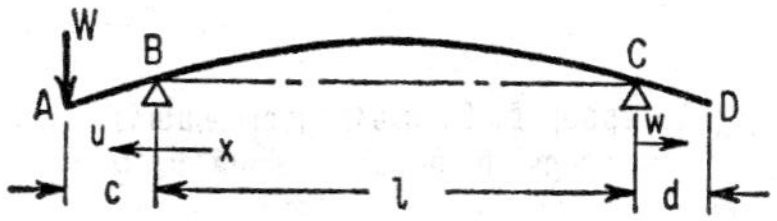 (6)	$R_1 = \frac{W(c + l)}{l}$ $R_2 = \frac{Wc}{l}$	$(A \text{ to } B)\ y = -\frac{Wu}{6EI}(3cu - u^2 + 2cl)$ $(B \text{ to } C)\ y = -\frac{Wcx}{6EIl}(l - x)(2l - x)$ $(C \text{ to } D)\ y = -\frac{Wclw}{6EI}$ $y_A = -\frac{Wc^2}{3EI}(c + l)$ $y_D = -\frac{Wcld}{6EI}$ Max $y = \frac{Wcl^2}{15.55EI}$ at $x = 0.42265L$

Table 15.1. Shear, Moment, and Deflection Formulas for Beams (Continued)

Loading, support, and reference number	Reactions R_1 and R_2, vertical shear V	Deflection y, maximum deflection, and end slope θ
Both ends overhanging supports, symmetrical overhanging loads 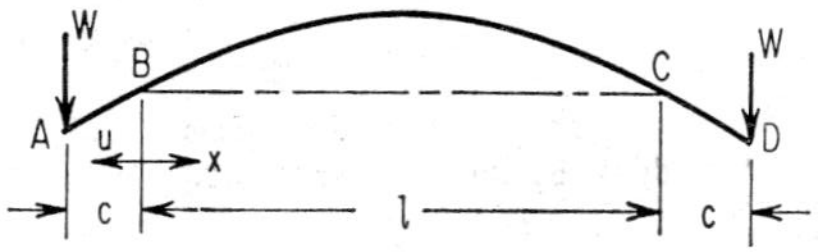 (7)	$R_1 = R_2 = W$	(A to B) and (C to D) $y = -\frac{Wu}{6EI}[3c(l+u) - u^2]$ (B to C) $y = -\frac{Wcx}{2EI}(l-x)$ $y_A = y_D = -\frac{Wc^2}{6EI}(2c+3l)$ Max $y = \frac{Wcl^2}{8EI}$
Cantilever Uniform load 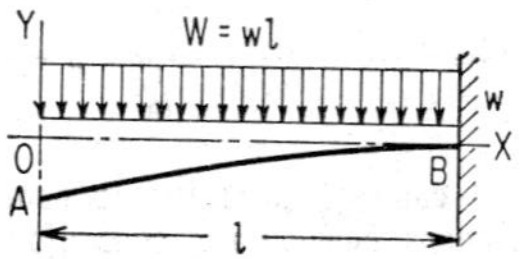 (8)	$R_2 = +W$ $V = -\frac{W}{l}x$	$y = -\frac{1}{24}\frac{W}{EIl}(x^4 - 4l^3x + 3l^4)$ Max $y = -\frac{1}{8}\frac{Wl^3}{EI}$ $\theta = +\frac{1}{6}\frac{Wl^2}{EI}$ at A

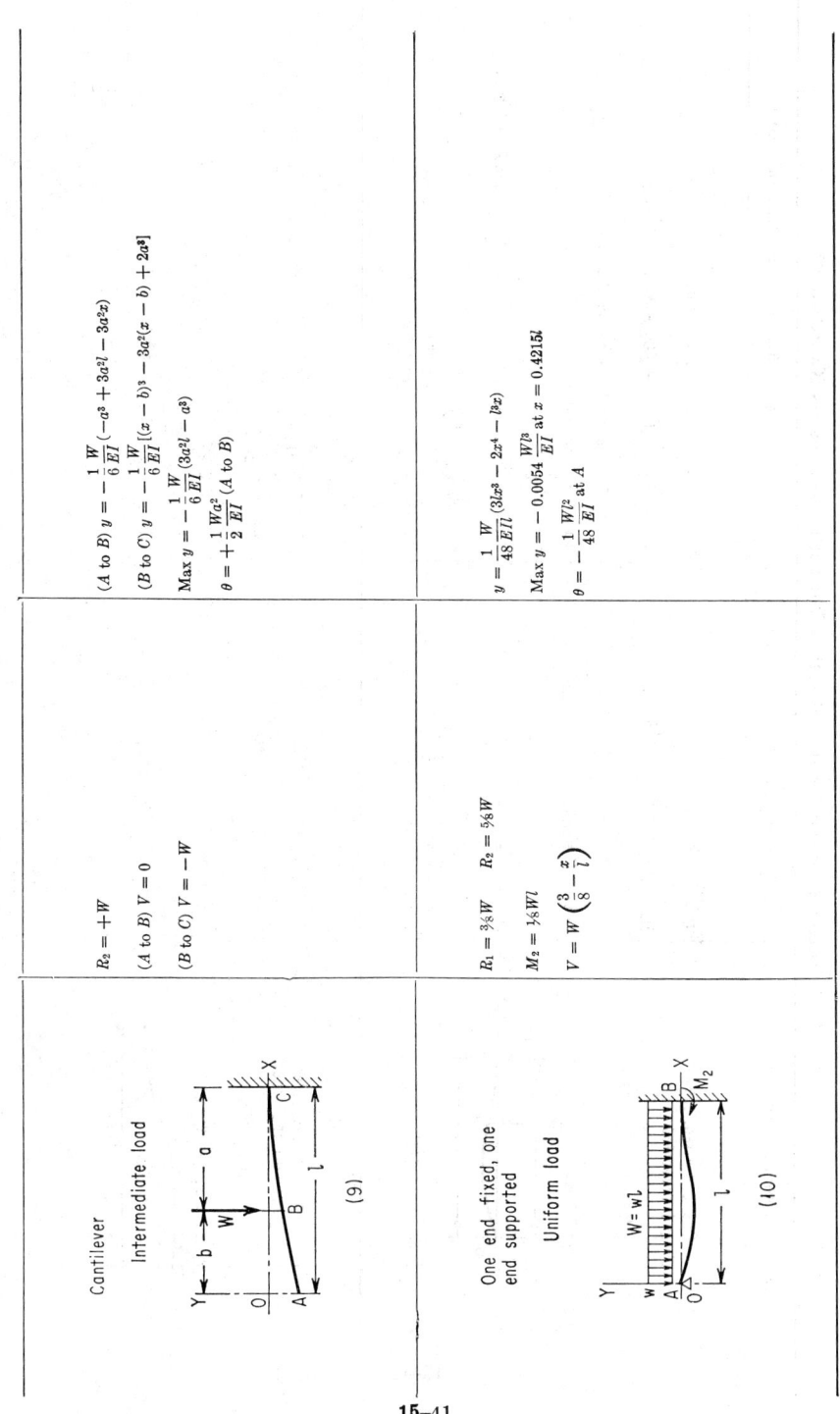

Cantilever Intermediate load 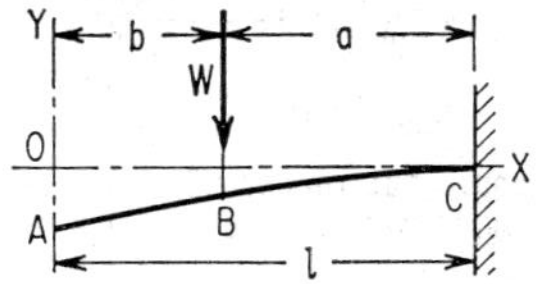 (9)	$R_2 = +W$ $(A \text{ to } B)\ V = 0$ $(B \text{ to } C)\ V = -W$	$(A \text{ to } B)\ y = -\frac{1}{6}\frac{W}{EI}(-a^3 + 3a^2l - 3a^2x)$ $(B \text{ to } C)\ y = -\frac{1}{6}\frac{W}{EI}[(x-b)^3 - 3a^2(x-b) + 2a^3]$ $\text{Max } y = -\frac{1}{6}\frac{W}{EI}(3a^2l - a^3)$ $\theta = +\frac{1}{2}\frac{Wa^2}{EI}\ (A \text{ to } B)$
One end fixed, one end supported Uniform load 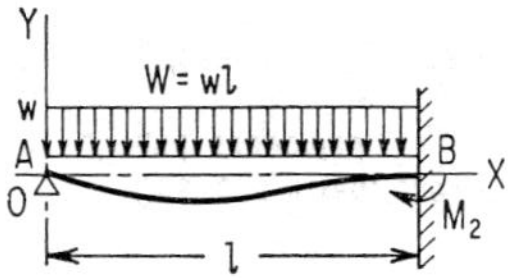 (10)	$R_1 = \frac{3}{8}W \qquad R_2 = \frac{5}{8}W$ $M_2 = \frac{1}{8}Wl$ $V = W\left(\frac{3}{8} - \frac{x}{l}\right)$	$y = \frac{1}{48}\frac{W}{EIl}(3lx^3 - 2x^4 - l^3x)$ $\text{Max } y = -0.0054\frac{Wl^3}{EI} \text{ at } x = 0.4215l$ $\theta = -\frac{1}{48}\frac{Wl^2}{EI} \text{ at } A$

Table 15.1. Shear, Moment, and Deflection Formulas for Beams (Continued)

Loading, support, and reference number	Reactions R_1 and R_2, vertical shear V	Deflection y, maximum deflection, and end slope θ
One end fixed, one end supported Intermediate load 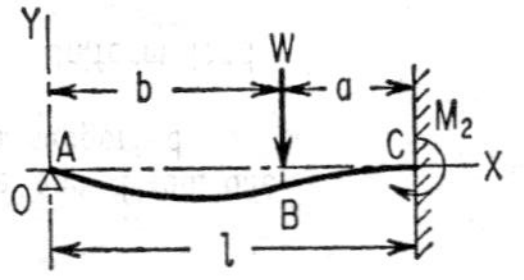 (11)	$R_1 = \frac{1}{2} W \left(\frac{3a^2l - a^3}{l^3}\right)$ $R_2 = W - R_1$ $M_2 = \frac{1}{2} W \left(\frac{a^3 + 2al^2 - 3a^2l}{l^2}\right)$ (A to B) $V = +R_1$ (B to C) $V = R_1 - W$	(A to B) $y = \frac{1}{6EI}[R_1(x^3 - 3l^2x) + 3Wa^2x]$ (B to C) $y = \frac{1}{6EI}\{R_1(x^3 - 3l^2x) + W[3a^2x - (x - b)^3]\}$ If $a < 0.586l$, max y is between A and B at: $x = l\sqrt{1 - \frac{2l}{3l - a}}$ If $a > 0.586l$, max y is at: $x = \frac{l(l^2 + b^2)}{3l^2 - b^2}$ If $a = 0.586l$, max y is at B and $= -0.0098\frac{Wl^3}{EI}$, max possible deflection $\theta = \frac{1}{4}\frac{W}{EI}\left(\frac{a^3}{l} - a^2\right)$ at A
Fixed at one end, free but guided at the other Uniform load 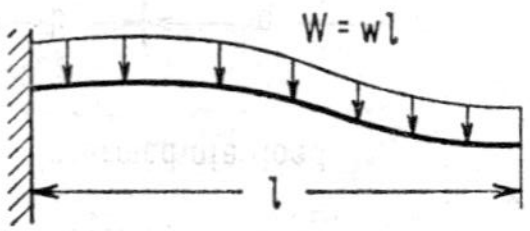 (12)	$R_1 = W$	$y = -\frac{Wx^2}{24EIl}(2l - x)^2$ Max $y = -\frac{Wl^3}{24EI}$

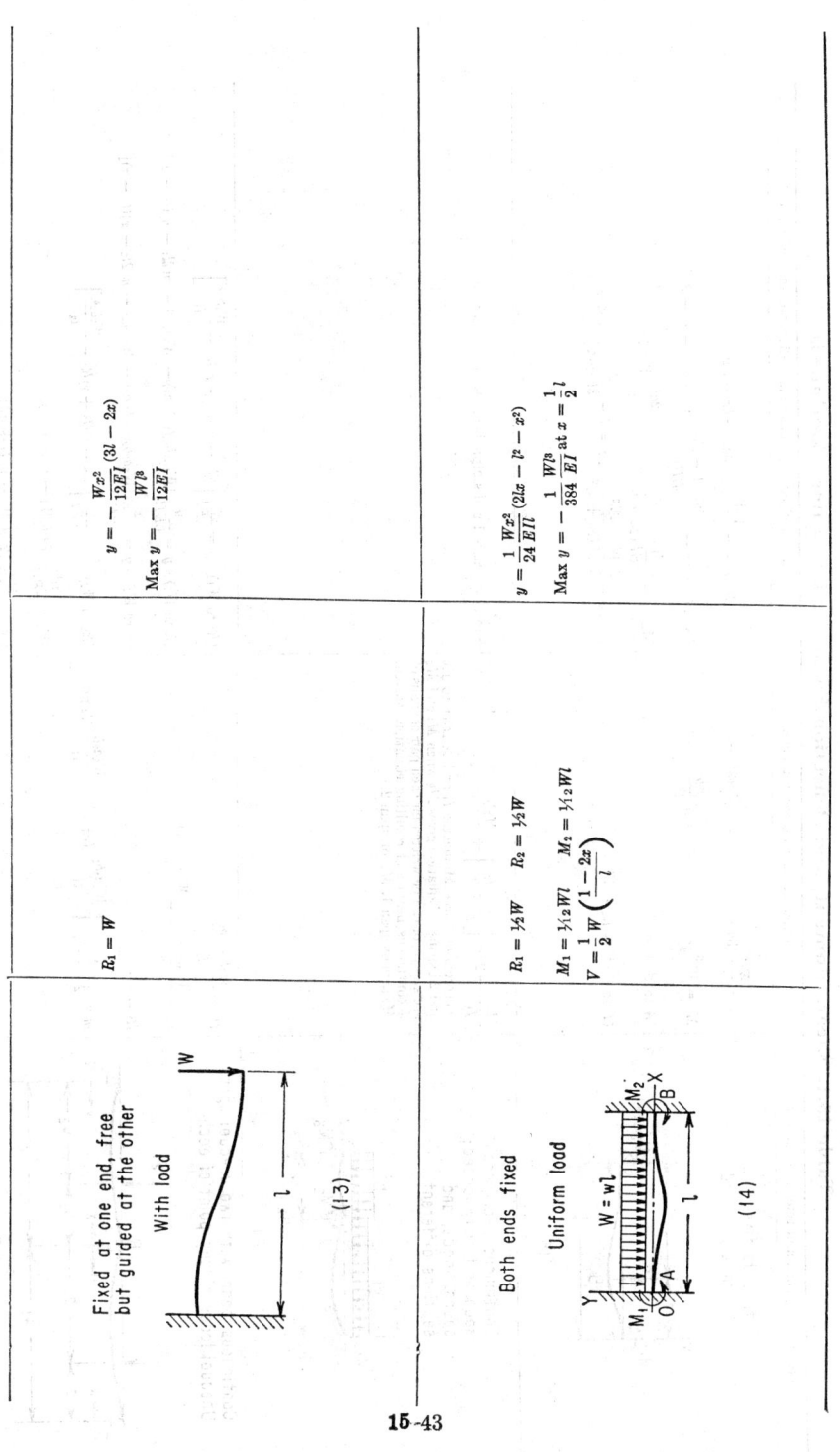

Fixed at one end, free but guided at the other With load 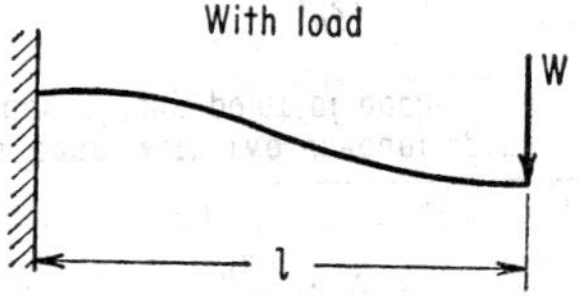 (13)	$R_1 = W$	$y = -\frac{Wx^2}{12EI}(3l - 2x)$ $\text{Max } y = -\frac{Wl^3}{12EI}$
Both ends fixed Uniform load 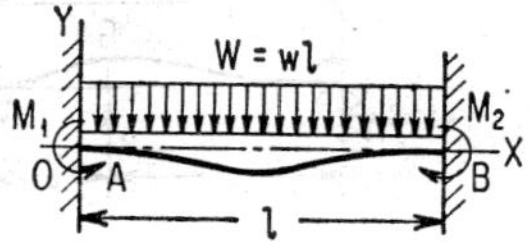 (14)	$R_1 = \frac{1}{2}W \qquad R_2 = \frac{1}{2}W$ $M_1 = \frac{1}{12}Wl \qquad M_2 = \frac{1}{12}Wl$ $V = \frac{1}{2}W\left(\frac{1-2x}{l}\right)$	$y = \frac{1}{24}\frac{Wx^2}{EIl}(2lx - l^2 - x^2)$ $\text{Max } y = -\frac{1}{384}\frac{Wl^3}{EI} \text{ at } x = \frac{1}{2}l$

Table 15.1. Shear, Moment, and Deflection Formulas for Beams (Continued)

Loading, support, and reference number	Reactions R_1 and R_2, vertical shear V	Deflection y, maximum deflection, and end slope θ
Both ends fixed Intermediate load 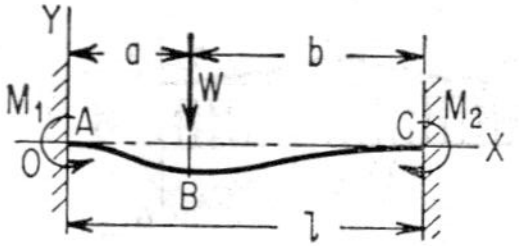 (15)	$R_1 = \frac{Wb^2}{l^3}(3a+b)$ $R_2 = \frac{Wa^2}{l^3}(3b+a)$ $M_1 = W\frac{ab^2}{l^2}$ $M_2 = W\frac{a^2b}{l^2}$ (A to B) $V = R_1$ (B to C) $V = R_1 - W$	(A to B) $y = -\frac{1}{6}\frac{Wb^2x^2}{EIl^3}(3ax + bx - 3al)$ (B to C) $y = -\frac{1}{6}\frac{Wa^2(l-x)^2}{EIl^3}[(3b+a)(l-x) - 3bl]$ Max $y = \frac{2}{3}\frac{W}{EI}\frac{a^3b^2}{(3a+b)^2}$ at $x = \frac{2al}{3a+b}$ if $a > b$ Max $y = \frac{2}{3}\frac{W}{EI}\frac{a^2b^3}{(3b+a)^2}$ at $x = l - \frac{2bl}{3b+a}$ if $a < b$
Continuous beam, each span uniformly loaded; spans, loads, and sections different 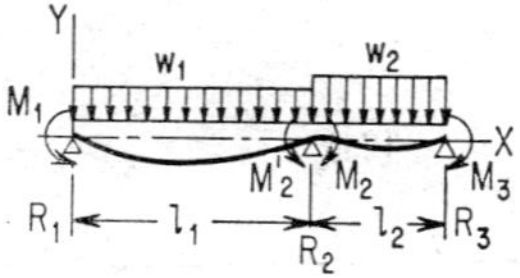 (16)	$\frac{M_1l_1}{I_1} + 2M_2\left[\frac{l_1}{I_1} + \frac{l_2}{I_2}\right] + \frac{M_3l_2}{I_2} = \frac{W_1l_1^3}{4I_1} + \frac{W_2l_2^3}{4I_2}$ (Theorem of Three Moments: I_1 and I_2 refer to 1st and 2d spans. Equation gives M_2 when M_1 and M_3 are known, or can be written for each pair of spans of a continuous beam and resulting equations solved. M_2 acts on span 1, M_2' on span 2.)	Superpose cases 1 and simple beam with end moment
Continuous beam with two unequal spans Unequal loads at any point of each 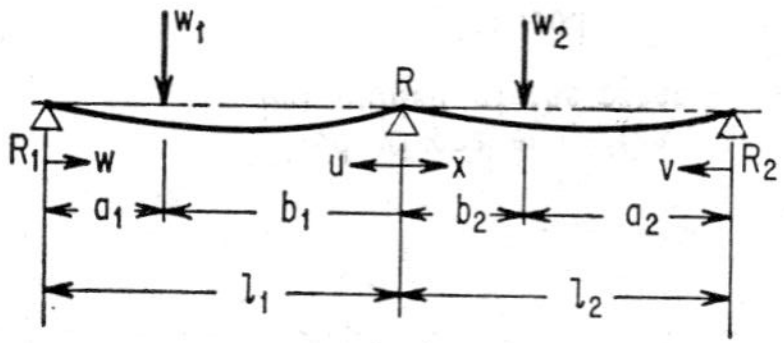 (17)	$R_1 = \frac{W_1b - m}{l_1}$ $R_2 = \frac{W_1a_1 + m}{l_1} + \frac{W_2a_2 + m}{l_2}$ $R^3 = \frac{W_2b_2 - m}{l_2}$ $m = \frac{1}{2(l_1 + l_2)}\left[\frac{W_1a_1b_1}{l_1}(l_1+a_1) + \frac{W_2a_2b_2}{l_2}(l_2+a_2)\right]$	(R_1 to W_1) $y = \frac{w}{6EI}\left[(l_1 - w)(l_1 + w)R_1 - \frac{W_1b_1^3}{l_1}\right]$ (R to W_1) $y = \frac{u}{6EIl_1}[W_1a_1b_1(l_1 + a_1) - W_1a_1u^2 - m(2l_1 - u)(l_1 - u)]$ (R to W_2) $y = \frac{x}{6EIl_2}[W_2a_2b_2(l_2 + a_2) - W_2a_2x^2 - m(2l_2 - x)(l_2 - x)]$ (R_2 to W_2) $y = \frac{v}{6EI}\left[(l_2 - v)(l_2 + v)R_2 - \frac{W_2b_2^3}{l_2}\right]$ $y_{W_1} = \frac{a_1b_1}{6EIl_1}[2a_1b_1W_1 - m(l_1 + a_1)]$ $y_{W_2} = \frac{a_2b_2}{6EIl_2}[2a_2b_2W_2 - m(l_2 + a_2)]$

15.13. PLATE THEORY[3,10]

(a) Fundamental Governing Equation

In deriving the first-order differential equation for a plate under action of a transverse load, the following assumptions are usually made:

1. The plate material is homogeneous, isotropic, and elastic.
2. The least lateral dimension (length or width) of the plate is at least ten times the thickness h.
3. At the boundary, the edges of the plate are unrestrained in the plane of the plate; thus the reactions at the edges are taken transverse to the plate.
4. The normal to the original middle surface remains normal to the distorted middle surface after bending.
5. Extensional strain in the middle surface is neglected.

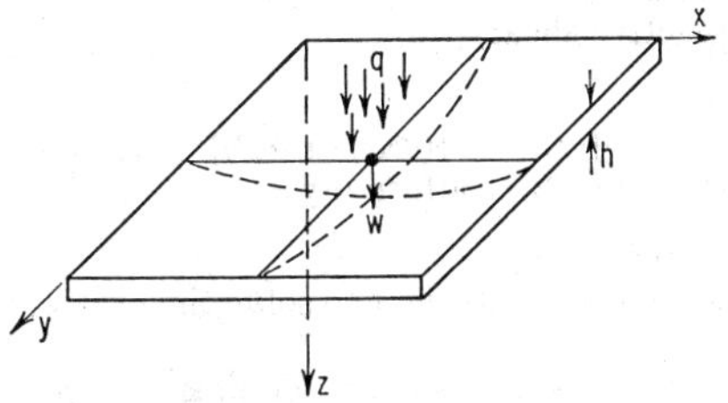

FIG. 15.31. Simply supported plate subjected to normal load q.

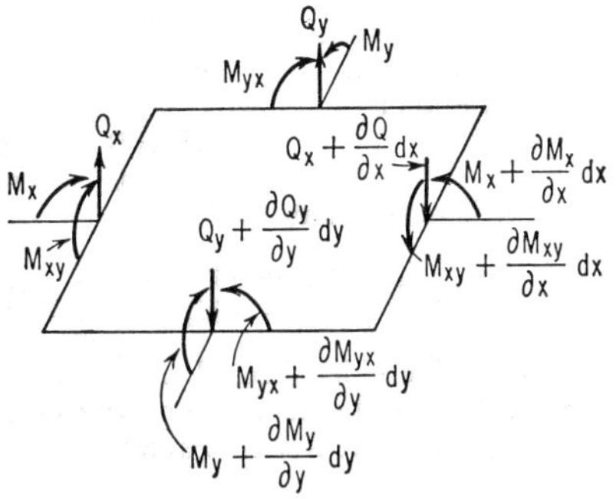

FIG. 15.32. Shears, twists, and moments on a plate element.

The deflected shape of a simply supported plate due to a normal load q s illustrated in Fig. 15.31, which also defines the coordinate system. The positive shears, twists, and moments which act on an element of the plate are depicted in Fig. 15.32.

Application of the equations of equilibrium and Hooke's law to the differential element leads to the following relations:

$$\begin{aligned} \partial M_x/\partial x + \partial M_{yx}/\partial y &= Q_x & M_{xy} &= D(1-\nu)(\partial^2 w/\partial x\,\partial y) = -M_{yx} \\ \partial M_y/\partial y + \partial M_{yx}/\partial x &= Q_y & M_x &= -D(\partial^2 w/\partial x^2 + \nu\partial^2 w/\partial y^2) \\ \partial Q_x/\partial x + \partial Q_y/\partial y &= -q & M_y &= -D(\partial^2 w/\partial y^2 + \nu\partial^2 w/\partial x^2) \end{aligned} \qquad (15.228)$$

where $D = Eh^3/[12(1-\nu^2)]$ = plate stiffness and is analogous to the flexural rigidity per unit width (EI) of beam theory.

Properly combining Eqs. (15.228) leads to the basic differential equation of plate theory

$$\nabla^4 w = \partial^4 w/\partial x^4 + 2(\partial^4 w/\partial^2 x\,\partial^2 y) + \partial^4 w/\partial y^2 = q/D \qquad (15.229)$$

This may be compared with the similar equation for beams

$$d^4 y/dx^4 = q/EI \qquad (15.230)$$

The solution of a specific plate problem involves finding a function w which satisfies Eq. (15.229) and the boundary conditions. With w known, the stresses may be evaluated by employing

$$\sigma_{\max} = \pm \frac{6M_x}{h^2}$$
$$\sigma_{\max} = \pm \frac{6M_y}{h^2} \qquad (15.231)$$
$$\tau_{\max} = \frac{6M_{xy}}{h^2}$$

(b) Boundary Conditions

The usual support conditions and the associated boundary conditions are as follows:
Simply supported plate:

$$w = 0 \quad \text{(zero deflection)}$$
$$M = 0 \quad \text{(zero moment)}$$

Built-in plate:

$$w = 0 \quad \text{(zero deflection)}$$
$$w' = 0 \quad \text{(zero slope)}$$

Free boundary:

$$M = 0 \quad \text{(zero moment)}$$
$$V = 0 \quad \text{(zero reactive shear)}$$

where it is noted that

$$\begin{aligned} V_x &= Q_x - \partial M_{xy}/\partial y = -D[\partial^3 w/\partial x^3 + (2-\nu)(\partial^3 w/\partial x\,\partial y^2)] \\ V_y &= Q_y - \partial M_{xy}/\partial x = -D[\partial^3 w/\partial y^3 + (2-\nu)(\partial^3 w/\partial y\,\partial x^2)] \end{aligned} \qquad (15.232)$$

where positive V is directed similar to positive Q.

Note that two boundary conditions are necessary and sufficient to solve the problem of bending for plates with transverse loads.

Example 1: For this problem the loading on a simply supported plate is taken to be distributed uniformly over the surface of that plate, and the coordinate axes are redefined as indicated in Fig. 15.33.

The deflection solution must satisfy Eq. (15.229) and the boundary conditions

$$w = 0 \quad \text{and} \quad \partial^2 w/\partial x^2 = 0 \quad \text{at } x = 0, a$$
$$w = 0 \quad \text{and} \quad \partial^2 w/\partial y^2 = 0 \quad \text{at } y = \pm b/2$$

A series solution of the form

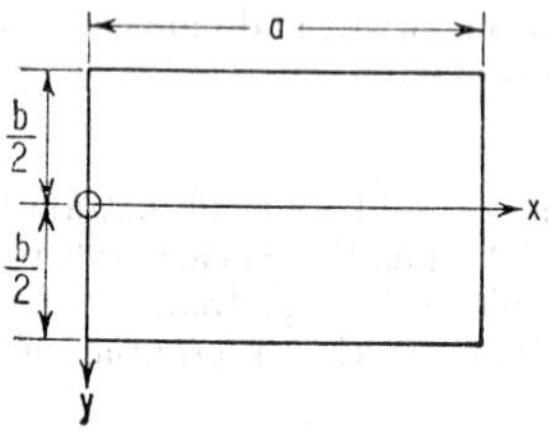

FIG. 15.33. Rectangular plate, simply supported.

$$w = \sum_{m=1}^{\infty} Y_m \sin(m\pi x/a) \qquad (15.233)$$

is assumed, and after manipulation, the following general expression results for the deflection surface:

$$w = \frac{4qa^4}{\pi^5 D} \sum_{m=1,3,\ldots}^{\infty} \frac{1}{m^5}\left(1 - \frac{\alpha_m \tanh \alpha_m + 2}{2\cosh \alpha_m}\cosh\frac{2\alpha_m y}{b} + \frac{\alpha_m}{2\cosh \alpha_m}\frac{2y}{b}\sinh\frac{2\alpha_m y}{b}\right)\sin\frac{m\pi x}{a} \qquad (15.234)$$

where $\alpha_m = m\pi b/a$. The maximum deflection occurs at $x = a/2$ and $y = 0$:

$$w_{\max} = \frac{4qa^4}{\pi^5 D}\sum_{m=1,3,\ldots}^{\infty}\frac{(-1)^{\frac{m-1}{2}}}{m^5}\left(1 - \frac{\alpha_m \tanh \alpha_m + 2}{2\cosh \alpha_m}\right) = \frac{\alpha q a^4}{D} \qquad (15.235)$$

where $\alpha = \alpha(b/a)$. The maximum moments $M_{x\,max}$ and $M_{y\,max}$ may be expressed by βqa^2 and $\beta_1 qa^2$, respectively, where β and β_1 are functions of b/a.

The final group of plate problems deals with edge reactions which are distributed along the edge and the concentrated forces at the corners to keep the corners of the plate from rising, as shown in Fig. 15.34.

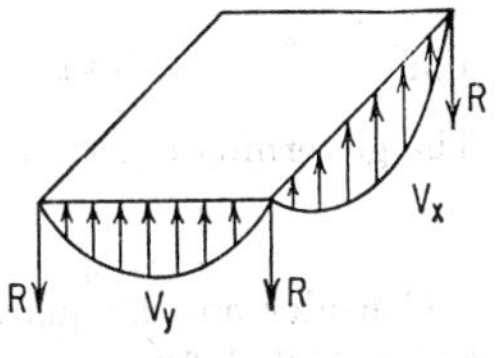

FIG. 15.34. Edge reactions of simply supported plate.

Note that the V_x and V_y are negative in sign, whereas R is positive and directed down in accordance with the adopted sign convention for plates.

The maximum value of V_x, V_y, and R may be expressed by δqa, $\delta_1 qa$, and ηqa, respectively, where δ, δ_1, and η have the same basic functional relation as the earlier defined coefficients.

Figure 15.35 presents curves of the important plate coefficients.

Example 2: The solution for the simple-support uniform-load problem as illustrated above is used as the basis for the solution for a built-in-edge problem. Transposing coordinates as in Fig. 15.36,

$$w = \frac{4qa^4}{\pi^5 D} \sum_{m=1,3,\ldots}^{\infty} \frac{(-1)^{\frac{m-1}{2}}}{m^5} \cos\frac{m\pi x}{a}\left(1 - \frac{\alpha_m \tanh \alpha_m + 2}{2\cosh \alpha_m}\cosh\frac{m\pi y}{a} + \frac{1}{2\cosh\alpha_m}\frac{m\pi y}{a}\sinh\frac{m\pi y}{a}\right) \qquad (15.236)$$

To satisfy the edge restrictions imposed by built-in supports, the deflection of a plate with moments distributed along the edges is superposed on this simple-support solution. These moments are then adjusted to satisfy the built-in boundary condition $\partial w/\partial n = 0$. The procedure finally results in a solution for the maximum deflection at the center of the plate of the form

$$w_{max} = \alpha' qa^4/D \qquad (15.237)$$

The coefficient α' can be evaluated as a function of b/a.

Similar expressions for the moments at the center of the plate and at two edges are as follows:

$$\text{Edge:}\quad M_x = -\beta' qa^2 \quad M_y = -\beta_1' qa^2 \qquad \text{Center:}\quad M_x = \gamma' qa^2 \quad M_y = \gamma_1' qa^2 \qquad (15.238)$$

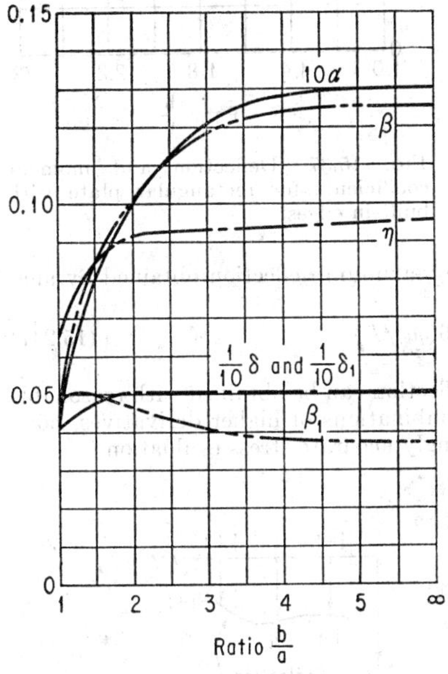

FIG. 15.35. Deflection and moment coefficients for rectangular plate on simple supports.

Figure 15.37 presents α', β', β_1', γ', γ_1', in graphical form.

Example 3: We shall use a numerical method to solve the simple-support uniform-load problem. The following expressions define the finite-difference form for the second-order partial derivatives in question (see Fig. 15.38).

$$\left.\frac{\partial^2 w}{\partial x^2}\right|_i = \frac{w_{i+1} - 2w_i + w_{i-1}}{h^2} \qquad \left.\frac{\partial^2 w}{\partial y^2}\right|_i = \frac{w_{i'+1} - 2w_i + w_{i'-1}}{k^2} \qquad (15.239)$$

For the grid, $h = k$,

$$\nabla^2 w = (1/h^2)(w_{i+1} + w_{i'+1} + w_{i-1} + w_{i'-1} - 4w_i) \tag{15.240}$$

Utilizing

$$\phi = (\nabla^2 w)D \tag{15.241}$$

and

$$\nabla^2 \phi = (1/h^2)(\phi_{i+1} + \phi_{i'+1} + \phi_{i-1} + \phi_{i'-1} - 4\phi_i) \tag{15.242}$$

The governing equation is

$$\nabla^2 \phi = q \tag{15.243}$$

Consider now a square plate coarsely divided into four segments. Then evidently $h = k = a/2$ and $w_{i+1} = w_{i'+1} = w_{i-1} = w_{i'-1} = 0$ since these points are on the boundary. Also $\nabla^2 w_{i+1} = \nabla^2 w_{i'+1} = 0$ and, using the last of the above equations, we find

$$\phi_{i+1} = \phi_{i'+1} = \phi_{i-1} = \phi_{i'-1} = 0$$

Therefore, $\nabla^2 \phi = -4\phi_i/h^2 = q$ (15.244)

Substituting back into the original equation yields

$$\nabla^2 w_i = \phi_i/D = -4w_i/h^2 = -qa^2/16D \tag{15.245}$$

$$w_i = (1/256)(qa^4/D) \tag{15.246}$$

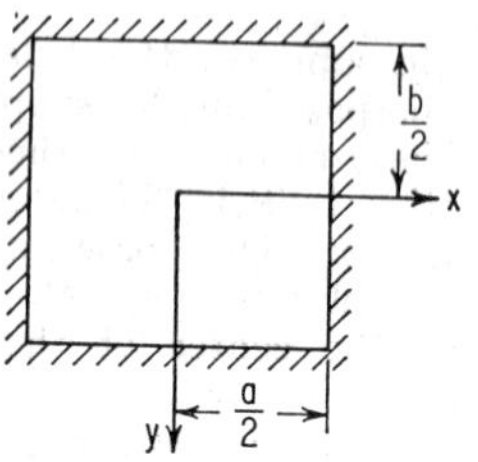

Fig. 15.36. Rectangular plate with built-in edges.

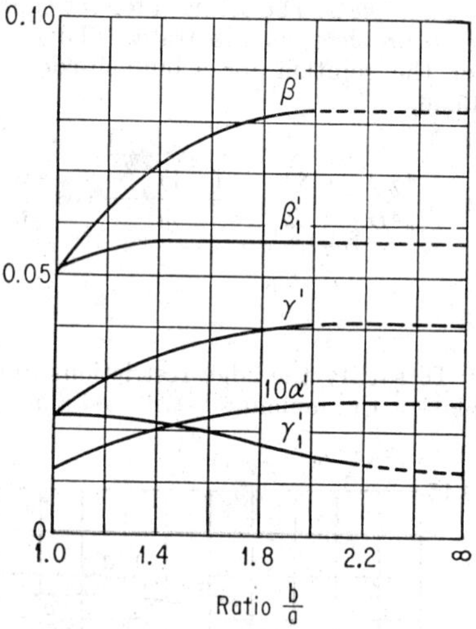

Fig. 15.37. Deflection and moment coefficients for rectangular plate with built-in edges.

This is approximately 3.5 per cent below the maximum deflection obtained by analytical means:

$$w_{max} = 0.00406(qa^4/D) \tag{15.247}$$

Thus, even with this crude grid, the center deflection can be obtained with reasonable accuracy. Since the stress is obtained by combinations of higher derivatives, however, a finer grid is required for a correspondingly accurate stress evaluation.

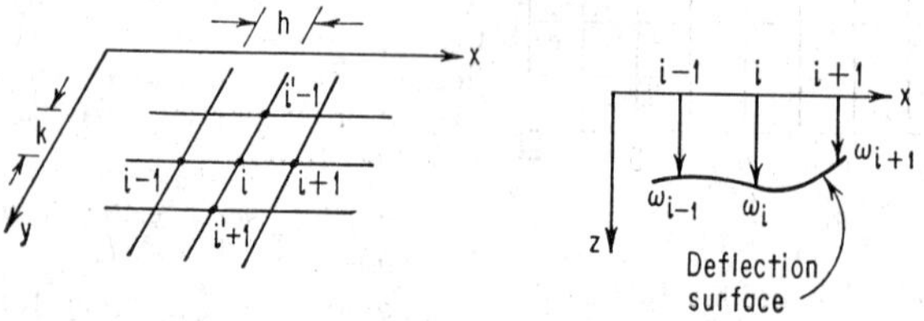

Fig. 15.38. Grid of rectangular plate, simply supported.

15.14. SHELL THEORY[3,10]

(a) Membrane Theory—Basic Equation

The general problem of determining stresses in shells may be subdivided into two distinct categories. The first of these is a moment-free *membrane* state of stress, which often predominates over a major portion of the shell. The second, associated with bending effects, is the *discontinuity stress*, which affects the shell for a limited segment in the vicinity of a load or profile discontinuity. The final solution often consists of the membrane solution, corrected locally in the regions of the boundaries for the discontinuity effects.

The *membrane solution* for shells in the form of surfaces of revolution, and loaded symmetrically with respect to the axis, often requires the following simplifying assumptions:

1. The thickness h of the shell is small compared with other dimensions of the shell, and the radii of curvature.
2. Linear elements which are normal to the undisturbed middle surface remain straight and normal to the deflected middle surface of the shell.
3. Bending stresses are small and can be neglected; so that only the direct stresses due to strains in the middle surface need be considered.

The basic element for a shell of revolution when subjected to axisymmetric loading is shown in Fig. 15.39. From conditions of static equilibrium the following equations may be derived:

$$(d/d\phi)(N_\varphi r_0) - N_\theta r_1 \cos\phi + Y r_1 r_0 = 0 \qquad (15.248)$$
$$N_\varphi r_0 + N_\theta r_1 \sin\phi + Z r_1 r_0 = 0 \qquad (15.249)$$

where Y and Z are the components of the external load parallel to the two-coordinate axis, respectively. Thus, in general, if these components of the load are known and the geometry of the shell is specified, the forces N_ϕ and N_θ can be calculated.

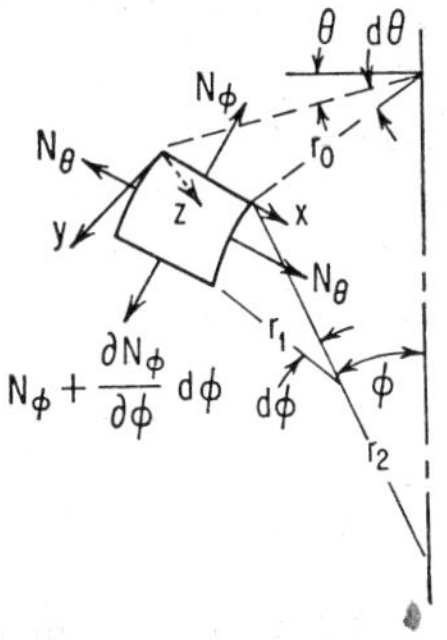

FIG. 15.39. Element of shell of revolution.

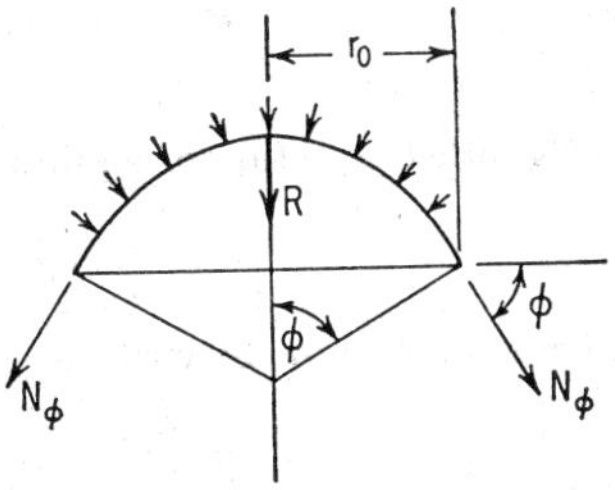

FIG. 15.40. Shell portion above parallel circle.

If now the conditions of static equilibrium are applied to a portion of a shell above a parallel circle instead of an element as indicated in Fig. 15.40 then the following equations result:

$$2\pi r_0 N_\phi \sin\phi + R = 0 \qquad (15.250)$$
$$N_\phi / r_1 + N_\theta / r_2 = -Z \qquad (15.251)$$

where R is the resultant of the total load on that part of the shell in question. Again these two equations suffice for the determination of the two forces N_ϕ and N_θ.

The associated principal stresses for these members are then

$$\sigma_\phi = N_\phi / h \quad \text{and} \quad \sigma_\theta = N_\theta / h \qquad (15.252)$$

To complete the consideration of membrane action for shells with axisymmetric loading, the concomitant displacements must be computed. The procedure indicated below applies for symmetrical deformations, in which the displacement (along a meridian) is indicated by v and the displacement in the radial direction is w (with an inward displacement taken as positive).

From geometry and Hooke's law,

$$dv/d\phi - v \cot \phi = f(\phi) \tag{15.253}$$

where

$$f(\phi) = (1/Eh)[N_\phi(r_1 + \nu r_2) - N_\theta(r_2 + \nu r_1)]$$

The solution of Eq. (15.253) is

$$v = \sin \phi\{\int[f(\phi)/\sin \phi]\, d\phi + C\} \tag{15.254}$$

The constant C is evaluated from the boundary conditions of the problem. The radial displacement is then determined from the following:

$$w = v \cot \phi - (r_2/Eh)(N_\theta - \nu N_\phi) \tag{15.255}$$

A similar analysis on a cylindrical shell without the restriction of symmetric loading will yield the basic equations

$$\partial N_x/\partial x + (1/r)(\partial N_{x\phi}/\partial \phi) = -X \tag{15.256}$$

$$\partial N_{x\phi}/\partial x + (1/r)(\partial N_\phi/\partial \phi) = -Y \tag{15.257}$$

$$N_\phi = -Zr \tag{15.258}$$

where the coordinate axes are redefined in Fig. 15.41 to be consistent with usual practice and Eqs. (15.254) and (15.255) are suitably modified.

(b) Example of Spherical Shell Subjected to Internal Pressure

For a hemispherical section of radius a, the forces N_ϕ distributed along the edges are required to maintain equilibrium. These may be determined from the previous equations. Now

$$2\pi a N_\phi \sin \phi + R = 0 \tag{15.259}$$

where

$$R = -\pi a^2 p \qquad \sin \phi = \sin (\pi/2) = 1$$

so that

$$N_\phi = pa/2 \tag{15.260}$$

From the other equation of equilibrium,

$$N_\phi/a + N_\theta/a = -Z = p \tag{15.261}$$

where

$$N_\theta = pa/2$$

The stresses may now be written

$$\sigma_\phi = \sigma_\theta = pa/2h \tag{15.262}$$

From loading symmetry, it is concluded that $v = 0$ and the increase in the radial direction is given by

$$w = (-a/Eh)(N_\theta - \nu N_\phi) \tag{15.263}$$

$$w = (-pa^2/2Eh)(1 - \nu) \tag{15.264}$$

(c) Example of Cylindrical Shell Subjected to Internal Pressure

Using a procedure similar to that outlined above, the following relationships are derived:

For open-ended cylinder:

$$\sigma_\phi = pa/h \tag{15.265}$$

$$w = -pa^2/Eh \tag{15.266}$$

For closed-ended cylinder:

$$\sigma_\phi = pa/h \tag{15.267}$$

$$\sigma_x = pa/2h \tag{15.268}$$

$$w = -(pa^2/Eh)(1 - \nu/2) \tag{15.269}$$

(d) Discontinuity Analysis

The *membrane solution* does not usually satisfy all "edge" conditions, and it is therefore often necessary to superpose a second, *edge-loaded* shell in order to obtain a *complete* solution.

To develop the general equations for the cylindrical shell with axisymmetric load, consider an element as shown in Fig. 15.42, where *bending moments are assumed acting.* The coordinate system is as defined in Fig. 15.41. All forces and moments are shown

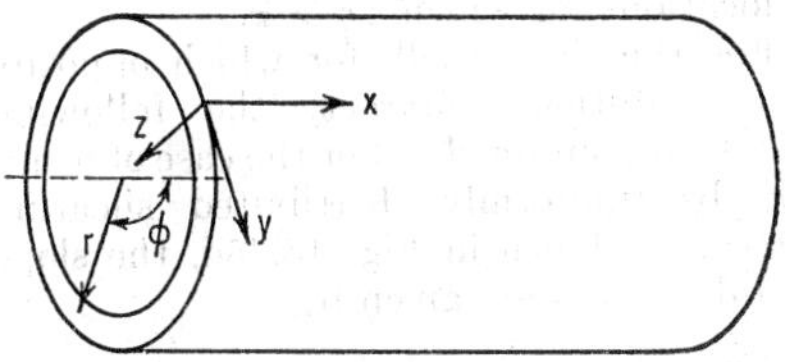

FIG. 15.41. Cylindrical shell.

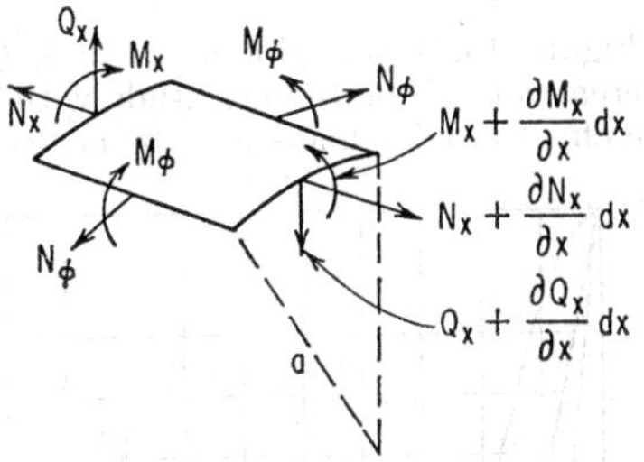

FIG. 15.42. Cylindrical-shell element.

in their positive directions. Referring to Fig. 15.42, Q_x is the shear force per unit length, M_x is the axial moment per unit length, M_ϕ is the circumferential moment per unit length, and N_x, N_ϕ are the normal forces defined in the discussion of membrane theory.

Applying the equations of equilibrium and Hooke's law and expressing the curvature as a function of moment (as was done in developing the plate equations), the following basic shell differential equation is obtained:

$$d^4w/dx^4 + 4\beta^4 w = Z/D \tag{15.270}$$

where

$$D = Eh^3/12(1 - \nu^2)$$

$$\beta^4 = \frac{3(1 - \nu^2)}{a^2h^2}$$

w = radial deflection of shell (positive inward)

The solution of the above equation in any particular case depends on the specific boundary conditions at the ends of the cylinder.

One very useful solution is for the case of a long cylinder, without radial pressure, subjected to uniformly distributed forces and moments along the edge, $x = 0$. The assumed positive directions of these loads are as shown in Fig. 15.43. The resultant expression for the deflection is

$$w = (e^{-\beta x}/2\beta^3 D)[\beta M_0(\sin \beta x - \cos \beta x) - Q_0 \cos \beta x] \tag{15.271}$$

The maximum deflection occurs at the loaded end and is evaluated as

$$w_{x=0} = -(1/2\beta^3 D)(\beta M_0 + Q_0) \tag{15.272}$$

The accompanying slope at the loaded end is

$$w'_{x=0} = (1/2\beta^2 D)(2\beta M_0 + Q_0) = (dw/dx)_{x=0} \tag{15.273}$$

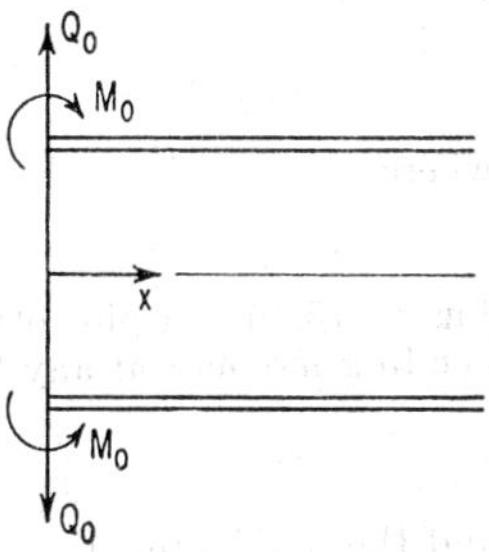

FIG. 15.43. Long-cylinder section.

The successive derivatives of the above expression for

deflection can be written in the following simplified form:

$$\begin{aligned} w &= -(1/2\beta^3 D)[\beta M_0\psi(\beta x) + Q_0\theta(\beta x)] \\ w' &= (1/2\beta^2 D)[2\beta M_0\theta(\beta x) + Q_0\phi(\beta x)] \\ w'' &= -(1/2\beta D)[2\beta M_0\phi(\beta x) + 2Q_0\xi(\beta x)] \\ w''' &= (1/D)[2\beta M_0\xi(\beta x) - Q_0\psi(\beta x)] \end{aligned} \tag{15.274}$$

where

$$\begin{aligned} \phi(\beta x) &= e^{-\beta x}(\cos \beta x + \sin \beta x) \\ \psi(\beta x) &= e^{-\beta x}(\cos \beta x - \sin \beta x) \\ \theta(\beta x) &= e^{-\beta x}\cos \beta x \\ \xi(\beta x) &= e^{-\beta x}\sin \beta x \end{aligned}$$

Figure 15.44 is a plot of ϕ, ψ, θ, ξ as a function of βx. Because each function decreases in absolute magnitude with increasing βx, in most engineering applications the effect of edge loads may be neglected at locations for which $\beta x > \pi$.

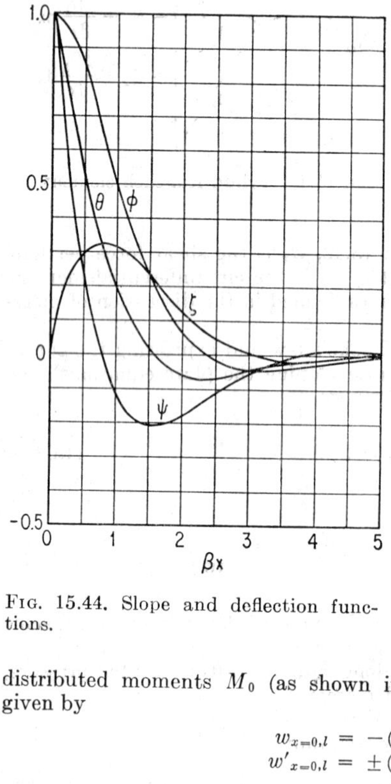

FIG. 15.44. Slope and deflection functions.

For the short shell, for which opposite end conditions interact, the following results are obtained. For the case of bending by uniformly distributed shearing forces, as shown in Fig. 15.45*a*, the slope and deflection are given by

$$w_{x=0,l} = -(2Q_0\beta a^2/Eh)\chi_1(\beta l) \tag{15.275}$$
$$w'_{x=0,l} = \pm(2Q_0\beta^2 a^2/Eh)\chi_2(\beta l) \tag{15.276}$$

where

$$\chi_1(\beta l) = \frac{\cosh \beta l + \cos \beta l}{\sinh \beta l + \sin \beta l}$$

$$\chi_2(\beta l) = \frac{\sinh \beta l - \sin \beta l}{\sinh \beta l + \sin \beta l}$$

For the case of bending by uniformly

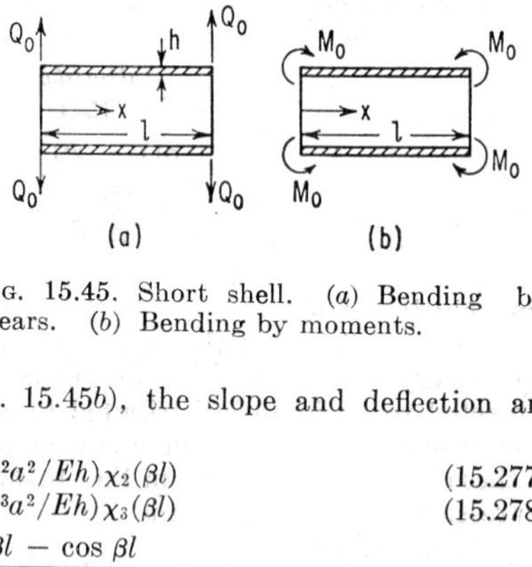

FIG. 15.45. Short shell. (*a*) Bending by shears. (*b*) Bending by moments.

distributed moments M_0 (as shown in Fig. 15.45*b*), the slope and deflection are given by

$$w_{x=0,l} = -(2M_0\beta^2 a^2/Eh)\chi_2(\beta l) \tag{15.277}$$
$$w'_{x=0,l} = \pm(4M_0\beta^3 a^2/Eh)\chi_3(\beta l) \tag{15.278}$$

where

$$\chi_3(\beta l) = \frac{\cosh \beta l - \cos \beta l}{\sinh \beta l + \sin \beta l}$$

Figure 15.46 is a plot of the functions χ_1, χ_2, and χ_3 as a function of βl. The axial bending moment at any location is given by

$$M_x = -Dw'' \tag{15.279}$$

and the maximum stress which occurs at the inside and outside surfaces of the shell

$$\sigma_{x,\text{bending}} = \pm(6M_x/h^2) \tag{15.280}$$

Likewise the shear force at any point is given by

$$Q_x = -Dw''' \tag{15.281}$$

and the associated maximum shear stress is $\tau_x = 3Q_x/2h$, which occurs midway between the inner and outer surfaces; the shear stress at the surface is zero.

Example: Examine the case of a long cylinder subjected to an internal pressure and fixed at the ends as depicted in Fig. 15.47*a*; axial pressure is taken to be zero.

The stress and deformation of this shell can be obtained by the superposition of two distinct problems, the membrane and edge-loaded cylinders. The first presupposes free ends and a membrane action as indicated in Fig. 15.47*b*. The built-in ends resist this membrane deflection at the edges through a system of forces Q_0 and moments M_0 which are required to enforce the boundary conditions of zero deflection and rotation, as shown in Fig. 15.47*c*.

The increase in the radius due to membrane action, as a result of pressure, is then obtained from the membrane solution

$$-w_p = \delta = pa^2/Eh \tag{15.282}$$

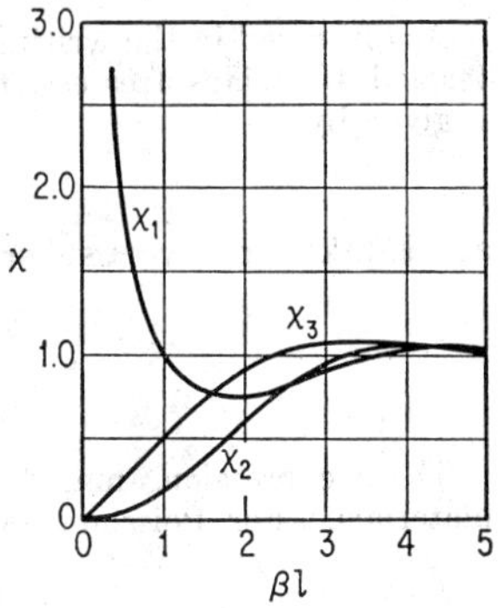

FIG. 15.46. Slope and deflection functions.

The boundary conditions for the edge-loaded problem, based on the actual built-in ends, become

$$w_x = \delta \quad \text{and} \quad w'_{x=0} = 0$$

Hence

$$\delta = (1/2\beta^3 D)(\beta M_0 + Q_0) \tag{15.283}$$

$$0 = (1/2\beta^2 D)(2\beta M_0 + Q_0) \tag{15.284}$$

Solving Eqs. (15.283) and (15.284), using Eq. (15.282)

$$M_0 = p/2\beta^2 \quad \text{and} \quad Q_0 = -p/\beta \tag{15.285}$$

Thus the *complete* solution for the deflection is

$$w = -(1/4\beta^4 D)[p\psi(\beta x) - 2p\theta(\beta x)] - pa^2/Eh \tag{15.286}$$

The axial stresses are given by

$$\sigma_x = \pm(6M_x/h^2) \tag{15.287}$$

where

$$M_x = -Dw'' = (p/2\beta^2)[\phi(\beta x) - 2\xi(\beta x)]$$

The mean circumferential stress can be evaluated from

$$\sigma_{\phi,\text{direct}} = -Ew/a \tag{15.288}$$

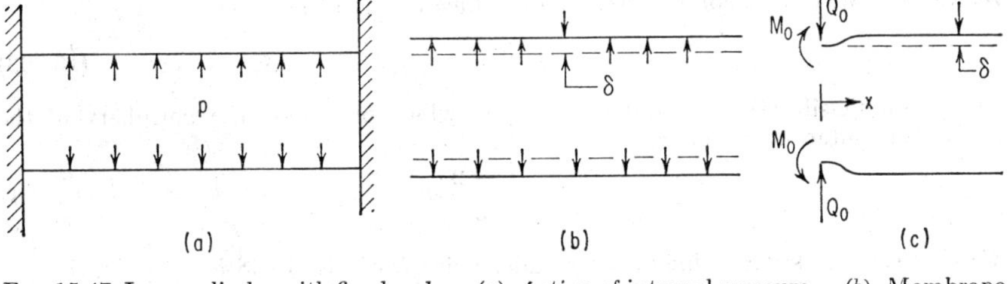

FIG. 15.47. Long cylinder with fixed ends. (*a*) Action of internal pressure. (*b*) Membrane action. (*c*) Discontinuity forces for boundary conditions.

and the added component of flexural stress due to the Poisson effect is

$$\sigma_{\phi,\text{bending}} = -\nu\sigma_{x,\text{bending}} \tag{15.289}$$

so that

$$\sigma_{\phi,\text{total}} = \sigma_{\phi,\text{direct}} + \sigma_{\phi,\text{bending}} \tag{15.290}$$

15.15. CONTACT STRESSES—HERTZIAN THEORY[2]

As discussed in the writings of Hertz (the contact stresses presented here are often termed Hertzian stresses), the maximum pressure q due to a compressive force P is given by

$$q = 3P/2\pi a^2 \tag{15.291}$$

and is taken to have a spherical distribution as shown in Fig. 15.48, where

$$a = \sqrt[3]{\frac{3P}{4}\frac{R_1R_2}{R_1+R_2}\left(\frac{1-\nu_1^2}{E_1}+\frac{1-\nu_2^2}{E_2}\right)} \tag{15.292}$$

These expressions may be simplified, if both spheres are composed of identical materials. For Poisson's ratio ν of approximately 0.3, which is common to steel,

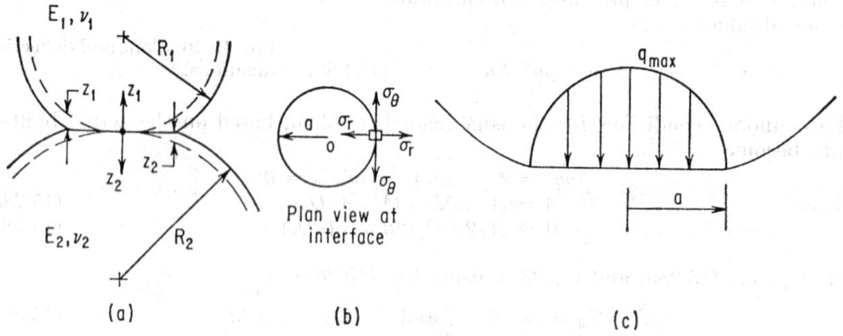

FIG. 15.48. Two spheres in contact. (a), (b), (c) Contact pressure distribution.

iron, aluminum, and most structural materials, there results

$$a = 1.11\sqrt[3]{\frac{P}{E}\frac{R_1R_2}{R_1+R_2}} \tag{15.292a}$$

and

$$q = 0.388\sqrt[3]{PE^2\left(\frac{R_1+R_2}{R_1R_2}\right)^2} \tag{15.292b}$$

The general stress levels in the spheres can now be presented based on the above relations. Maximum compressive stress, which occurs at point O,

$$\sigma_z = -q \tag{15.293}$$

Maximum tensile stress in radial direction, which occurs on the periphery of the surface of contact at radius a,

$$\sigma_r = \frac{1-2\nu}{3}q \tag{15.294}$$

Maximum shear stress, which occurs under point O of Fig. 15.48(b), at a depth

$$z_1 = 0.47a$$

Table 15.2. Contact Stresses[2,5,12]

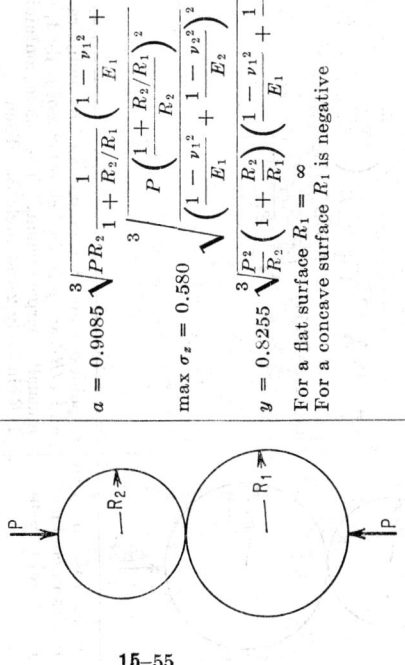

Sphere on a sphere, P = total load 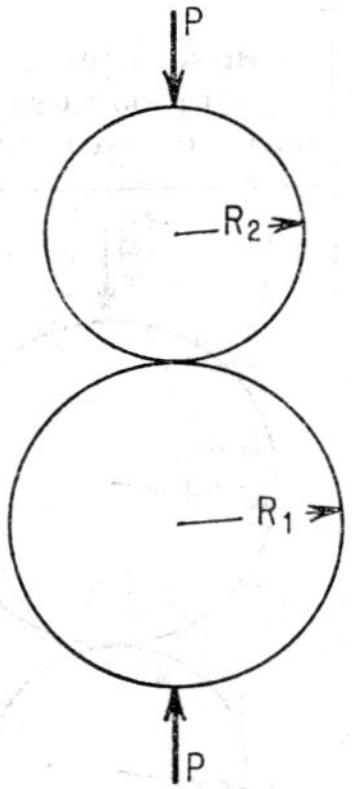	$a = 0.9085 \sqrt[3]{PR_2 \frac{1}{1 + R_2/R_1} \left(\frac{1 - \nu_1^2}{E_1} + \frac{1 - \nu_2^2}{E_2} \right)}$ $\max \sigma_z = 0.580 \sqrt[3]{\frac{P \left(\frac{1 + R_2/R_1}{R_2} \right)^2}{\left(\frac{1 - \nu_1^2}{E_1} + \frac{1 - \nu_2^2}{E_2} \right)^2}}$ $y = 0.8255 \sqrt[3]{\frac{P^2}{R_2} \left(1 + \frac{R_2}{R_1} \right) \left(\frac{1 - \nu_1^2}{E_1} + \frac{1 - \nu_2^2}{E_2} \right)^2}$ For a flat surface $R_1 = \infty$ For a concave surface R_1 is negative

y is the decrease in center-to-center distance between the two spheres.

Cylinder on cylinder axes, parallel, P = load/linear in.

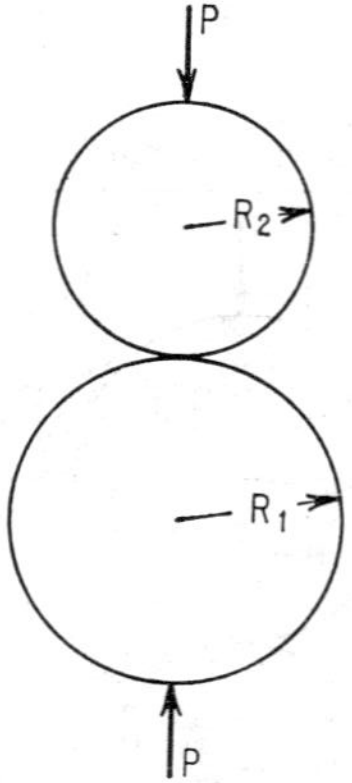

$$b = 1.13 \sqrt{PR_2 \frac{1}{1 + R_2/R_1}\left(\frac{1 - \nu_1^2}{E_1} + \frac{1 - \nu_2^2}{E_2}\right)}$$

$$\max \sigma_z = 0.564 \sqrt{\frac{(P/R_2)(1 + R_2/R_1)}{\frac{1 - \nu_1^2}{E_1} + \frac{1 - \nu_2^2}{E_2}}}$$

General case of two bodies in contact, P = total pressure

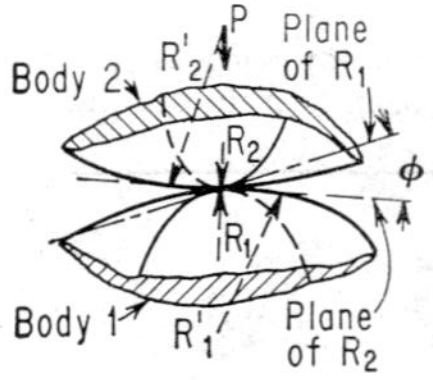

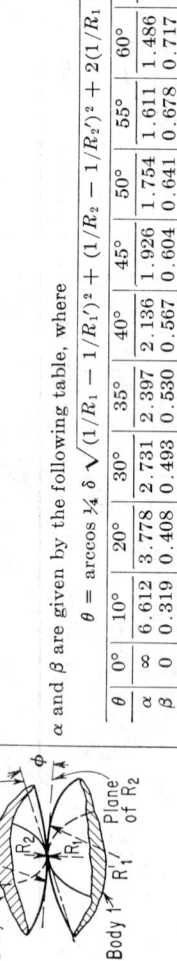

At point of contact minimum and maximum radii of curvature are R_1 and R_1' for body 1, R_2 and R_2' for body 2. Then $1/R_1$ and $1/R_1'$ are *principal curvatures* of body 1, and $1/R_2$ and $1/R_2'$ of body 2, and in each body the principal curvatures are mutually perpendicular. The plane containing curvature $1/R_1$ in body 1 makes with the plane containing curvature $1/R_2$ in body 2 the angle ϕ. Then

$\max \sigma_z = 1.5P/\pi cd$, $c = \alpha \sqrt[3]{P\delta/K}$, $d = \beta \sqrt[3]{P\delta/K}$, and $y = \lambda \sqrt[3]{P^2/K^2\delta}$, where $\delta = \dfrac{4}{1/R_1 + 1/R_2 + 1/R_1' + 1/R_2'}$

and $K = \dfrac{8}{3} \dfrac{E_1E_2}{E_2(1 - \nu_1^2) + E_1(1 - \nu_2^2)}$

α and β are given by the following table, where

$$\theta = \arccos \tfrac{1}{4} \delta \sqrt{(1/R_1 - 1/R_1')^2 + (1/R_2 - 1/R_2')^2 + 2(1/R_1 - 1/R_1')(1/R_2 - 1/R_2') \cos 2\phi}$$

θ	0°	10°	20°	30°	35°	40°	45°	50°	55°	60°	65°	70°	75°	80°	85°	90°
α	∞	6.612	3.778	2.731	2.397	2.136	1.926	1.754	1.611	1.486	1.378	1.284	1.202	1.128	1.061	1.00
β	0	0.319	0.408	0.493	0.530	0.567	0.604	0.641	0.678	0.717	0.759	0.802	0.846	0.893	0.944	1.00
λ	—	0.851	1.220	1.453	1.550	1.637	1.709	1.772	1.828	1.875	1.912	1.944	1.967	1.985	1.996	2.00

y is the decrease in center-to-center distance of the two cylinders. b is the half-width of the contact surface.

and is approximately equal to, and is in a plane inclined to the z axis,

$$\tau = \tfrac{1}{3}q \tag{15.295}$$

This latter stress is usually the governing criterion in the design for bodies in contact, fabricated from ductile materials. A compilation of important contact-stress cases is given in Table 15.2.

INELASTIC THEORY

15.16. CONCEPTS OF INELASTIC THEORY[6]

In the determination of the *actual carrying capacity* of a structure it is frequently necessary to investigate its behavior beyond its elastic limit. In most ductile structures the onset of yielding merely results in a gradual redistribution of load, such that the overstrained elements deform with little additional resistance, while the remaining elastic elements carry their full share of the load. This continuous redistribution

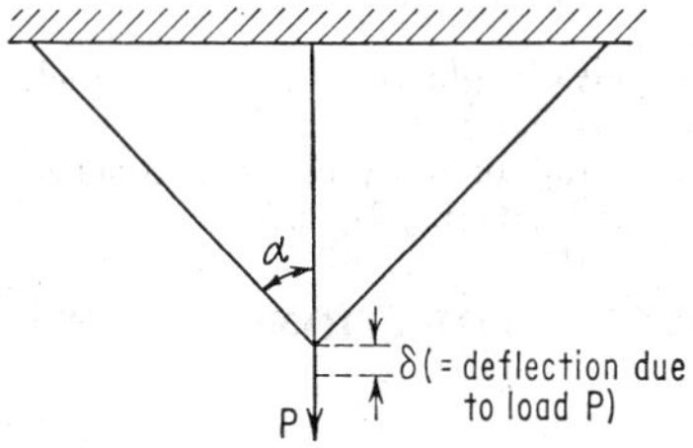

FIG. 15.49. Truss structure under load.

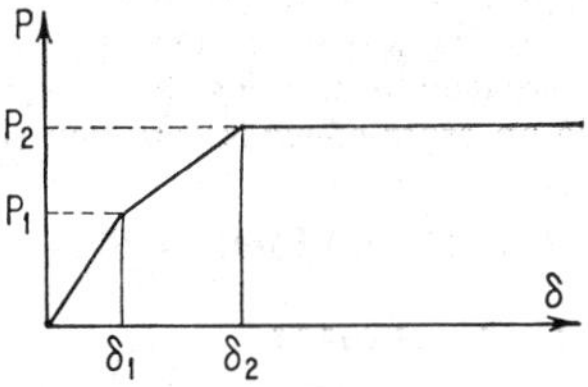

FIG. 15.50. Curve showing elastic limit and maximum carrying capacity.

takes place until no additional load can be carried by the structure and failure ensues. This maximum load is often greatly in excess of the load which signifies the onset of yield.

In addition to this important maximum load-carrying effect, it is often possible to create, through this inelastic behavior, a residual stress state, such that all future loading cycles tend to behave elastically, even though these are in excess of the original elastic loads. Finally, with the onset of yielding, many structures experience small but significant geometrical changes which tend to *stiffen* and *strengthen* them with respect to future loadings.

Consider the structure in Fig. 15.49. In this, each rod can carry a maximum load $P_{\max} = P_0$, beyond which it yields without any further load increase.

The structure load-deflection curve is given in Fig. 15.50. At deflection δ_1, load P_1, the central bar yields but the two outer bars remain elastic. When the load reaches P_2 the outer bars start to yield; so that the entire structure continues to deform without a further increase in load. Thus failure is said to occur at the load $P = P_2$, which is in excess of the *structural elastic limit* ($P = P_1$). In highly statically indeterminate structures the spread between the *elastic limit* and the *maximum carrying capacity* can be considerable.

If at some deflection between δ_1 and δ_2 the load is removed, the structure will unload elastically, and because there was a small amount of permanent strain in the central bar, the structure will be in a nonzero or *residual* stress state, even in the absence of an external load. Further positive reloading of the structure, less than the original external load, will unload the residual state and cause the structure to carry this new load *elastically;* only when the original load is exceeded will the structure again yield *inelastically*. This action is indicated in the load-deflection curve in Fig. 15.51.

This residual strengthening process is used in some restricted applications such as "surface hardening" and "autofrettage" of gun tubes. It offers a wide range of presently unexploited applications for the structural and machine designer.

The *theory of plasticity* treats these and many other complex problem types, with which are associated partial or full yielding of some of all of the basic *elemental units*.

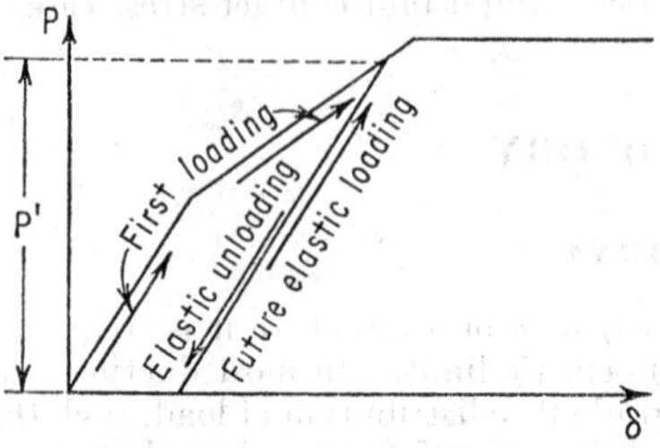

FIG. 15.51. Load-deflection curve.

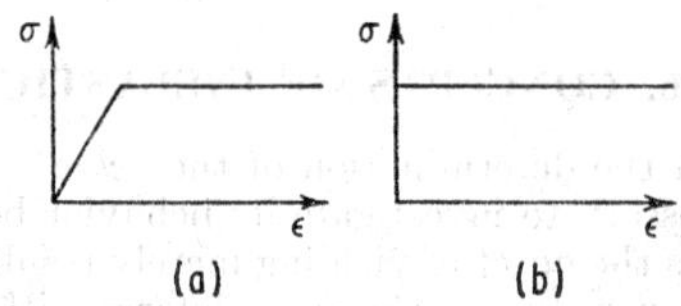

FIG. 15.52. Simple tensile test. (*a*) Elastic–ideally plastic. (*b*) Rigid–ideally plastic.

Consideration will be given here only to ductile ideally plastic materials, which are defined by the stress-strain curve indicated in Fig. 15.52*a*.

The treatment is adequate for many materials and in most cases also serves to provide an *order-of-magnitude* solution for the *strain-hardening material*.

15.17. FORMULATION OF IDEAL PLASTICITY EQUATIONS

(a) Introduction[6,7,8]

Neglecting geometry changes, the equilibrium, compatibility and strain-displacement, and yield criterion, which were established in conjunction with the elastic theory, apply to the plastic theory. The only new relationships that must be established are those which relate stress components and strain components (or strain increment). Although the yield criterion is not explicitly required in the *solution* of elasticity problems, it is required in plastic theory since it becomes necessary to distinguish between yielded and nonyielded elements.

(b) Stress-Strain (Increment) Relationships[6,7,8]

It is only necessary to modify the elastic theory by the introduction of a relationship between stress and strain (or strain increment). A significant and useful formulation for this was obtained by Saint-Venant, who suggested that there is a coincidence of principal axes of *strain increment* and stress. Levy and Von Mises first established these in equation form:

$$de_{ij} = \sigma_{ij}'\, d\lambda$$

where de_{ij} refers to the total strain increment and σ_{ij}' the deviatoric stress components; $d\lambda$ is a constant of proportionality. Reuss generalized this form to allow for the elastic as well as the plastic components of strain, which states that

$$de_{ij}^P = \sigma_{ij}'\, d\lambda \qquad (15.296)$$

the only change being that the total strain increment is replaced by the plastic strain increment. Considering the results of Bridgeman's experiments, which indicate that there is no change in volume due to plastic strain, or

$$de_{ii} = 0$$

the usual forms of the Levy-Mises-Reuss equations result

$$de_{ij}' = \sigma_{ij}'\,d\lambda + d\sigma_{ij}'/2G \tag{15.297}$$
$$de_{ii} = [(1-2\nu)/E]\,d\sigma_{ii} \tag{15.298}$$

where the primes indicate the deviatoric values of strain or stress; in particular

$$de_{ij}' = de_{ij} - \delta_{ij}\,de \tag{15.299}$$
$$de = \tfrac{1}{3}de_{ii} \tag{15.300}$$

A second type of stress-strain relationship, which is used to a much lesser degree, relates stress to strain (rather than strain increment) and is due to Hencky. These are often referred to as *finite* strain relationships. Although these lead to inconsistencies in the most general application, they often lead to reasonable approximations and more simple solutions. These relationships appear as

$$\epsilon_{ij}' = [(1+\phi)/2G]\sigma_{ij}' \tag{15.301}$$

with

$$\epsilon_{ij}^P = \frac{\phi}{2G}\sigma_{ij}'$$
$$\epsilon_{ii}^P = 0$$

where ϵ_{ij}^P is the plastic component of strain and ϕ is a proportionality factor.

In cartesian components these appear as

Levy-Mises:

$$\begin{aligned} de_x &= \sigma_x'\,d\lambda & d\gamma_{xy} &= 2\tau_{xy}\,d\lambda \\ de_y &= \sigma_y'\,d\lambda & d\gamma_{yz} &= 2\tau_{yz}\,d\lambda \\ de_z &= \sigma_z'\,d\lambda & d\gamma_{zx} &= 2\tau_{zx}\,d\lambda \end{aligned} \tag{15.302}$$

Reuss:

$$\begin{aligned} de_x' &= \sigma_x'\,d\lambda + d\sigma_x'/2G & d\gamma_{xy} &= 2\tau_{xy}\,d\lambda + d\tau_{xy}/G \\ de_y' &= \sigma_y'\,d\lambda + d\sigma_y'/2G & d\gamma_{yz} &= 2\tau_{yz}\,d\lambda + d\tau_{yz}/G \\ de_z' &= \sigma_z'\,d\lambda + d\sigma_z'/2G & d\gamma_{zx} &= 2\tau_{zx}\,d\lambda + d\tau_{zx}/G \end{aligned} \tag{15.303}$$

Hencky:

$$\begin{aligned} \epsilon_x' &= \frac{1+\phi}{2G}\sigma_x' & \gamma_{xy} &= \frac{1+\phi}{G}\tau_{xy} \\ \epsilon_y' &= \frac{1+\phi}{2G}\sigma_y' & \gamma_{yz} &= \frac{1+\phi}{G}\tau_{yz} \\ \epsilon_z' &= \frac{1+\phi}{2G}\sigma_z' & \gamma_{zx} &= \frac{1+\phi}{G}\tau_{zx} \end{aligned} \tag{15.304}$$

$$\epsilon_x^P + \epsilon_y^P + \epsilon_z^P = 0$$

15.18. LIMIT ANALYSIS

(a) Basic Theorems[7,8]

Limit analysis is principally used to evaluate the maximum load-carrying capacity of a structure. In the present treatment the material is taken as *elastic–ideally plastic* or *rigid–ideally plastic* (see Fig. 15.52). Further, changes in geometry (and the associated strengthening or instability effects) are usually neglected. Although none of the above restrictions need be imposed, the effects of their neglect can often be estimated, while their introduction leads to a particularly simple and useful theory.

As an example of these limitations consider the loading history of a plate under transverse loading (Fig. 15.53). The loading is assumed proportional; i.e., at any instant of time, the load intensity at every position is n times that which occurs at time t_0. The value of n, which is taken to increase monotonically, is bounded as follows: $0 \leq n \leq n_{max}$. Initially for sufficiently small n, the plate deforms elasticity as is shown in Fig. 15.53*a*. The plate is said to yield at the instant the elastic limit is reached at any position of the plate, and a "hinge" circle is said to form. That is, a "kink" is initiated in the plate at the value $n = n_{elastic\ limit}$.

However, since the plate is still statically indeterminate, the load can be increased until such time that the plate becomes a "mechanism" of sufficient "hinges" (kinks) such that any further load cannot be supported by the structure ($n = n_{limit}$). However, even at this point it is quite possible that changes in geometry will be of such a nature as to allow the plate to take additional loading due to "membrane" action. Thus, even though the plate can no longer support load in the manner for which it was initially designed, it can nevertheless support additional load. This load increase continues until such a time that rupture occurs (see Fig. 15.54).

Note that

$$n_{elastic} \leq n_{limit} \leq n_{max}$$

and that n_{limit} provides a conservative estimate of the carrying capacity of the structure and is often sufficiently in excess of $n_{elastic}$ to provide additional useful information. If required, detailed analysis can be used to obtain n_{max}.

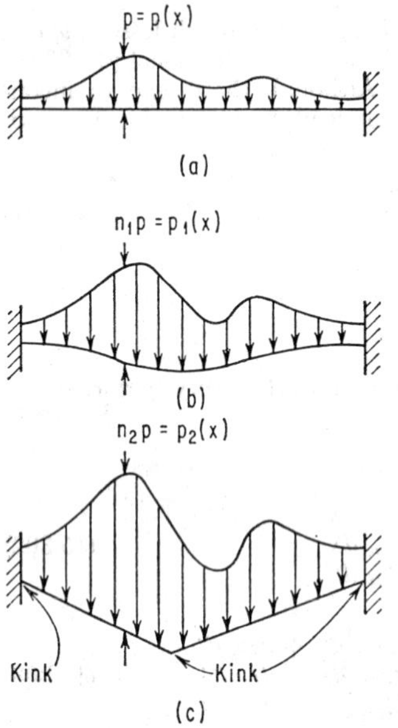

FIG. 15.53. Plate under transverse loading. (*a*), (*b*) Elastic. (*c*) Plastic.

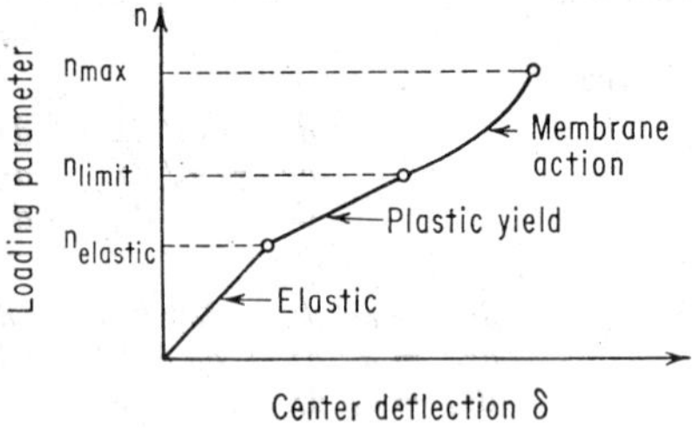

FIG. 15.54. Curve of loading parameter vs. center deflection.

A statically admissible stress field is defined by stress components which are everywhere in *equilibrium*, which satisfy the *external tractions*, and which nowhere violate the *yield criterion*.

A kinematically admissible velocity (or displacement) *field* is one which satisfies all the velocity constraints on the structure and at the same time provides that the external tractions do positive work. A strain-rate (strain) field can be associated with this velocity field through kinematic considerations, and a stress field through the yield criterion (or equivalently through the condition that the stress field maximizes the dissipation function $D = CQ_i{}^*\dot{q}_i{}^*$). In general there is no requirement that this associated stress field satisfy equilibrium.

The *first theorem of limit analysis* is that the limit load is the *largest* load for which a statically admissible stress state may be found. *All* statically admissible stress states furnish a lower bound on the limit load. The *second theorem of limit analysis* is that the limit load is the *smallest* load for which a kinematically admissible velocity field can be found. All kinematically admissible velocity fields furnish an upper bound in the limit load. To evaluate the *tractions* based on this upper bound it is necessary to equate the associated internal and external energies.

(b) Beam and Frame Theory[8]

In this application the bars or structure members are taken to be rigid, except at positions where the yield moment is exceeded and "kinks" form. It can be shown that this simplification leads to the same limit carrying capacity as would be obtained by considerations which include the elastic deformation of the structure. The basic limit-analysis theorems are usually modified and restated for specific applications to "bending" problems in the following manner.

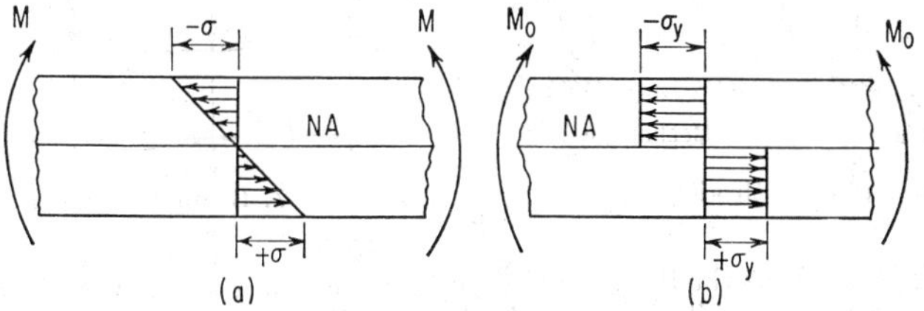

FIG. 15.55. Stress distributions. (*a*) Elastic stress distribution. (*b*) Fully plastic stress distribution.

The *yield moment* is the maximum moment that can be carried at any position of the structure. This can vary from point to point and is obtained from considerations of the fully plastic stress distribution (see Fig. 15.55).

Table 15.3 indicates the ratio of yield moment M_0 to the maximum elastic moment M_y, for a few typical cross sections.

Table 15.3. Ratio of Yield Moment to Maximum Elastic Moment for Various Sections and Comparison with Yield Moment for Rectangle

Section	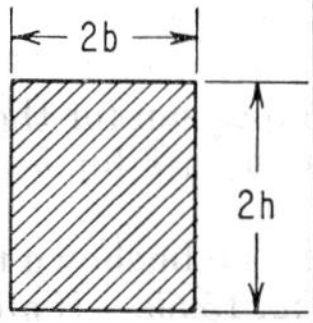	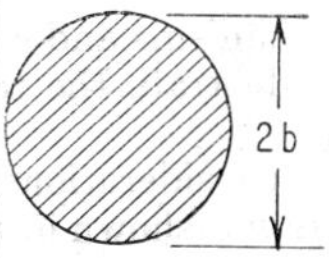	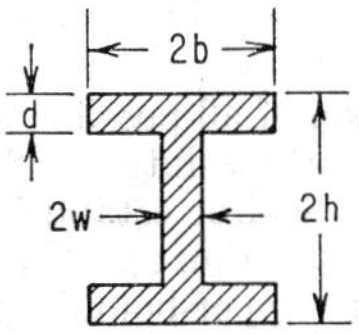	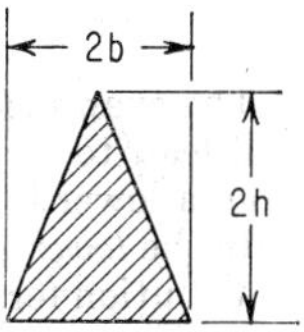
M_0/M_y	1.5	1.7	$1 \leq \dfrac{1 + wh/2bd}{1 + wh/3bd} \leq 1.5$ $(d = 0)$	2.34
$\dfrac{M_{0,\text{rectangle}}}{M_{0,\text{comparison section}}}$	1.0	1.5	$1 \leq \dfrac{M_{0\square}}{M_{0,I}} \leq \infty$	3/1.172

The yield hinge is an idealized hinge that occurs at the position where the yield moment exists, and which will permit rotation of any magnitude under constant moment M_0 but will not rotate under a lesser moment. Physically this phenomenon is envisioned as the formation of a "kink" in a member made up of non-strain-hardening material.

A *mechanism* is characterized by a combination of ideal yield hinges and rigid bars, for which an infinitesimal motion is possible (see Fig. 15.56).

A statically admissible bending-moment distribution is one which is in equilibrium both internally and with the external tractions, and nowhere exceeds the yield moment.

For example, the moment distribution in an elastic structure is statically admissible.

A kinematically admissible bending-moment distribution is one which is in equilibrium with external tractions, and one with which a mechanism may be associated. The moments must equal the yield moment at the hinges and be of algebraic sign consistent with the direction of rotation. The yield moment may be exceeded at nonhinge points. The special theorems of limit analysis of beams and frames follow.

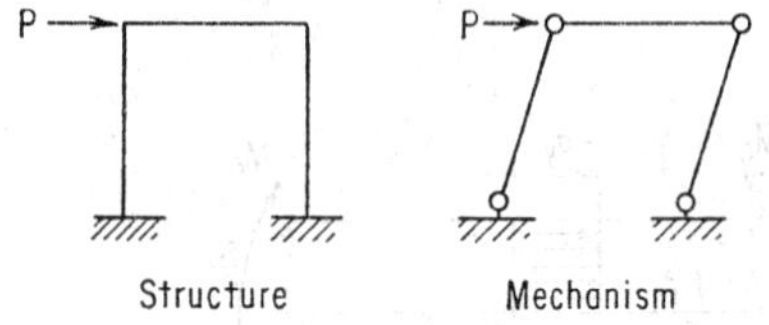

FIG. 15.56. Frame with yield hinges.

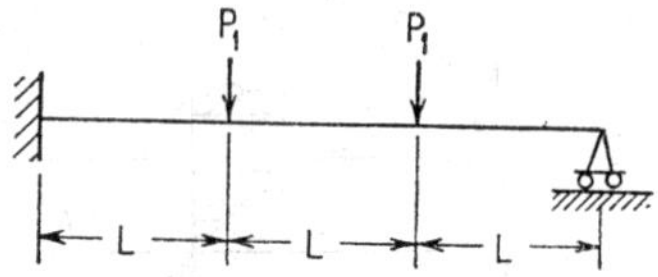

FIG. 15.57. Beam with loads P_1 applied.

The *first theorem* is that the limit load is the *largest* load for which a statically admissible bending-moment distribution may be found. The *second theorem* is that the limit load is the *smallest* load for which a kinematically admissible bending-moment distribution may be found.

Beam Example: It is desired to *bound* the maximum load P_1 that can exist on the structure shown in Fig. 15.57.

In the upper bound one possible mechanism is shown in Fig. 15.58. From virtual work,

$$\underbrace{P_1\delta + P_1(\delta/2)}_{\text{external work}} = \underbrace{M_0\theta + M_0(\theta + \theta/2)}_{\text{internal work}} \tag{15.305}$$

but $L\theta = \delta$, so that

$$P_1 = \tfrac{5}{3}M_0/L$$

From the second theorem,

$$P_{\text{lim}} \leq \tfrac{5}{3}M_0/L \tag{15.306}$$

The examination of all possible mechanisms and comparison for the lowest load P^* will lead to the exact solution. To illustrate the techniques for *bounding* the solution, a lower limit will be established.

In the *lower bound,* from equilibrium considerations it can be shown that the moment distribution for the upper-bound solution is shown in Fig. 15.59. If all external forces,

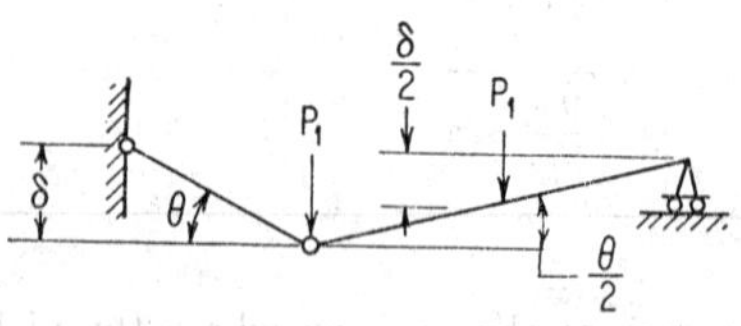

FIG. 15.58. Loaded structure with yield hinges.

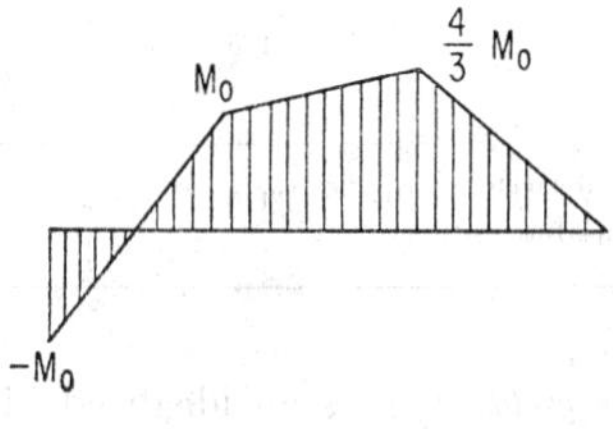

FIG. 15.59. Moment-distribution curve.

and hence internal moments, are multiplied by ¾, the yield moment is nowhere exceeded, and the distribution of Fig. 15.59 is statically admissible. Hence a lower bound on the limit is three-fourths that obtained for the upper bound:

$$P_{\text{lim}} \geq \tfrac{5}{4}M_0/L \tag{15.307}$$

Solution: From the above considerations,

$$\tfrac{1}{4} \leq P_{\text{lim}}L/5M_0 \leq \tfrac{1}{3}$$

For an error limited to 15 per cent,

$$P_{\text{lim}} = \tfrac{35}{24}M_0/L$$

Since the actual solution is

$$P_{\text{lim}} = \tfrac{32}{24}M_0/L$$

the actual error in the bounding solution does not exceed 9 per cent.

Frame Example: It is desired to examine the entire array of feasible mechanisms of the structure in Fig. 15.60 and thereby arrive at an *exact* determination of the limit load. Note that elementary considerations always lead to the result that no hinges need be considered except at loads and joints in concentrated loading problems.

From the second theorem of limit analysis the limit load is

$$P_{\text{lim}} = M_0/L$$

and the failure mechanisms can be either number 2 or 4.

(c) Plate Theory[8,11] (Circular Symmetric)

The *yield criterion and flow law* follows: in this it can be shown that the Tresca yield envelope, which will be used, reduces to that shown in Fig. 15.61, in terms of the nondimensional radial and tangential plate moments (m_r, m_ϕ). In this,

$$m_r = M_r/M_0 \tag{15.308}$$
$$m_\phi = M_\phi/M_0 \tag{15.309}$$

where M_r = radial moment/unit length
M_ϕ = tangential moment/unit length
M_0 = yield moment/unit length $= \sigma h^2$
$2h$ = plate thickness
σ = yield stress

The associated strain components are curvatures (K_r, K_ϕ) and changes of slope, in the event of curvature discontinuity (hinge circle). Based upon this an alternative and useful formulation of the associated flow law is that the curvature-rate vector is normal to the yield envelope, considering the m_r, K_r and m_ϕ, K_ϕ directions to be superposed. At the corners, the curvature-rate vector directions are not completely determined, as shown in Fig. 15.62. Each plate element is described by a point within or on the boundary of the yield envelope. The inside elements are elastic, and those on the boundary are plastic. Upon inspection Fig. 15.62 provides useful information. For instance, any element having $m_\phi = 1$, $0 \leq m_r \leq 1$ will have a radial curvature increment equal to zero and a tangential curvature increment which is proportional to m_ϕ.

The equilibrium equations and lower-bound solution follow. Although shear deformations are neglected, and correspondingly the shear force does not enter into the yield criteria and the associated flow law, it must be considered in the equilibrium equations. Based upon the circular plate element shown in Fig. 15.63, the equilibrium equations are

$$(d/dr)(rQ) - rp = 0 \tag{15.310}$$
$$(d/dr)(rM_r) - M_\phi + rQ = 0 \tag{15.311}$$

which reduces to the single equations

$$(d/dr)(rM_r) - M_\phi = -\int_0^r rp\, dr = -P(r)$$

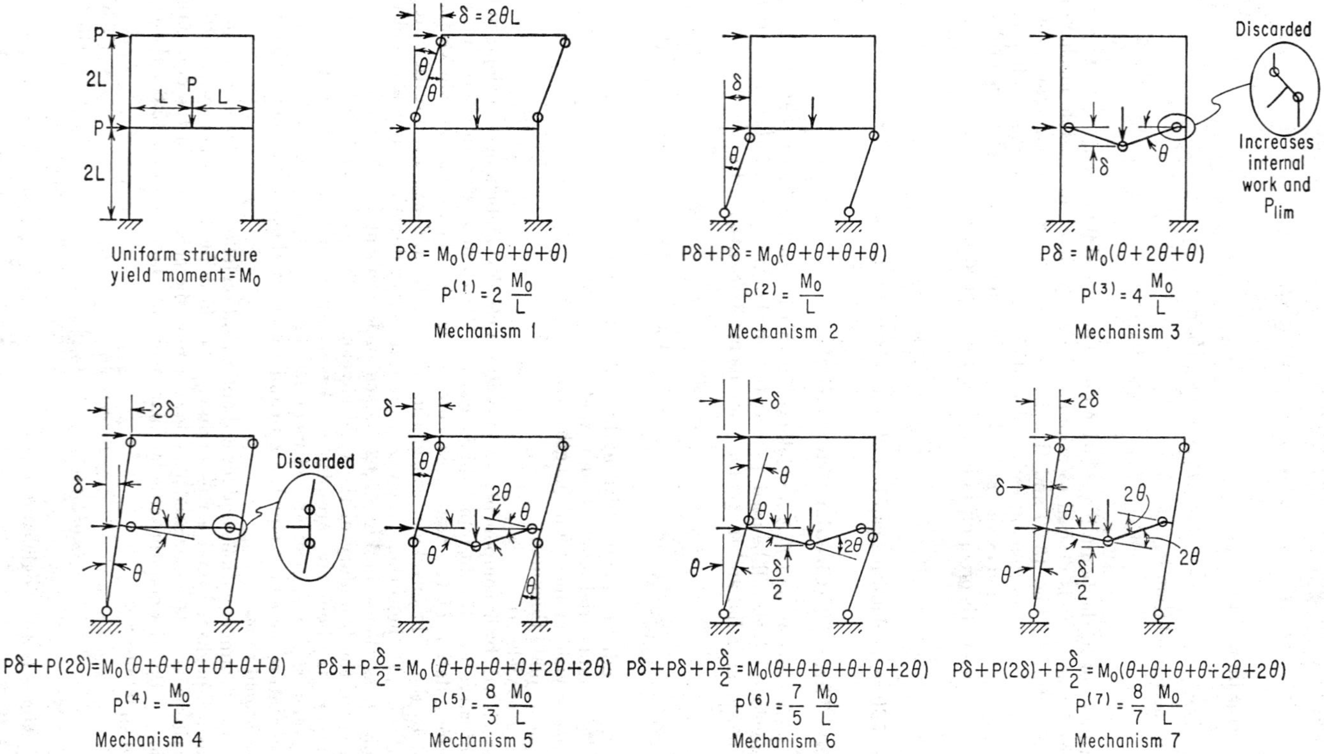

Uniform structure
yield moment = M_0
$P\delta = M_0(\theta+\theta+\theta+\theta)$
$P^{(1)} = 2\frac{M_0}{L}$
Mechanism 1
$P\delta + P\delta = M_0(\theta+\theta+\theta+\theta)$
$P^{(2)} = \frac{M_0}{L}$
Mechanism 2
$P\delta = M_0(\theta+2\theta+\theta)$
$P^{(3)} = 4\frac{M_0}{L}$
Mechanism 3
Discarded
Increases internal work and P_{lim}
$P\delta + P(2\delta) = M_0(\theta+\theta+\theta+\theta+\theta+\theta)$
$P^{(4)} = \frac{M_0}{L}$
Mechanism 4
Discarded
$P\delta + P\frac{\delta}{2} = M_0(\theta+\theta+\theta+\theta+2\theta+2\theta)$
$P^{(5)} = \frac{8}{3}\frac{M_0}{L}$
Mechanism 5
$P\delta + P\delta + P\frac{\delta}{2} = M_0(\theta+\theta+\theta+\theta+\theta+2\theta)$
$P^{(6)} = \frac{7}{5}\frac{M_0}{L}$
Mechanism 6
$P\delta + P(2\delta) + P\frac{\delta}{2} = M_0(\theta+\theta+\theta+\theta+2\theta+2\theta)$
$P^{(7)} = \frac{8}{7}\frac{M_0}{L}$
Mechanism 7

FIG. 15.60. Various mechanisms with yield hinges.

With the use of the basic yield criterion it is possible to solve the above equations for any specific yield regime. At regime A, for example,

$$M_\phi = M_r = M_0$$

which can exist only if $P = 0$. In regime AB, $M_\phi = M_0$ and

$$0 \leq \left[M_r = M_0 - \frac{1}{r} \int_0^r P(r)\, dr \right] \leq M_0$$

In regime BC, $M_r = M_\phi - M_0$ so that

$$0 \leq \left[M_\phi = M_0 (\ln r) - \int_0^r \frac{P(r)}{r}\, dr \right] \leq M_0$$
$$-M_0 \leq [M_r = -M_0 + M_\phi] \leq 0$$

Consideration of stress and moment-resultant discontinuities across possible *hinge circles* leads to the conclusion that M_r and Q must be continuous, but M_ϕ may be discontinuous across these hinge circles.

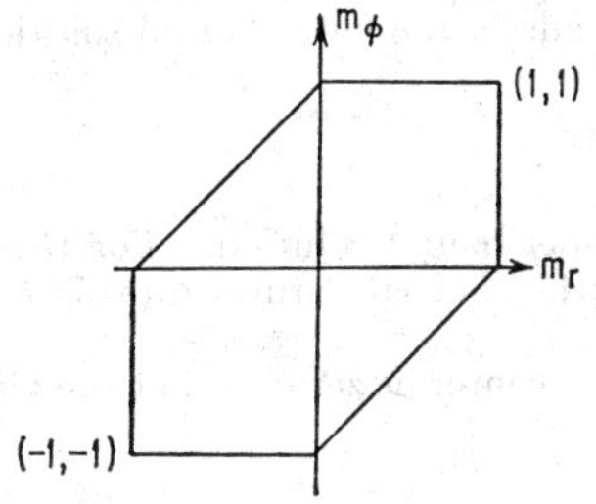

FIG. 15.61. Tresca yield envelope.

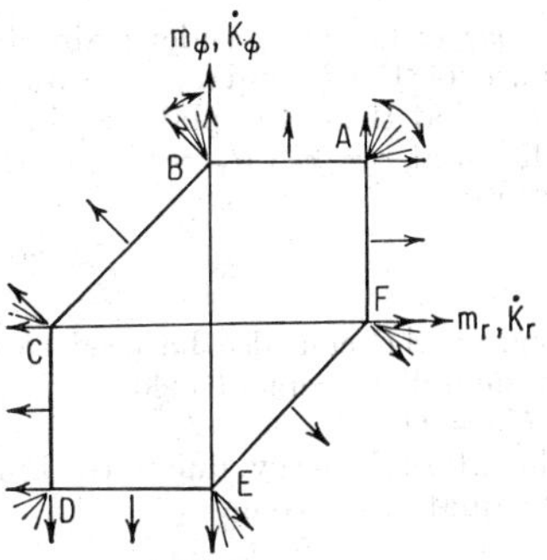

FIG. 15.62. Radial and tangential plate moments and associated strain components.

Arbitrarily separating the plate into one or more regimes [each of which satisfies the external traction $P(r)$, conditions across the hinge circle (boundary of regime), as well as inner and outer radius boundary conditions] will provide a lower-bound solution on the limit load. This limit load is defined by np where n is to be maximized and p defines the load distribution. The above results follow from the first theorem of limit analysis.

The relationships between curvature and transverse deflection are given by (upper-bound solution)

$$\dot{K}_\phi = -(1/r)(d\dot{w}/dr) \qquad (15.312)$$
$$K_r = -(d^2\dot{w}/dr^2) \qquad (15.313)$$

From Eqs. (15.312) and (15.313), it is seen that in regime A the requirement is that

$$d\dot{w}/dr \leq 0$$
$$d^2\dot{w}/dr^2 \leq 0$$

whereas in regime AB,

$$d\dot{w}/dr = \text{const}$$

and thus the deflection (rate) can be evaluated in each regime.

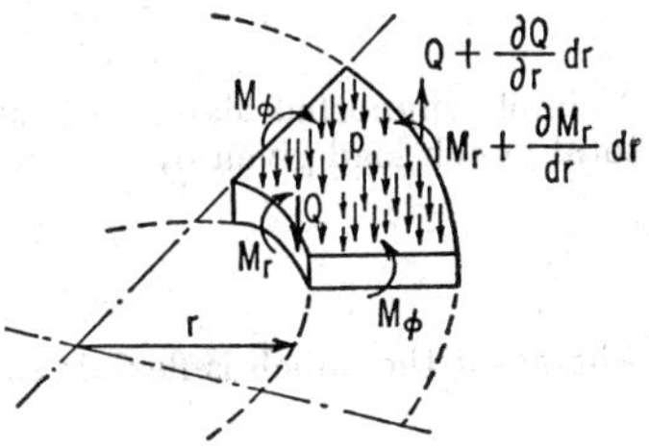

FIG. 15.63. Circular plate element.

The arbitrary selection of a displacement field, subject to externally prescribed displacement boundary conditions, leads to an upper-bound solution of the limit load. This follows from the second theorem of limit analysis. Actually to evaluate the limit load factor $n(p_{\text{lim}} = np)$ it is necessary to employ virtual-work equations and the associated moments from the yield criterion for each regime. This leads to the satisfaction and evaluation of the external tractions. Note that no requirement for the point-by-point satisfaction of equilibrium equations is imposed. This requirement would lead to the exact solution.

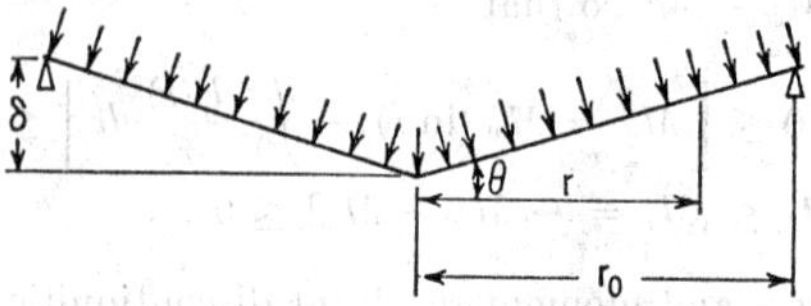

Fig. 15.64. Upper-bound displacement field on simply supported plate.

Example: Let us consider a simply supported circular plate under a uniform load Because of the boundary condition $M_r = 0$, and the center condition $M_r = M_\phi$, assume that the entire plate lies in regime AB. Solving for M_r, and introducing the condition at $r = r_0$ ($M_r = 0$) into the solution, leads to the lower-bound solution for pressure:

$$p_{\text{lim}}^{(L)} = 6M_0/r_0^2$$

Regime AB can also be used to obtain an upper-bound solution. For this case examine a displacement field as given in Fig. 15.64. The curvature (rate) is $\dot{K}_r = 0$ and $\dot{K}_\phi = (1/r)\dot{\theta}$.

The internal energy, due to the hinge circle at the center, is zero; so that the virtual-work equations become

$$2\pi \int_0^r \dot{K}_\phi M_\phi r\, dr = 2\pi \int_0^r \dot{\theta}(r_0 - r)pr\, dr \tag{15.314}$$

Thus the upper bound is

$$p_{\text{lim}}^{(U)} = 6M_0/r_0^2$$

Since the upper and lower bound coincide, the exact limit pressure is

$$p_{\text{lim}} = 6M_0/r_0^2 \tag{15.315}$$

From similar analyses, the limit pressure of a uniformly loaded clamped plate is given by

$$p_{\text{lim}} = 11.3M_0/r_0^2 \tag{15.316}$$

A simply supported plate, centrally loaded by a rigid punch of radius r, can support a total punch load given by

$$p_{\text{lim}} = \frac{2\pi M_0}{1 - r_1/r_0} \tag{15.317}$$

whereas if the punch is flexible so that it supplies its load uniformly

$$p_{\text{lim}} = \frac{2\pi M_0}{1 - \frac{2}{3}r_1/r_0} \tag{15.318}$$

A plate of arbitrary shape and fixity can carry a concentrated load equal to

$$p_{\lim} = 2\pi M_0 \tag{15.319}$$

at any position in that plate. This result holds whether the plate is simply supported, clamped, or in between, or whether the shape is circular, triangular, or any other, regardless of the position of load.

A simply supported plate of side $2L$ will support a uniform pressure of

$$p_{\lim} = 5.8(M_0/L^2) \pm 0.6(M_0/L^2) \tag{15.320}$$

where the tolerance represents the spread between the upper and lower bounds.

To aid in the evaluation of other shapes note that a clamped plate fully inscribed inside another clamped plate of a different shape will fully carry at least that load

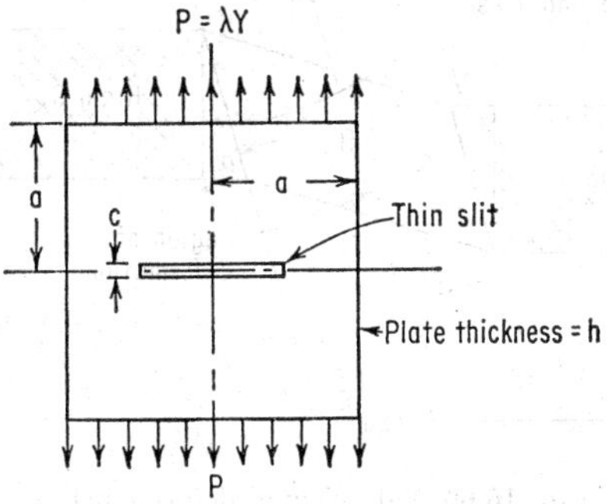

FIG. 15.65. Slab with thin slot.

which falls on it. That is, the square plate of side $2L$ can carry a pressure bounded by the limit pressure on the inscribed and circumscribed circular plates. Thus

$$p_{\lim} = 8.48(M_0/L^2) \pm 2.82(M_0/L^2) \tag{15.321}$$

Although the tolerance is broad, the approximate solution is expected to be close to the exact solution and adequate for most design purposes. It is intuitively anticipated a similar extension can be made to simple support boundaries, but presently this extension does not exist. Thus care should be used if such an assumption is made.

(d) Plane-stress Problem—Slab with Cutout

It is desired to find the maximum uniform edge load that can be carried by the slit plate shown in Fig. 15.65, in which λ is the *cutout factor*, Y is the yield stress, and $a \gg c > h$. In general it is necessary to determine the effective weakening of the plate due to a cutout, so that it can be properly reinforced. Note that, unless the plate is especially reinforced, $0 \leq \lambda \leq 1$. In this development, the pressure will be assumed uniformly distributed across the thickness h and the Tresca yield criterion will be used (slit width $= b$).

To obtain an upper bound, the assumed deformation model is as shown in Fig. 15.66, in which the side view of the slab is shown to illustrate the assumed deformation mode.

The assumed deformation field is summarized in Table 15.4. From this the internal dissipation rate is

$$D_i = \int_v \tau\gamma \, dv = 2YhUa(1 - b/a) \csc 2\alpha \tag{15.322}$$

and the external dissipation

$$D_e = (\lambda Y)(2ah)U \tag{15.323}$$

Thus

$$\lambda_{\text{kin}} = (1 - b/a) \csc 2\alpha \qquad 0 \leq \alpha \leq \pi/2 \tag{15.324}$$

the smallest λ_{kin} is associated with $\alpha = \pi/4$ so that

$$\lambda_{lim} \leq (1 - b/a)$$

To obtain a lower bound it is convenient to divide the slab into regions of constant stress and then satisfy equilibrium in each section, subject to allowable continuity and discontinuity requirements in stress across the boundaries. It can be shown that only the stress tangent to a boundary can be discontinuous; all others must be continuous.

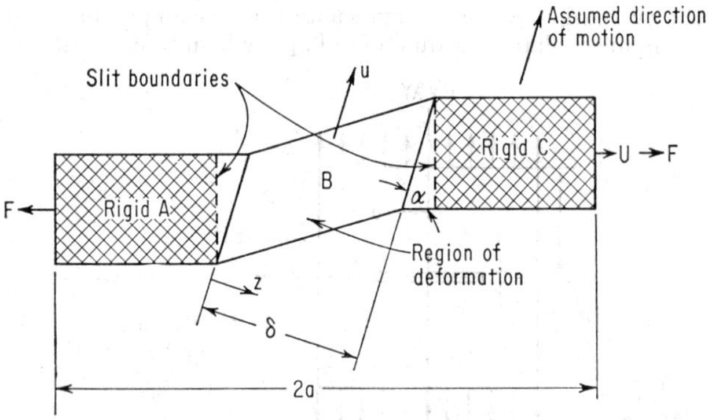

FIG. 15.66. Side view of deformed slab.

With this in mind the slab is divided as shown in Fig. 15.68. The typical elements are shown, along with their stresses, which are to be determined.

Note that there are two continuity conditions across each of three unique boundaries 1-2, 2-3, 2-4. Also there are six stress unknowns $(r_1, s_2, r_2, \theta, s_3, r_4)$ together with the cutout factor λ and the geometrical unknown x. The use of the six continuity conditions together with Mohr's circle will eliminate the six stress unknowns. It is thus

Table 15.4. Deformation Field

Regime	Total velocities	Shear strain
A	0	0
B	$(U \sec \alpha) \dfrac{z}{\delta}$	$\dfrac{U \sec \alpha}{\delta}$
C	$U \sec \alpha$	0

possible to express the stresses throughout the structure in terms of the unknowns x and λ. These stresses are written and then each regime is subjected to the yield criterion (the absolute value of each of the principal stresses and their difference τ shall not exceed the yield stress Y). The result is a number of inequalities for x and λ. Maximizing λ and eliminating x,

$$\lambda_{stat} = 1 - b/a$$

Correct Solution. Since the static and kinematic cutout factors are identical, the limit cutout factor becomes

$$\lambda_{\text{lim}} = 1 - b/a$$

which can be shown to hold also for the case where the ends are clamped, in which case boundary displacements rather than stresses are uniform.

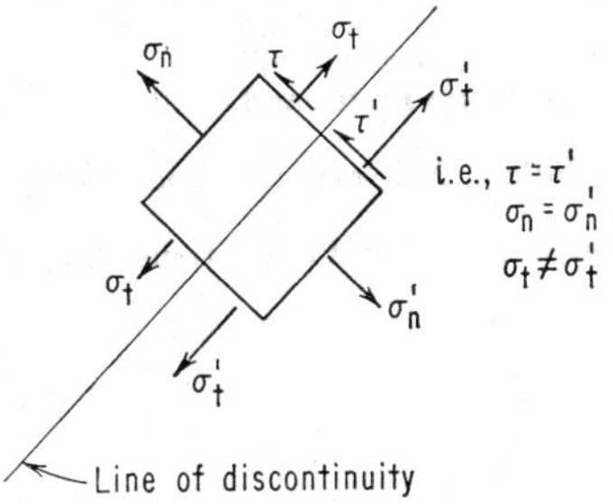

FIG. 15.67. Stress discontinuity region.

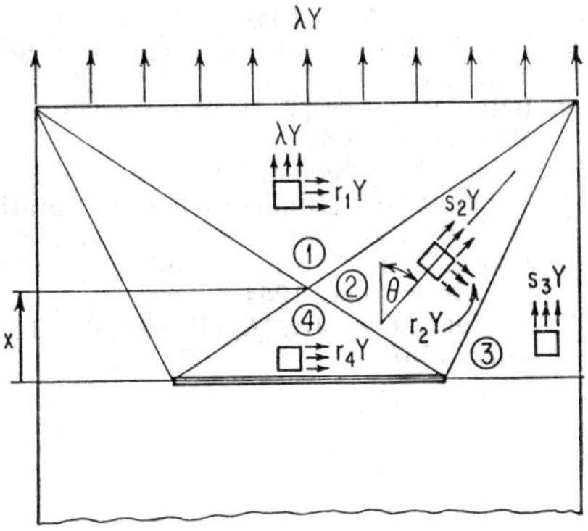

FIG. 15.68. Slab stress field.

Other Solutions. Considering the square slab of side $2A$, with a circular cutout of radius a, loaded similarly to the above problem, then the approximate cutout factor is as presented in Fig. 15.69.

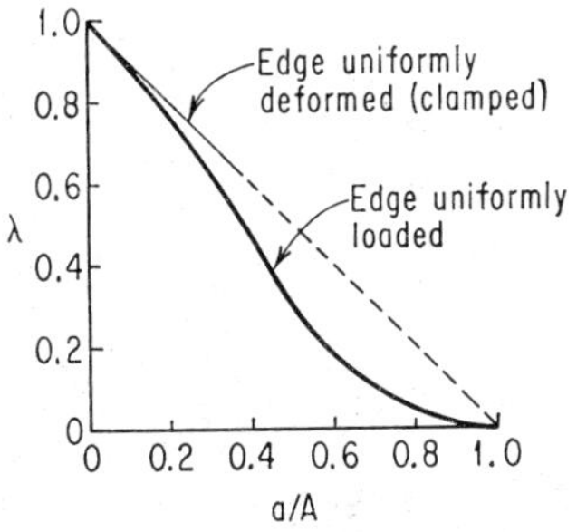

FIG. 15.69. Slab cutout factor as a function of radius.

In the case where the edge is clamped, so that the load may not be uniformly distributed (although the boundary displacement will be uniform), the approximate cutout factor becomes

$$\lambda = 1 - a/A$$

which is also shown in Fig. 15.69.

References

1. Timoshenko, S.: "Strength of Materials," 3d ed., Parts I and II, D. Van Nostrand Company, Princeton, N.J., 1955.
2. Timoshenko, S., and J. N. Goodier: "The Theory of Elasticity," 2d ed., McGraw-Hill Book Company, Inc., New York, 1951.

3. Timoshenko, S., and S. Woinowsky-Krieger: "Theory of Plates and Shells," 2d ed., McGraw-Hill Book Company, Inc., New York, 1959.
4. Sokolnikoff, I. S.: "Mathematical Theory of Elasticity," 2d ed., McGraw-Hill Book Company, Inc., New York, 1956.
5. Love, A. E. H.: "A Treatise on the Mathematical Theory of Elasticity," 4th ed., Dover Publications, Inc., New York, 1944.
6. Hill, R.: "The Mathematical Theory of Plasticity," Oxford University Press, London, 1950.
7. Prager, W., and P. G. Hodge: "Theory of Perfectly Plastic Solids," John Wiley & Sons, Inc., New York, 1951.
8. Hodge, P. G.: "Plastic Analysis of Structures," McGraw-Hill Book Company, Inc., New York, 1959.
9. Boley, B. A., and J. H. Weiner: "Theory of Thermal Stresses," John Wiley & Sons, Inc., New York, 1960.
10. Flugge, W.: "Stresses in Shells," Springer-Verlag OHG, Berlin, 1960.
11. Hodge, P. G.: "Limit Analysis of Rotationally Symmetric Plates and Shells," Prentice-Hall, Inc., Englewood Cliffs, N.J., 1963.
12. Roark, R. J.: "Formulas for Stress and Strain," 3d ed., McGraw-Hill Book Company, Inc., New York, 1954.
13. McConnell, A. J.: "Applications of Tensor Analysis," Dover Publications, Inc., New York, 1957.

Section 16

IMPACT

By

PHILIP BARKAN, Ph.D., *Senior Research Engineer, General Electric Company, Switchgear and Control Division, Philadelphia, Pa.*

CONTENTS

16.1. INTRODUCTION

Impact involves the study of the physical phenomena which attend the collision of bodies with initial relative velocity. Two extensive bibliographies[1,2] which cover abstracts of several thousand papers associated with the general subject and recent books[3,4] are recommended to the reader seeking more information than is covered in this section.

Impact phenomena are especially important to the machine designer since in nearly all systems the highest forces and greatest stresses arise as a consequence of impact. Many serious machine failures arise because impact forces were not properly recognized. Impact can be quite useful, since by means of impact it is possible to achieve many extreme short-duration effects. High force levels, rapid dissipation of energy, large accelerations and decelerations, and enormous power amplification may be so obtained from low-energy sources operating at low force levels.

(a) Nomenclature

A = cross-sectional area
a = rod diameter
B = function defined by Eqs. (16.10) and (16.11)
C = constant of proportionality
c = wave-propagation velocity
d = lateral dimension of a rod
e = coefficient of restitution
E = Young's modulus
F = force
f = wave function
G = Shear modulus
g = acceleration due to gravity or wave function
I = moment of inertia
K = net stiffness between centers of gravity of two impacting bodies
L = length of a body, measured in direction of impact
M, m = mass
n = number of a sequence or a constant
p = plastic stress, assumed independent of strain
P = gas pressure or dynamic stress
R, r = radius of curvature
T = kinetic energy
t = time
τ = kinetic energy
U = strain energy
U_s = strain energy density
u = displacement
v = velocity
V = volume
X, x = displacement or position
z = amplitude of bounce
α = deformation or approach distance between centers of gravity of impacting bodies.
γ = ratio of specific heat of a gas
δ = deflection of a beam
ϵ = strain
ζ = parameter involving strain rate and temperature
η = natural frequency
θ = absolute temperature
Λ = wavelength
λ = function defined by Eq. (16.22)
ν = Poisson's ratio
ρ = mass density
σ = stress
ϕ = force due to impact
Φ = function proportional to energy in a mode of vibration

Subscripts

c = common
e = externally applied
f = at end of impact

I = incident
m = maximum
0 = at beginning of impact
p = elastic limit
R = reflected
T = transmitted
t = at time t during impact
u = ultimate
y = yield

(b) Accuracy of Impact Calculations

Impact calculations suffer from several practical limitations which limit their value to establishing the approximate magnitude of the various phenomena involved and for providing some insight into the manner in which the various parameters affect the final result. In spite of these limitations available techniques provide a useful guide to the design of impacting systems and to the interpretation of final results. For precise data resort to experimental measurements is essential, employing such powerful tools as strain gages and high-speed photography.

The limitations to precise analysis include

1. The mathematics required for rigorous solution of impact problems is excessively complex, and approximate methods which omit one or more basic characteristics of the impact phenomenon are invariably employed.
2. The problem of adequately defining a complicated physical system such as a machine with discontinuities in both geometry and materials is formidable.
3. The properties of even conventional engineering materials are modified to some extent by the duration and rate of loading. These effects are neither well defined nor fully understood.

(c) Inadvertent Sources of Impact

Occasionally impact is overlooked in machine designs with unhappy results. It is therefore important for the designer to recognize and cope with possible conditions of severe impact. Typical sources include

1. Clearances as between cams and followers
2. Backlash or bearing clearances in mechanisms undergoing force or motion reversal
3. Mechanisms with kinematic discontinuities so that components with large relative velocities are engaged
4. High-speed systems which are nonlinearly elastic so that abrupt changes in stiffness occur with results similar to impact

(d) Design Factors to Be Considered in Minimizing Impact Effects[30]

1. Minimize the velocity of impact.
2. Minimize the mass of impacting bodies.
3. Design for minimum stiffness in the vicinity of the point of impact.
4. Design critical components for maximum strain-energy storage capacity within the tolerable maximum stress by the following means:
   (*a*) The maximum possible volume of materials should be stressed uniformly to the peak stress.
   (*b*) Other things being equal, select materials whose energy storage capacity per unit volume called modulus of resilience $\sigma_y^2/2E$, is the maximum, where σ_y = yield stress and E = Young's modulus.
5. Minimize sensitivity to local stress concentrations by these means:
   (*a*) Employ a ductile material with some capacity for plastic deformation to cope with localized stress concentrations.
   (*b*) Avoid surface irregularities, sharp discontinuities, and internal inhomogeneities in the design and manufacture of critical parts.

16.2. CLASSICAL THEORY OF IMPACT AND THE COEFFICIENT OF RESTITUTION

The simplest form of impact analysis, based on the laws of conservation of momentum and an empirical manifestation of the conservation of energy, can provide information on the net change in velocity of the center of mass of each body involved in the impact, the net impulse, and the energy-exchange processes accompanying the impact. This method can be employed when external forces either are not present during the period of impact or are negligibly small compared with the impact-induced forces.

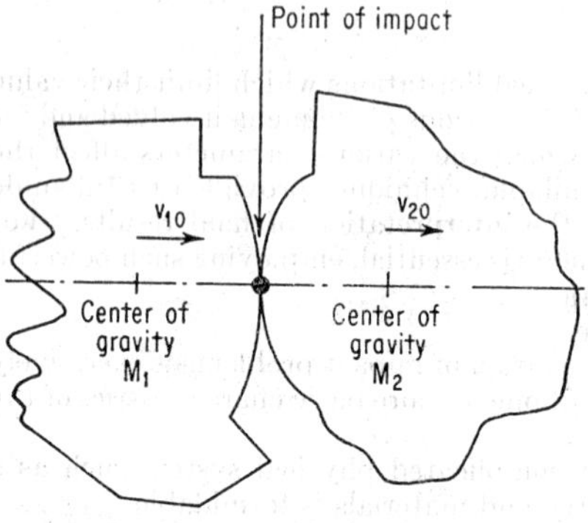

Fig. 16.1. Schematic representation of central impact. Initial impact velocity $= v_{10} - v_{20}$.

(a) Central Impact

In the simplest case two bodies collide on surfaces which are normal to the common line connecting their centers of mass (Fig. 16.1) and have velocity components only along this common line. With these restrictions no rotational or sliding effects occur.

Conservation of momentum requires that momentum immediately before and immediately after the impact be equal.

$$m_1 v_{10} + m_2 v_{20} = m_1 v_{1f} + m_2 v_{2f} \tag{16.1}$$

The second relationship involved can be regarded as a form of the conservation of energy. It recognizes that the total kinetic energy in the bodies after the impact must be equal to or less than the initial kinetic energy. The loss in kinetic energy is best expressed in terms of the coefficient of restitution (e), which Newton defined as

$$e = -\frac{v_{1f} - v_{2f}}{v_{10} - v_{20}} \tag{16.2}$$

It can then be shown that the kinetic energy loss is

$$T_{\text{loss}} = \frac{1 - e^2}{2} \frac{m_1 m_2}{m_1 + m_2} (v_{10} - v_{20})^2 \tag{16.3}$$

The coefficient of restitution is discussed in more detail at a later point in this section.

Simultaneous solution of Eqs. (16.1) and (16.2) results in the following dimensionless expressions for the terminal velocity of each of the impacting bodies:

$$\left.\begin{aligned} \frac{v_{1f} - v_{10}}{v_{10} - v_{20}} &= -\frac{(1 + e)(m_2/m_1)}{m_2/m_1 + 1} \\ \frac{v_{2f} - v_{20}}{v_{10} - v_{20}} &= \frac{1 + e}{1 + m_2/m_1} \end{aligned}\right\} \tag{16.4}$$

Figure 16.2, which is based on Eqs. (16.4), illustrates the effect of mass ratio and coefficient of restitution upon the velocities of two impacting bodies for the special case when the initial velocity of one of the masses is zero.

The net impulses produced by the impact upon the two masses are equal and opposite; their magnitude is obtained from conservation of momentum and Eqs. (16.4).

$$\left| \int \phi \, dt \right| = \left| (1 + e) \frac{m_1 m_2}{m_1 + m_2} (v_{10} - v_{20}) \right| \tag{16.5}$$

During the impact the total kinetic energy of the two bodies will decrease and strain energy will be developed within the impacting bodies. The maximum strain energy developed by the impact will occur at that instant when the centers of mass of the two impacting bodies share a common velocity. From conservation of momentum the common velocity is

$$v_c = \frac{m_1 v_{10} + m_2 v_{20}}{m_1 + m_2}$$

and the maximum possible strain energy is dictated by conservation of energy

$$U = \frac{m_1}{2} v_{10}^2 + \frac{m_2}{2} v_{20}^2 - \frac{m_1 + m_2}{2} v_c^2$$

$$= \frac{m_1 m_2}{m_1 + m_2} \frac{(v_{10} - v_{20})^2}{2} \qquad (16.6)$$

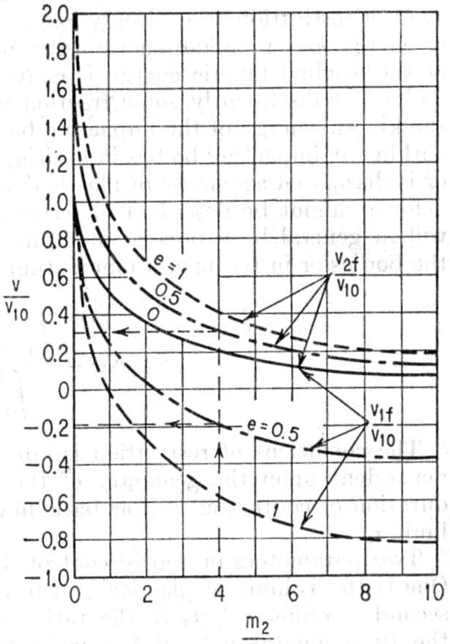

FIG. 16.2. Effect of coefficient of restitution and mass ratio upon velocity change when one mass is initially motionless.

Example:

$$\text{given } e = 0.5 \quad \frac{m_2}{m_1} = 4 \quad v_{20} = 0$$

$$\text{from graph } \frac{v_{2f}}{v_{10}} = 0.32; \quad \frac{v_{1f}}{v_{10}} = -0.18$$

(b) Most General Form of Classical Theory of Impact[38]

In its most general form, the calculation of the change in motion resulting from an impact between two bodies which are free of external constraints and external forces will require the designation of six quantities for each body:

Three components of translational velocity of the center of gravity of each body

Three components of angular velocity of each body about axes through the center of gravity of each body

Thus 12 equations are necessary to define the motions of the two bodies. Six of these equations are determined by the condition that the angular momentum of each body about any axis through the point of contact is unchanged, since the impact forces must act through this point. The seventh equation requires that the momentum of the entire system in the direction normal to the surface in contact is constant, since the normal forces acting on the two bodies are equal and opposite. The eighth equation involves the conservation of energy or the coefficient of restitution of the bodies, i.e., Eqs. (16.4) and (16.5). The last four equations consider conditions at the impact surface.

Case 1. If friction at the impact surface is negligible then two equations are employed for each body to state that the momentum of *each* body is unchanged in directions tangent to the plane of contact.

Case 2. If friction at the surface is sufficiently high (i.e., perfect roughness) so that no slippage or sliding occurs, then two equations are employed to state that the linear momentum of the system in mutually perpendicular directions parallel to the impact plane is unchanged. The last two equations then require that there is no relative motion between the two bodies in the plane of impact.

Case 3. If friction is intermediate between these two extreme cases, then as in case 2, the system invariance of linear momentum in the impact plane is still valid and this accounts for two of the equations. The last two equations equate the variation in linear momentum of each body along the direction of slip with the frictional impulse.

(c) The Coefficient of Restitution

The coefficient of restitution e is defined by Eq. (16.2) as the ratio (relative velocity between two bodies immediately after impact)/(relative velocity just prior to impact). However, it is most convenient for purposes of understanding to regard the coefficient of restitution as an energy-loss function, since all impacts are basically processes of energy exchange and energy transformation. When two bodies collide, a portion of the original kinetic energy is converted into strain energy within the impacting bodies. Subsequently some fraction of the strain energy is reconverted back into the kinetic energy of the impacting bodies. The remainder of the energy is trapped within the impacting bodies in exciting various modes of vibration within the bodies or is dissipated as energy of plastic deformation. Since the velocity of bodies which deform cannot be described in terms of a single value, the coefficient of restitution will in general be expressed in terms of the velocities of the centers of gravity of the bodies or in terms of a momentum averaged velocity of each body, which is

$$v_{av} = \frac{\int_0^L [dm(x)/dx]\, v(x)\, dx}{\int_0^L [dm(x)/dx]\, dx} \tag{16.7}$$

The coefficient of restitution is not a basic material property. Its magnitude is dependent upon the geometry of the bodies, the peak stresses produced, and the duration of contact as well as basic material properties such as E, ρ, and the elastic limit σ_p.

Two parameters primarily control the magnitude of the coefficient of restitution. One is the volume of plastically deformed material generated by the impact. The second parameter t_f/t_η is the ratio (duration of the impact period)/(the period of the fundamental natural frequency of the impacting bodies). As the ratio t_f/t_η becomes smaller, the vibrational modes of the bodies "trap" an increasingly larger proportion of the strain energy generated by the impact.

Coefficient of Restitution for Compact Bodies. For compact bodies such as spheres, the ratio t_f/t_η is quite large, and as a consequence the proportion of kinetic energy lost because of excitation of various modes of vibration is quite small. For such shapes the coefficient of restitution is primarily controlled by plastic deformation near the point of impact, or internal friction within the impacting bodies. The extent of plastic deformation is dependent upon the velocity of impact, and this effect accounts for the basic velocity sensitivity of the coefficient of restitution of spherelike bodies.

Typical coefficient-of-restitution data for the impact of spheres are shown in Fig. 16.3. Tests have shown that for spheres the effect of size upon the coefficient of restitution is rather small. For example, variation of about 100:1 in the weight of cast-iron balls (i.e., 5:1 in diameter) produces only 10 per cent variation in e. Bowden and Tabor[5] have derived an approximate relationship based on Hertz equations for deformation of spherical surfaces[6] and conservation of energy which accounts for the velocity variation of the coefficient of restitution of spheres. For spheres the relative velocities before and after impact are related by an equation of the form

$$v_f = k(v_0^2 - \tfrac{3}{8}v_f^2)^{3/8} \tag{16.8}$$

This nonlinear relationship yields values of $e = -v_f/v_0$ which correlate well in shape with experimental data such as Fig. 16.3. k is proportional to

$$\frac{\sigma_p^{5/8} r^{3/8}}{m^{1/8}} \left(\frac{E_1 + E_2}{E_1 E_2}\right)^{1/2}$$

Note that since for spheres $m \propto r^3$ the coefficient of restitution should be very nearly independent of size.

Coefficient of Restitution for Distributed Systems. For bodies such as rods, plates, or beams for which the period of impact tends to be short compared with the period of propagation of stress waves, the amount of energy remaining within the bodies after impact in the form of internal vibration or traveling stress waves is relatively large, and this factor has a predominant influence upon the coefficient of restitution. For such systems even if plastic deformation does not occur, a large proportion of the initial kinetic energy can be lost during the impact, and this is accounted for in trapped strain energy within the bodies.

When a relatively light, compact mass such as a sphere impacts a relatively massive rod, beam, or plate, the duration of impact is very short compared with the basic period of vibration of the resilient member and very long compared with the period of the fundamental mode of vibration of the sphere. The elastic vibration of the sphere then absorbs only a negligible fraction of the initial kinetic energy[31] and it is only necessary to calculate the energy absorbed by the body which the sphere strikes. Zener and Feshbach[21] showed that for such situations the strain

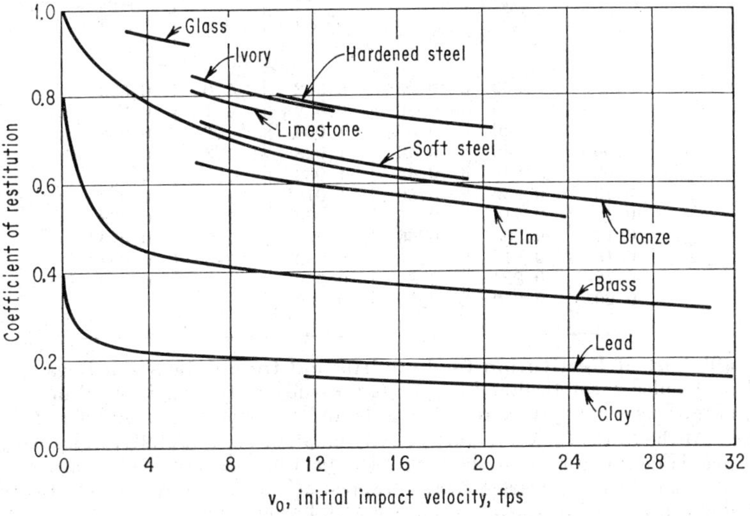

FIG. 16.3. Coefficient of restitution as a function of impact velocity for spheres of the same material.[3]

energy trapped within the body at the end of the impact can be accurately expressed in terms which do not require very precise estimates of the impact force-time characteristic during the impact. An approximation of this force-time history was made employing the Hertz methods of Art. 16.3, by assuming that the force history is dictated solely by strain conditions within the immediate region of contact. By normalizing this approximate interaction force in a form which is necessarily quite insensitive to the inaccuracy of the approximation, they calculated the energy remaining within a beam subject to such a short-duration force. Employing this technique, the coefficient of restitution for the impact of a relatively light sphere striking a relatively massive beam or plate was calculated to be

$$e = \frac{1 - B}{1 + B} \tag{16.9}$$

B is a function whose value is as follows:

1. For a sphere striking a simply supported beam:

$$B = (m/M)\, Q_1^{1/2}\, G(Q_1) \tag{16.10}$$

where $Q_1 = 1.087\ \eta_1 t_f$
$G(Q_1)$ = function defined by Table 16.1
m = mass of striking sphere
M = mass of beam
t_f = approximate duration of impact as calculated by Eq. (16.29)
η_1 = fundamental frequency of beam

The results of these calculations are plotted in Fig. 16.4*a* for a wide range of the parameter $(m/M)Q_1^{1/2}$.

2. Sphere striking a circular plate:

$$B = \frac{3m}{16\pi^2(1.087t_f)}\left[\frac{3(1-\nu^2)}{\rho E}\right]^{1/2} \tag{16.11}$$

Table 16.1. Table of Function $G(Q_1)$ for Calculation of Coefficient of Restitution for Impact of a Sphere on a Simple Hinged Beam[21]

Q_1	$G(Q_1)$	Q_1	$G(Q_1)$	Q_1	$G(Q_1)$
0	0.840	0.25	0.922	1.0	0.500
0.02	0.840	0.30	0.976	1.1	0.378
0.04	0.840	0.35	1.008	1.2	0.278
0.06	0.840	0.40	1.028	1.3	0.205
0.08	0.846	0.45	1.032	1.4	0.123
0.10	0.832	0.50	1.018	1.5	0.078
0.12	0.814	0.55	1.000	1.6	0.038
0.14	0.803	0.60	0.972	1.7	0.010
0.16	0.810	0.7	0.880	1.8	0.002
0.18	0.824	0.8	0.766	1.9	0.000
0.20	0.850	0.9	0.632		

Coefficient of Restitution for Longitudinal Impact between Bars. Sears[16] made a detailed theoretical and experimental study of the longitudinal impact steel bars of various lengths with rounded ends and reported the results shown in Fig. 16.4*b*. At the relatively low impact velocity employed in this study, losses associated with plastic deformation near the point of impact on the rounded end are relatively small compared with the strain energy trapped as traveling waves within the bodies. This factor accounts for the variation in *e* with bar length.

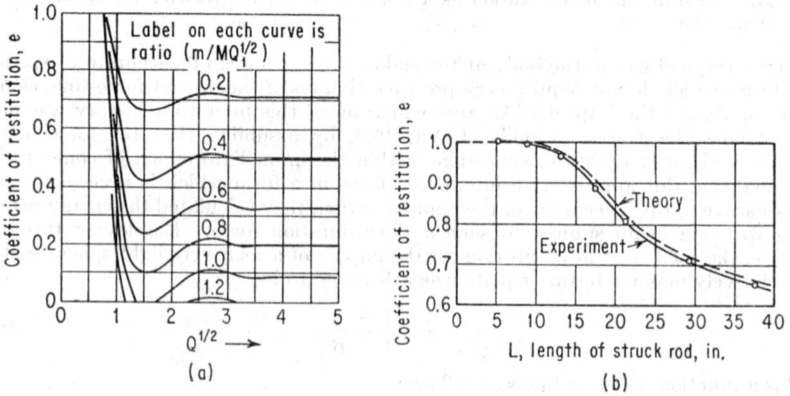

FIG. 16.4. Coefficient of restitution (*a*) of a sphere colliding with a beam pivoted at both ends,[21] (*b*) of originally stationary rod for the longitudinal impact of two ½-in.-diam alloy-steel bars (striker 5⅓ in. long, impact velocity 5 ips).[16]

16.3. SOLUTIONS OF EQUATIONS OF MOTION FOR IMPACT

Since the classical methods provide no information about the transient forces, stresses, and accelerations produced within the impacting bodies and tell nothing about the duration of the impact, it is necessary to solve the more basic equations of motion for such information. Because of the inherent complexity of these general solutions, it is necessary to exploit reasonable simplifications whenever the opportunity arises. Hence several different approaches are employed in this section, each with its pertinent field of application. The types of equations necessary to define the impacting systems adequately, and hence the solutions, depend largely upon the geometry of the bodies.

The first and simplest type of solution assumes that stresses are transmitted instantaneously to all points in the system. This quasi-static property is reasonably valid for one-degree-of-freedom systems in which the duration of impact is long compared with the time required for stress waves to propagate through the body.

Problems which can be solved in this way include collision of bodies in which the major contribution to the flexibility occurs in the vicinity of the point of impact and the major contribution to the inertia of the component is well behind the point of impact. This criterion is satisfied by compact bodies with curved surfaces near the region of impact and more complex geometries in which, by virtue of shape or material, the dominant source of flexibility is located in the immediate vicinity of the impact point. For such bodies it is possible to represent one region as having flexibility but negligible mass and a second region as having mass but infinite rigidity. These methods are described in Art. 16.3(a).

More precise theory, including the effects of both elasticity and inertia, gives rise to the concept of a finite velocity of propagation of stress waves. These considerations become necessary when the forces or stresses vary sufficiently rapidly during the time of propagation of these signals so that substantial deviation from quasi-static equilibrium occurs. Typical problems requiring this more rigorous approach include the longitudinal and transverse impact of long beams, and plates. Practical solutions employing this approach are restricted to simple geometries with uniform properties. These methods are covered in Art. 16.3(c).

Intermediate between bodies which are amenable to solution by these two methods are bodies which can be conveniently represented by multimass lumped systems, such as in Fig. 16.8. Many nonuniform geometries and nonlinear properties can be represented in this way. Although few general solutions are available, particular solutions are readily obtained on automatic computers. These methods are discussed in Art. 16.3(b).

(a) The Quasi-static Approach to Central Impact

The methods to be described here are most useful when impact is considered between bodies which satisfy the following criteria:

1. The period or duration of the impact is relatively long compared with the period of the fundamental natural frequency of the impacting bodies.
2. The stress-strain relationship is strain-rate-independent. To a reasonable approximation this can be assumed true for most common metals employed in machine work, *at stress levels below the elastic limit.*
3. Forces produced at the point of impact can be defined in terms of one-degree-of-freedom models.
4. The motion before and after the impact is adequately defined in terms of the translational motion of the center of gravity and rotational motion about the center of gravity of each body.
5. During impact the body deforms essentially as it would if deformed statically by the slow application of comparable forces between the centers of gravity of the two impacting bodies.

These restrictions allow the use of the simplest methods of impact, which are of considerable value in many machine-design applications.

Consider the collision of two masses with the following additional restrictions:

1. The line connecting the centers of gravity of the two bodies passes through the point of impact.

2. The center of the radius defining the curvature of the two surfaces must also be along this axis.

Unless these latter two conditions are met there will be both sliding at the surface of impact, requiring consideration of friction, and rotational effects.

Figure 16.5 shows a schematic diagram of a collision between two bodies satisfying these criteria. Positive motion is to the right, and external forces F_{e1} and F_{e2} drive the two bodies. The sign of velocities and forces is dictated by their direction. X_1 and X_2 define motions of centers of gravity of the two bodies. The equations of motion of each body during impact are

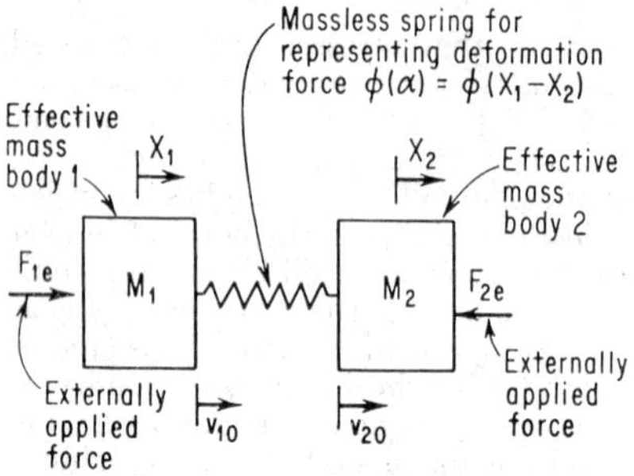

FIG. 16.5. Schematic representation of collision between two bodies satisfying quasi-static criteria. Positive direction is to right.

$$\ddot{X}_1 = F_{1e}/m_1 - \phi/m_1 \tag{16.12}$$

$$\ddot{X}_2 = F_{2e}/m_2 + \phi/m_2 \tag{16.13}$$

By subtracting $\ddot{X}_2$ from $\ddot{X}_1$ and defining $X_1 - X_2 = \alpha =$ deformation, the basic equation for this type of impact is obtained.

$$\frac{d^2\alpha}{dt^2} = \frac{d}{d\alpha}\frac{\dot{\alpha}^2}{2} = \frac{F_{1e}}{m_1} - \frac{F_{2e}}{m_2} - \frac{m_1 + m_2}{m_1 m_2}\phi(\alpha) \tag{16.14}$$

$\phi(\alpha)$ represents the reaction force produced at the surface of the impacting bodies. Even for perfectly elastic, constant-modulus materials $\phi(\alpha)$ will not in general be a linear function nor will it even be single-valued. The dissipation of kinetic energy normally attending the impact process necessitates that the force-deformation relationship $\phi(\alpha)$ exhibit a hysteresis effect. Equation (16.15) will be solved for general simple functions relating ϕ and α. The selection of the best relationship for any particular problem is a matter of engineering judgment.

The solution to Eq. (16.14) yields a relationship between α and $\dot{\alpha}$ at any instant t during the impact.

$$\dot{\alpha}_t^2 = \dot{\alpha}_0^2 - 2\frac{m_1 + m_2}{m_1 m_2}\int_0^{\alpha_t} \phi(\alpha)\, d\alpha - \left(\frac{F_{1e}}{m_1} - \frac{F_{2e}}{m_2}\right)\alpha \tag{16.15}$$

Solution of Eq. (16.15) with the terminal condition $\dot{\alpha}_t = 0$ yields Eq. (16.16) from which the value of α_m can be determined.

$$\frac{m_1 + m_2}{m_1 m_2}\int_0^{\alpha_m} \phi(\alpha)\, d\alpha - \left(\frac{F_{1e}}{m_1} - \frac{F_{2e}}{m_2}\right)\alpha_m = \dot{\alpha}_0^2 \tag{16.16}$$

Peak force is determined by substituting α_{max} in the known function $\phi(\alpha)$.

The order of magnitude of the impact duration can be estimated from Eq. (16.15) by the relationship

$$t_f = \int_0^{\alpha_m} d\alpha/\dot{\alpha} + \int_{\alpha_m}^{\alpha_f} d\alpha/\dot{\alpha} \tag{16.17}$$

Impact-force and Impact-period Formulas. The evaluation of peak force and impact duration depends finally upon the value of $\phi(\alpha)$, which will be a function of the properties of the materials involved, the geometry, and the stress levels produced.

The following compilation represents specific solutions to Eqs. (16.15), (16.16), and (16.17) for several functions $\phi(\alpha)$ which are of general interest. The force-deflection characteristics considered are illustrated in Fig. 16.6.

Linear Elastic Collision. If the geometry of a structure is such that a single-valued linear force-deflection relationship is valid, then $\phi = K\alpha$, $e = 1$, and solutions can be obtained either from Eq. (16.16) or by direct solution of Eq. (16.14). For compact bodies with high values of Young's modulus, impact force ϕ_m is normally quite large compared with external forces F_{1e} and F_{2e}. When forces F_{1e} and F_{2e} are negligible compared with ϕ_m, the solution to Eq. (16.14) yields an expression for peak force

$$\phi_m = (v_{10} - v_{20})\left(K \frac{m_1 m_2}{m_1 + m_2}\right)^{1/2} \tag{16.18}$$

Maximum deceleration of each mass is then

$$(\ddot{X}_1)_m = -\phi_m/m_1 \qquad (\ddot{X}_2)_m = +\phi_m/m_2 \tag{16.19}$$

and duration of impact

$$t_f = \pi \left[\frac{m_1 m_2}{K(m_1 + m_2)}\right]^{1/2} \tag{16.20}$$

For the special but important case where one of the masses is both infinitely large and stiff, Eqs. (16.18) and (16.20) reduce to

$$\phi_m = v_{10}(Km_1)^{1/2} \tag{16.18a}$$
$$t_f = \pi(m_1/K)^{1/2} \tag{16.20a}$$

When external forces F_{1e} and F_{2e} are significant with respect to ϕ_m,

$$\phi_m = K\{(F_{1e}/m_1 - F_{2e}/m_2)\eta^{-2} + [(F_{1e}/m_1 - F_{2e}/m_2)^2\eta^{-4} + \dot{\alpha}_0{}^2\eta^{-2}]^{1/2}\} \tag{16.21}$$

where

$$\eta = K\left(\frac{m_1 + m_2}{m_1 m_2}\right)^{1/2} \tag{16.21a}$$

Impact duration is determined by the relationship

$$\sin \eta t/(1 - \cos \eta t) = -(F_{1e}/m_1 - F_{2e}/m_2)(\dot{\alpha}_0\eta)^{-1} = \lambda \tag{16.22}$$

By entering Fig. 16.6, which is a plot of Eq. (16.22), with the argument λ, a value of ηt is obtained from which impact duration can be determined.

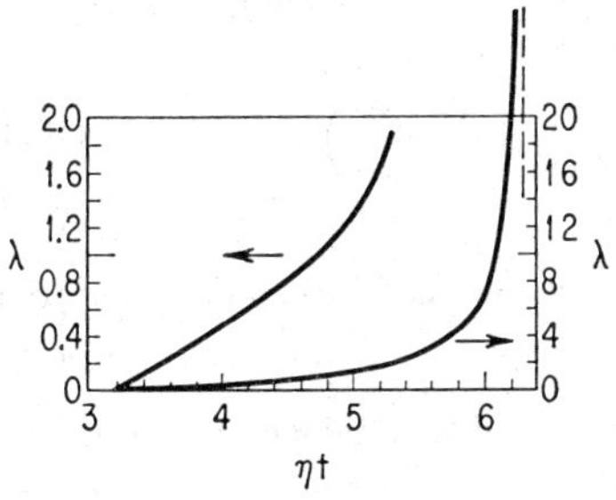

FIG. 16.6. Curve for the calculation of impact duration according to Eq. (16.22).

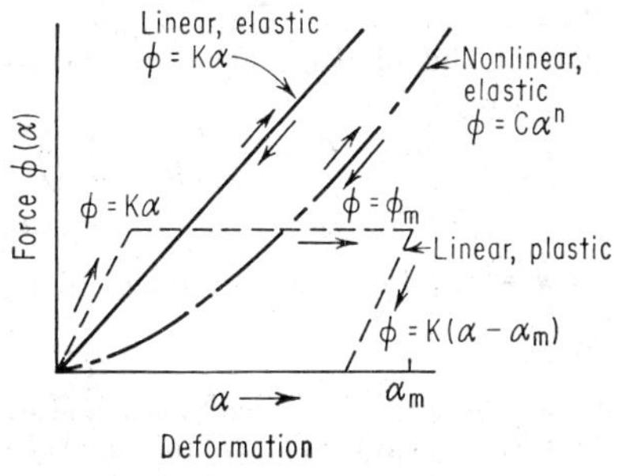

FIG. 16.7. Characteristic forces produced by impact deformation which are analyzed by the methods of Art. 16.3(a).

Linear Plastic Collision. For quasi-rigid bodies whose force-deflection is linear, plastic (e.g., the dashed curve, Fig. 16.7) deformation occurs until a prescribed value of peak force ϕ_m is reached. At force level ϕ_m plastic yielding begins. Further compression occurs plastically at constant force until relative motion between the bodies ceases. Relaxation occurs elastically, resulting in permanent deformation. For this case the coefficient of restitution is

$$e = -\frac{\phi_m}{\dot{\alpha}_0}\left(\frac{m_1 + m_2}{m_1 m_2}\frac{1}{K}\right)^{1/2} \tag{16.23}$$

When Eq. (16.23) gives values of $e > 1$, this indicates no plastic deformation and hence $e = 1$.

For the general case the impact duration is

$$t_f = \frac{\dot{\alpha}_0}{\phi_m} \frac{m_1 m_2}{m_1 + m_2} \left[1 - \left(\frac{\phi_m \eta}{K\dot{\alpha}_0}\right)^2\right] + \frac{\pi}{2\eta} + \frac{1}{\eta} \sin^{-1} \frac{\phi_m \eta}{K\dot{\alpha}_0} \tag{16.24}$$

where η is defined by Eq. (16.21*a*). In the limit, as the elastic intervals become negligibly short,

$$t_f = \frac{\dot{\alpha}_0}{\phi_m} \frac{m_1 m_2}{m_1 + m_2} \tag{16.25}$$

Nonlinear Elastic Collision. A general form of force-deflection relationship which is useful in representing many nonlinear systems is

$$\phi = C\alpha^n \tag{16.26}$$

(where C and n are constants). For the case where F_{1e} and F_{2e} are small compared with ϕ_m. Solution of Eqs. (16.12) and (16.13) yields

$$\text{Peak force } \phi_m = \left[\frac{\dot{\alpha}_0^2}{2}\left(\frac{m_1 m_2}{m_1 + m_2}\right)(n+1)C^{1/n}\right]^{n/(n+1)} \tag{16.27}$$

$$\text{Maximum compression } \alpha_m = \left[\frac{n+1}{C}\left(\frac{\dot{\alpha}_0^2}{2}\right)\frac{m_1 m_2}{m_1 + m_2}\right]^{1/(n+1)} \tag{16.28}$$

$$\text{Duration of impact } t_f = \frac{2\alpha_m}{\dot{\alpha}_0}\int_0^1 \frac{dz}{(1 - z^{n+1})^{1/2}} \tag{16.29}$$

Impact of Compact Bodies with Spherical Surfaces in the Vicinity of the Impact Region. In the case of impact of two compact bodies such as spheres, the dominant contribution to the flexibility of the system stems from the deformation in the immediate vicinity of the point of contact. Hertz[6] derived solutions for this case assuming perfect elasticity and ignoring plastic deformation. For the case of two spheres or comparable compact bodies with radii of curvature R_1, R_2 in the vicinity of the point of impact, solutions may be obtained from Eqs. (16.26) through (16.29), employing the following values for constants C and n:

$$C = \frac{4}{3\pi}\left(\frac{R_1 R_2}{R_1 + R_2}\right)^{1/2} \bigg/ \left(\frac{1 - \nu_1^2}{\pi E_1} + \frac{1 - \nu_2^2}{\pi E_2}\right) \tag{16.30}$$

$$n = 3/2$$

With $n = 3/2$, Eq. (16.29) reduces to the following expression for impact duration:

$$t_f = 2.943(\alpha_m/\dot{\alpha}_0) \tag{16.29a}$$

The case of a sphere impacting a massive plane is readily obtained: In the limit as m_1 and $R_1 \to \infty$, Eqs. (16.27) to (16.29) yield

$$\phi_m = (5\dot{\alpha}_0^2 m_2 C^{2/3}/4)^{3/5} \tag{16.27a}$$

$$\alpha_m = (5\dot{\alpha}_0^2 m_2/4C)^{2/5} \tag{16.28a}$$

where

$$C = \frac{4}{3\pi}(R_2)^{1/2}\left(\frac{1 - \nu_1^2}{\pi E_1} + \frac{1 - \nu_2^2}{\pi E_2}\right) \tag{16.30a}$$

The assumption of perfectly elastic behavior in such cases has been shown not to be completely valid.[5,7] Goldsmith and Lyman[7] have suggested that for more accurate work the assumption of perfect elasticity should not be used, but rather a quasi-statically measured force-indentation relationship should be employed. Correlations between force-indentation relationships obtained quasi-statically and under impact were found to be within 25 per cent.

Recognizing that the Hertz assumption of perfectly elastic behavior is not fully valid, completely plastic behavior has been assumed by some investigators of the impact of spherical surfaces.[5,8] Bowden and Tabor[5] analyzed the case for the impact of two masses with equal radii of curvature in the vicinity of the impact zone when the indentation is small compared with the curvature radius R.

Under such conditions the force is

$$\phi = \sigma_y R\alpha$$

$$\phi_m = \dot{\alpha}_0 \left(\sigma_y \pi R \frac{m_1 m_2}{m_1 + m_2}\right)^{1/2} \tag{16.31}$$

Impact duration is then

$$t_f = \frac{\pi}{2}\left(\frac{m_1 m_2}{m_1 + m_2}\frac{1}{\sigma_y \pi R}\right)^{1/2} \tag{16.32}$$

Neither assumption, perfect elasticity nor perfect plasticity, results in perfect correlation with test. Such discrepancies have been attributed by Bowden and Tabor[5] to strain-rate dependence in the material or they may possibly be due to the presence of shock-wave phenomena as studied by Bell.[10] (See Sec. 16.4*f*.)

Post-impact Behavior. How bodies behave subsequent to impact is a problem of frequent interest. As in Fig. 16.5, consider a pair of impacting bodies m_1 and m_2 under the influence of forces F_{1e} and F_{2e} which tend to force the two bodies together. For such a system, what is normally regarded as a single impact is in fact a series of impacts in rapid succession. This series of impacts is theoretically infinite in number but finite in total duration and involves the dissipation of enough kinetic energy to allow the two bodies to move together. From Eq. (16.3), the kinetic energy to be dissipated by the series of impacts is

$$\Delta T = \frac{1}{2}\frac{m_1 m_2}{m_1 + m_2}\dot{\alpha}_0^2 \tag{16.33}$$

The duration of the nth interval between impacts during which the bodies are separated is

$$t_n = 2e^n \dot{\alpha}_0 \left(\frac{F_{1e}}{m_1} - \frac{F_{2e}}{m_2}\right)^{-1} \qquad n = 1, 2, 3, \ldots \tag{16.34}$$

If the duration of impact is negligible compared to the bounce period, then the total time during which bouncing between the members will occur is

$$t_{\text{total}} = \sum_n t_n = \frac{2e}{1-e}\dot{\alpha}_0 \left(\frac{F_{1e}}{m_1} - \frac{F_{2e}}{m_2}\right)^{-1} \tag{16.35}$$

The amplitude of the nth bounce is

$$z_n = \frac{\dot{\alpha}_0^2}{2} e^{2n} \left(\frac{F_{1e}}{m_1} - \frac{F_{2e}}{m_2}\right)^{-1} = \frac{t_n^2}{8}\left(\frac{F_{1e}}{m_1} - \frac{F_{2e}}{m_2}\right) \tag{16.36}$$

The common velocity of the two bodies after the bouncing has ceased is given by Eq. (16.6) to a first approximation.

(b) Multi-degree-of-freedom Systems

Between the single-degree-of-freedom systems which were solved in Art. 16.3(a) and the continuum systems with an infinite number of degrees of freedom discussed in Art. 16.3(c) lies the important class of problems associated with a multiple but limited number of degrees of freedom. For many problems this approach offers a practical and reasonable approximation.

In representing complex mechanisms this approach enjoys the advantage of considerable versatility, particularly when computer facilities are available. However, this approach does not lend itself to great generalization because of the large number of variables involved, and specific solutions are normally developed. For linear cases, closed-form solutions are possible but may sometimes be too cumbersome for convenient application. The simplest multimass linear problem is of the type shown in Fig. 16.8, representing either the impact of a two-degree-of-freedom structure against an infinitely massive surface (Fig. 16.8*a*) or the impact of two one-degree-of-freedom

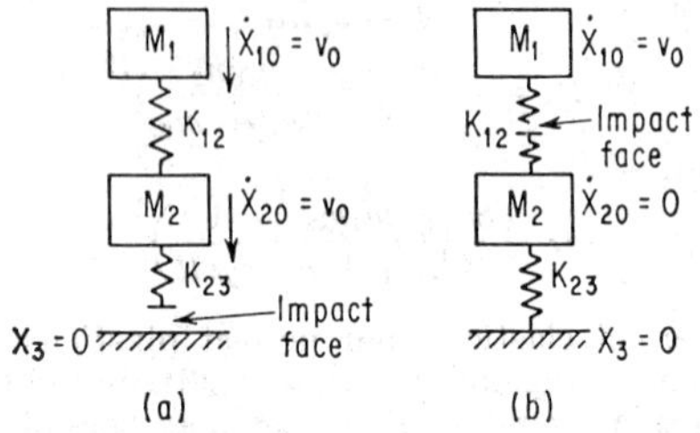

FIG. 16.8. Impact of two-degree-of-freedom systems.

systems, with one mass resiliently mounted to an infinitely massive plane as shown in Fig. 16.9*b*. The equations of motion for both systems are

$$\begin{aligned} m_1\ddot{X}_1 + K_{12}(X_1 - X_2) &= 0 \\ m_2\ddot{X}_2 - K_{12}(X_1 - X_2) + K_{23}X_2 &= 0 \end{aligned} \tag{16.37}$$

The two natural frequencies η_1, η_2 in terms of which the solutions may be expressed are obtained as the real positive roots from the following equations:

$$\eta_1^2 = \frac{H}{2J}\left[1 - \left(1 - \frac{4J}{H^2}\right)^{1/2}\right] \qquad \eta_2^2 = \frac{H}{2J}\left[1 + \left(1 - \frac{4J}{H^2}\right)^{1/2}\right] \tag{16.38}$$

where

$$H = \frac{m_1 + m_2}{K_{23}} + \frac{m_1}{K_{12}}$$

$$J = \frac{m_1 m_2}{K_{12}K_{23}}$$

The solutions to Eqs. (16.37) then depend upon the boundary conditions:

1. For the case described by Fig. 16.8*a* the initial conditions are

At $t = 0$:

$$\dot{X}_{10} = \dot{X}_{20} = v_0$$
$$X_1 = X_2 = 0$$

The solutions for displacement, velocity, and acceleration as functions of time are then

$$X_2 = \frac{F_{23}}{K_{23}} = \frac{v_0}{\eta_2^2 - \eta_1^2}\left(\frac{\beta^2 - \eta_1^2}{\eta_1}\sin \eta_1 t + \frac{\eta_2^2 - \beta^2}{\eta_2}\sin \eta_2 t\right) \tag{16.39a}$$

$$X_1 - X_2 = \frac{F_{12}}{K_{12}} = \frac{v_0}{\eta_2^2 - \eta_1^2}\frac{K_{23}}{M_2}\left(\frac{\sin \eta_1 t}{\eta_1} - \frac{\sin \eta_2 t}{\eta_2}\right) \tag{16.39b}$$

where

$$\beta^2 = K_{12}\frac{m_1 + m_2}{m_1 m_2} \tag{16.39c}$$

$$\dot{X}_2 = \frac{v_0}{\eta_1^2 - \eta_1^2}[(\beta^2 - \eta_1^2)\cos \eta_1 t + (\eta_2^2 - \beta^2)\cos \eta_2 t] \tag{16.39d}$$

$$\ddot{X}_2 = \frac{-v_0}{\eta_2^2 - \eta_1^2}[\eta_1(\beta^2 - \eta_1^2)\sin \eta_1 t + \eta_2(\eta_2^2 - \beta^2)\sin \eta_2 t] \tag{16.39e}$$

$$\dot{X}_1 = \dot{X}_2 + \frac{v_0}{\eta_2^2 - \eta_1^2}\frac{K_{23}}{m_2}[\cos \eta_1 t - \cos \eta_2 t] \tag{16.39f}$$

$$\ddot{X}_1 = \ddot{X}_2 - \frac{v_0}{\eta_2^2 - \eta_1^2}\frac{K_{23}}{m_2}[\eta_1 \sin \eta_1 t - \eta_2 \sin \eta_2 t] \tag{16.39g}$$

The duration of the impact, determined by equating Eq. (16.39a) to zero, is implicitly given by the following relationship:

$$\frac{\sin \eta_1 t_f}{\sin \eta_2 t_f} = \frac{-(\eta_2^2 - \beta^2)}{(\beta^2 - \eta_1^2)} \frac{\eta_1}{\eta_2} \tag{16.39h}$$

where graphical solution for t_f can be made. The terminal velocities $\dot{X}_{1f}$ and $\dot{X}_{2f}$ are found by substituting t_f in Eqs. (16.39c) and (16.39e). The coefficient of restitution is approximately

$$e = -\frac{m_1\dot{X}_{1f} + m_2\dot{X}_{2f}}{(m_1 + m_2)v_0} \tag{16.39i}$$

2. For the case described by Fig. 16.8b the initial conditions are

At $t = 0$:

$$X_{10} = X_{20} = 0$$
$$\dot{X}_{10} = v_0$$
$$\dot{X}_{20} = 0$$

and the solution to Eq. (16.35) is

$$\frac{F_{23}}{K_{23}} = X_2 = \frac{v_0\eta_1^2\eta_2^2(M_1/K_{23})}{\eta_1^2 - \eta_2^2}\left(\frac{\sin \eta_2 t}{\eta_2} - \frac{\sin \eta_1 t}{\eta_1}\right) \tag{16.39j}$$

$$\frac{F_{12}}{K_{12}} = X_1 - X_2 = \frac{-v_0}{\eta_1^2 - \eta_2^2}\left(\frac{\eta_2^2 - \gamma^2}{\eta_2}\sin \eta_2 t - \frac{\eta_1^2 - \gamma^2}{\eta_1}\sin \eta_1 t\right) \tag{16.39k}$$

where $\gamma^2 = \dfrac{2K_{12} + K_{23}}{M_2}$

$$\dot{X}_2 = \frac{v_0\eta_1^2\eta_2^2}{\eta_1^2 - \eta_2^2}\frac{M_1}{K_{23}}(\cos \eta_2 t - \cos \eta_1 t) \tag{16.39l}$$

$$\ddot{X}_2 = -\frac{v_0\eta_1^2\eta_2^2}{\eta_1^2 - \eta_2^2}\frac{M_1}{K_{23}}(\eta_2 \sin \eta_2 t - \eta_1 \sin \eta_1 t) \tag{16.39m}$$

$$\dot{X}_1 = \dot{X}_2 - \frac{v_0}{\eta_1^2 - \eta_2^2}[(\eta_2^2 - \gamma^2)\cos \eta_2 t - (\eta_1^2 - \gamma^2)\cos \eta_1 t] \tag{16.39n}$$

$$\ddot{X}_1 = \ddot{X}_2 + \frac{v_0}{\eta_1^2 - \eta_2^2}[\eta_2(\eta_2^2 - \gamma^2)\sin \eta_2 t - \eta_1(\eta_1^2 - \gamma^2)\sin \eta_1 t] \tag{16.39o}$$

The duration of impact, obtained by equating Eq. (16.39k) to zero, is implicitly defined by the equation

$$\frac{\sin \eta_1 t_f}{\sin \eta_2 t_f} = \frac{\eta_1}{\eta_2}\frac{\eta_2^2 - \gamma^2}{\eta_1^2 - \gamma^2} \tag{16.39p}$$

This solution is useful in evaluating the effect upon peak impact force F_{12} produced by the stiffness of the mounting K_{23} and for evaluating the peak force or shock transmitted to the foundation F_{23}. Further discussion of the problem of impact-induced shock is given in Sec. 6.

(c) Wave Phenomena in Impact

NOTE: A detailed discussion of longitudinal and transverse impact of taut cables is given in Sec. 31.

Consideration of the effects of stress-wave propagation during impact becomes necessary when the applied forces or stresses change significantly during the time required for the propagation of these signals throughout the system. Under such conditions substantial deviation from static equilibrium occurs. For example, at the beginning of impact remote portions of the body remain unstressed while large stresses develop near the point of impact and propagate at finite velocity.

As waves travel through a medium they alter both the stress magnitude and the velocity of the particles composing the medium. In most general terms plane waves are defined as *dilational* if they produce particle motion along the direction of propagation or *distortional* if they produce particle motion perpendicular to the direction of propagation. Different types of stress give rise to different waves with different velocities of propagation. For machine-design work it is convenient to limit consideration to the following types of waves.

1. Longitudinal waves which transmit compressive and tensile stresses are generated in such typical applications as pile driving, connecting rods in high-speed machines, pump rods in oil-well machinery, and elevator cables.

Longitudinal waves are a kind of plane dilational wave in that particle motion is along the direction of propagation. Although the velocity of propagation of longitudinal waves is somewhat dependent upon the frequency of the waves, the relatively low frequency waves of engineering interest propagate at the uniform velocity

$$c = (E/\rho)^{1/2}$$

2. Torsional waves transmitting shear stresses are encountered typically in (*a*) rapidly applied motions on helical springs and (*b*) long drill rods. Torsional waves are one example of distortional waves in that particle motion is normal to the direction of wave propagation. Plane torsional waves propagate at velocity

$$c = (G/\rho)^{1/2}$$

3. Flexural waves such as are encountered in the transverse impact of beams are among the most difficult to handle analytically, since they are multidimensional and because their velocity of propagation, even in elementary theory, is strongly dependent upon the frequency of the wave. Thus the ordinary case, in which the total wave is composed of many harmonics or frequency components, is strongly influenced by dispersive effects, as are discussed later. More detailed discussion of flexural waves is given by Kolsky.[11]

While the velocity of propagation of longitudinal waves is also dependent upon their frequency or wavelength, the effect is relatively small in the region of most practical importance. With flexural waves, however, the influence of wavelength is quite large in precisely the range of frequencies which is of interest. Figure 16.9 illustrates the manner in which the velocity of propagation of longitudinal and flexural waves is influenced by wavelength in magnesium rods. For small values of the parameter a/Λ, where a is the transverse dimension of the rod and Λ is the wavelength of a particular sinusoidal wave, it is seen that the velocity of propagation of longitudinal waves is essentially constant whereas long flexural waves are propagated at a velocity

$$c = 2\pi c_0 K/\Lambda$$

where K = radius of gyration of the beam cross section and $c_0 = (E/\rho)^{1/2}$

Wave Dispersion. Wave dispersion is one of the most important omissions in the approximate wave theory which is normally employed in engineering work. Because of wave dispersion a stress wave, as it propagates through a system, may become distorted in shape and attenuated in size, because of the following:

1. Internal friction within the propagating medium ordinarily tends more rapidly to damp out the higher-frequency components.

2. Large differences in propagation velocity are characteristic of different types of waves. Hence a combined stress composed of say shear and tension becomes distorted and spread out and may separate into completely separate signals.

3. The time-varying impact stress is composed of harmonic components of different amplitude and frequency; the velocity of propagation varies with the frequency and hence distorts the wave. This effect exists even for elastic materials with uniform properties and becomes of primary importance in materials exhibiting strain-rate sensitivity.

4. A dilational wave such as a longitudinal wave incident upon a free surface not only gives rise to a reflected dilational wave but also produces a distortional wave as well. Kolsky[11] also shows that, when an elastic wave reaches a slip-free

boundary, four waves are generated. Two of these waves are refracted into the second medium and two are reflected back. A more detailed consideration of such phenomena, which are prohibitively complex for most engineering applications, is described in ref. 11. Thus reflection from boundaries, free surfaces, or discontinuities gives rise to further distortion of the original wave.

In longitudinal impact dispersive effects are usually significant only in the vicinity of the point of impact, within the order of the lateral dimensions of the rod. Within this interval strong radial effects occur and plane-wave theory is not adequate. More modest radial effects occur even in the case of a uniform rod by virtue of Poisson's ratio, which causes lateral change and hence shear-stress waves as a consequence of a normal stress wave. For most engineering problems these effects are not predominant and the simple one-dimensional theory of plane waves is adequate. It is possible to regard the waves as essentially plane with substantial simplification of the theory provided that the duration of the pulse is such that $d/t_f c \ll 1$, where

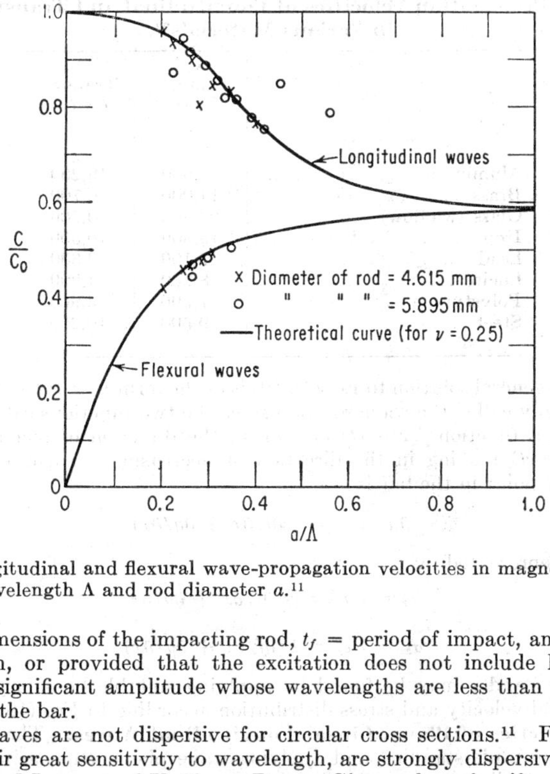

FIG. 16.9. Longitudinal and flexural wave-propagation velocities in magnesium rods as a function of wavelength Λ and rod diameter a.[11]

d = lateral dimensions of the impacting rod, t_f = period of impact, and c = velocity of propagation, or provided that the excitation does not include high-frequency harmonics of significant amplitude whose wavelengths are less than the transverse dimensions of the bar.

Torsional waves are not dispersive for circular cross sections.[11] Flexural waves, because of their great sensitivity to wavelength, are strongly dispersive.

Longitudinal Impact of Uniform Bars. Given a bar of uniform cross section composed of an isotropic material, with a known stress-strain relationship $\sigma(\epsilon)$, in which stresses are uniform in planes perpendicular to the boundaries. We ignore the secondary effects of Poisson's ratio which give rise to radial-shear and radial-inertia effects. As a result the equation formed will not involve any dispersive effects and waves will be propagated through the medium without change. It should be pointed out that real systems never undergo the perfectly abrupt discontinuities which are implied in the solution for the plane stress wave. The effects which arise from our assumption of perfect isotropy mean that these methods cannot be accurate when considering variations within the atomic dimensions or crystal-lattice dimensions of the medium. For the gross effects of engineering importance this simplification poses no serious difficulty. Let u equal the displacement of a point originally located at position x along the bar. Then the equation of motion is

$$\partial^2u/\partial t^2 = (1/\rho)(d\sigma/d\epsilon)(\partial^2u/\partial x^2) = c^2(\partial^2u/\partial x^2) \tag{16.40}$$

$$c = \text{velocity of propagation} = \sqrt{(d\sigma/d\epsilon)\rho^{-1}} \tag{16.40a}$$

In the case of linearly elastic waves $d\sigma/d\epsilon = E$, whence

$$c = (E/\rho)^{1/2} \tag{16.40b}$$

Any stress signal imposed at one point will be propagated at velocity c, along the bar. Table 16.2 summarizes propagation velocity for several common materials. For any material whose properties are not strongly strain-rate sensitive and for which the static stress-strain relationship is known, it is possible to determine with good accuracy the velocity of propagation.[9]

Table 16.2. Propagation Velocities of Longitudinal and Transverse Waves in Various Materials[22]

Material	Long., c_L, fps	Transv., c_T, fps
Aluminum	20,900	10,200
Brass	14,000	6,700
Glass (window)	22,300	10,700
Iron	19,500	10,500
Lead	7,100	2,300
Lucite	8,700	4,200
Polystyrene	7,500	3,900
Steel	19,500	10,200

St. Venant's general solution to Eq. (16.40) is of the form $u = f(x - ct) + g(x + ct)$. This is commonly called the wave solution since the two functions can be interpreted as two waves, a function $f(x - ct)$ moving in the direction of increasing x, and a function $g(x + ct)$ moving in the direction of decreasing x as in Fig. 16.8. The velocity of any point in the bar is

$$\partial u/\partial t = c(-\partial f/\partial x + \partial g/\partial x)$$

The strain is represented by

$$\epsilon_x = \partial u/\partial x = \partial f/\partial x + \partial g/\partial x \tag{16.41}$$

and the stress

$$\sigma_x = E\epsilon_x = E(\partial f/\partial x + \partial g/\partial x)$$

The functions f and g may be found by applying suitable boundary conditions to define the initial velocity and stress distribution according to Eq. (16.41).

Some Properties of Plane, Linear Longitudinal Waves. The following concepts, helpful in understanding and applying wave phenomena, arise directly from the solution of the linear wave equation.

1. *Particle velocity and wave velocity.* The velocity of wave propagation is a property of the medium [Eq. (16.40*b*)] and is distinct and different from the particle velocity. When a velocity change Δv is imposed on a linearly elastic bar, a stress wave of magnitude

$$\Delta\sigma = \Delta v\,(E\rho)^{1/2} = \Delta v\ \rho c \tag{16.42}$$

travels down the bar at velocity c and modifies the stress and velocity in the bar. If the cross-sectional area of the bar is A, the corresponding force is approximately $\Delta F = \Delta\sigma\, A$. The approximation arises from the fact that real plane waves are not fully uniform but are influenced by the boundaries of the body.[3,11]

2. *Effect of Stress Wave on Particle Motion.* (*a*) In a compressive stress wave the velocity change of the particles is in the same direction as the stress wave is propagated. (*b*) In a tension wave, conversely, the velocity change of the particles is in the direction opposite to the direction of wave propagation.

3. *Reflection of Waves.* *a. Reflection at a free end.* When an incident plane wave σ_I reaches a free end, a reflected wave σ_R is reflected back into the medium in order to satisfy the boundary condition that the total stress at a free boundary must be zero. Thus $\sigma_R + \sigma_I = 0$; hence a compression wave reaching a free end is reflected back as a tensile wave of equal and opposite magnitude. Corresponding to the reflected tensile wave, the particle velocity increases in the direction of the original incident compressive wave. An incident plane tension wave reaching a free end is covered by the converse of this description.

b. Reflection at a perfectly rigid fixed end. When an incident plane wave σ_I reaches a perfectly rigid, fixed end, the boundary condition requires that the total particle velocity must be zero. Thus if particle velocity v_I corresponds to σ_I then $v_I - v_R = 0$. Accordingly an incident wave is reflected without change in sign or magnitude, and as a result of superposition, the stress at the fixed boundary instantaneously doubles.

c. Reflection and refraction at a point of discontinuity. When an incident longitudinal plane wave σ_I passes through a discontinuity in the medium, either as a result of a change in material or to a first approximation as a result of a change in geometry, there will in general arise a reflected wave σ_R and a transmitted wave σ_T which must satisfy the following boundary conditions: (*a*) The force across the discontinuity must be continuous; thus

$$(\sigma_I + \sigma_R)A_1 - \sigma_T A_2$$

(*b*) The particle velocity must be continuous; thus

$$v_I + v_R = v_T$$

Combined with Eq. (16.42), these two conditions indicate that the magnitudes of these two waves are

$$\frac{\sigma_T}{\sigma_I} = \frac{2A_1\rho_2c_2}{A_1\rho_1c_1 + A_2\rho_2c_2} \qquad \frac{\sigma_R}{\sigma_I} = \frac{A_2\rho_2c_2 - A_1\rho_1c_1}{A_2\rho_2c_2 + A_1\rho_1c_1} \tag{16.43}$$

4. *Initial Velocity at an Impact Interface.* When two surfaces collide, the initial velocity at the common impact interface of both bodies is assumed to change instantaneously to a value $v = (v_{1i} - v_{2i})/2$. Subsequent reflections of the waves will, of course, modify this value in time.

5. *Energy in Wave.* The energy in a linear, longitudinal plane wave is composed of two equal parts: one half is the kinetic energy of the particles; the second half represents the strain energy of deformation. The total energy in a bar which is subjected for a time t to a uniform stress at one end is

$$T + U = (A/E)ct\sigma^2 \tag{16.44}$$

6. *Interference or Combining of Waves.* *a.* As long as the deformations within a body remain elastic, the principle of superposition can be used to determine the stress magnitude and particle velocity. Thus when several longitudinal waves of different sign and magnitude coincide in space and time at some point in the system, the net stress and velocity can be determined by simply adding the magnitudes of all the coincident waves with due regard for the sign of each wave, provided that the final stress does not exceed the elastic limit.

b. Waves moving in opposite directions can move through one another with no alteration provided that their combined stress magnitude does not exceed the elastic limit.

Figure 16.10 illustrates the superposition resulting from a decaying compression wave incident upon the free end of a bar, resulting in a reflected tension wave. The superposition of the two waves results in a net tensile stress within the body. This phenomenon accounts for scabbing-type failures which are encountered in violent impact of armor plate by projectiles.[4]

Wave Solution When Velocity History of One End of Bar Is Known. Solutions have been developed by DeJuhasz[12] and Marti[13] for the case where the velocity at one end of the bar is a known function of time. Such solutions are particularly useful in studying the surging phenomenon in helical springs. The known velocity of the moving end produces an initial stress wave which travels down the bar with a velocity

$$c = \sqrt{E/\rho}$$

and with a magnitude $\sigma = v\sqrt{E\rho}$. By suitably tracing the initial stress wave and its subsequent reflections from the boundaries, a solution can be developed for the total force-time history of the system. The original-impact-generated wave will travel down the bar until it reaches the boundary, where it will be reflected. The nature of the reflection dictates to a large degree the subsequent behavior of the bar.

In the case of a highly massive, rigid boundary, essentially perfect reflection occurs and the wave changes phase and moves in the reverse direction but is otherwise unmodified. At the point of reflection, an incident wave σ_I will add to its reflected image σ_R, giving rise to a stress $\sigma_I + \sigma_R$. In the case of the perfectly reflected wave

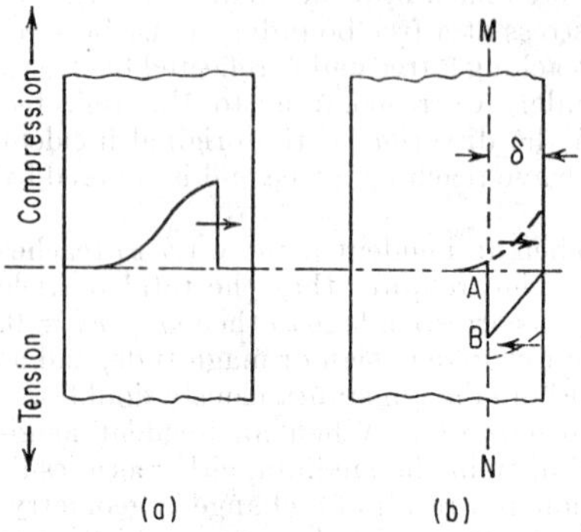

FIG. 16.10. Example of wave reflection and superposition near a free surface in a semi-infinite plate. (*a*) Before reflection. (*b*) Shortly after wave reflection.[4]

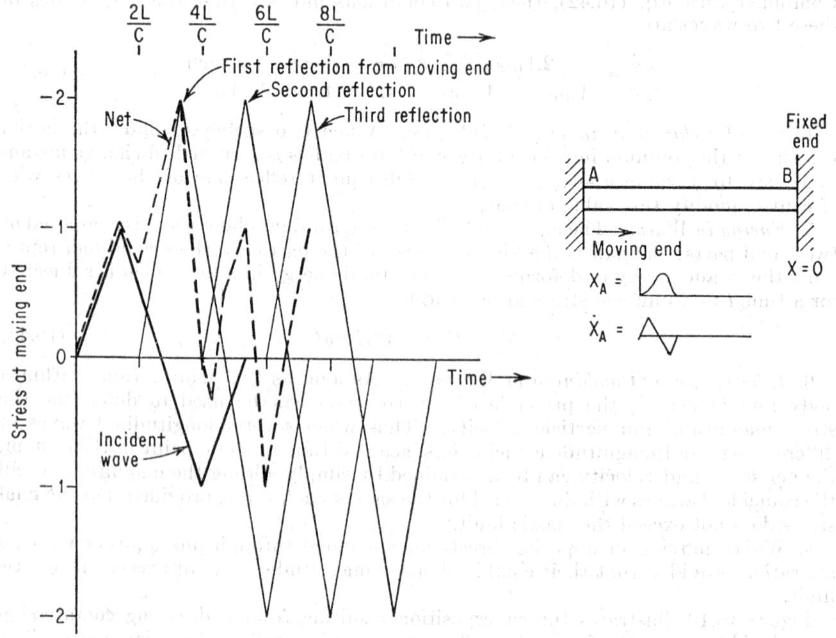

FIG. 16.11. Wave solution according to Eq. (16.45) for a fixed bar subjected to a known motion at its moving end.

$\sigma_I = \sigma_R$; thus perfect reflection involves doubling of the stress at the point of reflection. In every point in the system, the instantaneous stress will be equal to the sum of all stress waves, passing that point at that particular instant. If the wave travels at velocity c it will reach the fixed end at a time L/c and will return to its origin in a period $2L/c$, when it will be again reflected. As time progresses the number of traveling waves will correspondingly increase. At any time, the stress at the moving end of the bar will be

$$\sigma = (E\rho)^{1/2}(v + 2v_{t-2L/c} + 2v_{t-4L/c} + \cdots + 2v_{t-2nL/c}) \qquad (16.45)$$

The number of terms in the series at any instant is determined by the time the excitation has been applied. The number of waves n at any time is $n \leq ct/2L$ where n is a positive integer. At the fixed end the corresponding wave solution will be

$$\sigma = (E\rho)^{1/2}(2v_{t-L/c} + 2v_{t-3L/c} + \cdots + 2v_{t-(2n-1)L/c}) \qquad (16.46)$$

A graphical illustration of Eq. (16.45) is shown in Fig. 16.11. Many detailed solutions for both elastic and plastic behavior have been developed by DeJuhasz.[12]

Wave Solution When the Velocity History of Bar Is Not Known. The more general solution of impact involves situations where only the initial impact velocity is known, and the velocity history as well as the stress history are to be determined. An interesting problem of this type, shown in Fig. 16.12, involves the impact of a perfectly rigid mass M impacting one end of a long bar of mass ρAL and modulus E, which is rigidly clamped at its opposite end.[14,15]

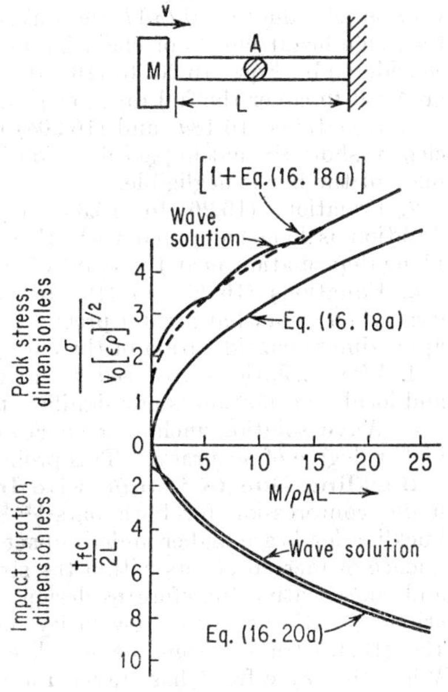

FIG. 16.12. Solution of impact of a finite mass and a fixed bar.[14]

For the first interval immediately following engagement ($0 \leq t \leq 2L/c$) a stress wave, whose magnitude decays as the mass decelerates, is generated at the moving end, travels down the bar, and is perfectly reflected from the fixed end and returns to the moving end. During the second interval ($2L/c \leq t \leq 4L/c$) and succeeding intervals the total stress at the moving end will be composed of multiple waves:

1. The wave instantaneously being generated by the velocity v of the mass and moving to the fixed end.
2. The wave generated at times $t - n \cdot 2L/c$ which has been reflected from the fixed end and now approaches the moving end.
3. The reflection of wave 2 from the moving end.

The sum of these waves gives rise to a decelerating force acting on the impacting mass which in turn modifies the generated wave.

The details of the calculation are treated in refs. 3 and 14. The significant results of the calculation are shown in Fig. 16.12, which gives values for the peak stress developed and the duration of the impact, as functions of the mass ratio $M/\rho AL$. For comparison there are also shown the results of the more approximate calculation based on Eqs. (16.18a) and (16.20a) which are approximately valid when the duration of the impact is long compared with the fundamental period ($=2L/c$) of longitudinal vibration of the bar, and which neglect the mass of the bar and consider only its stiffness. For this case $K = EA/L$ and Eq. (16.18a) gives

$$\sigma_m/v_i(E\rho)^{1/2} = (M/\rho AL)^{1/2}$$

From Eqs. (16.20a) and (16.40b) the duration of the impact is in dimensionless terms

$$ct_f/2L = (M/\rho AL)^{1/2}\pi/2$$

Figure 16.12 shows that contact duration can be calculated by either theory with little difference except at very small values of $M/\rho AL$. A more significant difference is noted for peak stress, showing the important effect of bar mass at small values of $M/\rho AL$ for which the duration of impact is not particularly long compared with the fundamental period of the bar. Such a condition violates a basic assumption involved in applying Eqs. (16.18a) and (16.20a), and a significant error arises. For very small values of $M/\rho AL$ the wave solution also is in error because it ignores the then significant effects of elasticity in the impacting mass and local deformation as considered by Eqs. (16.26) to (16.33). Thus several types of solution to this problem are used to cover the full range of possibilities:

1. Equations (16.18a) and (16.20a) may be used when $M/\rho AL \gg 1$ and L is sufficiently short so that impact duration is long compared with $2L/c$. For this case the mass of the bar is negligible.

2. Equations (16.26) to (16.33) may be used when $M/\rho AL \ll 1$ and the impact duration is long compared with $2L/c$. For this case the bar appears as a massive plane deformation near the point of contact dominates.

3. Equations (16.39i) to (16.39p), for a two-degree-of-freedom system, may be used when somewhat greater precision is desired than can be achieved by 1 or 2 since approximate consideration of the effects of bar mass can be included.

4. Wave solutions are used when impact duration is short compared with $2L/c$, and local deformation is not significant.

5. Wave solution including effects due to local deformation represent the next higher degree of accuracy. This problem was solved by Sears.[16]

Buckling Due to Compressive Impact Loading. Studies of impact loading under compression for both bars[23,24,25] and hollow cylinders[26] confirm that critical buckling loads are higher under impact than under quasi-static conditions, as a consequence of inertial effects within the struck rod. For design purposes, it is adequate and conservative therefore to design systems according to standard static-buckling criteria.[27] Under constant-velocity impact, a stress wave of magnitude defined by Eq. (16.42) travels from the struck end at wave velocity defined by Eq. (16.40a). When the wave front has traveled a critical distance along the bar from the struck end, instability or buckling will occur in this region, provided that the impact stresses and duration are sufficient. This phenomenon indicates that unlike static buckling a column can momentarily support a compressive stress of any magnitude imposed by the velocity of impact.[25] Thus the duration of the pulse is an important factor controlling buckling behavior.

(d) Transverse Impact of Beams

As in the previous section the method of solution of the problem of transverse impact is largely dictated by the ratio of the duration of the impact to the fundamental period of natural vibrations of the transverse beam. Where the impacting mass is sufficiently large compared with the effective mass of the beam and the stiffness of the beam is sufficiently low, the quasi-static techniques of Art. 16.3(a) may be applied to a reasonable approximation. At the other extreme the impacting mass is light compared with the beam and the impact duration is relatively short compared with the basic period of vibration of the beam. It is this more difficult problem which is considered here.

This section is limited then to the system shown in Fig. 16.13, involving the impact of a simply supported beam, struck at its center either by a sphere or by a similar compact, rigid mass. Unlike the problem of longitudinal impact, it is not possible

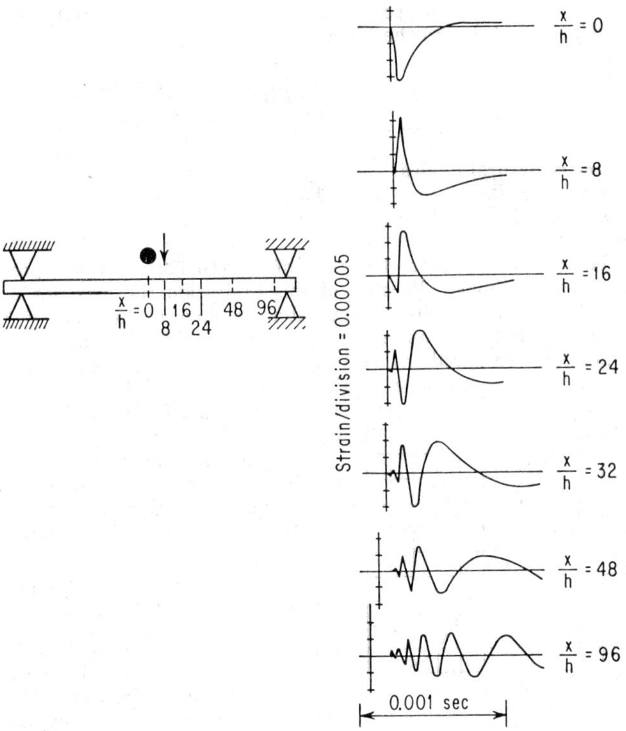

FIG. 16.13. Strain history in a simply supported beam of depth (h) impacted by a sphere. Strain measured on surface at varying distances (x) from point of impact. Note progressive distortion produced by dispersion.[32]

in transverse impact to ignore dispersive effects in considering the nature of stress-wave propagation. Some idea of the nature of dispersive effects in transverse impacts can be seen from the experimental data[32] of Fig. 16.14.

An approximate solution to the problem of transverse impact at the center of a simply supported beam was achieved by Timoshenko[17] in 1913 including the effects of local deformation at the point of contact as defined by the Hertz contact theory. This equation is based on the following assumptions:[18]

1. All the assumptions underlying the simple bending theory of beams.
2. All the assumptions underlying the theory of small-amplitude vibrations of beams.
3. The local strain distribution at the point of impact follows Hertz's law.[6] This implies that the fundamental period of vibration of the striking mass is short compared with the duration of contact between the striking mass and the beam.
4. The weight of the striking mass is negligible compared with the peak value of the impact force.

With these assumptions the following nonlinear integrodifferential equation defines the impact problem:

$$k[F(t)]^{2/3} = v_0 t - 1/m \int_0^t F(\tau)(t-\tau)\, d\tau - \sum_{n=1,3,5,\ldots} 1/M \int_0^t F(\tau) \sin \eta_n (t-\tau) d\tau \qquad (16.47a)$$

Only odd harmonics are present, consistent with the spatial symmetry of the problem. The central deflection δ due to the impact force is

$$\delta = \sum_{n=1,3,5,\ldots} \frac{1}{M} \int_0^t F(\tau) \frac{\sin \eta_n(t - \tau)\, d\tau}{\eta_n} \tag{16.47b}$$

where $\eta_n = n\eta_1$ = natural modes
m = mass of striker
M = effective mass of beam
k = constant

Because of the nonlinearity in the first term of Eq. (16.47*a*) solution is possible only by tedious numerical methods Because of this limitation the need for a more rapid calculation technique has resulted in several approximate solutions.[18,19,20]

An important assumption underlying all these methods is that the duration of the impact is short compared with the fundamental natural period of oscillation of the beam. All the assumptions of the Timoshenko solution are also pertinent. Based on refs. 18, 19, and 20, a method of calculation is as follows:

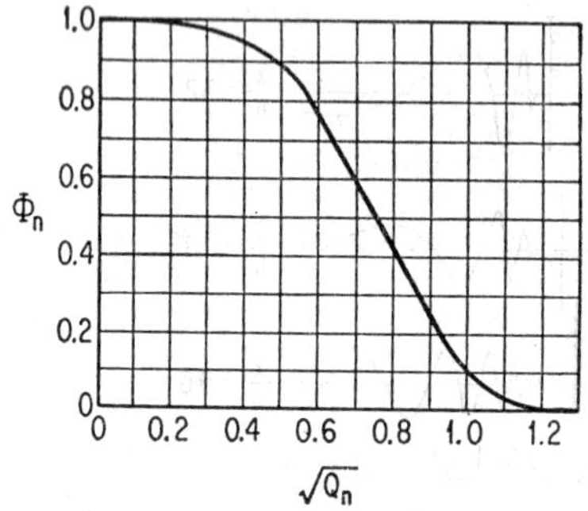

FIG. 16.14. Energy function for determining energy distribution in various modes of vibration in a beam subject to transverse impact.[20]

1. Calculate the natural frequencies of the beam. For a uniform-cross-section simply supported beam the natural frequencies of the symmetrically vibrating beam are

$$\eta_n^2 = n^2(EI\pi^4/L^4\rho A)^{1/2}$$
$$n = 1, 3, 5, \ldots$$

2. Calculate the approximate duration of the impact based on the local properties near the point of impact. If the impacting body and the beam are constituted of the same material, then the Hertz impact theory gives

$$t_f = 3.78 \left(\frac{m^2}{v_0 r} \frac{1 - \nu^2}{E^2}\right)^{1/5}$$

and let $t_l = 0.855t_f$. (This is an empirical relationship to improve accuracy.)

3. Calculate $Q_n = t_l\eta_n/2\pi$ and compute $\sqrt{Q_1}$, $\sqrt{Q_3}$, ... With these arguments, enter Fig. 16.14 and determine $\Phi_1, \Phi_3, \Phi_5, \ldots, \Phi_n$. Φ_n is proportional to the sum of potential and kinetic energies in the *n*th mode of vibration at the end of impact. It therefore follows that the fraction of the total energy involved in any particular mode of vibration is

$$\frac{T_k + U_k}{(T + U)_{\text{total}}} = \frac{\Phi_k}{\sum_{n=1,3,5,\ldots} \Phi_n}$$

This relationship then is of value in rapidly establishing whether higher modes will contribute significantly to the solution. With Φ_n evaluated, the coefficient of restitution e is determined by the following relationship:

$$e = \frac{1 - (m/M)\Sigma\Phi_n}{1 + (m/M)\Sigma\Phi_n} \tag{16.48}$$

This is similar to Eq. (16.9) but is somewhat more refined. Negative values of e imply the existence of subsequent subimpacts. The impact force experienced is then approximately

$$f(t) = \frac{\pi}{2} \frac{mv_0(1+e)}{t_l} \sin \frac{\pi t}{t_l} \tag{16.49}$$

The deflection δ of the beam can then be calculated by means of the two series solutions given below. Equation (16.50) is valid during the impact. Equation (16.51) is valid after the impact.

$$\delta_1 = \sum_{n=1,3,5,\ldots} \frac{\pi m v_0 (1+e) \sin (n\pi/2)}{2Mt_l} \left(\sin \frac{\pi t}{t_l} - \frac{\pi}{\eta_n t_l} \sin \eta_n t \right) \sin \frac{n\pi x}{L} \tag{16.50}$$

$$\delta_2 = \sum_{n=1,3,5,\ldots} \frac{\pi m v_0 (1+e) \sin (n\pi/2)}{2Mt_l} \left[\frac{\pi}{\eta_n t_l} \sin \eta_n t_l \cos \eta_n t - \frac{\pi}{\eta_n t_l} (\cos \eta_n t_l + 1) \sin \eta_n t \right] \sin \frac{n\pi x}{L} \tag{16.51}$$

By virtue of approximations made in the derivation of these equations, it will be found in general that the solution cited is more accurate for the interval after the impact than it is during the impact; i.e., Eq. (16.51) is more reliable than Eq. (16.50).

Strain. Strain in the beam will be affected in a most important way by the presence of higher modes of vibration in the beam. A strain relationship including effects of higher modes can be obtained by differentiating Eqs. (16.50) and (16.51) twice and employing the following relationship[33] for the strain ϵ_y in a beam at a distance y from the neutral axis:

$$\epsilon_y = y(\partial^2\delta/\partial x^2) \tag{16.52}$$

From Eqs. (16.50), (16.51), and (16.52) the strain relationship can be obtained:

$$\frac{\epsilon_1}{y} = \sum_{n=1,3,5,\ldots} -\left(\frac{n\pi}{L}\right)^2 \frac{\pi m v_0 (1+e) \sin (n\pi/2)}{2Mt_l} \left(\sin \frac{\pi t}{t_l} - \frac{\pi}{\eta_n t_l} \sin \eta_n t \right) \sin \frac{n\pi x}{L} \tag{16.53}$$

$$\frac{\epsilon_2}{y} = \sum_{n=1,3,5,\ldots} -\left(\frac{\eta\pi}{L}\right)^2 \frac{\pi m v_0 (1+e) \sin (\eta\pi/2)}{2Mt_l} \left[\frac{\pi}{\eta_n t_l} \sin \eta_n t_l \cos \eta_n t - \frac{\pi}{\eta_n t_l} (\cos \eta_n t_l + 1) \sin \eta_n t \right] \sin \frac{n\pi x}{L} \tag{16.54}$$

Because of the terms in n^2, Eqs. (16.53) and (16.54) will not converge so rapidly as Eqs. (16.50) and (16.52), demonstrating the importance of higher modes upon the strain. The higher modes are particularly sensitive to internal friction within the beam, and consequently the calculated strains may be somewhat excessive since friction has been neglected. Hoppmann's analysis[20] includes such effects.

Effect of Multiple Impacts on Maximum Possible Deflection.[19] The sign of the value of e calculated from Eq. (16.48) has an important physical significance. When e is positive and large the striker rebounds at a high speed, and it may be impossible for the beam to catch up with it. In this case the impact process consists of a single collision. Even if the beam should catch up with the rebounding striker, the second impact will not have any magnifying effect on the maximum deflection. Thus for positive e, maximum deflection occurs during or shortly after the first impact and before the second impact, should one occur.

For *negative* values of e the striker continues to move in its original direction after the first impact and will inevitably result in a second impact a short time later. The deflection resulting from this second impact can be no larger than

$$\delta_{n+1} \leq \delta_n(1 - e_1)$$

As an absolute upper bound, assuming an infinite number of impacts, each occurring at the most advantageous time, and assuming the same value of e calculated for the first impact, the limit in the maximum possible deflecting arising from the impact is

$$\delta_{\max} \leq \delta_1/(1 + e)$$

and as a realistic limit the following has been suggested:

$$\delta_{\max} \leq \delta_1(1 - e)$$

16.4. DYNAMIC BEHAVIOR OF MATERIALS

Several properties of engineering materials are modified as a consequence of impact-type rapid loading. For nonmetals such as plastics, rubbers, and other elastomers profound changes can occur as a consequence of rapid loading and these modified properties must be considered in even approximate calculation of dynamic behavior. For metals stressed below their elastic limit these effects are not normally so extreme, but under certain conditions significant deviations from quasi-static behavior can occur. Because of the incomplete state of our knowledge of these properties and the complexity of the applications, it is not generally possible to allow for these effects quantitatively. In general, stress limits for common metals under impact are approximately equivalent to quasi-static stress limits. However, in some cases this assumption may not be conservative. A qualitative understanding of these effects can be valuable as a guide to design and as an aid in the understanding and interpretation of some impact phenomena.

(a) Delayed Yield Phenomenon

Metals which exhibit well-defined static yield points also exhibit the characteristic of delayed yield when subjected to rapidly applied short-duration stresses. It is possible to subject low-carbon steels and iron (which exhibit well-defined yield points) to tensile stresses in excess of their static elastic limit without permanent deformation, provided that the time during which the stress exceeds the elastic limit is less than some well-defined limit. Tests on a 0.17 per cent carbon steel produced the relationships shown in Fig. 16.15 between the maximum stress and the cumulative time such a stress could be maintained without permanent deformation.[22] These times are so short at normal temperatures as to preclude their exploitation in repetitive operation, unless the cumulative effects are erased by heat-treatment. Temperature sensitivity of this phenomenon is also seen to be significant.

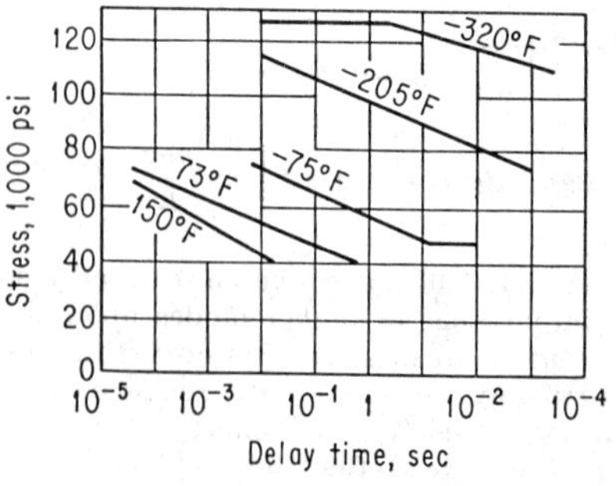

FIG. 16.15. Delay time for the initiation of yielding at rapidly applied stresses on 0.17 per cent carbon steel at various temperatures.[22]

Materials such as stainless steels, aluminum, and copper which do not exhibit well-defined yield points apparently do not exhibit the delayed yield phenomenon. Clark[22] reports that tests on such metals, including 18-8 stainless steel, normalized SAE 4130 steel, quenched and tempered 4130 steel, 24-ST aluminum alloy, and 75-ST aluminum alloy, all deformed plastically during the loading period with no evidence of delayed yield.

(b) Critical Impact Velocity[22]

A characteristic of importance in some applications is the existence of a "critical" impact velocity of metal. Wires or rods subjected to impact develop initial strains which are related to the difference between the velocity of propagation of strain in the metal and the velocity to which a point on the rod is impulsively subjected, as indicated by Eq. (16.42). Because $d\sigma/d\epsilon$ becomes smaller as σ is increased beyond the elastic limit, Eq. (16.40*a*) shows that high stresses propagate at lower velocities than low stresses. Thus a maximum velocity exists above which the large plastic strains, which are being generated, cannot propagate so rapidly as the end of the rod is being pulled, and fracture of the rod occurs almost immediately.

There exists then a critical velocity at which a tensile specimen will always fail under impact, independent of cross section of the specimen or the mass of the impacting mass, provided that the kinetic energy is at least sufficient to supply the energy absorbed by the necking-down process associated with tensile failures. This critical velocity is defined by Eq. (16.55).

$$v_{cr} = \int_0^{\epsilon_{ult}} [(1/\rho)(d\sigma/d\epsilon)]^{1/2}\, d\epsilon \qquad (16.55)$$

Table 16.3 shows results of impact experiments on critical velocity.

Table 16.3. Critical Impact Velocity and Tensile Properties of Common Materials[22]

Material	Ultimate strength, psi		Elongation in 8 in.		Critical velocity, fps	
	Static	Dynamic	Static	Dynamic*	Experimental	Theoretical
Ingot iron, annealed	37,100	57,400	25.7	16.2	100	†
SAE 1015, annealed	50,600	63,500	28.0	30.0	100	†
SAE 1022, cold-rolled	84,000	105,000	6.0	15.0	100	95
SAE 1040, annealed	78,050	91,800	20.4	20.7	200	†
SAE 1045, quenched and tempered	142,900	169,000	5.7	9.2	190	88
SAE 2345, quenched and tempered	145,250	175,250	8.4	14.1	200	161
SAE 4140, quenched and tempered	134,250	151,000	8.5	14.7	175	132
SAE 5150, quenched and tempered	139,000	148,100	8.5	13.3	170	159
Type 302 stainless steel	93,300	110,800	58.5	46.6	200	490
Copper, annealed	29,900	36,700	32.7	43.8	200	231
Copper, cold-rolled	45,000	60,000	2.5	10.7	50	42
2S aluminum, annealed	11,600	15,400	23.0	30.0	200	176
2S aluminum, ½ hard	17,200	22,100	4.6	7.0	110	36
24S-T aluminum alloy	65,150	68,600	11.3	13.5	200	290
Magnesium alloy (Dow J)	43,750	51,360	9.6	10.9	200	303

* Maximum percentage elongation up to the critical velocity.

† Existence of a yield point prevents computation of critical velocity. Values are given in feet per second.

(c) Transition from Ductile to Brittle Behavior

Ductility is normally regarded as a property of materials subject to stresses beyond their elastic limit. As such it might ordinarily be expected not to play too important a role in the behavior of machine components which are intended for long, repetitive life and are consequently designed to operate at stress levels well within their elastic limit. It might ordinarily be concluded then that the elastic limit or yield strength should be the all-important design criteria.

While, on a gross basis in ordinary machine design, yield strength is the important criterion, high local stress conditions in materials frequently arise which make ductility an important criterion for repetitive operations. Ductile behavior allows flow of the metal, providing a redistribution of the stress concentrations, and therefore helps to avoid failure due to local stress conditions. For impact applications the question of ductility takes on additional significance since, under impact, bodies are forced to absorb a definite amount of strain energy, for example, as is dictated by Eq. (16.7).

The significant distinctions between brittle and ductile failures are the amount of flow in response to shear stresses and the energy absorbed by the piece in order to produce fracture. Fractures are characterized as being brittle if failures involve relatively little flow of the metal and relatively little energy absorption or as ductile if considerable flow of metal and relatively greater energy are required to failure. Brittle behavior becomes more likely as temperature is reduced, the rate of loading is increased, and transverse or multiaxial stresses are increased, by shape and previous history.

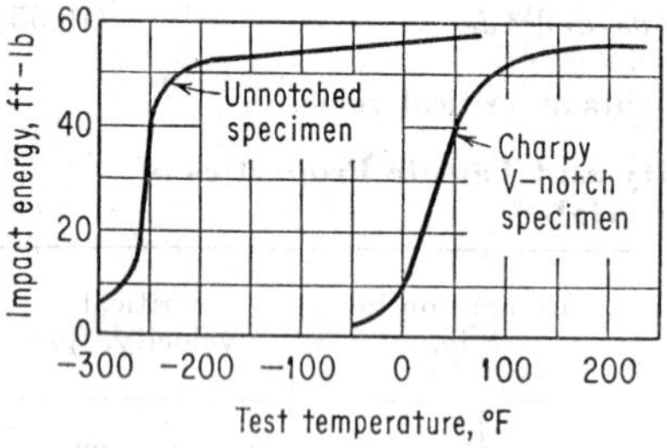

FIG. 16.16. Transition between ductile and brittle fracture in 0.1 per cent carbon-steel plate illustrating notch sensitivity.[39]

The presence of sharp corners, notches, or structural discontinuities produces complex multiaxial stresses, composed of tensile, compression, and shear components. The net result of these combined stresses is to increase the tensile stress necessary to produce failure.[27] Under such conditions, normally ductile materials can be made to fail in a brittle manner with a substantial reduction in the energy absorbed during failure.

For rapid loading the shape of a piece becomes important for still another reason since stress reflections occur at discontinuities which can give rise to instantaneous doubling of local stresses. The location and character of stress-wave reflections will determine location of peak stresses and determine preferred failure locations.

The temperature at which a particular specimen first demonstrates a change from ductile to brittle failure at a particular rate of loading is termed the *transition temperature.* The transition temperature is not a fundamental property of the material but at least for irons and ferritic steels is particularly sensitive to notch effects, as is evidenced by Fig. 16.16, and is quite sensitive to rate of strain. The existence of a high transition temperature in a metal intended for use at low temperatures should indicate the need for careful evaluation of the particular shape under the particular environmental conditions.

Increasing rate of strain also tends to produce a change from ductile to brittle fracture. Although ductile materials characteristically flow when subjected to shear stresses, with rapidly applied loads of short duration the shear stresses are not able to produce significant flow in the available time. Such materials then may fail in a brittle manner and at higher ultimate stresses, depending upon temperature and rate-of-strain conditions.

The following interesting relationship between temperature and rate of strain, as they influence the transition from ductile to brittle failure, has been proposed.[27]

$$\dot{\epsilon}_c/\dot{\epsilon}_0 = \exp\left[-(Q/R)(\theta_c^{-1} - \theta_0^{-1})\right]$$

where $\dot{\epsilon}_c$ and $\dot{\epsilon}_0$ are the minimum strain rates producing brittle fracture at temperatures θ_c and θ_0, respectively, and Q/R is an empirical constant related to the activation energy of the material.

The ultimate tensile strength σ_u in tension is also influenced by rate of strain and temperature. A parameter ζ derived from fundamental reasoning which appears successfully to relate these factors with ultimate strength is[27]

$$\sigma_u = \zeta^\gamma = \left(\frac{\dot{\epsilon} \exp Q/R\theta}{f_0}\right)^\gamma \tag{16.56}$$

where $f_0 = JQ/Nh$
Q = heat of activation, cal/g-atom (an empirical factor)
R = constant = 1.96
γ = empirical constant
N = Avogadro's number
J = mechanical equivalent of heat energy
h = Planck's constant
θ = temperature

Figure 16.17 shows a plot for the ultimate strength as a function of ζ for copper at temperatures between room temperature and 400°C. At higher temperatures correlation is not so successful. Comparable data for a typical steel are shown in Fig. 16.18.

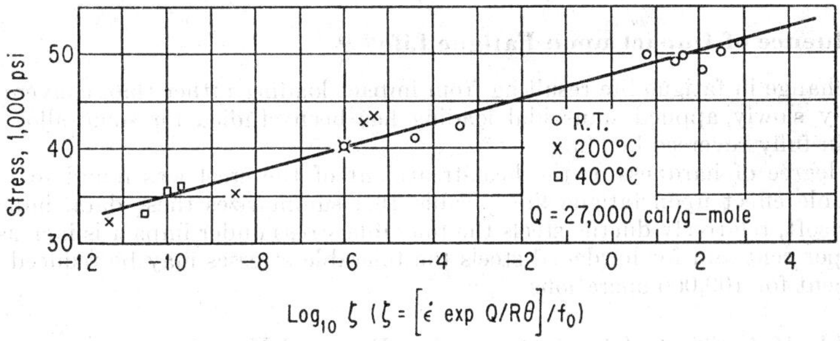

FIG. 16.17. Ultimate tensile strength of copper as a function of a parameter (ζ) including strain rate and temperature.[27]

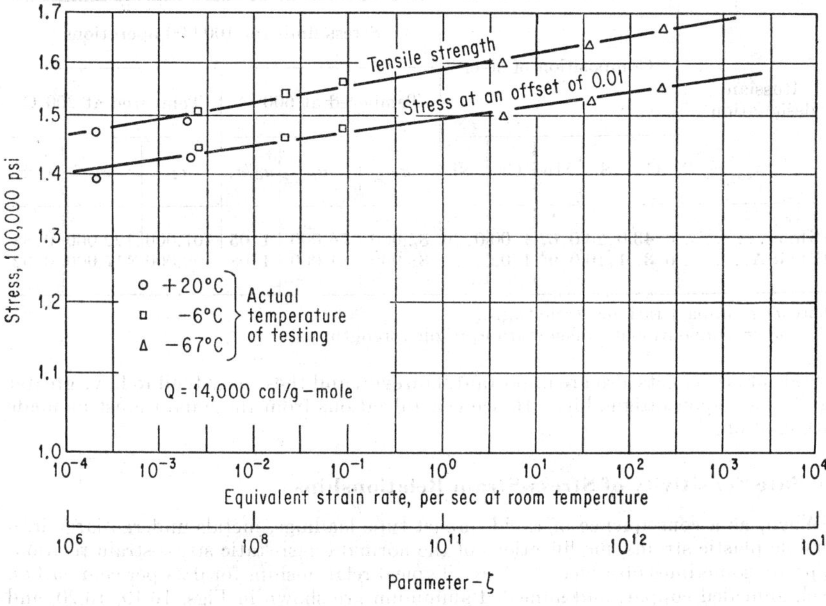

FIG. 16.18. Strain rate and temperature dependence of a typical steel. Ultimate tensile strength and plastic stress at a strain of 0.01 are plotted as functions of parameter ζ.[37]

These results show that only a limited amount of experimental data is necessary to obtain extensive information on the variation of ultimate strength over a wide range of strain rate and temperature.

(d) Effect of Residual Stresses

The importance of residual stresses upon impact behavior is not too well known. Little work has been done in this area and the results are not conclusive. Under normal static-loading conditions, time permits the leveling of high local residual stresses through plastic flow, provided severe restraints such as sharp notches are not acting. Under the conditions of impact loading, superposition of the impact stresses and residual stresses may result because time may be too short to allow plastic flow to attenuate and redistribute the peak residual stresses. Although validity of this theory is not yet proved, Miklowitz[36] has suggested that residual stresses may become important when high rates of loading are combined with low temperatures and notching. For such conditions the use of some method of stress relief is desirable.

(e) Influence of Impact upon Fatigue Life[28,29]

The change in fatigue life resulting from impact loading rather than conventional, relatively slowly applied sinusoidal loading has been studied for some alloy steels loaded in fully reversed bending.

The degree of hardness or the heat-treatment of the steel was found to have a measurable effect upon fatigue life. Table 16.4 summarizes these data, indicating that for soft, relatively ductile steels the tolerable stress under impact is increased at least 5 per cent and for hardened steels the tolerable stresses may be reduced up to 12 per cent for 100,000 operations.

Table 16.4. Effect of Impact-type Loading and Heat-treatment upon Fatigue Life of 10-mm-diameter Bars Loaded in Reverse Bending.[28,29] Dynamic Stresses Measured by Resistance-type Strain Gages

Russian designation	Composition of steel					Stress limit for 100,000 operations					
						Tempered at 600°C			Tempered at 200°C		
	C	S	Mn	Cr	Ni	σ_d	σ_s	σ_d/σ_s	σ_d	σ_s	σ_d/σ_s
45Kh.	0.43	0.26	0.67	1.00	0.29	82,000	76,000	1.05	107,000	122,000	0.88
30KhGSA.	0.3	1.10	0.95	1.02	. . . .	88,000	81,000	1.08	109,000	117,000	0.92

where σ_d = impact fatigue strength, psi
σ_s = conventional (quasi-static) fatigue strength, psi

Scale or size effects also are important, however, and these may tend to have greater significance upon fatigue life. Hence generalizations from these data must be made with caution.

(f) Rate Sensitivity of Stress-Strain Relationships

When, as a consequence of rapid impact-type loadings, metals undergo large irreversible plastic strains, modifications of the normal quasi-static stress-strain relationships are sometimes observed.[10,33,34,35] Typical relationships for 0.24 per cent carbon steel, annealed copper, and annealed aluminum are shown in Figs. 16.19, 16.20, and 16.21.

Bell[10] made a careful study of this behavior in annealed aluminum and copper. He showed that compression impact of long bars at velocities exceeding the critical impact velocity would produce this anomalous behavior, but only within a region restricted to the first one or two diameters near the impact surface. This behavior is

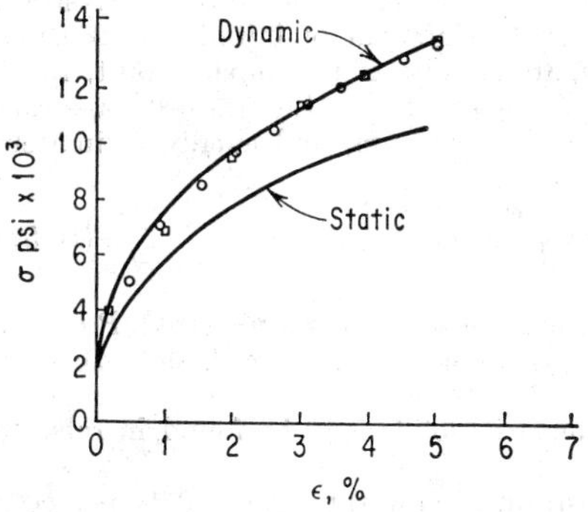

FIG. 16.19. A comparison of Bell's theoretical dynamic stress-strain curve in annealed aluminum, with the experimental (solid line) curve of ref. 35. The lower solid line is the static stress-strain curve.[10]

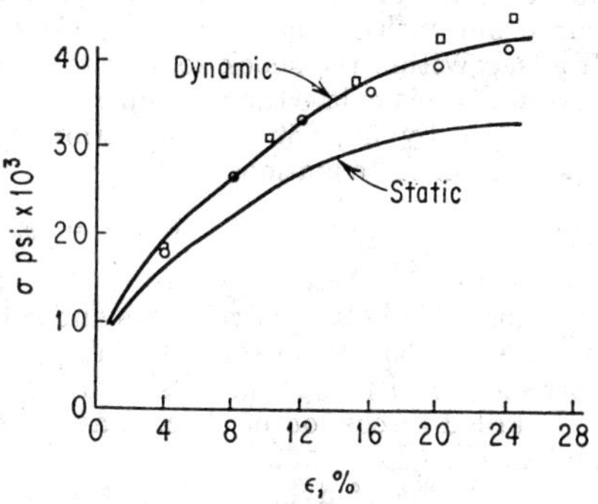

FIG. 16.20. A comparison of Bell's theoretical dynamic stress-strain curve in copper, with experimental data (solid line). The lower solid line is the static stress-strain curve.[10]

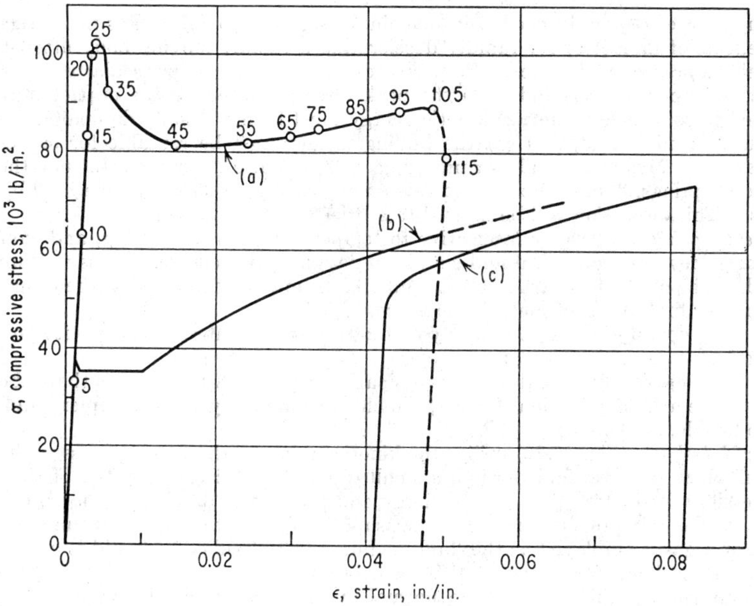

FIG. 16.21. Static and dynamic compressive stress-strain curves for a 0.24 per cent carbon steel, annealed at 1652°F. (a) Impact loading. (b) Static test of annealed specimen. (c) Static test of dynamically shocked specimen.

associated with the creation of a nondispersive shock front at the impact interface which ultimately develops into dispersive plastic waves as it propagates along the bar. In the region encompassed by the shock wave, modified stress-strain relationships are observed. Beyond the immediate vicinity of the impact zone (distances greater than

1½ to 2 diameters), the normal strain-rate-independent stress-strain relationships apply and can be used to determine the velocity of propagation with good accuracy, employing Eq. (16.38*a*).

The dynamic stress-strain curves for annealed aluminum and copper do not appear to be a measure of material properties in the same sense as the static-stress-strain curve but rather reflect the mechanics of the development of initial plastic wave fronts. Within the shock zone in the immediate vicinity of the impact point, Bell[10] was able to calculate the dynamically modified properties from the static-stress-strain properties by imposing conservation of mass, momentum, and energy ultimately resulting in the equations

$$P\epsilon_0/2 = \rho v_0^2/2 = \int_0^{\epsilon_0} \sigma_x \, d\epsilon_x \tag{16.57}$$

By equating the latter two terms and solving first for ϵ_0, the dynamic stress P can be obtained. Calculations employing this simple procedure produced the excellent correlations with experiment shown in Figs. 16.19 and 16.20.

Detailed discussion of shock wave phenomena in solids may be found in refs. 42 and 43.

Comparison of dynamic and static-stress-strain relationships for a 0.24 per cent carbon steel is shown in Fig. 16.21. This material exhibits both the delayed yield phenomenon and considerable difference between static and dynamic stresses in the plastic region. The elastic modulus appears to be independent of strain rate.

(g) Some Properties of Fluid and Elastomeric Buffers

Buffers are frequently needed in machine design to provide means for mitigating the effects of impact or for rapidly decelerating a rapidly moving body after it has traveled a prescribed distance. Properties of buffers which are particularly important for such applications include: (1) the peak force produced by the impact must be limited to some safe value, (2) kinetic energy must be irreversibly dissipated so that rebound is minimized, and (3) the buffing means must be reusable for successive operations. Means available for providing these important functions include the use of metal springs, fluid dashpots, and elastomers, each with its advantages and limitations. This discussion will be limited to a brief review of these means.

Metal springs are largely insensitive to temperature, rate of loading, and environment. They are economical and simple to design, and their behavior is well understood. The methods of Arts. 16.3(a) and 16.3(b) may be employed in their analysis. The major limitations of metal springs are:

1. They do not enjoy a particularly high ratio of energy-absorbing capacity per unit volume.
2. They have only a small energy-dissipating capacity, which means that rebound is a problem unless friction devices, which are inherently more complex, and less reproducible, are employed.

Fluid dashpots (Fig. 16.22) utilize the kinetic energy of the moving mass to drive a piston which forces a fluid through a small restricting orifice. The flow of the fluid irreversibly dissipates much of the energy and produces a decelerating force proportional to the fluid pressure produced. Dashpots may be either of the pneumatic type if air or a gas is employed as the fluid medium, or hydraulic if a liquid is employed.

Pneumatic dashpots are most effective when they can be operated in environments at pressures considerably higher than normal atmospheric pressure. When operating in a normal 14.7 psi atmosphere, their energy-absorbing capacity is relatively low per unit volume of dashpot, and they require high compression ratios to achieve even modest peak pressures. Pneumatic dashpots are rather critical in manufacture and in operation, require small restricting orifices, and are highly sensitive to leakage and stroke variations.

Because of compressibility effects, pneumatic dashpots pose mathematical difficulties, although the underlying theory is quite well understood. The differential equations of motion of a relatively rigid mass decelerated by a pneumatic dashpot

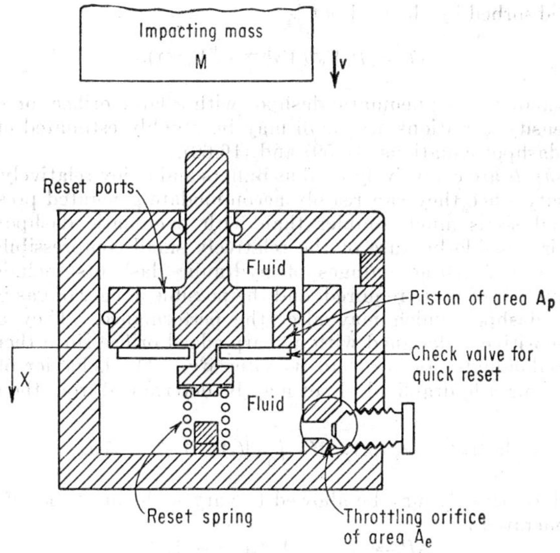

FIG. 16.22. Schematic drawing of fluid dashpot-type buffer.

containing an ideal gas are

$$\left.\begin{aligned}
&M\frac{d^2X}{dt^2} + (P - P_{\text{atm}})A_P = F_e\\
&V = V_0 - A_pX\\
&[A_e(RT_i 2g)^{1/2}]^{-1}\left(\frac{V}{P\gamma}\frac{dP}{dt} + \frac{dV}{dt}\right) = \left(\frac{\gamma}{\gamma - 1}\right)^{1/2} f\left(\frac{P}{P_{\text{atm}}}\right)\\
&f\left(\frac{P}{P_{\text{atm}}}\right) = \left\{\left(\frac{P_c}{P_{\text{atm}}}\right)^{(\gamma+1)/\gamma}\left(\frac{P_{\text{atm}}}{P}\right)^{2/\gamma}\left[\left(\frac{P}{P_c}\right)^{(\gamma-1)/\gamma} - 1\right]\right\}^{1/2}\\
&\text{where } P_c = \left(\frac{2}{\gamma+1}\right)^{\gamma/\gamma-1} P \quad \text{when} \left(\frac{2}{\gamma+1}\right)^{\gamma/\gamma-1} P > P_{\text{atm}}\\
&\text{or} \quad P_c = P_{\text{atm}} \quad \text{when} \left(\frac{2}{\gamma+1}\right)^{\gamma/\gamma-1} P < P_{\text{atm}}
\end{aligned}\right\} \tag{16.58}$$

and A_p = piston cross-sectional area
A_e = bleed-orifice effective area
M = mass of body being decelerated
V = volume of gas in dashpot
P = pressure
γ = ratio of specific heats of gas
R = gas constant for particular gas employed
X = stroke

These equations normally require solution by step-by-step numerical methods or by automatic computers. For crude approximations the limiting values for the behavior of a pneumatic dashpot must lie between the following limits:

1. With negligibly small leakage the following relationship holds approximately

$$PV^\gamma = \text{const}$$

and the energy absorbed by the dashpot is

$$\Delta T = P_0 V_0^{\gamma}(V_f^{1-\gamma} - V_0^{1-\gamma})$$

2. The performance of a pneumatic dashpot with a large orifice for which the gas pressure and density variations are small may be roughly estimated employing the incompressible-dashpot equations (16.59) and (16.60).

Hydraulic dashpots are extensively used as buffers and enjoy relatively high energy-absorbing capacity since they can readily accommodate generated pressures of 5 to 10,000 psi. Analysis is much simpler than with pneumatic dashpots and much tighter control is possible because of the relatively small compressibility of liquids compared with gases. Disadvantages of hydraulic dashpots include their cost, precision, and number of parts required, and the possible danger of gas-bubble formation within the dashpot which may alter the performance. They are relatively temperature-insensitive if designed with a sharp-edged orifice since then the density of the liquid predominates rather than its viscosity. The behavior of a relatively rigid mass impacting a hydraulic dashpot may be determined from the relationship

$$\ln (v_0/v_x) = (\rho A_p/M) \int_0^x [(A_p/A_e)^2 - 1]\, dx \qquad (16.59)$$

in which the orifice area A_e may be allowed to vary with the stroke of the dashpot. The pressure generated is

$$\Delta P = (\rho/2)[(A_p/A_e)^2 - 1]v^2 \qquad (16.60)$$

Rubber and Elastomeric Materials. For many applications the use of a rubber or similar elastomeric material is a good solution to the buffer problem. These materials excel in low cost, simplicity, reliability, and economy of space. Their use normally results in smaller deflections and hence higher impact forces than can be achieved with fluid dashpots.

Because of the complex properties of elastomers, practical solutions to the impact problem are greatly limited. In addition to sensitivity to rate of strain and temperature, elastomers are characterized by a rather long memory so that their behavior depends heavily upon their previous history, such as the magnitude of strain to which they have been previously subjected and the length of time they are allowed to "recover" following previous operations. In general the effect of frequent operations is to stiffen the elastomer since the long complex elastomer molecules are not allowed sufficient time to unwind. This effect also tends to reduce their energy-dissipating properties.

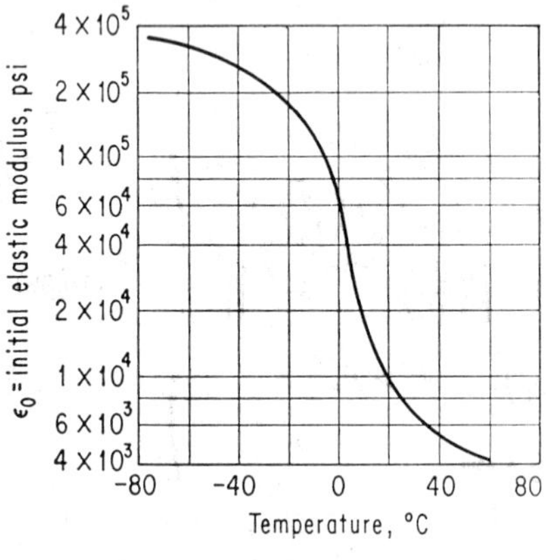

FIG. 16.23. Effect of temperature upon the dynamic modulus of a Buna N elastomer, $40/\text{sec} < \dot{\epsilon}_0 < 400/\text{sec}$.[41]

An additional complication arises because of the large strains to which elastomers are normally subjected in impact. Since the volume of the elastomer remains essentially constant, the elastomer must dilate radially, normal to the direction of applied strain. The dilation in turn is strongly influenced by the end conditions. Thus, if rigid surfaces are firmly bonded to the elastomer a higher effective modulus will be observed than if the engaging surfaces are lubricated, allowing radial dilation to occur at the ends. It has also been found[40] that when elastomer ends are rigidly bonded the response is sensitive to a "shape factor" which is the ratio of unstressed area normal to the direction of applied load divided by the lateral free or "bulge" area. When the ends are unconstrained the elastomer response is not strongly influenced by the shape factor. It is because of the large number of variables which markedly influence elastomer response that few data exist for predicting their impact behavior.

One attempt to determine the impact behavior of elastomers is described in ref. 41. The following information is drawn from that source.

The strong sensitivity to temperature and to rate of loading of the modulus of elasticity of an elastomer is shown in Figs. 16.23 and 16.24. Stress-strain curves are given in Fig. 16.24 for one sample of a Buna N buffer under dynamic and quasi-static

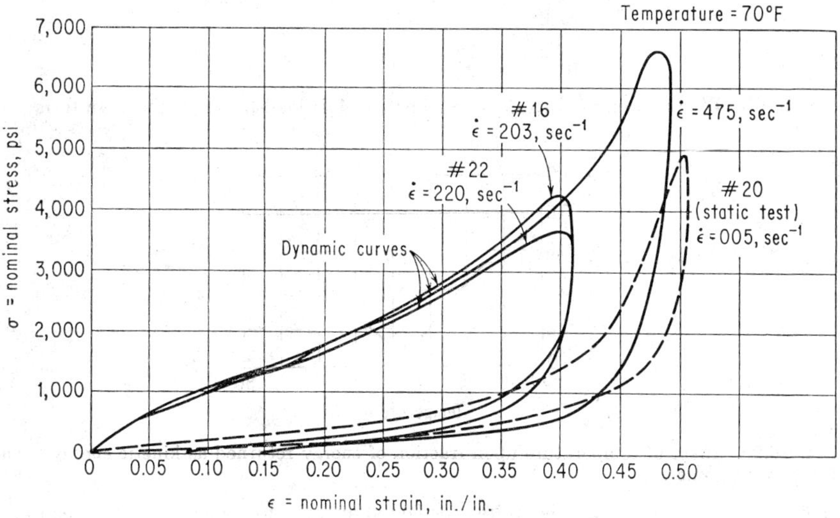

FIG. 16.24. Comparison of quasi-static and dynamic stress-strain curves for a Buna N elastomer, dynamic data from impact tests.[41]

Initial strain rate, sec^{-1}	Specific energy input, lb/in.^{3}	Approx. initial modulus, psi
475	1,244	
220	752	8,660
203	805	
0.0053	296	800

strain rates (strain rate $= v_0/L = \dot{\epsilon}$, where L = initial length of the rubber buffer). It is evident from these data that quasi-static data cannot be employed to predict dynamic behavior. If the initial slope of the stress-strain curve of Fig. 16.24 is taken as a measure of Young's modulus, it is seen that the modulus varies from 800 psi when $\dot{\epsilon} = 0.0053$/sec to 8,660 psi when $\dot{\epsilon}$ is in the range 200 to 500/sec. From these results, it is seen that a change in strain rate of 10^5 changes the apparent modulus by about a factor of 10.

The very large temperature sensitivity of Young's modulus as shown in Fig. 16.23 points up an important limitation in the use of elastomers for applications requiring precise response if large variations in ambient temperature can occur.

It was found possible to correlate several facets of compression impact behavior which are of engineering interest by restricting the problem to the case of one Buna N elastomer, with (1) ambient temperature of 70°F, (2) simple cylindrical samples of elastomer, (3) unbonded ends, (4) compression impact, (5) initial strain rates 40/sec < $\dot{\epsilon}$ < 400/sec, and (6) input energies into the elastomer limited to about 400 in.-lb/in.³ of elastomer, (7) fairly long intervals of the order of minutes between operations.

The following two design parameters were found to be adequate under these conditions to correlate the observed data:

$$U_s = \frac{1}{2g}\frac{W_1W_2}{W_1 + W_2}\frac{(v_{10} - v_{20})^2}{2AL} \qquad \text{in.-lb/in.}^3$$

$$\lambda = \frac{W_1W_2}{W_1 + W_2}\frac{L}{A} \qquad \text{lb/in.}$$

where W_1, W_2 are weights of impacting bodies, A is load-bearing area. In terms of

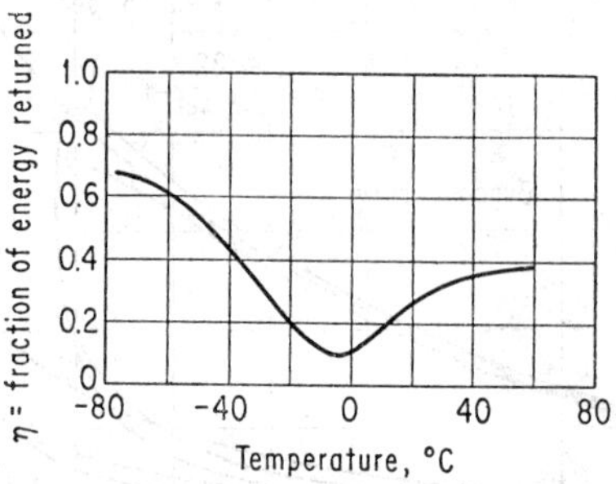

FIG. 16.25. Effect of temperature upon fraction of energy returned as kinetic energy for a Buna N elastomer.[41]

these two parameters, with the units specified above, the following empirical relationships were found to hold:

Duration of impact:

$$t_f = (\pi/E_0g)^{1/2}\lambda^{1/2}(U_s/100)^{-0.1}$$

here E_0 is the initial dynamic modulus as is given in Fig. 16.25.

Fraction of energy returned as kinetic energy after the impact:

$$\eta = 0.5(\lambda/U_s)^{0.2}$$

Maximum stress produced by the impact:

$$\sigma_{max} = 49U_s^{0.7}\lambda^{-0.1}$$

Maximum strain produced by the impact:

$$\epsilon_{max} = 0.026U_s^{0.38}\lambda^{0.12}$$

(h) Impact Properties of Nonmetals

Published information on the dynamic properties of such materials as thermoplastic and thermosetting compounds is extremely meager. In part this may be due to the wide and ever-increasing variety of such materials, their heterogeneous composition, and the extreme complexity of their properties. Fillers of diverse types ranging from wood to sand to fiber glass, combined with a resin binder, produce materials which are nonisotropic and which are sensitive to geometry and molding procedure. In addition, such materials are far more sensitive than metals to environmental conditions like strain rate, frequency of loading, and temperature. Because of these complications it is particularly difficult to make generalizations. It is well to keep in mind, however, that under impact-like conditions, published quasi-static data may bear little resemblance to the dynamic properties of plastic materials. As one example, the data shown in Fig. 16.26 is of interest. This curve shows the pronounced rate sensitivity of the energy absorbing capacity of specimens of one phenol-formaldehyde resin. Note that large and abrupt changes in energy absorption occur at impact velocities of 10^{-2} and 10^{2} ips. The trend is evidently strong toward increasingly brittle behavior as strain rate is increased. It should be expected that the apparent Young's modulus of such materials is also rate-sensitive, as is demonstrated for elastomers in Fig. 16.24. In their impact-fatigue properties plastic materials exhibit a pronounced sensitivity to the frequency of loading. As the frequency is increased, there is in general a reduction in the fatigue life. This is probably due to the long

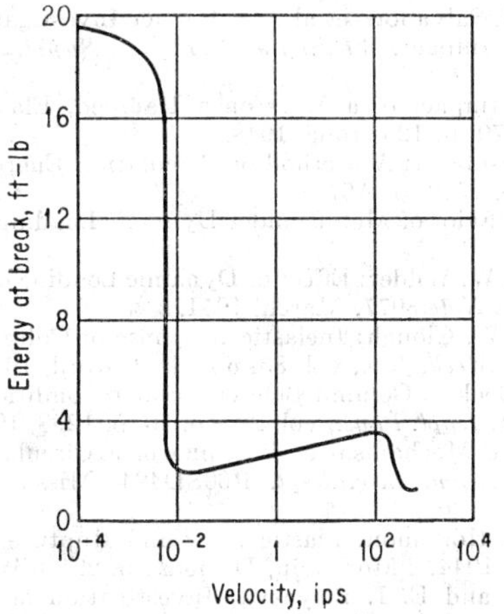

FIG. 16.26. The impact strength of phenol-formaldehyde resin as a function of the rate of application of load.[42]

relaxation time for residual stresses in plastics and to their high internal friction, which gives rise to thermal effects.

References

1. Brennan, J. N. (ed.): Bibliography on Shock and Shock Excited Vibrations, vols. I, II, *Engineering Research Bull.* 69, Pennsylvania State University, University Park, Pa., 1958.
2. Bennit, W. (ed.): Bibliography on Impact, *Gen. Elec. Rept.* R60SD302, Missile & Space Vehicle Dept., Philadelphia, Pa., 1960.
3. Goldsmith, W.: "Impact," Edward Arnold (Publishers) Ltd., London, 1960.
4. Rhinehart, J. S., and J. Pearson: "Behavior of Metals under Impulsive Loads," ASTM, Cleveland, 1954.
5. Bowden, F. P., and D. Tabor: "The Friction and Lubrication of Solids," Oxford University Press, Fair Lawn, N.J., 1954.
6. Love, A. E. H.: "Mathematical Theory of Elasticity," 4th ed., Dover Publications, Inc., New York, 1927.
7. Goldsmith, W., and P. T. Lyman: "The Penetration of Hard-steel Spheres into Plane Metal Surfaces," ASME paper 60-WA-15, *J. Appl. Mech.*
8. Crook, A. W.: A Study of Some Impacts between Metal Bodies by a Piezo-Electric Method, *Proc. Roy. Soc. (London) Series*, vol. A212, p. 377, 1952.
9. Bell, J. F.: Propagation of Large Amplitude Waves in Annealed Aluminum, *J. Appl. Phys.*, vol. 31, no. 2, pp. 277–282, February, 1960.
10. Bell, J. F.: Study of Initial Conditions in Constant Velocity Impact, *J. Appl. Phys.*, vol. 31, no. 12, pp. 2188–2195, December, 1960.
11. Kolsky, H.: "Stress Waves in Solids," Oxford University Press, Fair Lawn, N.J., 1953.
12. DeJuhasz, K. J.: Graphical Analysis of Impact of Bars above the Elastic Range, *J. Franklin Inst.*, vol. 248, nos. 1, 2, July and August, 1949.
13. Marti, W.: Vibrations in Valve Springs of Internal Combustion Engines, *Sulzer Tech. Rev. Switz.*, no. 2, 1936.

14. Timoshenko, S., and J. N. Goodier: "Theory of Elasticity," 2d ed., McGraw-Hill Book Company, Inc., New York, 1951.
15. Donnell, L. H.: Longitudinal Wave Transmission and Impact, *Trans. ASME*, vol. 2, p. 153, 1930.
16. Sears, J. E.: On the Impact of Bars with Rounded Ends, *Trans. Cambridge Phil. Soc.*, vol. 21, 1909–1911.
17. Timoshenko, S.: Zur Frage nach der Wirkung eines Stosse auf einer Balken, *Z. Math. Physik*, vol. 62, p. 193.
18. Lee, E. H.: The Impact of a Mass Striking a Beam, *J. Appl. Mech.*, vol. 62, p. A-129, December, 1940.
19. Krefeld, W. J., M. G. Salvadori, et al.: An Impact Investigation of Beams with Butt-welded Splices under Impact, *Welding J. N.Y. Res. Suppl.*, vol. 12, no. 7, pp. 372S–432S, 1947.
20. Hoppmann, W. H.: Impact of a Mass on a Damped, Elastically Supported Beam, *J. Appl. Mech.*, vol. 70, p. 125, June, 1948.
21. Zener, C., and H. Feshbach: A Method of Calculating Energy Losses during Impact, *J. Appl. Mech.*, June, 1939, p. A67.
22. Clark, D. S.: The Behavior of Metals under Dynamic Loading, *Metal Prog.*, November, 1953, p. 67.
23. Brooks, W. A., and I. W. Wilder: Effect of Dynamic Loading on Strength of an Inelastic Column, *NACA Tech Note* 3077, March, 1954.
24. Hartz, B. J., and R. W. Clough: Inelastic Response of Columns to Dynamic Loading, *Proc. ASCE, J. Eng. Mech. Div.*, vol. 83, no. Em2, April, 1957.
25. Gerard, G., and H. Becker: Column Behavior under Conditions of Compressive Stress Wave Propagation, *J. Appl. Phys.*, vol. 22, no. 10, p. 1298, 1951.
26. Coppa, A. P.: On the Mechanism of Buckling of a Circular Cylindrical Shell under Longitudinal Impact, *Gen. Elec. Rept.* R60SD494, Missile & Space Vehicle Dept., Philadelphia.
27. Zener, C., and J. H. Hollomon: Plastic Flow and Rupture in Metals, *Trans. ASM*, vol. 33, pp. 188–226, 1944. Morkovin, D.: *ibid.*, p. 221 (discussion).
28. Davidenkov, N. N., and E. I. Belyaeva: Investigation of Fatigue under Repeated Impact, *Brutcher Trans.* no. 4719, POB 157, Altadena, Calif., Russian original in *Metalloved. i Obrabotka Metal.*, no. 11, November, 1956.
29. Davidenkov, N. N., and E. I. Belyaeva: Study of Impact Fatigue Strength, *Brutcher Trans.*, no. 4349, Russian original in *Metalloved. i Obrabotka Metal.*, no. 9, September, 1958.
30. Welch, W. P.: "ASME Handbook—Metals Engineering—Design," Chap. 8, Impact Considerations in Design, McGraw-Hill Book Company, Inc., New York, 1953.
31. Rayleigh, J. W. S.: On the Production of Vibrations by Forces of Relatively Long Duration with Application to the Theory of Collisions, *Phil. Mag.*, ser. 6, vol. 11, 1906.
32. Dohrwend, C. O., D. C. Drucker, and P. Moore: Transverse Impact Transients, *Proc. SESA*, vol. 1, no. 2, pp. 1–11, 1944.
33. Campbell, J. D., and J. Duby: The Yield Behavior of Mild Steel in Dynamic Compression, *Proc. Roy. Soc. (London)*, vol. A236, 1956.
34. Maiden, C. J., and J. D. Campbell: The Static and Dynamic Strength of a Carbon Steel at Low Temperatures, *Phil. Mag.*, ser. 8, vol. 3, 1958.
35. Johnson, J. E., D. S. Wood, and D. S. Clark: *J. Appl. Mech.*, December, 1953, p. 523.
36. Osgood, W. R. (ed.): "Residual Stresses in Metals and Metals Construction," chapter by J. Miklowitz, The Effect of Residual Stresses on High-speed Impact Resistance, Reinhold Publishing Corporation, New York, 1954.
37. Zener, C., and J. H. Hollomon: Effect of Strain Rate upon Plastic Flow in Steel, *J. Appl. Phys.*, vol. 15, 1944.
38. Whittaker, E. T.: "A Treatise on the Analytical Dynamics of Particles and Rigid Bodies," Cambridge University Press, New York, 1937.
39. Davidenkov, N. N.: Allowable Working Stresses under Impact, *Trans. ASME*, vol. 56, 1934.
40. Cardillo, R. M., and D. F. Kruse: "Load Bearing Characteristics of Butyl Rubber," ASME paper 61-WA-335.
41. Barkan, P., and M. F. Sirkin: "Impact Behavior of Elastomers," ASME annual mtg. paper, 1962; published in *Trans. Soc. of Plastics Engrs.*, July, 1963, pp. 210–219.
42. Bradley, J. N.: "Shock Waves in Chemistry and Physics," John Wiley & Sons, Inc., New York, 1962.
43. Rice, M. H., R. G. McQueen, and J. M. Walsh: "Compression of Solids by Strong Shock Waves," appears in "Solid State Physics," vol. 6 (Seitz and Turnbull, eds.), Academic Press Inc., New York, 1958.
44. Maxwell, B.: Mechanical Properties of Plastic Dielectrics, *Elec. Mfg.*, vol. 58, p. 146, September, 1956.

Section 17

PROPERTIES OF ENGINEERING MATERIALS

By

THEODORE GELA, D.Sc., *Professor of Metallurgy, Stevens Institute of Technology, Hoboken, N.J.*

CONTENTS

17.1. MATERIAL-SELECTION CRITERIA IN ENGINEERING DESIGN

The selection of materials for engineering components and devices depends upon a knowledge of material properties and behavior in particular environmental states. Although a criterion for the choice of material in critically designed parts relates to the performance in a field test, it is usual in preliminary design to use appropriate data obtained from standardized tests. The following considerations are important in material selection:

1. Elastic properties
   a. Stiffness and rigidity
2. Plastic properties
   a. Yield conditions, stress-strain relations, and hysteresis

3. Time-dependent properties
   a. Elastic phenomenon (damping capacity), creep, relaxation, and strain-rate effect
4. Fracture phenomenon
   a. Crack propagation, fatigue, and ductile to brittle transition
5. Thermal properties
   a. Thermal expansion, thermal conductivity, and specific heat
6. Chemical interactions with environment
   a. Oxidation, corrosion, and diffusion

It is good design practice to analyze the conditions under which test data were obtained and to use the data most pertinent to anticipated service conditions.

The challenge that an advancing technology imposes on the engineer, in specifying treatments to meet stringent material requirements, implies a need for a basic approach which relates properties to structure in metals. As a consequence of the mechanical, thermal, and metallurgical treatments of metals, it is advantageous to explore, for example, the nature of induced internal stresses as well as the processes of stress relief. Better material performance may ensue when particular treatments can be specified to alter the structure in metals so that the likelihood of premature failure in service is lessened. Some of the following concepts are both basic and important:

1. Lattice structure of metals
   a. Imperfections, anisotropy, and deformation mechanisms
2. Phase relations in alloys
   a. Equilibrium diagrams
3. Kinetic reactions in the solid state
   a. Heat-treatment by nucleation and by diffusionless processes, precipitation hardening, diffusion, and oxidation
4. Surface treatments
   a. Chemical and structural changes in carburizing, nitriding, and localized heating
5. Metallurgical bonds
   a. Welded and brazed joints

17.2. STRENGTH PROPERTIES—TENSILE TEST AT ROOM TEMPERATURE

The yield strength determined by a specified offset, 0.2 per cent strain, from a stress-strain diagram is an important and widely used property for the design of statically loaded members exhibiting elastic behavior. This property is derived from a test in which the following conditions are normally controlled: surface condition of standard specimen is specified; load is axial; the strain rate is low, i.e., about 10^{-3} in./in./sec; and grain size is known. Appropriate safety factors are applied to the yield strength to allow for uncertainties in the calculated stress and stress-concentration factors and for possible overloads in service. Since relatively small safety factors are used in critically stressed aircraft materials, a proof stress at 0.01 per cent strain offset is used because this more nearly approaches the proportional limit for elastic behavior in the material. A typical stress-strain plot from a tensile test is shown in Fig. 17.1 indicating the elastic and plastic behaviors. In order to effect more meaningful comparisons in design strength properties among materials having different specific gravities, the strength property can be divided by the specific gravity, giving units of psi per pound per cubic inch.

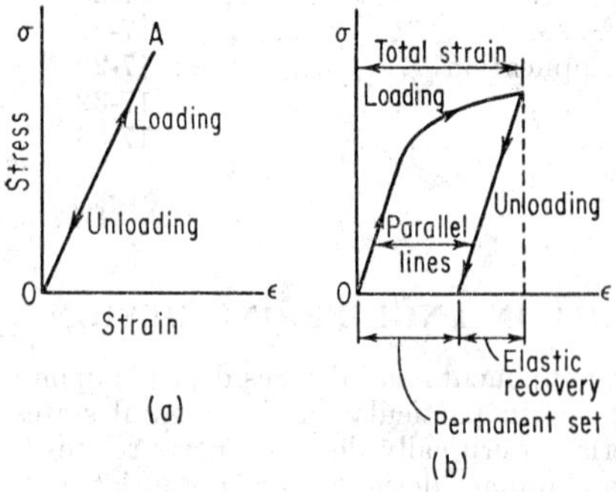

FIG. 17.1. Portions of tensile stress σ–strain E curves in metals.[1] (a) Elastic behavior. (b) Elastic and plastic behaviors.

The modulus of elasticity is a measure of the stiffness or rigidity in a material. Values of the modulus normally are not exactly determined quantities, and typical values are commonly reported for a given material. When a material is selected on the basis of a high modulus, the tendency toward whip and vibration in shaft or rod applications is reduced. These effects can lead to uneven wear. Furthermore the modulus assumes particular importance in the design of springs and diaphragms, which necessitate a definite degree of motion for a definite load. In this connection, selection of a higher-modulus material can lead to a thinner cross section.

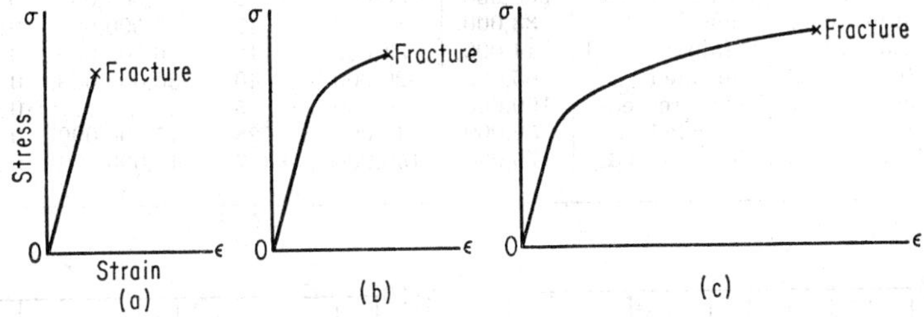

FIG. 17.2. The effects of treatments on tensile characteristics of a metal.[1] (*a*) Perfectly brittle (embrittled)—all elastic behavior. (*b*) Low ductility (hardened)—elastic plus plastic behaviors. (*c*) Ductile (softened)—elastic plus much plastic behaviors.

The ultimate tensile strength and the ductility, per cent elongation in inches per inch or per cent reduction in area at fracture are other properties frequently reported from tensile tests. These serve as qualitative measures reflecting the ability of a material in deforming plastically after being stressed beyond the elastic region. The strength properties and ductility of a material subjected to different treatments can vary widely. This is illustrated in Fig. 17.2. When the yield strength is raised by treatment to a high value, i.e., greater than two-thirds of the tensile strength, special concern should be given to the likelihood of tensile failures by small overloads in service. Members subjected solely to compressive stress may be made from high-yield-strength materials which result in weight reduction.

When failures are examined in statically loaded tensile specimens of circular section, they can exhibit a cup-and-cone fracture characteristic of a ductile material or on the other extreme a brittle fracture in which little or no necking down is apparent. Upon loading the specimen to the plastic region, axial, tangential, and radial stresses are induced. In a ductile material the initial crack forms in the center where the triaxial stresses become equally large, while at the surface the radial component is small and the deformation is principally by biaxial shear. On the other hand, an embrittled material exhibits no such tendency for shear and the fracture is normal to the loading axis. Some types of failures in round tensile specimens are shown in Fig. 17.3.

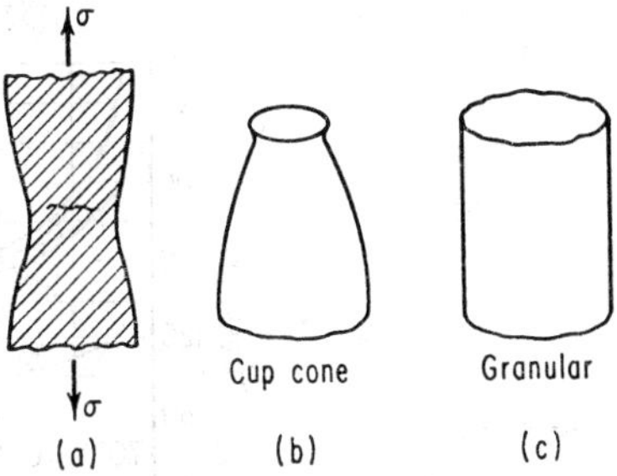

FIG. 17.3. Typical tensile-test fractures.[1] (*a*) Initial crack formation. (*b*) Ductile material. (*c*) Brittle material.

The properties of some wrought metals presented in Table 17.1 serve to show the significant differences relating to alloy content and treatment. Article 17.14 gives more information.

The tensile properties of metals are dependent upon the rate of straining, as shown for aluminum and copper in Fig. 17.4, and are significantly affected by the temperature as shown in Fig. 17.5. For high-temperature applications it is important to base

Table 17.1. Room-temperature Tensile Properties for Some Wrought Metals

Metal	Condition	Ultimate tensile σ_{ult}, psi	Yield strength σ_y, psi	% elongation	Modulus of elasticity, psi	Density, lb/cu in.
Aluminum (pure)..	Annealed	13,000	5,000	35	10,000,000	0.098
7075T6...........	Heat-treated	76,000	67,000	11	10,400,000	0.101
Copper (pure).....	Annealed	32,000	10,000	45	17,000,000	0.321
Cu-Be(2%).......	Heat-treated	200,000	150,000	2	19,000,000	0.297
Magnesium	Annealed	33,000	18,000	17	6,500,000	0.064
Mg-Al(8.5)A780A.	Strain relieved	44,000	31,000	18	6,500,000	0.065
Nickel (pure).....	Annealed	65,000	20,000	40	30,000,000	0.321
K Monel.........	Heat-treated	190,000	140,000	5	....	0.306
Titanium (pure)...	Annealed	59,000	40,000	28	15,000,000	0.163
Ti-Al(6)-V(14)α-β.	Heat-treated	170,000	150,000	7	15,000,000	0.163

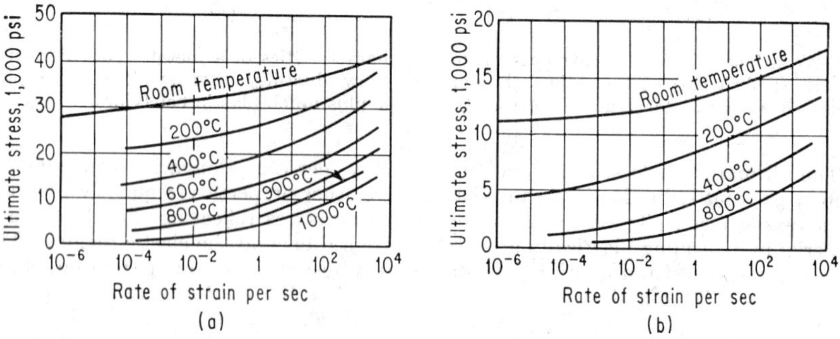

FIG. 17.4. Effects of strain rates and temperatures on tensile-strength properties of copper and aluminum.[1] (*a*) Copper. (*b*) Aluminum.

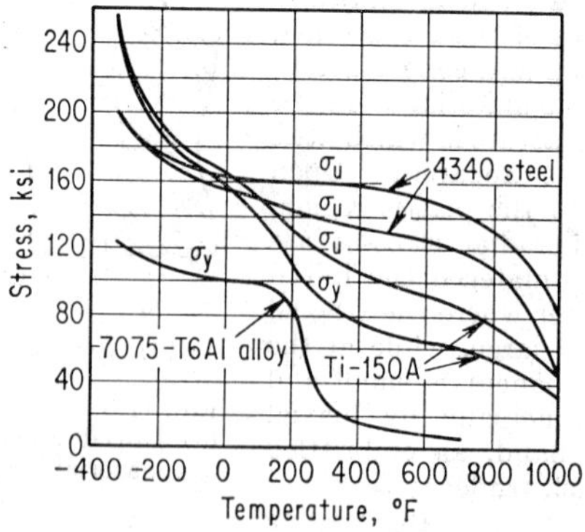

FIG. 17.5. Effects of temperatures on tensile properties.
σ_u = ultimate tensile strength
σ_y = yield strength

design on different criteria, notably the stress-rupture and creep characteristics in metals, both of which are also time-dependent phenomena. The use of metals at low temperatures requires a consideration of the possibility of brittleness, which can be measured in the impact test.

17.3. ATOMIC ARRANGEMENTS IN PURE METALS—CRYSTALLINITY

The basic structure of materials provides information upon which properties and behavior of metals may be generalized so that selection can be based on fundamental considerations. A regular and periodic array of atoms (in common metals whose atomic diameters are about one hundred-millionth of an inch) in space, in which a unit cell is the basic structure, is a fundamental characteristic of crystalline solids. Studies of these structures in metals lead to some important considerations of the behaviors

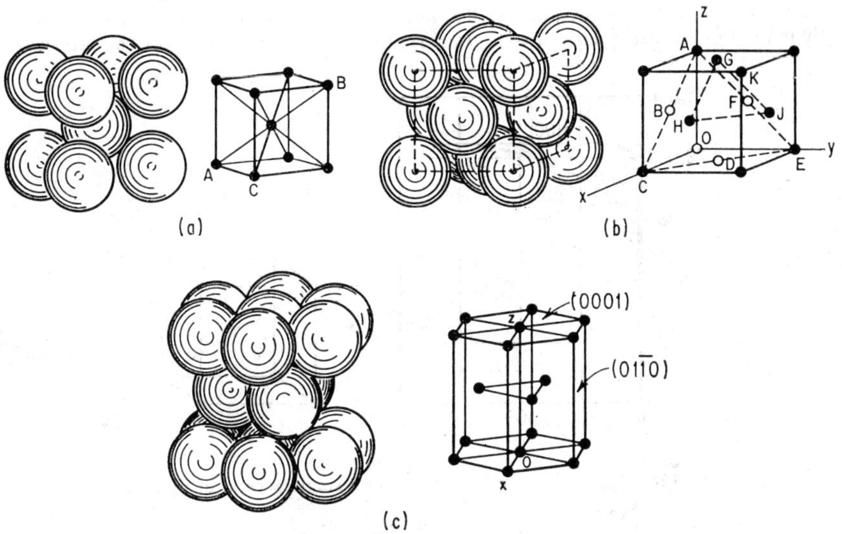

FIG. 17.6. Cell structure. (*a*) Body-centered cubic (b.c.c.) unit cell structure. (*b*) Face-centered cubic (f.c.c.) unit cell structure. (*c*) Hexagonal close-packed (h.c.p.) unit cell structure.

in response to externally applied forces, temperature changes, as well as applied electrical and magnetic fields.

The body-centered cubic (b.c.c.) cell shown in Fig. 17.6*a* is the atomic arrangement characteristic of αFe, W, Mo, Ta, βTi, V, and Nb. It is among this class of metals that transitions from ductile to brittle behavior as a function of temperature are significant to investigate. This structure represents an atomic packing density where about 66 per cent of the volume is populated by atoms while the remainder is free space. The elements Al, Cu, γFe, Ni, Pb, Ag, Au, and Pt have a closer packing of atoms in space constituting a face-centered cubic (f.c.c.) cell shown in Fig. 17.6*b*. Characteristic of these are ductility properties which in many cases extend to very low temperatures. Another structure, common to Mg, Cd, Zn, αTi, and Be, is the hexagonal close-packed (h.c.p.) cell in Fig. 17.6*c*. These metals are somewhat more difficult to deform plastically than the materials in the two other structures cited above.

It is apparent, from the atomic arrays represented in these structures, that the closest approach of atoms can vary markedly in different crystallographic directions. Properties in materials are anisotropic when they show significant variations in different

Table 17.2. Examples of Anisotropic Properties in Single Crystals

Property	Material and structure	Properties relation	Reference
Elastic modulus E in tension	αFe(b.c.c.)	$E_{\langle AB\rangle} \sim 2.2E_{\langle AC\rangle}$	Fig. 17.6*a*
Elastic modulus G in shear	Ag (f.c.c.)	$G_{\langle OC\rangle} \sim 2.3G_{\langle OK\rangle}$	Fig. 17.6*b*
Magnetization	αFe (b.c.c.)	Ease of magnetization	Fig. 17.7
Thermal expansion coefficient — α	Zn (h.c.p.)	$\alpha_{\langle OZ\rangle} \sim 4\alpha_{\langle OA\rangle}$	Fig. 17.6*c*

directions. Such tendencies are dependent on the particular structure and can be especially pronounced in single crystals (one orientation of the lattices in space). Some examples of these are given in Table 17.2. When materials are processed so that their final grain size is large (each grain represents one orientation of the lattices) or

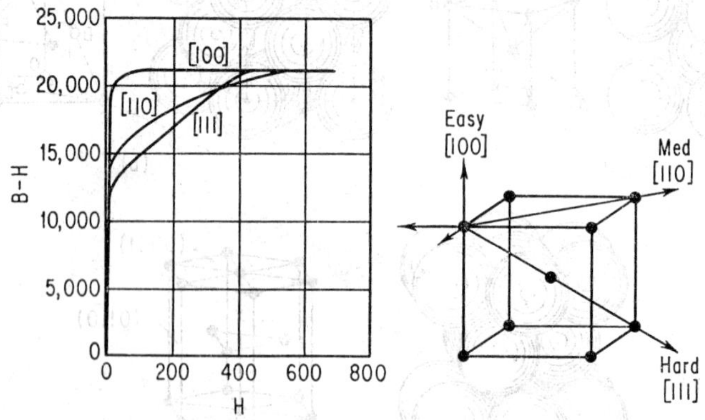

FIG. 17.7. Magnetic anisotropy in a single crystal of iron.[2]

$$I = \frac{B - H}{4\pi}$$

where I = intensity of magnetization
B = magnetic induction, gauss
H = field strength, oersteds

that the grains are preferentially oriented, as in extrusions, drawn wire, rolled sheet, sometimes in forgings and castings, special evaluation of anisotropy should be made. In the event that directional properties influence design considerations, particular attention must be given to metallurgical treatments which may control the degree of anisotropy. The magnetic anisotropy in a single crystal of iron is shown in Fig. 17.7.

17.4. PLASTIC DEFORMATION OF METALS

When metals are externally loaded past the elastic limit, so that permanent changes in shape occur, it is important to consider the induced internal stresses, property changes, and the mechanisms of plastic deformation. These are matters of practical consideration in the following: materials that are to be strengthened by cold work, machining of cold-worked metals, flow of metals in deep-drawing and impact extrusion operations, forgings where the grain flow patterns may affect the internal soundness, localized surface deformation to enhance fatigue properties, and cold working of some magnetic materials. Experimental studies provide the key by which important phenomena are revealed as a result of the plastic-deformation process. These studies indicate some treatments that may be employed to minimize unfavorable internal-stress distributions and undesirable grain-orientation distributions.

Plastic deformation in metals occurs by a glide or slip process along densely packed planes fixed by the particular lattice structure in a metal. Therefore, an applied load is resolved as a shear stress, on those particular glide elements (planes and directions) requiring the least amount of deformation work on the system. An example of this

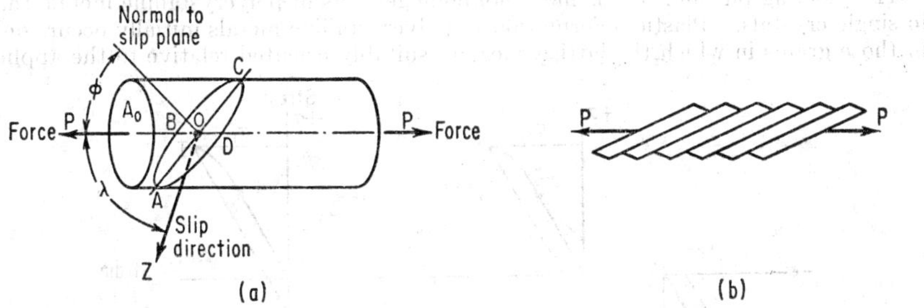

FIG. 17.8. Slip deformation in single crystals. (*a*) Resolved shear stress = $P/A_0 \cos\phi \cos\lambda$. *ABCD* is plane of slip. *OZ* is slip direction. (*b*) Sketch of single crystal after yielding.

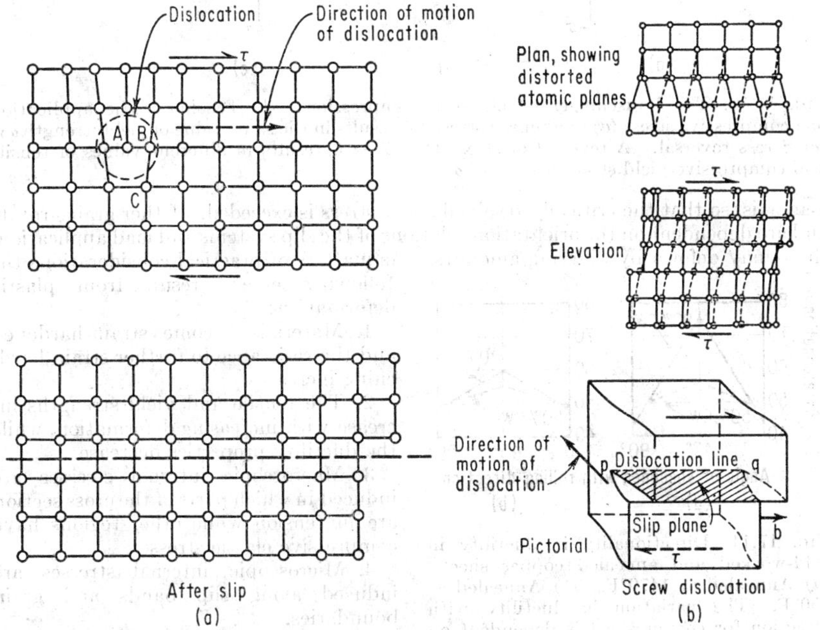

FIG. 17.9. Edge and screw dislocations as types of imperfections in metals.[2] (*a*) Edge dislocation. (*b*) Screw dislocation.

deformation process is shown in Fig. 17.8. Face-centered cubic structured metals, such as Cu, Al, and Ni, are more ductile than the hexagonal structured metals like Mg, Cd, and Zn at room temperature because in the f.c.c. structure there are four times as many possible slip systems as in a hexagonal structure. Slip is initiated at much lower stresses in metals than theoretical calculations based on a perfect array of atoms would indicate. In real crystals there are inherent structural imperfections termed dislocations (atomic misfits) as shown in Fig. 17.9, which account for the observed yielding phenomenon in metals. In addition, dislocations are made mobile by mechanical and thermal excitations and they can interact to result in strain hardening of metals by cold work. Strength properties can be increased while the ductility is decreased in those metals which are amenable to plastic deformation. Cold working of pure metals and single-phase alloys provides the principal mechanism by which these may be hardened.

The yielding phenomenon is more nonhomogeneous in polycrystalline metals than in single crystals. Plastic deformation in polycrystalline metals initially occurs only in those grains in which the lattice axes are suitably oriented relative to the applied

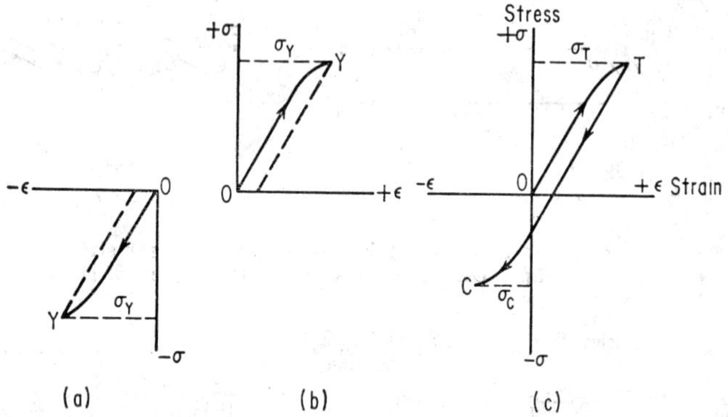

FIG. 17.10. The Bauschinger effect. (*a*) Compression. (*b*) Tension. The application of a compressive stress (*a*) or a tensile stress (*b*) results in the same value of yield strength σ_Y. (*c*) Stress reversal. A reversal of stress $O \rightarrow T \rightarrow C$ results in different values of tensile and compressive yield strengths; $\sigma_T \neq \sigma_C$.

load axis, so that the critically resolved shear stress is exceeded. Other grains rotate and are dependent on the orientation relations of the slip systems and load application; these may deform by differing amounts. As matters of practical considerations the following effects result from plastic deformation:

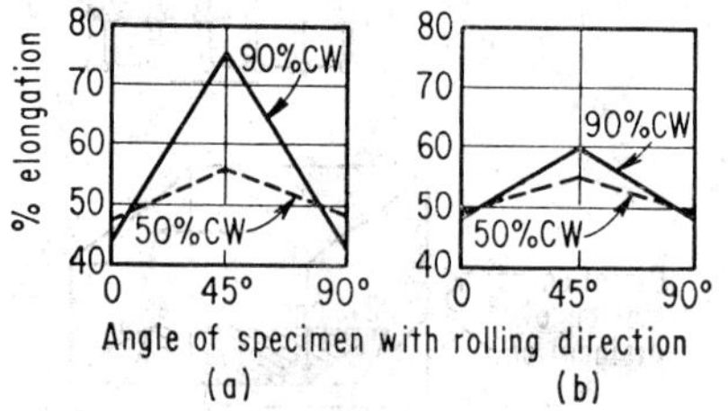

FIG. 17.11. Directionality in ductility in cold-worked and annealed copper sheet.[1] (*a*) Annealed at 1470°F. (*b*) Annealed at 750°F. The variation in ductility with direction for copper sheet is dependent on both the annealing temperature and the amount of cold work (per cent CW) prior to annealing.

1. Materials become strain-hardened and the resistance to further strain hardening increases.

2. The tensile and yield strengths increase with increasing deformation, while the ductility properties decrease.

3. Macroscopic internal stresses are induced in which parts of the cross section are in tension while other regions have compressive elastic stresses.

4. Microscopic internal stresses are induced along slip bands and grain boundaries.

5. The grain orientations change with cold work so that some materials may exhibit different mechanical and physical properties in different directions.

The Bauschinger effect in metals is related to the differences in the tensile and compressive yield-strength values, as shown at σ_T and σ_C in Fig. 17.10 when a ductile metal undergoes stress reversal. This change in polycrystalline metals is the result of the nonuniform character of deformation and the different pattern of induced macrostresses. These grains, in which the induced macrostresses are compressive, will yield at lower values upon the application of a reversed compressive stress because they are already part way toward yielding. This effect is encountered in cold-rolled metals where there is lateral contraction together with longitudinal elongation; this accounts for the decreased yield strength in the lateral direction compared with the increased longitudinal yield strength.

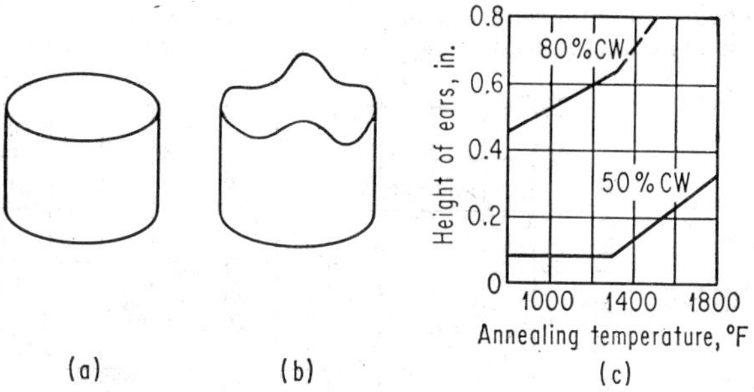

FIG. 17.12. The earing tendencies in cup deep-drawn from sheet. (*a*) Uniform flow, non-earing. (*b*) Eared cup, the result of nonuniform flow. (*c*) Height of ears in deep-drawn copper cups related to annealing temperature and amount of cold work.

The control of metal flow is important in deep-drawing operations performed on sheet metal. It is desirable to achieve a uniform flow in all directions. Cold-rolling sheet metal produces a structure in which the grains have a preferred orientation. This characteristic can persist, even though the metal is annealed (recrystallized), resulting in directional properties as shown in Fig. 17.11. A further consequence of this directionality, associated with the deep-drawing operation, is illustrated in Fig. 17.12. The important factors, involved with the control of earing tendencies, are the fabrication practices of the amount of cold work in rolling and duration and temperatures of annealing. When grain textural problems of this kind are encountered they can be studied by X-ray diffraction techniques and reasonably controlled by the use of optimum cold-working and annealing schedules.

17.5. PROPERTY CHANGES RESULTING FROM COLD-WORKING METALS

Cold-working metals by rolling, drawing, swaging, and extrusion is employed to strengthen them and/or to change their shape by plastic deformation. It is used principally on ductile metals which are pure, single-phase alloys and for other alloys which will not crack upon deformation. The increase in tensile strength accompanied by the decrease in ductility characteristic of this process is shown in Fig. 17.13. It is to be noted, especially from the yield-strength curve, that the largest rates of change occur during the initial amounts of cold reduction.

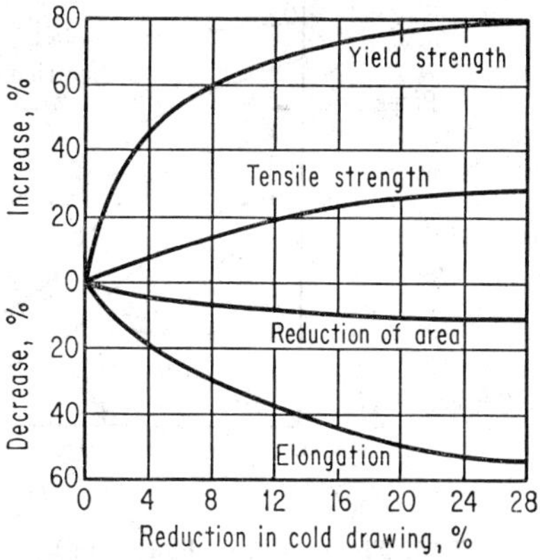

FIG. 17.13. Effect of cold drawing on the tensile properties of steel bars of up to 1 in. cross section having tensile strength of 110,000 psi or less before cold drawing.[3]

The variations in the macrostresses induced in a cold-drawn bar, illustrated in Fig. 17.14*a*, show that tensile stresses predominate at the surface. The equilibrium state of macrostresses throughout the cross section is altered by removing the surface layers in machining, the result of which may be warping in the machined part. It may be possible, however, to stress-relieve cold-worked metals, which generally have better machinability than softened (annealed) metals, by heating below the recrystallization temperature. A typical alteration in the stress distribution, shown in Fig. 17.14*b*, is achieved so that the warping tendencies on machining are reduced, without decreasing the cold-worked strength properties. This stress-relieving treatment may also inhibit season cracking in cold-worked brasses subjected to corrosive environments

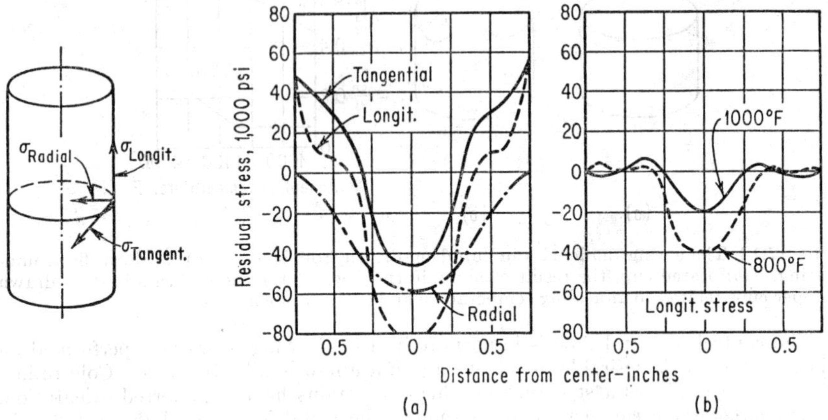

FIG. 17.14. Residual stress.[3] (*a*) In a cold-drawn steel bar 1½ in. in diameter 20 per cent cold drawn, 0.45C steel. (*b*) After stress-relieving bar.

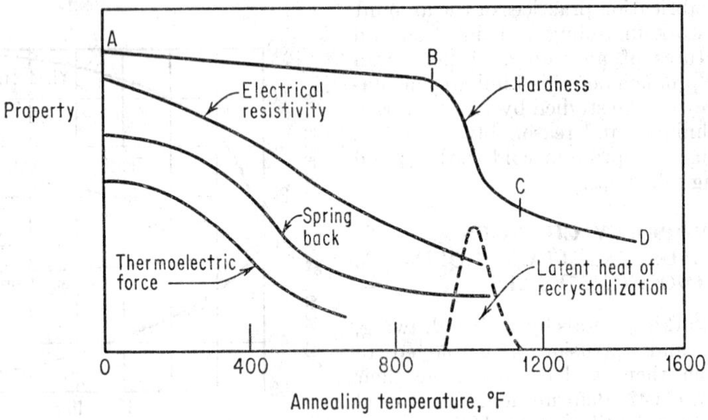

FIG. 17.15. The property changes in 95 per cent cold-worked iron with heating temperatures (1 hr). The temperature intervals:[4] $A \rightarrow B$, stress relieval, $B \rightarrow C$, recrystallization, and $C \rightarrow D$, grain growth, signify the important phenomena occurring.

containing amines. Since stressed regions, in a metal, are more anodic (i.e., go into solution more readily) than unstressed regions, it is often important to consider the relieving of stresses so that the designed member is not so likely to be subjected to localized corrosive attack.

Changes in electrical resistivity, elastic springback, and thermoelectric force, resulting from cold work, can be altered by a stress-relieval treatment, in a temperature range from *A* to *B*, as shown in Fig. 17.15. However, the grain flow pattern (preferred orientation) produced by cold working can be changed only by heating the metal to a temperature at which recrystallized stress-free grains will form.

Residual tensile stresses at the surface of a metal promote crack nucleation in the fatigue of metal parts. The use of a localized surface deformation treatment by shot peening, which induces compressive stresses in the surface fibers, offers the likelihood of improvement in fatigue and corrosion properties in alloys. Shot peening a forging flash line in high-strength aluminum alloys used in aircraft may also lessen the tendency toward stress-corrosion cracking. The effectiveness of this localized surface-hardening treatment is dependent on both the nature of surface discontinuities formed by shot peening and the magnitude of compressive stresses induced at the surface.

17.6. THE ANNEALING PROCESS

Metals are annealed in order to induce softening for further deformation, to relieve residual stresses, to alter the microstructure and, in some case after electroplating,

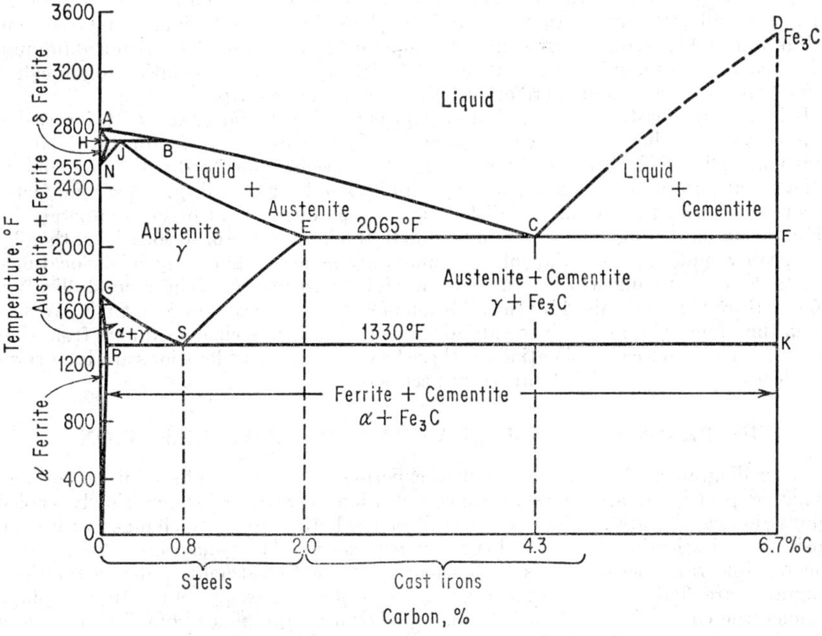

FIG. 17.16. The iron-carbon phase diagram.[4]

to expel, by diffusion, gases entrapped in the lattice. The process of annealing, the attaining of a strain-free recrystallized grain structure, is dependent mainly on the temperature, time, and the amount of prior cold work. The temperature indicated at *C* in Fig. 17.15 results in the complete annealing after 1 hr of the 95 per cent cold-worked iron. Heating beyond this temperature causes grains to grow by coalescence, so that the surface-to-volume ratio of the grains decreases together with decreasing the internal energy of the system. As the amount of cold work (from the originally annealed state) decreases, the recrystallization temperature increases and the recrystallized grain size increases. When a metal is cold-worked slightly (less than 10 per cent) and subsequently annealed, an undesirable roughened surface forms because of the abnormally large grain size (orange-peel effect) produced. These aspects of grain-size control in the annealing process enter in material specifications.

The annealing of iron-carbon-base alloys (steels) is accomplished by heating alloys of eutectoid and hypoeutectoid compositions (0.8 per cent C and less in plain carbon steels) to the single-phase region, austenite, as shown in Fig. 17.16 above the transition line *GS*; and for hypereutectoid alloys (0.8 to 2.0 per cent C) between the transition lines *SK* and *SE* in Fig. 17.16; followed by a furnace cool at a rate of about 25°F per hour to below the eulectoid temperature *SK*. In the annealed condition, a desirable distribution of the equilibrium phases is thereby produced. A control of the microstructure is manifested by this process in steels. Grain-size effects are principally controlled by the high-temperature treatment, grain sizes increasing with increasing temperatures, and in some cases minor impurity additions like vanadium inhibit grain coarsening to higher temperatures. These factors of grain-size control enter into the considerations of hardening steels by heat-treatment.

The control of the atmosphere in the annealing furnace is desirable in order to prevent gas-metal attack. Moisture-free neutral atmospheres are used for steels which oxidize readily. When copper and its alloys contain oxygen, as oxide, it is necessary to keep the hydrogen content in the atmosphere to a minimum. At temperatures lower than 900°F, the hydrogen should not exceed 1 per cent, and as the temperature is increased the hydrogen content should be reduced in order to prevent hydrogen embrittlement. In nickel and its alloys the atmosphere must be free from sulfur and slightly reducing by containing 2 per cent or more of CO. Some aluminum alloys containing magnesium are affected by high-temperature oxidation in annealing (and heat-treatment) and therefore require atmosphere control.

It is a characteristic property that strengths in all metals decrease with increasing temperatures. The coalescence of precipitate particles is one factor involved; so that material specifications for high-temperature use are concerned with alloy compositions that form particles having lower solubility and lower mobility. A second factor is concerned with the mobility of dislocations which increases at higher temperatures. Since strain hardening is reasoned to be due to the interaction of dislocations, then by the proper additions of solid-solution alloying elements that impede dislocations, resistance to softening will increase at the high temperature. The recrystallization temperature of iron is raised by the addition of 1 atomic per cent of Mn, Cr, V, W, Cb, Ta in the same order in which the atomic size of the alloying element differs from that of iron. The practical implications of these basic atomic considerations are important in selecting metals for high-temperature service.

17.7. THE PHASE DIAGRAM AS AN AID TO ALLOY SELECTION

Phase diagrams, which are determined experimentally and are based upon thermodynamic principles, are temperature-composition representations of slowly cooled alloys (annealed state). They are useful for predicting property changes with composition and selecting feasible fabrication processes. Phase diagrams also indicate the possible response of alloys to hardening by heat-treatment. Shown on these diagrams are first-order phase transitions and the phases present. In two-phase regions, the compositions of each phase are shown on the phase boundary lines and the relative amounts of each phase present can be determined by a simple lever relation at a given temperature.

The particular phases that are formed in a system are governed principally by the physical interactions of valency electrons in the atoms and secondarily by atomic-size factors. When two different atoms in the solid state exist on, or where one is in, an atomic lattice, the phase is a solid solution (e.g., γ austenite phase in Fig. 17.16) analogous to a miscible liquid solution. When the atoms are strongly electropositive and strongly electronegative to one another, an intermetallic compound is formed (e.g., Fe_3C, cementite). The two atoms are electronically indifferent to one another and a phase mixture issues (e.g., $\alpha + Fe_3C$) analogous to the immiscibility of water and oil.

The thermodynamic criteria for a first-order phase change, indicated by the solid lines on the phase diagram, are that, at the transition temperature, (1) the change in Gibbs's free energy for the system is zero; (2) there is a discontinuity in entropy (a latent heat of transformation and a discontinuous change in specific heat); and (3) there is a discontinuous change in volume (a dilational effect).

In the selection of alloys for sound castings, particular attention is given that part of the system where the liquidus line (*ABCD*) goes through a minimum. For alloys between the composition limits of $E \rightarrow F$, a eutectic reaction occurs at 2065°F such that

$$\text{liq } C = \gamma + Fe_3C$$

It is for this reason in the iron-carbon system that cast irons are classified as having carbon contents greater than 2 per cent. For purposes of controlling grain size, obtaining sound castings free from internal porosities (blowholes) and internal shrinkage cavities, and possessing good mold-filling characteristics, alloys and low-melting solutions are chosen near the eutectic composition (i.e., at *C*). Aluminum-silicon die-casting alloys have a composition of about 11 per cent silicon near the eutectic composition. Special considerations need be given to the properties and structures in cast irons because the Fe_3C phase is thermodynamically unstable and decomposition to graphite (in gray cast irons) may result.

The predominant phase-diagram characteristic in steels is the eutectoid reaction, in the solid state, along *GSE* where $\gamma_s = \alpha + Fe_3C$ (pearlite) at 1330°F. Steels are therefore classified as alloys in the Fe-C system having a carbon content less than 2.0 per cent C; and furthermore, according to their applications, compositions are designated as hypoeutectoid (C < 0.8 per cent), eutectoid (C = 0.8 per cent), and hypereutectoid (C < 2 per cent > 0.8 per cent). Since the slowly cooled room-temperature structures of steels contain a mechanical aggregate of the ferrite and Fe_3C-cementite phases, the property relations vary linearly as shown in Fig. 17.17. The ductility decreases with increasing carbon contents.

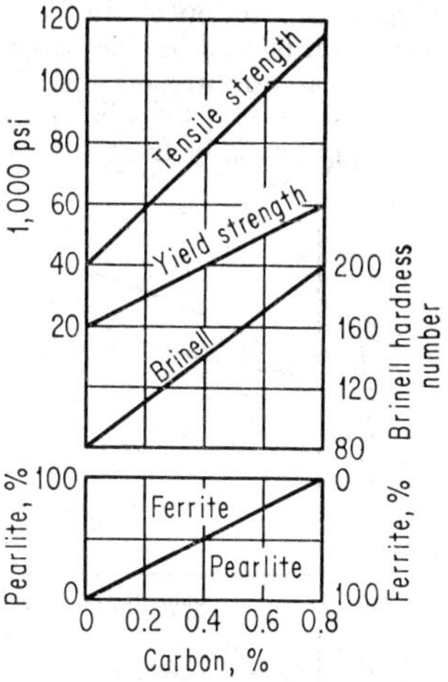

FIG. 17.17. Relation of mechanical properties and structure to carbon content of slowly cooled carbon steels.[4]

Some important characteristics of the equilibrium phases in steels are:

Phase	*Characteristics*
α ferrite	Low C solubility (less than 0.03%) body-centered cubic, ductile and ferromagnetic below 1440°F
Fe_3C, cementite	Intermetallic compound, orthorhombic, hard, brittle, and fixed composition at 6.7% C
γ austenite	Can dissolve up to 2% C in solid solution, face-centered cubic, nonmagnetic, and in this region annealing, hardening, forging, normalizing, and carburizing processes take place

Low-carbon alloys can be readily worked by rolling, drawing, and stamping because of the predominant ductility of the ferrite. Wires for suspension cables having a carbon content of about 0.7 per cent are drawn at about 1100°F (patenting) because of the greater difficulty, in room-temperature deformation, caused by the presence of a relatively large amount of the brittle Fe_3C phase.

Extensive substitutional solid-solution alloys form in binary systems when they have similar chemical characteristics and atomic diameters in addition to having the same lattice structure. Such alloys include copper-nickel (monel metal being a commercially useful one), chromium-molybdenum, copper-gold, and silver-gold (jewelry alloys). The phase diagram and the equilibrium-property changes for this system are shown in Fig. 17.18*a*. Each pure element is strengthened by the addition of the other whereby the strongest alloy is at an equal atom concentration. There are no first-order phase changes up to the start of melting (the solidus line *EHG*); so that these are not hardened by heat-treatment but only by cold work. The electrical conductivity decreases from each end of the composition axis. Because of the presence of but one phase, these alloys are selected for their resistance to electrochemical corrosion. High-temperature-service metals are alloys which have essentially a single-phase solid solution with minor additions of other elements to achieve specific effects.

Another important system is one in which there are present regions of partial solid solubility as shown in Fig. 17.18*b* together with equilibrium-property changes. An important consideration in the selection of alloys containing two or more phases is that galvanic-corrosion attack may occur when there exists a difference in the electromotive potential between the phases in the environmental electrolyte. Sacrificial galvanic protection of the base metal in which the coating is more anodic than the base metal is used in zinc-plating iron-base alloys (galvanizing alloys). The intimate mechanical mixture of phases which are electrochemically different may result in pitting corrosion, or even more seriously, intergranular corrosion may result if the alloy is improperly treated by causing localized precipitation at grain boundaries.

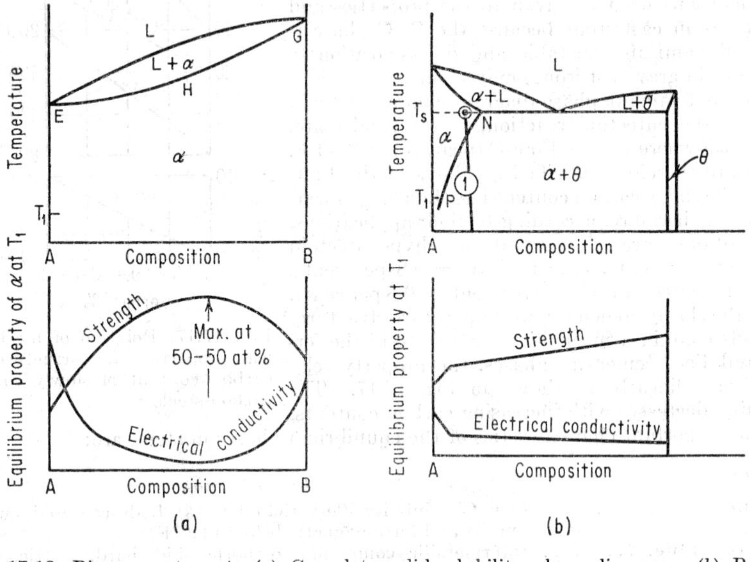

FIG. 17.18. Binary systems.[4] (*a*) Complete solid-solubility phase diagram. (*b*) Partial solid-solubility part of phase diagram. α is a substitutional solid solution, a phase with two different atoms on the same lattice. In the AlCu system θ is an intermediate phase (precipitant) having a composition nominally of $CuAl_2$.

Heat-treatment by a precipitation-hardening process is indeed an important strengthening mechanism in particular alloys such as the aircraft aluminum-base, copper-beryllium, magnesium-aluminum, and alpha-beta titanium alloys (Ti, Al, and V). In these alloys a distinctive feature is that the solvus line *NP* in Fig. 17.18*b* shows decreasing solid solubility with decreasing temperature. This in general is a necessary, but not necessarily sufficient, condition for hardening by precipitation since other thermodynamic conditions as well as coherency relations between the precipitated phases must prevail. The sequence of steps for this process is as follows: An alloy is solution heat-treated to a temperature T_s, rapidly quenched so that a metastable supersaturated solid solution is attained, and then aged at experimentally determined temperature-time aging treatments to achieve desired mechanical properties. This is the principal hardening process for those particular nonferrous alloys (including Inconels) which can respond to a precipitation-hardening process.

The engineer is frequently concerned with the strength-to-density ratio of materials and its variation with temperature. A number of materials are compared on this

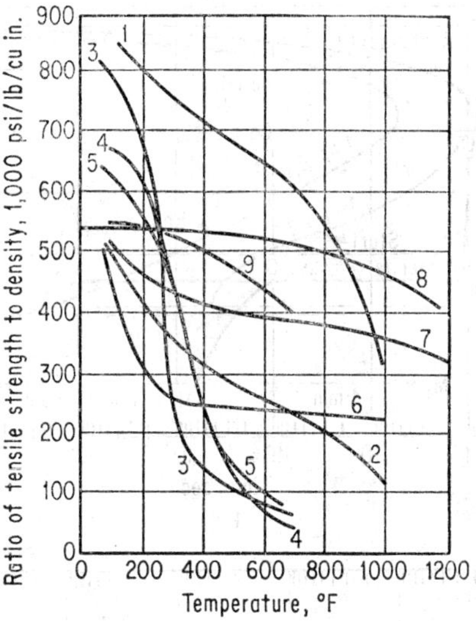

FIG. 17.19. Approximate comparison of materials on a strength-weight basis from room temperature to 1000°F.[6]

1. T_1 − 8Mn
2. 9990 T_1
3. 75S − T6Al
4. 24S − T4Al
5. AZ31A Mg
6. Annealed stainless steel
7. Half-hard stainless steel
8. Inconel X
9. Glass-cloth laminate

basis in Fig. 17.19 in which the alloys designated by curves 1, 3, 4, 5, and 8 are heat-treatable nonferrous alloys.

17.8. HEAT-TREATMENT CONSIDERATIONS FOR STEEL PARTS

The heat-treating process for steel involves heating to the austenite region where the carbon is soluble, cooling at specific rates, and tempering to relieve some of the stress which results from the transformation. Some important considerations involved in specifying heat-treated parts are strength properties, warping tendencies, mass effects (hardenability), fatigue and impact properties, induced transformation stresses, and the use of surface-hardening processes for enhanced wear resistance. Temperature and time factors affect the structures issuing from the decomposition of austenite; for a eutectoid steel (0.8 per cent C) they are:

Decomposition product from γ	Structure	Mechanism	Temp. range, °F
Pearlite........	Equilibrium ferrite + Fe_3C	Nucleation; growth	1300–1000
Bainite........	Nonequilibrium ferrite + carbide	Nucleation; growth	1000–450
Martensite.....	Supersaturated tetragonal lattice	Diffusionless	M_s(450 + lower)

The tensile strength of a slowly cooled (annealed) eutectoid steel containing a coarse pearlite structure is about 120,000 psi. To form bainite, the steel must be cooled rapidly enough to escape pearlite transformation and must be kept at an intermediate temperature range to completion, from which a product having a tensile

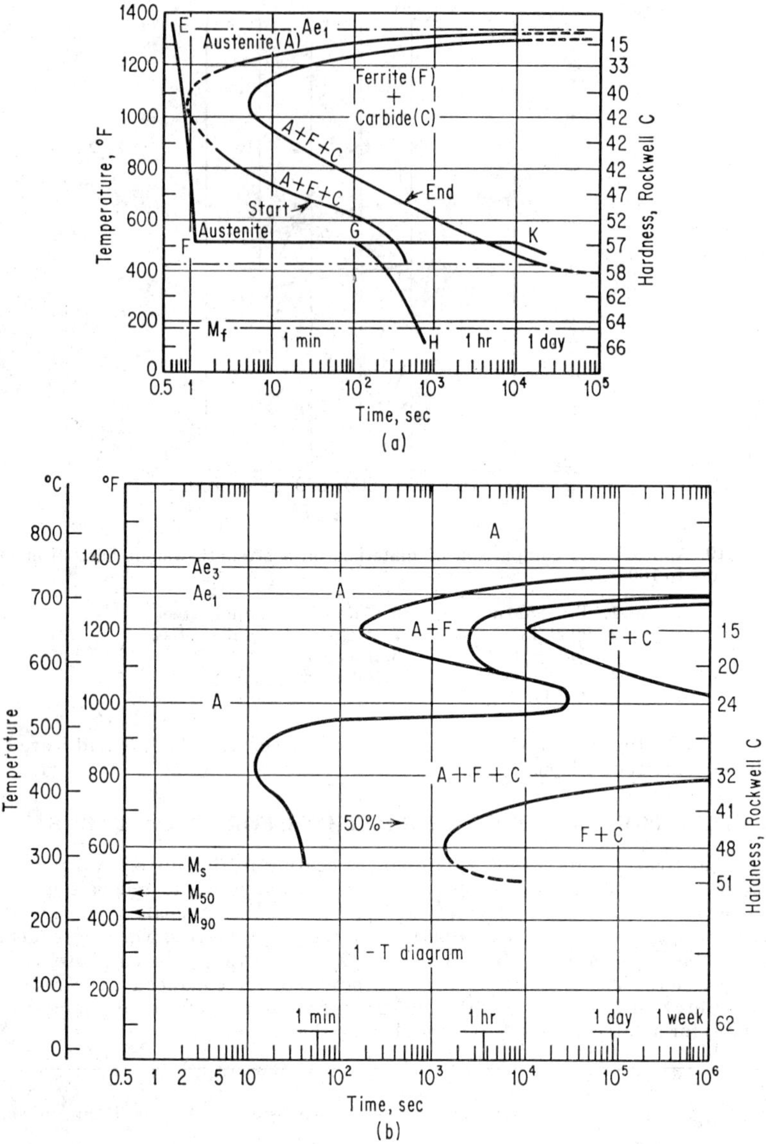

Fig. 17.20. Isothermal transformation diagram.[5] (*a*) Eutectoid carbon steel. (*b*) SAE 4340 steel.

M_S = start of martensite temperature
M_f = finish of martensite temperature
EFGH = martempering (follow by tempering)
EFK = austempering

A—Austenite; *F*—Ferrite; *C*—Carbide
M_s and M_f—temperatures for start and finish of Martensile transformation
M_{50} and M_{90}—temperatures for 50% and 90% Martensile transformation

strength of about 250,000 psi can be formed. Martensite, the hardest and most brittle product, forms independently of time by quenching rapidly enough to escape higher-temperature transformation products. The carbon atoms are trapped in the martensite, causing its lattice to be highly strained internally; its tensile strength is in excess of 300,000 psi. Isothermal transformation characteristics of all steels show the temperature-time and transformation products as in Fig. 17.20, where the lines indicate the start and end of transformation. On the temperature-time coordinates involved cooling curves can be superimposed, which show that, for a 1-in. round water-quenched specimen, mixed products will be present. The outside will be martensite and the middle sections will contain pearlite. Alloying elements are added to steels principally to retard pearlite transformation either so that less drastic quenching media can be used or to ensure more uniform hardness throughout. This retardation is shown in Fig. 17.20*b* for a SAE 4340 steel containing alloying additions of Ni, Cr, or Mo and 0.4 per cent C.

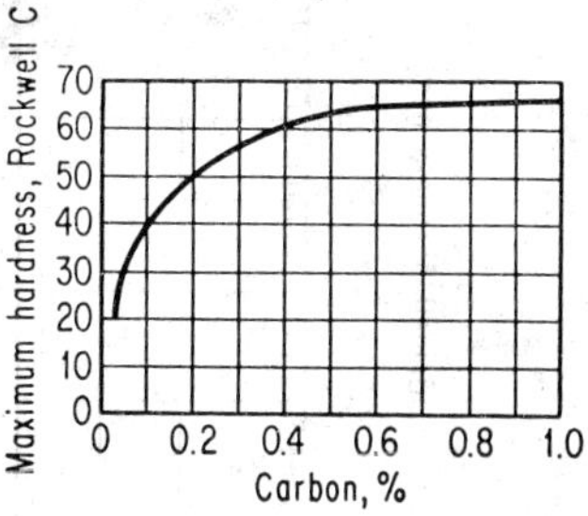

FIG. 17.21. Relation of maximum attainable hardness of quenched steels to carbon content.[7]

The carbon content in steels is the most significant element upon which selection for the maximum attainable hardness of the martensite is based. This relation is shown in Fig. 17.21. Since the atomic rearrangements involved in the transformation from the face-centered-cubic austenite to the body-centered-tetragonal martensite result in a volumetric expansion, on cooling, of about 1 per cent (for a eutectoid steel) nonuniform stress patterns can be induced on transformation. As cooling starts at the surface, by the normal process of heat transfer, parts of a member can be expanding, because of transformation, while further inward normal contraction occurs on the cooling austenite. The danger of cracking and distortion (warping) as a consequence of the steep thermal gradients and the transformation involved in hardening steels can be eliminated by using good design and the selection of the proper alloys. Where section size, time factors, and alloy content (as it affects transformation curves) permit, improved practices by martempering shown by *EFGH* in Fig. 17.20, followed by tempering or austempering shown by *EFK*, may be feasible and are worthy of investigation for the particular alloy used.

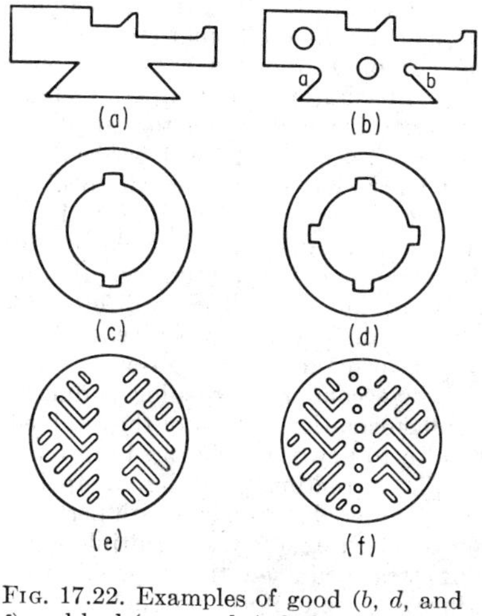

FIG. 17.22. Examples of good (*b*, *d*, and *f*) and bad (*a*, *c*, and *e*) designs for heat-treated parts.[8] (1) *b* is better than *a* because of fillets and more uniform mass distribution. (2) In *c*, cracks may form at keyways. (3) Warping may be more pronounced in *e* than in *f*, which are blanking dies.[8]

Uniform mass distribution, and the elimination of sharp corners (potential stress raisers) by the use of generous fillets, are recommended. Some design features pertinent to the elimination of quench cracks and the minimization of distortion by warping are illustrated by Fig. 17.22 in which *a*, *c*, and *e* represent poor designs in comparison with the suggested improvements apparent in *b*, *d*, and *f*.

Steels are tempered to relieve stresses, to impart ductility, and to produce a desirable microstructure by a reheating process of the quenched member. The tempering process is dependent on the temperature, time, and alloy content of the steel. Differ-

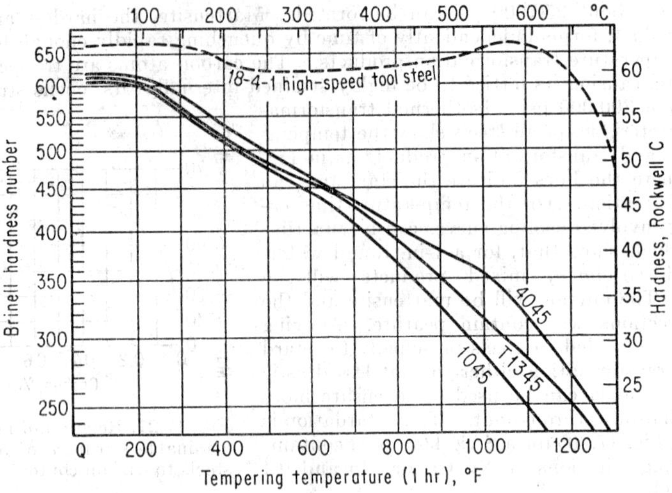

FIG. 17.23. Effect of tempering temperature on the hardnesses of SAE 1045, T1345, and 4045 steels. In the high-speed tool steel 18-4-1 secondary hardening occurs at about 1050°F.[9]

ent alloys soften at different rates according to the constitutionally dependent diffusional structure. The response to tempering for 1 hr for three different steels of the same carbon content is shown in Fig. 17.23. In addition, the tempering characteristics of a high-speed tool steel, 18 per cent W, 4 per cent Cr, 1 per cent V, and 0.9 per cent C, are shown to illustrate the secondary hardening at about 1050°F. The pronounced tendency for high-carbon steels to retain austenite on transformation normally has deleterious effects on dimensional stability and fatigue performance. In high-speed tool steel, the secondary hardening is due to the transformation of part of the retained austenite to newly transformed martensite. The structure contains tempered and untempered martensite with perhaps some retained austenite. Multiple tempering treatments on this type of steel produce a more uniform product.

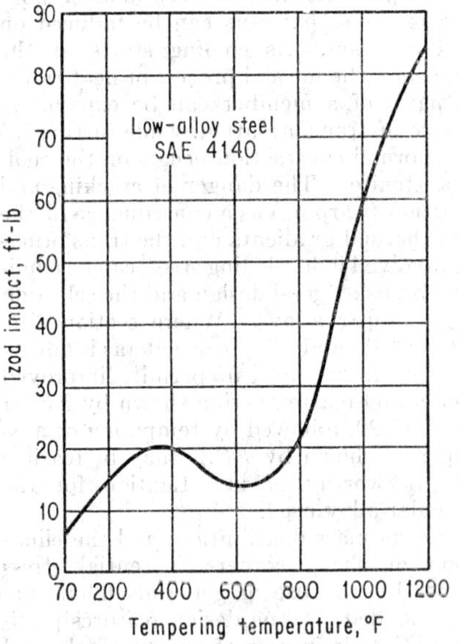

FIG. 17.24. In the tempering of this 4140 steel the notched-bar impact properties decrease in the range of 450 to 650°F.[9]

In low-alloy steels where the carbon content is above 0.25 per cent, there may be a tempering-temperature interval at about 450 to 650°F, during which the notch impact strength goes through a minimum. This is shown in Fig. 17.24 and is associated with the formation of an embrittling carbide network (ϵ carbide) about the martensite subgrain boundaries. Tempering is therefore carried out up to 400°F where the parts are to be used principally for wear resistance, or in the range of 800 to 1100°F where greater toughness is required. In the nomenclature of structural steels, adopted by the Society of Automotive Engineers and the American Iron and Steel Institute, the first two numbers designate the type of steel according to the principal alloying elements and the last two numbers designate the carbon content:

Series designation	*Types*
10xx	Nonsulfurized carbon steels
11xx	Resulfurized carbon steels (free-machining)
12xx	Rephosphorized and resulfurized carbon steels (free-machining)
13xx	Manganese 1.75%
23xx*	Nickel 3.50%
25xx*	Nickel 5.00%
31xx	Nickel 1.25%, chromium 0.65%
33xx	Nickel 3.50%, chromium 1.55%
40xx	Molybdenum 0.20 or 0.25%
41xx	Chromium 0.50 or 0.95%, molybdenum 0.12 or 0.20%
43xx	Nickel 1.80%, chromium 0.50 or 0.80%, molybdenum 0.25%
44xx	Molybdenum 0.40%
45xx	Molybdenum 0.52%
46xx	Nickel 1.80%, molybdenum 0.25%
47xx	Nickel 1.05%, chromium 0.45%, molybdenum 0.20 or 0.35%
48xx	Nickel 3.50%, molybdenum 0.25%
50xx	Chromium 0.25, 0.40, or 0.50%
50xxx	Carbon 1.00%, chromium 0.50%
51xx	Chromium 0.80, 0.90, 0.95, or 1.00%
51xxx	Carbon 1.00%, chromium 1.05%
52xxx	Carbon 1.00%, chromium 1.45%
61xx	Chromium 0.60, 0.80, or 0.95%, vanadium 0.12%, 0.10% min, or 0.15% min
81xx	Nickel 0.30%, chromium 0.40%, molybdenum 0.12%
86xx	Nickel 0.55%, chromium 0.50%, molybdenum 0.20%
87xx	Nickel 0.55%, chromium 0.50%, molybdenum 0.25%
88xx	Nickel 0.55%, chromium 0.50%, molybdenum 0.35%
92xx	Manganese 0.85%, silicon 2.00%, chromium 0 or 0.35%
93xx	Nickel 3.25%, chromium 1.20%, molybdenum 0.12%
94xx	Nickel 0.45%, chromium 0.40%, molybdenum 0.12%
98xx	Nickel 1.00%, chromium 0.80%, molybdenum 0.25%

* Not included in the current list of standard steels.

The most probable properties of tempered martensite for low-alloy steels fall within narrow bands even though there are differences in sources and treatments. The relations for these shown in Fig. 17.25 are useful in predicting properties to within approximately 10 per cent.

Structural steels may be specified by hardenability requirements, the H designation, rather than stringent specification of the chemistry. Hardenability, determined by the standardized Jominy end-quench test, is a measurement related to the variation in hardness with mass, in quenched steels. Since different structures are formed as a function of the cooling rate and the transformation is affected by the nature of the alloying elements, it is necessary to know whether the particular steel is shallow (*A*) or deeply hardenable (*C*), as in Fig. 17.26. The hardenability of a particular steel is a useful criterion in selection because it is related to the mechanical properties pertinent to the section size.

The selection of through-hardened steel based upon carbon content is indicated below for some typical applications.

Carbon range	Requirement	Approx tensile strength level, psi	Applications
Medium, 0.3 to 0.5%	Strength and toughness	150,000	Shafts, bolts, forgings, nuts
Intermediate, 0.5 to 0.7%	Strength	225,000	Springs
High, 0.8 to 1.0%	Wear resistance	Greater than 300,000	Bearings, rollers, bushings

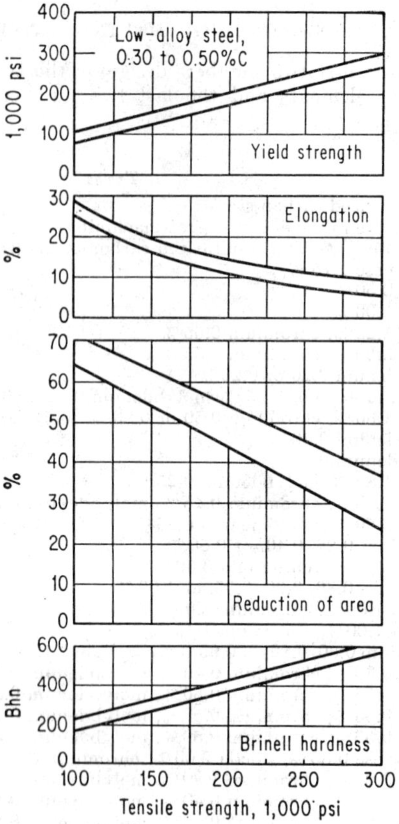

FIG. 17.25. The most probable properties of tempered martensite for a variety of low-alloy steels.[4]

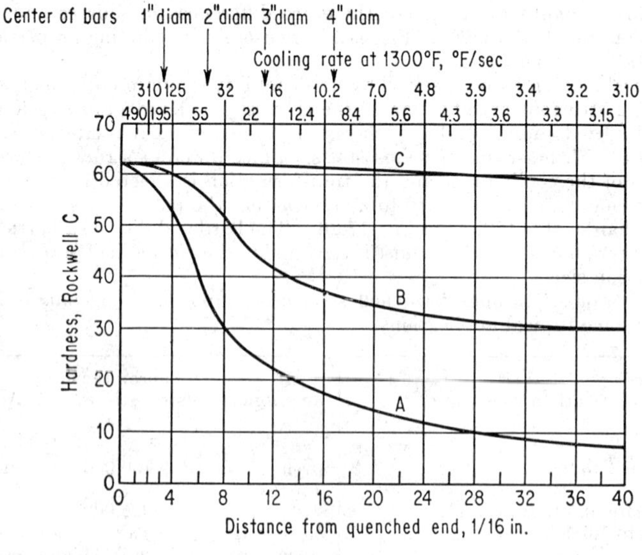

FIG. 17.26. Hardenability curves for different steels with the same carbon content.[7]

17.9. SURFACE-HARDENING TREATMENTS

The combination of high surface wear resistance and a tough-ductile core is particularly desirable in gears, shafts, and bearings. Various types of surface-hardening treatments and processes can achieve these characteristics in steels; the most important of these are the following:

Base metal	*Process*
Low C to 0.3%	Carburization—A carbon diffusion in the γ-phase region, with controlled hydrocarbon atmosphere or in a box filled with carbon. The case depth is dependent on temperature-time factors. Heat-treatment follows process
Medium C, 0.4 to 0.5%	Localized surface heating by induction or a controlled flame to above the Ac_3 temperature; quenched and tempered
Nitriding (Nitralloys, stainless steels)	Formation of nitrides (in heat-treated parts) in ammonia atmosphere at 950 to 1000°F, held for long times. A thin and very hard surface forms and there may be dimensional changes
Low C, 0.2%	Cyaniding—parts placed in molten salt baths at heat-treating temperatures; some limited carburization and nitriding occur for cases not exceeding 0.020 in. Parts are quenched and tempered

The carbon penetration in carburization is determined by temperature-time-distance relations issuing from the solution of diffusional equations where D, the diffusion coefficient, is independent of concentrations. These relations, shown in Fig. 17.27, permit the selection of a treatment to provide specific case depths. Typical applications are as follows:

Case depth, in.	*Applications (automotive)*
More than 0.020	Push rods, light-load gears, water pump shafts
0.020–0.040	Valve rocker arms, steering-arm bushings, brake and clutch pedal shafts
0.040–0.060	Ring gears, transmission gears, piston pins, roller bearings
Greater than 0.060	Camshafts

The heat-treatments used on carburized parts depend upon grain-size requirements, minimization of retained austenite in the microstructure, amount of undissolved carbide network, and core-strength requirements. As a result of carburization, the surface fiber stresses are compressive. This leads to better fatigue properties. This treatment, which alters the surface chemistry by diffusion of up to 1 per cent carbon in a low-carbon steel, gives better wear resistance because the surface hardness is treated for values above 60 Rockwell C, while the low-carbon-content core has ductile properties to be capable of the transmission of torsional or bending loads.

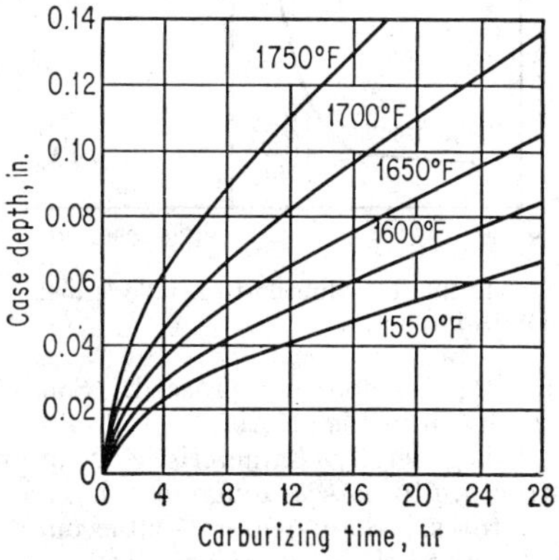

FIG. 17.27. Relation of time and temperature to carbon penetration in gas carburizing.[10]

Selection of the nitriding process requires careful consideration of cost because of the long times involved in the case formation. A very hard case having a hardness of about 70 Rockwell C ensures excellent wear resistance. Nitrided parts have good corrosion resistance and improved fatigue properties. Nitriding follows the finish machining and grinding operations, and many parts can be nitrided without great likelihood of distortion. Long service (several hundred hours) at 500°F has been attained in nitrided gears made from a chromium-base hot-work steel H11. Some typical nitrided steels and their applications are as follows:

Steel	Nitriding treatment		Case hardness, in.	Case depth Rockwell C	Applications
	Hr	°F			
4140	48	975	0.025–0.035	53–58	Gears, shafts, splines
4340	48	975	0.025–0.035	50–55	Gears, drive shafts
Nitralloy 135M	48	975	0.020–0.025	65–70	Valve stems, seals, dynamic faceplates
H11	70	960 + 980	0.015–0.020	67–72	High-temperature power gears, shafts, pistons

17.10. NOTCHED IMPACT PROPERTIES—CRITERIA FOR MATERIAL SELECTION

When materials are subject to high deformation rates and are particularly sensitive to stress concentrations at sharp notches, criteria must be established to indicate safe operating-temperature ranges. The impact test (Izod or Charpy V-notch) performed on notched specimens conducted over a prescribed temperature range indicates the likelihood of ductile (shear-type) or brittle (cleavage-type) failure. In this test the velocity of the striking head at the instant of impact is about 18 fps; so that the strain rates are several orders of magnitude greater than in a tensile test. The energy absorbed in fracturing a standard notched specimen is measured by the differences in potential energy from free fall of the hammer to the elevation after fracturing it. The typical effect of temperature upon impact energy for a metal which shows ductile and brittle characteristics is shown in Fig. 17.28. Interest centers on the transition temperature range and the material-sensitive factors such as composition, microstructure, and embrittling treatments. ASTM specifications for structural steel for ship plates specify a minimum impact energy at a given temperature, as, for example, 15 ft-lb at 40°F. It is desirable to use materials at impact energy levels and at service temperatures where crack propagation does not proceed. Some impact characteristics for construction steels are shown in Fig. 17.29.

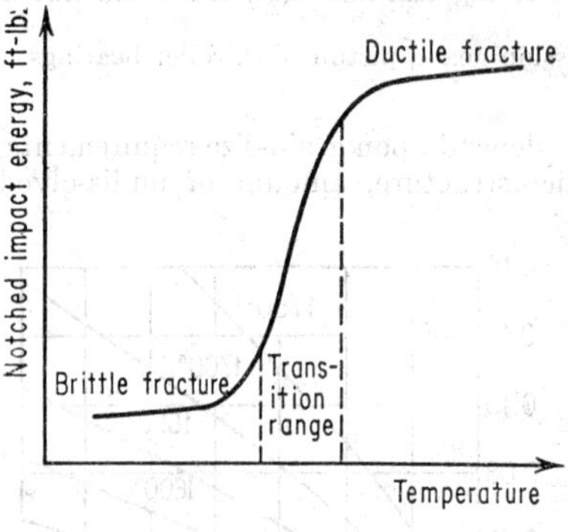

FIG. 17.28. The ductile-to-brittle transition in impact.

It is generally characteristic of pure metals which are face-centered cubic in lattice structure to possess toughness (have no brittle fracture tendencies) at very low temperatures. Body-centered-cubic pure metals, as well as hexagonal metals, do show ductile-to-brittle behavior. Tantalum, a body-centered-cubic metal, is a possible exception and is ductile in impact even at cryogenic temperatures. Alloyed metals do not follow any general pattern of behavior; some specific impact values for these are as follows:

Material	Charp V-notch at 80°F	Impact strength, ft-lb at −100°F
Cu-Be(2%) HT	5.4	5.5
Phosphor bronze (5% Sn):		
Annealed	167	193
Spring temper	46	44
Nilvar Fe-Ni (36%):		
Annealed	218	162
Hard	97	77
2024-T6 aluminum aircraft alloy, HT aged	12	12
7079-T6 aluminum aircraft alloy, HT aged	4.5	3.5
Mg-Al-Zn extruded	7.0	3.0

Austenitic stainless steels are ductile and do not exhibit transition in impact down to very low temperatures.

Some brittle service failures in steel structures have occurred in welded ships, gas-transmission pipes, pressure vessels, bridges, turbine generator rotors, and storage tanks. Serious consideration must then be given the effect of stress raisers, service temperatures, tempering embrittling structures in steels, grain-size effects, as well as the effects of minor impurity elements in materials.

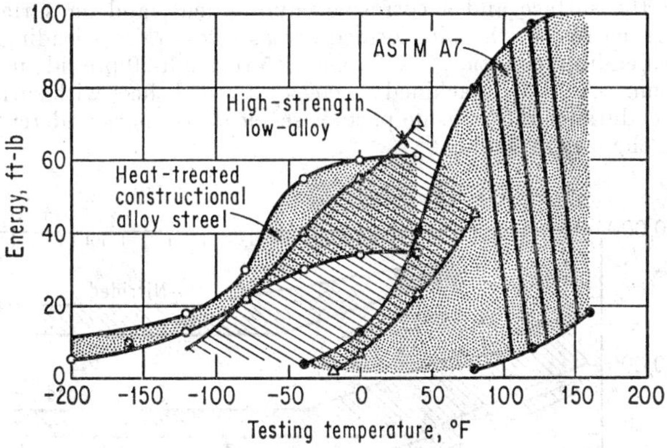

FIG. 17.29. Influence of testing temperature on notch toughness, comparing carbon steel of structural quality (ASTM A7) with high-strength low-alloy and heat-treated constructional alloy steel.[4] Charpy V-notch.

17.11. FATIGUE CHARACTERISTICS FOR MATERIALS SPECIFICATIONS

Most fatigue failures observed in service as well as under controlled laboratory tests are principally the result of poor design and machining practice. The introduction of potential stress raisers by inadequate fillets, sharp undercuts, and toolmarks at the surfaces of critically cyclically stressed parts may give rise to crack nucleation and propagation so that ultimate failure occurs. Particular attention should be given to material fatigue properties where rotating and vibrating members experience surface fibers under reversals of stress.

In a fatigue test, a highly polished round standard specimen is subjected to cyclic loading; the number of stress reversals to failure is recorded. For sheets, the standard specimen is cantilever supported. Failure due to tensile stresses usually starts at the surface. Typical of a fatigue fracture is its conchoidal appearance, where there is a smooth region in which the severed sections rubbed against each other and where the crack progressed to a depth where the load could no longer be sustained. From the fatigue data a curve of stress vs. number of cycles to failure is plotted. Note that there can be considerable statistical fluctuation in the results (about ±15 per cent variations in stress).

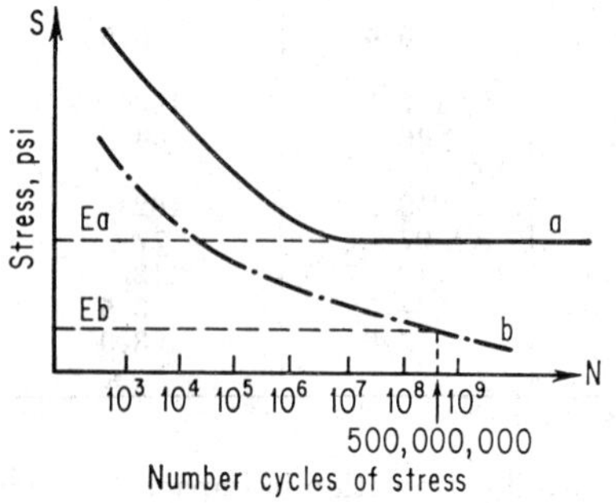

FIG. 17.30. Fatigue curves. (*a*) Most materials have an endurance limit E_a (asymptotic stress). (*b*) Endurance limit E_b (non-asymptotic) set at arbitrary value of *N*.

Two characteristics can be observed in fatigue curves with respect to the endurance limit shown by E_a and E_b in Fig. 17.30:

E_a, curve *A*, the asymptotic stress value, typical in most materials

E_b, curve *B*, a stress value taken at an arbitrary number of cycles; e.g., 500,000,000, typical in Al and Mg alloys

For design specifications, the endurance limit represents a safe working stress for fatigue. The endurance ratio is defined as the ratio of the endurance limit to the ultimate tensile strength. These values are strongly dependent upon the presence of notches on the surface and a corrosive environment, and on surface-hardening treatments. In corrosion, the pits formed act as stress raisers leading generally to greatly reduced endurance ratios. References 15 through 20 provide useful information on corrosion. A poorly machined surface or a rolled sheet with surface scratches evidences low endurance ratios, as do parts in service with sharp undercuts and insufficiently filleted changes in section.

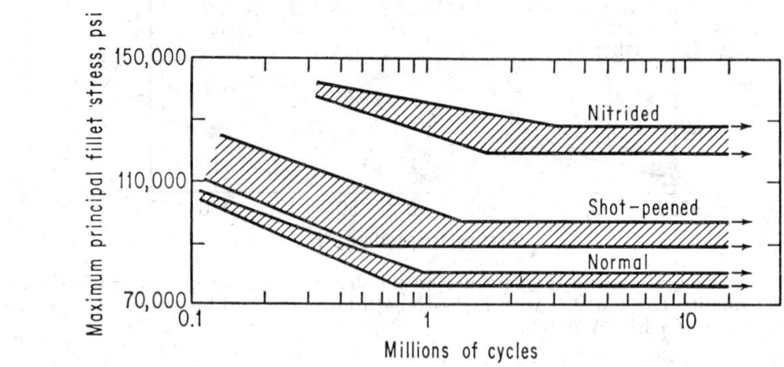

FIG. 17.31. Effects of surface-hardening treatments in improving the fatigue characteristics of a 4340 steel used as an aircraft crankshaft.[11]

Improvements in fatigue properties are brought about by those surface-hardening treatments which produce induced compressive stresses as in steels, nitrided (about 160,000 psi compressive stress) or carburized (about 35,000 psi compressive stress). An example of the effects of surface-hardening treatments on an aircraft crankshaft made from a 4340 hardened and tempered steel (30 Rockwell C) is shown in Fig. 17.31.

Metallurgical factors related to poorer fatigue properties are the presence of retained austenite in hardened steels, the presence of flakes or sharp inclusions in the microstructure, and treatments which induce preferential corrosive grain-boundary attack. When parts are quenched or formed so that surface tensile stresses are present, stress-relieval treatments are advisable.

17.12. MATERIALS FOR HIGH-TEMPERATURE APPLICATIONS

(a) Introduction

Selection of materials to withstand stress at high temperature is based upon experimentally determined temperature stress-time properties. Some useful engineering design criteria follow.

1. Dimensional change, occurring by plastic flow, when metals are stressed at high temperatures for prolonged periods of time, as measured by creep tests
2. Stresses that lead to fracture, after certain set time periods, as determined by stress-rupture tests, where the stresses and deformation rates are higher than in a creep test
3. The effect of environmental exposure on the oxidation or scaling tendencies
4. Considerations of such properties as density, melting point, emissivity, ability to be coated and laminated, elastic modulus, and the temperature dependence on thermal conductivity and thermal expansion

Furthermore, the microstructural changes occurring in alloys used at high temperatures are correlated with property changes in order to account for the significant discontinuities which occur with exposure time. As a result of these evaluations, special alloys that have been (or are being) developed are recommended for use in different temperature ranges extending to about 2800°F (refractory range). Vacuum or electron-beam melting and special welding techniques are of special interest here in fabricating parts.

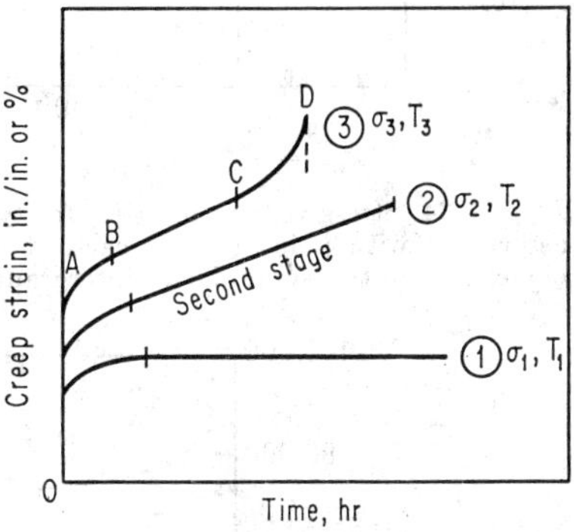

FIG. 17.32. Typical creep curves. At constant temperature, $\sigma_3 > \sigma_2 > \sigma_1$. At constant stress, $T_3 > T_2 > T_1$.

(b) Creep and Stress—Rupture Properties

In a creep test, the specimen is heated in a temperature-controlled furnace, an axial load is applied, and the deformation is recorded as a function of time, for periods of 1,000 to 3,000 hr. Typical changes in creep strain with time, for different conditions of stress and temperature, are shown in Fig. 17.32. Plastic flow creep, associated with the movement of dislocations by climb sliding of grain boundaries and the diffusion of vacancies, is characterized by:

1. *OA*, elastic extension on application of load
2. *AB*, first stage of creep with changing rate of creep strain
3. *BC*, second stage of creep, in which strain rate is linear and essentially constant
4. *CD*, third stage of increasing creep rate leading to fracture

Increasing stress at a constant temperature or increasing temperature at constant stress results in the transfers from the 1 to 2 to 3 curves in Fig. 17.32.

The engineering design considerations for dimensional stability are based upon

1. Stresses resulting in a second-stage creep rate of 0.0001 per cent per hour (1 per cent per 10,000 hr or 1 per cent per 1.1 years)
2. A second-stage creep rate of 0.00001 per cent per hour (1 per cent per 100,000 hr

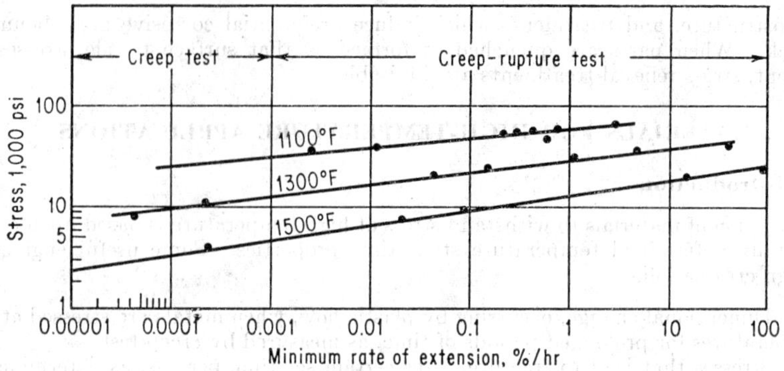

FIG. 17.33. Correlation of creep and rupture test data for type 316 stainless steel (18 Cr, 8 Ni, and Mo).[12]

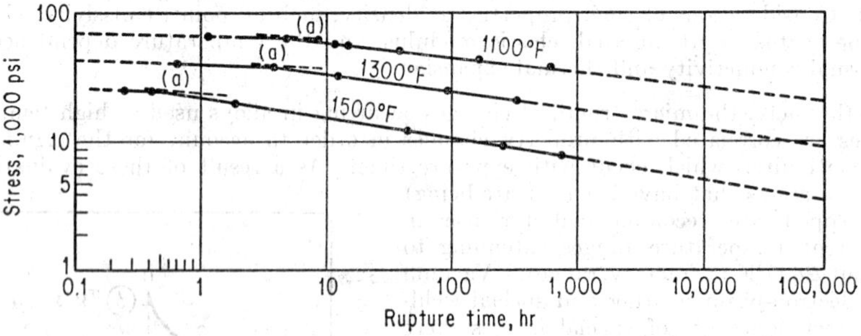

FIG. 17.34. Stress vs. rupture time for type 316 stainless steel.[12] The structural character associated with point (*a*), on each of the three relations, is that the mode of fracture changes from transgranular to intergranular.

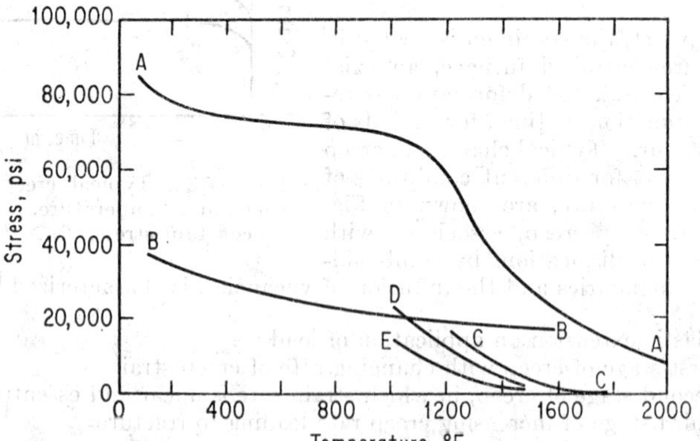

FIG. 17.35. Properties of type 316 stainless steel (18 Cr, 8 Ni, and Mo).[12]

A = short-time tensile strength
B = short-time yield strength, 0.2 per cent offset
C = stress for rupture 10,000 hr
D = stress for creep rate 0.0001 per cent per hr
E = stress for creep rate 0.00001 per cent per hr

or 1 per cent per 11 years), where weight is of secondary importance relative to long service life, as in stationary turbines

The time at which a stress can be sustained to failure is measured in a stress-rupture test and is normally reported as rupture values for 10, 100, 1,000, and 10,000 hr or more. Because of the higher stresses applied in stress-rupture tests, shown in Fig. 17.33, some extrapolation of data may be possible and some degree of uncertainty may ensue. Discontinuous changes at points *a* in the stress-rupture data shown in Fig. 17.34 are associated with a change from transgranular to intergranular fracture, and further microstructural changes can occur at increasing times. A composite picture of various high-temperature test results is given in Fig. 17.35 for a type 316 austenitic stainless steel.

(c) Material Selection

The aluminum and magnesium light alloys used in aircraft in their heat-treated condition have high-temperature applications limited to about 400°F. Low-density titanium alloys, of the alpha-beta heat-treatable type, have been used for aircraft gas-turbine compressor parts (where creep properties become important) in 600 to

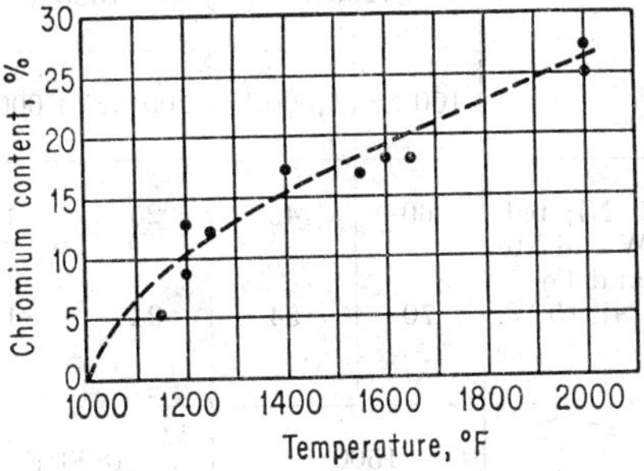

FIG. 17.36. The effect of chromium content on service temperature, without causing excessive sealing in alloys.[12]

1000°F applications. Low-alloyed steels are generally used safely to about 700°F in the high-temperature range. Above these temperatures, special alloys have been developed for the 1200 to 1800°F range and the 1800 to 2800°F range (the refractory alloys).

Alloys used in the temperature range of 1200 to 1800°F are essentially of the iron-, nickel-, and cobalt-base type and their structures, although predominantly of a single phase, may contain intermetallics and precipitating phases. Additions of chromium increase the oxidation resistance in a manner shown by Fig. 17.36. However, high-chromium alloys show depletion effects when used in a high vacuum. For nuclear applications, requiring low-neutron-absorption cross sections, alloys with high cobalt concentrations should be avoided. Some measured high-temperature stress-rupture properties for alloys used in this temperature range are listed in Table 17.3.

Applications of metals at the highest service temperatures call for the four refractory elements and their alloys. Each of these has a body-centered-cubic structure and a high melting point, oxidizes readily above 1200°F, and has the properties given in Table 17.4. The strength properties are affected by the impurities present (notably oxygen, carbon, and nitrogen), and therefore electron-beam melting under high vacuum is under investigation for attaining higher purities. Because of the oxidizing characteristics at high temperatures, protective coatings must be used.

Sheet products can be fabricated from Ta, Cb, and Mo by conventional means, since they have reasonably good ductilities at room temperature. However, tungsten is relatively brittle at room temperature and for this reason sheet-fabrication methods

Table 17.3. High-temperature Rupture Strength Properties for Some Superalloys[13]

Alloy	Principal alloy content, %	Rupture strengths, 1,000 psi				Typical applications
Fe base		1200°F		1500°F		
		100 hr	1,000 hr	100 hr	1,000 hr	
Incaloy 901	13 Cr; 43 Ni; 34 Fe, and Mo and Ti	94	78	24	15	Gas-turbine rotor disks
Refractory 26	18 Cr; 38 Ni; 20 Co, bal. Fe, and Mo and Ti	80	63	27	18	Gas-turbine parts, blading, bolting
Co base		1200°F		1500°F		
		100 hr	1,000 hr	100 hr	1,000 hr	
S816	20 Cr; 20 Ni; bal. Co and W and Mo and Cb and Fe	60	46	25	18	Jet-engine buckets
HS25	20 Cr; 10 Ni; 15 W, bal. Co	70	54	24	17	Jet-engine parts, sheet alloy
Ni base		1500°F		1800°F		
		100 hr	1,000 hr	100 hr	1,000 hr	
Inconel 713C	11.5 Cr; 74 Ni; Al and Mo and Cb	68	47	20	15	Jet-engine blades
Hastelloy R235	15.5 Cr; 10 Fe bal. Ni and Mo and Ti	40	30	8	5	Gas-turbine Jet-engine parts, sheet
Rene 41	19 Cr; 11 Co; 10 Mo; bal. Ni	45	29	11	. . .	Gas turbine, sheet, bolting

Table 17.4. Properties of Refractory Elements

Element	Melting point, °F	Density, lb cu in.	Recrystallization temp., °F	Young's modulus of elasticity at room temp., *E*, psi
Columbium, Cb.	4379	0.31	1785–2100	30,000,000
Molybdenum, Mo.	4730	0.369	2100–2200	47,000,000
Tantalum, Ta.	5425	0.60	2200–2400	27,000,000
Tungsten, W.	6170	0.697	2200–3000	50,000,000

require special development. Joining by fusion welding results in ductile joints for Ta and Cb, whereas for W and Mo grain-coarsening and cracking tendencies are factors of importance in attaining usable products.

High-temperature strength properties are substantially improved by alloying additions to the refractory elements. Tungsten-base alloys, some in the development stage, are used at the highest temperatures. Some typical rupture strength properties determined in heats are given in Table 17.5.

Table 17.5. Rupture Strength Properties[13]

Metal	Nominal alloy content, %	Rupture strength, 1,000 psi	
		2000°F	
Cb base FS82	33 Ta, 7.5 Zr	10 hr	100 hr
		25	18
Mo base	0.5 Ti	2000°F, 100 hr	2400°F, 100 hr
		34	10
		2500°F	
W base	2% ThO_2	10 hr	100 hr
		29	22

17.13. MATERIALS FOR LOW-TEMPERATURE APPLICATION

Materials for low-temperature application are of increasing importance because of the technological advances in cryogenics. The most important mechanical properties are usually strength and stiffness, which generally increase as the temperature is decreased. The temperature dependence on ductility is a particularly important criterion in design, because some materials exhibit a transition from ductile to brittle behavior with decreasing temperature. Factors related to this transition are microstructure, stress concentrations present in notches, and the effects of rapidly applied strain rates in materials. Mechanical design can also influence the tendency for brittle failure at low temperature, and for this reason, it is essential that sharp notches (which can result from surface-finishing operations) be eliminated and that corners at changes of section be adequately filleted.

Low-temperature tests on metals are made by measuring the tensile and fatigue properties on unnotched and notched specimens and the notched impact strength. Metals exhibiting brittle characteristics at room temperature, by having low values of per cent elongation and per cent reduction in area in a tensile test as well as low impact strength, can be expected to be brittle at low temperatures also. Magnesium alloys, some high-strength aluminum alloys in the heat-treated condition, copper-beryllium heat-treated alloys, and tungsten and its alloys all exhibit this behavior. At best, applications of these at low temperatures can be made only provided that they adequately fulfill design requirements at room temperature.

When metals exhibit transitions in ductile-to-brittle behavior, low-temperature applications should be limited to the ductile region, or where experience based on field tests is reliable, a minimum value of impact strength should be specified. The failure, by breaking in two, of 19 out of 250 welded transport ships in World War II, caused by the brittleness of ship plates at ambient temperatures, focused considerable attention on this property. It was further revealed in tests that these materials had Charpy V-notch impact strengths of about 11 ft-lb at this temperature. Design specifications for applications of these materials are now based on higher impact values. For temperatures extending from subatmospheric temperatures to liquid-nitrogen temperatures (−320°F), transitions are reported for ferritic and martensitic steels, cast steels, some titanium alloys, and some copper alloys.

Design for low-temperature applications of metals need not be particularly concerned with the Charpy V-notch impact values provided they can sustain some shear deformation and that tensile or torsion loads are slowly applied. Many parts are used successfully in polar regions, being based on material design considerations within the elastic limit. When severe service requirements are expected in use, relative to rapid rates of applied strain on notch-sensitive metals, particular attention is placed on selecting materials which have transition temperatures below that of the environment.

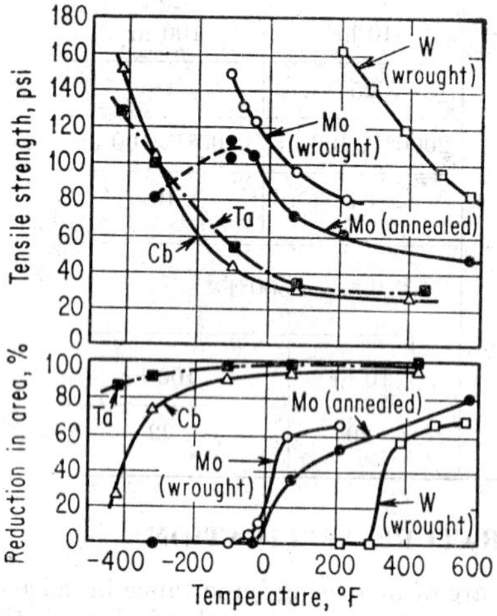

Fig. 17.37. Strength and ductility of refractory metals at low temperatures.[13]

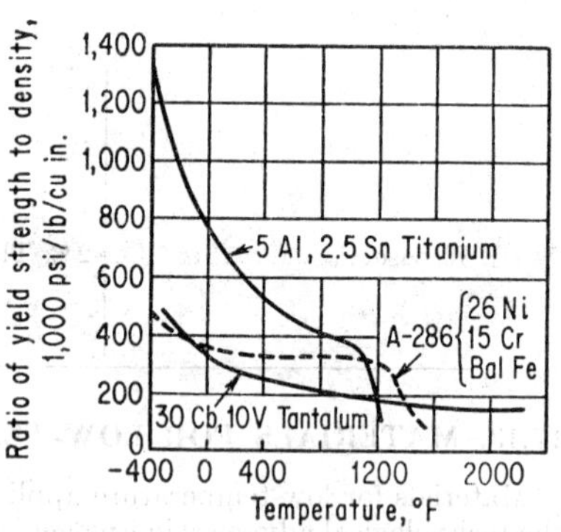

Fig. 17.38. Yield-strength-to-density ratios related to temperature for some alloys of interest in cryogenic applications.[13]

Some important factors related to the ductile-to-brittle transition in impact are the composition, microstructure, and changes occurring by heat-treatment, preferred directions of grain orientation, grain size, and surface condition.

The transition temperatures in steels are generally raised by increasing carbon content, by the presence of more than 0.05 per cent sulfur, and significantly by phosphorus at a rate of 13°F per 0.01 per cent P. Manganese up to 1.5 per cent decreases the transition temperature and high nickel additions are effective, so that in the austenitic stainless steels the behavor is ductile down to liquid-nitrogen temperatures. In high-strength medium-alloy steels it is desirable, from the standpoint of lowering the transition range, that the structure be composed of a uniformly tempered martensite, rather than containing mixed products of martensite and bainite or martensite and pearlite. This can be controlled by heat-treatment. The preferred orientation that can be induced in rolled and forged metals can affect notched impact properties; so that specimens made from the longitudinal or rolling direction have higher impact strengths than those taken from the transverse direction. Transgranular fracture is normally characteristic of low-temperature behavior of metals. The metallurgical factors leading to intergranular fracture, due to the segregation of

Table 17.6. Test Properties

Tensile Test Properties

Material	Tensile strength, 1,000 psi		Yield strength, 1,000 psi		Elongation, %		Modulus of elasticity, 100 psi	
	80°F	−100°F	80°F	−100°F	80°F	−100°F	80°F	−100°F
Beryllium copper 2% Be—heat-treated sheet...	189	194	154	170	3	3	19.1	19.1
Phosphor bronze 5% Sn; spring temp...........	98	107	90	97	7	11	16.5	17.1
Molybdenum sheet partly recrystallized..........	97	141	75	130	19	3.5	48.7	47.5
Tungsten wire, drawn....	214	...	196	...	2.6	....	56	
Tantalum sheet, annealed	55	71	36	66	27	25	28.2	28.3
Nilvar sheet 36 Ni; 74 Fe, rolled................	76	101	62	76	27	30	21.8	19.7

Sheet Fatigue Properties

Material	Fatigue strength at 2 × 10^7 cycles 1,000 psi		Endurance ratios	
	80°F	−100°F	80°F	−100°F
Beryllium copper..............	31	45	0.16	0.23
Molybdenum...................	46	65	0.59	0.47
Tantalum......................	35	41	0.64	0.58
Nilvar..........................	26	30	0.33	0.30

Charpy V-notch Impact Strength

Material	Impact strength, ft-lb	
	80°F	−100°F
Beryllium copper...............	5.4	5.4
Phosphor bronze...............	46	44
Nilvar..........................	97	77

embrittling constituents at grain boundaries, cause concern in design for low-temperature applications. In addition to the control of these factors for enhanced low-temperature use, it is important to minimize or eliminate notch-producing effects and stress concentration, by specifying proper fabrication methods and providing adequate controls on these, as by surface inspection.

Examples of some low-temperature properties of the refractory metals, all of which have body-centered-cubic structures, are shown in Fig. 17.37. The high ductility of tantalum at very low temperatures is a distinctive feature in this class that makes it attractive for use as a cryogenic (as well as a high-temperature) material. Based on the increase of yield-strength-to-density ratio with decreasing temperatures shown in Fig. 17.38, the three alloys of a titanium-base Al (5) Sn (2.5), an austenitic iron-base Ni (26), Cr (15) alloy A286, as well as the tantalum-base Cb (30), 10 V alloy, are also useful for cryogenic use.

Comparisons of the magnitude of property changes obtained by testing at room temperature and −100°F for some materials of commercial interest are shown in Table 17.6.

17.14. RADIATION DAMAGE[31]

A close relationship exists between the structure and the properties of materials Modification and control of these properties are available through the use of various metallurgical processes, among them the concept of nuclear radiation. Nuclear radiation is a process whereby an atomic nucleus undergoes a change in its properties brought about by interatomic collisions.

The energy transfer which occurs when neutrons enter a metal may be estimated by simple mechanics, the quantity of energy transferred being dependent upon the atomic mass. The initial atomic collision, or primary "knock-on" as it is called, has enough energy to displace approximately 1,000 further atoms, or so-called secondary knock-ons. Each primary or secondary knock-on must leave behind it a resulting vacancy in the lattice. The primaries make very frequent collisions because of their slower movement, and the faster neutrons produce clusters of "damage," in the order of 100 to 1,000 angstroms in size, which are well separated from one another.

Several uncertainties exist about these clusters of damage, and because of this it is more logical to speak of radiation damage than of point defects, although much of the damage in metals consists of point defects. Aside from displacement collisions, replacement collisions are also possible in which moving atoms replace lattice atoms. The latter type of collision consumes less energy than the former.

Another effect, important to the life of the material, is that of transmutation, or the conversion of one element into another. Due to the behavior of complex alloys, the cumulative effect of transmutation over long periods of time will quite often be of importance. U^{235}, the outstanding example of this phenomenon, has enough energy after the capture of a slow neutron to displace one or more atoms.

Moving charged particles may also donate energy to the valence electrons. In metals this energy degenerates into heat, while in nonconductors the electrons remain in excited states, and will sometimes produce changes in properties.

Figure 17.39 illustrates the effect of irradiation on the stress-strain curve of iron crystals at various temperatures.[32] In metals other than iron irradiation tends to produce a ferrous-type yield point and has the effect of hardening a metal. This hardening may be classified as a friction force and a locking force on the dislocations. Some factors of irradiation hardening are:

1. It differs from the usual alloy hardening in that it is less marked in cold-worked than annealed metals.
2. Annealing at intermediate temperatures may increase the hardening.
3. Alloys may exhibit additional effects, due for example to accelerated phase changes and aging.

The most noticeable effect of irradiation is the rise in transition temperatures of metals which are susceptible to cold brittleness. Yet another consequence of irradiation is the development of internal cracks produced by growth stresses. At high enough temperatures gas atoms can be diffused and may set up large pressures within the cracks.

Some other effects of irradiation are swelling, phase changes which may result in greater stability, radiation growth, and creep. The reader is referred to ref. 31 for an analysis of these phenomena.

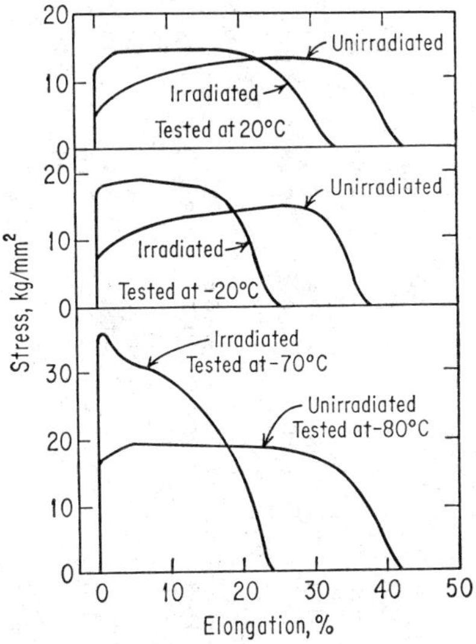

FIG. 17.39. Effect of irradiation on stress-strain curves of Fe single crystals tested at different temperatures. Irradiation dose 8×10^{17} thermal n/cm². (*Courtesy of ref. 31.*)

17.15. PRACTICAL REFERENCE DATA

Table 17.7 through 17.10 give various properties of commonly used materials. Figure 17.40 provides a hardness conversion graph for steel. References 21 through 30 yield more information.

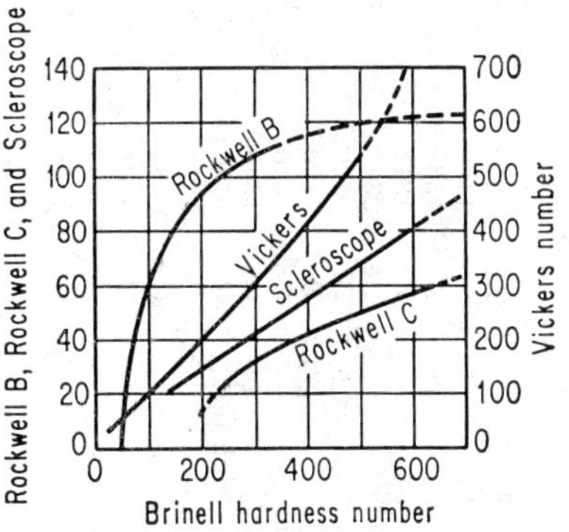

FIG. 17.40. Hardness conversion curves for steel.

Table 17.7. Physical Properties of Metallic Elements*

Element	Symbol	Melting point, °F	Boiling point, °F	Specific heat,[a] cal/g/°C	Thermal conductivity,[a] Btu/hr/sq ft/°F/ft	Density,[a] g/cm³	Modulus of elasticity in tension, million psi	Coefficient of linear thermal expansion,[a] μ in./in./°F	Electrical resistivity, microhm-cm	Crystal structure
Aluminum	Al	1220	4442	0.215	128.	2.70	9	13.1	2.65	f.c.c.
Antimony	Sb	1167	2516	0.049	10.8	6.62	11.3	4.7	39	Rhomb.
Arsenic	As	1503 (28 atm)	1135[b]	0.082	.....	5.72	.....	2.6	33.3	Rhomb.
Barium	Ba	1317	2980	0.068	.....	3.5	.....	....	....	b.c.c.
Beryllium	Be	2332	5020	0.45	84.4	1.85	42	6.4	4	h.c.p.
Bismuth	Bi	520	2840	0.029	4.8	9.80	4.6	7.4	107	Rhomb.
Boron	B	3690	......	0.309	.....	2.34	.....	4.6	10^{12}	Orthorhomb.
Cadmium	Cd	610	1409	0.055	53.	8.65	8	16.55	6.83	h.c.p.
Calcium	Ca	1540	2625	0.149	72.3	1.55	3.5	12.4	3.91	f.c.c.
Carbon (graphite)	C	6740[b]	8730	0.165	13.8	2.25	0.7	0.3 to 2.4	1375	Hexag.
Cerium	Ce	1479	6280	0.045	6.6	6.77	6	4.4	75	f.c.c.
Cesium	Cs	84	1273	0.048	.....	1.90	.....	54	20	b.c.c.
Chromium	Cr	3407	4829	0.11	40.3	7.19	36	3.4	12.9	b.c.c.
Cobalt	Co	2723	5250	0.099	41.5	8.85	30	7.66	6.24	h.c.p.
Columbium	Cb	4474	8901	0.065	31.5	8.57	.....	4.06	12.5	b.c.c.
Copper	Cu	1981	4703	0.092	226.	8.96	16	9.2	1.67	f.c.c.
Gallium	Ga	86	4059	0.079	19.4	5.91	.....	10	17.4	Orthorhomb.
Germanium[c]	Ge	1719	5125	0.073	33.7	5.32	.....	3.19	46×10^6	Diam. cubic
Gold	Au	1954	5380	0.031	171.	19.32	11.6	7.9	2.35	f.c.c.
Indium	In	313	3632	0.057	13.8	7.31	1.57	18	8.37	f.c.tetr.
Iridium	Ir	4449	9570	0.031	33.7	22.50	76	3.8	5.3	f.c.c.
Iron	Fe	2798	5430	0.11	43.3	7.87	28.5	6.53	9.71	b.c.c.
Lanthanum	La	1688	6280	0.048	8.	6.19	10.5	2.77	57	Hexag.
Lead	Pb	621	3137	0.031	20.	11.36	2	16.3	20.6	f.c.c.
Lithium	Li	357	2426	0.79	41.	0.534	.....	31	8.55	b.c.c.
Magnesium	Mg	1202	2025	0.245	88.5	1.74	6.35	15.05	4.45	h.c.p.
Manganese	Mn	2273	3900	0.115	.....	7.43	23	12.22	185	Complex cubic
Mercury	Hg	−37	675	0.033	4.7	13.55	.....	....	98.4	Rhomb.

Molybdenum	Mo	4730	10040	0.066	82.	10.22	47	2.7	5.2	b.c.c.
Nickel	Ni	2647	4950	0.105	53.	8.90	30	7.39	6.84	f.c.c.
Osmium	Os	4900	9950	0.031	.....	22.57	81	2.6	9.5	h.c.p.
Palladium	Pd	2826	7200	0.058	40.5	12.02	16.3	6.53	10.8	f.c.c.
Phosphorus (white)	P	112	536	0.177	.....	1.83	.....	70	10^{17}	Cubic
Platinum	Pt	3217	8185	0.0314	39.8	21.45	21.3	4.9	10.6	f.c.c.
Plutonium	Pu	1184	6000	0.033	4.8	19.00	14	30.55	141.4	Monoclinic
Potassium	K	147	1400	0.177	58.	0.86	.....	46	6.15	b.c.c.
Rhenium	Re	5755	10650	0.033	41.	21.04	66.7	3.7	19.3	h.c.p.
Rhodium	Rh	3571	8130	0.059	50.6	12.44	42.5	4.6	4.51	f.c.c.
Rubidium	Rb	102	1270	0.080	.....	1.53	.....	50	12.5	b.c.c.
Ruthenium	Ru	4530	8850	0.057	.....	12.20	60	5.1	7.6	h.c.p.
Selenium	Se	423	1265	0.084	.....	4.79	8.4	21	12	Hexag.
Silicon[c]	Si	2570	4860	0.162	48.2	2.33	16.35	1.6 to 4.1	10^5	Diam. cubic
Silver	Ag	1761	4010	0.056	242.	10.49	11	10.9	1.59	f.c.c.
Sodium	Na	208	1638	0.295	77.2	0.971	.....	39	4.2	b.c.c.
Strontium	Sr	1414	2520	0.176	.....	2.60	.....	.....	23	f.c.c.
Tantalum	Ta	5425	9800	0.034	31.3	16.60	27	3.6	12.45	b.c.c.
Tellurium	Te	841	1814	0.047	3.3	6.24	6	9.3	4.4×10^5	Hexag.
Thallium	Tl	577	2655	0.031	22.5	11.85	.....	16	18	h.c.p.
Thorium	Th	3182	7000	0.034	21.7	11.66	.....	6.9	13	f.c.c.
Tin	Sn	449	4120	0.054	36.2	7.30	6.3	13	11	Tetrag.
Titanium	Ti	3035	5900	0.124	9.8	4.51	16.8	4.67	42	h.c.p.
Tungsten	W	6170	10706	0.033	96.	19.30	50	2.55	5.65	b.c.c.
Uranium	U	2070	6904	0.028	17.1	19.07	24	3.8 to 7.8	30	Orthorhomb.
Vanadium	V	3450	6150	0.119	16.9	6.1	19	4.6	26	b.c.c.
Yttrium	Y	2748	5490	0.071	8.5	4.47	17	....	57	h.c.p.
Zinc	Zn	787	1663	0.092	65.	7.13	.....	22	5.92	h.c.p.
Zirconium	Zr	3366	6470	0.067	9.6	6.49	13.7	3.2	40	h.c.p.

* Courtesy of "Metals Handbook," vol. 1, 8th ed., American Society for Metals, Cleveland, 1961.

[a] Near 68°F (20°C).

[b] Sublimes—triple point at 2028 atm.

[c] Semiconductor.

Table 17.8. Typical Mechanical Properties at Room Temperature*

Metal	Modulus of elasticity E, million psi	Ultimate tensile strength σ_u, thousand psi	Yield strength σ_y, thousand psi	Endurance limit σ_{end}, thousand psi	Hardness, Brinell
Gray cast iron, ASTM 20, med. sec.	12	22	....	10	180
Gray cast iron, ASTM 50, med. sec.	19	53	....	25	240
Nodular ductile cast iron:					
Type 60-45-10	22–25	60–80	45–60	35	140–190
Type 120-90-02	22–25	120–150	90–125	52	240–325
Austenitic	18.5	58–68	32–38	32	140–200
Malleable cast iron, ferritic 32510	25	50	32.5	28	110–156
Malleable cast iron, pearlitic 60003	28	80–100	60–80	39–40	197–269
Ingot iron, hot rolled	29.8	44	23	28	83
Ingot iron, cold drawn	29.8	73	69	33	142
Wrought iron, hot rolled longit	29.5	48	27	23	97–105
Cast carbon steel, normalized 70000	30	70	38	31	140
Cast steel, low alloy, 100,000 norm. and temp	29–30	100	68	45	209
Cast steel, low alloy, 200,000 quench. and temp	29–30	200	170	85	400
Wrought plain C steel:					
C1020 hot rolled	29–30	66	44	32	143
C1045 hard. and temp. 1000°F	29–30	118	88	......	277
C1095 hard. and temp. 700°F	29–30	180	118	......	375
Low-alloy steels:					
Wrought 1330, HT and temp. 1000°F	29–30	122	100	......	248
Wrought 2317, HT and temp. 1000°F	29–30	107	72	......	222
Wrought 4340, HT and temp. 800°F	29–30	220	200	......	445
Wrought 6150, HT and temp. 1000°F	29–30	187	179	......	444
Wrought 8750, HT and temp. 800°F	29–30	214	194	......	423
Ultra high strength steel H11, HT	30	295–311	241–247	132	
300M HT and temp. 500°F	....	289	242	116	
4340 HT and temp. 400°F	30	287	270	107	
25 Ni Maraging	24	319	284		
Austenitic stainless steel 302, cold worked	28	110	75	34	240
Ferritic Stainless Steel 430, cold worked	29	75–90	45–80		
Martensitic Stainless Steel 410, HT	29	90–190	60–145	40	180–390
Martensitic Stainless Steel 440A, HT	29	260	240	......	510
Nitriding Steels, 135 Mod., hard. and temp. (core properties)	29–30	145–159	125–141	45–90	285–320
Nitiding Steels 5Ni-2A, hard. and temp	29–30	206	202	90	
Structural Steel	30	50–65	30–40	......	120
Aluminum Alloys, cast:					
195 SHT and aged	10.1	36	24	8	75
220 SHT	9.5	48	26	8	75
142 SHT and aged	10.3	28–47	25–42	9.5	75–110
355 SHT and aged	10.2	35–42	25–27	9–10	80–90
A13 as cast	10.3	39	21	19	

Table 17.8. Typical Mechanical Properties at Room Temperature* (Continued)

Metal	Modulus of elasticity E, million psi	Ultimate tensile strength σ_u, thousand psi	Yield strength σ_y, thousand psi	Endurance limit σ_{end}, thousand psi	Hardness, Brinell
Aluminum Alloys, wrought:					
EC ann.	10	12	4		
ECH 19, hard	10	27	24	7	
3003 H 18, hard	10	29	27	10	55
2024 H T (T3)	10.6	70	50	20	120
5052 H 38, hard	10.2	42	37	20	77
7075 HT (T6)	10.4	83	73	23	150
7079 HT (T6)	10.3	78	68	23	145
Copper alloys, cast:					
Leaded red brass BB11-4A	9–14.8	33–46	17–24	......	55
Leaded tin bronze BB11-2A	12–16	36–48	16–21	......	60–72
Yellow brass BB11-7A	12–14	60–78	25–40	......	80–95
Aluminum bronze BB11-9BHT	15	90	40	......	180
Copper alloys, wrought:					
Oxygen-free 102 ann	17	32–35	10	11	
Hard	....	50–55	45	13	
Beryllium copper, 172 HT	19	165–183	150–170	35–40	
Cartridge Brass, 260 hard	16	76	63	21	
Muntz metal, 280 ann	15	54	21		
Admiralty, 442 ann	16	53	22		
Manganese bronze, 675 hard	15	84	60		
Phosphor bronze, 521 spring	16	112	70		
Silicon bronze, 647 HT	18	100	88		
Cupro-Nickel, 715 hard	22	80	73		
Magnesium Alloys, cast:					
AZ63A, aged	6.5	30–40	14–19	11–15	55–73
AZ92A, aged	6.5	26–40	16–21	11–15	80–84
HK31A, T6	6.5	31	16	9–11	55
Magnesium Alloys, wrought:					
AZ61AF, forged	6.5	43	26	17–22	55
ZE-10A-H24	6.5	34–38	19–28	20–24	
HM31A-T5	6.5	42	33	12–14	
Nickel-alloy castings 210	21.5	45–60	20–30	......	80–125
Monel 411 cast	19	65–90	32–45	......	125–150
Inconel 610 cast	23	70–95	30–45	......	190
Nickel alloys, wrought:					
200 Spring	30	90–130	70–115		
Duranickel, 301 Spring	30	155–190			
K Monel, K500 Spring	26	145–165	130–180		
Titanium alloys, wrought, unalloyed	15–16	60–110	40–95	60–70	
5Al-2.5 Sn	16–17	115–140	110–135	95	
13V-11 Cr-3Al HT	14.5–16	190–240	170–220	50–55	
Zinc, wrought, comm. rolled	....	25–31	....	4.1	
Zirconium, wrought:	14	64	53		
Reactor grade	14	49	29		
Zircaloy 2	13.8	68	61		
Pure metals, wrought:					
Beryllium, ann	44	60–90	45–55		
Hafnium, ann	20	77	32		
Thorium, ann	10	34	26		
Vanadium, ann	20	72	64		
Uranium, ann	30	90	25		

Table 17.8. Typical Mechanical Properties at Room Temperature* (Continued)

Metal	Modulus of elasticity E, million psi	Ultimate tensile strength σ_u, thousand psi	Yield strength σ_y, thousand psi	Endurance limit σ_{end}, thousand psi	Hardness, Brinell
Precious metals:					
Gold, ann.	12	19	....	46	25
Silver, ann.	11	22	8	....	25–35
Platinum, ann.	21	17–26	2–5.5	....	38–52
Palladium, ann.	17	30	5	....	46
Rhodium, ann.	42	73	....	....	55–156
Osmium, cast	80	....	....	....	350
Iridium, ann.	74	....	....	....	170

Babbitt has a compressive elastic limit of 1.3 to 2.5 ksi and a Brinell hardness of 20.

Compressive yield strength of all metals, except those cold-worked = tensile yield strength.

Poisson's ratio is in the range 0.25 to 0.35 for metals.

Yield strength is determined at 0.2 per cent permanent deformation.

Modulus of elasticity in shear for metals is approximately 0.4 of the modulus of elasticity in tension E.

Compressive yield strength of cast iron 80,000 to 150,000 ksi.

* From *Materials in Design Engineering*, Materials Selector Issue, vol. 56, No. 5, 1962; courtesy of Reinhold Publishing Corporation, New York.

Table 17.9. Typical Properties of Refractory Ceramics and Cermets and Other Materials*

Type	Specific gravity	Melting point, °F	Mean specific heat, Btu/lb/°F	Mean coefficient of thermal expansion, μ in./in./°F	Mean thermal conductivity, Btu/hr/sq ft/°F/ft	Hardness, Mohs scale	Maximum service temperature (oxidiz.), °F
Alumina (99+) (Al_2O_3)	3.85	3725	0.23	4.3	10.7	9	3540
Beryllia (BeO)	3.0	4620	0.29	5.3	9.5	9	4350
Magnesia (MgO)	3.6	5070	0.26	7.8	1.47	6	4350
Thoria (ThO_2)	....	6000	....	5.28	0.0	7	4890
Zirconia (ZrO_2)	5.5–6	4710	0.16	3.06	0.53	7–8	4530
Quartzite (SiO_2)	2.65	2552	0.26	0.28	0.8		
Silicon carbide (Dens)(SiC)	3.2	......	0.33	2.17	25	9–10	3000
Boron carbide	2.5	......	....	1.73	16	....	1000
Titanium carbide (TiC)	6.5	......	....	4.3–7.5			
Tungsten carbide (WC)	14.3	......	....	2.5–3.9	26–50		
Boron nitride	2.05–2.15	4930	....	5.5	10–20		
Graphite	2.25	......	0.18	1.0–1.3	70–120	1.2	

* From *Materials in Design Engineering*, Materials Selector Issue, vol. 56, No. 5, 1962; courtesy of Reinhold Publishing Corporation, New York.

Table 17.10. Typical Properties of Plastics at Room Temperature

Type	Specific gravity	Coefficient of thermal expansion, 10^{-5}/°F	Thermal conductivity, Btu/hr/sq ft/°F/ft	Volume resistivity, ohm-cm	Dielectric strength [a], volts/mil	Modulus of elasticity in tension, 10^5 psi	Tensile strength, 10^3 psi
Acrylic, general purpose, type I	1.17–1.19	4.5	0.12	$>10^{15}$	450–530	3.5–4.5	6–9
Cellulose acetate, type I (med.)	1.24–1.34	4.4–9.0	0.1–0.19	10^{12}	250–600	....	2.7–6.5
Epoxy, general purpose	1.12–2.4	1.7–5.0	0.1–0.8	10^{13}	350–550	....	2–12
Nylon 6	1.13–1.14	4.6–5.4	0.1–0.14	10^{14}	420–485	2.5–3.4	10.2–12
Phenolic, type I (mech.)	1.31	3.3–4.4	....	1.7×10^{12}	350–400	4–5	6–9
Polyester, Allyl type	1.30–1.45	2.8–5.6	0.12	$>10^{13}$	330–500	2–3	4.5–7
Silicone, general (mineral)	1.80–2.0	2.8–3.2	0.09	$>10^{13}$	350–400	....	4.2
Polystyrene, general purpose	1.04–1.07	3.3–4.8	0.06–0.09	10^{18}	>500	4–5	5–8
Polyethylene, low density	0.92	8.9–11	0.19	10^{18}	480	0.22	1.4–2
Polyethylene, medium density	0.93	8.3–16.7	0.19	$>10^{15}$	480	....	2
Polyethylene, high density	0.96	8.3–16.7	0.19	$>10^{15}$	480	....	4.4
Polypropylene	0.89–0.91	6.2	0.08	10^{16}	769–820	1.4–1.7	5

[a] Short time.

References

1. Richards, C. W.: "Engineering Materials Science," Wadsworth Publ., San Francisco, 1961.
2. Barrett, C. S.: "Structure of Metals," 2d ed., McGraw-Hill Book Company, Inc., New York, 1952.
3. Sachs and Van Horn: "Practical Metallurgy," American Society for Metallurgy, 1940.
4. "Metals Handbook," vol. 1, 8th ed., American Society for Metals, Cleveland, 1961.
5. "Heat Treatment and Properties of Iron and Steel," *Natl. Bur. Std. (U.S.) Monograph* 18, 1960.
6. "Metals Handbook," 1954 Supplement, American Society for Metals, Cleveland,
7. *Natl. Bur. Std. (U.S.) Monograph* 18.
8. Palmer and Luersson: "Tool Steel Simplified," Carpenter Steel Co., 1948.
9. "Suiting the Heat Treatment to the Job," United States Steel Co.
10. "Metals Handbook," 1939 ed.
11. "Three Keys to Satisfaction," Climax Molybdenum Co., New York.
12. "Steels for Elevated Temperature Service," United States Steel Co.
13. *Metal Prog.*, vol. 80, nos. 4, 5, October, November, 1961.
14. Norton, J. T., and D. Rosenthal: *J. Am. Welding Soc.*, vol. 22, no. 2.
15. "ASME Handbook, Metals Engineering—Design," McGraw-Hill Book Company, Inc., New York, 1953.
16. "Symposium on Corrosion Fundamentals," A series of lectures presented at the University of Tennessee Corrosion Conference at Knoxville, The University of Tennessee Press, Knoxville, Tenn., 1956.
17. Evans, Ulich R.: "The Corrosion and Oxidation of Metals," St Martin's Press, Inc., New York, 1960.
18. Burns, R. M., and W. W. Bradley: "Protective Coatings for Metals," Reinhold Publishing Corporation, New York, 1955.
19. Bresle, Ake: "Recent Advances in Stress Corrosion," Royal Swedish Academy of Engineering Sciences, Stockholm, Sweden, 1961.
20. "ASME Handbook, Metals Engineering—Design," McGraw-Hill Book Company, Inc., New York, 1953.

Some suggested references recommended for the selections and properties of engineering materials are the following:

21. "Metals Handbook," American Society for Metals, vol. 1, 8th ed., Cleveland, 1961.
22. *Metals Progr.*, vol. 66, no. 1-A, July 15, 1954.
23. *Metals Progr.*, vol. 68, no. 2-A, Aug. 15, 1955.
24. Dumond, T. C.: "Engineering Materials Manual," Reinhold Publishing Corporation, New York, 1951.

25. "Steels for Elevated Temperature Service," United States Steel Co.
26. "Three Keys to Satisfaction," Climax Molybdenum Co., New York.
27. Zwikker, C.: "Physical Properties of Solid Materials," Interscience Publishers, Inc., New York, 1954.
28. Teed, P. L.: "The Properties of Metallic Materials at Low Temperatures," John Wiley & Sons, Inc., New York, 1950.
29. Hoyt, S. L.: "Metals and Alloys Data Book," Reinhold Publishing Corporation, New York, 1943.
30. *Materials in Design Engineering*, Materials Selector Issue, vol. 56, Reinhold Publishing Corporation, New York, 1962.
31. McLean, D.: "Mechanical Properties of Metals," pp. 363–382, John Wiley & Sons, Inc., New York, 1962.
32. Edmonson, B.: *Proc. Roy. Soc.* (*London*), *Ser. A*, vol. 264, p. 176, 1961.

Section 18

MECHANICAL DESIGN

By

SAUL FENSTER, Ph.D., *Chairman, Department of Mechanical Engineering, Fairleigh Dickinson University, Teaneck, N.J.*

HERBERT H. GOULD, M.M.E., *Engineer, Sperry Gyroscope Co., Great Neck, N.Y.*

CARL H. LEVINSON, B.S.M.E., *Reliability Engineer, Sperry Gyroscope Co., Great Neck, N.Y. (Systems Reliability)*

CONTENTS

DESIGN

SYSTEMS RELIABILITY ANALYSIS

DESIGN*

18.1. INTRODUCTION

In this section are discussed some of the concepts which must be considered in addition to nominal stress and nominal material strength before a machine member may be safely designed. This section is closely related to Properties of Engineering Materials, Sec. 17.

18.2. COMMON THEORIES OF FAILURE

(a) Maximum-shear Theory (Guest)[1,2]

Consider an element so oriented that its surface shearing stresses are zero; that is, the faces are subjected to principal stresses σ_1, σ_2, σ_3 only. In determining the values of the principal stress, it is important that the effects of impact or suddenly applied loads (Sec. 16), stress concentration and vibratory stresses (Sec. 6), be taken into account. The maximum-shear theory, applicable to ductile metals, states that when $|\sigma_1 - \sigma_2|$ or $|\sigma_2 - \sigma_3|$ or $|\sigma_3 - \sigma_1|$ (considering algebraic sign) is equal to or exceeds σ_{yp}, yielding will occur. σ_{yp} is the yield stress as obtained in a simple tension test.

In the biaxial case, $\sigma_3 = 0$, and the criteria become

$$|\sigma_1 - \sigma_2| = \sigma_{yp} \tag{18.1}$$

or

$$|\sigma_2| = \sigma_{yp} \tag{18.2}$$

or

$$|\sigma_1| = \sigma_{yp} \tag{18.3}$$

Thus, when σ_1 and σ_2 are of the same sign, Eq. (18.2) or (18.3) governs; that is, the largest principal stress is chosen. When σ_1 and σ_2 are of opposite sign, Eq. (18.1) should be used. The ultimate tensile strength σ_{ult} may be substituted for σ_{yp} if σ_{ult} is used as a failure criterion.

(b) Distortion-energy Theory (Shear-energy Theory, Hencky–Von Mises)[1,2]

This theory, applicable to ductile metals, states that yielding occurs when

$$(\sigma_1 - \sigma_2)^2 + (\sigma_2 - \sigma_3)^2 + (\sigma_3 - \sigma_1)^2 = 2(\sigma_{yp})^2 \tag{18.4}$$

(c) Maximum-stress Theory (Rankine)[1,2]

According to this theory, failure of brittle materials occurs when

$$\begin{array}{lcl} \sigma_1 = \sigma_{yp} & \text{or} & \sigma_1 = \sigma_{\text{ult}} \\ \sigma_2 = \sigma_{yp} & \text{or} & \sigma_2 = \sigma_{\text{ult}} \\ \sigma_3 = \sigma_{yp} & \text{or} & \sigma_3 = \sigma_{\text{ult}} \end{array} \tag{18.5}$$

where σ_{yp} and σ_{ult} are the respective values in tension or compression depending upon the type of loading.

* By S. Fenster and H. Gould

18.3. DESIGN STRESS

The design or allowable stress is generally defined as

$$\sigma_d = \frac{\sigma_{yp}}{\text{f.s.}} \tag{18.6}$$

or

$$\sigma_d = \frac{\sigma_{\text{ult}}}{\text{f.s.}} \tag{18.7}$$

where f.s. is the design factor of safety. σ_{yp} is used for ductile materials. σ_{ult} is used for brittle materials.

18.4. DESIGN STRESS–FAILURE RELATIONS

Design-stress relations corresponding to the various theories of failure cited above are obtained by modifying Eqs. (18.1) through (18.5):

(a) Maximum-shear Theory

$$\begin{aligned} |\sigma_1 - \sigma_2| &\leq \sigma_d \\ |\sigma_2 - \sigma_3| &\leq \sigma_d \\ |\sigma_3 - \sigma_1| &\leq \sigma_d \end{aligned} \tag{18.8}$$

(b) Distortion-energy Theory

$$(\sigma_1 - \sigma_2)^2 + (\sigma_2 - \sigma_3)^2 + (\sigma_3 - \sigma_1)^2 \leq 2(\sigma_d)^2 \tag{18.9}$$

(c) Maximum-stress Theory

$$\begin{aligned} \sigma_1 &\leq \sigma_d \\ \sigma_2 &\leq \sigma_d \\ \sigma_3 &\leq \sigma_d \end{aligned} \tag{18.10}$$

18.5. OTHER DESIGN CRITERIA

In addition to the design stress based upon σ_{yp} and σ_{ult}, other criteria are sometimes required. In the case of fatigue loading, the fatigue strength for a given number of cycles and an appropriate safety factor determine the design stress. Also when wear, creep, or deflections are to be limited to a prescribed value during the life of the machine element, the design stress can be based upon values different from σ_{yp} or σ_{ult}.

18.6. DESIGN FACTOR OF SAFETY

The magnitude of the design factor of safety, a number greater than unity, depends upon the application and the uncertainties associated with a particular design.[3] In the determination of the factor of safety, the following should be considered:

1. The possibility that failure of the machine element may cause injury or loss of human life
2. The possibility that failure may result in costly repairs
3. The uncertainty of the loads encountered in service
4. The uncertainty of material properties
5. The assumptions made in the analysis and the uncertainties in the determination of the stress-concentration factors and stresses induced by sudden impact and repeated loads
6. The knowledge of the environmental conditions to which the part will be subjected
7. The knowledge of stresses which will be introduced during fabrication (e.g., residual stresses[4]), assembly, and shipping of the part
8. The extent to which the part can be weakened by corrosion[4]

Many other factors obviously exist.

if the uncertainties are great enough to cause severe weight, volume, or economic penalties, testing and/or more thorough analyses should be performed rather than relying upon very large factors of safety.

Typical values of design safety factors are:[5]

f.s. = 1.25 to 1.5 for exceptionally reliable materials used under controllable conditions and subjected to loads and stresses that can be determined with certainty. Used almost invariably where low weight is a particularly important consideration.

f.s. = 1.5 to 2 for well-known materials under reasonably constant environmental conditions, subjected to loads and stresses that can be determined readily.

f.s. = 2 to 2.5 for average materials operated in ordinary environments and subjected to loads and stresses that can be determined.

f.s. = 2.5 to 3 for less tried as well as for brittle materials under average conditions of environment, load, and stress.

f.s. = 3 to 4 for untried materials used under average conditions of environment, load, and stress.

f.s. = 3 to 4 should also be employed with better-known materials that are to be used in uncertain environments or subjected to uncertain stresses.

f.s. = 2 for impact of very ductile materials where the small index of sensitivity results in low stress-concentration factors.

f.s. = 1.5 for less ductile materials where a higher sensitivity will provide a larger factor of stress concentration.

f.s. = 1.5 for design at higher temperatures, based on the creep strength of the material that will result in a permissible plastic deformation over a preestablished life period.

NOTE: 1. For repeated loads, the factors of safety established are acceptable but must be applied to the endurance limit rather than the yield strength of the material.

2. For castings, forgings, stampings, and welded components, factors of safety here used do not usually vary appreciably from those presented above.

3. Factors of safety to be used with standard design elements, commercially available, should be those recommended for them by reliable manufacturers and/or by established codes for design of machines.

4. Where higher factors of safety might seem desirable, a more thorough analysis should be undertaken before deciding upon their use.

18.7. TRUE FACTOR OF SAFETY

The true factor of safety, which may be defined in terms of load, stress, deflection, creep, wear, etc., is the ratio of the magnitude of any of the above parameters resulting in damage, to its actual value in service. For example:

$$\text{True factor of safety} = \frac{\text{maximum load part can sustain without damage}}{\text{maximum load part sustains in service}}$$

The true factor of safety is determined after a part is built and tested under service conditions.

18.8. STRESS CONCENTRATION[6,7]

Abrupt increases in local stress due to stress raisers, such as notches, holes, fillets, threads, shoulders, and scratches, are termed stress concentrations.

The theoretical (or elastic) stress-concentration factor is defined as

$$K_t \equiv \frac{\text{maximum stress at section}}{\text{average stress at section based upon net area}}$$

The theoretical stress-concentration factor is a function of geometry only and is determined from photoelastic studies, theory of elasticity, or actual strain measurement. K_t does not consider the mitigating effects of local yielding. Table 18.1 lists values of K_t.

Stress concentration should be considered with respect to its effect upon the strength reduction of the specimen. In statically loaded ductile materials, yielding at the points of high local stress tends to reduce the effect predicted by elastic theory, and it is found that the rupture strength of the material is essentially not affected by the stress raiser. Thus the stress-concentration factor is usually (and with caution) not considered in statically loaded ductile materials.

In statically loaded brittle materials, the effect of stress concentration must be considered cautiously.

Some designers[8] suggest that the full theoretical stress-concentration factor be used for statically loaded brittle materials. Others[9] suggest the use of the following expression to determine the effect of stress concentration in statically loaded members:

$$K' = 1 + q_s(K_t - 1) \tag{18.11}$$

where K' is the actual stress-concentration factor considering the effects of material structure and local yielding, and q_s is the index of sensitivity for static loading. Note that q_s is determined by test and defined by Eq. (18.11).

For ductile materials $q_s = 0$ and $K' = 1$, and we have the same result as before. For brittle materials such as hardened steels, $q_s = 0.15$ is recommended, while for very brittle materials such as quenched but not drawn steels, $q_s = 0.25$.[9] For cast iron, strength reduction associated with stress concentration is surprisingly low. This has been attributed to the fact that the internal stress raisers (due to the holes occupied by carbon) already produce the significant strength reduction. For this reason, if Eq. (18.11) is used, q_s may be taken as zero for cast iron.[9]

For materials subjected to impact, it is suggested that Eq. (18.11) be used with the modified values of q_s as follows:[9] for ductile materials, $q_s = 0.4$ to 0.6 with increasing brittleness, with $q_s = 1$ for hard and brittle materials. For cast iron, $q_s = 0.5$.

Because a great body of literature exists with regard to the theoretical stress-concentration factor many attempts, such as expressed by Eq. (18.11), have been made to utilize these data to the fullest extent in order to estimate the actual stress-concentration factors in static loading (and also fatigue loading). Because q_s is in actuality defined by K' and K_t and is by no means a precise function of material or shape, the experimental determination of the strength-reducing effects of stress concentration should be made wherever possible. Equation (18.11) should be used, therefore, only in the absence of better information.

In this connection, the factor of stress concentration at rupture K_r, which is effectively a factor of strength reduction, is defined as follows:[7]

$$K_r \equiv \frac{\text{computed stress at rupture for a specimen without stress raiser}}{\text{computed stress at rupture for specimen with stress raiser}}$$

Values of K_r are listed in Table 18.1. The stress-concentration factor for fatigue (or repeated) loading is defined as

$$K_f \equiv \frac{\text{endurance limit for specimen without stress raiser}}{\text{endurance limit for specimen with stress raiser}}$$

Values of K_f are given in Table 18.1. From Table 18.1 K_f is seen to be a function of material and shape. Tests indicate that K_f is smaller than K_t and that, as K_t increases, K_f/K_t appears to increase.

In the absence of experimental data, the following relationship has been used with some degree of success:

$$K_f = 1 + q(K_t - 1) \tag{18.12}$$

where q is the notch sensitivity and is defined by Eq. (18.12). Equation (18.12) is most successfully employed when q is known for the same material as the one for which the estimate is made, and provided further that the notch is not very sharp.[7]

Most data from which K_f is determined are based upon complete stress reversal. Considerably fewer data are available with respect to incomplete reversal of stress. It is generally recommended that the fatigue stress-concentration factor be applied only to the variable component of stress, although ref. 22 applies it to the mean stress also.

18.9. DESIGNING FOR REPEATED STRESS (FATIGUE)[1,10,11]

(a) Introduction

Experiments have shown that machine and structural elements will fail under cyclic load conditions even if the stresses are much smaller than the stresses which would cause failure in a static test. This phenomenon is called *fatigue* and is a very common cause of metal failure in service.

The history of a fatigue failure can be described by[12,13] the formation of a fatigue-crack nucleus, generally at a discontinuity or imperfection in the part; the propagation of the crack under repeated loading; rupture of part.

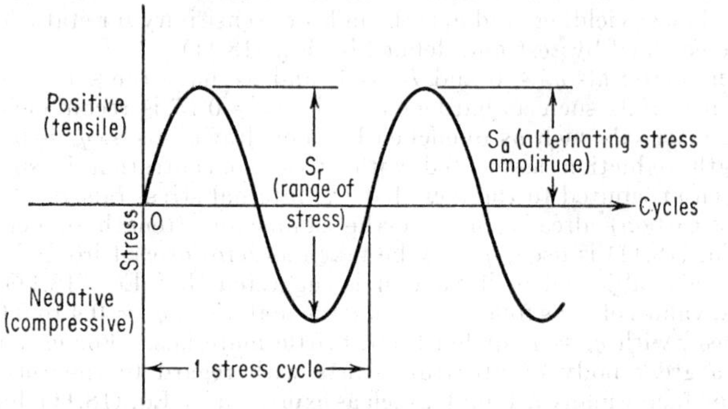

FIG. 18.1. Completely reversed stress.

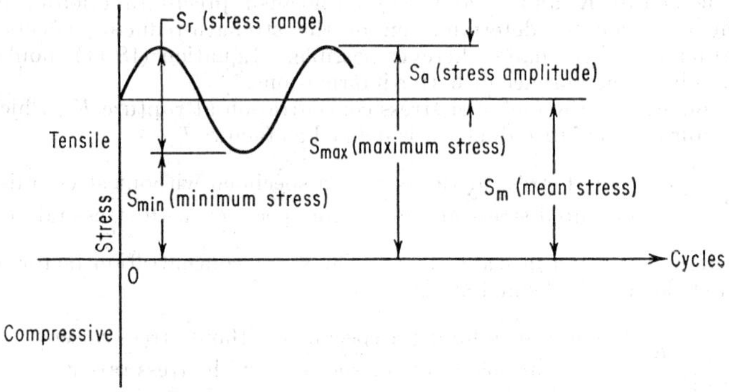

FIG. 18.2. Fluctuating tensile stress.

It is to be noted that, even with ductile materials, failure occurs without generally revealing plastic deformation.

In the design of a machine or structural element which will be subject to repeated loading, the objective is to ensure that the element will not fail during its contemplated life.

Test data for fatigue loading are generally obtained under the following conditions:

1. Completely reversed stress. All surface fibers of a specimen of symmetrical cross section are subjected to the maximum tensile and compressive stress by means such as a rotating-bending test. Figure 18.1 represents the variation of stress during such a test.

2. Oscillating stress and superimposed static stress (Fig. 18.2). A plate or sheet is repeatedly subjected to bending.

3. Oscillating direct axial stress. The specimen is subjected to axial push-pull

loading, which results in either tensile or compressive stress or both. The stresses are uniformly distributed over the cross section of the specimen.

4. Oscillating torsional (shear) stress. Repeated stress tests of specimen in torsion often provide data for completely reversed stressing. The significant stresses are maximum at the surface of the specimen and the stress pattern is more complex than in 2 or 3 above.

(b) Definitions of Terms Relative to Fatigue Testing and the Statistical Analysis of Fatigue Data*

Fatigue.† The process of progressive localized permanent structural change occurring in a material subjected to conditions which produce fluctuating stress and strains at some point or points and which may culminate in cracks or complete fracture after a sufficient number of fluctuations.‡

Fatigue Life N. The number of cycles of stress or strain of a specified character that a given specimen sustains before failure of a specified nature occurs.

The following definitions relating to fatigue tests and test methods apply to those cases where the conditions imposed upon a specimen result or are assumed to result in uniaxial principal stresses or strains which fluctuate in magnitude. Multiaxial stress, sequential loading, and random loading require more rigorous definitions which are at present beyond the scope of this section.

Nominal Stress S. The stress at a point calculated on the net cross section by simple elastic theory without taking into account the effect on stress produced by geometric discontinuities such as holes, grooves, and fillets.

Stress Cycle. The smallest segment of the stress-time function which is repeated periodically.

Maximum Stress S_{max}. The stress having the highest algebraic value in the stress cycle, tensile stress being considered positive and compressive stress negative. In this definition as well as in others that follow, the nominal stress is used most commonly.

Minimum Stress S_{min}. The stress having the lowest algebraic value in the cycle, tensile stress being considered positive and compressive stress negative.

Mean Stress (or Steady Component of Stress) S_m. The algebraic average of the maximum and minimum stresses in one cycle, that is,

$$S_m = \frac{S_{max} + S_{min}}{2}$$

Range of Stress S_r. The algebraic difference between the maximum and minimum stresses in one cycle, that is,

$$S_r = S_{max} - S_{min}$$

Stress Amplitude (or Variable Components of Stress) S_a. One-half the range of stress, that is,

$$S_a = \frac{S_r}{2} = \frac{S_{max} - S_{min}}{2}$$

Stress Ratio A or R. The algebraic ratio of two specified stress values in a stress cycle. Two commonly used stress ratios are:

(1) the ratio of the alternating stress amplitude to the mean stress, that is,

$$A = S_a/S_m$$

* From "Manual on Fatigue Testing" by the E-9 Committee of A.S.T.M. Reprinted by special permission of the American Society for Testing and Materials, 1916 Race St., Philadelphia 3, Pa.[14]

† The term fatigue in the materials-testing field has, in at least one case, glass technology, been used for static tests of considerable duration, a type of test generally designated as stress rupture.

‡ Fluctuations may occur in both stress and time (frequency) as in the case of "random vibration."

(2) the ratio of the minimum stress to the maximum stress, that is,

$$R = S_{min}/S_{max}$$

SN Diagram. A plot of S_{max}, S_{min} or S_a against the number of cycles to failure for a specified value of S_m, A, or R and for a specified probability of failure. For N, a log scale is almost always used. For S a linear scale is used most often, but a log scale is sometimes used.

Stress Cycles Endured n. The number of cycles of a specified character (that produce fluctuating stress or strain) which a specimen has endured at any time in its stress history.

Fatigue Strength at N Cycles, S_N. The hypothetical value of S_{max}, S_{min} or S_a obtained from an SN diagram, at which exactly N cycles of stress could be endured by a specimen from a given sample when tested at a specified value of S_m, A, or R.

NOTE: The value of S_N which is commonly found in the literature is the hypothetical value of S_{max} or S_a at which 50 per cent of the specimens of a given sample could survive N stress cycles in which $S_m = 0$. This is also known as the *median fatigue strength* for N cycles.

Fatigue Limit S_f. The limiting value of the median fatigue strength as N approaches infinity.

NOTE: Certain materials and environments preclude the attainment of a fatigue limit. Values tabulated as "fatigue limits" in the literature are frequently (but not always) values of S_N for 50 per cent survival at N cycles of stress in which $S_m = 0$.

Cycle Ratio C. The ratio of the number of stress cycles n of a specified character to the hypothetical fatigue life N, obtained from the SN diagram, for stress cycles of the same character, that is, $C = n/N$.

Theoretical Stress-concentration Factor, or Stress-concentration Factor K_t. The ratio of the greatest stress in the region of a notch or other stress concentrator as determined by the theory of elasticity (or by experimental procedures that give equivalent values) to the corresponding nominal stress.

NOTE: The theory of plasticity should not be used to determine K_t.

Fatigue Notch Factor K_f. The ratio of the fatigue strength of a specimen with no stress concentration to the fatigue strength at the same number of cycles with stress concentration for the same conditions.

NOTE: In specifying K_f it is necessary to specify the geometry and the values of S_{max}, S_m and N for which it is computed.

Fatigue Notch Sensitivity q. A measure of the degree of agreement between K_f and K_t for a particular specimen of a given size and material containing a stress concentrator of a given size and shape.

NOTE: A common definition of fatigue notch sensitivity is

$$q = \frac{K_f - 1}{K_t - 1}$$

in which q may vary between zero (where $K_f = 1$) and unity (where $K_f = K_t$).

Constant-life Fatigue Diagram. A plot (usually on rectangular coordinates) of a family of curves, each of which is for a single fatigue life N relating S_a, S_{max} and/or S_{min} to the mean stress S_m. The constant-life fatigue diagram is generally derived from a family of SN curves, each of which represents a different stress ratio A or R for a 50 per cent probability of survival.

Definitions Relating to Statistical Analysis of Fatigue Data

Median Fatigue Life. The middlemost of the observed fatigue-life values, arranged in order of magnitude, of the individual specimens in a group tested under identical conditions. In the case where an even number of specimens are tested, it is the average of the two middlemost values.

NOTE 1: The use of the sample median instead of the arithmetic mean (that is, the average) is usually preferred.

NOTE 2: In the literature, the abbreviated term "fatigue life" usually has meant the median fatigue life of the group. However, when applied to a collection of data without further qualification the term "fatigue life" is ambiguous.

Fatigue Life for p Per Cent Survival. An estimate of the fatigue life that p per cent of the population would attain or exceed at a given stress level. The observed value of the median fatigue life estimates the fatigue life for 50 per cent survival. Fatigue life for p per cent survival values, where p is any number, such as 95 or 90, may also be estimated from the individual fatigue-life values.

Median Fatigue Strength at N Cycles. An estimate of the stress level at which 50 per cent of the population would survive N cycles.

NOTE 1: The estimate of the median fatigue strength is derived from a particular point of the fatigue-life distribution, since there is no test procedure by which a frequency distribution of fatigue strengths at N cycles can be directly observed.

NOTE 2: This is a special case of the more general definition.

Fatigue Strength for p Per Cent Survival at N Cycles. An estimate of the stress level at which p per cent of the population would survive N cycles; p may be any number, such as 95 or 90.

NOTE: The estimates of the fatigue strengths for p per cent survival values are derived from particular points of the fatigue-life distribution since there is no test procedure by which a frequency distribution of fatigue strengths at N cycles can be directly observed.

Fatigue Limit for p Per Cent Survival. The limiting value of fatigue strength for p per cent survival as N becomes very large; p may be any number, such as 95 or 90.

SN Curve for 50 Per Cent Survival. A curve fitted to the median values of fatigue life at each of several stress levels. It is an estimate of the relationship between applied stress and the number of cycles to failure that 50 per cent of the population would survive.

NOTE 1: This is a special case of the more general definition.

NOTE 2: In the literature, the abbreviated term "*SN* curve" usually has meant either the *SN* curve drawn through the means (averages) or the medians (50 per cent values) for the fatigue-life values. Since the term "*SN* curve" is ambiguous, it should be used in technical papers only when adequately described.

SN Curve for p Per Cent Survival. A curve fitted to the fatigue life for p per cent survival values at each of several stress levels. It is an estimate of the relationship between applied stress and the number of cycles to failure that p per cent of the population would survive; p may be any number, such as 95 or 90.

NOTE: Caution should be used in drawing conclusions from extrapolated portions of the *SN* curves. In general, the *SN* curves should not be extrapolated beyond observed life values.

Response Curve for N Cycles. A curve fitted to observed values of percentage survival at N cycles for each of several stress levels, where N is a preassigned number such as 10^6 or 10^7. It is an estimate of the relationship between applied stress and the percentage of the population that would survive N cycles.

NOTE 1: Values of the median fatigue strength at N cycles and the fatigue strength for p per cent survival at N cycles may be derived from the response curve for N cycles if p falls within the range of the per cent survival values actually observed.

NOTE 2: Caution should be used in drawing conclusions from extrapolated portions of the response curves. In general, the curves should not be extrapolated to other values of p.

Population (or Universe). The hypothetical collection of all possible test specimens that could be prepared in the specified way from the material under consideration.

Sample. The specimens selected from the population for test purposes.

NOTE: The method of selecting the sample determines the population about which statistical inference or generalization can be made.

Group. The specimens tested at one time, or consecutively, at one stress level. A group may comprise one or more specimens.

Frequency Distribution. The way in which the frequencies of occurrence of members of a population, or a sample, are distributed according to the values of the variable under consideration.

Parameter. A constant (usually unknown) defining some property of the frequency distribution of a population, such as a population median or a population standard deviation.

Statistic. A summary value calculated from the observed values in a sample.

Estimation. A procedure for making a statistical inference about the numerical values of one or more unknown population parameters from the observed values in a sample.

Estimate. The particular value or values of a parameter for a given sample computed by an estimation procedure.

Point Estimate. The estimate of a parameter given by a single statistic.

Sample Median. The middle value when all observed values in a sample are arranged in order of magnitude if an odd number of samples are tested. If the sample size is even, it is the average of the two middlemost values. It is a point estimate of the population median, or 50 per cent point.

Sample Average (Arithmetic Mean). The sum of all the observed values in a sample divided by the sample size. It is a point estimate of the population mean.

Sample Variance s^2. The sum of the squares of the differences between each observed value and the sample average divided by the sample size minus 1. It is a point estimate of the population mean.

Sample Standard Deviation s. The square root of the sample variance. It is a point estimate of the population standard deviation, a measure of the "spread" of the frequency distribution of a population.

Sample Percentage. The percentage of observed values between two stated values of the variable under consideration. It is a point estimate of the percentage of the population between the same two stated values. (One stated value may be $-\infty$ or $+\infty$.)

Interval Estimate. The estimate of a parameter given by two statistics, defining the end points of an interval.

Confidence Interval. An interval estimate of a population parameter computed so that the statement "the population parameter lies in this interval" will be true, on the average, in a stated proportion of the times such statements are made.

Confidence Limits. The two statistics that define a confidence interval.

Confidence Level (or Coefficient). The stated proportion of the times the confidence interval is expected to include the population parameter.

Tolerance Interval. An interval computed so that it will include at least a stated percentage of the population with a stated probability.

Tolerance Limits. The two statistics that define a tolerance interval. (One value may be $-\infty$ or $+\infty$.)

Tolerance Level. The stated probability that the tolerance interval includes at least the stated percentage of the population. It is not the same as a confidence level, but the term *confidence level* is frequently associated with tolerance intervals.

Significant. Statistically significant. An effect or difference between populations is said to be present if the value of a test statistic is significant, that is, lies outside predetermined limits.

NOTE: An effect which is statistically significant may or may not have engineering significance.

Test Statistic. A function of the observed values in a sample that is used in a test of significance.

Test of Significance. A test of the hypothesis that the effect is not present. The rejection of the hypothesis indicates that the effect is present.

Significance Level. The stated probability (risk) that a given test of significance will reject the hypothesis that a specified effect is not present when the hypothesis is true.

(c) Influences on Fatigue Test Data

Fatigue test data, often characterized by considerable scatter, are influenced by the following variables:

1. Shape, size, and surface finish of specimen
2. Temperature at which test was performed

3. Type of loading imposed; the effect of combined steady and alternating stress
4. State, kind, and range of stress; the effect of combined shear, bending, and direct stresses
5. The effect of holes, fillets, notches, and other stress raisers
6. The effect of stress gradients

(d) Presentation of Fatigue Data

In *SN* curves (see Fig. 18.3), the stresses are completely reversed unless otherwise stated; i.e., the stress indicated on the ordinate equals one-half the stress range.

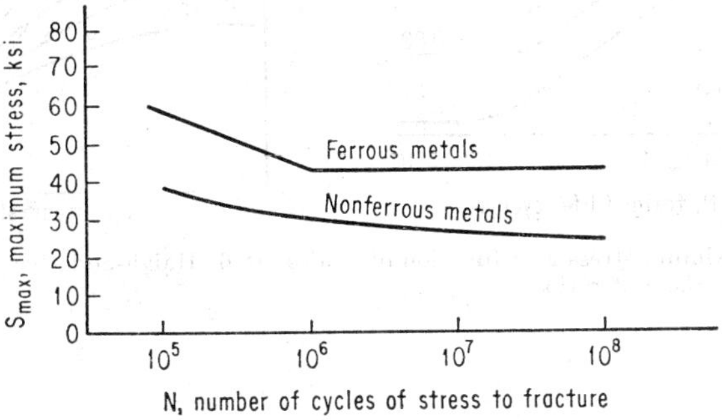

FIG. 18.3. *SN* curve.

The curve for ferrous metals will usually exhibit a characteristic knee and a horizontal asymptote defining this fatigue limit.

In the case of nonferrous metals, the curve is generally gradual and the true fatigue limit is not clearly defined. In such cases, it is common practice to define a "fatigue limit" at some arbitrary life such as 500,000,000 cycles of stress for aluminum alloys.

Available test data show that, for some materials, there is a definite correlation between fatigue limit for completely reversed stress and static tensile strength.

When an alternating stress is superimposed on a steady stress, the data are often represented by modified *SN* curves such as in Figs. 18.4 and 18.5.

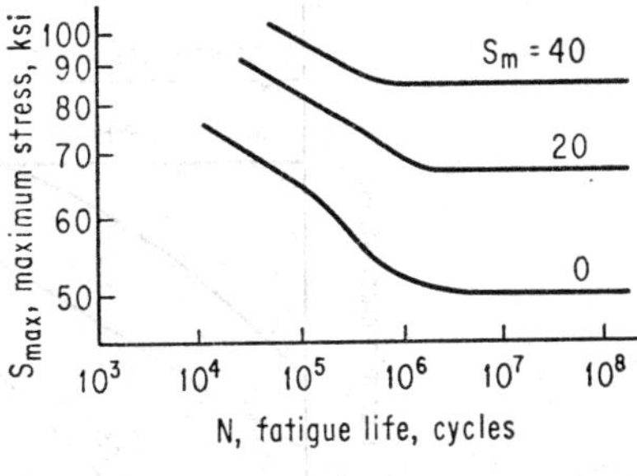

FIG. 18.4. Maximum stress as a function of fatigue life for constant mean stress.

In Fig. 18.4, curves for constant mean stress are given. The curve for $S_m = 0$ corresponds to a curve of the type shown in Fig. 18.3. In Fig. 18.5, curves for tests run under conditions of constant S_{min}/S_{max} are given.

In the *Haigh-Soderberg method*[14] (Fig. 18.6), the alternating stress amplitude S_a is plotted against the mean stress S_m, and the curves represent data for failure at a constant number of cycles.

The *modified Goodman diagram*[11] applies to failure at a constant number of cycles. This Goodman diagram (Fig. 18.7) represents actual test data (unlike other Goodman diagrams which assume a fatigue limit). The maximum or minimum stresses S_{max} or S_{min} are plotted against the mean stress S_m.

The *Goodman diagram*[2,11] (Fig. 18.8) represents the combined effect of steady and alternating stress on the fatigue limit. In constructing the diagram, it is assumed that the fatigue limit for complete reversal is one-third of the ultimate static tensile strength.

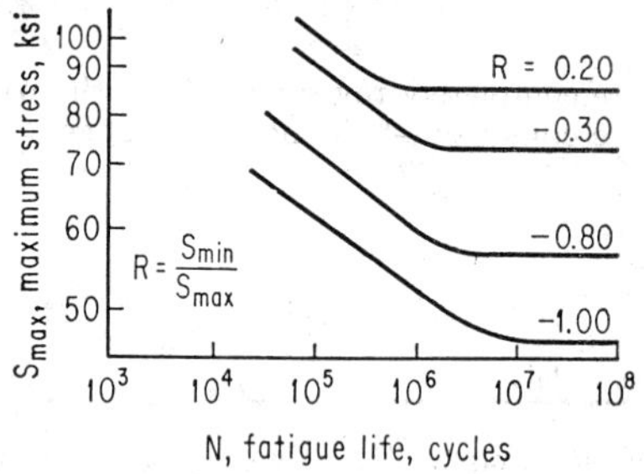

FIG. 18.5. Maximum stress as a function of fatigue life for constant ratios.

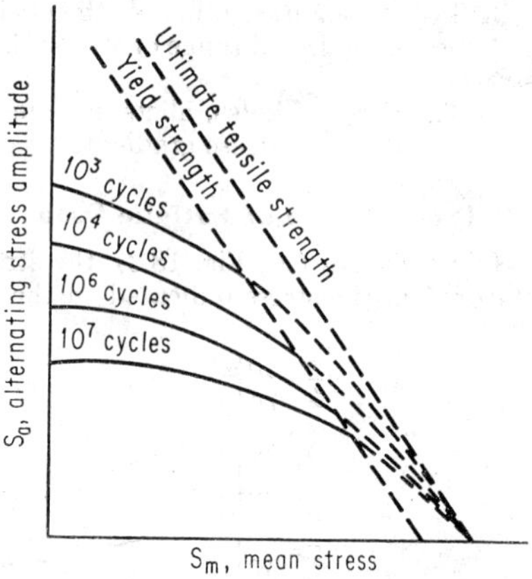

FIG. 18.6. Haigh-Soderberg representation.

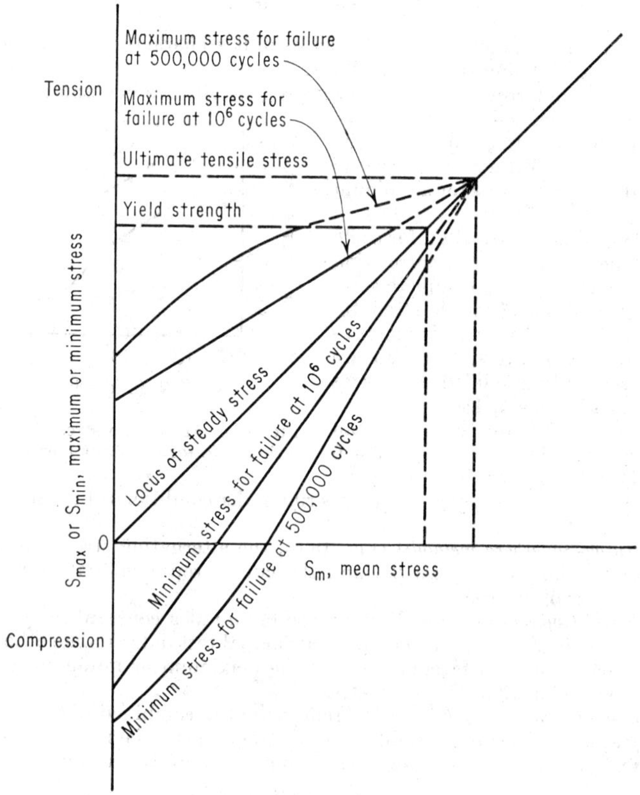

FIG. 18.7. Modified Goodman diagram.

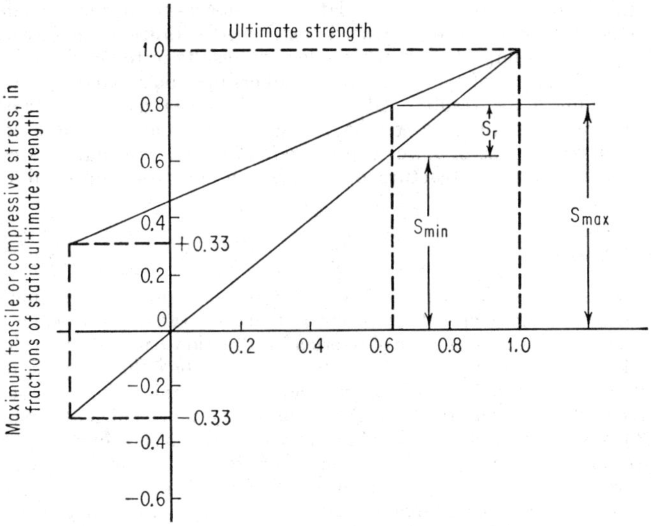

FIG. 18.8. Goodman diagram.

The Damage Line and Cumulative-damage Concepts.[15,16,17,18] For some materials under given types of loading it is possible to determine whether damage will be incurred if a part is subjected to a number of cycles at a stress above the fatigue limit. A damage line is obtained (Fig. 18.9) by subjecting specimens for n_1 cycles at stress levels S_{11}, S_{22}, . . . , above the fatigue limit, specimens for n_2 cycles at stress levels S_{21}, S_{22}, . . . , etc., and then running these specimens at the fatigue limit. The line drawn above the points representing specimens which have not failed subsequently at the fatigue limit is then the damage line.

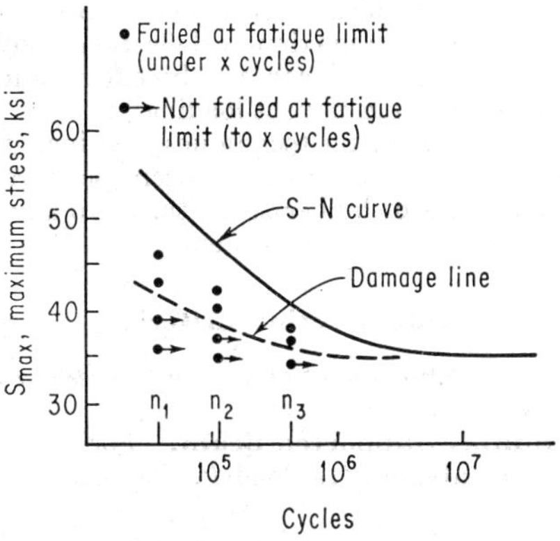

FIG. 18.9. The damage line.

The damage-line concept[15] represents an early effort to determine experimentally how the loading history affects the fatigue life of a specimen. A more recent concept, cumulative damage,[16] postulates that, if N_1, N_2, N_3, . . . , N_n are the number of cycles to failure at the corresponding stresses S_1, S_2, S_3, . . . , S_n, then the sum of the cycle ratios n_1/N_1, n_2/N_2, n_3/N_3, . . . , n_n/N_N, where n_1, n_2, n_3, . . . , n_n are the number of cycles applied at the corresponding stresses S_1, S_2, S_3, . . . , S_n, should equal unity. That is,

$$n_1/N_1 + n_2/N_2 + n_3/N_3 + \cdots + n_n/N_n = 1$$

It is to be noted that to date this concept is as often supported by experiment as not.[11]

18.10. LIGHTWEIGHT-STRUCTURE DESIGN[19,20]

(a) Introduction

While the need for lightweight structures for air and space-borne equipment is apparent, the practical advantages of reduced structure weight in more conventional designs should not be overlooked. Associated with lower-weight machine members is reduced inertia loading and often reduced cost, noise, vibration, and wear. Where weight reduction of an important machine element results in reduced inertia loading, the strength requirements of other machine members may be relaxed so that a general machine lightening may result. Additional advantages associated with lighter structures are increased payload weight (as in vehicles), and in some cases lower operating expenses because of lower energy requirements. In many cases the performance of a mechanical structure or device can only be improved by weight reduction.

(b) General Approach

Many areas can be explored in designing for light weight or redesigning existing machines. If a machine element has been designed on the basis of some peak load which rarely or never occurs, the use of overload protection (such as override clutches, see Sec. 27) should be investigated. Maximum loads can be reduced by use of cushioning devices. In some cases the load-carrying capacity per unit weight can be increased (or for a given capacity, the total weight reduced) by more efficient means of heat transfer. This is particularly applicable to brakes and clutches (see Sec. 28).

Efficient rearrangement of material can often result in increased load-carrying capacity or reduced weight for a machine of given loading requirements. The principle involved here is the removal of material from regions of zero or low stress and the addition of material to regions of high stress level, i.e., tending toward stress equalization or uniformity. Because bodies in bending, torsion, and buckling experience maximum stress at the outermost fibers, the removal of material from the central low-stress regions and the addition of material to the outer regions lead naturally to hollow structures such as hollow shafts and shells, to cellular structures such as honeycombs,[21] and to structural shapes such as I-beams.

In some applications, preloads can be effectively utilized to reduce the maximum tensile stress or maximum deflections, both of which can result in lighter structures. The use of preload can also be effective in reducing or eliminating backlash or play, thus minimizing impact loading.

Weight-reducing techniques also include the avoidance of such weakening effects as notches and other stress raisers (see Stress-concentration Factors), and the selection of materials of higher strength, more advantageous form (e.g., cast vs. welded), and more suitable heat-treatment.

(c) Material-selection Criteria

The establishment of specific *figures of merit* can be useful in comparing materials for a given application.

Simply Supported Beam. For a simply supported beam of span L, subjected to concentrated central load P, the maximum bending stress is given by

$$\sigma_{max} = M_{max}c/I = M_{max}/z = PL/4z \leq \sigma_d \qquad (18.13)$$

where M_{max} = maximum bending moment
z = section modulus = I/c
I = area moment of inertia
c = distance from neutral axis to outer fiber
σ_d = design stress

Assuming geometrically similar sections, a section factor S_1 may be defined such that

$$S_1 = A^{3/2}/z = \text{const} \qquad (18.14)$$

where A = cross-sectional area of beam

The required cross-sectional area of the beam may therefore be expressed

$$A = (M_{max}S_1/\sigma_d)^{2/3} \qquad (18.15)$$

For a given maximum moment (i.e., load and span), and since S_1 = constant, the required area thus varies as $(1/\sigma_d)^{2/3}$. The weight of the beam

$$W = \rho AL = \rho(M_{max}S_1/\sigma_d)^{2/3}L \qquad (18.16)$$

where ρ = weight per unit volume of beam material.

Thus beam weight varies as $\rho/\sigma_d^{2/3}$ for a given set of conditions and the figure of merit for beam weight for this application may be defined

$$F_{W,\sigma} = \rho/\sigma_d^{2/3} \qquad (18.17)$$

(Sim lar figures of merit for volume $1/\sigma^{2/3}$ and cost $\rho/\sigma^{2/3} \times$ cost per unit weight for the structural shape and material in question can also be developed.)

The parameter defined by Eq. (18.17) is based upon a "design-for-strength" criterion. Assume now that the deflection at mid-span for the centrally loaded simply supported beam is limited to some allowable deflection δ_{allow}, i.e., based upon a "design-for-rigidity" criterion,

$$\delta = PL^3/48EI \leqq \delta_{allow} \qquad (18.18)$$

where E = modulus of elasticity.

Employing a section factor S_2 defined by

$$S_2 = A^2/I \qquad (18.19)$$

the following expression for required cross-sectional area is derived:

$$A = (PL^3S_2/48E\delta_{allow})^{1/2} \qquad (18.20)$$

The weight of the beam is thus

$$W = \rho AL = \rho(PL^3S_2/48E\delta_{allow})^{1/2}L \qquad (18.21)$$

and a figure of merit for weight, based upon given δ_{allow}, P, L, and S_2, is defined

$$F_{W,\delta} = \rho/E^{1/2} \qquad (18.22)$$

(Similarly, figures of merit for volume and cost are $1/E^{1/2}$ and $\rho/E^{1/2} \times$ cost per unit weight, respectively.)

Tension or Short Compression Member. For a tension or short compression member, the required area for a given load is P/σ_d and the beam weight is $\rho LP/\sigma_d$ so that the figure of merit

$$F_W = \rho/\sigma_d \qquad (18.23)$$

Torsion Member. For a member subjected to torque T, the maximum shear stress is given by

$$\tau_{max} = Tc/J = T/Z_p \leqq \tau_d \qquad (18.24)$$

where J = polar moment of inertia

Z_p = polar section modulus J/c

Employing a section factor S_3 defined by

$$S_3 = A^{3/2}/Z_p \qquad (18.25)$$

the following expression for required cross-sectional area is obtained:

$$A = (S_3Z_p)^{2/3} = [S_3(T/\tau_d)]^{2/3} \qquad (18.26)$$

The weight of the bar is $\rho[S_3(T/\tau_d)]^{2/3}L$ and the figure of merit for weight is

$$F_{w,\tau} = \rho/\tau_d^{2/3} \qquad (18.27)$$

To derive a figure of merit under conditions of limited angle of twist, a section factor $S_4 = A^2/J$ should be used.

Column. The allowable column load may be written

$$P_{allow} = n\pi^2EI/L^2(\text{f.s.}) \qquad (18.28)$$

where f.s. = factor of safety and n depends upon the fixity of the column.

For a section factor $S_2 = A^2/I$, the column weight is

$$W = (L^2/\pi n^{1/2})[P(\text{f.s.})S_2]^{1/2}(\rho/E^{1/2}) \qquad (18.29)$$

and

$$F_w = \rho/E^{1/2} \qquad (18.30)$$

Note in the above examples that, where a stress criterion is used, the allowable stress selected depends upon the application (whether loading is static or impact, for example).

Table 18.1. Factors of Stress Concentration for Elastic Stress K_t, for Repeated Stress K_f, and for Rupture K_r*

Type of form irregularity or stress raiser	Stress condition	Manner of loading	Factor of stress concentration K_t, K_f, K_r, for various dimensions	
1. Two V notches in member of rectangular section 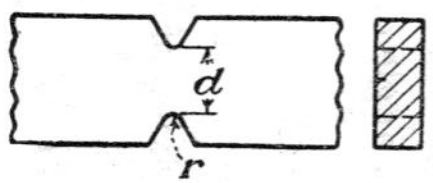	Elastic stress	Tension	$\frac{r}{d}$: 0.05, 0.10, 0.15, 0.20, 0.25, 0.30, 0.40, 0.50 K_t: 4.1, 3.0, 2.5, 2.2, 2.0, 1.9, 1.7, 1.55	(Refs. 1, 4)
		Bending	$\frac{r}{d}$: 0.025, 0.05, 0.10, 0.15, 0.20, 0.25, 0.30, 0.40, 0.50 K_t: 3.6, 2.9, 2.2, 1.9, 1.7, 1.6, 1.5, 1.4, 1.3	(Ref. 1)
	Static rupture	Bending	$D = 1\frac{1}{2}$, $d = 1$ (see table 1a below)	(Ref. 12)
2. Two U notches in member of rectangular section 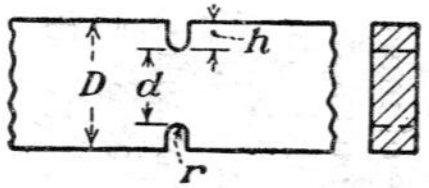	Elastic stress	Tension	See table 2a below	(Ref. 22) (Ref. 6)
		Bending	See table 2b below	(Ref. 6)

1a. Static rupture, bending, $D = 1\frac{1}{2}$, $d = 1$:

$\frac{r}{d}$	0	0.075	0.15
Plaster	2.10	1.72	1.46
Cast iron	1.38	1.23	

2a. Elastic stress, tension:

$\frac{h}{r}$ \ $\frac{r}{d}$	0.05	0.095	0.15	0.225	0.30	0.40	0.52	0.75
0.5	2.25	2.10	1.95	1.85	1.70	1.60	1.50	1.35
1.0	2.57	2.34	2.16	1.96	1.81	1.65	1.51	1.36
1.5	2.72	2.50	2.26	2.00	1.84	1.65	1.51	1.36
2.0	2.90	2.64	2.32	2.03	1.85	1.65	1.51	1.36
3.0		2.80	2.40	2.05	1.86	1.65	1.51	1.36
4.0		2.94	2.43	2.06	1.86	1.65	1.51	1.36

When $\frac{h}{r} = 1$ (semicircular notch) $k = 2.75 - 2.75\frac{h}{D} + 0.32\left(\frac{h}{D}\right)^2 + 0.68\left(\frac{h}{D}\right)^3$

2b. Elastic stress, bending:

$\frac{h}{r}$ \ $\frac{r}{d}$	0.05	0.10	0.20	0.30	0.50	0.75
0.5	1.90	1.75	1.54	1.40	1.27	1.17
1.0	2.20	1.86	1.59	1.45	1.30	1.18
1.5	....	1.91	1.60	1.45	1.30	1.18
2.0	....	1.94	1.60	1.46	1.30	1.18
3.0	....	2.00	1.61	1.47	1.30	1.18
4.0	....	2.05	1.62	1.47	1.30	1.18

* From Roark, Raymond J.: "Formulas for Stress and Strain," 3d ed., McGraw-Hill Book Company, Inc., New York, 1954.

3. One V notch in member of rectangular section

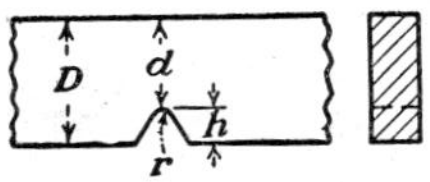

Stress	Loading	Stress concentration factor
Elastic stress	Tension	$K_t = 1 + 2\sqrt{\frac{h}{r}}$ (for h small compared with d) (Refs. 7, 8)
	Bending	$K_t = 1 + 2\sqrt{\frac{h}{r}}$ (for h small compared with d) (Refs. 7, 8)

Static rupture, Bending:

	Sharpness effect			Depth effect							Scale effect					
	$D = 1\frac{1}{2}''$, $h = \frac{1}{4}$ in.			$D = 1\frac{1}{2}$ in., $r = 0$							$\frac{D}{h} = 6$, $\frac{r}{h} = 0.1$					
	$r = 0$	0.03	0.15	$h =$ 0.02	$\frac{1}{8}$	$\frac{1}{4}$	$\frac{1}{2}$	$\frac{3}{4}$	1	1.21	$D =$ $1\frac{1}{2}$	1	$\frac{3}{4}$	$\frac{1}{2}$	$\frac{3}{8}$	$\frac{3}{16}$
Plaster	2.50	2.30	1.80	1.05	1.96	2.28	2.00	1.53	1.60	1.19						
G. Cast iron	1.56	1.54	1.43	1.28	1.51	1.59	1.55	1.53	1.34	1.14	1.24		1.19		1.20	1.15
W. Cast iron	1.33	1.66	1.40	1.03	1.22	1.21	1.37	1.08	1.08							
Ni. Cast iron	1.64	....	1.43	1.13	1.45	1.57	1.58	1.47	1.30	1.13	1.43		1.31		1.22	1.18
Al 112	1.77	....	1.39	....	1.37	1.61	1.53	1.82	1.30	1.02	1.39		1.33		1.18	0.94
Al 195 − T6	1.43	....	1.24								1.24		1.01		1.04	
Al 220 − T4	1.13	....	1.07													
Mag. Alloy	....	....	1.41								1.41		1.25		1.27	
Bakelite	....	....	1.13									1.13		1.15		

(Ref. 12)

4. One U notch in member of rectangular section

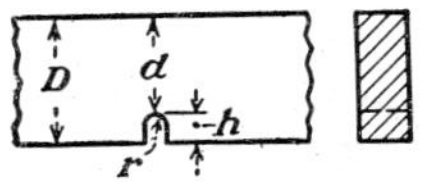

Elastic stress, Bending:

$\frac{r}{D} = 0.125$

$\frac{h}{r}$	0.667	1.33	2.00	2.67	3.33	4.00	4.67	5.33	6.00	6.67	7.34
K_t	1.96	2.04	1.96	1.92	1.84	1.76	1.65	1.56	1.46	1.35	1.06

(Ref. 12)

Static rupture, Bending:

$\frac{r}{D} = 0.125$

$\frac{h}{r}$	0.667	1.33	2.00	3.33	4.67	6.00
Plaster	1.49	1.57	1.67	1.55	1.42	1.25
Cast iron	1.28	1.37	1.42	1.30	1.17	1.10

(Ref. 12)

Table 18.1. Factors of Stress Concentration for Elastic Stress K_t, for Repeated Stress K_f, and for Rupture K_r (Continued)

Type of form irregularity or stress raiser	Stress condition	Manner of loading	Factor of stress concentration K_t, K_f, K_r for various dimensions
5. Circular hole in plate or rectangular bar 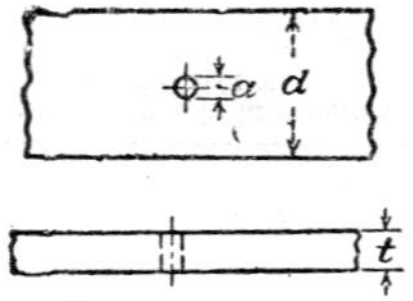	Elastic stress	Tension 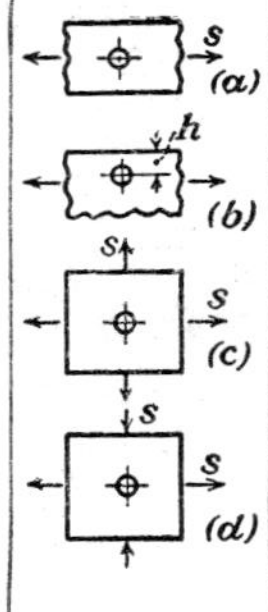	$K_t = 3 - 3.13\frac{a}{d} + 3.76\left(\frac{a}{d}\right)^2 - 1.71\left(\frac{a}{d}\right)^3$ (empirical formula, Ref. 22) (*a*) Uniaxial stress, hole central $K_t = \frac{3d}{a+d}$ (approximate formula, Ref. 20) (*b*) Uniaxial stress, hole near edge of wide plate $\frac{h}{a} = 0.67 \quad 0.77 \quad 0.91 \quad 1.07 \quad 1.29 \quad 1.56$ $K_t = 4.37 \quad 3.92 \quad 3.61 \quad 3.40 \quad 3.25 \quad 3.16$ (Ref. 21) (*c*) Biaxial stress, $\frac{a}{d}$ small $K_t = 2$ (*d*) Biaxial stress, $\frac{a}{d}$ small $K_t = 4$
		Bending 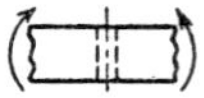	$K_t = \frac{(1+\nu)(5-\nu)}{3+\nu}$, $\frac{a}{d}$ small (Ref. 50)
	Repeated stress	Tension	$d = 1.50$, $t = 0.064$, $a = 0.0365$ to 0.20 Cold rolled hard steel strip, 1.4% ult. elongation $k = 2.15$ (Ref. 26)
		Bending	$d = \frac{1}{2}$ $t = 0.05$ $a = 0.055$ Material — K_f 1.20 per cent C steel (normalized) .... 1.25 0.52 per cent C steel (normalized) .... 1.31 0.37 per cent C steel (normalized) .... 1.22 Chr. Ni steel (3 heat treatments) .... 1.30, 1.53, 1.76 3.5 per cent Ni steel (2 heat treatments) .... 1.31, 1.38 Armco iron (0.02 per cent C) .... 1.30 0.49 per cent C steel .... 1.12 (Ref. 17)
	Static rupture	Bending	Beams, $d = 1$ $t = 1\frac{1}{2}$ $a = \frac{1}{16}$ $a = \frac{1}{2}$ Plaster 1.33 1.43 Cast iron 1.03 1.11 Circular flat plate, dia. = 8.4, $t = 0.375$. $a = \frac{1}{4}$ $a = \frac{1}{2}$ Plaster 1.5 .... Cast iron ... 1.19 (Ref. 12)

6. Elliptical hole in plate 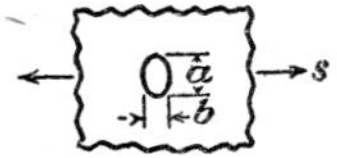	Elastic stress	Tension	$K_t = 1 + 2\frac{a}{b}$ (for wide plate) (Ref. 7)
7. Circular hole with bead in wide plate 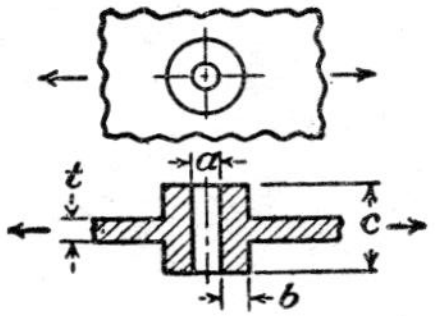	Elastic stress	Tension	(Bead area) $A_b = b(c - t)$ (Hole area) $A_h = at$ $\frac{A_b}{A_h} = 0.1 \quad 0.2 \quad 0.3 \quad 0.4 \quad 0.5$ $K_t = 2.53 \quad 2.17 \quad 1.90 \quad 1.69 \quad 1.53$

Table 18.1. Factors of Stress Concentration for Elastic Stress K_t, for Repeated Stress K_f, and for Rupture K_r (Continued)

Type of form irregularity or stress raiser	Stress condition	Manner of loading	Factor of stress concentration K_t, K_f, K_r for various dimensions
8. Square shoulder with fillet in rectangular bar 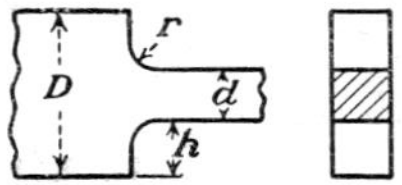	Elastic stress	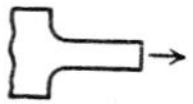 Tension	See table (a) below (Ref. 6)
	Elastic stress	 Bending	See table (b) below (Ref. 6)
	Repeated stress	 Bending	See table (c) below (Ref. 17)

(a) Elastic stress, tension

$\frac{h}{r}$ \ $\frac{r}{d}$	0.05	0.10	0.20	0.27	0.50	1.0
0.5	1.70	1.60	1.53	1.47	1.39	1.21
1.0	1.93	1.78	1.67	1.59	1.42	1.22
1.5		1.89	1.72	1.65	1.43	1.23
2.0		1.95	1.80	1.70	1.44	1.23
3.5		2.10	1.93	1.78	1.47	1.24

(Ref. 6)

(b) Elastic stress, bending

$\frac{h}{r}$ \ $\frac{r}{d}$	0.05	0.10	0.20	0.27	0.50	1.0
0.5	1.61	1.49	1.39	1.34	1.22	1.07
1.0	1.91	1.70	1.48	1.38	1.22	1.08
1.5	2.00	1.73	1.50	1.39	1.23	1.08
2.0		1.74	1.52	1.39	1.23	1.09
3.5		1.76	1.54	1.40	1.23	1.10

(Ref. 6)

(c) Repeated stress, bending

Material	K_f		
1.20 per cent C steel (normalized)	1.23		
0.52 per cent C steel (normalized)	1.23		
0.37 per cent C steel (normalized)	1.03		
Chr. Ni steel (3 Heat Treatments)	1.26	1.48	1.46
3.5 per cent Ni steel (2 Heat Treatments)	1.25	1.10	
Armco iron (.02 per cent C)	1.08		
0.49 per cent C steel (normalized)	1.14		

(Ref. 17)

9. Square or filleted corner in tension 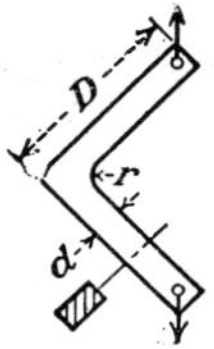	Elastic stress	As shown	$\frac{D}{d} = 5.5$ $\frac{r}{d} = 0.125$ 0.15 0.20 0.25 0.30 0.40 0.50 0.70 1.00 $K_t = 2.50$ 2.30 2.03 1.88 1.70 1.53 1.40 1.26 1.20 (Ref. 12)
	Static rupture	As shown	$D = 8\frac{1}{4}$ $d = 1\frac{1}{2}$ $\frac{r}{d} = 0$ 0.393 0.914 Plaster 2.60 1.54 1.28 Cast iron 1.56 1.02 1.00 (Ref. 12)
10. Square or filleted corner in compression 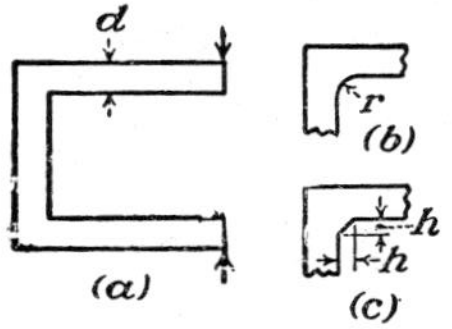	Static rupture	As shown	Reinforced concrete (d = 6 in. and 3 in.) Condition (a) sharp corner (b) fillet, $\frac{r}{d} = \frac{1}{3}$ (c) chamfered, $\frac{h}{d} = \frac{1}{3}$ $K_r =$ 1.00 1.02 1.06 (Ref. 12)

Table 18.1. Factors of Stress Concentration for Elastic Stress K_t, for Repeated Stress K_f, and for Rupture K_r (Continued)

Type of form irregularity or stress raiser	Stress condition	Manner of loading	Factor of stress concentration K_t, K_f, K_r for various dimensions
11. Square shoulder with fillet in circular shaft	Elastic stress	Tension	Approximately same as Case 8
		Bending	Approximately same as Case 8
		Torsion	See table below (Refs. 2, 3)
	Repeated stress	Tension	See table below (Ref. 25)
		Bending	See table below (Ref. 10)
		Torsion	Approximately same as bending (Ref. 10)

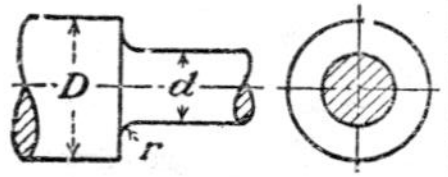

Elastic stress, Torsion:

$\frac{D}{d}$ \ $\frac{r}{d}$	0.005	0.01	0.02	0.03	0.04	0.06	0.08	0.10	0.12
2.00	....	3.0	2.25	2.00	1.82	1.65	1.51	1.44	1.39
1.33	....	2.7	2.16	1.91	1.76	1.60	1.48	1.40	1.35
1.20	3.00	2.5	2.00	1.75	1.62	1.50	1.40	1.34	1.30
1.09	2.20	1.88	1.53	1.40	1.30	1.20	1.16	1.15	1.15

(Refs. 2, 3)

Repeated stress, Tension:

Material	D	d	$\frac{r}{d}$	K_f	$\frac{r}{d}$	K_f
0.065 % C steel.....	0.57	0.295	0	1.56	.21	1.39
0.331 % C steel.....	0.57	0.295	0	1.82	.21	1.39
0.446 % C steel.....	0.57	0.295	0	1.67	.21	1.41
0.645 % C steel.....	0.57	0.295	0	2.08	.21	1.47

(Ref. 25)

Repeated stress, Bending:

Material	D	d	$\frac{r}{d}$	K_f
.57 % C steel (H.T.)..........	....	0.080	0.15	1.03
.57 % C steel (H.T.)..........	....	0.410	0.15	1.50
.57 % C steel (H.T.)..........	....	2.13	0.15	1.75
.30 % C steel..................	....	0.37	0.21	1.13
.30 % C steel..................	....	0.37	0.053	1.59
.49 % C steel (H.T.)..........	0.40	0.275	0	2.04
.46 % C steel (H.T.)..........	2.0	1.0	0.188	1.35
Alloy steel (3.5 % Ni; 0.8 % Cr.)	....	0.30	0.267	1.21
Alloy steel (3.5 % Ni; 0.8 % Cr.)	....	0.30	0.062	2.17

(Ref. 10)

Static rupture

Tension

$\frac{D}{d} = 0.875$ and 0.50; $\frac{r}{d} = 0$

K_r for cast iron = 1 (Ref. 12)

Bending

$\frac{D}{d}$ \ $\frac{r}{d}$		0	0.015	0.021	0.031	0.042	0.062	0.083	0.093	0.125	0.166	0.186	0.25	0.375	
Plaster...	$\frac{4}{2}$	1.84	1.80		1.80		1.72		1.50	1.37		1.44	1.21	1.13	(Ref. 13)
Plaster...	$\frac{4}{3}$	1.82		1.53		1.55		1.38			1.26		1.19		(Ref. 13)
Cast iron.	$\frac{1}{0.5}$	1.00													(Ref. 12)

($\frac{D}{d}$ values represent actual dimensions)

Torsion

Material	D	d \ $\frac{r}{d}$	0	0.062	0.073	
Plaster....................	4	2	1.30	1.13		(Ref. 13)
Plaster....................	1.10	0.823	0.87		0.78	(Ref. 12)
Cast iron....................	0.65	0.50	1.00		1.03	(Ref. 12)
Ni cast iron....................	0.65	0.50	0.95		0.80	(Ref. 12)
Al 112....................	0.475	0.331	0.91		0.76	(Ref. 12)
Al 195–T6....................	0.475	0.331	0.87		0.87	(Ref. 12)
Al 220–T4....................	0.475	0.331	0.94		0.86	(Ref. 12)

Table 18.1. Factors of Stress Concentration for Elastic Stress K_t, for Repeated Stress K_f, and for Rupture K_r (Continued)

Type of form irregularity or stress raiser	Stress condition	Manner of loading	Factor of stress concentration K_t, K_f, K_r for various dimensions
12. U notch in circular shaft	Elastic stress	Tension	Approximately same as Case 2 (Ref. 1)
		Bending	Approximately same as Case 2 (Ref. 1)
		Torsion	$K_t = \frac{(D-d+2r)(d+2r)^2 + 4r^2(D-d-2r)}{2rD(d+4r)}$ For semicircular notch ($d = D - 2r$) $K_t = \frac{2D}{D+2r}$ (Refs. 2, 3)
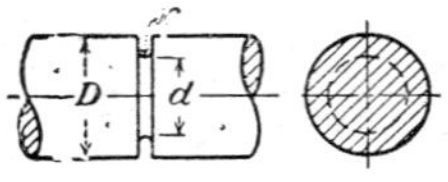	Repeated stress	Bending	See table below
	Static rupture	Tension	See table below

Repeated stress — Bending:

Material	D	h	r	K_f	
6130 steel, 0.29 per cent C. (H.T.)	0.48	0.015	0.015	2.20	(Ref. 14)
6130 steel, 0.29 per cent C. (H.T.)	0.716	0.156	0.125	1.37	(Ref. 14)
0.10 per cent C steel	0.3	0.008	0.002	1.21	(Ref. 14)
	0.6	0.008	0.002	1.07	(Ref. 14)
0.62 per cent C steel	0.34	0.012	0.006	1.78	(Ref. 10)
Chr Ni steel {3.5 per cent Ni, 0.8 per cent Cr (H.T.)}	0.60	0.008	0.002	2.00	(Ref. 10)
0.2 per cent C cast steel, annealed	0.30	0.008	0.002	1.31	(Ref. 10)

Static rupture — Tension:

Material	D	h	r	K_r
Cast iron	0.6	0.052	0.03	1.01
Cast iron	0.4	0.035	0.02	1.18
Nickel cast iron	0.6	0.052	0.03	1.06
Nickel cast iron	0.4	0.035	0.02	1.01
Cast magnesium	0.475	0.04	0.02	1.16
Al 112	0.35	0.035	0.02	1.24
Al 195–T6	0.35	0.035	0.02	0.81
Al 220–T4	0.35	0.035	0.02	0.76

(Ref. 12)

13. V notch in circular shaft

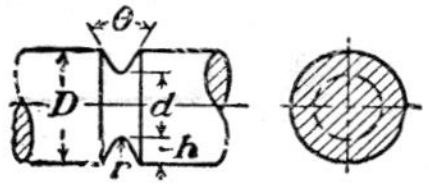

Elastic stress

Tension

$\frac{r}{d}$ =	0.05	0.10	0.15	0.20	0.30	0.40	0.50
K_t =	3.4	2.5	2.1	1.9	1.62	1.5	1.4

(Refs. 1, 4)

Bending

$\frac{r}{d}$ =	0.05	0.10	0.15	0.20	0.30	0.40	0.50
K_t =	2.55	2.0	1.74	1.6	1.43	1.31	1.25

(Refs. 1, 4)

For $\frac{h}{d}$ small, $K_t = \dfrac{\left(1 + 2\sqrt{\frac{h}{r}}\right)}{\left(1 + \sqrt{\frac{h}{r}}\right)}$ times value given below for K_t in torsion (Ref. 8)

Torsion

θ \ $\frac{h}{r}$	0.5	1	3	5	9
0°	1.85	2.01	2.66	3.23	4.54
60°	1.84	2.00	2.54	3.06	3.90
90°	1.81	1.95	2.40	2.64	3.12
120°	1.66	1.75	1.95	2.06	2.13

(Ref. 8)

Repeated stress

Bending

Material	θ \ $\frac{h}{r}$	1	4	7	∞		
0.33 per cent C steel	72°	1.16	1.31	1.42		($h < 0.0025$) $D = 0.45$	(Ref. 8)
0.33 per cent C steel	$63\frac{1}{2}$°	1.45	1.90	2.20		($h = 0.005$ to 0.045) $D = 0.45$	(Ref. 8)
Normalized ext. mag Al alloy	60°		1.50			($h = 0.036$ to 0.040, $D = 0.5$)	(Ref. 14)
1050 steel (H.T.)	60°		2.37			($h = 0.036$ to 0.040, $D = 0.5$)	(Ref. 14)
6130 (.29 per cent C) steel (H.T.)	60°		2.62			($h = 0.036$ to 0.040, $D = 0.5$)	(Ref. 14)
25 S Al	60°		1.80			($h = 0.036$ to 0.040, $D = 0.5$)	(Ref. 14)
Cast Al 220–T4*	72°				0.81	($h = 0.05$, $D = 0.40$)	(Ref. 12)
Cast iron	72°				0.90	($h = 0.05$, $D = 0.40$)	(Ref. 12)
Ext. Al 17 ST	60°	1.67					(Ref. 15)

* Tested to 100,000,000 cycles.

Static rupture

Tension

$\theta = 72°$ $\frac{h}{D} = 0.87$

Material	$\frac{h}{r}$ = ∞	1.72
Cast iron	1.10	1.10
Ni cast iron	1.05	1.04
Al 112	1.09	1.24
Al 195 — T6	0.82	0.81
Al 220 — T4	0.81	0.76
Mag — H1	1.23	1.16

(Ref. 12)

Table 18.1. Factors of Stress Concentration for Elastic Stress K_t, for Repeated Stress K_f, and for Rupture K_r (Continued)

Type of form irregularity or stress raiser	Stress condition	Manner of loading	Factor of stress concentration K_t, K_f, K_r for various dimensions
14. Radial hole in circular shaft	Elastic stress	Tension	Approximately same as Case 5a
		Bending	Approximately same as Case 5a
		Torsion	For $\frac{a}{d}$ very small, $K_t = 4$
	Repeated stress	Bending	See table below
	Static rupture	Bending	See table below
		Torsion	See table below

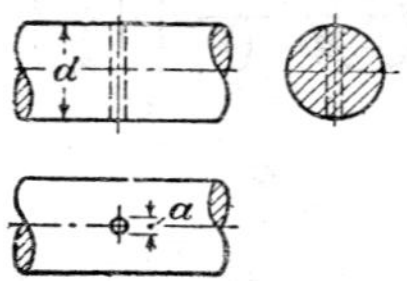

Repeated stress — Bending

Material	$\frac{a}{d}$ \ d	0.10	0.50	1	3	0.273	2.13	0.30	0.35	
0.45 per cent C steel	0.0625		1.34	1.55	1.88					(Ref. 9)
0.45 per cent C steel	0.250	1.1	1.36	1.40	1.55					(Ref. 9)
0.57 per cent C steel (H.T.)	0.150					1.40	2.22			(Ref. 9)
Armco iron	0.183							1.38		(Refs. 17, 10)
0.49 per cent C steel (H.T.)	0.183							1.27		(Refs. 17, 10)
0.52 per cent C steel (H.T.)	0.183							1.34		(Refs. 17, 10)
Cyclops metal (annealed)	0.183							1.05		(Refs. 17, 10)
Cast iron	0.157								1.15	(Refs. 16, 10)

Static rupture — Bending

Material	$\frac{a}{d}$ \ d	2	3
Plaster	0.0625	1.84	1.65
Plaster	0.125	1.50	1.70
Plaster	0.250	1.37	1.46

(Ref. 13)

Static rupture — Torsion

Material	$\frac{a}{d}$ \ d	2	0.82	0.5	0.331
Plaster	0.0625	1.86			
Plaster	0.125	1.89			
Plaster	0.250	2.12			
Plaster	0.076		1.17		
Cast iron	0.125			1.22	
Ni cast iron	0.125			1.11	
Al 112	0.20				1.14
Al 195–T6	0.20				1.06
Al 220–T4	0.20				1.05

(Ref. 13)

15. Screw thread

D = over-all diameter

h = depth of thread

Repeated stress — Tension

$D = 0.375$

Material	Whitworth	U.S. Standard	Rolled	
0.065 per cent C steel....	1.35			(Ref. 25)
0.331 per cent C steel....	1.41			
0.446 per cent C steel....	1.51			
0.645 per cent C steel....	1.45			
0.300 per cent C steel....	1.76	2.84	2.15	(Ref. 27)
S.A.E. 2320 Ni steel H.T.	3.32	3.85		

Repeated stress — Bending

Material	$D = \frac{1}{4}''$, $h = 0.0232$, 28 threads to inch		$D = \frac{3}{8}''$, $h = 0.0271$, 24 threads to inch	
	Continuous thread	Single notch	Continuous thread	Single notch
Tie rod steel 0.43 per cent C	1.18	4.25	1.38	2.42

(Ref. 14)

16. Keyway in circular shaft

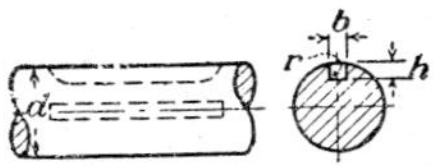

Sled-runner Type (a)

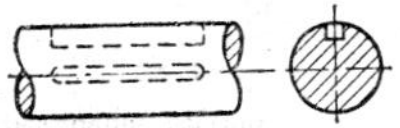

Profile Type (b)

Elastic stress — Torsion

For $r = 0$, theoretical $K_t = \infty$.
Tests to determine elastic failure gave $k = 1.30$ (Refs. 18, 19)

Repeated stress — Bending

Material	(a)	(b)
Chr Ni steel (H.T.)............	1.35	1.74
Medium C steel (normalized).....	1.11	1.35

(Ref. 23)

Repeated stress — Torsion

Type (b), $d = 0.4$, $h = 0.036$

Material	$b = 0.109$	$b = 0.055$
0.65 per cent C steel............	1.27	1.27
Armco iron....................	1.14	1.14

(Ref. 18)

Static rupture — Bending

Material	Type (a)	Type (b)
Plaster........................	1.08	1.28

(Ref. 13)

Static rupture — Torsion

Material	Type (a)	Type (b)
Plaster........................	1.44–1.29	1.68
Cast iron......................	1.17	....

(Refs. 12, 13)

17. Circular thickening (assumed rigid) in wide plate

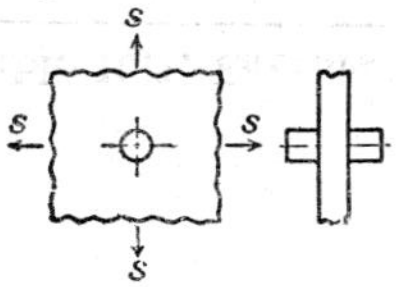

Elastic stress — Tension

For either uniaxial or biaxial (equal) stress, $K_t = 1.54$ (Ref. 24)

Table 18.1. Factors of Stress Concentration for Elastic Stress K_t, for Repeated Stress K_f, and for Rupture K_r (Continued)

Type of form irregularity or stress raiser	Stress condition	Manner of loading	Factor of stress concentration K_t, K_f, K_r for various dimensions
(Additional data) Two V-notches in member of rectangular section	Elastic stress	Bending (normal to plane of d)	$\frac{r}{d}$: 0.01, 0.0125, 0.025, 0.05, 0.10 K_t: 4.0, 3.6, 2.7, 2, 1.5 (Ref. 28)
		Torsion	$K_t = 1 + 0.47 \frac{d}{r}$ (Ref. 28)
(Additional data) 5. Circular hole in plate or rectangular bar	Elastic stress	Tension (load applied through close-fitting pin)	$\frac{a}{d}$: 0.10, 0.15, 0.20, 0.25, 0.30, 0.40, 0.50, 0.60, 0.70 K_t: 9, 7, 5.5, 4.5, 4.0, 3.0, 2.7, 2.2, 2.0 (approximate values, K_t increases with looseness of fit and with proximity of hole to end of plate) (Ref. 29)
	Repeated stress	Tension	(see below)
		Bending	(see below)

Repeated stress, Tension:

Material	Specimen	$\frac{a}{d}$	K_f	
Steel, t.s. 84,000 lb. per sq. in.	0.87 in. wide × 0.65, hole as bored	0.23	1.61	tension ¼s to s, 10,000,000 cycles (Ref. 30)
Same	Hole polished, edges sharp	0.23	1.49	
Same	Hole polished, edges rounded	0.23	1.42	
Steel, .2 per cent C, t.s. 61,800 lb. per sq. in.	8 × ¾ in. plate, two 1-in. rivet holes		2.13, 1.43	tension 0 to s, 2,000,000 cycles. First values comparison small polished specimens, second values comparison unpierced plate.
Silicon steel, t.s. 80,800	Same		2.33, 1.49	
Nickel steel, t.s. 99,000	Same		3.03, 1.60	

(Ref. 31)

Repeated stress, Bending:

Material	Specimen	$\frac{a}{d}$	K_f
Alloy steel H.T., t.s. 100,000 lb. per sq. in.	0.75 in. wide, hole as bored	0.16	1.78
Same	Hole edges rounded	0.16	1.63
Same	Hole edges rounded and pressed	0.16	1.40

(Ref. 30)

(Additional data)

Case	Stress	Loading	Material	D	d	$\frac{r}{d}$	K_f	Ref.
11. Square shoulder with fillet in circular shaft	Elastic stress	Bending	$K_t = 1 + 0.375\sqrt{\frac{d}{r}}$ (approximate formula)					(Ref. 32)
		Torsion	$K_t = 1 + 0.188\sqrt{\frac{d}{r}}$ (approximate formula)					(Ref. 32)
	Repeated stress	Bending	Steel *A*, low alloy, t.s. 78,400 lb. per sq. in.	0.5	0.25	0.08	2.20	(Ref. 33)
			Steel *B*, low alloy, t.s. 70,800	0.5	0.25	0.08	2.08	(Ref. 33)
			Steel *C*, low alloy, t.s. 91,300	0.5	0.25	0.08	2.10	(Ref. 33)
			Steel, 1.42 per cent C, t.s. 73,800	3.5	1.60	0	2.30	(Ref. 34)
			Same	1.94	1.60	0	1.90	(Ref. 34)
			Steel 1.44 per cent C, t.s. 87,000	....	0.35	0	2.10	(Ref. 35)
		Torsion	Steel S.A.E. 1020 hot rolled, t.s. 63,300 lb. per sq. in.	0.75	0.375	0.0053	1.15	(Ref. 36)
			Steel S.A.E. 3140 hot rolled, t.s. 115,000	0.75	0.3	0.027	1.54	(Ref. 36)
			Same	0.75	0.3	0.067	1.57	(Ref. 36)
			Steel S.A.E. 3140 heat treated, t.s. 162,000	0.75	0.3	0.0087	1.51	(Ref. 36)
			Steel *N*	0.552	0.394	0.17	0.96	(Ref. 37)
			Same	0.552	0.394	0.03	0.99	(Ref. 37)
			Steel *E*	0.552	0.394	0.17	1.11	(Ref. 37)
			Same	0.552	0.394	0.03	1.66	(Ref. 37)
			Cr-Ni-W steel	0.99	0.55	0.14	1.10	(Ref. 38)
			Same	2.12	1.18	0.10	1.17	(Ref. 38)
			Same	3.19	1.77	0.11	1.03	(Ref. 38)

Table 18.1. Factors of Stress Concentration for Elastic Stress K_t, for Repeated Stress K_f, and for Rupture K_r (Continued)

Type of form irregularity or stress raiser	Stress condition	Manner of loading	Factor of stress concentration K_t, K_f, K_r for various dimensions
(Additional data) 13. V-notch in circular shaft	Elastic stress	Bending	$K_t = 1 + 2\sqrt{\frac{h}{r}}$ (approximate formula, $\frac{h}{d}$ small) (Ref. 32)
		Torsion	$K_t = 1 + \sqrt{\frac{h}{r}}$ (approximate formula, $\frac{h}{d}$ small) (Ref. 32)
	Repeated stress	Tension	See table below (Ref. 39)
		Bending	See table below (Ref. 40, 41, 45)
		Torsion	See table below (Ref. 45)

Repeated stress, Tension — $D = 0.25$, $d = 0.200$, $r = 0.01$ (Ref. 39)

Material	θ	$\frac{h}{r}$ = 2.5
Cast iron, grey, t.s. 22,000 lb. per sq. in.	60°	1.08
Cast iron, Ni. Moly., t.s. 53,400	60°	1.21
Cast iron, Ni-Cr-Cu, t.s. 46,000	60°	1.31
Cast iron, Ni-Cr-Cu, t.s. 31,600	60°	1.42

Repeated stress, Bending

Material	θ	h = 0.025, $\frac{h}{r}$ = 2.5	h = 0.0035, $\frac{h}{r}$ = 0.45	h = 0.0035, $\frac{h}{r}$ = 0.83	h = 0.025, $\frac{h}{r}$ = 3.2	h = 0.0035, $\frac{h}{r}$ = 5.83	h = 0.05, $\frac{h}{r}$ = 6.4	h = 0.025, $\frac{h}{r}$ = 8.33	Dimensions
Steel, S.A.E. 2330 H.T., t.s. 128,700 lb. per sq. in.	60°	2.55							$D = 0.3$, $d = 0.25$, $r = 0.01$ (Ref. 40)
Steel S.A.E. 4130 normal, t.s. 76,200	60°	1.72							
Steel S.A.E. 4130 H.T., t.s. 139,500	60°	2.87							
Steel S.A.E. 4130 H.T., t.s. 199,300	60°	2.15							
Ni-Cr steel, anneal., t.s. 82,300	60°	0.83							
Ni-Cr steel, cold dr., t.s. 132,500	60°	1.40							
Alcoa 27 S.T. H.T., t.s. 60,000	60°	2.14							
Steel 0.84–0.86 per cent C, H.T. 360 Brinell	55°		1.60		2.65		2.65		$D = 0.35$ (Ref. 41)
Same	75°					2.33			
Steel 0.16 per cent C	55°				1.52				
Same	75°			1.08		1.41			
Steel 0.91 per cent C, t.s. 225,000	60°							2.22	$D = 0.25$, $d = 0.20$, $r = 0.003$ (Ref. 45)
Steel 1.04 per cent C, t.s. 237,000	60°							2.04	
Cr-Van steel t.s. 237,000	60°							3.70	
Si-Man steel t.s. 236,000	60°							3.23	
Beryllium bronze t.s. 166,000	60°							1.89	

Repeated stress, Torsion — (Ref. 45)

Material	θ	$\frac{h}{r}$ = 8.33	
Steel 0.91 per cent C, t.s. 225,000 lb. per sq. in.	60°	1.28	$D = 0.25$
Steel 1.04 per cent C, t.s. 237,000	60°	1.30	$d = 0.20$
Cr-Van. steel, t.s. 237,000	60°	1.75	$r = 0.003$
Si-Man steel, t.s. 263,000	60°	2.19	torsion 0 to τ
Beryllium bronze, t.s. 166,000	60°	1.15	

	Static rupture	Tension, compression, and torsion	see table below (Ref. 39)

Material	K_r for tension	K_r for compression	K_r for torsion	
Cast iron *A*	0.91	0.77	0.81	$\theta = 60°$, $D = 0.625$
Cast iron *B*	1.03	0.87	0.95	$d = 0.50$, $r = 0.01$
Cast iron *C*	1.05	0.73	0.92	$\frac{h}{r} = 6.25$
Cast iron *D*	0.98	0.56	0.88	

(cast irons described under "Repeated stress, Bending," above.) (Ref. 39)

(*Additional data*) 14. Radial hole in circular shaft	Repeated stress	Bending	see table below

Material	$\frac{a}{d}$ \ d	0.4	0.35	
Steel *A* (Described under 11 above)	0.1	2.16		(Ref. 33)
Steel *B* (Described under 11 above)	0.1	2.10		(Ref. 33)
Steel *C* (Described under 11 above)	0.1	2.06		(Ref. 33)
Steel *D* (Described under 11 above)	0.1	2.54		(Ref. 33)
Cast iron *A* (Described under 13 above)	0.1	1.12		(Ref. 39)
Cast iron *B* (Described under 13 above)	0.1	1.50		(Ref. 39)
Cast iron *C* (Described under 13 above)	0.1	1.18		(Ref. 39)
Cast iron *D* (Described under 13 above)	0.1	1.71		(Ref. 39)
Steel 0.84–0.86 per cent C (Described under 13 above)	0.134		2.07	(Ref. 41)

		Torsion	see table below

Material	$\frac{a}{d}$ \ d	0.4	0.38	0.394	0.55	1.18	1.77	
Steel S.A.E. 1020, hot rolled, t.s. 63,300 lb. per sq. in.	0.1	1.31						(Ref. 36)
Same	0.175	1.43						(Ref. 36)
Same	0.250	1.62						(Ref. 36)
Steel S.A.E. 3140, hot rolled, t.s. 115,000	0.1	2.00						(Ref. 36)
Same	0.25	2.25						(Ref. 36)
Rail steel, 0.78 per cent C, t.s. 133,000	0.1	1.64						(Ref. 36)
Same	0.25	2.26						(Ref. 36)
Steel S.A.E. 3140 H.T., t.s. 162,000	0.095		1.87					(Ref. 36)
Steel *N*	0.15			1.39				(Ref. 37)
Steel *V*	0.15			1.31				(Ref. 37)
Steel *E*	0.15			1.66				(Ref. 37)
Cr-Ni-W steel	0.14				1.60			(Ref. 38)
Same	0.10					1.88		(Ref. 38)
Same	0.11						1.74	(Ref. 38)
Cast iron, alloy, t.s. 43,900	0.10	1.07						(Ref. 42)
Cast iron *A* (Described under 13 above)	0.10	1.37						(Ref. 39)
Cast iron *B* (Described under 13 above)	0.10	1.38						(Ref. 39)
Cast iron *C* (Described under 13 above)	0.10	1.20						(Ref. 39)
Cast iron *D* (Described under 13 above)	0.10	1.20						(Ref. 39)

Table 18.1. Factors of Stress Concentration for Elastic Stress K_t, for Repeated Stress K_f, and for Rupture K_r (Continued)

Type of form irregularity or stress raiser	Stress condition	Manner of loading	Factor of stress concentration K_t, K_f, K_r for various dimensions
(Additional data) 16. Key way in circular shaft	Elastic stress	Torsion	$K_t = 1 + \sqrt{\frac{h}{r}}$ (approximate formula for any longitudinal groove; r = min. corner radius) (Ref. 32)
	Repeated stress	Bending	See sub-table below
18. Gear tooth	Elastic stress	Bending plus some compression	For 14.5° pressure angle: $K_t = 0.22 + \left(\frac{t}{r}\right)^{0.2} \left(\frac{t}{h}\right)^{0.4}$ For 20° pressure angle: $K_t = 0.18 + \left(\frac{t}{r}\right)^{0.15} \left(\frac{t}{h}\right)^{0.45}$ K_t = (max s_t by photoelastic analysis) ÷ $\left(\text{calculated max } s_t = \frac{6Ph}{bt^2} - \frac{P \tan \phi}{bt}\right)$ **(Ref. 43)**
19. Press-fitted collar on circular shaft	Repeated stress	Bending	See sub-table below

16. Key way in circular shaft — Repeated stress, Bending:

Material	Type (a)	Type (b)	
Steel 0.84–0.86 per cent, 360 Brinell		1.35	(Ref. 41)
Steel, low carbon, t.s. 60,000 lb. per sq. in.	1.25	1.71 (ends rounded as to width)	(Ref. 30)

18. Gear tooth:

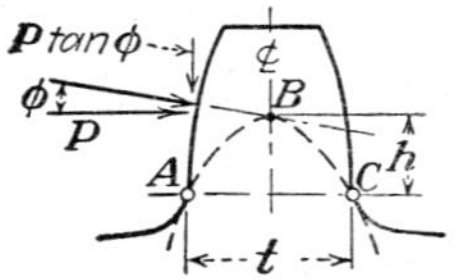

A and *C* are points of tangency of the inscribed parabola *ABC* with tooth profile, b = tooth width normal to plane of figure, r = minimum radius of tooth fillet.

19. Press-fitted collar on circular shaft — Repeated stress, Bending:

Material	Shaft diameter	Collar fit or pressure	K_f (initial cracking)	K_f (break off)	
Steel 0.84–0.86 per cent C, H.T. 360 Brinell	0.25	<12,000 psi	....	1.97	(Ref. 41)
Steel S.A.E. 1045 H.T., t.s. 88,800 lb. per sq in.	2.00	0.003–0.004 in.	4.3	2.65	(Ref. 44)
Same, surface metallized	2.00	0.003–0.0073	1.64	1.64	(Ref. 44)
Same, surface flame hardened	2.00	0.003–0.0063	1.38	.....	(Ref. 44)
Steel 0.42 per cent C, t.s. 73,800	2.50	not given	....	1.50	(Ref. 34)
Same	1.60	16,000	....	2.00	(Ref. 34)
Same	1.60	90	....	1.40	(Ref. 34)
Steel 0.57 per cent C, t.s. 93,000	0.55	9,000	....	1.80	(Ref. 35)
Same	0.55	17,000	....	1.80	(Ref. 35)
Same	0.55	30,000	....	1.90	(Ref. 35)
Same, surface rolled	0.55	30,000	....	1.10	(Ref. 35)
Same, collar grooved near shaft	0.55	30,000	....	1.40	(Ref. 35)
Same, surface rolled, collar grooved	0.55	30,000	....	1.00	(Ref. 35)
Same, collar edge rounded 0.04 in. rad	0.55	30,000	....	1.80	(Ref. 35)
Same, collar edge rounded 0.12 in. rad	0.55	30,000	....	1.60	(Ref. 35)
Same, edge bearing only	0.55	30,000	....	1.70	(Ref. 35)
Cr-Ni steel H.T., t.s. 150,000	0.55	13,000	....	2.60	(Ref. 35)
Low-carbon steel, t.s. 62,000	0.55	17,000	....	1.80	(Ref. 35)
Steel 0.47 per cent C, t.s. 87,000	0.55	13,000	....	1.70	(Ref. 35)

			Steel S.A.E. 1045 hot rolled	1.75	......	3.44 2.82 (compared 1.5 in. pol. spec.)	(Ref. 46)
			Steel S.A.E. 1045 H.T.	2.00	......	4.30 2.64 (compared 2.0 in. pol. spec.)	
			Steel S.A.E. 1045 as forged	7.00	......	1.93 1.59 (comp. 6 in. smooth turn. spec.)	
						3.67 3.00 (comp. 0.3 in. pol. spec.)	
			Steel S.A.E. 1045 hot rolled, tapered from 1.89 to 1.69 on 0.75 in. rad.	1.89	......	2.95 2.48 (comp. 1.5 in. pol. spec.)	
			Steel S.A.E. 1045 as forged, tapered from 7.625 to 6.8 on 2.5 in rad.	7.625	......	1.46 1.25 (comp. 6 in. smooth turn. spec.)	
			Steel S.A.E. 1045 as forged, tapered from 8.125 to 6.77 on 5 in. radius	8.125	......	1.30 1.09 (comp. 6 in. smooth turn spec.)	
						2.44 2.06 (comp. 0.3 in. pol. spec.)	
			Steel S.A.E. 1045 H.T., shaft grooved next collar, groove radius 0.156 in.	2.00	......	1.81 1.56 (comp. 2 in. pol. spec.)	

			Material	Surface finish: Polished	Surface finish: Mill scale		
20. Surface roughness	Repeated stress	Tension	Steel, 0.2 per cent C, t.s. 61,800 lb. per sq. in.	1	1.48	tension 0 to s, 2,000,000 cycles	(Ref. 31)
			Silicon steel, t.s. 80,800	1	1.56		
			Ni. steel, t.s. 99,000	1	1.89		

				Surface finish								
			Material	Highly polished	Fine emery	Coarse emery	Ground	Smooth file	Coarse file	Smooth turned	Rough turned	
		Bending	Steel, 0.49 per cent C, H.T. 197 Brinell	1	1.06		1.13			1.19	1.21	(Ref. 30)
			Steel, 0.33 per cent C	1	1.02	1.06		1.08	1.24	1.14		
			Steel *A*, t.s. 55,200 lb. per sq. in.	1							1.21	(Ref. 48; see also Ref. 49)
			Steel *B*, t.s. 65,300	1							1.18	
			Steel *C*, t.s. 73,800	1							1.27	
			Steel *D*, t.s. 78,200	1							1.27	
			Steel *E*, t.s. 81,300	1							1.24	
			Steel, annealed, t.s. 60,000	1							1.19	
			Steel, annealed, t.s. 90,000	1							1.22	
			Steel, H.T., t.s. 140,000	1							1.31	

Table 18.1. Factors of Stress Concentration for Elastic Stress K_t, for Repeated Stress K_f, and for Rupture K_r (Continued)

Type of form irregularity or stress raiser	Manner of loading	$\sqrt{h/r}$ \ $\sqrt{c/r}$	1	2	3	4	5	6	6.5	7.5
			Factor of elastic stress concentration K_t for various dimensions (Ref. 51)							
(Additional data) 1 and 2	Tension	0.5	1.45	1.75	1.85	1.90	1.95	1.98	1.99	
Two notches in member of rectangular section		1	1.49	2.26	2.62	2.75	2.85	2.87	2.89	
		2	1.50	2.50	3.30	3.80	4.20	4.40	4.50	
		3	1.51	2.60	3.60	4.40	5.00	5.40	5.60	
		3.5	....	2.63	3.65	4.55	5.26	5.75	6.00	
r = notch radius,	Bending	0.5	1.35	1.65	1.80	1.85	1.90	1.93	1.95	1.98
h = notch depth,		1	1.37	1.80	2.26	2.60	2.70	2.80	2.83	2.85
c = half net width d		2	1.40	1.97	2.60	3.15	3.60	3.90	4.05	4.21
		3	1.42	1.99	2.63	3.40	3.88	4.42	4.65	5.02
		3.5	....	....	....	3.45	4.00	4.60	4.85	5.25
(Additional data) 3 and 4	Tension	0.5	1.24	1.50	1.68	1.80	1.86	1.90	1.91	1.92
One notch in member of rectangular section		1	1.25	1.55	1.87	2.18	2.40	2.57	2.62	2.70
r = notch radius,		2	1.26	1.57	1.98	2.40	2.81	3.14	3.32	3.61
h = notch depth,		3	....	1.58	2.00	2.48	2.92	3.37	3.61	3.91
c = net width d		3.5	....	....	....	2.50	2.98	3.43	3.63	4.00
	Bending	0.5	1.37	1.65	1.82	1.90	1.93	1.95	1.96	1.98
		1	1.41	1.83	2.25	2.51	2.68	2.77	2.80	2.86
		2	1.45	1.91	2.50	3.05	3.50	3.80	3.91	4.11
		3	1.46	1.94	2.55	3.20	3.83	4.32	4.50	4.86
		3.5	....	....	2.58	3.25	3.88	4.48	4.69	5.10
(Additional data) 6	Tension	0.5	1.80	1.92	1.96	1.98				
Elliptical hole in plate		1	2.20	2.70	2.85	2.90				
r = end radius of hole,		2	2.42	3.60	4.20	4.45				
h = half long axis a,		3	2.50	3.88	4.85	5.50				
c = distance end of hole to edge of plate		3.5	2.52	4.00	5.08	5.87				
	Bending	1	1.80	1.91	1.98	1.99				
		2	2.20	2.70	2.83	2.90				
		3	2.35	3.20	3.58	3.76				
		4	2.48	3.55	4.11	4.45				
		5	2.49	3.74	4.60	5.05				
		6	2.50	3.90	4.87	5.50				
		7	2.51	4.00	5.18	5.90				
(Additional data) 12 and 13	Tension	0.5	1.36	1.75	1.86	1.91	1.96	1.98	1.99	
Circumferential notch in circular shaft		1	1.41	1.99	2.44	2.65	2.80	2.88	2.90	
r = notch radius,		2	1.43	2.06	2.87	3.49	3.85	4.25	4.40	
h = notch depth,		3	....	2.11	3.00	3.77	4.40	5.13	5.43	
c = half net diameter d		3.5	....	....	3.07	3.85	4.60	5.40	5.80	
	Bending	0.5	1.35	1.68	1.80	1.87	1.91	1.95	1.98	
		1	1.40	1.80	2.20	2.48	2.62	2.76	2.80	
		2	1.42	1.86	2.40	2.91	3.30	3.80	4.05	
		3	....	1.89	2.48	3.06	3.62	4.30	4.70	
		3.5	....	....	2.49	3.10	3.70	4.48	4.89	
	Torsion	1	1.25	1.42	1.60	1.74	1.80	1.85	1.90	
		2	1.25	1.48	1.72	1.95	2.16	2.40	2.50	
		3	....	1.50	1.75	2.00	2.32	2.62	2.85	
		4	....	....	....	....	2.40	2.75	3.00	

21. V-notch (Case 13) in hollow circular shaft.
b = distance central axis to root of notch,
c = distance inner wall to root of notch,
h = notch depth,
r = notch radius

	$\sqrt{h/r}$	0.5				1				2				3			
	$\sqrt{c/r}$ \ $\sqrt{b/r}$	1	2	4	6	1	2	4	6	1	2	4	6	1	2	4	6
Tension	1	1.34	1.68	1.77	1.79	1.35	1.87	2.10	2.15	1.36	1.94	2.30	2.39	1.37	2.00	2.35	2.43
	2	1.35	1.75	1.89	1.90	1.38	2.05	2.55	2.62	1.38	2.16	3.10	3.37	1.39	2.20	3.25	3.61
	3	1.36	1.76	1.91	1.94	1.39	2.09	2.66	2.80	1.40	2.24	3.40	3.85	1.40	2.25	3.66	4.30
	4	1.36	1.77	1.94	1.97	1.40	2.11	2.68	2.86	1.40	2.25	3.52	4.04	1.42	2.28	3.86	4.68
Bending	1	1.24	1.54	1.75	1.78	1.25	1.64	2.05	2.10	1.25	1.65	2.18	2.30	1.25	1.68	2.25	2.38
	2	1.24	1.60	1.80	1.89	1.25	1.74	2.35	2.55	1.25	1.76	2.75	3.16	1.25	1.77	2.85	3.35
	3	1.24	1.63	1.86	1.92	1.25	1.76	2.45	2.65	1.25	1.77	2.90	3.50	1.25	1.80	3.00	3.77
	4	1.25	1.65	1.90	1.94	1.25	1.80	2.48	2.70	1.25	1.80	2.95	3.60	1.25	1.81	3.10	4.00

	$\sqrt{h/r}$	1				2				3				4			
	$\sqrt{c/r}$ \ $\sqrt{b/r}$	2	4	6	7.5	2	4	6	7.5	2	4	6	7.5	2	4	6	7.5
Torsion	1	1.25	1.30	1.35	1.37	1.25	1.31	1.36	1.38	1.25	1.31	1.37	1.39				
	2	1.34	1.52	1.55	1.58	1.35	1.70	1.72	1.74	1.35	1.70	1.73	1.76				
	3	1.35	1.73	1.77	1.79	1.40	1.83	2.00	2.06	1.42	1.90	2.08	2.13	1.42	1.91	2.09	2.14
	4	1.36	1.74	1.81	1.83	1.41	1.95	2.13	2.20	1.43	2.05	2.30	2.40	1.43	2.07	2.35	2.50
	5	1.37	1.75	1.85	1.88	1.42	2.01	2.24	2.33	1.43	2.20	2.35	2.60	1.44	2.25	2.60	2.75
	6.5	1.39	1.77	1.88	1.91	1.43	2.10	2.35	2.48	1.44	2.30	2.63	2.80	1.44	2.37	2.75	2.95
	7.5	1.40	1.78	1.89	1.92	1.43	2.15	2.42	2.55	1.44	2.35	2.70	2.85	1.44	2.45	2.82	3.04

References for Table 18.1

1. Peterson, R. E., and A. M. Wahl: Two and Three-dimensional Cases of Stress Concentration, and Comparison with Fatigue Tests, *J. Appl. Mech.*, vol. 3, no. 1, p. A-15, 1936.
2. Jacobsen, L. S.: Torsional Stresses in Shafts Having Grooves or Fillets, *J. Appl. Mech.*, vol. 2, no. 4, p. A-154, 1935.
3. Sonntag, R.: Zur Torsion von runden Wellen mit veränderlichem Durchmesser, *Z. Angew. Math. Mechanik*, vol. 9, p. 1, 1929.
4. Neuber, H. P.: Elastische strenge Lösungen zur Kerbwirkung bei Scheiben und Umdrehungskörpern and Ein neuer Ansatz zur Lösung raumlicher Probleme der Elastizitätstheorie, *Z. Angew. Math. Mechanik*, vol. 13, p. 439, 1933; vol. 14, p. 203, 1934.
5. Timoshenko, S., and W. Diet: Stress Concentration Produced by Holes and Fillets, *Trans. ASME*, vol. 47, p. 199, 1925.
6. Frocht, M.: Factors of Stress Concentration Photoelastically Determined, *J. Appl. Mech.*, vol. 2, no. 2, p. A-67, 1935.
7. Inglis, C. E.: Stresses in a Plate Due to the Presence of Cracks and Sharp Corners, *Engineering (London)*, vol. 95, p. 415, 1913.
8. Thomas, W. N.: The Effect of Scratches and Various Workshop Finishes upon the Fatigue Strength of Steel, *Engineering (London)*, vol. 116, p. 449, 1923.
9. Peterson, R. E.: Model Testing as Applied to Strength of Materials, ASME paper APM 55-11, *J. Appl. Mech.*, vol. 1, no. 2, 1933.
10. Peterson, R. E.: Stress Concentration Phenomena in Fatigue of Materials, ASME paper APM 55-19, *J. Appl. Mech.*, vol. 1, no. 4, 1933.
11. Trinks, W., and J. H. Hitchcock: Strength of Roll Necks, R. P. 55-5, *Trans. ASME*, vol. 55, p. 67, 1933.
12. Roark, R. J., R. S. Hartenberg, and R. Z. Williams: Influence of Form and Scale on Strength, *Univ. Wis., Eng. Expt. Sta. Bull.*, 1938.
13. Private communication.
14. Moore, R. R.: Effect of Grooves, Threads and Corrosion upon the Fatigue of Metals, *ASTM Proc.*, vol. 26, part II, p. 255, 1926.
15. Templin, R. L.: The Fatigue Properties of Light Metals and Alloys, *ASTM Proc.*, vol. 33, part II, p. 364, 1933.
16. Moore, H. F., S. H. Lyons, and N. P. Inglis: Tests of the Fatigue Strength of Cast Iron, *Univ. Ill., Eng. Expt. Sta. Bull.* 164, 1927.
17. Moore, H. F., and T. M. Jasper: An Investigation of the Fatigue of Metals, *Univ. Ill., Eng. Expt. Sta. Bull.* 152, 1925.
18. Gough, H. F.: British Aero. Research Committee Reports, vol. II, p. 488, 1924–1925.
19. Moore, H. F.: The Effect of Keyways on the Strength of Shafts, *Univ. Ill., Eng. Expt. Sta. Bull.* 42, 1910.
20. Coker, E. G.: Photoelastic and Strain Measurements of the Effects of Circular Holes on the Distribution of Stress in Tension Members, *Engineering (London)*, vol. 109, p. 259.
21. Jeffery, J. B.: Plane Stress and Plane Strain in Bipolar Coordinates, *Proc. Roy. Phil. Soc. (London)*, vol. A221, p. 265, 1921.
22. Wahl, A. M., and R. Beeuwkes: Stress Concentration Produced by Holes and Notches, *Trans. ASME*, vol. 56, p. 617, 1934.
23. Peterson, R. E.: Fatigue of Shafts having Keyways, *ASTM Proc.*, vol. 32, part II, p. 413, 1932.
24. Palmblad, E.: Störung der gleichmässigen Spannungsverteilung in einer Scheibe durch Verdickungen, Doctor's dissertation, Darmstadt.
25. Stanton, T. E., and L. Bairstow: On the Resistance of Iron and Steel to Reversals of Direct Stress, *Inst. Civil Eng. (British)*, vol. 166, part IV, p. 78, 1905–1906.
26. Haigh, B. P.: Report of Committee on Complex Stress Distribution, Reports of British Association for Advancement of Science, 1922–1924.
27. Moore, H. F., and P. E. Henwood: The Strength of Screw Threads under Repeated Tension, *Univ. Ill., Eng. Expt. Sta. Bull.* 264, 1934.
28. Lee, G. H.: The Influence of Hyperbolic Notches on the Transverse Flexure of Elastic Plates, ASME paper A-53, *J. Appl. Mech.*, vol. 7, no. 2, June, 1940.
29. Frocht, M. M., and H. N. Hill: Stress Concentration Factors around a Central Circular Hole in a Plate Loaded through a Pin in Hole, *J. Appl. Mech.*, vol. 7, no. 1, p. A-5, March, 1940.
30. "Prevention of Fatigue of Metals," Batelle Mem. Inst., John Wiley & Sons, Inc., New York, 1941.
31. Wilson, W. M., and F. P. Thomas: Fatigue Tests of Riveted Joints, *Univ. Ill., Eng. Expt. Sta. Bull.* 302, 1938.

32. Wilson, W. K.: "Practical Solution of Torsional Vibration Problems," 2d ed., vol. II, John Wiley & Sons, Inc., New York, 1941.
33. Collins, W. L., and T. J. Dolan: Physical Properties of Four Low-alloy High-strength Steels, *ASTM Proc.*, vol. 38, part II, p. 157, 1938.
34. Peterson, R. E., and A. M. Wahl: Fatigue of Shafts at Fitted Members, with a Related Photoelastic Analysis, ASME paper A-1, *J. Appl. Mech.*, vol. 2, no. 1, March, 1935.
35. Thum, A., and F. Wunderlich: Der Einfluss von Einspann und Kraffangriffsstellen auf die Dauerhaltbarkeit der Konstruktionen, *VDI Zeitschrift*, vol. 77, no. 31, Aug. 5, 1933.
36. Dolan, T. J.: The Combined Effect of Corrosion and Stress Concentration at Holes and Fillets in Steel Specimens Subjected to Reversed Torsional Stresses, *Univ. Ill., Eng. Expt. Sta. Bull.* 293, 1937.
37. Armbruster, E.: Einfluss der Oberflachenbeschaffenheit auf den Spannungsverlauf und die Schwingungsfestigkeit, Vereines Deutscher Ing. Verlag, 1931.
38. Mailander, R., and W. B. Bauersfeld: Einfluss der Probengrosse und Probenform auf die Schwingungsfestigkeit von Stahl, *Technische Mitteilungen Krupp*, vol. 2, p. 143, December, 1934.
39. Collins, W. L., and J. O. Smith: The Notch Sensitivity of Alloyed Cast Irons Subjected to Repeated and Static Loads, *ASTM Proc.*, vol. 42, part II, 1942.
40. Oberg, T. T., and J. B. Johnson: Fatigue Properties of Metals Used in Aircraft Construction at 3450 and 10600 Cycles, *ASTM Proc.*, vol. 37, part II, p. 195, 1937.
41. Lea, F. C.: The Effect of Discontinuities and Surface Conditions on Failure under Repeated Stress, *Engineering*, vol. 144, July 23, 1937.
42. Draffin, J. O., and W. L. Collins: The Mechanical Properties of a High Strength Cast Iron, *ASTM Proc.*, vol. 39, part II, p. 589, 1939.
43. Dolan, T. J., and E. L. Broghamer: A Photo-elastic Study of Stresses in Gear Tooth Fillets, *Univ. Ill., Eng. Expt. Sta. Bull.* 335, 1942.
44. Horger, O. J., and T. V. Buckwalter: Fatigue Strength of 2-inch Diameter Axles with Surfaces Metal Coated and Flame Hardened, *ASTM Proc.*, vol. 40, part II, p. 733, 1940.
45. Johnson, J. B.: Fatigue Characteristics of Helical Springs, *Iron Age*, vol. 133, Mar. 15 and 22, 1934.
46. Horger, O. J., and T. V. Buckwalter: Improving Engine Axles and Piston Rods, *Metal Progr.*, vol. 39, no. 2, February, 1941.
47. Horger, O. J., and H. R. Neifert: Effect of Surface Conditions on Fatigue Properties, "The Surface Treatment of Metals," American Society of Metals, 1940.
48. Horger, O. J.: Fatigue Strength of Members as Influenced by Surface Conditions, *Product Eng.*, November and December, 1940; January, 1941.
49. Karpov, A. V.: Fatigue Problems in Structural Design, *Metals and Alloys*, vol. 10, November and December, 1939.
50. Dumont, C.: Stress Concentration around an Open Circular Hole in a Plate Subjected to Bending Normal to the Plane of the Plate, *NACA Tech. Note* 740, 1939.
51. Neuber, H.: "Theory of Notch Stresses," J. W. Edwards, Publishers, Incorporated, Ann Arbor, Mich., 1946.
52. Peterson, R. E.: "Stress Concentration Design Factors," John Wiley & Sons, Inc., New York, 1953.

SYSTEMS RELIABILITY ANALYSIS*

18.11. INTRODUCTION

Ever since man began building machines and systems, he has been concerned not only with achieving designs capable of functioning to specified accuracies and performance levels but also in assuring operation for long periods of time without failure or relatively free of malfunctions and the associated costly maintenance. When rotating machinery was first developed, long life was a primary consideration, and the subjects of wear, lubrication, fatigue stress, and margin of safety were duly considered as part of good design practice. Reliability was considered only in terms of long operating life. The concept of reliability has been broadened and changed in the past two decades, however, and long operating life is not today considered the only criterion in

* By Carl H. Levinson.

assuring reliable design. Indeed, much of the equipment being built today fails long before it would wear out.

One of the primary reasons for this change in attitude is that modern-day equipment and systems are far more complex than those of 100 or even 10 years ago, and the trend is toward increasing complexity. In addition, the applications have changed so that the designer is not always striving for long life. Indeed, many of the very complex systems now being designed are intended to operate for only a short period of time, with a primary requirement being that no failure occurs during this short operating period. Consequently, designing for long operating life will not be here considered. This subject has been given ample treatment in the literature. Wear-out will be considered only from the standpoint of determining adequate systems-maintenance procedures.

The primary concern here will be with those principles and procedures applicable for determining the probability of successful operation, for a short period of time, of a complex system operated prior to wear-out. The development of this reliability concept and its outstanding applications today may be found in a large variety of military and space projects.

The subject of reliability has an extensive literature, and the reader is referred to references 23 through 32 for more information.

(a) Definition of Reliability

The reliability of a simple device or complex system is defined as the probability that the device will perform its intended design function for a specified period of time under specified environmental conditions.

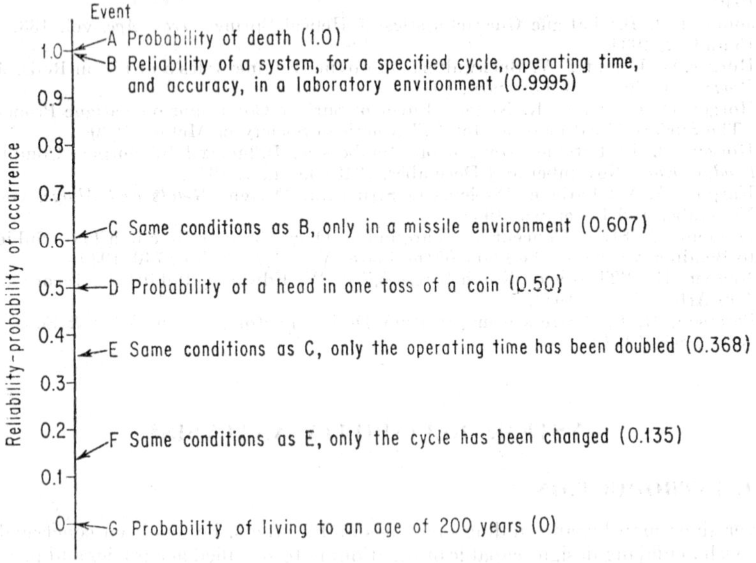

FIG. 18.10. Reliabilities of various events.

From this definition, the basic factors associated with reliability are: (1) operating cycle or function, (2) tolerance or accuracy of the output function, (3) operating time, and (4) ambient environmental conditions.

The mere specification of a reliability number is totally meaningless unless the above factors are also specified, as illustrated by the following simple examples.

1. Cycle: A gas bearing may be highly reliable when operated continuously, but if

the operating cycle is changed to on-off operation with reversals in direction of rotation, the reliability will certainly be lower.

2. Accuracy: A missile may have a very high probability of reaching its target within 50 miles, but a very low probability of reaching this same target within 1 mile.

3. Operating time: The longer a system is operated, the more opportunity it has for failure, and consequently its reliability decreases with time.

4. Environment: An inertial guidance system may be highly reliable when operated in a laboratory environment but totally unreliable in a missile application where it is subjected to severe vibration and acceleration.

Reliability is therefore the probability of the success of an event, and this event (successful equipment operation) must be accurately defined.

Since reliability is a probability, the reliability of any device will be a number in the range from 0 to 1. A reliability of 1 means absolute certainty that the event will occur; a reliability of 0 means absolute certainty that the event will not occur. Figure 18.10 illustrates the reliabilities of various events.

18.12. DERIVATION OF THE RELIABILITY FUNCTION

A large number (so as to arrive at statistically valid results) of devices N_0 are placed on test, with a specified operating cycle (condition 1) in a specified environment (condition 4). After t hr of operation (condition 3), all devices are functionally tested, and based upon the specified accuracy (condition 2), N devices are still operating properly. Consequently, for the specified conditions 1, 2, and 4, the reliability for an operating time of t hr is

$$R = N/N_0 \tag{18.31}$$

If this test is continued, more and more devices will fail; that is, N will decrease, and consequently the reliability will be some function of the operating time t.

The instantaneous rate at which the devices fail λ is called the failure rate (or percentage failures per unit time period). If after t hr of operation, N devices are still operating, and in the time interval from t to $t + dt$, dN devices fail, then the per cent change in population in time interval dt

$$f(t) = \lambda = -\frac{dN/dt}{N}$$

Rearranging and integrating

$$-\int_0^t \lambda\, dt = \int_{N_0}^{N} dN/N$$

Substituting Eq. (18.32) we find the probability of successful operation for t hr

$$R = e^{-\int_0^t \lambda\, dt} \tag{18.32}$$

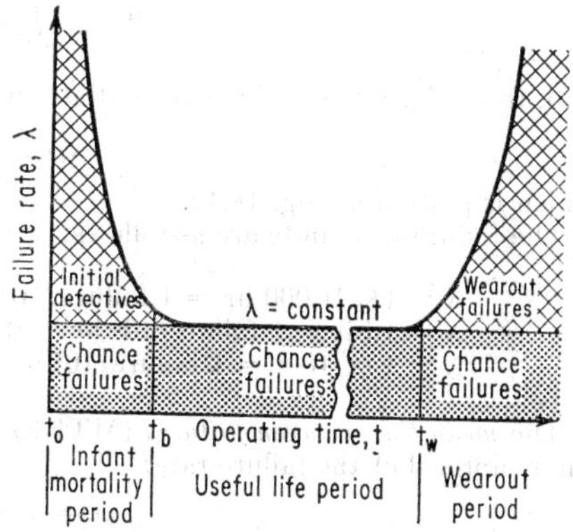

FIG. 18.11. Failure rate–life characteristic curve.

(a) Determination of the Failure Rate

At this point, the problem of determining the reliability of the device reduces to a determination of its failure rate as a function of time, i.e., $f(t) = \lambda$. This determination must be an empirical one, for it is very difficult and usually impossible to arrive at an answer based upon theoretical considerations.

Figure 18.11 illustrates a very typical failure rate–life characteristic curve, based upon empirical test observation for many types of devices and equipment.

The *infant mortality period* will be discussed first. In the first few hours of equipment operation, the failure rate is high because of initial defectives. These early failures are attributed to inherently weak units resulting from poor quality of materials and defects introduced during the manufacturing process. Rigorous quality control of procured materials and during the manufacturing process can minimize this type of failure. Most new equipments are "burned in" for a certain time period after manufacture (debugging period), in order to weed out initial defectives and assure delivery to the customer of an equipment operating in the region beyond its infant-mortality period.

The *wear-out period* is characterized by a sharp increase in the failure rate due to the detrimental effects of age on the equipment. Preventive-maintenance procedures, requiring repair or replacement of parts prior to wear-out, are a requisite to assuring that the equipment will not be operated in this high-failure-rate period.

The *useful-life period* is of greatest interest, for it is in this period that the equipment will be operated; the equipment will be burned in so as to preclude initial defectives, operated during its useful-life period, and taken out of service or repaired so as to avoid the high-failure-rate period associated with wear-out.

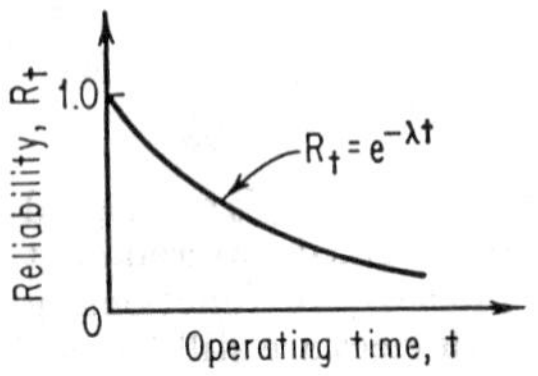

FIG. 18.12. Exponential reliability function.

During the useful-life period, the failure rate λ is constant. Failures that occur during this time period are purely of a random nature, as, for example, failure caused by a speck of dirt which could at any time work its way into a bearing; the initial defectives have already been weeded out prior to time t_b, and the effects of age will not be manifest until time t_w; consequently any failures which occur are unpredictable and of a random nature. If we have no prior reason to suspect that these random causes have a higher probability of occurrence at one time than another, then the probability of their occurrence at any time is the same, and we are led to the conclusion that $f(t) = \lambda$ is constant for all times in the interval t_b to t_w.

The cumulative probability of survival as a function of time for an equipme ' of constant failure rate λ is now derived from Eq. (18.32)

$$f(t) = \int_0^t \lambda \, dt = \lambda t$$

and, from Eq. (18.32), the exponential reliability function reduces to

$$R = e^{-\lambda t} \tag{18.33}$$

which is plotted as Fig. 18.12.

The failure-rate units are as follows:

$$\lambda = 1\%/1{,}000 \text{ hr} = 1 \text{ failure}/10^5 \text{ hr} = 10 \times 10^{-6} \text{ failure/hr/part}$$
$$\lambda = 1 \text{ bit} = 1 \text{ failure}/10^8 \text{ hr} = 0.01 \times 10^{-6} \text{ failure/hr/part}$$
$$\lambda = 1 \times 10^{-6} = 1 \text{ failure}/10^6 \text{ hr} = 1 \times 10^{-6} \text{ failure/hr/part}$$

The *mean time between failures* (MTBF) of an equipment is designated as m and is the reciprocal of the failure rate:

$$m = 1/\lambda \qquad \text{hr}$$

If a number of equipments are placed on test and operated within their useful-life period, then, on the average, one failure will be observed for every m hr of operating time for all equipments:

$$m = \frac{\text{total operating time for all equipments}}{\text{number of failures}}$$

The MTBF is *not* the wear-out life; in many cases it is much longer than the wear-out life.

Table 18.2. The Exponential Distribution $R = e^{-\lambda t}$

Failure rate λ: Failures /hr $\times 10^6$	Failure rate λ: Failures per 1,000 hr, %	Failure rate λ: Bits	Mean life or MTBF $m = 1/\lambda$, hr	Probability of success (reliability R), % after time period of: 5 hr	10 hr	20 hr	50 hr	100 hr	200 hr	500 hr	1,000 hr	2,000 hr	5,000 hr
1	0.1	100	1,000,000	.....	.....	.....	.....	99.99	99.98	99.95	99.90	99.80	99.50
2	0.2	200	500,000	.....	.....	.....	99.99	99.98	99.96	99.90	99.80	99.60	99.00
3	0.3	300	333,333	.....	.....	.....	99.99	99.97	99.94	99.85	99.70	99.40	98.51
4	0.4	400	250,000	.....	.....	.....	99.98	99.96	99.92	99.80	99.60	99.20	98.01
5	0.5	500	200,000	.....	.....	99.99	99.98	99.95	99.90	99.75	99.50	99.00	97.53
6	0.6	600	166,670	.....	.....	99.99	99.97	99.94	99.88	99.70	99.40	98.80	97.04
7	0.7	700	142,860	.....	.....	99.99	99.97	99.93	99.86	99.65	99.30	98.60	96.56
8	0.8	800	125,000	.....	.....	99.99	99.96	99.92	99.84	99.60	99.20	98.41	96.07
9	0.9	900	111,100	.....	.....	99.99	99.96	99.91	99.82	99.55	99.10	98.21	95.59
10	1	1,000	100,000	.....	99.99	99.98	99.95	99.9	99.8	99.50	99.0	98.0	95.1
20	2	2,000	50,000	99.99	99.98	99.96	99.90	99.8	99.6	99.0	98.0	96.1	90.5
30	3	3,000	33,333	99.99	99.97	99.94	99.85	99.7	99.4	98.5	97.0	94.2	86.1
40	4	4,000	25,000	99.98	99.96	99.92	99.80	99.6	99.2	98.0	96.1	92.3	81.9
50	5	5,000	20,000	99.98	99.95	99.90	99.75	99.5	99.0	97.5	95.1	90.5	77.9
60	6	6,000	16,667	99.97	99.94	99.88	99.70	99.4	98.8	97.0	94.2	88.7	74.1
70	7	7,000	14,286	99.97	99.93	99.86	99.65	99.3	98.6	96.6	93.2	86.9	70.5
80	8	8,000	12,500	99.96	99.92	99.84	99.60	99.2	98.4	96.1	92.3	85.2	67.0
90	9	9,000	11,111	99.96	99.91	99.82	99.55	99.1	98.2	95.6	91.4	83.6	63.8
100	10	10,000	10,000	99.95	99.9	99.80	99.50	99.0	98.0	95.1	90.5	81.9	60.7
200	20	20,000	5,000	99.90	99.8	99.6	99.0	98.0	96.1	90.5	81.9	67.0	36.8
300	30	30,000	3,333	99.85	99.7	99.4	98.5	97.0	94.2	86.1	74.1	54.9	22.3
400	40	40,000	2,500	99.80	99.6	99.2	98.0	96.1	92.3	81.9	67.0	44.9	13.5
500	50	50,000	2,000	99.75	99.5	99.0	97.5	95.1	90.5	77.9	60.7	36.8	8.2
600	60	60,000	1,667	99.70	99.4	98.8	97.0	94.2	88.7	74.1	54.9	30.1	5.0
700	70	70,000	1,429	99.65	99.3	98.6	96.6	93.2	86.9	70.5	49.7	24.7	3.0
800	80	80,000	1,250	99.60	99.2	98.4	96.1	92.3	85.2	67.0	44.9	20.2	1.8
900	90	90,000	1,111	99.55	99.1	98.2	95.6	91.4	83.6	63.8	40.7	16.5	1.1
1×10^3	100	1×10^5	1,000	99.50	99.0	98.0	95.1	90.5	81.9	60.7	36.8	13.5	0.7
2×10^3	200	2×10^5	500	99.0	98.0	96.1	90.6	81.9	67.0	36.8	13.5	1.8	
3×10^3	300	3×10^5	333	98.5	97.0	94.2	86.1	74.1	54.9	22.3	5.0	0.25	
4×10^3	400	4×10^5	250	98.0	96.1	92.3	81.9	67.0	44.9	13.5	1.8	0.03	
5×10^3	500	5×10^5	200	97.5	95.1	90.5	77.9	60.7	36.8	8.2	0.7		
6×10^3	600	6×10^5	167	97.0	94.2	88.7	74.1	54.9	30.1	5.0	0.25		
7×10^3	700	7×10^5	143	96.6	93.2	86.9	70.5	49.7	24.7	3.0	0.09		
8×10^3	800	8×10^5	125	96.1	92.3	85.2	67.0	44.9	20.2	1.8	0.03		
9×10^3	900	9×10^5	111	95.6	91.4	83.6	63.8	40.7	16.5	1.1	0.01		
1×10^4	1,000	1×10^6	100	95.1	90.5	81.9	60.7	36.8	13.5	0.7			
2×10^4	2,000	2×10^6	50.0	90.5	81.9	67.0	36.8	13.5	1.8				
3×10^4	3,000	3×10^6	33.3	86.1	74.1	54.9	22.3	5.0	0.25				
4×10^4	4,000	4×10^6	25.0	81.9	67.0	44.9	13.5	1.8	0.03				
5×10^4	5,000	5×10^6	20.0	77.9	60.7	36.8	8.2	0.7					
6×10^4	6,000	6×10^6	16.7	74.1	54.9	30.1	5.0	0.25					
7×10^4	7,000	7×10^6	14.3	70.5	49.7	24.7	3.0	0.09					
8×10^4	8,000	8×10^6	12.5	67.0	44.9	20.2	1.8	0.03					
9×10^4	9,000	9×10^6	11.1	63.8	40.7	16.5	1.1	0.01					
10×10^4	10,000	10×10^6	10.0	60.7	36.8	13.5	0.7						

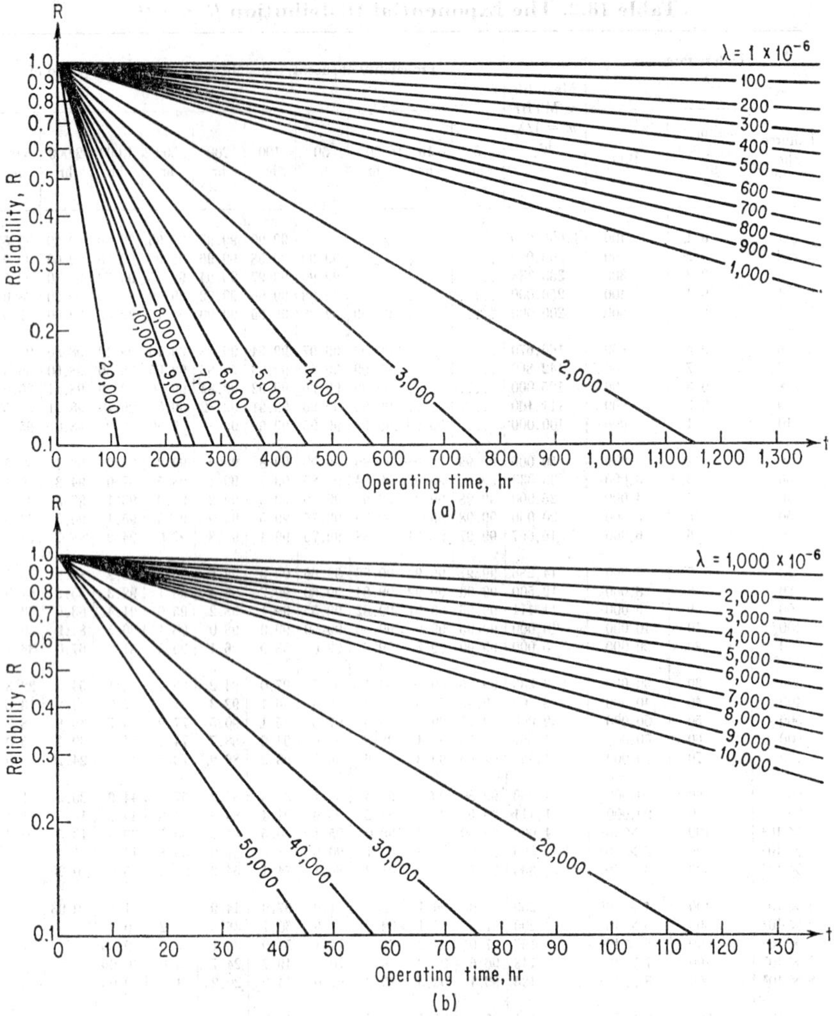

FIG. 18.13. Semilog plot of reliability as a function of operating time for different values of λ. (*a*) From 0 to 1,300 hr. (*b*) From 0 to 130 hr.

Table 18.2 is a tabulation of R for different values of λ and t and Fig. 18.13 is a plot of Table 18.2. When λt is small (less than 0.03) the following useful approximation can be made: $e^{-\lambda t} \cong 1 - \lambda t$. This expression is correct to three significant figures.

18.13. SYSTEM RELIABILITY

(a) Block Diagrams

The first step in determining the reliability of a system is to draw a block diagram of the system.

Figure 18.14*a* is a reliability block diagram of a simple system composed of two components. The components are shown in series to indicate that both must operate for the system to function.

Figure 18.14*b* is a reliability block diagram of a simple system composed of two redundant elements, either of which must operate for the system to function. The elements are shown in parallel to indicate that the system can operate if one component fails. In the usual redundant configuration, both elements are identical. Primary operation utilizes element *A* with *B* on a standby basis. *B* is switched into the system should *A* fail.

Figure 18.14*c* illustrates a more typical system configuration employing both series and redundant elements.

This system can be considered as a series system with elements *A*, *L*, *D*, *E*, *M*, and *K*. Element *L* is a single-unit redundant configuration wherein component *C* is

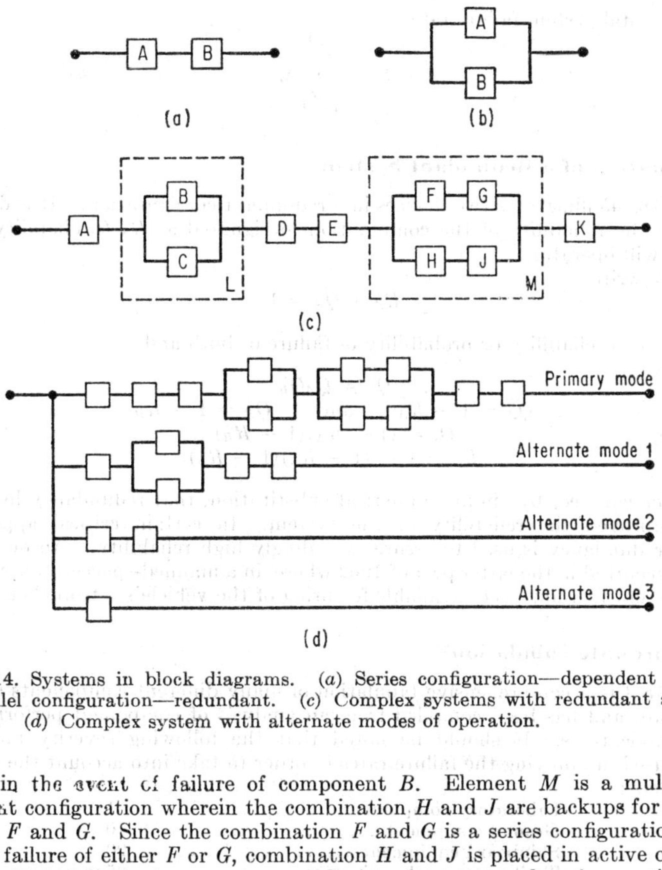

FIG. 18.14. Systems in block diagrams. (*a*) Series configuration—dependent elements. (*b*) Parallel configuration—redundant. (*c*) Complex systems with redundant and series elements. (*d*) Complex system with alternate modes of operation.

utilized in the event of failure of component *B*. Element *M* is a multiple-unit redundant configuration wherein the combination *H* and *J* are backups for the combination *F* and *G*. Since the combination *F* and *G* is a series configuration, in the event of failure of either *F* or *G*, combination *H* and *J* is placed in active operation.

Figure 18.14*d* illustrates a system employing alternate modes of operation. The alternate modes are usually less complex, less accurate, and more reliable than the primary mode.

(b) Evaluation of a Series System

From probability theory, the probability of simultaneous occurrence of *k* mutually exclusive events is the product of the probabilities of occurrence for each event.

Each event in this case is the failure-free operation of each system element, and the probabilities are the individual element reliabilities.

The reliability of the system is the product of the reliabilities of the individual elements. The series system reliability (Fig. 18.14*a*) composed of k elements

$$R_S = R_A R_B R_C \cdots R_K \tag{18.34}$$

Since the reliability of the ith element

$$R_i = e^{-\lambda_i t}$$

substituting in Eq. (18.34),

$$R_S = e^{-\lambda_A t} e^{-\lambda_B t} e^{-\lambda_C t} \cdots e^{-\lambda_k t} = e^{-\lambda_s t}$$

where the total system failure rate

$$\lambda_s = \sum_{i=A}^{i=k} \lambda_i$$

(c) Evaluation of a Redundant System

Figure 18.14*b* illustrates two devices in a redundant configuration. It is desired to determine the reliability of the combination, designated as R_S (probability that at least one will operate).

We may write

$$R_S + Q_S = 1$$

where Q_S = unreliability or probability of failure of both and

$$Q_S = Q_A Q_B$$

but $\quad Q_A = 1 - R_A \quad$ and $\quad Q_B = 1 - R_B$

Therefore, $\quad Q_s = (1 - R_A)(1 - R_B)$

and

$$R_S = 1 - (1 - R_A)(1 - R_B) \tag{18.35}$$

The reader can see, by simple numerical substitution, that redundancy has significantly improved the reliability of the system. In certain critical applications, multiple redundancy is used to assure exceedingly high reliability. An outstanding example occurred in the latter part of 1962 where, in a manned-spacecraft application, eight different methods were available for firing of the vehicle's retrorockets.

(d) Failure-rate Tabulation[32]

Table 18.3 is a comprehensive tabulation of many different components and their failure rates and has been compiled from an analysis of component performance in actual applications. It should be noted that the following severity factors K_F must be used in applying the failure rates in order to take into account the effects of environment:

Laboratory computer	1
Ground equipment	10
Shipboard equipment	20
Trailer-mounted equipment	25
Rail-mounted equipment	30
Aircraft equipment (bench test)	50
Missile equipment (bench test)	75
Aircraft equipment (in flight)	100
Missile equipment (in flight)	1,000

Therefore, the failure rate

$$\lambda = \lambda_G K_F \tag{18.36}$$

Table 18.3. Generic Failure-rate Distributions λ_G

Component or part	Upper extreme	$\lambda_G/10^6$ hr, mean	Lower extreme
Absorbers, r-f	1.20	0.687	0.028
Accelerometers	7.5	2.8	0.35
Accelerometers, strain gage	21.4	8.0	1.00
Accumulator	19.3	7.2	0.40
Actuators	13.7	5.1	0.35
Actuators, booster servo	33.6	12.5	0.86
Actuators, sustainer servo	33.6	12.5	0.86
Actuators, small utility	9.6	3.6	0.17
Actuators, large utility	18.5	6.9	0.60
Adapters, bore-sight	6.53	2.437	0.01
Adapters, wave-guide	9.31	3.475	0.139
Alternators	2.94	0.7	0.033
Antennas	3.52	2.0	0.48
Antenna drives	10.04	5.7	1.36
Attenuators	1.30	0.6	0.15
Base castings	0.70	0.175	0.015
Baffles	1.3	1.0	0.12
Batteries, chargeable	14.29	1.4	0.5
Batteries, one shot	300 cycles	30 cycles	10 cycles
Bearings	1.0	0.5	0.02
Bearings, ball, high-speed heavy-duty	3.53	1.8	0.072
Bearings, ball, low-speed light-duty	1.72	0.875	0.035
Bearings, rotary, sleeve-type	1.0	0.5	0.02
Bearings, rotary, roller	1.0	0.5	0.02
Bearings, translatory, sleeve shaft	0.42	0.21	0.008
Bellows	4.38	2.237	0.090
Bellows, motor in excess of 0.5 in. stroke	5.482	2.8	0.113
Bellows, null-type	5.879	3.0	0.121
Blowers	3.57	2.4	0.89
Boards, terminal	1.02	0.0626	0.01
Bolts, explosive	400 cycles	40 cycles	10 cycles
Brackets, bore-sight	0.05	0.0125	0.003
Brackets, mounting	0.05	0.0125	0.003
Brackets, miscellaneous	0.55	0.1375	0.034
Bracket assemblies	7.46	2.1	0.94
Brushes, rotary devices	1.11	0.1	0.04
Bulbs, temperature	3.30	1.0	0.05
Bumpers, ring assembly	0.073	0.0375	0.002
Bumper ring supports (bracket)	2.513	1.2875	0.052
Bushings	0.08	0.05	0.02
Buzzers	1.30	0.60	0.05
Cabinet assemblies	0.330	0.03	0.003
Cable assemblies	0.170	0.02	0.002
Cams	0.004	0.002	0.001
Circuit breakers	0.04	0.1375	0.045
Circuit breakers, thermal	0.50	0.3	0.25
Clamshell, plug-in assemblies	0.70	0.175	0.10
Clutches	1.1	0.04	0.06
Clutches, magnetic	0.93	0.6	0.45
Clutches, slip	0.94	0.3	0.07
Connectors, electrical	0.47/pin	0.2/pin	0.03/pin
Connectors, AN type	0.385/pin	0.2125/pin	0.04/pin
Counters	5.25	4.2	3.5
Counterweights, large	0.545	0.3375	0.13
Counterweights, small	0.03	0.0125	0.005
Coolers	7.0	4.20	1.40

Table 18.3. Generic Failure-rate Distributions λ_G (Continued)

Component or part	Upper extreme	$\lambda_G/10^6$ hr, mean	Lower extreme
Couplers, directional	3.21	1.6375	0.065
Couplers, rotary	0.049	0.025	0.001
Couplings, flexible	1.348	0.6875	0.027
Couplings, rigid	0.049	0.025	0.001
Covers, bore-sight adapter	0.347	0.1837	0.02
Covers, dust	0.01	0.006	0.002
Covers, protective	0.061	0.038	0.015
Crankcases	1.8	0.9	0.10
Cylinders	0.81	0.007	0.005
Cylinders, hydraulic	0.12	0.008	0.005
Cylinders, pneumatic	0.013	0.004	0.002
Delay lines, fixed	0.25	0.1	0.08
Delay lines, variable	4.62	3.00	0.22
Diaphragms	9.0	6.00	0.10
Differentials	0.168	0.04	0.012
Diodes	1.47	0.2	0.16
Disconnects, quick	2.1/pin	0.4/pin	0.09/pin
Drives, belt	15.0	3.875	0.142
Drives, direct	5.26	0.4	0.33
Drives, constant-speed, pneumatic	6.2	2.8	0.3
Driving-wheel assemblies	0.1	0.025	0.02
Ducts, blower	1.3	0.5125	0.21
Ducts, magnetron	3.0	0.075	0.04
Dynamotors	5.46	2.8	1.15
Fans, exhaust	9.0	0.225	0.21
Filters, electrical	3.00	0.345	0.140
Filters, light	0.80	0.20	0.12
Filters, mechanical	0.8	0.3	0.045
Fittings, mechanical	0.71	0.1	0.04
Gaskets, cork	0.077	0.04	0.003
Gaskets, impregnated	0.225	0.1375	0.05
Gaskets, monel mesh	0.908	0.05	0.0022
Gaskets, O-ring	0.03	0.02	0.01
Gaskets, phenolic	0.07	0.05	0.01
Gaskets, rubber	0.03	0.02	0.011
Gages, pressure	7.8	4.0	0.135
Gages, strain	15.0	11.6	1.01
Gears	0.20	0.12	0.0118
Gearboxes, communications	0.36	0.20	0.11
Gears, helical	0.098	0.05	0.002
Gears, sector	1.8	0.9125	0.051
Gears, spur	4.3	2.175	0.087
Gear trains (communications)	1.79	0.9	0.093
Generators	2.41	0.9	0.04
Gimbals	12.0	2.5	1.12
Gyros	7.23	4.90	0.85
Gyros, rate	11.45	7.5	3.95
Gyros, reference	25.0	10.0	2.50
Hardware, miscellaneous	0.121	0.087	0.0035
Heaters, combustion	6.21	4.0	1.112
Heater elements	0.04	0.02	0.01
Heat exchangers	18.6	15.0	2.21
Hoses	3.22	2.0	0.05
Hoses, pressure	5.22	3.9375	0.157
Housings	2.05	1.1	0.051
Housings, cast, machined bearing surface	0.91	0.4	0.016

Table 18.3. Generic Failure-rate Distributions λ_G (Continued)

Component or part	Upper extreme	$\lambda_G/10^6$ hr, mean	Lower extreme
Housings, cast, tolerances 0.001 in. or wider	0.041	0.0125	0.0005
Housings, rotary	1.211	0.7875	0.031
Insulation	0.72	0.50	0.011
Iris, wave-guide	0.08	0.0125	0.003
Jacks	0.02	0.01	0.002
Joints, hydraulic	2.01	0.03	0.012
Joints, mechanical	1.96	0.02	0.011
Joints, pneumatic	1.15	0.04	0.021
Joints, solder	0.005	0.004	0.0002
Joints, solder	0.08	0.04	0.02
Lamps	35.0	8.625	3.45
Lines and fittings	7.80	0.02	0.05
Motors	7.5	0.625	0.15
Motors, blower	5.5	0.2	0.05
Motors, electrical	0.58	0.3	0.11
Motors, hydraulic	7.15	4.3	1.45
Motors, servo	0.35	0.23	0.11
Motors, stepper	0.71	0.37	0.22
Mounts, vibration	1.60	0.875	0.20
Orifices, bleeds fixed	2.11	0.15	0.01
Orifices, variable area	3.71	0.55	0.045
Pins, grooved	0.10	0.025	0.006
Pins, guide	2.60	1.625	0.65
Pistons, hydraulic	0.35	0.2	0.08
Pumps	24.3	13.5	2.7
Pumps, engine-driven	31.3	13.5	3.33
Pumps, electric drive	27.4	13.5	2.9
Pumps, hydraulic drive	45.0	14.0	6.4
Pumps, pneumatic driven	47.0	14.7	6.9
Pumps, vacuum	16.1	9.0	1.9
Regulators	5.54	2.14	0.70
Regulators, flow and pressure	5.54	2.14	0.70
Regulators, helium	5.26	2.03	0.65
Regulators, liquid oxygen	7.78	3.00	0.96
Regulators, pneumatic	6.21	2.40	0.77
Relays, general-purpose	0.48/cs	0.25/cs	0.10/cs
Resistors, carbon deposit	0.57	0.25	0.11
Resistors, fixed	0.07	0.03	0.01
Resistors, precision tapped	0.292	0.125	0.041
Resistors, WW, accurate	0.191	0.091	0.052
Resolvers	0.07	0.04	0.02
Rheostats	0.19	0.13	0.07
Seals, rotating	1.12	0.7	0.25
Seals, sliding	0.92	0.3	0.11
Sensors, altitude	7.50	3.397	1.67
Sensors, beta-ray	21.30	14.00	6.70
Sensors, liquid-level	3.73	2.6	1.47
Sensors, optical	6.66	4.7	2.70
Sensors, pressure	6.6	3.5	1.7
Sensors, temperature	6.4	3.3	1.5
Servos	3.4	2.0	1.1
Shafts	0.62	0.35	0.15
Shields, bearing	0.14	0.0875	0.035
Shims	0.015	0.0012	0.0005
Snubbers, surge dampers	3.37	1.0	0.3
Springs	0.221	0.1125	0.004
Springs, critical to calibration	0.42	0.22	0.009

Table 18.3. Generic Failure-rate Distributions λ_G (Continued)

Component or part	Upper extreme	$\lambda_G/10^6$ hr, mean	Lower extreme
Springs, simple return force	0.022	0.012	0.001
Starters	16.1	10.0	3.03
Structural sections	1.35	1.0	0.33
Suppressors, electrical	0.95	0.3	0.10
Suppressors, parasitic	0.16	0.09	0.02
Switches	0.14/cs	0.5/cs	0.009/cs
Synchros	0.61	0.35	0.09
Synchros, resolver	1.94	1.1125	0.29
Tachometers	0.55	0.3	0.25
Tanks	0.27	0.15	0.083
Thermisters	1.40	0.6	0.20
Thermostats	0.14	0.06	0.02
Timers, electronic	1.80	1.2	0.24
Timers, electromechanical	2.57	1.5	0.79
Timers, pneumatic	6.80	3.5	1.15
Transducers	45.0	30.0	20.0
Transducers, pressure	52.2	35.0	23.2
Transducers, strain gage	20.0	12.0	7.0
Transducers, thermister	28.00	15.0	10.0
Transformers	0.02	0.2	0.07
Transistors	1.02	0.61	0.38
Tubes, electron, commercial, single-diode	2.20	0.80	0.24
Turbines	16.67	10.0	3.33
Valves	8.0	5.1	2.00
Valves, ball	7.7	4.6	1.11
Valves, blade	7.4	4.6	1.08
Valves, bleeder	8.94	5.7	2.24
Valves, butterfly	5.33	3.4	1.33
Valves, bypass	8.13	5.88	1.41
Valves, check	8.10	5.0	2.02
Valves, control	19.8	8.5	1.68
Valves, dump	19.0	10.8	1.97
Valves, four-way	7.22	4.6	1.81
Valves, priority	14.8	10.3	7.9
Valves, relief	14.1	5.7	3.27
Valves, reservoir	10.8	6.88	2.70
Valves, shutoff	10.2	6.5	1.98
Valves, selector	19.7	16.0	3.70
Valves, sequence	81.0	4.6	2.10
Valves, spool	9.76	6.9	2.89
Valves, solenoid	19.7	11.0	2.27
Valves, three-way	7.41	4.6	1.87
Valves, transfer	1.62	0.5	0.26
Valves, vent and relief	15.31	5.7	3.41

18.14. COMPONENT LIFE EXPECTATION[31]

Table 18.5 gives a comprehensive tabulation of life expectancies for many different components. Systems-maintenance procedures should be based upon these life expectancies. Where maintenance is not practical, such as in space-vehicle applications, alternate-mode operation with appropriate switching must be provided. This is necessary if the design cannot be changed so as to utilize a longer-life component. It should be noted that the life expectancy severity factors K_L from Table 18.4 must be used in determining component life expectancies in order to take into account the effects of the environment. Therefore, the wear-out life

$$t_W = t_G K_* \qquad (18.37)$$

Table 18.4. Life-expectancy Severity Factor K_L

Installation environment	All equipment	Electronic and electrical equipment	Electro-mechanical equipment	Dynamic mechanical equipment
Satellites	2.50	2.60	2.40	2.10
Laboratory computer	1.00	1.00	1.00	1.00
Bench test	0.54	0.55	0.51	0.50
Ground	0.30	0.31	0.26	0.25
Shipboard	0.19	0.21	0.17	0.15
Aircraft	0.16	0.18	0.14	0.12
Missiles	0.15	0.17	0.13	0.11

Table 18.5. Generic Life-expectancy Distributions t_G

Component or part	Upper extreme	Mean $t_G/10^6$	Lower extreme
Accelerometers	0.052 hr,	0.02 hr,	0.001 hr,
	0.5 cycle	0.1 cycle	0.02 cycle
Accumulators	0.1 hr,	0.06 hr,	0.02 hr,
	0.01 cycle	0.008 cycle	0.002 cycle
Actuator, electric counter		0.1 cycle	
Actuator, linear		0.2 cycle,	
		0.0004 hr	
Actuator, rotary	10.0 cycles	0.50 cycle	0.001 cycle
Air-conditioning unit	0.0017 hr	0.004 hr,	0.007 hr
		0.15 cycle	
Alternators	0.01 hr	0.009 hr	0.001 hr
Amplifier, signal transistor, a-c		0.004 hr	
Antenna drivers	0.013 hr	0.008 hr	0.001 hr
Antennas, plasma sheath	0.001 hr	0.000050 hr	0.000040 hr
Antenna switch		0.04 hr	
Attenuators	0.01 hr	0.005 hr,	0.0025 hr
		0.0036 hr	
Auxiliary power units	0.001 hr	0.000500 hr	0.0000013 hr
Batteries, chargeable	0.005 cycle	0.002 cycle	0.00003 cycle
Battery, lead-acid		0.0015 cycle	
Battery, primary type	0.016 hr	0.12 hr	0.0008 hr
Beacon, S-band radar		0.040 hr	
Bearings, dry	0.0012 hr	0.0005 hr	0.00001 hr
Bearings, lubricated	0.02 hr	0.007 hr	0.002 hr
Bearings, ball	0.016 hr	0.006 hr	0.0005 hr
Bearings, ball, midget		0.008 hr	
Bearings, ball, precision		0.04 hr	
Bearings, ball, turbine	0.001 hr	0.00065 hr	0.00005 hr
Bearings, clutch release		3.0 cycles	
Bearings, heavy-duty, lub	0.003 hr	0.002 hr	0.0001 hr
Bearings, light-duty, lub	0.01 hr	0.006 hr	0.001 hr
Bearings, precision, lub	0.05 hr	0.020 hr	0.005 hr
Bearings, rotary, roller, lub		0.052 hr	
Bearings, stagger roller		0.016 hr	
Bearings, tracker roller		0.6 cycle	
Bellows, plastic	0.050 cycle	0.01 cycle	0.001 cycle
Bellows, steel and beryllium copper	10.0 cycles	0.1 cycle	0.001 cycle

Table 18.5. Generic Life-expectancy Distributions t_G (Continued)

Component or part	Upper extreme	Mean $t_G/10^6$	Lower extreme
Bellows, aluminum and magnesium	0.1 cycle	0.01 cycle	0.001 cycle
Blowers	0.008 hr	0.004 hr	0.001 hr
Blowers, vane, axial	0.005 hr	0.0025 hr	0.001 hr
Brake, assembly	...	0.020 hr	
Brushes, rotary device	0.01 hr	0.003 hr	0.001 hr
Buckets, turbine wheel	...	0.0065 hr	
Buzzers	1.0 cycle, 0.0025 hr	0.1 cycle, 0.001 hr	0.01 cycle, 0.0005 hr
Cameras (slit)	0.002 hr	0.001 hr	0.0005 hr
Cams	100.0 cycles	1.0 cycle	0.1 cycle
Capacitor, ceramic	0.02 hr	0.016 hr	0.002 hr
Capacitor, ceramic variable	...	0.016 hr	
Capacitors, electrolytic	0.012 hr	0.01 hr	0.004 hr
Cells, solar	0.02 hr	0.003 hr	0.0005 hr
Choppers	0.008 hr	0.005 hr	0.0012 hr
Chopper, synchro	...	0.026 hr	
Chopper, microsignal	...	0.004 hr	
Circuit breakers	0.05 cycle	0.035 cycle	0.01 cycle
Clutch, high-speed backstopping	...	0.019 hr	
Clutch, precision indexing	...	80.0 cycles	
Coatings, vitrous ceramic	0.0006 hr	0.00045 hr	0.0001 hr
Commutators	0.014 hr	0.006 hr	0.001 hr
Compressors (lub. bearings)	0.02 hr	0.007 hr	0.002 hr
Compressor diaphragm (oil-free)	...	0.003 hr	
Connector, electrical	0.024 hr, 0.0005 cycle	0.019 hr, 0.0001 cycle	0.012 hr, 0.00005 cycle
Contactors	0.1 cycle	0.05 cycle	0.02 cycle
Contactors, power	...	0.05 cycle	
Contactor, rotary power	...	0.1 cycle	
Control package, pneumatic	...	0.02 hr	
Converter, analog	0.004 hr	0.003 hr	0.002 hr
Counters to digital	0.6 cycle	0.3 cycle	0.1 cycle
Counters, electronic	100.0 cycles	60.0 cycles	30.0 cycles
Counter, Geiger	...	0.0034 hr	
Counter, heavy-duty	...	0.2 cycle	
Counter, magnetic	...	400.0 cycles	
Cylinder, piston 250 psi, air	50.0 cycles	30.0 cycles	12.0 cycles
Delay lines, fixed	0.025 hr	0.005 hr	0.01 hr
Delay lines, variable	...	0.010 hr	
Detection mechanism, star wheel	...	0.10 cycle	
Detectors, micrometeorite	0.05 hr	0.01 hr	0.001 hr
Diaphragms, Teflon	0.01 cycle	0.005 cycle	0.001 cycle
Differentials	...	1.0 cycle	
Digital telemetry extraction equipment	...	0.007 hr	
Diodes, semicondutor	...	0.20 hr	
Drive assembly	...	0.016 hr	
Drive, belt	...	0.002 hr	
Drives, direct lub bearing	0.02 hr	0.01 hr	0.002 hr
Drive rotary solenoid	...	5.0 cycles	

Table 18.5. Generic Life-expectancy Distributions t_G (Continued)

Component or part	Upper extreme	Mean $t_G/10^6$	Lower extreme
Equalizer, pressure-bellows type	....	100.0 cycles	
Fastener, Nylatch	....	0.06 cycle	
Fasteners, socket head	0.45 cycle	0.4 cycle	0.3 cycle
Fastener, threaded (bolt and nut)	....	8.0 cycles	
Filter, electrical	0.02 hr	0.012 hr	0.008 hr
Filter, pneumatic	....	0.040 hr	
Gaskets, rubber nonworking	0.044 hr	0.035 hr	0.026 hr
Gaskets, phenolic	0.088 hr	0.035 hr	0.018 hr
Gages, pressure	10.0 cycles	0.1 cycle	0.001 cycle
Gages, electric field and ion	....	0.04 hr	
Gage, gas density	....	0.04 hr	
Gear boxes, communications	0.02 hr	0.005 hr	0.002 hr
Gear head	....	0.001 hr	
Gears, steel	10.0 cycles	1.0 cycle	0.1 cycle
Gear trains, communications	0.0025 hr	0.0018 hr	0.001 hr
Generators	0.02 hr	0.01 hr	0.002 hr
Generator, brushless synchronous motor	....	0.010 hr	
Generators, d-c	0.02 hr	0.006 hr	0.002 hr
Generator, phase	....	0.03 hr	
Generators, reference	0.02 hr	0.008 hr	0.0015 hr
Generator, solid propellant	....	0.00004 sec	
Gyros, rate	0.040 hr	0.001 hr	0.0001 hr
Gyros, reference	0.002 hr	0.001 hr	0.0002 hr
Heater elements	0.016 hr	0.012 hr	0.001 hr
Hoses, flex	0.8 cycle	0.5 cycle	0.25 cycle
Hoses, plastic, metal-braided	0.5 cycle	0.22 cycle	0.05 cycle
Indicator, elapsed time	....	0.01 cycle	
Inductor	....	0.15 hr	
Inertial reference package	....	0.02 hr	
Integrating hydraulic package	....	0.02 hr	
Inverters, 400 cps	....	0.04 hr	
Jack, tip	....	0.0001 cycle	
Joints, mechanical	1.0 cycle	0.1 cycle	0.0025 cycle
Joint, rotary	....	0.016 hr	
Junction box	....	0.04 hr	
Lamps	0.025 hr	0.02 hr	0.000005 hr
Lines and fittings	0.02 hr	0.007 hr	0.003 hr
Meters, electrical	....	1.0 cycle,	
	0.0135 hr	0.009 hr	0.005 hr
Meter, relay	20.0 cycles	15.0 cycles	10.0 cycles
Monitor, speed (engine)	....	0.01 hr	
Monitor, cosmic-ray	....	0.04 hr	
Motors (lub. bearings)	0.02 hr	0.007 hr	0.002 hr
Motors, blower (lub. bearings)	0.02 hr	0.007 hr	0.002 hr
Motors, electrical (lub. bearings)	0.02 hr	0.01 hr	0.001 hr
Motor, electrical, a-c	....	0.03 hr	

Table 18.5. Generic Life-expectancy Distributions t_G (Continued)

Component or part	Upper extreme		Mean $t_G/10^6$		Lower extreme	
Motors, electrical, d-c	200.0	cycles	100.0	cycles	0.4	cycle
Motors, electrical, d-c torque	...		200.0	cycles		
Motor, electrical, d-c subminiature reversible	...		0.003	hr		
Motor, electrical, d-c nonferrous rotor	...		0.002	hr		
Motor, hydraulic	...		0.01	hr		
Motors, servo	0.02	hr	0.008	hr	0.001	hr
Multiplexer	...		0.012	hr		
Pistons, hydraulic	0.1	cycle	0.05	cycle	0.01	cycle
Power unit, electrohydraulic	...		0.002	hr		
Potentiometers	0.03	hr	0.02	hr	0.00025	hr
Pump	...		0.005	hr		
Pumps, engine-driven	0.005	hr	0.0035	hr	0.002	hr
Pumps, electric-driven	0.001	hr	0.0005	hr	0.00015	hr
Pumps, ion	0.2	hr	0.02	hr	0.002	hr
Pump, pneumatic-driven	...		0.003	hr		
Pump, variable-displacement (hyd.)	...		0.001	hr		
Pump, variable displacement (hyd.) miniature	0.0005	hr	0.00025	hr	0.00003	hr
Pumps, vane, miniature	...		0.002	hr		
Recorders, video-tape	0.005	hr	0.002	hr	0.001	hr
Rectifiers	0.04	hr	0.021	hr	0.003	hr
Regulator, flow and pressure	...		0.001	hr		
Regulator, pressure pneumatic	...		0.04	hr		
Regulator, gas pressure	...		0.0015	hr		
Regulators, voltage	0.04	hr	0.02	hr	0.005	hr
Relays, general-purpose	0.7	cycle	0.2	cycle	0.02	cycle
Relays, heavy-duty	0.2	cycle	0.145	cycle	0.02	cycle
Relays, sensitive	0.3	cycle	0.2	cycle	0.02	cycle
Resistors, carbon deposit	0.05	hr	0.015	hr	0.012	hr
Resistors, composition	0.04	hr	0.014	hr	0.011	hr
Resistors, variable, composition	0.017	hr	0.016	hr	0.014	hr
Resolver	...		0.002	hr		
Rheostat	...		0.03	hr		
Ring, seal	...		0.001	min		
Rotor	...		0.01	hr		
Seals, mechanical	0.01	hr	0.007	hr	0.003	hr
Seals, electronic	...		0.01	hr		
Sensor, temperature	...		0.01	hr		
Sensors, pressure differential, bellows type	1.0	cycle	0.35	cycle	0.01	cycle
Servo	0.065	hr	0.04	hr	0.02	hr
Socket, electron tube	...		0.02	hr		
Solar collector	...		0.0035	hr		
Solar cells	...		0.000720	hr		
Solenoids	100.0	cycles	60.0	cycles	20	cycles
Spark plug	...		0.0025	hr		
Switches	0.05	cycle	0.03	cycle	0.01	cycle
Switch, cam	...		0.01	cycle		
Switch, miniature	...		0.2	cycle		

Table 18.5. Generic Life-expectancy Distributions t_G (Continued)

Component or part	Upper extreme		Mean $t_G/10^6$		Lower extreme	
Tachometers	0.016	hr	0.01	hr	0.005	hr
Thermostats	0.5	cycle	0.25	cycle	0.02	cycle
Timers, electronic	0.02	hr	0.01	hr	0.005	hr
Timer, pneumatic	....		0.0025	hr		
Timer, high-precision totalizer	....		2.0	cycles		
Timer, elementary	....		0.04	hr		
Transducer	....		0.01	hr		
Transducer, potentiometer	....		0.015	hr		
Transducer, strain gage	....		0.02	hr		
Transducer, temperature	....		0.005	hr		
Transducer, low-pressure	....		10.0	cycles		
Transducer, power	....		0.02	hr		
Transducers, pressure	0.1	cycle	0.05	cycle	0.02	cycle
Transformers	0.03	hr	0.01	hr	0.004	hr
Transistors	....		0.2	hr		
Tube, electron, receiving	....		0.009	hr		
Tubes, electron, power	0.01	hr	0.007	hr	0.003	hr
Turbines (limited by bearings)	0.015	hr	0.011	hr	0.008	hr
Turbines, gas	0.1	hr	0.05	hr	0.007	hr
Valves	0.2	cycle	0.15	cycle	0.06	cycle

CONDITIONS:

1. Assume continuous operation.
2. Assume a theoretical laboratory computer element or component in order to comply with general trend data for actual usage element or component. Equipment that is normally not used in laboratory computers can be visualized in such an application for purposes of arriving at a generic condition.
3. Assume no adjustments other than automatic.
4. The mean and extremes do not necessarily apply to statistical samples but encompass application and equipment-type variations, also.
5. The values should be applied at operating installation conditions only and no comparisons made at the generic level.

References

1. Marin, J.: "Significance of Material Properties in Design for Fatigue Loading," Machine Design Reprint, Penton Publishing Company, Cleveland, 1957.
2. Timoshenko, S., and G. H. MacCullough: "Elements of Strength of Materials," 2d ed., D. Van Nostrand Company, Inc., Princeton, N.J., 1940.
3. Powell, H. R.: "The Impact of Reliability on Design," ASME paper 60-MD-2, May, 1960.
4. Horger, O. J. (ed.): "ASME Handbook, Metals Engineering—Design," McGraw-Hill Book Company, Inc., New York, 1953.
5. Shigley, J. E.: "Machine Design," McGraw-Hill Book Company, Inc., New York, 1956.
6. Peterson, R. E.: "Stress Concentration Design Factors," John Wiley & Sons, Inc., New York, 1953.
7. Roark, R. J.: "Formulas for Stress and Strain," 3d ed., McGraw-Hill Book Company, Inc., New York, 1954.
8. Peterson, R. E.: Application of Stress Concentration Factors in Design, *Proc. SESA*, vol. 1, no. 1, 1943.
9. Maleev, V. L.: "Machine Design," International Textbook Company, Scranton, Pa., 1946.
10. Lessells, J. M.: "Strength and Resistance of Metals," John Wiley & Sons, Inc., New York, 1954.
11. Grover, H. J., S. A. Gordon, and L. R. Jackson: Fatigue of Metals and Structures, Bureau of Naval Weapons, Navweps 00-25-534, June, 1960.

12. Lipson, C.: Why Machine Parts Fail, *Machine Design*, May, June, July, August, September, October, November, December, 1960.
13. Bennett, J. A., and G. W. Quick: Mechanical Failures of Metals in Service, *Natl. Bur. Std. (U.S.), Circ.* 550, September, 1954.
14. American Society for Testing and Materials, "Manual on Fatigue Testing," Special Technical Publication 91.
15. French, H. J.: Fatigue and the Hardening of Steels, *Trans. Am. Chem. Soc. Steel Treatment*, vol. 21, 1933.
16. Miner, M. A.: Cumulative Damage in Fatigue, *J. Appl. Mech.*, vol. 12, no. 3, September, 1945.
17. Gatts, R. R.: Application of a Cumulative Damage Concept to Fatigue, *Trans. ASME, J. Basic Eng.*, December, 1961.
18. Mains, R. M.: "A Generalization of Cumulative Damage," ASME paper 59-Met-1, April, 1959.
19. Niemann, G.: "Maschinenelemente," vol. 1, Springer-Verlag OHG, Berlin, 1960.
20. Shanley, F. R.: "Weight-strength Analysis of Aircraft Structures," 2d ed., Dover Publications, Inc., New York, 1960.
21. "Honeycomb Sandwich Design," Hexcel Products, Inc., Berkeley, Calif.
22. Langer, B. F.: Application of Stress Concentration Factors, *Bettis Tech. Rev.* WAPD-BT-18, April, 1960.
23. Lloyd, David K., and Myron Lipow: "Reliability: Management, Methods and Mathematics," Prentice-Hall, Inc., Englewood Cliffs, N.J., 1962.
24. Bazovsky, Igor: "Reliability; Theory and Practice," Prentice-Hall, Inc., Englewood Cliffs, N.J., 1961.
25. Chwick, A.: "Elementary Reliability for Design Engineers," Sperry Gyroscope Co., 1960.
26. *Proceedings Eighth National Symposium on Reliability and Quality Control*, sponsored by IRE, EIA, ASQC, AIEE, January, 1962.
27. Technical Publications Committee (ed.): "Research and Development Reliability," American Society for Quality Control, 1961.
28. "Reliability Engineering—Statistical Concepts and Reliability Analysis—A Brochure of the Material Presented in Films MN 8770c and MN 8770d," Government Printing Office, Washington, D.C., 1962.
29. "Reliability—Fundamental Concepts Part I—A Brochure of the Material Presented in Film MN 8770a," Government Printing Office, Washington, D.C.
30. "Reliability Engineering—Fundamental Concepts Part II—A Brochure of the Material Presented in Film MN 8770b," Government Printing Office, Washington, D.C.
31. Earles, D., M. Eddins, and D. Jackson: "A Theory of Component Part Life Expectancies," 8th National Symposium on Reliability and Quality Control, 1962.
32. "Generic Failure Rates," Martin Company Report, Baltimore, Md.

Section 19

FABRICATION PRINCIPLES

By

E. T. FORTINI, M. S., *Staff Engineer, Compugraphic Corp., Reading, Mass., and Lecturer in Mechanical Engineering, Lincoln College, Northeastern University (Dimensions for Tolerances)*

CARL H. RINGE, Dipl. Ing., *Research Engineer, Manager, Patent Dept., R. Hoe and Co., New York, N.Y. (Designing for Manufacturing and Materials)*

CONTENTS

DIMENSIONS FOR TOLERANCES

DESIGNING FOR MANUFACTURING AND MATERIALS

DIMENSIONS FOR TOLERANCES*

19.1. DEFINITIONS AND GENERAL CONSIDERATIONS

(a) Terminology and Notation

Dimension. A physical property as modulus of elasticity, density, etc., but more commonly a measure of length or angle. x_a denotes any value of a dimension having a maximum limit A and minimum a. Other dimensions are denoted by suitable subscripts, usually taken from the beginning of the alphabet.

Table 19.1. Clearance Fits

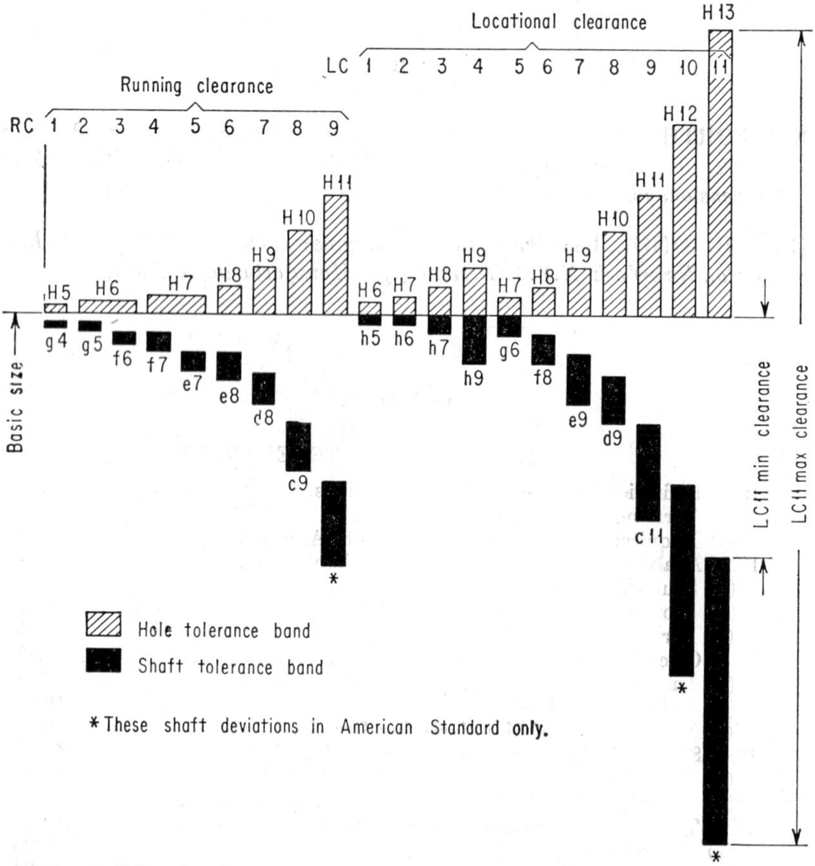

Running and sliding fits are intended to provide similar running performance with suitable lubrication allowance throughout the range of sizes. Clearances of RC1 and RC2 fits increase more slowly with diameter than other classes so that accurate location is maintained even at the expense of free relative motion

ASA class

RC1 *Close sliding fits.* For accurate location of parts which must assemble without perceptible play

* By E. T. Fortini.

Table 19.1. Clearance Fits (Continued)

RC2	*Sliding fits.* For accurate location but with greater maximum clearance than RC1. Parts made to this fit move and turn easily but are not intended to run freely. Larger sizes may seize with small temperature changes
RC3	*Precision running fits.* About the closest fit expected to run freely. Intended for precision work at slow speeds and light journal pressures. Not suitable where appreciable temperature differences are likely
RC4	*Close running fits.* Intended for precision machinery with moderate surface speeds and journal pressures where accurate location and minimum play are desired
RC5, RC6	*Medium running fits.* For higher running speeds or heavy journal pressures, or both
RC7	*Free running fits.* For use where accuracy is not essential, or where large temperature variations are likely, or under both these conditions
RC8, RC9	*Loose running fits.* For use where materials made to commercial tolerances (such as cold-rolled shafting and tubing) are involved

Locational clearance fits are intended for parts which are normally stationary but which can be freely assembled or disassembled

ASA class	
LC1, LC2, LC3, LC4	*Snug location fits.* Clearance ranges from zero in the maximum metal condition to some small amount in the minimum metal condition depending on the tolerance grades involved
LC5, LC6, LC7	*Medium location fits.* For use where accurate location is necessary but with greater ease of assembly than probable with LC1 to LC4
LC8, LC9, LC10, LC11	*Loose location fits.* Intended for use where freedom of assembly is of prime importance

Tolerance. A measure of the permissible difference between maximum and minimum dimension limits. This difference allows for inaccuracies of manufacture and dimensional changes in operation. t_a is the tolerance of dimension x_a, t_b of x_b, etc. Two possible interpretations of tolerances and dimensions are shown in Fig. 19.1.

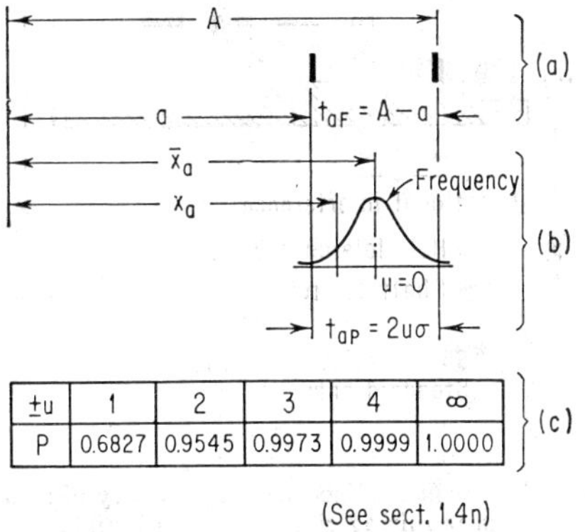

±u	1	2	3	4	∞
P	0.6827	0.9545	0.9973	0.9999	1.0000

FIG. 19.1. Two definitions of length dimensions. (*a*) Defined at limits only. (*b*) Defined by a frequency distribution. The probability that a dimension will not differ from its mean value $\bar{x}_a$ by more than $\pm\mu\sigma$ is tabulated at (*c*); σ is the standard deviation of all sources of inaccuracy.

Dimension Condition. A variable condition controlled by two or more dimensions as clearance in a journal bearing. y_w denotes any value of a dimension condition between a maximum limit W and minimum w. Other dimension conditions are denoted by suitable subscripts, usually taken from the end of the alphabet.

Table 19.2. Transition Fits

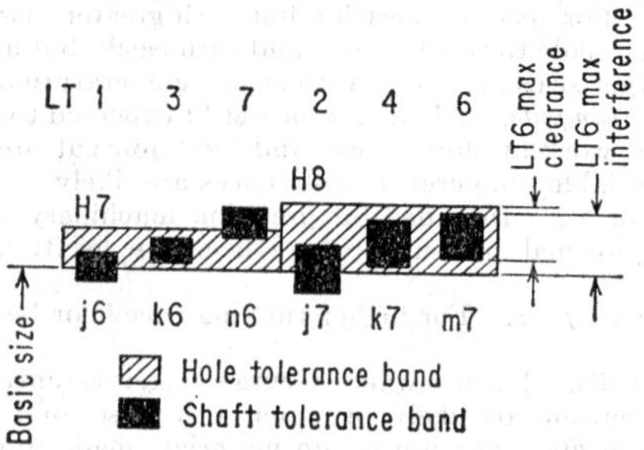

Transition fits are a compromise between clearance and interference fits for applications where accuracy of location is important. A small amount of either clearance or interference is permissible

ASA class	
LT1, LT2	*Push fits.* These fits allow assembly by hand or with light blows
LT3, LT4	*Wringing fits.* For use where less freedom of rotation is wanted than probable with LT1 or LT2
LT6, LT7	*Tight fits.* These fits will usually result in light interference with only a small chance of clearance

Table 19.3. Interference Fits

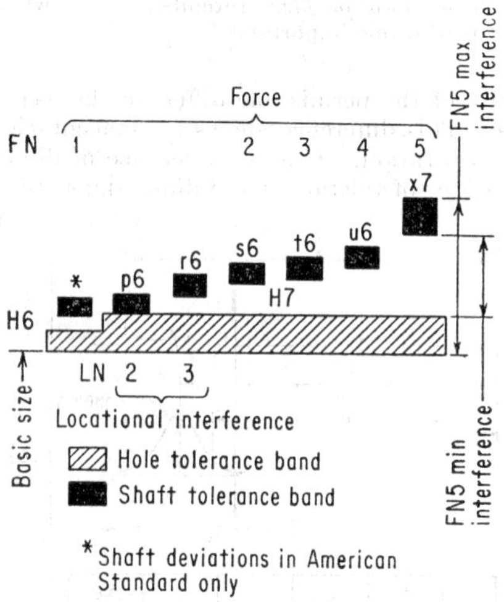

ASA class	
LN2, LN3	*Locational interference fits.* Used where accuracy of location is of prime importance and for parts requiring rigidity and alignment with no special requirement for bore pressure. These fits are not designed to transmit frictional loads from one part to another by virtue of tightness of fit
FN1	*Light drive fits.* Require light assembly forces and produce more or less permanent assemblies. Suitable for thin sections or long fits with cast-iron external members
FN2	*Medium drive fits.* Suitable for ordinary steel parts, or for shrink fits on thin sections. About the tightest fit that can be used with high-grade cast-iron members
FN3	*Heavy drive fits.* For heavy steel parts or shrink fits in medium sections
FN4, FN5	*Force fits.* Suitable for parts which can be highly stressed, or for shrink fits where the heavy pressing forces required are impractical

Variation of a Dimension Condition. A measure of the difference between maximum and minimum limits of a dimension condition. v_w is the variation of y_w, v_u of y_u, etc. Any v is generally a function of dimensions and their tolerances, or in certain cases of tolerances only.

Mechanism Error. The deviation from an ideal position of a mechanism output member. ϵ_1 denotes the error for an output position of mechanism 1, ϵ_2 for mechanism 2, etc.

Train Output Error. The deviation from an ideal position of the output member of a train of mechanism. ϵ_A denotes the error for an output at terminal A, ϵ_B for terminal B, etc.

System Positional Error. The deviation from an ideal performance of a system comprising two or more trains of mechanism. ϵ_{I} denotes an error for an aspect of system I, ϵ_{II} for system II, etc.

Additional subscripts F and P are applied as appropriate to symbols of tolerance, variation of dimension condition, mechanism error, train output error, and system positional error. For example, t_{aF} indicates a tolerance with fixed value; t_{aP} indicates a tolerance with a probability distributed value. Distributions found in engineering practice are usually normal.

(b) Requirements for and Limitations to Accuracy

The functioning of a product may involve mechanical, electrical, magnetic, thermal, optical, or other physical effects. These primary effects must be translated into a set of design requirements capable of being met by dimensional specifications. For instance, the operating requirements of a linkage regulating air-fuel mixture can be specified by defining allowable limits for positions of the output terminal with respect to positions of input terminals; operation of a high-speed gearset may be specified in terms of profile errors, misalignments of axes of rotation, elastic deflections of shafts, etc.

Many dimensional requirements can be met by selective or special assembly practices. The degree of interchangeable manufacture necessary or convenient will therefore influence the development of the design and its manufacturing specifications. The appropriate degree of interchangeable manufacture depends on (1) rate and quantity of manufacture, (2) available facilities, (3) extent of quality control, and (4) service requirements.

Parts are interchangeable if, from a group conforming to the same limits of size and material properties as specified by the product drawings, any one part can be used in place of any other with equal assurance that the part will assemble and function satisfactorily. *Universal interchangeability* implies complete assurance regardless of where parts are made so long as they conform to the drawing. Parts have *local interchangeability* when tooling and inspection gages control assembly and functional requirements. Parts made with *limited interchangeability* require selection or fitting at assembly and may not be suitable for service replacement.

In establishing conditions controlling operation and assembly the designer may take the following approaches: (1) assign fixed limits of variability to design conditions and then develop controlling dimensional specifications that will assure conformance within these limits, (2) assume that design requirements may deviate from ideal values in terms of probability distributions, or (3) use a combination of the fixed-limit and probability approaches. Whatever approach is taken, the aim is to obtain highest product quality for a given cost, or conversely, minimum cost for a given product quality. Techniques for realizing optimum results have been explored[1,2] but are not generally adaptable to ordinary design and production practices.

Limitations of Accuracy. Size and other dimensional properties of materials cannot be produced to exact values, and once produced do not necessarily retain manufactured values. Dimensional changes result from fluctuations of temperature and humidity (in the case of hygroscopic materials such as nylon), the action of external forces, the relief of residual stresses, changes in the unstable constituents following heat-treatment of metals, and deterioration of surfaces because of wear and corrosion. The engineering tolerance allowed a dimension cannot be less than the probable variability due to all sources of inaccuracies throughout the expected useful life of a part. However, merely because a dimension has a small tolerance does not necessarily assure dimensional *accuracy*. A dimension is accurate when its ideal value is not exceeded by a specified amount as allowed by its tolerance. But if the ideal value is not within the tolerance zone defined by the limits, then the dimension will be inaccurate, regardless of the precision demanded in manufacturing the part. The purpose of a dimension study is to determine the ideal values of dimensions controlling operation and assembly conditions and to assign tolerances so that the limits of the dimension conditions are not exceeded by amounts specified by the designer.

19.2. ANALYSIS OF ERRORS IN TRAINS OF MECHANISM

(a) Sources and Types of Errors

Errors originating in individual mechanisms are propagated through trains of mechanism to system output terminals. Each mechanism in the system can be the source of more than one type of error. These types are classified as follows:

Structural error, the error resulting from the use of an approximate instead of an exact mechanism. For example, a gear ratio of 6.13:1 may be an ideal requirement,

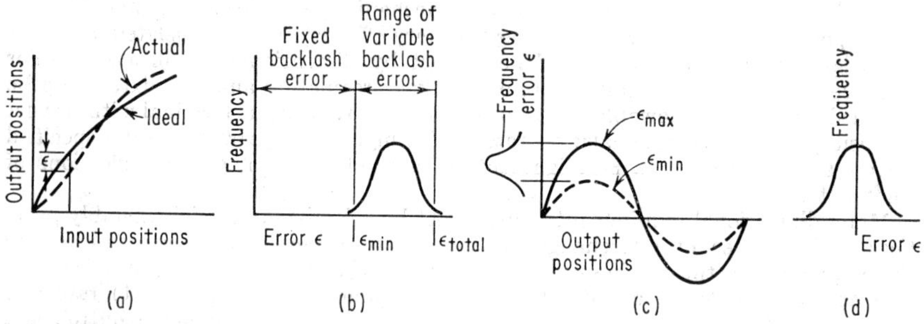

FIG. 19.2. Examples of common types of mechanism errors. (*a*) Structural. (*b*) Backlash. (*c*) Cyclic. (*d*) Random.

but because of design limitations the approximate ratio of 6:1 must be used instead. The difference between ideal and approximate ratios results in a structural error. Structural errors of variable-aspect mechanisms will not necessarily be uniform throughout the operating range of the mechanism, as shown in Fig. 19.2*a*.

Minimum Backlash Error. Many mechanisms have clearances necessary for assembly and operation, for example, clearance between gear teeth, or the tongue and groove of an Oldham's coupling. The smallest necessary clearance will result in a minimum backlash error. For many mechanical systems, backlash is the major source of error and special means are often used to reduce or eliminate it entirely.

Variable backlash error, due to the additional clearance caused by tolerances. If the tolerances controlling the amount of additional clearance are normally distributed, then values of additional backlash will also be normally distributed. Figure 19.2*b* shows total backlash as a combination of fixed and probable values.

Systematic or cyclic error, usually associated with the eccentricity of rotating elements, the wobble of gear faces, etc. The amplitude of cyclic errors may be normally distributed as shown in Fig. 19.2*c*.

Random error produced by errors in the location or profile of driving surfaces. Errors of this type are generally assumed to be normally distributed as shown in Fig. 19.2*d*.

Transient or short-period error results from vibrations, shock, foreign particles lodged between driving surfaces, etc. The character of these errors may be complex and unpredictable.

Errors may be identified by type and source using the following scheme:

Type of error	Mechanism 1	2	···	n
1. Structural	ϵ_{11}	ϵ_{21}	···	ϵ_{n1}
2. Minimum backlash	ϵ_{12}	ϵ_{22}		·
3. Variable backlash	ϵ_{13}	ϵ_{23}		·
4. Systematic or cyclic	ϵ_{14}	ϵ_{24}		·
5. Random	ϵ_{15}	ϵ_{25}		·
6. Transient or short-period	ϵ_{16}	ϵ_{26}	···	ϵ_{n6}

(b) Input-Output Relationships

Mechanical systems may be analyzed as mechanisms in parallel and series combinations. (Some examples of mechanisms are a cam and follower, a Scotch yoke, a gear differential, a universal joint.) A *simple mechanism* has one input and one output terminal. Figure 19.3*a* represents a system of simple mechanisms. For mechanism 6, the position of the input terminal Y_5 is related to the position of the output terminal Y_6 by the *mechanism function* $Y_6 = f(Y_5)$. Couplings comprise a special class of

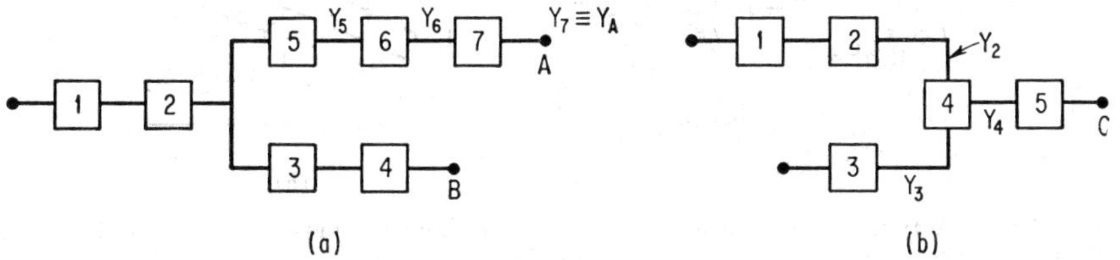

FIG. 19.3. Block diagrams for systems of mechanisms. (*a*) System consisting of simple mechanisms only. (*b*) System with a compound mechanism (mechanism 4).

mechanisms; the mechanism function of a coupling designated as the *i*th mechanism in a train is $Y_i = Y_{i-1}$. Machines are constructed so that the output member of one simple mechanism and the input member of the next mechanism in train are on a common part. A *compound mechanism* has more than one input and/or one output terminal. (A gear differential is a familiar compound mechanism.) Referring to the system represented by Fig. 19.3*b*, mechanism 4 is compound; its mechanism function is $Y_4 = f(Y_2, Y_3)$.

Each mechanism in a system is a possible source of any type of error as already described. An error of some type k originating in the jth mechanism will be propagated through a series of mechanisms to some system output terminal G contributing to the train output error a value $M_{jG}\epsilon_{jk}$. The *train propagation factor* M_{jG} is the product of the absolute values of the derivatives of the mechanism functions between mechanism j and the terminal G. For example, referring to Fig. 19.3*a*,

$$M_{5A} = |dY_6/dY_5| \cdot |dY_7/dY_6|$$

In calculating train propagation factors for systems having a compound mechanism, partial derivatives must be taken with respect to the appropriate input terminal. Referring to Fig. 19.3*b*, $M_{1C} = |dY_2/dY_1| \cdot |\partial Y_4/\partial Y_2| \cdot |dY_5/dY_4|$.

How errors combine at an output terminal will depend on the character of each ϵ_{jk}. For most engineering practice it is good enough to assume that an error will either have a fixed or a normally distributed value. Thus all errors originating in a mechanism j can be grouped into two components ϵ_{jF} and ϵ_{jP}. Any ϵ_{jF} can be the algebraic sum of two or more types of fixed errors. The sum of all fixed errors propagated to an output terminal G has itself a fixed value

$$\epsilon_{GF} = \Sigma M_{jG}\epsilon_{jF} \tag{19.1}$$

Errors with normally distributed probability values will result in a normally distributed error component at terminal G having the value

$$\epsilon_{GP} = [\Sigma(M_{jG}\epsilon_{jP})^2]^{1/2} \tag{19.2}$$

where ϵ_{jP}^2 may be the sum of the squares of two or more normally distributed errors originating in mechanism j. The combination of ϵ_{GF} and ϵ_{GP} thus defines the error at an output terminal G.

Relationships between train output and system errors must be examined on an individual basis. Referring to Fig. 19.4*a*, if there were no errors at the output terminals, the wire would be wound on the bobbin as represented by the solid line of Fig. 19.4*b*. Because of inaccuracies, the position of the wire with respect to the bobbin

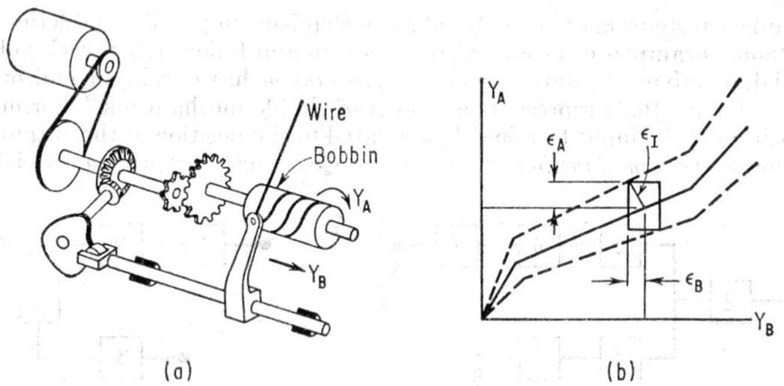

FIG. 19.4. A typical mechanical system. (*a*) Schematic representation. (*b*) Diagram defining system error.

can deviate from its ideal by an amount within the broken lines. The relationship of ϵ_A and ϵ_B thus defines the allowable magnitude of system error ϵ_I for one output position.

(c) Partitionment of Errors

Equations (19.1) and (19.2) make it possible to partition an allowable system error among contributing mechanisms. However, the error allotted any mechanism cannot be assigned only on the basis of these equations. Mechanism errors contribute to both the malfunctioning of the mechanism itself and the performance of the system. For instance, the requirements for quiet operation of a gearset may be more stringent than those for system accuracy.

Error values can be assigned by repeated trials until Eqs. (19.1) and (19.2) are satisfied. A direct approach, based on the assumption that all $M\epsilon^2$ terms for unassigned error values are equal, leads to the equations

$$\epsilon_{jF} = \epsilon_{GF}'/M_j^{1/2}\Sigma(M^{1/2}) \tag{19.3}$$

$$\epsilon_{jP} = \epsilon_{GP}'/M_j^{1/2}(\Sigma M)^{1/2} \tag{19.4}$$

When all errors are initially unassigned, $\epsilon_{GF}' \equiv \epsilon_{GF}$ and $\epsilon_{GP}' \equiv \epsilon_{GP}$; otherwise ϵ_{GF} and ϵ_{PF} must be suitably adjusted by subtracting $M\epsilon$ or $(M\epsilon)^2$ terms from ϵ_{GF} or ϵ_{GP} in order to obtain numerical values for ϵ_{GF}' or ϵ_{GP}'. Moreover, the sums $\Sigma(M^{1/2})$ and ΣM in Eqs. (19.3) and (19.4) should include train propagation factors only for mechanisms having unassigned error values.

19.3. CALCULATION OF FUNCTIONAL DIMENSIONS

(a) Direct Relationships

The limits of dimensions controlling operation and assembly are determined on the basis of relationships between mean dimension values, tolerances, dimension conditions, and the variation of dimension conditions. Direct relationships are deduced from the geometry of design without the need for performing mathematical operations. Problems involving direct relationships can take so many forms that no general procedure can include all cases. Some examples are given in ref. 3. Because of the frequency with which *length fits* and the *eccentric rotation of shafts* occur in practice, these problems are discussed as follows.

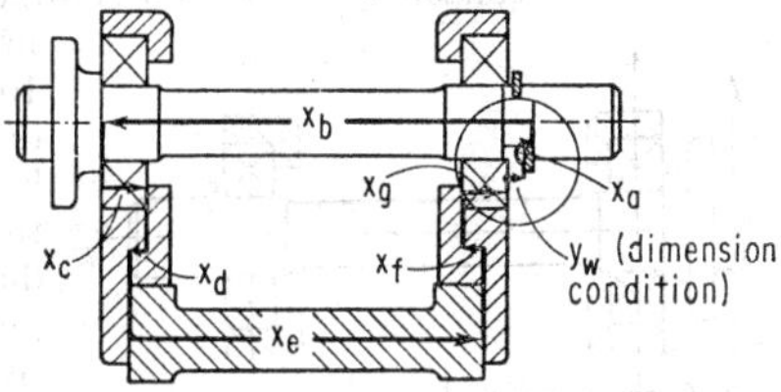

FIG. 19.5. Dimension loop controlling a length fit.

Length Fits. An example of an assembly fit controlled by length dimensions is shown in Fig. 19.5. By definition, the vector of dimension condition is in the plus direction. If $(+\bar{x})$ denotes the mean value of a dimension in the plus direction and $(-\bar{x})$ in the minus direction, then

$$\bar{y}_w + \Sigma(+\bar{x}) = \Sigma(-\bar{x})$$

The condition of least clearance is ordinarily the basis for assigning values to dimensions; hence

$$w + v/2 + \Sigma(+\bar{x}) = \Sigma(-\bar{x}) \tag{19.5}$$

If tolerances are assumed to have values only at extreme limits

$$v = v_F = \Sigma t_F \tag{19.6}$$

On the other hand, if tolerances are assumed to be normally distributed,

$$v = v_P = (\Sigma t_P{}^2)^{1/2} \tag{19.7}$$

Equations (19.5) to (19.7) provide the basis for calculations. When standard or commercial parts are involved in a dimensional relationship, their mean values and

(a) Limit Method

$\bar{x}$	t_F, mils	$(+\bar{x})$	$(-\bar{x})$
$\bar{x}_a$	2	0.120	
$\bar{x}_b$	4		$\bar{x}_b$
$\bar{x}_c$	6	0.500	
$\bar{x}_d$	4		0.250
$\bar{x}_e$	6	2.750	
$\bar{x}_f$	4		0.250
$\bar{x}_g$	6	0.500	
v_F	32		
	$v_F/2$	0.016	
	w	0.010	
	Σ	3.896	$\bar{x}_b + 0.500$
		$\bar{x}_b$	3.396

(b) Probability Method

$\bar{x}$	t_P, mils	$t_P{}^2$	$(+\bar{x})$	$(-\bar{x})$
$\bar{x}_a$	2	4	0.120	
$\bar{x}_b$	4	16		$\bar{x}_b$
$\bar{x}_c$	6	36	0.500	
$\bar{x}_d$	4	16		0.250
$\bar{x}_e$	6	36	2.750	
$\bar{x}_f$	4	16		0.250
$\bar{x}_g$	6	36	0.500	
	$v_P{}^2$	160		
		$v_P/2$	0.006	
		w	0.010	
		Σ	3.886	$\bar{x}_b + 0.500$
			$\bar{x}_b$	3.386

FIG. 19.6. Work sheets for length-dimension calculations.

tolerances are first entered onto a work sheet as shown in Fig. 19.6. Next, tolerances are assigned to the dimensions of all other parts—subject to whatever restrictions are imposed on v by the limits of fit. Mean dimensions for all new parts *except one* are scaled from layouts and entered on the work sheet. Following the form of the work sheet, the mean value of the unassigned dimension is calculated. The work sheet will then record the mean value and tolerance of each dimension involved in the relationship.

Eccentric rotation of shafts is a common dimension condition which usually results in cyclic mechanism errors. Referring to Fig. 19.7, sources of eccentricity at bearing 1 are designated by e_{11}, e_{12}, etc., and at bearing 2 by e_{21}, e_{22}, etc. These sources may be ascribed to out of roundness, eccentricity between shaft diameters at mounting and bearing positions, play in housing and bore fits, etc. The resulting fixed value of eccentricity at mounting position A,

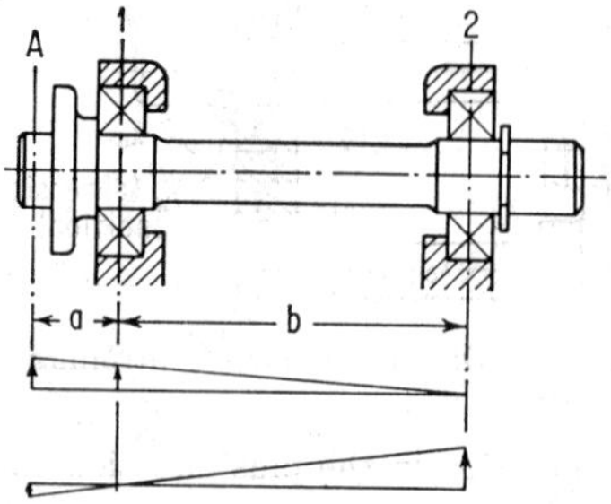

FIG. 19.7. Bearing eccentricities reflected to shaft mounting position.

$$e_{AF} = \frac{a+b}{b} e_{1F} + \frac{a}{b} e_{2F}$$

e_{1F} represents the algebraic sum of all eccentricity sources at bearing 1, and e_{2F} at bearing 2. When sources of eccentricity are assumed to have normally distributed values, the resultant value at the mounting position will also be normally distributed,

$$e_{AP} = \left[\left(\frac{a+b}{b}\right)^2 e_{1P}^2 + \left(\frac{a}{b}\right)^2 e_{2F}^2\right]^{1/2}$$

e_{1P}^2 and e_{2P}^2 represent the sums of the squares of contributing values at bearings 1 and 2, respectively.

(b) Indirect Relationships

The friction shoe shown in Fig. 19.8 illustrates a situation in which the variation of a dimension condition is not obvious from the geometry of design but must be deduced by mathematical operation. In problems of this sort it is first necessary to find some function

$$\bar{y}_w = f(\bar{x}_a, \bar{x}_b, \ldots) \tag{19.8}$$

For the illustrative example this function is furnished by the law of cosines. When the tolerances have fixed values, the variation of dimension condition will also have a fixed value

$$v_{wF} = D_a t_{aF} + D_b t_{bF} + \cdots \tag{19.9}$$

When tolerances are normally distributed, so is the variation of dimension condition,

$$v_{wP} = (D_a^2 t_{aP}^2 + D_b^2 t_{bP}^2 + \cdots)^{1/2} \tag{19.10}$$

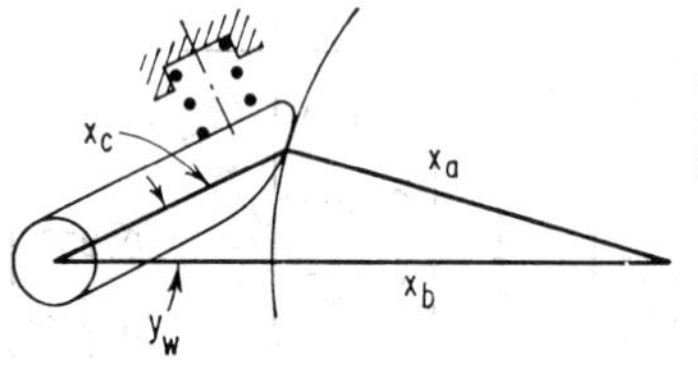

FIG. 19.8. Indirect dimensional relationship. Variation of the dimension condition must be derived from Eq. (19.9) or (19.10) by operating on the equation shown.

$$\bar{y}_w = \cos^{-1} \frac{\bar{x}_b^2 + \bar{x}_c^2 - \bar{x}_a^2}{2\bar{x}_b\bar{x}_c}$$

In both equations $D_a = |\partial y_w/\partial x_a|$, $D_b = |\partial y_w/\partial x_b|$, etc.

Mean values for the dimension condition and all dimensions except one are scaled from the design layout; the mean value of the remaining dimension is calculated using an equation corresponding to Eq. (19.8). If either limit of the dimension condition is a controlling factor, then Eq. (19.8) can be modified by substituting $\bar{y} = W - v/2$ or $\bar{y} = w + v/2$; v in either case is calculated from Eq. (19.9) or (19.10) as appropriate. Tolerances satisfying Eqs. (19.9) and (19.10) may be determined by repeated trials. A more direct approach is possible by assuming all Dt^2 terms for unassigned tolerances are equal so that

$$t_{jF} = v_{wF}'/D_j^{1/2}\Sigma(D^{1/2}) \quad (19.11)$$
$$t_{jP} = v_{wP}'/D_j^{1/2}(\Sigma D)^{1/2} \quad (19.12)$$

When all tolerances are initially unassigned, $v_{wF}' \equiv v_{wF}$ and $v_{wP}' \equiv v_{wP}$; otherwise v_{wF} and v_{wP} terms must be suitably adjusted by subtracting Dt or D^2t^2 terms. In addition the sums $\Sigma(D^{1/2})$ and ΣD should include partial derivatives only for dimensions having unassigned tolerance values.

(c) Dimensioning of Manufacturing Drawings

Drawings of complex parts include many dimensional specifications only some of which are derived by a dimension study. Dimensions that control operating and assembly conditions are called *functional;* other dimensions are called *constructional.* Any dimension, whether functional or constructional, will either define the size and shape of a geometrical feature or locate these features with respect to suitable datum planes. In good production design, datums used for functional dimensions also serve manufacturing purposes. There is, however, no general assurance that datums selected for functional considerations will be suitable for manufacturing purposes. With this in mind, the best practice is to dimension end-product drawings so that these drawings conform most directly to the functional relationships derived from dimension studies. If these dimensions are inappropriate for manufacturing purposes, production engineers can specify alternate relationships on process drawings. It may even be desirable to introduce special surfaces into the design of parts for production use. The substitution of secondary dimensional relationships in place of functional dimensions will, however, carry the penalty of reduced manufacturing tolerances.

19.4. STANDARD LIMITS AND FITS FOR CYLINDRICAL PARTS

(a) Technical Features of the American Standard

ASA B4.1-1955 is an inch-measure version of the ISA tolerance system.[4] American, British, and Canadian versions are in exact agreement with respect to hole and shaft deviations for diameters to 20 in. An explanation of the symbolism representing a typical class of fit is shown in Fig. 19.9. As this symbolism has international currency engineering data concerning the application and selection of cylindrical fits can be drawn from foreign as well as domestic sources.

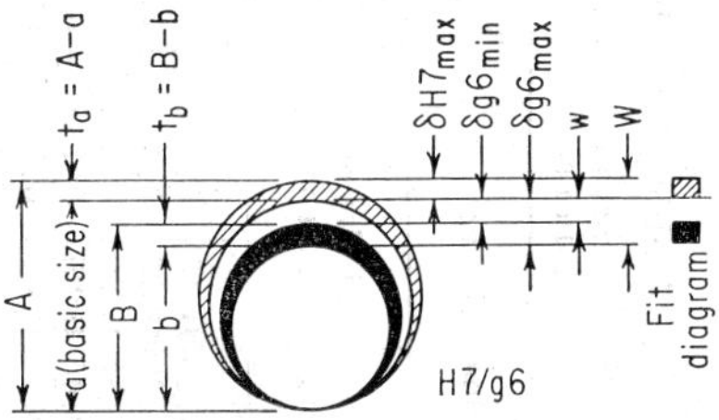

FIG. 19.9. Elements of a typical basic hole fit. $\delta H7_{\max}$, $\delta g6_{\max}$, and $\delta g6_{\min}$ are deviations used to calculate diameters A, B, and b.

Tables of deviations published in the American and other national standards permit limiting diameters to be calculated for any association of hole and shaft classes. By associating holes with limits over, and shafts with limits under, the basic size, clearance fits are obtained. Other associations result in transition or interference fits. A fit is *basic hole, basic shaft,* or *composite* according to whether the basic size corresponds to the minimum hole diameter, the maximum shaft diameter, or neither. A discussion concerning the relative merit and application of basic hole and basic shaft fits is given in ref. 5. The majority of engineering requirements are satisfied by basic hole fits.

(b) Selection of Basic Hole Fits

Tables 19.1, 19.2, and 19.3 describe the basic hole fits recommended in the American Standard. A limited number of these fits will usually do for a particular type of

Table 19.4. Clearance Fits Limits and Deviations

Hole and shaft deviations and limits of clearance are in thousandths of an inch. Data to horizontal lines are in accordance with the recommendations of British-Canadian-American conferences

Range of diameters, in.		H5 hole deviation	RC1 close sliding				H6 hole deviation	LC1 snug location				RC2 sliding				RC3 precision running			
Over	To		g4 shaft deviations		Limits of clearance			h5 shaft deviations		Limits of clearance		g5 shaft deviations		Limits of clearance		f6 shaft deviations		Limits of clearance	
0.04	0.12	+0.20	−0.10	−0.25	0.10	0.45	+0.25	0	−0.20	0	0.45	−0.10	−0.30	0.10	0.55	−0.3	−0.55	0.3	0.8
0.12	0.24	+0.20	−0.15	−0.30	0.15	0.50	+0.30	0	−0.20	0	0.50	−0.15	−0.35	0.15	0.65	−0.4	−0.7	0.4	1.0
0.24	0.40	+0.25	−0.20	−0.35	0.20	0.60	+0.40	0	−0.25	0	0.65	−0.20	−0.45	0.20	0.85	−0.5	−0.9	0.5	1.3
0.40	0.71	+0.30	−0.25	−0.45	0.25	0.75	+0.40	0	−0.30	0	0.70	−0.25	−0.55	0.25	0.95	−0.6	−1.0	0.6	1.4
0.71	1.19	+0.40	−0.30	−0.55	0.30	0.95	+0.50	0	−0.40	0	0.90	−0.30	−0.70	0.30	1.20	−0.8	−1.3	0.8	1.8
1.19	1.97	+0.4	−0.4	−0.7	0.4	1.1	+0.6	0	−0.4	0	1.0	−0.4	−0.8	0.4	1.4	−1.0	−1.6	1.0	2.2
1.97	3.15	+0.5	−0.4	−0.7	0.4	1.2	+0.7	0	−0.5	0	1.2	−0.4	−0.9	0.4	1.6	−1.2	−1.9	1.2	2.6
3.15	4.73	+0.6	−0.5	−0.9	0.5	1.5	+0.9	0	−0.6	0	1.5	−0.5	−1.1	0.5	2.0	−1.4	−2.3	1.4	3.2
4.73	7.09	+0.7	−0.6	−1.1	0.6	1.8	+1.0	0	−0.7	0	1.7	−0.6	−1.3	0.6	2.3	−1.6	−2.6	1.6	3.6
7.09	9.85	+0.8	−0.6	−1.2	0.6	2.0	+1.2	0	−0.8	0	2.0	−0.6	−1.4	0.6	2.6	−2.0	−3.2	2.0	4.4
9.85	12.41	+0.9	−0.8	−1.4	0.8	2.3	+1.2	0	−0.9	0	2.1	−0.8	−1.7	0.8	2.9	−2.5	−3.7	2.5	4.9
12.41	15.75	+1.0	−1.0	−1.7	1.0	2.7	+1.4	0	−1.0	0	2.4	−1.0	−2.0	1.0	3.4	−3.0	−4.4	3.0	5.8
15.75	19.69	+1.0	−1.2	−2.0	1.2	3.0	+1.6	0	−1.0	0	2.6	−1.2	−2.2	1.2	3.8	−4.0	−5.6	4.0	7.2
19.69	30.09	+1.2	−1.6	−2.5	1.6	3.7	+2.0	0	−1.2	0	3.2	−1.6	−2.8	1.6	4.8	−5.0	−7.0	5.0	9.0
30.09	41.49	+1.6	−2.0	−3.0	2.0	4.6	+2.5	0	−1.6	0	4.1	−2.0	−3.6	2.0	6.1	−6.0	−8.5	6.0	11.0
41.49	56.19	+2.0	−2.5	−3.7	2.5	5.7	+3	0	−2.0	0	5.0	−2.5	−4.5	2.5	7.5	−8	−11	8	14
56.19	76.39	+2.5	−3.0	−4.6	3.0	7.1	+4	0	−2.5	0	6.5	−3.0	−5.5	3.0	9.5	−10	−14	10	18
76.39	100.90	+3.0	−4.0	−6.0	4.0	9.0	+5	0	−3.0	0	8.0	−4.0	−7.0	4.0	12.0	−12	−17	12	22
100.90	131.90	+4.0	−5.0	−7.5	5.0	11.5	+6	0	−4.0	0	10.0	−5.0	−9.0	5.0	15.0	−16	−22	16	28
131.90	171.90	+5.0	−6.0	−9.0	6.0	14.0	+8	0	−5.0	0	13.0	−6.0	−11.0	6.0	19.0	−18	−26	18	34
171.90	200.00	+6	−8	−12	8	18	+10	0	−6	0	16	−8	−14	8	24	−22	−32	22	42

Table 19.4. Clearance Fits Limits and Deviations (Continued)

Range of diameters, in.		H7 hole deviation	LC2 snug location				LC5 medium location				RC4 close running				RC5 medium running			
Over	To		h6 shaft deviations		Limits of clearance		g6 shaft deviations		Limits of clearance		f7 shaft deviations		Limits of clearance		e7 shaft deviations		Limits of clearance	
0.04	0.12	+0.4	0	−0.25	0	0.65	−0.10	−0.35	0.10	0.75	−0.3	−0.7	0.3	1.1	−0.6	−1.0	0.6	1.4
0.12	0.24	+0.5	0	−0.30	0	0.80	−0.15	−0.45	0.15	0.95	−0.4	−0.9	0.4	1.4	−0.8	−1.3	0.8	1.8
0.24	0.40	+0.6	0	−0.40	0	1.00	−0.20	−0.60	0.20	1.20	−0.5	−1.1	0.5	1.7	−1.0	−1.6	1.0	2.2
0.40	0.71	+0.7	0	−0.40	0	1.10	−0.25	−0.65	0.25	1.35	−0.6	−1.3	0.6	2.0	−1.2	−1.9	1.2	2.6
0.71	1.19	+0.8	0	−0.50	0	1.30	−0.30	−0.80	0.30	1.60	−0.8	−1.6	0.8	2.4	−1.6	−2.4	1.6	3.2
1.19	1.97	+1.0	0	−0.6	0	1.6	−0.4	−1.0	0.4	2.0	−1.0	−2.0	1.0	3.0	−2.0	−3.0	2.0	4.0
1.97	3.15	+1.2	0	−0.7	0	1.9	−0.4	−1.1	0.4	2.3	−1.2	−2.4	1.2	3.6	−2.5	−3.7	2.5	4.9
3.15	4.73	+1.4	0	−0.9	0	2.3	−0.5	−1.4	0.5	2.8	−1.4	−2.8	1.4	4.2	−3.0	−4.4	3.0	5.8
4.73	7.09	+1.6	0	−1.0	0	2.6	−0.6	−1.6	0.6	3.2	−1.6	−3.2	1.6	4.8	−3.5	−5.1	3.5	6.7
7.09	9.85	+1.8	0	−1.2	0	3.0	−0.6	−1.8	0.6	3.6	−2.0	−3.8	2.0	5.6	−4	−5.8	4	7.6
9.85	12.41	+2.0	0	−1.2	0	3.2	−0.7	−1.9	0.7	3.9	−2.5	−4.5	2.5	6.5	−5	−7.0	5	9.0
12.41	15.75	+2.2	0	−1.4	0	3.6	−0.7	−2.1	0.7	4.3	−3.0	−5.2	3.0	7.4	−6	−8.2	6	10.4
15.75	19.69	+2.5	0	−1.6	0	4.1	−0.8	−2.4	0.8	4.9	−4.0	−6.5	4.0	9.0	−8	−10.5	8	13.0
19.69	30.09	+3.0	0	−2.0	0	5.0	−0.9	−2.9	0.9	5.9	−5.0	−8.0	5.0	11.0	−10	−13.0	10	16.0
30.09	41.49	+4.0	0	−2.5	0	6.5	−1.0	−3.5	1.0	7.5	−6.0	−10.0	6.0	14.0	−12	−16.0	12	20.0
41.49	56.19	+5	0	−3	0	8	−1.2	−4.2	1.2	9.2	−8	−13	8	18	−16	−21	16	26
56.19	76.39	+6	0	−4	0	10	−1.2	−5.2	1.2	11.2	−10	−16	10	22	−20	−26	20	32
76.39	100.90	+8	0	−5	0	13	−1.4	−6.4	1.4	14.4	−12	−20	12	28	−25	−33	25	41
100.90	131.90	+10	0	−6	0	16	−1.6	−7.6	1.6	17.6	−16	−26	16	36	−30	−40	30	50
131.90	171.90	+12	0	−8	0	20	−1.8	−9.8	1.8	21.8	−18	−30	18	42	−35	−47	35	59
171.90	200.00	+16	0	−10	0	26	−1.8	−11.8	1.8	27.8	−22	−38	22	54	−45	−61	45	77

Table 19.4. Clearance Fits Limits and Deviations (Continued)

Range of diameters, in.		H8 hole deviation	LC3 snug location				LC6 medium location				RC6 medium running				H9 hole deviation	LC4 snug location			
Over	To		h7 shaft deviations		Limits of clearance		f8 shaft deviations		Limits of clearance		e8 shaft deviations		Limits of clearance			h9 shaft deviations		Limits of clearance	
0.04	0.12	+0.6	0	−0.4	0	1.0	−0.3	−0.9	0.3	1.5	−0.6	−1.2	0.6	1.8	+1.0	0	−1.0	0	2.0
0.12	0.24	+0.7	0	−0.5	0	1.2	−0.4	−1.1	0.4	1.8	−0.8	−1.5	0.8	2.2	+1.2	0	−1.2	0	2.4
0.24	0.40	+0.9	0	−0.6	0	1.5	−0.5	−1.4	0.5	2.3	−1.0	−1.9	1.0	2.8	+1.4	0	−1.4	0	2.8
0.40	0.71	+1.0	0	−0.7	0	1.7	−0.6	−1.6	0.6	2.6	−1.2	−2.2	1.2	3.2	+1.6	0	−1.6	0	3.2
0.71	1.19	+1.2	0	−0.8	0	2.0	−0.8	−2.0	0.8	3.2	−1.6	−2.8	1.6	4.0	+2.0	0	−2.0	0	4.0
1.19	1.97	+1.6	0	−1.0	0	2.6	−1.0	−2.6	1.0	4.2	−2.0	−3.6	2.0	5.2	+2.5	0	−2.5	0	5
1.97	3.15	+1.8	0	−1.2	0	3.0	−1.2	−3.0	1.2	4.8	−2.5	−4.3	2.5	6.1	+3.0	0	−3.0	0	6
3.15	4.73	+2.2	0	−1.4	0	3.6	−1.4	−3.6	1.4	5.8	−3.0	−5.2	3.0	7.4	+3.5	0	−3.5	0	7
4.73	7.09	+2.5	0	−1.6	0	4.1	−1.6	−4.1	1.6	6.6	−3.5	−6.0	3.5	8.5	+4.0	0	−4.0	0	8
7 09	9.85	+2.8	0	−1.8	0	4.6	−2.0	−4.8	2.0	7.6	−4.0	−6.8	4.0	9.6	+4.5	0	−4.5	0	9
9.85	12.41	+3.0	0	−2.0	0	5.0	−2.2	−5.2	2.2	8.2	−5	−8.0	5	11	+5	0	−5	0	10
12.41	15.75	+3.5	0	−2.2	0	5.7	−2.5	−6.0	2.5	9.5	−6	−9.5	6	13	+6	0	−6	0	12
15.75	19.69	+4.0	0	−2.5	0	6.5	−2.8	−6.8	2.8	10.8	−8	−12.0	8	16	+6	0	−6	0	12
19.69	30.09	+5.0	0	−3.0	0	8.0	−3.0	−8.0	3.0	13.0	−10	−15.0	10	20	+8	0	−8	0	16
30.09	41.49	+6.0	0	−4.0	0	10.0	−3.5	−9.5	3.5	15.5	−12	−18.0	12	24	+10	0	−10	0	20
41.49	56.19	+8	0	−5	0	13	−4.0	−12.0	4.0	20.0	−16	−24	16	32	+12	0	−12	0	24
56.19	76.39	+10	0	−6	0	16	−4.5	−14.5	4.5	24.5	−20	−30	20	40	+16	0	−16	0	32
76.39	100.90	+12	0	−8	0	20	−5.0	−17.0	5.0	29.0	−25	−37	25	49	+20	0	−20	0	40
100.90	131.90	+16	0	−10	0	26	−6.0	−22.0	6.0	38.0	−30	−46	30	62	+25	0	−25	0	50
131.90	171.90	+20	0	−12	0	32	−7.0	−27.0	7.0	47.0	−35	−55	35	75	+30	0	−30	0	60
171.90	200.00	+25	0	−16	0	41	−7	−32	7	57	−45	−70	45	95	+40	0	−40	0	80

Table 19.4. Clearance Fits Limits and Deviations (Continued)

Range of diameters, in.		H9 hole deviation	LC7 medium location				RC7 free running				H10 hole deviation	LC8 loose location				RC8 loose running			
Over	To		e9 shaft deviations		Limits of clearance		d8 shaft deviations		Limits of clearance			d9 shaft deviations		Limits of clearance		c9 shaft deviations		Limits of clearance	
0.04	0.12	+1.0	−0.6	−1.6	0.6	2.6	−1.0	−1.6	1.0	2.6	+1.6	−1.0	−2.0	1.0	3.6	−2.5	−3.5	2.5	5.1
0.12	0.24	+1.2	−0.8	−2.0	0.8	3.2	−1.2	−1.9	1.2	3.1	+1.8	−1.2	−2.4	1.2	4.2	−2.9	−4.0	2.8	5.8
0.24	0.40	+1.4	−1.0	−2.4	1.0	3.8	−1.6	−2.5	1.6	3.9	+2.2	−1.6	−3.0	1.6	5.2	−3.0	−4.4	3.0	6.6
0.40	0.71	+1.6	−1.2	−2.8	1.2	4.4	−2.0	−3.0	2.0	4.6	+2.8	−2.0	−3.6	2.0	6.4	−3.5	−5.1	3.5	7.9
0.71	1.19	+2.0	−1.6	−3.6	1.6	5.6	−2.5	−3.7	2.5	5.7	+3.5	−2.5	−4.5	2.5	8.0	−4.5	−6.5	4.5	10.0
1.19	1.97	+2.5	−2.0	−4.5	2.0	7.0	−3	−4.6	3	7.1	+4.0	−3	−5.5	3	9.5	−5	−7.5	5	11.5
1.97	3.15	+3.0	−2.5	−5.5	2.5	8.5	−4	−5.8	4	8.8	+4.5	−4	−7.0	4	11.5	−6	−9.0	6	13.5
3.15	4.73	+3.5	−3.0	−6.5	3.0	10.0	−5	−7.2	5	10.7	+5.0	−5	−8.5	5	13.5	−7	−10.5	7	15.5
4.73	7.09	+4.0	−3.5	−7.5	3.5	11.5	−6	−8.5	6	12.5	+6.0	−6	−10.0	6	16.0	−8	−12.0	8	18.0
7.09	9.85	+4.5	−4.0	−8.5	4.0	13.0	−7	−9.8	7	14.3	+7.0	−7	−11.5	7	18.5	−10	−14.5	10	21.5
9.85	12.41	+5	−4.5	−9.5	4.5	14.5	−8	−11.0	8	16.0	+8	−7	−12	7	20	−12	−17	12	25
12.41	15.75	+6	−5.0	−11.0	5.0	17.0	−10	−13.5	10	19.5	+9	−8	−14	8	23	−14	−20	14	29
15.75	19.69	+6	−5.0	−11.0	5.0	17.0	−12	−16.0	12	22.0	+10	−9	−15	9	25	−16	−22	16	32
19.69	30.09	+8	−6.0	−14.0	6.0	22.0	−16	−21.0	16	29.0	+12	−10	−18	10	30	−20	−28	20	40
30.09	41.49	+10	−7.0	−17.0	7.0	27.0	−20	−26.0	20	36.0	+16	−12	−22	12	38	−25	−35	25	51
41.49	56.19	+12	−8	−20	8	32	−25	−33	25	45	+20	−14	−26	14	46	−30	−42	30	62
56.19	76.39	+16	−9	−25	9	41	−30	−40	30	56	+25	−16	−32	16	57	−40	−56	40	81
76.39	100.90	+20	−10	−30	10	50	−40	−52	40	72	+30	−18	−38	18	68	−50	−70	50	100
100.90	131.90	+25	−12	−37	12	62	−50	−66	50	91	+40	−20	−45	20	85	−60	−85	60	125
131.90	171.90	+30	−14	−44	14	74	−60	−80	60	110	+50	−25	−55	25	105	−80	−110	80	160
171.90	200.00	+40	−14	−54	14	94	−80	−105	80	145	+60	−25	−65	25	125	−100	−140	100	200

Table 19.4. Clearance Fits Limits and Deviations (Concluded)

Range of diameters, in.		H11 hole deviation	LC9 loose location				RC9 loose running				H12 hole deviation	LC10 loose location				H13 hole deviation	LC11 loose location			
Over	To		c11 shaft deviations		Limits of clearance		Shaft deviations*		Limits of clearance			Shaft deviations*		Limits of clearance			Shaft deviations*		Limits of clearance	
0.04	0.12	+2.5	−2.5	−5.0	2.5	7.5	−4.0	−5.6	4.0	8.1	+4	−4.0	−8.0	4.0	12.0	+6	−5	−11	5	17
0.12	0.24	+3.0	−2.8	−5.8	2.8	8.8	−4.5	−6.0	4.5	9.0	+5	−4.5	−9.5	4.5	14.5	+7	−6	−13	6	20
0.24	0.40	+3.5	−3.0	−6.5	3.0	10.0	−5.0	−7.2	5.0	10.7	+6	−5.0	−11.0	5.0	17.0	+9	−7	−16	7	25
0.40	0.71	+4.0	−3.5	−7.5	3.5	11.5	−6.0	−8.8	6.0	12.8	+7	−6.0	−13.0	6.0	20.0	+10	−8	−18	8	28
0.71	1.19	+5.0	−4.5	−9.5	4.5	14.5	−7.0	−10.5	7.0	15.5	+8	−7.0	−15.0	7.0	23.0	+12	−10	−22	10	34
1.19	1.97	+6	−5	−11	5	17	−8	−12.0	8	18.0	+10	−8	−18	8	28	+16	−12	−28	12	44
1.97	3.15	+7	−6	−13	6	20	−9	−13.5	9	20.5	+12	−10	−22	10	34	+18	−14	−32	14	50
3.15	4.73	+9	−7	−16	7	25	−10	−15.0	10	24.0	+14	−11	−25	11	39	+22	−16	−38	16	60
4.73	7.09	+10	−8	−18	8	28	−12	−18.0	12	28.0	+16	−12	−28	12	44	+25	−18	−43	18	68
7.09	9.85	+12	−10	−22	10	34	−15	−22.0	15	34.0	+18	−16	−34	16	52	+28	−22	−50	22	78
9.85	12.41	+12	−12	−24	12	36	−18	−26	18	38	+20	−20	−40	20	60	+30	−28	−58	28	88
12.41	15.75	+14	−14	−28	14	42	−22	−31	22	45	+22	−22	−44	22	66	+35	−30	−65	30	100
15.75	19.69	+16	−16	−32	16	48	−25	−35	25	51	+25	−25	−50	25	75	+40	−35	−75	35	115
19.69	30.09	+20	−18	−38	18	58	−30	−42	30	62	+30	−28	−58	28	88	+50	−40	−90	40	140
30.09	41.49	+25	−20	−45	20	70	−40	−56	40	81	+40	−30	−70	30	110	+60	−45	−105	45	165
41.49	56.19	+30	−25	−55	25	85	−50	−70	50	100	+50	−40	−90	40	140	+80	−60	−140	60	220
56.19	76.39	+40	−30	−70	30	110	−60	−85	60	125	+60	−50	−110	50	170	+100	−70	−170	70	270
76.39	100.90	+50	−35	−85	35	135	−80	−110	80	160	+80	−50	−130	50	210	+125	−80	−205	80	330
100.90	131.90	+60	−40	−100	40	160	−100	−140	100	200	+100	−60	−160	60	260	+160	−90	−250	90	410
131.90	171.90	+80	−50	−130	50	210	−130	−180	130	260	+125	−80	−205	80	330	+200	−100	−300	100	500
171.90	200.00	+100	−50	−150	50	250	−150	−210	150	310	+160	−90	−250	90	410	+250	−125	−375	125	625

* Shaft deviations not in conformance to recommendations of British-Canadian-American conferences.

product. Where special requirements prevail—especially for basic shaft and composite fits—the tables of hole and shaft deviations given in ASA B4.1-1955 can be used.

Each ASA recommended fit has a class symbol (RC1, LC3, etc.) in addition to the symbols previously described. Names designating recommended fits in Tables 19.1, 19.2, and 19.3 are from the American Standard or follow customary usage where the standard is not specific. The same name is applied to more than one class of fit if a particular type of fit is available in more than one tolerance grade. Diagrams included in the tables indicate relative fit conditions. Tolerance grades associated with each class of fit are explicit in the ISA symbolism. For instance, H7/g6 indicates a fit having a hole with a grade 7 tolerance and a shaft with a grade 6 tolerance. Numerical values of tolerances and the manufacturing processes which are normally expected to produce work to the various tolerance grades are given in Tables 19.7 and 19.8.

Table 19.5. Transition Fits Limits and Deviations

Hole and shaft deviations and limits of fits are in thousandths of an inch. Data are in accordance with the recommendations of British-Canadian-American conferences

Range of diameters, in.		H7 hole deviation	LT1 push fit				LT3 wringing fit				LT7 tight fit			
Over	To		j6 shaft deviations		Limits of fit		k6 shaft deviations		Limits of fit		n6 shaft deviations		Limits of fit	
0.04	0.12	+0.4	+0.15	−0.1	−0.15	+0.5	....	....	....	....	+0.5	+0.25	−0.5	+0.15
0.12	0.24	+0.5	+0.20	−0.1	−0.20	+0.6	....	....	....	....	+0.6	+0.30	−0.6	+0.20
0.24	0.40	+0.6	+0.30	−0.1	−0.30	+0.7	+0.5	+0.1	−0.5	+0.5	+0.8	+0.40	−0.8	+0.20
0.40	0.71	+0.7	+0.30	−0.1	−0.30	+0.8	+0.5	+0.1	−0.5	+0.6	+0.9	+0.50	−0.9	+0.20
0.71	1.19	+0.8	+0.30	−0.2	−0.30	+1.0	+0.6	+0.1	−0.6	+0.7	+1.1	+0.60	−1.1	+0.20
1.19	1.97	+1.0	+0.4	−0.2	−0.4	+1.2	+0.7	+0.1	−0.7	+0.9	+1.3	+0.7	−1.3	+0.3
1.97	3.15	+1.2	+0.4	−0.3	−0.4	+1.5	+0.8	+0.1	−0.8	+1.1	+1.5	+0.8	−1.5	+0.4
3.15	4.73	+1.4	+0.5	−0.4	−0.5	+1.8	+1.0	+0.1	−1.0	+1.3	+1.9	+1.0	−1.9	+0.4
4.73	7.09	+1.6	+0.6	−0.4	−0.6	+2.0	+1.1	+0.1	−1.1	+1.5	+2.2	+1.2	−2.2	+0.4
7.09	9.85	+1.8	+0.7	−0.5	−0.7	+2.3	+1.4	+0.2	−1.4	+1.6	+2.6	+1.4	−2.6	+0.4
9.85	12.41	+2.0	+0.7	−0.6	−0.7	+2.6	+1.4	+0.2	−1.4	+1.8	+2.6	+1.4	−2.6	+0.6
12.41	15.75	+2.2	+0.7	−0.7	−0.7	+2.9	+1.6	+0.2	−1.6	+2.0	+3.0	+1.6	−3.0	+0.6
15.75	19.69	+2.5	+0.8	−0.7	−0.8	+3.2	+1.8	+0.2	−1.8	+2.3	+3.4	+1.8	−3.4	+0.7

Range of diameters, in.		H8 hole deviation	LT2 push fit				LT4 wringing fit				LT6 tight fit			
Over	To		j7 shaft deviations		Limits of fit		k7 shaft deviations		Limits of fit		m7 shaft deviations		Limits of fit	
0.04	0.12	+0.6	+0.3	−0.1	−0.3	+0.7	....	....	....	....	+0.55	+0.15	−0.55	+0.45
0.12	0.24	+0.7	+0.4	−0.1	−0.4	+0.8	....	....	....	....	+0.70	+0.20	−0.70	+0.50
0.24	0.40	+0.9	+0.4	−0.2	−0.4	+1.1	+0.7	+0.1	−0.7	+0.8	+0.80	+0.20	−0.80	+0.70
0.40	0.71	+1.0	+0.5	−0.2	−0.5	+1.2	+0.8	+0.1	−0.8	+0.9	+1.00	+0.30	−1.00	+0.70
0.71	1.19	+1.2	+0.5	−0.3	−0.5	+1.5	+0.9	+0.1	−0.9	+1.1	+1.10	+0.30	−1.10	+0.90
1.19	1.97	+1.6	+0.6	−0.4	−0.6	+2.0	+1.1	+0.1	−1.1	+1.5	+1.4	+0.4	−1.4	+1.2
1.97	3.15	+1.8	+0.7	−0.5	−0.7	+2.3	+1.3	+0.1	−1.3	+1.7	+1.7	+0.5	−1.7	+1.3
3.15	4.73	+2.2	+0.8	−0.6	−0.8	+2.8	+1.5	+0.1	−1.5	+2.1	+1.9	+0.5	−1.9	+1.7
4.73	7.09	+2.5	+0.9	−0.7	−0.9	+3.2	+1.7	+0.1	−1.7	+2.4	+2.2	+0.6	−2.2	+1.9
7.09	9.85	+2.8	+1.0	−0.8	−1.0	+3.6	+2.0	+0.2	−2.0	+2.6	+2.4	+0.6	−2.4	+2.2
9.85	12.41	+3.0	+1.0	−1.0	−1.0	+4.0	+2.2	+0.2	−2.2	+2.8	+2.8	+0.8	−2.8	+2.2
12.41	15.75	+3.5	+1.2	−1.0	−1.2	+4.5	+2.4	+0.2	−2.4	+3.3	+3.0	+0.8	−3.0	+2.7
15.75	19.69	+4.0	+1.3	−1.2	−1.3	+5.2	+2.7	+0.2	−2.7	+3.8	+3.4	+0.9	−3.4	+3.1

Table 19.6. Interference Fits Limits and Deviations

Hole and shaft deviations and limits of interference are in thousandths of an inch. Data to horizontal lines are in accordance with the recommendations of British-Canadian-American conferences

Range of diameters, in.		H6 hole deviation	FN1 light drive				H7 hole deviation	LN2 locational interference				LN3 locational interference			
Over	To		Shaft deviations*		Limits of interference			p6 shaft deviations		Limits of interference		r6 shaft deviations		Limits of interference	
0.04	0.12	+0.25	+0.50	+0.3	0.05	0.50	+0.4	+0.65	+0.4	0.0	0.65	+0.75	+0.5	0.1	0.75
0.12	0.24	+0.30	+0.60	+0.4	0.10	0.60	+0.5	+0.80	+0.5	0.0	0.80	+0.90	+0.6	0.1	0.90
0.24	0.40	+0.40	+0.75	+0.5	0.10	0.75	+0.6	+1.00	+0.6	0.0	1.00	+1.20	+0.8	0.2	1.20
0.40	0.56	+0.40	+0.80	+0.5	0.10	0.80	+0.7	+1.10	+0.7	0.0	1.10	+1.40	+1.0	0.3	1.40
0.56	0.71	+0.40	+0.90	+0.6	0.20	0.90	+0.7	+1.10	+0.7	0.0	1.10	+1.40	+1.0	0.3	1.40
0.71	0.95	+0.5	+1.1	+0.7	0.2	1.1	+0.8	+1.3	+0.8	0.0	1.3	+1.7	+1.2	0.4	1.7
0.95	1.19	+0.5	+1.2	+0.8	0.3	1.2	+0.8	+1.3	+0.8	0.0	1.3	+1.7	+1.2	0.4	1.7
1.19	1.58	+0.6	+1.3	+0.9	0.3	1.3	+1.0	+1.6	+1.0	0.0	1.6	+2.0	+1.4	0.4	2.0
1.58	1.97	+0.6	+1.4	+1.0	0.4	1.4	+1.0	+1.6	+1.0	0.0	1.6	+2.0	+1.4	0.4	2.0
1.97	2.56	+0.7	+1.8	+1.3	0.6	1.8	+1.2	+2.1	+1.4	0.2	2.1	+2.3	+1.6	0.4	2.3
2.56	3.15	+0.7	+1.9	+1.4	0.7	1.9	+1.2	+2.1	+1.4	0.2	2.1	+2.3	+1.6	0.4	2.3
3.15	3.94	+0.9	+2.4	+1.8	0.9	2.4	+1.4	+2.5	+1.6	0.2	2.5	+2.9	+2.0	0.6	2.9
3.94	4.73	+0.9	+2.6	+2.0	1.1	2.6	+1.4	+2.5	+1.6	0.2	2.5	+2.9	+2.0	0.6	2.9
4.73	5.52	+1.0	+2.9	+2.2	1.2	2.9	+1.6	+2.8	+1.8	0.2	2.8	+3.5	+2.5	0.9	3.5
5.52	6.30	+1.0	+3.2	+2.5	1.5	3.2	+1.6	+2.8	+1.8	0.2	2.8	+3.5	+2.5	0.9	3.5
6.30	7.09	+1.0	+3.5	+2.8	1.8	3.5	+1.6	+2.8	+1.8	0.2	2.8	+3.5	+2.5	0.9	3.5
7.09	7.88	+1.2	+3.8	+3.0	1.8	3.8	+1.8	+3.2	+2.0	0.2	3.2	+4.2	+3.0	1.2	4.2
7.88	8.86	+1.2	+4.3	+3.5	2.3	4.3	+1.8	+3.2	+2.0	0.2	3.2	+4.2	+3.0	1.2	4.2
8.86	9.85	+1.2	+4.3	+3.5	2.3	4.3	+1.8	+3.2	+2.0	0.2	3.2	+4.2	+3.0	1.2	4.2
9.85	11.03	+1.2	+4.9	+4.0	2.8	4.9	+2.0	+3.4	+2.2	0.2	3.4	+4.7	+3.5	1.5	4.7

Table 19.6. Interference Fits Limits and Deviations (Continued)

Range of diameters, in.		H6 hole deviation	FN1 light drive				H7 hole deviation	LN2 locational interference				LN3 locational interference			
Over	To		Shaft deviations*		Limits of interference			p6 shaft deviations		Limits of interference		r6 shaft deviations		Limits of interference	
11.03	12.41	+1.2	+4.9	+4.0	2.8	4.9	+2.0	+3.4	+2.2	0.2	3.4	+4.7	+3.5	1.5	4.7
12.41	13.98	+1.4	+5.5	+4.5	3.1	5.5	+2.2	+3.9	+2.5	0.3	3.9	+5.9	+4.5	2.3	5.9
13.98	15.75	+1.4	+6.1	+5.0	3.6	6.1	+2.2	+3.9	+2.5	0.3	3.9	+5.9	+4.5	2.3	5.9
15.75	17.72	+1.6	+7.0	+6.0	4.4	7.0	+2.5	+4.4	+2.8	0.3	4.4	+6.6	+5.0	2.5	6.6
17.72	19.69	+1.6	+7.0	+6.0	4.4	7.0	+2.5	+4.4	+2.8	0.3	4.4	+6.6	+5.0	2.5	6.6
19.69	24.34	+2.0	+9.2	+8	6.0	9.2	+3	+5.5	+3.5	0.5	5.5	+9.0	+7	4	9.0
24.34	30.09	+2.0	+10.2	+9	7.0	10.2	+3	+5.5	+3.5	0.5	5.5	+9.0	+7	4	9.0
30.09	35.47	+2.5	+11.6	+10	7.5	11.6	+4	+7.0	+4.5	0.5	7.0	+11.5	+9	5	11.5
35.47	41.49	+2.5	+13.6	+12	9.5	13.6	+4	+7.0	+4.5	0.5	7.0	+11.5	+9	5	11.5
41.49	48.28	+3.0	+16.0	+14	11.0	16.0	+5	+9.0	+6.0	1.0	9.0	+15.0	+12	7	15.0
48.28	56.19	+3	+18.0	+16	13	18.0	+5	+9	+6	1	9	+15	+12	7	15
56.19	65.54	+4	+20.5	+18	14	20.5	+6	+11	+7	1	11	+20	+16	10	20
65.54	76.39	+4	+24.5	+22	18	24.5	+6	+11	+7	1	11	+20	+16	10	20
76.39	87.79	+5	+28.0	+25	20	28.0	+8	+14	+9	1	14	+25	+20	12	25
87.79	100.90	+5	+31.0	+28	23	31.0	+8	+14	+9	1	14	+25	+20	12	25
100.90	115.30	+6	+34	+30	24	34	+10	+18	+12	2	18	+31	+25	15	31
115.30	131.90	+6	+39	+35	29	39	+10	+18	+12	2	18	+31	+25	15	31
131.90	152.20	+8	+50	+45	37	50	+12	+24	+16	4	24	+38	+30	18	38
152.20	171.90	+8	+55	+50	42	55	+12	+24	+16	4	24	+38	+30	18	38
171.90	200.00	+10	+66	+60	50	66	+16	+30	+20	4	30	+50	+40	24	50

* Shaft deviations not in conformance to recommendations of British-Canadian-American conferences.

Table 19.6. Interference Fits Limits and Deviations (Continued)

Range of diameters, in.		H7 hole deviation	FN3 heavy drive				FN4 force				FN5 heavy force			
Over	To		t6 shaft deviations		Limits of interference		u6 shaft deviations		Limits of interference		x7 shaft deviations		Limits of interference	
0.04	0.12	+0.4	......	......	......	......	+0.95	+0.7	0.3	0.95	+1.3	+0.9	0.5	1.3
0.12	0.24	+0.5	......	......	......	......	+1.20	+0.9	0.4	1.20	+1.7	+1.2	0.7	1.7
0.24	0.40	+0.6	......	......	......	......	+1.60	+1.2	0.6	1.60	+2.0	+1.4	0.8	2.0
0.40	0.56	+0.7	......	......	......	......	+1.80	+1.4	0.7	1.80	+2.3	+1.6	0.9	2.3
0.56	0 71	+0.7	......	......	......	......	+1.80	+1.4	0.7	1.80	+2.5	+1.8	1.1	2.5
0.71	0.95	+0.8	......	......	......	......	+2.1	+1.6	0.8	2.1	+3.0	+2.2	1.4	3.0
0.95	1.19	+0.8	+2.1	+1.6	0.8	2.1	+2.3	+1.8	1.0	2.3	+3.3	+2.5	1.7	3.3
1.19	1.58	+1.0	+2.6	+2.0	1.0	2.6	+3.1	+2.5	1.5	3.1	+4.0	+3.0	2.0	4.0
1.58	1.97	+1.0	+2.8	+2.2	1.2	2.8	+3.4	+2.8	1.8	3.4	+5.0	+4.0	3.0	5.0
1.97	2.56	+1.2	+3.2	+2.5	1.3	3.2	+4.2	+3.5	2.3	4.2	+6.2	+5.0	3.8	6.2
2.56	3.15	+1.2	+3.7	+3.0	1.8	3.7	+4.7	+4.0	2.8	4.7	+7.2	+6.0	4.8	7.2
3.15	3.94	+1.4	+4.4	+3.5	2.1	4.4	+5.9	+5.0	3.6	5.9	+8.4	+7.0	5.6	8.4
3.94	4.73	+1.4	+4.9	+4.0	2.6	4.9	+6.9	+6.0	4.6	6.9	+9.4	+8.0	6.6	9.4
4.73	5.52	+1.6	+6.0	+5.0	3.4	6.0	+8.0	+7.0	5.4	8.0	+11.6	+10.0	8.4	11.6
5.52	6.30	+1.6	+6.0	+5.0	3.4	6.0	+8.0	+7.0	5.4	8.0	+13.6	+12.0	10.4	13.6
6.30	7.09	+1.6	+7.0	+6	4.4	7.0	+9.0	+8	6.4	9.0	+13.6	+12	10.4	13.6
7.09	7.88	+1.8	+8.2	+7	5.2	8.2	+10.2	+9	7.2	10.2	+15.8	+14	12.2	15.8
7.88	8.86	+1.8	+8.2	+7	5.2	8.2	+11.2	+10	8.2	11.2	+17.8	+16	14.2	17.8
8.86	9.85	+1.8	+9.2	+8	6.2	9.2	+13.2	+12	10.2	13.2	+17.8	+16	14.2	17.8
9.85	11.03	+2.0	+10.2	+9	7.0	10.2	+13.2	+12	10.0	13.2	+20.0	+18	16.0	20.0

Table 19.6. Interference Fits Limits and Deviations (Concluded)

Range of diameters, in.		H7 hole deviation	FN3 heavy drive				FN4 force				FN5 heavy force			
Over	To		t6 shaft deviations		Limits of interference		u6 shaft deviations		Limits of interference		x7 shaft deviations		Limits of interference	
11.03	12.41	+2.0	+10.2	+9	7.0	10.2	+15.2	+14	12.0	15.2	+22.0	+20	18.0	22.0
12.41	13.98	+2.2	+11.4	+10	7.8	11.4	+17.4	+16	13.8	17.4	+24.2	+22	19.8	24.2
13.98	15.75	+2.2	+13.4	+12	9.8	13.4	+19.4	+18	15.8	19.4	+27.2	+25	22.8	27.2
15.75	17.72	+2.5	+13.6	+12	9.5	13.6	+21.6	+20	17.5	21.6	+30.5	+28	25.5	30.5
17.72	19.69	+2.5	+15.6	+14	11.5	15.6	+23.6	+22	19.5	23.6	+32.5	+30	27.5	32.5
19.69	24.34	+3	+20.0	+18	15	20.0	+27.0	+25	22	27.0	+38	+35	32	38
24.34	30.09	+3	+22.0	+20	17	22.0	+32.0	+30	27	32.0	+43	+40	37	43
30.09	35.47	+4	+27.5	+25	21	27.5	+37.5	+35	31	37.5	+54	+50	46	54
35.47	41.49	+4	+30.5	+28	24	30.5	+43.5	+40	36	43.5	+64	+60	56	64
41.49	48.28	+5	+38.0	+35	30	38.0	+53.0	+50	45	53.0	+75	+70	65	75
48.28	56.19	+5	+43	+40	35	43	+63	+60	55	63	+85	+80	75	85
56.19	65.54	+6	+49	+45	39	49	+74	+70	64	74	+106	+100	94	106
65.54	76.39	+6	+54	+50	44	54	+84	+80	74	84	+126	+120	114	126
76.39	87.79	+8	+65	+60	52	65	+95	+90	82	95	+148	+140	132	148
87.79	100.90	+8	+75	+70	62	75	+105	+100	92	105	+168	+160	152	168
100.90	115.30	+10	+86	+80	70	86	+126	+120	110	126	+190	+180	170	190
115.30	131.90	+10	+96	+90	80	96	+146	+140	130	146	+210	+200	190	210
131.90	152.20	+12	+108	+100	88	108	+168	+160	148	168	+232	+220	208	232
152.20	171.90	+12	+128	+120	108	128	+188	+180	168	188	+262	+250	238	262
171.90	200.00	+16	+150	+140	124	150	+210	+200	184	210	+316	+300	284	316

Table 19.7. Nominal Tolerance Grades for Common Manufacturing Processes

Grade

4 *Machine lapping.* Very little stock removed so that prior operations must also be carefully controlled
5 *Surface grinding. Fine boring* using precision boring machines with diamond or tungsten-carbide tools
6 *Honing* of relatively short holes. *Cylindrical grinding,* external and internal. *Broaching* of holes. *Reaming* small holes with lengths to 10 diameters
7 *Honing* of relatively long holes. *Turning* and *boring* using precision equipment. *Surface broaching. Hobbing*
8 *Reaming* large or long holes. *Slotting. Automatic screw machining*
9 *Milling* with careful workmanship. *Vertical* and *horizontal boring* of large parts. *Powder-metal parts,* important diameters
10 *Milling,* general workmanship. *Vertical* and *horizontal boring* of large parts. *Capstan* and *turret-lathe* length dimensions
11 *Drilling, rough turning,* and *boring*
12 *Extruded sections* of brass. *Piercing* and *blanking* holes in sheet metal to 1/16 in. thick
13 *Extruded sections* of light alloys. *Piercing* and *blanking* holes in sheet metal 1/16 to 3/32 in. thick
14 *Die casting* in zinc, dimensions not across parting line. *Piercing* and *blanking* holes in sheet metal 3/32 to 1/8 in.
15 *Blanking* and *forming* operations on stamped parts, *die casting* in light alloys, dimensions not across parting line. *Plaster-mold casting. Forging* length dimensions
16 *Sand casting. Flame cutting. Shearing* of plates

Table 19.8. Tolerance Values for International Tolerance Grades
Values are in thousandths of an inch. Data to line in accordance with recommendations of British-Canadian-American conferences

Range of diameters, in.		Tolerance grades												
Over	To	4	5	6	7	8	9	10	11	12	13	14*	15*	16*
0.04	0.12	0.15	0.20	0.25	0.4	0.6	1.0	1.6	2.5	4	6	10	16	25
0.12	0.24	0.15	0.20	0.30	0.5	0.7	1.2	1.8	3.0	5	7	12	18	30
0.24	0.40	0.15	0.25	0.40	0.6	0.9	1.4	2.2	3.5	6	9	14	22	35
0.40	0.71	0.20	0.30	0.40	0.7	1.0	1.6	2.8	4.0	7	10	16	28	40
0.71	1.19	0.25	0.40	0.50	0.8	1.2	2.0	3.5	5.0	8	12	20	35	50
1.19	1.97	0.3	0.4	0.6	1.0	1.6	2.5	4.0	6	10	16	25	40	60
1.97	3.15	0.3	0.5	0.7	1.2	1.8	3.0	4.5	7	12	18	30	45	70
3.15	4.73	0.4	0.6	0.9	1.4	2.2	3.5	5.0	9	14	22	35	50	90
4.73	7.09	0.5	0.7	1.0	1.6	2.5	4.0	6.0	10	16	25	40	60	100
7.09	9.85	0.6	0.8	1.2	1.8	2.8	4.5	7.0	12	18	28	45	70	120
9.85	12.41	0.6	0.9	1.2	2.0	3.0	5	8	12	20	30	50	80	120
12.41	15.75	0.7	1.0	1.4	2.2	3.5	6	9	14	22	35	60	90	140
15.75	19.69	0.8	1.0	1.6	2.5	4.0	6	10	16	25	40	60	100	160
19.69	30.09	0.9	1.2	2.0	3.0	5.0	8	12	20	30	50			
30.09	41.49	1.0	1.6	2.5	4.0	6.0	10	16	25	40	60			
41.49	56.19	1.2	2.0	3	5	8	12	20	30	50	80			
56.19	76.39	1.6	2.5	4	6	10	16	25	40	60	100			
76.39	100.90	2.0	3.0	5	8	12	20	30	50	80	125			
100.90	131.90	2.5	4.0	6	10	16	25	40	60	100	160			
131.90	171.90	3.0	5.0	8	12	20	30	50	80	125	200			
171.90	200.00	4	6	10	16	25	40	60	100	160	250			

* In British Standard B.S. 1916 only

Table 19.9. Surface-quality Values
Solid bar indicates most common range

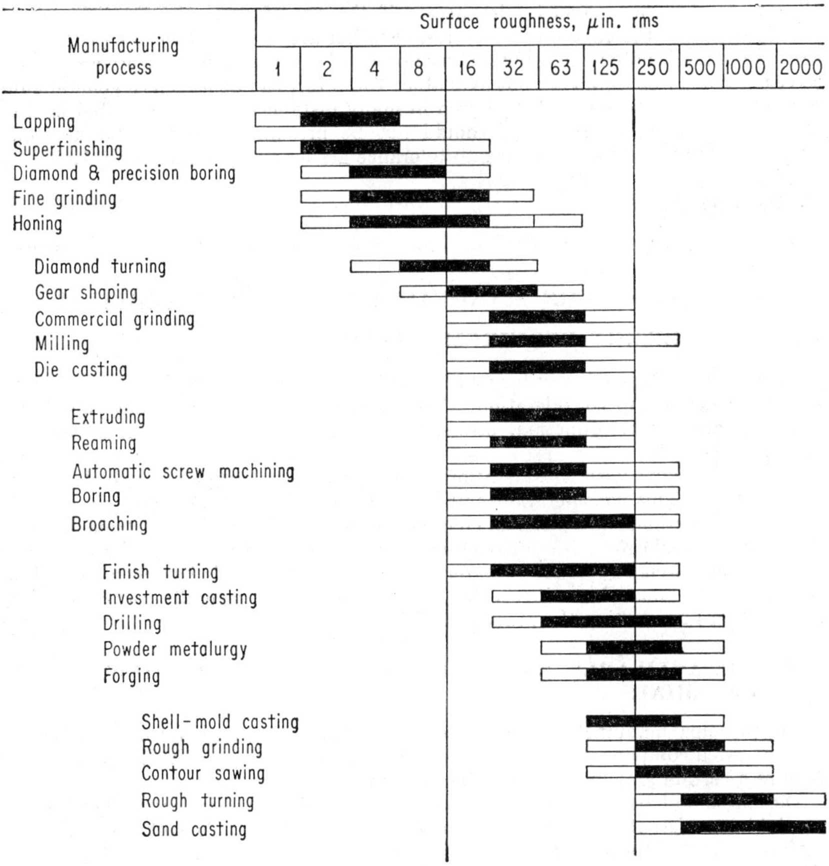

Final selection depends on the actual limits of fit as shown in Tables 19.4, 19.5, and 19.6. These tables are arranged by hole grades and include deviations from which shaft and hole limits are calculated. A typical calculation follows for an H7/g6 fit with a 1.3750 basic size.

	H7 hole		g6 shaft	
Basic size	1.3750	1.3750	1.3750	1.3750
Deviations	0	+0.0010	−0.0004	−0.0010
Limits of diameter	1.3750	1.3760	1.3746	1.3740

19.5. TOLERANCE GRADES AND SURFACE QUALITY

(a) Tolerances

Every machine tool and production process has inherent limitations as to its ability for producing accurate work. These limitations depend on the size, material, and shape of the pieces being made, as well as the machine or process itself. There is, depending on individual circumstances, a characteristic variability that must be allowed for by the assigned tolerances. If the use of particular facilities can be anticipated, and the capabilities of these facilities are known, then design tolerances can be made to match the inherent variability of the machines and processes themselves. When exact data are not available, Table 19.7 can be used as a general guide. Finer tolerance grades may be realized in many instances than those listed in Table 19.7; finer tolerances, however, should not be presumed without careful study.[6] Values of tolerances corresponding to tolerance grades are given in Table 19.8.

(b) Surface Quality

Table 19.9 shows ranges of surface quality for common manufacturing processes.

DESIGNING FOR MANUFACTURING AND MATERIALS*

19.6. BASIC DESIGN REQUIREMENTS[7–12]

To obtain the best solution for a given design problem, various determining factors should be taken into consideration. The most important of these may be divided into two groups: (1) function, stress, and weight; (2) manufacturing, cost, and maintenance.

All design begins with consideration of the function or purpose for which a part is being made. This consideration embraces the principle of maximum simplicity in view of all functional and physical characteristics, and this leads to a stress investigation. A material can usually be selected in conjunction with deciding the significance of component weight. When a material or group of materials has been singled out, the factors of manufacturing, cost, and maintenance generally provide enough information for selecting the proper material.

19.7. THE AVAILABLE MANUFACTURING PROCESS IN VIEW OF THE MATERIALS

The principal manufacturing processes are casting, forging, cold forming, and joining prefabricated components by welding, soldering, and brazing. Obviously most parts require a certain amount of machining for finishing.

Design should be based upon the most economical production method or methods available consistent with other limitations or conditions. To suit the production method as well as the other basic requirements, the material must be properly selected. The most economical production method will result in a minimum number of separate operations in machining, forming, molding, casting, or assembling. The specification of finish and accuracy should not be more stringent than necessary for function and must be capable of achievement with the production methods contemplated.

(a) Design with Regard to Material

All materials have inherent characteristics which must be taken into account for the most satisfactory design solution.

Cast Iron. Various types of cast iron are available: gray cast iron, white cast iron, malleable or ductile iron, Meehanite. For these materials the ultimate strength and yield point in compression are several times greater than in tension. Gray cast iron should be selected for parts which are primarily subjected to compression forces, and the design of those parts can be arranged accordingly (Fig. 19.10). Where rigidity is required, proper stiffening by ribs under compression can be easily attained (Fig. 19.11).

The microstructure of cast iron shows the typical discontinuities associated with the distribution of graphite within the ferrite base metal. This explains its vibration-damping characteristics, which can be used advantageously in many designs. Fur-

* By Carl H. Ringe.

thermore, cast iron has a low sensitivity to stress concentration resulting from notch effects as in keyways.

In designing for cast iron, the shrinkage which occurs during solidification and cooling must be considered. The material should be distributed as uniformly as possible. Where sections of greatly differing thickness meet, stress concentrations resulting from the material contraction during cooling occur and there is a strong likelihood of material failure. Thus the arrangement of thin ribs or webs between thicker flanges constitutes poor design. Tapering of casting walls (draft) for ease of

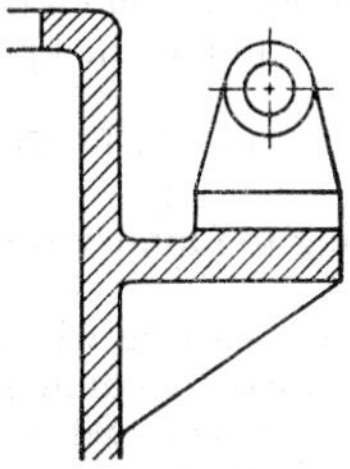

FIG. 19.10. Brackets.

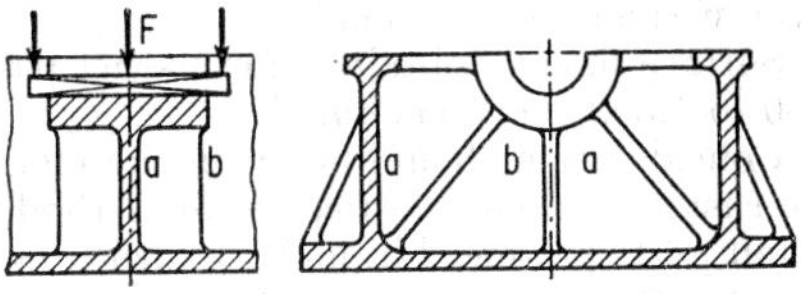

FIG. 19.11. Examples of proper stiffening.

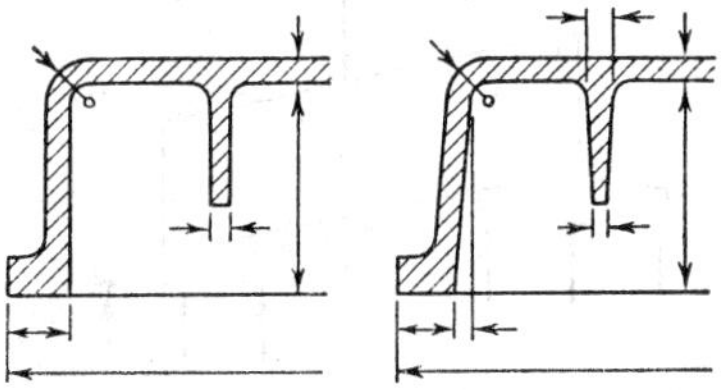

FIG. 19.12. Draft.

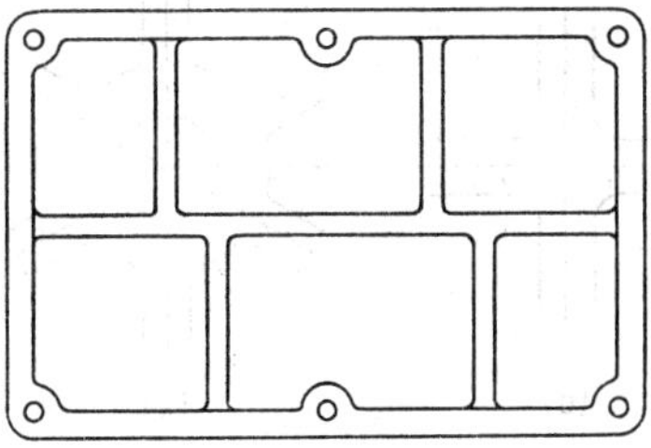

FIG. 19.13. Ribs on a flat plate.

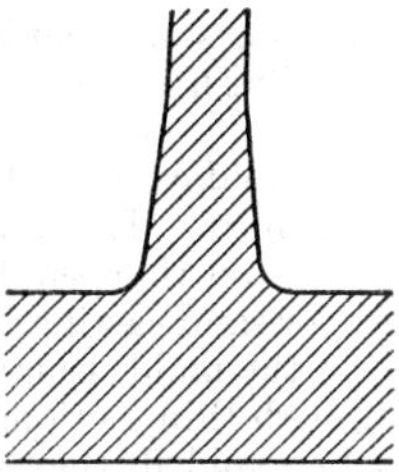

FIG. 19.14. Transition from wall to plane.

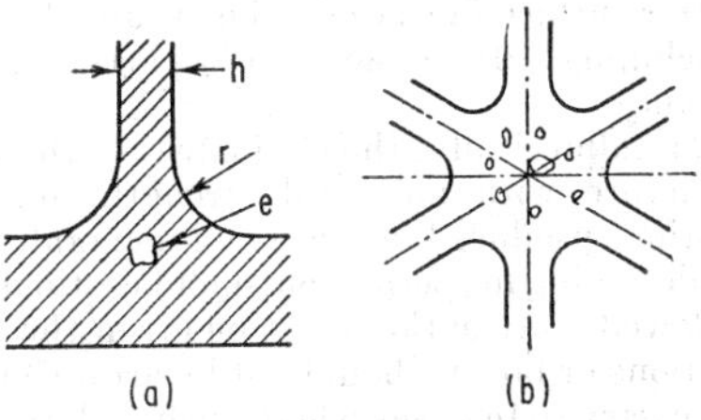

FIG. 19.15. Casting imperfections. (*a*) Cavity at *e*. (*b*) Shrinkage voids.

molding also serves a useful purpose with regard to joining sections of different thicknesses. The molten metal flows more easily from the larger section into the smaller section, thus decreasing the stress at the point of juncture (see Fig. 19.12). A good example of stiffening ribs for a flat plate is shown in Fig. 19.13. Another example showing a proper transition from wall to flange in a T-shaped casting is shown in Fig. 19.14. Local accumulation of material at intersections leads to formation of cavities caused by material shrinkage. The narrower sections solidify faster, drawing some of the still liquid material from the heavier portion of the intersection. The consequences can be seen in Fig. 19.15*a* and *b*. Figure 19.16*a* and *b* indicates ways in which these faults can be avoided.

Several types of cast iron are available. Great hardness and wear resistance can be obtained by chilling (white iron); however, white iron is brittle and not shock-resistant.

By proper heat-treatment of gray iron having low silicon content, castings can be produced with increased tensile strength and ductility. Cast irons with such characteristics are known as *malleable iron* and *ductile iron*. These types are suitable for members stressed in tension or bending (Fig. 19.17). The ability of malleable and ductile iron to absorb shock exceeds that of steel; the impact resistance is high, the notch sensitivity low. However, the design considerations which pertain to cast iron must not be neglected, e.g., size of corner fillets, transitions from a thick to a thin section.

The practical limits for section thickness of sand castings are: for small parts ⅛ in.; medium-sized parts ¼ in.; heavy sections three-fourths maximum thickness of solid sections for malleable iron, practically about 4 in.

Attainable tolerances for cast-iron parts, as cast, are $\frac{3}{64}$ to $\frac{1}{32}$ in./ft of length, even if the parts are small.

Surface finishes obtained for as-cast parts vary according to foundry practice from 250 to 1,000 μin. (very rough).

Consideration should be given the method of molding and casting when designing for cast iron. The most common method is sand casting for castings of all sizes. For small parts produced in high quantity, permanent molds are used. Since molds and cores of these permanent molds are made of fine-grain cast iron, closer tolerances and

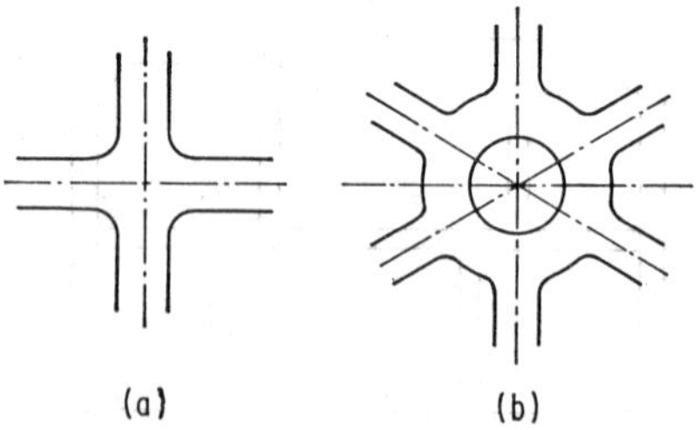

FIG. 19.16. Properly designed intersections.

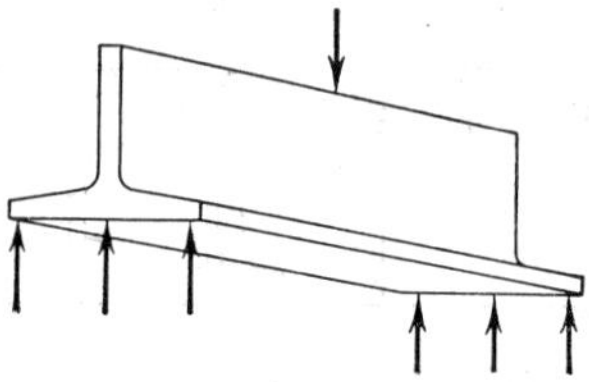

FIG. 19.17. Tension in plate—compression in rib.

better surface finishes can be obtained, and often the surface does not require further machining. Permanent-mold parts have an as-cast surface-finish range from 250 to 32 μin.

Metallurgically, the chill effect of the metallic permanent mold causes a finer grain structure with increased strength, uniformity, and soundness. The strength of castings produced in this way is generally 10 to 20 per cent greater than sand castings. In designing for permanent-mold castings, the general rules for sand castings should be followed. Attention must be given the flow of the metal in the mold; isolated heavy sections or bosses should not be arranged so that filling through thin sections becomes necessary. In cases where such isolated sections cannot be avoided, they should be placed at the mold parting line, or heavy ribs should be provided to assure full flow to the isolated sections. These precautions serve to prevent shrinkage cracking and internal stresses.

Metal cores must be withdrawn before complete solidification occurs in order to prevent shrink cracking. If the casting design does not permit this withdrawal or if the casting is large, sand cores rather than metal cores are used. This process is known as *semipermanent-mold casting*.

Nonferrous Metals. The nonferrous metals can be grouped in two principal classes: (1) copper-alloy metals, usually having a higher specific gravity, and (2) lightweight metals and their alloys.

The metals in group 1 are recommended when it is impractical, impossible, or more costly to obtain desired shapes and properties by other means. The commonly used alloys permit casting or irregular and often complex external and internal contours. The lightweight metals such as zinc, aluminum, magnesium, and their alloys can also be formed by the various casting processes.

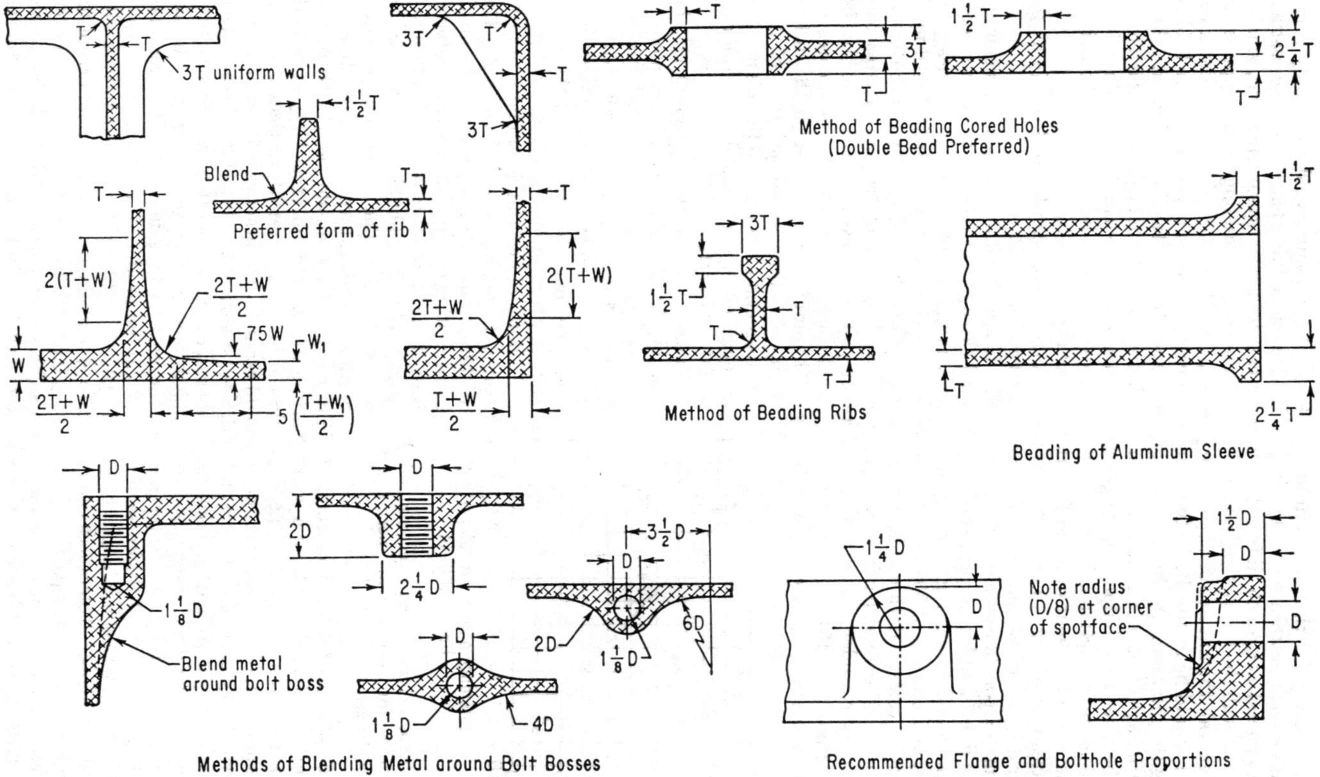

FIG. 19.18. Typical design details of aluminum-alloy castings.

Sand casting for the metals in groups 1 and 2 is recommended when only a few castings of one pattern are to be produced. Design requirements for these castings are in general not much different from good cast-iron practices. However, smaller minimum wall thicknesses can be obtained than in cast-iron castings. In copper-alloy castings wall thicknesses of 3⁄32 in. are possible, except in cases of extended flat areas (which should be avoided). Minimum thicknesses of aluminum parts can be held to 1⁄8 in.; the minimum thickness for magnesium and most of its alloys is 5⁄32 in. Since the shrinkage factor of the nonferrous metals is greater than that of cast iron, attention must be given this characteristic in design. This includes consideration of directional solidification, reduction of abrupt changes in cross sections, liberal fillets and provision for the best placement of gates and risers, thus avoiding internal shrinkage cavities and stresses (see Fig. 19.18). Tolerances of sand castings can be held to 3⁄32 in. for copper alloys and 1⁄32 in. for aluminum and magnesium alloys.

Permanent-mold castings not exceeding 48 in. in any direction offer a much smoother surface finish and higher precision than do sand castings. However, the process does not lend itself to casting small quantities. A range of 200 pieces, if the design is simple, and upward of 500 pieces, is generally considered the lower limit, based upon production costs. Both molds and cores are made of cast iron. The mold consists in most cases of several parts which swing or slide for rapid operation. The casting design must consider core withdrawal before complete solidification, in order to obviate shrink cracking. If the casting is too large or the shapes are intricate, sand cores are usually necessary, as in cast-iron molding. The improvements in strength, surface finish, and dimensional accuracy of nonferrous-metal parts realized by this method are even more apparent than in the case of cast-iron parts.

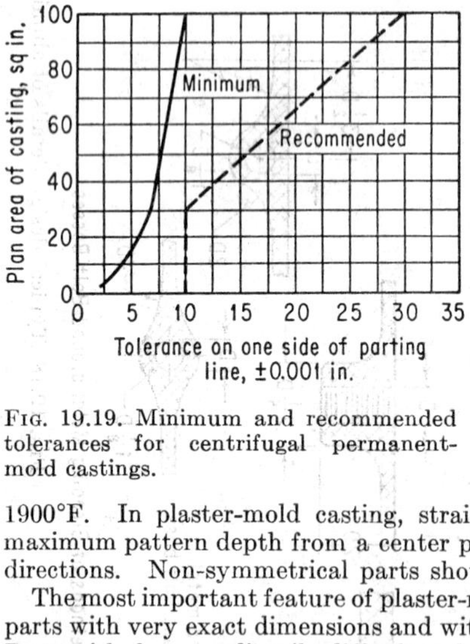

Fig. 19.19. Minimum and recommended tolerances for centrifugal permanent-mold castings.

Dimensional tolerances generally may be held in the fixed cavity of the mold to within ±0.020 in. and across the parting line to within ±0.030 in. On parts of smaller sizes and of simple design, these tolerances may be reduced to 0.010 and 0.020 in., respectively (see Fig. 19.19).

Plaster-mold casting is in many respects related to permanent-mold casting. The limitations of the process lie in the maximum casting size and the melting temperature of the material to be cast. The maximum mold size should not exceed 24 by 36 by 12 in. The maximum practical casting temperature is approximately 1900°F. In plaster-mold casting, straight parting lines should be provided, and maximum pattern depth from a center parting line should not exceed 1½ in. in both directions. Non-symmetrical parts should not have a greater depth than 2 in.

The most important feature of plaster-mold casting is the practicability of producing parts with very exact dimensions and with as-cast surface finishes of 30 μin. or better. Parts with shapes ordinarily difficult and expensive to machine can be readily produced to such exacting dimensions that further operations are often unnecessary. The low thermal conductivity of plaster permits the casting of thin walls. In areas not exceeding 2 in.^{2}, a wall thickness of 0.040 in. can be produced; for larger areas up to 4 to 6 in.^{2}, a practical minimum thickness is 0.062 in. Blade sections can be cast with knife-sharp edges. Tolerances can be held to ±0.005 in./in. of dimension, except for dimensions crossing the parting line, where ±0.010 in./in. is the tolerance. Plaster-mold casting has been used for alloys such as beryllium copper with tensile strengths up to 170,000 psi and hardness up to 46 Rockwell C, silicon bronzes, nickel bronzes, nickel copper, high-grade tin bronzes, and silicon-aluminum bronzes.

Centrifugal casting is performed by true centrifugal, semicentrifugal, and centrifuge casting methods. In true centrifugal casting, the mold revolves outside the axis of rotation, usually in a horizontal position. This method is used for castings which are internally cylindrical. For securing the necessary balance, the outside of the cylindrical body must generally be symmetrical around the axis. Flanges and small bosses are essentially the only possible modifications. The need for radii adjoining flanges, fillets, etc., is not so pronounced as in sand castings, but some fillet radii always should be specified in order to avoid sharp corners. The minimum wall thickness of a tubular casting can be held to approximately 0.25 in. Maximum thicknesses of 5 in. are obtainable. Centrifugal casting produces a dense product of great uniformity, in which impurities such as dirt, sand, or slag are forced to the inside of the casting because of their lower density, where they can be completely eliminated. This casting process is used in the manufacture of pressuretight parts and for components subject to high loads and stresses. Physical properties approach those of forgings, and in many cases, parts can be produced by the centrifugal casting of metals which are

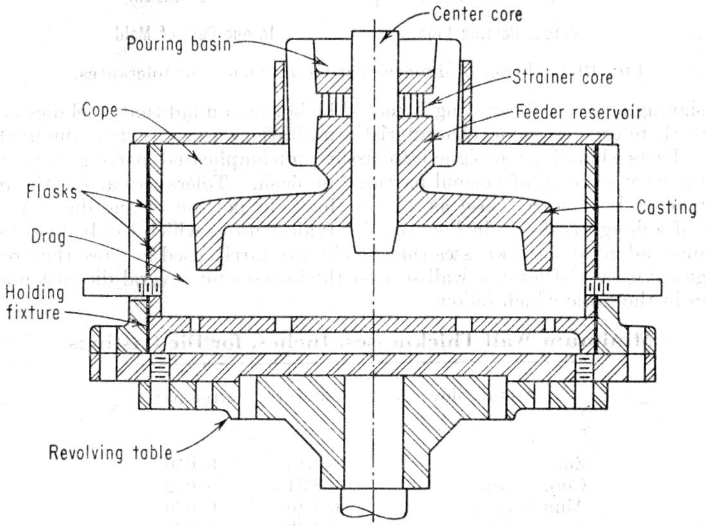

FIG. 19.20. Semicentrifugal casting.

extremely difficult or impossible to forge. The casting centrifugal force varies within the range of 50 to 100 times the acceleration of gravity.

The semicentrifugal method is almost identical to true centrifugal casting. The difference is in that the mold axis is vertical and that a center core is used (see Fig. 19.20). A small-diameter center bore should not be cored because the metal in the center has a poorer quality than that at the periphery. Otherwise, all kinds of cored center bores can be produced in wheel-type castings of superior quality which are very difficult or even impossible to cast by other methods.

Centrifugal forces also are used in the centrifuge casting method. Nonsymmetrical parts which cannot be rotated about their own axis can be arranged in multiples on a vertical "tree." Design of such parts should follow sand-casting procedures closely although the parts should be designed properly for the centrifuge casting method in order to obtain the maximum benefit of the uniformly dense metal structure, which in the case of cast iron approximates steel quality. It is important that the complete mold is balanced, and parts must be arranged so that a directional solidification of the metal in the mold can be obtained. This can sometimes be provided by adding material in the center portion.

Generally, all metals can be cast by this method. The properties of sand castings made by this method are up to 10 per cent better than those made by stationary casting, whereas as much as 40 per cent improvement can be realized by using permanent-mold practices.

Minimum wall thicknesses are approximately 0.25 in. and maximum thicknesses may not exceed 5 in. Metal allowances for machining can be kept to ⅛ to $\frac{3}{16}$ in. Surface roughness is the same as for stationary casting.

The die-casting process is used where relatively large quantities of parts are produced. A wide range of basic metals and their alloys can be advantageously used, ranging from zinc alloys with casting temperatures up to 750 to 800°F to copper-base alloys with casting temperatures up to 1900°F. The maximum size of die castings is 24 in. in any direction, with occasional exceptions.

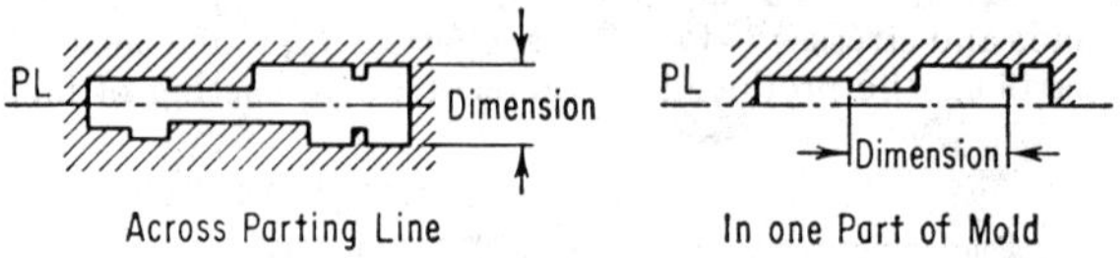

FIG. 19.21. Types of dimensional requirements for tolerances.

In designing parts for die casting it should be borne in mind that steel dies are used and that therefore the same characteristic limitations exist as in permanent-mold casting. Parts should be designed to assure uncomplicated parting lines and to permit the easy removal of normal parting-line flash. Tolerances across the parting line can never be held as close as within the solid portion of the die. A typical example of a design is shown in Fig. 19.21. Wall sections which can be held to small dimensions and greater thicknesses than 0.5 in. are rarely used because they result in shrinkage cavities. Minimum wall-section thicknesses for several die-cast materials are given in the table which follows.

Minimum Wall Thicknesses, Inches, for Die Castings

Metal alloy	Small parts	Large parts
Zinc	0.015	0.050
Copper base	0.031	0.062
Aluminum	0.040	0.080
Magnesium	0.040	0.062

Cored holes should not be deeper than about five times the diameter for zinc alloys and about three times the diameter for aluminum, magnesium, and copper-base alloys. Minimum core diameters can be assumed to $\frac{3}{32}$ in. for aluminum and magnesium, $\frac{3}{16}$ in. for copper-base alloys, and $\frac{1}{32}$ in. for zinc alloys.

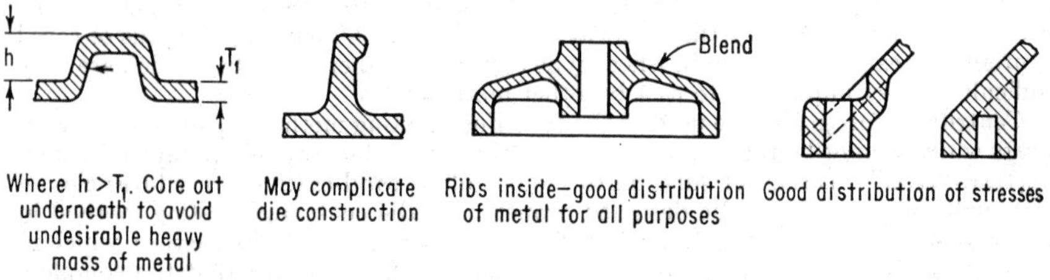

FIG. 19.22. Design guide lines.

Undercuts and recesses should be avoided if possible; external threads over ¾ in. diameter and coarser than 24 pitch can be cast but usually add substantially to the die cost. Internal undercuts are very difficult to cast and should be avoided. Large flat surfaces are not easy to obtain without imperfections. Generous drafts should be provided for better removal of the parts from the die. This also applies to properly arranged ribs, which are useful in distributing stress. Sharp corners must be avoided; fillets which are too large, however, may create shrinkage voids. Figure 19.22 presents some design guide lines.

As-cast tolerances can be held to an approximate minimum of ± 0.002 in./in. within the solid die for the light metals and zinc, 0.005 in. for copper-base alloys. The tolerances across the parting line are normally twice the values indicated.

Steel Castings. Steel castings differ from gray-iron castings in various aspects. Steel in its liquid form has poor fluidity relative to other metals. Therefore, wall and section thickness cannot be made less than ¼ in. The volumetric contraction which occurs during solidification is comparatively high and must be considered in the design. The stresses which result when this contraction effect is overlooked in design often cannot be entirely removed by normalizing. A properly designed steel casting can be lighter in weight and stronger than castings made of other metals. Cast steel has excellent impact resistance and does not show directional properties as does wrought steel.

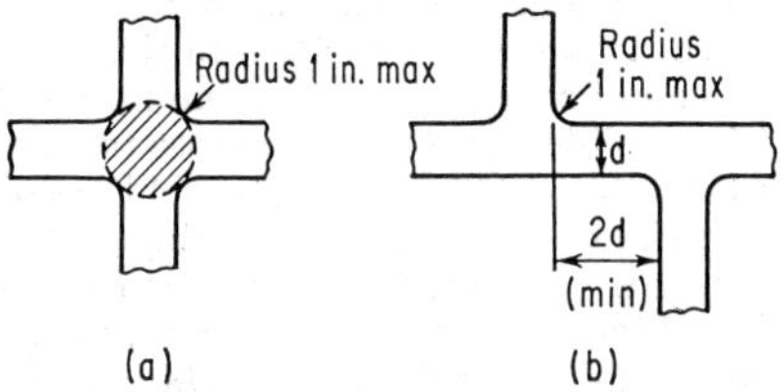

Fig. 19.23. Steel castings. Mass of metal at junction.

In designing for steel casting, the general design considerations such as uniform thicknesses, gradual sections, and avoidance of sharp corners must be given attention.

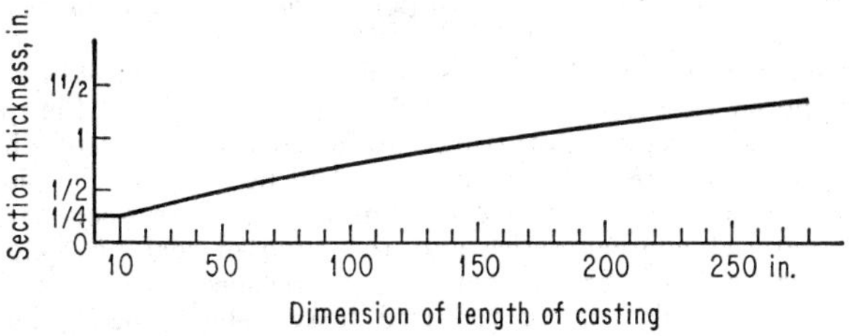

Fig. 19.24. Minimum recommended thicknesses.

Complicated designs do not lend themselves to a one-piece casting; it is better to divide such parts into simple sections which can be joined at assembly. Accumulation of material in areas where ribs or sections are joined must be avoided; such junctions tend to have shrinkage cavities which initiate cracks. Fillet radii must therefore not be too large (see Fig. 19.16*a* and *b* and Fig. 19.23*a* and *b*).

Tolerances for parts which are not be be machined may be taken from the table below, where D is the longest dimension of the casting (in inches). Minimum thicknesses for sections may be taken from Fig. 19.24. Steel castings of intricate design can be advantageously produced by the investment-casting process. All types of alloy steel as well as essentially all metals which can be melted are being cast by this process. In general, the higher grades of alloy steel including stainless steels are easier to cast than carbon steel. Heat-treatments for metals which can be hardened or annealed will not result in warping of the parts. The process is best suited for small- to medium-sized parts usually not exceeding approximately 15 lb.

	Under 12 in.	12–36 in.	36–120 in.
Average..........	0.06–0.006D	0.06–0.006D	0.08–0.006D
Minimum........	0.03–0.004D	0.04–0.004D	0.06–0.004D

Parts cast by this process have very good finishes, generally 70 to 80 μin.; with plastic patterns, 10 to 20 μin. can be obtained. Thus further machining can in many instances be eliminated, except for the reaming of bores, threading, etc. For dimensions over 1 in., the as-cast tolerance can be held to ±0.005 in./in. Under 1 in., ±0.003 in./in. can be held. Uniform sections from 1⁄32 to 1⁄4 in. are ideal; heavy sections exceeding 1⁄2 in. and not larger than 1 in. should be avoided. Strength and stiffness should be achieved by the use of ribs instead. Where sections vary in thickness or in areas of transition to ribs, the transition should be gradual. Thin edges at the end of a tapered section can be cast to a minimum thickness of 0.012 to 0.015 in.; it is possible in exceptional cases to cast a trailing edge of 0.005 in. on a thin section. Edge radii of 0.005 in. can be held and fillets should be at least 1⁄16, preferably 1⁄8 in., if possible. Contrary to all other casting methods, a draft on walls or sections is not required. Through holes can be cast, preferably not exceeding a depth about three times their diameter. Blind holes should be avoided. Complicated internal passages are difficult to mold and should be avoided. Flat, uninterrupted areas, larger than approximately 10 to 12 in.^{2}, are difficult to produce without imperfections and are preferably designed with small ribs or steps. Fine serrations or external threads may be produced by the frozen-mercury process but should be avoided and left for machining.

Characteristics and Influence of Hardening and Annealing Steels. Wrought carbon steels and alloy steels are used for members subjected to higher stresses. Internal stresses in these materials are induced by several operations: cold working, heat-treatment, including welding and electroplating. Attention must be paid to removing or at least reducing these stresses where they are not desirable. In other instances stresses are intentionally induced in order to increase the strength (as in the prestressing of beams).

All steels are more or less sensitive to stress concentrations caused by notch effects. Many of these notch effects can be avoided by proper design. Sharp corners and inadequate fillets and transitions contribute to fatigue failure. Holes in plates, keyways in shafts, and notches in elements stressed in bending and torsion are typical stress raisers and must be carefully considered in design. Stress-concentration factors depend upon material characteristics and the type of loading. Generally these factors are higher for hard, brittle materials, decreasing with increasing ductility. Stress concentration is highest for repeated heavy shock loads, decreasing with decreasing shock intensity and reaching a minimum when the load is more or less static. Material sensitivity is expressed by a sensitivity factor which is discussed in Sec. 18.

One of the most important factors in machine structure is the existence of residual stresses. These stresses occur in most members and are often detrimental. They may be related to the negative factors of (1) warping and distortion after heat-treatment and machining, (2) cracks due to quenching and grinding operations, (3) stress-corrosion season cracking, (4) cracking in cold drawing, and (5) premature failure of tools, dies, and welded members. Residual stresses are originated in three basic ways: (1) cold working, i.e., shot peening, rolling, or cold drawing; (2) heating or cooling, i.e., heat-treatment, surface hardening, and welding; and (3) electrodeposition as in chrome plating. Residual stresses are considered as internal body stresses and are of two types: one type is distributed over appreciable areas, and the other varies within and around the grain.

It has been found by X-ray measurements that understressing or overstressing a material can increase its endurance limit by as much as 35 per cent. Also, it has been shown that existing residual stresses in a member will fade under repeated stressing.

Recent work indicates that surface conditions have a significant effect on fatigue strength and a negligible effect on the static tensile strength of a material. If the ultimate is to be obtained from an alloy, the surfaces of mechanical parts must be treated with care in formation and use. This does not mean that the best finish (low surface roughness) is always applied, but rather the most suitable and economical methods of surface condition should be adopted. In some cases better results are obtained with a relatively rough surface given the proper surface treatment than by fine finishing or polishing. Among the surface treatments which markedly affect the fatigue strength are nitriding, carburizing, flame or induction hardening, shot peening, sand blasting, and cold rolling.

Where deflections, bending or torsional, must be kept within given limits, all steels are of equal value for design consideration because the modulus or elasticity for all steels is approximately the same. Certain alloy steels, preferably of the Cr-Ni and Cr-Mo classes, excel in toughness. These are successfully used for gears, for example. Other alloy steels are used for tools and dies, particularly owing to their ability to withstand shock loads, their wear resistance, and their ability to withstand hardening-process distortion. Hot-rolled steels can generally be better formed by the various processes than can cold-rolled steel. The latter has the advantage of a higher ratio of yield strength to tensile strength but has in many instances internal stresses stemming from the cold-working process, which may cause trouble, as, for example, where these stresses induce warping in long keyways.

Fatigue resistance of steels, especially alloy steels, can be improved by hardening and annealing. These processes improve the tensile strength, although to some degree at the expense of ductility. Not all steels, especially plain-carbon steels, can be improved by hardening. Alloy steels can be heat-treated and hardened to good advantage. The notch sensitivity usually increases with hardening. Hardening cracks can be avoided by proper design such as avoidance of abrupt transitions. Fatigue resistance can be improved by adequate design, especially by avoiding buildup of stress concentrations rather than using high-grade steels and heat-treatments. Various methods of heat-treatment and hardening which can be applied to hardenable steels corresponding to the desired result, depth of hardened zone over an annealed tough core, through hardening, etc., are available.

Because of their elastic nature, steels do not absorb shock as do cast iron or certain light metals. In systems where vibrations can result in fatigue-induced failures or excessive stress caused by resonance, consideration must be given the proper distribution of design elements in critical areas to prevent resonance.

Plastics. Plastics can be classified in two groups, thermosetting and thermoplastic. To the former group belong the phenolic resins, mostly reinforced by additive materials, such as fibrous substances. These resins are preferably used for producing laminations, which may or may not be postformed by applying moderate heat and pressure. The design for these plastics can exploit their toughness, wear resistance (silent gears), and light weight. Such parts do not require lubrication and must not be lubricated by oil or grease. Shock absorption is good, but parts must be well supported except in cases where loads are light. Parts are easily machined. In cases where pressure molding is economically justified, the reinforcing substances can be arranged in the direction of the analyzed stresses (Fig. 19.25).

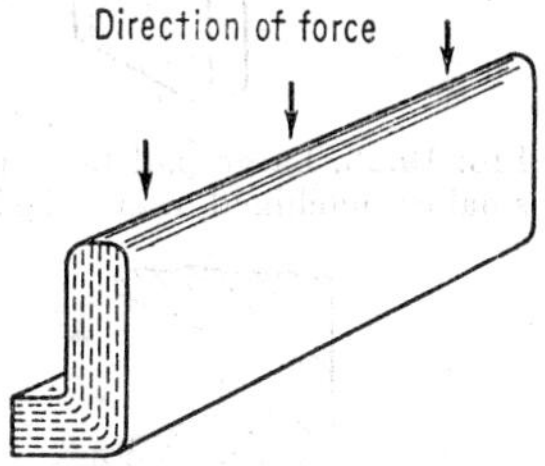

FIG. 19.25. Example of laminated thermosetting, postformed; so-called cheek woods in cutting cylinders of rotary-printing-machine folders.

The variety of materials of the thermoplastic group is very great. These materials usually are molded by pressforming under heat from raw materials in pelletized or powder form. They are also available in extruded shapes, such as bars, rods, tubing, squares, or in sheet form (maximum thickness of slabs usually not exceeding 2 in.). They have a homogeneous structure, are easy to machine, have high dielectric strength, corrosion and chemical resistance, excellent surface condition, and stability of form. In designing plastic parts, it is of paramount importance to keep wall thickness as uniform as possible, particularly for laminated thermosetting materials. Nonuniformity of sections may result in warpage or nonuniform shrinkage.

In compression molding, a desirable wall thickness is generally 1/16 in. Thick sections are more difficult to mold. Large, unsupported walls should be reinforced by ribs for strength and warpage resistance. Sufficient taper or draft on all surfaces normal to the parting line of the mold must be taken into consideration. Average draft values range from ½ to a maximum of 5°. Generous fillets and radii add strength and facilitate molding. Metallic inserts, such as pins, hubs (in gears or wheels), lugs, and bushings, are easily molded into the plastic. Sufficient material is required around the inserts to withstand shrinkage upon cooling without cracking. Material wall thicknesses around inserts vary with the diameter of the insert and with the type of plastic and approximates ½ to 1 times the diameter. Parting lines should be carefully considered in design so that flash can be removed with a minimum of finishing. Straight parting lines are preferred where a design modification cannot be made for permitting molding in one plane.

Tolerances on dimensions parallel to the parting line can be held to an approximate minimum of ±0.003 in., whereas the minimum tolerance on dimensions normal to the parting line is not better than +0.010, −0.000 in. Thermosetting plastics provide the greatest accuracy and stability of dimension. Closer tolerances require machining or grinding.

As-molded surface finish is generally good. Special surfaces such as high gloss or satin finishes are not difficult to obtain.

(b) Design for Manufacturing

Although almost all materials can be machined, certain general design rules should be adhered to: Large pads or flat areas should always be interrupted (see Fig. 19.26). Improved joint rigidity can be similarly achieved (see Fig. 19.27). Bosses are

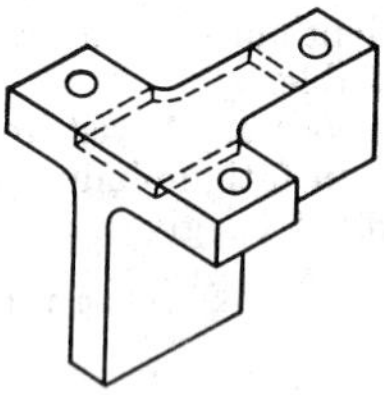

FIG. 19.26. Large pad broken up into three smaller machined areas (dashed lines).

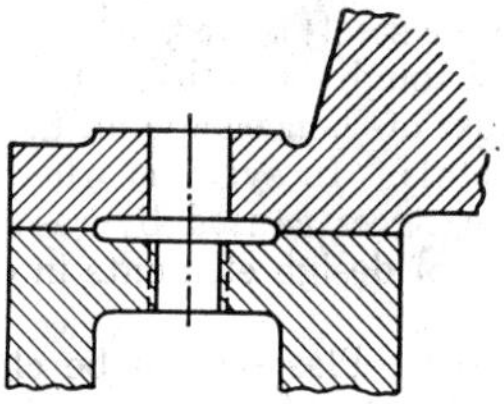

FIG. 19.27. Improved joint rigidity.

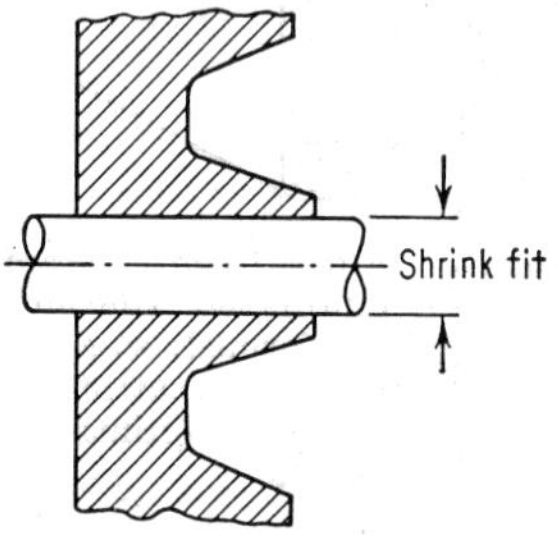

FIG. 19.28. Stress concentration in shrink fit reduced by hub design.

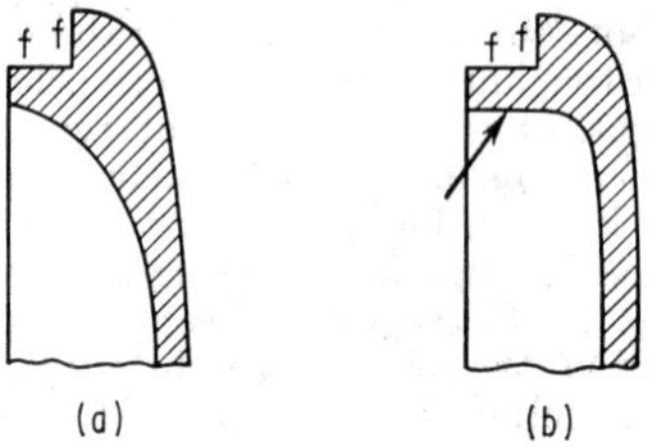

FIG. 19.29. Provision for chucking for turning seating areas on a cover. (*a*) Without. (*b*) Improved with facility for chucking.

generally better than spot-faced areas, especially where stress-raising sharp corners should be avoided. In general, it is better to avoid designs in which stresses can build up. For example, shrink-fit stress concentration can be materially reduced by making the gripping part conical (see Fig. 19.28). Provision should be made for auxiliary pads or bosses on parts for clamping on machine-tool tables. This also is true for chucking of pieces to be turned (see Fig. 19.29). Adequate dimensions must be provided for the tools to cut free at the end of a machined area; otherwise toolmarks can result which lead to fatigue failure.

Whenever it is possible to preform pieces by hot forming (other than casting), the strength of the material can be improved. One typical example is the die-forged head of a machine screw with its characteristic grain flow. Free forging, use of setting tools, is costly and merely used for single parts. Where the cost of dies can be justified by quantity, this manufacturing process can be advantageously applied for both reducing subsequent finishing costs and improving strength.

Practically all steels and the forgeable light-metal alloys can be chipless formed by die forging. Good forging design is influenced to a considerable extent by the manufacturing process. A knowledge of the forging process, die-construction requirements, and the forgeability of the alloy will help to develop sound basic design. One of the most important rules in connection with design for forging refers to draft angles. A draft angle ranging from 7 to 10° should be used for all steel alloys and carbon steels; for hammer forgings of copper alloys, an angle of 3 to 7° is recommended, and for press forging, 1 to 5°. For aluminum and magnesium and their alloys, a common draft angle for both hammer and press forgings of 5 to 7° is used; in some cases (provision of

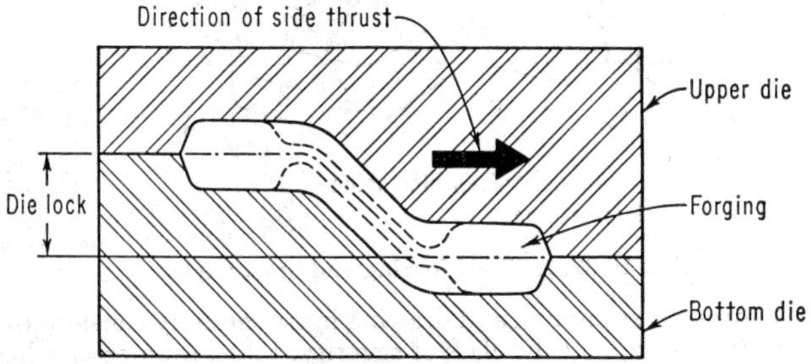

FIG. 19.30. Example of irregular or dislocated halves—impractical. Side thrust makes it impossible to hold the dies in match accurately.

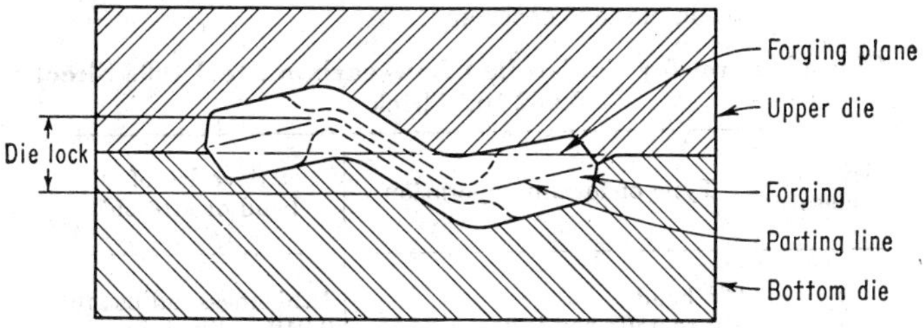

FIG. 19.31. Example of offset lever. The best method is to incline the forging with respect to the forging plane.

gripper dies and stripper lugs in mechanical upsetting machines is necessary) a draft as low as $\frac{1}{4}$° may be permissible. In general, the design of die-forged parts must consider the flow of the material. Therefore, corners should be generously rounded and fillet radii should be larger than in castings.

Another important factor in design for forging is proper parting-line determination. Straight parting lines are preferred; irregular parting lines should be avoided. Grain flow, trimming, draft requirement, and die costs also influence the design. Undercuts anywhere in the forging are impossible because the die halves must be permitted to separate and the forging must be removable from the die. Normally the parting line is at the center of a web which surrounds the piece. If the web is less than $\frac{3}{16}$ in., the parting line should coincide with one surface of the web. Irregular parting lines necessitate costly counterlocks or locking pins in the dies, and when the die starts to wear, irregular or dislocated halves occur (Fig. 19.30). An example of an offset lever is shown in Fig. 19.31, where side-thrust forces are avoided by inclining the forging in the die, so that the forging plane and the parting line are substantially parallel. Thus a counterlock is unnecessary (Fig. 19.31). In many cases, similar to the one shown in Fig. 19.32, favorable draft-angle solutions can be attained at the same time.

Cold forming of metal parts having uniform wall thickness, usually not exceeding 0.050 in., can be performed by either bending, braking, rolling, drawing, or pressing. The materials available for these processes are in sheet or strip form, either cold rolled or hot rolled, and include low-carbon steels, certain stainless steels, most ductile copper-base alloys, and aluminum and magnesium and their alloys. Profiled sections in straight form for an almost unlimited number of shapes can be made. Radii should not be smaller than 0.010 in., and in cases of quarter-hard cold-rolled steel not less than the thickness of the sheet when bent parallel to the rolling direction, half the thickness of the sheet for bends perpendicular to the rolling direction.

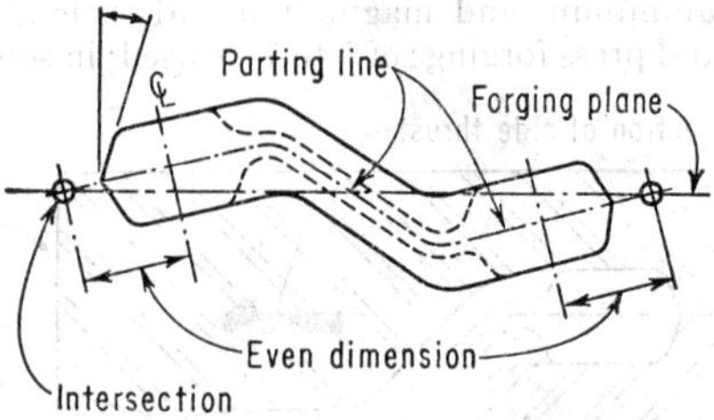

FIG. 19.32. Inclined forging and draft angle.

Design for die forming, stamping, and drawing must carefully consider the material, particularly the elongation, and the class of severity, taken from established tables of the severity classification in comparison with similar existing parts. The severity classifications can be taken from the table below. In all cases of stamping and drawing it is recommended that prototypes be made before actual production so that design modifications can be made if necessary.

Severity Classification for Cold-rolled Low-carbon Steel 1008 Sheet up to 0.062 in. Thick

Severity class	Type of forming operation	Severity of bend or stretch*
1	90° bend	1 thickness radius min
2	Up to 180° bend, drawing	0.010 radius min
3	Drawing	10–20%
4	Drawing	20–30%
5R	Drawing	30–35%
5K	Drawing with possible buckling	30–35%
6K	Drawing with possible buckling	35–40%

* The values shown are based on the greatest percentage stretch of the metal as measured between lines which before forming are marked on the blank as a grid pattern of 1-in. squares.

The usual dimensional tolerances are $\pm 1/32$ in.

For the mass production of parts with punched holes, notches, shallow draws, etc., progressive die sets can be designed and arranged for automated production from coiled strip. In many instances, parts can be redesigned from casting to stamping with considerable cost reduction for medium production runs of from 10,000 to 50,000 pieces.

Many complicated machine parts which can only be cast become easier to manufacture by assembly of prefabricated components. It is often better (from the point of view of strength and weight) to divide the design into simpler segments which may be made of cast steel, forged or machined bar stock, angle bars, or the like, which then can be welded, brazed, soldered, or otherwise joined. Strong plastic adhesives (epoxy resins) sometimes provide an adequate bonding for fabricating strong lightweight structures.

When the weight of machine parts is a critical requirement, advantage can be taken of high-strength material which can be preformed by forging or casting into components or which is available in prefabricated shapes eventually to be joined by welding. The same pertains to machine elements which are exposed to corrosive conditions or to high temperatures. In such cases, use can be made of laminations or combinations of corrosion- (or heat-) resistant but weak materials with high-strength materials with inadequate corrosion- or heat-resisting properties.

(c) Design Considerations for Production

In addition to the mechanical and technological considerations, every design must be viewed in the light of production requirements. Tooling should be taken into account, whether this relates to accessibility for machining, to clearance and adequate runouts for machine tools, or to jigs and fixtures. The latter frequently influence the design with respect to the machinability of the material and to specific requirements for mass production. Since machine elements and detail parts invariably work with or are fitted with other component members, these machine elements must be manufactured to proper tolerances, neither too close nor too loose. Consideration must be given the compatibility of tolerance with material. This is also true of surface finish. Finishing processes in general require designs which are consistent with the pertinent operations such as grinding and polishing; otherwise costly hand finishing becomes necessary. Finishing requirements should not be exaggerated.

Hardening processes and the behavior of various hardenable materials must be carefully considered in the design. Warpage cannot usually be avoided, and therefore, the design must be made in such way that warpage in certain directions can be anticipated and provision made for easily correcting the deformation by a final machining operation, e.g., grinding.

With respect to costs, emphasis is placed upon simplicity, functionality, and facility of manufacture.

References

1. Evans, D. H.: Optimum Tolerance Assignment to Yield Minimum Manufacturing Cost, *Bell System Tech. J.*, vol. 37, no. 2, pp. 461–484, March, 1958.
2. Pike, E. W., and T. R. Silverberg: Assigning Tolerances for Maximum Economy, *Machine Design*, vol. 25, no. 9, pp. 139–146, September, 1953.
3. Fortini, E. T.: Dimension Control in Design, *Machine Design*, vol. 28, no. 7, pp. 82–87, Apr. 5, 1956; no. 8, pp. 112–117, Apr. 19; no. 9, pp. 92–95, May 3; no. 10, pp. 79–84, May 17; no. 12, pp. 113–118, June 14; no. 13, pp. 82–87, June 28; no. 14, pp. 110–115, July 12, 1956.
4. "ISA Tolerance System," *ISA Bulletin* 25, International Federation of the National Standardizing Associations, Basel, Switzerland, January, 1941.
5. Leinweber, P.: "Passung und Gestaltung," Springer-Verlag OHG, Berlin, 1941.
6. Conway, H. G.: "Engineering Tolerances," Pitman Publishing Corporation, New York, 1948.
7. Maleev, V. L., and J. B. Hartman: "Machine Design," International Textbook Co., Scranton, 1954.
8. Niemann, G.: "Maschinenelemente," vols. 1 and 2, Springer-Verlag OHG, Berlin, 1958–1960.
9. Stulen, F. B., H. N. Cummings, and N. C. Schulte: Preventing Fatigue Failures, *Machine Design*, vol. 33, 1961, May 11, p. 191.
10. Bolz, R. W.: "Production Processes," vols. 1 and 2, Penton Publishing Co., Cleveland, 1949.
11. Lehr, E.: "Spannungsverteilung in Konstruktion Selementen," Verlag VDI, Berlin, 1934.

Section 20

STANDARDS FOR MECHANICAL ELEMENTS

By

JOHN J. VIEGAS, M.M.E., *Assistant to the Engineering Vice President, R. Hoe and Company, New York, N.Y.*

CONTENTS

20.1. SCREW-THREAD SYSTEMS

(a) Introduction

The most commonly known threads used in this country today are the following:

1. United States thread or Sellers' profile thread introduced by Sellers in 1864
2. V thread
3. American National screw-thread system developed in 1933
4. Unified system—called the Unified and American Standard—was accepted in 1948. Recommended for all new designs
5. British Standard Whitworth (B.S.W.)
6. British Standard Fine (B.S.F.)
7. Whitworth truncated or American truncated
8. Whitworth—developed as a war standard in 1944
9. British Association Standard (B.A.)
10. International metric standard—uses the metric system—European
11. Unified miniature screw thread
12. Microscopic-objective screw thread

(b) Classification

Screws for transmitting power are commonly termed power screws. Screw fastenings are generally categorized as to function, i.e., machine screws, cap screws, setscrews, bolts.

(c) Unified Thread Standardization and Unification

Essentially, only one screw-thread form is now being used for fasteners in this country; it is the *unified thread form*[1] (Fig. 20.1), which is based on earlier standards.

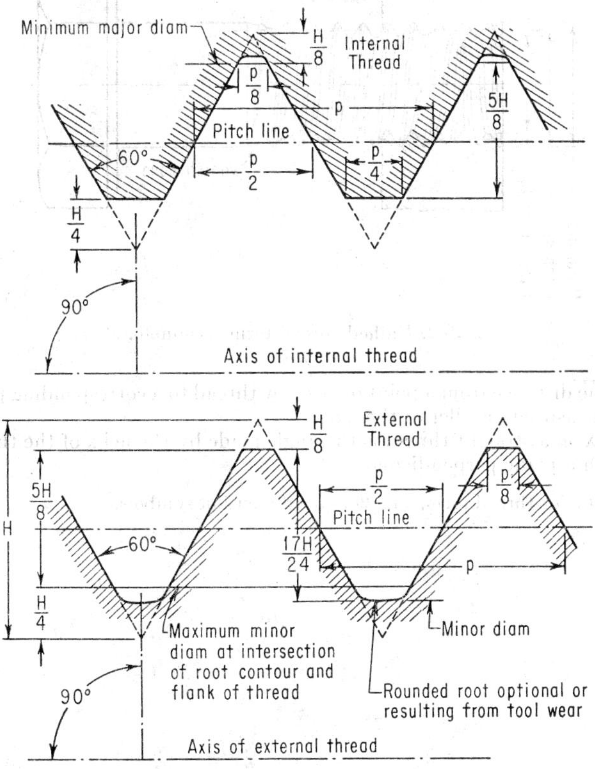

FIG. 20.1. Unified thread form.[1]

This new standard is additional to earlier American standards.[2] Its outstanding characteristic is general interchangeability of threads.

Set up by the standardization committees from Canada, Great Britain, and the United States in 1948, the unified system is herein referred to as the *Unified and American Standard* and is identified by the letters UN. Briefly, the standard comprises the following categories:

Diameter-Pitch Combinations. Coarse series (UNC and NC), fine series (UNF and NF), extra-fine series (UNEF and NEF), miniature (UNM), constant-pitch series (4UN, 6UN, 8UN, 12UN, 16UN, 20UN, 28UN, 32UN).

Tolerance Classes. External (1A, 2A, 3A, 2, 3); internal (1B, 2B, 3B, 2, 3).

(d) **Unified Thread Form Terminology*** (Fig. 20.2)

Major diameter D_m of a threaded screw or nut is the largest diameter of the screw thread.

Minor diameter D_r is the smallest diameter of the screw thread.

Pitch diameter E, also called effective diameter, is the diameter of an imaginary cylinder whose surface would pass through the thread profiles at such points as to make the width of the groove equal to one-half the basic pitch. On a perfect thread, this occurs at the point where the widths of thread and groove are equal.

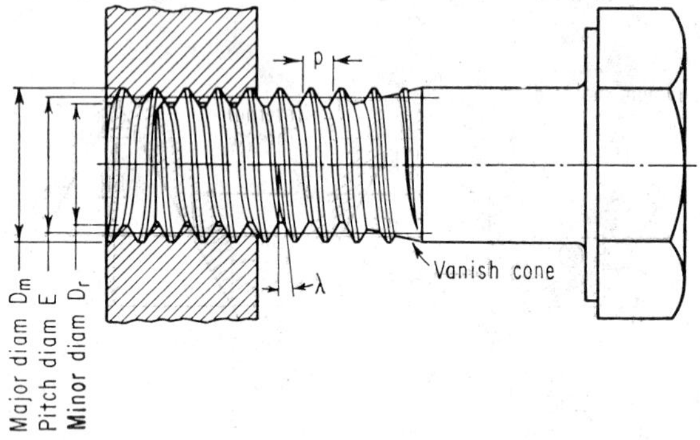

Fig. 20.2. Unified thread form terminology.

Pitch p is the distance from a point on a screw thread to a corresponding point on the next thread measured parallel to the axis.

Lead angle λ on a straight thread is the angle made by the helix of the thread at the pitch line with a plane perpendicular to the axis.

* See Fig. 20.1 for thread form, Fig. 20.3 for dimension symbols.

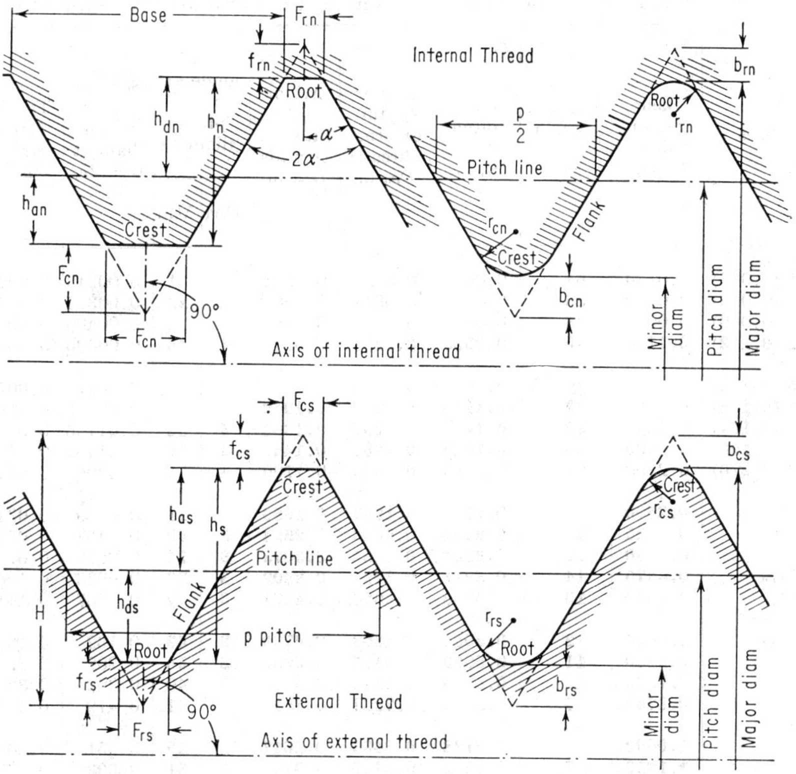

FIG. 20.3. Unified thread symbol application.

n = threads per inch
p = pitch
$F_{cn} = p/4 = 0.250$
$F_{rn} = F_{cs} = 0.125$
H = 0.866025-height of sharp V-thread
$f_{rn} = f_{cs} = H/8 = 0.10825p$-truncation of internal thread root and external thread crest
$b_{rs} = H/6 = 0.14434p$-truncation of external thread root
$\frac{3}{16}H = 0.16238p$-half addendum of external thread
$f_{cn} = H/4 = 0.21651p$-truncation of internal thread crest
$h_{as} = \frac{3}{8}H = 0.32476p$-addendum of external thread
$h_n = h_e = \frac{5}{8}H = 0.54127p$-height of internal thread and depth of thread engagement
$h_s = \frac{17}{24}H$ = 0.61343-height of external thread
$h_b = 2h_{as} = \frac{3}{4}H = 0.649519p$-twice the external thread addendum
$\frac{11}{12}H = 0.79386p$-difference between maximum major pitch diameters of internal thread
$2h_n = 1\frac{1}{4}H = 1.08253p$-double height of internal thread
$1\frac{5}{12}H = 1.22687p$-double height of external thread

(e) Diameter-Pitch Selection[3]

Coarse-thread Series (UNC), Table 20.1. Used where rapid assembly or disassembly is required. Applicable with cast and malleable iron, soft metals, and plastics.

Fine-thread Series (UNF), Table 20.2. For application, where thread engagement or wall thickness is limited. Normally not recommended for use in soft metals or plastics.

Table 20.1. Unified and American Coarse-thread Series UNC and NC[1]

Sizes	Basic major diam D_m, in.	Threads per inch n	Basic pitch diam E, in.	Minor diam external threads D_{rb}, in.	Minor diam internal threads D_{rn}, in.	Lead angle at basic pitch diam λ		Section at minor diam at $D - 2h_b$, in.^{2}	Tensile stress area, in.^{2}
						Deg	Min		
1 (0.073)	0.0730	64	0.0629	0.0538	0.0561	4	31	0.00218	0.00263
2 (0.086)	0.0860	56	0.0744	0.0641	0.0667	4	22	0.00310	0.00370
3 (0.099)	0.0990	48	0.0855	0.0734	0.0764	4	26	0.00406	0.00487
4 (0.112)	**0.1120**	**40**	**0.0958**	**0.0813**	**0.0849**	**4**	**45**	0.00496	0.00604
5 (0.125)	0.1250	40	0.1088	0.0943	0.0979	4	11	0.00672	0.00796
6 (0.138)	**0.1380**	**32**	**0.1177**	**0.0997**	**0.1042**	**4**	**50**	0.00745	0.00909
8 (0.164)	**0.1640**	**32**	**0.1437**	**0.1257**	**0.1302**	**3**	**58**	0.01196	0.0140
10 (0.190)	**0.1900**	**24**	**0.1629**	**0.1389**	**0.1449**	**4**	**39**	0.01450	0.0175
12 (0.216)	0.2160	24	0.1889	0.1649	0.1709	4	1	0.0206	0.0242
1/4	**0.2500**	**20**	**0.2175**	**0.1887**	**0.1959**	**4**	**11**	0.0269	0.0318
5/16	**0.3125**	**18**	**0.2764**	**0.2443**	**0.2524**	**3**	**40**	0.0454	0.0524
3/8	**0.3750**	**16**	**3.3344**	**0.2983**	**0.3073**	**3**	**24**	0.0678	0.0775
7/16	**0.4375**	**14**	**0.3911**	**0.3499**	**0.3602**	**3**	**20**	0.0933	0.1063
1/2	**0.5000**	**13**	**0.4500**	**0.4056**	**0.4167**	**3**	**7**	0.1257	0.1419
9/16	**0.5625**	**12**	**0.5084**	**0.4603**	**0.4723**	**2**	**59**	0.162	0.182
5/8	**0.6250**	**11**	**0.5660**	**0.5135**	**0.5266**	**2**	**56**	0.202	0.226
3/4	**0.7500**	**10**	**0.6850**	**0.6273**	**0.6417**	**2**	**40**	0.302	0.334
7/8	**0.8750**	**9**	**0.8028**	**0.7387**	**0.7547**	**2**	**31**	0.419	0.462
1	**1.0000**	**8**	**0.9188**	**0.8466**	**0.8647**	**2**	**29**	0.551	0.606
1 1/8	**1.1250**	**7**	**1.0322**	**0.9497**	**0.9704**	**2**	**31**	0.693	0.763
1 1/4	**1.2500**	**7**	**1.1572**	**1.0747**	**1.0954**	**2**	**15**	0.890	0.969
1 3/8	**1.3750**	**6**	**1.2667**	**1.1705**	**1.1946**	**2**	**24**	1.054	1.155
1 1/2	**1.5000**	**6**	**1.3917**	**1.2955**	**1.3196**	**2**	**11**	1.294	1.405
1 3/4	**1.7500**	**5**	**1.6201**	**1.5046**	**1.5335**	**2**	**15**	1.74	1.90
2	**2.0000**	**4 1/2**	**1.8557**	**1.7274**	**1.7594**	**2**	**11**	2.30	2.50
2 1/4	**2.2500**	**4 1/2**	**2.1057**	**1.9774**	**2.0094**	**1**	**55**	3.02	3.25
2 1/2	**2.5000**	**4**	**2.3376**	**2.1933**	**2.2294**	**1**	**57**	3.72	4.00
2 3/4	**2.7500**	**4**	**2.5876**	**2.4433**	**2.4794**	**1**	**46**	4.62	4.93
3	**3.0000**	**4**	**2.8376**	**2.6933**	**2.7294**	**1**	**36**	5.62	5.97
3 1/4	**3.2500**	**4**	**3.0876**	**2.9433**	**2.9794**	**1**	**29**	6.72	7.10
3 1/2	**3.5000**	**4**	**3.3376**	**3.1933**	**3.2294**	**1**	**22**	7.92	8.33
3 3/4	**3.7500**	**4**	**3.5876**	**3.4433**	**3.4794**	**1**	**16**	9.21	9.66
4	**4.0000**	**4**	**3.8376**	**3.6933**	**3.7294**	**1**	**11**	10.61	11.08

Bold type indicates unified threads, UNC.

Table 20.2. Unified and American Fine-thread Series, UNF and NF[1]

Sizes	Basic major diam D_m, in.	Threads per inch n	Basic pitch diam E, in.	Minor diam external threads D_{rb}, in.	Minor diam internal threads D_{rn}, in.	Lead angle at basic pitch diam λ Deg	Lead angle at basic pitch diam λ Min	Section at minor diam at $D - 2h_b$, in.[2]	Tensile stress area, in.[2]
0 (0.060)	0.0600	80	0.0519	0.0447	0.0465	4	23	0.00151	0.00180
1 (0.073)	0.0730	72	0.0640	0.0560	0.0580	3	57	0.00237	0.00278
2 (0.086)	0.0860	64	0.0759	0.0668	0.0691	3	45	0.00339	0.00394
3 (0.099)	0.0990	56	0.0874	0.0771	0.0797	3	43	0.00451	0.00523
4 (0.112)	0.1120	48	0.0985	0.0864	0.0894	3	51	0.00566	0.00661
5 (0.125)	0.1250	44	0.1102	0.0971	0.1004	3	45	0.00716	0.00830
6 (0.138)	0.1380	40	0.1218	0.1073	0.1109	3	44	0.00874	0.01015
8 (0.164)	0.1640	36	0.1460	0.1299	0.1339	3	28	0.01285	0.01474
10 (0.190)	**0.1900**	**32**	**0.1697**	**0.1517**	**0.1562**	**3**	**21**	0.0175	0.0200
12 (0.216)	0.2160	28	0.1928	0.1722	0.1773	3	22	0.0226	0.0258
1/4	**0.2500**	**28**	**0.2268**	**0.2062**	**0.2113**	**2**	**52**	0.0326	0.0364
5/16	**0.3125**	**24**	**0.2854**	**0.2614**	**0.2674**	**2**	**40**	0.0524	0.0580
3/8	**0.3750**	**24**	**0.3479**	**0.3239**	**0.3299**	**2**	**11**	0.0809	0.0878
7/16	**0.4375**	**20**	**0.4050**	**0.3762**	**0.3834**	**2**	**15**	0.1090	0.1187
1/2	**0.5000**	**20**	**0.4675**	**0.4387**	**0.4459**	**1**	**57**	0.1486	0.1599
9/16	**0.5625**	**18**	**0.5264**	**0.4943**	**0.5024**	**1**	**55**	0.189	0.203
5/8	**0.6250**	**18**	**0.5889**	**0.5568**	**0.5649**	**1**	**43**	0.240	0.256
3/4	**0.7500**	**16**	**0.7094**	**0.6733**	**0.6823**	**1**	**36**	0.351	0.373
7/8	**0.8750**	**14**	**0.8286**	**0.7874**	**0.7977**	**1**	**34**	0.480	0.509
1	**1.0000**	**12**	**0.9459**	**0.8978**	**0.9098**	**1**	**36**	0.625	0.663
1 1/8	**1.1250**	**12**	**1.0709**	**1.0228**	**1.0348**	**1**	**25**	0.812	0.856
1 1/4	**1.2500**	**12**	**1.1959**	**1.1478**	**1.1598**	**1**	**16**	1.024	1.073
1 3/8	**1.3750**	**12**	**1.3209**	**1.2728**	**1.2848**	**1**	**9**	1.260	1.315
1 1/2	**1.5000**	**12**	**1.4459**	**1.3978**	**1.4098**	**1**	**3**	1.521	1.581

Bold type indicates unified threads, UNF.

Extra-fine-thread Series (UNEF), Table 20.3. Extensively used in aircraft, missile, and other allied fields where requirements of threaded thin walls, minimum thread depth of nuts and coupling flanges, and maximum number of threads required within a given length of thread engagement are important.

Miniature[4] **Threads** (UNM). Primarily used on instrument parts and miniature mechanisms.

Table 20.3. Unified and American Extra-fine-thread Series, UNEF and NEF[1]

Sizes	Basic major diam D_m, in.	Threads per inch n	Basic pitch diam E, in.	Minor diam external threads D_{rb}, in.	Minor diam internal threads D_{rn}, in.	Lead angle at basic pitch diam λ		Section at minor diam at $D - 2h_b$, in.[2]	Tensile stress area, in.[2]
						Deg	Min		
12 (0.216)	0.2160	32	0.1957	0.1777	0.1822	2	55	0.0242	0.0270
1/4	0.2500	32	0.2297	0.2117	0.2162	2	29	0.0344	0.0379
5/16	0.3125	32	0.2922	0.2742	0.2787	1	57	0.0581	0.0625
3/8	0.3750	32	0.3547	0.3367	0.3412	1	36	0.0878	0.0932
7/16	**0.4375**	**28**	**0.4143**	**0.3937**	**0.3988**	**1**	**34**	0.1201	0.1274
1/2	**0.5000**	**28**	**0.4768**	**0.4562**	**0.4613**	**1**	**22**	0.162	0.170
9/16	0.5625	24	0.5354	0.5114	0.5174	1	25	0.203	0.214
5/8	0.6250	24	0.5979	0.5739	0.5799	1	16	0.256	0.268
11/16	0.6875	24	0.6604	0.6364	0.6424	1	9	0.315	0.329
3/4	**0.7500**	**20**	**0.7175**	**0.6887**	**0.6959**	**1**	**16**	0.369	0.386
13/16	**0.8125**	**20**	**0.7800**	**0.7512**	**0.7584**	**1**	**10**	0.439	0.458
7/8	**0.8750**	**20**	**0.8425**	**0.8137**	**0.8209**	**1**	**5**	0.515	0.536
15/16	**0.9375**	**20**	**0.9050**	**0.8762**	**0.8834**	**1**	**0**	0.598	0.620
1	**1.0000**	**20**	**0.9675**	**0.9387**	**0.9459**	**0**	**57**	0.687	0.711
1 1/16	1.0625	18	1.0264	0.9943	1.0024	0	59	0.770	0.799
1 1/8	1.1250	18	1.0889	1.0568	1.0649	0	56	0.871	0.901
1 3/16	1.1875	18	1.1514	1.1193	1.1274	0	53	0.977	1.009
1 1/4	1.2500	18	1.2139	1.1818	1.1899	0	50	1.090	1.123
1 5/16	1.3125	18	1.2764	1.2443	1.2524	0	48	1.208	1.244
1 3/8	1.3750	18	1.3389	1.3068	1.3149	0	45	1.333	1.370
1 7/16	1.4375	18	1.4014	1.3693	1.3774	0	43	1.464	1.503
1 1/2	1.5000	18	1.4639	1.4318	1.4399	0	42	1.60	1.64
1 9/16	1.5625	18	1.5264	1.4943	1.5024	0	40	1.74	1.79
1 5/8	1.6250	18	1.5889	1.5568	1.5649	0	38	1.89	1.94
1 11/16	1.6875	18	1.6514	1.6193	1.6274	0	37	2.05	2.10
1 3/4	**1.7500**	**16**	**1.7094**	**1.6733**	**1.6823**	**0**	**40**	2.19	2.24
2	**2.0000**	**16**	**1.9594**	**1.9233**	**1.9323**	**0**	**35**	2.89	2.95

Bold type indicates unified threads, UNEF.

Uniform-pitch Series. 4UN, 6UN, 8UN, 12UN, 16UN, 20UN, 28UN, and 32UN; see ref. 1.

(f) Unified Thread Class Selection

The required assembly fit is obtained by selecting the thread class for each component. The class dictates the amount of manufacturing tolerance and allowance permitted (Fig. 20.4). The letter A designates tolerances and allowances applicable to externally threaded fasteners. The letter B identifies internally threaded numbers.

Classes 1A/1B. Loose fit, quick and easy assembly.

Classes 2A/2B. Commercially manufactured in the majority of threaded fasteners. The allowance of this class minimizes galling, minimizes seizure in high-temperature applications, and allows for plating.

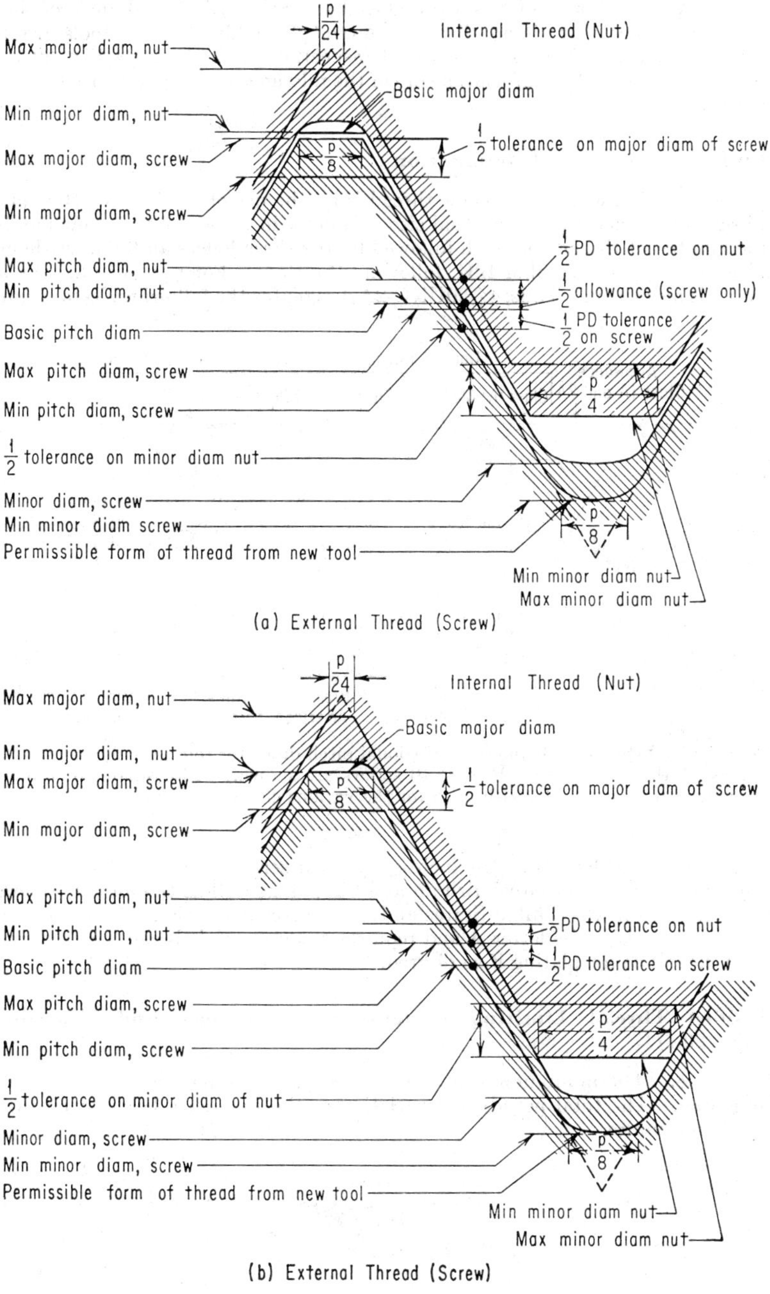

FIG. 20.4. Basic unified thread form.

Classes 3A/3B. For use where close fit and accuracy of lead and angle of thread are required. Since there is no allowance intended, highly accurate tools are used in its production together with controlled inspection methods. See Fig. 20.4, for example, showing the physical relationship of the tolerance as designated by the different classes.[5]

(g) Unified Screw Thread Designation

The method of designating a screw thread is by the use of the initial letters of the thread series preceded by the nominal size (diameter in inches or the screw number) and the number of threads per inch followed by the thread class and, if a left-handed thread is required, the symbol L.H. following the class. For right-handed threads, no symbol is necessary. Thus, ¼-20UNC-2A designates the following: ¼ in. nominal

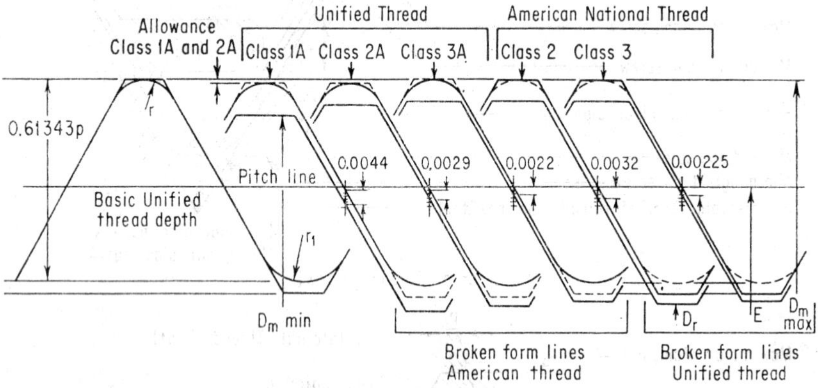

FIG. 20.5. Unified thread and American National pitch diameter tolerance figures.[5] (Various classes of thread for one-half of ¾-10 screws are shown.) Pitch-diameter tolerance figures shown are one-half of ¾-10 screw for the various classes of thread.

diameter having 20 threads to the inch is a unified series, therefore interchangeable with the thread system of Great Britain and Canada as well as the former American Standard threads in the United States and is in the coarse series. 2A signifies it is an external thread by the letter A and requires a commercial thread which provides sufficient clearance for plating if so desired.

(h) Unified Thread Limiting Dimensions—Allowances and Tolerances[6] (Fig. 20.4)

Allowance. The minimum clearance or maximum interference.
Tolerance. The total permissible variation or the difference between maximum and minimum limits.

Table 20.4. American National Coarse-thread Series[1]

Identification		Screw basic diameters			Major screw diam tolerances			Max minor screw diam	Max minor nut diam	Minor nut diam tolerance	Nut pitch diam tolerances			
Size	Threads per in. n	Major diam D	Pitch diam E	Minor diam K	Class 1	Class 2	Classes 3 and 4				Class 1	Class 2	Class 3	Class 4
1	64	0.0730	0.0629	0.0527	0.0052	0.0692	0.0038	0.0538	0.0623	0.0062	0.0026	0.0019	0.0014	
2	56	0.0860	0.0744	0.0628	0.0056	0.0820	0.0040	0.0641	0.0737	0.0070	0.0028	0.0020	0.0015	
3	48	0.0990	0.0855	0.0719	0.0062	0.0946	0.0044	0.0734	0.0841	0.0077	0.0031	0.0022	0.0016	
4	40	0.1120	0.0958	0.0795	0.0068	0.1072	0.0048	0.0813	0.0938	0.0089	0.0034	0.0024	0.0017	
5	40	0.1250	0.1088	0.0925	0.0068	0.1202	0.0048	0.0943	0.1062	0.0083	0.0034	0.0024	0.0017	
6	32	0.1380	0.1177	0.0974	0.0076	0.1326	0.0054	0.0997	0.1145	0.0103	0.0038	0.0027	0.0019	
8	32	0.1640	0.1437	0.1234	0.0076	0.1586	0.0054	0.1257	0.1384	0.0028	0.0038	0.0027	0.0019	
10	24	0.1900	0.1629	0.1359	0.0092	0.1834	0.0066	0.1389	0.1559	0.0110	0.0046	0.0033	0.0024	
12	24	0.2160	0.1889	0.1619	0.0092	0.2094	0.0066	0.1649	0.1801	0.0092	0.0046	0.0033	0.0024	
¼	20	0.2500	0.2175	0.1850	0.0102	0.2428	0.0072	0.1887	0.2060	0.0101	0.0051	0.0036	0.0026	0.0013
5⁄16	18	0.3125	0.2764	0.2403	0.0114	0.3043	0.0082	0.2443	0.2630	0.0106	0.0057	0.0041	0.0030	0.0015
⅜	16	0.3750	0.3344	0.2938	0.0126	0.3660	0.0090	0.2983	0.3184	0.0111	0.0036	0.0045	0.0032	0.0016
7⁄16	14	0.4375	0.3911	0.3447	0.0140	0.4277	0.0098	0.3499	0.3721	0.0119	0.0070	0.0049	0.0036	0.0018
½	13	0.5000	0.4500	0.4001	0.0148	0.4896	0.0104	0.4056	0.4290	0.0123	0.0074	0.0052	0.0037	0.0019
9⁄16	12	0.5625	0.5084	0.4542	0.0158	0.5513	0.0112	0.4603	0.4850	0.0127	0.0079	0.0056	0.0040	0.0020
⅝	11	0.6250	0.5660	0.5069	0.0170	0.6132	0.0118	0.5135	0.5397	0.0131	0.0085	0.0059	0.0042	0.0021
¾	10	0.7500	0.6850	0.6201	0.0184	0.7372	0.0128	0.6273	0.6553	0.0136	0.0092	0.0064	0.0045	0.0023
⅞	9	0.8750	0.8028	0.7307	0.0200	0.8610	0.0140	0.7387	0.7689	0.0142	0.0100	0.0070	0.0049	0.0024
1	8	1.0000	0.9188	0.8376	0.0222	0.9848	0.0152	0.8466	0.8795	0.0148	0.0111	0.0076	0.0054	0.0027
1⅛	7	1.1250	1.0322	0.9394	0.0248	1.1080	0.0170	0.9497	0.9858	0.0154	0.0124	0.0085	0.0059	0.0030
1¼	7	1.2500	1.1572	1.0644	0.0248	1.2330	0.0170	1.0747	1.1108	0.0154	0.0124	0.0085	0.0059	0.0030
1⅜	6	1.3750	1.2667	1.1585	0.0290	1.3548	0.0202	1.1705	1.2126	0.0180	0.0145	0.0101	0.0071	0.0036
1½	6	1.5000	1.3917	1.2835	0.0290	1.4798	0.0202	1.2955	1.3376	0.0180	0.0145	0.0101	0.0071	0.0036
1¾	5	1.7500	1.6201	1.4902	0.0338	1.7268	0.0232	1.5046	1.5551	0.0216	0.0169	0.0116	0.0082	0.0041
2	4½	2.0000	1.8557	1.7113	0.0368	1.9746	0.0254	1.7274	1.7835	0.0241	0.0184	0.0127	0.0089	0.0044
2¼	4½	2.2500	2.1057	1.9613	0.0368	2.2246	0.0254	1.9774	2.0335	0.0241	0.0184	0.0127	0.0089	0.0044
2½	4	2.5000	2.3376	2.1752	0.0408	2.4720	0.0280	2.1933	2.2564	0.0270	0.0204	0.0140	0.0097	0.0048
2¾	4	2.7500	2.5876	2.4252	0.0408	2.7220	0.0280	2.4433	2.5064	0.0270	0.0204	0.0140	0.0097	0.0048
3	4	3.0000	2.8376	2.6752	0.0408	2.9720	0.0280	2.6933	2.7564	0.0270	0.0204	0.0140	0.0097	0.0048
3¼	4	3.2500	3.0876	2.9252	0.0408	3.2220	0.0280	2.9433	3.0064	0.0270	0.0204	0.0140	0.0097	0.0048
3½	4	3.5000	3.3376	3.1752	0.0408	3.4720	0.0280	3.1933	3.2564	0.0270	0.0204	0.0140	0.0097	0.0048
3¾	4	3.7500	3.5876	3.4252	0.0408	3.7220	0.0280	3.4433	3.5064	0.0270	0.0204	0.0140	0.0097	0.0048
4	4	4.0000	3.3876	3.6752	0.0408	3.9720	0.0280	3.6933	3.7564	0.0270	0.0204	0.0140	0.0097	0.0048

(i) American National Thread Standard[2]

Also called American Standard or National Standard, this standard was formerly known as the United States thread or Sellers profile and provided the foundation to its successor the Unified and American Standard. It is interchangeable with the Unified and American Standard.

The American Standard consisted of the following: *Three thread series:*

1. Coarse series (Table 20.4). General engineering work for rapid assembly.
2. Fine series (Table 20.5).
3. Special-pitch series. 8, 12, 16; see ref. 2.

Table 20.5. American National Fine-thread Series[17]

Identification		Screw basic diam				Major screw diam tolerances		Max minor screw diam		Max minor nut diam		Nut pitch diam tolerances		
Size	Threads per in., n	Major diam, classes 2 and 3, D_m	Major diam, class 4, D_m	Pitch diam, classes 2 and 3	Pitch diam, class 4	Classes 2 and 3	Class 4	Classes 2 and 3	Class 4	Classes 2 and 3	Class 4	Class 2	Class 3	Class 4
0	80	0.0600	......	0.0519	......	0.0034	......	0.0447	......	0.0514	......	0.0017	0.0013	
1	72	0.0730	......	0.0640	......	0.0036	......	0.0560	......	0.0634	......	0.0018	0.0013	
2	64	0.0860	......	0.0759	......	0.0038	......	0.0668	......	0.0746	......	0.0019	0.0014	
3	56	0.0990	......	0.0874	......	0.0040	......	0.0771	......	0.0856	......	0.0020	0.0015	
4	48	0.1120	......	0.0985	......	0.0044	......	0.0864	......	0.0960	......	0.0022	0.0016	
5	44	0.1250	......	0.1102	......	0.0046	......	0.0971	......	0.1068	......	0.0023	0.0016	
6	40	0.1380	......	0.1218	......	0.0048	......	0.1073	......	0.1179	......	0.0024	0.0017	
8	36	0.1640	......	0.1460	......	0.0050	......	0.1299	......	0.1402	......	0.0025	0.0018	
10	32	0.1900	......	0.1697	......	0.0054	......	0.1517	......	0.1624	......	0.0027	0.0019	
12	28	0.2160	......	0.1928	......	0.0062	......	0.1722	......	0.1835	......	0.0031	0.0022	
¼	28	0.2500	0.2500	0.2268	0.2270	0.0062	0.0062	0.2062	0.2062	0.2173	0.2173	0.0031	0.0022	0.0011
5/16	24	0.3125	0.3125	0.2854	0.2857	0.0066	0.0066	0.2614	0.2614	0.2739	0.2739	0.0033	0.0024	0.0012
⅜	24	0.3750	0.3750	0.3479	0.3482	0.0066	0.0066	0.3239	0.3238	0.3364	0.3364	0.0033	0.0024	0.0012
7/16	20	0.4375	0.4375	0.4050	0.4053	0.0072	0.0072	0.3762	0.3672	0.3906	0.3906	0.0036	0.0026	0.0013
½	20	0.5000	0.5000	0.4675	0.4678	0.0072	0.0072	0.4387	0.4387	0.4531	0.4531	0.0036	0.0026	0.0013
9/16	18	0.5625	0.5625	0.5264	0.5267	0.0082	0.0082	0.4943	0.4943	0.5100	0.5100	0.0041	0.0030	0.0015
⅝	18	0.6250	0.6250	0.5889	0.5892	0.0082	0.0082	0.5568	0.5568	0.5725	0.5725	0.0041	0.0030	0.0015
¾	16	0.7500	0.7500	0.7094	0.7098	0.0090	0.0090	0.6733	0.6733	0.6903	0.6903	0.0045	0.0032	0.0016
⅞	14	0.8750	0.8750	0.8286	0.8290	0.0098	0.0098	0.7874	0.7874	0.8062	0.8062	0.0049	0.0036	0.0018
1	14	1.0000	1.0000	0.9536	0.9540	0.0098	0.0008	0.9124	0.9124	0.9312	0.9312	0.0049	0.0036	0.0018
1⅛	12	1.1250	1.1250	1.0709	1.0714	0.0112	0.0112	1.0228	1.0228	1.0438	1.0438	0.0056	0.0040	0.0020
1¼	12	1.2500	1.2500	1.1959	1.1964	0.0112	0.0112	1.1478	1.1478	1.1688	1.1688	0.0056	0.0040	0.0020
1⅜	12	1.3750	1.3750	1.3209	1.3214	0.0112	0.0112	1.2728	1.2728	1.2938	1.2938	0.0056	0.0040	0.0020
1½	12	1.5000	1.5000	1.4459	1.4464	0.0112	0.0112	1.3978	1.3978	1.4188	1.4188	0.0056	0.0040	0.0020

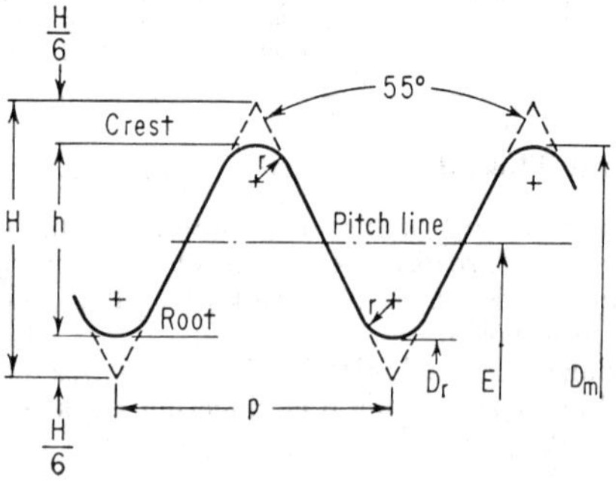

FIG. 20.6. British Standard Whitworth thread.[5]

$$E = D - h = D - 0.64033p$$
$$H = 0.96049p$$
$$h = 0.64033p$$
$$r = 0.13733p$$

Four classes are provided:
Class 1. Used where clearance is not objectional.
Class 2. Used in commercial items and is most commonly specified.
Class 3. Used in high-quality equipment.
Class 4. For unusual requirements, selective fit.

(j) American National Thread Form[2]

This screw thread is substantially identical with the unified thread form (Fig. 20.1). The formulas for its proportions are given in Fig. 20.3.

(k) British Standard Whitworth Thread (B.S.W.)

These threads used in Great Britain employ a 55° included angle (Fig. 20.6). Basic sizes are listed in Table 20.6.

Table 20.6. British Standard Whitworth Threads[16]

Nominal diam, in.	Threads per in.	Pitch, in.	Depth of thread, in.	Major diam, in.	Effective* diam, in.	Minor diam, in.	Cross-sectional area at bottom of thread, in.^{2}
⅛	40	0.025 00	0.0160	0.1250	0.1090	0.0930	0.0068
3/16	24	0.041 67	0.0267	0.1875	0.1608	0.1341	0.0141
¼	20	0.050 00	0.0320	0.2500	0.2180	0.1860	0.0272
5/16	18	0.055 56	0.0356	0.3125	0.2769	0.2413	0.0457
⅜	16	0.062 50	0.0400	0.3750	0.3350	0.2950	0.0683
7/16	14	0.071 43	0.0457	0.4375	0.3918	0.3461	0.0941
½	12	0.083 33	0.0534	0.5000	0.4466	0.3932	0.1214
9/16	12	0.083 33	0.0534	0.5625	0.5091	0.4557	0.1631
⅝	11	0.090 91	0.0582	0.6250	0.5668	0.5086	0.2032
¾	10	0.100 00	0.0640	0.7500	0.6860	0.6220	0.3039
⅞	9	0.111 11	0.0711	0.8750	0.8039	0.7328	0.4218
1	8	0.125 00	0.0800	1.0000	0.9200	0.8400	0.5542
1⅛	7	0.142 86	0.0915	1.1250	1.0335	0.9420	0.6969
1¼	7	0.142 86	0.0915	1.2500	1.1585	1.0670	0.8942
1½	6	0.166 67	0.1067	1.5000	1.3933	1.2866	1.300
1¾	5	0.200 00	0.1281	1.7500	1.6219	1.4938	1.753
2	4.5	0.222 22	0.1423	2.0000	1.8577	1.7154	2.311
2¼	4	0.250 00	0.1601	2.2500	2.0899	1.9298	2.925
2½	4	0.250 00	0.1601	2.5000	2.3399	2.1798	3.732
2¾	3.5	0.285 71	0.1830	2.7500	2.5670	2.3840	4.464
3	3.5	0.285 71	0.1830	3.0000	2.8170	2.6340	5.449
3¼	3.25	0.307 69	0.1970	3.2500	3.0530	2.8560	6.406
3½	3.25	0.307 69	0.1970	3.5000	3.3030	3.1060	7.577
3¾	3	0.333 33	0.2134	3.7500	3.5366	3.3232	8.674
4	3	0.333 33	0.2134	4.0000	3.7866	3.5732	10.03
4½	2.875	0.347 83	0.2227	4.5000	4.2773	4.0546	12.91
5	2.75	0.363 64	0.2328	5.0000	4.7672	4.5344	16.15
5½	2.625	0.380 95	0.2439	5.5000	5.2561	5.0122	19.73
6	2.5	0.400 00	0.2561	6.0000	5.7439	5.4878	23.65

* Basic pitch diameter.

(l) British Standard Fine Thread (B.S.F.)[7]

This supplements the Whitworth standards where finer pitches are required. The thread form is the same as the Whitworth (Fig. 20.6). See Table 20.7 for bolt sizes.

Table 20.7. British Standard Fine Threads[16]

Nominal diam, in.	Threads per in.	Pitch, in.	Depth of thread, in.	Major diam, in.	Effective* diam, in.	Minor diam, in.	Cross-sectional area at bottom of thread, in.[2]
3/16	32	0.031 25	0.0200	0.1875	0.1675	0.1475	0.0171
7/32	28	0.035 71	0.0229	0.2188	0.1959	0.1730	0.0235
1/4	26	0.038 46	0.0246	0.2500	0.2254	0.2008	0.0317
9/32	26	0.038 46	0.0246	0.2812	0.2566	0.2320	0.0423
5/16	22	0.045 45	0.0291	0.3125	0.2834	0.2543	0.0508
3/8	20	0.050 00	0.0320	0.3750	0.3430	0.3110	0.0760
7/16	18	0.055 56	0.0356	0.4375	0.4019	0.3663	0.1054
1/2	16	0.062 50	0.0400	0.5000	0.4600	0.4200	0.1385
9/16	16	0.062 50	0.0400	0.5625	0.5225	0.4825	0.1828
5/8	14	0.071 43	0.0457	0.6250	0.5793	0.5336	0.2236
3/4	12	0.083 33	0.0534	0.7500	0.6966	0.6432	0.3249
13/16	12	0.083 33	0.0534	0.8125	0.7591	0.7057	0.3911
7/8	11	0.090 91	0.0582	0.8750	0.8168	0.7586	0.4520
1	10	0.100 00	0.0640	1.0000	0.9360	0.8720	0.5972
1 1/8	9	0.111 11	0.0711	1.1250	1.0539	0.9828	0.7586
1 1/4	9	0.111 11	0.0711	1.2500	1.1789	1.1078	0.9639
1 3/8	8	0.125 00	0.0800	1.3750	1.2950	1.2150	1.159
1 1/2	8	0.125 00	0.0800	1.5000	1.4200	1.3400	1.410
1 5/8	8	0.125 00	0.0800	1.6250	1.5450	1.4650	1.686
1 3/4	7	0.142 86	0.0915	1.7500	1.6585	1.5670	1.928
2	7	0.142 86	0.0915	1.0000	1.9085	1.8170	2.593
2 1/4	6	0.166 67	0.1067	2.2500	2.1433	2.0366	3.258
2 1/2	6	0.166 67	0.1067	2.5000	2.3933	2.2866	4.106
2 3/4	6	0.166 67	0.1067	2.7500	2.6433	2.5366	5.054
3	5	0.200 00	0.1281	3.0000	2.8719	2.7438	5.913
3 1/4	5	0.200 00	0.1281	3.2500	3.1219	2.9938	7.039
3 1/2	4.5	0.222 22	0.1423	3.5000	3.3577	3.2154	8.120
3 3/4	4.5	0.222 22	0.1423	3.7500	3.6077	3.4654	9.432
4	4.5	0.222 22	0.1423	4.0000	3.8577	3.7154	10.84
4 1/4	4	0.250 00	0.1601	4.2500	4.0899	3.9298	12.13

NOTE: It is recommended that for larger diameters in this series four threads per inch be used.
* Basic pitch diameter.

(m) British Association Thread (B.A.) (Fig. 20.7)

This type is recommended by the British Standards for screws below 1/4 inch diameter. This thread has an included angle of 47 1/2°. Used in watchmaking and instrument work. Basic sizes of bolts are listed in Table 20.8.

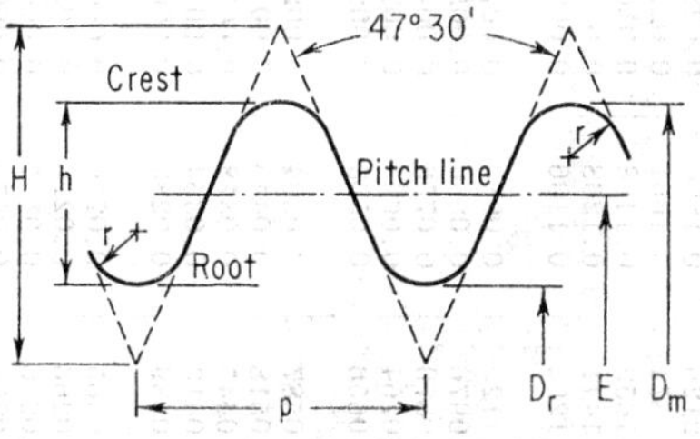

FIG. 20.7. British Association thread.[18]

$$E = D - h = D - 0.60p$$
$$H = 1.13634p$$
$$h = 0.60p$$
$$r = 0.1818p$$

(n) International Metric Thread Standard

This thread form (Fig. 20.8 and data of Table 20.9) is similar to the American National Standard in that the included angle is 60°. A rounded root profile is recommended.

(o) Löwenherz Thread[8]

Based on the metric system, this thread is used for the fine threads of measuring instruments (see Fig. 20.9). Crest and root flats resemble the Unified and American

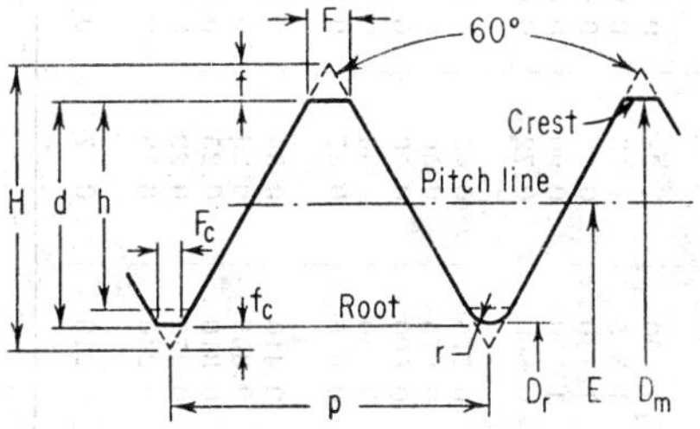

FIG. 20.8. International 60° metric thread.[19]

$$E = D - h = D - 0.64952p$$
$$F = 0.1250p \qquad F_c = 0.0625p$$
$$G = 0.10825p \qquad G_c = 0.05412p$$
$$H = 0.86603p$$
$$h = 0.64952p \qquad d = 0.70364p$$
$$r = 0.054p \text{ (optional)}$$

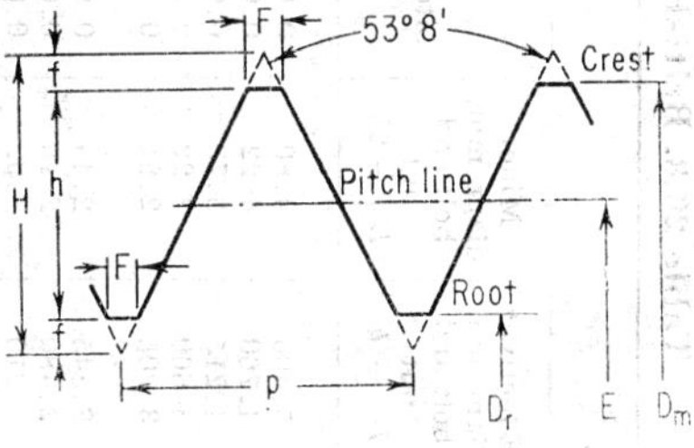

FIG. 20.9. Löwenherz thread.[8]

$$E = D - h = D - 0.75p$$
$$F = 0.125p$$
$$H = p$$
$$h = 0.75p$$

Standard. Since both the German (DIN) and the French metric thread systems carry diameters down to 1 mm with other details varying slightly, the Löwenherz has been superseded.

(p) Dardelet Thread[8] (Fig. 20.10)

This type is a self-locking thread with an included angle of 29°. When the nut is tightened, a wedging action due to the 6° angle results, locking both bolt and nut. Note the inclined crests on the nut and inclined flats at the roots of the bolt.

Table 20.8. British Association Threads[18]

No.	Metric dimensions					Approx. in equivalents						
	Pitch p, mm	Depth of thread $0.6p$, mm	Major diam, mm, bolt and nut D_m	Effective* diam, mm, bolt and nut $D_m - h$	Minor diam, mm, bolt and nut $D_m - 2h$	Approx pitch, in.	Approx threads per in.	Depth of thread, in.	Major diam, in., bolt and nut	Effective* diam, in., bolt and nut	Minor diam, in., bolt and nut	Approx area at bottom of thread, in.^{2}
0	1.0000	0.600	6.00	5.400	4.80	0.03937	25.4	0.0236	0.2362	0.2126	0.1890	0.0281
1	0.9000	0.540	5.30	4.760	4.22	0.03543	28.2	0.0213	0.2087	0.1874	0.1661	0.0217
2	0.8100	0.485	4.70	4.215	3.73	0.03189	31.3	0.0191	0.1850	0.1659	0.1468	0.0169
3	0.7300	0.440	4.10	3.660	3.22	0.02874	34.8	0.0173	0.1614	0.1441	0.1268	0.0126
4	0.6600	0.395	3.60	3.205	2.81	0.02598	38.5	0.0156	0.1417	0.1262	0.1106	0.0096
5	0.5900	0.355	3.20	2.845	2.49	0.02323	43.1	0.0140	0.1260	0.1120	0.0980	0.0075
6	0.5300	0.320	2.80	2.480	2.16	0.02087	47.9	0.0126	0.1102	0.0976	0.0850	0.0057
7	0.4800	0.290	2.50	2.210	1.92	0.01890	52.9	0.0114	0.0984	0.0870	0.0756	0.0045
8	0.4300	0.260	2.20	1.940	1.68	0.01693	59.1	0.0102	0.0866	0.0764	0.0661	0.0034
9	0.3900	0.235	1.90	1.665	1.43	0.01535	65.1	0.0093	0.0748	0.0656	0.0563	0.0025
10	0.3500	0.210	1.70	1.490	1.28	0.01378	72.6	0.0083	0.0669	0.0587	0.0504	0.0020
11	0.3100	0.185	1.50	1.315	1.13	0.01220	81.9	0.0073	0.0591	0.0518	0.0445	0.0016
12	0.2800	0.170	1.30	1.130	0.96	0.01102	90.7	0.0067	0.0512	0.0445	0.0378	0.0011
13	0.2500	0.150	1.20	1.050	0.90	0.00984	102.0	0.0059	0.0472	0.0413	0.0354	0.0010
14	0.2300	0.140	1.00	0.860	0.72	0.00905	110.0	0.0055	0.0394	0.0339	0.0283	0.0006
15	0.2100	0.125	0.90	0.775	0.65	0.00827	121.0	0.0049	0.0354	0.0305	0.0256	0.0005
16	0.1900	0.115	0.79	0.675	0.56	0.00748	134.0	0.0045	0.0311	0.0266	0.0220	0.0004
17	0.1700	0.100	0.70	0.600	0.50	0.00669	149.0	0.0039	0.0276	0.0236	0.0197	0.0003
18	0.1500	0.090	0.62	0.530	0.44	0.00590	169.0	0.0035	0.0244	0.0209	0.0173	0.0002
19	0.1400	0.085	0.54	0.455	0.37	0.00551	182.0	0.0033	0.0213	0.0179	0.0146	0.0002
20	0.1200	0.070	0.48	0.410	0.34	0.00472	213.0	0.0028	0.0189	0.0161	0.0134	0.0001
21	0.1100	0.065	0.42	0.355	0.29	0.00433	232.0	0.0026	0.0165	0.0140	0.0114	0.0001
22	0.1000	0.060	0.37	0.310	0.25	0.00394	256.0	0.0024	0.0146	0.0122	0.0098	0.00008
23	0.0900	0.055	0.33	0.275	0.22	0.00354	286.0	0.0022	0.0130	0.0108	0.0087	0.00006
24	0.0800	0.050	0.29	0.240	0.19	0.00315	322.0	0.0020	0.0114	0.0094	0.0075	0.00004
25	0.0700	0.040	0.25	0.210	0.17	0.00276	357.0	0.0016	0.0098	0.0083	0.0067	0.00004

* Basic pitch diameter.

Table 20.9. International 60° Metric Threads

Millimeter sizes						Equivalent sizes, in.			
Diam of screw D	Pitch p	Depth of thread $0.6495p$ (h)	Pitch diam $D_m - h$ (E)	Minor diam $D_m - 2h$ (D_r)	Approx threads per in. n	Depth of thread $0.6495p$ (h)	Major diam D_m	Pitch diam $D_m - h$ (E)	Minor diam $D - 2h$ (D_r)
3	0.50	0.32475	2.67525	2.350	50.8	0.01279	0.11811	0.10532	0.09253
3.5	0.60	0.38970	3.11030	2.721	42.3	0.01534	0.13780	0.12246	0.10712
4	0.70	0.45465	3.54535	3.091	36.3	0.01790	0.15748	0.13958	0.12168
4.5	0.75	0.48712	4.01288	3.526	33.9	0.01918	0.17716	0.15798	0.13880
5	0.80	0.51960	4.48040	3.961	31.8	0.02046	0.19685	0.17639	0.15593
5.5	0.90	0.58455	4.91545	4.331	28.2	0.02301	0.21654	0.19353	0.17052
6	1.00	0.64950	5.35050	4.70	25.4	0.02557	0.23622	0.21065	0.18508
7	1.00	0.64950	6.35050	5.70	25.4	0.02557	0.27559	0.25002	0.22445
8	1.25	0.81188	7.18813	6.38	20.3	0.03196	0.31496	0.28300	0.25104
9	1.25	0.81188	8.18813	7.38	20.3	0.03196	0.35433	0.32237	0.29041
10	1.5	0.97425	9.026	8.05	16.9	0.03836	0.39370	0.35534	0.31698
11	1.5	0.97425	10.026	9.05	16.9	0.03836	0.43307	0.39471	0.35635
12	1.5	0.97425	11.026	10.05	16.9	0.03836	0.47244	0.43408	0.39572
12	1.75	1.13662	10.863	9.73	14.5	0.04475	0.47244	0.42769	0.38294
14	2.0	1.2990	12.701	11.40	12.7	0.05114	0.55118	0.50004	0.44890
16	2.0	1.2990	14.701	13.40	12.7	0.05114	0.62992	0.57878	0.52764
18	2.5	1.62375	16.376	14.75	10.2	0.06393	0.70866	0.64473	0.58080
20	2.5	1.62375	18.376	16.75	10.2	0.06393	0.78740	0.72347	0.65954
22	2.5	1.62375	20.376	18.75	10.2	0.06393	0.86614	0.80221	0.73828
22	3.0	1.9485	20.052	18.10	8.5	0.07671	0.86614	0.78943	0.71272
24	3.0	1.9485	22.052	20.10	8.5	0.07671	0.94488	0.86817	0.79146
26	3.0	1.9485	24.052	22.10	8.5	0.07671	1.02362	0.94691	0.87020
27	3.0	1.9485	25.052	23.10	8.5	0.07671	1.06299	0.98628	0.90957
28	3.0	1.9485	26.052	24.10	8.5	0.07671	1.10236	1.02565	0.94894
30	3.5	2.27325	27.727	25.45	7.3	0.08950	1.18110	1.09160	1.00210
32	3.5	2.27325	29.727	27.45	7.3	0.08950	1.25984	1.17034	1.08084
33	3.5	2.27325	30.727	28.45	7.3	0.08950	1.29921	1.20971	1.12021
34	3.5	2.27325	31.727	29.45	7.3	0.08950	1.33858	1.24908	1.15958
36	4.0	2.5980	33.420	30.80	6.4	0.10228	1.41732	1.31504	1.21276
38	4.0	2.5980	35.402	32.80	6.4	0.10228	1.49606	1.39378	1.29150
39	4.0	2.5980	36.402	33.80	6.4	0.10228	1.53543	1.43315	1.33087
40	4.0	2.5980	37.402	34.80	6.4	0.10228	1.57480	1.47252	1.37024
42	4.5	2.92275	39.077	36.15	5.6	0.11507	1.65354	1.53847	1.42340
44	4.5	2.92275	41.077	38.15	5.6	0.11507	1.73228	1.61721	1.50214
45	4.5	2.92275	42.077	39.15	5.6	0.11057	1.77165	1.65658	1.54151
46	4.5	2.92275	43.077	40.15	5.6	0.11057	1.81102	1.69595	1.58088
48	5.0	3.2475	44.752	41.50	5.1	0.12785	1.88976	1.76191	1.63406
50	5.0	3.2475	46.752	43.50	5.1	0.12785	1.96850	1.84065	1.71280
52	5.0	3.2475	48.752	45.50	5.1	0.12785	2.04724	1.91939	1.79154
56	5.5	3.57225	52.428	48.86	4.6	0.14064	2.20472	2.06408	1.92344
60	5.5	3.57225	56.428	52.86	4.6	0.14064	2.36220	2.22156	2.08092
64	6.0	3.8970	60.103	56.21	4.2	0.15342	2.51968	2.36626	2.21284
68	6.0	3.8970	64.103	60.21	4.2	0.15342	2.67716	2.52374	2.37032
72	6.5	4.22175	67.778	63.56	3.9	0.16621	2.83464	2.66843	2.50222
76	6.5	4.22175	71.778	67.56	3.9	0.16621	2.99212	2.82591	2.65970
80	7.0	4.5465	75.454	70.91	3.6	0.17900	3.14960	2.97060	2.79160

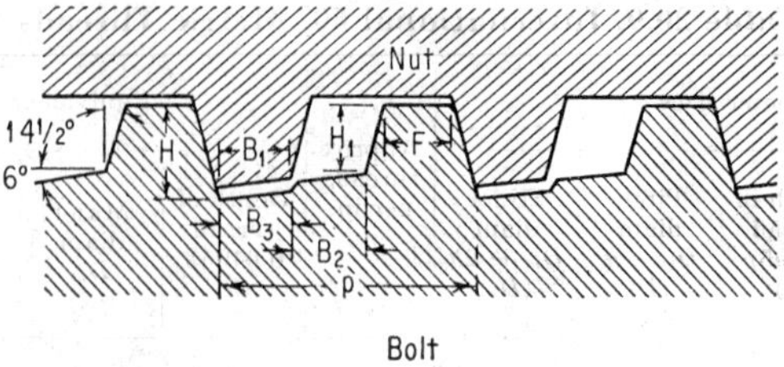

FIG. 20.10. Dardelet relieved thread (unlocked).[8]

(q) Lok-Thread[9]

This has an included thread angle of 60° (Fig. 20.11) and is seen to be similar to the Unified and American Standard. The flat at the root of the screw is a 6° inclined

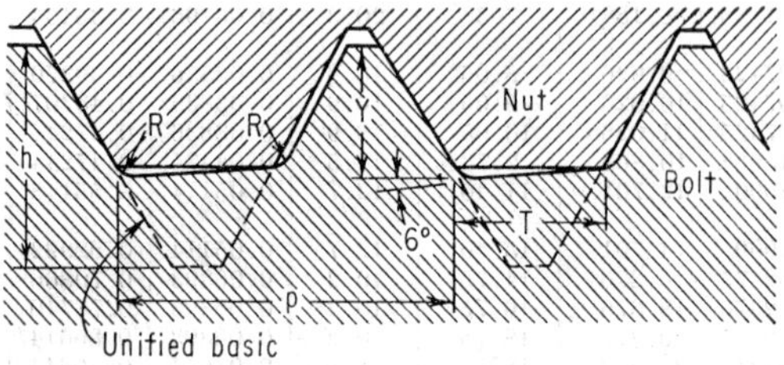

FIG. 20.11. Lok-Thread.[9]

spiral. Locking is obtained by the interference between the crest of the internal thread and the tapered flats of the external thread.

Recommended Thread Depths

For cap screws and bolts, 70 per cent thread:

$$\text{Thread depth } d = 0.7x \text{ (Unified and American depth)}$$
$$\text{Width of root } T = 0.5p \qquad R = 0.137p$$

For studs, 60 per cent thread:

$$\text{Thread depth } d = 0.6x \text{ (Unified and American depth)}$$
$$\text{Width of root } T = 0.425p \qquad R = 0.137p$$

(r) Aero-Thread[10] (Fig. 20.12)

This type is particularly applicable where soft and light internally threaded parts such as aluminum or magnesium alloys are used, as in the aircraft field, with a high-strength steel screw. Wear of the light-alloy thread is prevented by a spring-shaped insert, usually made of phosphor bronze.

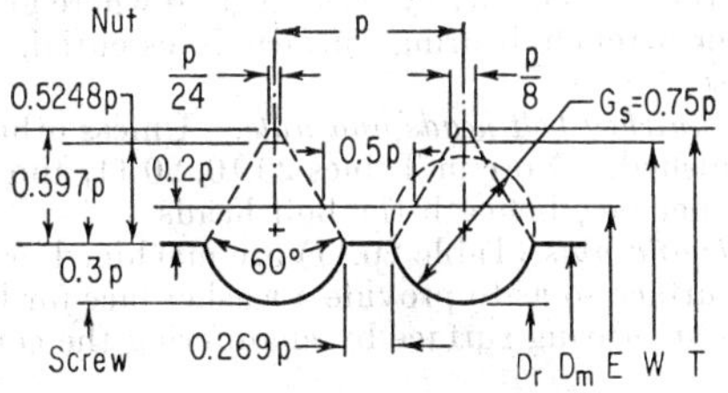

FIG. 20.12. Aero-Thread.[10] (*Courtesy of Aircraft Screw Products Co., Inc.*)

D = major diameter, screw, also minor diameter of tapped hole
E = pitch diameter = $D + 0.4p$
K = minor diameter = $D - 0.6p$
W = major diameter, insert = $D + 1.0495p$
G_s = diameter of thread-form circle = $0.75p$
T = major diameter of tap = $D + 1.194p$ and minimum = $D + 1.122p$

(s) Microscopic-objective Screw Thread[11]

Based on the Whitworth thread, except truncated, this thread is used for mounting the microscope objective to the nosepiece. This thread is recommended for other optical assemblies of microscopes, such as photomicrographic equipment. It has become universally accepted as a standard for microscopic-objective and nosepiece threads.[12]

For data on hose-coupling threads, fire-hose-coupling threads, and threads for electric sockets and lamp bases, see refs. 13, 14, and 15, respectively.

20.2. BOLTS, NUTS, AND SCREWS

(a) Introduction

Screw fasteners are generally classified according to function, style, and application such as bolts, nuts, machine screws, and cap screws.

Proportions of these fasteners have been standardized by the American Standards Association and are summarized herein.

(b) Bolts and Nuts

Bolts, used in through holes with mating nut, are shown with leading dimensions in Tables 20.10, 20.11, and 20.12.

Terms Applied to Square and Hexagon Bolts and Nuts

Series Selection. Square and hexagon-head bolts and nuts[20] are supplied in two series of sizes, known as "regular" series and "heavy" series. Both series refer to the bolt-head or nut proportions. The heavy series is used where greater bearing surface is required or a greater wrench bearing surface is essential. Regular-series bolt heads are for general use.

Finish Selection. Unfinished bolt heads and nuts. Unless otherwise specified, only the thread portion is finished. Note in Tables 20.10, 20.11, and 20.12 the unfinished bolts do not have a washer face beneath the bolt heads.

Semifinished bolt heads and nuts (Table 20.11) are machined or otherwise formed or treated on the bearing surface so as to provide a washer face for bolt heads and either a washer face or a circular bearing surface by chamfering the corners for nuts.

Table 20.10. Regular Unfinished Square Bolts[20]

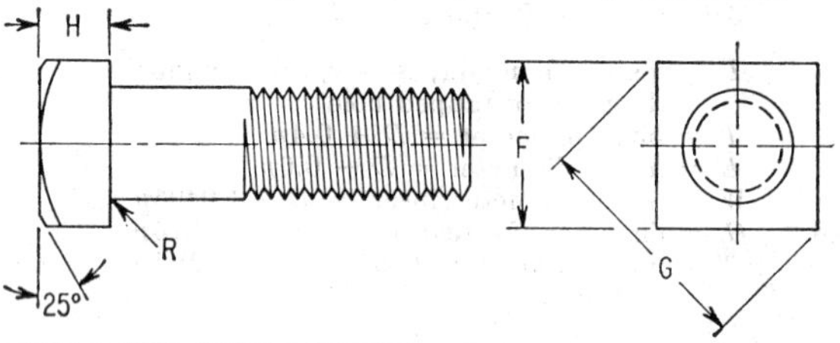

Nominal size or basic major diam of thread		Body diam, max	Width across flats F			Width across corners G		Height H			Radius of fillet R
			Max (basic)		Min	Max	Min	Nom	Max	Min	Max
1/4	0.2500	0.280	3/8	0.3750	0.362	0.530	0.498	11/64	0.188	0.156	0.031
5/16	0.3125	0.342	1/2	0.5000	0.484	0.707	0.665	13/64	0.220	0.186	0.031
3/8	0.3750	0.405	9/16	0.5625	0.544	0.795	0.747	1/4	0.268	0.232	0.031
7/16	0.4375	0.468	5/8	0.6250	0.603	0.884	0.828	19/64	0.316	0.278	0.031
1/2	0.5000	0.530	3/4	0.7500	0.725	1.061	0.995	21/64	0.348	0.308	0.031
5/8	0.6250	0.675	15/16	0.9375	0.906	1.326	1.244	27/64	0.444	0.400	0.062
3/4	0.7500	0.800	1 1/8	1.1250	1.088	1.591	1.494	1/2	0.524	0.476	0.062
7/8	0.8750	0.938	1 5/16	1.3125	1.269	1.856	1.742	19/32	0.620	0.568	0.062
1	1.0000	1.063	1 1/2	1.5000	1.450	2.121	1.991	21/32	0.684	0.628	0.062
1 1/8	1.1250	1.188	1 11/16	1.6875	1.631	2.386	2.239	3/4	0.780	0.720	0.125
1 1/4	1.2500	1.313	1 7/8	1.8750	1.812	2.652	2.489	27/32	0.876	0.812	0.125
1 3/8	1.3750	1.469	2 1/16	2.0625	1.994	2.917	2.738	29/32	0.940	0.872	0.125
1 1/2	1.5000	1.594	2 1/4	2.2500	2.175	3.182	2.986	1	1.036	0.964	0.125
1 5/8	1.6250	1.719	2 7/16	2.4375	2.356	3.447	3.235	1 1/32	1.132	1.056	0.125

Bolt is not finished on any surface.

Minimum thread length shall be twice the diameter plus 1/4 in. for lengths up to and including 6 in. and twice the diameter plus 1/2 in. for lengths over 6 in.

Thread shall be coarse-thread series, class 2A.

Table 20.11. Hexagon Bolts[20]

Dimensions of Regular Semifinished Hexagon Bolts

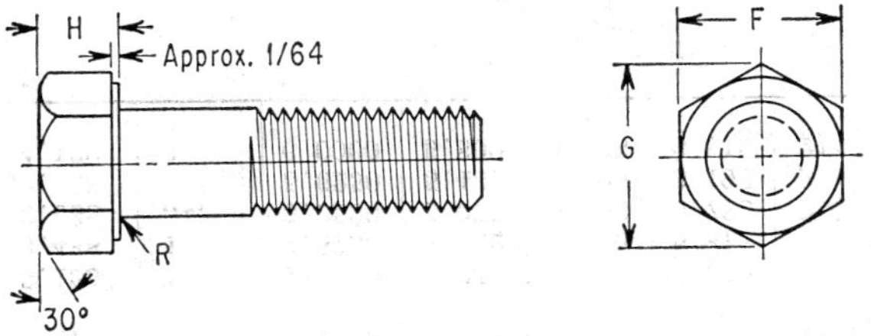

Dimensions of Regular Hexagon Bolts

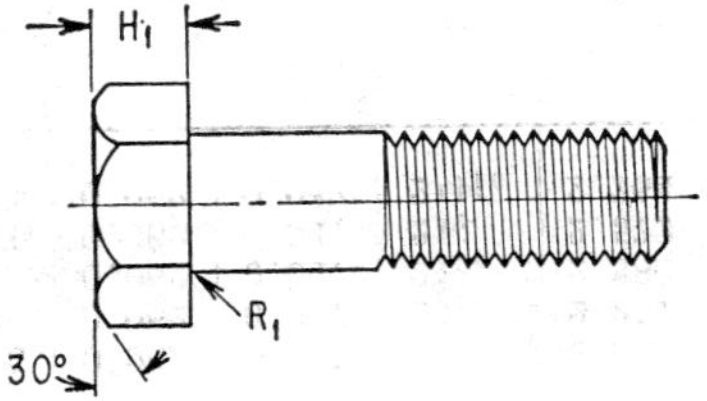

Nominal size or basic major diam of thread		Body diam, max	Width across flats F			Width across corners G		Semifinished height H			Radius of fillet R		Regular height H_1			Radius of fillet R_1
			Max (basic)		Min	Max	Min	Nom	Max	Min	Max	Min	Nom	Max	Min	Max
1/4	0.2500	0.280	7/16	0.4375	0.425	0.505	0.484	5/32	0.163	0.150	0.031	0.016	11/64	0.188	0.150	0.031
5/16	0.3125	0.342	1/2	0.5000	0.484	0.577	0.552	13/64	0.211	0.195	0.031	0.016	7/32	0.235	0.195	0.031
3/8	0.3750	0.405	9/16	0.5625	0.544	0.650	0.620	15/64	0.243	0.226	0.031	0.016	1/4	0.268	0.226	0.031
7/16	0.4375	0.468	5/8	0.6250	0.603	0.722	0.687	9/32	0.291	0.272	0.031	0.016	19/64	0.316	0.272	0.031
1/2	0.5000	0.530	3/4	0.7500	0.725	0.866	0.826	5/16	0.323	0.302	0.031	0.016	11/32	0.364	0.302	0.031
5/8	0.6250	0.675	15/16	0.9375	0.906	1.083	1.033	25/64	0.403	0.378	0.031	0.016	27/64	0.444	0.378	0.062
3/4	0.7500	0.800	1 1/8	1.1250	1.088	1.299	1.240	15/32	0.483	0.455	0.047	0.031	1/2	0.524	0.455	0.062
7/8	0.8750	0.938	1 5/16	1.3125	1.269	1.516	1.447	35/64	0.563	0.531	0.047	0.031	37/64	0.604	0.531	0.062
1	1.0000	1.063	1 1/2	1.5000	1.450	1.732	1.653	39/64	0.627	0.591	0.047	0.031	43/64	0.700	0.591	0.062
1 1/8	1.1250	1.188	1 11/16	1.6875	1.631	1.949	1.859	11/16	0.718	0.658	0.062	0.047	3/4	0.780	0.658	0.125
1 1/4	1.2500	1.313	1 7/8	1.8750	1.812	2.165	2.066	25/32	0.813	0.749	0.062	0.047	27/32	0.876	0.749	0.125
1 3/8	1.3750	1.469	2 1/16	2.0625	1.994	2.382	2.273	27/32	0.878	0.810	0.062	0.047	29/32	0.940	0.810	0.125

Table 20.11. Hexagon Bolts[20] (Continued)

Nominal size or basic major diam of thread		Body diam, max	Width across flats F			Width across corners G		Semifinished height H			Radius of fillet R		Regular height H_1			Radius of fillet R_1
			Max (basic)		Min	Max	Min	Nom	Max	Min	Max	Min	Nom	Max	Min	Max
1½	1.5000	1.594	2¼	2.2500	2.175	2.598	2.480	15/16	0.974	0.902	0.062	0.047	1	1.036	0.902	0.125
1⅝	1.6250	1.719	2 7/16	2.4375	2.356	2.815	2.686	1	1.038	0.962	0.062	0.047	1 1/16	1.100	0.962	0.125
1¾	1.7500	1.844	2⅝	2.6250	2.538	3.031	2.893	1 3/32	1.134	1.054	0.062	0.047	1 5/32	1.196	1.054	0.125
1⅞	1.8750	1.969	2 13/16	2.8125	2.719	3.248	3.100	1 5/32	1.198	1.114	0.062	0.047	1 7/32	1.260	1.114	0.125
2	2.0000	2.094	3	3.0000	2.900	3.464	3.306	1 7/32	1.263	1.175	0.062	0.047	1 11/32	1.388	1.175	0.125
2¼	2.2500	2.375	3⅜	3.3750	3.262	3.897	3.719	1⅜	1.423	1.327	0.062	0.047	1½	1.548	1.327	0.188
2½	2.5000	2.625	3¾	3.7500	3.625	4.330	4.133	1 17/32	1.583	1.479	0.062	0.047	1 21/32	1.708	1.479	0.188
2¾	2.7500	2.875	4⅛	4.1250	3.988	4.763	4.546	1 11/16	1.744	1.632	0.062	0.047	1 13/16	1.869	1.632	0.188
3	3.0000	3.125	4½	4.5000	4.350	5.196	4.959	1⅞	1.935	1.815	0.062	0.047	2	2.060	1.815	0.188
3¼	3.2500	3.438	4⅞	4.8750	4.712	5.629	5.372	2	2.064	1.936	0.062	0.047	2 3/16	2.251	1.936	0.188
3½	3.5000	3.688	5¼	5.2500	5.075	6.062	5.786	2⅛	2.193	2.057	0.062	0.047	2 5/16	2.380	2.057	0.188
3¾	3.7500	3.938	5⅝	5.6250	5.437	6.495	6.198	2 5/16	2.385	2.241	0.062	0.047	2½	2.572	2.241	0.188
4	4.0000	4.188	6	6.0000	5.800	6.928	6.612	2½	2.576	2.424	0.062	0.047	2 11/16	2.764	2.424	0.188

Table 20.12. Finished Hexagon Bolts[20]

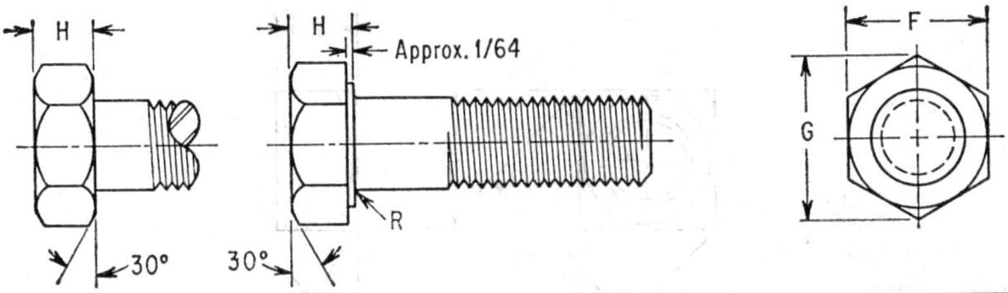

Nominal size or basic major diam of thread		Body diam, min (max equal to nominal size)	Width across flats F			Width across corners G		Height H			Radius of fillet R	
			Max (basic)		Min	Max	Min	Nom	Max	Min	Max	Min
1/4	**0.2500**	**0.2450**	**7/16**	**0.4375**	**0.428**	**0.505**	**0.488**	**5/32**	**0.163**	**0.150**	**0.023**	**0.009**
5/16	**0.3125**	**0.3065**	**1/2**	**0.5000**	**0.489**	**0.577**	**0.557**	**13/64**	**0.211**	**0.195**	**0.023**	**0.009**
3/8	**0.3750**	**0.3690**	**9/16**	**0.5625**	**0.551**	**0.650**	**0.628**	**15/64**	**0.243**	**0.226**	**0.023**	**0.009**
7/16	**0.4375**	**0.4305**	**5/8**	**0.6250**	**0.612**	**0.722**	**0.698**	**9/32**	**0.291**	**0.272**	**0.023**	**0.009**
1/2	**0.5000**	**0.4930**	**3/4**	**0.7500**	**0.736**	**0.866**	**0.840**	**5/16**	**0.323**	**0.302**	**0.023**	**0.009**
9/16	**0.5625**	**0.5545**	**13/16**	**0.8125**	**0.798**	**0.938**	**0.910**	**23/64**	**0.371**	**0.348**	**0.041**	**0.021**
5/8	**0.6250**	**0.6170**	**15/16**	**0.9375**	**0.922**	**1.083**	**1.051**	**25/64**	**0.403**	**0.378**	**0.041**	**0.021**
3/4	**0.7500**	**0.7410**	**1 1/8**	**1.1250**	**1.100**	**1.299**	**1.254**	**15/32**	**0.483**	**0.455**	**0.041**	**0.021**
7/8	**0.8750**	**0.8660**	**1 5/16**	**1.3125**	**1.285**	**1.516**	**1.465**	**35/64**	**0.563**	**0.531**	**0.062**	**0.047**
1	**1.0000**	**0.9900**	**1 1/2**	**1.5000**	**1.469**	**1.732**	**1.675**	**39/64**	**0.627**	**0.591**	**0.062**	**0.047**
1⅛	1.1250	1.1140	1 11/16	1.6875	1.631	1.949	1.859	11/16	0.718	0.658	0.125	0.110
1¼	1.2500	1.2390	1⅞	1.8750	1.812	2.165	2.066	25/32	0.813	0.749	0.125	0.110
1⅜	1.3750	1.3630	2 1/16	2.0625	1.994	2.382	2.273	27/32	0.878	0.810	0.125	0.110
1½	1.5000	1.4880	2¼	2.2500	2.175	2.598	2.480	15/16	0.974	0.902	0.125	0.110
1⅝	1.6250	1.6130	2 7/16	2.4275	2.356	2.815	2.686	1	1.038	0.962	0.125	0.110
1¾	1.7500	1.7380	2⅝	2.6250	2.538	3.031	2.893	1 3/32	1.134	1.054	0.125	0.110
1⅞	1.8750	1.8630	2 13/16	2.8125	2.719	3.248	3.100	1 5/32	1.198	1.114	0.125	0.110
2	2.0000	1.9880	3	3.0000	2.900	3.464	3.306	1 7/32	1.263	1.175	0.125	0.110
2¼	2.2500	2.2380	3⅜	3.3750	3.262	3.897	3.719	1⅜	1.423	1.327	0.188	0.173
2½	2.5000	2.4880	3¾	3.7500	3.625	4.330	4.133	1 17/32	1.583	1.479	0.188	0.173
2¾	2.7500	2.7380	4⅛	4.1250	3.988	4.763	4.546	1 11/16	1.744	1.632	0.188	0.173
3	3.0000	2.9880	4½	4.5000	4.350	5.196	4.959	1⅞	1.935	1.815	0.188	0.173

Bold type indicates unified thread.

Table 20.13. Regular Square Nuts[20]

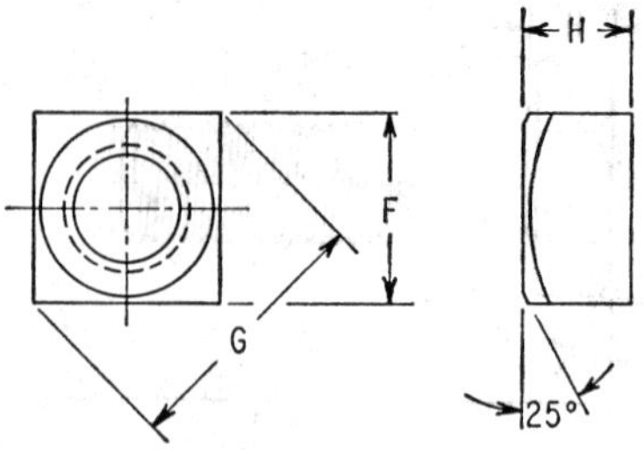

Nominal size or basic major diam of thread		Width across flats F Max (basic)		Width across flats F Min	Width across corners G Max	Width across corners G Min	Thickness H Nom	Thickness H Max	Thickness H Min
1/4	0.2500	7/16	0.4375	0.425	0.619	0.584	7/32	0.235	0.203
5/16	0.3125	9/16	0.5625	0.547	0.795	0.751	17/64	0.283	0.249
3/8	0.3750	5/8	0.6250	0.606	0.884	0.832	21/64	0.346	0.310
7/16	0.4375	3/4	0.7500	0.728	1.061	1.000	3/8	0.394	0.356
1/2	0.5000	13/16	0.8125	0.788	1.149	1.082	7/16	0.458	0.418
5/8	0.6250	1	1.0000	0.969	1.414	1.330	35/64	0.569	0.525
3/4	0.7500	1 1/8	1.1250	1.088	1.591	1.494	21/32	0.680	0.632
7/8	0.8750	1 5/16	1.3125	1.269	1.856	1.742	49/64	0.792	0.740
1	1.0000	1 1/2	1.5000	1.450	2.121	1.991	7/8	0.903	0.847
1 1/8	1.1250	1 11/16	1.6875	1.631	2.386	2.239	1	1.030	0.970
1 1/4	1.2500	1 7/8	1.8750	1.812	2.652	2.489	1 3/32	1.126	1.062
1 3/8	1.3750	2 1/16	2.0625	1.994	2.917	2.738	1 13/64	1.237	1.169
1 1/2	1.5000	2 1/4	2.2500	2.175	3.182	2.986	1 5/16	1.348	1.276
1 5/8	1.6250	2 7/16	2.4375	2.356	3.447	3.235	1 27/64	1.460	1.384

Finished bolt heads and nuts (Table 20.12) refers to quality of manufacture and closeness of tolerance and does not indicate that surfaces are necessarily machined.

Nut styles are shown with leading dimensions in Tables 20.13 through 20.16, inclusive.

Stud Bolts. To date there is no American Standard for stud bolts; however, the Industrial Fasteners Institute[31] has published a recommended standard. This standard has four basic classes of studs.

Class 1. Tap end
Class 2. Double end
Class 3. Bolt studs
Class 4. Continuous thread

The first two classes are then subdivided into four types:

a. Unfinished stud
b. Optional, full or undersized body
c. Full-sized body
d. Close-tolerance body

For stud proportions in general usage today, see ref. 21.

Table 20.14. Hexagon and Hexagon Jam Nuts[20]

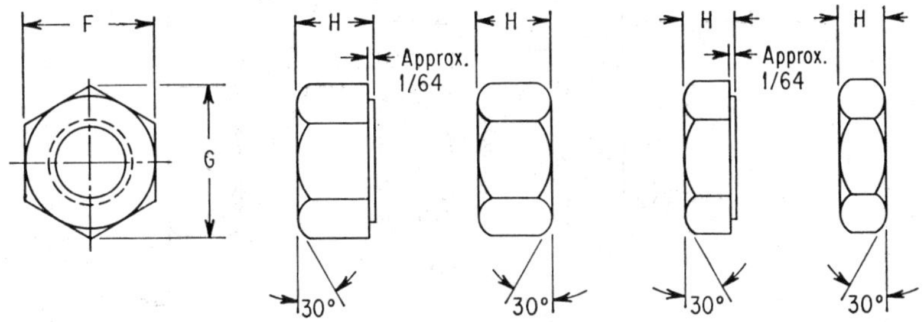

Nominal size or basic major diam of thread		Width across flats *F*			Width across corners *G*		Finished and regular semi-finished hexagon nuts thickness *H*			Finished and regular semi-finished jam nuts thickness *H*		
		Max (basic)		Min	Max	Min	Nom	Max	Min	Nom	Max	Min
1/4	**0.2500**	**7/16**	**0.4375**	**0.428**	**0.505**	**0.488**	**7/32**	**0.226**	**0.212**	**5/32**	**0.163**	**0.150**
5/16	**0.3125**	**1/2**	**0.5000**	**0.489**	**0.577**	**0.557**	**17/64**	**0.273**	**0.258**	**3/16**	**0.195**	**0.180**
3/8	**0.3750**	**9/16**	**0.5625**	**0.551**	**0.650**	**0.628**	**21/64**	**0.337**	**0.320**	**7/32**	**0.227**	**0.210**
7/16	**0.4375**	**11/16**	**0.6875**	**0.675**	**0.794**	**0.768**	**3/8**	**0.385**	**0.365**	**1/4**	**0.260**	**0.240**
1/2	**0.5000**	**3/4**	**0.7500**	**0.736**	**0.866**	**0.840**	**7/16**	**0.448**	**0.427**	**5/16**	**0.323**	**0.302**
9/16	**0.5625**	**7/8**	**0.8750**	**0.861**	**1.010**	**0.982**	**31/64**	**0.496**	**0.473**	**6/16**	**0.324**	**0.301**
5/8	**0.6250**	**15/16**	**0.9375**	**0.922**	**1.083**	**1.051**	**35/64**	**0.559**	**0.535**	**3/8**	**0.387**	**0.363**
3/4	**0.7500**	**1 1/8**	**1.1250**	**1.088**	**1.299**	**1.240**	**41/64**	**0.665**	**0.617**	**27/64**	**0.446**	**0.398**
7/8	**0.8750**	**1 5/16**	**1.3125**	**1.269**	**1.516**	**1.447**	**3/4**	**0.776**	**0.724**	**31/64**	**0.510**	**0.458**
1	**1.0000**	**1 1/2**	**1.5000**	**1.450**	**1.732**	**1.653**	**55/64**	**0.887**	**0.831**	**35/64**	**0.575**	**0.519**
1 1/8	1.1250	1 11/16	1.6875	1.631	1.949	1.859	31/32	0.999	0.939	39/64	0.639	0.579
1 1/4	1.2500	1 7/8	1.8750	1.812	2.165	2.066	1 1/16	1.094	1.030	23/32	0.751	0.687
1 3/8	1.3750	2 1/16	2.0625	1.994	2.382	2.273	1 11/64	1.206	1.138	25/32	0.815	0.747
1 1/2	1.5000	2 1/4	2.2500	2.175	2.598	2.480	1 9/32	1.317	1.245	27/32	0.880	0.808
1 5/8	1.6250	2 7/16	2.4375	2.356	2.815	2.686	1 25/64	1.429	1.353	29/32	0.944	0.868
1 3/4	1.7500	2 5/8	2.6250	2.538	3.031	2.893	1 1/2	1.540	1.460	31/32	1.009	0.929
1 7/8	1.8750	2 13/16	2.8125	2.719	3.248	3.100	1 39/64	1.651	1.567	1 1/32	1.073	0.989
2	2.0000	3	3.0000	2.900	3.464	3.306	1 23/32	1.763	1.675	1 3/32	1.138	1.050
2 1/4	2.2500	3 3/8	3.3750	3.262	3.897	3.719	1 59/64	1.970	1.874	1 13/64	1.251	1.155
2 1/2	2.5000	3 3/4	3.7500	3.625	4.330	4.133	2 9/64	2.193	2.089	1 29/64	1.505	1.401
2 3/4	2.7500	4 1/8	4.1250	3.988	4.763	4.546	2 23/64	2.415	2.303	1 37/64	1.634	1.522
3	3.0000	4 1/2	4.5000	4.350	5.196	4.959	2 37/64	2.638	2.518	1 45/64	1.763	1.643

Bold type indicates unified threads.

Table 20.15. Finished Hexagon Slotted Nuts[20]

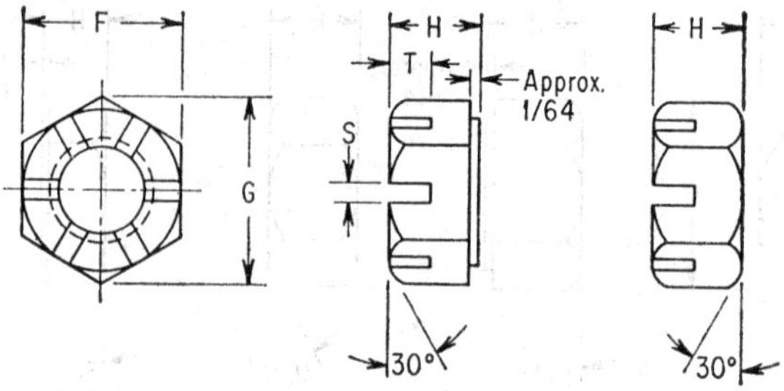

Nominal size or basic major diam of thread		Width across flats F			Width across corners G		Thickness H			Slot	
		Max (basic)		Min	Max	Min	Nom	Max	Min	Width S	Depth T
1/4	**0.2500**	**7/16**	**0.4375**	**0.428**	**0.505**	**0.488**	**7/32**	**0.226**	**0.212**	0.078	0.094
5/16	**0.3125**	**1/2**	**0.5000**	**0.489**	**0.577**	**0.557**	**17/64**	**0.273**	**0.258**	0.094	0.094
3/8	**0.3750**	**9/16**	**0.5625**	**0.551**	**0.650**	**0.628**	**21/64**	**0.337**	**0.320**	0.125	0.125
7/16	**0.4375**	**11/16**	**0.6875**	**0.675**	**0.794**	**0.768**	**3/8**	**0.385**	**0.365**	0.125	0.156
1/2	**0.5000**	**3/4**	**0.7500**	**0.736**	**0.866**	**0.840**	**7/16**	**0.448**	**0.427**	0.156	0.156
9/16	**0.5625**	**7/8**	**0.8750**	**0.861**	**1.010**	**0.982**	**31/64**	**0.496**	**0.473**	0.156	0.188
5/8	**0.6250**	**15/16**	**0.9375**	**0.922**	**1.083**	**1.051**	**35/64**	**0.559**	**0.535**	0.188	0.219
3/4	**0.7500**	**1 1/8**	**1.1250**	**1.088**	**1.299**	**1.240**	**41/64**	**0.665**	**0.617**	0.188	0.250
7/8	**0.8750**	**1 5/16**	**1.3125**	**1.269**	**1.516**	**1.447**	**3/4**	**0.776**	**0.724**	0.188	0.250
1	**1.0000**	**1 1/2**	**1.5000**	**1.450**	**1.732**	**1.653**	**55/64**	**0.887**	**0.831**	0.250	0.281
1 1/8	1.1250	1 11/16	1.6875	1.631	1.949	1.859	31/32	0.999	0.939	0.250	0.344
1 1/4	1.2500	1 7/8	1.8750	1.812	2.165	2.066	1 1/16	1.094	1.030	0.312	0.375
1 3/8	1.3750	2 1/16	2.0625	1.994	2.382	2.273	1 11/64	1.206	1.138	0.312	0.375
1 1/2	1.5000	2 1/4	2.2500	2.175	2.598	2.480	1 9/32	1.317	1.245	0.375	0.438
1 5/8	1.6250	2 7/16	2.4375	2.356	2.815	2.686	1 25/64	1.429	1.353	0.375	0.438
1 3/4	1.7500	2 5/8	2.6250	2.538	3.031	2.893	1 1/2	1.540	1.460	0.438	0.500
1 7/8	1.8750	2 13/16	2.8125	2.719	3.248	3.100	1 39/64	1.651	1.567	0.438	0.562
2	2.0000	3	3.0000	2.900	3.464	3.306	1 23/32	1.763	1.675	0.438	0.562
2 1/4	2.2500	3 3/8	3.3750	3.262	3.897	3.719	1 59/64	1.970	1.874	0.438	0.562
2 1/2	2.5000	3 3/4	3.7500	3.625	4.330	4.133	2 9/64	2.193	2.089	0.562	0.688
2 3/4	2.7500	4 1/8	4.1250	3.988	4.763	4.546	2 23/64	2.415	2.303	0.562	0.688
3	3.0000	4 1/2	4.5000	4.350	5.196	4.959	2 37/64	2.638	2.518	0.625	0.750

Bold type indicates unified threads.

Table 20.16. Regular Hexagon and Hexagon Jam Nuts[20]

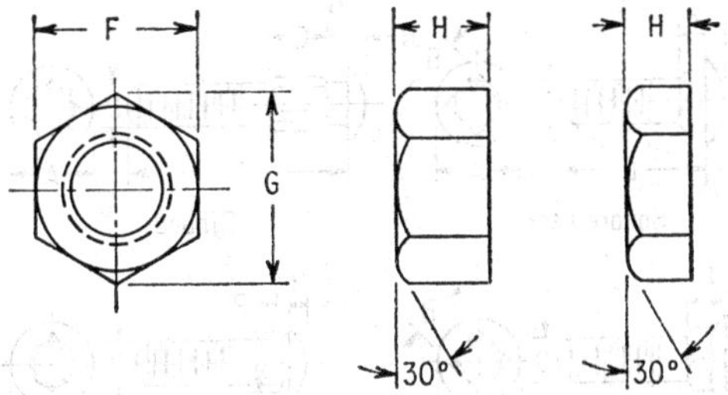

Nominal size or basic major diam of thread		Width across flats F			Width across corners G		Thickness regular nuts H			Thickness regular jam nuts H		
		Max (basic)		Min	Max	Min	Nom	Max	Min	Nom	Max	Min
1/4	0.2500	7/16	0.4375	0.425	0.505	0.484	7/32	0.235	0.203	5/32	0.172	0.140
5/16	0.3125	9/16	0.5625	0.547	0.650	0.624	17/64	0.283	0.249	3/16	0.204	0.170
3/8	0.3750	5/8	0.6250	0.606	0.722	0.691	21/64	0.346	0.310	7/32	0.237	0.201
7/16	0.4375	3/4	0.7500	0.728	0.866	0.830	3/8	0.394	0.356	1/4	0.269	0.231
1/2	0.5000	13/16	0.8125	0.788	0.938	0.898	7/16	0.458	0.418	5/16	0.332	0.292
9/16	0.5625	7/8	0.8750	0.847	1.010	0.966	1/2	0.521	0.479	11/32	0.365	0.323
5/8	0.6250	1	1.0000	0.969	1.155	1.104	35/64	0.569	0.525	3/8	0.397	0.353
3/4	0.7500	1 1/8	1.1250	1.088	1.299	1.240	21/32	0.680	0.632	7/16	0.462	0.414
7/8	0.8750	1 5/16	1.3125	1.269	1.516	1.447	49/64	0.792	0.740	1/2	0.526	0.474
1	1.0000	1 1/2	1.5000	1.450	1.732	1.653	7/8	0.903	0.847	9/16	0.590	0.534
1 1/8	1.1250	1 11/16	1.6875	1.631	1.949	1.859	1	1.030	0.970	5/8	0.655	0.595
1 1/4	1.2500	1 7/8	1.8750	1.812	2.165	2.066	1 3/32	1.126	1.062	3/4	0.782	0.718
1 3/8	1.3750	2 1/16	2.0625	1.994	2.382	2.273	1 13/64	1.237	1.169	13/16	0.846	0.778
1 1/2	1.5000	2 1/4	2.2500	2.175	2.598	2.480	1 5/16	1.348	1.276	7/8	0.911	0.839

Round-head Bolts. Carriage bolts, used where the head bears against wood, are illustrated in Table 20.17. Threads are Unified National coarse class 2A. Minimum thread length $= 2D + \frac{1}{4}$ for bolts up to 6 in. long, $2D + \frac{1}{2}$ for bolts over 6 in. long.

Styles and leading dimensions of countersunk, buttonhead, and step bolts are given in Table 20.18.

Elevator bolts[22] and plow bolts[23] are similar to carriage bolts.[22]

Track bolts[24] are specials, designed to meet the requirements of the railways.

Table 20.17. Carriage Bolts

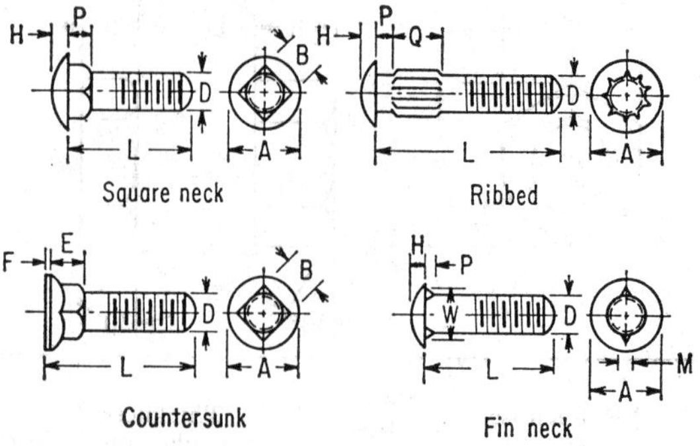

Nominal diam of bolt *D*	Countersunk: Diam of head, max, *A*	Countersunk: Feed thickness *F*	Countersunk: Depth of square and countersink, max, *E*	Countersunk: Width of square, max, *B*	Fin neck: Diam of head, max, *A*	Fin neck: Height of head, max, *H*	Fin neck: Depth of fins, max, *P*	Fin neck: Distance across fins, max, *W*	Fin neck: Thickness of fins, max, *M*
No. 10	0.520	0.016	0.250	0.199	0.469	0.114	0.088	0.395	0.098
¼	0.645	0.016	0.312	0.260	0.594	0.145	0.104	0.458	0.114
5/16	0.770	0.031	0.375	0.324	0.719	0.176	0.135	0.551	0.145
⅜	0.895	0.031	0.437	0.388	0.844	0.208	0.151	0.645	0.161
7/16	1.020	0.031	0.500	0.452	0.969	0.239	0.182	0.739	0.192
½	1.145	0.031	0.562	0.515	1.094	0.270	0.198	0.833	0.208
⅝	1.400	0.031	0.687	0.642					
¾	1.650	0.047	0.812	0.768					

Nominal diam of bolt *D*	Diam of head, max, *A*	Height of head, max, *H*	Square neck: Depth of square, max, *P*	Square neck: Width of square, max, *B*	Ribbed neck: Ribs below head *P*, $L \leqq$ ⅞	Ribbed neck: Ribs below head *P*, $L \geqq 1$	Ribbed neck: Length of ribs *Q*, $L \leqq$ ⅞	Ribbed neck: Length of ribs *Q*, $L = 1$, $L =$ 1⅛	Ribbed neck: Length of ribs *Q*, $L \geqq$ 1¼	Number of ribs
No. 10	0.469	0.114	0.125	0.199	0.031	0.063	0.188	0.313	0.500	9
¼	0.594	0.145	0.156	0.260	0.031	0.063	0.188	0.313	0.500	10
5/16	0.719	0.176	0.187	0.324	0.031	0.063	0.188	0.313	0.500	12
⅜	0.844	0.208	0.219	0.388	0.031	0.063	0.188	0.313	0.500	12
7/16	0.969	0.239	0.250	0.452	0.031	0.063	0.188	0.313	0.500	14
½	1.094	0.270	0.281	0.515	0.031	0.063	0.188	0.313	0.500	16
⅝	1.344	0.344	0.344	0.642	0.094	0.094	0.188	0.313	0.500	19
¾	1.594	0.406	0.406	0.768	0.094	0.094	0.188	0.313	0.500	22
⅞	1.844	0.469	0.469	0.895						
1	2.094	0.531	0.531	1.022						

Table 20.18. Countersunk, Buttonhead, and Step Bolts

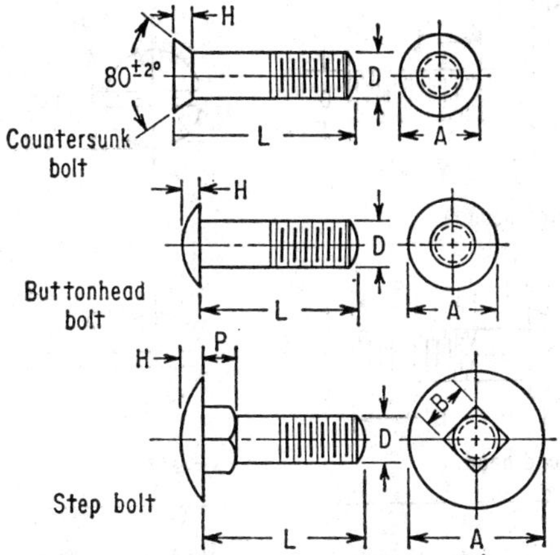

Countersunk bolt				Buttonhead		Step bolt			
Nominal diam of bolt *D*		Head diam, max, *A*	Head depth *H*	Head diam, max, *A*	Head height, max, *H*	Head diam, max, *A*	Head height, max, *H*	Depth of square, max, *P*	Width of square, max, *B*
10	24	.....	.....	0.469	0.114	0.656	0.114	0.125	0.199
¼	20	0.493	0.140	0.594	0.145	0.844	0.145	0.156	0.260
5⁄16	18	0.618	0.176	0.719	0.176	1.031	0.176	0.187	0.324
⅜	16	0.740	0.210	0.844	0.208	1.219	0.208	0.219	0.388
7⁄16	14	0.803	0.210	0.969	0.239	1.406	0.239	0.250	0.452
½	13	0.935	0.250	1.094	0.270	1.594	0.270	0.281	0.515
⅝	11	1.169	0.313	1.344	0.344				
¾	10	1.402	0.375	1.594	0.406				
⅞	9	1.637	0.438	1.844	0.469				
1	8	1.869	0.500	2.094	0.531				
1⅛	7	2.104	0.563						
1¼	7	2.337	0.625						
1⅜	6	2.571	0.688						
1½	6	2.804	0.750						

Table 20.19. Machine-screw Heads[25,32]

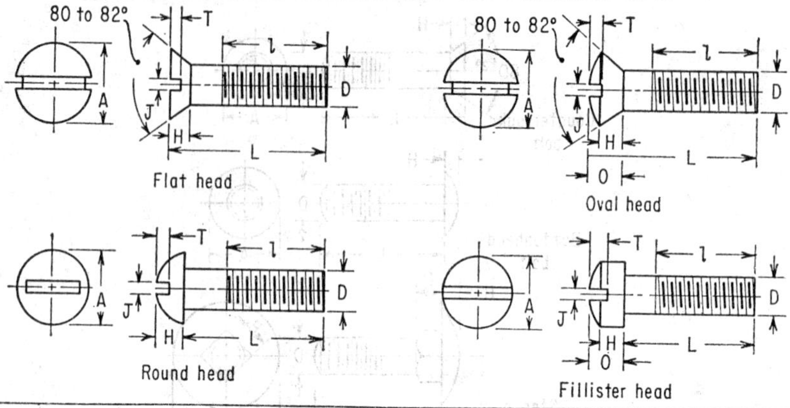

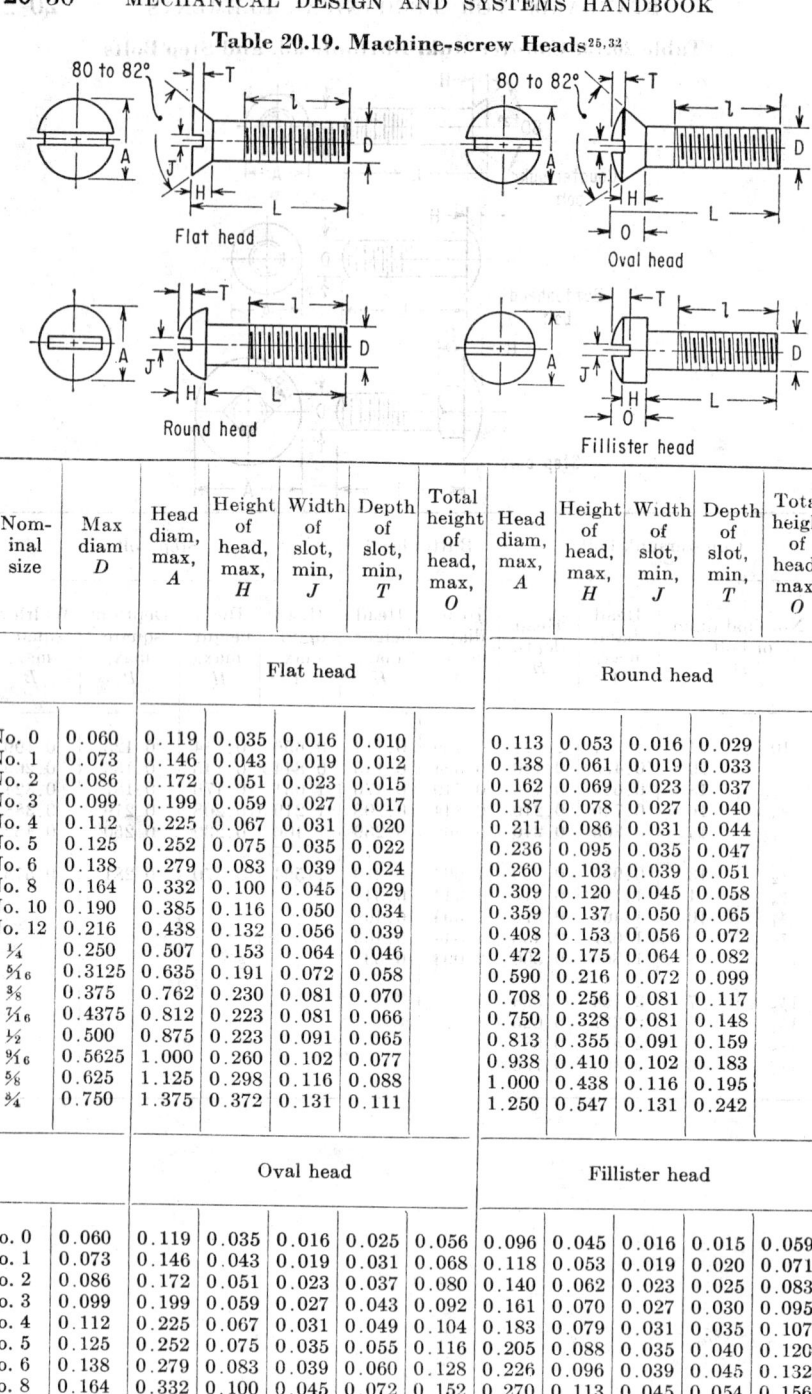

Nominal size	Max diam *D*	Head diam, max, *A*	Height of head, max, *H*	Width of slot, min, *J*	Depth of slot, min, *T*	Total height of head, max, *O*	Head diam, max, *A*	Height of head, max, *H*	Width of slot, min, *J*	Depth of slot, min, *T*	Total height of head, max, *O*
		Flat head					Round head				
No. 0	0.060	0.119	0.035	0.016	0.010		0.113	0.053	0.016	0.029	
No. 1	0.073	0.146	0.043	0.019	0.012		0.138	0.061	0.019	0.033	
No. 2	0.086	0.172	0.051	0.023	0.015		0.162	0.069	0.023	0.037	
No. 3	0.099	0.199	0.059	0.027	0.017		0.187	0.078	0.027	0.040	
No. 4	0.112	0.225	0.067	0.031	0.020		0.211	0.086	0.031	0.044	
No. 5	0.125	0.252	0.075	0.035	0.022		0.236	0.095	0.035	0.047	
No. 6	0.138	0.279	0.083	0.039	0.024		0.260	0.103	0.039	0.051	
No. 8	0.164	0.332	0.100	0.045	0.029		0.309	0.120	0.045	0.058	
No. 10	0.190	0.385	0.116	0.050	0.034		0.359	0.137	0.050	0.065	
No. 12	0.216	0.438	0.132	0.056	0.039		0.408	0.153	0.056	0.072	
¼	0.250	0.507	0.153	0.064	0.046		0.472	0.175	0.064	0.082	
5⁄16	0.3125	0.635	0.191	0.072	0.058		0.590	0.216	0.072	0.099	
⅜	0.375	0.762	0.230	0.081	0.070		0.708	0.256	0.081	0.117	
7⁄16	0.4375	0.812	0.223	0.081	0.066		0.750	0.328	0.081	0.148	
½	0.500	0.875	0.223	0.091	0.065		0.813	0.355	0.091	0.159	
9⁄16	0.5625	1.000	0.260	0.102	0.077		0.938	0.410	0.102	0.183	
⅝	0.625	1.125	0.298	0.116	0.088		1.000	0.438	0.116	0.195	
¾	0.750	1.375	0.372	0.131	0.111		1.250	0.547	0.131	0.242	
		Oval head					Fillister head				
No. 0	0.060	0.119	0.035	0.016	0.025	0.056	0.096	0.045	0.016	0.015	0.059
No. 1	0.073	0.146	0.043	0.019	0.031	0.068	0.118	0.053	0.019	0.020	0.071
No. 2	0.086	0.172	0.051	0.023	0.037	0.080	0.140	0.062	0.023	0.025	0.083
No. 3	0.099	0.199	0.059	0.027	0.043	0.092	0.161	0.070	0.027	0.030	0.095
No. 4	0.112	0.225	0.067	0.031	0.049	0.104	0.183	0.079	0.031	0.035	0.107
No. 5	0.125	0.252	0.075	0.035	0.055	0.116	0.205	0.088	0.035	0.040	0.120
No. 6	0.138	0.279	0.083	0.039	0.060	0.128	0.226	0.096	0.039	0.045	0.132
No. 8	0.164	0.332	0.100	0.045	0.072	0.152	0.270	0.113	0.045	0.054	0.156
No. 10	0.190	0.385	0.116	0.050	0.084	0.176	0.313	0.130	0.050	0.064	0.180

Table 20.19. Machine-screw Heads[25,32] (Continued)

Nominal size	Max diam D	Oval head: Head diam, max, A	Oval head: Height of head, max, H	Oval head: Width of slot, min, J	Oval head: Depth of slot, min, T	Oval head: Total height of head, max, O	Fillister head: Head diam, max, A	Fillister head: Height of head, max, H	Fillister head: Width of slot, min, J	Fillister head: Depth of slot, min, T	Fillister head: Total height of head, max, O
No. 12	0.216	0.438	0.132	0.056	0.096	0.200	0.357	0.148	0.056	0.074	0.205
¼	0.250	0.507	0.153	0.064	0.112	0.232	0.414	0.170	0.064	0.087	0.237
5/16	0.3125	0.635	0.191	0.072	0.141	0.290	0.518	0.211	0.072	0.110	0.295
⅜	0.375	0.762	0.230	0.081	0.170	0.347	0.622	0.253	0.081	0.133	0.355
7/16	0.4375	0.812	0.223	0.081	0.174	0.345	0.625	0.265	0.081	0.135	0.368
½	0.500	0.875	0.223	0.091	0.176	0.354	0.750	0.297	0.091	0.151	0.412
9/16	0.5625	1.000	0.260	0.102	0.207	0.410	0.812	0.336	0.102	0.172	0.466
⅝	0.625	1.125	0.298	0.116	0.235	0.467	0.875	0.375	0.116	0.193	0.521
¾	0.750	1.375	0.372	0.131	0.293	0.578	1.000	0.441	0.131	0.226	0.612

Edges of head on flat- and oval-head machine screws may be rounded.

Radius of fillet at base of flat- and oval-head machine screws shall not exceed twice the pitch of the screw thread.

Radius of fillet at base of round- and fillister-head machine screws shall not exceed one-half the pitch of the screw thread.

All four types of screws in this table may be furnished with cross-recessed heads.

Fillister-head machine screws in sizes No. 2 to ⅜ in., inclusive, may be furnished with a drilled hole through the head along a diameter at right angles to the slot but not breaking through the slot.

(c) Machine Screws (Tables 20.19 to 20.21)

Machine screws are classified according to head style. From the tables note that the No. 0 series has a body diameter of 0.060 in. and successive numbers add 0.013 in. to the diameter; e.g.,

$$\text{body diameter of No. 6 screw} = 0.060 + 0.013(6) = 0.060 + 0.078 = 0.138$$

Machine screws are regularly supplied with plain sheared ends, not pointed. On machine screws up to 2 in. long, complete threads extend to within two threads of the bearing surface of the head; long screws have a minimum complete thread length of 1¾ in.

Table 20.20. Machine-screw Heads—Pan, Hexagon, Truss, and 100° Flat Heads[25,32]

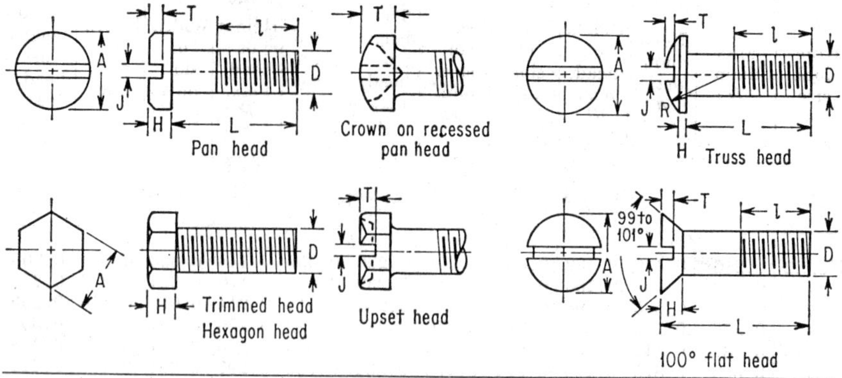

Nominal size	Max diam *D*	Head diam, max, *A*	Height of slotted head, max, *H*	Width of slot, min, *J*	Depth of slot, min, *T*	Radius *R*	Height of recessed head, max, *O*	Head diam, max, *A*	Height of head, max, *H*	Width of slot, min, *J*	Depth of slot, min, *T*
		Pan head						Hexagon head			
No. 2	0.086	0.167	0.053	0.023	0.023	0.035	0.062	0.125	0.050		
No. 3	0.099	0.193	0.060	0.027	0.027	0.037	0.071	0.187	0.055		
No. 4	0.112	0.219	0.068	0.031	0.030	0.042	0.080	0.187	0.060	0.031	0.025
No. 5	0.125	0.245	0.075	0.035	0.032	0.044	0.089	0.187	0.070	0.035	0.030
No. 6	0.138	0.270	0.082	0.039	0.038	0.046	0.097	0.250	0.080	0.039	0.033
No. 8	0.164	0.322	0.096	0.045	0.043	0.052	0.115	0.250	0.110	0.045	0.052
No. 10	0.190	0.373	0.110	0.050	0.050	0.061	0.133	0.312	0.120	0.050	0.057
No. 12	0.216	0.425	0.125	0.056	0.060	0.078	0.151	0.312	0.155	0.056	0.077
1/4	0.250	0.492	0.144	0.064	0.070	0.087	0.175	0.375	0.190	0.064	0.083
5/16	0.3125	0.615	0.178	0.072	0.092	0.099	0.218	0.500	0.230	0.072	0.100
3/8	0.375	0.740	0.212	0.081	0.113	0.143	0.261	0.562	0.295	0.081	0.131
		Truss head						100° flat head			
No. 2	0.086	0.194	0.053	0.023	0.022	0.129					
No. 3	0.099	0.226	0.061	0.027	0.026	0.151					
No. 4	0.112	0.257	0.069	0.031	0.030	0.169		0.225	0.048	0.031	0.017
No. 5	0.125	0.289	0.078	0.035	0.034	0.191					
No. 6	0.138	0.321	0.086	0.039	0.037	0.211		0.279	0.060	0.039	0.022
No. 8	0.164	0.384	0.102	0.045	0.045	0.254		0.332	0.072	0.045	0.027
No. 10	0.190	0.448	0.118	0.050	0.053	0.283		0.385	0.083	0.050	0.031
No. 12	0.216	0.511	0.134	0.056	0.061	0.336					
1/4	0.250	0.573	0.150	0.064	0.070	0.375		0.507	0.110	0.064	0.042
5/16	0.3125	0.698	0.183	0.072	0.085	0.457		0.635	0.138	0.072	0.053
3/8	0.375	0.823	0.215	0.081	0.100	0.538		0.762	0.165	0.081	0.064
7/16	0.4375	0.948	0.248	0.081	0.116	0.619					
1/2	0.500	1.073	0.280	0.091	0.131	0.701					
9/16	0.5625	1.198	0.312	0.102	0.146	0.783					
5/8	0.625	1.323	0.345	0.116	0.162	0.863					
3/4	0.750	1.573	0.410	0.131	0.182	1.024					

Radius of fillet at base of truss- and pan-head machine screws shall not exceed one-half the pitch of the screw thread.

Truss-, pan-, and 100° flat-head machine screws may be furnished with cross-recessed heads.

Hexagon-head machine screws are usually not slotted; the slot is optional. Also optional is an upset-head type for hexagon-head machine screws of sizes 4, 5, 8, 12, and 1/4 in.

Table 20.21. Machine-screw Heads—Binding Head[25,32]

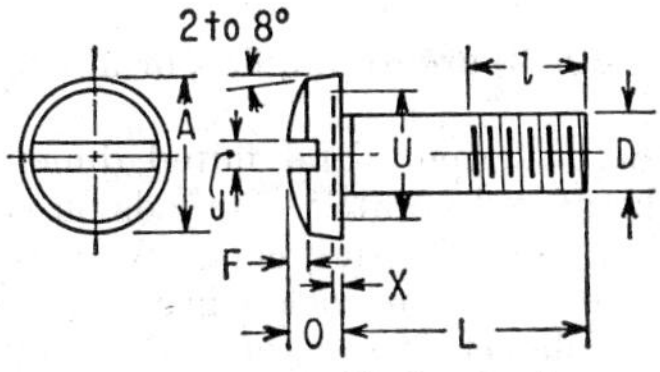

Binding head

Nominal size	Max diam D	Head diam, max, A	Total height of head, max, O	Width of slot, min, J	Depth of slot, min, T	Height of oval, max, F	Diam of undercut,* min, U	Depth of undercut, min, X
No. 2	0.086	0.181	0.046	0.023	0.024	0.018	0.124	0.005
No. 3	0.099	0.208	0.054	0.027	0.029	0.022	0.143	0.006
No. 4	0.112	0.235	0.063	0.031	0.034	0.025	0.161	0.007
No. 5	0.125	0.263	0.071	0.035	0.039	0.029	0.180	0.009
No. 6	0.138	0.290	0.080	0.039	0.044	0.032	0.199	0.010
No. 8	0.164	0.344	0.097	0.045	0.054	0.039	0.236	0.012
No. 10	0.190	0.399	0.114	0.050	0.064	0.045	0.274	0.015
No. 12	0.216	0.454	0.130	0.056	0.074	0.052	0.311	0.018
1/4	0.250	0.513	0.153	0.064	0.088	0.061	0.360	0.021
5/16	0.3125	0.641	0.193	0.072	0.112	0.077	0.450	0.027
3/8	0.375	0.769	0.234	0.081	0.136	0.094	0.540	0.034

Binding-head machine screws may be furnished with cross-recessed heads.
* Use of undercut is optional.

Screw threads of machine screws are unified or American National coarse- or fine-thread series, class 2 or 2A, supplied with a naturally bright finish, not heat-treated.

Machine-screw Heads. Four head styles are available, namely, screwdriver slot (Tables 20.19 to 20.21), two styles of cross recess,[25] and the "clutch head,"[25] not shown.

(d) Cap Screws[27]

Slotted-head styles (Table 20.22) are round, flat, or fillister, in screw diameters from $\frac{1}{4}$ to $1\frac{1}{2}$ in.

Socket-head Cap Screw. Hexagon and fluted dimensions are given in Table 20.23. Cap screws have chamfered points. Threads on cap screws are Unified and American coarse, fine, or eight-thread series, class 2A for plain (uncoated screws).

Minimum thread length for fluted or socket-head cap screw, National coarse, is $l = 2D \times \frac{1}{4}$ in. or $\frac{1}{2}L$, whichever is greater. Thread length for fine threads is $l = 1\frac{1}{2}D + \frac{1}{2}$ or $\frac{3}{8}L$, whichever is greater. Screw thread is class 3A; body is made of high-grade alloy steel, hardened by oil quenching and tempered to 36 to 43 Rockwell C.

Table 20.22. Slotted-head Cap Screws[27]

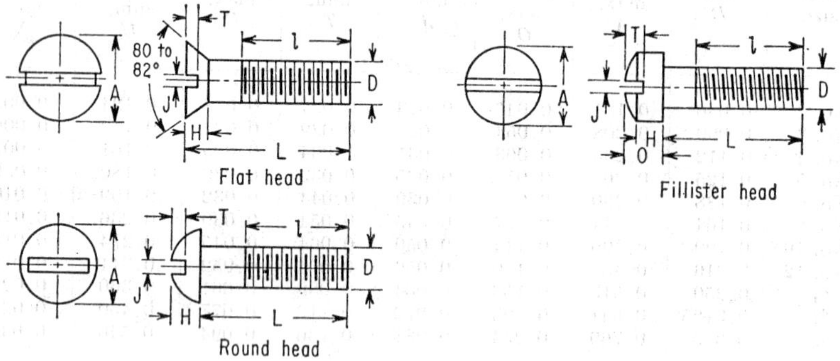

Nominal size (body diam, max) *D*	Width of slot, min, *J*	Fillister head				Flat head			Round head		
		Head diam, max, *A*	Height of head, max, *H*	Total height of head, max, *O*	Depth of slot, min, *T*	Head diam, max, *A*	Height of head, average, *H*	Depth of slot, min, *T*	Head diam, max, *A*	Height of head max, *H*	Depth of slot, min, *T*
$\frac{1}{4}$	0.064	0.375	0.172	0.216	0.077	0.500	0.140	0.046	0.437	0.191	0.097
$\frac{5}{16}$	0.072	0.437	0.203	0.253	0.090	0.625	0.176	0.057	0.562	0.246	0.126
$\frac{3}{8}$	0.081	0.562	0.250	0.314	0.113	0.750	0.210	0.069	0.625	0.273	0.138
$\frac{7}{16}$	0.081	0.625	0.297	0.368	0.133	0.8125	0.210	0.069	0.750	0.328	0.167
$\frac{1}{2}$	0.091	0.750	0.328	0.412	0.148	0.875	0.210	0.069	0.812	0.355	0.179
$\frac{9}{16}$	0.102	0.812	0.375	0.466	0.169	1.000	0.245	0.080	0.937	0.410	0.208
$\frac{5}{8}$	0.116	0.875	0.422	0.521	0.190	1.125	0.281	1.092	1.000	0.438	0.220
$\frac{3}{4}$	0.131	1.000	0.500	0.612	0.233	1.375	0.352	0.115	0.125	0.547	0.277
$\frac{7}{8}$	0.147	1.125	0.594	0.720	0.264	1.625	0.423	0.139			
1	0.166	1.312	0.656	0.802	0.292	1.875	0.494	0.162			
$1\frac{1}{8}$	0.178	.....	.....	.....	.....	2.062	0.529	0.173			
$1\frac{1}{4}$	0.193	.....	.....	.....	.....	2.312	0.600	0.197			
$1\frac{3}{8}$	0.208	.....	.....	.....	.....	2.562	0.665	0.220			
$1\frac{1}{2}$	0.240	.....	.....	.....	.....	2.812	0.742	0.244			

Table 20.23. Socket-head Cap Screws[28]

Nominal size	Body diam, max, *D*	Head diam, max, *A*	Head height, max, *H*	Head side height, max, *S*	Hexagon*	Fluted socket			
					Socket width across flats, min, *J*	Number of flutes	Socket diam minor, min, *K*	Socket diam major, min, *M*	Width of socket land, max, *N*
No. 0	0.060	0.096							
No. 1	0.073	0.118							
No. 2	0.0860	0.140	0.086	0.0803	1/16	6	0.063	0.073	0.016
No. 3	0.0990	0.161	0.099	0.0923	5/64	6	0.080	0.097	0.022
No. 4	0.1120	0.183	0.112	0.1043	5/64	6	0.080	0.097	0.022
No. 5	0.1250	0.205	0.125	0.1163	3/32	6	0.096	0.113	0.025
No. 6	0.1380	0.226	0.138	0.1284	3/32	6	0.096	0.113	0.025
No. 8	0.1640	0.270	0.164	0.1522	1/8	6	0.126	0.147	0.032
No. 10	0.1900	5/16	0.190	0.1765	5/32	6	0.161	0.186	0.039
No. 12	0.2160	11/32	0.216	0.2005	5/32	6	0.161	0.186	0.039
1/4	0.2500	3/8	1/4	0.2317	3/16	6	0.188	0.219	0.050
5/16	0.3125	7/16	5/16	0.2894	7/32	6	0.219	0.254	0.060
3/8	0.3750	9/16	3/8	0.3469	5/16	6	0.316	0.377	0.092
7/16	0.4375	5/8	7/16	0.4046	5/16	6	0.316	0.377	0.092
1/2	0.5000	3/4	1/2	0.4620	3/8	6	0.383	0.460	0.112
9/16	0.5625	13/16	9/16	0.5196	3/8	6	0.383	0.460	0.112
5/8	0.6250	7/8	5/8	0.5771	1/2	6	0.506	0.601	0.138
3/4	0.7500	1	3/4	0.6920	9/16	6	0.531	0.627	0.149
7/8	0.8750	1 1/8	7/8	0.8069	9/16	6	0.600	0.705	0.168
1	1.0000	1 5/16	1	0.9220	5/8	6	0.681	0.797	0.189
1 1/8	1.1250	1 1/2	1 1/8	1.0372	3/4	6	0.824	0.966	0.231
1 1/4	1.2500	1 3/4	1 1/4	1.1516	3/4	6	0.824	0.966	0.231
1 3/8	1.3750	1 7/8	1 3/8	1.2675	3/4	6	0.824	0.966	0.231
1 1/2	1.5000	2	1 1/2	1.3821	1	6	1.003	1.271	0.298

* Maximum socket depth *T* should not exceed three-fourths of minimum head height *H*.

Head chamfer angle *E* is 28 to 32°, the edge between flat and chamfer being slightly rounded.

Screw point chamfer angle 35 to 40°, the chamfer extending to the bottom of the thread. Edge between flat and chamfer is slightly rounded.

Hexagon-head cap screws are shown and dimensions given in Table 20.12.

(e) Shoulder Screws[25]

Sometimes termed *stripper bolts,* these are used as pivots for oscillating or rotating parts.

(f) Setscrews

These are made with either of three head styles:

1. Square head[27]
2. Headless,[27] with a slot for a screwdriver
3. Socket head,[28] or fluted key

Square-head setscrews (Table 20.24) are threaded the entire length of the body. Point styles are illustrated and dimensions are given in Table 20.25.

Table 20.24. Square-head Setscrews[27]

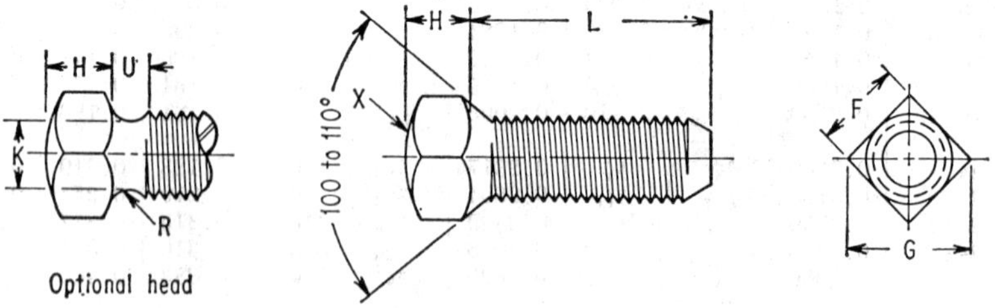

Nominal size		Width across flats F Max	Width across flats F Min	Width across corners G Min	Height of head H Nom	Height of head H Max	Height of head H Min	Diam of neck relief K Max	Diam of neck relief K Min	Radius of head X Nom	Radius of neck relief R Max	Width of neck relief U Max
No. 10	0.190	0.1875	0.180	0.247	9/64	0.148	0.134	0.145	0.140	15/32	0.027	0.083
No. 12	0.216	0.216	0.208	0.292	5/32	0.163	0.147	0.162	0.156	35/64	0.029	0.091
1/4	0.250	0.250	0.241	0.331	3/16	0.196	0.178	0.185	0.170	5/8	0.032	0.100
5/16	0.3125	0.3125	0.302	0.415	15/64	0.245	0.224	0.240	0.225	25/32	0.036	0.111
3/8	0.3750	0.375	0.362	0.497	9/32	0.293	0.270	0.294	0.279	15/16	0.041	0.125
7/16	0.4375	0.4375	0.423	0.581	21/64	0.341	0.315	0.345	0.330	1 3/32	0.046	0.143
1/2	0.500	0.500	0.484	0.665	3/8	0.389	0.361	0.400	0.385	1 1/4	0.050	0.154
9/16	0.5625	0.5625	0.545	0.748	27/64	0.437	0.407	0.454	0.439	1 13/32	0.054	0.167
5/8	0.6250	0.625	0.606	0.833	15/32	0.485	0.452	0.507	0.492	1 9/16	0.059	0.182
3/4	0.750	0.750	0.729	1.001	9/16	0.582	0.544	0.620	0.605	1 7/8	0.065	0.200
7/8	0.875	0.875	0.852	1.170	21/32	0.678	0.635	0.731	0.716	2 3/16	0.072	0.222
1	1.000	1.000	0.974	1.337	3/4	0.774	0.726	0.838	0.823	2 1/2	0.081	0.250
1 1/8	1.125	1.125	1.096	1.505	27/32	0.870	0.817	0.989	0.914	2 13/16	0.092	0.283
1 1/4	1.250	1.250	1.219	1.674	15/16	0.966	0.908	1.064	1.039	3 1/8	0.092	0.283
1 3/8	1.376	1.375	1.342	1.843	1 1/32	1.063	1.000	1.159	1.134	3 7/16	0.109	0.333
1 1/2	1.500	1.500	1.464	2.010	1 1/8	1.159	1.091	1.284	1.259	3 3/4	0.109	0.333

Table 20.25. Square-head Setscrew Points[27]

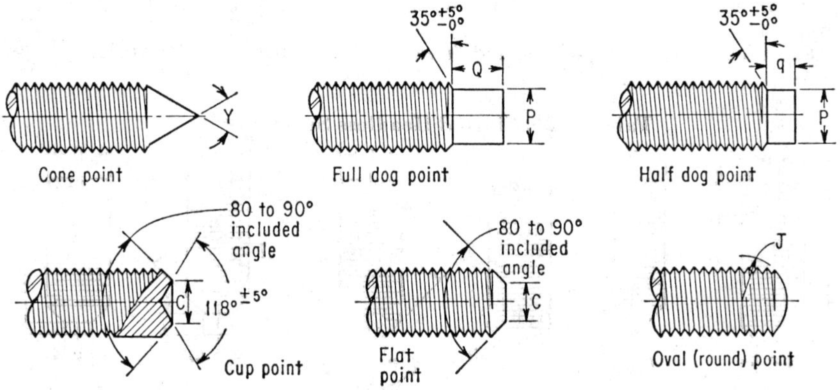

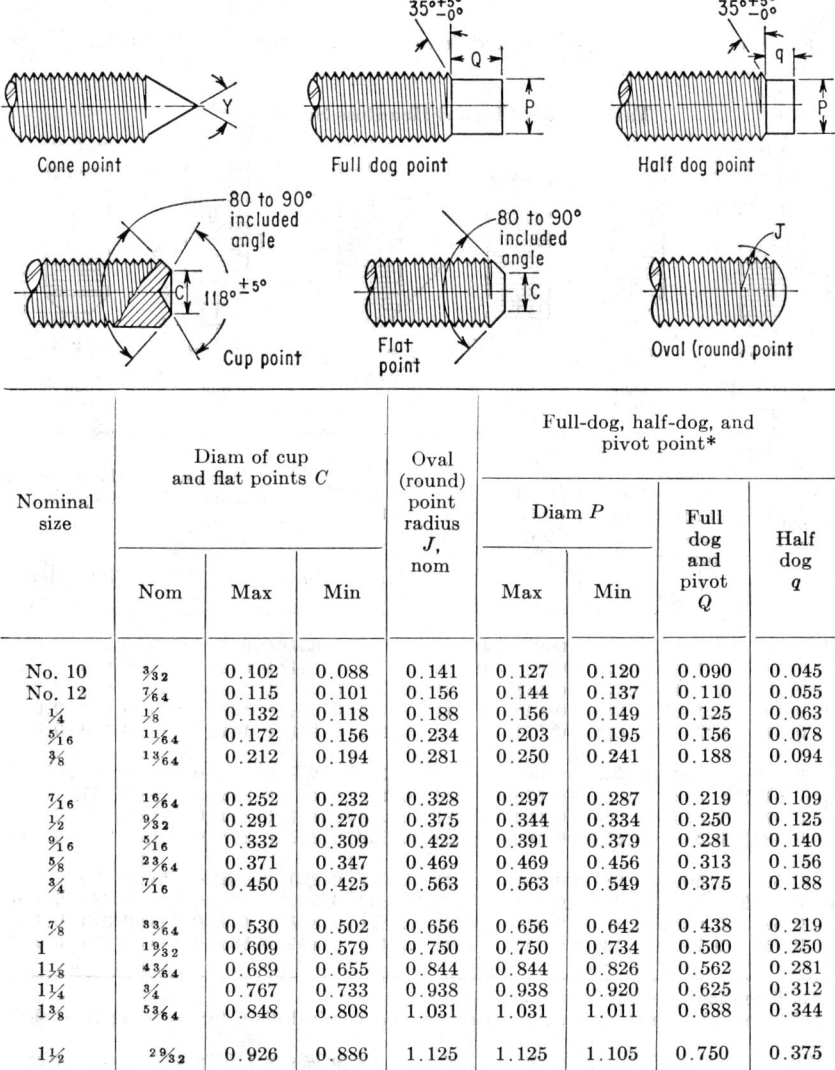

Nominal size	Diam of cup and flat points *C*			Oval (round) point radius *J*, nom	Full-dog, half-dog, and pivot point*			
					Diam *P*		Full dog and pivot *Q*	Half dog *q*
	Nom	Max	Min		Max	Min		
No. 10	3/32	0.102	0.088	0.141	0.127	0.120	0.090	0.045
No. 12	7/64	0.115	0.101	0.156	0.144	0.137	0.110	0.055
1/4	1/8	0.132	0.118	0.188	0.156	0.149	0.125	0.063
5/16	11/64	0.172	0.156	0.234	0.203	0.195	0.156	0.078
3/8	13/64	0.212	0.194	0.281	0.250	0.241	0.188	0.094
7/16	15/64	0.252	0.232	0.328	0.297	0.287	0.219	0.109
1/2	9/32	0.291	0.270	0.375	0.344	0.334	0.250	0.125
9/16	5/16	0.332	0.309	0.422	0.391	0.379	0.281	0.140
5/8	23/64	0.371	0.347	0.469	0.469	0.456	0.313	0.156
3/4	7/16	0.450	0.425	0.563	0.563	0.549	0.375	0.188
7/8	33/64	0.530	0.502	0.656	0.656	0.642	0.438	0.219
1	19/32	0.609	0.579	0.750	0.750	0.734	0.500	0.250
1 1/8	43/64	0.689	0.655	0.844	0.844	0.826	0.562	0.281
1 1/4	3/4	0.767	0.733	0.938	0.938	0.920	0.625	0.312
1 3/8	53/64	0.848	0.808	1.031	1.031	1.011	0.688	0.344
1 1/2	29/32	0.926	0.886	1.125	1.125	1.105	0.750	0.375

All dimensions are given in inches.

* Pivot points are similar to full-dog point except that the point is rounded by a radius equal to *J*.

Where usable length of thread is less than the nominal diameter, half-dog point shall be used.

When length equals nominal diameter or less, $Y = 118° \pm 2°$; when length exceeds nominal diameter, $Y = 90° \pm 2°$

Headless setscrews are threaded the entire length (Table 20.26). Threads are coarse or fine, class 2 or 2A. Slotted-type setscrews are regularly furnished casehardened.

Dimensions of keys for *hexagon and fluted sockets* are given in Table 20.27.

Table 20.26. Slotted Headless Setscrews[27]

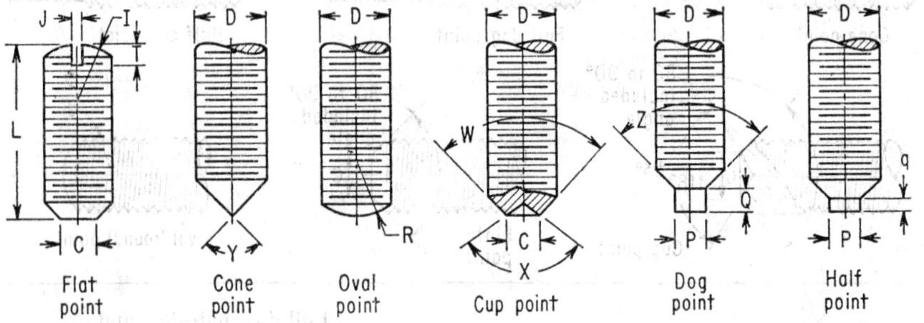

Nominal size D		Radius of headless crown I	Width of slot J	Depth of slot T	Oval-point radius R	Diam of cup and flat points C		Diam of dog point P		Length of dog point*	
						Max	Min	Max	Min	Fill Q	Half q
5	0.125	0.125	0.023	0.031	0.094	0.067	0.057	0.083	0.078	0.060	0.030
6	0.138	0.138	0.025	0.035	0.109	0.074	0.064	0.092	0.087	0.070	0.035
8	0.164	0.164	0.029	0.041	0.125	0.087	0.076	0.109	0.103	0.080	0.040
10	0.190	0.190	0.032	0.048	0.141	0.102	0.088	0.127	0.120	0.090	0.045
12	0.216	0.216	0.036	0.054	0.156	0.115	0.101	0.144	0.137	0.110	0.055
¼	0.250	0.250	0.045	0.063	0.188	0.132	0.118	0.156	0.149	0.125	0.063
5/16	0.3125	0.313	0.051	0.078	0.234	0.172	0.156	0.203	0.195	0.156	0.078
⅜	0.375	0.375	0.064	0.094	0.281	0.212	0.194	0.250	0.241	0.188	0.094
7/16	0.4375	0.438	0.072	0.109	0.328	0.252	0.232	0.297	0.287	0.219	0.109
½	0.500	0.500	0.081	0.125	0.375	0.291	0.270	0.344	0.344	0.250	0.135
9/16	0.5625	0.563	0.091	0.141	0.422	0.332	0.309	0.391	0.379	0.281	0.140
⅝	0.625	0.625	0.102	0.156	0.469	0.371	0.347	0.469	0.456	0.313	0.156
¾	0.750	0.750	0.129	0.188	0.563	0.450	0.425	0.563	0.549	0.375	0.188

All dimensions given in inches.

* Where usable length of thread is less than the nominal diameter, half-dog point shall be used.

When L (length of screw) equals nominal diameter or less = 118° ± 2°; when L exceeds nominal diameter, $Y = 90° \pm 2°$.

Point angles, W = 80 to 90°; $X = 118° \pm 5°$; Z = 100 to 110°.

Allowable eccentricity of dog-point axis with respect to axis of screw shall not exceed 3 per cent of nominal screw diameter with maximum 0.005 in.

A setscrew is often employed to provide frictional resistance in order to prevent relative motion, e.g., between a shaft and a hub. Setscrews are made with heads and headless. Headed setscrews have square heads. A setscrew with an unprotected head on a moving machine member represents a safety hazard. Headless setscrews are manufactured to accommodate either a hexagonal- or spline-shaped wrench (hollow setscrews) or a screwdriver blade. The disadvantage of the cone and cup types is associated with the shaft burr which may form, making removal of parts difficult. For this reason, a flat or conical seat may be machined into the shaft. Setscrews with dog and cone points and sometimes oval points are spotted in holes. Flat (used on hardened steels) and cup points (used with softer materials) are generally used without spotting. Setscrews usually have coarse threads and hardened points (which may be knurled to prevent loosening under vibration).

(g) Tapping Screws[29,30]

Tapping screws tap their own mating threads in holes in materials such as metals and plastics. Tapping screws are of two types, thread-forming (displaces the material), type A, B, BP, and C, U, or thread-cutting (actual removal of material), type D, F, G, T, BF, BG, and BT.

Thread-forming Screws

Type A. Spaced thread with gimlet point provides a strong joint in light sheet metal up to 18 gage.

Type B. Spaced thread with blunt point. Used for light and heavy sheet metal, nonferrous castings, plastics, or soft metals. Drives faster than any screw except type A and U.

Type BP. Spaced-thread screw same as type B but has a cone point. Used in assemblies where holes are misaligned.

Type C. Used where the use of a machine screw is preferred. Threads are the same pitch as standard machine screws. Useful when chips from machine-screw thread-cutting screws are objectionable. The finer pitch and more engaged thread surface offers an increased functional resistance to loosening.

Type U. Multiple-thread, blunt point, metallic-drive screw threads have a large helix angle, resulting in good holding power, and must be hammered or forced into the work. Used for permanent fastening, since it is difficult to remove.

Thread-cutting Screws

Types D, F, G, T. Threads approximate machine-screw threads, with blunt point. Front end has tapered entering threads for easy starting. Type T has same usage as D, except wide flute provides more chip clearance. Type G is used where less driving torque than type D is required. Type F replaces type C where a low driving torque is required; however, chip space of the flutes is not suitable for deep penetration. The above types are suitable for use with aluminum, zinc, die castings, carbon- and stainless-steel sheets and shapes, cast iron, brass, and plastics.

Type BF, BG, *and* BT. Spaced threads blunt point same as type B. Used in plastics and die castings.

Table 20.27. Fluted and Hexagon Socket-headless Setscrews[28]

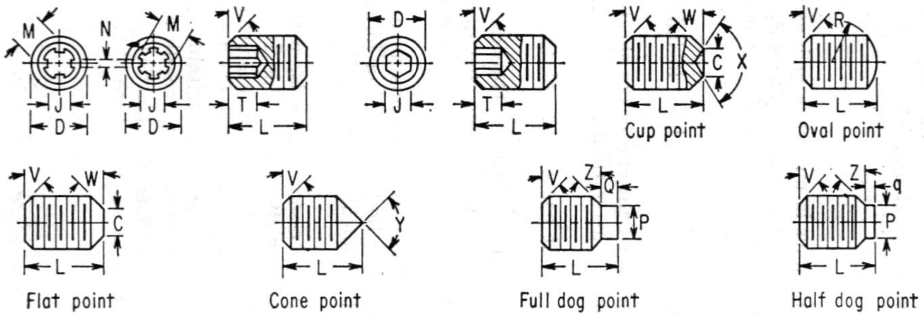

Screw size nominal diam D	Fluted and hexagon socket types										Hexagon type		Fluted type*					
	Cup- and flat-point diameter C		Oval-point radius R	Cone-point angle Y		Dog point				Key engagement† min	Socket width across flats J		Socket diam, minor, J		Socket diam, major, M		Socket land width N	
				118 ± 2° for these lengths and under	90 ± 2° for these lengths and over	Diam P		Full Q	Half q									
	Max	Min				Max	Min				Max	Min	Max	Min	Max	Min	Max	Min
No. 0	0.033	0.027	3/64	1/16	5/64	0.040	0.037	0.030	0.015	0.022	0.0285	0.028	0.026	0.0255	0.035	0.034	0.012	0.0115
No. 1	0.040	0.033	0.055	5/64	3/32	0.049	0.045	0.037	0.019	0.028	0.0355	0.035	0.026	0.0255	0.035	0.034	0.012	0.0115
No. 2	0.047	0.039	1/16	3/32	7/64	0.057	0.053	0.043	0.022	0.028	0.0355	0.035	0.038	0.0375	0.050	0.049	0.017	0.016
No. 3	0.054	0.045	5/64	7/64	1/8	0.066	0.062	0.050	0.025	0.040	0.051	0.050	0.038	0.0375	0.050	0.049	0.017	0.016
No. 4	0.061	0.051	0.084	1/8	5/32	0.075	0.070	0.056	0.028	0.040	0.051	0.050	0.051	0.050	0.062	0.061	0.014	0.013

No. 5	0.067	0.057	3/32	1/8	3/16	0.083	0.078	0.06	0.03	0.050	0.0635	1/16	0.053	0.052	0.071	0.070	0.022	0.021
No. 6	0.074	0.064	7/64	1/8	3/16	0.092	0.087	0.07	0.035	0.050	0.0635	1/16	0.056	0.055	0.079	0.078	0.023	0.022
No. 8	0.087	0.076	1/8	3/16	1/4	0.109	0.103	0.08	0.04	0.062	0.0791	5/64	0.082	0.080	0.098	0.097	0.022	0.021
No. 10	0.102	0.088	9/64	3/16	1/4	0.127	0.120	0.09	0.045	0.075	0.0947	3/32	0.098	0.096	0.115	0.113	0.025	0.023
No. 12	0.115	0.101	5/32	3/16	1/4	0.144	0.137	0.11	0.055	0.075	0.0947	3/32	0.098	0.096	0.115	0.113	0.025	0.023
1/4	0.132	0.118	3/16	1/4	5/16	5/32	0.149	1/8	1/16	0.100	0.1270	1/8	0.128	0.126	0.149	0.147	0.032	0.030
5/16	0.172	0.156	15/64	5/16	3/8	13/64	0.195	5/32	5/64	0.125	0.1582	5/32	0.163	0.161	0.188	0.186	0.039	0.037
3/8	0.212	0.194	9/32	3/8	7/16	1/4	0.241	3/16	3/32	0.150	0.1895	3/16	0.190	0.188	0.221	0.219	0.050	0.048
7/16	0.252	0.232	21/64	7/16	1/2	19/64	0.287	7/32	7/64	0.175	0.2207	7/32	0.221	0.219	0.256	0.254	0.060	0.058
1/2	0.291	0.270	3/8	1/2	9/16	11/32	0.334	1/4	1/8	0.200	0.2520	1/4	0.254	0.252	0.298	0.296	0.068	0.066
9/16	0.332	0.309	27/64	9/16	5/8	25/64	0.379	9/32	9/64	0.200	0.2520	1/4	0.254	0.252	0.298	0.296	0.068	0.066
5/8	0.371	0.347	15/32	5/8	3/4	15/32	0.456	5/16	5/32	0.250	0.3155	5/16	0.319	0.316	0.380	0.377	0.092	0.089
3/4	0.450	0.425	9/16	3/4	7/8	9/16	0.549	3/8	3/16	0.300	0.3780	3/8	0.386	0.383	0.463	0.460	0.112	0.109
7/8	0.530	0.502	21/32	7/8	1	21/32	0.642	7/16	7/32	0.400	0.5030	1/2	0.509	0.506	0.604	0.601	0.138	0.134
1	0.609	0.579	3/4	1	1 1/8	3/4	0.734	1/2	1/4	0.450	0.5655	9/16	0.535	0.531	0.631	0.627	0.149	0.145
1 1/8	0.689	0.655	27/32	1 1/8	1 1/4	27/32	0.826	9/16	9/32	0.450	0.5655	9/16	0.604	0.600	0.709	0.705	0.168	0.164
1 1/4	0.767	0.733	15/16	1 1/4	1 1/2	15/16	0.920	5/8	5/16	0.500	0.6290	5/8	0.685	0.681	0.801	0.797	0.189	0.185
1 3/8	0.848	0.808	1 1/32	1 3/8	1 5/8	1 1/32	1.011	11/16	11/32	0.500	0.6290	5/8	0.744	0.740	0.869	0.865	0.207	0.203
1 1/2	0.926	0.886	1 1/8	1 1/2	1 3/4	1 1/8	1.105	3/4	3/8	0.600	0.7540	3/4	0.828	0.824	0.970	0.966	0.231	0.227
1 3/4	1.086	1.039	1 5/16	1 3/4	2	1 5/16	1.289	7/8	7/16	0.800	1.0040	1	1.007	1.003	1.275	1.271	0.298	0.294
2	1.244	1.193	1 1/2	2	2 1/4	1 1/2	1.474	1	1/2	0.800	1.0040	1	1.007	1.003	1.275	1.271	0.298	0.294

All dimensions are in inches.

* The number of flutes for setscrews Nos. 0, 1, 2, 3, 5, and 6 is four. The number of flutes for Nos. 4, 8, and larger is six.

† These dimensions apply to cup- and flat-point screws one diameter in length or longer. For screws shorter than one diameter in length, and for other types of points, socket to be as deep as practicable.

20.3. POWER - TRANSMITTING SCREW THREADS[37]

(a) Types

Acme screw thread
Stub Acme or 29° stub Acme, form 1 and form 2 stub
60° stub
Modified square screw thread
Buttress

(b) Acme Screw Thread[33] (Fig. 20.13)

Developed for the purpose of producing transverse motions, Acme screw thread is Standardized for two applications, namely, (1) general purpose and (2) centralizing.

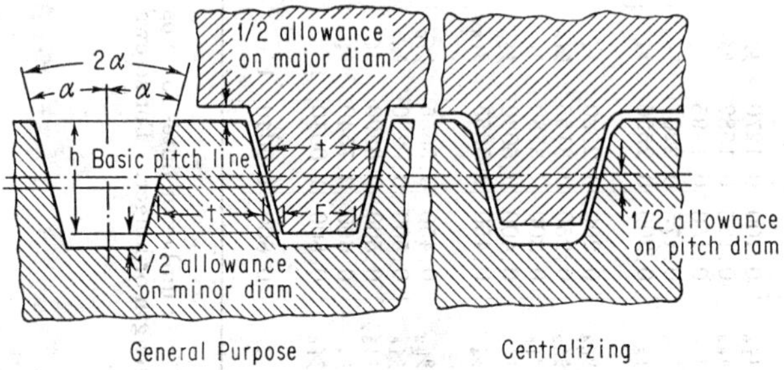

FIG. 20.13. Acme thread forms.

$\alpha = 14°30'$
$h = 0.5p$
$t = 0.5p$
$F = 0.3707p$
p = pitch, in.

1. General-purpose thread has clearance on all diameters for free movement. Class 2G is preferred for gene al-purpose assemblies. Classes 3G and 4G are used where backlash and end play are undesirable. Table 20.28 lists the recommended diameter-pitch Acme threads.

2. Centralizing threads have a limited clearance at the major diameter, resulting in controlled concentricity of screw and nut. External threads must have the crest corners clarified 45° × $p/15$ maximum.

Classes. For classes 2C, 3C, 4C the fillet at the minor diameter is equal to $0.1p$ or less. For classes 5C, 6C the minimum fillet is $0.07p$ and maximum fillet is $0.1p$. The major diameter of all classes may have a fillet not greater than $0.06p$. Class 2C provides the maximum end play or backlash.

Table 20.28. Acme and Stub Acme Thread Series[38]

Nominal size	Threads per in. $1/p$	Acme threads: Basic height of thread h	Acme threads, general-purpose (all classes) and centralizing classes 2C, 3C, and 4C: Basic major diam D_m	Acme threads, general-purpose (all classes) and centralizing classes 2C, 3C, and 4C: Helix angle at basic pitch diam λ	Acme threads, centralizing classes 5C and 6C: Basic major diam D_m'	Acme threads, centralizing classes 5C and 6C: Helix angle at basic pitch diam λ	Stub Acme threads: Basic height of thread h'	Stub Acme threads: Helix angle at basic pitch diam
1/4	16	0.03125	0.2500	5° 12′	......	.....	0.01875	4° 54′
5/16	14	0.03571	0.3125	4° 42′	......	.....	0.02143	4° 28′
3/8	12	0.04167	0.3750	4° 33′	......	.....	0.02500	4° 20′
7/16	12	0.04167	0.4375	3° 50′	......	.....	0.02500	3° 41′
1/2	10	0.05000	0.5000	4° 3′	0.4823	4° 13′	0.03000	3° 52′
5/8	8	0.06250	0.6250	4° 3′	0.6052	4° 12′	0.03750	3° 52′
3/4	6	0.08333	0.7500	4° 33′	0.7284	4° 42′	0.05000	4° 20′
7/8	6	0.08333	0.8750	3° 50′	0.8516	3° 57′	0.05000	3° 41′
1	5	0.10000	1.0000	4° 3′	0.9750	4° 10′	0.06000	3° 52′
1 1/8	5	0.10000	1.1250	3° 33′	1.0985	3° 39′	0.06000	3° 25′
1 1/4	5	0.10000	1.2500	3° 10′	1.2220	3° 15′	0.06000	3° 4′
1 3/8	4	0.12500	1.3750	3° 39′	1.3457	3° 44′	0.07500	3° 30′
1 1/2	4	0.12500	1.5000	3° 19′	1.4694	3° 23′	0.07500	3° 12′
1 3/4	4	0.12500	1.7500	2° 48′	1.7169	2° 52′	0.07500	2° 43′
2	4	0.12500	2.0000	2° 26′	1.9646	2° 29′	0.07500	2° 22′
2 1/4	3	0.16667	2.2500	2° 55′	2.2125	2° 58′	0.10000	2° 50′
2 1/2	3	0.16667	2.5000	2° 36′	2.4605	2° 39′	0.10000	2° 32′
2 3/4	3	0.16667	2.7500	2° 21′	2.7085	2° 23′	0.10000	2° 18′
3	2	0.25000	3.0000	3° 19′	2.9567	3° 22′	0.15000	3° 12′
3 1/2	2	0.25000	3.5000	2° 48′	3.4532	2° 51′	0.15000	2° 43′
4	2	0.25000	4.0000	2° 26′	3.9500	2° 28′	0.15000	2° 22′
4 1/2	2	0.25000	4.5000	2° 8′	4.4470	2° 10′	0.15000	2° 6′
5	2	0.25000	5.0000	1° 55′	4.9441	1° 56′	0.15000	1° 53′

For general-purpose and centralizing classes 2C, 3C, and 4C, basic pitch diameter $= D_m - h$; basic minor diameter $= D_m - 2h$, $h = p/2$.

For centralizing classes 5C and 6C, basic pitch diameter $= D_m - h$; basic minor diameter $= D_m - 2h$.

For stub Acme, basic pitch diameter $= D_m - h'$; basic minor diameter $= D_m - 2h'$.

(c) Stub Acme Screw Thread (Fig. 20.14)[34]

The stub Acme, or 29° stub, thread is used where a coarse-pitch thread of shallow depth is required. Like the Acme thread, it has an angle of 29° between flanks.

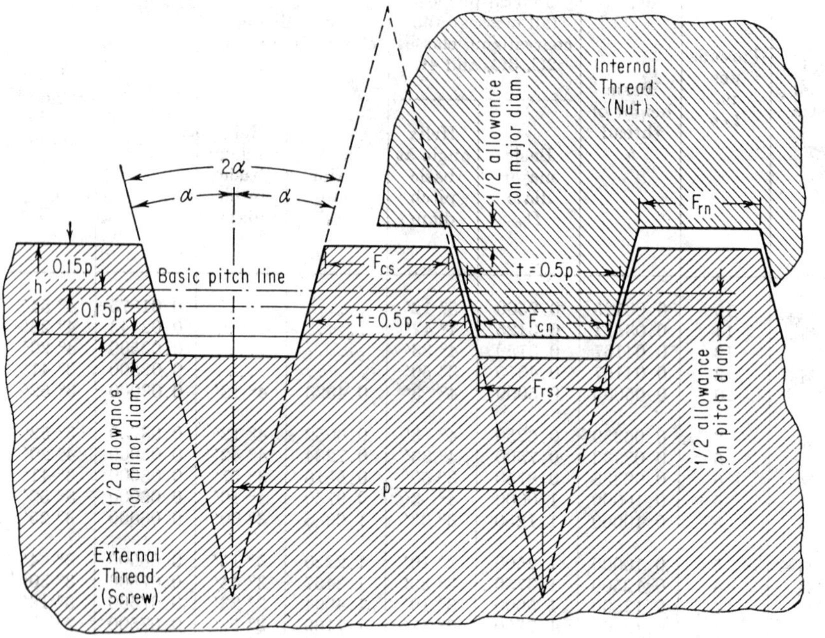

FIG. 20.14. 29° stub Acme thread. Stub Acme thread used in place of the standard Acme form when shallow depth is required.

2α = 29°
α = 14°30′
p = pitch
n = number of threads per inch
N = number of turns per inch
h = $0.3p$, basic height of thread
F_{cn} = $0.4224p$ = basic width of flat of crest of internal thread
F_{cs} = $0.4224p$ = basic width of flat of crest of external thread
F_{rn} = $0.4224p - 0.259 \times$ (major-diameter allowance on internal thread)
F_{rs} = $0.4224p - 0.259 \times$ (minor-diameter allowance on external thread − pitch-diameter allowance on external thread)

The stub or height is $0.3p$. Only one class is provided, class 2G (general-purpose) (Table 20.28).

Alternate stub Acme threads[34] (Fig. 20.15) are provided where design requirements necessitate their use.

(d) 60° Stub Acme Thread*[,35]

The angle between the flanks of the thread is 60°. The threads are truncated top and bottom. Basic dimensions are given in Table 20.29.

* In accordance with standard practice, this thread is designated as follows, for example, "1⅛-9 special form, 60° or 10° thread," followed by all limits of size.

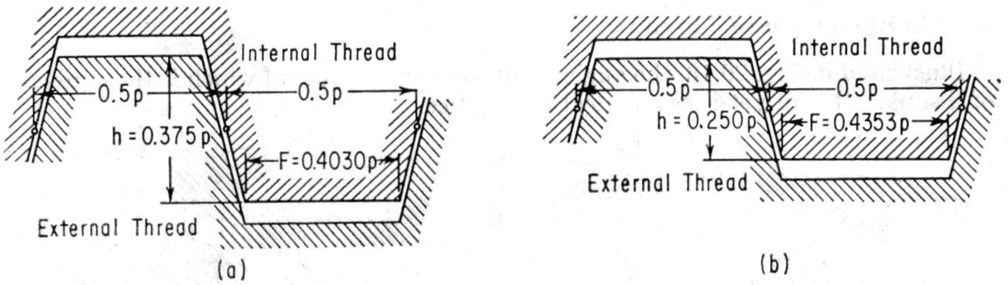

FIG. 20.15. Modified stub Acme thread with basic height of 0.250 pitch. (*a*) Modified stub Acme thread with basic height of 0.375 pitch (form 1). (*b*) Modified stub Acme thread with basic height of 0.250 pitch (form 2).

Table 20.29. 60° Stub Threads[34]

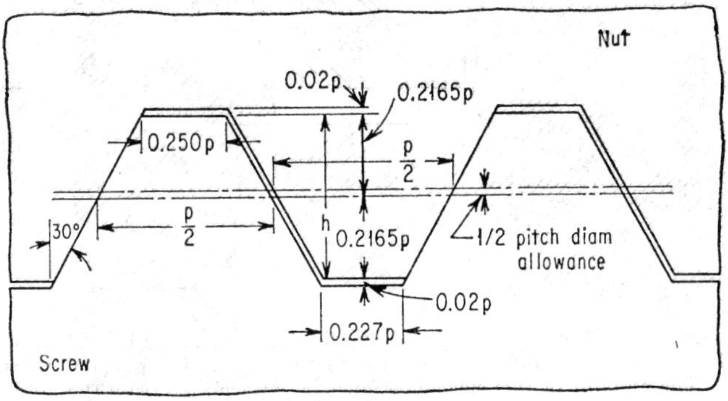

Threads per in.	Pitch p, in.	Depth of thread (basic) $h = 0.433p$, in.	Total depth of thread $(h + 0.02p)$, in.*	Thread thickness (basic) $t = 0.5p$, in.	Width of flat, in. Crest of screw (basic) $F = 0.250p$	Width of flat, in. Root of screw $F_c = 0.227p$
16	0.06250	0.0271	0.0283	0.0313	0.0156	0.0142
14	0.07143	0.0309	0.0324	0.0357	0.0179	0.0162
12	0.08333	0.0361	0.0378	0.0417	0.0208	0.0189
10	0.10000	0.0433	0.0453	0.0500	0.0250	0.0227
9	0.11111	0.0481	0.0503	0.0556	0.0278	0.0252
8	0.12500	0.0541	0.0566	0.0625	0.0313	0.0284
7	0.14286	0.0619	0.0647	0.0714	0.0357	0.0324
6	0.16667	0.0722	0.0755	0.0833	0.0417	0.0378
5	0.20000	0.0866	0.0906	0.1000	0.0500	0.0454
4	0.25000	0.1083	0.1133	0.1250	0.0625	0.0567

* A clearance of at least $0.02p$ is added to h to produce extra depth, thus avoiding interference with threads of mating part at minor or major diameters.

(e) Modified Square Thread[35]

Illustrated in Fig. 20.16, the angle of 10° results in a thread which is the equivalent of a square thread yet is economical to produce.

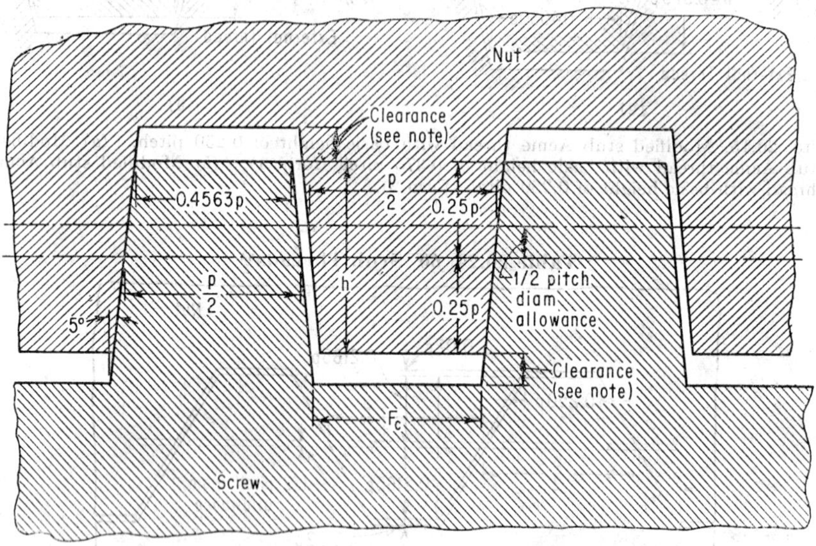

FIG. 20.16. 10° modified square thread. *Note:* A clearance should be added to h to produce extra depth, thus avoiding interference with threads of mating parts at major and minor diameters. The amount of this clearance must be determined from the application.

p = pitch
h (basic depth of thread) $= 0.5p$
H (total depth of thread) $= 0.5p +$ clearance
t (thickness of thread) $= 0.5p$
F_c (flat at root of thread) $= 0.4563p - (0.17 \times \text{clearance})$
F (basic width of flat at crest of thread) $= 0.4563p$

(f) Buttress Thread[36] (Fig. 20.17)

Used in specially designed components involving high stresses in one direction only, along the thread axis. As the thrust side of the thread is made very nearly perpendicular to the thread axis, the radial component of the thrust is reduced to a minimum, making this form of thread particularly suitable for tubular members which have to be screwed together. Applications include breech mechanisms of large guns and airplane-propeller hubs.

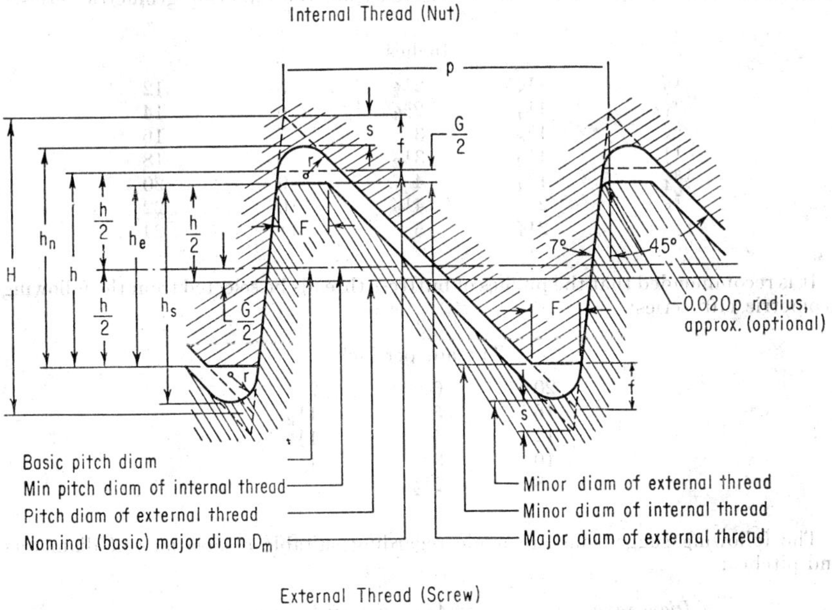

FIG. 20.17. Form of buttress thread having 7° pressure flank and 45° clearance flank.

Nominal major diameter D
Height of sharp V-thread $H = 0.89064p$
Basic height of thread $h = 0.6p$
Root radius $r = 0.07141p$
Root truncation $s = 0.08261p$
Depth of engagement $h_e = h - G/2$
Crest truncation $f = 0.14532p$
Crest width $F = 0.16316p$
Major diameter of internal thread (nut) $D_n = D + 0.12542p$
Minor diameter of external thread (screw) $K_s = D - 1.32542p - G$
Height of thread of internal thread (nut) $h_n = 0.66271p$
Height of thread of external thread (screw) $h_s = 0.66271p$

Pressure flank angle measured in an axial plane is 7° from the normal to the axis, and the trailing flank angle is 45°. There is no standard series, inasmuch as this thread applies mainly to specially designed components. However, it is recommended that the nominal major diameters be selected from the following geometric series:[36]

Inches

1/2	1 1/8	2 1/2	5 1/2	12
9/16	1 1/4	2 3/4	6	14
5/8	1 3/8	3	7	16
11/16	1 1/2	3 1/2	8	18
3/4	1 3/4	4	9	20
7/8	2	4 1/2	10	22
1	2 1/4	5	11	24

It is recommended that the pitches of buttress threads be selected from the following geometric (10) series:

Threads per inch

20	6	2
16	5	1 1/2
12	4	1 1/4
10	3	1
8	2 1/2	

The following suggestions are made regarding suitable associations of diameters and pitches:

Diam range, in.	*Associated pitches, threads per in.*
From 1/2 to 11/16	20, 16, 12
Over 11/16 to 1	16, 12, 10
Over 1 to 1 1/2	16, 12, 10, 8, 6
Over 1 1/2 to 2 1/2	16, 12, 10, 8, 6, 5, 4
Over 2 1/2 to 4	16, 12, 10, 8, 6, 5, 4
Over 4 to 6	12, 10, 8, 6, 5, 4, 3
Over 6 to 10	10, 8, 6, 5, 4, 3, 2 1/2, 2
Over 10 to 16	10, 8, 6, 5, 4, 3, 2 1/2, 2, 1 1/2, 1 1/4
Over 16 to 24	8, 6, 5, 4, 3, 2 1/2, 2, 1 1/2, 1 1/4, 1

20.4. WASHERS

(a) Usage

Washers[39] under the heads of screws or bolts are used for the following purposes: to prevent the loosening of the associated screw, bolt, or nut; to distribute the compressive stress over areas larger than that of the bolt head or nut; to distribute thrust loads; between rotating parts to reduce wear; and to prevent damage to finishes by bolt head or nut.

(b) Plain Washers

Type A washer (Table 20.30) is selected by the choosing of the desired inside diameter, outside diameter, and thickness. *Type* B (Table 20.31) is selected by the nominal screw or bolt size and appropriate thickness, narrow, regular, or wide. Metal washers are normally made from SAE 1060 steel or equivalent, heat-treated to a hardness of 45 to 53 Rockwell C.

Table 20.30. Type A Plain Washers[39]

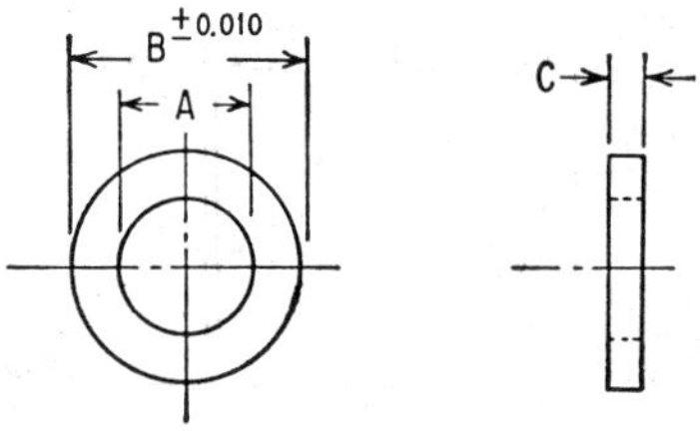

ID*	OD†	Thickness			ID*	OD†	Thickness		
		Nom	Max	Min			Nom	Max	Min
5/64	3/16	0.020	0.025	0.016	5/8	1 1/2	0.109	0.132	0.086
3/32	7/32	0.020	0.025	0.016	5/8	2 1/8	0.134	0.160	0.108
3/32	1/4	0.020	0.025	0.016	21/32	1 5/16	0.095	0.121	0.074
1/8	1/4	0.022	0.028	0.017	11/16	1 1/2	0.134	0.160	0.108
1/8	5/16	0.032	0.040	0.025	11/16	1 3/4	0.134	0.160	0.108
5/32	5/16	0.035	0.048	0.027	11/16	2 3/8	0.165	0.192	0.136
5/32	3/8	0.049	0.065	0.036	13/16	1 1/2	0.134	0.160	0.108
11/64	13/32	0.049	0.065	0.036	13/16	1 3/4	0.148	0.177	0.122
3/16	3/8	0.049	0.065	0.036	13/16	2	0.148	0.177	0.122
3/16	7/16	0.049	0.065	0.036	13/16	2 7/8	0.165	0.192	0.136
13/64	15/32	0.049	0.065	0.036	15/16	1 3/4	0.134	0.160	0.108
7/32	7/16	0.049	0.065	0.036	15/16	2	0.165	0.192	0.136
7/32	1/2	0.049	0.065	0.036	15/16	2 1/4	0.165	0.192	0.136
15/64	17/32	0.049	0.065	0.036	15/16	3 3/8	0.180	0.213	0.153
1/4	1/2	0.049	0.065	0.036	1 1/16	2	0.134	0.160	0.108
1/4*	9/16	0.049	0.065	0.036	1 1/16	2 1/4	0.165	0.192	0.136
1/4*	9/16	0.065	0.080	0.051	1 1/16	2 1/2	0.165	0.192	0.136
17/64	5/8	0.049	0.065	0.036	1 1/16	3 7/8	0.238	0.280	0.210
9/32	5/8	0.065	0.080	0.051	1 3/16	2 1/2	0.165	0.192	0.136
5/16	3/4	0.065	0.080	0.051	1 1/4	2 3/4	0.165	0.192	0.136
5/16	7/8	0.065	0.080	0.051	1 5/16	2 3/4	0.165	0.192	0.136
11/32	11/16	0.065	0.080	0.051	1 3/8	3	0.165	0.192	0.136
3/8	3/4	0.065	0.080	0.051	1 7/16	3	0.180	0.213	0.153
3/8	7/8	0.083	0.104	0.064	1 1/2	3 1/4	0.180	0.213	0.153
3/8	1 1/8	0.065	0.080	0.051	1 9/16	3 1/4	0.180	0.213	0.153
13/32	13/16	0.065	0.080	0.051	1 5/8	3 1/2	0.180	0.213	0.153
7/16	7/8	0.083	0.104	0.064	1 11/16	3 1/2	0.180	0.213	0.153
7/16	1	0.083	0.104	0.064	1 3/4	3 3/4	0.180	0.213	0.153
7/16	1 3/8	0.083	0.104	0.064	1 13/16	3 3/4	0.180	0.213	0.153
15/32	59/64	0.065	0.080	0.051	1 7/8	4	0.180	0.213	0.153
1/2	1 1/8	0.083	0.104	0.064	1 15/16	4	0.180	0.213	0.153
1/2	1 1/4	0.083	0.104	0.064	2	4 1/4	0.180	0.213	0.153
1/2	1 3/8	0.083	0.104	0.064	2 1/16	4 1/4	0.180	0.213	0.153
17/32	1 1/16	0.095	0.121	0.074	2 1/8	4 1/2	0.180	0.213	0.153
9/16	1 1/4	0.109	0.132	0.086	2 3/8	4 3/4	0.220	0.248	0.193
9/16	1 3/8	0.109	0.132	0.086	2 5/8	5	0.238	0.280	0.210
9/16	1 7/8	0.109	0.132	0.086	2 7/8	5 1/4	0.259	0.310	0.228
19/32	1 3/16	0.095	0.121	0.074	3 1/8	5 1/2	0.284	0.327	0.249
5/8	1 3/8	0.109	0.132	0.086					

All dimensions are given in inches.

* Tolerance is ±0.005 in. for inside diameters up to 7/32 in., inclusive, and also for the two 1/4 in. inside diameters marked with an asterisk (*). Tolerance for all other inside diameters larger than 7/32 in. is ±0.010 in.

† Tolerance is ±0.010 in. for all outside diameters.

Table 20.31. Type B Plain Washers[39]

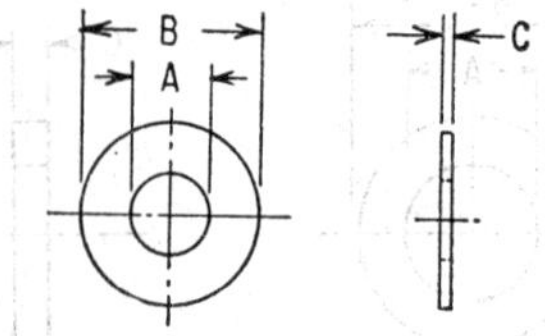

Screw or bolt size	Series	ID A Nom	OD B Nom	Thickness C Nom	Thickness C Max	Thickness C Min
0.060 (No. 0)	Narrow		1/8	0.025	0.028	0.022
	Regular	0.068	3/16	0.025	0.028	0.022
	Wide		1/4	0.025	0.028	0.022
0.073 (No. 1)	Narrow		5/32	0.025	0.028	0.022
	Regular	0.084	7/32	0.025	0.028	0.022
	Wide		9/32	0.032	0.036	0.028
0.086 (No. 2)	Narrow		3/16	0.025	0.028	0.022
	Regular	0.094	1/4	0.032	0.036	0.028
	Wide		11/32	0.032	0.036	0.028
0.099 (No. 3)	Narrow		7/32	0.025	0.028	0.022
	Regular	0.109	5/16	0.032	0.036	0.028
	Wide		13/32	0.040	0.045	0.036
0.112 (No. 4)	Narrow		1/4	0.032	0.036	0.028
	Regular	0.125	3/8	0.040	0.045	0.036
	Wide		7/16	0.040	0.045	0.036
0.125 (No. 5)	Narrow		9/32	0.032	0.036	0.028
	Regular	0.141	13/32	0.040	0.045	0.036
	Wide		1/2	0.040	0.045	0.036
0.138 (No. 6)	Narrow		5/16	0.032	0.036	0.028
	Regular	0.156	7/16	0.040	0.045	0.036
	Wide		9/16	0.040	0.045	0.036
0.164 (No. 8)	Narrow		3/8	0.040	0.045	0.036
	Regular	0.188	1/2	0.040	0.045	0.036
	Wide		5/8	0.063	0.071	0.056
0.190 (No. 10)	Narrow		13/32	0.040	0.045	0.036
	Regular	0.208	9/16	0.040	0.045	0.036
	Wide		47/64	0.063	0.071	0.056
0.216 (No. 12)	Narrow		7/16	0.040	0.045	0.036
	Regular	0.240	5/8	0.063	0.071	0.056
	Wide		7/8	0.063	0.071	0.056
1/4	Narrow		1/2	0.063	0.071	0.056
	Regular	0.281	47/64	0.063	0.071	0.056
	Wide		1	0.063	0.071	0.056
5/16	Narrow		5/8	0.063	0.071	0.056
	Regular	0.344	7/8	0.063	0.071	0.056
	Wide		1 1/8	0.063	0.071	0.056
3/8	Narrow		47/64	0.063	0.071	0.056
	Regular	0.406	1	0.063	0.071	0.056
	Wide		1 1/4	0.100	0.112	0.090
7/16	Narrow		7/8	0.063	0.071	0.056
	Regular	0.480	1 1/8	0.063	0.071	0.056
	Wide		1 15/32	0.100	0.112	0.090
1/2	Narrow		1	0.063	0.071	0.056
	Regular	0.540	1 1/4	0.100	0.112	0.090
	Wide		1 3/4	0.100	0.112	0.090
9/16	Narrow		1 1/8	0.063	0.071	0.056
	Regular	0.604	1 15/32	0.100	0.112	0.090

Table 20.31. Type B Plain Washers[39] (Continued)

Screw or bolt size	Series	ID A	OD B	Thickness C		
		Nom	Nom	Nom	Max	Min
	Wide		2	0.100	0.112	0.090
5/8	Narrow		1 1/4	0.100	0.112	0.090
	Regular	0.666	1 3/4	0.100	0.112	0.090
	Wide		2 1/4	0.160	0.174	0.146
3/4	Narrow		1 3/8	0.100	0.112	0.090
	Regular	0.812	2	0.100	0.112	0.090
	Wide		2 1/2	0.160	0.174	0.146
7/8	Narrow		1 15/32	0.100	0.112	0.090
	Regular	0.938	2 1/4	0.160	0.174	0.146
	Wide		2 3/4	0.160	0.174	0.146
1	Narrow		1 3/4	0.100	0.112	0.090
	Regular	1 1/16	2 1/2	0.160	0.174	0.146
	Wide		3	0.160	0.174	0.146
1 1/8	Narrow		2	0.100	0.112	0.090
	Regular	1 3/16	2 3/4	0.160	0.174	0.146
	Wide		3 1/4	0.160	0.174	0.146
1 1/4	Narrow		2 1/4	0.160	0.174	0.146
	Regular	1 5/16	3	0.160	0.174	0.146
	Wide		3 1/2	0.250	0.266	0.234
1 3/8	Narrow		2 1/2	0.160	0.174	0.146
	Regular	1 7/16	3 1/4	0.160	0.174	0.146
	Wide		3 3/4	0.250	0.266	0.234
1 1/2	Narrow		2 3/4	0.160	0.174	0.146
	Regular	1 9/16	3 1/2	0.250	0.266	0.234
	Wide		4	0.250	0.266	0.234
1 5/8	Narrow		3	0.160	0.174	0.146
	Regular	1 3/4	3 3/4	0.250	0.266	0.234
	Wide		4 1/4	0.250	0.266	0.234
1 3/4	Narrow		3 1/4	0.160	0.174	0.146
	Regular	1 7/8	4	0.250	0.266	0.234
	Wide		4 1/2	0.250	0.266	0.234
1 7/8	Narrow		3 1/2	0.250	0.266	0.234
	Regular	2	4 1/4	0.250	0.266	0.234
	Wide		4 3/4	0.250	0.266	0.234
2	Narrow		3 3/4	0.250	0.266	0.234
	Regular	2 1/8	4 1/2	0.250	0.266	0.234
	Wide		5	0.250	0.266	0.234
2 1/4	Narrow		4	0.250	0.266	0.234
	Regular	2 3/8	5	0.250	0.266	0.234
	Wide		5 1/2	0.375	0.393	0.357
2 1/2	Narrow		4 1/2	0.250	0.266	0.234
	Regular	2 5/8	5 1/2	0.375	0.393	0.357
	Wide		6	0.375	0.393	0.357
2 3/4	Narrow		5	0.250	0.266	0.234
	Regular	2 7/8	6	0.375	0.393	0.357
	Wide		6 1/2	0.375	0.393	0.357
3	Narrow		5 1/2	0.375	0.393	0.357
	Regular	3 1/8	6 1/2	0.375	0.393	0.357
	Wide		7	0.375	0.393	0.357

All dimensions are given in inches.

The inside diameter has a tolerance of −0.005 in. for sizes 0.060 to 0.190 in., inclusive; −0.010 for sizes 0.216 to 7/8 in., inclusive; and ±0.010 in. for all other sizes. The outside diameter has a tolerance of ±0.005 for narrow series sizes 0.060 to 5/16 in., inclusive; regular series sizes 0.060 to 0.216 in., inclusive; and wide series sizes 0.060 to 0.164 in., inclusive. All other outside diameters have a tolerance of ±0.010 in.

(c) Helical-spring Lock Washers

These have the standard dimensions listed in Table 20.32. Spring lock washers are made from carbon steel and stainless steel, types 302 and 420, aluminum zinc alloy, phosphorus or silicon bronze, and K monel. Consult ref. 40 for specifications on plating and heat-treatment of the various materials. Spring lock washers come in four series: (1) light, (2) medium, (3) heavy, and (4) extra heavy. Designation is made by the nominal size and the series, e.g., ⅜ light.

(d) Tooth Lock Washers[40,41]

Table 20.33 illustrates the four basic series: internal-light and heavy, external, countersunk external, and internal-external. Designation is made by the nominal size, description, and type, e.g., ⅜ in. internal tooth, type A.

20.5. PINS

For more on pins, see Sec. 27.

Table 20.32. Helical-spring Lock Washers

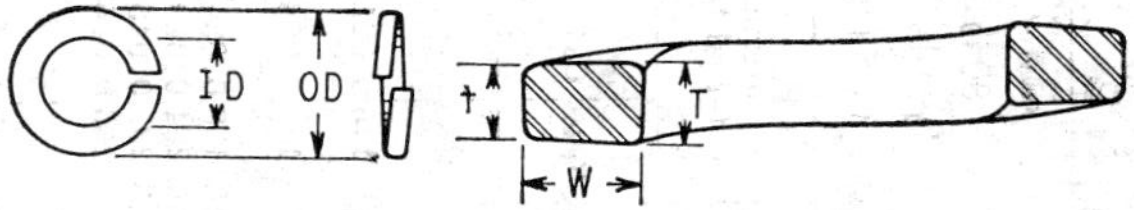

Nominal size	ID, min	Clearance of nominal bolt size		Light		Medium		Heavy		Extra heavy		OD, max			
		Min	Max	Width W	Thickness $\frac{T+t}{2}$	Width W	Thickness $\frac{T+t}{2}$	Width W	Thickness $\frac{T+t}{2}$	Width W	Thickness $\frac{T+t}{2}$	Light	Medium	Heavy	Extra heavy
No. 2	0.088	0.002	0.011	0.030	0.015	0.035	0.020	0.040	0.025	0.053	0.027	0.165	0.175	0.185	0.211
No. 3	0.102	0.002	0.011	0.035	0.020	0.040	0.025	0.047	0.031	0.062	0.034	0.188	0.198	0.212	0.242
No. 4	0.115	0.003	0.012	0.035	0.020	0.040	0.025	0.047	0.031	0.062	0.034	0.202	0.212	0.226	0.256
No. 5	0.128	0.003	0.012	0.040	0.025	0.047	0.031	0.055	0.040	0.079	0.045	0.225	0.239	0.255	0.303
No. 6	0.141	0.003	0.013	0.040	0.025	0.047	0.031	0.055	0.040	0.079	0.045	0.237	0.251	0.267	0.315
No. 8	0.168	0.004	0.014	0.047	0.031	0.055	0.040	0.062	0.047	0.096	0.057	0.280	0.296	0.310	0.378
No. 10	0.194	0.004	0.015	0.055	0.040	0.062	0.047	0.070	0.056	0.112	0.068	0.323	0.337	0.353	0.437
No. 12	0.221	0.005	0.016	0.062	0.047	0.070	0.056	0.077	0.063	0.130	0.080	0.364	0.380	0.394	0.500
1/4	0.255	0.005	0.017	0.107	0.047	0.109	0.062	0.110	0.077	0.132	0.084	0.489	0.493	0.495	0.539
5/16	0.319	0.006	0.020	0.117	0.056	0.125	0.078	0.130	0.097	0.143	0.108	0.575	0.591	0.601	0.627
3/8	0.382	0.007	0.023	0.136	0.070	0.141	0.094	0.145	0.115	0.170	0.123	0.678	0.688	0.696	0.746
7/16	0.446	0.008	0.026	0.154	0.085	0.156	0.109	0.160	0.133	0.186	0.143	0.780	0.784	0.792	0.844
1/2	0.509	0.009	0.029	0.170	0.099	0.171	0.125	0.176	0.151	0.204	0.162	0.877	0.879	0.889	0.945
9/16	0.573	0.010	0.032	0.186	0.113	0.188	0.141	0.193	0.170	0.223	0.182	0.975	0.979	0.989	1.049
5/8	0.636	0.011	0.035	0.201	0.126	0.203	0.156	0.210	0.189	0.242	0.202	1.082	1.086	1.100	1.164
11/16	0.700	0.012	0.038	0.216	0.138	0.219	0.172	0.227	0.207	0.260	0.221	1.178	1.184	1.200	1.266
3/4	0.763	0.013	0.041	0.233	0.153	0.234	0.188	0.244	0.226	0.279	0.241	1.277	1.279	1.299	1.369
13/16	0.827	0.014	0.044	0.249	0.168	0.250	0.203	0.262	0.246	0.298	0.261	1.375	1.377	1.401	1.473
7/8	0.890	0.015	0.047	0.264	0.179	0.266	0.219	0.281	0.266	0.322	0.285	1.470	1.474	1.504	1.586
15/16	0.954	0.016	0.050	0.277	0.191	0.281	0.234	0.298	0.284	0.345	0.308	1.562	1.570	1.604	1.698
1	1.017	0.017	0.053	0.289	0.202	0.297	0.250	0.319	0.306	0.366	0.330	1.656	1.672	1.716	1.810
1 1/16	1.081	0.018	0.056	0.301	0.213	0.312	0.266	0.338	0.326	0.389	0.352	1.746	1.768	1.820	1.922
1 1/8	1.144	0.019	0.059	0.314	0.224	0.328	0.281	0.356	0.345	0.411	0.375	1.837	1.865	1.921	2.031
1 3/16	1.208	0.020	0.062	0.324	0.234	0.344	0.297	0.373	0.364	0.431	0.396	1.923	1.963	2.021	2.137
1 1/4	1.271	0.021	0.065	0.336	0.244	0.359	0.312	0.393	0.384	0.452	0.417	2.012	2.058	2.126	2.244
1 5/16	1.335	0.022	0.068	0.346	0.254	0.375	0.328	0.410	0.403	0.472	0.438	2.098	2.156	2.226	2.350
1 3/8	1.398	0.023	0.071	0.356	0.264	0.391	0.344	0.427	0.422	0.491	0.458	2.183	2.253	2.325	2.453
1 7/16	1.462	0.024	0.074	0.366	0.273	0.406	0.359	0.442	0.440	0.509	0.478	2.269	2.349	2.421	2.555
1 1/2	1.525	0.025	0.077	0.375	0.282	0.422	0.375	0.458	0.458	0.526	0.496	2.352	2.446	2.518	2.654

Table 20.33. Dimensions of Tooth Lock Washers[40,42]

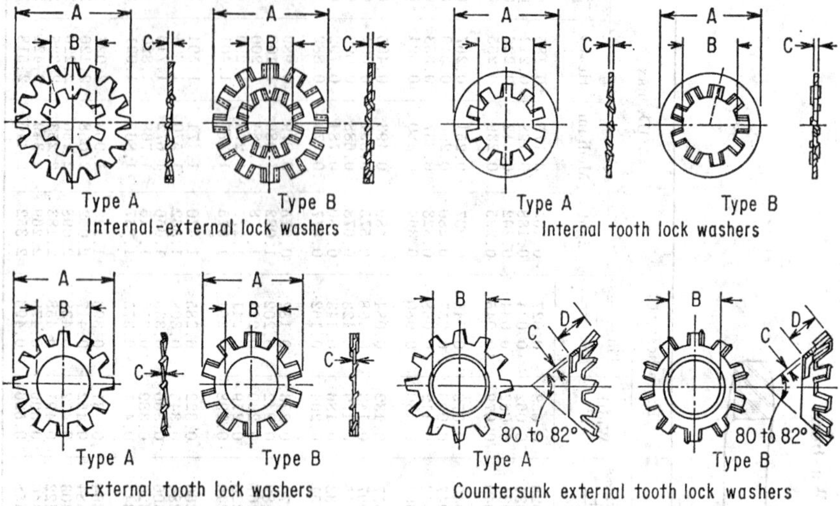

Nominal size	Internal and external tooth				Heavy B_{min}	Internal tooth		Countersunk external tooth			Internal-external tooth		
	B_{min}	C_{max}	Internal tooth A_{max}	External tooth A_{max}		C_{max}	A_{max}	B_{min}	C_{max}	D_{max}	B_{min}	C_{max}	A
No. 2	0.089	0.015	0.200										
No. 3	0.102	0.019	0.232										
No. 4	0.115	0.019	0.270	0.290	.....	.....	.....	0.113	0.019	0.065	0.115	0.021	0.610
No. 5	0.129	0.021	0.280										
No. 6	0.141	0.021	0.295	0.320	.....	.....	.....	0.140	0.021	0.092	0.141	0.028	0.690
No. 8	0.168	0.023	0.340	0.381	.....	.....	.....	0.167	0.021	0.105	0.168	0.034	0.760
No. 10	0.195	0.025	0.381	0.410	.....	.....	.....	0.195	0.025	0.099	0.204	0.040	0.900
No. 12	0.221	0.025	0.410	0.475	.....	.....	.....	0.220	0.025	0.128	0.231	0.045	0.985
1/4	0.256	0.028	0.478	0.510	0.256	0.045	0.536	0.255	0.025	0.128	0.256	0.045	1.070
No. 16	.....	.....	.....	.....	.....	.....	.....	0.273	0.028	0.147			
5/16	0.320	0.034	0.610	0.610	0.320	0.050	0.607	0.318	0.028	0.192	0.320	0.050	1.155
3/8	0.384	0.040	0.692	0.694	0.384	0.050	0.748	0.383	0.034	0.255	0.384	0.050	1.260
7/16	0.448	0.040	0.789	0.760	0.448	0.067	0.858	0.448	0.045	0.270	0.448	0.055	1.315
1/2	0.512	0.045	0.900	0.900	0.512	0.067	0.924	0.512	0.045	0.304	0.512	0.067	1.620
9/16	0.576	0.045	0.985	0.985	0.576	0.067	1.034	.....	.....	.....	0.576	0.067	1.830
5/8	0.640	0.050	1.071	1.070	0.640	0.067	1.135	.....	.....	.....	0.640	0.067	1.975
11/16	0.704	0.050	1.166	1.155									
3/4	0.769	0.055	1.245	1.260	0.768	0.084	1.265						
13/16	0.832	0.055	1.315	1.315									
7/8	0.894	0.060	1.410	1.410	0.894	0.084	1.447						
1	1.019	0.067	1.637	1.620									
1 1/8	1.144	0.067	1.830										
1 1/4	1.275	0.067	1.975										

(a) Dowel Pins

Dowel pins are used primarily to preserve alignment or to retain parts in a fixed position. Dowel-pin size is governed by its application. The general rule is to use dowel pins of the same size as the screws used in fastening the work. The length of the dowel pin should be approximately one and one-half to two times its diameter. Good practice, when using hardened dowel pins in soft parts, calls for a reamed hole 0.001 in. smaller than the pin. Tables 20.34 to 20.36 list pin dimensions.

Table 20.34. Hardened and Ground Dowel Pins[43]

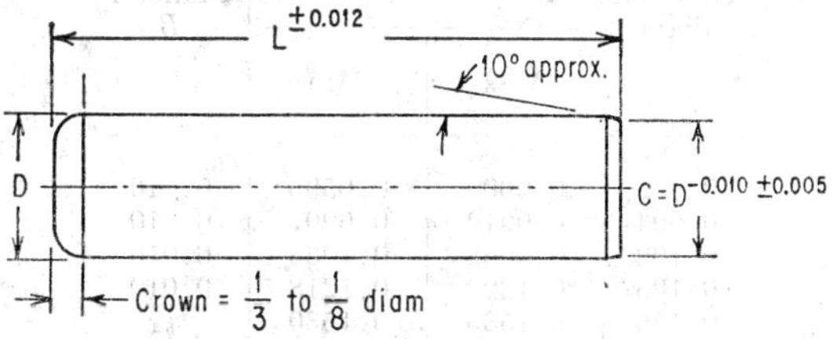

Length L	Nominal diam. D									
	1/8	3/16	1/4	5/16	3/8	7/16	1/2	5/8	3/4	7/8
	Diam. standard pins ±0.0001									
	0.1252	0.1877	0.2502	0.3127	0.3752	0.4377	0.5002	0.6252	0.7502	0.8752
	Diam. oversize pins ±0.0001									
	0.1260	0.1885	0.2510	0.3135	0.3760	0.4385	0.5010	0.6260	0.7510	0.8760
1/2	X	X	X	X						
5/8	X	X	X	X						
3/4	X	X	X	X	X					
7/8	X	X	X	X	X	X				
1	X	X	X	X	X	X				
1 1/4		X	X	X	X	X	X	X		
1 1/2		X	X	X	X	X	X	X	X	
1 3/4		X	X	X	X	X	X	X	X	
2		X	X	X	X	X	X	X	X	X
2 1/4				X	X	X	X			
2 1/2				X	X	X	X	X	X	X
3							X	X	X	X
3 1/2							X	X		
4							X	X	X	X
4 1/2								X	X	X
5									X	X
5 1/2									X	X

The letter X indicates range of diameters for corresponding length.
All dimensions are given in inches.
These pins are extensively used in the tool and machine industry and a machine reamer of nominal size may be used to produce the holes into which these pins tap or press fit. They must be straight and free from any defects.

Table 20.35. Dimensions of Ground Dowel Pins, Not Hardened[43]

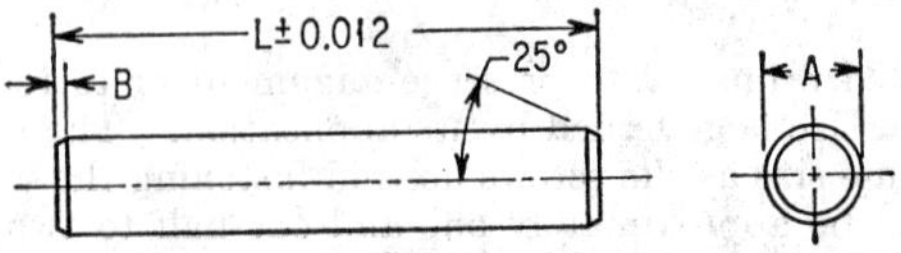

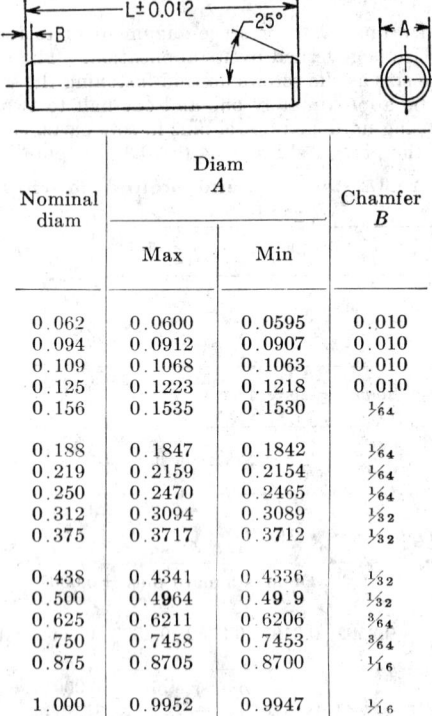

Nominal diam	Diam *A*		Chamfer *B*
	Max	Min	
0.062	0.0600	0.0595	0.010
0.094	0.0912	0.0907	0.010
0.109	0.1068	0.1063	0.010
0.125	0.1223	0.1218	0.010
0.156	0.1535	0.1530	1/64
0.188	0.1847	0.1842	1/64
0.219	0.2159	0.2154	1/64
0.250	0.2470	0.2465	1/64
0.312	0.3094	0.3089	1/32
0.375	0.3717	0.3712	1/32
0.438	0.4341	0.4336	1/32
0.500	0.4964	0.49 9	1/32
0.625	0.6211	0.6206	3/64
0.750	0.7458	0.7453	3/64
0.875	0.8705	0.8700	1/16
1.000	0.9952	0.9947	1/16

All dimensions are given in inches.
Maximum diameters are graduated from 0.0005 on 1/16-in. pins to 0.0028 on 1-in. pins under the minimum commercial bar stock sizes.

(b) Taper Pins

These are made in standard sizes with a taper of 1/4 in./ft and in length increments of 1/4 in. For standard dimensions see Table 20.37.

(c) Grooved Pins

These are resistant to loosening from shock or vibration. Longitudinal grooves pressed into the cylindrical body deform the pin stock outward. When forced into a drilled hole of proper size, a locking fit is obtained. The various types of grooved pins are:

Type A. The full-length taper grooved pin replaces the taper dowel pin in many applications.

Type B. The half-length taper grooved pin, often used as a dowel pin or stop, has tapering grooves and body on only half of the pin length.

Type C. The full-length constant-diameter pin has straight grooves along its length and a pilot at one end to facilitate easy assembly. Particularly applicable where longitudinal stress due to vibration and shock is severe.

Type F. The full-length constant-diameter pin is similar to type C except it has a pilot at both ends. This pin has applications similar to type C.

Type D. The half-length reverse-taper grooved pin has applications similar to those of type B. Note the pin diameter is constant but stepped.

Type E. The center-grooved half-length pin has oval grooves and body. Used as a fulcrum bolt, T-handle, and hinge.

For dimensions, see ref. 43.

Table 20.36. Dimensions of Straight Pins[43]

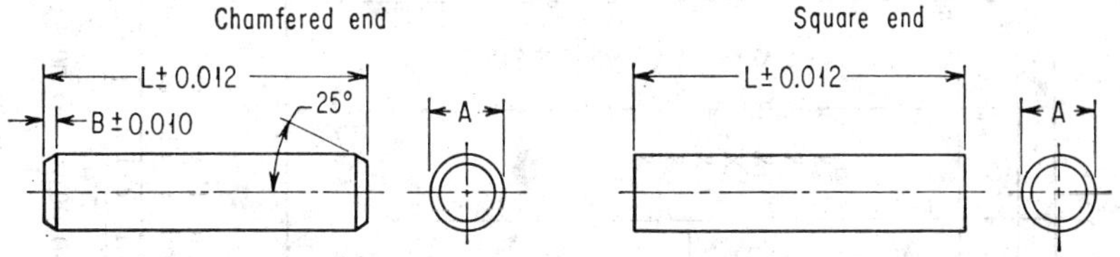

Nominal diam	Diam A		Chamfer B
	Max	Min	
0.062	0.0625	0.0605	0.015
0.094	0.0937	0.0917	0.015
0.109	0.1094	0.1074	0.015
0.125	0.1250	0.1230	0.015
0.156	0.1562	0.1542	0.015
0.188	0.1875	0.1855	0.015
0.219	0.2187	0.2167	0.015
0.250	0.2500	0.2480	0.015
0.312	0.3125	0.3095	0.030
0.375	0.3750	0.3720	0.030
0.438	0.4375	0.4345	0.030
0.500	0.500	0.4970	0.030

All dimensions are given in inches.

These pins must be straight and free from burrs or any other defects that will affect their serviceability.

(d) Spring-type Straight Pins[43,44]

Driven into a hole, it exerts continuous spring pressure against the sides of the hole, preventing loosening by vibration. No reaming is required. Construction of the pin permits using a screw through the pin in some applications. It can be used as a replacement for stop pins, setscrews, taper pins, hinge pins, dowel pins, grooved pins, or clevis pins. One important advantage of this type of pin is that it is easily removed and replaced with little loss in holding power. The characteristics are shown in Table 20.38.

Table 20.37. Dimensions of Taper Pins[43]

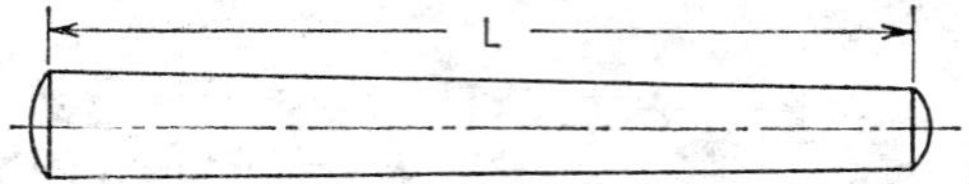

Number	7/0	6/0	5/0	4/0	3/0	2/0	0	1	2	3	4	5	6	7	8	9	10
Size (large end)	0.0625	0.0780	0.0940	0.1090	0.1250	0.1410	0.1560	0.1720	0.1930	0.2190	0.2500	0.2890	0.3410	0.4090	0.4920	0.5910	0.7060
Length, L																	
0.375	X	X															
0.500	X	X	X	X	X	X	X										
0.625	X	X	X	X	X	X	X										
0.750	...	X	X	X	X	X	X	X	X	X							
0.875	...	...	...	...	X	X	X	X	X	X							
1.000	...	...	X	X	X	X	X	X	X	X	X	X					
1.250	...	...	...	...	...	X	X	X	X	X	X	X	X				
1.500	...	...	...	...	...	...	X	X	X	X	X	X	X				
1.750	...	...	...	...	...	...	...	X	X	X	X	X	X				
2.000	...	...	...	...	...	...	...	X	X	X	X	X	X	X	X		
2.250	...	...	...	...	...	...	...	...	X	X	X	X	X	X	X		
2.500	...	...	...	...	...	...	...	...	X	X	X	X	X	X	X		
2.750	...	...	...	...	...	...	...	...	...	X	X	X	X	X	X	X	
3.000	...	...	...	...	...	...	...	...	...	X	X	X	X	X	X	X	
3.250	...	...	...	...	...	...	...	...	...	...	...	...	X	X	X	X	
3.500	...	...	...	...	...	...	...	...	...	...	...	...	X	X	X	X	X
3.750	...	...	...	...	...	...	...	...	...	...	...	...	X	X	X	X	X
4.000	...	...	...	...	...	...	...	...	...	...	...	...	X	X	X	X	X
4.250	...	...	...	...	...	...	...	...	...	...	...	...	...	...	X	X	X
4.500	...	...	...	...	...	...	...	...	...	...	...	...	...	...	X	X	X
4.750	...	...	...	...	...	...	...	...	...	...	...	...	...	...	X	X	X
5.000	...	...	...	...	...	...	...	...	...	...	...	...	...	...	X	X	X
5.250	...	...	...	...	...	...	...	...	...	...	...	...	...	...	...	X	X
5.500	...	...	...	...	...	...	...	...	...	...	...	...	...	...	...	X	X
5.750	...	...	...	...	...	...	...	...	...	...	...	...	...	...	...	X	X
6.000	...	...	...	...	...	...	...	...	...	...	...	...	...	...	...	X	X

All dimensions are given in inches.
Standard reamers are avilable for pins given above the line.
Pins Nos. 11 (size 0.8600), 12 (size 1.032), 13 (size 1.241), and 14 (1.523) are special sizes—hence their lengths are special.
To find small diameter of pin, multiply the length by 0.02083 and subtract the result from the large diameter.

Types	Commercial type	Precision type
Sizes	7/0 to 14	7/0 to 10
Tolerance on diameter	+0.0013, −0.0007	(+0.0013, −0.0007)
Taper	¼ in./ft	¼ in./ft
Length tolerance	(±0.030)	(±0.030)
Concavity tolerance	None	0.0005 up to 1 in. long 0.001 1 1/16 to 2 in. long 0.002 2 1/16 and longer

Table 20.38. Spring-type Straight Pins[43,44]

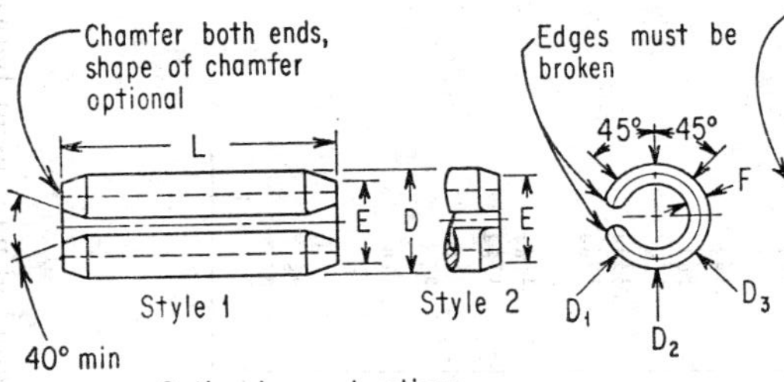

Optional constructions slotted

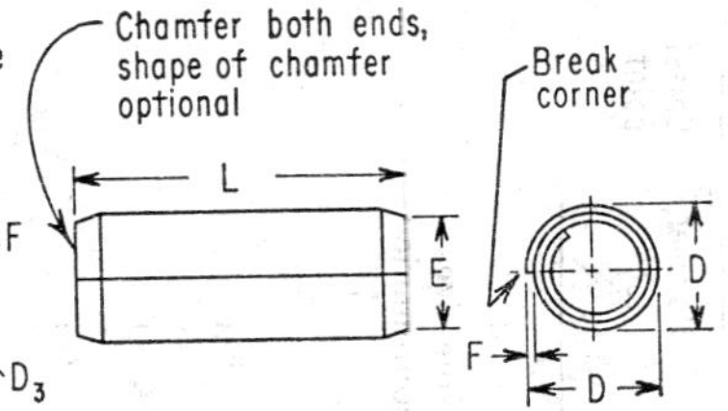

Coiled

Nominal pin size	D[a] Slotted Series A[b] and B[b] Max	D[a] Slotted Series A[b] and B[b] Min[c]	D[a] Coiled Series A Max	D[a] Coiled Series A Min	D[a] Coiled Series B Max	D[a] Coiled Series B Min	D[a] Coiled Series C Max	D[a] Coiled Series C Min	E, slotted and coiled, series A, B, and C, max	F Slotted Series A, nominal	F Slotted Series B Style I nominal	F Slotted Series B Style II nominal	F Coiled Series A, nominal	F Coiled Series B, nominal	F Coiled Series C, nominal	Double shear strength, lb,[e] slotted and coiled Series A, min	Series B, min	Series C, min	Recommended hole size Slotted Series A and B Max	Slotted Series A and B Min	Coiled Series A, B, and C Max	Coiled Series A, B, and C Min
1/32	.....	.....	.....	.....	0.035	0.033	.....	.....	0.029[d]	.....	.....	.....	.....	0.003	.....	.....	75	.....	.....	.....	0.0325	0.0310
3/64	.....	.....	.....	.....	0.052	0.049	.....	.....	0.045[d]	.....	.....	.....	.....	0.004	.....	.....	170	.....	.....	.....	0.0485	0.0470
0.052	.....	.....	.....	.....	0.057	0.054	.....	.....	0.050[d]	.....	.....	.....	.....	0.004	.....	.....	230	.....	.....	.....	0.0535	0.0520
1/16	0.069	0.066	0.070	0.066	0.071	0.067	0.072	0.067	0.059	0.012	.....	0.006	0.007	0.005	0.003	425	300	160	0.065	0.062	0.065	0.061
5/64	0.086	0.083	0.086	0.082	0.087	0.083	0.088	0.083	0.075	0.018	.....	0.008	0.007	0.005	0.003	650	480	260	0.081	0.078	0.081	0.077
3/32	0.103	0.099	0.103	0.098	0.104	0.099	0.105	0.099	0.091	0.022	0.012	0.012	0.010	0.007	0.005	1000	690	370	0.097	0.094	0.097	0.093
7/64	0.118	0.113	0.118	0.113	0.119	0.114	0.120	0.114	0.106	0.022	.....	0.018	0.010	0.007	0.005	1410	940	510	0.112	0.109	0.112	0.108
1/8	0.135	0.131	0.136	0.130	0.137	0.131	0.138	0.131	0.122	0.028	0.012	0.018	0.014	0.010	0.007	1840	1000	660	0.129	0.125	0.129	0.124
9/64	0.149	0.145	0.151	0.145	0.152	0.146	0.153	0.146	0.136	0.028	.....	0.022	0.014	0.010	0.007	2200	1550	830	0.141	0.140	0.144	0.139
5/32	0.167	0.162	0.168	0.161	0.170	0.163	0.171	0.163	0.152	0.032	0.018	0.022	0.017	0.011	0.007	2880	1750	1040	0.160	0.156	0.160	0.155
3/16	0.199	0.194	0.202	0.194	0.204	0.196	0.206	0.196	0.182	0.040	0.022	0.028	0.020	0.015	0.010	4140	2500	1500	0.192	0.187	0.192	0.185
7/32	0.232	0.226	0.235	0.226	0.238	0.229	0.240	0.229	0.214	0.048	0.028	0.032	0.024	0.017	0.011	5640	3760	2040	0.224	0.219	0.224	0.217
1/4	0.264	0.258	0.268	0.258	0.270	0.260	0.272	0.260	0.245	0.048	0.028	0.032	0.028	0.020	0.015	7360	4600	2660	0.256	0.250	0.256	0.248
5/16	0.328	0.321	0.340	0.327	0.341	0.327	0.342	0.327	0.306	0.062	.....	0.040	0.032	0.024	0.017	11500	7670	4160	0.318	0.312	0.318	0.308
3/8	0.392	0.385	0.407	0.391	0.408	0.391	0.409	0.391	0.368	0.077	.....	0.048	0.040	0.028	0.020	16580	11040	6000	0.382	0.375	0.382	0.368
7/16	0.456	0.448	0.475	0.457	0.476	0.457	0.478	0.457	0.430	0.077	.....	0.048	0.047	0.036	0.024	20000	15020	8160	0.445	0.437	0.445	0.429
1/2	0.527	0.513	0.542	0.522	0.543	0.522	0.545	0.522	0.490	0.094	.....	0.062	0.055	0.040	0.028	25800	19600	10640	0.510	0.500	0.510	0.490

[a] Maximum D shall be checked by a "go" ring gage.
[b] Series designation applies to stock thickness, A being heaviest.
[c] Minimum D shall be the average of the D_1, D_2, and D_3 diameters.
[d] Series B coiled.
[e] Applies to pins made from SAE 1070 to 1095 steel and SAE 51410 or AISI 420 corrosion-resistant steel. SAE 30302 stainless steel has a minimum shear strength equal to 85 per cent of values shown for coiled pins.

(e) Cotter Pins

These are used with threaded members as a positive locking device (see Table 20.39.)

Table 20.39. Dimensions of Cotter Pins[43]

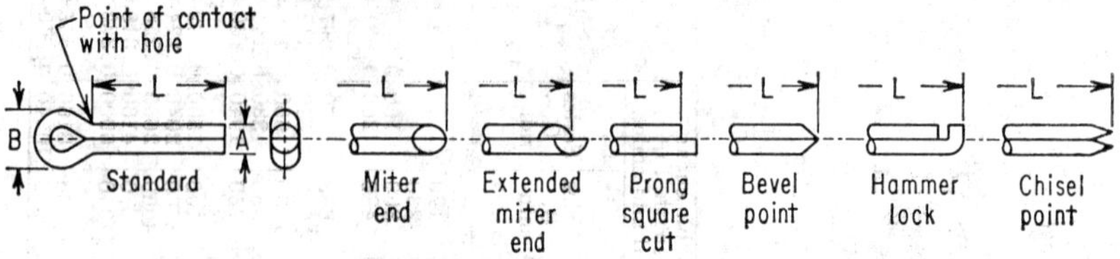

Nominal diam	Diam A		Outside eye diam B, min	Hole sizes recommended
	Max	Min		
0.031	0.032	0.028	1/16	3/64
0.047	0.048	0.044	3/32	1/16
0.062	0.060	0.056	1/8	5/64
0.078	0.076	0.072	5/32	3/32
0.094	0.090	0.086	3/16	7/64
0.109	0.104	0.100	7/32	1/8
0.125	0.120	0.116	1/4	9/64
0.141	0.134	0.130	9/32	5/32
0.156	0.150	0.146	5/16	11/64
0.188	0.176	0.172	3/8	13/64
0.219	0.207	0.202	7/16	15/64
0.250	0.225	0.220	1/2	17/64
0.312	0.280	0.275	5/8	5/16
0.375	0.335	0.329	3/4	3/8
0.438	0.406	0.400	7/8	7/16
0.500	0.473	0.467	1	1/2
0.625	0.598	0.590	1 1/4	5/8
0.750	0.723	0.715	1 1/2	3/4

20.6. KEYS AND SPLINES

(a) Introduction

Keys and splines are used to prevent relative circumferential movement between rotating parts and respective shafts. Except for extremely heavy loads, the key is sized according to the shaft diameter rather than to the torsional loads developed.

(b) Sunk and Feather Keys

The "*sunk*" *key* (Table 20.40) is fitted in a groove in the shaft and projects into a keyway in the hub.

Table 20.40. Square and Flat Plain Parallel Stock Keys[48]

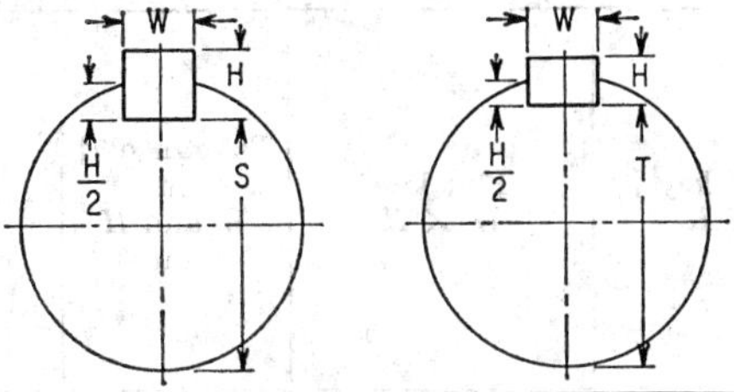

Shaft diam	Square key W × H	Flat key W × H	Tolerance*,† on W and H (minus)	Bottom of keyseat to opposite side of shaft	
				Square key S	Flat key T
1/2	1/8 × 1/8	1/8 × 3/32	0.0020	0.430	0.445
9/16	1/8 × 1/8	1/8 × 3/32	0.0020	0.493	0.509
5/8	3/16 × 3/16	3/16 × 1/8	0.0020	0.517	0.548
11/16	3/16 × 3/16	3/16 × 1/8	0.0020	0.581	0.612
3/4	3/16 × 3/16	3/16 × 1/8	0.0020	0.644	0.676
13/16	3/16 × 3/16	3/16 × 1/8	0.0020	0.708	0.739
7/8	3/16 × 3/16	3/16 × 1/8	0.0020	0.771	0.802
15/16	1/4 × 1/4	1/4 × 3/16	0.0020	0.796	0.827
1	1/4 × 1/4	1/4 × 3/16	0.0020	0.859	0.890
1 1/16	1/4 × 1/4	1/4 × 3/16	0.0020	0.923	0.954
1 1/8	1/4 × 1/4	1/4 × 3/16	0.0020	0.986	1.017
1 3/16	1/4 × 1/4	1/4 × 3/16	0.0020	1.049	1.081
1 1/4	1/4 × 1/4	1/4 × 3/16	0.0020	1.112	1.144
1 5/16	5/16 × 5/16	5/16 × 1/4	0.0020	1.137	1.169
1 3/8	5/16 × 5/16	5/16 × 1/4	0.0020	1.201	1.232
1 7/16	3/8 × 3/8	3/8 × 1/4	0.0020	1.225	1.288
1 1/2	3/8 × 3/8	3/8 × 1/4	0.0020	1.289	1.351
1 9/16	3/8 × 3/8	3/8 × 1/4	0.0020	1.352	1.415
1 5/8	3/8 × 3/8	3/8 × 1/4	0.0020	1.416	1.478
1 11/16	3/8 × 3/8	3/8 × 1/4	0.0020	1.479	1.542
1 3/4	3/8 × 3/8	3/8 × 1/4	0.0020	1.542	1.605
1 13/16	1/2 × 1/2	1/2 × 3/8	0.0025	1.527	1.590
1 7/8	1/2 × 1/2	1/2 × 3/8	0.0025	1.591	1.654
1 15/16	1/2 × 1/2	1/2 × 3/8	0.0025	1.655	1.717
2	1/2 × 1/2	1/2 × 3/8	0.0025	1.718	1.781
2 1/16	1/2 × 1/2	1/2 × 3/8	0.0025	1.782	1.843
2 1/8	1/2 × 1/2	1/2 × 3/8	0.0025	1.845	1.908
2 3/16	1/2 × 1/2	1/2 × 3/8	0.0025	1.909	1.971
2 1/4	1/2 × 1/2	1/2 × 3/8	0.0025	1.972	2.034
2 5/16	5/8 × 5/8	5/8 × 7/16	0.0025	1.957	2.051
2 3/8	5/8 × 5/8	5/8 × 7/16	0.0025	2.021	2.114
2 7/16	5/8 × 5/8	5/8 × 7/16	0.0025	2.084	2.178
2 1/2	5/8 × 5/8	5/8 × 7/16	0.0025	2.148	2.242
2 5/8	5/8 × 5/8	5/8 × 7/16	0.0025	2.275	2.368
2 3/4	5/8 × 5/8	5/8 × 7/16	0.0025	2.402	2.495

Table 20.40. Square and Flat Plain Parallel Stock Keys[48] (Continued)

Shaft diam	Square key $W \times H$	Flat key $W \times H$	Tolerance*,† on W and H (minus)	Bottom of keyseat to opposite side of shaft	
				Square key S	Flat key T
2⅞	¾ × ¾	¾ × ½	0.0025	2.450	2.575
2¹⁵⁄₁₆	¾ × ¾	¾ × ½	0.0025	2.514	2.639
3	¾ × ¾	¾ × ½	0.0025	2.577	2.702
3⅛	¾ × ¾	¾ × ½	0.0025	2.704	2.829
3¼	¾ × ¾	¾ × ½	0.0025	2.831	2.956
3⅜	⅞ × ⅞	⅞ × ⅝	0.0030	2.880	3.005
3⁷⁄₁₆	⅞ × ⅞	⅞ × ⅝	0.0030	2.944	3.069
3½	⅞ × ⅞	⅞ × ⅝	0.0030	3.007	3.132
3⅝	⅞ × ⅞	⅞ × ⅝	0.0030	3.140	3.259
3¾	⅞ × ⅞	⅞ × ⅝	0.0030	3.261	3.386
3⅞	1 × 1	1 × ¾	0.0030	3.309	3.434
3¹⁵⁄₁₆	1 × 1	1 × ¾	0.0030	3.373	3.498
4	1 × 1	1 × ¾	0.0030	3.437	3.562
4¼	1 × 1	1 × ¾	0.0030	3.690	3.815
4⁷⁄₁₆	1 × 1	1 × ¾	0.0030	3.881	4.006
4½	1 × 1	1 × ¾	0.0030	3.944	4.069
4¾	1¼ × 1¼	1¼ × ⅞	0.0030	4.042	4.229
4¹⁵⁄₁₆	1¼ × 1¼	1¼ × ⅞	0.0030	4.232	4.420
5	1¼ × 1¼	1¼ × ⅞	0.0030	4.296	4.483
5¼	1¼ × 1¼	1¼ × ⅞	0.0030	4.550	4.733
5⁷⁄₁₆	1¼ × 1¼	1¼ × ⅞	0.0030	4.740	4.927
5½	1¼ × 1¼	1¼ × ⅞	0.0030	4.803	4.991
5¾	1½ × 1½	1½ × 1	0.0030	4.900	5.150
5¹⁵⁄₁₆	1½ × 1½	1½ × 1	0.0030	5.091	5.341
6	1½ × 1½	1½ × 1	0.0030	5.155	5.405

* Stock keys are applicable to the general run of work and the tolerances have been set accordingly. It is understood that these keys are to be cut from cold-finished stock and are to be used without machining. They are not intended to cover the finer applications where a closer fit may be required.

† These tolerances are *negative* and represent the maximum allowable variation *below* the exact nominal size. For example, the standard stock square key for a 2-in. shaft has a maximum size of 0.500 × 0.500 in. and a minimum size of 0.4975 × 0.4975 in.

Feather keys differ from sunk keys in that longitudinal movement between shaft and hub is permitted as illustrated in Fig. 20.18.

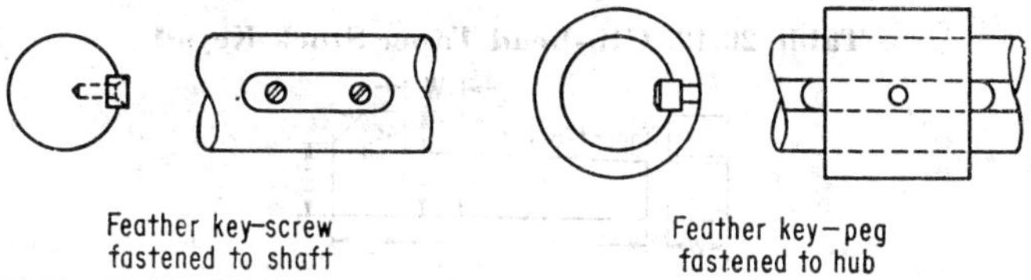

Fig. 20.18. Keys.

Some of the key forms most commonly used are:

1. *Plain Parallel Key.*[45] Square keys have sides equal to approximately one-fourth the shaft diameter. Both square and flat keys should fit tightly on all four sides to prevent any tendency to rock. Standard dimensions of square and flat keys are listed in Table 20.40.

2. *Plain Taper Key.*[46] One surface of this square- or rectangular-section key is tapered, resulting in a wedging action. Refer to Table 20.41 for standard dimensions.

Table 20.41. Square and Flat Plain Taper Stock Keys[45]

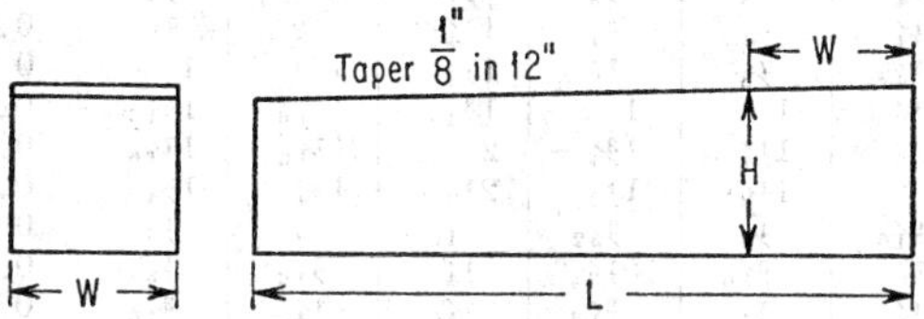

Shaft diam, incl	Square type		Flat type		Tolerance*	
	Max width *W*	Height at large end† *H*	Max width *W*	Height at large end† *H*	On width (minus)	On height (plus)
1/2 – 9/16	1/8	1/8	1/8	3/32	0.0020	0.0020
5/8 – 7/8	3/16	3/16	3/16	1/8	0.0020	0.0020
15/16–1 1/4	1/4	1/4	1/4	3/16	0.0020	0.0020
1 5/16 –1 3/8	5/16	5/16	5/16	1/4	0.0020	0.0020
1 7/16 –1 3/4	3/8	3/8	3/8	1/4	0.0020	0.0020
1 13/16–2 1/4	1/2	1/2	1/2	3/8	0.0025	0.0025
2 5/16 –2 3/4	5/8	5/8	5/8	7/16	0.0025	0.0025
2 7/8 –3 1/4	3/4	3/4	3/4	1/2	0.0025	0.0025
3 3/8 –3 3/4	7/8	7/8	7/8	5/8	0.0030	0.0030
3 7/8 –4 1/2	1	1	1	3/4	0.0030	0.0030
4 3/4 –5 1/2	1 1/4	1 1/4	1 1/4	7/8	0.0030	0.0030
5 3/4 –6	1 1/2	1 1/2	1 1/2	1	0.0030	0.0030

All dimensions given in inches.

* Not intended to cover the finer applications where a closer fit may be required.

† This height of the key is measured at the distance *W*, equal to the width of the key, from the large end.

3. *Gib-head Key.*[47] Used where the small end of the key is not accessible. The gib provides a means for removal as well as a driving head when assembling; refer to Table 20.42.

Table 20.42. Gib-head Taper Stock Keys[56]

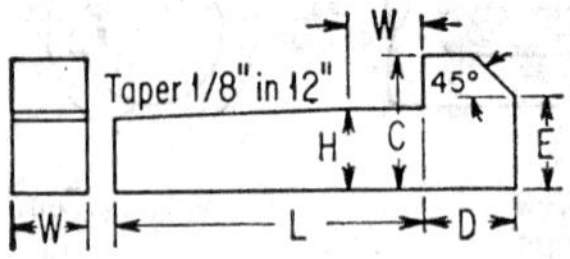

Type of key	Shaft diam, incl	Key		Gib head			Tolerances	
		W	*H**	*C*	*D*	*E*	Width (minus)	Height (plus)
Square type	1/2 - 9/16	1/8	1/8	1/4	7/32	5/32	0.0020	0.0020
	5/8 - 7/8	3/16	3/16	5/16	9/32	7/32	0.0020	0.0020
	15/16-1 1/4	1/4	1/4	7/16	11/32	11/32	0.0020	0.0020
	1 5/16 -1 3/8	5/16	5/16	9/16	13/32	13/32	0.0020	0.0020
	1 7/16 -1 3/4	3/8	3/8	11/16	15/32	15/32	0.0020	0.0020
	1 13/16-2 1/4	1/2	1/2	7/8	19/32	5/8	0.0025	0.0025
	2 5/16 -2 3/4	5/8	5/8	1 1/16	23/32	3/4	0.0025	0.0025
	2 7/8 -3 1/4	3/4	3/4	1 1/4	7/8	7/8	0.0025	0.0025
	3 3/8 -3 3/4	7/8	7/8	1 1/2	1	1	0.0030	0.0030
	3 7/8 -4 1/2	1	1	1 3/4	1 3/16	1 3/16	0.0030	0.0030
	4 3/4 -5 1/2	1 1/4	1 1/4	2	1 7/16	1 7/16	0.0030	0.0030
	5 3/4 -6	1 1/2	1 1/2	2 1/2	1 3/4	1 3/4	0.0030	0.0030
Flat type	1/2 - 9/16	1/8	3/32	3/16	1/8	1/8	0.0020	0.0020
	5/8 - 7/8	3/16	1/8	1/4	3/16	5/32	0.0020	0.0020
	15/16-1 1/4	1/4	3/16	5/16	1/4	3/16	0.0020	0.0020
	1 5/16 -1 3/8	5/16	1/4	3/8	5/16	1/4	0.0020	0.0020
	1 7/16 -1 3/4	3/8	1/4	7/16	3/8	5/16	0.0020	0.0020
	1 13/16-2 1/4	1/2	3/8	5/8	1/2	7/16	0.0025	0.0025
	2 5/16 -2 3/4	5/8	7/16	3/4	5/8	1/2	0.0025	0.0025
	2 7/8 -3 1/4	3/4	1/2	7/8	3/4	5/8	0.0025	0.0025
	3 3/8 -3 3/4	7/8	5/8	1 1/16	7/8	3/4	0.0030	0.0030
	3 7/8 -4 1/2	1	3/4	1 1/4	1	13/16	0.0030	0.0030
	4 3/4 -5 1/2	1 1/4	7/8	1 1/2	1 1/4	1	0.0030	0.0003
	5 3/4 -6	1 1/2	1	1 3/4	1 1/2	1 1/4	0.0030	0.0030

All dimensions in inches.
* This height of the key is measured at a distance *W* from the gib head.

(c) Woodruff Key[48]

This key cannot move axially but can adjust itself to the taper, if any, in the hub keyway. The key is tightly fitted in a keyway formed in the shaft. Table 20.43 lists the standard key dimensions and Table 20.44 the keyslot data.

Refer to references for the following key types: round end,[49] pin,[50] beveled or Barth,[51] Kennedy,[55] Lewis,[51] and Sec. 27.

Table 20.43. Woodruff Keys[48]

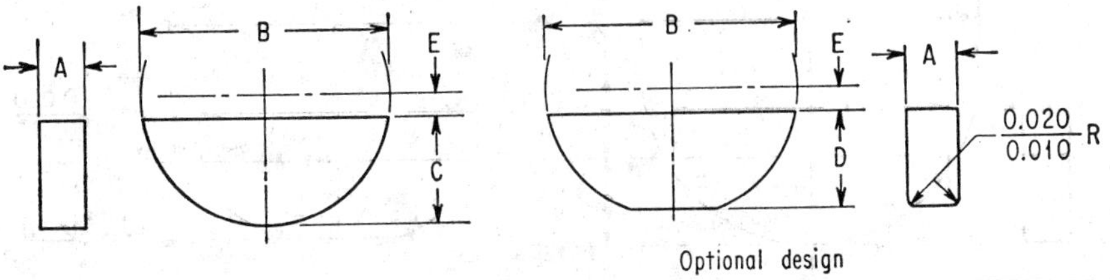

American Standard No.*	Old standard No.†	SAE nominal size	Width A	Diam B	Heights C	Heights D	Heights E	Key area at shear line
			+0.001 −0.000	+0.000 −0.010	+0.000 −0.005	+0.000 −0.006	Nominal	
202	201	1/16 × 1/4	0.0625	0.250	0.109	.....	1/64	0.0145
202½	206	1/16 × 5/16	0.0625	0.312	0.140	.....	1/64	0.0184
302½	207	3/32 × 5/16	0.0938	0.312	0.140	.....	1/64	0.0264
203	211	1/16 × 3/8	0.0625	0.375	0.172	.....	1/64	0.0225
303	212	3/32 × 3/8	0.0938	0.375	0.172	.....	1/64	0.0328
403	213	1/8 × 3/8	0.1250	0.375	0.172	.....	1/64	0.0420
204	1	1/16 × 1/2	0.0625	0.500	0.203	0.194	3/64	0.0296
304	2	3/32 × 1/2	0.0938	0.500	0.203	0.194	3/64	0.0434
404	3	1/8 × 1/2	0.1250	0.500	0.203	0.194	3/64	0.0512
305	4	3/32 × 5/8	0.0938	0.625	0.250	0.240	1/16	0.0523
405	5	1/8 × 5/8	0.1250	0.625	0.250	0.240	1/16	0.0716
505	6	5/32 × 5/8	0.1563	0.625	0.250	0.240	1/16	0.0871
605	61	3/16 × 5/8	0.1875	0.625	0.250	0.240	1/16	0.0105
406	7	1/8 × 3/4	0.1250	0.750	0.313	0.303	1/16	0.0884
506	8	5/32 × 3/4	0.1563	0.750	0.313	0.303	1/16	0.1086
606	9	3/16 × 3/4	0.1875	0.750	0.313	0.303	1/16	0.1279
806	91	1/4 × 3/4	0.2500	0.750	0.313	0.303	1/16	0.1623
507	10	5/32 × 7/8	0.1563	0.875	0.375	0.365	1/16	0.1294
607	11	3/16 × 7/8	0.1875	0.875	0.375	0.365	1/16	0.1531
707	12	7/32 × 7/8	0.2188	0.875	0.375	0.365	1/16	0.1813
807	A	1/4 × 7/8	0.2500	0.875	0.375	0.365	1/16	0.1976
608	13	3/16 × 1	0.1875	1.000	0.438	0.428	1/16	0.1781
708	14	7/32 × 1	0.2188	1.000	0.438	0.428	1/16	0.2100
805	15	1/4 × 1	0.2500	1.000	0.438	0.428	1/16	0.2320
1008	B	5/16 × 1	0.3125	1.000	0.438	0.428	1/16	0.2811
609	16	3/16 × 1⅛	0.1875	1.125	0.484	0.475	5/64	0.2007
709	17	7/32 × 1⅛	0.2188	1.125	0.484	0.475	5/64	0.2320
809	18	1/4 × 1⅛	0.2500	1.125	0.484	0.475	5/64	0.2622
1009	C	5/16 × 1⅛	0.3125	1.125	0.484	0.475	5/64	0.3193
610	19	3/16 × 1¼	0.1875	1.250	0.547	0.537	5/64	0.2284
710	20	7/32 × 1¼	0.2188	1.250	0.547	0.537	5/64	0.2608
810	21	1/4 × 1¼	0.2500	1.250	0.547	0.537	5/64	0.2955
1010	D	5/16 × 1¼	0.3125	1.250	0.547	0.537	5/64	0.3621
1210	E	3/8 × 1¼	0.3750	1.250	0.547	0.537	5/64	0.4243
811	22	1/4 × 1⅜	0.2500	1.375	0.594	0.584	3/32	0.3259
1011	23	5/16 × 1⅜	0.3125	1.375	0.594	0.584	3/32	0.4003
1211	F	3/8 × 1⅜	0.3750	1.375	0.594	0.584	3/32	0.4705
812	24	1/4 × 1½	0.2500	1.500	0.641	0.631	7/64	0.3562
1012	25	5/16 × 1½	0.3125	1.500	0.641	0.631	7/64	0.4384
1212	G	3/8 × 1½	0.3750	1.500	0.641	0.631	7/64	0.5166

* Numbers of standard shank-type cutters, and expected numbers in future revision of Woodruff-key standard.
† SAE listing of manufacturers' part numbers.

Table 20.43. Woodruff Keys[48] (Continued)

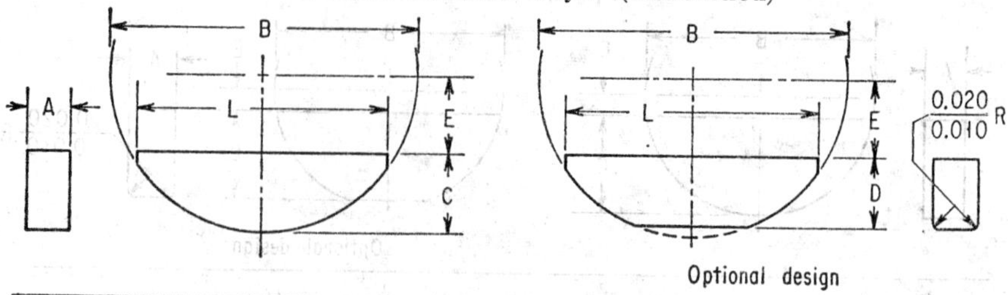

American Standard No.*	Old standard No.†	SAE nominal size	Width A +0.001 −0.000	Diam B +0.000 −0.010	Heights C +0.000 −0.005	Heights D +0.000 −0.006	Heights E Nominal	Length L +0.000 −0.010	Key area at shear line
	126	3/16 × 2 1/8	0.1875	2.125	0.406	0.396	21/32	1.380	0.2578
	127	1/4 × 2 1/8	0.2500	2.125	0.406	0.396	21/32	1.380	0.3437
	128	5/16 × 2 1/8	0.3125	2.125	0.406	0.396	21/32	1.380	0.4296
	129	3/8 × 2 1/8	0.3750	2.125	0.406	0.396	21/32	1.380	0.4833
617	26	3/16 × 2 1/8	0.1875	2.125	0.531	0.521	17/32	1.723	0.3222
817	27	1/4 × 2 1/8	0.2500	2.125	0.531	0.521	17/32	1.723	0.4178
1017	28	5/16 × 2 1/8	0.3125	2.125	0.531	0.521	17/32	1.723	0.5062
1217	29	3/8 × 2 1/8	0.3750	2.125	0.531	0.521	17/32	1.723	0.5868
	Rx	1/4 × 2 3/4	0.2500	2.750	0.594	0.584	25/32	2.000	0.5000
	Sx	5/16 × 2 3/4	0.3125	2.750	0.594	0.584	25/32	2.000	0.6286
	Tx	3/8 × 2 3/4	0.3750	2.750	0.594	0.584	25/32	2.000	0.6943
	Ux	7/16 × 2 3/4	0.4375	2.750	0.594	0.584	25/32	2.000	0.8253
	Vx	1/2 × 2 3/4	0.5000	2.750	0.594	0.584	25/32	2.000	0.9094
822	R	1/4 × 2 3/4	0.2500	2.750	0.750	0.740	5/8	2.317	0.5718
1022	S	5/16 × 2 3/4	0.3125	2.750	0.750	0.740	5/8	2.317	0.7071
1222	T	3/8 × 2 3/4	0.3750	2.750	0.750	0.740	5/8	2.317	0.8319
1422	U	7/16 × 2 3/4	0.4375	2.750	0.750	0.740	5/8	2.317	0.9499
1622	V	1/2 × 2 3/4	0.5000	2.750	0.750	0.740	5/8	2.317	1.0606
1228	30	3/8 × 3 1/2	0.3750	3.500	0.938	0.927	13/16	2.880	1.0781
1428	31	7/16 × 3 1/2	0.4375	3.500	0.938	0.927	13/16	2.880	1.2371
1628	32	1/2 × 3 1/2	0.5000	3.500	0.938	0.927	13/16	2.880	1.3905
1828	33	9/16 × 3 1/2	0.5625	3.500	0.938	0.927	13/16	2.880	1.5368
2028	34	5/8 × 3 1/2	0.6250	3.500	0.938	0.927	13/16	2.880	1.6755
2228	35	11/16 × 3 1/2	0.6875	3.500	0.938	0.927	13/16	2.880	1.8062
2428	36	3/4 × 3 1/2	0.7500	3.500	0.938	0.927	13/16	2.880	1.9281

Material: Carbon steel or alloy heat-treated steel as specified.

Carbon-steel keys to be 0.30 carbon min, with hardness of 10 Rockwell C min.

Alloy-steel keys to be SAE 2330 or 8630 steel, heat-treated to a hardness of 40 to 50 Rockwell C; or other alloy steels having equal physical properties at the same hardness.

Alloy heat-treated keys are marked with depressions on the top to distinguish them from carbon-steel keys.

* Numbers of standard arbor-type cutters, and expected numbers in future revision of Woodruff-key standard.

† SAE listing of manufacturers' part numbers.

Table 20.44. Woodruff Keyslot and Keyway Dimensions[56]

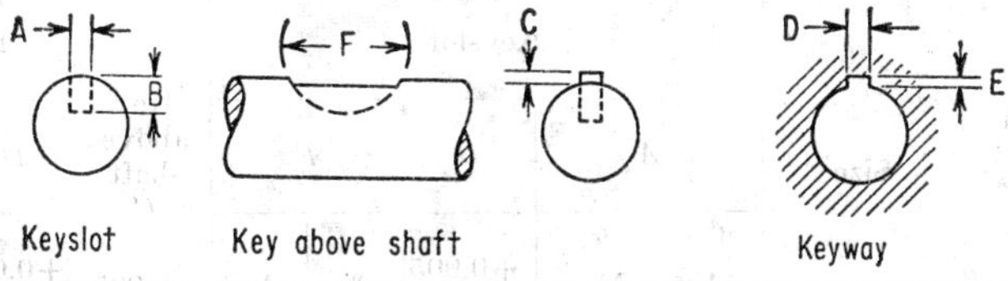

American Standard No.	Old standard* No.	Size	Keyslot					Key above shaft C	Keyway	
			A		B	F			D	E
			Min	Max	+0.005 −0.000	Min	Max	±0.005	+0.002 −0.000	+0.005 −0.000
202†	201	1/16 × 1/4	0.0615	0.0630	0.0728	0.255	0.260	0.0312	0.0635	0.0372
202½	206	1/16 × 5/16	0.0615	0.0630	0.1038	0.317	0.322	0.0312	0.0635	0.0372
302½	207	3/32 × 5/16	0.0928	0.0943	0.0882	0.317	0.322	0.0469	0.0948	0.0529
203	211	1/16 × 3/8	0.0615	0.0630	0.1358	0.380	0.385	0.0312	0.0635	0.0372
303	212	3/32 × 3/8	0.0928	0.0943	0.1202	0.380	0.385	0.0469	0.0984	0.0529
403	213	1/8 × 3/8	0.1240	0.1255	0.1045	0.380	0.385	0.0625	0.1260	0.0685
204	1	1/16 × 1/2	0.0615	0.0630	0.1668	0.510	0.515	0.0312	0.0635	0.0372
304	2	3/32 × 1/2	0.0928	0.0943	0.1511	0.510	0.515	0.0469	0.0948	0.0529
404	3	1/8 × 1/2	0.1240	0.1255	0.1355	0.510	0.515	0.0625	0.1260	0.0685
305	4	3/32 × 5/8	0.0928	0.0943	0.1981	0.635	0.640	0.0469	0.0948	0.0529
405	5	1/8 × 5/8	0.1240	0.1255	0.1825	0.635	0.640	0.0625	0.1260	0.0685
505	6	5/32 × 5/8	0.1553	0.1568	0.1669	0.635	0.640	0.0781	0.1573	0.0841
605	61	3/16 × 5/8	0.1863	0.1880	0.1513	0.635	0.640	0.0937	0.1885	0.0997
406	7	1/8 × 3/4	0.1240	0.1255	0.2455	0.760	0.765	0.0625	0.1260	0.0685
506	8	5/32 × 3/4	0.1553	0.1568	0.2299	0.760	0.765	0.0781	0.1573	0.0841
606	9	3/16 × 3/4	0.1863	0.1880	0.2143	0.760	0.765	0.0937	0.1885	0.0997
806	91	1/4 × 3/4	0.2487	0.2505	0.1830	0.760	0.765	0.1250	0.2510	0.1310
507	10	5/32 × 7/8	0.1553	0.1568	0.2919	0.887	0.892	0.0781	0.1573	0.0841
607	11	3/16 × 7/8	0.1863	0.1880	0.2763	0.887	0.892	0.0937	0.1885	0.0997
707	12	7/32 × 7/8	0.2175	0.2193	0.2607	0.887	0.892	0.1093	0.2198	0.1153
807	A	1/4 × 7/8	0.2487	0.2505	0.2450	0.887	0.892	0.1250	0.2510	0.1310
608	13	3/16 × 1	0.1863	0.1880	0.3393	1.012	1.017	0.0937	0.1885	0.0997
708	14	7/32 × 1	0.2175	0.2193	0.3237	1.012	1.017	0.1093	0.2198	0.1153
808	15	1/4 × 1	0.2487	0.2505	0.3080	1.012	1.017	0.1250	0.2510	0.1310
1008	B	5/16 × 1	0.3111	0.3130	0.2768	1.012	1.017	0.1562	0.3135	0.1622
609	16	3/16 × 1 1/8	0.1863	0.1880	0.3853	1.137	1.142	0.0937	0.1885	0.0997
709	17	7/32 × 1 1/8	0.2175	0.2193	0.3697	1.137	1.142	0.1093	0.2198	0.1153
809	18	1/4 × 1 1/8	0.2487	0.2505	0.3540	1.137	1.142	0.1250	0.2510	0.1310
1009	C	5/16 × 1 1/8	0.3111	0.3130	0.3228	1.137	1.142	0.1562	0.3135	0.1622
610	19	3/16 × 1 1/4	0.1863	0.1880	0.4483	1.265	1.270	0.0937	0.1885	0.0997
710	20	7/32 × 1 1/4	0.2175	0.2193	0.4327	1.265	1.270	0.1093	0.2198	0.1153
810	21	1/4 × 1 1/4	0.2487	0.2505	0.4170	1.265	1.270	0.1250	0.2510	0.1310
1010	D	5/16 × 1 1/4	0.3111	0.3130	0.3858	1.265	1.270	0.1562	0.3135	0.1622
1210	E	3/8 × 1 1/4	0.3735	0.3755	0.3545	1.265	1.270	0.1875	0.3760	0.1935
811	22	1/4 × 1 3/8	0.2487	0.2505	0.4640	1.390	1.395	0.1250	0.2510	0.1310

Table 20.44. Woodruff Keyslot and Keyway Dimensions[56] (Continued)

American Standard No.	Old standard* No.	Size	Keyslot A Min	Keyslot A Max	Keyslot B +0.005 −0.000	Keyslot F Min	Keyslot F Max	Key above shaft C ±0.005	Keyway D +0.002 −0.000	Keyway E +0.005 −0.000
1011	23	5/16 × 1 3/8	0.3111	0.3130	0.4328	1.390	1.395	0.1562	0.3135	0.1622
1211	F	3/8 × 1 3/8	0.3735	0.3755	0.4015	1.390	1.395	0.1875	0.3760	0.1935
812	24	1/4 × 1 1/2	0.2487	0.2505	0.5110	1.515	1.520	0.1250	0.2510	0.1310
1012	25	5/16 × 1 1/2	0.3111	0.3130	0.4798	1.515	1.520	0.1562	0.3135	0.1622
1212	G	3/8 × 1 1/2	0.3735	0.3755	0.4485	1.515	1.520	0.0875	0.3760	0.1935
	126	3/16 × 2 1/8	0.1863	0.1880	0.3073	2.125	2.135	0.0937	0.1885	0.0997
	127	1/4 × 2 1/8	0.2487	0.2505	0.2760	2.125	2.135	0.1250	0.2510	0.1310
	128	5/16 × 2 1/8	0.3111	0.3130	0.2448	2.125	2.135	0.1562	0.3135	0.1622
	129	3/8 × 2 1/8	0.3735	0.3755	0.2135	2.125	2.135	0.1875	0.3760	0.1935
617‡	26	3/16 × 2 1/8	0.1863	0.1880	0.4323	2.125	2.135	0.0937	0.1885	0.0997
817	27	1/4 × 2 1/8	0.2487	0.2505	0.4010	2.125	2.135	0.1250	0.2510	0.1310
1017	28	5/16 × 2 1/8	0.3111	0.3130	0.3698	2.125	2.135	0.1562	0.3135	0.1622
1217	29	3/8 × 2 1/8	0.3735	0.3755	0.3385	2.125	2.135	0.1875	0.3760	0.1935
	Rx	1/4 × 2 3/4	0.2487	0.2505	0.4640	2.750	2.760	0.1250	0.2510	0.1310
	Sx	5/16 × 2 3/4	0.3111	0.3130	0.4328	2.750	2.760	0.1562	0.3135	0.1622
	Tx	3/8 × 2 3/4	0.3735	0.3755	0.4015	2.750	2.760	0.1875	0.3760	0.1935
	Ux	7/16 × 2 3/4	0.4360	0.4380	0.3703	2.750	2.760	0.2187	0.4385	0.2247
	Vx	1/2 × 2 3/4	0.4985	0.5005	0.3390	2.750	2.760	0.2500	0.5010	0.2560
822	R	1/4 × 2 3/4	0.2487	0.2505	0.6200	2.750	2.760	0.1250	0.2510	0.1310
1022	S	5/16 × 2 3/4	0.3111	0.3130	0.5888	2.750	2.760	0.1562	0.3135	0.1622
1222	T	3/8 × 2 3/4	0.3735	0.3755	0.5575	2.750	2.760	0.1875	0.3760	0.1935
1422	U	7/16 × 2 3/4	0.4360	0.4380	0.5263	2.750	2.760	0.2187	0.4385	0.2247
1622	V	1/2 × 2 3/4	0.4985	0.5005	0.4950	2.750	2.760	0.2500	0.5010	0.2560
1228	30	3/8 × 3 1/2	0.3735	0.3755	0.7455	3.500	3.510	0.1875	0.3760	0.1935
1428	31	7/16 × 3 1/2	0.4360	0.4380	0.7143	3.500	3.510	0.2187	0.4385	0.2247
1628	32	1/2 × 3 1/2	0.4985	0.5005	0.6830	3.500	3.510	0.2500	0.5010	0.2560
1828	33	9/16 × 3 1/2	0.5610	0.5630	0.6518	3.500	3.510	0.2812	0.5635	0.2872
2028	34	5/8 × 3 1/2	0.6235	0.6255	0.6205	3.500	3.510	0.3125	0.6260	0.3185
2228	35	11/16 × 3 1/2	0.6860	0.6880	0.5893	3.500	3.510	0.3437	0.6885	0.3497
2428	36	3/4 × 3 1/2	0.7485	0.7505	0.5580	3.500	3.510	0.3750	0.7510	0.3810

Width *A*. Dimensions shown are set with the maximum keyslot width as that figure which will receive a key with the greatest amount of looseness permissible to assure the key sticking in the slot.

Minimum keyslot width is that figure permitting the largest shaft distortion acceptable when assembling maximum key in minimum keyslot.

B, *C*, and *E*. Dimensions to be taken at side intersection.

* SAE listing of manufacturers' part numbers.

† Numbers of standard shank-type cutters, and expected numbers in future revision of Woodruff-key standard.

‡ Numbers of standard arbor-type cutters, and expected numbers in future revision of Woodruff keyslot standard.

(d) Splines[55]

Splines are multiple keys cut in a shaft and hub to prevent relative motion and are used in preference to multiple keys for the transmission of power. Types of splines commonly used are the parallel-side or square and the more recent involute spline.[53] Table 20.45 lists maximum dimensions for the square-side splines. The involute spline is similar in form to involute gears and has a pressure angle of 30° and a stub

Table 20.45. Dimensions of Spline Fittings for All Fits[53,54]

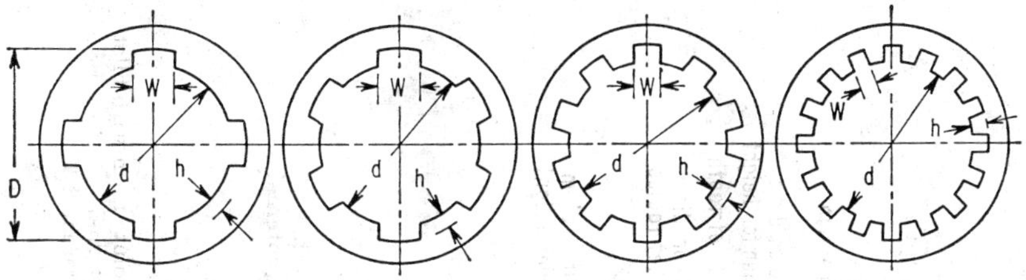

Nominal diam	4-spline		6-spline		10-spline		16-spline	
	D max[1]	W max[2]	D max[1]	W max[2]	D max[1]	W max[2]	D max[1]	W max[2]
3/4	0.750	0.181	0.750	0.188	0.750	0.117		
7/8	0.875	0.211	0.875	0.219	0.875	0.137		
1	1.000	0.241	1.000	0.250	1.000	0.156		
1 1/8	1.125	0.271	1.125	0.281	1.125	0.176		
1 1/4	1.250	0.301	1.250	0.313	1.250	0.195		
1 3/8	1.375	0.331	1.375	0.344	1.375	0.215		
1 1/2	1.500	0.361	1.500	0.375	1.500	0.234		
1 5/8	1.625	0.391	1.625	0.406	1.625	0.254		
1 3/4	1.750	0.422	1.750	0.438	1.750	0.273		
2	2.000	0.482	2.000	0.500	2.000	0.312	2.000	0.196
2 1/4	2.250	0.542	2.250	0.563	2.250	0.351		
2 1/2	2.500	0.602	2.500	0.625	2.500	0.390	2.500	0.245
3	3.000	0.723	3.000	0.750	3.000	0.468	3.000	0.294
3 1/2	.....	.....	.....	.....	3.500	0.546	3.500	0.343
4	.....	.....	.....	.....	4.000	0.624	4.000	0.392
4 1/2	.....	.....	.....	.....	4.500	0.702	4.500	0.441
5	.....	.....	.....	.....	5.000	0.780	5.000	0.490
5 1/2	.....	.....	.....	.....	5.500	0.858	5.500	0.539
6	.....	.....	.....	.....	6.000	0.936	6.000	0.588

1 Tolerance allowed of −0.001 in. for shafts 3/4 to 1 3/4 in., inclusive; of −0.002 for shafts 2 to 3 in., inclusive; −0.003 in. for shafts 3 1/2 to 6 in., inclusive, for 4-, 6-, and 10-spline fittings; tolerance of −0.003 in. allowed for all sizes of 16-spline fittings.

2 Tolerance allowed of −0.002 in. for shafts 3/4 in. to 1 3/4 in., inclusive; of −0.003 in. for shafts 2 to 6 in., inclusive, for 4-, 6-, and 10-spline fittings; tolerance of −0.003 allowed for all sizes of 16-spline fittings.

tooth with a height equal to one-half that used in gear practice. Nomenclature and basic formulas are given in Table 20.46.

For complete calculation of any pitch, refer to Tables 20.47 and 20.48, which are for one to two diametral (1/2) pitch. These tables can be considered the masters for all other tables in the standard.[54]

Table 20.46. Basic Formulas for Involute Splines[53]

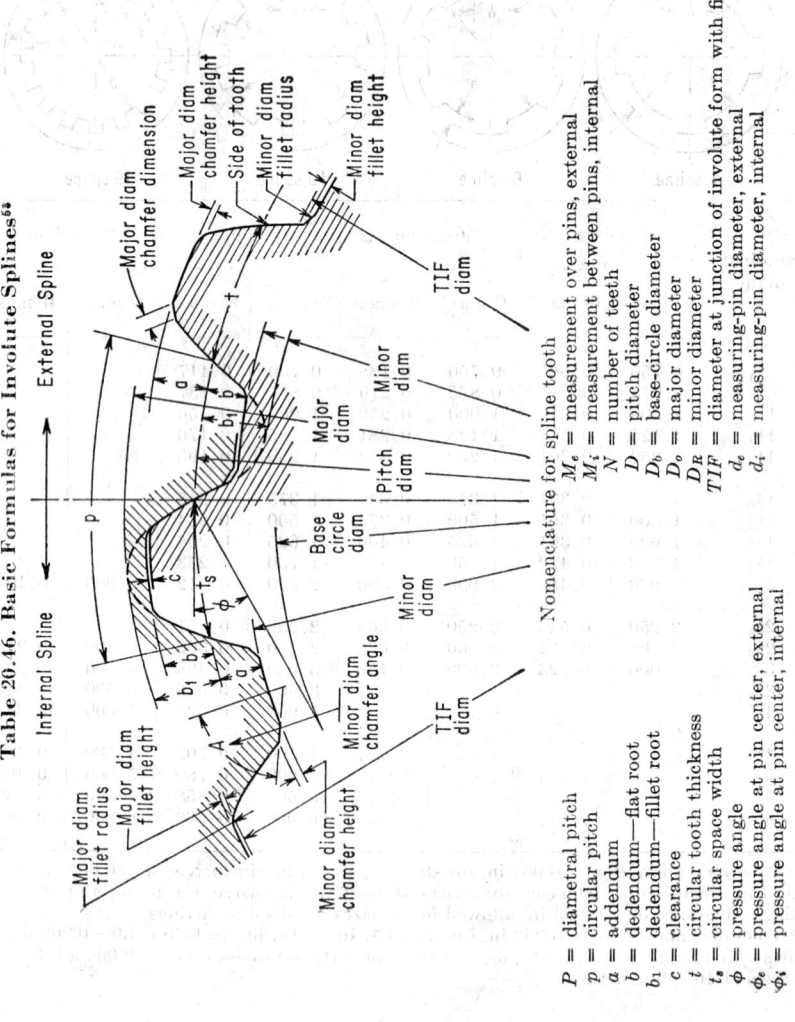

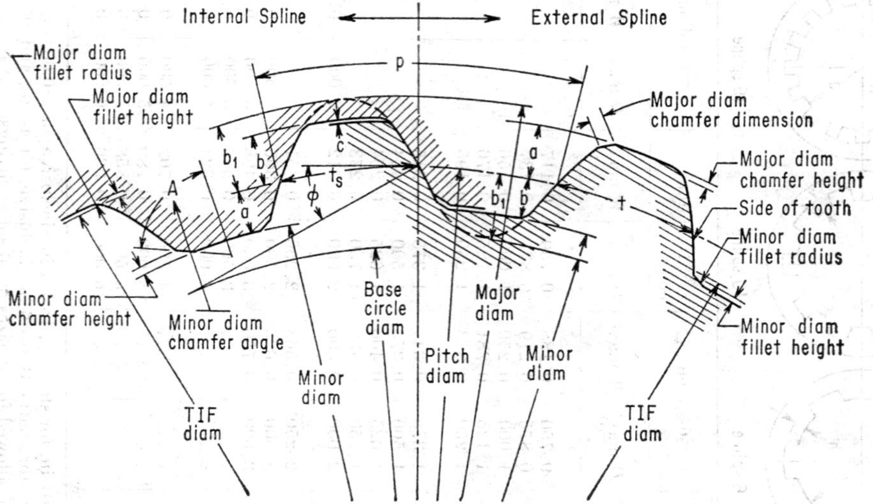

Nomenclature for spline tooth

P = diametral pitch
p = circular pitch
a = addendum
b = dedendum—flat root
b_1 = dedendum—fillet root
c = clearance
t = circular tooth thickness
t_s = circular space width
ϕ = pressure angle
ϕ_e = pressure angle at pin center, external
ϕ_i = pressure angle at pin center, internal

M_e = measurement over pins, external
M_i = measurement between pins, internal
N = number of teeth
D = pitch diameter
D_b = base-circle diameter
D_o = major diameter
D_R = minor diameter
TIF = diameter at junction of involute form with fillet
d_e = measuring-pin diameter, external
d_i = measuring-pin diameter, internal

Flat and fillet root		Fillet root only		
		Diametral pitch	$\frac{1}{2}$–$1\frac{1}{24}$	$1\frac{6}{32}$–$4\frac{8}{96}$
Pitch diam $D = \frac{N}{P}$	Major diam (external) $D_o = \frac{N+1}{P}$	Major diam (internal)	$D_o = \frac{N+1.8}{P}$	$D_o = \frac{N+1.8}{P}$
Circular pitch $p = \frac{3.141593}{P}$	TIF diam (internal) $= \frac{N+1}{P}$	Minor diam (external)	$D_R = \frac{N-1.8}{P}$	$D_R = \frac{N-2}{P}$
Circular tooth thickness $t = \frac{1.570796}{P}$	Minor diam (minor diam fits only) $D_R = \frac{N-1}{P}$			
Addendum $a = \frac{0.500}{P}$	TIF diam (external) $= \frac{N-1}{P}$	Dedendum (internal)	$b_1 = \frac{0.900}{P}$	$b_1 = \frac{0.900}{P}$
Dedendum $b = \frac{0.600}{P} + 0.002$		Dedendum (external)	$b_1 = \frac{0.900}{P}$	$b_1 = \frac{1.000}{P}$

For major-diameter fits the dedendum of the internal spline is the same as the addendum, and for minor-diameter fits the dedendum of the external spline is the same as the addendum.

Table 20.47. 1/2 Diametral Pitch Internal Involute Splines[53]

Pressure angle 30°						Addendum (basic) 0.5000				Circular pitch 3.1416, Measuring-pin diam 1.4400					
Internal and external			Internal												
			Major-diam fit			Flat root side fit				Fillet root side fit				All fits	
						Major-diam fillet		Major diam							
N	Pitch diam ref	Base-circle diam ref	Major diam basic	TIF diam	Minor diam	Radius	Height	Full[a] dedenum	Short[b] dedenum	TIF diam	Minor diam	Major diam	Fillet radius	Measurement between pins	Space width min effective 1.5708
(1)	(2)	(3)	(4)	(5)	(6)	(7)	(8)	(9A)	(9B)	(10)	(11)	(12)	(13)	(14)	(15)
Recommended tolerance........			+0.0007 + L^c −0.0000	Min	+0.0050 −0.0000	Approx	Max	+0.1050 −0.0000	+0.0007 + L^c −0.0007 + L	Min	+0.0050 −0.0000	+0.1050 −0.0000	Min[d]	Max	Dimensional max[e]
6	6.0000	5.1962	7.0000	6.897	5.2062	0.145	0.052	7.2040	7.0000	7.0000	5.2062	7.8000	0.166	4.4022	1.5755
7	7.0000	6.0622	8.0000	7.881	6.0722	0.160	0.060	8.2040	8.0000	8.0000	6.0722	8.8000	0.194	5.2320	1.5756
8	8.0000	6.9282	9.0000	8.870	7.0400	0.170	0.066	9.2040	9.0000	9.0000	7.0400	9.8000	0.208	6.4043	1.5756
9	9.0000	7.7942	10.0000	9.860	8.0000	0.178	0.071	10.2040	10.0000	10.0000	8.0000	10.8000	0.220	7.2707	1.5757
10	10.0000	8.6603	11.0000	10.852	9.0000	0.184	0.075	11.2040	11.0000	11.0000	9.0000	11.8000	0.230	8.4055	1.5757
11	11.0000	9.5263	12.0000	11.845	10.0000	0.189	0.079	12.2040	12.0000	12.0000	10.0000	12.8000	0.238	9.2955	1.5757
12	12.0000	10.3923	13.0000	12.839	11.0000	0.193	0.081	13.2040	13.0000	13.0000	11.0000	13.8000	0.245	10.4063	1.5758
13	13.0000	11.2583	14.0000	13.834	12.0000	0.196	0.084	14.2040	14.0000	14.0000	12.0000	14.8000	0.249	11.3129	1.5758
14	14.0000	12.1244	15.0000	14.830	13.0000	0.199	0.086	15.2040	15.0000	15.0000	13.0000	15.8000	0.253	12.4070	1.5759
15	15.0000	12.9904	16.0000	15.826	14.0000	0.201	0.088	16.2040	16.0000	16.0000	14.0000	16.8000	0.257	13.3258	1.5759
16	16.0000	13.8564	17.0000	16.823	15.0000	0.203	0.090	17.2040	17.0000	17.0000	15.0000	17.8000	0.260	14.4073	1.5759
17	17.0000	14.7224	18.0000	17.820	16.0000	0.205	0.091	18.2040	18.0000	18.0000	16.0000	18.8000	0.262	15.3357	1.5760
18	18.0000	15.5885	19.0000	18.818	17.0000	0.207	0.093	19.2040	19.0000	19.0000	17.0000	19.8000	0.265	16.4078	1.5760
19	19.0000	16.4545	20.0000	19.815	18.0000	0.208	0.094	20.2040	20.0000	20.0000	18.0000	20.8000	0.267	17.3437	1.5761
20	20.0000	17.3205	21.0000	20.813	19.0000	0.209	0.095	21.2040	21.0000	21.0000	19.0000	21.8000	0.269	18.4081	1.5761

21	21.0000	18.1865	22.0000	21.811	20.0000	0.210	0.096	22.2040	22.0000	22.0000	20.0000	22.8000	0.270	19.3499	1.5761
22	22.0000	19.0526	23.0000	22.810	21.0000	0.212	0.097	23.2040	23.0000	23.0000	21.0000	23.8000	0.272	20.4085	1.5762
23	23.0000	19.9186	24.0000	23.808	22.0000	0.212	0.097	24.2040	24.0000	24.0000	22.0000	24.8000	0.273	21.3552	1.5762
24	24.0000	20.7846	25.0000	24.807	23.0000	0.213	0.098	25.2040	25.0000	25.0000	23.0000	25.8000	0.274	22.4088	1.5763
25	25.0000	21.6506	26.0000	25.805	24.0000	0.214	0.099	26.2040	26.0000	26.0000	24.0000	26.8000	0.275	23.3598	1.5763
26	26.0000	22.5167	27.0000	26.804	25.0000	0.215	0.100	27.2040	27.0000	27.0000	25.0000	27.8000	0.276	24.4089	1.5763
27	27.0000	23.3827	28.0000	27.803	26.0000	0.215	0.100	28.2040	28.0000	28.0000	26.0000	28.8000	0.277	25.3637	1.5764
28	28.0000	24.2487	29.0000	28.802	27.0000	0.216	0.101	29.2040	29.0000	29.0000	27.0000	29.8000	0.278	26.4092	1.5764
29	29.0000	25.1147	30.0000	29.801	28.0000	0.217	0.101	30.2040	30.0000	30.0000	28.0000	30.8000	0.279	27.3669	1.5764
30	30.0000	25.9808	31.0000	30.801	29.0000	0.217	0.102	31.2040	31.0000	31.0000	29.0000	31.8000	0.280	28.4094	1.5765
31	31.0000	26.8468	32.0000	31.800	30.0000	0.218	0.102	32.2040	32.0000	32.0000	30.0000	32.8000	0.281	29.3699	1.5765
32	32.0000	27.7128	33.0000	32.799	31.0000	0.218	0.103	33.2040	33.0000	33.0000	31.0000	33.8000	0.282	30.4097	1.5766
33	33.0000	28.5788	34.0000	33.798	32.0000	0.219	0.103	34.2040	34.0000	34.0000	32.0000	34.8000	0.282	31.3725	1.5766
34	34.0000	29.4449	35.0000	34.798	33.0000	0.219	0.103	35.2040	35.0000	35.0000	33.0000	35.8000	0.283	32.4097	1.5766
35	35.0000	30.3109	36.0000	35.797	34.0000	0.219	0.104	36.2040	36.0000	36.0000	34.0000	36.8000	0.284	33.3748	1.5767
36	36.0000	31.1769	37.0000	36.796	35.0000	0.220	0.104	37.2040	37.0000	37.0000	35.0000	37.8000	0.284	34.4100	1.5767
37	37.0000	32.0429	38.0000	37.796	36.0000	0.220	0.104	38.2040	38.0000	38.0000	36.0000	38.8000	0.285	35.3770	1.5768
38	38.0000	32.9090	39.0000	38.795	37.0000	0.220	0.105	39.2040	39.0000	39.0000	37.0000	39.8000	0.286	36.4101	1.5768
39	39.0000	33.7750	40.0000	39.795	38.0000	0.221	0.105	40.2040	40.0000	40.0000	38.0000	40.8000	0.286	37.3787	1.5768
40	40.0000	34.6410	41.0000	40.794	39.0000	0.221	0.105	41.2040	41.0000	41.0000	39.0000	41.8000	0.286	38.4104	1.5769
41	41.0000	35.5070	42.0000	41.794	40.0000	0.221	0.106	42.2040	42.0000	42.0000	40.0000	42.8000	0.287	39.3805	1.5769
42	42.0000	36.3731	43.0000	42.794	41.0000	0.222	0.106	43.2040	43.0000	43.0000	41.0000	43.8000	0.287	40.4106	1.5770
43	43.0000	37.2391	44.0000	43.793	42.0000	0.222	0.106	44.2040	44.0000	44.0000	42.0000	44.8000	0.287	41.3821	1.5770
44	44.0000	38.1051	45.0000	44.793	43.0000	0.222	0.106	45.2040	45.0000	45.0000	43.0000	45.8000	0.288	42.4107	1.5770
45	45.0000	38.9711	46.0000	45.793	44.0000	0.222	0.106	46.2040	46.0000	46.0000	44.0000	46.8000	0.288	43.3835	1.5771
46	46.0000	39.8372	47.0000	46.792	45.0000	0.222	0.107	47.2040	47.0000	47.0000	45.0000	47.8000	0.288	44.4108	1.5771
47	47.0000	40.7032	48.0000	47.792	46.0000	0.223	0.107	48.2040	48.0000	48.0000	46.0000	48.8000	0.288	45.3849	1.5772
48	48.0000	41.5692	49.0000	48.792	47.0000	0.223	0.107	49.2040	49.0000	49.0000	47.0000	49.8000	0.288	46.4111	1.5772
49	49.0000	42.4352	50.0000	49.791	48.0000	0.223	0.107	50.2040	50.0000	50.0000	48.0000	50.8000	0.289	47.3860	1.5772
50	50.0000	43.3013	51.0000	50.791	49.0000	0.223	0.107	51.2040	51.0000	51.0000	49.0000	51.8000	0.289	48.4113	1.5773

[a] Intended for cutting by a generating process.

[b] If this dimension is used, the dimension in column (24), Table 20.48, should be decreased by twice the amount of maximum dimensional tooth clearance and the chamfer applied.

[c] $L = 0.0001 \times$ diam [column (4)].

[d] Represents minimum allowable radius of curvature and is based on 75 per cent of the full tangent radius for maximum depth.

[e] Allowable errors[53] except lead have been added to the machining tolerance in computing the maximum space width. When allowances for lead errors must be made, add 60 per cent of the lead error to this dimension.

Table 20.48. ½ Diametral Pitch External Involute Splines[53]

½ diametral pitch, pressure angle 30°						Addendum (basic) 0.5000						Circular pitch 3.1416, measuring-pin diam 1.9200					
External																	
	Major-diam fit[a]					Flat roof side fit			Fillet root[b]			Dimensions for all fits					
N	Major diam			Major diam chamfer		Minor diam	Minor diam fillet		Major[c] diam	Minor diam	Fillet rad	TIF diam	Measurement over pins			Tooth thickness	
																Class A	Class B	Class C
																Max effective		
	Class I	Class II	Class III	Dim	Ht		Rad	Ht					Class A	Class B	Class C	1.5703	1.5723	1.5753
(1)	(16)	(17)	(18)	(19)	(20)	(21)	(22)	(23)	(24)	(25)	(26)	(27)	(28)	(29)	(30)	(31)	(32)	(33)
Recommended tolerance	+0.0000 −0.0019 + J[f]	+0.0000 −0.0006 + J[f]	+0.0009 + J[f] −0.0000	Approx	Min	+0.0000 −0.1050	Approx	Max	+0.0000 −0.0100	+0.0000 −0.1050	Min[e]	Max	Min	Min	Min	Min dimensional[d]		
6	6.9985	6.9999	7.0015	0.228	0.091	4.7960	0.170	0.168	7.0000	4.2000	.....	5.2062	8.8590	8.8622	8.8661	1.5654	1.5679	1.5709
7	7.9985	7.9999	8.0016	0.230	0.094	5.7960	0.150	0.150	8.0000	5.2000	.....	6.0722	9.6747	9.6780	9.6819	1.5653	1.5678	1.5708
8	8.9985	8.9999	9.0017	0.230	0.095	6.7960	0.140	0.137	9.0000	6.2000	0.379	7.0400	10.8869	10.8903	10.8914	1.5653	1.5678	1.5708
9	9.9985	9.9999	10.0018	0.232	0.097	7.7960	0.130	0.127	10.0000	7.2000	0.373	8.0000	11.7460	11.7494	11.7535	1.5653	1.5678	1.5708
10	10.9985	10.9999	11.0019	0.233	0.098	8.7960	0.120	0.119	11.0000	8.2000	0.367	9.0000	12.9065	12.9100	12.9143	1.5652	1.5677	1.5707
11	11.9985	11.9999	12.0020	0.234	0.099	9.7960	0.116	0.113	12.0000	9.2000	0.362	10.0000	13.7924	13.7959	13.8002	1.5652	1.5677	1.5707
12	12.9985	12.9999	13.0021	0.234	0.100	10.7960	0.110	0.107	13.0000	10.2000	0.357	11.0000	14.9215	14.9250	14.9294	1.5652	1.5677	1.5707
13	13.9985	13.9999	14.0022	0.235	0.100	11.7960	0.108	0.103	14.0000	11.2000	0.353	12.0000	15.8253	15.8289	15.8332	1.5651	1.5676	1.5706
14	14.9985	14.9999	15.0023	0.235	0.101	12.7960	0.105	0.099	15.0000	12.2000	0.350	13.0000	16.9328	16.9365	16.9409	1.5651	1.5676	1.5706
15	15.9985	15.9999	16.0024	0.236	0.102	13.7960	0.102	0.096	16.0000	13.2000	0.348	14.0000	17.8500	17.8537	17.8581	1.5651	1.5676	1.5706
16	16.9985	16.9999	17.0025	0.236	0.102	14.7960	0.100	0.093	17.0000	14.2000	0.345	15.0000	18.9421	18.9458	18.9503	1.5650	1.5675	1.5705
17	17.9985	17.9999	18.0026	0.236	0.103	15.7960	0.100	0.091	18.0000	15.2000	0.343	16.0000	19.8691	19.8729	19.8774	1.5650	1.5675	1.5705
18	18.9985	18.9999	19.0027	0.237	0.103	16.7960	0.100	0.089	19.0000	16.2000	0.341	17.0000	20.9497	20.9534	20.9579	1.5650	1.5675	1.5705
19	19.9985	19.9999	20.0028	0.237	0.103	17.7960	0.100	0.087	20.0000	17.2000	0.339	18.0000	21.8845	21.8883	21.8928	1.5649	1.5674	1.5704
20	29.9985	20.9999	21.0029	0.237	0.104	18.7960	0.100	0.085	21.0000	18.2000	0.338	19.0000	22.9559	22.9597	22.9643	1.5649	1.5674	1.5704

21	21.9985	21.9999	22.0030	0.237	0.104	19.7960	0.100	0.083	22.0000	19.2000	0.336	20.0000	23.8970	23.9009	23.9054	1.5648	1.5673	1.5703
22	22.9985	22.9999	23.0031	0.237	0.104	20.7960	0.100	0.082	23.0000	20.2000	0.334	21.0000	24.9612	24.9650	24.9696	1.5648	1.5673	1.5703
23	23.9985	23.9999	24.0032	0.238	0.104	21.7960	0.100	0.081	24.0000	21.0000	0.333	22.0000	25.9075	25.9114	25.9160	1.5648	1.5673	1.5703
24	24.9985	24.9999	25.0033	0.238	0.105	22.7960	0.100	0.080	25.0000	22.2000	0.332	23.0000	26.9657	26.9696	26.9743	1.5647	1.5672	1.5702
25	25.9985	25.9999	26.0034	0.238	0.105	23.7960	0.100	0.079	26.0000	23.2000	0.331	24.0000	27.9164	27.9203	27.9250	1.5647	1.5672	1.5702
26	26.9985	26.9999	27.0035	0.238	0.105	24.7960	0.100	0.078	27.0000	24.2000	0.330	25.0000	28.9698	28.9737	28.9784	1.5647	1.5672	1.5702
27	27.9985	27.9999	28.0036	0.239	0.105	25.7960	0.100	0.077	28.0000	25.2000	0.330	26.0000	29.9240	29.9279	29.9326	1.5646	1.5671	1.5701
28	28.9985	28.9999	29.0037	0.239	0.105	26.7960	0.100	0.076	29.0000	26.2000	0.329	27.0000	30.9732	30.9771	30.9818	1.5646	1.5671	1.5701
29	29.9985	29.9999	30.0038	0.239	0.106	27.7960	0.100	0.075	30.0000	27.2000	0.328	28.0000	31.9307	31.9347	31.9394	1.5646	1.5671	1.5701
30	30.9985	30.9999	31.0039	0.239	0.106	28.7960	0.100	0.074	31.0000	28.2000	0.327	29.0000	32.9763	32.9802	32.9849	1.5645	1.5670	1.5700
31	31.9985	31.9999	32.0040	0.239	0.106	29.7960	0.100	0.074	32.0000	29.2000	0.327	30.0000	33.9365	33.9405	33.9452	1.5645	1.5670	1.5700
32	32.9985	32.9999	33.0041	0.239	0.106	30.7960	0.100	0.073	33.0000	30.2000	0.326	31.0000	34.9789	34.9828	34.9876	1.5644	1.5669	1.5699
33	33.9985	33.9999	34.0042	0.239	0.106	31.7960	0.100	0.072	34.0000	31.2000	0.326	32.0000	35.9415	35.9455	35.9503	1.5644	1.5669	1.5699
34	34.9985	34.9999	35.0043	0.239	0.106	32.7960	0.100	0.072	35.0000	32.2000	0.325	33.0000	36.9814	36.9854	36.9902	1.5644	1.5669	1.5699
35	35.9985	35.9999	36.0044	0.239	0.106	33.7960	0.100	0.071	36.0000	33.2000	0.325	34.0000	37.9461	37.9501	37.9549	1.5643	1.5668	1.5698
36	36.9985	36.9999	37.0045	0.240	0.106	34.7960	0.100	0.071	37.0000	34.2000	0.324	35.0000	38.9835	38.9875	38.9923	1.5643	1.5668	1.5698
37	37.9985	37.9999	38.0046	0.240	0.107	35.7960	0.100	0.070	38.0000	35.2000	0.324	36.0000	39.9503	39.9543	39.9591	1.5643	1.5668	1.5698
38	38.9985	38.9999	39.0047	0.240	0.107	36.7960	0.100	0.070	39.0000	36.2000	0.323	37.0000	40.9854	40.9894	40.9942	1.5642	1.5667	1.5697
39	39.9985	39.9999	40.0048	0.240	0.107	37.7960	0.100	0.069	40.0000	37.2000	0.323	38.0000	41.9539	41.9579	41.9627	1.5642	1.5667	1.5697
40	40.9985	40.9999	41.0049	0.240	0.107	38.7960	0.100	0.069	41.0000	38.2000	0.323	39.0000	42.9874	42.9914	42.9962	1.5642	1.5667	1.5697
41	41.9985	41.9999	42.0050	0.240	0.107	39.7960	0.100	0.069	42.0000	39.2000	0.322	40.0000	43.9572	43.9612	43.9661	1.5641	1.5666	1.5696
42	42.9985	42.9999	43.0051	0.240	0.107	40.7960	0.100	0.068	43.0000	40.2000	0.322	41.0000	44.9888	44.9928	44.9927	1.5641	1.5666	1.5696
43	43.9985	43.9999	44.0052	0.240	0.107	41.7960	0.100	0.068	44.0000	41.2000	0.322	42.0000	45.9603	45.9643	45.9692	1.5641	1.5666	1.5696
44	44.9985	44.9999	45.0053	0.240	0.107	42.7960	0.100	0.068	45.0000	42.2000	0.322	43.0000	46.9903	46.9943	46.9992	1.5640	1.5565	1.5695
45	45.9985	45.9999	46.0054	0.240	0.107	43.7960	0.100	0.067	46.0000	43.2000	0.321	44.0000	47.9630	47.9670	47.9719	1.5640	1.5665	1.5695
46	46.9985	46.9999	47.0055	0.240	0.107	44.7960	0.100	0.067	47.0000	44.2000	0.321	45.0000	48.9916	48.9957	49.0005	1.5639	1.5664	1.5694
47	47.9985	47.9999	48.0056	0.240	0.107	45.7960	0.100	0.067	48.0000	45.2000	0.320	46.0000	49.9654	49.9694	49.9743	1.5639	1.5664	1.5694
48	48.9985	48.9999	49.0057	0.240	0.107	46.7960	0.100	0.066	49.0000	46.2000	0.320	47.0000	50.9929	50.9969	51.0018	1.5639	1.5664	1.5694
49	49.9985	49.9999	50.0058	0.240	0.107	47.7960	0.100	0.066	50.0000	47.2000	0.320	48.0000	51.9676	51.9717	51.9766	1.5638	1.5663	1.5693
50	50.9985	50.9999	51.0059	0.240	0.107	48.7960	0.100	0.066	51.0000	48.2000	0.319	49.0000	52.9940	52.9981	53.0030	1.5638	1.5663	1.5693

[a] Measurement over pins for class A is recommended [column (28)], but if tighter fits are required, class B [column (29)] may be used.

[b] This may be used for a major-diameter fit by using dimension in column (16), (17), or (18) instead of that in column (24).

[c] When column (9B), Table 20.47, is used for the internal spline, reduce this dimension as covered in Table 20.47, note *b*.

[d] Allowable errors,[53] except lead, have been added to the machining tolerance in computing the minimum tooth thickness. When allowances for lead errors must be made, subtract 60 per cent of the lead error from this dimension.

[e] Represents minimum allowable radius of curvature, and is based on 75 per cent of the full tangent radius for maximum depth.

[f] $J = 0.0002 \times$ diam [column (24)].

(e) Involute Serrations

Involute serrations are multiple keys in the general form of internal and external involute gear teeth, intended for parts which are to be permanently fitted together.

There are three basic classes of fit: loose, close, and press. The fit is obtained by varying the external serration. The pitches included in the standard are 10/20, 16/32, 24/48, 32/64, 40/80, 48/96, 64/128, 80/160, 128/256 (Table 20.49).

For calculating the dimensions of any serration size, Table 20.50 may be used.

20.7. WIRE AND SHEET-METAL GAGES

Introduction. In the United States today, two wire gages are most commonly used.

(a) United States Steel Wire Gage[57]

Also called *steel wire gage,* this is used for practically all steel wire in the United States. See column (1), Table 20.51.

(b) American Wire Gage

Also called Brown & Sharpe gage, this is used to identify copper and aluminum wire as well as sheets of copper, aluminum, and other nonferrous metals. See column (2), Table 20.51.

Other gages in use are listed in Table 20.51 but are not recommended for new design specifications.

(c) Sheet-metal Gage

The United States Standard and the Manufacturers Standard gage for sheet metal are listed in columns (6) and (7). The gage values listed in Table 20.51 are currently used and accepted. Where tolerances are important, however, it is recommended that the decimal equivalent in parentheses follow the gage number.

Table 20.49. Tooth Dimensions of Involute Serrations[53,56]

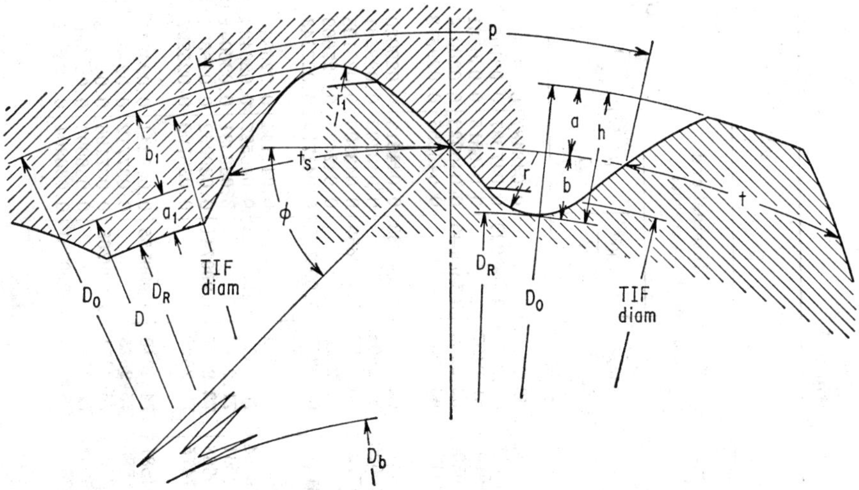

Nomenclature for involute serrations

P = diametral pitch
p = circular pitch
a = addendum, external
a_1 = addendum, internal
b = dedendum, external
b_1 = dedendum, internal
h = total depth
t = circular tooth thickness
t_s = circular space width
ϕ = pressure angle at pitch line (45°)

N = number of teeth
D = pitch diameter
D_b = base circle diameter
D_o = major diameter
D_R = minor diameter
d = measuring-pin diameter
TIF = diameter at junction of involute form with fillet
r, r_1 = approximate radius of fillet

M_e = measurement over pins, external serration
M_i = measurement between pins, internal serration

Diametral pitch P	External addendum dedendum $a = b$	Internal serration		Circular pitch p	Effective space min t_s	Effective tooth thickness max			Pin diam d
		Addendum a_1	Dedendum b_1			Class of fit A	Class of fit B	Class of fit C	
1/2	0.5000	0.3000	0.7000	3.1416	1.7708	1.7703	1.7718	1.7738	1.9200
10/20	0.0500	0.0300	0.0700	0.3142	0.1771	0.1766	0.1781	0.1801	0.1920
16/32	0.0313	0.0188	0.0438	0.1963	0.1107	0.1102	0.1117	0.1137	0.1200
24/48	0.0208	0.0125	0.0292	0.1309	0.0738	0.0733	0.0748	0.0768	0.0800
32/64	0.0156	0.0094	0.0218	0.0982	0.0553	0.0548	0.0563	0.0583	0.0600
40/80	0.0125	0.0075	0.0175	0.0785	0.0443	0.0438	0.0453	0.0473	0.0480
48/96	0.0104	0.0063	0.0145	0.0654	0.0369	0.0364	0.0379	0.0399	0.0400
64/128	0.0078	0.0047	0.0109	0.0491	0.0277	0.0272	0.0287	0.0307	0.0300
80/160	0.0063	0.0038	0.0088	0.0393	0.0221	0.0216	0.0231	0.0251	0.0240
128/256	0.0039	0.0023	0.0055	0.0245	0.0138	0.0133	0.0148	0.0168	0.0150

Table 20.50. Basic Data for Involute Serrations[53]

½P			All pitches				½P			All pitches			
Internal and external		External measure over pins	Internal and external		Internal	External	Internal and external		External measure over pins	Internal and external		Internal	External
N	D_b		$\cos \frac{90°}{N}$	$\frac{d}{D_b}$	$\frac{t_s}{D}$	$\frac{\pi}{N}$	N	D_b		$\cos \frac{90°}{N}$	$\frac{d}{D_b}$	$\frac{t_s}{D}$	$\frac{\pi}{N}$
6	4.242641	9.1631		0.452548	0.295133	0.523599	56	39.597980	59.2492		0.048487	0.031621	0.056100
7	4.949747	9.9666	0.974928	0.387899	0.252971	0.448799	57	40.305086	66.2273	0.999620	0.047637	0.031067	0.055116
8	5.656854	11.1820		0.339411	0.221350	0.392699	58	41.012193	61.2497		0.046815	0.030531	0.054165
9	6.363961	12.0330	0.984808	0.301699	0.196756	0.349066	59	41.719300	62.2286	0.999646	0.046022	0.030014	0.053247
10	7.071068	13.1949		0.271529	0 177080	0 314159	60	42 426407	63.2502		0.045255	0.029513	0.052360
11	7.778175	14.0749	0.989821	0.246845	0.160982	0.285599	61	43.133514	64.2297	0.999668	0.044513	0.029030	0.051502
12	8.485281	15.2042		0.226274	0.147567	0.261799	62	43.840620	65.2506		0.043795	0.028561	0.050671
13	9.192388	16.1038	0.992709	0.208868	0.136215	0.241661	63	44.547727	66.2308	0.999689	0.043100	0.028108	0.049867
14	9.899495	17.2114		0.193949	0.126486	0.224399	64	45.254834	67.2510		0.042426	0.027669	0.049087
15	10.606602	18.1251	0.994522	0.181019	0.118053	0.209440	65	45.961941	68.2319	0.999708	0.041774	0.027243	0.048332
16	11.313708	19.2171		0.169706	0.110675	0.196350	66	46.669047	69.2514		0.041141	0.026830	0.047600
17	12.020815	20.1413	0.995734	0.159723	0.104165	0.184800	67	47.376154	70.2328	0.999725	0.040527	0.026430	0.046889
18	12.727922	21.2215		0.150849	0.098378	0.174533	68	48.083261	71.2518		0.039931	0.026041	0.046200
19	13.435029	22.1541	0.996584	0.142910	0.093200	0.165347	69	48.790368	72.2337	0.999741	0.039352	0.025664	0.045530
20	14.142136	23.2253		0.135764	0.088540	0.157080	70	49.497475	73.2521		0.038790	0.025297	0.044880
21	14.849242	24.1645	0.997204	0.129300	0.084324	0.149600	71	50.204581	74.2346	0.999755	0.038244	0.024941	0.044248
22	15.556349	25.2284		0.123422	0.080491	0.142800	72	50.911688	75.2525		0.037712	0.024594	0.043633
23	16.263456	26.1731	0.997669	0.118056	0.076991	0.136591	73	51.618795	76.2354	0.999768	0.037196	0.024258	0.043036
24	16.970563	27.2311		0.113137	0.073783	0.130900	74	52.325902	77.2528		0.036693	0.023930	0.042454
25	17.677670	28.1804	0.998027	0.108612	0.070832	0.125664	75	53.033009	78.2362	0.999781	0.036204	0.023611	0.041888

26	18.384776	29.2334		0.104434	0.068108	0.120830	76	53.740115	79.2531		0.035728	0.023300	0.041337
27	19.091883	30.1865	0.998308	0.100566	0.065585	0.116355	77	54.447222	80.2369	0.999792	0.035264	0.022997	0.040800
28	19.798990	31.2354		0.096975	0.063243	0.112200	78	55.154329	81.2534		0.034811	0.022703	0.040277
29	20.506097	32.1918	0.998533	0.093631	0.061062	0.108331	79	55.861436	82.2376	0.999802	0.034371	0.022415	0.039767
30	21.213203	33.2372		0.090510	0.059027	0.104720	80	56.568542	83.2536		0.033941	0.022135	0.039270
31	21.920310	34.1965	0.998717	0.087590	0.057123	0.101342	81	57.275649	84.2383	0.999812	0.033522	0.021862	0.038785
32	22.627417	35.2387		0.084853	0.055338	0.098175	82	57.982756	85.2539		0.033113	0.021595	0.038312
33	23.334524	36.2006	0.998867	0.082282	0.053661	0.095200	83	58.689863	86.2389	0.999821	0.032714	0.021335	0.037851
34	24.041631	37.2401		0.079861	0.052082	0.092400	84	59.396970	87.2541		0.032325	0.021081	0.037400
35	24.748737	38.2042	0.998993	0.077580	0.050594	0.089760	85	60.104076	88.2395	0.999829	0.031945	0.020833	0.036960
36	25.455844	39.2414		0.075425	0.049189	0.087266	86	60.811183	89.2544		0.031573	0.020591	0.036530
37	26.162951	40.2074	0.999099	0.073386	0.047859	0.084908	87	61.518290	90.2401	0.999837	0.031210	0.020354	0.036110
38	26.870058	41.2425		0.071455	0.046600	0.082673	88	62.225397	91.2546		0.030856	0.020123	0.035700
39	27.577164	42.2103	0.999189	0.069623	0.045405	0.080554	89	62.932503	92.2406	0.999844	0.030509	0.019897	0.035299
40	28.284271	43.2435		0.067882	0.044270	0.078540	90	63.639610	93.2548		0.030170	0.019676	0.034907
41	28.991380	44.2129	0.999266	0.066227	0.043190	0.076624	91	64.346717	94.2411	0.998951	0.029838	0.019459	0.034523
42	29.698485	45.2444		0.064650	0.042162	0.074800	92	65.053824	95.2550		0.029514	0.019248	0.034148
43	30.405592	46.2153	0.999333	0.063146	0.041181	0.073060	93	65.760931	96.2416	0.999857	0.029197	0.019041	0.033781
44	31.112698	47.2453		0.061711	0.040245	0.071400	94	66.468037	97.2552		0.028886	0.018838	0.033421
45	31.819805	48.2175	0.999391	0.060340	0.039351	0.069813	95	67.175144	98.2421	0.999863	0.028582	0.018640	0.033069
46	32.526912	49.2461		0.059028	0.038496	0.068295	96	67.882251	99.2554		0.028284	0.018446	0.032725
47	33.234019	50.2195	0.999442	0.057772	0.037677	0.066842	97	68.589358	100.2426	0.999869	0.027993	0.018256	0.032388
48	33.941125	51.2468		0.056569	0.036892	0.065450	98	69.296464	101.2556		0.027707	0.018069	0.032057
49	34.648232	52.2213	0.999486	0.055414	0.036139	0.064114	99	70.003571	102.2430	0.999874	0.027427	0.017887	0.031733
50	35.355339	53.2474		0.054306	0.035416	0.062832	100	70.710678	103.2557		0.027153	0.017708	0.031416
51	36.062446	54.2230	0.999526	0.053241	0.034722	0.061600							
52	36.769553	55.2481		0.052217	0.034054	0.060415							
53	37.476659	56.2245	0.999561	0.051232	0.033411	0.059275							
54	38.183766	57.2487		0.050283	0.032793	0.058178							
55	38.890873	58.2260	0.999592	0.049369	0.032196	0.057120							

Constants:

$\pi = 3.141593$

$\cos 45° = 0.707107$

$\text{inv } 45° = 0.214602$

Table 20.51. Wire and Sheet-metal Gages

No. of wire gage	Steel wire gage (U.S.)* (1)	American Wire or Brown & Sharpe gage (2)	Birmingham or Stub's iron wire gage (3)	Stub's steel wire gage (4)	No. of wire gage	Stub's steel wire gage	Music or piano wire gage (5)	U.S. Standard (6)	Manufacturers Standard for sheet steel nominal (7)
7/0	0.4900	....	......	.....	51	0.066	.....	0.500	
6/0	0.4615	0.5800	......	.....	52	0.063	0.004	0.469	
5/0	0.4305	0.5165	0.5000	.....	53	0.058	0.005	0.438	
4/0	0.3938	0.4600	0.4540	.....	54	0.055	0.006	0.406	
3/0	0.3625	0.4096	0.4250	.....	55	0.050	0.007	0.375	
2/0	0.3310	0.3648	0.3800	.....	56	0.045	0.008	0.344	
1/0	0.3065	0.3249	0.3400	.....	57	0.042	0.009	0.312	
1	0.2830	0.2893	0.3000	0.227	58	0.041	0.010	0.281	
2	0.2625	0.2576	0.2840	0.219	59	0.040	0.011	0.266	
3	0.2437	0.2294	0.2590	0.212	60	0.039	0.012	0.250	0.2391
4	0.2253	0.2043	0.2380	0.207	61	0.038	0.013	0.234	0.2242
5	0.2070	0.1819	0.2200	0.204	62	0.037	0.014	0.219	0.2092
6	0.1920	0.1620	0.2030	0.201	63	0.036	0.016	0.203	0.1943
7	0.1770	0.1443	0.1800	0.199	64	0.035	0.018	0.188	0.1793
8	0.1620	0.1285	0.1650	0.197	65	0.033	0.020	0.172	0.1644
9	0.1483	0.1144	0.1480	0.194	66	0.032	0.022	0.156	0.1495
10	0.1350	0.1019	0.1340	0.191	67	0.031	0.024	0.141	0.1345
11	0.1205	0.0907	0.1200	0.188	68	0.030	0.026	0.125	0.1196
12	0.1055	0.0808	0.1090	0.185	69	0.029	0.029	0.109	0.1046
13	0.0915	0.0720	0.0950	0.182	70	0.027	0.031	0.0938	0.0897
14	0.0800	0.0641	0.0830	0.180	71	0.026	0.033	0.0781	0.0747
15	0.0720	0.0571	0.0720	0.178	72	0.024	0.035	0.0703	0.0673
16	0.0625	0.0508	0.0650	0.175	73	0.023	0.037	0.0625	0.0598
17	0.0540	0.0453	0.0580	0.172	74	0.022	0.039	0.0562	0.0538
18	0.0475	0.0403	0.0490	0.168	75	0.020	0.041	0.0500	0.0478
19	0.0410	0.0359	0.0420	0.164	76	0.018	0.043	0.0438	0.0418
20	0.0348	0.0320	0.0350	0.161	77	0.016	0.045	0.0375	0.0359
21	0.0317	0.0285	0.0320	0.157	78	0.015	0.047	0.0344	0.0329
22	0.0286	0.0253	0.0280	0.155	79	0.014	0.049	0.0312	0.0299
23	0.0258	0.0226	0.0250	0.153	80	0.013	0.051	0.0281	0.0269
24	0.0230	0.0201	0.0220	0.151	...	.....	0.055	0.0250	0.0239
25	0.0204	0.0179	0.0200	0.148	...	.....	0.059	0.0219	0.0209
26	0.0181	0.0159	0.0180	0.146	...	.....	0.063	0.0188	0.0179
27	0.0173	0.0142	0.0160	0.143	...	.....	0.067	0.0172	0.0164
28	0.0162	0.0126	0.0140	0.139	...	.....	0.071	0.0156	0.0149
29	0.0150	0.0113	0.0130	0.134	...	.....	0.075	0.0141	0.0135
30	0.0140	0.0100	0.0120	0.127	...	.....	0.080	0.0125	0.0120
31	0.0132	0.00893	0.0100	0.120	...	.....	0.085	0.0109	0.0105
32	0.0128	0.00795	0.0090	0.115	...	.....	0.090	0.0102	0.0097
33	0.0118	0.00708	0.0080	0.112	...	.....	0.095	0.00938	0.0090
34	0.0104	0.00630	0.0070	0.110	...	.....	0.100	0.00859	0.0082
35	0.0095	0.00561	0.0050	0.108	...	.....	0.106	0.00781	0.0075
36	0.0090	0.00500	0.0040	0.106	...	.....	0.112	0.00703	0.0067
37	0.0085	0.00445	......	0.103	...	.....	0.118	0.00664	0.0064
38	0.0080	0.00396	......	0.101	...	.....	0.124	0.00625	0.0060
39	0.0075	0.00353	......	0.099	...	.....	0.130		
40	0.0070	0.00314	......	0.097	...	.....	0.138		
41	0.0066	0.00280	......	0.095	...	.....	0.146		
42	0.0062	0.00249	......	0.092	...	.....	0.154		
43	0.0060	0.00222	......	0.088	...	.....	0.162		
44	0.0058	0.00198	......	0.085	...	.....	0.170		
45	0.0055	0.00176	......	0.081	...	.....	0.180		
46	0.0052	0.00157	......	0.079					
47	0.0050	0.00140	......	0.077					
48	0.0048	0.00124	......	0.075					
49	0.0046	0.00111	......	0.072					
50	0.0044	0.00099	......	0.069					

* Also known as Washburn and Moen, American Steel and Wire Co., and Roebling wire gages. A greater selection of sizes is available and is specified by what are known as split gage numbers. They can be recognized by the ¼ and ½ fractions which follow the gage number; i.e., 4¼, 4½, 4¾. The decimal equivalents of split gage numbers are in the Steel Products Manual entitled "Wire and Rods, Carbon Steel" published by the American Iron and Steel Institute, New York, N.Y.

(d) **Twist-drill Gage**[58] is given by decimal-size drill diameters ranging from 0.0135 to 0.2280, inclusive (Table 20.52).

Table 20.52. Twist-drill and Steel Wire Gage

No.	Size, in.	No.	Size, in.	No.	Size, in.	No.	Size, in.	No.	Size, in.	No.	Size, in.
1	0.2280	14	0.1820	27	0.1440	40	0.0980	53	0.0595	67	0.0320
2	0.2210	15	0.1800	28	0.1405	41	0.0960	54	0.0550	68	0.0310
3	0.2130	16	0.1770	29	0.1360	42	0.0935	55	0.0520	69	0.0292
4	0.2090	17	0.1730	30	0.1285	43	0.0890	56	0.0465	70	0.0280
5	0.2055	18	0.1695	31	0.1200	44	0.0860	57	0.0430	71	0.0260
6	0.2040	19	0.1660	32	0.1160	45	0.0820	58	0.0420	72	0.0250
7	0.2010	20	0.1610	33	0.1130	46	0.0810	59	0.0410	73	0.0240
8	0.1990	21	0.1590	34	0.1110	47	0.0785	60	0.0400	74	0.0225
9	0.1960	22	0.1570	35	0.1100	48	0.0760	61	0.0390	75	0.0210
10	0.1935	23	0.1540	36	0.1065	49	0.0730	62	0.0380	76	0.0200
11	0.1910	24	0.1520	37	0.1040	50	0.0700	63	0.0370	77	0.0180
12	0.1890	25	0.1495	38	0.1015	51	0.0670	64	0.0360	78	0.0160
13	0.1850	26	0.1470	39	0.0995	52	0.0635	65	0.0350	79	0.0145
								66	0.0330	80	0.0135

20.8. STEEL PIPE AND FITTINGS

(a) Pipe-thread Types

1. Taper pipe threads for general use (NPT)
2. Internal straight threads in pipe couplings (NPSC)
3. Taper pipe threads for railing joints (NPTR)
4. Straight pipe threads for mechanical joints (NPSM, NPSL, NPSH)
5. Dryseal pressuretight joints
6. Aeronautical

(b) American Standard Pipe-thread Form

The basic form known as the "American Standard taper pipe-thread form"[59] is shown in Fig. 20.19. American Standard taper pipe threads are designated by specifying in sequence the nominal size, number of threads per inch, and the symbols for thread series and form, e.g., (1) 3/8-18 NPT LH, (2) 1/8-27 NPSC.

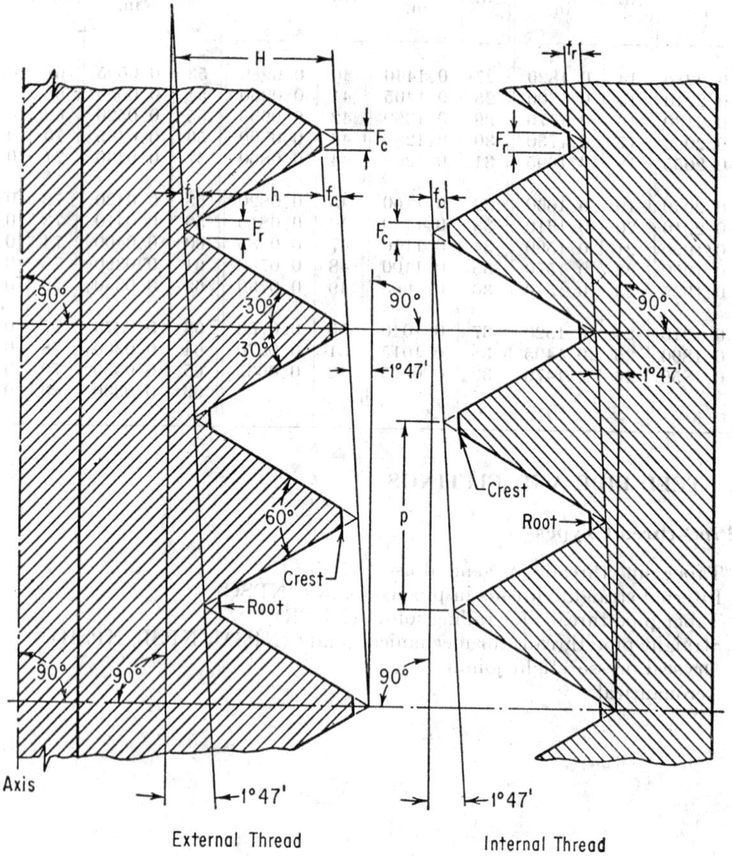

FIG. 20.19. Basic form of American Standard taper pipe thread.

H = 0.866025p = height of 60° sharp V-thread
h = 0.800000p = height of thread on product
p = $1/n$ = pitch (measured parallel to axis)
n = number of threads per inch
f_c = depth of truncation at crest
f_r = depth of truncation at root
F_c = width of flat at crest
F_r = width of flat at root

NOTE: For a symmetrical straight screw thread, $H = \cot \alpha/2n$. For a symmetrical taper screw thread, $H = (\cot \alpha - \tan^2 \beta \tan \alpha)/2n$; so that the exact value for an American Standard taper pipe thread is $H = 0.865743p$ as against $H = 0.866025p$, the value given above. For an eight-pitch thread, which is the coarsest standard taper pipe-thread pitch, the corresponding values of H are 0.108218 and 0.108253 in., respectively, the difference being 0.000035 in. This difference being too small to be significant, the value of $H = 0.866025p$ continues in use for threads of 3/4 in. or less taper per foot on the diameter.

(c) Pipe-thread Symbols

N = American (National) Standard
P = pipe
T = taper
C = coupling
S = straight
M = mechanical
L = locknut
H = hose coupling
R = railing fitting

Unless otherwise specified right-hand threads are obtained.

(d) Taper Pipe-thread Formula

The taper of the thread is 1:16 or 0.750 in./ft measured on the diameter along the axis. Basic taper pipe-thread dimensions, obtained by substituting into the following formulas, are given in Table 20.53.

$$L_2 = (0.80D + 6.8)(1/n) = (0.80D + 6.8)p$$

where L_2 = length of engagement
D = outside diameter of pipe
n = threads per inch

This formula determines directly the length of effective thread engagement, which includes two usable threads slightly imperfect.

The basic pitch diameters of the taper thread are determined by the following formulas:*

$$E_0 = D - (0.05D + 1.1)(1/n)$$
$$E_1 = E_0 + 0.0625L_1$$

where D = outside diameter of pipe
E_0 = pitch diameter of threaded end of pipe
E_1 = pitch diameter of thread at large end of internal thread
L_1 = normal engagement by hand between external and internal threads
n = threads per inch

* For the ⅛-27 and ¼-18 sizes, E_1 approx $= D - (0.05D + 0.827)p$.

Table 20.53. American Standard Taper Pipe Thread, NPT[59,64]

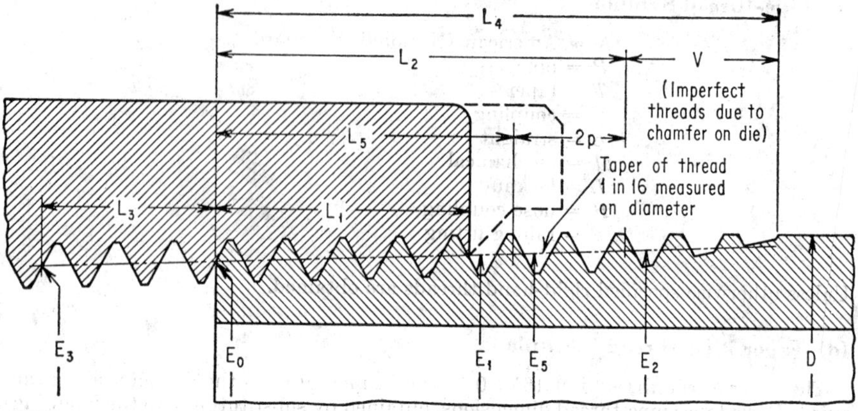

Nominal pipe size	OD of pipe D	Threads per in. n	Pitch of thread p	Pitch diam at beginning of external thread E_0	Handtight engagement: Length L_1, In.	Handtight engagement: Length L_1, Threads	Handtight engagement: Diam E_1	Effective thread, external: Length L_2, In.	Effective thread, external: Length L_2, Threads	Effective thread, external: Diam E_2
1/16	0.3125	27	0.03704	0.27118	0.160	4.32	0.28118	0.2611	7.05	0.28750
1/8	0.405	27	0.03704	0.36351	0.1615	4.36	0.37360	0.2639	7.12	0.38000
1/4	0.540	18	0.05556	0.47739	0.2278	4.10	0.49163	0.4018	7.23	0.50250
3/8	0.675	18	0.05556	0.61201	0.240	4.32	0.62701	0.4078	7.34	0.63750
1/2	0.840	14	0.07143	0.75843	0.320	4.48	0.77843	0.5337	7.47	0.79179
3/4	1.050	14	0.07143	0.96768	0.339	4.75	0.98887	0.5457	7.64	1.00179
1	1.315	11½	0.08696	1.21363	0.400	4.60	1.23863	0.6828	7.85	1.25630
1¼	1.660	11½	0.08696	1.55713	0.420	4.83	1.58338	0.7068	8.13	1.60130
1½	1.900	11½	0.08696	1.79609	0.420	4.83	1.82234	0.7235	8.32	1.84130
2	2.375	11½	0.08696	2.26902	0.436	5.01	2.29627	0.7565	8.70	2.31630
2½	2.875	8	0.12500	2.71953	0.682	5.46	2.76216	1.1375	9.10	2.79062
3	3.500	8	0.12500	3.34062	0.766	6.13	3.38850	1.2000	9.60	3.41562
3½	4.000	8	0.12500	3.83750	0.821	6.57	3.88881	1.2500	10.00	3.91562
4	4.500	8	0.12500	4.33438	0.844	6.75	4.38712	1.3000	10.40	4.41562
5	5.563	8	0.12500	5.39073	0.937	7.50	5.44929	1.4063	11.25	5.47862
6	6.625	8	0.12500	6.44609	0.958	7.66	6.50597	1.5125	12.10	6.54062
8	8.625	8	0.12500	8.43359	1.063	8.50	8.50003	1.7125	13.70	8.54062
10	10.750	8	0.12500	10.54531	1.210	9.68	10.62094	1.9250	15.40	10.66562
12	12.750	8	0.12500	12.53281	1.360	10.88	12.61781	2.1250	17.00	12.66562
14 OD	14.000	8	0.12500	13.77500	1.562	12.50	13.87262	2.2500	18.90	13.91562
16 OD	16.000	8	0.12500	15.76250	1.812	14.50	15.87575	2.4500	19.60	15.91562
18 OD	18.000	8	0.12500	17.75000	2.000	16.00	17.87500	2.6500	21.20	17.91562
20 OD	20.000	8	0.12500	19.73750	2.125	17.00	19.87031	2.8500	22.80	19.91562
24 OD	24.000	8	0.12500	23.71250	2.375	19.00	23.86094	3.2500	26.00	23.91562

Table 20.53. American Standard Taper Pipe Thread, NPT[59,64] (Continued)

Nominal pipe size	Wrench makeup length for internal thread: Length L_3, In.	Length L_3, Threads	Diam E_3	Vanish thread V, In.	Vanish thread V, Threads	Overall length external thread L_4	Nominal perfect external threads: Length L_5*	Nominal perfect external threads: Diam E_5	Height of thread h	Increase in diam per thread $0.0625/n$
1/16	0.1111	3	0.26424	0.1285	3.47	0.3896	0.1870	0.28287	0.02963	0.00231
1/8	0.1111	3	0.35656	0.1285	3.47	0.3924	0.1898	0.37537	0.02963	0.00231
1/4	0.1667	3	0.46697	0.1928	3.47	0.5946	0.2907	0.49556	0.04444	0.00347
3/8	0.1667	3	0.60160	0.1928	3.47	0.6006	0.2967	0.63056	0.04444	0.00347
1/2	0.2143	3	0.74504	0.2478	3.47	0.7815	0.3909	0.78286	0.05714	0.00446
3/4	0.2143	3	0.95429	0.2478	3.47	0.7935	0.4029	0.99286	0.05714	0.00446
1	0.2609	3	1.19733	0.3017	3.47	0.9845	0.5089	1.24543	0.06957	0.00543
1 1/4	0.2609	3	1.54083	0.3017	3.47	1.0085	0.5329	1.59043	0.06957	0.00543
1 1/2	0.2609	3	1.77978	0.3017	3.47	1.0252	0.5496	1.83043	0.06957	0.00543
2	0.2609	3	2.25272	0.3107	3.47	1.0582	0.5826	2.30543	0.06957	0.00543
2 1/2	0.2500	2	2.70391	0.4337	3.47	1.5712	0.8875	2.77500	0.100000	0.00781
3	0.2500	2	3.32500	0.4337	3.47	1.6337	0.9500	3.40000	0.100000	0.00781
3 1/2	0.2500	2	3.82188	0.4337	3.47	1.6837	1.0000	3.90000	0.100000	0.00781
4	0.2500	2	4.31875	0.4337	3.47	1.7337	1.0500	4.40000	0.100000	0.00781
5	0.2500	2	5.37511	0.4337	3.47	1.8400	1.1563	5.46300	0.100000	0.00781
6	0.2500	2	6.43047	0.4337	3.47	1.9462	1.2625	6.52500	0.100000	0.00781
8	0.2500	2	8.41797	0.4337	3.47	2.1462	1.4625	8.52500	0.100000	0.00781
10	0.2500	2	10.52969	0.4337	3.47	2.3587	1.6750	10.65000	0.100000	0.00781
12	0.2500	2	12.51719	0.4337	3.47	2.5587	1.8750	12.65000	0.100000	0.00781
14 OD	0.2500	2	13.75938	0.4337	3.47	2.6837	2.0000	13.90000	0.100000	0.00781
16 OD	0.2500	2	15.74688	0.4337	3.47	2.8837	2.2000	15.90000	0.100000	0.00781
18 OD	0.2500	2	17.73438	0.4337	3.47	3.0837	2.4000	17.90000	0.100000	0.00781
20 OD	0.2500	2	19.72188	0.4337	3.47	3.2837	2.6000	19.90000	0.100000	0.00781
24 OD	0.2500	2	23.69688	0.4337	3.47	3.6837	3.0000	23.90000	0.100000	0.00781

* The length L_5 from the end of the pipe determines the plane beyond which the thread form is imperfect at the crest. The next two threads are perfect at the root. At this plane the cone formed by the crests of the thread intersects the cylinder forming the external surface of the pipe.

It should be noted that, in special applications as in high-pressure work, longer thread engagement is used; thus E_1, the pitch diameter shown in Table 20.53, is maintained and the pitch diameter E_0 at the end of the pipe is made smaller.

(e) Pipe-coupling Threads

The internal straight threads in pipe couplings[60] are straight or parallel threads of the same thread form as the American taper pipe thread. This thread is normally specified where a tight low-pressure joint is required with an American Standard external taper pipe thread.

(f) Railing Joints

These require a more rigid joint obtained by using a shortened external thread, permitting the use of the larger end of the pipe thread. The form of thread is the same as the form of the American Standard taper pipe thread (Fig. 20.19). Straight pipe threads[59] are based on the pitch diameter of the American Standard pipe thread at the gaging notch, E_1 of Table 20.53.

(g) Dryseal Pipe Thread[61]

Dryseal threads are used for both external and internal threads where a pressuretight joint is required and the use of a sealer is objectionable. The external thread is tapered while the internal threads may be tapered or straight. The crest and root truncations are modified American Standard pipe threads, in most cases, to produce a pressuretight joint.

(h) Aeronautical Taper Pipe Thread[62]

As given in Military Specifications MILP-7105, these are identical in all respects with American Standard taper pipe threads except for E_3 and L_3 dimensions in the 2½ and 3 sizes. The importance of the Aeronautical pipe-thread specifications lies in the gaging system, which provides a closer dimensional control than gages used for American Standard taper pipe threads or American Standard Dryseal pipe threads.

(i) Pipe Designation

Commercial sizes of wrought iron and steel pipe[68] are known by their nominal inside diameter from ⅛ to 12 in. Above 12 in. in diameter, the outside diameter is referred to. All classes of pipe of a given nominal size have the same outside diameter; the extra thickness is reflected on the inside diameters.

The former designation, standard, extra strong, and double extra weight pipe, has been replaced by a simpler and more versatile system. The new standard uses pipe-thickness schedules.[63] It gives a larger selection of wall thicknesses with the flexibility of adding new wall thicknesses (Table 20.54).

(j) Pipe Wall Thickness[64]

Minimum wall thickness is determined by the following formula:

$$t_m = \frac{PD}{2S} + C \quad \text{or} \quad \frac{D}{2}\left(1 - \frac{S - P}{S + P}\right) + C$$

where t_m = minimum pipe wall thickness, in.*
P = maximum internal pressure, psig
D = outside diameter of pipe, in.
S = allowable stress, psi
C = thickness allowance for corrosion, threading
C = 0.15 for cast iron
C = 0.05 for threaded steel or wrought iron ⅜ in. and smaller
$C = 0.8/n$ for threaded steel or wrought iron ½ in. or larger

* For pipe sizes 4 in. or greater, t_m = 10 per cent × ID.

Table 20.54. Dimensions of Welded and Seamless Steel Pipe[65]

Nom diam, in.	Schedule	OD, in.	Wall thickness, in.	ID, in.	Internal area, in.²
1/8	40 (S)	0.405	0.068	0.269	0.0568
	80 (X)		0.095	0.215	0.0363
1/4	40 (S)	0.540	0.088	0.364	0.1041
	80 (X)		0.119	0.302	0.0716
3/8	40 (S)	0.675	0.091	0.493	0.1909
	80 (X)		0.126	0.423	0.1405
1/2	40 (S)	0.840	0.109	0.622	0.3039
	80 (X)		0.147	0.546	0.2341
	160		0.187	0.466	0.1706
	(XX)		0.294	0.252	0.0499
3/4	40 (S)	1.050	0.113	0.824	0.5333
	80 (X)		0.154	0.742	0.4324
	160		0.219	0.612	0.2942
	(XX)		0.308	0.434	0.1479
1	40 (S)	1.315	0.133	1.049	0.8643
	80 (X)		0.179	0.957	0.7193
	160		0.250	0.815	0.5217
	(XX)		0.358	0.599	0.2818
1 1/4	40 (S)	1.660	0.140	1.380	1.496
	80 (X)		0.191	1.278	1.283
	160		0.250	1.160	1.057
	(XX)		0.382	0.896	0.6305
1 1/2	40 (S)	1.900	0.145	1.610	2.036
	80 (X)		0.200	1.500	1.767
	160		0.281	1.338	1.406
	(XX)		0.400	1.100	0.9503
2	40 (S)	2.375	0.154	2.067	3.356
	80 (X)		0.218	1.939	2.953
	160		0.344	1.687	2.235
	(XX)		0.436	1.503	1.774
2 1/2	40 (S)	2.875	0.203	2.469	4.788
	80 (X)		0.276	2.323	4.233
	160		0.375	2.125	3.547
	(XX)		0.552	1.771	2.464
3	40 (S)	3.500	0.216	3.068	7.393
	80 (X)		0.300	2.900	6.605
	160		0.438	2.624	5.408
	(XX)		0.600	2.300	4.155
3 1/2	40 (S)	4.000	0.226	3.548	9.887
	80 (X)		0.318	3.364	8.888
4	40 (S)	4.500	0.237	4.026	12.73
	80 (X)		0.337	3.826	11.50
	120		0.438	3.624	10.32
	160		0.531	3.438	9.283
	(XX)		0.674	3.152	7.803
5	40 (S)	5.563	0.258	5.047	20.01
	80 (X)		0.375	4.813	18.19
	120		0.500	4.563	16.35
	160		0.625	4.313	14.61
	(XX)		0.750	4.063	12.97
6	40 (S)	6.625	0.280	6.065	28.89
	80 (X)		0.432	5.761	26.07
	120		0.562	5.501	23.77
	160		0.719	5.187	21.13
	(XX)		0.864	4.897	18.83

Table 20.54. Dimensions of Welded and Seamless Steel Pipe[65] (Continued)

Nom diam, in.	Schedule	OD, in.	Wall thickness, in.	ID, in.	Internal area, in.²
8	20	8.625	0.250	8.125	51.85
	30		0.277	8.071	51.16
	40 (S)		0.322	7.981	50.03
	60		0.406	7.813	47.94
	80 (X)		0.500	7.625	45.66
	100		0.594	7.437	43.44
	120		0.719	7.187	40.57
	140		0.812	7.001	38.50
	(XX)		0.875	6.875	37.12
	160		0.906	6.813	36.46
10	20	10.75	0.250	10.250	82.52
	30		0.307	10.136	80.69
	40 (S)		0.365	10.020	78.85
	60 (X)		0.500	9.750	74.66
	80		0.594	9.562	71.81
	100		0.719	9.312	68.11
	120		0.844	9.062	64.50
	140 (XX)		1.000	8.750	60.13
	160		1.125	8.500	56.75
12	20	12.75	0.250	12.250	117.86
	30		0.330	12.090	114.80
	(S)		0.375	12.000	113.10
	40		0.406	11.938	111.93
	(X)		0.500	11.750	108.43
	60		0.562	11.626	106.16
	80		0.688	11.374	101.61
	100		0.844	11.062	96.11
	120 (XX)		1.000	10.750	90.76
	140		1.125	10.500	86.59
	160		1.312	10.126	80.53
14 OD	10	14.00	0.250	13.500	143.14
	20		0.312	13.376	140.52
	30 (S)		0.375	13.250	137.89
	40		0.438	13.124	135.28
	(X)		0.500	13.000	132.67
	60		0.594	12.812	128.92
	80		0.750	12.500	122.72
	100		0.938	12.124	115.45
	120		1.094	11.812	109.58
	140		1.250	11.500	103.87
	160		1.406	11.188	98.31
16 OD	10	16.00	0.250	15.500	188.69
	20		0.312	15.376	185.69
	30 (S)		0.375	15.250	182.65
	40 (X)		0.500	15.000	176.72
	60		0.656	14.688	169.44
	80		0.844	14.312	160.88
	100		1.031	13.938	152.58
	120		1.219	13.562	144.46
	140		1.438	13.124	135.28
	160		1.594	12.812	128.92
18 OD	10	18.00	0.250	17.500	240.53
	20		0.312	17.376	237.13
	(S)		0.375	17.250	233.71
	30		0.438	17.124	230.00

Table 20.54. Dimensions of Welded and Seamless Steel Pipe[65] (Continued)

Nom diam, in.	Schedule	OD, in.	Wall thickness, in.	ID, in.	Internal area, in.^{2}
18 OD	(X)		0.500	17.000	226.98
	40		0.562	16.876	223.68
	60		0.750	16.500	213.83
	80		0.938	16.124	204.19
	100		1.156	15.688	193.30
	120		1.375	15.250	182.65
	140		1.562	14.876	173.81
	160		1.781	14.438	163.72
20 OD	10	20.00	0.250	19.500	298.65
	20 (S)		0.375	19.250	291.04
	30 (X)		0.500	19.000	283.53
	40		0.594	18.812	277.95
	60		0.812	18.376	265.21
	80		1.031	17.938	252.72
	100		1.281	17.438	238.83
	120		1.500	17.000	226.98
	140		1.750	16.500	213.83
	160		1.969	16.062	202.62
24 OD	10	24.00	0.250	23.500	433.74
	20 (S)		0.375	23.250	424.56
	(X)		0.500	23.000	415.48
	30		0.562	22.876	411.01
	40		0.688	22.624	402.00
	60		0.969	22.062	382.28
	80		1.219	21.562	365.15
	100		1.531	20.938	344.32
	120		1.812	20.376	326.92
	140		2.062	19.876	310.28
	160		2.344	19.312	292.92
30 OD	10	30.00	0.312	29.376	677.76
	(S)		0.375	29.250	671.62
	20 (X)		0.500	29.000	660.52
	30		0.625	28.750	649.18

S = Wall thickness formerly designated, "standard weight."
X = Wall thickness formerly designated, "extra strong."
XX = Wall thickness formerly designated, "double extra strong."

(k) Pipe-fitting Dimensions

For dimensions of fittings refer to Tables 20.55 through 20.60.

Table 20.55. General Dimensions of Class 125 Flanged Fittings Straight Sizes[66,37]

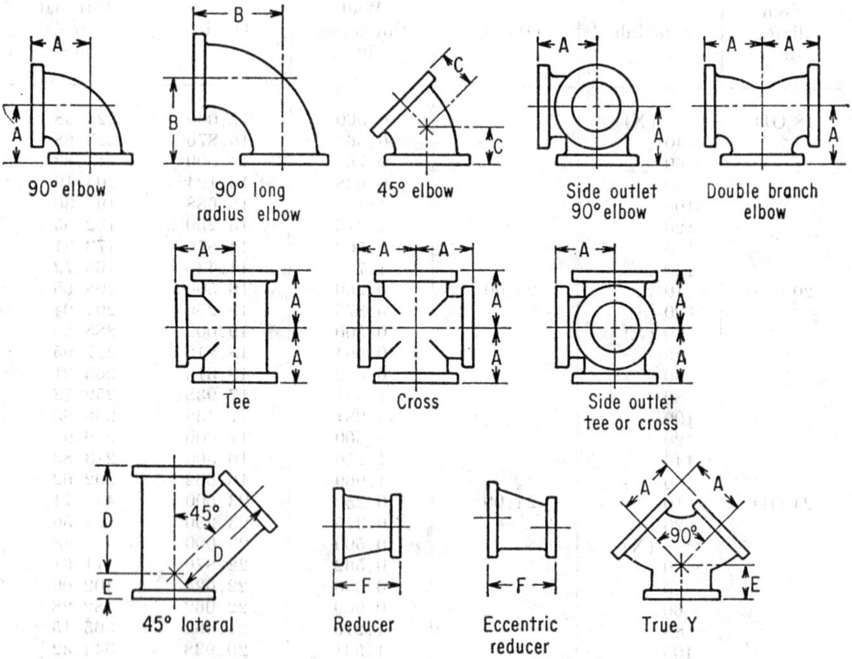

Nominal pipe size	ID of fittings	Center to face 90° elbow, tees, crosses, true Y, and double-branch elbow *A*	Center to face 90° long radius elbow *B*	Center to face 45° elbow *C*	Center to face lateral *D*	Short center to face true Y and lateral *E*	Face to face reducer *F*	Diam of flange	Thickness of flange, min	Wall thickness
1	1	3½	5	1¾	5¾	1¾	....	4¼	7/16	5/16
1¼	1¼	3¾	5½	2	6¼	1¾	....	4⅝	½	5/16
1½	1½	4	6	2¼	7	2	....	5	9/16	5/16
2	2	4½	6½	2½	8	2½	5	6	⅝	5/16
2½	2½	5	7	3	9½	2½	5½	7	11/16	5/16
3	3	5½	7¾	3	10	3	6	7½	¾	⅜
3½	3½	6	8½	3½	11½	3	6½	8½	13/16	7/16
4	4	6½	9	4	12	3	7	9	15/16	½
5	5	7½	10¼	4½	13½	3½	8	10	15/16	½
6	6	8	11½	5	14½	3½	9	11	1	9/16
8	8	9	14	5½	17½	4½	11	13½	1⅛	⅝
10	10	11	16½	6½	20½	5	12	16	1 3/16	¾
12	12	12	19	7½	24½	5½	14	19	1¼	13/16
14	14	14	21½	7½	27	6	16	21	1⅜	⅞
16	16	15	24	8	30	6½	18	23½	1 7/16	1
18	18	16½	26½	8½	32	7	19	25	1 9/16	1 1/16
20	20	18	29	9½	35	8	20	27½	1 11/16	1⅛
24	24	22	34	11	40½	9	24	32	1⅞	1¼
30	30	25	41½	15	49	10	30	38¾	2⅛	1 7/16
36	36	28	49	18	....	....	36	46	2⅜	1⅝
42	42	31	56½	21	....	....	42	53	2⅝	1 13/16
48	48	34	64	24	....	....	48	59½	2¾	2

Table 20.56. Center to Contact Surface of American Standard Steel Flanged Fittings[66,67]

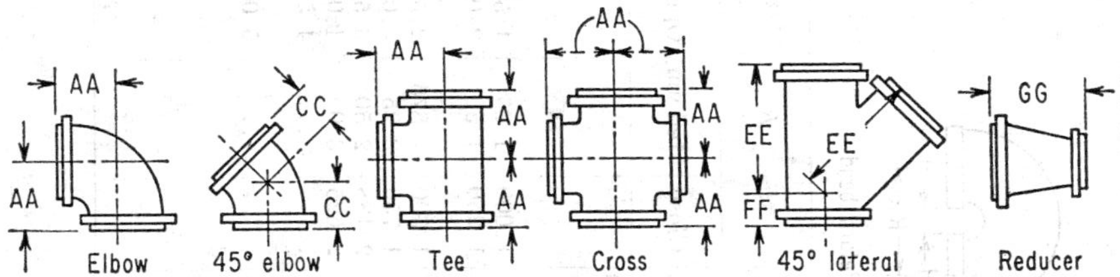

Nom pipe size	400-lb standard					600-lb standard					900-lb standard					1,500-lb standard				
	AA	CC	EE	FF	GG	AA	CC	EE	FF	GG	AA	CC	EE	FF	GG	AA	CC	EE	FF	GG
½	For sizes below 4 in., use the dimensions of 600-lb fittings					3¼	2	5¾	1¾	5	For sizes below 3 in., use the dimensions of 1,500-lb fittings					4¼	3			
¾						3¾	2¼	6¾	2	5						4½	3¼			
1						4¼	2¼	7¼	2¼	5						5	3½	9	2½	5
1¼	....	....	....	....	....	4½	2¾	8	2½	5	....	....	....	....	....	5½	4	10	3	5¾
1½	....	....	....	....	....	4¾	3	9	2¾	5	....	....	....	....	....	6	4¼	11	3½	6¼
2	....	....	....	....	....	5¾	4¼	10¼	3½	6	....	....	....	....	....	7¼	4¾	13¼	4	7¼
2½	....	....	....	....	....	6½	4½	11½	3½	6¾	....	....	....	....	....	8¼	5¼	15¼	4½	8¼
3	....	....	....	....	....	7	5	12¾	4	7¼	7½	5½	14½	4½	7¾	9¼	5¾	17¼	5	9¼
3½	....	....	....	....	....	7½	5½	14	4½	7¾										
4	8	5½	16	4½	8¼	8½	6	16½	4½	8¾	9	6½	17½	5½	9¼	10¾	7¼	19¼	6	10¾
5	9	6	16¾	5	9¼	10	7	19½	6	10¼	11	7½	21	6½	11¼	13¼	8¾	23¼	7½	13¾
6	9¾	6¼	18¾	5¼	10	11	7½	21	6½	11¼	12	8	22½	6½	12¼	13⅞	9⅜	24⅞	8⅛	14½
8	11¾	6¾	22¼	5¾	12	13	8½	24½	7	13¼	14½	9	27½	7½	14¾	16⅜	10⅞	29⅞	9⅛	17
10	13¼	7¾	25¾	6¼	13¼	15½	9½	29½	8	15¾	16½	10	31½	8½	16¾	19	12	36	10¼	20¼
12	15	8¾	29¾	6½	15¼	16½	10	31½	8½	16¾	19	11	34½	9	17¾	22¼	13¼	40¾	12	23
14 OD	16¼	9¼	32¾	7	16½	17½	10¾	34½	9	17¾	20¼	11½	36½	9½	19	24¾	14¼	44	12½	25¾
16 OD	17¾	10¼	36¼	8	18½	19½	11¾	38½	10	19¾	22¼	12½	40¾	10½	21	27¼	16¼	48¼	14¾	28¼
18 OD	19¼	10¾	39¼	8½	19½	21½	12¼	42	10½	21¾	24	13¼	45½	12	24½	30¼	17¾	53¼	16½	31½
20 OD	20¾	11¼	42¾	9	21	23½	13	45½	11	23¾	26	14½	50¼	13	26½	32¾	18¾	57¾	17¾	34
24 OD	24½	12¾	50¼	10½	24½	27½	14¾	53	13	27¾	30½	18	60	15½	30½	38¼	20¾	67¼	20½	39¾

All dimensions in inches.

Table 20.57. Dimensions of American 150-lb Standard Malleable-iron Screwed Fittings (Straight Sizes)[66]

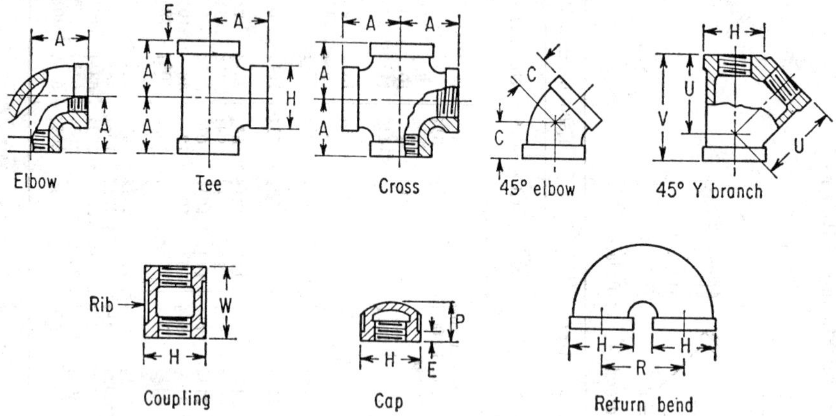

Size	A	H	E	C	V	U	W	P	R		
									Close	Medium	Open
⅛	0.69	0.693	0.200	....	....	....	0.96				
¼	0.81	0.844	0.215	0.73	....	....	1.06				
⅜	0.95	1.015	0.230	0.80	1.93	1.43	1.16				
½	1.12	1.197	0.249	0.88	2.32	1.71	1.34	0.87	1.000	1.25	1.50
¾	1.31	1.458	0.273	0.98	2.77	2.05	1.52	0.97	1.250	1.50	2.00
1	1.50	1.771	0.302	1.12	3.28	2.43	1.67	1.16	1.500	1.875	2.50
1¼	1.75	2.153	0.341	1.29	3.94	2.92	1.93	1.28	1.750	2.25	3.00
1½	1.94	2.427	0.368	1.43	4.38	3.28	2.15	1.33	2.188	2.50	3.50
2	2.25	2.963	0.422	1.68	5.17	3.93	2.53	1.45	2.625	3.00	4.00
2½	2.70	3.589	0.478	1.95	6.25	4.73	2.88	1.70	.....	.....	4.50
3	3.08	4.285	0.548	2.17	7.26	5.55	3.18	1.80	.....	.....	5.00
3½	3.42	4.843	0.604	2.39	....	....	3.43	1.90			
4	3.79	5.401	0.661	2.61	8.98	6.97	3.69	2.08			
5	4.50	6.583	0.780	3.05	....	....	....	2.32			
6	5.13	7.767	0.900	3.46	....	....	....	2.55			

All dimensions in inches.

Table 20.58. Dimensions of American 125- and 250-lb Standard Cast-iron Screwed Fittings (Straight Sizes)*,[66]

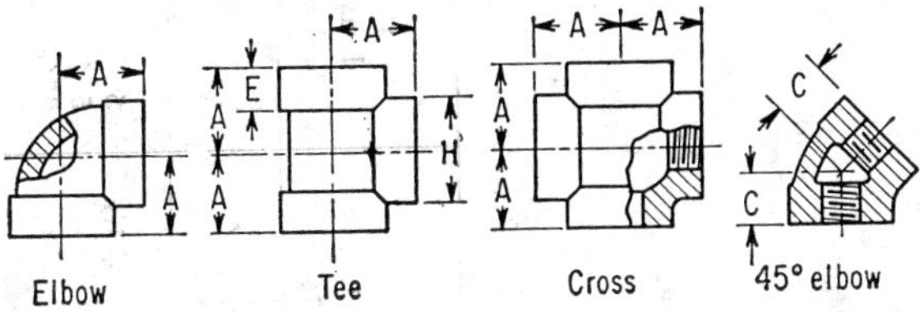

Size	125 lb				250 lb			
	A	*H*	*E*	*C*	*A*	*H*	*E*	*C*
¼	0.81	0.93	0.38	0.73	0.94	1.17	0.49	0.81
⅜	0.95	1.12	0.44	0.80	1.06	1.36	0.55	0.88
½	1.12	1.34	0.50	0.88	1.25	1.59	0.60	1.00
¾	1.13	1.63	0.56	0.98	1.44	1.88	0.68	1.13
1	1.50	1.95	0.62	1.12	1.63	2.24	0.76	1.31
1¼	1.75	2.39	0.69	1.29	1.94	2.73	0.88	1.50
1½	1.94	2.68	0.75	1.43	2.13	3.07	0.97	1.69
2	2.25	3.28	0.84	1.68	2.50	3.74	1.12	2.00
2½	2.70	3.86	0.94	1.95	2.94	4.60	1.30	2.25
3	3.08	4.62	1.00	2.17	3.38	5.36	1.40	2.50
3½	3.42	5.20	1.06	2.39	3.75	5.98	1.49	2.63
4	3.79	5.79	1.12	2.61	4.13	6.61	1.57	2.81
5	4.50	7.05	1.18	3.05	4.88	7.92	1.74	3.19
6	5.13	8.28	1.28	3.46	5.63	9.24	1.91	3.50
8	6.56	10.63	1.47	4.28	7.00	11.73	2.24	4.31
10	8.08	13.12	1.68	5.16	8.63	14.37	2.58	5.19
12	9.50	15.47	1.88	5.97	10.00	16.84	2.91	6.00

All dimensions in inches.

* This applies to elbows and tees only.

Table 20.59. Dimensions of American Standard Cast-iron Screwed Drainage Fittings*,[66]

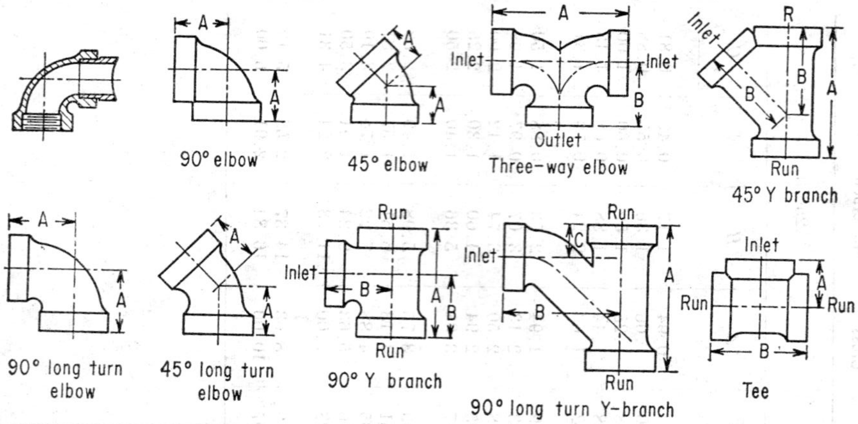

Size, in.	90° elbows	45° elbows	90° long-turn elbows	45° long-turn elbows	Three-way elbows		Tees		90° Y branches		90° long-turn Y branches			45° Y branches	
	A	*A*	*A*	*A*	*A*	*B*	*A*	*B*	*A*	*B*	*A*	*B*	*C*	*A*	*B*
1¼	1¾	1⅜	2¼	1¾	4½	2¼	1¾	3½	3¾	2¼	4¾	3⅝	1 3/16	5	3¼
1½	2 3/16	1 7/16	2½	1⅞	5¼	2⅝	2 3/16	4⅜	4¼	2½	5⅜	4⅛	1¼	5½	3⅝
2	2⅜	1¾	3 1/16	2¼	6¼	3⅛	2 5/16	4⅝	5 3/16	3 1/16	7 1/16	5 7/16	1⅝	6 7/16	4 5/16
2½	2 13/16	2 1/16	3 11/16	2⅝	7⅜	3 11/16	2 13/16	5⅝	6 5/16	3 11/16	8¼	6¼	2	7⅞	5⅜
3	3 3/16	2⅜	4¼	2⅞	8⅝	4 5/16	3 3/16	6⅜	7¼	4¼	9 13/16	7½	2 5/16	9	6 3/16
4	3 11/16	2¾	5 3/16	3 7/16	10⅜	5 3/16	4	8	8¾	5 3/16	13¾	9⅞	2⅞	10⅞	7 11/16
5	4½	3 3/16	6⅛	4⅛	12¼	6⅛	4⅝	9¼	10 5/16	6⅛	15¾	12¼	3½	12 15/16	9 3/16
6	5 3/16	3½	7⅛	4⅞	14¼	7⅛	5 3/16	11⅜	11 15/16	7⅛	18 11/16	14 9/16	4⅛	14¾	10¾
8	6½	4 3/16	9	6	....	....	6½	13	15 5/16	9	24 9/16	19 5/16	5¼	18 13/16	13 9/16

All dimensions in inches.
* Crane Co.

Table 20.60. Soldered-joint Fittings—Dimensions of Elbows, Tees, and Crosses[66]

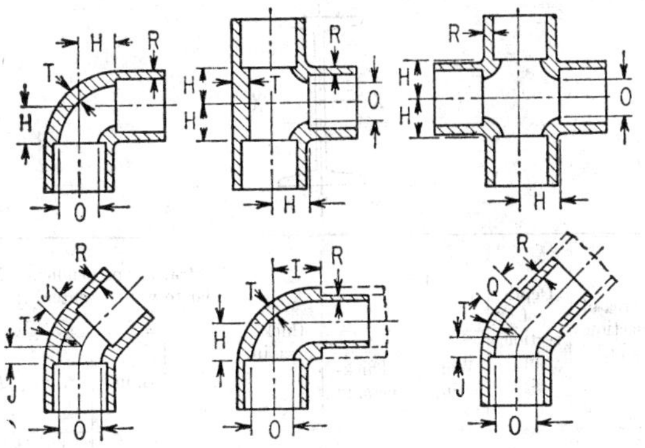

Nominal size	Cast brass							Wrought metal (*T* and *R*)
	H	*I*	*J*	*Q*	*O*	*T*	*R*	
1/4	1/4	3/8	3/16	1/4	0.31	0.08	0.048	0.030
3/8	5/16	7/16	3/16	5/16	0.43	0.08	0.048	0.035
1/2	7/16	9/16	3/16	5/16	0.54	0.09	0.054	0.040
3/4	9/16	11/16	1/4	3/8	0.78	0.10	0.060	0.045
1	3/4	7/8	5/16	7/16	1.02	0.11	0.066	0.050
1 1/4	7/8	1	7/16	9/16	1.26	0.12	0.072	0.055
1 1/2	1	1 1/8	1/2	5/8	1.50	0.13	0.078	0.060
2	1 1/4	1 3/8	9/16	3/4	1.98	0.15	0.090	0.070
2 1/2	1 1/2	1 5/8	5/8	7/8	2.46	0.17	0.102	0.080
3	1 3/4	1 7/8	3/4	1	2.94	0.19	0.114	0.090
3 1/2	2	2 1/8	7/8	1 1/8	3.42	0.20	0.120	0.100
4	2 1/4	2 3/8	15/16	1 1/4	3.90	0.22	0.132	0.110
5	3 1/8	.....	1 7/16	....	4.87	0.28	0.168	0.125
6	3 5/8	.....	1 5/8	....	5.84	0.34	0.204	0.140

20.9. STRUCTURAL SECTIONS[68,69]

The most economical form of beam cross section for vertical loading only is a beam of I section. Beams designed to resist bending should have the greatest depth of material in the direction of force applied, as indicated by the formula for section modulus.

This section includes pertinent data of standard sections useful to the designer as well as the detailer of fabricated steel.

Table 20.61. Wide-flange Beams and Columns

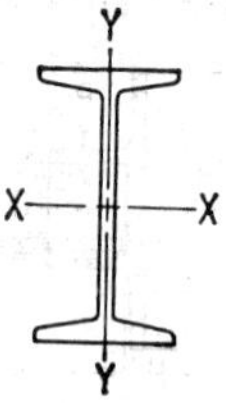

Nominal size, in.	Weight /ft, lb	Area of section, in.²	Depth of section, in.	Flange		Web thickness, in.	Neutral axis perpendicular to web at center			Neutral axis parallel to web at center		
				Width, in.	Thickness, in.		I, in.⁴	S, in.³	r, in.	I, in.⁴	S, in.³	r, in.
36 × 16½	300	88.17	36.72	16.65	1.680	0.945	20,290	1,105	15.17	1,225	147.1	3.73
	280	82.32	36.50	16.59	1.570	0.885	18,819	1,031	15.12	1,127	135.9	3.70
	260	76.56	36.24	16.55	1.440	0.845	17,233	951	15.00	1,020	123.3	3.65
	245	72.03	36.06	16.51	1.350	0.802	16,092	892	14.95	944	114.4	3.62
	230	67.73	35.88	16.47	1.260	0.765	14,988	835	14.88	870	105.7	3.59
36 × 12	194	57.11	36.48	12.11	1.260	0.770	12,103	663	14.56	355	58.7	2.49
	182	53.54	36.32	12.07	1.180	0.725	11,281	621	14.52	327	54.3	2.47
	170	49.98	36.16	12.02	1.100	0.680	10,470	579	14.47	300	50.0	2.45
	160	47.09	36.00	12.00	1.020	0.653	9,738	541	14.38	275	45.9	2.42
	150	44.16	35.84	11.97	0.940	0.625	9,012	502	14.29	250	41.8	2.38
33 × 15¾	240	70.52	33.50	15.86	1.400	0.830	13,585	811	13.88	874	110.2	3.52
	220	64.73	33.25	15.81	1.275	0.775	12,312	740	13.79	782	99.0	3.48
	200	58.79	33.00	15.75	1.150	0.715	11,048	669	13.71	691	87.8	3.43
33 × 11½	152	44.71	33.50	11.56	1.055	0.635	8,147	486	13.50	256	44.3	2.39
	141	41.51	33.31	11.53	0.960	0.605	7,442	446	13.39	229	39.8	2.35
	130	38.26	33.10	11.51	0.855	0.580	6,699	404	13.23	201	35.0	2.29
30 × 15	210	61.78	30.38	15.10	1.315	0.775	9,872	649	12.64	707	93.7	3.38
	190	55.90	30.12	15.04	1.185	0.710	8,825	586	12.57	624	83.1	3.34
	172	50.65	29.88	14.98	1.065	0.655	7,891	528	12.48	550	73.4	3.30
30 × 10½	132	38.83	30.30	10.55	1.000	0.615	5,753	379	12.17	185	35.1	2.18
	124	36.45	30.16	10.52	0.930	0.585	5,347	354	12.11	169	32.3	2.16
	116	34.13	30.00	10.50	0.850	0.564	4,919	327	12.00	153	29.2	2.12
	108	31.77	29.82	10.48	0.760	0.548	4,461	299	11.85	135	25.8	2.06
27 × 14	177	52.10	27.31	14.09	1.190	0.725	6,728	492.8	11.36	518.9	73.7	3.16
	160	47.04	27.08	14.02	1.075	0.658	6,018	444.0	11.31	458.0	65.3	3.12
	145	42.68	26.88	13.96	0.975	0.600	5,414	402.9	11.26	406.9	58.3	3.09
27 × 10	114	33 53	27.28	10.07	0.932	0.570	4,080	299.2	11.03	149.6	29.7	2.11
	102	30.01	27.07	10.02	0.827	0.518	3,604	266.3	10.96	129.5	25.9	2.08
	94	27.65	26.91	9.99	0.747	0.490	3,266	242.8	10.87	115.1	23.0	2.04
24 × 14	160	47.04	24.72	14.09	1.135	0.656	5,110	413.5	10.42	492.6	69.9	3.23
	145	42.62	24.49	14.04	1.020	0.608	4,561	372.5	10.34	434.3	61.8	3.19
	130	38.21	24.25	14.00	0.900	0.565	4,009	330.7	10.24	375.2	53.6	3.13
24 × 12	120	35.29	24.31	12.08	0.930	0.556	3,635	299.1	10.15	254.0	42.0	2.68
	110	32.36	24.16	12.04	0.855	0.510	3,315	274.4	10.12	229.1	38.0	2.66
	100	29.43	24.00	12.00	0.775	0.468	2,987	248.9	10.08	203.5	33.9	2.63
24 × 9	94	27.63	24.29	9.06	0.872	0.516	2,683	220.9	9.85	102.2	22.6	1.92
	84	24.71	24.09	9.02	0.772	0.470	2,364	196.3	9.78	88.3	19.6	1.89
	76	22.37	23.91	8.98	0.682	0.440	2,096	175.4	9.68	76.5	17.0	1.85
21 × 13	142	41.76	21.46	13.13	1.095	0.659	3,403	317.2	9.03	385.9	58.8	3.04
	127	37.34	21.24	13.06	0.985	0.588	3,017	284.1	8.99	338.6	51.8	3.01
	112	32.93	21.00	13.00	0.865	0.527	2,620	249.6	8.92	289.7	44.6	2.96
21 × 9	96	28.21	21.14	9.03	0.935	0.575	2,088	197.6	8.60	109.3	24.2	1.97
	82	24.10	20.86	8.96	0.795	0.499	1,752	168.0	8.53	89.6	20.0	1.93
21 × 8¼	73	21.46	21.24	8.29	0.740	0.455	1,600	150.7	8.64	66.2	16.0	1.76
	68	20.02	21.13	8.27	0.685	0.430	1,478	139.9	8.59	60.4	14.6	1.74
	62	18.23	20.99	8.24	0.615	0.400	1,326	126.4	8.53	53.1	12.9	1.71
18 × 11¾	114	33.51	18.48	11.83	0.991	0.595	2,033	220.1	7.79	255.6	43.2	2.76
	105	30.86	18.32	11.79	0.911	0.554	1,852	202.2	7.75	231.0	39.2	2.73
	96	28.22	18.16	11 75	0.831	0.512	1,674	184.4	7.70	206.8	35.2	2.71

Table 20.61. Wide-flange Beams and Columns (Continued)

Nominal size, in.	Weight /ft, lb	Area of section, in.²	Depth of section, in.	Flange		Web thickness, in.	Neutral axis perpendicular to web at center			Neutral axis parallel to web at center		
				Width, in.	Thickness, in.		I, in.⁴	S, in.³	r, in.	I, in.⁴	S, in.³	r, in.
18 × 8¾	85	24.97	18.32	8.83	0.911	0.526	1,429	156.1	7.57	99.4	22.5	2.00
	77	22.63	18.16	8.78	0.831	0.475	1,286	141.7	7.54	88.6	20.2	1.98
	70	20.56	18.00	8.75	0.751	0.438	1,153	128.2	7.49	78.5	17.9	1.95
	64	18.80	17.87	8.71	0.686	0.403	1,045	117.0	7.46	70.3	16.1	1.93
18 × 7½	60	17.64	18.25	7.56	0.695	0.416	984	107.8	7.47	47.1	12.5	1.63
	55	16.19	18.12	7.53	0.630	0.390	889	98.2	7.41	42.0	11.1	1.61
	50	14.71	18.00	7.50	0.570	0.358	800	89.0	7.38	37.2	9.9	1.59
16 × 11½	96	28.22	16.32	11.53	0.875	0.535	1,355	166.1	6.93	207.2	35.9	2.71
	88	25.86	16.16	11.50	0.795	0.504	1,222	151.3	6.87	185.2	32.2	2.67
16 × 8½	78	22.92	16.32	8.58	0.875	0.529	1,042	127.8	6.74	87.5	20.4	1.95
	71	20.86	16.16	8.54	0.795	0.486	936	115.9	6.70	77.9	18.2	1.93
	64	18.80	16.00	8.50	0.715	0.443	833	104.2	6.66	68.4	16.1	1.91
	58	17.04	15.86	8.46	0.645	0.407	746	94.1	6.62	60.5	14.3	1.88
16 × 7	50	14.70	16.25	7.07	0.628	0.380	655	80.7	6.68	34.8	9.8	1.54
	45	13.24	16.12	7.03	0.563	0.346	583	72.4	6.64	30.5	8.7	1.52
	40	11.77	16.00	7.00	0.503	0.307	515	64.4	6.62	26.5	7.6	1.50
	36	10.59	15.85	6.99	0.428	0.299	466	56.3	6.49	22.1	6.3	1.45
14 × 16	426	125.2	18.69	16.69	3.033	1.875	6,610	707.4	7.26	2,359	282.7	4.34
	398	116.9	18.31	16.59	2.843	1.770	6,013	656.9	7.17	2,169	261.6	4.31
	370	108.8	17.94	16.47	2.658	1,655	5,454	608.1	7.08	1,986	241.1	4.27
	342	100.6	17.56	16.36	2.468	1.545	4,911	559.4	6.99	1,806	220.8	4.24
	314	92.30	17.19	16.23	2.283	1.415	4,399	511.9	6.90	1,631	201.0	4.20
	287	84.37	16.81	16.13	2.093	1.310	3,912	465.5	6.81	1,466	181.8	4.17
	264	77.63	16.50	16.02	1.938	1.205	3,526	427.4	6.74	1,331	166.1	4.14
	246	72.33	16.25	15.94	1.813	1.125	3,228	397.4	6.68	1,226	153.9	4.12
	237	69.69	16.12	15.91	1.748	1.090	3,080	382.2	6.65	1,174	147.7	4.11
	228	67.06	16.00	15.86	1.688	1.045	2,942	367.8	6.62	1,124	141.8	4.10
	219	64.36	15.87	15.82	1.623	1.005	2,798	352.6	6.59	1,073	135.6	4.08
	211	62.07	15.75	15.80	1.563	0.980	2,671	339.2	6.56	1,028	130.2	4.07
	202	59.39	15.63	15.75	1.503	0.930	2,538	324.9	6.54	979	124.4	4.06
14 × 16	193	56.73	15.50	15.71	1.438	0.890	2,402	310.0	6.51	930	118.4	4.05
	184	54.07	15.38	15.66	1.378	0.840	2,274	295.8	6.49	882	112.7	4.04
	176	51.73	15.25	15.64	1.313	0.820	2,149	281.9	6.45	837	107.1	4.02
	167	49.09	15.12	15.60	1.248	0.780	2,020	267.3	6.42	790	101.3	4.01
	158	46.47	15.00	15.55	1.188	0.730	1,900	253.4	6.40	745	95.8	4.00
	150	44.08	14.88	15.51	1.128	0.695	1,786	240.2	6.37	702	90.6	3.99
	142	41.85	14.75	15.50	1.063	0.680	1,672	226.7	6.32	660	85.2	3.97
	320	94.12	16.81	16.71	2.093	1.890	4,141	492.8	6.63	1,635	195.7	4.17
14 × 14½	136	39.98	14.75	14.74	1.063	0.660	1,593	216.0	6.31	567	77.0	3.77
	127	37.33	14.62	14.69	0.998	0.610	1,476	202.0	6.29	527	71.8	3.76
	119	34.99	14.50	14.65	0.938	0.570	1,373	189.4	6.26	491	67.1	3.75
	111	32.65	14.37	14.62	0.873	0.540	1,266	176.3	6.23	454	62.2	3.73
	103	30.26	14.25	14.57	0.813	0.495	1,165	163.6	6.21	419	57.6	3.72
	95	27.94	14.12	14.54	0.748	0.465	1,063	150.6	6.17	383	52.8	3.71
	87	25.56	14.00	14.50	0.688	0.420	966	138.1	6.15	349	48.2	3.70
14 × 12	84	24.71	14.18	12.02	0.778	0.451	928	130.9	6.13	225.5	37.5	3.02
	78	22.94	14.06	12.00	0.718	0.428	851	121.1	6.09	206.9	34.5	3.00
14 × 10	74	21.76	14.19	10.07	0.783	0.450	796	112.3	6.05	133.5	26.5	2.48
	68	20.00	14.06	10.04	0.718	0.418	724	103.0	6.02	121.2	24.1	2.46
	61	17.94	13.91	10.00	0.643	0.378	641	92.2	5.98	107.3	21.5	2.45
14 × 8	53	15.59	13.94	8.06	0.658	0.370	542	77.8	5.90	57.5	14.3	1.92
	48	14.11	13.81	8.03	0.593	0.339	484	70.2	5.86	51.3	12.8	1.91
	43	12.65	13.68	8.00	0.528	0.308	429	62.7	5.82	45.1	11.3	1.89
14 × 6¾	38	11.17	14.12	6.77	0.513	0.313	385	54.6	5.87	24.6	7.3	1.49
	34	10.00	14.00	6.75	0.453	0.287	339	48.5	5.83	21.3	6.3	1.46
	30	8.81	13.86	6.73	0.383	0.270	289	41.8	5.73	17.5	5.2	1.41
12 × 12	190	55.86	14.38	12.67	1.736	1.060	1,892	263.2	5.82	589.7	93.1	3.25
	161	47.38	13.88	12.51	1.486	0.905	1,541	222.2	5.70	486.2	77.7	3.20
	133	39.11	13.38	12.36	1.236	0.755	1,221	182.5	5.59	389.9	63.1	3.16

Table 20.61. Wide-flange Beams and Columns (Continued)

Nominal size, in.	Weight /ft, lb	Area of section, in.²	Depth of section, in.	Flange		Web thickness, in.	Neutral axis perpendicular to web at center			Neutral axis parallel to web at center		
				Width, in.	Thickness, in.		I, in.⁴	S, in.³	r, in.	I, in.⁴	S, in.³	r, in.
12 × 12 (cont'd)	120	35.31	13.12	12.32	1.106	0.710	1,071	163.4	5.51	345.1	56.0	3.13
	106	31.19	12.88	12.23	0.986	0.620	930	144.5	5.46	300.9	49.2	3.11
	99	29.02	12.75	12.19	0.921	0.580	858	134.7	5.43	278.2	45.7	3.09
	92	27.06	12.62	12.15	0.856	0.545	788	125.0	5.40	256.4	42.2	3.08
	85	24.98	12.50	12.10	0.796	0.495	723	115.7	5.38	235.5	38.9	3.07
	79	23.22	12.38	12.08	0.736	0.470	663	107.1	5.34	216.4	35.8	3.05
	72	21.16	12.25	12.04	0.671	0.430	597	97.5	5.31	195.3	32.4	3.04
	65	19.11	12.12	12.00	0.606	0.390	533	88.0	5.28	174.6	29.1	3.02
12 × 10	58	17.06	12.19	10.01	0.641	0.359	476	78.1	5.28	107.4	21.4	2.51
	53	15.59	12.06	10.00	0.576	0.345	426	70.7	5.23	96.1	19.2	2.48
12 × 8	50	14.71	12.19	8.07	0.641	0.371	394	64.7	5.18	56.4	14.0	1.96
	45	13.24	12.06	8.04	0.576	0.336	350	58.2	5.15	50.0	12.4	1.94
	40	11.77	11.94	8.00	0.516	0.294	310	51.9	5.13	44.1	11.0	1.94
12 × 6½	36	10.59	12.24	6.56	0.540	0.305	280	45.9	5.15	23.7	7.2	1.50
	31	9.12	12.09	6.52	0.465	0.265	238	39.4	5.11	19.8	6.1	1.47
	27	7.97	11.95	6.50	0.400	0.240	204	34.1	5.06	16.6	5.1	1.44
10 × 10	112	32.92	11.38	10.41	1.248	0.755	718.7	126.3	4.67	235.4	45.2	2.67
	100	29.43	11.12	10.34	1.118	0.685	625.0	112.4	4.61	206.6	39.9	2.65
	89	26.19	10.88	10.27	0.998	0.615	542.4	99.7	4.55	180.6	35.2	2.63
	77	22.67	10.62	10.19	0.868	0.535	457.2	86.1	4.49	153.4	30.1	2.60
	72	21.18	10.50	10.17	0.808	0.510	420.7	80.1	4.46	141.8	27.9	2.59
	66	19.41	10.38	10.11	0.748	0.457	382.5	73.7	4.44	129.2	25.5	2.58
	60	17.66	10.25	10.07	0.683	0.415	343.7	67.1	4.41	116.5	23.1	2.57
	54	15.88	10.12	10.02	0.618	0.368	305.7	60.4	4.39	103.9	20.7	2.56
	49	14.40	10.00	10.00	0.558	0.340	272.9	54.6	4.35	93.0	18.6	2.54
10 × 8	45	13.24	10.12	8.02	0.618	0.350	248.6	49.1	4.33	53.2	13.3	2.00
	39	11.48	9.94	7.99	0.528	0.318	209.7	42.2	4.27	44.9	11.2	1.98
	33	9.71	9.75	7.96	0.433	0.292	170.9	35.0	4.20	36.5	9.2	1.94
10 × 5¾	29	8.53	10.22	5.79	0.500	0.289	157.3	30.8	4.29	15.2	5.2	1.34
	25	7.35	10.08	5.76	0.430	0.252	133.2	26.4	4.26	12.7	4.4	1.31
	21	6.19	9.90	5.75	0.340	0.240	106.3	21.5	4.14	9.7	3.4	1.25
8 × 8	67	19.70	9.00	8.28	0.933	0.575	271.8	60.4	3.71	88.6	21.4	2.12
	58	17.06	8.75	8.22	0.808	0.510	227.3	52.0	3.65	74.9	18.2	2.10
	48	14.11	8.50	8.11	0.683	0.405	183.7	43.2	3.61	60.9	15.0	2.08
	40	11.76	8.25	8.07	0.558	0.365	146.3	35.5	3.53	49.0	12.1	2.04
	35	10.30	8.12	8.02	0.493	0.315	126.5	31.1	3.50	42.5	10.6	2.03
	31	9.12	8.00	8.00	0.433	0.228	109.7	27.4	3.47	37.0	9.2	2.01
8 × 6½	28	8.23	8.06	6.54	0.463	0.285	97.8	24.3	3.45	21.6	6.6	1.62
	24	7.06	7.93	6.50	0.398	0.245	8.25	20.8	3.42	18.2	5.6	1.61
8 × 5¼	20	5.88	8.14	5.27	0.378	0.248	69.2	17.0	3.43	8.5	3.2	1.20
	17	5.00	8.00	5.25	0.308	0.230	56.4	14.1	3.36	6.7	2.6	1.16

Flanges of wide-flange beams and columns are not tapered, have constant thickness.
Lightweight beams for each nominal size, and beams with depth in even inches, are most usually stocked.
Designation of wide-flange beams is made by giving nominal depth and weight, thus, 8WF40.

Table 20.62. American Standard I-beams

Depth of beam, in.	Weight ft, lb	Area of section, in.	Width of flange, in.	Thickness of web, in.	Neutral axis perpendicular to web at center			Neutral axis coincident with center line of web		
					I, in.^{4}	r, in.	z in.^{3}	I, in.^{4}	r, in.	z, in.^{3}
24	120.0	35.13	8.048	0.798	3010.8	9.26	250.9	84.9	1.56	21.1
	105.9	30.98	7.875	0.625	2811.5	9.53	234.3	78.9	1.60	20.0
	100.0	29.25	7.247	0.747	2371.8	9.05	197.6	48.4	1.29	13.4
	90.0	26.30	7.124	0.624	2230.1	9.21	185.8	45.5	1.32	12.8
	79.9	23.33	7.000	0.500	2087.2	9.46	173.9	42.9	1.36	12.2
20	95.0	27.74	7.200	0.800	1599.7	7.59	160.0	50.5	1.35	14.0
	85.0	24.80	7.053	0.653	1501.7	7.78	150.2	47.0	1.38	13.3
	75.0	21.90	6.391	0.641	1263.5	7.60	126.3	30.1	1.17	9.4
	65.4	19.08	6.250	0.500	1169.5	7.83	116.9	27.9	1.21	8.9
18	70.0	20.46	6.251	0.711	917.5	6.70	101.9	24.5	1.09	7.8
	54.7	15.94	6.000	0.460	795.5	7.07	88.4	21.2	1.15	7.1
15	50.0	14.59	5.640	0.550	481.1	5.74	64.2	16.0	1.05	5.7
	42.9	12.49	5.500	0.410	441.8	5.95	58.9	14.6	1.08	5.3
12	50.0	14.57	5.477	0.687	301.6	4.55	50.3	16.0	1.05	5.8
	40.8	11.84	5.250	0.460	268.9	4.77	44.8	13.8	1.08	5.3
	35.0	10.20	5.078	0.428	227.0	4.72	37.8	10.0	0.99	3.9
	31.8	9.26	5.000	0.350	215.8	4.83	36.0	9.5	1.01	3.8
10	35.0	10.22	4.944	0.594	145.8	3.78	29.2	8.5	0.91	3.4
	25.4	7.38	4.660	0.310	122.1	4.07	24.4	6.9	0.97	3.0
8	23.0	6.71	4.171	0.441	64.2	3.09	16.0	4.4	0.81	2.1
	18.4	5.34	4.000	0.270	56.9	3.26	14.2	3.8	0.84	1.9
7	20.0	5.83	3.860	0.450	41.9	2.68	12.0	3.1	0.74	1.6
	15.3	4.43	3.660	0.250	36.2	2.86	10.4	2.7	0.78	1.5
6	17.25	5.02	3.565	0.465	26.0	2.28	8.7	2.3	0.68	1.3
	12.5	3.61	3.330	0.230	21.8	2.46	7.3	1.8	0.72	1.1
5	14.75	4.29	3.284	0.494	15.0	1.87	6.0	1.7	0.63	1.0
	10.0	2.87	3.000	0.210	12.1	2.05	4.8	1.2	0.65	0.82
4	9.5	2.76	2.796	0.326	6.7	1.56	3.3	0.91	0.58	0.65
	7.7	2.21	2.660	0.190	6.0	1.64	3.0	0.77	0.59	0.58
3	7.5	2.17	2.509	0.349	2.9	1.15	1.9	0.59	0.52	0.47
	5.7	1.64	2.330	0.170	2.5	1.23	1.7	0.46	0.53	0.40

Table 20.63. American Standard Channels

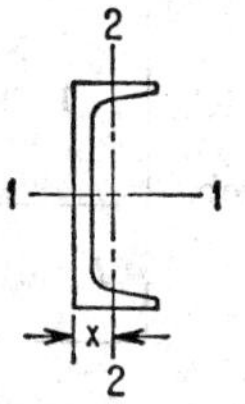

Depth of channel, in.	Weight ft, lb	Area of section, in.²	Width of flange, in.	Thickness of web, in.	Axis 1-1			Axis 2-2	x, in.
					I, in.⁴	r, in.	S, in.³	r, in.	
15	50.0	14.64	3.716	0.716	401.4	5.24	53.6	0.87	0.80
	40.0	11.70	3.520	0.520	346.3	5.44	46.2	0.89	0.78
	33.9	9.90	3.400	0.400	312.6	5.62	41.7	0.91	0.79
12	30.0	8.79	3.170	0.510	161.2	4.28	26.9	0.77	0.68
	25.0	7.32	3.047	0.387	143.5	4.43	23.9	0.79	0.68
	20.7	6.03	2.940	0.280	128.1	4.61	21.4	0.81	0.70
10	30.0	8.80	3.033	0.673	103.0	3.42	20.6	0.67	0.65
	25.0	7.33	2.886	0.526	90.7	3.52	18.1	0.68	0.62
	20.0	5.86	2.739	0.379	78.5	3.66	15.7	0.70	0.61
	15.3	4.47	2.600	0.240	66.9	3.87	13.4	0.72	0.64
9	20.0	5.86	2.648	0.448	60.6	3.22	13.5	0.65	0.59
	15.0	4.39	2.485	0.285	50.7	3.40	11.3	0.67	0.59
	13.4	3.89	2.430	0.230	47.3	3.49	10.5	0.67	0.61
8	18.75	5.49	2.527	0.487	43.7	2.82	10.9	0.60	0.57
	13.75	4.02	2.343	0.303	35.8	2.99	9.0	0.62	0.56
	11.5	3.36	2.260	0.220	32.3	3.10	8.1	0.63	0.58
7	14.75	4.32	2.299	0.419	27.1	2.51	7.7	0.57	0.53
	12.25	3.58	2.194	0.314	24.1	2.59	6.9	0.58	0.53
	9.8	2.85	2.090	0.210	21.1	2.72	6.0	0.59	0.55
6	13.0	3.81	2.157	0.437	17.3	2.13	5.8	0.53	0.52
	10.5	3.07	2.034	0.314	15.1	2.22	5.0	0.53	0.50
	8.2	2.39	1.920	0.200	13.0	2.34	4.3	0.54	0.52
5	9.0	2.63	1.885	0.325	8.8	1.83	3.5	0.49	0.48
	6.7	1.95	1.750	0.190	7.4	1.95	3.0	0.50	0.49
4	7.25	2.12	1.720	0.320	4.5	1.47	2.3	0.46	0.46
	5.4	1.56	1.580	0.180	3.8	1.56	1.9	0.45	0.46
3	6.0	1.75	1.596	0.356	2.1	1.08	1.4	0.42	0.46
	5.0	1.46	1.498	0.258	1.8	1.12	1.2	0.41	0.44
	4.1	1.19	1.410	0.170	1.6	1.17	1.1	0.41	0.44

Table 20.64. Standard Angles, Equal Legs

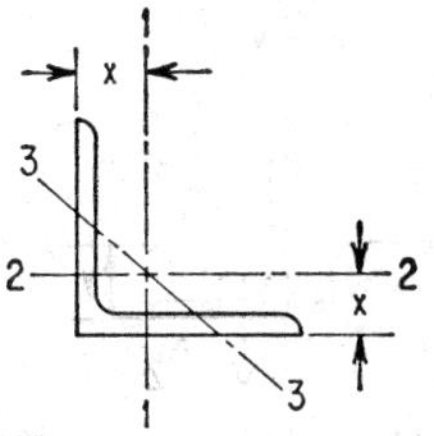

Size, in.	Weight/ft, lb	Area of section, in.^{2}	Axis 1-1 and axis 2-2				Axis 3-3, r min, in.
			I, in.^{4}	r, in.	S, in.^{3}	x, in.	
8 × 8 × 1⅛	56.9	16.73	98.0	2.42	17.5	2.41	1.56
1	51.0	15.00	89.0	2.44	15.8	2.37	1.56
⅞	45.0	13.23	79.6	2.45	14.0	2.32	1.57
¾	38.9	11.44	69.7	2.47	12.2	2.28	1.57
⅝	32.7	9.61	59.4	2.49	10.3	2.23	1.58
½	26.4	7.75	48.6	2.50	8.4	2.19	1.59
6 × 6 × 1	37.4	11.00	35.5	1.80	8.6	1.86	1.17
⅞	33.1	9.73	31.9	1.81	7.6	1.82	1.17
¾	28.7	8.44	28.2	1.83	6.7	1.78	1.17
⅝	24.2	7.11	24.2	1.84	5.7	1.73	1.18
½	19.6	5.75	19.9	1.86	4.6	1.68	1.18
⅜	14.9	4.36	15.4	1.88	3.5	1.64	1.19
5 × 5 × ⅞	27.2	7.98	17.8	1.49	5.2	1.57	0.97
¾	23.6	6.94	15.7	1.51	4.5	1.52	0.97
⅝	20.0	5.86	13.6	1.52	3.9	1.48	0.98
½	16.2	4.75	11.3	1.54	3.2	1.43	0.98
⅜	12.3	3.61	8.7	1.56	2.4	1.39	0.99
4 × 4 × ¾	18.5	5.44	7.7	1.19	2.8	1.27	0.78
⅝	15.7	4.61	6.7	1.20	2.4	1.23	0.78
½	12.8	3.75	5.6	1.22	2.0	1.18	0.78
⅜	9.8	2.86	4.4	1.23	1.5	1.14	0.79
¼	6.6	1.94	3.0	1.25	1.1	1.09	0.80
3½ × 3½ × ½	11.1	3.25	3.6	1.06	1.5	1.06	0.68
⅜	8.5	2.48	2.9	1.07	1.2	1.01	0.69
¼	5.8	1.69	2.0	1.09	0.79	0.97	0.69
3 × 3 × ½	9.4	2.75	2.2	0.90	1.1	0.93	0.58
⅜	7.2	2.11	1.8	0.91	0.83	0.89	0.58
¼	4.9	1.44	1.2	0.93	0.58	0.84	0.59
2½ × 2½ × ½	7.7	2.25	1.2	0.74	0.72	0.81	0.49
⅜	5.9	1.73	0.98	0.75	0.57	0.76	0.49
¼	4.1	1.19	0.70	0.77	0.39	0.72	0.49
2 × 2 × ⅜	4.7	1.36	0.48	0.59	0.35	0.64	0.39
¼	3.19	0.94	0.35	0.61	0.25	0.59	0.39
⅛	1.65	0.48	0.19	0.63	0.13	0.55	0.40
1¾ × 1¾ × ¼	2.77	0.81	0.23	0.53	0.19	0.53	0.34
⅛	1.44	0.42	0.13	0.55	0.10	0.48	0.35
1½ × 1½ × ¼	2.34	0.69	0.14	0.45	0.13	0.47	0.29
⅛	1.23	0.36	0.08	0.47	0.07	0.42	0.30
1¼ × 1¼ × ¼	1.92	0.56	0.08	0.37	0.09	0.40	0.24
⅛	1.01	0.30	0.04	0.38	0.05	0.36	0.25
1 × 1 × ¼	1.49	0.44	0.04	0.29	0.06	0.34	0.20
⅛	0.80	0.23	0.02	0.30	0.03	0.30	0.20

Table 20.65. Selected Standard Angles, Unequal Legs

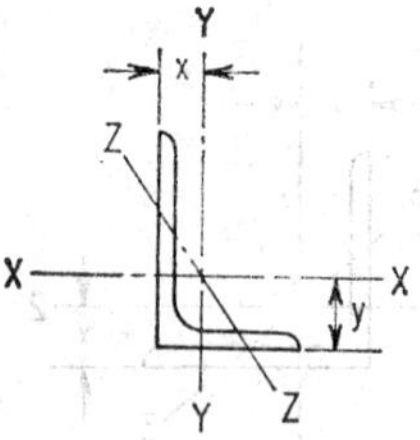

Size, in.	Thickness, in.	Weight/ft, lb	Area of section, in.²	Axis X-X				Axis Y-Y				Axis Z-Z
				I, in.⁴	*S*, in.³	*r*, in.	*y*, in.	*I*, in.⁴	*S*, in.³	*r*, in.	*x*, in.	*r*, in.
8 × 6	1	44.2	13.00	80.8	15.1	2.49	2.65	38.8	8.9	1.73	1.65	1.28
	¾	33.8	9.94	63.4	11.7	2.53	2.56	30.7	6.9	1.76	1.56	1.29
	½	23.0	6.75	44.3	8.0	2.56	2.47	21.7	4.8	1.79	1.47	1.30
	7/16	20.2	5.93	39.2	7.1	2.57	2.45	19.3	4.2	1.80	1.45	1.31
8 × 4	1	37.4	11.00	69.6	14.1	2.52	3.05	11.6	3.9	1.03	1.05	0.85
	¾	28.7	8.44	54.9	10.9	2.55	2.95	9.4	3.1	1.05	0.95	0.85
	½	19.6	5.75	38.5	7.5	2.59	2.86	6.7	2.2	1.08	0.86	0.86
	7/16	17.2	5.06	34.1	6.6	2.60	2.83	6.0	1.9	1.09	0.83	0.87
7 × 4	⅞	30.2	8.86	42.9	9.7	2.20	2.55	10.2	3.5	1.07	1.05	0.86
	¾	26.2	7.69	37.8	8.4	2.22	2.51	9.1	3.0	1.09	1.01	0.86
	½	17.9	5.25	26.7	5.8	2.25	2.42	6.5	2.1	1.11	0.92	0.87
	⅜	13.6	3.98	20.6	4.4	2.27	2.37	5.1	1.6	1.13	0.87	0.88
6 × 4	⅞	27.2	7.98	27.7	7.2	1.86	2.12	9.8	3.4	1.11	1.12	0.86
	¾	23.6	6.94	24.5	6.3	1.88	2.08	8.7	3.0	1.12	1.08	0.86
	½	16.2	4.75	17.4	4.3	1.91	1.99	6.3	2.1	1.15	0.99	0.87
	⅜	12.3	3.61	13.5	3.3	1.93	1.94	4.9	1.6	1.17	0.94	0.88
	5/16	10.3	3.03	11.4	2.8	1.94	1.92	4.2	1.4	1.17	0.92	0.88
6 × 3½	½	15.3	4.50	16.6	4.2	1.92	2.08	4.3	1.6	0.97	0.83	0.76
	⅜	11.7	3.42	12.9	3.2	1.94	2.04	3.3	1.2	0.99	0.79	0.77
	¼	7.9	2.31	8.9	2.2	1.96	1.99	2.3	0.85	1.01	0.74	0.78
5 × 3½	¾	19.8	5.81	13.9	4.3	1.55	1.75	5.6	2.2	0.98	1.00	0.75
	½	13.6	4.00	10.0	3.0	1.58	1.66	4.1	1.6	1.01	0.91	0.75
	¼	7.0	2.06	5.4	1.6	1.61	1.56	2.2	0.83	1.04	0.81	0.76
5 × 3	½	12.8	3.75	9.5	2.9	1.59	1.75	2.6	1.1	0.83	0.75	0.65
	⅜	9.8	2.86	7.4	2.2	1.61	1.70	2.0	0.89	0.84	0.70	0.65
	¼	6.6	1.94	5.1	1.5	1.62	1.66	1.4	0.61	0.86	0.66	0.66
4 × 3½	⅝	14.7	4.30	6.4	2.4	1.22	1.29	4.5	1.8	1.03	1.04	0.72
	½	11.9	3.50	5.3	1.9	1.23	1.25	3.8	1.5	1.04	1.00	0.72
	⅜	9.1	2.67	4.2	1.5	1.25	1.21	3.0	1.2	1.06	0.96	0.73
	¼	6.2	1.81	2.9	1.0	1.27	1.16	2.1	0.81	1.07	0.91	0.73
4 × 3	⅝	13.6	3.98	6.0	2.3	1.23	1.37	2.9	1.4	0.85	0.87	0.64
	½	11.1	3.25	5.1	1.9	1.25	1.33	2.4	1.1	0.86	0.83	0.64
	¼	5.8	1.69	2.8	1.0	1.28	1.24	1.4	0.60	0.90	0.74	0.65
3½ × 3	½	10.2	3.00	3.5	1.5	1.07	1.13	2.3	1.1	0.88	0.88	0.62
	¼	5.4	1.56	1.9	0.78	1.11	1.04	1.3	0.59	0.91	0.79	0.63
3½ × 2½	½	9.4	2.75	3.2	1.4	1.09	1.20	1.4	0.76	0.70	0.70	0.53
	¼	4.9	1.44	1.8	0.75	1.12	1.11	0.78	0.41	0.74	0.61	0.54
3 × 2½	½	8.5	2.50	2.1	1.0	0.91	1.00	1.3	0.74	0.72	0.75	0.52
	⅜	6.6	1.92	1.7	0.81	0.93	0.96	1.0	0.58	0.74	0.71	0.52
	¼	4.5	1.31	1.2	0.56	0.95	0.91	0.74	0.40	0.75	0.66	0.53
3 × 2	½	7.7	2.25	1.9	1.0	0.92	1.08	0.67	0.47	0.55	0.58	0.43
	3/16	3.07	0.90	0.84	0.41	0.97	0.97	0.31	0.20	0.58	0.47	0.44
2½ × 2	⅜	5.3	1.55	0.91	0.55	0.77	0.83	0.51	0.36	0.58	0.58	0.42
	3/16	2.75	0.81	0.51	0.29	0.79	0.76	0.29	0.20	0.60	0.51	0.43
2 × 1½	¼	2.77	0.81	0.32	0.24	0.62	0.66	0.15	0.14	0.43	0.41	0.32
	⅛	1.44	0.42	0.17	0.13	0.64	0.62	0.09	0.08	0.45	0.37	0.33
1¾ × 1¼	¼	2.44	0.69	0.20	0.18	0.54	0.60	0.09	0.10	0.35	0.35	0.27
	⅛	1.23	0.36	0.11	0.09	0.56	0.56	0.05	0.05	0.37	0.31	0.27

Table 20.66. Carnegie Zees

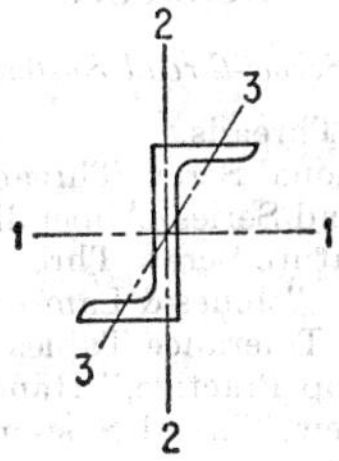

Size			Weight/ft, lb	Area of section, in.²	Axis 1-1			Axis 2-2			Axis 3-3
Depth, in.	Flanges, in.	Thickness, in.			*I*, in.⁴	*r*, in.	*S*, in.³	*I*, in.⁴	*r*, in.	*S*, in.³	*r* min, in.
6⅛	3⅝	½	21.1	6.19	34.4	2.36	11.2	12.9	1.44	3.8	0.84
6	3½	⅜	15.7	4.59	25.3	2.35	8.4	9.1	1.41	2.8	0.83
5	3¼	½	17.9	5.25	19.2	1.91	7.7	9.1	1.31	3.0	0.74
5⅛	3⅜	7/16	16.4	4.81	19.1	1.99	7.4	9.2	1.38	2.9	0.77
5 1/16	3 5/16	⅜	14.0	4.10	16.2	1.99	6.4	7.7	1.37	2.5	0.76
5	3¼	5/16	11.6	3.40	13.4	1.98	5.3	6.2	1.35	2.0	0.75
4 1/16	3⅛	½	15.9	4.66	11.2	1.55	5.5	8.0	1.31	2.8	0.67
4⅛	3 3/16	⅜	12.5	3.66	9.6	1.62	4.7	6.8	1.36	2.3	0.69
4 1/16	3⅛	5/16	10.3	3.03	7.9	1.62	3.9	5.5	1.34	1.8	0.68
4	3 1/16	¼	8.2	2.41	6.3	1.62	3.1	4.2	1.33	1.4	0.67
3	2 11/16	½	12.6	3.69	4.6	1.12	3.1	4.9	1.15	2.0	0.53
3	2 11/16	⅜	9.8	2.86	3.9	1.16	2.6	3.9	1.17	1.6	0.54
3	2 11/16	¼	6.7	1.97	2.9	1.21	1.9	2.8	1.19	1.1	0.55

Table 20.67. Gages for Angles, Inches

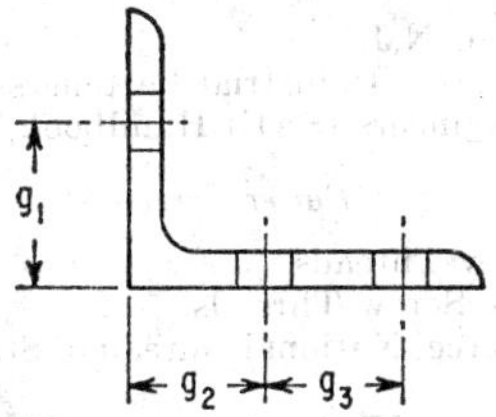

Leg	8	7	6	5	4	3½	3	2½	2	1¾	1½	1⅜	1¼	1
g_1	4½	4	3½	3	2½	2	1¾	1⅜	1⅛	1	⅞	⅞	¾	⅝
g_2	3	2½	2¼	2										
g_3	3	3	2½	1¾										
Max rivet	1⅛	1	⅞	⅞	⅞	⅞	⅞	¾	⅝	½	⅜	⅜	⅜	¼

References

Screw-thread Systems

1. ASA B1.1-1960, Unified Screw Threads.
2. ASA B1.1-1960, American National Screw Threads, Appendix I.
3. ASA B1.1-1960, Standard Thread Series, Appendix C.
4. ASA B1.10-1958, Unified Miniature Screw Threads.
5. "Thread Elements and Formulas," Jones & Lamson Machine Company, Springfield, Vt.
6. ASA B1.1-1960, Allowance and Tolerance Tables, Appendix E.
7. "British Standards for Workshop Practice," Handbook 2.
8. "Machinery's Handbook," Screw Thread Systems, The Industrial Press, New York, 1960.
9. Trade name, patented by Lock Thread Corp.
10. Trade name, patented by Aircraft Screw Products Company.
11. ASA B1.11-1958, Microscopic Objective Threads.
12. "Screw Thread Standards for Federal Services," part III, p. 36, U.S. Department of Commerce, National Bureau of Standards Handbook H28, 1957.
13. ASA B33.1-1947, Hose Coupling Screw Threads.
14. ASA B26-1953, Fire Hose Coupling Screw Threads.
15. "Screw Thread Standards for Federal Services," U.S. Department of Commerce, National Bureau of Standards Handbook H28, 1957.
16. Kent: "Mechanical Engineers' Handbook," 12th ed., Carmichael, Design and Production, Sec. 10, John Wiley & Sons, Inc., New York, 1951.
17. Marks, Lionel S.: "Mechanical Engineers' Handbook" (Baumeister rev.), 6th ed., Sec. 8, McGraw-Hill Book Company, Inc., New York, 1958.
18. British Standard 93.
19. British Standard 1095, 1943.

Bolts, Nuts, Screws

20. ASA B18.2-1960, Square and Hexagon Bolts and Nuts.
21. "Machinery Handbook," Screw Thread Systems, The Industrial Press, New York, 1960.
22. ASA B18.5-1959, Round Head Bolts.
23. ASA B18.9-1958, Plow Bolts.
24. ASA B18.10-1952, Track Bolts and Nuts.
25. ASA B18.6-1960, Machine Screws.
26. Kent, "Mechanical Engineers Handbook," 12th ed., Carmichael, Design and Production, pp. 10-32 through 10-51, John Wiley & Sons, Inc., New York.
27. ASA B18.6.2-1960, Hexagon, Slotted Cap Screws; Square, Slotted Headless Set Screws.
28. ASA B18.3-1954, Socket Head Cap and Set Screws.
29. ASA B18.6.4-1960, Slotted and Recessed Head Tapping Screws and Metallic Drive Screws.
30. Parker Kalon Corp., Clifton, N.J.
31. Bolt, Nut and Rivet Standards, Industrial Fasteners Institute, Cleveland, Ohio.
32. "Society of Automotive Engineers (SAE) Handbook," 1957.

Power Screws

33. ASA B1.5-1952, Acme Screw Threads.
34. ASA B1.9-1953, Stub Acme Screw Threads.
35. U.S. Department of Commerce, National Bureau of Standards Handbook H28, part III, p. 46, 1957.
36. U.S. Department of Commerce, National Bureau of Standards Handbook H28, part III, p. 29, 1957.
37. Laughner, Vallory H., and Augustus D. Hargan: "Handbook of Fastening and Joining Metal Parts," p. 8, McGraw-Hill Book Company, Inc., New York, 1956.
38. Kent: "Mechanical Engineers' Handbook," 12th ed., p. 10-03, Power, John Wiley & Sons, Inc., New York.

Washers

39. ASA B27.2-1958, Plain Washers.
40. ASA B27.1-1958, Lock Washers.
41. Laughner, Vallory H., and Augustus D. Hargan: "Handbook of Fastening and Joining Metal Parts," McGraw-Hill Book Company, Inc., New York, 1956
42. "Machinery's Handbook," The Industrial Press, New York, 1960

Pins

43. ASA B5.20-1958, Machine Pins.
44. Spring Type Straight Pin, trade name, Rollpin, patented by Elastic Stop Nut Corp.

Keys and Splines

45. Laughner, Vallory H., and Augustus D. Hargan: "Handbook of Fastening and Joining Metal Parts," Table 11.1, McGraw-Hill Book Company, Inc., New York, 1956.
46. Table 11.2, see ref. 45.
47. Table 11.3, see ref. 45.
48. ASA B17.F-1955, Woodruff Keys, Keyslots and Cutters.
49. Table 11.4, see ref. 45.
50. Phelan, Richard M.: "Fundamentals of Mechanical Design," McGraw-Hill Book Company, Inc., New York, 1957.
51. Kent: "Mechanical Engineers' Handbook," chap. 11, p. 15, 12th ed., John Wiley & Sons, Inc., New York, 1950.
52. "Machinery's Handbook," The Industrial Press, New York, 1960.
53. ASA B5.15-1960, Involute Splines, Serrations and Inspection.
54. LeGrand, Rupert (ed.): "The New American Machinists' Handbook," chap. 4, p. 114, McGraw-Hill Book Company, Inc., New York, 1955.
55. "SAE Handbook," p. 626, 1957.
56. LeGrand, Rupert (ed.): "The New American Machinists' Handbook," McGraw-Hill Book Company, Inc., New York, 1955.

Wire and Sheet-metal Gages

57. ref. 56, p. 28-4.
58. ASA B5.12-1940, Twist Drills.

Pipe

59. ASA B2.1-1960, Pipe Threads.
60. ASA B33.1-1960, Hose Coupling Screw Threads.
61. ASA B2.2-1960, Dryseal Pipe Threads.
62. ref. 56, Sec. 10-65.
63. ASA B36.10-1959, Wrought Steel and Wrought Iron Pipe.
64. "Machinery's Handbook," Bursting Pressures of Pipe, The Industrial Press, New York, 1962. 16th ed.
65. "American Petroleum Institute Handbook."
66. ASA B16-1960, Pipe Flanges and Fittings.
67. "Machinery's Handbook," 16th ed., p. 1969, The Industrial Press, New York, 1962.

Structural Sections

68. AISC Handbook, "Manual of Steel Construction," 6th ed., American Institute of Steel Construction, New York, 1963.
69. "Bethlehem Steel Handbook," Bethlehem, Pa.

General References

70. AF-NAVY: Air Material Command, Wright Patterson Air Force Base, Dayton, Ohio.
71. Federal: Business Service Center, General Services Administration, Washington 25, D.C.
72. IFI: Industrial Fasteners Institute, 1517 Terminal Tower, Cleveland 13, Ohio.
73. MIL: Commanding Officer, Naval Aviation Supply Depot, 700 Robbins Ave., Philadelphia 11, Pa. Att: C D S.
74. NAS: National Aircraft Standards Committee, National Standards Association, 610 Washington Loan & Trust Bldg., Washington 4, D.C.
75. *Machine Design:* The Fastener's Book, Sept. 29, 1960, Sec. 2, Penton Publishing Company, Cleveland, Ohio.

Section 21

BOLTED JOINTS

By

EUGENE I. RADZIMOVSKY, Ph.D., *Professor of Mechanical Engineering, University of Illinois, Urbana, Ill.*

CONTENTS

21.1. GENERAL

The joining of removable machine and structural elements is commonly accomplished by means of threaded bolts and screws. External threads, such as those found on screws, bolts, and pipes, are usually formed by cutting or rolling a helicoidal groove on a screw blank, bolt, or rod. Internal threads are generally produced by cutting. Cold hardening causes rolled threads to be generally stronger than threads formed by cutting.[24]

Sections 20 and 26 deal with proportions of threaded fasteners and thread strength, respectively. References 14 and 15 concern the physical and mechanical properties of threaded fasteners.

A fine thread is especially suited to joints subject to vibration. It offers greater strength at the thread root, may be used to provide fine adjustment, and is recommended for tapping hard nuts. The strength of fine threads is greater for sizes 1 in. and under, increasing as the size decreases.

A coarse thread is desirable where the hole is tapped in a weaker material as, for example, cast iron, aluminum, and magnesium alloys. It is stronger in sizes 1 in. and over and is especially suited for tapping in brittle materials.

Threads are scored by repeated unscrewing and tightening of steel nuts. After 50 tightenings, the torsional resistance is approximately 100 per cent greater than for the first tightening. After 200 tightenings, it is approximately 150 per cent greater.

Three basic forms of standard bolts are shown in Fig. 21.1. The through bolt (Fig. 21.1*a*) is least expensive but requires that both ends of the bolt be accessible. The tap bolt and stud (Fig. 21.1*b* and *c*, respectively) require strong threaded sections.

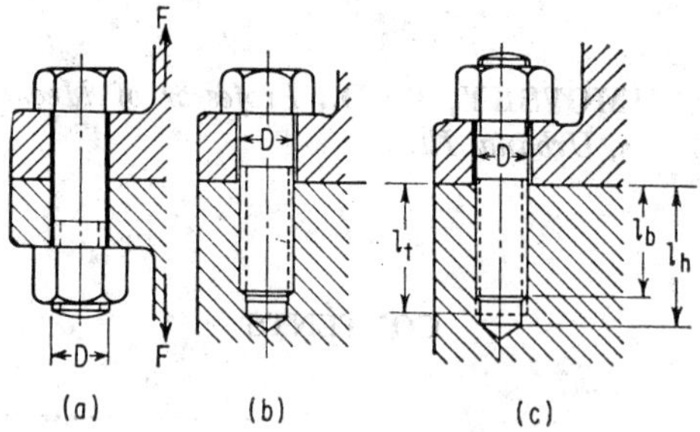

FIG. 21.1. Bolts. (*a*) Through. (*b*) Tap. (*c*) Stud.

An important disadvantage of the tap bolt is that wear may occur in the tapped material, which is not easily replaced.

The required length of a tapped hole depends upon the material. For a steel bolt or stud it is common practice to use the following dimensions. Referring to Fig. 21.1*c*:

$$\text{Length of threaded bolt portion} = l_b = C_1 D, \text{ in.}$$
$$\text{Length of tap} = l_t = C_1 D + \tfrac{1}{8} \text{ to } \tfrac{1}{4} \text{ in.}$$
$$\text{Length of blind hole} = l_h = C_1 D + \tfrac{1}{4} \text{ to } \tfrac{1}{2} \text{ in.}$$

where C_1 = 1.0 for steel, 1.5 for cast iron and bronze, and 2.0 for aluminum.

An oversize tap drill should be used for softer metals or the tap will roll the material, forming a deeper thread.

21.2. MATERIALS AND STRENGTH

(a) General

In addition to strength, materials used for threaded fasteners must often meet specifications related to corrosion resistance, magnetic properties, electrical conductivity, thermal conductivity, and cost.

The majority of bolts, however, are made of steel. Standard specifications of steels used for threaded fasteners cover a very broad range of mechanical properties. Table 21.1 lists properties of various SAE grades of steel. Commercial screws and bolts are made of low-carbon steels: SAE 1010, 1018, 1025, 1030, and 1040. Cap screws have been made of C1335, 3135, 4047, and 8635, and studs of 1035 to 1045. Heat-treatable alloy steels are the 4700, 8600, and 8700 series. Automatic screw machines use the free-cutting 1100 series. AISI 430 has been popular as a corrosion-resistant metal. High-strength bolts with hardened washers having the same dimensions as standard (ASTM A325) are being used in large structures.

Table 21.1. Properties of ASTM and SAE Grades of Steel Bolts and Cap Screws

SAE grade	Steel designation	Bolt size diam, in.	Yield strength σ_y, psi	Ultimate tensile strength (min) σ_u, psi	Hardness (Brinell)	Comments
0	Low-carbon	All sizes	....	....	....	General use
1	ASTM A307 low-carbon commercial	All sizes	....	55,000	207 max	Cold or hot heading
2	Low-carbon bright finish 0.28C, 0.04P, max 0.05S	Up to ½ incl. Over ½ to ¾ incl.	55,000 52,000	69,000 64,000	241 max	Cold-headed product. Automotive applications
3	Medium-carbon 0.28 to 0.55C, max 0.04P, max 0.05S. May be aged to 700°F to suit	Up to ½ incl. Over ½ to ⅝ incl.	85,000 80,000	110,000 100,000	207–269	Cold worked by heading or roll threading. Used for high fatigue strength
4	Commercial	Over ¾ to 1½ incl.	28,000	55,000	207 max	
5	ASTM A325 medium-carbon 0.28 to 0.55C, max 0.04P, max 0.05S, quenched and tempered at 800°F	Up to ¾ incl. Over ¾ to 1 incl. Over 1 up 1½ incl.	85,000 78,000 74,000	120,000 115,000 105,000	241–302 235–302 223–285	Used where high preload is necessary
6	Special medium-carbon, oil-quenched and tempered at 800°F min	Up to ⅝ incl. Over ⅝ to ¾ incl.	110,000 105,000	140,000 133,000	285–331 269–331	Used where higher strength than grade 5 is required
7	Medium-carbon fine-grain alloy 0.28 to 0.55C, max 0.04P, max 0.05S, oil-quenched and tempered at 800°F min	Up to 1½ incl.	105,000	133,000	269–331	Roll-threaded after heat-treatment for improved fatigue strength
8	Same as grade 7 but higher strength	Up to 1½ incl.	120,000	150,000	302–352	

The pulling of a hardened-steel threaded mandrel through a nut provides a stripping test for nuts known as "proof strength." This strength is slightly less than the yield strength.

References 20 and 21 compare riveted and bolted joints and indicate that, with a significant clearance hole, the clamping force is the most important factor in the fatigue strength of the joint. They also indicate that, because of a somewhat better controlled preload, bolted joints have a stronger fatigue strength than riveted ones. In some cases the joint strength may be increased by application of high-strength materials for bolts or rivets.

Standard aircraft bolts have an ultimate tensile and shear strength of 160,000 and 95,000 psi, respectively. High-strength aircraft bolts have been developed having strength of 230,000 and 130,000 psi, respectively. Titanium bolts[19] are stronger and 43 per cent lighter in weight than high-strength aircraft bolts.

Table 21.2 presents properties of nonferrous fastener materials.[15]

Table 21.2. Nonferrous Fastener Materials[15]

Material	Tensile strength, psi	Yield strength, 0.5% elongation, psi	Rockwell hardness
Aluminum:			
2024-T4	55,000 min 60,000 avg	50,000	54–63 B
2011-T3	55,000 avg	48,000	50–60 B
1100	13,000 min 16,000 avg	5,000	20–25 F
Brasses:			
Yellow brass	60,000 min 72,000 avg	43,000	57–65 B
Free-cutting brass	50,000 min 55,000 avg	25,000	53–59 B
Commercial bronze	45,000 min 50,000 avg	40,000	40–50 B
Naval bronze, composition A	55,000 min 65,000 avg	33,000	45–50 B
Naval bronze, composition B	50,000 min 60,000 avg	31,000	45–50 B
Copper	35,000–45,000	10,000–37,000	40–80 F
Silicon bronze:			
High-silicon, type A	70,000 min 80,000 avg	38,000	74–80 B
Low-silicon, type B	70,000 min 76,000 avg	35,000	67–75 B
Silicon-aluminum	80,000 min 85,000 avg	42,000	76–82 B
Nickel and high nickel:			
Monel	82,000 min 97,000 avg	60,000	90 B
Nickel	68,000–82,000	20,000–65,000	75–86 B
Inconel	80,000–120,000	25,000–70,000	72 B to 24 C
Stainless steel (AISI):			
Type 302	90,000–124,000	35,000–116,000	89 B to 25 C
Type 303	90,000–124,000	35,000–116,000	89 B to 25 C
Type 304	85,000–112,000	30,000–92,000	87–98 B
Type 305	80,000–110,000	30,000–95,000	84–97 B
Type 309	100,000–120,000	40,000–100,000	94 B to 23 C
Type 310	100,000–120,000	40,000–100,000	94 B to 23 C
Type 316	90,000–115,000	30,000–95,000	89–99 B
Type 317	90,000–115,000	30,000–95,000	89–99 B
Type 321	90,000–112,000	30,000–92,000	89–98 B
Type 347	90,000–112,000	35,000–95,000	89–98 B
Type 410	75,000–190,000	40,000–140,000	81 B to 42 C
Type 416	75,000–190,000	40,000–140,000	81 B to 42 C
Type 430	70,000–90,000	40,000–183,000	77–90 B

(b) High Temperature

Three factors affecting the performance of a bolted assembly at high working temperature are the rupture strength of the locknut, relaxation occurring in the locknut and bolt, and thermal expansion.[17] The first two are predicted by test and the third can be avoided by selection of materials having similar expansion rates. For high temperatures the following are recommended material choices:

60–400°F	Aluminum alloy 7075-T6, and steel SAE 4140 and 8740
400–800°F	Alloy steels, 300–400 series, stainless steels, and titanium
Up to 900°F	Martensitic chromium steels, and 17-7PH
900°F	Superalloy 8A and 286 Inconel

Notch effects cannot be predicted except by test. High-temperature bolts have all thread dimensions reduced by 0.003 in. from standard dimensions to eliminate galling of the bolt thread in the nut, and the shape of the nut is an important factor in high-temperature service. Above 500°F bolts are not reusable.

21.3. BOLTS SUBJECTED TO STEADY LOADS

If a bolt is loaded by a steady force (time-invariant) the influence of local stress and stress concentration in the threaded portion of the bolt may be disregarded if the stress is less than the ultimate stress. In a ductile material, a readjustment of local stress will occur as a consequence of material yielding.

(a) Axial Load Only

Consider a bolt loaded by an external load only, e.g., the threaded portion of a hook used on a hoist. Only an axial load acts; i.e., no twisting moment exists in the bolt because of the initial turning of the nut. The stress area

$$A_s = \pi D_s^2/4 = F_e/\sigma_d \tag{21.1}$$

where $D_s = (D_m + D_r)/2$ = diameter corresponding to stress area, in.
D_m = major or nominal diameter, in.
D_r = root diameter, in.
$\sigma_d = \sigma_y/N$ = allowable nominal bolt stress, psi
σ_y = yield strength of bolt material, psi (see Table 21.1 for carbon steel)
N = factor of safety = 2 to 3 for carbon steels = 1.5 to 3 for alloy steels
F_e = external bolt load, lb

The strength contributed by the threads is small in comparison with other unknowns. The value of N depends on the degree of accuracy with which the magnitudes of load and mechanical properties of bolt material are known, as well as upon the quality of workmanship. For inferior workmanship, the allowable stress in Eq. (21.1) must be decreased. The allowable stress should be lowered for screws having a nominal diameter of ½ in. or smaller. The allowable nominal stress should be also lowered when a number of bolts in the joint work together and an uneven load distribution among the individual bolts is possible. For additional simplicity and safety, the root area may be used in lieu of the stress area.

(b) Tension Bolt Stresses

The permissible magnitude of the external load per bolt may be reasonably expressed by

$$F_e = C_2(A_s)^{1.42} = \sigma_w A_s \tag{21.2}$$

where F_e = external load per bolt, lb
C_2 = an empirical constant
σ_w = working stress, psi = $C_2(A_s)^{0.42}$
A_s = bolt stress area, in.^{2}

Equation (21.2) applies for bolts of ¾ in. diameter and over made of steel containing 0.08 to 0.25 per cent carbon.

For bolts 2 in. and smaller, C_2 = 5,000 for carbon steel of ultimate strength

60,000 psi, increasing in proportion to the ultimate strength to 15,000 for alloy-steel bolts. For bronze bolts, $C_2 = 1{,}000$.

For bolts 2 in. and larger, C_2 is increased by 40 per cent.

ASME Boiler and Pressure Vessel Code[2] recommends a method by which bolt stress is approximately determined by the bolt material regardless of bolt size, for assemblies where leak prevention as well as strength must be considered. The code lists the values for permissible stresses for common bolt materials at temperatures from −20 to +1100°F. The maximum axial load acting on the bolt is assumed to be greater than the externally applied load plus the initial tightening force necessary to prevent leakage under the operating conditions. This axial load is assumed to be also larger than the initial tightening force required to "seat" the contact surfaces of the joint without the external load. The code contains data concerning the properties of gasket materials necessary for calculating these initial tightening forces.

(c) Bolts Subjected to Initial Tightening

When the nut shown in Fig. 21.1 is tightened, the friction between the nut and screw threads causes a twisting moment in the bolt which results in principal tensile and shear stresses.

Because of the many factors influencing this moment, it is convenient in practice to use the following simple relationship for the twisting moment M:

$$M = C_3 F_i D_m \qquad \text{in.-lb} \tag{21.3}$$

where C_3 = an experimentally determined coefficient
F_i = tightening-up load, lb
D_m = bolt nominal diameter, in.

Lacking experimental data, the following average values of C_3 may be used for conventional bolt threads:[3]

Well-lubricated, smooth surfaces, $C_3 = 0.10$
Unlubricated, smooth surfaces, $C_3 = 0.12$
Well-lubricated, surfaces not smooth, $C_3 = 0.13$–0.15
Unlubricated, surfaces not smooth, $C_3 = 0.18$–0.20

The values of C_3 given are for unplated steel bolts with the threads cut (not rolled or ground).

Wrench torque overcomes frictional resistance between the nut and bolt, and between the nut and supporting surface; therefore, the wrench torque may be 50 to 100 per cent greater than moment M.

In particular cases where the bolts are highly stressed, it is desirable to remove the torsional moment by turning the nut back through a small angle after it has been tightened. If this is done after the nut is locked (by a pin driven through the nut and bolt), the initial bolt tension is unaffected.

In general, the influence of the torsional stress on the bolt strength may be taken into account by decreasing the allowable nominal stress σ_d by 25 to 30 per cent. Equation (21.1) then takes the form

$$A_s = \pi D_s^2/4 = F_i/0.75\sigma_d \tag{21.4}$$

where F_i = initial tightening induced by the nut, lb

In metal-to-metal surfaces, the tightening force should be somewhat greater than F_e. If $F_e > F_i$, then F_e should be used in Eq. (21.4).

Equation (21.4) should be employed only for screws with nominal diameters ¾ in. or larger. For diameters less than ¾ in., the tightening moment depends to such a great extent upon the judgment and experience of the mechanic that the calculation of the actual combined stress in the thread is almost impossible.

The tightening moment may be established by means of a torque wrench, by measurement of the actual elongation of a through bolt, or by inducing a predetermined strain through temperature control. These methods may be employed to achieve the desirable initial tightening and to predict, if only approximately, the nominal stresses. The approximations arise in the unknown friction between the nut and the supporting surface.

Special care must be exercised if consistent results are to be achieved when using the torque wrench. Threads must be in a good condition and the fit must be such that the nut may be screwed up snug by finger pressure alone. Good graphite grease is suggested. Where faulty lubrication occurs, the variation in bolt tension may be as high as 10:1.[23]

(d) Combined Initial-tightening Load and Applied Load

In the discussion which follows, reference is made to Fig. 21.2, in which a gasketed bolted joint is shown. The bolt has been tightened to load F_i, and an external load F_e is applied to the joint. In order to size the bolt properly, it is necessary to calculate the resultant bolt load

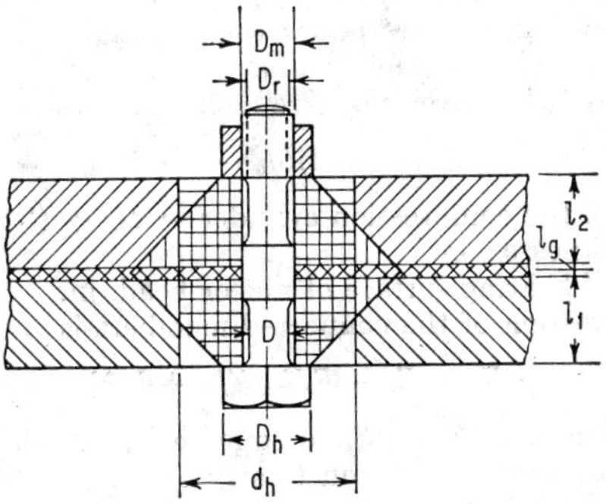

FIG. 21.2. Assembly of two plates and a gasket.

Before application of the external load, the bolt is in tension because of the tightening load F_i. F_e therefore causes an increase in the bolt deformation. The gasket (and other connected parts) undergo a decrease in deformation because of the compression relief associated with F_e.

The change of bolt (and gasket) length δ associated with the application of external load F_e (considering only bolt and gasket)

$$\delta = \frac{F_e}{E_g A_g / l_g + E_b A_b / l_b} \tag{21.5}$$

where E_g = modulus of elasticity of gasket, psi
E_b = modulus of elasticity of bolt, psi
l_g = gasket thickness, in.
l_b = head to nut bolt length, in.
A_g = gasket load-carrying area, in.^{2}
A_b = bolt-cross-sectional area, in.^{2}

The resultant bolt load is therefore

$$F_t = \frac{\delta E_b A_b}{l_b} + F_i = F_e \frac{E_b A_b / l_b}{E_g A_g / l_g + E_b A_b / l_b} + F_i$$

$$= F_e \frac{k_b}{k_b k_g} + F_i = K F_e + F_i \tag{21.6}$$

where k_g = gasket stiffness, lb/in.
k_b = bolt stiffness, lb/in.

The magnitude of K in Eq. (21.6) lies between zero and unity. Where soft gasketing material is used, the gasket stiffness may become so small in comparison with the bolt stiffness as to cause K to approach unity. A hard thin gasket of large area results in a high gasket stiffness with corresponding diminution of K. The gasket elasticity approaches zero as its thickness diminishes, and therefore k_g is infinite in the absence of a gasket. In this case, $K = 0$.

Equation (21.6) may be generalized to include any connected members

$$F_t = F_e(k_b/k_m k_b) + F_i \tag{21.7}$$

where k_m is the resultant stiffness of the connected members (including gasket, if any), defined by

$$1/k_m = 1/k_1 + 1/k_2 + 1/k_3 + \cdots + 1/k_n \tag{21.8}$$

where $k_1, k_2, k_3, \ldots, k_n$ are the stiffness constants EA/l of the connected members.

If the stiffness constant of a flange is to be evaluated, the value of A used should be the effective cross-sectional area of the flange portions undergoing deformation. These cross sections can be represented by compression cones shown in Fig. 21.2, which intersect the bearing surfaces under the nut and head at approximately 45° angles. The stiffness of a double cone can be determined approximately by replacing the cones with a hollow cylinder of effective cross-sectional area.

$$A_c = (\pi/4)(d_h^2 - D_m^2) \tag{21.9}$$

where d_h = diameter of the compression cylinder = $D_h + (l_1 + l_2)/2$, in.
l_1, l_2 = flange thicknesses, in.
D_h = diameter of bearing surface of nut or head, in.

The resultant compression of the connected members is

$$F_g = F_i - F_e[k_m/(k_b + k_m)] \tag{21.10}$$

If F_e becomes so large as to cause F_g to equal zero, i.e., the joint is unloaded, the joint members separate, and the bolt must carry the entire load F_e.

Representative values of K are given in Table 21.3.

Table 21.3. Stiffness of Bolted Assembly

Type of joint	Ratio $\frac{k_g}{k_b + k_g}$	Ratio $\frac{k_b}{k_b + k_g}$
Soft packing with studs	0.00	1.00
Soft packing with through bolts	0.25	0.75
Asbestos gasket	0.40	0.60
Soft-copper gasket with long through bolts	0.50	0.50
Hard-copper gasket with long through bolts	0.75	0.25
Very rigid metal-to-metal joint with long through bolts	1.00	0.00

When the assembly of machine members is complex in form (e.g., the bolted assembly of a connecting rod), the stiffness constants must be determined experimentally.

(e) Bolts Subjected to External Shear Load

When small transverse forces act on the bolted assembly, fitted bolts in reamed holes may be used. For larger forces, taper bolts (Fig. 21.3a) with tan α = 1/20 to 1/10 are used. However, it is usually better to relieve the transverse force by the use of special members, such as the cylindrical key in Fig. 21.3b. Preloading bolts so that the frictional resistance between the joints exceeds the shear load is another commonly used method. In this case the bolt load is equal to the initial tightening load only, since the local stresses due to the pressure between the hole wall and the bolt shank do not exist. Dowel pins in reamed holes are also used to take shear. In addition, a knurled bolt body has been successfully used.

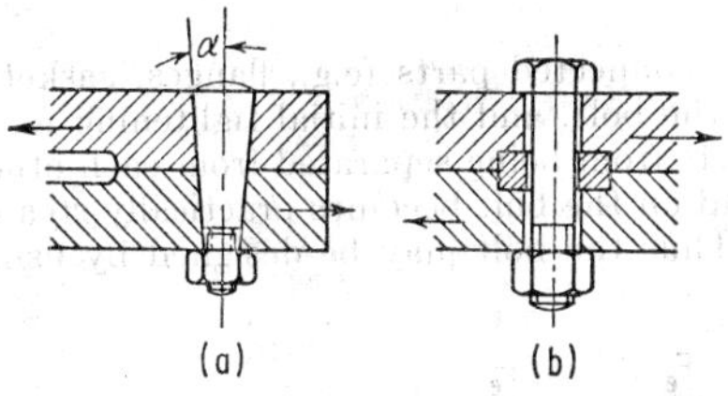

FIG. 21.3. Bolted assemblies for transverse force. (a) Taper bolt. (b) Cylindrical key.

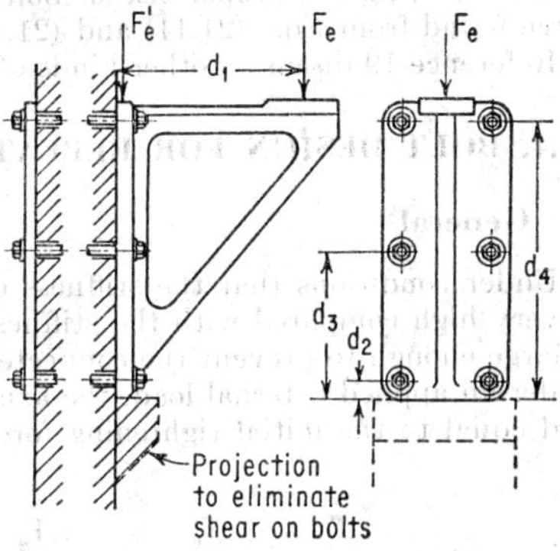

FIG. 21.4. Bracket subjected to eccentric load.

(f) Bolted Joints with Eccentric Load

In Fig. 21.4 a cast-iron bracket, which is connected by six bolts to a wall, is shown. An external force F_e applied at the distance d_1 from the wall may be replaced by moment $M_e = F_e d_1$ and force F_e' ($F_e' = F_e$). The moment tends to rotate the bracket clockwise about its lower edge, while the force F_e' tends to move the bracket downward along the wall surface. The projection shown prevents downward motion of the bracket so that no direct shear load acts on the bolts.

When the bracket flange is heavy and the bolts are relatively long, as in the case presented in the figure, it is reasonable to assume that the wall and the bracket flange are very rigid when compared with the bolts. In this case, as long as the individual bolt load is less than its initial tightening force, the load on each bolt is equal to the tightening load. In practice, all bolts are of the same size and are tightened with approximately the same initial force

$$F_i = \frac{C_4 F_e d_1}{2(d_2 + d_3 + d_4)} \tag{21.11}$$

where C_4 is a coefficient larger than unity to assure that the moment produced by bolt tightening is larger than the external moment M_e, and the bracket does not separate from the wall. This coefficient C_4 may be from 1.2 to 2 depending upon the accuracy with which the external force is known, the uniformity of initial tightening among the individual bolts, and the type of external load F_e. Larger values for C_4 should be used for variable external loads. If the initial tightening force is determined from Eq. (21.11) the bolt size can be selected by using Eq. (21.4).

Bolts in a bolted connection of this kind should not be subjected to direct shear. If there is no projection in the wall preventing the bracket from sliding, the frictional force between the wall and the bracket surfaces must be larger than the external force which tends to move the bracket along the wall. Thus

$$F_i = C_5 F_e / \mu n \tag{21.12}$$

where μ = coefficient of friction between wall and bracket surfaces
n = number of bolts in connection
C_5 = a coefficient larger than unity to provide a frictional force larger than external force

This coefficient C_5 may be between 1.5 and 3 depending upon the accuracy with which the values of external loads and coefficient of friction are known, and the uniformity of tightening of bolts. The larger values for C_5 should be used in cases when F_e is variable or when vibrations may occur.

For selecting the proper size of bolts, Eq. (21.4) is again applicable. The larger force found from Eqs. (21.11) and (21.12) must be substituted in Eq. (21.4).

Reference 19 discusses other kinds of eccentric loads.

21.4. BOLT DESIGN FOR REPEATED LOADING

(a) General

Under conditions that the stiffness of the connected parts (e.g., flanges, gaskets) is very high compared with the stiffness of the bolt, and the initial tightening force is large enough to prevent the connected parts from being separated from each other under an applied external load, the axial load on the bolt becomes practically steady and equal to the initial tightening force. Thus the bolt may be designed by using

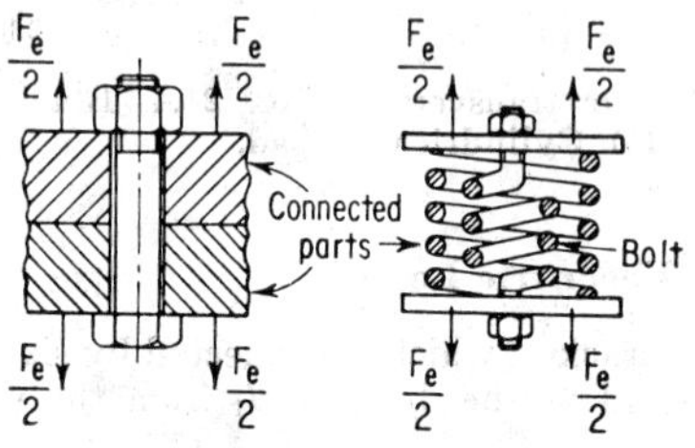

FIG. 21.5. Bolted joint (F_e = external load). (*a*) Bolted assembly. (*b*) Elastic equivalent.

the methods for Art. 21.2, even if the bolt connection is actually subjected to periodically changing loads. However, these methods become unsatisfactory if the compressive stiffness of the connected members is of the same order as the tensile stiffness of the bolt and a variable external force exists.

The tightened bolted assembly can be represented schematically as a system comprised of two springs (Fig. 21.5). One of these springs represents the connected parts (including gasket) and is compressed.

(b) Bolt Loads

The additional symbols defined below pertain to the discussion which follows:

F_v = additional tension force acting on the bolt due to external force F_e, lb

$F_{i(c)}$ = critical tightening load, the smallest magnitude of F_i necessary to prevent separation of the connected parts when F_e is applied, lb

$R = F_{e\min}/F_{e\max}$ = range ratio of the load cycle, the algebraic ratio of the externally applied load

K = stress-concentration factor (effective), the ratio of the fatigue strength of the polished specimen containing no discontinuities to the strength of the given member of the same material and having the same critical cross-sectional area. In determining the factor K, the same loading-range ratio R must be applied to both the polished specimen and the member in question

K' = thread-concentration factor (effective)

K'' = fillet-concentration factor (effective)

K_{-1} = stress-concentration factor (effective) for alternate loading ($R = -1$)

F_t = total axial load on bolt, lb

σ_R = endurance limit, for a steel specimen with a notch representing threaded part of bolt, for an asymmetrical loading cycle with range ration R, psi

σ_{-1} = endurance limit for bolt material for symmetrical reversed cycle of stress, as obtained from tests of polished notch-free specimens, psi

σ = nominal bolt stress, psi

$\sigma_m = (\sigma_{max} + \sigma_{min})/2$ = nominal stress at the critical bolt cross section due to steady (mean) component of the load, psi

$\sigma_v = (\sigma_{max} - \sigma_{min})/2$ = nominal stress due to the alternate component of the load, psi

τ = nominal shearing stress due to steady twisting moment, psi

The critical tightening load

$$F_{i(c)} = F_e[k_m/(k_b + k_m)] \tag{21.13}$$

In order to avoid impact conditions, which will occur if the joint is opened by F_e, and to ensure against leakage, the initial tightening force

$$F_i = C_6 F_{i(c)} = C_6 F_e[k_m/(k_b + k_m)] \tag{21.14}$$

where C_6 usually falls between 1.2 and 1.5 but can be greater depending upon the type of assembly and upon the accuracy with which F_e, k_m, and k_b can be determined.

The additional force F_v acting on the bolt as a result of external force F_e

$$F_v = F_e[k_b/(k_b + k_m)] \tag{21.15}$$

The total axial force acting on the bolt

$$\begin{aligned} F_{tmin} &= F_i + F_{vmin} = C_6 F_{emax}[k_m/(k_b + k_m)] + F_{emin}[k_b/(k_b + k_m)] \\ F_{tmax} &= F_i + F_{vmax} = C_6 F_{emax}[k_m/(k_b + k_m)] + F_{emax}[k_b/(k_b + k_m)] \end{aligned} \tag{21.16}$$

The range ratio of cycle of the load on the bolt

$$R = \frac{F_{tmin}}{F_{tmax}} = \frac{C_6 F_{emax} k_m + F_{emin} k_b}{F_{emax}(C_6 k_m + k_b)} \tag{21.17}$$

If the external force varies from zero to F_e, then $F_{tmin} = F_i$, and Eq. (21.17) reduces to

$$R = C_6 k_m/(C_6 k_m + k_b) \tag{21.18}$$

(c) Stress Concentration and Local Stresses

Loaded threads experience a complex combination of nonuniform bending, shear, and compressive stress. The effect of this combined stress is summarized in the effective stress-concentration factors K which will be applied. Any fluctuating load (as well as corresponding nominal stress) without impact can be interpreted as a combination of a steady (mean) load ($R = +1$) and an alternate load ($R = -1$). For static loading, the factor $K_{+1}(R = +1)$ can be assumed equal to 1. In the case of alternate loading, the approximate values of thread effective stress-concentration factor K_{-1}' are shown in Table 21.4. Values in Table 21.4 refer to bolts with the

Table 21.4. Thread Effective Stress-concentration Factor K_{-1}' for Alternately Loaded Bolt[3]

Thread form	Medium-carbon steels	Heat-treated alloy steels
Whitworth thread and Unified thread with rounded root..................	3.2–3.8	5.4–6.0
Metrical standard DIN and OST.......	4.4–5.0	5.6–6.4
Sellers (American National) form with flat root..........................	5.0–5.8	6.4–7.2

threads formed by cutting. When the threads are not heat-treated after forming and are formed either by rolling or by rolling after initial cutting, the values for K_{-1}' are lower than those given in the table. The percentage decrease in the values shown is medium-carbon steel 15, ductile alloy steels 20, and hard alloy steels 35. The lower range values in Table 21.4 are for softer and more ductile materials.

Nonuniformity of the stress distribution also occurs at the junction between the shank and the head and at other locations where the diameter changes. Figure 21.6 presents fillet stress-concentration factors K_{-1}''.

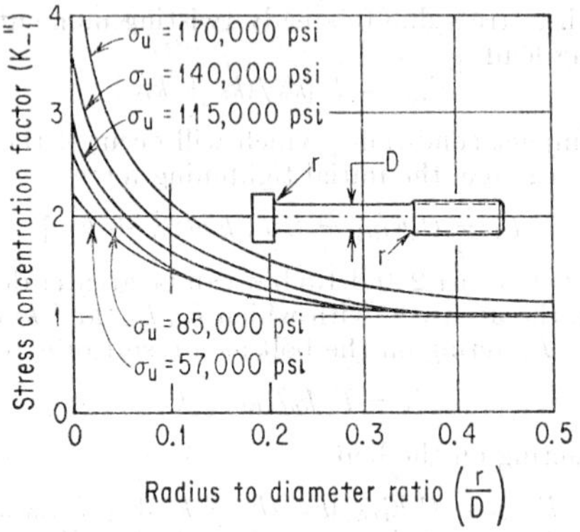

FIG. 21.6. Bolt fillet stress-concentration factors K_{-1}'' (σ_u = ultimate tensile stength).

Using the Goodman diagram for an asymmetrical loading cycle, the endurance limit for the given nominal stress range is given by[3]

$$\sigma_R = \sigma_m + \sigma_{-1}/K_{-1} \tag{21.19}$$

If only the tension forces act on the bolt, the factor of safety is determined by[3]

$$n = \frac{\sigma_R}{\sigma_{\max}} = \frac{\sigma_m + \sigma_{-1}/K_{-1}}{\sigma_m + \sigma_v}$$

(d) Nut Loosening by Vibration[22]

In a bolted joint, the nut is under compressive loading while the bolt is loaded in tension. Increasing the bolt load results in an increase in longitudinal strain and a small decrease in bolt diameter because of the Poisson's ratio effect. Corresponding to the increased tensile bolt load is an increased compressive nut load which causes a slight increase in the nut diameter. The decreased bolt diameter and increased nut diameter (nut flattening) cause a radial slippage of the engaged thread. It is not purely radial because of the circumferential component of the normal thread pressure.

The net effect of each vibration cycle is a small amount of loosening called *frictional ratchet*. Where the applied load variation and a relatively soft gasket material result in large bolt-load variation, special care must be exercised to prevent nut loosening. To reduce the possibility of nut loosening by vibration, the amount of stretch in tightening should be as great as possible.

Since threaded fasteners lose their effectiveness and purpose when they become loosened by vibration, continuous use, or other reasons, auxiliary locking devices have been established. These devices are of two types, friction and positive engagement. Figure 21.7 shows such devices. *a*, *d*, *e*, *i*, and *j* rely on friction, and *b*, *c*, *f*, *g*, and *h* are of positive engagement. The most positive device is *c*, where a cotter pin is placed through the bolt and nut. The most recent positive fastening device is a self-aligning nut that reduces the stress concentration which would be present in the initial thread as shown in Fig. 21.7*c*. Thread inserts have some advantages over other types. This is discussed in ref. 27. For a discussion on locknut devices as shown in Fig. 21.7, ref. 27 is recommended. For a guide to locknut performance of critical application, aircraft specifications AN-N-5 and AN-N-10 are recommended.

The trend to higher-strength fasteners is discussed in ref. 28. Tensile strengths of over 100,000 psi are not uncommon.

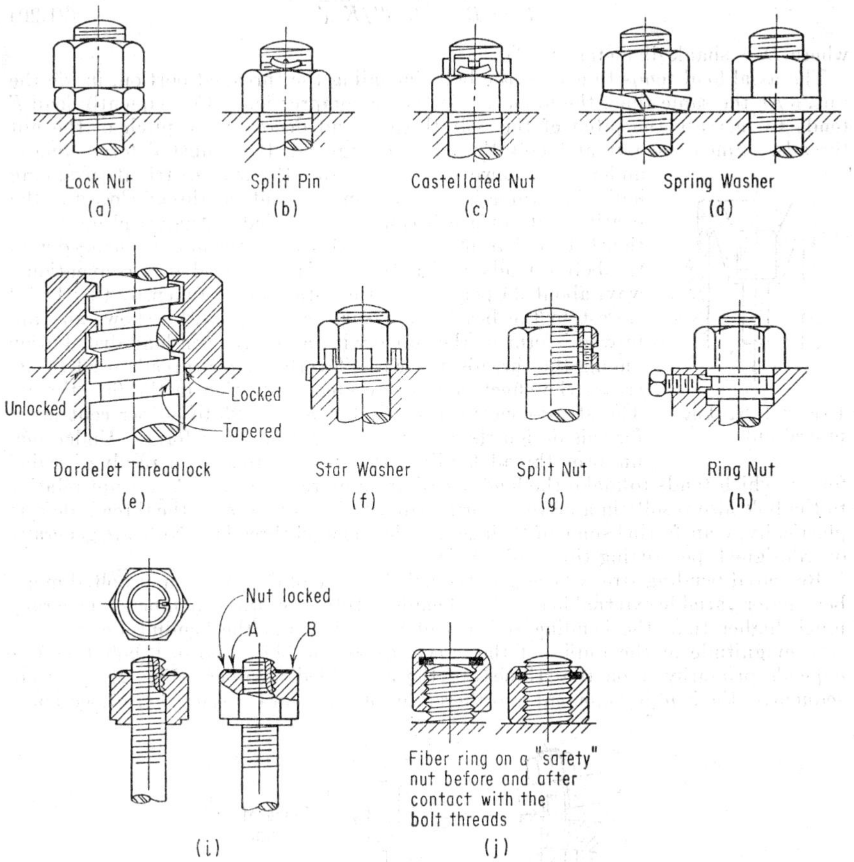

FIG. 21.7. Locking devices.

21.5. IMPROVING BOLT PERFORMANCE AND RELIABILITY

In some cases the reliability of a bolted assembly can be increased considerably without changing the basic diameter of the bolt or increasing the size or weight of the entire assembly. This can be accomplished by (1) decreasing the stress-concentration factor in the threads and other critical cross sections of the bolt, (2) decreasing the longitudinal stiffness of the bolt, (3) improving the load distribution among the individual threads of the nut, (4) reducing bending stresses in the bolt arising from the deformation of the assembled members, and (5) controlling the bolt preload. The stress-concentration factor can be decreased by using rolled threads and by increasing the fillet radii at the junction between the shank and the head and at other places where the diameter changes. Longitudinal bolt stiffness can be decreased by decreasing the shank cross-section area. This method is limited because of the accompanying increase in shank stress. Theoretically, the highest bolt reliability is achieved when the safety factor at the critical section in the threaded portion of the bolt and that of the critical section of the shank are equal. For cases in which the external load acting on the assembly changes from zero to maximum, and the bolt is unloaded from the torsional moment, the safety factors in the two critical cross sections become equal when[3,6]

$$D = D_r \sqrt{K_{-1}''/K_{-1}'} \tag{21.20}$$

where D = shank diameter, in.

The axial load tends to extend the bolt, including the threaded portion, inside the nut. At the same time, the nut is subjected to compression. Consequently load F tends to increase the pitch of the bolt threads and decrease the pitch of the nut threads. Since the nut and bolt threads are engaged, they must deform equally under load, however. As a result the load distribution among individual threads becomes uneven and the thread closest to the bearing nut surface becomes overloaded. Analytical investigations[5] show that, if six or more threads in the nut are engaged with the bolt threads, and if the nut is supported in a conventional way, about 35 per cent of the total load is taken by the thread closest to the bearing nut surface. Figure 21.8 shows an "improved" form of the nut[5,7] with a nearly even load distribution among the threads and with "overlapping" threads used to decrease the effect of the stress concentration in the first thread. The stress-concentration factor K_{-1}' is 25 to 30 per cent lower for this design than for the conventional nut form. Under nonuniform thread loading, the nut expands as a result of radial forces, which tends to make the load distribution more uniform. A soft nut relative to the bolt also results in a more uniform load distribution because the threads deform plastically, transferring some of their load to less loaded threads. Nuts are generally overdesigned, permitting this choice.

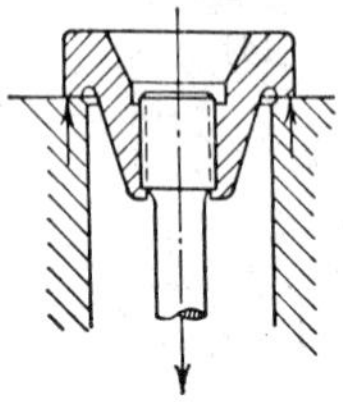

Fig. 21.8. Improved nut.

Repeated bending stresses occur in the bolt because of deformation of bolted members under variable external load. The bending stiffness of these members is generally much higher than the bending stiffness of the bolts, and the bending deformation (the magnitude of the radius of the curvature of the bolt axis) of a bolt therefore depends primarily upon the stiffness of the assembled members (Fig. 21.9). Consequently, the bending moment and therefore the stress in the critical cross section of the bolt (usually in the threaded portion) depend upon the moment of inertia of the shank cross section, and it is highly desirable to decrease this moment of inertia as much as possible. The actual radius of curvature cannot be readily calculated in most cases. It is therefore difficult to consider the nominal bending stress when determining the safety factor analytically. However, its influence on the strength of the assembly may be rather significant.[30] The cross section of the shank shown in Fig. 21.9*b* is a good design because the bending stresses in the bolt cross sections and the longitudinal stiffness of the bolt shank are reduced if the bending moment acts in one definite plane ($x - x'$) relative to the bolt.[3,6]

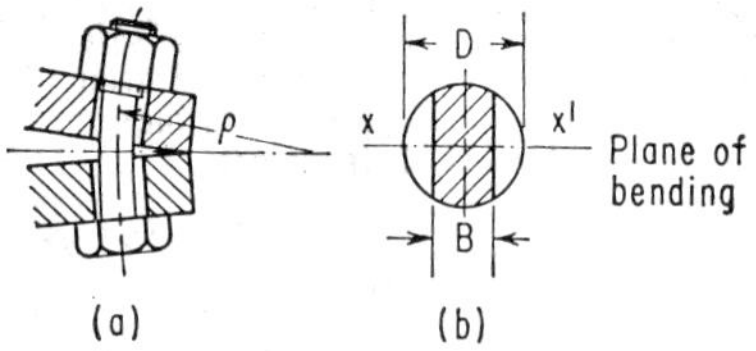

FIG. 21.9. Bolts in bending. (*a*) Deformation in assembly. (*b*) Bolt shank cross section for reducing bending moment in the bolt.

In general, if the load producing the deflection of an assembly is a variable load, any precaution which tends to decrease the bending stresses in the critical sections of the bolt may considerably increase the reliability of a joint. Control of preload is quite significant in bolt reliability. Preloading should be maintained throughout the life of the assembly. An assembly tends to relax or reduce its initial preload and torsional and shear stress in service. This is due to surface high points and flow of metal at these points of high stress and the crushing of bolted material (especially when hard fasteners are used on softer bolted materials). Adequate base area is necessary. As mating surfaces adjust themselves, relaxation reaches a stable value lower than the initial preload. Since the amount of relaxation is fixed, this factor becomes more important in short stiff bolts than in longer, more resilient ones.

Preloading can generally reduce the problem of stress changes in bolts so that fatigue is not significant. When excessive amplitude of bolt stress change exists, preloading may not prevent loosening. In such cases locking devices are necessary. Increasing the thread length somewhat increases the fatigue strength and makes a bolt more resistant to loosening.

All locking devices used to combat the loosening effect of vibrations and shock loads are based on increasing frictional drag between contacting surfaces. Liquid threadlocking agents have been applied on thread surfaces to develop different locking strengths. Thermosetting plastics[16] have also been used.

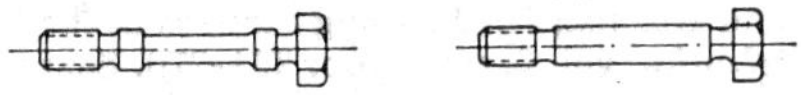

FIG. 21.10. Bolts designed for shock and creep. (*a*) Shock. (*b*) Creep.

Figure 21.10 shows bolts designed for shock and creep. Both have a smooth change in cross section. Figure 21.10*a* for shock design has short lengths of normal shank diameter provided for location in boltholes. Also, the shank of the bolt is reduced to the core diameter of the thread to provide better resilience. Figure 21.10*b* shows a bolt designed for creep. The large shank diameter results in lower stresses and thus lower creep rates than the previous bolt.

References

1. Seaton, A. E., and H. M. Routhwaite: "Marine Engineer's Pocket Book," 16th ed. Charles Griffin & Company, Ltd., London, 1914.
2. ASME Boiler and Pressure Vessel Code, Sec. VIII, 1952 ed.
3. Radzimovsky, E. I.: Bolt Design for Repeated Loading, *Machine Design*, November, 1952.
4. Rotscher, F.: "Maschinenelemente," Springer-Verlag OHG, Berlin, 1929.
5. Maduschka, L.: *Forsch. Gebiete Ingenieurw.*, vol. 7, no. 6, 1936.
6. Radzimovsky, E. I.: "Schraubenverbindungen bei veränderlicher Belastung," Manu Verlag, Augsburg, 1949.
7. Wiegand, H., and B. Haas: "Berechnung und Gestaltung Schraubenverbindungen," Springer-Verlag OHG, Berlin, 1940.
8. Almen: On the Strength of Highly Stressed Dynamically Loaded Bolts and Studs, *Diesel Power*, vol. 24, August, 1946.
9. Dolan, T. J., and J. H. McClow: The Influence of Bolt Tension and Eccentric Tensile Loads on the Behavior of a Bolt Joint, *Proc. SESA*, vol. 8, no. 1, 1950.
10. Field, J. E.: Fatigue Strength of Screw Threads, *Engineer*, vol. 198, July, 1954.
11. Boomsma, M.: Loosening and Fatigue Strength of Bolted Joints, *Engineer*, vol. 200, August, 1955.
12. Stewart, W. S.: Properties of Preloaded Steel Bolts, *Prod. Eng.*, vol. 24, November, 1953.
13. Boomsma, M.: Strength Calculation of Bolted Joints, *Engineer*, vol. 203, May, 1957.
14. "ASME Handbook, Metals Engineering—Design," p. 126, McGraw-Hill Book Company, Inc., New York, 1953.
15. "Machine Fasteners Book," Penton Publishing Company, Cleveland, 1960.

16. Krieble, R. H.: New Developments in Thread Locking Agents, *Machine Design*, vol. 29, no. 19, pp. 129–134, Sept. 19, 1957.
17. Selwones, R. S., and R. A. Degen: High Temperature Design of Bolted Assemblies, *Prod. Eng.*, vol. 28, no. 12, pp. 79–83, Sept. 30, 1957.
18. Viglione, J.: Strength of Titanium Bolts, *Prod. Eng.*, vol. 26, no. 4, pp. 129–133, April, 1955.
19. Shigley, J. E.: "Machine Design," McGraw-Hill Book Company, Inc., New York, 1956.
20. Baron, F., and E. A. Larson: Comparative Behavior of Bolted and Riveted Joints, *Trans. ASME*, ser. E 26, pp. 285–290, June, 1959.
21. Carter, J. W., K. E. Lenzen, and L. T. Wyly: Fatigue in Riveted and Bolted Single Lap Joints, *Proc. ASCE*, vol. 85, no. ST3, pp. 7–28, March, 1959.
22. Goodier, J. N., and R. J. Sweeney: Loosening by Vibration of Threaded Fastenings, *Mech. Eng.*, vol. 67, pp. 798–802, December, 1945.
23. Almen, J. O.: Tightening Is a Vital Factor in Bolt Endurance, *Machine Design*, vol. 16, pp. 158–162, February, 1944.
24. Almen, J. O.: Fatigue Durability of Prestressed Screw Threads, *Prod. Eng.*, vol. 22, pp. 153–156, April, 1951.
25. Ollis, R., Jr.: Self Aligning Nuts, *Machine Design*, June 21, 1962, pp. 176–179.
26. Wolfe, P. C.: Threaded Inserts, *Elec. Mfg.*, January, 1954, pp. 120–123.
27. Feroni, C. C.: Fundamentals of Selecting Lock Nuts, *Prod. Eng.*, December, 1953, pp. 177–179.
28. Stewart, W. C.: Mechanical Fasteners, *Machine Design*, September, 1954, pp. 220–224.
29. Vallance, A., and V. L. Doughtie: "Design of Machine Elements," 3d ed., McGraw-Hill Book Company, Inc., New York, 1951.
30. Radzimovsky, E. I., and Kasuba, R.: Bending Stresses in the Bolts of a Bolted Assembly, *Experimental Mechanics*, vol. 2, no. 9, pp. 264–270, September, 1962.

Section 22

RIVETS AND RIVETED JOINTS

By

MARSHALL HOLT, Ph.D., *Chief, Mechanical Testing Division, Aluminum Company of America, Alcoa Research Laboratories, New Kensington, Pa.*

CONTENTS

22.1. INTRODUCTION

A rivet is a fastening device which is secured by distortion or upsetting of the shank and ends. This definition is used to differentiate rivets from threaded fasteners, nails, and screws, but it also includes upset pins with enlarged ends. Rivets have been made of many materials; soft steel and aluminum alloy are the most common.

Figure 22.1 shows some types of cold-formed fastening devices. Figures 22.1*a* and *b* show a *solid rivet* and a *tubular* rivet, respectively. Tubular rivets can be used to advantage in thin material which would buckle under the action of solid rivets of small diameter. Since the strength of the tubular shank may be limited by the buckling strength of the thin wall, the strength of such rivets should be determined by test rather than computation. In driving, a mandrel should be forced into the rivet to expand it (called *setting*) into close contact with the pieces being jointed before the ends are flared to form the heads. Split or bifurcated rivets and eyelets are similar. An eyelet has a through hole and is applied where great strength is not required. Figure 22.1*c* shows a staked joint which is often used as a connection to thin parts. *Metal stitching* is the stapling of sheet metal and other softer materials with hard wire. Production has been as high as 300 stitches per minute with $\frac{1}{16}$-in. steel plate. Figure 22.1*d* shows a flat-clinch type and ref. 23 provides strength and pitch information.

For those special cases where both sides of the work are not accessible, the rivet must be inserted and driven from the same side. Several varieties of "blind rivets" have been developed. The manufacturer's recommendations for hole size and grip length should be strictly followed. Many of the rivets, in addition to their blind-driving characteristics, offer the additional advantage that they can be driven rapidly by only one workman. Blind rivets are usually of a tubular nature. Driving may be accomplished by drawing a preset mandrel either completely or partially through the tube, thus upsetting the end. In the latter case, the connection can be made pressuretight. An explosive force, detonated by heat, may also be used.

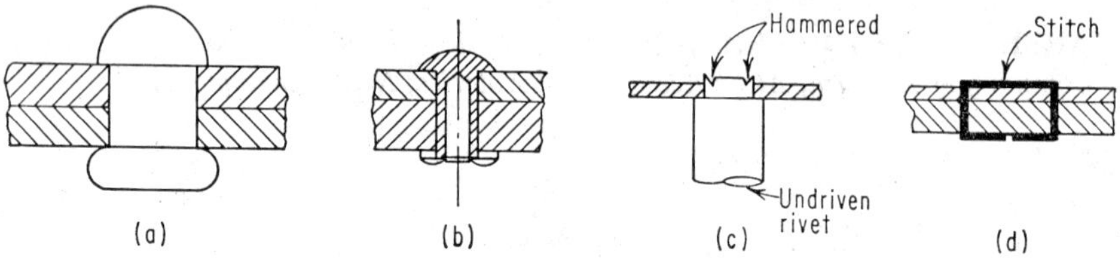

FIG. 22.1. Types of rivets and fasteners. (*a*) Solid rivet. (*b*) Tubular rivet. (*c*) Staking. (*d*) Metal stitching.

22.2. GENERAL COMMENTS

(a) Head Forms and Driving Pressure

The shapes and proportions of many of the heads commonly used in various applications are shown in Fig. 22.2.

The tinners' rivets are similar to flat-head rivets.[22] The 100° flat-top countersunk head is used almost exclusively in aircraft and the cone point, No. 9, is described in a tentative specification of the American Institute of Bolt, Nut and Rivet Manufacturers and is used extensively for structural-size rivets of high-strength materials. The annular point, No. 10, was developed by the builders of the aluminum bridge at Arvida, Canada, and is covered by a British standard for large aluminum-alloy rivets. It shows considerable advantage for upsetting high-strength materials.

The effort required to form the driven head depends on the shape and diameter of the head, the material, and the temperature at the time of driving.

The force required for cold forming buttonheads on mild-steel rivets can be estimated on the basis of a stress of 150,000 psi and the diameter of the head. For example, the forming of a buttonhead on a $\frac{3}{4}$-in.-diameter rivet requires a force of about 200,000 lb. For hot driving (1600 to 1900°F), the driving pressures are only about 70 per cent as great. The pressures required to form heads on rivets of other materials can be found in the same way.

The high pressure required to develop some head forms may result in bulging or even fracturing of the edge of the workpiece or buckling of the workpiece between rivets, especially in the case of thin material. The feasibility of some combinations of conditions should be evaluated by test. Undersized hammers should be avoided because they tend to peen the rivet rather than upset it. Special squeeze riveters have been built capable of cold-driving 3-in.-diameter steel rivets. Since the force varies roughly as the square of the diameter of the rivet head, a small driven head such as a buttonhead may have an advantage over a mushroom head. The cone-point head can be formed with from 20 to 50 per cent of the force required to form the buttonhead.

In many cases, the malleability of the rivet can be increased and the driving force reduced by hot driving. Care must be taken to control the temperature properly so that there will be no adverse effect on the strength or other characteristics of the material. Rivets of materials subject to natural age hardening, such as aluminum alloy 2024, require less force if driven immediately after quenching or if refrigerated between the quenching and driving operations to arrest the age hardening.

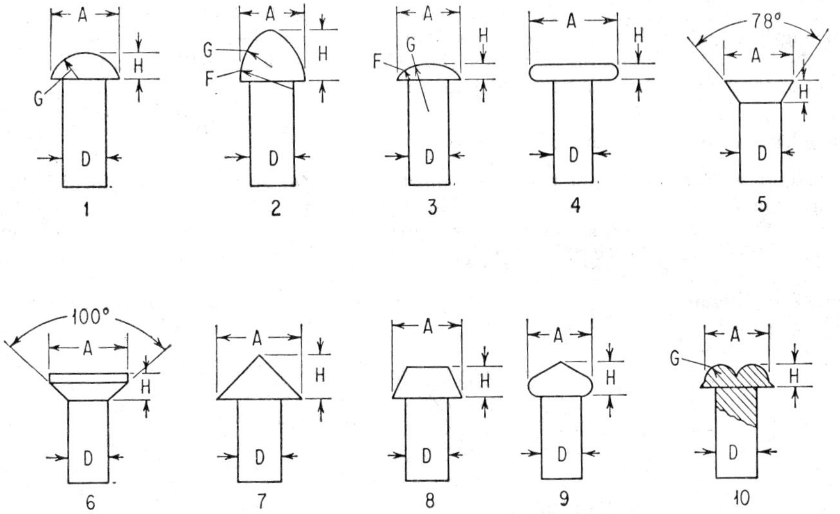

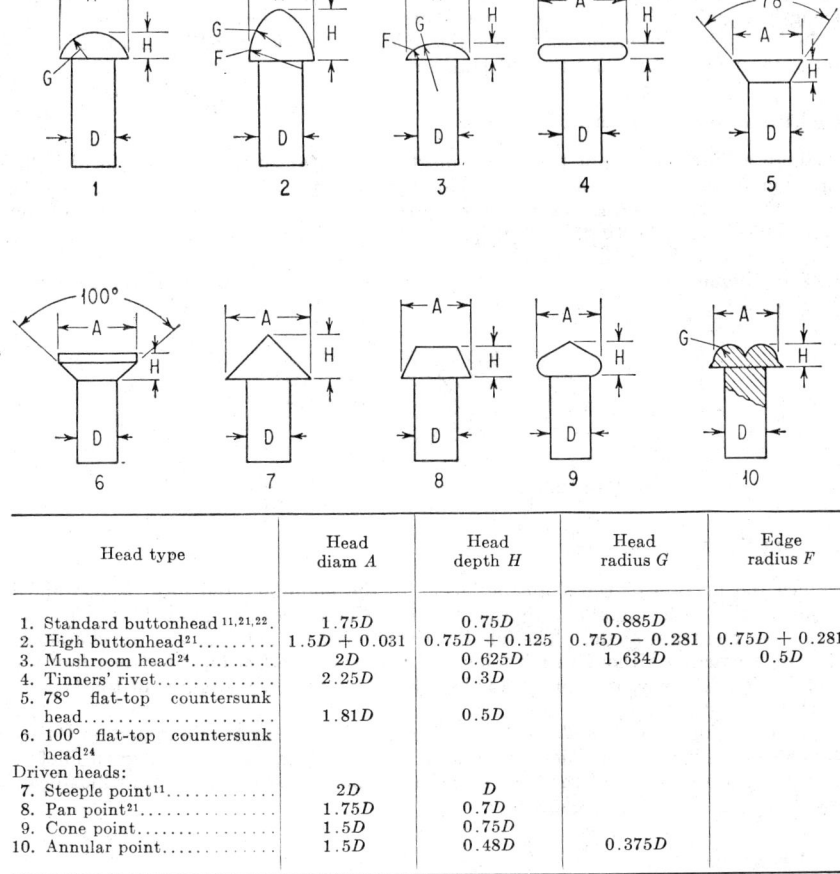

Head type	Head diam A	Head depth H	Head radius G	Edge radius F
1. Standard buttonhead[11,21,22]	$1.75D$	$0.75D$	$0.885D$	
2. High buttonhead[21]	$1.5D + 0.031$	$0.75D + 0.125$	$0.75D - 0.281$	$0.75D + 0.281$
3. Mushroom head[24]	$2D$	$0.625D$	$1.634D$	$0.5D$
4. Tinners' rivet	$2.25D$	$0.3D$		
5. 78° flat-top countersunk head	$1.81D$	$0.5D$		
6. 100° flat-top countersunk head[24]				
Driven heads:				
7. Steeple point[11]	$2D$	D		
8. Pan point[21]	$1.75D$	$0.7D$		
9. Cone point	$1.5D$	$0.75D$		
10. Annular point	$1.5D$	$0.48D$	$0.375D$	

FIG. 22.2. Common types of heads for rivets.

Tests have shown that rather small heads are sufficient to develop the full tensile strength of the rivet shank, and heads of even insignificant size are sufficient to develop the shear strength under double-shear loading. Therefore, where ease of driving is important and where large, fully formed heads of the button or brazier type are not required for the sake of appearance or other reasons, heads of smaller size and simpler shape should be satisfactory. Even mildly upset pins will be satisfactory for many applications.[1] Obviously, if the rivet must resist tensile loads, the heads must be of such diameter that they will not pull through the holes, and of such height that a ring will not be sheared off.

Shear cracks in heads resulting from overdriving should be avoided, but with some materials, cracks may develop in driven heads not yet completely formed. Tests indicate that these cracks do not adversely affect the static tensile strength or the shear strength under either static or cyclic loading. The only valid objection to such head cracks would seem to be a poor appearance.[2]

Reliability was not impaired according to tests of structural-steel rivets subjected to tensile loading, even though some of the specimens were intentionally the product of malpractices and would not pass ordinary inspection.[3] Even the rivets heated to such a high temperature that they were considered badly burned developed tensile strengths equal to that of the rods from which they were made. On the other hand, in the case of heat-treatable materials, the strictest attention must be paid to the control of the temperature and time at temperature.

(b) Flush-riveting Dimpled Sheets

Where a flush surface of thin parts must be maintained, as in the skin of high-performance aircraft, the sheets are sometimes dimpled, as shown in Fig. 22.3, to accommodate the countersunk head of the rivet and to increase the strength of the joint. In aircraft riveting, as opposed to most structural riveting, the countersunk head is the manufactured one and the driven head is usually flat. In order to avoid cracked dimples, it may be necessary to increase the temperature of the sheets to be dimpled.[4]

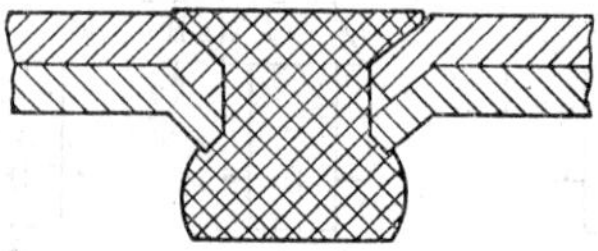

FIG. 22.3. Riveted joint in dimpled sheet.

(c) Folds under Heads

Since the portion of the shank adjacent to the manufactured head is restrained against swelling, the metal in a thoroughly upset shank folds down against the head to form a flat reentrant angle. The fold has the appearance and many of the characteristics of a crack. Although such folds seem to have no effect on the strength of the rivet in double shear, it is probable that they promote rivet failures of the type in which the heads "pop off."

The formation of folds under the manufactured heads can be eliminated by (1) use of small clearance in the holes, (2) avoiding overdriving, (3) use of a small fillet or chamfer at the junction of the head and shank, and (4) use of a rivet set which bears heavily at the axis of the head.

(d) Hole Clearance

A factor which may have considerable influence upon the strength and behavior of a rivet is the degree to which the shank is upset to fill the hole. For practical reasons, the holes must be larger than the rivet shank and the greater the clearance the greater is the possibility of bending the shank and producing eccentric heads. The best clearance is the smallest one that permits easy insertion of the rivet. For cold-driven rivets, the hole diameter need not be more than about 4 per cent greater than the nominal diameter of the rivet, while for hot-driven rivets in structural work a clearance of $\frac{1}{16}$ in. is commonly used. Table 22.1 gives the cross-sectional areas of rivets expanded 4 per cent over the nominal diameter, of rivets filling holes with a clearance of $\frac{1}{16}$ in., and of standard rivet-hole sizes for aircraft work.[5] The lengths of undriven rivets depend on the rivet diameter, kind of head, plate thickness being riveted, and the hole diameter.

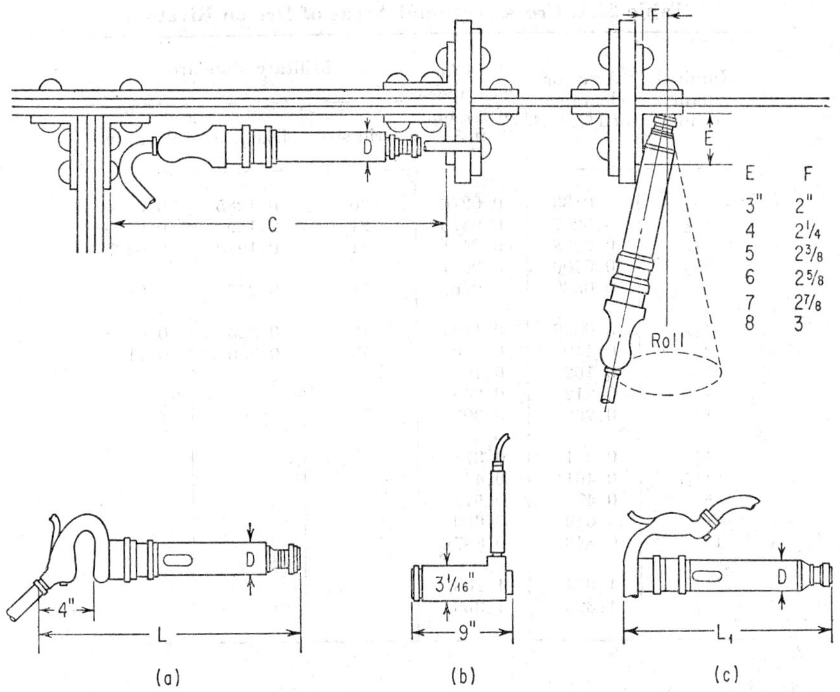

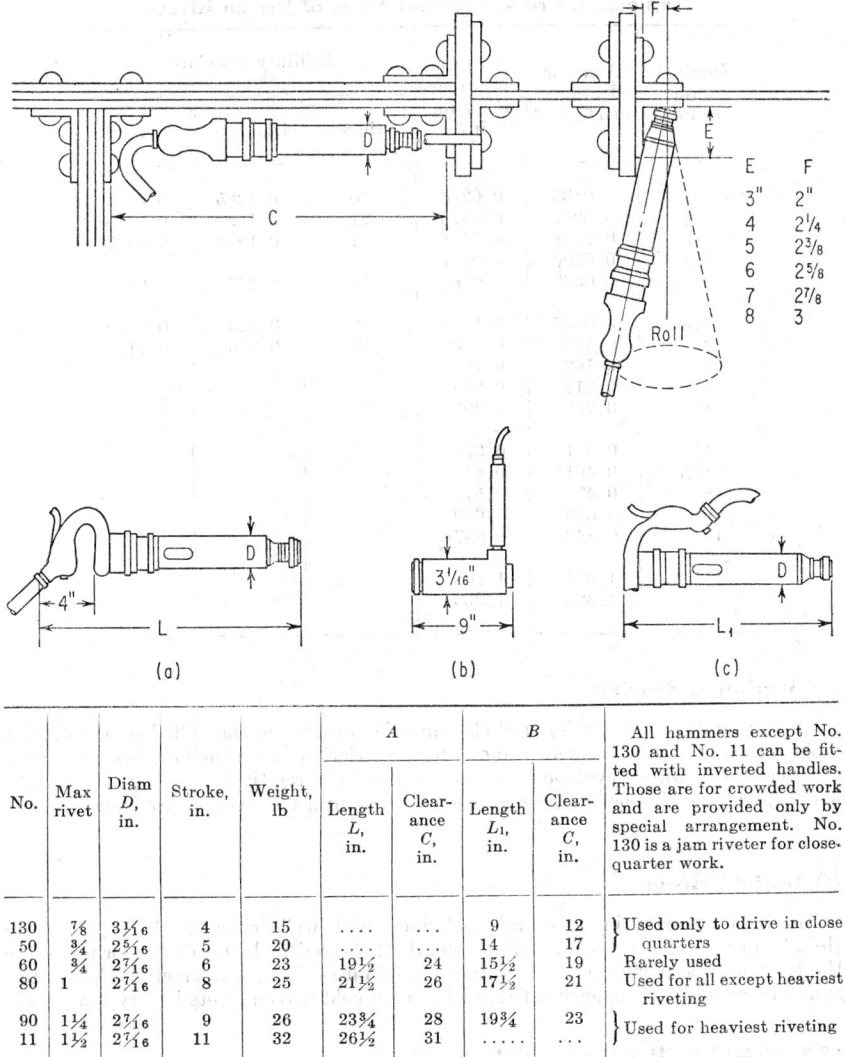

No.	Max rivet	Diam D, in.	Stroke, in.	Weight, lb	*A* Length L, in.	*A* Clearance C, in.	*B* Length L_1, in.	*B* Clearance C, in.	All hammers except No. 130 and No. 11 can be fitted with inverted handles. Those are for crowded work and are provided only by special arrangement. No. 130 is a jam riveter for close-quarter work.
130	7/8	3 1/16	4	15	....	...	9	12	Used only to drive in close quarters
50	3/4	2 5/16	5	20	....	...	14	17	
60	3/4	2 7/16	6	23	19½	24	15½	19	Rarely used
80	1	2 7/16	8	25	21½	26	17½	21	Used for all except heaviest riveting
90	1¼	2 7/16	9	26	23¾	28	19¾	23	Used for heaviest riveting
11	1½	2 7/16	11	32	26½	31	.....	...	

(*Reproduced by permission of "AISC Manual."*)

FIG. 22.4. Erection clearances for inserting and driving rivets. If hammer can be rolled, easier driving and more symmetrical heads are obtained. To permit this, distance F must be as given here and field rivets must have a perfect stagger with shop rivets. (*a*) Standard open-handle riveter. (*b*) Jam riveter No. 130. (*c*) Inverted-handle riveter.

Table 22.1. Cross-sectional Areas of Driven Rivets

Nominal diam d, in.	Area for clearance of 4%, in.^{2}	Area for clearance of 1/16 in., in.^{2}	Military standard[5]		
			Drill size	Diam, in.	Area, in.^{2}
1/8	0.0133	0.0276	30	0.1285	0.0130
5/32	0.0207	0.0376	21	0.159	0.0200
3/16	0.0298	0.0491	11	0.191	0.0287
7/32	0.0406	0.0621			
1/4	0.0530	0.0767	F	0.257	0.0519
5/16	0.0828	0.1104	P	0.323	0.0819
3/8	0.119	0.150	W	0.386	0.117
7/16	0.162	0.196			
1/2	0.212	0.249			
9/16	0.268	0.307			
5/8	0.331	0.371			
11/16	0.401	0.442			
3/4	0.477	0.518			
7/8	0.649	0.690			
1	0.848	0.887			
1 1/8	1.074	1.108			
1 1/4	1.325	1.353			

(e) Minimum Spacing

The minimum rivet spacing and clearance in corners are controlled by the driving tools available. Small rivets impose few restrictions on spacing because special tools can be readily developed to accommodate the relatively small driving loads. The controlling dimensions usually used in structural engineering are shown in Fig. 22.4.

(f) Initial Tension

The initial tension in structural-steel rivets with well-driven heads may be more than 70 per cent[3] of the yield-point strength of the rod if the heads are formed while the rivet is at a cherry-red temperature. The initial tension in rivets with countersunk or flat heads is somewhat less, and that in cold-driven rivets is very low.

22.3. STRENGTH OF RIVETED JOINTS

(a) Static Strength

Under conditions approaching the ultimate load of a joint in which the rivets are subjected to double shear, as indicated in Fig. 22.5, experience and test results

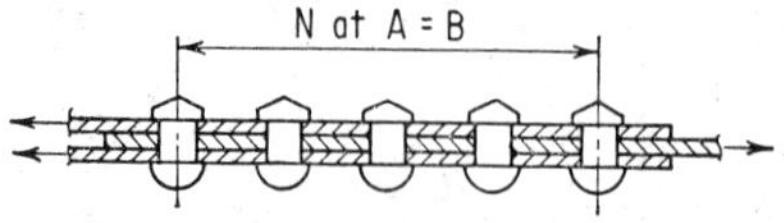

FIG. 22.5. Rivets stressed in double shear.

indicate that the load can be assumed uniformly distributed among the rivets. Failure may occur by one or a combination of processes; i.e., the rivets may shear or the plates may fail in tension or in bearing. The bearing failure of the plates may take the form of bulging in the region adjacent to the rivet or splitting out to the edge of the plate, or shearing of a portion of the plate to the edge on longitudinal planes tangent to the rivet at the ends of the transverse diameter, as shown in Fig. 22.6.

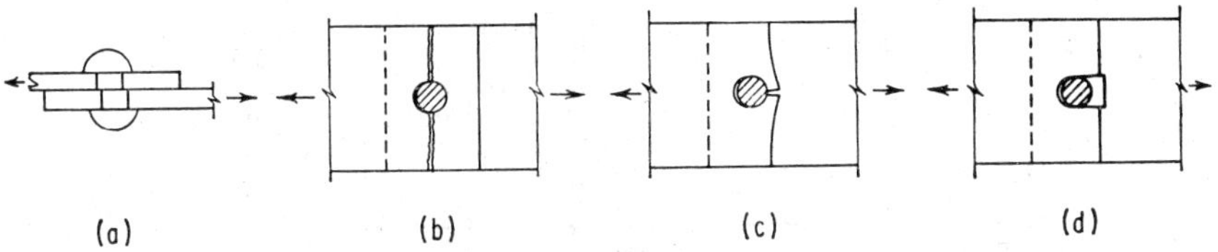

FIG. 22.6. Possible modes of failure of a riveted joint. (*a*) Shear in rivet. (*b*) Tension in plate. (*c*) Tearing of plate behind hole. (*d*) Shearing of plate behind hole.

The strength of the joint is essentially equal to the smallest of the following quantities:

$$P_s = nNA_R\tau_u \tag{22.1}$$
$$P_t = W_n t\sigma_u \tag{22.2}$$
$$P_b = NDt\beta_u \tag{22.3}$$

in which P_s = shear strength of all the rivets, lb
P_t = tensile strength of the plates, lb
P_b = bearing strength of the plates, lb
n = number of planes on which shearing failure must occur, 1 for single shear or 2 for double shear
N = number of rivets in the connection
A_R = cross-sectional area of the driven rivet shank, in.^{2}
W_n = least net width of the joint, in.
t = thickness of the plate, in.
D = diameter of the driven rivet shank, in.
τ_u = shear strength of rivet material, psi
σ_u = tensile strength of plate material, psi
β_u = bearing strength of plate material, psi (function of the edge distance and of the ratio of the rivet diameter to the thickness of the material being joined)

Table 22.2a. Strength of Steel Rivet Materials

ASTM spec.	Description of material	Ultimate tensile strength σ_u, psi	Yield point σ_y, psi	Elongation in 8 in., %
A31*......	Boiler rivet steel and rivets, grade A	45,000–55,000	23,000	27
	Boiler rivet steel and rivets, grade B	58,000–68,000	29,000	22
A141......	Structural rivet steel	52,000–62,000	28,000	24
A195*.....	High-strength structural rivet steel	68,000–82,000	38,000	20
A406......	High-strength structural-alloy rivet steel	68,000–82,000	50,000	20
A131......	Structural steel for ships	55,000–65,000	30,000	23
A152......	Wrought-iron rivets and rivet rounds	47,000	28,000	22–28

NOTE: For zinc coatings (hot-dip) on iron and steel hardware see ASTM A153*.
* Approved as American Standard by the American Standards Association.

Table 22.2b. Shear Strengths of Aluminum-alloy Rivets

Specification	Alloy and temper	Ultimate shear strength τ_u, psi*
QQ-A-430	2017-T4	33,000
	2024-T4	37,000
	2117-T4	26,000
	6053-T61	20,000
	6061-T6	25,000
	7075-T6	42,000
MIL-R-12221A(CE)	7277-T6	35,000

* Based on tests of undriven rivets, applicable for finished rivets in lieu of tensile tests.

Table 22.3a. Tensile Properties of Structural Steels (¾-in.-thick Plate)

Material	Tensile properties			
	Specification	Ultimate strength σ_u, psi	Yield point σ_y, psi	Elongation in 8 in., %
Carbon steel, grade C	ASTM A283	55,000–65,000	33,000	23
Low-alloy steel	ASTM A242	70,000	50,000	18
Silicon steel	ASTM A94	80,000–95,000	45,000	16
Nickel steel	ASTM A8	90,000–115,000	55,000	14

When the three values of strength computed by Eqs. (22.1), (22.2), and (22.3) are equal, the joint is said to have balanced design and is equally likely to fail by each of the three modes. The efficiency of the joint is the ratio of the smallest of these three loads to the computed strength of the gross section of the members being joined.

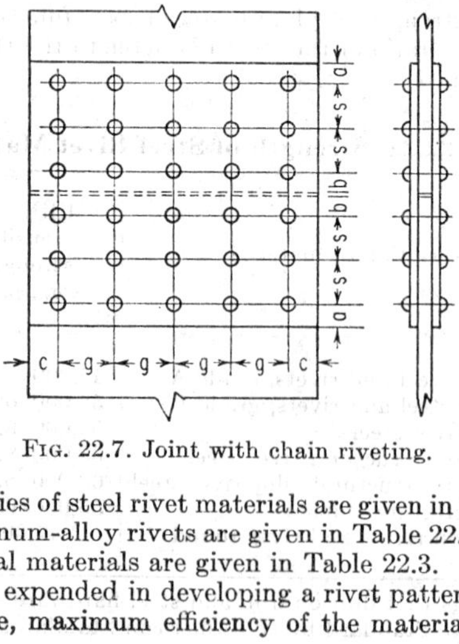

Fig. 22.7. Joint with chain riveting.

Ultimate tensile properties of steel rivet materials are given in Table 22.2*a*. Ultimate shear strengths of aluminum-alloy rivets are given in Table 22.2*b*. Values of strength of a number of structural materials are given in Table 22.3.

Much effort has been expended in developing a rivet pattern which leads to a balanced design and, hence, maximum efficiency of the materials. Most of the effort

Table 22.3b. Minimum Tensile Properties of Aluminum-alloy Plate (ASTM B209)

Alloy	Temper	Thickness, in.	Ultimate tensile strength σ_u, psi	Yield strength σ_y, psi	Elongation in 2 in., %
1100	H14	0.250–0.499	16,000	14,000	6
		0.500–1.000	16,000	14,000	10
3003	H14	0.250–0.499	20,000	17,000	8
		0.500–1.000	20,000	17,000	10
3004	H34	0.250–1.000	32,000	25,000	5
Alclad 2014	T42	0.250–0.499	57,000	34,000	15
		0.500–1.000	58,000*	34,000*	15
	T62	0.250–0.499	64,000	57,000	8
		0.500–1.000	67,000*	59,000*	6
		1.001–1.500	67,000*	59,000*	4
2024	T42	0.250–0.499	64,000	38,000	12
		0.500–1.000	62,000	38,000	8
Alclad 2024	T42	0.250–0.499	62,000	38,000	12
		0.500–1.000	62,000*	38,000*	8
6061	T6	0.250–0.499	42,000	35,000	10
		0.500–1.000	42,000	35,000	9
		1.001–2.000	42,000	35,000	8
Alclad 6061	T6	0.250–0.499	38,000	32,000	10
		0.500–1.000	42,000*	35,000*	9
		1.001–2.000	42,000*	35,000*	8

* The tension test specimen from plate 0.500 in. and thicker is machined from the core and does not include the cladding alloy.

has concerned the subject of net width of the joint, W_n. In joints with a single row of rivets and those with chain riveting (Fig. 22.7), the net width is simply the gross width less the product of the hole diameter and the number of holes per row. In

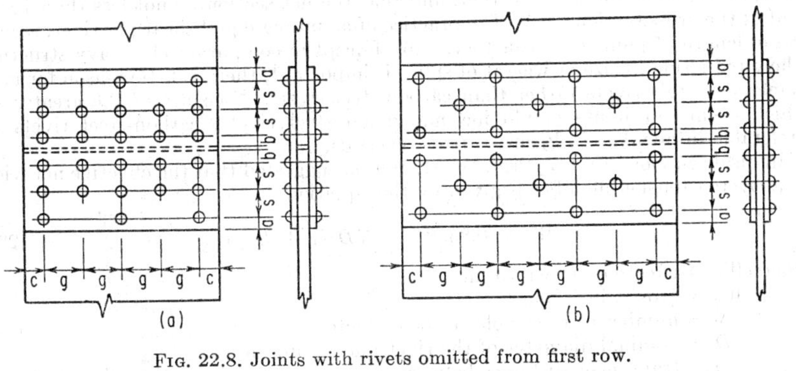

Fig. 22.8. Joints with rivets omitted from first row.

joints with staggered riveting or with some rivets omitted from the first row (Fig. 22.8), the spacing of the rows must be considered in determining the net width.

The rules for determining the net width, as given in several of the specifications for structures, are as follows:[6]

1. In the case of a chain of holes extending across a part in any diagonal or zigzag line, the net width of the part shall be obtained by deducting from the gross width the sum of the diameters of all holes in the chain, and adding, for each gage space in the chain, the quantity

$$b = s^2/4g \tag{22.4}$$

where s = longitudinal spacing (pitch), in., of any two successive holes
g = transverse spacing (gage), in., of the same two holes

The critical net section of the part is obtained from the chain which gives the least net width.

2. For angles, the gross width shall be the sum of the widths of the legs less the thickness. The gage for holes in opposite legs shall be the sum of the gages from the back of the angle less the thickness. The gage distances commonly used are shown in Table 22.4.

Table 22.4. Usual Gages for Angles, Inches

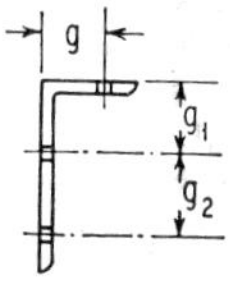	Leg	8	7	6	5	4	$3\frac{1}{2}$	3	$2\frac{1}{2}$	2	$1\frac{3}{4}$	$1\frac{1}{2}$	$1\frac{3}{8}$	$1\frac{1}{4}$	1
	g	$4\frac{1}{2}$	4	$3\frac{1}{2}$	3	$2\frac{1}{2}$	2	$1\frac{3}{4}$	$1\frac{3}{8}$	$1\frac{1}{8}$	1	$\frac{7}{8}$	$\frac{7}{8}$	$\frac{3}{4}$	$\frac{5}{8}$
	g_1	3	$2\frac{1}{2}$	$2\frac{1}{4}$	2										
	g_2	3	3	$2\frac{1}{2}$	$1\frac{3}{4}$										

3. In computing the net area, the diameter of a rivet hole shall be taken as ⅛ in. greater than the nominal diameter of the rivet.

An analysis of much of the test data available in 1934, on the static strength of riveted joints in steel members, together with the results of an extensive series of tests undertaken in connection with the design of the San Francisco-Oakland Bay Bridge, concluded:[7] (1) Nothing is gained by an attempt to detail a tension member with a critical net area greater than about 75 per cent of the gross area. (2) There is no justification for elaborate formulas to calculate the effect of rivet stagger on net section. (3) Joints should be as compact as practicable (optimum results will probably be obtained by full-row riveting with a gage of about 4.5 rivet diameters and a pitch of 3.5 to 4 rivet diameters. (4) Statements 1 to 3 lead to the suggestion that the allowable loads on riveted tension members be based on, and expressed in terms of, stress in the gross section; and that the net section be not less than 75 per cent of the gross section. (5) The practice of assuming equal shear per rivet, regardless of length of joint, is satisfactory. (6) Except in comparatively heavy structures where reduction in size or weight of splice is important, there is little reason for using manganese-steel rivets rather than carbon-steel rivets. Because of the greater slip which occurs in joints employing manganese-steel rivets, carbon-steel rivets are preferable in members subject to stress reversals.

In his discussion of ref. 7, Prof. W. M. Wilson suggested that the effective net width of a riveted tension member is given by the equation

$$W_n = 0.85(W_g - ND)(1 + D/g) \tag{22.5}$$

where W_n = effective net width, in.
W_g = gross width, in.
N = number of rivet holes to be deducted
D = nominal diameter of the rivet plus ⅛ in., in.
g = transverse distance between rivets or two times the edge distance,* whichever is greater, in.

His analysis of data involving 25 rivet patterns showed that this formula yielded

* In a discussion of test specimens or narrow joints, this is not to be confused with the use of the expression in the discussion of bearing strength.

net widths within 7 per cent of the value deduced from the test results, whereas one difference of 15 per cent was obtained with the rule involving Eq. (22.4).

In another discussion of ref. 7, Hartmann and Holt present an analysis of data for 98 steel specimens and 39 aluminum-alloy specimens and suggest the addition of a statement similar to the following:

"The transverse distance between rivets in the outside row shall not exceed eight times the nominal diameter of the rivet, and the edge distance for the end rivet in the outside row shall not exceed four times the nominal diameter of the rivet."

This discussion of net width is applicable to joints which are transverse to the load line. The use of diagonal joints, however (Fig. 22.9), could result in structures of greater efficiency, as in the case of spirally wound pipe. The number of rivets to be considered in determining the net width may be relatively small. Results of tests on such a joint which developed an efficiency of 86 per cent are reported in ref. 8.

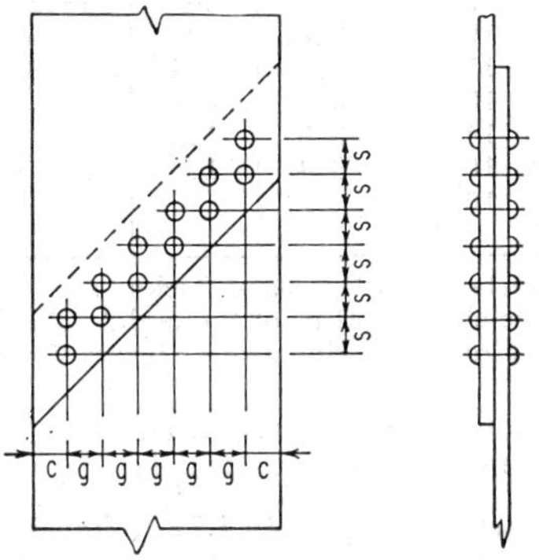

FIG. 22.9. Diagonal rivet pattern.

For high-performance structures, there may be some question as to the value of τ_u to be used in Eq. (22.1). Reference 8 reports an average shear strength of 41,250 psi for 12 undriven steel rivets tested in single shear and only 37,750 psi for 12 other rivets from the same lot tested in double shear. It is to be expected that the strength of hot-driven rivets would be a little higher and that of cold-driven rivets considerably higher than the strengths of these undriven rivets. References 9 and 10 employ a safety factor for steel rivets in shear of about 3 based on the ultimate strength. References 15 and 16 for aluminum-alloy rivets in shear use a factor of safety of about $2\frac{1}{2}$.

For general purposes, a value of shear strength equal to about 60 per cent of the tensile strength can be used; however, ref. 11 sets the allowable stress for shear equal to 80 per cent of the allowable stress in tension.

When a rivet is used in conjunction with a relatively thin plate, the plate seems to cut into the rivet and cause an apparent lowering of the shear strength of the rivet. The associated reduction in shear strength depends upon whether the rivet is in single or double shear. Table 22.5 shows the per cent reduction in shear strength of aluminum-alloy rivets resulting from the use of thin plates and shapes. These reductions are applicable to the average shear strengths. This type of behavior is not recognized in the specifications of steel structures.

A small amount of data on aluminum-alloy rivets at elevated temperatures[12] indicates that the shear strength is affected by changes in temperature in the same way that the tensile strength of the material is affected.

The bearing strength of parts being joined is greatly influenced by the distance (parallel to the load line) between the center of the loaded rivet (or pin) and the free edge of the plate (or specimen) when that distance is less than about two times the diameter of the rivet. The ASME Boiler and Pressure Vessel Code[11] permits the use of allowable stress values for bearing equal to 1.60 times the allowable stress value

Table 22.5. Percentage Reduction in Shear Strength of Aluminum-alloy Rivets Resulting from Their Use in Thin Plates and Shapes[15,16]

Ratio* D/t	Loss in double shear†	Ratio* D/t	Loss in double shear†	Ratio* D/t	Loss in		Ratio* D/t	Loss in	
					Single shear	Double shear		Single shear	Double shear
1.5	0	2.2	9.1	2.9	0	18.2	3.5	2.0	26.0
1.6	1.3	2.3	10.4	3.0	0	19.5	3.6	2.4	27.3
1.7	2.6	2.4	11.7	3.1	0.4	20.8	3.7	2.8	28.6
1.8	3.9	2.5	13.0	3.2	0.8	22.1	3.8	3.2	29.9
1.9	5.2	2.6	14.3	3.3	1.2	23.4	3.9	3.6	31.2
2.0	6.5	2.7	15.6	3.4	1.6	24.7	4.0	4.0	32.5
2.1	7.8	2.8	16.9						

* Ratio of the rivet diameter D to the plate thickness t. The thickness used is that of the thinnest plate in a single-shear joint or of the middle plate in a double-shear joint.

† The percentage loss of strength in single shear is zero for D/t less than 3.0.

for tension, when the edge distance e is at least $2D$. Tests on a wide variety of aluminum alloys[13] indicate the following relations:

For $e = 1.5D$, Bearing strength $= (1.5 \text{ to } 1.7) \times$ (tensile strength)
For $e = 2.0D$, Bearing strength $= (1.8 \text{ to } 2.4) \times$ (tensile strength)
For $e = 4.0D$, Bearing strength $= (2.4 \text{ to } 3.4) \times$ (tensile strength)

In ref. 13, bearing yield strength is defined as the stress corresponding to a deformation of the hole equal to 2.0 per cent of the initial diameter, and the following ratios between bearing and tensile yield strengths for aluminum alloys are given:

For $e = 1.5D$, Bearing yield strength $= (1.4 \text{ to } 2.1) \times$ (tensile yield strength)
For $e = 2.0D$, Bearing yield strength $= (1.5 \text{ to } 2.5) \times$ (tensile yield strength)
For $e = 4.0D$, Bearing yield strength $= (1.6 \text{ to } 2.6) \times$ (tensile yield strength)

(b) Distribution of Load under Working Conditions

In the preceding discussion, it has been considered that, at failure, the load is uniformly distributed among the rivets. As pointed out in a discussion of ref. 14 by A. E. R. de Jonge, all the analyses made for determining the distribution of load among the rivets under working conditions lead to the conclusion that these loads are not uniformly distributed and that the end rivets resist a relatively large percentage of the load.

The analysis developed in ref. 14 leads to the load-distribution factors given in Table 22.6 for certain special cases based on rivets and plates of the same material. In a discussion of ref. 14, Hill and Holt present experimental and calculated distribution factors for double-strap butt joints between aluminum-alloy and steel plates given in Table 22.7. In every case involving three or more rows of rivets, the end rivets support more than their share of working loads.

(c) Fatigue Strength

The nonuniform distribution of load among rivets is given special consideration in structures subject to fatigue, which requires an understanding of the entire previous history of the joint, including the surface condition of the plates, rivet-driving temperature, and previous loading.

Tests on joints in aluminum-alloy plates indicate that the shear fatigue strengths

Table 22.6. Values of Distribution Coefficients C
Diameter of rivets = ⅞ in.

No. of rivets	Width of strip, in.	Pitch of rivets, in.	Thickness of plate t, in.		Stress factor k, in./kip	Distribution coefficients				% over-stress
			t_1	t_2		C	C_1	C_2	C_3	
					Lap Joints					
3	3	2.5	0.5		1/8200	0.333	0.371	0.257	......	11.4
	3	5.0	0.5		1/8200	0.333	0.395	0.210	......	18.6
	3	2.5	1.0		1/6400	0.333	0.351	0.298	......	5.4
	3	5.0	1.0		1/6400	0.333	0.365	0.270	......	9.6
4	3	2.5	0.5		1/8200	0.250	0.327	0.173	......	30.8
	3	5.0	0.5		1/8200	0.250	0.368	0.132	......	47.1
	3	2.5	1.0		1/6400	0.250	0.288	0.212	......	15.2
	3	5.0	1.0		1/6400	0.250	0.316	0.184	......	26.4
5	3	2.5	0.5		1/8200	0.200	0.312	0.140	0.096	56.0
	3	5.0	0.5		1/8200	0.200	0.3618	0.1095	0.0574	80.9
	3	2.5	1.0		1/6400	0.200	0.2570	0.1705	0.1450	28.5
	3	5.0	1.0		1/6400	0.200	0.2955	0.1495	0.110	47.7
6	3	2.5	0.5		1/8200	0.1667	0.305	0.128	0.067	83.0
	3	5.0	0.5		1/8200	0.1667	0.3596	0.1036	0.0368	115.6
	3	2.5	1.0		1/6400	0.1667	0.241	0.149	0.110	44.5
	3	5.0	1.0		1/6400	0.1667	0.286	0.135	0.079	71.5
					Butt Joints					
2	3	2.5	0.5	0.375	1/6180	0.250	0.268	0.232	......	7.2
	3	5.0	0.5	0.375	1/6180	0.250	0.277	0.223	......	10.8
	*	*	0.75	0.375	1/7150	0.250	0.250	0.250	......	0.0
	3	2.5	0.75	0.5	1/7645	0.250	0.262	0.238	......	4.8
	3	5.0	0.75	0.5	1/7645	0.250	0.268	0.232	......	7.2
	*	*	1.00	0.5	1/8200	0.250	0.250	0.250	......	0.0
3	3	2.5	0.75	0.5	1/7645	0.1667	0.2052	0.1252	0.1696	23.1
	3	5.0	0.75	0.5	1/7645	0.1667	0.2237	0.1003	0.1760	34.2
	3	2.5	1.00	0.5	1/8200	0.1667	0.1860	0.1280	0.1860	11.6
	3	5.0	1.0	0.5	1/8200	0.1667	0.1981	0.1039	0.1981	18.8
	3	2.5	1.0	0.75	1/7200	0.1667	0.2019	0.1363	0.1618	21.1
	3	5.0	1.0	0.75	1/7200	0.1667	0.2211	0.1152	0.1637	32.6

* Any value.

of driven rivets are a much higher percentage of the ultimate strength than are the tensile fatigue strengths of rolled plates with either open holes or holes filled with idle rivets. Thus the designer of riveted joints subject to cyclic loading is especially concerned with the plates or shapes, rather than with the rivets.

For members subject to reversal of stress, except for wind loads, ref. 6 provides the following rule:

To the net total compressive stress, and to the net total tensile stress, add arithmetically 50 per cent of the smaller of these two; proportion the connected material and the connecting rivets, bolts, pins, or welds for each of the two increased stresses thus separately obtained at the unit stresses prescribed (for rivets, 15,000 psi shear, 44,000 psi bearing under double shear, 35,000 psi bearing under single shear, supple-

Table 22.7. Percentage of Total Load Carried by Each Rivet in a Large Butt Joint

At normal working load

Specimen No.	Thickness of main plates, in.	Thickness of cover plates, in.	Kind of rivets	Determination	Aluminum main plate rivets Nos.			Steel main plate rivets Nos.				
					1 and 5	2 and 4	3	1	2	3	4	5
1	¾	⅜	Steel	Experimental	34	13	6	17	8.5	12.5	27.5	34.5
				Calculated	31.5	14	9	17	9	10	19.5	44.5
2	1	½	Steel	Experimental	29.5	14	13	23	14.5	16	22	24
					25.5*	19	11	16	17	16	19	32
					25.5†	18	13	18.5	9.5	13.5	23.5	35
				Calculated	28	16	12	17	12	13	20	38
3	1	½	Aluminum	Experimental	24.5	19	13	19.5	10	19	21.5	30
				Calculated	24.5	18	15	17	15	16	21	31

* Specimen straightened before test.
† Specimen loaded in compression.

mentary provisions of June, 1960). If 100,000 reversals are expected throughout the life of the building, the stresses in the connected material and in the connecting rivets, bolts, pins, or welds shall not exceed 75 per cent of those specified. Sharp notches, copes, and other sudden changes of cross section shall be particularly avoided in and adjacent to such connections.

References 9 and 10 provide that the connections of members subject to alternating (cyclic) loadings occurring in succession during one passage of the live load shall be proportioned for the sum of the net alternating stresses.

References 15 and 16 provide the following rule for members subjected to cyclic loading:

Riveted members designed in accordance with the other requirements of the specifications and constructed so as to be free of severe reentrant corners and other unusual stress raisers require no further consideration of cyclic loadings for numbers of cycles less than 300,000 cycles for structures of 6061-T6 and 100,000 cycles for structures of 2014-T6.

When greater numbers of repetitions of some particular loading cycle are expected during the life of a structure, the calculated net section tensile stresses for the loading in question shall not exceed the values given by the curves in Figs. 22.10 and 22.11.

The following points are worthy of note:

1. The most severe combination of loadings for which a structure is designed (dead load, maximum live load, maximum impact, maximum wind, etc.) rarely occurs in actual service and is of little or no interest from the standpoint of fatigue.
2. The loading of most interest from the fatigue standpoint is the steady dead load with a superimposed, repetitive applied live load under normal operating conditions.
3. The number of cycles of load encountered in structures is usually small compared with those encountered in fatigue problems involving machine parts.
4. Careful attention to details of design and fabrication will greatly enhance resistance to fatigue. A structural fatigue failure is usually at a point of stress concentration where the state of stress could have been improved with little or no added expense.

Test results[17] show that the geometry of a joint is more important in governing the fatigue strength than is the choice of alloy. Stress concentrations should be alleviated by avoiding sharp reentrant corners and by using generous fillets and smooth transitions. Secondary bending should be minimized.

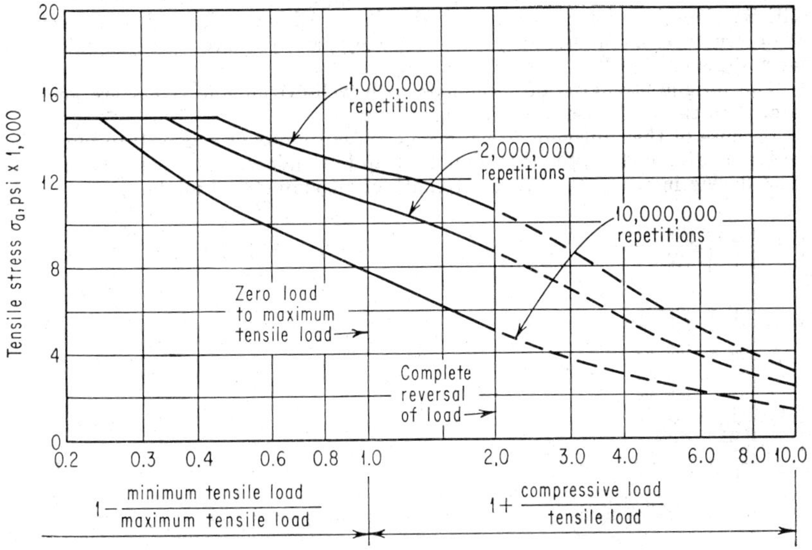

FIG. 22.10. Allowable tensile stresses on net section for various numbers of repetitions of load application, aluminum alloy 6061-T6.[15]

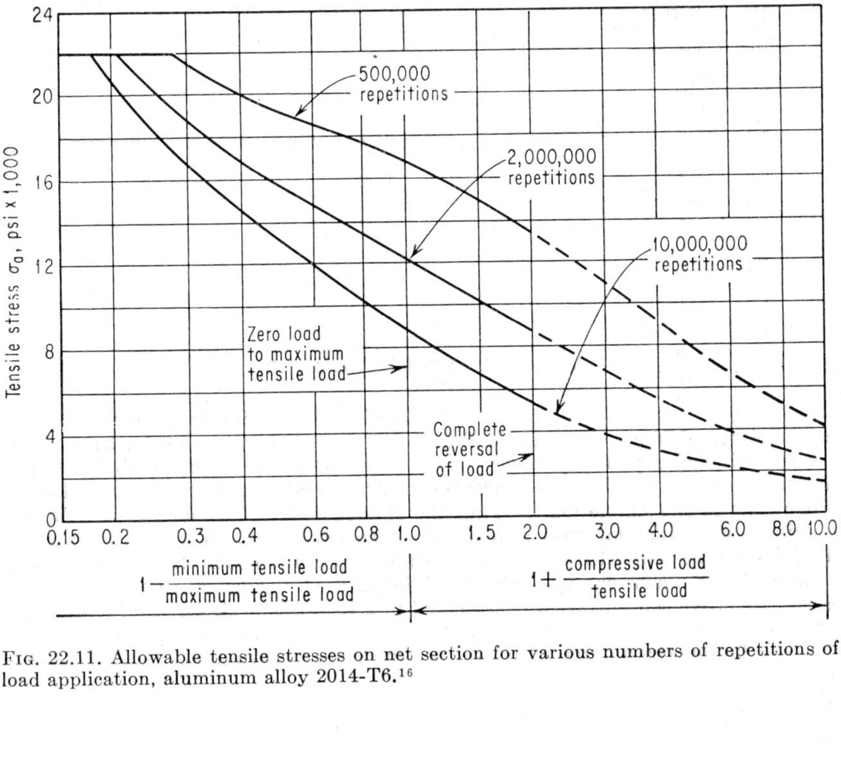

FIG. 22.11. Allowable tensile stresses on net section for various numbers of repetitions of load application, aluminum alloy 2014-T6.[16]

Reference 18 describes fatigue tests which support the design procedures in the ASCE Specifications for Structures of Aluminum Alloys. Although the specimens were of structural aluminum alloys, the following principles are equally applicable to other materials:

1. The fatigue strengths of specimens with well-driven idle rivets are generally greater than those of specimens with open holes.

2. Because of their unsymmetrical geometry, lap joints or single-strap butt joints have lower fatigue strengths than double-strap butt joints. Increasing the stiffness against flexing increases the fatigue strength of unsymmetrical joints.

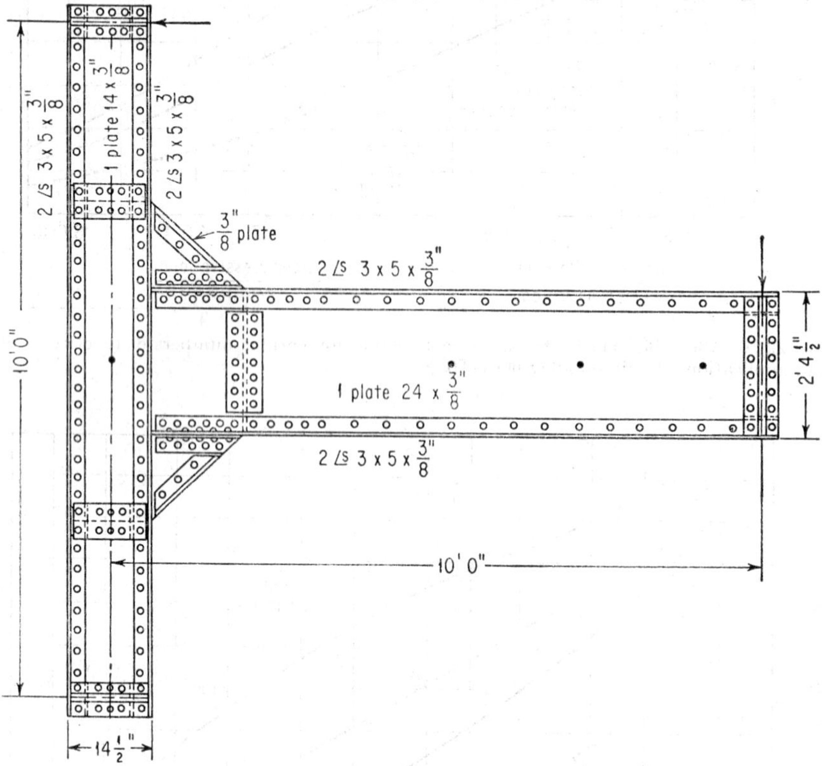

FIG. 22.12. Rigid joint (no slip).[20]

The fatigue strength of a butt joint of balanced static design with a computed gross-area efficiency of 76 per cent was found to be less than that of a simple butt joint with a computed efficiency of only 68 per cent.[18] This difference can be attributed to the fact that the computed stress concentration was greater in the joint of balanced design than in the simple joint. Thus the differences in stress-concentration factors associated with different rivet spacings must be recognized. The fatigue strength of a joint will be controlled by the concentrated stress and not by the average stress.

Reference 18 points out that the ASCE design curves (Figs. 22.10 and 22.11) are based on a stress-concentration factor equal to 2.4. Fatigue strengths of joints with larger values of the stress-concentration factor are obtained by multiplying the allowable fatigue stress derived from the design curves by the ratio of 2.4 to the stress-concentration factor of the joint.

(d) Fretting

Under cyclic loading, the small relative movement of the parts of a joint may develop fretting or galling of the surfaces. Ordinarily fretting acts as a very serious stress raiser. Applications of paint or sealing compounds on the faying surfaces may reduce the frictional forces between parts, but at the same time they may serve a useful purpose in reducing the tendency of the surfaces to fret. The small amount of data available indicates that such treatments of the faying surfaces have no deleterious effect on the fatigue strengths of joints.

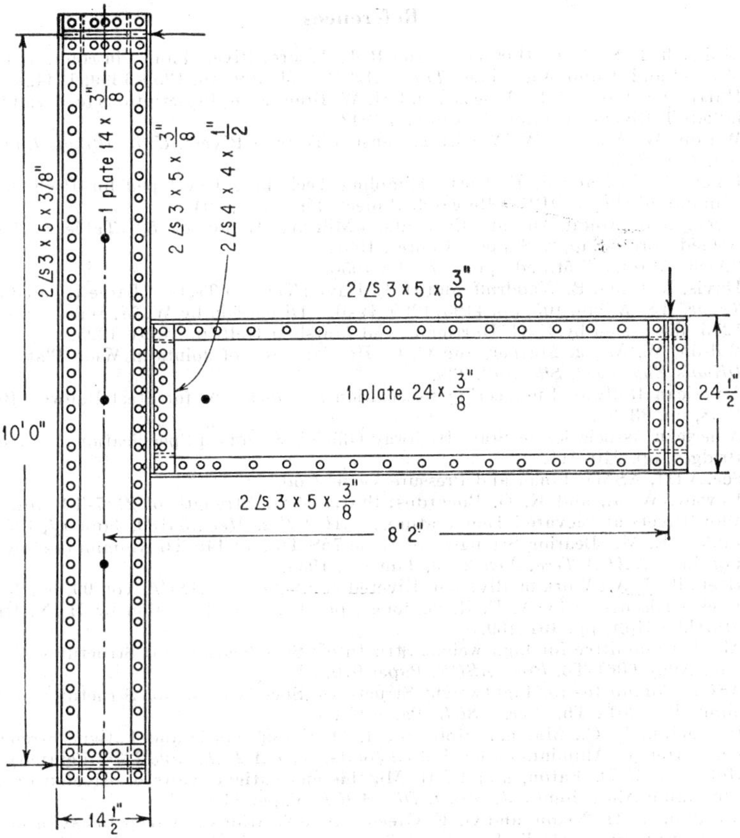

FIG. 22.13. Joint (significant amount of slip).[20]

(e) Creep

Although the high local stresses may be reduced by creep deformations, the fatigue life of a joint may still be reduced by the interaction of the mechanisms of fatigue and creep. The creep strength and stress-rupture strength are not affected by the stress concentration and interaction of fatigue mechanisms.[19] It is improbable that highly localized creep deformations in a joint would affect the behavior of the structure as a whole.

(f) Slip

Wilson and Moore[20] describe a series of tests designed to determine whether serious error is introduced into computations for stresses in steel frames by the assumption that the joints are perfectly rigid. They found that connections of the type used for the specimens shown in Fig. 22.12 were so rigid that they could be considered perfectly rigid. The connections of the type used for the specimens shown in Fig. 22.13 cannot be considered perfectly rigid.

Additional material related to structural specifications may be found in refs. 10, 11, 15, and 16.

References

1. Moisseiff, L. S., E. C. Hartmann, and R. L. Moore: Riveted and Pin-connected Joints of Steel and Aluminum Alloys, *Trans. ASCE*, vol. 109, pp. 1359–1399, 1944.
2. Hartmann, E. C., C. F. Wescoat, and M. W. Brennecke: Prescriptions for Head Cracks in 24S-T Rivets, *Aviation*, November, 1943.
3. Wilson, W. M., and W. A. Oliver: Tension Tests of Rivets, *Univ. Illinois Eng. Expt. Sta. Bull.* 210.
4. Finch, D. M., and J. E. Dorn: Dimpling Technique Developed for High Strength Aluminum Alloys, OPRD Research Project, February, 1945.
5. Strength of Metal Aircraft Elements, "Military Handbook 5," Table 8.1.1.1.1(d), Armed Forces Supply Support Center, 1961.
6. "AISC Manual," 5th ed., pp. 277–315, 1956.
7. Davis, R. E., G. B. Woodruff, and H. E. Davis: Tension Tests of Large Riveted Joints, *Trans. ASCE*, vol. 105, pp. 1193–1299, 1940. Discussion by W. M. Wilson, pp. 1264–1275; discussion by E. C. Hartmann and Marshall Holt, pp. 1291–1295.
8. Wilson, W. M., J. Mather, and C. O. Harris: Tests of Joints in Wide Plates, *Univ. Illinois Eng. Expt. Sta. Bull.* 239.
9. American Railway Engineering Association Specifications for Steel Railway Bridges, 1938, pp. 33–39.
10. American Association of State Highway Officials Standard Specifications for Highway Bridges, pp. 210–211.
11. Sec. VIII, ASME Boiler and Pressure Vessel Code.
12. Dewalt, W. J., and K. O. Bogardus: Static Shear Strength of 2117-T4 Aluminum-alloy Rivets at Elevated Temperatures, *NACA Res. Mem.* 55130, January, 1956.
13. Finley, E. M.: Bearing Strengths of Some 75S-T6 and 14S-T6 Aluminum-alloy Hand Forgings, *NACA Tech. Note* 2883, January, 1953.
14. Hrennikoff, A.: Work of Rivets in Riveted Joints, *Trans. ASCE*, vol. 99, pp. 437–449, 1934. Discussion by A. E. R. de Jonge, pp. 474–484; discussion by H. N. Hill and Marshall Holt, pp. 464–469.
15. ASCE Committee for Lightweight Structures: Specifications for Structures of Aluminum Alloy 6061-T6, *Proc. ASCE*, Paper 970.
16. ASCE Committee for Lightweight Structures: Specifications for Structures of Aluminum Alloy 2014-T6, *Proc. ASCE*, Paper 971.
17. Hartmann, E. C., Marshall Holt, and I. D. Eaton: Static and Fatigue Strengths of High-strength Aluminum-alloy Bolted Joints, *NACA Tech. Note* 2276, February, 1951.
18. Holt, M., I. D. Eaton, and R. B. Matthiesen: Fatigue Tests of Riveted or Bolted Aluminum Alloy Joints, *J. Struct. Div. ASCE*, Paper 1148.
19. Mordfin, L., H. Nixon, and G. E. Greene: Investigations of Creep Behavior of Structural Joints under Cyclic Loads and Temperatures, *NASA* Tech. Note D-181.
20. Wilson, W. M., and Herbert F. Moore: Tests to Determine the Rigidity of Riveted Joints in Steel Structures, *Univ. Illinois Eng. Expt. Sta. Bull.* 104.
21. Large Rivets (½ in. Nominal Diameter and Larger), ASA B18.4-1960, American Society of Mechanical Engineers, New York.
22. Small Solid Rivets, ASA B18.1-1955, American Society of Mechanical Engineers, New York.
23. Laughner, V. H., and A. D. Hargan: "Handbook of Fastening and Joining Metal Parts," McGraw-Hill Book Company, Inc., New York, 1956.
24. Military Standard MS 20426, Procurement Specification Mil-R-5674 B.
25. Vallance, A., and V. L. Doughtie: "Design of Machine Members," 3d ed., McGraw-Hill Book Company, Inc., New York, 1951.

Section 23

SHRINK- AND PRESS-FITTED ASSEMBLIES

By

STANLEY J. BECKER, M.S., *Engineer, Advanced Techniques, Re-entry Systems Department, General Electric Corporation, Philadelphia, Pa.*

CONTENTS

23.1. INTRODUCTION

Shrink and *press fits* are permanent connections which operate by interference of materials. They differ in method of assembling. A press fit or force fit is obtained by forcing a shaft into a smaller hole. A shrink fit is produced by heating the member having the hole and allowing it to cool to the shaft and ambient temperature. The same effect is sometimes utilized when the shaft is initially subcooled. Both methods may be used for holding a hub and shaft together. The holding power (torsional and axial) of a shrink joint is considerably greater than that of a press joint with the same interference. Fits are classified in Sec. 20. However, the shrink fit may also be applied to compounded cylinders, designed to withstand high internal pressures with a minimum use of material. In the earlier portion of this section the theoretical relationships of the compounded-cylinders design will be emphasized. Joint designs represent simpler cases and utilize the fundamentals shown.

23.2. SHRINK FITTING

The elastic design of cylinders which sustain an internal pressure greater than one-tenth of the allowable tensile stress requires the use of the thick-cylinder formulas originally developed by Lamé.[1] The designer may frequently find formulas in use for thick cylinders or for shrink fitting which appear to be modifications of the original Lamé formulas. Often these are Lamé's formulas combined with an end condition of closure or nonclosure and a yield condition to provide some equivalent, but not real, stress value. For a clear understanding of design, it is best to give separate consideration to stress, end closure, and yield condition.

(a) Nomenclature

$\sigma_t, \sigma_r, \sigma_z$ = tangent, radial, and axial normal stresses, respectively, positive if tension

$\epsilon_t, \epsilon_r, \epsilon_z$ = tangent, radial, and axial normal strains, respectively, positive if extension

p_j = pressure acting at typical radius j

p_j' = passive interference pressure at radius j after first subassembly

p_j'' = passive interference pressure at radius j after second subassembly

$a, b, \ldots, f, g, h, \ldots$ = typical boundary radii

δ_j = diametral interference at typical radius j

$\Delta_1\delta_j$ = change in measured δ_j after first subassembly

$\Delta_2\delta_j$ = change in measured δ_j after second subassembly

ν = Poisson's ratio

E = Young's modulus at operating temperature

E' = Young's modulus at ambient temperature

α = thermal coefficient of expansion

ΔT = temperature change

u = radial outward deflection at radius r

r = arbitrary radius

C_1, C_2 = constants of integration

τ_{fg} = shear yield stress in cylinder with inside boundary radius f, outside boundary radius g. Shear yield stress is taken at 57.7 per cent of tensile yield stress

(b) Basic Equations

Lamé's equations can be developed directly from a fundamental law, a mathematical relationship, and two assumed physical relationships.

Fundamental Law. Equilibrium of forces on an elementary volume of the cylinder (Fig. 23.1). Since every radius exhibits the same force conditions, equilibrium leads to the stress equation

$$\sigma_t - \sigma_r - r(d\sigma_r/dr) = 0 \tag{23.1}$$

Mathematical Relationship—Compatibility (Fig. 23.2). By definition $\epsilon_r = du/dr$. If a circle of radius r is expanded to radius $r + u$, then its new length is $2\pi(r + u)$ while its original length was $2\pi r$. Thus its length has increased by $2\pi u$ and the tangent strain along the circle is $2\pi u/2\pi r = u/r$. Since ϵ_r and ϵ_t are both defined by the same quantity u, they must be related (i.e., compatible). The compatibility relationship is

$$\epsilon_t - \epsilon_r + r(d\epsilon_t/dr) = 0 \tag{23.2}$$

The *assumed physical relationships* hold reasonably well for most materials and most cylinders.

1. *Hooke's Law.* This linear relationship between stress and strain is based upon experimental evidence provided by simple tension and compression tests on many specimens. Within the elastic limit, it is a good approximation for some materials such as mild steel, and a poor approximation for others. In either case, it is the simplest relationship available for analysis which approximates the true state of affairs.

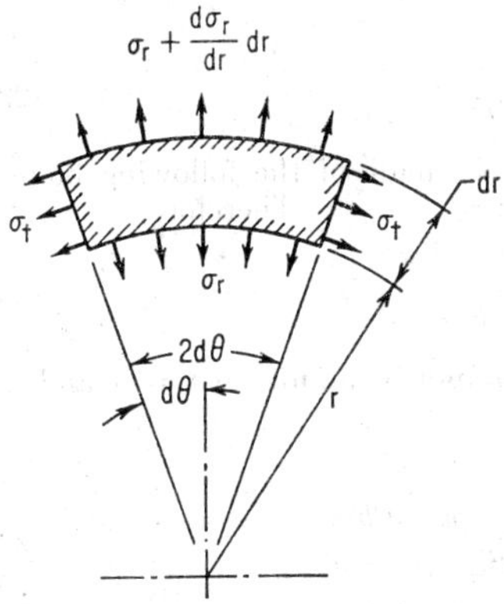

FIG. 23.1. Elementary cylindrical volume under radially symmetric forces.

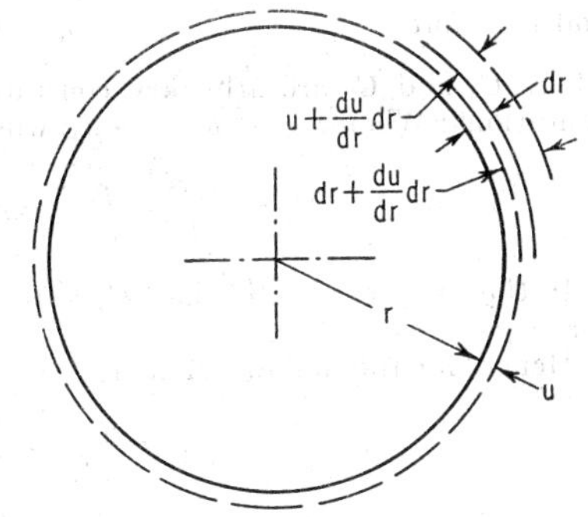

FIG. 23.2. Strains under radial expansion.

$$\epsilon_r = \frac{[dr + (du/dr)dr] - dr}{dr} = \frac{du}{dr}$$

$$\epsilon_t = \frac{2\pi(r + u) - 2\pi r}{2\pi r} = \frac{u}{r}$$

Taking into account the effect of lateral contraction of the material, the three principal strains can be expressed in terms of the three principal stresses in the following form of Hooke's law:

$$E\epsilon_r = \sigma_r - \nu(\sigma_t + \sigma_z) \qquad (23.3)$$

$$E\epsilon_t = \sigma_t - \nu(\sigma_z + \sigma_r) \qquad (23.4)$$

$$E\epsilon_z = \sigma_z - \nu(\sigma_r + \sigma_t) \qquad (23.5)$$

2. The second assumed relationship is *generalized plane strain;* that is, ϵ_z is a constant all across any given cross section of the cylinder. Its value depends on the axial and radial load and the particular cross section in question. This relationship is very nearly true for most cylinders at distances from an end closure or a very sudden change in section greater than about three times the mean value of average radius and wall thickness.

(c) Lamé's Equations for Cylinders

Solving Eq. (23.5) for σ_z and applying the result to Eqs. (23.3) and (23.4) yields

$$E\epsilon_r = (1 - \nu^2)\sigma_r - \nu(1 + \nu)\sigma_t - \nu E\epsilon_z \qquad (23.6)$$

and

$$E\epsilon_t = (1 - \nu^2)\sigma_t - \nu(1 + \nu)\sigma_r - \nu E\epsilon_z \qquad (23.7)$$

By inserting Eqs. (23.6) and (23.7) into the compatibility equation [Eq. (23.2)] and using the generalized plane strain relationship that ϵ_z is independent of r and therefore has no derivative with respect to r, the following stress equation results:

$$(1 - \nu^2)r(d\sigma_t/dr) - \nu(1 + \nu)r(d\sigma_r/dr) = (1 + \nu)(\sigma_r - \sigma_t) \qquad (23.8)$$

From the equilibrium equation,

$$\sigma_r - \sigma_t = -r(d\sigma_r/dr) \qquad \text{and} \qquad d\sigma_t/dr = r(d^2\sigma_r/dr^2) + 2(d\sigma_r/dr)$$

Inserting these into Eq. (23.8) and simplifying the result,

$$d^2\sigma_r/dr^2 + (3/r)(d\sigma_r/dr) = 0$$

or, in another form,

$$(1/r^3)\{(d/dr)[r^3(d\sigma_r/dr)]\} = 0 \tag{23.9}$$

This integrates immediately to

$$\sigma_r = C_1 + C_2/r^2$$

and therefore

$$\sigma_t = C_1 - C_2/r^2 \tag{23.10}$$

where C_1 and C_2 are arbitrary constants. Use is made of the following boundary conditions: When $r = a$, $\sigma_r = -p_a$; when $r = b$, $\sigma_r = -p_b$. Therefore,

$$C_1 = \frac{p_a a^2 - p_b b^2}{b^2 - a^2} \qquad C_2 = -\frac{a^2 b^2}{b^2 - a^2}(p_a - p_b) \tag{23.11}$$

In the case of the solid shaft $C_2 = 0$, since σ_r cannot be infinite at $r = 0$, and therefore $C_1 = -p_b$.

Hence, for the hollow cylinder,

$$\sigma_r = \frac{p_a a^2 - p_b b^2}{b^2 - a^2} - \frac{p_a - p_b}{r^2}\frac{a^2 b^2}{b^2 - a^2} \tag{23.12}$$

$$\sigma_t = \frac{p_a a^2 - p_b b^2}{b^2 - a^2} + \frac{p_a - p_b}{r^2}\frac{a^2 b^2}{b^2 - a^2} \tag{23.13}$$

and, for the solid shaft,

$$\sigma_r = \sigma_t = -p_b \tag{23.14}$$

Since $\epsilon_t = u/r$, from Eq. (23.7),

$$u = (r/E)[(1 - \nu^2)\sigma_t - \nu(1 + \nu)\sigma_r - \nu E\epsilon_z] \tag{23.15}$$

or

$$u = \frac{(1 + \nu)(1 - 2\nu)}{E}\frac{p_a a^2 - p_b b^2}{b^2 - a^2}r + \frac{1 + \nu}{E}\frac{p_a - p_b}{r}\frac{a^2 b^2}{b^2 - a^2} - \nu r\epsilon_z \tag{23.16}$$

From Eq. (23.10), it follows that $\sigma_t + \sigma_r = 2C_1$, a constant. From Eq. (23.5), since ϵ_z is also constant for a given cross section, it follows that σ_z is constant for a section. This has not been previously assumed. Hence σ_z is known once the axial force is known; that is, σ_z = axial force per unit cross section. It follows from $\epsilon_t = u/r$[Eq. (23.4) and Eqs. (23.10) and (23.11)] that

$$u = \frac{1 - \nu}{E}\frac{p_a a^2 - p_b b^2}{b^2 - a^2}r + \frac{1 + \nu}{E}\frac{p_a - p_b}{r}\frac{a^2 b^2}{b^2 - a^2} - \frac{\nu r}{E}\sigma_z \tag{23.17}$$

Equation (23.16) will often prove more useful than the usual alternative form, Eq. (23.17).

(d) The Compound Cylinder (Fig. 23.3)

From the form of Eqs. (23.12) and (23.13), it can be seen that the largest magnitude of stress difference, $\sigma_t - \sigma_r$, in the thick hollow cylinder appears at the bore of the cylinder and decreases rapidly with increasing radius. Because of this unequal distribution of stress along a radius, the material in a thick hollow cylinder is not efficiently utilized in resisting internal pressure. The situation can be improved by proper prestressing of the material.

One method of prestressing a thick cylinder against an internal pressure is by constructing the cylinder of two or more concentric cylinders with an interference fit between the component cylinders. This places an external pressure on the innermost cylinder which tends to balance the internal pressure load on this component. At the same time internal pressure acts on the outermost cylinder before any pressure load is put on the assembly, thus increasing the maximum internal pressure on the outermost component. Such an assemblage of concentric cylinders is known as a *compound cylinder*.

If two adjacent components with the same elastic modulus of such a compound cylinder are considered, with inside radius f, interference radius g, and outer radius h, and with pressures acting at these radii with values p_f, p_g, and p_h, respectively the

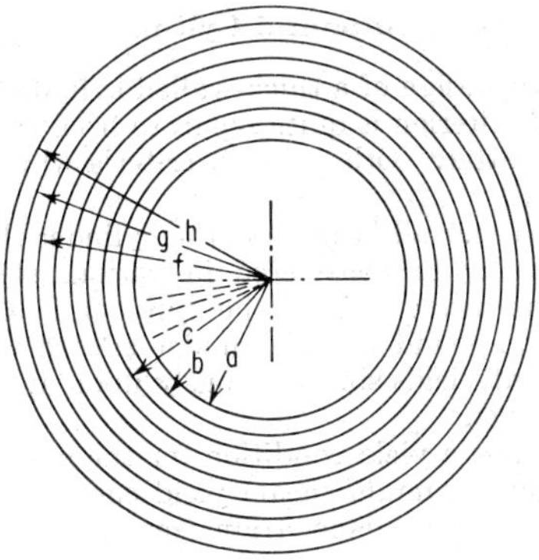

FIG. 23.3. The compound cylinder.

outward radial deflection at the outer radius of the inner cylinder from Eq. (23.16) is

$$u = \frac{(1+\nu)(1-2\nu)}{E}\frac{p_f f^2 - p_g g^2}{g^2 - f^2} g + \frac{1+\nu}{E}\frac{p_f - p_g}{g}\frac{f^2 g^2}{g^2 - f^2} - \nu g \epsilon_z$$

The outward radial deflection at the inner radius of the outer cylinder is

$$u = \frac{(1+\nu)(1-2\nu)}{E}\frac{p_g g^2 - p_h h^2}{h^2 - g^2} g + \frac{1+\nu}{E}\frac{p_g - p_h}{g}\frac{g^2 h^2}{h^2 - g^2} - \nu g \epsilon_z$$

The difference between these two deflections must equal one-half the diametral interference.

The diametral interference at g is therefore

$$\delta_g = \frac{4g(1-\nu^2)}{E}\left[\frac{p_g - p_h}{1-(g/h)^2} - \left(\frac{f}{g}\right)^2 \frac{p_f - p_g}{1-(f/g)^2}\right] \tag{23.18}$$

δ_g is not changed by changes in pressure loads at any internal or external radius of this subassembly or any other larger subassembly or the entire assembly of the compound cylinder.[2] Therefore, the values of p_f, p_g, and p_h in Eq. (23.18) can be either the passive interference load or the active external load plus interference load. In particular, for the two-component assembly with radii a, b, and c (in order of magnitude), the passive interference pressure at b is

$$p_b = \frac{E\delta_b}{4b(1-\nu^2)}\frac{[1-(a/b)^2][1-(b/c)^2]}{1-(a/c)^2} \tag{23.19}$$

The case of a cylindrical sleeve of outside radius b, placed over a solid shaft of radius a, with diametral interference δ_a, will be considered separately. For this case Eqs. (23.14) and (23.15) result, for the solid shaft, in

$$u = (r/E)[(1+\nu)(1-2\nu)p_a + \nu E \epsilon_z]$$

which when evaluated at $r = a$, deducting the value from that of Eq. (23.16) at $r = a$, results in one-half of δ_a. Thus

$$\delta_a = \frac{4a(1 - \nu^2)}{E} \frac{p_a - p_b}{1 - (a/b)^2} \tag{23.20}$$

or

$$p_a - p_b = \frac{E}{4a(1 - \nu^2)} \delta_a \left[1 - \left(\frac{a}{b}\right)^2\right] \tag{23.21}$$

(e) Balanced Design for the Compound Cylinder

A criterion for the elastic design of a thick-walled cylinder is quite properly chosen as that of yielding. This criterion is quite separate from that of rupture and should not be confused with the latter. Other design criteria which may be considered are fatigue and creep.

Since yielding does not occur all at once in the cylinder throughout its wall thickness, the yield criterion may be considered in various stages.[9] For basic elastic design, initial yield at the inside wall of the cylinder is a necessary criterion. A balanced design of the compound cylinder under such a criterion is one in which every cylindrical component yields simultaneously upon application of internal pressure on the whole assemblage.

The yield criterion requires a yield condition for the materials used. To this end the Tresca (also called the maximum shear) yield condition is selected, modified so that it will match, for the case of plane strain (zero third principal strain), the Von Mises (also called the distortion energy or octahedral shear stress) yield condition. To use this condition, it is required that the yield shear stress is taken to be 57.7 per cent of the tensile yield and that the maximum difference of principal stress at any point in the material is $\sigma_t - \sigma_r$.

Stress geometry requires that the maximum shear stress at a point be one-half of the maximum difference of principal stresses at the point. The modified Tresca condition thus cannot satisfy one-dimensional tensile or compressive stress test data but because of its construction is quite satisfactory for design of thick cylindrical pressure vessels.

For a typical cylindrical component with inner radius f and outer radius g, using Eqs. (23.10) and (23.11), the maximum shear stress, which must occur at the smallest radius, is

$$\tau_{fg} = \left(\frac{\sigma_t - \sigma_r}{2}\right)_{r=f} = \frac{C_2}{f^2} = \frac{p_f - p_g}{1 - (f/g)^2} \tag{23.22}$$

From Eqs. (23.18) and (23.22), it follows that the diametral interference necessary to achieve the balanced design is given by

$$\delta_g = \frac{4g(1 - \nu^2)}{E} \left[\tau_{gh} - \left(\frac{f}{g}\right)^2 \tau_{fg}\right] \tag{23.23}$$

The pressure at f, from Eq. (23.22),

$$p_f = \tau_{fg}[1 - (f/g)^2] + \tau_{gh}[1 - (g/h)^2] + p_h \tag{23.24}$$

It is desired to adjust g so that p_f will be maximized with all other quantities unaltered. This requires that $\partial p_f/\partial g = 0$, or that

$$\tau_{fg}(f/g)^2 = \tau_{gh}(g/h)^2 \tag{23.25}$$

or

$$g = (\tau_{fg}/\tau_{gh})^{1/4}(fh)^{1/2} \tag{23.26}$$

Since $\partial^2 p_f/\partial g^2$ is readily shown to be negative for any radius g, Eqs. (23.25) and (23.26) are a maximizing condition for p_f.

(f) Corrections for Operating Temperature and Assembly Order

The general method of economical balanced design of compound cylinders can be achieved by repeated and successive use of Eqs. (23.22), (23.23), and (23.25). However, the values of interference given by Eqs. (23.18) and (23.23) are those of the completely disassembled compound cylinder. Since it is usually preferable to manufacture the interferences at various stages of fabrication, proper account must be made of this modification. Also the operating temperature may not be the same as the fabrication temperature and proper account must be made of this effect on the interference values. (The optimum design radii are so slightly affected by the latter change that it is not worthwhile to alter the basic radii for operating-temperature considerations.)

In placing larger cylinders over smaller subassemblies, the growth of the outside diameter of the last subassembly due to that assembly must be added to the free-state (disassembled) interference at the same diameter in order to maintain the correct interference pressures.[2] Likewise, in placing smaller cylinders inside subassemblies, the contraction of the inside diameter of the last subassembly should be added to the free-state (disassembled) interference at the same diameter in order to maintain the correct interference pressure.

If the operating temperature is ΔT above ambient or fabrication temperature, and if α is the coefficient of expansion of the material used, then, at ambient, all dimensions including the interferences are smaller by a factor $(1 - \alpha\,\Delta T)$. The pressures and stresses should be multiplied by $(E'/E)(1 - \alpha\,\Delta T)$ in going from operating to ambient conditions, where E' is Young's modulus at ambient conditions while E is Young's modulus at operating conditions.

Assume that the calculated interferences of disassembly have been corrected to ambient temperature and measured at this temperature. Suppose that a four-cylinder assembly is to be manufactured with successive radii $e < f < g < h < j$ and suppose that the middle subassembly $f < g < h$ is to be assembled first with design interference at g as given by Eq. (23.18) or (23.23) and corrected to ambient conditions. From Eq. (23.16), the additional diametral interference necessary at f at the ambient temperature to maintain the proper operating interference pressure is

$$\Delta_1\delta_f = \frac{4f(1 - \nu^2)}{1 - (f/g)^2}\frac{p_g'}{E'} \tag{23.27}$$

where p_g' is the passive interference pressure at g at ambient temperature caused by the first subassembly [and calculated from Eq. (23.18)].

Similarly, the additional diametral interference at h to be accounted for is

$$\Delta_1\delta_h = \frac{(g/h)^2 4h(1 - \nu^2)}{1 - (g/h)^2}\frac{p_g'}{E'} \tag{23.28}$$

at ambient temperature.

Suppose that the first subassembly is followed by slipping the subassembly over cylinder ef with corrected interference. This will cause a passive interference pressure at f at ambient conditions of p_f'', and the total corrected diametral interference at h at ambient conditions will be

$$\Delta_2\delta_h = \frac{4(1 - \nu^2)}{E'}\left[\frac{hp_g'(g/h)^2}{1 - (g/h)^2} + \frac{hp_f''(f/h)^2}{1 - (f/h)^2}\right] \tag{23.29}$$

However, after the second subassembly, the ambient interference pressure at g, using Eq. (23.12) and superposition, is

$$p_g'' = p_g' + p_f''\frac{(f/g)^2 - (f/h)^2}{1 - (f/h)^2} \tag{23.30}$$

If Eq. (23.30) is solved for p_g' and the result applied to Eq. (23.29), it is easily shown that the correction to the interference at h is

$$\Delta_2\delta_h = \frac{(g/h)^2 4h(1 - \nu^2)}{1 - (g/h)^2} \frac{p_g''}{E'} \qquad (23.31)$$

which has the same form as Eq. (23.28) with p_g'' substituted for p_g'.

The passive interference pressure of the first subassembly at ambient temperature, from Eq. (23.18) with $p_f' = p_h' = 0$, is

$$p_g' = \frac{E'}{1 - \nu^2} (1 - \alpha\,\Delta T) \frac{\delta_g}{4g} \frac{[1 - (f/g)^2][1 - (g/h)^2]}{1 - (f/h)^2} \qquad (23\text{-}32)$$

The factor $(1 - \alpha\,\Delta T)$ is used here because it is applied to δ_g in actual measurement but not to g, which is kept at nominal value as measured.

For passive interference pressure at operating condition of this same subassembly at g, the factor $E'(1 - \alpha\,\Delta T)$ in Eq. (23.32) is changed to E.

For the next subassembly, with $p_e'' = p_h'' = 0$, the passive interference pressure at f at ambient temperature is

$$p_f'' = \frac{E'}{1 - \nu^2} (1 - \alpha\,\Delta T) \frac{\delta_f + \Delta_1\delta_f}{4f} \frac{[1 - (e/f)^2][1 - (f/h)^2]}{1 - (e/h)^2} \qquad (23.33)$$

where $(1 - \alpha\,\Delta T)\Delta_1\delta_f$ is the correction to be made to δ_f for the previous subassembly and is given by Eq. (23.27).

The method of calculation can be extended for any order of assemblage.

(g) Unbalanced Design, Creep, Fatigue, Partial Yielding, Autofrettage

The analysis of the balanced design ignores certain inelastic effects which may make an unbalanced design preferable. For example, a design which causes a particular component of an assembly to operate above its fatigue limit for the anticipated number of cycles during a lifetime, even though below the yield point, would not be preferred to one utilizing more material in an unbalanced design, but below the fatigue limit. Furthermore, the effect of creep may destroy the balance of an original design, thereby reducing its strength. The latter may be especially true where heavy shrink stresses are involved. Even if the assembly is fully loaded over a short portion of its lifetime and unloaded during a major portion of its lifetime, creep properties of the materials involved may govern the design. In this connection, it is noted that creep acts to lower the design interference pressures, thereby decreasing the support and strength of the innermost cylindrical components. The design interference pressure may therefore be governed by permissible creep limits rather than by the yield strength.

It is not the purpose here to enter into a full discussion of fatigue and creep properties of simple or compound cylinders. Work on these subjects is not extensive and much research is in progress. Rather, reference is made to a few representative published articles.[3,4,5,6]

Because the balanced design is based upon initial yield criteria, it is not directly related to either full yielding or bursting of such pressure vessels. Some theoretical work has been done on these subjects for simple cylinders (refs. 7 and 8, for example) but little on compound cylinders. Full yielding is treated incidental to partial yielding of compound cylinders under restricted conditions in ref. 9, and the possibility of increasing the strength of such compound cylinders by a measured amount of partial yielding (autofrettaging) has been investigated, all as an extension of the previous theory. It has been found that the ideal autofrettage pressure, to be used as a test pressure, is given by the formula [Eq. (10) of ref. 9]

$$p_a = (\tau_{ab+} - \tau_{ab-})[1 - (a/l)^2] \qquad (23.34)$$

provided that certain conditions, given below, are met. In this formula, p_a is the internal pressure that, when released, will cause the compound cylinder to be on the verge of reversed yielding at the inside radius a, *and no other place*, provided that the Tresca yield condition, as given by Eq. (23.22), is satisfied on load application and that the materials exhibit no strain hardening. l is the outside radius of the assembly, τ_{ab+} is the shear yield stress for the inner cylinder under load (57.7 per cent of tensile yield), and τ_{ab-} is the corresponding shear yield stress on reversed yielding. τ_{ab-} is negative to τ_{ab+} and usually $\tau_{ab-} = -\tau_{ab+}$; however, the formula permits different conditions of yielding on stress reversal.

(h) Unequal Tube Lengths

The development previously presented is essentially one-dimensional, in that all effects depend solely on a radius location as the one independent space coordinate. Recently more attention has been given to the effects of variations along the axial length of the cylinders. In particular, attention has been focused on the problem of pressure and stress concentrations at compound cylinder tube ends which are short of the ends of the rest of the cylinder. The same problem occurs where a cylindrical hub is shrunk onto a solid shaft of greater length (Fig. 23.4). The nature of the problem is indicated in ref. 10, where the problem is treated by means of thin-shell theory

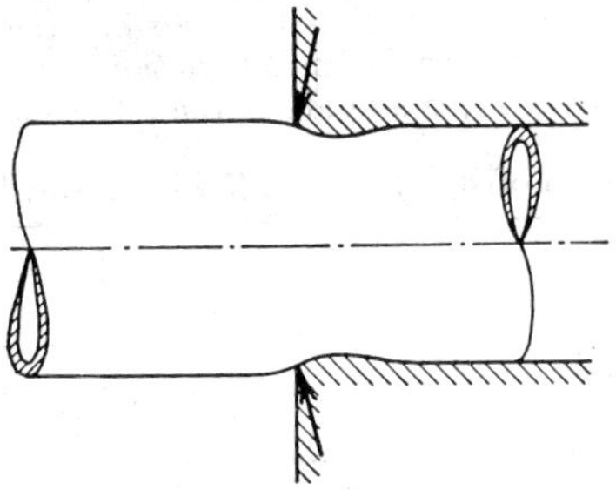

FIG. 23.4. Pinch effect at a discontinuity of an interference fit.

(with application to thin shells). The solution indicates that the interference pressure becomes a concentrated force per unit circumference at such short ends. Concentrated forces are permitted in thin-shell theory just as in beam theory.

In refs. 11 and 12, the effect of unequal tube lengths is studied in greater detail without restriction to thin-shell theory. There is no contradiction in the results from that of the previous reference.

Because of this concentration of forces, design modifications have been suggested to alleviate the condition. Essentially, these efforts have been to make the ends of the shorter tubes more flexible.

23.3. PRESS AND SHRINK JOINTS

(a) Description

The term "joints" applied to a mechanism denotes a connection between two members, capable of power transmission. A press joint or a shrink joint refers specifically to a connection between a shaft and tube, or between two tubes, or between two shafts connected by a tube. Such a joint is an interference fit depending solely upon the resulting friction for its ability to transmit power. There are two design questions associated with this joint:

1. Is there enough friction present to transmit the necessary power without slippage?
2. Is there enough strength in the structural members to carry both the stresses caused by the interference and the additional stresses caused by power transmission?

The first question, which is basic to the design, is easily answered. The second, somewhat more complex, can be answered most easily if the effects of interference and power transmission are considered separately and then superposed, together with other effects to be discussed shortly.

(b) Power vs. Friction

There are two types of power that can be transmitted by a press or shrink joint: torsional and axial. The holding ability is

$$T = 2\pi f p_b L b^2 \quad \text{for torsion} \tag{23.35}$$

and

$$F = 2\pi f p_b L b \quad \text{for axial force} \tag{23.36}$$

where T = torque
F = force
L = interference length
b = interference radius
f = coefficient of friction
p_b = interference pressure

Equation (23.36) is approximate since it assumes a constant value of p_b. The axial force, however, will cause p_b to vary along the length of the joint because of the Poisson effect of the axial stresses. This is of no practical consequence, however, because of the uncertainty in the coefficient of friction.

Table 23.1, from ref. 16 via 15, is a guide to the selection of coefficients of friction.

Table 23.1. Coefficients of Friction for Pressed and Shrunk Assemblies

	Coefficient of friction		
	Min	Max	Avg
a. Assembly of 62 press fits; lower values due to hub yielding as a result of too large fit allowance	0.030	0.250	0.086
b. Pressing four steel axles of 5¼ in. diam and 175 Brinell hardness into a cast-steel spider	0.100	0.140	0.115
Pressing off the pressed-on spider	0.120	0.150	0.137
Pressing off the shrunk-on spider	0.160	0.170	0.165
c. Pressing 15 axles of 6$\frac{1}{16}$ in. diam into gears, both of which were steel having a hardness of 170 Brinell	0.075	0.250	0.150
d. Pressing steel gears off steel shafts having tapered fits, where the mean diameter was 6½ in.:			
When the parts were pressed together	0.150	0.210	0.177
When the parts were shrunk together	0.210	0.230	0.220

Equations (23.35) and (23.36), together with a knowledge of the coefficient of friction, are sufficient to answer the first design question, provided that b and p_b are known. These can be selected, essentially on the basis of Lamé's equations and the theory of the compound cylinder. The stresses caused by power transmission should be superposed on the interference stresses. (Also stresses caused by discontinuity, as indicated in refs. 10, 11, and 12, should be included.)

(c) Interference, Interference Pressure, and Interference Stresses

In the case of joint interference, unlike that of the compound cylinder, the condition of the interference is essentially plane stress. Equations (23.12) and (23.13) remain valid, since they are independent of the value of σ_z or ϵ_z, but Eq. (23.17) with σ_z set equal to zero will be found more useful than Eq. (23.16). If the joint members all have the same elastic modulus then Eqs. (23.18), (23.19), (23.20), and (23.21) remain valid with the substitution of unity for the factor $1 - \nu^2$ wherever it appears so that plane stress conditions are maintained. In the present notation, considering only interference pressures, these equations become

$$\delta_b = \frac{4b}{E} p_b \frac{1 - (a/c)^2}{[1 - (a/b)^2][1 - (b/c)^2]} \quad \text{for a hollow shaft} \tag{23.18a}$$

$$p_b = \frac{E\delta_b}{4b} \frac{[1 - (a/b)^2][1 - (b/c)^2]}{1 - (a/c)^2} \quad \text{for a hollow shaft} \tag{23.19a}$$

$$\delta_b = \frac{4b}{E} p_b \frac{1}{1 - (b/c)^2} \quad \text{for a solid shaft} \tag{23.20a}$$

$$p_b = (E\delta_b/4b)[1 - (b/c)^2] \quad \text{for a solid shaft} \tag{23.21a}$$

Where the mating parts of a press or shrink joint are made of materials with different elastic moduli, direct application of Eq. (23.17) with $\sigma_z = 0$ results in

$$\delta_b = 2p_b b \left[\frac{1}{E_{ab}} \left(\frac{b^2 + a^2}{b^2 - a^2} - \nu_{ab} \right) + \frac{1}{E_{bc}} \left(\frac{c^2 + b^2}{c^2 - b^2} + \nu_{bc} \right) \right] \quad \text{for a hollow shaft} \tag{23.37}$$

where E_{ab} and ν_{ab} are Young's modulus and Poisson's ratio, respectively, in the shaft with inside radius a and interference radius b, and E_{bc} and ν_{bc} are Young's modulus and Poisson's ratio, respectively, in the hub with interference radius b and outside radius c.

Hence the interference pressure is

$$p_b = \frac{\delta_b/2b}{\dfrac{1}{E_{ab}} \left(\dfrac{b^2 + a^2}{b^2 - a^2} - \nu_{ab} \right) + \dfrac{1}{E_{bc}} \left(\dfrac{c^2 + b^2}{c^2 - b^2} + \nu_{bc} \right)} \quad \text{with a hollow shaft} \tag{23.37a}$$

Similarly, for a solid shaft, using Eq. (23.5) with $\sigma_z = 0$, Eqs. (23.14) and (23.16) for the solid shaft, and Eq. (23.17) for the hub, with $\sigma_z = 0$, the interference relations become the same as given by Eqs. (23.37) and (23.37a), but with $a = 0$. Thus for a solid shaft

$$\delta_b = 2p_b b \left[\frac{1 - \nu_{ab}}{E_{ab}} + \frac{1}{E_{bc}} \left(\frac{c^2 + b^2}{c^2 - b^2} + \nu_{bc} \right) \right] \tag{23.37b}$$

The stresses caused by the interference are found by application of Eqs. (23.12), (23.13), and (23.14):

$$\sigma_r = -\frac{p_b}{1 - (a/b)^2} \left[1 - \left(\frac{a}{r} \right)^2 \right] \quad \text{for a hollow shaft} \tag{23.12a}$$

$$\sigma_t = -\frac{p_b}{1 - (a/b)^2} \left[1 + \left(\frac{a}{r} \right)^2 \right] \quad \text{for a hollow shaft} \tag{23.13a}$$

$$\sigma_r = -\frac{p_b}{1 - (b/c)^2} \left[\left(\frac{b}{r} \right)^2 - \left(\frac{b}{c} \right)^2 \right] \quad \text{for the hub} \tag{23.12b}$$

$$\sigma_t = \frac{p_b}{1 - (b/c)^2} \left[\left(\frac{b}{r} \right)^2 + \left(\frac{b}{c} \right)^2 \right] \quad \text{for the hub} \tag{23.13b}$$

$$\sigma_t = \sigma_r = -p_b \quad \text{for the solid shaft} \tag{23.14}$$

(d) Stresses Due to Power Transmission in Combination with Interferences

Superposed upon the interference stresses and pressures must be those stresses caused by power transmission, bending, and centrifugal action. The latter is discussed in Sec. 27. The true stress picture associated with power transmission is quite complex because load transfer from the shaft to the hub does not occur at one position but rather all along the length of the interference. However, it is conservative and simplifying to assume this transfer occurs at one position along the length of the interference.

Employing this assumption in the case of only axial force and interference, appeal is made to St. Venant's principle; so that at an axial distance of about $c - b$ in the hub from the point of load transfer, and still within the zone of effect of the interference, the axial shear stresses have become essentially dispersed to give a resultant uniform axial normal stress of $F/\pi(c^2 - b^2)$ in the hub. This stress is simply treated as the third principal stress, the other two being σ_r and σ_t as determined by the interference fit. Similarly for the shaft, the axial principal stress is taken as $F/\pi(b^2 - a^2)$. Such axial forces may affect the interference fit, although this effect is not calculated. Of course, other sources of axial stress, such as shaft bending and discontinuity effects, must be superposed.

A similar treatment and use of St. Venant's principle are used for torsional power transmission, but in this case the effect of torsion is to rotate the directions of the principal stresses so that σ_t and σ_z are no longer principal values. Furthermore, since maximum torsional shear stress occurs at the outside while maximum tangential interference stress is at the inside surface of either mating part, it can be shown that it is possible for the maximum principal stress to occur at any point along the radius of either mating part (with the sole exception of the solid shaft), depending on the geometry, interference, and torque. The formulas to be used are Eqs. (23.13a), (23.13b), (23.14), and the following:

$$\tau = 2\,Tr/\pi(b^4 - a^4) \qquad \text{in the shaft} \tag{23.38}$$

$$\tau = 2\,Tr/\pi(c^4 - b^4) \qquad \text{in the hub} \tag{23.39}$$

and principal stresses at any point are net σ_r (including centrifugal action) and

$$\tfrac{1}{2}(\sigma_t + \sigma_z) \pm \sqrt{\left(\frac{\sigma_t - \sigma_z}{2}\right)^2 + \tau^2} \tag{23.40}$$

where σ_t = total tangent stress due to all sources (pressure, discontinuity effects, centrifugal)

σ_z = total axial stress due to all sources (axial force, bending of shaft as a beam, discontinuities)

For further aid in design, assistance in the design of power transmitting joints can be obtained from the charts of ref. 15. It should be remembered, however, that the total stress in these components is made up of stress due to interference, power transmission, bending, discontinuity effects (as outlined in refs. 10, 11, and 12), and centrifugal action, if any.

(e) Centrifugal Action on Shafts and Disks

The study of centrifugal effects on rotating parts appears to be a subject quite different from that of shrunk or press fits. However, in the case of shafts or flat disks, the mathematical treatment is so similar to that which has preceded that it is quite natural to discuss the theory in this section.

Equation of Motion. The same conditions exist as were used in writing Eq. (23.1), except that the volume element now has a constant acceleration of $\omega^2\rho = v^2/\rho$ toward the center of rotation, where ω = angular velocity, v = tangent linear velocity, and ρ = radius of rotation.

It is assumed that the center of rotation coincides with the geometric center of the shaft or disk, so that $\rho = r$. In this case Newton's second law of motion ($F = ma$) becomes

$$\sigma_t - \sigma_r - r(d\sigma_r/dr) = (\gamma/g)\omega^2 r^2 \tag{23.41}$$

where γ = weight density of the rotating part

g = acceleration of gravity

Hooke's Law under Plane Stress. Since most problems of this type occur under conditions of essentially plane stress or one on which a known axial stress can be superposed upon a system with zero axial stress, Eqs. (23.3) and (23.4) can be used in the simplified form

$$E\epsilon_r = \sigma_r - \nu\sigma_t \tag{23.42}$$

and

$$E\epsilon_t = \sigma_t - \nu\sigma_r \tag{23.43}$$

while Eq. (23.5) loses all utility.

Compatibility. Equation (23.2) remains valid. In a development quite smiliar to that of Eq. (23.9), the following differential equation can be constructed:

$$(1/r^3)(d/dr)[r^3(d\sigma_r/dr)] = -(3 + \nu)(\gamma/g)\omega^2 \tag{23.44}$$

Note that this equation is linear in σ_r. It is therefore permissible to separate the static effects of pressure entirely from the dynamic system and superimpose them later, since the static effects are simply part of the homogeneous solution to this equation. If conditions of generalized plane strain had been assumed ($\epsilon_z = 0$) instead of plane stress ($\sigma_z = 0$), the resulting equation of motion would have contained the factor $(3 - 2\nu)/(1 - \nu)$ in place of $(3 + \nu)$ on the right-hand side, and otherwise would be identical to Eq. (23.44).

The general solution to Eq. (23.44) (homogeneous solution plus particular integral) is

$$\sigma_r = C_1 + C_2/r^2 - [(3 + \nu)/8](\gamma/g)\omega^2 r^2 \tag{23.45}$$

For the solid shaft or disk, C_2 must be zero for stress to be finite, and if c is the outside radius of shaft, disk, or hub of a press- or shrink-fitted assembly, then $\sigma_r = 0$ at this radius. Thus for a system with a solid shaft, and with the same elastic constants for hub and shaft, the stresses due solely to centrifugal action are given by (with $r \leq c$)

$$\sigma_r = [(3 + \nu)/8](\gamma/g)\omega^2(c^2 - r^2) \qquad \text{with solid shaft} \tag{23.46}$$

and, from Eq. (23.41),

$$\sigma_t = (\gamma\omega^2/8g)[(3 + \nu)c^2 - (1 + 3\nu)r^2] \qquad \text{with solid shaft} \tag{23.47}$$

If the shaft is hollow with inside radius a, then C_2 is not zero and the boundary conditions for centrifugal stresses are $\sigma_r = 0$ at both $r = a$ and $r = c$. This results in stresses due solely to centrifugal action which are given by ($a \leq r \leq c$)

$$\sigma_r = [(3 + \nu)/8](\gamma/g)\omega^2(c^2 + a^2 - r^2 - a^2c^2/r^2) \tag{23.48}$$

and, from Eq. (23.41),

$$\sigma_t = (\gamma\omega^2/8g)[(3 + \nu)(c^2 + a^2 + a^2c^2/r^2) - (1 + 3\nu)r^2] \tag{23.49}$$

For generalized plane strain, the factor $(3 + \nu)$ in Eqs. (23.46) through (23.49) is replaced by $(3 - 2\nu)/(1 - \nu)$ and the factor $(1 + 3\nu)$ is replaced by $(1 + 2\nu)/(1 - \nu)$.

This system of stresses is to be superposed upon the stresses due to interference and other causes. In particular, note that σ_r at the interference radius b is always positive and therefore tensile and thus reduces the interference pressure.

Further discussion of this subject will be found in refs. 14 and 15. The former contains a description of an interesting and useful method for extending the analysis of the flat disk to one of variable thickness. Essentially, the method consists of approximating the variable thickness by a set of concentric rings, each of uniform thickness, and then imposing the condition that the radial force (not stress) per unit circumference shall be continuous across the mating boundaries. The same technique can be applied, where necessary, to the compound cylindrical vessel.

The radial deflections because of centrifugal action can be found from Eq. (23.43) [or Eq. (23.7) for the case of generalized plane strain] since $\epsilon_t = u/r$. Thus, for centrifugal action alone,

$$u = (r/E)(\gamma/8g)\omega^2[(3 + \nu)(1 - \nu)(c^2 + a^2) + (3 + \nu)(1 + \nu)(a^2c^2/r^2) - (1 - \nu^2)r^2] \tag{23.50}$$

For generalized plane strain, the factor $3 + \nu$ in Eq. (23.50) is replaced by $(3 - 2\nu)/(1 - \nu)$ and $1 - \nu^2$ is replaced by $1 - [2\nu^2/(1 - \nu)]$, while $-\nu r\epsilon_z$ is added to the total radial deflection.

(f) Design

Figure 23.5 shows an excellent design for shrink or press fit. Internal recessing provides better symmetry of the shrink-fit stresses and prevents a fulcrum in imperfect fits. Diameters of shaft are reduced at ends of hub, reducing the stress concentration and shrinkage allowance at these points. If a key is added to the fit, the

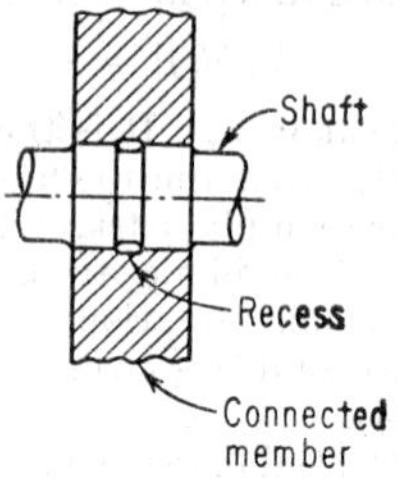

FIG. 23.5. Excellent press-fit design.

shaft seat diameter may be further increased and a more gradual step down to the shaft diameter is provided.[17]

References

1. Lamé and Clapeyron: Mémoire sur l'équilibre interieur des corps solides homogènes, *Mémoires présentés par divers savans*, vol. 4, 1833.
2. Becker, S. J., and L. Mollick: The Theory of the Ideal Design of a Compound Vessel, *Trans. ASME*, vol. 82, ser. B, pp. 136–142, May, 1960.
3. Morrison, J. L. M., B. Crossland, and J. S. C. Perry: Fatigue under Triaxial Stress, Development of a Testing Machine and Preliminary Results, *Proc. Inst. Mech. Engrs. (London)*, vol. 170, pp. 697–712, 1956.
4. Davis, E. A.: Relaxation of a Cylinder on a Rigid Shaft, *J. Appl. Mech.*, vol. 27, ser. E, no. 1, pp. 41–44, March, 1960.
5. Davis, E. A.: Relaxation of Stress in a Heat-exchanger Tube of Ideal Material, *Trans. ASME*, vol. 75, no. 1, pp. 381–385, April, 1952.
6. Finnie, Iain: Steady State Creep of a Thick-walled Cylinder under Combined Axial Load and Internal Pressure, *Trans. ASME*, vol. 82, ser. D, no. 3, pp. 689–694, September, 1960.
7. Svensson, N. L.: The Bursting Pressure of Cylindrical and Spherical Vessels, *J. Appl. Mech.*, vol. 25, pp. 89–96, 1958.
8. Marin, J., and F. P. J. Rimrott: Design of Thick-walled Pressure Vessels Based on the Plastic Range, *Welding Research Council Bulletin Series, Bull.* 41, July, 1958 (Welding Research Council of the Engineering Foundation; other pertinent references are listed in the bibliography of this bulletin).
9. Becker, S. J.: An Analysis of the Yielded Compound Cylinder, *Trans. ASME*, vol. 83, ser. B, pp. 43–49, February, 1961.
10. Friedrich, C. M., and S. J. Becker: A Thin Cylinder under Semi-infinite Outward Restraint, *Bettis Tech. Rev.*, WAPD-BT-9, pp. 78–83, August, 1958 (available from the Office of Technical Services, Department of Commerce, Washington 25, D.C.).
11. Sparenberg, J. A.: On a Shrink-fit Problem, *Appl. Sci. Res.*, Sec. A, vol. 7, pp. 109–120, 1958.
12. Severn, R. T.: Shrink-fit Stresses between Tubes Having a Finite Interval of Contact, *Quart. J. Mech. Appl. Math.*, vol. 12, pp. 82–88, February, 1959.
13. "Manual of Standard and Recommended Practice Wheel and Axle Manual," Association of American Railroads.
14. Timoshenko, S.: "Strength of Materials," part II, D. Van Nostrand Company, Inc., Princeton, N.J., 1958.
15. Horger, O. J.: "ASME Handbook, Metals Engineering—Design," Press-fitted Assembly, pp. 178–189, McGraw-Hill Book Company, Inc., New York, 1953.
16. Baugher, J. W.: Transmission of Torque by Means of Press and Shrink Fits, *Trans. ASME*, vol. 53, paper MSP-53-10, pp. 85–92, 1931.
17. Baugher, J. W.: Internal Stress and Fracture of Metals, *Metallurgia*, March, 1946.

Section 24

PRESSURE COMPONENTS (SEALS)

By

JOHN W. AXELSON, Ph.D., *Section Chief, Packings Development, Johns Manville Corp., Research Center, Manville, N.J.*

F. B. PINTARD, *Research Engineer, Johns Manville Corp., Research Center, Manville, N.J.*

CONTENTS

24.1. INTRODUCTION

A breakdown of pressure components into specific categories is difficult because of their great variety, but this section will be devoted entirely to seals—static and dynamic. The static seal will include those normally referred to as gaskets as well as other types which serve the same purpose and, in some cases, have additional uses. Dynamic seals are those which function between moving components, although the seal itself or part of the seal may be stationary.

The discussion of these seals will cover their use from a design viewpoint, not the design of the seal so much, but rather the design and characteristics of the component parts in order to best utilize the seal. A number of important factors in design which affect performance are discussed. These should aid in the design of the proper machine component and in the selection of the best seal for the application.

For more information than is here supplied on seal materials, see ref. 1.

STATIC SEALS

24.2. COMPRESSION TYPE

(a) General

Compression-type static seals are those which are installed between the surfaces to be sealed, and are preloaded to cause them to conform to the sealing surfaces and thereby eliminate or at least reduce leakage. When one of these seals, referred to as gaskets, is in use, certain forces are always present, and it is a balance of these forces, in combination with the gasket and joint characteristics, which determines the resultant sealability. A diagrammatic representation of these forces is shown in Fig. 24.1.

The mechanics of this type of system can be explained by the use of a simple equation:

$$F_b = N_b \sigma_b A_b \tag{24.1}$$

where F_b = total bolt load, lb
N_b = number of bolts
σ_b = bolt stress, psi
A_b = stress area (mean of pitch and root area) per bolt, in.^{2}

Very often the number of bolts is selected on the basis of aesthetic reasoning, and it is little wonder that some gasketed joints give trouble. A later section presents a method for calculating the number of bolts to use in a particular joint.

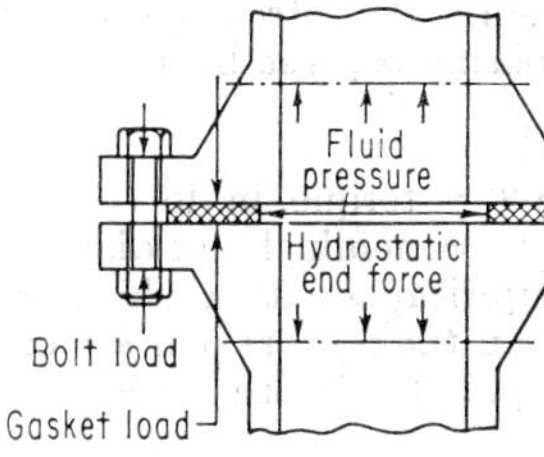

FIG. 24.1. Forces on a gasketed joint.

In almost all cases in the design of a gasketed joint, the total bolt load required F_b can be calculated. From this, the load per bolt or the product $\sigma_b A_b$ can be calculated by dividing the total bolt load by the number of bolts. If the maximum allowable stress σ_b for the particular bolting material is known, the stress area A_b can be calculated and the size of the required bolt determined. Until quite recently the root diameter of the bolts was generally used as the stress area, but a stress area based upon a diameter that is the mean of the pitch and root diameters has recently been adopted by ASA (Standard B1.1-1960).

When a joint or joint design is already available, the calculations can be made in the reverse manner. The stress area of the bolts is multiplied by the maximum allowable stress for the bolting material and by the number of bolts to give the total bolt load F_b.

The maximum allowable stress and the yield stress for standard bolting materials are given in Table 24.1. These are established by the American Society of Mechanical

Table 24.1. Bolting Materials and Allowable Stresses

Type of metal	ASTM specification	Nominal composition	Max allowable design stress, psi		Min yield stress, psi
Carbon steel	A261-BD	0.55 C	16,250		75,000
Carbon steel	A307-B	....	7,000		
Carbon steel	A325	0.30 C	18,750		77,000, 1–1½ in. 55,000, 1¾–3 in.
Low-alloy steel	A193-BA	....	16,250		
Low-alloy steel	A193-BB	....	18,750		
Low-alloy steel	A193-BC	....	20,000		
Low-alloy steel	A193-B5	5 Cr, ½ Mo	20,000		75,000– 90,000
Chrome steel	A193-B6	12 Cr	20,000		80,000–120,000
Low-alloy steel	A193-B7	1 Cr, 0.2 Mo	20,000		65,000–115,000
Low-alloy steel	A193-B7a	1 Cr, 0.6 Mo	20,000		95,000–120,000
Low-alloy steel	A193-B14	1 Cr, 0.3 Mo-V	20,000		105,000–120,000
Low-alloy steel	A193-B16	1 Cr, ½ Mo-V	20,000		85,000–105,000
Chrome-nickel steel	B8, B8C, B87	18 Cr, 8 Ni	15,000		30,000–100,000
Low-alloy steel	A354-BB	....	18,750		75,000– 80,000
Low-alloy steel	A354-BC	....	20,000		95,000–105,000
Low-alloy steel	A354-BD	....	20,000		120,000
			100°F	400°F	
Aluminum	B211-GS11A, T6	....	7,000	2,750	35,000
Aluminum	B211-CG42A, T4	....	8,000	3,650	40,000
Aluminum	B211-CS41A, T6	....	10,800	2,350	55,000
Copper	B-12	Copper	2,000	1,600	10,000
Copper	B98-A, C, and D	Copper-silicon	3,000	....	15,000
Copper	B98-B	Copper-silicon	2,400-11,000	....	12,000– 55,000
Nickel	B160	Low-carbon Nickel	2,000	1,900	10,000
Nickel	B164	Nickel-copper	4,900– 8,000	4,000–7,200	25,000– 40,000
Nickel	B166	Nickel-chromium Iron	5,900– 6,900	5,300–6,300	30,000– 35,000

Engineers, but other organizations such as the Society of Automotive Engineers may set up different values for the same or special materials. Under no circumstances should the yield stress be exceeded, and many times, especially with larger bolts, even the maximum design stress cannot be utilized.

Two methods of estimating the bolt load actually applied to a bolted joint are by the use of a torque wrench and a table or formula and by the use of an empirical equation. Bolt stresses as a function of torque are given in Table 24.2. A general relationship is given below.

$$F_b = N_b T / 0.2 D_b \tag{24.2}$$

where T = torque, in.-lb
D_b = nominal bolt diameter, in.

An empirical equation which can be used when a torque wrench is not available is based on the normal pull a mechanic would put on a standard open-end wrench for the required bolt size.

$$F_b = 16{,}000 D_b N_b \tag{24.3}$$

It should be understood that neither of these equations gives a very accurate determination of the bolt load, but often nothing more is available. Torques, especially, are misleading because the stress developed is so dependent on the friction factors present. Test data of applied torque vs. bolt stress calculated from bolt elongation show a band variation of about ± 40 per cent. This is largely attributed to the variation in the torque readings. The only accurate way to measure bolt load is by the measurement of bolt elongation using an extensometer or strain gages.

As can be seen from Fig. 24.1, the bolt load is applied to the gasket during installation, and therefore it is equal to the gasket load at that time. The gasket stress is this value divided by the gasket area which is in contact with the sealing faces. When a gasketed joint is placed in service so that an internal pressure is present, a new force, the hydrostatic end force, comes into play. This force can be calculated from the known pressure and the area over which it is effective. A number of experiments with different types of gaskets of different size have conclusively shown that the effective area at the leakage pressure is essentially the area based on the mid-diameter of the gasket.

$$F_h = P_i A_m \tag{24.4}$$

where F_h = hydrostatic end load, lb
P_i = internal pressure, psi
$A_m = (\pi/4) D_m^2$ = effective hydrostatic pressure area, in.^{2}
D_m = mid-diameter of gasket, in.

The effect of this hydrostatic end force is highly dependent upon the particular assembly. This is one of the reasons a gasketed joint design should include an ample factor of safety. In a rigid assembly where a constant load is maintained on the joint, the hydrostatic end force balances some of the initial load. When the two become equal even the best gasketed joint will leak.

In a rigid assembly where a constant load is maintained on the gasket rather than on the joint (constant strain), the hydrostatic end load adds directly to the initial load. Most gasketed assemblies are somewhere between these two extremes, but no one has yet been able to calculate exactly where, because of other factors such as flange rotation, gasket stress relaxation, and bolt creep.

A factor in all seal applications which most designers do not usually consider is the amount of leakage which can be tolerated. This will vary from a relatively large amount which can be expected in certain oven or furnace seals to the molecular-diffusion magnitude of leakage which is almost too much in certain radioactive or toxic-fluid applications. The time element also becomes important in establishing a criterion of leakage because it may change radically, especially if there are temperature and pressure cycles involved.

Table 24.2. Torque vs. Bolt Stress and Bolt Load

Nominal diam	Bolt stress, psi											
	7,500		15,000		30,000		45,000		60,000		90,000	
	Torque, ft-lb	Load, lb	Torque, ft-lb	Load, lb	Torque, ft-lb	Load, lb	Torque, ft-lb	Load, lb	Torque, ft-lb	Load, lb	Torque, ft-lb	Load, lb
National Coarse-thread Series												
1/4	1	240	2	480	4	950	6	1,430	8	1,900		
5/16	2	390	4	780	8	1,570	12	2,350	16	3,130		
3/8	3	600	6	1,160	12	2,320	18	3,480	24	4,640		
7/16	5	800	10	1,590	20	3,180	30	4,770	40	6,360		
1/2	8	1,060	15	2,120	30	4,250	45	6,370	60	8,500		
9/16	12	1,360	23	2,720	45	5,450	68	8,170	90	10,900		
5/8	15	1,690	30	3,380	60	6,770	90	10,200	120	13,500		
3/4	25	2,510	50	5,010	100	10,000	150	15,000	200	20,000		
7/8	40	3,460	80	6,920	160	13,800	240	20,800	320	27,700		
1	62	4,540	123	9,080	245	18,200						
1 1/8	98	5,720	195	11,400	390	22,900						
1 1/4	137	7,260	273	14,500	545	29,100						
1 3/8	183	8,650	365	17,300	730	34,600						
1 1/2	219	10,500	437	21,100	875	42,100						
1 3/4	390	14,200	775	28,500	1,550	56,900						
2	563	18,700	1,125	37,500	2,250	74,900						
8-thread Series												
1	...	......	.....	......	245	18,200	368	27,200	490	36,300		
1 1/8	...	......	.....	......	355	23,700	533	35,500	710	47,400		
1 1/4	...	......	.....	......	500	30,000	750	44,900	1,000	59,900		
1 3/8	...	......	.....	......	680	37,000	1,020	55,400	1,360	73,900		
1 1/2	...	......	.....	......	800	44,700	1,200	67,000	1,600	89,400		

Table 24.2. Torque vs. Bolt Stress and Bolt Load (Continued)

Nominal diam	Bolt stress, psi											
	7,500		15,000		30,000		45,000		60,000		90,000	
	Torque, ft-lb	Load, lb	Torque, ft-lb	Load, lb	Torque, ft-lb	Load, lb	Torque, ft-lb	Load, lb	Torque, ft-lb	Load, lb	Torque, ft-lb	Load, lb
8-thread Series (Continued)												
1⅝	...	......	.....	......	1,100	53,200	1,650	79,800	2,200	106,300		
1¾	...	......	.....	......	1,500	62,400	2,250	93,600	3,000	124,800		
1⅞	...	......	.....	......	2,000	72,300	3,000	108,500	4,000	144,600		
2	...	......	.....	......	2,200	83,000	3,300	124,500	4,400	166,000		
2¼	...	......	.....	......	3,180	106,600	4,770	159,800	6,360	213,100		
2½	...	......	.....	......	4,400	133,100	6,600	199,600	8,800	266,100		
2¾	...	......	.....	......	5,920	162,500	8,880	243,700	11,840	325,000		
3	...	......	.....	......	7,720	194,900	11,580	292,300	15,440	389,700		
National Fine-thread Series												
¼	...	......	.....	......	6	1,090	9	1,630	12	2,120	18	2,930
5/16	...	......	.....	......	12	1,740	18	2,600	24	3,480	36	5,210
⅜	...	......	.....	......	22	2,630	31	3,940	44	5,260	66	7,880
7/16	...	......	.....	......	32	3,560	50	5,070	64	7,100	96	10,700
½	...	......	.....	......	50	4,790	76	7,100	100	9,600	150	14,200
9/16	...	......	.....	......	72	6,080	109	9,120	144	12,200	216	18,200
⅝	...	......	.....	......	98	7,670	148	11,500	196	15,300	294	23,000
¾	...	......	.....	......	170	11,200	255	16,800	340	22,300	510	33,500
⅞	...	......	.....	......	270	15,300	404	22,800	540	30,500	810	45,800
1	...	......	.....	......	390	19,900	587	29,800	780	39,700	1,170	59,600
1⅛	...	......	.....	......	574	25,600	860	38,500	1,148	51,300	1,722	76,900
1¼	...	......	.....	......	790	32,200	1,180	48,300	1,580	64,300	2,370	96,500
1⅜	...	......	.....	......	1,044	42,400	1,565	59,000	2,088	78,800	3,132	118,200
1½	...	......	.....	......	1,358	47,400	2,060	71,000	2,716	94,800	4,074	142,200

(b) ASME Code

The American Society of Mechanical Engineers' Code for Unfired Vessels[2] presents the most commonly used design methods for gasketed joints. Although these design calculations have been used for about 20 years, it is doubtful whether the intricacies required to obtain some of the values are worthwhile because of the lack of real data and the number of other factors of unknown value which enter into any static-sealing problem.

The code makes use of two basic equations to calculate bolt load, with the larger calculated load being used for design.

$$W_{m1} = H + H_p = 0.785G^2P + (2b \times 3.14GmP) \qquad (24.5)$$
$$W_{m2} = H_y = 3.14bGy \qquad (24.6)$$

where W_{m1} = required bolt load for maximum operating or working conditions, lb
W_{m2} = required initial bolt load at atmospheric-temperature conditions without internal pressure, lb
H = total hydrostatic end force, lb ($0.785\ G^2P$)
H_p = total joint-contact-surface compression load, lb
H_y = total joint-contact-surface seating load, lb
G = diameter at location of gasket load reaction; generally defined as follows:
When $b_0 < \frac{1}{4}$ in., G = mean diameter of gasket contact face, in.
When $b_0 > \frac{1}{4}$ in., G = outside diameter of gasket contact face less $2b$, in.
P = maximum allowable working pressure, psi
b = effective gasket or joint-contact-surface seating width, in.
$2b$ = effective gasket or joint-contact-surface pressure width, in.
b_0 = basic gasket seating width per table*; the table defines b_0 in terms of flange finish and type of gasket, usually from one-half to one-fourth gasket contact width
m = gasket factor per table*; the table shows m for different types and thicknesses of gaskets ranging from 0.5 to 6.5
y = gasket or joint-contact-surface unit seating load, psi, per table*, which shows values from 0 to 26,000 psi

In Eq. (24.5) use is made of an m value which provides a margin of safety to be applied when hydrostatic end force becomes a determining factor. This value is very difficult to determine experimentally since it is not a constant. Equation (24.6) assumes that a certain unit stress is required on a gasket to make it conform to the sealing surfaces and be effective. The y or yield-stress values given in the code were apparently based upon experience and judgment but are very difficult to obtain experimentally. Other factors considered by the code relate to the surface finish and the effective gasket width to be used with these finishes. No design criteria are given for the spacing of the bolts, an item of great importance.

(c) Simplified Design

A much simpler method of calculation has been found to be very effective.[3] This is based on the stress required to seat the gasket and the hydrostatic end force developed in the joint. Basically, the equations do the same thing as the code but they are simplified by using the full gasket contact width regardless of the flange width or the surface finish of the sealing faces.

The method of calculation depends upon whether a joint is being designed or whether a gasket is being determined for a joint already in existence.

In all cases it is necessary to select the type of gasket required based on temperature, pressure, fluid, flange finish, etc. These are described in numerous articles and in manufacturers' catalogues and literature. After the desired gasket has been selected, the minimum seating stress, as given in Table 24.3, is used to calculate the total bolt

* Tables UA-47.1 and 47.2 in ASME Code for Unfired Pressure Vessels.[2]

Table 24.3. Minimum Seating Stresses for Gaskets

Material	Thickness	Minimum seating stresses, psi
Rubber sheet:		
75 durometer	1/32 in. and up	200
60 durometer	1/32 in. and up	175
Compressed asbestos	1/64 in.	3,500
	1/32 in.	2,500
	1/16 in.	1,800
	1/8 in.	1,400
Rubberized cloth	2 ply	2,500
	3 ply	2,100
	4 ply	1,800
Vegetable-fiber sheets	All	750
Teflon:		
Virgin	1/64 in.	14,000
	1/32 in.	6,500
	1/16 in.	3,700
	1/8 in.	1,600
Glass-filled	1/64 in.	14,000
	1/32 in.	11,000
	1/16 in.	6,000
	1/8 in.	3,000
Asbestos cloth, impregnated	3/32 in.	1,600
Flat metals:		
Aluminum	All	20,000
Copper		45,000
Carbon steel		70,000
Monel		85,000
Stainless		95,000
Corrugated sheet metal:		
Lead	All	500*
Aluminum		1,000
Copper		3,500
Monel		4,500
Stainless		6,000
Profile solid metal:		
Aluminum	All	25,000†
Copper		35,000
Carbon steel		55,000
Monel		65,000
Stainless		75,000
Corrugated sheet with asbestos:		
Aluminum	All	2,000*
Copper		2,500
Carbon steel		3,000
Monel		3,500
Stainless		4,000
Plain metal jacketed:		
Lead	All	500
Aluminum		2,500
Copper		4,000
Carbon steel		6,000
Monel		7,500
Stainless		10,000
Spiral wound	All	3,000–30,000

* Based on total projected area.

† Based on actual contact area (0.010-in.-width profiles).

load required by Eq. (24.7).

$$F_b = S_g A_g \tag{24.7}$$

where S_g = gasket seating stress, psi
A_g = gasket contact area, in.^{2}

This equation will be satisfactory for the lower pressures in the smaller sizes since hydrostatic end force is not a major factor. No limits can be set to specify these values of size and pressure; so another calculation is required:

$$F_b \geqq K P_t A_m \tag{24.8}$$

where P_t = test pressure, psi
A_m = effective hydrostatic area based on gasket mid-diameter, in.^{2}
K = safety factor

Safety factors are based upon the joint conditions and the operating conditions but not on surface finish or the type of gasket. For minimum-weight applications where all installation factors (bolt lubrication, tension, parallel surfaces, etc.) are carefully controlled, low-temperature applications, and applications where adequate proof pressure is applied the K factor is 1.2 to 1.4. For most normal designs where weight is not a major factor, vibration is moderate, and temperatures do not exceed 700 to 800°F, the K factor is 1.5 to 2.5. Use the high end of the range where bolts are not lubricated. For cases of extreme fluctuations in pressure, temperature, or vibration; where no test pressure is applied; or where uniform bolt tension is difficult to ensure; K factor is 2.6 to 4.0.

In essence, these factors are similar to the m values in the code. As noted, the total bolt load F_b calculated by Eq. (24.7) should always be equal to or greater than the value given by Eq. (24.8). If it is not, it should be increased. In most cases this can be done safely without changing the gasket because most gaskets can withstand at least four times the minimum gasket seating stress without sustaining any damage. However, in some cases it may be necessary to change the gasket dimensions or even to change the type of gasket in order to meet the requirement of Eq. (24.8).

(d) Rule of Thumb

A rule of thumb for determining the range of usage of different types of gasket materials is based on the product of temperature in degrees Fahrenheit and pressure in psi with a maximum temperature being specified. Values are given in Table 24.4.

Table 24.4. *PT* Values for Gasket Materials

Material	Max $P_i T$*	Max temp, °F
Rubber	15,000	300
Vegetable fiber	40,000	250
Rubberized cloth	125,000	400
Compressed asbestos	250,000	850
Metals	>250,000	Depends on metal

* P_i = pressure, psi
T = temperature, °F

(e) Flange Width

There are no known methods of calculating the optimum flange width for gaskets, and the widths are usually picked on a basis of common-sense judgment based upon the following factors:

1. Most gasket have a minimum flange width based upon their strength and fabrication limitations. These can range from a minimum of about 1/32 in. for a rubber or solid metal O-ring to 3/8 in. for a corrugated, jacketed metal gasket.

2. The maximum flange width will usually be limited by the bolt load available or by the practicability of manufacture.

3. The optimum flange width is that width which will give an effective seal with a minimum bolt load. It can be arrived at by proper consideration of the seating stress, hydrostatic end force, and compression characteristics of the gasket.

(f) Flange Rotation

Flange rotation is a factor which is present in all flanged joints, but it becomes important only in large sizes, thin flanges, and at relatively high pressures. Generally, if the proper gasket is selected and the joint is well designed, flange rotation is not important because it has very little effect on maintaining a seal. When it does become a problem, some means to prevent excessive rotation such as a full-face gasket or a limiting device outside the bolts are usually incorporated into the gasket. The rotation is caused by the bolt load being applied a great distance from the center line of the flange and the effect of the uneven hydrostatic stresses over the flange width.

(g) Gasket Thickness

Gasket thickness is another factor which is difficult to specify. Generally, the thinnest gasket consistent with the requirements should be used. These requirements are determined by the limits of manufacture, flange surface finish, flange flatness, flange parallelism, and compressibility required. Rubber and compressed-sheet gaskets are often used in $\frac{1}{32}$-in. thicknesses when the flanges have a good finish and are parallel; they are seldom used above $\frac{1}{8}$-in. thickness. Some gaskets like the spiral-wound asbestos-steel types are not generally available in thicknesses less than $\frac{1}{8}$ in.

(h) Compressibility and Recovery

For years compressibility and recovery values have been specified for gasket materials without any real correlations between these values and the performance of the material. Actually, compression and recovery values were for quality-control purposes in specifications for materials which had been found to work in a certain application. When a new material came along, the compressibility and recovery probably had to be in a certain range for it to get further consideration. This further study consisted of other tests more related to performance or even an actual application test. Many materials were then discarded despite their successful compressibility and recovery results. It is probable that as many good materials are discarded because of their compression and recovery values as are obtained after additional application tests.

Knowledge of gasket compressibility is most important in two particular types of use. First, if the mating surfaces have a major deficiency in surface finish, flatness, or parallelism, a certain amount of compressibility is necessary in order to obtain contact between the gasket and flanges over their entire area. Second, many joints require a definite spacing (e.g., metal-to-metal connections and parts with critically aligned components) and the gasket must therefore have enough compressibility to allow the joint to be made up metal-to-metal with the available bolt load. It must not be so compressible that the seating stress is not obtained on the gasket before the load shifts to the metal-to-metal joint.

Recovery of a gasket material can be related to sealability in long-time service or where flange separation can occur. The recovery as usually measured, however, has no real value because it is taken from full to zero (or essentially zero) load and any joint would undoubtedly leak long before this recovery could occur. If recovery were a straight-line function, these figures might still be of some use but usually most of the recovery occurs at the very low loads which cannot be allowed in actual service.

Figure 24.2 gives some typical compression curves for a few types of gasket materials. It should be understood that different formulations of the same types can give widely different values.

(i) Sealability

Sealability is the primary property of a gasket material. Since there is no general, all-inclusive definition of sealability, there is no general method of measuring it. Sealability is usually empirically determined; that is, an actual or simulated joint is set up under a certain loading and tested for sealability under certain conditions of temperature, pressure, time, and fluid. Many times a compression machine is used to provide a more accurate load although it does not duplicate the load behavior

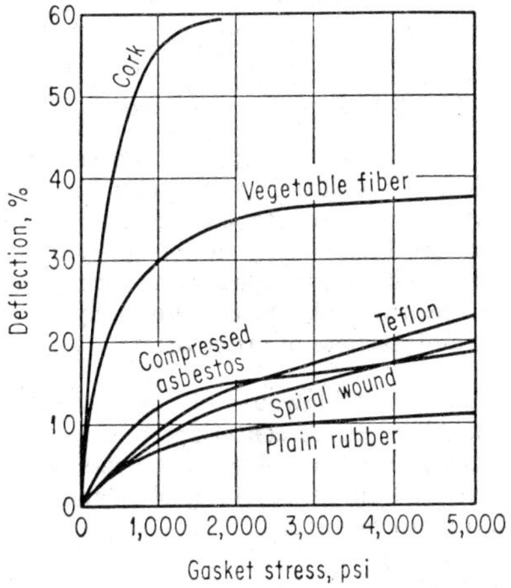

FIG. 24.2. Compression of gasket materials.

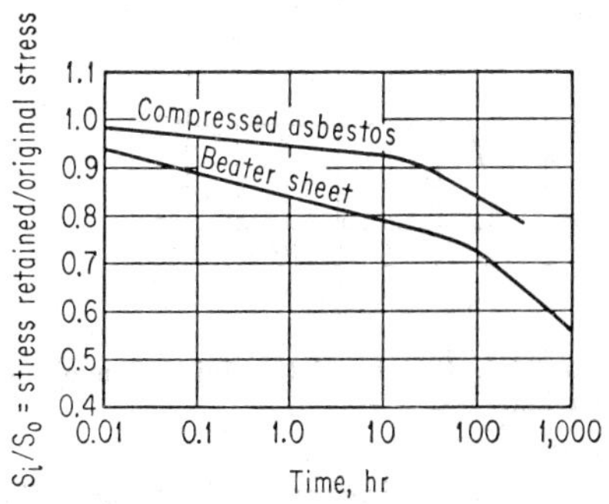

FIG. 24.3. Stress relaxation of gasket materials

in an actual joint. A standardized sealability test procedure (D2025-62T) has been published by ASTM and SAE.

Probably one of the most important properties of a gasket material is stress relaxation. Although there is undoubtedly a correlation between recovery and stress relaxation, most of the measured recovery is in the area of very low stress, whereas the stress relaxation is usually measured at stresses 50 per cent of the full stress or higher. Although data are very limited, Fig. 24.3 shows the type of plot obtained with two different types of materials. ASTM has recently adopted standard D2139-62T for this characteristic.

(j) Bolt Spacing

Bolt spacing is very important in gasketed joints, but little attention is paid to it and there are no good rules. Generally, the total load required is calculated and either the number of bolts is selected from an appearance or manufacturing viewpoint or the size of bolt is determined by appearance or flange width. This situation led Roberts[4] to develop the following approximate technique.

Based upon the assumption that a flange is analogous to an elastic beam with point loading representing the bolt loads, equations were derived to show the load midway between the two bolts. It was further assumed that this load should be 95 per cent of the load at the bolts, and a curve was derived plotting the relationship between the ratio of bolt spacing to flange thickness and a function S defined as follows:

$$S = tb_f E_f / db_g E_g \tag{24.9}$$

where t = gasket thickness, in.
b_f = flange width, in.
b_g = gasket width, in.
d = flange thickness, in.
r = bolt spacing, in.
E_g = modulus of elasticity of gasket
E_f = modulus of elasticity of flange material

This plot is shown in Fig. 24.4 where q is the load midway between two bolts and q_0 is the load at the bolts. When the factor r/d has been determined, the value of r is calculated and the number of bolts obtained by dividing the spacing into the bolt-circle circumference.

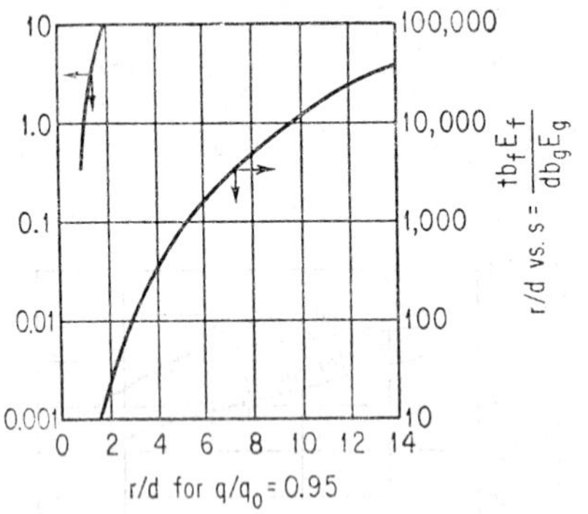

FIG. 24.4. Plot for determining flange bolt spacing.

(k) Confined vs. Unconfined

The primary example of the confined gasket is the static O-ring seal. Here it is very necessary to confine the O-ring because it is usually a relatively thick gasket with high elongation. However, it is desirable also to confine many other types of gaskets in order to:

1. Prevent blowout
2. Allow easier centering
3. Ensure controlled compression to prevent damage to the gasket and misalignment of parts

The major disadvantages of confined gaskets are:
1. It is more difficult to produce the desired surface finish.
2. Gasket width must be reduced for the same overall flange width.
3. Flange thickness may have to be increased.
4. Incorrect gasket thickness will result in too much or too little load on the gasket.
5. Bolt elasticity has no effect on maintaining tightness of joint if stress relaxation occurs in the gasket.

Although the disadvantages outnumber the desirable characteristics, there are many times when the confined gasket joint is very advantageous.

24.3. PRESSURE-ACTUATED TYPES

(a) Theory and Design

One of the troublesome aspects in the use of standard flat-faced gaskets is the difficulty of sealing high pressures. This is caused by the hydrostatic end force which tends to separate the flanges, thereby decreasing the flange load on the gasket. The same net effect occurs if the gasket or the joint tends to relax during service.

At high pressures a compensating action can be obtained if the pressure can be utilized to force the gasket into sealing contact with the flanges. This can be accomplished in several ways, depending upon whether the gasket is sealing in a radial or axial direction.

Probably the most common type of gasket which may be pressure-actuated is the O-ring. When an O-ring is installed in a groove between two flanges, a certain amount of compression or gasket load is supplied to give at least an initial seal. Let us assume that the flanges are brought together metal-to-metal. As pressure is applied to this joint, it causes the O-ring to expand outward against the outside diameter of the groove and at the line between the flanges. With an increase in internal pressure the flanges may tend to separate but the O-ring is forced into this opening and thereby seals it off. This will continue until the pressure becomes so high or the opening between the flanges becomes so great that the O-ring is extruded out between the flanges and failure occurs.

The same principle applies to other types of gaskets such as the delta, lens, and new ring joint, where radial movement of the gasket causes it to seal the opening between the two flanges.

The other type of pressure-actuated gasket is represented by the Bridgman joint, bellows gasket, perforated metal O-ring, and other specialty types such as the Skinner, Cadillac, and Bar-X seals. In the case of the Bridgman joint the method of sealing is the same as for the O-ring, delta, etc., but the forces are applied in an axial direction. However, the sealing is accomplished by sealing off a space between two mating parts by wedging the gasket in between.

With the other types which seal by the action of pressure in the axial direction, the sealing is conventional except that the internal pressure gives the ultimate gasket

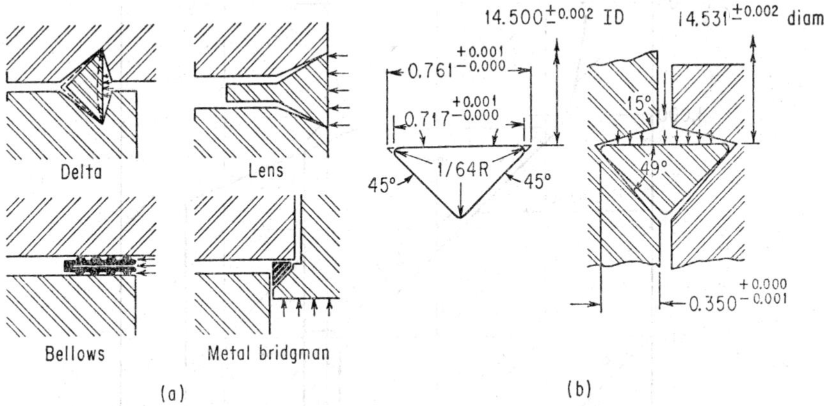

FIG. 24.5. (*a*) Pressure-actuated gaskets. (*b*) Delta gasket design.

load. The primary factor in design is to provide as large an area as possible for the pressure actuation and as small an area as possible for the sealing area. This results in a high stress at the sealing surface and thereby promotes good sealing ability. The design and mechanics of pressure-actuated joints are given in Fig. 24.5*a* and *b*.

24.4. DIAPHRAGM SEALS

(a) Theory and Design

Diaphragm seals form a continuous membrane across the fluid path and therefore prevent leakage past the plane of the diaphragm. They are used exclusively in valves and pumps where relative motion between two components is quite small and in a reciprocating action. In addition to creating a seal across the flow path of the fluid, the diaphragm serves as a gasket around its perimeter and center which serve as the points of attachment to the equipment.

In a valve, the diaphragm center is attached to the valve stem or valve-operating mechanism so that it flexes and allows reciprocating motion of the valve-stem assembly as the valve is opened or closed. At the extremes of motion the diaphragm is extended close to its maximum without causing excessive tensile stress. The action in a diaphragm pump is essentially the same. The outer periphery of the diaphragm is attached between the bonnet and the body of the valve or between the halves of the pump as with a standard gasket.

The design of a diaphragm seal is relatively simple since the major requirement is that the diaphragm material have the flex life, ease of flexing, and pressure resistance required. The gasket requirements must also be met and the basic design criteria for gaskets can be applied.

Although many diaphragm materials are used, the most common are rubber, rubberized cloth with and without wire reinforcement, Teflon, polyethylene, and beryllium bronze.

DYNAMIC SEALS

24.5. General

Dynamic seals are required to seal four types of shaft motion: rotary, reciprocating, helical, and swinging-rotary. There are a wide variety of materials and designs from which to choose. Before the selection of the proper packing can be made many

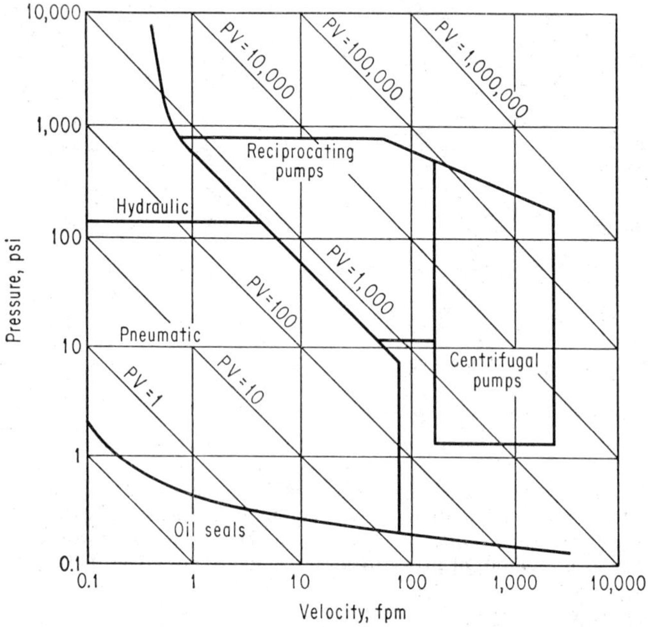

FIG. 24.6. *PV* application ranges for dynamic seals.

factors must be considered: cost, equipment requirements necessary for the installation, service conditions, and performance level required. The types of dynamic seals or packing are classified: (1) compression types, sometimes referred to as mechanical packings or jam type; (2) automatic or pressure-actuated, almost always molded or machined; (3) oil seals; (4) mechanical seals, including face seals; (5) labyrinth or positive-clearance seal; and (6) piston rings, both metallic and nonmetallic.

The selection of the type of dynamic seal required for a particular application can be guided by estimating the severity of the pressure and velocity conditions in terms of a *PV* factor, which is defined as the product of the fluid pressure in psi and the surface velocity in fpm. The *PV* parameter is an approximate measure of the frictional heat generated between the packing and the shaft. Figure 24.6 is a plot of the *PV* ranges for various common dynamic-seal applications.

24.6. COMPRESSION TYPES

(a) General

There are many types of compression packings and some variations in installation but all function by restricting the axial flow of the medium being sealed to an acceptable rate. Whenever there is considerable relative motion with this type of packing, there must always be some leakage to provide lubrication and prevent temperature buildup. Figure 24.7 illustrates the basic installation of a compression packing.

The packing may be considered as providing an adjustable labyrinth which expands radially against the shaft because of a compressive force applied at one end, either from applied gland load or from the pressure of the fluid being sealed. When fluid flow or leakage occurs, there is a drop in fluid pressure from 100 per cent at the packing ring at the bottom of the stuffing box to almost zero per cent at the packing ring next to the gland. The leakage rate is dependent upon the ability of the packing to maintain its sealing force against the shaft under the operating conditions. With

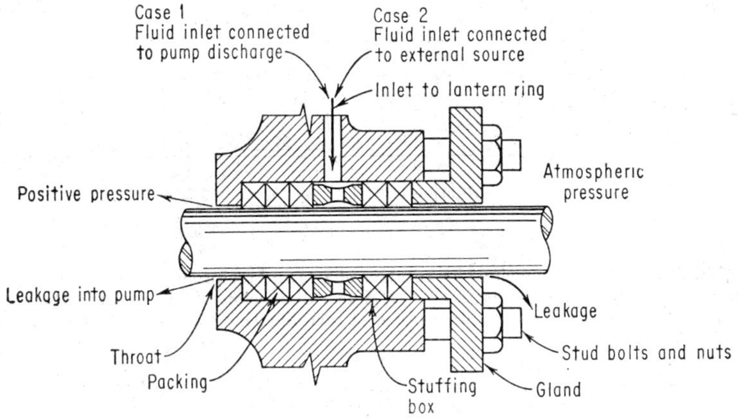

FIG. 24.7. Typical stuffing box with lantern ring.

fluctuating fluid pressures, compression packings do not function satisfactorily without gland adjustment. Although the packing can be seated effectively for a high fluid pressure, an abrupt pressure drop will cause the leakage to be reduced and the packing will overheat; or if effective sealing is accomplished at the low pressure, the leakage will be excessive at the high pressure.

(b) Design

The stuffing box shown in Fig. 24.8*a* and *b* and Table 24.5 for rotating and reciprocating shaft dimensions is based upon the recommendations of the Mechanical Packings Association.[5] These dimensions are intended for soft packings to approximately 1,500 psi and are independent of shaft speed. The shaft clearance at the throat can vary between 0.008 and 0.030 in. dependent upon the pressure sealed. Gland clearance at the inside diameter and outside diameter should be held to a minimum to prevent extrusion of the packing because of the relative motion. The inside diameter clearance is usually between 0.020 and 0.075 in.; clearance at the outside diameter can be held at 0.010 to 0.046 in. These clearances provide gland alignment and should not permit shaft interference at the inside diameter. The recommended dimensions will also accommodate molded-type packings.

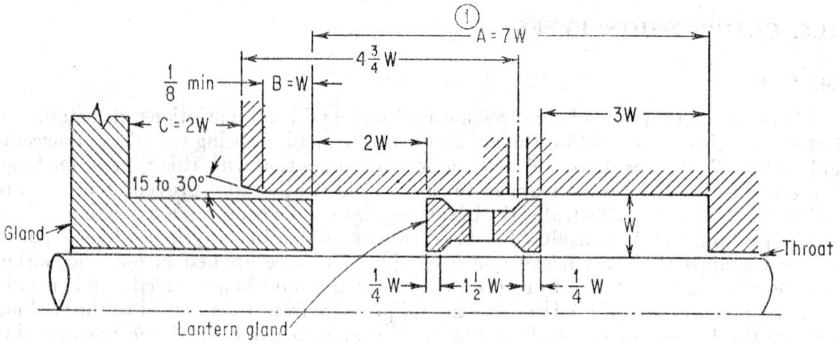

(a)

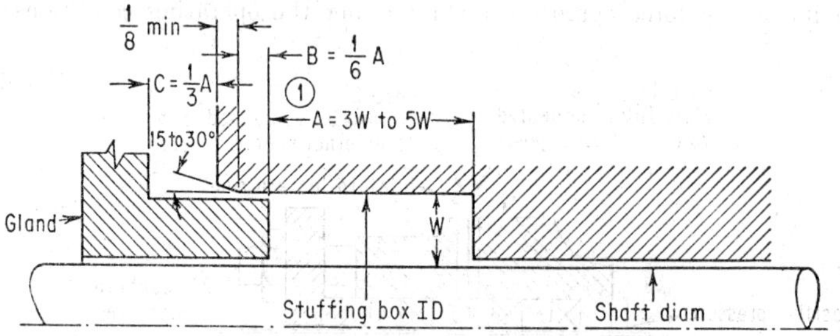

(b)

FIG. 24.8. Stuffing-box dimensions for rotating shafts and reciprocating rods. (a) Stuffing-box dimensions for rotating shafts ①. A depth dimension of 7W is used where lantern gland is used. (b) Stuffing-box dimensions for reciprocating rods ①. A = total depth of packing.

Table 24.5. Packing Width vs. Shaft Diameter

Rotary shafts

Shaft diam, in.	Packing width, in. W
⅝ to and including 1⅛	$\frac{5}{16}$
Over 1⅛ to and including 1⅞	⅜
Over 1⅞ to and including 3	½
Over 3 to and including 4¾	⅝
Over 4¾ to and including 12	¾

Reciprocating rods

Rod diam, in.	Packing width, in.	Dash No.
¼ to and including 1¼	¼	8-24
Over 1¼ to and including 2½	$\frac{5}{16}$	25-35
Over 2½ to and including 3¾	⅜	36-46
Over 3¾ to and including 5½	$\frac{7}{16}$	49-55
Over 5½ to and including 15	½	56-80

Soft-compression-type packings are usually made in straight lengths which are cut to form rings that fit the stuffing-box dimensions. The ratio of shaft to bore dimensions should conform to the dimensions given in order to minimize uneven density of the packing from inside to outside diameter. If the width W of the packing is too large, the inside diameter of the formed ring will bulge and the outside diameter will neck down, resulting in high shaft friction while still tending to leak along the bore.

In the following discussion numerous references are made to the excellent work of Denny and Turnbull.[6]

(c) Number of Rings, Stress Decay, and Pressure Drop

When a packing is installed, it is seated by tightening the gland stud bolts and thus applying an axial force on the packing. The axial stress results in a radial stress which determines the sealing effectiveness of the packing, the magnitude of the radial stress depending upon the elasticity of the packing. In a packing set, the greatest axial stress exists on the ring next to the gland. Each succeeding ring is subject to less axial stress, as shown in Fig. 24.9 for a six-ring set.

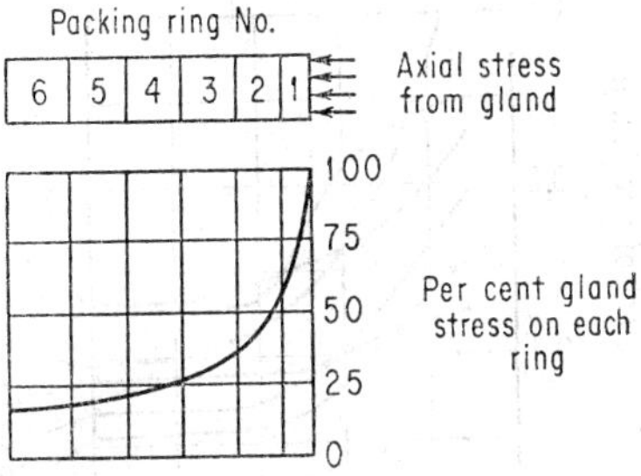

FIG. 24.9. Stress decay through a set of packing.

This loss in stress transmission through the packing set is principally caused by the friction between the stuffing-box bore and the packing outside diameter and, to some extent, the friction between the shaft and the inside diameter of the packing. The rougher the bore, the greater the proportion of gland force taken up by the last ring installed. There are no fixed axial-stress values required to seat compression packing, but the following estimates may be useful as a guide:

Type of packing	*"Estimated" min axial stress, psi, required to seat packings*
Teflon-impregnated braided asbestos	200
Plaited asbestos, lubricated	250
Plastic	200
Braided vegetable fiber, lubricated	160
Braided metallic	400–500

The fluid pressure distribution along a set of packing can be illustrated as shown in Fig. 24.10.

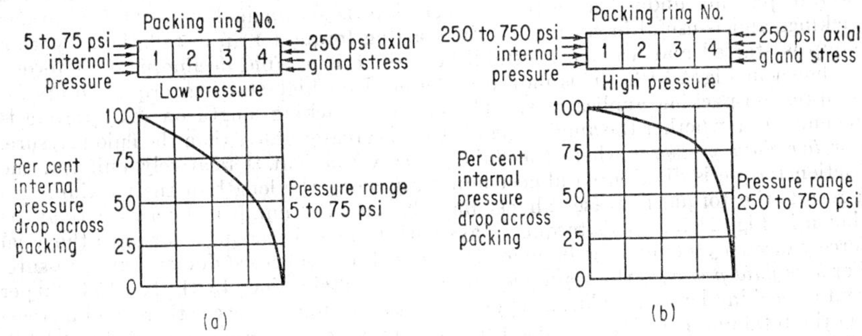

FIG. 24.10. Fluid pressure drop vs. packing length. (*a*) Low pressure; (*b*) high pressure.

On the basis of the foregoing, it might appear that a packing set would require two or possibly three rings only because the two rings next to the gland do approximately 75 per cent of the sealing. However, additional rings are required to compensate for packing wear, to help throttle pressure, to aid in spreading out the shaft wear, and in some cases, to protect the sealing rings from abrasive media. Suggested number of packing rings to be used are: for a stuffing box without lantern ring, five or six rings; for a stuffing box with lantern ring, three rings inside, two outside; for smothering-type gland (used with hazardous fluids), six, eight, or ten rings depending on service. Greater numbers are often used in high-pressure installations.

(d) Power Requirements

The power requirement of a compression packing is influenced by many variables such as packing coefficient of friction, packing length, initial gland-compression load

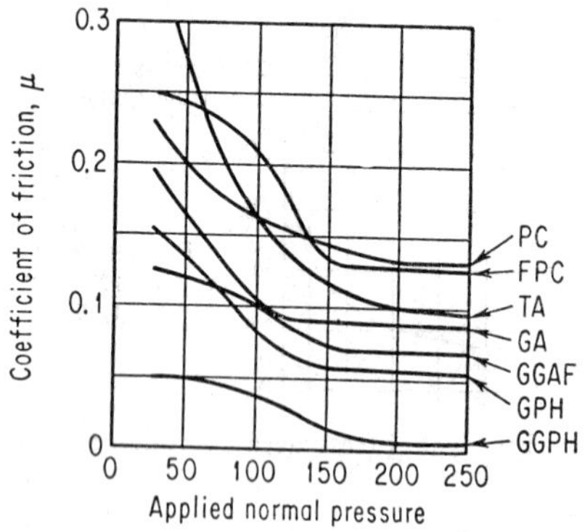

FIG. 24.11. Packing coefficient of friction.
TA = Teflon-impregnated plaited asbestos
GA = graphite-plaited asbestos
$GGAF$ = graphite-greasy asbestos fiber
GPH = graphite-plaited hemp
$GGPH$ = graphited-greasy plaited hemp
PC = plain plaited cotton
FPC = Teflon-impregnated plaited cotton

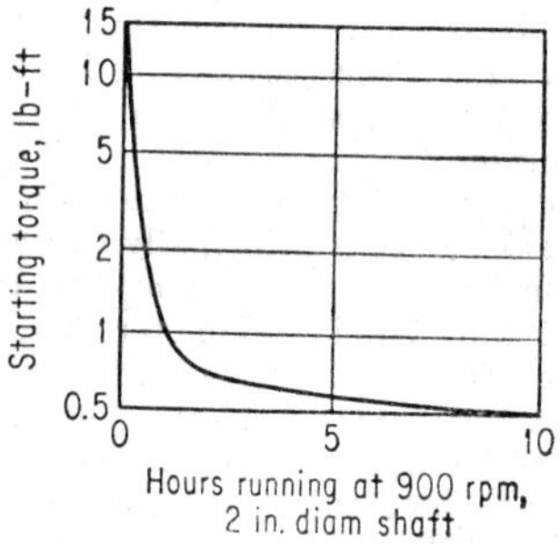

FIG. 24.12. Effect of running time on friction torque. Four rings of Teflon-impregnated plaited cotton. Applied gland pressure, 215 psi. Water pressure, 75 psi. Curve calculated from coefficient of friction.

on packing, fluid pressure, shaft diameter, shaft speed, shaft finish, leakage rate, and lubrication.

The packing coefficient of friction as determined by the pull required to move a piece of packing under load across a steel plate is shown in Fig. 24.11 for various packings and is useful in calculating the starting torque for an installation. Once a pump is started and leakage is established, there is a rapid decline in shaft torque as shown in Fig. 24.12. This indicates the need for high-starting-torque motors for compression-packing applications. The effect of packing length on shaft torque is dependent on whether the applied gland stress is greater or less than the fluid pressure. For *low fluid pressures* where the fluid pressure gradient is relatively uniform, the friction torque is distributed almost uniformly along the length of the packing. As a result, the torque increases almost linearly with the number of packing rings, as shown in Fig. 24.13. This torque decays with time and packing wear until the axial force caused by gland load becomes less than the axial force due to fluid pressure. For *high fluid pressures* the sealing is automatically maintained by the last 10 to 20 per cent of packing length. Thus most of the friction torque will occur there, and increasing the total number of rings will not result in a marked increase in torque (Fig. 24.13). This torque will tend to remain constant until the packing no longer seals effectively.

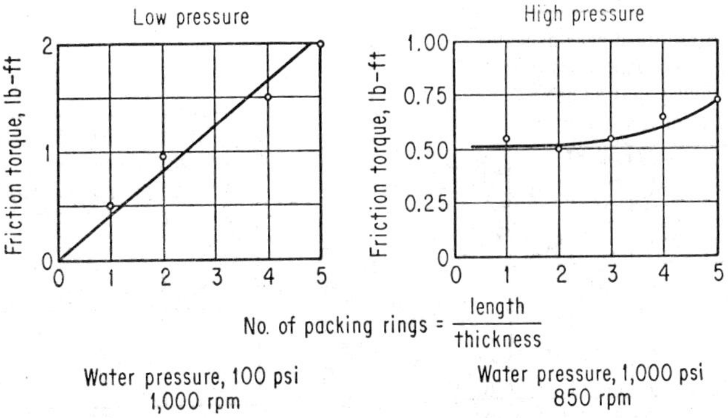

FIG. 24.13. Variation of friction torque with number of packing rings. Graphite-greasy plaited hemp.

The initial compression load on the packing directly affects the initial starting torque and will be a determining factor with respect to friction torque as long as the gland load exceeds the fluid-pressure load on the packing. When the fluid-pressure load is greater, the gland load does not affect friction torque, which increases in proportion to fluid pressure. Figure 24.14 confirms these effects.

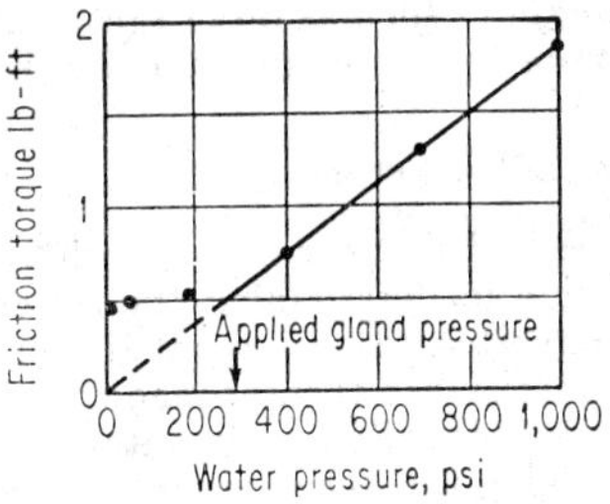

FIG. 24.14. Effect of water pressure on friction torque. Average of two glands, Teflon-impregnated cotton, 850 rpm.

Shaft speed and diameter directly determine the amount of work done and are therefore basic in calculating the power requirements. Shaft finish also affects the power requirements because the coefficient of friction is a function of surface finish. The effect of shaft speed on torque is complex, and little information is available at high speeds but the data plotted in Fig. 24.15 give an indication of results at slower speeds. Data at 3,500 rpm show the average friction torque was 1.3 ft-lb for ⅝-in.-square plastic packing in a 300-hr test against water at 300°F and 150 psi with a 3¼-in. shaft. The leakage rate has also been found to be independent of shaft speed.

The leakage rate is a determining factor in providing lubrication and influencing the power requirements as shown in Fig. 24.16. In laboratory tests, a 5-hp electric motor driving a 2-in. shaft was readily stalled when packing leakage dried up. Figure 24.17

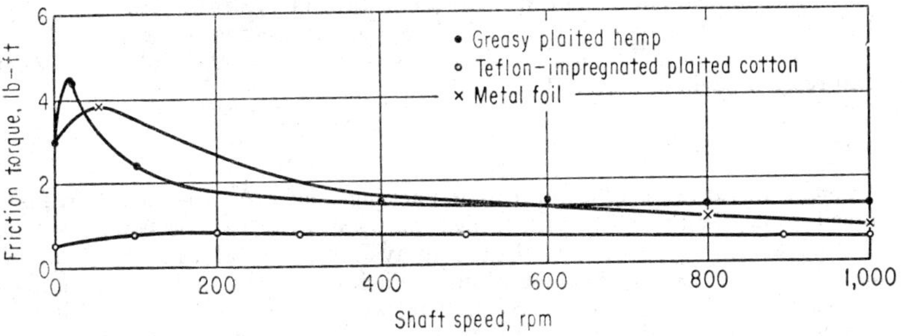

FIG. 24.15. Typical torque vs. rpm values. Three rings, applied gland pressure 200 psi, water pressure 75 psi.

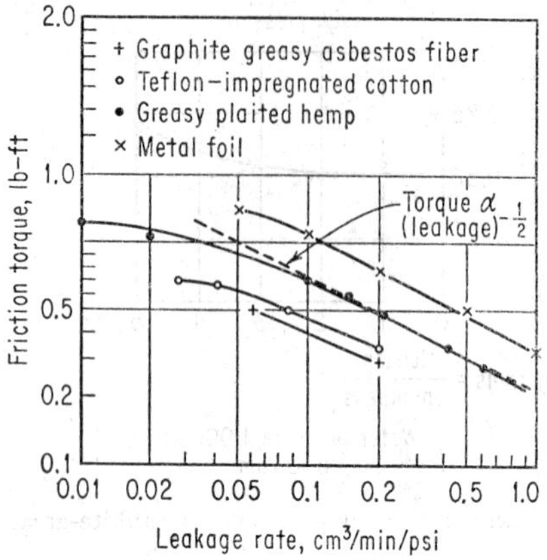

FIG. 24.16. Friction torque vs. leakage rate. Three rings of packing, 1,000 rpm.

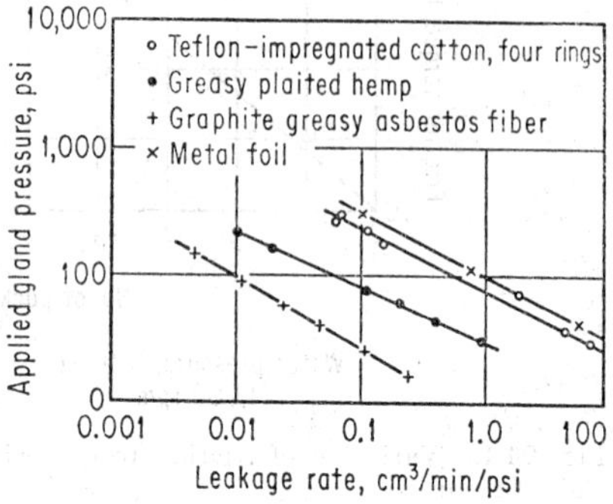

FIG. 24.17. Effect of gland load on leakage rate. Shaft speed, 1,000 rpm.

shows how the rate of leakage is controlled by gland pressure when it exceeds fluid pressure.

In addition to the lubrication provided by the leakage of the sealed fluid, a lubricating fluid such as oil or grease may be introduced through the lantern gland and this more desirable lubricant can reduce the power requirements.

(e) *PV* Factor

The *PV* factor (pressure times shaft linear velocity) referred to previously is a useful guide in determining the selection of a compression packing. This factor may be converted to heat input if a packing coefficient of friction is estimated. A packing selection can then be made on the basis of the heat resistance of the packing materials available. The temperature condition of the application must also be considered in making this selection. The heat generated in a stuffing box can be expressed as

$$Q = PLdV\mu/J \tag{24.10}$$

where Q = quantity of heat generated, Btu/min
P = fluid pressure to be sealed, psi
L = depth of packing, in.
d = shaft diameter, in.
V = shaft velocity, fpm = $(\pi d/12)$ rpm
μ = packing coefficient of friction
J = mechanical equivalent of heat, a constant of 778 ft-lb/Btu

The chart in Fig. 24.18 in which the *PV* factor is automatically established by the rpm, shaft diameter in inches, and fluid pressure in psi will be helpful in the selection of a compression packing material.

(f) Wear

One of the advantages of compression packings, as compared with lip seals or mechanical seals, is the ease with which packing wear can be compensated for with gland take-up. The amount of take-up available is established by gland clearance and the compressibility of the packing. Packing life can be increased and shaft wear decreased by the use of hardened shafting with a 20 rms or better finish. Hard chrome-plating shafts and finishing to 10 rms improves shaft and packing wear. It also reduces the possibility of shaft corrosion, which often is the cause of premature packing failure. Another common practice is to provide shaft sleeves in the stuffing-box area that may be replaced when worn so that new packings will operate against optimum shaft conditions. It is good practice to have a shaft hardness in excess of 400 Brinell; for metallic packings a shaft hardness of 500 Brinell or better is recommended.

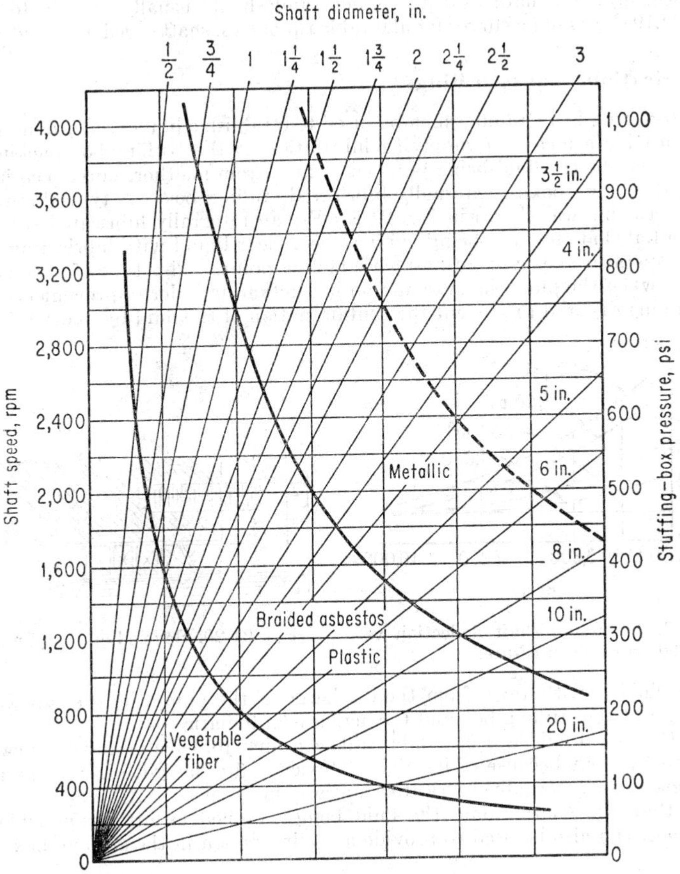

FIG. 24.18. *PV* chart.

(g) Shaft Finish

In addition to reducing wear, a good shaft finish of 16 to 20 rms will improve the sealing performance of packings. This is especially important in reciprocating service where any shaft imperfections would establish a leakage channel across the sealing surface of the packing.

(h) Misalignment and Shaft Deflection

Misalignment between shaft and bore makes the sealing problem more difficult, and it should be held to an acceptably low value. If a shaft runs true but is located eccentrically to the bore, most compression packings can accommodate this quite readily up to about 0.010 in. The effect is most harmful at high pressures because excessive clearance is available on one side of the shaft, and extrusion of the packing may occur. The misalignment resulting in a runout condition or eccentric motion of the shaft about its center line is the most difficult for the packing to seal against. This results in a continuous pounding of the packing during each revolution or stroke. The packing has to be extremely resilient to withstand this action and still seal effectively. Runout may be caused by faulty installation, out-of-round shaft, imbalance, or inadequate bearing support and should usually be held to 0.003 in. Figure 24.19 shows some curves for allowable runout vs. shaft speed and fluid pressure.

(i) Lubrication—Lantern Rings

A lantern ring is an annular member used to establish a liquid seal around the shaft and to provide a means of supplying lubrication to the stuffing-box packing. The lubricant may be the fluid being pumped or fluid from another source, i.e., integrally lubricated stuffing box or externally lubricated stuffing box. A typical stuffing box with lantern ring was shown in Fig. 24.7. For an integrally lubricated stuffing box, it is essential that the fluid being pumped be a clear liquid with lubricating qualities and a temperature below its atmospheric boiling point. This type of arrangement is necessary when the pump has vacuum on its suction, in order to prevent the packing from running dry and to prevent the contamination of the fluid by another lubricant.

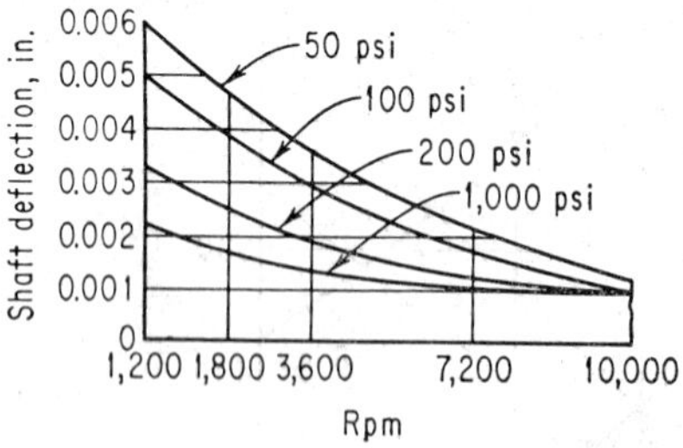

FIG. 24.19. Maximum shaft deflection vs. shaft speed and fluid pressure.

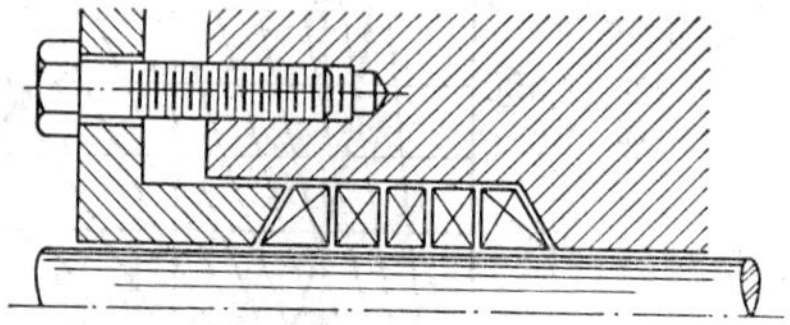

FIG. 24-20. Beveled-end stuffing box.

Since the fluid is withdrawn from the discharge of the pump, its pressure is always higher than at the stuffing box and the action is automatic.

When the pump fluid is a poor lubricant, contains abrasive material, or cannot be allowed to leak out because of its highest toxicity, fire hazard, high temperature or corrosiveness, an externally lubricated lantern ring is used. A lubricant should be selected that will contaminate the fluid being pumped as little as possible. This arrangement can also be used to provide a cooling action in the stuffing box.

(j) Beveled Boxes

Although square-end stuffing boxes are generally used for compression packings, some stuffing boxes are made with beveled ends as illustrated in Fig. 24.20. The theory is that the beveled ends tend to force the packing against the shaft where the sealing takes place. However, it also tends to force the packing into the shaft and follower clearance, thus promoting extrusion, and results in greater shaft and packing wear.

(k) Spring Followers

Spring followers are sometimes used to provide a follow-up on gland load to compensate for the decay of gland stress because of packing wear and loss of impregnation. Laboratory experiments have shown that, although this technique prevents increase of leakage rate with time, the friction torque remains very high, and wear of the packing becomes excessive. Spring followers should be used only in special low-speed applications.

(1) Dynamic Seals for Reciprocating Service

Compression packings are used in reciprocating service. Many of the installation recommendations for rotary application, such as shaft finish, runout, and lubrication, apply equally in this case. In reciprocating service the packing is almost invariably subjected to fluctuating pressure, from atmospheric to pump output pressure. As a result, leakage is intermittent and shaft friction varies during a pressure stroke. Because the gland load on the packing is adjusted to seal maximum pump pressure, during the low-pressure part of the stroke the packing is not lubricated by the fluid being pumped. If heavy leakage were permitted, packing erosion would result in premature failure; hence it is usually necessary to keep a positive gland load on the packing all the time. Outside lubrication such as an oil drip on a horizontal shaft is commonly employed to ensure adequate lubrication. Temperature buildup caused by the packing is usually not a problem because packing contact is not concentrated and speeds are relatively slow. The maximum shaft speed recommended is 250 fpm

24.7. AUTOMATIC OR PRESSURE-ACTUATED

(a) Theory

In automatic or pressure-actuated packings, fluid pressure automatically supplies the seating force required by the packing to effect a seal. There are two general classifications: (1) lip packings such as cups, U-cups, and V-rings; and (2) squeeze types such as O-rings. They all rely upon built-in interference to establish an initial low-pressure seal; as fluid pressure increases the packing is forced tighter against the surfaces to be sealed, as illustrated in Fig. 24.21.

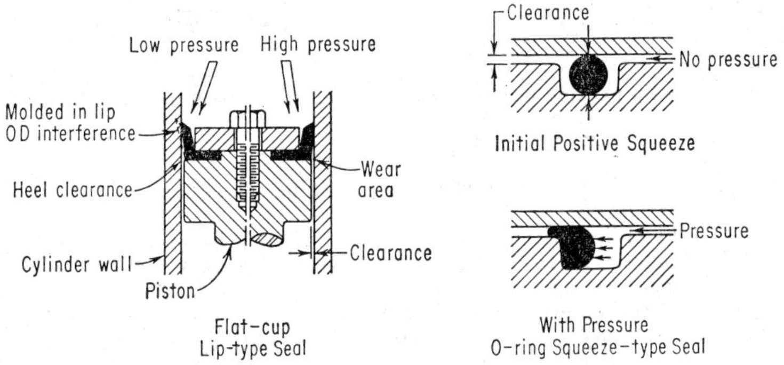

FIG. 24.21. Typical pressure-actuated packing installations.

Pressure-actuated packings are generally used to seal reciprocating motion, but occasionally for slow-speed rotary motion. These packings seal very effectively so that there is a greater temperature rise than with compression packings where a certain amount of leakage is allowed. In reciprocating service the heat tends to dissipate because the packing is continuously in contact with a new surface which has been lubricated and cooled. This is not the case in rotary service where the heat developed within the packing under pressure will destroy the packing unless some heat-removing means is provided.

Most pressure-actuated packings are installed with no provision for packing take-up. Therefore, in order to ensure long life, wear must be held to a minimum; one of the best methods to accomplish this is to have a good shaft finish of 16 rms or better. This will also reduce frictional heat, which shortens seal life. In hydraulic service, the fluid will usually supply sufficient lubrication but in pneumatic service external lubrication usually must be provided. In any installation the clearance between the sealed members is critical. The packing under pressure, if improperly supported, tends to extrude into the clearance and wear away. The higher the pressure, the more critical the clearance becomes; therefore, it is always advisable to hold clearance to a minimum. Back-up washers are necessary in some applications. These packings should never be installed over sharp edges or come into contact with them during operation.

(b) Lip-type Packings

These packings are generally made of leather, rubber, or fabric-reinforced rubber with the choice of material depending on the service conditions. A properly selected and installed lip-type pressure-actuated packing will seal with almost no leakage. They are sensitive to pressure and will wipe lubrication from the shaft more effectively as pressure is increased. Friction on a pressure stroke varies proportionately

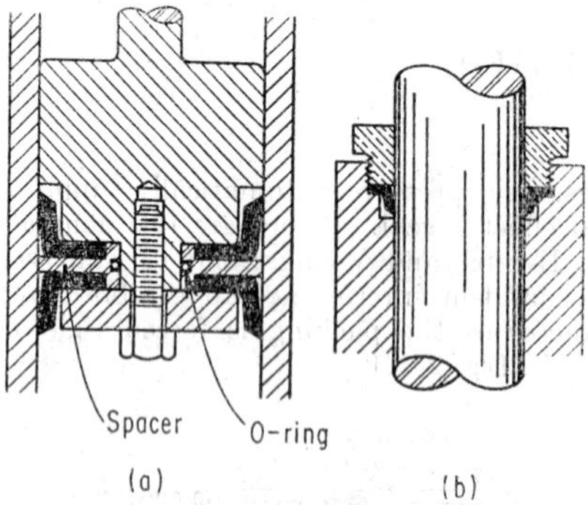

FIG. 24.22. Single-lip packing installations. (*a*) Piston assembly. (*b*) Internally threaded design, hat gasket shaft assembly.

FIG. 24.23. U-cup packing installation.

with the pressure. On the return stroke the packing relaxes and friction drops to a low level.

The oldest type of lip packing is a flat cup with a single lip on the outer circumference. This is sometimes classified as unbalanced packing. When the lip is on the inner circumference, it is referred to as a hat or flange gasket. The flat cup is usually used to seal a piston or other driving or driven members, while the hat gasket is almost always stationary and seals a shaft or ram. Both types seal across the base as a static seal or gasket, while the lip is pressure-actuated. Typical installations are shown in Fig. 24.22*a* and *b*.

Figure 24.22*a* illustrates an application where the cup flange is sealed by a known amount of compression as determined by the dimensions of the spacer boss. The spacer also prevents transfer of load between cups. There are numerous designs for flat-cup installations each of which has certain functional or cost advantages. For example, an all-compound or homogeneous flat cup is not clamped across the base but is permitted to float and depends upon fluid pressure to create a base seal. This results in a simplified design of minimum no-pressure friction. The design should be such that the cup base is not overstressed, or the base of the lip or heel will be forced into contact with the cylinder walls to create high friction and cause rapid wear. Overcompression also causes the lip to tilt inward, resulting in loss of lip interference. Clearance must be provided at the base of the cup to prevent binding action that might result from swelling of the packing. The hat gasket (Fig. 24.22*b*) has application where space is limited and functions best up to about 5 in. inside diameter. Single-lip packings may be installed with a spring expander or contractor against the sealing lip in order to maintain contact with the cylinder wall or shaft.

Table 24.6. Homogeneous U-cap Packing—Nominal Sizes

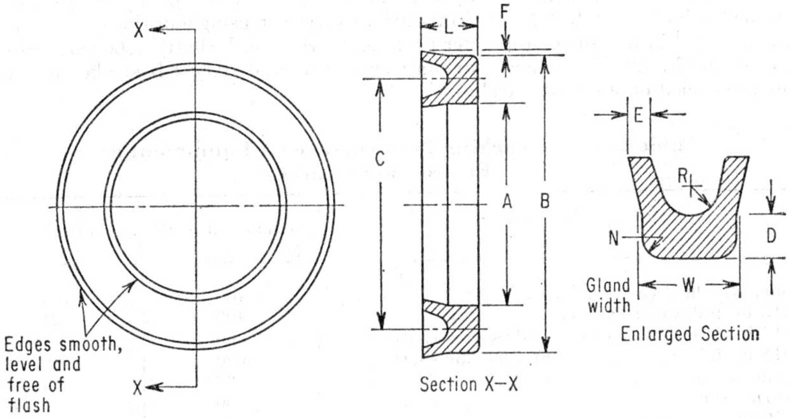

Dash No.	Dimensions, width and length, in.	Nominal size, in.	
		ID	OD
8-24	¼	¼–1¼	¾–1¾
25-35	5/16	1¼–2½	1⅞–3⅛
36-40	⅜	2½–3	3¼–3¾

The U-cup and V-ring are considered balanced packings because they have a sealing lip on the inside and outside diameters. They are completely automatic, seal with low friction, and require no gland adjustment. They can be used to seal either a shaft or single- and double-acting pistons.

Figure 24.23 illustrates installations of a U-cup with a pedestal ring used to center the packing and hold it in position. The tongue should allow clearance for the lip, both at the sides and at the tip. The corners on the tongue should be rounded so as not to cut the cup. Some compression of the cup base is desirable, but this should not be excessive or lip distortion will occur. About 1/64-in. compression is usually satisfactory. Table 24.6 gives recommended flange widths for various sizes of U-cups.

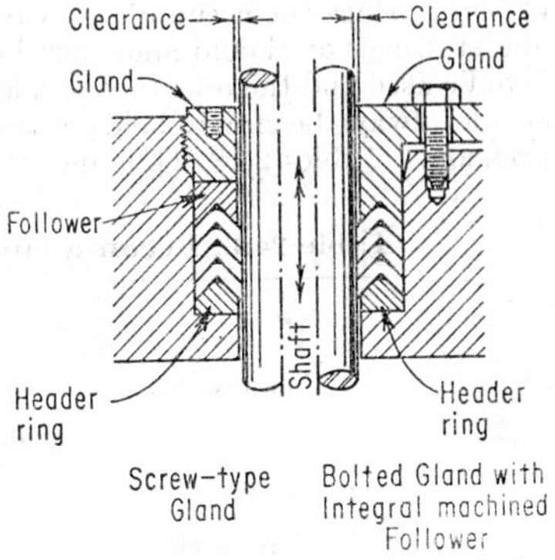

FIG. 24.24. V-ring packing installation.

(c) V-rings

The most popular of the multiple-lip seals or nested sets are V-rings, although there are other types which can be used to advantage in many applications. Properly installed, V-rings will seal considerable pressure with long life and low friction. They are a balanced packing that may be used to seal a shaft and single- or double-acting piston. Typical installations are as shown in Fig. 24.24.

Table 24.7 illustrates the importance of good design and realistic operating conditions relative to the performance of a set of typical all-compound lip packing as obtained in laboratory tests. The standard test was run using four rings of 3 in. inside diameter by 3¾ in. outside diameter packing, 20 rms finish shaft, 3,000 psi hydraulic fluid at 160°F, 0.005-in. follower to shaft clearance, and sixty 1-ft strokes/min with pressure cycled on alternate strokes.

Table 24.7. Lip-packing Performance vs. Equipment and Environment Variables

Test variable	Leakage rate, %	Life,* %
None, standard test	100	100
0.015-in. follower clearance	400	24
0.015-in. follower clearance and 32 rms shaft	360	21
0.015-in. follower clearance and 60 rms shaft	620	8
32 rms shaft	166	60
120 rms shaft	500	6
0.010-in. undersize shaft	300	24
6,000 psi	515	4
2-ring set	220	31
3-ring set	101	79
200°F	180	11
250°F	286	8

* Life based on number of cycles before leakage rate became excessive.

In some installations, such as control valves, spring loading is used (with the spring located under the male adapter) to maintain the lip interference of a V-ring set. This is especially effective with an inelastic low-friction material such as Teflon. The spring load can be calculated to provide the exact value necessary to effect a seal without creating high friction. This will compensate for some lip wear. The design of V-rings is fairly well established. They are made to standard dimensions, as are the header and follower adapter rings required to make them function. A tolerance of ±0.010 in. is permitted in the height of each V-ring and adapter; so that in a set of four V-rings the accumulated variance could be 0.060 in. The design of the gland and stuffing box should allow for the maximum stack height if a nonadjustable gland is to be used and the tolerance should be compensated for by the use of shims. V-rings should always be snugly held by the gland, with leather V-ring requiring slight compression. Tables 24.8, 24.9, and 24.10 provide data on V-rings.

Table 24.8. Recommended Number of V-rings vs. Pressure

Pressure, psi	No. of V-rings*		
	Leather	Homogeneous	Fabricated
Up to 500	3	3	3
500– 1,500	4	4	4
1,500– 3,000	4	5	4
3,000– 5,000	5	5	5
5,000–10,000	5	...	6
10,000 and over	6		

* For solid rings.

Table 24.9. Homogeneous, Leather, and Fabricated V-ring Packings—Nominal Sizes

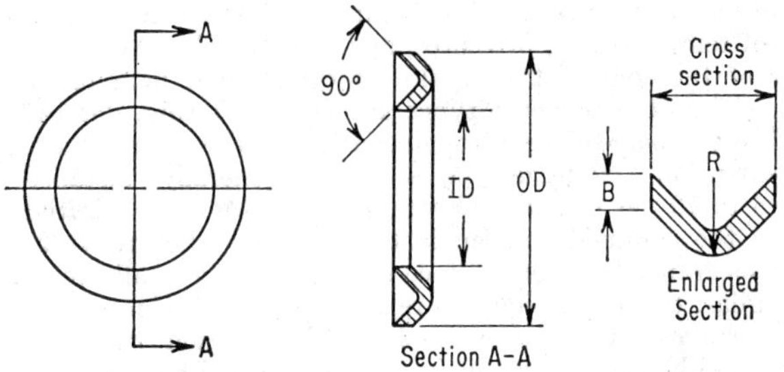

Dash No.	Gross section *W*, in.	Nominal size		*B*, in. −0.010 in.	Stack height, in., 3-ring set*
		ID	OD		
8-24	1/4	1/4–1 1/4	3/4–1 3/4	0.083	5/8
25-35	5/16	1 1/4–2 1/2	1 7/8–3 1/8	0.140	55/64
36-46	3/8	2 1/2–3 3/4	3 1/4–4 1/2	0.156	31/32
49-55	7/16	4 –5 1/2	4 7/8–6 3/8	0.197	1 5/32
56-80	1/2	5 1/2–15	6 1/2–16	0.197	1 7/32

* Includes adapters.

Table 24.10. Dimensions for V-ring Adapters

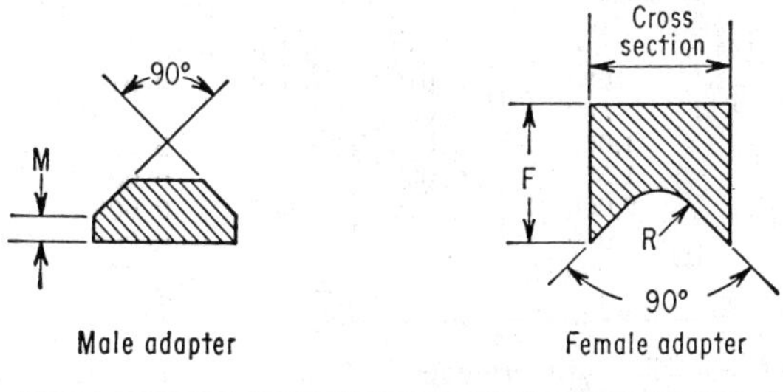

Dash No.	Cross section	Dimensions, in.		
		F	*M*	*R*
8-24	1/4	1/4	1/8	1/16
25-35	5/16	5/16	1/8	7/64
36-46	3/8	3/8	1/8	1/8
49-55	7/16	7/16	1/8	5/32
56-80	1/2	1/2	1/8	5/32

(d) Squeeze-type Packings

This style of packing is the simplest sealing device, and it will seal effectively over a wide range of temperatures, pressures, and fluids. The O-ring is the simplest and most popular of the squeeze-type seals; its design and installation are fairly well standardized. The other squeeze types are D-ring, T-ring, delta-ring, and X-ring, each designed to overcome some of the limitations of the O-ring, such as spiral failure, friction, or extrusion. Squeeze-type seals are generally installed in rectangular grooves with 5 to 10 per cent initial compression to establish initial sealing. Fluid pressure forces the confined ring against one side of the groove and mating part, sealing off the clearance between. Since the functioning of all squeeze-type packings is basically similar, this discussion will relate to the more commonly used O-ring.

O-rings are primarily used to seal reciprocating motion, but they are being more widely used to seal helical motion (valve stems) and, with special design, to seal low-speed and low-pressure rotary motion. The three types of packing grooves are:

1. Rectangular—most common for dynamic applications.
2. V—recommended for low temperature where they increase squeeze. High friction and wear are an objection.
3. Undercut or dovetail—used for slow-speed reciprocating motion. Reduce friction and tendency to blow out. Have concentrated wear and high cost. Useful for valve-seat seals.

Figure 24.21 illustrates the operation of O-rings.

Anti-extrusion backup rings are used mainly to raise the pressure range by preventing extrusion of the O-ring into the clearance. They are sometimes employed to eliminate excessive clearance beyond the values recommended in Table 24.11.

Table 24.11. Recommended Clearances for O-rings

Operating pressure, psi	Clearance, in.		
	For 70 Shore A	For 80 Shore A	For 90 Shore A
0	0.010	0.010	0.010
250	0.010	0.010	0.010
500	0.008	0.010	0.010
1,000	0.005	0.008	0.010
1,500	0.003	0.005	0.008
2,000	.....	0.004	0.005
3,000	.....	0.003	0.004
5,000	.....	.....	0.003

The softer O-rings will seal more effectively and have less friction at zero pressure. Operating friction at pressure is influenced by the amount of O-ring surface making contact, as illustrated by Fig. 24.25.

There are two types of recommended groove designs for O-rings. One is for military aircraft, MIL-R-5514-D; and the other, for industrial applications, is based upon the military specification MIL-R-5514.

Tables of standard size and installation data are available from government specifications and O-ring manufacturers.

Limitations on surface-finish requirements are given below.

For military applications:

1. Diameter over which packing must slide, 16 μin. maximum.
2. Groove root diameter, 32 μin. maximum.
3. O-ring groove sides without backup rings, 32 μin. maximum.
4. O-ring groove sides with backup rings, 62 μin. maximum.

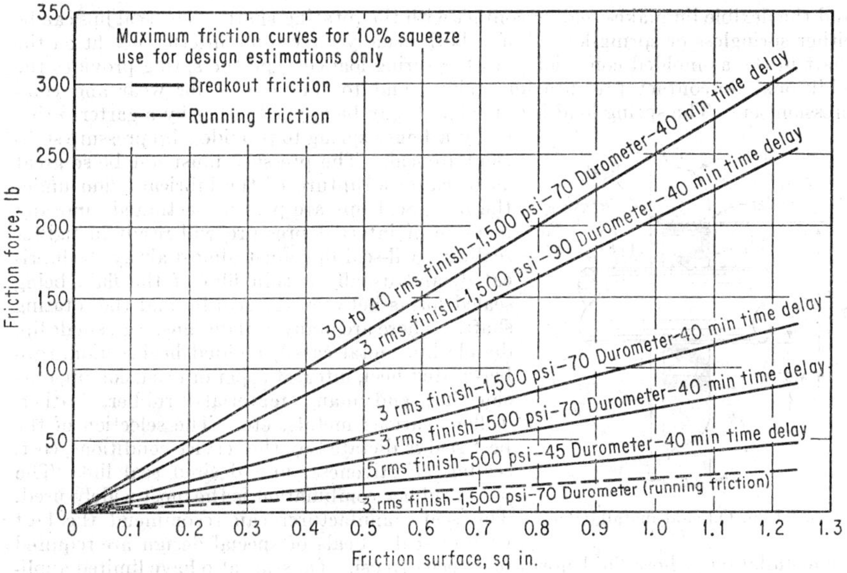

FIG. 24.25. O-ring friction. Maximum friction curves for 10 per cent squeeze. Use for design estimations only. Solid line shows breakout friction. Dashed line shows running friction.

For commercial applications:

1. Diameter over which O-ring must slide, 16 μin. maximum.
2. O-ring groove root diameter, 64 to 125 μin.
3. O-ring groove side, with or without backup rings, 64 to 125 μin.

If possible, toolmarks should run in the same direction as O-ring movement to promote better sealing and reduce wear. Honed and hardened steel that has been hard-nickel plated is best. Hard chrome plating is good but increases friction. Aluminum alloy, bronze, brass, monel, and soft stainless steels are not recommended for long life. They have limited low-pressure application.

The design of O-rings for rotary applications up to 600 FPM has been established with good lubrication and cooling. For successful rotary usage O-rings should be installed so that they are in circumferential compression rather than in tension (developed in stretching around a shaft) in order to eliminate the "Joule effect" common to some rubbers. Design tables are available that give about 5 per cent peripheral compression by having the O-ring inside diameter larger than the shaft it seals and the groove width equal to the O-ring cross section. Shaft speed should be limited as follows:

Shaft diam, in.	*Max speed, fpm*
$\frac{1}{8}$–$\frac{9}{32}$	350
$\frac{3}{8}$–$\frac{11}{16}$	450
$\frac{3}{4}$–$1\frac{1}{4}$	600

24.8. OIL SEALS

(a) Theory

An oil seal, sometimes referred to as a radial positive-contact seal, is generally used to seal off lubricant flow along a rotating shaft as illustrated in Fig. 24.26.

The oil seal has two essential elements, a rigid heel and a flexible lip. The heel is a force fit into the bore to seal off outside-diameter leakage and prevent seal rotation, and the flexible lip makes sealing contact with the rotating shaft. The seal lips can be either springless or spring-loaded, and both usually have some interference fit on the shaft in the as-molded condition. In the spring-loaded type the spring provides the additional lip-contact pressure for sealing, and to compensate for wear and compression set. The spring load is critical and may be applied either by a garter spring or by a finger spring to provide a lip pressure sufficient to seal. The pressure must not be so great as to cause a rupture of the lubricant film under the lip. Seal lips are pressure-actuated, and any increase of internal pressure will result in tighter sealing. Oil-seal lips must almost always be lubricated, and usually a thin film of the fluid being sealed forms between the seal lip and the rotating shaft. There are many seal designs, e.g., single lip, double lip, metal cased, molded heel section, rubber-coated heel, integral wiper or exclusion lip, outside lip; and many materials: rubber, leather, Teflon, various metals, etc. The selection of the best seal is dictated by the service conditions, cost, sealing effectiveness, and desired seal life. The metal-rubber combination is the most widely used. The seal manufacturer can recommend the best type of seal. Seals of special design are required for installations where fluid pressure exceeds 10 psi. Oil seals also have limited application for reciprocating and oscillating motion.

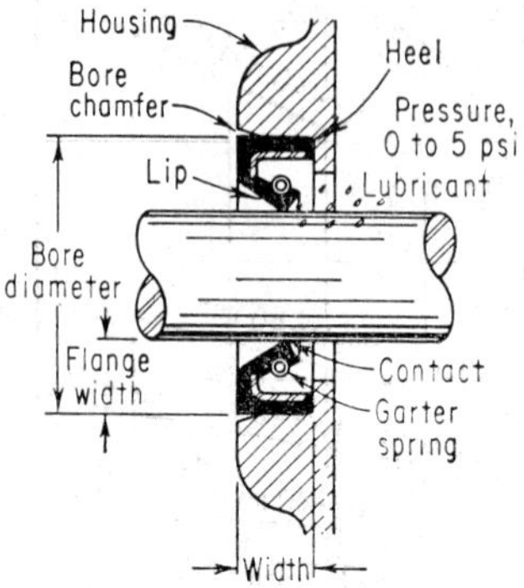

FIG. 24.26. Oil-seal installation.

(b) Design

The determination of flange width and seal width (bore depth) is important in order that seals of standard dimension can be used without going to the expense of securing special designs required by poorly dimensioned recesses. A square cross section is best for designing any seal. Table 24.12 presents suggested oil-seal cross-section dimensions as a function of shaft diameters where seal flange and seal width are equal.

Table 24.12. Oil-seal Dimensions

Shaft diam, in.	*Seal flange and width, in.*
To ½	¼
9/16– 1	5/16
1 1/16– 2	⅜
2 1/16– 5	½
5 1/16– 8	⅝
8 1/16–16	¾
16 1/16–24	¾ and 1
Over 24	1 and over

(c) Power or Torque Requirements

Generally the power absorption of an oil seal is not great enough to be significant in a normal application. The peak torque usually occurs at startup, especially after a lengthy shutdown period. Once operating speed is reached and the lip lubricating film established, the torque will be low. For installations where torque may be a factor, such as in instrumentation, seals with minimum lip load can be designed. Torque will be appreciable if there is any fluid pressure; this usually requires special design consideration. Typical oil-seal torque values are presented in Table 24.13.

The horsepower requirement may be calculated as follows:

$$\text{hp} = 9.917 \times 10^{-7} NT \qquad (24.11)$$

where N = shaft rpm and T = seal torque, oz-in.

Table 24.13. Oil-seal Torque

Shaft Diameter vs. Seal Torque

Shaft diam., in.	*Torque, oz-in.*
5/8	8–10
7/8	10–20
1	15–30
2	60–90
3	90–120

Seal Torque vs. Pressure*

Pressure, psi	*Torque, oz-in.*
Atm	10
5	17
10	23
15	50

* 7/8-in.-diameter shaft at 1,750 rpm.

The following torque limits for oil seals established by Naval Ordinance, MIL Specification 13882, is a useful guide in estimating power requirements:

	Max torque, oz-in.
A limitation (low-friction seals)	$4.8\pi D^2$
B limitation (normal-friction seals)	$10.4\pi D^2$

where D = shaft diameter, in.

(d) Performance Variables

Among the important factors which influence seal performance are:

1. Shaft speed in terms of feet per minute rubbing speed.
2. Temperature. For a discussion dealing with the problems of dynamic O-ring seals and glands operating at high temperatures see ref. 18.
3. Shaft finish.
4. Fluid pressure.
5. Severity of fluid sealed and lubricity.
6. Shaft concentricity and runout.

An oil seal can give satisfactory performance with some of the factors at a relatively severe level but will have short life if all factors are adverse. The shaft speed limit for leather seals is about 2,000 fpm, and for rubber-lip seals about 3,000 fpm (although when conditions are favorable the latter has been increased to 4,000 rpm or higher). The temperature limit is dependent upon the heat resistance of the lip material.

A finely finished shaft is required for maximum seal life; 10 to 20 rms is the general recommendation but 10 to 16 rms is preferred. Laboratory tests show that sealing effectiveness is not improved with better than a 10 rms finish, and seal life has actually been shortened with a 2 rms finish. It has been suggested that, with a superfinish, a lubricating film cannot be maintained. Tool finish marks sometimes create a spiral lead for leakage; if they do exist, the direction of shaft rotation should be in the direction that would direct the flow inward.

A shaft hardness of from 40 to 50 Rockwell C will promote good performance by limiting shaft wear. Soft metal shafts such as brass or aluminum should be provided with a hardened sleeve or ring to contact the seal lip. Since seal lips are readily pressure-actuated, a means of preventing pressure buildup should be provided, especially at high speeds, unless the seal is designed for a pressure application. If the fluid being sealed is a poor lubricant, such as air, auxiliary lubrication must be provided.

The built-in seal lip interference plus spring load will compensate to some extent for a shaft that is off-center but true-running. Runout, however, whether from an out-of-round shaft or shaft whip, in combination with high speed, will tend to throw the lip off the shaft and permit leakage. Table 24.14 will act as a guide to indicate maximum misalignment and runout.

Table 24.14. Shaft Speed vs. Shaft Misalignment for Oil Seals

Shaft speed, rpm	Shaft to bore misalignment (static measurements) in., max	Shaft runout and whip (dynamic measurement) T.I.R.,* in., max
0– 800	0.010	0.010
800–2,000	0.005	0.005
2,000–4,000	0.005	0.003

* T.I.R. = total indicator reading.

24.9. MECHANICAL SEALS

(a) Theory

A mechanical or face seal is defined as a mechanism which seals a rotating shaft in a plane vertical to the shaft. This is accomplished by placing the sealing surfaces perpendicular to the shaft with one face attached to the rotating shaft and the other fastened to the stationary housing. These seals are also referred to as radial or end-face seals. There are innumerable designs ranging from the simple inexpensive unit on the automobile water pump to the expensive elaborate seal used in pipeline pumping stations. They all function by maintaining the two sealing faces in contact so no leakage can occur. The complexity of a mechanical seal usually increases with pressure and shaft size.

(b) Design

In order to function properly a mechanical seal must have its two mating faces in intimate contact to prevent leakage. This requires that the faces be flat and parallel. Flatness can be assured during manufacture by proper machining and polishing techniques. Parallelism can be accomplished to a degree during the installation, but complete parallelism is never attained and the design must incorporate features which will ensure parallel sealing surfaces both initially and during service. This is most readily accomplished by providing that at least one of the faces be free to move so it can align itself parallel to the other. If a member is located against a flexible member, such as a resilient gasket or a spring, it will be free to adjust itself within limits, as required.

In addition to creating a seal across the mating faces, one of the members must be sealed on the shaft to prevent leakage at that point. This is usually accomplished by an O-ring, V-rings, or a flexible bellows which allow the part to move in any plane. The other member usually also has a certain freedom of movement by having a resilient member between it and the part to which it is fastened. To create the axial movements necessary to keep the sealing faces in contact, one or more springs are usually used behind one of the members.

Although there are many mechanical seal designs, the following breakdown will cover most of the basic types. The simplest type is a seal with a spring-loaded member mounted to a shaft with an O-ring seal and which rotates against a face which is machined into the housing of the unit. This can be accomplished only where the housing is made of a material such as cast iron which provides a good seal surface. Two of the major problems connected with this type of seal are the difficulty of machining and polishing a flat surface in the unit, and the difficulty and cost of repair if excessive wear occurs. These can be overcome by installing a separate wear ring in the unit, and this mechanical seal then becomes representative of most seals on the market.

One of the major differences between high- and low-pressure seals is the unbalanced area on which the pressure acts to augment the spring pressure which holds the sealing faces together. For low pressures, up to about 100 psi, an unbalanced seal is usually satisfactory. In this type no effort is made to counteract the effect of the fluid pressure holding the sealing faces together. However, at higher pressures this force would become excessive and cause high friction with resultant heat build-up, wear, and early failure. To counteract this, the seal is balanced, or an area is provided on the movable sealing face for a force to be generated in the direction opposite to the main pressure

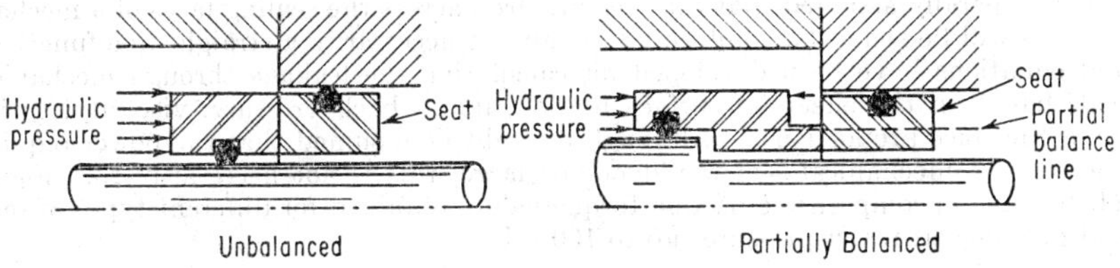

FIG. 24.27. Unbalanced and balanced seal faces.

force so that the net force on the faces is satisfactory. Complete balance is never desired because some force, in addition to the springs, must be provided to prevent the higher pressures from forcing the faces apart with resultant catastrophic leakage. A properly designed balanced seal will have the unit pressure on the sealing faces created by the fluid load equal to about 70 per cent of the fluid pressure. Balance and unbalance of seal faces are illustrated in Fig. 24.27.

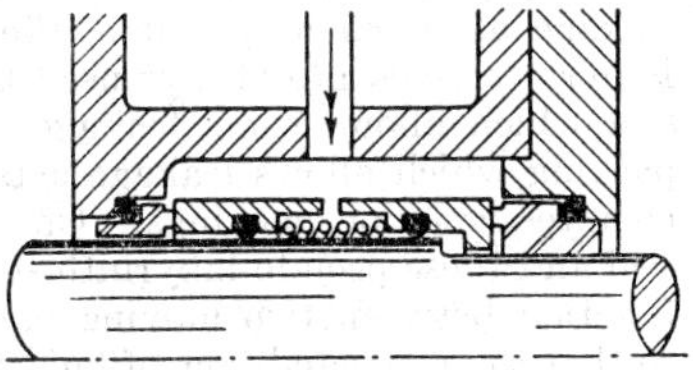

FIG. 24.28. Balanced double seal.

In cases where leakage of the fluid being sealed cannot be tolerated or where the fluid is a slurry or has poor lubricating qualities, double seals are used as shown in Fig. 24.28. These function somewhat like a stuffing box with a lantern ring, since an auxiliary fluid is used. It is prevented from entering the mechanism by one seal and from leaking to the atmosphere by the other seal. The auxiliary fluid is usually chosen on the basis of its lubricating qualities and its noncontaminating characteristics.

Although most seals are internal, or seal inside the unit, external seals can also be provided. These require that both sealing faces be sealed against their locating members, but this is usually done in all seals anyway.

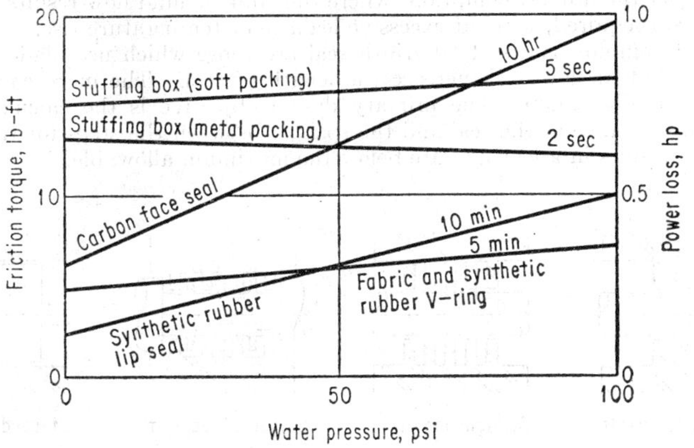

FIG. 24.29. Friction torques for various packings. 2-in.-diameter shaft. $N = 1{,}600$ rpm.

The materials used in mechanical seals are dictated by the conditions of operation. They must be resistant to the fluid being handled and able to withstand the temperatures involved. The seal faces are usually most critical and consist of a wearable surface such as a carbon ring rotating against a metal such as bronze, cast iron, or a hard surface such as Stellite or Comoloy. Elonka[7] lists a large number of materials that are used for sealing faces in different fluids.

It is generally assumed that the pressure drop across the sealing faces of a mechanical seal will be somewhat uniform although not necessarily a straight-line function,[8] and equations have been developed for calculating the leakage through mechanical seals based on the effects of surface tension, liquid back pressure, viscosity, spring force, and face proportions.[9] Although it might be assumed that the power requirements for a mechanical seal would not be large, this is not necessarily so. Figure 24.29 presents comparative friction torque values obtained for different types of seals and packings at water pressures up to 100 psi.

24.10. LABYRINTH SEALS

(a) Theory

The basic theory of a labyrinth seal is that a pressure drop will accompany the flow of a fluid through a passage. The longer the passage and the more turbulent the flow, the greater the drop in pressure will be. Since the primary purpose is to limit leakage as much as possible, the cross-sectional flow area should be a minimum. Labyrinth seals are always used between moving parts, although in essence a static seal which allows some leakage is also a labyrinth seal. Similarly, any dynamic packing which allows leakage acts as a labyrinth seal and this especially applies to compression-type packings. One manner to differentiate between a true labyrinth seal and these pseudo-labyrinth seals is to define a labyrinth seal as one "designed to function between two moving parts by having a fixed clearance between the parts so that the restricted flow of fluid through this clearance creates the required pressure drop with a tolerable loss of fluid." This definition also excludes the segmented carbon rings used in many installations.

(b) Design

There are innumerable labyrinth-seal designs, depending upon fluid, temperature, pressure, type of motion, speed, allowable leakage, space, and material requirements. The simplest type, of course, is a close-fitting bushing around the shaft. This could be used under the mildest conditions where the laminar fluid flow results in the small pressure drop required, without excessive leakage or temperature rise.

The most common types of labyrinth seal are those which use blades or steps to promote turbulent flow and increase the length of seal. The more common types are shown in Fig. 24.30. The primary design objective is the calculation of the length and clearance of the seal and the spacing between the steps for the staggered type which results in a leakage rate below the maximum allowable.

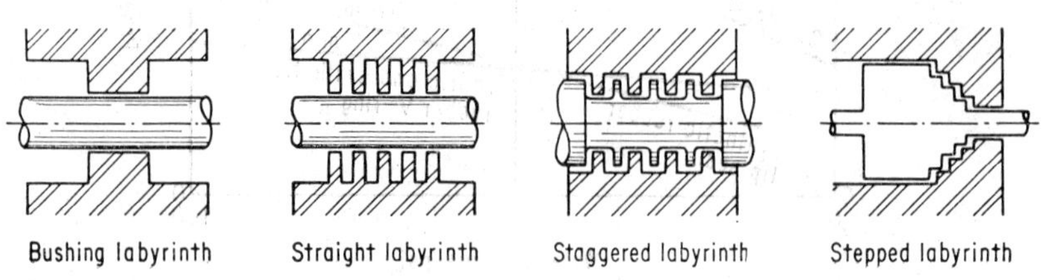

FIG. 24.30. Labyrinth seals.

(c) Bushing Labyrinth

The bushing labyrinth presents the simplest type, but even so, complications arise when the relative motion and possible eccentricity are considered. Standard fluid-flow theories are the basis for the calculations and were used by Tao and Donovan[10] in their work in which a general equation relating the friction factor to the Reynolds number was derived:

$$F = C/(N_{\mathrm{Re}})n \tag{24.12}$$

For laminar flow through a small annulus with no relative motion between adjoining surfaces, the values of C and n were found to be 170 and 1.03, respectively. For

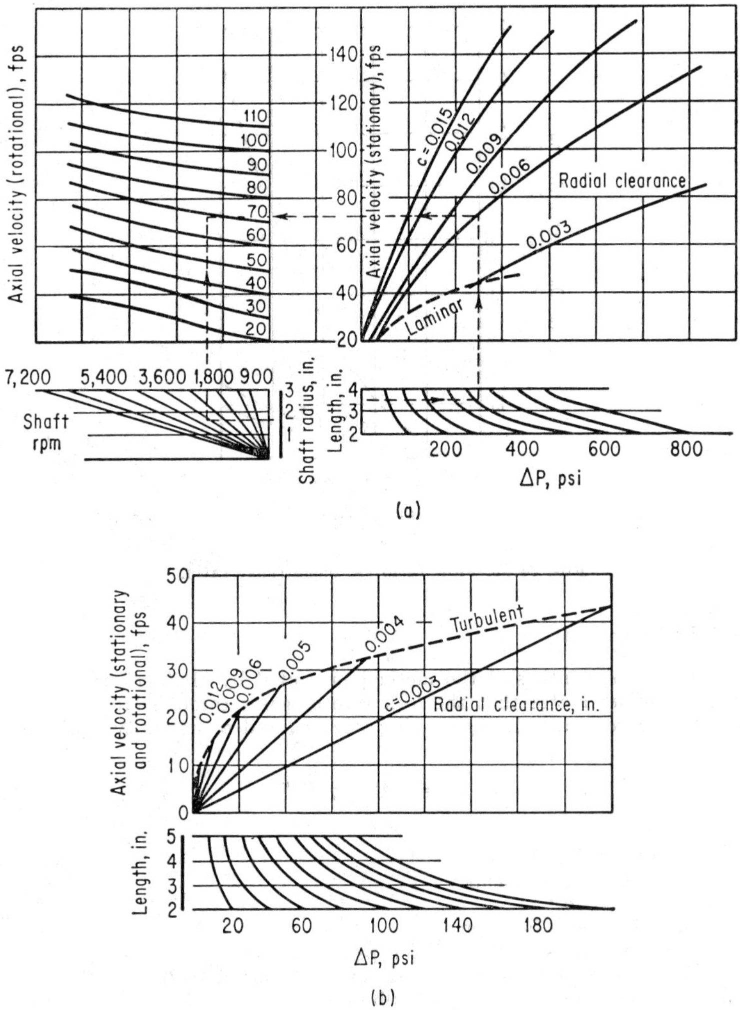

FIG. 24.31. Leakage past a labyrinth seal. (*a*) For turbulent flow of water at 68°F. (*b*) For laminar flow of water at 68°F.

turbulent flow the values were 0.316 and 0.21, and in both cases the equation applies for both concentric and eccentric parts.

When relative motion is considered, a number of variations must be made. By the application of these to a specific system Tao and Donovan developed nomographs for laminar and turbulent flows showing the relationship between pressure drop, radial clearance, sleeve length, shaft radius, and shaft speed for water at 68°F with concentric motion. By using a relationship between eccentricity and the ratio of eccentric to concentric flow, it is a simple matter to calculate the flow with eccentric motion. Reproductions of the nomographs are shown in Fig. 24.31*a* and *b*. The axial velocity is read directly from the charts, and when multiplied by the annulus area, the amount of flow with concentric motion is obtained. This value multiplied by the ratio of eccentric to concentric flow from Fig. 24.32 will give the eccentric-flow value.

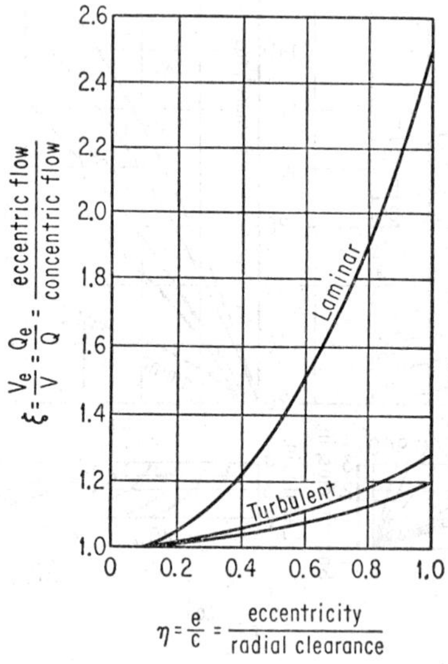

FIG. 24.32. Flow ratio vs. eccentricity ratio.

(d) Straight Labyrinth

For a straight labyrinth seal the method presented by Heffner[11] gives the best correlation. Heffner states that the leak rate of gases through labyrinth seals is analogous to the flow rate through a single orifice, and a similar type of equation can be used.

$$M = \alpha A \phi P (g/RT)^{1/2} \tag{24.13}$$

where M = leakage rate, lb/sec
α = contraction coefficient
A = flow area, in.^{2}
ϕ = pressure-ratio function
P = absolute pressure at seal inlet, psi
g = gravitational constant, (lb mass)(in.)/(lb force)(sec^2)
R = gas constant, (lb force)(in.)/(lb mass)(°R)
T = absolute temperature, °R

In order to use this equation effectively, proper expressions must be found for α and

ϕ. α can be determined experimentally for a single-blade seal using the equation

$$M = \alpha AP(g/RT)^{1/2}[2K/(K-1)]^{1/2}r^{1/K}(1 - r^{K-1/K})^{1/2} \tag{24.14}$$

where K = gas specific-heat ratio
r = pressure ratio P_a/P
P_a = absolute pressure at seal outlet, psi

The correlation between single-blade data and multiple-blade seals can be made by means of Reynolds number:

$$N_{Re} = (W/\alpha)[2Ph/\mu(T)^{1/2}]\alpha \tag{24.15}$$

where h = radial clearance, in.

The pressure-ratio function is defined by the following equation from Egli:[12]

$$\phi = \left[\frac{1 - r^2}{N + (2/K)\ln(1/r)}\right]^{1/2} \tag{24.16}$$

where N = number of blades

If W is now defined as the specific leakage rate in (lb mass)(°R$^{1/2}$)(sec)(lb force) or $W = G(T^{1/2})/P$ the basic leakage-rate equation can be rewritten

$$W/\alpha = \phi(g/R)^{1/2} \tag{24.17a}$$

Heffner's experimental data did not quite correlate with Eq. (24.17*a*); so it is rewritten

$$W/\alpha = F(\phi) \tag{24.17b}$$

A plot of this equation based upon experimental data will allow the determination of W/α for a specific value of ϕ calculated from Eq. (24.16). The value for W/α is then substituted in Eq. (24.15), which had previously been plotted for the experimental data. A value of α which will fit the experimental plot can then be determined and the specific leakage rate W can be calculated with W/α and α known.

Although the above summary of Heffner's work is probably not in enough detail to work out a design of a labyrinth seal, it does give an indication of the method and the complexity of the problem.

Equations (24.18) and (24.19), which will give at least an estimate of the number of rings, or the leakage rate for labyrinth seals, are given by Peickii and Christensen.[13]

$$N = \frac{(40P - 2{,}600)(W/A)}{540(W/A) - P} \tag{24.18}$$

where N = number of rings
W = permissible leakage, lb/sec
A = cross-sectional flow area $C\pi D$, in.^{2}
C = clearance, in.
D = diameter, in.
P = absolute pressure, psi

It is necessary to calculate N for the inlet and outlet pressures for a constant W/A and then subtract the two values to obtain the number of rings required to control the leakage at a definite rate with the specified inlet and outlet pressures.

The leakage rate can be calculated from the following equation if the number of rings is known:

$$W = 25KA\left\{\frac{P_1/V_1 - [1 - (P_2/P_1)^2]}{N - \ln(P_2/P_1)}\right\}^{1/2} \tag{24.19}$$

where W = flow rate, lb/hr
V_1 = initial specific volume, cu ft/lb
K = experimental coefficient

For interlocking labyrinths $K \cong 55$ and is independent of clearance in the usual range.

For noninterlocking types, K varies with the ratio of labyrinth spacing to radial clearance such that $K = 100$ for a ratio of 5 and $K = 60$ for a ratio of 50.

Additional labyrinth-seal theory and data may be found in refs. 16 and 17.

References

1. Mantell, C. L. (ed.): "Engineering Materials Handbook," Sec. 40, McGraw-Hill Book Company, Inc., New York, 1958.
2. "Rules for Construction of Unfired Pressure Vessels," 1952 ed., American Society of Mechanical Engineers.
3. Whalen, J. J.: How to Select the Right Gasket Material, *Prod. Eng.*, Oct. 3, 1960, pp. 52–56.
4. Roberts, I.: Gaskets and Bolted Joints, *J. Appl. Mech.*, vol. 17, pp. 169–179, June, 1950.
5. "Handbook of Mechanical Packings and Gasket Materials," Mechanical Packing Association, 1960.
6. Denny, D. F., and D. E. Turnbull: "Sealing Characteristics of Stuffing-box Seals for Rotating Shafts," The Institution of Mechanical Engineers, advance copy, 1959.
7. Elonka, S.: Mechanical Seals, *Power*, March, 1956, pp. 109–132.
8. Greiner, H. F.: Rotating Seals for High Pressure, *Prod. Eng.*, February, 1956, pp. 140–143.
9. Mayer, E.: Leakage and Wear in Mechanical Seals, *Machine Design*, Mar. 3, 1960, pp. 106–113.
10. Tao, L. N., and W. E. Donovan: Through-flow in Concentric and Eccentric Annuli of Fine Clearance with and without Relative Motion of Boundaries, *Trans. ASME*, vol. 77, no. 8, pp. 1291–1299, November, 1955.
11. Heffner, F. E.: A General Method for Correlating Labyrinth Seal Leak-rate Data, *J. Basic Eng.*, June, 1960, pp. 265–275.
12. Egli, A.: The Leakage of Steam through Labyrinth Seals, *Trans. ASME*, vol. 57, pp. 115–122, 1935.
13. Peickii, V. L., and D. A. Christensen: How to Choose a Dynamic Seal, *Prod. Eng.*, Mar. 20, 1961, pp. 57–69.
14. The Seals Book, *Machine Design*, Jan. 19, 1961.
15. Williams, B. G.: Mechanical Seals—A Survey of Present Day Practice—part 2, *Chem. & Process Eng.*, April, 1955, pp. 124–126.
16. Zabriskie, W., and B. Sternlicht: Labyrinth-seal Leakage Analysis, *J. Basic Eng.*, vol. 81, ser. D, no. 3, September, 1959.
17. Kearton, W. J.: The Flow of Air through Radial Labyrinth Glands, *Proc. Inst. Mech. Engrs. (London)*, vol. 169, 1955.
18. McCuistion, T. J.: O-ring and Gland Design for High Temperature Seals, *Prod. Eng.*, January, 1956, pp. 151–155.

Section 25

WELDING, BRAZING, SOLDERING, AND ADHESIVE BONDING

By

GEORGE F. KAPPELT, *Director, Engineering Laboratories, Bell Aerosystems Company, Buffalo, N.Y.*

CONTENTS

25.1. WELDING

(a) Introduction

Advances in welding, brazing, soldering, and adhesive bonding during the past decade have permitted consideration of these fastener methods in machine design. By use of these fasteners, cast or machined components assembled by bolting, screwing, riveting, or doweling can often be replaced by standard structural shapes cut to size. A welded base structure is shown in Fig. 25.1. As shown, extensive use can be made of standard steel shapes in lieu of special castings. But the use of such construction often requires new perspectives of designers and engineers.[1,32,33,34,35] It is misleading to consider welding as merely one of the various methods of fastening component parts together. In order to employ welding successfully as a method of joining machine components, the machine must be designed for welding. To duplicate a design fabricated by other fastener methods is unsatisfactory and will be wasteful in both the weight (size, thickness, reinforcement) and fabrication costs. With welded construction it is often possible to use steel plates, bars, angles, channels, etc., and steel castings as required to meet the actual stress and service requirements, in lieu of the use of ordinary cast-iron castings. Such standard steel shapes are usually three to six times as strong as standard iron castings. One major design consideration that must be recognized in using the welding processes is that welding is positive in action and that the joints so made are *rigid* joints. This contrasts with other joining processes where slippage or movement between parts always occurs. Thus a design that functions satisfactorily when riveted may be completely unsatisfactory when welded.

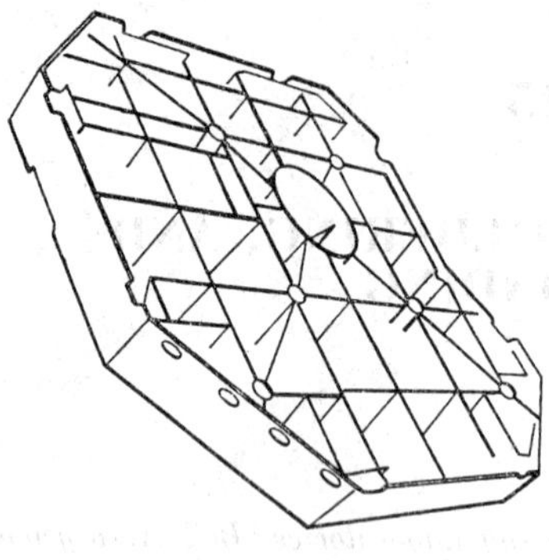

FIG. 25.1. Welded base structure.

Discussed herein are the design parameters affecting welding processes and the results that can be expected from the use of these processes. Intentionally omitted are the process details of the various fasteners under discussion. Process information is found in the specialized handbooks devoted to those subjects.[2,3,4,5,6] Process descriptions herein have been limited to those factors required to clarify the design factors presented.

(b) The Joining Process

Of the 37 different welding processes (ref. 5, p. 4) available, most machine fastener requirements can be met by three, namely, arc and resistance welding and brazing. Within these three categories, shielded metal arc, submerged arc, shielded or unshielded stud welding, seam, spot, and projection welding, and furnace or induction brazing are most widely used. All these processes are characterized by the fact that the metal components being joined have either been molten during the joining process or have been at a temperature approaching the melting point of the base material. The heat associated with these processes has an effect on design and must be considered as discussed later. Soldering is generally limited to nonstructural applications. Dip and torch methods are the standard methods. Soldering is a low-temperature operation (420 to 450°F) which has little if any significant effect on the properties of the material being joined. Adhesive bonding is generally the lowest-temperature joining process and as such rarely, if ever, raises the temperature of the surfaces being joined to much above 300°F. Many of the adhesives, specifically designed for fastener use, are room-temperature curing.[7] Advances in materials for adhesive bonding indicate that this method of joining of machine parts is in its infancy and will become more significant in future years.

The design factor dictating the selection of fusion welding over all other welding processes is the type of load the joint carries. A tension joint dictates the use of a fusion weld; such joints as shear, compression, and scarf permit joining by any of the methods under discussion in this chapter.

The design processes for welded *construction* are standard[8], using known mathematical formulas. But the design process for the *joint* involves many factors unique to the processes and must be clearly and carefully thought through.

The designers' specifications as to the type of weld (fusion or resistance) desired are expressed on the drawing through the use of the AWS weld symbol (ref. 2, Sec. 33). In its basic form it appears as a broken arrow with symbol elements as shown (Figs. 25.2 and 25.3). The symbols used are shown in Fig. 25.3. Welding should not be specified on drawings except through the use of the AWS welding symbols. They are precise, eliminate discussions with manufacturing and inspection personnel, and assure the design engineer that the joint will be fabricated in accordance with his wishes. Working stresses need not be reduced to take into consideration defective materials or workmanship when the desired parameters are specified by means of the welding symbol.

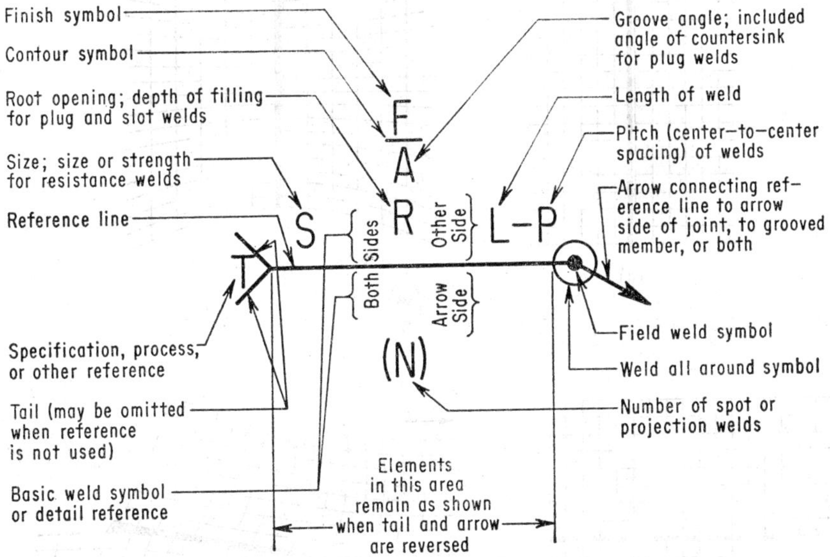

FIG. 25.2. Standard location of elements of a welding symbol.

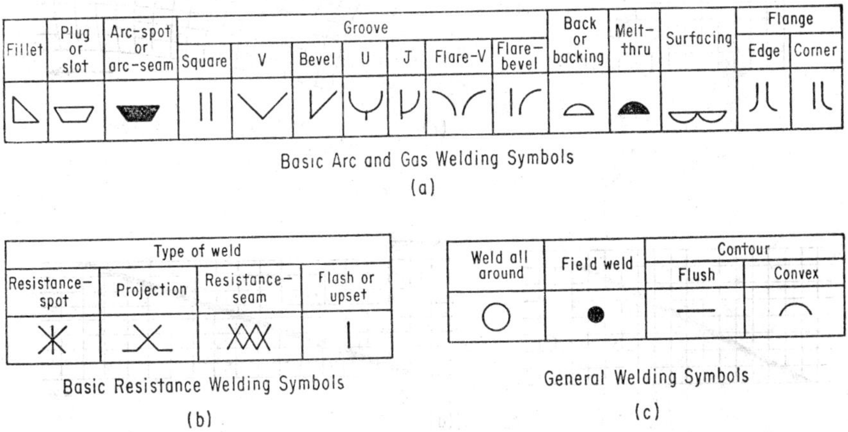

Fillet
Plug or slot
Arc-spot or arc-seam
Groove
Square
V
Bevel
U
J
Flare-V
Flare-bevel
Back or backing
Melt-thru
Surfacing
Flange
Edge
Corner
Basic Arc and Gas Welding Symbols
(a)
Type of weld
Resistance-spot
Projection
Resistance-seam
Flash or upset
Basic Resistance Welding Symbols
(b)
Weld all around
Field weld
Contour
Flush
Convex
General Welding Symbols
(c)

FIG. 25.3. Welding symbols.

The abnormalities of welded construction that must be considered are:

1. Undesirable mechanical properties in the weld- and heat-affected zone
2. Thermal stresses
3. Limited knowledge of fatigue characteristics

These factors are related to the fact that the metal in the joint has been molten during the welding process and has not received mechanical work subsequent to solidification.

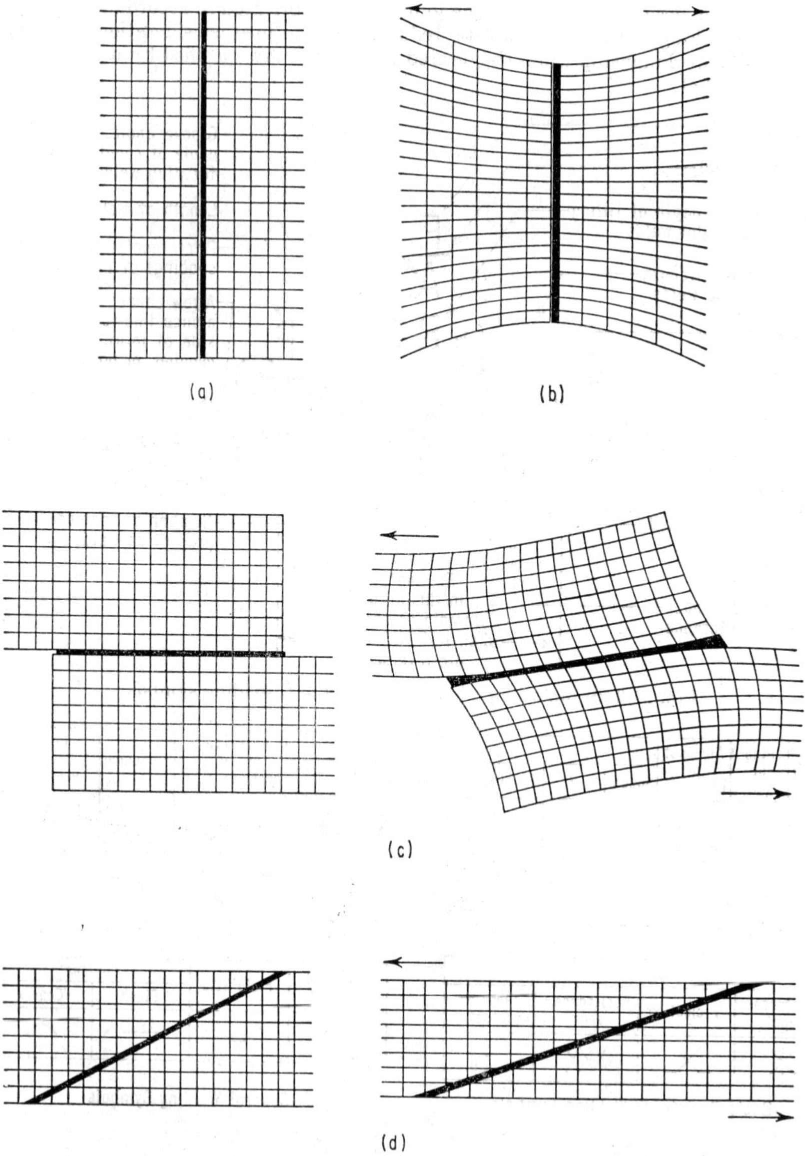

FIG. 25.4. Stress distribution within joints. (*a*) Butt joint is depicted here in two dimensions as it might look before (left) and after (right) being subjected to an applied load. (*b*) Arrows indicate the direction of stress; the constriction in the center, the nonuniform distribution of stress across the area of the joint. (*c*) Lap-joint adhesive under an applied load is strained chiefly at both ends. One component of the stress results in a shearing motion, which tends to slide the two portions of the joint over each other; a second component acts to force the joint open. (*d*) Scarf joint is far stronger than either a butt joint or a lap joint (see illustrations above). This is a type of joint in which the two adherent materials are tapered off so that they stretch uniformly throughout the length of the joint when subjected to an applied load.

On the positive side, welding is the most economical method of joining tension members. At least a 20 per cent reduction in net cross section can be realized in welded construction by the elimination of rivet or other fastener holes. Further reductions in weight can be experienced in the final design because welded construction does not require heavy end-connection details. Ease of fabrication should also be considered, especially in machines where such items as pressure vessels are used. With any other type of construction a secondary sealing operation, which at best is highly unreliable, would be required. Generally, fusion-welding equipment is the lowest-cost joining equipment that can be used for assembly purposes.

In addition to stresses caused by weld-metal solidification, welded joints are subjected to the forces produced by external loads. The stress concentration caused by external loads on three common types of joints is shown in Fig. 25.4.[9]

Note that the single lap joint is the poorest design. In a lap joint the weld metal is strained at each end by shearing moments (tending to cause the two parts to slide over one another), and by the bending moment developed (which tends to tear the weld metal). Thus it can be seen that the strength of a lap joint is proportional to the width of the joint but does not increase linearly with increasing overlap. It is affected also by the thickness of the members. The tension joint shows a nonuniform stress across the length of the joint, with the edges again carrying the major load. The scarf joint shows the most uniform stress distribution, both within the weld metal and in the adjoining base metal. Uneven concentrations of stress will cause the joint to fail under load at a value lower than that calculated for the joint. These theoretical facts must be considered in many of the joint designs later discussed.

(c) Welding Processes, Selection Considerations

Spot, roll, seam, flash, and upset are the most common resistance-welding processes used in machine construction. All are characterized by the fact that the heat for coalescence is obtained from electrical resistance at the interface of the pieces being joined. For process and equipment details see ref. 2 (Chaps, 30, 31, and 32) and ref. 10.

In gas welding, the heat is generated by the combustion of a fuel gas (principally acetylene and hydrogen) with air or oxygen.

In arc welding, heat is generated by the resistance to flow of an electric current across an air gap. Filler metal may or may not be added to the joint. Metal arc, MIG (metal arc in inert atmosphere), and TIG (tungsten arc in inert atmosphere) are the most commonly used processes.

The advantages and disadvantages associated with these processes are shown in Table 25.1.

Table 25.1. Welding Processes Compared

Process	Advantages	Disadvantages
Resistance welding	Very flexible Economical Joins three or more pieces with one weld Properties of joint independent of process Can join any combination of metals that alloy together	Poor in fatigue Requires expensive equipment Must be able to reach both sides of the joint
Gas welding	Good control of weld metal Heat input anneals weld metal and heat-affected zone	High cost Low production rate Higher heat input; steep temperature gradient More warpage, distortion, etc. Material thicknesses must be nearly equal
Arc welding	High production rate Minimum distortion Low annealing of base metal	Stress concentrations at edge of welds, undercutting, overlaps, etc., likely

Selection of the actual process depends upon (1) the alloy being welded, (2) metal thickness, (3) desired properties in the joint, (4) production rate, (5) quality standards, (6) number of units to be fabricated, (7) amount and type of tooling available.

(d) Resistance Welding

Material Weldability. The basic theory and uses of resistance welding are given in ref. 38. The characteristics of the spot weld formed when joining different materials together are shown in Fig. 25.5. This table shows that it is physically possible

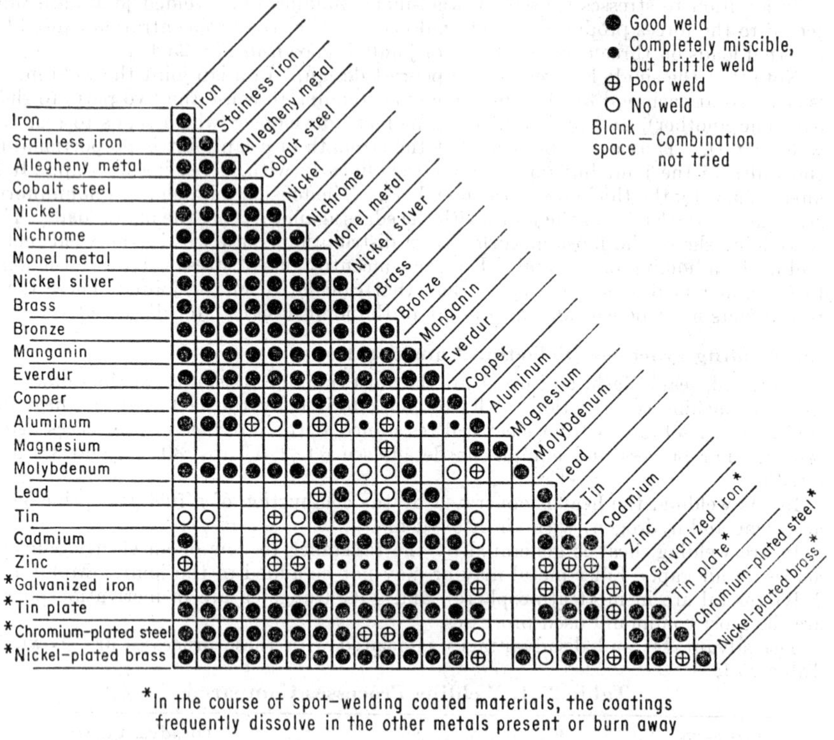

FIG. 25.5. Will they weld? Characteristics of spot welding of some 250 combinations of metals. This study was made by Lawrence Furguson of the Bell Telephone Laboratories, Inc. As in spot welding of steel, an essential detail is the choice of a suitable electrode tip. For welding high-resistance materials such as iron, nickel, nichrome, and nickel-silver, electrode tips of copper are suitable. Where greater strength or longer life is required, however, a copper-rich copper-tungsten alloy is advisable. For low-resistance materials such as copper or aluminum, tungsten tips are the best, as they do not easily alloy or weld at ordinary welding temperatures.

to weld a wide variety of metals and alloys to one another. In actual practice, spot properties, corrosion resistance, and other factors must be taken into account. Recommendations for a variety of current aluminum, magnesium, steel, and nickel alloys are shown in Figs. 25.6 and 25.7.

When two or more sheets of unequal thickness are to be welded, the maximum ratio between the thickness of the two sheets is given in Fig. 25.8. For welding three or more sheets, the minimum thickness of a thinner outer sheet is given in Fig. 25.9. Combinations of more than three thicknesses can be welded, but the welding technique needs to be developed for each application and therefore should be avoided if possible.

Basic alloy designation	1100-0, 1100-H14, 5052-0, 5052-H34 aluminum	2024-0 bare aluminum	2024-0 clad aluminum	2024-T3 clad aluminum	6061 (all tempers) aluminum	7075-T6 bare aluminum	7075-T6 clad aluminum	2014 clad aluminum	AZ31A magnesium
1100-0, 1100-H14, 5052-0, 5052-H34, aluminum......	G	P	F	N	P	P	N	P	
2024-0 bare aluminum........	P	N	F	F	P	N	F	N	
2024-0 clad aluminum........	F	F	G	N	P	F	N	F	
2024-T3 clad aluminum.......	N	F	N	G	P	F	G	F	
6061 (all tempers) aluminum..	P	P	P	P	G	P	P	P	
7075-T6 bare aluminum.......	P	N	F	F	P	N	F	N	
7075-T6 clad aluminum.......	N	F	N	G	P	F	G	F	
2014 clad aluminum..........	P	N	F	F	P	N	F	G	
AZ31A magnesium..........									G

Code:
G = good
F = fair
P = poor
N = not recommended

FIG. 25.6. Weldability of various aluminum and magnesium alloys.

	Inconel	Monel	K monel	1010 through 1035	1040 and above	4130	4135	8630	18-8 stainless*
Inconel............	G	N	N	N	N	N	N	N	N
Monel.............	N	F	P	N	N	N	N	N	N
K monel..........	N	P	F	N	N	N	N	N	N
1010 through1035 ..	N	N	N	G	F	F	F	F	P
1040 and above.....	N	N	N	F	F	F	F	F	P
4130..............	N	N	N	F	F	G	F	G	P
4135..............	N	N	N	F	F	F	G	F	P
8630..............	N	N	N	F	F	G	F	G	P
18-8 stainless*......	N	N	N	P	P	P	P	P	G

Code:
G = good
F = fair
P = poor
N = not recommended
* 18-8 welded to carbon or alloy steel may produce a brittle nugget unless postheating is used.

FIG. 25.7. Weldability of various steels and nickel alloys.

Steels and nickel alloys		All other materials	
Thinner sheet	Max ratio	Thinner sheet	Max ratio
0.020 and under	1:2	0.020 and under	1:1.5
0.021 and over	1:3	0.021 and over	1:2.5

FIG. 25.8. Resistance-welding thickness limitations, two sheets.

Steels and nickel alloys		All other materials	
Thinner outer sheet	% of total thickness	Thinner outer sheet	% of total thickness
0.020 and under 0.021 and over	33⅓ 25	0.020 and under 0.021 and over	40 33⅓

FIG. 25.9. Resistance-welding thickness limitations, three or more sheets.

Spot Pattern. The following spot patterns are most commonly used:

1. Single rows with the spacing according to structural requirements. A type 1 joint is not as reliable as a type 2 joint because of the possibility of progressive failure (see Fig. 25.10). More than three welds in tandem are of no structural value for loading in a type 1 welded joint.

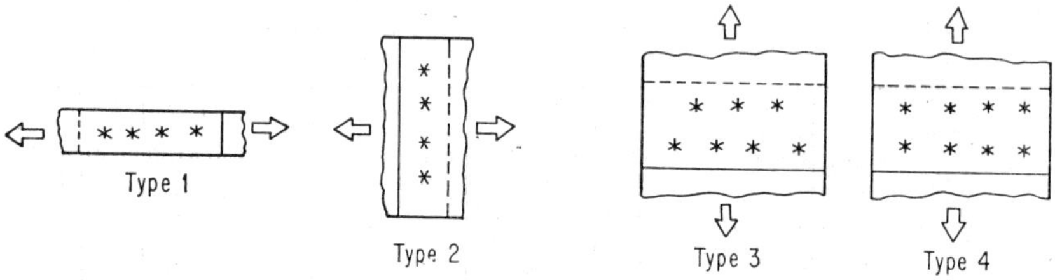

FIG. 25.10. Typical spot patterns.

2. Double rows either staggered or in line. Type 4 joints are somewhat more efficient for the loadings indicated in Fig. 25.10 than type 3 joints, provided the minimum weld spacing is maintained.

Spot Location. Spots may be located as desired provided the minimum spacing shown in Fig. 25.11 is maintained. The values shown in Fig. 25.11 are based on a control thickness which is equal to one-half the total thickness of the combinations being considered. Closer spot spacings than the minimum shown in the table may cause partial short circuiting of the welding current through the adjacent spots, producing a weld of smaller diameter and lower shear strength.

Loading of Spot-welded Joints. Spot-welded joints should be designed to transmit shear or compression loads only. The following limitations must be observed in spot-welded joints:

1. Avoid welds in tension or those subject to undue vibration.
2. Avoid designs where considerable "tension-field" action is developed even though the tension load is below the static strength of the joint.
3. Single isolated welds shall not be used in structural assemblies, as spot welds provide little resistance to torsional stresses.
4. A minimum of two spot welds should always be used.

The shear strength of single spot welds in a joint consisting of two equal thicknesses of material is shown in Table 25.2. As noted, single spot welds cannot be used in actual designs. When two or more spot welds are used, the shear strengths shown in Table 25.2 must be reduced by at least 10 per cent to allow for shunting action (flow of a portion of the current through the already made spot weld which reduces the size of the second and subsequent welds). In large-assembly fixture-type spot-welding machines, the actual shear strength is determined on the basis of tests run on actual parts. In general, the test values will approach those shown in Table 25.2.

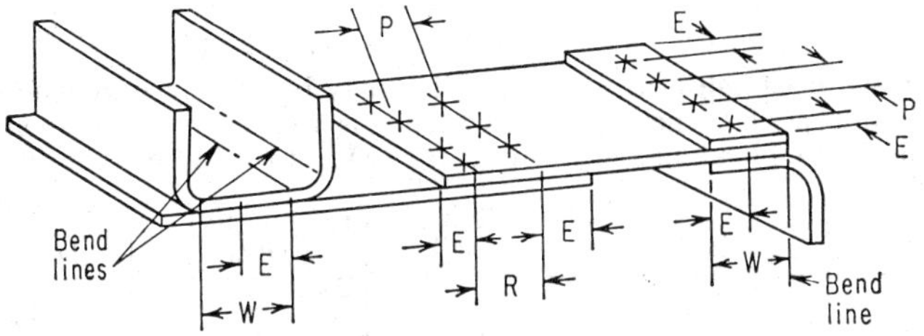

Min values for spot location

Control thickness (½ total thickness)	Steel and nickel alloys					Other material				
	P min	*E* min	*E* min[a]	*R* min	*W* min	*P* min	*E* min	*E* min[a]	*R* min	*W* min
0.010	0.25	0.18	0.12	0.22	0.36	0.31	0.18	0.12	0.27	0.36
0.015	0.30	0.18	0.12	0.26	0.36	0.31	0.18	0.12	0.27	0.36
0.020	0.38	0.22	0.12	0.33	0.44	0.38	0.18	0.12	0.33	0.36
0.025	0.44	0.25	0.14	0.38	0.50	0.38	0.22	0.14	0.33	0.44
0.030	0.50	0.25	0.14	0.43	0.50	0.38	0.25	0.16	0.33	0.50
0.035	0.50	0.25	0.16	0.43	0.50	0.40	0.25	0.16	0.34	0.50
0.040	0.62	0.25	0.18	0.54	0.50	0.40	0.25	0.18	0.38	0.50
0.045	0.75	0.25	0.20	0.65	0.50	0.46	0.28	0.20	0.40	0.56
0.050	0.88	0.31	0.20	0.76	0.62	0.50	0.30	0.20	0.43	0.60
0.055	0.94	0.31	0.22	0.81	0.62	0.50	0.33	0.22	0.43	0.66
0.060	0.98	0.31	0.22	0.85	0.62	0.50	0.36	0.22	0.43	0.72
0.065	1.03	0.31	0.22	0.89	0.62	0.50	0.38	0.22	0.43	0.76
0.070	1.12	0.31	0.24	0.97	0.62	0.54	0.38	0.24	0.47	0.76
0.075	1.20	0.31	0.24	1.04	0.62	0.58	0.38	0.25	0.50	0.76
0.080	1.25	0.38	0.24	1.08	0.76	0.62	0.38	0.25	0.54	0.76
0.085	1.30	0.38	0.25	1.12	0.76	0.62	0.40	0.25	0.54	0.80
0.090	1.34	0.38	0.25	1.16	0.76	0.62	0.44	0.25	0.54	0.88
0.095	1.38	0.38	0.25	1.19	0.76	0.66	0.44	0.25	0.57	0.88
0.100	1.44	0.38	0.25	1.25	0.76	0.73	0.44	0.28	0.63	0.88
0.105	1.48	0.38	0.28	1.28	0.76	0.78	0.44	0.28	0.67	0.88
0.110	1.54	0.40	0.28	1.33	0.80	0.84	0.46	0.28	0.73	0.92
0.115	1.60	0.40	0.28	1.38	0.80	0.90	0.46	0.28	0.78	0.92
0.120	1.64	0.40	0.28	1.42	0.80	0.94	0.48	0.28	0.81	0.96
0.125	1.68	0.44	0.28	1.45	0.88	1.00	0.50	0.28	0.87	1.00

[a] Nonstructural only.

FIG. 25.11. Design limitations, spot locations.

Joint Design and Equipment Limitations. It is of the utmost importance to provide for accessibility of welding apparatus in all resistance-welding applications. The high pressure and current involved require heavy machine structures, and the need for water cooling the electrodes further complicates the accessibility problem. Wherever possible, the design should be such that simplified electrode components can be used with a minimum of offsetting.

Figure 25.12 illustrates typical machine-joint designs, classified as to desirability.

(e) Fusion Welding

General Design Considerations. Welding may be employed in the fabrication of practically all types of structures. Gas welding is more expensive than arc welding and may cause more buckling and warping when used to weld very thin flat sections

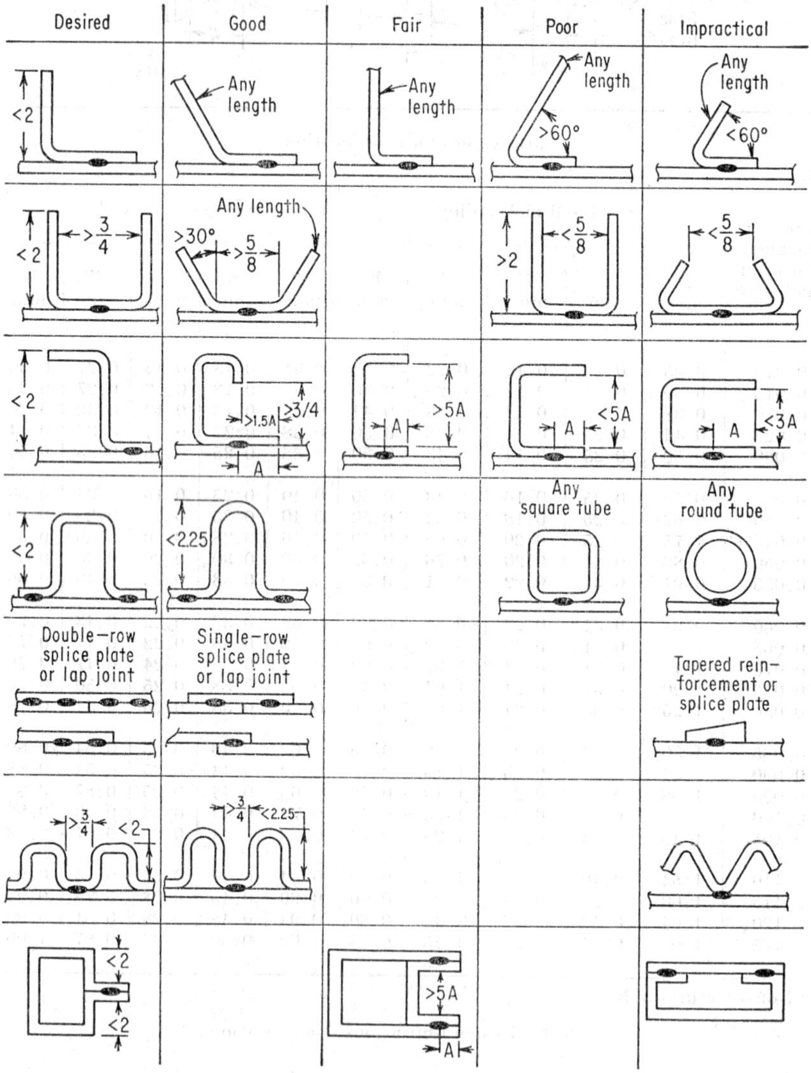

FIG. 25.12. Spot weldability of typical machine joints.

unless excessive clamping is employed. Gas welding is usually specified for metals between 0.010 and 1/16 in. in thickness, stainless steel frequently being an exception. Arc welding is used in applications where these limitations apply, and in general where high production rates are involved since arc welding is a lower-production-cost process.

Table 25.2. Single-spot Shear Strength

Alloy thickness	Low-carbon steel <70,000 psi	Low-carbon steel >70,000	Stainless steel, tensile strength 70/90,000 psi	Stainless steel, tensile strength 90/150,000 psi	Stainless steel, tensile strength <150,000 psi	Magnesium AZ-31	Nickel	Inconel
0.010	130	180	150	170	210	100	135	175
0.021	320	440	370	470	500	176	350	545
0.031	570	800	680	800	930	272	760	920
0.040	920	1,200	1,000	1,270	1,400	344		
0.050	1,350	.....	1,450	1,700	2,000	432		
0.062	1,850	.....	1,950	2,400	2,900	544	2,400	2,750
0.078	2,700	.....	2,700	3,400	4,000	660		
0.094	3,450	.....	3,550	4,200	5,300	770	3,600	4,400
0.109	4,150	.....	4,200	5,000	6,400	925		
0.125	5,000	.....	5,000	6,000	7,600	1,075	5,600	6,400

Reference 10 summarizes all the recommended welding practices (for automotive welding design), showing specifics in welding-process selection.

The following general rules apply:

1. Metals being joined must have approximately the same melting point.

2. Welded joints have the same strength as the annealed material of the parts welded if the welding metal is of the same analysis as the base metal. When tensile tested to failure, the joint will usually break at the edge of the weld in the base-metal area annealed by the welding heat rather than at the weld. Welded fittings may be heat-treated after welding to remove cooling strains and overheated spots in the fitting. Welded alloy steels should be heat-treated after welding if possible.

3. Materials of equal thicknesses are best suited for welding. Where the thickness ratio is 1:1.5 or less, the materials are considered to be of equal heating value and can be welded easily.

4. Joints prepared for welding should have the edges beveled on one or both pieces to allow space for filling in the welding material, which should be of nearly the same analysis as the welded pieces (see Table 25.3).

Table 25.3. Weld-rod Types for Various Base Alloys

Base metal	*Weld rod*
Steel castings	Steel
Steel pipe	Steel
Steel plate	Steel
Steel sheet	Steel or bronze
High-carbon steel	Steel
Wrought iron	Steel
Galvanized iron	Steel
Cast iron, gray	Cast iron or bronze
Cast iron, malleable	Bronze
Cast-iron pipe	Cast iron
Chrome-nickel steel castings	Base metal or 25-12 chrome-nickel steel
Chrome-nickel steel sheet	Columbium stainless steel or base metal
Chromium steel	Columbium stainless steel or base metal

5. Single-plate gussets, welded on one side only, should have the weld carried around the end to provide positive anchoring of the gusset end.

6. Welded joints should not be designed to take bending loads unless supported by auxiliary means such as pinning, riveting, telescoping, or other positive support.

7. Avoid joints between thin and thick sections. Materials where the thickness ratio is greater than 1:1.5 are considered to be of unequal heating value and are difficult to weld. A maximum ratio of 1:3 is normally considered the limit. An exception to this exists where one side of the joint has a greater thickness but smaller mass.

For instance, a ⅛-in. continuous sheet can be welded to a 1-in. cube because the block, once heated, will not dissipate the heat so rapidly as the continuous sheet containing the greater area.

8. Avoid the convergence of more than six members to form a welded cluster. The repeated application of heat on the small area where all members meet weakens the base metal adjacent to the remelted weld metal.

9. Do not use tin-lead solder in the proximity of a joint. Repair of the joint might cause contamination of the weld metal with subsequent embrittlement and cracking of the affected area.

10. Do not weld a brazed joint. The reverse practice of brazing a welded joint is permissible.

11. Avoid a joint which requires welding at the inside junction of two members forming an acute angle.

12. Avoid joints in parts which are heat-treated before welding, especially to a tensile strength in excess of 125,000 psi.

13. Avoid washers welded to lugs or other parts of fittings.

14. Do not weld parts previously spot-welded because of the probability of inclusion of flux or flux-removal fluids in the spot-welded seam.

15. Avoid designs which require welding into deep pockets, which are difficult to handle because of blowback or arc-over. The maximum condition should be a corner where three plates intersect at right angles to one another. The corner should be left open if possible.

16. Avoid welding parts having widely different melting points.

Detail Fusion-welding Design Considerations. Three factors influence the length of fusion welds, namely, strength requirements, design of the parts, and distortion of the parts and possible resultant cracking of the weld. Depending upon these factors, welds may be either continuous, intermittent, or tack welds.

Continuous welds are used whenever strength requirements are high, or where a liquid- or gastight joint is required. They are costly because of the postwelding straightening operation usually required to eliminate distortion caused by the heat of the welding operation. Intermittent welds are used on long joints where strength and rigidity requirements are not exacting enough to warrant the extra cost and weight of a continuous weld.

There are five fundamental types of welded joints, namely, butt, lap, corner, edge or flange, and tee. The more common types of welding applied to these joints are given in Fig. 25.13. Recommendations on structural efficiency are given in Fig. 25.14.

In all types of butt welds, the strength characteristics may be improved by welding both sides. However, welding both sides is not recommended on steel or aluminum alloys in gages less than 0.125, or on any gage of magnesium alloy.[30]

Fittings should be made simply, with the smallest practicable number of component parts to reduce welding to a minimum. Single-piece forgings or castings are preferable to built-up welded fittings.

When welding sheet or tube to fittings, the thinner stock is frequently burned away before the thicker material is brought to the fusing point. Furthermore, finish machining of a heavy fitting included in a welded assembly cannot be completed because excessive warpage or shrinkage has left insufficient material for the final operation. It is therefore necessary to rough-machine all heavy fittings prior to gas welding to eliminate excess material, reducing the area to be welded to a thickness not to exceed three times the thickness of the attaching members. When such precautions are exercised, it may be machined to within $\frac{1}{32}$ in. of its final thickness before welding.

The joints shown in Fig. 25.15 represent good practice because equal gages of material are welded. The fitting in the area to be joined is first beveled at 45°.

A bead around a welded fitting helps eliminate warpage and cracking by permitting expansion and contraction (see Fig. 25.16).

If the members to be joined are of the same diameter, they should be welded at a 30° angle scarf joint (Fig. 25.17). A tubular liner should be used, extending at least 1 in. or $1\frac{1}{4}D$, whichever is greater, beyond the welded joint. This not only prevents

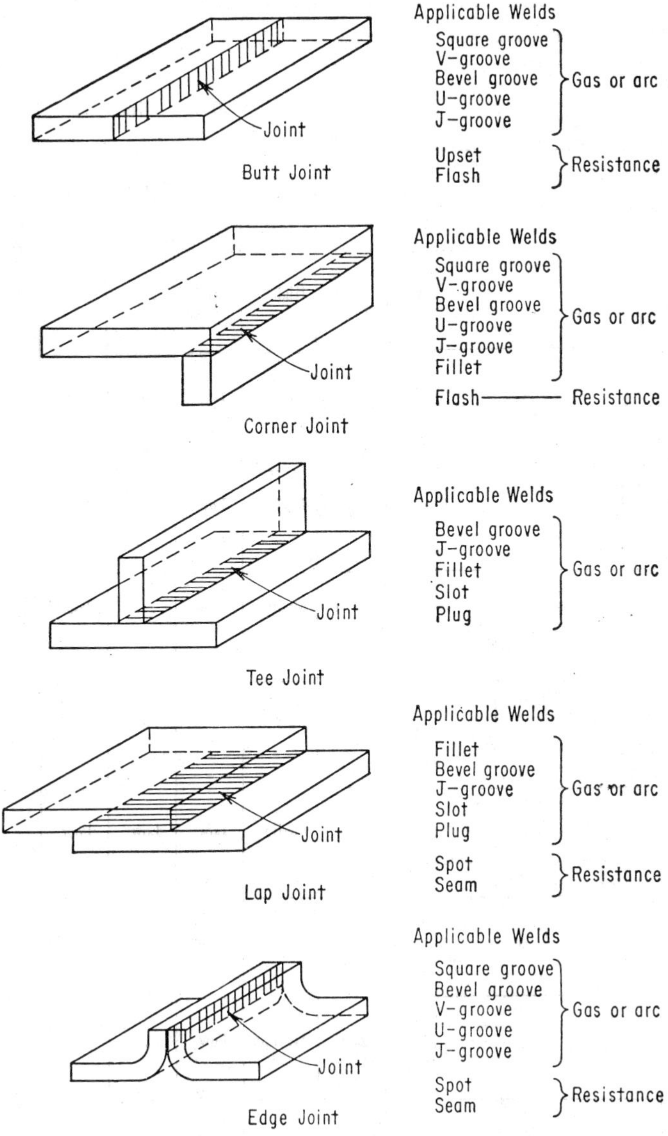

FIG. 25.13. Welds applicable to basic joints.

reduction of diameter when the joint is under tension but also serves to align the tube ends for welding. All liners must fit snugly inside the tubular members.

Tubes of different diameters should be telescoped one inside the other. The tubing may be swaged or a sleeve may be used to ensure a snug fit. The female tube (and sleeve, if used) must be scarfed or fishmouthed (see Fig. 25.18).

Truss joints generally involve one continuous member, to which are welded vertical and/or diagonal members. The axes of several members must intersect at a common

Type of weld	Illustration	Sheet gage	Efficiency in			Fatigue resistance	Application
			Shear	Tension	Compression		
Square butt welded one side	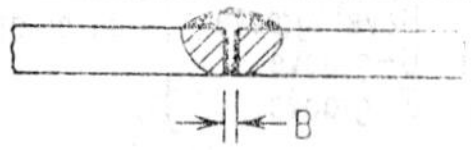	Up to 0.124	High	High	High	Fair	
Single V welded one side	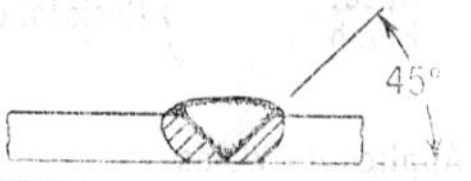	0.125 and over	High	High	High	Good	All general applications. The choice depends upon the material, gage, loading, fatigue requirements, etc.
Single bevel welded one side	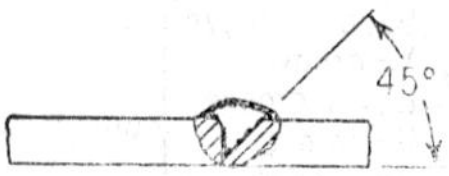	0.125 and over	High	High	High	Good	
Double V welded both sides	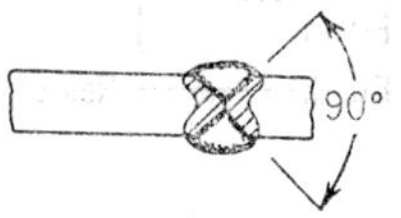	0.375 and over	High	High	High	Good	
Open-square groove corner	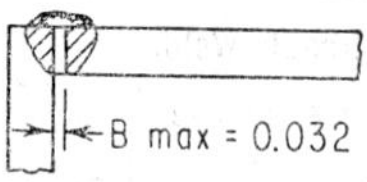	Up to 0.065	Medium	Low	High	Poor	
Single V corner fillet	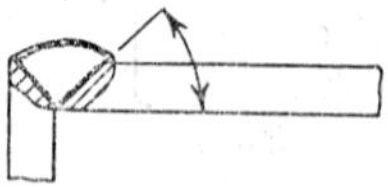	0.065 and over	Medium	Low	....	Poor	Closed structures and heavy-gage tanks
Outside or inside single fillet	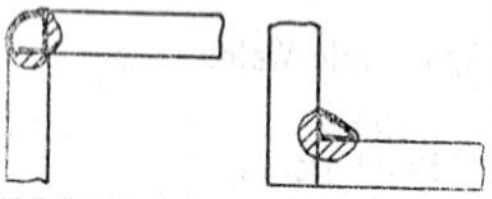	Up to 0.094	High	Low	....	Poor	
Single fillet tee	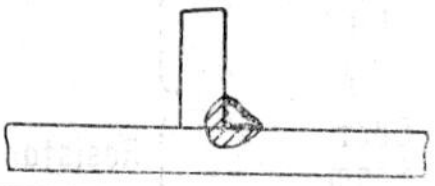	Any gage	Medium	Low	Low	Poor	
Double fillet tee	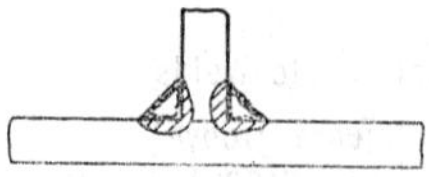	Any gage	High	High	High	Good	General purposes
Edge fillet weld	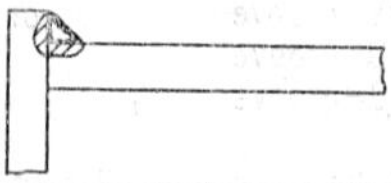	Any gage	Medium	Low	....	Fair	
Double-lap fillet		Up to 0.250	High	High	....	Fair	Closed structure and heavy-gage tanks
Flange weld	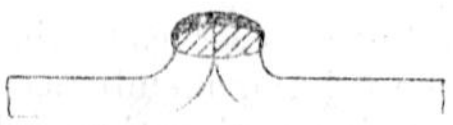	Up to 0.081	Medium	Low	....	Poor	Tanks and closed nonstructural parts

Flange height	Material gage
4 × material gage	0.040 or less
3 × material gage	0.041–0.081

Root Opening B

0.064 gage and less, no gap

0.064–0.124 gage, gap = ½ sheet thickness

FIG. 25.14. Structural efficiency of various types of welds.

point to avoid eccentric loading. The fixity and tensile strength of such joints are increased by the use of gussets (see Fig. 25.19). They also permit gradual introduction of the stress into the joint.

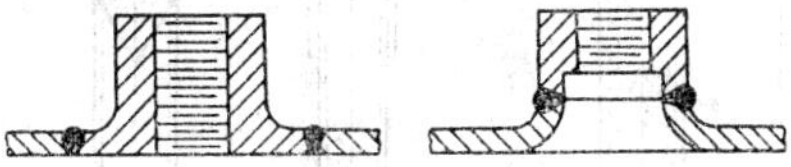

FIG. 25.15. Preferred fitting joints.

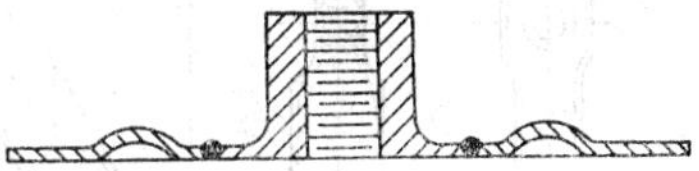

FIG. 25.16. Method of stress relief around a fitting.

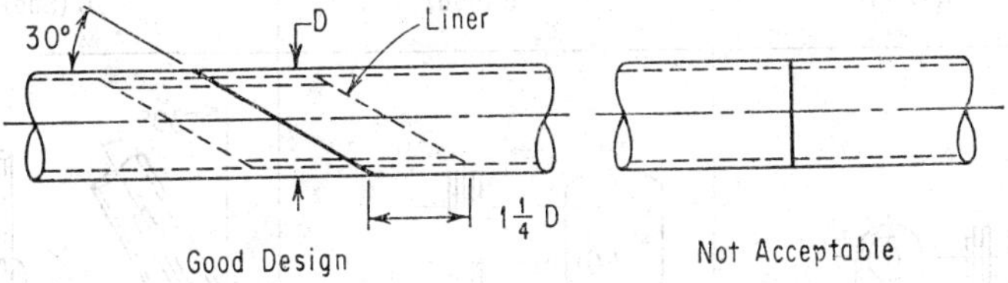

FIG. 25.17. Tubular joint splices, equal-size tubes.

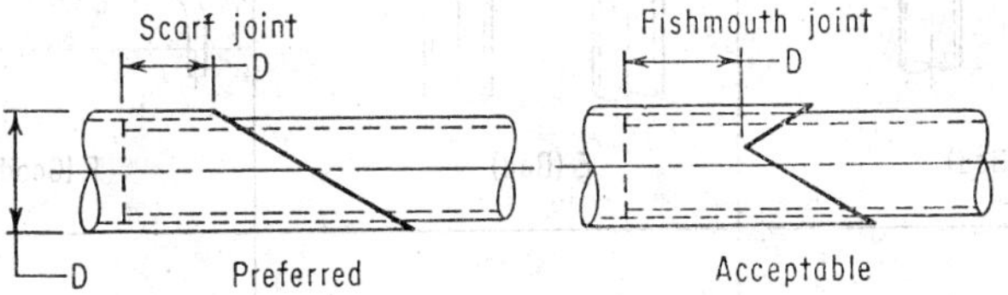

FIG. 25.18. Tubular joint splices, unequal-size tubes.

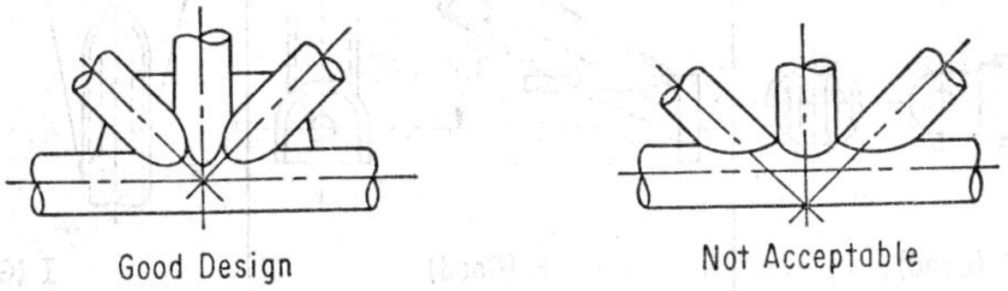

FIG. 25.19. Tubular truss joints, acceptable and unacceptable.

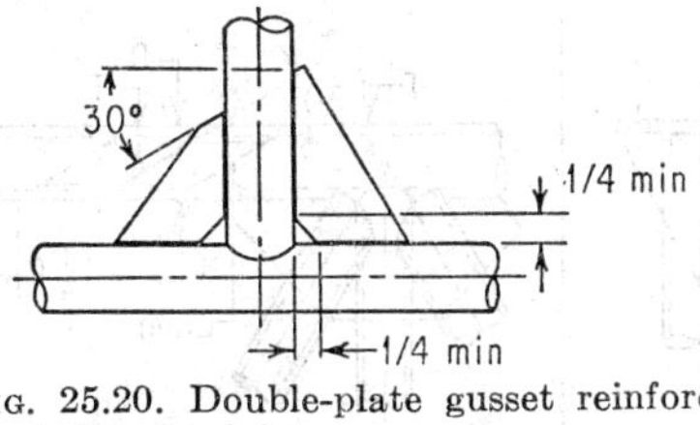

FIG. 25.20. Double-plate gusset reinforcement, tubular joints.

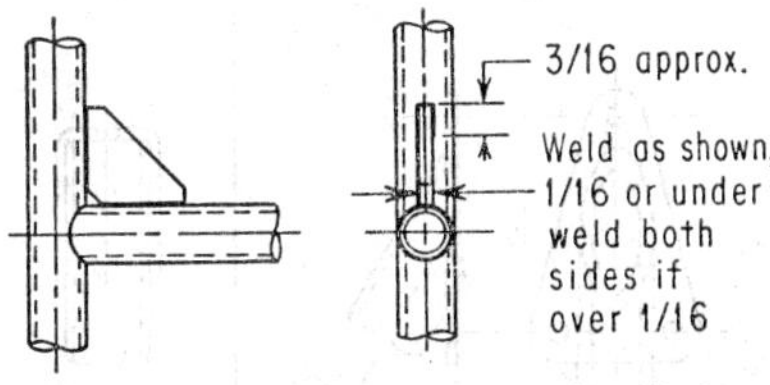

FIG. 25.21. Single-plate gusset reinforcement, tubular joints.

Because of stress concentration, welds in slotted tubes should not end diametrically opposite one another. A minimum diagonal of 30° across the tube should be maintained (see Fig. 25.20).

Single-plate gussets 1/16 in. thick and under should be welded on one side only and have the weld carried around the end to provide positive anchoring of the gusset plate (see Fig. 25.21). Gussets over 1/16 in. in thickness should be welded on both sides.

Splicing of tubes can often be done conveniently near a truss joint, which tends to strengthen the splice.

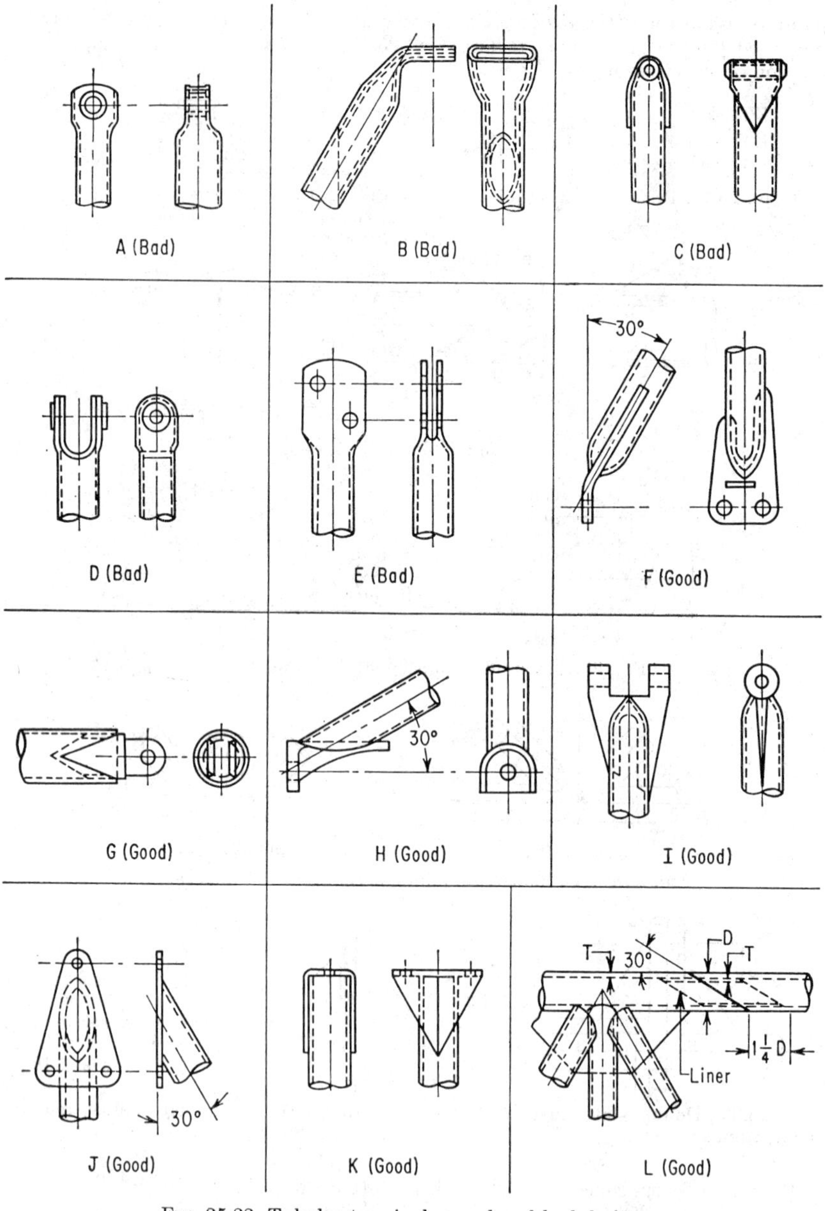

FIG. 25.22. Tubular terminals, good and bad designs.

Lugs may be welded to any point on a member if the load on the lug is of low magnitude. High-stress lugs should be located at truss points. Washers or bushings may be welded to lugs to increase bearing area when weight is critical, but this practice *must not* be followed on highly stressed fittings.

End fittings designed for tubular members may be forgings flash-welded to the tubes where large quantities are involved. For highly stressed members of limited quantity, the fitting may be a forging or high-tensile steel casting jointed to the tube by a fishmouth weld or similar method. Abrupt changes in section must be avoided because of the probability of high stress concentrations.

Figure 25.22 illustrates both good and poor end fitting design.

Figure 25.23 illustrates methods of welding a tube to a sheet or two sheets together. In the latter case, except for capping a tube or for a low-stressed joint, the tube should be inserted into a hole punched in the sheet.

Plug or slot welds may be employed for attaching two sheets or plates at specific points. A certain degree of local warping usually accompanies this process (see Fig. 25.24). Plug welds are often used in lieu of spot welds to assemble plate stock too thick to be spot-welded with available machines.

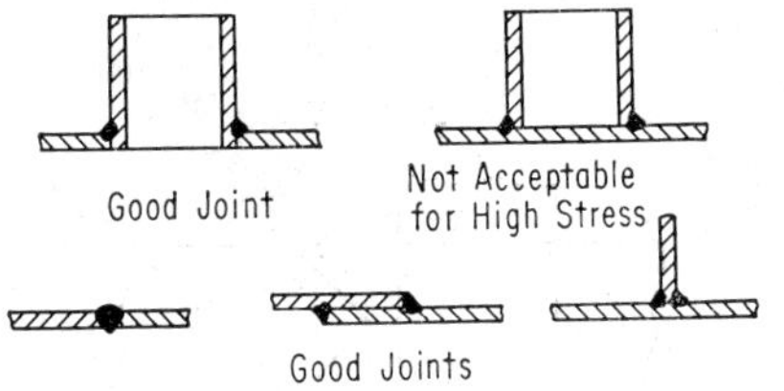

FIG. 25.23. Recommended sheet and plate joints.

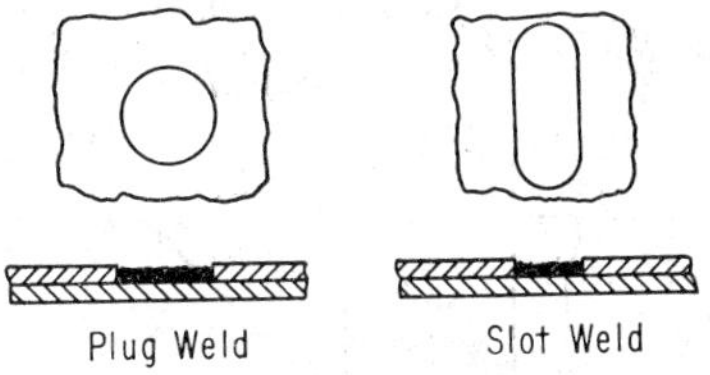

FIG. 25.24. Recommended plug and slot welds.

Weldability of Materials. The relative welding qualities of similar and dissimilar metals, and types of welding which may be employed are indicated in Fig. 25.25. A complete discussion of weldability of steels is given in ref. 36 and of nonferrous metals in ref. 37.

Basic alloy designation	1100 and 3003	5052	5053	6061	AZ61X	M1	1010, 1020, 1025	2330	4130	4140	4340	18-8
1100 and 3003	B E 1											
5052	B E 1	B E 1										
5053	B E 1	B E 1	B E 1									
6061	B E 1	B E 1	B E 1	B E 1								
AZ61X					E 1							
M1					E 1	E 1						
1010, 1020, 1025							A C 1					
2330							A C 2	A C 2				
4130							A C 4	A C 4	A C 4			
4140							A C 4	A C 5	A C 5	A C 5		
4340							A C 5	A C 5	A C 5	A C 5	A C 5	
18-8							A C 1	A C 1	A C 1	A C 2	A C 2	A E 1

The necessity for heat-treatment after welding should be judged on the basis of the specific design. Welding may be done after heat-treatment, provided the reduced strength in the weld area is taken into consideration.

Hydrogen welding may be used on any steel fitting where deemed expedient.

Code:

A = oxyacetylene
B = oxyhydrogen
C = metallic arc
E = Heliarc

1 = good
2 = fair
3 = poor
4 = heat-treat after welding
5 = special methods, such as preheating and controlled cooling, required

FIG. 25.25. Weldability of various alloys.

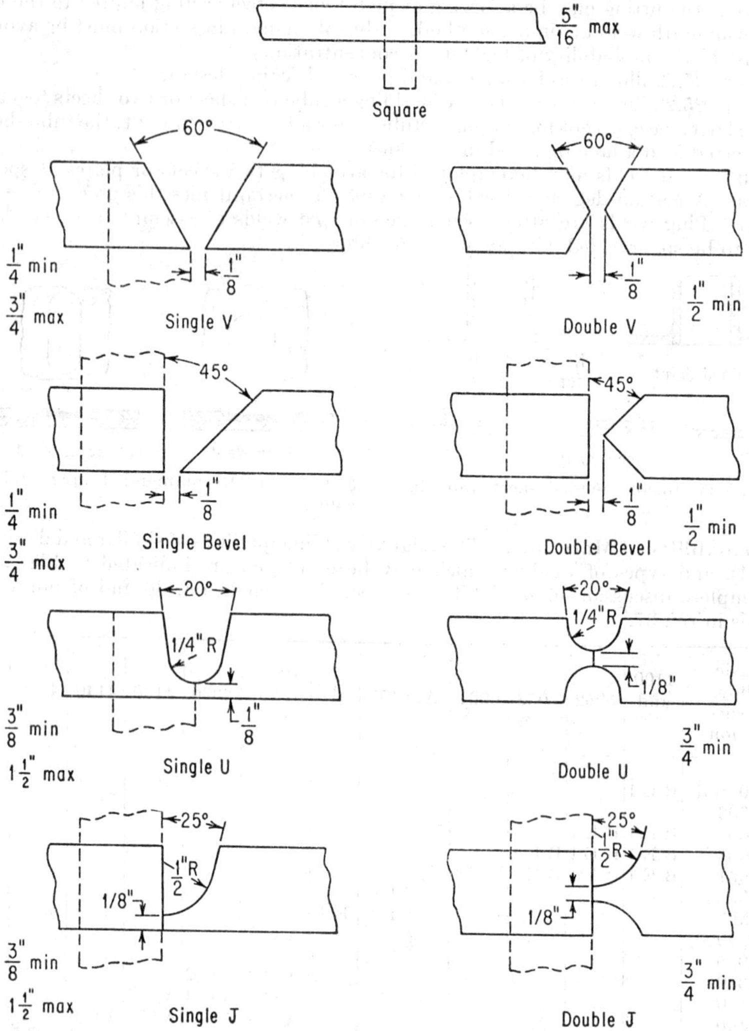

FIG. 25.26. Types of joints.

(f) Metallic-arc Welding

This method is used for most steel applications over 0.062 in. or aluminum over 0.091 in. Joint recommendations are given in Fig. 25.26.

Design Considerations for Magnesium and High-nickel Alloys. In addition to the general and detail design considerations listed above, the following specific considerations should be observed:

With Magnesium Alloys. Heliarc welding should be employed on sheets and plates ranging from 0.030 to 1.00 in. in thickness. A backing plate is desirable to minimize "drop-through" and distortion due to heat dissipation.

All magnesium joints should be stress-relieved.

Desirable joint designs for magnesium alloys are illustrated in Fig. 25.27. For monel, nickel, and Inconel, filler metal should be the same as the base alloy, or the electrode design for the alloy.[11] Such joints are equal in corrosion resistance and strength to the parent metal, requiring no aftertreatment in heavy gages. Fabricated assemblies in thin gages should be stress-relieved after welding. The high coefficient of thermal expansion makes it necessary to use heavier clamping pressure to keep parts from buckling than is used with low-alloy steel. Recommended methods of making joints in these alloys are shown in Fig. 25.28 and ref. 11.

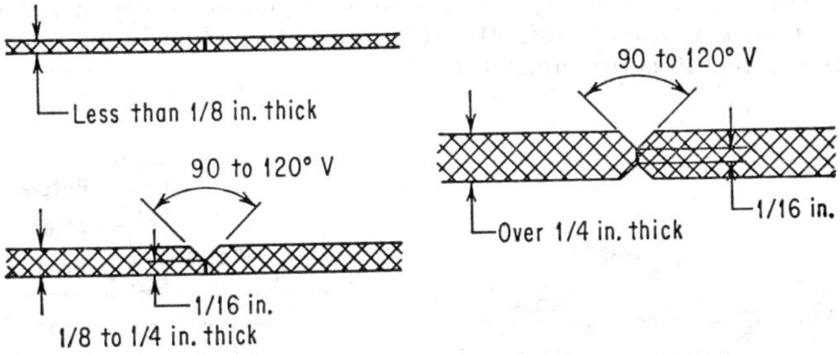

FIG. 25.27. Recommended joint chamfer for magnesium alloys.

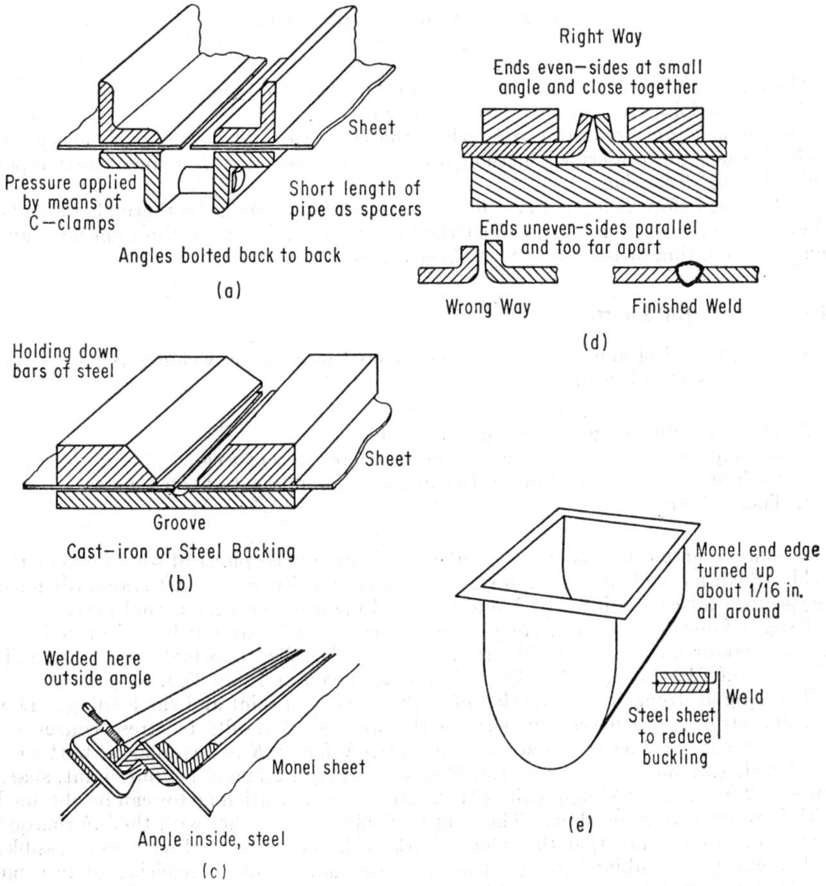

FIG. 25.28. Typical clamping fixtures needed for high-nickel alloys.

(g) Distortion and Heat Flow

Distortion associated with a given material and welding process is a function of joint design. This phenomenon is due to the unequal cooling rate which exists between the materials being welded and the materials being used for the weld (bead). The effect of unequal cooling rate upon two welded structures is shown in Fig. 25.29.

If a complete mathematical approach is desired, a general theory of welding deformation and stress is treated in ref. 39. If further information is desired on residual welding stresses, ref. 40 is recommended.

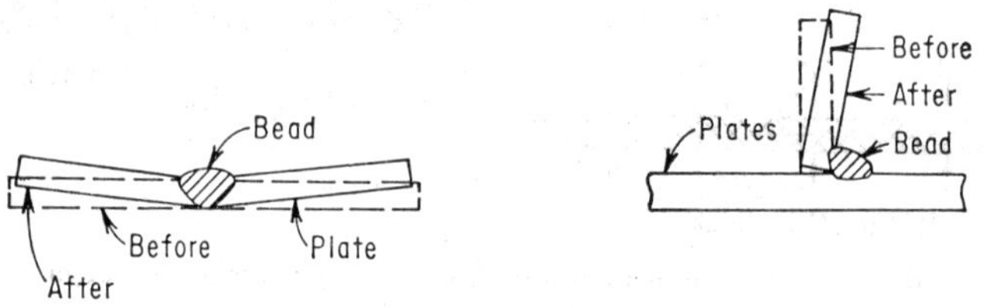

FIG. 25.29. Distortion after cooling.

The following rules should be observed to minimize distortion:

Use U or J joints in preference to V or bevel.

Use double joints (front and back) where thickness warrants instead of a single joint. In both cases the prescribed method uses less weld metal; this reduces heat input and therefore distortion.

Preheating before welding serves to reduce distortion by reducing thermal gradients. Metal thickness and joint type affect the flow of heat away from the weld bead and therefore resulting distortion. This effect is shown in Fig. 25.30[12].

(h) Unfavorable Factors

The major undesirable factors to be considered in selecting welding as opposed to other joinint methods are:

1. The possibility of generating inferior welds
2. Incomplete knowledge of the fatigue characteristics of welded joints
3. Undesirable mechanical properties in the weld and heat-affected zones
4. Thermal stresses

A nondestructive test is often specified to evaluate the physical soundness of the weld. Such methods include X-ray, Magnaflux, and Zyglo. Exact standards must be specified for each joint and should be based upon tests on the actual part.

Fatigue failures show after some time has passed and may not be recognized as a limiting factor until many units are in service. Detailed information on fatigue is found in refs. 12 through 25. Typical data are given in Table 25.4.

The fatigue strength is a function of both the type of joint and the loading. Low working stresses are necessary because the weld joint results in stress concentration. Accepted values of the stress-concentration factor K are as follows: butt weld machined, 1.2; fillet weld at toe, 1.5; fillet weld at end, 2.7; and tee butt joint, sharp corners, 2.0. More desirable values of the stress-concentration factor can be obtained with improved workmanship. The data in Table 25.4 together with the information mentioned above show that the smooth butt welds should be used wherever possible.

The effect of combined loads (static plus dynamic) must be considered in some machine components as shown in Fig. 25.31. The equation for the straight line relating the stready stress S_0 and the variable stress KS_V and joining the variable working stress S_wV and the steady working stress S_{w0} is

$$KS_V/S_{wV} + S_0/S_{w0} = 1$$

By substitution into the above formula, and using the preceding formulas to solve for the length of weld needed to carry the stress calculated, a determination can be made as to whether there is sufficient joint area to produce the length of weld required for safe operation.

Table 25.4. Relative Fatigue Strength of Various Welded Joints

Type of joint	Notes	Endurance limit, tons/in.^{2}	
		Tension/ compression	Tension only
	Weld machine flush	±9.0	7.0 ± 7.0
	As welded	±7.0	5.75 ± 5.75
	Incomplete root fusion	±5.0	4.0 ± 4.0
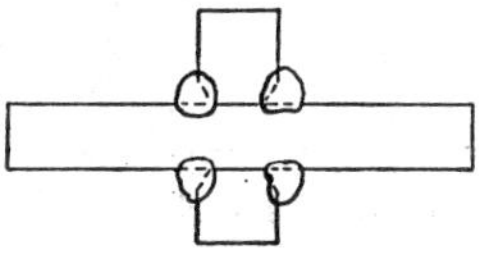	Complete root fusion	±5.5	4.75 ± 4.75
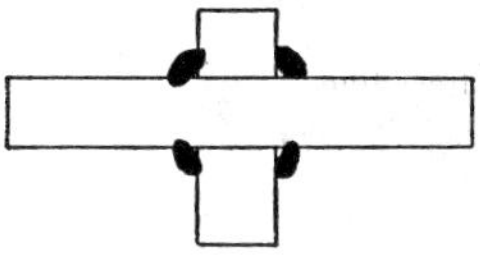	Incomplete root fusion	±3.7	3.5 ± 3.5
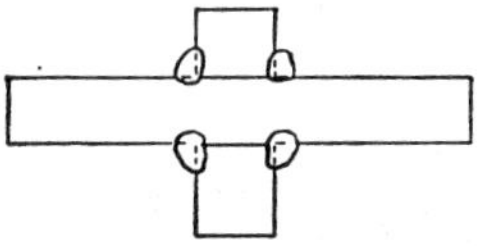	No joint preparation	±2.25	2.0 ± 2.0

(i) Stress Calculations[26,31]

The basic formulas for determining shear and normal stresses in welded joints are given in Fig. 25.32.

Heat flow	Sheet thickness, in.	Thermal severity no.
Two way 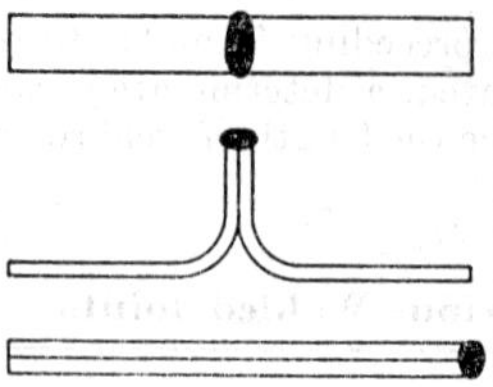	Both ¼	2
	¼–½	3
	¼–¾	4
	Both ½	4
	½–1	6
	1	8
	1–2	12
Three way 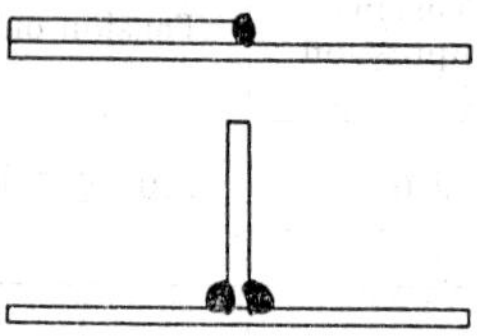	Both ¼	3
	½	6
	1	12
	2	24
	¼ + ½ + ½	5
	½ + 1 + 1	10
	2 + 1 + 1	16
Four way 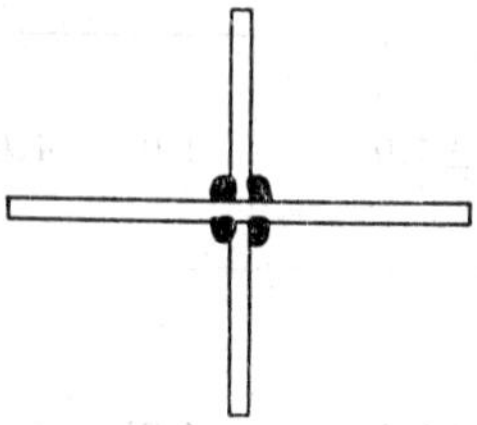	Both ¼	4
	½	8
	1	16
	2	32
	¼ + ½ + ½ + ½	7
	½ + ½ + 1 + 1	12

FIG. 25.30. Heat-flow paths in typical joints and nominal thermal severity.

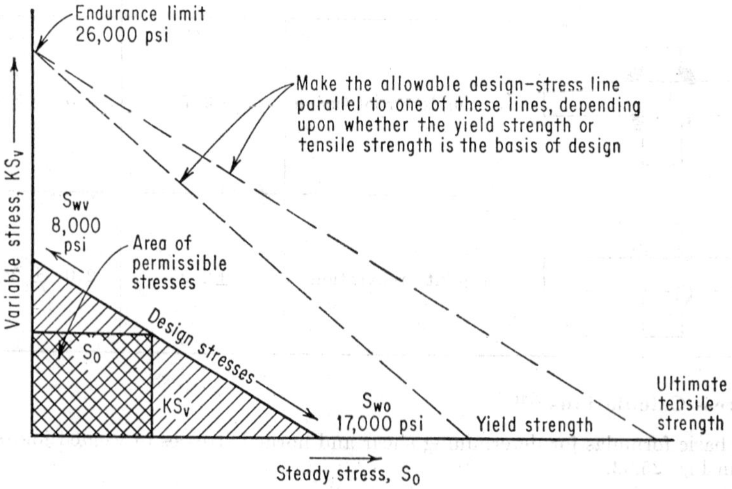

FIG. 25.31. Variable and steady relationship.

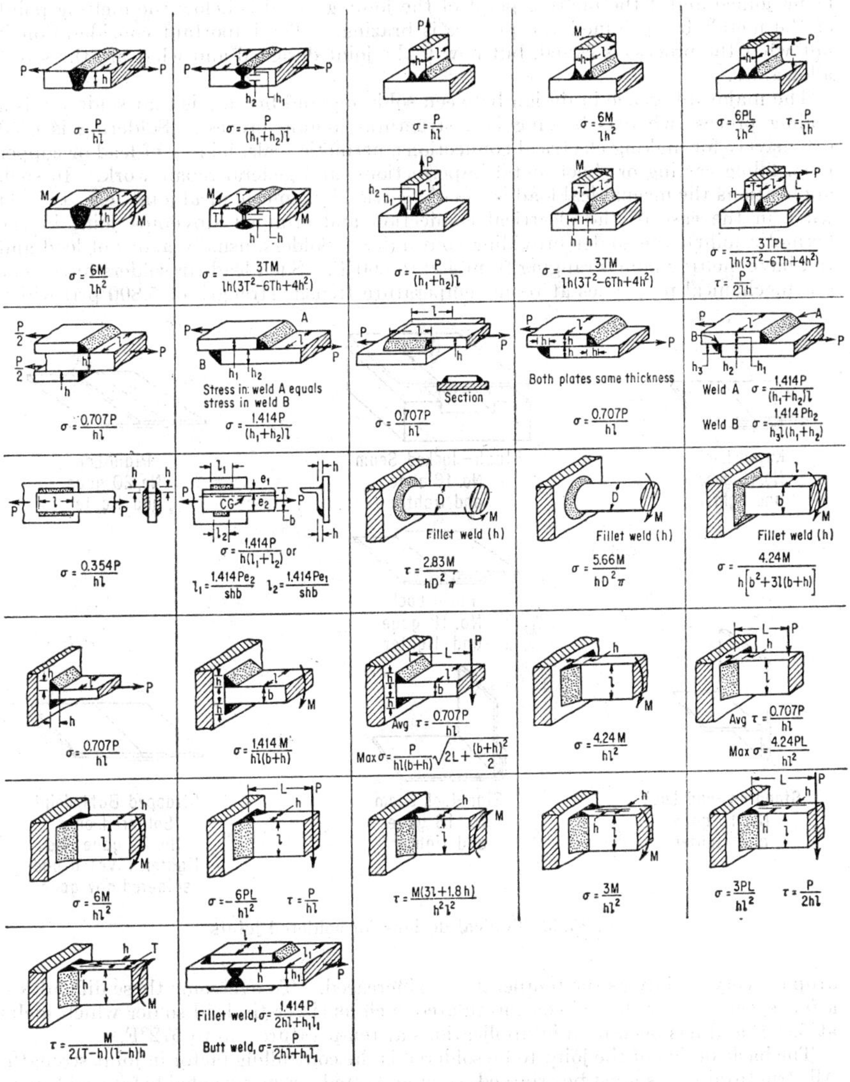

FIG. 25.32. Stress calculations.

α = normal stress, psi
τ = shear stress, psi
M = bending moment, in.-lb
P = external load, lb
L = linear distance, in.
h = size of weld, in.
l = length of weld, in.

25.2. BRAZING AND SOLDERING

(a) Design

There is no firm definition of soldering and brazing. If the joining metal melts below 800°F the operation is soldering; if above 800°F, it is brazing. Another definition states that if there is marked diffusion of the joining metal into the metals being joined and if the melting point of the joining metal is below the melting point of the metals being joined, the process is brazing. The important consideration is not what the process is called, but how is the joint designed and what are the stress allowables.

The major difference in design between soldering and brazing is that soldering is a sealing process whereas brazing is a structural joining process. Soldering is used extensively for making electrical connections, hermetic seals, joining of lead or copper pipe, filling casting or sheet-metal imperfections, and general repair work. In such applications the mechanical load is usually carried by a mechanical joint (e.g., twisted wires in the case of the electrical connection and crimped stovepipe joint in the hermetic joint), the solder providing continuity. Solders, usually alloys of lead and tin, have melting points ranging from 450 to 600°F. Such lead-tin solders have very low mechanical properties at room temperature (tensile strength of 5,800 psi) which

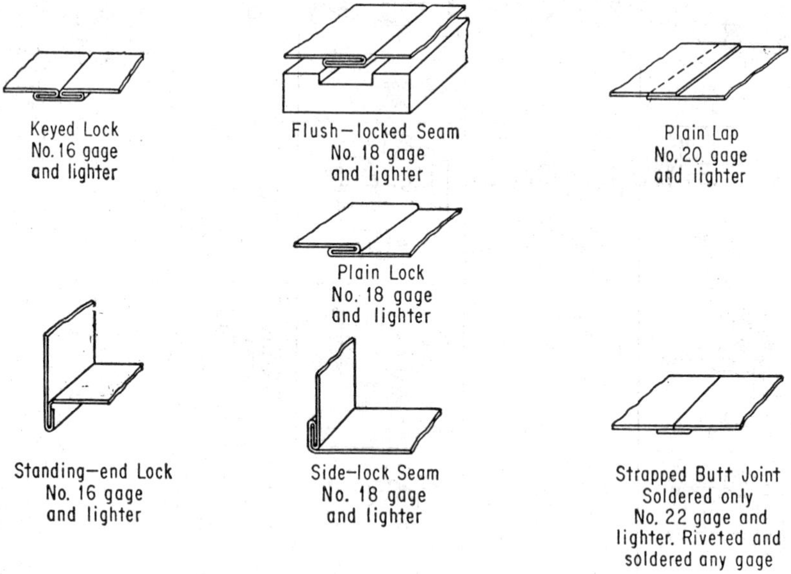

FIG. 25.33. Typical designs for soldered joints.

drop off very rapidly as the temperature is increased. To overcome these difficulties, a few special solders have been formulated such as silver-tin-lead solder which melts at 589°F and has been used in applications at temperatures up to 572°F.

The basic design of the joint to be soldered is the controlling factor in joint strength. All structural joints must be crimped, riveted, bolted, or spot-welded before soldering. Any thickness or section can be soldered provided the base metal can be heated to a temperature above the flow temperature of the solder. In general, butt welds should be avoided because of the low tensile strength of solders. If such joints are absolutely necessary, a reinforcement strap should be used. Lap joints are in more common usage because they can be fitted more easily and achieve greater strength by extending the lap area. The area of the overlap is a function of application. Long overlaps should be avoided because it is difficult to assure full flow of the solder in the joint area. Voids caused by such poor design practice can be determined only by nondestructive inspection techniques such as X-ray (which is not usually used in commercial practice). The standard practice is to make the width of the overlap equal to 1½ times the thickness of the thinner base-metal member. Such joints should not be subjected to bending loads because the solder will fail rapidly under the tension loads applied (see Fig. 25.4 for loading distribution). In overlap tubular members, the overlap should not be greater than the diameter of the smaller tube.

For solder to flow, the surface of the metal being joined must be chemically clean and free of oxides, grease, dirt, and other contaminants. Such dirt is usually removed by solvent cleaning and then chemical etching by means of a flux. To assure the flow of solder into the joint, the clearance between members should be held within 0.001 to 0.010 in. Joints with 0.003 in. of clearance offer the best combination of strength, ease of deposit, and capillary action to draw the solder into the joint. Typical solder-joint applications are shown in Fig. 25.33.

Brazing is similar to soldering in that good joint design is similar for both. They both require joints having capillary crevices into which the solder or braze alloy flows. Advances in braze alloys and heating methods during the last few years have advanced

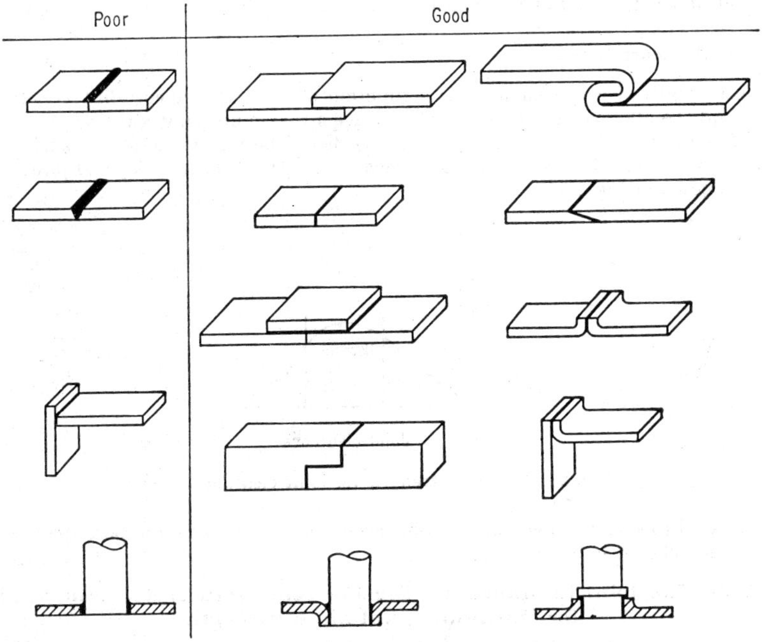

Fig. 25.34. Typical designs for brazed joints.

the importance of brazing to the point where brazing rivals welding in importance. With brazing it is possible to produce joints of low cost, ductility, and high strength. Most metals can be joined either to themselves or to other metals. The principal exception is the exclusion of aluminum and magnesium from dissimilar-metal combinations.

From a design point of view, brazing offers minimum joint distortion, good appearance as brazed, absence of weld spatter or other deposits (which require removal), adaptability to mechanization, and ability to perform operations with lower labor skill.

The disadvantages include the following: loss of joint strength at lower temperatures than welded joints; the color of the braze alloy may not match the color of the alloy being joined; the braze alloy may have lower corrosion resistance than the base alloy and therefore need protection (paint, etc.) if the machine is to be used in a corrosion area.

The braze joints used are similar in principle to those specified for soldering except that thicker base-metal sections are often employed. Braze joints include lap, scarf, butt, tee, and combinations thereof. Typical joints are shown in Fig. 25.34.

The lap joint is preferred when the joint must be leakproof or is subjected to high loads. Here, also, the strength of the joint can be varied by increasing or decreasing the area of the lap. The lap joint is preferred because it permits the easy assembly of members and the maintenance of proper capillary clearances through inexpensive jigging. For tubular lap joints, correct fit tolerances must be achieved through control of the diametric tolerances. To assure correctness of fit, the machining of one or both members to a tolerance closer than that obtained with "as-drawn" tubing must usually be specified.

Square butt joints can be successfully used if care is taken to specify that the joint components are machined square to provide uniform joint clearance between members. Square butt joints are usually specified for applications where double-lap joints cannot be used. The scarf butt joint is a variation of the butt joint offering a stronger, more ductile, but more costly joint.

(b) Joint Clearances

The strength of a brazed joint is a function of joint clearance, joint type, braze alloy, and the alloys being joined. To achieve a maximum joint strength, the thickness of the braze alloy, in a joint as brazed, should be thin enough to assure failure in the base alloy. The relationship of shear strength (of braze alloy) to joint clearance is shown in Fig. 25.35.

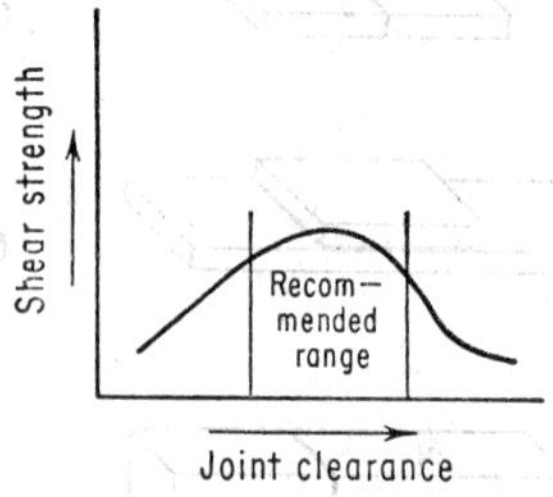

Fig. 25.35. Shear strength vs. joint clearance.

As related to specific braze alloys, joint clearances (at brazing temperature) recommended are shown in Table 25.5.

Table 25.5. Joint Clearances (at Brazing Temperature) Recommended for Maximum Joint Shear Strength

Braze alloy *ASTM* B260–56T	*Joint clearance*
B Al Si group..........	0.006–0.010 for laps under ¼ in.
	0.010–0.025 for laps over ¼ in.
B Cw P group..........	0.001–0.005
B Ag group............	0.002–0.005
B Cw Ag group.........	0.002–0.005
B Cw	0.000–0.002
B Cw Zn group..........	0.002–0.005
B Mg..................	0.004–0.010
B Ni Cr...............	0.002–0.005

If the joint clearance is too small or too large, the braze alloy will not flow into the joint and it will not develop full strength.

Composition and uses of braze alloys are given in ASTM Specification B260-56T. For machine-component assembly, the copper- and silver-base alloys are usually used. A shear strength of 15,000 psi can be used in calculating joint strength with either copper braze or silver braze when the joint does not exceed 0.010 in. in thickness.[27] Where copper braze alloy is used, brazed assemblies may be heat-treated after brazing without affecting the characteristics of the joint.

Brazing may be used where the gages are too thin for welding or where it is desirable to avoid forging and machining costs by simultaneously brazing a number of simple fabricated details into an assembly.

25.3. ADHESIVE BONDING

(a) Design

Adhesives are now used in joining of metals for structural and nonstructural applications. The generally accepted theories of the adhesive bond relate to mechanical adhesion and chemical adhesion.[6] The quantities fundamental to problems of adhesion are:[28]

The surface energy γ_{LA} of a liquid against air (the energy required to make a free surface of a unit area α starting from the bulk liquid). The work of cohesion of the liquid is sometimes referred to as $2\gamma_{LA}$.

The energy difference between unit area α free solid plus bulk liquid and unit area α solid wetted by the liquid, which energy difference equals $\gamma_{SA} - \gamma_{SL}$.

The energy of adhesion W gained when two free unit areas (2α) of liquid and solid surface are brought into contact:

$$W = \gamma_{LA} + \gamma_{SA} - \gamma_{SL}$$

No theory has yet been established which describes cohesion and adhesion phenomena quantitatively. Most of the information available on adhesives and adhesion is the result of experiment.

Some of the possible advantages of bonding over other metal-fastening methods are:[29]

Uniform stress distribution with resultant increased service life
Reduced weight
Smooth surfaces and contours

Table 25.6. Adhesive Bonding

Type of adhesive	Applications	Remarks
Animal glues	Wood	Natural—fungus attack under high temperature and humidity
Vegetable glues	Paper	
Casein	Wood and paper	
Sodium silicate	Glass and paper	Labels, etc.
Sodium silicate + fuller's earth	Cement for electric heaters	
Natural and synthetic rubbers	Rubber, felt, paper to metal and plastic	Lower-strength material
Phenol-formaldehyde	Wood, rubber, plastics	No good for metals, glass
Phenol, acetals or polyamide	Metals	
Urea-formaldehyde	Wood, rubber, plastics	No good for metals, glass
Resorcinols	Porous materials	
Alkyd resins	Good, porous materials	
Vinyl resins (acetate and butyral)	Metal and glass	
Acrybis	General	Pressure-sensitive adhesive
Cellulose derivatives	General	(i.e., Duco cement)
Melamine resins	Porous materials	
Polyester resins	Glass, fiber, etc.	
Polyurethane resins	Metals and glass	
Epoxy resins	....	Most important last 5 years. No solvent, no shrinking, good strength, bonding copper foil to plastic, printed circuits
Silicone	....	Heat resistance

Table 25.7. Metal-to-metal Adhesives[6]

Type	Cure	Pressure	Properties	Comments
Epoxy	Catalyzed at room temp. Elevated-temp. cure for high-temp. service	Contact	Up to 8,000 psi tensile strength with no significant change up to 180°F	
Epoxy phenolics	Temp, 325–350°F Time, 40 min to 1 hr	From few pounds to 100 psi	Particularly useful for bonding stainless steel, titanium, beryllium, and other high-temp. alloys. Service temp. from −67 to 500°F plus. Lap shear strength *a.* Aluminum to aluminum—2,500–3,000 psi at room temp. *b.* Stainless steel to stainless steel—3,100–3,800 psi at room temp.	Suitable for bonding materials exposed to high temperatures for extended periods
Epoxy polyamides	Range from approximately 3–5 days at room temp. to 3–5 min at 400°F	....	Bonds exhibit good shear, peel, and impact strength but are more subject to deformation or "creep" at elevated temperatures than unmodified epoxies. Degree of resistance to heat, chemicals, oils, and solvents decreases with increase of polyamide content. Typical lap shear values for aluminum-to-aluminum bonds run between 3,000 and 4,000 psi at room temperature with good strength retention down to −70°F and approximate reduction of shear values by 33–50 per cent at temperatures between 200 and 250°F. Higher percentage of polyamide resin in mixture will increase peel but lower shear strength	Suitable for bonding a wide variety of rigid and flexible materials, including metals, rubbers, glass, ceramics, and most plastics, particularly nylon

Ability to join very thin metals
May not require high temperatures for joining
Ability to join dissimilar metals
No need for perforating the metals
Can provide leakproof joints
Can incorporate vibration damping and insulation properties
Can provide corrosion resistance
May reduce production costs

Some of the significant limitations of bonding metals are: (1) the maximum operating temperature of a bonded machine tool is determined by the adhesive; (2) the best synthetic-resin adhesives available have a maximum operating temperature of about 500°F (adhesives are now being developed based on ceramic and other inorganic-type materials which should extend the useful temperature range); (3) tooling costs for bonding may be excessive when only a few assemblies are made; (4) structural adhesives generally have high shear and tensile strength, but their resistance to peel and

Table 25.8. Test Data on Selected Adhesives

Type	Temp., °F	Adhesive A*	Adhesive B*	Adhesive C*	Adhesive D*	Adhesive E*
Standard-temp. shear strength, psi	72 to 76	3,700	2,700	3,300	3,700	5,400
Elevated-temp. shear strength, psi	178 to 182	4,000	1,900	1,900	2,100	2,700
Low-temp. shear strength, psi	−65 to −70	3,400	4,500	No test	2,200	2,800
Standard-temp. impact strength, ft-lb	72 to 76	7.9	16.1	.....	45.9	65.9
Low-temp. impact strength, ft-lb	−65 to −70	8.4	0.7	.....	0.9	6.1
Shear strength after 30 days salt spray exposure (panel), psi	Specification QQ-M-51	3,200	1,800	2,500	3,000	5,100
Shear strength after 30 days immersion in tap water, psi	72 to 76	3,800	2,200	2,300	2,800	5,300
Shear strength after 7 days immersion in ethylene glycol (AN-E-2), psi	72 to 76	4,000	3,100	3,300	3,100	4,100
Shear strength after 7 days immersion in anti-icing fluid (AN-F-13), psi	72 to 76	3,900	3,000	3,000	3,100	4,300
Shear strength after 7 days immersion in hydraulic oil (AN-O-366), psi	72 to 76	3,900	2,800	3,400	4,000	4,800
Shear strength after 7 days immersion in hydrocarbon fluid (AN-F-42, type II), psi	72 to 76	3,300	2,500	1,700	3,600	4,400
Peel strength, lb/lin in	Room temp.	5.67	49.87	.....	61.81	27.07
Tension (24ST Al alloy to 24ST Al alloy), psi	Room temp.	7,700	1,500	1,200	2,500	6,500
Tension (18:8 steel to 18:8 steel), psi	Room temp.	8,300	1,500	700	3,300	6,200
Tension (Alclad 24ST Al alloy to 18:8 steel), psi	Room temp.	4,900	1,300	800	3,000	7,800
Salt-spray specimen exposure for 30 days (1-in. strips cut from panel specimen), psi	Specification QQ-M-51	1,100	14	0	0	3,900

* Adhesive A. High-temperature-setting adhesive in powder form assumed to be a formulation based on epichlorohydrin resins.

Adhesive B. High-temperature-setting formulation of a thermosetting resin and synthetic rubber (believed to be neoprene).

Adhesive C. Phenolic nylon. Neoprene rubber primer plus nylon-fiber tape impregnated with phenolic nylon neoprene rubber.

Adhesive D. Nitrate rubber phenolic.

Adhesive E. Redux. High-temperature-setting two-component modified vinyl-resin formulation.

cleavage stresses is relatively poor; and (5) no reliable nondestructive testing technique exists which will locate unbonded areas or those areas where a very weak bond exists.

From an application standpoint, adhesives are generally classified as structural and nonstructural. Structural adhesives are those which produce high-strength joints used in high-stress areas. The nonstructural adhesives are used in areas where stresses are low because of either small applied loads or large bond areas in relation to the load being applied. Chemically, adhesives are divided into thermosetting and thermoplastic resins, elastomeric materials, and organic and inorganic materials. A comprehensive list of adhesives and their uses is given in Table 25.6. The most important commercial adhesives by generic classification and significant properties are listed in Table 25.7. A compilation of design test data on a selected number of structural adhesives is given in Table 28.8.

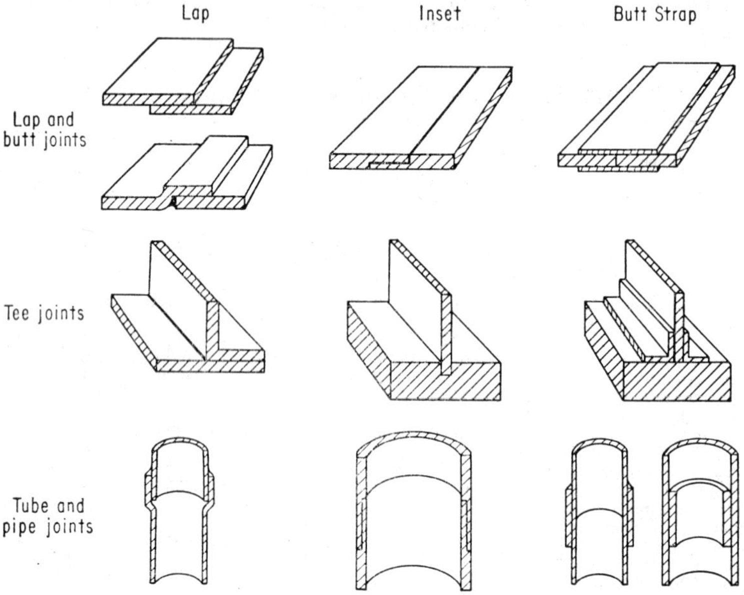

FIG. 25.36. Joint designs for adhesive bonding.

Surface treatment of metals to be bonded is an important factor in obtaining strong joints. In designing joints to be bonded, the high shear strength and relatively low peel strength of adhesively fastened metal systems must be taken into account. Lap joints are the most common type of adhesive joint employed in industry. Recommended joint designs for adhesive bonding are shown in Fig. 25.36. The thickness of the adhesive is a contributing factor in bonded joint strength. Generally the bond strength of the more rigid adhesives used for structural applications decreases as the thickness of the bond line increases.

At higher temperatures the variation in strength with bond thickness becomes less pronounced. For flexible adhesives, bond strength may increase as the thickness of the adhesive bond line increases, as shown in Table 25.9. Note that this is opposite to the effect obtained in brazed and soldered joints (see Fig. 25.35). Reference 41 presents interesting theoretical information with regard to adhesive joints.

Table 25.9. Effect of Adhesive Thickness on Shear Impact Strength

Type adhesive	Adhesive thickness, mils	Shear impact strength, ft-lb
Buna N vinyl	1.1	1.6
	2.9	3.0
	4.4	3.7
	13.9	5.2
	18.2	8.2
	21.8	9.2
	34.5	10.3
Buna N phenolic	1.0	2.6
	2.0	2.8
	4.0	4.1
	8.5	5.2
	25.0	8.9
	30.6	9.4
Vinyl butyral	1.3	0.7
	2.5	1.3
	15.7	4.1
	28.0	5.3
Thiokol	2.1	5.5
	4.7	7.0
	7.5	7.5
	14.5	10.3
	30.9	12.2
	46.0	15.9
Resorcinol (rubber-casein primer)	2.2	7.0
	5.0	6.0
	8.0	6.1
	17.8	5.7
	23.0	5.8
	33.0	6.8
Resorcinol (buna N vinyl primer)	4.9	5.2
	14.2	5.2
	23.4	4.3
	34.8	3.9

References

1. Ericson, R. G.: "Reduction in Manufacturing Costs on Medium to Light Machinery through Greater Use of Steel and Improved Fabrication Methods," ASME paper 61-WA-260.
2. "Welding Handbook," 4th ed., American Welding Society, New York, 1960.
3. "Procedure Handbook of Arc Welding, Design and Practice," The Lincoln Electric Company, Cleveland, Ohio, 1945.
4. American Welding Society: "Resistance Welding, Theory and Use," Reinhold Publishing Corporation, New York, 1956.
5. Rossi, B. E.: "Welding Engineering," McGraw-Hill Book Company, Inc., New York. 1954.
6. Guttmann, Werner H.: "Concise Guide to Structural Adhesives," Reinhold Publishing Corporation, New York, 1961.
7. Technical Data Sheets, Nos. 4, 6, 8, 9, and 10, Loctite Sealants, American Sealants Company, Hartford 11, Conn.

8. Marks, Lionel S. (ed.): "Mechanical Engineers' Handbook," 6th ed., sec. 5, McGraw-Hill Book Company, Inc., New York, 1958.
9. *Sci. Am.*, April, 1962, p. 124.
10. "Recommended Practices for Automotive Welding," AWS Automotive Welding Committee, American Welding Society, New York, 1962.
11. Fusion Welding of Nickel and High Nickel Alloys, International Nickel Company, *Tech. Bull.* T2.
12. Harman, C. R.: "Handbook for Welding Design," vol. I, Pitman Publishing Corporation, New York, 1956.
13. Wilson, W. M.: Fatigue of Structural Joints, *J. Am. Welding Soc.*, vol. 29, no. 3, p. 204, March, 1950.
14. Adams Lecture, American Welding Society, 1949.
15. Fatigue Tests of Riveted Joints, *Univ. Illinois Bull.* 302, 1938.
16. Fatigue Tests of Butt Welds in Structural Steel Plates, *Univ. Illinois Bull.* 310, 1938.
17. Fatigue Tests of Connection Angles, *Univ. Illinois Bull.* 317, 1939.
18. Fatigue Tests of Welded Joints in Structural Plates, *Univ. Illinois Bull.* 327, 1941.
19. Fatigue Tests of Commercial Butt Welds in Structural Steel Plates, *Univ. Illinois Bull.* 344, 1943.
20. Fatigue Tests of Fillet Welds and Plug-weld Connections in Steel Structural Members, *Univ. Illinois Bull.* 350, 1944.
21. Rate of Propagation of Fatigue Cracks in 12 × ¾ inch Steel Plates with Severe Geometrical Stress Risers, *Univ. Illinois Bull.* 371, 1947.
22. Flexural Fatigue Strength of Steel Beams, *Univ. Illinois Bull.* 377, 1948.
23. Fatigue Strength of Fillet-weld, Plug-weld, and Slot Weld Joints Connecting Steel Structural Members, *Univ. Illinois Bull.* 380, 1949.
24. The Fatigue Strength of Various Details Used for the Repair of Bridge Members, *Univ. Illinois Bull.* 382, 1949.
25. The Fatigue Strength of Various Types of Butt Welds Connecting Steel Plates, *Univ. Illinois Bull.* 384, 1949.
26. Jennings, C. H.: Welding Design, *Welding J. (N.Y.)*, vol. 15, no. 10, pp. 58–70, 1936.
27. MIL-HDBK-5 (Military Handbook), March, 1959, Armed Forces Supply Center, Washington 25, D.C.
28. de Bruyne, N. A., and R. Houwink: "Adhesion and Adhesives," Elsevier Publishing Company, New York, 1951.
29. Epstein, G.: "Adhesive Bonding for Metals," Reinhold Publishing Corporation, New York, 1954.
30. Koopman, K. H.: Elements of Joint Design for Welding, *J. Am. Welding Soc.*, June, 1958, pp. 579–588.
31. Blodgett, O. W.: A Simplified Approach to Calculating Design Stresses, *J. Am. Welding Soc.*, December, 1958, pp. 1182–1192.
32. DeWitt, E. J.: Welding of Small Subassemblies in a Program of All Welded Machines, *J. Am. Welding Soc.*, January, 1939, pp. 27–31.
33. Mikulak, J.: Design of Welded Machining, *J. Am. Welding Soc.*, January, 1945, pp. 13–21.
34. Snyder, G. L.: Design Considerations for Welded Machining Parts, *J. Am. Welding Soc.*, February, 1946, pp. 105–118.
35. Ideas for Designing Machining Parts, *J. Am. Welding Soc.*, 1951, pp. 269–270.
36. Stout, R. D., and W. D. Doty: "Weldability of Steels," Welding Research Council, New York, 1953.
37. West, E. G.: "The Welding of Non-ferrous Metals," Chapman & Hall, Ltd., London, 1951.
38. Eickner, H. W., et al.: The Effect of Temperature from −70 to 600°F on Strength of Adhesive Bonded Lap Shear Specimens of Clad 24ST3 Aluminum Alloy and Cotton-and Glass-fabric Plastic Laminates, *NACA Tech. Note* 2717, June, 1952.
39. Okerblom, N. O.: "The Calculations of Deformations of Welded Metal Structures," Her Majesty's Stationery Office, London, 1958.
40. Gunnert, R.: "Residual Welding Stresses," Almqvist & Wiksell, Stockholm, 1955.
41. Bikerman, J. J.: "The Science of Adhesive Joints," Academic Press Inc., New York. 1961.

Section 26

POWER SCREWS

By

KARL BRUNELL, B.M.E., *Research Associate, Princeton Laboratory, American Can Co., Princeton, N.J.*

CONTENTS

Power screws are used to translate rotary motion into uniform longitudinal motion. Common applications include jacks, valves, machine tools, and presses. Screws also permit very accurate position adjustment because they provide a high reduction ratio from rotational to longitudinal displacement.

26.1. THREADS

(a) V-threads

The efficiency of a power screw depends upon the profile angle of the thread: the larger the angle, the lower the efficiency. V-threads are therefore not well suited for transmission of great loads. They are generally used only where accuracy of adjustment and low production cost are required and the power demands are quite small.

(b) Square Threads

The square thread (Fig. 26.1) has the greatest efficiency (zero profile angle). However, it is costly to manufacture, because it cannot be cut with dies, and it is difficult to engage with a moving split nut as is sometimes required.

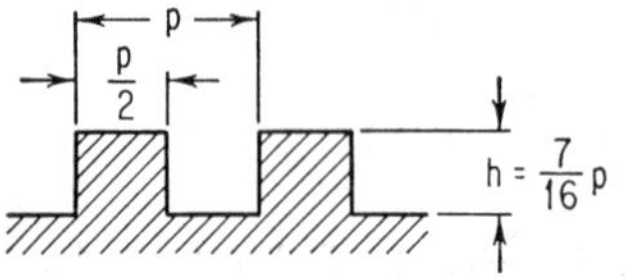

FIG. 26.1. Square thread.

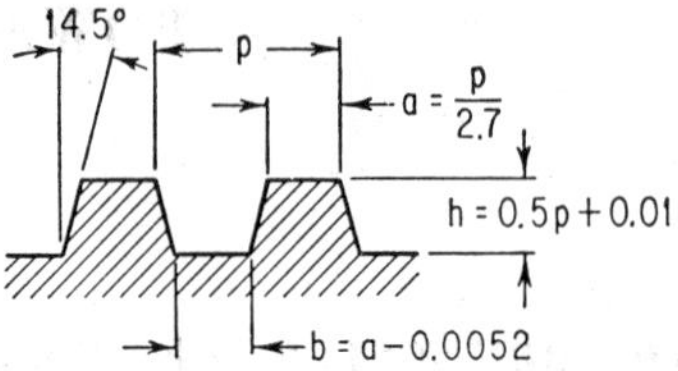

FIG. 26.2. Acme thread, profile angle $\beta = 14.5°$.

(c) Acme Threads

The Acme thread (Fig. 26.2) is often used in order to overcome the difficulties associated with the square thread.

While its efficiency is lower than that of a square thread, it has the advantage that lost motion resulting from manufacturing tolerances or wear can be taken out by using a split nut.

(d) Buttress Threads

Where unidirectional power transmission is required and the nut returns with little or no load, the buttress thread (Fig. 26.3) can be used. Because its square face is used for power transmission, it has the square-thread efficiency but slightly lower manufacturing cost.

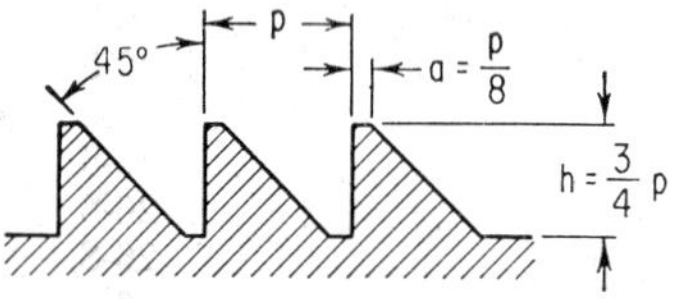

FIG. 26.3. Buttress thread.

(e) Multiple Threads

Two or more parallel threads can be used to reduce the ratio of screw rotation to nut displacement. This reduces mechanical advantage but increases efficiency because of increased helix angle.

(f) Ball-round Threads

To reduce friction, special threads using a ball between screw and nut are used· The efficiency of such a screw is very high, on the order of 90 per cent. The grooves are semicircular in cross section and the nut has 1½ to 2½ rows of bearing balls. The nut has a return groove so that the ball at the end of the nut returns to the starting thread in the nut.

A special development is the Eaton overrunning ball screw. The screw has helical threads formed to fit balls of a size depending upon the loading. The balls are captured in a retainer and thus their spacing never changes. The nut has a number of annular grooves corresponding to the rings of balls running on the screw. Thus when the screw is turned, the balls, their retainer, and the nut are displaced, but turning of the nut will not cause any displacement of the screw. When the ball retainer hits stops provided at either end of the screw, it locks against the stops in the direction of rotation. Screw and retainer then turn as one unit and the annular grooves in the nut are merely ball races. This overrunning action will persist until the rotation of the screw is reversed.[1]

26.2. FORCES

The pitch of a thread p is the distance from a point on one thread to the corresponding point on an adjacent thread regardless of whether the screw has a single or multiple thread. The displacement d_s of the nut or screw resulting from one full turn of either is the lead l. Thus for a multiple thread of m threads,

$$l = mp \tag{26.1}$$

and the displacement for n revolutions of either screw or nut

$$d_s = nl = nmp \tag{26.2}$$

The angle of the thread at its mean diameter with respect to a normal to the screw axis is the helix angle α. To determine the force P required to overcome a certain load Q it is necessary to observe the relation of the load direction with respect to the displacement direction. If the load opposes the direction of motion (Fig. 26.4) the force required to overcome it, neglecting friction, is given by

$$P_0 = Q \tan \alpha \qquad \text{(see Fig. 26.5)} \tag{26.3}$$

However, friction displaces the normal N (Fig. 26.6) by angle ϕ, and P_1 is given by

$$P_1 = Q \tan(\alpha + \phi) = Q \frac{\tan \alpha + \tan \phi}{1 - \tan \alpha \tan \phi} \tag{26.4}$$

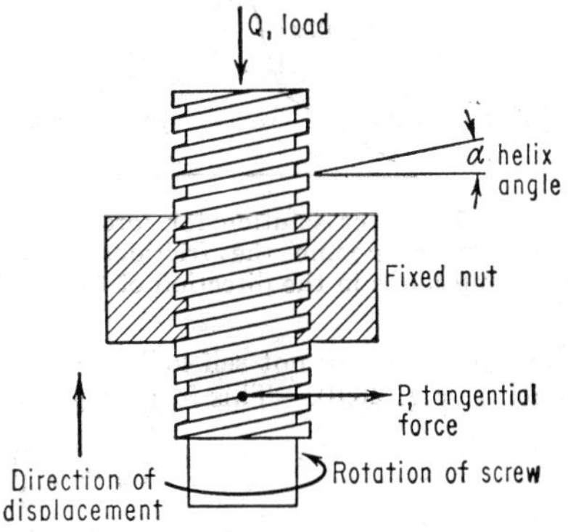

FIG. 26.4. Force P required to overcome load Q.

Replacing $\tan \phi$ with μ, the coefficient of friction,

$$P_1 = Q \frac{\tan \alpha + \mu}{1 - \mu \tan \alpha} \tag{26.5}$$

Because of the thread profile angle β, the resultant R should be replaced by $R/\cos \beta$. This affects only the friction terms since friction gave rise to R to begin with; hence

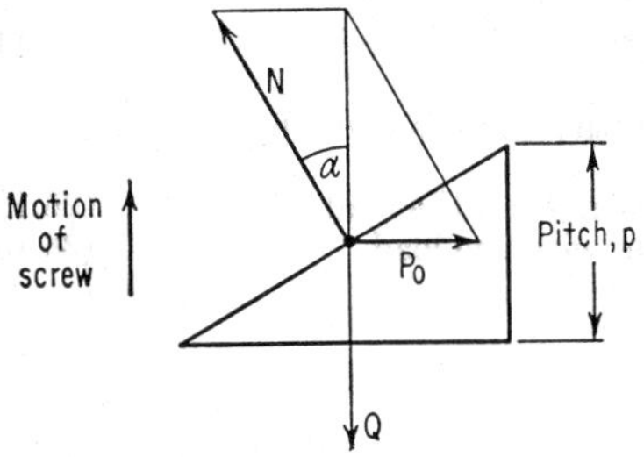

FIG. 26.5. Vector solution for force P_0 required to overcome load Q, neglecting friction.

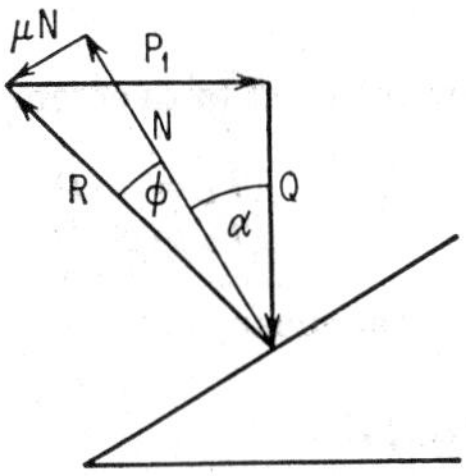

FIG. 26.6. Vector solution for force P_1 required to overcome load Q, including frictional force μN.

these must be divided by $\cos \beta$. Thus

$$P_1 = Q \frac{\mu \sec \beta + \tan \alpha}{1 - \mu \sec \beta \tan \alpha} \tag{26.6}$$

When a load is applied to the screw in Fig. 26.4 but P is removed, friction alone keeps the screw from turning and moving in the direction of Q. Hence, when motion in the direction of the load is required, the applied force need only be large enough to overcome the friction (Fig. 26.7).

$$P_2 = Q \tan (\phi - \alpha) \tag{26.7}$$

For a profile angle β

$$P_2 = Q \frac{\mu \sec \beta - \tan \alpha}{1 + \mu \sec \beta \tan \alpha} \tag{26.8}$$

When motion of the screw or nut may be caused by the load applied to either, the

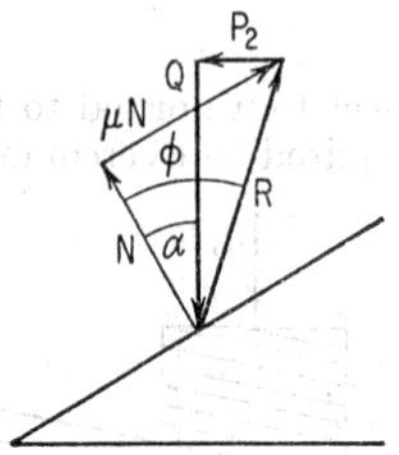

FIG. 26.7. Vector solution for force P_2 required to overcome the frictional force, when motion is in the direction of the load.

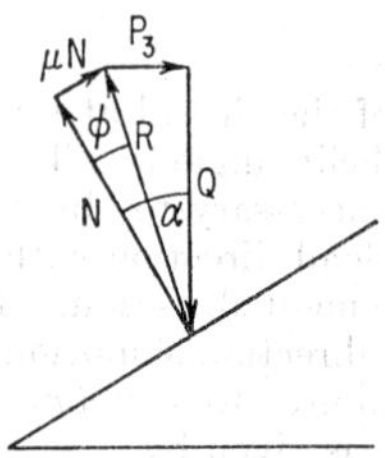

FIG. 26.8. Force P_3 required to prevent motion of screw or nut due to an applied load.

screw and nut are not self-locking and a tangential force will be required to prevent motion (Fig. 26.8). This force is given by

$$P_3 = Q \frac{\tan \alpha - \mu \sec \beta}{1 + \mu \sec \beta \tan \alpha} \tag{26.9}$$

When $P_3 = 0$ the friction force will just cancel the tangential force produced by the load, and the screw will be self-locking. Setting $P_3 = 0$, the helix angle α at which the screw is self-locking is given by

$$\tan \alpha = \mu \sec \beta \tag{26.10}$$

Note that, for the helix angle at the equilibrium point when the screw is just self-locking, force P_1, to move against the load, will be

$$P_1 = Q \tan 2\alpha \tag{26.11}$$

The torque required to overcome the load Q on the screw is given by

$$T_1 = \tfrac{1}{2} D_s P_1 \quad \text{for motion against the load} \tag{26.12}$$

$$T_2 = \tfrac{1}{2} D_s P_2 \quad \text{for motion with the load} \tag{26.13}$$

where D_s is the mean diameter of the screw.

26.3. FRICTION

From Eq. (26.10), it can be seen that the helix angle at which a screw will be self-locking depends upon the coefficient of friction as well as the thread profile. For an average coefficient of friction $\mu = 0.150$, the helix angle must be at least 9° for square and buttress and 10° for Acme threads. These values allow for a small (½°) margin of safety so that the screw will not keep turning under load for kinetic coefficient of friction slightly less than 0.150 if the applied force P is removed. Coefficients of friction are given in Table 26.1.[2]

Table 26.1. Coefficient of Friction μ

Screw	Nut			
	Steel	Brass	Bronze	Cast iron
Steel, dry	0.15–0.25	0.15–0.23	0.15–0.19	0.15–0.25
Steel, machine oil	0.11–0.17	0.10–0.16	0.103–0.15	0.11–0.17
Bronze	0.08–0.12	0.04–0.06	....	0.06–0.09

The effect of friction in bearings and thrust collars, which must always be used either on the nut or the screw depending upon application, was not included in the preceding considerations. The force necessary to overcome these frictional forces must be determined separately and added to Eqs. (26.6), (26.7), and (26.8).

Where two surfaces are in sliding contact, good design practice requires that nut and screw are made of different materials in order to reduce both wear and friction. Because the nut is usually smaller and easier to replace than the screw, it is made of the softer material, usually high-grade bronze or brass where loads are light.

Workmanship also has an important effect upon friction. A 30- to 50-μin. finish will result in a coefficient of friction about one-third lower than a finish of 100 to 125 μin.[3]

26.4. DIFFERENTIAL AND COMPOUND SCREWS

The displacement of the nut or screw depends upon the pitch. From Eqs. (26.1) and (26.2),

$$d_s = nmp \tag{26.2}$$

where n is the number of turns of the screw and m the number of threads. If small displacement per turn is required, m must equal 1 and $d = p$. However, very small displacements require a very small pitch, and this results in a weak thread. This difficulty can be overcome to some extent by using a differential screw.

When a fast motion is required, a multiple thread of $m = 2$, 3, or more can be used. However, machining is expensive because each thread must be separately cut. Also, since the helix angle on multiple-thread screws is quite large, the screw will not be self-locking. Here, the remedy may be the use of a compound screw.

(a) Differential Screw

The differential screw has two threads in series. Both are of the same hand but different pitch. Every revolution of the screw will move the two nuts toward or away

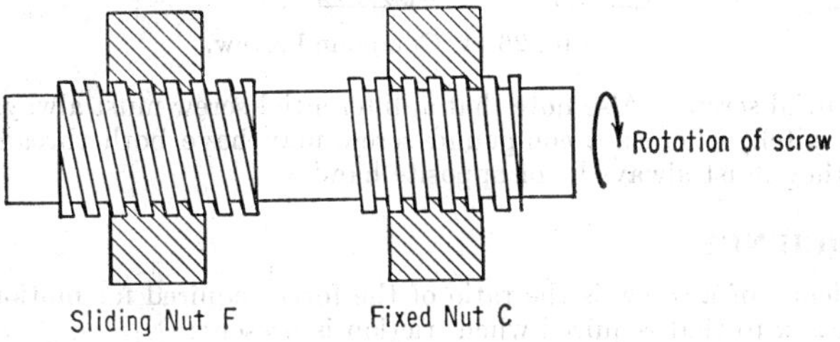

FIG. 26.9. Differential screw.

from each other by an amount equal to the difference in pitch. For the arrangement in Fig. 26.9 the nuts C and F will separate, and with the coarser pitch at the fixed nut

$$d_s = p_C - p_F \tag{26.14}$$

If the fixed nut has the finer pitch, then, for the same sense of rotation for the screw,

$$-d_s = p_F - p_C \tag{26.15}$$

that is, the nuts will move toward each other. Another arrangement (Fig. 26.10)

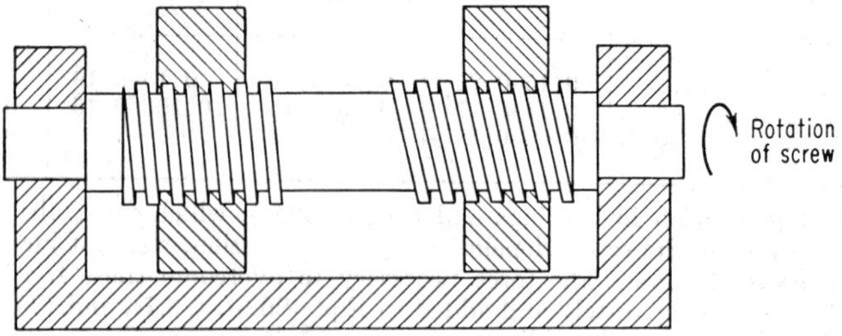

FIG. 26.10. Differential screw.

shows the nuts F and C approaching each other. Their relative displacement

$$d_R = p_C - p_F \tag{26.16}$$

If the rotation of the screw is reversed, the nuts will separate:

$$-d_R = p_F - p_C \tag{26.17}$$

(b) Compound Screws

If as in the arrangements in Fig. 26.11 the threads are of opposing hands, the result will be a compound screw. The displacement will then be the sum of the two pitches.

$$d_s = p_C + p_F \tag{26.18}$$

Note that for the arrangement in Fig. 26.10 interchange of coarse and fine pitch between the nuts will not result in a reversal of the displacement direction as it would

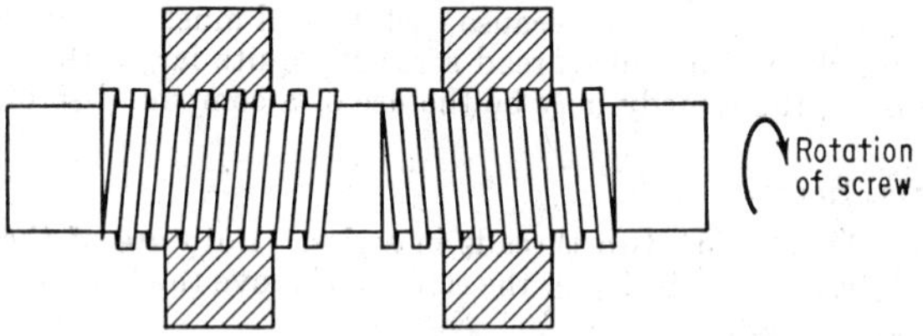

FIG. 26.11. Compound screw.

for a differential screw. Also note that a differential screw must always have threads of different pitch, whereas a compound screw may have both threads of the same pitch, but they must always be of opposite hand.

26.5. EFFICIENCY

The efficiency of a screw is the ratio of the force required for motion against load without friction to that required when friction is present:

$$e_1 = \frac{P_0}{P_1} = \frac{\tan\alpha\,(1 - \mu \sec\beta \tan\alpha)}{\tan\alpha + \mu \sec\beta} \tag{26.19}$$

$$e_1 = \frac{\cos\beta - \mu \tan\alpha}{\cos\beta + \mu \cot\alpha} \tag{26.20}$$

Thus the efficiency of a square thread ($\beta = 0$),

$$e_{\text{sq. thread}} = \frac{1 - \mu \tan \alpha}{1 + \mu \cot \alpha} \tag{26.21}$$

From Eq. (26.20) it is seen that max e occurs when $\alpha = 45°$. The maximum attainable efficiency

$$e_{\max} = \frac{\cos \beta - \mu}{\cos \beta + \mu} \tag{26.22}$$

To obtain the greatest efficiency, μ should be as small as possible.

When the motion occurs in the direction of the load, it is meaningless to speak of efficiency unless it is proposed to convert an axial load into a tangential force with a non-self-locking screw, a most unlikely design application. When the screw is self-locking, efficiency must be considered negative because a tangential force is required to do what the axial load cannot accomplish by itself. Zero efficiency then represents the equilibrium condition given by Eqs. (26.9) and (26.10). However, the efficiencies are required to consider differential and compound screws, and mathematical expression can be derived provided the foregoing is kept in mind. When the screw is not self-locking the efficiency is given by

$$e_3 = \frac{P_3}{P_0} = \frac{\cos \beta - \mu \cot \alpha}{\cos \beta + \mu \tan \alpha} \tag{26.23}$$

The 100 per cent efficient case occurs when P_3 equals P_0, to keep the nut or screw from overrunning. Zero efficiency represents the equilibrium case when screw and nut are just self-locking. For the latter, setting Eq. (26.23) equal to zero,

$$\cos \beta = \mu \cot \alpha \tag{26.24}$$

or

$$\tan \alpha = \mu \sec \beta \tag{26.25}$$

For a self-locking nut, the negative efficiency is given by

$$e_2 = 1 - \frac{P_0}{P_0 + P_2} = 1 - \frac{\cos \beta + \mu \tan \alpha}{\mu(\tan \alpha + \cot \alpha)} \tag{26.26}$$

For the equilibrium case of self-locking, when $e = 0$,

$$\tan \alpha = \mu \sec \beta \tag{26.27}$$

For a differential screw, note that, regardless of rotational direction, one thread will always cause displacement against the load, while the other causes displacement with the load. For the screw in Fig. 26.9, nut F may move with the load with respect to the screw thread, and the nut at C against the load. Reversal of rotation reverses the motions as well. Therefore, the efficiency for a differential screw must be checked for both directions of rotation. The efficiency is given by

$$e_4 = e_1 e_2 \tag{26.28}$$

when both threads are self-locking. Care must be taken to substitute the proper values of α, μ, and β for each thread into the respective efficiency relationships. When only one thread is self-locking,

$$e_5 = e_1 e_3 \tag{26.29}$$

for one direction of motion, and e_4 from Eq. (26.28) for the other. When neither thread is self-locking, Eq. (26.29) applies for either direction.

While the efficiency for a differential screw is quite low for self-locking threads, it can be improved by making one or both threads overrunning. With proper design the tangential force P_3 resulting from a load Q on one thread is not large enough to cause motion against the load on the other thread; so that, even if either thread is separately not self-locking, the combined result in a differential screw produces the desired self-locking feature.

Since the threads are of opposite hand in a compound screw, the displacement with respect to the load is the same for both threads. The efficiency is therefore the product of the efficiency of each thread.

$$e_6 = e_{1,\text{thread}1} e_{1,\text{thread}2} \tag{26.30}$$

26.6. DESIGN CONSIDERATIONS

(a) Thread Bearing Pressure

Design of power screws, as all other screws, is based upon the assumption that the load is uniformly distributed over all threads. This assumption is not true. When the screw is in tension and the nut in compression only the first threads will carry the load. The other threads merely maintain contact with one another. This is especially true when the threads are new and not worn. After the threads have undergone some plastic and elastic deformation, some of the load will be carried by the other threads too, but the load distribution nevertheless is far from uniform.

When both screw and nut are in tension or compression, the load is distributed among all threads in engagement but the distribution again is not uniform. It depends upon the total number of threads in engagement. For three threads, the distribution for a screw in tension is $\frac{2}{3}Q$ from the first to the second thread and $\frac{1}{3}Q$ from the second to the third thread. The distribution in the nut is the same in reverse order.[4] This condition can be alleviated by using a nut of variable cross section. The outside of the nut is parabolic; uniform pressures produced in such nuts greatly reduce wear.[5] To simplify design, however, the assumption of uniformly distributed load is usually made, allowance being made by selecting low values of bearing pressure. The required number of threads is then given by

$$Q = 0.785(D^2 - d^2)p_b N \tag{26.31}$$

where D is the outside diameter of the screw, d the inside diameter of the nut (root diameter of the screw), and N the number of threads required. Values for p_b may be taken from Table 26.2.[4,6]

Table 26.2. Safe Bearing Pressures

Screw	Nut	Safe bearing pressure p_b, psi	Speed range
Steel.	Bronze	2,500–3,500	Low speed
Steel.	Bronze	1,600–2,500	Not over 10 fpm
	Cast iron	1,800–2,500	Not over 8 fpm
Steel.	Bronze	800–1,400	20–40 fpm
	Cast iron	600–1,000	20–40 fpm
Steel.	Bronze	150–240	50 fpm and over

(b) Tensile and Compressive Stresses

The screw tensile stress is based upon the cross-sectional area at the root diameter. A factor of 4 to 5 is recommended for stress concentrations in the absence of fillets at the root of the thread.

Compressive stresses also cause stress concentrations, but these are not so dangerous and may be neglected.

If the screw is longer than six times its root diameter it should be treated as a short column.

(c) Shear Stresses

The shear produced at the thread root in a direction parallel to the screw axis is generally not dangerous.

The number of threads in the nut, where shear occurs at the major diameter, is usually controlled by wear considerations.

(d) Deflection

An applied turning moment produces torsional shear, the effect of which should be considered. It must also be borne in mind that torsional, bending, and axial deflections are not negligible. A torsional deflection of 1° causes a pitch variation of 0.00005 in. For a precision screw, where the pitch tolerance may be ±0.0002, this represents a 25 per cent deviation. When the screw is a beam, the bending deflection will cause the threads on the compression side to move together and those on the tension side to separate. When the effect of all three deflections is added, the resulting deflection may be quite large (0.008 to 0.015 in. is not unusual on heavily loaded screws) and must not be neglected lest undue wear result.[3]

(e) Collar Friction

The frictional forces associated with thrust collars must be determined separately and added to the frictional forces arising from thread loads in order to determine the torque required to turn the screw or nut. Since power screws almost always move with higher than negligible speed, the equations used to determine the static-friction forces at the head of a screw or nut fastening should not be used. A useful relationship is $pv = 20{,}000$, where p is the allowable pressure, psi, and v the velocity at the mean diameter of the collar, fpm. The bearing pressure thus determined should not exceed the value associated with load Q over the thrust-collar area. Coefficients of collar friction are found in Table 26.3.[2]

(f) Efficiency

Advantageous use of Eq. (26.10) should be made in order to obtain the most efficient thread for a self-locking screw. The helix angle so derived will always be the most efficient. When the screw can be overrunning, the most efficient thread will have a helix angle as close to 45° as possible.

Table 26.3. Coefficients of Friction for Thrust Collars

Material	Running	Starting
Soft steel on cast iron	0.12	0.17
Hard steel on cast iron	0.09	0.15
Soft steel on bronze	0.08	0.10
Hard steel on bronze	0.06	0.08

References

1. *Engineering Forum*, vol. 21, July, 1960, Eaton Mfg. Co., Cleveland, Ohio.
2. Ham, C. W., and D. G. Ryan: *Univ. Illinois Bull.*, vol. 29, no. 81, June, 1932.
3. Hieber, G. E.: Power Screws, *Machine Design*, November, 1953.
4. Maleev, V. L., and J. B. Hartman: "Machine Design," 3d ed., p. 385, International Textbook Company, Scranton, Pa., 1954.
5. Timoshenko, S., and J. M. Lessels: "Applied Elasticity," Westinghouse Press, East Pittsburgh, Pa., 1925.
6. Vallance, A., and V. L. Doughtie: "Design of Machine Members," 3d ed., p. 169, McGraw-Hill Book Company, Inc., New York, 1951.
7. Faires, V. M.: "Design of Machine Elements," 3d ed., The Macmillan Company, New York, 1955.
8. Niemann, G.: "Maschinen Elemente," p. 161, Springer-Verlag OHG, Berlin, 1960.
9. Spotts, M. F.: "Design of Machine Elements," 2d ed., Prentice-Hall, Inc., Englewood Cliffs, N.J., 1953.

Section 27

SHAFTS, COUPLINGS, KEYS, ETC.

By

M. SADOWY, Ph.D., *Professor of Mechanical Engineering, Marquette University, Milwaukee, Wis.*

CONTENTS

Preliminary Note. The reader is referred to Sec. 20 of this handbook for a compilation of standards dealing in part with the subject matter of this chapter.

27.1. SHAFTING

(a) Introduction

The term *shaft* applies to rotating machine members used for power or torque transmission, in which case the shaft is subject to torsion and bending; and to stationary and rotating members, called *axles*, which carry rotating elements, in which cases the shaft is subjected primarily to bending.

Short axles are called *bolts* and short shafts for torque transmission are called *spindles*. According to the shape of the shaft, straight shafts and irregular shafts (crankshafts) can be distinguished (see Fig. 27.1).

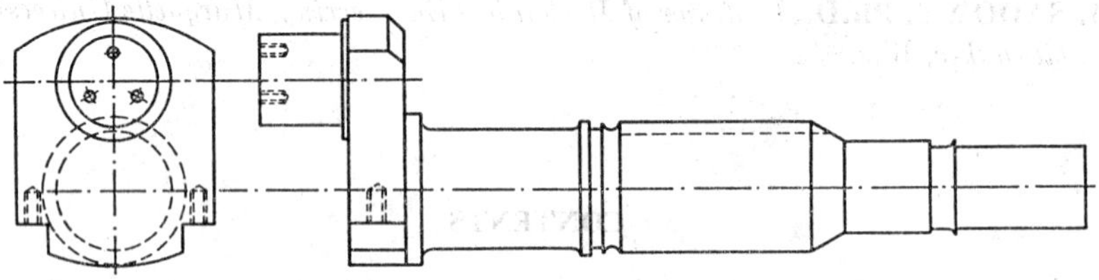

FIG. 27.1. One-piece crankshaft with side crank.

According to their position within the system, shafts are referred to as transmission or line shafts, which are relatively long shafts which transmit torque from motor to machine. Countershafts are short shafts between driven motor and driven machine or line shaft. Head shafts or stub shafts are shafts directly connected to the motor.

The transmission of motion or power through an angle can be accomplished without gear trains, chains, or belts by using flexible shafting. Such shafting is fabricated by building up on a single central wire one or more superimposed layers of coiled wire.[1]

Regardless of design requirement, care must be taken to reduce the stress concentration in notches, keyways, etc. Proper consideration of notch sensitivity may improve the strength more significantly than material consideration.

High-speed shafts require not only higher shaft stiffness but also stiff bearing supports, machine housing, etc. High-speed shafts must be carefully checked for static and dynamic unbalance and for first- and second-order critical speeds.

The lengths of journals, clutches, pulleys, and hubs should be viewed critically because these very strongly influence the overall assembly length. Pulleys, gear coupling, etc., should be placed as close as possible to the bearing supports in order to reduce the bending stresses.

Depending on application, shafts may be designed on the basis of strength and rigidity. In the first case, the maximum allowable stress is specified. In the second case, the maximum allowable deformation is specified.

(b) Stresses

In the process of designing, the complexity of stresses is caused by torque transmitted to the shaft, bending of the shaft due to its weight or load, and axial forces imparted to the shaft. The three basic types or cases will be considered.

Case 1 considers pure torque. For a shaft transmitting power P_0 (hp) at a rotational speed N (rpm), the transmitted torque T (lb-in.) and its relation to the nominal shear stress τ_{nom} (psi) and polar section modulus Z_p (in.^{3}) are given by

$$T = 63{,}000P_0/N \tag{27.1}$$

and

$$T = \tau_{\text{nom}}Z_p \tag{27.2}$$

For a hollow circular shaft the nominal shear stress

$$\tau_{nom} = (16/\pi d_0^3 B)T \tag{27.3}$$

where $B = 1/(1 - \alpha^4)$
$\alpha = d_i/d_0$
d_0 = outside diameter
d_i = inside diameter

Shear stress in noncircular shafts may be calculated from relations found in Table 27.1 and Sec. 14. Torsion and design of crankshafts are found in refs. 2, 3, and 4.

Case 2 considers simple bending. For a given bending moment M(lb-in.), rectangular (equatorial) section modulus Z (in.^{3}), and nominal stress in bending σ_{nom},

$$M = \sigma_{nom} Z \tag{27.4}$$

For a circular shaft,

$$\sigma_{nom} = (32/\pi d_0^3)BM \tag{27.5}$$

For the equation above, B for a circular shaft is equal to 1.

Case 3 considers torsion and bending in a circular shaft. When a shaft is subjected to torsion and bending, the induced stresses may be larger than the direct stress due to T and M alone. The maximum values of σ and τ must be considered:

$$\sigma_{max} = (32/\pi d_0^3)BM_{eq} \tag{27.6}$$
$$\tau_{max} = (16/\pi d_0^3)BT_{eq} \tag{27.7}$$

where the "equivalent" bending moment and "equivalent" torque are given by

$$M_{eq} = \tfrac{1}{2}(M + \sqrt{M^2 + T^2})$$

and

$$T_{eq} = \sqrt{M^2 + T^2}$$

(c) Design

Circular Shaft. The diameter for a solid shaft subjected to a pure torque is given by solving for d in Eq. (27.3) and substituting $\tau_{nom} = \tau_d$. Then τ_d = design shear stress, τ_μ = ultimate stress, τ_e = endurance limit, fs = safety factor, k = stress-concentration factor, $\tau_d = \tau_\mu/fs$, and $\tau_d = \tau_e/k(fs)$. For steady loads, design shear stress τ_d for estimating shaft diameters can be taken as 4,000 psi approximately for main power shafting, 6,000 psi for line shafts with pulleys, and 8,000 psi for short and small shafts.

For short solid shafts with heavy transverse shear only, the shaft diameter may be calculated from

$$d_0 = \sqrt{1.7V/\tau_d} \tag{27.8}$$

where V = maximum transverse shearing load.

ASME Transmission Design Code.[5,6,7] This method considers various loading conditions and is based upon the maximum-shear-stress theory of failure for ductile shear strength. Many designers prefer to use both the maximum-shear- and maximum-normal-stress theories of failure and to work with the larger stress values obtained. Distinction between ductile and brittle materials is somewhat arbitrary; frequently an elongation of 5 per cent is considered to represent the point of division. For circular shafting subject to torsion and bending,

$$\sigma_{max} = (16/\pi d_0^3)B\sqrt{(K_M M)^2 + (K_T T)^2} \tag{27.9}$$

where σ_{max} = maximum combined (induced) stress
K_T, K_M = combined shock and fatigue factors to be applied in every case to the computed torque and bending moment, respectively, and T and M are computed according to the loading case

From Eq. (27.9) the shaft diameter is given by

$$d_0^3 = (16/\pi\sigma_d)B\sqrt{(K_M M)^2 + (K_T T)^2} \tag{27.10}$$

Table 27.1. Shaft Data

Cross section of shaft	Equivalent length of circular bar of diameter D	Maximum shear stress
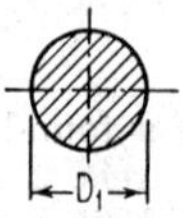	$L = L_1 D^4 \left[\frac{1}{D_1^4}\right]$	$\tau = \frac{5{\cdot}1\,T}{D_1^3}$ at periphery
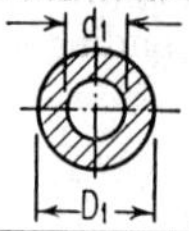	$L = L_1 D^4 \left[\frac{1}{D_1^4 - d_1^4}\right]$	$\tau = \frac{5{\cdot}1\,D_1 T}{(D_1^4 - d_1^2)}$ at periphery
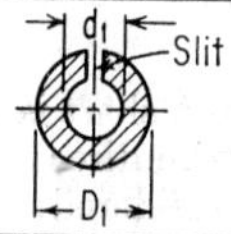	$L = L_1 D^4 \left[\frac{1{\cdot}5}{(D_1+d_1)(D_1-d_1)^3}\right]$	$\tau = \frac{7{\cdot}6\,T}{(D_1+d_1)(D_1-d_1)^2}$
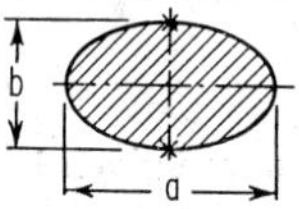	$L = L_1 D^4 \left[\frac{(a^2+b^2)}{2a^3 b^3}\right]$	$\tau = \frac{5{\cdot}1\,T}{ab^2}$ at X
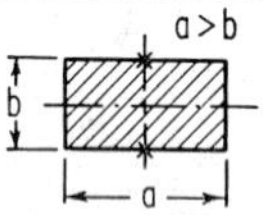	$L = L_1 D^4 \frac{(a^2+b^2)}{3{\cdot}1 a^3 \cdot b^3}$	$\tau = \left[\frac{15a+9b}{5a^2 \cdot b^2}\right] T$ at X
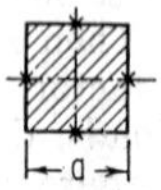	$L = L_1 D^4 \left[\frac{1}{1{\cdot}43 a^4}\right]$	$\tau = \frac{4{\cdot}8\,T}{a^3}$ at X
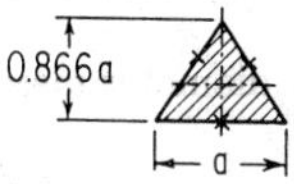	$L = L_1 D^4 \left[\frac{4.53}{a^4}\right]$	$\tau = \frac{20\,T}{a^3}$ at X
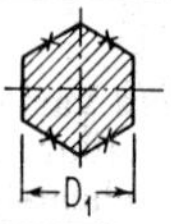	$L = L_1 D^4 \left[\frac{1}{1{\cdot}18 D_1^4}\right]$	$\tau = \frac{5{\cdot}3\,T}{D_1^3}$ at X
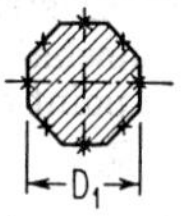	$L = L_1 D^4 \left[\frac{1}{1{\cdot}10 D_1^4}\right]$	$\tau = \frac{5{\cdot}4\,T}{D_1^3}$ at X

For brittle circular shafts, the maximum-normal-stress theory applies, and with

$$\sigma_{max} = (16/\pi d_0^3)B[K_M M + \sqrt{(K_M M)^2 + (K_T T)^2}] \quad (27.11)$$

the shaft diameter may be calculated from

$$d_0^3 = (16/\pi \sigma_d)B[K_M M + \sqrt{(K_M M)^2 + (K_T T)^2}] \quad (27.12)$$

For suggested values of K_M and K_T, and for the maximum stresses σ_d, see Tables 27.2 and 27.3.

Table 27.1. Shaft Data (Continued)

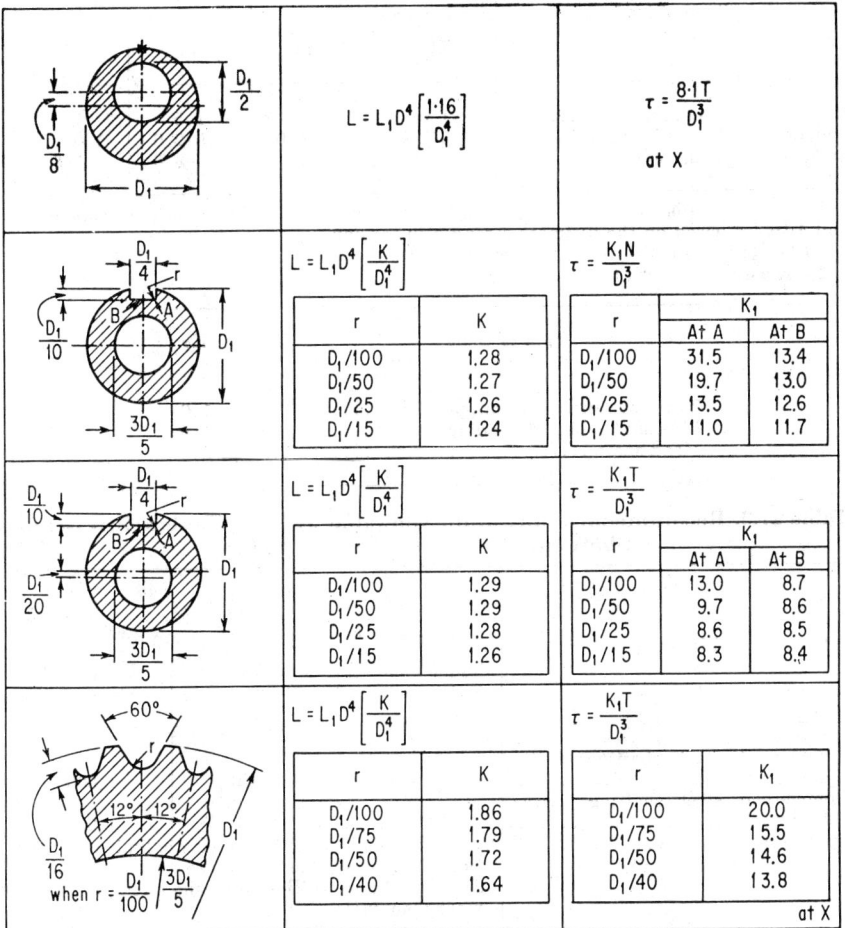

Cross section of shaft	Equivalent length of circular bar of diameter D	Maximum shear stress

Row 1

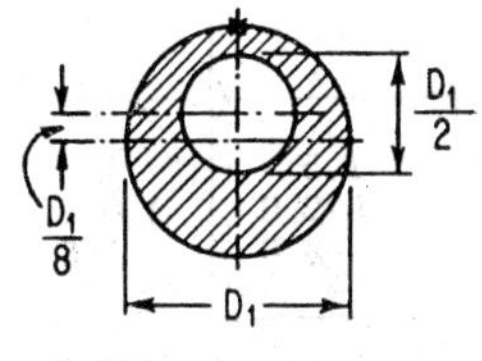

Equivalent length: $L = L_1 D^4 \left[\frac{1.16}{D_1^4}\right]$

Maximum shear stress: $\tau = \frac{8.1T}{D_1^3}$ at X

Row 2

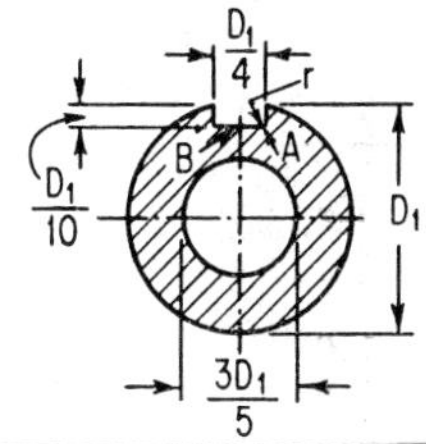

Equivalent length: $L = L_1 D^4 \left[\frac{K}{D_1^4}\right]$

r	K
$D_1/100$	1.28
$D_1/50$	1.27
$D_1/25$	1.26
$D_1/15$	1.24

Maximum shear stress: $\tau = \frac{K_1 N}{D_1^3}$

r	K_1	
	At A	At B
$D_1/100$	31.5	13.4
$D_1/50$	19.7	13.0
$D_1/25$	13.5	12.6
$D_1/15$	11.0	11.7

Row 3

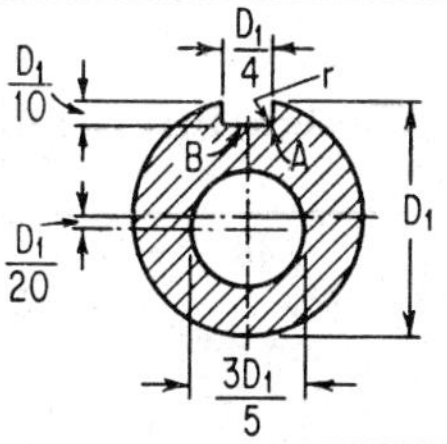

Equivalent length: $L = L_1 D^4 \left[\frac{K}{D_1^4}\right]$

r	K
$D_1/100$	1.29
$D_1/50$	1.29
$D_1/25$	1.28
$D_1/15$	1.26

Maximum shear stress: $\tau = \frac{K_1 T}{D_1^3}$

r	K_1	
	At A	At B
$D_1/100$	13.0	8.7
$D_1/50$	9.7	8.6
$D_1/25$	8.6	8.5
$D_1/15$	8.3	8.4

Row 4

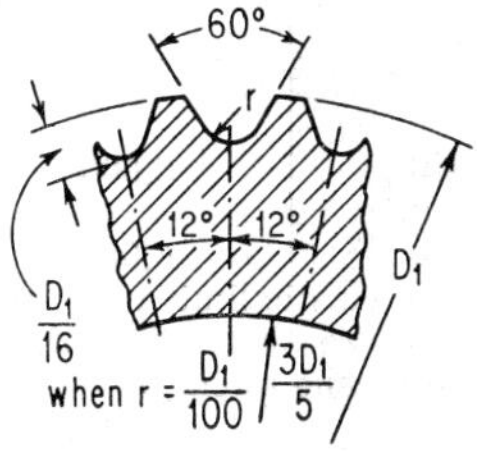

Equivalent length: $L = L_1 D^4 \left[\frac{K}{D_1^4}\right]$

r	K
$D_1/100$	1.86
$D_1/75$	1.79
$D_1/50$	1.72
$D_1/40$	1.64

Maximum shear stress: $\tau = \frac{K_1 T}{D_1^3}$

r	K_1
$D_1/100$	20.0
$D_1/75$	15.5
$D_1/50$	14.6
$D_1/40$	13.8

at X

L = length of solid circular shaft of diameter D having the same torsional rigidity as length L_1 of the actual shaft

T = torque transmitted by shaft

The equations for d are simplified in the case for simple bending, pure torsion, and also for solid shafts ($B = 1$). The Transmission Designing Code may be used in cases where loading conditions cannot be precisely accounted for, e.g., when the fluctuating components of the bending moment and/or torque are not known.

Westinghouse Code.[9] For more specific evaluations of load conditions, shape, and surface quality of shaft, and shaft material, the equations of the Transmission Shafting Code are modified by using more specific values of T and M, as well as K_M and K_T, based upon considerations of the average (av) and range (r) components of the torque and bending loads, and on the modification of these components by the stress-concentration factor k_t for torsion and k_b for bending (see Sec. 18 and ref. 8).

Table 27.2. Shock and Fatigue Factors

Type of load	Stationary shaft		Rotating shaft	
	K_M†	K_T‡	K_M	K_T
Gradually applied, steady*..........	1.0	1.0	1.5	1.0
Suddenly applied, minor shocks......	1.5–2.0	1.5–2.0	1.5–2.0	1.0–1.5
Suddenly applied, heavy shocks.....	....	....	2.0–3.0	1.5–3.0

* For static loads on this basis, a solid shaft will have *fs* of 2⅓ based on the elastic limit, and 4 to 4.5 based on the ultimate strength in tension.
† K_M, associated with bending moment.
‡ K_T, associated with torque.

Type of shaft	K_M	K_T
Transmission shaft, torque only........	1.0	1.0
Line shaft, limited bending.............	1.0	1.5
Head shafts..........................	1.0	2.5

Table 27.3. Recommended Maximum Allowable Working (Design) Stresses for Shafts under Various Types of Loads

Material	Type of load		
	Simple bending stress, psi	Pure torsion stress, psi	Combined stress, psi
Commercial steel shafting (without/keyways)..................	16,000	8,000	8,000
Commercial steel shafting (with/keyways)........................	12,000	6,000	6,000
Special steels......................	60% of the elastic limit in tension but not more than 36% of the ultimate tensile strength	30% of the elastic limit in tension but not more than 18% of the ultimate tensile strength	

According to the Westinghouse Code,

$$T = T_{av} + k_t T_r \tag{27.13}$$

$$M = M_{av} + k_b M_r \tag{27.14}$$

$$K_M = \frac{M_{av} + (\sigma_{Y.P.}/\sigma_e)k_0 M_r}{M_{av} + k_t M_r} \tag{27.15}$$

$$K_T = \frac{T_{av} + (\sigma_{Y.P.}/\sigma_e)k_t T_r}{T_{av} + k_t T_r} \tag{27.16}$$

The equation for maximum combined stress in ductile materials can be written

$$\sigma_{max} = (16/\pi d_0^3 B)\sqrt{K_M^2(M_{av} + k_b M_r)^2 + (T_{av} + k_t T_r)^2} \tag{27.17}$$

(d) Rigidity

When machine shafts are designed for rigidity, a maximum allowable deformation is usually specified. In such cases, the shaft size is determined by rigidity rather than by strength, and the usual procedure is to compute deformation and then check for strength. Deformation may be considered for bending and/or torsion (in some cases for buckling also). Deformation will be discussed here for static and quasi-static loads only (for dynamic loads, see Secs. 6, 10, and 16).

When designing for shafts in bending, the permissible value of shaft deflection depends upon the particular application and service intended. Some acceptable values are: for transmission shafts 0.01 in./ft length; shafts carrying gears 0.005 in./f (f = width of the gear face in inches); shafts supported on plain bearings $0.0015L$ (L = distance from load point to center of the bearing in inches).

Shaft-deflection analysis is treated in Sec. 15.

When circular shafts are in torsion, the shaft twist or angular deformation θ in radians is due to the torque (in.-lb). Then

$$\theta = TL/I_pG \tag{27.18}$$

where L = shaft length, in.
G = modulus of rigidity in shear, psi
I_p = polar moment of inertia of the shaft section

Some indication of permissible values for shaft twist: transmission shafts, 1° for a length of 20 d, or 0.1°/ft; machine shafts, from 6 min/ft for steady loads to 3 min/ft for suddenly changing loads; milling cutter spindles 1° at the cutter.

The torsional rigidity (stiffness) of a shaft is defined by

$$k = T/\theta = GI_p/L \tag{27.19}$$

For more complicated shaft shapes the equivalent length L_e is introduced as the length of constant-diameter shaft having the stiffness of the original shaft. See Table 27.1 and ref. 10.

(e) Shaft Materials

Ordinary shafts are made of medium machinery steel (open-hearth steel with 0.15 to 0.40 per cent carbon). Turned and ground shafts are usually made of shafting steel (similar to SAE 1015 with higher Mn, Ph, and C content), special-order higher-carbon-content steel, or stainless steel. For high-quality shafts requiring greater strength, alloy steels (Ni alloy, Ni-Cr alloy, and Cr-Va alloy) are used. Alloy-steel shafts are heat-treated. Properties of some shafting materials are given in Table 27.4.

Most commercial material is hot-rolled and sometimes cold-finished (cold drawing and cold rolling). Larger-diameter shafts, special-purpose shafts (including connecting flanges), and most hollow shafts are forged.

Shaft stiffness is a function of shaft dimensions, and E and G. Since low-priced steels have approximately the same moduli as high-quality (alloy) steels, there is no need to use high-quality steel if stiffness is the only criterion. High-quality steel, however, has advantages if, for example, higher strength is required.

(f) Shaft Vibration

Because of higher machine speeds, increasing size of mechanical systems, and the increasing stringency of system performance requirements, it is frequently necessary to determine not only the first natural frequency but higher natural frequencies as well. Sometimes the corresponding amplitudes of vibration are also required. The analytical and numerical techniques for analyzing shaft vibration are discussed in ref. 10 and Sec. 6. Consideration will be given here to some of the practical factors which influence shaft vibration.

Bearings constitute the end conditions for static and dynamic deflection of shafting systems. Assumption of a simple support is justified only if the bearings adjust

Table 27.4. Common Uses of Steel

Automotive

	Automobile, light truck, tractor				Heavy truck	Diesel engine
	Arms and knuckles		Axles and shafts		Crank-shafts	Crank-shafts
Tensile strength, σ_{ult} psi	125,000–165,000	150,000–200,000	150,000–200,000	175,000–225,000	140,000–170,000	100,000–125,000
Mn	1340	....	1330	1340		
NiCr	3130–35–40	....	3130–45			
Mo	4047–53	....	4063–68			
Cr Mo	4135–42	....	4140–45–50	....	4142	
NiCrMO	....	4337–40	....	4340–45	4340	
NiMo	....	....	4640			
Cr	....	....	....	....	5145	5046–5145
NiCrMo	8640–42	....	8640–50	8653		
	8740–42	....	8740–50			
	....	9840	....	9840–45		

Aircraft Shafts

	Carburized	Heat-treated
Crankshafts	3310, 9310	3140, 4337, 4340
Gears and shafts	8615, 8620, 9310, 9315	4340, 6150

themselves to shaft deformation. Fixed bearings change the end conditions from "simple support" to "fixed ended," thereby increasing the shaft natural frequency. Lateral elasticity of bearings and insufficient stiffness of the machine housing tend to produce the condition of a "shaft on elastic supports." If the elasticity is known, the change in natural frequency can be calculated. Because of nonlinear-spring characteristics of bearings, this type of calculation will not always produce a completely realistic result. In ref. 11 it is suggested that the Rayleigh method be applied to the problem of bearing deformation. It becomes even more difficult to account for the end conditions when damping within the bearing must be considered, and also in the case of multiple-damped elastic supports. Forced lateral vibrations of shafts on damped flexible end supports are treated in ref. 12. For more complex systems, computer techniques have been developed.[13,14]

Shrink fits may introduce an additional stiffness in the affected section of the shaft This effect, however, is difficult to account for analytically in an accurate manner. Gyroscopic effects in larger rotors cause an apparent increase in shaft stiffness by creating a reacting moment which opposes shaft deflection and tends to increase the natural frequency of the rotating shaft.[16,18]

27.2. KEYS AND KEYWAYS

(a) Fundamentals

The primary function of a key is to transmit torque between a shaft and the mating machine element, usually a hub. Keys prevent relative rotational motion of the joined elements and in most cases also prevent relative axial motion except in the cases of feather and spline keys.

The distribution of forces on the surfaces of a key is complex and depends upon the key fit in the shaft and hub. The force distribution is not uniform over the key surfaces. Because of uncertainties in analysis, a sufficiently high factor of safety should be employed in calculating key dimensions. This factor of safety may be taken as 1.5 for a steady torque, and it should be increased up to 2.5 for a strong fluctuating torque. Practical considerations require that the minimum hub length be 1.2 shaft diameters for good grip and to prevent rocking on the shaft. This length is also recommended as the minimum length of the key. A light press fit between the shaft and the hub should be used in order to prevent rocking and eccentricity (except with feather keys).

(b) Prestressed Keys

The standard key taper is ⅛ in./ft. The key seat in the shaft is plain and the seat in the hub is tapered and slightly more shallow to provide for fitting.

In the most usual case of a "top-fitting" key (see Fig. 27.2*a*) compressive stresses exist on the upper and lower surfaces (*a-b* and *c-d*) and a clearance may exist as shown. For "top and side" fitting keys, an additional side fit of the key is required.

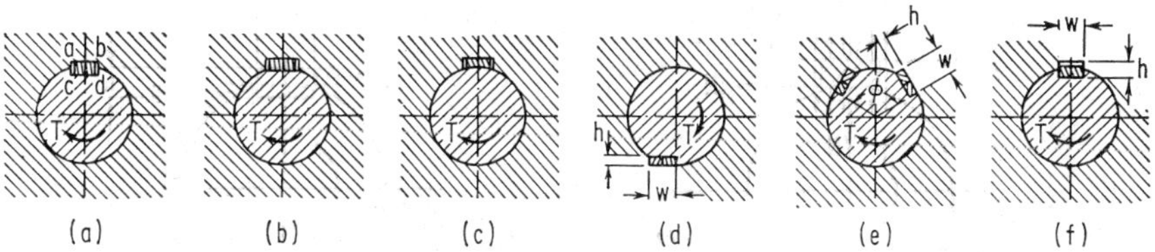

FIG. 27.2. Sunk key installations. (*a*) Top fitting key. (*b*) Flat. (*c*) Saddle. (*d*) Single tangential. (*e*) Double tangential. (*f*) Parallel.

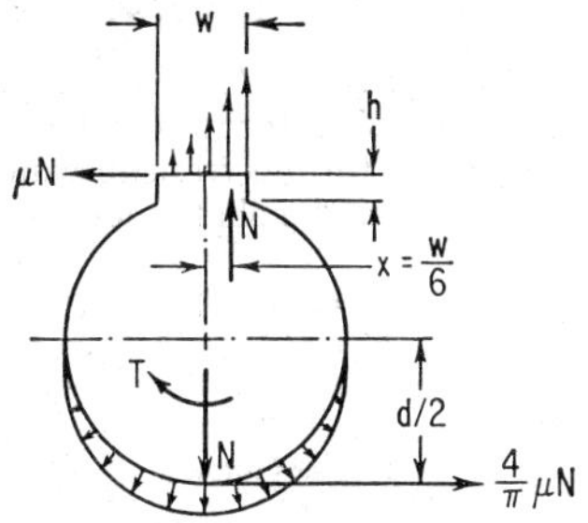

FIG. 27.3. Top fitting key—torque transmission.

The mechanism of torque transmission is indicated in Fig. 27.3.[4,19] The equations which follow are based upon torque transmission by friction and it is therefore advisable also to check shaft bearing stress and key shear stress. The normal force

$$N = WL\sigma/2 \tag{27.20}$$

The force necessary to drive the key home

$$Q = N(\tan \alpha + 2\mu) \tag{27.21}$$

The transmitted torque

$$T \leqq (WL\sigma_d/4)(W/3 + 1.44\mu d) \tag{27.22}$$

where L = effective key length
σ_d = design stress
α = taper
μ = coefficient of friction = 0.1 for greased keys, 0.15 for dry keys

Flat keys (Fig. 27.2*b*) and *saddle keys* (Fig. 27.2*c*) are used for transmission of small steady torques in cases where shaft weakening cannot be tolerated. Under excessive load, they tend to loosen, rotate, and damage the shaft.

The torque transmission occurs because of friction between key and hub, key and shaft, and shaft and hub (Fig. 27.4). With $N = WL\sigma$, considering the equilibrium condition for moments of forces μN and $(4/\pi)\mu N$, the transmitted torque is given by

$$T \leqq 1.14Nd \tag{27.23}$$

or

$$T \leqq 1.14L\sigma_d\mu \tag{27.24}$$

with $d \approx 3W_1$ and $\mu \approx 0.15$,

$$T \leqq 0.5W^2L\sigma_d \tag{27.25}$$

With *tangential keys* (Fig. 27.2*d*) torque is transmitted by means of compressive stress alone. For each key connection, two mating taper keys are used as shown. The key seats in the shaft and in the hub are *not* tapered. A single tangential key transmits torque in one direction only. For intermittent torque transmission double

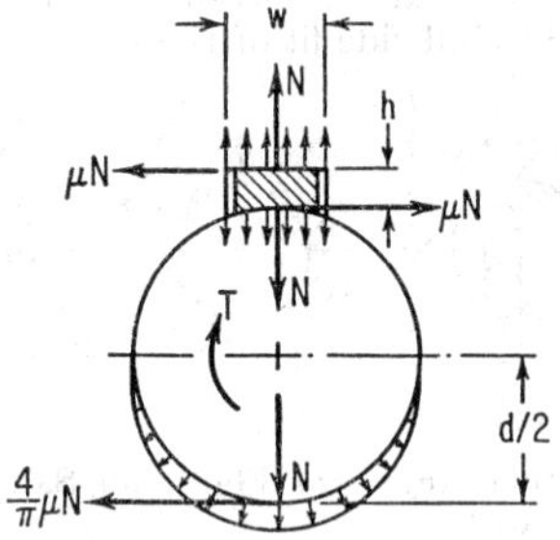

FIG. 27.4. Saddle key—torque transmission.

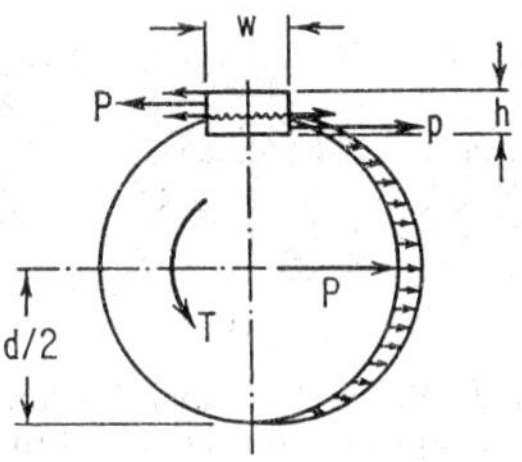

FIG. 27.5. Parallel key—torque transmission

keying of shafts is used (Fig. 27.2*d*); $\phi = 120$ to $135°$. Tangential keys are suitable for transmission of large and oscillating torques such as those associated with flywheels. They are more effective in torque transmission then square keys of equal cross section. However, they weaken the shaft more severely than do square keys. With

$$T = N(d/2) \tag{27.26}$$

and

$$N = hL\sigma \tag{27.27}$$

the transmitted torque

$$T \leqq \tfrac{1}{2}dhL\sigma_d \tag{27.28}$$

(c) Nonprestressed Keys

A *parallel key* is shown in Fig. 27.2*f*. These keys are made square ($w = h$) up to $w = 1\frac{1}{2}$ in. and $w = \frac{1}{4}d_{\text{shaft}}$. For larger key width w, the key height h is usually smaller than w. The depth of the key seat in the shaft and in the hub usually is equal to $\frac{1}{2}h$. Upper and lower surfaces of the key are parallel. The key is fitted on the sides of the key seat and may have clearance on the top. Torque is transmitted through the sides and no axial forces can be resisted. Neglecting the friction between contacting surfaces, and considering the transmission of torque by means of compressive and shear stresses, the transmitted torque (Fig. 27.5) is

$$T \leqq dwL\tau_d \quad \text{for shear} \tag{27.29}$$

and

$$T \leqq \tfrac{1}{4}dhL\sigma_d \quad \text{for compression} \tag{27.30}$$

or, for $d \cong 4w$ and $h \cong \frac{3}{2}w$,

$$T \leqq 2w^2L\tau_d \quad \text{for shear} \tag{27.31}$$

$$T \leqq 6w^2L\sigma_d \quad \text{for compression} \tag{27.32}$$

Woodruff keys (Fig. 27.6) are used for light or medium torque transmission, especially in machine-tool and automotive engineering, usually in shafts up to 2½ in. in diameter. They offer the advantage of permitting easy removal of pulleys from shafts and of preventing tipping of the keys. They adjust themselves easily to a

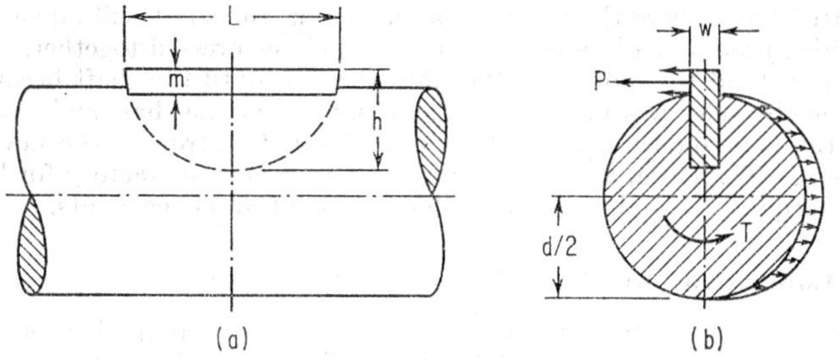

FIG. 27.6. Woodruff key. (*a*) Installation. (*b*) Torque transmission.

tapered-hub key seat and may also be used for angular location of parts on tapered shaft ends. Because of the relatively deep key seats, a disadvantage associated with Woodruff keys is shaft weakening. Two keys should therefore be used for longer hubs. Woodruff keys should not be used as sliding keys.

The torque transmitted can be calculated (Fig. 27.6*b*) from

$$T \leqq \tfrac{1}{2} d h b \tau_b \qquad \text{for shear} \qquad (27.33)$$

and

$$T \leqq \tfrac{1}{2} d w m \sigma_d \qquad \text{for compression} \qquad (27.34)$$

Sliding and *guiding* keys (feather keys) (Fig. 27.7) prevent only relative rotational motion between hub and shaft but permit axial motion. The tight-fit key is fastened either to the shaft or hub, usually by means of a countersunk screw or riveted pin. The pressure p on the bearing faces of the key should be less than 1,000 psi if the hub is to slide under load.

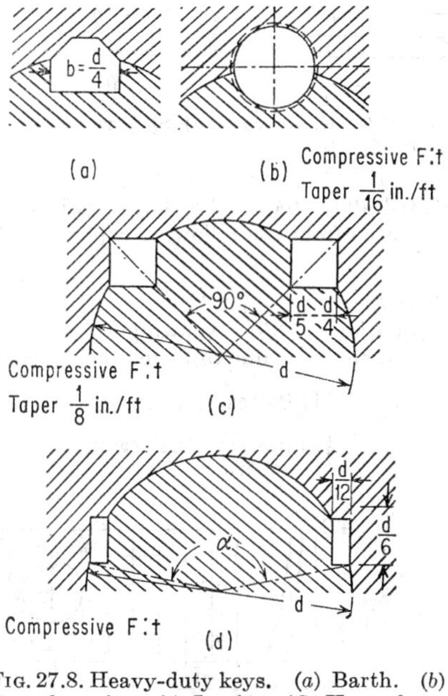

FIG. 27.8. Heavy-duty keys. (*a*) Barth. (*b*) Round or pin. (*c*) Lewis. (*d*) Kennedy.

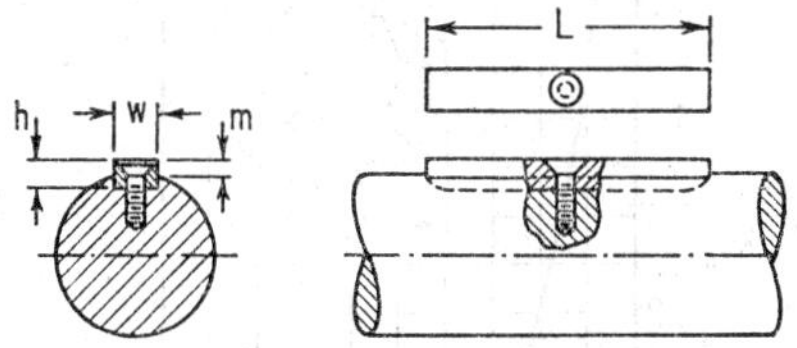

FIG. 27.7 Sliding and guiding keys.

(d) Heavy-duty Keys

The Barth key (Fig. 27.8*a*) utilizes the torque to force the key into the key seat, thereby inducing a compressive stress. This key does not require a tight fit.

Other heavy-duty keys are the round or pin key (Nordberg key) (Fig. 27.8*b*), the Lewis key (Fig. 27.8*c*), and the Kennedy key (Fig. 27.8*d*).

(e) Effect of Elasticity

With a tapered key and fitted key, the shaft has a tendency to twist in the hub during torque fluctuations. The torque distribution is approximately the same as with the fitted key; where the torque is a maximum, the shaft will move slightly in the hub, which produces a gradual wear of the surfaces pressed together. The wear will therefore extend the motion farther into the hub until the shaft becomes loose. A good fit on the sides will limit the torsional motion of the shaft and will lengthen the life of this key connection. If the torque fluctuates strongly the key will ultimately loosen. Therefore, a taper rectangular key is not satisfactory for heavy and fluctuating torque. A nonlinear-spring effect exists at all keyed joints.

(f) Stress Concentration

Stress concentration in keyways is especially pronounced at the keyway ends, and fatigue cracks usually start there. The size of the fillet radius r strongly influences the value of the stress-concentration factor K_t, and rectangular keyways should be avoided, especially if the shaft is already highly stressed. For keyways in bending,. Hetenyi[20] finds in photoelastic tests that

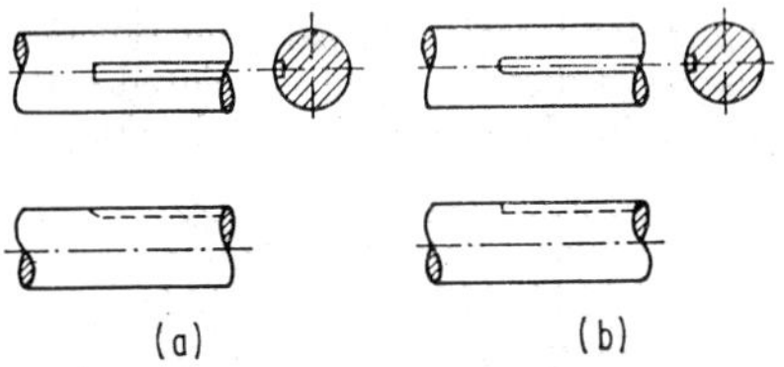

FIG. 27.9. Keyways. (*a*) Sled runners. (*b*) Profiled.

K_t = 1.38 for sled-runner keyways (Fig. 27.9*a*).
K_t = 1.79 for profiled keyways (Fig. 27.9*b*).

According to Peterson[8,21] these values may be somewhat low, especially for heat-treated alloy steels. For keyways in torsion, the stress-concentration factor can be calculated from the following expression due to Neuber:[22]

$$K_{ts} = 1 + \sqrt{t/r}\,C \tag{27.35}$$

where $C = 2(R - T)3R$

Equation (27.35) is shown graphically in Fig. 27.10*a*. where the results of a mathematical analysis (due to Leven, see Peterson[8]) is shown for comparison. Figure 27.10*b*

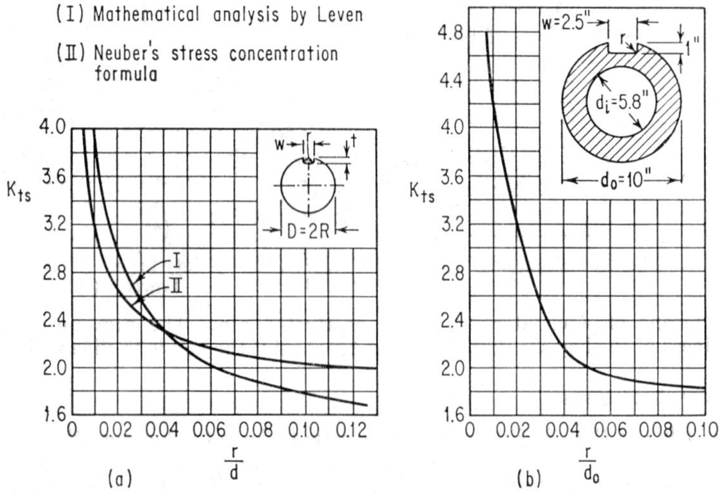

FIG. 27.10. Stress-concentration values for shafts. (*a*) Solid. (*b*) Hollow.

shows stress-concentration values found by the plaster-model method for hollow shafts in torsion.[21]

27.3. SPLINES

(a) Introduction

Splines may be considered multiple keys, integral with the shaft. Compared with keys, splines have the advantage of self-centering. For a given torque transmission, a splined shaft is weakened less than a keyed shaft. Thus a splined shaft can transmit higher torque than a keyed shaft of comparable dimension.

Parallel-side splines[25] are available as four-, six-, ten-, and sixteen-spline fits (see Fig. 27.11 for a four-spline fit).

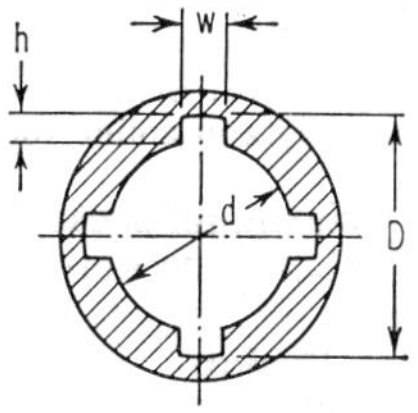

FIG. 27.11. Parallel spline—four spline.

Involute splines[23,24] have the same general form as involute gear teeth, except that the pressure angle is 30° and the depth of tooth is half the depth of a standard gear tooth. Involute splines are stronger; there is no danger of excessive stress concentration, and a better surface quality can be achieved (for manufacturing and accuracy check, see ref. 24). ASA standards provide 15 diametral pitches, from ½ to 48⁄16, with the number of teeth varying within each group from 6 to 50. Splines are made with *flat root* (Fig. 27.12*a*) and with *fillet root* (Fig. 27.12*b*). *Ball-bearing* splines are discussed in ref. 25.

Three types of fits are provided (see Fig. 27.12) by varying the *outside diameter* of the external spline, the *tooth thickness* (used mainly for fillet root splines), and the *root diameter* of the internal spline. Within each type of fit the following three classes may be discerned:

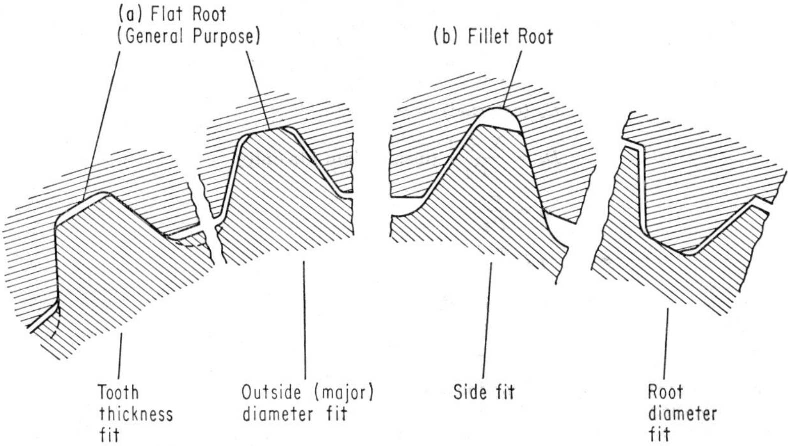

FIG. 27.12. Involute splines.

1. *Sliding fit*
2. *Close fit,* close on either the outside diameter, inside diameter, or tooth sides
3. *Press fit,* with interference on the outside diameter, inside diameter, or tooth sides

(b) Stresses

A splined shaft is weaker in torsion than in bending. The stiffness of a splined shaft is usually based upon root diameter d, so that the equivalent length

$$L_e = L(d_e^4/d^4) \tag{27.36}$$

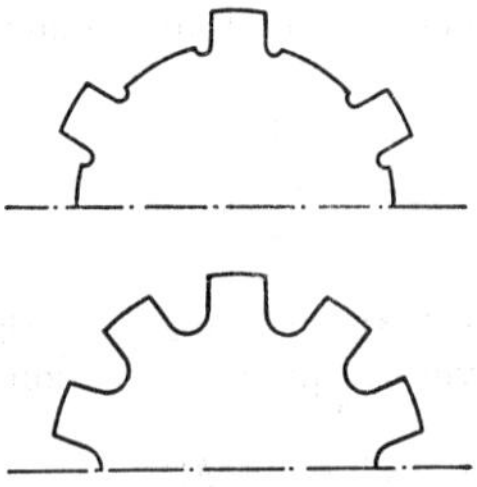

FIG. 27.13. Stress-relieving fillets in spline roots.

The strength of a straight splined shaft is based upon the equivalent shaft diameter $d_e = D - h$.[4] The torque-carrying capacity for parallel-side splines is described in ref. 24.

$$T = i(D_m/2)Lsp \tag{27.37}$$

where i = number of splines
D_m = $(D + d)/2$, mean diameter, in.
L = effective spline length, in.
s = contact distance, in. = $\frac{1}{2}(D - d) - 2h$
p = permissible side pressure, psi; p is limited by material properties and stress concentration; suggested value $p = 4 \times$ permissible nominal shear stress

Maximum shear stress may be calculated by utilizing Neuber's formula; see Eq. (27.38). Stress relieving of the fillets at the spline roots (Fig. 27.13) is advisable.

(c) Design

The following equations relate to parallel-side and involute splines.[19] For relations specifically for involute splines see ref. 4 and Table 27.5.

Table 27.5. Basic Formulas for Involute Splines

D = pitch diam
D_o = outside diam
D_i = inside diam
a = addendum
N = number of teeth
b = dedendum
P_d = diametral pitch
p = circular pitch
t = circular tooth thickness

Flat and fillet roots:

$$D = \frac{N}{P_d}$$

$$D_o \text{ (external)} = \frac{N+1}{P_d}$$

$$D_i = \frac{N+1}{P_d} \text{ (inside-diam fits only)}$$

$$p = \frac{\pi}{P_d}$$

$$t = \frac{p}{2}$$

$$a = 0.500\, P_d$$

$$b = \frac{0.600}{P_d} + 0.002$$

For outside-diam fits internal spline $a = b$
For inside-diam fits external spline $a = b$
Flat roots only:

Pitch	From 2.5/5 through 12/24	From 16/32 through 48/96
D_o (internal)	$\frac{N+1.8}{P_d}$	$\frac{N+1.8}{P_d}$
D_i (external)	$\frac{N-1.8}{P_d}$	$\frac{N-2}{P_d}$
b (internal)	$\frac{0.900}{P_d}$	$\frac{0.900}{P_d}$
b (external)	$\frac{0.900}{P_d}$	$\frac{1.000}{P_d}$

For splines in shear, the force P acting on a spline at the mean spline diameter D_m is given by

$$P = 2T/D_m = 4T/(D + d) \tag{27.38}$$

and

$$P = Lbik\tau_d \tag{27.39}$$

where b = spline width measured at the base, in.
k = coefficient accounting for nonuniform load distribution; suggested value $k = 0.75$[19]

For splines in bending, given the bending moment

$$M_b = P[(D - d)/2] \tag{27.40}$$

and the rectangular section modulus

$$Z = Lb^2k/6 \tag{27.41}$$

the bending stress

$$\sigma = 3P(D - d)/Lb^2ik = 12T(D - d)/(D + d)Lbik \tag{27.42}$$

For bearing stresses with the nominal bearing stress twice the nominal stress in compression σ,

$$P = (D - d)Lik\sigma \tag{27.43}$$

(d) Involute Serrations

Involute serrations are involute splines with a pressure angle of 45° and a much finer pitch. All the definitions related to involute splines apply to involute serrations, and the same symbols are used for both. Involute serrations are used in a pitch range of $^{10}/_{20}$ to $^{128}/_{256}$ and a pitch-diameter range from 0.1 to 10.0 in.

Involute serrations are mainly used for close fits, with no provision for sliding. Other classes of fits may be employed, however. Compared with involute splines, the teeth of involute serrations are shallower, stronger, have less radial depth of contact, and frequently offer manufacturing advantages. Under the same load conditions, the contact pressure, radial forces, and resistance to sliding are greater in serrations than in splines. Finer pitches in the case of serrations result in a greater number of teeth and therefore provide a wider range of index positions.

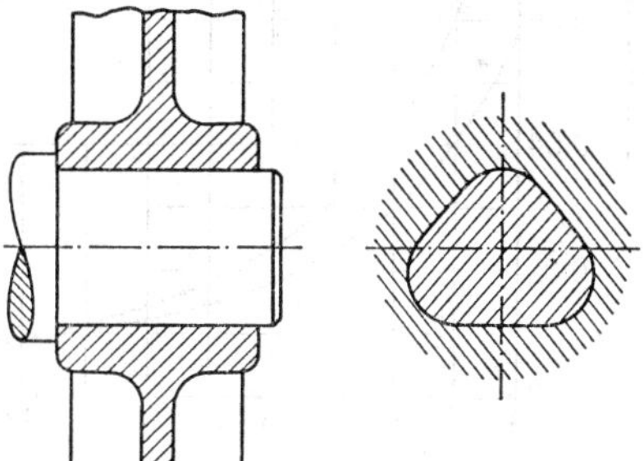

FIG. 27.14. Tri-Lobe spiral shaft.

(e) Miscellaneous

Another type of form-locked shaft connection, the Tri-Lobe spline shaft connection,[26] is shown in Fig. 27.14.

27.4. SNAP RINGS (RETAINING RINGS)

Snap rings are used to prevent axial motion of mating concentric parts, e.g., shafts and bearing sleeves. They are made of hardened steel wire and can be round, square, and rectangular in cross section. Snap rings may be used in expansion (Fig. 27.15*a*, *external snap ring* expanded on a shaft) or in contraction (Fig. 27.15*b*, *internal snap ring* contracted in a groove).

The stress distribution within the snap ring can be calculated according to the theory of curved beams in bending. For calculation of approximate ring deflection δ, the following formula has been suggested for a square-cross-section snap ring.[28]

$$\sigma = SEK/h \tag{27.44}$$

where h = ring thickness (for circular cross section substitute the cross-section diameter d for h)
E = modulus of elasticity
K = constant to be taken from Fig. 27.16 (K is a function of h, or d, and the mean diameter D of the "free" snap ring)

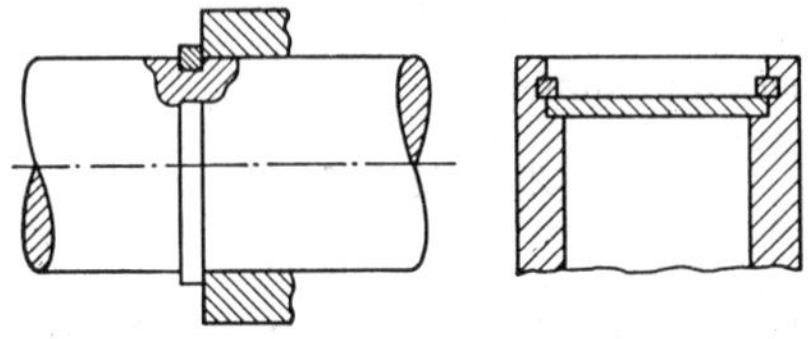

FIG. 27.15. Snap rings. (*a*) External. (*b*) Internal.

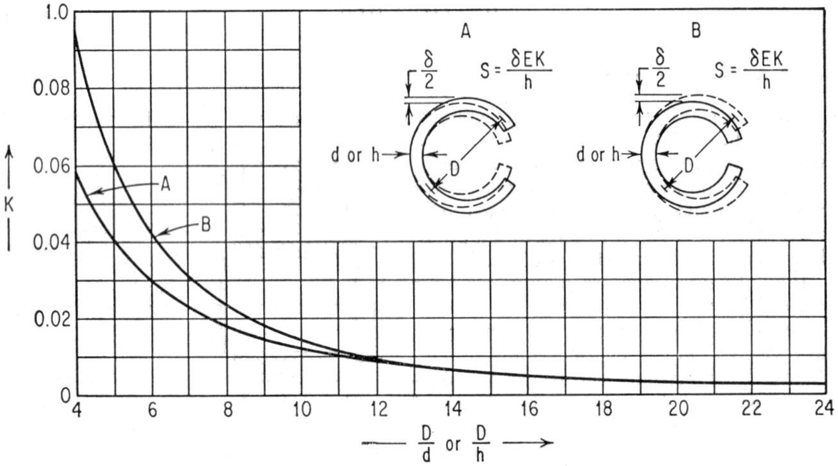

FIG. 27.16. Constant table for snap rings.

It is suggested that the maximum stress σ should not exceed 200,000 psi for snap rings to be deflected without permanent set, and 260,000 where permanent set may be acceptable.

27.5. COUPLINGS

(a) Introduction

In contrast to clutches, couplings do not permit, during shaft rotation, disengagement of the two shafts which they connect. The selection of couplings is based upon the following considerations:

Loading. Maximum load, steady and vibratory load, torque transmitted.

Misalignment. Maximum parallel and angular misalignment, and often ability to compensate for axial displacement.

Stiffness. Especially in the case of flexible couplings, the maximum permissible deflection across the coupling due to static and dynamic loads, and also the ability to provide damping and detuning effects.

Most couplings have not been standardized, and specific information regarding a particular coupling is therefore best obtained from the manufacturer. It is current practice for manufacturers to quote a single value of coupling stiffness. For couplings with variable stiffness, the stiffness characteristics should be experimentally determined.[17]

In the design of couplings, it is important to consider:

Compactness. The total length of the assembly.

Weight. A factor in static and dynamic loading of the shafting system.

Static and dynamic balancing, the latter being especially recommended for high-speed application.

Installation and maintenance, requiring accessibility without shafting-system disassembly.

(b) Rigid Couplings

While rigid couplings offer the advantage of design simplicity, they do not provide compensation for misalignment. This type of coupling may be employed in low-rotational-speed applications and in cases where only small misalignments are anticipated. *Sleeve-type* couplings (Fig. 27.17*a*) are fitted over both shaft ends, connected

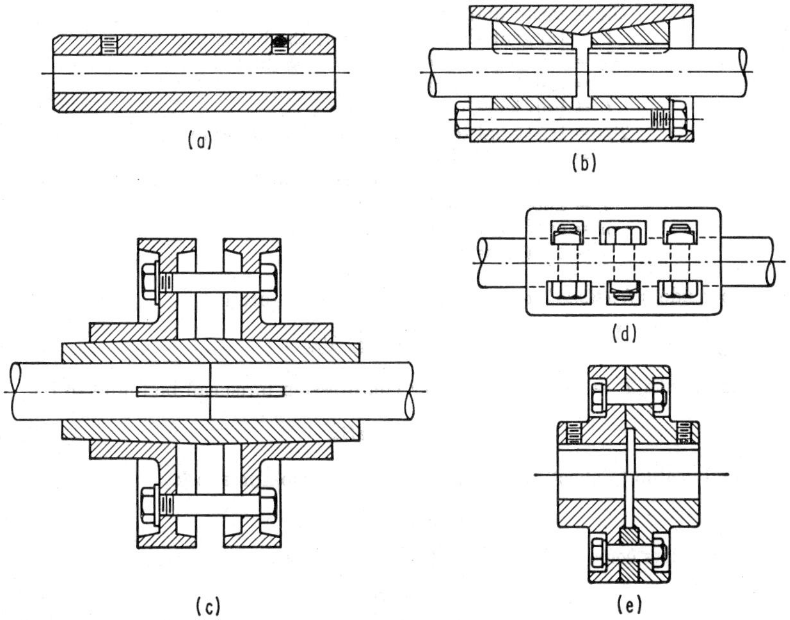

FIG. 27.17. Rigid couplings. (*a*) Sleeve type. (*b*) Ring compression. (*c*) Flange compression. (*d*) Clamp type. (*e*) Flange type.

by means of keys, taper pins, or setscrews. Sleeve couplings are generally used with small-diameter shafts, except in some marine installations utilizing heavy-duty couplings in which special hydraulic sleeve-expanding devices are provided to disconnect the sleeves. The difficulties associated with disconnecting sleeve couplings can be overcome for small-diameter shafts by employing *compression* couplings. The *ring compression* coupling, shown in Fig. 27.15*b*, represents a simple and inexpensive solution and is in use for shaft diameters up to 5⁵⁄₁₆ in. Likewise, *flange compression* couplings (keyed or keyless) have the advantage of simple adjustability and low cost (Fig. 27.15*c*). The number and size of the bolts used depend upon the size of the coupling. Outside dimensions are the same as those for flange-faced couplings (discussed later). Generally in compression couplings $L = 4d$; other dimensions can be calculated from strength considerations.

Clamp-type couplings (see Fig. 27.17*d*) are split longitudinally, and both parts of the coupling are bolted together. Torque is generally transmitted through a key. Flange-type couplings (see Fig. 27.17*e*) transmit torque by virtue of friction between flange faces, and shearing stress in the bolts or pins. Where friction is employed, bolts are required to provide the normal force associated with sufficient friction. Should the available frictional torque be exceeded by the load torque, slipping will occur and the coupling will rely upon bolt shear stress. Flange-faced couplings are generally used for heavy loads and in permanent installations. They have the advantage of being shorter and lighter than many other heavy-duty couplings. Flange couplings are not standardized, except for integrally forged hydroelectric unit couplings which have been standardized for optimum weight and strength.

In the following equations which relate to flange-faced couplings the following symbols are used:

d = shaft diameter
d_b = bolt diameter
D = diameter of bolt circle
D_h = hub diameter
D_m = mean diameter at which F acts
F = tangential force acting at the bolts
F_0 = compressive force in each of the bolts
i = number of bolts
L = hub length
T_{fr} = frictional torque at the flange interface
T = transmitted torque
Z_0 = polar sectional modulus of the shaft
τ = allowable shear stress
μ = coefficient of friction
σ = compressive stress

Case 1 considers *frictional contact.* The transmitted torque

$$T = F_0 i D_m/2 \leqq T_{fr} \tag{27.45}$$

Furthermore, with $T = Z_0\tau$, the necessary compression force can be found to be

$$F_0 = 2Z_0\tau/\mu i D_m \tag{27.46}$$

Here it is advisable to check the strength of bolts in compression:

$$\sigma = 4F_0/d^2$$

Case 2 considers *bolts in shear.* The tangential force can be expressed as

$$F = 2T/D \tag{27.47a}$$

or

$$F = i\pi d_b^2\tau/4 \tag{27.47b}$$

and the shear stress in bolts

$$\tau = 8T/i\pi d_b^2 D \leqq \tau_{\text{design}} \tag{27.48}$$

Since there are no standards available, empirical values may be used for preliminary calculations. Reference 4 suggests

$d_b = 0.5d/\sqrt{i}$ $D = 2(d + 1)$ $D_h = 1.5d + 1$ $i = 0.5d_b + 3$ $L = 1.25d + 0.75$

(c) Flexible Couplings[31–33,41]

Fexible couplings provide compensation for misalignment of shaft ends by virtue of angular, lateral, and longitudinal (axial) flexibility. Generally, flexibility can be achieved by:

Loose fitting of hard-coupling members (for low-speed application only)

Close fitting of hard-coupling members and provision for lubricated sliding motion between these parts

Employing resilient metallic or nonmetallic members

Couplings in which flexibility is obtained by the first two means are frequently considered rigid or only kinematically flexible. Such couplings are not capable of compensating for dynamic disturbances. By way of contrast, couplings employing resilient materials not only possess kinematic flexibility, but also are able to withstand shock and vibration during rotation and also to change the dynamic characteristics of the shafting system (detuning action).

For the dynamics of high-speed couplings see refs. 34 to 40.

Flexible Couplings Employing Rigid Members. The *double-slider (Oldham)* coupling (Fig. 27.18) may be used for shafts with lateral misalignment amounting to as much as 5 per cent of the shaft diameter with considerable axial play and with

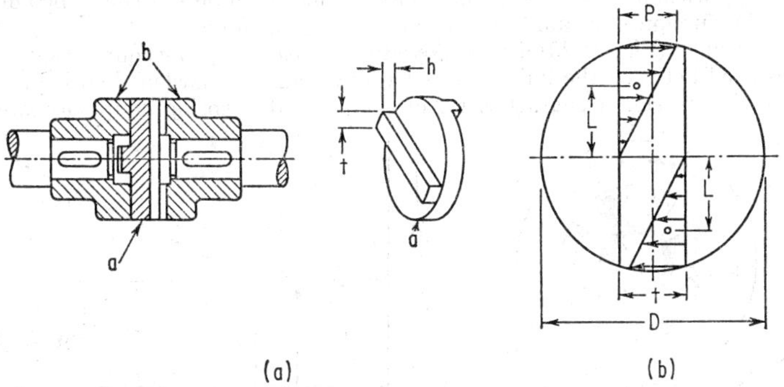

Fig. 27.18. Double-slider (Oldham) coupling. (*a*) Installation. (*b*) Pressure distribution.

relatively small angular misalignment (1°). The tongues of the centerpiece are perpendicular to one another and are free to slide in the grooves in both hubs. Lubrication and cleanliness of the sliding surfaces are important.

The design of the Oldham coupling is based upon an allowable pressure of the sliding surfaces $p = 1{,}000$ psi. The pressure distribution is shown in Fig. 27.18*b*. According to ref. 4, the force F due to the total pressure on each side of the tongue

$$F = (1/4)pDh \tag{27.49}$$

from which the torque transmitted by both sides of the tongue

$$T = 2FL = (1/6)pD^2h \tag{27.50}$$

and the horsepower transmitted at a rotational speed of n rpm

$$P = Tn/63{,}030 = pD^2hn/378{,}180 \tag{27.51}$$

where h = axial dimension of the contact area (approximately the height of the tongue in Fig. 27.16*a*)

D = outside diameter of the centerpiece

F = force due to the total pressure on each side of the tongue

L = distance of the pressure-area centroid from the center line (Fig. 27.16*a*)

Suggested values for preliminary calculations: $D \cong (3 \text{ to } 4)d_{\text{shaft}}$ and the width of the tongue $t = 0.45d_{\text{shaft}}$.

Gear-tooth couplings (Fig. 27.19) are double-engagement type and may be used in the case of angular and parallel misalignment. The use of gears improves the accuracy of the fits and results in smoother and quieter operation. The coupling consists of two externally geared hubs (a) and an internally geared sleeve (b). The sleeve gear meshes with the hub gear and the coupling acts as a spline drive. Clearance between shaft ends should be provided.

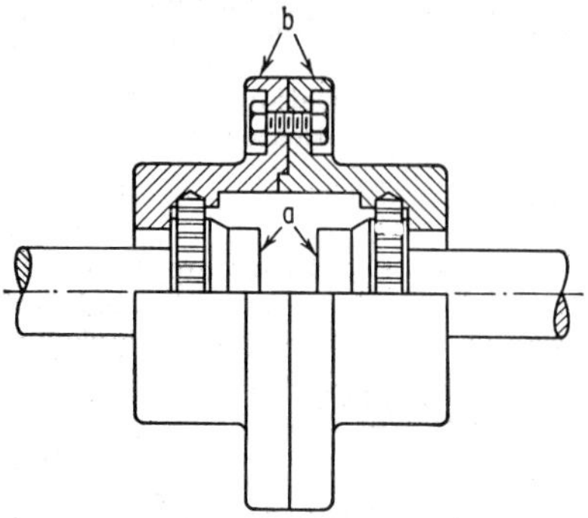

FIG. 27.19. Gear-tooth coupling.

Flexible Couplings Employing Resilient Members. The ability of a coupling to compensate for fluctuating torsional load depends primarily upon the dynamic characteristics of the resilient member employed. The design and selection of the coupling will therefore be based upon strength considerations as well as the effective spring constant of the coupling elements. Applications of spring elements as resilient members are described in refs. 17 and 19. Discussion of a number of resilient elements follows:

Helical springs (Fig. 27.20) are employed in *peripheral-spring* couplings, the springs being inserted between the hubs of the coupling (against hardened spring seats). A representative torsional characteristic of nonpreloaded spring couplings is given in Fig. 27.20b. Because of a certain amount of backlash, the coupling is capable of detuning action. For *calculation for strength* the effective load F can be calculated from

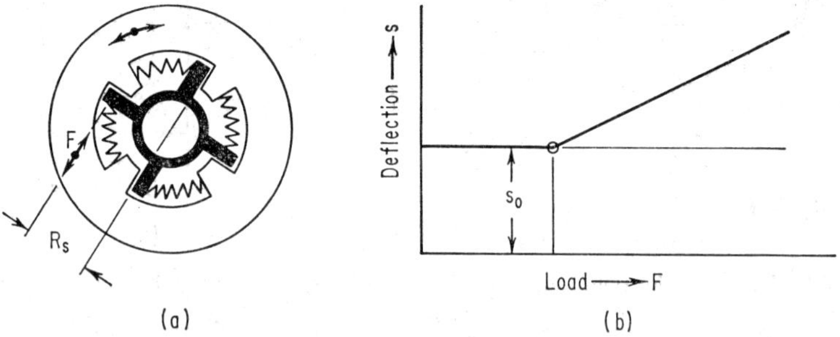

FIG. 27.20. Peripheral-spring coupling. (a) Installation. (b) Torsion characteristic of nonprestressed spring coupling.

$$F = (T/R_s)\gamma \qquad (27.52)$$

where T = transmitted torque
R_s = effective spring radius
γ = load factor ($\gamma > 1$)

The *torsional spring constant* of the coupling

$$k_s = T/\theta = kR_s^2 i \qquad (27.53)$$

where i = number of active springs
k_s = compression stiffness of each spring
R_s = effective spring radius
θ = relative angular displacement of the hub

Spoked couplings (Fig. 27.21) permit angular and transverse misalignment. The spokes are usually fixed at one end and may be fixed, pivoted, or free at the other end. Loading of a spoke is similar to the loading of a cantilever beam. Angular displacement changes the effective length of the spoke, which in turn changes the spring constant and the effect force $F[F = T/i(R + L)]$. Both can be controlled, e.g., by varying the slope α in Fig. 27.21*b*. A typical load-deflection curve is shown in Fig. 27.21*c*.

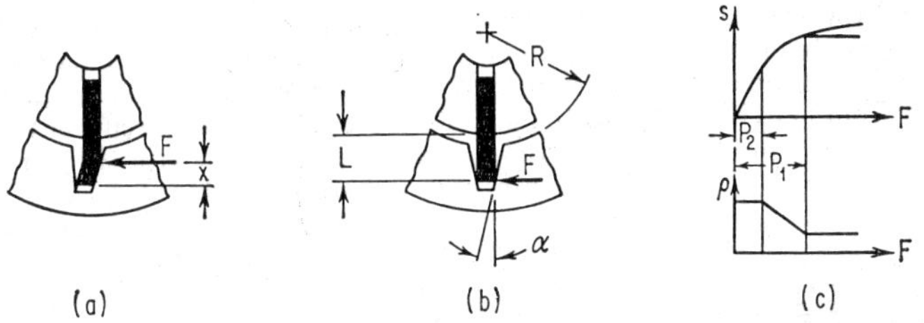

FIG. 27.21. Spoked coupling. (*a*) Angular misalignment. (*b*) Slope. (*c*) Deflection curve.

In *Bibby* couplings (Fig. 27.22) each of the hubs is provided with teeth, and both hubs are coupled together by a continuous strip spring. The teeth are flared and the stiffness of the coupling spring changes with the deflection or load. These couplings allow axial, transverse, and angular misalignment and have a detuning ability in case of vibratory disturbances. Generally, sections of the coupling spring are subjected to bending; shear may prevail under impact loading. The force F may be calculated for n rpm from

$$F = \frac{63{,}000 P_0 \gamma}{R_s i n} \qquad (27.54)$$

where P_0 = power
γ = loading coefficient (1 to 3, in special cases up to 6)
R_s = effective spring radius
i = number of springs

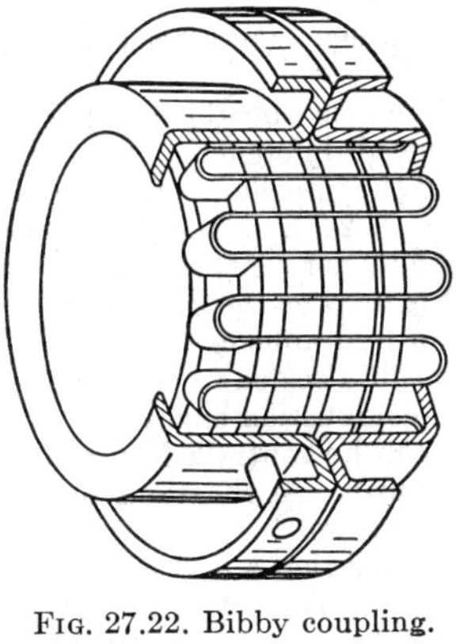

FIG. 27.22. Bibby coupling.

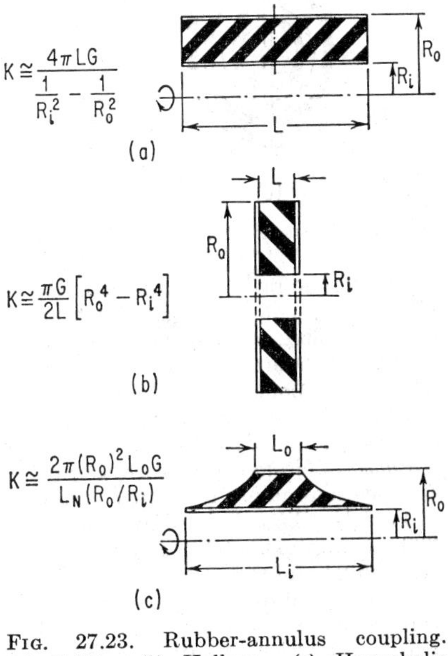

FIG. 27.23. Rubber-annulus coupling. (*a*) Solid. (*b*) Hollow. (*c*) Hyperbolic contour.

Resilient-material Flexible Couplings.[42] These couplings offer the advantages of simplicity of design and low cost. Torque is transmitted through resilient elements (elastomers) generally exposed to shear and usually permitting larger misalignments.

Because of strength considerations, the application of rubber elements is limited mainly to lighter loads. Owing to the nonlinear characteristics of rubber elements, this type of flexible coupling can also be used for detuning purposes. Approximate formulas for spring constants of some simply shaped rubber elements are given in Fig. 27.23.

(d) Hydraulic Couplings[43]

Hydraulic couplings operate on hydrokinetic principles. These couplings consist of two radially vaned rotors, the first of which (impeller) is connected to the prime mover, and the second (runner) to the output shaft. When the impeller rotates, the liquid (oil) located between the vanes of the impeller flows radially outward under the influence of centrifugal force, as indicated by the arrows in Fig. 27.24*a*, and into the runner. From here it returns to the impeller, thus closing the circuit of the vortex of liquid. The drag produced on the runner causes it to rotate with the impeller. The impeller acts as a hydrodynamic pump, the runner as a turbine.

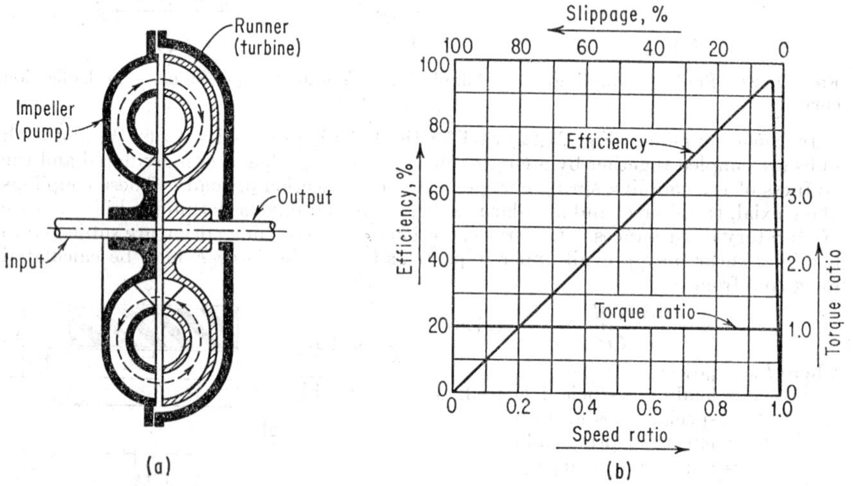

FIG. 27.24. Hydraulic coupling. (*a*) Installation. (*b*) Efficiency.

The torque ratio of a fluid coupling is 1:1, which means that the coupling does not aid the driver in overcoming high torques. Any speed difference at the output side of the coupling is due to slip between the runner and the impeller. This slip increases nearly in proportion to the torque transmitted (Fig. 27.24*b*). Slip of 1 to 3 per cent, which corresponds to transmission efficiency of 97 to 99 per cent, is usual. Usually a mineral oil (viscosity of 180 to 200 SSU at 130°F) is used as the driving liquid.

Two different designs with respect to the filling of hydraulic couplings are:

Constant-filling design with a constant volume of fluid sealed in the rotating parts (application mainly to electric motors and internal-combustion engines).

Variable-filling design in which the fluid volume can be controlled by levers (up to 300 hp, approximately) or by pumps (up to 3,000 hp). Applications include fans, centrifugal and reciprocating pumps, agitators, and mixers.

With regard to torsional vibrations, it is recommended that a hydraulic coupling be considered as having zero torsional stiffness. For calculations which include damping effects, the viscous friction within the coupling should be considered and the damping values for the coupling should be determined experimentally or obtained from the manufacturer.

27.6. PINS

Pins are employed when parts are to be accurately aligned. Figure 27.25 shows installation of dowel, tapered pins, and tapered dowel pins. These are in sequence of increasing strength requirements. Cylindrical dowel pins are preferred in tool and gage work. For more stringent requirements regarding accuracy of alignment and

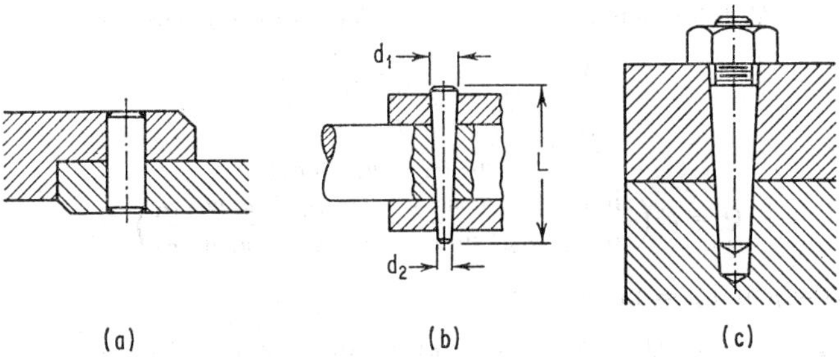

FIG. 27.25. Pins. (*a*) Dowel. (*b*) Taper. (*c*) Taper dowel.

for parts which must be taken apart frequently, tapered dowel pins (with a threaded larger-diameter end) are advisable.

27.7. JOINTS

(a) Cotter Joints

A *cotter* is a form of key. In cotter joints, the connected parts are subjected to either shear or compression; the cotter itself is subjected to shear at two cross sections.

Prestressed cotter joints (Fig. 27.26*a*) can withstand alternating loads. Only unidirectional loads can be resisted by a nonprestressed cotter joint (Fig. 27.26*b*).

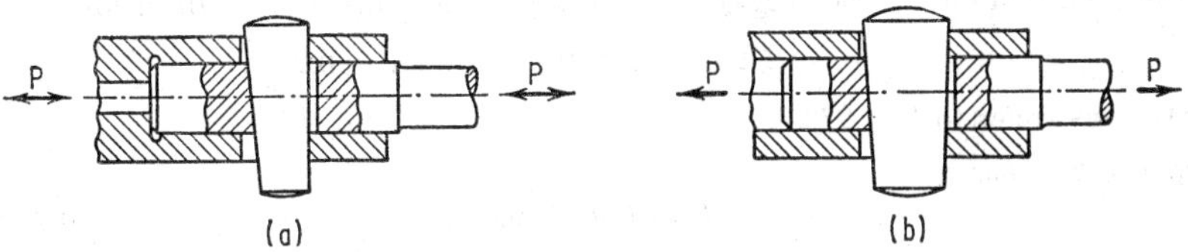

FIG. 27.26. Cotter joints. (*a*) Prestressed. (*b*) Nonprestressed.

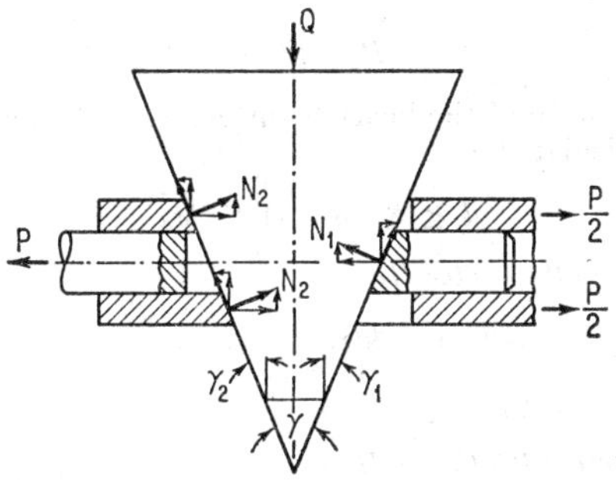

FIG. 27.27. Cotter-pin force analysis.

Taper on the cotter should be properly selected in order to prevent loosening. For fluctuating loads, a taper of ½ in./ft is normally sufficient. The taper may be increased if additional locking of the cotter is provided, e.g., by a setscrew. Additional locking devices are especially advisable for vibration applications. For machine tapers and self-holding key drives, see Sec. 20.

Geometry and Force Analysis. The equation of equilibrium for a symmetrical nonprestressed cotter (Fig. 27.27) is written

$$Q = 2N_2(\sin \alpha_2 + \mu_2 \cos \alpha_2) + N_1(\sin \alpha_1 + \mu_1 \cos \alpha_1) \tag{27.55}$$

Since

$$N_1 = \frac{P}{\cos \alpha_1 - \mu_1 \sin \alpha_1} \tag{27.56}$$

$$N_2 = \frac{P}{2(\cos \alpha_2 - \mu_2 \sin \alpha_2)} \tag{27.57}$$

$$Q = P\left(\frac{\sin \alpha_2 + \mu_2 \cos \alpha_2}{\cos \alpha_2 - \mu_2 \sin \alpha_2} + \frac{\sin \alpha_1 + \mu_1 \cos \alpha_1}{\cos \alpha_1 - \mu_1 \sin \alpha_1}\right) \tag{27.58}$$

For $\mu_1 = \mu_2 = \mu$ and $\mu = \tan \rho$,

$$Q = P[\tan(\alpha_2 \pm \rho) + \tan(\alpha_1 \pm \rho)] \tag{27.59}$$

The positive sign applies if the driven motion of the cotter and Q have the same direction, the negative sign if the directions are opposite.

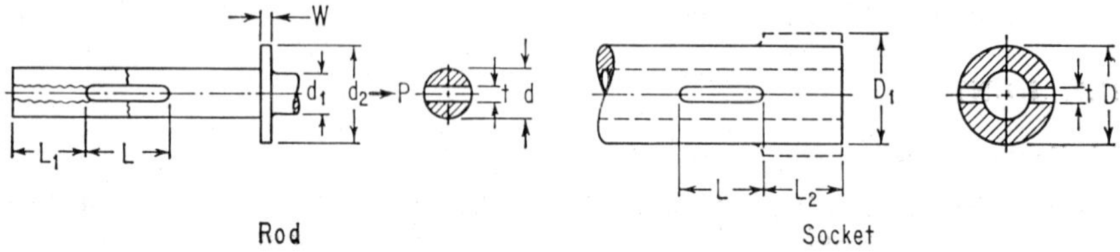

FIG. 27.28. Stressed area of rod and socket.

The condition for self-locking, $Q \leqq 0$, for a one-sided cotter ($\alpha_2 = 0$) yields

$$\alpha_1 = \alpha_2 \leqq 2\rho \qquad \mu \cong 0.1 \qquad \tan \alpha \cong 0.2$$

Stresses (refer to Fig. 27.28).

Rod in Tension

$$P = (\pi d_1^2/4)\sigma_d \tag{27.60}$$

or

$$d_1 \geqq \sqrt{4P/\pi\sigma_d} = 1.13\sqrt{P/\sigma_d} \tag{27.61}$$

Crushing of the Cotter Bearing Surface, at the Rod End

$$P \leqq td\sigma_b \tag{27.62}$$

(where the ultimate strength of the bearing surface $\sigma_b = 2\sigma$) and $t \cong (\frac{1}{3}$ to $\frac{1}{4})d$, the rod diameter can be calculated as

$$d \geqq \sqrt{(3 \text{ to } 4)P/2\sigma_d} \tag{27.63}$$

Rod End Tearing across the Cotter Hole

$$P = (\pi d^2/4 - td)\sigma \tag{27.64}$$

Shearing Rod End

$$P = 2L_1 d\iota_d \tag{27.65}$$

Socket End Tearing across the Cotter Hole

$$P = [(\pi/4)(D^2 - d^2) - (D - d)t]\sigma \tag{27.66}$$

or

$$\sigma = \frac{P}{(D - d)[(\pi/4)(D + d) - t]} \leqq \sigma_d \tag{27.67}$$

It is advisable to check whether an increase in D is required.

Crushing of the Bearing Surface in the Socket End

$$P = [(D - d)t]\sigma_{\text{crushing}} \tag{27.68}$$

Note that $\sigma_{crushing} = 2\sigma_{ult}$. Check for possible need to increase D.

Shearing Socket End

$$P \leqq 2[L_2(D - d)\tau_d] \tag{27.69}$$

Double Shearing of the Cotter

$$P \leqq 2th_{cotter}\tau_d \tag{27.70}$$

Crushing the Collar of the Rod

$$P = 2\pi(d_2^2 - d^2)\sigma_d \tag{27.71}$$

Shearing of the Collar of the Rod

$$P \leqq d^2\pi w\tau_d \tag{27.72}$$

Other Dimensions. Considering the cotter loading condition, and assuming $d_1 = \frac{3}{4}d$, the following dimensions can be derived:

$$L_{cotter} = (1.3 \text{ to } 1.5)d \qquad \text{and} \qquad h_{cotter} = 1.3d$$

Further recommended dimensions:

$$\begin{aligned} L_1 &= L_2 = (0.7 \text{ to } 1.0)L_{cotter} \\ D &= (0.8 \text{ to } 0.9)D_1 \\ D_1 &= 2d \\ w &= \tfrac{1}{4}d \\ d_2 &= 1.3d \end{aligned}$$

(b) Friction Joints

Connections between hub and shaft can be achieved by means of friction in a cylindrical or conical fit. Typical examples of friction hub-shaft fits are shown in Fig. 27.29.

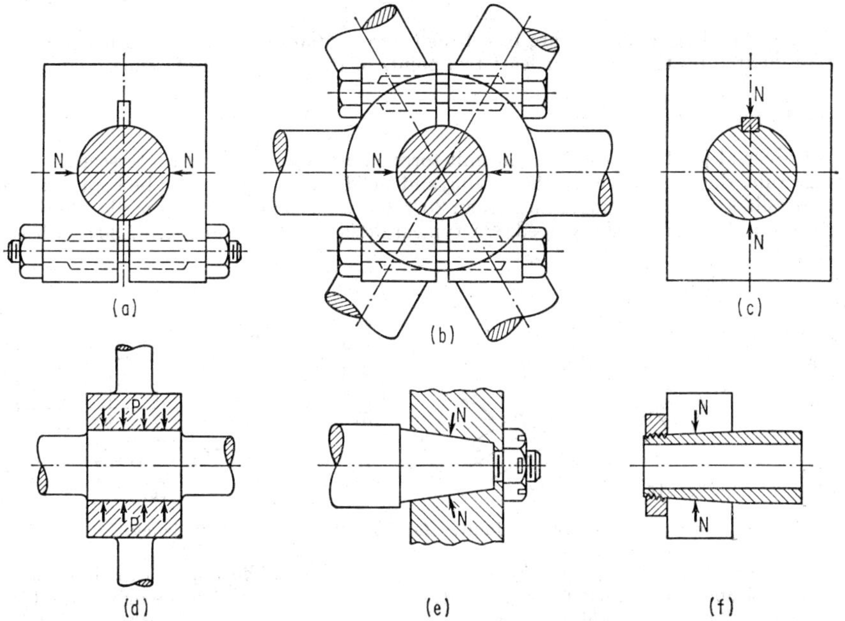

Fig. 27.29. Friction hub. (*a*) Split hub. (*b*) Split hub. (*c*) Solid hub with taper key (*d*) Press fit. (*e*) Conical fit with taper fit. (*f*) Conical fit with nut.

Assuming a uniform pressure distribution at the hub-shaft interface, the retaining force may be expressed

$$R = \Sigma N\mu \geqq F_t = 2T/d \tag{27.73}$$

where $N = p(\pi dL)$ and μ = coefficient of friction; $\mu = 0.13$ for steel on cast iron, 0.33 for steel on cast iron or steel in shrink joints in which cold welding occurs,[4] and as much as 0.65 for steel and cast iron if the surfaces are treated with carborundum.[24]

Calculations of forces and stress are identical to pressure-vessel calculations. The bursting force F is given by

$$F = dLp \tag{27.74}$$

and

$$F = F_t/\mu\pi = 2T/\pi\mu d \tag{27.75}$$

The average normal stress in section *a-a* (hub stress)

$$\sigma_{av} = F/2(D - d)L \tag{27.76}$$

If *taper* keys are used in order to originate the normal force N, the retaining friction force R can be expressed as

$$R = 2F_k\mu \tag{27.77}$$

or, from Eq. (27.73),

$$R \geqq F_t = 2T/d \tag{27.78}$$

where the force of the key F_k can be calculated from the force Q necessary to drive home the key:

$$Q = F_k(\tan\alpha + 2\mu) \tag{27.79}$$

and, from Eq. (27.77),

$$Q = (T/d\mu)(\tan\alpha + 2\mu) \tag{27.80}$$

The driving force Q is limited by key deformation. Thus

$$Q \leqq bh\sigma_{d,\text{key}} \tag{27.81}$$

The preceding equations are based upon the assumption that the normal force N is originated by the key at at least two points of the shaft circumference.

(c) Universal Joints[44–47]

Universal joints are used to transmit torque between shafts with intersecting axes. As a rule, the angle between the driving and the driven shafts should be kept as small as possible (5 to 15°, seldom more than 30°), but an exception can be made where low speeds and torques are involved. A typical design is shown in Fig. 27.30*a* with the

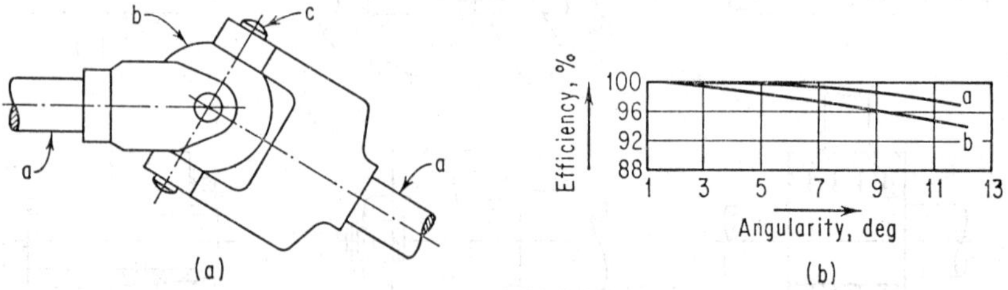

FIG. 27.30. Universal joint. (*a*) Installation. (*b*) Efficiency.

hubs (*a*) and the inner yoke (*b*) made of cast iron, and the pins (*c*) made of steel. Figure 27.30*b* shows efficiency as a function of angularity between the driving and the driven shafts. The angularity between shafts also causes a change in the relationship between the angular speed of the driven and driving shafts during each revolution and thus prohibits the use of universal joints of *this type* where close conformity between the angular speeds of both shafts is required. Uniformity of angular motion can be restored if the two shafts are connected by an intermediate shaft and two universal joints, provided that both shafts make the same angle with the intermediate shaft, and the forks on the intermediate shaft are in the same plane (for the most common case of parallel driving and driven shafts). If the intermediate shaft can telescope, the driving and driven shafts can be moved longitudinally and laterally, and independently of one another, e.g., many machine-tool drives.

Single-joint velocity drives are discussed in ref. 46.

27.8. RIGID-BODY CLUTCHES

(a) Jaw Clutch

Positive-engagement clutches (Fig. 27.31) with spiral jaws are made for driving in one direction only (positive-engagement clutches with square jaws are designed for driving in both directions). Design of jaw clutches should be based on strength considerations, and it is advisable to check jaws for shearing and for crushing at the sides. Approximate principal dimensions for cast-iron jaw clutches are shown in ref. 4.

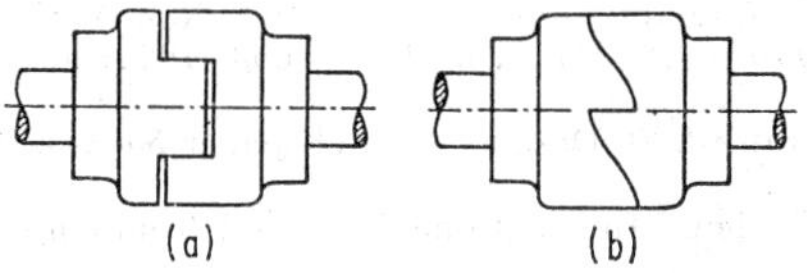

FIG. 27.31. Positive clutch. (*a*) Square jaw. (*b*) Spiral jaw.

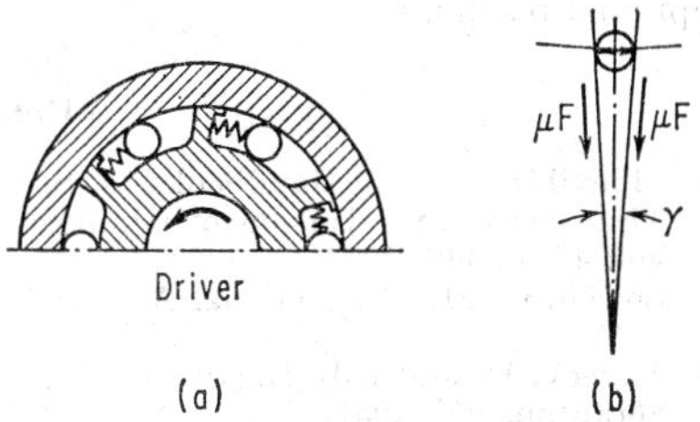

FIG. 27.32. Roller clutch.

(b) Unidirectional Clutches

Freewheeling or overrunning clutches are employed when torque must be transmitted in one direction only or where the driven member is to be permitted to "overrun" the driver. They can be considered a simple and inexpensive substitute for a differential drive and are also used in multiple-speed drives where they automatically disengage the lower-speed unit when the higher-speed unit is engaged. When one of the members of the overrunning clutch is fixed, the other member can rotate only in one direction and the clutch becomes "backstopping," e.g., in gear reducers where reverse motion must be excluded.

The most common designs utilize the wedging principle (Fig. 27.32). Generally, the construction consists of a shell-shaped outer member and an inner member with wedge-shaped pockets, or flat cam profiles, around its periphery; the coupling between the two members is provided by the locking action of rolls. Generally, instantaneous wedging action at any position without backlash or slip is required. The wedge-shaped pocket may have the disadvantages of expensive manufacturing costs and of the relatively small number of rolls employed.

For proper locking action, the condition for self-locking, $\alpha \leqq 2\phi$, must be satisfied. Here α = the angle between tangents to the cam contour and to the roller surface at the point of contact, and $\phi = \tan^{-1} \mu$ (μ = coefficient of friction). To check the strength of the roller the crushing force

$$F_c = F_t/\tan \alpha = 2T/D \tan \alpha \tag{27.82}$$

where T = the torque transmitted and D = the effective diameter of the rollers. For more than three rollers, uniform distribution of load is difficult to achieve without special precautions in regard to accuracy of manufacture. A more detailed force analysis of this type of clutch is found in ref. 27 (vol. 2).

27.9. SETSCREWS

The setscrew diameter d is related to shaft diameter D by the following empirical relationship:[29]

$$d = D/8 + 5/16 \quad \text{in.} \tag{27.83}$$

The maximum safe holding force is given by[30]

$$F = 2{,}500d^{2.31} \tag{27.84}$$

A factor of safety of 1.5 to 2 is recommended.[4] If a lower-grade steel is used, the holding power is approximately one-sixth of the values given. The holding power of headless setscrews is still lower.[4]

It should be noted that setscrew holding power is related to the initial torque of the screw, the shaft diameter, the setscrew-point size, the shaft material, the setscrew material, vibration of the shaft, etc. Smaller cup points have been developed with improved results.[48]

References

1. "Flexible Shafting Handbook," S. S. White Company Industrial Division, New York.
2. Timoshenko, S.: Torsion of Crankshafts, *Trans. ASME*, vol. 44, p. 653, 1922; also vol. 45, p. 449, 1923.
3. Lowell, C. M.: "A Rational Approach to Crankshaft Design," ASME paper 55-A-57, 1955.
4. Maleev, V., and J. B. Hartman: "Machine Design," International Textbook Company, Scranton, Pa., 1957.
5. ASME Code, ASME Standard B-17e-1927, reaffirmed 1947.
6. Marks, Lionel S. (ed.): "Mechanical Engineers' Handbook," 6th ed., McGraw-Hill Book Company, Inc., New York, 1958.
7. Spotts, M. S.: "Design of Machine Elements," Prentice-Hall, Inc., Englewood Cliffs, N.J., 1955.
8. Peterson, R. E.: "Stress Concentration Design Factors," John Wiley & Sons, Inc., New York, 1953.
9. Roark, R. J.: "Formulas for Stress and Strain," 3d ed., McGraw-Hill Book Company, Inc., New York, 1954.
10. Wilson, W. K.: "Practical Solution of Torsional Vibrations," 3d ed., vol. 1, Frequency Calculations, John Wiley & Sons, Inc., New York, 1956.
11. Kimball, A. L.: "Vibration Prevention in Engineering," John Wiley & Sons, Inc., New York, 1932.
12. Miller, D. F.: Forced Lateral Vibration of Beams on Damped Flexible End Supports, *J. Appl. Mech.*, no. 52-A-23, pp. 167–172, June, 1953.
13. Koenig, E. C.: "Analysis for Calculating Lateral Vibration Characteristics for Rotary Systems with a Number of Flexible Supports," part 1, ASME paper 61-APMW-16A.
14. Guenther, T. G., and D. C. Lovejoy: "Analysis for Calculating Lateral Vibration Characteristics for Rotary Systems with a Number of Flexible Supports," part II, ASME paper 61-APMW-16B.
15. MacDuff, John N., and John R. Curreri: "Vibration Control," McGraw-Hill Book Company, Inc., New York, 1958.
16. Den Hartog, J. P.: "Mechanical Vibrations," 4th ed., McGraw-Hill Book Company, Inc., New York, 1956.
17. Nestorides, E. T.: "A Handbook of Torsional Vibrations," Cambridge University Press, New York, 1958.
18. Phelan, R. M.: "An Introduction to the Dynamics of Shaft Systems," Pennsylvania State University, Seminar on High Spaced Flexible Couplings, University Park, June 10–13, 1962.
19. Kotchina, N. I., W. S. Lomiakov, and others: "Machine Elements" (Russian), Gos Nautchno-Techn. Isdat., Moscow, 1958.
20. Hetenyi, M. I.: The Application of Hardening Resins to Three Dimensional Photoelastic Studies, *J. Appl. Phys.*, vol. 10, 1939.

21. Peterson, R. E.: Fatigue of Shafts Having Keyways, *ASTM Proc.*, vol. 32, part II, p. 413, 1932.
22. Neuber, H.: "Kerspannungslehre," Springer-Verlag OHG, Berlin, 1939.
23. Seely and Dolan: Stress Concentration at Fillets, Holes, and Keyways as Found by the Plaster-model Method, *Univ. Illinois Eng. Expt. Sta. Bull.*, no. 276, June, 1935.
24. "SAE Handbook," Society of Automotive Engineers, Inc., New York, 1957.
25. Dudley, D. W.: How to Design Involute Splines, *Prod. Eng.*, Oct. 28, 1957.
26. Rowland, D. R.: Matching Ball Bearing Splines to Torque, Radial-load, and Life Requirements, *Machine Design*, July 20, 1961.
27. Niemann: "Maschinenelemente," Springer-Verlag OHG, Berlin, 1960.
28. "Mechanical Springs," Associated Spring Corp., Bristol, Conn.
29. Vallance, A., and V. L. Doughtie: "Design of Machine Members," 3d ed., McGraw Hill Book Company, Inc., New York, 1951.
30. Pinkney, B. H. D.: *Machinery*, Oct. 15, 1914.
31. Spector, L. F.: Design Guide–Flexible Couplings, *Machine Design*, Oct. 30, 1958.
32. Conway, H. G.: Couplings of Parallel Shafts, Products Design Handbook Issue, *Prod. Eng.*, mid-October, 1955.
33. Pampel, W. P.: "Kupplungen," vol. I, VEB Verlag Technik, Berlin, 1958.
34. "High Speed Gear Coupling Development," Pennsylvania State University Seminar on High Speed Flexible Couplings, University Park, Pa., June 10–13, 1962.
35. Allen, E. A.: "A Coupling Design to Minimize Dynamic Instabilities of High Speed Systems," Pennsylvania State University Seminar on High Speed Couplings, University Park, Pa., June 10–13, 1962.
36. Grundtner, R. R.: "The Torsionally Soft Coupling and Its Application," Pennsylvania State University Seminar on High Speed Flexible Couplings, University Park, Pa., June 10–13, 1962.
37. Thoma, F. A.: "Coupling Induced Shaft Vibration," Pennsylvania State University Seminar on High Speed Flexible Couplings, University Park, Pa., June 10–13, 1962.
38. "Some Design Problems Associated with a New High Speed Flexible Disk Type Coupling," Pennsylvania State University Seminar on High Speed Flexible Couplings, University Park, Pa., June 10–13, 1962.
39. Ho, J. Y. L.: "Dynamic Balance of Gear Type Flexible Coupling," Pennsylvania State University Seminar on High Speed Flexible Couplings, University Park, Pa., June 10–13, 1962.
40. Rothfuss, N. B., and C. B. Gibbons: "Contoured Flexible Diaphragm Type High Speed Couplings," Pennsylvania State University Seminar on Flexible High Speed Couplings, University Park, Pa., June 10–13, 1962.
41. Gensheimer, J. R.: How to Design Flexible Couplings, *Machine Design*, Sept. 14, 1961.
42. Larsen, P. J.: Special Report on Engineering Elastomers, *Machine Design*, Jan. 8, 1962.
43. Gibson, W. B.: Fluid Coupling, *Machine Design*, Mar. 31, 1960.
44. Rzeppa, A. H.: University Joint Drives, *Machine Design*, April, 1953.
45. Rosenburg, R. M.: "On the Dynamical Behavior of Rotating Shafts Driven by Universal (Hooke) Couplings," ASME paper 57-A-1, 1957.
46. Saari, O.: How to Obtain Useful Speed Variations with Universal Joints, *Machine Design*, October, 1954.
47. Mabie, H. H.: Constant Velocity Joints, *Machine Design*, May, 1948.
48. Gates, C. S.: Allenpoint Cup Point, Allen Manufacturing Co., Hartford, Conn.

Section 28

FRICTION CLUTCHES AND BRAKES

By

Z. J. JANIA, B.S., *Senior Research Engineer, Applied Research Office, Ford Motor Company, Dearborn, Mich.* *(Friction Clutches)*

DAVID SINCLAIR, Ph.D., *Research Physicist, Johns-Manville Research and Engineering Center, Manville, N.J.* *(Friction Brakes)*

CONTENTS

FRICTION CLUTCHES

FRICTION BRAKES

FRICTION CLUTCHES*

Acknowledgment. The author wishes to thank D. Zawada of the Applied Mathematics Section, Research and Engineering Staff, Ford Motor Company, for his advice and assistance in obtaining analog-computer solutions to the problems illustrating the dynamic behavior of the systems discussed in this section.

28.1. INTRODUCTION

The operation of friction clutches and brakes is dependent upon the phenomenon of friction. The reader is referred to Sec. 12 for a more detailed coverage of friction than is here presented. Those areas of friction which are of particular importance to the study of clutches and brakes will be discussed in the following paragraphs.

28.2. FRICTION AND FRICTION MATERIALS

(a) Effect of Temperature on Friction

Because the interface of rubbing clutch and brake surfaces is often at high temperatures, and because of the dependence of their operation upon friction, it is important to examine, if only briefly, the effect of temperature upon friction.

Friction of clean metals is not markedly influenced by temperature unless the temperature is so high as to cause appreciable thermal softening. At high temperatures, therefore, there is a decrease of frictional force, and friction may fall to a very low value when temperature achieves a value sufficiently close to the melting point of the material.

In the case of contaminated surfaces the effect of temperature is quite complex, but in general, friction increases with temperature because of the decomposition and breakdown of surface films.

(b) Effect of Sliding Velocity on Friction

In general, the effect of sliding velocity is to decrease the frictional force acting on two sliding bodies. A distinction must be made here between (1) dry surfaces, (2)

* By Z. J. Jania.

full fluid-film lubrication, and (3) boundary lubrication. In case 1, static frictional force is in most cases higher than kinetic friction, which may remain constant, increase, or decrease with sliding velocity. The decrease in frictional force with sliding velocity is most common in practice, the variation being asymptotic, tending to some constant value at a large enough sliding velocity. At some sufficiently large value of sliding velocity, an increase of frictional force may occur.

In general, boundary-lubricated surfaces show the same behavior. The frictional force always decreases with sliding velocity to a certain point at which it begins to increase, probably because of the establishment of at least partial hydrodynamic film.

In the case of full fluid-film lubrication, friction, being viscous in nature, always increases with increasing velocity.

(c) Dynamic Friction Characteristics

The relationship between the coefficient of friction and sliding velocity has not received enough attention to produce sufficient data for design use. This relationship can be controlled to a certain degree by changing not only the composition of the material itself but also the type of lubricant, the additives used, and their concentration.

Figure 28.1 shows several recordings of individual engagement cycles of a clutch tested on an inertia dynamometer. Each cycle consists of decelerating a flywheel of a given moment of inertia from certain maximum speed to zero by the test clutch, one side of which is grounded. Some recordings clearly show the effect of negative slope of friction speed characteristics—the system becomes oscillatory—the amplitude of the oscillations increasing with time near the end of the cycle.

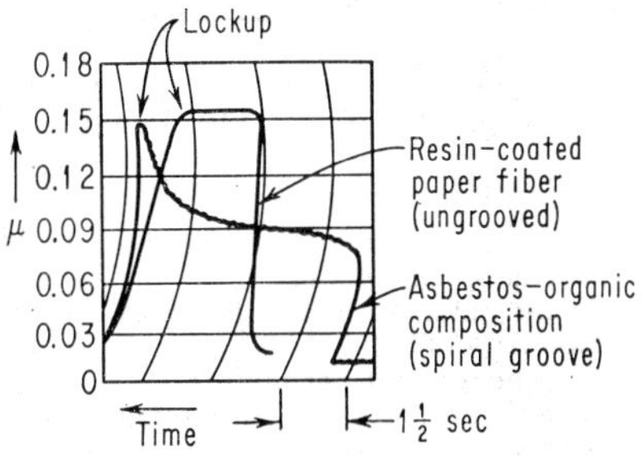

FIG. 28.1. Engagement curves of materials. Inertia dynamometer tests. (*Courtesy of Raybestos—Manhattan, Inc.*)

Figure 28.1 also shows the characteristic engagement curves of a paper-type material which are very desirable from the standpoint of system stability. The curves show a slightly positive slope, indicating that a dynamical system employing a clutch using this type of material will be essentially stable. Note that this implies that kinetic friction is higher than static friction, clearly indicated in the figure.

Figures 28.2 and 28.3 show the variation of the coefficient of friction (average values) with number of engagements and temperature and pressure, respectively. Note the decrease in friction with increasing unit pressure.

(d) Effect of Grooving

The effective coefficient of friction in a lubricated clutch will be affected markedly by grooves on the plates of the clutch. Grooving of the plates coated with friction material is usually accomplished by either machining or molding during the manufacturing process. The most commonly used forms of grooving patterns are spiral, single lead; radial; and "waffle iron" (90°).

28.3. TORQUE EQUATIONS

Friction clutches are used in power-transmission systems (1) to transmit the desired torque from one point in the mechanism to the other and simultaneously to limit this torque to a desired value and (2) to couple the prime mover to the load in a manner which allows the acceleration of the load to be controlled.

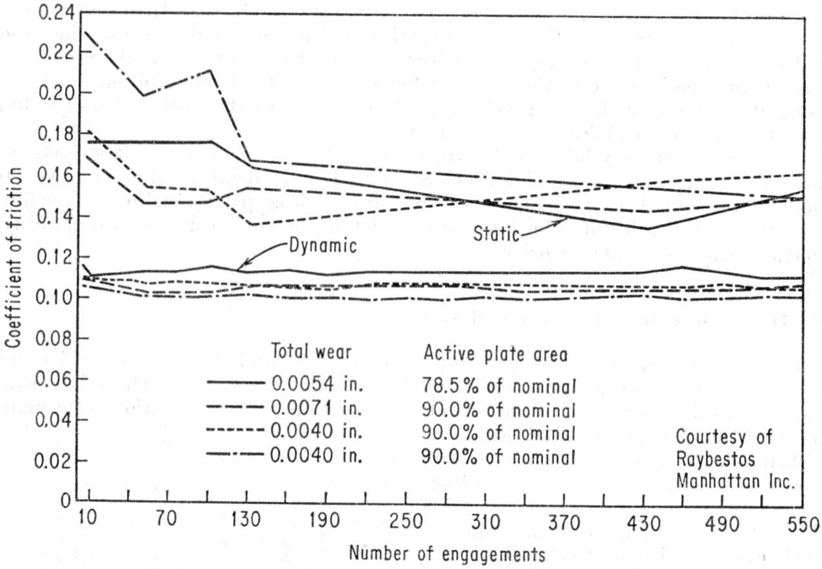

Fig. 28.2. Coefficient of friction vs. number of engagements.
Inertia dynamometer.
Single plate, two active faces, 5⁵⁄₁₆ by 4⁷⁄₆₄ in. diameter, 118 psi gross unit pressure.
Esso AQ-ATF-198-A oil.
One engagement per minute.
Material, asbestos, organic composition, grooved.
(*Courtesy of Raybestos—Manhattan, Inc.*)

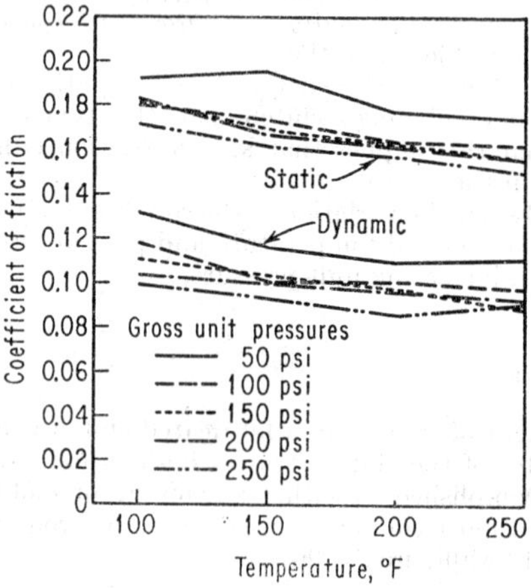

Fig. 28.3. Friction vs. temperature and pressure.
Inertia dynamometer.
Single plate, two active faces, 5.51 by 4.10 in. diameter.
Amoco AQ-ATF-795-A oil,
Material, asbestos, organic composition, grooved.
Active plate area 90 per cent of nominal.
(*Courtesy of Raybestos—Manhattan, Inc.*)

Consider the arrangement shown in Fig. 28.4, in which it is desired to determine the torque M due to frictional forces existing at the interface of members A and B, and due to the axial force P.

The normal load acting on the differential strip of width dl is $2\pi rp\,dl$, where p is the pressure. Since $dr = dl \sin \alpha$, the axial load supported by the strip

$$dP = 2\pi rp\,dr \tag{28.1}$$

so that

$$P = 2\pi \int_{R_1}^{R_2} pr\,dr \tag{28.2}$$

The frictional force acting on the strip dl

$$\mu\,dP = 2\pi\mu pr\,dl \tag{28.3}$$

where μ = coefficient of friction.

The associated moment

$$dM = 2\pi\mu pr^2 \csc \alpha\,dr \tag{28.4}$$

and the total frictional moment

$$M = 2\pi \csc \alpha \int_{R_1}^{R_2} \mu pr^2\,dr \tag{28.5}$$

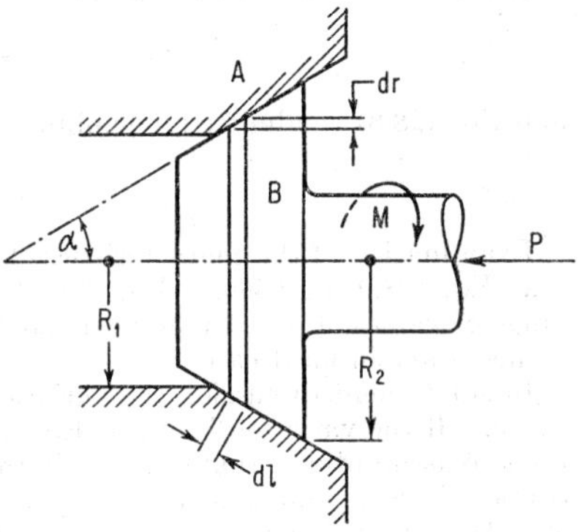

FIG. 28.4. Friction clutch.

Before Eq. (28.2) can be integrated, the dependence of μ and p on r and the relationship connecting μ and p must be known.

Variation of μ with r. If friction is independent of sliding velocity, μ is independent of r. If, as is commonly the case in practice, friction is a decreasing function of linear sliding velocity v, the relationship can be written

$$\mu = f(v) = f(\omega_r r) \tag{28.6}$$

where ω_r is the relative angular velocity between the members A and B of the clutch in Fig. 28.4.

In case μ varies linearly with v, expression (28.6) can be written

$$\mu = \mu_0 - mv = \mu_0 - m\omega_r r \tag{28.7}$$

where μ_0 = static coefficient of friction

m = slope of friction speed characteristic of the interface (depends on material combination)

Variation of p with r. There is, in any practical situation, considerable uncertainty as to the distribution of axial load P over the contact area. Because of the difficulty of obtaining perfect geometrical fit (especially in the case of cone clutches) between the two surfaces, and particularly with relatively hard materials such as sintered metals and ceramics, contact between the surfaces is far from uniform and pressure distribution highly irregular. When dealing with resilient friction materials such as granulated cork or paper, the pressure is reasonably independent of r, and the assumption of uniform contact is safe.

Wear. The rate at which material is removed is dependent upon the pressure and the sliding velocity. Little information is available to indicate the nature of this dependence, except for the correlations similar to those given below, gained from experience.[3]

$pv \leqq 83{,}000$ for continuous-load application and good heat dissipation, as in an oil bath.

$pv \leqq 55{,}000$ for intermittent-load application, relatively long periods of rest, and poor heat dissipation.

$pv \leq 28{,}000$ for continuous-load application and poor heat dissipation.

An assumption is often made that rate of wear is uniform over the area of the frictional surfaces, that is,

$$pr = C \text{ const} \tag{28.8}$$

Relationship between μ and p. As previously indicated, μ may be a function of p. If p is invariant with r, the expression for the clutch frictional moment will not be affected by the variation of μ with p. However, the design of the clutch (i.e., the number of plates required, etc.) will be different for different values of p.

If p varies with r and μ is a function of p as in

$$\mu = \mu_p - ap \tag{28.9}$$

and Eq. (28.8) is admitted, one obtains

$$\mu = \mu_p - aC/r \tag{28.10}$$

Examination of the characteristics of those materials in which μ varies with p shows that Eq. (28.9) and hence Eq. (28.10) can be expected to hold only over a limited range of values of p. A nonlinear representation of μ as a function of p will probably be necessary in most cases.

In subsequent discussion the variation of friction with sliding velocity is of importance. If the variation is linear, Eq. (28.7) may be used, and depending on which of the two assumptions, that of the uniform rate of wear or of the constancy of p over the frictional area, is taken as the basis upon which Eqs. (28.2) and (28.5) are integrated, one obtains two expressions for the frictional moment of the clutch.

p invariant with r, $\mu = \mu_0 - mr\omega_r$:

$$M = \frac{2nP\mu_0}{3\sin\alpha}\frac{R_2^3 - R_1^3}{R_2^2 - R_1^2} - \frac{nmP}{2\sin\alpha}(R_2^2 + R_1^2)\omega_r \tag{28.11}$$

Uniform rate of wear, $pr = C_3$, $\mu = \mu_0 - mr\omega_r$:

$$M = \frac{Pn\mu_0}{2\sin\alpha}(R_1 + R_2) - \frac{nmP}{3\sin\alpha}\frac{R_2^3 - R_1^3}{R_2 - R_1}\omega_r \tag{28.12}$$

where n = number of pairs of frictional surfaces in contact.

It is seen that Eqs. (28.11) and (28.12) have the form

$$M = M_0 - \alpha\omega_r \tag{28.13}$$

where M_0 is recognized as the static torque capacity of a friction clutch

$$M_0 = \frac{2nP\mu_0}{3\sin\alpha}\frac{R_2^3 - R_1^3}{R_2^2 - R_1^2}$$

or

$$M_0 = \frac{nP\mu_0}{2\sin\alpha}(R_1 + R_2) \tag{28.14}$$

depending on the assumption made in integrating Eqs. (28.2) and (28.5).

Simple calculation will show that for clutches so designed that R_2 is not very much larger than R_1 both equations (28.11) and (28.12) give values of M which, for all practical purposes, are identical. In general, M obtained from Eq. (28.12) has a slightly lower value than that given by Eq. (28.11).

Cone Clutches. Equations (28.11) and (28.12) can be used as they stand for computation of M.

Disk Clutches. In this case $\alpha = \pi/2$, $\sin\alpha = 1$, and Eqs. (28.11) and (28.12) become, respectively,

$$M = \tfrac{2}{3}nP\mu_0\frac{R_2^3 - R_1^3}{R_2^2 - R_1^2} - \tfrac{1}{2}nmP(R_2^2 + R_1^2)\omega_r \tag{28.15}$$

$$M = \tfrac{1}{2}nP\mu_0(R_1 + R_2) - \tfrac{1}{3}nmP\frac{R_2^3 - R_1^3}{R_2 - R_1}\omega_r \tag{28.16}$$

Multiple-disk Clutches. In the case of multiple-disk clutches Eqs. (28.11) through (28.14) require correction in that not all the disks in the assembly are subject to the same externally applied axial clamping force P_0.

Referring to Fig. 28.5, individual disks are in equilibrium under the action of three axial forces P_N, P_{N-1}, and the force originating at the hub spline because of friction between the disk and the hub as a consequence of torque transmitted by that disk, T_{N-1}/r_1.

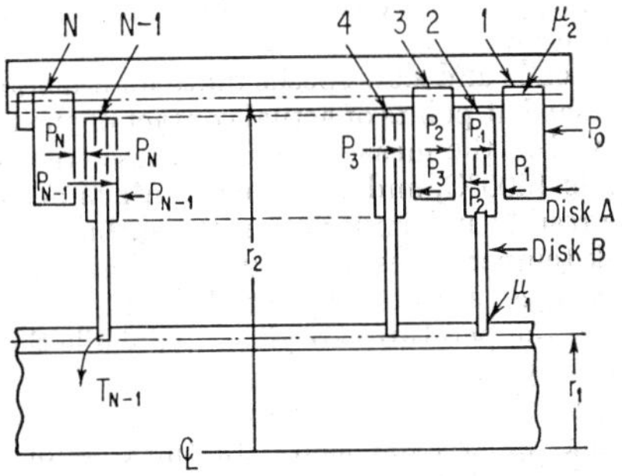

FIG. 28.5. Multiple-disk clutch.

Considering disk 1, the following relationship holds between forces acting on this disk when the clutch transmits full rated torque:

$$P_1 = P_0 - T_1\mu_2/r_2 \tag{28.17}$$

where T_1 = frictional moment transmitted by disk 1
μ_2 = coefficient of friction between disk 1 and its housing
r_2 = radius, as indicated in Fig. 28.5

But

$$T_1 = kP_1 \tag{28.18}$$

where, for simplicity, k may be taken from Eq. (28.14).

$$k = \tfrac{1}{2}\mu_0(R_1 + R_2)$$

for $\alpha = \pi/2$ and $n = 1$.

Thus

$$P_1 = \frac{1}{1 + k\mu_2/r_2} \tag{28.19}$$

Considering now the conditions under which disk 2 remains in balance one obtains

$$P_2 = \frac{1 - k\mu_1/r_1}{1 + k\mu_1/r_1} P_1 = \frac{1}{1 + k\mu_2/r_2} \frac{1 - k\mu_1/r_1}{1 + k\mu_1/r_1} P_0 \tag{28.20}$$

where μ_1 = coefficient of friction between disk 2 and the hub
r_1 = radius, as indicated in Fig. 28.5

Extending this procedure to include all disks in the assembly and expressing axial forces in terms of P_0, one can write

$$\frac{P}{P_0} = \frac{1}{(N-1)(1+a)} \left\{ \left(1 + \frac{1-b}{1+b}\right) \sum_{n=1}^{j} \left[\frac{(1-b)(1-a)}{(1+b)(1+a)}\right]^{n-1} \right\} \tag{28.21}$$

where $a = k\mu_2/r_2$
$b = k\mu_1/r_1$
$j = (N-1)/2$
N = total number of disks (both sets)

Knowing P_0, P is computed through Eq. (28.21) and then inserted into Eqs. (28.11) through (28.14) to compute the value of the total frictional moment of the clutch.

28.4. THERMAL PROBLEMS IN CLUTCH DESIGN

When one solid body slides over another, most of the work done against frictional forces opposing the motion will be liberated as heat at the interface. Consequently the temperature of the rubbing surfaces will increase and may reach values high enough to destroy the clutch.

Even under moderate loads and slow speeds quite high instantaneous values of surface temperature are reached at points of actual contact of the two surfaces. Thermoelectric methods employed in the measurement of these transient local temperatures indicate values as high as 1000°C, and their duration has been observed to be of the order of 10^{-4} sec. The results are similar when the surfaces are lubricated with mineral oil except that peak temperatures are reduced. Also, it has been observed that, when large quantities of heat are liberated at the rubbing interface during a short time interval, high surface temperatures are generated while the bulk of the two bodies in contact remains relatively cool.

Most premature clutch failures can be attributed to excessive surface temperatures generated during slipping under axial load. In the case of metallic clutch plates, high temperatures existing at the rubbing interface may cause the individual plates to be welded together. When nonmetallic or semimetallic plates are used with steel or cast-iron separating plates, the friction material usually disintegrates because of high temperatures, and if the temperature penetration into the bulk of friction coating is appreciable, the material will be completely removed from the steel core of the plate.

Other effects of high surface temperatures are distortion of the shape of the plates, surface cracks in solid metallic plates caused by thermal stresses, appreciable fluctuation in the value of the friction coefficient ("fading"), and in the case of lubricated clutches, oxidation of oil resulting in the formation of deposits on the working surfaces and in grooves.

In order to design a clutch that is satisfactory from the thermal point of view, it is necessary to know or estimate the value of surface temperature considered safe, and the maximum value of surface temperature likely to occur in a given assembly operating under known conditions of loading.

Because the required information is not readily obtained, it is probably best to perform a carefully executed series of tests. Direct measurement of surface temperatures in a clutch assembly presents great difficulties, however, especially when the friction material is a poor electrical conductor. Furthermore, experimentation with different clutch designs working under different conditions is rather expensive and time-consuming.

If a model is created, based on some simplifying assumptions, surface temperature of the clutch plates sliding relative to each other may be easily computed. The calculated values may then be compared with the safe temperature values usually supplied by friction-material manufacturers. Complete correlation cannot be expected if this procedure is followed, but in most situations a good criterion for establishing the size of the total frictional area of the clutch is thus obtained.

The imperfection with which the model represents the actual system should be borne in mind. The results obtained analytically should be regarded as an approximation. Even though the calculated values of the surface temperature of a clutch plate may differ significantly from the actual transient local values, the equations express the manner in which temperature will vary with system parameters and thus will show which of them should be altered to obtain desired change in temperature.

One widely used method of evaluating the thermal capacity of a friction clutch consists of calculating the average rate of energy dissipation per unit area of the clutch frictional surface during the slip period. The answer, usually stated in terms of hp/in.^{2} (often of the order of 0.3 to 0.5 hp/in.^{2}), is compared with a similar figure computed for some other successfully operating design and is adjusted accordingly by variation of plate area and number of plates in the assembly. This method affords a quick and rough estimation of the extent of thermal loading to which the active surfaces are subjected and can serve as a very approximate basis for comparing two similar assemblies operating under identical conditions. This method completely neglects such factors of utmost importance in heat transfer as conductivity, specific heat, and plate thickness which have large effects on the surface temperature.

The temperature of the clutch plates will be evaluated by solving the Fourier conduction equation, and the necessary assumptions discussed below.

1. The heat flux generated at the rubbing interface is uniformly distributed over the total frictional area of the clutch; i.e., the rate of energy dissipation is a function of time only. This assumption is based upon the premise that the rate of wear of the rubbing surfaces and not the intensity of pressure is uniformly distributed over the area of the plate.

Assumption 1 is justified as follows: First, in a properly designed system, the rate of heat dissipation is a function of time—it decreases from an initial maximum value at the beginning of the slip period to zero at the end, i.e., at the time when the two shafts are effectively coupled by the clutch. Secondly, most clutches are so designed that the ratio of the outside diameter to the inside diameter of the plate is not very much greater than unity (about 1.3) and, therefore, the effect of the heat-flux variation with radius (if any) on plate temperature will be small. Finally, the manner in

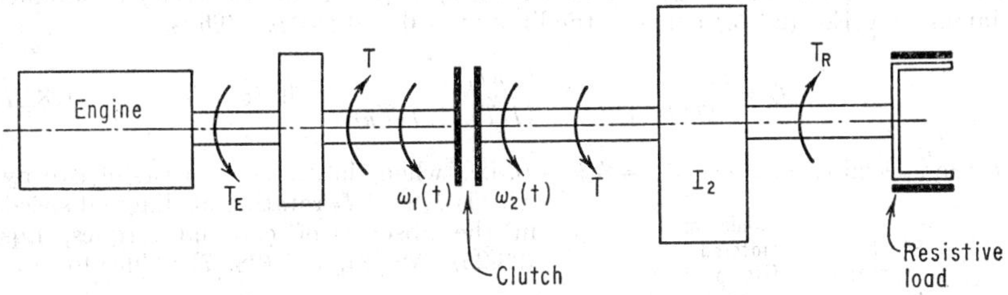

FIG. 28.6. Basic power-transmission system.

which the thermal parameters affect the surface temperature is not greatly influenced by the flux distribution.

2. Heat transfer from the clutch plates to the surroundings during slipping is negligible.

This assumption is well justified when dealing with slip periods of short duration, of the order of several seconds or less. Also, in most clutches the edge area of the plate is much smaller than the area receiving heat; thus the heat flow in the radial direction is small and can be neglected in comparison with axial direction.

3. The effect of the lubricating oil is neglected.

The effect of the lubricant is difficult to account for in the analysis, mainly because, during the time interval in which the plates move closer together under the action of the axial force, lubrication changes from hydrodynamic to boundary while the oil is squeezed out of the space between the plates. The presence of oil generally lowers surface temperature considerably, depending on the type of lubricant. Under boundary lubrication the temperature can be expected to be between 30 to 60 per cent lower than in the case of dry friction.

(a) Derivation of Temperature Equation

First to be considered is the form of the expression for the rate of heat dissipation to be used in subsequent analysis. Consider a basic power-transmission system shown in Fig. 28.6.

T_E is the engine output torque, T the clutch frictional torque, T_R the resistive load torque, and $\omega_1(t)$ and $\omega_2(t)$ are angular velocities of engine and load sides of clutch, respectively. All torques acting on the system will, for simplicity, be assumed constant.

It is easily shown that, when the compliance of the system is neglected,

$$\omega_1(t) = \frac{T_E - T}{I_1} t + \Omega_1 \qquad \text{rad/sec} \tag{28.22}$$

$$\omega_2(t) = \frac{T - T_R}{I_2} t + \Omega_2 \qquad \text{rad/sec} \tag{28.23}$$

where Ω_1 and Ω_2 are initial velocities of the two shafts.

The rate at which energy is dissipated in the clutch during slipping is given by

$$q_0(t) = T[\omega_1(t) - \omega_2(t)] \qquad \text{lb-ft/sec} \tag{28.24}$$

$$q_0(t) = T\left[\frac{I_2T_E + I_1T_R - T(I_1 + I_2)}{I_1I_2} t + \Omega_1 - \Omega_2\right] \tag{28.25}$$

The duration of the slip period is given by

$$t_0 = \frac{I_1I_2(\Omega_1 - \Omega_2)}{T(I_1 + I_2) - (I_2T_E + I_1T_R)} \qquad \text{sec} \tag{28.26}$$

The total amount of energy dissipated during slipping (in one cycle) is obtained by integrating Eq. (28.25) between the limits $t = 0$ and $t = t_0$. Thus

$$Q = \frac{T(\Omega_1 - \Omega_2)^2 I_1I_2}{2T(I_1 + I_2) - (I_2T_E + I_1T_R)} \qquad \text{lb-ft} \tag{28.27}$$

In the special case, when $T_R = T_E = 0$, i.e., when the system consists of two flywheels I_1 and I_2 rotating at different speeds in the absence of external torques, Eqs. (28.25), (28.26), and (28.27) reduce to

$$q_0(t) = \left(\Omega_1 - \Omega_2 - T\frac{I_1 + I_2}{I_1I_2} t\right) T \qquad \text{lb-ft/sce} \tag{28.28}$$

$$t_0 = \frac{I_1I_2(\Omega_1 - \Omega_2)}{T(I_1 + I_2)} \qquad \text{sce} \tag{28.29}$$

$$Q = \frac{I_1I_2(\Omega_1 - \Omega_2)^2}{2(I_1 + I_2)} \qquad \text{lb-ft} \tag{28.30}$$

There is considerable experimental evidence to suggest that the rate of heat dissipation varies approximately linearly with time and thus has the general form

$$q_0(t) = -At + B \tag{28.31}$$

which is identical with Eqs. (28.25) and (28.28).

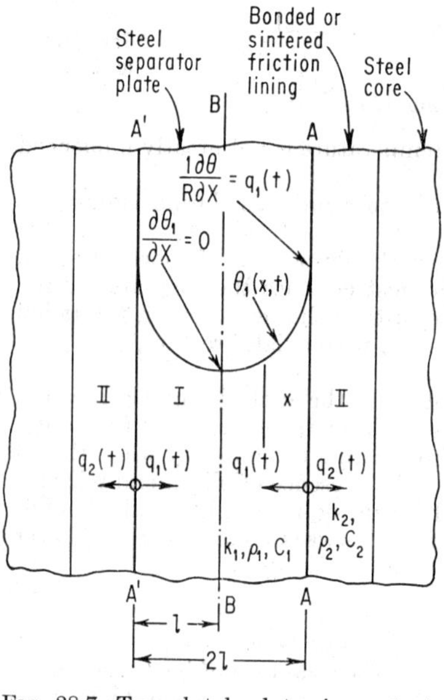

FIG. 28.7. Two clutch plates in contact under the action of the axial clamping force.

Figure 28.7 shows the two clutch plates in contact under the action of the axial clamping force. Surface *A-A*, where rubbing occurs, can be thought of as a plane source of heat producing $q(t)$, Btu/sec/ft². At the rubbing interface the surface temperature of both plates I and II must be the same. The rates of heat flow into I and II will not, in general, be equal but will depend on the thermal properties of the two materials in contact.

While the manner of division of the total flux is yet to be determined, the mathematical form of solution of the heat-conduction equation can be written

$$q_s(t) = m(-at + b) = -\alpha_s t + \beta_s \tag{28.32}$$
$$q_f(t) = (1 - m)(-at + b) = -\alpha_f t + \beta_f \tag{28.33}$$

where m is as yet undetermined and $q_s(t)$ = heat flux entering steel plate, $q_f(t)$ = heat flux entering friction lining. The heat-diffusion equation written for the steel plate is[4,5,6]

$$\partial^2\theta_s/\partial x^2 = C_s R_s(\partial\theta_s/\partial t) \tag{28.34}$$

where $C_s = c_s \rho_s$
$R_s = 1/k_s$
θ_s = temperature of the steel plate over ambient temperature
c_s = specific heat of steel
k_s = thermal conductivity of steel
ρ_s = density of steel

Using Eq. (28.32) together with the boundary conditions,

At $x = 0$ $\qquad -(1/R_s)(\partial\theta_s/\partial x) = q_s(t)$
At $x = l$ $\qquad \partial\theta_s/\partial x = 0$

From the solution of Eq. (28.34), the surface temperature of a clutch plate as a function of time is determined.

$$\theta_s(0,t) = \zeta\left(\frac{t}{l\sqrt{R_s C_s}} + \frac{l^3\sqrt{R_s C_s}}{3} - \frac{2l\sqrt{R_s C_s}}{\pi^2}\sum_{n=1}^{\infty}\frac{1}{n^2}\exp - \frac{n^2\pi^2}{l^2 R_s C_s}t\right) - \nu\left[\frac{t^2}{2l\sqrt{R_s C_s}} + \frac{l\sqrt{R_s C_s}}{3} - \frac{l^3(C_s R_s)^{3/2}}{45} + \frac{2l^3(R_s C_s)^{3/2}}{\pi^4}\sum_{n=1}^{\infty}\frac{1}{n^4}\exp - \frac{n^2\pi^2}{l^2 R_s C_s}t\right] \tag{28.35}$$

where $\zeta = \beta_s\sqrt{R_s/C_s}$
$\nu = \alpha_s\sqrt{R_s/C_s}$

For very large values of $l\sqrt{R_s C_s}$ or very small values of t, Eq. (28.35) is not convenient for computation of the surface temperature. It can be shown that, for very large plate thickness, such that reflection from the opposite side can be neglected,

$$\theta_s(0,t)_{l\to\infty} = 2\zeta\sqrt{t/\pi} - \tfrac{4}{3}(\nu t^{3/2}/\sqrt{\pi}) \tag{28.36}$$

In this case, the maximum temperature occurs at the time $t_m = \frac{1}{2}t_0$ and its value is $\sqrt{2}$ times that at $t = t_0$, i.e., the temperature at the end of engagement period.

The value of m in Eqs. (28.32) and (28.33) is determined by considering that the temperature at the rubbing interface ($x = 0$) must be the same for both the steel plate and the friction lining, i.e.,

$$\theta_s(0,t) = \theta_f(0,t) \tag{28.37}$$

The result, obtained by using any one of the temperature expressions, is

$$m = \frac{\sqrt{R_f/C_f}}{\sqrt{R_s/C_s} + \sqrt{R_f/C_f}} \tag{28.38}$$

It can be seen from Eq. (28.38) that the thermal properties and the densities of both mating materials are important in determining the value of the surface temperature.

(b) Application

Equation (28.38) rewritten in more familiar notation

$$m = \frac{\sqrt{c_f \rho_f k_f}}{\sqrt{c_f \rho_f k_f} + \sqrt{c_s \rho_s k_s}} \tag{28.39}$$

shows that the total heat flux divides equally between the two contacting plates when the products of their thermal properties and densities are equal.

From Eqs. (28.32), (28.33), (28.35), and (28.39) it can be shown that the temperature at the interface is proportional to

$$\frac{1}{\sqrt{c_f \rho_f k_f} + \sqrt{c_s \rho_s k_s}}$$

High values of thermal conductivity, specific heat, and high density will decrease the temperature at the rubbing interface. These properties, however, are not the only criteria to be used in the selection of materials when designing a clutch.

Other factors, such as coefficient of friction, fading, crushing strength, wear characteristics, and conformability, must also be considered when designing a friction clutch. Other criteria are discussed in the section on temperature flashes.

In practice, many clutches, especially those of multiple-disk type, are so designed that the thickness of the plate $2l$ does not exceed 0.10 in. When l(0.05 in.) is substituted in Eq. (28.35) the constant in the exponent is

$$n^2\pi^2/l^2R_sC_s = 71.3n^2 \qquad \text{for steel}$$

Thus, even with $t = 0.1$ and $n = 1$, $e^{-7.13} = 9 \times 10^{-4}$ and for all practical purposes the contribution of the infinite series in n can be neglected. Thus

$$\theta_s(0,t) = -\frac{\nu t^2}{2l\sqrt{R_sC_s}} + \left(\frac{\zeta}{l\sqrt{R_sC_s}} - \frac{\nu l\sqrt{R_sC_s}}{3}\right)t + \left[\frac{\zeta l\sqrt{R_sC_s}}{3} + \frac{\nu l^3(R_sC_s)^{3/2}}{45}\right] \tag{28.40}$$

In order to find the maximum surface temperature of the plates, we differentiate Eq. (28.40) and set $\partial\theta_s/\partial t = 0$. The time of incidence of the maximum surface temperature is thus given by

$$t_m = \zeta/\nu - l^2R_sC_s/3 \qquad \text{sec} \tag{28.41}$$

or, since $\zeta/\nu = t_0$,

$$t_m = t_0 - l^2R_sC_s/3 \qquad \text{sec} \tag{28.42}$$

Substituting the value of t_m from Eq. (28.41) into Eq. (28.40) the maximum surface temperature is found to be (for small l)

$$\theta_s \max = \frac{1}{2}\frac{\zeta^2}{\nu l\sqrt{R_sC_s}} + \frac{7}{90}\nu l^3(R_sC_s)^{3/2} \tag{28.43}$$

A detailed examination of Eq. (28.43) will yield interesting and useful results depending on the nature of the system in which the clutch is used. Thus, if the system consists of two flywheels rotating at different speeds in the absence of external torques, and the purpose of the clutch is to transfer energy from one flywheel to the other (such systems form the basis of various forms of inertia starters), Eq. (28.30) shows that the total amount of energy dissipated in the clutch during the process of coupling is independent of the clutch frictional torque T. On the other hand, the duration of the slip period t_0, according to Eq. (28.29), is inversely proportional to T. The question may be raised, how does the change in t_0, caused by varying T, affect the maximum surface temperature of the plates? It is to be understood that the variation of clutch frictional torque is effected by varying the axial clamping force F only, and that clutch dimensions, number of plates, and their frictional and thermal properties are held constant.

This problem may arise when it is desired to adapt a clutch of certain thermal and mechanical capacity to a system demanding a higher-capacity clutch, and the deficiency is to be compensated for by increased pressure on the frictional surfaces.

It can be shown that the amount of energy absorbed by the steel plate per unit surface area is given by

$$Q_s = \bar{\mu}\delta(Q/A) \tag{28.44}$$

where Q = defined in Eq. (28.30)

$$\bar{\mu} = \frac{\sqrt{R_f/C_f}}{\sqrt{R_f/C_f} + \sqrt{R_s/C_s}}$$

$\delta = 1.285 \times 10^{-3}$ (conversion factor from lb-ft to Btu)

A = total frictional area of clutch, ft^2

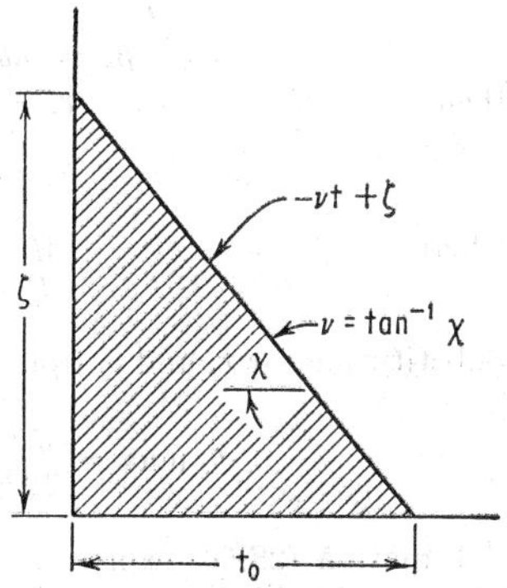

FIG. 28.8. Area under $-\nu t + \zeta$ plot which is proportional to Q_s.

Also, it is not difficult to verify that Q_s will be proportional to the shaded area in Fig. 28.8, and since Q_s is independent of either T or t_0, we may write

$$\zeta t_0 = K_1 = \text{const} \tag{28.45}$$

where $K_1 = 2Q_s\sqrt{R_s/C_s}$

Now

$$\zeta/t_0 = \nu \tag{28.46}$$

Then, from Eqs. (28.45) and (28.46),

$$\zeta = K_1/t_0 \tag{28.47}$$

$$\nu = K_1/t_0^2 \tag{28.48}$$

$$\zeta^2/\nu = K_1 \tag{28.49}$$

but $t_0 = K_2/T$ where $K_2 = I_1I_2(\Omega_1 - \Omega_2)/(I_1 + I_2)$. Substituting for t_0 into Eq. (28.48),

$$\nu = (K_1/K_2)T^2 \tag{28.50}$$

Substitution of ζ^2/ν from Eq. (28.49) and ν from Eq. (28.50) into Eq. (28.43) results in

$$\theta_s \max = \frac{Q_s}{lC_s} + \frac{7}{45}\frac{Q_sT^2l^3R_s^2C_s(I_1 + I_2)^2}{I_1^2I_2^2(\Omega_1 - \Omega_2)^2} \tag{28.51}$$

Although the contribution of the second term in Eq. (28.51) to the value of θ_s max becomes appreciable only at very high values of T, it will be observed that surface temperature is proportional to the square of the clutch frictional torque. Thus, in the case when the clutch is used to start or stop pure inertia loads, too small and *not* too large values of t_0 will tend to damage the clutch.

Another much more frequently used system is one in which external driving and resistive torques, as well as the clutch frictional torques, act on the rotating inertias. In such systems, Eqs. (28.25), (28.26), and (28.27) would be used to calculate the rate of energy dissipation, duration of slip period, and the total amount of energy dissipated per cycle. It will be observed that, in the present case, Q is no longer independent of T and that both $t_0 \to \infty$ and $Q \to \infty$ when $T = (I_2T_E + I_1T_R)/(I_1 + I_2)$.

We again wish to examine the effect of varying clutch frictional torque on maximum surface temperature when the change in T is brought about by varying axial force on the clutch plates only.

From Eqs. (28.25) and (28.27), it follows that

$$\alpha_s = \frac{\bar{\mu}\delta}{A}\,\frac{T^2(I_1 + I_2) - T(I_2T_E + I_1T_R)}{I_1I_2} \qquad \text{Btu/sec}^2\text{ft}^2 \tag{28.52}$$

$$\beta_s = (\bar{\mu}\delta/A)T(\Omega_1 - \Omega_2) \qquad \text{Btu/sec/ft}^2 \tag{28.53}$$

then

$$\zeta_s = (\bar{\mu}\delta/A)T(\Omega_1 - \Omega_2)\sqrt{R_s/C_s} \tag{28.54}$$

$$\nu_s = (\bar{\mu}\delta/A)\frac{T^2 - MT}{P}\sqrt{R_s/C_s} \tag{28.55}$$

where

$$M = (I_1T_E + I_1T_R)/(I_1 + I_2)$$
$$P = I_1I_2/(I_1 + I_2)$$

Substituting for ζ_s and ν_s from Eqs. (28.54) and (28.55) into Eq. (28.43), one obtains

$$\theta_s \max = \frac{\delta\bar{\mu}T(\Omega_1 - \Omega_2)^2P}{2Al(T - M)C_s} + \frac{7\bar{\mu}\,\delta T(T - M)l^3R_s{}^2C_s}{90PA} \tag{28.56}$$

Equation (28.56) expresses analytically what may, perhaps, have been arrived at by some qualitative speculation, namely, that when driving and resistive torques are present in the system shown in Fig. 28.6, maximum surface temperature will reach very high values when the frictional clutch torque approaches the value of the driving torque. The same equation also shows that, when $T \gg (I_2T_E + I_1T_R)/(I_1 + I_2)$ (i.e., when t_0 is very small), the contribution of the second term to the value of θ_s max may become appreciable; so that, in spite of decreased slip period, the surface temperature of clutch plates may begin to increase.

(c) Variation of Maximum Surface Temperature with Plate Thickness

If in the previously discussed system it is desired to lower the value of the maximum surface temperature by properly designing the friction elements, two methods are generally available: (1) increase in the total clutch frictional area A by increasing the number of plates or their size, and (2) increase in the thickness of the steel separator plate.

In applying the first of these methods, Eq. (28.56) shows that the maximum surface temperature is inversely proportional to the total frictional area A. Thus it follows that virtually any desired value of maximum surface temperature may be obtained by assigning proper value to A. The addition of one or more sets of plates, however, will generally be more expensive than the increase in thickness of the steel separator plates in an already designed clutch.

A general criterion for the maximum useful plate thickness when heated on both sides is based upon plotting the nondimensional Fourier number $\mathfrak{F} = t_0/l^2CR$ as a function of plate thickness. It can be shown that any increase beyond the value corresponding approximately to $1/\mathfrak{F} = 1.75$ ceases to be significant.

$$l_{\max} = \sqrt{1.75t_0/C_sR_s} = \sqrt{1.75t_0k_s/c_s\rho_s} \tag{28.57}$$

where $l_{\max} = \frac{1}{2}$ maximum useful plate thickness when heated on both sides.

In the case when the plate is heated on one side only, Eq. (28.57) gives the actual maximum useful plate thickness.

(d) Optimum Radial Dimensions of a Clutch Plate

The optimum dimensions of a clutch plate are dependent on the criterion used to define the meaning of the word "optimum." If the criterion is formulated on the basis that the ratio of frictional torque to the clutch-plate surface temperature should be a maximum one can obtain the relationship between the inside and outside radii as follows:

$$T \propto R_o + R_i \tag{28.58}$$

where T = torque
R_o = outside radius of plate
R_i = inside radius of plate

The surface temperature is inversely proportional to the area of the plate

$$\theta_s \propto [1/(R_o^2 - R_i^2)] \tag{28.59}$$

$$T/\theta_s = (1 - k^2)(1 + k) \tag{28.60}$$

where $$k = R_i/R_o$$

from which (T/θ_s) is a maximum when $k = \frac{1}{3}$ or

$$R_i = \tfrac{1}{3}R_o \tag{28.61}$$

(e) Flash Temperatures

Very high values of transient surface temperature occur at the interface between two sliding bodies, these temperature flashes occurring at points of intimate contact between the two surfaces.

The effective or microscopic contact between the two surfaces is confined to surface asperities, and it can be shown that the relationship between the effective and the nominal or macroscopic area of contact is

$$A'/A = p/p_m \tag{28.62}$$

where p_m = mean yield pressure of the softer material
p = nominal pressure (average pressure)
A' = effective frictional area over which contact actually occurs

If contact occurs at n spots, and if these spots are assumed to be circular in shape and equal in size

$$A' = \pi n r^2 \tag{28.63}$$

where r = radius of the spot.

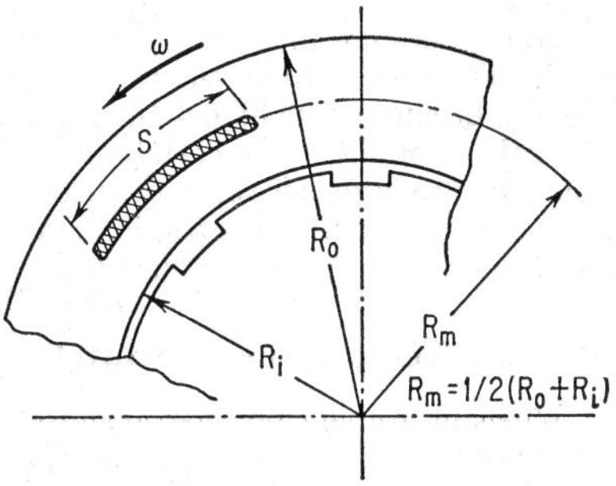

FIG. 28.9. A portion of a clutch plate showing the distance covered by an isolated hot spot.

When relative motion exists between the surfaces, it has been observed that a number of luminous points appear at the rubbing interface and that the position of these spots changes rapidly as points of intimate contact wear away and new asperities come into contact. Referring to Fig. 28.9, which represents a portion of a clutch plate, let S be the distance covered by such an isolated hot spot in the small time interval t_s. Then

$$t_s = S/\omega R_m \tag{28.64}$$

where ω = angular velocity
R_m = mean radius of the plate

Also $$S \propto r \tag{28.65}$$

If a clutch with only one frictional surface is considered, the rate at which heat enters each spot is given by

$$q_r = \frac{\delta T \omega}{\pi n r^2} \tag{28.66}$$

Because t_s is very small, q_r will remain essentially constant. Also, the effect of heat reflection from the other side of the plate can be neglected. This allows the use of the simplest expression for the surface temperature, which is derived for the case of a semi-infinite solid receiving heat at a constant rate.

$$\theta_f(t) \propto \sqrt{R_s/C_s}\, q_r \sqrt{t_s} \tag{28.67}$$

From $T = \mu p A R_m$, $A \propto R_m^2$, and Eqs. (28.62) through (28.67), the following expression is obtained:

$$\theta_f(t) \propto \sqrt{R_s/C_s}\,(\mu p_m{}^{3/4} R_m p^{1/4} \omega^{1/2}/n^{1/4}) \tag{28.68}$$

Equation (28.68) indicates that, in order to decrease the value of flash temperature occurring at the areas of intimate contact, the following points should be considered when selecting friction material and clutch-plate size:

1. Soft friction materials having good surface conformability are preferable.
2. The coefficient of friction should be low. Since, in order to minimize clutch size, high values of μ are usually preferred, a material having sharp fade characteristics at high temperatures should be used.
3. Contact between the mating plates should be as uniform as possible; i.e., the number of contact spots n should be as large as possible.
4. For a clutch of given nominal frictional area A, the mean radius of the plate should be kept small. Since, for the plate of a given outside diameter, the frictional torque of the clutch is proportional to R_m, a compromise decision regarding the value of R_m should be made which takes into account both flash temperatures and the value of the frictional torque desired.

The foregoing analysis to a certain extent explains why substitution of hard sintered-bronze plates for those coated with such materials as cork or wood flour and asbestos mixtures does not always result in solving the high-temperature problems in a poorly designed clutch. Even though, according to Eq. (28.35), the use of sintered bronze should, because of its higher thermal conductivity, result in lower surface temperature than that obtained with cork lining, this advantage is offset by high values of surface-temperature flashes if the contact between the rubbing plates is not uniform. Therefore, when using metallic friction materials, great care should be taken in the preparation and finish of working surfaces.

(f) Bulk Temperature of Clutch Assembly

When in a given power-transmission system the clutch is cycled continuously, it is necessary to have some means of computing the value of the average temperature at which the assembly will operate. This temperature is based upon radiative and convective heat transfer.

Problems encountered in analysis include the following:

1. Although most clutch assemblies are similar in outside appearance, i.e., cylindrical, the details of their design vary widely and may, in many cases, influence the process of heat transfer considerably.

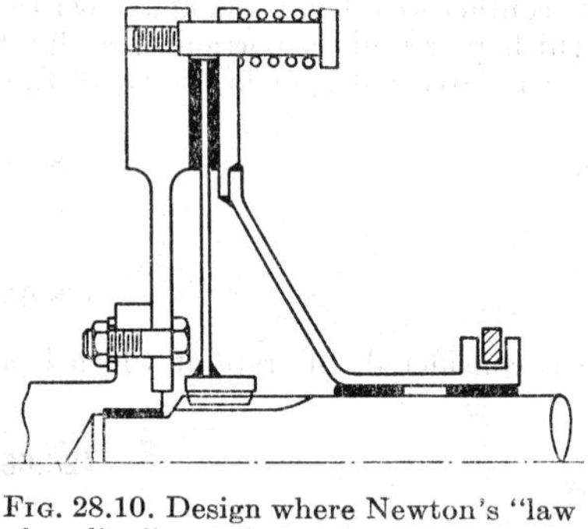

FIG. 28.10. Design where Newton's "law of cooling" may be used.

2. Heat-transfer data, which could be directly applied to conditions under which the clutch is required to perform, are not available (e.g., shape, large diameter/length ratio, oil-mist atmosphere, end effects).
3. When formulas and methods applicable to simple and idealized cases are used to describe complicated systems errors result, mainly because many of the so-called "constants" rarely remain truly constant, and also some of the effects which can be considered negligible in idealized cases may become quite prominent in an actual system.

Two approaches are possible toward the estimation of the bulk temperature of the clutch, depending on its construction. When the design is such as shown in Fig. 28.10, i.e., when the heat generated at the rubbing surfaces is transmitted directly by conduction through a homogeneous material to the outer surfaces and from there transferred to the surroundings by the mechanism of forced convection, Newton's "law of cooling" may be used. This method is strictly correct only when the temperature everywhere within the body, including the surface, is the same and the surface coefficient of heat transfer is a constant independent of temperature.

In practice, the above conditions are approximated if the thermal conductivity of the material in question is high, when temperature does not vary too rapidly, and when the relative size of the body is small.

When a solid body of weight W, specific heat c, having an exposed surface area A cools slowly from an initial temperature θ'', the temperature at any subsequent time t is given by

$$\theta - \theta_a = (\theta'' - \theta_a) \exp(-\alpha t) \tag{28.69}$$

where θ_a = ambient temperature
$\alpha = Ah/Wc$
h = surface coefficient of heat transfer

The temperature rise per cycle is

$$\theta = \theta'' - \theta' = \delta Q/Wc \qquad °\text{F} \tag{28.70}$$

where θ' = temperature at the beginning of the slip period (or cycle)
θ'' = temperature at the end of the slip period

It is assumed that the clutch attains the temperature θ'' instantaneously, which is approximately correct when the time during which it is allowed to cool is large in comparison with the time required for the heat to diffuse throughout the bulk of the plates.

If t_0 is the equal time interval between successive applications of the clutch, the temperatures θ' and θ'' are given by

First cycle:

$$\theta' = \theta_a$$
$$\theta'' = \theta_a + \Delta\theta$$

Second cycle:

$$\theta_2' = \theta_a + \Delta\theta \exp(-\alpha t_c)$$
$$\theta_2'' = \theta_a + \Delta\theta[1 + \exp(-\alpha t_c)]$$

nth cycle:

$$\theta_n' = \theta_a + \Delta\theta[\exp(-\alpha)t_c + \exp(-2\alpha)t_c + \cdots + \exp(-n\alpha t_c)]$$
$$= \theta_a + \Delta\theta \frac{\exp[-(n-1)\alpha t_c] - 1}{\exp(-\alpha t_c) - 1} \exp(-\alpha t_c)$$

$$\theta_n'' = \theta_a + \Delta\theta[1 + \exp(-\alpha t_c) + \exp(-2\alpha t_c) + \cdots + \exp(-n\alpha t_c)]$$
$$= \theta_a + \Delta\theta \frac{\exp(-n\alpha t_c) - 1}{\exp(-\alpha t_c) - 1}$$

The steady-state values of θ_n' and θ_n'' are obtained by making $n \to \infty$. Thus

$$\theta_{ss}' = \theta_a + \Delta\theta \frac{\exp(-\alpha t_c)}{1 - \exp(-\alpha t_c)} \tag{28.71}$$

$$\theta_{ss}'' = \theta_a + \Delta\theta \frac{1}{1 - \exp(-\alpha t_c)} \tag{28.72}$$

Before Eqs. (28.71) and (28.72) can be used, the value of h must be determined. This can be approximately obtained by the use of the charts given in Figs. 28.11 and 28.12, taken from refs. 7 and 8. Strictly speaking, the data contained in Fig. 28.11 pertain to rotating cylinders in which the length/diameter ratio is large, so that end effects are negligible, which is not quite the case when dealing with clutches. Figure 28.12 represents the results of experiments designed to establish the heat transfer in rotating electrical machines. Figure 28.13 gives those properties of air which are needed to evaluate Re and Nu.

The other method of evaluating the average temperature of the clutch assembly may be applied in cases where the preceding exponential development cannot be used. Consider that the clutch shown in Fig. 28.14 (disregard, for this purpose, the presence of cooling oil) is intended to be cycled continuously and at a regular time interval of 60 sec duration. The clutch cylinder rotates at a constant speed of 4,000 rpm and is

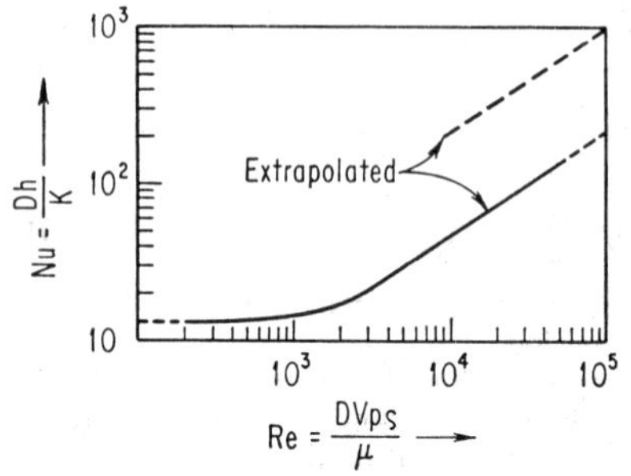

FIG. 28.11. Correlation of Nu (Nusselt number) vs. Re (Reynolds number) for a rotating cylinder without cross flow in air.
D = diameter
h = coefficient of heat transfer
k = thermal conductivity of air
Vp = peripheral velocity
s = specific weight of air
μ = viscosity of air

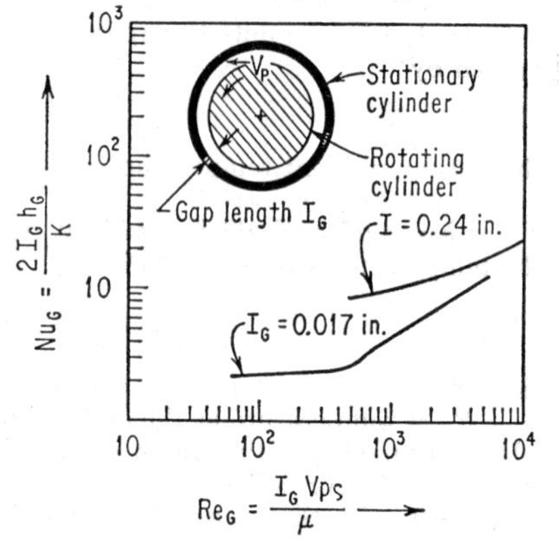

FIG. 28.12. Correlation of Nu_G (gap Nusselt number) vs. Re_G (gap Reynolds number) for a pair of coaxial cylinders. Inner cylinder rotating in air without axial flow (heat transfer from rotor surface to stator surface). Note curves show heat transfer across gap from rotating to stationary cylinder.

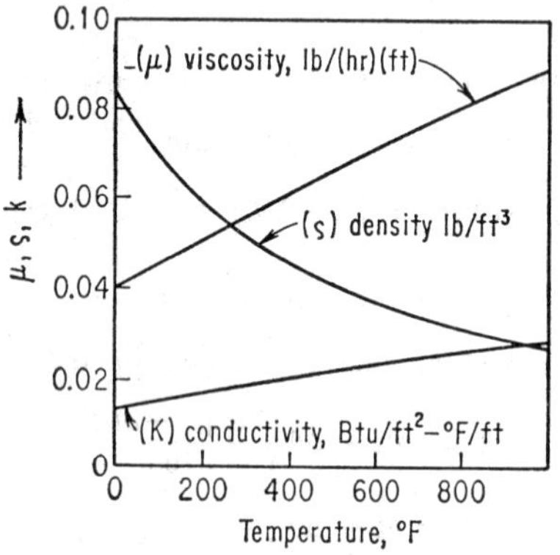

FIG. 28.13. Properties of air.

completely enclosed in a cylindrical housing made of good heat-conducting material. The ambient temperature outside the housing is 100°F. Assume that the average rate of energy dissipation at the rubbing surfaces during the slip period is 0.5 hp/in.² per cycle and that $t_0 = 0.8$ sec. With $A = 117$ in.², the total amount of energy liberated per cycle in the form of heat is 33.1 Btu. Thus in the steady state the average rate of heat rejection from the clutch assembly is $q = 33.1/60 = 0.55$ Btu/sec.

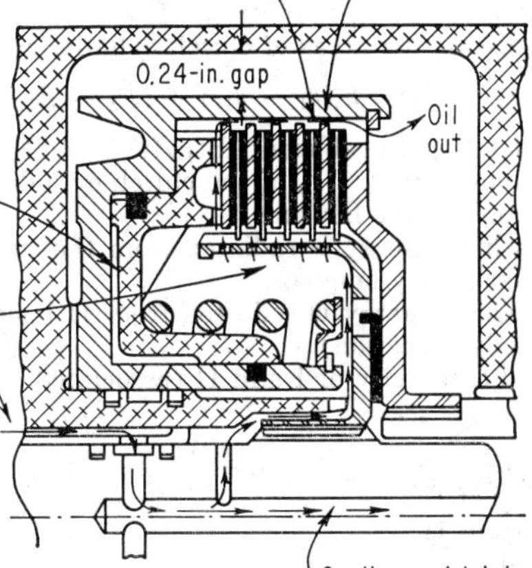

FIG. 28.14. Design where Newton's "law of cooling" cannot be used.

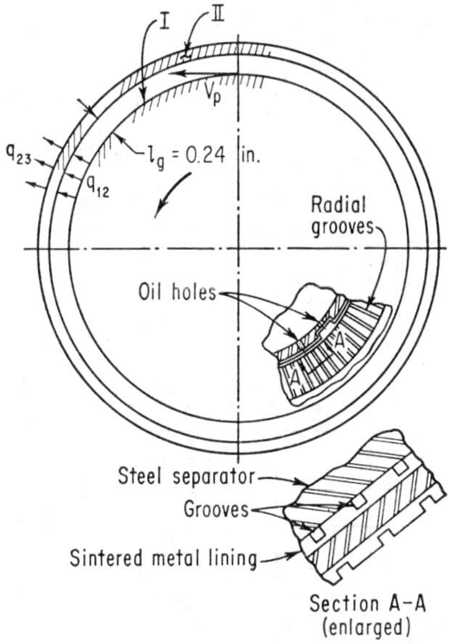

FIG. 28.15. Heat dissipation from outer surface of clutch cylinder for clutch shown in Fig. 28.14.
I = clutch cylinder outer surface
II = housing
III = surrounding air

Because of the high thermal conductivity of the housing material and its small wall thickness, the temperature drop across it will be neglected in the foregoing.

Referring to Fig. 28.15, the heat-dissipating areas of the housing and of the clutch cylinder are not equal,

$$q_{12} = q/A_{\text{I}} = 0.55/0.44 = 1.25 \text{ Btu/sec/ft}^2$$
$$q_{23} = q/A_{\text{II}} = 0.55/0.52 = 1.06 \text{ Btu/sec/ft}^2$$

where q_{12} = rate of heat flow per ft² from I to II
q_{23} = rate of heat flow per ft² from II to III
$A_{\rm I}$ and $A_{\rm II}$ = heat-dissipating areas of clutch cylinder and housing, respectively

Now

$$q_{12} = h_g(\theta_{\rm I} - \theta_{\rm II}) + (E_{\rm I} - E_{\rm II})\epsilon_c \qquad (28.73)$$

where h_g = gap heat-transfer coefficient, Btu/ft²/sec²/°F
$\theta_{\rm I}$ = surface temperature, °F, of the clutch cylinder
$\theta_{\rm II}$ = temperature of the housing
$E_{\rm I}$ = rate at which energy is radiated from the surface of I at a temperature $\theta_{\rm I}$ °F (black-body emission)
$E_{\rm II}$ = rate at which energy is reflected from II back to I at a temperature $\theta_{\rm II}$ (black-body reflection)
ϵ_c = combined emissivity of I and II (associated with geometry of system)

$$\epsilon_c = \frac{\epsilon_{\rm I}\epsilon_{\rm II}}{\epsilon_{\rm I}(A_{\rm I}/A_{\rm II}') + \epsilon_{\rm II} - \epsilon_{\rm I}\epsilon_{\rm II}(A_{\rm I}/A_{\rm II}')}$$

$\epsilon_{\rm I}$ = emissivity of I
$\epsilon_{\rm II}$ = emissivity of II
$A_{\rm I}$ = outside area of I
$A_{\rm II}'$ = inside area of II

The rate of heat flow from II to III is given by

$$q_{23} = h_c(\theta_{\rm II} - \theta_{\rm III}) + (E_{\rm II} - E_{\rm III})\epsilon_{\rm II} \qquad (28.74)$$

where h_c = coefficient of heat transfer due to natural (or forced) convection, Btu/ft²/sec/°F
$\theta_{\rm III}$ = ambient temperature, °F
$E_{\rm III}$ = rate of radiant-heat reflection from the surroundings

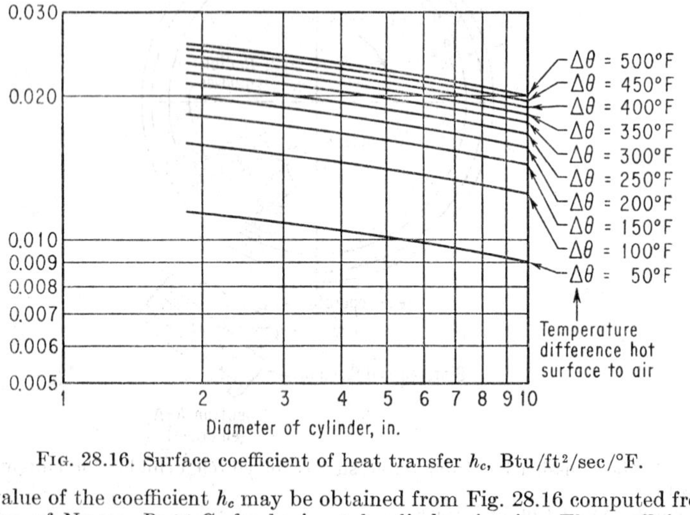

FIG. 28.16. Surface coefficient of heat transfer h_c, Btu/ft²/sec/°F.

The value of the coefficient h_c may be obtained from Fig. 28.16 computed from the correlation of Nu vs. $Pr \times Gr$ for horizontal cylinders in air. The coefficient h_g is obtained from Fig. 28.12. The values of black-body radiant-heat transfer are plotted in Fig. 28.17 in a manner designed to facilitate computation. Values of $\epsilon_{\rm I}$ and $\epsilon_{\rm II}$ may be obtained from refs. 4 and 5.

All five quantities, h_g, h_c, $E_{\rm I}$, $E_{\rm II}$, and $E_{\rm III}$, are, for any physical configuration, functions of temperatures $\theta_{\rm I}$, $\theta_{\rm II}$, and $\theta_{\rm III}$. The simplest method of solution is outlined below:

1. With $\theta_{\rm III}$ known, assume a series of several values of $\theta_{\rm II}$ and calculate corresponding values of q_{23}. Plot these values on a temperature scale as a function of $\theta_{\rm II}$.

The resulting graph is shown in Fig. 28.18, curve *A*. The intersection of this curve (point *X*) with the horizontal line q_{23} = 1.06 Btu/ft²/sec determines the value of θ_{II}.

2. The same procedure is applied in order to determine the surface temperature of the clutch cylinder θ_I. The point of intersection of curve *B* with the horizontal line q_{12} = 1.25 Btu/ft²/sec determines the surface temperature of the rotating clutch cylinder, which in this case is 537°F.

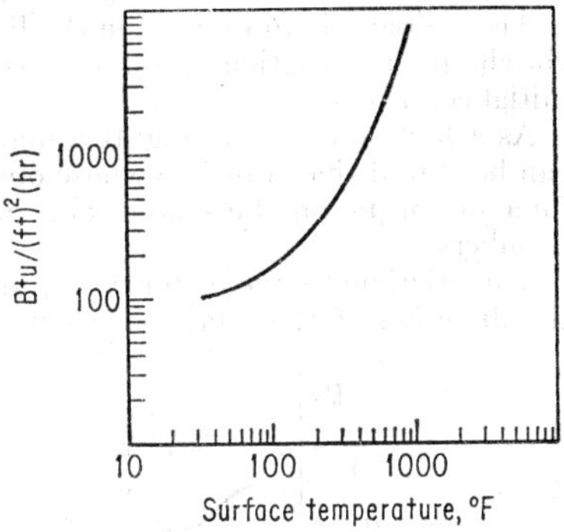

FIG. 28.17. Power radiated (Btu/ft²/hr) by a black-body surface.

With this value of the surface temperature of the clutch cylinder, the temperature existing at the rubbing surfaces of the plates during the slip period would be several hundred degrees higher, and thus steps must be taken to provide added cooling of the clutch assembly. In the case of lubricated clutches, such as shown in Fig. 28.14, this is most effectively accomplished by providing sufficient amounts of cooling oil with passages arranged so that the flow is directed at the surfaces where heat is generated. The flow of oil should be continuous, and the clutch so designed that the oil is in direct contact with the plates prior to, during, and after the slip period. In a heavy-duty clutch, sintered-metal friction plates would probably be used and, in this case, radial grooves, such as shown in Fig. 28.15, can be most easily and inexpensively provided.

Heat transfer from the clutch assembly by conduction to other system components (shafts, gears, etc.) has been neglected. This would probably amount to some 10 to 15 per cent of the average rate of heat generation; so that the calculated value of θ_I = 537°F may be too high.

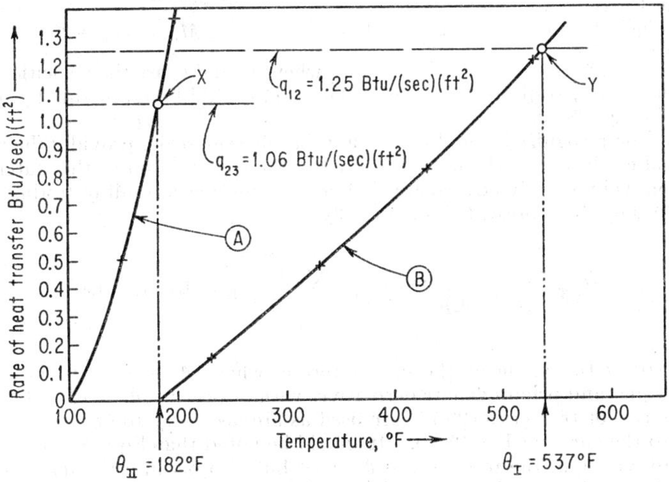

FIG. 28.18. Plot of values of q from assumed values of θ in order to find θ_I and θ_{II}.

28.5. DYNAMICS OF SYSTEMS EMPLOYING FRICTION CLUTCHES[9,10,11]

A rigorous dynamic analysis of a mechanical power-transmission system containing a friction clutch presents a very difficult problem in derivation and solution of the differential equations describing the motion of the system. This is because these systems are usually characterized by several degrees of freedom and are generally nonlinear. Because simple and general analytical techniques for dealing with nonlinear differential equations of higher order than the second are not available, there will generally be some sacrifice of mathematical rigor and elegance in the analysis. Reference 58 provides more information.

The problem is to determine the behavior of a system following application of the clutch, given a friction clutch of certain specified characteristics, and given a set of initial conditions.

As a first step in arriving at a mathematical representation of a friction clutch it can be stated that a device whose operation depends upon friction exerts a frictional force or torque on the system only when relative motion exists between contacting members.

This frictional force or torque is, in general, a function of both the magnitude and the direction of the relative velocity, and it may also be made time-dependent by introducing external control.

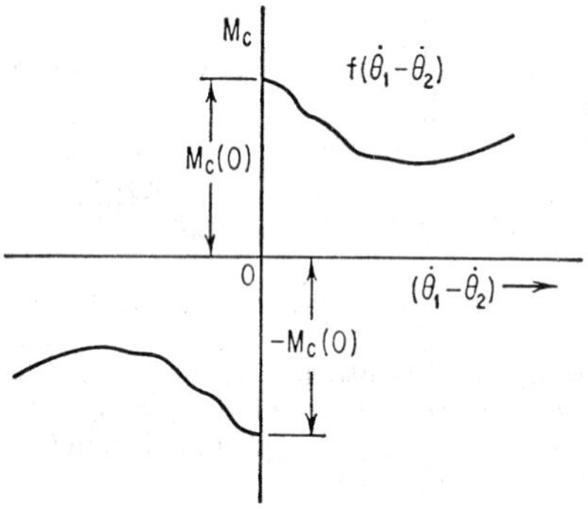

FIG. 28.19. Curve of $M_c = f(\dot\theta_1 - \dot\theta_2)$ for a clutch in which the torque is not externally controlled.

Thus a friction clutch may be characterized

$$M_c = f(\Delta\dot\theta, t) \tag{28.75}$$

where M_c = frictional torque
$\Delta\dot\theta$ = relative angular velocity
t = time

The variation in the friction coefficient with variables such as temperature, humidity, duty cycle, and loading history is slow in comparison with the frequencies encountered in vibrating mechanical systems.

For a clutch in which the torque is not externally controlled,

$$M_c = f(\dot\theta_1 - \dot\theta_2) \tag{28.76}$$

where $\dot\theta_1$ and $\dot\theta_2$ are the velocities of input and output plates, respectively. This relationship could be represented graphically as shown in Fig. 28.19.

It should be recognized that Fig. 28.19, to a different scale, provides the variation of frictional coefficient of the materials used in the construction of the clutch. This information will normally be available in graphical form as a result of testing. Equation (28.76) may be represented analytically:

$$M_c = \frac{\dot\theta_1 - \dot\theta_2}{|(\dot\theta_1 - \dot\theta_2)|} M_c(0) + \sum_{n=1}^{\infty} \frac{1}{n!} (\dot\theta_1 - \dot\theta_2)^n M_c^{(n)}(0) \tag{28.77}$$

where $M_c(0)$ is the value of the clutch torque when $(\dot\theta_1 - \dot\theta_2) = 0$ (i.e., "static" friction torque) and primes denote derivatives with respect to $(\dot\theta_1 - \dot\theta_2)$. In practice, as many terms of the series (28.77) are used as are necessary to obtain an adequate "fitting" to the curve in Fig. 28.19. It should be noted that Eq (28.77) is nonlinear not only by virtue of the terms $(\dot\theta_1 - \dot\theta_2)^{2,3,\ldots,n}$ but also because the term $M_c(0)$ is a function of the relative velocity $(\dot\theta_1 - \dot\theta_2)$.

Since the torque-speed curve of a clutch may be characterized over the whole or part of the range by negative slope, it is of interest to see what effect this may have on the behavior of the dynamical system of which the clutch is a part.

Consider a simple mechanical oscillator consisting of mass m, spring of constant k, and a damper of constant α. The equation of motion is

$$m\ddot{x} + \alpha\dot{x} + kx = 0 \tag{28.78}$$

Multiply Eq. (28.78) by x and integrating from 0 to τ

$$\left[\frac{m(\dot{x})^2}{2}\right]_0^\tau + \left[\frac{kx^2}{2}\right]_0^\tau = -\alpha \int_0^\tau (\dot{x})^2\,dt \tag{28.79}$$

The sum of the left-hand side of Eq. (28.79) represents the variation of the total energy, kinetic plus potential, during this time. It can be seen that this variation may be either positive or negative depending upon the sign of α. If α is positive, the variation in the total energy is negative, i.e., the system loses energy as $\tau \to \infty$. If α is negative the system gains energy as $\tau \to \infty$. In a system without a source of energy this is not physically realizable. α represents the slope of the friction-speed characteristic of the damper.

A power-transmission system to be considered here consists of three essential elements: (1) a source of energy (in the form of some type of prime mover on the input to the clutch), (2) a clutch controlling the flow of power, and (3) the load on the output side of the clutch.

Since the presence of inertia and compliance is inevitable in mechanical systems, the entire system may become oscillatory following the application of the clutch. If the clutch characteristic has anywhere a portion with negative slope, these oscillations will tend to build up as suggested by Eq. (28.78) until limited by some system nonlinearity. In this case, the nonlinearity which eventually limits the growth of the oscillation and causes the amplitude to decrease to zero in a properly designed system is the term

$$\frac{\dot{\theta}_1 - \dot{\theta}_2}{|(\dot{\theta}_1 - \dot{\theta}_2)|} M_c(0)$$

in Eq. (28.77).

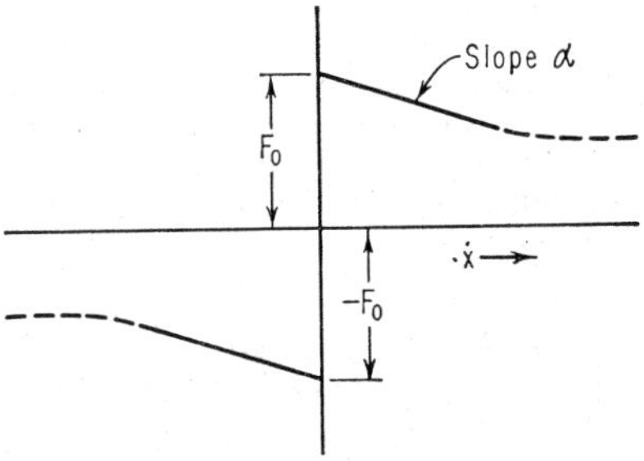

FIG. 28.20. Characteristics of damper of mechanical oscillator.

Consider now a simple mechanical oscillator, letting the damper have a characteristic shown in Fig. 28.20.

This characteristic has been constructed by retaining only the first and the second terms of Eq. (28.77), i.e.,

$$F = (\dot{x}/|\dot{x}|)F_0 - \alpha\dot{x} \tag{28.80}$$

Equation (28.78) is now written

$$\ddot{x} + (1/m)[(\dot{x}/|\dot{x}|)F_0 - \alpha\dot{x}] + \omega^2 x = 0 \tag{28.81}$$

where $\omega = \sqrt{k/m}$ (natural frequency of the oscillator).

If oscillations are admitted they will be of the form $x = A(t)\sin(\omega t + \Phi)$. If the rate of variation of $A(t)$ is small in comparison with ω, we find

$$d[A(t)]/dt = -(1/2\pi\omega m)\left[\int_0^{\pi/2} (F_0 - \alpha A\omega\cos\psi)\cos\psi\,d\psi - \int_{\pi/2}^{3\pi/2} (-F_0 - \alpha A\omega\cos\psi)\cos\psi\,d\psi - \int_{3\pi/2}^{2\pi} (F_0 - \alpha A\omega\cos\psi)\cos\psi\,d\psi\right] \tag{28.82}$$

Integration yields

$$d[A(t)]/dt = \alpha A/2m - 2F_0/\pi\omega m \tag{28.83}$$

Equation (28.83), a linear differential equation of the first order, is solved to obtain

$$A(t) = F_0/\pi\alpha\omega + (A_0 - F_0/\alpha\pi\omega)\exp(\alpha/2m) \tag{28.84}$$

where A_0 = initial amplitude [value of $A(t)$ at $t = 0$].

The damper characteristic given by Eq. (28.80) is a mathematical fiction used to obtain Eq. (28.84). A more realistic case would be represented by the dashed extension of the curves in Fig. 28.20. Thus Eq. (28.84) should be used with the restriction that $|A_0| < |4F_0/\alpha\pi\omega|$, which implies that the case of $|\alpha\dot{x}| > |F_0|$ is inadmissible.

Examining Eq. (28.84) in the light of restrictions just imposed, it is seen that the amplitude of the oscillation decreases exponentially despite the fact that the damper is characterized by negative slope.

When $F_0 = 0$, the amplitude increases exponentially, e.g.,

$$A(t)\Big|_{F_0=0} = A_0 \exp\left[(\alpha/2m)t\right] \tag{28.85}$$

which shows that it is the nonlinearity of the frictional characteristic of the damper which has the effect of positive damping.

(a) Transient Analysis of Mechanical Power Transmission Containing a Friction Clutch

A mechanical power-transmission system to be discussed in this section consists of a prime mover, a friction clutch, and a load. It will also be characterized by the presence of a multiplicity of energy-storage elements in the form of inertias, compliances, and dissipative elements. Included will be various nonlinear and linear coupling elements.

Backlash and hysteresis in various coupling elements will not be considered, but otherwise no restriction is made as to the linearity of the components present in the system, with the exception that inertia is assumed to remain invariant.

The following method of analysis is general enough to permit extension to systems employing any number of energy-storage elements and any number of dissipative elements.

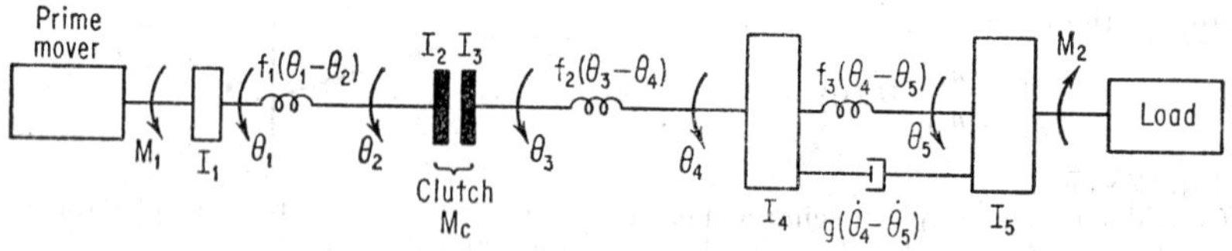

FIG. 28.21. Mechanical power-transmission system with a friction clutch.

Consider a system shown in Fig. 28.21.

I_1 = moment of inertia of prime mover
I_2 = moment of inertia of input plate of clutch
I_3 = moment of inertia of output plate of clutch
I_4, I_5 = moment of inertia of load side of clutch
M_1 = driving torque
M_c = clutch frictional torque
M_2 = load torque
$f_1(\theta_1 - \theta_2), f_2(\theta_3 - \theta_4), f_3(\theta_4 - \theta_5)$ = equivalent characteristics of various shafts; torque as a function of the angle of twist
C_1 = coefficient of viscous resistance corresponding to $\dot{\theta}_1$
C_2 = coefficient of viscous resistance corresponding to $\dot{\theta}_2$
C_3 = coefficient of viscous resistance corresponding to $\dot{\theta}_3$
C_4 = coefficient of viscous resistance corresponding to $\dot{\theta}_4$
C_5 = coefficient of viscous resistance corresponding to $\dot{\theta}_5$
$g(\dot{\theta}_4 - \dot{\theta}_5)$ = characteristic of a damper placed in the system as shown in Fig. 28.21; torque as a function of relative velocity $(\dot{\theta}_4 - \dot{\theta}_5)$

Having described the lumped parameters of the transmission system, consider next the characteristics of the prime mover and the load. In the general case, torques M_1 and M_2 can be considered functions of time, angular velocity, and angular displacement.

In practice, the torque pulsations of multicylinder engines are small in comparison with the average value of torque and can be neglected in the present analysis.

If the torque M_1 is considered a function of speed only, it can be represented analytically by an infinite series.

$$M_1(\dot{\theta}_1) = M_1(0) + \sum_{n=1}^{\infty} (1/n!)(\dot{\theta}_1)^n M_1^{(n)}(0) \tag{28.86}$$

where $M_1(0)$ is the value of $M_1(\dot{\theta}_1)$ at $\dot{\theta}_1 = 0$.

Generally, the load torque M_2 can be considered a function of both speed $\dot{\theta}_5$ and the angular position of the shaft θ_5 and can be represented by the sum of two infinite series of the same form

$$M_2(\dot{\theta}_5,\theta_5) = M_2(0) + \sum_{n=1}^{\infty} (1/n!)(\dot{\theta}_5)^{(n)} M_2^{(n)}(0) + \bar{M}_2(0) + \sum_{r=1}^{\infty} (1/r!)(\theta_5)^r \bar{M}_2^{(r)}(0) \tag{28.87}$$

In certain situations it may be more convenient to replace the second term on the right-hand side of Eq. (28.87) with a Fourier series, e.g., if $M_2(\dot{\theta}_5,\theta_5)$ is periodic in θ_5.

In Eq. (28.87) $M_2(0)$ is the value of the load torque at $\dot{\theta}_5 = 0$ and $\bar{M}_2(0)$ is the value of the load torque at $\theta_5 = 0$.

For a mechanical system with k degrees of freedom the form of Lagrange's equations will be suited for deriving the equations of motion if the system of Fig. 28.21 is written

$$(d/dt)(\partial T/\partial \dot{q}_k) - \partial T/\partial q_k + \partial V/\partial q_k + \partial F/\partial \dot{q}_k = Q_k \tag{28.88}$$

where T = total kinetic energy of the system
V = total potential energy of the system
F = dissipation function
Q_k = generalized force
q_k = generalized position coordinate
$\dot{q}_k$ = generalized velocity coordinate
k = 1, 2, 3, . . . number of independent coordinates

The computation of T is normally quite straightforward and needs no special explanation. In the computation of V should be included all elastic forces whether internal or external that are functions of displacement, and all constant forces. All external and internal forces and moments that are functions of velocities must be included in the computation of F. Q_k includes all external forces and moments which are functions of time

The coordinates used below are:

Velocities	Positions	k
$\dot{q}_1, \dot{\theta}_1$	q_1, θ_1	1
$\dot{q}_2, \dot{\theta}_2$	q_2, θ_2	2
$\dot{q}_3, \dot{\theta}_3$	q_3, θ_3	3
$\dot{q}_4, \dot{\theta}_4$	q_4, θ_4	4
$\dot{q}_5, \dot{\theta}_5$	q_5, θ_5	5

In the case considered presently $Q_k = 0$, since it was assumed that no time-dependent torques act on the system. The potential energy stored in a single elastic element whose torque-deflection characteristic is given by $f(\theta_1 - \theta_2)$ can be expressed as

$$V = \int_0^{(\theta_1-\theta_2)} f(\theta_1 - \theta_2)d(\theta_1 - \theta_2) \tag{28.89}$$

Similarly the dissipation function or the energy dissipated in a damper whose torque-velocity characteristic is $g(\dot\theta_1 - \dot\theta_2)$ is given by

$$F = \int_0^{(\dot\theta_1-\dot\theta_2)} g(\dot\theta_1 - \dot\theta_2)d(\dot\theta_1 - \dot\theta_2) \tag{28.90}$$

T, V, and F may now be established:

$$T = \tfrac{1}{2}[I_1(\dot\theta_1)^2 + I_2(\dot\theta_2)^2 + I_3(\dot\theta_3)^2 + I_4(\dot\theta_4)^2 + I_5(\dot\theta_5)^2] \tag{28.91}$$

$$V = \int_0^{(\theta_1-\theta_2)} f_1(\theta_1 - \theta_2)d(\theta_1 - \theta_2) + \int_0^{(\theta_3-\theta_4)} f_2(\theta_3 - \theta_4)d(\theta_3 - \theta_4)$$
$$+ \int_0^{(\theta_4-\theta_5)} f_3(\theta_4 - \theta_5)d(\theta_4 - \theta_5) + \int_0^{\theta_5} \left[M_2(0) + \sum_{r=1}^{\infty} (1/r!)(\theta_5)^r \bar{M}_2^{(r)}(0) \right] d\theta_5 \tag{28.92}$$

Note that the part of the load torque M_2 which is a function of position θ_5 has been included in the computation of V.

In cases dealing with linear springs the first three terms of Eq. (28.92) become

$$\tfrac{1}{2}k_1(\theta_1 - \theta_2)^2 + \tfrac{1}{2}k_2(\theta_3 - \theta_4)^2 + \tfrac{1}{2}k_3(\theta_4 - \theta_5)^2$$

where k_1, k_2, and k_3 are spring constants.

The dissipation function for this system is given by

$$F = \tfrac{1}{2}[C_1(\dot\theta_1)^2 + C_2(\dot\theta_2)^2 + C_3(\dot\theta_3)^2 + C_4(\dot\theta_4)^2 + C_5(\dot\theta_5)]$$
$$+ \int_0^{(\dot\theta_2-\dot\theta_3)} M_c(\dot\theta_2 - \dot\theta_3)d(\dot\theta_2 - \dot\theta_3) + \int_0^{(\dot\theta_4-\dot\theta_5)} g(\dot\theta_4 - \dot\theta_5)d(\dot\theta_4 - \dot\theta_5)$$
$$- \int_0^{\dot\theta_1} M_1(\dot\theta_1)d\dot\theta_1 + \int_0^{\dot\theta_5} M_2(\dot\theta_5,\theta_5)d\dot\theta_5 \tag{28.93}$$

In computing F the driving torque $M_1(\dot\theta_1)$ supplies energy to the system. Therefore, the associated dissipation function is negative.

It should be noted that dissipation due to load torque $M_2(\dot\theta_5,\theta_5)$ is obtained by integration with respect to $\dot\theta_5$ only.

In order to obtain the dynamical equations of motion for the system, Eqs. (28.91), (28.92), and (28.93) are substituted into Eq. (28.75) and indicated partial differentiation performed in order for $k = 1$, 2, etc.

The result is

$$I_1\ddot\theta_1 + (\partial/\partial\theta_1) \int_0^{(\theta_1-\theta_2)} f_1(\theta_1 - \theta_2)d(\theta_1 - \theta_2) + C_1\dot\theta_1 - M_1(\dot\theta_1) = 0$$

$$I_2\ddot\theta_2 + (\partial/\partial\theta_2) \int_0^{(\theta_1-\theta_2)} f_1(\theta_1 - \theta_2)d(\theta_1 - \theta_2)$$
$$+ (\partial/\partial\dot\theta_2) \int_0^{(\dot\theta_2-\dot\theta_3)} M_c(\dot\theta_2 - \dot\theta_3)d(\dot\theta_2 - \dot\theta_3) + C_2\dot\theta_2 = 0$$

$$I_3\ddot\theta_3 + (\partial/\partial\theta_3) \int_0^{(\theta_3-\theta_4)} f_2(\theta_3 - \theta_4)d(\theta_3 - \theta_4)$$
$$+ (\partial/\partial\dot\theta_3) \int_0^{(\dot\theta_2-\dot\theta_3)} M_c(\dot\theta_2 - \dot\theta_3)d(\dot\theta_2 - \dot\theta_3) + C_3\dot\theta_3 = 0$$

$$I_4\ddot{\theta}_4 + (\partial/2\theta_4)\int_0^{(\theta_3-\theta_4)} f_2(\theta_3 - \theta_4)d(\theta_3 - \theta_4) + (\partial/\partial\theta_4)\int_0^{(\theta_4-\theta_5)} f_3(\theta_4 - \theta_5)d(\theta_4 - \theta_5) + (\partial/\partial\dot{\theta}_4)\int_0^{(\dot{\theta}_4-\dot{\theta}_5)} g(\dot{\theta}_4 - \dot{\theta}_5)d(\dot{\theta}_4 - \dot{\theta}_5) + C_4\dot{\theta}_4 = 0$$

$$I_5\ddot{\theta}_5 + (\partial/\partial\theta_5)\int_0^{(\theta_4-\theta_5)} f_3(\theta_4 - \theta_5)d(\theta_4 - \theta_5) + (\partial/\partial\dot{\theta}_5)\int_0^{(\dot{\theta}_4-\dot{\theta}_5)} g(\dot{\theta}_4 - \dot{\theta}_5)d(\dot{\theta}_4 - \dot{\theta}_5) + C_5\dot{\theta}_5 + M_2(\dot{\theta}_5,\theta_5) = 0 \quad (28.94)$$

Equations (28.94) apply only during the transient, i.e., only when $(\dot{\theta}_2 - \dot{\theta}_3) \neq 0$. When $(\dot{\theta}_2 - \dot{\theta}_3) = 0$ and the clutch is engaged permanently a different set of equations must be derived to represent the system.

In deriving the equations of motion, the coefficient of the highest derivative, no matter how small, must not be neglected or the resulting equations will not properly represent the system.

Equations (28.94) represent a starting point in the analysis of power transmissions utilizing friction clutches when the various functions describing system parameters are known. If it is desired to obtain explicit solutions for the various speeds, shaft deflections, and torques existing in the system and their variation in time, because of the complexity of the mathematics recourse must be made to analog or digital computers.

In order to gain some approximate information regarding the behavior of the system, Eqs. (28.94) can be "linearized" and some useful techniques normally applicable to sets of linear differential equations can be applied with the understanding that only a limited amount of information can be obtained in this way.

If the system does not contain any highly nonlinear springs (or couplings), or if these can be represented by means of piecewise linear functions, the functions $f_1(\theta_1 - \theta_2)$, etc., can be considered linear of the form $k_1(\theta_1 - \theta_2)$. The same pertains to various dissipation functions, which can also be made representable by linear functions or combinations of linear functions.

It is often useful to determine, even approximately, the natural frequencies of the system. A method of calculating these is given below and is based upon the assumption that Eqs. (28.94) can be linearized.

Equations (28.94) are rewritten, omitting all dissipation terms and all external torques acting on the system and considering all springs to be linear.

$$\begin{aligned} I_1\ddot{\theta}_1 + k_1(\theta_1 - \theta_2) &= 0 \\ I_2\ddot{\theta}_2 - k_1(\theta_1 - \theta_2) &= 0 \\ I_3\ddot{\theta}_3 + k_2(\theta_3 - \theta_4) &= 0 \\ I_4\ddot{\theta}_4 - k_2(\theta_3 - \theta_4) + k_3(\theta_4 - \theta_5) &= 0 \\ I_5\ddot{\theta}_5 - k_3(\theta_4 - \theta_5) &= 0 \end{aligned} \quad (28.95)$$

When the expression for the clutch torque is omitted, as it is in Eqs. (28.95), there is no coupling between the input and output sides of the clutch and the transmission system consists of two dynamically independent subsystems. Since the presence of the dissipation term which acts as a coupling term between the two subsystems will not greatly influence their natural frequencies, they remain essentially independent during the transient. The situation changes after the engagement of the clutch is completed since then $\dot{\theta}_2 = \dot{\theta}_3$ and I_2 and I_3 form one flywheel.

In order to obtain the natural frequencies of the system, solutions of the form $\theta_k = A_k \sin(\omega t + \phi_k)$ are sought, where A_k is the amplitude and ϕ_k is the phase angle of the k'th harmonic, with $\omega = 2\pi f$ the natural angular frequency of the system. Substituting θ_k for θ_1 and θ_2 in Eqs. (28.95), a set of equations is obtained

$$\begin{aligned} (k_1 - \lambda I_1)A_1 - k_1A_2 &= 0 \\ -k_1A_1 + (k_1 - \lambda I_2)A_2 &= 0 \end{aligned} \quad (28.96)$$

where $\lambda = \omega^2$, which results in a quadratic equation in λ which, when solved, yields the roots

$$\lambda_1 = \omega_1^2 = 0$$

$$\lambda_2 = \omega_2^2 = k_1 \frac{I_1 + I_2}{I_1 I_2}$$

The natural frequency of the input side of the clutch is given by

$$f_2 = \frac{1}{2\pi} \sqrt{k_1 \frac{I_1 + I_2}{I_1 I_2}} \tag{28.97}$$

Similarly,

$$f_3 = \frac{1}{2\pi} \left(\frac{a + \sqrt{a^2 - 4b}}{2} \right)^{1/2} \tag{28.98}$$

$$f_4 = \frac{1}{2\pi} \left(\frac{a - \sqrt{a^2 - 4b}}{2} \right)^{1/2} \tag{28.99}$$

where

$$a = \frac{k_2(I_3 I_5 + I_4 I_5) + k_3(I_3 I_5 + I_3 I_4)}{I_3 I_4 I_5}$$

$$b = \frac{k_2 k_3 (I_3 + I_4 + I_5)}{I_3 I_4 I_5}$$

Equations (28.94) can be useful in another way to gain more information about the dynamic behavior of the system during the transient following the application of the clutch.

Earlier, a simple mechanical oscillator was analyzed from the standpoint of stability, and in this connection, a question may well be asked how will a more complicated system behave under similar conditions.

For the purpose of the present discussion it is more convenient to take as a model a simpler system shown in Fig. 28.22.

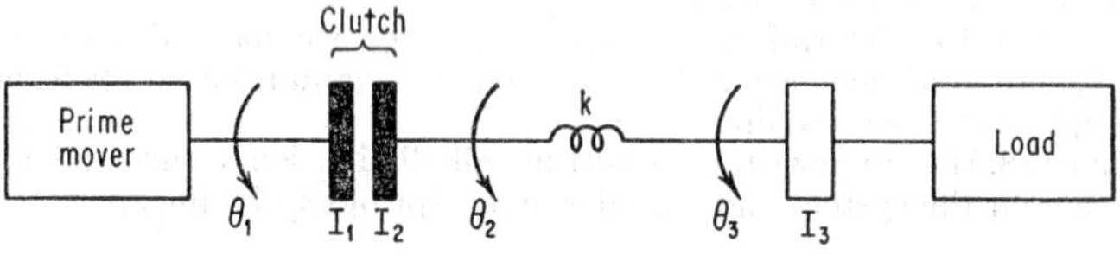

FIG. 28.22. Simplified power-transmission system.

The stiffness of the shaft connecting the prime mover to the input member of the clutch is made very large in comparison with the shaft connecting the clutch and the load. Such a system approximately represents a power transmission consisting of an engine, a clutch, flexible coupling of stiffness k, a gearbox, and a load. It will be assumed that the prime mover, the clutch, and the load have torque-speed characteristics as shown in Fig. 28.23.

Equations (28.77), (28.86), and (28.87) yield, respectively,

Clutch torque $$M_c = M_c(0) \frac{(\dot\theta_1 - \dot\theta_2)}{|(\dot\theta_1 - \dot\theta_2)|} + \alpha(\dot\theta_1 - \dot\theta_2) \tag{28.100}$$

Driving torque $$M_1 = M_1(0) + a\dot\theta_1 \tag{28.101}$$

Load torque $$M_2 = M_2(0) + b\dot\theta_3 \tag{28.102}$$

where there is no restriction placed on the algebraic sign of α, a, and b.

Furthermore, if the discussion is limited to apply only for values of $0 < (\dot\theta_1 - \dot\theta_2) < \Omega$ where Ω is the value of the relative velocity of clutch plates corresponding to some

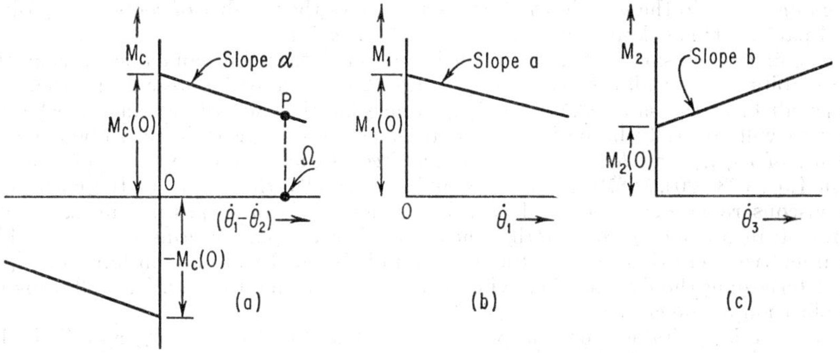

FIG. 28.23. Torque-speed characteristics for clutch, prime mover, and load shown in Fig. 28.22. (*a*) Clutch characteristics. (*b*) Prime-mover characteristics. (*c*) Load characteristics.

point P on the torque-speed characteristic of the clutch in Fig. 28.23*a*, a set of equations of motion for the system can be obtained which will be linear in the interval specified. With these restrictions, the expression for the clutch torque becomes

$$M_c = M_c(0) + \alpha(\dot\theta_1 - \dot\theta_2) \tag{28.103}$$

and the equations of motion are written

$$\begin{aligned} I_1\ddot\theta_1 + c_1\dot\theta_1 - [M_1(0) + a\dot\theta_1] + [M_c(0) + \alpha(\dot\theta_1 - \dot\theta_2)] &= 0 \\ I_2\ddot\theta_2 + c_2\dot\theta_2 + k(\theta_2 - \theta_3) - [M_c(0) + \alpha(\dot\theta_1 - \dot\theta_2)] &= 0 \\ I_3\ddot\theta_3 + c_3\dot\theta_3 - k(\theta_2 - \theta_3) + [M_2(0) + b\dot\theta_3] &= 0 \end{aligned} \tag{28.104}$$

Expansion of the determinant resulting from Eq. (28.104) leads to

$$\begin{aligned} p^2\{I_1I_2I_3p^4 + [I_1(I_2R_3 + I_3R_2) + R_1I_2I_3]p^3 + [I_1(I_2k + R_2R_3 + I_3k) \\ + R_1(I_2R_3 + I_3R_2) - \alpha^2I_3]p^2 + [I_1k(R_2 + R_3) + (I_2k + R_2R_3 + I_3k) \\ + \alpha R_3]p + k[R_1(R_2 + R_3) + \alpha]\} = 0 \end{aligned} \tag{28.105}$$

where $p = d/dt$

$$R_1 = c_1 - a + \alpha \tag{28.106}$$
$$R_2 = c_2 + \alpha \tag{28.107}$$
$$R_3 = c_3 + b \tag{28.108}$$

If the system represented by Eqs. (28.104) is to be stable, the roots of Eq. (28.105) must have all their real parts negative. The presence of two equal roots $p^2 = 0$ merely signifies that the system is capable of rotation in space without executing any oscillations, and is of no consequence in this discussion.

In general, in order to discover whether or not an algebraic equation of the form

$$a_0p^n + a_1p^{n-1} + a_2p^{n-2} + \cdots + a_n = 0 \tag{28.109}$$

possesses roots having positive real parts, an array of coefficients is formed. The coefficients are first written in two rows:

I	a_0	a_2	$a_4 \cdots$	etc.
II	a_1	a_3	$a_5 \cdots$	etc.

By cross multiplication of the first column with each succeeding column in turn a further $n - 1$ rows are formed:

III	$b_1 = (a_1a_2 - a_0a_3)$	$b_2 = (a_1a_4 - a_0a_5) \cdots$	etc.
IV	$c_1 = (b_1a_3 - a_1b_2)$	$c_2 = (b_1a_5 - a_1b_3) \cdots$	etc.

Equation (28.109) contains no roots with positive real parts if, and only if, all terms in the first column of the array (i.e., a_0, a_1, b_1, c_1, etc.) are positive for positive a_0, or negative for negative a_0. The number of times a change in sign takes place in going from one term to the next in the first column gives the number of roots with positive real parts. This is known as *Routh's stability criterion.*

The first conclusion that can be made is that a necessary but not sufficient condition for stability is that all coefficients of p in Eq. (28.109) must be present and that they must all have the same algebraic sign. Since inertia and stiffness in a mechanical system will generally be positive, this statement imposes certain restrictions on the values of R_1, R_2, and R_3 in Eq. (28.105), and consequently on the values of α, a, and b in Eqs. (28.100), (28.101), and (28.102). Normally the values of the coefficients of viscous resistance c_1, c_2, and c_3 will be made as small as possible in an efficient power-transmission system; so that the three other dissipation terms a, α, and b which characterize the prime mover, the clutch, and the load will play an important part in determining the dynamic behavior of the system during the transient following the application of the clutch.

Since it is desirable from the viewpoint of stability that R_1, R_2, and R_3 in Eq. (28.105) be positive the following statements can be made:

1. The output torque of the prime mover should decrease with speed.
2. The torque-speed characteristic of the clutch should be a curve having positive slope.
3. The torque-speed characteristic of the load should be a curve having positive slope, but the value of the load torque at any particular speed within the range of operation should not exceed that of the static clutch torque, $M_c(0)$.

In most cases of practical importance statements 1 and 3 above are satisfied, but the case of a friction clutch exhibiting a torque-speed characteristic with positive slope is rare. In a friction clutch, the choice of friction materials, the preparation of surfaces, lubrication, temperature, and pressure on the frictional surfaces will all influence the torque-speed characteristic, which in some cases can be made, if not increasing with speed, at least of zero slope over most of the operating range.

If the transmission system is such that the vibrations of the drive line are particularly disturbing during the transient caused by the application of the clutch, some form of tuned and damped vibration absorber is necessary if the change in the torque-speed characteristic of the clutch is inadmissible. Drive-line vibration frequency caused by the action of a friction clutch usually falls within the lower end of audio spectrum and may also have values low enough to be quite unpleasant to a human being. Examples of this are clutch "chatter" in various forms of motor vehicles when starting from rest and the noise occurring during the shift from one geared ratio to the next in automatic transmissions.

In order to develop criteria for the design of a damped vibration absorber to be used with a friction clutch, consider a system shown in Fig. 28.24, which is essentially the

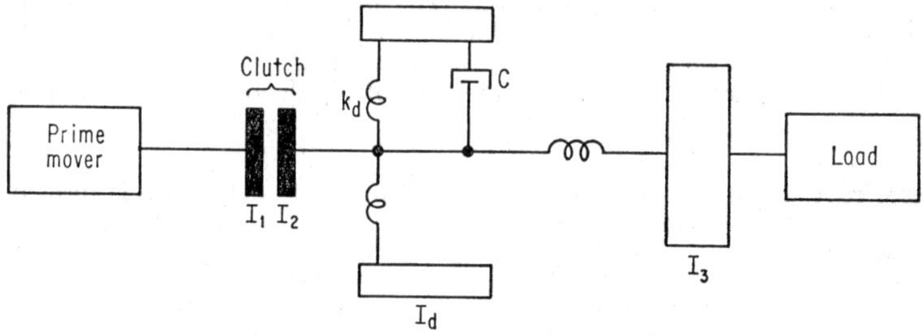

FIG. 28.24. Transmission system shown in Fig. 28.22 with damped vibration absorber added.

same as that shown in Fig. 28.22, the difference being that another oscillating system consisting of a small flywheel has been flexibly attached to the clutch plate and a viscous damper connected between the clutch plate and the absorber inertia.

If, as is usually the case with mechanical power transmission, $I_3 \gg I_2$, the natural frequency of the system on the output side of the clutch is very closely given by $\sqrt{k/I_2}$. The output end of the clutch can thus be represented as in Fig. 28.25.

I_2 = moment of inertia of driven clutch plate(s)
c = viscous damping coefficient of the absorber
α = slope of the torque-speed characteristic of the clutch, equivalent dynamically to viscous damping coefficient
k_1 = spring constant of the main drive shaft
k_d = spring constant of the absorber

The total kinetic energy of the system

$$T = \tfrac{1}{2}I_2(\dot{\theta}_2)^2 + \tfrac{1}{2}I_d(\dot{\theta}_d)^2 \tag{28.110}$$

Total potential energy of the system

$$V = \tfrac{1}{2}k_1(\theta_2)^2 + \tfrac{1}{2}k_d(\theta_2 - \theta_d)^2 \tag{28.111}$$

Dissipation function

$$F = \tfrac{1}{2}\alpha(\dot{\theta}_2)^2 + \tfrac{1}{2}c(\dot{\theta}_2 - \dot{\theta}_d)^2 \tag{28.112}$$

Substituting Eqs. (28.110), (28.111), and (28.112) into Eq. (28.88) yields, in the absence of any external forces,

$$\begin{aligned} I_2\ddot{\theta}_1 + k_1\theta_1 + k_d(\theta_1 - \theta_d) + \alpha\dot{\theta}_1 + c(\dot{\theta}_1 - \dot{\theta}_d) &= 0 \\ I_d\ddot{\theta}_d - k_d(\theta_1 - \theta_d) - c(\dot{\theta}_1 - \dot{\theta}_d) &= 0 \end{aligned} \tag{28.113}$$

Equations (28.113) describe the motion of the system of Fig. 28.25.

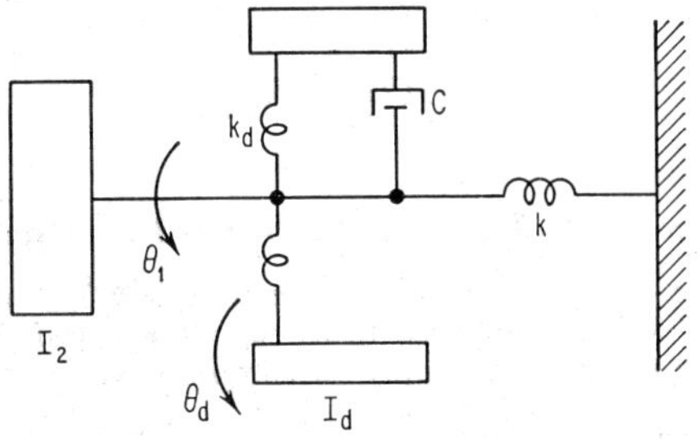

FIG. 28.25. Output end of clutch shown in Fig. 28.24.

Making α negative in Eqs. (28.113) and expanding the determinant associated with Eqs. (28.113), an equation in p is obtained:

$$I_1I_dp^4 + [c(I_1 + I_d) - \alpha I_d]p^3 + [k_1I_d + k_d(I_1 + I_d) - c\alpha]p^2 + (k_1c - k_d\alpha)p + k_1k_d = 0 \tag{28.114}$$

Applying Routh's criterion to Eq. (28.114) yields the follo ving array of coefficients:

I I_2I_d
II $c(I_2 + I_d) - I_d\alpha$
III $[c(I_2 + I_d) - I_d\alpha][k_2I_2 + k_2(I_2 + I_d) - c\alpha] - I_2I_d(k_1c - k_2\alpha)$
IV $[c(I_2 + I_d) - I_d\alpha][k_1I_d + k_2(I_2 + I_d) - c\alpha](k_1c - k_2\alpha) - [c(I_2 + I_d) - I_d\alpha]^2k_1k_2 - I_2I_d(k_1c - k_2\alpha)^2$

Since $I_2I_d > 0$, all the subsequent rows must be positive if the system is to be dynamically stable. This affords a means of obtaining a range of values for the viscous damping coefficient of the vibration absorber c which will result in stable operation.

In order for all the terms in the array to be positive all the coefficients of all powers of p in Eq. (28.114) must be positive. This gives a very rough indication of the range of acceptable values of c.

$$c > \frac{I_d \alpha}{I_2 + I_d}$$
$$c > \frac{k_d}{k_1} \alpha \qquad (28.115)$$
$$c < \frac{k_1 I_d + k_d(I_2 + I_d)}{\alpha}$$

The vibration absorber must also be proportioned so that the natural frequency of the inertia I_d (the damper flywheel) is the same as that of I_2. This imposes a second condition

$$k_1/I_2 = k_d/I_d \qquad (28.116)$$

In cases where I_3 in Fig. 28.24 is not very large in comparison with I_2, for the natural frequencies to be the same,

$$\frac{k_d}{I_d} = k_1 \frac{I_2 I_3}{I_2 + I_3} \qquad (28.117)$$

Substituting Eq. (28.116) into Eq. (28.115) yields

$$c > \frac{I_d}{I_2 + I_d} \alpha$$
$$c > \frac{k_d}{k_1} \alpha \qquad (28.118)$$
$$c < \frac{(2I_2 + I_d)k_d}{\alpha}$$

(b) Simplified Approach to Dynamic Problems Involving Friction Clutches

In designing power-transmission systems containing friction clutches situations frequently arise in which it is required to calculate energy dissipation in the clutch during the engagement period or simply to compute the inertial torques acting on the system during a transition from one ratio to the other. These problems can be greatly simplified if the compliance is neglected.

This procedure is justified in cases where the natural frequencies of the system are relatively high and the resulting amplitude of oscillation small in comparison with the changes in velocities which take place as a consequence of the application of the clutch.

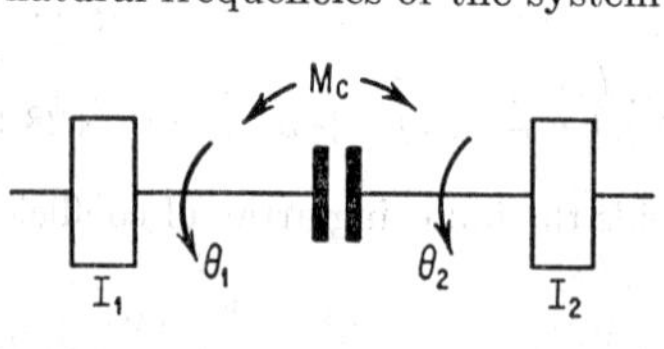

FIG. 28.26. Simple two-inertia system.

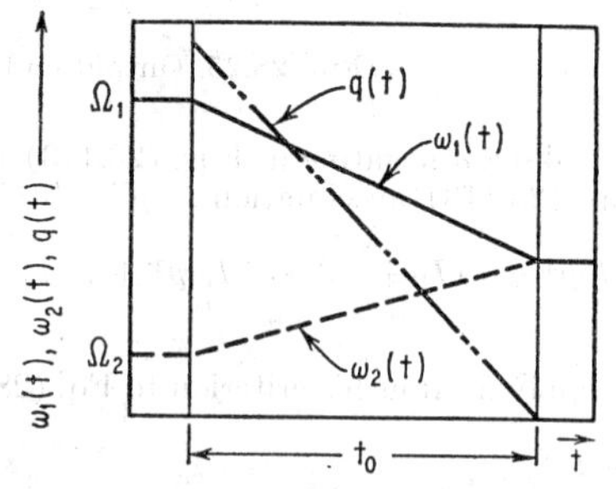

FIG. 28.27. Plot showing time t_0, which torque will be exerted, and the rate of energy dissipation.

Consider a simple two-inertia system shown in Fig. 28.26. Such systems frequently form the basis of various forms of inertia starters.

It is convenient here to use velocities as coordinates rather than angular displacements, because the compliance is neglected.

The coefficient of friction will be assumed to be a constant, independent of speed. Assume also that no external torques act on the system.

In Fig. 28.26, the two flywheels initially rotate at two different angular velocities Ω_1 and Ω_2. Let the clutch be instantaneously applied at time $t = 0$ and let the torque exerted by it on the two flywheels be M_c = constant. This torque will act on the system only as long as there exists a difference in speeds between the two shafts, as shown in Fig. 28.27.

Equations for speed, rate of heat dissipation, etc., are therefore valid for time $0 < t < t_0$, where t_0 is the time required to couple the two shafts.

The equations of motion for the two sides of the clutch are

$$I_1(d\omega_1/dt) = -M_c \qquad (28.119)$$
$$I_2(d\omega_2/dt) = M_c \qquad (28.120)$$

where $\omega_1 = \dot{\theta}_1$ and $\omega_2 = \dot{\theta}_2$, the angular velocities of I_1 and I_2, respectively. Integrating Eqs. (28.119) and (28.120) and applying the conditions at $t = 0$, $\omega_1(t) = \Omega_1$, and $\omega_2(t) = \Omega_2$ yields

$$\omega_1(t) = -(M_c/I_1)t + \Omega_1 \qquad (28.121)$$
$$\omega_2(t) = (M_c/I_2)t + \Omega_2 \qquad (28.122)$$

The relative velocity of I_1 with respect to I_2 is given by

$$\omega_1(t) - \omega_2(t) = \omega_r(t) = -M_c \frac{I_1 + I_2}{I_1 I_2} t + \Omega_1 - \Omega_2 \qquad (28.123)$$

The rate at which energy is dissipated in the clutch during the engagement period is given by $q(t) = T\omega_r(t)$.

$$q(t) = M_c \left(\Omega_1 - \Omega_2 - M_c \frac{I_1 + I_2}{I_1 I_2} t \right) \quad \text{lb-in./sec} \qquad (28.124)$$

The duration of the engagement period, found by considering that when $t = t_0$, $\omega_1(t) = \omega_2(t)$, is given by

$$t_0 = \frac{I_1 I_2 (\Omega_1 - \Omega_2)}{M_c (I_1 + I_2)} \quad \text{sec} \qquad (28.125)$$

The total energy dissipated is obtained by integration of Eq. (28.124):

$$Q = \int_0^{t_0} q(t)dt = \frac{I_1 I_2 (\Omega_1 - \Omega_2)^2}{2(I_1 + I_2)} \quad \text{lb-in.} \qquad (28.126)$$

The efficiency with which energy is transferred from I_1 to I_2 during the clutching operation is

$$\eta = 1 - Q/E_0 \qquad (28.127)$$

where E_0 = kinetic energy of the system at a time $t = 0$. Thus

$$\eta = \frac{(I_1\Omega_1 + I_2\Omega_2)^2}{(I_1 + I_2)(I_1\Omega_1^2 + I_2\Omega_2^2)} \qquad (28.128)$$

If $\Omega_2 = 0$, this expression reduces to

$$\eta = \frac{I_1}{I_1 + I_2}$$

and if $I_1 = I_2 = I$ and $\Omega_2 \neq 0$

$$\eta = \frac{(\Omega_1 + \Omega_2)^2}{2(\Omega_1^2 + \Omega_2^2)}$$

The case where the frictional clutch torque is a function of relative velocity of clutch plates is analyzed next assuming negligible compliance and the dependence of friction on velocity represented by a piecewise linear function, shown in Fig. 28.28.

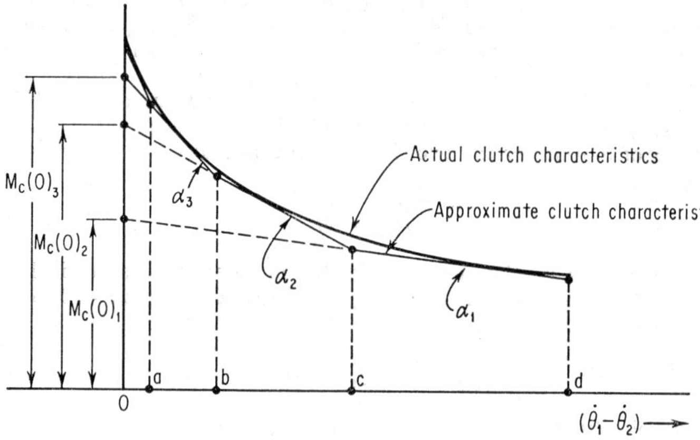

FIG. 28.28. Actual and approximate clutch characteristics.

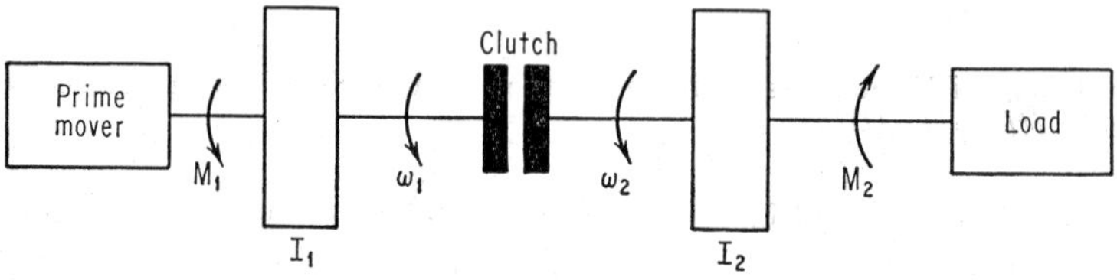

FIG. 28.29. Two-inertia transmission system.

For the system shown in Fig. 28.29, the equations of motion are

$$I_1\dot\omega_1 - \alpha_1\omega_1 + \alpha_1\omega_2 = M_1 - M_c(0)_1 \tag{28.129}$$
$$I_2\dot\omega_2 - \alpha_1\omega_2 + \alpha_1\omega_1 = M_c(0)_1 - M_2 \tag{28.130}$$

In Eqs. (28.129) and (28.130), α is the slope of the clutch frictional characteristic corresponding to the interval $c < (\omega_1 - \omega_2) < d$, and $M_c(0)$ is the point on the M_c axis obtained by intersection of the line of slope α_1 with the axis.

Solution of Eqs. (28.129) and (28.130) yields

$$\omega_1(t) = \frac{(A_1\beta^2 + B_1\beta + C_1)\exp(\beta t) - \beta(C_1t + B_1) - C_1}{I_1I_2\beta^2} \tag{28.131}$$

$$\omega_2(t) = \frac{(A_2\beta^2 + B_2\beta + C_1)\exp(\beta t) - \beta(C_1t + B_2) - C_1}{I_1I_2\beta^2} \tag{28.132}$$

where
$$\begin{aligned} A_1 &= I_1I_2\omega_1(0) \\ B_1 &= I_2[M_1 - M_c(0)_1] - \alpha_1[I_1\omega_1(0) + I_2\omega_2(0)] \\ C_1 &= M_2 - M_1 \\ \beta &= \alpha_1(I_1 + I_2)/I_1I_2 \\ A_2 &= I_1I_2\omega_2(0) \\ B_2 &= I_1[M_c(0)_1 - M_2] - \alpha_1[I_1\omega_1(0) + I_2\omega_2(0)] \end{aligned} \tag{28.133}$$

In Eqs. (28.133) $\omega_1(0)$ and $\omega_2(0)$ are the values of the angular velocities $\omega_1(t)$ and $\omega_2(t)$ at $t = 0$, i.e., at point (d) in Fig. 28.28.

Equations (28.131) and (28.132) are used to calculate the time which must elapse for the two velocities to reach the value $\omega_1(t) - \omega_2(t) = \omega_r(c)$ (where ω_r denotes relative velocity), the point on the $\omega_1(t) - \omega_2(t)$ axis in Fig. 28.28 at which the slope of the clutch characteristic changes discontinuously. This value of time is given by

$$t_c = \frac{1}{\beta} \log_e \frac{\omega_r(c) I_1 I_2 \beta + (B_1 + B_2)}{(A_1 - A_2)\beta + (B_1 - B_2)} \tag{28.134}$$

The computation is carried out along the line of slope α_2 with α_2 substituted for α_1, $\omega_1(t_c)$ and $\omega_2(t_c)$ for $\omega_1(0)$ and $\omega_2(0)$, respectively, and $M_c(0)_2$ for $M_c(0)_1$ in Eqs. (28.133). The process is repeated at points (b), (a), and (0), each time using appropriate constants in Eqs. (28.133) and finding the time the system takes to arrive at the appropriate point on the $\omega = \omega_1 - \omega_2$ axis in Fig. 28.28.

The rate at which the energy is dissipated in the clutch when operating along the linear segment of its characteristic is given by

$$q_0(t) = \{M_c(0) - \alpha[\omega_1(t) - \omega_2(t)]\}[\omega_1(t) - \omega_2(t)] \tag{28.135}$$

Since the coefficient of friction is in this case a variable depending on the relative speed between clutch plates, it is of interest to examine the conditions under which engagement of the clutch can occur.

To this end, Eqs. (28.131) and (28.132) are differentiated, and by setting $t = 0$, expressions for accelerations of the two

$$\dot{\omega}_1(0) = \frac{M_1 - [M_2 - \alpha(\Omega_1 - \Omega_2)]}{I_1} \tag{28.136}$$

$$\dot{\omega}_2(0) = \frac{M_1 - M_2}{I_2} - \frac{M_1 - [M_2 - \alpha(\Omega_1 - \Omega_2)]}{I_2} \tag{28.137}$$

Examining Eqs. (28.136) and (28.137) it can be noted that:

1. Most desirable condition for fast and positive engagement occurs when the following inequality is satisfied:

$$M_1 < \{M_c(0)_1 - \alpha_1[\omega_1(0) - \omega_2(0)]\} \tag{28.138}$$

2. When

$$M_1 = \{M_c(0)_1 - \alpha_1[\omega_1(0) - \omega_2(0)]\} \tag{28.139}$$

engagement will be slower but will be accomplished in a finite length of time depending on the value of $M_1 - M_2$. $M_1 - M_2$ must be positive in order that engagement may be accomplished.

The rate at which $\omega_2(t)$ will begin to increase immediately after the clutch is applied is determined by the quantity $(M_1 - M_2)/I_2$.

The increase of speed of the output side of the clutch decreases the relative velocity of clutch plates, thus increasing clutch torque. Any increase in clutch torque above the value given by Eq. (28.139) will cause the deceleration of input shaft. Thus the effect of the initial increase in speed of the load side of clutch is cumulative and leads to positive engagement.

3. When the prime-mover output torque is greater than the clutch torque at the instant of clutch application, i.e., when

$$M_1 > \{M_c(0)_1 - \alpha_1[\omega_1(0) - \omega_2(0)]\} \tag{28.140}$$

engagement is possible only when the following inequality is satisfied:

$$\frac{M_1 - \{M_c(0)_1 - \alpha_1[\omega_1(0) - \omega_2(0)]\}}{I_1} < \frac{\{M_c(0)_1 - \alpha_1[\omega_1(0) - \omega_2(0)]\} - M_2}{I_2} \tag{28.141}$$

Nevertheless, the analysis provides a basis for continued study of friction clutches and associated control components pertinent to any specific design through the use of modern computing equipment which makes it possible to include in the analysis the dependence of friction on speed, temperature, etc., while simultaneously greatly reducing the time required for computation.

28.6. ELECTROMAGNETIC FRICTION CLUTCHES

A method widely used for controlling friction clutches depends on the electromagnetic force of attraction between two magnetized iron parts. The axial clamping force on the friction disks is thus a function of exciting current, and current may be made time-dependent, thus controlling the torque transmitted through the clutch. Electromagnetic clutches are frequently used in applications requiring close control of acceleration of rotating masses because the exciting current is easily and accurately controlled.

Electromagnetic clutches can be operated dry or in oil and can be of single- and multiple-disk design. The clutch may be so designed that the magnetic flux either (1) passes through the disk stack or (2) does not pass through the stack. The latter case provides more flexibility in design since the disk stack can then be formed and friction materials selected without regard to the geometry of the magnetic circuit and the magnetic properties of the disk materials. Figure 28.30 shows one example of such a clutch.

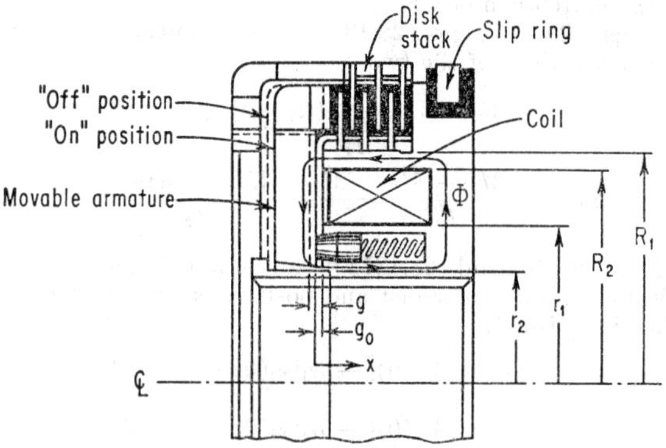

FIG. 28.30. Electromagnetic clutch.

In the following analysis, it will be assumed that effects of saturation, leakage flux, flux fringing, eddy currents, and hysteresis are all negligible, and that the ampere-turns required to maintain the flux in the iron parts of the magnetic circuit are very small compared with the ampere-turns required to maintain the field across the air gap.

Let α = mechanical friction coefficient (armature parts)
d = value of x at which spring force = 0
E = voltage applied to the coil
N = number of turns in the coil
i = coil current, amp
μ = magnetic permeability of air (rationalized mks)
Φ = magnetic flux, webers
λ = flux linkages, weber-turns
L = inductance of the coil, henrys
R = resistance of the coil, ohms
x = displacement of armature, m
g = maximum air-gap length, m
g_0 = minimum air-gap length, m
m = mass of the armature, kg
k = return spring elastance, newton/m.

R_1, R_2, r_1, r_2 are defined in Fig. 28.30.

Under the assumptions made above (using rationalized mks units),

$$\Phi = \frac{\mu A_1 A_2 N i}{(g - x)(A_1 + A_2)} \quad \text{webers} \tag{28.142}$$

where $A_1 = \pi(R_1^2 - R_2^2)$
$A_2 = \pi(r_1^2 - r_2^2)$
also

$$\lambda = N\Phi \tag{28.143}$$

The self-inductance of the coil can be defined as

$$L = L_0/(g - x) \quad \text{henrys} \tag{28.144}$$

where

$$L_0 = \mu A_1 A_2 N^2/(A_1 + A_2) \tag{28.145}$$

The magnetic energy stored in the field existing in the air gap is given by

$$W_m = \tfrac{1}{2}Li^2 \quad \text{joules} \tag{28.146}$$

From Eq. (28.146) the electromagnetic force acting on the armature is

$$F_m = \partial W_m/\partial x = -[L_0 i^2/2(g - x)^2] \quad \text{newtons} \tag{28.147}$$

Equation (28.147) shows that the force, and also the torque transmitted by the clutch, are very sensitive to change in the coil current and to the difference $(g - x)$. Whereas in the case of the hydraulically actuated clutch the position of the actuating piston relative to the main body of the clutch is immaterial when computing torque transmitted, the minimum distance of the armature from the body of the electromagnetic clutch is an important parameter.

The applied coil voltage is given by

$$E = Ri + d\lambda/dt \tag{28.148}$$

However, λ is a function of x and i, and both x and i are functions of time. Therefore,

$$E = R_i + (dL/dx)(dx/dt) + L(di/dt) \tag{28.149}$$

Summing all forces acting on the armature and performing the differentiation indicated by Eq. (28.149) yields

$$M(d^2x/dt^2) + \alpha(dx/dt) - k(d + x) + L_0 i^2/2(g - x)^2 = 0 \tag{28.150}$$

$$-[L_0 i/(g - x)^2](dx/dt) + [L_0/(g - x)](di/dt) + R_i = E \tag{28.151}$$

Equations (28.150) and (28.151) describing dynamic behavior of the clutch system are nonlinear and cannot be solved by employing elementary methods. However, they provide a reasonable basis for studying the response of the system to the control voltage E using an analog computer, in which case the nonlinearity of the magnetic circuit (saturation and hysteresis) can also be taken into account when formulating system equations.

It is difficult to calculate complete response of the system from the time the voltage E is applied across the coil, but when it is realized that the clutch begins to transmit torque only after the armature has been attracted and the clamping force applied to the disk stack, Eqs. (28.150) and (28.151) can be simplified considerably and can be used to examine the factors affecting the torque capacity of the clutch.

After the motion of the armature has ceased $d^2x/dt^2 = dx/dt = 0$ and $g - x = g_0$ and Eqs. (28.150) and (28.151) become, respectively,

$$-k(d + g - g_0) + L_0 i^2/2(g_0)^2 = F_c \tag{28.152}$$

$$L(di/dt) + Ri = E \tag{28.153}$$

where $L = L_0/g_0$
F_c = net axial clamping force

Solution of Eq. (28.153) yields

$$i(t) = E/R - (E/R - I_0) \exp[-(R/L)t] \tag{28.154}$$

where I_0 is the value of i at the time $t = 0$, i.e., at the time when $x = g - g_0$ and armature motion ceased.

Substituting for $i(t)$ from Eq. (28.154) into (28.152) the response of the axial clamping force to the sudden application of control voltage E is obtained:

$$F_c = [L_0/2(g_0)^2]\{(E/R)^2 - 2(E/R)[(E/R) - I_0] \exp[-(R/L)t] + [(E/R) - I_0]^2 \exp[(2R/L)t]\} - K(d + g - g_0) \quad (28.155)$$

For sufficiently large values of t, the steady-state value of the clamping force on the disk stack is given by

$$F_c = [L_0/2(g_0)^2](E/R) - K(d + g - g_0) \quad (28.156)$$

To disengage the clutch, the control voltage E is reduced to zero, and similar procedure leads to the expression for F_c

$$F_c = \frac{L_0}{2(g_0)^2}\left(\frac{E}{R + r}\right)^2 \exp\left[-\frac{2(R + r)}{L}t\right] - K(d + g - g_0) \quad (28.157)$$

Equation (28.157) shows how the clamping force F_c acting on the disk stack decreases with time following disappearance of the control voltage E at a time $t = 0$. In this equation, r is any external resistance through which the coil discharges and which may be added to decrease the disengagement time by decreasing the time constant of the system.

If the coefficient of friction of the disk surfaces is a reasonably invariant quantity Eqs. (28.155), (28.156), and (28.157) also show how the clutch torque varies with time. In practice, these equations may represent clutch torque variation with time to a very good approximation in the case of dry clutches, but when dealing with oil-lubricated clutches additional factors must be considered which pertain to the mechanical-design details of friction disks.

Frictional torque characteristics of magnetic clutches have been studied experimentally and the results of these investigations discussed in Ref. 13.

The value of frictional torque developed by the clutch and its variation with time depends not only on the magnetizing current but also on the kind of disks and disk material used. Similarly the decrease of the clutch torque besides being a function of time such as indicated by Eq. (28.157) also depends on adhesion between the disks, which in turn is a function of disk design and material.

FRICTION BRAKES*

Acknowledgments. I am greatly indebted to my colleagues at Johns Manville for their cooperation; to Ruth Keusseff and H. G. Koch for reading the manuscript; to C. L. Meserve, R. J. Taylor, and J. F. Herr for supplying data on friction materials; to W. F. Gulick for his assistance in tests and computations; and to P. J. Reardon (now with Princeton-Pennsylvania Accelerator, Princeton, N.J.) for his collaboration on the theory of frictional vibrations.

28.7. FRICTION MATERIALS

(a) Types and Physical Properties

The most common type of friction material is "organic," a heterogeneous mixture of rubber and/or resin binders, carbon, sulfur, asbestos fiber, catalysts, metal particles, friction particles, mineral fillers, etc. These and other ingredients are needed to provide the great variety of properties which brake linings must possess. Among these are[14] (1) appropriate coefficient of friction, (2) uniform coefficient over wide temperature range—low fade, (3) low wear rate, (4) good water resistance and quick

* By David Sinclair.

recovery after immersion, (5) low "moisture sensitivity," (6) low noise, (7) resistance to oil, (8) adequate mechanical strength, (9) absence of abrasive action on the brake drum, (10) good bonding properties to the brake shoe, (11) absence of odor, (12) low shrinkage or swelling, (13) low cost.

Table 28.1 lists the physical properties of three important types.[15]

Table 28.1. Properties of Friction Materials

Type	Woven lining (flexible)	Molded lining (sheeter molded)	Molded block (rigid molded)
Compressive strength, psi	10,000–15,000	10,000–18,000	10,000–15,000
Shear strength, psi	3,500– 5,000	8,000–10,000	4,500– 5,500
Tensile strength, psi	2,500– 3,000	4,000– 5,000	3,000– 4,000
Modulus of rupture, psi	4,000– 5,000	5,000– 7,000	5,000– 7,000
Hardness, Brinell	*	*	20–28
Specific gravity	2.0–2.5†	1.7–2.0	2.1–2.5
Thermal expansion, in./in./°F (mean to 350°F)	*	1.5×10^{-5}	1.3×10^{-5}
Thermal conductivity, Btu in./hr/ft²/°F (125°F mean)	5.5	3	3.5–4
Principal type of service	Industrial	Automotive (passenger)	Automotive (truck and bus) and railway‡
Max continuous operating temperature, °F	400–500	500	750
Max rubbing speed, fpm	7,500	5,000	7,500
Max continuous operating pressure, psi	50–100	100	150
Avg. friction coefficient (350°F, 50 psi, 600 fpm)	0.45	0.47	0.40–0.45

* Inhomogeneous.
† High wire-mesh content.
‡ See refs. 16 and 17.

A more general list of types is as follows:[18–20] (1) woven; (2) folded and compressed; (3) molded, (*a*) dry mix (lining and block), (*b*) extruded, wet mix, (*c*) sheeter, laminated wet mix, (*d*) wire-back, (*e*) millboard; (4) resilient (for operation in oil).

The more important ingredients and their function are as follows:

1. Asbestos fiber imparts heat resistance and mechanical strength. (The oriented fiber in sheeter molded lining probably accounts for its higher mechanical strength. See Table 28.1.)
2. Friction particles provide wear resistance and high friction. They may consist of (*a*) hard-rubber dust (20 to 80 mesh) heavily loaded with inorganic fillers (barytes) to reduce plasticity, (*b*) fully cured resins such as cashew nutshell oil, (*c*) crushed coal and coke (low friction).
3. Bonding agents—resin, rubber, or a combination (*a*) thermosetting phenolic resins provide mechanical strength and resistance to oil and heat; (*b*) natural or synthetic rubber produces a more flexible material; (*c*) rubber-resin combinations may provide the best features of both.
4. Vulcanizing agents for rubber—sulfur, litharge, zinc oxide, accelerators.
5. Carbon black may be used to increase tensile strength or as a coloring agent.
6. Graphite or lead powder reduces drum scoring and friction.
7. Brass chips improve wear resistance and reduce drum scoring.

Inorganic-metallic[19] and cerametallic[21] friction materials are less common. They are durable and have high heat resistance, high heat capacity, and high heat dissipation. They are often used in the form of buttons in conjunction with organic materials for heavy-duty service.

(b) The Causes of Friction

The laws of friction of metals and polymers apply in general to brake linings and other friction materials. The performance of brake linings is similar in many respects to that of metals. One reason is that the mating surface against which the brake lining rubs is always a metal, usually the cast iron of the brake drum, and the two surfaces act together to determine the coefficient of friction and its characteristics.

Five causes of brake-lining friction should be considered:

1. The lifting over surface asperities, discarded by proponents of the adhesion theory, should be considered because adhesion does not explain all frictional properties of brake linings.[22]

2. Welding of junctions, which may occur between metals, does not occur between the brake lining and drum. Nevertheless, adhesion and shearing of junctions occur, because the materials of the lining and the drum often become smeared over each other. The real area of contact varies markedly with the condition of the lining surfaces. Examination with the optical microscope shows that the pits and irregularities of the surface become filled with wear particles, and at high temperatures the entire surface becomes glazed with partially decomposed binder, drum metal, etc. As with metals, surface contamination has a marked effect on friction. It is always necessary to run in a brake lining before a representative coefficient of friction can be obtained under any given set of conditions.[23]

3. "Plowing," which may be considered shearing below the surface, helps to account for the high wear rates when the lining is abrasive.

4. The elastic-plastic properties of the bulk material below the surface contribute to sliding friction, because the surface irregularities (always large compared with the molecular dimension) "work" the material below the surface.[24]

5. Triboelectrification accompanies the rubbing of brake linings as well as most other solids. Some observers believe this is a cause of friction.[25] Observations of frictional-electric potentials accompanying frictional vibrations indicate that triboelectrification is an effect rather than a cause of friction. Fluctuating potentials of a few millivolts, accompanied by d-c potentials of a few hundredths or tenths of a volt, occurred across the brake-lining samples. The fluctuating potentials appeared to be caused by slight changes in the electrical capacity, resulting from changes in spacing as the sample was lifted over surface asperities or over wear particles. If electrical potentials add to the friction, it was reasoned that large applied potentials should have an observable effect on the coefficient of friction, but no such effect could be found.

(c) The Effect of Sliding Speed on Friction

When measured at constant temperature, the coefficient of friction of most brake linings decreases somewhat with increasing sliding velocity. The static coefficient is usually about 10 per cent higher than the initial sliding coefficient. The greater static friction is the primary cause of brake squeal. The variation of coefficient depends upon the temperature and pressure during a particular test, and on the previous history of the lining.

Figure 28.31 shows the performance of a dry clutch facing, loaded at 35 psi, when tested at three temperatures on a friction test machine.[26] The rubbing velocity was increased from 0.77 fps to 43 fps during a period of 6 min and then reduced during an equal period. Other materials show less variation than that in Fig. 28.31, depending on the composition and method of manufacture. Some materials, such as transmission band linings run in oil, show a slight increase in coefficient at very low velocities just after sliding begins.[27]

(d) The Effect of Temperature and Pressure on Friction

High temperatures, induced by frictional work, cause a marked decrease in the coefficient of friction of brake linings, called "fade." Fade is responsible for the great loss of effectiveness of self-actuating brakes during severe use, and for the frequent imbal-

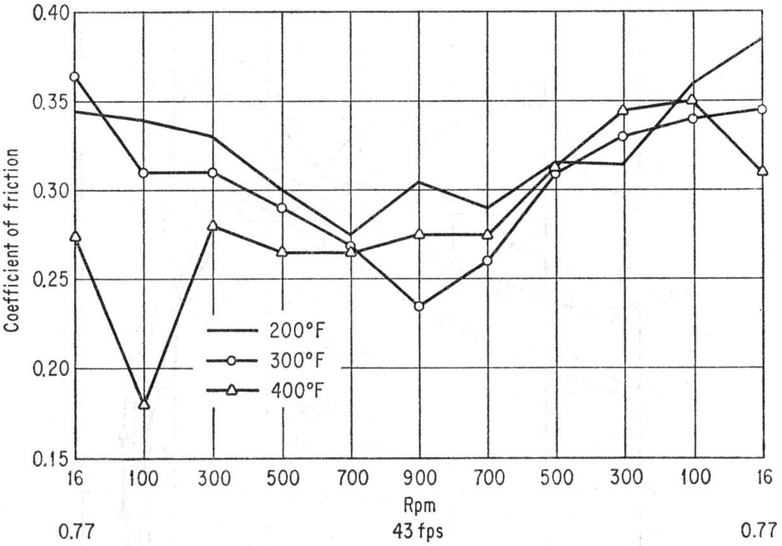

FIG. 28.31. Coefficient of friction of dry clutch facing versus rubbing velocity at three temperatures, 35 psi, 0.77 fps to 43 fps to 0.77 fps.

ance among the brakes in an automobile. In continuous tests of brake linings in the laboratory the coefficient tends to increase with increasing temperature up to about 500°F. In road tests, which are intermittent, fade usually begins at lower temperatures, but only after an elapsed time of several minutes. The response of the lining is delayed and depends on both the instantaneous temperature and the previous temperature history.

The effects of pressure and temperature are closely interrelated, because pressure affects the temperature at a given rubbing velocity, as well as the rate of wear and the smearing of the softer components, which in turn affect the coefficient of friction. Figure 28.32 shows continuous test[26] measurements, on a typical molded brake lining, of the variation of coefficient with temperature and pressure.[28,29] The rubbing velocity was 20 fps and the duration of a run at a given pressure was 15 min, with one measurement each minute.

Brake linings usually fade less and less after a series of tests until a comparatively stable condition is reached. This results from the baking and curing of the outer layers of the lining during use. The curing progresses into the lining with prolonged use, which accounts for the marked effect of the previous history on performance.

(e) The Effect of Moisture on Friction

The coefficient of friction of all brake linings is more or less affected by moisture. When a brake lining is thoroughly wet it becomes ineffective; the lining is lubricated by a film of water of low coefficient which, combined with the behavior of self-actuating brakes, causes almost complete loss of brakes. The "wet" coefficient also decreases markedly with velocity,[30] sometimes leading to brake noise.

The coefficient of a lining slightly damp is often much higher than when dry. This condition, known as moisture sensitivity, can develop during the drying of a wet lining or from exposure to a humid atmosphere. This accounts for the occasional locking of self-actuating brakes when first applied in humid weather. Moisture sensitivity is more common in dry-mix linings, which are more porous than wet-mix linings. The moisture sensitivity is greater when the brake drum is worn smooth than when new and comparatively rough.

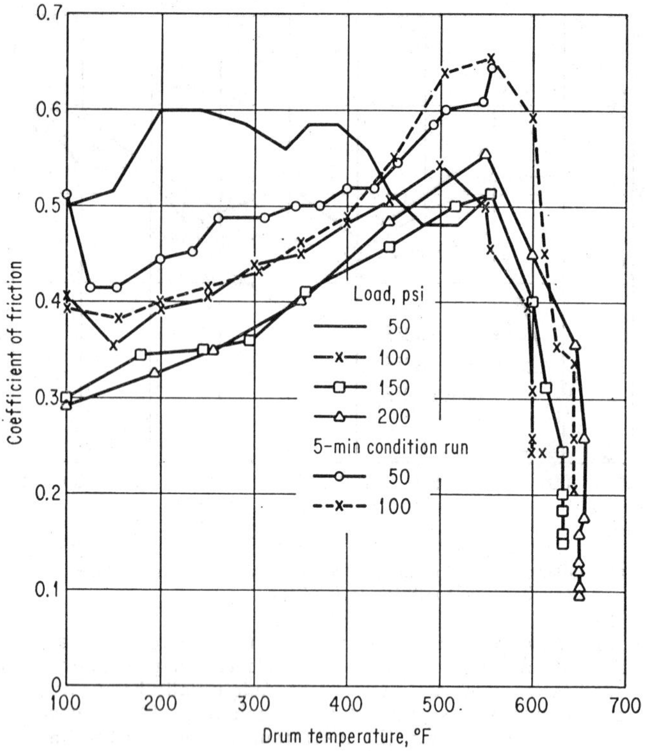

Fig. 28.32. Coefficient of friction of molded brake lining versus temperature at various pressures.

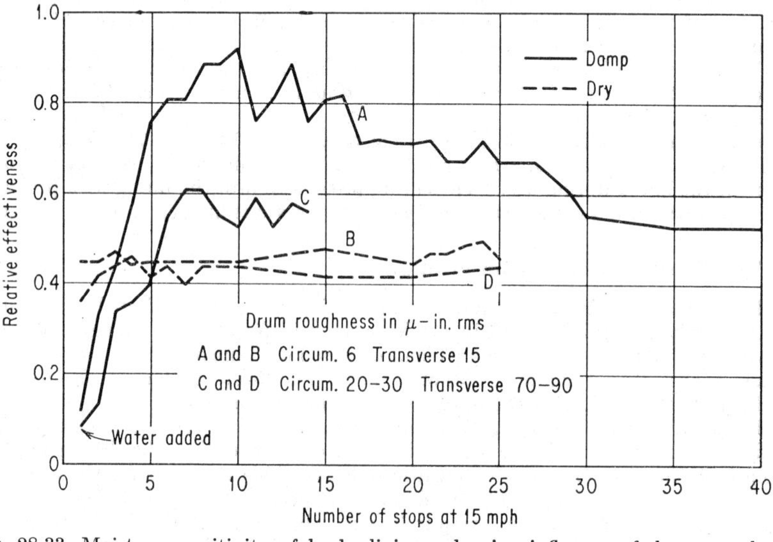

Fig. 28.33. Moisture sensitivity of brake linings, showing influence of drum roughness, micro-inches rms.

Figure 28.33 shows typical measurements of moisture sensitivity made on a standard self-actuating brake, equipped with a dry-mix lining, mounted in an inertia dynamometer. The ordinates show the relative effectiveness and the abscissas, the number of stops from 15 mph. The dashed curves show that the effectiveness with a dry lining is practically independent of drum roughness. When water was introduced between the rubbing surfaces, the effectiveness fell and then rose to a high value, particularly with the smooth drum (curve *A*). After 45 stops, the lining dried out, and the effectiveness returned to the original value.

(f) The Wear of Materials

Wear not only has an important effect on the life of the lining and drum but strongly affects the coefficient of friction. Rising temperatures tend to reduce friction while wear tends to restore friction. This applies particularly to the extreme outer surface where the temperatures are often excessive.[31,32] A moderate amount of wear is therefore desirable to clean the lining and drum surfaces.[28]

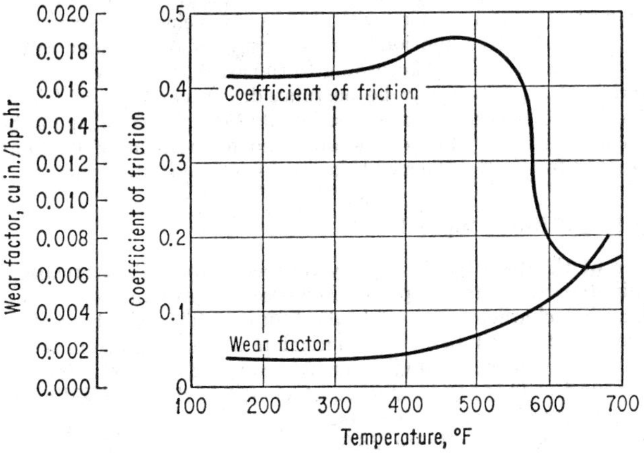

FIG. 28.34. Rate of wear and coefficient of friction of truck brake block versus temperature, 10½ fps and 50 psi on 1-in.-square sample.

Figure 28.34 shows results obtained from a laboratory wear-test machine on a 1-in. square sample of a truck brake block run at 10½ fps and 50 psi. The abscissas show the temperature in degrees Fahrenheit and the ordinates show the coefficient of friction and the wear factor in cubic inches per horsepower hour. The rate of wear is seen to increase rapidly at high temperature where the coefficient drops sharply. The coefficient will recover substantially when the test is stopped and resumed at low temperature.

28.8. FRICTIONAL VIBRATIONS—BRAKE SQUEAL

(a) The Cause of Brake Squeal

Brake squeal is a form of intermittent motion which frequently occurs when two solid bodies are rubbed together. In brakes, a high-pitched squeal is often drum vibration, and the low-pitched groan and chatter are vibration of the "backing plate" or "spider" to which the brake shoes are attached. The cause is the decrease of friction with increasing velocity which produces a negative damping factor causing relaxation oscillations.[33] It may also result from the wedging action of a self-actuating brake, causing vibration of the shoes and, hence, variation of the load in resonance with some natural frequency of the system.

Intermittent motion, commonly called "stick-slip," has been studied with metal sliding on metal[34-36] and with polymers sliding on metal.[37] Similar stick-slip motion has been observed with brake lining sliding on metal, usually cast iron.[30,38] The primary cause of stick-slip is the decrease in coefficient of friction as the sliding velocity is increased from zero to a very small value. This change is actually continuous,[36] but it occurs over so short a velocity range that the coefficient usually appears to fall instantly from its static value to its sliding value.

The static coefficient is often called the "breakaway" coefficient. It is this sudden "breakaway" which may set a massive elastic member into vibration. At the instant of breakaway, an unbalanced force (which may amount to 50 lb in a brake) suddenly appears and produces high acceleration. The vibration may then be sustained and amplified by the energy supplied by the driving mechanism. The further decrease of friction which often occurs at higher velocities can be a contributing cause of brake squeal.

The increase of coefficient with decreasing velocity, particularly the greater static coefficient relative to sliding coefficient, is believed to be caused by a time lag in adhesion between the rubbing bodies. During sliding, the time of contact between the areas of real contact increases with decreasing velocity, causing an increase in coefficient. When relative motion ceases, the coefficient rises rapidly, but not instantly, to the static value. According to the asperity theory, a similar time lag in the interlocking of asperities may be postulated. A variation of friction with time of contact has been observed in some experiments.[23,39] No time lag has been observed with friction materials.[30] Most of the observed phenomena of frictional vibrations can be accounted for by the theory of damped frictional vibrations.

(b) Reduction of Brake Squeal

Alteration of Friction by Boundary Lubrication. Theoretically, brake squeal can be eliminated by means of positive or viscous damping which counteracts the negative damping characteristics of brake linings. This has been demonstrated in the laboratory, but found to be impractical in an automotive brake. A sample of brake lining, run in a mechanical vibrator, was coated with a boundary layer of a polar organic compound. This gave a static coefficient less than the sliding coefficient and caused the sliding coefficient to increase with increasing velocity. Figure 28.35 shows the damping of frictional vibrations by lubrication with soap and glycerine,[30] and Fig. 28.36 shows the variation of coefficient with velocity when dry and lubricated. Similar results were obtained with lithium stearate[30] and other polar compounds[40] and with Teflon.[41] In practice, damping materials of this nature cannot withstand the severe abrasion and high temperature in a brake; so that they become either volatilized or burned away.

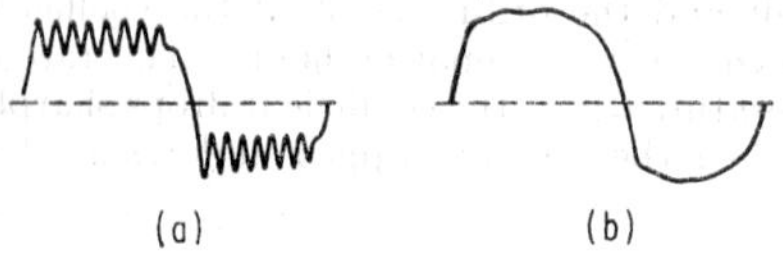

FIG. 28.35. Damping of frictional vibrations by lubrication with soap and glycerin.

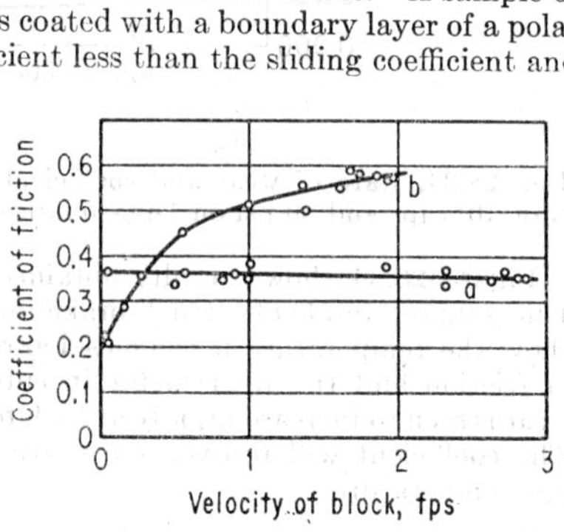

FIG. 28.36. Variation of coefficient of friction with velocity. (a) Lubricated. (b) Dry.

The frictional characteristics of a lining (or drum) can sometimes be altered so as to reduce noise, by grinding the surfaces. The partially faded or overcured surface of a worn lining may have an unusually large difference between the static and sliding coefficients. This difference can be reduced by grinding a fresh surface. Regrinding the highly polished or oxidized surface of a worn drum may have the same beneficial effect. Noise may sometimes be reduced by cleaning the wear dust from the drum and lining and from within the brake. In general, however, the benefit is not permanent and noise may again occur when the linings and drums become repolished, or from unfavorable atmospheric conditions, or from accidental mechanical changes, etc.

Mechanical Alterations. Brake noise or any frictional vibration is highly dependent upon the characteristics of the mechanical system: its mass and stiffness, the natural damping,* the load, the rubbing velocity, etc.

The method of manufacture of the drum has a considerable influence on squeal. A two-piece riveted drum provides its own damping, while a one-piece or solidly welded drum is much more easily excited. Proper adjustment of the brake will sometimes reduce or eliminate noise. However, whether or not a well-seated and adjusted lining will reduce noise depends upon many factors: the condition of the lining and drum surfaces, the resiliency of the drum and backing plate, the ambient temperature, humidity, dust, etc.

Backing-plate chatter may be eliminated by loading the shoes or by stiffening the backing plate or making it more flexible.

(c) Theory of Damped Frictional Vibrations

The equations of frictional vibration without damping show that stick-slip motion occurs whenever the static coefficient is higher than the sliding coefficient.[30,34,42] They also show that constant (Coulomb) friction does not produce any damping. Vibration at the natural frequency of the system takes place about the mean stressed position which balances the friction force.[30] The energy dissipated in friction is supplied by the driving mechanism, but the vibration is the same as a free vibration without friction. This means that constant sliding friction less than static friction also provides no damping. Sufficient viscouslike friction from some source must be present to counteract the effect of the higher static coefficient.

The equations which follow are based upon two assumptions: (1) the difference between the static and sliding coefficients is a constant, and (2) the sliding coefficient does not vary with velocity. These assumptions are approximately correct for dry friction materials sliding on metal, although they may be inadequate for metal on metal.[41] For friction materials, the static coefficient varies relatively little with velocity.

The equations herein given are a modification of Blok's.[43] They show that a higher static coefficient is a necessary but not a sufficient condition for frictional vibration when viscous damping is present. When the applied velocity (for example, the velocity of the rotating drum relative to the fixed lining) exceeds a critical value, only transient vibration occurs. The value of the critical velocity varies with the mechanical properties of the system; load, mass, stiffness, and viscous damping.

The following equation applies equally well during the slip and during damped continuous motion in which transient vibration and subsequent uniform motion occur. It does not, of course, apply during the stick portion of the cycle.

$$m\ddot{x} + R\dot{x} + kx = F_0 - F + kvt \qquad (28.158)$$

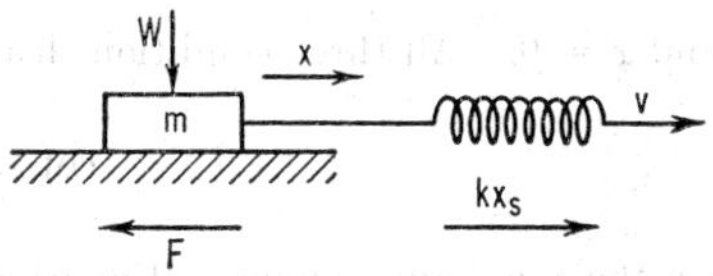

FIG. 28.37. Schematic diagram of brake system.

Referring to Fig. 28.37, m is the mass of the moving member; k is the stiffness of the mechanical system supporting the moving member; R is the viscous damping constant (from whatever source); F_0 and F are the static- and sliding-friction forces (product of coefficient and load, $L = W + mg$); and v is the constant applied velocity. The displacement of the moving member is x, $\dot{x}(dx/dt)$ its velocity, and $\ddot{x}(d^2x/dt^2)$ its acceleration. Slip begins at $t = 0$ and $x = 0$.

Equation (28.158) may be rewritten in terms of two dimensionless ratios;

$$B = R/2\sqrt{km}$$

* Positive or viscouslike damping arising from internal solid friction, air resistance, etc.

the ratio of the viscous resistance of the system to the viscous resistance of a critically damped system, and $E = v\sqrt{km}/(F_0 - F)$. With these substitutions, Eq. (28.158) becomes

$$\ddot{x} + 2B\omega_n\dot{x} + \omega_n^2 x = v\omega_n(1/E + \omega_n t) \tag{28.159}$$

The solution is

$$x = \exp(-B\omega_n t)[(v/\omega_n)(2B - 1/E)\cos\omega t + (v/\omega)(2B^2 - B/E - 1)\sin\omega t] + (v/\omega_n)(1/E - 2B) + vt \tag{28.160}$$

where the undamped natural frequency $\omega_n = \sqrt{k/m}$ and the damped natural frequency $\omega = \omega_n\sqrt{1 - B^2}$.

When stick-slip motion occurs, the moving member comes to rest and sticks to the fixed base at some time during the vibration cycle. It then remains at rest until the spring force again equals the static-friction force, when the slip is repeated. If it does not come to rest, the exponential factor in Eq. (28.160) approaches zero and uniform sliding occurs at the velocity v.

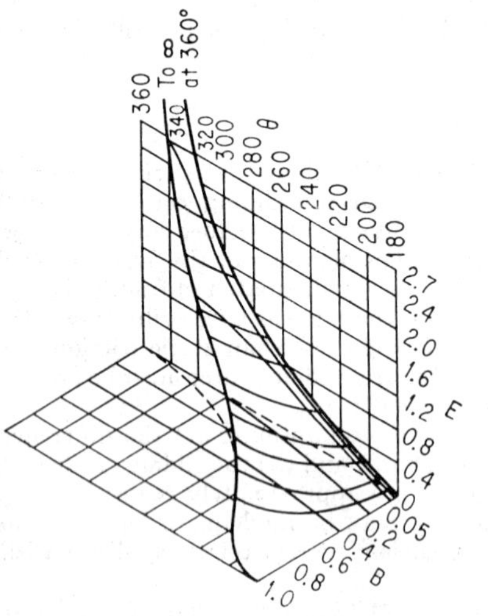

FIG. 28.38. Stick-slip surface.

The condition for zero velocity $\dot{x}$ is found by equating the derivative of Eq. (28.160) to zero, yielding

$$\exp(B\theta/\sqrt{1 - B^2}) = \cos\theta + [(B - 1/E)/\sqrt{1 - B^2}]\sin\theta \tag{28.161}$$

where θ, the "stopping" angle, is the value of ωt at which the block stops during the damped vibration cycle.

The variables B, E, and θ define the "stick-slip surface" (Fig. 28.38). All points on this surface satisfy Eq. (28.161) and each point defines a trio of values of B, E, and θ which must occur simultaneously when stick-slip motion occurs. The range of values is B: 0 (no damping) to 1 (critical damping), E: 0 to ∞, θ: π to 2π. This means that the slip always lasts at least one-half a damped vibration cycle.

The stick-slip surface is bounded by three edges. Along one edge $E = 0$ and along the second edge $B = 0$. The third edge defines maximum values B_m, E_m, and θ_m beyond which stick-slip motion cannot occur. Along this edge,

$$(\partial E/\partial\theta)_B = 0 \qquad (\partial B/\partial\theta)_E = 0$$

and $\ddot{x} = 0$. All three conditions lead to the equation

$$\cot\theta_m = \frac{B_m - E_m}{\sqrt{1 - B_m^2}} \tag{28.162}$$

for the maximum values. Equation (28.162) applies only at the boundary, while Eq. (28.161) applies over the whole surface including the boundary.

Table 28.2 lists a set of values of B_m, E_m, and θ_m which satisfy Eq. (28.162).

Figure 28.39 shows a plot of these values of B_m and E_m, the projection of the maximum boundary on the BE plane. Figure 28.40 shows a plot of B vs. θ at constant E, together with the projection of the maximum boundary on the $B\theta$ plane. It is evident from Fig. 28.40 that when B is small and E large (the usual condition) the slip may last almost a full vibration cycle. That is why stick-slip vibrations often look like and sound like a complete harmonic vibration; the stick time is only a small portion of the cycle.

Table 28.2. Boundary Values of B_m, E_m, and θ_m

Where $1 \geq B \geq 0$
$0 \leq E \leq \infty$
$\pi \leq \theta \leq 2\pi$

B_m	E_m	θ_m
1.000	0.000	180°00′
0.999	>0	182°34′
0.990	>0	188°06′
0.975	>0	192°55′
0.950	>0	198°16′
0.900	>0	205°50′
0.800	0.007	217°08′
0.700	0.020	226°25′
0.600	0.045	235°14′
0.500	0.081	244°10′
0.400	0.137	254°00′
0.300	0.226	265°30′
0.200	0.378	280°18′
0.100	0.691	300°42′
0.050	1.092	316°12′
0.030	1.491	325°36′
0.010	2.700	339°36′
0.000	∞	360°00′

The curves show that increasing either B or E while keeping the other constant causes the system to approach and finally cross the boundary line into the uniform-motion region. B can be increased independently of E only by increasing the damping, which may not be feasible. E can be increased independently of B by increasing the velocity v or by decreasing $F_0 - F$ by reducing the load or the difference between the static and sliding coefficients. By means of such changes, a critical value is reached (the boundary) beyond which stick-slip motion quickly ceases.

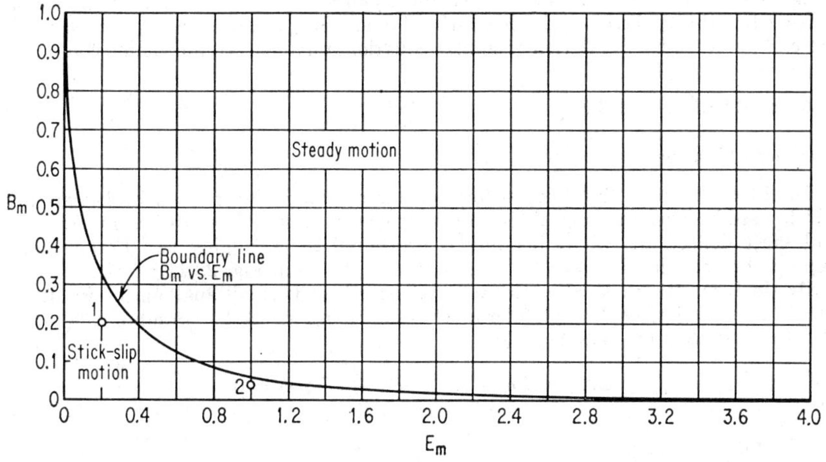

FIG. 28.39. Boundary line B_m vs. E_m.

E may also be increased by increasing $\sqrt{km}$, which causes a proportionate decrease in B, and conversely. Stick-slip can be eliminated by either increasing or decreasing $\sqrt{km}$, depending upon the initial values of B and E. If, for example, the initial values of B and E are each 0.2 (Fig. 28.39, point 1) either a twelvefold increase or a fourfold decrease in $\sqrt{km}$ will place the system in the uniform-motion region. If $B = 0.04$ and $E = 1.0$ initially (Fig. 28.39, point 2), a twofold increase in $\sqrt{km}$, compared with a fifteenfold decrease, will produce uniform motion.

In most practical systems, where the damping factor B is small and $\sqrt{km}$ is usually large, it is more advantageous to increase $\sqrt{km}$ than to decrease it. This can be done by increasing either the mass or the stiffness or both. If both are increased proportionately, stick-slip can be eliminated without any change in the natural frequency $\sqrt{k/m}$.

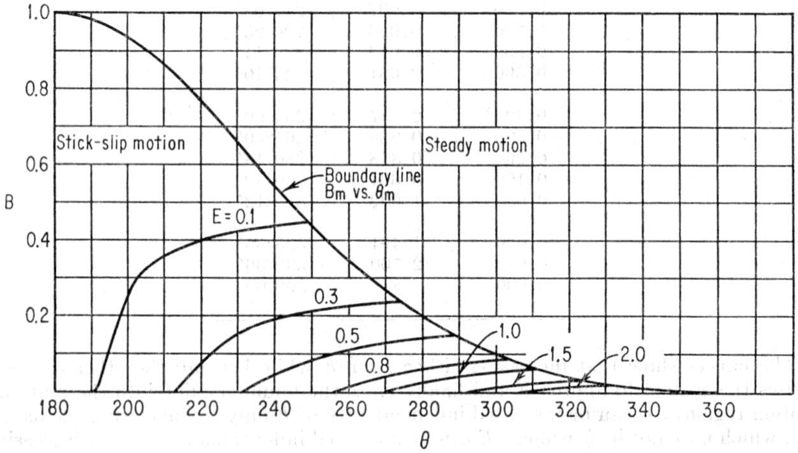

FIG. 28.40. B vs. θ at constant E and boundary line B_m vs. θ_m.

It is well known that a critical velocity often exists above which stick-slip motion does not occur.[44] As mentioned earlier, brake noise usually develops near the end of a stop when the applied velocity is relatively low. Singh[45] gives values of the critical velocity obtained by simulation of similar equations on an electronic analog computer.

28.9. SIMPLE BRAKES

(a) Block Brakes

The block brake is the simplest form of brake. Single-lever block brakes are not widely used because the normal braking force results directly in shaft bearing pressure and shaft deflection. Shaft bending is prevented by arranging two brake blocks opposite one another in a double block brake as shown in Fig. 28.41.

Single-lever Block Brake (Case I). Shoe subtends small angle on brake drum. Line of action F passes through fulcrum O as shown in Fig. 28.42. Summing moments about O and equating to zero yields

$$\Sigma M_0 = Pa - Nb = Pa - (F/\mu)b = 0$$

or

$$F = \mu(a/b)P \tag{28.163}$$

The braking torque

$$T = FD/2 = \mu aDP/2b \tag{28.164}$$

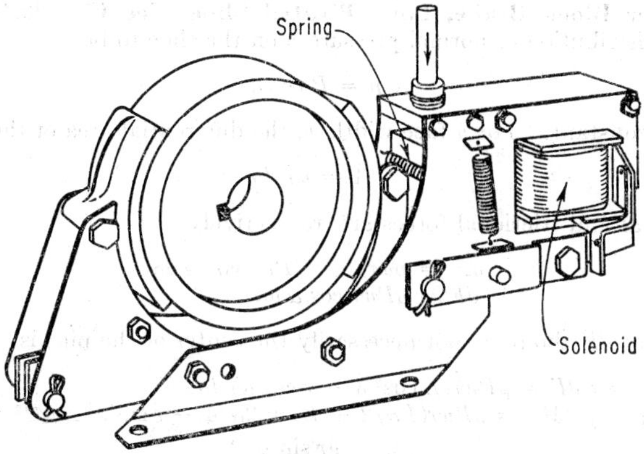

FIG. 28.41. Double block brake.

where N = normal reaction between drum and shoe
P = applied load
$F = \mu N$ = friction force
μ = coefficient of friction

The above equations apply to drum rotation in either direction.

Single-lever Block Brake (Case II). Shoe subtends small angle on brake drum. Line of action F passes a distance e below fulcrum O. For counterclockwise drum rotation,

$$\Sigma M_0 = Pa - (F/\mu)b + Fe = 0$$

or

$$F = \frac{Pa}{b/\mu - e} \qquad (28.165)$$

$$T = \frac{Pa}{b/\mu - e}\frac{D}{2} \qquad (28.166)$$

The frictional force in this case helps to apply the brake. The brake is therefore "self-energizing."

If $e > b/\mu$, the brake is self-locking, and some force P is necessary to disengage the brake. For clockwise drum rotation,

$$F = \frac{Pa}{b/\mu + e} \qquad (28.167)$$

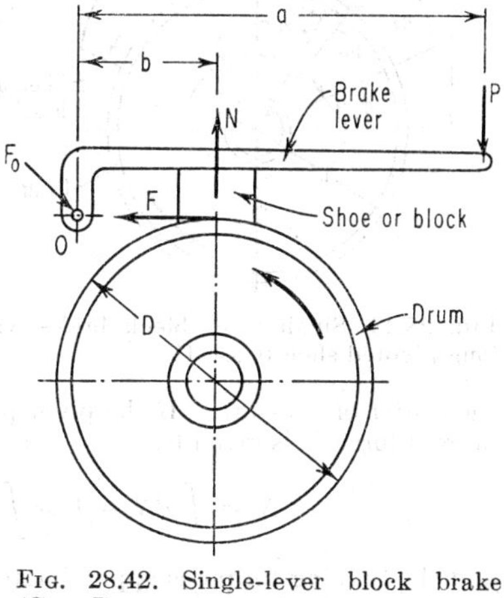

FIG. 28.42. Single-lever block brake (Case I).

Single-lever Block Brake (Case III). Shoe subtends small angle on brake drum. Line of action F passes a distance e above fulcrum O. For counterclockwise drum rotation,

$$\Sigma M_0 = Pa - (F/\mu)b - Fe = 0 \qquad (28.168)$$

$$F = \frac{Pa}{b/\mu + e}$$

For clockwise drum rotation,

$$F = \frac{Pa}{b/\mu - e} \qquad (28.169)$$

Single-lever Block Brake, Long Pivoted Shoe (See Fig. 28.43) (Case IV). Assume the distribution of normal pressure p on the shoe to be

$$p = P \cos \alpha$$

where P is a constant. For a face width w, the differential area of the shoe

$$dA = wr\, d\alpha$$

and the normal and frictional forces are, respectively;

$$dN = pwr\, d\alpha = Pwr \cos \alpha\, d\alpha \tag{28.170}$$
$$dF = \mu Pwr \cos \alpha\, d\alpha \tag{28.171}$$

The moment of dF about A (not necessarily the center of the pin) is

$$dM_A = e\, dF = \mu Pwr(L \cos^2 \alpha - r \cos \alpha)\, d\alpha$$
$$M_A = \int dM_A = \mu Pwr(L\alpha/2 + L \sin 2\alpha/4 - r \sin \alpha)_{-\phi/2}^{\phi/2} = 0 \tag{28.172}$$

and

$$L = \frac{4r \sin \phi/2}{\phi + \sin\phi} \tag{28.173}$$

where L is the distance from the drum center to the line of action of F. Point A is the center of pressure. If the pivot pin is located at A, the shoe will not tip. The normal force N is given by

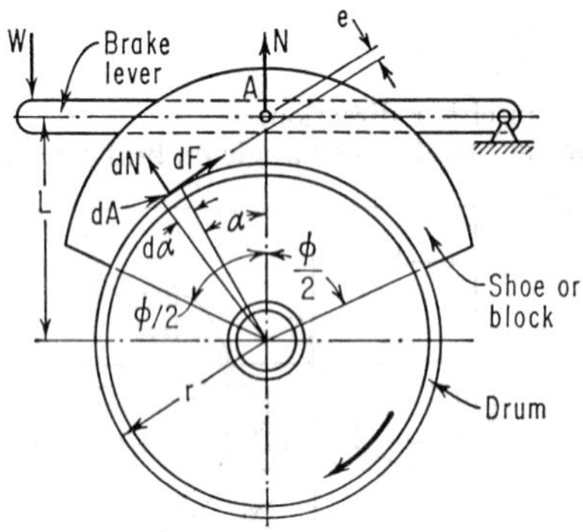

FIG. 28.43. Single-lever block brake with long pivoted shoe (Case IV).

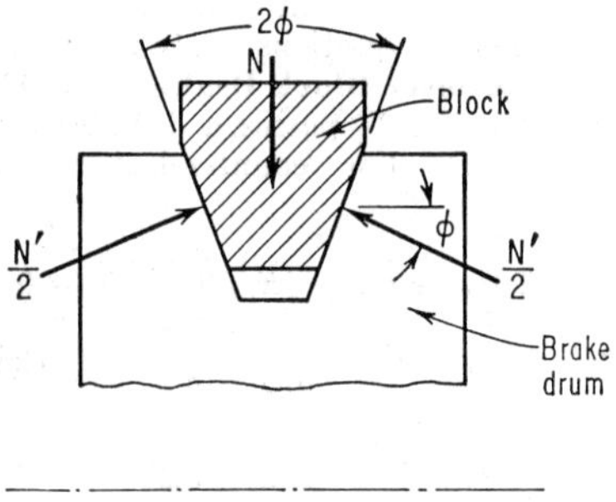

FIG. 28.44. Grooved wheel brake.

$$N = \int dN = Pwr \int_{-\phi/2}^{\phi/2} \cos \alpha\, d\alpha = 2Pwr \sin \phi/2 \tag{28.174}$$

and the braking torque is approximately

$$T = \int r\, dF = \int \mu pwr^2\, d\alpha = \mu Pwr^2 \int \cos \alpha\, d\alpha = \mu Nr \tag{28.175}$$

(b) Grooved Wheel Brake

If the brake drum is grooved as shown in Fig. 28.44, the application of normal force N will result in forces normal to the groove side N'.

$$N' = N/\sin \phi \tag{28.176}$$

(c) Band Brakes

In a band brake, braking action is obtained by pulling a band tightly against the drum. The braking force F_t is defined as the difference between the tensions F_1 and

F_2 at the two ends of the band. Thus

$$F_t = F_1 - F_2 \tag{28.177}$$

Referring to Fig. 28.45, summation of the horizontal and vertical components of the forces acting on a differential element of band length yields, respectively,

$$\mu F_N - dF \cos (d\theta/2) = 0$$

and

$$F_N - (2F + dF) \sin (d\theta/2) = 0$$

where F_N = force between band and drum
F = band tension
μF_N = frictional force
θ = angle of contact

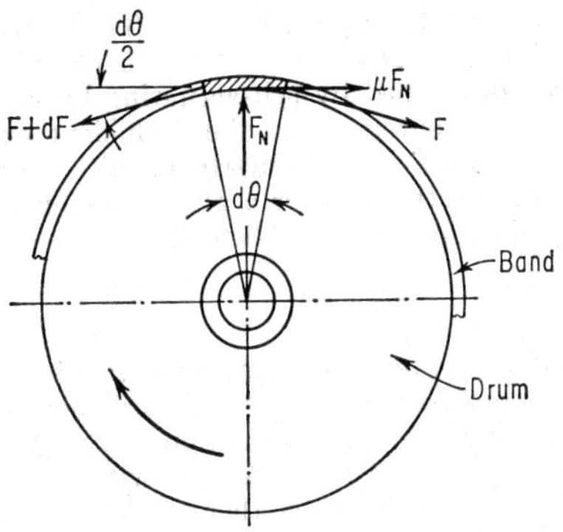

FIG. 28.45. Band tension.

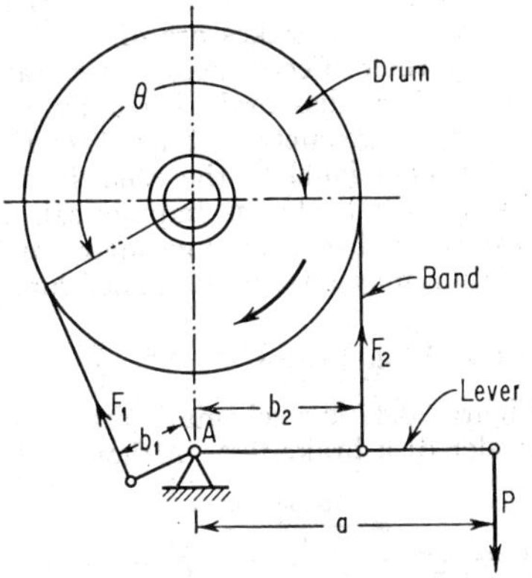

FIG. 28.46. Differential band brake.

Making the small-angle approximations $\sin (d\theta/2) = d\theta/2$ and $\cos (d\theta/2) = 1$ and eliminating F_N from the equilibrium equations yields

$$\mu F\, d\theta - dF = 0$$

Integrating,

$$\int_{F_2}^{F_1} dF/F = \mu \int_0^\theta d\theta$$

Thus

$$\frac{F_1}{F_2} = e^{\mu\theta} \qquad \text{or} \qquad F_1 = \frac{F_t e^{\mu\theta}}{e^{\mu\theta} - 1} \tag{28.178}$$

Figure 28.46 shows a differential band brake. Equating to zero the moments acting on the lever about point A,

$$M_A = Pa + F_1 b_1 - F_2 b_2 = 0$$

and substituting Eq. (28.178), the solution for P is

$$P = \frac{F_t(b_2 - e^{\mu\theta} b_1)}{a(e^{\mu\theta} - 1)} \tag{28.179}$$

Normal operation of the arrangement shown in Fig. 28.46 requires that $b_2 > b_1 e^{\mu\theta}$. If $b_2 = b_1 e^{\mu\theta}$, the brake is self-locking. If $b_2 < b_1 e^{\mu\theta}$, a force must be applied in the opposite direction in order to permit drum rotation. Similar analyses are used for other combinations of direction of P, fulcrum location, and direction of drum rotation.

For $b_1 = 0$ and clockwise drum rotation

$$P = \frac{F_t b_2}{a(e^{\mu\theta} - 1)} \tag{28.180}$$

and the arrangement is termed a simple band brake.

28.10. ANALYSIS OF SELF-ACTUATING BRAKES

Automotive brakes commonly used in the United States are self-actuating or "self-energizing" brakes. They utilize wedging action to cause the normal force exerted by the lining on the drum to increase nonlinearly with the coefficient of friction. The "caliper" disk brake, commonly used in Europe, is not self-actuating. The normal force is independent of the coefficient of friction and depends only on the mechanical design. Examples are the Girling, Dunlop, and Lockheed brakes.[46]

Self-actuating brakes are of two general types, disk and drum. Both types are rated by their effectiveness, the ratio of friction force to applied force.* The effectiveness of the disk brake is readily calculated, but that of the drum brake is complicated by its geometry. Since the friction force developed by self-actuating brakes is not proportional to the coefficient of friction, measurement of the coefficient on self-actuating brakes is impractical. The drum brake is particularly unsuited for this purpose, because its geometry and hence its effectiveness for a given coefficient of friction may alter with applied load because of drum and shoe distortion.

(a) The Disk Brake

Figure 28.47 is a schematic diagram of one of the six ball-and-ramp systems of the Chrysler disk brake.[47,48] Although this brake is obsolete, it serves well to illustrate the principles of self-actuation. The system acts like two rigid wedges with negligible friction between them. When the load L is applied, the wedges separate, forcing the linings against the upper and lower "drums," assumed rigid. For the direction of rotation shown, the lower stop prevents rotation of the disks. Since the system is symmetrical, its behavior is identical in reverse.

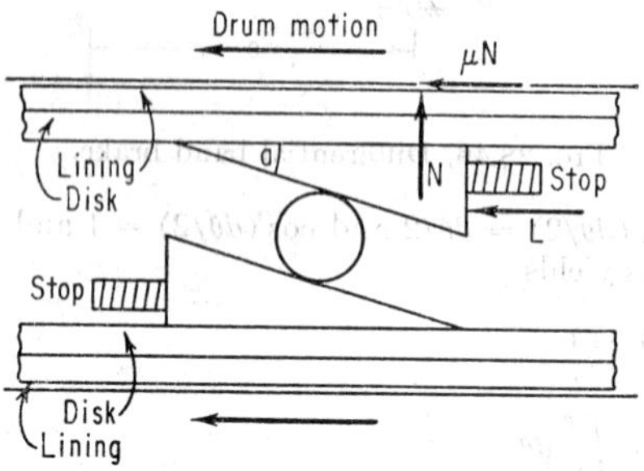

FIG. 28.47. One of the six ball-and-ramp systems of a disk brake.

Because of the wedging action, the normal force N exerted by the upper disk against the drum by applied load L is

$$N = L \cot a \tag{28.181}$$

where a is the wedge or ramp angle. Because of the motion of the drum, the friction force μN on the lining is added to the applied load L, thereby increasing the normal force and in turn the friction force. The force equation then becomes

$$N = (\mu N + L) \cot a \tag{28.182}$$

where N is now the increased normal force and μ is the coefficient of friction. Solving Eq. (28.182) for N yields

$$N = \frac{L \cot a}{1 - \mu \cot a} \tag{28.183}$$

so that the friction force is

$$F = \frac{\mu L \cot a}{1 - \mu \cot a} \tag{28.184}$$

* The effectiveness is often defined as the ratio of vehicle deceleration to hydraulic pressure. This quantity is directly proportional to the effectiveness as defined above, the proportionality factor varying with type of brake and weight of vehicle.

The factor $\mu \cot a$ is the actuation factor A. Its significance can be demonstrated as follows: Let L' be the load required to produce a given friction force without self-actuation and L, as before, be the load sufficient to produce the same friction force with self-actuation. Then the actuation factor is found to be the fractional decrease in load due to self-actuation:

$$\frac{L' - L}{L'} = \frac{\mu N}{\mu N + L} = \mu \cot a \tag{28.185}$$

since from Eq. (28.182) $L' = \mu N + L$. $\cot a$ is the actuation constant, a purely geometric factor.

The actuation constant Q is the reciprocal of the locking coefficient. When the coefficient of friction μ is such as to cause the denominator in Eq. (28.184) to equal zero, the friction force becomes infinite and the brake will lock. For usual coefficients, the friction force increases rapidly with increasing coefficient. For example, an increase in coefficient from 0.40 to 0.49 will double the friction force (Fig. 28.58, curve A).

In this disk brake, the load is applied circumferentially through two symmetrically placed hydraulic cylinders at radius r_1 from the brake center, while the average lining radius is r_2. The friction force [Eq. (28.184)] is therefore reduced by the ratio r_1/r_2.

Equation (28.184) gives the friction force on the lining of the upper disk only. An equal friction force is exerted on the lower lining.

The total friction force produced by the brake, referred to a single cylinder load L*, is

$$F = 2\frac{r_1}{r_2}\frac{\mu L \cot a}{1 - \mu \cot a} \tag{28.186}$$

and the brake effectiveness is

$$\frac{F}{L} = 2\frac{r_1}{r_2}\frac{\mu \cot a}{1 - \mu \cot a} \tag{28.187}$$

Expressions of the type of Eq. (28.187) will hereafter be referred to as "actuation equations."

Equation (28.187) neglects the force required to overcome the retraction springs and rolling friction, which must be measured for each type of brake, whether disk or drum. In the disk brake considered, a load of about 140 lb is required to place the linings just in contact with the drum.

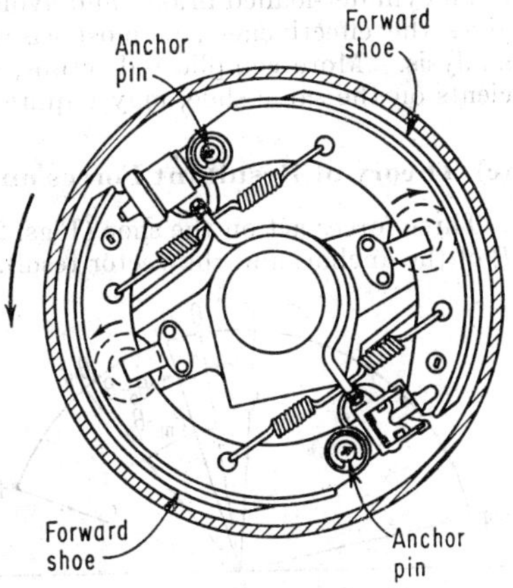

FIG. 28.48. Fixed-anchor internal drum brake.

(b) The Internal Drum Brake

Self-actuating drum brakes are of three types: (1) fixed anchor (Fig. 28.48), (2) movable or link anchor (Fig. 28.49), and (3) floating or sliding anchor (Fig. 28.50). Each brake contains two shoes. Both shoes may be self-actuating, called *forward shoes*, commonly used on front wheels; or one shoe may be anti-self-actuating, called the *drag shoe*, commonly used on rear wheels. In passenger car brakes, the load is applied through hydraulic wheel cylinders and is given by the product of the fluid pressure and wheel-cylinder area. In many truck brakes, the load is applied through cams and must be measured.

* Or, referred to unit fluid pressure. This provides the same basis of comparison for the brakes considered, all of which have two wheel cylinders.

Several methods are in use for calculating brake effectiveness. They usually assume negligible drum and shoe distortion, an assumption valid for light loads only. They also assume a uniform coefficient of friction over the surface of the lining, valid for most practical purposes. The equations derived herein are based upon a graphical method described by Fazekas[49] and others.[50–52]

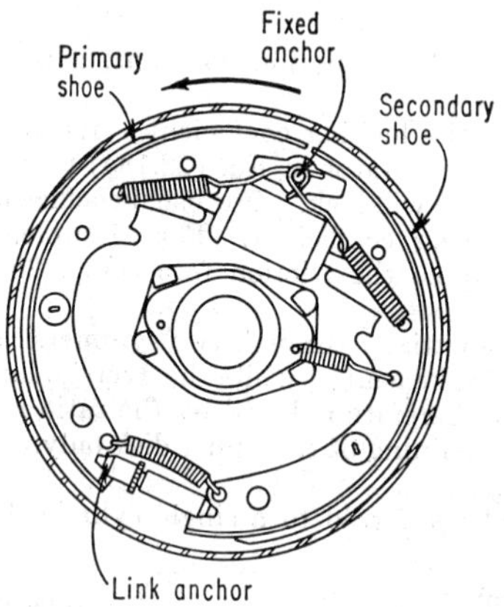

FIG. 28.49. Movable- or link-anchor internal drum brake.

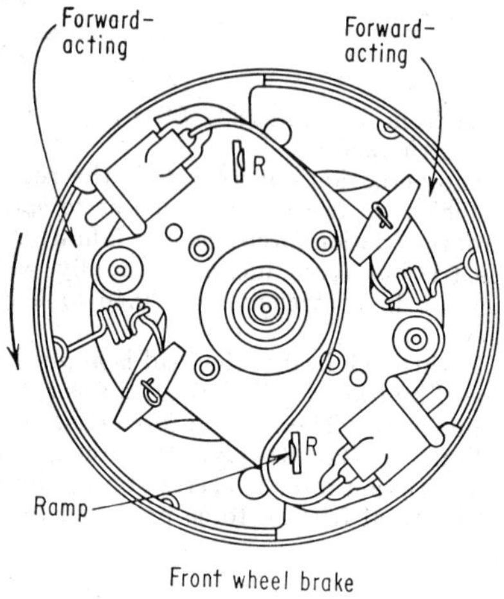

FIG. 28.50. Floating- or sliding-anchor internal drum brake.

Accurate methods involve integration of the friction and normal forces on an element of lining. Some methods[53,54] require the calculation of a definite integral for each type of brake and for each change of dimensions. The present method applies to all cylinder-loaded brakes and avoids special integration. The actuation equation gives the effectiveness, in most cases more quickly and accurately than graphical analysis. More complicated setups, such as two pieces of lining of different coefficients on the same shoe, may require special integration and graphical construction.

(c) Theory of Resultant Forces on Shoe

Three forces act on the shoe (Figs. 28.55 to 28.57), the applied load L, the reaction R of the anchor, and the vector resultant reaction V of the friction and normal forces exerted by the drum through the lining. The force V acts through the so-called "drag point" or "center of pressure" (CP). In most practical cases, the CP lies outside the drum of any internal curved-shoe brake. With negligible distortion, the pressure distribution is harmonic. With moderate distortion, it is practically uniform over the lining.

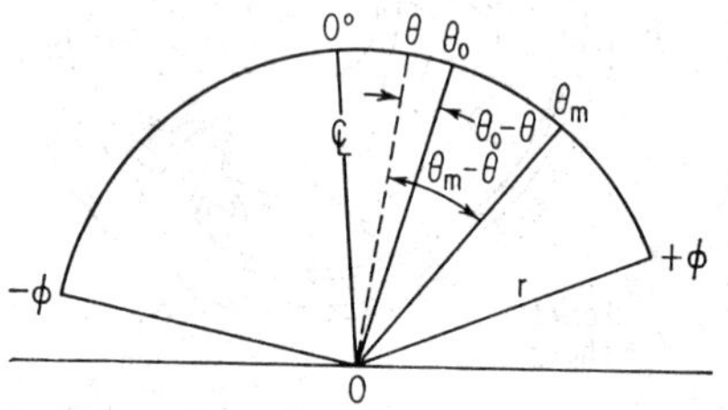

FIG. 28.51. Schematic diagram of the circular surface of the lining of angular arc 2ϕ.

Moment of Friction Forces about Center of Drum. Figure 28.51 is a schematic diagram of the circular surface of the lining of angular arc 2ϕ. The origin of coordinates is the center of the brake. Let p_m be the normal (radial) pressure at θ_m, locating the line of maximum pressure (LMP), θ_m being measured from the center line of the lining. The pressure at an angle θ from the center line is

$$p = p_m \cos (\theta_m - \theta) \qquad (28.188)$$

The normal force on an element of area of lining is $pwr\,d\theta$, where w is the width of lining and r is the drum radius. The corresponding friction force in the direction of rotation of the drum is $\mu pwr\,d\theta$ and the moment of this force about the center is $\mu pwr^2\,d\theta$, where μ is the coefficient of friction.

Integrating these moments over the lining gives the total friction moment

$$Fr = \int_{-\phi}^{+\phi} \mu p_m w r^2 \cos(\theta_m - \theta)\,d\theta$$

$$Fr = 2\mu p_m w r^2 \cos\theta_m \sin\phi \tag{28.189}$$

Resultant Normal Force. The normal forces $pwr\,d\theta$ acting together produce a vector resultant normal force N along a drum radius at some angle θ_0 from the center line of the lining. The component of the pressure p in the direction θ_0 is $p\cos(\theta_0 - \theta)$; so that the component of the normal force in this direction is $pwr\cos(\theta_0 - \theta)\,d\theta$.

Integrating these forces over the lining gives the resultant normal force

$$N = p_m wr \int_{-\phi}^{+\phi} \cos(\theta_m - \theta)\cos(\theta_0 - \theta)\,d\theta$$

$$N = p_m wr[\cos\theta_m \cos\theta_0(\phi + \sin\phi\cos\phi) + \sin\theta_m \sin\theta_0(\phi - \sin\phi\cos\phi)] \tag{28.190}$$

Relation of θ_0 to θ_m. By definition, the resultant normal force perpendicular to the direction θ_0 is zero. The component of the normal force perpendicular to θ_0 is $pwr\sin(\theta_0 - \theta)\,d\theta$. Integrating these forces over the lining gives

$$\int_{-\phi}^{+\phi} \cos(\theta_m - \theta)\sin(\theta_0 - \theta)\,d\theta = 0$$

$$\cos\theta_m \sin\theta_0(\phi + \sin\phi\cos\phi) - \sin\theta_m\cos\theta_0(\phi - \sin\phi\cos\phi) = 0$$

Transposing and dividing gives

$$\tan\theta_m = \tan\theta_0 \frac{\phi + \sin\phi\cos\phi}{\phi - \sin\phi\cos\phi} \tag{28.191}$$

Center-of-pressure Circle—CP Locus. The resultant friction force μN acts through the CP at right angles to N. The moment of μN about the center of the drum $\mu Nc = Fr$, where c is the distance of the CP from the center of the drum. Using the values for Fr and N from Eqs. (28.189) and (28.190) together with Eq. (28.191),

$$c = \frac{Fr}{\mu N} = \frac{2r\sin\phi\cos\theta}{\phi + \sin\phi\cos\phi} \tag{28.192}$$

This is the polar equation of a circle[55] with the origin on its circumference, its center at point $d/2$, 0, and of diameter

$$d = \frac{2r\sin\phi}{\phi + \sin\phi\cos\phi} \tag{28.193}$$

The relationship between d/r and ϕ is shown in Fig. 28.52, curve A.

Relation of Center of Pressure to Line of Maximum Pressure. This relationship is given in Eq (28.191) which, employing Eq. (28.193), may be written

$$\tan\theta_0 = \tan\theta_m[1 - (d/r)\cos\phi] \tag{28.194}$$

The relationship between $\tan\theta_m/\tan\theta_0$ and ϕ is shown in Fig. 28.53.

(d) Effect of Moderate Distortion of Drum and Shoe—Uniform Pressure

Assume the CP is initially at the center line of the lining, i.e., $\theta_0 = 0$, for the usual coefficient of friction. This gives the minimum variation of pressure over the lining

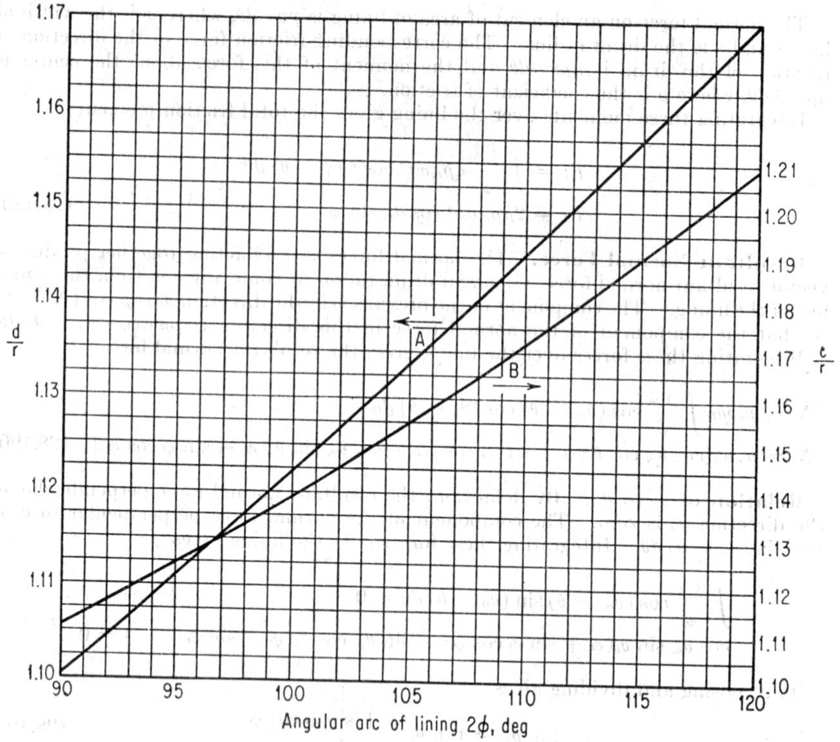

FIG. 28.52. Relationship between d/r and ϕ.

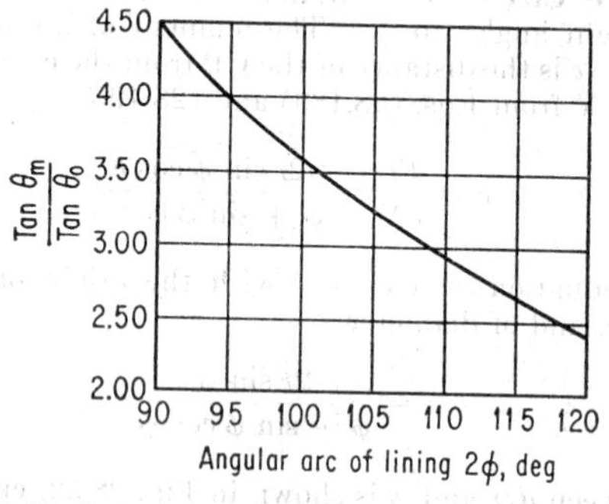

FIG. 28.53. Relationship between $\tan \theta_m / \tan \theta_0$ and ϕ.

and consequently the most uniform heating and wear, and maximum effectiveness. In well-adjusted brakes, the wear is often uniform, i.e., the pressure is uniform over the lining. This means that the CP remains at the center line after moderate distortion. This will not necessarily be the case if the coefficient changes considerably, since, in general, the CP is at the center line for only one coefficient of friction. With uniform pressure p,

$$Fr = \int_{-\phi}^{+\phi} \mu p w r^2 \, d\theta = 2\mu p w r^2 \phi$$

The component of p in the direction θ_0 is $p \cos \theta$, since $\theta_0 = 0$. Therefore,

$$N = \int_{-\phi}^{+\phi} pwr \cos \theta \, d\theta = 2pwr \sin \phi$$

The distance of the CP from the center of the drum

$$c = \frac{Fr}{\mu N} = \frac{2\mu pwr^2\phi}{2\mu pwr \sin \phi} = \frac{r\phi}{\sin \phi} \qquad (28.195)$$

This is a circle of radius c with its center at the drum center. The function c/r vs. 2ϕ is shown plotted in Fig. 28.52, curve B.

(e) Effect of Lining Insert of Different Coefficient of Friction

Equation (28.189) gives the braking torque of a one-piece lining. The braking torque of a two-piece lining is given by

$$T = \mu_1 p_m wr^2[\cos \theta_m(\sin \phi + \sin \phi_1) - \sin \theta_m(\cos \phi_1 - \cos \phi)] + \mu_2 p_m wr^2[\cos \theta_m(\sin \phi - \sin \phi_1) + \sin \theta_m(\cos \phi_1 - \cos \phi)] \qquad (28.196)$$

Here μ_1 is the coefficient of the insert, μ_2 that of the main segment of lining, and ϕ_1 is the angular position of the junction measured from the center line of the total lining. When $\mu_1 = \mu_2$, Eq. (28.196) is identical to Eq. (28.189).

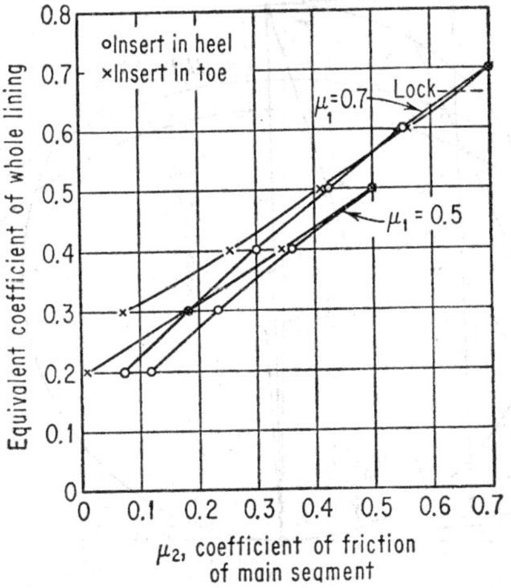

FIG. 28.54. Coefficient of friction of whole lining vs. coefficient of friction of main segment.

Equating the torque of Eqs. (28.189) and (28.196) gives

$$\mu = \frac{1}{2 \sin \phi} [\mu_1(\sin \phi + \sin \phi_1) + \mu_2(\sin \phi - \sin \phi_1) - (\mu_1 - \mu_2) \tan \theta_m(\cos \phi_1 - \cos \phi)] \qquad (28.197)$$

Equation (28.197) is solved by finding pairs of values of μ_1 and μ_2 which make the right side equal to some chosen value of μ, the coefficient of a one-piece lining, the equivalent coefficient of a two-piece lining.

Equation (28.197) assumes harmonic-pressure distribution, valid provided the compressibilities of the two linings are equal, or nearly so. When Eq. (28.197) is satisfied, the LMP of the composite lining necessarily coincides with that of the single lining. Figure 28.54 shows the results of computations on a 4-in. insert in a 12-in. lining in a sliding anchor brake. The insert of higher coefficient $\mu_1 = 0.5$ or 0.7 has more effect in raising the equivalent coefficient when in the toe than when in the heel. The points for $\mu_1 = 0.6$ lie midway between and are omitted to avoid overlapping.

(f) The Actuation Equation—Effectiveness

The above relations, together with the known geometry of a given brake, suffice to obtain the actuation equation and hence the effectiveness, usually without graphical construction. Either the LMP (θ_m) or the CP (θ_0) can always be found from the geometry of the brake, and the other found from Fig. 28.53.

Fixed-anchor brakes are the most stable. Each shoe has a separate anchor (Fig. 28.48) which allows rotation about a fixed axis. The actuation equation for the forward shoe, anchored at the trailing end, is similar to Eq. (28.187). The actuation equation for the drag shoe, anchored at the leading end, is obtained by reversing the sign of the actuation factor. The anchored end is herein called the *heel* and the loaded end the *toe*.

Figure 28.55 is a schematic diagram showing the lining of a forward shoe of an 11-in.-diameter brake and the angles and dimensions required to obtain the actuation

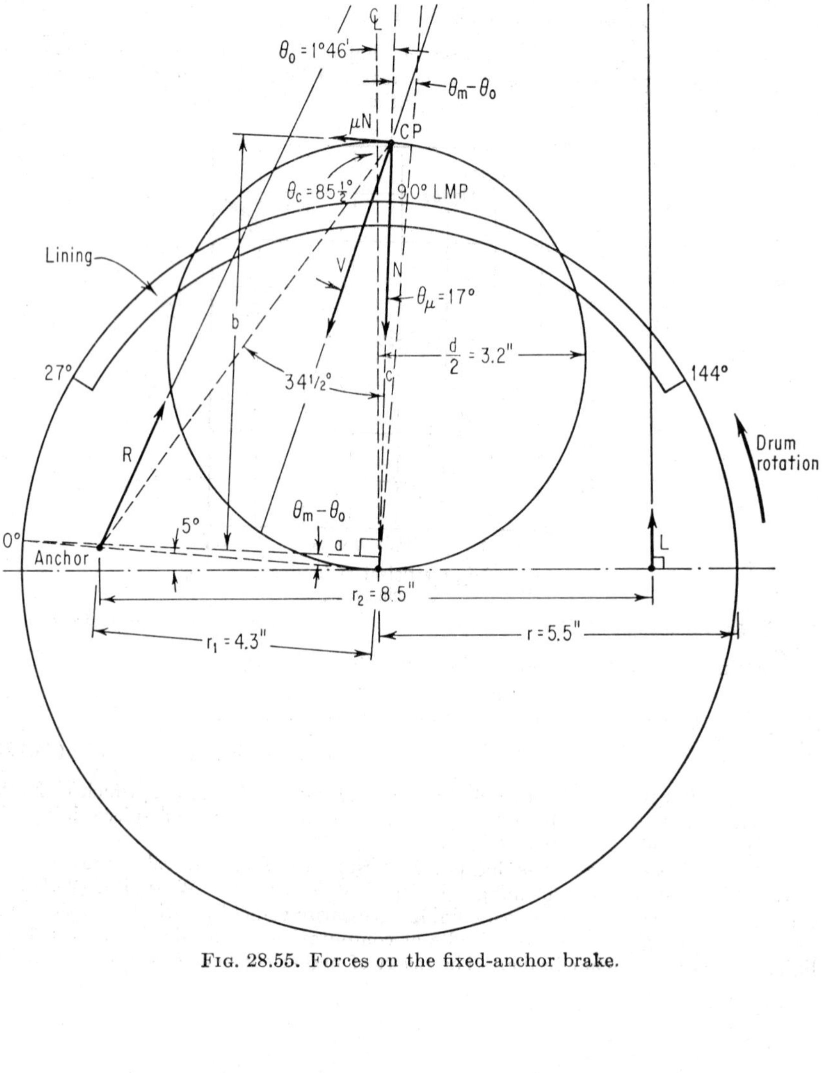

FIG. 28.55. Forces on the fixed-anchor brake.

equation. The drum reaction V is drawn through the CP, making the friction angle $17° = \arctan 0.3$ with the normal force N. The anchor reaction R is drawn through the anchor and the intersection of V with the load L (off the diagram). The magnitudes of R and V may be obtained by constructing a force triangle on L.

The angles θ_m and θ_0 are measured from the lining center line. All other lining angles are measured from the anchor line, a line from the center of the brake to the anchor. From Fig. 28.52 for $2\phi = 117°$, the CP-circle diameter $d = 6.4$ in. for no drum distortion. The CP circle is drawn as defined by Eq. (28.193), with its center on the center line of the lining at $85\frac{1}{2}°$.

In a fixed-anchor brake, the LMP necessarily lies on a line at right angles to the anchor line.[49] Therefore, $\theta_m = 90 - 85\frac{1}{2}° = 4\frac{1}{2}°$. From Fig. 28.53 $\theta_0 = 1°46'$.

The CP lies on the CP circle at its intersection with the drum radius inclined at $1°46'$ to the center line of the lining. The pressure distribution on this lining is thus very nearly symmetrical.

Since the LMP is fixed in a fixed-anchor brake, the location of the CP is independent of the coefficient of friction and depends only on the length and position of the lining. This explains its greater stability. In link- and sliding-anchor brakes, the CP and the LMP both move as the coefficient of friction varies.

It is convenient to write the actuation equation in terms of the ratio of the braking torque to the applied torque about the anchor:

$$\frac{Fr}{Lr_2} = \frac{\mu Nc}{Na - \mu Nb} = \frac{\mu(c/a)}{1 - \mu(b/a)} \tag{28.198}$$

where μ = coefficient of friction
a = moment arm of the normal force N about the anchor
b = moment arm of the friction force μN about the anchor
c = moment arm of the friction force μN about the center of the brake
r = radius of the brake drum
r_2 = moment arm of the load L about the anchor

The ratio b/a is the actuation constant Q. The factor $\mu(b/a)$ is the actuation factor A. The shoe locks when the drum reaction V passes through the anchor,* i.e., when $\mu = a/b = \tan \angle(V - N) = \tan 34\frac{1}{2}° = 0.687$, in this case.

In a drag shoe, the algebraic sign of the actuation factor is changed. It cannot lock and its effectiveness is less than if non-self-actuating.

Figure 28.55 illustrates the graphical method. The quantities needed to obtain the actuation equation may be measured on a full-scale drawing, or they may be obtained analytically. From Fig. 28.55,

$$\begin{aligned} a &= r_1 \cos(\theta_m - \theta_0) \\ b &= c - r_1 \sin(\theta_m - \theta_0) \\ c &= d \cos \theta_0 \end{aligned} \tag{28.199}$$

where r_1 is the length of the anchor line. These quantities can all be obtained from Figs. 28.52 and 28.53 and the geometry of the brake and lining, without graphical construction.

The effectiveness of two forward shoes of the 11-in. brake, referred to a single cylinder load L, is plotted in Fig. 28.58, curve B. For any coefficient, the effectiveness is a maximum when the CP lies on the center line of the lining. The magnitude and distribution of pressure on the lining can be calculated from Eqs. (28.188), (28.189), (28.198), and (28.199).

The link-anchor brake, exemplified by the Bendix Duo-Servo brake[56] (Fig. 28.49), has one movable-anchor shoe, called the primary, and one fixed-anchor shoe, called the secondary. The load is applied near the toe of the primary and the heel of the secondary through hydraulic cylinders placed back to back. The heel of the primary

* Neglecting the force of the retraction springs. In ref. 48, the left side of Fig. 9 illustrates the locking condition, although the condition is not so defined. The drum reaction must lie inside the anchor if the shoe is to return when the load is removed—at least, as long as the drum is rotating.

is connected to the toe of the secondary through a rigid, hinged link. The end of the link attached to the primary is the link anchor which can move toward or away from the drum as the primary rotates about it.

Figure 28.56 is a schematic diagram of the primary in the 12-in. Bendix brake. The forces shown are those exerted on the primary shoe and lining. The drum rotation moves the primary out of contact with the fixed anchor. The load L acts at a

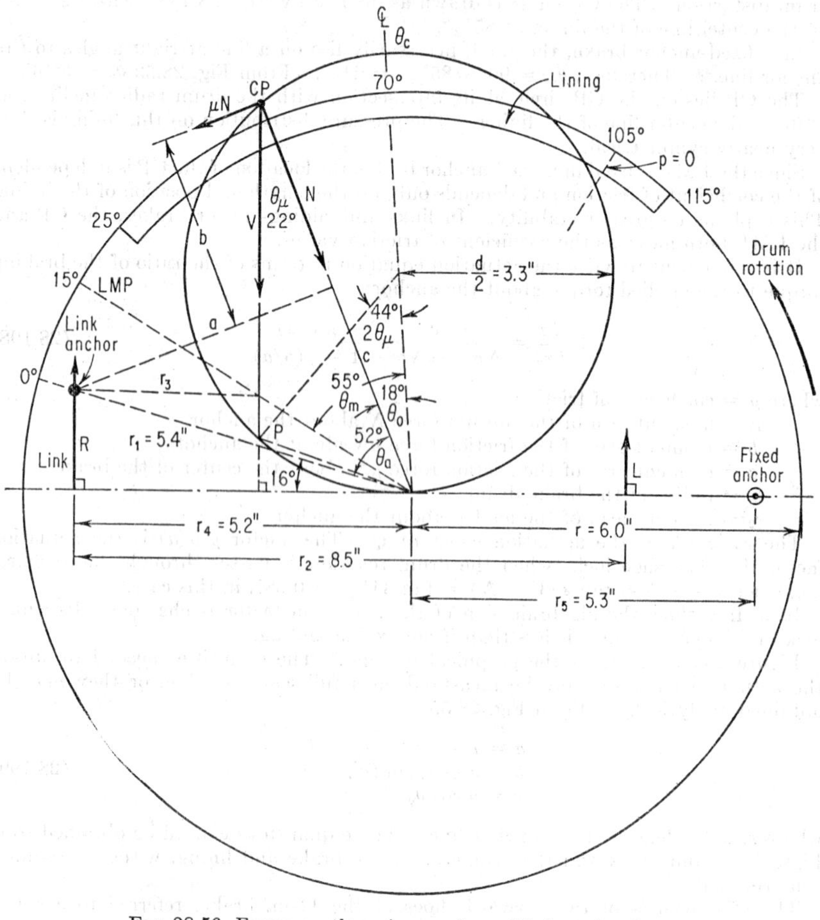

FIG. 28.56. Forces on the primary shoe of link-anchor brake.

right angle to the center line of the brake. The reaction R through the link is, for all practical purposes, parallel to L. Therefore, $V = R + L$. The moments about the link anchor $Vr_3 = Lr_2$. These two equations give the magnitude of V and R. The pressure distribution is harmonic, but the LMP is not known a priori as for a fixed anchor.

In the graphical method, the CP is found by drawing a line parallel to L through a point P located on the CP circle at an angular distance $2\theta_\mu$ from the lining center line. This line also intersects the CP circle at the CP. Since an inscribed angle is measured by half the intercepted arc, this line makes the angle θ_μ with the drum radius through the CP, thus defining the position and direction of V and N.

Analytically, from Fig. 28.56,

$$\theta_a = 74 - \theta_\mu = 52° \tag{28.200}$$
$$\theta_0 = \theta_c - \theta_a = \theta_c + \theta_\mu - 74° = 18° \tag{28.201}$$

where θ_a = angular position of the CP measured from the anchor line
θ_μ = friction angle, arctan μ
θ_c = angular position of the lining

The LMP (θ_m) is found from Eq. (28.201) and Figs. 28.52 and 28.53 to be 15° from the anchor line, i.e., 10° beyond the heel of the lining. At 90° from this, or 10° in from the toe, the pressure is zero. The remainder of the toe is lifted off the drum.

The pressure distribution is extremely asymmetrical with the lining in this position. For $\mu = 0.4$, the pressure distribution would be symmetrical if the lining were moved 18° toward the link anchor. In the symmetrical-pressure position, the effectiveness is 10 per cent higher than in the position of Fig. 28.56.

The actuation equation is, as before,

$$\frac{Fr}{Lr_2} = \frac{\mu(c/a)}{1 - \mu(b/a)} \tag{28.202}$$

From Fig. 28.56,

$$\begin{aligned} a &= r_1 \sin \theta_a \\ b &= c - r_1 \cos \theta_a \\ c &= d \cos \theta_0 \end{aligned} \tag{28.203}$$

Since θ_0 and θ_a are both functions of μ, the effectiveness is a complicated function of μ. In general, both the numerator and denominator of the right-hand side of Eq. (28.202) are quadratic functions of μ. This shoe cannot lock, since the drum reaction V cannot pass through the anchor; the quadratic denominator of the right-hand side of Eq. (28.202) has no real roots. The secondary shoe, however, can lock.

In this case, the friction force in the primary can be found by a simpler method without using the actuation equation. Since V is normal to the brake center line, the moment of V about the center of the brake is $Vc \sin \theta_\mu = (Lr_2/r_3)c \sin \theta_\mu$, which in turn is equal to Fr.

The secondary is analyzed as is any other fixed-anchor shoe. The drum rotation forces the secondary against the fixed anchor. Two loads are exerted on the secondary, the reaction R of the primary and the applied load L, both opposite in direction to that shown in Fig. 28.56. The total torque exerted by these two loads about the fixed anchor is equal to that exerted by V.

The total torque T applied to the secondary about the fixed anchor is therefore

$$T = L(r_2/r_3)(r_5 + c \sin \theta_\mu) \tag{28.204}$$

where r_2 = moment arm of L about the link anchor
r_3 = moment arm of V about the link anchor $= r_4 - c \sin \theta_\mu$
r_4 = distance from the link anchor to the center of the brake
r_5 = distance from the fixed anchor to the center of the brake

The actuation equation is

$$\frac{Fr}{T} = \frac{\mu(c/a)}{1 - \mu(b/a)} \tag{28.205}$$

where a, b, and c have the values for a fixed-anchor brake [Eq. (28.199)].

For the values given in Fig. 28.56, $T = 22.8L$, which may be compared with $Lr_2 = 8.5L$ for the primary. The secondary lining commonly has a lower coefficient but greater arc length than the primary. Since this brake is symmetrical, except for the linings, its behavior is identical in reverse, the primary and secondary becoming interchanged.

The nearly threefold greater braking torque exerted by the secondary results in an asymmetrically loaded and distorted drum. The pressure on the secondary is three to four times that on the primary and the rate of wear correspondingly greater.

The combined effectiveness of the two shoes of the 12-in. Duo-Servo brake is shown plotted in Fig. 28.58, curve *C*.

The floating-shoe or sliding-anchor brake[49] is the least stable. It has two separately loaded shoes (Fig. 28.50). The heel of each shoe rotates and slides on a ramp inclined at about 8° to the anchor line. Figure 28.57 is a schematic diagram of one forward shoe in a 12-in.-diameter brake. The rotation of the drum causes the heel to slide away from the center of the brake toward the drum. The pressure distribution is harmonic but neither the CP nor the LMP is known a priori.

The CP varies with the coefficient of friction both between drum and lining and between heel and ramp. The LMP is very sensitive to the ramp coefficient, which partially explains the instability. In Fig. 28.57, the anchor reaction is drawn to correspond to a ramp coefficient of 0.1, i.e., assuming good lubrication. The anchor reaction makes the angle 5°43′ = arctan 0.1, with a normal to the ramp.

In the graphical construction, the drum reaction V is drawn through the intersection of R and L and the point on the CP circle at the angular distance $2\theta_\mu$ from the lining center line. (For a drag shoe, this point is located on the other side of the center line from that shown in Fig. 28.57.) The other quantities are then obtained by the methods just described.

Analytically, from Fig. 28.57,

$$\theta_a = 155 - \theta_\mu - \theta_V \tag{28.206}$$
$$\theta_0 = \theta_a - \theta_c \tag{28.207}$$

The angle θ_V between the drum reaction and the brake center line may be obtained as follows:

As is shown in Fig. 28.57,

$$\tan \theta_V = \frac{y - y_2}{l + r_3} \tag{28.208}$$

where $y = r_2 \tan \theta_R + r_1 \sin \theta_A$

$y_2 = x \sin B = d \sin \theta_\mu \sin B$

$l = x \cos B = d \sin \theta_\mu \cos B$

$B = \theta_c + \theta_\mu - 65°$

θ_R is the angle between the anchor reaction and the brake center line.

The angle θ_A between the anchor line and the brake center line is 25° in the brake considered. The angle of the CP, θ_0, is obtained from Eqs. (28.206), (28.207), and (28.208). θ_m is obtained from Figs. 28.52 and 28.53. The actuation equation is given by Eqs. (28.202) and (28.203).

Computations for the conditions shown in Fig. 28.57, lining coefficient = 0.4, ramp coefficient = 0.1, give the LMP at θ_m = 6°40′. For a ramp coefficient = 0.2, θ_m = 18°45′ and for a ramp coefficient = 0.4, θ_m = 39°. The effectiveness of two forward shoes for various lining coefficients and a ramp coefficient of 0.1 is shown in Fig. 28.58, curve *D*. If the ramp becomes roughened or is not lubricated, the shoe may stick and the lining make contact with the drum only at the toe. The performance of this brake would be improved by mounting a roller on the heel.

The locking coefficient can be measured with a protractor on the graphical construction. The shoe locks when V coincides with R, i.e., $\theta_V = \theta_R$, and the CP lies on the CP circle at its intersection with R.

The locking coefficient μ_l decreases with increasing ramp coefficient. It is calculated by equating the normal moment Na to the friction moment $\mu_l Nb$ when locking occurs. Substitution of the values for a and b [Eq. (28.203)] into the equation $Na = \mu_l Nb$

$$\mu_l^2(1 - X) + \mu_l \cot Y - X = 0 \tag{28.209}$$

where $X = (r_1/d)(\cos \theta_c + \cot Y \sin \theta_c)$

$Y = 155° - \theta_R - \theta_c$

The following values for μ_l were calculated from Eq. (28.209): ramp coefficient 0.1, μ_l = 0.67; ramp 0.4, μ_l = 0.60; ramp 1.0, μ_l = 0.53. The variation of locking coefficient adds to the instability.

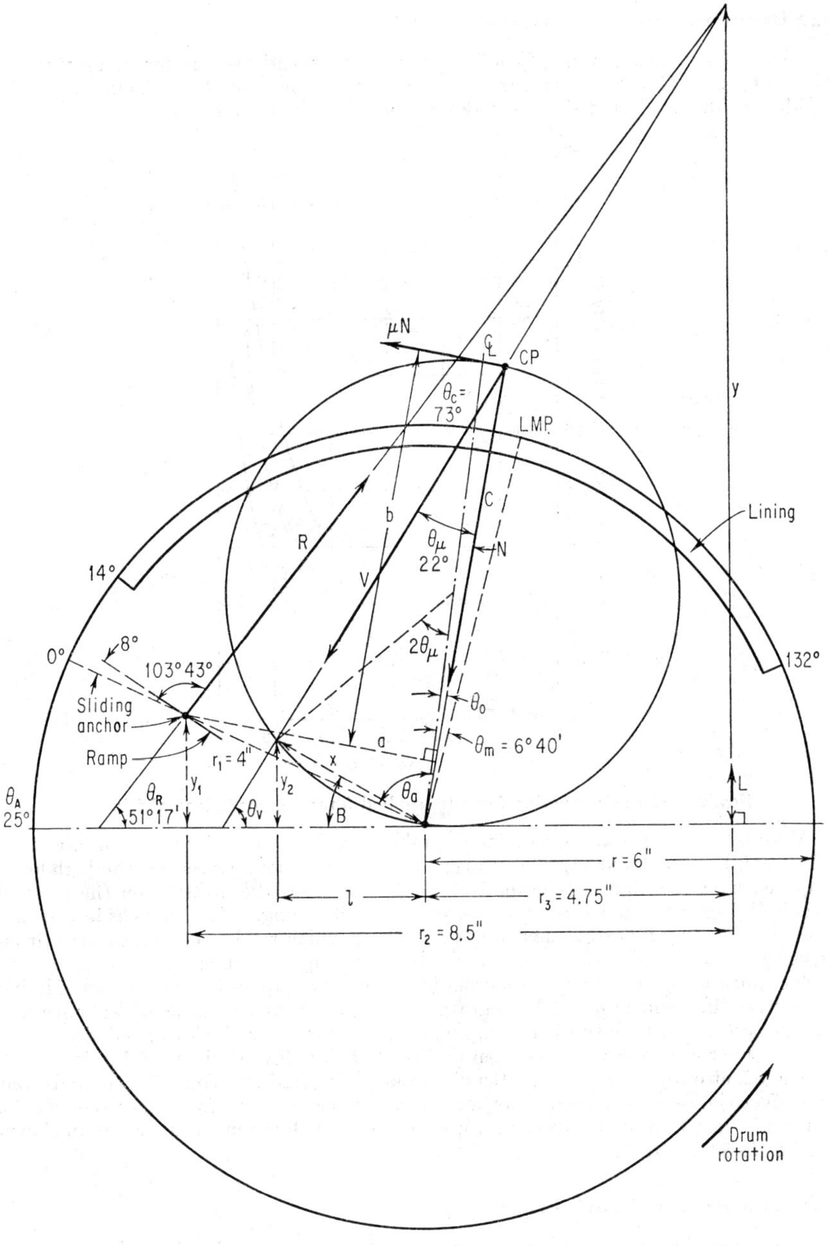

FIG. 28.57. Forces on the sliding-anchor brake.

(g) Drum and Shoe Distortion

Distortion, which can greatly alter the magnitude and distribution of pressure on the lining and, to some extent, the effectiveness, depends upon the applied load, the rigidity of the drum and shoe, and the compressibility of the lining.

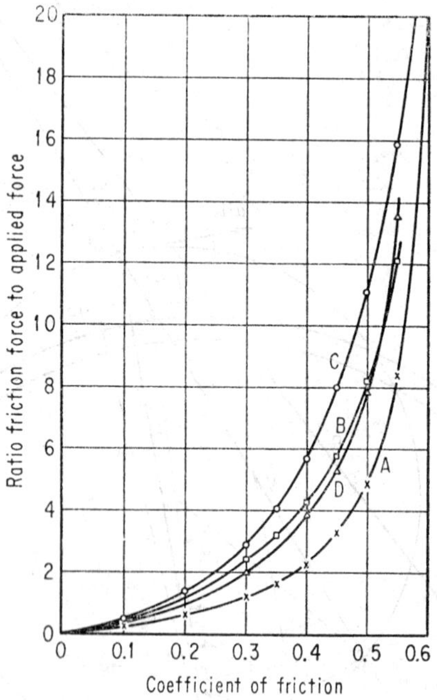

FIG. 28.58. Ratio, friction force to applied force vs. coefficient of friction.

Moderate distortion produces the desirable condition of uniform pressure, and its effect can be calculated [Art. 28.10(d)]. Severe distortion, caused by the high loads required by a low lining coefficient, markedly increases the pressure on the toe and heel and decreases the pressure on the center of the lining. The changes in pressure are minimized by a flexible shoe and a soft lining which tend to conform to the drum when it distorts. The effect on brake performance must be measured.

Computation of lining pressure for a distorted drum required measurement of lining compressibility and numerical integration of the moment equations of Vallance and Doughtie[53] and Chase.[54] The computed effectiveness for the distorted drum was about 20 per cent greater than when undistorted, for all coefficients of friction.

The effect of distortion on the Bendix brake is reported by Winge.[57] He finds that the effectiveness is increased more for lower coefficients than for higher coefficients. The wide variation of effectiveness with coefficient is thus somewhat less than shown in Fig. 28.58, curve *C*.

(h) Cam Brakes—Truck Brakes

Cam-operated drum brakes have one forward and one drag shoe loaded by a double cam[48] mounted on a single pivot and actuated through a lever arm. The load and hence the effectiveness of cam brakes cannot be calculated, because the rotation of the drum pushes the drag shoe against the cam, thereby increasing the load, and tends to pull the forward shoe away from the cam, thereby decreasing the load. The amount of this effect varies with the coefficients of friction between lining and drum and between cam and shoe, the clearance in the cam pivot, the compressibility of the lining, and the distortion of the drum and shoes. The usual result is that the effectiveness of the two shoes is about equal.

Figure 28.59 shows the forces and lining pressure distribution in a Timken 14½-in.-diameter cam-and-roller-type brake. A similar brake is the cam-and-plate type.

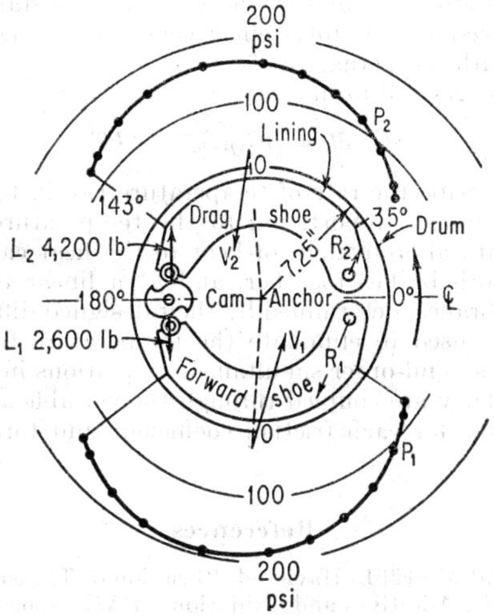

FIG. 28.59. Forces and lining pressure distribution in a Timken brake.

28.11. TORQUE TRANSFER IN AUTOMOTIVE BRAKES

The braking torque in an automobile equipped with self-actuating drum brakes has been observed to vary among the four brakes. Recorded tests[58] in a fully instrumented car show that the individual brakes seldom exert their designed relative braking torque because the coefficient of friction is seldom the same in each. Fade, a decrease in coefficient of friction following temperatures of 500 to 800°F or more, and recovery, following lowered temperatures and wear, cause erratic and often large changes in the coefficient. Self-actuation serves to magnify the changes in coefficient to cause large changes in the effectiveness of individual brakes. The results[58] show that braking torque is continually being transferred in large amounts from one or more brakes to the others.

Torque transfer between right and left brakes produces "erratics" readily detected by the driver. Torque transfer between front and rear is more common because front brakes, designed to be more effective than rear brakes, heat and fade more rapidly. The tests[58] showed that the driver cannot detect front-rear torque transfer.

Similar results were obtained in tests on a specially constructed inertia dynamometer[59] equipped with two brakes. Simulation of torque transfer in a dual-brake dynamometer is necessary since fade in a car causes periodic overloading and overheating of all the brakes, one or more at a time. Some brake linings exhibit "antifade," an increase in coefficient with increasing temperature. In that case, torque tends to pile up on whichever brake happens to get hotter first, so that it either locks or is overloaded to destruction of the lining. None of the above effects can be observed in a single-brake dynamometer,[58] the standard test machine.[60,61]

A mathematical analysis has been given[62] for the case of one front and one rear brake acting together. A series of high-speed fade stops at constant deceleration is simulated by a continuous drag test at constant velocity and power. This preserves the essential condition of constant total torque of both brakes acting together.

The equation

$$a(d\mu/dt) - b\mu = -T \tag{28.210}$$

where μ is the coefficient of friction and T is the temperature, describes the decrease in coefficient with increase in temperature after a time delay. Without the observed time delay, brakes would fail almost immediately, since the temperature often reaches 800°F in the first fade stop. Values of the arbitrary constants a and b are chosen to simulate the observed loss of total effectiveness resulting from the decrease in coefficient combined with self-actuation.

Two equations of the general form

$$dT/dt = BC(\mu/\mu^*) - kT \tag{28.211}$$

one for each brake, describe the rate of temperature rise in terms of the heat input, less the heat loss—assumed proportional to the temperature. The constant B is the ratio of the mechanical equivalent of heat to the heat capacity, C the constant power expanded by both brakes together, and μ^* a linear function of the friction coefficients of the two brakes, determined by their designed difference in effectiveness.

Equation (28.210) is used to eliminate the temperature from the two equations (28.211) to obtain two second-order simultaneous equations in the friction coefficient, one for each brake. They are coupled through the variable μ^*. The solutions show different rates of change for each friction coefficient and torque transfer similar to that observed.

References

1. Rodgers, John J., and Merrill L. Haviland: "Friction of Transmission Clutch Materials as Affected by Fluids, Additives and Oxidation," SAE paper 194A, June 5, 1960.
2. Twiss, Sumner B., and Paul R. Basford: "Properties of Friction Materials," parts I and II, ASME paper 57-SA-97, *Trans. ASME,* February, 1958.
3. Spotts, M. F.: "Design of Machine Elements," 3d ed., Prentice-Hall, Inc., Englewood Cliffs, N.J., 1961.
4. Jakob, Max: "Heat Transfer," vols. I and II, John Wiley & Sons, Inc., New York, 1949, 1957.
5. MacAdams, W. H.: "Heat Transmission," 3d ed., McGraw-Hill Book Company, Inc., New York, 1954.
6. McLachlan, N. W.: "Transform Calculus and Complex Variable Theory," Cambridge University Press, New York, 1953.
7. Gazley, Carl, Jr.: "Heat Transfer Characteristics of the Rotational and Axial Flow between Coaxial Cylinders," ASME paper 56-A-128, *Trans. ASME,* January, 1958.
8. Kays, W. M., and I. S. Bjorklund: "Heat Transfer from a Rotating Cylinder with and without Crossflow," ASME paper 56-A-71, *Mech. Eng.,* 1957.
9. Andronow, A. A., and C. E. Chaikin: "Theory of Oscillations," Princeton University Press, Princeton, N.J., 1949.
10. MacLachlan, N. W.: "Ordinary Nonlinear Differential Equations," 2d ed., Oxford University Press, New York, 1956.
11. Timoshenko, S.: "Vibration Problems in Engineering," 3d ed., D. Van Nostrand Company, Inc., Princeton, N.J., 1955.
12. Friction Clutch Transmissions, *Machine Design,* Nov. 13, 27; Dec. 11, 25, 1958.
13. Nitsche, C.: "Electro Magnetic Multi-disk Clutches," ASME paper 58-5A-61, *Trans. ASME,* May, 1957.
14. Whitehouse, A. A. K.: Friction Materials, *Res. Appl. Ind.,* vol. 12, pp. 206–11, June, 1959.
15. As manufactured and tested by Johns Manville. Unpublished reports.
16. Willaman, P. O.: The Composition Shoe Contribution to Modern Railroad Service, for Presentation before the Air Brake Association, Chicago, Ill., Sept. 16, 1958. Braking with Composition Shoes, *Railway Locomotives and Cars,* November, 1956, pp. 68–69.
17. Cabble, G. M., Jr.: Technique in the Use of Composition Shoes, for Presentation at the Air Brake Association, Chicago, Ill., Sept. 12, 1960.
18. Shearer, Andrew W.: Friction Materials—Today and Tomorrow, *Auto. Inds.,* Apr. 1, 1958, p. 62.
19. Salter, Elwin J.: Selecting Friction Materials, *Mater. Design Eng.,* October, 1957, pp. 130–137.

20. Halstead, R. T.: The Fundamentals of Asbestos Friction Materials, *Paper Trade J.*, vol. 118, pp. 30–33, Mar. 2, 1944.
21. DuBois, W. H.: Cerametallic Friction Material, SAE National Tractor Meeting, Sept. 10–13, 1956.
22. Bikerman, J. J.: Surface Roughness and Sliding Friction, *Rev. Mod. Phys.*, vol. 16, pp. 53–68, January, 1944.
23. Spurr, R. T.: Frictional Behavior of a Simple Rheological Material, *J. Appl. Phys.*, vol. 32, p. 1450, August, 1961.
24. Flom, D. G.: Rolling Friction of Polymeric Materials, *J. Appl. Phys.*, vol. 32, p. 1426, August, 1961; vol. 31, p. 306, 1960.
25. Sohl, G. W., J. Gaynor, and S. M. Skinner: Electrical Effects Accompanying the Stick-slip Phenomenon of Sliding of Metals on Plastics and Lubricated Surfaces, *Trans. ASME*, vol. 79, p. 1963, November, 1957.
26. Brake Lining Quality Control Test Procedure, "1961 SAE Handbook," p. 745.
27. Smith, G. R., V. J. Jandasek, S. R. Sprague, and R. B. Singer: "A New Concept of Measuring Friction Materials," paper 363A, SAE 1961 summer meeting.
28. Rabins, M. J., and R. J. Harker: "The Dynamic Frictional Characteristics of Molded Friction Materials," paper 60-WA-35, ASME 1960 winter annual meeting.
29. Oetzel, J. George: Performance Characteristics of Molded and Woven Friction Materials, *Machine Design*, vol. 23, p. 128, Aug. 3, 1961.
30. Sinclair, D.: Frictional Vibrations, *J. Appl. Mech.*, vol. 22, p. 207, June, 1955.
31. Newcomb, T. P.: Transient Temperatures in Brake Drums and Linings, *Proc. Inst. Mech. Engrs. Auto. Div.*, no. 7, p. 227, 1958–1959.
32. Toghill, E. C., and H. T. Angus: Cast Iron Brake Drums, *Auto. Engr.*, October, 1953, p. 409.
33. Dudley, B. R., and H. W. Swift: Frictional Relaxation Oscillations, *Phil. Mag.*, vol. 40. p. 849, August, 1949.
34. Bowden, F. P., and D. Tabor: "The Friction and Lubrication of Solids," Oxford University Press, New York, 1950.
35. Morgan, F., M. Muskat, and D. W. Reed: Studies in Lubrication. Friction Phenomena and the Stick-slip Process, *J. Appl. Phys.*, vol. 12, p. 743, October, 1941.
36. Sampson, J. B., F. Morgan, and D. W. Reed: Studies in Lubrication. Friction Behavior during the Slip Portion of the Stick-slip Process, *J. Appl. Phys.*, vol. 14, p. 689, December, 1943.
37. James, D. I.: Measurement of Friction between Rubber-like Polymers and Steel, *J. Sci. Instr.*, vol. 38, p. 294, July, 1961.
38. Basford, P. R., and S. B. Twiss: Properties of Friction Materials, *Trans. ASME*, vol. 80, pp. 402, 407, February, 1958.
39. Rabinowicz, E.: The Nature of the Static and Dynamic Coefficients of Friction, *J. Appl. Phys.*, vol. 22, p. 1373, November, 1951.
40. Merchant, M. E.: Characteristics of Polar and Non-polar Lubricant Additives under Stick-slip Conditions, *Lubrication Eng.*, vol. 2, p. 56, 1946.
41. Rabinowicz, E.: A Study of the Stick-slip Process, "Friction and Wear," Elsevier Publishing Company, New York, 1959.
42. Den Hartog, J. P.: "Mechanical Vibrations," 4th ed., McGraw-Hill Book Company, Inc., New York, 1956.
43. Blok, H.: Fundamental Mechanical Aspects of Boundary Lubrication, *SAE J.* (*Trans.*), vol. 46, p. 54, February, 1940.
44. Dokos, S. J.: Sliding Friction under Extreme Pressures, *J. Appl. Mech.*, vol. 13, p. A-148, June, 1946, discussion, vol. 14, p. A-68, March, 1947.
45. Singh, B. R.: Study of Critical Velocity of Stick-slip Sliding, *Trans. ASME* (B), vol. 82, p. 393, November, 1960.
46. Brakes, *Auto. Engr.*, Nov. 25, 1953; November, 1956; November, 1958.
47. Rodger, W. R.: Disk Brakes, *Machine Design*, vol. 22, p. 148, July, 1950.
48. Frazee, Irving, and Earl L. Bedell: "Automotive Brakes and Power Transmission Systems," chap. 1, American Technical Society, Chicago, Ill., 1956.
49. Fazekas, G. A. G.: Graphical Shoe-brake Analysis, *Trans. ASME*, vol. 79, p. 1322, August, 1957; Some Basic Properties of Shoe Brakes, *J. Appl. Mech.*, vol. 25, p. 7, March, 1958.
50. Acres, F. A. S.: Some Problems in the Design of Braking Systems, *Inst. Auto. Engs. J.*, vol. 15, pp. 33–35, 41–49, November, 1946.
51. Robinson, J. G.: Brake Design Considerations, *Auto. Engr.*, vol. 49, p. 340. September, 1959.
52. Oldershaw, R. M., and A. F. Prestidge: Brake Design Considerations, *Auto. Engr.*, vol. 50, p. 157, April, 1960.

53. Vallance, A., and V. L. Doughtie: "Design of Machine Members," McGraw-Hill Book Company, Inc., New York, 1943.
54. Chase, T. P.: Passenger Car Brake Performance, *SAE J.*, vol. 56, p. 51, October, 1948
55. Middlemiss, Ross R.: "Analytic Geometry," 2d ed., p. 204, McGraw-Hill Book Company, Inc., New York, 1955.
56. Lupton, Clark R.: Drum Brakes, *Machine Design*, vol. 22, p. 146, July, 1950.
57. Winge, J. L.: "Instrumentation and Methods for the Evaluation of Variables in Brakes," paper 361B, SAE summer meeting, 1961.
58. Sinclair, D., and D. H. Wright: "Recording Brake Performance in Road and Dynamometer Tests," SAE National Passenger Car Body and Materials Meeting, Detroit, Mar. 6, 1958.
59. Sinclair, D., and W. F. Gulick: "The Dual-brake Inertia Dynamometer—A New Tool for Brake Testing," Paper 488C, SAE National Automobile Week, Detroit, Mar. 12–16, 1962; *SAE Trans.*, vol. 71, 1963.
60. "Brake Test Code-Dynamometer," 1961 SAE Handbook, p. 749.
61. Report of SAE Brake Subcommittee No. 3, Construction and Operation of Brake Testing Dynamometers, *SAE Trans.*, vol. 61, pp. 329–346, 1953.
62. Sinclair, D., Torque Transfer in Automotive Brakes, *Proc. Inst. Mech. Engrs. Auto. Div.*, 1963.

Section 29

BELTS

By

W. H. BAIER, Ph.D., *Senior Research Engineer, IIT Research Institute, Technology Center, Chicago, Ill.*

CONTENTS

29.1. INTRODUCTION

Belt drives, employing flat, V-, or other belt cross sections, are used in the transmission of power between shafts which may be located some distance apart. There are basically two types of belting: friction and positive drive.

Friction belting transmits power by means of friction between belt and pulley or sheave; hence because of slippage, the velocity ratio between shafts is not constant. The positive drive "timing" belt has teeth formed in one face which mesh with toothed or grooved pulleys.

A number of belt materials are currently in use. Among these are leather, natural and synthetic rubber, cotton, canvas, nylon, and animal hair. Shaft center-line position may be fixed or may vary.

Belting is generally operated at speeds less than 6,000 fpm. Flat woven or synthetic endless belts are used at speeds up to 18,000 fpm.

29.2. FLAT BELTS

(a) Forces

In transmitting power from one shaft to another by means of a flat belt and pulleys, the belt must have an initial tension T_0. When power is being transmitted, the tension T_1 in the tight side exceeds the tension T_2 in the slack side (Fig. 29.1).

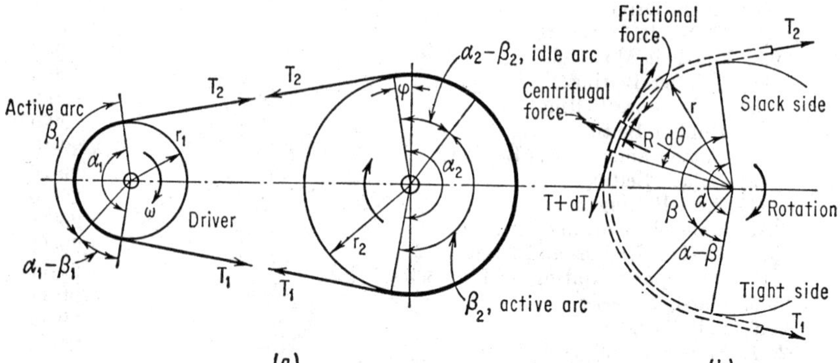

FIG. 29.1. Action of belts on pulleys.

The tension in a belt transmitting power is determined by considering belt deformation under load. For leather belts the stress-strain relationship is not linear. If it is assumed that the total stretch of the belt is the same when it is transmitting power as it is when it is at rest, T_1, T_2, and T_0 are related by the following equations:

for vertical (and short horizontal) belts,

$$T_1^{1/2} + T_2^{1/2} = 2T_0^{1/2} \tag{29.1}$$

and for horizontal belts,

$$T_1^{1/2} + T_2^{1/2} = 2T_0^{1/2} + \frac{w^2E_1C^2A^{1/2}}{24}\left(\frac{1}{T_1^2} + \frac{1}{T_2^2} - \frac{2}{T_0^2}\right) \tag{29.2}$$

where w is the weight of the belt, lb/ft (Sec. 29.3), E the elastic constant for leather determined[1] to be in the range 860 to 900 (psi)$^{1/2}$, A the cross-sectional area, in.^{2}, and C the center distance, in. The last term in Eq. (29.2) is introduced by the catenary effect resulting from belt sag. For a given T_1 the resulting T_2 is higher for a horizontal belt.

For belts having a linear stress-strain relation the above equations become:

for vertical (and short horizontal) belts,

$$T_1 + T_2 = 2T_0 \tag{29.3}$$

and for horizontal belts,

$$T_1 + T_2 = 2T_0 + \frac{w^2EC^2A}{24}\left(\frac{1}{T_1^2} + \frac{1}{T_2^2} - \frac{2}{T_0^2}\right) \tag{29.4}$$

where E is the modulus of elasticity of the belt material, psi.

The horsepower transmitted by the belt is given by

$$\text{hp} = \frac{(T_1 - T_2)V}{33{,}000} \tag{29.5}$$

where V is the belt velocity, fpm.

(b) Action of Belt on Pulley

Because the tension T_1 in the tight side of the belt is greater than that in the slack side T_2, the belt material undergoes a change in strain as it passes around the belt. The belt has a small relative motion, called *creep*, with respect to the pulley to compensate for these different strains. Except when the drive delivers maximum power, this creep occurs only on the so-called active arc β of the pulley (see Fig. 29.1). The active arc increases as the effective tension $(T_1 - T_2)$ increases until the active arc equals the entire arc of contact.

The forces acting on the belt are shown in Fig. 29.1*b*. Summing forces in the tangential and normal directions and substituting and integrating the tangential forces over the active arc, 0 to β, yields

$$(T_1 - wv^2/g)/(T_2 - wv^2/g) = e^{\mu\beta} \tag{29.6}$$

where wv^2/g accounts for centrifugal effects, v is the belt velocity, fps, and μ is the coefficient of friction (Table 29.1). The output of the belt is shown to decrease with increasing belt velocities as a result of centrifugal effects. Equation (29.6) indicates that, at some particular speed, the output drops to zero. Effects other than those considered allow the belt to transmit power even at these speeds and the equation therefore gives conservative results, especially in the case of heavy belts.

Flat belts in line-shaft and machine-belting service are generally operated in the 1,000 to 4,500 fpm range. While operation in the 3,500 to 4,500 fpm range results in the highest horsepower capacity per dollar expended for belts and pulleys, drive efficiency is less than that obtained when operating at lower speed. Total annual cost of lower-speed (1,000 to 3,000 fpm) drives is in most cases lower.

To utilize the weight of the belt in maintaining tension (by increasing the arc of contact) the driving member should rotate so that the tension side of the belt is at the bottom in a horizontal drive. When the drive is vertical, steeply inclined, or horizontal with short center-line distance, no such advantage can be gained from the belt weight.

A number of methods are in use to increase contact pressure or angle of contact to enable the belt to transmit a greater effective tension. One such method utilizes an idler pulley between driver and driven pulleys. While effective tension is increased, belt life is reduced because of the reversed flexing introduced by the idler.

In a vertical drive, the belt tension at the top of the belt on each side of the pulley is greater than the tension at the bottom, by an amount equal to the weight of the freely hanging belt. It is therefore preferable to have the larger of the two pulleys at the bottom, so that the loss of tension may be offset by an increased contact arc.

(c) Belt Stresses

Figure 29.2 shows the tensile stresses acting in an operating belt. Note the location of the maximum stress.

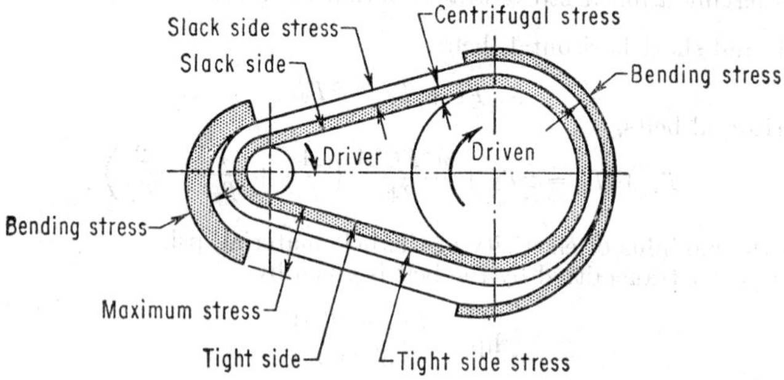

FIG. 29.2. Belt stresses.

(d) Belt Geometry

Open and Crossed Belts. The length of belt L required for an open drive (Fig. 29.3) is given by the series

$$L = 2C + (\pi/2)(D_1 + D_2) + (D_1 - D_2)^2/4C + \cdots \qquad (29.7)$$

where C is the center distance, and D_1, D_2 are the diameters of the large and small

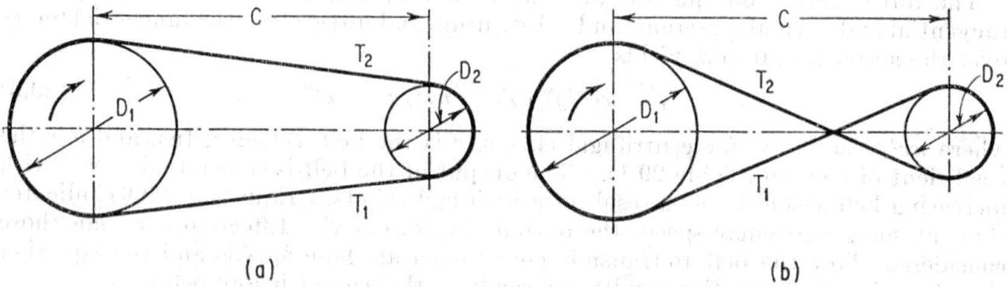

FIG. 29.3. Belt drives.

pulleys, respectively. The length of a crossed belt is given by the series

$$L = 2C + (\pi/2)(D_1 + D_2) + (D_1 + D_2)^2/4C + \cdots \qquad (29.8)$$

For rough calculations, the first two terms provide sufficient accuracy. For increased accuracy, D_1 and D_2 should be taken as the pulley diameter plus one belt thickness. For large center distances, it is evident that the same belt may be used for either an open or a crossed drive. Center distance for an open belt is given by

$$4C = b + [b^2 - 2(D_1 - D_2)^2]^{1/2} \qquad (29.9)$$

where $b = L - (\pi/2)(D_1 + D_2)$. The above discussion neglects the effects of belt sag.

With drives employing a short center distance and a high speed ratio, the angle of contact on the small pulley is decreased and therefore the capacity of the drive is reduced.

To provide reasonable values of effective tension in drives having angles of contact less than 165° on the small pulley requires high belt tensions. Contact angles less than 155° are never recommended.

Cone Pulleys. Cone or stepped pulleys (Fig. 29.4) may be used to obtain a variable velocity ratio. Generally, the velocity ratios are chosen so that the driven velocities increase in a geometric ratio from step to step. That is, if the angular velocity of the first pulley is a, that of the second is ka, of the third k^2a, etc. In practice, k is in the range 1.25 to 1.75. The value[4] for machine tools is about 1.2.

For crossed belts, for a constant belt length, $D_{11} + D_{21} = D_{1i} + D_{2i}$, where D_{11}, D_{21} are the diameters of the first set of pulleys and D_{1i}, D_{2i} are the diameters of any succeeding pair. In addition, for the velocity ratio n_{11}/n_{2i}, where n_{11} is the angular velocity of the driver shaft and n_{2i} is the angular velocity of the ith driven pulley, the diameters D_{1i} and D_{2i} must satisfy the ratio $D_{2i}/D_{1i} = n_{11}/n_{2i}$. These two relations must be solved simultaneously to determine D_{1i}, D_{2i}.

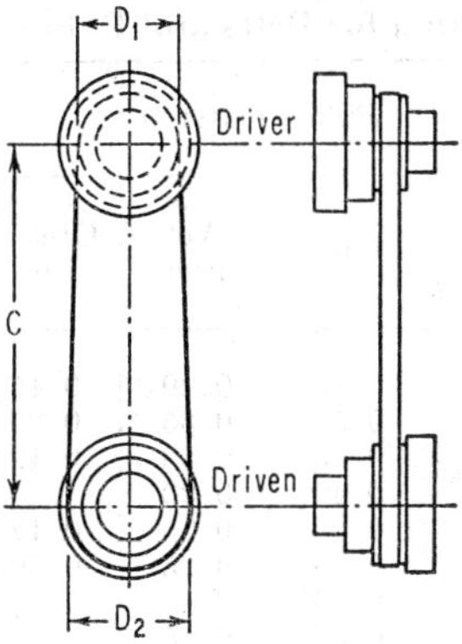

FIG. 29.4. Cone or stepped pulleys.

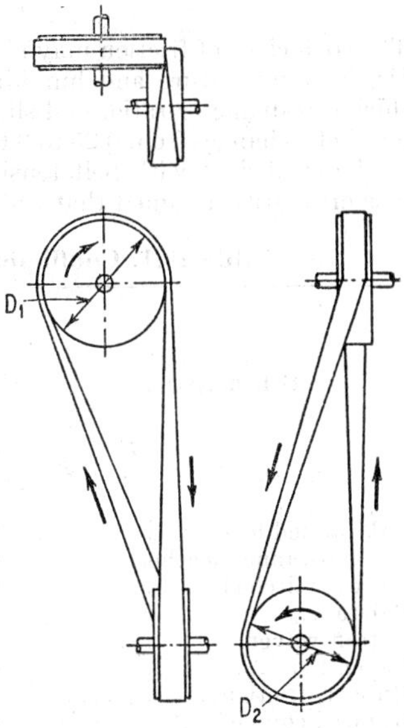

FIG. 29.5. Quarter-turn drive

A more complex relation exists for open belts. If the required diameter ratio is substituted in the open-belt-length equation, the result is

$$L = 2C + (\pi/2)(1 + n_{11}/n_{2i})D_{1i} + (1 - n_{11}/n_{2i})^2(D_{1i}^2/4C) + \cdots \quad (29.10)$$

This equation is solved for D_{1i} and D_{2i} obtained from the required ratio of diameters. Graphical methods for solution of this equation are described in books on mechanisms.

An alternative approximate method for open belts, accurate to four decimal places, follows. Let $m = 1.58114C - D_0$, where D_0 is the diameter of either pulley when $D_2/D_1 = 1$. Then $(D_1 + m)^2 + (k^n D_1 + m)^2 = 5L^2$, where k is the geometric ratio between steps. This equation is solved for D_1 after D_0, C, and k have been chosen. C is the center distance, L the belt length.

Quarter-turn Drive. The length of belt L for a quarter-turn drive (Fig. 29.5) is given by

$$L = (\pi/2)(D_1 + D_2) + (C^2 + D_1^2)^{1/2} + (C^2 + D_2^2)^{1/2} \quad (29.11)$$

Arc of Contact. Arc of contact for an open drive is

$$\theta = \pi \pm \sin^{-1}(D_1 - D_2)/C \quad (29.12)$$

where the plus sign corresponds to the larger pulley and the minus sign to the smaller pulley. Arc of contact for both pulleys in a crossed drive is

$$\theta = \pi + \sin^{-1}(D_1 + D_2)/C \tag{29.13}$$

(e) Coefficient of Friction

The coefficient of friction between belt and pulley depends upon the materials involved and the condition of the surfaces. Coefficients of friction, for design purposes, for a number of belt and pulley combinations are given in Table 29.1. Barth has shown that the coefficient varies with velocity. Experimental data for leather belts on iron pulleys gave the equation

$$\mu = 0.54 - 140/(500V) \tag{29.14}$$

The coefficient of friction may also be influenced by the cleanliness of the surrounding air, its temperature and humidity, the area of surface contact, the belt-to-pulley-thickness-diameter ratio, and slip. The coefficient rises rapidly after installation of a new belt; changes from 0.25 to 0.65 in a short time have been observed. The variation of the coefficient with belt tension has not been determined. For design purposes, it is ordinarily assumed that the coefficient remains constant.

Table 29.1. Coefficients of Friction μ for Belts and Pulleys

Belt material	Pulley material					
	Iron, steel	Wood	Paper	Wet iron	Greasy iron	Oily iron
Oak-tanned leather	0.25	0.30	0.35	0.20	0.15	0.12
Mineral-tanned leather	0.40	0.45	0.50	0.35	0.25	0.20
Canvas stitched	0.20	0.23	0.25	0.15	0.12	0.10
Balata	0.32	0.35	0.40	0.20		
Cotton woven	0.22	0.25	0.28	0.15	0.12	0.10
Camel hair	0.35	0.40	0.45	0.25	0.20	0.15
Rubber, friction	0.30	0.32	0.35	0.18		
Rubber, covered	0.32	0.35	0.38	0.15		
Rubber on fabric	0.35	0.38	0.40	0.20		

Belt dressings or cork inserts tend to raise these values.

(f) Creep and Slip

The tension in a belt as it passes around the driver pulley increases from slack-side tension to the tight-side tension. At the driven pulley, the reverse is true. Consequently, the belt elongates as it passes around the driver and shortens at the driven. A greater length of belt leaves the driver than is delivered by the driven pulley; hence the belt must move relative to the surface of both pulleys. This motion, caused by elongation, is called *creep*.

At low values of effective tension, creep occurs over only a small arc of the pulley. This arc, called the *active arc of contact,* increases with increasing effective tension until the belt creeps over the entire pulley. If the effective tension is further increased, the belt begins to slide on the pulley. The latter motion is called *slip*.

A creep velocity of 2 per cent of belt velocity is considered reasonable. Slip usually begins at a creep velocity of about 1 to 3 per cent of belt velocity. The equation for creep velocity in a leather belt is

$$v = (V_1/2)(\sigma_1^{1/2} - \sigma_2^{1/2})/(830 + \sigma_2^{1/2}) \tag{29.15}$$

where σ_1 and σ_2 are the unit stresses (psi) in the tight and slack sides, respectively, and V_1 is the tight-side velocity.

(g) Angle Drives

Open or crossed-belt drives are used to greatest advantage where center distances are large. In many instances, however, space is at a premium and power must be transmitted between shafts whose center lines are not parallel. Shafts which lie at an angle to each other may lie in the same plane, in which case they would intersect if extended, or they may not lie in the same plane. The most common arrangements are parallel and right-angle or quarter-turn shafting.

Parallel shafting on close centers can be connected by means of the nonreversible drive shown in Fig. 29.6. The effect is similar to a crossed drive, the shafts rotating in opposite directions. The drive of Fig. 29.6 is nonreversible because pulley B does not deliver the belt in the plane of guide pulley C. A reversible drive can be made, however, but the guide pulleys will no longer have a common shaft.

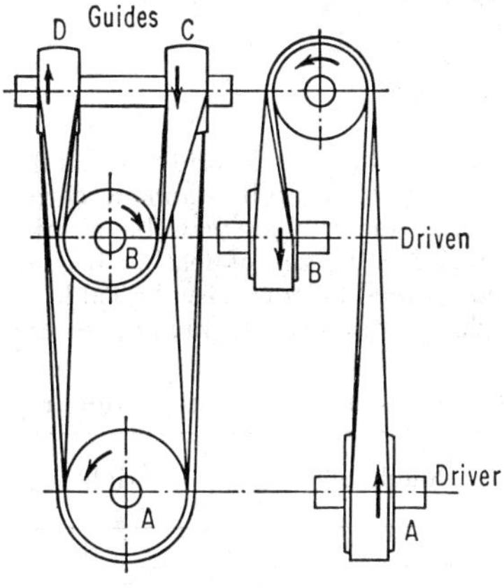

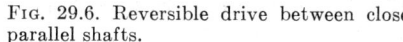

FIG. 29.6. Reversible drive between close parallel shafts.

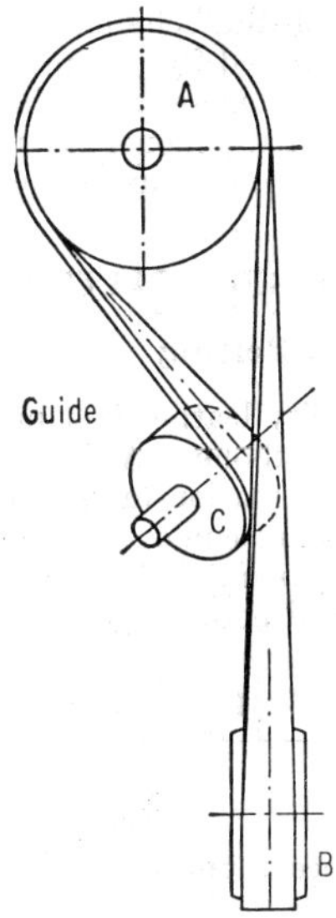

FIG. 29.7. Reversible quarter-turn drive, one guide pulley.

One of the most common angular drives is the quarter turn, in which the driven pulley is mounted at right angles to the driver. As the belt leaves one pulley, it must be turned so as to run onto the other pulley correctly. When rotation is in one direction, guides are usually not required. Guide pulleys are required for reversible operation.

To overcome the belt distortion introduced in quarter-turn drives, the belt is turned through a 180° angle so that, at the joint, the grain side is adjacent to the flesh side. With this construction, wear is equalized on the faces and edges of the belt. An alternative method of combating belt distortion is to make one edge longer than the other. For heavy belts, this method is preferred.

A reversible one-pulley quarter-turn drive is shown in Fig. 29.7. A two-guide-pulley arrangement, not shown, has an advantage in that the belt leaves the pulley in a straighter condition, reducing side stretch. The possibility of the belt's rubbing on itself is eliminated, and tension adjustment is possible if one guide pulley is made movable.

If the shafts intersect in a horizontal plane, the guide pulleys may have to be carried on a vertical axis. In this case, they are called *mule pulleys* and must be heavily crowned to hold the belt. For long belts, crowning will not hold the belt on the pulley and flanges must be provided. The flanges must be made with a reentrant fillet to reduce the tendency for the belt to ride upon the flanges. A guide plate mounted on the shaft adjacent to the pulley may be substituted for the flange. The flange or plate should be regarded as a safety measure; the crowning should be sufficient to keep the belt on the pulley under ordinary circumstances.

(h) Flat-belt Fasteners

Belts may be made endless by splicing and cementing, or the ends may be joined by means of belt fasteners. The strength of a properly made splice is equal to that of the belt, while the use of a fastener decreases belt capacity from 10 to 50 per cent dependent upon the fastener used. Fasteners offer a more convenient method of joining belt ends, however, when it becomes necessary to take up stretch or repair a damaged section.

Fasteners are made in four general classes: laces, hooks, plates, and the pressure type in which the belts ends are held together. Laces are made of rawhide or wire. Hooks are formed from wire and plates are of solid or hinged construction.

Metallic fasteners are fabricated from ordinary steel for normal applications. Stainless steel and monel are used for highly corrosive or abrasive conditions; Everdur (copper, silicon, and manganese alloy) where electrical static protection is required.

29.3. FLAT-BELT MATERIALS

(a) Leather Belting

High-quality leather belt material is obtained from the hide behind the shoulders near the backbone. Although most leather belting is vegetable-tanned, chrome-tanned, combination-tanned belting, rawhide, and semirawhide can be obtained. Leather belting comes in sizes shown in Table 29.2. Weight of leather belting is about 0.035 $lb/in.^3$ Horsepower capacity for oak-tanned belts is given in Table 29.2.

Oak-tanned belts are preferred in dry areas, chrome-tanned and rawhide in damp

Table 29.2. Horsepower per Inch of Width, Oak-tanned Leather Belting

Belt speed, fpm		Single ply		Double ply			Triple ply	
		11/64 medium	13/64 heavy	18/64 light	20/64 medium	23/64 heavy	30/64 medium	34/64 heavy
1,000		1.8	2.1	2.6	3.1	3.6	4.1	4.5
2,000		3.5	4.1	4.9	6.0	6.9	8.1	8.9
3,000		5.2	5.9	7.2	8.7	10.0	11.6	12.8
4,000		6.4	7.4	9.0	10.9	12.6	14.5	16.0
5,000		7.4	8.4	10.3	12.5	14.3	16.5	18.2
6,000		7.8	8.9	10.9	13.2	15.2	17.6	19.3
	rpm							
Min pulley diam,* in.	0–2,500	2½	3	4	5	8	16	20
	2,500–4,500	3	3½	4½	6	9	18	22
	4,000–6,000	3½	4	5	7	10	20	24

National Industrial Leather Assoc.

* Belts to 8 in. wide. For wider belts add 2 in. to minimum pulley diameter for single- and double-ply belts. Add 4 in. for triple-ply.

locations. In damp or oily areas, waterproof belts should be used instead of non-treated belts.

Leather belt ends may be joined by cement joints, which, when made correctly, have a joint strength equal to that of the leather; leather laced and riveted joints have a capacity of one-third to two-thirds that of the belt, respectively; wire-laced joints have 85 to 90 per cent of that strength.

Oak-tanned leather belting should not be used at temperatures greater than 110°F. Chrome or retanned leather can be used at higher temperatures. Total losses encountered from belt flexing and windage are less than 2 per cent of the power transmitted.

Leather belts in which the surface is ribbed or treated to increase friction at the pulley are available in mineral-tanned or oak-tanned form. Such belts are not recommended for installations where the belt is permitted a momentary slip.

In addition to flat types, leather belts are obtainable as V-belts, laminated V-belts, block V-belts, and round and oval, solid or plied belting.

(b) Nylon-core Belting

Nylon-core belting is made with leather or rubberized friction material on both outside plies or with leather on the pulley side and textile on the outside. The former type is reversible. Lightweight belts without leather or rubber facing are available. Nylon-core belting has assumed a major role in flat-belt transmissions. The largest drive reported to date is 6,000 hp. Speeds to 10,000 fpm or higher are utilized. Quarter-turn drives should be avoided.

Nylon-core belting has the ability to run at high speeds, to operate around very small pulleys, and to absorb shock. In addition, it possesses high strength, low initial tension, and predictable elastic qualities. Its tensile strength of 35,000 psi is about eight times that of leather. It is advantageous to use the greatest possible thickness of nylon consistent with recommended pulley diameters, and the narrowest possible width. Nylon-core belts can be operated from −20 to 160°F or higher. Friction materials available have resistance to oil, gasoline, and static. Center distances to 75 ft are possible. For information regarding the design of a nylon-core belt, see refs. 8, 9, and 10.

(c) Rubber Belting

Flat rubber belts are made up of plies consisting of fabric or cord strength members impregnated with vulcanized-rubber or synthetic-rubber compounds which resist moisture. The loading-carrying member may be prestretched cotton, rayon, nylon, or steel cable. Rubber belts are available in endless form or may be spliced by vulcanizing. They should not be operated in the presence of grease or oil. For such service neoprene-impregnated belts are required. Belts employing cord construction permit the use of smaller-diameter pulleys and are less extensible. They must be used in endless form, however, unless an oil-field-type clamp is used. Rubberized-fabric belts permit use of metal end fasteners. Neoprene-covered belts may be used at temperatures to 250°F. Static-resistant belts are available. Dust-jacketed belts are available to exclude destructive vapors or liquids and to provide wear resistance at the edges.

Initial tensions should be in the 15 to 25 lb/ply/in. width range. Generally, belts are made 1 per cent undersize from measured length, 1½ per cent for severe loading. For humid conditions this should be reduced 50 per cent. Provision for 2 to 4 per cent additional take-up during operation should be made to allow for stretch or manufacturing variations. Ultimate tensile strength is 280 to 600 (or more) lb/ply/in. width. Commercial belt widths run from 1 to 60 in. A disadvantage of rubberized belting is the decrease in coefficient of friction which results when the rubber is worn away to the glazed fabric below.

In practice, high-speed belting covers the range 4,000 to 5,500 fpm, medium speed 3,000 to 4,000 fpm, and low speed 2,500 to 3,000 fpm.

A number of special treatments are available. Neoprene-coated belts 3 to 11 plies thick are available in widths to 48 in. Tension members are hard-woven light duck molded with neoprene between plies. Hycar belts are similar but have higher resistance to animal, vegetable, or mineral oils or greases.

(d) Cotton and Canvas Belting

Fabric belts of this type are made up of layers of woven material held together by gums, rubbers, or synthetic resins which also serve to protect the fabric. Stitched canvas belts are inexpensive belts which need little protection against oil or moisture. Friction coefficient is low (0.15 to 0.22 against a steel pulley); hence pulley bearing pressure will be high. Resistance to high temperature is good. Bituminous-coated belts resist moisture and are recommended for rough usage. Combination leather and cotton belts consist of a single or double ply of oak or chrome leather cemented to a woven cotton backing.

Initial tension should be 20 to 25 lb/in. width/ply.

Because of the lower coefficient of friction, the largest possible pulleys which do not produce excessive belt speeds should be used to reduce bearing loads. Cotton and canvas belts are limited to belt velocities of 5,000 to 6,000 fpm.

(e) Balata Belting

Balata belts are made from closely woven duck of high tensile strength impregnated with balata gum obtained from South American trees. Balata is tough, stretches very little, is waterproof, resistant to aging, and affected by mineral oil but not by animal oils or humidity. Maximum allowable ambient temperature is 100 to 120°F. Initial tension is ordinarily in the range 22 to 25 lb/ply/in. width. Balata belting can be laced by metal fasteners or made endless. Its horsepower capacity is similar to that of canvas stitched or rubber belts. The coefficient of friction remains high even under damp conditions.

The specific weight of balata and other belts in lb/in.^{3} is as follows: leather ranges from 0.035 to 0.045, while canvas is 0.044, rubber 0.041, balata 0.040, single-woven cotton belt 0.042, and double-woven cotton belt 0.045.

(f) Steel Belting

Thin steel belts are generally used in installations where belt speed is high. Initial tension is generally high because of the lower coefficient of friction. Stretch is low because of the high elastic modulus of steel. Steel belts provide very high power transmission for a given belt cross section. Belt material is similar to that used for clock springs. Carbon content is high; the material is drawn and rolled, and ground to size. It is available in widths from ½ to 3 in. in 0.01-in. thickness. Tensile strength is in excess of 300,000 psi. Belt material has rounded edges to reduce the hazard while running.

Joints may be riveted or silver-soldered. Rivets can be phosphor bronze or cold-drawn iron wire. Joints are about 80 per cent efficient, failure occurring by shear. Silver-soldered butt joints, with 60° beveled edges, which can be used with small pulleys, are about 65 per cent efficient, rupture occurring in the adjoining belting, which is softened by the joining process. Cover-plate joints are bent to conform to the smallest pulley.

Hampton,[3] using 0.01- by 0.75-in. steel belting on cork pulleys, found that horsepower transmitted increased in proportion to slip velocity until the tension in the slack side approached that due to centrifugal effects, at which point slip velocity increased rapidly without a corresponding increase in horsepower transmitted. This breakaway point occurs at higher horsepowers for higher belt speeds because effective belt pull for a given horsepower is lower at higher speeds. The power in the range before slip occurred is given by

$$\text{hp} = 0.17 T_1^{0.65} V_s \qquad (29.16)$$

where V_s is the slip velocity and T_1 the tight-side tension.

Friction coefficient f rises with slip velocity but drops with increasing belt speed. For the belts of the previous reference,

$$f = V_s^{1.37} \times 3.2 \times 10^5/(p^{0.5} \times V_d^{1.5}) \qquad (29.17)$$

where p is the unit pressure on the driver pulley face, psi, and V_d the belt velocity, fpm. Drive efficiency is over 98 per cent. Slip velocities to approximately 3 fpm are considered reasonable. Young[4] showed that the horsepower which can be transmitted by a steel belt running on wood pulleys is

$$\text{hp} = 1.51p^{0.81}/1{,}000V_d^{0.78} \qquad (29.18)$$

29.4. FLAT-BELT WORKING STRESSES

(a) Design Stresses and Loads

Allowable tension for flat-belt materials is customarily expressed in terms of lb/in. width or lb/ply/in. width rather than in terms of allowable stress, psi. Ultimate strength for oak-tanned leather belts is in the range 3,000 to 4,500 psi; for chrome-tanned leather, 4,000 to 5,500 psi; for rubber, balata, cotton, or canvas belts, 900 to 1,500 psi; and for nylon-core belts, 35,000 psi.

As belt thickness and number of plies increase, recommended maximum allowable tension, lb/in. width, increases. For leather belting particularly, however, recommended maximum allowable stress, psi, and maximum allowable tension per ply, lb/ply/in. width, decrease as a result of increased bending stresses at the pulleys.

Recommended maximum values of allowable tension (lb/in. width) for oak-tanned leather belts range from 108 for single-ply medium to 275 for triple-ply heavy; for rubber, canvas, and balata, 32-oz duck, they range from 75 for 3-ply to 300 for 12-ply. Recommended maximum stress for oak-tanned leather belts ranges from 620 psi for single-ply medium to 520 psi for triple-ply heavy. For rubber, balata, or canvas belts the corresponding value is 500 psi.

(b) Horsepower Ratings of Flat Belts

The Goodyear Company has developed the following empirical relation relating service life I to d, the small pulley diameter; l, the belt length; S, the belt speed; N, the belt thickness; and T, the tight-side tension:

$$I = (k \times d^{5.35} \times l)/(S^{0.5} \times N^{6.27} \times T^{4.12}) \qquad (29.19)$$

where k is a constant dependent upon the particular belt characteristics. From Eq. (29.19) it is seen that a reduction in small pulley diameter, an increase in belt thickness, or an increase in tight-side tension have the greatest effect in decreasing belt life. A 50 per cent decrease in pulley diameter will decrease drive life to $\frac{1}{32}$ its original value; a 10 per cent increase in tension will decrease belt life 60 per cent.

The difference in horsepower recommendations among several manufacturers for a given type of belt may result from a difference in choice of service life. In addition some manufacturers base their recommended horsepower for smaller belts on a shorter service life because ordinarily the user does not expect so long a life from such drives.

29.5. FLAT-BELT DESIGN

Design techniques for all types of flat belts are basically similar; however, variations are encountered as outlined in this section. Current design is generally based on empirical data concerning the load-carrying capacity of a given belt. These recommended horsepowers are chosen to provide reasonable life under most circumstances. As may be deduced from Eq. (29.19), belt horsepower capacity is a function of desired service life, pulley diameters, belt length, belt speed, belt thickness, and tight-side tension. In addition, such factors as service conditions and contact angle must be included.

Existing practice is such that the effect of some factors such as belt length is ignored. In the case of leather belts, that of contact angle is neglected, while for nylon belts, no account is taken of service conditions. These omissions are permissible because of their comparatively minor effects or because recommended horsepower values are somewhat conservative.

In the design of a flat-belt drive, the diameters of the driver and driven pulleys, center distance, horsepower, and belt speed are first determined. Then, in accordance

Table 29.3. Horsepower per Inch of Width, Rubber, Balata, Cotton Duck, and Canvas Stitched Belts

35-oz silver hard duck, ply	Small pulley diam, in.	Belt speed, fpm			
		1,000	2,000	4,000	6,000
4	4	1.07	1.87	2.75	2.31
	8	1.98	3.52	5.72	6.49
	12	2.20	4.18	7.04	8.47
	14+	2.20	4.29	7.59	9.24
6	8	1.87	3.41	5.17	4.84
	12	2.86	5.06	8.36	9.46
	16	3.30	6.05	10.01	11.77
	28+	3.30	6.49	11.33	13.86
8	14	2.86	5.17	8.14	8.36
	20	4.07	7.37	12.21	13.97
	26	4.40	8.58	14.52	17.49
	36+	4.40	8.58	15.18	18.37
10	24	4.18	7.48	11.88	12.76
	30	5.39	9.68	17.16	18.81
	36	5.50	10.67	17.82	21.45
	48+	5.50	10.78	18.92	22.99
32-oz duck:					
4	4	0.9	1.2	2.2	1.8
	8	1.5	2.8	4.5	5.1
	12	1.8	3.5	5.8	7.0
	18+	1.8	3.6	6.3	7.7
6	8	1.4	2.5	3.9	3.5
	12	2.3	4.1	6.6	7.5
	16	2.7	4.9	8.1	9.6
	26+	2.7	5.3	9.4	11.5
8	16	2.5	4.4	7.0	7.4
	22	3.5	6.3	10.3	12.1
	28	3.6	6.9	11.5	13.9
	36+	3.6	7.1	12.5	15.3
10	28	3.7	6.7	10.9	12.2
	34	4.5	8.6	14.4	17.3
	40	4.5	8.9	15.2	18.4
	44+	4.5	8.9	15.7	19.2

Other sizes available.

For woven cotton belts, multiply corresponding values for 32-oz duck belts by the following factors:

	Single	Double	Triple
Light	0.96	1.45	2.00
Medium	1.10	1.60	2.15
Heavy	1.30	1.80	2.30

with the recommendations of Table 29.2 for leather belting; Table 29.3 for rubber, balata, woven cotton, or canvas stitched belting; and Table 29.4 for nylon belting, a belt of proper thickness is chosen. From these tables, the rated horsepower/in. width (hp_R) is also obtained.

Table 29.4. Horsepower Capacity per Inch of Width, Nylon-core Belts

Nylon-core thickness, in., approx	Small pulley diam, in.	Belt speed, fpm									
		1,000	2,000	3,000	4,000	5,000	6,000	7,000	8,000	9,000	10,000
0.020	½	0.15	0.20	0.30	0.40	0.50					
	1	0.20	0.30	0.50	0.62	0.81					
	2	0.30	0.55	0.85	1.12	1.45					
0.030	1	0.55	1.0	1.62	2.02	2.62	3.25	3.37	3.5	3.62	3.7
	4	1.12	2.25	3.12	4.0	5.25	6.0	6.5	6.7	6.8	6.9
	8	1.3	2.37	3.62	4.7	5.7	6.7	7.5	8.0	8.1	8.2
0.060	2	1.5	2.75	4.37	5.02	6.7	8.0	9.2	10.0	10.5	10.6
	4	1.87	4.6	6.0	6.5	10.0	11.7	13.2	13.7	15.2	15.4
	12	2.75	5.0	8.0	10.0	12.5	15.0	17.5	18.7	20.0	20.2
0.090	4	3.75	5.62	9.0	11.7	13.7	17.0	18.7	21.2	22.5	23.0
	8	4.12	6.87	10.5	13.7	17.5	20.0	22.5	25.0	27.6	28.0
	20	4.5	8.12	13.0	16.0	21.2	25.0	27.5	31.2	33.7	35.0
0.120	8	5.5	9.5	14.5	18.7	22.7	27.7	32.5	35.5	40.0	42.0
	15	6.35	11.87	18.7	23.8	29.0	35.0	37.0	40.0	44.0	46.0
	40	7.0	13.0	20.0	25.5	32.5	37.5	42.5	45.0	48.0	50.0

Published data vary. Other recommendations range from 75 to 100 per cent of these values.

To account for variations from the normal conditions assumed in Tables 29.2, 29.3, and 29.4 for (hp_R), these values must be modified by service factors to obtain the belt design horsepower per in. width (hp_D). Recommendations regarding appropriate service factors vary for different types of belting as listed below.

1. Leather

F, special operating conditions (Table 29.5)
M, motor and starting method (Table 29.5)
P, small pulley diameter (Table 29.5)

Accordingly, for leather belts,

$$hp_D = hp_R(P/F \times M) \tag{29.20}$$

2. Rubber, cotton, canvas, balata

K_1, arc of contact (Table 29.6)
F, special operating conditions (Table 29.5)
M, motor and starting method (Table 29.5)

Accordingly, for such belts,

$$hp_D = hp_R(K_1/F \times M) \tag{29.21}$$

3. Nylon core

K_1 arc of contact (Table 29.6)

Accordingly, for nylon-core belts,

$$hp_D = hp_R(K_1) \tag{29.22}$$

When the allowable design horsepower per in. width (hp_D) is established, the necessary belt width t is obtained from

$$t = (hp_T)/(hp_D) \tag{29.23}$$

where hp_T is the horsepower being transmitted by the belt. For more information on flat-belt design, see refs. 5, 6, and 7.

Table 29.5. Correction Factors for Belt Drives

Factor	Condition	Factor F
F, special operating conditions	Oily, wet, dusty	1.35
	Vertical drive	1.2
	Jerky loads	1.2
	Shock and reversing	1.4
	Method	Factor M
M, motor and starting method	Squirrel cage, compensator start	1.5
	Squirrel cage, line start	2.0
	Slip ring and high start torque	2.5
	Diam, in.	Factor P
P, small pulley diameter	0–4	0.5
	4.5–8	0.6
	9–12	0.7
	13–16	0.8
	17–30	0.9
	30+	1.0

Table 29.6. Arc Correction Factor K_1

Arc of contact, deg*	*Factor*
180	1.00
174	0.99
168	0.97
162	0.96
156	0.94
150	0.92
144	0.90
138	0.88
132	0.87
126	0.85
120	0.83
114	0.80
108	0.78
102	0.75
96	0.72
90	0.69

* Small sheave.

29.6. BELT FATIGUE

From Fig. 29.2, which represents the tensile-stress distribution for an operating belt, it is apparent that a given section of the belt undergoes cyclic stress loading.

This type of loading results in fatigue failures, and several theories have evolved to relate the number of belt cycles to failure, that is, the life of the belt, to various stress parameters.

29.7. FLAT-BELT PULLEYS AND IDLERS

(a) Pulleys

As the belt passes around the pulley, it flexes and bending stresses are induced. As a result of the reduced radius of curvature, bending stresses are higher for small pulleys. Because of the reduced fatigue stresses, a large pulley results in an increased belt life. Pulleys should not be so large, however, as to cause excessive belt speeds.

Table 29.2 lists minimum pulley sizes for leather belting. If pulley diameters less than those shown are used, belt life will be greatly reduced as a result of the higher bending stresses and greater elongation of the outer belt fiber.

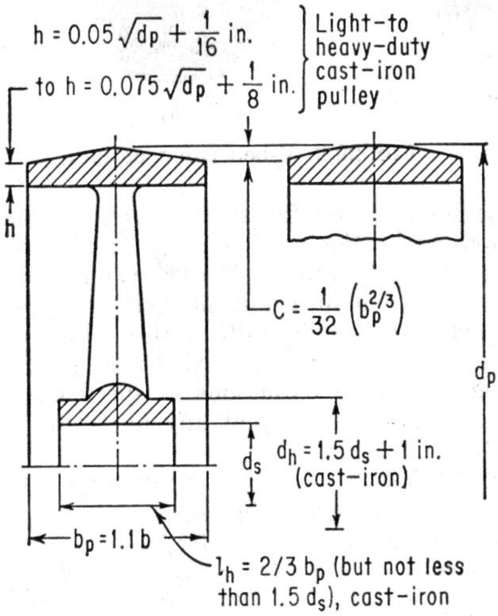

FIG. 29.8. Crowned pulley proportions.

In order to aid in keeping the belt on the pulley face it is recommended that the face be crowned as shown in Fig. 29.8, which also shows approximate pulley proportions. An alternate form of crowning consists of making the central half of the pulley face cylindrical and incorporating a slight straight or curved taper in the face at either side of the cylindrical portion.

(b) Idlers

Idlers are always flat-faced and located on the slack side of the belt next to the small pulley, whether the small pulley is driving or driven.

Only where the drive is steady and the small pulley is driven may the idler be fixed.

Spring-loaded idlers used as gravity idlers have the disadvantage of relieving the spring tension on the belt when the slack increases upon application of load.

The clearance between the small pulley and the idler should not exceed 2 in., and the idler should be positioned so that the belt wrap around the small pulley is 225 to 245°.

Idlers should be well balanced.

29.8. V-BELTS

(a) Introduction

The cross section of a V-belt is such (see Fig. 29.9) that the belt tension forces the belt into the groove. This wedging action between belt and groove effectively increases the coefficient of friction, permitting the drive to operate with reduced angles of contact and low initial tension.

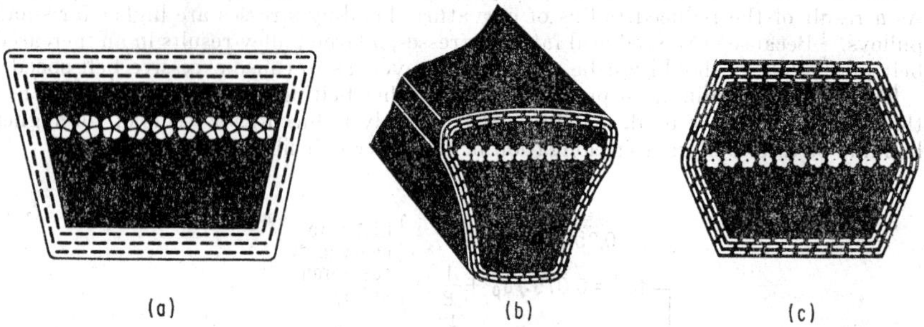

FIG. 29.9. V-belt construction.

The V-belt is particularly suited for small center distances requiring no idler. V-belts may be used for speed ratios as high as 10:1 and at belt speeds to 7,000 fpm.

V-belts offer the following advantages: Power output can be increased by use of multiple belts; failure of one belt in a multiple-belt drive will not necessarily cause machine stoppage; V-belts are made endless, thus eliminating splicing problems; V-belts offer a more positive drive because of reduced slippage; because V-belts may be operated over small pulleys, large reductions in speed can be realized in a single drive; the V-belt drive may be inclined at any operating angle, slack side top or bottom.

Construction features of V-belts vary among manufacturers. A typical modern belt consists of three major components: the uppermost or tension section of synthetic or natural rubbers, which is free to bend; the load-carrying member composed of nylon, rayon, or cotton cords (for severe operating conditions, steel cords may be used); and the compression or cushion section of blended rubber. The belt is encased in a rubber-impregnated canvas cover.

V-belts are made in standard letter-designated sizes listed in Table 29.7.

Table 29.7. Properties of V-belts

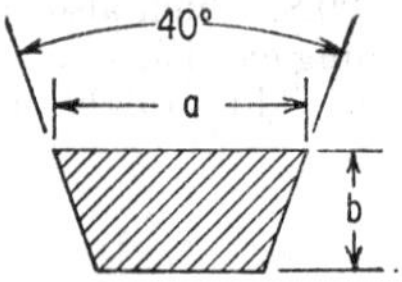

Belt section	Width a, in.	Thickness b, in.	Min sheave diam, in.	Hp, one or more belts
A	½	11/32	3.0	To 25
B	21/32	7/16	5.4	4–75
C	⅞	17/32	9.0	1½–150
D	1¼	¾	13.0	5–250
E	1½	1	21.6	15 and higher

The included angle of the groove usually lies between 30 and 38°. The sheave groove angle should be made less than that of the belt cross section so that the belt wedges to increase the effective friction. The sheave groove angle should not be too much smaller than the belt angle, however, or the force required to pull the belt from the groove as it leaves the sheave will be too great.

(b) Forces

Forces normal to the sides of the grooves act on a V-belt as shown in Fig. 29.10. Referring to Fig. 29.10, the normal force on the groove face

$$P_n = P/2 \sin \beta \tag{29.24}$$

The tractive force

$$F = 2\mu P_n = 2\mu P/2 \sin \beta = \mu_e P \tag{29.25}$$

where $P = T_1 + T_2$, the sum of the tight- and slack-side tensions
μ = coefficient of friction
μ_e = effective coefficient of friction = $\mu/\sin \beta$

The relation between tight and slack tensions [Eq. (29.6)] is applicable to V-belts when μ_e replaces μ in that equation.

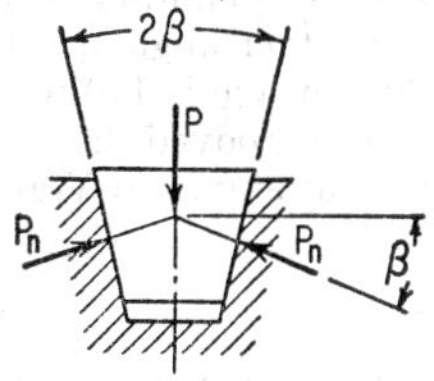

FIG. 29.10. V-belt forces.

(c) Geometry

Center Distance. Drive center distance is generally larger than the larger sheave diameter but less than the sum of the larger and smaller sheave diameters. Because the life of the V-belt is substantially shortened by slack-side shock and vibration, long center distances, in excess of 2½ to 3 times the large sheave diameter, are usually not recommended for V-belts.

Provision should be made to permit a small increase or decrease in the center distance to accommodate the nearest standard size of belt.

Larger center distances may be used with link-type V-belts which, because of better balance, are subject to less vibration.

The drive should make provision for the adjustment of center distance above and below the nominal value for belt take-up after stretching and installation, respectively. Take-up allowance should be at least 2½ per cent of the belt length for belts up to 200 in. long, slightly less for longer belts. The application allowance should be at least equal to the belt thickness, and up to three times this long for heavier sections and longer belts.

In multiple-belt drives, all belts should elongate equally in order to assure an equal distribution of load. When one belt breaks, all the belts should be replaced in order that all the belts will have nearly equal properties.

The adjustable sheave in a multiple-belt drive should permit an increase in center distance of at least 5 per cent above the nominal value and should also permit a decrease in this distance by an amount at least equal to the belt thickness to permit installation.

(d) Sheaves

Sheaves are generally made of wood, stamped sheet steel, or cast iron. For higher speeds, cast-steel sheaves are used.

A minimum clearance of ⅛ in. is required between the inner surface or base of the belt and the bottom of the sheave groove. This clearance prevents the belt from bottoming as it becomes narrower from wear.

It is important that sheaves be statically balanced for applications up to 5,000 fpm and dynamically balanced for higher speeds.

Sheaves are available which permit the groove width to be adjusted, thus varying the effective pitch diameter of the sheave and permitting moderate changes in the speed ratio.

The diameter of the small sheave should not be less than the values given in Table 29.7.

Sheaves should be as large as possible; however, if possible, the belt speed should not exceed 5,000 fpm.

(e) V-flat Drives

Where the speed ratio is 3:1 or more and the center distance is short (equal to or slightly less than the diameter of the large sheave), it is possible to develop the required tractive force and maintain efficiency while omitting the grooving from the face of the large sheave. In this case, the belt contact is at its inner surface and the cost of cutting the grooves of the large sheave is eliminated. The maximum tractive power of the belts is obtained when the contact angle on the larger sheave is in the range 240 to 250° and that on the smaller sheave is 110 to 120°.

When power must be transmitted to grooved sheaves from both the top and bottom of the belt, double V-belts, made to fit standard V-grooves, are used (see Fig. 29.9*c*).

(f) Quarter-turn V-belt Drives

Quarter-turn drives may be used with V-belts. The horsepower rating of a V-belt used in this arrangement is 75 per cent of its rating in a straight drive under the same load conditions.

If the sheaves are relatively small, and take-up can be provided on one of the shafts, an idler may be omitted. The minimum recommended center distance in this case is $6(D + b)$ where D is the large sheave diameter and b is the belt width.

An idler is necessary to keep the belt on the sheaves when a larger speed ratio is used. The center distance between the idler and the small sheave should not be less than $8b$.

For information involving V-belt drives see refs. 11 to 13.

29.9. V-BELT DESIGN

Current design techniques for V-belts are based on empirical data concerning the load-carrying capacity of a given belt. These recommended horsepowers are chosen to provide reasonable life under most circumstances. Relations similar to that of Eq. (29.19) for flat belts may be developed for V-belts, relating horsepower capacity to service life, diameters, belt length, speed and thickness, and tight-side tension. In addition, such factors as operating conditions and arc of contact must be considered.

In the design of a V-belt drive, the diameters of the driver and driven pulleys, center distance, horsepower, and belt speed are first determined. Then, in accordance with Tables 29.7 and 29.8, a belt of proper cross section is chosen. From these tables, the rated horsepower per belt (hp_R) is also obtained. To account for variations from the normal conditions assumed in Table 29.8, these values must be modified by service factors to obtain the belt design horsepower per belt (hp_D). These factors are listed below.

K_1, arc of contact (Table 29.6)
K_2, belt-length correction (Table 29.9)
F, special operating conditions (Table 29.5)

Table 29.8. Horsepower Ratings, Standard and Premium-quality V-belts

Belt section	Rating diam, in.	Belt speed, fpm: 1,000 S	1,000 P	2,000 S	2,000 P	3,000 S	3,000 P	4,000 S	4,000 P	5,000 S	5,000 P	6,000 S	6,000 P
A	2.0	.....	0.01										
	3.0	0.66	0.90	1.01	1.38	1.12	1.60	0.93	1.51	0.38	1.04	.....	0.12
	4.0	.....	.....	.....	.....	.....	.....	.....	.....	.....	.....	1.29	2.78
	5.0	1.17	1.61	2.03	2.80	2.64	3.73	2.96	4.39	2.91	4.59	2.43	4.38
B	3.0	.....	0.06										
	4.0	0.95	1.22	1.35	1.73	1.33	1.77	0.79	1.27	.....	0.11		
	5.0	1.44	1.92	2.33	3.13	2.80	3.86	2.76	4.06	2.10	3.60	0.68	2.38
	6.0	1.77	2.39	2.99	4.06	3.78	5.26	4.07	5.92	3.73	5.93	2.65	5.17
	7.0	2.01	3.78	3.46	4.72	4.49	6.26	5.01	7.25	4.90	7.59	4.05	7.17
C	6.0	1.84	2.28	2.66	3.25	2.72	3.36	1.87	2.50	.....	0.48		
	8.0	2.96	3.90	4.90	6.48	6.09	8.21	6.36	8.97	5.52	8.57	3.34	6.80
	10.0	3.64	6.74	6.25	8.42	8.11	11.1	9.06	12.9	8.89	13.4	7.38	12.6
	12.0	4.09	7.72	7.15	9.72	9.46	13.1	10.9	15.4	11.1	16.7	10.1	16.5
D	10.0	4.14	4.93	6.13	7.08	6.55	7.46	5.09	5.83	1.35	1.82		
	12.0	5.71	7.23	9.26	11.7	11.2	14.3	11.4	15.0	9.18	13.3	4.27	8.77
	14.0	6.82	8.87	11.5	15.0	14.6	19.4	15.8	21.6	14.8	21.5	11.0	18.6
	16.0	7.66	10.1	13.2	17.4	17.1	23.0	19.2	26.5	19.0	27.6	16.0	26.0
	17.0	8.01	10.6	13.9	18.4	18.1	24.5	20.6	28.5	20.7	30.0	18.1	29.0
E	15.0	7.94	9.82	12.7	15.6	15.3	18.8	15.1	19.1	11.6	15.9	4.20	8.70
	18.0	9.92	12.7	16.7	21.4	21.2	27.5	23.0	30.7	21.5	30.5	16.1	26.2
	21.0	11.3	14.8	19.5	25.6	25.4	33.8	28.6	39.1	28.6	41.0	24.5	38.8
	24.0	12.4	16.4	21.6	28.7	28.6	38.5	32.9	45.4	33.8	48.8	30.9	48.2
	28.0	13.4	18.0	23.7	31.9	31.8	43.2	37.1	51.6	39.1	56.6	37.2	57.6

S, standard; P, premium.

Table 29.9. Belt-length Correction Factor K_2

Length factor	Nominal belt length, in.: A belts	B belts	C belts	D belts	E belts
0.85	Up to 35	Up to 46	Up to 75	Up to 128	
0.90	38–46	48–60	81–96	144–162	Up to 195
0.95	48–55	62–75	105–120	173–210	210–240
1.00	60–75	78–97	128–158	240	270–300
1.05	78–90	105–120	162–195	270–330	330–390
1.10	96–112	128–144	210–240	360–420	420–480
1.15	120 up	158–180	270–300	480	540–600
1.20	....	195 up	330 up	540 up	660

Accordingly, design horsepower for V-belts,

$$hp_D = hp_R(K_1K_2/F) \tag{29.26}$$

where the rated horsepower per belt is based on the rating diameter.

When the allowable design horsepower per belt (hp_D) is established, the number of belts n required is obtained from

$$n = (hp_T)/(hp_D) \tag{29.27}$$

where hp_T is the horsepower being transmitted by the drive.

29.10. WEDGE V-BELTS

The wedge V-belt (Fig. 29.9*b*) is a V-belt which is truncated for about 15 per cent of the belt height. The belt has increased capacity per unit cross section when compared with a standard V-belt, by reason of a concave side wall which, when the belt is bent around a sheave, permits the belt to expand and grip the sheave evenly, and a crowned top which serves to keep the top from dishing in and causing a nonuniform tension in the load-carrying members.

29.11. POSITIVE-DRIVE BELTS

Positive-drive belts do not depend upon friction for their power-transmitting properties. One side of such belts is formed into teeth which mesh with similar teeth in the mating sheave to obtain positive drive. Speed is uniform, there being no chordal rise and fall of pitch line.

The positive-drive belt shown in Fig. 29.11 utilizes continuous, helically wound, brass-plated high-carbon steel, high-tenacity-rayon (special conditions), or stainless-

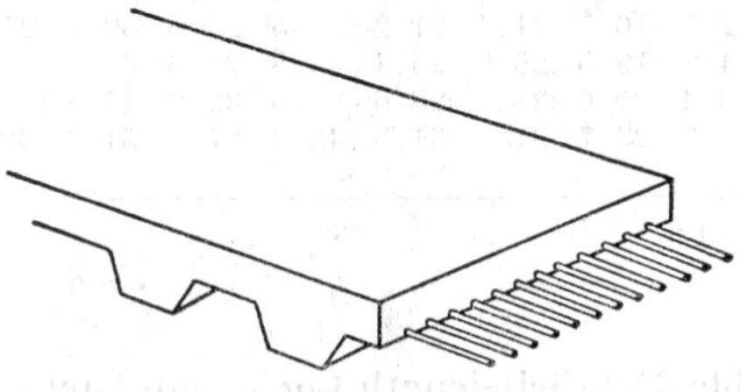

FIG. 29.11. Positive-drive belt construction.

steel (wet conditions) cable as the load-carrying element of the belt. The load-carrying member is encased by a neoprene backing member. Other backing is available for severe oil conditions. The teeth, molded integral with the backing, are of a shear-resistant, moderately hard rubber or a neoprene compound. Belt tooth strength exceeds the tensile strength of the steel tension member when six or more teeth are in contact with a pulley. The tooth root line coincides with the center line of the tension member so that tooth spacing is not altered by flexing. The nylon fabric covering the tooth surface protects the teeth from wear. While power is ordinarily transmitted by means of the teeth, it may also be transmitted by frictional means from the back of the belt. Such belts have been used on drives transmitting up to 600 hp at speeds of 100 to 10,000 fpm.

The advantages claimed for positive-drive belting include elimination of slippage, and lower belt tension, resulting in reduced bearing loads, smaller space requirements, constant angular velocity, and high mechanical efficiency with low heat output and low backlash.

Positive-drive pulleys must be run with belts of the same pitch. Pulley pitch diameter, which coincides with that of the belt, is greater than the face diameter of the pulley. Positive-drive belts generate a small, inherent side thrust; hence at least one flanged pulley (usually the smaller for economy reasons) is required to keep the belt on the pulley. Both pulleys should be flanged when the center distance is greater than eight times the diameter of the smaller pulley for horizontal drives. Generating hobs and shaper cutters suitable for forming pulley teeth are available from belt manufacturers.

Small pulleys generally employ integral hubs, while larger pulleys are used with bushings. Speeds in excess of 6,000 fpm require the pulleys to be dynamically balanced; in excess of 7,000 fpm, they must be made of ductile iron instead of cast iron. Stock pulleys, generally of cast iron, are statically balanced. Aluminum and steel stock pulleys are available.

Stock neoprene-backed belts can be operated at −30 to above 185°F. Special belt materials are available for extreme temperatures (−65 to 260°F) and for caustic environments. Oil-resistant, static-conductive, and nonmarking belts may be ordered. A limited number of special-pitch belts are available from individual manufacturers.

Positive belt drives do not require an initial tension. A larger, ⅞- or 1¼-in.-pitch, belt can even be installed slightly slack unless shock loads are high. Pulley alignment should be checked with a straightedge. On long center drives, the belt should be sufficiently taut to avoid contact between teeth on tension and slack sides. As a rule of thumb, pulley diameter should not be less than belt width. In installing, the belt should not be pried over the pulley flange.

29.12. PIVOTED MOTOR BASES

With small-diameter pulleys necessary with high-speed motors and short center-distance drives, it is impractical to maintain belt tension by belt sag, and pivoted

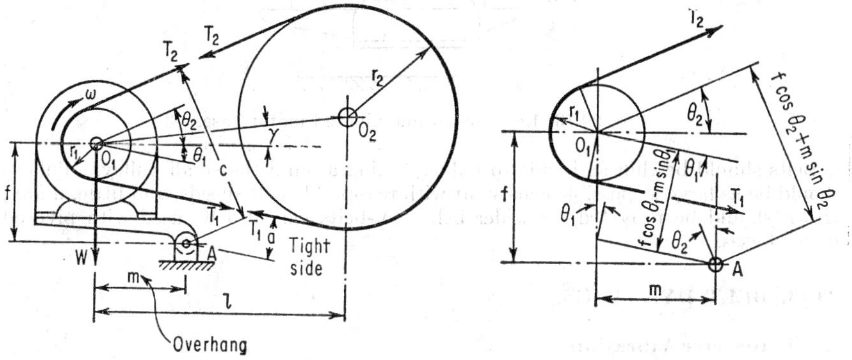

FIG. 29.12. Gravity-type pivoted motor base.

motor bases and idler pulley drives are employed. Pivoted motor bases which automatically maintain correct belt tension are of two types, gravity and reaction torque.

The gravity type (Fig. 29.12) utilizes part or all of the weight of the driving motor to maintain belt tension. In horizontal drives, with the driven pulley above the driver, motor weight keeps the belt in tension as it stretches because of the torque transmitted. Vertical drives are also used.

Reaction-torque motor bases are of all-steel construction. The standard base (Fig. 29.13) is designed to function satisfactorily only with the tight side nearest the pivot axis. Special bases are available for reverse-direction drives. Reaction torque on the motor stator tends to rotate the stator about the pivot axis in the opposite direction to the belt pull, thus tightening the belt.

Vertical drives with the motor mounted above the driven pulley are called *down drives*. A counterweight is utilized to balance the motor weight.

The spring-type motor base maintains belt tensions by means of springs built into the motor base.

Belt tension in a gravity-type pivoted motor base drive (Fig. 29.12) is found by taking moments about the pivot A. The resulting moment equation,

$$Wm = T_1 a + T_2 c$$

is solved simultaneously with the equation for tension ratio, $T_1/T_2 = e^{\mu\beta}$, to find the belt tensions. Maximum power, which can be transmitted when $T_2 = 0$, is

$$\text{hp} = mWNd_1/126{,}000a$$

where N is the motor rpm and all dimensions are in inches. During operation, only about 75 per cent of this horsepower can be transmitted by the drive. For ordinary oak-tanned-leather belt drives values of $T_1/T_2 = 5$ at 180° arc of contact and $T_1/T_2 = 3$ at 120° are recommended. For high-capacity belts these values may be increased to 6.5 and 3.5 respectively. Centrifugal effects in the belt are usually neglected because the motor weight maintains belt pulley contact. From Fig. 29.12, the moment arms a and c are given by $a = f \cos \theta_1 - m \sin \theta_1 - r_1$, and $c = f \cos \theta_2 + m \sin \theta_2 + r_1$.

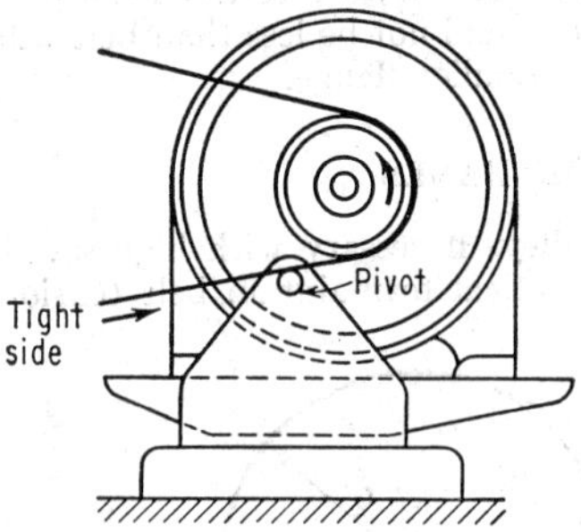

FIG. 29.13. Reaction-torque pivoted motor base.

Belts should be thin and wide to reduce flexing around the small pulley. Pulleys should be as large as possible consistent with reasonable belt speeds, and little, if any, crown should be provided for wider belts. V-belts can also be used with pivoted motor bases.

29.13. BELT DYNAMICS

(a) Transverse Vibration

When the belt is excited by forces which have a frequency corresponding to its transverse natural frequency, it may undergo serious flapping vibrations. It is accordingly desirable to determine the natural frequency of the belt to avoid such operation. In the treatment below, it is assumed that the flexural rigidity of the belt is negligible.

The velocity of sound u, which is equal to the velocity of propagation of a disturbance along the belt, is given by

$$u = (Pg/\rho A)^{1/2} \quad \text{in./sec} \tag{29.28}$$

where P is the tension in the belt, lb, g the gravitational constant, in./sec², ρ the density of the belt, lb/in.³, and A the cross-sectional area, in.²

The velocity v of the belt is

$$v = 2\pi(RN/60) \quad \text{in./sec} \tag{29.29}$$

where R and N are the radius and rpm of the sheave. The time for a disturbance to travel up and down the belt is $t = 2L_S u/(u^2 - v^2)$ sec, where L_S is the distance between points of tangency of the belt, so that the frequency of the fundamental mode is

$$F_1 = 60/t = 60(u^2 - v^2)/2L_S u \qquad \text{cycles/min} \tag{29.30}$$

In addition to the fundamental mode, the belt may execute higher modes of vibration whose frequency is given by

$$F_i = mF_1 \tag{29.31}$$

where $m = 1, 2, 3$, etc. It should be noted that the natural frequency of the slack side differs from that of the tight side and that this frequency differs with belt speed.

To account for its flexural rigidity, the belt is assumed to be a simply supported beam in lateral vibration whose natural frequency

$$F_2 = 30(m^2/L_S^2)(EIg/A)^{1/2} \qquad \text{cycles/min} \tag{29.32}$$

where $m = 1, 2, 3$, etc. For a steel flat belt this reduces to

$$F_2 = 5.5 \times 10^6 bm^2/L_S^2 \tag{29.33}$$

where b is the belt thickness.

The approximate natural frequency F of a belt having flexural rigidity is given by

$$F^2 = F_1^2 + F_2^2 \tag{29.34}$$

The effect of flexural rigidity, even for a steel belt, is generally negligible.

Excitation of lateral vibrations in a belt drive can arise as a joint, debris, or thickened area of the belt passes over a pulley. The excitation frequency in this case is obtained by dividing the linear velocity of the belt by the total length of the belt, or

$$F_e = 2\pi R N_R / L_t \qquad \text{cycles/min} \tag{29.35}$$

(b) Torsional Rigidity of Belt Drives

The torsional rigidity of a belt drive is the ratio of the moment applied to one pulley to the angular deflection undergone by the pulley. The torsional rigidity for the open drive may be written

$$C_R = \frac{R^2\beta}{eL_S + (R + r)/3} \qquad \text{in.-lb/rad} \tag{29.36}$$

$$C_r = \frac{r^2\beta}{eL_S + (R + r)/3} \qquad \text{in.-lb/rad} \tag{29.37}$$

where C_R, C_r are the torsional rigidities referred to the larger and smaller pulleys, respectively, $e = 1/AE$ where A is the cross-section area, E the modulus of elasticity of the belt, L_S the distance between points of tangency of the belt

$$L_S = [C^2 - (R - r)^2]^{1/2}$$

and R and r are the radii of the large and small pulley, respectively. The empirical factor $(R + r)/3$ is included to account for stretching effects as the belt passes around the pulley and may be omitted in approximate calculations. β is a constant whose value depends on the relation between steady state and vibratory torques imposed. If the magnitude of the vibratory force is less than the magnitude of the steady-state force in the slack side, $\beta = 2$. If the vibratory torque is large enough that the force in the slack side is negligible throughout the cycle, $\beta = 1$. If the force in the slack side is zero for ½ cycle, $\beta = 1.5$ (approximately). For steady-state conditions, with no vibratory torque, $\beta = 2$.

The belt may be converted to an equivalent length of either drive or driven shafting for frequency calculations. This value is, when referred to the larger pulley,

$$L_E = (\pi/32)(GD^4/\beta EAR^2)L_S \tag{29.38}$$

or, when referred to the smaller pulley,

$$L_e = (\pi/32)(Gd^4/\beta EAr^2)L_S \tag{29.39}$$

where G is the torsional rigidity of the shaft and D and d are the diameters of large and small pulleys, respectively.

Example: Calculate the torsional rigidity of a steel belt 1.0 in. wide and 0.004 in. thick; diameter of both pulleys is 6 in. and center distance is 7 in. With these values, $e = 1/(1.0 \times 0.004) \times 30 \times 10^6 = 250/30 \times 10^6$.

$$C_r = C_R = \frac{2 \times 9 \times 30 \times 10^6}{250 \times 7 + (3 + 3)/3} = 240{,}000 \text{ in.-lb/rad}$$

The value of E for leather is usually taken as 25,000 psi.

When calculating the fundamental frequency of a multimass system in which a belt drive is included, it is usually assumed that this mode of vibration has a node in the belt, whose stiffness is low compared with that of other members of the system.

Excitations at this frequency can be caused by sheave unbalance or a bent or misaligned shaft. Excitations at twice this frequency are produced by out-of-round sheaves or by shaft misalignment. Excitations at ½, 1, and 1½ times this frequency arise when the belt is driven by four-stroke-cycle internal-combustion engines.

References

1. Barth, C. G.: The Transmission of Power by Leather Belting, *Trans. ASME*, vol. 31, pp. 29–103, 1909.
2. "Handbook of Power Transmission," Flat Belting, p. 11, Publication S-5119, The Goodyear Tire and Rubber Co., Akron, Ohio, 1954.
3. Hampton, F. G., C. F. Leh, and W. E. Helmick: An Experimental Investigation of Steel Belting, *Mech. Eng.*, vol. 42, pp. 369–378, July, 1920.
4. Young, G. L., and G. V. D. Marx: Performance Tests of Steel Belts with Compressed Spruce Pulleys, *Mech. Eng.*, vol. 45, pp. 246–247, 1923.
5. "Modern Leather Belting," p. 10, National Industrial Leather Association, New York, 1957.
6. "Power Transmission Machinery," p. 318, Engineering Catalog D56, Dodge Manufacturing Corp., Mishawaka, Ind., 1956.
7. Engineering Guide for Flat Transmission Belts and Belting, U.S. Rubber International, *Bull.* 142, New York, 1958.
8. Page-Lon Catalog 611, Page Belting Company, Concord, N.H.
9. Nycor-M Catalog 31-10M, L. H. Shingle Co., Worcester, Mass.
10. Foulds Vitalastic Horsepower Tables, I, Foulds and Sons, Inc., Hudson, Mass.
11. The Dayton Handbook of V-belt Drive Design and Selection Catalog 280-B, The Dayton Rubber Co., Melrose Park, Ill., 1961.
12. Power Transmission Equipment, Catalog 61, Browning Manufacturing Co., Maysville, Ky., 1960.
13. Engineering Design Data, Positive Drive Belts, Catalog S-5188, The Goodyear Tire and Rubber Company, Akron, Ohio, 1959.

Section 30

POWER-TRANSMISSION CHAINS

By

G. V. TORDION, Dipl. Ing., *Director, Mechanical Engineering Department, Université Laval, Faculté des Sciences, Quebec, Canada*

CONTENTS

30.1. INTRODUCTION

Power-transmission chains are primarily of two kinds, roller and silent. They transmit power in a positive manner through sprockets rotating in the same plane. This fairly old transmission element (sketches of its design have been found in notebooks of Leonardo da Vinci) has been improved in the course of time to a very high standard of precision and quality.

Chain transmission is positive and there is no slip present as in belt drives. Large center distances can be dealt with more easily, with fewer elements and in less space than with gears. Chain drives have high efficiency. No initial tension is necessary and shaft loads are therefore smaller. The only maintenance required, after a careful alignment of elements, is lubrication. Chains as well as sprockets are thoroughly standardized in ASA standard B.29.1. The primary specifications of the standard single-strand roller chain are given in Table 30.1.

Table 30.1. American Standard Roller Chain

Pitch, in.	ASA No.	Weight, lb/ft	Avg ultimate strength, lb	Roller width, in.	Roller diam, in.	Side plate	
						Thickness, in.	Height, in.
1/4	25	0.085	925	1/8	0.130	0.03	0.23
3/8	35	0.22	2,100	3/16	0.200	0.050	0.36
1/2	41	0.28	2,000	1/4	0.306	0.050	0.39
1/2	40	0.41	3,700	5/16	5/16	0.060	0.46
5/8	50	0.68	6,100	3/8	0.400	0.080	0.59
3/4	60	0.96	8,500	1/2	15/32	0.094	0.68
1	80	1.70	14,500	5/8	5/8	0.125	7/8
1 1/4	100	2.70	24,000	3/4	3/4	0.156	1 5/32
1 1/2	120	4.00	34,000	1	7/8	0.187	1 13/32
1 3/4	140	5.20	46,000	1	1	0.218	1 5/8
2	160	6.80	58,000	1 1/4	1 1/8	0.250	1 7/8
2 1/4	180	9.10	76,000	1 13/32	1 13/32	0.281	2 1/8
2 1/2	200	10.80	95,000	1 1/2	1 9/16	0.312	2 5/16

30.2. NOMENCLATURE

p = pitch, in.
T = chain tension, lb
T_c = chain tension due to centrifugal forces, lb
w = chain weight per unit length, lb/in.
$m = w/g$ = chain mass per unit length, lb-sec²/in.²
D_1, D_2 = sprocket diameters, in.
N_1, N_2 = sprocket-teeth numbers
n_1, n_2 = revolutions per minute of the sprockets
$\rho = n_2/n_1$ = speed ratio
V = chain speed, in./sec in equations, fpm in specifications
$V_0 = \sqrt{T/m}$ = velocity of propagation of transverse waves, in./sec
$L = L'p$ = chain length, in.
$C = C'p$ = center distance of the sprockets, in.
a = strand length between seated rollers, in.
h = chord length of the strand between seated rollers, in.
s = strand maximum sag, in.
γ = pressure angle
$\alpha_1 = 180°/N_1$ and $\alpha_2 = 180°/N_2$
ω_n = natural circular frequency of vibrations, rad/sec
E = Young's modulus, lb/in.²
f = roller width, in.

30.3. DESIGN OF ROLLER CHAINS (FIG. 30.1)

The speed ratio is given by $\rho = n_2/n_1 = N_1/N_2$. For one-step transmission, it is recommended that $\rho < 7$. Values between 7 and 10 may be used at low speeds (<650 fpm). The minimum wrap angle of the chain on the smaller sprocket is 120°. A smaller angle may be used on idler sprockets used for adjustment of the chain slack, where the center distance is not adjustable. Horizontal drive is recommended, in which case the power should be transmitted by the upper strand. Preferred inclined-drive arrangements are shown in Fig. 30.2. Vertical drives should be used with idlers to prevent the chain from sagging and to avoid disengagement from the lower sprocket. When running outside the chain, idlers should be located near the smaller sprocket.

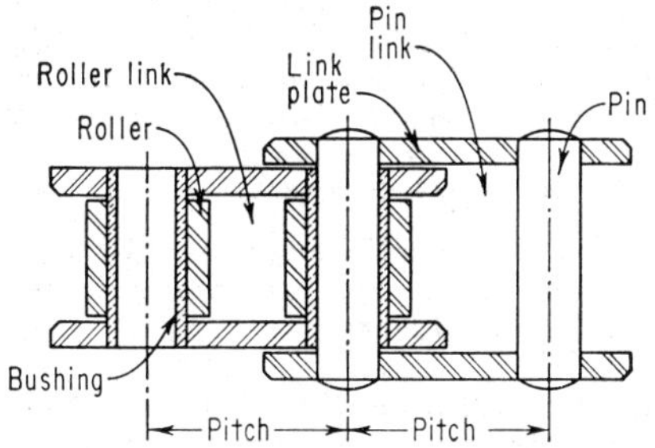

FIG. 30.1. Roller chain.

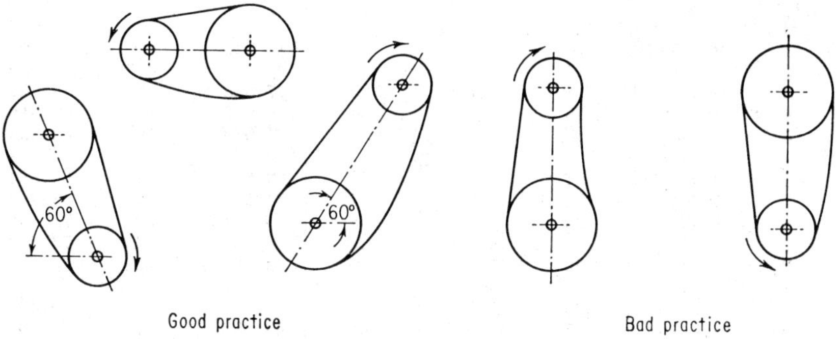

FIG. 30.2. Chain-drive arrangements.

(a) Chain Length

The chain length L' in pitch numbers, for a given distance C' in pitch numbers, is exactly (Fig. 30.3)

$$L' = 2C' \cos \epsilon + \frac{N_1 + N_2}{2} + \frac{(N_2 - N_1)\epsilon}{\pi} \tag{30.1}$$

where

$$\epsilon = \sin^{-1} \frac{1/\sin \alpha_2 - 1/\sin \alpha_1}{2C'} \tag{30.2}$$

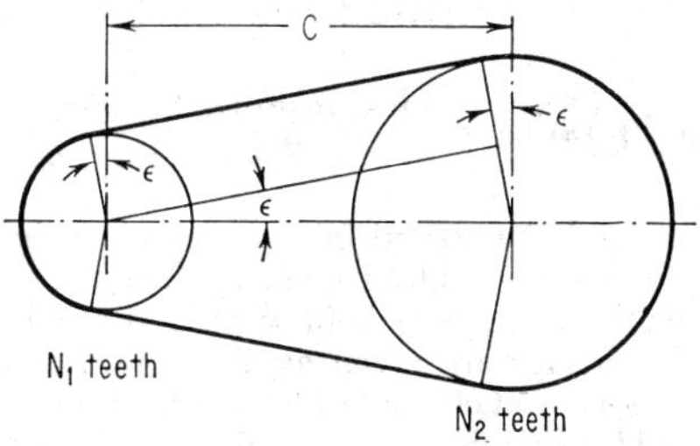

FIG. 30.3. Angle ϵ used in chain-length calculation.

Since ϵ is generally small, it is sufficient in most cases to use the approximation $\epsilon = (N_2 - N_1)/(2C')$. Then

$$L' = 2C' + \frac{N_1 + N_2}{2} + \frac{S}{C'} \tag{30.3}$$

where $S = [(N_2 - N_1)/(2\pi)]^2$. The values of S are tabulated in Table 30.2. The actual chain length is $L = L'p$, and the actual sprocket center distance $C = C'p$. It is recommended that C' lie between 30 and 50 pitches. If the center distance is not given, the designer is free to fix C' and to calculate L' from Eq. (30.3). The nearest larger (preferably even) integer L' should be chosen. In this case an offset link is avoided. With L' an integer, the center distance becomes exactly

$$C' = e/4 + \sqrt{(e/4)^2 - S/2} \tag{30.4}$$

with $e = L' - (N_1 + N_2)/2$. C' should be decreased by about 1 per cent to provide slack in the nondriving chain strand. For horizontal drive, this will result in a chain sag of some 2 per cent of the strand length.

Table 30.2. Values of S for Different $N_2 - N_1$

$N_2 - N_1$	S	$N_2 - N_1$	S	$N_2 - N_1$	S	$N_2 - N_1$	S
1	0.02533	16	6.4846	31	24.342	46	53.599
2	0.10132	17	7.3205	32	25.938	47	55.955
3	0.22797	18	8.2070	33	27.585	48	58.361
4	0.40528	19	9.1442	34	29.282	49	60.818
5	0.63326	20	10.132	35	31.030	50	63.326
6	0.91189	21	11.171	36	32.828	51	65.884
7	1.2412	22	12.260	37	34.677	52	68.493
8	1.6211	23	13.400	38	36.577	53	71.153
9	2.0518	24	14.590	39	38.527	54	73.863
10	2.5330	25	15.831	40	40.528	55	76.624
11	3.0650	26	17.123	41	42.580	56	79.436
12	3.6476	27	18.466	42	44.683	57	82.298
13	4.2808	28	19.859	43	46.836	58	85.211
14	4.9647	29	21.303	44	49.039	59	88.175
15	5.6993	30	22.797	45	51.294	60	91.189

The chain speed is $V = \pi n_1 D_1/12$ fpm, or approximately $V = n_1 p N_1/12$ fpm. The sprocket diameters are $D_i = p/\sin \alpha_i$ in., $i = 1$ or 2.

(b) Ratings

The transmitted horsepower is related to the required horsepower rating by the following:

$$\text{Required hp rating} = \frac{\text{hp transmitted} \times \text{service factor}}{\text{multiple-strand factor}} \tag{30.5}$$

where the service factors and the strand factors are given in Tables 30.3 and 30.4. The ratings are given in extensive tables in ref. 2. They are partially reproduced in graphical form in Fig. 30.4. For every chain pitch and number of teeth in the small sprocket, the horsepower ratings are given as a function of rpm. The service life expectancy is approximately 15,000 hr. The lubrication conditions are: type I, manual lubrication; type II, drip lubrication from a lubricator; type III, bath lubrication or disk lubrication; type IV, forced-circulation lubrication.

The rating graphs have the peculiar "tent form." At lower speeds chain fatigue is responsible for the chain failures. From a certain speed the roller impact resistance becomes responsible for roller breakage, and the rated horsepower falls rapidly until, at the ultimate speed, failure is caused by joint galling.[3] Even with the new reliable ratings, careful consideration should be given the application. The design is a compromise between life and cost. A large pitch provides more bearing area than a

Table 30.3. Service Factor[2]

Type of driven load	Type of input power		
	Internal-combustion engine with hydraulic drive	Electric motor or turbine	Internal-combustion engine with mechanical drive
Smooth	1.0	1.0	1.2
Moderate shock	1.2	1.3	1.4
Heavy shock	1.4	1.5	1.7

Table 30.4. Strand Factor[2]

No. of strands	*Multiple-strand factor*
2	1.7
3	2.5
4	3.3

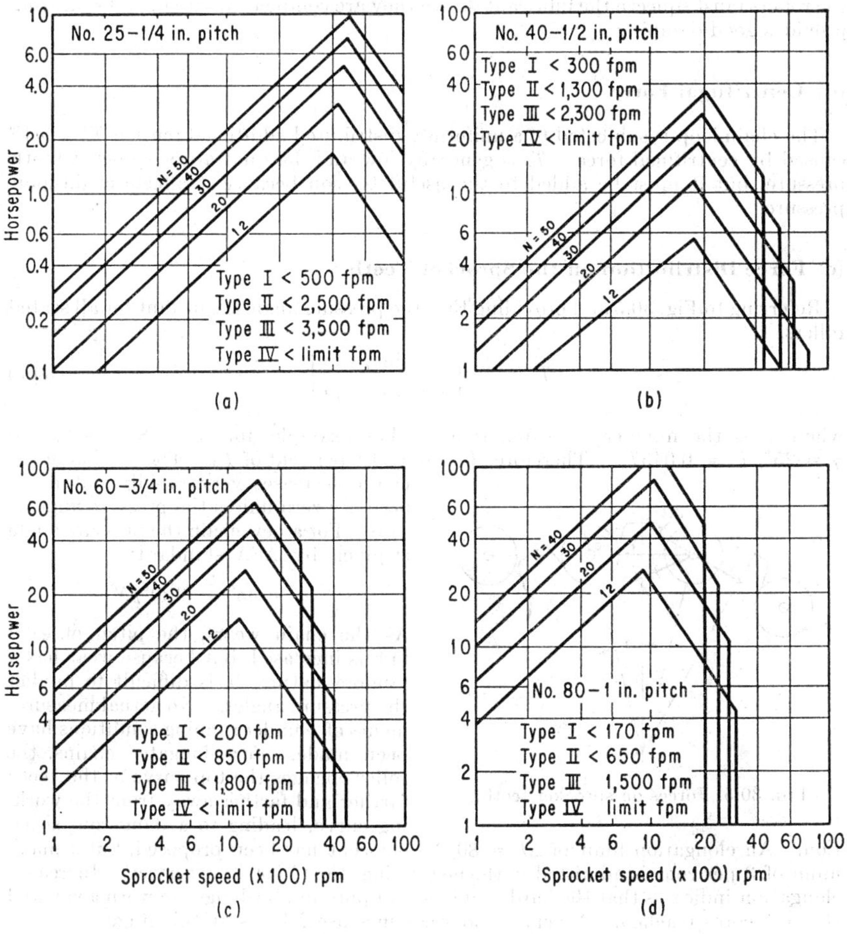

FIG. 30.4. Horsepower ratings vs. sprocket speed.

small one for the same load but requires fewer teeth on the sprocket. This leads to strong chordal action and thus to large dynamic effects, which in turn results in premature wear. A multiple strand with smaller pitch is preferable in such cases. Reference 23 gives more information.

(c) Roller Impact Velocity

The roller impact velocity is given by $\omega p \sin(2\alpha + \gamma)$. How much of the chain mass is involved in the impact is unknown. It is sometimes assumed[1,5] that a single link of mass mp is involved. In this case the kinetic energy is $\frac{1}{2}mp\omega^2p^2 \sin^2(2\alpha + \gamma)$. The strain energy is $\frac{1}{2}P\Delta$ (P = impact force, Δ = local deformation). Using the approximation[5] $\Delta = 3P/fE$, the strain energy becomes $3P^2/2fE$ where f is the roller width. If the kinetic energy is completely transformed into strain energy, the impact force

$$P = (2\pi V/N_1)\sqrt{wpfE/3g}\,\sin^2(2\alpha + \gamma) \tag{30.6}$$

A sprocket having a small number of teeth, running at high speeds, leads to roller breakage. Lubricant viscosity also plays an important part in roller fatigue because the rollers must squeeze the lubricant before they are completely seated, and squeezing provides good damping.

(d) Centrifugal Force

The chain loop is subjected to a uniformly distributed additional tension $T_c = mV^2$ caused by centrifugal force. T_c is generally low and does not affect sprocket tooth pressure, but it must be added to the useful tension because it affects chain-joint pressure.

(e) Force Distribution on the Sprocket Teeth

Referring to Fig. 30.5 and providing that the pressure angle is constant for all seated rollers,

$$T_n = T_0\left[\frac{\sin\gamma}{\sin(2\alpha + \gamma)}\right]^n \tag{30.7}$$

where n is the number of seated rollers. For example, for $N = 18$, $n = 6$, and $\gamma = 25°$, $T_6 = 0.045T_0$. Therefore, T_6 is only 4.5 per cent of T_0. The tension in the chain decreases very rapidly but never becomes zero unless the pressure angle is zero. For a new chain the pressure angle is given by ASA standard:

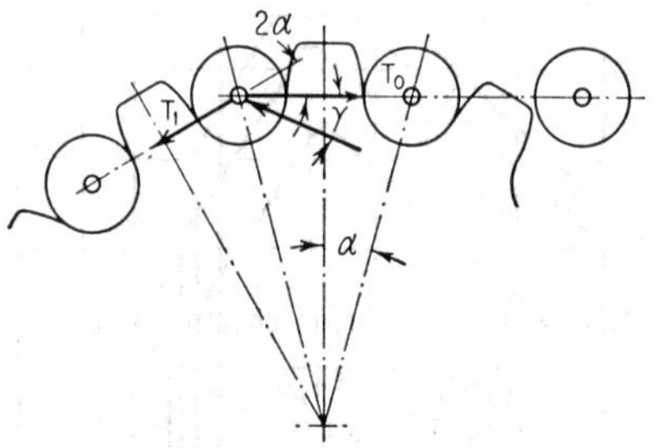

FIG. 30.5. Forces on sprocket teeth.

$$\gamma = 35° - (120°/N)$$

As the chain wears, the pitch of each link is increased, but because wear has a random nature, it is difficult to predict the pressure angles. No actual measurements of γ under running conditions have been made. For elongated chains, the roller engages the top curve of the teeth farther and farther away from the working curve, leading to a dangerous situation. An elongation limit of $\Delta L = 80/N_2$ per cent has been proposed, but a maximum of 3 per cent is admitted in the new ratings for 15,000 hr of service. Increased elongation indicates that the hardened layers of pins and bushings are worn away and the soft cores reached. A very rapid wear increase follows at this stage.

30.4. SYSTEM ANALYSIS

(a) Chordal Action

Chordal action, or polygonal effect, is the variation of the velocity of the driven sprocket which results because the actual instantaneous velocity ratio is not constant but a function of the sprocket angular position $\rho = d\varphi_2/d\varphi_1 = f(\varphi_1)$. Therefore, we have the angular-displacement equation $\varphi_2 = \int f(\varphi_1)\, d\varphi_1 +$ constant, the velocity equation $\dot{\varphi}_2 = \rho(\varphi_1)\dot{\varphi}_1$, and the acceleration equation $\ddot{\varphi}_2 = \rho(\varphi_1)\ddot{\varphi}_1 + (d\rho/d\varphi_1)\dot{\varphi}_1^2$. From the last equation, it is seen that, even in the case of constant driving speed $\dot{\varphi}_1$, the angular acceleration of the driven sprocket is not zero. This produces dynamic loads, vibrations, and premature chain wear. The analysis of chordal action is made under the assumption that the chain strand acts like a connecting rod in a four-bar linkage. Experiments at low speeds[7] yield excellent agreement with the theoretical results. At higher speeds the chain vibrations are sufficient to alter the static behavior of the chordal action. Nevertheless this action still remains a main source of excitation. Analysis shows[1,7] that the speed variation is maximum for a strand length equal to an *odd multiple of half pitches*. For this case,

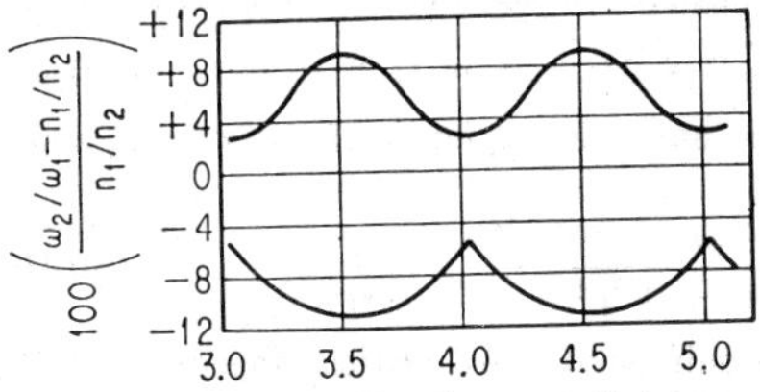

FIG. 30.6. Speed variation vs. center distance in pitches.

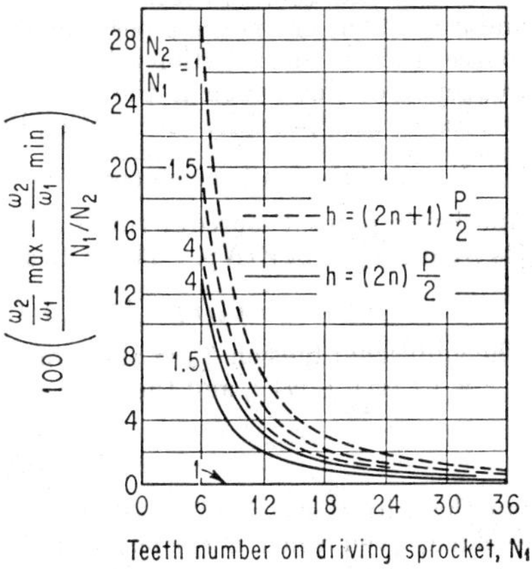

FIG. 30.7. Speed fluctuation vs. number of teeth on driving sprocket.

$$\left(\frac{\Delta\omega}{\omega}\right)_{\max} = \frac{N_2}{N_1}\left(\frac{\tan\alpha_2}{\sin\alpha_1} - \frac{\sin\alpha_2}{\tan\alpha_1}\right) \tag{30.8}$$

If the strand length is equal to an *even number of half pitches*, the speed variation is close to a minimum:

$$\left(\frac{\Delta\omega}{\omega}\right)_{\min} = \frac{N_2}{N_1}\left(\frac{\sin\alpha_2}{\sin\alpha_1} - \frac{\tan\alpha_2}{\tan\alpha_1}\right) \tag{30.9}$$

For $N_1 = N_2$, and an even number of half pitches, the speed variation is zero. Generally, for a fractional number of half pitches the speed variation fluctuates between maximum and minimum (Fig. 30.6). For large N_2/N_1 the values of $\Delta\omega/\omega$ in both cases approach the same limit (Fig. 30.7). There is a considerable advantage in having as many teeth as possible on the small sprocket. Chordal action in general may be ignored if the minimum number of sprocket teeth is 15.

(b) Transverse Vibrations of Chains

The power-transmitting strand may be considered as a tight string with a uniformly distributed mass. The small, transverse, undamped free vibrations $y(x,t)$ are described by the differential equation[10]

$$(V^2 - V_0^2)(\partial^2 y/\partial x^2) + 2V(\partial^2 y/\partial x\,\partial t) + \partial^2 y/\partial t^2 = 0 \tag{30.10}$$

which has the solution

$$y = C_1 \cos[\omega t + \omega x/(V_0 - V) + \phi_1] + C_2 \cos[\omega t - \omega x/(V_0 + V) + \phi_2] \tag{30.11}$$

The natural frequency, for the boundary conditions $y = 0$ at $x = 0$ and $x = a$, is

$$\omega_n = (n\pi V_0/a)[1 - (V/V_0)^2] \qquad n = 1, 2, 3, \ldots \tag{30.12}$$

To each ω_n corresponds a mode of vibration. For strings with continuously distributed mass, the number of modes is infinite. For a chain, there is a limit to this number because the strand is composed of a finite number of rollers s. The string approximation is therefore unsuitable for n close to s, and for small s in general. Considering the chain as a tight string of zero mass with point masses m_0 at every pitch distance, the frequency at zero speed

$$\omega_n = (n\pi V_0/a)(2s/n\pi) \sin [n\pi/(2s + 2)] \qquad n = 1, 2, \ldots, s \tag{30.13}$$

For $s \to \infty$ Eq. (30.13) becomes identical with Eq. (30.12) with $V = 0$. For $s = 19$ and $s = 44$ the errors are 5 per cent and 2.2 per cent, respectively. One important source of vibration excitation is the chordal action discussed above. The fundamental frequency of this action is the tooth-engagement circular frequency $(2\pi V)/p$; harmonics are also possible. Neglecting harmonics, the critical chain speeds

$$V_{\text{crit}} = (aV_0/np)[\sqrt{1 + (pn/a)^2} - 1] \cong nV_0p/2a \qquad n = 1, 2, 3, \ldots \tag{30.14}$$

Another important source of vibration excitation is the runout of the shaft or sprocket. Its frequency is $2\pi V/Np$. The critical speed attributable to eccentricity is therefore

$$V_{\text{crit}} = (aV_0/npN)[\sqrt{1 + (Npn/a)^2} - 1] \qquad n = 1, 2, 3, \ldots \tag{30.15}$$

The additional tension $T_c = mV^2$ changes the natural frequency of the strand, since $V_0^2 = (T/m) + mV^2$, and therefore

$$\omega_n = \frac{n\pi}{a} \frac{T/m}{[(T/m) + V^2]^{1/2}} \tag{30.16}$$

but the critical speeds [Eq. (30.14)] remain unchanged. All critical speeds must be avoided, even when a resonance does not result in infinite vibration amplitudes; damping is always present and reduces resonant amplitudes. For the computation of transverse amplitudes as a function of $V \neq V_{\text{crit}}$, see ref. 10. Resonance may be remedied by varying tight-strand tension and length, and the possible use of guides or idlers.

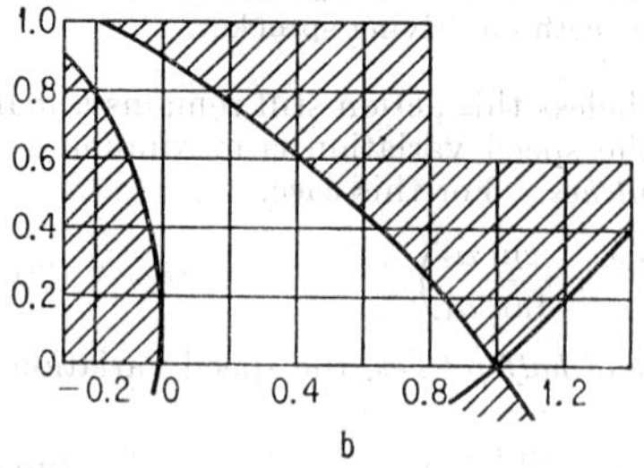

FIG. 30.8. Plot of the dimensionless parameters b and q of the Mathieu differential equation.

The traveling curve of the slack strand under the action of gravity is the catenary of a heavy chain in equilibrium. The catenary is disturbed by the chordal action at the ends. The stability criteria are formulated in ref. 14, where the problem is reduced, after linearization, to a Mathieu differential equation of the type

$$\ddot{y} + (b - 2q \cos 2t)y = 0$$

The dimensionless parameters

$$b = [(\omega_n/V)^2 - (7.5/a)^2](p/\pi)^2 \qquad \text{and} \qquad q = 2(\omega_n^2/g)(p^2/D_1) \tag{30.17}$$

are shown in Fig. 30.8. The shaded regions represent unstable running conditions and must be avoided if possible. The slack strand ω_n can be determined either experimentally or from Table 30.5. The given values are valid for a horizontal arrange-

Table 30.5. Values of ω_n

h/a	0.95	0.96	0.97	0.98	0.99	0.998
s/h	0.140	0.125	0.110	0.088	0.060	0.030
$\omega_n^2 a/g$	10.3	14.2	18	20.4	29	91.5

ment of the suspension points (seated rollers) and small s/h, when the catenary of the heavy chain can be approximated by a parabola. The values were experimentally verified to within 5.7 per cent.

(c) Longitudinal Vibrations of the Tight Strand

The chain is elastic in tension also, and the cushioning of shocks between the driving and driven shafts is one of the advantages of chain transmissions. The elongation of a chain under 50 per cent of ultimate load has been measured with great care.[11] A statistical average for all pitches is 0.081 in./ft. Not more than 5 per cent of chains of a given pitch will vary from this average by more than 0.014 in./ft. The values of the elastic constant given in Table 30.6 are based upon an assumed linear relationship between load and elongation.

Table 30.6. Values of the Elastic Constant K

p, in.	3/8	1/2	5/8	3/4	1	1 1/4	1 1/2	1 3/4	2
$K/10^6$ lb-in./in.	0.163	0.269	0.420	0.604	1.075	1.68	2.41	3.3	4.3

The natural frequency at zero speed is

$$\omega_n = (\pi n/a)\sqrt{K/m} \qquad n = 1, 2, 3, \ldots \tag{30.18}$$

while the critical speeds due to the tooth-engagement frequency become

$$V_{\text{crit}} = (np/2a)\sqrt{K/m} \qquad n = 1, 2, 3, \ldots \tag{30.19}$$

(d) Equivalent Stiffness of Chain for Torsional-vibration Calculations

The arrangement to be considered is that of a chain drive with two sprockets. The torsional stiffness depends upon whether the shaft of the first or second sprocket is taken as a reference value. With reference to the first shaft, the torsional stiffness of the tight strand is $K_1 = (K/a)R_1^2$ lb-in./rad. With reference to the second shaft, it is $K_2 = (K/a)R_2^2 = K_1(R_2/R_1)^2$ lb-in./rad. In both equations K is the tensile elastic constant of the chain. The equivalent shaft length in terms of a reference diameter d_0 and shear modulus G psi is

$$L_{1,\text{eq}} = \pi d_0^4 G/32K_1 \quad \text{in.} \qquad \text{or} \qquad L_{2,\text{eq}} = \pi d_0^4 G/32K_2 \quad \text{in.} \tag{30.20}$$

Further development follows the standard torsional-vibration procedure. For sufficiently high torsional amplitudes, however, the tight strand can become slack during a half cycle of vibration, introducing a nonlinear element into the computations. A theory taking this phenomenon in account has not yet been developed.

30.5. SILENT OR INVERTED-TOOTH CHAIN

Silent chains consist of inverted-toothed links alternately assembled on articulating-joint parts. There are two types of chain, (1) flank contact and (2) heavy-duty chordal-action-compensating.

The *flank-contact-type chains*[19,20] are shown in Fig. 30.9*a*. The chain width may be as great as 16 times the pitch, and speed-reduction ratios as high as 8 are employed. The speed limit is about 6,000 rpm for 3/8-in.-pitch chain and 900 rpm for 2-in.-pitch chain. The minimum and maximum number of sprocket teeth suggested is 17 and 150, respectively. The sprockets have a pitch diameter approximately equal to the outside diameter.

These chains are subject to the same speed variations and chordal rise and fall due to the polygonal chordal action as are roller chains. In general they are suggested for slow to moderate speed and 20,000 or more hours of life expectancy. They have an approximate ultimate tensile strength of 12,500 × pitch × width.

Heavy-duty chordal-action-compensating-type chains are designed to obtain the optimum in link strength and to eliminate the effects of chordal action. Figure 30.9*b* shows how the chain pitch line is maintained in a constant tangential relation to the sprocket pitch circle. This characteristic is obtained by generating the link profile for conjugate action with the sprocket tooth and/or by special articulating-joint design. These chains are used for a life requirement of less than 20,000 hr and/or widely varying load conditions. They have an approximate ultimate tensile strength of 20,000 × pitch × width.

For further information on silent-tooth chains see refs. 19 to 22.

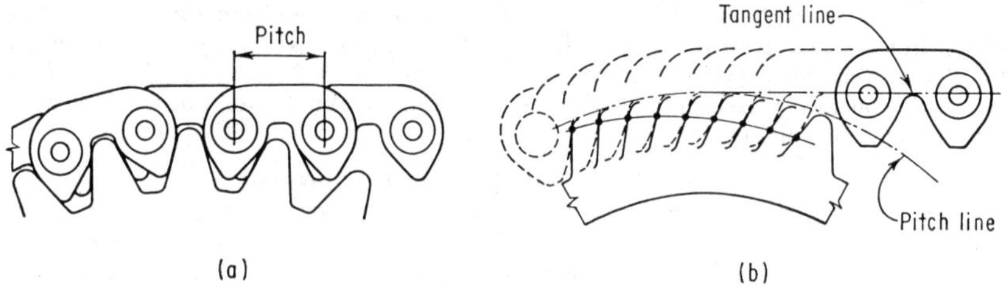

FIG. 30.9. Types of silent chain. (*a*) Flank-contact. (*b*) Chordal-action-compensating.

References

1. Binder, R. C.: "Mechanics of the Roller Chain Drive," Prentice-Hall, Inc., Englewood, Cliffs, N.J., 1956.
2. "New Horsepower Ratings of American Standard Roller Chains," proposed by ARSCM, Diamond Chain Company, Inc., Indianapolis, 1960.
3. Frank, J. F., and C. O. Sundberg: New Roller Chain Horsepower Ratings, *Machine Design*, July 6, 1961.
4. Kuntzmann, P.: "Roller Chain Drives" (French), Dunod, Paris, 1961.
5. Niemann, G.: "Machine Design" (German), vol. 2, Springer-Verlag OHG, Berlin, 1960.
6. Jackson and Moreland: "Design Manual for Roller and Silent Chain Drives," prepared for ARSCM, 1955.
7. Bouillon, G.: "Polygonal Action, Theoretical and Experimental Study," M.Sc. Thesis, Laval University, 1961.
8. Morrison, R.: Polygonal Action in Chain Drives, *Machine Design*, vol. 24, no. 9, September, 1952.
9. Mahalingham, S.: Polygonal Action in Chain Drives. *J. Franklin Inst.*, vol. 265, no. 1, January, 1958.
10. Mahalingham, S.: Transverse Vibrations of Power Transmission Chains, *Brit. J. Appl. Phys.*, vol. 8, April, 1957.
11. Whitney, L. H., and P. M. MacDonald: Elastic Elongation of Chains, *Prod. Eng.*, February, 1952.
12. Germond, H. S.: "Wear Limits for Roller and Silent Chain Drives," ASME paper 60-WA-8, 1960.
13. Schakel, R. A., and C. O. Sundberg: "Proven Concepts in Oil Field Roller Chain Drive Selection," ASME paper 57-PET-24.
14. Ignatenko, V. V.: Variable Forces in the Strand of a Chain Drive (Russian), *Bull. Inst. Higher Education*, no. 4, 1961.
15. "Roller Chains and Sprockets," Catalog 8 ACME Chain Corp., Holyoke, Mass.
16. "Stock Power Transmission and Conveyor Products," Catalog 760, Diamond Chain Co., Inc., Indianapolis, Ind.
17. Radzimovsky, E. I.: Eliminating Pulsations in Chain Drives, *Prod. Eng.*, July, 1955.
18. Hofmeister, W. F., and H. Klaucke: Dynamic Check Point Way to Longer Chain Life, *Iron Age*, August, 1956.
19. American Standards Association: ASA B29.1-1963, Transmission Roller Chains and Sprocket Teeth (SAE SP-69).
20. American Standards Association: ASA B29.3-1954, Double Pitch Power Transmission Chains and Sprockets (SAE SP-69).
21. American Standard Association: ASA B29.2-1957, Inverted Tooth (Silent) Chains and Sprocket Teeth (SAE SP-68).
22. American Standard Association: ASA B29.9-1958, Small Pitch Silent Chains and Sprocket Tooth Form (Less Than 3/8 Inch Pitch) (SAE TR-96).
23. Rudolph, R. O., and P. J. Imse: Designing Sprocket Teeth, *Machine Design*, Feb. 1. 1962, pp. 102–107.

Section 31

CABLE DYNAMICS

By

FRIEDRICH O. RINGLEB, Ph.D., *Staff Physicist, Naval Air Engineering Laboratory, Philadelphia, Pa., and Professor of Mechanics, University of Delaware, Newark, Del.*

CONTENTS

31.1. BASIC CONCEPTS

(a) Introduction

Cable or wire rope has been applied in various mechanical systems, i.e., hoists, elevators, instruments, and especially aircraft arresting gear. In some rare cases it has been used as a means for power transmission. Wire-rope manufacturers' catalogues contain much valuable information.[15]

A cable or rope is defined as a flexible, axially symmetric structure consisting of one or more filaments of equal or different homogeneous materials. The length of this structure is assumed to be large compared with its largest diameter.

For power transmission the rope wrap tension is given by Eq. (31.2). The coefficient of friction for wire rope in a cast-iron or steel groove $\mu = 0.1$. Unless stated otherwise, all the material of this section refers to cable or wire rope.

(b) Cable Data

Wire rope is made up of 7, 19, or 37 wires twisted into a strand. A number of strands (usually 6, 7, or 8) are then twisted helically about a fiber core. The strands are laid around the core either to the right or to the left. The resulting cable is accordingly designated as a right- or left-lay cable. Cores are generally hemp or plastic. The fiber core acts as an elastic support for the strands and is intended to prevent excessive contact stresses between the wires.[16]

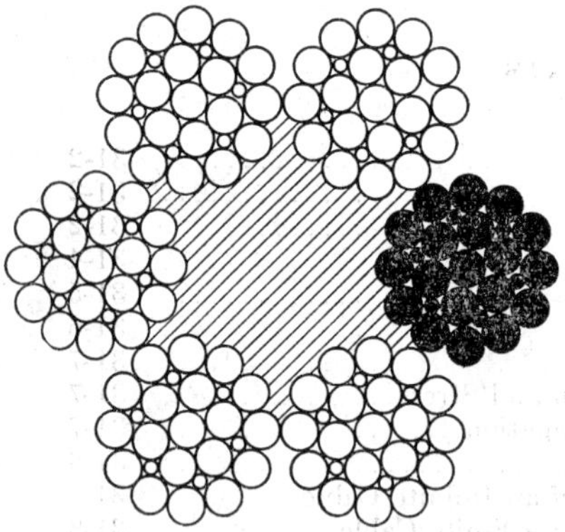

FIG. 31.1. Cross section of a 6 × 19 steel wire cable with fiber core and filler wires.

Although steel is most popular, wire rope is made of many kinds of material such as copper, bronze, stainless steel, and wrought iron.[20] A standard steel wire cable consisting of six strands of wire wound around a core, each strand containing 19 wires, is shown in Fig. 31.1. The characteristics of this 6 × 19 wire rope having 6 filler wires are presented in Table 31.1. The 6 × 17, 6 × 19, 6 × 37, and 8 × 19 wire ropes have wire diameters d_w equal to $d_r/9$, $d_r/16$, $d_r/32$, and $d_r/19$ where d_r is the rope diameter. The design factors are flexibility, wear, strength, core strength, and corrosion resistance. Flexibility is obtained by using a large number of small-diameter wires. Large wires give better wear resistance, and strength depends on the wires and the structure. Corrosion by moisture and stray electric currents is generally inhibited by proper lubrication of the wire rope.

Data for the same type of cable produced by different manufacturers can differ considerably.

Minimum factors of safety based on breaking strength are: hoists and cranes 4 to 7, mine shaft 4 to 8, elevators 7 to 12, depending on risk. These factors of safety are based upon favorable conditions of environment and lubrication of the rope and depend on the loads, acceleration, speed, attachments, number, size, arrangements of sheaves and drums, corrosion, abrasion, rope length, and accuracy of stress investigation. A factor of safety of 1.2 is used in the design of Navy arresting gear under controlled design conditions (discussed later) and limited life. The breaking strength is 80 per cent or more of the total strength of the wires in the rope. The modulus of elasticity (Table 31.1) is less than half that of wire material because the rope is composed of twisted wires which act as helical springs. As an approximation for plow steel, the breaking strength $F_u = 75{,}000 d_r^2$ and the rope weight is $1.6 d_r^2$ lb/ft. Friction loss of 5 per cent is assumed in hoisting practice, and 3 per cent is used for very flexible rope with extra large pulleys.

Table 31.1. Data for 6 × 19 Wire Rope

Diam, in.	Max pitch, in.	Metallic area, ft^2	Weight per ft, lb/ft	Breaking strength F_u, lb	Breaking stress σ_u, $lb/ft^2 \times 10^{-6}$
5/8	4.0	0.00106	0.63	40,000	38.2
11/16	4.5	0.00119	0.71	46,000	38.6
7/8	5.7	0.00210	1.23	73,000	34.8
1	6.5	0.00274	1.60	95,000	34.5
1 1/8	7.3	0.00346	2.03	119,000	34.3
1 1/4	8.1	0.00428	2.50	146,000	34.1
1 3/8	8.9	0.00519	3.03	175,000	33.7
1 1/2	9.8	0.00617	3.60	208,000	33.7
1 5/8	10.6	0.00725	4.23	242,000	33.4
1 3/4	11.4	0.00838	4.90	280,000	33.4
1 7/8	12.2	0.00962	5.63	314,000	32.6
2	13.0	0.01094	6.40	350,000	32.0

Mass density (average) $\rho = 18.1$ lb-sec²/ft⁴.
Modulus of elasticity (average) $E = 15.9 \times 10^8$ lb/ft².

The following formulas are useful for structural computations. It is assumed that the center of a wire of diameter $2r$ forms a cylindrical helix with the coordinates

$$\xi = R\cos\varphi \qquad \eta = R\sin\varphi \qquad \zeta = (h/2\pi)\varphi$$

in a rectangular x, y, z coordinate system. R is the radius of the circular cylinder, h the pitch of the helix, and φ a variable parameter. The pitch angle α is determined by

$$\cos\alpha = \frac{R}{\sqrt{R^2 + (h/2\pi)^2}} \qquad \sin\alpha = \frac{h}{2\pi\sqrt{R^2 + (h/2\pi)^2}}$$

If the parameter φ is replaced by the arc length

$$s = \varphi\sqrt{R^2 + (h/2\pi)^2}$$

of the helix its equations can be written

$$\begin{aligned} \xi &= R\cos[(s/R)\cos\alpha] \\ \eta &= R\sin[(s/R)\cos\alpha] \\ \zeta &= s\sin\alpha \end{aligned}$$

The surface of the wire is described by a circle of radius r within a plane which is perpendicular to the tangent of the helix and which moves so that the center of the circle describes the helix. A radius r of the circle which connects a point x, y, z of the wire surface with its center ξ, η, ζ forms with the principal normal of the helix at ξ, η, ζ an angle t which is used as a second parameter for the representation of the surface of the wire. Then the equations of the wire surface as functions of the parameters s and t are[24]

$$\begin{aligned} x &= (R - r\cos t)\cos[(s/R)\cos\alpha] + r\sin t\sin\alpha\sin[(s/R)\cos\alpha] \\ y &= (R - r\cos t)\sin[(s/R)\cos\alpha] - r\sin t\sin\alpha\cos[(s/R)\cos\alpha] \\ z &= s\sin\alpha + r\sin t\cos\alpha \end{aligned}$$

Setting z = constant one obtains s as a function of t. The first two of these equations then represent the contour of a cross section through the wire which is perpendicular to the cylinder axis. For the pitch angle $\alpha = 90°$ this curve is a circle of radius r. For decreasing pitch angles the shape of the cross section is first approximately an ellipse. It finally approaches two concentric circles for $\alpha = 0°$. In between the shape of the cross section is similar to a bent ellipse.

In a corresponding way the geometry of a strand of wires whose center forms a helix can be studied.

The ultimate tension of a cable decreases if it is used statically in a state of bending. Table 31.2 shows the ultimate tension of a 6 × 19 plow-steel wire rope of various diameters bent 180° over a sheave of various diameters. The ultimate tension of the straight cable is approximately given by

$$T_{us} = 83{,}000d^2 \qquad \text{lb}$$

The tensile strength of the single wire is 230,000 psi.

Table 31.2. Strength of Ropes on Sheaves

Diam d of rope, in.	Diam D of sheave, in.					
	10		14		18	
	T_u, lb	T_u/T_{us}	T_u, lb	T_u/T_{us}	T_u, lb	T_u/T_{us}
5/8	28,170	87.4	39,910	96.0	30,700	95.3
3/4	39,840	80.3	45,450	91.5	45,840	92.5
7/8	51,470	81.9	56,300	89.5	58,390	92.9
1	65,000	79.3	71,670	87.5	75,780	92.5
1 1/4	99,070	75.8	106,320	81.3	110,630	84.7
3/4 (worn)	35,830	83.5	35,300	82.3	34,700	80.9

The column T_u/T_{us} is the ratio of the ultimate tension of the bent rope and the straight rope.[23]

(c) Accessories

The use of spliced rope should be avoided. Many kinds of connectors have been used. The best is the rope socket (efficiency 100 per cent) in which wires are opened

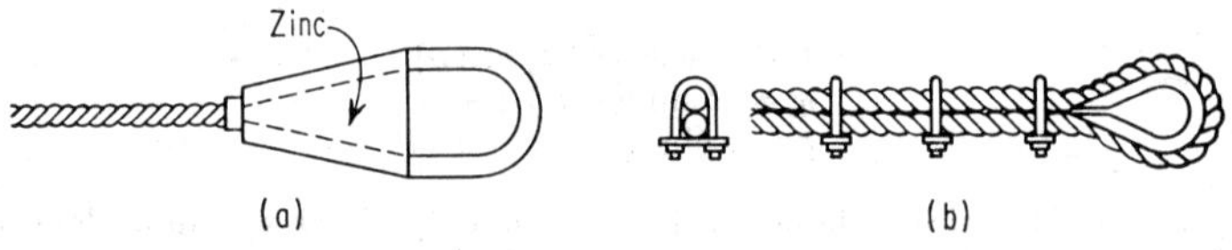

Fig. 31.2. Rope connections. (a) Rope socket. (b) Thimble and clamps.

and high-grade zinc poured on (Fig. 31.2a). Figure 31.2b shows a thimble and clamp which has an approximate efficiency of 75 per cent. Woven wire has been used in lieu of clamps. Because of stress concentration thimbles and other splices cannot strengthen filler rope.

Drums and sheaves should be of largest practical diameter. Where feasible, a grooved drum is desirable and grooves should provide ample clearance between windings. The winding of one layer of rope on another should be avoided, since this considerably reduces rope life. Long ropes may run as high as 4,000 fpm. Reverse bending of wire rope should also be avoided. Figure 31.3 shows recommended sheave and drum proportions. Sheaves for rope are made similar to those of chains. The

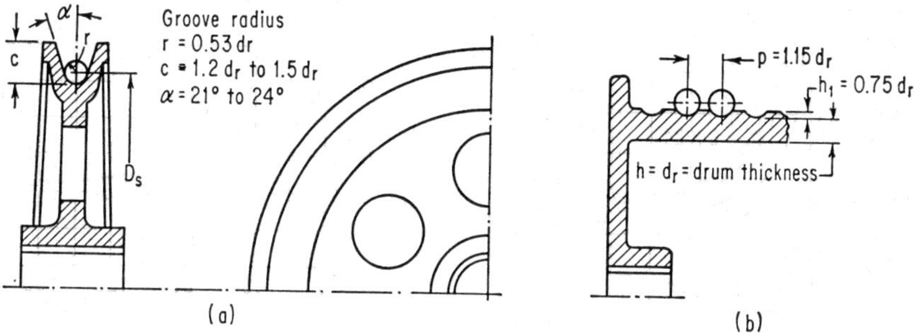

FIG. 31.3. Sheave and drum proportions. (*a*) Sheave (arms may also be of elliptical cross section). (*b*) Drum.

life of the rope can be lengthened by use of an insert at bottom of the sheave groove. The insert may be made of leather, hard rubber, wood, etc. Small drums on hoists are made plain.

The drum thickness should be made equal to the rope diameter d_r and checked for strength. Ropes 6 × 7, 6 × 19, 6 × 37, and 8 × 19 have a minimum sheave or drum diameter D_s equal to $42d_r$, $24d_r$, $18d_r$, and $20d_r$, respectively. If possible, twice these values are recommended for longer rope life.

(d) Elementary Stresses

The basic stresses in a wire rope are (1) tension due to primary loads and dynamic loads and (2) tension due to bending or wrapping around drum and sheaves. The fatigue condition associated with all must be considered. Bending is often the predominant load.

The *tension* load is comprised of the initial load and the inertia load. In hoisting and elevator services the acceleration may be as high as 10 ft/sec². The stress wave, considered later, should also be included.

The *tension bending stress* is given by

$$\sigma_t = d_w E/D_s \qquad \text{psi} \qquad (31.1)$$

The pressure between the rope and sheave (if assumed uniformly distributed)

$$P_i = 2T/d_r D_s \qquad \text{psi} \qquad (31.2)$$

where T is the tension in the rope, lb. This pressure may cause high compressive stresses in the wires.

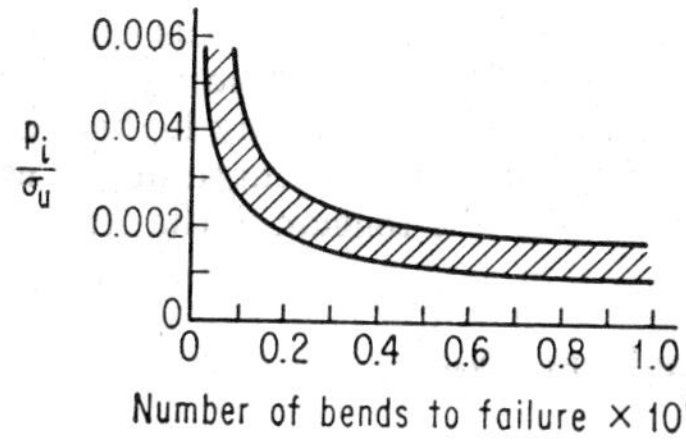

FIG. 31.4. Relationship between fatigue life and sheave pressure.

A fatigue failure may first appear as a few broken wires on the rope surface. Figure 31.4 shows the results of tests of P_i/σ_u as a function of the number of bends to failure for various kinds of rope. Here σ_u denotes the ultimate stress of the wire (psi).* Note that the flexing and unflexing of the rope as it passes over a sheave count as a single bend. The critical stress in the rope is due to tension in bending and the contact stress of the wires. In addition, fretting and corrosion initiate and accelerate fatigue damage.

* This is the customary way to represent such measurements. It is proposed to replace the ultimate stress of the wire by the ultimate stress in the cable.

Figure 31.5 represents the case where the bearing pressure is rather small and the number of bends large. This is the case of true fatigue. Figure 31.6 shows the result of measurements in the case of extremely high bearing pressures and accordingly small number of bends to failure. This case is of importance for the use of cables at aircraft

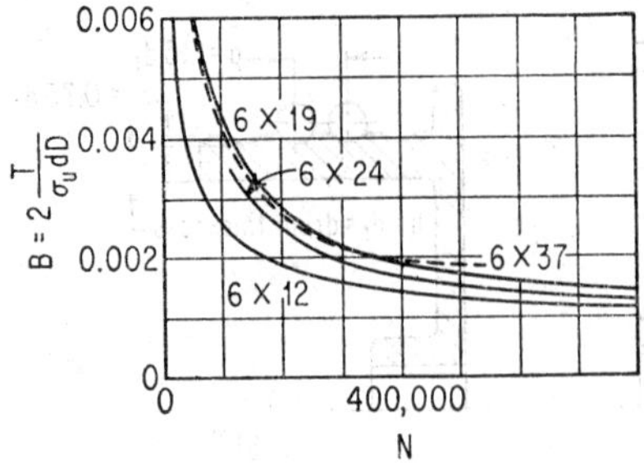

FIG. 31.5. Relation between fatigue and number of bends with small bearing pressure.

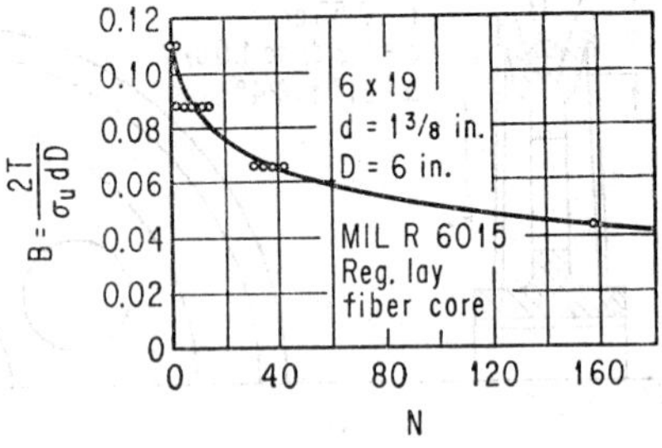

FIG. 31.6. Relation between fatigue and number of bends with extremely high bearing pressures. (*After Robert F. Barthelemy, Naval Air Engineering Laboratory, Philadelphia, Pa.*)

arresting gears [see Art. 31.5(*d*)]. The deck pendant which is engaged by the hook of the landing airplane, for instance, is bent over a rather small radius at high impact loads.

(e) Equivalence to an Elastic String

The dynamics of a moving cable is treated in the following as that of an equivalent homogeneous elastic string. The initial mass density ρ_0 (lb sec²/ft⁴) of the cable is defined by the equation

$$\rho_0 q = W/g \tag{31.3}$$

where W is the weight per unit length of cable (lb/ft), g the acceleration due to gravity, and q the cross-sectional area of the densest material of the cable, measured normal to the longitudinal axis of the cable (ft²). In the case of a cable of uniform material the mass density, defined in this way, is equal to the mass density of the material. In the case of a steel wire cable with fiber core the mass density of the (equivalent homogeneous) cable is slightly higher than that of steel.

If l_0 is the initial length of the cable (ft) under zero tension and Δl_0 its elongation (ft) under the tension T the strain is defined by

$$\epsilon_0 = \Delta l_0 / l_0 \tag{31.4}$$

and the stress

$$\sigma = T/q \qquad \text{lb/ft}^2 \tag{31.5}$$

Under the assumption that Hooke's law holds, according to which stress and strain are proportional, the modulus of elasticity E (lb/ft²) of the cable (with respect to the cross-sectional area q) is defined by

$$Eq = T/\epsilon_0 \tag{31.6}$$

In the case of a cable of uniform material the modulus of elasticity of the cable is equal to the modulus of elasticity of its material.

If l is the length of the cable under the initial stress $\sigma_0 = T_0/q$ and Δl its elongation under the stress $\sigma = T/q$ then

$$[(\sigma - \sigma_0)/E](1 + \sigma_0/E) = \Delta l/l \tag{31.7}$$

Usually (at least in the case of a steel wire cable with fiber core) σ_0/E is negligibly small compared with unity and therefore

$$(\sigma - \sigma_0)/E = \Delta l/l \qquad \text{approx} \tag{31.8}$$

It is assumed that the cross-sectional area q, as defined before, is a constant and does not change with the stress. The Poisson ratio, therefore, is neglected in the following. A logical consequence of this simplification is that the mass density depends on the stress according to the formula

$$\rho_0/\rho = 1 + \sigma/E \tag{31.9}$$

where ρ = mass density under stress, lb sec²/ft⁴. In general σ/E is negligibly small compared with unity so that approximately $\rho = \rho_0$ constant. A prestressed wire rope rather accurately satisfies Hooke's law. Otherwise sufficiently accurate results can be obtained by considering the modulus of elasticity as a piecewise constant variable during a dynamic process. The case where E is a function of σ has been studied by T. von Kármán and others.[18]

31.2. LONGITUDINAL MOTION OF A CABLE

(a) Equations of Longitudinal Motion and Stress

The longitudinal motion of a cable can be described by an equation of the form[12]

$$x = f(s,t) \tag{31.10}$$

The cable is situated on and moves along a straight line, the x axis. s is a parameter, characterizing the mass points of the cable. x is the coordinate of the mass point s at the time t. If the function $f(s,t)$ satisfies the condition $f(s,0) = s$ then s is the x coordinate of a cable point at time $t = 0$.

If the motion along the x axis occurs under the influence of a tension distribution along the cable only, then the function [Eq. (31.10)] describing the motion satisfies the condition

$$\partial^2 x/\partial t^2 = c^2(\partial^2 x/\partial s^2) \tag{31.11}$$

where c is the longitudinal wave velocity.

$$c^2 = (1 + \sigma_0/E)(E/\rho) \tag{31.12}$$

or approximately

$$c^2 = E/\rho \tag{31.13}$$

if σ_0/E is negligibly small compared with unity.

The local stress σ at a cable point s at time t is given by

$$(\sigma - \sigma_0)/E = (1 + \sigma_0/E)(\partial x/\partial s - 1) \tag{31.14}$$

(b) General Solution and Stress Propagation

The general solution of Eq. (31.11) is

$$x = F(s + ct) + G(s - ct) \tag{31.15}$$

where $F(\zeta)$ and $G(\zeta)$ are arbitrary twice-differentiable functions of the argument $\zeta = s + ct$, respectively, $\zeta = s - ct$. If s is the x coordinate of any cable point at time $t = 0$ the general solution can be written in the form

$$x = \phi(s + ct) - \phi(s - ct) + s \tag{31.16}$$

(ϕ is an arbitrary function).

The velocity u at time t of a cable particle, characterized by its initial x coordinate at time $t = 0$, is determined by

$$u/c = \phi'(s + ct) + \phi'(s - ct) \tag{31.17}$$

where $\phi'(\zeta) = d\phi/d\zeta$. The stress σ at any point s at any time t is determined by

$$(\sigma - \sigma_0)/E = (1 + \sigma_0/E)[\phi'(s + ct) - \phi'(s - ct)] \tag{31.18}$$

A function $x = F(s + ct)$ represents at time $t = 0$ a longitudinal distortion $x = F(s)$ of the x axis. With variable time this distortion moves along the x axis with the velocity c (the longitudinal wave velocity). Therefore, the general solution of Eq. (31.15) is the superposition of two longitudinal waves, moving in opposite directions with the velocity c. Accordingly two stress waves move in opposite directions with the velocity c.

(c) Longitudinal Impact

If an end point O of an infinitely long cable, initially at rest, suddenly moves with a constant velocity v_0 in longitudinal direction, the stress σ induced at the moving end is constant and determined[5] by

$$(\sigma - \sigma_0)/E = (1 + \sigma_0/E)(v_0/c) \tag{31.19}$$

The longitudinal wave velocity c is determined by Eq. (31.12). At time t the end point moved from O to P where $OP = v_0t$ and the stress σ propagated from O to R where $OR = -ct$ (Fig. 31.7). Beyond R the cable is at rest and has the initial stress σ_0.

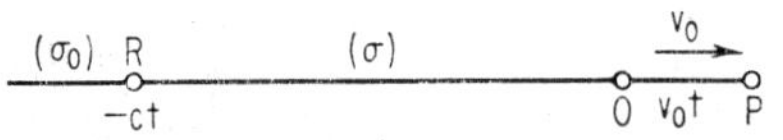

FIG. 31.7. Longitudinal impact.

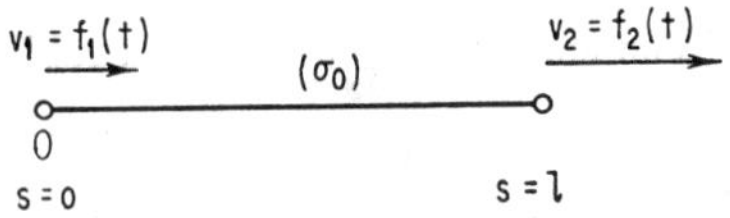

FIG. 31.8. Longitudinal cable motion.

The work done at time t by moving the end point O of the cable is

$$H = q\sigma v_0 t \qquad \text{ft-lb} \tag{31.20}$$

The kinetic energy of the moving cable part

$$H_k = q\rho ct(v_0^2/2) \tag{31.21}$$

where v_0 is the initial velocity, ft/sec. The strain energy

$$H_\sigma = (\sigma + \sigma_0)qv_0t/2 \tag{31.22}$$

Also $H = H_k + H_\sigma$ and

$$H_\sigma/H_k = (\sigma + \sigma_0)/(\sigma - \sigma_0) \tag{31.23}$$

In the case of zero initial stress $H_\sigma = H_k$.

(d) Longitudinal Motion and Stress of an Infinite Cable

If the end point O of an infinitely long cable, initially at rest, moves with the velocity $v_0 = f_0(t)$ for $t > 0$ [$f_0(t) = 0$ for $t \leq 0$] then the velocity u of the cable particle s at time t is determined[12] by

$$u/c = (1/c)f_0(s/c + t) \tag{31.24}$$

and the stress σ by

$$(\sigma - \sigma_0)/E = (1 + \sigma_0/E)(u/c) \tag{31.25}$$

(e) Longitudinal Motion and Stress of a Finite Cable

A finite cable is initially situated between point $O(s = 0)$ and $s = l$ with constant initial stress and is at rest. It is assumed that point O moves with the velocity $v_1 = f_1(t)$ and point $s = l$ with the velocity $v_2 = f_2(t)$ for $t > 0$ (Fig. 31.8). Then the velocity u of any cable point s at time t is determined by ref. 12,

$$u/c = \phi'(s + ct) + \phi'(s - ct) \tag{31.26}$$

and the stress by

$$(\sigma - \sigma_0)/E = (1 + \sigma_0/E)[\phi'(s + ct) - \phi'(s - ct)] \tag{31.27}$$

where the function $\phi'(\zeta)$ is completely defined by the conditions

$$\begin{aligned} \phi'(\zeta) &= 0 \qquad \text{for } \zeta \leqq 0 \\ \phi'(\zeta) + \phi'(-\zeta) &= (1/c)f_1(\zeta/c) \\ \phi'(l + \zeta) + \phi'(l + \zeta) &= (1/c)f_2(\zeta/c) \qquad \text{for } \zeta > 0 \end{aligned} \tag{31.28}$$

(f) Variable Initial Conditions

If $u_0(s)$ is the longitudinal velocity and $\sigma_0(s)$ the local stress of a cable point with the coordinate $x_0 = x_0(s)$ at time $t = 0$ (s being a variable parameter) then for time $t \geqq 0$

$$x = \tfrac{1}{2}\left[x_0(s + ct) + x_0(s - ct) + 1/c\int_{s-ct}^{s+ct} u_0(\zeta)\,d\zeta\right] \tag{31.29}$$

$$\frac{u}{c} = \frac{1}{2}\left[\frac{\sigma_0(s + ct) - \sigma_0(s - ct)}{E} + \frac{u_0(s + ct) + u_0(s - ct)}{c}\right] \tag{31.30}$$

$$\frac{\sigma}{E} = \frac{1}{2}\left[\frac{\sigma_0(s + ct) + \sigma_0(s - ct)}{E} + \frac{u_0(s + ct) - u_0(s - ct)}{c}\right] \tag{31.31}$$

(d'Alembert).

31.3. LONGITUDINAL INTERACTION BETWEEN INELASTIC MASSES AND CABLES

(a) Motion of an Inelastic Mass by Means of a Cable

A mass m is attached to the end point $x_0 = 0$ of a cable of length l with zero initial stress. The other end point suddenly moves longitudinally with the constant velocity v_0. Assuming that the developing stress in the cable is approximately constant along its length (but variable with time) the x coordinate of the mass at time t

$$x = v_0[t - (l/\omega)\sin\omega t] \tag{31.32}$$

and the stress in the cable is given by

$$\sigma/E = v_0/c\sqrt{m/m_c}\sin\omega t \tag{31.33}$$

where

$$\omega^2 = Eq/lm \tag{31.34}$$

and m_c denotes the mass of the total cable.

The exact solution is obtained by stepwise application of the result of Art. 31.2(*e*). For details see ref. 12.

(b) Prescribed Acceleration of a Mass by Means of a Cable

A mass m is attached at point $s = 0$ to a cable which is situated between $s = 0$ and $s = l$ and which has zero initial stress. The velocity of the mass, to be obtained from the prescribed acceleration, is denoted by $u = u(t)$, and the unknown velocity of the cable end point $s = l$ by $v = v(t)$. Application of the result of Art. 31.2(*e*) yields

$$\phi'(\zeta) = \tfrac{1}{2}[u/c + (m/qE)\dot{u}] \tag{31.35}$$

$$\phi'(-\zeta) = \tfrac{1}{2}[u/c - (m/qE)\dot{u}] \tag{31.36}$$

where $\zeta = ct$. The unknown function $v(t)$ is determined by

$$v/c = \phi'(l + \zeta) + \phi'(l - \zeta) \tag{31.37}$$

It is assumed that the prescribed acceleration of the mass starts at time $t = l/c$. The graphical solution of the problem to determine $v(t)$ is shown in Fig. 31.9. The variable $\frac{1}{2}(m/qE)\dot{u}$ is plotted as function of $\zeta = ct$ and graphically integrated starting from $\zeta = l$. The value $\frac{1}{2}u/c$ is plotted as a function of ζ. The sum of both curves gives $2\phi'$ for $\zeta > 0$. The difference gives $2\phi'$ for $\zeta < 0$. From Eq. (31.37) ratio v/c is obtained.

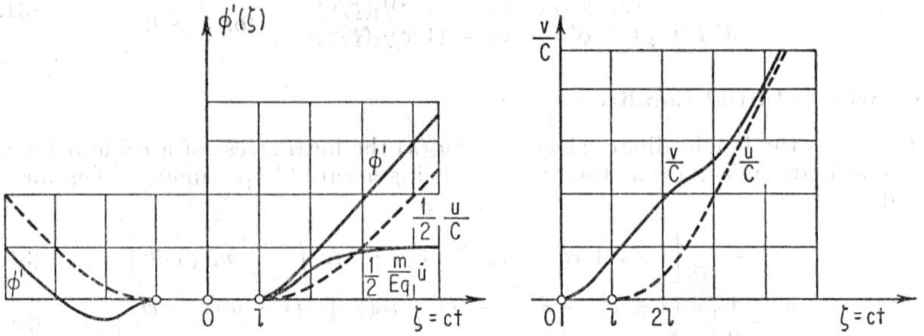

FIG. 31.9. Prescribed acceleration of a mass by means of a cable.

(c) Influence of a Mass between Two Cables on the Stress Propagation

In general, two different cables, situated along the x axis, are distinguished by the indices 1 and 2. For instance, the moduli of elasticity are E_1 and E_2, respectively. Between the two cables and attached to them is the inelastic mass m so that cable 1 is located on the right and cable 2 on the left side of the mass. The initial stresses produced by a tension T_0 in both cables are denoted by

$$\sigma_{01} = T_0/q_1 \qquad \sigma_{02} = T_0/q_2 \tag{31.38}$$

It is assumed that the free end point of cable 1 suddenly moves with the constant velocity v_0 along the x axis and that the stress ratios σ/E produced by this motion in the cables are negligibly small compared with unity. In the case of an infinitely long cable 2 the mass m then moves with the velocity v given as a function of t by

$$v/v_0 = (A/B)(1 - e^{-Bt}) \tag{31.39}$$

where

$$A_1 = \frac{2q_1\sqrt{E_1\rho_1}}{m} \qquad B = \frac{q_1\sqrt{E_1\rho_1} + q_2\sqrt{E_2\rho_2}}{m} \tag{31.40}$$

The stresses induced in the cables at the right and the left side of m are, respectively,

$$\sigma_r = \sigma_{01} + (2v_0 - v)\sqrt{E_1\rho_1} \tag{31.41}$$

$$\sigma_l = \sigma_{02} + v\sqrt{E_2\rho_2} \tag{31.42}$$

valid as long as there are no stress reflections returning from the cable end point.

(d) Two Cables Attached to Each Other

In the special case $m = 0$ one obtains

$$\frac{T_r - T_0}{T_1 - T_0} = \frac{2}{1 + (q_1/q_2)\sqrt{E_1\rho_1/E_2\rho_2}} \tag{31.43}$$

where T_1 is the tension in cable 1 due to longitudinal impact with velocity v_0. This can also be written

$$\frac{T_r - T_0}{T_1 - T_0} = \frac{2}{1 + (\mu_1/\mu_2)} \tag{31.44}$$

where μ_1 and μ_2 are the masses of the two cables through which a tension wave propagates per unit time.

If the second cable is replaced by an elastic link at which the tension in the first cable is measured by a strain gage, its diameter should be chosen so that $\mu_1 = \mu_2$. For stress propagation through three cables see ref. 12.

31.4. TWO-DIMENSIONAL MOTION OF A CABLE

(a) Equations of Motion

It is assumed that the initial position of a cable at time $t = 0$ is the straight position along the x axis of a rectangular x, y coordinate system and that its initial stress σ_0 = constant. The x coordinate of a cable point in the initial position is denoted by s. Then the motion of the cable is generally described by two functions

$$x = f(s,t) \qquad y = g(s,t) \tag{31.45}$$

satisfying the initial conditions

$$x = f(s,0) = s \qquad y = g(s,0) = 0 \tag{31.46}$$

If the motion of the cable occurs under the influence of a tension distribution only, the following equations hold:

$$\frac{\partial^2 x}{\partial t^2} = \frac{E}{\rho}\frac{\partial}{\partial s}\left(\frac{\sigma}{E}\cos\theta\right) \qquad \frac{\partial^2 y}{\partial t^2} = \frac{E}{\rho}\frac{\partial}{\partial s}\left(\frac{\sigma}{E}\sin\theta\right) \tag{31.47}$$

$$\frac{\partial x}{\partial s} = \frac{1+\sigma/E}{1+\sigma_0/E}\cos\theta \qquad \frac{\partial y}{\partial s} = \frac{1+\sigma/E}{1+\sigma_0/E}\sin\theta \tag{31.48}$$

where θ is the angle between a cable element and the x axis and $\sigma = T/q$ the local stress. θ is the angle between the direction corresponding to increasing s values and the positive x axis measured positive in counterclockwise direction. For the functions σ and θ of s and t the following equations result from Eqs. (31.47) and (31.48):

$$\begin{aligned}\frac{\partial^2}{\partial t^2}\left(\frac{1+\sigma/E}{1+\sigma_0/E}\cos\theta\right) &= \frac{E}{\rho}\frac{\partial^2}{\partial s^2}\left(\frac{\sigma}{E}\cos\theta\right)\\ \frac{\partial^2}{\partial t^2}\left(\frac{1+\sigma/E}{1+\sigma_0/E}\sin\theta\right) &= \frac{E}{\rho}\frac{\partial^2}{\partial s^2}\left(\frac{\sigma}{E}\sin\theta\right)\end{aligned} \tag{31.49}$$

(Compare refs. 2, 3, and 7.)

(b) Wave Velocities

If a cable element moves so that θ is constant, Eqs. (31.49) show that the stress σ propagates with velocity

$$c = [(1 + \sigma_0/E)(E/\rho)]^{1/2} \tag{31.50}$$

with respect to s. This is the longitudinal wave velocity which reduces to

$$c = \sqrt{E/\rho} \tag{31.51}$$

if σ_0/E is negligibly small compared with unity.

In the case where a cable element moves under constant stress σ, Eqs. (31.49) show that a θ value propagates with velocity

$$\bar{c} = \left(\frac{1+\sigma_0/E}{1+\sigma/E}\frac{\sigma}{\rho}\right)^{1/2} \tag{31.52}$$

with respect to s. This velocity is the transverse wave velocity which reduces to

$$\bar{c} = \sqrt{\sigma/\rho} \tag{31.53}$$

if σ/E and σ_0/E are negligibly small compared with unity.

(c) Oblique Impact

If the end point O of a straight infinitely long cable with initial stress σ_0 suddenly moves with constant velocity v_0 in a given direction at an impact angle $\beta > 0$ (Fig. 31.10), the cable moves at any time in two straight parts, namely, the segment PQ in the direction β with the velocity v_0 and the segment RQ in the direction from R to Q with a velocity u, called the *particle velocity*. The part beyond R is at rest. The stress induced in both segments is equal and constant during the motion. The velocity of the kink in the cable at point Q, moving toward R, in the following is denoted by ω.

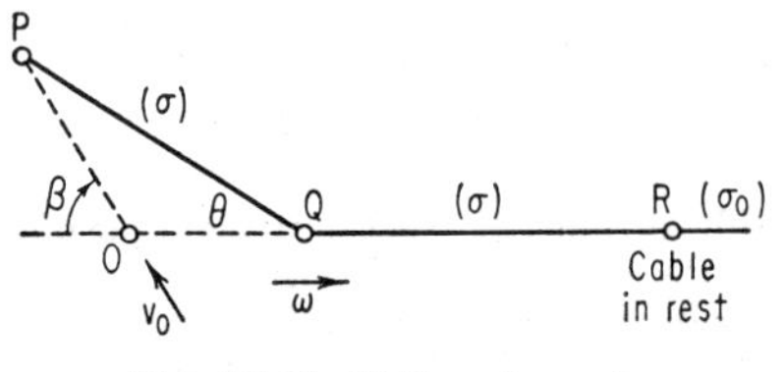

FIG. 31.10. Oblique impact.

The impact velocity v_0, the impact angle β, and the impact stress σ are related by

$$\left(\frac{v_0}{c_0}\right)^2 + 2\frac{v_0}{c_0}\left[\left(\frac{\sigma}{E}\right)^{1/2}\left(1+\frac{\sigma}{E}\right)^{1/2} - \frac{\sigma-\sigma_0}{E}\right]\cos\beta$$
$$= 2\frac{\sigma-\sigma_0}{E}\left(\frac{\sigma}{E}\right)^{1/2}\left(1+\frac{\sigma}{E}\right)^{1/2} - \left(\frac{\sigma-\sigma_0}{E}\right)^2 \qquad (31.54)$$

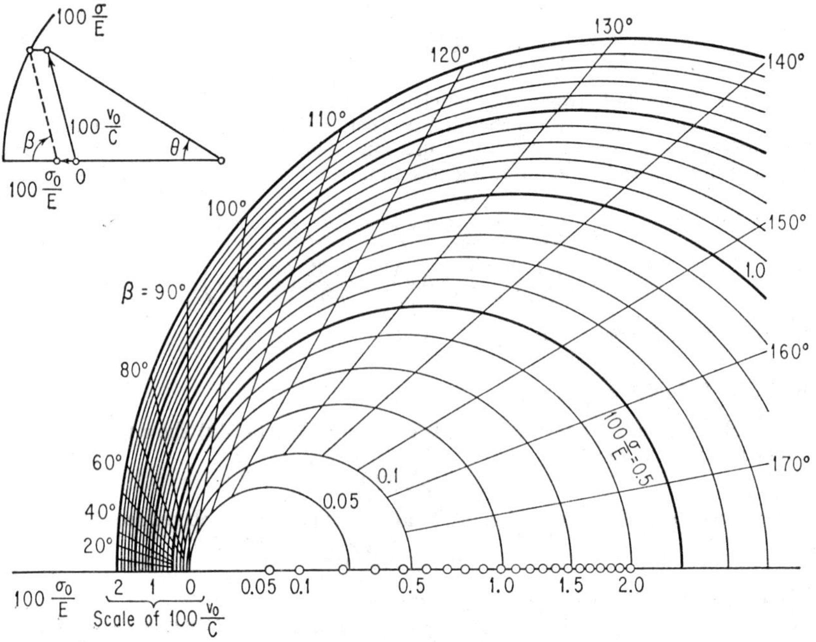

FIG. 31.11. Evaluation of the oblique impact formula.

(F. O. Ringleb, 1957) called in the following the *oblique impact formula*, where

$$c_0 = \sqrt{\frac{E}{\rho_0}}$$

and ρ_0 is the mass density at zero stress. The particle velocity u is given by

$$\frac{u}{c} = \frac{(\sigma - \sigma_0)/E}{1 + \sigma_0/E} \tag{31.55}$$

and the kink velocity c_0 by

$$\omega/c = \bar{c}/c - (u/c)(1 - \bar{c}/c) \tag{31.56}$$

where $\bar{c}$ is the transverse wave velocity (ft/sec). The kink angle θ is determined by

$$\cot\theta = \frac{\omega/v_0 + \cos\beta}{\sin\beta} \qquad \sin\theta = \frac{v_0/c}{\bar{c}/c}\sin\beta \tag{31.57}$$

See refs. 12 and 13.

These relations can be simplified as before if σ/E is negligibly small compared with unity. In particular one obtains in the case of transverse (perpendicular) impact, the approximation[6,7,11]

$$(\sigma - \sigma_0)/E = (\tfrac{1}{2})^{2/3}(v_0/c_0)^{4/3} \tag{31.58}$$

Figure 31.11 serves to evaluate numerically approximately the oblique impact of Eq. (31.54). The small figure in the upper left corner describes the application. Figure 31.12 shows the transverse impact stresses for various initial stresses, and Fig. 31.13 presents a comparison of theory and measurement.

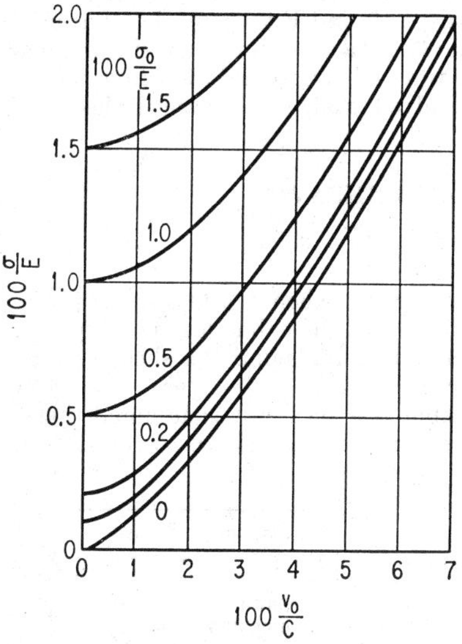

FIG. 31.12. Transverse impact stress for various initial stresses.

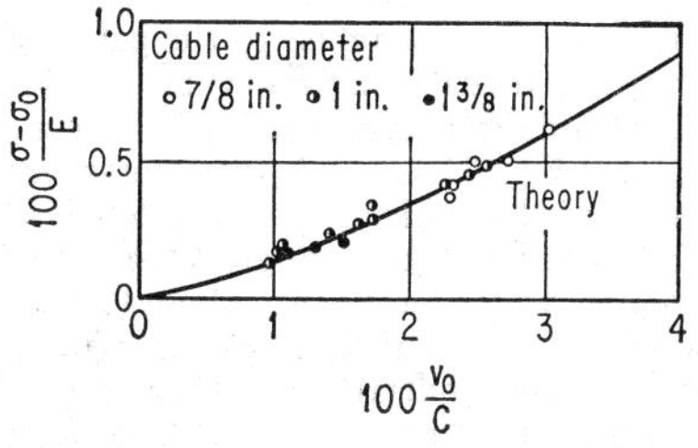

FIG. 31.13. Comparison of transverse impact stress theory and measurement.

(d) Energy Relations

Denoting the kinetic energy of the moving cable segment PQ (Fig. 31.10) by H_1, the kinetic energy of the segment QR by H_2, and the energy stored in the cable when raising its stress from σ_0 to σ by H_σ, the following relations are satisfied:[13]

$$H_1/H_0 = (\bar{c}/c)(v_0/c)^2 \tag{31.59}$$
$$H_2/H_0 = (1 - \bar{c}/c)(u/c)^2 \tag{31.60}$$
$$H_\sigma/H_0 = 2(\bar{c}/c)^2(u/c)(1 + u/c) - (u/c)^2 \tag{31.61}$$

where

$$H_0 = \rho q c t(c^2/2) \tag{31.62}$$

In the case of transverse impact,

$$H_1 : H_2 : H_\sigma = 2:1:1 \tag{31.63}$$

(e) Reflection and Interaction of Waves

The reflection and interaction of longitudinal and transverse waves can be studied using the oblique impact relation [Eq. (31.54)]. For instance, in order to determine the effect of a wave approaching a point S of a moving cable, one has to think oneself moving with point S so that the two cable segments ending in S appear to be at rest. The arriving wave produces in general oblique impacts at the end points S with the same stress resulting near S in both cable segments, unless S moved with the longitudinal wave velocity (the speed of sound in the cable). Of great value for the simplification of the interactions is the following result: A stress propagates over a kink with small angle θ (say less than 20°) approximately without disturbance.[10]

(f) Characteristics Theory of Cable Motion

If point O (Fig. 31.10) moves along any given path with a given time-dependent velocity, the resulting cable motion can theoretically be determined by application of the oblique impact formula, approximating the given path by polygons along which the cable end point moves with stepwise constant velocity.

This procedure is related to the characteristics theory of the moving cable.[19,4,9] If instead of $\partial x/\partial t$ and $\partial y/\partial t$ the velocity components U and V tangential and normal to the cable are introduced into Eqs. (31.47) and (31.48) a set of four quasilinear equations results for the unknown functions U, V, θ, and σ of s and t and their first derivatives which is totally hyperbolic and can be solved using the classical theory of characteristics.

31.5 INTERACTION BETWEEN CABLES AND SHEAVES

(a) Cable Impact at a Sheave

A longitudinal impact of a cable at a sheave occurs when a stress wave passes through the cable and over the sheave. Assuming that the cable does not slip over the sheave the effect of the impact can be computed, replacing the sheave by an equivalent mass attached to the cable (compare Art. 31.3).

In order to study a transverse impact of a cable at a sheave it is assumed that a cable of infinite length and initial stress σ_0 passes over the sheave S and terminates at point O (Fig. 31.14). A transverse impact at point O of the cable of constant velocity v_0 produces a kink wave at angle θ_1 and an impact tension σ_1 which propagate along the cable. It is assumed that the sheave turns with the constant circumferential velocity u_1 (the particle velocity in the cable associated with stress σ_1) before the transverse impact wave arrives at the sheave. This will be approximately valid if the mass of the sheave is sufficiently small. Then the stress σ_2 caused by the impact of the transverse wave at the sheave is given by

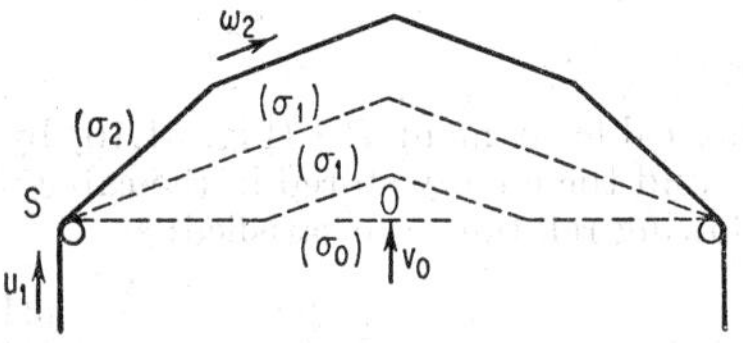

FIG. 31.14. Cable impact at a sheave.

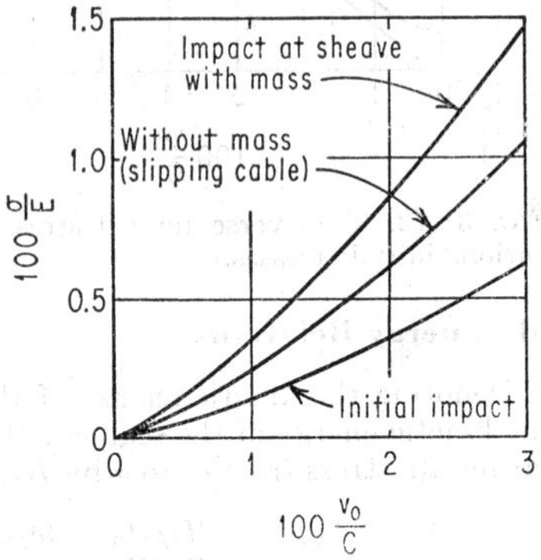

FIG. 31.15. Cable impact at a sheave with and without mass.

$$\frac{\sigma_2 - \sigma_1}{E} = \sqrt{\left(\frac{v_0}{c}\right)^2 + \left(\frac{\omega_2}{c}\right)^2 + 2\frac{v_0}{c}\frac{\omega_2}{c}\sin\theta_1} - \frac{\omega_2}{c} - \frac{u_1}{c} \tag{31.64}$$

where

$$\omega_2/c = \bar{c}_2/c = \sqrt{\sigma_2/E} \tag{31.65}$$

This shows that the stress σ_2 caused by the impact of the transverse wave at the sheave is approximately the same as the stress produced by an oblique impact at a cable with the initial stress σ_0 under the impact angle $\beta = 90° - \theta_1$ with the velocity v_0.

In the case of a massless (or frictionless) sheave Eq. (31.65) must be replaced by

$$\frac{\sigma_2 - \sigma_1}{E} = \frac{1}{2}\left[\sqrt{\left(\frac{v_0}{c}\right)^2 + \left(\frac{\omega_2}{c}\right)^2 + 2\frac{v_0}{c}\frac{\omega_2}{c}\sin\theta_1} - \frac{\omega_2}{c} - \frac{u_1}{c}\right] \tag{31.66}$$

because the stress σ_2 spreads at the moment of impact over both segments of the cable separated by the sheave. The curves of Fig. 31.15 represent, for an initial stress $\sigma_0 = 0$, the initial impact stress ratio σ_1/E and the stress ratio σ_2/E resulting from Eqs. (31.64) and (31.65). For the numerical computation it should be noted that the unknown stress σ_2 is contained also in the expression ω_2 which occurs on the right side of these equations.[12,14]

(b) Slack in a Pulley

Figure 31.16 shows a set of five sheaves on a movable crosshead and a set of four fixed sheaves. A cable is reeved over the two sets of sheaves as shown. One cable end point is anchored; the other is assumed to move suddenly with constant velocity u,

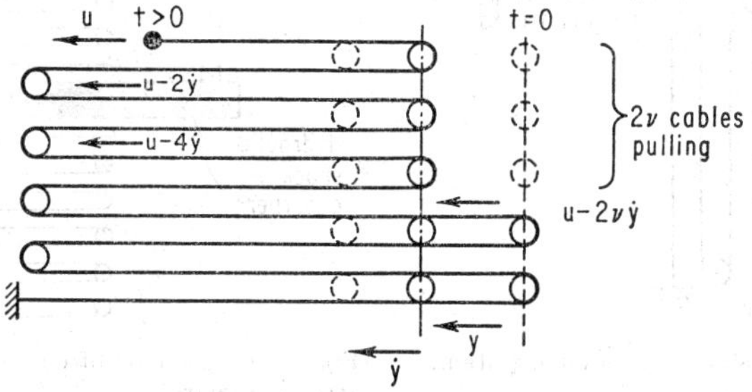

FIG. 31.16. Slack in a pulley.

producing a longitudinal impact at the cable. This arrangement explains the effect of a longitudinal impact on a common pulley. It is assumed that the crosshead carries a mass and that in general a force F acts on it. The impact stress propagates along the cable and produces a force on the crosshead when it has passed over the first crosshead sheave. In general the crosshead will move if the impact stress has passed a sufficient number of crosshead sheaves so that the total tension is larger than the force F acting against the crosshead. In general the moving crosshead will produce slack in the cable segments through which the impact stress did not propagate. The figure represents a moment where the first three pairs of cable segments are under stress while the others are still in battery position. The slack is picked up because of the propagating stress. From the formula for reverse impact (oblique impact under the angle $\beta = 180°$) it follows that the slack pickup velocity[14]

$$u_\nu = u - 2\nu\dot{y} \tag{31.67}$$

where ν is the number of crosshead sheaves under stress and $\dot{y}$ the velocity of the crosshead.

At the moment when all slack is picked up at the last sheave (which usually takes a considerable time) the cable can have a large speed which then will result in a strong longitudinal impact with high stress when the motion is stopped by the anchor.

(c) Slackless Reeving Systems

In order to avoid high longitudinal impact stresses due to slack pickup slackless reeving systems have been developed.[12] Figure 31.17 represents a system where independently moving pairs of sheaves are incorporated between the moving crosshead and the fixed sheaves. A stress which enters the cable at the upper right side of this system quickly propagates across the fixed sheaves because of the short cable length involved and therefore almost simultaneously reaches the crosshead sheaves. Other combinations are of course possible.

(d) Aircraft Arresting Gear

The typical example of a modern dynamic use of cables is the aircraft arresting gear[1,8,14,17] as shown in Fig. 31.18. The tail hook of the airplane engages a cable (the

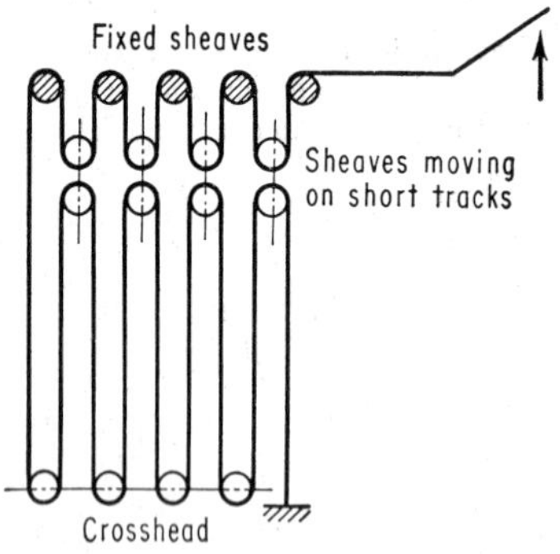

FIG. 31.17. A slackless reeving system.

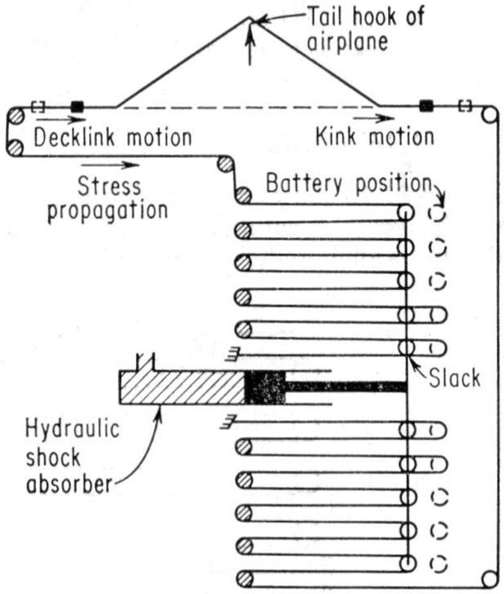

FIG. 31.18. Scheme of conventional aircraft arresting gear.

"deckpendent" in Navy notation) which is connected by two links to the "purchase cable." The latter is reeved over sheaves guiding it to the arresting engine, a hydraulic shock absorber, consisting of a movable crosshead with two sets of sheaves, of two sets of fixed sheaves, of a hydraulic cylinder with piston, and of an automatically controlled orifice. Other designs having slackless reeving systems have been developed.

31.6. SPECIAL ROPE DRIVES

Rope drives consisting of manila or cotton rope on multiple-grooved pulleys may often be a most economical form of drive over long distances and for large amounts of power. The velocity should be quite high, and 5,000 fpm is a good speed for economy.

Maximum permissible working load for manila rope for hoisting purposes is $200d_r^2$, lb (400 to 800 fpm), $400d_r^2$ (150 to 300 fpm), and $1{,}000d_r^2$ below 100 fpm. Reference 22 shows data for manila rope. Nylon rope is also used. It has from 1½ to 2 times tensile strength and 3 times elasticity and gives greater resistance to fatigue and surface abrasion than manila rope. For power transmission the rope wrap tension is shown by Eq. (29.6). The coefficient of friction for hemp rope on a cast-iron pulley is $\mu = 0.2$ and on a wood pulley $\mu = 0.4$.

References

1. Ayre, R. S., and J. I. Abrams: Dynamic Analysis and Response of Aircraft Arresting Systems, *Proc. ASCE, Eng. Mech. Div.*, April, 1958.
2. Carrier, G. F.: On the Non-linear Vibration Problem of the Elastic String, *Quart. Appl. Math.*, vol. 3, pp. 157–165, 1945.

3. Carrier, G. F.: A Note on the Vibrating String, *Quart. Appl. Math.*, vol. 7, pp. 97–101, 1949.
4. Cole, J. C., C. B. Dougherty, and J. H. Huth: Constant Strain Waves in Strings, *J. Appl. Mech.*, vol. 20, no. 4, pp. 19–22, 1953.
5. de Saint Venant, B.: Choc longitudinal de deux barres elastiques, *Compt. Rend.*, vol. 66, pp. 650–653, 1868.
6. Flügge, W., and H. Köller: Ausbreitung von Stosswellen im gespannten Seil, *Jahrbuch der deutschen Luftfahrtforschung*, 1941.
7. Goldsmith, W.: "Impact," Edward Arnold (Publishers) Ltd., London, 1960.
8. Kaufman, W. J.: Recovery Equipment Study and Proposed Mark 8 Arresting Gear Program, *Naval Air Material Center Rept.* M-6070, 1956.
9. Marble, F. E.: The Motion of a Finite Elastic Cable, *North American Instrument, Inc., Rept.*, Altadena, Calif., 1954.
10. Neidhardt, G. L., N. F. Eslinger, and F. Sasaki: An Analytical Approach to the Alleviation of Dynamic Tensions in Aircraft Arresting Gear Cables, *WADC Tech. Rept.* 58-217, Astia no. AD155542.
11. Ringleb, F. O.: Dynamics of a Moving Cable, *Naval Air Material Center, Rept.* M-4812, Philadelphia, Pa., 1948.
12. Ringleb, F. O.: Cable Dynamics, *Naval Air Material Center, Rept.* NAEF-Eng-6169, Philadelphia, Pa.
13. Ringleb, F. O.: Motion and Stress of an Elastic Cable due to Impact, *J. Appl. Mech.*, vol. 24, no. 3, pp. 417–425, September, 1957.
14. Ringleb, F. O.: Basic Problems in the Dynamics of the Aircraft Arresting Gear, from "A Decade of Basic and Applied Science in the Navy," pp. 59–75, Office of Naval Research, 1957.
15. John A. Roebling' Sons Corp. Handbook.
16. Starkey, W. L., and H. A. Cress: An Analysis of Critical Stresses and Mode of Failure of a Wire Rope, *Trans. ASME, J. Eng. Ind.*, vol. 81, ser. B, no. 4, p. 307, November, 1959.
17. Tuman, C.: "High Velocity Engagement of Arresting Wires," Naval Air Missile Test Center, Point Mugu, Calif., 1954.
18. von Kármán, T., and P. Duwez: The Propagation of Plastic Deformation in Solids, *J. Appl. Phys.*, vol. 21, pp. 987–994, 1950.
19. Wen-Hsiung Li: Elastic Flexible Cable in Plane Motion under Tension, *J. Appl. Mech.*, vol. 81, ser. E, pp. 589–593, December, 1959. Compare also the literature on textiles.
20. Simplified Practice Recommendations 198, CO 1C 614, U.S. Department of Commerce, 1950.
21. Drucker and Tasbau: A New Design Criterion for Wire Rope, *Trans. ASME*, 66 A-33 T 945.
22. U.S. Government Spec. TR 601A, Nov. 26, 1935.
23. Skillman, E.: Some Tests of Steel Wire Rope on Sheaves, Technological Papers of the Bureau of Standards, Department of Commerce, no. 229, Mar. 2, 1923.
24. Shitkow, D. G., and I. T. Pospechow: "Drahtseile," German ed., VEB Verlag Technik, Berlin, 1957.

Section 32

GEARING

By

W. A. TUPLIN, D. Sc., M. I. Mech. E., *Professor of Applied Mechanics, University of Sheffield, Sheffield, England* (*Gear Design*)

GEORGE W. MICHALEC, M.S., *Head Special Products Dept., General Precision, Inc., Pleasantville, N.Y.* (*Precision Gearing*)

CONTENTS

GEAR DESIGN

PRECISION GEARING

GEAR DESIGN*

32.1. INTRODUCTION[16,17,18]

Tooth gearing is used for one or more of three principal purposes.

1. Transmission of motion between parts which have to move in accurately timed relation, without necessarily involving a considerable amount of power.
2. Transmission of power between two shafts which are required to run at different speeds.
3. Transmission of power between two shafts that are not in the same straight line.

The operation of determining appropriate manufacturing dimensions for the common types of gear for normal purposes can be reduced to simple routine. The essential feature of rational gear design is that production of the gears shall not demand the use of any nonstandard tool. Fortunately the adaptability of the involute-gear system is such that insistence on the use of only standard tools is not a perceptible restriction, and it is adopted here.

The design procedures outlined here assume the use of cutters corresponding to a basic rack with 20° flank angle and working depth equal to 0.636 times the normal pitch, and these dimensions represent what is now an almost international standard. Figure 32.1 presents the basic dimensions for a number of standard gear systems.

32.2. NOTATION

(a) General Symbols

Where small letters and large letters are given, or where a symbol is repeated with the subscripts P and G, the former refers to the pinion and the other refers to the gear.

A = cone distance
$A_C A_L A_q$ = dynamic-load factors (Figs. 32.24 and 32.25)
B = designed normal backlash
c = clearance
C = designed center distance
C_B = meshing center distance
d, D = mean diameter = blank diameter − working depth
d_b, D_b = base diameter
d_B, D_B = meshing diameter
D_c = diameter of cutter
d_F, D_F = reference diameter
d_0, D_0 = blank diameter
d_R, D_R = root diameter
d_u = diameter of measuring roller
D_t = throat diameter of worm wheel
$e = C - \frac{1}{2}(d_f + D_f)$
E = Young's modulus
F = face width
g = (revolutions in life of gear, relative to axes of mating gears) × (number of mating gears)
h = height setting of caliper
$i(d/d_b)$ = involute function of angle whose secant is d/d_b
$= [(d-d_b)/d_b] [(d - d_b)/(d + d_b/q)]^{1/2}$

* By W. A. Tuplin.

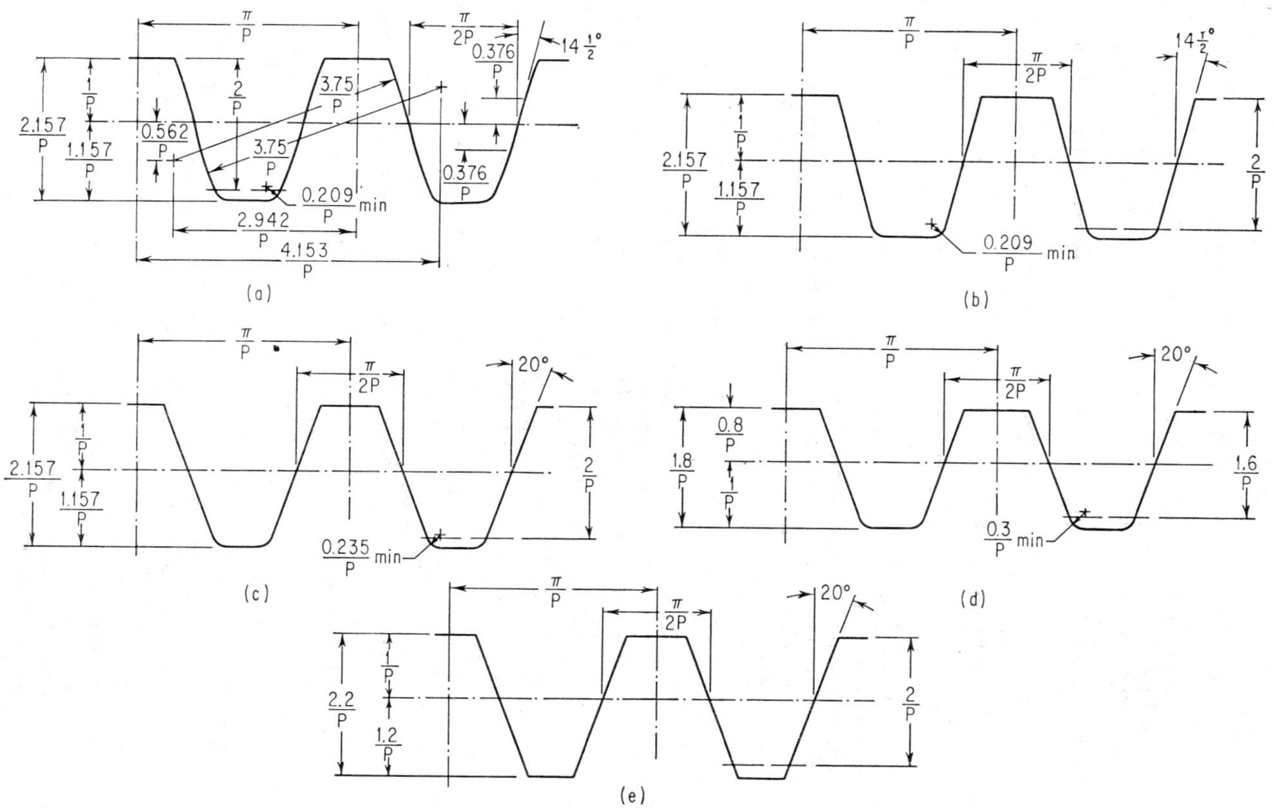

FIG. 32.1. Gear-tooth systems. (*a*) Basic rack for the 14½° composite system (ASA standard). (*b*) Basic rack for the 14½° full-depth involute system (ASA standard). (*c*) Basic rack for the 20° full-depth involute system (ASA standard). (*d*) Basic rack for the 20° stub involute system (ASA standard). (*e*) Basic rack for the 20° involute fine-pitch system (ASA standard).

J_{Ry} = repetition factor for bending stress (values in Fig. 32.22)
J_{Rz} = repetition factor for surface stress (values in Fig. 32.22)
k = normal backlash with full-depth teeth (Eq. 32.30)
l = lead of tooth helix
l_e = lead of cutter guides
m_G = gear ratio $= N_G/N_P$
$n_P n_G$ = rotational speed, rpm, relative to a common perpendicular to the axis of the gears
$N_P(N_W)N_G$ = number of teeth (or number of threads in worm), negative for an internal gear
$N_{Pe}N_{Ge}$ = equivalent number of spur-gear teeth. $N_{Pe} = t \sec^3 \psi_1 \sec \gamma_1$.

$$N_{Ge} = T \sec^3 \psi_2 \sec \gamma_2$$

P = diametral pitch
p_x = axial pitch
p_n = normal pitch of conjugate standard rack
p_{nb} = normal base pitch of gear and generating cutter ($p_{nb} = p_n \cos \phi_{nf}$)
p_t = transverse pitch at some circle defined or implied
r, R = radius
σ_y = bending stress (see Table 32.7)
σ_{yb} = basic allowable bending stress for the material concerned (see Table 32.7)
σ_z = contact stress (see Table 32.7)
σ_{zb} = basic allowable contact stress for the material concerned (see Table 32.7)
σ_s = contact stress in crossed-axis gear (see Table 32.7)
σ_{sb} = basic allowable contact stress for the material of a crossed-axis gear (see Table 32.9)
t_{nc} = thickness setting of caliper
t = normal arc thickness of tooth
t_t = transverse arc thickness of tooth
v = speed of tooth at mean circle relative to axes of mating gears
v_{fm} = v, fpm
v_{nb} = velocity of involute tooth in direction of normal at any point
v_s = sliding (or rubbing) speed, fpm
v_{se} = endwise sliding velocity, i.e., along a helical line on the tooth
v_{su} = "up-and-down" sliding velocity, i.e., in the direction of a line of intersection of the tooth flank and a transverse plane
V_1V_2 = blank-diameter factors
V_C = $(V_1 + V_2)$ for given value of C
W = axial load, lb
w_d = dynamic load, lb/in. width of tooth
w_g = gross tooth load, lb/in. width
Y = strength factor
z = basic backlash factor
β = lubrication factor (Table 32.10)
γ = meshing-cone angle
γ_0 = blank-cone angle
γ_R = root-cone angle
Δ = transmission error (Table 32.13)
η = force angle or pressure angle
κ = zone factor (Table 32.12)
λ = lead angle ($= 90° - \psi$)
Σ = shaft angle
ϕ_x = axial flank angle
ϕ_n = normal flank angle
ϕ_{nF} = normal flank angle at reference circle, i.e., the normal flank angle of the basic rack
ψ = helix angle or spiral angle at reference circle

(b) Meanings of Subscripts

A = allowable
b = basic
B = meshing
c = cutter
D = driving gear
F = reference cylinder
G = gear wheel
i = internal gear
n = normal section
o = outer
P = pinion
r = radial
R = root of tooth
s = sliding (or rubbing)
t = transverse section, or throat of worm wheel
u = measuring roller
y = bending stress
z = surface stress
1 = pinion
2 = wheel

32.3. COMPARISON WITH OTHER DRIVES

For many power-transmission purposes, a belt made of leather or other flexible material may be used successfully in conjunction with suitable pulleys. Because the load is transmitted between pulleys and belts by virtue of friction, and along the belt itself by virtue of the tensile strength of the belt, the dimensions of a belt drive must be large, if any considerable torque is to be handled. A gear drive has, by comparison, a considerable advantage in the way of compactness, because the load is transmitted positively through metal members.

Because it relies upon friction, a friction belt drive cannot be used where accurate timing of the driving and driven shafts is required.

A chain drive has characteristics that lie between those of the belt drive and those of the gear drive. It provides a positive drive and can be arranged with smaller dimensions in general than those of the corresponding belt drive, although it cannot reasonably give such a small center distance as is possible with the gear drive for the same duty. The chain drive can be used only when the driving and driven shafts are parallel or nearly so. It also requires means for adjustment to take out slack which always tends to develop owing to wear in the various pins.

32.4. RUNNING OF GEARS

(a) Noise

The most frequent objection to toothed gearing has been its tendency to produce objectionable noise at high speeds. The only way in which the objection can be overcome is to manufacture the gears to high standards of accuracy in respect of tooth spacing and tooth shape. Modern gear-production equipment produces such accurate work that tooth gearing can be made to give satisfactory service at any speeds.

A pair of gears which are objectionably noisy are not necessarily inefficient or subjected to excessive stresses by the conditions which lead to the noise.

(b) Mounting

The accurate meshing of gears is dependent not only upon the accuracy of the gears themselves but also upon that of the mounting. This depends upon the accuracy and rigidity of bearings and of the bearing housings. Furthermore, the application of load to the gears inevitably causes some deformation of all the loaded parts, and if the smoothest possible running is to be secured, the gears must be made to dimensions differing slightly from the theoretical ideal in ways that anticipate the result of such deformation. This cannot be accurately predicted, and adjustment may have to be made after the gears have been mounted and used under load.

(c) Lubrication

The load capacity of a pair of gears is dependent not only upon the materials and surface finish of the gear teeth themselves but also upon the load capacity of the lubricant. In any type of gear, there is a certain amount of sliding between the tooth surfaces when the gears are running, and unless a suitable lubricant is present, the surfaces would quickly be destroyed, even under very light load.

The choice of a suitable lubricant and the adoption of a reliable method of applying it to the gears are therefore important points in design and operation of gear drives.

32.5. TYPES OF GEARS

Classification of Gear Types According to Shaft Position (see Table 32.1)

Table 32.1. Ranges of Application of Different Classes of Gear
(For general guidance)

Shaft position	Gear ratio	Tooth speed, fpm*	Type of gear	Max wheel torque, lb in.
Parallel.	. . . .	Up to 1,000	Spur or helical	60×10^6
	Up to 10	Up to 4,000	Helical or profile-ground spur	1×10^6
		Over 1,000	Helical	40×10^6
	Up to 10	Up to 4,000	Internal (straight or helical)	0.5×10^6
	Over 10	. . . .	Multistage spur or helical	60×10^6
Intersecting.	Up to 7	Up to 500	Bevel or spiral bevel	0.3×10^6
		Up to 10,000	Spiral bevel	
	7 to 100	. . . .	(Bevel or worm) and (spur or helical)	6×10^6
	Over 25	. . . .	Worm and (spur or helical)	6×10^6
Perpendicular and nonintersecting	Up to 50	Up to 4,000†	Worm	2.5×10^6
	Over 50	Up to 4,000†	Worm and (spur or helical)	6×10^6
Nonparallel and nonintersecting	Up to 20	Up to 4,000†	Crossed helical	1.0×10^6
	Up to 50	Up to 4,000†	Worm (shaft angles between 80 and 100°)	1.0×10^6
	Up to 7	Up to 10,000	Hypoid	0.3×10^6

* Relative to a common perpendicular to the shaft axes.
† Sliding speed, fpm.

Parallel shafts. Spur, helical, and double-helical gears

Shafts not parallel and with nonintersecting center lines. Crossed-helical gears, wormgears, hypoid gears

Shafts not parallel but with intersecting center lines. Straight bevel gears, spiral bevel gears

(a) Spur Gears

The general form of a spur gear is cylindrical with teeth parallel to the center line, as shown in Fig. 32.2. Any section of a spur gear on a plane perpendicular to its center line is the same as any other section. This is consequently the simplest possible type of gear.

(b) Helical Gears

This is a development of the spur gear, the general form being cylindrical but the teeth being of spiral formation. Any section of the gear on a plane perpendicular to its axis is the same as any other section, except for angular position. Successive plane sections perpendicular to the center line have the same shape rotated through successively greater angles. In a pair of helical gears for connecting parallel shafts, the hand of the helices for one gear must be opposite to the hand of the helices for the other gear and the axial pitches must be numerically equal.

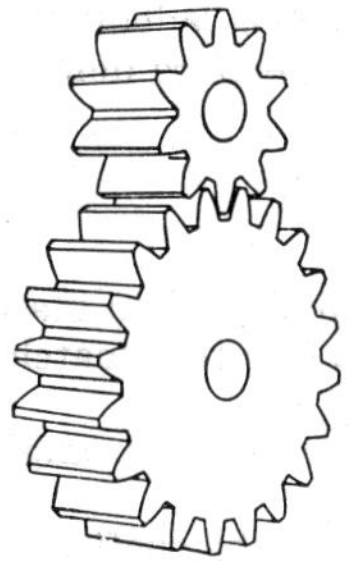

FIG. 32.2. Spur gears.

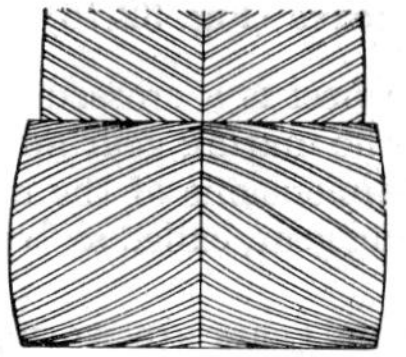

FIG. 32.3. Continuous herringbone gears.

(c) Double-helical Gears

A pair of double-helical or herringbone gears, shown in Fig. 32.3, is equivalent to two pairs of single-helical gears. One half of either gear is identical with the other half of the same gear, except that the hand of the helix is reversed. The object of the double-helical gear is to balance out the end thrusts that are induced in single-helical gears when transmitting load.

(d) Rack Gears

A rack, shown in Fig. 32.4, is produced by cutting equally spaced identical straight grooves in one face of a rectangular prism.

The rack is of limited importance in practice because it cannot be used in any application where continuous movement in one direction is required. It is important, however, in studying the process of "generation" of gear teeth because it represents the simplest possible case.

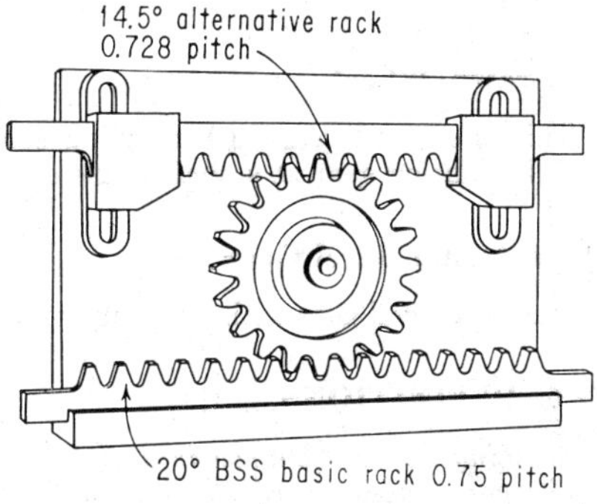

FIG. 32.4. Alternative racks for a 20-tooth gear.

If the direction of motion of a rack is not at right angles to its teeth, it is described as a "helical rack."

(e) Internal Gears

The rack represents the transition between the external gear and the internal gear. The mating gear to an internal gear is an external gear placed inside it. The internal gear and pinion have certain advantages over external gears but are used only to a limited extent, because it is usually difficult to provide adequate bearing support for both gears. Furthermore, the cutting of accurate internal gear teeth is more difficult than that of external teeth.

(f) Crossed Helical Gears (Fig. 32.13)

Helical gears meshed with their axes not parallel (or "crossed") are called "crossed helical gears." Contact between any pair of teeth at any instant is limited to a very small area, and the relative sliding is of a nature that makes adequate lubrication difficult. The materials of the gears and their lubricant therefore demand specially careful selection.

(g) Wormgears (Fig. 32.15)

A worm wheel is a crossed helical gear with rim section shaped to match the root cylinder of the mating gear and with teeth generated by a hob that matches the form and major dimensions of the mating gear.

The mating gear, called a "worm," is usually a helical gear with helix angle exceeding 45°.

The mating gears are wormgears. Only in rare cases is the shaft angle different from 90°, and the departure never exceeds a few degrees.

(h) Bevel Gears (Fig. 32.16*a* and *b*)

These are basically of truncated conical form and are made to mesh with intersecting axes. The teeth may be straight or curved. The shaft angle may have any value from a few degrees to 90°.

(i) Hypoid Gears (Fig. 32.17)

These are curved-tooth bevel gears designed and made to work with axes at right angles and with a center distance much smaller than the diameter of the larger gear, but not zero.

(j) Crown Wheel

This is a bevel gear with meshing cone angle equal to 90°. Such a gear meshes with its mating gear at a shaft angle exceeding 90°.

(k) Internal Bevel Gear

This is a bevel gear with meshing cone angle greater than 90°. Such gears are rare.

32.6. DEFINITIONS

Meshing Circles (Diameters d_B,D_B). A pair of spur gears meshing with parallel shafts corresponds to a pair of smooth cylinders on the same center lines. If those cylinders touch along a line parallel to the center lines, and if their diameters are proportional to the numbers of teeth in the spur gears, then they form the same type of connection between the shafts as do the spur gears.

The smooth cylinders corresponding to spur gears are regarded as the "meshing cylinders." The end view of the two pitch cylinders shows two circles in contact and these are called the "meshing circles" (see Fig. 32.5).

The sizes of the meshing circles corresponding to a given pair of spur gears are determined by the following:

1. The sum of their radii is equal to the center distance.
2. The ratio of their radii is the ratio of the numbers of teeth in the gears.

This means that the meshing diameter of the smaller gear is twice the center distance divided by 1 plus the ratio of the larger number of teeth to the smaller one.

The meshing circles do not bear any essential relation to the tip circles or root circles of the gears. In fact, if the center distance of the gears is altered (and this is permissible in most circumstances), the meshing circles are altered in size, despite the fact that the gears themselves have not changed at all.

An involute gear itself has no meshing circle. During the process of generating the

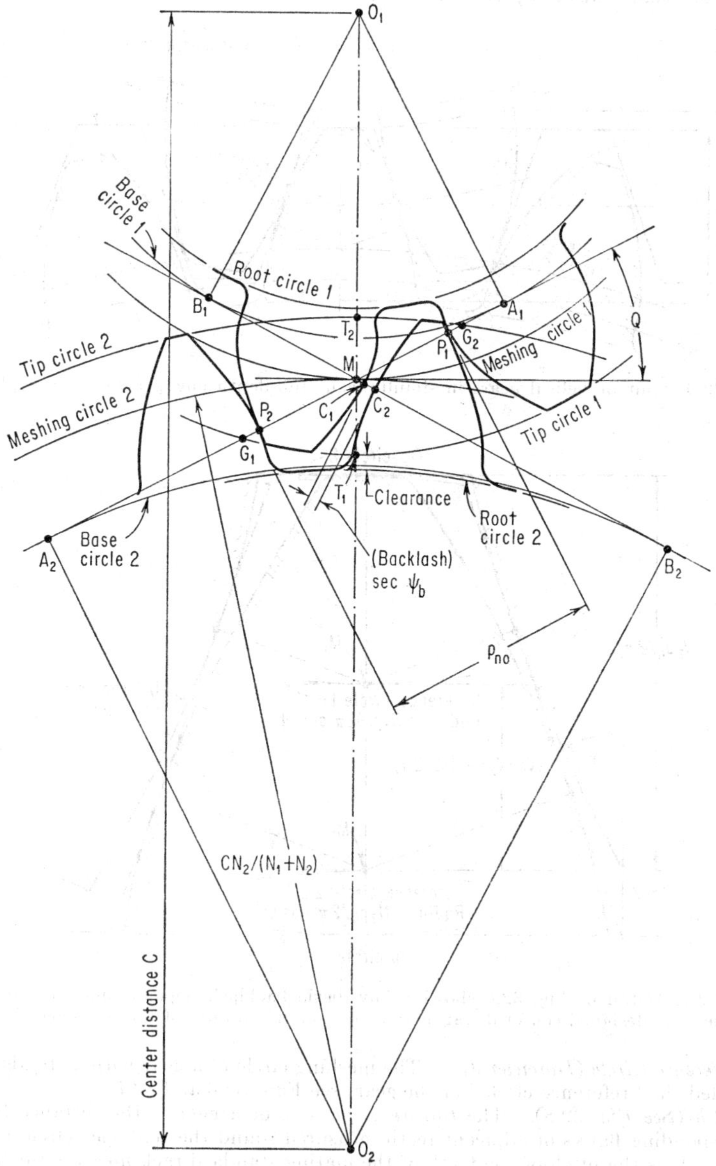

FIG. 32.5. Transverse section of helical gears meshed with axes parallel. Mesh point M divides O_1O_2 in proportion to the numbers of teeth in the gears. G_1G_2 = path of contact = part of common base tangent A_1A_2 intercepted by tip circles.

teeth it had a meshing circle with the generating cutter, and this is often the same as the meshing circle that is created when the gear meshes with its mating gear, but there is no necessity for this.

In other types of gear, e.g., the cycloidal gear, there is an identifiable "pitch circle," and when two such gears are meshed together they can operate correctly only at one particular center distance, which is equal to the sum of the radii of their pitch circles.

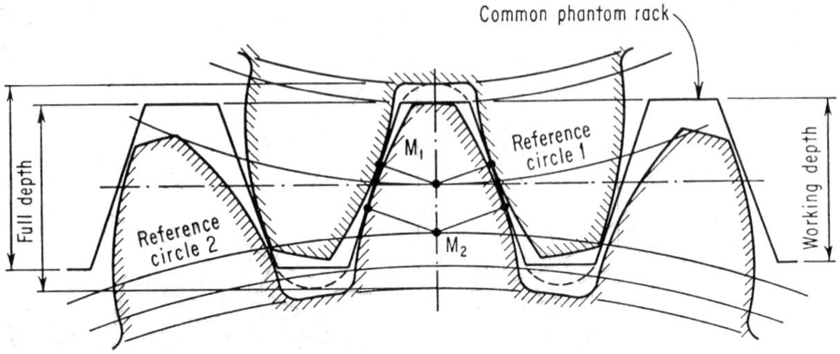

FIG. 32.6. Spur or helical gears in simultaneous full-depth engagement with a phantom rack.

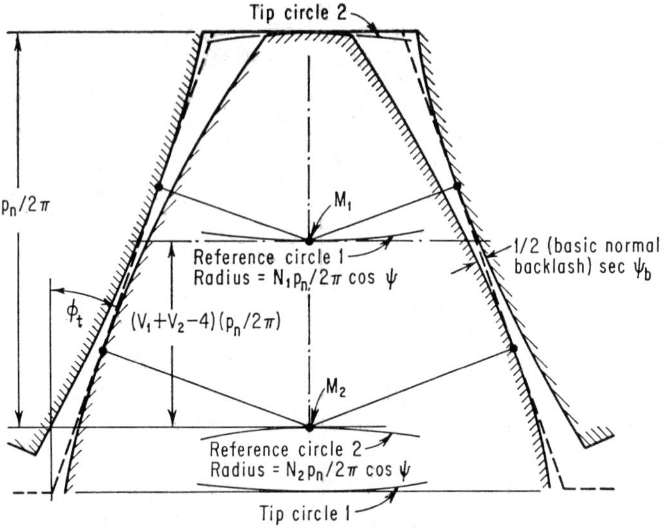

FIG. 32.7. Detail of Fig. 32.6 showing how basic backlash depends upon the difference between the designed center distance and the sum of the radii of the reference circles.

Reference Circle (*Diameter* d_F). The meshing circle of a gear with a standard rack is called the "reference circle" of the gear (see Figs. 32.6 and 32.7).

Pitch (*See Fig.* 32.8). The *transverse pitch* p_t of a gear is the distance between corresponding flanks of adjacent teeth measured round the reference circle (see Fig. 32.6). It is the pitch of the teeth of the mating standard rack measured in a transverse plane, i.e., perpendicular to the axis of the gear.

The *normal pitch* p_n is the pitch of the mating standard rack measured at right angles to its tip lines.

The *axial pitch* p_x is the distance between corresponding flanks of adjacent teeth of the gear, measured on a line parallel to the axis.

The *normal base pitch* p_{nb} is the distance measured from any point on any tooth flank along the normal to the flank to its point of intersection with the corresponding flank of an adjacent tooth. It applies only to involute teeth and has the same value everywhere on every gear generated by any one rack-shaped cutter with flat-flanked teeth.

Total Depth. The total depth of a tooth is half the difference between the tip diameter and the root diameter.

Working Depth. The working depth of the teeth in either gear of a mating pair is the sum of the tip radii minus the center distance.

Clearance. The clearance is the difference between the total depth and the working depth.

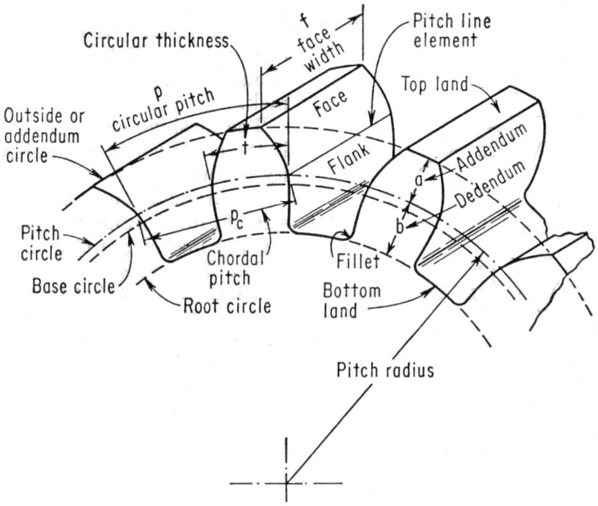

FIG. 32.8. Gear-teeth definitions.

Path of Contact. This is the line upon which the point of contact of any two teeth always lies. There may be two pairs of teeth in contact at the same time, in which case the two points of contact are different points on the path of contact.

The path of contact must pass through the point of contact of the meshing circles. This is the "mesh" or "pitch" point.

In involute gearing the two common tangents to the base circles of a mating pair contain the paths of contact for the two sets of mating flanks. The paths lie within the tip circles.

Line of Force (or "Pressure"). This is the line along which acts the force exerted by one tooth on the mating tooth. If the gears give constant-velocity transmission, the line of pressure passes through the mesh point. In involute gearing, the line of force is the same for all points of contact. It is therefore the same line as the path of contact.

Force Angle (or Pressure Angle) (η). This is the angle between the line of force and the common tangent to the meshing circles at the mesh point.

Uniform Velocity Transmission. The condition that a pair of tooth profiles shall transmit uniform velocity in all angular positions is that the common normal at every point of contact shall pass through the mesh point.

In the special case of a rack with flat-flanked teeth meshing with a gear, this condition reduces to a very simple form. If a straight line is drawn through the mesh point perpendicular to the rack tooth flank that is making contact with the gear, that line must be the common normal for all points of contact, because it is the only line (in the particular transverse plane considered) that lies perpendicular to any of those flanks and also passes through the pitch point.

For this reason the flat-flanked rack tooth form is the easiest of all to study from the point of view of tooth contact. It also has many practical advantages, and that is why the involute tooth form is almost universal.

32.7. GENERATION OF GEARS

Consider a flat-flanked rack pressed into a cylindrical gear blank made of plastic material that the rack can mold into shape by mere pressure as in Fig. 32.9. Suppose that the rack is moving with uniform velocity in a direction parallel to the plane containing the tips of the teeth and that the plastic blank is rotated at uniform velocity while in mesh with the rack about its axis parallel to the teeth of the rack.

The meshing circle of rack and blank is the circle whose circumferential velocity is equal to the velocity of the rack.

The meshing line of the rack is the straight line parallel to its tips and touching the meshing circle. From this definition of meshing circle and meshing line it will be realized that the meshing line of the rack has no fixed position in relation to the tips and roots of the teeth. Usually it is about halfway down the depth of the tooth, but there is no necessity for this. In fact it need not be within the depth of tooth at all.

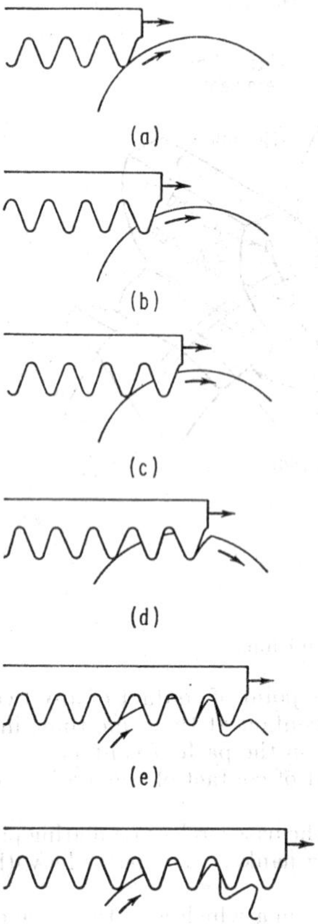

FIG. 32.9. Involute-gear generation.

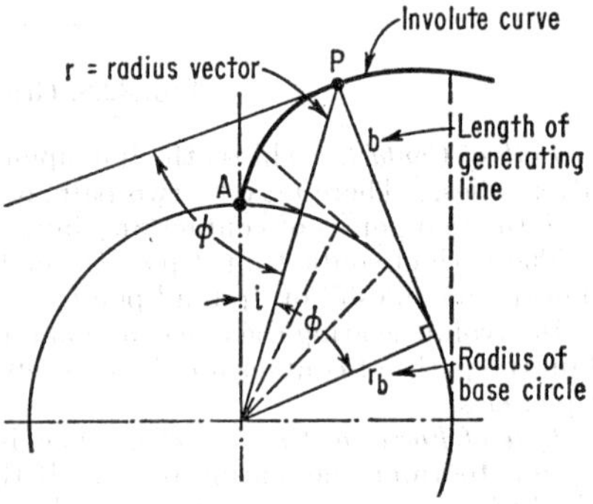

FIG. 32.10. Involute-curve generation.

It is found that the teeth generated by the rack in the blank are of involute form. The involute is a curve whose shape can be demonstrated by unwrapping a string from a circle which is called the *base circle*.

Referring to Fig. 32.10, the length of generating line b = length of arc from origin of involute (A). That is,

$$b = r_b(\phi + i)$$

Also $$b = r_b \tan \phi$$

whence $$i = \tan \phi - \phi = \text{“involute function” of } \phi \qquad (32.1)$$

Also $$r/r_b = \sec \phi$$

or $$\phi = \sec^{-1} (r/r_b)$$

Hence $$i = \text{involute function of } \sec^{-1} (r/r_b) \qquad (32.2)$$

which may be written $i(r/r_b)$. An approximation close enough for all practical purposes is

$$i(r/r_b) = [(r - r_b)/r_b][(r - r_b)/(r + r_b/8)]^{1/2} \qquad (32.3)$$

This is the angle between the radius of the base circle at the origin of the involute and the radius vector of any length r.

The flank angle of the rack is the angle between the flank of the tooth and a line perpendicular to the tip surface of the rack. The base circle of the involutes generated is the circle whose radius is equal to the radius of the meshing circle multiplied by the cosine of the flank angle of the rack.

These radii may be expressed in slightly different ways by reference to the pitch of the rack teeth, which is the distance between adjacent similar flanks measured parallel to the tip surface. The "base pitch" of the rack is the distance between adjacent similar flanks measured perpendicular to the flanks.

The diameter of the meshing circle is equal to the pitch of the rack multiplied by the number of teeth in the gear divided by π. The diameter of the base circle of the involutes is equal to the base pitch of the rack multiplied by the number of teeth in the gear divided by π.

(a) Types of Generation Processes

Rack-generating Process. Here the cutter is in the form of a rack which is reciprocated parallel to the length of its teeth in order to give a cutting action. The cutter is also moved transversely in unison with rotation of the gear blank in order to give a generating action.

The rack-generation process is used for production of straight-tooth spur gears, single-helical gears, and double-helical gears.

Generation by Pinion-type Cutter. In this process the cutter is in the form of a spur gear modified to provide cutting relief. One set of flanks forms part of a helical gear having a right-hand helix of small helix angle, and the other set of flanks forms part of a corresponding helical gear having a left-hand helix.

The outside of the cutter is also tapered to provide relief at the tips. The profiles are precision-ground, and the cutter is sharpened when required by the simple operation of grinding the front face. There is a change in tooth profile when the cutter is sharpened, but the profiles are parts of the original involutes, and the cutter can produce the same gears when worn as when new.

This process has the advantage that it can be used for generating internal gears.

Gear Generation by Hobbing. The hob is in the form of a worm whose thread section is very nearly that of the basic rack, having relieved cutting edges.

In use, the hob is rotated at a suitable cutting speed and the blank is also rotated at the speed which the finished gear would need to have to mesh with a worm rotating at the speed of the hob.

Rotation of the hob in conjunction with appropriate rotation of the blank gives the required generating action, and with straight-sided hob teeth, involute teeth are generated in the gear. The hob is traversed parallel to the axis of the blank in order to cover the whole face width.

The hobbing process is applied to the production of spur and helical gears, and owing to its continuous nature (there is no change in speed of any part of the machine from the start to the finish of the cut), it gives the highest degree of accuracy possible in machine-cut gears.

(b) Generation of Worm Wheels

The only practicable method of generating the teeth in a worm wheel is to use a hob of the same general dimensions as the worm which will ultimately mesh with the wheel.

The hob is rotated at a suitable cutting speed and the wheel blank at the speed that the finished worm wheel would have to mesh with a worm running at the speed of the hob.

(c) Form Cutting of Bevel-gear Teeth

In the Gleason bevel-gear-planing process, a planing tool is reciprocated along a slide which is mounted in such a way that it can pivot in any direction about the apex of the gear. As the tool reciprocates along a line passing through the apex, the slide is slowly moved under the control of a former plate on which presses a roller carried by the slide. The shape of the former plate is that which the gear tooth would have if it extended out to the roller.

The path of the tool is always a straight line lying in the required surface of the tooth whose shape is therefore produced by form planing.

(d) Generation of Bevel-gear Teeth

Bevel-gear teeth may be generated by a planing process in which a straight-edged planing tooth represents one flank of a tooth of the imaginary crown wheel. This process would be impossible with any crown-wheel tooth form that is not straight-sided. The section of any bevel-gear tooth becomes smaller in passing across the face width toward the apex. If, however, the profile is a straight line the section at the small end of the tooth is simply a portion of that at the large end. So a straight cutting edge can represent, at any point in the face width, a flat-flanked crown-wheel tooth.

The straight-edged planing tool reciprocates along the straight line in such a way that it sweeps out the surface of one flank of a tooth of the imaginary crown wheel. The full depth of the planing tool may be used in cutting the large end of the gear tooth, but only part of it is used at the small end. While the tool reciprocates to perform the cutting operation, the slide upon which it is carried is rotated about the apex of the gear to correspond to the rotation of the imaginary crown wheel. At the same time the gear blank itself is rotated at the speed which it would need to have to mesh with the rotating imaginary crown wheel. This "rolling" motion causes the straight-edged planing tool to generate a curved gear tooth which will mesh accurately with a crown wheel having teeth of straight profile.

(e) Generation of Spiral-bevel-gear Teeth

In the Gleason process for generating spiral bevel gears, the cutter is in the form of a disk carrying a number of axially projecting blades whose cross section represents the tooth of an imaginary crown wheel with spiral teeth. Each blade has only one cutting edge, alternate blades cutting one flank of the gear tooth while the other set of alternate blades cuts the other flank. Continuous rotation of the disk causes an endless series of blades to follow each other through the tooth space.

The cutter disk is mounted in a head which can rotate about an axis passing through the apex in a way that corresponds to rotation of an imaginary crown wheel. Simultaneously, the gear blank is rotated about its axis in a way that would cause it to mesh correctly with the rotating imaginary crown wheel. Thus the cutter blades generate a tooth space of spiral form, and with an indexing operation for each tooth required in the blank, they complete the cutting of a spiral gear that would mesh accurately with the imaginary crown wheel. The mating gear is generated in the same way, and as each of these gears would then mesh with the imaginary crown wheel, they mesh accurately with each other.

The tooth produced by this process is of a curved spiral type owing to the circular path of the cutter blades. There is no particular operational advantage in the curved spiral, the essential for smooth running being that the length of the tooth should not lie perpendicular to the direction of its motion.

Spiral-bevel-gear teeth may also be cut with "continuous indexing" by hobbing (Klingelnberg) or on the Mammano principle using a cutter mounted on a disk substantially as in the Gleason process but with the difference that, as in hobbing helical gears, the work and the cutter are rotated in timed relation to each other. The Klingelnberg process uses tapered hobs and the basic form of the tooth spiral is an involute rather than a circular arc. The exact nature of the tooth spiral is kinematically unimportant; whatever the particular process naturally produces is, in general, as good as anything else.

(f) Hypoid Gears

By using an appropriately modified machine, the teeth of a spiral bevel pinion may be cut in such a way that it may mesh with a spiral bevel wheel while the axes of the pair do not intersect each other. Such a combination is a pair of "hypoid gears." An immense amount of time is commonly spent on calculating gear dimensions and machine settings for gears of this type, but even so, the best contact conditions for any design can be found only by repeated trial.

(g) Trochoid

Consideration of the generating action shows that not all the profile of any generated tooth is of involute form. There is a part near the root which is of a form called *trochoid.* That part of the tooth can never make useful contact with any mating tooth, but its shape and its degree of finish affect the fatigue strength of the tooth.

(h) Interference

If the tips of the rack teeth project too far inside the base circle of the gear which is being generated, the trochoidal part of the tooth becomes extended so far that it cuts away part of the involute profile previously generated. This effect is called *interference.* Not only does it reduce the useful part of the involute profile, but it produces a tooth which is "undercut" and therefore mechanically weak. This interference occurs if the effective part of the rack tooth projects inside the reference circle by an amount greater than the radius of the meshing circle multiplied by the square of the sine of the flank angle of the rack teeth.

(i) Standard Rack-tooth Form

An AGMA standard basic rack (Fig. 32.11) has a flank angle of 20° and the total tooth depth is 0.716 times the pitch of the rack. The working depth of the tooth is 0.636 times pitch and the difference between full depth and working depth is the clearance.

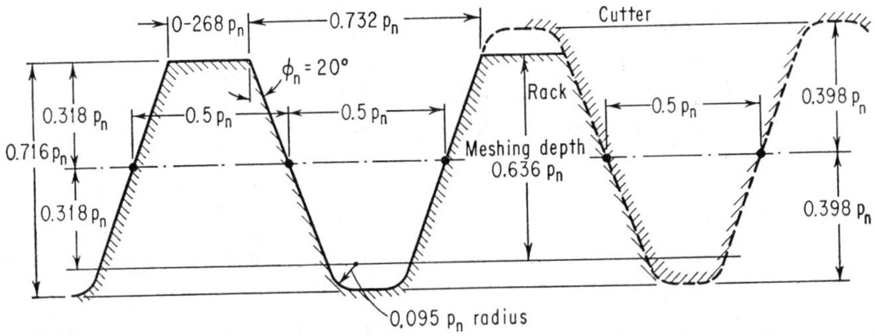

FIG. 32.11. Normal section of standard basic rack.

The greater part of the basic rack tooth profile within the clearance is of nearly semicircular form and the clearance part of the profile of a gear of any number of teeth is also of approximately semicircular form. The greater the fillet radius, the greater the fatigue strength of the tooth.

(j) Undercutting and Sharp Pointing

A natural carry-over from old cycloidal gear practice was to make the difference between the blank diameter of a gear and its reference diameter equal to some arbitrary multiple of the pitch of the generating cutter. This restriction led to difficulties in gears with small numbers of teeth but the practice could be justified for teeth cut to shape by formed cutters.

This restriction can be abandoned in the case of gears cut by a generating process (as is now almost universal practice), and if the full advantage is taken of the adaptability of the involute system it becomes unnecessary ever to use a nonstandard cutter for any ordinary application of involute gears. This is done by admitting in the design of any gear with a given number of teeth generated by a given cutter any blank diameter within the range that avoids undercutting on the one hand and sharp pointing on the other.

32.8. GEOMETRY OF GEARS

(a) Involute Helicoid Gears

The base pitch p_{nb} of an involute gear or generating cutter is the distance from any point on any tooth flank along the normal to the surface at that point to either of the adjacent similar flanks which are intersected normally by the normal already mentioned.

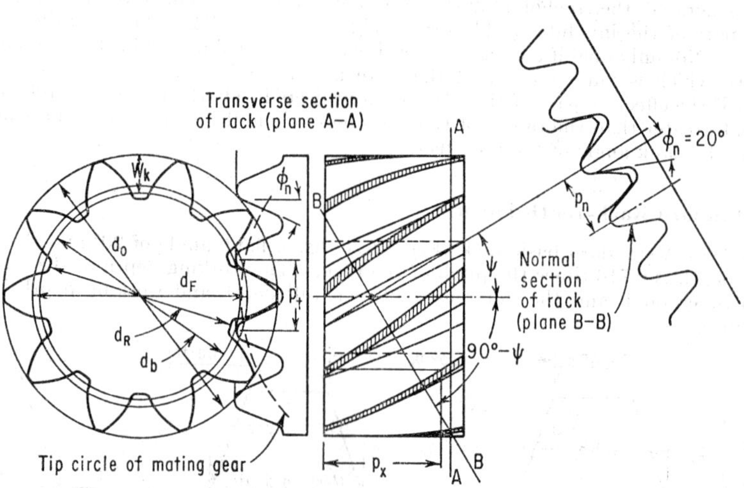

FIG. 32.12. Helical gears, fundamentals.

The base pitch is one of the fundamentals of an involute gear. An involute generating cutter imposes its own normal base pitch on every gear that it generates, and that is why any pair of them can be correctly meshed together.

Another fundamental dimension of a helical gear (see Fig. 32.12) is the axial pitch p_x which is the pitch measured along any line parallel to the axis. The axial pitch is the measure of the helical characteristic of the gear.

The angle between the intersection of any tooth flank with any coaxial cylinder of diameter d is the "helix angle" ψ_d at that cylinder; it is given by

$$\tan \psi_d = \pi d / N p_x \tag{32.4}$$

where N is the number of teeth. The normal pitch p_n at diameter d is given by

$$p_n = p_x \sin \psi_d \tag{32.5}$$

The transverse pitch p_t at diameter d is given by

$$p_t = \pi d / N \tag{32.6}$$

These three pitches are connected by the relation

$$1/p_n^2 = 1/p_t^2 + 1/p_x^2 \tag{32.7}$$

The diameter of the cylinder at which the normal pitch has any given value p_n is

$$d = (N/\pi) p_x p_n / \sqrt{(p_x^2 - p_n^2)} \tag{32.8}$$

The essential condition for correct meshing of a pair of involute helical gears is that they have a common normal base pitch; this is ensured by using the same cutter to generate their teeth. If they are to mesh with full-width contact they must have a common axial pitch and be set to run with axes parallel.

Straight-tooth spur gears are helical gears in which p_x is infinite.

A tooth may be intersected at any point on a cylinder of diameter d by various planes containing the perpendicular from the point to the axis. Such a plane:

1. Perpendicular to the axis is "transverse."
2. Containing the axis is "axial."
3. Normal to the tooth helix at the point is "normal."

The angles made by the intersections of these planes with the normal to the axis are

1. Transverse flank angle ϕ_t
2. Axial flank angle ϕ_x
3. Normal flank angle ϕ_n

Some relations between them and the helix angle at the point are

$$\tan \phi_n = \tan \phi_t \cos \psi \tag{32.9}$$
$$\tan \phi_n = \tan \phi_x \sin \psi \tag{32.10}$$
$$\sec \phi_t = d/d_b \tag{32.11}$$
$$\tan \psi = \pi d / N p_x \tag{32.12}$$

Hence the flank angles at any point can be determined in terms of d, d_b, p_x, and N.

If the flanks of the teeth of a rack are flat, the teeth generated by a cutter of the same form as the rack are "involute helicoid." This means that the section of each tooth by a transverse plane (i.e., perpendicular to the axis of the gear) is part of an involute and that sections of the teeth by different transverse planes are identical except for orientation, the angular displacement being equal to one revolution multiplied by the ratio of the separation of the planes to the lead of the tooth helices. The lead is equal to the number of teeth in the gear multiplied by the pitch p_x of the teeth in the section of the rack by a plane containing the axis of the gear.

A characteristic of the flat-flanked rack is that the distance p_{nb} from any point on a flank to the foot of the perpendicular from that flank to the next similar flank is the same everywhere. This characteristic is reproduced, with the same normal base pitch p_{nb} on every gear generated by any cutter corresponding to the rack. The fundamental dimensions of an involute helicoid gear are its normal base pitch p_{nb}, its axial pitch p_x, and the diameter d_b of its base circle given by

$$d_b = (N/\pi) p_x p_{nb} / \sqrt{(p_x^2 - p_{nb}^2)} \tag{32.13}$$

The normal to an involute helicoid tooth at any point on it is a tangent to the base cylinder and is inclined to the direction of the axis at an angle equal to the complement of the helix angle ψ_b at the base cylinder given by

$$\tan \psi_b = \pi d_b / N p_x \qquad (32.14)$$

Every tangent plane to the base cylinder intersects one set of tooth flanks in straight lines inclined at ψ_b to the direction of the axis and the other set in curves that are of no importance here—as no contact normally occurs on them.

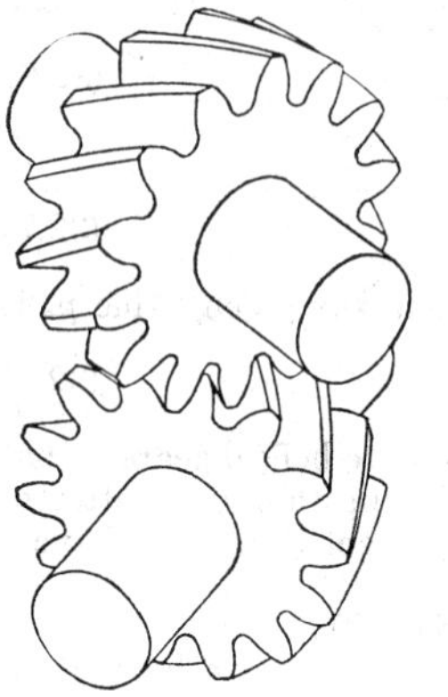

FIG. 32.13. Crossed-axis helical gears.

Meshing of Involute Helicoid Gears (General). Two involute helicoid gears of common normal base pitch may be set in mesh and their axes are not (in general) parallel. There cannot (in general) be any common tangent plane to the base cylinders. For each pair of sets of mating flanks there is, however, a straight line that touches the base cylinders and is perpendicular, at each point of contact with them, to the direction of the base helix of a tooth at that point. Contact between the flanks concerned is limited to their points of intersection with this common tangent line to the base cylinders.

These are crossed-helical gears (Fig. 32.13).

(b) Meshing of Involute Helicoid Gears (Special).

If two involute helicoid gears have a common axial pitch of opposite hands and a common normal base pitch, they will mesh with parallel axes and there are two common tangent planes to their base cylinders. Each plane corresponds to two sets of mating flanks which make contact on their common straight lines of intersection with the tangent plane.

The rectangular part of the common tangent plane defined by its intersections with the tip cylinders of the gears and with the end planes of the gears is the "zone of contact." If the width of the gears is an integral multiple of the axial pitch of the teeth or if the other side of the rectangular zone of contact is an integral multiple of the transverse base pitch p_{tb} of the teeth (the spacing of the teeth on a circular section of the base cylinder), then the total length of the contact lines is the area of the contact zone divided by the normal base pitch p_{nb}. Otherwise the total length varies over a narrow range with the phase of engagement.

(c) Center Distance (*C*)

Every standard basic rack has associated with it a "meshing depth" equal to some arbitrary fraction of the pitch of the teeth, and in normal circumstances every pair of gears is intended to mesh at such a center distance (more accurately "axis distance") that the tip circles of the gears overlap by a distance equal to the "meshing depth" that corresponds to the pitch of the basic rack and to its tooth proportions, i.e.,

$$C = \tfrac{1}{2}(d_{01} + d_{02}) - 2p_n/\pi \qquad (32.15)$$

where $2p_n/\pi$ is the "meshing depth" of the teeth.

By definition the meshing circles touch each other whatever the center distance may be; in general the reference circles do not touch each other although if there is no close restriction on center distance involute gears may be designed so that, when they are meshed at the designed center distance, the reference circles do touch. In this case the basic backlash is zero and so its calculation is unnecessary but the point is of no practical importance.

The working center distance of involute gears may differ from the designed center distance by about 2 per cent or by about a quarter of the working depth, whichever is the less, without adverse effect on the meshing of the gears, although the backlash varies with change in center distance.

Increase in center distance of a pair of spur or helical gears has the effect of reducing the overlap of the tip cylinders of the gears and therefore the width of the rectangular contact zone. If this becomes too small, there will be phases of engagement in which no loaded tooth flank intersects the zone, edge/flank contact of the teeth then replacing the flank/flank contact that is necessary for correct tooth action.

There is no approach to this limit in rationally designed gears working at their designed center distance with ordinary machining tolerance.

(d) Blank Diameter of Spur or Helical Gear

In the following it is assumed that teeth are to be generated by a rack-shaped cutter of normal pitch p_n and normal flank angle 20° set to produce teeth of axial pitch p_x with meshing depth $(2p_n/\pi)$ and full depth at least $(2.25p_n/\pi)$. Alternatively use may be made of any other type of generating cutter that will produce such teeth. All standard cutters will do so, in some cases with slight variations in root dimensions.

The blank diameter of any such gear (No. 1) with N_1 teeth is

$$d_0 = (p_n/\pi)(N_1 \sec \psi_1 + V_1) \tag{32.16}$$

where ψ_1 is the nominal helix angle (at the reference cylinder) defined by

$$p_n/p_{x1} = \sin \psi_1$$

and V_1 is a number that may have any value that is

$$\begin{array}{ll} \text{(1) greater than} & 1 + 20/N_1 \sec^3 \psi_1 \\ \text{and (2) less than} & 4 - 10/N_1 \sec^3 \psi_1 \end{array} \tag{32.17}$$

The corresponding expressions, with suffix 2 instead of 1, apply to the mating gear (No. 2).

If the gears are to mesh correctly together, it is essential that $p_{nb1} = p_{nb2}$, and this is ensured by cutting all the teeth with the same cutter.

If the gears are to mesh correctly together with axes parallel, then, in addition, it is essential that $p_{x1} = -p_{x2}$, i.e., that the gears have a common axial pitch but with opposite hands of helix.

If both gears have straight teeth, then

$$p_{x1} = p_{x2} = \infty$$

and they are to be meshed together with axes parallel.

Values of V. In the very old conventional system of gear design

$$V_P = V_G = 2$$

In a more recent system

$$V_P + V_G = 4$$

and this has the advantage that, even when N_P, N_G, p_n, p_{x1}, and p_{x2} are fixed, there is latitude in blank diameter although not in center distance.

In a rational system, at the design stage, $V_P + V_G$ = any value within the range defined by those permitted for V_P and V_G by Eqs. (32.17).

If the teeth of each gear of a pair are cut to exactly the standard depth with an accurate cutter and are then meshed at the usual designed center distance given by Eq. (32.15), there will be some backlash between the teeth unless $V_P + V_G = 4$, when the backlash is zero.

As backlash is necessary or desirable in most power-transmitting gears, most gears with $V_P + V_G = 4$ have teeth cut to a depth slightly greater than standard depth in order that there shall be backlash at the designed center distance. If $(V_P + V_G)$ is not equal to 4, it is necessary to calculate the backlash that full-depth teeth will have and to make appropriate allowance for it. The conventional form of this calculation is so labored as to deter some designers from departing from $V_P + V_G = 4$, and some of the adaptability of the involute system is thereby rejected.

(e) Backlash B

The backlash between the teeth of a pair of meshed gears is the minimum distance between any pair of matable teeth when they are held as far apart as is permitted by contact between other pairs of teeth. Backlash may be measured either by a feeler gage or by holding one gear stationary and repeatedly rotating the other about its axis between the limits imposed by the teeth and noting the range of reading of a dial indicator set with the axis of its plunger normal to the surface of a tooth at its point of contact with the stylus.

This is "normal backlash." The desired or "designed" normal backlash is denoted by B.

The "basic normal backlash" k is that between teeth of exactly full depth meshed to standard depth, and its value is given by Eq. (32.30).

(f) Normal Arc Thickness of Tooth

The thickness of a helical gear tooth may be measured at its intersection with any cylinder coaxial with the base cylinder. Here attention is confined to measurement at the "reference cylinder," which is the one at which the normal pitch and normal flank angle of the gear teeth are the same as those of the standard basic rack corresponding to the gear-generating tool. The difference in radius between that cylinder and the tip cylinder is $\frac{1}{2}Vp_n/\pi$.

(g) Tooth Thickness

Tooth thickness is usually measured by means of a "gear-tooth caliper." The designed distance t_{nc} between the jaws when touching a tooth at the reference cylinder is the chord of the arc whose length is t and radius is $\frac{1}{2}(d_0 - Vp_n/\pi)\sec^2\psi$. This is

$$t_{nc} = t - t^3/6(d_0 - Vp_n/\pi)^2 \sec^4\psi \tag{32.18}$$

The last term is usually negligible.

The value of t is the thickness of the basic rack tooth at diameter $d_0 - Vp_n/\pi$ less half the difference between the desired backlash and the basic backlash measured in the direction of the chord of the gaged arc. Thus

$$t = \tfrac{1}{2}p_n + (V - 2)(p_n/\pi)\tan\phi_n - \tfrac{1}{2}(B - k)\sec\phi_n$$

If $\phi_n = 20°$ this becomes

$$t_{nc} = t = (0.843 + 0.364V)(p_n/\pi) - 0.532(B - k) \tag{32.19}$$

The "height setting" is the difference between the tip radius and the reference radius plus the height of the arc, i.e.,

$$h = \tfrac{1}{2}V(p_n/\pi) + t_{nc}^2/4(d_0 - Vp_n/\pi)\sec^2\psi \tag{32.20}$$

Tooth thickness may also be deduced from measurement over rollers of known diameter resting in opposite tooth spaces. For simplicity in calculation, the roller diameter should be

$$d_u = (p_n \cos\phi_n)(1 - t/p_n) \tag{32.21}$$

The dimension over two such rollers in opposite tooth spaces is

$$s_u = d_u + N/\pi(1/p_n^2 - 1/p_x^2)^{1/2} \tag{32.22}$$

If the number of teeth is odd, the measurement may be taken over one such roller and the tip of the tooth opposite to the space in which the roller lies. The measurement in this case is

$$s_u = \tfrac{1}{2}[d_u + d_0 + N/\pi(1/p_n^2 - 1/p_x^2)^{1/2}] \qquad (32.23)$$

Tooth thickness may also be deduced from "base-tangent" measurement over any convenient number (q) of teeth by means of a micrometer or a vernier gage. This dimension is given by

$$s_q = p_{nb}\{q - 1 + Nt_{n1}[1/L^2 + (1/\pi d_1)^2]^{1/2} + (N/\pi)i(d_1/d_b)\}$$

where t_{n1} is the normal tooth thickness at diameter d_1 and

$$i(d_1/d_b) = [(d_1 - d_b)/d_b][(d_1 - d_b)/(d_1 + d_b/8)]^{1/2} \qquad (32.23a)$$

The thickness of a straight bevel-gear tooth may be measured by the application of a gear-tooth caliper to the large end of the tooth. Because the jaws bear on an edge, this is less satisfactory than for spur or helical gears, and there is a similar difficulty with spiral bevel teeth. High-precision measurement of bevel-gear-tooth thickness is hardly possible with the gear still on the tooth-generating machine. A meshing test is the usual way of determining what adjustment of thickness is necessary. This also applies to wormgears.

Note: Equation (32.21) represents a particular example of the general problem of calculating what diameter of roller will sit with its axis touching the cylinder of any diameter d_2 in a helical gear of base diameter d_b, normal base pitch p_{nb}, number of teeth N, and with normal tooth thickness t_1 at the cylinder of any diameter d_1 where the normal pitch is p_{n1}.

The roller diameter d_{u2} is given by

$$d_{u2} = p_{nb}\{1 - t_1/p_{n1} + (N/\pi)[i(d_2/d_b) - i(d_1/d_b)]\}$$

where $i(d/d_b)$ is given by Eq. (32.23a).

If $(d_2 - d_1)$ is less than $\tfrac{1}{2}(d_1 - d_b)$, then the value of d_{u2} is obtainable rather more conveniently by using

$$d_{u2} = p_{nb}\{1 - t_1/p_{n1} + (N/\pi)[(d_2 - d_1)/d_b(d_2 + d_1)][(d_1 + d_2 - 2d_b)(d_1 + d_2 + 2d_b)]^{1/2}\}$$

(h) Crossed-helical Gears

Helical gears designed to mesh with axes not parallel ("crossed") are called *crossed-helical gears* (see Fig. 32.13). There can be no common tangent plane to the base cylinders but there are two straight lines that touch both base cylinders and are perpendicular to the base helices at the contact points. Each line corresponds to one combination of sets of tooth flanks which it intersects normally; it is the common normal at every possible point of contact of those flanks and is the path of contact on those flanks.

In the conventionally restricted system of gear design these lines intersect the common perpendicular to the axes of the gears. By abandoning this restriction, design procedure is simplified.

The ratio of the diameters of a pair of parallel-shaft gears is approximately the same as the reciprocal of the velocity ratio. In crossed-helical gears there is no essential relation between those ratios, but it happens that power loss by tooth friction is minimized by designing the gears so that the ratios are nearly equal. This is not practicable for high velocity ratios as the pinion would be excessively small, and it is here that crossed-helical gears (and the wormgears derived from them) have a great advantage.

To offset this, the power loss by tooth friction in crossed-axis gears is higher than in parallel-shaft gears, and the sliding conditions on the teeth demand special care in choosing the materials of the gears and the lubricant for them.

The shaft angle associated with any pair of crossed-helical gears is defined by refer-

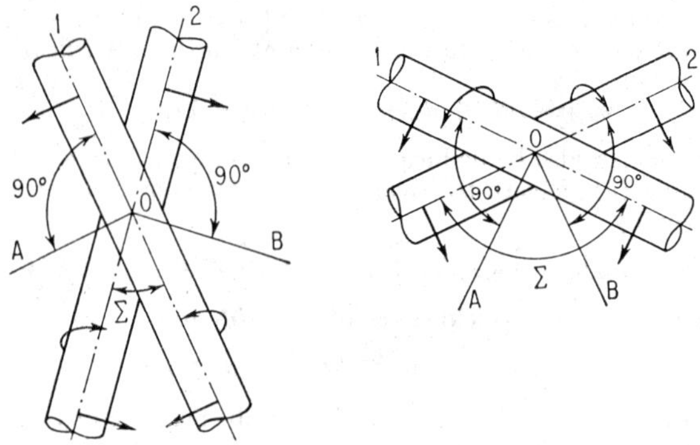

FIG. 32.14. Shaft angle (or axis angle) of crossed-helical gears. Common direction of tooth helices at the mesh point must lie with the smaller angle *AOB*.

ence to the directions of motion of the nearest lines on the shafts when they are viewed along the common perpendicular to their axes (see Fig. 32.14). The shaft angle Σ is the angle across which the directions of motion are opposed.

Mating crossed-helical gears normally have helices of common hand. If the shaft angle is less than about 20°, it is possible to use crossed-helical gears of opposite hands, but this is unusual except in the particular case of parallel shafts, where it is normal.

The helix angle of a helical gear driven by meshing with its mating gear should not exceed 60°. If it exceeds the complement of the angle of friction between the teeth, the gear itself cannot be rotated by tooth pressure of the mating gear.

(i) Wormgears (Fig. 32.15)

If the smaller gear of a crossed-helical pair is made of much harder material than the mating gear, the original very narrow track of contact on each tooth of the latter is

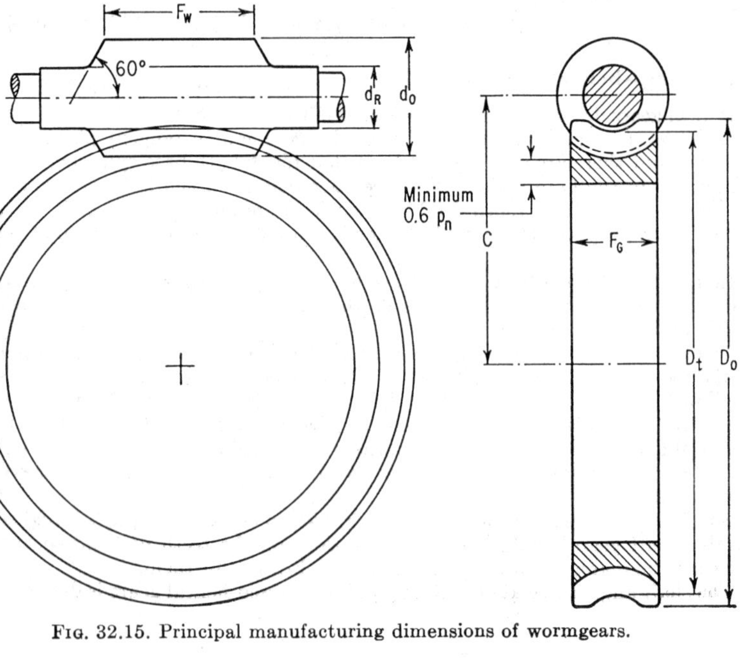

FIG. 32.15. Principal manufacturing dimensions of wormgears.

broadened by prolonged running under load into a band representing successive positions of a line of contact. The load capacity is then much greater than it was in the original condition.

By cutting the teeth of the softer gear by means of a generating cutter of substantially the same dimensions as the harder gear the effect of "running in" is anticipated and line contact is obtained right from the start.

A gear specially cut in this way is a "worm wheel" and the smaller (and harder) mating gear is a "worm" whose teeth are called "threads." Many thread forms have been used for worms but for all-round economy none surpasses the involute helicoid derived from the same basic rack form as that used for spur and helical gears. A worm may reasonably be defined as a helical gear of which the axial pitch is less than 1½ times the normal base pitch or, in other words, of which the helix angle at the base cylinder exceeds about 45°.

The geometrical form of worm-wheel teeth is far less easily analyzed than that of an involute helicoid gear, but in practice this is not detrimental. Unlike other forms of gear, wormgearing requires a hob for each design of worm; it is therefore specially important to minimize the number of different designs of worm, and this can be done only by creating standard dimensions for worms in a series sufficiently closely spaced to cover all reasonably foreseeable requirements. Such worms should naturally have threads capable of generation by standard helical-gear-generating tools with normal pitches taken so far as possible from Table 32.2. By taking full advantage of the fact that a worm wheel that is to mesh with a defined worm at a defined center distance may have any number of teeth within a definable range usually of three consecutive

Table 32.2. Standard Pitches

Circular pitch, in.	Diametral pitch, in.	Module, mm	Circular pitch, in.	Diametral pitch, in.	Module, mm	Circular pitch, in.	Diametral pitch, in.	Module, mm
0.1855	16.933	1.5	0.618	5.080	5	1.396	2.25	11.286
0.1875	16.755	1.516	**0.625**	**5.026**	5.053	1.484	2.117	12
0.1963	**16**	1.587	**0.628**	**5**	5.077	**1.5**	**2.094**	12.127
0.2164	14.514	1.75	0.680	4.618	5.5	**1.571**	**2**	12.701
0.2244	**14**	1.814	0.6875	4.569	5.558	1.608	1.954	13
0.247	12.700	2	0.742	4.233	6	1.625	1.933	13.138
0.250	**12.566**	2.021	**0.75**	**4.189**	6.064	1.732	1.814	14
0.262	**12**	2.118	**0.7854**	**4**	6.350	**1.75**	**1.795**	14.148
0.278	11.288	2.25	0.8125	3.867	6.569	**1.795**	**1.75**	14.512
0.309	10.160	2.5	0.866	3.628	7	1.855	1.693	15
0.3125	10.053	2.527	**0.875**	**3.590**	7.074	**1.875**	**1.676**	15.159
0.314	**10**	2.540	**0.898**	**3.5**	7.260	1.978	1.587	16
0.340	9.236	2.75	0.9375	3.351	7.580	**2**	**1.571**	16.170
0.371	8.466	3	0.990	3.175	8	**2.094**	**1.5**	16.930
0.375	**8.377**	3.032	**1.000**	**3.142**	8.085	2.226	1.412	18
0.393	**8**	3.177	**1.047**	**3**	8.465	**2.25**	**1.396**	18.191
0.433	7.257	3.5	1.113	2.822	9	2.474	1.270	20
0.4375	7.181	3.537	1.125	2.792	9.095	**2.5**	**1.257**	20.212
0.449	**7**	3.630	**1.142**	**2.75**	9.233	2.513	1.25	20.317
0.495	6.350	4	1.237	2.54	10	2.75	1.142	22.233
0.5	**6.283**	4.042	**1.25**	**2.513**	10.106	**3**	**1.047**	24.254
0.524	**6**	4.236	**1.257**	**2.5**	10.163	**3.142**	**1**	25.400
0.557	5.644	4.5	1.361	2.309	11			
0.5625	5.585	4.547	**1.375**	**2.285**	11.117			

The pitches printed in heavy type are preferred.

numbers, the minimum number of standard worms to cover every center distance and every velocity ratio in the usual industrial range can be discovered. By using the principle of geometrical similarity the number of different designs of standard worm for each number of threads and for each of the necessary standard normal pitches (Table 32.3) can be reduced to five, as shown in Table 32.4.

Table 32.3. Normal Pitches p_n of Worms

	0.500	1.000	2.000
	0.5236	1.0472	2.094
	0.5625	1.125	2.250
0.297	0.594	1.1875	2.375
0.3142	0.6283	1.2561	2.513
0.3401	0.6803	1.3606	2.721
0.3491	0.6981	1.3963	2.792
0.375	0.750	1.500	3.000
0.406	0.8125	1.625	
0.4375	0.875	1.750	
0.469	0.9375	1.875	

Table 32.4. Proportions of Standard Worms

N_P (number of threads)	$A = d_R/p_n$	1.468	1.868	2.268	2.668	3.068	3.468	3.868	4.268	4.668	5.068	5.468
1	$B = l/d_R$	0.69	0.54	0.45	0.38	0.33						
	$K = l/p_n$	1.014	1.008	1.008	1.005	1.003						
	$D = d_0/p_n$	2.900	3.300	3.700	4.100	4.500						
	$E = r_c/p_n$	40	40	40	40	40						
2	B	1.44	1.11	0.91	0.77	0.66						
	K	2.12	2.08	2.064	2.04	2.032						
	D	2.900	3.300	3.700	4.100	4.500						
	E	40	40	40	40	40						
3	B	2.36	1.77	1.41	1.18	1.02						
	K	3.465	3.300	3.204	3.150	3.120						
	D	2.900	3.300	3.700	4.100	4.500						
	E	10	10	10	40	40						
4	B	3.68	2.57	2.01	1.65	1.41						
	K	5.400	4.800	4.550	4.400	4.320						
	D	2.800*	3.300	3.700	4.100	4.500						
	E	3	5.3	10	10	10						
5	B	......	3.70	2.73	2.17	1.83	1.59					
	K	......	6.900	6.200	5.800	5.600	5.500					
	D	......	3.200*	3.700	4.100	4.500	4.900					
	E	......	3	5.3	5.3	10	10					
6	B	......	......	3.72	2.86	2.35	2.00	1.74				
	K	......	......	8.448	7.650	7.200	6.900	6.720				
	D	......	......	3.600*	4.100	4.500	4.900	5.300				
	E	......	......	3	5.3	5.3	5.3	10				
7	B	......	......	......	3.77	2.97	2.48	2.12	1.87			
	K	......	......	......	10.080	9.100	8.575	8.190	7.980			
	D	......	......	......	4.000*	4.500	4.900	5.300	5.700			
	E	......	......	......	3	5.3	5.3	5.3	5.3			
8	B	......	......	......	......	3.78	3.05	2.57	2.25	1.97		
	K	......	......	......	......	11.600	10.560	9.920	9.600	9.200		
	D	......	......	......	......	4.400*	4.900	5.300	5.700	6.100		
	E	......	......	......	......	3	3	5.3	5.3	5.3		
10	B	......	......	......	......	......	......	3.84	3.20	2.75	2.42	2.17
	K	......	......	......	......	......	......	14.850	13.600	12.800	12.250	11.880
	D	......	......	......	......	......	......	5.200*	5.700	6.100	6.500	6.900
	E	......	......	......	......	......	......	3	3	5.3	5.3	5.3

Each worm is an involute helical gear with $V = 3.142$. Where the value of D is marked *, the value of d_z is less than normal for a worm of the same root diameter by $0.1\ p_n$ in order to provide adequate crest width.

The quantity $r_c = Ep_n$ is the radius of curvature of the circular-arc profile of a milling cutter (of diameter in the range 4 to $10p_n$) which will produce worm threads very closely approximating the required involute helicoid form.

For design purposes the essential dimensions of a worm are its root diameter (which is the most important single measure of its strength and rigidity) and its lead, which is the measure of its size as a kinematic mate for the worm wheel.

Wormgears have the advantage that they permit a high velocity ratio to be attained in a single pair of gears with a diameter ratio of about 5, although the consequently high ratio of sliding speed to circumferential speed of the worm wheel makes for rather low mechanical efficiency at high velocity ratios. The highest possible efficiency at any given velocity ratio demands the smallest worm diameter consistent with load capacity. For the usual steel/bronze combination of materials the best root diameter for the worm is about one-fifth to one-quarter of the center distance of the gears. This may be used as the starting point of design of wormgears for any specified purpose; once a standard list of worms has been prepared, design of wormgears is extremely simple.

The range of worms exemplified by the dimensions given in Table 32.4 have blank diameters equal to 2.9, 3.4, 3.9, etc., times the normal pitch; with standard depth of thread, this means that root diameters are 1.468, 1.968, 2.463, etc., times the normal pitch. The value of V is 3.142 in all cases.

In certain worms with lead angles near to 45° these proportions would mean an undesirably narrow crest for the worm thread. To avoid this the blank diameter is reduced by 0.1 times the normal pitch without change in any other diameter; such a worm is said to be "topped" and every one in Table 32.4 is identified by an asterisk.

The throat diameter of a wormwheel may be given any value

(1) greater than $[N_G + 0.2 + 15/(N_G - 10)]l/3.16N_P$
and (2) less than $[N_G + 4.5 - 15/(N_G - 10)]l/3.16N_P$

The outside diameter of the worm wheel may exceed this by any amount up to about $0.5p_n$.

Worm-wheel flanks have been made in many varieties of cross-sectional form but there is little to be gained by departing from the simple rectangle with a circular-arc throat of radius equal to the standard clearance plus half the root diameter of the worm.

The maximum effective face width of a worm wheel clearly cannot exceed the blank diameter of the worm with which it meshes, and it is in general more severely limited by consideration of worm-wheel tooth form. Rather arbitrarily the maximum useful width is here taken to be $p_n + 0.33d_0$.

There is similarly a minimum desirable face width for the worm, and this is taken to be $4.5p_n$ or, for fast-running single-thread worms, $4p_x$. This exception is made in order to minimize vibration due to unbalance of the worm.

(j) Bevel Gears (Figs. 32.16*a* and *b*)

Bevel gearing is similar to wormgearing in that, while the tooth-generating cutter is of simple geometrical form, bevel gear teeth (and especially spiral bevel gear teeth) themselves are of such complicated form that exact analysis would be burdensome.

A pair of bevel gears is kinematically equivalent to a pair of cones with coincident apices and in straight-line contact along a common generator of their surfaces. The cone angles are equal to the "meshing cone angles" γ_P and γ_G of the gears and their sum is the angle between the center lines of the shafts. The cones can also make rolling contact with a disk centered at their common apex and touching each of them along their common straight-line generator. The disk may be regarded as a cone of 90° cone angle and the corresponding gear, a bevel gear of 90° meshing cone angle, is a "crown wheel."

The large ends of the teeth of a bevel gear are defined by the "back cone" generated by a straight line perpendicular to the intersecting generator of the meshing cone. The distance from the "back cone" to the apex of the meshing cone is the "cone distance" A common to the two gears of a mating pair. The apices of the meshing cones of

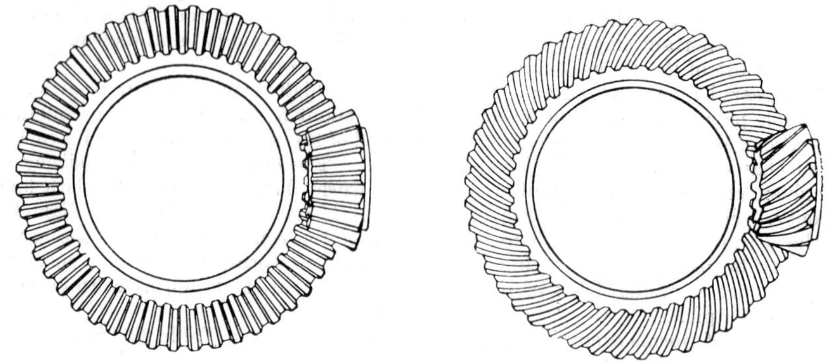

(a) (b)

FIG. 32.16. (a) Straight bevel gear. (b) Spiral bevel gear.

correctly mounted bevel gears coincide and the back cones have a common generator at the mesh point.

Bevel-gear teeth are usually cut so that the apex of the root cone coincides with that of the meshing cone. Because the radius of the fillets at the roots of the teeth cannot be conveniently made to differ across the face width of the gear, the clearance between tips and roots is more easily made uniform. For this reason the angle γ_0 of the tip cone is made equal to the difference between the shaft angle Σ and the root angle γ_R of the mating gear. In consequence, the apex of the tip cone lies between the body of the gear and the apex of the root cone.

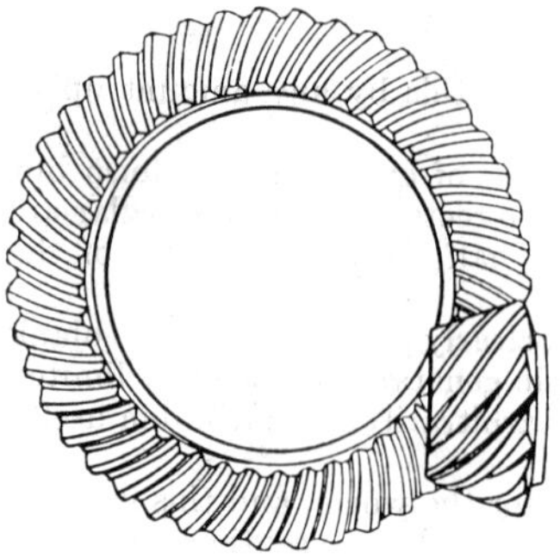

FIG. 32.17. Hypoid gear.

If the tooth flanks of the crown wheel intersect the meshing plane in straight lines they are planes inclined to the meshing plane at the complement of the standard flank angle. If the tooth flanks intersect the meshing plane in curved lines, each has a straight-line intersection with any plane perpendicular to the meshing plane and to its intersection with the tooth flank.

Although bevel-gear teeth and crown-wheel teeth are necessarily tapered, each flank may be swept by a straight-edged cutter blade and so the flat-flanked teeth of the *BS* rack have an analogy in crown-wheel tooth flanks that are either flat or are generated by a curved movement of a straight line. The crown wheel bears the same relation to bevel gears as the rack bears to spur and helical gears.

A bevel gear has no specific pitch in the ordinary sense, but the spacing of the teeth at their intersection with the large end of either meshing cone is the nominal pitch. In design there is no need to make this nominal pitch coincide with any standard pitch, and this slightly simplifies design procedure. The normal pitch must be chosen with regard for the normal pitches for which available cutters have been designed.

32.9. DETAILED DIMENSIONS OF GEARS

In this section "gear design" means the determination of manufacturing dimensions of a pair of gears of specified type that shall have a specified velocity ratio with about ±3 per cent tolerance and shall mesh at a specified center distance or, in the case of bevel gears, at a specified cone distance. The major dimensions are thus fixed, and the problem is to determine the detail dimensions required in manufacturing suitable gears without using any nonstandard tool.

Table 32.5 gives guidance in tentatively selecting the number of teeth in the pinion, and Table 32.6 defines a limitation that it is desirable to observe in the interest of keeping down noise. Table 32.2 gives a list of pitches that may be regarded as standard.

Table 32.5. Suggested Number of Teeth in Pinion*

Underriding minimum 10 $\cos^3 \psi \cos \gamma$ in every case

Type of gear	Both gears of steel not casehardened	Either gear of cast iron	Both gears of steel casehardened, flame hardened, or induction hardened
Spur or helical	$17 + 30/m_G$	$15 + 24/m_G$	$10 + 15/m_G$
Spur or helical pinion to mesh with internal gear	$10 + 15/m_G$ but not less than $10/(m_G - 1)$		
Bevel or hypoid	$10 + 30/m_G$	$9 + 24/m_G$	$7 + 15/m_G$
Crossed helical†	.....	$5 + 15/m_G$	$5 + 15/m_G$
Worm	Either of the whole numbers nearest to $(30 + C)/m_G$ or to $(20 + C)/m_G$ for service with heavy overloads. In all cases between $2C/m_G$ and $20C/m_G$ (C = center distance, in.)		

* See also Table 32.6.

† Limited to material combinations in Table 32.9.

Table 32.6. Minimum Desirable Number of Teeth in Pinion

This is ascertained by dividing $d_P{}^2 n_P$ by the number given in the table for the gear-finishing process to be employed.

Type of gear	Method of finishing teeth				
	Form cut	Recipro-cation generation	Hobbing or shaving	Profile grinding	Face-mill generation
Straight spur	250	750	1,200	3,000	
Helical	400	2,500	10,000	10,000	
Crossed helical and worm	400	2,500	3,000	10,000	
Internal (straight)	750	1,500	2,500	5,000	
Internal (helical)	1,200	6,000	10,000	10,000	
Bevel (straight)	200	750			
Spiral bevel or hypoid	.....	1,500	3,000	3,000	3,000

N_P should preferably be less than $480{,}000/d_P n_P$ or greater than $750{,}000/d_P n_P$.

(a) Spur and Helical Gears

In this discussion spur gears are considered helical gears with zero helix angle.

The helix angle ψ of helical gears that are to mesh with axes parallel may have any value between about 5 and 45°. Where there is no restriction the angle ψ may be taken as about 30°. Where it is important to minimize the axial thrust developed by the gears, smaller angles down to 5° may be used, provided that $\sin \psi$ is not less than (p_n/face width).

If the teeth are to be generated by a pinion-type cutter, the helix angle of the gear is determined by

$$\sin \psi = (\text{normal pitch of cutter})/(\text{axial pitch of cutter})$$

If the teeth are to be cut by hobbing or by rack-shaped cutters reciprocating on slides of adjustable inclination, there is no manufacturing restriction on helix angle except that values between 0 and 1° may be difficult to achieve by hobbing because of limitations in the machine.

Let it be required to determine dimensions of gears to work with axes parallel at center distance C and to run at a maximum sustained pinion speed of n_P rpm with a velocity ratio of $m_G \pm 3$ per cent.

The mean diameters are approximately

$$d_P = 2C/(1 + m_G) \qquad d_G = 2m_GC/(1 + m_G) \tag{32.24}$$

Table 32.5 suggests a value of N_P, and on the basis of the intended gear-cutting process, Table 32.6 shows whether any modification should be made.

Then p_n is (tentatively) the nearest standard pitch to $(\pi d_P \cos \psi)/N_P$ where ψ is the tentatively assumed helix angle.

Then p_x is the nearest low-factored number of inches to $p_n/\sin \psi$. [By "low-factored" is meant a number (ignoring the decimal point) equal to the product of two numbers neither of which has a prime factor exceeding 100.]

Helix angle ψ is now determined by $\sin \psi = p_n/p_x$.

The (possibly revised) number (N_P) of teeth in the pinion is the nearest whole number to

$$(\pi d_P \cos \psi)/p_n$$

Of the numbers without prime factors exceeding 100, take N_G as the nearest to m_GN_P. (The existence of high common factors in N_P and N_G is not now regarded as objectionable.) Let

$$N_{Pe} = N_P \sec^3 \psi \tag{32.25}$$

and

$$N_{Ge} = N_G \sec^3 \psi \tag{32.26}$$

$$\left.\begin{aligned} \text{Maximum permitted value of } V_P &= 4 - 10/N_{Pe} \\ \text{Maximum permitted value of } V_G &= 4 - 10/N_{Ge} \\ \text{Minimum permitted value of } V_P &= 1 + 20/N_{Pe} \\ \text{Minimum permitted value of } V_G &= 1 + 20/N_{Ge} \end{aligned}\right\} \tag{32.27}$$

From these the maximum and minimum permitted values of $V_P + V_G$ are determined by addition. Let

$$V_C = 2\pi C/p_n + 4 - (N_P + N_G) \sec \psi \tag{32.28}$$

If this lies within the permitted range of $V_P + V_G$, then the assumed values of N_P, N_G, p_n, and ψ may be accepted. If not, then some change must be made, e.g., to N_G within the permitted range of m_G with or without a change in N_P.

The procedure of using Eqs. (32.25) to (32.27) is repeated with the revised values.

Then V_P and V_G may be given any of their permitted values that add up to V_C. Provided that there are no special circumstances to prevent it, give V_P its nearest permitted value to $\frac{1}{2}V_C + 10(1/N_{Pe} - 1/N_{Ge})$ provided that $V_C - V_P$ is then a permitted value for V_G. If it is not, give V_G its nearest permitted value, and then $V_P = V_C - V_G$.

$$\left.\begin{aligned} d_0 &= (p_n/\pi)(N_P \sec \psi + V_P) \\ D_0 &= (p_n/\pi)(N_G \sec \psi + V_G) \end{aligned}\right\} \tag{32.29}$$

As a check, calculate $\frac{1}{2}(d_0 + D_0) - 2p_n/\pi$. This should be equal to C; any small difference may be corrected by subtracting twice its value from D_0. Any big difference shows an error in the arithmetic.

$$\text{Lead of teeth of pinion } l_P = N_Pp_x$$
$$\text{Lead of teeth of pinion } l_G = N_Gp_x$$

The helices must be of opposite hand. Preferred directions of axial thrust may help to decide which hand to adopt for one of the gears.

Note: No matter how the numerical value of N_Pp_x is "rounded off" it should be made clear on drawings that l_G is exactly (N_G/N_P) times l_P.

Basic normal backlash (always positive for a pair of external gears) is given by

$$k = (2.58z^2C\cos^2\psi)/[1 + z(2 + 4\cos^2\psi) - z^2(20\cos^4\psi - 1)] \qquad (32.30)$$

where $z = (V_P + V_G - 4)/(N_P + N_G)\sec\psi$.

Gear-tooth caliper settings to give normal backlash B at center distance C:
Pinion thickness $= t_{ncP}$, height $= h_p$

$$t_{ncP} = (0.843 + 0.364V_P)(p_n/\pi) - 0.532(B - k) \qquad (32.31)$$
$$h_P = \tfrac{1}{2}V_P(p_n/\pi) + \tfrac{1}{4}(t_{ncP}\cos\psi)^2/(d_0 - V_Pp_n/\pi) \qquad (32.32)$$

Gear thickness $= t_{ncW}$, height $= h_W$

$$t_{ncG} = (0.843 + 0.364V_G)(p_n/\pi) - 0.532(B - k) \qquad (32.33)$$
$$h_G = \tfrac{1}{2}V_Wp_n/\pi + \tfrac{1}{4}(t_{cnW}\cos\psi)^2/(D_0 - V_Wp_n/\pi) \qquad (32.34)$$

Check:

$$B = k + [p_n - (t_{ncP} + t_{ncG})]/1.064 + 0.109(V_P + V_G - 4)p_n$$

These tooth thicknesses are attained by sinking the cutter to the appropriate depths in the gear blanks. In some circumstances the resulting tip-and-root clearance when the gears are meshed is less than the standard $0.08p_n$. There is no harm in this except that a clearance less than about $0.02p_n$ is undesirable. To avoid this condition each blank diameter should be reduced by $0.37(k - B) - 0.03p_n$ if this is a positive quantity. If such reduction is made, then h_P and h_G must each be reduced by half the reduction in diameter.

Possible Numbers of Teeth in a Given Blank. Inversion of the principle expressed in Eqs. (32.16) and (32.17) defines the maximum and minimum permissible numbers of teeth that may be cut in a given blank with a specified axial pitch by a specified standard cutter. Let $\cos\psi = (1 - p_n^2/p_x^2)^{1/2}$.

The number of teeth N must be not less than

$$[\pi d_0/p_n - 4 + (10\cos^2\psi)/(\pi d_0/p_n - 4)]\cos\psi$$

and not greater than $[\pi d_0/p_n - 1 - (20\cos^2\psi)/(\pi d_0/p_n - 1)]\cos\psi$ (32.35)

Neither these formulas nor the limits of V in Eq. (32.17) are rigorous and if, for example, Eq. (32.35) suggests that the maximum permissible value of N is 25.9, it is probable that 26 is permissible. To clarify such points as this, and for general reassurance, results from Eq. (32.35) are preferably checked by Eqs. (32.16) and (32.17).

(b) Internal Gears

Much of the geometry of an internal gear is the same as that of the corresponding external gear except that the radius is negative instead of positive.

Thus for a pinion and a mating internal gear, generated by a 20° cutter with working depth $2p_n/\pi$,

$$d_0 = (p_n/\pi)(N_P\sec\psi + V_P)$$
$$D_0 = (p_n/\pi)(N_G\sec\psi - V_{Gi})$$
$$C = -[\tfrac{1}{2}(d_0 - D_0) - 2p_n/\pi] = \tfrac{1}{2}(D_0 - d_0) + 2p_n/\pi$$

Details of tooth design for a pinion/internal gear combination are subject to restrictions additional to those which apply to external gears. For example, the tips of the teeth of the pinion tend to foul those of the internal gear near the line of intersection of the tip cylinders of the gears. For teeth produced by standard cutters with working depth equal to $2p_n/\pi$ this difficulty is avoided if the difference between the numbers of teeth in the gears exceeds 10. Similarly the difference between the numbers of teeth in an internal gear and the pinion-type cutter that generates it should exceed 14 if this type of interference is to be avoided.

The form of the blank for the internal gear must leave clearance for the "runout" of the cutter beyond the end of the face width. For a pinion-type cutter, this runout may be as little as 0.2 in. Internal teeth may be generated by a special kind of hob; this needs much more clearance for "runout."

The procedure for determining detail dimensions for a pinion and internal gear to work together at a specified center distance with a specified gear ratio and maximum pinion speed is the same as that detailed for external gears except that in Eq. (32.24) the quantity $(1 + m_G)$ is replaced by $(m_G - 1)$.

Then, corresponding to Eq. (32.27), we have

$$\begin{aligned} \text{Maximum permitted value of } V_P &= 4 - 10/N_{Pe} \\ \text{Maximum permitted value of } V_{Gi} &= 3 - 20/N_{Ge} \\ \text{Minimum permitted value of } V_P &= 1.5 + 20/N_{Pe} \\ \text{Minimum permitted value of } V_{Gi} &= 2 \end{aligned} \tag{32.36}$$

From these the maximum and minimum permitted values of $V_P + V_{Gi}$ are determined by addition. Let

$$V_C = (N_G - N_P) \sec \psi + 4 - 2\pi C/p_n$$

If this is less than $4 + [(N_G - N_P) \sec \psi]/25$ and also lies within the permitted range of $V_P + V_{Gi}$ then the proposed values of N_P, N_G, p_n, and ψ may be adopted. If not, the conditions must be achieved by some change in the tentative selections, e.g., in N_G or N_P.

The procedure down to Eq. (32.36) is repeated with the revised values.

Give V_P its nearest permitted value to 3, provided that $V_C - V_P$ is then a permitted value for V_{Gi}. If it is not, give V_{Gi} its nearest permitted value to this and take $V_P = V_C - V_{Gi}$.

$$\begin{aligned} d_0 &= (p_n/\pi)(N_P \sec \psi + V_P) \\ D_0 &= (p_n/\pi)(N_G \sec \psi - V_{Gi}) \end{aligned}$$

As a check, calculate $\frac{1}{2}(D_0 - d_0) + 2p_n/\pi$. Any small difference between this and C is corrected by subtracting twice its value from D_0. Any big difference shows an error in arithmetic.

$$l_p = N_P p_x \qquad l_G = N_G p_x$$

The helices must be of common hand.

NOTE: It should be made clear on drawings that l_G is exactly (N_G/N_P) times l_P.

Basic normal backlash (always negative for a pinion/internal gear pair) is given by

$$k = -(2.58z^2C \cos^2 \psi)/[1 - z(2 + 4 \cos^2 \psi) + z^2(20 \cos^4 \psi - 1)]$$

where $z = (V_P + V_{Gi} - 4)/(N_G - N_P) \sec \psi$. Gear-tooth caliper settings are given by Eqs. (32.31) to (32.34) except that in the expression for h_G the sign preceding $\frac{1}{4}$ is reversed.

(c) Crossed-helical Gears

Let it be required to determine dimensions of gears to work with a shaft angle Σ at a center distance C and to run at a maximum sustained pinion speed of n_P rpm with a velocity ratio of $m_G \pm 3$ per cent.

There is no essential restriction on the ratio d_P/d_G, but unless some restriction peculiar to the particular problem forbids, tentatively take

$$d_P = 2C/(1 + m_G) \qquad d_G = m_G d_P$$

Then the approximate blank diameters and root diameters are

$$\begin{aligned} d_0 &= d_P(15 + 7m_G)/(15 + 5m_G) \\ d_R &= d_P(15 + 2.5m_G)/(15 + 5m_G) \\ D_0 &= d_G(17 + 5m_G)/(15 + 5m_G) \\ D_R &= d_G(12.5 + 5m_G)/(15 + 5m_G) \end{aligned}$$

Ascertain whether these diameters are acceptable for clearing adjacent components and for adequate bore diameters. If they are, take $\psi_1 = \psi_2 = \frac{1}{2}\Sigma$. If not, select values of d and D (noting that $d + D = 2C$) that produce satisfactory blank diameters and root diameters.

Determine approximate values of ψ_1 and ψ_2 from

$$\tan \psi_1 = (m_G d/D - \cos \Sigma)/\sin \Sigma \tag{32.37}$$
$$\tan \psi_2 = (D/m_G d - \cos \Sigma)/\sin \Sigma \tag{32.38}$$

The cotangent of ψ for the driven gear must not be less than the coefficient of friction between the teeth. Check $\psi_1 + \psi_2 = \Sigma$. Then

$$p_{n1} = 2\pi C/N_P[\sec \psi_1 + m_G \sec (\Sigma - \psi_1)]$$

Starting from the value of N_P suggested by Table 32.5, find one that brings p_{n1} within 5 per cent of a standard pitch and within the limitation of Table 32.6. If none can be found with the originally selected value of ψ_1, then other values of ψ_1 may be tried.

When suitable values of N_P and p_{n1} have been found, take N_G as the nearest whole number to $m_G N_P$ and recalculate ψ_1 and ψ_2 in Eqs. (32.37) and (32.38) with m_G replaced by N_G/N_P. If there is a standard pitch within 5 per cent of p_{n1}, adopt it as p_n. If not, change N_G (within the tolerance on m_G) or ψ_1.

Take p_{x1} as the next convenient dimension above $p_n \csc \psi_1$, avoiding numbers with prime factors over 100. Similarly take p_{x2} as the next convenient dimension below $p_n \csc \psi_2$. (Here p_n is the adopted standard pitch.)

Recalculate ψ_1 and ψ_2 by

$$\sin \psi_1 = p_n/p_{x1} \qquad \sin \psi_2 = p_n/p_{x2}$$

Check $\psi_1 + \psi_2 = \Sigma \pm 10'$. The meshing helix angles are given by

$$\left.\begin{aligned} \tan \psi_{m1} &= (\sin \Sigma)/(\cos \Sigma + p_{x1}/p_{x2}) \\ \tan \psi_{m2} &= (\sin \Sigma)/(\cos \Sigma + p_{x2}/p_{x1}) \end{aligned}\right\} \text{not required in design}$$

Determine $N_{Pe} = N_P \sec^3 \psi_1$ and $N_{Ge} = N_G \sec^3 \psi_2$.

Determine the maximum and minimum permitted values for V_P and V_G and hence the permitted range of $(V_P + V_G)$ [see Eq. (32.27)]. Let

$$V_C = 2\pi C/p_n - N_P \sec \psi_1 - N_G \sec \psi_2 + 4$$

Then proceed as for helical gears to Eq. (32.30). Check that the values of d_0, D_0, d_R, and D_R are acceptable for clearances.

$$l_1 = N_P p_{x1} \qquad l_2 = N_G p_{x2}$$

The basic normal backlash for crossed-helical gears is zero if the sum of the helix angles of the gears is equal to the shaft angle. This is the normal aim in design, but a small discrepancy may be produced by adjustment of axial pitches so that they are factorable numbers. On this account

$$k = (V_P + V_G - 4)(p_n/\pi) \cot \eta_n(\cos \eta_n)(\Sigma - \psi_P - \psi_G)/(\tan \psi_P + \tan \psi_G)$$

where η_n is the normal flank angle of the imaginary rack of engagement of the gears. For this purpose $\eta_n = \phi_n = 20°$ and $\Sigma - \psi_P - \psi_G$ is in radians.

The expression for k may be slightly simplified thus

$$k = (V_P + V_G - 4)p_n(\Sigma - \psi_P - \psi_G)'/4{,}200(\tan \psi_P + \tan \psi_G) \tag{32.39}$$

where $\Sigma - \psi_P - \psi_G$ is now expressed in minutes of angle.

Gear-tooth caliper settings are determined as in Eqs. (32.31) to (32.34).

The paths of contact of the two pairs of mating flanks are limited by the tip cylinders of the gears, provided that the face widths are great enough. On this account the minimum desirable face width of a crossed-helical gear is about

$$0.9p_n[(V' + 0.5) \sin \psi + 1.2 \cos \psi - \Omega(V_P + V_G - 4)/(\tan \psi_P + \tan \psi_G) \cos \psi]$$

where ψ is the helix of the gear under consideration, V' means V or $(4 - V)$, whichever is greater, $\Omega(V_P + V - 4)$ means $(V_P + V_G - 4)$ or $(4 - V_P - V_G)$, whichever is positive, and all angles are positive.

(d) Wormgears

The general problem is to determine dimensions of wormgears that shall work at a specified center distance C in. with a specified velocity ratio m_G at a maximum sustained worm speed of n_P rpm.

The number N_P of worm threads may be either of the whole numbers nearest to (1) $(30 + C)/m_G$ for normal industrial purposes or (2) $(20 + C)/m_G$ for drives with recurrent short-period overloads, and in every case N_P should be between $2C/m_G$ and $20C/m_G$ and should not be less than $C^2n/6{,}000(m_G^2 + 25)$.

N_G is either of the whole numbers nearest to m_GN_P. If there is a common factor in N_P and N_G the teeth should be marked so that the original meshing may be repeated after any dismantling.

Approximate root diameter of a worm made in one piece with its shaft, as is usual, is

$$d_R = 0.1 + C(0.1 + 5/N_G) \tag{32.40}$$

or $1.8C/(1 + m_G)$, whichever is greater, or as near to this as may be permitted by any restriction on the diameter of the worm wheel.

This suffices for all ordinary loading conditions. Exceptionally high short-period overloads may, however, demand a larger root diameter to avoid overstress.

The root diameter of a worm that is to be fitted to a separate shaft should be such as to leave a minimum thickness of metal of $0.3(1 + \sin\lambda)p_n$ in. outside the keyway.

Approximate lead of worm threads is, tentatively,

$$l_1 = (2C - d_R)N_P/0.316(N_G + 3)$$

In the part of Table 32.4 concerned with the selected value of N_P locate the two nearest values of B to l_1/d_R. Let these be B_1 and B_2 with corresponding values K_1 and K_2 of K.

Then p_n is the standard normal pitch (in Table 32.3) that lies between l_1/K_1 and l_1/K_2. If there is more than one such pitch, select the one that lies nearest to whichever of l_1/K_1 and l_1/K_2 corresponds to the value of B that is nearer to l_1/d_R.

In Table 32.4 locate the value of A that is nearest to [d_R from Eq. (32.40)]/(p_n from Table 32.3).

In the vertical column headed by the value of A previously determined and in the horizontal group associated with the selected value of N_P are the ratios of dimensions of the appropriate worm. The actual dimensions are $p_{nb} = p_n \cos 20°$.

$$d_R = Ap_n \qquad d_0 = Dp_n \qquad l = Kp_n \qquad p_x = l/N_P$$

and the radius of curvature of the cutter or grinding wheel for the worm is Ep_n.

It should be confirmed that d_0 is acceptable for clearance. If the gears will be subjected to unusually heavy short-period loading it should be confirmed that d_R is adequate to resist them. For normal service conditions this check is unnecessary as gears capable of taking the sustained load are, with the proportions recommended here, capable of withstanding short-period overloads.

Throat diameter of worm wheel

$$D_t = 2C - d_R - 0.16p_n$$

Blank diameter of worm wheel

$$D_0 = D_t + 0.5p_n$$

or as near to this as clearances permit.

As a check, let $W = N_PD_t/0.316l$. Then the numbers of teeth that the worm wheel may have lie between $W - 0.2 - 15/(W - 12)$ and $W - 4.5 + 15/(W - 12)$.

Confirm that this range contains a value of N_G that places N_G/N_P within the acceptable range of m_G.

Face width of worm $= 4.5p_n$ or, in the case of a single-thread worm, $4p_x$.

Maximum useful face width of worm wheel $= p_n + d_0/3$. Minimum desirable face width of worm wheel $= lp_n/1.7d_R$ or $D_t(1 + l/4d_R)/12$, whichever is the greater.

Caliper settings for worm $t_{ne} = 0.557p_n$. $h = 0.4p_n$ or $0.35p_n$ if the value of D in Table 32.4 is marked *.

Lead angle λ_F at the reference cylinder (diameter $d_F = d_R + 0.432p_n$) is derived from

$$\sin \tfrac{1}{2}\lambda_F = [(p_x - p_n)/2p_x]^{1/2}$$

Lead angle λ at mid-depth of thread is derived from

$$\cot \lambda = 1.57(d_R + d_0)/l$$

Table 32.7. Basic Allowable Stresses for Spur, Helical, and Bevel Gears

For Forged Steels

Basic allowable surface stress σ_{zb}	Basic allowable bending stress σ_{yb}	Steel	Heat-treatment	Brinell hardness	Ultimate tensile strength, psi	Approx. relative cost/lb
1,450*	18,000†	1045	Normalized	145	90,000	10
1,800	20,000	1335	Normalized	180	100,000	17
2,100*	21,000†	1060	Normalized	210	110,000	11
3,000	30,000	E9310	Heat-treated before finish cutting	250	125,000	35
4,400	41,000	E9310	Heat-treated after finish cutting	360	180,000	35
8,500	27,000	4140	Nitrided after finish cutting	270§	135,000§	70
10,000	34,000‡	4320	Casehardened after cutting	200§	100,000§	45
11,000	43,000‡	4820	Casehardened after cutting	250§	125,000§	45

For Other Materials

Basic allowable surface stress σ_{zb}	Basic allowable bending stress σ_{yb}	Material	Condition	Brinell hardness	Ultimate tensile strength, psi.	Relative cost/lb
700	7,000	Phosphor bronze	Sand cast	69	27,000	100
850	8,500	(12% tin)	Chill cast	82	34,000	100
1,000	5,800	Cast iron	Average	165	27,000	8
1,450	10,000	Cast iron	High grade	220	50,000	10
1,450	16,000	Cast steel (0.4% carbon)	As cast	145	80,000	20

* Flame hardening or induction hardening multiplies this by 2.

† Flame hardening multiplies this by 0.8.

‡ Very smooth root finish multiplies this by 1.8 if the fillet surface is *not* ground.

§ Refers to the core as distinct from the hard case.

Table 32.8. Combinations of Materials for Spur, Helical, and Bevel Gears

Combination	Steel	Heat-treatment	Relative cost/lb	Relative load capacity	Tooth-finishing process
1	P 1060	Normalize	11	1	Teeth cut after heat-treatment
	G 1045	Normalize	10		
2	P E9310	Harden to 250 Brinell	35	1.5	Teeth cut after heat-treatment
	G 1060	Normalize	11		
3	P 1060	Flame harden or induction harden	11	2	Finished teeth hardened in these ways usually have very little hardening deformation
	G 1045	Flame harden or induction harden	10		
4	P E9310	Harden to 400 Brinell	35	2.2	Hardening deformation usually necessitates subsequent finishing by grinding or lapping
	G E9310	Harden to 250 Brinell	35		
5	P 4140	Nitride	70	4	Negligible deformation produced by nitriding
	G 4140	Nitride	70		
6	P 4320	Caseharden	45	4.5	Quiet running at high speed demands grinding or lapping after hardening
	G 4320	Caseharden			
7	P 4820	Caseharden	45	5	Quiet running at high speed demands grinding or lapping after hardening
	G 4820	Caseharden			

Base diameter of worm is

$$d_b = 1/\pi[(1.064/N_P p_n)^2 - 1/l^2]^{1/2}$$

Base lead angle λ_b is given by

$$\cot \lambda_b = \pi d_b / l$$

Unless external considerations forbid, right-hand worm threads should be used.

(e) Spiral Bevel Gears

Let it be required to determine dimensions of a pair of spiral bevel gears to work at a shaft angle Σ with cone distance A (Fig. 32.18) and velocity ratio $m_G \pm 3$ per cent and to have teeth generated by the face-milling (Gleason) process.

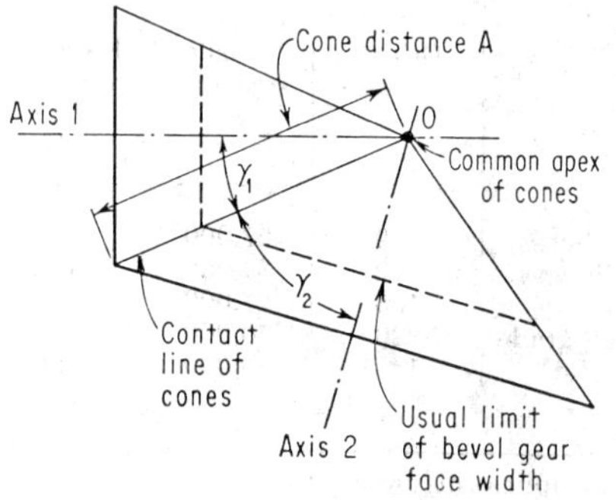

FIG. 32.18. Spiral bevel gears at a shaft angle Σ.

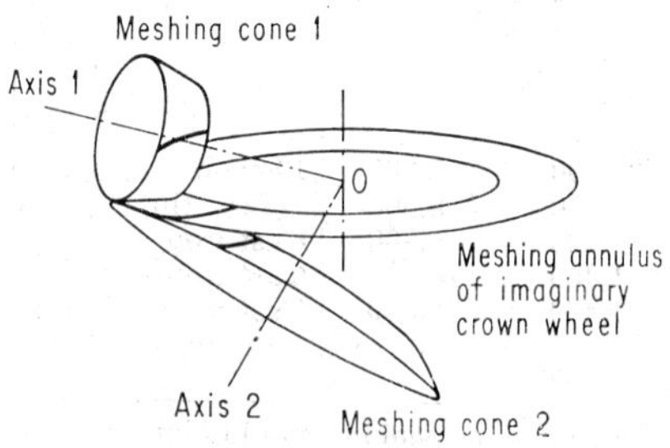

FIG. 32.19. Meshing cones of a pinion and gear with imaginary crown wheel.

Such dimensions as normal pitch, spiral angle, and blank diameter apply to the large ends of the teeth.

Determine approximate meshing cone angles

$$\tan \gamma_P = (\sin \Sigma)/(\cos \Sigma + m_G)$$
$$\tan \gamma_G = (\sin \Sigma)/(\cos \Sigma + 1/m_G)$$

Check $\gamma_P + \gamma_G = \Sigma$.

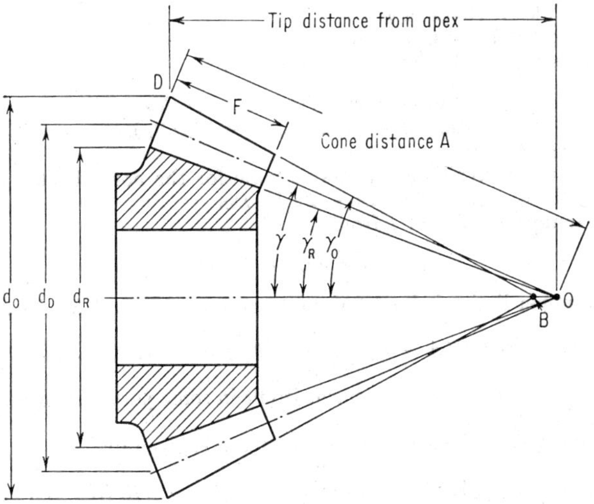

FIG. 32.20. Principal manufacturing dimension of a bevel gear. Apex of tip cone is at B because generator DB of this cone is parallel to that of the root cone of the mating gear. Meshing cone and root cone converge at point O, which coincides, in operation, with corresponding point of mating gear.

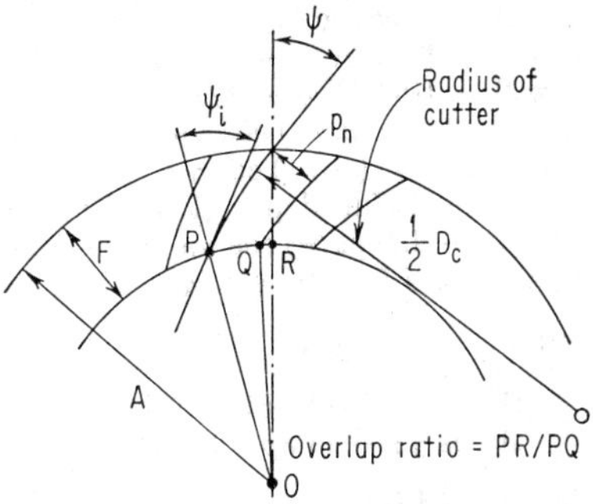

FIG. 32.21. Meshing plane of basic crown wheel corresponding to spiral bevel gear teeth generated by face-mill cutter.

The meshing cones of the pinion and the gear are in every case identical with the "reference cones" which are the meshing cones with the basic crown wheel (Fig. 32.19). Hence, from Fig. 32.20,

$$d_F = 2A \sin \gamma_P \qquad D_F = 2A \sin \gamma_G$$

The face width F is taken as $0.3A$.

The spiral angle ψ (at the reference circles) is first tentatively assumed to be 30° (see Fig. 32.21).

The tentative value of N_P is that derived from Table 32.5 consistently with the restrictions of Table 32.6. Then the normal pitch at the reference circles is

$$p_n = (2\pi A \sin \gamma_P \cos \psi)/N_P \tag{32.41}$$

This should be between $A/12$ and $A/6$ and should not be greater than half the face width. It need not be any standard pitch.

In order to comply with these restrictions it may be necessary to adopt a different value of N_P and also perhaps of ψ.

N_G is the nearest number to $m_G N_P$.

Revised values of γ_1 and γ_2 are now obtained from

$$\tan \gamma_P = \sin \Sigma/(\cos \Sigma + N_G/N_P)$$
$$\tan \gamma_G = \sin \Sigma/(\cos \Sigma + N_P/N_G)$$

Check $\gamma_P + \gamma_G = \Sigma$.

Of the diameters of available cutters suitable for the normal pitch calculated in Eq. (32.41) let the one nearest to $4A - 2F$ be D_C.

Again fix a tentative value of ψ from

$$\sin \psi = (2A - F)/D_C$$

but impose an upper limit of about 35° on ψ and a minimum defined by

$$\tan \psi = [2A(A - F)/F(2A - F)] \tan (450 \sin \gamma_P/N_P)^\circ$$

If these restrictions are found to be inconsistent in some particular case, the value of N_P should be changed.

The spiral angle ψ_i at the small end of the teeth is given by

$$\sin \psi_i = [A/(A - F)] \sin \psi - F(2A - F)/D_C(A - F)$$

Using the adopted value of ψ, determine p_n again by the formula in Eq. (32.41). The numbers of equivalent spur-gear teeth are given by

$$N_{Pe} = N_P \sec^3 \psi \sec \gamma_P$$
$$N_{Ge} = N_G \sec^3 \psi \sec \gamma_G$$
$$V_P = 2 + 10(1/N_{Pe} - 1/N_{Ge})$$
$$V_G = 2 - 10(1/N_{Pe} - 1/N_{Ge})$$
$$d_0 = d_F + V_P(p_n/\pi) \cos \gamma_P$$
$$D_0 = D_F + V_G(p_n/\pi) \cos \gamma_G$$
$$d_R = d_0 - 2 \text{ (full depth of tooth) } \cos \gamma_P$$
$$= d_0 - 4.5(p_n/\pi) \cos \gamma_P$$
$$D_R = D_0 - 4.5(p_n/\pi) \cos \gamma_G$$

Tip distances from apex:

Pinion $\qquad A \cos \gamma_P - (V_P p_n/2\pi) \sin \gamma_P$

Gear $\qquad A \cos \gamma_G - (V_G p_n/2\pi) \sin \gamma_G$

Root cone angles are determined by

Pinion $\qquad \sin \gamma_{RP} = d_R/2A$

Gear $\qquad \sin \gamma_{RG} = D_R/2A$

Blank cone angles are determined by

Pinion $\qquad \gamma_{0P} = \Sigma - \gamma_{RG}$

Gear $\qquad \gamma_{0G} = \Sigma - \gamma_{RP}$

Caliper settings for backlash B (in the system of design described here

$$V_P + V_G = 4$$

and so the basic backlash $k = 0$):

Pinion $\quad t_{ncP} = (0.843 + 0.364V_P)(p_n/\pi) - 0.532B$

$$h_P = \tfrac{1}{2}V_P(p_n/\pi) + \tfrac{1}{4}(t_{ncP}\cos\psi)^2/(d_0 \sec\gamma_P - V_P p_n/\pi)$$

Gear $\quad t_{ncG} = (0.843 + 0.364V_G)(p_n/\pi) - 0.532B$

$$h_G = \tfrac{1}{2}V_G p_n/\pi + \tfrac{1}{4}(t_{ncG}\cos\psi)^2/(D_0 \sec\gamma_G - V_G p_n/\pi)$$

32.10. GEARS UNDER LOADING

(a) Load Capacity

Although there are many formulas for the load capacity of gears and although most of them include numerical values, and algebraic expressions that may give the impression of high precision, the allowable values of stresses calculated in any particular way can be decided (or estimated) only on the basis of experience with many gears of many different designs for many duties. Even then, every estimate of load capacity has an uncertainty of at least 10 per cent in each direction. It is based on the assumption of the use of sound material in gears accurately mounted, adequately lubricated, and fully protected against dust, dirt, and excessively high temperature. Any default in any of these respects may severely reduce the load capacity of the gears.

It may be emphasized that the possible limitations on the load capacity of any pair of gears are more numerous than the following analysis suggests. Formulas for load capacity of gears are largely empirical.

(b) Gear Materials

Apart from crossed-helical gears and wormgears, gears are usually made of a type of steel. Although a great many varieties of steel have been used, all requirements are met by the materials noted in Table 32.7. Although these are not the only steels that can be used, caution should be exercised before departing from this list. It is especially important to note that the load capacity of a steel as a gear material cannot be reliably assessed from its hardness or from its tensile strength. These qualities are necessary, but equally important is compatibility with the mating material while the two slide together in loaded line contact.

If it is known that a particular pair of steels work together satisfactorily as gear materials the basic allowable surface stress σ_{zb} for each material may be assessed on the basis of its hardness by comparison with similarly hard steels noted in Table 32.7.

The basic allowable bending stress σ_{yb} may be assessed on the basis of ultimate tensile strength by comparison with similarly strong steels noted in Table 32.7.

Combination 1 in Table 32.8 should be the first tentative choice. If it is impossible or inconvenient to find room for the gears that are necessary to transmit the load in these materials, combination 2 should be tried, and so on.

The ductility of 1045 steel and of 1060 steel is a valuable and insufficiently appreciated means of offsetting small errors in manufacture and mounting. Steels of lower carbon content have even greater ductility but nevertheless should not be used as gear materials because their resistance to sliding under load is likely to be inadequate.

(c) Stresses in Gear Teeth

Photoelastic examination of a typical gear-tooth model shows that the stresses just below the contact line on the tooth flank are very much higher than those induced at the root fillets by the bending effect of the tooth load. The tensile stress at the root fillet is more destructive than the shear stresses produced by line contact, but nevertheless the safe loading of a gear tooth is usually determined by contact stress rather than by bending stress. Use of the value of N_P suggested in Table 32.5 ensures this in most circumstances except where a casehardened steel gear is intended for only a short life. All such cases should be specially examined in this respect.

The nominal gear-tooth load is equal to the external torque divided by the radius of

Table 32.9. Basic Allowable Contact Stresses (σ_{sb}) for Materials of Crossed-axis Gears (Including Wormgears)

Material ref.	Material of wheel	Material of pinion (or worm)				
		1	2	3	4	5
1	Casehardened and ground steel 4320	700†	......	......	......	500*
2	Nickel-chrome steel E9310	......	......	......	......	450*
3	0.5% carbon steel 1060	......	......	......	......	400*
4	0.4% carbon steel 1045	......	......	......	......	400*
5	Gray cast iron	500*	450*	400*	400*	700*
6	Sand-cast phosphor bronze	700	700	700	700	700*
7	Chill-cast phosphor bronze	900	900	900	800	800*
8	Centrifugally cast phosphor bronze	1,200	1,100	1,100	1,000	1,000*

* Sliding speed should not exceed 500 fpm.

† Only for high-precision crossed-helical gears smoothly finished and run in with antiweld oil.

the meshing circle. The nominal load divided by the tooth width gives the nominal tangential loading per inch of width, denoted by w_t.

An effect of dimensional errors in the teeth is, however, to induce "dynamic load"; the dynamic load per inch width is denoted by w_d. The gross load per inch width is $w_g = w_t + w_d$. This load induces a maximum contact stress σ_z in the teeth, a maximum bending stress σ_{yP}, in the pinion teeth, and a maximum bending stress σ_{yG} in the gear teeth.

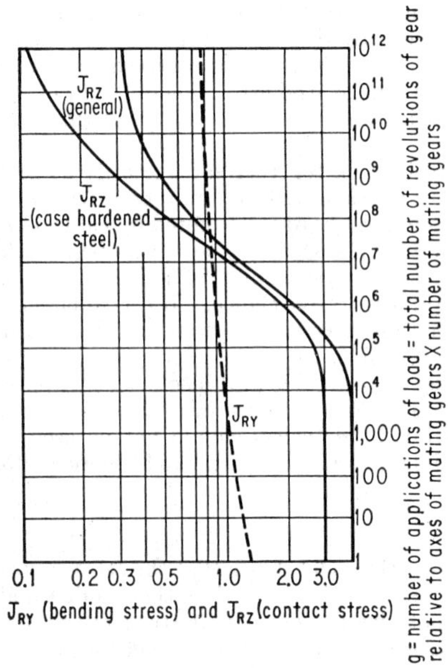

FIG. 32.22. Stress repetition factors.

The ratio of σ_z to the basic allowable stress σ_{zbP} for the material of the pinion gives an estimate of the number of load applications that a pinion tooth may be expected to withstand so far as contact stress is concerned. Similar comparisons lead to figures for life expectation of each gear for contact stress and bending stress.

In the case of wormgears and crossed-helical gears the contact stress is denoted by σ_s and is compared with σ_{sb}, which is the basic allowable stress for the material primarily under consideration in "crossed-axis" conditions. The basic allowable contact stresses σ_{sb} for materials of crossed-axis gears are given in Table 32.9.

(d) Limitations on Tooth Loading

The allowable load per inch width of a gear tooth is w_A and is the lowest value imposed by (1) contact stress, (2) bending stress (both applied g times during the life of the gear), and (3) load capacity of the lubricant at the sliding speed concerned.

The allowable contact stress is $J_{Rz}\sigma_{zb}$ where J_{Rz} is a "repetition factor" determinable from Fig. 32.22 if g is known or assumed. The value of g is usually equal to the number of gears mating with the one concerned multiplied by the number of revolutions (relative to a common perpendicular to its axis and that of the mating gear) made by the gear during its life. The value of σ_{zb} is determined from Table 32.7.

The mating gears have a common contact stress at any instant but their allowable contact stresses are (in general) different, the lower allowable stress governing under limitation (1). It is $J_{RzP}\sigma_{zbP}$ or $J_{RzG}\sigma_{zbG}$, whichever is the less.

There is a corresponding limitation by allowable bending stresses, with the difference that the maximum bending stress is not (in general) the same for the pinion as for the wheel.

Limitation (3) by the lubricant depends on the maximum sliding speed v_s of the teeth and the rotational speed of the pinion.

(e) Maximum Sliding Speed

For use in connection with questions associated with lubrication of the gears and with the coefficient of friction between the teeth, it is necessary to know the approximate value of the maximum sliding speed of the teeth.

For spur or helical gears with axes parallel, the maximum sliding speed is given by

$$v_{su} = 1.7 d_P n_P (1/N_P + 1/N_G) \cos \psi \qquad \text{fpm (approx.)}$$

or $0.22 V p_n (n_P + n_G)$.

For bevel gears the same expression applies except that N_P is multiplied by sec γ_P and N_G by sec γ_G.

In gears working with crossed axes inclined at angle Σ there is also an endwise sliding of the teeth at velocity

$$v_{se} = 0.26(d_P^2 n_P^2 + d_G^2 n_G^2 - 2 d_P n_P d_G n_G \cos \Sigma)^{1/2} \qquad \text{fpm}$$

where

$$d_P = (N_P \sin \Sigma)/\pi[(\cos \Sigma)/p_{xP} + 1/p_{xG}]$$

$$d_G = (N_G \sin \Sigma)/\pi[(\cos \Sigma)/p_{xG} + 1/p_{xP}]$$

For hypoid gears,

$$v_{se} = 0.52 C n_G \sec \psi_1$$

The resultant maximum sliding velocity v_s is the greater of the velocities v_{su} and v_{se}, plus one-third of the smaller.

(f) Limitation by Lubricant

The consequences of metallic contact between metal surfaces sliding under load increase in severity with increased sliding speed and load.

There is evidence that the load required to produce scuffing of given surfaces with a given sliding speed diminishes with rise of temperature of the surfaces. This implies that, for mating gear teeth of given dimensions, the scuffing load falls with rise in sliding speed and also with rise in rotational speed of the gear because the more frequently load is applied to a tooth, the higher its temperature must be to dissipate heat at the increased rate. A factor in dissipation of the heat is the thermal conductivity of the material of the gear; this suggests some discrimination in favor of bronze in comparison with steel. Examination of general practice and research results make it possible to suggest a rule for guidance. Here it is recommended that, for a steel/steel combination of gears, σ_z should not exceed

$$(\sigma_{zbP}/100) + \beta[10^{10}/(v_s + 100)(n_p + 10{,}000)]$$

where v_s (fpm) is the maximum sliding speed on the teeth, n_p (rpm) is the higher rotational speed of the two gears, and β is a factor given in Table 32.10.

The thickness of the oil film between loaded surfaces such as gear teeth depends on the velocities of the contact line over the surfaces and on the viscosity of the oil

Table 32.10. Lubrication Factor β

Materials	Lubricant	β	
		Axes parallel or intersecting	Axes crossed
Casehardened* steel/casehardened* steel....	Mineral oil	2.5	1
Other steel/other steel....................	Mineral oil	1.5	0
Steel/bronze...............................	Mineral oil	4	2
Casehardened* steel/casehardened* steel, very smoothly finished..................	Antiscuffing oil†	5	2
Other steel/other steel, very smoothly finished..............................	Antiscuffing oil†	3	0

* This includes nitriding.
† At least for the running-in period.

immediately ahead of the contact line. If the velocities are low, production of an oil film requires that the viscosity of the oil be high.

The viscosity (Saybolt Universal) of the oil *at its working temperature* should not be less than about

$$80 + [80/(v + 50)]A^2$$

where v = 0.7 × tooth speed, fpm, relative to a common perpendicular to the axes or sliding speed, fpm, whichever is higher
A = tensile yield stress, tons/in.^{2} of the softer material of the mating gears

(g) Allowable Gross Tooth Load per Inch Width (w_{gA})

w_{gA} is the smaller of the quantities w_{Az} and w_{Ay} determined by the formulas in Table 32.11. If the value of N_P is in accordance with Table 32.5, the limitation will be w_{Az} and there is no need to calculate w_{Ay}, unless the required life of the gears is short.

From this allowable gross load the dynamic load due to tooth errors must be subtracted, leaving the tooth load (per inch width) corresponding to the external (useful) torque applied to the gears.

The dynamic load produced by random transmission errors is independent of the external torque on the gears. The dynamic load produced at and near resonance with periodic transmission errors is, however, limited by tooth separation to a value associated with the external load in a way that is different for helical teeth from what it is for straight teeth.

(h) Dynamic Load

The load induced in the teeth by any random error of pitch depends on (1) the magnitude of the error (denoted by Δ_c), (2) the elasticity (or "compliance" denoted by H) of the meshing teeth and the bodies of the gears, and (3) the ratio of the time in which the error takes effect to the natural period of rotational vibration of the gears by reason of the elasticity of the teeth.

The quantity Δ_c (here called the "transmission error") is the greatest difference of the spacing errors of any pair of adjacent teeth of the pinion and of any pair of mating teeth of the gear.

A corresponding quantity Δ_q is the amplitude of periodic transmission error at tooth-engagement frequency.

The expressions in Table 32.13 may be used for estimating purposes where no more detailed information is available.

Table 32.11. Allowable Total Load per Inch Width

Type of gear	w_{Az}	w_{Ay}
Spur or helical	$\left\{\begin{array}{l} J_{RzP}\sigma_{zbP} \\ \text{or } J_{RzG}\sigma_{zbG} \\ \text{or } 10^{10}\beta/(v_s + 100)(n_p + 10{,}000) + \sigma_{zbP}/100 \\ \text{whichever is least} \end{array}\right\} \kappa d_P/(1 + 1/m_G)$	$\left\{\begin{array}{l} J_{RyP}\sigma_{ybP}Y_P \\ \text{or } J_{RyG}\sigma_{ybG}Y_G \\ \text{whichever is the less} \end{array}\right\} p_n/\pi$ where Y_P and Y_G are from Fig. 32.25
Pinion and internal gear	As above but $m_G = -N_G/N_P$	As above but with special values of Y_P and Y_G
Straight bevel or spiral bevel	As above but with $(1 + 1/m_G)$ replaced by $0.9(A - F)/A[\cos \gamma_P + (\cos \gamma_G)/m_G]$	$[0.9(A - F)/A]$ times what is given for spur and helical gears
Worm wheel	$\left\{\begin{array}{l} J_{RzG}\sigma_{sbG} \\ \text{or } 10^{10}\beta/(v_s + 100)(n_P + 10{,}000) + \sigma_{zbP}/100 \\ \text{whichever is the less} \end{array}\right\} D_G/4$	$J_{RyG}\sigma_{ybG}(Y_G + 3F_G/d_{oP})(p_n/\pi)$
Cross-helical wheel	$\left\{\begin{array}{l} J_{RzG}\sigma_{sbG} \\ \text{or } 10^{10}\beta/(v_s + 100)(n_P + 10{,}000) + \sigma_{zbP}/100 \\ \text{whichever is least} \end{array}\right\} D_G^2/80F_G$	$J_{RyG}\sigma_{ybG}Y_G p_n^2(\cos \psi_G)/5F_G$

NOTE: Values of J_{Rz} and J_{Ry} are obtained from Fig. 32.22.
Values of σ_{yb} and σ_{zb} are in Table 32.7 and of σ_{sb} in Table 32.9.
Values of κ are given in Table 32.12. Values of Y are obtained from Fig. 32.25 (special note about internal gear).

Table 32.12. Zone Factor κ

Grade of accuracy	κ		
	Helical teeth	Straight teeth	
		Max $(V_1 + V_2)$	Min $(V_1 + V_2)$
High precision	0.24	0.15	0.12
High-class commercial	0.22	0.14	0.11
Average commercial	0.20	0.13	0.10

Table 32.13. Approximate Transmission Errors in Millionths of an Inch

Grade of accuracy	Δ_c	Δ_q	
		Straight teeth	Helical teeth
High precision	$200(1 + p_t)$	$400p_n$	$200p_n$
High-class commercial	$400(1 + p_t)$	$600p_n + 0.07w_t$	$400p_n$
Average commercial	$600(1 + p_t)$	$800p_n + 0.07w_t$	$600p_n$

w_t is the load per inch of face width corresponding to the external torque.

(i) Critical Speed

Critical Speed of Spur and Helical Gears. Let

m_i = (internal diameter of the rim of the gear)/D

$B = 1/(1 - m_i^4)$

$G = (1 + m_i^2)^{1/2}$

H = compliance of stressed parts in./lb/in. width

$$= [0.7 \sec \psi + 0.011(G_P + m_G G_G)^2(B_P + B_G/m_G^2)] \times 10^{-6} \tag{32.42}$$

n_{PC} = critical rotational speed, rpm, of pinion

$$= (560/N_P d_P)[(B_P + B_G/m_G^2)/H]^{1/2} \tag{32.43}$$

The dynamic load is determined by use of factors A_C, A_q, and A_L given by Figs. 32.23 and 32.24 for every value of n_P/n_{PC}.

The above expressions for spur and helical gears may be taken to apply also to bevel gears, B and G having been given the values corresponding to the transverse plane at the mid-point of the face width.

Critical Speed of Wormgears. The tangential loads on the two members of a mating pair of crossed-axis gears are, in general, different. In the following details regarding wormgearing, attention is confined to the tangential load on the worm wheel which meshes with the worm in substantially the same way as a helical gear meshes with a rack.

The compliance includes the elements which constitute the compliances of a helical gear and a rack and an additional one associated with the flexibility of the worm shaft. The quantities G_P and B_G in Eq. (32.42) are each unity.

The compliance associated with the flexibility of the worm shaft is

$$H_w = SF_G[8d_P^2 + 1.33S^2(\tan^2 \lambda + \tan^2 \phi_n \sec^2 \lambda)]/\pi E d_{PR}^4 \tag{32.44}$$

where d_P = diameter at mid-depth of worm threads
d_{PR} = root diameter of worm
E = modulus of elasticity for the material of the worm ($E = 30 \times 10^6$ for steel)
S = span of worm-shaft bearings
λ = lead angle of worm at diameter d_P
ϕ_n = normal flank angle of worm threads

In wormgears of average proportions with m_G not less than about 7, the value of this compliance for a steel worm is about

$$H_w = (3.5 + 10 \tan^2 \lambda) \times 10^{-6} \text{ in./lb/in. face width of the worm wheel}$$

The critical worm-shaft speed is

$$n_{PC} = 4.8(d_G/N_P)[F_G(1/m_G^2 I_P + 1/I_G)/(H + H_w)]^{1/2} \quad (32.45)$$

where I = moment of inertia about axis of rotation. For a steel ring of width F and diameters d_1 and d_2 the moment of inertia is

$$I = F(d_1^4 - d_2^4)/14{,}000 \text{ lb in. sec}^2 \quad (32.46)$$

The dynamic load on the worm-wheel teeth per inch width is given in the following paragraphs, but H is replaced by $(H + H_w)$.

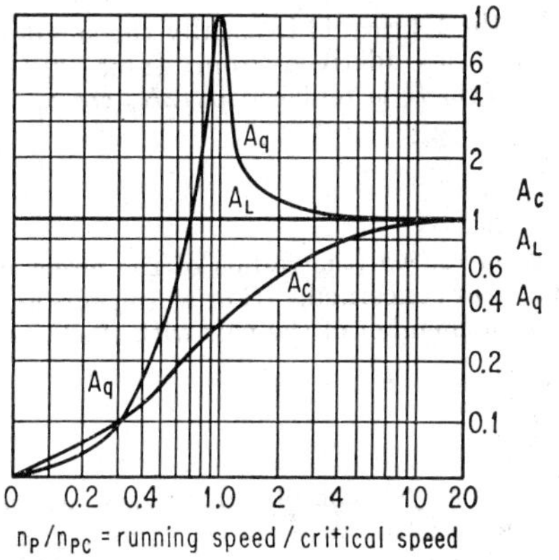

FIG. 32.23. Dynamic-load factors for helical teeth.

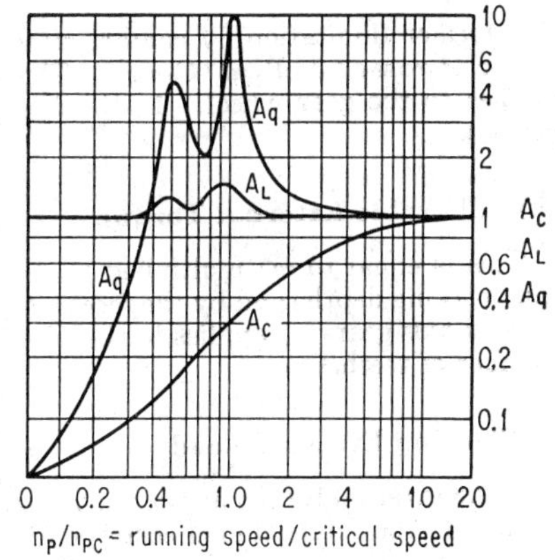

FIG. 32.24. Dynamic-load factors for straight teeth.

Critical Speed of Crossed-helical Gears. For crossed-helical gears the normal compliance per pound of normal tooth load is about

$$H_n = [0.7 + 0.8 \times 10^{-6} d_p^4 \cos^2 \psi_P (1/I_P + m_G^2/I_G) \times (G_P + m_G G_G \cos \psi_P \sec \psi_G)^2] 10^{-6}$$

The critical speed of the pinion shaft is

$$n_{PC} = 4.8(d_p \cos \psi_P/N_P)[(1/I_P + m_G^2/I_G)/H_n]^{1/2} \quad \text{rpm}$$

(j) Dynamic-load Factors A_C, A_L, and A_q

These are determined on the basis of (n_P/n_{PC}) from Fig. 32.23 for helical teeth or Fig. 32.24 for straight teeth.

Transmission errors Δ_c and Δ_q are estimated from Table 32.13.

The factor A_c is associated with the random transmission error Δ_c and rises from zero at zero speed to unity at infinite speed.

The factor A_q is associated with transmission error Δ_q at tooth-engagement frequency. In the absence of damping and backlash it would be infinite at the critical speed. Damping limits A_q to about 10. Tooth separation (if the teeth have the normal amount of backlash) limits this dynamic load to an amount equal to the mean load in the case of helical teeth (for which A_L accordingly is equal to unity) or to about 1½ times the mean load in the case of straight teeth, with a lower peak at half the critical speed. So A_L for straight teeth varies with the speed.

The allowable load per inch width w_A corresponding to external torque (applies to worm wheel in wormgears) is given by

$$w_A = (w_{gA} - A_c\Delta_c/H)/(1 + A_L) \quad \text{or} \quad (w_{gA} - A_c\Delta_c/H - A_q\Delta_q/H) \qquad (32.47)$$

whichever is the greater.

This suffices for most purposes, but if further investigation is necessary, then w_A may be defined in more detail by stating that it is the smaller of the quantities w_{tz} and w_{ty} whose values are given by

$$w_{tz} = (w_{Az} - A_c\Delta_c/H)/(1 + A_L) \quad \text{or} \quad (w_{Az} - A_c\Delta_c/H - A_q\Delta_q/H)$$

whichever is the greater

and $$w_{ty} = (w_{Ay} - A_c\Delta_c/H)/(1 + A_L) \quad \text{or} \quad (w_{Ay} - A_c\Delta_c/H - A_q\Delta_q/H)$$

whichever is the greater

The former (limitation by contact stress or by the lubricant) is usually the smaller.

For Crossed-helical Gears. The quantity w_A (derived from Table 32.11) applies to the gear wheel. In the formulas above, H is replaced by $2H_n(\cos\psi_P + \cos\psi_G)$.

(k) Allowable External Torque

This is equal to w_A multiplied by the *meshing radius* (the radius of the circle at mid working depth) and by the *face width* of the gear concerned. (For wormgears and crossed-axis gears w_A applies to the wheel, i.e., the member of the pair with the larger number of teeth.)

(l) Varying External Load

If the external torque applied to a gear varies in a cycle completed in every revolution, then the external load corresponds to the maximum torque in the cycle because the same teeth are subjected to that torque in every revolution. (This may not be true of the mating gear, but the effect of this may be ignored.)

If the external torque applied to a gear varies in a cycle composed of maximum torque m_0 for fraction b_0 of the cycle and torques m_1, m_2, etc., for fractions b_1, b_2, etc., of the cycle then the duty may be taken as equivalent to the application of the maximum torque m_0 for a fraction of the total number of load cycles equal to

$$b_0 + (m_1/m_0)^3 b_1 + (m_2/m_0)^3 b_2 + \cdots \qquad (32.47a)$$

For an idler gear, in which each tooth is subjected to alternating loads in opposite directions, the number of load cycles is equal to the number of revolutions, but because the bending stresses are reversed in each cycle, the basic allowable bending stress is 0.67 times that given in Table 32.7.

32.11. ESTIMATING MAJOR DIMENSIONS OF GEARS

Preceding articles show how to estimate the probable life of a pair of gears of known dimensions subjected to known loading.

The converse operation, that of fixing dimensions of gears so that they may perform a specified duty, is not straightforward inasmuch as every problem of this nature has many different and equally good solutions. In principle a set of dimensions must be assumed and the expected life of such gears estimated. If the result is not satisfactory a modified set of dimensions must be assumed and the process repeated.

To afford some guidance for the first assumed set of dimensions, some *approximate* formulas are given below on the basis of a life expectation of 26,000 hr.

For small adjustments in load capacity, the linear dimensions of geometrically similar gears are approximately proportional to the cube root of the load capacity.

The recommendations concerning face width are shown in Table 32.14.

Table 32.14. Recommendations about Face Width

Type of gear	Face width		
	First approximation	Upper limit	Lower limit
Spur	$2dm_G/(m_G + 1)$	$2d$	$D/20$
Single helical	$2dm_G/(m_G + 1)$	$2d$	p_x and $D(1 + 2 \tan \psi)/20$
Double helical	$2dm_G/(m_G + 1)$	$2.5d$	$2p_x$ and $D/20$
Crossed helical	....	....	$3.5p_n \sin \psi$ and $D(1 + 2 \tan \psi)/20$
Spur or helical pinion and internal	$0.6d$	$0.8d$	p_x and $D(1 + 2 \tan \psi)/20$
Worm	....	$6p_x$	$4.5p_n$ and $4p_x$
Worm wheel	....	$p_n + 0.33d_0$	$lp_n/1.7d_R$ and $D_t(1 + l/4d_R)/12$
Bevel	$0.3A$	$0.3A$	$2p_t$
Spiral bevel and hypoid	$0.3A$	$0.3A$	$p_n \csc \psi$

Helical Gears in 1060/1045 Steel. Allowable torque (lb in.) on the pinion shaft is

$$200d_P^{1.75}Fm_G/(1 + m_G) \log_{10} (10 + n_P) \quad (32.48)$$

If, for example, it is decided that $F = 2d_Pm_G/(1 + m_G)$, the value of d_P necessary for a pinion-shaft torque T_P lb in. is about

$$d_P = [(T_P/400\, m_G^2)(1 + m_G)^2 \log_{10} (10 + n_P)]^{1/2.75} \quad (32.49)$$

This diameter applies to the combination of 1060 steel for the pinion and 1045 steel for the gear. If this diameter, or the associated dimensions, are unacceptably large in a ratio S, the necessary correction can be applied by using instead of 1060/1045 steel the combination in Table 32.8 for which the relative load capacity is the next higher than $S^{2.75}$.

Alternatively, if S is less than about 1.5, the desired increase in load capacity may be attained by multiplying F by S, without change in diameters, but F should not exceed about $2.2d_P$.

Spur Gears in 1060/1045 Steel. The procedure for helical gears applies equally to spur gears except that 200 is to be replaced by 150 in Eq. (32.48) and 400 by 300 in Eq. (32.49).

Spiral Bevel Gears (for 90° Shaft Angle). For 1060/1045 steel gears, with face width equal to 0.3 times the cone distance, the value of d_P necessary for a pinion-shaft torque T_P lb in. is about

$$d_P = \{[T_P/(22\, m_G - 5.5)] \log_{10} (10 + n_P)\}^{1/2.75} \quad (32.50)$$

As for helical and spur gears, the corresponding value for other material combinations may be estimated by dividing T_P by the "relative load capacity" factor in Table 32.8.

For Straight-tooth Bevel Gears. Proceed as for spiral bevel gears, but multiply 5.5 by 0.7.

Wormgears (Steel/Bronze). For a casehardened-steel worm meshing with a phosphor-bronze wheel of width equal to 0.16 times the meshing diameter d_G, the value of d_G necessary for a worm-wheel shaft torque of T_G lb in. with n_P between 10 and 10,000 is about

$$d_G = [T_G/(85 - 25/m_G - 19 \log_{10} n_P)]^{1/2.5} \tag{32.51}$$

NOTE: For this preliminary purpose, a torque T_1 for a period of H_1 hr may be taken as equivalent to a torque of about $T_1[(1{,}000 + H_1)/27{,}000]^{1/3}$ for 26,000 hr.

32.12. SERVICE FACTORS

Very often the gear designer is given the nominal power capacity of the motor or engine that drives the gears and economic design of the gears requires him to discover or to assume for what proportions of the life of the gears they will actually be required to transmit the power or certain fractions of the power.

To this end he may employ "service factors," some of which represent little more than guided guesswork. With persistence, it is usually possible to evaluate any particular duty rationally, but it is certainly easier to use a "service factor" and it can often be done without serious disadvantage.

32.13. EXAMPLES

(a) Design of Helical Gears

Let it be required to determine dimensions of a pair of helical gears to perform for 5 years a duty involving the transmission of

1,000 hp for 20 per cent of the time
800 hp for 30 per cent of the time
400 hp for 30 per cent of the time
Zero for 20 per cent of the time

while reducing speed from 6,000 to 1,200 rpm.

The total numbers of revolutions made during the 5 years are

$$\text{Pinion} = 5 \times 365 \times 24 \times 60 \times 6{,}000 = 1.6 \times 10^{10}$$
$$\text{Wheel} = 0.32 \times 10^{10}$$

The equivalent numbers of load cycles may be estimated by applying Eq. (32.47*a*) to the nominal tooth loading in the first instance, although it must later be applied to the sum of nominal tooth load and dynamic tooth load.

Equivalent number of load cycles of pinion at 1,000 hp is

$$1.6 \times 10^{10}[0.2 + (800/1{,}000)^3 \times 0.3 + (400/1{,}000)^3 \times 0.3 + 0] = 0.6 \times 10^{10}$$

Equivalent number of load cycles of the wheel is

$$0.6 \times 10^{10} \times 1{,}200/6{,}000 = 0.12 \times 10^{10}$$

Each of these is equivalent to a running time of $0.6 \times 10^{10}/6{,}000 \times 60 = 16{,}700$ hr.

The note at the end of Art. 32.11 shows that the equivalent power for 26,000 hr is about $1{,}000[(1{,}000 + 16{,}700)/27{,}000]^{1/3} = 870$ hp.

The pinion shaft torque corresponding to this is

$$T_P = (870/6{,}000)63{,}000 = 9{,}150 \text{ lb in.}$$

Hence from Eq. (32.49) the necessary diameter of pinion for 1060/1045 steel helical gears with face width equal to $2 \times 5/(1 + 5) = 1.67$ times the pinion diameter is about $d_P = [(9{,}150/400 \times 5^2)6^2 \log_{10} 6{,}010]^{1/2.75} = 5.75$. Then

$$F = 2 \times 5.75 \times 5/(1 + 5) = 9.6$$

Hence $d_P{}^2 n_P = 5.75^2 \times 6{,}000 = 200{,}000$, and so from Table 32.5 the minimum desirable value of N_P if the teeth are to be cut by hobbing is $200{,}000/10{,}000 = 20$. Also, from the note following Table 32.5, N_P should lie outside the range

$$450{,}000/5.75 \times 6{,}000 = 13 \text{ to } 700{,}000/5.75 \times 6{,}000 = 20$$

Hence take N_P not less than 20.

Assuming helix angle $\psi = 30°$ the value of p_n corresponding to $d_P = 5.75$ is

$$(5.75\pi \cos 30°)/N_P = 15.6/N_P$$

If $N_P = 25$ the value of p_n is 0.625 and $p_t = 0.72$.

(b) Calculation of Load Capacity

Assuming the "high-precision" grade of manufacture we find from Table 32.13 that

$$\Delta_c = 200(1 + 0.72) = 345 \times 10^{-6} \text{ in.}$$
$$\Delta_q = 200 \times 0.625 = 125 \times 10^{-6} \text{ in.}$$

Assuming the gears to be in the form of solid cylinders the "internal diameter of rim" for Eq. (32.42) is zero for both gears and so

$$B_P = B_G = 1 \qquad \text{and} \qquad G_P = G_G = 1$$

From Eq. (32.42),

$$H = [0.7 \sec 30° + 0.011(1 + 5)^2(1 + 1/5^2)] \times 10^{-6} = 0.88 \times 10^{-6}$$

From Eq. (32.43),

$$n_{PC} = (560/25 \times 5.75)[(1 + 1/5^2)/0.88 \times 10^{-6}]^{1/2} = 4{,}250 \text{ rpm}$$

Hence $n_P/n_{PC} = 6{,}000/4{,}250 = 1.41$. From Fig. 32.24,

$$A_c = 0.4 \qquad A_q = 2$$

The maximum sliding speed is about

$$0.22 V p_n (n_P + n_G)$$

where V is the greater of V_P and V_G. Taking V_P as 2.5, this gives

$$v_s = 0.22 \times 2.5 \times 0.625 \times 7{,}200 = 2{,}500 \text{ fpm}$$

From Table 32.7,

For 1060 steel, $\sigma_{zb} = 2{,}100$, $\sigma_{yb} = 21{,}000$.

For 1045 steel, $\sigma_{zb} = 1{,}450$, $\sigma_{yb} = 18{,}000$.

From Table 32.10 for unhardened steel/unhardened steel very smoothly finished and lubricated with antiscuffing oil for the running-in period, the value of β is 3. Hence, from Table 32.11, we find for helical gears that a limitation in connection with w_{Az} is $10^{10} \times 3/(2{,}500 + 100)(6{,}000 + 10{,}000) + 2{,}100/100 = 742$.

From Fig. 32.22, we find for 0.6×10^{10} cycles and 0.12×10^{10} cycles

$$J_{RzP} = 0.39 \qquad J_{RzG} = 0.45$$

Hence

$$J_{RzP}\sigma_{zbP} = 0.39 \times 2{,}100 = 820$$
$$J_{RzG}\sigma_{zbG} = 0.45 \times 1{,}450 = 650$$

Of the three relevant quantities the least is 650.

From Table 32.12, we find $\kappa = 0.24$ for high-precision helical gears. Hence, from Table 32.12,

$$w_{Az} = 650 \times 0.24 \times 5.75 \times 5/6 = 745$$

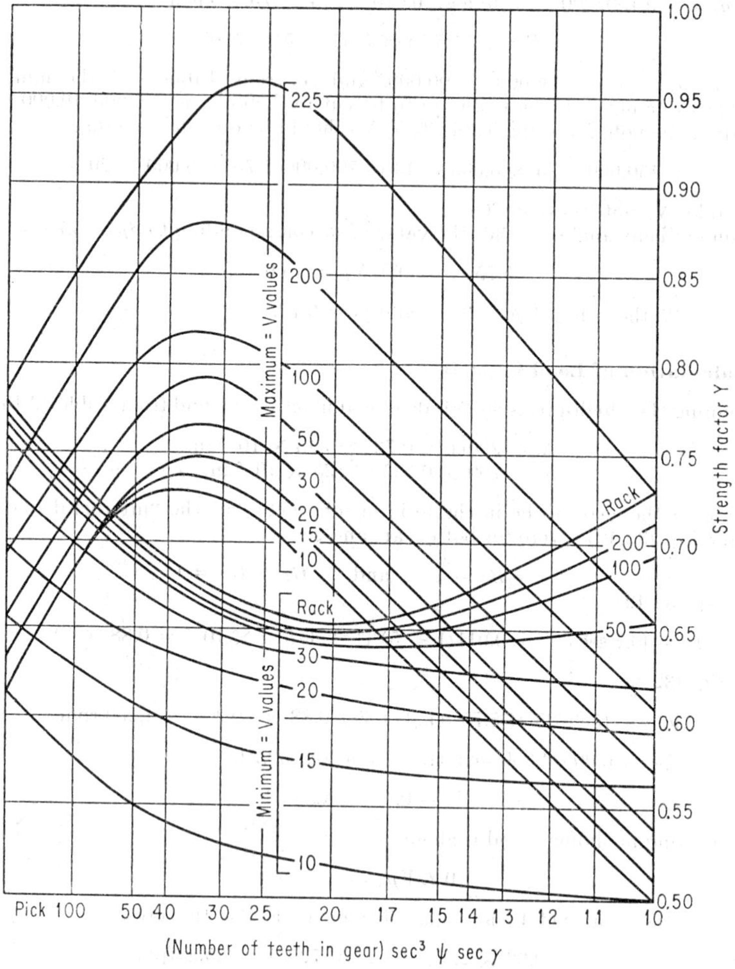

FIG. 32.25. Strength factors. For a pinion/(internal) gear combination, the strength factors are $1 + (2N_P/N_G)^2$ times as great as those of the combination of the same pinion with a rack.

To evaluate w_{Ay} we find from Fig. 32.22 that

$$J_{RyP} = 0.8 \qquad J_{RyG} = 0.8$$

The equivalent numbers of spur-gear teeth are

$$25 \sec^3 30° = 38 \qquad \text{and} \qquad 125 \sec^3 30° = 190$$

From Fig. 32.25, the strength factors for teeth with maximum V values are

$$Y_P = 0.86 \qquad \text{and} \qquad Y_G = 0.7$$

For minimum V values

$$Y_P = 0.67 \qquad \text{and} \qquad Y_G = 0.72$$

Taking the mean values,

$$Y_P = 0.76 \quad \text{and} \quad Y_G = 0.71$$

The quantities in Table 32.11 to be compared in estimating w_{A_y} are therefore

$$0.8 \times 21{,}000 = 16{,}800 \quad \text{and} \quad 0.8 \times 18{,}000 = 14{,}400$$

Hence $$w_{A_y} = 14{,}400 \times 0.625/\pi = 2{,}860$$

Hence w_{gA} (the lesser of w_{A_z} and w_{A_y}) is equal to 745.

Now $$A_c\Delta_c/H = 0.4 \times 345/0.88 = 157$$
$$A_q\Delta_q/H = 2 \times 125/0.88 = 284$$

For helical teeth $A_L = 1$. From Eq. (32.47), $w_A = (745 - 157)/2 = 294$ or $(745 - 157 - 284) = 304$, whichever is greater, i.e., $w_A = 304$.

Allowable external torque $= w_A(\frac{1}{2}d)F = 304 \times \frac{1}{2} \times 5.75 \times 9.6 = 8{,}700$ lb in., whereas 9,150 lb in. is required, and increase of the face width to 10 in. suffices to achieve this.

Hence the duty may be performed by helical gears of the following dimensions:

	Pinion	Gear
Mean diam.	5.75	28.8
Working face width	10	
Normal pitch p_n	0.625	
Helix angle ψ	30°	
Axial pitch (p_n csc ψ)	1.25	
Numbers of teeth	25	125
Center distance	17.3 (approx)	
Material of pinion	1060 steel normalized	
Material of gear	1045 steel normalized	

32.14. POWER LOSS IN GEAR TRAINS

In any pair of gears transmitting power there is a loss of power associated with friction between the teeth, friction in the bearings, turbulence in the oil, and windage.

(a) Friction between the Teeth

This loss may be expressed as the product of the tangential load on the teeth of the driven gear, the speed of the teeth of that gear relative to a common perpendicular to the axes, and a "loss factor" L.

1. For spur and helical gears with parallel axes,

$$L_a = (1/N_P + 1/N_G)/5 \sec \psi \tag{32.52}$$

(for an internal gear, N_G is negative).

2. For bevel gears,

$$L_d = [(\cos \gamma_P)/N_P + (\cos \gamma_G)/N_G]/5 \sec \psi \tag{32.53}$$

3. For crossed-axis gears,

$$L_c = \mu v_s/(\text{speed of driven teeth relative to perpendicular to axes}) \tag{32.54}$$

where μ is the coefficient of friction for the materials concerned at sliding speed v_s.

For values of v_s up to 10,000:

Steel/bronze: $\mu_1 = 0.1 - 0.04 (\log_{10} v_s)^{1/2}$ (approx.)
Cast iron/bronze: $\mu = 1.2\mu_1$
Cast iron/cast iron: $\mu = 1.35\mu_1$
Steel/steel: $\mu = 2\mu_1$

(b) Friction in the Bearings

The frictional torque on a sleeve bearing of normal proportions under normal loading at ordinary speeds is about

$$(\text{Journal load})(\text{journal diam})/400$$

This also applies (approximately) to a ball bearing or roller bearing if for "journal diam" is substituted the diameter of the smaller running track.

(c) Turbulence in the Oil

The power loss by turbulence in an oil bath cannot be usefully evaluated by any simple formula. It may be assumed to be of the same order as the power loss in the bearings.

(d) Windage

The windage loss is negligible at all ordinary speeds.

The power loss in oil sprayed onto the teeth is about

$$[\text{Weight (lb) of oil fed per min}](v_{fm}/60{,}000)^2\text{hp}$$

where v_{fm} is the speed, fpm, of the teeth relative to a common perpendicular to the axes.

The number of pounds per minute of oil fed is usually about five times the number of horsepower lost by friction.

(e) Temperature Rise of Oil in Gearbox

The power dissipated by friction and oil turbulence in a gear unit is converted to heat which raises the temperatures of the oil and the metal until the rate of heat dissipation from the surface of the gearbox is equal to the rate of heat generation.

In normal circumstances the oil temperature may be allowed to reach 200°F. Temperatures higher than 200°F tend to raise questions about the chemical stability of the oil, and this point should be discussed with the manufacturer of the oil.

Heat dissipation from the outer surfaces of a gearbox corresponds to about 0.0013 hp/ft^2 of exposed area/°F of difference in temperature between the surface of the gearbox and the local atmosphere.

If the power transmitted by a gear unit is large in relation to its size, natural cooling may be inadequate to avoid excessive temperature rise. In such a case, fan cooling of the gearbox may suffice to keep the temperature of the oil within safe limits. Otherwise artificial cooling may be introduced by (1) giving the gearbox radiating fins, (2) cooling the gearbox by a fan, (3) cooling the gearbox by an internal pipe conveying cold water, or (4) pumping oil to and from a separate oil tank with or without a cooling element in the tank.

32.15. BEARING LOADS

Accuracy of manufacture and meshing is the most important single factor in the performance of gears in service. Rigidity of gears, shaft bearings, and gearboxes is therefore important, and with this in mind, rational design of mountings demands an estimate of the loads imposed by gears on the bearings that support them.

Assuming the ideal uniform distribution of load across the width of the teeth of a gear, bearing loads are satisfactorily estimated on the assumption that the load corresponding to the external torque on a spur, helical, or bevel gear is concentrated at a point at the mid working depth of the tooth, at the mid-plane of the width of the teeth, and in the common plane of the shaft axes. In crossed-helical gears the point may be considered to be on the common perpendicular to the axes.

If the point P is distant $\frac{1}{2}d$ from the axis of the gear and the external torque is T, then, for a helical gear, through the point P there are forces (1) $2T/d$ perpendicular to the axis and to the perpendicular from P onto it, (2) $(2T/d) \tan \psi$ parallel to the axis, and (3) $(2T/d) \tan \phi_n \sec \psi$ perpendicular to both (1) and (2).

For a bevel gear, the forces are (1) $2T/d$ as before, (2) $(2T/d) \tan \psi$ in a direction inclined at angle γ to the axis, and in the common plane of the axes of the gears, and (3) $(2T/d) \tan \phi_n \sec \psi$ in a direction inclined at angle $(90° - \gamma)$ to the axis, and in the common plane of the axes of the gears.

Application of the principles of elementary mechanics shows the directions in which these forces act and it is then a simple matter to calculate the bearing loads corresponding to them. A point to be noted particularly is that the force b on a helical gear not only induces an axial bearing load (i.e., an "end thrust") of the same amount but also induces equal and opposite journal loads equal in magnitude to ($d/2 \times$ bearing span) times itself.

This last fact suggests that the journal bearings for a gear that is narrow in relation to its diameter (e.g., a worm wheel) should not necessarily be placed as close as possible to it. Apart from this case and that of an "overhung" gear, bearings should be placed as close as possible to the gear that they support.

32.16. EPICYCLIC GEAR TRAINS

An epicyclic gear train is one in which at least one common perpendicular to the axes of a pair of gears has rotation relative to the observer. A simple epicyclic gear train is shown in Fig. 32.26 in which O is the center of the fixed gear C, O' is the center of gear B, and A is the arm which connects B to C. This mechanism is termed "epicyclic" because a point on gear B will describe an epicyclic curve as B rotates about fixed gear C.

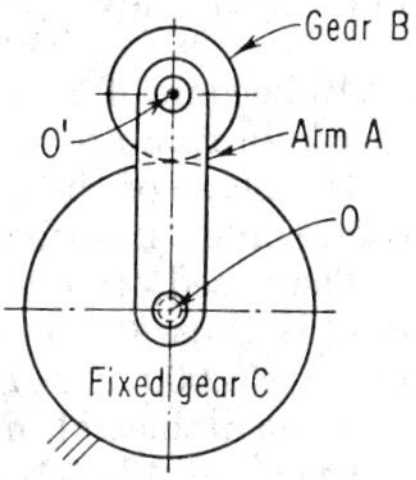

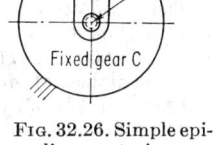

FIG. 32.26. Simple epicyclic gear train.

In order to determine the number of revolutions of B about O' for each revolution of A about O, the motion may be divided as follows:

1. Assume C is free to rotate and let A, B, and C be welded together as the entire assembly makes one positive (clockwise) revolution about O.

2. With arm A temporarily fixed, let gear C make one revolution counterclockwise (negative), thus making the net motion of C about O, for steps 1 and 2, zero. Gear B, which is free to move about O', will therefore rotate C/B revolutions clockwise.

By adding the motions of each member as shown in Table 32.15, total revolutions may be obtained.

Table 32.15. Number of Revolutions for a Simple Epicyclic Gear Train Shown in Fig. 32.26

	Motion of A	Motion of B	Motion of C
Step 1 (motion with arm A)	+1	+1	+1
Step 2 (motion relative to arm A)	+0	$+C/B$	−1
Total motion	+1	$1 + C/B$	0

Table 32.16. Number of Revolutions for a Compound Epicyclic Gear Train Shown in Fig. 32.27

	A	B	C	D	F
Step 1 (motion with arm A)......	+1	+1	+1	+1	+1
Step 2 (motion relative to arm A)..	0	$+F/D$	$-(F/D)(B/C)$	$+F/D$	−1
Total motion..................	+1	$1+F/D$	$1-FB/DC$	$1+F/D$	0

Thus the ratio of the angular velocity of gear B to A is $(1 + C/B)/1$ and the motion of B is in the same direction as A, as indicated by the algebraic sign of the velocity ratio. If an idler gear is interposed between B and C or if C is an internal gear, the velocity ratio will be negative; that is, B will rotate oppositely to A.

For the compound epicyclic gear train shown in Fig. 32.27 the procedure is similar and Table 32.16 is used to determine the total motions of the members.

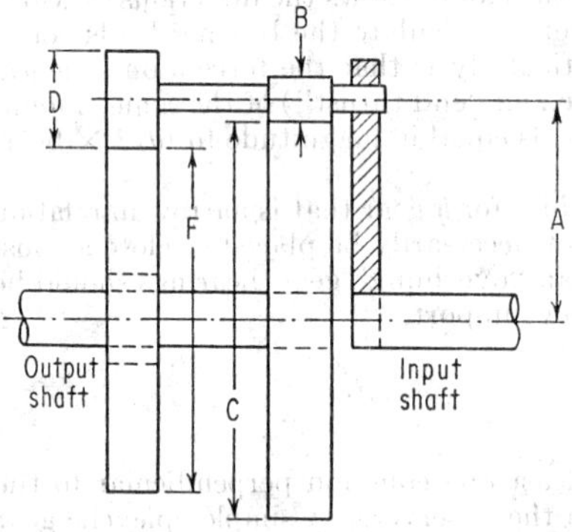

FIG. 32.27. Epicyclic gear train.

32.17. POWER LOSSES IN EPICYCLIC GEAR TRAINS

The following example will serve to demonstrate the importance of power-loss considerations in epicyclic gear trains. Referring to Fig. 32.27, consider a gear train described by the following data: $A = 7$ in., $B = 3$ in., $C = 11$ in., $D = 4$ in., $F = 10$ in., horsepower to be transmitted = 1, input shaft speed = 1,200 rpm. The values in the last line of Table 32.16 become $A = 1$, $B = 3.5$, $C = 0.318$, $D = 3.5$, and $F = 0$. The corresponding speeds in rpm are 1,200, 4,200, 382, 4,200, and 0.

The plane of the axes of gears B and C is fixed in relation to A, which runs at 1,200 rpm, while the speed of B (and D) is 4,200 rpm and that of C is 382 rpm.

The speed of B relative to the plane that contains its axis and that of the mating gear is thus $4{,}200 - 1{,}200 = 3{,}000$ rpm, and the speed of the teeth through that plane is $3{,}000 \times 3\pi/12 = 2{,}350$ fpm.

If the diametral pitch of the teeth is 10, there are 30 teeth in B and 110 in C.

From Eq. (32.52) the loss factor is therefore $(1/30 + 1/110)/5 = 1/118$.

The tooth load on B is equal to

$$\text{Output torque/radius of } C = (1/382) \times 63{,}000/5.5 = 30 \text{ lb}$$

Hence the power loss in tooth friction between B and C is

$$30 \times 2{,}350 \times (1/118) \times 1/33{,}000 = 0.018 \text{ hp}$$

By similar procedure the power loss in tooth friction between D and F is found to be 0.015 hp, giving a total of 0.033 hp, i.e., a loss equal to about 3.3 per cent of the input power. To this must be added the losses in oil turbulence and bearing friction.

PRECISION GEARING*

Introduction. The need for dimensional accuracy in gear trains increases as pitch-line velocities become greater and as requirements for angle-true motion become more severe. Of these two, the problem of wear and strength failure due to high

* By George W. Michalec.

pitch-line velocities is treated in the foregoing portions of this section. Here consideration is limited to evaluating errors in output motions of gear systems. The problem of accurate transformation of motion is of concern principally in the application of small, lightly loaded, fine-pitch gearing to controls and instruments. The design of this type of gearing demands an approach essentially different from that required by power gearing. Accordingly, the concern here is on a survey of factors influencing the ability of gear trains to transmit accurate motion, on the recognition of error sources and how their limits are specified, and on the presentation of formulas suitable for design calculations.

32.18. REFERENCE INFORMATION

(a) Notation

Letter symbols found in the text are listed below. Units assigned to each symbol satisfy the needs for consistency and utility. However, as other systems of units are possible, due allowance must be given when reference is made to other works. Besides the subscripts appearing in the list below, additional subscripts are used throughout the text: P and G denote pinion and gear; 1, 2, . . . , j, . . . , n indicate the number position of a mesh in a train of n meshes.

B = backlash, in.
B_A = backlash allowance, in.
B_V = variable backlash, or backlash tolerance, in.
C = center distance, in.
D = pitch diameter, in.
E = gear or gear-mounting error
E_p = position error, rad
E_r = runout error of mounting components, in.
E_{si} = supplementary intermediate-frequency error, in.
E_{tc} = total composite error, in.
E_{tt} = tooth-to-tooth composite error, in.
e = eccentricity, in.
K_m = change factor for measurement over pins, dimensionless
M = train ratio, dimensionless
m = mesh ratio, dimensionless
N = number of teeth, dimensionless
N_c = undercut number, dimensionless
P = pitch
P_d = diametral pitch
P_c = circular pitch
R = pitch radius, in.
t = tolerance
t_R = tolerance of meshing radius, in.
t_C = tolerance of center distance, in.
γ = phase angle, rad
Δr = change in meshing radius, in.
ΔM = change in measurement over pins, in.
δ = deviation from a nominal dimension, in.
δ_R = deviation from nominal pitch radius, in.
δ_C = center distance allowance, in.
ϵ = mesh or train error, rad
ϵ_{bc} = constant-backlash mesh error, rad
ϵ_{bv} = variable-backlash mesh error, rad
ϵ_{tr} = transmission mesh error, rad
ϵ_T = train error, rad
$\epsilon_{T_{bc}}$ = constant-backlash component of train error, rad
$\epsilon_{T_{bv}}$ = variable-backlash component of train error, rad
$\epsilon_{T_{tr}}$ = transmission component of train error, rad
θ = shaft angle, rad

θ_i = input shaft angle, rad
θ_o = output shaft angle, rad
θ_w = shaft angle for worst-error position radian
φ = pressure angle, degrees

(b) Terminology

A general knowledge of elementary gear theory is prerequisite to the understanding of this article; hence no presentation of basic nomenclature is offered. However, a few terms need definition, if only for the purpose of avoiding possible ambiguous interpretation, and to make the text self-contained. Terms peculiar to this article are defined where they are introduced. Figure 32.28 illustrates most of the common gear dimensions referred to in the text. Diametral pitch is one term that cannot be given geometrical interpretation, although it is an index of tooth size. Diametral pitch equals the number of teeth per inch of pitch diameter $P_d = N/D$. Circular pitch is another index of gear size, but less commonly used; it is equal to the spacing between adjacent tooth profiles measured linearly along the pitch circle, $P_c = \pi D/N$. Diametral and circular pitch are related by the formula $P_c = \pi/P_d$.

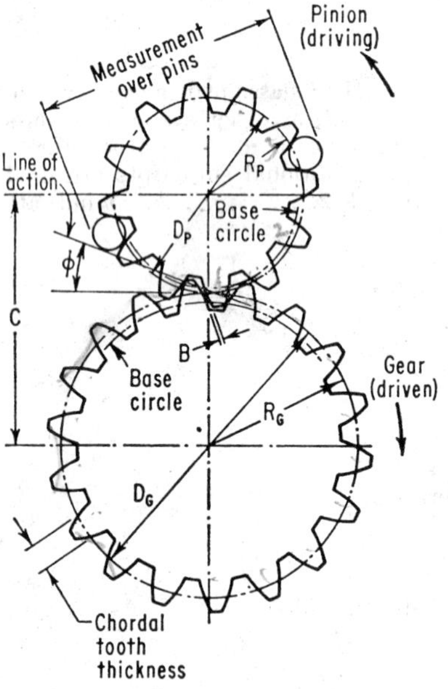

FIG. 32.28. Gear symbols and terms.

(c) Design Standards

The main sources of information for calculating gear dimensions, wear and strength, etc., are the publications of the American Standards Association (ASA) and the American Gear Manufacturers Association (AGMA). Table 32.17 lists standards especially important for designers of precision gear trains. Note that some standards have both an ASA and an AGMA number and may therefore be obtained from either organization.

Table 32.17. Gear Standards

Title	ASA No.	AGMA No.
Tooth Proportions for Coarse-pitch Involute Spur Gears	....	201.02
Zerol Bevel Gear System	....	202.02
Fine Pitch Straight Bevel Gears	B6.8-1950	206.03
20-degree Involute Fine-pitch System for Spur and Helical Gears	B6.7-1956	207.04
System for Straight Bevel Gears	B6.13-1955	208.01
System for Spiral Bevel Gears	....	209.02
Design for Fine-pitch Wormgearing	B6.9-1956	374.03
Gear Specification Manual for Spur, Helical, and Herringbone Gears	....	390.01
Inspection of Coarse-pitch Spur and Helical Gears	....	231.03
Pin Measurement Tables for Involute Spur Gears	....	231.51
Inspection of Fine Pitch Gears	B6.11-1956	236.04
Gear Tolerances and Inspection	B6.6-1946	

(d) Gear-specification Drawings

Where practical, gear-specification drawings should conform to a standard format. Figure 32.29 shows a specimen drawing of a fine-pitch precision spur gear. Other recommendations for drawing formats will also be found in the standards of the ASA and AGMA.

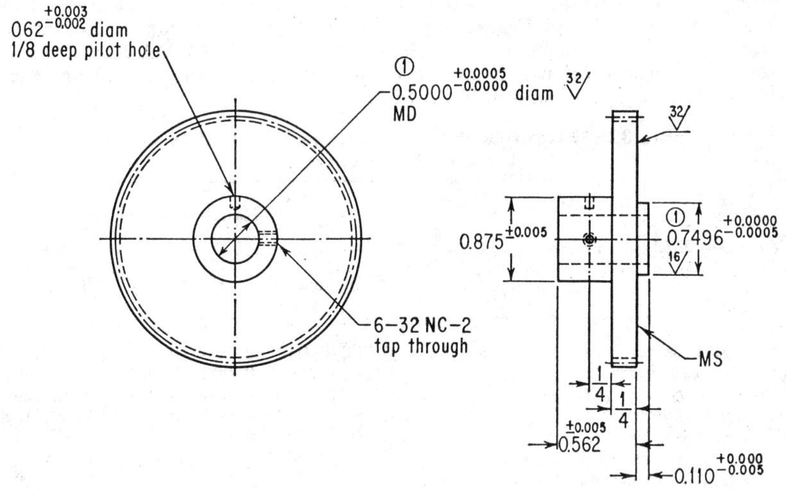

Fig. 32.29. Specimen drawing of a light-duty fine-pitch spur gear.

1. Diameters marked ① to be concentric within 0.0005 TIR.
2. Lateral runout of registering surface *RS* not to exceed 0.001 TIR.
Convexity of *RS* not to exceed 0.0005.
Concavity of *RS* not to exceed 0.001.
3. Surface finish[63]; RMS unless otherwise stated.
4. Tolerance on fractional dimensions ± 1/64.
5. Material, AISI 416 CRS. Hardness, 28 to 33 Rockwell C. Finish, clear passivate

Spur-gear Data

No. of teeth	80	
Diametral pitch	32	
Pressure angle	20°	
Pitch diam (standard nom.)	2.5000	
Tooth form	ASA B6.7-1950	
Addendum	0.0313	
Dedendum	0.0395	
Clearance	0.0083	
Outside diam	2.562	+0.000 −0.002
Max outside diam runout	0.0015	
Measurement over 2 pins	2.5750	+0.0000 −0.0018
Diam of pins	0.0540	
Testing radius	1.2493	+0.0000 −0.0012
Max total composite error	0.0005	
Max tooth-to-tooth composite error	0.0003	
AGMA quality No	12-C	
Surface finish of active profile	16 microin. max RMS	

32.19. TYPES OF GEARS AND FACTORS AFFECTING THEIR CHOICE

(a) Classifications

A simple scheme for classifying gears consists of three groupings: commercial, precision, and ultraprecision. Gears with tolerances measured in thousandths and ten-thousandths of an inch fall into the precision classifications. Table 32.18 shows comparative tolerance ranges for these three groupings. The system of classes defined in AGMA 390.01 for fine-pitch gearing is considerably more detailed. Also, extensive use is still made of the former system of classification given in AGMA 236.04. A comparison of the classification system used in this article with the current and former AGMA systems is to be found in Table 32.19.

Table 32.18. Comparative Tolerances for Quality Grades
Values in inches except as noted

Measurement	Range of values		
	Commercial	Precision	Ultraprecision
Tooth thickness (unilateral)	0.002–0.010	0.0005–0.002	0.0002 max
Total composite error (max)	0.0015–0.010	0.0003–0.0015	0.00025 max
Tooth-to-tooth composite error (max)	0.001–0.005	0.0002–0.001	0.0002 max
Linear position error	0.001–0.005	0.0005–0.001	0.0001–0.0005
Profile deviation (±)	0.001–0.005	0.0005 max	0.0001 max
Bore or journal diameter	0.001–0.005	0.0005 max	0.0002 max
Tooth surface finish (rms, μ in.)	63–250	8–32	16 max

Table 32.19. Approximate Correspondence of Gear Precision Classifications

In this article	AGMA 390.01 class	AGMA 236.04
Commercial	3	
	4	
	5 or 6	Commercial 1
	6 or 7	Commercial 2
	8	Commercial 3
	9	Commercial 4
Precision	10 or 11	Precision 1
	12	Precision 2
	13 or 14	Precision 3
Ultraprecision	15	
	16	

Although precision gears are made in all pitches and sizes, most are fine pitch and for light duty. Nevertheless, many coarse-pitch power gears are made to precision specifications.

(b) Types of Gears

Spur gears offer the greatest opportunity for precision because their tooth form is a basic involute, the pressure angle constant, and tooth dimensions identical in all planes of rotation. With these simple characteristics specification and control in fabrication can be direct and exact; complete and careful inspection is possible. The simple, straight spur gear should be used where maximum precision is desired.

Helical gears are second to spur gears in simplicity and attainable precision. They are equivalent to spur gears except for the additional complication of the helix angle, which is a further source of error in manufacture and inspection and results in helical gears being more sensitive to mounting inaccuracies than spur gears. Except where power and speeds are high, most precision-gear applications do not require the special features of the helical gear.

Crossed-helical gears, sometimes referred to as spiral gears, are used on nonparallel, nonintersecting shafts. Although the precision of crossed-helical gears can be identical with that of helical gears, the resulting quality of operation will be lower because of heavy sliding on point contact. For this reason, crossed-helical gears are not recommended for precision power transmission.

Worms and wormgears are less suitable for precision applications than spur or helical gears. However, worms can be made to high precision as they have simple straight-sided tooth forms which can be accurately produced by thread milling or grinding. Inspection and control of the simple tooth form and tooth spacing can be done to high accuracy. On the other hand, because the wormgear tooth form is not an involute it cannot be checked with readily available standard master gears. Since the wormgear has a self-generated conjugate-type gear form, its accuracy is directly related to the hob used in its fabrication. The gear pair will therefore be conjugate only if the worm is identical to the hob that generated the wormgear. Moreover, conjugate action is possible only if the center distance of the pair is identical to that used in generating the wormgear. Axial alignment of the worm along the wormgear axis must be perfect for true conjugate action. In spite of these difficulties, high load capacity, high reduction, and suitability for right-angle transmission warrant the use of worm meshes in certain high-precision applications. These include gear-generating machines, indexing tables, and optical devices. However, worm meshes cannot generally compete with spur gears for precision and so should be avoided unless their special features outweigh inherent disadvantages.

Bevel gears have even greater precision limitations than wormgears because both members of a pair may be equally inexact. Unlike involute gearing, the tooth form is conjugate for the ideal cone distance only. The tapered tooth form makes measurement and control difficult. There are no standard masters for inspection. Because bevels must be developed in matched pairs, individual gears will not be interchangeable.

Special gear types such as face, side-worm, and tapered gears also have tooth forms conjugate only to the generating tooth form, and at the generating center distance. This inherent restriction affects accuracy of operation, limits interchangeability, and causes difficulties in manufacture and inspection, thus making their choice for precision applications inadvisable.

(c) Pressure Angle and Pitch

For fine-pitch precision gears, the 20° pressure angle is favored over 14½°. This is a result of having tooth profiles formed by involutes farther out from the critical base-circle point. As a result specific sliding has a lower average value because mating involutes are better matched. Tooth profiles are formed in the generating process by increased number of shorter, straight-cut flats and will therefore be more symmetrical. For higher pressure angles there is less undercutting on gears with low numbers of teeth, and less lost motion caused by radial looseness upon reversal of drive.

There are certain disadvantages to having a larger pressure angle. An increase in pressure angle from 14½ to 20° results in a 40 per cent increase in the radial component of the gear force. This greater load makes devices using 20° pressure angle gears more sensitive to deflection, wear, and stress failures. The most significant disadvantage of higher pressure angle is the effect on center-distance change and backlash. Gears with 20° pressure angle teeth have 40 per cent more backlash than those with 14½° pressure angle teeth for the same increase in center distance.

For gears to function with high accuracy the pressure angle must be accurate. Any difference between the pressure angles of driving and driven gears will result in a non-uniform velocity of the output gear even when the input gear has a uniform velocity.

To minimize pressure-angle deviation due to inaccuracies in manufacture, it is preferable for mating high-precision gears to be made using the same or matched cutting tools.

Pitch. Precision gears fall into two categories: those which transmit power as well as motion, and those which transmit motion only. For power precision gears, pitch choice is determined by tooth strength and wear-rating considerations. Use of the finest pitch allowable on this basis requires careful study to be certain that tooth deflection does not significantly contribute to position error. Power precision gears are usually coarser than 20 diametral pitch.

Coarse-pitch gears are those having diametral pitches less than 20. Medium-pitch gears range from 20 to 48; fine-pitch gears are from 48 to 120. From 120 and finer are the less frequent ultrafine pitches, with 200 diametral pitch as the usual upper limit, although many cycloidal watch gears are of still finer pitches (200 to 400).

For gears not required to transmit appreciable power, 32 to 64 diametral pitch is the optimum range. In this range, choice of the highest pitch possible results in a slight increase in contact ratio, hence more tooth-to-tooth error averaging. The result is smoother and more accurate motion transmission. Also greater precision and surface-finish quality are possible with smaller teeth because of lighter manufacturing operations.

Diametral pitches finer than 64 are generally avoided. Fragile teeth cause manufacturing and inspection problems that increase markedly as pitch becomes finer. Contact ratio is more difficult to maintain because of the proportionately greater effect of change in center distance, outside-diameter variation, and tip rounding.

(d) Standard and Modified Tooth Forms

Difficulty in achieving precision with gears having standard tooth forms rises sharply when pinions have low numbers of teeth. Not only is the involute profile more sensitive to manufacturing inaccuracies, but where undercutting occurs, contact ratio is reduced, thus decreasing the benefits of tooth-to-tooth error averaging. Undercutting can be avoided by designing gears with more teeth than the undercut number N_c. For gears with American standard spur-gear tooth forms (ASA B6.1-1932), N_c is 20 when the pressure angle is 20° and 36 when the pressure angle is 14½°. However, for 20° fine-pitch spur and helical gears conforming to ASA B6.7-1956, clearance is a function of pitch. Therefore, the minimum number of teeth required to avoid undercutting is also a function of pitch. Minimum and recommended values are listed in Table 32.20.

Table 32.20. Number of Teeth for 20° Fine-pitch Gears, ASA B6.7-1956

P_d	N_c	N	P_d	N_c	N
20	21.2	25	72	23.0	26
24	21.3	25	80	23.3	27
32	21.6	25	96	23.8	27
48	22.2	26	120	24.6	28
64	22.7	26	200	27.4	31

N_c is the calculated minimum number of teeth to avoid undercutting.
N is the recommended number for best precision.

When low numbers of teeth are required, undercutting should be avoided by enlarging the pinion. To allow for an enlarged pinion, the addendum of the mating gear should be reduced if center distance is maintained, or the center distance increased if the mating gear is to be of standard proportions. By enlarging the pinion, tooth profiles are shifted outward so that the pinion has the characteristics of a gear with one or more additional teeth. Besides eliminating undercutting, the tooth profile permits more accurate fabrication. For power gears the enlarged pinion has the further advantage of improved strength and wear characteristics.

(e) Materials; Surface Quality

The initial achievement of high dimensional accuracy in a gear and the subsequent ability to retain accuracy in operation for a desired period depends in part on the choice of materials, the treatment given to these materials, and the initial surface quality of tooth profiles. Recommendations for pinion and gear combinations are given in Table 32.21. An indication of the inherent difficulty in manufacturing various materials to precision quality gears is found in Table 32.22.

Table 32.21. Combinations of Materials for Lightweight Fine-pitch Gearsets*

Unlubricated		Lubricated	
Pinion	Gear	Pinion	Gear
Sinite D-10S	Anodized aluminum	Stainless steel	Anodized aluminum
Sinite D-10S	Stainless steel	Aluminum	Aluminum
Nylasint 64	Anodized aluminum	Stainless steel	Aluminum
Nylasint 66	Anodized aluminum	Anodized aluminum	Anodized aluminum
Nylon 3001	Anodized aluminum		
Stainless steel	Anodized aluminum		

Sinite D-10S, impregnated with Dow Corning silicon oil.
Aluminum type 2024-T3.
Stainless steel type 416 25 to 36 Rockwell C.
* From Benson, *Prod. Eng.*, Feb. 27, 1961, p. 37. Choices listed in descending order of wear resistance. Based on tests of 48-diametral-pitch gears, 48-tooth pinions, and 97-tooth gears.

Table 32.22. Relative Rating of Various Materials for Ease of Attaining Quality (In Descending Order)

Teeth generated by cutting	Teeth generated by grinding
Bronze alloys (free cutting)	Hardened-tool-steel alloys
Brasses (free cutting)	Hardened-carbon-steel alloys
Aluminum, hard alloys (2024 and 2075 types)	Hardened-stainless-steel alloys
Steels, mild	Semihard-steel alloys
Steels, alloy	Stainless-steel alloys
Stainless steels, 400 series	Soft steels
Stainless steels, 300 series	Bronzes
Plastics, (phenolic, nylon, etc.)	Brasses
	Aluminum alloys

Close tolerances require comparable values of surface quality for tooth profiles. Surface roughness can represent a significant part of the tolerance and contribute toward early wear, noise, etc. Recommended values of maximum roughness for tooth profiles are listed in Table 32.23.

In order to improve wear qualities and for the protection of surfaces against corrosion, gears may be finished by plating or by a surface-conversion process. As most finishes alter dimensions, their application must be carefully considered with regard to the degree of precision expected of the gear. A guide to the application of common finish treatments is given in Table 32.24. In some instances, protective finishes are required for gear bodies, but not for the teeth, which may be sufficiently protected by a film of lubricant. When this is so, and if the finish treatments would limit the attainment of the required degree of accuracy, then the teeth and other closely toleranced features can be machined after the blank has been finish treated.

Table 32.23. Tooth Surface Finish Micro-inches

Gear quality	Steel	Nonferrous metals
Commercial	64–125	64–125
Precision	16–32	16–64
High precision	8–16	16–32
Ultraprecision	4–8	8–16

Table 32.24. Effect of Surface Treatments on Gear Dimensions

Surface treatment	Dimensional change, in.	Recommended practice
Electroplate	+0.00005 to +0.0002	
Anodize (aluminum)	0 to ±0.0002	Avoid for high-precision gears
Iridite	0 to +0.00005	
Chromium-alloy coating	+0.00005 to +0.0001	Avoid for ultraprecision
Passivation (stainless steel)	No change	Recommended

32.20. SPECIFICATIONS FOR PRECISION SPUR GEARS

(a) Introduction

The inherent accuracy of a precision gear depends on the faithfulness of the teeth and on the dimensions of the gear blank controlling the location of the teeth. In addition to the dimensions of the gear itself, the action of a gear in operation is contingent with the design of the train in which the gear is mounted. In this article data are given concerning the specification of gear features having an effect on the inherent accuracy of a gear. The following article is concerned with the more inclusive problem of gear-train accuracy.

(b) Critical Reference Surfaces and Gear-blank Design

Three types of critical reference surfaces are involved in the specification and manufacture of precision gears: mounting surfaces *MS*, generating mounting surfaces *GMS*, and registering surfaces *RS*.

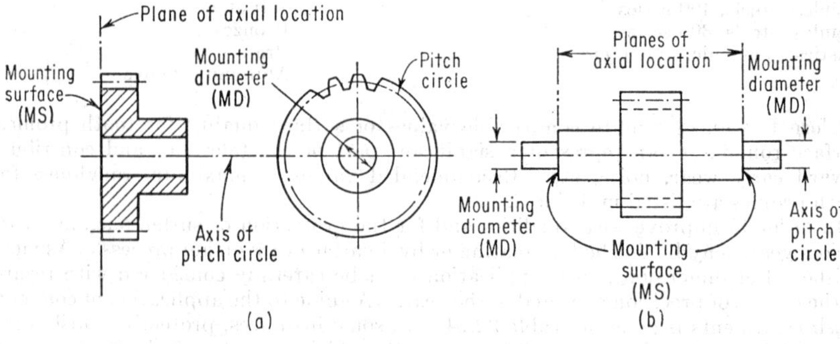

FIG. 32.30. Examples of typical mounting surfaces *MS* and mounting diameters *MD*. (*a*) Gear with bore. (*b*) Gear with integral shaft.

Mounting surfaces are illustrated in Fig. 32.30. These position a gear with respect to the axis of its pitch circle, and to a plane or planes defining its axial location. A mounting surface can be a bore, shaft journal, shoulder between two diameters, etc. For precision gears it is important to designate mounting surfaces, identifying them on drawings with the symbols *MS*. When the mounting surface is a diameter, it is often referred to as *mounting diameter,* in which case the symbol *MD* is used instead of *MS*.

Gear teeth should preferably be formed with the blank located from the same surfaces used for its assembly mounting. The adoption of other than mounting surfaces during manufacture is undesirable for gears of high precision. When it is not convenient or possible to use mounting surfaces, then alternate surfaces are designated. These alternate surfaces are called generating mounting surfaces; examples are shown

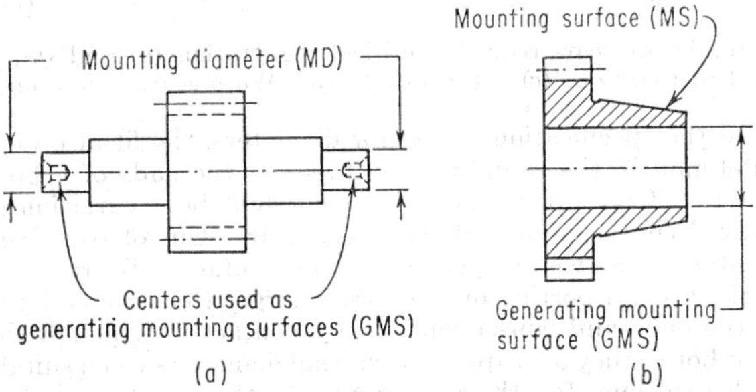

FIG. 32.31. Examples of surfaces other than assembly mounting surfaces used for mounting gears during generation of teeth. (*a*) Gear with integral shaft and centers. (*b*) Gear mounting on tapered surface.

in Fig. 32.31. The penalty paid for the employment of alternate surfaces is the additional tolerances between the mounting surfaces and the alternate generating mounting surfaces. Recommended values for these tolerances are given in Table 32.25.

Table 32.25. Tolerances for Relationship of Mounting Surface *MS* and Generating Mounting Surface *GMS*

Quality	Runout between *GMS* and *MS*		Parallelism of *GMS* and *MS*, in./in. . . in. max
	Shaft centers, in.	Concentric diameters, in.	
Commercial........	0.0008	0.001	0.0010. . .0.0030
Precision..........	0.0001	0.0002	0.0002. . .0.0005
Ultraprecision......	0.00005	0.00005	0.0001. . .0.0003

Locations of mounting surfaces perpendicular to the axis of the pitch circle are not critical for all types of gears. In Fig. 32.32, for example, the axial location of the teeth on the spur gear with respect to the mounting surface does not require accurate registration either for generating the teeth or for the subsequent mounting of the gear. Certain types of gears do require that the teeth be accurately located from a registering surface; examples are the bevel and wormgears shown in Fig. 32.32.

The quality of critical reference surfaces depends on the control of various characteristics as listed in Table 32.26. Size of the bore or external diameters is perhaps the most important. Both the diameter of the gear and the diameter of its mating part must be closely dimensioned so as to assure a satisfactory condition of fit in the assembled condition.

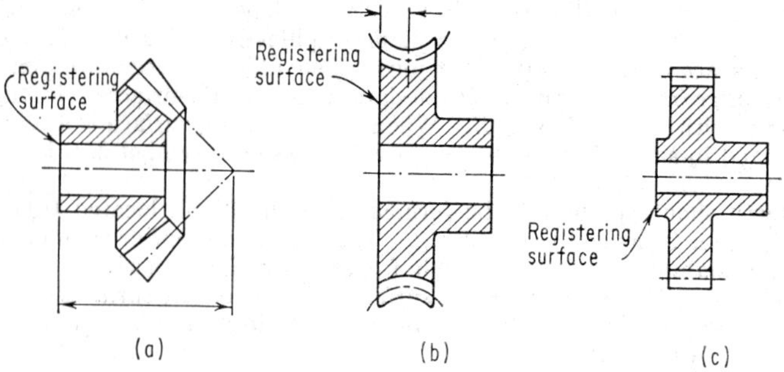

FIG. 32.32. Examples of gears requiring critical registration (*a*) and (*b*), and a gear not requiring critical registrations (*c*). (*a*) Bevel. (*b*) Wormgear. (*c*) Spur.

In addition to the specification of size for diameters, the fit also depends on other conditions. Bellmouth, the rounding and taper at the ends of a bore, reduces the useful bore length. Taper, the amount of consistent bore variation along the axis, results in localized contact and cocking of the gear. Out of roundness causes false diameter measurement as well as poor mounting contact. Barrel is a special condition in which the central portion of the bore is larger than the ends; it can result in early wear of the ends and consequent sloppy contact. Although these conditions are described for bores, they also apply to external diameters when suitable changes are made in the descriptions for their different effects on external than on internal diameters.

Table 32.26. Gear-blank Tolerances for Critical Surfaces

	Commercial	Precision	Ultraprecision
Diameters:			
Tolerance, in.	0.0005	0.0002	0.0001
Taper, in./in.	0.001	0.0005	0.0002
Out of roundness, TIR, in.	0.0003	0.0001	0.00005
Bellmouth, % of total length	50	20	5
Barrel, % of total length	50	20	10
Surface roughness, μin. rms	64–125	16–32	4–8
Flat surfaces:			
Convexity, in./in.	0.001	0.003	0.0001
Concavity, in./in.	0.001	0.0005	0.0002
Parallelism, in./in.	0.002	0.001	0.0005
Lateral runout, TIR/in. radius	0.001	0.0003	0.0002
Surface roughness, μin. rms	64–250	16–64	8–16

Ideally, critical reference surfaces in planes of rotation used for mounting and registering are flat planes perpendicular to the axis of mounting diameters. For these surfaces, convexity and concavity refer to departures from flat to convex and concave, parallelism to the departure from parallel of two nominally parallel mounting surfaces, lateral runout to the wobble of a surface in the plane of rotation measured as the total indicator reading at the farthest distance on the surface from the axis of rotation.

Outside-diameter runout does not directly affect gear function if it does not cause binding at the tightest point of gear mesh or if it does not otherwise affect gear accuracy. It is measured as the total indicator reading between the axis of the mounting diameter and the outside diameter of the gear blank. Recommended values are given in Table 32.27.

Table 32.27. Typical Tolerance Values for Outside-diameter Runout
Maximum runout, in., total indicator reading

Diametral pitch	Precision	Ultraprecision
Coarser than 1	0.010	
1–10	0.005	0.002
11–20	0.003	0.0015
21–31	0.002	0.0010
32–63	0.0015	0.0008
64–99	0.0010	0.0005
100–125	0.0005	0.0003
126 and finer	0.0005	0.0002

(c) Gear Measurement

Assurance of quality in precision gears depends on the method of measurement used in their manufacture and inspection. The three common methods are the measurement of chordal tooth thickness, overpin measurement, and the testing-radius measurement.

The measurement of chordal tooth thickness is not applicable to precision gears because of obvious limitation to gears of coarse pitch and because of the incomplete information yielded by such measurements.

Overpin measurements are occasionally used for precision gearing although gaging is awkward for extremely fine pitches and a single measurement may not be representative for the entire gear. Moreover, overpin measurements suffer the disadvantage of not always being sufficiently accurate when applied to small gears of fine pitch because of the effects of surface finish and gaging pressure. In common with measurements of chordal tooth thickness, overpin measurements do not indicate certain important properties such as concentricity of teeth with the center line of mounting surfaces, or the accuracy of the tooth forms. However, it is common practice to give dimensions or overpin measurements if only as reference because this method of measurement offers so handy a means for checking gear size. Data for the calculation of overpin measurements are given in AGMA 231.51.

The method of measurement most suitable for precision gears is the testing-radius measurement using a variable center distance fixture with master gear as shown in Fig. 32.33. Instruments of this sort can be used with gears of any pitch but are generally limited to gears with pitch diameters less than 6 in. The gear being tested is rotated while intimately meshed with a master gear. Gear errors have the effect of varying the meshing radius of the test gear. The movement of the center of the test gear is continuously recorded for a full rotation or more of the test gear. An idealized recording is shown in Fig. 32.34. Any difference between the recording and the straight line representing a perfect gear can be attributed to several kinds of tooth errors: pitch-line runout, tooth-thickness variation, profile error, etc. For this reason, the result of a testing-radius measurement is called a *composite check*. From the recording it is possible to evaluate total composite error, tooth-to-tooth composite error, and size variation.

(d) Total Composite Error; Tooth-to-tooth Composite Error

Total composite error E_{tc} is defined as the maximum variation in center distance indicated by the recording of the testing-radius measurement. This error is the combined effect of pitch circle and lateral runout, pitch and profile errors, and tooth-thickness variation.

Pitch-circle runout is a single-cycle variation caused by eccentricity between the center of rotation and the center about which the teeth were generated. For preci-

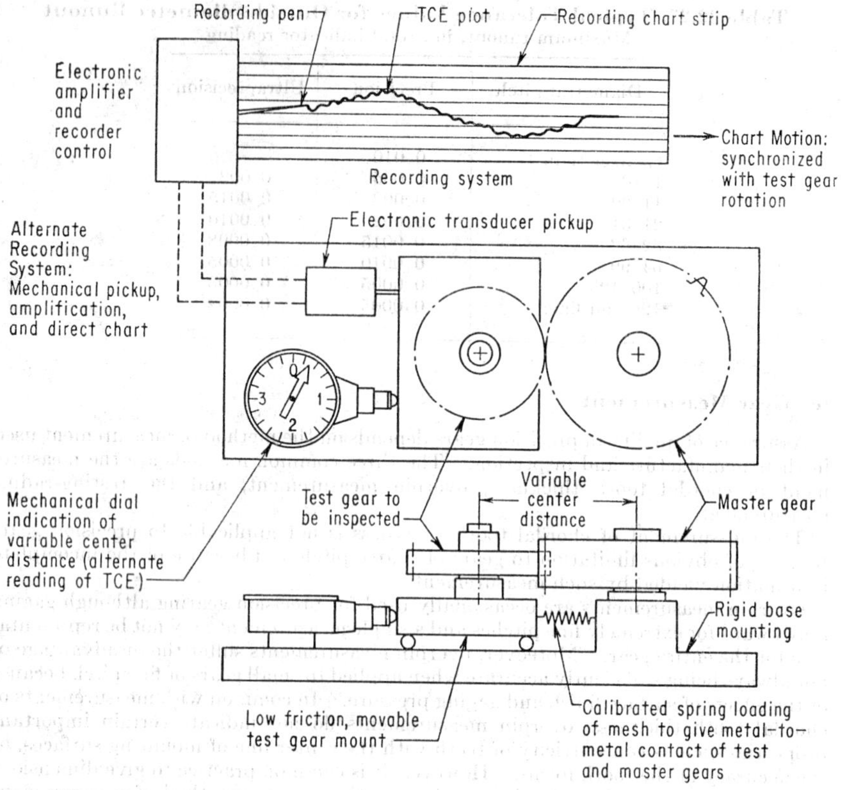

Fig. 32.33. Essential features of variable-center-distance gear rolling inspection device.

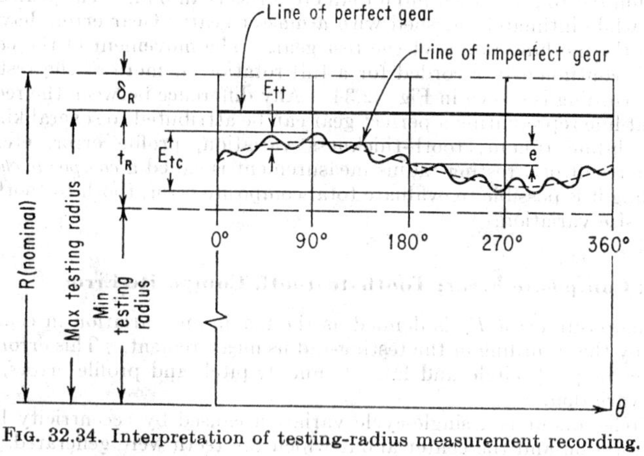

Fig. 32.34. Interpretation of testing-radius measurement recording.

sion-quality gears, pitch-circle runout ranges from 0.0001 to 0.0010 in. regardless of pitch and pitch diameter. It may be controlled by specifying maximum allowable values or, more preferably, by the inclusive specification of total composite error.

Lateral runout, the side wobble of a gear relative to the plane of rotation, is a one-cycle variation and has the effect of causing the pitch circle to run out. Because it varies as the cosine of the equivalent angle of wobble, it is a secondary effect contributing but little to the value of total composite error. Lateral runout is controlled by gear-blank tolerances; see Table 32.26.

Tooth-to-tooth composite error E_{tt} is caused by variations in tooth thickness and pitch and by profile errors. Because total composite error includes tooth-to-tooth composite error, the inclusive specification of total composite error is often all that is required.

A concise listing of ranges for total and tooth-to-tooth composite errors for commercial, precision, and ultraprecision quality gears is to be found in Table 32.28. Table 32.29, taken from AGMA standards, lists tolerances according to AGMA quality numbers. Because extensive reference is made in the literature and by gear manufacturers to former AGMA quality classes, Table 32.30 lists these former values of composite errors and shows correspondence to current AGMA quality grades.

Table 32.28. Typical Value for E_{tc}, E_{tt}, and E_{si}

Values in inches

Gear quality	E_{tc}	E_{tt}	E_{si}
Commercial	0.0020	0.0010	0.0020
	0.0015	0.0007	0.0015
Precision	0.0010	0.0004	0.0005
	0.0005	0.0003	0.0003
Ultraprecision	0.00025	0.0002	0.0002
	0.00015	0.0001	0.0002
	0.00010	0.0001	0.0001

E_{tc} = total composite error
E_{tt} = tooth-to-tooth composite error
E_{si} = supplementary intermediate-frequency error

(e) Position Error E_p

Also referred to as index error, accumulative error, total spacing error, and angular error, position error is caused by displacements of tooth profiles from their ideal positions. Although tooth profiles should be concentric with the axis of rotation and equally spaced along the pitch circle, in practice there will be small deviations from the ideal. The contributors of position errors are the error functions associated with pitch-circle runout and tooth-to-tooth composite error. In combining these error functions, the separate components can either reinforce or diminish each other. It is thus difficult to be able to predict with assurance what the maximum value of position error will be. Except for gears having wrong numbers of teeth, an integration over one revolution is zero.

The error function associated with pitch-circle runout $e \sin \theta$ can generally be inferred from the recording of a testing-radius measurement as indicated on Fig. 32.34. However, the behavior of the error function due to tooth-to-tooth composite error is difficult to measure. Conventional testing-radius measurement instruments are sensitive only to tooth-to-tooth errors between adjacent teeth. A buildup or cancellation of position error over several teeth is generally not detectable. The error function for the buildup and cancellation of the tooth-to-tooth component of position error supplements the error function due to pitch-circle runout; its frequency is inter-

Table 32.29. AGMA Tolerances for Total Composite Error and Tooth-to-tooth Composite Error (AGMA 390.01)

Coarse-pitch Gear Tolerances, in Ten-thousandths of an Inch

AGMA quality No.	Normal diametral pitch	Total composite error E_{tc}						
		Pitch diam, in.						
		1	3	6	12	25	50	100
8	2	...	...	...	80	94	111	135
	4	...	46	52	58	68	79	93
	8	35	38	42	46	52	60	70
	16–19.99	27	30	34	37	44	50	58
9	2	...	...	...	57	66	78	95
	4	...	33	37	42	48	54	66
	8	...	27	30	33	38	43	50
	16–19.99	19	22	24	27	31	36	42
10	2	...	...	...	40	48	58	69
	4	...	23	26	29	34	39	47
	8	...	19	21	24	26	30	35
	16–19.99	14	15	17	19	22	26	30
11	4	...	17	18	21	24	28	34
	8	...	14	15	17	19	22	26
	16–19.99	10	11	12	14	16	18	21
12	4	...	12	13	15	17	20	25
	8	...	10	11	12	13	15	18
	16–19.99	7	8	9	10	11	13	16

Fine-pitch Gear Tolerances, in Inches

AGMA quality No.	No. of teeth N and pitch diam, D	Diametral pitch, P_d	Tooth-to-tooth composite error, E_{tt}	Total composite error E_{tc}
5	Up to 20 T (inclusive)	20–80	0.0037	0.0052
	Over 20 T; to 1.999 in.	20–32	0.0027	0.0052
	Over 20 T; 2–3.999 in.	20–24	0.0027	0.0061
	Over 20 T; 4 in. and over	20	0.0027	0.0072
6	Up to 20 T	20–200	0.0027	0.0037
	Over 20 T; to 1.999 in.	20–48	0.0019	0.0037
	Over 20 T; 2–3.999 in.	20–32	0.0019	0.0044
	Over 20 T; 4 in. and over	20–24	0.0019	0.0052
7	Up to 20 T	20–200	0.0019	0.0027
	Over 20 T; to 1.999 in.	20–100	0.0014	0.0027
	Over 20 T; 2–3.999 in.	20–48	0.0014	0.0032
	Over 20 T; 4 in. and over	20–40	0.0014	0.0037
8	Up to 20 T	20–200	0.0014	0.0019
	Over 20 T; to 1.999 in.	20–200	0.0010	0.0019
	Over 20 T; 2–3.999 in.	20–100	0.0010	0.0023
	Over 20 T; 4 in. and over	20–64	0.0010	0.0027
9	Up to 20 T	20–200	0.0010	0.0014
	Over 20 T; to 1.999 in.	20–200	0.0007	0.0014
	Over 20 T; 2–3.999 in.	20–200	0.0007	0.0016
	Over 20 T; 4 in. and over	20–120	0.0007	0.0019
10	Up to 20 T	20–200	0.0007	0.0010
	Over 20 T; to 1.999 in.	20–200	0.0005	0.0010
	Over 20 T; 2–3.999 in.	20–200	0.0005	0.0012
	Over 20 T; 4 in. and over	20–200	0.0005	0 0014

Table 32.29. AGMA Tolerances for Total Composite Error and Tooth-to-tooth Composite Error (AGMA 390.01) (Continued)

AGMA quality No.	No. of teeth N and pitch diam, D	Diametral pitch, P_d	Tooth-to-tooth composite error E_{tt}	Total composite error E_{tc}
11	Up to 20 T	20–200	0.0005	0.0007
	Over 20 T; to 1.999 in.	20–200	0.0004	0.0007
	Over 20 T; 2–3.999 in.	20–200	0.0004	0.0009
	Over 20 T; 4 in. and over	20–200	0.0004	0.0010
12	Up to 20 T	20–200	0.0004	0.0005
	Over 20 T; to 1.999 in.	20–200	0.0003	0.0005
	Over 20 T; 2–3.999 in.	20–200	0.0003	0.0006
	Over 20 T; 4 in. and over	20–200	0.0003	0.0007
13	Up to 20 T	20–200	0.0003	0.0004
	Over 20 T; to 1.999 in.	20–200	0.0002	0.0004
	Over 20 T; 2–3.999 in.	20–200	0.0002	0.0004
	Over 20 T; 4 in. and over	20–200	0.0002	0.0005
14	Over 20 T	20–200	0.00019	0.00027
	Over 20 T; to 1.999 in.	20–200	0.00014	0.00027
	Over 20 T; 2–3.999 in.	20–200	0.00014	0.00032
	Over 20 T; 4 in. and over	20–200	0.00014	0.00037
15	Up to 20 T	20–200	0.00014	0.00019
	Over 20 T; to 1.999 in.	20–200	0.00010	0.00019
	Over 20 T; 2–3.999 in.	20–200	0.00010	0.00023
	Over 20 T; 4 in. and over	20–200	0.00010	0.00027
16	Up to 20 T	20–200	0.00010	0.00014
	Over 20 T; to 1.999 in.	20–200	0.00007	0.00014
	Over 20 T; 2–3.999 in.	20–200	0.00007	0.00016
	Over 20 T; 4 in. and over	20–200	0.00007	0.00019

Table 32.30. Composite Errors for Former and Present AGMA Quality Classes

Class per AGMA 236.04	E_{tc}, in.	E_{tt}, in.	Present AGMA quality No.*
Commercial 1	0.006	0.002	5 or 6
Commercial 2	0.004	0.0015	6 or 7
Commercial 3	0.002	0.0010	8
Commercial 4	0.0015	0.0007	9
Precision 1	0.0010	0.0004	10 or 11
Precision 2	0.0005	0.0003	12
Precision 3	0.00025	0.0002	13 or 14

* Per AGMA 390.01.

mediate between that of two successive teeth and the frequency of a single revolution, which is the frequency of the pitch-circle runout error function. Thus the error component due to tooth-to-tooth composite error can be called a supplementary intermediate-frequency error E_{si}; typical values are given in Table 32.28.

As all precision gears do not require the same degree of linearity of action, no one value for position error applies to all gears within a given quality range. Accordingly, Table 32.31 lists several grades of precision error typical for precision-quality and ultraprecision-quality gears. Because linearity of action may not be sufficiently important in some cases to warrant the cost of control, the specification of precision error may be omitted if the indirect control exerted by total composite error is considered adequate.

Table 32.31. Typical Position-error Values

Quality and grade designation	$\pm E_p$, sec of arc	$\pm E_p$, rad
Precision *A*	60	0.00029
Precision *B*	40	0.00019
Precision *C*	30	0.00015
Precision *D*	25	0.00012
Ultraprecision *E*	20	0.00010
Ultraprecision *F*	15	0.00007
Ultraprecision *G*	10	0.00005
Ultraprecision *H*	5	0.00002

(f) Size Variation; Inherent Backlash

Although the backlash of a gear appears only when mated with another in an assembly, part is directly identifiable in the individual gears before they are mated. Thus it is possible to consider the *inherent backlash* of a single gear independent of assembly. Inherent backlash is attributable to size variation, tooth errors, and pitch-circle runout. The total backlash for an assembled pair is due also to center-distance allowance and tolerance, mounting errors, and bearing play. Calculation of the effect of total backlash on the output of a train of gears is treated in the next article.

Size variation is the departure from ideal tooth thickness because of allowance and tolerance. Allowance is the intentional thinning of teeth to provide clearance at assembly so as to avoid possible interference due to bearing runout and differential thermal expansion, and to allow for lubricant.

Table 32.32. Gear Size—Typical Backlash Allowance B_A and Tolerance B_V Ranges

Units for B_A and B_V, in.

Gear quality	Range of P_d	Range of B_A	Range of B_V
Commercial	Up to 10	−0.005 to −0.010	−0.005 to −0.015
	11 to 19	−0.002 to −0.005	−0.003 to −0.010
	20 to 50	−0.001 to −0.003	−0.002 to −0.005
	51 to 80	−0.001 to −0.002	−0.001 to −0.003
Precision	Up to 20	−0.001 to −0.002	−0.001 to −0.002
	20 to 50	−0.0005 to −0.001	−0.0005 to −0.001
	51 to 100	−0.0002 to −0.0005	−0.0002 to −0.0005
	Above 100	−0.0000 to −0.0003	−0.0002 to −0.0004
Ultraprecision	Up to 20	−0.001 to −0.002	−0.0005 to −0.001
	20 to 50	−0.0005 to −0.001	−0.0002 to −0.0005
	51 to 100	−0.0000 to −0.0005	−0.0001 to −0.0003
	Above 100	−0.0000 to −0.0002	−0.0001 to −0.0002

Nominal gear dimensions and the distance between gear centers are calculated according to the rules of the gear system used. Actual dimensions are obtained by modifying these nominal values, providing for allowance and tolerance. Minimum backlash for a pair of mated gears can be obtained by either increasing the center distance or decreasing tooth thickness. If the first course is chosen, size variation will be attributable only to tooth-thickness tolerance. The gears will have no allowance as the maximum tooth thickness will equal the nominal value. The minimum tooth thickness possible will then be determined only by the tooth-thickness tolerance necessary for fabrication. This tolerance is always unilateral so as to provide more backlash. If the second course is chosen and the minimum center distance is a nominal value, then size variation will consist of both allowance and tolerance.

The relationship between backlash and a change in meshing radius Δr of a testing measurement is

$$B = 2 \tan \varphi(\Delta r) \tag{32.55}$$

A similar relationship between backlash and the change in overpin measurements ΔM is

$$B = \Delta M / K_m \tag{32.56}$$

Values of K_m are tabulated in AGMA 231.51.

A concise listing of recommended ranges for tooth-thickness allowance and tolerance is given in Table 32.32. The more detailed AGMA backlash allowances and tolerances for coarse- and fine-pitch gears are given in Table 32.33.

Table 32.33. AGMA Backlash Allowances and Tolerances for Coarse- and Fine-pitch Gears (AGMA 390.01)

Backlash Tolerances for Coarse-pitch Gearing, in Inches Measured in Normal Plane

Center distance	Normal diametral pitches				
	0.5–1.99	2–3.49	3.5–5.99	6–9.99	10–19.99
Up to 5	....	....	....	....	0.005–0.015
Over 5 to 10	....	....	....	0.010–0.020	0.010–0.020
Over 10 to 20	....	....	0.020–0.030	0.015–0.025	0.010–0.020
Over 20 to 30	....	0.030–0.040	0.025–0.030	0.020–0.030	
Over 30 to 40	0.040–0.060	0.035–0.045	0.030–0.040	0.025–0.035	
Over 40 to 50	0.050–0.070	0.040–0.055	0.035–0.050	0.030–0.040	
Over 50 to 80	0.060–0.080	0.045–0.065	0.040–0.060		
Over 80 to 100	0.070–0.095	0.050–0.080			
Over 100 to 120	0.080–0.110				

Backlash Tolerances for Fine-pitch Gearing, in Inches Measured in Normal Plane

Backlash designation	Normal diametral-pitch range	Tooth thinning to obtain backlash (per gear)	Resulting approx. backlash (per pair)
A	20– 45	0.0020	0.0040–0.0080
	46– 70	0.0015	0.0030–0.0070
	71– 90	0.0010	0.0020–0.0055
	91–200	0.0008	0.0015–0.0030
B	20– 60	0.0010	0.0020–0.0040
	61–120	0.0008	0.0015–0.0030
	121–200	0.0005	0.0010–0.0020
C	20– 60	0.0005	0.0010–0.0020
	61–120	0.0004	0.0008–0.0015
	121–200	0.0003	0.0006–0.0010
D	20– 60	0.0003	0.0006–0.0010
	61–120	0.0002	0.0004–0.0007
	121–200	0.0001	0.0002–0.0004
E	All pitches	Do not provide tooth thinning	0.0000–0.0002

32.21. PRECISION GEAR TRAINS*

(a) Train Output Error ϵ_T

The diagram of Fig. 32.35 shows a simple gear train. Errors originating at each mesh contribute toward the train output error depending on the characteristics of the errors and the mesh ratios. For any mesh its ratio is defined by

$$m = N_P/N_G \tag{32.57}$$

In precision gear trains three kinds of mesh errors must be considered: constant backlash errors ϵ_{bc}, variable backlash errors ϵ_{bv}, and transmission errors ϵ_{tr}. The units of these errors are radians. A particular kind of error originating at some mesh j

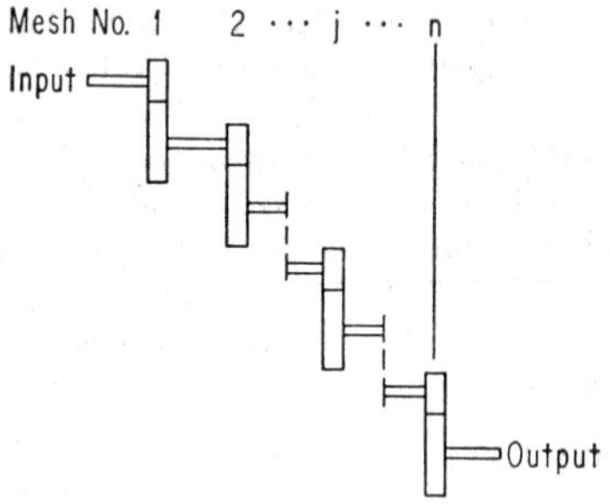

FIG. 32.35. Diagram of gear train.

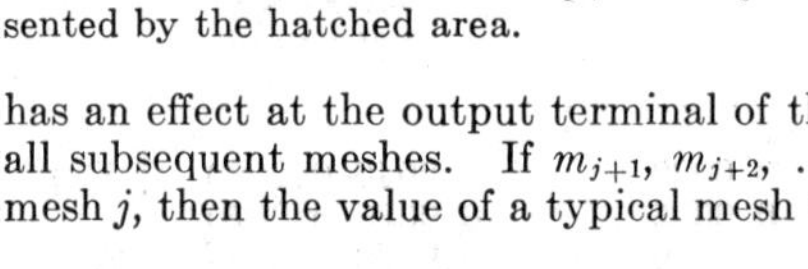

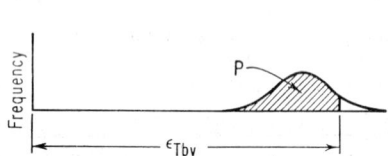

FIG. 32.36. Relationship of ϵ_{Tbv} with its probability of occurrence P. The total area under the curve is unity; P is represented by the hatched area.

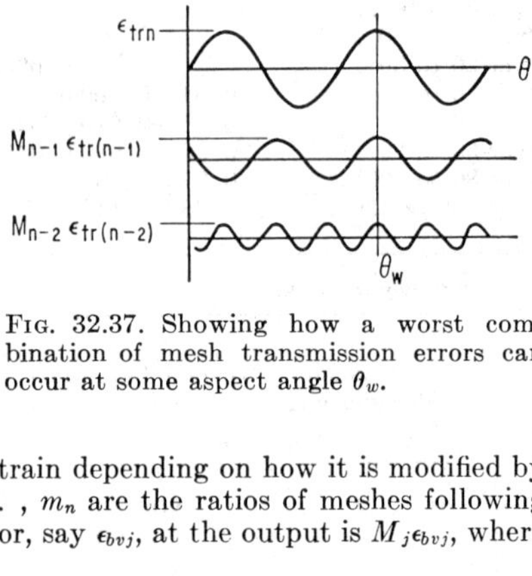

FIG. 32.37. Showing how a worst combination of mesh transmission errors can occur at some aspect angle θ_w.

has an effect at the output terminal of the train depending on how it is modified by all subsequent meshes. If $m_{j+1}, m_{j+2}, \ldots, m_n$ are the ratios of meshes following mesh j, then the value of a typical mesh error, say ϵ_{bvj}, at the output is $M_j\epsilon_{bvj}$, where

$$M_j = m_{j+1}m_{j+2} \cdots m_n \tag{32.58}$$

In combining constant backlash errors originating at each mesh, the effective errors at the output can ordinarily be assumed to add linearly. Hence the constant backlash component of train error

$$\epsilon_{Tbc} = M_1\epsilon_{bc1} + M_2\epsilon_{bc2} + \cdots + \epsilon_{bcn} \tag{32.59}$$

In combining variable backlash errors it is not reasonable to assume that the actual value occurring in every mesh will be the maximum allowable. If the magnitudes of

* This portion contributed by E. T. Fortini, Compugraphic Corporation, Reading, Mass.

these errors are assumed to have distributions approximating the normal, then the variable backlash component of train error

$$\epsilon_{Tbv} = (M_1{}^2\epsilon_{bv1}{}^2 + M_2{}^2\epsilon_{bv2}{}^2 + \cdots + \epsilon_{bvn}{}^2)^{1/2} \tag{32.60}$$

Figure 32.36 illustrates the relationship between the value of ϵ_{Tbv} and its probability of occurrence. The value of this probability is related to the probabilities of mesh errors. If all variable backlash mesh errors were normally distributed so that the probability of each one of these being no greater than ϵ_{bv} were P, then the variable component of backlash error would also be normally distributed with the probability of ϵ_{Tbv} not being exceeded being P. Even when the distributions of the mesh errors are not normal, the distribution of the train error will be very close to normal if its

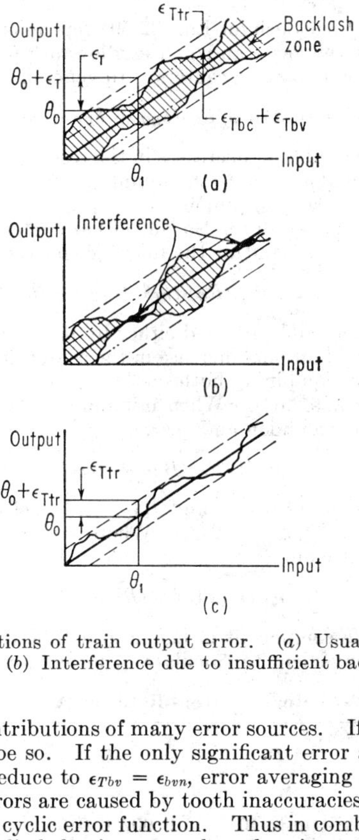

FIG. 32.38. Various conditions of train output error. (*a*) Usual combination of backlash and transmission errors. (*b*) Interference due to insufficient backlash. (*c*) Transmission error with no backlash.

value depends on the contributions of many error sources. If one or two error sources dominate, this will not be so. If the only significant error at the train output were ϵ_{bvn}, Eq. (32.60) would reduce to $\epsilon_{Tbv} = \epsilon_{bvn}$, error averaging having no effect.

Mesh transmission errors are caused by tooth inaccuracies and pitch-circle runout. The effect of runout is a cyclic error function. Thus in combining mesh transmission errors, their possible cyclic behaviors must be taken into account. A transmission error originating at mesh j will have a frequency at the output M_j times greater than the frequency of the train output shaft. At some aspect of the train it is possible that a worst combination of mesh transmission errors can occur as indicated in Fig. 32.37. For such an event,

$$\epsilon_{Ttr} = M_1\epsilon_{tr1} + M_2\epsilon_{tr2} + \cdots + \epsilon_{trn} \tag{32.61}$$

If, however, train accuracy can be based on an average error expectation, the occasional worst combination not considered,

$$\epsilon_{Ttr} = (M_1{}^2\epsilon_{tr1}{}^2 + M_2{}^2\epsilon_{tr2}{}^2 + \cdots + \epsilon_{trn}{}^2)^{1/2} \tag{32.62}$$

The qualifications cited for Eq. (32.60) apply as well to Eq. (32.62). The two equations above set limiting values for the transmission component of train error. In some cases it may be prudent to assume the actual value of ϵ_{Ttr} will fall between the values calculated by the two equations.

The train output error function, as shown in Fig. 32.38*a*, consists of a backlash zone bounded by the oscillating transmission error curves. Total backlash depends, therefore, on ϵ_{Ttr} as well as ϵ_{Tbc} and ϵ_{Tbv}. If the transmission error curves were to overlap as illustrated in Fig. 32.38*b*, then a condition of interference would be indicated. By designing the gear train so that backlash is eliminated, the only component of train output error would be ϵ_{Ttr} as shown in Fig. 32.38*c*.

(b) Evaluation of Mesh Backlash Errors ϵ_{bc} and ϵ_{bv}

Constant components of backlash for Eq. (32.59) depend on the minimum backlash allowed for in the size of the gears and/or the allowance for backlash provided for in the dimensioning of gear mounting centers. In general,

$$\epsilon_{bc} = (2 \tan \varphi / R_G)(\delta_{RG} + \delta_{RP} + \delta_C) \tag{32.63}$$

where δ_{RG} and δ_{RP} are the minimum decreases in the testing-radius measurement from nominal of the gear and pinion, and δ_C the minimum increase in the center distance from nominal between the gear and pinion.

Variable components of backlash for Eq. (32.60) are determined by the tolerances for testing-radius measurements and/or the tolerance for center distance:

$$\epsilon_{bv} = (2 \tan \varphi / R_G)(t_{RP}^2 + t_{RG}^2 + t_C^2)^{1/2} \tag{32.64}$$

Constant backlash is generally provided either by intentional thinning of the gear and pinion teeth or by an intentional increase in the center distance between gear and pinion, and only rarely by combining both methods. It is convenient, therefore, to have a special form for Eq. (32.63). When minimum backlash is provided only by the backlash allowance in the pinion and gear,

$$\epsilon_{bc} = (1/R_G)(B_{AG} + B_{AP}) \tag{32.65}$$

Values of the backlash allowances B_{AG} and B_{AP} can be selected on the basis of the recommendations in Table 32.32 or 32.33. Where minimum backlash is provided by center-distance allowance only,

$$\epsilon_{bc} = (2 \tan \varphi / R_G)\delta_C \tag{32.66}$$

In either case the variable component of backlash will be the same as given by Eq. (32.64). If it is more convenient to calculate ϵ_{bv} using values of backlash tolerance

Table 32.34. Ranges of Typical Center-distance Allowances and Tolerances

Gear quality	P_d	δ_C	t_C
Commercial.	Up to 10	0.002–0.010	0.010–0.020
	11 to 19	0.002–0.005	0.006–0.014
	20 to 50	0.001–0.003	0.004–0.010
	51 to 80	0.001–0.002	0.002–0.006
Precision.	Up to 20	0.002–0.005	0.002–0.004
	20 to 50	0.001–0.002	0.001–0.003
	51 to 100	0.0003–0.0005	0.0006–0.002
	Above 100	0.0–0.0005	0.0004–0.001
Ultraprecision.	Up to 20	0.001–0.003	0.001–0.002
	20 to 50	0.0005–0.001	0.0006–0.002
	51 to 100	0–0.0005	0.0004–0.001
	Above 100	0–0.0002	0.0001–0.0004

from Table 32.32 or 32.33, then

$$\epsilon_{bv} = (1/R_G)(4 \tan^2 \varphi\, t_C^2 + B_{VG}^2 + B_{VP}^2)^{1/2} \tag{32.67}$$

Table 32.35 gives recommended values for center-distance allowance and tolerance. Note that these recommendations refer to the actual distance between gear mounting centers. Both δ_C and t_C should be realistic estimates of all sources of fixed error, that is, error components that do not vary as the gears rotate. Besides dimensions locating bores on the supporting structure or housing, sources of fixed errors include eccentricities in the stationary races of ball bearings, adapters, etc.

(c) Evaluation of Mesh Transmission Errors ϵ_{tr}

Transmission errors are attributable to deviations in true positions of both pinion and gear. Certain of these errors originate in the mounting of the gears; others are inherent in the gears themselves. As total composite error provides an inclusive measurement of inherent gear error, it is convenient to use this and the supplementary intermediate-frequency error in evaluating that part of mesh transmission error inherent in the gears.

An approximation of the error function due to total composite error is $\frac{1}{2}E_{tc} \sin \theta$. To this must be added the supplementary intermediate-frequency error function $E_{si} \sin (n\theta + \gamma_1)$ and the error function due to the runout of bearing mounting components $\frac{1}{2}E_r \sin (\theta + \gamma_2)$. Angles γ_1 and γ_2 represent phasing differences between the three components of ϵ_{tr}. The method appropriate for combining the three error functions depends on the assumptions allowed concerning the phasing differences and the probability distributions of the functions of E_{tc}, E_{si}, and E_r.

A conservative assumption is that there will be phase matching of the error functions and that the probabilities of E_{tc}, E_{si}, and E_r are fractiles of the normal distribution. These assumptions lead to the equation

$$\epsilon_{tr} = \pm(\tan \varphi/R_G)(E_{tcG}^2 + E_{rG}^2 + 4E_{siG}^2 + E_{tcP}^2 + E_{rP}^2 + 4E_{siP}^2)^{1/2} \tag{32.68}$$

in which the units of all E's are inches. Values of E_{tcG} and E_{tcP} are part of the accuracy specifications given on the gear and pinion drawings. Values for E_{rG} and E_{rP} depend on the eccentricities of mounting components. Probable maximum values for these must be evaluated from an analysis of the design and dimensioning of component parts. Values for E_{siP} and E_{siG} substituted in the above equation must be in terms of changes in meshing radius.

Supplementary intermediate-frequency errors E_{siG} and E_{siP} are often small enough so that they can be neglected. When there is an inclusive position-error specification E_p (see recommended values in Table 32.32), the above equation can be modified so as to read

$$\epsilon_{tr} = \pm(1/R_G)(E_{pG}^2 R_G^2 + E_{rG}^2 \tan^2 \varphi + E_{pP}^2 R_P^2 + E_{rP}^2 \tan {}^2\varphi)^{1/2} \tag{32.69}$$

Because the units of position errors for the above equations are radians, each E_p is multiplied by the radius of the corresponding pinion or gear to obtain an equivalent linear distance at the pitch circle.

(d) Approaches to Optimum Accuracy

Ideally, tolerances affecting the accuracy of a precision gear train should be assigned so that, for a given train output error, cost is a minimum—or conversely, for a given cost, train output error is a minimum. This is not easy to do. Necessary cost-tolerance data are seldom available; even when they are, the effort required for calculation may not be justified. However, optimum accuracy in reducing gear trains of conventional design can be approached in three ways. First, each mesh reduction should be as high as practical, especially the reduction of the last mesh. Secondly, the diameters of gears should be as large as possible, in particular the gear at the train output. Thirdly, values of ϵ allowed at each mesh should be in some relationship to their corresponding values of M. The basis for the three approaches to optimum accuracy is evident in the equations relating allowances and tolerances to mesh error, and mesh errors to a component of train error.

The important term in Eqs. (32.59) to (32.62) is the effective mesh error $M\epsilon$. For example, in a train of four meshes, each mesh having a 5:1 reduction, the effective mesh errors are $0.008\epsilon_1$, $0.04\epsilon_2$, $0.2\epsilon_3$, ϵ_4. The most critical error is that of the output mesh because it is not modified by a subsequent mesh reduction. Assigning the same

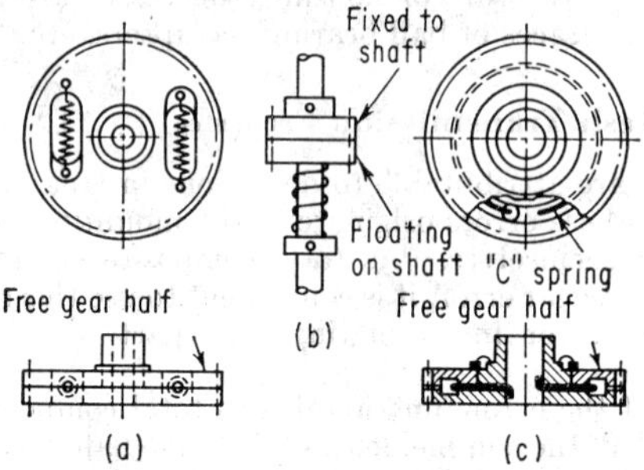

FIG. 32.39. Types of spring-loaded anti-backlash gears.[15] (a) Extension- or compression-spring type. (b) Torsion-spring type. (c) C-type spring.

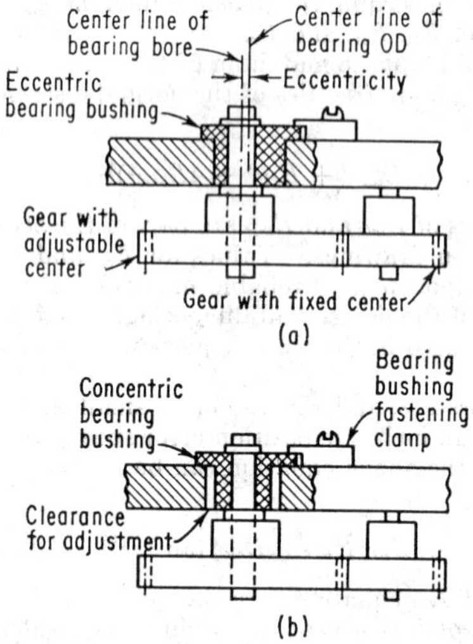

FIG. 32.40. Two types of adjustable-gear-center designs.[15] (a) Eccentric bearing. (b) Floating bearing.

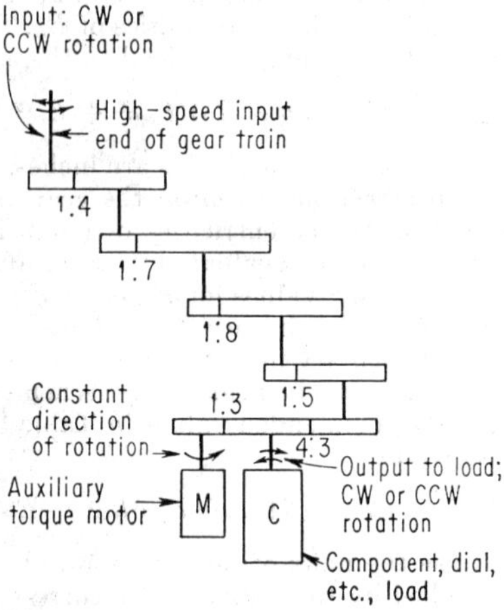

FIG. 32.41. Schematic diagram of a gear train using an auxiliary torque motor to reduce backlash.[15]

value to each ϵ would not make sense because the effect of the mesh errors at the input end is small compared with that at the output—yet the cost of both ϵ_1 and ϵ_4 would be about the same. Suppose that, by doubling the allowable value of ϵ_1 its cost were cut by \$10 and that by spending this \$10 on ϵ_4 its value could be reduced by 20 per cent. The net result would be an improvement in train accuracy at no additional cost. Thus a reducing gear train will approach optimum accuracy when $\epsilon_n < \epsilon_{n-1} < \epsilon_{n-2}$ and so on until any further increase will cause a deterioration of function for reasons other than train accuracy. The amount by which any ϵ should differ from ϵ_n can be based on some reasonable assumption, for instance, that $\epsilon_n = M_j^{1/2}\epsilon_j$.

By making each mesh as large as possible, the number of meshes, hence the number of error sources, will be a minimum. If in order to obtain a desired train value it is convenient to have unequal mesh reductions, then the largest reductions should occur closest to the output because each reduction modifies all previous mesh errors. For example, in a gear train having three reductions, 5:1, 4:1, and 3:1 in that order, the corresponding effective mesh errors are $0.083\epsilon_1$, $0.33\epsilon_2$, and ϵ_3. Changing the order of the reductions to 3:1, 4:1, and 5:1 results in effective mesh errors $0.05\epsilon_1$, $0.2\epsilon_2$, and ϵ_3.

Equations (32.63) to (32.69) show that mesh accuracy varies inversely as the radius of the driven gear R_G. A gear train of small-diameter gears is inherently less capable of accuracy than one with larger gears. As the last mesh is most critical, any increase or decrease in R_{Gn} will have a considerable effect on train error.

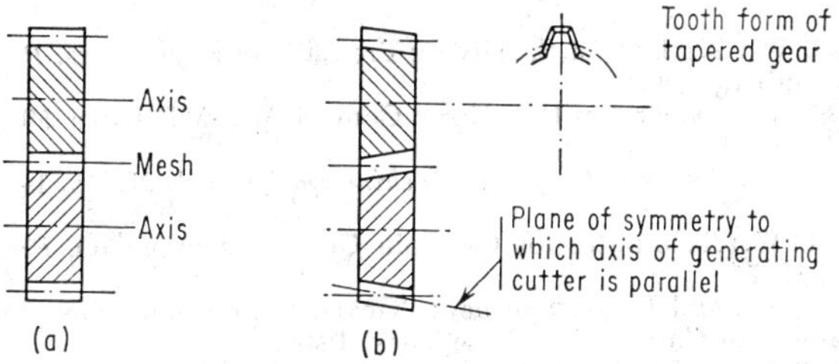

FIG. 32.42. Tapered gears contrasted to normal spur gears.[15] (*a*) Normal spur gear. (*b*) Tapered gear mesh.

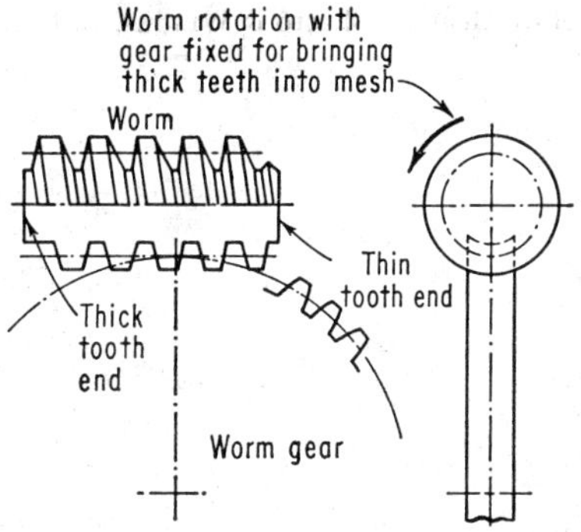

FIG. 32.43. Variable-tooth-thickness worm for eliminating backlash.[15]

Another approach to optimum accuracy is to eliminate all or part of the backlash component of train error. Various methods for achieving this are shown in Figs. 32.39 to 32.43. Antibacklash spring-loaded gears are suitable for light-duty instrument and control gear trains. These gears eliminate all backlash and allow manufacture to liberal tolerances. With adjustable gear centers only the constant component of train backlash ϵ_{Tbc} can be eliminated by adjusting for no backlash at the condition of tightest mesh. Gear trains with adjustable gear centers can be designed to transmit substantial torque. Because of the skill required for adjusting gear centers, assembly and field replacement is apt to be troublesome. Selective assembly techniques and the use of matching gear pairs also have limited application because of assembly and field replacement difficulties. Methods such as those shown in Figs. 32.41 to 32.43 are used only in special circumstances.

References

1. Dudley, D. W.: "Gear Handbook," McGraw-Hill Book Company, Inc., New York, 1962.
2. Vogel, W. F.: "Involutometry and Trigonometry," Michigan Tool Co., Detroit, Mich., 1945.
3. Candee, Allan H.: "Introduction to the Kinematic Geometry of Gear Teeth," Chilton Company, Philadelphia, 1961.
4. Chironis, Nicholas P.: New AGMA Classification System for Gears, *Prod. Eng.*, July 10, 1961.
5. Dean, Paul M.: Effects of Size on Gear Design Calculations, *Prod. Eng.*, April, 1954.
6. Dudley, D. W.: "Practical Gear Design," McGraw-Hill Book Company, Inc., New York, 1954.
7. Merritt, H. E.: "Gears," Pitman Publishing Corporation, New York, 1954.
8. Michalec, G. W.: Critical Criteria for Precision Gears, *Elec. Mfg.*, vol. 66, no. 1, July, 1960.
9. Michalec, G. W.: Methods of Specifying Precision of Spur Gears, *Prod. Eng.*, vol. 27, no. 12, November, 1956.
10. Michalec, G. W.: "Gear Position Error Control," ASME paper 59-A-21, December, 1959.
11. Michalec, G. W.: Precision Gearing, *Machine Design*, vol. 27, pp. 1–4, January–April, 1955.
12. Michalec, G. W.: Tolerances for Over-pin Gear Measurements, *Prod. Eng.*, vol. 28, no. 5, May, 1957.
13. Parkinson, A. C., and H. W. Downey: "Gears, Gear Production and Measurement," Pitman Publishing Corporation, New York, 1948.
14. Steeds, W.: "Involute Gears," Longmans, Green & Co., Inc., New York, 1948.
15. Truxal, John C. (ed.): "Control Engineers' Handbook," chap. 13, McGraw-Hill Book Company, Inc., New York, 1958.
16. Tuplin, W. A.: "Gear Design," The Industrial Press, New York, 1963.
17. Tuplin, W. A.: "Gear Load Capacity," Sir Isaac Pitman & Sons, Ltd., London, 1962.
18. Tuplin, W. A.: "Involute Gears," Chatto & Windus, Ltd., London, 1962.

Section 33

SPRINGS*

By

WILLIAM R. JOHNSON, B.S., *Assistant Director of Research and Development, Associated Spring Corp., Bristol, Conn.*

CONTENTS

33.1. INTRODUCTION

A spring is a device which stores energy or provides a force over a distance by elastic deflection. Energy may be stored in a compressed gas or liquid, or in a solid that is bent, twisted, stretched, or compressed. The energy is recoverable by the elastic return of the distorted material. Spring structures are characterized by their ability to withstand relatively large deflections elastically. The energy stored is measured by the volume of elastically distorted material, and in the case of metal springs, the volume of distorted material is limited by the spring configuration and the stress-carrying capacity (elastic limit) of the most highly strained portion. For this reason, most commercial spring materials come from the group of high-strength materials, including high-carbon steel, cold-rolled and precipitation-hardening stainless and non-ferrous alloys, and a few specialized nonmetallics such as laminated fiberglass. Thus spring design is basically an analysis of machine elements which undergo relatively large elastic deflections and are made of materials capable of withstanding high stress levels without yielding.

Units. The basic units used in the text are the pound, the inch, and the second.

Symbols. Following are the letter symbols for equations and their corresponding units. Where a symbol can have more than one interpretation its meaning is made explicit in each section where it is used.

b = breadth or width, in.
C = spring index, dimensionless
d = wire or bar diameter, in.
D = diameter, in.
E = modulus of elasticity, psi
f = frequency of vibration, cps
g = gravitational constant, 386.4 in./sec/sec
G = modulus of elasticity in shear, psi
h = height, in.
k = rate, lb/in. or in.-lb/rad
K = factor, used variously
ln = natural logarithm
L = length, in.
M = moment of force, in.-lb
N = number of active coils in helically wound springs, dimensionless
n = quantity or count, dimensionless
p = pitch, in.
P = force, lb
r = radius, in.
R = radius, in.
t = thickness, in.
T = number of turns, dimensionless
U = energy, in.-lb
v = velocity, in./sec
V = volume, in.^{3}
δ = linear deflection, in.
θ = angular deflection, rad
μ = Poisson's ratio, dimensionless
γ = density, lb/in.^{3}
σ = normal stress, psi
τ = shear stress, psi
ϕ = angle, deg

The following terms are peculiar to spring engineering. Definitions of general terms are not given.

Rate (see Art. 33.2). Also referred to as scale, gradient, and load factor.

Spring index, applies to helical-type springs. For springs wound with circular wire, spring index $C = D/d$; for springs wound with rectangular wire $C = D/t$ if t is in the radial direction, or $C = D/b$ if b is in the radial direction. See Figs. 33.8 and 33.13*b*.

Active coils, coils free to deflect under load.

Inactive coils, coils not free to deflect under load.

Total coils, the sum of active and inactive coils.

Free length or height, a length or height measurement of a spring in the free (unloaded) condition.

Solid length or height, a length or height measurement of a compression spring when no further deflection is possible.

Solid stress, the stress corresponding to deflection to solid length or height.

Set, a permanent distortion resulting when a spring is stressed beyond the elastic limit.

Initial tension, the force holding coils closed in the free or unloaded condition of helical extension or torsion springs.

Sizes (Table 20.51). Heavy steel wire springs are specified according to Washburn and Moen (Roebling) wire gage. However, sizes that are multiples of 1/32 in. can usually be obtained; sizes 3/8 in. and larger come in multiples of 1/32 in. up to 5/8 in. and in multiples of 1/16 in. for larger sizes. Nonferrous round wire is specified according to the American wire gage (Brown and Sharpe). Spring plates are made in accordance with Birmingham wire gage and also in thicknesses varying in 1/32-in. increments from 3/16 to 1/2 in. thick. Precision spring wires are usually specified in thousandths of an inch.

33.2. SPRING CHARACTERISTICS

The load-deflection curve is the most fundamental of all spring properties. For a load-deflection curve of the form $P = f(\delta)$ the force P will cause a linear deflection δ in the direction of P; when the load is a moment of force M, the load-deflection curve has the form $M = f(\theta)$ and M is in the direction of θ. Four types of load-deflection curve are represented in Fig. 33.1*a* to *d*.

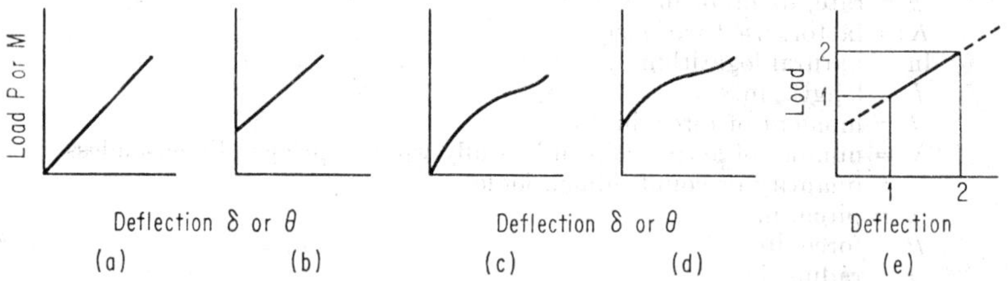

FIG. 33.1. Types of load-deflection curve. (*a*) Linear, no initial load. (*b*) Linear, initial load. (*c*) Nonlinear, no initial load. (*d*) Nonlinear, initial load. (*e*) Operating range.

(a) Rate

By definition, rate is the slope of the load-deflection curve

$$k = dP/d\delta \quad \text{or} \quad k = dM/d\theta \tag{33.1}$$

When the load-deflection curve is linear, rate is constant so that

$$k = \Delta P/\Delta\delta \quad \text{or} \quad k = \Delta M/\Delta\theta \tag{33.2}$$

For a linear load-deflection curve with no initial load, Eq. (33.2) reduces to

$$k = P/\delta \quad \text{or} \quad k = M/\theta \tag{33.3}$$

(b) Operating Range

Many springs operate over only a part of their useful load-deflection curve. The operating range is defined by the limits of load and deflection as shown in Fig. 33.1*e*.

(c) Energy Storage

The energy stored by deflecting a spring over its operating range,

$$U = \int_{\delta_1}^{\delta_2} P\,d\delta \quad \text{or} \quad U = \int_{\theta_1}^{\theta_2} M\,d\theta \tag{33.4}$$

For any linear load-deflection curve,

$$U = \frac{(P_2 + P_1)(\delta_2 - \delta_1)}{2} \quad \text{or} \quad U = \frac{(M_2 + M_1)(\theta_2 - \theta_1)}{2} \tag{33.5}$$

If in addition to the load-deflection curve being linear, the initial load is zero, Eq. (33.5) reduces to

$$U = P\delta/2 \qquad \text{or} \qquad U = M\theta/2 \tag{33.6}$$

(d) Stress

Values of maximum allowable load are determined by equations relating stress and load. Calculated stresses differ from true stresses because of the approximate nature of the stress formulas, because these formulas do not take into account residual stresses induced in manufacturing, and because of the difficulty in evaluating stress concentrations. For these reasons, limits for calculated stresses are not always based merely on the permissible stress values for spring materials as determined by standard tests. Rather, the amount by which calculated stresses can approach permissible values for spring materials often depends on factors devised from tests.

Residual stresses exist. They may be either beneficial or detrimental, depending upon the direction of the applied service stresses. Favorable residual stresses may be deliberately added to the system already present from the cold-forming operations.

As a general rule, plastic flow caused by a forming system which is in the same direction as the service load will result in a favorable residual-stress system, one in which the pertinent residual stresses are opposite in sign to the service stresses. These residual stresses may be both tension and shear, and calculations must be made on a combined-stress basis. For example, a helical spring to be used in compression will be able to support a higher load elastically if it has been previously "set out" or overloaded into the plastic region by a compression load. If, however, it has been stretched out previously (to meet a blueprint dimension), the residual-stress system is unfavorable, and the elastic limit in compression will be below expectations unless the spring is again stress-relieved. Similarly, if a flat spring is formed by bending pretempered material, the residual stresses will be favorable to loads tending to close up the bend and unfavorable to loads opening the bend. A cold-wound torsion spring is best stressed when its coils tend to wind down onto an arbor and unfavorably stressed when the coils unwind.

For best spring design it is thus necessary to predict the probable magnitude and direction of the residual stresses and add them to the service stresses. In extreme cases of unfavorable residual stresses, hardening the spring after forming may be necessary. However, it is always best to attempt to design so that residual stresses can be utilized effectively; if not, the spring manufacturer must have the necessary design information so that he can stress-relieve the springs at the highest possible temperature.

The simple spring formulas do not take into account the residual-stress system which is always present in cold-formed springs. Most stress calculations are therefore directed toward determining relative rather than absolute stress figures. Charts of allowable design stresses take into account the residual-stress factor and a safety factor.

Conditions of stress concentrations introduced by notches, holes, and curvature not otherwise included in stress equations should be included in stress calculations.

(e) Volume Efficiency[2,7]

The volume efficiency of a spring depends on the amount of energy that can be stored per unit volume of active spring material when the spring is loaded so that stress is the maximum allowable. For springs with load-deflection curves of the type shown in Fig. 33.1*a* and subject to a static load, the resilience

$$U/V = K(\max \sigma)^2/E \qquad \text{or} \qquad U/V = K(\max \tau)^2/G \tag{33.7}$$

V is the volume of active spring material. Values of K for common spring types are given in Table 33.1; K is an index of volume efficiency.

Table 33.1. Values of K

Spring type	K
1. Helical extension and compression, round wire:	
C-3	1/5.4
C-6	1/4.7
C-9	1/4.5
C-12	1/4.3
2. Helical torsion, round wire	1/8
3. Helical torsion, rectangular wire	1/6
4. Cantilever leaf, uniform rectangular section	1/18
5. Triangular cantilever leaf, rectangular section	1/6
6. Spiral spring, rectangular section	1/6
7. Torsion bar, circular section	1/4
8. Belleville spring	1/40 to 1/10

33.3. SPRINGS IN COMBINATION

Springs are often used in combination because of space limitations, because a combination may be more efficient than a single equivalent spring or because a combination will give a load-deflection curve or dynamic characteristics not possible with a single spring of standard design.

For simplicity, the examples shown are of combinations having the fewest number of springs; these examples can be easily generalized to apply to larger combinations.

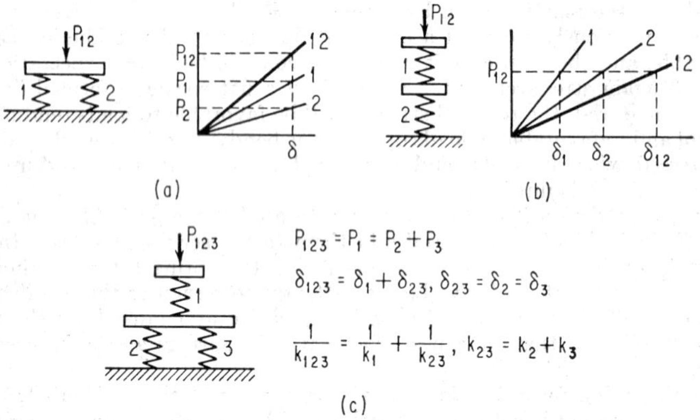

FIG. 33.2. Spring combination. (*a*) Parallel springs, linear load deflection. $P = 0$ where $\delta = 0$. (*b*) Series springs, linear load deflection. $P = 0$ where $\delta = 0$. (*c*) Springs in series-parallel combinations.

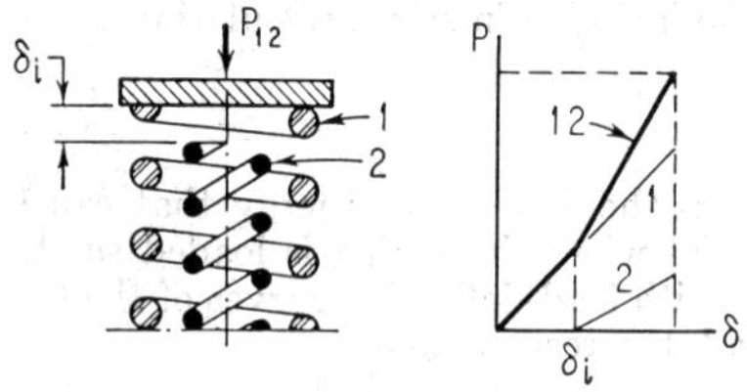

FIG. 33.3. Nest of helical springs. Inner spring shorter than outer results in abrupt change in load-deflection curve. Note that springs are wound opposite in hand to prevent tangling.

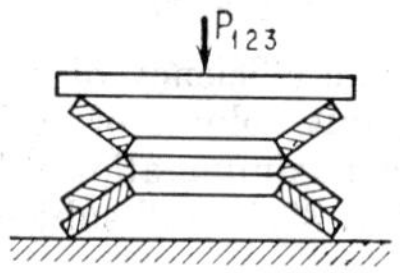

FIG. 33.4. Nest of Belleville springs.

(a) Series and Parallel Combinations

Figure 33.2 shows schematically how springs are combined in parallel and series. For the parallel combination,

$$P_{12} = P_1 + P_2 \qquad \delta_{12} = \delta_1 = \delta_2 \qquad k_{12} = k_1 + k_2 \tag{33.8}$$

For the series combination,

$$P_{12} = P_1 = P_2 \qquad \delta_{12} = \delta_1 + \delta_2 \qquad 1/k_{12} = 1/k_1 + 1/k_2 \tag{33.9}$$

A parallel-series combination is shown in Fig. 33.2*c*.

When the operating range of a combination is within the operating range of each component spring, the resulting load-deflection curve will not have any abrupt changes. However, for springs combined as in Fig. 33.3, the load-deflection curve will change abruptly when the inner spring begins to deflect. Similarly, the load-deflection curve for the combination of Belleville springs shown in Fig. 33.4 will change abruptly when the single spring has deflected to the flat condition.

(b) Energy Storage

The energy stored within the operating range of a combination of springs may be determined from the load-deflection characteristics of the combination or by adding together the energy stored by the separate springs.

33.4. SPRING MATERIALS AND OPERATING CONDITIONS

In this section properties of some common spring metals are given. Data applicable only to specific types of springs are given in the sections devoted to these springs. For further data consult refs. 1, 3, and 4.

(a) Common Spring Metals

Designations for and properties of common spring metals are given in Table 33.2.

(b) Strength of Spring Wire and Flat Stock

Figure 33.5 shows the dependence of tensile strength on wire diameter. Figure 33.6 and Table 33.3 give recommended working stresses for flat-spring materials.

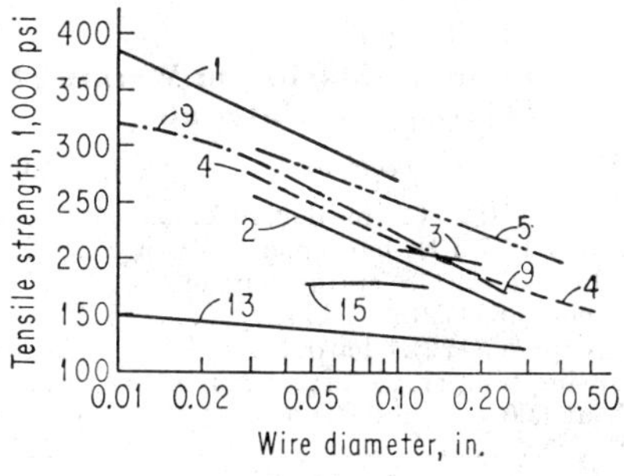

FIG. 33.5. Tensile strength of spring wires vs. diameter.[3] Numbers correspond to listing in Table 33.2. (1) Music wire ASTM A228. (2) Hard-drawn ASTM A227. (3) Valve spring ASTM A230. (4) Oil-tempered ASTM A229. (5) Chrome vanadium AISI 6150. (9) Stainless steel AISI 302. (13) Phosphor bronze ASTM B159. (15) Beryllium copper ASTM B197.

Table 33.2. Common Spring Metals

Common name, specification	E 10^6 psi	G 10^6 psi	γ, density lb/in.³	Max service temp., °F	Electrical conductivity*	Principal characteristics
High-carbon steels:						
1. Music wire, ASTM A228	30	11.5	0.283	250	7	High strength, excellent fatigue life
2. Hard drawn, ASTM A227	30	11.5	0.283	250	7	General-purpose use, poor fatigue life
3. Valve spring, ASTM A230	30	11.5	0.283	300	7	For infinite fatigue life at average stress range
4. Oil tempered, ASTM A229	30	11.5	0.283	300	7	For moderate stress; no impact or shock
Alloy steels:						
5. Chrome vanadium, AISI 6150	30	11.5	0.283	425	7	High strength, resists shock, good fatigue life. AISI 9260 used as less expensive substitute for AISI 6150
6. Chrome silicon, AISI 9254	30	11.5	0.283	475	5	
7. Silicon manganese, AISI 9260	30	11.5	0.283	450	4.5	
Stainless steels:						
8. Martensitic, AISI 410, 420	29	11	0.280	500	2.5	Unsatisfactory for subzero applications
9. Austenitic, AISI 301, 302	28	10	0.286	600	2	Good strength at moderate temperatures, low creep
10. Precipitation hardening, 17-7PH	29.5	11	0.286	700	2	Long life under extreme service conditions
11. High temperature, A286	29	10.4	0.290	950	2	For elevated-temperature applications
Copper-base alloys:						
12. Spring brass, ASTM B134	15	5.5	0.308	200	26	Low cost, high conductivity; poor mechanical properties
13. Phosphor bronze, ASTM B159	15	6.3	0.320	200	18	Able to withstand repeated flexures. Popular alloy
14. Silicon bronze, ASTM B99	15	6.4	0.308	200	7	Used as less expensive substitute for phosphor bronze
15. Beryllium copper, ASTM B197	19	6.5	0.297	400	21	High elastic and fatigue strength; hardenable
Nickel-base alloys:						
16. Inconel	31	11	0.307	600	1.5	Good strength, high corrosion resistance
17. Inconel X	31	11	0.298	1100	1	Precipitation hardening; for high temperatures
18. Ni-Span C	27	9.6	0.294	200	1.6	Constant modulus over a wide temperature range

* Electrical conductivity given as a per cent of the conductivity of the International Annealed Copper Standard.

Table 33.3. Recommended Maximum Working Stresses for Flat-spring Materials[4]

Material, specifications	1,000 *psi* σ
Stainless steel, AISI 302, hard	130
Stainless steel 17-7 PH, hardened	160
Phosphor bronze (521),* spring temper	70
Beryllium copper (172),* ½ hard	90
Beryllium copper (172),* hard	110
Inconel, spring temper	90
Inconel X at 500°	100
At 700°	90
At 900°	75

* Numbers are of the Copper and Brass Research Association (CABRA) standard alloys.

Data apply to springs subject to static and low-cycle duty at normal temperatures. Recommendations as to the use of the data are given under specific spring types.

(c) Cyclic Loading

Fatigue failures usually start at a surface defect or point of stress concentration. Metals hardened by heat-treatment show typical fatigue fractures. Some hard-drawn wires develop long longitudinal cracks which are often mistaken for metallurgical seams; hence tests of springs made from hard-drawn wire can be misleading if not properly monitored, as a crack can grow over several hundred thousand cycles without external sign of failure.

It is difficult to apply to springs much of the data obtained from the usual reverse-bending and torsional-fatigue tests. There are numerous reasons for this difficulty. The chief reasons are: Pressure of residual stresses is inherent in springs but not found in test specimens; the hardness at which metals are used in springs makes them more sensitive to stress concentrations than the same metals fatigue-tested at lower hardnesses; most springs are not subject to complete stress reversals as is the case with test specimens.

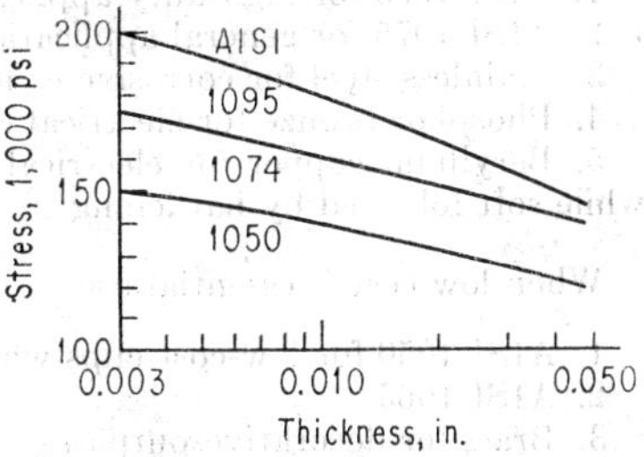

FIG. 33.6. Recommended working stresses for pretempered carbon steels.

The difference between stresses at the extremes of the operating range is called the *stress range*. Allowable stress ranges must be evaluated by using Goodman or Soderberg diagrams, *SN* curves, etc. Because of the variables involved in cyclic loading, the number of cycles to failure between similar springs may differ by a factor of 5.

Springs in cyclic service may also fail by setting, particularly when highly stressed, as small amounts of plastic flow undetectable in a single cycle can accumulate over many operations.

(d) Selection of Materials

When fatigue resistance is important and wire sizes small, use:

1. Music wire where space is limited and stresses high
2. Stainless steel when corrosive conditions exist or temperatures are as high as 600°F
3. Oil-tempered or hard-drawn wire when space is available and costs must be kept low
4. Chrome-vanadium steel when temperature approaches 400°F
5. Phosphor bronze when electrical conductivity is important

When fatigue resistance is important and wire sizes above ⅛ in., use:

1. Valve-spring wire for critical applications
2. Oil-tempered or hard-drawn wire when space is available and costs must be kept low
3. Chrome-vanadium steel when temperatures approach 400°F
4. Chrome-silicon steel when temperatures are as high as 500°F
5. Stainless steel when corrosive conditions exist
6. Phosphor bronze when electrical conductivity is important

When resistance to set in static loads is important, use:

1. Music wire where space is limited and wire size small
2. Oil-tempered wire, lower stress than music wire

3. Hard drawn wire where low cost is essential
4. Chrome-vanadium steel when temperatures are as high as 400°F
5. Chrome-silicon steel when temperatures are as high as 500°F
6. Stainless steel for corrosion resistance and temperatures approaching 600°F
7. Inconel, A286 for temperatures 500 to 800°F
8. Inconel X for temperatures above 800°F
9. Beryllium copper when electrical conductivity is important and stresses are high

When fatigue resistance or high resistance to setting is important use (flat springs):

1. AISI 1095 for high-duty applications
2. AISI 1075 for general application
3. Stainless steel for corrosive conditions
4. Phosphor bronze for electrical conductivity
5. Beryllium copper for electrical conductivity and shapes that demand forming while soft followed by hardening

When low cost is essential use:

1. AISI 1050 for low-cost clips where volume is substantial
2. AISI 1065
3. Brass for decorative purposes

(e) Tolerances

The designer of springs, like any other machine designer, should not call for dimensional restrictions more severe than needed. Often the exact dimensions of the spring may not be critical, only its load-deflection characteristics. Any spring to be used in large quantities should be carefully designed with this in mind; the problem will be simplified if adequate space is left for the spring at an early stage in the mechanical design.[13]

Allowable variations in commercial spring-wire sizes are given in Table 33.4. Tolerance on free length, coil diameter, spring rate, load, and squareness can be calculated from the following relations, used in conjunction with Table 33.5:

$$\begin{aligned} \text{Free-length tolerance} &= \pm T_1\, CL \\ \text{Coil-diameter tolerance} &= \pm T_2 D \\ \text{Spring-rate tolerance} &= \pm (T_3 + T_4)k \\ \text{Load tolerance} &= \pm (T_4 + T_5)P \\ \text{Squareness tolerance} &= \psi,\ \text{deg} \end{aligned}$$

Table 33.4. Allowable Variations in Commercial Spring-wire Sizes

Material	ASTM Specification No.	Wire diam, in.	Permissible variation, in.
Hard-drawn spring wire	A227	0.028–0.072	±0.001
Chrome-vanadium spring wire	A231	0.073–0.375	±0.002
Oil-tempered wire	A229	0.376 and over	±0.003
Music wire	A228	0.026 and under	±0.0003
		0.027–0.063	±0.0005
		0.064 and over	±0.001
Carbon-steel valve spring wire	A230	0.093–0.148	±0.001
		0.149–0.177	±0.0015
Chrome-vanadium valve spring wire	A232	0.178–0.250	±0.002

Table 33.5. Constants for Computing Tolerances of Commercial Springs

L	T_1	d	T_2	T_4	N	T_3	L/δ	T_5	$D^3/2Ld$	ψ, deg
0.4	0.0092	0.010	0.0485	0.0490	2	0.0560	1.1	0.0200	0.5	1.62
0.7	0.0073	0.020	0.0367	0.0435	3	0.0480	1.5	0.0269	0.7	1.72
1.0	0.0063	0.030	0.0314	0.0405	4	0.0430	2	0.0353	1	1.84
1.5	0.0054	0.040	0.0278	0.0385	5	0.0395	3	0.0512	2	2.09
2	0.00477	0.060	0.0238	0.0360	6	0.0368	4	0.0675	3	2.24
3	0.00405	0.080	0.0210	0.0342	8	0.0330	5	0.083	4	2.36
4	0.00362	0.100	0.0194	0.0330	10	0.0303	6	0.098	5	2.45
5	0.00332	0.150	0.0164	0.0307	15	0.0260	8	0.129	6	2.54
6	0.00306	0.200	0.0147	0.0293	20	0.0233	10	0.159	8	2.66
7	0.00288	0.300	0.0125	0.0273	25	0.0214	13	0.203	10	2.76
8	0.00274	0.400	0.0113	0.0260	30	0.0200	16	0.247	12	2.85
10	0.00250	0.500	0.0103	0.0250	40	0.0180	20	0.303	15	2.96

(f) Springs at Reduced and Elevated Temperatures

The mechanical properties of spring metals vary with temperature. For most spring metals E and G decrease almost linearly with temperature within their useful ranges of application. One exception is the family of nickel-base alloys which includes Elinvar and Ni-Span C whose composition is specifically designed to confer a near zero temperature coefficient of spring rate over a limited temperature range around room temperature. This anomalous behavior is caused by a gradual and reversible loss of ferromagnetic effect as the material is raised in temperature toward its Curie point. Once the Curie point is neared (at temperatures above 200°F), the temperature coefficient of rate is comparable with that of other spring alloys.

Reduced Temperature. The body-centered-cubic materials such as spring steel are normally considered to be brittle at subzero temperatures, and the plain-carbon and low-alloy steels are not generally recommended for such service. However, a spring structure is designed to be resilient, and thus the spring configuration may protect the spring material from brittle failure. In special cases of low-temperature application where a large quantity of springs is required and the cost must be low, ordinary spring materials should not be disregarded without a service test. Otherwise, the face-centered-cubic materials, such as copper-base, nickel-base, or austenitic iron-base alloys, should be specified.

Elevated Temperature. Springs used at temperatures above room temperature must be designed with allowance for the rate change associated with modulus decrease, if accuracy of rate is essential.

The type of failure characteristic of springs at elevated temperature is setting, loss of load. The setting of a spring is the result of two types of flow—plastic and anelastic. Plastic flow of the spring material is simply the result of exceeding the elastic limit or yield point at the service temperature. This level limits the allowable stress at any given temperature, regardless of length of time under load. The loss of load caused by plastic flow is not recoverable.

In addition, a spring under load for a period of time will lose more load or set more, by a process called *anelastic flow*. This flow is time- and temperature-dependent and may be recoverable when the load is removed. The net effect in spring service is that a spring will lose load or relax in service at a rate which is stress-, time-, and temperature-dependent. If the spring is unloaded at temperature, the load loss may be partially recoverable. The complete relaxation curve for a spring loaded to a constant deflection at elevated temperature consists of a very rapid first stage characterized by plastic flow caused by the lower yield point at elevated temperature, and a second stage of anelastic flow at a rate that is dependent on the log of the time under load. There has been no evidence of a third stage or that relaxation ever stops, although the rate is decreasing. If the spring is loaded under constant-load (creep) conditions, instead of constant-deflection (relaxation) conditions, the rate of load loss is far higher, particularly at longer test times.[5]

The accurate and efficient design of a high-temperature spring is a difficult undertaking. Assuming that operating temperatures and spring-loading conditions are well known (and they seldom are), the designer must know what relaxation he can accept within his specifications of load loss and time of service, for zero change is close to unobtainable. The more stable the spring must be and the longer the time of service, the lower the design stress must be. On the other hand, if the service is intermittent, with periods of no load at temperature, the spring may recover a portion of its load loss in the "rest" periods and relax far less than might be anticipated. Because of these variables, data plots can only indicate the worst service condition, that of constant loading for the times and temperatures indicated.

If the spring requirements for minimum relaxation are severe, heat setting may be used. This is a process for inducing favorable residual stresses during manufacture by preloading the spring at higher than service temperatures. Such a process can contribute markedly to the apparent relaxation resistance of a spring, but since the heat setting is actually an anelastic-flow process, its effects are recoverable. In other words, a heat-set spring will grow and lose its stability if it is unloaded at the service temperature. Furthermore, if the heat-setting operation is improperly carried out, its effects may be transient, and the spring will behave unpredictably.

(g) Constancy of Rate

Some springs are such that their rate is constant over the entire usable range of deflection. For example, a helical extension spring has an essentially constant rate after the initial tension has been exceeded and before the yield point is reached. A torsion spring exhibits a constant rate within the elastic range, because the decrease of diameter as the spring winds down on an arbor is compensated for by the increase in number of coils. In both examples cited, minor deviations from a straight-line rate may be caused by end or loop deflection.

A helical compression spring will show a constant rate only in the central portion of its load-deflection plot. Initially, nonuniform compression of the end coils appears to produce a lower rate than calculated, and as the spring approaches solid, the rate increases. A constant-rate design should utilize only the center 60 per cent of the total deflection. If the compression spring must have a constant rate over a major portion of its deflection range, certain special techniques can achieve this at added cost. The closed-end coils may be soldered or brazed together to reduce the effect of early end closure. A more exact solution is to braze an insert into the dead coils. The most accurate spring is made by introducing slits in tubing to form either a continuous helix or a series of cantilever sections. This is a very expensive solution.

Some springs are deliberately designed with a nonconstant rate, e.g., conical springs and volute springs. These are useful in shock-absorbing operations because their rate continues to increase with deflection. No helical compression spring can be designed with a decreasing rate.[6]

A few specialized spring designs may have a rate close to zero, a feature desirable in counterbalance springs and springs used in seals. Belleville springs of special configurations, the constant-force spring and a buckling column, fall in this category. Belleville springs can also be designed with a decreasing rate or even a negative rate over a given deflection range, so that snap-through action will occur. Buckling flat springs, which will snap through, are also a common spring application of negative rate.

The torque output of a motor spring decreases with each turn, and the rate of change depends upon the spring design and manufacturing technique. A constant-force spring can be made to have a torque output that is nearly constant over the useful range of the spring.

33.5. HELICAL COMPRESSION SPRINGS

(a) Principal Characteristics

The load-deflection curve is of the type shown in Fig. 33.1*a*. For springs wound from circular wire (see Fig. 33.7),

$$P = \delta G d^4/8ND^3 \qquad \tau = 8K_A PD/\pi d^3 \tag{33.10, 33.11}$$

For springs wound from rectangular wire,

$$P = K_1 \delta G b t^3/ND^3 \qquad \tau = K_A K_2 PD/bt^2 \tag{33.12, 33.13}$$

Values for stress factor K_A are given in Table 33.6. Values for the shape factors K_1 and K_2 are given in Table 33.7.[7]

Table 33.6. Stress Factors for Helical Springs

C	3	4	5	6	8	10	12	16
K_A	1.58	1.40	1.31	1.25	1.18	1.15	1.12	1.09
K_B	1.33	1.23	1.17	1.14	1.10	1.08	1.07	1.05
K_C	1.29	1.20	1.15	1.12	1.09	1.07	1.05	1.04

Table 33.7. Shape Factors K_1 and K_2 for Helical Compression Springs Wound from Rectangular Wire[6]

b/t	1.00	1.50	1.75	2.00	2.50	3.00	4.00	6.00
K_1	0.180	0.250	0.272	0.292	0.317	0.335	0.358	0.381
K_2	2.41	2.16	2.09	2.04	1.94	1.87	1.77	1.67

Types of ends for helical compression springs are shown in Fig. 33.8. *Open ends not ground* are satisfactory only if accuracy of load is not important. Springs of this type tangle easily when loose packed for shipment or storage. *Open ends ground* also tangle easily and are generally specified only where it is necessary to obtain as many

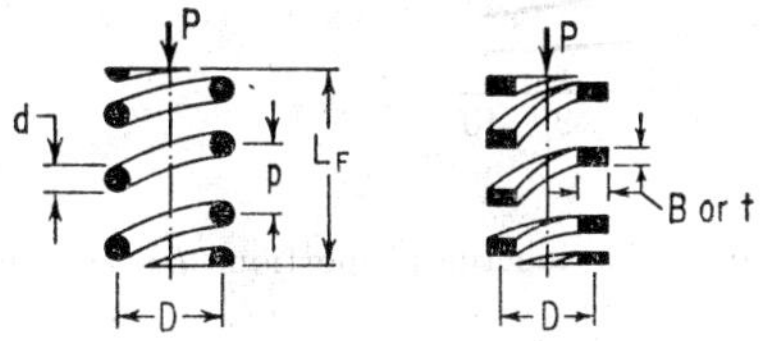

FIG. 33.7. Helical compression springs. L_F is the free length. For rectangular wire $b > t$; $b = t$ if the wire has a square section.

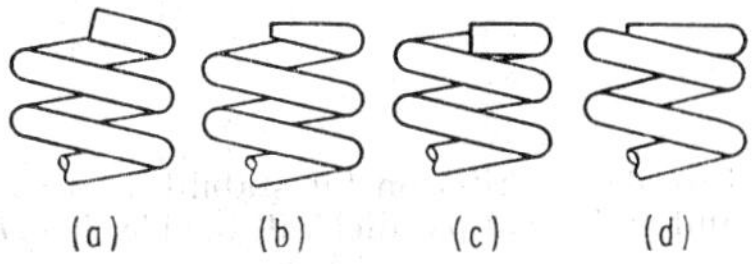

FIG. 33.8. Types of ends for helical compression spring. (*a*) Open, not ground. (*b*) Open and ground. (*c*) Closed, not ground. (*d*) Closed and ground.

coils as possible in a limited space. *Closed ends* not ground are preferred for wire sizes less than 1⁄32 diameter or thickness, or if $C > 13$. Closed ends ground are preferred for wire sizes greater than 1⁄32 diameter or thickness and if $C < 13$. Relationships between types of ends and spring measurements are given in Table 33.8.[1]

Table 33.8. Helical Compression Spring Measurements for Common Types of Ends

Type of ends	L_F	N_T	L_S
Open, not ground.	$pN + d$	N	$dN + d$
Open and ground.	pN	N	dN
Closed, not ground.	$pN + 3d$	$N + 2$	$dN + 3d$
Closed and ground.	$pN + 2d$	$N + 2$	$dN + 2d$

L_F = free length, N_T = total number of coils, L_S = solid height.
Table applies to springs wound from rectangular wire if b or t as appropriate is substituted for d.

(b) Lateral Loading

Occasionally an application requires lateral as well as axial loading of a helical compression spring. Although the axial rate of a spring is constant for all loads, the lateral rate varies with compression. An adjustable spring-rate system can thus be designed to take advantage of this fact.

The combined stress of a laterally loaded spring

$$\tau = \tau_A(1 + F_L/D + P_L/PD) \qquad (33.14)$$

where τ_A is the axial stress, P the axial load, L the length of the spring under load, and F_L and P_L the lateral deflection and lateral load, respectively.

(c) Buckling and Squareness in Helical Springs

If the free length of a compression spring is more than four times the spring diameter, its stability under load may become critical, and the spring may buckle as a column. The lateral stability of a spring depends on the ratio of its free length to mean diameter L_F/D. A slender spring will buckle under load when the ratio of deflection to free length S/L_F exceeds a critical value. Figure 33.9 can be used to

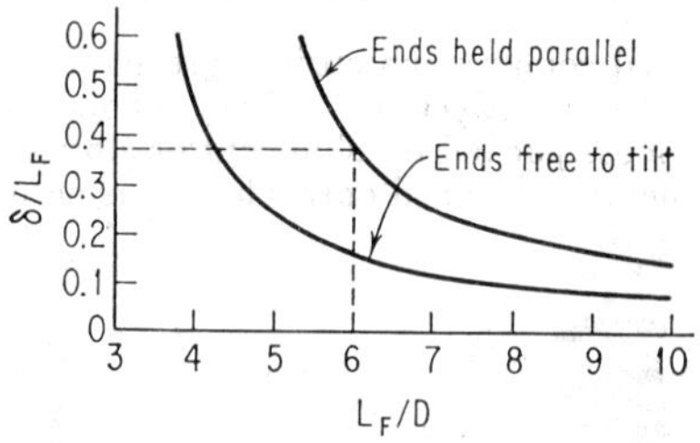

FIG. 33.9. Criterion for stability. Example: A spring having the proportions $L_F/D > 6$ and ends held parallel will buckle if $\delta/L_F > 3.7$.[6]

evaluate lateral stability. If possible, the spring should be so designed that it will not buckle under service conditions; otherwise the spring must be guided within a tube or on a rod. The friction between spring and guides will interfere somewhat with the accuracy of the load-deflection curve and reduce the spring endurance limit (see Table 33.9).

Torsion springs, if long and slender, may also buckle. This may sometimes be prevented by properly clamping the ends or by using initial tension between coils; if the spring still buckles, guides or arbors must be used.

A coiled compression spring does not exert force directly along its axis; the loading is usually eccentric. If the spring must be square under load, as where uniform pressure must be exerted on a valve in guides, special design considerations are used. It is difficult to design a spring which will be square when unloaded and square throughout its loading cycle. Squareness under a specific load is often achieved at the expense of squareness at other deflections. Practical experience in spring design is far ahead of theory in this area at present.

The effect of eccentric loading is to cause an increase in torsional stress on one side of a compression spring and a decrease in stress on the other side.

An approximate relationship for the load eccentricity

$$e = 1.12R(0.504/N + 0.121/N^2 + 2.06/N^3) \qquad (33.15)$$

Equation (33.14) indicates that the eccentricity is especially important when there are few active coils since

$$\frac{\text{Stress in eccentrically loaded spring}}{\text{Stress in spring under pure axial load}} = 1 + \frac{e}{R} \qquad (33.16)$$

where e is a measure of load displacement from the spring axis. However, this stress correction is usually ignored unless the number of active turns is two or less. Since the eccentricity varies with load, these calculations are only approximate.

(d) Change in Diameter during Deflection

A helical spring will change in diameter during deflection, which may be significant if the spring is functioning under severe space restrictions. The maximum increase in diameter of a compression spring deflected to solid height is given by

$$\Delta D_{max} = 0.05\,\frac{p^2 - d^2}{D} \qquad (33.17)$$

(e) Surge and Vibration

When a spring is velocity loaded, maximum stress is limited not necessarily by total deflection but by a surge wave which travels through the spring. Stress due to surging because of a velocity of loading v is independent of spring dimensions:

$$\tau = v\,\sqrt{2\gamma G/g} \qquad (33.18)$$

For a steel spring, $\tau = 131v$.[8] (33.19)

Surge waves can travel through a spring at a frequency characteristic of the spring. The fundamental frequency of a helical compression spring wound from circular wire,

$$f = (Kd/9ND^2)\,\sqrt{gG/\gamma} \qquad (33.20)$$

For springs with one end fixed and one free, $K = \frac{1}{2}$; $K = 1$ for springs with both ends fixed. For a steel spring,

$$f = 13{,}900Kd/ND^2 \qquad (33.21)$$

To avoid vibration, the fundamental frequency of a spring should be at least twelve times the frequency of the operating cycle.[7]

(f) Limiting-stress Values for Static and Low-cycle Duty

The value of τ calculated by Eqs. (33.11) or (33.13) must be less than the permissible limiting-stress value of the spring material by a suitable margin of safety. Values of permissible limiting stress are determined by multiplying the tensile strength obtained from Fig. 33.5 by the factors given in Table 33.9. Preferred practice is to design for deflection to the solid height.

Table 33.9. Factors for Determining Limiting Stress Values for Helical Compression Springs

Material	*Factor*
Music wire	0.45
Valve spring	0.45
AISI 6150	0.45
Oil tempered	0.40
Hard drawn	0.40
Phosphor bronze	0.35
Beryllium copper	0.40
Stainless steel	0.35

(g) Cyclic-loading Stress Charts

The chart shown in Fig. 33.10 provides an accurate basis for the design of helical compression springs wound from music wire, valve spring wire, and stainless-steel wire. If the number of stress cycles, wire size, and material are given:

1. Draw a vertical line C from number of stress cycles to the line A or A'.
2. Draw a horizontal line D.
3. Draw an oblique line E intersecting line B at the given wire diameter.

The intersection of any combination of lines F and G with line E will define an allowable combination of maximum and minimum torsional-stress values. The chart may be used in the reverse order.

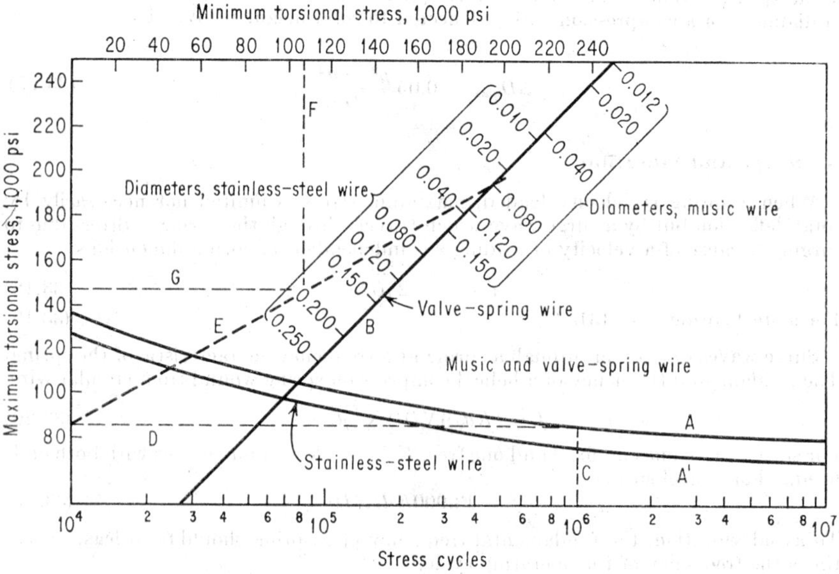

FIG. 33.10. Cyclic-loading stress chart for helical springs, with diameter in inches.

33.6. HELICAL EXTENSION SPRINGS (Fig. 33.11)

(a) Principal Characteristics

For close-wound springs the load-deflection curve is of the type shown in Fig. 33.1*b* and

$$P = P_i + \delta G d^4/8ND^3 \qquad \tau = \tau_i + 8K_A PD/\pi d^3 \tag{33.22}$$

P_i is the initial tension wound into the spring and τ_i the corresponding torsional stress,

$$P_i = \pi\tau_i d^3/8K_A D \tag{33.23}$$

Recommended values of τ_i depend on the spring index (see Fig. 33.12). The factor K_A is also a function of spring index and is given in Table 33.6.

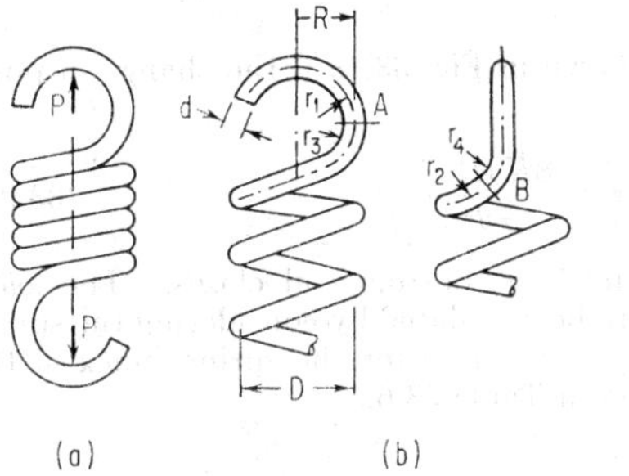

FIG. 33.11. Helical extension springs. (*a*) Close wound. (*b*) Open wound.

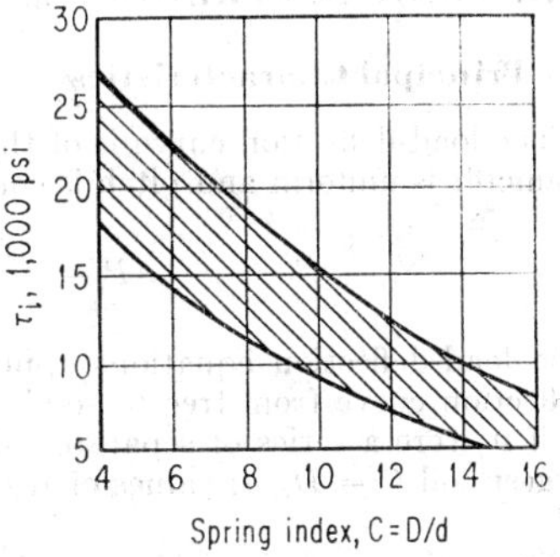

FIG. 33.12. Ranges of recommended values for stress due to initial tension in close-wound helical extension springs.[9]

For open-wound springs τ_i and P_i are zero and the load-deflection curve is of the type shown in Fig. 33.1*a*.

(b) Stresses in Hooks[6]

Critical stresses may occur at sections A and B as shown in Fig. 33.11. At section A the stress is due to bending, and at section B the stress is due to torsion:

$$\sigma = (32PR/\pi d^3)(r_1/r_3) \qquad \tau = (16PR/\pi d^3)(r_2/r_4) \tag{33.24}$$

Recommended practice is that $r_4 > 2d$. By winding the last few coils with a smaller diameter than the main part of the spring so that R is reduced, the stresses in hooks can be minimized.

(c) Limiting-stress Values for Static and Low-cycle Duty

Values of τ calculated by Eqs. (33.22) and (33.24) must be less than the permissible limiting stress of the spring by a suitable margin of safety. Values of permissible limiting stress are determined by multiplying the value of tensile strength obtained from Fig. 33.5 by the multiplying factor given in Table 33.10. Bending stresses in hooks calculated by Eq. (33.24) must not exceed the values in Fig. 33.5 by a suitable margin of safety.

Table 33.10. Factors for Determining Limiting-stress Values for Helical Extension Springs to Be Used in Eq. (33.24)

Material	*Factor*
Music wire	0.36
Valve spring	0.36
AISI 6150	0.36
Oil-tempered	0.32
Hard-drawn	0.32
Phosphor bronze	0.28
Beryllium copper	0.32
Stainless steel	0.28

(d) Cyclic-loading Stress Chart

Figure 33.10 applies to helical extension springs if the stress values shown on the chart are reduced by 20 per cent.

33.7. CONICAL SPRINGS (Fig. 33.13*a*)

(a) Principal Characteristics

The load-deflection curve is of the type shown in Fig. 33.1*a* if the change in pitch diameter is uniform and pitch is constant:

$$P = \frac{\delta G d^4}{8 N D_2^3} \qquad \tau = \frac{8 K_A D_2 P}{\pi d^3} \tag{33.25}$$

The load-deflection equations apply only until the bottom coil closes. The load-deflection curve from free to solid height can be calculated by considering the spring as if it were a series of separate coils. K_A is evaluated for the spring index of the largest coil $C = D_2/d$; values of K_A are given in Table 33.6.

(b) Limiting-stress Values

For static and cyclic loading, limiting-stress values may be determined on the same basis as for helical compression springs.

33.8. HELICAL TORSION SPRINGS (Fig. 33.13*b*)

(a) Principal Characteristics

For open-wound springs the load-deflection curve is of the type shown in Fig. 33.1*a*. For springs wound from circular wire,

$$M = \theta E d^4/64ND \qquad \sigma = 32K_B M/\pi d^3 \tag{33.26}$$

For springs wound from rectangular wire,

$$M = \theta E b t^3/12\pi ND \qquad \sigma = 6K_C M/bt^2 \tag{33.27}$$

Deflection is given in radians to facilitate the calculation of energy storage. However, it is often convenient to express deflection in number of turns $T = \theta/(2\pi)$. Stress factors K_B and K_C are given in Table 33.6.

For close-wound springs there may be an initial turning moment due to friction between the coils. Friction between coils will affect rate and cause considerable hysteresis in a complete deflection cycle.

(b) Changes of Dimension with Deflection

As a helical torsion spring winds up, pitch diameter decreases and number of coils increases. After T turns the pitch diameter $D_T = DN/(T + N)$ and the number of coils $N_T = T + N$; D and N are the pitch diameter and number of coils before windup.

(c) Limiting-stress Values for Static and Low-cycle Duty

The value of σ calculated by Eq. (33.26) or (33.27) can be as high as 90 per cent of the tensile strength if the residual stress is favorable. If unfavorable, use 50 per cent.

(d) Cyclic-loading Stress Chart

Figure 33.10 applies to helical torsion springs if the stress values shown on the chart are increased by 75 per cent. See Art. 33.5(*g*) for use of chart.

33.9. GARTER SPRINGS (Fig. 33.13c)

(a) Principal Characteristics

Garter springs are special forms of helical extension springs with ends fastened together to form garterlike rings. The inside diameter D_u of an unmounted spring is

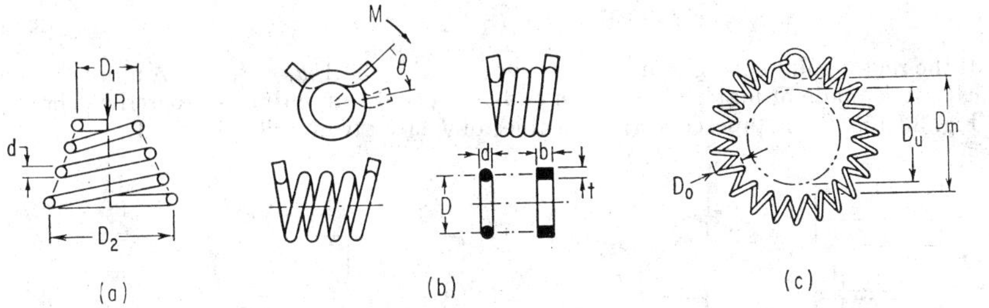

FIG. 33.13. Wound springs. (a) Conical spring. (b) Helical torsion. (c) Garter spring.

made to assemble over a larger mounting diameter D_m so that the circumferential load P_c psi exerted by the spring per inch of circumference,

$$P_c = 4/\pi D_m[P_i + k\pi(D_m - D_u)] \tag{33.28}$$

(b) Application

Garter springs are used in mechanical seals for shafting and to hold circular segments in place.

33.10. LEAF SPRINGS

The term leaf spring as it is used here applies to springs whose load-deflection and load-stress formulas are based on beam equations. Load-deflection curves are of the type shown in Fig. 33.1*a*.

(a) Single-leaf Cantilever Springs (Fig. 33.14)

For springs of rectangular section,

$$P = \delta KEbt^3/4L^3 \qquad \sigma = 6PL/bt^2 \tag{33.29}$$

If the rectangular section is uniform as shown in Fig. 33.14(*a*), $K = 1$. A more economical use is made of material where a cantilever leaf spring is tapered as shown in Fig. 33.4(*b*). For tapered springs, values of K are given in Table 33.11.

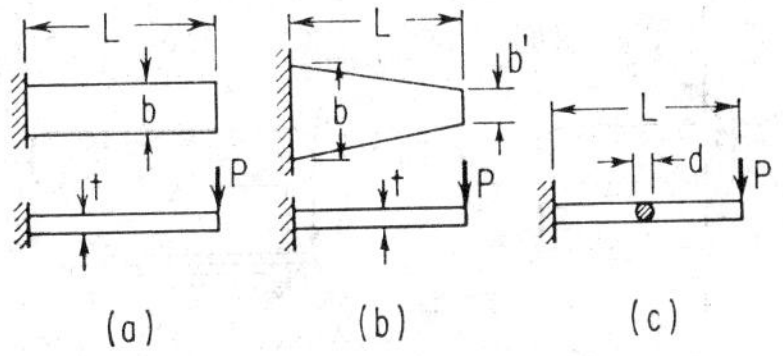

FIG. 33.14. Single-leaf cantilever spring. (*a*) Uniform cantilever section. (*b*) Tapering. uniform rectangular. (*c*) Uniform circular section.

Table 33.11. Values of *K* for Tapered Leaf Springs[7]

b'/b	0	0.1	0.2	0.3	0.4	0.5	0.6	0.7	0.8	0.9	1.0
K	1.50	1.39	1.32	1.25	1.20	1.16	1.12	1.09	1.05	1.03	1.00

For cantilever springs made from circular wire,

$$P = 3\pi\delta Ed^4/64L^3 \qquad \sigma = 32PL/\pi d^3 \tag{33.30}$$

(b) Single-leaf End-supported Springs (Fig. 33.15)

For springs of rectangular section,

$$P = 4\delta KEbt^3/L^3 \qquad \sigma = 3PL/2bt^2 \tag{33.31}$$

If the rectangular spring is uniform as shown in Fig. 33.15*a*, $K = 1$. A more economical use is made of material when an end-supported leaf spring is tapered as shown in Fig. 33.15*b*. For tapered springs, values of K are given in Table 33.11.

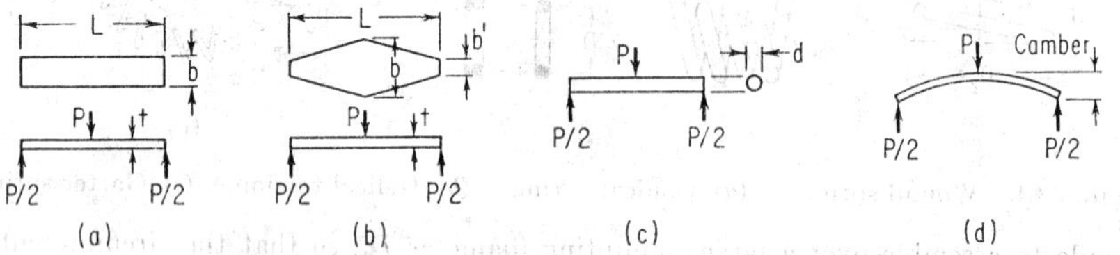

FIG. 33.15. Single-leaf end-supported springs. (*a*) Uniform rectangular section. (*b*) Tapering rectangular section. (*c*) Uniform circular section. (*d*) Elliptical type.

For end-supported springs made from round wire,

$$P = 3\pi\delta Ed^4/4L^3 \qquad \sigma = 32PL/\pi d^3 \tag{33.32}$$

Springs initially bowed as shown in Fig. 33.15*d* are called *elliptical springs*. Camber at no load is often made so the spring will be flat under load.

(c) Multiple-leaf Springs (Fig. 33.16)

The formulas for multiple-leaf springs are based on those for equivalent single-leaf springs. For example, the single-leaf cantilever spring shown in Fig. 33.16*a* is equivalent to a multiple-leaf cantilever spring for which $b = nw$, n being the number of leaves.

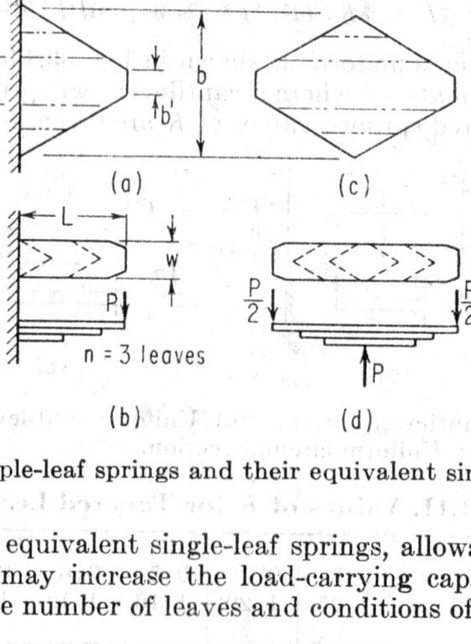

FIG. 33.16. Multiple-leaf springs and their equivalent single-leaf spring.

In using equations for equivalent single-leaf springs, allowance must be made for interleaf friction, which may increase the load-carrying capacity by from 2 to 12 per cent depending on the number of leaves and conditions of lubrication.

(d) Limiting-stress Values for Static and Low-cycle Duty

Values of calculated stress must be less than the permissible limiting stress of the spring material by a suitable margin of safety. Values of permissible limiting stress can be taken from Fig. 33.6 and Table 33.3 for springs formed of pretempered flat stock and loaded so that residual and load stresses are not additive. For materials hardened after forming, values of permissible stress should be reduced by 20 per cent.

(e) Cyclic Loading

Maximum and minimum values of stress for cyclic loading must fall within the permissible limits defined by a Goodman or similar diagram. The design chart shown in Fig. 33.17 applies to carbon-steel springs.

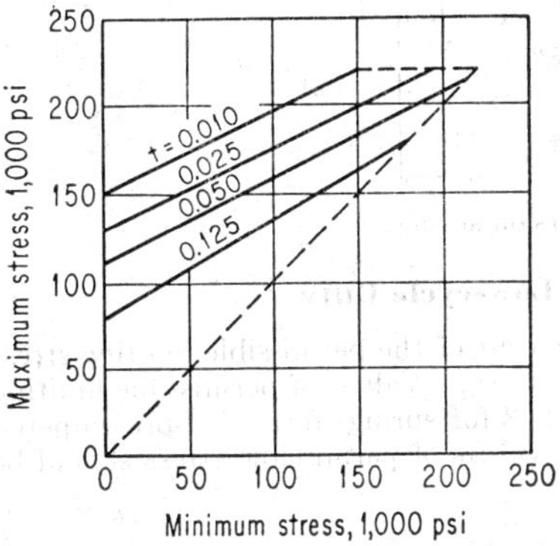

FIG. 33.17. Design chart for cyclically loaded flat steel springs, applies to 0.65 to 0.80 carbon steel at 45 to 48 Rockwell C.[6]

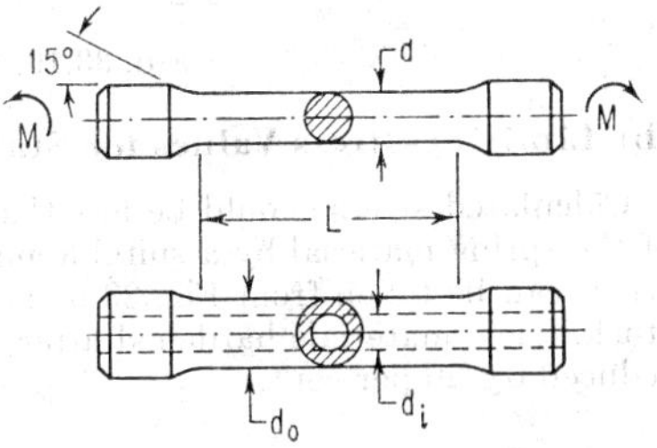

FIG. 33.18. Torsion-bar springs.

33.11. TORSION - BAR SPRINGS (Fig. 33.18)

(a) Principal Characteristics

The load-deflection curve is of the type shown in Fig. 33.1*a*. For torsion bars with hollow circular sections,

$$M = \pi\theta G(d_0^4 - d_i^4)/32L \qquad \tau = 16Md_0/\pi(d_0^4 - d_i^4) \tag{33.33}$$

For torsion bars with solid circular sections $d_i = 0$ and $d_0 = d$. Torsion-bar ends should be stronger than the body to prevent failure at the ends.

(b) Material

For highly stressed applications, torsion bars are generally made of silicomanganese. Fatigue resistance is increased by shot peening. Presetting increases strength in the direction of preset but should not be used if bars are loaded in both directions.

(c) Application

Torsion-bar springs are used where highly efficient energy-storage devices are required. The most familiar applications are in automotive equipment.

33.12. SPIRAL TORSION SPRINGS (Fig. 33.19)

(a) Principal Characteristics

The load-deflection characteristic is of the type shown in Fig. 33.1*a* and

$$M = \theta E b t^3/12L \qquad \sigma = 6M/bt^2 \tag{33.34}$$

The length of active material $L \approx \pi n(D_0 + D_i)/2$; n is the number of coils. Deflection in Eq. (33.34) is expressed in radians to facilitate the calculation of energy storage. The corresponding number of rotations $T = \theta/(2\pi)$.

Stresses at the spring ends may exceed the values of stress calculated by Eq. (33.34) depending on the design of the end and its treatment in manufacture.

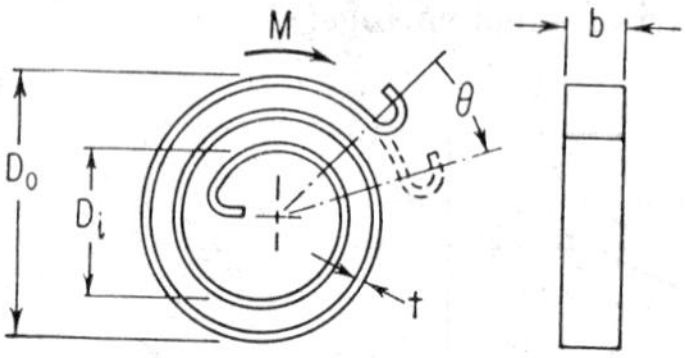

FIG. 33.19. Spiral torsion spring.

(b) Limiting-stress Values for Static and Low-cycle Duty

Calculated stress should be less than 50 per cent of the permissible limiting stress of the spring material by a suitable margin of safety. Values of permissible limiting stress can be taken from Fig. 33.6 and Table 33.3 for springs formed of pretempered stock. For materials hardened after forming, values of permissible stress should be reduced by 20 per cent.

(c) Application

Spiral torsion springs are widely used for locks in automobiles and to maintain pressure on carbon brushes of electric motors and generators.

33.13. CLOCK SPRINGS (Fig. 33.20*a*)

(a) Principal Characteristics

The clock spring is a special form of the spiral torsion spring. However, Eq. (33.34) applies only to maximum torque when the spring is fully wound. Figure 33.20*b*

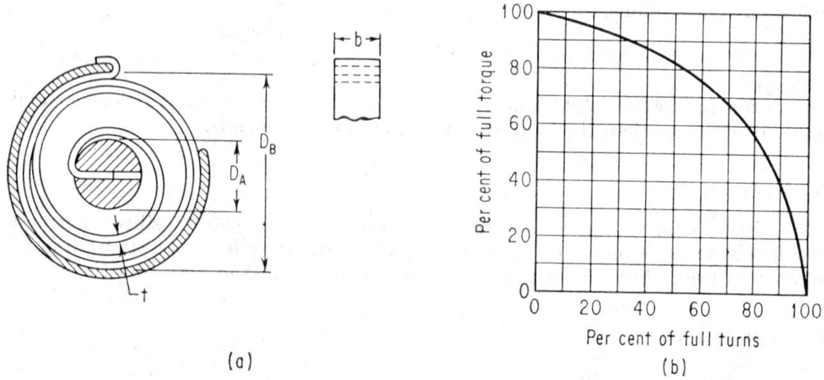

FIG. 33.20. Clock spring. (*a*) Orientation. (*b*) Load-deflection curve.

indicates the form of a typical load-deflection curve. To deliver the maximum number of turns, the space occupied by the spring coil should be half the space available in the drum so that

$$Lt = (\pi/8)(D_B^2 - D_A^2) \tag{33.35}$$

in which case the number of initial coils n_A wound tight on the arbor, and the number of rotations T that can be made by the arbor before the spring is unwound solid against the drum,

$$n_A = \frac{-D_A + \sqrt{\frac{1}{2}(D_A^2 + D_B^2)}}{2t} \qquad T = \frac{-D_A - D_B + \sqrt{2(D_A^2 + D_B^2)}}{2.55t} \tag{33.36}$$

The arbor diameter D_A should be from $15t$ to $25t$. The ratio of length to thickness L/t is generally between 3,000 and 4,000. If an extremely large number of operating cycles are desired, L/t should be less than 1,000.

(b) Materials and Recommended Working Stresses

Equation (33.34) is used for calculation of maximum stress when the spring is tightly wound on its arbor so that M is a maximum. Clock springs are usually made from cold-rolled blue tempered spring steel AISI 1074, or cold-rolled blue tempered and polished clock spring steel AISI 1095. Recommended maximum values of working stress for these materials depend on desired life. They may be as high as 50 per cent above the values from Fig. 33.6.

33.14. CONSTANT-FORCE SPRINGS

Constant-force springs are made from flat stock that has been given a constant natural curvature R by prestressing. If, however, R_n increases during extension, the rate decreases and vice versa. The principle of the constant-force spring is applied to extension and motor types.

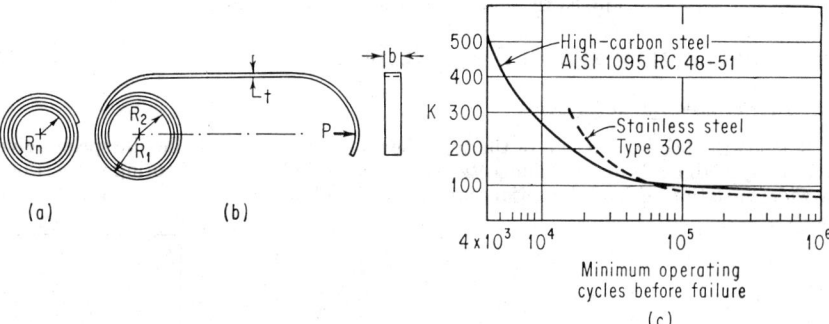

FIG. 33.21. Constant-force spring. (*a*) Unmounted coil. (*b*) Extension type. (*c*) Stress factors.[1]

(a) Extension Type (Fig. 33.21*a*,*b*)

In its relaxed condition the spring will form a tightly wound spiral. An extension spring is mounted on a spool of radius $R_2 > R_n$. The force P is virtually constant; it will vary somewhat as the coil unwinds and the radius R decreases:

$$P = (Ebt^3/24)(2/R_1R_n - 1/R_1) \qquad R_1 = R_2 + nt \tag{33.37}$$

where n is the number of coils still wound on the roller. R_2 should be about $1.2R_n$.

The length of prestressed extension springs should be

$$L = \delta_{max} + 10R \tag{33.38}$$

A stress factor K relates spring material, proportions, load, and number of operations[1]

$$P = Kbt$$

Values of K for materials commonly used for constant-force springs are given in Fig. 33.21c. Recommended practice is that $50 < b/t < 200$, $b/t = 100$ being a good median value.

Constant-force extension springs are used where large deflections and a zero rate are required as in counterbalancing devices.

(b) Motor Type (Fig. 33.22)

A constant-force spring reverse-wound on the larger of two spools will wind up on the smaller spool, driving the larger spool with a torque approximated by the equation

$$M = ER_Abt^3/24(1/R_A + 1/R_B)^2 \tag{33.39}$$

Spring load and stress are related by the equation

$$\sigma = 2Et(1/R_n + 1/R_A) \tag{33.40}$$

Recommended practice is that $50 < b/t < 200$, $b/t = 100$ being a good median value.

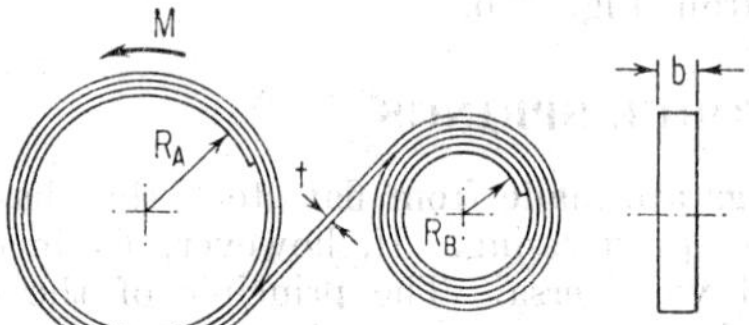

FIG. 33.22. Motor-type constant-force spring.

33.15. BELLEVILLE SPRINGS (Fig. 33.23a)

(a) Principal Characteristics

The load-deflection curve is of the type shown in Fig. 33.1c. Belleville springs give relatively high loads for small deflections. By a suitable choice of spring proportions a wide variation of load-deflection curves can be obtained as shown in Fig. 33.23b.

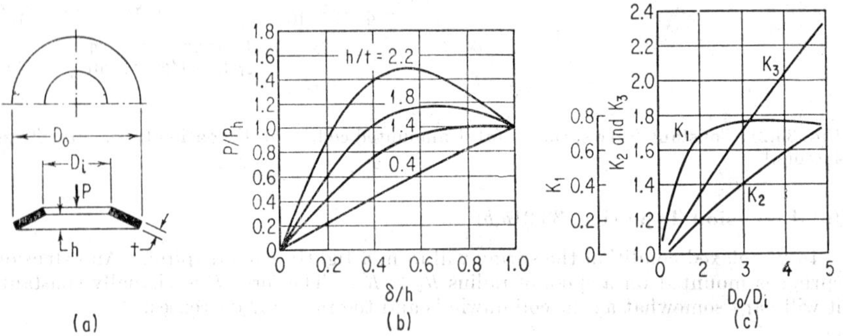

FIG. 33.23. Belleville spring. (a) Orientation. (b) Load-deflection curve P/P_h is the ratio of any P to the load P_h when deflected to the flat condition. (c) Factors K_1, K_2, and K_3 for spring equation.[11]

Rate is almost constant when $h/t = 0.4$ and close to zero for a large part of the deflection range when $h/t = 1.4$. When $h/t > 2.8$ a Belleville spring becomes unstable and will snap over. The general equations for Belleville springs are[10]

$$P = \frac{4E\delta}{(1-\mu^2)K_1D_0^2}\left[(h-\delta)\left(h-\frac{\delta}{2}\right)t + t^3\right] \tag{33.41}$$

$$\sigma = \frac{4E\delta}{(1-\mu^2)K_1D_0^2}\left[K_2\left(h-\frac{\delta}{2}\right) + K_3t\right] \tag{33.42}$$

The factors K_1, K_2, and K_3 are all functions of D_0/D, as shown in Fig. 33.23c.

Simpler equations are possible if springs are designed on the basis of selected proportions. The ratio h/t as shown in Fig. 33.23b determines load-deflection characteristics. The ratio of diameters D_0/D_i is usually restricted by assembly requirements; one of the ratios 1.2, 1.8, 2.4, or 3.0 will ordinarily be found suitable. If the most efficient use of spring material is a factor, then D_0/D_i should be 1.8. For springs designed to deflect to the flat positions so that σ is a solid stress,

$$P = \frac{\sigma^2(1-\mu^2)K_4D_0^2}{4E} \qquad t = \left[\frac{\sigma(1-\mu^2)}{4E}\right]^{1/2} K_5D_0 \tag{33.43}$$

For steel springs $\mu = 0.3$ and $E = 30 \times 10^6$ psi; hence

$$P = 75.8 \times 10^{-10}K_4\sigma^2D_0^2 \qquad t = 87.0 \times 10^{-6}K_5\sigma^{1/2}D_0 \tag{33.44}$$

Factors K_4 and K_5 are given in Table 33.12.

Table 33.12. Factors K_4 and K_5 for Simplified Belleville-spring Equations

D_o/D_i	K	$h/t = 0.4$	$h/t = 1.4$	$h/t = 1.8$	$h/t = 2.2$
1.2	K_4	0.475	0.068	0.043	0.029
	K_5	0.770	0.346	0.290	0.248
1.8	K_4	0.681	0.103	0.064	0.043
	K_5	1.024	0.465	0.388	0.336
2.4	K_4	0.598	0.089	0.056	0.038
	K_5	1.039	0.467	0.391	0.338
3.0	K_4	0.480	0.075	0.048	0.033
	K_5	0.990	0.452	0.381	0.328

(b) Materials and Recommended Working Stresses

The value of stress calculated by Eqs. (33.42) and (33.43) is a compressive stress at the upper edge of the inside diameter. In cyclic service, cracks may start at this edge but will usually not progress. Actual fatigue failures start as cracks at the outer diameter in a zone stressed in tension. For essentially static service where failure should originate at the upper edge of the inside diameter, conditions of prestress and plastic flow may invalidate stress calculations. It is therefore often possible to allow calculated stress values to exceed limiting values for the material. However, this practice is not recommended without careful engineering study. A conservative basis for design is to limit the values of calculated stress to the permissible-stress value of the material for the condition of loading.

(c) Application

Belleville springs are used where a high load with a relatively small deflection is required or where their unique load-deflection characteristics can be used to advantage. Belleville springs are used for preloading bearings in spindles, as pressure disks for power brakes, and in buffer assemblies for absorbing impact loads.

33.16. WAVE-WASHER SPRINGS (Fig. 33.24*a*)

(a) Principal Characteristics

The load-deflection curve is of the type shown in Fig. 33.1*a* and

$$P = 16K\delta Ebt^3n^4/\pi^3D^3 \qquad \sigma = 3\pi PD/4bt^2n^2 \tag{33.45}$$

The factor K is plotted in Fig. 33.24*b*. The number of waves n can be three or more. Usual practice is to have three, four, or six waves.

Allowance must be made in the assembly of wave washers for the slight decrease in diameters from the loaded to the unloaded condition.

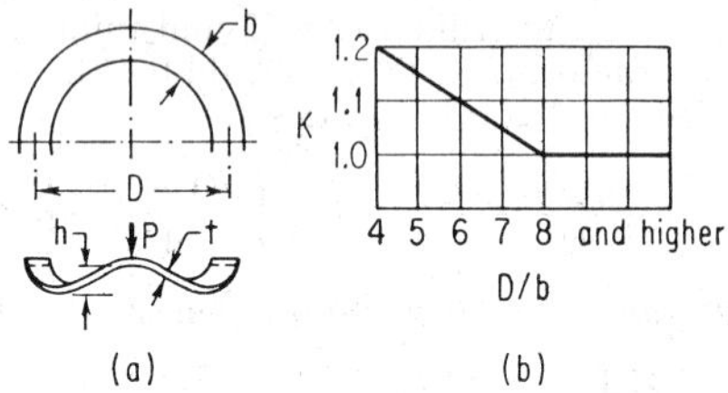

FIG. 33.24. Wave-washer spring. (*a*) Orientation. (*b*) Factor K for wave-washer spring equation.[6]

(b) Limiting-stress Values for Static and Low-cycle Duty

Wave washers are similar to leaf springs; hence the recommendations in Art. 33.10(*d*) apply. Maximum stress should be based on deflection to the solid height, in which case

$$\sigma = 12KhEtn^2/\pi^2D^2 \tag{33.46}$$

(c) Application

Wave-washer springs are used because of their compact form where a static load or small deflection range is required. A common use for wave washers is to assemble them between parts mounted on a shaft in order to compensate for variations in assembly clearances.

33.17. NONMETALLIC SYSTEMS

(a) Elastomer Springs

An elastomer is rubber or a rubberlike material which can be stretched to at least twice its free length at normal temperatures and will quickly return to its original length when the load is released. The ability to undergo very large deformations makes elastomers ideally suited to applications where energy absorption is important. Because of their high damping properties elastomers are used for resilient mountings in applications requiring vibration isolation. The stress-strain characteristics of elastomers are not linear; hence load-deflection curves are of the type shown in Fig. 33.1*c*.

Elastomers can be applied to a variety of spring designs. Only the simplest types of elastomer springs are considered here. The load-deflection curves for these simple types are nearly linear for limited deflections. For more complete information on elastomer springs, see refs. 7, 11, and 12.

Compression Cylinders (Fig. 33.25*a*). Principal formulas are

$$P = \pi\delta ED^2/4h \qquad \sigma = 4P/\pi D^2 \tag{33.47}$$

The value of E is not constant but depends on the relative area available for bulging as measured by the form factor.[11]

$$K = [(\pi/4)D^2]/\pi Dh = D/4h$$

The relationship between E and K is given by Fig. 33.26*a*. Recommended practice is for $\delta < 0.2h$ to avoid creep.

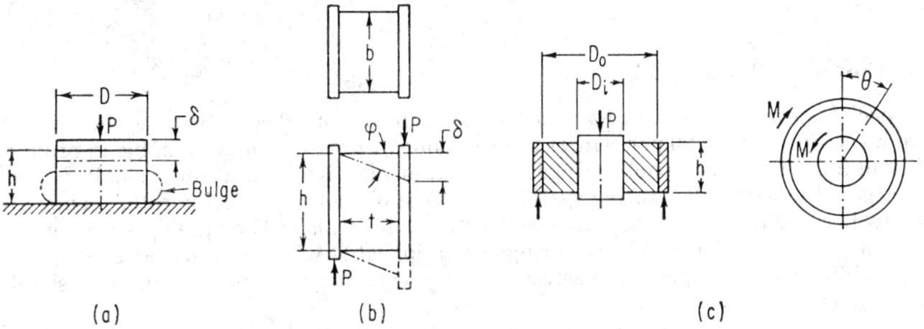

FIG. 33.25. Elastomer springs. (*a*) Compression cylinder. (*b*) Cylindrical bushing—force applied for direct shear loading—applied torque. (*c*) Shear sandwich.

Shear Sandwiches (Fig. 33.25*c*). For $\phi < 20°$ the load-deflection curve is nearly linear. Principal formulas are

$$P = \delta bhG/t \qquad \tau = P/bh \tag{33.48}$$

Recommended practice is that $t < h/4$ or $t < b/4$, whichever is the least value.

Cylindrical Bushings (Fig. 33.25*b*). Cylindrical bushings can be loaded in either direct shear or torsion. For direct-shear loading,

$$\delta = P \ln (D_0/D_i)/2\pi hG \qquad \tau = P/\pi hD \tag{33.49}$$

Equation (33.49) is sufficiently accurate for practical purposes if $P/(\pi hGD_0) < 0.4$. Shear stress calculated by Eq. (33.49) refers to the bond between the elastomers and the surface at D_i. For torsion loading

$$\theta = (M/\pi hG)(1/D_i^2 - 1/D_0^2) \qquad \tau = 2M/\pi hD_i^2 \tag{33.50}$$

Equation (33.50) is valid for $\theta < 0.7$ rad (40°). Shear stress calculated by Eq. (33.50) is for the bond between the elastomer and the surface at D_i.

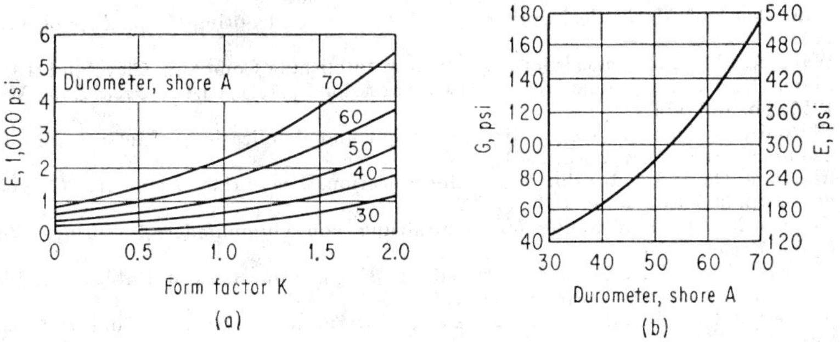

FIG. 33.26. Natural-rubber charts. (*a*) Modulus of elasticity vs. form factor K. (*b*) G and E vs. durometer.

Mechanical Properties and Recommended Working Stresses. Both E and G depend on hardeners; values for static loading are shown in Fig. 33.26*b* to have different values for dynamic loading.

For compression cylinders subject to static or low-cycle loading, maximum recommended compressive stress is 700 psi. Lower values should be used for cyclic duty.

For shear sandwiches and cylindrical bushings loaded in direct shear, recommended maximum design stress based on the bond strength is 35 psi; for cylindrical bushings loaded in torsion, recommended design stress based on the bond strength is 50 psi. Durometer hardness of 30 to 60 is usual for elastomer springs subject to shear loading.

(b) Compressible Liquids

Special spring devices utilize the compressibility of silicone-base liquids, which can be compressed to 90 per cent of their volume at pressures maintainable in production units. These compact springs will sustain high loads in a much smaller space than possible with other spring systems, because of their high spring rate. They are costly because of the necessity of excellent pressure seals. They are also sensitive to temperature variations. These springs are applicable where space is at a premium, as in aircraft landing gear, presses, and dies. Operating frequency generally should be kept below 1 cps.

Compressible-liquid springs may also be used as shock absorbers with excellent damping qualities attained by utilizing dashpot action.

(c) Compressible Gases

The air springs in use in many shock-absorbing applications utilize gas compressibility.[14] Many applications take advantage of the increasing rate of such a spring as the deflection continues. The spring rate can be changed at will by intentionally altering the gas pressure, which also affords a method of adjusting the loaded height of the spring system. There are problems associated with the design of a leakproof flexible gas container and with increasing temperatures. Generally spring rates are low, deflections high, and considerable space is needed.

References

1. Mechanical Drawing Requirements for Springs, Military Standard, MIL-STD-29 Sept. 23, 1958.
2. Maier, Karl W.: Springs That Store Energy Best, *Prod. Eng.*, Nov. 10, 1958, pp. 71–75.
3. Milck, John T.: Materials for Springs, *Mater. Design Eng.*, August, 1959, pp. 115–126.
4. Carson, Robert W.: Flat Spring Materials, *Prod. Eng.*, June 11, 1961, pp. 68–80.
5. Crooks, R. D., and W. R. Johnson: The Performance of Springs at Temperatures above 900°F., *Trans. SAE*, vol. 69, pp. 325–330, 1961.
6. "Handbook of Mechanical Spring Design," Associated Spring Corp., Bristol, Conn., 1958.
7. Wahl, A. M.: "Mechanical Springs," Penton Publishing Company, Cleveland, 1944.
8. Maier, Karl W.: Dynamic Loading of Compression Springs, part II, *Prod. Eng.*, March, 1955, pp. 162–174.
9. Chironis, Nicholas P. (ed.): "Spring Design and Application," McGraw-Hill Book Company, Inc., New York, 1961.
10. Almen, J. O., and A. Laszlo: The Uniform Section Disc Spring, *Trans. ASME*, vol. 58, no. 4, pp. 305–314, May, 1936.
11. Gobel, E. F.: "Berechnung und Gestaltung von Gummifedern," Springer-Verlag OHG, Berlin, 1955.
12. McPherson, A. T., and A. Klemin (eds.): "Engineering Uses of Rubber," Reinhold Publishing Corporation, New York, 1956.
13. Maleev and Hartman: "Machine Design," International Textbook Company, Scranton, Pa., 1954.
14. Deist: Air Springs Cushion the Ride, *Mech Eng.*, June, 1958, p. 61.

Section 34

DAMPERS

By

ANTONIO F. BALDO, Ph.D., *Associate Professor of Mechanical Engineering, The City University of New York, New York, N.Y.*

CONTENTS

34.1. INTRODUCTION

In vibrating mechanical systems, forces occur which tend to dissipate some of the energy of vibration. Such a system is said to be *damped.* These damping forces may be due to natural causes inherent in the mechanical system and its components, or they may be due to intentional constructions in the system. Natural and intentional damping may also occur together.

Component damping (material damping, internal friction, internal damping, or hysteretic damping) is related to the energy dissipation in a volume of material. It is associated with the internal micro- and macrostructure of the material.

System damping is related to energy dissipation due to interaction of components by various physical phenomena. Important system types are (1) structural systems, with energy dissipation occurring at joints, interfaces, or fasteners; (2) electromechanical systems, with energy dissipation occurring because of electrical or electromagnetic coupling (i.e., magnetic hysteresis and eddy currents); (3) hydromechanical systems, with energy dissipation occurring because of fluid motion through restricted passages.

(a) Mathematical Models for Damped Vibrations

Nearly all descriptions of damping are derived from the single-degree-of-freedom system (Fig. 34.1) with a damper in parallel with a spring. The linear system is important because of its susceptibility of mathematical solution. The governing equation is

$$M\ddot{x} + f(\dot{x}) + kx = Fe^{i\omega t} \tag{34.1}$$

where $f(\dot{x})$ is the damping term.

A few of the more widely encountered types of damping terms are as follows:

Viscous Damping (Linear System). The damping force is proportional to and in opposite phase with the velocity of the vibrating mass.

$$f(\dot{x}) = c\dot{x} \tag{34.2}$$

where c = coefficient of viscous damping.

Hysteretic Damping (Linear System). The damping force is proportional to the displacement of the mass but is in opposite phase with its velocity.

$$f(\dot{x}) = (h/w)\dot{x} \tag{34.3}$$

where h = coefficient of hysteretic damping.

Dry Friction or Coulomb Damping (Nonlinear System). The damping force is nominally a constant and is independent of the position or velocity of the mass.

$$f(\dot{x}) = \pm c \tag{34.4}$$

Hydraulic Damping (Nonlinear System). The damping force is proportional to the square of the velocity and in opposite phase with it.

$$f(\dot{x}) = c'(\dot{x})^2 \tag{34.5}$$

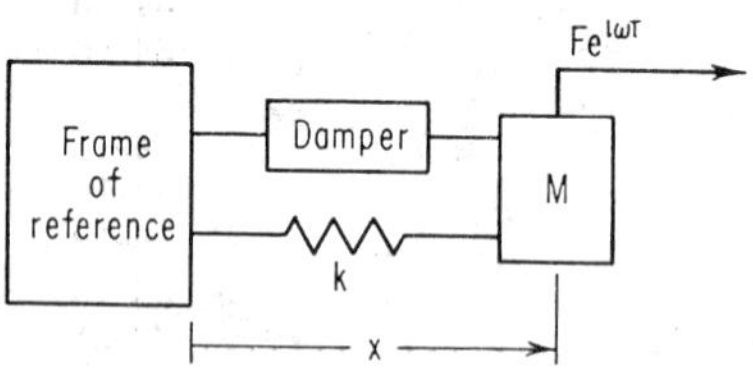

FIG. 34.1. Single-degree-of-freedom system.

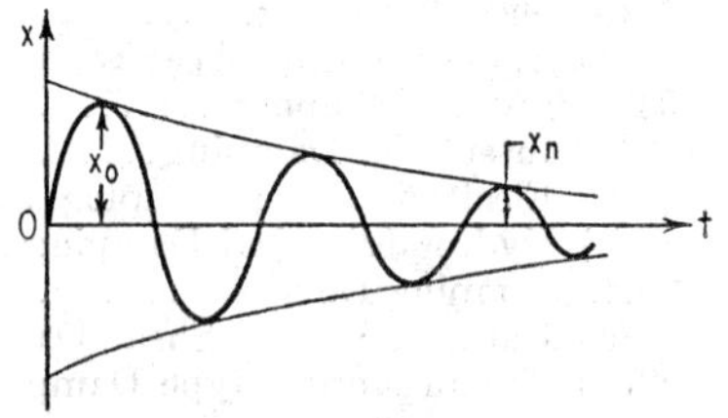

FIG. 34.2. Relative decrease in amplitude of a viscously damped system.

(b) Measures of Damping

Logarithmic Decrement δ. A viscously damped system, or one in which the energy loss per cycle is proportional to the square of the amplitude or stress, has linear damping. In such a case, the relative decrease in amplitude is constant from cycle to cycle as shown in Fig. 34.2.

$$\delta = -(1/n)\ln(x_n/x_0) \tag{34.6}$$

where x = amplitude
n = number of cycles between x_0 and x_n

Amplification Factor A

$$A = x_{\text{res}}/x_{\text{st}} \tag{34.7}$$

where x_{res} = vibration amplitude at resonance
x_{st} = amplitude of a constant sinusoidal excitation force at zero frequency

Equivalent Dashpot Constant, or Equivalent Viscous Damping c_{eff}. Very often in computations, a nonlinear damping term is replaced by an equivalent viscous-damping term. The linearized system can then be treated rather easily.

$$c_{\text{eff}} = F'/(\omega_N x_{\text{res}}) \tag{34.8}$$

where c_{eff} = effective viscous-damping coefficient
F' = amplitude of excitation force
ω_N = circular frequency at resonance
x_{res} = amplitude at resonance

Quality Factor Q

$$Q = 2\pi W/\Delta W \tag{34.9}$$

where W = stored energy per cycle
ΔW = energy dissipated per cycle

Specific Damping Energy D. For viscous and hysteretic damping, respectively, the vibration equation can be written in terms of complex stiffness

$$M\ddot{x} + (k + i\omega c)x = Fe^{i\omega t} \tag{34.10}$$

and

$$M\ddot{x} + (k + ih)x = Fe^{i\omega t} \tag{34.11}$$

then

$$D = 2\pi W'(E''/E') \tag{34.12}$$

where W' = stored energy per unit volume
E'' = imaginary portion of the complex elastic modulus (i.e., stress divided by the component of strain 90° out of phase with the strain)
E' = real portion of the complex elastic modulus (i.e., stress divided by the component of strain in phase with the stress)

Damping Ratio r. With enough damping, free vibrations are prevented from occurring, and an initially disturbed system returns to equilibrium in a single motion. The threshold of such a state is called *critical damping*, and the critical damping coefficient has the value $c_c = 2M\omega_N$, where ω_N = the circular natural frequency. The damping ratio is defined as the ratio of actual damping coefficient to the critical damping coefficient:

$$r = c/c_c \tag{34.13}$$

(c) Relationship among Definitions

$$r = 1/2Q = \delta/2\pi = E'/2E'' = 1/2A$$

34.2. COMPONENT DAMPING

(a) Structural Metals and Nonmetals

Only crudely approximate mathematical analyses of structural-material damping exist, especially for stress levels of engineering interest. Description of damping properties involves energy-dissipation terms.

The absolute damping energy dissipated

$$D_0 = D_d V_0 \left[\int_0^1 \frac{D}{D_d} \frac{d(V/V_0)}{d(\sigma/\sigma_d)} d(\sigma/\sigma_d) \right] = D_d V_0 \alpha = D_d V_0 \frac{D_a}{D_d} \tag{34.14}$$

where D_0 = total damping energy dissipated by the whole specimen, in.-lb/cycle
D_a = average unit damping energy, determined by dividing the total damping energy D_0 by the volume V_0, in.-lb/in.^{3}/cycle
D = specific damping energy of material at any stress σ in a component under nonuniform stress. This energy is proportional to the area within the

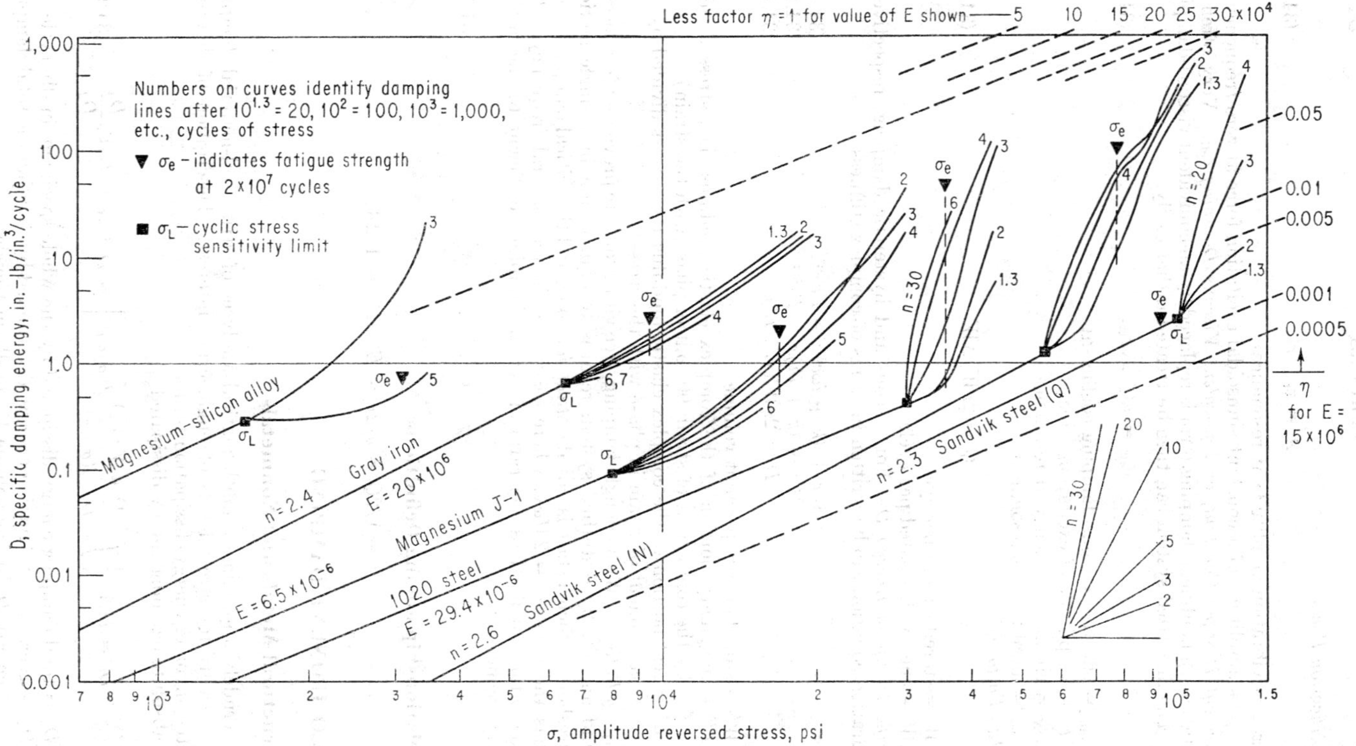

FIG. 34.3. Specific damping energy of various materials as a function of amplitude of reversed stress and number of fatigue cycles.

stress-strain hysteresis loop of a uniformly loaded cubic inch of the material, in.-lb/in.³/cycle

D_d = specific damping energy at the maximum stress $\sigma_d (0 < D < D_d)$, in.-lb/in.³/cycle

V = volume of specimen subjected to a stress less than σ, in.³

V_0 = total effective volume of specimen contributing to the dissipation of damping energy and subjected to a stress between 0 and σ_d, in.³

σ = amplitude of reversed stress at any point in a specimen $(0 < \sigma < \sigma_d)$, psi

σ_d = maximum induced stress, psi

α = damping-energy integral

In the low- to intermediate-stress range, the specific damping energy for many materials obeys the simple relation

$$D = J\sigma^n \tag{34.15}$$

where n ranges from 2 to 3 as shown in Fig. 34.3 and Table 34.1.

The ratio of damping energy to strain energy

$$\eta_s = \frac{D_0}{2\pi W_0} = \frac{D_0}{2\pi \dfrac{\sigma_d^2}{2E} V_0 \displaystyle\int_0^1 \left(\frac{\sigma}{\sigma_d}\right)^2 \frac{d(V/V_0)}{d(\sigma/\sigma_d)}\, d(\sigma/\sigma_d)} = \frac{D_0}{2\pi \dfrac{\sigma_d^2}{2E} V_0 \beta} \tag{34.16}$$

where η_s = specimen loss factor, dimensionless

W_0 = strain energy at maximum stress, in.-lb/in.³

E = conventional modulus of elasticity, psi

β = strain-energy integral

Conversion from specimen properties to material properties can be made by multiplying by β/α, i.e., $\eta = \eta_s(\beta/\alpha)$; see Table 34.2.

(b) Viscoelastic Materials

If a stress is applied to a viscoelastic material, the ensuing strain does not appear immediately. The delay is attributable to the viscous behavior of the material. Assuming sinusoidal loading, the situation (see Fig. 34.4) may be represented as

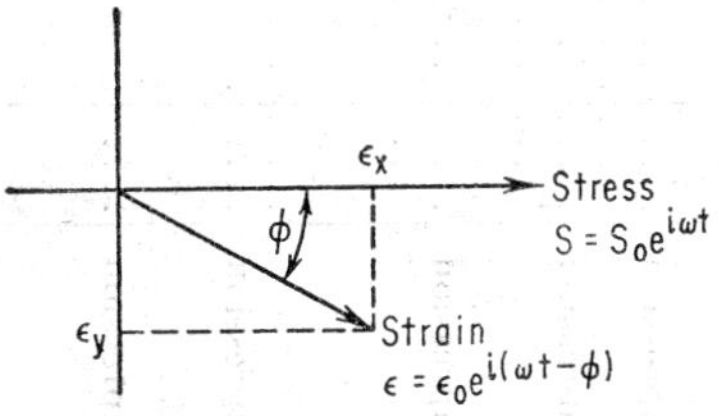

FIG. 34.4. Strain from a stress applied to a viscoelastic material.

follows: Energy storage (elastic) is due to component ϵ_x in phase with the stress, and energy dissipation (viscous) is due to ϵ_y. If a complex Young's modulus is defined as

$$E^* = S/\epsilon = (S_0/\epsilon_0)e^{i\phi} = E' + jE'' \tag{34.17}$$

then the loss factor (loss tangent) of the material becomes

$$\eta = E''/E' \tag{34.18}$$

For viscoelastic materials subjected only to shear, the loss factor is written

$$\beta = G''/G' \tag{34.19}$$

where G'' = imaginary portion of complex shear modulus

G' = real portion of complex shear modulus

Table 34.1. Damping Capacity of Common Materials[2]

Material	Mode of stress	Reversed stress, ±psi														
		200	400	600	800	1000	1500	2000	4000	6000	8000	10,000	15,000	20,000	30,000	40,000
SAE 1025 steel	Tension-compression	...	...	...	...	...	...	...	...	...	0.026	0.050	0.17	0.40	1.35	10
	Shear	...	...	...	...	...	...	...	0.026	0.089	0.21	0.41	1.38	3.3		
SAE 4130 steel	Tension-compression	...	...	...	...	...	...	...	...	...	...	0.043	0.145	0.3	1.1	2.8
	Shear	...	...	...	...	...	...	...	...	0.044	0.102	0.20	0.68	1.60	5.4	
Monel (67% Ni, 30% Cu, 1.4% Fe, 1% Mn)	Tension-compression	...	...	...	...	...	...	...	...	...	...	0.04	0.135	0.32	1.1	2.6
	Shear	...	...	...	...	...	...	...	...	0.041	0.097	0.19	0.64	1.52	5.1	
Magnesium alloy J-1 (6.5% Al, 1% Zn, 0.2% Mn, bal. Mg)	Tension-compression	...	...	...	...	...	...	...	0.035	0.12	0.28	0.55				
	Shear	...	...	...	...	...	...	0.031	0.25	0.84						
Cast iron (Meehanite GA)	Shear	...	...	...	...	...	...	...	0.083	0.28	0.66	1.3	4.4	10.4		
Duralumin 17 S-T	Tension-compression	...	...	...	...	...	...	...	...	...	...	0.115	0.272	0.575	1.77	
Wood—birch, along grain (sp. gr. 0.74, moisture 10.9%)	Tension-compression	...	...	0.003	0.008	0.015	0.051	0.12	0.40	.096						
Compressed plywood—birch 1/48-in. ply thickness (sp. gr. 0.87)	Tension-compression	...	...	0.004	0.010	0.019	0.064	0.15	0.51	1.22						
Paper-base phenolic grade XX	Tension-compression	...	...	0.005	0.0085	0.015	0.04	0.075	0.19							
Methyl methacrylate (Plexiglass)	Tension-compression	0.011	0.053	0.13	0.26	0.43										
Cellulose acetate (FM 6)	Tension-compression	0.022	0.05	0.23	0.40	0.68	1.1									
Polystyrene (Luxtrex)	Tension-compression	...	0.01	0.025	0.045	0.08	0.20	0.40								

D = energy absorbed, in.-lb/in.^{3} of stressed material per cycle of stress.

Table 34.2. Expressions and Values for α and β/α for Various Stress-distribution and Damping Functions[1]

	Type of specimen and loading	Volume-stress function V/V_0	Dimensionless damping energy integral α for various damping functions: General case $D = f(\sigma)$	For special case $D = J\sigma^n$: For any value of n	For special case $D = J\sigma^n$: $n = 2.4$	For special case $D = J\sigma^n$: $n = 8$	Dimensionless strain energy integral β	β/α if $n = 8$
1	Tension-compression member	....	1	1	1	1	1.0	1
2	Cylindrical torsion member or rotating beam	$\left(\frac{\sigma}{\sigma_d}\right)^2$	$\left(1 + \frac{\sigma_d}{2D_0}\frac{dD_0}{d\sigma_d}\right)^{-1}$	$\frac{2}{n+2}$	0.45	0.20	0.5	2.5
3	Rectangular beam under uniform bending	$\frac{\sigma}{\sigma_d}$	$\left(1 + \frac{\sigma_d}{D_0}\frac{dD_0}{d\sigma_d}\right)^{-1}$	$\frac{1}{n+1}$	0.29	0.11	0.33	3.0
4	Cylindrical beam under uniform bending	$\frac{2}{\pi}\left[\frac{\sigma}{\sigma_d}\sqrt{1 - \left(\frac{\sigma}{\sigma_d}\right)^2} + \sin^{-1}\left(\frac{\sigma}{\sigma_d}\right)\right]$		$\frac{1}{\sqrt{\pi}}\frac{2}{n+2}\frac{\Gamma[(n+1)/2]}{\Gamma[(n+2)/2]}$	0.21	0.055	0.24	4.5
5	Diamond beam under uniform bending	$2\frac{\sigma}{\sigma_d} - \left(\frac{\sigma}{\sigma_d}\right)^2$	$\left(1 + \frac{2\sigma_d}{D_0}\frac{dD_0}{d\sigma_d} + \frac{\sigma_d^2}{2D_0}\frac{d^2D_0}{d\sigma_d^2}\right)^{-1}$	$\frac{2}{n^2+3n+2}$	0.13	0.022	0.17	7.7
6	Rectangular beam having bending moment shown: $M_x = \frac{x}{L}M_0$	$\frac{\sigma}{\sigma_d}\left(1 - \log_e\frac{\sigma}{\sigma_d}\right)$	$\left(1 + \frac{3\sigma_d}{D_0}\frac{dD_0}{d\sigma_d} + \frac{\sigma_d^2}{D_0}\frac{d^2D_0}{d\sigma_d^2}\right)^{-1}$	$\frac{1}{(n+1)^2}$	0.088	0.012	0.11	9.1
7	Rectangular beam having bending moment shown: $M_x = \left(\frac{x}{L}\right)^2 M_0$	$2\sqrt{\frac{\sigma}{\sigma_d}} - \frac{\sigma}{\sigma_d}$	$\left(1 + \frac{5\sigma_d}{D_0}\frac{dD_0}{d\sigma_d} + \frac{2\sigma_d^2}{D_0}\frac{d^2D_0}{d\sigma_d^2}\right)^{-1}$	$\frac{1}{2n^2+3n+1}$	0.051	0.0065	0.067	10
8	Tuning fork in bending	$K\frac{\sigma}{\sigma_d}\left(1 - \log_e\frac{\sigma}{\sigma_d}\right)$	$K\left(1 + \frac{3\sigma_d}{D_0}\frac{dD_0}{d\sigma_d} + \frac{\sigma_d^2}{D_0}\frac{d^2D_0}{d\sigma_d^2}\right)^{-1}$	$\frac{K}{(n+1)^2}$ For $K = 0.8$.......	0.091	0.0099	0.089	9.0

NOTE: $\beta/\alpha = 1$ for all cases if $n = 2$.

See Tables 34.3, 34.4, 34.5.

Table 34.3. Dynamic-property Data for Wide Range of Temperature and Frequency[3]

Material	Max loss factor β_{max}	Position of β_{max}		Shear* modulus G_1 at β_{max}, psi
		Frequency cps	Temp, °C	
Plasticized polyvinyl acetate	2.6	50	5	2,180
Polystyrene	>2.0	>2,000	~140	14,500
Polyisobutylene	>2.0	>6,000	25	290
Polysulfide rubber (Thiokol RD)	1.9	7	5	10,100
Polyvinyl chloride	1.8	20	92	3,330
Buna N (type B-1), vulcanized	1.5	4,000	20	1,450
Polymethyl methacrylate	1.5	1,200	142	14,500
Plasticized PVC (Koroseal)	1.45	660	50	3,470
Polyester	1.1	200	108	10,100
Polytetrafluoroethylene (Teflon-TFE)	1.0	400	23	5,800
Hard rubber	1.0	40	60	10,100
Nitrile rubber	0.8	1,800	20	15,900
Urethane rubber (Shore 80 A)	0.8	30	−8	2,320
Filled rubber	0.5	2,500	22	33,300
Urethane rubber (Shore 94 A)	0.4	140	−23	43,500
Aquaplas (water-soluble)	0.38	1,000	40	24,200

* To get tensile modulus E_1, multiply G_1 figures by 3.

Table 34.4. Dynamic-property Data Available at Low Frequencies Only[3]

Material	Max loss factor β_{max}	Temp range, °C (for $\beta \geq 0.75\ \beta_{max}$)	Test frequency, cps	Shear* modulus G_1 at β_{max}, psi
Polyvinylcarbazol	>1.6	200 to 220	10.0	1,200
Polyvinyl-*n*-butyl ether	>1.6	−40 to −20	0.8	1,500
Polyvinyl-isobutyl ether	>1.6	−15 to 10	1.2	1,500
Polyvinyl-tert-butyl ether	>1.6	70 to 90	1.7	1,500
Butyl rubber (unvulcanized)	1.2	−20 to 20	1.0	15,000
Neoprene GN-50 (vulcanized)	0.5	−20 to 5	~150.0	5,800
Natural rubber (high mol. wt.)	~0.5	−70 to −50	<10.0	1,500
GR-S rubber	~0.5	−60 to −40	<10.0	7,200
Polytrifluorochloroethylene (e.g., Fluorothene)	0.42	90 to 120	3.3	11,600
Polyvinylfluoride	0.36	20 to 60	1.7	5,800
Polyethylene	0.23	30 to 80	12.0	2,900
Buna N (B-5), carbon-filled, vulcanized	0.15	0 to 10	50.0	29,000

* To get tensile modulus E_1, multiply G_1 figures by 3.

Table 34.5. Dynamic-property Data Available at Room Temperature Only[3]

Material	Max loss factor β_{max}	Frequency for β_{max} cps	Loss factor at five frequencies, cps					Test temp, °C	Shear* modulus G_1 at β_{max}, psi
			1	10	100	1,000	10,000		
Polysulfide rubber (Thiokol H-5)	5.00	1,000	0.17	0.50	1.20	5.00	......	25 ± 50	1,500
Butyl rubber (Enjay 9–262–4)	4.02	3,100	....	......	0.40	0.92	>2.0	21	1,500
Urethane rubber (Disogrin IDSA 9250)	2.59	3,000	....	......	0.10	0.13	~0.50	25	2,900
Butyl rubber (Enjay 9–262–1)	2.56	3,000	....	......	0.55	0.84	>1.80	21	2,900
(Enjay 9–26–3)	2.20	3,000	....	......	0.45	0.68	>1.40	24	2,900
Polyvinyl butyral	2.0	2	....	0.14	0.06	0.06	......	30 ± 70	29,000
3M tape adhesive, No. 466	1.82	1,000	....	~1.00	1.17	1.82	~0.80	25	360
Butyl gum (type U-50), vulcanized	1.80	10,000	....	......	......	1.20	1.80	20	1,000
Buna N (type B-5), carbon-filled, vulcanized	1.60	10,000	....	0.40	0.75	1.20	1.60	20	1,500–4,400
GR-S rubber	1.60	10,000	....	0.20	0.50	1.00	1.60	20	1,500
Urethane rubber (82% Pb powder filler)	1.40	3,000	....	......	0.40	0.49	......	25	2,500
Fluoro rubber (3M No. IF4)	1.30	4,000	....	~0.30	0.46	0.79	......	45 ± 55	17
Neoprene GRT	1.18	1,775	....	......	0.60	1.01	......	20	220
Polyvinyl chloride acetate	1.14	100	...	......	1.14	0.73	0.90	40 ± 20	580
Neoprene (type GN-50), vulcanized	1.10	10,000	....	......	......	0.70	1.10	20	290
Buna N (type B-O), unvulcanized	1.00	1,000	....	0.50	0.65	1.00	......	15	1,200
Hycar 1014	0.76	1,850	....	......	0.40	0.63	......	20	200
3M damping tape adhesive, No. 435	0.73	4,000	....	~0.50	0.58	0.67	~0.70	10 ± 50	37
Neoprene (type CG-1), vulcanized	0.60	10,000	....	......	......	0.02	0.60	20	4,350
Fluorosilicone rubber (Silastic LS53)	0.56	4,000	....	~0.10	0.16	0.46	......	30 ± 70	87
Polysulfide rubber (Thiokol ST)	0.53	1,300	....	......	0.40	0.51	......	20	160
Urethane rubber (Disogrin IDSA 7560)	0.51	4,250	....	......	0.10	0.22	......	25	1,500
GR-S rubber (23.5% styrene)	0.47	1,700	....	......	~0.30	0.39	......	20	200
(3.0% stryene)	0.38	1,400	....	......	~0.25	0.37	......	20	13
GR-S (type S-50) carbon-filled vulcanized	0.35	10,000	....	......	......	0.20	0.35	20	435
Silicone gum (Linde No. Y-1032)	0.33	400	....	~0.15	0.20	0.40	......	60 ± 40	2.5
Natural gum rubber (Type N-1)	0.30	1,000	0.04	0.08	0.14	0.30	......	20	150
GR-S rubber (Krylene)	0.30	10,000	0.11	0.12	0.13	0.15	0.3	20	100
Hevea rubber, filled	0.25	10,000	0.11	0.11	0.11	0.14	0.25	20	870
Vulcanized	0.20	10,000	0.03	0.03	0.03	0.05	0.20	20	870
Natural rubber (tire-tread stock)	0.20	500	0.16	0.18	0.20	....	......	25 ± 50	150

* To get tensile modulus E_1, multiply G_1 figures by 3.

(c) Damping of Flexural Plate Vibrations Using Viscoelastic Adhesives

The differential equation for damped flexural motion of thin elastic plates may be written

$$B(1 + j\eta_B)(\partial^4 z/\partial x^4) + m(\partial^2 z/\partial t^2) = f(x,t) \tag{34.20}$$

where z = displacement of plate
B = flexural rigidity of cross section
m = mass per unit length, x direction
x = direction of propagation of a straight-crested flexural wave
η_B = internal loss factor of the material

Internal or viscoelastic losses in the structure are represented by the complex flexural rigidity $B(1 + j\eta_B)$.

For a homogeneous plate, Young's modulus is used to evaluate the internal loss factor η_B. For a composite plate of elastic and viscoelastic laminae, an effective loss factor η can be substituted by determining the total flexural rigidity. This is done by assuming each layer to be an ideal elastic material, and then replacing the real moduli by complex quantities.

Damping in a composite plate results from extension and shear of the individual layers. The three-layer plate is the simplest configuration in which both these effects occur. Total flexural rigidity is given in the following. For multilayer plates, total flexural rigidity may be computed in a manner similar to the three-layer case.

In the three-layer plate, all laminae are assumed to have the same flexural motion (i.e., the wavelength of flexural vibration is the same in all three layers). The laminae

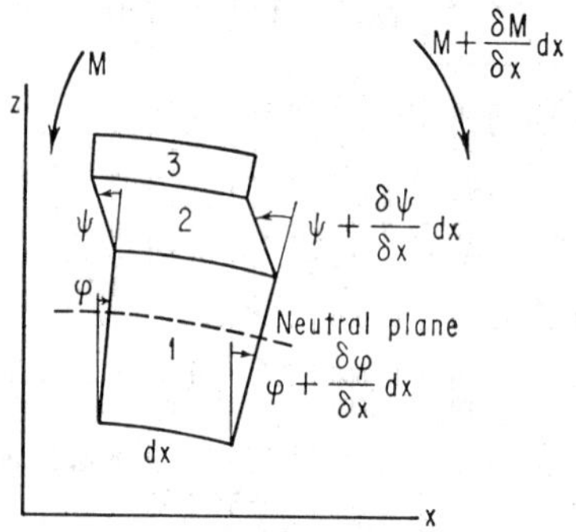

FIG. 34.5. Element of a three-layer plate in flexural vibration.

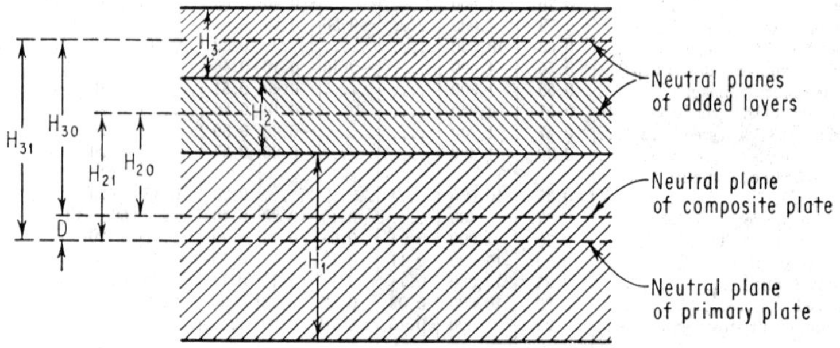

FIG. 34.6. Dimensions used in analysis of the three-layer plate in flexural vibration.[1]

thicknesses are small relative to the flexural, longitudinal, and shear wave-motion wavelengths (Figs. 34.5 and 34.6). The total flexural rigidity takes the form

$$\begin{aligned} B = {} & K_1(H_1^2/12) + K_2(H_2^2/12) + K_3(H_3^2/12) - K_2(H_2^2/12)\partial\psi/\partial\phi + K_1D^2 \\ & + K_2(H_{21} - D)^2 + K_3(H_{31} - D)^2 \\ & - [(K_2/2)(H_{21} - D) + K_3(H_{31} - D)]H_2(\partial\psi/\partial\phi) \end{aligned} \qquad (34.21)$$

where K_i = extensional stiffness of a unit length of the ith layer = E_ih_i
E_i = Young's modulus of the ith layer
H_i = thickness of the ith layer
H_{21}, H_{31}, D = dimensions as shown in Fig. 34.6

The quantities D and $\partial\psi/\partial\phi$ take the form

$$D = \frac{K_2(H_{21} - H_{31}/2) + (K_2H_{21} + K_3H_{31})(G_2C_f/K_3H_{2\omega})}{K_1 + K_2/2 + (K_1 + K_2 + K_3)(G_2C_f/K_3H_{2\omega})} \qquad (34.22)$$

and

$$\frac{\partial\psi}{\partial\phi} = \frac{1}{H_2}\,\frac{K_1H_{31} + K_2(H_{31} - H_{21})}{K_1 + K_2/2 + (K_1 + K_2 + K_3)(G_2C_f/K_3H_{2\omega})} \qquad (34.23)$$

where G_2 = shear modulus of middle layer
C_f = speed of propagation of flexural waves
ω = circular frequency of particular wave being considered

Substitution of imaginary components of Young's moduli in Eq. (34.21) yields the imaginary component of B'' of the flexural rigidity, while substitution of real components yields B'. The loss factor η_B equals B''/B'.

(d) Practical Considerations

In comparison with the extensional stiffness of the base plate, the extensional stiffness of the adjoining layer can be considered very small. The stiffness of the outer, or constraining, layer is also relatively small, being at most one-fourth to one-fifth that of the base plate ($K_2 \ll K_1$ and $K_3^2 \ll K_1^2$). The more common treatments are discussed briefly below.

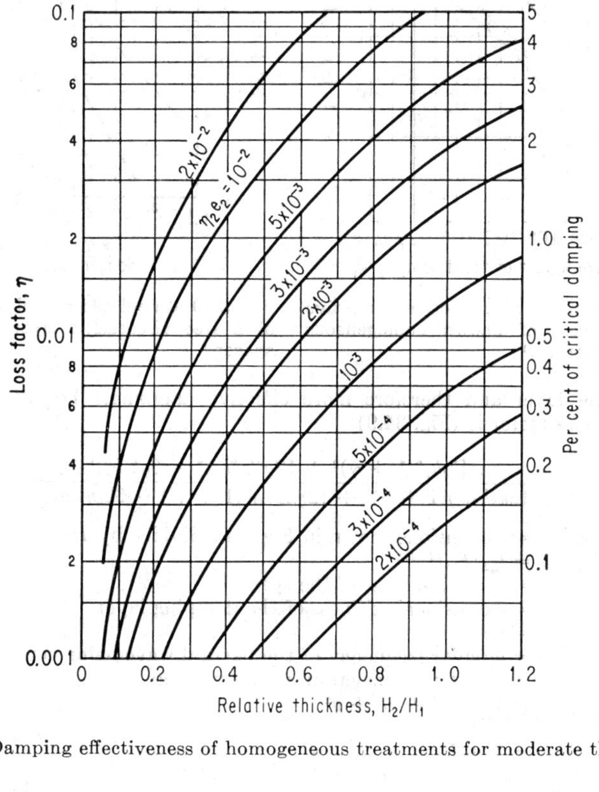

FIG. 34.7. Damping effectiveness of homogeneous treatments for moderate thicknesses.[1]

Homogeneous Damping Treatments. For a single layer applied directly to a plate (sound-deadening mastics) extensional damping predominates. The composite loss factor is approximately

$$\eta = \frac{\eta_2 e_2 h_2(3 + 6h_2 + 4h_2^2)}{1 + e_2 h_2(3 + 6h_2 + 4h_2^2)} \qquad (34.24)$$

where $e_2 = E_2/E_1$ = Young's modulus of the damping layer divided by Young's modulus of the base plate

η_2 = loss factor of the applied layer

$h_2 = H_2/H_1$ = applied layer thickness divided by base-plate thickness

The charts given in Figs. 34.7 and 34.8 present useful information for evaluating and designing damping treatments of moderate thickness ratios, based upon Eq. (34.24).

Spaced Homogeneous Treatments. Interposing a light shear-stiff spacer between the damping layer and the base plate results in greater relative extension of

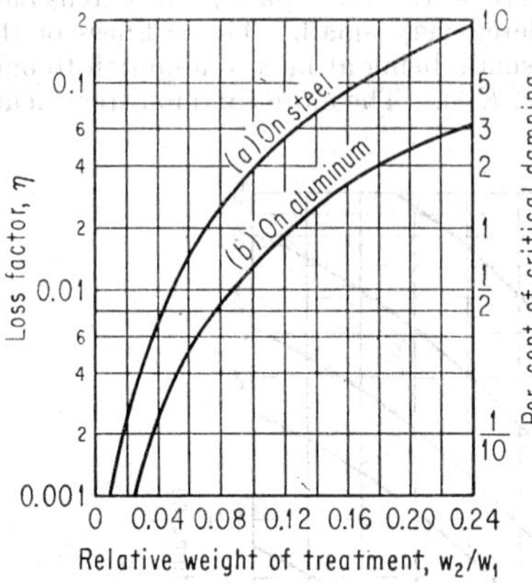

FIG. 34.8. Damping as a function of relative weight for best of known homogeneous damping materials.[1]

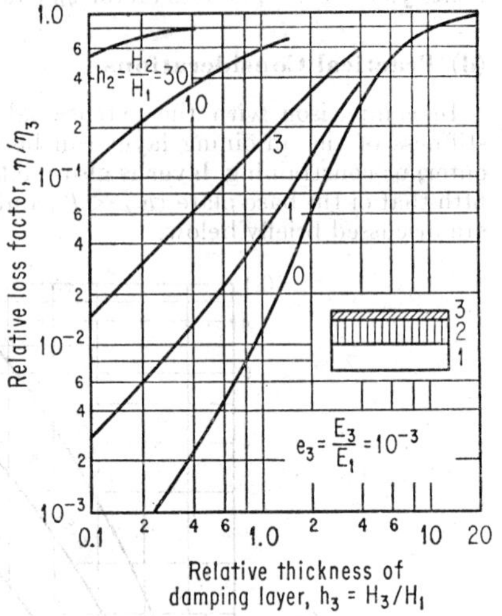

FIG. 34.9. Calculated damping performance of a free viscoelastic layer with an ideal spacer.[1]

the damping layer and therefore more effective damping. For a nondissipative spacer, the loss factor is (Fig. 34.9)

$$\eta = \frac{\eta_3[k_3h_3^2(1+\gamma)^2 + 12g^2k_3^2h_{31}^2 - 6gk_2h_{21}h_{31}]}{(1+12k_2h_{21}^2)(1+\gamma)^2 + 12gk_3h_{31}^2(1+\gamma') - 6k_2h_{21}h_{31}(1+\gamma')} \tag{34.25}$$

where h_3, h_{21}, h_{31} = dimensions shown in Fig. 34.6 divided by H_1
$g = G_2/K_3H_2P^2$
G_2 = shear modulus of the spacer
K_3 = extensional stiffness of the damping layer
$P = \omega/c_f$
ω = circular frequency of the flexural wave being considered
c_f = speed of propagation of flexural waves
$(1+\gamma)^2 = 1 + \eta_3^2 + 2g + g^2$
$(1+\gamma') = 1 + \eta_3^2 + g$
$2h_{21} = 1 + h_2$
$2h_{31} = 1 + 2h_2 + h_3$
$k_2 = e_2h_2$
$k_3 = e_3h_3$

(e) Shear Damping

Adding a "constraining" layer atop the damping layer causes shear as well as extensional distortion in the damping layer. On an equal-weight basis, treatments using shear distortion are likely to provide more effective damping than those using only extensional distortion. The loss factor may be written

$$\eta_s = \frac{\beta Y(1-\epsilon)g(1+k_s)}{1 + [2 + Y(1-\epsilon)]g(1+k_3) + [1 + Y(1-\epsilon)](1+\beta^2)g^2(1+k_3)^2} \tag{34.26}$$

where β = loss factor for the shear damping layer
$g(1 + k_3)$ = modified shear parameter
$\epsilon = (k_2/gk_3)(h_{21}/2h_{31})$

The quantity Y is called the *geometrical parameter* and, for the case where the extensional stiffness of the damping layer is strictly negligible, can be expressed as

$$Y = \frac{3e_2h_3(1 + 2h_2 + h_3)^2}{(1 + e_3h_3)(1 + e_3h_3^3)} \tag{34.27}$$

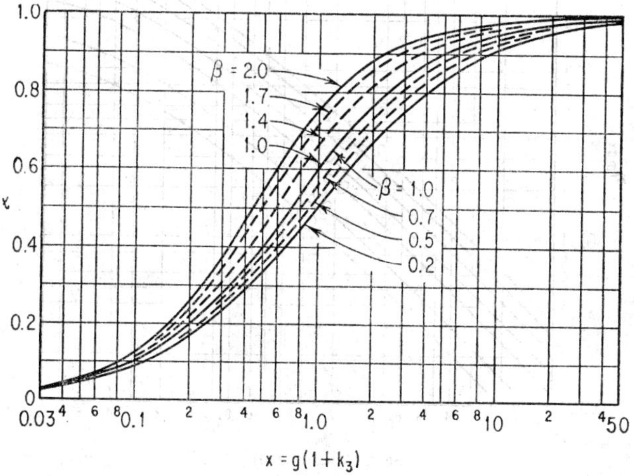

FIG. 34.10. Flexural-rigidity factor as a function of shear parameter and loss factor.[1]

For maximum shear damping, Eq. (34.27) and the following equations and graphs are useful for determining the optimum thickness H_2 of the damping layer.

$$g = \frac{G_2}{K_3H_2\omega}\sqrt{\frac{B_1}{m_1}}\sqrt{\frac{(1 + \epsilon_3h_3^3)(1 + \alpha Y)}{1 + d_2h_2 + d_3h_3}} \tag{34.28}$$

where $B_1 = \frac{1}{12}E_1H_1^3$ = flexural rigidity of base plate
m_1 = mass per unit area of base plate
d_2, d_3 = densities of layers 2 and 3
α = quantity shown in Fig. 34.10

$$g_{opt}(1 + k_3) = \frac{1}{\sqrt{1 + Y}\,\sqrt{1 + \beta^2}} \tag{34.29}$$

$$\eta_{max} = \frac{\beta Y}{(2 + Y) + 2\sqrt{1 + Y}\,\sqrt{1 + \beta^2}} \tag{34.30}$$

$$\frac{\eta}{\eta_{max}} = \frac{2(1 + N)(g/g_{opt})}{1 + 2N(g/g_{opt}) + (g/g_{opt})^2} \tag{34.31}$$

$$N = \frac{1 + \frac{1}{2}Y}{\sqrt{1 + Y}\,\sqrt{1 + \beta^2}} \tag{34.32}$$

$$H_{2opt} = (8{,}600/f_0)(G_2/E_1)(1/g_{opt})(1/e_3h_3) \tag{34.33}$$

where f_0 = mid-frequency of the range of interest
G_2 = shear modulus of damping layer at f_0, and at a representative temperature of the range of interest

Values for η_{max}, N, and η/η_{max} are given in Figs. 34.11, 34.12, and 34.13 in graphical form.

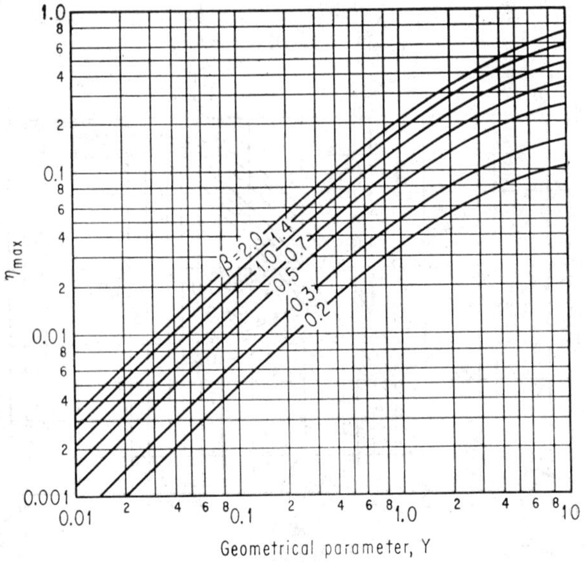

FIG. 34.11. Maximum shear damping as function of geometrical parameter and loss factor of material.[1]

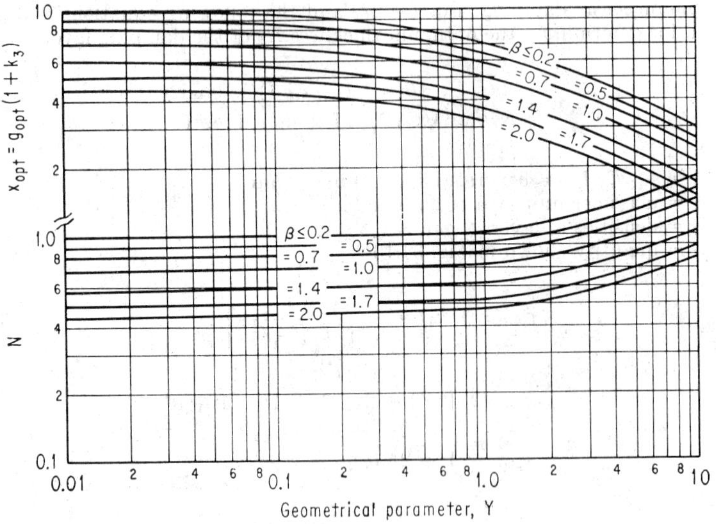

FIG. 34.12. Optimum shear parameter and parameter N as functions of Y and β.[1]

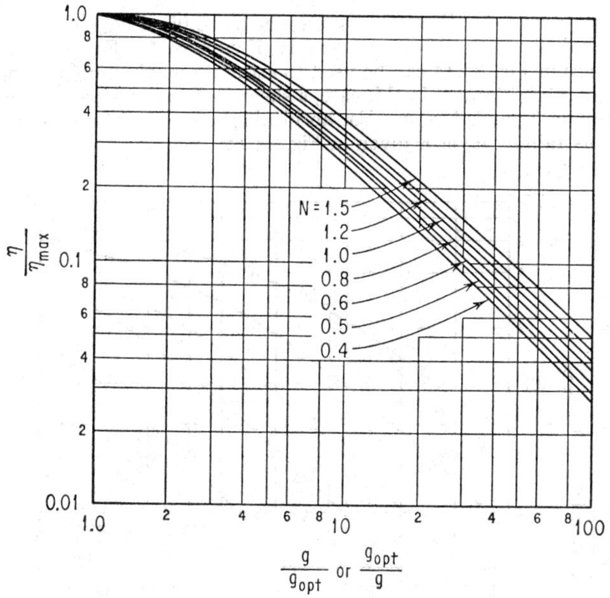

FIG. 34.13. Reduction in damping when operating shear damping treatments at non-optimum values of the shear parameter.[1]

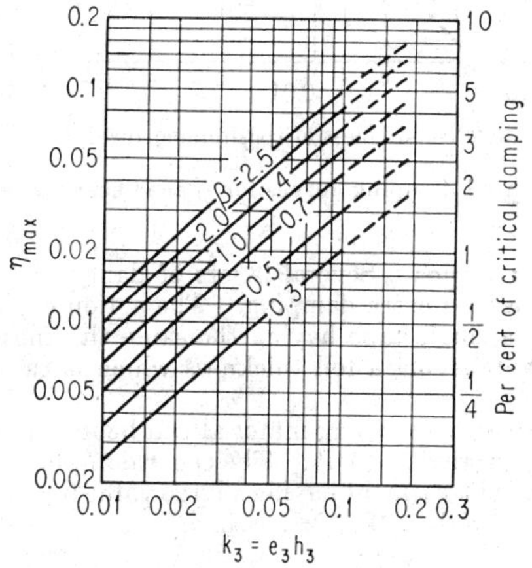

FIG. 34.14. Maximum damping for relatively thin damping tapes as a function of the foil stiffness.[1]

(f) Applications of Shear Damping

Damping Tape with Thin Damping Layer. The common damping tape is constructed of a thin damping layer, usually an adhesive, surmounted by a relatively thin, stiff metal foil. Damping effectiveness is primarily a function of the relative foil thickness.

In many instances, tape dimensions are such that $0.05 < 2h_2 + h_3 < 0.15$; so that the "geometrical parameter" becomes simply $Y \simeq 3.5e_3h_3$. For most tape applications, the curves in Fig. 34.14 can be used to estimate maximum damping.

Spaced Damping Tape. In this case, a shear-stiff spacer separates the base plate from the damping-layer-foil tape. For an ideal spacer, η_{max} as a function of relative stiffness of constraining layer is given in Fig. 34.15.

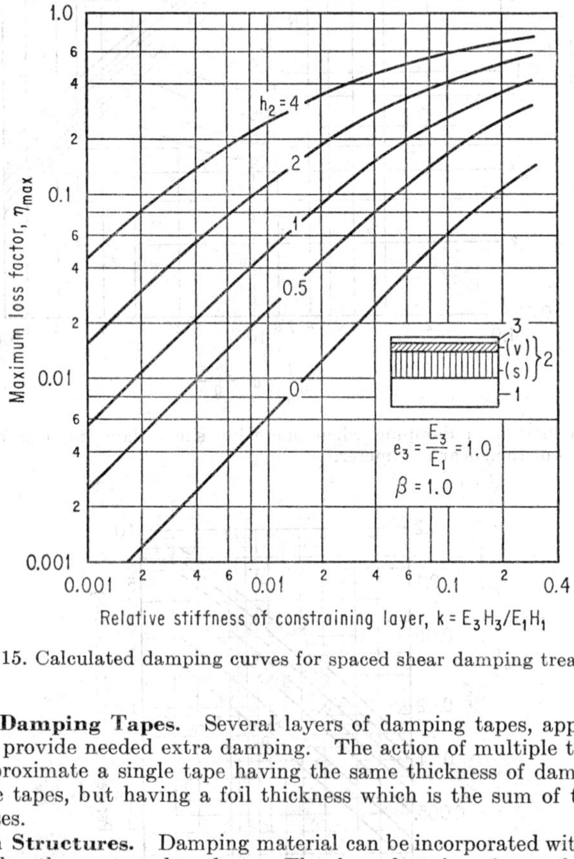

FIG. 34.15. Calculated damping curves for spaced shear damping treatments.[1]

Multiple Damping Tapes. Several layers of damping tapes, applied one atop another, can provide needed extra damping. The action of multiple tapes has been found to approximate a single tape having the same thickness of damping material as one of the tapes, but having a foil thickness which is the sum of the individual foil thicknesses.

Sandwich Structures. Damping material can be incorporated within structural elements, rather than external to them. The shear damping of a sandwich structure is equivalent to an applied treatment having a large value for the "geometrical parameter" Y. Loss factors as high as 0.3 are possible.

34.3. SYSTEM DAMPING

(a) Interfacial Damping

Two cases may be distinguished for damping at a joint. In the first case relative gross motion occurs between the adjacent members. In the second case the exciting force does not produce overall motion, but portions of the joint do slip past one another because of elastic deformations.

In the case of relative gross motion, treating the members as rigid is usually justified. Determination of the relative motion occurring at a joint is often difficult because of the nonlinear mathematical formulations, and direct integration of differential equations is abandoned for special techniques. The equations for some special situations are given below.

Damping of Supported Beams with Fixed Ends. In studying the damping at beam supports it is convenient to investigate the fundamental mode of vibration. To do this, a shape for the mode is assumed. If the damping is viscous, and there is

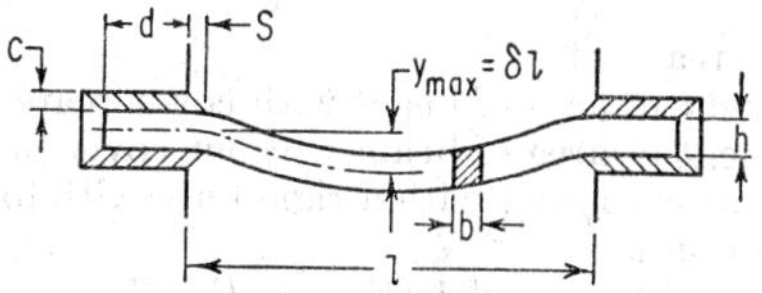

FIG. 34.16. Damping of supporting beams with fixed ends.

only axial deformation of the beam end with the support as shown in Fig. 34.16, the energy dissipation per cycle is given by

$$D_v = (bhlE)(\pi C^2 \delta^4) \frac{s \sinh 2r(d/l) + r \sin 2s(d/l)}{4[\cosh 2r(d/l) + \cos 2s(d/l)] + 4[r \sinh 2r(d/l) - s \sin 2s(d/l)]} \tag{34.34}$$

where d = depth of beam into the support
b = width of beam
h = height of beam
l = length of span
E = Young's modulus
C = a constant depending on the mode shape assumed
$\delta = y_{max}/l$
y_{max} = maximum transverse deflection
$r = \sqrt{el^2(1 + m)/ch}$
$e = \sqrt{(G_1^2 + 4G_2^2)/E}$
$s = \sqrt{el^2(1 - m)/ch}$
$m = G_1/\sqrt{G_1^2 + G_2^2}$
G_1 = real component of viscoelastic modulus
G_2 = imaginary component of viscoelastic modulus
c = thickness of viscoelastic bond

If the ends are considered rigid, the expression is

$$D_v = (bhlE)(\pi C^2 \delta^4) \frac{(le/c)(d/l)(h/l)n}{2[(le/c)(d/l)n] + 2[(le/c)(d/l)m + h/l]^2} \tag{34.35}$$

where $n = 2G_2/\sqrt{G_1^2 + 4G_2^2}$

In the analysis of joints where there is no overall motion, Coulomb's law of friction is assumed to hold, and inertia forces are neglected. A slipped region starts, grows, and stops until the variable external load reaches its maximum value, after which it decreases as the exciting force diminishes. Variation of Poisson's ratio between material of shaft and bushing is neglected, and damping pressure is assumed to be uniform. Figure 34.17 illustrates some simple interfacial slip joints.

The energy dissipated for the joint shown in Fig. 34.17*a* during each loading cycle is

$$D_0 = \frac{2}{3} \frac{1 - 2k}{k_s} \frac{(F_{max} - F_{min})^3}{q} \tag{34.36}$$

where k_s = stiffness of shaft
k_B = stiffness of bushing
$k = k_s/(k_s + k_B)$
q = shear force per unit length at which slip occurs (i.e., for circular shaft $q = 2\pi R\mu p$, where R = radius of shaft and p = damping pressure)

For a joint in pure flexure (thin cover plate), the energy dissipation per cycle is

$$D_0 = \frac{2}{3}\frac{hc^2}{EI_0}\frac{(M_{\max} - M_{\min})^3}{q} \tag{34.37}$$

where h = depth of beam
$c = hA/2I$
A = cover-plate area
I = second moment of area of beam with cover plate
I_0 = second moment of area of beam without cover plate

For an elastic plate in press contact with a rigid base with force parallel to the interface, the energy absorption is

$$D_0 = \frac{1}{6}\frac{1}{k}\frac{(F_{\max} - F_{\min})^3}{q} \tag{34.38}$$

(b) Electrodynamic Damping

Electrodynamic damping is obtained when a short-circuited electrical conductor is made to move in a magnetic field. An induced current appears in the conductor.

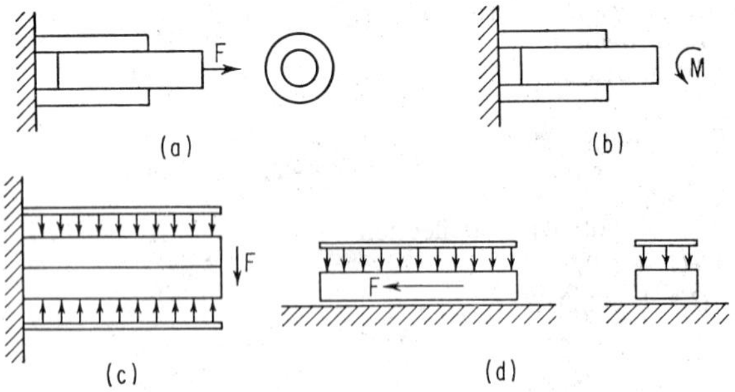

FIG. 34.17. Some simple interfacial slip joints.

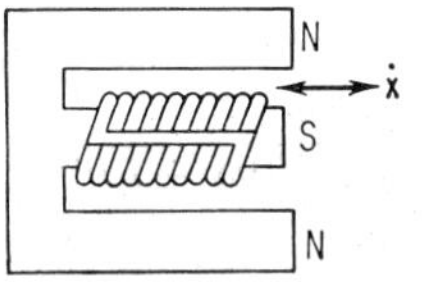

FIG. 34.18. Electrodynamic damper with induced current.

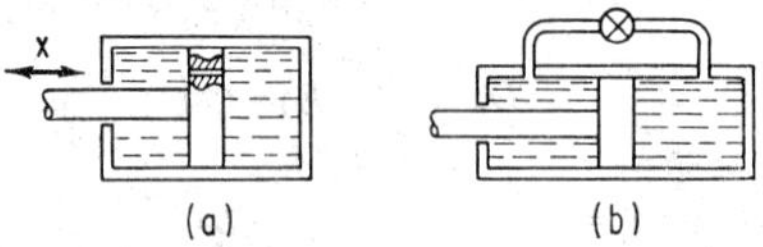

FIG. 34.19. Hydromechanical damping with incompressible fluid.

The conductor experiences a force which is proportional to but opposite to the velocity (see Fig. 34.18). The damping force is expressed by

$$F = \frac{H^2Al \times 10^{-9}}{\rho} = \frac{H^2V \times 10^{-9}}{\rho} \tag{34.39}$$

where F = force
H = average magnetic-field strength
A = cross-sectional area of coil wire
l = length of wire
V = volume of wire
ρ = specific resistance

The damping constant ($F = c\dot{x}$) is therefore

$$c = \frac{H^2V \times 10^{-9}}{\rho\dot{x}} \tag{34.40}$$

(c) Hydromechanical Damping

Incompressible Fluids (Fig. 34.19). A general form of the damping force may be written

$$F = k(\dot{x})^n \tag{34.41}$$

where $1 \leq n \leq 2$. In terms of the pressure drop, the equation becomes

$$\Delta p = f(l/d)(\dot{x}^2/2g) \tag{34.42}$$

where f = a dimensionless friction coefficient which is a function of Reynolds number
l = length of duct
d = diameter of duct

At low speeds in which fully developed laminar flow is achieved Eq. (34.42) assumes the Poiseuille form

$$\Delta p = (32\eta l/d^2)\dot{x} \tag{34.43}$$

where η viscosity of the liquid.

Compressible Fluids. The principle of damping in this case is similar to that for incompressible fluids. However, the fluid also acts as a spring. The total behavior can be imagined as a spring in parallel with a damper. Assuming adiabatic compression and an ideal-gas behavior, the effective spring constant of the gas can be evaluated from

$$k = p\gamma A^2/V \tag{34.44}$$

where p = gas pressure
$\gamma = C_p/C_v$
C_p = specific heat at constant pressure
C_v = specific heat at constant volume
V = volume
A = cross-sectional area of duct

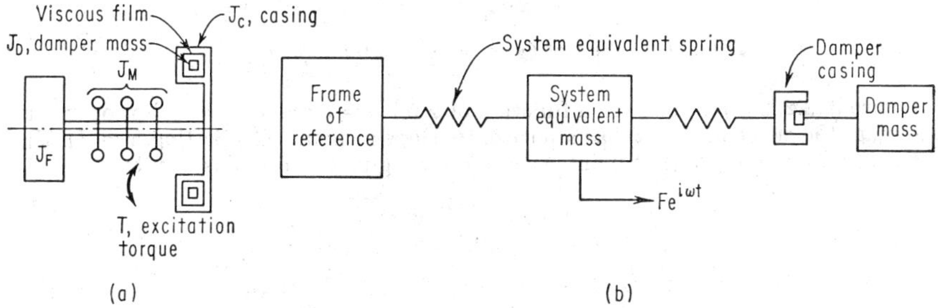

FIG. 34.20. Untuned viscous shear damper. (a) Torsional system. (b) Translational system.

34.4. DAMPING DEVICES

(a) Untuned Viscous Shear Damper (Fig. 34.20)

This damper (also viscous-fluid damper, untuned damped vibration absorber) consists of an annular mass enclosed by an annular ring. A viscous fluid fills the volume between mass and casing. Since there is no elastic connection between the casing, which is attached to the torsionally vibrating system, and the mass, the damper is untuned. Reduction in amplitudes and a lowering of the natural frequency are obtained without the introduction of additional resonances.

Reduction of a complex system to an equivalent two-mass system yields a useful design approximation. Letting $J_F = \infty$, the equivalent system is shown in Fig. 34.21*a* and *b*. The impedance equations for steady state can be written

$$(-\omega^2 J_D + j\omega c)\theta_1 - (j\omega c)\theta_2 = 0 \tag{34.45}$$

$$-(j\omega c)\theta_1 + [K_2 - \omega^2(J_E + J_c) + j\omega c]\theta_2 = T_E \tag{34.46}$$

where J_D = mass moment of inertia of damper
J_E = equivalent mass moment of inertia of torsionally vibrating system
J_c = mass moment of inertia of damper casing
ω = circular natural frequency of vibrating system with damper
θ_1 = steady-state vibration amplitude at damper mass
θ_2 = steady-state vibration amplitude at damper casing

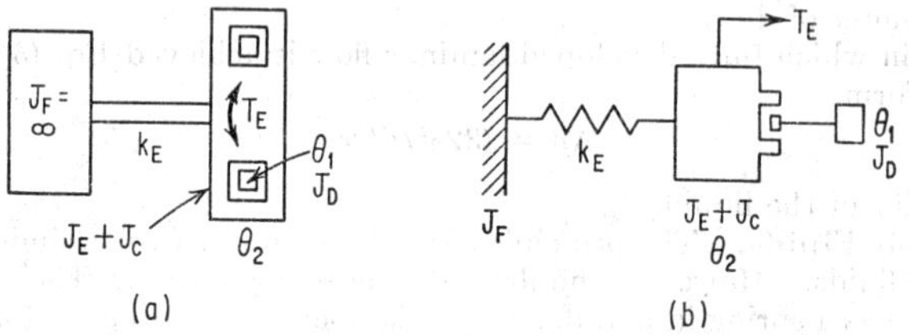

FIG. 34.21. Equivalent two-mass system.

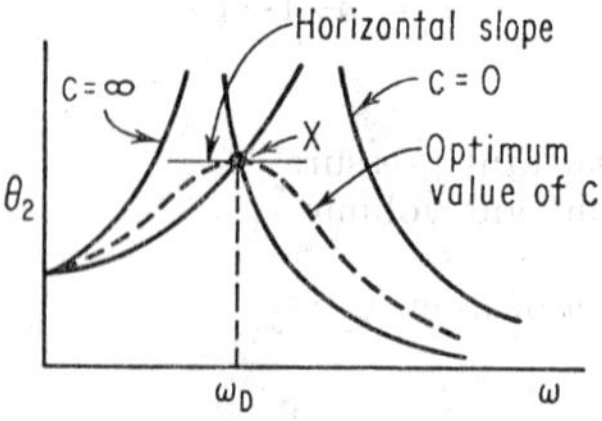

FIG. 34.22. Plot of θ_2 vs. ω for $c = 0$ and $c = \infty$.

Figure 34.22 is a plot of θ_2 as a function of ω for $c = 0$ and $c = \infty$. For optimum damping, the curve of θ vs. ω has a horizontal slope at point X. Setting $\mu = J_D/J_E$, and solving for conditions at point X, the following useful formulas result:

$$\omega_D/\omega_2 = \sqrt{2/(2 + \mu)} \tag{34.47}$$

$$\theta_2/(T_0/K_E) = M = (2 + \mu)/\mu \tag{34.48}$$

$$c/(2J_D\omega_2) = \gamma = 1/\sqrt{2(1 + \mu)(2 + \mu)} \tag{34.49}$$

$$|\theta_{SH}/\theta_2| = |\theta_{SH}/(T_0/K_E)|/M = \sqrt{(1 + \mu)/(2 + \mu)} \tag{34.50}$$

where ω_2 = circular natural frequency of torsionally vibrating system without damper
ω_D = circular frequency corresponding to point X in Fig. 34.22
c = optimum value of coefficient of viscous damping
T_0 = excitation torque
θ_{SH} = $|\theta_1 - \theta_2|$
M = dynamic magnifier

Knowledge of such items as θ_2, T_0, K_E, and ω_2 permits the determination of μ, from which J_D and c may be calculated.

In reciprocating engines, μ tends to fall in the range 0.4 to 1.0. Also J_C/J_D usually lies between 0.35 and 0.8. The dimensions are so proportioned that sufficient surface is provided for proper heat dissipation. Determination of the necessary fluid viscosity is based upon a modified value of c which accounts for varying shear rates on the lateral and peripheral surfaces of the damper.

(b) Slipping-torque-type Dampers

In this type of damper, relative motion between a shaft-fixed hub and a damping mass occurs only when the relative acceleration of the two exceeds a predetermined value. An effective change of natural frequency occurs as the damping mass "locks" and "unlocks" from the hub during each oscillation. Dissipation of energy by damping occurs during the intervals of relative motion.

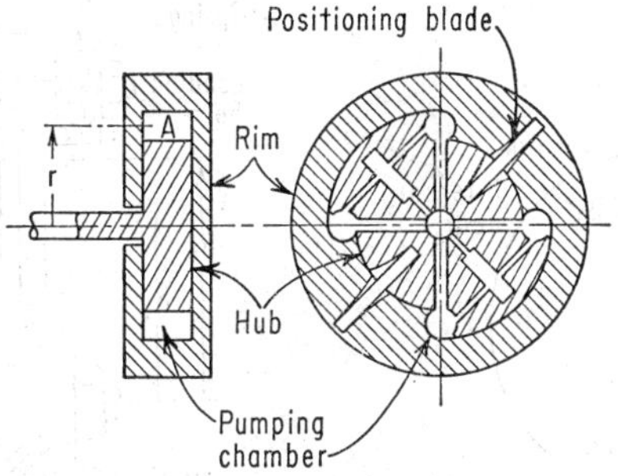

FIG. 34.23. The Sandner damper (pumping-chamber type).

An analysis of the input and dissipated energy can be made by assuming continual slip, sinusoidal motion of the hub, and a linear time variation of the damper mass. The maximum energy dissipated is

$$U_{D\max} = (4/\pi)\omega^2\theta_1{}^2 J_R \tag{34.51}$$

where ω = circular natural frequency of vibrating system without damper
θ_1 = permissible oscillation amplitude at damper hub
J_R = moment of inertia of damper

Energy input $U_{\text{in}} = \pi T_0 \theta_1$, where T_0 is the peak value excitation torque at the hub. Equating input and dissipated energy yields the useful design equation

$$J_R = \pi^2 T_0 / 4\omega^2\theta_1 \tag{34.52}$$

The Sandner Damper (Pumping-chamber Type). A rim and side plates, which act as the damper mass, and a hub are arranged to form internal cavities at certain points of the interfaces. The cavities act as pumping chambers. Oil passes through radial passages starting at the hub, moves to the pumping chambers, returns to the hub to pass through spring-loaded relief valves, and then is discharged into some convenient space. The slipping torque is accurately set by adjustment of the relief valves.

The damper moment of inertia is determined from Eq. (34.52). Maximum torque at which slipping occurs is calculated from

$$T_{R\max} = (\sqrt{2}/\pi)\omega^2\theta_1 J_R \tag{34.53}$$

Determination of the relief-valve spring pressure can be calculated from

$$p = \frac{T_{R\max}}{2rA}$$

where A is the cross-sectional area of one of the pumping chambers and r is the mean radius to the area A as shown in Fig. 34.23.

The Sandner Damper (Gear-wheel Type). In this type of damper, a rim is cut with gear teeth on its inner surface. Pinions mesh with this gear and are enclosed in special recesses in the hub. Passages connect opposite sides of each recess to centrally located relief valves.

The rim tends to rotate the pinions because of its inertia torque. Actual rotation takes place beyond a certain critical value as predetermined by the relief-valve adjusted pressure. Oil is passed from one gear chamber to the next as the pinions rotate, as shown in Fig. 34.24. Calculations for this damper are the same as for the pumping-chamber type. However, the pressure is computed as $p = T_{R\max}/8rA$, where A is the effective area of the gear-pump recess.

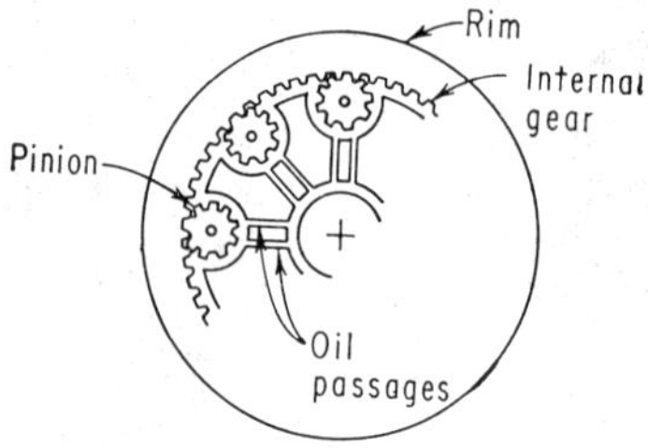

FIG. 34.24. The Sandner damper (gear-wheel type).

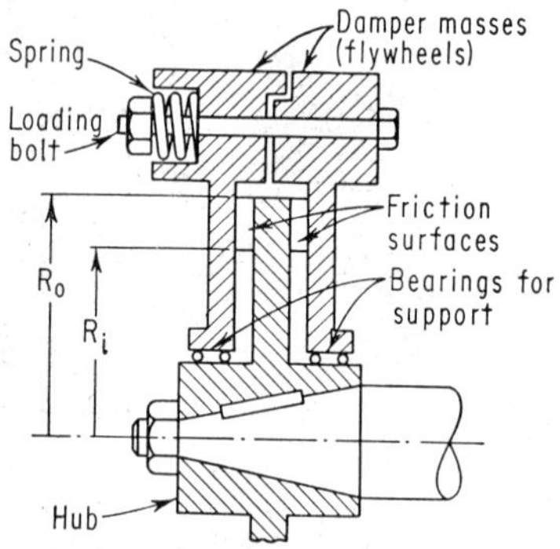

FIG. 34.25. The Lanchester damper (semi-dry-friction type).

The Lanchester Damper (Semi-dry-friction Type). In this type of damper a hub, fixed coaxially to a vibrating shaft, carries friction plates on an annulus near its rim. Pressing against the friction surfaces is the damper mass consisting of two flywheels and loading bolts. The damper mass lies coaxial with the shaft but is coupled to it only through the friction surfaces as shown in Fig. 34.25.

The moment of inertia J_{2R} of the two flywheels is calculated using Eq. (34.52). Maximum torque is computed from $T_{R\max} = (\sqrt{2}/\pi)\omega^2\theta_1 J_{2R}$ and is equated to the friction torque T_f. Spring load can then be computed from

$$T_f = \tfrac{4}{3}\mu P(R_o^3 - R_i^3)/(R_o^2 - R_i^2) \tag{34.54}$$

where μ equals coefficient of friction, R_i and R_o are the inner and outer radii of the friction surfaces, respectively, and P is the total spring load.

References

1. Ruzicka, J. E. (ed.): "Structural Damping," Applied Mechanics Division of ASME, December, 1959. Contains colloquium papers, which contain extensive bibliographies.
2. Yorgiadis, A.: Damping Capacity of Materials, *Prod. Eng.*, November, 1954, pp. 164–170.
3. Ungar, E. E., and D. Kent Hatch: Your Selection Guide to High-damping Materials, *Prod. Eng.*, Apr. 17, 1961, pp. 44–56.
4. Van Santen, G. W.: "Mechanical Vibrations," Philips' Technical Library, Elsevier Publishing Company, Houston, Tex., 1953.
5. Nestorides, E. J.: "A Handbook on Torsional Vibration," Cambridge University Press, New York, 1958.
6. Ungar, E. E.: A Guide to Designing Highly Damped Structures . . . Using Layers of Viscoelastic Material, *Machine Design*, Feb. 14, 1963, pp. 162–168.

Section 35

ELECTROMECHANICAL COMPONENTS

By

JEROME LEIGHT, MEE, *Chief, Systems Engineering Section, Systems Management Division, Kollsman Instrument Corp., Elmhurst, N.Y.*

STANLEY SPORN, MSEE, *Supervisor, Accelerometer Section, American Bosch Arma Corporation, Arma Division, Garden City, N.Y.*

CONTENTS

35.1. INTRODUCTION

The successful implementation of good control-system concepts rests on the proper selection and application of components. Although major emphasis is frequently placed on stability analysis, the more serious development problems generally are a result of improper component selection.

The designer must understand the system requirements and environment in which it must operate. In addition to satisfying these requirements, the choice of components must also be guided by considerations of cost, weight, and volume, and compatibility (e.g., power requirements, reliability, etc).

Having selected the component type, the choice of a particular model must be made. A rapid method of converging on the best model is to select one more or less arbitrarily, under the constraint of the factors cited above and previous experience. System accuracy is then evaluated and the desired component parameters quickly become apparent. As a result of the accuracy analysis it may become necessary to select another model or even to modify the initial choice of component type.

It is the intent of this section to provide the designer with information to select the optimum component type for his application, information to determine the system error directly and indirectly attributable to the component, and information to apply the component properly. For more on components see ref. 17.

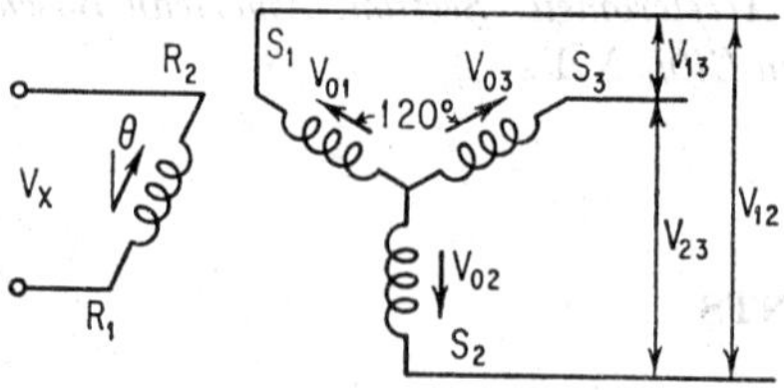

FIG. 35.1. Schematic of a synchro transmitter showing voltages induced in the stator windings. R = rotor terminals. S = stator terminals.

35.2. TRANSDUCERS

(a) Synchros, General

Application. Synchros are inherently suited for angle-transmission systems because they are capable of continuous rotation and their balanced three-wire transmission is relatively unaffected by changes of frequency, reference voltage, temperature, and cable capacitance.

Operation. The transmitter synchro converts angular data from shaft position to three-wire electrical signals. The excitation voltage is applied to the rotor. The voltage induced in each stator winding is proportional to the cosine of the angle between the rotor and stator winding.

From Fig. 35.1 the following relationships can be derived:

$$V_{01} = KV_x \cos(\theta + 60) = KV \sin(\theta + 150) \tag{35.1}$$

$$V_{02} = KV_x \cos(\theta + 180) = KV \sin(\theta - 90) \tag{35.2}$$

$$V_{03} = KV_x \cos(60 - \theta) = KV \sin(\theta + 30) \tag{35.3}$$

where V = voltage across any points

V_x = excitation voltage on rotor

K = voltage transformation, rotor to stator

$$V_{13} = K\sqrt{3}\, V_x \sin\theta \tag{35.4}$$

$$V_{23} = K\sqrt{3}\, V_x \sin(\theta + 120) \tag{35.5}$$

$$V_{12} = K\sqrt{3}\, V_x \sin(\theta + 240) \tag{35.6}$$

When connected to the stator of a receiver (or differential) the current in each transmitter winding is the same as the current in corresponding receiver windings. The resultant flux field in the receiver (or differential) therefore has the same angular position as the rotor of the transmitter. When a differential synchro is imposed between a transmitter and receiver it serves to rotate the field induced in the receiver relative to the angular position of the rotor of the transmitter.

The basic difference between the operation of control synchros and torque synchros is the manner in which a null is achieved. In the case of control synchros, the receiver (*CT*) shaft is positioned to null the voltage induced in the rotor. The null occurs when the rotor of the *CT* is displaced 90° from the rotor of the transmitter. In the case of torque synchros, the rotor of the receiver (*TR*) is energized by the same voltage as the transmitter. When the rotors are in correspondence, the receiver stator voltages are equal to those induced in the transmitter. Then the current in the stator coils will be nulled, a magnetic field is not established, and no torque is exerted on the rotors. It should be noted that in the above discussion a condition of null rather than

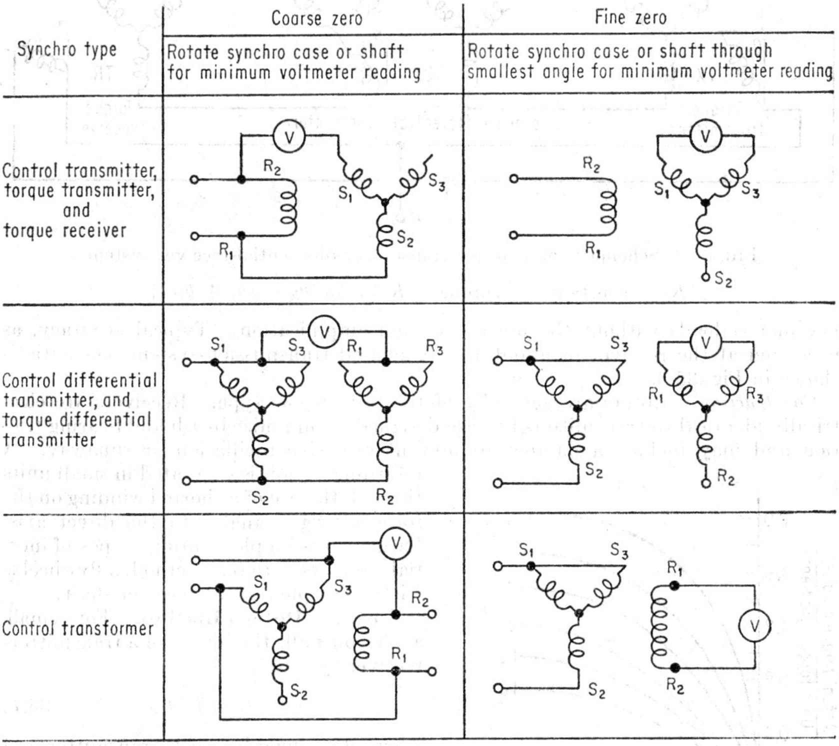

Coarse zero
Fine zero
Synchro type
Rotate synchro case or shaft for minimum voltmeter reading
Rotate synchro case or shaft through smallest angle for minimum voltmeter reading
Control transmitter, torque transmitter, and torque receiver
Control differential transmitter, and torque differential transmitter
Control transformer
V
R1
R2
R3
S1
S2
S3

FIG. 35.2. Zeroing synchros.

a zero was indicated. A null refers to a zero of only the information-bearing portion of the signal. In addition to information, the signal consists of quadrature (90° phase-shifted fundamental) and harmonic terms.

Construction. The construction of a synchro is similar to a miniature wound-rotor induction motor. The stator is a three-phase winding, usually *Y*-connected, set into skewed slots to minimize slot effect.

The rotor is single-phase, two-pole, either salient or nonsalient construction. The windings are brought out to slip rings to permit continuous rotation. In standard synchros, the bearings are designed for continuous operation at speeds up to 1,200 rpm.

Synchro size is designated by two digits which represent the maximum diameter (to the next higher tenth if not an integral number of tenths) in tenths of an inch.

The accuracy of standard synchro types is more or less independent of size, but special high-accuracy devices have been produced in the larger sizes.

The range of impedance increases with size since there is more room for windings and a larger mass for heat dissipation.

The location of the synchro electrical zero in accordance with BUORD specifications is shown in Fig. 35.2.

(b) Torque-type Synchros

Application. Application of torque-type synchros is found in the self-contained angle-transmission system used for positioning dials, pointers, and other relatively low inertia loads without the use of external amplification. Typical accuracy, as measured at the receiver, is about 1°. A general transmission-system schematic is shown in Fig. 35.3.

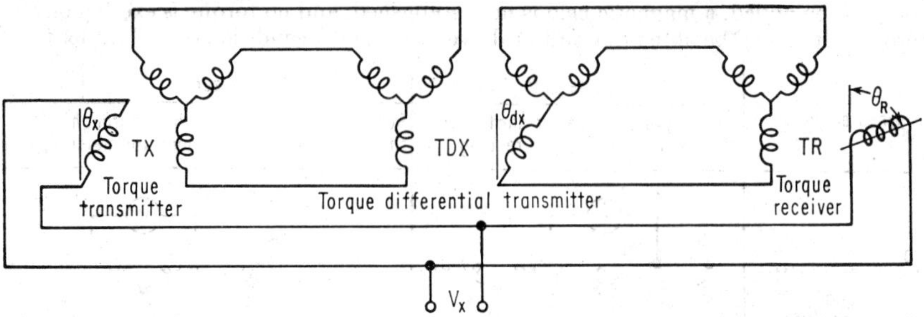

FIG. 35.3. Schematic of a torque transmitter-differential-receiver system.

$$K_1 = \text{constant} \qquad \text{Torque} = K_1 V_x \sin\left[\theta_R - (\theta_x + \theta_{Dx})\right]$$

Construction. Rotors are generally of the salient-pole type. Receivers are electrically identical to transmitters but are designed to minimize brush and bearing friction and may include a damper to prevent excessive oscillation or runaway. A retarding action is generated in small units through the use of a shorted winding on the rotor at right angles to the direct axis. Larger units employ various types of inertial dampers. In one method, a flywheel is friction-coupled to the receiver shaft.

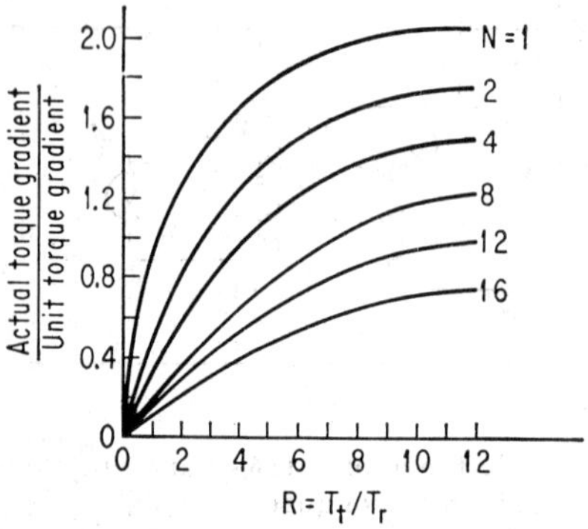

FIG. 35.4. Torque-type synchro actual torque gradient.

Torque Determination. For small angles off null, the torque of a transmitter-receiver system

$$T = T_r'\theta \tag{35.7}$$

where θ = angle between transmitter and receiver rotors

T_r' = torque gradient of the receiver

T_r' is determined for the general case where several receivers are following one transmitter, by

$$T_r' = 2RT_r/(N + R) \tag{35.8}$$

where T_r = receiver-unit torque gradient, which is the torque gradient of the receiver when excited by an identical transmitter

N = number of receivers

R = ratio T_t/T_r

T_t = transmitter-unit torque gradient, which is the torque gradient of a synchro system using components electrically identical to the transmitter

Figure 35.4 is a graph of T_r'/T_r as a function of R with N as parameter.

Errors. The angular deviation between the transmitter position and the position of the receiver constitutes the synchro error. The inherent synchro errors are:

1. Friction. Bearing and brush friction usually produce up to 1° error.
2. Electrical. Caused by physical design limitations and manufacturing tolerances resulting in nonsinusoidal flux distribution and out-of-round stators and eccentric rotors. The error thus produced is small relative to the friction error.
3. Mechanical. (See Sec. 19 and references to specific components listed in the index.)

Errors due to load and conditions of use are results of the effects of load friction and load inertia.

The load-friction-error effect

$$\epsilon = F/T_r' \tag{35.9}$$

where F = friction torque. Friction should be determined at extremes of the required temperature range.

In general the load inertia should be limited to a small fraction of the receiver inertia in order to avoid excessive oscillations. If the system viscous damping coefficient f is known, then the damping ratio

$$\xi = f/2\sqrt{1/T_r'J} \tag{35.10}$$

where J is the combined inertia of receiver and load. For good performance ξ should not be less than 0.4. A torque data-transmission system is relatively inexpensive and reliable; however, accuracy is poor and may be further degraded by increased friction at temperature extremes.

(c) Control-type Synchros

Application. The angle-transmission system is used in conjunction with a follow-up servo. Typical system accuracy, as measured at the controlled shaft, is about 10 arc minutes. An order-of-magnitude improvement of accuracy as measured at the output shaft can be achieved by reduction gearing from the controlled shaft to the components positioned by the servo, since the error at the controlled shaft is therefore divided by the gear ratio. However, backlash errors must be added. Synchronism is maintained by the use of a coarse synchro.

Figure 35.5 shows two data-transmission systems. The configuration of Fig. 35.5*a* is used more frequently. However, when the servoamplifier is located near the

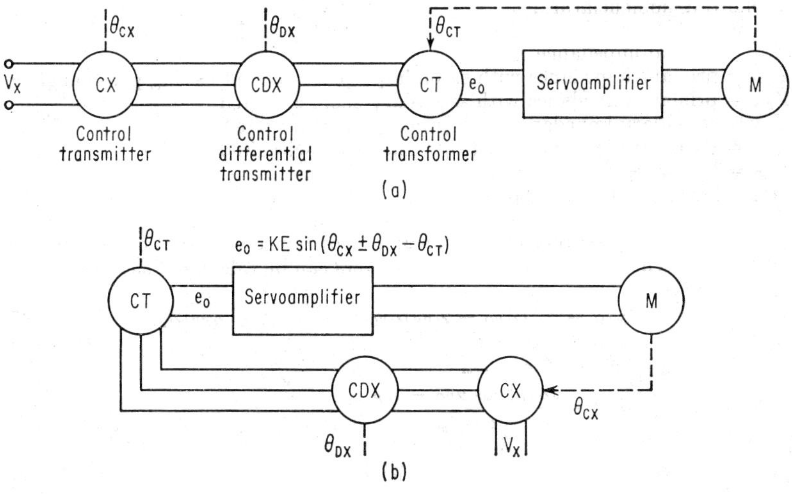

FIG. 35.5. Control-type synchros data-transmission systems. (*a*) Amplifier located near receiver. (*b*) Amplifier located near transmitter.

input shaft, it is advantageous to use the system shown in Fig. 35.5*b* in order to avoid long cable runs in the null circuit.

Construction. Control transformers employ nonsalient pole rotors to produce a uniform reluctance around the air gap and therefore an output voltage which varies sinusoidally with angle from the null position. For small errors the gradient is thus proportional to the error, and the stator impedance with open-circuited rotor does not change with rotor position. The impedance of the *CT* is generally higher than that of the *CX* so that several *CT*'s can be driven from one *CX*.

Differential generators closely resemble control transformers in construction except that three coils are wound on the rotor displaced from one another by 120°.

The output error voltage of a synchro system is

$$e_0 = V_x K \sin \epsilon_e \tag{35.11}$$

where ϵ_e is the angular error between the commanded and the actual position of the *CT* rotor.

The value of K is supplied by the manufacturer for the condition of design load on the transmitter. The actual output voltage can be determined for any condition

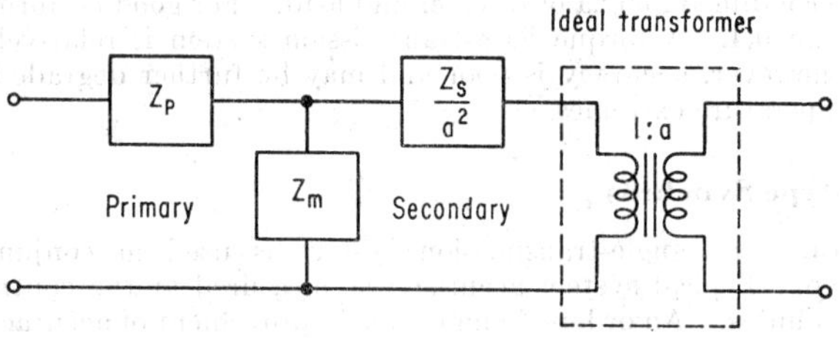

FIG. 35.6. Synchro equivalent circuit.

of load impedance, number of *CT*'s, and temperature effects on winding impedance, etc., by using the general equivalent circuit of Fig. 35.6.

The rotor and stator impedances are supplied by the manufacturer. For generality, they will be here referred to as primary and secondary impedances in accordance with the following nomenclature:

Z_m = mutual impedance
Z_{PO} = primary impedance with secondary open-circuited having phase angle θ_{PO}
Z_{PS} = primary impedance with secondary short-circuited
Z_{SO} = secondary impedance with primary open-circuited
Z_{SS} = secondary impedance with primary short-circuited
R_P = resistance of primary
R_S = resistance of secondary

These impedance measurements are made in accordance with the conventions shown in Fig. 35.7. The equivalent circuit parameters can be calculated as follows:

$$Z_m/a = \sqrt{(Z_{PO} - Z_{PS})Z_{SO}} = \sqrt{(Z_{SO} - Z_{SS})Z_{PO}} \tag{35.12}$$

$$Z_P = Z_{PO} - Z_m \tag{35.13}$$

$$Z_S = Z_{SO} - Z_m \tag{35.14}$$

$$|Z_m| = \frac{|Z_{PO}| \cos \theta_{PO} - R_P}{\cos \theta_m} \tag{35.15}$$

Therefore

$$a = \frac{|Z_m|}{Z_m/a} \tag{35.16}$$

$$Z_m = (Z_m/a)a \tag{35.17}$$

Errors. The error is defined as the angular deviation between the commanded position and the actual position of the receiver rotor.

Manufacturers generally specify electrical error as either maximum error spread or maximum error from electrical zero, as shown in Fig. 35.8. The causes of electrical errors and their characteristics are as follows: nonsinusoidal flux distribution, $6n$

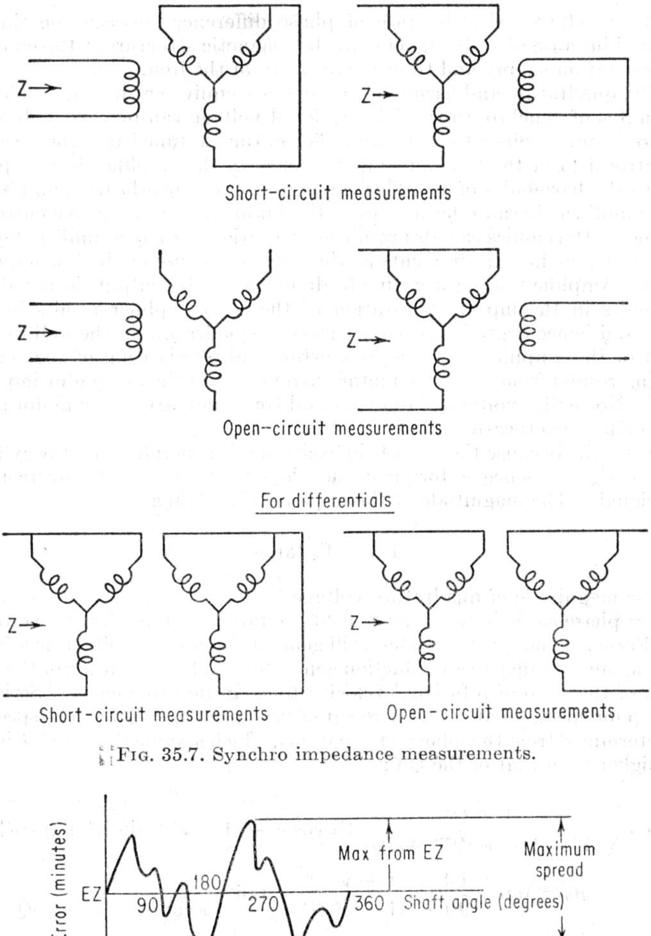

FIG. 35.7. Synchro impedance measurements.

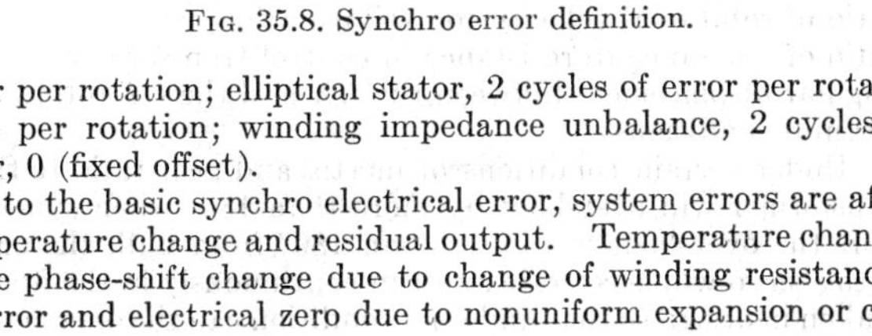

FIG. 35.8. Synchro error definition.

cycles of error per rotation; elliptical stator, 2 cycles of error per rotation; eccentric rotor, 1 cycle per rotation; winding impedance unbalance, 2 cycles per rotation; elliptical rotor, 0 (fixed offset).

In addition to the basic synchro electrical error, system errors are affected by such factors as temperature change and residual output. Temperature change can produce output-voltage phase-shift change due to change of winding resistance, and change of electrical error and electrical zero due to nonuniform expansion or contraction.

Synchro friction can produce a servo static error. However, this error is generally negligible in a well-designed servo.

Mechanical errors are the same as described under Torque Synchros.

Residual-voltage (Null) Output Error. If $V_\omega \angle\theta$ is defined as the output voltage of fundamental frequency, where θ is the phase of the output for large displacements from the null position, then when the rotor is adjusted for $V_\omega \angle\theta = 0$, the output will contain harmonics of ω and signals of fundamental frequency whose phase is $\theta \pm 90°$ (quadrature).

Quadrature voltage exists because of phase difference between the three stator windings and because of eddy currents in the magnetic structure. Harmonics result from the excitation supply and from nonlinearity of the iron.

Typically, quadrature and harmonic voltages are equivalent in magnitude to about 3 to 5 minutes of synchro error. This residual voltage can be extremely serious in some servosystems. Since the servoamplifier output is tuned to resonance with the motor control field at the fundamental frequency ω, the amplifier load impedance is very low to the harmonics of ω. This can cause serious overheating and burnout of transistor amplifiers because harmonics in the input produce excessive current in the output stage. Harmonics can also result in saturation of early amplifier stages, while not being apparent in measurements at the output because of the low impedance to harmonics. Amplifiers using negative feedback from the output do not degenerate the harmonics in the input. Saturation of the servoamplifier results in a loss of sensitivity and hence degraded servo accuracy. Quadrature in the null can produce saturation of the amplifier, a servo bias error, and overheating of the servomotor. Overheating results from excessive motor current due to the low motor impedance at standstill. Normally, voltage of fundamental frequency drives the motor and hence does not produce overheating.

The bias results because the main field excitation is generally not exactly 90° out of phase with $V_\omega \angle\theta$; hence a torque is developed which must be overcome by an in-phase signal. The magnitude of the in-phase bias voltage

$$V_{\text{in}} = V_q \tan\alpha \tag{35.18}$$

where V_q = magnitude of quadrature voltage
α = phase angle between main field excitation and quadrature voltage

Speed Error. The synchro system will generate an output voltage as a function of speed in a manner similar to an induction generator. The magnitude of the voltage is a function of the ratio of rotational velocity to excitation frequency. Serious errors will develop for speeds of about 20 per cent of synchronous speed. The speed voltage may be determined from the following equations, which assume that the *CT* impedance is much higher than that of the *CX*:

$$|V_0| = KV_x \sqrt{\frac{1 + Q^2}{[1 - (1 - \nu^2)Q^2]^2 + 4Q^2} \{[\nu\cos\sigma - (1 - \nu^2)Q\sin\sigma]^2 + \sin^2\sigma\}}$$

$$\beta - \beta_0 = \tan^{-1}\frac{1}{Q}\frac{1 + (1 + \nu^2)Q^2}{1 + (1 - \nu^2)Q^2} + \tan^{-1}\frac{1}{\nu\cot\sigma - (1 - \nu^2)Q} \tag{35.19}$$

where V_0 = magnitude of speed voltage
$\beta - \beta_0$ = phase angle of V_0 relative to phase angle at standstill
ν = ratio of rotational velocity to excitation frequency
Q = ratio of reactance to resistance in control transformer
σ = angular displacement between the *CX* and the *CT* rotors (nominally 90° for null conditions)

Runaway. Under certain conditions of inertia and gear ratio to the servomotor, the control transformer will overshoot by more than 180° when synchronizing from large offsets. If the overshoot is greater than 180° from null, the error signal will reverse; hence the servo will accelerate and run continuously in the original direction. Care must be taken in design so that under no condition can the overshoot exceed 180°. Ninety degrees is a reasonable design objective.

Reaction Torque. The effect of synchro reaction torques must be considered in sensitive applications such as gyro systems.

Phase Shift. The phase shift through the synchro system must be considered for the following reasons:

1. The phase of the control-field excitation should be 90° from the main field in order to obtain maximum torque and minimize the biasing effect of quadrature.
2. Systems using generator damping and those having signals injected in the null line as shown in Fig. 35.9 will suffer excessive quadrature unless the phase of the *CT* output is shifted such that $\gamma \approx \rho$.

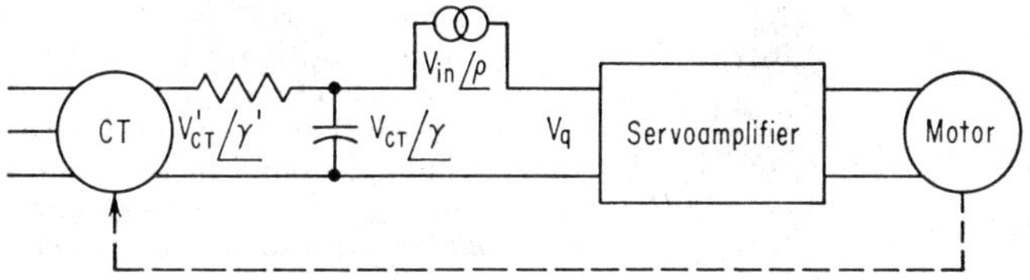

FIG. 35.9. Phase-shifting *CT* output.

Under null conditions and for small $(\gamma - \rho)$ angles,

$$V_q = V_{in}(\gamma - \rho) \tag{35.20}$$

The seriousness of the phase difference between γ and ρ can be illustrated by a simple example. Assuming V_{in} = 5 volts and $\gamma - \rho$ = 0.1 rad, the resulting quadrature is 500 mv, more than enough to saturate most systems.

Multispeed System. The accuracy of synchro systems can be increased by the use of coarse and fine synchros. Two problems must be considered in this type of system: transfer of control and false synchronization. Control of the servo should be maintained by the highest-speed synchro until its error falls between 45 and 90°. Control is then transferred to the next highest speed synchro. Transfer circuits are generally constructed using solid-state diodes and taking advantage of their breakdown characteristics.

The problem of false synchronization develops when the speed of the high-speed unit is an even multiple of the low-speed unit. Thus at the 180° position (unstable

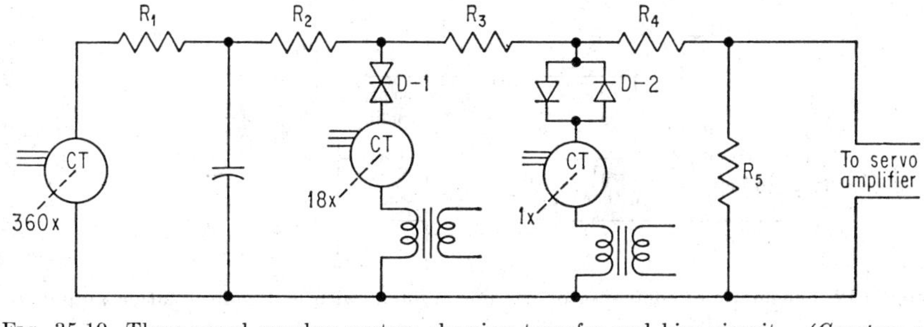

FIG. 35.10. Three-speed synchro system showing transfer and bias circuit. (*Courtesy of Arma Div., American Bosch Arma Corp.*)

null) of the low-speed unit, the high-speed unit will be at a stable null position. The system may therefore synchronize 180° from the command position. The problem is solved by adding a "stick-off" voltage in series with the low-speed synchro equal to its output when the high-speed synchro is at 90°. The low-speed synchro case is then rotated relative to its zero position until the stick-off voltage is "bucked out." Thus at the correct servo null the offset synchro output cancels the "stick-off" voltage while adding to it at the 180° position. The unstable null of the low-speed synchro is thus shifted by an angle equivalent to 180 high-speed degrees. Unstable nulls now occur simultaneously for both units, and the system will synchronize properly.

A three-speed system including transfer and stick-off voltage is shown in Fig. 35.10.

(d) Resolvers

Application. Included among resolver applications are analog-computer computation of trigonometric functions (typical accuracy 0.1 per cent) and angle transmission (typical accuracy 3 minutes).

Construction. Resolver construction is similar to that of a control transformer except that both rotor and stator have two windings in space quadrature. The winding distribution is designed to yield an accurate sinusoidal voltage output as a function of rotor position. Figure 35.11 shows a cutaway view of a resolver.

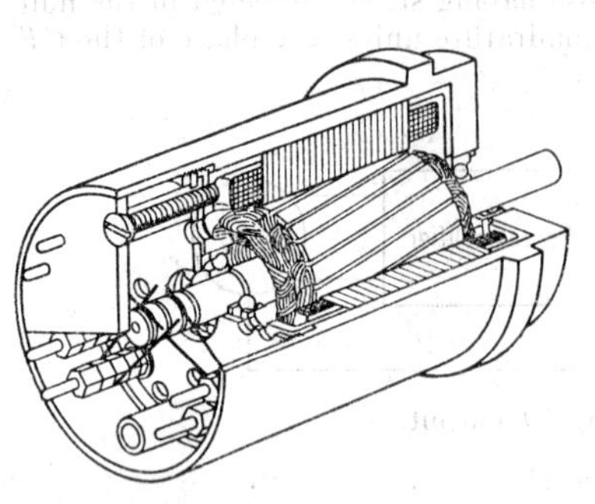

FIG. 35.11. Resolver. (*Courtesy of Reeves Instrument Company, New York.*)

Some resolvers are equipped with an auxiliary winding, wound in the same slots as, and similar to, the stator winding. These components are called *winding-compensated resolvers*. The auxiliary winding is used in conjunction with an isolation amplifier as shown in Fig. 35.12 to compensate for the effects of temperature and frequency changes.

The operation of the feedback loop is such that an error voltage is developed if V_c/R_2 is not equal to V_I/R_1. The ratio of R_2/R_1 is usually adjusted so that, at maximum coupling, the output voltage of the rotor winding is equal to the input voltage.

Various resolver computations are shown in Table 35.1.

Errors. Resolver accuracy is specified by (1) conformity error, the deviation between actual and theoretical sinusoidal output as a percentage of the maximum output measured at nominal excitation voltage with one primary winding excited and the other shorted with the rotor open-circuited, and (2) interaxis error, the angular deviation of the windings from 90°. The specification refers to stator as well as rotor windings.

Inherent electrical errors, which consist of conformity and interaxis errors, result from rotor and stator ellipticity and eccentricity, and imperfect winding and flux

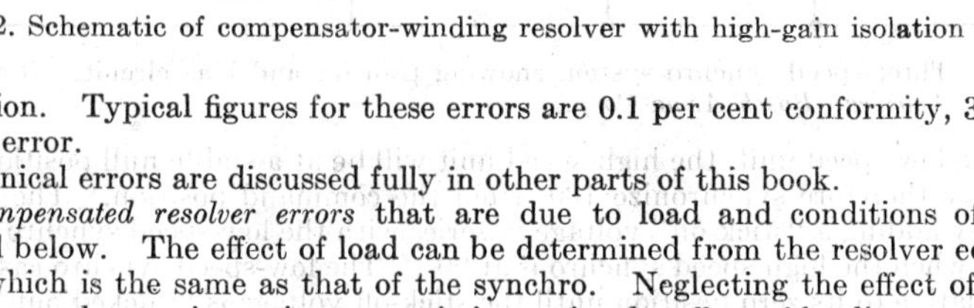

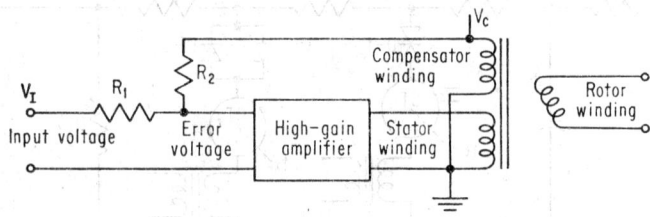

FIG. 35.12. Schematic of compensator-winding resolver with high-gain isolation amplifier.

distribution. Typical figures for these errors are 0.1 per cent conformity, 3 minutes interaxis error.

Mechanical errors are discussed fully in other parts of this book.

Noncompensated resolver errors that are due to load and conditions of use are described below. The effect of load can be determined from the resolver equivalent circuit, which is the same as that of the synchro. Neglecting the effect of primary inductance and assuming unity transmission ratio, the effect of loading can be calculated from

$$\frac{V_0}{V_x} = \frac{Z_L}{Z_{\text{in}} + Z_L + R_p + R_S + jX_S + iX_p} \tag{35.21}$$

Table 35.1. Various Resolver Computations

Computation	Schematic	Vector diagram
Resolution of vector into components	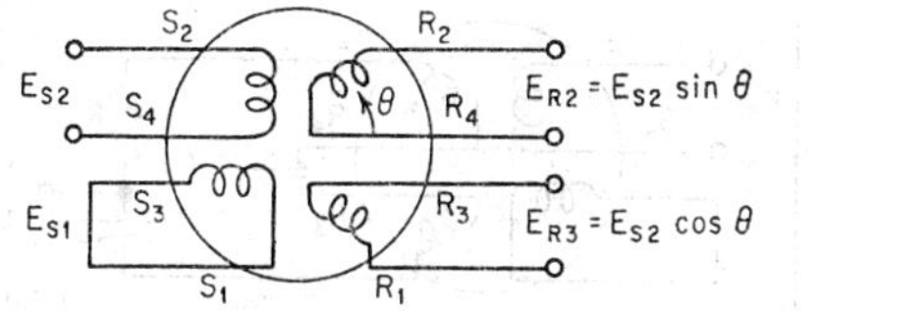	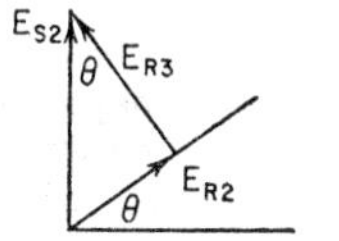
Rotation of coordinates	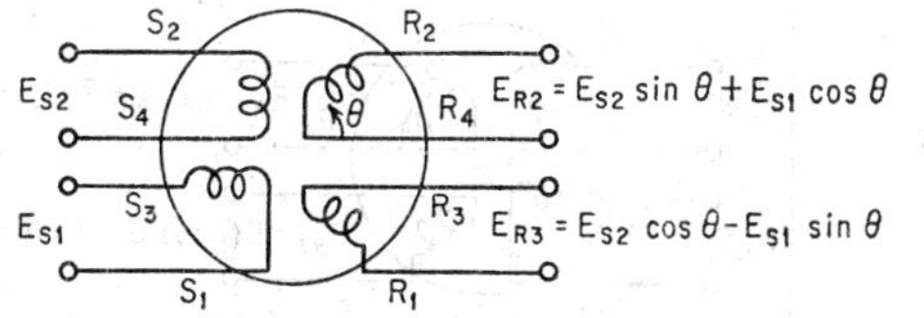	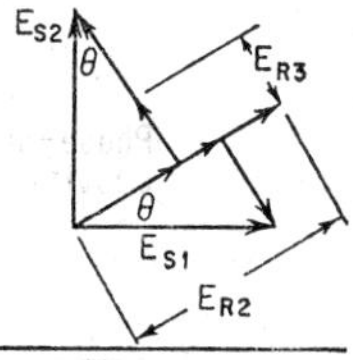
Vector addition	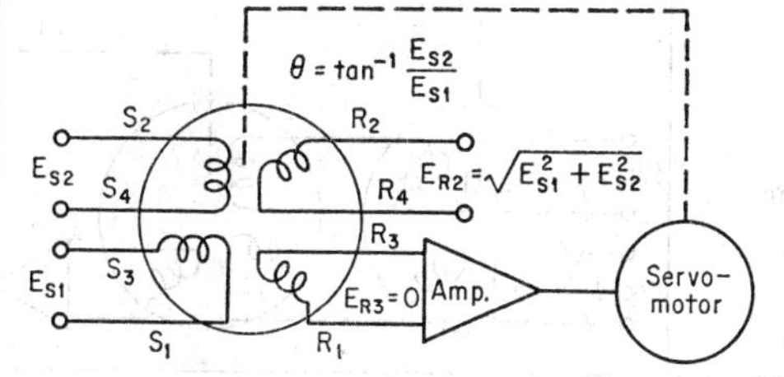	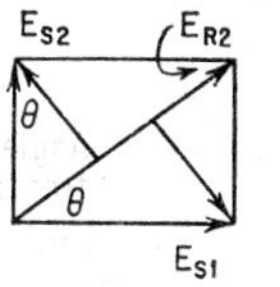 $E_{S2} \cos \theta = E_{S1} \sin \theta$
Angle computation	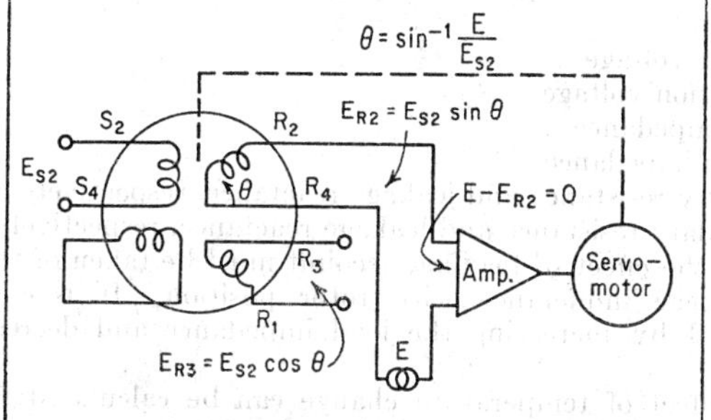	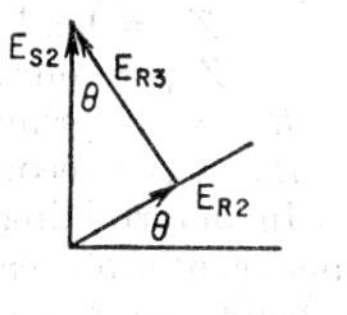
Secant computation	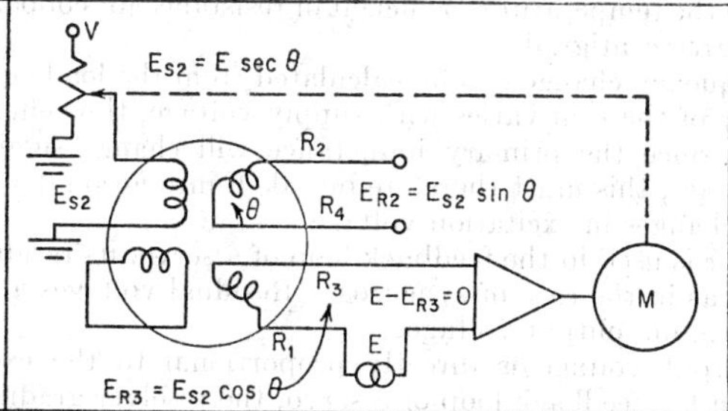	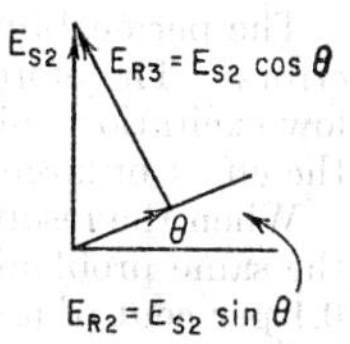

Table 35.1. Various Resolver Computations (*Continued*)

Computation	Schematic
Phase-shift circuit	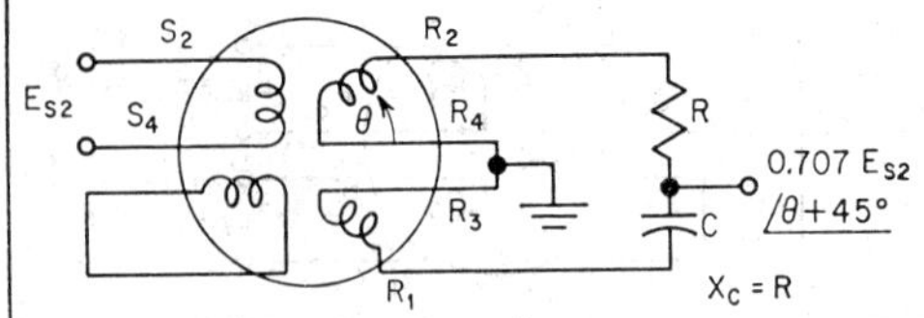
Phase-shift circuit	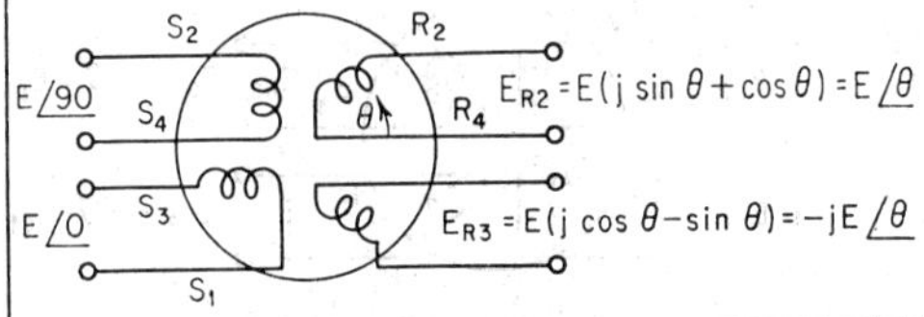
Angle transmission	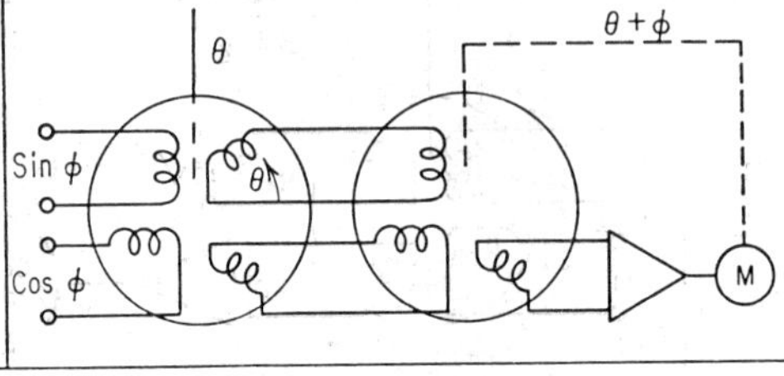

where V_0 = output voltage
V_x = excitation voltage
Z_L = load impedance
Z_{in} = source impedance
R_p, X_p = primary resistance and leakage reactance, respectively
R_S, X_S = secondary resistance and leakage reactance, respectively

In determining the effect of loading, account must be taken of the variation (±3 per cent) of leakage inductance with rotor position. It is clear that loading effects are reduced by increasing the load impedance and decreasing the source impedance.

The principal effect of temperature change can be calculated from the loading equation above and the temperature coefficient of resistance for copper (approximately 0.4 per cent per degree centigrade).

The effect of frequency change can be calculated from the load equation above.

The permeability of the iron varies with supply voltage, thus effecting conformity errors. Therefore, since the primary inductance will change significantly at very low excitation voltages, this must therefore be taken into account when determining the effect of large changes in excitation voltage.

When the resolver is used in the feedback loop of a servo, its residual voltages pose the same problems as in the case of synchros. Residual voltages are in the order of 0.1 per cent of maximum output voltage.

The resolver output voltage is directly proportional to the excitation voltage. Hence when used in the feedback loop of a servo, the resolver gradient and therefore servo gain is directly proportional to resolver excitation. For small excitation voltages, servo performance will be sluggish and static errors large. The servo may become unstable for large excitation voltages. Figure 35.13 shows a method of compensating for this effect when high servo performance must be maintained over a wide range of excitation voltage. The resolver gradient

$$\partial E_{R3}/\partial\theta = -(E_{S2}\sin\theta + E_{S1}\cos\theta) \tag{35.22}$$

It is thus seen that the magnitude of the resolver gradient is equal to E_{R2}. Therefore, if E_{R2} is used to bias an amplifier such that its gain is proportional to $1/E_{R2}$ the servo-loop gain will be constant and independent of excitation voltage.

Compensated resolver errors that are due to load and conditions of use are described below. The effect of winding compensation on errors due to load, temperature, and frequency can be determined approximately from Eq. (35.21) as before, except that the source impedance Z_{in} and stator impedance can be assumed zero, since these are common to both compensator and stator circuits. The errors attributable to the above effects are thus greatly reduced.

The effect of voltage-level errors is greatly reduced in the compensated case because the feedback circuit compensates for changes of iron permeability, since the output of the compensator winding must match the input to the isolation amplifier.

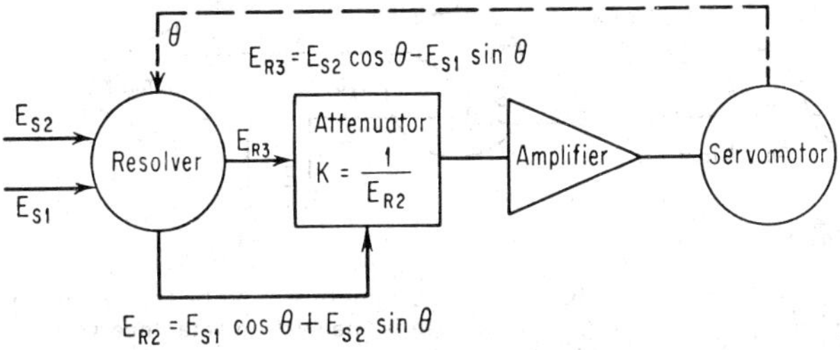

FIG. 35.13. Equalizing servo gain.

Residual-voltage and gradient-variation effects are the same as for the uncompensated case.

Application Notes. In order to minimize the variation of primary impedance with rotor position and to cancel stray flux in quadrature with the excitation axis, the primary windings should be excited from low-impedance sources, or short-circuited if unused.

Since the transformation ratio of many resolvers is specified under conditions of a nominal load, this load should be used to obtain specified results. Unused secondary windings should be terminated with the nominal impedance.

(e) Induction Potentiometer

Application. Induction potentiometers are used for analog multiplication and division with an accuracy of approximately 0.2 per cent, and as feedback elements in servomechanisms.

Construction. The construction of an induction potentiometer is similar to that of a resolver except that the rotor and stator have one computing winding each, which are distributed to obtain an accurate linear output as a function of shaft position. Since the induction potentiometer is designed for continuous rotation, the output is linear with shaft position only within $\pm 45°$ from null position.

Some induction potentiometers have auxiliary windings on both the rotor and the stator. These windings serve to maintain output impedance approximately independent of shaft position and to reduce the effect of air-gap asymmetry. The auxiliary stator winding is permanently short-circuited. The auxiliary rotor winding should be terminated in the same load as the computing winding, which, for best results, should be the standard load recommended by the manufacturer.

Errors. Induction-potentiometer error is the deviation from the correctly scaled linear output as a function of shaft position (see Fig. 35.14).

Inherent electrical error, manifested as a departure from perfect linearity, is caused by imperfect flux and winding distribution.

Mechanical errors are discussed fully in other parts of this book.

Errors due to load and conditions of use include output scaling (potentiometer gradient affected by temperature changes, frequency changes, and voltage level as in an uncompensated resolver), and residual-voltage and phase-shift errors (present in all induction devices, handled in the same manner as resolvers and synchros).

Advantages and Disadvantages. As compared with resistance potentiometers, the induction potentiometer offers the following advantages: longer life, higher-speed operation, electrically isolated output, higher input to output impedance ratio, and infinite resolution.

Its disadvantages include the variation of scale factor with temperature, frequency and voltage level, higher phase shift, and difficulty of trimming to desired gradient

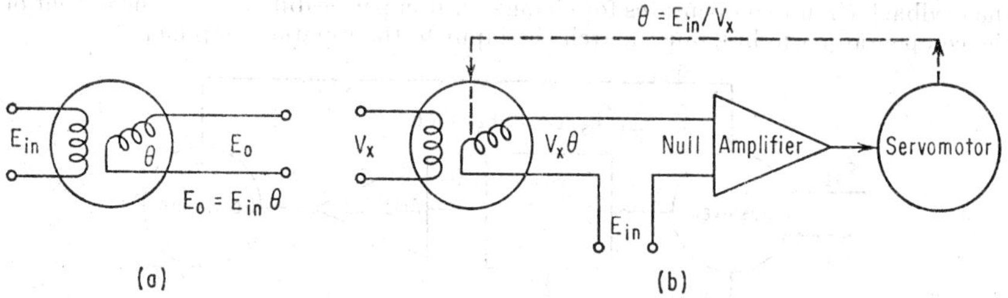

FIG. 35.14. Computing with induction potentiometers. Note that angle transmission is similar to division except that V_x in that case is the system reference voltage. (*a*) Multiplication. (*b*) Division.

(f) Precision Potentiometer

Application. The precision potentiometer can perform functions similar to those of the induction potentiometer, in addition to many others. These functions include those requiring specially shaped functions or d-c excitation. Precision potentiometers can be obtained with 0.05 per cent or better linearity.

Construction. The principal components of the potentiometer are the resistance element, wiper, ball-bearing-supported shaft, slip rings, and case. Resistance elements are available in two basic forms: wire and deposited film. Wire potentiometers are of either the slide-wire or wire-wound type. Film types use deposited carbon, metal, or conducting plastic.

Errors. Potentiometer error is the deviation between the actual and theoretical output at any angular position. The mechanical errors are discussed fully in other parts of this book. The inherent electrical error sources are principally linearity or conformity error, resolution, and noise.

Linearity Error. The linearity error generally specified by manufacturers is termed *independent linearity.* This may be defined as the maximum deviation from a straight reference line whose slope and position are chosen to minimize the deviations over the actual effective electrical rotation. This is represented by Fig. 35.15.

In applying a potentiometer with a given independent linearity tolerance to a computing problem, the slope and zero of the reference line must be adjusted for the best error split. This is usually accomplished with trim potentiometers at both excitation points, as shown in Fig. 35.16. Using the potentiometer represented by Fig. 35.15 as an example, it can be seen that the excitation voltage applied to the lower end must be below zero by an amount equal to the voltage at point *a*, in order to set the zero of the *best reference line* properly. To set the slope of the reference line, the voltage applied to the high end should equal V_{in}, plus V_{in} minus the voltage at point *b*, if the actual electrical travel equals the specified value. If the electrical travel deviates (as it generally does) from the specified or theoretical value the potentiometer excitation should be increased or decreased in the same ratio.

Potentiometer linearity is sometimes specified as *zero-based linearity*. This definition deviates from independent linearity in that the origin of the best reference line is at the zero output, zero function angle point. The actual output at this point need not be zero but must be within the specified tolerance of zero. The system error may be adjusted to the specified zero-based linearity error by the use of one trim potentiometer to set the slope as described above in the discussion of independent linearity.

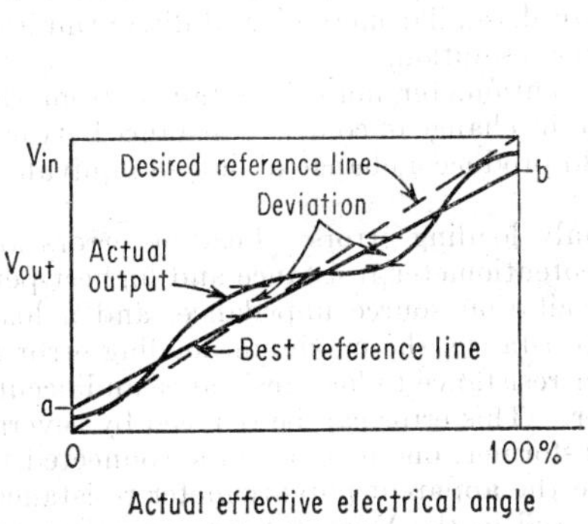

FIG. 35.15. Independent linearity.

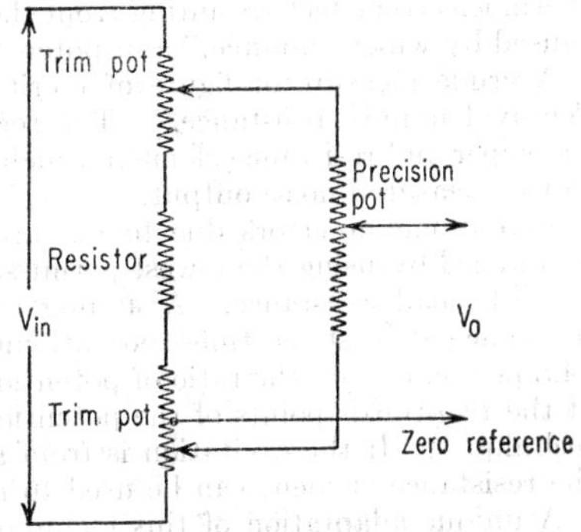

FIG. 35.16. Trimming a precision potentiometer.

When linearity is described as *terminal linearity,* no trimming is required to achieve the specified accuracy. Terminal linearity differs from the foregoing in that the reference line is drawn through points located at zero angle, zero output, and at 100 per cent output, 100 per cent of specified or theoretical electrical angle (Fig. 35.16). The error at all points including the end points must be within the specified tolerance.

In summary, only in the case of terminal linearity are the slope and zero of the reference line completely specified. In the case of independent linearity the slope and zero of the reference line are not related to specified points. It is simply a line drawn through the error curve so as to minimize deviations.

Potentiometers whose linearity is described as terminal linearity can be expected to be more costly than similar units of the same numerical linearity specification but described as independent linearity. Since the reference line must be drawn through fixed points in the case of terminal linearity, peak plus and minus errors cannot be averaged. In addition, included in the specified linearity error is the error due to the deviation of the actual electrical travel from the specified value. That this is important is demonstrated by the fact that a 1° deviation in electrical travel from the specified value will change the slope of a 350° potentiometer by almost 0.3 per cent. This error alone is greater than that specified for many precision computing problems. Care should be taken to check with the manufacturer when using a potentiometer whose linearity is specified as terminal. It should be verified that the error due to the deviation of electrical travel has been included since there appears to be some lack of standardization on this point.

Resolution. Resolution refers to the smallest change of shaft position which will produce an output-voltage change. In the case of wire-wound units, it is generally specified as the reciprocal of the number of turns in per cent, or as the number of degrees of shaft rotation required for a change of output. In the case of film potentiometers, the resolution figure is virtually zero. However, it is measurable, because of granularity of the film, wiper "stiction," and hysteresis.

The resolution of a potentiometer limits the maximum attainable linearity. The resolution error, however, is included in the linearity specification. The resolution error becomes important as a separate consideration when using a potentiometer as the feedback element of a servo. In that case the servo "stiction" error should be greater than one-half the potentiometer resolution in order to avoid hunting between voltage steps. This is because an input voltage to the servo which lies between steps cannot be exactly nulled out.

Noise. Noise refers to spurious signals that may appear in the voltage output during potentiometer operation. Noise is principally due to contamination between the wiper and resistance element. Other sources may include the triboelectric effect (voltage due to friction between the moving wiper and the resistance element), the thermoelectric effect (resulting from the use of dissimilar metals), and discontinuities caused by wiper "bounce," and potentiometer resolution.

A crude measure or figure of merit of potentiometer noise is obtained from the "equivalent noise resistance." This refers to the change of contact resistance between the wiper and resistance element which would produce a change of output equivalent to the measured noise output.

Potentiometer errors due to use are mainly loading errors. Loading errors are minimized by using the lowest permissible potentiometer resistance and highest permissible load resistance. Assuming zero excitation source impedance, and a load resistance at least ten times potentiometer resistance, the maximum loading error is 14.8 per cent times the ratio of potentiometer resistance to load resistance and occurs at the two-thirds points of the potentiometer. This error can be reduced by several techniques. If the excitation is from a transformer, one or more taps connected to the resistance element can be used to reduce the apparent potentiometer resistance.

A unique adaptation of this technique is used in the Vernistat a-c potentiometer manufactured by the Perkin-Elmer Co. A tapped autotransformer is used in conjunction with a resistance element which is automatically switched between taps as the potentiometer shaft is rotated.

Other techniques for reducing loading errors consist of shaping the potentiometer by varying the resistance element or by the use of shunts across selected taps.

35.3. Motor Devices for Servosystems

(a) Servomotors

Three basic categories are used to fulfill the requirements of a given servo application. They are (1) type, i.e., a-c, d-c, or torquer; (2) size, i.e., torque and power output required; (3) damping, i.e., viscous damping, inertial damping, and network damping.

Type. A-C motors are used for most instrument servo applications since a-c transducers are the most commonly used, and the electronics required consist of a simple a-c amplifier.

D-C motors are generally used in higher-power control amplifications since they develop considerably more power and operate at higher efficiency than a-c motors of identical size and weight. D-C motors have the disadvantage of relatively high starting torque required because of brush friction. A more complex electronics system consisting of an a-c preamplifier and/or a chopper-stabilized d-c amplifier is generally required to avoid excessive error due to amplifier drift. Sparking at the brushes usually results in excessive radio-frequency interference.

Torque motors of the direct-coupled type are used primarily in inertial applications such as stable platforms. In these applications the feedback element, generally a pendulum or gyro, is referenced to inertial space. Since the rotor of the torquer is coupled direct, rather than through gearing, it has the advantage of not requiring any acceleration in space. The torquer must supply only the torque required to overcome friction between the stabilized and support members plus any static load unbalance.

Size. The motor must be capable of delivering the required torque at the required speed. The torque that the motor must deliver is the sum of:

1. Friction torque (T_f, lb-ft)
2. Static-load unbalance torque (T_u, lb-ft)

3. Torque required to accelerate load (T_a, lb-ft)
4. Externally applied torques (T_e, lb-ft)

Thus the total torque

$$T_R = T_f + T_u + T_a + T_e \tag{35.23}$$

If the maximum torque and the speed are T_R and ω_R, respectively, the maximum required motor power is

$$P_r = T_R\omega_R/550 \quad \text{hp} \tag{35.24}$$

or

$$P_r = 1.36T_R\omega_R \quad \text{watts} \tag{35.25}$$

Damping. There are two basic methods of damping. One method provides damping as a function of error signal and is derived from electrical networks. The other type of damping is a function of the output and is derived from mechanical or electromechanical components. These generally provide damping as a function or either output rate or output acceleration. The components which produce damping as a function of output rate are tachometer generators whose output is fed back in opposition to the input, or viscous dampers in which a disk rotates in a fluid or magnetic field exerting a torque on the output shaft proportional to speed. Components which produce damping as a function of acceleration are called *inertial dampers.* The principal components of inertial dampers are a flywheel which rotates freely relative to the output shaft, and a low-inertia disk which rotates with the output shaft. The disk exerts a torque on the flywheel through viscous or magnetic coupling as a function of their speed difference. Thus when the output shaft is turning at constant speed the flywheel will rotate at the same speed as the disk and exert no reaction torque on the output shaft. An acceleration of the output will produce a speed difference, and the flywheel will then exert a torque proportional to that difference.

Control systems actuated by d-c motors or torquers are generally damped by compensation networks or separate tachometer generators. A-C motors, however, can be obtained with integral tachometers, viscous dampers, or inertial dampers. Significant advantages can be realized by using integral dampers if the application is compatible with their use. An integral device generally costs less and takes up less volume than separate motors and dampers. The following discussion is intended to provide guidance in the selection of the method of damping (see Table 35.2).

The method of damping is determined by the requirements and characteristics of the system. The primary consideration is the required system velocity constant. Generators and viscous dampers produce velocity lags; hence the maximum attainable velocity constant is low (about 100). Using inertial dampers, a velocity constant of up to about 1,000 can be attained. However, there are certain advantages of generator or viscous dampers which make their use desirable if a high velocity constant is not required. Backlash between the motor and feedback element is critical in an inertially damped servo. The servo will theoretically always oscillate within the backlash region unless damped out by static friction. A rule of the thumb to follow to eliminate backlash oscillation is that the total backlash should not exceed twice the total servo dead zone. In a generator or viscous damped servo, oscillations due to backlash are extremely unlikely with the gear meshes generally used. Another feature of inertial damping which may be undesirable in certain applications is overshoot during synchronization. The servo must overshoot since the flywheel initially turns in the same direction as the motor, and therefore its torque adds to the effect of system inertia in producing an overshoot. The damping action of the flywheel occurs when the motor reverses, thereby decreasing the second overshoot. Viscous or generator damped servos need not have an overshoot. The choice between generator and viscous damping must be made on the basis of the flexibility required. Generator dampers can be easily adjusted for any value required. Viscous dampers are not easily adjustable but have the advantage of requiring no electrical connections, and not introducing quadrature and harmonics into the servo null.

Standard viscous and inertial dampers are designed for good servo performance under conditions of negligible load inertia and load torque. If load torque and inertia are not small, the parameters of the dampers must be specially designed for the particular application.

Table 35.2. Motor Classifications

Motor	Block diagram	Transfer function
Standard a-c	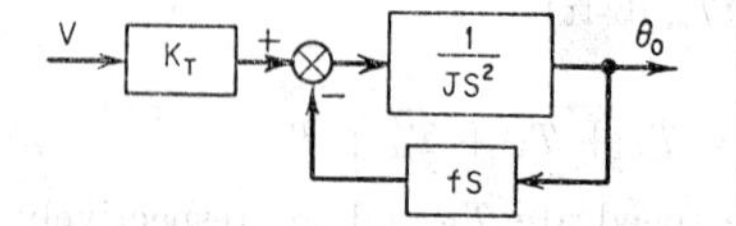	$\frac{\theta_o}{V} = \frac{\frac{K_T}{f}}{S\left(\frac{J}{f}S+1\right)}$
Viscous-damped a-c	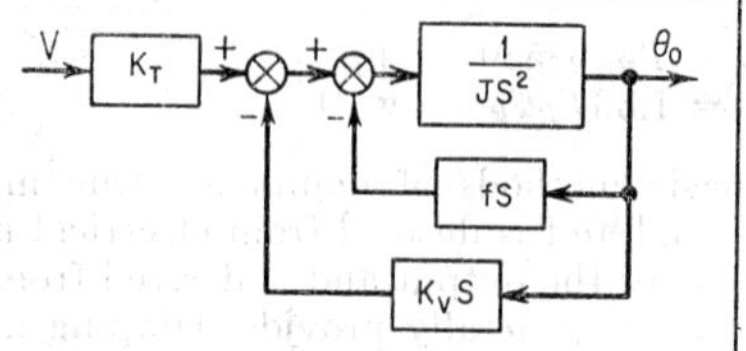	$\frac{\theta_o}{V} = \frac{\frac{K_T}{f+K_v}}{S\left(\frac{JS}{f+K_v}+1\right)}$
Generator-damped a-c	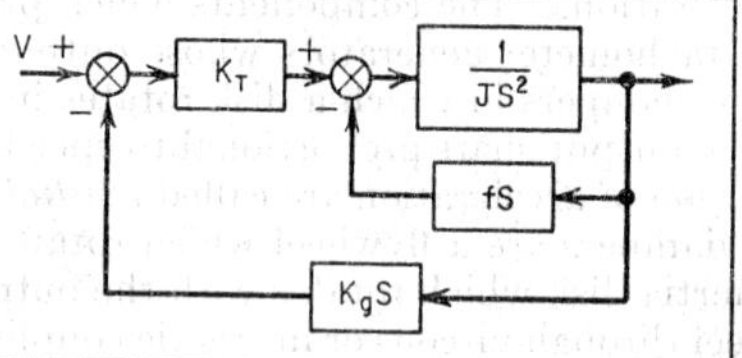	$\frac{\theta_o}{V} = \frac{\frac{K_T}{f+K_gK_T}}{S\left(\frac{JS}{f+K_gK_T}+1\right)}$
Inertial-damped a-c	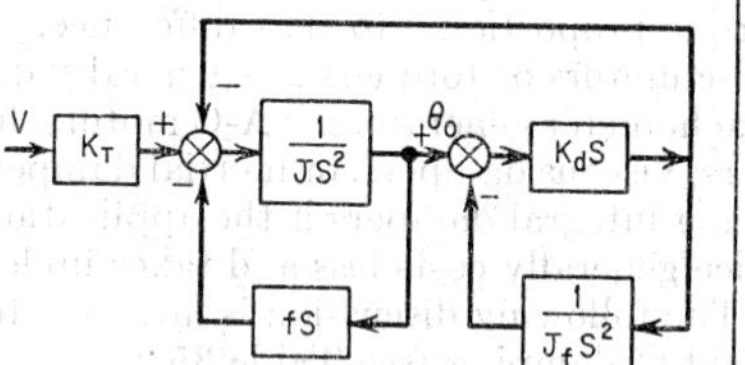	$\frac{\theta_o}{V} = \frac{K_T}{\frac{JS^2}{\frac{J_f}{K}S+1}+JS^2+fs}$ $\frac{\theta_o}{V} = \frac{\frac{K_T}{f}\left(\frac{J_f}{K_d}S+1\right)}{S\left(\frac{JJ_f}{K_df}S^2+\frac{JK_d+J_fK_d+J_ff_2}{K_df}S+1\right)}$ $\frac{\theta_o}{V} \approx \frac{\frac{K_T}{f}\left(\frac{J_f}{K_d}S+1\right)}{S\left\{\left[\left(\frac{J_f}{K_d}+\frac{J_f}{f}\right)S+1\right]\left[\frac{J}{K_d+f}S+1\right]\right\}}$ When $J_m \ll J$
D-C separately by excited or permanent magnet field shunt-wound (includes D-C torquers)	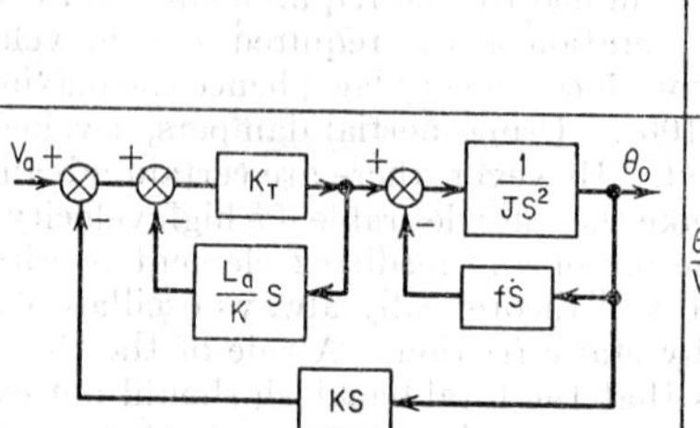	$\frac{\theta_o}{V_a} = \frac{\frac{K_T}{f}}{S\left[\frac{JL_aK_T}{fK}S^2+\left(\frac{J}{f}+\frac{L_aK_T}{K}\right)+\left(1+KK_T\right)\right]}$
D-C constant armature current	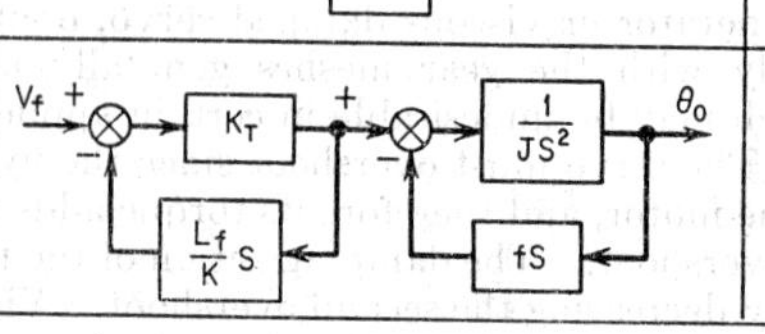	$\frac{\theta_o}{V_f} = \frac{\frac{K_T}{f}}{S\left(\frac{K_TL_fS}{K}+1\right)\left(1+\frac{J}{f}S\right)}$

f = viscous friction of motor plus load (torque/speed)
J = inertia of motor plus load
J_f = inertia of flywheel
J_m = inertia of motor
K_d = viscous coupling between motor and flywheel (torque/speed)
K_g = generator constant (volts/speed)
$K = K_TR_r$ = torque gradient measured at stall (torque/amp)
K_T = torque gradient measured at stall (torque/volts)
K_v = viscous friction due to damping (torque/speed)
L_a = inductance of armature
L_f = inductance of field
R_r = either armature or field resistance

If the servo band-pass requirements cannot be attained with output function damping, a standard servometer should be employed and network dampers used. Network damping can provide any velocity constant required, approaching infinity with the use of an integrating circuit. However, careful consideration must be given the effect of harmonics and quadrature in the null signal, since these spurious signals can easily cause amplifier saturation and bias errors. Lead networks, for example, decrease signal-to-noise ratio, since they attenuate the fundamental while having little effect on harmonics.

(b) Tachometers

Tachometers are used to derive signals which are proportional to the rate of change of their shaft angular position. These signals can be used for servomechanism damping or for computation. If the tachometer output signal is used directly, the derivative of the shaft angular position is obtained. When the output is used in the feedback loop, the integral of the input can be obtained.

At zero speed the tachometer output signal is not zero but contains both harmonic and fundamental frequency components. Residual voltage is due to imperfect materials and machining. Imperfections in the rotor circuit give rise to a residual voltage which varies with rotor position, while the residual voltages due to stator imperfections are independent of rotor position. The residual voltage varies directly with the excitation voltage and is slightly affected by temperature and conditions of mounting which deform the case.

The voltage gradient is not constant but in fact decreases with speed. This is due to the reduction of main flux caused by voltage drops in the primary winding, since the rotor, and hence the stator current, is directly proportional to speed.

The phase angle of the output voltage varies with temperature, primary voltage, and frequency since these factors affect the winding resistance, primary inductance, and leakage reactance.

The nominal voltage gradient of the tachometer is specified for a standard load. Variation of the load or the standard conditions of temperature, primary voltage, or frequency will change the scale factor.

Most a-c tachometers used for control, such as damping and integrating tachometers, are of the drag-cup type. They are similar in construction but differ in certain details such as drag-cup material, winding turns, temperature compensation, and machining tolerances. These differences give rise to differences in noise and nonlinearity levels. The cost of tachometers varies greatly according to the control placed on these errors. Table 35.3 is designed to aid the designer to select the tachometer quality required for his particular application, by describing the effect of each error on the tachometer application.

35.4. PRECISION ACCELEROMETERS

(a) Introduction

Until the advent of inertial navigation, accelerometers and accelerometer-like devices received wide use in two areas. For vibration measurement (covering the broad spectral range from the low seismic through the structural mechanical spectrum), pendulum, crystal, and strain-gage devices have enjoyed a virtual monopoly, and the present industrial and commercial literature is quite complete with discussions of their theory and use.[1]

For indication of the magnitude and direction of gravity, pendulums of various types monopolize the field.[2]

The introduction of the inertially guided long-range ballistic missile and the closely allied area of launching space vehicles have spurred the development of rugged accelerometers which are extremely accurate (accuracy measured in the 1/20,000 class or better).

Table 35.3. Effect of Tachometer Errors

Function	Block diagram	Error	Effect
Damping	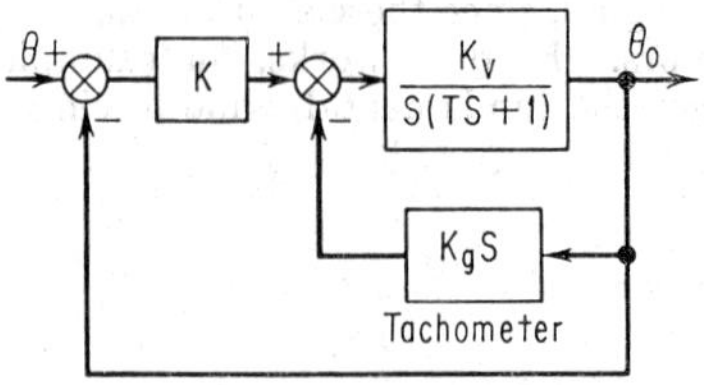	Residual voltage:	
		Harmonics	Amplifier saturation
		Quadrature	Amplifier saturation, motor overheating
		In phase	Servo bias error = (V/K) rad where V = in-phase zero-speed voltage and K = transducer constant, volts/rad
		Linearity error	Negligible effect
		Phase-angle variation from nominal	Amplifier saturation and loss of motor torque except at low speed
		Scaling variation from nominal	Generally minor effect except under the most extreme conditions of temperature and excitation-voltage variation
Differentiation	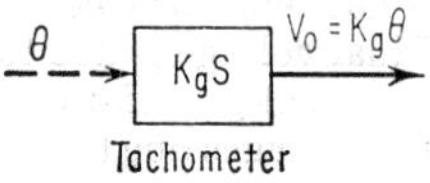	Residual voltage:	
		Harmonics Quadrature	Effects of harmonics and quadrature depend on how the tachometer output is used. When used as the input to a servo the effect is similar to the damping application
		In phase	Represents a fixed error in the magnitude of the derivative
		Linearity error	Specified as a percentage of the speed-sensitive output of 6/5 of the calibration speed
		Phase angle variation from nominal	Produces a scale error = $1 - \cos\phi$ where ϕ is the phase shift from the reference phase. When used as the input to a servo the effect on amplifier gain and motor torque is similar to the damping application
		Scaling variation from nominal	Produces a percentage error of actual output
Integration	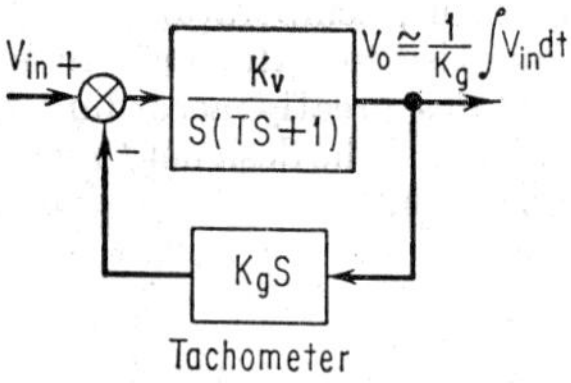	Residual voltage:	
		Harmonics	Amplifier saturation
		Quadrature	Amplifier saturation, motor overheating
		In phase	Produces an output of the integrator = $1/K_g \int_0^T V dt$, where V = in-phase zero-speed voltage; K_g = tachometer gradient
		Linearity error	Since linearity error is specified as a percentage of the speed-sensitive voltage at maximum rated speed, the error in the integral is a function of the magnitude of the input (integrand). For example, with a constant input voltage V_{in} the percentage error in the integral = % linearity error × $V_{in,\,max}/V_{in}$ where $V_{in,\,max}$ is the input voltage corresponding to maximum tachometer speed
		Phase-angle variation from nominal. Scaling variation from nominal	Same as effect on differention

(b) Theory

The following rules describe what accelerometers measure:

Rule 1. As is postulated and enlarged upon in Einstein's theory of relativity, equipment that responds to gravitational effects responds in an equivalent manner to the inertial effects of linear accelerations. Thus gravitation and inertial forces are inseparably mixed in the output of any accelerometer and cannot be distinguished in a measurement unless additional information is furnished.

Rule 2. The output of any accelerometer with respect to an earth-centered inertial frame may be approximated by

$$\mathbf{A}_{\text{indicated}} = -\mathbf{A}_{\text{true}} + \mathbf{g} \qquad (35.26)$$

where $\mathbf{A}_{\text{indicated}}$ = indicated acceleration (instrument output)

$\mathbf{A}_{\text{true}}$ = true acceleration with respect to earth-centered inertial system

$\mathbf{g}$ = gravitational field vector

$\mathbf{A}_{\text{true}}$ includes any coordinate-induced centrifugal and/or Coriolis accelerations.

Because the meaning of Eq. (35.26) sometimes tends to be clouded in its use and interpretation, we shall abstract its derivation, following precisely the elegant derivation demonstrated in ref. 3.

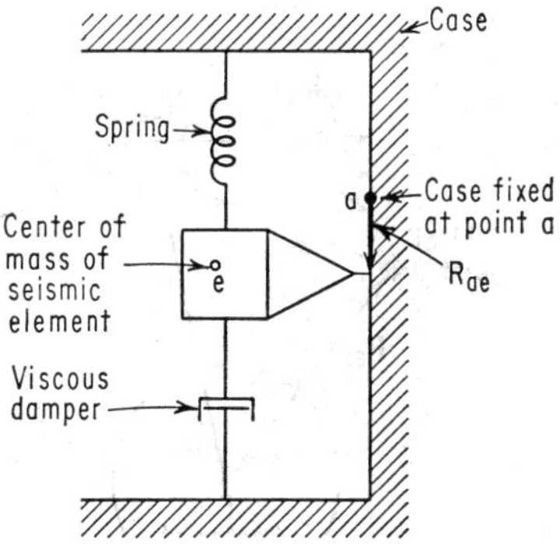

FIG. 35.17. Accelerometer mass system.

For the accelerometer mass element of Figs. 35.17 and 35.18, the force equation is

$$M_e[\ddot{\mathbf{R}}_{Ie}]_I = -k\mathbf{R}_{ae} - c[\dot{\mathbf{R}}_{ae}]_a + M_e[\mathbf{G}_{Ee} + \mathbf{G}_{Me} + \mathbf{G}_{Se} + \cdots] \qquad (35.27)$$

Buoyant force is omitted, as perfect flotation reduces the effective mass to zero, thus making accelerometer output zero.

M_e = mass of seismic element

$\mathbf{R}_{Ie}$ = displacement vector of mass center e relative to arbitrary inertial point I

$[\ddot{\mathbf{R}}_{Ie}]_I$ = acceleration of e with respect to an inertial frame centered at I

k = spring constant

$\mathbf{R}_{ae}$ = displacement vector of e relative to a reference point a in the accelerometer case

c = viscous damping coefficient

$\mathbf{G}_{Ee}, \mathbf{G}_{Me}, \mathbf{G}_{Se}$ = gravitational fields at point e due to earth, moon, and sun, respectively

But

$$\mathbf{R}_{Ie} = \mathbf{R}_{Ia} + \mathbf{R}_{ae} \qquad (35.28)$$

where $\mathbf{R}_{Ia}$ = displacement vector of point a relative to arbitrary inertial point I

and

$$\mathbf{R}_{Ia} = \mathbf{R}_{IA} + \mathbf{R}_{Aa} \qquad (35.29)$$

where $\mathbf{R}_{IA}$ = displacement vector of center of mass A of vehicle carrying the accelerometer relative to I

$\mathbf{R}_{Aa}$ = displacement vector of accelerometer case relative to point A (see Fig. 35.18)

Also

$$M_A[\ddot{\mathbf{R}}_{IA}]_I = \mathbf{L} + \mathbf{D} + \mathbf{T} + \cdots + M_A[\mathbf{G}_{EA} + \mathbf{G}_{MA} + \mathbf{G}_{SA} + \cdots] \qquad (35.30)$$

where M_A = mass of vehicle

$\mathbf{L}, \mathbf{D}, \mathbf{T}$ = lift, drag, and thrust acting on vehicle

Differentiating Eqs. (35.28) and (35.29), recognizing that $[\dot{\mathbf{R}}_{Aa}]_A$ and $[\ddot{\mathbf{R}}_{Aa}]_A$ are zero because the accelerometer case is assumed fixed to the vehicle, and substituting in Eq. (35.27) along with Eq. (35.30) gives

$$[\ddot{\mathbf{R}}_{ae}]_a + (c/M_e)[\dot{\mathbf{R}}_{ae}]_a + (k/M_e)\mathbf{R}_{ae} = -(\mathbf{L} + \mathbf{D} + \mathbf{T})/M_A - \mathbf{W}_{IA} \times (\mathbf{W}_{IA} \times \mathbf{R}_{Aa}) - [\dot{\mathbf{W}}_{IA}]_I \times \mathbf{R}_{Aa} - \mathbf{W}_{Ia} \times (\mathbf{W}_{Ia} \times \mathbf{R}_{ae}) - [\dot{\mathbf{W}}_{Ia}]_I \times \mathbf{R}_{ae} - 2W_{Ia} \times [\dot{\mathbf{R}}_{ae}]_a \quad (35.31)$$

where it is considered that the gravitational fields are homogeneous over the volume of the vehicle such that $\mathbf{G}_{Ee} + \mathbf{G}_{Me} + \mathbf{G}_{Se} + \cdots = \mathbf{G}_{EA} + \mathbf{G}_{MA} + \mathbf{G}_{SA} + \cdots$.

Under these circumstances, the direct effect of the gravity fields vanishes; only nonfield forces remain.

Equation (35.31) explains the statement, often seen in the literature, that an accelerometer measures only nonfield forces.

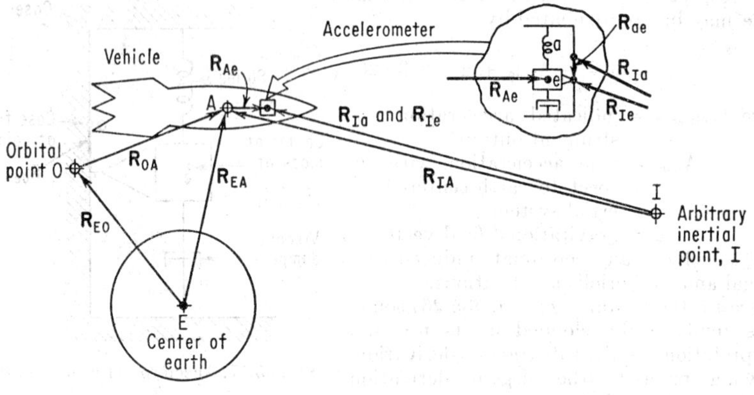

FIG. 35.18. Vehicle-borne accelerometer geometry.

Conversion to Earth-centered Inertial Frame. Consider that

$$\mathbf{R}_{IA} = \mathbf{R}_{IE} + \mathbf{R}_{EA} \quad (35.32)$$

Since the earth is in orbit, its motion is due to gravitational fields alone, and

$$[\ddot{\mathbf{R}}_{IE}]_I = \mathbf{G}_{ME} + \mathbf{G}_{SE} + \cdots \quad (35.33)$$

Taking the second derivative of Eq. (35.32), recognizing that

$$[\ddot{\mathbf{R}}_{EA}]_I = [\ddot{\mathbf{R}}_{EA}]_{I_E} \quad (35.34)$$

where I_E = frame centered at E and inertially nonrotating, and substituting along with Eq. (35.30), into Eq. (35.31) gives

$$[\ddot{\mathbf{R}}_{ae}]_a + (c/M_e)[\dot{\mathbf{R}}_{ae}]_a + [(k/M_e)\mathbf{R}_{ae}] = -[\ddot{\mathbf{R}}_{EA}]_{I_E} + \mathbf{G}_{EA} + (\mathbf{G}_{MA} - \mathbf{G}_{ME}) + (\mathbf{G}_{SA} - \mathbf{G}_{SE}) + \cdots \quad (35.35)$$

[terms in W_{IA} and W_{Ia} from Eq. (35.31)].

The two gravitational-difference terms have magnitudes of the order of $10^{-7}g_E$, where g_E is the value of gravity at the earth's surface. Taking the steady state of Eq. (35.35), we have Eq. (35.26).

Conversion to Inertial Frame in Orbit around the Earth

$$\mathbf{R}_{EA} = \mathbf{R}_{EO} + \mathbf{R}_{OA} \quad (35.36)$$

Since point O is in orbit, by definition

$$[\ddot{\mathbf{R}}_{EO}]_{I_E} = \mathbf{G}_{EO} \quad (35.37)$$

Also

$$[\ddot{\mathbf{R}}_{EA}]_{I_E} = \mathbf{G}_{EO} + [\ddot{\mathbf{R}}_{OA}]_{I_o} \quad (35.38)$$

where I_o = frame centered at O and inertially nonrotating

Since

$$[\ddot{\mathbf{R}}_{oA}]_{I_o} = [\ddot{\mathbf{R}}_{oA}]_{I_E} \quad (35.39)$$

Substituting Eq. (35.38) into Eq. (35.39),

$$[\ddot{\mathbf{R}}_{ae}]_a + (c/M_e)[\dot{\mathbf{R}}_{ae}]_a + (k/M_e)\mathbf{R}_{ae} = -[\ddot{\mathbf{R}}_{OA}]_{I_0} + (\mathbf{G}_{EA} - \mathbf{G}_{EO}) \qquad (35.40)$$

+ gravitational-difference terms).
− [terms in W_{IA} and W_{Ia} from Eq. (35.31)].
$\mathbf{G}_{EA} - \mathbf{G}_{EO}$ *may not necessarily be negligible*

Conversion to Vehicle in Orbit. If the vehicle itself is in orbit, we can consider point O in the vehicle and $[\ddot{\mathbf{R}}_{OA}]_{I_0} = 0$. Then Eq. (35.40) holds, if $[\ddot{\mathbf{R}}_{OA}]_{I_0}$ is set equal to zero. The term $\mathbf{G}_{EA} - \mathbf{G}_{EO}$ is the "gravity-gradient" term, which has been discussed with increasing frequency in the recent literature.[4,5,6]

Model Equation. The output of a theoretically ideal accelerometer is perfectly linear, the constant of proportionality being a tabulated value, invariant with time and environment. Because instrument engineering permits the ideal to be only approximated in practice, the predominant linear term is modified by "higher-order effects." It is possible to handle these effects in two relatively distinct manners. Recognizing that, in general, there exists some well-defined functional relationship between the output of any usable accelerometer and the input to be measured, one could assume such a relationship (adding some dependence on cross acceleration and any other pertinent variable such as temperature or time, for further generality). Because a natural process cannot generate derivatives of infinite order, one can write the acceleration-dependent function as a MacLaurin series in the two (or more) variables, e.g., direct-axis acceleration and cross-axis acceleration. To assure more rapid convergence it may sometimes be convenient to add other functions to the power series described, thus obtaining greater accuracy with fewer terms. This is especially pertinent when certain "anomalous effects" such as "dead-band," "threshold," "resolution," and "hysteresis" are applicable and are to be described.

The previous *ad hoc* approach suffers from the disadvantage that no hint is given as to how many higher-order terms must be considered for a given accuracy. No knowledge (except by direct experiment) of the nominal relative magnitudes of the power-series coefficients is provided. By studying the physical process inherent in the accelerometer mechanism one can often construct an appropriate mathematical model for the accelerometer which can be correlated, term by term, with the series description. Hopefully, proper substitution of the accelerometer parameters gives a good approximation to the power-series coefficients which are measured during a linearity experiment.[7] This latter approach is a necessity if a quantitative engineering design is to be accomplished because otherwise no guide for properly proportioning the physical parameters (in the light of the inevitable design compromises) exists. Thus a mathematical description can be given to the output of an accelerometer by the following:

$$\theta_0 = K_0 + K_1 a + K_2 a^2 + K_3 a^3 + \cdots = \sum_n K_n a^n \qquad (35.41)$$

where θ_0 = accelerometer output
K_0 = accelerometer fixed bias
K_1 = accelerometer scale factor
K_2 = second-order nonlinearity
K_3 = third-order nonlinearity

Cross-axis coupling and other similar effects have been omitted. They add other terms to the series of Eq. (35.41) and help define a set of "coupling coefficients."

One measure of quality for an accelerometer is the smallness of the terms K_n, $n \geq 2$, compared with K_1. However, a corollary measure, that of the constancy of the K_n, $K_n \geq 0$, appears just as important. This is true because it is possible to provide vibration isolation against rectification effects and otherwise to compensate for the higher-order effects, provided that they are small enough and that their values are known with sufficient exactness.

We can logically define an accelerometer linearity test as an experiment designed to enable a precise determination to be made of the values of the K_n. Because an accelerometer is constructed from a finite number of parts and because "lumped-constant" performance is usually a sufficiently good description, we infer that many of the higher-order terms in Eq. (35.41) must be describable as functions of the lower terms. This is because the finite number of parts demands description by only a finite number of independent parameters. Stated more succinctly, there is only a finite number of independent parameters which may be adjusted to change performance. In general then, the K_n's are interrelated.

It is important to determine this interrelationship in order to limit any uncertainties in anticipated accelerometer performance. In fact, a measure of quality, in the sense of "uniformity in manufacture," is given by noting the value of n above which we cease to have independent parameters. The lower this value of n, the higher the chance of manufacturing uniformity because the fewer will be the number of critical adjustments.

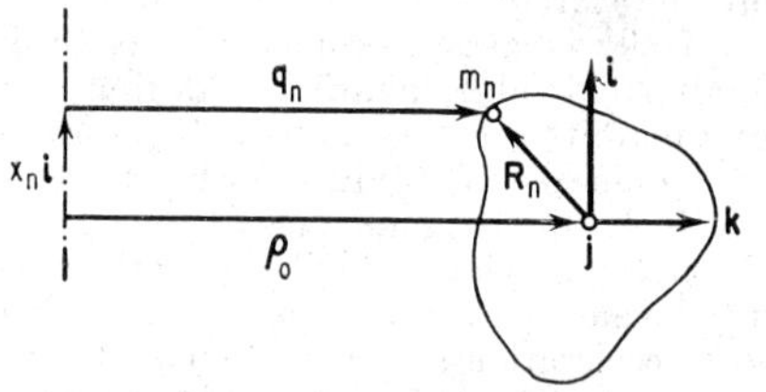

FIG. 35.19. Seismic mass on centrifuge.

Centrifuge Testing for Linearity. For centrifuge testing, Eq. (35.41) may have to be modified because the acceleration field possesses cylindrical symmetry with a significant g gradient for an accelerometer of finite dimensions.

In Fig. 35.19, M_n is a typical elementary mass which goes to make up the total seismic mass of the accelerometer. The origin of the i, j, k coordinate system is at the center of mass of the seismic element. Then, following a procedure similar to that of ref. 6,

$$\rho_0\mathbf{k} + \mathbf{R}_n - \mathbf{q}_n - X_n\mathbf{i} = 0 \tag{35.42}$$

$$\mathbf{q}_n = \rho_0\mathbf{k} + \mathbf{R}_n + X_n\mathbf{i} \tag{35.43}$$

$$\mathbf{a}_n = -w^2 q_n \tag{35.44}$$

$$\mathbf{F}_n = \omega^2 m_n \mathbf{q}_n \tag{35.45}$$

$$\mathbf{T}_n = \mathbf{R}_n \times \mathbf{F}_n = \omega^2 m_n \mathbf{R}_n \times \mathbf{q}_n \tag{35.46}$$

Case I. Accelerometers Working by Torque Detection

$$T = \sum_n T_n = \omega^2 \sum_n M_n\mathbf{R}_n \times (\rho_0\mathbf{k} + \mathbf{R}_n - X_n\mathbf{i}) \tag{35.47}$$

$$T = \omega^2 \sum_n \rho_0 m_n \mathbf{R}_n \times \mathbf{k} - \omega^2 \sum_n m_n X_n \mathbf{R}_n \times \mathbf{i} \tag{35.48}$$

$$T = \omega^2 \rho_0 \left(\sum_n m_n \mathbf{R}_n\right) \times \mathbf{k} - \omega^2 \sum_n m_n X_n \mathbf{R}_n \times \mathbf{i} \tag{35.49}$$

but $\sum_n m_n\mathbf{R}_n = 0$ defines the mass center; so

$$T = -\omega^2 \sum_n m_n X_n \mathbf{R}_n \times \mathbf{i} \tag{35.50}$$

$$T_x = \mathbf{T}\cdot\mathbf{i} = -\omega^2 \sum_n m_n X_n \mathbf{R}_n \times \mathbf{i}\cdot\mathbf{i} \tag{35.51}$$

but

$$\mathbf{R}_n \times \mathbf{i}\cdot\mathbf{i} = \mathbf{R}_n\cdot\mathbf{i}\times\mathbf{i} = 0 \tag{35.52}$$

So

$$T_x = 0$$

$$T_y = \mathbf{T}\cdot\mathbf{j} = -\omega^2 \sum_n m_n X_n \mathbf{R}_n \times \mathbf{i}\cdot\mathbf{j} \tag{35.53}$$

$$\mathbf{R}_n \times \mathbf{i}\cdot\mathbf{j} = \mathbf{R}_n\cdot\mathbf{i}\times\mathbf{j} = \mathbf{R}_n\cdot\mathbf{k} = Z_n \tag{35.54}$$

$$T_y = -\omega^2 \sum_n m_n X_n Z_n = -\omega^2 I_{xz} \tag{35.55}$$

where $I_{xz} \triangleq \sum_n m_n X_n Z_n$ = product of inertia.

$$T_z = \mathbf{T} \cdot \mathbf{k} = -w^2 \sum_n m_n X_n \mathbf{R}_n \times \mathbf{i} \cdot \mathbf{k} \tag{35.56}$$

$$\mathbf{R}_n \times \mathbf{i} \cdot \mathbf{k} = \mathbf{R}_n \cdot \mathbf{i} \times \mathbf{k} = -\mathbf{R}_n \cdot \mathbf{j} = -y_n \tag{35.57}$$

$$T_z = +\omega^2 \sum_n m_n X_n Y_n = \omega^2 I_{xy} \tag{35.58}$$

where $I_{xy} \triangleq \sum_n m_n X_n Y_n$ = product of inertia.

Equations (35.50) to (35.58) indicate that spurious torques, due to the centrifuge acceleration gradient, can appear as a function of the product of inertias with respect to axes through the seismic center of mass.

However, torque T_y (direct-axis torque) does not affect the higher-order terms (to first order), but rather it causes a virtual change to the accelerometer scale factor since its magnitude is proportional to ω^2 and hence to a/ρ_0.

Case II. Accelerometers Working by Force Detection

From Eq. (35.35),

$$\mathbf{F} = \sum_n \mathbf{F}_n = \omega^2 \sum_n m_n \mathbf{q}_n \tag{35.59}$$

$$\mathbf{F} = \omega^2 \sum_n m_n(\rho_0 \mathbf{k} + \mathbf{R}_n - X_n \mathbf{i}) \tag{35.60}$$

$$\mathbf{F} = \omega^2 \rho_0 \sum_n m_n \mathbf{k} \tag{35.61}$$

Since

$$\sum_n m_n \mathbf{R}_n = 0 \qquad \sum_n m_n X_n \mathbf{i} = 0$$

define the mass center, then

$$F_x = F_y = 0 \tag{35.62}$$

$$F_z = \omega^2 \rho_0 M \tag{35.63}$$

where M = total seismic mass. It should be noted that a longer centrifuge arm (ρ_0 large) tends to reduce the effect of the gradient and hence the precision required in removing the gradient effect by computation.

Measurement of K_1 and K_3. Of the many techniques available for measuring K_1 and K_3, one which has met with considerable success is described as follows: The basic measurement consists of applying known accelerations in the plus and minus directions. By subtracting the results, the even terms are suppressed, thus revealing the odd terms K_1 and K_3. As shown in the references,[7] it is possible to reduce the linearity run for the odd coefficients to the following form:

$$K_{1c} a_2 + K_{3c} a_2^3 = N_2 \tag{35.64}$$

$$K_{1c} a_4 + K_{3c} a_4^3 = N_4 \tag{35.65}$$

$$\cdots \qquad \cdots \qquad \cdot$$

$$\cdots \qquad \cdots \qquad \cdot$$

$$K_{1c} a_m + K_{3c} a_m^3 = N_m \tag{35.66}$$

where K_1 and K_3 are as previously defined, subscript c referring to centrifuge measurement, a_m = known acceleration applied to accelerometer, $2g$, $4g$, $6g$, etc., through perhaps $20g$ in the standard test, and N_m = processed accelerometer output in the sense of Eq. (35.41) and the references.

Equations (35.64) to (35.66) constitute an overdetermined situation in variables K_{1c} and K_{3c}. Although standard least-squares techniques could be utilized to remedy the situation by extracting the "best" K_{1c} and K_{3c}, there is no a priori knowledge that these values would be the "best" approximation to the true value obtainable from the experiment. For example, the least-squares technique has the defect that it weighs relatively large errors much more heavily than small errors. What is really needed is a means for displaying the consistency of the data in such a manner as not only to indicate "the general state of agreement but also, if some of the data depart markedly from the general consensus, to indicate these readily."[8] A recommended technique is a special case of a method first suggested by R. A. Beth in 1938 which has become known as the "isometric consistency chart."[8]

The method becomes apparent when it is recognized that Eqs. (35.64) to (35.66) are linear in the variables K_{1c} and K_{3c}. Each equation defines a straight line in these variables with slope determined by the test g level. K_{1c} and K_{3c} would be indicated by the unique intersection of these 10 lines. In actuality, of course, experimental errors lead to a variety of intersections, the "tightness" of their distribution being an excellent measure of the dispersion of the measurement.

In computing a_m in Eq. (35.41), it is necessary to have a very precise value for acceleration. Because centrifuge speed is measured with respect to the earth and because the earth rotates in space, it can be shown[7] that a Coriolis acceleration component must be added to the centrifugal acceleration such that the formula for acceleration produced by a level centrifuge is

$$a = -R_c S(S \pm 2\Omega \sin \lambda) \tag{35.67}$$

where R_c = centrifuge radius
Ω = earth's angular rate
λ = latitude angle at location of centrifuge
S = spin speed of centrifuge with respect to earth

For the case of a 100-in. arm it is seen that, if the earth's rate correction were omitted, acceleration errors of 1 part in 20,000 would be produced at the $2g$ level, and at the $10g$ level the error would be 1 part in 100,000.

The measurement of K_2, as shown in ref. 7, is accomplished by adding the results of a plus and minus g test. The odd terms are suppressed, thus enabling the direct computation of K_2 to be made at each test g level. Thus

$$K_2 = \frac{(\Delta f^+ + \Delta f^-) - (K_0^+ + K_0^-)}{2a_m^2} \tag{35.68}$$

This technique has the added advantage of asymptotically canceling out the size effect. Also, since K_2 is very small, nominally zero, the precision needed in the knowledge of a_m is low.

Testing in the Earth's g Field. Although it is theoretically possible to use the one g input of the earth's field to measure all the K_n's of Eq. (35.41), experience indicates that the precision needed in the angle-setting device and in the auxiliary equipment is greater than can be obtained in practice. As a result, the earth's g field is mainly used for calibration of K_0 and K_1 by the following equations:

$$K_0 + K_2 = \frac{\theta_0(+1g) + \theta_0(-1g)}{2} \tag{35.69}$$

$$K_1 + K_3 = \frac{\theta_0(+1g) - \theta_0(-1g)}{2} \tag{35.70}$$

where $\theta_0(+1g)$ = accelerometer output with $+1g$ input
$\theta_0(-1g)$ = accelerometer output with $-1g$ input

Estimates of K_2 and K_3 are known from other sources (possibly the centrifuge).

The utility of Eqs. (35.69) and (35.70) is that they may be used to recalibrate an accelerometer in the field, provided that g at the calibration location is known with sufficient precision.

(c) Integrating Accelerometers

Because implementation of inertial navigation requires the integration of accelerometer signals to provide the wanted velocity and position information, modern technique has searched for ways to construct accelerometers which incorporate within their mechanism one or two stages of integration. These devices have come to be known as "velocity meters" or "distance meters," respectively.

Since most modern system computers are of the digital type, one further restriction is usually placed on the accelerometer component; its output must be digital.

As examples, we shall outline briefly several velocity-meter instruments which have demonstrated their performance in a variety of inertial systems.

Vibrating-string Accelerometers. In a single-string single-mass system, a string under tension will vibrate when excited. A piano string vibrates when struck, a guitar string vibrates when plucked, and a violin string vibrates when bowed. The frequency of vibration is a function of the dimensions of the string, the string material, and the string tension.

If a string is suspended vertically, clamped at its high end with a mass attached at the other end, its resonant frequency can be varied by varying the mass. This varies the tension on the string. The frequency of vibration F is directly proportional to the square root of the tension of the string T and inversely proportional to the square root of the mass of the string m and its length L:

$$F = K\sqrt{T/mL} \tag{35.71}$$

The string tension is equal to the initial tension T_0 due to the earth's gravitational field, and any other forces

$$T = T_0 + Ma$$

where Ma = inertial reaction force.

From this it can be seen that, for any particular string and mass, the frequency of vibration can be varied by changing the acceleration applied to the sensitive mass.

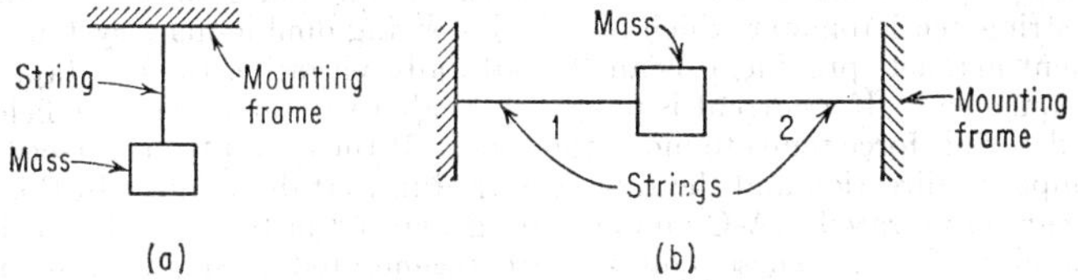

FIG. 35.20. Vibrating-string accelerometer schematics. (*a*) Single-mass system. (*b*) Double-string single-mass system.

If we consider the acceleration as variable, the frequency-tension relationship can be expressed as a Taylor-series expansion about zero (MacLaurin series, see Fig. 35.20*a*).

$$F = K(T_0 + Ma)^{1/2} = F(0) + \frac{F'(0)a}{1!} + \frac{F''(0)a^2}{2!} + \frac{F'''(0)a^3}{3!} + \cdots \tag{35.72}$$

or

$$F = K_0 + K_1a + K_2a^2 + K_3a^3 + \cdots \tag{35.73}$$

See Eq. (35.41).

Arma* Vibrating-string Accelerometer. This instrument provides an inherently digital output with a minimum of auxiliary electronics. When a, the additional acceleration, is zero, the bias K_0 is the initial frequency. With applied acceleration, there is a linear change in frequency and other higher-order terms.

In a double-string instrument, the higher-order terms in the Taylor-series expansion can be reduced in magnitude by using two strings along one axis with the mass between them (see Fig. 35.20*b*). The coefficients approach zero as the order approaches infinity and the series converges. If we express the frequency of each string as a series, we have

$$F_1 = K_{01} + K_{11}a + K_{21}a^2 + K_{31}a^3 + \cdots \quad \text{for } T = T_0 + M_a \tag{35.74}$$
$$F_2 = K_{02} - K_{12}a + K_{22}a^2 - K_{32}a^3 + \cdots \quad \text{for } T = T_0 - M_a \tag{35.75}$$

* Division of American Bosch Arma Corp.

Since the K_1 terms are dominant, the frequency of one string will increase while the other will decrease. Taking the differences of these two frequencies, we have

$$F_1 - F_2 = (K_{01} - K_{02}) + (K_{11} + K_{12})a + (K_{21} - K_{22})a^2 + (K_{31} + K_{32})a^3 + \cdots \quad (35.76)$$

If the two strings are closely matched, all the even terms in the difference-frequency equation become negligibly small and only the odd terms remain. Since the series converges rapidly, the only nonlinear term of any significance for most applications is the K_3 term. This method is useful in reducing the effect of the higher-order terms.

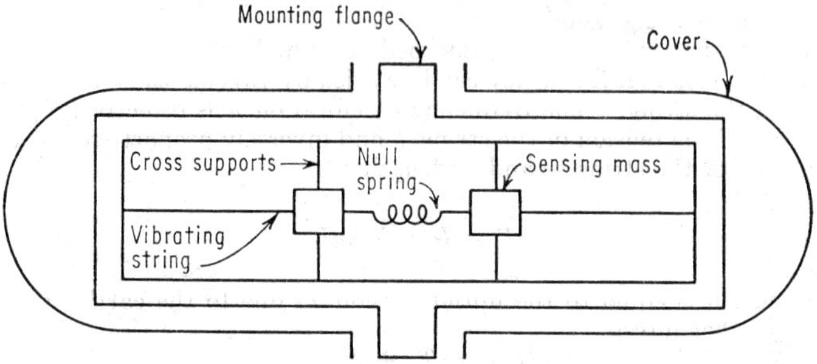

FIG. 35.21. Arma vibrating-string accelerometer.

In order to reduce the effect of accelerations not directed along the axis of the tapes, it becomes necessary to restrain the masses. This is accomplished by cross-support tapes placed perpendicular to the sensitive axis. Figure 35.21 shows the basic Arma vibrating-string accelerometer which is a double-string double-mass system.

Permanent magnets provide a means for laterally vibrating the end tapes at their natural frequencies. If a current is passed through a wire in a magnetic field, a force is produced. This force tends to move the tape. If the current is reversed, the force is in the opposite direction and the tape is moved in that direction. In this manner, the tapes can be vibrated. A-C voltages are generated in the tapes by their motion in the magnetic fields. These voltages are regenerated through stable high-gain amplifiers and returned to the tapes to provide the energy necessary to sustain oscillation. The tape is therefore part of a tank circuit which oscillates at its natural frequency.

The dynamics of the vibrating tapes in their respective magnetic fields provided by the permanent magnets have properties which are completely analogous to those of a

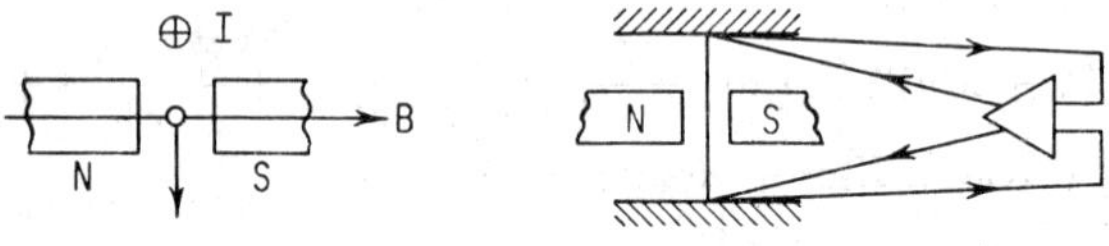

FIG. 35.22. Magnetic field for *VSA*.

parallel resonant electric circuit. This feature permits the design of an appropriately controlled feedback amplifier to maintain a constant amplitude of oscillation (see Fig. 35.22). During operation the two masses are free to move along the sensitive axis and are restrained in motion along the cross sensitive axis by the suspension system. An acceleration along the axis of the end tape causes the masses to increase the tension of one tape and decrease the tension of the other tape by approximately an equal amount. Since the frequency of tape oscillation is proportional to the square root of tension, the instantaneous frequency difference of the tape oscillations is a measure of the acceleration. The integral of acceleration is velocity, and by integrating (summing the total number of $F_1 - F_2$ cycles) over any time interval, the velocity at that time is obtained. A further integration will give the distance traveled. Using three accelerometers of known orientation the position and velocity at any point in space can be determined. Knowing the position and velocity of the vehicle, the computer in a ballistic missile, for example, can determine the proper engine cut-off time to allow the payload to fall in a ballistic trajectory toward a preselected target.

Because of the square-root relationship, the frequency difference is only approximately proportional to acceleration. The small effects (mainly the K_3 terms) introduced by higher-order terms in an actual accelerometer can be compensated for in the computer.

Pendulous Integrating Gyro Accelerometer. In the gyro-type accelerometer[1] (Fig. 35.23) the unbalance weight is mounted in such a manner that it will produce a torque about the output axis OA which is proportional to the acceleration. This torque causes motion about the output axis which is detected by the output-axis

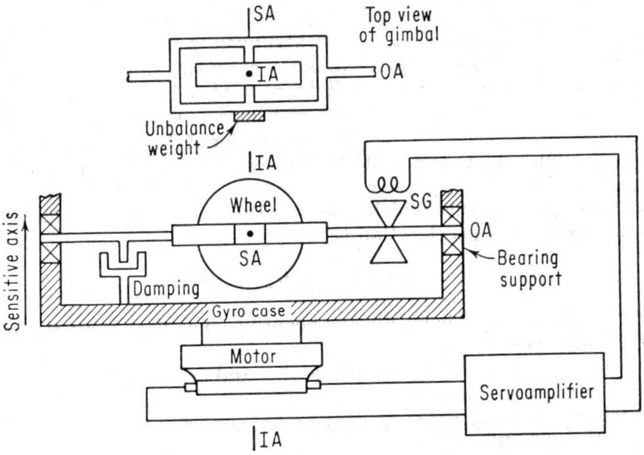

FIG. 35.23. Pendulous integrating gyro accelerometer.

pickoff SG. A servoamplifier accepts the error signal and, by means of a servomotor, drives the gyro about its input axis with a rate which nulls the error.

In the steady state, the condition of torque balance about the output axis demands that

$$[Ml]a = H\omega_{IA} \tag{35.77}$$

where Ml = pendulosity
a = input acceleration
H = wheel momentum
ω_{IA} = input axis rate supplied by servomotor

From Eq. (35.77) we see that the scale factor involves H directly, and the wheel frequency (speed) must therefore be controlled with sufficient precision to accommodate the desired precision of scale factor. Because the gyro is inherently a good rate detector, special techniques must be used when precision centrifuging a gyro accelerometer.

The output of the instrument is obtained by a digital pickoff on the servomotor shaft. Since servomotor speed is proportional to acceleration, servomotor angle is proportional to velocity. Thus, like the vibrating-string accelerometer, the output is directly a velocity in digital form.

Although Eq. (35.77) indicates a perfectly linear output, certain effects combine in a manner to make a more accurate description conform to Eq. (35.41) in which higher-order terms are present. These effects are primarily due to certain dynamic-balance conditions.

(d) Analog Output Devices

Unlike the accelerometers previously described which have outputs in a digital form directly representing velocity, many useful accelerometers have outputs which are inherently analog in nature. As such, their accuracy and utility in a modern system are limited. The force-feedback type of instrument can be implemented with an analog-to-digital conversion system which forms an integral part of a force-feedback loop.

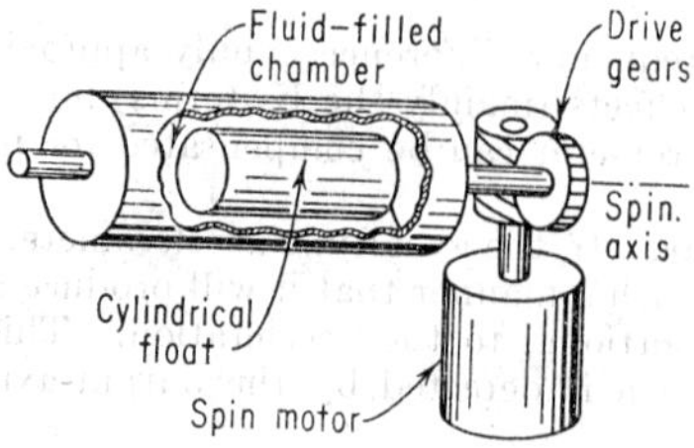

FIG. 35.24. Cutaway view of the horizontal integrating accelerometer. Test mass flotation as a way of reducing the suspension friction, provided float is overbuoyant.

Sperry* Horizontal Integrating Accelerometer. In this device the seismic mass is in the form of an overbuoyant cylindrical float contained within a fluid-filled cylinder, as shown in Fig. 35.24. When the outer cylinder is spun about the longitudinal axis, the fluid is centrifuged to the outer diameter of the containing cylinder so that the overbuoyant (lighter) float is brought into alignment with the spin axis. The float has no contact with the remainder of the structure, being restrained by the viscous fluid. Thus the friction level can be made to approach zero, as in the case of a fluid-supported gyroscope.

The equation of performance which describes the longitudinal motion of the float in response to a steady axial acceleration is

$$ma = k\mu V_F \tag{35.78}$$

where m = effective mass of float, always negative since float is overbuoyant
a = steady acceleration
k = a constant depending on float and chamber dimensions
μ = absolute viscosity of fluid
V_F = relative velocity between float and chamber

Integrating Eq. (35.78),

$$x = \int_0^T V_F\,dt = m/k\mu \int_0^T a\,dt \tag{35.79}$$

Thus the float relative displacement is proportional to velocity. The device is inherently integrating. Displacement can be read out by a variety of pick-offs; optical, capacitor bridge, and inductive pick-offs, as well as those employing the resistance of fluid between either end and the float, have all been employed. Aside from the disadvantage of the analog readout, the constant case spin which must be provided produces a serious life limitation over other accelerometer types.

Torque-balanced Pendulum. Here, the sensing mass is suspended pendulously, as either a normal or inverted pendulum so that it is free to move in one direction but is restricted to as great an extent as possible in the other two directions. Forces applied by the suspension in the direction of free motion (sensitive axis) are made as small as possible. Any inertial force imposed on the mass along the sensitive axis is balanced by a force generated by a torquer winding. The relationship between current and force (with a fixed magnetic field) becomes the means by which acceleration can be read out. Displacement of the mass, caused by an unbalance between inertial- and torquer-produced forces, is sensed by a position pick-off, which controls the torquer current by means of an amplifier in such a way that the seismic mass is nulled with respect to the instrument case. Viscous damping against vibration can be supplied by filling the case with a silicone damping fluid.

A diagrammatic view of a unique accelerometer employing these principles is shown in Fig. 35.25. This device, designed at the Jet Propulsion Laboratory by Dr. A. R. Johnston, employs a fused-quartz suspension with an optical pick-off to produce a

* Division of Sperry Rand Corp.

rather small unit for which unusual stability of bias with time (due to the stability of quartz) is claimed.

Pulse Torquing. In an attempt to overcome the limitations of the analog output in a modern system, an interesting approach to digitalization has been used. This method, known as pulse torquing, is ideally suited to the needs of the electrical force-rebalance type of instrument. The principle depends upon the ability to generate a sequence of absolutely constant torque impulses of variable repetition rate. If this can be done, the average value of torque produced (proportional to acceleration) is determined strictly by the repetition rate (frequency of the torque impulses). Thus

$$T_{av} = 1/\tau \int_0^\tau T_i(t)\, dt = (N/\tau)|T_i| \tag{35.80}$$

where T_{av} = average torque
$T_i(t)$ = torque impulse as a function of time
τ = τth second or interval of time starting from zero
N = number of identical torque impulses generated up to time τ
$|T_i|$ = magnitude of absolutely constant torque impulses

From Eq. (35.80) it is seen that average torque is indeed proportional to the number of impulses per unit of time. The output of such an accelerometer is then somewhat

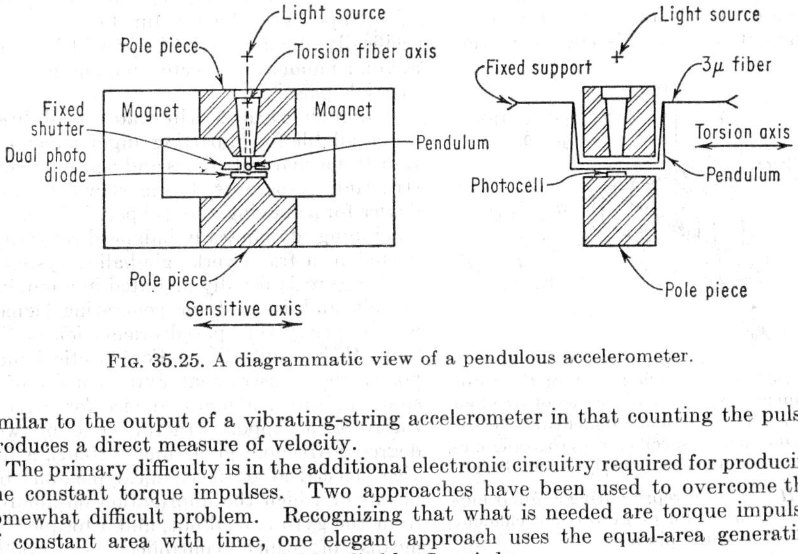

Fig. 35.25. A diagrammatic view of a pendulous accelerometer.

similar to the output of a vibrating-string accelerometer in that counting the pulses produces a direct measure of velocity.

The primary difficulty is in the additional electronic circuitry required for producing the constant torque impulses. Two approaches have been used to overcome this somewhat difficult problem. Recognizing that what is needed are torque impulses of constant area with time, one elegant approach uses the equal-area generating capabilities of magnetic material implied by Lenz's law:

$$e = -k(d\phi/dt) \tag{35.81}$$

and

$$\int_0^\tau e\, dt = -k(\phi_A - \phi_B) \tag{35.82}$$

where e = voltage
k = constant
$d\phi/dt$ = rate of change of flux
τ = time for completing one complete magnetization cycle
ϕ_A and ϕ_B = constant flux values for the extremes of the hysteresis loop for some heavily saturating material, such as the toroidal cores used in magnetic amplifiers

From Eq. (35.82) we see that if ϕ_A and ϕ_B or their difference are kept constant, an equal-area voltage impulse is generated. The voltage impulse can be converted to torque by suitable electronics.

The second method for generating equal-area torque impulses recognizes that one area-determining coordinate, time, can be generated with excellent absolute precision. The problem of generating the second coordinate as a voltage is approached by means of a suitable regulator.

Until the stable Zener diode was introduced, this method was not very practical. However, even the Zener-diode approach requires complicated electronic conversion circuits.

When comparisons of pulse torquing systems are made with other precision accelerometers, one finds that serious compromise(s) may have to be made in the areas of reliability, size, weight, and power. Nevertheless, development of such instruments has proceeded to the point where various manufacturers can provide them for specialized applications in the area of inertial guidance.

The method of pulse torquing may be universally applied to any force rebalance accelerometer and is the basis for such devices as the MIT accelerometer designated PIPA (pulse integrating pendulous accelerometer).

35.5. GYROSCOPES

(a) Introduction

The function of a gyroscope in modern instrumentation is to provide a direct link to a stable, convenient coordinate system—inertial space, the reference formed by the remote "fixed" stars. The property of a gyroscope from which its function derives is inherent in Newton's laws of motion; specifically, an untorqued body will have its angular-momentum vector remain fixed in inertial space.

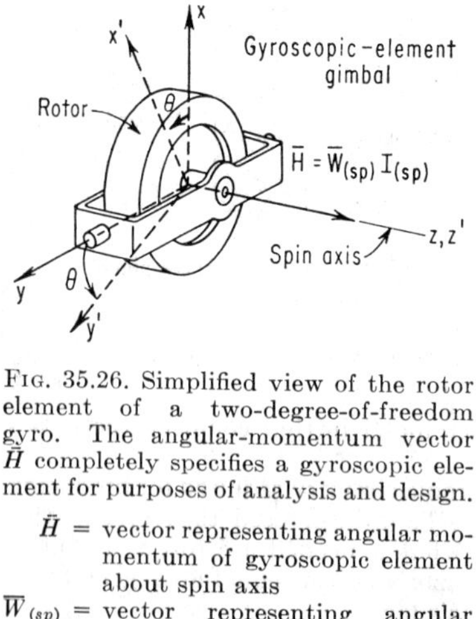

FIG. 35.26. Simplified view of the rotor element of a two-degree-of-freedom gyro. The angular-momentum vector $\bar{H}$ completely specifies a gyroscopic element for purposes of analysis and design.

$\bar{H}$ = vector representing angular momentum of gyroscopic element about spin axis

$\bar{W}_{(sp)}$ = vector representing angular velocity of rotor about spin axis = W_{Z^1}

$I_{(sp)}$ = moment of inertia of rotor about spin axis = I_{Z^1}

By definition, a gyroscopic element is a mechanical system with the following properties: (1) System contains a rotor spinning about an axis of symmetry. (2) Rotor is spinning at constant speed; i.e., $W_{(sp)}$ is constant.

Although various alternate possibilities are available for imparting angular momentum to mechanical bodies and hence for constructing gyroscopes, the most widely used device for achieving gyro properties is that of a spinning, dynamically balanced rotor supported in a framework (gimbaling system).

The gyro is usually mounted in a moving vehicle and is used for generating signals which specify the spatial orientation of the body to be controlled. For practical purposes, then, instrument gyros consist of a spinning rotor, a motor device for driving the rotor, a support structure (one- or two-degree-of-freedom gimbaling system), a device (electrical or mechanical pick-off) by means of which the information obtainable from the gyro is read out, and a torquer by means of which controlled, measurable torque may be imparted to the rotor assembly.

Examples of the instrument use of a gyro are a stable vertical carried on a ship or an airplane, a gyrocompass, and an inertial guidance system.

(b) Two-degree-of-freedom Gyro

Consider Fig. 35.26, a simplified view of the rotor element of Fig. 35.27. The primed coordinate system rotates with the wheel, whereas the unprimed system is fixed to the inner-rotor support gimbal. Further, the primed system is chosen to be along the principal axis of inertia of the wheel.

Under these circumstances, the vector equation

$$T = dH/dt \tag{35.83}$$

where T = applied torque
H = angular momentum

reduces to the Euler equations

$$\begin{aligned} T_{x'} &= I_{x'}(d\omega_{x'}/dt) - (I_{y'} - I_{z'})\omega_{y'}\omega_{z'} \\ T_{y'} &= I_{y'}(d\omega_{y'}/dt) - (I_{z'} - I_{x'})\omega_{z'}\omega_{x'} \\ T_{z'} &= I_{z'}(d\omega_{z'}/dt) - (I_{x'} - I_{y'})\omega_{x'}\omega_{y'} \end{aligned} \tag{35.84}$$

where $I_{x'}, I_{y'}, I_{z'}$ = rotor moments of inertia about principal axes x', y', z'
$\omega_{x'}, \omega_{y'}, \omega_{z'}$ = inertial-referenced angular velocity of the rotor about axes x', y', z'
x', y', z' = rotating coordinate system fixed to wheel
x, y, z = coordinate system fixed to gimbal

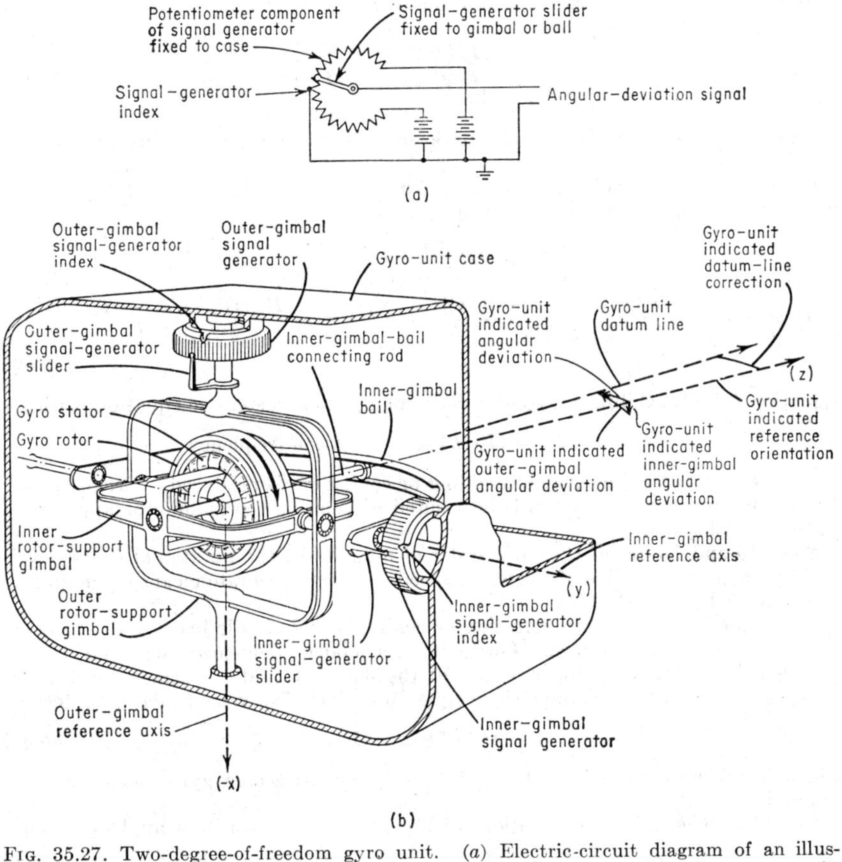

FIG. 35.27. Two-degree-of-freedom gyro unit. (*a*) Electric-circuit diagram of an illustrative signal generator. (*b*) General features of a two-degree-of-freedom gyro unit. z = spin reference axis. y = sensitive axis, inner. x = sensitive axis, outer. Notes: (1) Signal generator indexes show positions of sliders when output signals are zero. (2) The indicated datum-line correction is the angle through which the case would have to be rotated in order to bring the datum line into coincidence with the reference orientation. (3) Gyro units with the general features shown in the illustrative diagram of this figure are commonly used in automatic pilots for aircraft.

Transforming the primed coordinate system to the more convenient gimbal-referenced system we have

$$\begin{aligned}
T_{x'} &= T_x \cos\theta + T_y \sin\theta \\
T_{y'} &= -T_x \sin\theta + T_y \cos\theta \\
\omega_{x'} &= \omega_x \cos\theta + \omega_y \sin\theta \\
\omega_{y'} &= -\omega_x \sin\theta + \omega_y \cos\theta \\
T_{z'} &= T_z \\
\omega_{z'} &= \omega_z
\end{aligned} \tag{35.85}$$

Substituting Eq. (35.85) into Eq. (35.84) and letting $I_{x'} = I_{y'}$, which makes the rotor a figure of revolution as is the case for a practical gyro, we have

$$\begin{aligned}
T_x &= I_{x'}(d\omega_x/dt) + I_{z'}\omega_y\omega_z \\
T_y &= I_{x'}(d\omega_y/dt) - I_{z'}\omega_x\omega_z \\
T_z &= I_{z'}(d\omega_z/dt)
\end{aligned} \tag{35.86}$$

Considering ω_z = a constant (rotor up to speed) and letting $I_{z'}\omega_z = H_z$ (wheel angular momentum),

$$\begin{aligned}
T_x &= I_{x'}(d\omega_x/dt) + H_z\omega_y \\
T_y &= I_{x'}(d\omega_y/dt) - H_z\omega_x \\
T_z &= 0
\end{aligned} \tag{35.87}$$

Equations (35.87) constitute the basic differential equations of the two-degree-of-freedom gyro.

Gyro Precession and Nutation. Taking the Laplace transform of Eq. (35.87) (assuming zero initial conditions) yields

$$\begin{aligned}
T_x(s) &= I_{x'}s\omega_x(s) + H_z\omega_y(s) \\
T_y(s) &= -H_z\omega_x(s) + I_{x'}s\omega_y(s)
\end{aligned} \tag{35.88}$$

and $$\omega_x(s) = \frac{(1/H_z)[sT_x(s)/\omega_n - T_y(s)]}{s^2/\omega_n^2 + 1} \qquad \omega_y(s) = \frac{(1/H_z)[sT_y(s)/\omega_n + T_x(s)]}{s^2/\omega_n^2 + 1} \tag{35.89}$$

Equation (35.89) exhibits a second-order undamped characteristic equation, with

$$\omega_n = H_z/I_{x'} = (I_{z'}/I_{x'})\omega_z = \text{nutation frequency}$$

because of the three-dimensional nodding or coning effect produced. Because there is no damping, the nutation will be excited as soon as the spinmotor approaches normal wheel speed.

Since $I_{z'}$ and $I_{x'}$ are of the same order of magnitude, the nutational frequency is high and a sizable fraction of the wheel speed. A more accurate means of computing ω_n is outlined in Eq. (35.93).

As will be discussed later, nutation or coning is undesirable because of the effective drift rate it can produce. Fortunately, a modern fluid-suspended gyro has its nutation damped out by the viscosity of the flotation fluid. Mathematically, this changes the form of the characteristic equation to include a velocity damping term

$$s^2 + 2\xi\omega_n s + 1 \tag{35.90}$$

which is the characteristic factor of a two-degree-of-freedom gyro including nutational damping.

Consider the steady-state solutions of Eq. (35.89), with constant applied torques.

$$\omega_x = -T_y/H_z \qquad \omega_y = T_x/H_z \tag{35.91}$$

Thus a steady torque about y produces a steady drift about x and a steady torque about x produces a steady drift about y.

These torque-induced drifts are the gyro precession effect. They exhibit cross-coupling behavior (torque in x produces drift in y).

Because the function of a gyro is to retain a fixed (or follow a desired dynamic) direction in inertial space, and because torque produces a precessional drift rate, we are immediately introduced to the importance of the torque to angular-momentum ratio.

Undesired torques, from *whatever* cause, show up as *undesired* drift. The gyro designer, then, strives toward maximizing the H/T ratio in the presence of varying environmental conditions which produce a subtle array of disturbing torques. Some of these torques are listed in Table 35.4.

Table 35.4. Some Causes of Gyro Drift

Item	*Comments*
Constant torque, deg/hr	Independent of acceleration
Unbalance torque, deg/hr/g	Proportional to acceleration
Anisoelastic torque, deg/hr/g^2	Exists when strain vector is not parallel to force vector
Elastic restraint torque, deg/hr	Due to some springy member (possibly lead in wires to spin motor)
Friction torque, deg/hr	Minimized by flotation and jeweled pivots or flexible torsion wire suspension
Reaction torque, deg/hr	Caused by low-impedance load on gyro pick-off
Magnetic torque, deg/hr	This is a form of constant torque. Could be caused by interaction of earth's field with magnetic material in gyro element
Thermal fluid torques, deg/hr/g	Caused by convection currents in suspension fluid from nonhomogeneous heating
Random torque	Cause generally unknown—exercise chief limit on performance of high-precision gyro

(c) The Block Diagrams for a Floated Two-degree-of-freedom Gyro and a Single-degree-of-freedom Gyro

In an endeavor to bound frictional uncertainty torques and their resultant drift, in most precision gyros constructed today, the rotor element is floated in a dense fluid. When properly adjusted, the rotor weight is supported almost completely by the buoyant force of the displaced fluid. The adjustment of buoyancy is made by controlling the temperature of the fluid (which determines the fluid density). The temperature is adjusted until the fluid density matches the average "float" density so that the "float" hangs suspended in the fluid at neutral buoyancy. Under these conditions, substantially zero bearing loads exist so that comparatively delicate jewel bearings (pivots) or slender torsion wires may be used, with the result that vanishingly small friction torque uncertainties are realized. Flotation also allows the unit to withstand violent shocks and strong vibration without damage.

Basic Differential Equations Including Case Motion and Fluid Damping. Starting with Eq. (35.87) we may write (assuming orthogonal gimbals, for simplicity)

$$
\begin{aligned}
k_x(\theta_{cx} - \theta_{gx}) + f_x(\dot\theta_{cx} - \dot\theta_{gx}) + T_x &= I_x\ddot\theta_{gx} + H_z\dot\theta_{gy} \\
k_y(\theta_{cy} - \theta_{gy}) + f_y(\dot\theta_{cy} - \dot\theta_{gy}) + T_y &= I_y\ddot\theta_{gy} - H_z\dot\theta_{gx}
\end{aligned}
\tag{35.92}
$$

where θ_{gx}, θ_{gy} = displacement of gyro float with respect to inertial space
θ_{cx}, θ_{cy} = displacement of outer case with respect to inertial space
k_x, k_y = spring restraints as might be caused by "flex" harnesses bringing power to the spin motor, or by torsion wires, if this type of suspension is employed
f_x, f_y = viscous damping coefficients about x and y
I_x = total inertia about x = wheel inertia + gimbal, etc.
I_y = total inertia about y = wheel inertia + gimbal, etc.

Taking the Laplace transform of Eq. (35.92) and letting

$$
\begin{aligned}
\epsilon_x &= \theta_{cx} - \theta_{gx} \\
\epsilon_y &= \theta_{cy} - \theta_{gy}
\end{aligned}
$$

$$
\begin{aligned}
(k_x + sf_x)\epsilon_x(s) + T_x(s) &= s^2I_x\theta_{gx}(s) + sH_z\theta_{gy} \\
(k_y + sf_y)\epsilon_y(s) + T_y(s) &= s^2I_y\theta_{gy}(s) - sH_z\theta_{gx}
\end{aligned}
\tag{35.93}
$$

From Eq. (35.93), we can draw the block diagram as shown in Fig. 35.28.

We note, as expected, that this gyro provides two inertially referenced error signals ϵ_x and ϵ_y. In this sense, we may say that the two-degree-of-freedom gyro functions as a "two-axis space synchro."

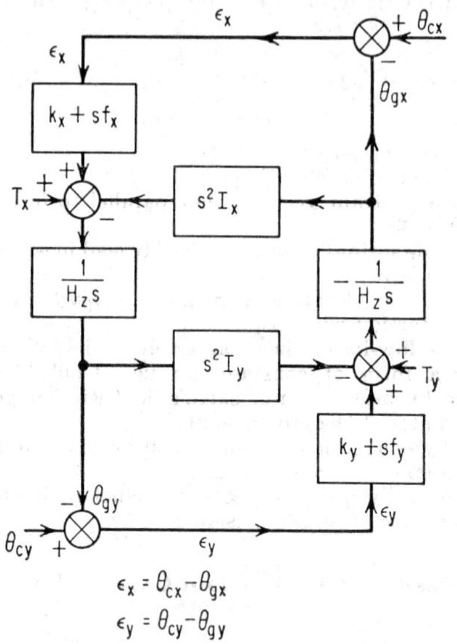

FIG. 35.28. Block diagram for a two-degree-of-freedom gyro.

This representation (Fig. 35.28) puts into sharp focus the "closed-loop" cross-coupling effects which tend to cause nutation, as well as give rise to the precessional effects.

In some applications, we want the precessional relationships as given by Eq. (35.91) to be exact, whereas Fig. 35.28 indicates some dependence on ϵ_x and ϵ_y (via the elastic and viscous effects). In reality, these effects will not materially change Eq. (35.91) because the gyro is invariably used with a follow-up servo (space-stabilization servo) in a manner which nulls ϵ_x and ϵ_y. In view of the fact that the servo frequency response is usually much below the nutational frequencies, ϵ_x and ϵ_y "exist" for these higher frequencies, and the viscous effect is useful in damping nutation.

(d) Specialization to a Single-degree-of-freedom Gyro by the Introduction of Constraints

Figure 35.29 shows a single-degree-of-freedom gyro unit derived from the two-degree-of-freedom gyro (Fig. 35.27). We note that the x axis is no longer free but is fixed to the case.

With the equation of constraint, $\theta_{cx} = \theta_{gx}$ or $\epsilon_x = 0$, Eq. (35.93) becomes

$$T_{IA}(s) = s^2 I_{IA}\theta_{gIA}(s) + sH_z\theta_{gOA} \qquad (35.94)$$
$$(k_{OA} + sf_{OA})\epsilon_{OA}(s) + T_{OA}(s) = s^2 I_{OA}\theta_{gOA}(s) - sH_z\theta_{gIA} \qquad (35.95)$$

where $X \to IA$ = input axis
$Y \to OA$ = output axis
$Z \to SRA$ = spin rotation axis

as defined in Fig. 35.29.

Since $\epsilon_{OA}(s) = \theta_{cOA}(s) - \theta_{gOA}(s)$, Eqs. (35.94) and (35.95) can be rewritten

$$T_{OA}(s) + (k_{OA} + sf_{OA})\theta_{cOA}(s) = (s^2 I_{OA} + sf_{OA} + k_{OA})\theta_{gOA}(s) - sH_z\theta_{gIA}(s) \qquad (35.96)$$

Equation (35.94) is of interest in defining the normal forces acting on the gimbal pivots as well as indicating a "gyro-stabilization" torque which helps keep a platform fixed in inertial space, independent of servo action.

From Eq. (35.96), we can draw the block diagram shown in Fig. 35.30.

From Fig. 35.30, note that, unlike the two-degree-of-freedom unit (Fig. 35.28), the angular-momentum vector is coerced in the single-degree-of-freedom gyro; i.e., we must have some motion about OA if a readout is to be obtained. This characteristic can lead, as is seen below, to some subtle drift effects. Control of this characteristic is obtained by using the gyro in conjunction with a high-performance platform servo (usually gearless). Further specialization can be made, as shown in the following discussion.

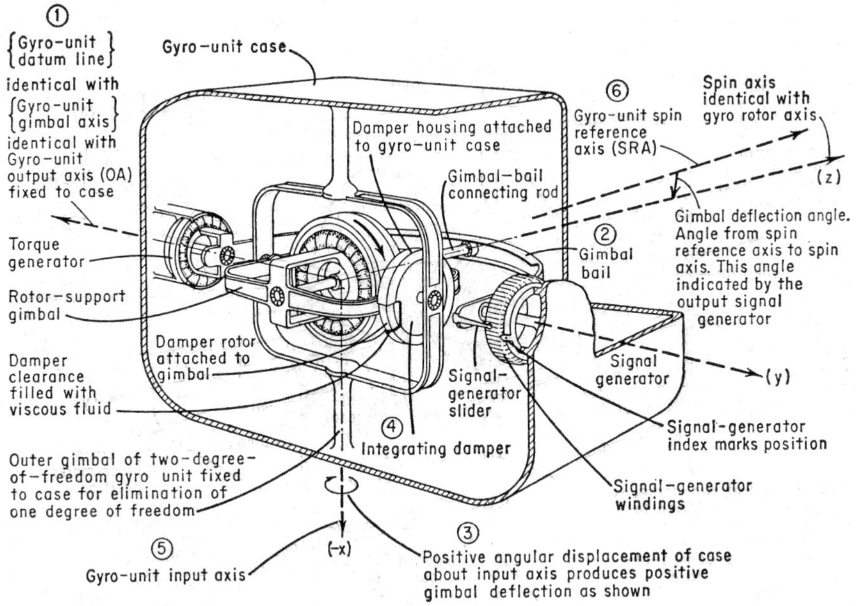

FIG. 35.29. Single-degree-of-freedom gyro unit with integrating damper.

1. The datum line for the single-degree-of-freedom gyro unit is fixed to the case at right angles (along inner gimbal reference axis) to the datum line for the two-degree-of-freedom gyro unit.

2. In an actual unit the gimbal bail is not required and would not be used. The signal generator and the torque generator would act directly on the single gimbal. The unnecessarily complicated form of this diagram is used to illustrate the relationship between two-degree-of-freedom gyro units and single-degree-of-freedom gyro units.

3. Positive rotations about case axes are related to the positive direction of the axis by the right-hand-screw rule.

4. The action of the viscous liquid in the damper clearance space causes a torque opposing angular velocity of the gimbal with a torque proportional to the magnitude of the gimbal angular velocity with respect to the unit case.

5. Gyro-unit input axis $(-x)$ (IA): Fixed to case at right angles to the gimbal axis. Its positive sense is determined by the right-hand-screw rule when the output axis is turned toward the spin reference axis.

6. Gyro-unit spin reference axis (SRA): Fixed to the case at right angles to the input axis and the output axis. The spin reference axis is identical with the spin axis when the output signal is zero.

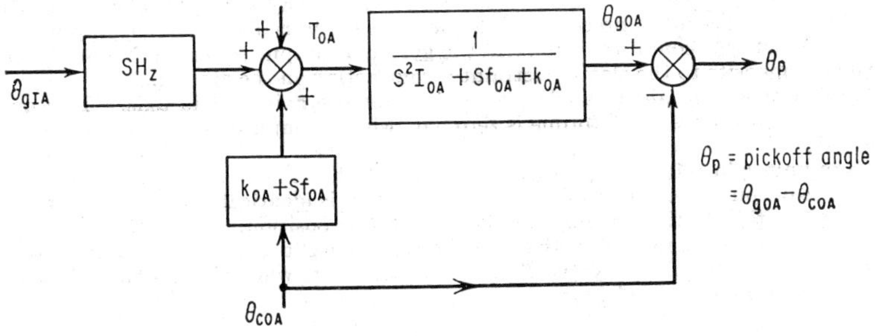

FIG. 35.30. Block diagram for a single-degree-of-freedom gyro.

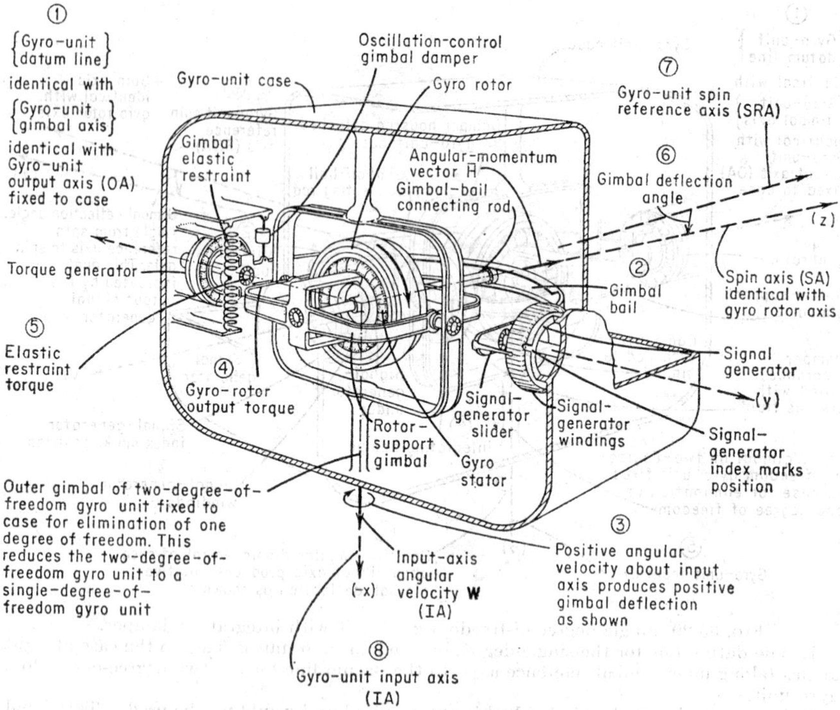

FIG. 35.31. Rate gyro-based on the single-degree-of-freedom gyro unit with elastic restraint and oscillation control damper.

1. The datum line for the single-degree-of-freedom gyro unit is fixed to the case at right angles (along inner gimbal reference axis) to the datum line for the two-degree-of-freedom gyro unit.

2. In an actual unit the gimbal bail is not required and would not be used. The signal generator and the torque generator would act directly on the single gimbal. The unnecessarily complicated form of this diagram is used to illustrate the relationship between two-degree-of-freedom gyro units and single-degree-of-freedom gyro units.

3. Positive rotations about case axes are related to the positive direction of the axis by the right-hand-screw rule.

4. The gyro-rotor output torque tends to turn the angular-momentum vector toward alignment with the input angular-velocity vector.

5. The elastic-restraint torque due to gimbal deflection always tends to cause the spin axis to have a precessional angular velocity about the input axis with the same direction as the direction in which the case is being rotated. Steady-state equilibrium is reached when the elastic-restraint precessional angular velocity is equal to the input angular velocity. When this condition is fulfilled the gyro unit output torque acting on the gimbal is equal and opposite to the elastic-restraint torque acting on the gimbal.

6. Gimbal deflection angle: angle from spin reference axis to spin axis. This angle is zero when the elastic restraint torque is zero, i.e., when no input angular velocity is applied to the case.

7. Gyro-unit spin reference axis (*SRA*): Fixed to the case at right angles to the input axis and the output axis. The spin reference direction is identical with the direction of the spin axis when the elastic restraint is in its zero torque position.

8. Gyro-unit input axis (*IA*): Fixed to the case at right angles to the gimbal axis; its positive sense is determined by the right-hand-screw rule when the gimbal axis is turned toward the spin reference axis.

(e) The Rate Gyro

Figure 35.31 shows a single-degree-of-freedom gyro unit with an introduced spring restraint so that it will function as a rate gyro or "space tachometer." To develop its performance, we start with Fig. 35.30 and, for simplicity, neglect cross-coupling inputs (case motion θ_{COA} as well as noncoincidence of x and IA due to θ_{gOA}).

From Fig. 35.32 the equation of performance is

$$\theta_{gOA} = \frac{H_z/k_{OA}}{s^2/\omega_n^2 + (2\xi/\omega_n)s + 1}\,\omega_{gIA} \tag{35.97}$$

Note the importance of the natural frequency parameter ω_n and the damping factor ξ. For frequencies above ω_n, the gyro ceases to perform as a rate gyro. To keep ω_n high, high k_{OA} and low I_{OA}, high spring constant, and low inertia are required. High spring constant results in low sensitivity [Eq. (35.97)], however. This, then, is the design compromise—a relatively high natural frequency invariably means a relatively low sensitivity. Some attention must also be paid to obtain a relatively constant damping ratio ξ over a wide range of temperature, without heating.

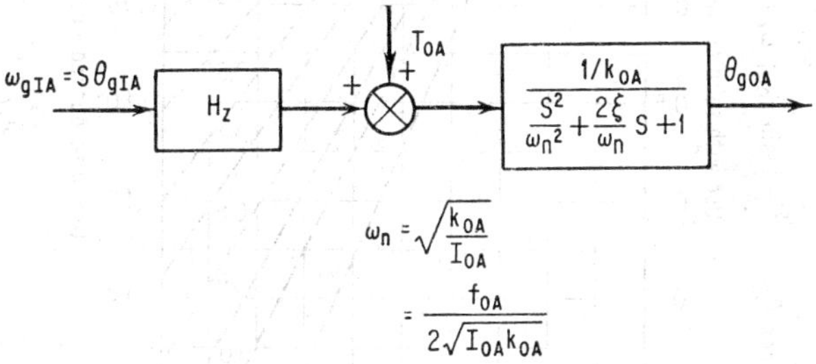

FIG. 35.32. Block diagram of rate gyro.

Rate gyros have been used to provide rate-of-turn information in automatic pilots, fire-control systems, oscillation dampers, missile stabilization, and telemetering systems.

Because, for the rate gyro, the gimbal angle θ_{gOA} must increase as the input increases, the *effective* input axis (which is tied to the gimbal, Fig. 35.31) also displaces. This introduces a gyro cross-coupling effect, which may be important; i.e., the gyro will respond to case angular velocity about the spin reference axis *SRA* when it is at the same time subjected to an angular velocity about the input axis.

The factor by which the case-referenced input rate must be multiplied before using Eq. (35.97) is shown in Fig. 35.33, which effectively summarizes the magnitude error of the cross-coupling effect.

(f) The Rate-integrating Gyro (Draper Gyro)

Figure 35.29 shows (pictorially) a single-degree-of-freedom gyro unit with an introduced integrating damper and negligible spring restraint; so that it will function as a rate-integrating gyro or "space synchro." To develop its performance, we start with Fig. 35.30 and, for simplicity, neglect cross-coupling effects.

A condition for the rate-integrating gyro is that k_{OA} approaches zero. From Fig. 35.30 and the above condition, the block diagram given in Fig. 35.34 can be drawn.

From Fig. 35.34, the equation of performance is

$$\theta_{gOA} = \frac{H_z/f_{OA}}{\tau S + 1}\,\theta_{gIA} \tag{35.98}$$

Note the importance of the H/f ratio (gain of the gyro) as well as the gyro time constant τ. To keep τ small, high damping (large f_{OA}) is required, but large f_{OA}

results in low sensitivity. This is the design compromise—a relatively high sensitivity (gain) invariably means a relatively long time constant. Various manufacturers supply these gyros with H/f ratios varying from about 0.5 to 20, and τ varying from 0.0015 to 0.005 sec.

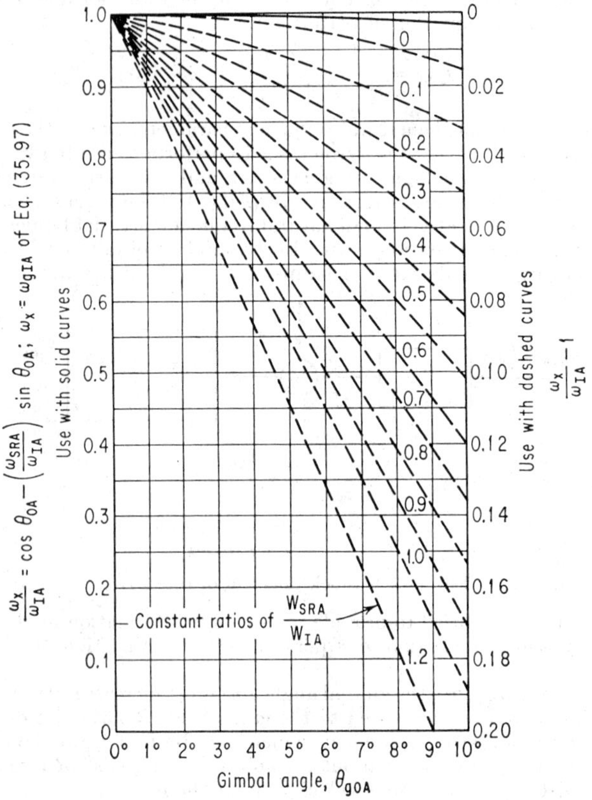

FIG. 35.33. Angular velocity–voltage sensitivity performance ratio as a function of gimbal angle and angular-velocity ratio for the single-axis rate gyro unit.

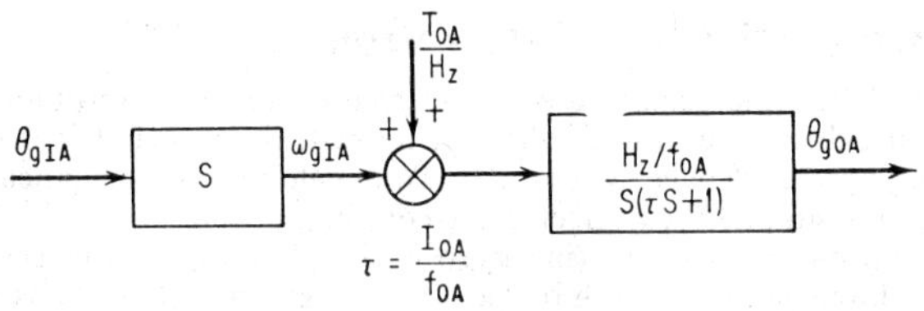

FIG. 35.34. Block diagram of rate-integrating gyro.

Single-axis integrating gyros for actual use differ greatly in design from the construction shown in Fig. 35.29. A more realistic functional diagram is shown in Fig. 35.35.

The integrating gyro is sometimes used as a rate gyro. This is done by feeding back the output signal θ_{gOA} to the torquer T_{OA} and measuring torquer current as the rate indicator, as shown in Fig. 35.36.

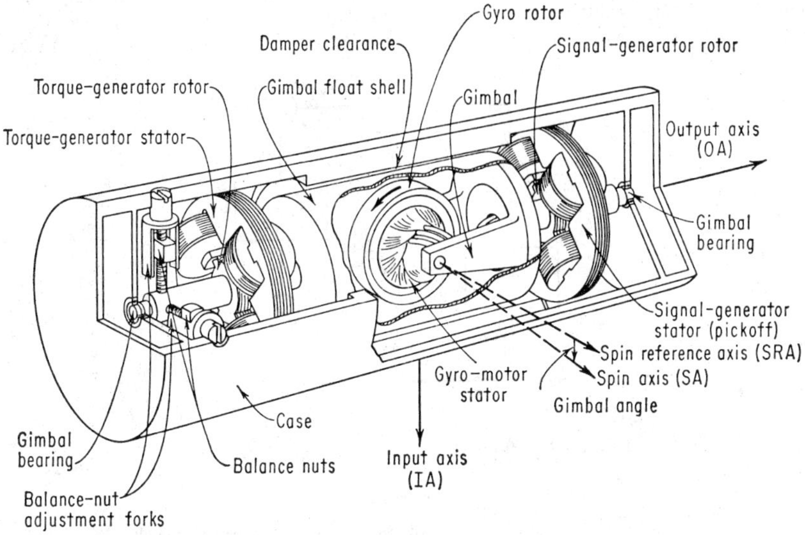

FIG. 35.35. Pictorial diagram for the single-axis integrating gyro unit (single-degree-of-freedom). Note: The volume between the outside of the float and the inside of the case is completely filled with a fluid having the property of Newtonian viscosity. That is, the coefficient of viscosity is independent of shear gradient.

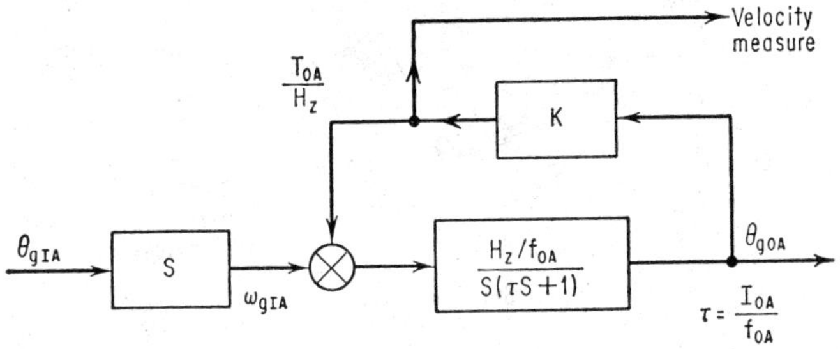

FIG. 35.36. Rate-integrating gyro converted to rate-gyro duty by use of a torque feedback loop.

The equation of performance is

$$\frac{k\theta_{gOA}}{\omega_{gIA}} = \frac{1}{s^2/\omega_n^2 + (2\xi/\omega_n)s + 1} \tag{35.99}$$

where $\omega_n = \sqrt{kH_z/I_{OA}}$

$\xi = f_{OA}/2\sqrt{I_{OA}kH_z}$

Comparing Eq. (35.99) with (35.97), rate gyro performance is recognized. Here, however, θ_{gOA} can be extremely small while sensitivity is high because of the gain k. Thus this way of constructing a rate gyro bypasses some of the geometric cross coupling shown in Fig. 35.33.

Rate-integrating gyros have been used in fire-control systems, inertial-guidance systems, rate measurement, and flight-control systems for navigation.

(g) The Free Gyro

Free gyros are two-degree-of-freedom units that are employed in applications where information relating to the displacement of a vehicle from some specific reference is desired. This is the normal two-degree-of-freedom function previously

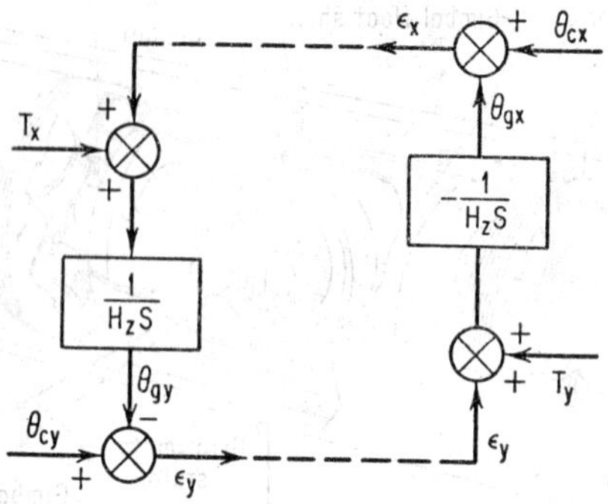

FIG. 35.37. Block diagram of an "ideal" free gyro.

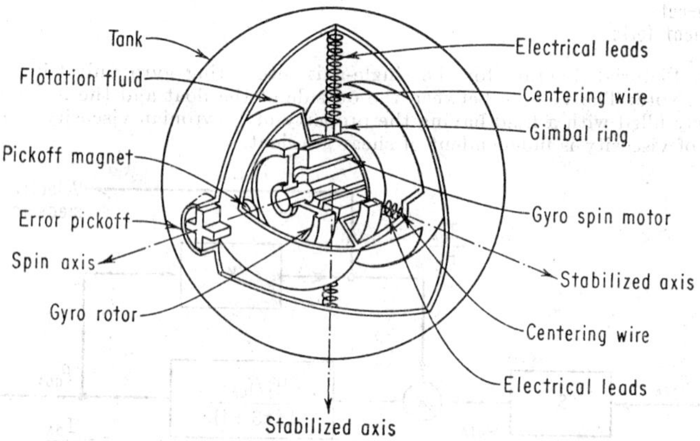

FIG. 35.38. Two-degree-of-freedom gyro typical of those used in inertial platforms. (*Developed by Arma Div., American Bosch Arma Corp., Garden City, N.Y.*)

described. However, the block diagram of Fig. 35.28 can be materially simplified, if the following assumptions are made:

1. Viscous and elastic restraints are negligible.
2. Gimbal inertial torques are negligible.

Under these circumstances, the block diagram of an "ideal" free gyro is shown in Fig. 35.37.

A realistic functional diagram of a two-degree-of-freedom unit which approaches an "ideal" free gyro is shown in Fig. 35.38.

(h) The Undamped Single-axis Gyro

It is possible (and in some respects desirable) to solve the problem of reducing friction uncertainties in a manner different from the fluid suspension previously discussed. One could, for example, employ a liquid- or gas-bearing technique, each of which is generally recognized to have lower-uncertainty torques (by several orders of magnitude) than the best ball or roller bearings.

Gyros with hydrostatic gas output-axis bearings have been produced and exhibit a number of the unique features to be expected from such a device:

1. Relatively low drift (inertial quality)
2. Extremely low damping so that special care (several cascaded lead networks) must be taken in the platform servo design
3. A large amount of "gyro stabilization" as given by Eq. (35.94)
4. Relative freedom from critical-temperature requirements

A functional diagram of an output-axis air-bearing gyro developed by the U.S. Army Ballistic Missile Agency is shown in Fig. 35.39.

The block diagram for an undamped gyro is shown in Fig. 35.40 and is derived from Fig. 35.30 by allowing $k_{OA} \to 0$ and $f_{OA} \to 0$.

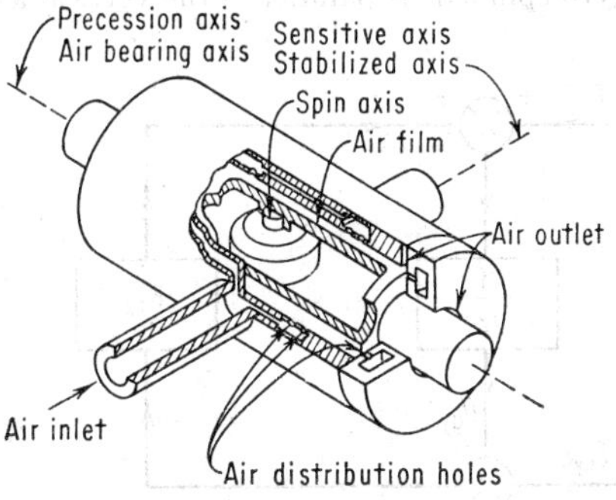

FIG. 35.39. Cutaway drawing of a single-axis gyro developed by U.S. Army Ballistic Missile Agency, Redstone Arsenal, Ala. (Suspension of the output axis is by a unique air bearing. A thin film of compressed air separates the "floated" output axis from the housing. The air gap between float and housing is about 1.5 thousandths of an inch. The distribution holes feeding air into the gap are a few thousandths of an inch in diameter.) (*Courtesy of ref.* 9.)

(i) Gyro Applications

Directional Gyro. A directional gyro is used for "memorizing" or "smoothing" a vehicle heading indication. It is usually a two-degree-of-freedom unit, one degree of freedom permitting the spin axis to be maintained horizontal and the second degree of freedom permitting the given azimuth reference to be maintained independent of vehicle motion.

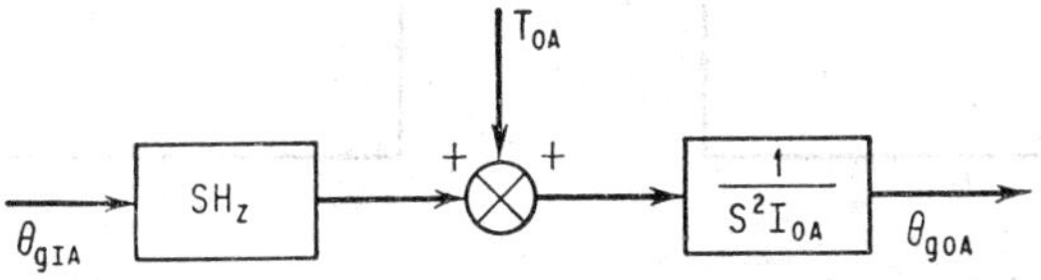

FIG. 35.40. Block diagram for a single-degree-of-freedom gyro (undamped).

A simplified block diagram of these two functions may be obtained from Fig. 35.37 as shown in Fig. 35.41. Thus

θ_{gy} = gyro angle driven to vertical by torquer's servo action
k_{Tx} = torquer gradient in x

θ_{cy} could be obtained independently by pendulum referenced to inner gimbal.

θ_{gx} = gyro angle driven to heading by torquer's servo action
k_{Ty} = torquer gradient in y

θ_{cx} could be obtained independently by compass (if a north indication is desired). The memory function is achieved by removing the θ_{cx} signal. The azimuth channel now indicates a heading referenced to the last azimuth input as modified by the gyro drift caused by the T_y uncertainty torques.

Vertical Gyro. A vertical gyro is used in conjunction with a pendulous element to give a continuous indication of "true" vertical (to within a tolerance) in a moving vehicle. It is usually a two-degree-of-freedom unit aligned so that one degree of freedom is parallel to the fore-aft axis of the vehicle, the other degree of freedom indicating relative displacement about a transverse axis. Common nomenclature has established that the fore-aft information is designated *longitudinal* or *roll*, whereas the transverse information is designated *lateral* or *pitch*. In normal operation, the gyro spin axis is parallel to the vertical axis.

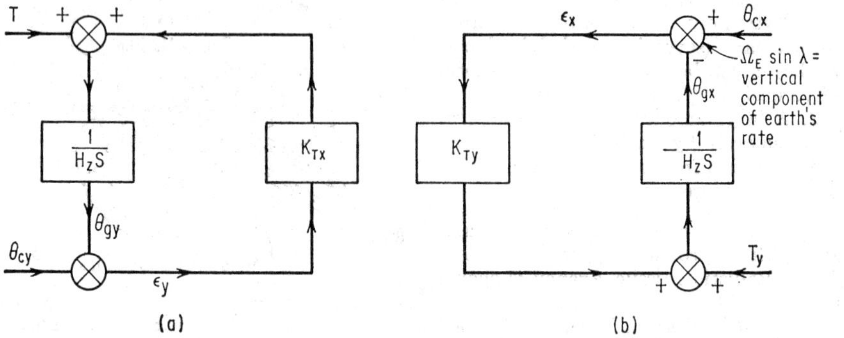

FIG. 35.41. Functional block diagram of directional gyro loops. (*a*) Erection channel (closed loop). (*b*) Azimuth channel.

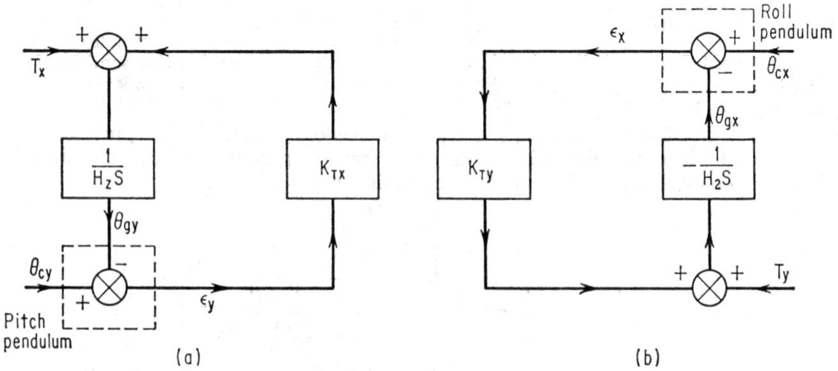

FIG. 35.42. Block diagram of the direction loops of a gyro vertical. (*a*) Pitch pendulum. (*b*) Roll pendulum.

Since the spin axis of a gyroscope will tend to maintain a fixed orientation in inertial space, the gyro will tilt relative to the earth because of earth's rotation in inertial space.

The erection loops of a vertical gyro are entirely similar to the directional gyro loops shown in Fig. 35.41 and repeated in Fig. 35.42 for clarity, with a slight redefinition of θ_{cy}, θ_{gy}, θ_{cx}, θ_{gx}.

From Fig. 35.42,

$$\begin{aligned}\theta_{gy} &= \frac{1}{\tau_x S + 1}\,\theta_{cy} + \frac{1/k_{T_x}}{\tau_x S + 1}\,T_x \\ \theta_{gx} &= \frac{1}{\tau_y S + 1}\,\theta_{cx} + \frac{1/k_{T_y}}{\tau_y S + 1}\,T_y\end{aligned} \tag{35.100}$$

where $\tau_x = H_z/k_{T_y}$

$\tau_y = H_z/k_{T_x}$

We may consider the earth's rate input as supplying a drift torque, say T_y, even though it is the coordinate system that moves and not the gyro.

From Eq. (35.100),

$$
\begin{aligned}
[\theta_{gx}] \text{ due to earth's rate} &= \lim_{s\to 0} S \frac{1/k_{T_y}}{\tau_y S + 1} \frac{H_z \Omega_E \cos\lambda}{S} \\
&= H_z / k_{T_y} \Omega_E \cos\lambda \\
&= \tau_x \Omega_E \cos\lambda
\end{aligned}
\tag{35.101}
$$

For the case of a 200-sec time constant at 45° latitude, the westward tilt is

$$
\begin{aligned}
[\theta_{gx}] &= 200 \text{ sec} \times \tfrac{1}{4} \text{ minute of arc per second} \times 0.707 \\
&= 35.35 \text{ minutes of arc}
\end{aligned}
$$

This large error cannot be tolerated in most cases. Although the time constant can be reduced, this makes the gyro more susceptible to short-term acceleration disturbances. On many moderate-accuracy gyro verticals the erection-loop time constant is chosen as a compromise between acceleration errors and earth-rotation errors. In addition, switching out the pendulum input (and going on gyro memory) during accelerations larger than a preset amount (such as in a turn) is also employed. The tilt caused by earth's rotation is often called the *latitude error*.

An exact means for removing latitude error would be to compute the equivalent earth's rate torque and feed these as a boost signal into the T_x and T_y inputs of Fig. 35.42.

Considering a vehicle with a heading angle A, the appropriate torques are

$$
\begin{aligned}
T_y &= T_{\text{roll}} = H_z \Omega_E \cos\lambda \sin A \\
T_x &= T_{\text{pitch}} = H_z \Omega_E \cos\lambda \cos A
\end{aligned}
\tag{35.102}
$$

The analog computer used can obtain sin A and cos A from a resolver on the heading reference of the vehicle.

Single-degree-of-freedom gyros can also be used to construct vertical references. Also, accelerometers rather than pendulums can be used as gravity sensors. When more detailed considerations of system errors are made, one is naturally led to the use of the famous Schuler tuning principle to obtain a precision vertical on maneuvering vehicles.

(j) Gyro Drift

Gyro drift may be described as an undesired motion of the precessional element with respect to inertial space.

Undesired (or uncompensated) torques, from *whatever* cause, show up as gyro drift. Several causes for gyro drift are listed in Table 35.4, together with their classification, in terms of acceleration sensitivity.

Table 35.5. Accuracy Requirements of Gyroscopes[9]

Application	*Drift requirement, deg/hr*
Rate gyroscope for fire and flight control	10–150
Directional gyroscope and gyro horizon for flight indication and automatic-pilot applications	1–15
Marine gyro vertical	1–5
Polar (high-altitude) directional gyroscope (compensated for earth rotation; used in grid navigation)	0.1–1
Gyrocompass (ship-borne)	0.03–0.4
Gyroscope for aircraft inertial navigator	0.0005–0.1

All disturbing torques must be kept within moderately small limits; maximum values depending on accuracy requirements and the angular momentum of the gyro. A more appropriate figure of merit would be the $H/T_{\text{disturbance}}$ ratio, which must be maximized for a given gyro configuration. To scale the magnitude of torque effect properly, which is important in a modern precision gyro, it is appropriate to mention that torque is measured in *dyne-centimeter* units, corresponding to drift rates in thousandths of a degree per hour!

In gyros employing a floated element, shifts of the center of mass (displacement of center of mass from center of support) or unbalance of the floating part become one of the most important (if not *the* most important) reason for drift. Mass shifts can be caused by improper factory adjustment, dimensional changes of the float or gimbal assembly, and/or relative displacement of parts making up the spin motor itself.

The most critical item is usually the gyro spin motor, being made of a relatively large number of parts of different materials, having different thermal characteristics important because of the temperature rise in operation due to heater and spin-motor power. Thus we have the rotor, the laminated stator, insulation material, and copper for windings.

Heating and cooling (operation and shutdown) create slight strains (which relieve with time or do not repeat) which correspond to minute mass shifts resulting in a disturbance torque which is sensitive to acceleration (or position in the earth's gravity field).

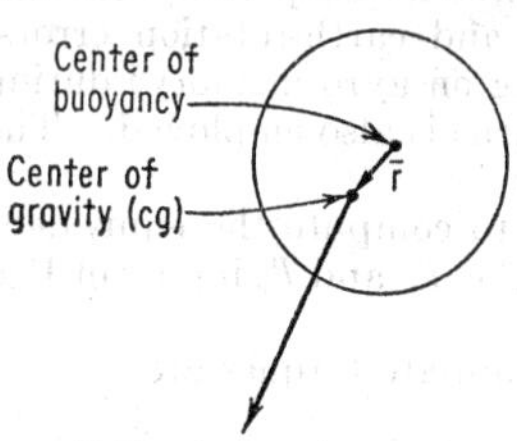

FIG. 35.43. Gyro float with mass unbalance. $\bar{r}$ = vector separation of center of gravity and center of buoyancy.

In some gyro systems, the shutdown problem is avoided by always keeping the gyros hot, flotation fluid liquid, from the moment they leave the factory, during shipment, and up until the moment of final use. This technique is both troublesome and expensive and can be justified only for extremely precise special-purpose applications (usually military) in which the extra performance is traded off against the extra expense.

A preliminary evaluation of a precision gyro can be made by measuring the change in drift between a number of operating and shutdown cycles.

Table 35.5 indicates typical drift-rate ranges for several applications.

Acceleration-sensitive Drifts. Consider the gyro float, shown in Fig. 35.43, in which an unbalance is shown by noncoincidence of the center of gravity and the center of buoyancy.

We are interested in computing the torque due to the acceleration-induced force F. For generality, we assume that the gyro float structure (motor, wheel bearings, etc.) is nonrigid and not equally stiff in all directions (anisoelastic). Under these conditions,

$$\mathbf{T} = \mathbf{r} \times \mathbf{F} \tag{35.103}$$

$$\mathbf{F} = m\mathbf{a} \tag{35.104}$$

$$\mathbf{r} = (x_0 + ma_x/k_x)\mathbf{i} + (y_0 + ma_y/k_y)\mathbf{j} + (z_0 + ma_z/k_z)\mathbf{k} \tag{35.105}$$

where x_0, y_0, z_0 = initial unbalance due to factory misadjustment or quasi-static shift from the last temperature shutdown

k_x, k_y, k_z = equivalent structure spring constant in the x, y, z directions, respectively

a_x, a_y, a_z = component accelerations along the reference coordinates

m = mass point at center of mass

$$\delta_x = ma_x/k_x \qquad \delta_y = ma_y/k_y \qquad \delta_z = ma_z/k_z$$

Combining Eqs. (35.103), (35.104), and (35.105),

$$\begin{aligned} T_x &= m(y_0a_z - z_0a_y) + m^2a_ya_z(1/k_y - 1/k_z) \\ T_y &= m(z_0a_x - x_0a_z) + m^2a_xa_z(1/k_z - 1/k_x) \\ T_z &= m(x_0a_y - y_0a_x) + m^2a_xa_y(1/k_x - 1/k_y) \end{aligned} \tag{35.106}$$

Equation (35.106) indicates that the equivalent drift-producing torques are composed of two distinct components: a "mass-unbalance" component due to the x_0, y_0, z_0 shift and an "anisoelastic" component due to the nonidentity of k_x, k_y, and k_z.

Anisoelastic Rectification. Because the "anisoelastic" torque is a function of the product of two acceleration components, rectification torques can exist in the presence of vibration. The nonrigid gyro assembly will exhibit resonances under vibration and will have a transmissibility curve with at least two peaks, one depending on the structural stiffness in the spin-axis direction and one depending on the stiffness in the transverse axis (assuming symmetry in the transverse directions).

Because of the resonance effects, $a_{x,y,z}$ and δ_x, δ_y, δ_z are computed as follows:

$$\begin{aligned} a_x &= a_{xc}\frac{1+(2\xi_x/\omega_x)s}{s^2/\omega_x^2+(2\xi_x/\omega_x)s+1} & \delta_x &= \frac{ma_{xc}/k_x}{s^2/\omega_x^2+(2\xi_x/\omega_x)s+1} \\ a_y &= a_{yc}\frac{1+(2\xi_y/\omega_y)s}{s^2/\omega_y^2+(2\xi_y/\omega_y)s+1} & \delta_y &= \frac{ma_{yc}/k_y}{s^2/\omega_y^2+(2\xi_y/\omega_y)s+1} \\ a_z &= a_{zc}\frac{1+(2\xi_z/\omega_z)s}{s^2/\omega_z^2+(2\xi_z/\omega_z)s+1} & \delta_z &= \frac{ma_{zc}/k_z}{s^2/\omega_z^2+(2\xi_z/\omega_z)s+1} \end{aligned} \tag{35.107}$$

where a_{xc}, a_{yc}, a_{zc} = case accelerations
ω_x, ω_y, ω_z = resonance frequencies
s = Laplace-transform operator
ξ_x, ξ_y, ξ_z = damping factors

Assuming

$$\begin{aligned} a_{xc} &= A_x \sin \omega t \\ a_{yc} &= A_y \sin (\omega t + \phi_y) \\ a_{zc} &= A_z \sin (\omega t + \phi_z) \end{aligned} \tag{35.108}$$

It can be shown by substituting Eqs. (35.108) and (35.107) in Eq. (35.106) that T_x and T_y are (T_z spin-axis torque assumed constant, although these effects could cause difficulty in synchronous operation under marginal torque conditions)

$$\begin{aligned} T_x &= (m^2A_yA_z\mathfrak{Z}_y\mathfrak{Z}_z/2)\cos\psi(1/k_y - 1/k_z) \\ &\quad - m^2A_yA_z\mathfrak{Z}_y\mathfrak{Z}_z\omega \sin\psi(\xi_z/\omega_z k_y + \xi_y/\omega_y k_z) \\ T_y &= (m^2A_xA_z\mathfrak{Z}_x\mathfrak{Z}_z/2)\cos\rho(1/k_z - 1/k_x) \\ &\quad - m^2A_xA_z\mathfrak{Z}_x\mathfrak{Z}_z\omega \sin\rho(\xi_x/k_z\omega_x + \xi_z/k_x\omega_z) \end{aligned} \tag{35.109}$$

where ψ and ρ are functions of ϕ_y, ϕ_z and the dynamic phase shift due to the transmissibilities

$$\begin{aligned} \psi &= \phi_y + \theta_y - \phi_z - \theta_z \\ \rho &= \phi_z + \theta_z - \theta_x \end{aligned} \tag{35.110}$$

where θ_y, θ_z are transmissibility phase shifts.

$$\begin{aligned} \mathfrak{Z}_x &= \left|\frac{1}{s^2/\omega_x^2+(2\xi_x/\omega_x)s+1}\right|_{s=j\omega} \\ \mathfrak{Z}_y &= \left|\frac{1}{s^2/\omega_y^2+(2\xi_y/\omega_y)s+1}\right|_{s=j\omega} \end{aligned} \tag{35.111}$$

When the applied vibration frequency is close to either of the two resonant frequencies, maximum torque is developed.

It is interesting to note from Eq. (35.109) that two torque components are produced, one due to the difference in elasticities and one due to the sum of elasticities. The first is the expected "anisoelastic" effect while the second is the "cylindrical"[18] effect, since it can be shown to exist for the cylindrical-motion effects implied by Eq. (35.108) and the definitions of ψ and ρ.

The error at zero frequency (Fig. 35.44) is referred to as the "anisoelastic coefficient" of the gyro. Typically, peaks occur in the 500 to 3,000 cps region and the "drift amplification" (determined by the Q of the resonant peaks) can be in the neighborhood of 30 to 50 times the "anisoelastic coefficient."

Geometric Effects. Recent investigations have shown the existence of a variety of related drift-producing effects which have their origin in motion of the gyro input axis (the effect described in Fig. 35.33 being a specialized case).

The coning error is a geometric effect which would be present even in a theoretically perfect single-degree-of-freedom gyro. The gyro is responding to an actual input rate which is caused by the coning motion. It can be shown that the existence of the real rate detected is very closely related to the well-known fact that a sequence of finite angular motions are noncommutative, and indeed, the "coning effect" is often described in the literature as the "noncommutative effect."

If a rigid body (the gyro) undergoes a sequence of finite rotations about body fixed axes, the final orientation is not a unique function of the angles alone but depends on

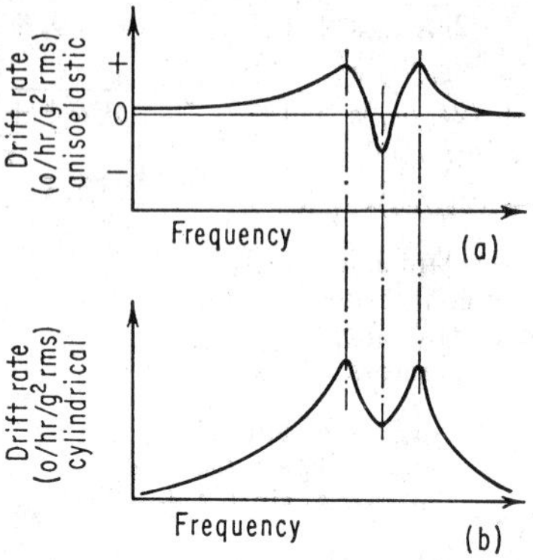

FIG. 35.44. Gyro anisoelastic and cylinder drift vs. frequency.

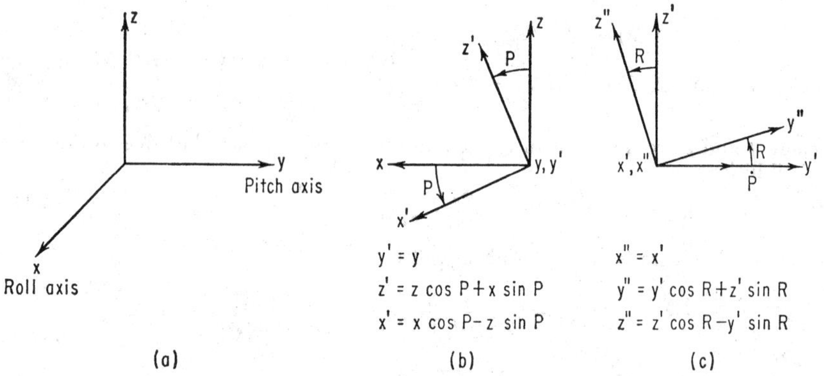

FIG. 35.45. Rotations of a rigid body to illustrate coning. (*a*) Unrotated. (*b*) Rotated in pitch. (*c*) Rotated in roll.

the order in which the rotations occur. Thus, if an ordered sequence of rotations is made about two different body axes approximating conical motion of some third axis, the final orientation will not, in general, repeat the original orientation, even though the net rotation about each axis is zero. The drift due to nutation, mentioned previously, stems from this cause.

Because the single-degree-of-freedom gyro, by its very nature, requires a motion (however small) of the precessional element, the single-degree-of-freedom gyro is more susceptible to coning effects than the two-degree-of-freedom gyro. If a theoretical advantage can be claimed for one type of gyro vs. the other type, it is clearly rooted in the lesser susceptibility of two-degree-of-freedom units to coning errors.

In practice, the effect is noticeable in strap-down gyro applications where the gyro is subjected to full vehicle angular motions. It is also noticeable during vibration testing at shake-table resonant frequencies where the shaker has a rocking motion. Even in a platform application, if phase-coherent errors occur because of any cause such as platform vibration, a cumulative drift can occur, which increases approximately linearly with time.

Consider Fig. 35.45, a rigid body with body axes x, y, z. As a means for gaining an insight into the effect, assume P and R rates about the pitch and roll axes. We then compute the rate coupled into z. Let

$$\begin{aligned} P &= p \sin \omega t \\ R &= r \sin (\omega t + \phi) = r \sin \omega t \cos \phi + r \cos \omega t \sin \phi \end{aligned} \qquad (35.112)$$

then

$$\begin{aligned} \dot{P} &= p\omega \cos \omega t \\ \dot{R} &= r\omega \cos \phi \cos \omega t - r\omega \sin \phi \sin \omega t \end{aligned} \qquad (35.113)$$

$$\omega_{z''} = -\dot{P} \sin R \qquad (35.114)$$

Substituting Eqs. (35.112) and (35.113) in Eq. (35.114) and for small angles,

$$\begin{aligned} \sin R &\approx R \\ \omega_{z''} &\approx -p\omega \cos \omega t (r \cos \phi \sin \omega t + r \sin \phi \cos \omega t) \end{aligned} \qquad (35.115)$$

Integrating Eq. (35.115) over a complete cycle,

$$\omega_{z''\text{av}} \approx -\tfrac{1}{2} pr\omega \sin \phi \qquad (35.116)$$

Thus, if the coning motion is a small wobble (the most practical situation), the resultant drift rate is given by Eq. (35.116).

More detailed analysis shows that the angle through which the body has rotated is equal to the area intercepted by the coning axis (z'' axis in Fig. 35.45) on a sphere of unit radius.

References

1. Canfield, E. B.: Accelerometers and Their Characteristics, *Elec. Mfg.*, November, 1959.
2. "Encyclopaedia Britannica."
3. Wrigley, W., R. B. Woodbury, and J. Hovorka: "Inertial Guidance," MIT, Institute of Aeronautical Sciences, 1957.
4. Schindler, G. M.: On Satellite Vibrations, *Am. Rocket Soc. J.*, May, 1959.
5. Roberson, R. E.: Gravitational Torque on a Satellite Vehicle, *J. Franklin Inst.*, January, 1958.
6. Nidley, R. A.: Gravitational Torque on a Satellite of Arbitrary Shape, *Am. Rocket Soc. J.*, February, 1960.
7. Sporn, S. R.: Use of a Centrifuge for the Precision Measurement of Accelerometer Characteristics, *Trans. ASME, J. Eng. Ind.*, May, 1961.
8. Cohen, E. R., K. M. Crowe, and J. W. M. Dumond: "Fundamental Constants of Physics," Interscience Publishers, Inc., New York, 1957.
9. Savet, Paul H. (ed.): "Gyroscopes," McGraw-Hill Book Company, Inc., New York, 1961.
10. Draper, C. S., W. Wrigley, and L. R. Grohe: "The Floating Integrating Gyro and Its Application to Geometrical Stabilization Problems on Moving Bases," MIT, Institute of Aeronautical Sciences, SMF Fund Paper FF-13.
11. "Gyros, Technical Information for Engineers," Kearfott Company, Inc.
12. Mueller, F. K.: The How and Why of Inertial Guidance, *Missiles Rockets*, Feb. 9, 1959.
13. Fellows, William E.: "The Performance of Gyroscopes in a Vibration Environment," ASME winter annual meeting, New York, 1960.
14. Stewart, R. M.: Some Effects of Vibration and Rotation on the Drift of Gyroscopic Instruments, *Am. Rocket Soc. J.*, January, 1959.
15. Culver, W. H.: Nuclear Gyros, *Am. Rocket Soc. J.*, June, 1962.
16. Goodman, L. E., and A. R. Robinson: Effects of Finite Rotation and Gyro Sensing Devices, *J. Appl. Mech.*, June, 1958.
17. Truxal, J. C.: "Control Engineers' Handbook," McGraw-Hill Book Company, Inc., New York, 1958.

Section 36

PNEUMATIC COMPONENTS

By

CHARLES B. SCHUDER, M.S.M.E., *Director Systems Analysis Group, Research Dept., Fisher Governor Company, Marshalltown, Iowa*

CONTENTS

36.1. INTRODUCTION

Pneumatic components can be built up into a wide variety of systems offering a high degree of dependability and safety along with relatively low cost. Their desirability is evidenced by the continued popularity of pneumatic systems in the process-control field where the feature of economical energy storage is of major importance.

Systems composed of nozzle-flapper amplifiers, pneumatic relays, and volume-restriction networks can exhibit high gain, high power, and a variety of desirable dynamic characteristics.

Most of the advantages of pneumatic systems arise from the inherent compressibility of air, a factor which unfortunately contributes to the time lag of the components. In spite of this drawback, systems can be designed with excellent dynamic characteristics, often superior to hydraulic and electrical systems in the same application class. Intelligent design requires a knowledge of feedback-control-system theory and also a good understanding of the static and dynamic characteristics of the components.

36.2. BASIC RELATIONSHIPS

The flow of compressible fluids through orifices, relay valves, and other similar restrictions is described by a reversible adiabatic process from the inlet up to the vena contracta. From this point of minimum area to a downstream point where pressure recovery has occurred the process is irreversible.

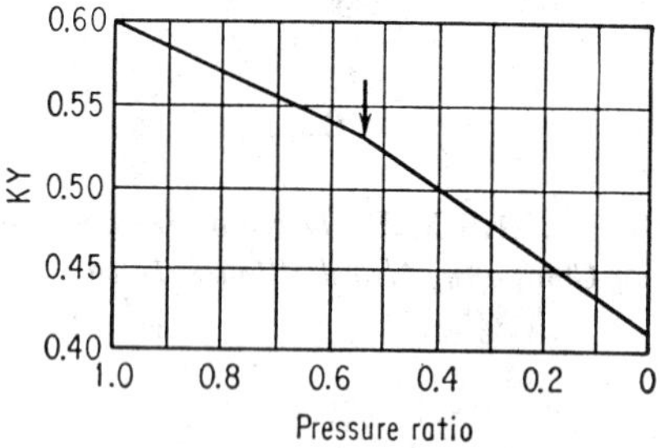

FIG. 36.1. Effect of pressure ratio on KY for a sharp-edged orifice with air.[6]

Of the various formulas proposed for flow-rate calculations based on this overall loss, the form given by Perry[6] is probably the most useful one for design purposes.

$$W = KYA[2gw(P_1 - P_2)]^{1/2} \tag{36.1}$$

where P_1 = upstream pressure, lb/ft² abs
P_2 = downstream pressure, lb/ft² abs
W = weight rate of flow, lb/sec
K = discharge coefficient, including the velocity-of-approach factor
Y = expansion factor
w = upstream specific weight, lb/ft³
g = acceleration of gravity, 32.2 ft/sec²
A = area of restriction, ft²

For air at 70°F the relationship reduces to

$$W = 0.048KYA[P_1(P_1 - P_2)]^{1/2} \tag{36.2}$$

The equations are usable in both the critical and subcritical regions, with P_2 being used as the actual downstream pressure, not limited to $0.53P_1$. In pneumatic components the ratio of restriction area to pipe area is small so that the product KY is a function only of P_2/P_1 and the ratio of specific heats. Figure 36.1 gives values of KY for air flow through a sharp-edged orifice. The slope in the subcritical region is inversely proportional to k, the ratio of specific heats, so that the curve can be used for fluids other than air.[4] The work of Cunningham[5] indicates that k has practically no effect on the slope in the critical region.

As the configuration of the restriction departs from the "sharp-edged" concept the KY curve starts at higher values because of an increase in K and the slope in both flow regions becomes steeper because of less opportunity for radial expansion at the vena contracta. Since the designs of restrictions vary considerably it is usually necessary to obtain the KY curve experimentally, although some data are available in the literature.[7,20,21]

Data collected by Turnquist[26] for wide-open globe-type valves indicate that the expansion factor could be calculated for all pressure ratios by the relationship

$$Y = \frac{1.2 + (P_2/P_1)}{2.2}$$

The flow of compressible fluids in tubes is characterized by a region of laminar flow and a region of turbulent flow with a transition between the two occurring at Reynolds numbers between 2,000 and 4,000. In the turbulent range at high Reynolds numbers and low pressure drops the flow varies nearly with the square root of the pressure drop but reaches limiting values as the pressure drop is increased (ref. 1, chap. 17). In most pneumatic devices the steady-state flow through tubing is zero so that an analysis can be made on the basis of small pressure drops. This permits the assumption of laminar flow, which gives a linear relationship between flow and pressure drop.

For laminar flow of air through a capillary tube (ref. 1, Eq. 8-10) in the range $N_r < 2{,}000$, $(P_1 - P_2)/P_1 \ll 1$,

$$W = \pi w D^4(P_1 - P_2)/128\mu L \tag{36.3}$$

where W = weight rate of flow, lb/sec
D = inside diameter of tube, ft
μ = viscosity, (lb-sec)/ft², 0.37×10^{-6} for air at 60°F
L = tube length, ft
P_1 = upstream pressure, lb/ft² abs
P_2 = downstream pressure, lb/ft² abs
w = specific weight, lb/ft³ (mean)
N_r = Reynolds number, $\rho UD/\mu$
ρ = density, slugs/ft³ = (w/g)
U = velocity, ft/sec

Laminar flow of air through a needle valve is derived from ref. 1, Eq. (6-3), and a force balance; see also ref. 3. With $b \ll D_m$ and ϕ small, $N_r < 2{,}000$, $(P_1 - P_2)/P_1 \ll 1$,

$$W = \pi w D_m b^3(P_1 - P_2)/12\mu L \tag{36.4}$$

where W, μ, P_1, P_2, and w are as in Eq. (36.3) and L, b, and D_m are as shown in Fig. 36.2 measured in feet. The length to use in Reynolds number is $2b$. Typical dimensions for valves used in pneumatic circuits in process-control equipment are $L = \frac{1}{8}$ in., $D_m = \frac{1}{16}$ in., $\phi = 1°$, and $b = 0.001$ in., which gives laminar flow over the usual pressure drops encountered.

The rate of pressure change within a volume is derived from conservation of mass, PV^k = constant, $C^2 = KgRT$ [Eq. (17-21), ref. 1], and the perfect gas law. It is given by

$$dP/dt = WC^2/gV \tag{36.5}$$

for a reversible-adiabatic process. For PV = constant we obtain $dP/dt = WC^2/gVk$ for an isothermal process

where P = pressure, lb/ft²
W = net flow into the volume, lb/sec
t = time, sec
C = velocity of sound (1,120 ft/sec for air at 60°F)
k = ratio of specific heats c_p/c_v (1.4 for air)
g = acceleration of gravity, 32.2 ft/sec²
V = volume, ft³

For most applications the isothermal relationship is realistic only at very low frequencies, probably less than 0.01 cps. For pneumatic devices we are usually concerned with frequencies above 0.1 cps, so that the reversible-adiabatic relationship is applicable and will be used throughout this section except as noted. The relationship (36.5) is not affected by the irreversible expansion which occurs at the entrance to the volume considered.

The transfer function relating pressure to flow from Eq. (36.5) is

$$\mathcal{L}P/\mathcal{L}W = C^2/gVS = Z_v \tag{36.6}$$

In cases where the volume V is free to change with pressure as in Fig. 36.3 a correction must be made for the apparent increase in volume as seen by the flow source. This can be derived from conservation of mass [Eq. (36.5)], PV^k = constant,

$$C^2 = KgRT$$

(Eq. 17-21, ref. 1), and the perfect gas law. In this case the equivalent volume V to use in Eq. (36.6) is

$$V = V_m + 1.4PA^2/K_s \tag{36.7}$$

where V_m = mean volume, ft³
P = mean pressure, lb/ft² abs
A = area, ft²
K_s = spring rate, lb/ft

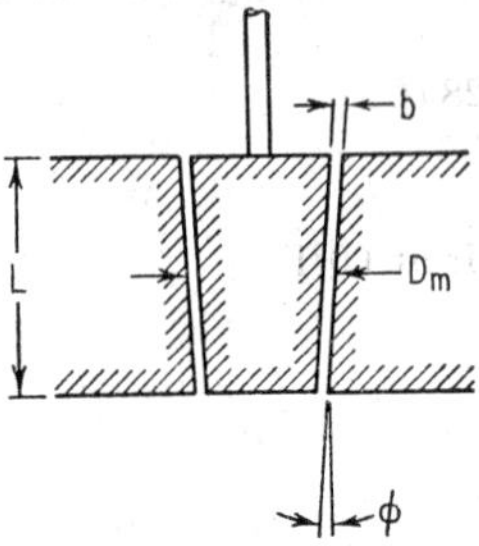

FIG. 36.2. Needle-valve nomenclature.

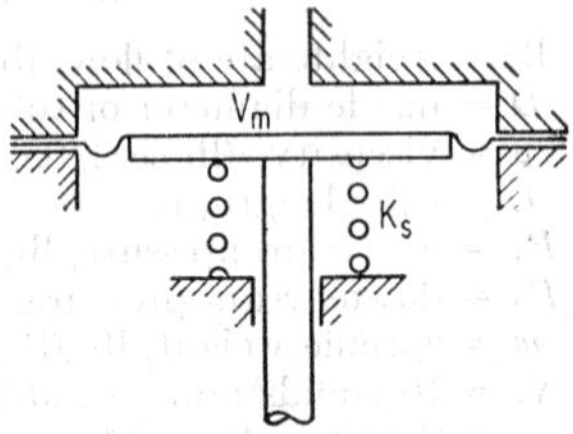

FIG. 36.3. Spring-loaded volume.

This correction for spring loading should not be overlooked in design since the volume given by Eq. (36.7) is often many times larger than that given by V_m alone.

If the output pressure acts on an inertia load the "equivalent-volume" concept is not applicable. Here the mean volume should be used with the load velocity shown fed back through the load ram area to a flow-summing point. An example of this approach is given in Art. 36.9.

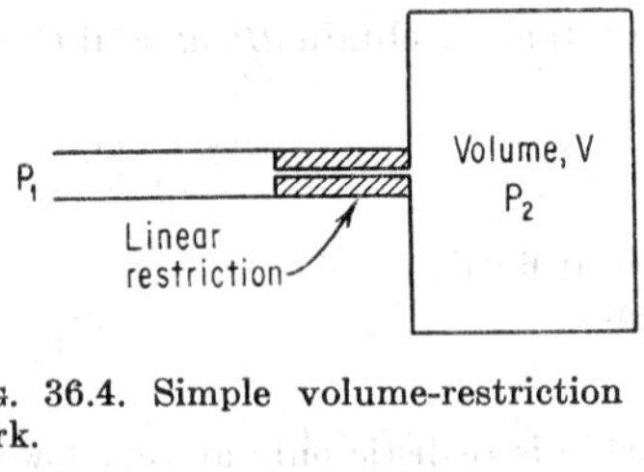

FIG. 36.4. Simple volume-restriction network.

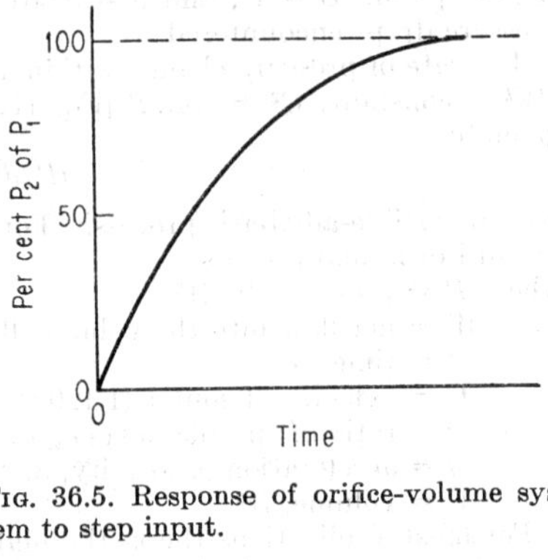

FIG. 36.5. Response of orifice-volume system to step input.

36.3. SIMPLE NETWORKS

The majority of pneumatic systems are made up of simple networks involving a volume and either one or two restrictions. The static and dynamic characteristics of these networks depend upon the linearity of the restriction and whether or not a steady flow is established through the device.

(a) Case 1. Single Linear Restriction, No Steady Flow

The combination shown in Fig. 36.4 is commonly used as a feedback element for integral or derivative action with a needle valve or a capillary tube for the restriction.

The transfer function derived from conservation of mass [Eq. (36.5)] and

$$dw/dt = g/C^2\, dP/dt$$

(ref. 12) is

$$\mathcal{L}P_2/\mathcal{L}P_1 = 1/(\tau S + 1) \tag{36.8}$$

where τ is $128\mu LgV/w\pi D^4C^2$ for the capillary tube and $12\mu LgV/w\pi C^2D_mb^3$ for a needle valve. Nomenclature is from Eqs. (36.3) through (36.5).

(b) Case 2. Single Nonlinear Restriction, No Steady Flow

If a restriction such as a sharp-edged orifice is used in the network of Fig. 36.4 the system will be nonlinear and cannot be linearized even with the assumption of small deviations. This is the only case that we shall consider where it is not possible to write a transfer function. We can, of course, obtain time-domain solutions for a specified disturbance. If, for example, a small step change is made in P_1, we have from Eqs. (36.1) and (36.5) and also conservation of mass

$$t = [P_1^{1/2} - (P_1 - P_2)^{1/2}]V/wKAC^2$$

with P_1 being the magnitude of the step and P_2 the deviation of the pressure in the volume from its initial value. This solution is plotted in Fig. 36.5. Note that P_2 becomes equal to P_1 in a finite length of time rather than approaching it asymptotically. This is the result of an "apparent time constant" approaching zero as the pressure difference approaches zero. This feature makes nonlinear restrictions undesirable in networks with no established steady flow.

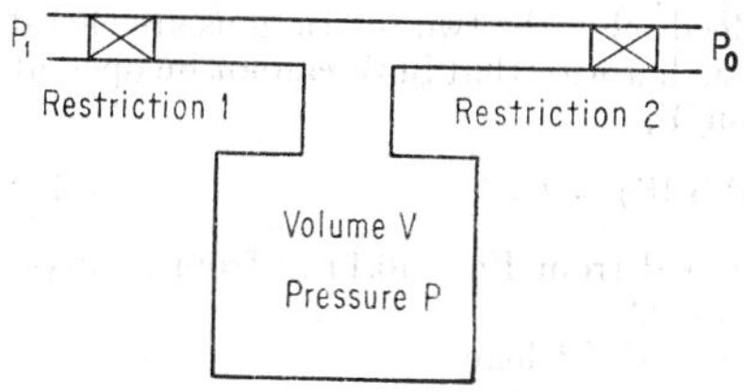

FIG. 36.6. Double-restriction network.

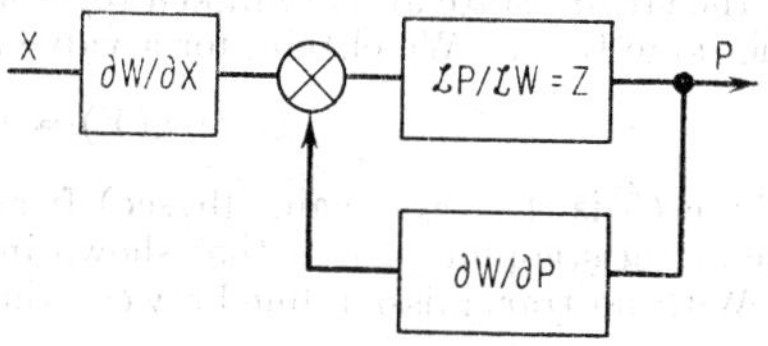

FIG. 36.7. Block diagram of double-restriction network.

(c) Case 3. Two Linear or Nonlinear Restrictions, Established Steady Flow

Figure 36.6 shows the general case of a single volume with two restrictions. The restrictions may be fixed or variable but in all cases a steady flow exists through the device. The output of the system is P and the input or disturbance may be P_1, P_0, or a change in either restriction. The transfer function for any input X is obtained from the block diagram in Fig. 36.7 [since system flow $\bar{W} = f(P,X)$] as

$$\mathcal{L}P/\mathcal{L}X = (\partial W/\partial X)(\mathcal{L}P/\mathcal{L}W)/[1 + (\mathcal{L}P/\mathcal{L}W)(\partial W/\partial P)]$$

$\partial W/\partial X$ is the change in flow produced by a change in input with P held constant. Likewise $\partial W/\partial P$ is the flow change produced by a change in P with the input or disturbance held constant.

Letting $G = (\partial W/\partial X)/(\partial W/\partial P)$ we obtain the general transfer function

$$\mathcal{L}P/\mathcal{L}X = \frac{G}{[1/Z(\partial W/\partial P)] + 1} \tag{36.9}$$

G is the steady-state change in P produced by the change in the input X. It is best obtained from the slope dP/dX of the calculated characteristic curve. This curve is normally calculated from static considerations and is therefore an "isothermal" gain.

The "reversible adiabatic" gain is slightly higher with a maximum difference of 16 per cent occurring with critical drop across both restrictions. This "reversible adiabatic" gain is given by $(\partial W/\partial X)/(\partial W/\partial P)$ with these quantities evaluated as shown below.

$(\partial W/\partial P)$ is a function of the operating point of the system and can be obtained by differentiation of the basic flow equations. Formulas for $(\partial W/\partial P)$ for various flow regimes have been calculated from modifications of the basic equations and are given in Table 36.1. All symbols in the table refer to the steady-state operating conditions about which the deviations occur. W is in lb/sec and pressures are in lb/ft² absolute. Table 36.1 is derived by differentiating Eq. (36.2) for critical drop. For subcritical drop see ref. 12.

Z is $\mathcal{L}P/\mathcal{L}W$. For the simple volume termination $Z = Z_v = C^2/gVS$ giving for a transfer function

$$\mathcal{L}P/\mathcal{L}X = G/(\tau S + 1) \tag{36.10}$$

with $\tau = gV/C^2(\partial W/\partial P)$.

The calculation of Z for a transmission line between the output volume and the load volume is given in Art. 36.10.

If the output pressure is used to drive an inertia load provisions must be made to show negative feedback from the load flow. The summing point shown in Fig. 36.7 may be used although it is sometimes more convenient to evaluate the overall gain $\partial P/\partial X$ and introduce the load flow through $1/(\partial W/\partial X)$ to a displacement summing point. The $\partial W/\partial X$ can be evaluated as W/X from the steady-state air-consumption curve.

(d) Case 4. Two Linear or Nonlinear Restrictions, No Steady Flow

In the slide valve or forward path of a nonbleed relay the two restrictions are closed in the steady state and are linked together in such a way that both cannot be open at the same time. We obtain, for a valve motion Y,

$$\mathcal{L}(P)/\mathcal{L}(Y) = C_1(\mathcal{L}P/\mathcal{L}W) = C_1 Z \tag{36.11}$$

where C_1 is the valve gain (lb/sec)/ft as obtained from Eq. (36.1) or from a valve-characteristic plot such as that shown in Fig. 36.22.

With no transmission line between the valve and the load

$$\mathcal{L}P/\mathcal{L}Y = C_1 C^2/gVS \tag{36.12}$$

If the valve is driving an inertia load or appreciable leakage occurs in the neutral position, the above transfer functions should be broken down so that C_1 appears as a separate block preceding a flow summing point. This point can then be used to pick up negative feedback from the load flow or from $\partial W/\partial P$ due to leakage flow.

36.4. NOZZLE-FLAPPER SYSTEMS

Figure 36.8 shows a common system used to convert mechanical motion into a pneumatic pressure. The gain of the device is high, often above 10,000 psi/in., and the load on the input element is insignificant, usually less than 0.01 oz/psi output. Figure 36.9 shows the steady-state characteristics and Fig. 36.10 the time constants of a typical system with $P_s = 35$ psia, $D_1 = 0.010$ in., $D_2 = 0.020$ in., and $V = 1$ in.³ (no transmission line). The curve of pressure vs. flapper displacement is calculated from

$$X = \frac{D_1^2 Y_1 K_1}{4 D_2 Y_2 K_2}\left[\frac{P_s(P_s - P)}{P(P - P_0)}\right]^{1/2} \tag{36.13}$$

The nomenclature is from Fig. 36.8 with Y as the expansion factor and K as the flow coefficient. The dimensions are in inches with pressures in pounds per square inch absolute. Eq. (36.1) gives the flow across the first restriction and letting

$$Q = 47{,}000W$$

(volumetric flow, standard cubic feet per hour). The general transfer function is given by Eq. (36.9) and, for the volume load shown, by Eq. (36.10).

Since both G and τ vary with flapper position it is advisable to follow the nozzle flapper with a medium- to high-gain relay so that the operation of the nozzle flapper will be restricted to a short "straight-line" portion of the characteristic curve.

In process-control equipment the primary orifice diameter is usually between 0.010 and 0.020 in. and the nozzle diameter is 0.020 to 0.060 in. Smaller primary orifice dimensions present problems in plugging and produce excessive time constants, as

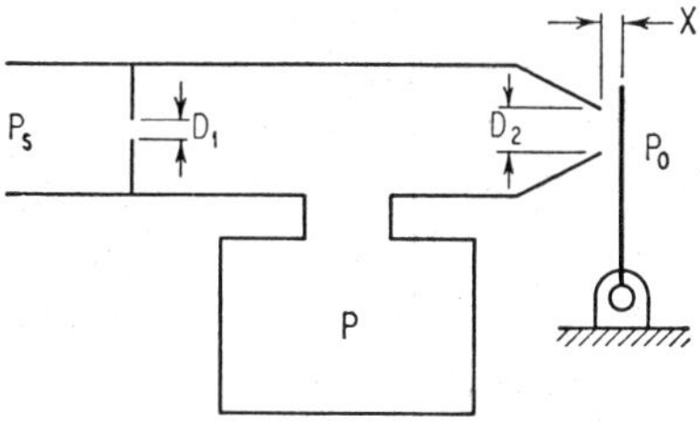

FIG. 36.8. Nozzle-flapper system.

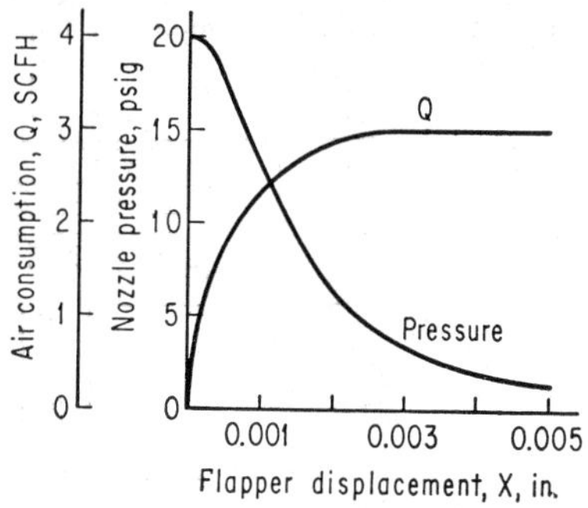

FIG. 36.9. Typical steady-state characteristics of the nozzle flapper.

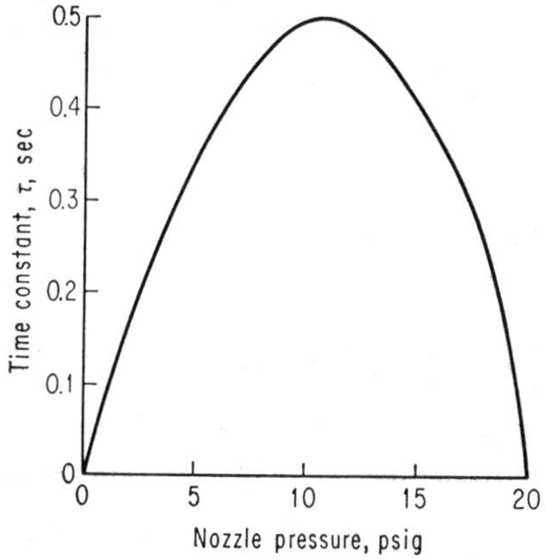

FIG. 36.10. Effect of operating point on nozzle-flapper time constant.

indicated by Eq. (36.10). Small nozzle diameters result in low gain in the low-pressure range and limit the minimum pressure that can be obtained. Equation (36.13) is based on the assumption that the minimum pressure is governed by the nozzle-flapper clearance area, not by the nozzle area. A high ratio D_2/D_1 can produce high gain but nozzle-flapper alignment becomes critical. Large primary orifice diameters result in low gain unless a corresponding increase is made in nozzle diameter. High-frequency flapper oscillation limits the maximum nozzle size that can be used. These oscillations can be minimized by reducing nozzle pressure or diameter, increasing flapper spring rate, or decreasing the flapper mass. A study of this problem in connection with hydraulic systems has been made by Feng.[7]

36.5. EJECTOR SYSTEMS

Occasionally it becomes desirable to omit a power relay and work a nozzle-flapper system over most of its range. In this situation an ejector system can be used to minimize the nonlinearities of the basic system. Figure 36.11 shows a typical design. The geometry roughly follows recommendations for fluid-handling ejectors[8,9] except that no attempt is made to recover pressure in the discharge section. These recommendations show discharge throat diameters two to three times the supply nozzle diameter with the nozzle positioned about one nozzle diameter past the start of the converging section. For proper operation the output connection must be in the general area shown but the nozzle flapper can be some distance away if necessary.

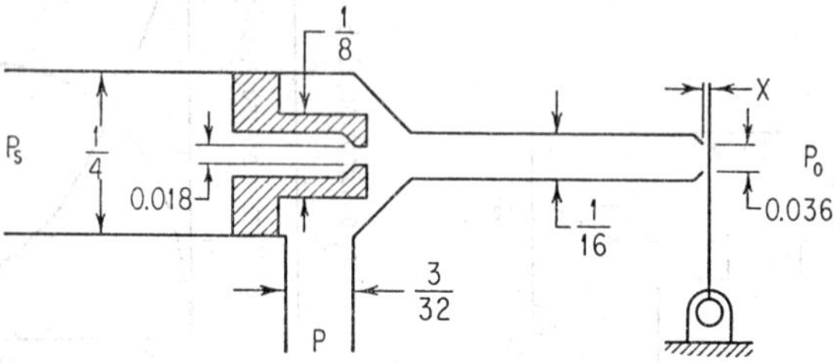

FIG. 36.11. Typical ejector design.

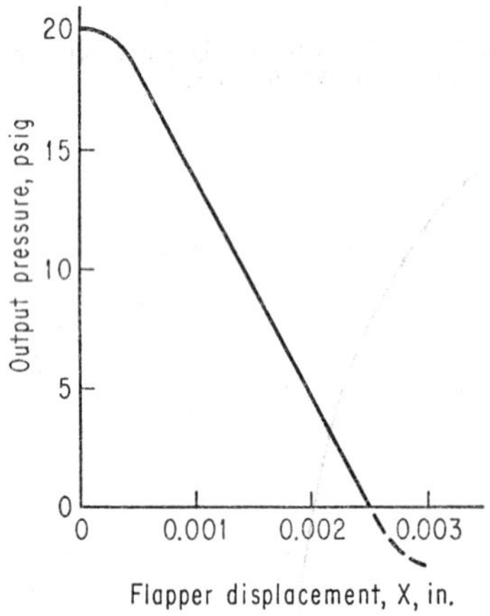

FIG. 36.12. Ejector characteristic.

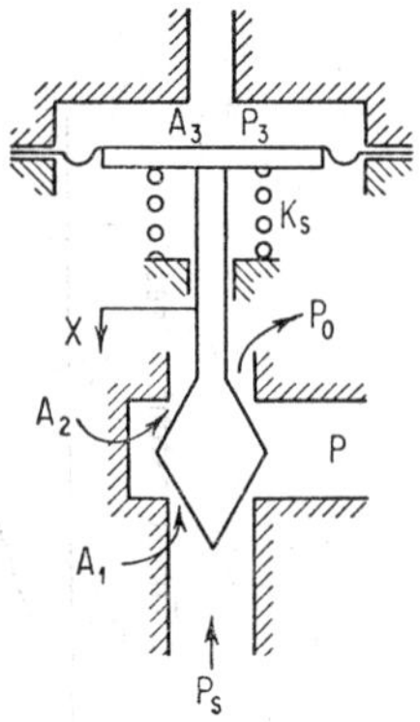

FIG. 36.13. Bleed relay.

Figure 36.12 illustrates the excellent linearity that can be obtained by properly positioning the supply restriction in the converging section. Varying the location of this restriction can produce characteristics between those of the basic nozzle-flapper system and the ejector. This becomes necessary when negative load pressures are to be avoided. Analytical relationships for the static and dynamic characteristics of these systems are not available, but laboratory tests indicate time constants on the order of two or three times those of a basic nozzle-flapper system with the same size restrictions.

36.6. BLEED RELAYS

Figure 36.13 shows a typical design for a bleed relay. The input can be a pressure on the diaphragm as shown or a direct mechanical input. A port size of 0.060 in. is typical when the device is used as a first-stage motion-pressure transducer. For a power relay 1/8- to 3/16-in. ports are common, with the valve travel being about one-half the port diameter. The characteristic curve is obtained by first relating the port-area ratio A_2/A_1 to the motion input X or pressure input ($X = A_3P_3/K_s$). The area ratio is then related to the output pressure P by

$$\frac{A_2}{A_1} = \frac{Y_1K_1}{Y_2K_2}\left[\frac{P_s(P_s - P)}{P(P - P_0)}\right]^{1/2} \tag{36.14}$$

The nomenclature is from Fig. 36.13 with Y as the expansion factor and K as the flow coefficient. Areas are in square inches with pressures in pounds per square inch absolute. The steady-state air consumption is obtained by using Eq. (36.1) across either restriction, with the flow in standard cubic feet per hour given by $47{,}000W$. Figures 36.14 and 36.15 show the characteristics typical of the $\frac{1}{16}$-in. port designs with a supply pressure of 35 psia and a volume load of 1 in.^{3}. The seat-to-seat distance L is sometimes made adjustable so that the gain can be varied. The magnitude of gain variation obtained by this method is limited by the effect of L on linearity, air consumption, and the time constant.

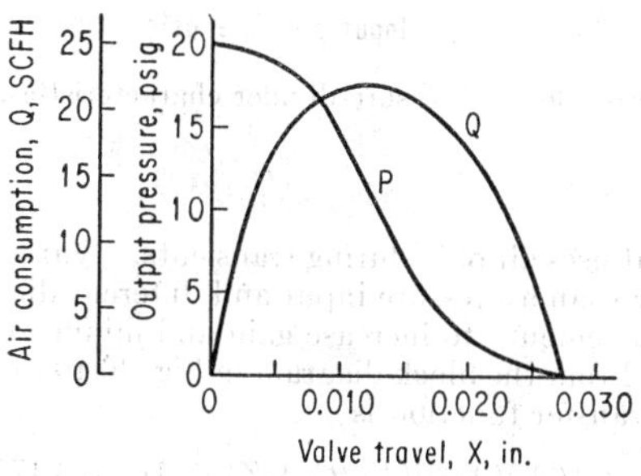

FIG. 36.14. Typical bleed-relay characteristics.

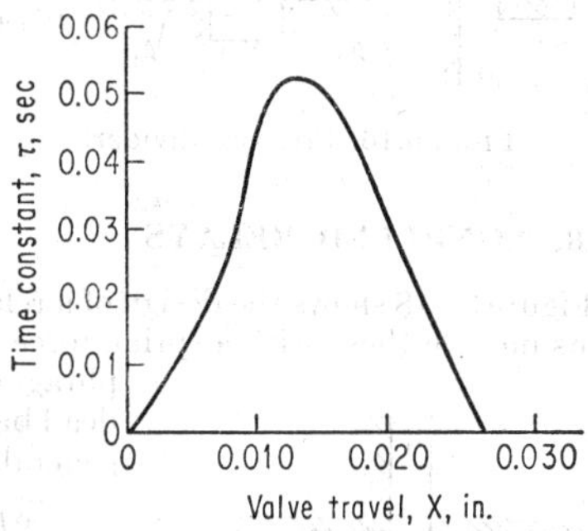

FIG. 36.15. Effect of valve position on bleed-relay time constant.

Where the bleed relay is used without a transmission line its transfer function is given from Eq. (36.10) by

$$\mathcal{L}P/\mathcal{L}X = G/(\tau S + 1) \tag{36.15}$$

with $G = dP/dX$ from the calculated characteristic curve and $\tau = gV/C^2(\partial W/\partial P)$. For a pressure input the time constant is the same and G is dP/dP_3. Where the output goes to a transmission line the general relationship (36.9) must be used, with Z obtained from the characteristics of the line and the load.

The bleed relay has no dead band and therefore can be designed with a small diaphragm offering a high input impedance (low equivalent volume). The input impedance is also affected by the spring rate [see Eq. (36.7)] so that high gains obtained by low spring rates increase the load on the preceding stage. Since there is no internal feedback the time constant, for a fixed pressure range and load, is determined by the port size—increasing the port size decreases the time constant but increases the steady-state air consumption.

36.7. PRESSURE DIVIDER

The three-way valve may also be used as a pressure divider as shown in Fig. 36.16. The usual application is in the feedback path of a controller to vary the closed-loop gain. As in the case of the bleed relay the dynamic response is characterized by a first-order lag with the time constant equal to $gV/C^2(\partial W/\partial P)$. In the subcritical region the gain dP/dP_1 is obtained by plotting the characteristic curve from Eq. (36.14),

$$P = \tfrac{1}{2}\{(-P_1/C_1 - P_0) + [(P_1/C_1 - P_0)^2 + 4P_1^2/C_1]^{1/2}\} \tag{36.16}$$

where $C_1 = (A_2Y_2K_2/A_1Y_1K_1)^2$.

Over the usual pressure ranges used in process instrumentation the ratio Y_2K_2/Y_1K_1 remains very close to unity, and in spite of the apparent nonlinearity of relationship (36.16) the gain is essentially constant. Figure 36.17 is a plot of Eq. (36.16) for an area ratio of 1.11 showing a nearly constant gain of 0.54.

The way in which the time constant and air consumption vary with valve position is the same as for the bleed relay, with τ and Q becoming smaller as P_1 is increased.

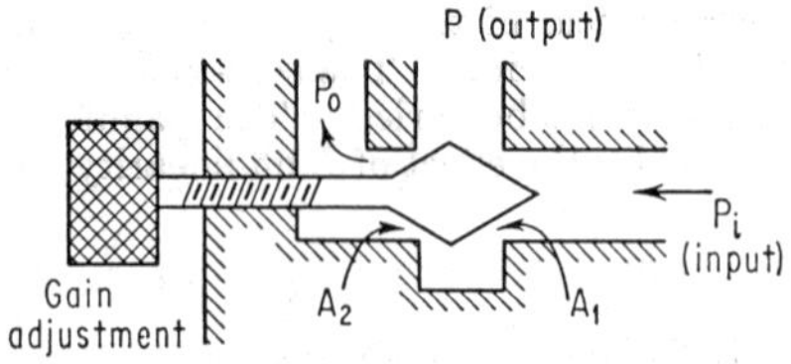

FIG. 36.16. Pressure divider.

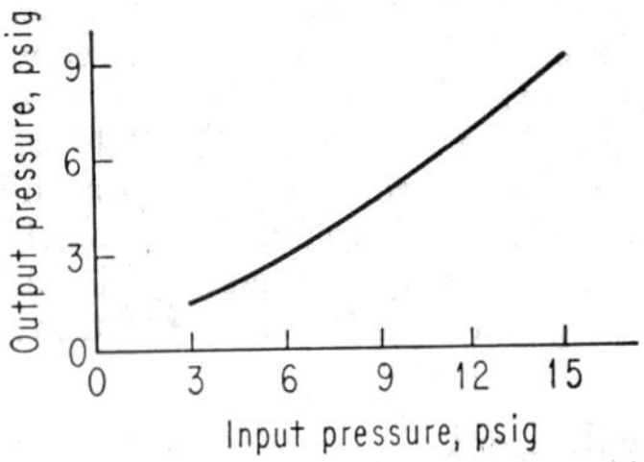

FIG. 36.17. Pressure-divider characteristics.

36.8. NONBLEED RELAYS

Figure 36.18 shows the design of a relay that uses air only during transients. Variations include those with a spring force rather than a pressure input and different diaphragm arrangements to increase gain and minimize dead band. From the block diagram in Fig. 36.19 the generalized transfer function is

$$\mathcal{L}P_2/\mathcal{L}P_1 = (A_1/A_2)/[(K_s/C_1A_2Z) + 1] \qquad (36.17)$$

where K_s = main spring rate, lb/ft
A_1 = input area, ft²
A_2 = feedback area, ft²
Z = $\mathcal{L}P_2/\mathcal{L}W$, sec/ft²
C_1 = valve gain, (lb/sec)/(ft lift)

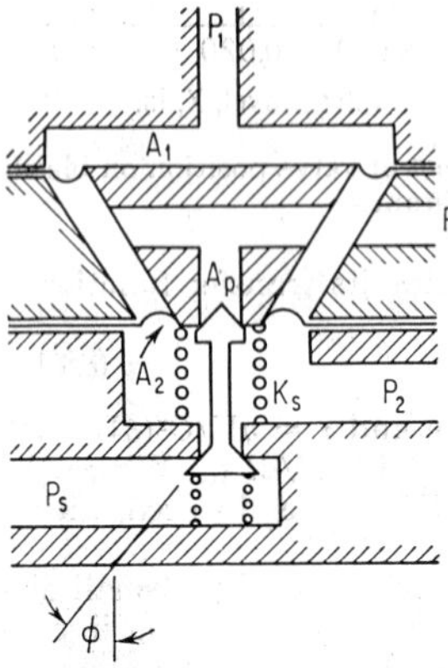

FIG. 36.18. Nonbleed relay.

The value of C_1 can be obtained from Eq. (36.1) or (36.2) with $A = \pi D \sin \phi$. For a pure volume load, substituting Eq. (36.6) in Eq. (36.17), the transfer function becomes

$$\mathcal{L}P_2/\mathcal{L}P_1 = (A_1/A_2)/[(K_sgVS/C^2C_1A_2) + 1] \qquad (36.18)$$

C_1 will vary with P_2 and will, in general, be different for exhaust and supply. This results in two 22½° asymptotes on a "Bode" plot rather than the usual single 45° break.

The gain is seen to be dependent only on the area ratio A_1/A_2 with the time constant proportional to $VK_s/A_2\,D \sin \phi$. The dead band, as a fractional part of the input signal from a force balance, is

$$\text{Dead band} = [A_p(P_s - P_0) + F_s]/A_1(\Delta P_1) \qquad (36.19)$$

where ΔP_1 is the input signal range and F_s is the force exerted by the inner valve spring with the system in a neutral position. The effect of dead band is minimized by keeping the ratio A_1/A_p as large as practical, and the steady-state error is eliminated by unavoidable (and sometimes intentional) leakage across the valve ports.

Flow-closed valves have a tendency to sustain high-frequency oscillations unless damping is provided. The normal input system provides some damping (see Art. 36.7), but unless other damping is provided the port size and pressure drop are limited.

In the low-pressure range, ⅛-in. ports are common; ⅜-in. ports with 100 psi drop have been used in conjunction with air or dry-friction dampers.

Because of the internal feedback in these relays they are characterized by relatively low gain (usually less than 5) and small time constants. Variations in the design to produce high gain (Fig. 36.20) give a corresponding increase in the time constant as indicated by the transfer function.

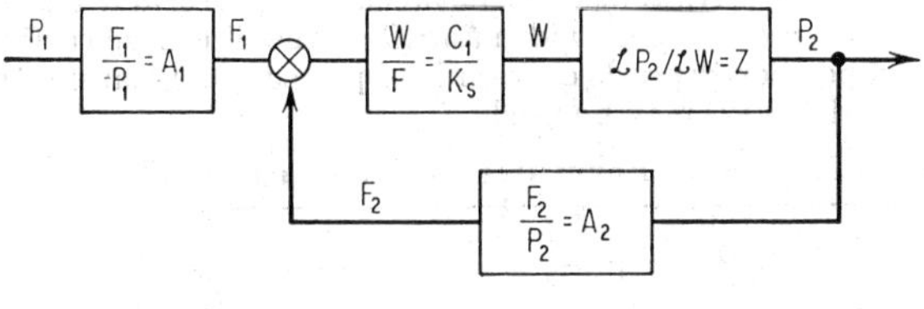

FIG. 36.19. Block diagram of nonbleed relay.

Another variation includes an exhaust port bypass of such size that the exhaust port never opens on normal signal amplitudes. The transfer function for small-amplitude signals is given by Eq. (36.9), with feedback added, and for large signals is approximated by the relationships developed for the nonbleed relay. Some designs omit the exhaust end of the inner valve in favor of a small fixed restriction. It should be noted that the input to a nonbleed relay appears as a spring-loaded diaphragm only when the relay time constant is much larger than the time constant of the device driving the relay. Where this is not the case the flow utilized in moving the relay must be fed back to the flow-summing point in the preceding stage.

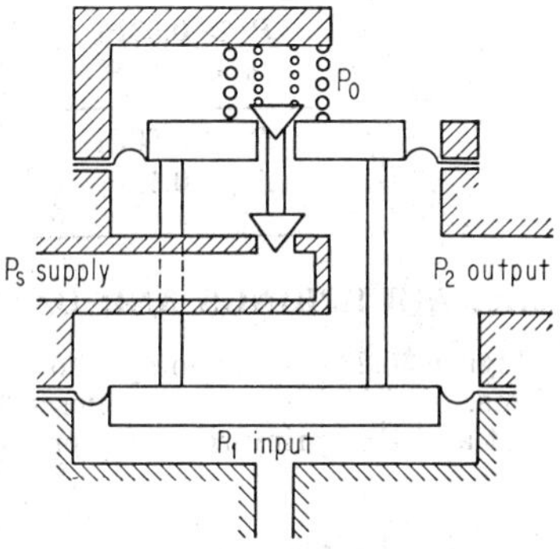

FIG. 36.20. High-gain nonbleed relay.

36.9. SLIDE VALVES

Slide valves of the spool or plate type as used in hydraulic systems have found limited application in pneumatic devices because of the high power loss due to leakage and the problem of providing adequate lubrication.

An analysis of a pneumatic system utilizing a sliding-plate valve is given by Shearer.[19] Figure 36.21 shows the block diagram used for the linearized analysis. K_1 is the valve gain, in.³/sec-in. Using D as the Laplace transform variable the blocks $(1/K_3)(1/D)$ are the Laplace transform of pressure with respect to flow, and k_2 is the partial derivative of flow with respect to load pressure. The load pressure

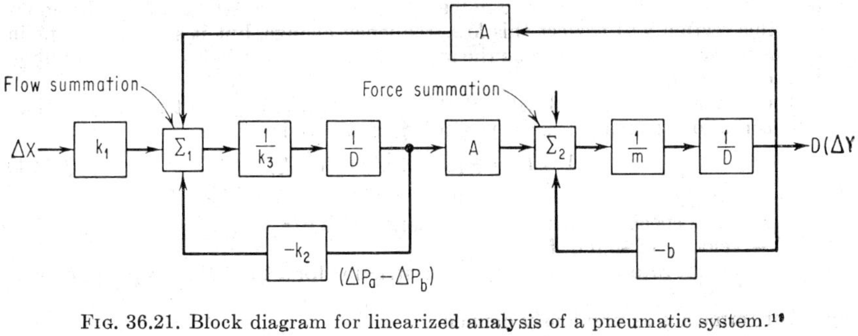

FIG. 36.21. Block diagram for linearized analysis of a pneumatic system.[19]

acts through the ram area A on an inertia load of mass m and damping coefficient b. The load velocity $D(\Delta Y)$ is considered to be fed back through the ram area to the flow-summing point to account for the flow utilized in moving the load.

In making the analysis the only information needed on the slide valve is the value of k_1 and k_2. These may be obtained from Eq. (36.1) and Table 36.1 for a given operating point or plotted for a variety of operating conditions as shown in Fig. 36.22. k_1 is proportional to the vertical distance between the curves, and $1/k_3$ is given by the slope of the curves. w is the port width and C_d is the discharge coefficient.

Where the load position depends only on load pressure the transfer functions given by Eqs. (36.11) and (36.12) may be used.

Table 36.1. Formulas for Evaluating $\partial W/\partial P$

Flow regime	$\partial W/\partial P$
$P < 0.53P_1$ $P_0 < 0.53P$	$\dfrac{0.86W}{P}$
$P > 0.53P_1$ $P_0 > 0.53P$	$\dfrac{W(2P - P_1)}{2(P_1 - P)P} + \dfrac{W[1 + 0.4(P_0/P)]}{2.8(P - P_0)}$
$P > 0.53P_1$ $P_0 < 0.53P$	$\dfrac{W(2P - P_1)}{2(P_1 - P)P} + \dfrac{0.86W}{P}$
$P < 0.53P_1$ $P_0 > 0.53P$	$\dfrac{W[1 + 0.4(P_0/P)]}{2.8(P - P_0)}$

36.10. AIR SPRINGS AND DAMPERS

Figure 36.23 shows a component used as a damping device in various control mechanisms. With the restriction closed the rate of the air spring alone, from PV^k = constant and the spring-rate relationship, is

$$K_a = 1.4A^2P/V \tag{36.20}$$

where K_a = air spring rate, lb/ft
A = diaphragm area, ft²
P = pressure, lb/ft² abs
V = chamber volume, ft³

The rate is nonlinear, increasing with P. It is linearized by using a mean pressure for P and restricting the analysis to a narrow range of pressures around this mean. At very low frequencies the process becomes isothermal and we would use 1.0 in place of the 1.4. Consequently the system shows a higher spring rate at high frequencies than it does at low frequencies. The break point associated with this change could be determined by a heat-transfer analysis but is probably <0.01 cps. The equations in this section will neglect this low-frequency change, but it should be kept in mind that for certain configurations the steady-state characteristics will depend upon the isothermal case.

For the system in Fig. 36.23 without the mechanical spring the transfer function, from conservation of mass, Equations (36.3) and (36.5), is

$$\mathcal{L}Y/\mathcal{L}F = (\tau_1 S + 1)/(wA^2S/C_1) \tag{36.21}$$

where $\tau_1 = Vg/C_1C^2$
w = mean specific weight of air in chamber, lb/ft³
C = velocity of sound, ft/sec
C_1 = weight rate of flow through the restriction for a 1 lb/ft² pressure difference, lb/sec

and other terms are as previously defined.

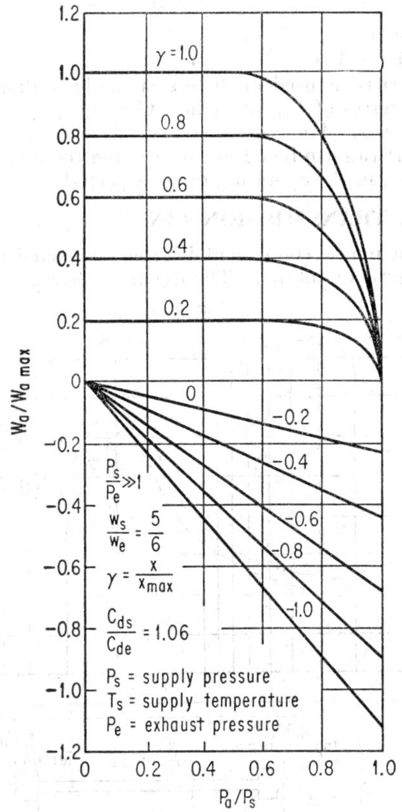

FIG. 36.22. Computed dimensionless pressure-flow curves of ideal sliding-plate valve.[19] $W_{a,\max} = 0.53(P_s/\sqrt{T_s})(0.89w_sX_{\max})$

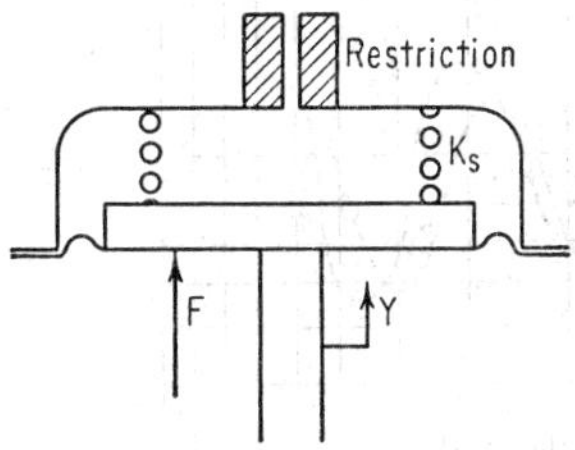

FIG. 36.23. Pneumatic damping device.

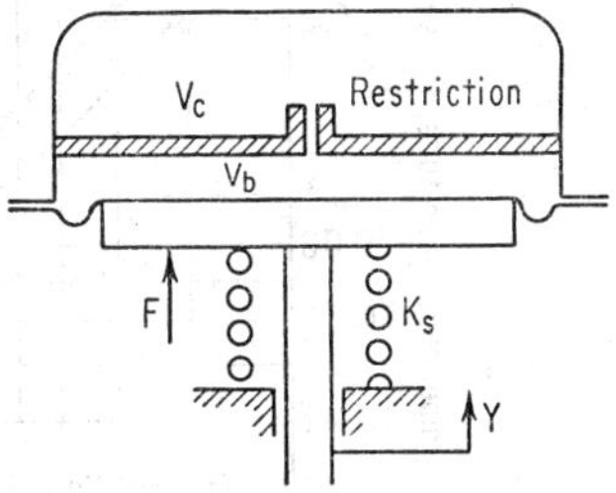

FIG. 36.24. Pressure-loaded damping device.

Adding the mechanical spring gives the transfer function

$$\mathcal{L}Y/\mathcal{L}F = (1/K_s)(\tau_1S + 1)/[(\tau_1 + \tau_2)S + 1] \qquad (36.22)$$

where K_s is the mechanical spring rate in lb/ft and $\tau_2 = wA^2/K_sC_1$.

For the arrangement shown in Fig. 36.24 the transfer function is

$$\mathcal{L}Y/\mathcal{L}F = \frac{[1/(K_s + J)](\tau_2S + 1)}{[(K_s\tau_2 + J\tau_1)/(K_s + J)]S + 1} \qquad (36.23)$$

where $\tau_1 = gV_c/C_1C^2$
$\tau_2 = \tau_1 V_b/(V_b + V_c)$
$J = C^2 wA^2/g(V_b + V_c)$

In practice the restrictions used in these systems have diameters on the order of 0.02 to 0.06 in. and lengths of $\frac{1}{16}$ to $\frac{1}{4}$ in. Consequently they are linear only for small pressure amplitudes. For large pressure differences a crude approximation of the dynamic characteristics can be obtained by using the relationships given with C_1 evaluated as $dW/d(\Delta P)$ in the operating range expected.

36.11. PNEUMATIC TRANSMISSION LINES

The output of a pneumatic component is often connected to the load volume by means of a length of pipe or tubing. The dynamic characteristics of the line affect

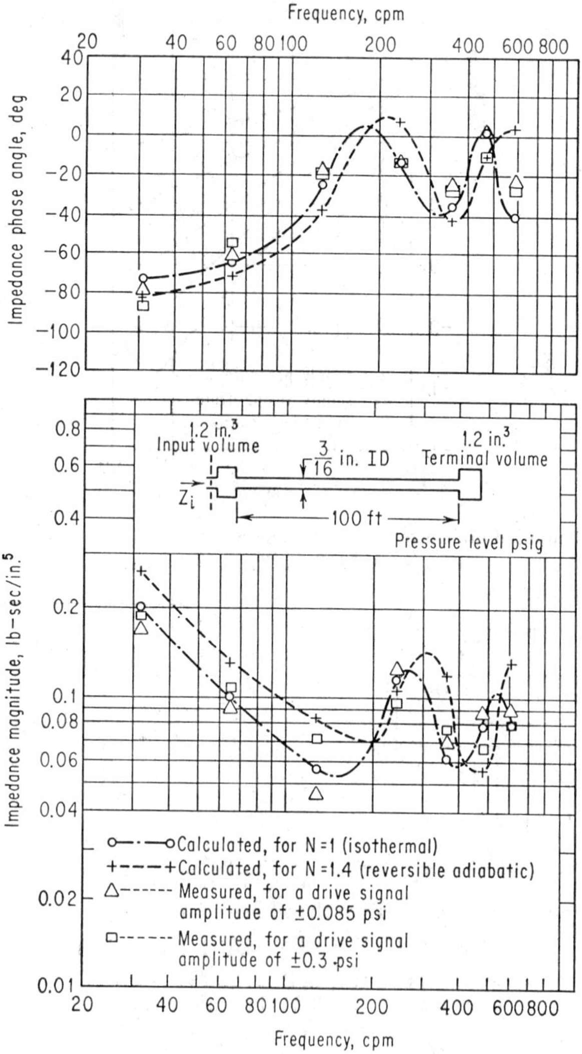

FIG. 36.25. Driving-point impedance of 100 ft of $\frac{3}{16}$-in. inside-diameter copper tubing.

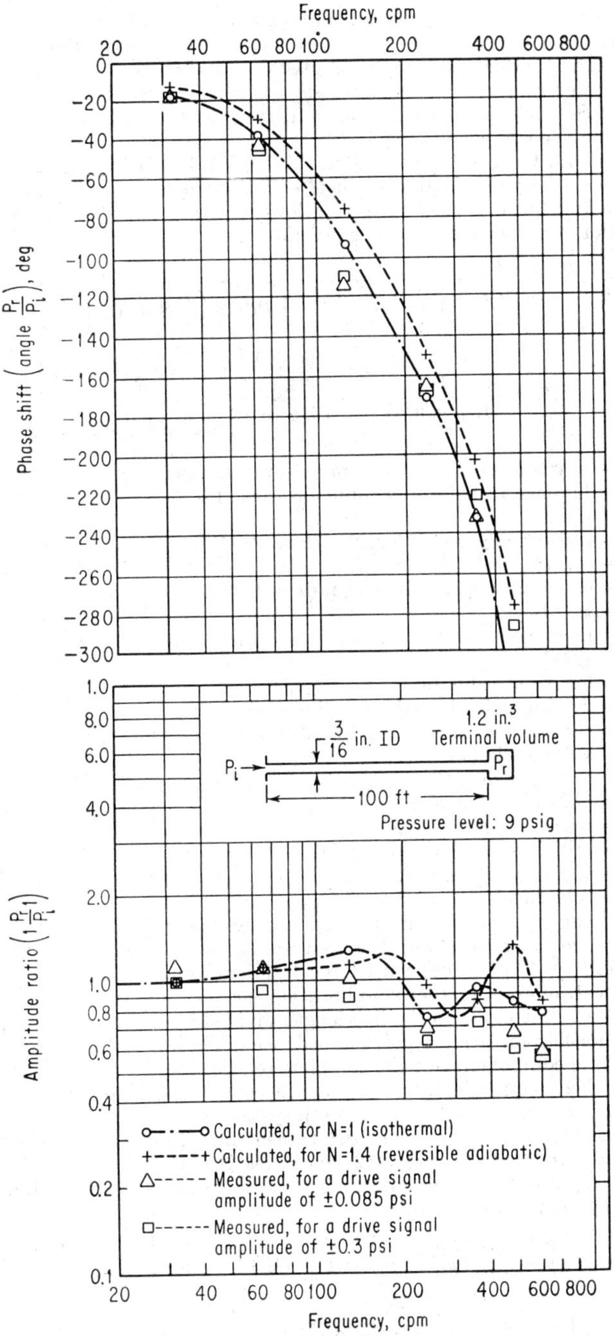

FIG. 36.26. Output/input frequency response of 100 ft of 3/16-in. inside-diameter copper tubing.

the performance of both the component and the overall system. A treatment of transmission-line dynamics is complicated by the fact that the line's resistance, capacitance, and inductance are distributed over its length.

Two transfer functions are of interest: $\mathcal{L}P_2/\mathcal{L}P_1$ for the input/output characteristics and $\mathcal{L}P_1/\mathcal{L}W_1$ (Z_i, driving-point impedance) to describe the loading effects of the line upon the component.

Rohmann and Grogan[10] have treated this subject by analogy with electrical transmission lines. The application of their equations to a typical system is shown in Figs. 36.25 and 36.26. They give, as a low-frequency approximation to their complete solution, the following transfer functions:

$$\frac{P_r}{P_i} = \frac{1}{1 + \left(\frac{2g_0 + 1}{2}\right)\left[RCl^2S\left(\frac{L}{R}S + 1\right)\right]} \tag{36.24}$$

$$Z_i = \frac{1}{(1 + g_0)ClS}\,\frac{1 + \frac{2g_0 + 1}{2}\left[RCl^2S\left(\frac{L}{R}S + 1\right)\right]}{1 + \frac{1}{6}\frac{3g_0 + 1}{g_0 + 1}\left[RCl^2S\left(\frac{L}{R}S + 1\right)\right]} \tag{36.25}$$

where P_r = Laplace transform of terminal pressure, psi
P_i = Laplace transform of input pressure, psi
$g_0 = C_r/Cl$
R = resistance per unit length, (lb-sec/in.^{5})/in.
$= (1.2)8\mu/\pi r^4$
C = capacitance per unit length, (in.^{5}/lb)/in.
$= \pi r^2/np$
L = inertance per unit length, (lb-sec^2/in.^{5})/in.
$= \delta/\pi r^2$
l = length of tube, in.
Z_i = driving-point impedance, (lb-sec)/in.^{5}
C_r = capacitance of receiver, in.^{5}/lb, (V/np)
I_i = volumetric flow rate, in.^{3}/sec
δ = mass density, (lb-sec^2)/in.^{4}
n = polytropic exponent
V = receiver volume, in.^{3}
p = initial system pressure, psia
μ = viscosity, (lb-sec)/in.^{2}
r = tube inside radius, in.

The input/output relationship is adequate for the majority of applications since the dynamics of the process and the final control element will restrict the frequency ranges of interest. For lines with large terminal volumes the relationship in Art. 36.3(a), case 1, offers a further simplification for very low frequencies. In all cases there is a dead time equal to the line length divided by the velocity of sound.

The relationship for driving-point impedance might be used in a preliminary study of a pneumatic component but the complete relationship should be used for accurate work since the frequencies involved in the component loop may be considerably higher than those of interest in the overall system.

The transient response of pneumatic transmission lines has been investigated by Schuder and Binder.[11] Figure 36.27 indicates the ways in which these lines can respond to step inputs. In practice, most long lines are well damped, and their step response may be approximated by

$$\begin{aligned} (p - p_o)/(p_m - p_o) &= 0 && \text{for } 0 < t < L/c \\ (p - p_o)/(p_m - p_o) &= 1 - Ge^{-tM} && \text{for } t \geq L/c \end{aligned} \tag{36.26}$$

where $G = \dfrac{[1 + R/(\rho\theta)]}{\alpha[(Q/aL + 1)\sin\alpha + (Q/aL)\cos\alpha]}$
$M = \frac{1}{2}(\theta - R/\rho)$
α = solution of $\alpha \tan \alpha = aL/Q$, rad (from Fig. 36.28)

Q = receiver volume, ft³
a = cross-sectional area of tube, ft²
L = tube length, ft
t = time, sec
$\theta = [(R/\rho)^2 - (2\alpha c/L)^2]^{1/2}$
ρ = air density, slugs/ft³
R = resistance, (lb-sec)/ft⁴, $32\mu/d^2$
c = velocity of sound, ft/sec
d = tube inside diameter, ft
μ = viscosity, slugs/(ft-sec)
p = receiver pressure at time t, lb/ft²
p_0 = initial tubing and receiver pressure, lb/ft²
p_m = step pressure applied at input, lb/ft²

References 10 and 11 cover only a small portion of the work that has been done in this field. The additional references 13 through 18 cover other approaches to the problem and contain considerable useful test data.

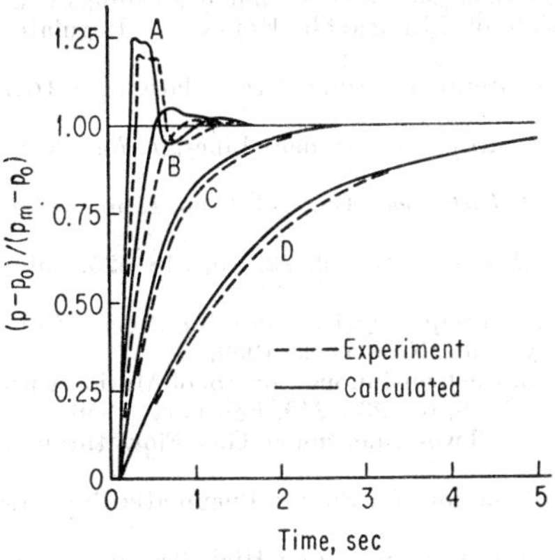

FIG. 36.27. Response of pneumatic transmission lines to step inputs.

Curve	L, ft	Q, in.³	OD, in.
A	100	91.2	⅜
B	100	294.0	⅜
C	100	91.2	¼
D	100	294.0	¼

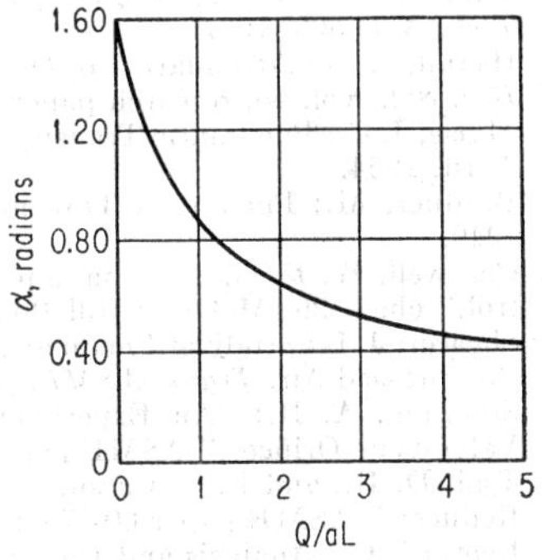

FIG. 36.28. Solution of $\alpha \tan \alpha = aL/Q$.

36.12. OTHER COMPONENTS

For Bourdon tubes, bellows, diaphragms, etc., in regard to hysteresis linearity and other design consideration, see refs. 27 through 34. In addition, most of this material is fairly well covered in manufacturers' catalogues. References 19 and 20 discuss pneumatic slide valves; ref. 21 shows data of pneumatic valve unbalance and discharge coefficients; ref. 25 shows some component analysis; ref. 23 indicates the dynamic characteristics of orifice-tube volume systems. All refs. 19 through 25 have system discussions.

References

1. Binder, R. C.: "Fluid Mechanics," Prentice-Hall, Inc., Englewood Cliffs, N.J., 1955.
2. Gibson, J. E., and F. B. Tuteur: "Control System Components," p. 439, McGraw-Hill Book Company, Inc., New York, 1958.

3. Blackburn, J. F., G. Reethof, and J. L. Shearer: "Fluid Power Control," John Wiley & Sons, Inc., New York, 1960.
4. "Fluid Meters, Their Theory and Application," p. 79, American Society of Mechanical Engineers, New York, 1959.
5. Cunningham, R. G.: Orifice Meters with Supercritical Compressible Flow, *Trans. ASME*, vol. 73, p. 625, July, 1951.
6. Perry, J. A., Jr.: Critical Flow through Sharp-edged Orifices, *Trans. ASME*, vol. 71, p. 757, October, 1949.
7. Feng, Tsun-Ying: Static and Dynamic Characteristics of Flapper-nozzle Valves, *Trans. ASME, Basic Eng.*, vol. 81, p. 275, 1959.
8. Elrod, H. G., Jr.: Theory of Ejectors, *J. Appl. Mech.*, September, 1945.
9. Keenan, J. H., and E. P. Neumann: A Simple Air Ejector, *J. Appl. Mech.*, June, 1942.
10. Rohmann, C. P., and E. C. Grogan: On the Dynamics of Pneumatic Transmission Lines, *Trans. ASME*, vol. 79, pp. 853–867, 1957.
11. Schuder, C. B., and R. C. Binder: The Response of Pneumatic Transmission Lines to Step Inputs, *Trans. ASME, J. Basic Eng.*, vol. 81, 1959.
12. Schuder, C. B.: Valve Positioner Design by Dynamic Analysis, *J. Instr. Soc. Am.*, paper 24-SL-61, 1961.
13. Eckman, D. P., and L. Guess: Pneumatic Transmission of Instrument Readings over Long Distances, Brown Instrument Division of Minneapolis-Honeywell Regulator Company, *Bull.* B-59-2.
14. Wildhack, W. A.: Pressure Drop in Tubing in Aircraft Instrument Installations, *NACA Tech. Note* 593, 1937.
15. Iberall, A. S.: Attenuation of Oscillatory Pressures in Instrument Lines, *J. Res. Natl. Bur. Std.*, vol. 45, research paper 2115, 1951.
16. Moise, J. C.: Pneumatic Transmission Lines, *J. Instr. Soc. Am.*, vol. 1, no. 4, pp. 35–40, April, 1954.
17. Bradner, M.: Pneumatic Transmission Lag, *Instruments*, vol. 22, pp. 618–625, July, 1949.
18. Caldwell, W. I., G. A. Coon, and L. M. Zoss: "Frequency Response for Process Control," chap. 26, McGraw-Hill Book Company, Inc., New York, 1959.
19. Shearer, J. L.: Study of Pneumatic Processes in the Continuous Control of Motion with Compressed Air, *Trans. ASME*, parts I, II, vol. 78, pp. 233–249, February, 1956.
20. Stenning, A. H.: "An Experimental Study of Two-dimensional Gas Flow through Valve-type Orifices," ASME paper 54-A-45.
21. Tsai, D. H., and E. C. Cassidy: "Dynamic Behavior of a Simple Pneumatic Pressure Reducer," ASME paper 60-Wa-186.
22. Reethof, G.: Analysis and Design of a Servomotor Operating on High Pressure Compressed Gas, *Trans. ASME*, vol. 79, pp. 875–879, May, 1957.
23. Benedict, R. P.: "The Response of a Pressure-sensing System," ASME paper 59-A-289.
24. Rivard, J., and J. Pembleton: "High-temperature Pneumatics—Its Use and Control," Joint Automatic Control Conference, paper ISA-2-60, 1960.
25. Finegan, F. J., Jr.: "High-performance Pneumatic Controllers," Joint Automatic Control Conference, paper ISA-13-60, 1960.
26. Turnquist, R.: "Comparison of Gas Flow Formulas for Control Valve Sizing," ISA paper 17-SL-61.
27. Newell, Floyd B.: "Diaphragm Characteristics, Designs and Terminology," American Society for Mechanical Engineers, New York.
28. Dressler, R. F.: "Bending and Stretching of Corrugated Diaphragms," ASME paper n58-A-62 for meeting Nov. 30–Dec. 5, 1958.
29. Liu, F. F.: "Dynamic Response Behavior of Diaphragms in Relation to Driving Function and Surrounding Media," ASME paper n58-A-224 for meeting, Nov. 30–Dec. 5, 1958.
30. Tueda, Masasuke: Mathematical Theories of Bourdon Pressure Tubes and Bending of Curved Pipes, *Mem. Coll. Eng. Univ. Kyoto*, vol. 8, 1934–1935.
31. Seegers, H.: Precision Bourdon-tube Gauge, *Instr. Control Systems*, vol. 34, no. 2, pp. 234–236, February, 1961.
32. Buffenmyer, W. L.: Selecting Bourdon-tube Gauges, *Instr. Control Systems*, vol. 34, no. 2, pp. 238–241, February, 1961.
33. Turner, C. E.: Stress and Deflection Studies of Flat-plate and Toroidal Expansion Bellows, Subjected to Axial, Eccentric or Internal Pressure Loading, *J. Mech. Eng. Sci.*, vol. 1, no. 2, pp. 130–143, September, 1959.
34. Matheny, James D.: Bellows Spring Rate for Seven typical convolution Shapes, *Machine Design*, Jan. 4, 1962.

Section 37

HYDRAULIC COMPONENTS

By

G. REETHOF, Sc.D., *Manager, Reliability Engineering, Large Jet Engine Department, General Electric Co., Cincinnati, Ohio*

CONTENTS

37.1. INTRODUCTION

Hydraulic components find their widest application as the actuation (power-output) elements of power-control systems. The more important advantages of hydraulic systems are summarized as follows:

1. High-pressure hydraulic power can be generated efficiently, with pump efficiencies of 92 per cent common.

2. Hydraulic components are comparatively light in weight compared with equivalent mechanical and electrical components because the highly stressed structures of the hydraulic system make very efficient use of structural material. Hydraulic pumps and motors with a power density of less than 1 lb/hp are common. This light weight is made possible by the high pressures now available from commercially available pumps. Hydraulic systems operating at 3,000 psi are quite common and higher-pressure systems are readily available.

3. As seen by the load, the hydraulic actuator is extremely stiff compared with an equivalent pneumatic or electrical system. The hydraulic approach permits the maintenance of load position against significant and varying load forces, with lower loop gain and higher response speed.

4. Hydraulic actuation offers the highest torque (or force) to inertia ratio in comparison with most mechanical, pneumatic, and electrical systems. This property, coupled with the incompressible nature of the medium, results in exceptionally fast response and high power output.

Some of the disadvantages of hydraulic systems follow:

1. Most hydraulic systems use organic-base fluids which present serious fire and explosion hazards because of fluid spillage from leaks in piping or seals as well as breaks. The wide use of hydraulics in the die-casting and machine-tool field has resulted in the search for nonflammable hydraulic fluids, with good lubricity and no toxicity. Although good progress has been made, no fully satisfactory fluid is available at this time. The article on fluid properties will discuss some of the fluids and their relative fire-hazard problems.

2. Associated with the fire-hazard problem is the inherent difficulty of preventing leaks in normal usage and the subsequent "messiness" of the hydraulic system. This problem can be avoided at the expense of maintainability by all-welded plumbing.

3. High-speed-of-response hydraulic systems are sensitive to solid contaminants in the fluid which interfere with the smooth and proper function of the control valves and have been known to cause either poor operation or outright failure. Cleanliness in manufacture, assembly, normal operation, and service is an absolute necessity to assure good reliability. Adequate provision for filtering is only one of the necessities of good hydraulic design.

Even a cursory review of the design requirements for such systems as high-speed steel rolling-mill tension controls, jet-engine controls, or radar-antenna drives conveys the need for the building of comprehensive and representative analytical models to assure satisfactory dynamic response and adequate stability throughout widely varying operating conditions and realistic tolerances.

The purpose of this section is to provide the designer with necessary basic building blocks to permit the preparation of an analytical model of the hydraulic system for experimentation and design specification. For illustrations of present design practice of commercially available hydraulic components, see refs. 12 and 40.

37.2. PROPERTIES OF HYDRAULIC FLUIDS

(a) Density and Related Properties

Density ρ is defined as the mass per unit of volume.

Specific weight W is defined as the weight per unit of volume.

Specific gravity σ is the ratio of the density of the substance in question to that of water at 60°F.

The petroleum industry uses a measure of relative density called "API gravity." API gravity in terms of specific gravity is given by Eq. (37.1).

$$\text{Degrees API} = \frac{141.5}{\sigma 60°\text{F}/60°\text{F}} - 131.5 \tag{37.1}$$

$\sigma 60°\text{F}/60°\text{F}$ represents the specific gravity of the substance at 60°F relative to water at 60°F.

Specific gravity in terms of degrees API is given by

$$\sigma 60°\text{F}/60°\text{F} = \frac{141.5}{\text{deg API} + 131.5} \tag{37.2}$$

The density of a liquid is a function of both pressure and temperature. At any one temperature a good approximation is given by Eq. (37.3)

$$\rho = \rho_0(1 + aP - bP^2) \tag{37.3}$$

where P is pres[illegible]nd ρ_0 is the density at the reference condition. Empirical cons[illegible] a and b are functions of temperature. Typical values for hydraulic oils a[illegible]°F are[1]

$$a = 4.38 \times 10^{-6}\ \text{in.}^2/\text{lb}$$
$$b = 5.65 \times 10^{-11}\ \text{in.}^4/\text{lb}^2$$

The decrease of density with increasing temperature (thermal expansion) is approximated by Eq. (37.4).

$$\rho = \rho_1[1 - \alpha(T - T_1)] \tag{37.4}$$

where α is called the cubical-expansion coefficient. This linear approximation is accurate within 0.5 per cent for most hydraulic fluids over temperature ranges of 500°F.

Compressibility K is a most important property to the designer of fast-response hydraulic systems. It is defined as the unit rate of change of density with change in pressure.

For small pressure changes,

$$K \triangleq (1/\rho)(d\rho/dP) \tag{37.5}$$

The *bulk modulus* β of a fluid is the reciprocal of compressibility,

$$\beta \triangleq \rho(dP/d\rho) \tag{37.6}$$

From Eq. (37.3), compressibility and bulk modulus are, respectively,

$$K = \frac{a - 2bP}{1 + aP - bP^2} \tag{37.7}$$

$$\beta = \frac{1 + aP - bP^2}{a - 2bP} \tag{37.8}$$

Typical values of β are given in the table of fluid properties (Table 37.1).

The computed values of bulk moduli for a given fluid should be used with great caution in the analysis of hydraulic control systems. The effective compressibility as seen by the flow source must be considered. Entrained air and transmission conduit compliance can drastically reduce the bulk modulus of the pure liquid.

(b) Viscosity

A true fluid is commonly defined as matter which is unable to store energy in shear; it is not elastic in shear. Therefore, any shear force applied to a fluid film will result in a finite shear rate. A Newtonian fluid is one for which shear rate is proportional to the shear stress. A constant of proportionality μ, the absolute viscosity, is defined by the following expression:

$$\mu = \tau/(du/dx) \tag{37.9}$$

where τ is the shear stress, du the incremental change in velocity resulting from the shear stress, and x the direction of the shear stress.

Most hydraulic fluids behave like Newtonian liquids up to shear rates of about 1,000,000 psi/sec. However, some of the highly compounded high-temperature hydraulic fluids may lose as much as 40 per cent viscosity as the long-chain molecules are oriented or sheared.[2] This loss is temporary in some cases, with the original viscosity restored after an appreciable lapse of time. In other cases the loss is permanent.

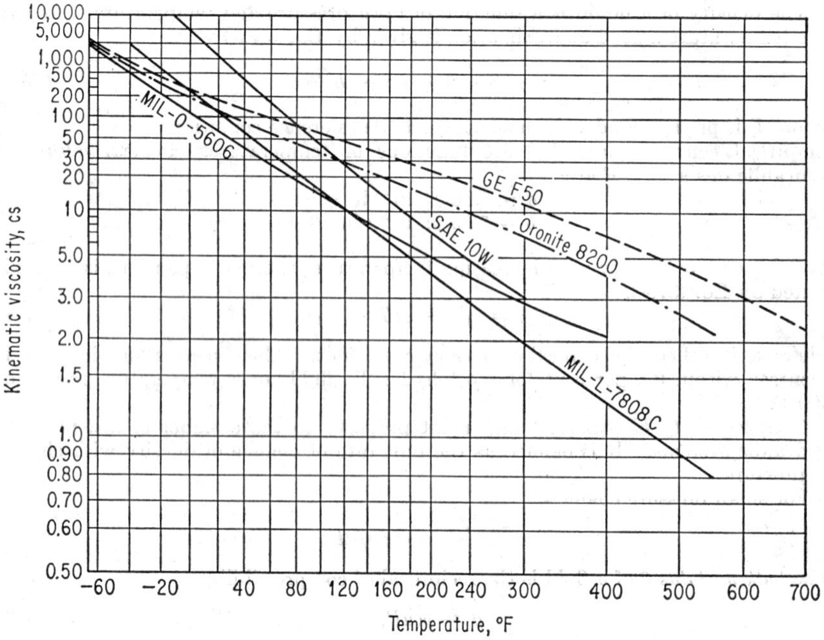

FIG. 37.1. Viscosity-temperature curves for various hydraulic fluids.

Since several units of viscosity are in use, they should be carefully defined:

Reyn. A very large inconvenient unit in the English system.

$$1 \text{ Reyn} \equiv 1 \text{ lb sec/in.}^2$$

Centipoise (*cp*) (metric system). One centipoise is the viscosity of a fluid such that a force of 1 dyne will give two parallel surfaces 1 cm² area, 1 cm apart, a velocity of 0.01 cm/sec. The centipoise is thus 0.01 dyne-sec/cm².

Centistoke (*cs*). Absolute viscosity divided by density is defined as kinematic viscosity. The common units are the stoke and the centistoke corresponding to the poise and the centipoise, respectively, divided by the density in consistent units. The centistoke is thus 0.01 cm²/sec.

Saybolt Universal Seconds. The Saybolt viscosimeter is commonly used to determine the viscosity of petroleum products. The time required for 60 ml of the sample to flow through an 0.176-cm-diameter and 1.225-cm-long tube is measured and designated SSU.

The following lists various conversion factors: 1 lb sec/in.^{2} = 68,747.2 poises, 1 lb sec/ft^2 = 478.8 poises, 1 in.^{2}/sec = 6.4516 stokes, 1 ft^2/sec = 229.03 stokes, 1 dyne sec/cm^2 = 1 poise, and 1 cm^2/sec = 1 stoke.

Effect of Pressure. The viscosity of liquids increases with pressure in a manner approximated as follows:

$$\log_{10} \mu/\mu_0 = cP \tag{37.10}$$

where c is a constant which for petroleum products at room temperature is approximately[3]

$$c = 7 \times 10^{-4} \text{ in.}^2/\text{lb}$$

Effect of Temperature. The viscosity of hydraulic fluids decreases markedly with increasing temperature. Of the many empirical formulas which have been proposed to describe the variation, Eq. (37.11) has found broadest acceptance.

$$\mu_t = \mu_0 e^{-\lambda(T-T_0)} \tag{37.11}$$

where T is the temperature °F, T_0 is the reference temperature, and λ is the temperature coefficient of viscosity.

For most petroleum-base oils an empirical formula called the *Walther formula* is a better approximation.[4] It is valid for relatively pure oils in the temperature range −20 to +160°F. The ASTM charts[5] are based on this approximation. Figure 37.1 gives the viscosity-temperature characteristics of several typical hydraulic fluids on the ASTM chart. The temperature coefficients of viscosity of several typical hydraulic fluids are given in Table 37.1.

Pour Point.[6] The pour point is the temperature at which a fluid will no longer pour from a standard container when tested according to a standard ASTM procedure. It is considerably below the temperature which can be considered a practical lower limit for use in hydraulic systems. A viscosity in excess of 2,500 cs is not generally practical for hydraulic-system use.

(c) Chemical Properties

Thermal stability relates to the fact that some hydraulic fluids when heated to high temperatures either decompose to form gaseous, liquid, or solid products or polymerize to form gels, varnishes, or even cokes.

Oxidative stability relates to the reaction of the hydraulic fluid with the oxygen of either the atmospheric air, the dissolved air, or other oxidizing agents. Since the products of oxidation are often acidic in nature, corrosion problems may result. Sludges and varnishes which result can clog filters, freeze sleeve valves, and foul orifices. Oxidation problems arise primarily in high-temperature applications.

Hydrolytic stability relates to the reaction of the hydraulic fluid with free water.

Fire Safety. Standard tests have been devised to rate the relative "fire safety" of various fluids.[7,8]

The *flash point* is the temperature at which sufficient vapors are evolved from the fluid in a heated cup to cause a transient flame when a pilot flame is brought into the test area.

The *fire point* is the temperature at which the flame above a test cup will be self-sustaining.

The *autogenous ignition* temperature is considerably above the fire point and is the temperature at which a liquid droplet will ignite upon contact with heated air.

Compatibility of hydraulic fluid relates to the property of the fluid to either be affected or affect surrounding metallic and nonmetallic materials.

Toxicity. Several of the special hydraulic fluids for use in high-temperature applications contain special additives which may be injurious when inhaled as vapor.

(d) Thermal Properties

The specific heat is the amount of heat which must be supplied to a unit mass to raise its temperature one degree.

For most fluids the specific heat increases with temperature. The specific heats of pure petroleum-base oils are well approximated by a Bureau of Standards equation[9]

$$C = (1/\sqrt{\sigma})(0.388 + 0.00045T) \tag{37.12}$$

where C = specific heat, Btu/(lb)(°F)
σ = specific gravity at 60°F/60°F
T = temperature, °F

The thermal conductivity for petroleum-base oils

$$k = (0.813/\sigma)[1 - 0.0003(T - 32)] \quad \text{Btu/(hr)(ft}^2\text{)(°F)(in.)} \tag{37.13}$$

(e) Surface Properties

Two areas involving surface energies which are of interest to the designer of hydraulic equipment are foaming and boundary lubrication. A foam is an emulsion of gas bubbles in a liquid. Since foam is many times more compressible as the gas-free liquid, the presence of foam in a servosystem will drastically decrease the bulk modulus of the mixture depending on the volumetric proportions of foam and liquid in the chamber. Developers of hydraulic fluids frequently add antifoaming agents.[6]

Friction and wear between metal surfaces in sliding are strongly affected by the molecular structure of the fluid-metal interfaces. Boundary lubrication relates to physicochemical relations which occur in very thin films. "Oiliness" is sometimes defined as a property of a fluid which will give low coefficients of friction to two sliding surfaces (see Sec. 12).

(f) Choice of Hydraulic Fluid

System performance, both steady and transient, is affected by fluid properties as follows:[4,10]

Viscosity affects damping effects, pipe flow, lubrication, leakage, motor and pump efficiencies.

Density affects orifice flow, acoustic effects, pump and motor efficiency.

Specific heat and thermal conductivity combined with viscosity and density affect temperature rise and heat rejection.

Compressibility is all-important in determining transmission characteristics, stability and response of closed-loop control systems, and pressure pulsations from pumps.

Vapor pressure and gas solubility affect cavitation effects and the associated pump-filling problem as well as the potential effects on compressibility.

Hydraulic-system life and reliability are closely associated with such fluid properties as:

Lubricity. Boundary lubrication affects wear in pumps and motors, and friction in sliding valves.

Thermal stability, where poor performance results in high acidity, gas evolution, solid-particle formation, gum formation, varnish formation, depending upon fluid and temperature level.

Compatibility. Poor performance may result in seal deterioration and other side effects.

Not to be overlooked should be *hydrolytic stability, oxidative stability,* and *flammability* for consideration of *fire safety.*

Certain of the disadvantages of the types of fluids listed in Table 37.1[4] are as follows:

1. Petroleum-base oils have poor viscosity-temperature characteristics, are volatile and highly flammable.

2. Esters of dibasic organic acids have poor low-temperature viscosity, poor thermal stability, and poor compatibility with elastomers and other organic materials.

3. Phosphate esters are good in many respects for aircraft and industrial applications, but thermal stability is poor above 250°F.

4. Polyglycols have poor thermal stability above 400°F and their decomposition

Table 37.1. Properties of Hydraulic Fluids

Type	Light turbine	Phosphate ester	Chlorinated hydrocarbon	SAE 30	Petroleum base MIL-H −5606 A	Phosphate ester
Use	Petroleum industry	Industrial fire resistant	Industrial fire resistant	Mobile petroleum	Aircraft missile	Aircraft missile
Typical commercial	DTE light	Pydraul F9	Aroclor 1242	…	Esso Univis. J43	Skydrol 500A
Specific gravity, 60°F/60°F	0.865	1.20	1.32	0.887	0.848	1.086
Expansion coefficient α, 1/°F	4.21×10^{-4}	4×10^{-4}	…	4.21×10^{-4}	0.5×10^{-3} 2130/−65°F	
Viscosity, cs:						
−40°F	…	…	…	…	500	
0°F	…	18,000/20°F	2,000/30°F	…	1	240
100°F	32	47	18	120	14.3	11.5
210°F	5.2	6.3	2.0	12	5.1	3.92
400°F	…	2.5/300°F	…	…	1.9	
550°F						
700°F						
Viscosity-temperature coefficient, λ, 1/°F	1.85×10^{-2}	…	…	…	1.04×10^{-2}	
Pour point, °F	0	−5	2	0	−90	−85
Fire point, °F	445	675	None	495	235	425
Flash point, °F	400	430	348	435	225	360
Autogenous ignition temperature, °F	…	1100	…	…	700	>1100
Bulk modulus $\times 10^{-3}$ psi	≈250	387	…	≈250	270	387
Thermal conductivity, Btu/(hr)(ft²)(°F/ft)	…	0.065/70°F	0.053	0.082	0.0615	0.0365
Specific heat, Btu/(lb) (°F)	…	0.33	0.285/60°F	…	0.5/100°F	0.42/68°F

Table 37.1. Properties of Hydraulic Fluids (Continued)

Type	Halogenated silicone	Silicone ester	Diester MIL-L 7808 C	Poly-phenyl ether	NAK 77	Mercury	JP 6
Use	Aircraft missile	Aircraft missile	Jet-engine lubrication	Aircraft nuclear	Aircraft nuclear	...	Jet fuel
Typical commercial	G.E. F50	Oronite 8200					
Specific gravity, 60°F/60°F	1.03	0.93	0.93	1.18	0.87	12.76/662°F, 13.55/68°F	0.75 to 0.84
Expansion coefficient α, 1/°F	...	0.445×10^{-3} 2400/−65°F	0.44×10^{-3}				
Viscosity, cs:							
−40°F	934	630	1920	...	...	...	5.22
0°F	287	195	230	...	...	0.1305	2.6
100°F	60	32.5	14.2	609	6.7	0.1087	0.97
210°F	21.2	11.3	3.6	6.0	0.5	0.0935	0.53
400°F	6.5	3.8	1.2	1.3	0.4	0.0797	
550°F	3.5	2.2	0.8	0.9	0.3	0.0745	
700°F	2.2	...	...	0.5	0.2	0.0710	
Viscosity-temperature coefficient, λ, 1/°F							
Pour point, °F	−100	<−100	<−75	+5	+10		
Fire point, °F	>650	450	485	...	NA		
Flash point, °F	>525	390	430	...	NA		
Autogenous ignition temperature, °F	900	760	485	465	NA	...	449
Bulk modulus $\times 10^{-3}$ psi	150/75°F, 30/700°F	218	290/0°F, 60/440°F	...	...	2,600/90°F, 1,400/600°F	170/100°F, 80/200°F, 30/400°F
Thermal conductivity, Btu/(hr)(ft²)(°F/ft)	...	0.070	0.084/200°F	...	...	4.7/50°F, 8.1/600°F	0.087/0°F, 0.082/210°F
Specific heat, Btu/(lb)(°F)	0.37/75°F, 0.51/500°F	0.47/68°F	0.35/65°F, 0.60/400°F	...	...	0.033 2240°F(Tc)* −38°F(FP)† 674.4°F(BP)‡	0.43/0°F, 0.53/210°F

* Tc—critical temperature. † FP—freezing point. ‡ BP—boiling point.

is catalyzed by some metals. They are incompatible with even a trace of hydrocarbon, this incompatibility resulting in the formation of gummy precipitates.

5. Fluorcarbons and chlorinated hydrocarbons have poor viscosity-temperature characteristics, do not accept additives, and are heavy.

6. Silicones have poor lubricity and are unresponsive to additives.

7. Polysiloxanes and orthosilicate esters have poor hydraulytic stability and are as flammable as petroleum-base oils.

8. Phenyl ethers have poor low-temperature characteristics limiting their use to above 0°F.

9. Liquid metals such as NAK 77 (a eutectic of 77 per cent Na, 23 per cent K) and mercury have poor lubricity, present serious handling problems. Mercury is highly toxic, NAK 77 highly flammable in air.

37.3. FUNDAMENTAL RELATIONSHIPS IN HYDRAULIC FLOW

(a) Introduction

In the design of such hydraulic components as pumps, motors, valves, and conduits the observation and proper application of certain basic relationships will assure sound fundamental design. These concepts are[4,9,12]

1. Conservation of mass (continuity)
2. Conservation of momentum
3. Conservation of energy

Conservation of mass requires that the rate of mass flow into a defined volume equals the rate of mass flow out plus the rate at which mass accumulates within the control volume. Thus (see Fig. 37.2)

$$\int_{A_s} \rho V_n dA_s + d/dt \int_V \rho \, dv = 0 \tag{37.14}$$

where V_n = velocity normal to control surface
A_s = control surface
v = control volume

Conservation of momentum requires that the net rate of outflow of momentum in a specific direction x plus the rate at which x momentum accumulates within the control volume is equal to the force applied to the control volume in the x direction.

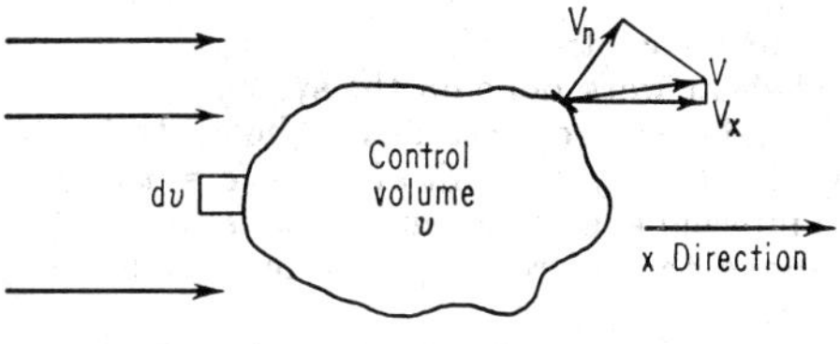

FIG. 37.2. Control volume in a flow field.

Thus

$$\sum F_x = d/dt \int_v \rho V_x \, dv + \int_{A_s} \rho V_x V_n \, dA \tag{37.15}$$

Example 1:[4] How is the pressure difference $P_1 - P_2$ related to the rate of change of flow of a frictionless incompressible fluid in a uniform tube of length L and area A?

Conservation of mass requires that

$$\rho Q_1 = \rho Q_2 \qquad \text{or} \qquad Q_1 = Q_2 = Q \qquad (\rho = \text{const}) \tag{37.16}$$

where Q = volume flow, in.³/sec

Conservation of momentum requires that

$$P_1A - P_2A = \rho QV - \rho QV + (d/dt)(\rho VAL) \tag{37.17}$$

Thus

$$P_1 - P_2 = (\rho L/A)(dQ/dt) \tag{37.18}$$

Example 2: One-dimensional, frictionless, incompressible, streamline flow of a fluid (Fig. 37.3).

Conservation of mass demands that

$$V\,dA + A\,dV = 0 \tag{37.19}$$

Conservation of momentum demands that

$$-g\rho\,dZ - dP = \rho V^2(dA/A) + \rho V\,dV \tag{37.20}$$

Combining Eqs. (37.19) and (37.20) yields Euler's equation

$$V\,dV + dP/\rho + g\,dZ = 0 \tag{37.21}$$

Since ρ is constant, Euler's equation can be integrated yielding Bernoulli's equation

$$V^2/2g + P/\rho g + Z = \text{const} \tag{37.22}$$

Conservation of energy requires that the increase in internal energy of a system of fixed identity is equal to the work done on the system plus the heat added to the system. Applied to a control volume fixed in space and not subject to acceleration,

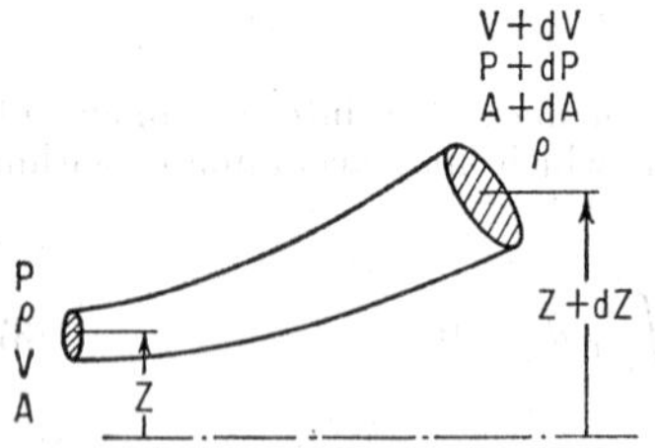

FIG. 37.3. One-dimensional, frictionless, steady incompressible, streamline flow with area change.

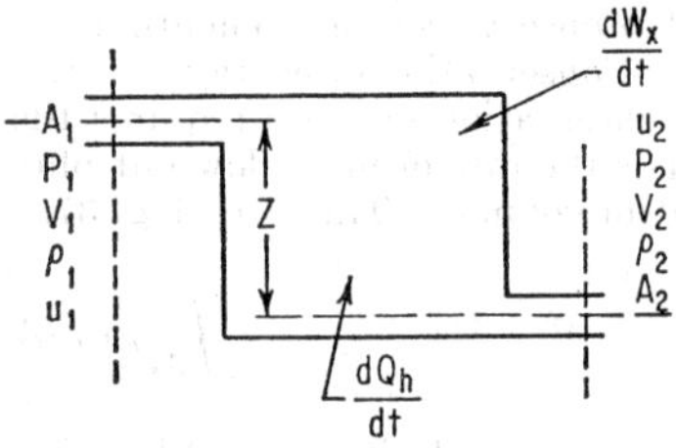

FIG. 37.4. Conservation of energy applied to a simple control volume.

conservation of energy yields (see Fig. 37.4)

$$dQ_h/dt + dW_x/dt - dE/dt + \int_{A_s} (P/\rho + e)\rho V_n\,dA \tag{37.23}$$

where Q_h = heat flow to the control volume
W_x = shaft and shear work done on the system
E = total internal energy of fluid inside the control volume
e = total internal energy per unit of mass ($e = u + gZ + V^2/2$)
u = intrinsic internal energy per unit of mass of fluid
Z = height above the reference point, in. (gZ is the potential energy per unit of mass)
P = pressure on an element of area at the surface of the control volume

(b) Orifice Flow

The design of valves for control and regulation purposes and the design of pumps and motors require the analysis of flow through rounded and sharp-edged orifices.

Consider the case of *frictionless flow through a nozzle or orifice* as illustrated in Fig. 37.5. At condition 1 the velocity is negligible. From Bernoulli's equation and ρAV = constant,

$$\rho Q = A_2\sqrt{2\rho(P_1 - P_2)} \tag{37.24}$$

The discharge coefficient C_d is defined

$$C_d \triangleq A_2/A_0 \tag{37.25}$$

where A_0 is the nominal area of the constriction. The discharge coefficient varies from 0.6 to 1.0 depending upon the geometry of the upstream passage. With this definition the mass-flow equation becomes

$$\rho Q = C_d A_0 \sqrt{2\rho(P_1 - P_2)} \qquad (37.26)$$

The fact that friction phenomena from viscous effects are neglected requires that the Reynolds number of the flow at section 2 must be in excess of a certain critical value which varies from 260 for the sharp-edged annular orifice to 100,000 for certain circular orifices in pipes. The Reynolds number, a dimensionless quantity, is defined by

$$\mathrm{Re} \triangleq \rho V D_e / \mu$$

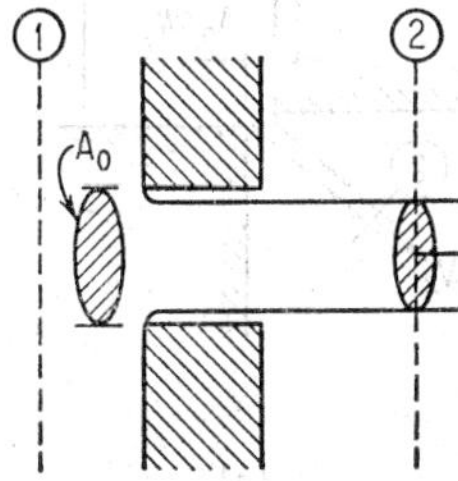

FIG. 37.5. Frictionless flow through a nozzle or orifice.

where D_e is the equivalent hydraulic diameter of the constriction, μ is the absolute viscosity, V is the velocity, and ρ is the density. For Re larger than the critical value the discharge coefficient C_d is constant and thus independent of Re. For Re less than the critical value C_d is a function of both geometry and Re, as demonstrated in the following paragraphs.

The Sharp-edged Circular Orifice. Orifice meters for flow measurement commonly consist of sharp-edged circular orifices in pipes with pressure taps on each side

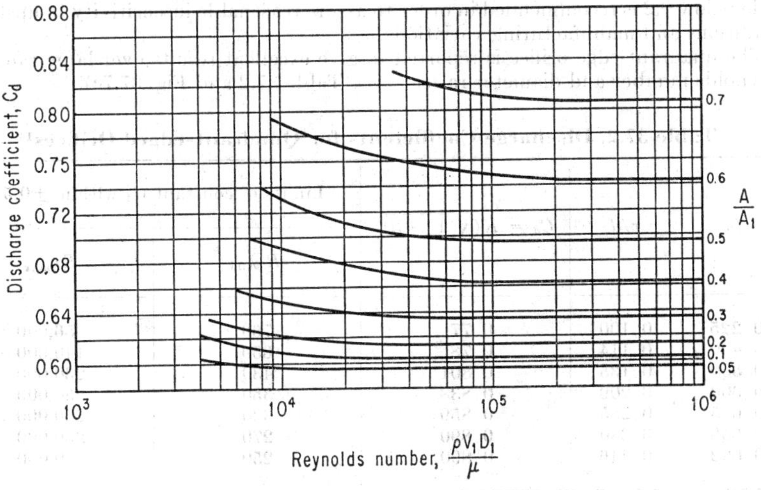

FIG. 37.6. Data for knife-edged orifices.

of the orifice plate. Sharp-edged circular orifices are also commonly used in flow-control valves as fixed restrictions. The effect of upstream pipe diameter is commonly included as the ratio of orifice diameter to pipe diameter β. The mass-flow equation thus becomes

$$\rho Q = C_d A_0 \sqrt{\frac{2\rho(P_1 - P_2)}{1 - \beta^4}} \qquad (37.27)$$

For a circular orifice,

$$\mathrm{Re} = 4\rho Q / \pi \mu D \qquad (37.28)$$

Figure 37.6 presents data for knife-edged orifices[13] with 45° chamfers. As a properly constructed measuring device, the sharp-edged orifice will give a measurement accuracy of about ±0.5 per cent. As a control device, the sharp-edged circular orifice

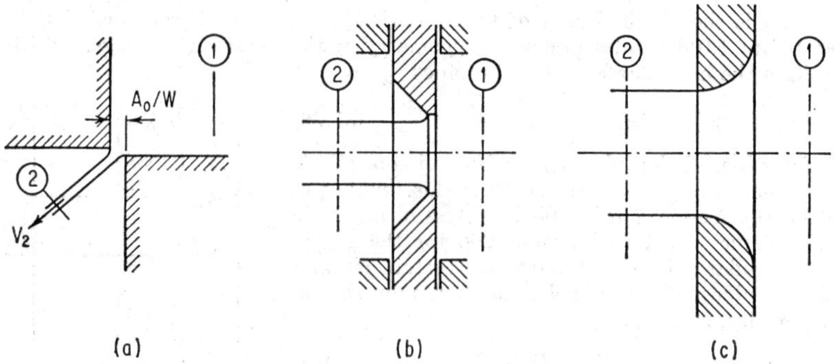

FIG. 37.7. Flow through nozzles and orifices. (*a*) Sharp-edged annular orifice. (*b*) Sharp-edged orifice. (*c*) Quadrant-edged orifice.

can be expected to give the most reproducible results as compared with other orifice configurations (see Fig. 37.7*b*).

The thick-plate circular orifice, particularly in the smaller sizes, and for diameter-to-length ratio of unity, gives a 30 per cent spread in discharge coefficient at any one Reynolds number for Reynolds numbers from 100 to 6,000. A diameter-to-length ratio below 0.2 is recommended in order to assure reasonable insensitivity to upstream conditions and manufacturing variations.

The quadrant-edge orifice is reported to give excellent results over large ranges of Reynolds number and diameter ratios[14] (see Table 37.2 and Fig. 37.7*c*).

Table 37.2. Discharge Coefficients for Quadrant-edged Orifices[14]

β	r/d	$C_d = K(\sqrt{1-\beta^4})$	Limits of constant C_d within $\pm 0.5\%$	
			$R_{D_{min}}$	$R_{D_{max}}$
0.225	0.100	0.77	760	56,000
0.400	0.114	0.78	650	140,000
0.500	0.135	0.801	430	230,000
0.600	0.209	0.838	350	250,000
0.625	0.285	0.859	330	250,000
0.656	0.380	0.890	270	250,000
0.692	0.446	0.890	250	250,000

The Sharp-edged Annular Orifice. For sliding-type control valves with sharp edges (see Fig. 37.7*a*) the discharge coefficient C_d is reasonably constant

$$C_d = 0.60 \text{ to } 0.65$$

above the critical value of the Reynolds number of 300. Below the critical value of Re, the discharge coefficient increases to about 0.95, then decreases almost linearly to Re = 0 (see Fig. 37.8). With decreasing pressure drop the peak value of C_d tends to decrease. This phenomenon becomes pronounced for ΔP below about 150 psi.[4]

Poppet-type valves show a similar variation of discharge coefficient with Re. Experiments[15] show very good agreement with Fig. 37.8 over pressure-drop ranges of 500 to 200 psi and 45° poppet angles. The geometry of the valve and the valve chamber should be considered in applying any data, as the effect may be large for special cases.

The "Flapper-valve" Type of Annular Orifice. The "flapper valve" has been for many years and is today frequently employed in control valves. Figure 37.9*a* shows the sharp-edged orifice. Figure 37.9*b* is the more practical chamfered and flat design. Experiments for a two-dimensional glass-sided model gave discharge coefficients of 0.65 for Reynolds numbers in excess of 600 for the sharp-edged orifices as shown in Fig. 37.10. Figure 37.11 presents the results for several flats and chamfers as shown in Fig. 37.9*b*.[7,16] Outside chamfer angles α can vary from 30 to 60° without materially affecting the result. The results indicate that internally chamfered orifices of the flapper type can have discharge coefficients up to 0.90 which remain constant for Reynolds numbers in excess of 600, where Re is defined by

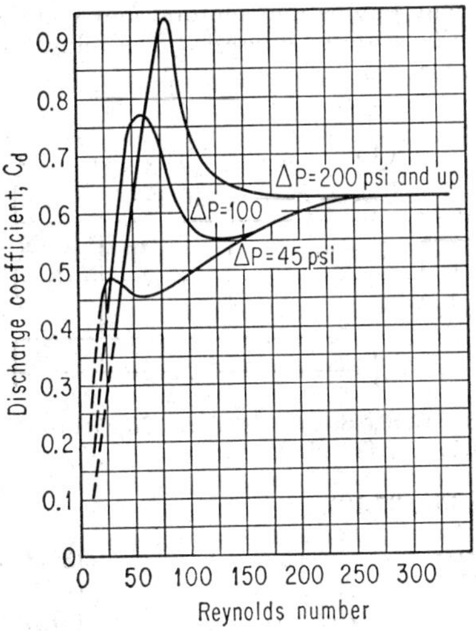

FIG. 37.8. Orifice coefficients for annular sharp-edged orifices as a function of orifice Reynolds number.

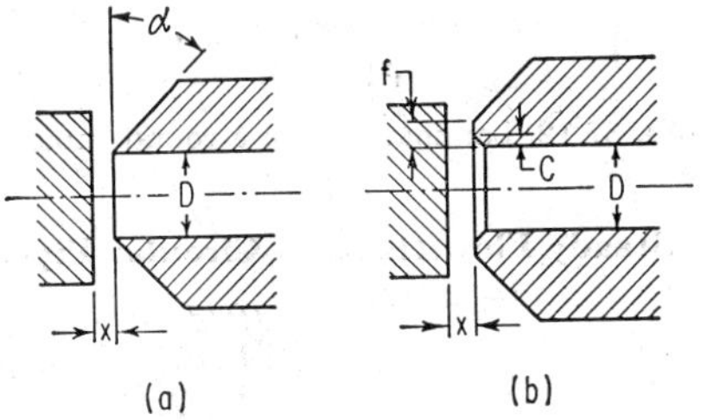

FIG. 37.9. Schematic diagrams of flapper orifice configurations.

$$\mathrm{Re} = \rho V x/\mu \tag{37.29}$$

where $V = \sqrt{2(\Delta P)/\rho}$ (37.30)

x = flapper opening, in.

ΔP = pressure drop, lb/in.²

ρ = density of fluid, lb-sec/in.⁴

V = average velocity, in./sec

since

$$Q = C_d A \sqrt{2(\Delta P)/\rho} \tag{37.31}$$

and

$$A = \pi D x \tag{37.32}$$

$$\mathrm{Re} = Q/\pi D C_d \mu \tag{37.33}$$

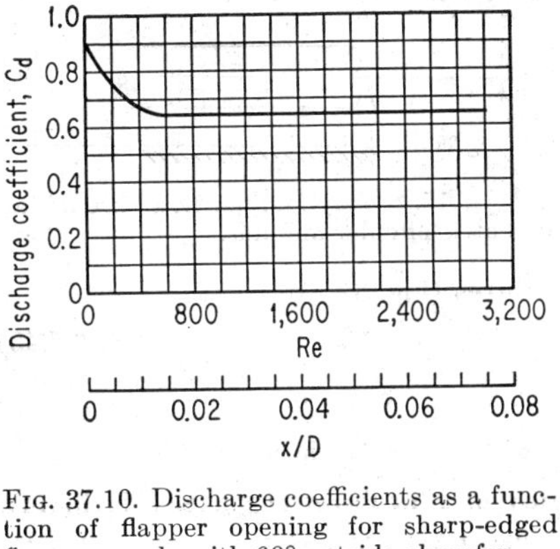

FIG. 37.10. Discharge coefficients as a function of flapper opening for sharp-edged flapper nozzle with 60° outside chamfer.

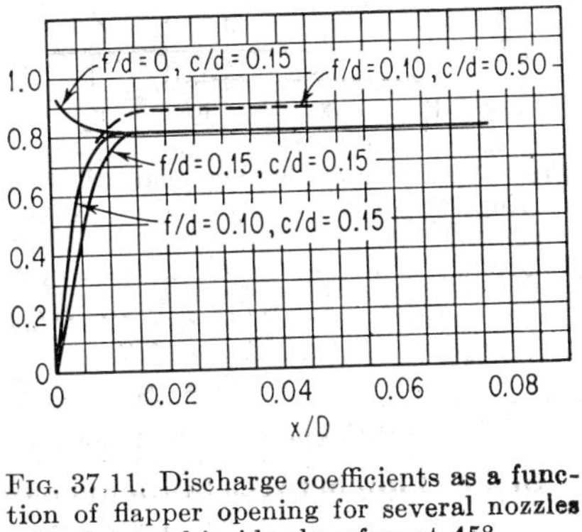

FIG. 37.11. Discharge coefficients as a function of flapper opening for several nozzles with flats and inside chamfers at 45°.

(c) Laminar Flow

In the design of such hydraulic components as certain closely fitted valves, the pistons and valve plates of positive-displacement pumps and motors, and certain seals, the equations of fully developed laminar flow are of interest. The following cases assume constant fluid viscosity and density.[4]

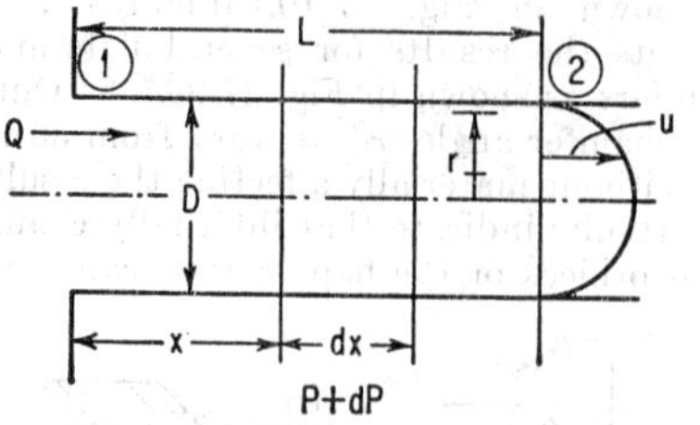

FIG. 37.12. Laminar flow in a round tube.

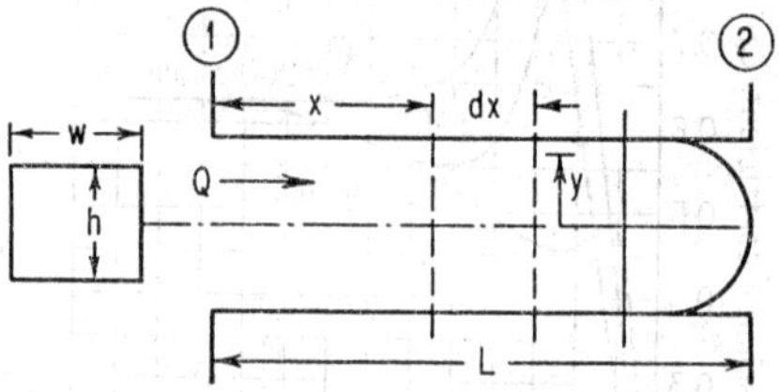

FIG. 37.13. Steady laminar flow through stationary flat plates.

Steady Flow through Circular Pipes (see Fig. 37.12).

$$dP/dx = -128\,\mu Q/\pi D^4 \tag{37.34}$$
$$dP/dx = -32\,\mu V_{av}/D^2 \tag{37.35}$$
$$Q = (\pi D^4/128\,\mu L)(P_1 - P_2) \tag{37.36}$$
$$u = 2V_{av}(1 - 4r^2/D^2) \tag{37.37}$$
$$V_{av} = Q^4/\pi D^2 \tag{37.38}$$

If the friction factor f is defined by

$$\Delta P = f\rho(L/D)(V^2/2) \tag{37.39}$$
$$f \triangleq 2D/\rho V^2(dP/dx) \tag{37.40}$$

then for laminar flow the friction factor is given by Eq. (37.41).

$$f = 64/\text{Re} \tag{37.41}$$

The distance from the entrance of a tube to the length at which fully developed laminar flow is obtained is given by

$$L/D = 0.058\,\text{Re} \qquad \text{where Re} = \rho VD/\mu < 2{,}000 \tag{37.42}$$

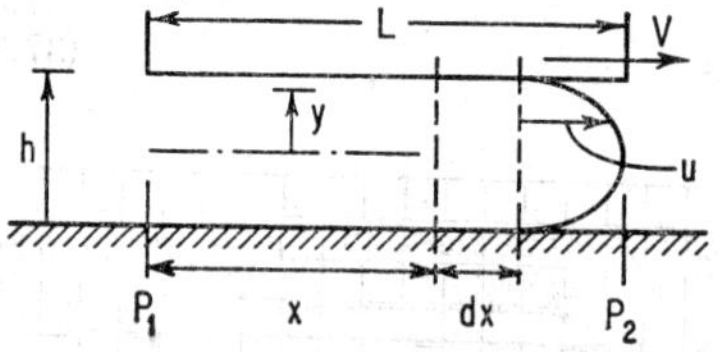

FIG. 37.14. Steady laminar flow through moving flat plates.

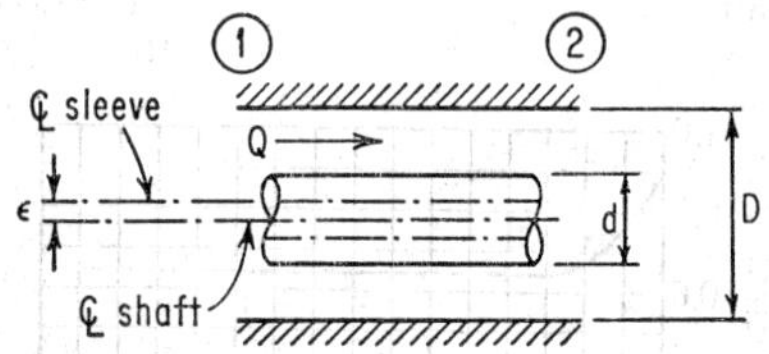

FIG. 37.15. Steady laminar flow through an eccentric circular annulus.

Steady Flow through Stationary Flat Plates (Fig. 37.13).

$$dP/dx = -12\,\mu Q/wh^3 = -12\,\mu V_{av}/h^2 \tag{37.43}$$
$$V_{av} = Q/hw \tag{37.44}$$
$$Q = (wh^3/12\,\mu L)(P_1 - P_2) \qquad \text{for } w \gg h \tag{37.45}$$

where h is the thickness of the passage, w is the width, and L is the length of the passage.

Steady Flow between Stationary and Moving Flat Parallel Plates (Fig. 37.14).

$$Q/w = Vh/2 - (h^3/12\,\mu)(dP/dx) \qquad \text{if } w \gg h \tag{37.46}$$

where Q is rate of flow through plates relative to fixed surface.

$$dP/dx = (P_1 - P_2)/L \tag{37.47}$$
$$Q/w = Vh/2 - (h^3/12\,\mu)[(P_1 - P_2)/L] \tag{37.48}$$

The force on the upper plate

$$F_u = Lw\{u(V/h) + h/2[(P_1 - P_2)/L]\} \tag{37.49}$$

The force on the lower plate

$$F_1 = Lw\{-uV/h + (h/2)[(P_1 - P_2)/L]\} \tag{37.50}$$

Steady Flow through an Eccentric Circular Annulus (Fig. 37.15).

$$Q = (\pi Dh^3/12\,\mu L)(P_1 - P_2)[1 + 1.5(\epsilon/h)^3] \tag{37.51}$$

where ϵ is the eccentricity and $h = (D - d)/2$.

(d) Flow with Varying Viscosity and Density

The assumption of constant density and constant viscosity will not fully explain the behavior of such devices as pistons moving in cylinders or valve plates rotating against cylinder blocks. Since work is done on a particle of fluid in viscous shear as it passes through a passage, the temperature of the fluid will increase for the adiabatic process. Experiments[17] have verified the following relationships and design curves.

Consider that fluid properties as a function of temperature are defined by

$$\rho = \rho_1(1 - \alpha\,\Delta T) \tag{37.52}$$
$$E = E_1 + C_v\,\Delta T \tag{37.53}$$

where C_v is the specific heat, in. lb/lb°F

$$\mu = \mu_1 e^{-\lambda}(\Delta T) \tag{37.54}$$

The temperature gradient is approximated by the following equation, assuming no side leakage:

$$dT/dx = (1/\lambda)[K_2/(1 + K_2x)] \tag{37.55}$$

where

$$K_2 \triangleq 2Vx\mu_1/h^2C_v \tag{37.56}$$

The gradient is largest at $x = 0$, the leading edge of the slider, since the fluid there is coolest and therefore most viscous. At $x = 0$,

$$\left.\frac{dT}{dx}\right|_{x=0} = \frac{2V\mu_1}{h^2C_v} \tag{37.57}$$

The pressure distribution is given by

$$p = P_m - \frac{3\alpha C_v}{2x^2}\left(\ln\frac{1 + K_2x}{1 + K_2x_m}\right)^2 \tag{37.58}$$

where the subscript m refers to conditions for the maximum pressure point or where

$$dp/dx = 0 \tag{37.59}$$

Integrating the local pressure p over the area of the pad results in the expression for the load-carrying capacity of the parallel slider

$$F = K_3\left[\frac{2 + K_2L}{K_2L}\ln(1 + K_2L) - 2\right] \tag{37.60}$$

For small values of K_2L (up to approximately 0.1) the simplified force equation may be written

$$F = (K_3/6)[(K_2L)^2 - (K_2L)^3] \tag{37.61}$$

where

$$K_3 \triangleq 3\alpha C_vLw/2\lambda^2 \tag{37.62}$$

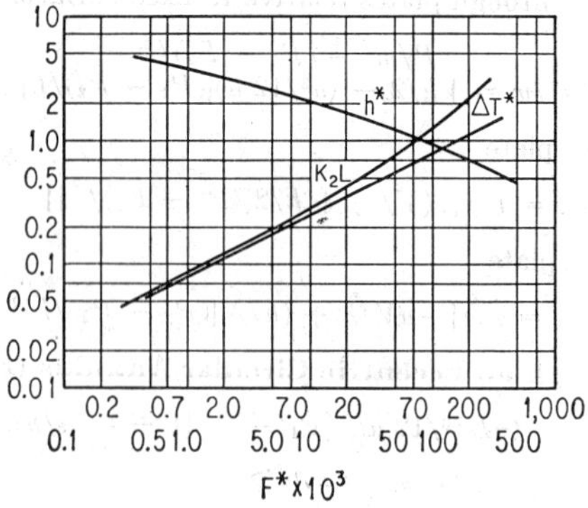

FIG. 37.16. Normalized force vs. normalized clearances; normalized temperature rise and length for the parallel slider.[4]

For design purposes plots of normalized forces, clearance and temperatures will be found useful. The normalized parameters are defined by

$$F^* \triangleq F/F_1 \qquad (37.63)$$

$$h^* = 1/\sqrt{{}_2KL} \qquad (37.64)$$

$$\Delta T^* = \ln\,(1 + K_2L) \qquad (37.65)$$

$$\Delta T^* = \lambda\,\Delta T_L \qquad (37.66)$$

Figure 37.16 provides a cross plot of the normalized quantities.

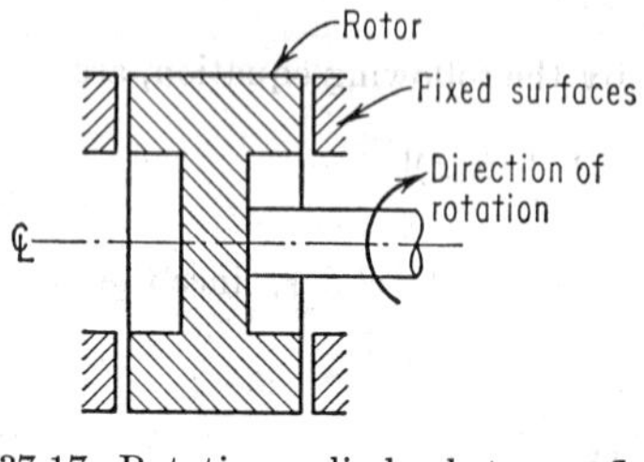

FIG. 37.17. Rotating cylinder between flat side plates.

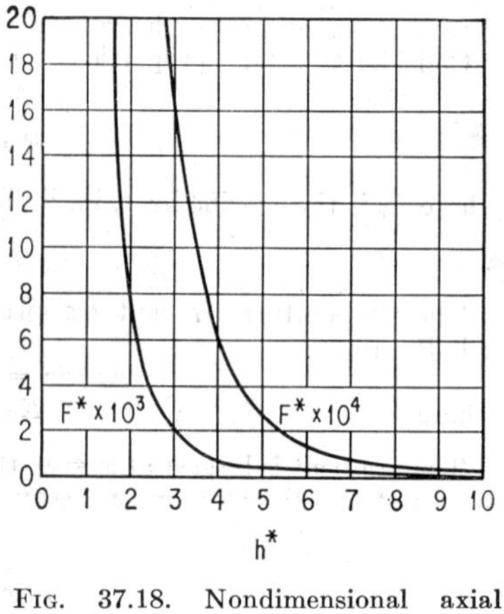

FIG. 37.18. Nondimensional axial centering force vs. eccentricity for the rotating cylinder between side plates.[4]

The centering force on a rotating cylinder is of interest in gear and vane pump design. Figure 37.17 schematically shows a typical example, and Fig. 37.18 is the plot of the nondimensional centering force vs. the eccentricity in the clearance space.

37.4. HYDRAULIC POWER GENERATION

Two fundamentally different types of hydraulic pumps are available:

1. Positive-displacement pumps of the piston, vane, or gear type used to generate relatively high pressures at relatively low flows. Clean fluids of good lubricity and adequate viscosity are commonly used.

2. Hydrokinetic pumps of the axial-flow or centrifugal type used to generate large volume flows at relatively low pressures. Centrifugal pumps will tolerate such low lubricity fluids as liquid hydrogen, liquid metals, and jet fuels, and fluids with considerable contaminant content.

(a) **Positive-displacement Pumps**

Gear pumps are used extensively for jet fuels, lubricating oils, and other applications where pressures up to 1,500 psi suffice. Gear pumps are fixed-displacement pumps; i.e., delivery per revolution cannot be changed over large ranges with good retention of efficiency.

Vane pumps find broad use in such applications as roadworking machinery, machine-tool application, and many other uses where pressures do not exceed 2,000 psi. Variable-displacement vane pumps are available but pressures rarely exceed 1,500 psi.

High-pressure generation of fluid power in the 3,000 to 5,000 psi range can be accomplished by the *piston* pump. Either axial-piston or radial-piston pumps, fixed and variable delivery, are available.

Efficiency. The power required to drive a pump of specific delivery and pressure rise and the heat rejected into the fluid during the pumping process are a function of pump efficiency.

The volumetric efficiency η_v of a positive-displacement machine is defined as the ratio of the actual delivery to the displacement or the theoretical delivery:

$$\eta_v = Q_A/Q_T = (Q_T - Q_L)/Q_T = 1 - Q_L/Q_T \tag{37.67}$$

where Q_A = actual delivery
Q_T = theoretical delivery
Q_L = internal leakage

For most well-designed pumps the dominant leakage is laminar in nature, and therefore loss due to cavitation at the inlet, and orifice-type backflow can be assumed negligible. The leakage flow Q_L can be expressed in terms of a nondimensional parameter[4] $\Delta P D_p/2\pi\mu$ and a slip coefficient C_S, where ΔP = pressure drop, D_p = displacement per revolution at zero pressure drop across pump, μ = viscosity of fluid in clearance passages.

The mechanical or torque efficiency η_m is defined as the theoretical torque T_P required to drive a pump at a particular speed N, pressure rise ΔP, and delivery Q_A divided by the actual torque T_A.

$$\eta_m = T_P/T_A = T_P/(T_P + T_F) \tag{37.68}$$

where T_F is the friction drag torque.

The applied torque T_A is

$$T_A = \Delta P D_p/2\pi + C_d D_p \mu N + C_f(\Delta P D_p/2\pi) + T_c \tag{37.69}$$

The theoretical torque is given by

$$T_P = \Delta P D_p/2\pi \tag{37.70}$$

C_d is the viscous drag coefficient (dimensionless).

C_f is the drag coefficient resulting from pressure-dependent and speed-independent friction sources such as bearings and seals.

T_c is the remaining frictional drag, which is independent of speed and pressure drop. It is the result of tight seals, poorly lubricated rubbing surfaces, etc.

Substituting expression (37.69) into (37.68) yields

$$\eta_m = \frac{1}{1 + C_d(\mu N/\Delta P) + C_f + T_c/D_p\Delta P} \tag{37.71}$$

C_d is inversely proportional to the first power of typical pump clearances. T_c and C_f tend to be proportional to the size of the pump.

The overall pump efficiency is equal to the product of volumetric and mechanical efficiencies. The volumetric efficiency is

$$\eta_v = 1 - C_s(\Delta P/\mu N) \tag{37.72}$$

It can be shown that the slip (or leakage) coefficient C_s is proportional to the cube of the typical clearance of a pump. The pump efficiency is given by

$$\eta_p = \eta_v \eta_m \tag{37.73}$$

From Eqs. (37.71) and (37.72), Eq. (37.73) becomes

$$\eta_p = \frac{1 - C_s(\Delta P/\mu N)}{1 + C_d(\mu N/\Delta P) + C_f + T_c/D_p\,\Delta P} \tag{37.74}$$

The efficiency of a pump is thus determined by C_s, C_d, C_f, and $\mu N/\Delta P$ if T_c is considered negligible. Table 37.3 gives typical values for several hydraulic pumps.[18]

Table 37.3. Hydraulic Pump Constants

Hydraulic unit	D, in.³/rev	C_d	C_s	C_f	T_c, in.-lb
Piston pump	3.60	16.8×10^4	0.15×10^{-7}	0.045	0
Vane pump	2.865	7.3×10^4	0.477×10^{-7}	0.212	0
Spur-gear pump	2.965	10.25×10^4	0.48×10^{-7}	0.179	0
Internal-gear pump	2.965	9.77×10^4	1.02×10^{-7}	0.045	0
Internal-gear pump	1.800	5.06×10^4	1.6×10^{-7}	0.075	13

The volume under compression in the pump as seen by the high-pressure circuit is of interest to the system designer. Typical values for fixed- and variable-delivery pumps are as follows:

Fixed displacement

$$\frac{\text{Volume under compression}}{\text{Displacement per revolution}} = 1.2 \text{ to } 1.5$$

Variable delivery

$$\frac{\text{Volume under compression}}{\text{Displacement per revolution}} = 1.8 \text{ to } 3.0$$

(b) Variable-delivery Pressure-compensated Pump

Position servosystems requiring only intermittent actuation of the output member commonly use a variable-delivery pump with the displacement of the pump controlled

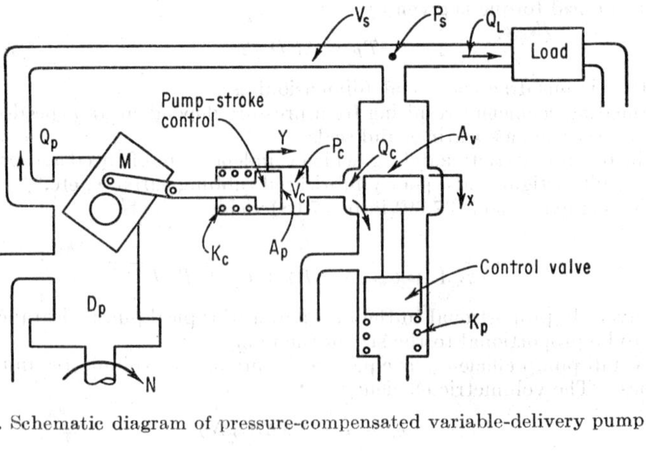

FIG. 37.19. Schematic diagram of pressure-compensated variable-delivery pump circuit.

so as to maintain a constant system supply pressure. Use of a fixed-delivery pump would require sizing to provide sufficient flow to the load for the maximum load velocity requirement. Thus for normal standby and load holding of a multiple servosystem the flow in excess of normal leakage would be bypassed through a throttling valve. Under such conditions the bypass system causes almost all the fluid power to be converted to a temperature rise of the fluid. This objectionable heat input can be drastically reduced by automatically adjusting the delivered flow from the pump to maintain the desired pressure.

The variable-delivery pressure-compensated pump is shown schematically in Fig. 37.19. The control valve compares the force associated with actual system pressure

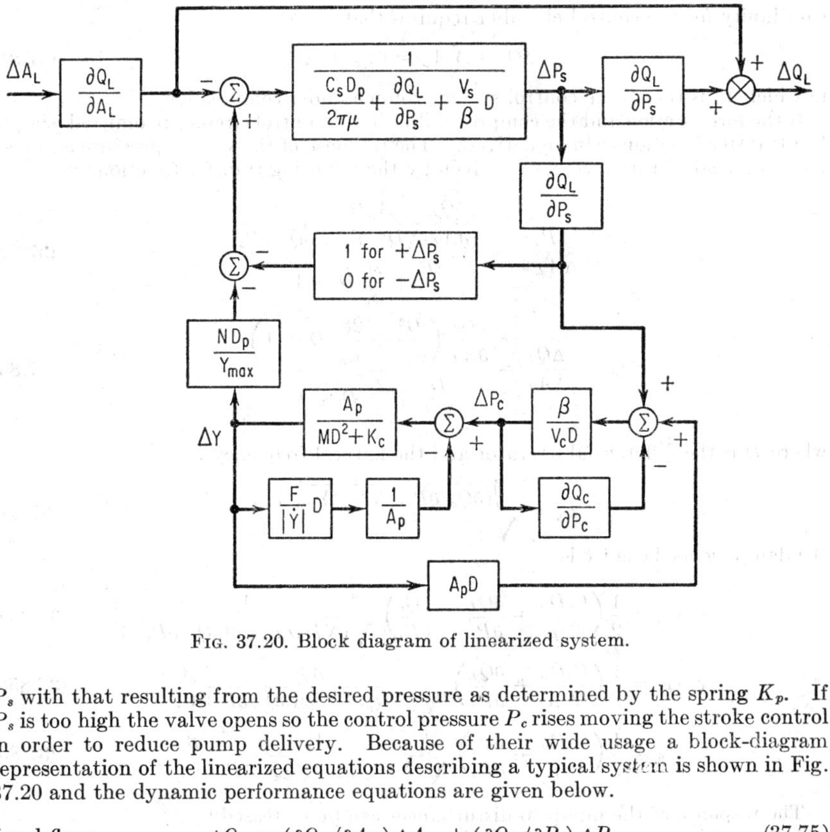

FIG. 37.20. Block diagram of linearized system.

P_s with that resulting from the desired pressure as determined by the spring K_p. If P_s is too high the valve opens so the control pressure P_c rises moving the stroke control in order to reduce pump delivery. Because of their wide usage a block-diagram representation of the linearized equations describing a typical system is shown in Fig. 37.20 and the dynamic performance equations are given below.

Load flow: $\Delta Q_L = (\partial Q_L/\partial A_L)\,\Delta A_L + (\partial Q_L/\partial P_s)\,\Delta P_s$ (37.75)

Pump flow: $\Delta Q_p = (ND_p/Y_{\max})\,\Delta Y - (C_sD_p/2\pi\mu)\,\Delta P_s$ (37.76)

Control-valve flow: $\Delta Q_c = (\partial Q_c/\partial P_s)\,\Delta P_s - (\partial Q_c/\partial P_c)\,\Delta P_c$ (37.77)

where A_L = load-orifice area
Q_P = valve flow delivered by pump
Y = nondimensional valve displacement
Q_c = control flow

Continuity demands that

$$\Delta Q_L - \Delta Q_p + \Delta Q_c = (-V_s/\beta)\,\Delta \dot{P}_s \tag{37.78}$$

where V_s is the system volume under compression and β is the bulk modulus. The force equation for the stroke control is given by

$$M\,\Delta \ddot{Y} + (F/|\dot{Y}|)\,\Delta \dot{Y} + K_c\,\Delta Y = \Delta P_c A_p \tag{37.79}$$

where M = mass of stroke control
K_c = effective spring rate in pressure control
A_p = control-piston area

The pressure- and flow-sensitive reaction forces of the variable-delivery mechanism for certain yoke and wobble plate pumps are given by an equation of the form of Eq. (37.80) where C_2 is lumped with K_c. The reaction force from pump-stroke control is

$$F_r = C_1 P_s - C_2 Y - C_3 \tag{37.80}$$

Continuity for the control chamber requires that

$$\Delta Q_c + \dot{Y} A_p = (V_c/\beta)\,\Delta \dot{P}_c \tag{37.81}$$

in which V_c is the stroke-control system volume under compression.

If the force section and the compressibility in the control circuit are omitted, simplified transfer functions can be derived. The response of the system pressure and load flow to a load disturbance ΔA_L is given by the following transfer functions:[11]

$$\frac{\Delta P_s}{\Delta A_L} = \frac{-\dfrac{\partial Q_L}{\partial A_L}\dfrac{A_p D}{ND_p/Y_{\max}}\dfrac{1}{\partial Q_c/\partial P_s}}{\dfrac{D^2}{\omega_n^2} + \dfrac{2\zeta}{\omega_n}D + 1} \tag{37.82}$$

$$\frac{\Delta Q_L}{\Delta A_L} = \frac{\dfrac{\partial Q_L}{\partial A_L}\left(\dfrac{D^2}{\omega_n^2} + \dfrac{2\zeta_0}{\omega_n}D + 1\right)}{\dfrac{D^2}{\omega_n^2} + \dfrac{2\zeta}{\omega_n}D + 1} \tag{37.83}$$

where D is the differential operator and the natural frequency is

$$\omega_n = \sqrt{\frac{(\partial Q_c/\partial P_s)(ND_p/Y_{\max})}{A_p}\frac{\beta}{V_s}} \tag{37.84}$$

the damping coefficient ζ is

$$\zeta_{+\Delta P_s} = \frac{1}{2}\left(\frac{C_s D_p}{2\pi\mu} + \frac{\partial Q_L}{\partial P_s} + \frac{\partial Q_c}{\partial P_s}\right)\sqrt{\frac{A_p}{(ND_p/Y_{\max})(\partial Q_c/\partial P_s)}\frac{\beta}{V_s}} \tag{37.85a}$$

$$\zeta_{-\Delta P_s} = \frac{1}{2}\left(\frac{C_s D_p}{2\pi\mu} + \frac{\partial Q_L}{\partial P_s}\right)\sqrt{\frac{A_p}{(ND_p/Y_{\max})(\partial Q_c/\partial P_s)}\frac{\beta}{V_s}} \tag{37.85b}$$

$$\zeta_0 = \frac{1}{2}\left(\frac{C_s D_p}{2\pi\mu} + \frac{\partial Q_c}{\partial P_s}\right)\sqrt{\frac{A_p}{(ND_p/Y_{\max})(\partial Q_c/\partial P_s)}\frac{\beta}{V_s}} \tag{37.86}$$

The response of the pump to disturbances can be increased by:

1. Increasing the control-valve gain $\partial Q_c/\partial P_s$.
2. Increasing the pump gain $ND_p/Y_{\max}$.
3. Increasing the bulk modulus of the oil.
4. Decreasing control piston area.
5. Volume under compression in systems circuit.

The damping coefficient, Eqs. (37.85*a*) and (37.85*b*), shows that increases in pump leakage, load flow pressure sensitivity, and control-valve gain will increase damping whereas pump system gain increases $(\partial Q_c/\partial P_s)(ND_p/Y_{\max})(1/A_p)$ and system compliance increases β/V_s will decrease damping.

For negative values of ΔP_s, system leakage through the control valve is much reduced, as seen from the schematic in Fig. 37.19 and the block diagram in Fig. 37.20. Thus for negative values of ΔP_s, Eq. (37.85*b*) should be used.

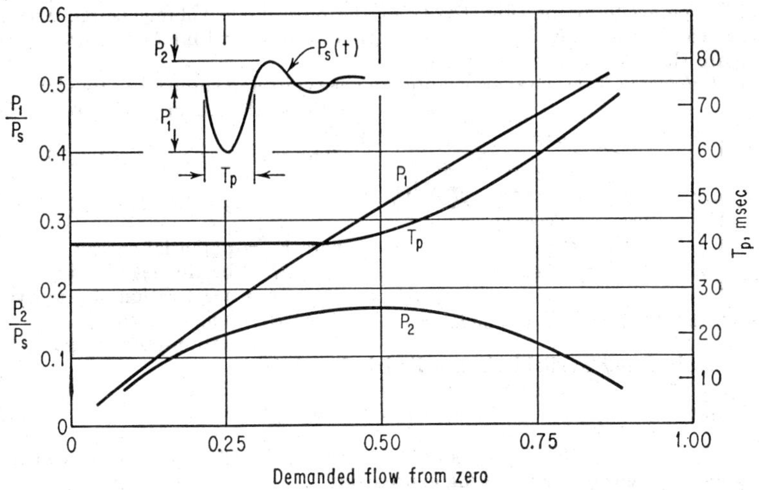

FIG. 37.21. Response characteristics of typical pressure-compensated variable-delivery pumps.

Figure 37.21 is a normalized curve for the step response of a typical variable-delivery pump with a purely resistive load. The disturbance in each case is a step change in ΔA_L.

System Considerations. The stability of the variable-delivery servo-type pump (either pressure-controlled or electrohydraulically controlled stroke mechanisms) is sensitive to many design parameters. The more important causes of instability are:

1. Operation near cutoff can be shown to be the least-damped condition.
2. Servo-valve dead band, because of the integrator action of the control, will cause hunting.
3. High friction level in the stroking mechanism with high gain control will cause instability.
4. Backlash in the stroking mechanism can induce instability.
5. Piston pumps develop pressure pulsations in the output line which, because of the nonlinear nature of the system, can induce subharmonic oscillations which with low system damping may result in sustained oscillations.
6. The frequency of the pressure pulsations in the output line, if too close to the characteristic frequencies of the servo valve or cam plate system, can be troublesome.

37.5. HYDRAULIC POWER STORAGE

The basic intentional power-storage element in hydraulic systems is the accumulator which may be either the spring-loaded or gas-loaded ram (Fig. 37.22*a* and *b*), the

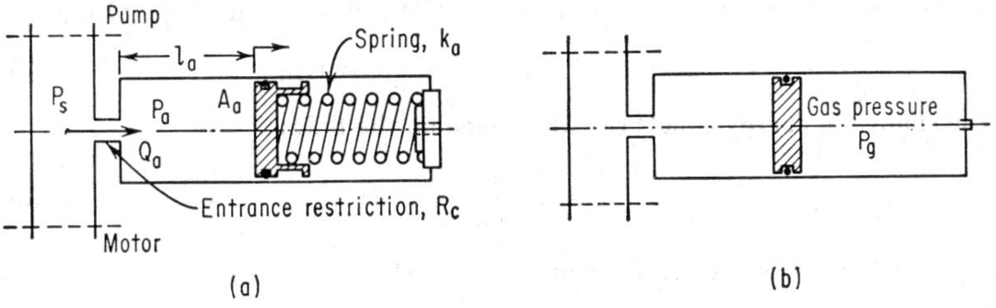

FIG. 37.22. Ram-type accumulator. (*a*) Spring-loaded type. (*b*) Air-loaded type.

bladder-type accumulator (Fig. 37.23), or the pure liquid volume (surge tank). In addition to the intentional power storage there are several inherent power storages. There are three types of power storage:

Potential energy E_p is the energy stored in reservoirs and surge tanks by virtue of the height of the liquid level above a datum line.

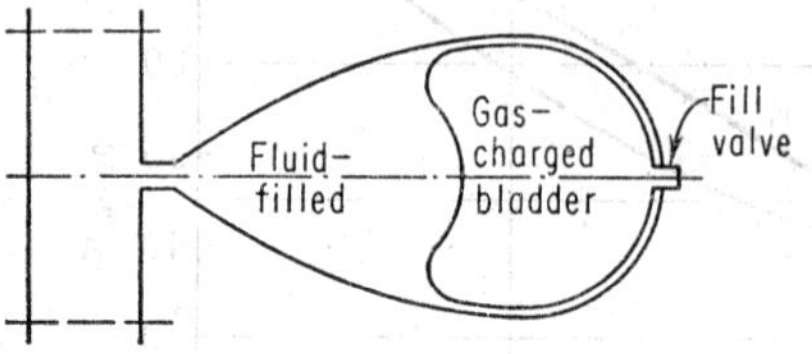

FIG. 37.23. Bladder-type accumulator.

The energy at any one level is

$$E_p = Mg(L - L_0) \tag{37.87}$$

where E_p = potential energy
M = total mass of fluid
g = acceleration of gravity
L = height of liquid
L_0 = datum-point height

The total energy available in a cylindrical tank is

$$E_{pt} = Mg[(L - L_0)/2] \tag{37.88}$$

Kinetic energy E_k, the energy stored in the moving fluid in conduits and large actuators, is given by the following expression:

$$E_k = \tfrac{1}{2}\rho V_1^2 A_1^2 \int_0^L dl/A(l) \tag{37.89}$$

where A_1 and V_1 are the area and velocity at station 1. $A(l)$ is the functional relationship between A and length l.

If the velocity is not constant across a tube diameter, such as is the case for highly laminar flow, Eq. (37.89) must also contain the integral of velocity across the radius R. Consider the case of fully developed laminar flow for which

$$V = V_{max}(1 - r^2/R^2) \tag{37.90}$$

The double integral in Eq. (37.91)

$$E_k = \rho \int_0^L \int_0^A (V^2/2)\, dA\, dl \tag{37.91}$$

when evaluated between limits, yields

$$E_k = \rho V_{max}^2(LA/4) \tag{37.92}$$

Elastic energy E_e is the energy stored by virtue of fluid elasticity and the elasticity of an enclosure such as a conduit or accumulator.[19]

When an element of fluid undergoes an increase in pressure its volume decreases. From Eq. (37.3),

$$\rho = \rho_0(1 + aP - bP^2) \tag{37.93}$$

Neglecting the second-order term in Eq. (37.3) and employing Eq. (37.6),

$$d\rho = \rho_0(dP/\beta) \qquad (\text{thus } a \equiv 1/\beta) \tag{37.94}$$

The change in energy equals the work done on the fluid:

$$E_e = M \int_{v_1}^{v_2} - P\, dv \tag{37.95}$$

Since $V = 1/\rho$ = specific volume and

$$dv = -v_0(dP/\beta) \tag{37.96}$$

the elastic energy of fluid compression becomes

$$E_e = (M/2\rho_0\beta)(P_2^2 - P_1^2) \tag{37.97}$$

for a pressure change of $P_2 - P_1$.

For the elastic thin-walled pipe[20] the stored energy is

$$E_e = (M/2\rho_0\beta_e)(P_2^2 - P_1^2) \tag{37.98}$$

where

$$1/\beta_e = 1/\beta + 1/[E(P_0/P_1 - 1)] \tag{37.99}$$

Accumulators such as the added-volume type or surge tank, the spring-loaded or air-loaded ram type, or the bladder type are used to provide system compensation because they reduce system natural frequency or provide lead-lag effects.[2,4]

A simple example of a spring-loaded accumulator is given to illustrate the method (Fig. 37.22*a*).

The flow across the laminar restriction

$$\Delta Q_a = (\Delta P_s - \Delta P_a)(1/R_c) \tag{37.100}$$

where R_c is the laminar resistance to flow.

The flow into the accumulator

$$\Delta Q_a = (A_a/k_a)D\,\Delta P_a \tag{37.101}$$

for which k_a is the spring rate of the spring-loaded accumulator.

Eliminating ΔP_a between equations yields

$$\Delta Q_a = \frac{A_a^2/k_a}{(A_a^2R_c/k_a)D + 1} D(\Delta P_s) \tag{37.102}$$

This first-order lag can be shown to be in series with the load compressibility flow[4] and thus presenting the possibility of lead compensation to permit added system stability.

37.6. HYDRAULIC POWER TRANSMISSION

(a) Introduction

The connection between the power-generating, power-controlling, and power-utilization devices requires the transmission of flows and pressures through the transmission lines. In hydraulic systems the major proportion of power is transmitted as flow. Because the change in fluid power requires pressure changes, transmission of pressure signals becomes an all-important consideration in system design to assure a dynamically stable and responsive system. The problem of hydraulic-power transmission can thus be divided into two parts:

1. Steady flow through pipes, tubes, and fittings
2. Dynamic response of the hydraulic transmission line

(b) Steady Flow through Pipes and Fittings

Flow in Tubing. The tubing used in normal hydraulic practice is made of drawn-aluminum alloy, copper, or stainless steel of very smooth surface.

For Reynolds numbers below the lower critical value of 2,000, the friction factor is given by

$$f = 64/\text{Re} \tag{37.103}$$

where $\text{Re} = \rho VD/\mu$.

For Reynolds numbers above the upper critical value of 4,000, the resistance law for pipes is given by the Karman-Prandtl equation

$$1/\sqrt{f} = 2.0 \log_{10} (\text{Re}\sqrt{f}) - 0.8 \tag{37.104}$$

This relationship agrees exactly with experiment for ultrasmooth pipes with roughness ratios k/D less than 0.00001.

Figure 37.24 is the Stanton diagram, which presents the friction factor as a function of Reynolds number.[12] For standard drawn tubing an absolute roughness of

$$k = 0.00006 \text{ in.}$$

should be used.

The pressure drop in a tube of length L and diameter D is given by

$$\Delta P = f(L/D)(\rho V^2/2) \tag{37.105}$$

Hydraulic circuits are characterized by a large number of bends in tubing and fittings of various types such as elbows, tees, unions, contractions, and enlargements. To

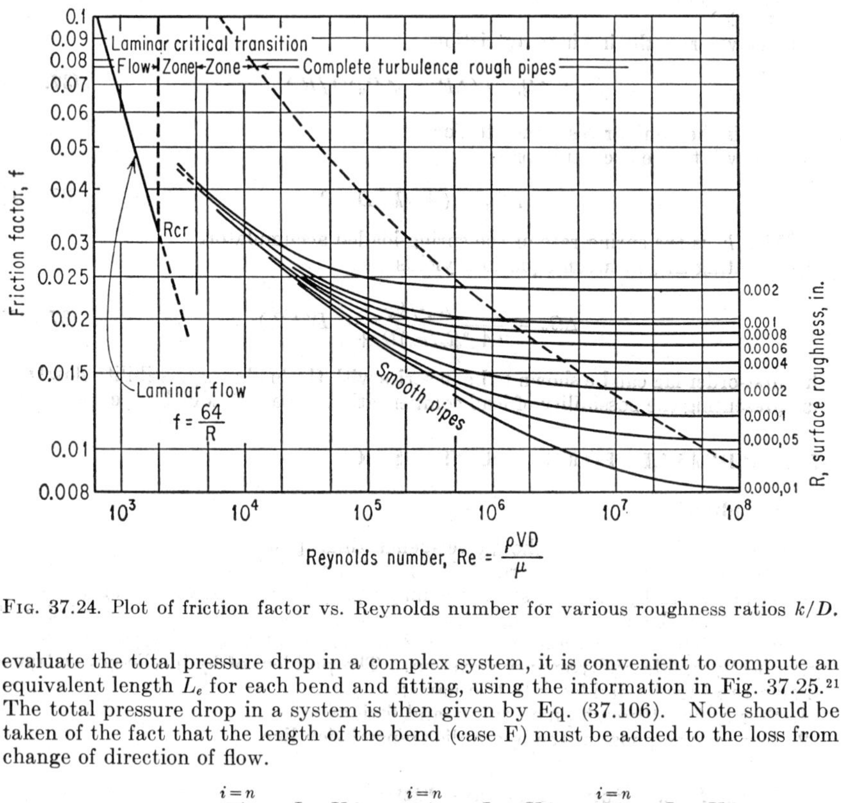

FIG. 37.24. Plot of friction factor vs. Reynolds number for various roughness ratios k/D.

evaluate the total pressure drop in a complex system, it is convenient to compute an equivalent length L_e for each bend and fitting, using the information in Fig. 37.25.[21] The total pressure drop in a system is then given by Eq. (37.106). Note should be taken of the fact that the length of the bend (case F) must be added to the loss from change of direction of flow.

$$\Delta P(\text{system}) = \underbrace{\sum_{i=1}^{i=n}}_{\text{tubing}} f_i \frac{L_i}{D_i}\frac{\rho V_i^2}{2} + \underbrace{\sum_{i=1}^{i=n}}_{\text{fittings}} f_i \frac{L_{ei}}{D_i}\frac{\rho V_i^2}{2} + \underbrace{\sum_{i=1}^{i=n}}_{\text{bends}} f_i \frac{L_{ei}}{D_i}\frac{\rho V_i^2}{2} \tag{37.106}$$

Since the coefficient of energy loss $L_e f/D$ is inversely proportional to fluid density, the plots must be corrected for wide departures from the reference values of the aircraft hydraulic oil (An-0-366b) at 75°F ($W = 0.0312$ lb/in.³).

In many systems flexible hosing is used. Where actual pressure-drop test data are not available from the manufacturer, Fig. 37.26 can be used as a guide in estimating the effect of roughness ratios on terminal-friction factors. For orifice data, see Art. 37.1.

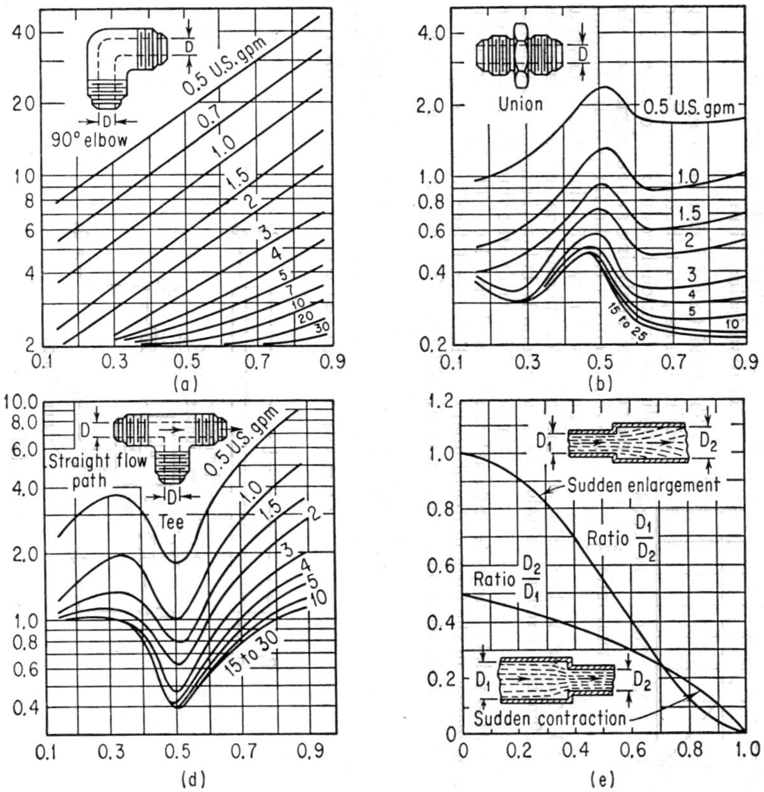

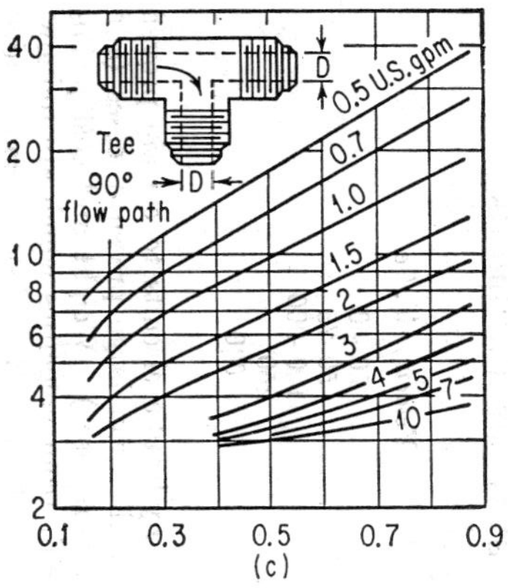

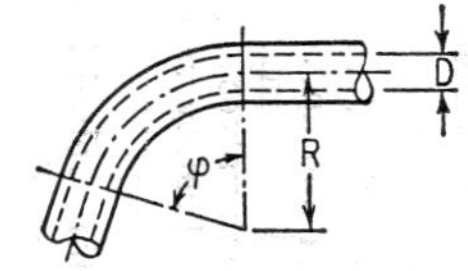

Coefficient of energy loss due to change in direction of flow				
Angle of bend, φ deg	Ratio, R/D			
	2	4	8	12
45	0.13	0.12	0.19	0.25
90	0.19	0.16	0.27	0.35
120	0.21	0.17	0.30	0.39

(f)

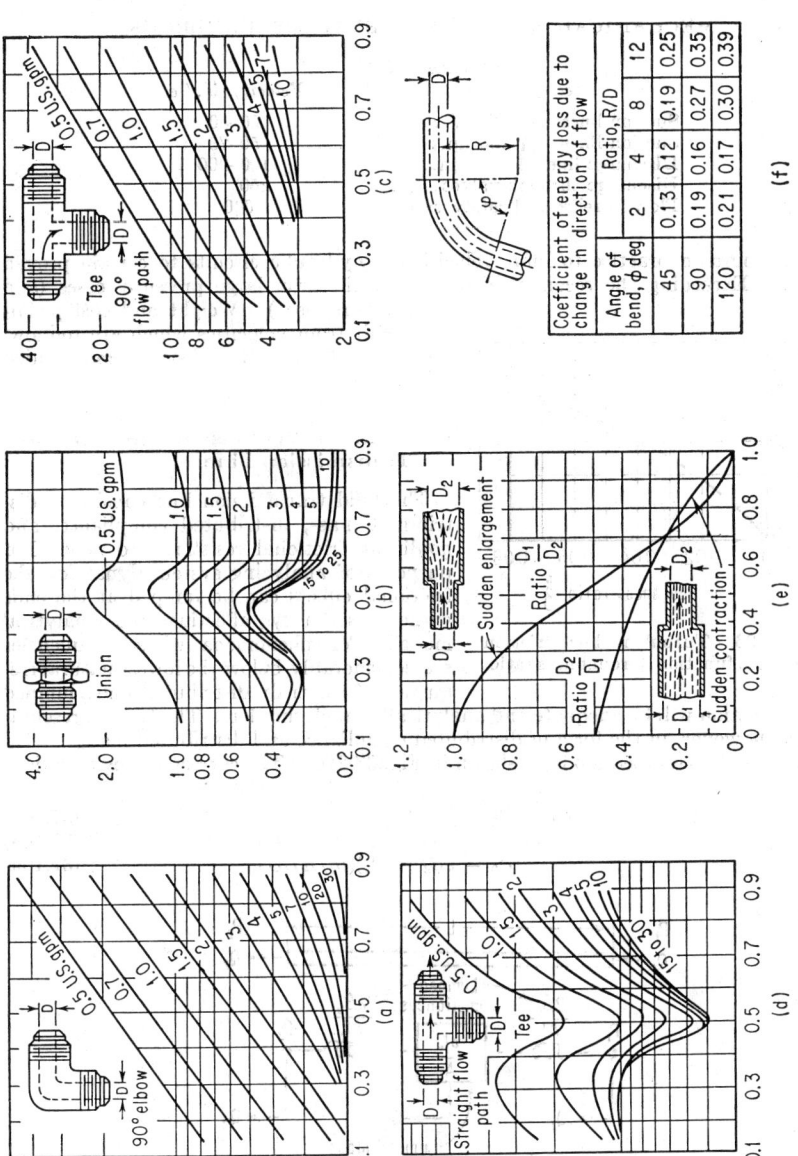

FIG. 37.25. Plots of coefficient of energy loss $L_c f/D$.

Table 37.4

Area of system	*Velocity, in./sec*
Suction lines, ¼ to 1 in.	20–50
Suction lines, 1¼ and up	50–75
Discharge lines, ¼ to 2 in.	100–200
Discharge through valves	250
Flow in relief and safety valves	1,000

It is common practice to limit velocities in hydraulic circuits to certain proved ranges. Exceeding the recommended values will mean large pressure losses (and temperature rises). Weight and cost penalties result from velocities which are too low, however. Table 37.4 lists recommended velocities.

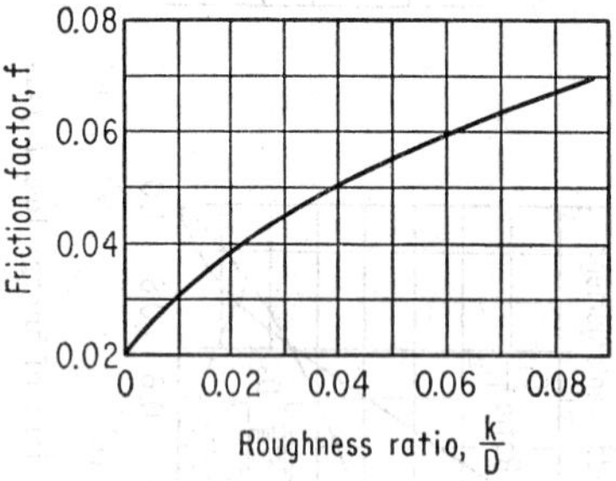

FIG. 37.26. Turbulent-flow friction factors as a function of roughness ratio.

(c) Dynamic Response of the Hydraulic Transmission Line

The fluid transmission line consists of distributed mass, distributed compliance, and nonlinear frictional resistance to flow. For purposes of analyzing system dynamics, the frictional effects can be lumped at the ends without seriously affecting the analytical model. For phenomena whose frequencies are considerably below the lowest natural frequency of the truly distributed line a lumped representation will be adequate and analytically much simpler. The true representation is, however, in the form of distributed compliance and distributed mass.

The simplest lumped model is shown in Fig. 37.27*a*. The corresponding equations are

$$P_1 - P_2 = \rho LDQ \tag{37.107}$$

$$P_1 = 2(Q_1 - Q)(\beta_e/ALD) \tag{37.108}$$

$$P_2 = 2(Q - Q_2)(\beta_e/ALD) \tag{37.109}$$

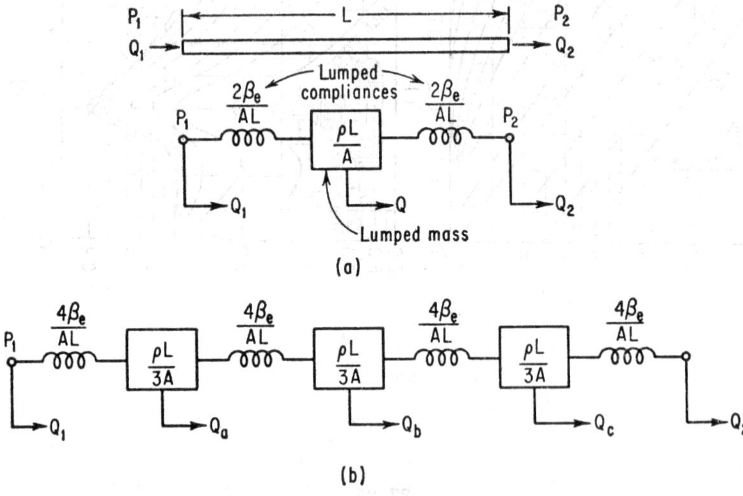

FIG. 37.27. Transmission lines. (*a*) One-lump line. (*b*) Three-lump line.

These fundamental equations can be rearranged into the more convenient form[20]

$$P_1 = G_{1a}Q_1 + G_{1b}Q_2 \tag{37.110}$$
$$P_2 = G_{2a}Q_1 + G_{2b}Q_2 \tag{37.111}$$

where
$$G_{1a} = \frac{\beta_e}{ALD}\,\frac{1 + \rho L^2D^2/2\beta_e}{1 + \rho L^2D^2/4\beta_e} \tag{37.112}$$

$$G_{2a} = \frac{\beta_e}{ALD}\,\frac{1}{1 + \rho L^2D^2/4\beta_e} \tag{37.113}$$

$$G_{1b} = -G_{2a} \tag{37.114}$$
$$G_{2b} = -G_{1a} \tag{37.115}$$
transfer coefficients of pipe impedance

where β_e = equivalent bulk modulus is given by Eq. (37.116)

$$\frac{1}{\beta_e} = \frac{1}{\beta} + \frac{2}{E(D_0/D_i - 1)} \tag{37.116}$$

β = fluid bulk modulus
A = cross-sectional area of pipe
D = differential operator
E = modulus of elasticity of conduit material

The input impedance of a transmission line is given by Eq. (37.117), the output impedance by Eq. (37.118).

$$Z_1 = G_{1a} + G_{1b}G_{2a}\frac{1}{Z_2 - G_{2b}} \tag{37.117}$$

$$Z_2 = G_{2b} + G_{2a}G_{1b}\frac{1}{Z_1 - G_{1a}} \tag{37.118}$$

Frequency-response representations can be obtained by the usual substitution of $j\omega$ for the differential operator D. The accuracy of the lumped representation can be

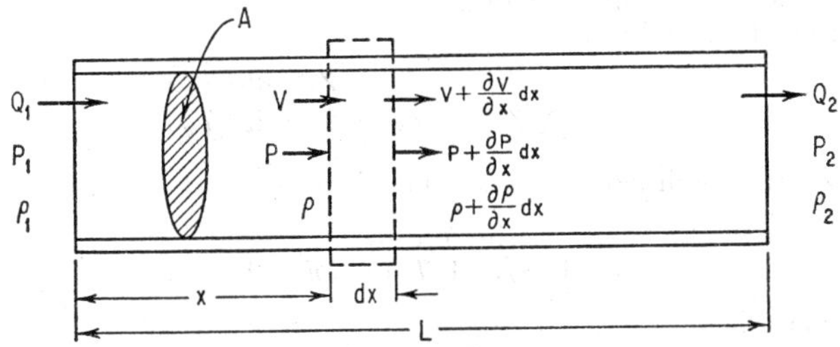

FIG. 37.28. Distributed-parameter transmission line.

checked at a particular frequency range by comparing gain and phase relationships with the more accurate distributed parameter equations. For this purpose it is advisable to expand the exponentials of the distributed cases in appropriate series.

The compliance can be further broken down, more closely approaching the distributed case as shown in Fig. 37.27b, but the ensuing analysis becomes cumbersome.

The distributed-case analysis is based upon the continuity and momentum equations, employing the conventions of Fig. 37.28.

The continuity equation becomes

$$-\partial V/\partial x = (1/\beta_e)(\partial P/\partial t) \tag{37.119}$$

The momentum equation becomes

$$-\partial P/\partial x = \rho(\partial V/\partial t) \tag{37.120}$$

These two equations are known as the *wave equations*. Their solution in the form of hyperbolic functions is given by

$$P_2 = (\cosh \Gamma)P_1 - (Z_s \sinh \Gamma)Q_1 \tag{37.121}$$
$$Q_2 = -[(1/Z_s) \sinh \Gamma]P_1 + (\cosh \Gamma)Q_1 \tag{37.122}$$
$$P_1 = (\cosh \Gamma)P_2 + (Z_s \sinh \Gamma)Q_2 \tag{37.123}$$
$$Q_1 = [(1/Z_s) \sinh \Gamma]P_2 + (\cosh \Gamma)Q_2 \tag{37.124}$$

where

$$\Gamma = L\sqrt{\rho/\beta_e}\, D \tag{37.125}$$
$$Z_s = \text{characteristic line impedance } \sqrt{\rho\beta_e^2/A} \tag{37.126}$$

It should be noted that

$$T_p = L\sqrt{\rho/\beta_e} \tag{37.127}$$

is the propagation time for a wave to travel the length of the tube L.

Equations (37.121) through (37.124) can be rewritten in the form of Eqs. (37.110) and (37.111) with the coefficients now given by

$$G_{1a} = Z_s \cosh \Gamma / \sinh \Gamma \tag{37.128}$$
$$G_{2a} = Z_s / \sinh \Gamma \tag{37.129}$$
$$G_{2b} = -G_{1a} \tag{37.130}$$
$$G_{1b} = -G_{2a} \tag{37.131}$$

The impedance equations for input and outlet can be derived for the specific case using Eqs. (37.117) and (37.118) and the appropriate terms from Eqs. (37.128) through (37.131). For the specific case of a resistive load at the outlet (downstream end) the inlet impedance can be shown to be[22]

$$Z_1 = Z_s \frac{(Z_2 + Z_s)e^{\Gamma} + (Z_2 - Z_s)e^{-\Gamma}}{(Z_2 + Z_s)e^{\Gamma} - (Z_2 - Z_s)e^{-\Gamma}} \tag{37.132}$$

Since $\Gamma = L\sqrt{\rho/\beta_e}\, D$, the substitution of the complex frequency $j\omega$ for D permits the separation into real and imaginary parts, the derivation of the amplitude and phase relationships as a function of the frequency ω.

The amplitude is given by

$$|Z_1| = Z_s \sqrt{\frac{Z_2^2 + (Z_s^2 - Z_2^2)\sin^2 T_p\omega}{Z_s^2 - (Z_s^2 - Z_2^2)\sin^2 T_p\omega}} \tag{37.133}$$

When $Z_s > Z_2$, the amplitude is a maximum when

$$\sin^2 T_p\omega = 0$$

or

$$\omega = (2n - 1)(\pi/2)(1/T_p) \qquad n = 1, 2, 3, \ldots$$

and a minimum when

$$\sin^2 T_p\omega = 0$$

or

$$\omega = (n - 1)\pi(1/T_p) \qquad n = 1, 2, 3, \ldots$$
$$|Z_1|_{\max} = Z_s^2/Z_2$$
$$|Z_1|_{\min} = Z_2$$

When $Z_s < Z_2$ the impedance is a minimum when

$$\omega = (2n - 1)(\pi/2)(1/T_p) \qquad n = 1, 2, 3, \ldots$$

and a maximum when

$$\omega = (n - 1)\pi(1/T_p) \qquad n = 1, 2, 3, \ldots$$

Then

$$|Z_1|_{\min} = Z_s^2/Z_2$$

and

$$|Z_1|_{\max} = Z_2$$

The phase angle is given by

$$\phi_1 = \tan^{-1}[(Z_s/Z_2)\tan T_p\omega] - \tan^{-1}[(Z_2/Z_s)\tan T_p\omega] \tag{37.134}$$

Reference 22 shows good experimental agreement with the wave-equation solution in the frequency response of a 68-ft-long 1-in.-diameter hydraulic transmission line. Some of the conclusions were:

1. Pipe friction had negligible effect.
2. Radial tube vibration had negligible effect.
3. Longitudinal tube vibration had marked effect on the frequency response in certain frequency ranges.
4. Mean fluid velocity (usually lower than 200 in./sec), being much lower than sonic velocities (for JP4, 46,800 in./sec), can be neglected in line-performance studies.

37.7. HYDRAULIC POWER CONTROL

(a) Introduction

There are essentially two methods of controlling the flow of hydraulic power to a load:

1. By varying some characteristic of the pump (power generator) so that the rate at which fluid energy is generated is controlled in a definable manner.
2. By throttling the fluid power in a single- or multiple-orifice valve, thereby effecting control by predictable flow restrictions. Some of the fluid power is thus changed into heat, resulting in a fluid-temperature rise.

In addition to the flow characteristics, the flow- and pressure-induced forces on the valve components must be described. Understanding the nature of these forces is important in the design of pressure and flow regulators as well as the specification and analysis of the behavior of the flow-valve actuators such as torque motors.

(b) Flow-valve Configurations

The hydraulic control valve is an assembly of flow-controlling elements which operate in a prescribed and predictable manner. Several common constructions are described as follows:

The Spool Valve. The spool valve is one of the most common types. Figure 37.29 is the two-way application, Fig. 37.30 the three-way, and Fig. 37.31 the more complex but most versatile four-way valve system. Each system permits the driving of a load.

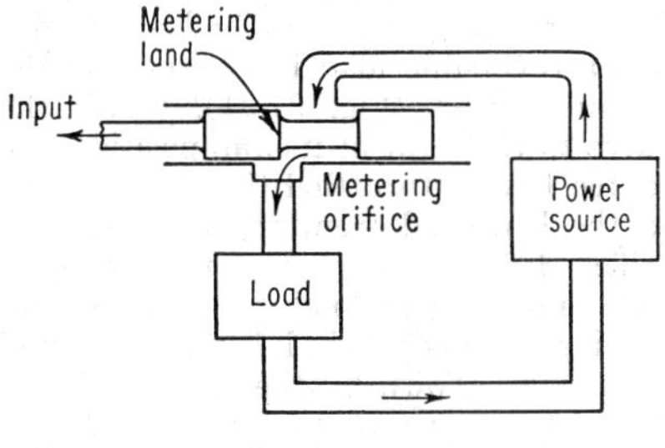

FIG. 37.29. Two-way spool valve.

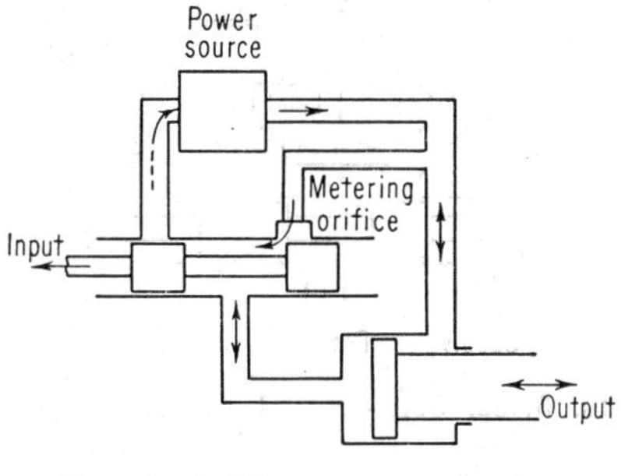

FIG. 37.30. Three-way spool valve.

The *two-way valve* introduces a restriction in the line to the load which reduces a part of the supply pressure and dissipates some of the supply power. Thus the fluid power to the load can be controlled. With a simple restriction, the direction of flow to the load cannot be changed.

The *three-way valve* (Fig. 37.30) is the simplest configuration which permits load reversal. The differential-area double-acting piston exposes its smaller area to the supply pressure and the larger (usually double) area to the valve-controlled pressure.

The exhaust port permits fluid to escape to the return line so that the supply pressure can push the actuating piston to the left.

If the distance between the metering edges on the spool and the sleeve are identical, the valve is said to be of the *closed-center* and zero-lapped type. If the distance between the metering edges on the spool is larger than those on the sleeve, the valve is defined as an *open-center* (or underlapped) valve. With the open-center valve there is a steady flow of fluid from the source to the return line for valve positions in the underlap region. The closed-center and overlapped valve has the reverse geometry with

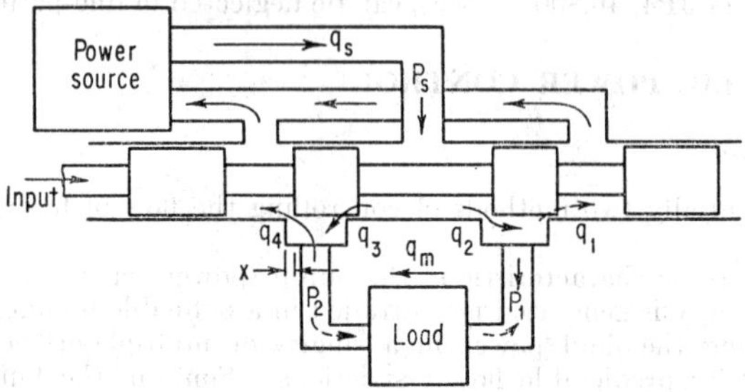

FIG. 37.31. Four-way spool valve, closed-center type.

the load sealed from both source and return over the overlap region of the valve. In practice the unavoidable clearance between valve and sleeve will permit some small flow to occur.

The three-way valve permits an interesting variation in the power source. The source can be a *constant-flow source* working into the underlapped valve, with the valve operating primarily in the underlap region and the pressure in the source built up to meet load-force power requirements. Beyond the underlap region the valve functions as a two-way valve, with the load velocity determined by the source flow. The source can be a *constant-pressure source* with a closed-center valve. The flow from the source will vary depending upon load-velocity demands, load direction and leakage.

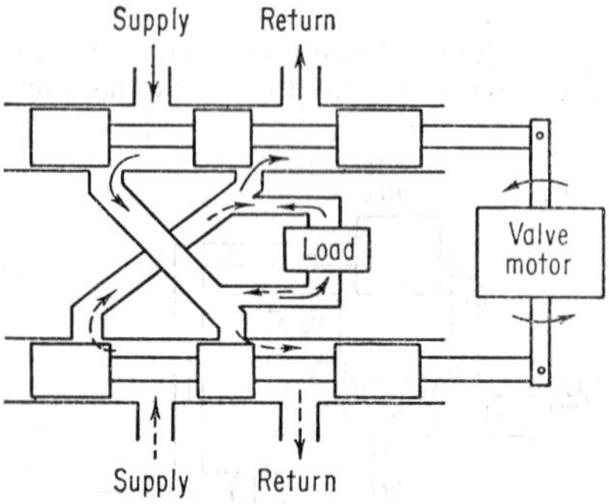

FIG. 37.32. Split-type closed-center four-way valve.

The most common method of controlling a reversible load is by means of the *four-way valve* (Fig. 37.31). With the valve to the right of its null position, the source flow is throttled through orifice 2 to the load and from the load through orifice 4 to return. The reverse flows from and to the load are attained by moving the spool to the left of the centered position as shown by the dotted arrows. The four-way valve, similar to the three-way valve, is used as either an underlapped valve with a constant-flow source (power steering) or a closed-center valve with a constant-pressure source. If a single hydraulic power supply has to provide power to several valve-motor-load systems, the closed-center constant-pressure design is used with either a variable-delivery pump or a constant-delivery pump and bypassing pressure regulator. Power dissipation, load cycles, and cost consideration dictate which system is used.

The three-way valve has several limitations compared with the four-way valve:

1. It is difficult to apply to a rotary motor.

2. The volume under compression compared with an equivalent four-way system is larger, thus tending to give poorer dynamic response in a closed-loop system.

3. Under certain conditions there is a lower load stiffness.
4. It tends to be more nonlinear in its flow-pressure characteristics.

The prime advantage of the three-way valve is its simplicity. There is only one critical axial port dimension rather than three for the single-spool four-way valve.

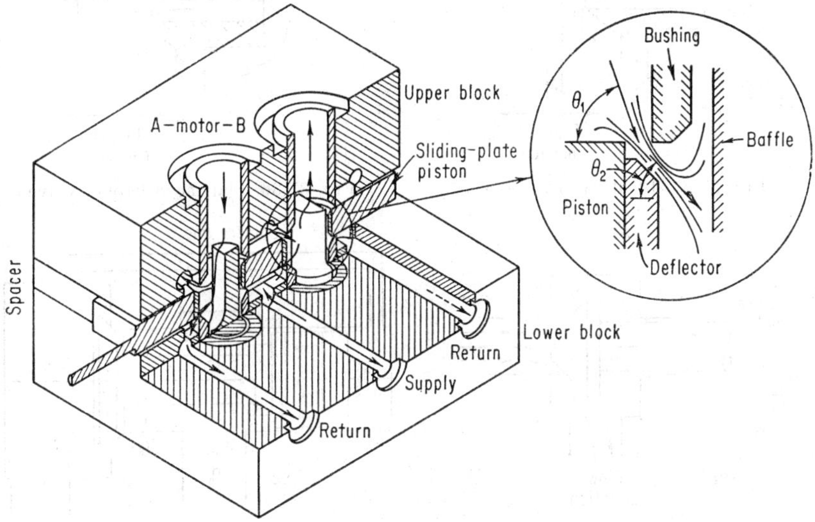

FIG. 37.33. Flat-plate valve.

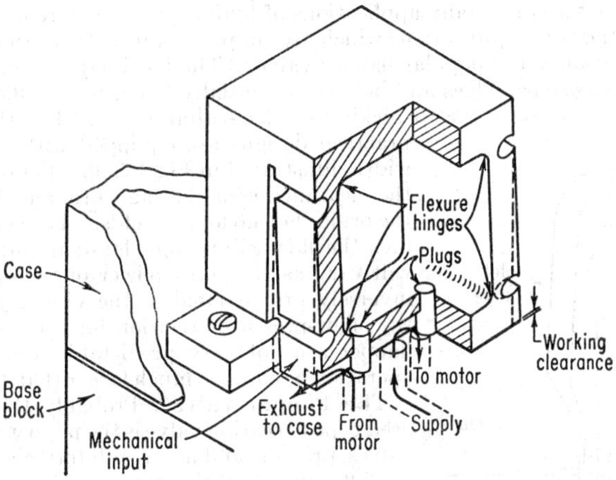

FIG. 37.34. Suspension valve.

It is possible to construct the four-way valve out of two three-way valves which are mechanically linked as shown in Fig. 37.32. The dimensional stability of this construction does not compare favorably with the solid-spool construction.

Other Sliding-type Valves. Many attempts have been made to overcome some of the manufacturing difficulties of the sleeve-type four-way valve by unwrapping the inaccessible sleeve parts and constructing flat plate valves.[4,23] The flat-plate valve is illustrated in Fig. 37.33. A modification of the flat-plate valve is the suspension valve shown in Fig. 37.34.

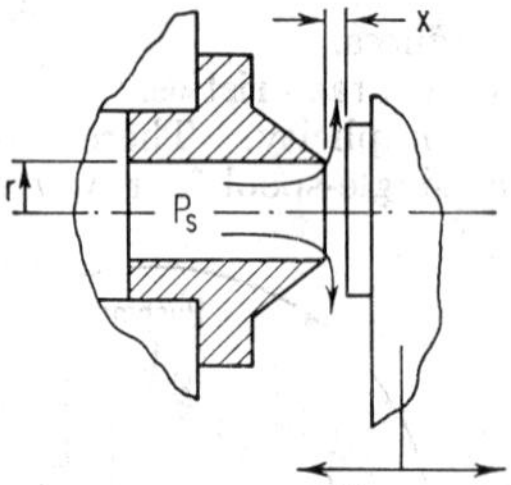

FIG. 37.35. Ideal flapper valve.

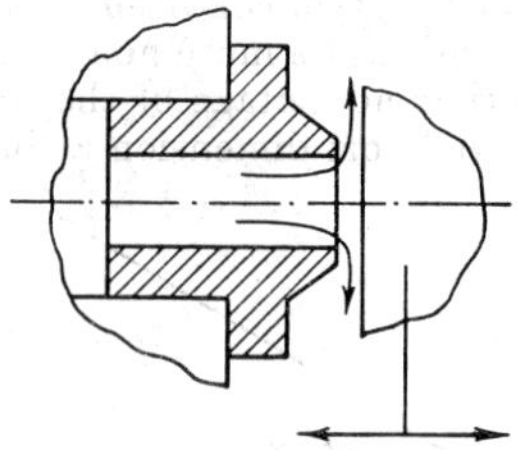

FIG. 37.36. Flat-faced flapper valve.

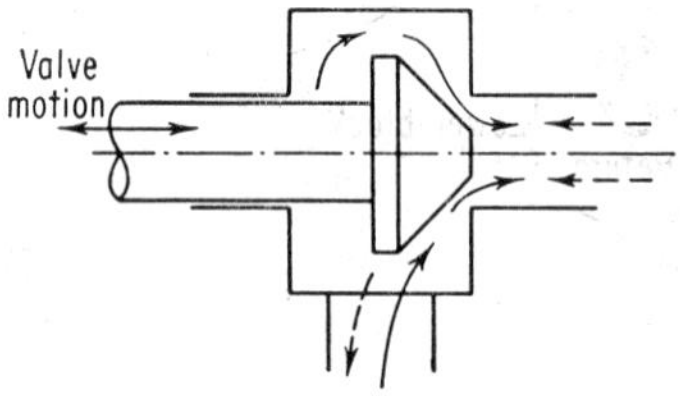

FIG. 37.37. Poppet valve.

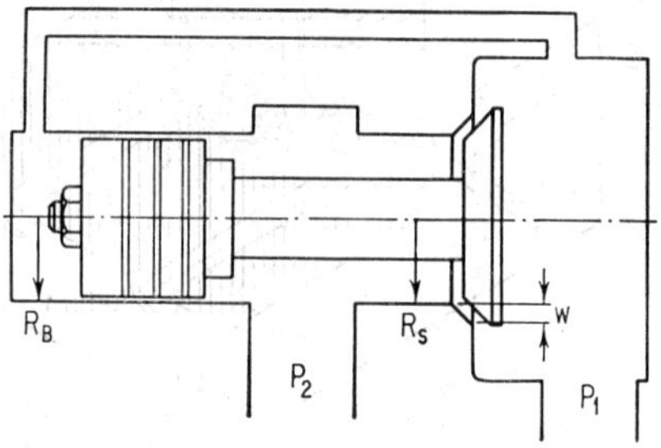

FIG. 37.38. Balanced inverted poppet valve. For approximate balance $R_B = R_s + \frac{1}{2}W$.

The Flapper Valve. Many applications of hydraulic controls require pilot valves (for the first stage of amplification) which are simple and insensitive to contaminants. One mechanization is the popular flapper valve. The ideal flapper valve, illustrated in Fig. 37.35, has sharp edges and behaves essentially like a pure orifice. The ideal flapper valve is too sensitive to nicking of the sealing edges. For this reason the practical designs are equipped with a face of finite width, illustrated in Fig. 37.36. For openings where the clearance is of the same order as the seat width, the orifice law no longer holds. For very small openings the flow will become laminar and thus the flow will vary as the third power of the opening and be inversely proportional to the viscosity. In general the flow-pressure relationships of the flat-faced flapper are not very predictable unless the normal operation opening is much larger than the seat width.

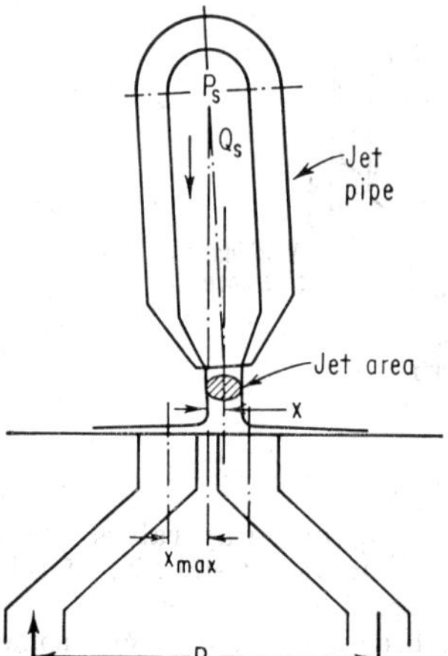

FIG. 37.39. Jet-pipe valve.

The Poppet Valve. Probably the most widely used type of seating valve is the poppet valve. Relief valves, pressure and flow regulators, check valves, and shutoff valves use the poppet valve extensively. A typical unbalanced poppet is shown in Fig. 37.37. The inverted and pressure-balanced valve is shown in Fig. 37.38. With a sharp-edged seat it behaves like a pure orifice. Near the cutoff point, similar to the flapper, the valve is highly nonlinear, since a particular design must have a seat of finite width, usually 0.002 to 0.005 in.

Jet-pipe Valve. The process-control field has long used the jet-pipe valve. It is a hydrokinetic valve in that the jet-velocity impingement on the receiver (Fig. 37.39) creates the pressure by momentum recovery. The flow pattern is highly complex and subject to wide variation depending upon fabrication details.[24] The valve, since it contains neither closely fitted sliding surfaces nor small orifices, is remarkably insensitive to dirt. The flow-induced forces and flow characteristics are empirically determined. Recent test results are given later.

(c) Pressure-Flow Relationships

Introduction. The pressure-flow-displacement relationships for many of the configurations mentioned in the previous paragraphs will be derived, based upon the following assumptions:

1. The fluid is nonviscous and incompressible. For very small valve openings of flappers and poppet valves and low pressure drops and for flow through clearance spaces in spool valves, these assumptions should be examined as the case demands. The assumption of incompressibility inside the valve, for the flow relationships, results in negligible errors.
2. The fluid source is ideal. This assumption demands a constant-pressure-source (CP) supply fluid to the valve and a constant pressure at the intake regardless of flow demands. The constant-flow source (CQ) similarly provides a constant flow at the valve intake port regardless of pressure demands. The idealized characteristics should thus be corrected for excessive line drops and regulation droops.
3. The geometry of the valve is ideal. It is assumed that the metering edges of the valve are sharp and that the clearances between spool and sleeve are zero.
4. Steady-state conditions prevail inside the valve. For a stable valve and system the assumption is sound. For valve squeal and chatter the characteristics no longer hold. Valve stability will be discussed in Art. 37.7(d).

The volume flow through an orifice, derived previously, is given by

$$Q = C_d A_0 \sqrt{2(\Delta P)/\rho} \tag{37.135}$$

The discharge coefficient C_d is very nearly constant for Reynolds numbers larger than 260 ($\text{Re} = 2Vx\rho/\mu$)

where A_0 = nominal area for annular opening ($A_0 = \pi D_0 x_0$) in.^{2}
D_0 = annular mean diameter, in.
x = width of annular slit, in.
ΔP = pressure drop

For the sharp-edged orifice it is assumed that $C_d = 0.625$.

Caution should be exercised in interpreting test data. If the valve saturates, meaning that an appreciable portion of the pressure drop occurs in the valve chambers and passages, rather than across the orifice, the analysis is not valid. Similarly, if the opening approaches the clearance, the relationships are in error.

The characteristics of the following cases will be given:

1. The symmetrical four-way valve with constant-pressure source, (*a*) and (*b*) four arms variable, (*c*) two arms variable
2. The symmetrical four-way valve with constant-flow source, (*a*) four arms variable, (*b*) two arms variable
3. The three-way valve with constant-pressure source, (*a*) two arms variable, (*b*) one arm variable
4. The three-way valve with constant-flow source, (*a*) two arms variable, (*b*) one arm variable
5. The partially underlapped symmetrical case of the four-way valve with the constant-pressure source
6. Asymmetric cases

For details of the derivations the reader is referred to ref. 4, chap. 7. The plots of these equations are useful in understanding valve characteristics. The slopes and spacing of the individual curves, the *valve gains*, are of paramount interest to the system designer and are therefore presented in analytical form.

Case 1a. Symmetrical Closed-center Four-way Valve with Constant-pressure Source (Rectangular Ports) (see Fig. 37.31). The dimensionless equation relating flow to pressure to valve displacement is

$$P_m = 1 - 2Q_m^2/y^2 \tag{37.136}$$

where

$$P_m = (P_1 - P_2)/P_s = \text{motor pressure} \tag{37.137}$$

$$Q_m = q_m/G\sqrt{P_s} = \text{flow to motor} \tag{37.138}$$

$$y = x/x_0 = \text{displacement where } x \text{ is valve opening} \tag{37.139}$$

$$x_0 = \text{maximum valve opening}$$

$$G = g/y = \text{unit valve conductance} \tag{37.140}$$

with $g = AC_d\sqrt{2/\rho}$ (A is the port area corresponding to x) where G is the conductance for $y = 1$ (the valve fully open).

Equation (37.136) is represented by a set of parabolas and is plotted in Fig. 37.40*a*.

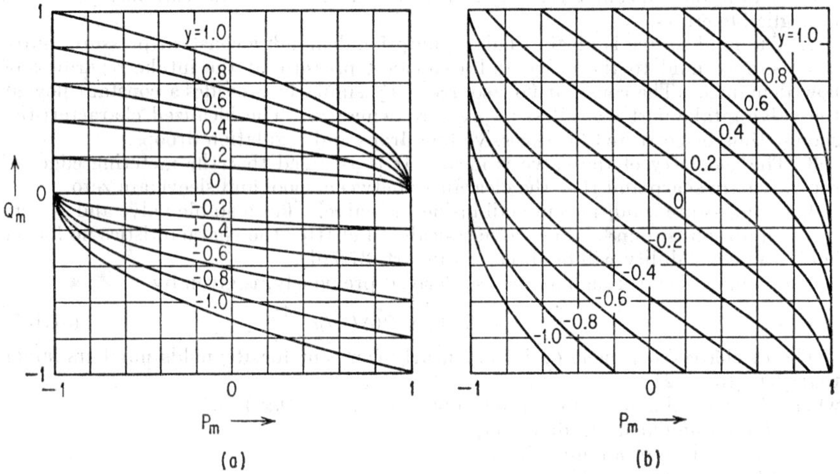

FIG. 37.40. Dimensionless pressure-flow valve characteristics. (*a*) Closed-center four-way valve, constant pressure supply. (*b*) Open-center four-way valve, constant pressure supply.

The flow sensitivity is given by k_{q0}, the spacing of the curves at constant load pressure.

$$k_{q0} = \partial q_m/\partial x \Big|_{p_m = \text{const}} \tag{37.141}$$

The pressure sensitivity is given by k_{p0}, the slope of the curves at constant flow to the load,

$$k_{p0} = \partial p_m/\partial x \Big|_{q_m = \text{const}} \tag{37.142}$$

Thus, near the origin

$$k_{q0} = (G/x_0)\sqrt{P_s/2} \tag{37.143}$$

k_{p0} for the ideal case is infinite; for practical cases it is possible to attain values of 10^6 psi/in. for supply pressures in the 1,000 to 3,000 psi range.

Case 1b. Symmetrical Open-center Four-way Valve with Constant-pressure Source. If the distance between the lands on the spool in Fig. 37.31 is less than the distance between the lands on the sleeve by an amount $2x_0$ (x_0 being the so-called underlap or open center) the valve becomes the underlapped valve. For operation in the underlap region the valve characteristics are given by (see Fig. 37.40*b*)

$$Q_m^2 = 1 + y^2 - 2yP_m - (1 - y^2)\sqrt{1 - P_m^2} \qquad (37.144)$$

The curves in Fig. 37.40*b* are remarkably straight, parallel, and evenly spaced, indicating a linear device over most of its operating range. The standby power loss is very high relative to the power supplied to the load. If the operation of the valve is not limited to the underlap region, case 5 results.

For this case, the values of valve gain are

$$k_{q0} = 2(G/x_0)\sqrt{P_s/2}$$
$$k_{p0} = 2(P_s/x_0)$$

Case 1c. Symmetrical Four-way Valve with Constant-pressure Source with the Two Upstream Arms Fixed and the Two Downstream Arms Variable (Twin Flapper-nozzle System) (see Fig. 37.41). The dimensionless equations relating flow, pressure, and geometry are very complicated for the total valve. A

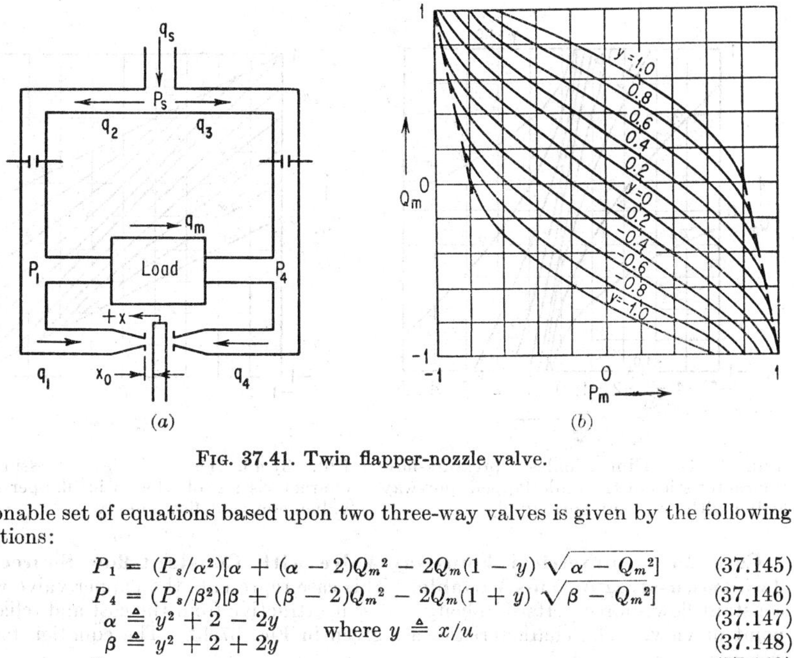

FIG. 37.41. Twin flapper-nozzle valve.

reasonable set of equations based upon two three-way valves is given by the following equations:

$$P_1 = (P_s/\alpha^2)[\alpha + (\alpha - 2)Q_m^2 - 2Q_m(1 - y)\sqrt{\alpha - Q_m^2}] \qquad (37.145)$$
$$P_4 = (P_s/\beta^2)[\beta + (\beta - 2)Q_m^2 - 2Q_m(1 + y)\sqrt{\beta - Q_m^2}] \qquad (37.146)$$
$$\alpha \triangleq y^2 + 2 - 2y \qquad (37.147)$$
$$\beta \triangleq y^2 + 2 + 2y \qquad (37.148)$$
$$\text{where } y \triangleq x/u$$
$$P_m = p_1 - p_4 \qquad (37.149)$$
$$Q_m = q_m/G\sqrt{P_s} \qquad (37.150)$$

Figure 37.41*b* is a plot of the valve characteristics.

The valve gains at the origin are

$$k_{q0} = (G/x_0)\sqrt{P_s/2} \qquad (37.151)$$
$$k_{p0} = P_s/x_0 \qquad (37.152)$$

Case 2a. Symmetrical Four-way Valve with Constant-flow Source and Four Arms Variable. With a constant-flow source the supply pressure will adjust itself to meet load demands. Thus at no load the power loss is dictated by the orifice sizing in relation to the constant flow. Systems incorporating this design must contain supply-pressure-limiting means such as relief valves for protection. The flow-pressure-displacement characteristics are given by Eq. (37.153), Fig. 37.42.

$$P_m(1 - y^2)^2 = (1 + Q_m^2)y - Q_m(1 + y^2) \quad (37.153)$$

where

$$Q_m \triangleq q_m/q_s \quad (37.154)$$

$$P_m \triangleq p_m g_0^2/q_s^2 \quad (37.155)$$

where g_0 is the conductance for $y = 0$.

For $q_m = 0$, $y = 0$ the supply pressure P_{s0} is of interest because it indicates the standby power loss. If x_0 is the motion within the underlapped range, P_{s0} is given by

$$P_{s0} = \tfrac{1}{2}(q_s/g_0)^2 \quad (37.156)$$

thus

$$P_m = p_m/2P_{s0} \quad (37.157)$$

The flow sensitivity

$$k_{q0} = q_s/x_0 \quad (37.158)$$

The pressure sensitivity

$$k_{p0} = q_s^2/G^2x_0 \quad (37.159)$$

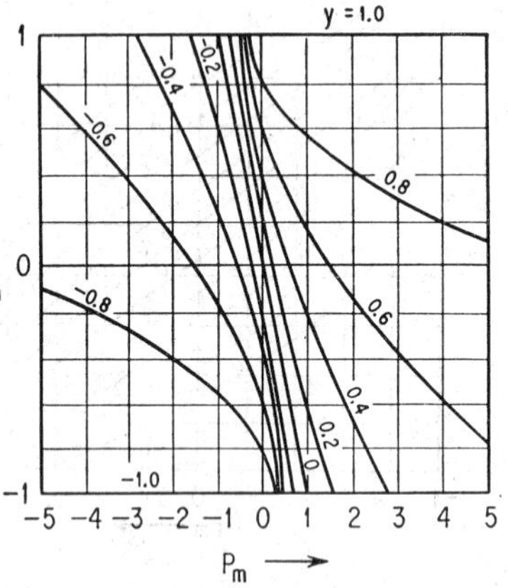

FIG. 37.42. Dimensionless pressure-flow characteristics of the underlapped four-way valve with constant flow supply.

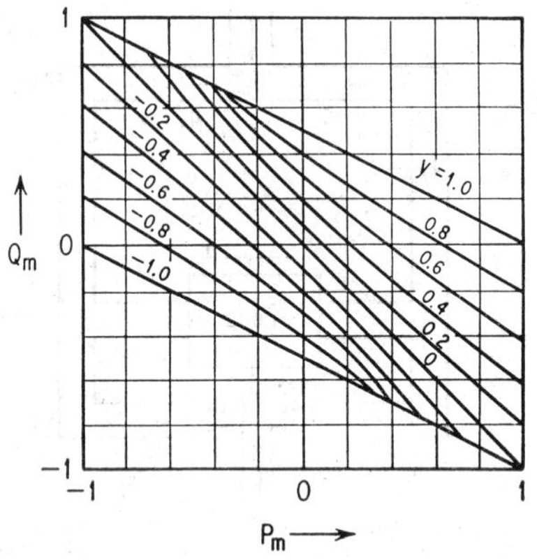

FIG. 37.43. Dimensionless pressure-flow characteristics of the twin flapper-nozzle valve, constant flow supply.

Case 2*b*. Symmetrical Four-way Valve with Constant-flow Source and Two Down-stream Arms Variable. This case represents the flapper valve with a constant-flow source. Its simplicity makes it attractive from the cost and reliability point of view. The characteristics are given in Fig. 37.43. The equation for this case is

$$P_m(1 - y^2)^2 = [1 + (P_m + 2Q_m)^2]y - (P_m + 2Q_m)(1 + y^2) \quad (37.160)$$

The flow sensitivity

$$k_{p0} = q_s/x_0 \quad (37.161)$$

The pressure sensitivity

$$k_{p0} = q_s^2/2G^2x_0 \quad (37.162)$$

Case 3*a*. Underlapped Three-way Valve with Constant-pressure Source and Both Arms Variable (Fig. 37.44*a*). The characteristics here given will be valid within the underlap region (Fig. 37.44*b*).

$$x = \pm x_0$$

The governing relationships are

$$Q_m = (1 + y)\sqrt{1 - P_1} - (1 - y)\sqrt{P_1} \tag{37.163}$$

where $Q_m = q_m/G\sqrt{P_s}$ $\quad P_1 = p_1/P_s$ $\quad y = x/x_0$

Rearrangement yields

$$P_1 = \frac{(y+1)^2(y^2+1) - 2yQ_m{}^2 - (1-y^2)Q_m\sqrt{2(1-y)^2 - Q_m{}^2}}{2(y^2+1)^2} \tag{37.164}$$

The flow sensitivity

$$k_{q0} = (2G/x_0)\sqrt{P_s/2} \tag{37.165}$$

The pressure sensitivity

$$k_{p0} = P_s/x_0 \tag{37.166}$$

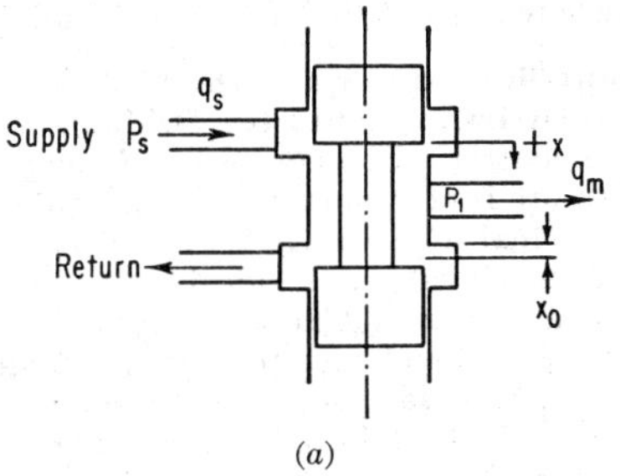

(a)

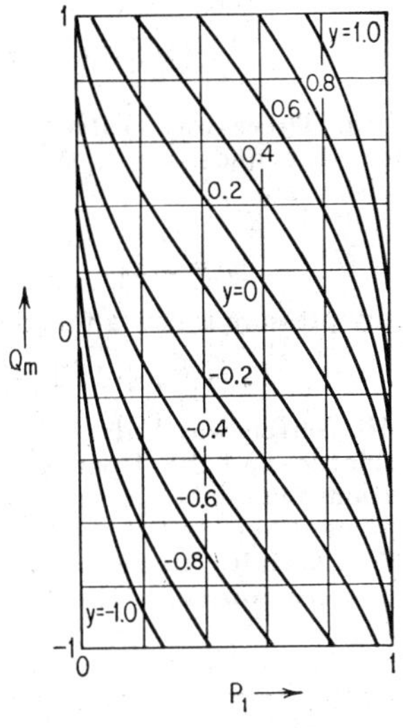

(b)

FIG. 37.44. Three-way underlapped valve.

Case 3b. Three-way Valve with Constant-pressure Source and the Downstream Arm Variable. The single-jet flapper is shown in Fig. 37.45a. While its simplicity recommends it, experience has shown that its characteristics are sensitive to variations in supply pressure. The governing equations are

$$Q_m = \sqrt{1 - P_1} - (1 - y)\sqrt{P_1} \tag{37.167}$$

$$P_1 = \frac{2y - y^2(1 - Q_m{}^2) - 2(1-y)Q_m\sqrt{2 - 2y + y^2 - Q_m{}^2}}{(2 - 2y + y^2)^2} \tag{37.168}$$

Figure 37.45b gives the valve characteristics.

The flow sensitivity

$$k_{q0} = (G/x_0)\sqrt{P_s/2} \tag{37.169}$$

The pressure sensitivity

$$k_{p0} = P_s/x_0 \tag{37.170}$$

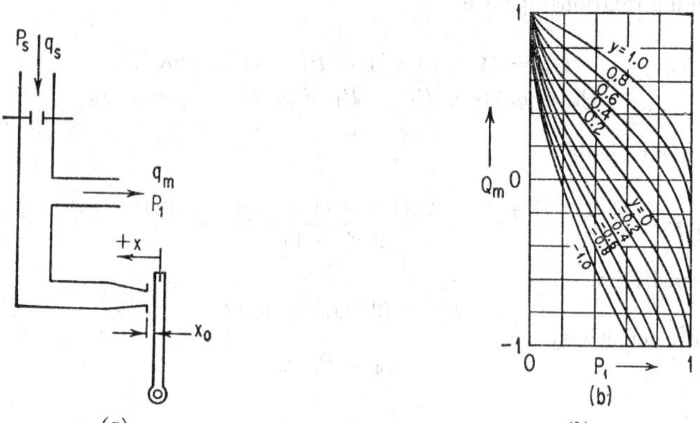

(*a*) (*b*)

FIG. 37.45. Single-jet flapper-nozzle valve.

Case 4*a*. Three-way Valve with Constant-flow Source and Two Arms Variable. The underlapped three-way valve is shown in Fig. 37.44. The flow characteristics are

$$Q_m = 1 - (1 - y)\sqrt{P_m} \tag{37.171}$$

where $y = x/x_0 \qquad P_m = p_1/(q_s^2/G^2) \qquad Q_m = q_m/q_s$

The supply pressure is now a function of P_m and y.

$$P_s = P_m + 1/(1 + y)^2 \qquad \text{where } P_s = P_s/(q_s^2/G^2) \tag{37.172}$$

Case 4*b*. Three-way Valve with Constant-flow Source and the Downstream Arm Variable and No Upstream Arm Restriction (Quarter Bridge). This configuration represents the simplest possible control. The equations are identical to those of case 4*a* except that the supply pressure is now equal to the load pressure.

Case 5. Partially Underlapped Symmetrical Four-way Valve with Constant-pressure Source. Case 1*a*, the closed-center valve in its ideal form, assumes no leakage from clearance between spool and sleeve and metering edge rounding. All practical designs will have clearances in excess of 0.0002 in. (radial) and an equivalent radius on the metering edge. The metering-edge radius, in particular, will increase with use because of erosion from fluid and contaminant flow, and crushing from hard foreign particles. It is important to evaluate carefully the valve gain near the closed position. k_{p0} is usually less than 10^6 psi/in. The valve characteristics of the actual valve are thus similar to an open-center valve near the null position. They approach closed-center characteristics for openings two to three times larger than the valve clearance or 0.0004 to 0.0006 in. Depending upon the full travel of the valve, the underlap region may occupy an appreciable portion of the valve travel. A typical case is illustrated in Fig. 37.46. The characteristics should be compared with those of Fig. 37.40*a* and *b*. For a

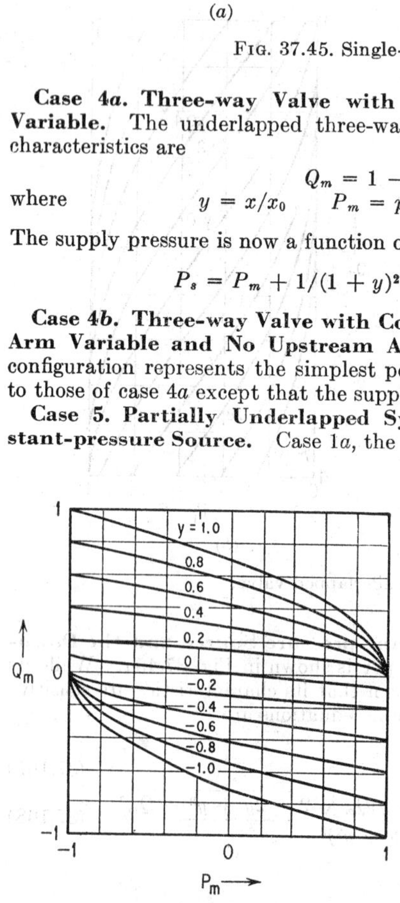

FIG. 37.46. Dimensionless pressure-flow characteristics of the partially underlapped (20 per cent) four-way valve with constant pressure supply.

particular case the valve characteristics can be constructed by appropriate approximations using Figs. 37.46 and 37.40.

Unsymmetrical Cases. The nonlinear characteristics of the zero-lapped valve and particularly its high gain near the null position make the design of stable systems difficult. The underlapped valve has been shown to result in more stable systems by virtue of its lower and more constant gains as well as inherently higher damping

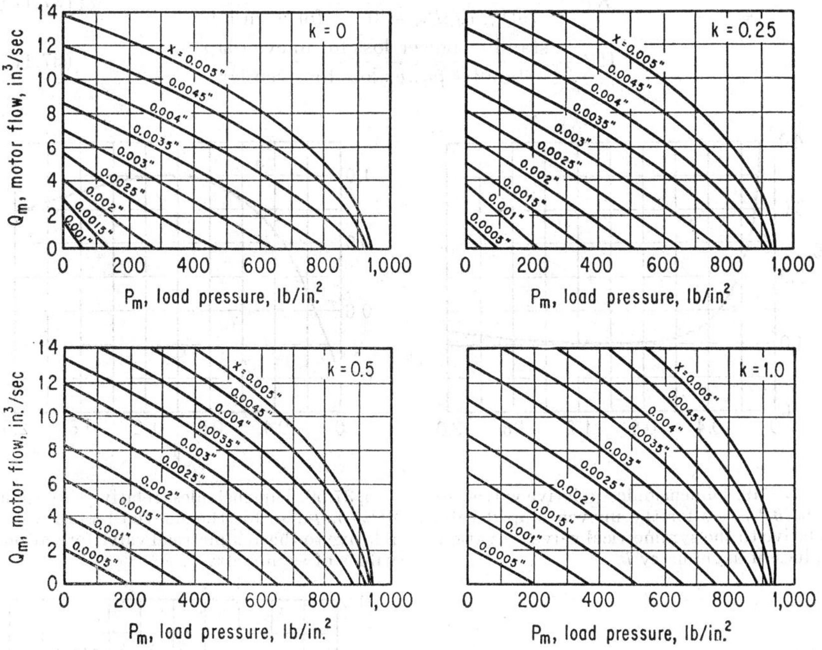

FIG. 37.47. Flow-pressure characteristics of unevenly underlapped four-way valves with constant supply pressure.

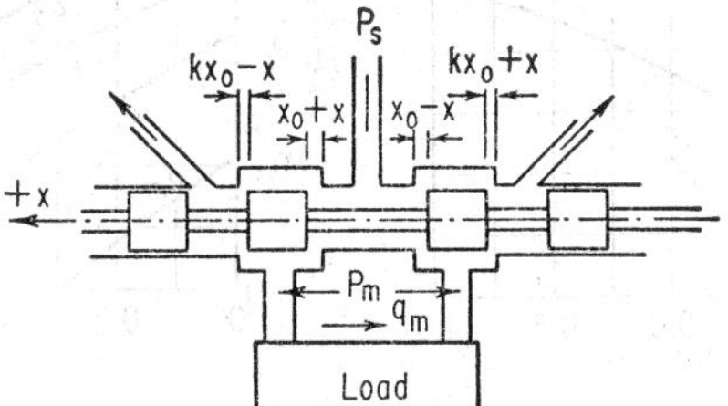

FIG. 37.48. Schematic drawing of unevenly underlapped four-way valve.

coefficients. The high standby power loss of this valve has led to studies of unsymmetrically lapped valves which could combine the good features of both types, namely, low standby power loss and good damping near the null. Another reason for opening the supply pressure to the load ports while keeping the return ports closed at center is the need for cooling flow through the actuator in high-temperature applications. Figure 37.47 shows four cases of a specific valve design[25] in which $P_s = 950$ psi, $x_0 = 0.005$ in., and W = port width = 1.0 in. with varying unsymmetrical underlap. The term k is the ratio of return-port underlap to pressure-port underlap as shown in Fig. 37.48; thus $k = 1$ represents a symmetrically lapped valve.

The relative effects upon valve gain and standby power loss of the uneven underlap are shown in Figs. 37.49 to 37.51, in which

$$C_1 = \frac{(\partial Q_m/\partial P_m)x = 0 \quad \text{for uneven lap}}{(\partial Q_m/\partial P_m)x = 0 \quad \text{for even lap}} \tag{37.173}$$

$$K_1 = \frac{(\partial Q_m/\partial x)P_m = 0 \quad \text{for uneven lap}}{(\partial Q_m/\partial x)P_m = 0 \quad \text{for even lap}} \tag{37.174}$$

$$\text{HP} = \frac{\text{standby power loss for uneven lap}}{\text{Standby power loss for even lap}} \tag{37.175}$$

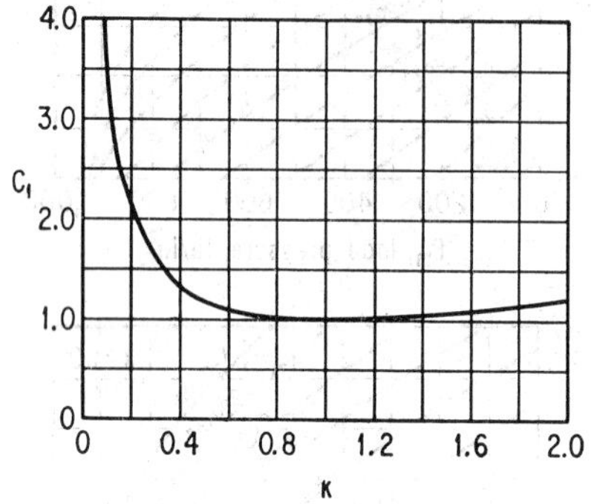

FIG. 37.49. Dimensionless relative valve gain $\partial Q_m/\partial P_m|_{x=0}$ for the uneven lapped valve relative to the symmetrical valve for various values of asymmetry k.

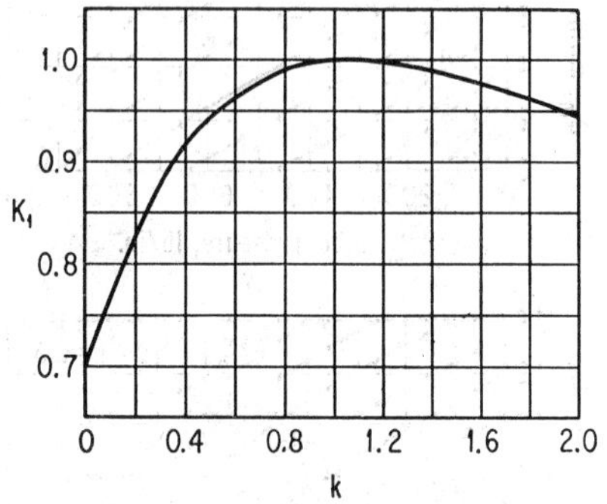

FIG. 37.50. Dimensionless relative valve gain $\partial Q_m/\partial x|_{P_m=0}$ for the uneven lapped valve relative to the symmetrical valve for various values of asymmetry k.

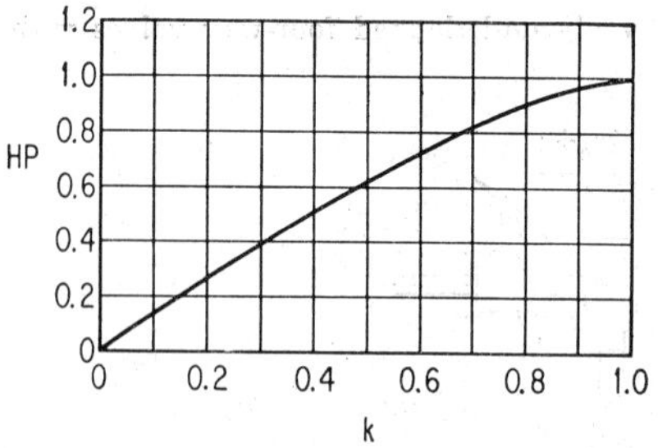

FIG. 37.51. Ratio of standby power loss of unevenly lapped valve over standby power loss of symmetrical valve for various values of asymmetry k.

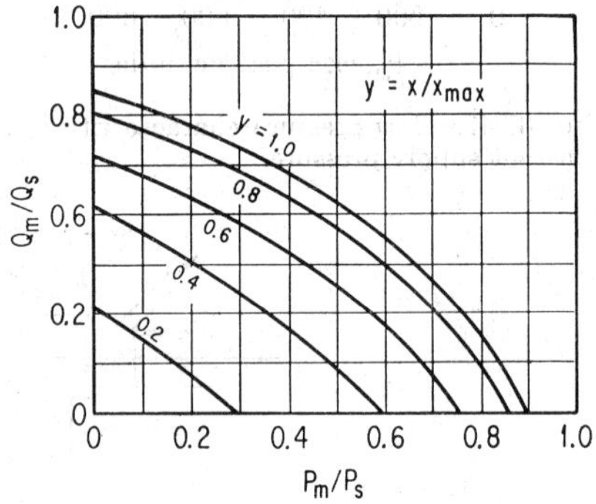

FIG. 37.52. Dimensionless pressure-flow characteristics of the jet-pipe valve.

The Jet-pipe Valve. The jet-pipe valve differs from those previously described in that it is more akin to a hydrokinetic class of devices (see Fig. 37.39). The momentum of the fluid against the two receiver holes is regulated by controlling the jet direction relative to the receiver block. Pressure recoveries up to 90 per cent can be achieved with well-designed receivers. Jet-pipe flow is complex three-dimensional flow which has defied useful analysis to date. Experience indicated that:

1. The receiver holes should be about twice the jet diameter.

2. The jet pipe should be separated from the receiver block by at least two jet diameters. Separation in excess of two jet diameters has little effect on the valve characteristics.

The flow-pressure characteristics are given by Fig. 37.52 and are seen to be similar to the characteristics of the four-way flapper valve (Fig. 37.42).

(d) Steady-state Forces in Flow-control Valves

The ability to position the spool valve accurately in its sleeve, the flapper valve relative to its plug, and the jet-pipe valve relative to its receiver is the heart of good fluid-power control-system design. The forces acting on the valve elements can be divided as follows: flow induced, pressure unbalance induced, inertia, viscous drag, and nonviscous drag.

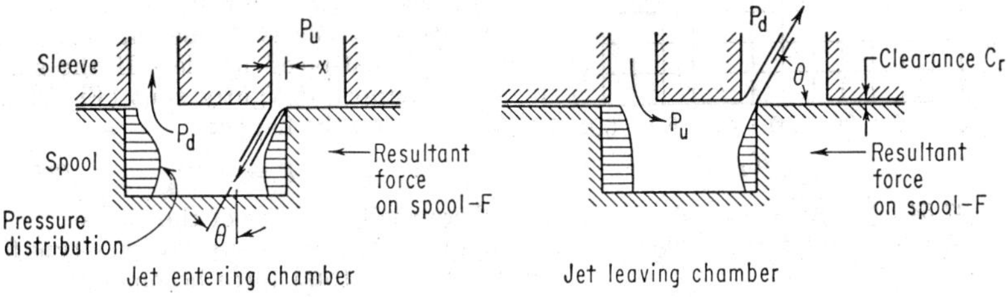

FIG. 37.53. Origin of the flow-induced force on the spool-type valve.

For *spool-type valves,* the flow-induced forces generally represent the major portion of the force to be overcome by the valve actuator. The origin of the force is best explained by studying the pressure distribution on the walls of the spool lands. Figure 37.53 shows that flow either into the spool or out of the spool at the metering edge results in a flow-induced force causing the valve to close.

The force can be shown to be

$$F = \sqrt{\rho} \cos \theta Q \sqrt{P_u - P_d} \tag{37.176}$$

where θ = angle between jet and valve-chamber wall
P_u = upstream pressure
P_d = downstream pressure

Angle θ is equal to 69° subject to the following limitations:

1. The fluid is nonviscous and incompressible.

2. The peripheral width of the orifice (into the paper) is large compared with the axial length (two-dimensional flow).

3. Flow is irrotational (potential flow) ahead of the orifice.

4. The upstream chamber angles are 90° (within 10°).

5. Valve clearance is an order of magnitude smaller than valve opening x.

6. Orifice edges are sharp.

For the case of finite clearance between valve spool and valve sleeve C_r the efflux angle is effected as shown in Fig. 37.54. The effect of rounded corners is similar because for small openings θ will become less than 69°, resulting in an increased flow force.

For petroleum-base fluids with a specific gravity of 0.85, square sharp corners, and clearance small relative to opening, the following useful relations result for a single land:

$$F = 0.0032Q \sqrt{P_u - P_d} \tag{37.177}$$

or

$$F = 4.8 \times 10^{-5} Q^2/A \tag{37.178}$$

$$F = 0.215A(P_u - P_d) \tag{37.179}$$

These flow forces can be substantially reduced or even reversed with proper design of the valve to direct the efflux momentum. The design considerations are rather complex in that both upstream and downstream chamber designs affect the efflux angle. In principle, the flow is to be guided with retention of jet velocity until the jet leaves at an angle θ of 90°.[4]

Similar fundamental considerations hold for the *poppet valve*. Yet the resulting forces on a poppet valve have been found to be difficult to predict because of the effect of fluid shear forces on the poppet cone in addition to the complex flow pattern in the downstream chamber.[15] Guiding the flow as illustrated in Fig. 37.55 has been found to reduce the flow-induced closing forces and result in improved performance of pressure regulators.

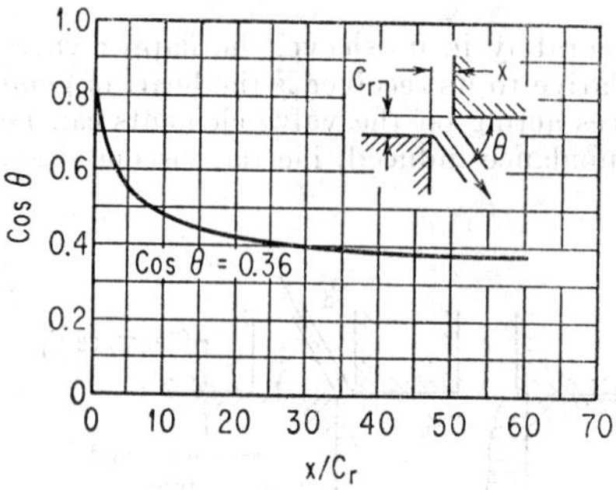

FIG. 37.54. Effect of radial clearance between spool and sleeve on efflux charge θ.

The flapper-nozzle-valve forces are best studied for two configurations, the ideal sharp-edged nozzle flapper and the more common flat-faced nozzle. If the sharp-edged flapper is tightly shut (Fig. 37.56) the force on the flapper

$$F = \pi r^2 P_s \tag{37.180}$$

where r is the radius of the nozzle.

As the flapper opens and orifice-type flow occurs, the total force becomes the static pressure P_1 times the nozzle area plus the momentum transferred to the flapper. For small values of x/r, it can be shown that the force varies parabolically with opening x (ref. 4)

$$F = \pi r^2 P_s[1 + (2C_d x/r)^2] \tag{37.181}$$

In the limit all the momentum is transferred to the flapper.

$$F = 2\pi r^2 P_s \tag{37.182}$$

Design realities do not permit either the manufacture or maintenance of a sharp-edged

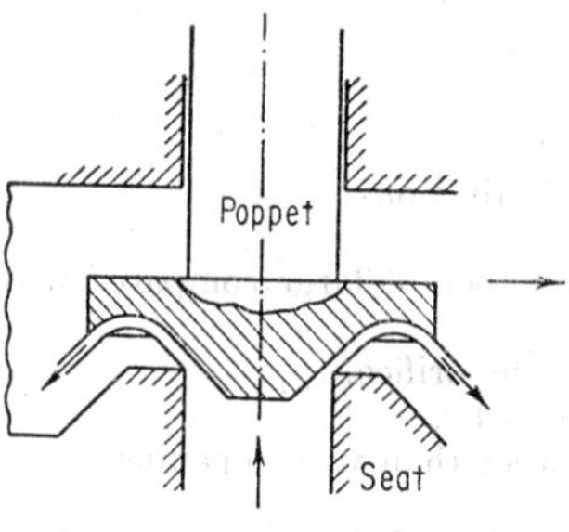

FIG. 37.55. Flow-force-compensated poppet valve.

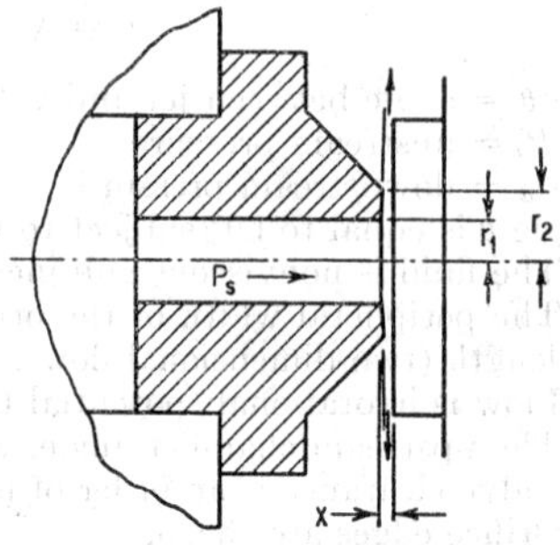

FIG. 37.56. Flat-faced flapper-nozzle valve.

flapper-nozzle unit. Many flapper valves are therefore made with relatively wide flats. For very small values of $x/(r_2 - r_1)$ (see Fig. 37.56), the flow is essentially laminar and proportional to x^3 as given by Eq. (37.183). The limiting value of $x/(r_2 - r_1)$ is determined by the onset of separation. A good estimate can be made by limiting Re at r_1 to less than 2,000.

$$q = \frac{\pi x^3 P_s}{6\mu \ln (r_2/r_1)} \tag{37.183}$$

The force is given by

$$F = \frac{\pi(r_2^2 - r_1^2)P_s}{2 \ln r_2/r_1} \tag{37.184}$$

The pressure distribution across the flat may result in a force considerably larger than would be obtained from the sharp-edged orifice. For values of $x/(r_2 - r_1)$ larger than the critical value, flow separation occurs, making analytical predictions impossible.

(e) Transient Flow Forces in Flow-control Valves

The rate of change of flow into and out of valve chambers results in forces on the moving valve element. These forces are usually proportional to the valve velocity and thus similar to damping forces. Since the force may either aid or oppose the valve velocity, the damping may be negative or positive. If the valve, as is usually the case, is part of a spring-mass system (torque motor or valve actuator), negative damping will contribute to dynamic instability.

The detailed derivation of the transient flow forces can be found in ref. 4. The total flow-induced force on a spool with constant pressure drop across the land is given by

$$F = -[2C_d^2W \cos \theta(P_u - P_d)]x - [C_dW \sqrt{2(P_u - P_d)\rho}\, L](dx/dt) \tag{37.185a}$$

for the configuration of Fig. 37.57. The first term on the right-hand side is similar to a spring force tending to close the valve; the second term, since it opposes motion in

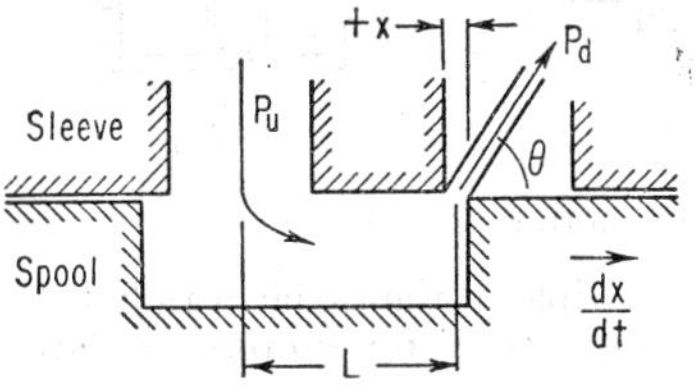

FIG. 37.57. Positively damped valve port.

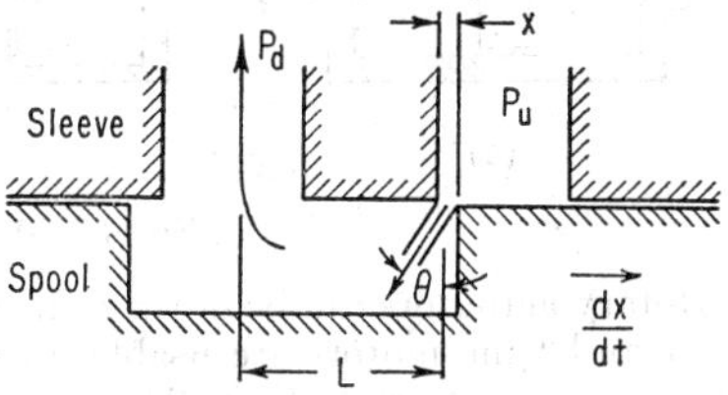

FIG. 37.58. Negatively damped valve port.

the direction of opening, is like a positive damping force. If the flow is reversed as shown in Fig. 37.58, the spring force still tends to close the valve, but the damping force is now in the direction of motion and tends to open the valve, thus producing a negative damping effect

$$F = -[2C_d^2W \cos \theta(P_u - P_d)]x + [C_dW \sqrt{2(P_u - P_d)\rho}\, L](dx/dt) \tag{37.185b}$$

The porting design can thus be arranged to give positive damping to assure stability.

However, even valves with positively damped ports can be unstable when installed in a specific system. Such instabilities are usually associated with acoustic resonances between the transmission line, the valve, and the load.

Article 37.6(c) discusses various methods of analyzing and simulating the resonant transmission line. A detailed discussion is contained in ref. 4.

37.8. HYDRAULIC MOTORS

(a) Introduction

Hydraulic motors are the output devices of the hydraulic system. They convert the controlled fluid power into mechanical position (either rotary or linear) for the position servo, or controlled power for the power control in the form of torque at angular velocity (or force at linear velocity) (see Fig. 37.59). The input impedance and load sensitivity are the characteristics of importance.

Several types of motors (actuators) are commonly used:

Linear actuators, or ram-type motors, are shown in Fig. 37.60*a*, *b*, and *c*. Figure 37.60*a* is the single-acting ram, which is rather uncommon because it relies on the load force for return.

Figure 37.60*b* is the double-acting differential-area ram used commonly with three-way valves such that the small area is constantly at supply pressure and the area ratio is 2:1. For applications with four-way valves the area ratio is usually much less than 2:1 and is determined by the structural requirements of the actuator rod.

Figure 37.60*c* is the double-acting equal-area ram. In each case differential pressures across the ports result in force on the piston which is transmitted through the rod to the load. Pressures vary from a few hundreds to 5,000 psi and up.

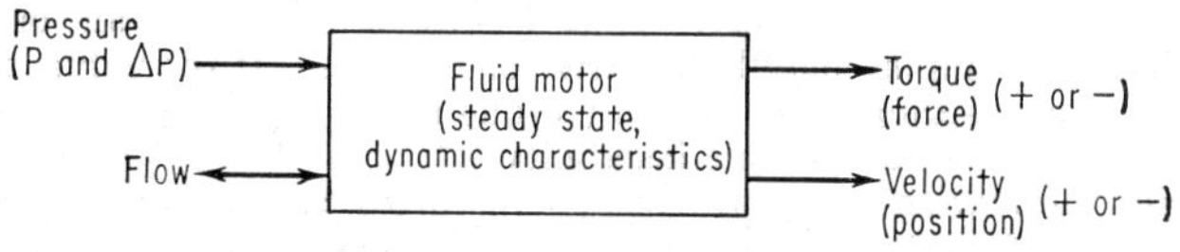

FIG. 37.59. Block diagram of motor.

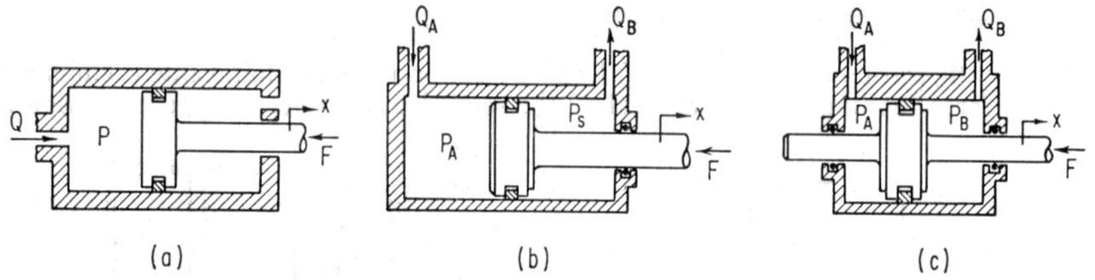

FIG. 37.60. Ram-type motors.

Rotary actuators can be of the gear, vane, or multiple-piston (axial or radial) types. Gear and vane motors are usable up to 1,000 psi, with piston motors available for hydraulic pressures up to 5,000 psi.

Hydraulic motors are used for both steady and intermittent duty.

(b) Characteristics of Motors

The steady-state characteristics of motors can be expressed in a manner similar to those of the pump as developed in Art. 37.4. The torque output of the motor is

$$T_{AM} = \Delta P D_m/2\pi - C_d D_m \mu N_m - C_f(\Delta P D_m/2\pi) - T_c \qquad (37.186)$$

The flow to the motor is given by

$$Q_{AM} = D_m N_m + (2\pi C_s D_m/\mu)\,\Delta P - Q_r \qquad (37.187)$$

Certain piston motors have additional retarding torques and leakage flows proportional to the sum of the port pressures. Appropriate terms can be added to Eqs. (37.186) and (37.187)

where T_{AM} = shaft torque
ΔP = pressure drop
D_m = motor ideal displacement
C_f = drag coefficient dependent on pressure
T_c = friction torque dependent on neither speed nor pressure from shaft seals and bearings
Q_{AM} = flow to the motor, volume per revolution
N_m = motor speed
C_s = leakage coefficient
Q_r = flow-loss rate from cavitation effects

The overall efficiency η_m, torque efficiency η_{mm}, and volumetric efficiency η_{vm} are given by Eqs. (37.188) and (37.189).

$$\eta_m = \frac{1 + C_s(\Delta P/\mu N) - Q_r/2\pi D_m N}{1 - C_d(\mu N/\Delta P) - C_f - T_c/D_m \,\Delta P} \tag{37.188}$$

$$\eta_m = \eta_{vm}/\eta_{mm} \tag{37.189}$$

The steady-state characteristics can be shown on torque-speed diagrams (Fig. 37.61) for further clarification.[20]

The dynamic characteristics of motors are described by several design parameters: volume under compression V_m, inertia relative to torque J_m, mechanical compliance of output shaft K_m, friction as a function of speed only C_d, friction as a function of pressure only C_f, and nonviscous friction T_c.

The volume under compression as seen by the load circuit for fixed-displacement gear, vane, and piston motors varies between 1.2 and 1.5 times the displacement (volume per revolution).

The inertia of axial-piston motors is extremely low relative to the torque at the maximum pressure differential at no load. Typical values[26] for axial-piston motor at $\Delta P = 3{,}000$ psi:

$$T_m/J_m = 100{,}000 \text{ 1/sec}^2$$
$$T_m{}^2/J_m = 4 \times 10^7 \text{ lb-in./sec}^2$$

Vane motors at $\Delta P = 1{,}000$ psi:

$$T_m/J_m = 200{,}000 \text{ 1/sec}^2$$
$$T_m{}^2/J_m = 7.5 \times 10^7 \text{ lb-in./sec}^2$$

Values for gear-type motors are similar in magnitude to vane motors.

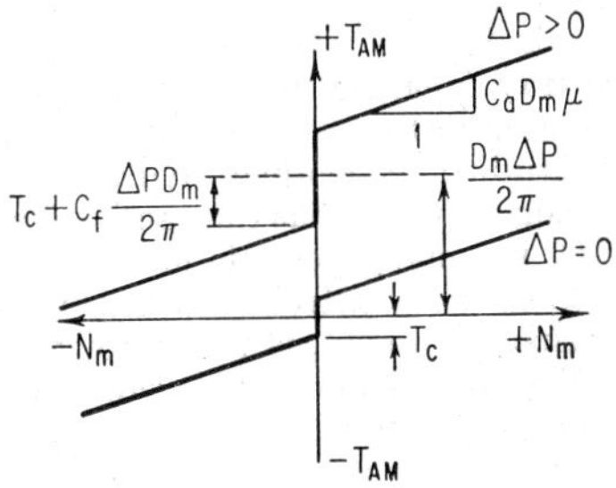

FIG. 37.61. Torque-speed characteristics of positive-displacement motors.

The compliance of the output member is a result of the mechanical connections between the pistons and the shaft. Flexibility of connecting rods, universal joints, wobble plates, etc., results in shaft motion with torque change and no flow to the backlash in the output drive. For reversals in direction this phenomenon becomes important and must be considered in the systems simulation.

Friction as a function of speed results in damping in the closed-loop position-controlled system.

Friction as a function of pressure only can be considered as a reduction in the effective displacement of the motor.

The nonviscous friction term T_c resulting from seals is a major obstacle to achieving accurate performance and stability. T_c is a strongly dependent upon the average pressure of the ports to the motor. Thus the lower the average pressure, the lower T_c.

A further problem exists in the large difference between starting torque and running torque for some motors. Test data[27] show that a straight-line approximation for the pressure drop required to start a hydraulic motor as a function of downstream port pressure is valid:

$$\Delta P = A + BP_d \tag{37.190}$$

B varies from 0.220 to 0.550 depending upon the motor design; A varies from 20 to 50 psi for rotary motors with displacement ranging from 0.065 to 0.375 in³/rev. Ram-type motors depending upon the rod-seal design may have considerably higher breakaway pressure differentials ranging into several hundred psi.

The no-load running torque can also be represented by a straight-line approximation. The intercept A is the same as for the previous case, but the slope is generally one-quarter to one-third of the starting P/P_d. Certain balanced-vane-type motors of special design exhibit exceedingly low starting pressure drops with P/P_d equal to almost zero. Special attention to this phenomenon in the design of motors can thus result in good starting performance. Figure 37.61 is somewhat of an idealization in that it does not recognize the starting-torque problem. Under normal operating conditions pump pressure fluctuations and normal closed-loop system dither may reduce the difference between starting and running torque somewhat. For initial system design, consideration must be given this phenomenon or unstable performance (limit cycle) may result.

37.9. HYDRAULIC CONTROL SYSTEMS

(a) Introduction

In this article will be described, in a very limited and restricted manner, several typical hydraulic control circuits by means of block diagrams and transfer functions.

The greatest care must be taken in the assumptions made in deriving basic system equations. The common practice of complete equation linearization has led to many disappointments, since only small perturbations about a well-defined operating point can be analyzed. The approach of incremental linearization about many operating points has met with some success. However, such violent nonlinearities as dead band and backlash cannot thus be treated with the linear approach if the dead band or backlash constitutes an appreciable portion of the excursion. The first two examples portray linearized approaches, and the third example indicates the use of the describing function in the treatment of certain typical nonlinearities. If large excursions are to be studied in complex hydraulic systems containing substantial nonlinearities the analog computer has time and again proved to be the ideal design-analysis tool.[28,29,30,31]

Of the many possible methods of hydraulic control, two basic systems stand out:

The valve-controlled system wherein some characteristic of the valve, the power-dissipative device, is varied to control the fluid power supplied to the load.

The pump-controlled system wherein some characteristic of the pump is varied which will result in a control of the fluid energy generated.

Both types are used in high-performance hydraulic servomechanisms. For very large systems involving several hundred horsepower, the pump-controlled system incorporates a valve-controlled system as a preliminary stage in the actuation of the variable-delivery mechanism of the pump. Generally speaking, valve-controlled systems are used for hydraulic power levels up to about 20 hp. For the larger systems, pump-controlled approaches have found preference because of their lower standby heat rejection.

(b) Valve-controlled Circuits[32]

The most widely used method of hydraulic control consists of a four-way control valve of either the overlapped or underlapped type as described in Art. 37.7 and a ram- or rotary-type motor as described in Art. 37.8. The system is shown in Fig. 37.62. The following assumptions are made in the analysis:[4]

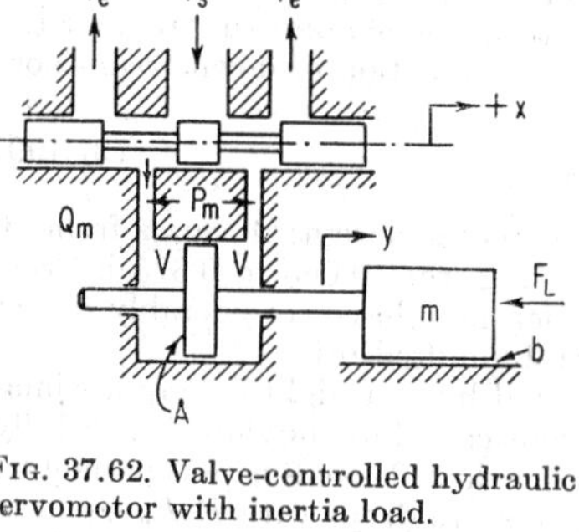

FIG. 37.62. Valve-controlled hydraulic servomotor with inertia load.

1. The control valve is ideal, edges are sharp, and radial clearances are zero between spool and sleeve.
2. Flow through orifices is simple-type orifice flow with no viscous effects.
3. All flow passages between valve and motor are short so that fluid-mass effects can be neglected.
4. Friction losses in lines are negligible.
5. Leakage flow in the motor is laminar.
6. Supply characteristics are unaffected by valve demands (pressure remains constant or flow remains constant, respectively).

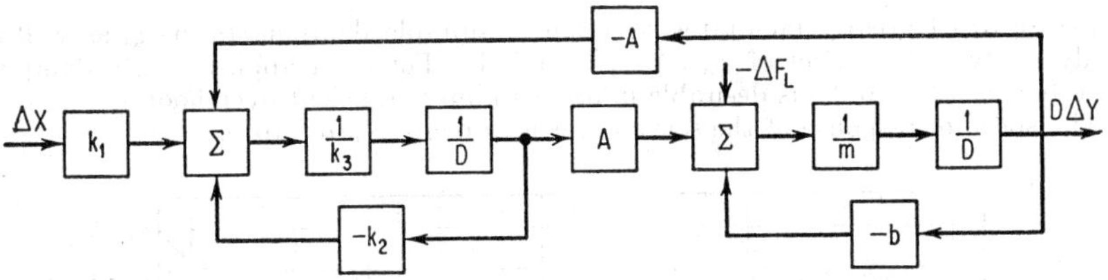

FIG. 37.63. Graph of servomotor damping ratio as a function of the dimensionless parameters π_a and π_b.

7. Exhaust pressure is zero.
8. Motion of the ram is restricted to small excursions from its center position, and the valve characteristics can be assumed linear for small load-pressure, load-flow, and valve-travel changes. The block diagram shown in Fig. 37.63 represents the following equation:

$$\left(\frac{k_3 m}{k_2 b + A^2} D^2 + \frac{k_2 m + k_3 b}{k_2 b + A^2} D + 1\right) \Delta \dot{Y} = \frac{k_1 A(\Delta X) - (k_2 + k_3 D)(\Delta F_L)}{k_2 b + A^2} \tag{37.191}$$

where X = valve travel
Y = ram travel
m = load inertia
b = load viscous damping
A = ram area
k_1 = valve gain, $\partial Q_m/\partial x \big|_{P_m=\text{const}}$
Q_m = volume flow to motor
P_m = ram pressure drop
C_1 = valve gain, $\partial Q_m/\partial P_m \big|_{x=\text{const}}$
$k_2 = C_1 + C_2$
C_2 = laminar-flow coefficient for leakage past ram $\partial Q_e/\partial P_m$
$k_3 = V/2\beta + k_e$
k_e = coefficient of elasticity of lines and motor
F_L = external load force
D = differential operator
V = volume under compression in each side

Equation (37.191) can be put into the form of Eq. (37.192):

$$(D^2/\omega_{ns}^2 + 2\xi_s D/\omega_{ns} + 1)\,\Delta \dot{Y} = k_{ss}\,\Delta X - k_{sL}(1 + k_3 D/k_2)\,\Delta F_L \tag{37.192}$$

where

$$\omega_{ns} = \sqrt{\frac{k_2 b + A^2}{k_3 m}} \quad \text{natural frequency of servomotor} \tag{37.193}$$

$$\xi_s = \frac{k_2 m + k_3 b}{2\sqrt{k_3 m(k_2 b + A^2)}} \quad \text{servomotor damping ratio} \tag{37.194}$$

$$k_{ss} = \frac{k_1 A}{k_2 b + A^2} \quad \text{servomotor steady-state gain} \tag{37.195}$$

$$k_{sL} = \frac{k_2}{k_2 b + A^2} \quad \text{servomotor steady-state load sensitivity} \tag{37.196}$$

Equation (37.194) is represented in graphical form in Fig. 37.64 where

$$\pi_a \triangleq (k_2/A)\sqrt{m/k_3} \tag{37.197}$$

$$\pi_b \triangleq (b/A)\sqrt{k_3/m} \tag{37.198}$$

$$\xi_s = \frac{\pi_a + \pi_b}{2\sqrt{\pi_a \pi_b + 1}} \tag{37.199}$$

Figure 37.64 greatly facilitates selection of suitable damping ratios ξ_s as well as ready study of the effect of k_2, k_3, m, b, and A. For most applications a damping ratio between 0.6 to 0.8 is desirable in order to limit transient overshoot.

The speed of response of the servomotor is strongly related to ω_{ns}.

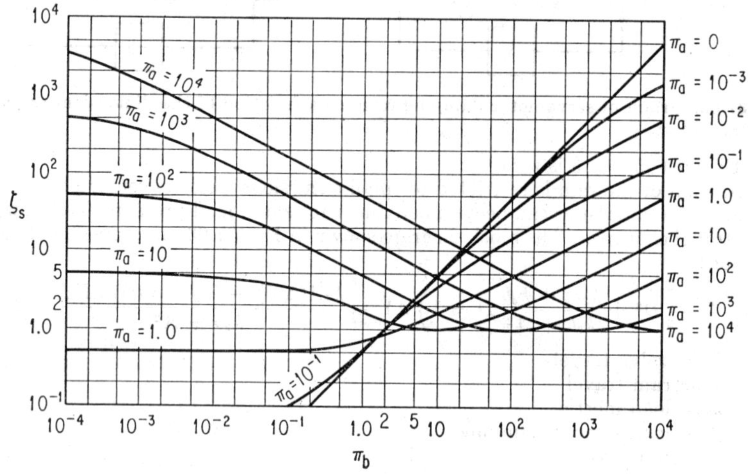

FIG. 37.64. ζ_s vs. π_a and π_b for a valve-controlled hydraulic servomotor.

$$\zeta_s = \frac{\pi_a + \pi_b}{2\sqrt{\pi_a \pi_b + 1}} \qquad \pi_a = (k_2/A)\sqrt{m/k_3} \qquad \pi_b = (b/A)\sqrt{k_3/m}$$

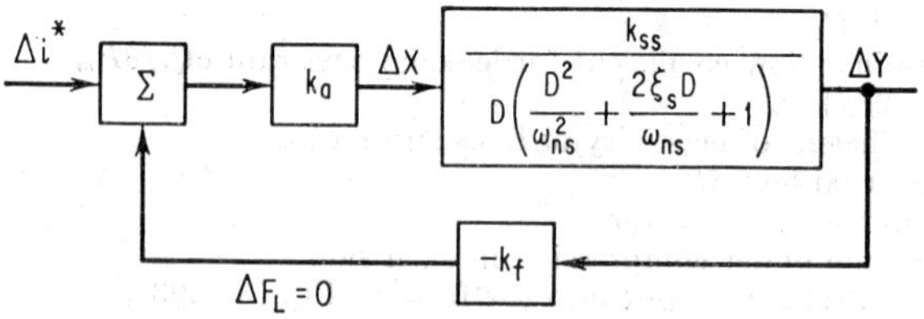

FIG. 37.65. Operational block diagram of valve-controlled servomotor.

If the system is overdamped ($\xi_s > 1$), the speed of response τ_1 is given by

$$\tau_1, \tau_2 = \frac{k_2 m + k_3 b}{2(k_2 b + A^2)} \left[1 \pm \sqrt{1 - \frac{4k_3 m(k_2 b + A^2)}{(k_2 m + k_3 b)^2}}\right] \tag{37.200}$$

where the left-hand side of Eq. (37.191) now becomes $(\tau_1 D + 1)(\tau_2 D + 1)$.

If the loop is closed between load position and valve actuator input such as shown in Fig. 37.65 the response equation becomes[4]

$$(D^3/\omega_{ns}^2 + 2\xi_s D^2/\omega_{ns} + D + k_a k_{ss} k_f)\,\Delta Y = k_{ss} k_a\,\Delta i^* - k_{sL}(1 + k_3 D/k_2)\,\Delta F_L \tag{37.201}$$

For the steady-state case the terms containing D go to zero so that steady-state sensitivities are given by

$$(\Delta Y)_{ss} = (1/k_f)(\Delta X)_{ss} - (k_2/k_a k_1 k_f A)(\Delta F_L)_{ss} \tag{37.202}$$

where k_f = position-feedback gain
k_a = valve-actuator gain

The validity of this simplified approach must be carefully examined. If a pressure-compensated variable-delivery pump is used as a fluid power source, such as is described in Art. 37.4, the sensitivity of the source-to-flow demand variations from the valve and their effect on supply pressure and flow dynamics must be examined. Of particular interest are the pump damping coefficient and natural frequency. A low damping coefficient (below 0.5) and a pump natural frequency close to the valve and load natural frequency ω_{ns} require closer study.

In certain jet-engine applications the ram motor is removed from the valve by as much as 20 ft. If the frequency of the standing wave and its harmonics in the transmission line even approaches the system frequency ω_{ns} the transmission-line dynamics such as described in Art. 37.6(*c*) must be considered to assure a stable system.

Of prime importance is the full understanding of the nature of the load. Mechanical compliance between ram and inertia and inertia against ground and the nonviscous friction in the load are among the considerations which must be carefully examined to assure validity of the analysis.

(c) Pump-displacement-controlled Variable-speed Drive[4]

For high-power applications, a variable-speed output positive-displacement hydraulic transmission is used. This transmission consists of a variable-displacement hydraulic pump driven by a constant-speed power source and transmission lines to

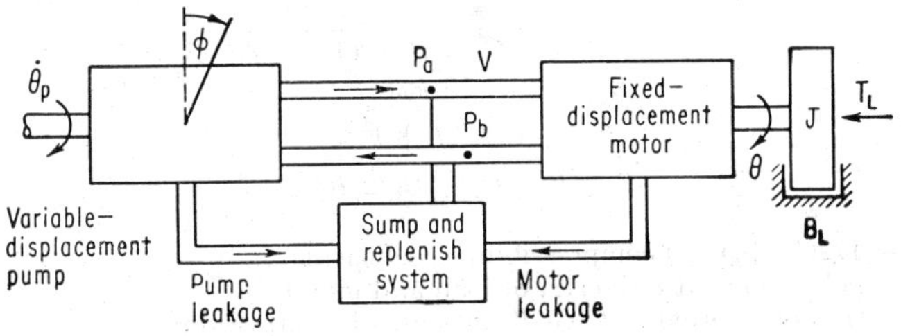

FIG. 37.66. Pump-displacement-controlled servomotor.

and from the fixed-displacement hydraulic motor which drives an inertial load with some viscous damping (see Fig. 37.66).

The following assumptions are made in the simplified analysis:

1. Pressures P_a and P_b are uniform throughout the lines and chambers in the pump and motor. Short lines are thus assumed between pump and motor so that fluid-inertia effects and wave phenomena are neglected.
2. Leakage in pump and motor is considered to consist of three parts:
   a. Internal laminar leakage between P_a and P_b.
   b. External laminar leakage from P_a to case.
   c. External laminar leakage from P_b to case.
3. Cavitation phenomena are considered negligible.
4. Operation of the system is such that relief valves between lines *a* and *b* never open.
5. The replenishing system assures that lines *a* and *b* remain filled.
6. Motor friction torque is proportional to $(P_a - P_b)$ and motor speed.
7. Motor-to-load connection is rigid.

The operational block diagram of the complete system is given in Fig. 37.67.

For small changes of all variables from the condition of zero motor speed, the system equation is

$$\left[\frac{k_3J}{(C_i + C_e/2)B + (1 - C_{fm})D_m^2}D^2 + \frac{(C_i + C_e/2)J + k_3B}{(C_i + C_e/2)B + (1 - C_{fm})D_m^2}D + 1\right](\Delta\theta_m)$$
$$= \frac{D_m(1 - C_f)k_p\theta_p(\Delta\phi) - [(C_i + C_e/2) + k_3D]\,\Delta T_L}{(C_i + C_e/2)B + (1 - C_{fm})D_m^2} \quad (37.203)$$

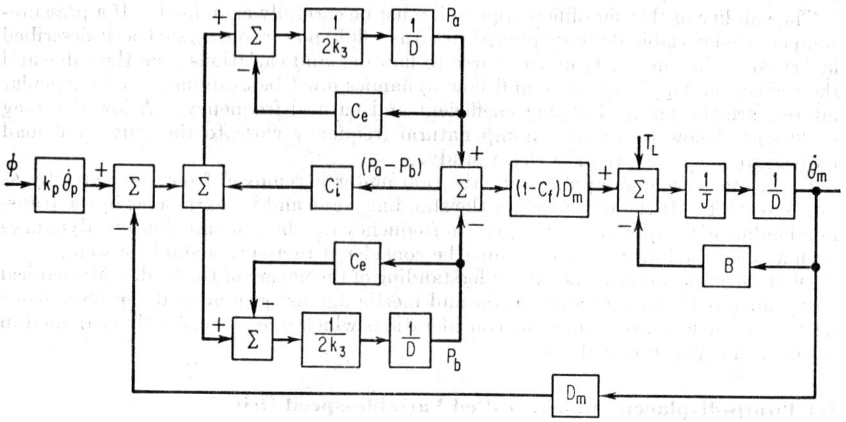

FIG. 37.67. Operational block diagram of pump-controlled servomechanism.

ξ and ω_n are given by Eqs. (37.204) and (37.205):

$$\omega_n = \sqrt{\frac{(C_i + C_e/2)B + (1 - C_f)D_m^2}{k_3 J}} \tag{37.204}$$

$$\xi = \frac{(C_i + C_e/2)J + k_3 B}{2\sqrt{k_3 J[(C_i + C_e/2)B]}} \tag{37.205}$$

where $k_3 = 1/2(k_e + V/\beta)$ compressibility coefficient
J = polar moment of inertia of load and motor
C_i = sum of internal leakage coefficients of pump and motor
C_e = sum of external leakage coefficients of pump and motor
B = sum of motor and load viscous friction coefficient
C_{fm} = motor coefficient of friction due to pressure
D_m = motor displacement
θ_m = motor angular displacement
$\dot\theta_p$ = pump shaft speed
k_p = displacement coefficient of pump
ϕ = pump stroke angle
T_L = load torque
Δ = incremental change in variable
k_e = elasticity coefficient of pipe
V = volume under compression in each line and motor

Equation (37.203) then becomes

$$[D^2/\omega_n^2 + (2\xi/\omega_n)D + 1]\dot\theta_m = k_{\mathrm{I}}\phi - k_{\mathrm{II}}[(k_3/k_2)D + 1]T_L \tag{37.206}$$

where $\dot\theta_m$ = motor speed

$$k_{\mathrm{I}} = \frac{(1 - C_{fm})D_m k_p \dot\theta_p}{k_2 B + (1 - C_{fm})D_m^2} \tag{37.207}$$

$$k_2 = C_i + C_e/2 \tag{37.208}$$

$$k_{\mathrm{II}} = \frac{k_2}{k_2 B + (1 - C_{fm})D_m^2} \tag{37.209}$$

The equations given do not hold for reversals of motor speed because the pressure-dependent friction torque reverses direction, introducing a gross nonlinearity into the analysis.

(d) Nonlinearities in Hydraulic Power Control Systems

Although linear models of hydraulic systems do provide a good understanding of the relative effects of the various parameters on dynamic performance, detailed system design must include consideration of the inherent nonlinearities to assure a good understanding of large signal response and stability. Probably the most widely used method of design analysis of nonlinear systems utilizes the analog computer.[28,29]

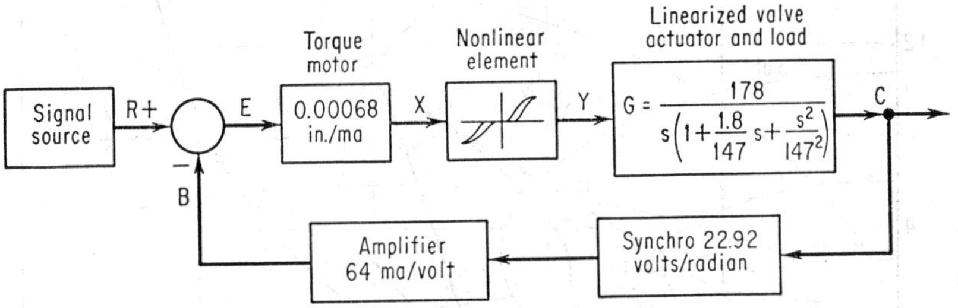

Fig. 37.68. Block diagram of system employing hydraulic servo valve with adjustable dead zone.

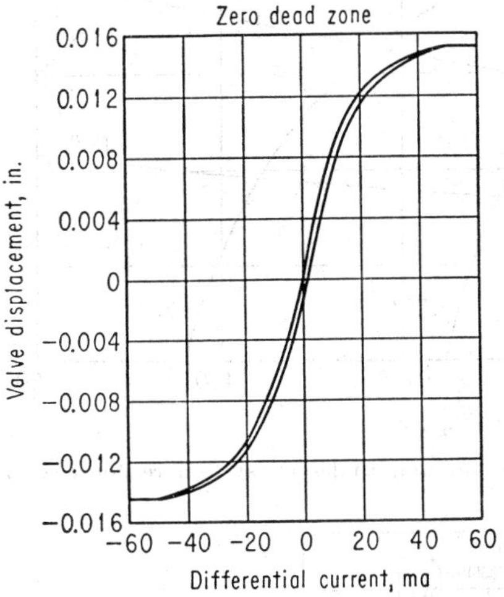

Fig. 37.69. Valve displacement vs. input differential current of torque motor.

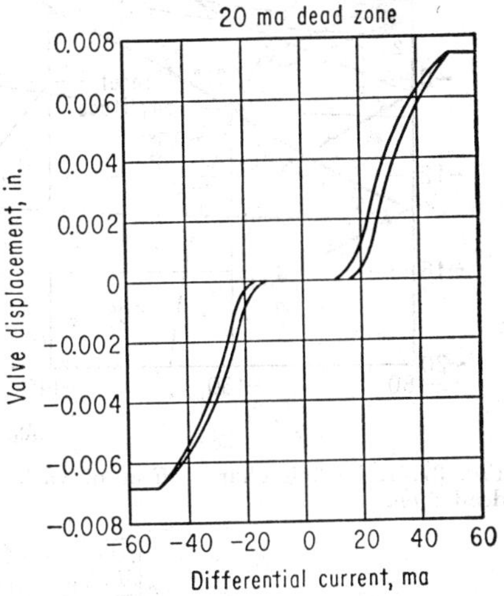

Fig. 37.70. Valve displacement vs. input differential current of torque motor.

There are, however, pencil-and-paper methods which permit the study of grossly nonlinear systems with some facility. The purpose of this section is to demonstrate the use of the describing-function technique to the design of a nonlinear electrohydraulic system,[33] and describe the nonlinear problem in general. The describing-function technique is amply described in the literature.[34,35,36,37,38]

The system considered here consists of an electric-torque-motor-driven two-stage hydraulic four-way servo valve. There is unity feedback between the first and second stages. The actuator drives an inertia load. The system loop is closed by comparing the load position with the input signal (see Fig. 37.68).

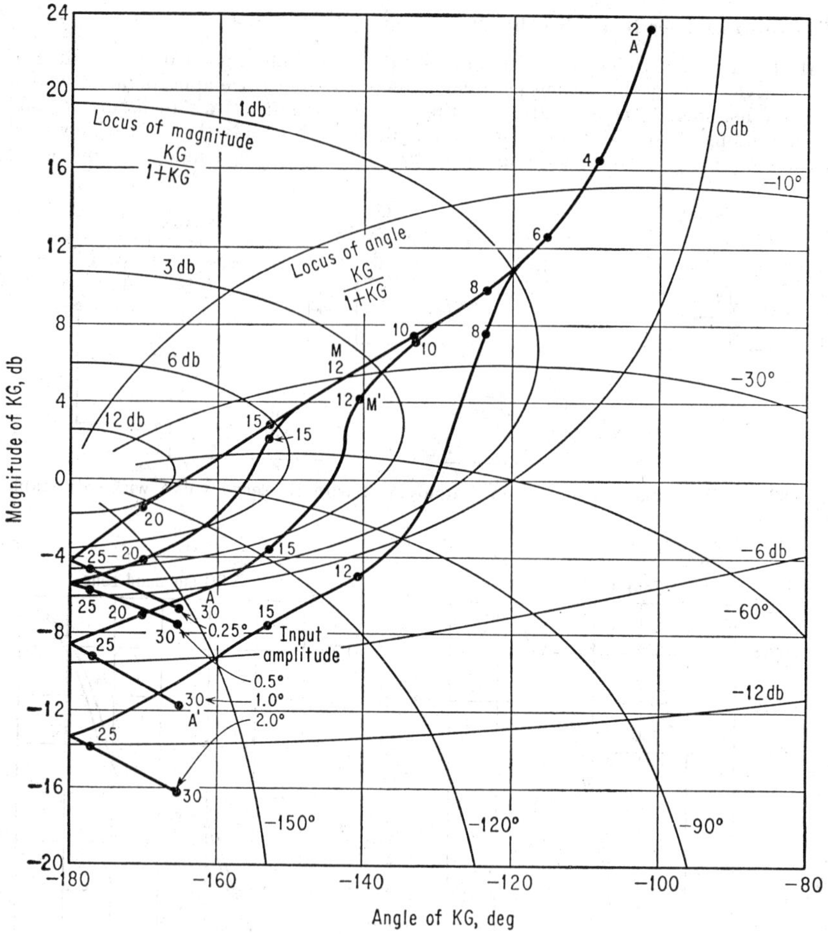

FIG. 37.71. Nichols chart—effect of various input amplitudes on system response zero dead zone.

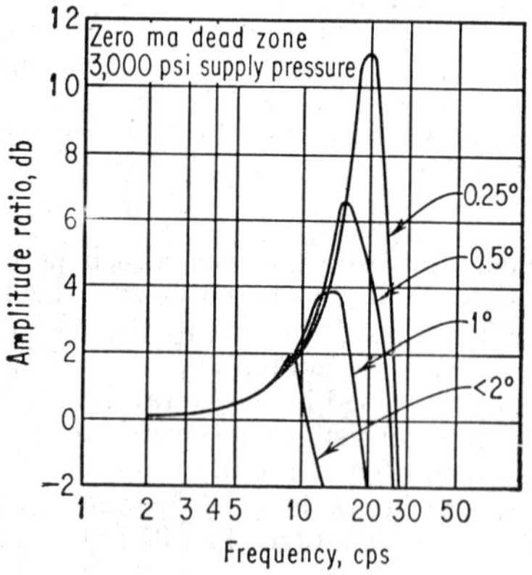

FIG. 37.72. Closed-loop amplitude response as a function of input amplitude.

The nonlinearities considered here are:

1. Torque motor and valve hysteresis
2. Torque motor and valve limiting
3. Torque motor and valve dead zone
4. Valve flow gain change as a function of torque motor current (but not load pressure)

The torque motor-valve characteristics including hysteresis and limiting, and with no dead zone, are shown in Fig. 37.69. Figure 37.70 shows the case with dead zone. The frequency response of the system with no dead zone but with limiting is shown on the Nichols chart in Fig. 37.71. The curve marked ($<0.25°$) indicates linear behavior. Obviously, saturation effects lower the effective system gain with increasing amplitude as evidenced by the lower amplitude ratios at lower frequencies. If the amplitude

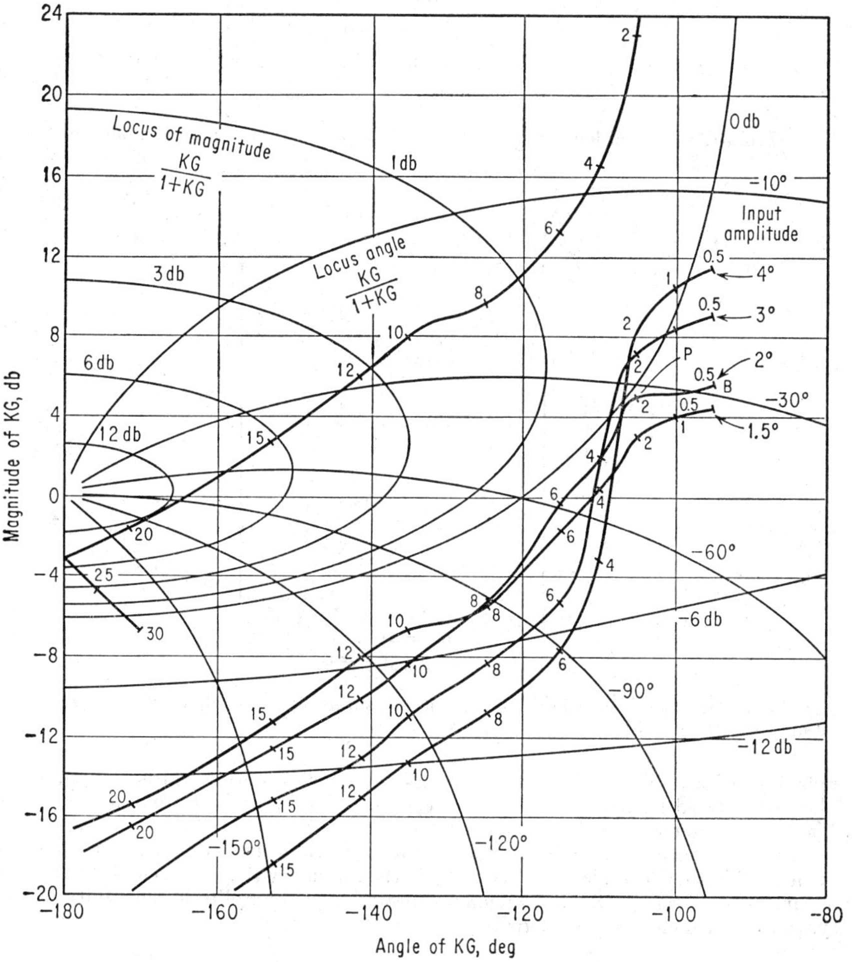

FIG. 37.73. Nichols chart—effect of various input amplitudes on system response.

ratio vs. frequency plot is constructed (Fig. 37.72) the expected behavior is further clarified.

With dead zone included, the Nichols chart is shown in Fig. 37.73 with the linear characteristics indicated. For very small input signals, the output will now be zero until the dead-zone limit is exceeded. The gross departure from linear behavior is

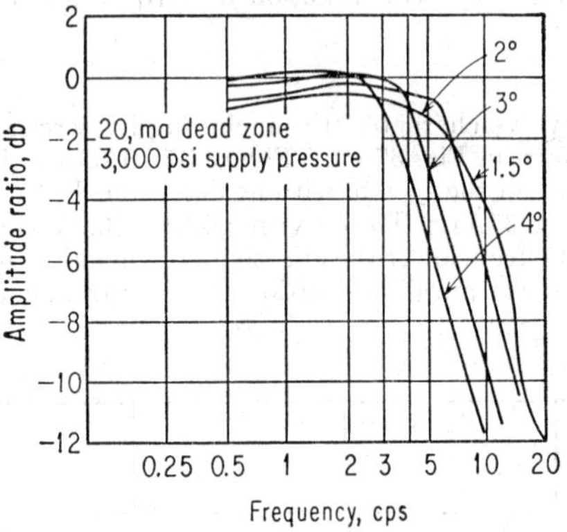

Fig. 37.74. Closed-loop amplitude response as a function of input amplitude.

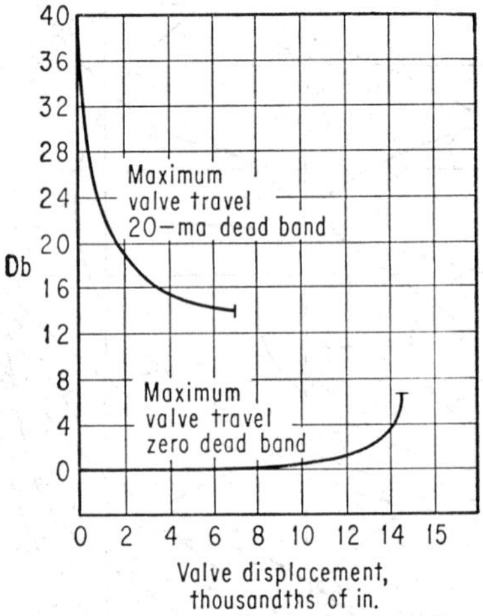

Fig. 37.75. Describing functions from static measurements of nonlinearities for zero and 20 ma dead zone.

apparent even for large signals. Figure 37.74 is the amplitude ratio vs. frequency plot constructed from Fig. 37.73. The corresponding describing functions given in Fig. 37.75 were used for the construction of Figs. 37.71 and 37.73.

The importance of recognizing nonlinearities in specifying system performance is evident from this example. The concept of the frequency response is arrived at from linear-system theory. Realities demand that the signal-amplitude sensitivity of the system performance be considered.

References

1. Dow, R. B., and C. E. Fink: Computation of Some Physical Properties of Lubricating Oils of High Pressures, *J. Appl. Phys.*, vol. 11, pp. 353–357, May, 1940.
2. Klaus, E. E., and M. R. Fenske: Some Viscosity-shear Characteristics of Lubricants, *Lubrication Eng.*, March–April, 1955, pp. 101–108.
3. "Viscosity and Density of Over 40 Lubricating Fluids of Known Composition of Pressures up to 150,000 psi and Temperatures up to 425°F," vols. I, II, ASME Research Report, 1953.
4. Blackburn, J. F., G. Reethof, and J. L. Shearer: "Fluid Power Control," John Wiley & Sons, Inc., New York, 1960.
5. ASTM D 341–43, "Standard Viscosity-temperature Charts for Liquid Petroleum Products."
6. ASTM D 97–47, "Cloud and Pour Points."
7. ASTM D 92–52, "Flash and Fire Points by Means of Cleveland Open Cup."
8. Setchkin test.
9. Hunsaker, J. C., and B. G. Rightmire: "Engineering Applications of Fluid Mechanics," McGraw-Hill Book Company, Inc., New York, 1947.
10. Leslie, R. L.: "Hydraulic Fluids—Today and Tomorrow," ASME paper 60-MD-6, May 23–26, 1960.
11. Taplin, L. B.: "Dynamic Performance of Variable Delivery, Constant Pressure Pumps," presented before SAE-AEC meeting, Apr. 17, 1961, Detroit, Mich.
12. Ernst, Walter: "Oil Hydraulic Power and Its Industrial Applications," 2d ed., McGraw-Hill Book Company, Inc., New York, 1960.
13. Grace, H. P., and C. E. Lapple: Discharge Coefficients of Small-diameter Orifices and Flow Nozzles, *Trans. ASME*, vol. 73, pp. 639–647, July, 1951.
14. Jorissen, A. L.: Discharge Measurements at Low Reynolds Numbers—Special Devices, *Trans. ASME*, February, 1956, pp. 365–368.
15. Stone, J. A.: "Discharge Coefficients and Steady State Flow Forces for Poppet Valves," ASME 59–HYD–18, 1959.
16. Merrill, R. F.: "Effect of Nozzle Geometry on Flapper Valve Characteristics," S. B. Thesis, M.E. Dept., MIT, Jan. 16, 1956.
17. Reethof, G., C. Goth, and H. Kord: "Thermal Effects in the Flow of Fluids between Parallel Flat Plates in Relative Motion," American Society of Lubrication Engineers paper 58 AM 4A–1.
18. Wilson, W. E.: "Performance Criteria for Positive-displacement Pumps of Fluid Rotors," ASME paper 48–SA–14.
19. Reethof, G.: "On the Dynamics of Pressure Controlled Hydraulic Systems," ASME paper 54–SA–7, June, 1954.
20. Shearer, J. L. (ed.): "Handbook of Fluid Dynamics," McGraw-Hill Book Company, Inc., New York, 1961.
21. Sverdrup, N. N.: Calculating the Energy Losses in Hydraulic Systems, *Prod. Eng.*, April, 1951, p. 146.
22. Regetz, John D., Jr.: An Experimental Determination of the Dynamic Response of a Long Hydraulic Line, *NASA Tech. Note* D–576, December, 1960.
23. Lee, S. Y.: Contributions to Hydraulic Control—6, New Valve Configurations for High-performance Hydraulic and Pneumatic Systems, *Trans. ASME*, vol. 76, pp. 905–911, August, 1954.
24. Reid, K. N., Jr.: "Some Aspects of Jet Pipe Valves," special notes prepared for UCLA 1961 summer course on Fluid Power Control.
25. Tsun-Ying, Feng: "Characteristics of Unevenly Underlapped Four-way Hydraulic Servo Valves," ASME paper 56–A–140, November, 1956.
26. Taylor, L. D.: Hydraulic Motors for Automatic Control, *Prod. Eng.*, 1954.
27. Bowman, R. J.: "No Load Starting Characteristics of Fluid Motors," B. S. Thesis, M.E. Dept., MIT, June, 1955.
28. Shearer, J. L.: Proportional Control of Rate-type Servomotors, *Trans. ASME*, vol. 76, pp. 889–903, August, 1954.
29. Reethof, G.: "The Analog Computer as a Design Tool in the Study of Velocity-control Hydraulic System," Proceedings of the 11th National Conference on Industrial Hydraulics, Chicago, 1955.
30. Royle, J. K.: "Inherent Nonlinear Effects in Hydraulic Control Systems with Inertia Loading," Institute of Mechanical Engineers (London), January, 1959.
31. Eckman, D. P., C. K. Taft, and R. H. Schuman: "Electrohydraulic Servo Mechanism with an Ultrahigh Frequency Response," ASME paper 56–IRD–8, Mar. 26, 1956.

32. Shearer, J. L.: Dynamic Characteristics of Valve-controlled Hydraulic Servomotors, *Trans. ASME*, vol. 76, August, 1959.
33. Scherba, M. B., and A. J. Gregory: "The Effect of Certain Nonlinearities on the Dynamic Performance of Valve Controlled Hydraulic Servomechanisms," Proceedings National Conference on Industrial Hydraulics, 1957.
34. Zaborszky, J., and H. J. Harrington: Generalized Charts of the Effects of Nonlinearities in Electrohydraulic Control Valves, *Trans. AIEE*, vol. 76, part I, pp. 191–198, May, 1957.
35. Zaborszky, J., and H. J. Harrington: A Describing Function for Multiple Nonlinearities Present in Electrohydraulic Control Valves, *Trans. AIEE*, vol. 76, part I, pp. 183–190, May, 1957.
36. Johnson, E. C.: "Sinusoidal Analysis of Feedback-control Systems, Containing Nonlinear Elements," *Trans. AIEE*, vol. 71, part II, Applications and Industry, pp. 169–181, 1952.
37. Truxal, J. C.: "Control Engineers' Handbook," Sec. 2, p. 69, McGraw-Hill Book Company, Inc., New York, 1958.
38. Vande Vegte, J.: "Frequency Response Analysis of a Hydraulic Servomechanism with Several Nonlinearities," ASME paper 60-WA-45, 1960.
39. Moody, L. F.: Friction Factors for Pipe Flow, *Trans. ASME*, vol. 66, p. 671, 1944.
40. Fluid Power Book, Power Components, *Machine Design*, Penton Publishing Co., Cleveland, Ohio, Dec. 12, 1963.

INDEX